WHAT DO *YOU* WANT FROM TECHNOLOGY?

When HRW decided to develop technology products, we started by asking teachers what they needed in the classroom. We polled thousands of educators, and their ideas and advice became our starting point.

The first thing we learned is that most technology products available today share a common flaw — they aren't integrated into your curriculum. Most of them were created by people who know a great deal about RAM and ROM, but not very much about teaching the other three Rs. The result for teachers: frustration and hours of planning time wasted trying to mix and match.

Because HRW has been creating educational curriculum since 1866, we know that the focus of the classroom is the student, not the technology. That's why we've organized our new technology materials around your curriculum, and that's why — before we tell you how fast it runs or how many images you can access in 30 seconds — we want to show you the instructional content and value that characterize every product in the new *HRW Multimedia Curriculum Systems.* Look for this special logo to identify these curriculum-based products and to remind you that, at HRW, technology is not an extra — it's part of an integrated approach to classroom management and instruction.

The *Holt Biology Videodiscs* program is one of these new products, and this is your opportunity to preview it for 30 days.

Like all *HRW Multimedia Curriculum Systems* materials, the *Holt Biology Videodiscs* program —

- *Requires less planning time because we've already matched its content to fit your curriculum goals and sequence*
- *Addresses the diversity of learning styles found in today's classroom*
- *Offers greater flexibility with both English and Spanish soundtracks*
- *Correlates directly to* Holt Biology: Visualizing Life, ©1994; Modern Biology, ©1993, 1991, 1989; *or* Biology Today, ©1991

PLUS

- *A separate* Teacher's Correlation Guide *for each HRW biology text makes it even easier to integrate the program with the instructional materials you're now using*
- *Optional* Management Software System *(for Macintosh®) saves you planning and presentation time*

That's our story. For a glimpse of the future that's ready for the classroom today, please let us send *Holt Biology Videodiscs* to preview for 30 days. Just complete the attached order form and either fax or mail it to us today.

William A. Talkington
President
Holt, Rinehart and Winston

P.S.— This extensive multimedia resource is also available for CD-ROM and is called *Concepts of Biology.* Call **1-800-HRW-9799** for more information.

**HOLT
RINEHART
WINSTON**

224S95B15

BIOLOGY

Visualizing Life

Your Evaluation Checklist

Fulfilling the Promise—Biology Education in the Nation's Schools, a report by the National Research Council, tells you what to look for when choosing a biology text.

☒ Appropriateness of students' level and interest

See the many high-interest lessons and features throughout the text.
- Biology and You, page 10
- The Puzzle of Sickle Cell Anemia, page 189
- Human Impact on Ecosystems, page 282
- Testing for HIV! Who's at Risk? page 364
- Exploration of an Arthropod, page 470
- Tour of a Frog, page 563
- Cholesterol and Your Health, page 726

☒ Factual accuracy

Read George Johnson's statements on accuracy, pages 14T–15T.

☒ Incorporation of current conceptual understanding and new subject matter

See the Table of Contents, pages 4T–13T, for important highlights.

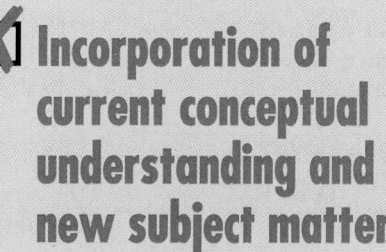

☒ Logical coherence

Read George Johnson's Philosophy on pages 14T–19T, and see the Author's Rationale section on the first page of every chapter.

☒ Adequate but not encyclopedic coverage

See the *Biology: Visualizing Life* Table of Contents on pages 4T–13T and what students really need to know on pages 18T–19T.

☒ Representation of biology as an experimental subject

See how *Biology: Visualizing Life* effectively links lecture and lab with its unique approach to the lab on pages 20T–21T.

☒ Clarity in explanation and effectiveness of illustrations

Look at the careful integration of all illustrations throughout the text. Note the extended captions (such as on pages 101, 140, 190, 225, 575, 580, 612) that aid in the visual presentation of concepts.

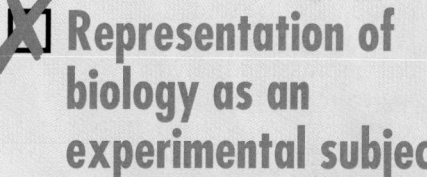

Author

George B. Johnson
Professor of Biology
Washington University
St.Louis, MO

For permission to reprint copyrighted material, grateful acknowledgment is made to the following sources:
National Academy Press, Washington, DC:
From *Fulfilling the Promise: Biology Education in the Nation's Schools*, 1990. Copyright © 1990 by the National Academy of Sciences.

Acknowledgments

Design Development
Foca, Inc.
New York, NY

Contributing Writers
Gary J. Brusca, Ph.D.
Department of Biological Sciences
Humboldt State University
Arcata, CA

Dorothy L. Domm
Fort Worth, TX

Salvatore Tocci
Science Department
East Hampton High School
East Hampton, NY

Regina Abernathy
Publisher, *Multicultural Education Review*
Karamu House
Cleveland, OH

Jo Arnett
San Jose, CA

Tracey Cohen
Science Writer
Highland Park, NJ

Rhoda Elovitz
Writer
Austin, TX

Jacquelyn Jarzem, Ph.D.
Instructor
Austin Community College
Austin, TX

Thomas R. Koballa, Jr, Ph.D.
Science Education Department
University of Georgia
Athens, GA

Glenn Leto
Biology Teacher
Barrington High School
Barrington, IL

Ronnee Yashon
Biology Teacher
Lakeview High School
Chicago, IL

Christina Zikos
WARD'S Natural Science Establishment, Inc.
Rochester, NY

Lab Reviewers
Don Chmielowiec
Alex Molinich
George Nassis
Laboratory Investigations
WARD'S Natural Science Establishment, Inc.
Rochester, NY

Kenneth Rainis
Safety
WARD'S Natural Science Establishment, Inc.
Rochester, NY

Reviewers
Hugh C. Allen
Biology Teacher
American High School
Dade County, FL

David Anderson
Biology Teacher
Miami Northwestern High School
Miami, FL

Carol Baskin, Ph.D.
School of Biological Science
University of Kentucky
Lexington, KY

Lowell Bethel, Ph.D.
Science Education Center
University of Texas
Austin, TX

Thomas G. Betz, M.D., M.P.H.
Austin/Travis County Health Department
Austin, TX

Mark W. Bierner, Ph.D.
Department of Biology
Southwest Texas State University
San Marcos, TX

Barry Bogin, Ph.D.
Professor of Anthropology
University of Michigan
Dearborn, MI

Beverly J. Bradley, Ph.D., R.N., C.H.E.S.
Supervisor of School Health Programs
San Francisco Unified School District
San Francisco, CA

Steve Bratteng
Austin, TX

Linda Butler, Ph.D.
Lecturer
University of Texas
Austin, TX

Diane Calabrese, Ph.D.
PAPILLONS: Diversified Endeavors
Columbia, MO

Thomas J. Conley
Biology Teacher/Science Chairman
Parkway West High School
Ballwin, MO

Mary Coyne, Ph.D.
Professor, Department of Biological Sciences
Wellesley College
Wellesley, MA

Joe Crim, Ph.D.
Professor of Zoology
University of Georgia
Athens, GA

John Delaney
Biology Teacher
Harlandale High School
San Antonio, TX

Andrew A. Dewees, Ph.D.
Department of Biological Sciences
Sam Houston State University
Huntsville, TX

Claudia Dickerson
Biology Teacher
Oliver Wendell Holmes High School
San Antonio, TX

Robert Dimmick
Biology Teacher
Harlandale High School
San Antonio, TX

Tommy C. Douglas, Ph.D.
Graduate School of
Biomedical Sciences
University of Texas Health
Science Center at Houston
Houston, TX

Marvin Druger, Ph.D.
Departments of Biology
and Science Teaching
Syracuse University
Syracuse, NY

John Edwards, Ph.D.
Department of Zoology
University of Washington
Seattle, WA

Richard Farrar
Lead Science Teacher
Northern High School
Accident, MD

Carl Gans, Ph.D.
Department of Biology
University of Michigan
Ann Arbor, MI

Cynthia Hanes R.D., L.D.
Austin Regional Clinic
Austin, TX

Denny O. Harris, Ph.D.
School of Biological
Sciences
University of Kentucky
Lexington, KY

Joseph M. Hepfinger
Biology Teacher
Webster Groves
High School
Webster Groves, MO

Agnes Higbie
Biology Teacher
Fulton Junior High School
Indianapolis, IN

Vivian Huang
Biology Teacher
Skyline High School
Oakland, CA

Andrea Huvard, Ph.D.
Department of Biology
California Lutheran
University
Thousand Oaks, CA

Duane E. Jeffery
Department of Zoology
Brigham Young University
Provo, UT

Deborah L. Jensen
Biology Teacher/depart-
ment Chairperson
Oak Ridge High School
Conroe, TX

Janet Jones
Biology Teacher
Sullivan High School
Chicago, IL

Gayle Karriker
Biology Teacher/Science
Chairperson
Olympic High School
Charlotte, NC

Karen Martin, Ph.D.
Natural Sciences Division
Pepperdine University
Malibu, CA

Emily Mims
Biology Teacher
Edison High School
San Antonio, TX

Patricia Mokry
Biology Teacher
Westlake High School
Austin, TX

Ted Molskness, Ph.D.
Oregon Regional Primate
Research Center
Beaverton, OR

Jane Moncure
Biology Teacher
East Mecklenburg High
Charlotte, NC

Betty K. Moore
Biology Teacher
East Mecklenburg High
Charlotte, NC

David Moury, Ph.D.
Department of EPO Biology
University of Colorado
Boulder, CO

Dorothy Ngongang
Biology Teacher
Providence Senior High
School
Charlotte, NC

Martin Nickels, Ph.D.
Anthropology Program
Illinois State University
Normal, IL

Celia Rainwater
Biology/Anatomy Teacher
Clark High School
San Antonio, TX

Peter Raven, Ph.D.
Director
Missouri Botanical Garden
St. Louis, MO

Patricia Recker
Biology Teacher
John Jay High School
San Antonio, TX

Linda Fox Simmons
Nutrition & Health Training
Alternatives
Austin, TX

Marian Smith, Ph.D.
Biology Department
Southern Illinois University
Edwardsville, IL

Scott Spear, M.D.
University of Texas
Student Health Center
Austin, TX

George F. Spiegel, Jr.
Biology Instructor
Austin Community College
Austin, TX

Susan Talkmitt
Biology Teacher
Monterey High School
Lubbock, TX

William Thwaites, Ph.D.
Biology Department
San Diego State University
San Diego, CA

David Zeigler, Ph.D.
Pembroke State University
Pembroke, NC

Page 561 contains infor-
mation on the biology
of this salamander and
other amphibians.

The leg bone's
connected to the
hip bone? Check it
out on page 624.

Biology
Visualizing Life

Table of Contents

Unit 1 Study of Life

Section 1.1
Science literacy is tied to personal decision making.

Energy is a unifying scientific theme. For others, see page 16.

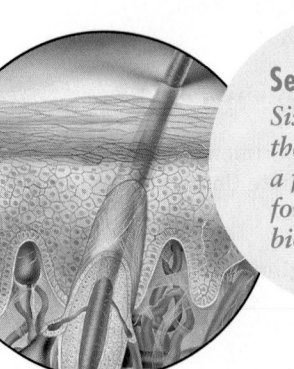

Section 1.3
Six essential themes provide a framework for learning biology.

For the role of cells in living things, including your skin, see page 46.

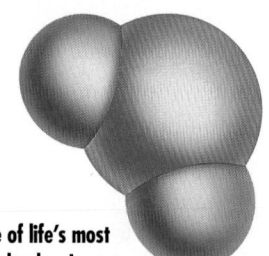

Water is one of life's most important molecules. Learn about it on page 48.

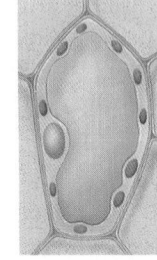

How will the loss of water from a cell affect this sunflower? See page 75.

What do enzymes have to do with the color of this cat's fur? For an answer, see page 93.

Unit 2 Continuity of Life

To learn about Mendel's studies of garden peas, see page 119.

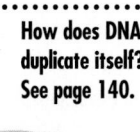

How does DNA duplicate itself? See page 140.

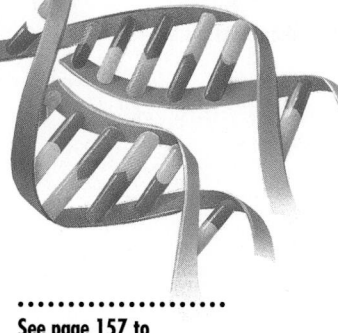

See page 157 to learn about how the *Xenopus* frog is used in genetic engineering studies.

Chapter 9 highlights the story of evolution and natural selection.

How does natural selection result in the various forms of life on Earth? See page 186.

Table of Contents

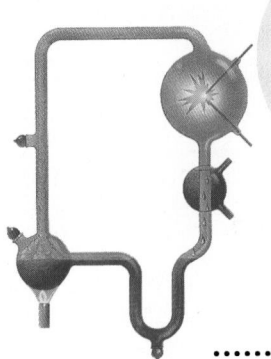

To learn about the role of primates in human evolution, see page 221.

Unit 3
Environmental issues are presented early in the course.

Unit 3 The Environment

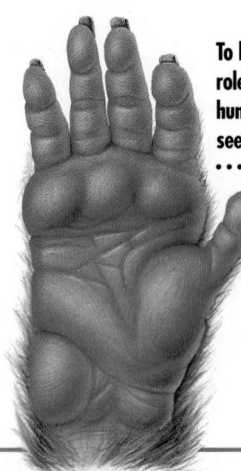

How do organisms interact in an ecosystem? See page 248.

See page 275 to read about how these mussels were affected by the removal of their major predator.

Chapter 13
stresses the importance of ecology.

Chapter 14
gives practical strategies for solving environmental problems.

To learn how you can help an ecosystem, see page 307.

Unit 4 Diversity of Life

See page 325 to learn about how living things are classified.

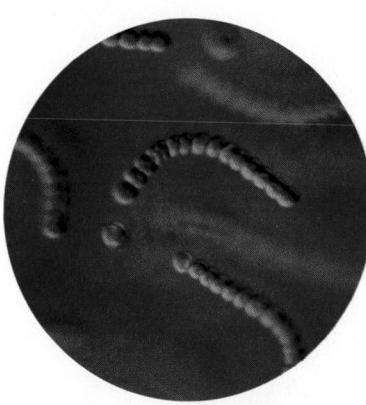

To learn about organisms such as these blue-green bacteria, see page 342.

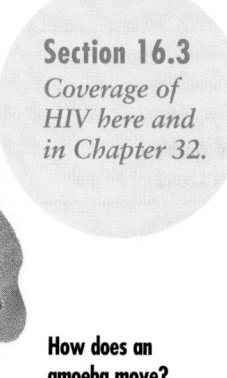

Section 16.3
Coverage of HIV here and in Chapter 32.

How does an amoeba move? See page 368.

Study page 385 to learn about the biology of fungi such as this mushroom.

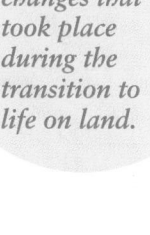

Chapter 18
covers the evolutionary changes that took place during the transition to life on land.

Section 19.2
Classification is presented on the basis of evolutionary relationships.

How do water and nutrients move through plants? See page 410.

Table of Contents

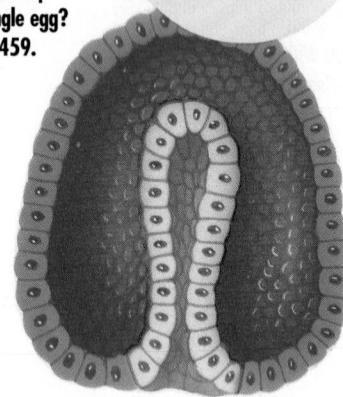

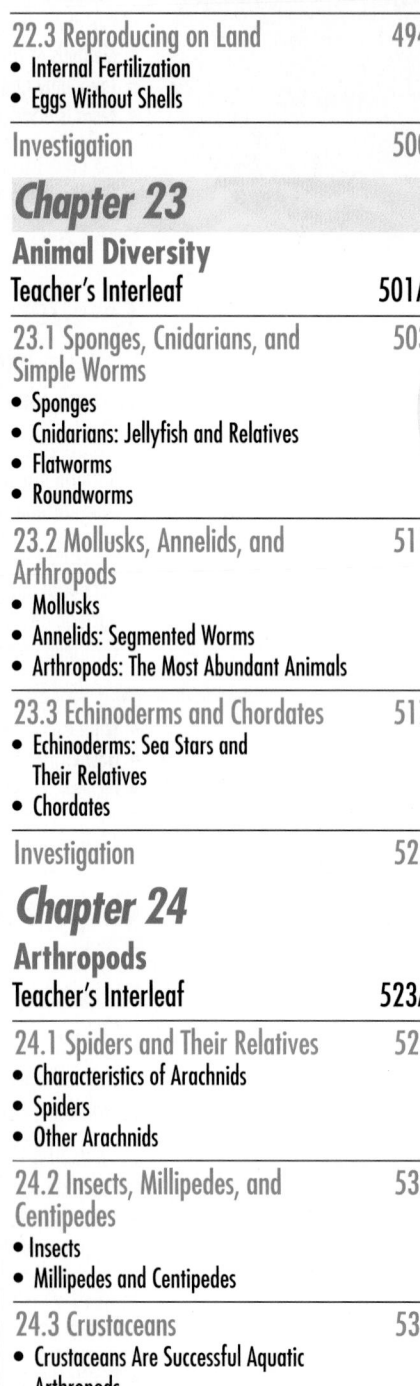

See page 525 to read about this tarantula.

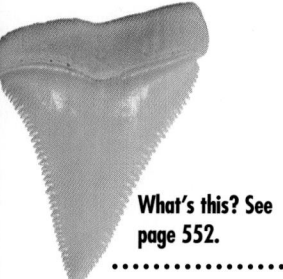

What's this? See page 552.

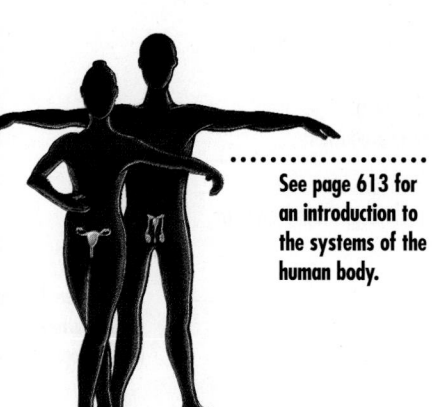

To learn about the biology of this lizard, see page 573.

Unit 6 Human Life

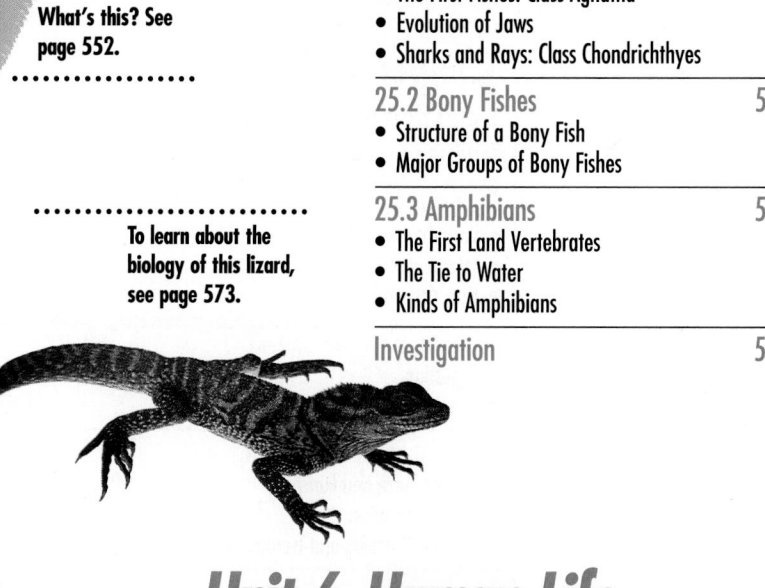

See page 613 for an introduction to the systems of the human body.

Chapter 28 *covers the fundamentals students need to understand drug addiction in Chapter 30.*

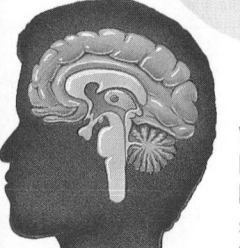

What role does the brain play in the body's hormone system. See page 671.

Table of Contents

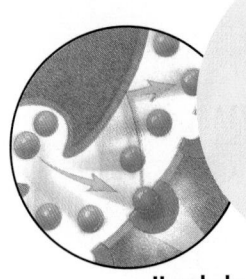
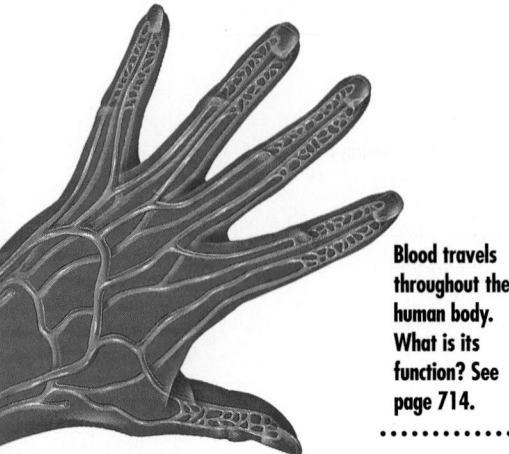
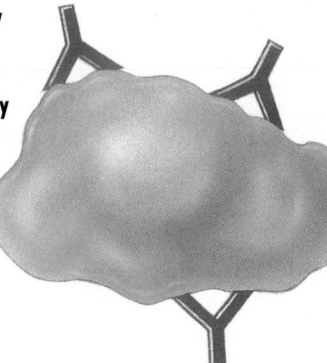
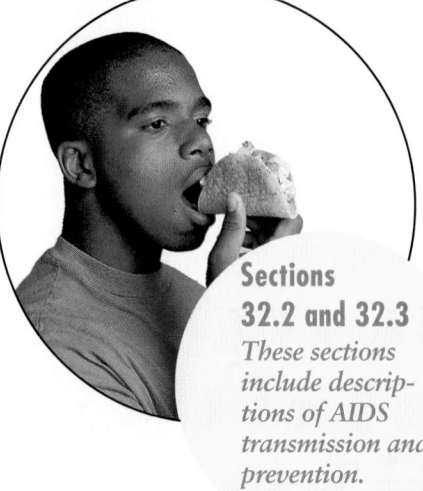

Features

Science in Action

Science in Action
These features provide accounts of real people in their professions, or advancements in biological research.

To learn what inspired Diana Punales-Morejon to become a genetic counselor see page 146.

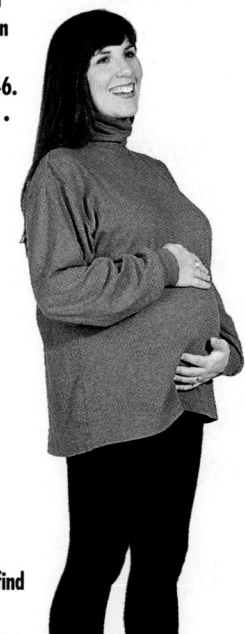

On page 792, you can find out how this woman is helping to ensure that her baby will be healthy.

These high school students are helping preserve the environment. To find out what you can do, see page 304.

Table of Contents

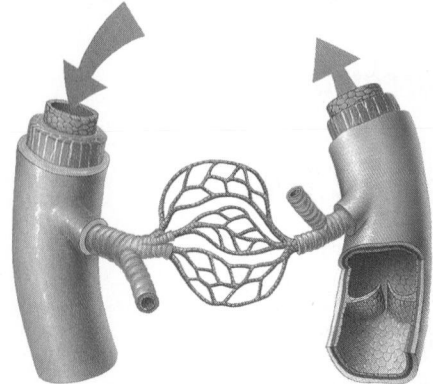

Journeys

Journeys
Students get a close-up view of an organism or see the visual summary of an evolutionary story.

See page 371 for a close-up look at this aquatic organism.

On page 715, you can study the details of how blood flows from arteries to veins.

Get up-close and personal with a jellyfish on page 506.

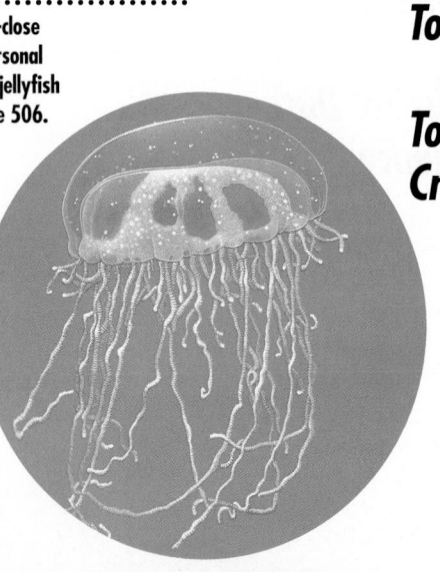

Page 587 explores the biology of the cedar waxwing.

12T

Science, Technology, and Society

Discoveries in Science

Breaking new ground

in curriculum and instruction

"In designing a course, we must identify the central concepts and principles that every high school student should know and pare from the curriculum everything that does not explicate and illuminate these relatively few concepts."

Fulfilling the Promise—Biology Education in the Nation's Schools

Why I Wrote This Book

No one who teaches biology today can fail to be aware of the current clamor for better ways to teach science in our classrooms. As science becomes more important to our everyday lives, we see national studies repeatedly indicating that today's students are not learning the science they will need to know to live in the twenty-first century. This concern of educators has led to a flurry of proposals for improving how science is taught, from Project 2061's suggestion that the different sciences be integrated into a single curriculum, to the National Research Council's suggestion that we get away from encyclopedic detail and focus more on teaching concepts. It is in response to these concerns that this book was written. In writing *Biology: Visualizing Life*, I set out to accomplish five things.

George Johnson

❶ Get the Science Right

Working as a research scientist for some 24 years with over 50 research papers to my credit has given me the background and experience needed to get the science right. The experience of writing two successful college biology textbooks has given me the experience to write a high school text that is authoritative, accurate, and up-to-date.

❷ Focus on Concepts

The most common complaint about high school texts is that they are too long and are filled to the brim with detail that is too much for any student. What you don't hear as often, but can easily demonstrate to yourself by opening any of the texts, is that these texts are hard to learn from. Just turn, for example, to the chapter on photosynthesis. It's not just that they present too much detail, the basic concepts that the details are intended to elaborate are not themselves presented clearly.

In writing *Biology: Visualizing Life*, I have stressed at every stage a clear presentation of ideas and concepts, rather than a listing of detailed information. In every chapter, I try to present how things work and why things happen the way they do, rather than naming parts and stressing definitions.

Focus on Concepts
A focus on themes is a fundamental part of a conceptual approach to biology. The Themes Trace shows how the six themes, introduced in Chapter 1, become focal points for students as they review their knowledge.

Themes Trace

Theme	Chapter 1	2	3	4	5	6	7	8	9	10	11	12	13	14	15	16	17	18	19	20	21	22	23	24	25	26	27	28	29	30	31	32	33	34
Interacting Systems				x	x			x		x	x	x		x		x	x	x		x			x						x	x		x	x	x
Energy and Life		x	x	x										x	x	x	x	x	x	x		x			x	x	x							
Evolution				x	x	x	x	x		x	x		x	x	x	x	x	x	x	x	x	x	x	x	x	x	x	x			x	x		x
Patterns of Change		x			x	x	x		x	x	x	x	x	x	x	x			x	x	x			x	x	x			x	x	x	x		x
Scale and Structure				x		x		x		x	x	x		x		x	x	x		x		x		x	x	x	x		x	x	x	x	x	x
Stability				x	x	x	x	x		x				x								x					x	x	x	x	x		x	

❸ Write Well

In teaching biology for more than 20 years, I have come to the unpleasant but unavoidable conclusion that today's students can't read as well as those in the past. Whether this is the result of substituting television for reading as entertainment during childhood, or for some other reason, the fact is that students today neither read nor write with facility. It is the job of the text to bridge this gap. What I have done is to teach with analogies rather than detail whenever possible. When the student confronts the complex problem of how a repressor protein prevents RNA polymerase from reading a gene, the text points out that the locations of the DNA overlap where these two proteins attach, and that "two people can't sit in the same chair at the same time." Simple analogies are very effective teaching tools.

❹ Relate to the Student

Biology: Visualizing Life was written in the firm belief that you cannot teach a student who is not paying attention. Most of my students have been college freshmen, and from all those years of fun and frustration, one lesson has emerged clearly: Students will not learn what I am trying to teach them if they see no relationship between their lives and what is going on in my classroom. This simple fact has colored every aspect of *Biology: Visualizing Life*. There is no concept or idea presented without emphasizing why it is important to students.

❺ Teach Visually

Today's students, raised on video screens and computers have had much more exposure to learning via visual presentations than ever before. We now know that many students learn most easily through visual presentations, and that is how we must teach them. Illustrations do not supplement the text, but rather are active partners with it. This lively presentation not only captivates student interest and makes the task of learning seem less intimidating, it is actually a better way to teach.

Relate to the Student
Students are shown throughout the text actively engaged in everyday biology.

Teach Visually
Illustrations and photographs are an integral part of the delivery of scientific information, not simply decorative elements. Visual learning is enhanced, and students become eager to master the content.

Interacting Systems

How would an ecosystem be affected if its nitrogen-fixing bacteria were destroyed?

Nitrogen-fixing bacteria enrich the soil

Because nitrogen is a component of proteins, plants cannot photosynthesize, grow, or reproduce without it. Much of the nitrogen available for plants is produced by nitrogen-fixing bacteria. Nitrogen-fixing bacteria transform atmospheric nitrogen, which cannot be absorbed by plants, into ammonia, a nitrogen compound that plants can absorb. No other organisms have this ability.

Nitrogen-fixing bacteria are found in the soil, in aquatic ecosystems, and within the roots of some plant species. The legumes, plants including the peas and beans, contain nitrogen-fixing bacteria in swellings on their roots. These bacteria enable legumes to grow in nitrogen-poor soils where other plants cannot.

Figure 16.6
The young woman below is eating yogurt, while the young man on the right is eating olives. Yogurt and olives are made when the bacteria indicated are added during the food manufacturing process.

Bacteria are used to manufacture food and drugs

Like the young woman and man in **Figure 16.6**, when you eat yogurt or olives, you are eating foods that are the product of bacterial decomposition. Humans have learned that decomposition is occasionally beneficial because it adds flavor to food. For example, bacteria convert cabbage and cucumbers into tangy sauerkraut and pickles.

Modern technology is taking advantage of the enormous genetic diversity among the bacteria. Using genetic engineering technology, biologists can now "reprogram" bacteria to manufacture any protein for which a gene has been isolated. For example, bacteria now produce most of the insulin needed by diabetics in the United States. Before genetic engineering, insulin had to be isolated from the pancreases of animals killed in slaughterhouses. Other drugs and products produced by genetically engineered bacteria are described in Chapter 8.

Streptococcus thermophilus and *Lactobacillus bulgaricus*

Leuconostoc mesenteroide and *Lactobacillus plantarum*

Yogurt

Bacteria and Viruses **347**

As shown in Figure 26.15, birds have two major types of feathers. Most of a bird's feathers are **contour feathers**.

These feathers cover the body of the bird and give the wings and tail their shape. Contour feathers also insulate against heat loss. Fine **down feathers** growing underneath or among the contour feathers are specialized for insulation. The down feathers of eider ducks are used in sleeping bags because they are a lightweight, effective insulation.

Figure 26.15
a Contour feathers and down feathers have different functions. Down feathers serve primarily as insulation. Contour feathers provide insulation, steering, balance, and coloration.

Bird skeletons are lightweight
The skeleton of a bird is adapted for flight. The bones are thin and hollow. Many are reinforced by internal struts, like the wings of an airplane. The sternum is large and has a keel, providing solid anchorage for some of the large flight muscles.

Vane

Shaft

b The individual filaments, or barbs, of a contour feather are linked together by hooked barbules to form a continuous surface, thereby decreasing wind resistance.

Barbules

Barbs

Quill

Bird wings are modified forelimbs. The bones of the forelimbs fully support and move the wings. The finger bones are very tiny, but the arm and hand bones are long, providing strength and enabling complex movements of the wings. As you learned in Chapter 9, the bones of a bird's wing are homologous to the bones of your arm and hand.

Birds are endothermic and active
Flight is an energy-demanding activity. Birds, like mammals, are endothermic. Endothermy enables birds to meet the energetic demands of flight. In addition, birds have a four-chambered heart with separate circulatory loops to the lungs and to the body. Therefore, oxygen-rich blood is rapidly delivered to tissues where it is needed, without mixing with deoxygenated blood. Bird respiration is very efficient because their system of air sacs permits air to flow in only one direction through the lungs, as explained in Chapter 22. You can read more about birds in the *Tour of a Bird* on page 587.

c Most feathers are shed every year during molting. Birds clean their feathers by bathing in either water or dust to get rid of dirt and parasites.

d Down feathers trap air, helping maintain the bird's constant, high body temperature.

Reptiles, Birds, and Mammals **585**

What This Text Is Like

These five goals, taken together, describe not so much what I did in writing *Biology: Visualizing Life*, but rather describe how I set out to do it, the tools I used as I sat down to write. What sort of a book is *Biology: Visualizing Life?*

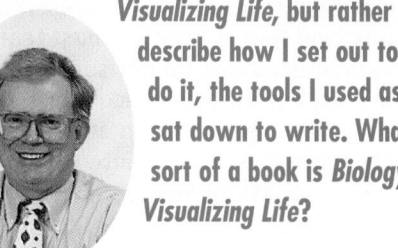

Table of Contents

My first act was to attack the traditional table of contents, a straightjacket that forces the teaching of biology into overwhelming detail and away from the core concepts that we should be teaching students. Traditionally, biology is taught from a phylogenetic point of view, starting with the little stuff. After learning a little about cells, students are taken on a long tour of the biological kingdoms, passing through detailed chapters on the different kinds of animals and other organisms. It is not by accident that these books are overloaded with detail—detail is the very essence of what the first year course has become.

A Conceptual Approach

In *Biology: Visualizing Life*, we abandon traditional approaches and focus instead on key concepts—on those ideas that explain biology—that tell us why and how rather than what. You can see the result of this conceptual approach in the early chapters of the text. Instead of starting the text with a student-terrifying chapter on chemistry, *Biology: Visualizing Life* integrates the chemistry into the flow of the text, teaching only as much as students need to know in order to understand what they are reading.

Coverage of the Animal Kingdom

The focus on concepts can be clearly seen in Unit 5, which is devoted to the animal kingdom. Instead of a deadly progression of chapters, devoted to each phylum, a sure way to kill the joy of biology, the unit opens with Chapter 21, which describes the evolutionary journey that produced animals. Special features called Explorations introduce the animals in terms of a series of key adaptions. This organization serves to highlight both the progressive nature of animal evolution and the important role of certain key elements of body architecture. Each advance is coupled to a pedagogical overview of the phylum that represents this advance— Exploration of a Snail, etc.—so that the student may put the information into perspective. The intent is to minimize the presentation of detail, focusing instead on the process of evolution and recounting how it has produced the diversity we see today. If we are to succeed in substituting concepts and principles for encyclopedic detail, then

Journeys ▪ Journeys ▪ Journeys ▪ Journeys ▪ Journeys ▪

Tour of a Lizard

The largest group of reptiles today is the lizards. This gecko is a very fast runner. Most living lizards are small; few are bigger than a squirrel.

All reptiles have a tough, dry, scaly skin. The outer layer of scales is shed periodically.

Most lizards have external ears; snakes do not.

Most lizards have four legs, but some are legless. These lizards are not snakes because they lack the "floating jaws" that allow snakes to swallow large prey.

A gecko has adhesive toe pads that enable it to walk upside down and on vertical surfaces.

Many lizards drop their tails when seized by a predator; the tail regrows in about seven weeks.

Modern lizards evolved late in the age of dinosaurs.

Most lizards lay leathery, shelled eggs that do not dry out when exposed to air.

576 Chapter 26

▲ *A Conceptual Approach*
In covering the phyla, representative organisms are used in illustrating adaptive characteristics.

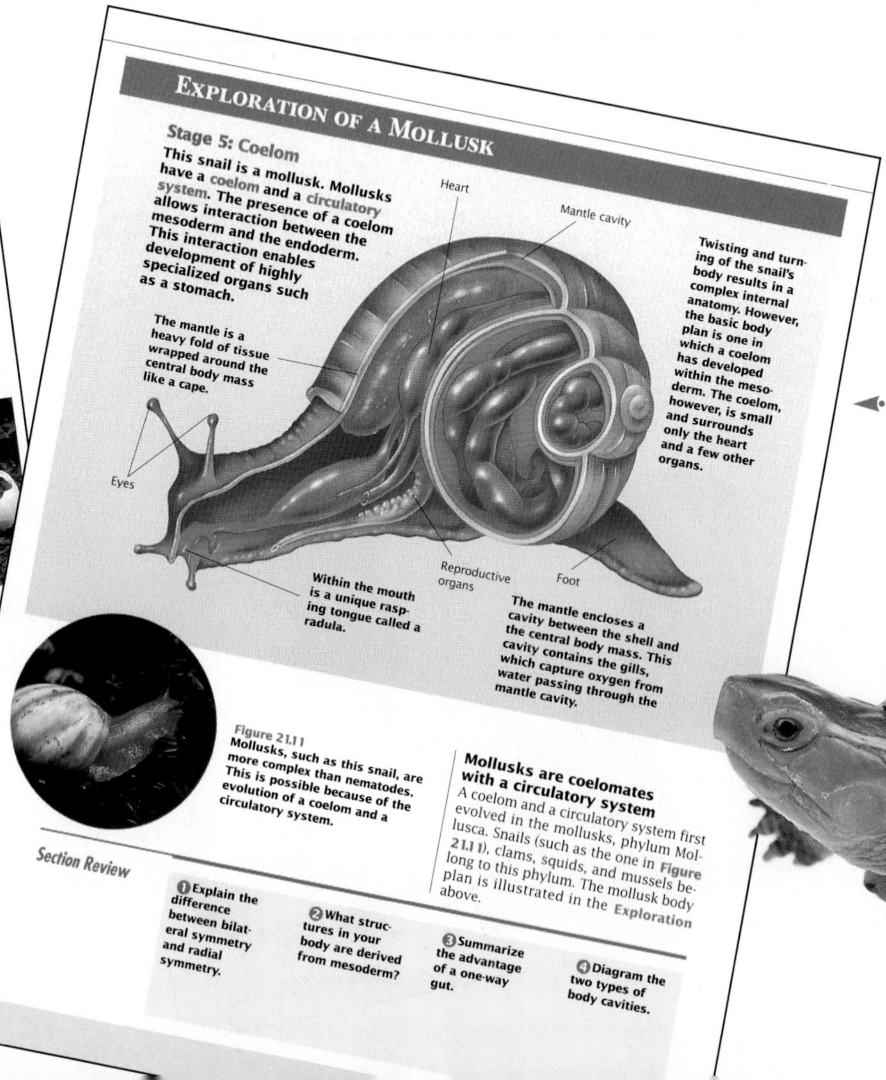

EXPLORATION OF A MOLLUSK

Stage 5: Coelom

This snail is a mollusk. Mollusks have a coelom and a circulatory system. The presence of a coelom allows interaction between the mesoderm and the endoderm. This interaction enables development of highly specialized organs such as a stomach.

The mantle is a heavy fold of tissue wrapped around the central body mass like a cape.

Heart

Mantle cavity

Twisting and turning of the snail's body results in a complex internal anatomy. However, the basic body plan is one in which a coelom has developed within the mesoderm. The coelom, however, is small and surrounds only the heart and a few other organs.

Eyes

Within the mouth is a unique rasping tongue called a radula.

Reproductive organs

Foot

The mantle encloses a cavity between the shell and the central body mass. This cavity contains the gills, which capture oxygen from water passing through the mantle cavity.

Figure 21.11
Mollusks, such as this snail, are more complex than nematodes. This is possible because of the evolution of a coelom and a circulatory system.

Mollusks are coelomates with a circulatory system
A coelom and a circulatory system first evolved in the mollusks, phylum Mollusca. Snails (such as the one in **Figure 21.11**), clams, squids, and mussels belong to this phylum. The mollusk body plan is illustrated in the **Exploration** above.

Section Review

❶ Explain the difference between bilateral symmetry and radial symmetry.

❷ What structures in your body are derived from mesoderm?

❸ Summarize the advantage of a one-way gut.

❹ Diagram the two types of body cavities.

this principle—how evolution explains the animals we see—is surely one of the most important to convey.

Chapter 22 continues the evolutionary journey, considering key adaptations involved in the invasion of the land from the sea. In each case, the solution says a great deal about the relation between structure and function, and the possibility of different successful solutions to the same adaptive problem. The focus here is on how adaptation has driven progressive evolutionary change, a lesson that permits students to see themselves as part of a larger evolutionary picture. If students can come to view themselves as the product of a long evolutionary history, they will gain a sense of themselves as being part of life's tapestry, not apart from it. This lesson is far more important than memorizing the structural details of animals.

Coverage of Environmental Issues

Another clear example of the focus on concepts is seen in the organization of Unit 3, which is devoted to ecology. First, note where it is—up front, and not at the tail end of the book! Ecology is often skipped in high school courses for the simple reason that it comes at the end of the course. In today's world, with the need to understand environmental issues, that omission is a tragedy. Besides, it is difficult to teach animal and plant diversity properly without ecology. What is diversity, but the product of evolution and ecology? In *Biology: Visualizing Life*, ecology is presented early in the sequence, where it makes the most sense.

Now take a look at the three chapters of Unit 3. They are not the traditional "levels of organization" chapters (populations, ecosystems, biomes). Instead, Chapter 12 focuses immediately on ecosystems, the basic unit of the environment. Students learn how ecosystems work—how they cycle nutrients and process energy, and how climate shapes who and what lives where. Chapter 13 focuses on the key issue in environmental science today—how ecosystems respond to disruption. The biodiversity and complex relationships built over centuries of coevolution tend to make ecosystems more stable, while humans today are disrupting natural ecosystems. The principles students learn here will guide their thinking about ecological issues for years to come, a very important tool in our rapidly changing world. Chapter 14 provides a careful and honest evaluation of where we are now, and what is being done to save our environment. The approach is positive and pragmatic. There are few issues that students will encounter in biology that deserve more careful treatment.

Coverage of the Animal Kingdom
The overview of animals includes significant milestones in the evolution of the animal body.

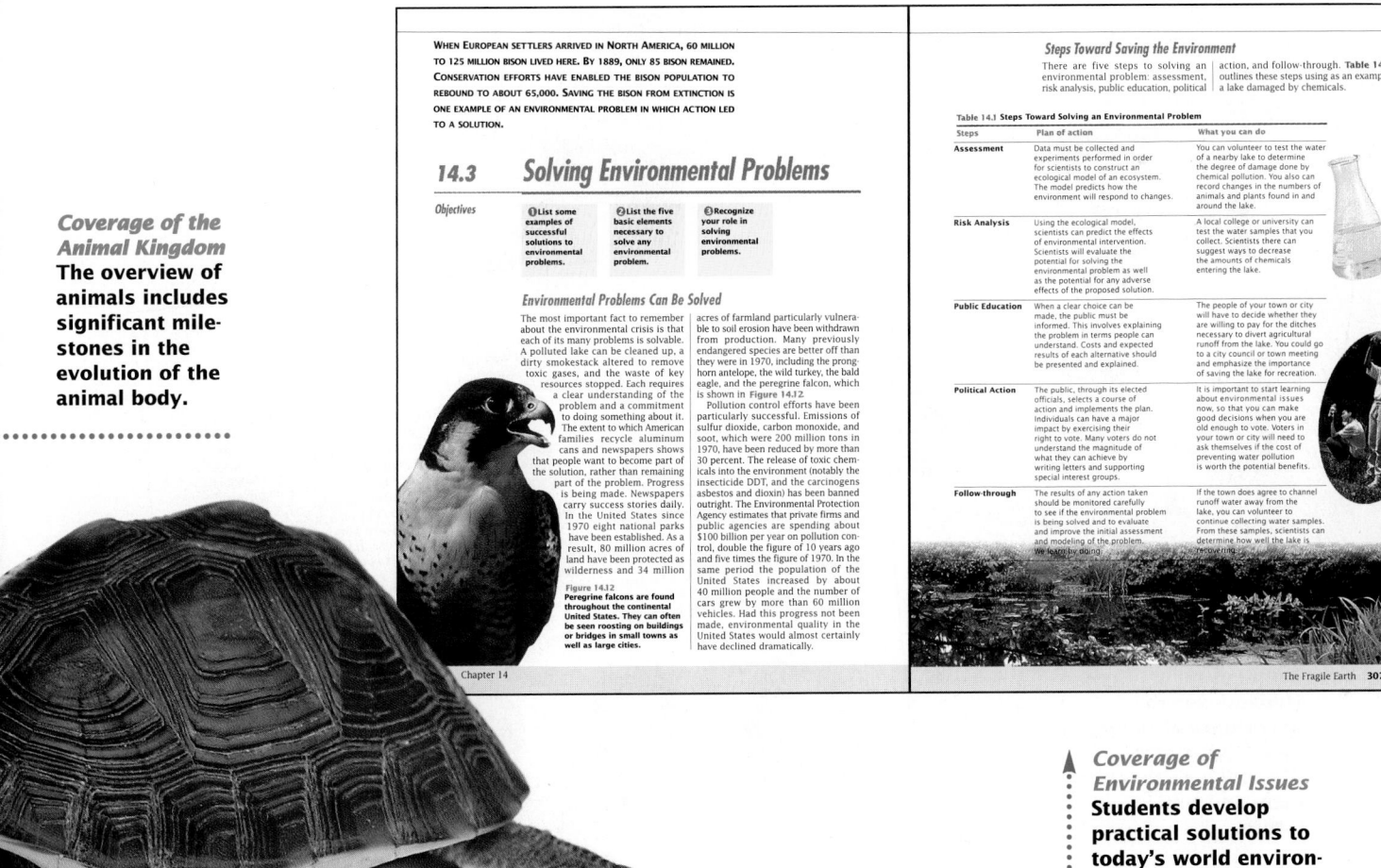

Coverage of Environmental Issues
Students develop practical solutions to today's world environmental problems using a systematic method of analysis.

17T

What Students Need to Know

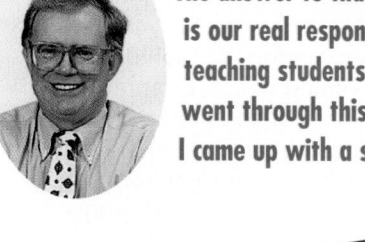

At some point, anyone who writes a text needs to sit down and ask a simple question without regard to curricula or state requirements: What things do students really need to know? The answer to that question is our real responsibility in teaching students. When I went through this exercise, I came up with a short but powerful list—things that impact students in ways that change their lives. The list included teaching students why smoking cigarettes will kill them, how to avoid contracting AIDS, why drugs are dangerous, and how their bodies work. You will find all these topics in *Biology: Visualizing Life*, and will find many others that directly impact the lives of students. This approach isn't taken for entertainment, but because I regard it as my responsibility to tell students about the science behind these issues, for this knowledge will affect their futures.

Coverage of Adolescent Issues

To get some sense of what I mean, look at Chapter 30, which is devoted to drugs. Almost every high school text has such a chapter, but this one is different in that it does not devote the bulk of its space to a boring litany of drug types or preaching about the evils of drug abuse. Instead, it focuses on explaining just how drugs work to create addicts. Most high school texts devote one paragraph to this subject. Yet addiction is a major problem in this country. In my teaching I have found that if I explain the mechanism of addiction to students so that they see it is a simple cause-and-effect process, then I don't need to preach to them, because the danger is so obvious. The mechanism of addiction is not difficult to explain. The basic idea of lowering the numbers of drug receptors within nerve synapses as a compensation for increased levels of neurotransmitters is not difficult to

Biology and Medicine

Stop for a moment and think about the headlines you read in the newspaper this week, the news stories you heard on the radio or TV. Mixed in with bad news and politics is news that modern science is exploring new ways to cure genetically inherited disorders like cystic fibrosis and muscular dystrophy. The fight against infectious diseases like AIDS and malaria, and ways to protect the body from cancer and heart disease, affects everyone, as shown in Figure 1.2. The most direct impact of biology on our lives is in medicine, where scientific advances are improving health and health care every day.

Figure 1.2
AIDS is one of the most devastating diseases of the 1990s. A knowledge of biology helps us to understand what causes AIDS and how the virus that causes it is spread.

a The virus that causes AIDS is small but deadly. Because it may take years for the symptoms of infection to appear, people can spread the virus unknowingly.

b Anyone can get AIDS. Knowing how the virus is transmitted can be the first step in controlling its spread.

6 Chapter 1

Biological knowledge is used in stopping infectious diseases

At the beginning of this century, the infectious diseases flu, tuberculosis, and pneumonia were the top three causes of death in the United States. The flu epidemic of 1918 killed 22 million people worldwide in 18 months! Now, thanks to intense biological research, deaths from these three diseases are far less common. Because of the discovery of antibiotics like penicillin and the development of vaccines that prevent infection, these diseases will probably not be the major causes of death in the United States in coming years.

The battle against disease is not over. More than 1 million people are likely to die of malaria this year alone. Spread by mosquitoes, this disease is prevalent in tropic regions—more than 250 million people suffer from malaria at any one time. Almost every child under the age of five who contracts malaria dies. The organism that causes malaria has a very complex life cycle, and is therefore difficult to eliminate. Using modern genetic engineering techniques, scientists are trying to design a vaccine that will attack at a critical stage of the organism's life cycle. Also, new ways to control the large mosquito population are being researched.

New strains of bacteria causing tuberculosis, the fatal lung disease, have recently arisen that are resistant to today's antibiotics.

Other scientists are working to find a cure for AIDS, a fatal disease caused by a virus that destroys the body's ability to defend itself from infections. Because the virus changes so quickly, normal vaccines don't work. No one yet knows what the solutions to these problems will be, but many approaches are being explored.

c This child has AIDS, which she acquired before her birth. Scientists predict that 40 million people world-wide will be infected with the virus that causes AIDS by the year 2000.

Biological knowledge is used in curing genetic disorders

Among the most disheartening medical problems are fatal disorders that result from inherited defective genes. Two common fatal human genetic disorders are cystic fibrosis and muscular dystrophy. Cystic fibrosis kills 1 in 1,800 Caucasian children. Their lungs are clogged with mucus because of a defect in a single gene. Muscular dystrophy kills about 1 in 10,000 humans. Their muscles waste away because of a defect in another gene.

Until recently, such genetic disorders were almost always fatal, and no cure was known. In 1990, biologists trying a new approach used genetic engineering techniques to transfer copies of a normally functioning gene from a healthy individual into a patient with defective copies of that gene. Gene transfer therapy offers the first hope that cures for previously fatal genetic disorders will be found. Progress with the new therapy is reported almost daily. A major effort is now underway to catalog every gene in the human body, a program that may soon open the avenue to curing many other genetic disorders.

Teen Focus ••••••••••••••••••••
Students see the importance of biology by studying topics that will directly affect them now and in the future.

Biology and You

Much of what you will learn in this course will give you information you will need in making critical personal decisions like the ones contemplated in **Figure 1.4**.

Should I smoke? Biologists now know how cigarette smoking causes lung cancer—chemicals in the smoke enter the cells of your lungs and damage them, and cancer results. Most people who die of lung cancer today are smokers. Smoking cigarettes is really a form of suicide. You'll learn a great deal about the dangers of smoking in Chapter 30.

How can I avoid developing heart disease? Biologists have learned that diet and exercise have a major influence over whether you are likely to have a heart attack or stroke. You will learn about what you can do to prevent heart disease in Chapter 31.

What are the risks in taking drugs? Most mind-altering drugs are addictive and thus very harmful. This includes the nicotine in cigarettes and the alcohol in wine, beer, and spirits. Biologists have learned much about the physical basis of drug dependency, and about the far-reaching effects drugs can have. You will learn about drug abuse and safe drug use in Chapter 30.

How does our species reproduce? The biology of reproduction is now well known, and because you may someday plan to raise a family of your own, you need to gain a clear understanding of these critical processes. You will learn about human reproduction in Chapter 34.

Figure 1.4
Asking questions is the first step to finding answers. A knowledge of biology will help answer many of the questions that these young adults have.

"Many of my friends smoke and they say smoking is all right as long as you quit when you're older. I wonder if that is true?"

"My uncle died three years ago at 45 of a heart attack. Will I have heart problems too?"

"I see other kids using drugs. They say they can stop whenever they want to. I wonder if that's true?"

"My sister is going to have a baby. What's going to happen to her?"

Section Review

❶ List three

teach or understand. Few students can read this chapter and contemplate taking drugs, and that decision is based on their own reasoning process—not on the acceptance of my opinions on the subject. The same approach is taken in discussing cigarette smoking and cancer. Students don't encounter preaching, but rather they read a clear description of how chemicals in cigarette smoke cause mutations at random in DNA. They learn that when key growth regulating genes are hit by these mutations, cancer results. Smoking is like going into a dark room and letting someone shoot a gun at you—you might not be hit at first, but how many times would you let them shoot? Every cigarette is one more bullet. Because the presentation of how cancer works is very concrete, grounded in basic concepts that students can grasp, not one fails to get the point.

A New Beginning

I took time from a busy teaching schedule to write this book, because I felt it was important to make a start at improving how we teach biology. Anyone who looks at this book will think of things I could have done differently. I hope you will tell me about them by writing to me care of Holt, Rinehart and Winston. No single person can get the job done alone, and I certainly appreciate any help that experience and interest can provide. Thank you in advance.

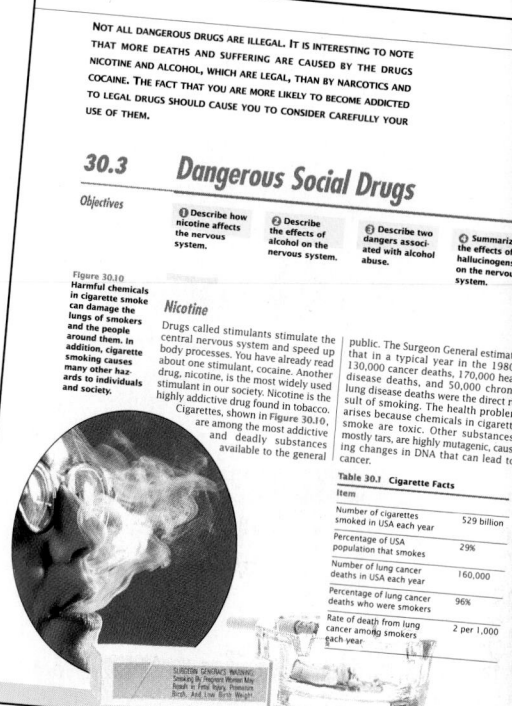

Smoking
The hazards of smoking are covered in relation to mutations of DNA.

Drugs
The biological basis of addiction is the core of the drug chapter.

Breaking new ground

to develop the link between process and communication skills

"They must be taught by a process that engages all students in examining why they believe what they believe. That requires building slowly, with ample time for discussion with peers and with the teacher. In science, it also means observation and experimentation, not as an exercise in following recipes, but to confront the essence of the material."

Fulfilling the Promise—Biology Education in the Nation's Schools

Biology: Visualizing Life gives you the tools you need to teach scientific processes and bridge the knowledge gap between lecture and the lab. Using Vee Diagram Report Sheets, you can be assured that your time spent in lab is effective in reinforcing fundamental science skills.

Using Gowin's Vee to Link Lecture to the Laboratory

Gowin's Vee is a scaffold device that illustrates why knowledge is the end result of the process of inquiry. You can use the Gowin's Vee to make the connection between lecture and lab, guide science fair research, and make oral reports and lab reports easier to understand. It also helps students decipher professional research journals. The diagram at right shows the common elements of the Vee.

Lab Reporting

You can use the Vee with all Investigations in the text. Vee report sheets are included as an option for all Laboratory Investigations in the Laboratory Manual for *Biology: Visualizing Life. Students fill out the left side and center portions of the Vee as a prelab activity. The right side of the Vee is completed after the Investigation is*

complete. The completed Vee serves as a comprehensive assessment of students' understanding and findings from the Investigation.

The Vee on the right summarizes an Investigation concerning osmosis. The information on the point of the Vee gives an "event-sense" focus to the study. An event-sense is the recognition of the things that are happening, whether in an investigation or in a research article. Methodology is presented on the right side of the Vee. That knowledge is guided by the concepts on the left side of the Vee. The knowledge on the right side is interpreted to finally answer the focus questions.

Benefits of Vee Use

One of the most valuable results of using Gowin's Vee is that students come to know the tentativeness of scientific truths. They see that knowledge claims are indeed claims and not conclusions.

The text Investigations provide fundamental hands-on experiences requiring limited equipment.

20T

Definitions of Vee Parts

Concept Statements

A concept statement demonstrates one's knowledge and understanding of a fundamental idea on which the Investigation is based.

Subject Area

relates the concept statements and concepts and explains why a phenomenon occurs.

Focus Question

describes the objects and main event of the Investigatin and indicates the kinds of records that will be collected. This question can usually be constructed from reading the introduction and objectives of the Investigation. The question may include some of the key concepts of the study and may provide some indication of the research procedures to be used.

Knowing side

Subject Area
Cell theory

Concept Statements
1. Homeostasis, or a steady state of balance, of a cell depends on movement of materials in and out of a cell.
2. Molecules constantly move at random from areas of greater to areas of lesser concentration until they reach equilibrium.
3. Osmosis is the diffusion of water.
4. Cells have membranes.
5. The cell is the basic unit of structure and function of living things.

Concepts/Vocabulary
homeostasis, molecular motion, diffusion, osmosis, membrane permeability, molecular size, equilibrium, starch, iodine, membrane pore-size, observation, description

Focus Questions
Do cell membranes as represented by the plastic model allow substances to pass in and out of a cell with equal ease? Do chemical substances move from areas of high concentration to areas of low concentration as they pass in and out of a cell membrane model?

New Focus Questions
How does the cell prevent small, toxic molecules from entering?

Materials/Procedures
Starch and iodine solutions on either side of a model membrane are observed within a beaker, test tube, and plastic bag.

Doing side

Value Claim
Cells can be selective in absorbing and excreting particles to maintain homeostasis, which helps them stay alive.

Knowledge Claims
1. Cell membranes are selectively permeable to molecules of a certain size.
2. Chemical substances move from areas of greater to areas of lesser concentration.

Records

	Before	After
Color of starch inside bag (cell)		
Color of starch outside bag		

Value Claim

describes the significance of the knowledge gained in the Investigation.

Knowledge Claims

are the answers to the focus questions and may be the basis of new questions that lead to further investigation. It is here that the student makes statements concerning what is claimed to be known as results.

Records

are the data collected during the Investigation and should be presented in an organized fashion, such as tables, charts, and graphs.

Concepts/Vocabulary

Concepts are the ideas, signs, or symbols that delineate the scope of the Investigation. Vocabulary includes any new terms used in the Investigation.

Materials

are the subjects or things that allow one to perform the Investigation procedure.

Procedure

is the process that occurs to collect data on the Investigation.

New Focus Questions

are derived from the Knowledge Claims. They describe the next stage in the Investigation.

Assessing Student-Constructed Vee Reports

Comprehensive criteria for assessing student-constructed vees.
The higher the score, the more complete/correct the vee part.

Focus Question

0 No question is identified.

1 A question is identified, but does not focus upon the objects and the major event or the conceptual side of the Vee.

2 A focus question is identified; includes concepts, but does not suggest materials or procedure, or the wrong materials and procedures are identified in relation to the rest of the laboratory exercise.

3 A clear focus question is identified; includes concepts to be used and suggests the procedure and materials.

Material/Procedure

0 No materials and procedures are identified.

1 The materials and procedure are identified, but are inconsistent with the focus question.

2 The materials and procedure are identified, and are consistent with the focus question.

3 Same as above, but also suggests what records will be taken.

Concepts/Concept Statements

0 No conceptual side is identified.

1 A few concepts are identified, and no concept statements or a concept written is really the knowledge claim sought in the laboratory exercise.

2 Concepts and at least one type of concept statement are identified.

3 Concepts and two types of concept statements are identified.

4 Concepts, two types of concept statements, and all relevant vocabulary are identified.

Records

0 No records or observations are identified.

1 Records are identified but are inconsistent with the focus question or the major event.

2 Records or observations are identified, but not both.

3 Records are identified for the major event; but observations are inconsistent with the intent of the focus question.

4 Records are identified for the major event and observations are consistent with the focus question and the grade level and ability of the student.

Knowledge Claim

0 No knowledge claim is identified.

1 A knowledge claim is unrelated to the left-hand side of the Vee.

2 A knowledge claim that includes a concept that is used in an improper context, or any generalization that is inconsistent with the records.

3 A knowledge claim that includes the concepts from the focus question and is derived from the records.

4 Same as above, but the knowledge claim leads to a new focus question.

New Focus Question

0 No new focus question is given.

1 A new focus question consistent with the knowledge claim is identified.

Value Claim

0 No value claim is given.

1 A claim consistent with the significance of the Investigation describing the usefulness of the knowledge claim for pure or applied scientific endeavors.

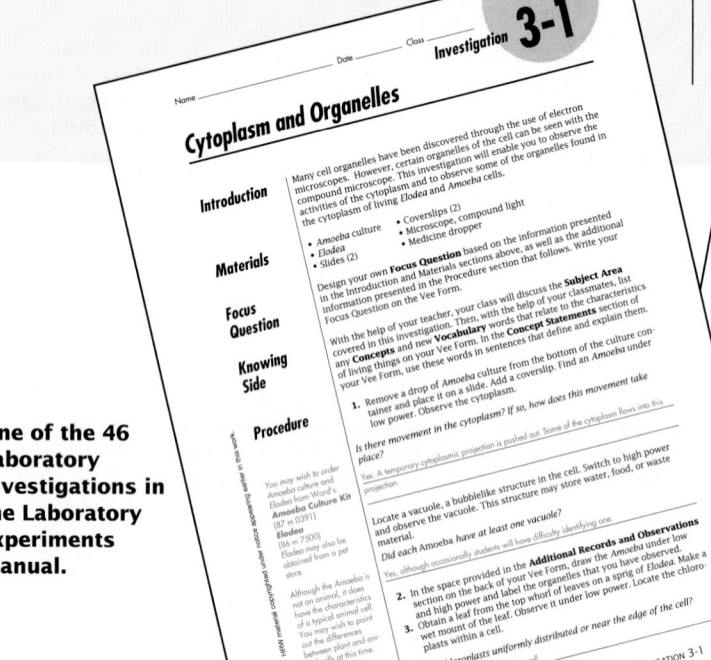

One of the 46 Laboratory Investigations in the Laboratory Experiments manual.

One of the 34 Focus Activities in the Laboratory Experiments manual.

Process Skill Matrix

The process skills highlighted in each in-text Investigation are highlighted below.
You can see all skills are covered from simple observation through experimental design.

Skill	1	2	3	4	5	6	7	8	9	10	11	12	13	14	15	16	17	18	19	20	21	22	23	24	25	26	27	28	29	30	31	32	33	34
Observing		x	x	x	x	x				x	x	x	x	x	x		x		x		x	x	x	x	x	x	x	x		x		x	x	x
Analyzing Data	x	x	x	x	x	x				x	x	x	x	x	x		x		x	x	x		x		x	x						x	x	x
Relating			x											x			x				x	x				x	x		x	x		x		x
Inferring								x	x	x	x	x	x	x	x		x				x	x					x		x	x		x	x
Evaluating					x				x				x												x									
Identifying				x												x			x				x	x	x	x	x		x			x		
Designing Experiments	x				x								x					x													x			
Demonstrating				x																	x										x			
Collecting Data	x	x	x	x	x	x	x			x	x	x	x	x	x	x	x	x		x	x			x	x						x	x	x	x
Measuring					x								x	x							x													
Calculating	x					x							x			x					x											x		
Comparing & Contrasting	x		x	x			x	x		x	x	x	x		x	x				x	x	x	x		x	x						x	x	x
Classifying																x								x										
Modeling							x		x							x					x										x			x
Hypothesizing	x		x										x					x	x									x		x				
Testing Hypotheses	x		x							x			x					x	x												x			
Organizing Data	x		x	x	x	x				x	x	x	x				x	x				x	x	x							x	x	x	
Predicting	x					x				x				x				x			x				x						x			

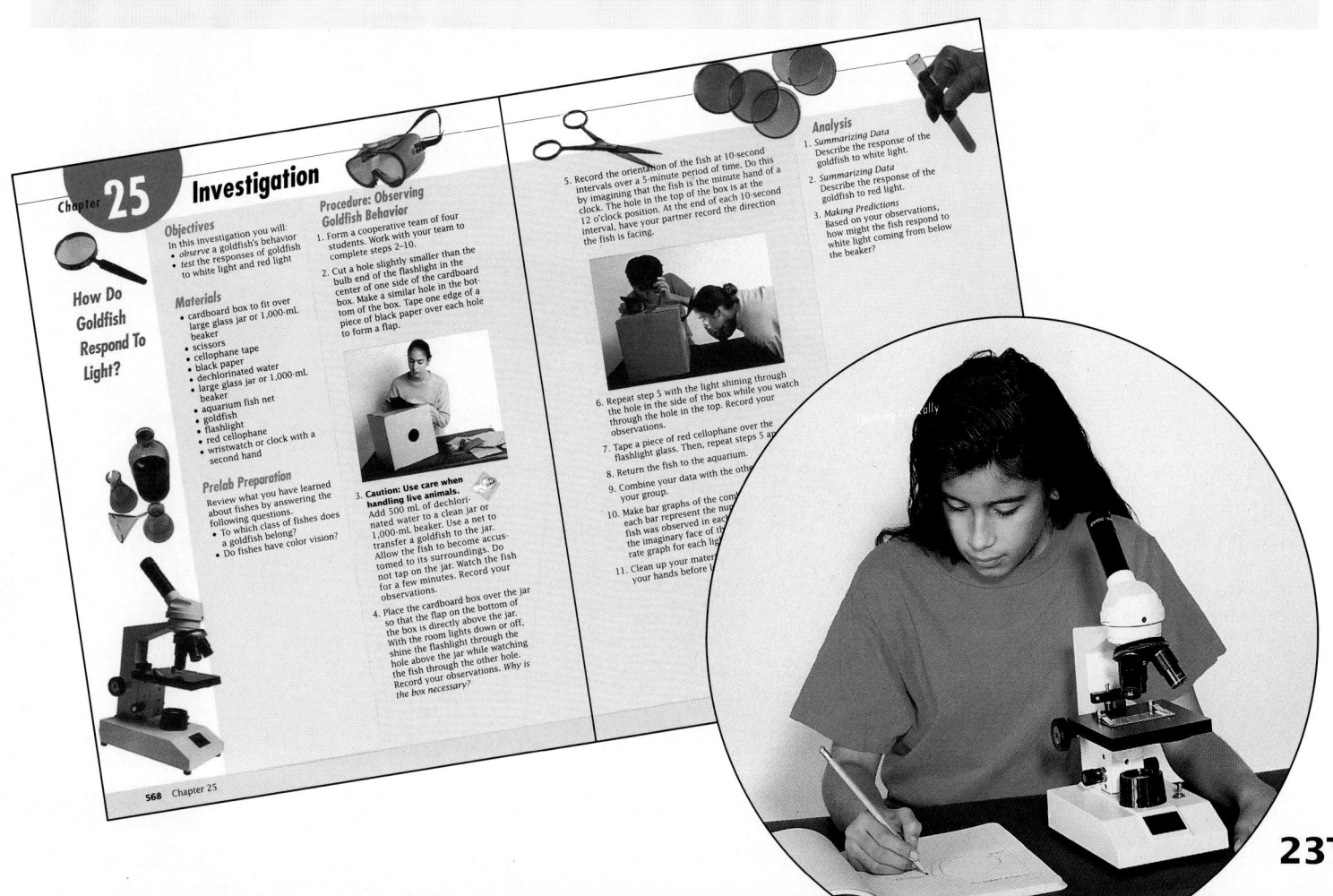

Chapter 25 — Investigation: How Do Goldfish Respond To Light?

Objectives
In this investigation you will:
- *observe* a goldfish's behavior
- *test* the responses of goldfish to white light and red light

Materials
- cardboard box to fit over large glass jar or 1,000-mL beaker
- scissors
- cellophane tape
- black paper
- dechlorinated water
- large glass jar or 1,000-mL beaker
- aquarium fish net
- goldfish
- flashlight
- red cellophane
- wristwatch or clock with a second hand

Prelab Preparation
Review what you have learned about fishes by answering the following questions.
- To which class of fishes does a goldfish belong?
- Do fishes have color vision?

Procedure: Observing Goldfish Behavior
1. Form a cooperative team of four students. Work with your team to complete steps 2–10.
2. Cut a hole slightly smaller than the bulb end of the flashlight in the center of one side of the cardboard box. Make a similar hole in the bottom of the box. Tape one edge of a piece of black paper over each hole to form a flap.
3. **Caution: Use care when handling live animals.** Add 500 mL of dechlorinated water to a clean jar or 1,000-mL beaker. Use a net to transfer a goldfish to the jar. Allow the fish to become accustomed to its surroundings. Watch the fish for a few minutes. Record your observations.
4. Place the cardboard box over the jar so that the flap on the bottom of the box is directly above the jar. With the room lights down or off, shine the flashlight through the hole above the jar while watching the fish through the other hole. Record your observations. *Why is the box necessary?*
5. Record the orientation of the fish at 10-second intervals over a 5-minute period of time. Do this by imagining that the fish is the minute hand of a clock. The hole in the top of the box is at the 12 o'clock position. At the end of each 10-second interval, have your partner record the direction the fish is facing.
6. Repeat step 5 with the light shining through the hole in the side of the box while you watch through the hole in the top. Record your observations.
7. Tape a piece of red cellophane over the flashlight glass. Then, repeat steps 5 and 6.
8. Return the fish to the aquarium.
9. Combine your data with the other groups in your group.
10. Make bar graphs of the combined data. Let each bar represent the number of times the fish was observed in each direction, using the imaginary face of the clock. Make a separate graph for each light.
11. Clean up your materials and wash your hands before leaving the lab.

Analysis
1. *Summarizing Data* Describe the response of the goldfish to white light.
2. *Summarizing Data* Describe the response of the goldfish to red light.
3. *Making Predictions* Based on your observations, how might the fish respond to white light coming from below the beaker?

568 Chapter 25

Breaking new ground

to deliver instruction using various learning modalities

Biology: Visualizing Life gives you all the support you need to address various learning styles with its comprehensive support package.

Visual Learners

Visual learners require extensive instruction with physical models to form mental images of concepts. The unique illustration program for *Biology: Visualizing Life* puts a wealth of physical models at your fingertips.

Auditory Learners

Auditory Learners benefit from the *Biology: Visualizing Life* Audiocassettes. The tapes focus on delivering instruction using the auditory modality combined with a focus on the visual presentation of the text.

Visual Learners
Concepts are presented through graphics in a way that always relates the parts and structural details to the whole of the organism or system. Students can build relationships because they can see them.

Figure 28.9

a The sensory neurons in this person's hand react to stimuli such as heat and initiate nerve signals that are sent to the spine and brain.

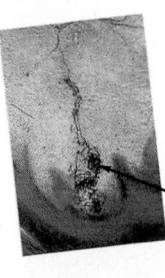

b After the brain has received the information, it sends a signal back to the motor neurons to react in a particular way.

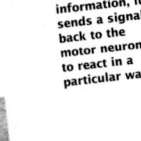

c In the event of a dangerous situation where fast action is critical, information is sent directly to the spine (shown here in cross section) and immediately back to the motor neurons. The shortened path allows for a faster reflex response.

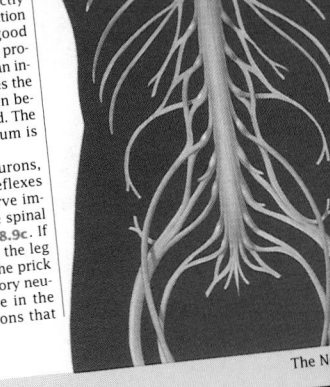

The Peripheral Nervous System

All of the nervous system outside the spinal cord and brain is known as the **peripheral nervous system**. It carries all the messages sent back and forth between the central nervous system and the rest of the body. The peripheral nervous system has two main types of neurons: sensory neurons and motor neurons.

Sensory neurons relay signals to the central nervous system
Sensory neurons tell the central nervous system what is happening. They carry nerve impulses from sense organs to the central nervous system, as shown in **Figure 28.9a**.

Sense organs are organs that react to changes inside and outside the body. They detect many different things, including changes in blood pressure, strain on ligaments, and smells in the air. Sense organs include complex organs, such as eyes and ears. Your skin has many small structures called sensory receptors, which enable you to sense pressure and temperature.

Motor neurons deliver information to muscles and glands

Motor neurons are partners of sensory neurons. Motor neurons carry information from the central nervous system to a muscle or gland, as shown in **Figure 28.9b**. They act on the information delivered by the sensory neurons. If your eyes see a runaway truck speeding toward you, the central nervous system sends messages through motor neurons to glands that secrete adrenaline. The adrenaline increases your heartbeat and breathing rate. The central nervous system also sends messages through motor neurons to many muscles, which contract and get the body out of there—fast!

In each segment of the spine, sensory nerves go into the cord and motor nerves come out of it. The motor nerves control most of the muscles below the head. This is why injuries to the spinal cord often paralyze the lower part of the body. A muscle is paralyzed and cannot move if its motor neurons are damaged.

Reflexes enable you to act quickly

All animals have the ability to act particularly quickly at times of danger. These sudden, involuntary movements are reflexes. A **reflex** produces a rapid motor response to a stimulus because the sensory neuron synapses directly to a motor neuron. The escape reaction of a cockroach is a reflex. A good example in humans is a reflex that protects the eye. If anything, such as an insect or a cloud of dust, approaches the eye, the eyelid blinks closed even before we realize what has happened. The reflex occurs before the cerebrum is aware the eye is in danger.

Because they involve few neurons, reflexes are very fast. Many reflexes never reach the brain. The nerve impulse travels only as far as the spinal cord, as indicated in **Figure 28.9c**. If you step on something sharp, the leg jerks away from the danger. The prick causes nerve impulses in sensory neurons. These neurons synapse in the spinal cord with motor neurons that cause the leg to pull away.

Language-Deficient Learners

Language-deficient learners benefit from the vocabulary development strategies in the Annotated Teacher's Edition and in the Review Guide. The Audiocassettes can be successfully used with language-deficient students because the instruction focuses on an audio narrative coupled with the illustration presentation in the text. Audio with graphics is an excellent way to develop specific vocabulary and language skills.

Multisensory Learners

Multisensory learners benefit from the many individual and group strategies in the Annotated Teacher's Edition. Each Chapter's objectives have been distilled to a set of core objectives that are reasonable knowledge goals for students. The strategies utilize all learning modalities along with peer tutoring to provide an optimal and efficient instructional mix.

Kinesthetic and Tactile Learners

Kinesthetic and tactile learners benefit from the many demonstrations throughout the lesson plan sections of the Annotated Teacher's Edition. The wealth of laboratory opportunities in the *Biology: Visualizing Life* program text and Laboratory Manual provide the concrete hands-on experiences these students need for effective learning.

Investigations

The Investigations and Focus Activities from the text and Laboratory Experiments manual provide 114 hands-on experiences for students.

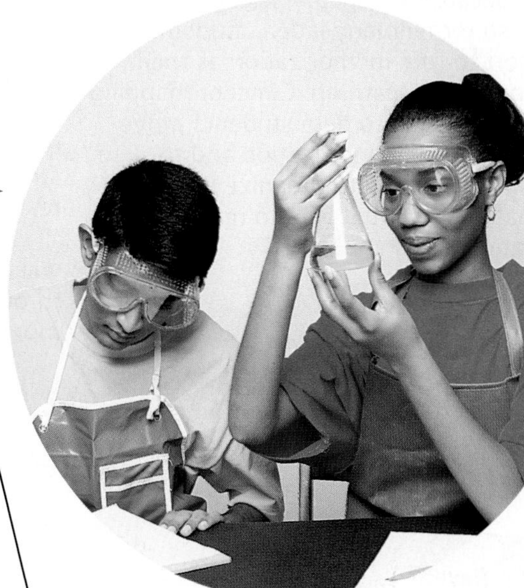

Audio Tapes

Reviewing key concepts and important illustrations using audio prompts helps students with reading problems and language deficiencies. The tapes provide audio review in English and Spanish.

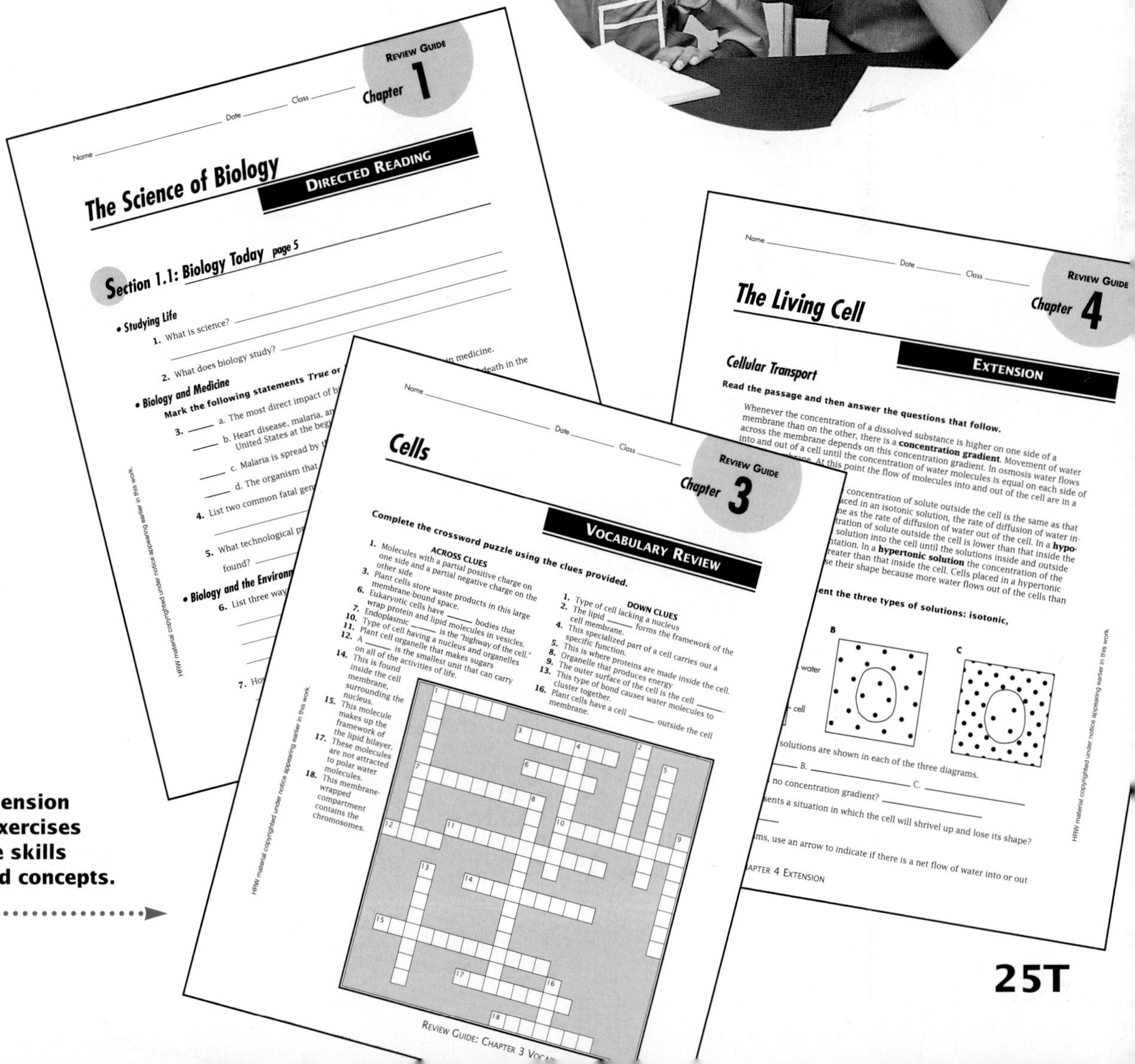

Review Guide

Reading comprehension and vocabulary exercises develop language skills necessary to build concepts.

Concept Maps

Concept-Mapping as an Effective Teaching and Study Strategy

Many secondary teachers focus so intently on teaching that they relieve their students of any responsibility for learning. Are your students responsible for their learning? Do they know *how* to learn?

Because secondary science courses are so terminology laden, students resort to rote memorization as their approach to learning. Concept-mapping is a technique to help students move beyond rote memorization and to teach them how to organize important ideas in ways that lead to meaningful relationships.

Every Chapter Review in *Biology: Visualizing Life* has a concept-mapping exercise. In addition, there are many concept-mapping strategies suggested throughout the lesson plan portion of the Annotated Teacher's Edition. This technique is so powerful in increasing retention that we have made it an important part of the Portfolio Assessment options for each chapter.

Introducing Concept-Mapping to the Class

Have students read the concept-mapping appendix in the back of their text. Let students know that you will expect them to learn and use this strategy. Be prepared for some initial resistance.

If students understand that they will be graded on concept maps throughout the course, they will be more likely to learn the technique.

Making Concept Maps

Drawing concept maps involves the following steps:

1

Start with a list of concepts or ideas to be mapped.This list does not have to be complete, but it should be complete enough to allow you to choose the main idea of the map.

2

Look through the list to identify the concept words that directly relate to the main idea. Place these words below the main idea. Continue this procedure with all the words in your list and any supporting ideas until they are all placed in order of priority under the main idea.

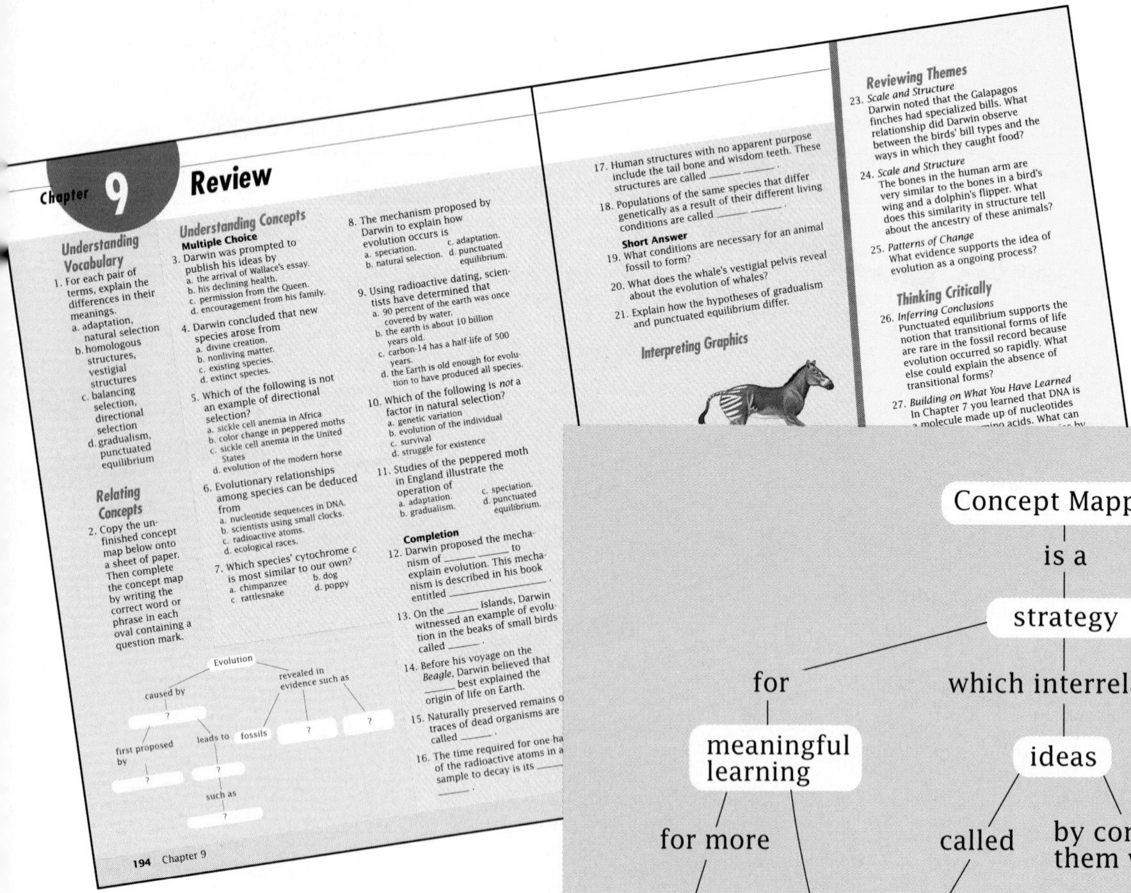

Chapter 9 Review

Understanding Vocabulary

1. For each pair of terms, explain the differences in their meanings.
   a. adaptation, natural selection
   b. homologous structures, vestigial structures
   c. balancing selection, directional selection
   d. gradualism, punctuated equilibrium

Relating Concepts

2. Copy the unfinished concept map below onto a sheet of paper. Then complete the concept map by writing the correct word or phrase in each oval containing a question mark.

Evolution — caused by — ? — first proposed by — ?; ? — leads to — ? — such as — ?; Evolution — revealed in evidence such as — fossils, ?, ?

Understanding Concepts

Multiple Choice

3. Darwin was prompted to publish his ideas by
   a. the arrival of Wallace's essay.
   b. his declining health.
   c. permission from the Queen.
   d. encouragement from his family.
4. Darwin concluded that new species arose from
   a. divine creation.
   b. nonliving matter.
   c. existing species.
   d. extinct species.
5. Which of the following is not an example of directional selection?
   a. sickle cell anemia in Africa
   b. color change in peppered moths
   c. sickle cell anemia in the United States
   d. evolution of the modern horse
6. Evolutionary relationships among species can be deduced from
   a. nucleotide sequences in DNA.
   b. scientists using small clocks.
   c. radioactive atoms.
   d. ecological races.
7. Which species' cytochrome *c* is most similar to our own?
   a. chimpanzee b. dog
   c. rattlesnake d. poppy
8. The mechanism proposed by Darwin to explain how evolution occurs is
   a. speciation. c. adaptation.
   b. natural selection. d. punctuated equilibrium.
9. Using radioactive dating, scientists have determined that
   a. 90 percent of the earth was once covered by water.
   b. the earth is about 10 billion years old.
   c. carbon-14 has a half-life of 500 years.
   d. the Earth is old enough for evolution to have produced all species.
10. Which of the following is *not* a factor in natural selection?
    a. genetic variation
    b. evolution of the individual
    c. survival
    d. struggle for existence
11. Studies of the peppered moth in England illustrate the operation of
    a. adaptation. c. speciation.
    b. gradualism. d. punctuated equilibrium.

Completion

12. Darwin proposed the mechanism of ______ ______ to explain evolution. This mechanism is described in his book entitled ______.
13. On the ______ Islands, Darwin witnessed an example of evolution in the beaks of small birds called ______.
14. Before his voyage on the *Beagle*, Darwin believed that ______ best explained the origin of life on Earth.
15. Naturally preserved remains o[...] traces of dead organisms are called ______.
16. The time required for one-ha[...] of the radioactive atoms in a sample to decay is its ______ ______.
17. Human structures with no apparent purpose include the tail bone and wisdom teeth. These structures are called ______ ______.
18. Populations of the same species that differ genetically as a result of their different living conditions are called ______ ______.

Short Answer

19. What conditions are necessary for an animal fossil to form?
20. What does the whale's vestigial pelvis reveal about the evolution of whales?
21. Explain how the hypotheses of gradualism and punctuated equilibrium differ.

Interpreting Graphics

Reviewing Themes

23. *Scale and Structure* Darwin noted that the Galapagos finches had specialized bills. What relationship did Darwin observe between the birds' bill types and the ways in which they caught food?
24. *Scale and Structure* The bones in the human arm are very similar to the bones in a bird's wing and a dolphin's flipper. What does this similarity in structure tell about the ancestry of these animals?
25. *Patterns of Change* What evidence supports the idea of evolution as a ongoing process?

Thinking Critically

26. *Inferring Conclusions* Punctuated equilibrium supports the notion that transitional forms of life are rare in the fossil record because evolution occurred so rapidly. What else could explain the absence of transitional forms?
27. *Building on What You Have Learned* In Chapter 7 you learned that DNA is a molecule made up of nucleotides [...] amino acids. What can [...]

194 Chapter 9

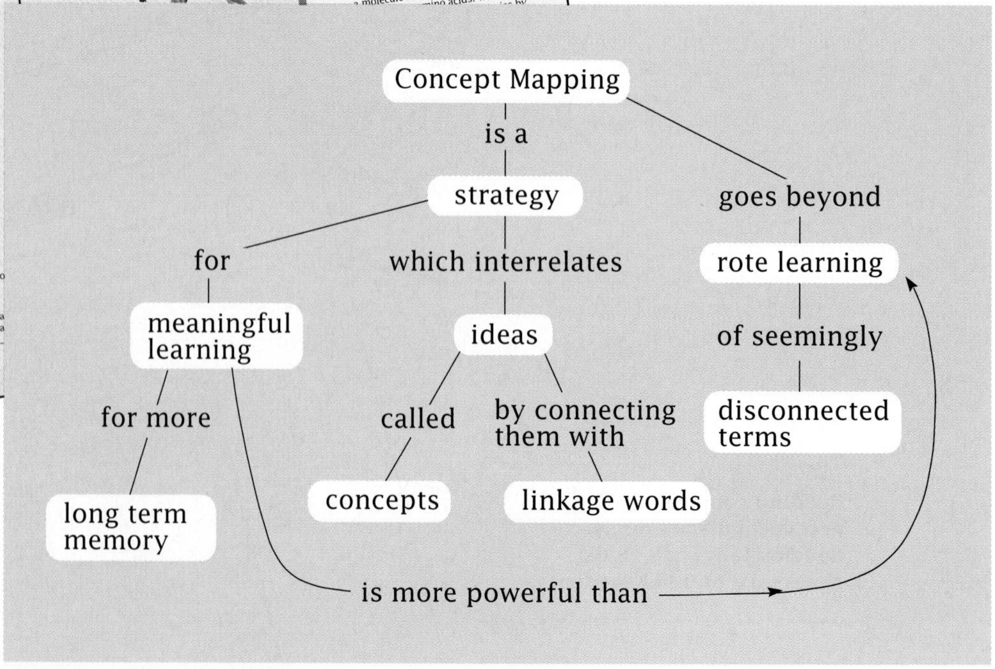

A concept map of the idea of concept-mapping. This map shows all the characteristics of a good map—hierarchy of concepts, adequate links, branching, and cross-links.

❸

Use lines to connect the concepts, based on relationships that link them.

❹

Use connecting words to label the linking lines so that the relationship between any two concepts is a clear and complete thought.

❺

Look for all possibilities to add cross-links to the map. Cross-links show how deeply one understands the relationships among concepts.

Evaluating Concept Maps

When working with students to evaluate maps, emphasize that there is not a single correct map for an idea. Each map should be evaluated on the following criteria.

❶

Appropriate labeling of concepts

Concepts should be succinctly represented—no longer than three words. The arrangement of concepts should be hierarchical from general to specific. Concepts should not be repeated on a map. A repeating concept should be represented by a cross-link.

❷

Appropriate labeling of linking words

Does the map show a clear distinction between concepts and links? Is the link between two concepts meaningful? Does it correctly represent the relationship? Can complete ideas be traced through several links?

❸

Adequate branching

Students' first maps generally look similar to outlines and as such are linear in form without adequate branching. Provide incentives for students to turn in highly branched maps so that they will work hard to develop this skill.

❹

Cross linking of concepts

The best maps show sufficient cross-links among concepts. Cross-links show that students know how multiple ideas are connected.

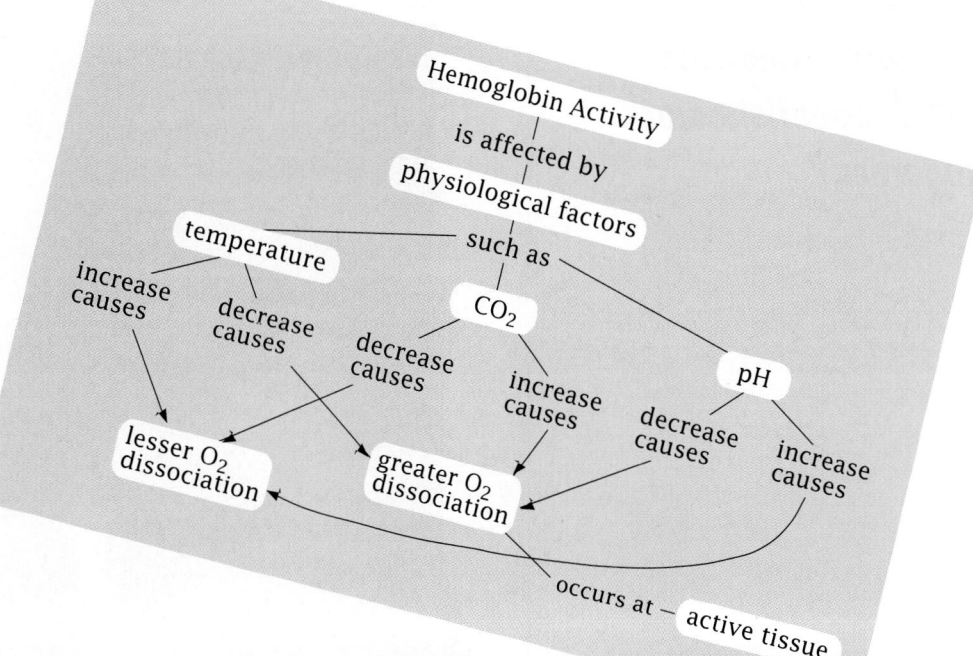

Two maps depicting information on hemoglobin. Note their differences in construction. The map above is too linear and there are no cross-links. Note that some of the concepts are not really concepts, others repeat or are not succinctly stated. The map on the right shows good form.

Breaking new ground

in peer tutoring

with effective

cooperative

learning

strategies

"One of the true benefits of cooperative learning is that it models the real scientific experience in which scientists work together."

Biology: Visualizing Life Cooperative Learning Strategies

① In the Lab

The in-text Investigations, Focus Activities, and Laboratory Investigations are excellent group work situations. Use of the Vee Form Report Sheets (as described on pages 20T–21T) provides a realistic model for the way scientists work in a research situation.

② In the Classroom

Your Annotated Teacher's Edition has one formal group activity per chapter. Each ATE activity is broken down into the following areas:

Timing
When to use the activity in your teaching plan.

Group size
Use small groups to assure that everyone has accountability in the activity.

In the Lab
Working through a Vee report from an Investigation can be an excellent peer learning opportunity.

Concept-Mapping
Students enjoy building concept maps in groups. They find it fun, like putting together a jigsaw puzzle.

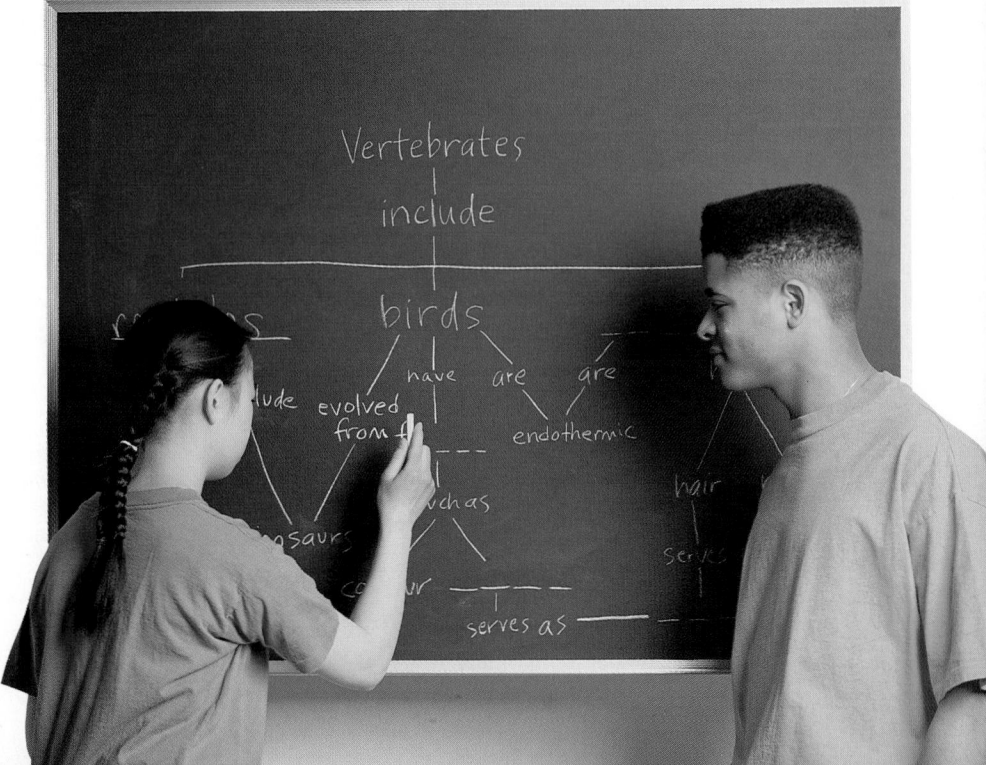

Outcome

Students need to understand the Objective and assess the outcome of the activity.

Individual Accountability

Each group member must have a specific responsibility that is established in the beginning of the activity and has a purpose in the intended learning of all group members.

Concept-Mapping

Throughout your Annotated Teacher's Edition you'll find prompts for the development of concept maps as assignments. The Chapter Review concept-mapping section can be used as a group activity as well.

Science, Technology, and Society Features

The goal of these features is to stimulate class discussion and debate. The class can be divided into teams to do further research on an issue or set up a role-playing situation. Each team presents their particular point of view. Because these situations are realistic, they become models for issues students will face as voting citizens.

Science, Technology, and Society Features
Thinking Critically and Acting on the Issue items from these features make excellent group projects where students learn to work cooperatively.

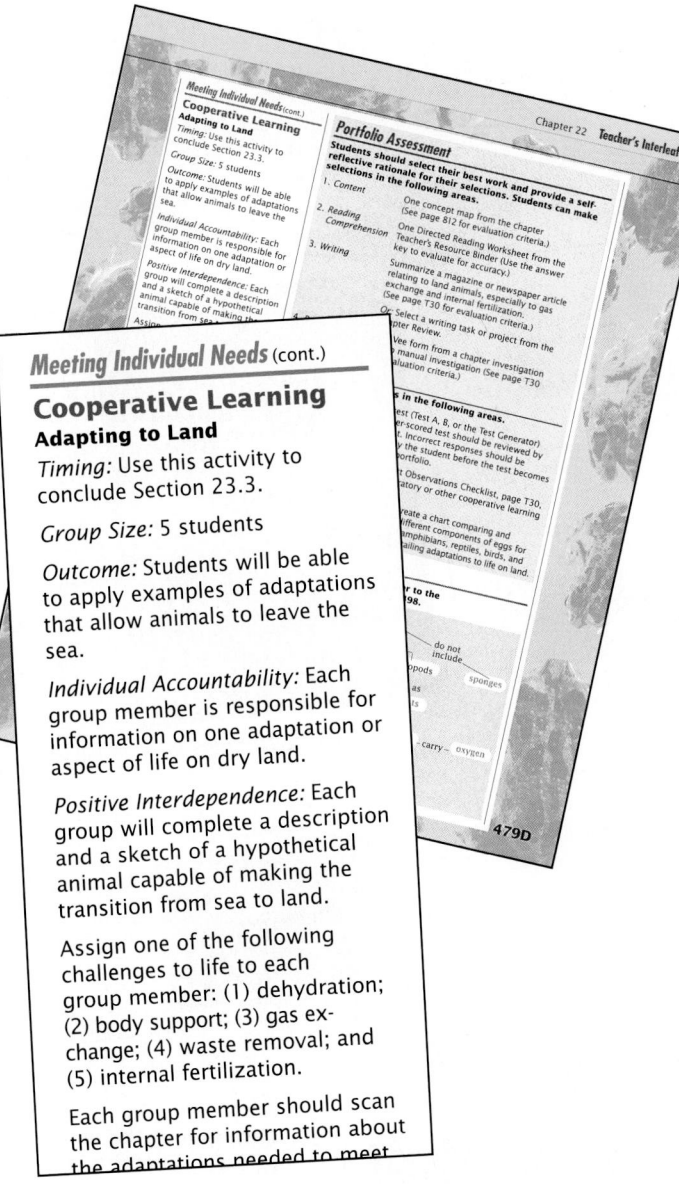

Cooperative Learning Activities **are in the Teacher's Interleaf section for each chapter.**

Meeting Individual Needs (cont.)

Cooperative Learning

Adapting to Land

Timing: Use this activity to conclude Section 23.3.

Group Size: 5 students

Outcome: Students will be able to apply examples of adaptations that allow animals to leave the sea.

Individual Accountability: Each group member is responsible for information on one adaptation or aspect of life on dry land.

Positive Interdependence: Each group will complete a description and a sketch of a hypothetical animal capable of making the transition from sea to land.

Assign one of the following challenges to life to each group member: (1) dehydration; (2) body support; (3) gas exchange; (4) waste removal; and (5) internal fertilization.

Each group member should scan the chapter for information about the adaptations needed to meet

Breaking new ground

to present the cultural diversity in science

Biology: Visualizing Life recognizes the influences of various cultures on the accumulation of scientific knowledge. This program gives you the necessary tools to link science concepts to cultural events and provides role models in science for students of all cultural backgrounds.

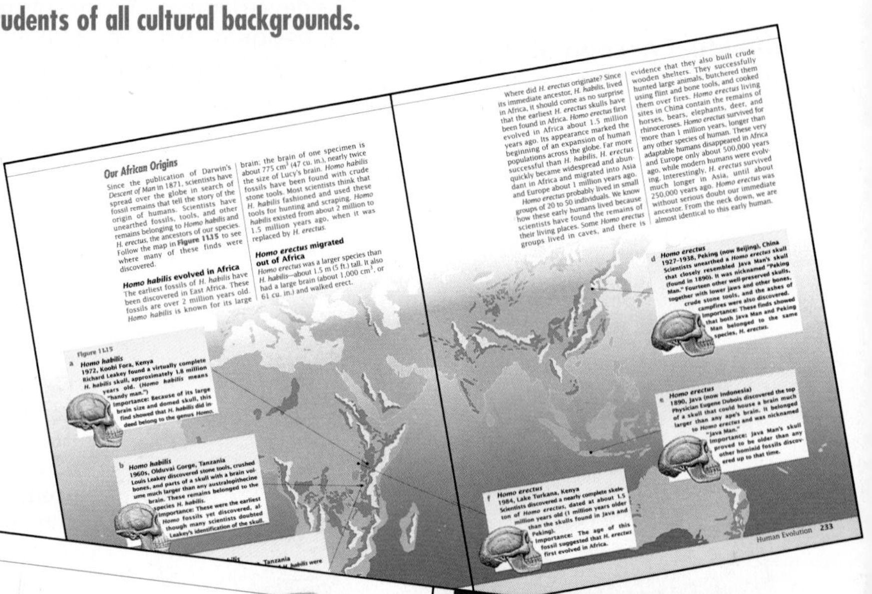

"Today's classrooms are places of diversity— populated by students from a variety of ethnic backgrounds and cultures."

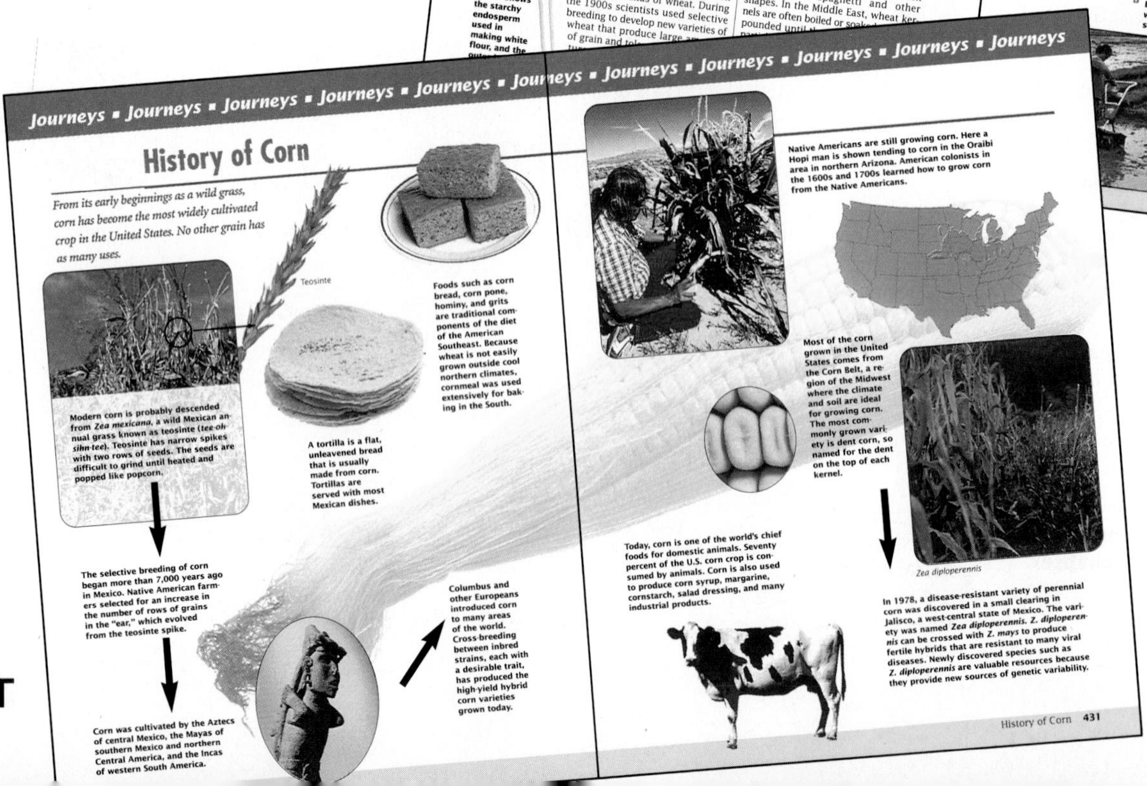

Text

Throughout the text cultural diversity is stressed in relation to a biological concept.

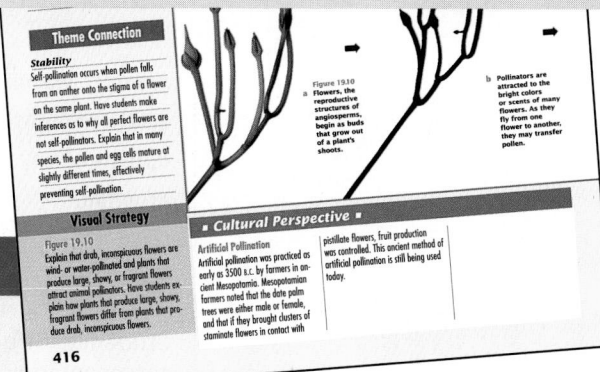

Cultural Perspective
These annotations can be used to show students that biological knowledge comes from people of all cultures.

■ Cultural Perspective ■

Kola Nuts

Kola nuts are cultivated in the West Indies, West Africa, and South Africa. They are the fruits of several types of evergreen trees. People who live in African countries call the nuts guru or goora nuts and chew them like gum. Kola nuts are also used to make cola soft drinks and medicines. The nuts contain caffeine and theobromine. Caffeine has a stimulating effect and assists in combating fatigue.

Discoveries in Science
These features highlight role models from all cultures, to show students that scientific knowledge is a collective and cooperative effort.

Multicultural Lesson Plan

Exploring Differences in Languages
(Dealing with Ethnic Diversity)

Preparation: If a foreign exchange student is available, arrange for him/her to visit the biology class in order to help you and the students compare names of animals in different languages. As an alternative, bilingual students, a foreign-language teacher, or students taking foreign-language classes can assume this role. Be sure to have several foreign language-English dictionaries available, especially a Latin-English one. Field

Multicultural Lesson Plan
To enhance the study of cultural diversity in relation to biology, optional multicultural lesson plans are provided.

Science in Action
Personal profiles often highlight cultural challenges faced by individuals in pursuing their careers.

Breaking new ground

to assist you in thoroughly assessing student progress— portfolio assess- ment options

Biology: Visualizing Life gives you the most comprehensive array of assessment options of any high school biology program.

Portfolio Assessment

Student portfolios should contain a representative sampling of a student's work, providing a realistic look at a student's efforts, progress, and achievement. Portfolio assessment must include:

1. student-selected items
2. guidelines for the selection of items to be included
3. evaluation guidelines for items
4. evidence of student self-reflection
5. ongoing use as the assessment method

Using these criteria, you can select from options listed for each chapter and include tasks of your own design.

Chapter Tests

These printed tests are available in your Teacher's Resource Binder. Each test exists in two formats: A and B. Test A is all multiple choice and can be used easily with data-card scoring machines. Test B includes short-answer and essay formats. Both tests are objective-based and include graphics to ensure accurate evaluation of course objectives.

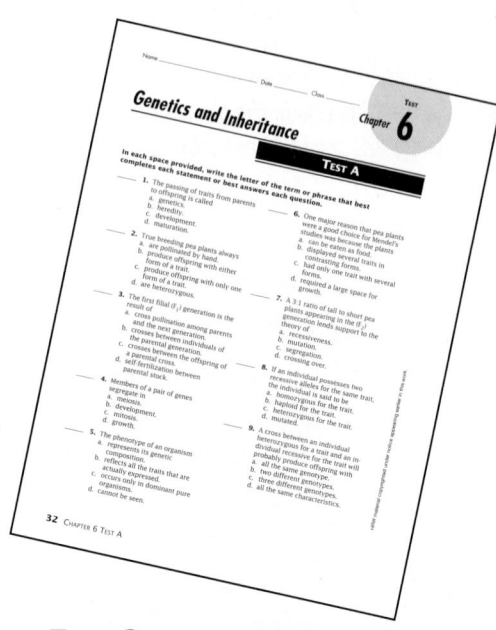

Test Generator

The test generator teacher utility program provides quick and easy access to a bank of over 2,000 test items for building customized tests. This program is available for Macintosh and IBM compatible PCs. It includes graphics to fully assess students' visual comprehension of biological concepts.

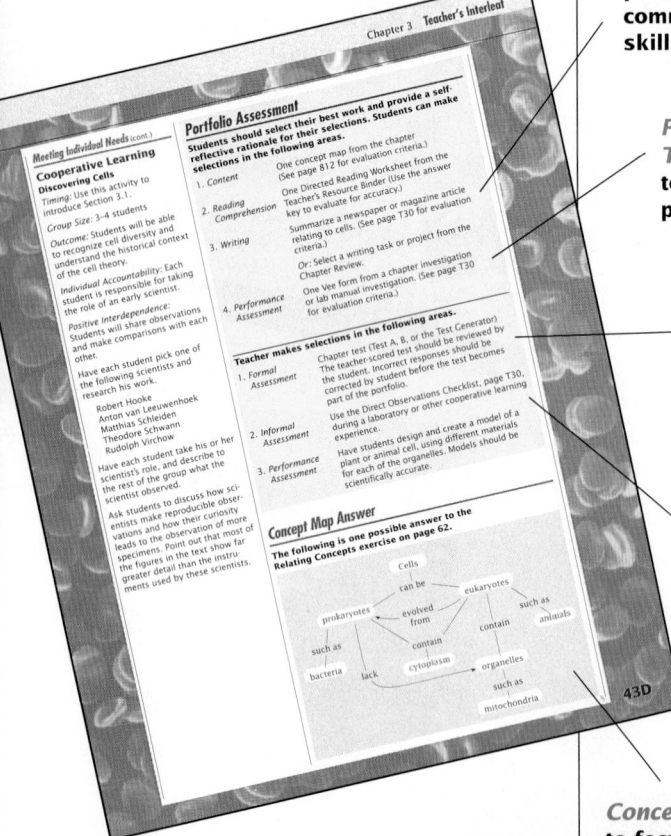

Writing
prompts to show communication skills.

Performance Tasks
to assess process skills.

Formal Assessment
to assess content knowledge.

Informal Assessment
to assess stu- dents' work prac- tices using the Direct Observations Checklist.

Concept Maps
to focus on text theme and relation- ships.

Performance Tasks Scoring Rubric

6 Points, Experienced Level
Work is superior with no serious inaccuracies, or any inaccuracies noted are minor.

Conceptual understanding: clearly demonstrated throughout the performance of the task

Reasoning: logical throughout
Communication: exemplary

5 Points, Competent Level
Work is effective with no serious inaccuracies and few minor inaccuracies.

Conceptual understanding: apparent throughout student's work
Reasoning: generally logical
Communication: effective

4 Points, Intermediate Level
Work is satisfactory with few serious inaccuracies and some minor ones, or many minor inaccuracies

Conceptual understanding: somewhat apparent.
Reasoning: somewhat weak in logic
Communication: satisfactory

3 Points, Transitional Level
Work is marginal with some serious inaccuracies and some minor ones.

Conceptual understanding: not always apparent
Reasoning: not always logical
Communication: marginal

2 Points, Beginning Level
Work is unsatisfactory with serious inaccuracies throughout.

Conceptual understanding: lacking
Reasoning: very weak
Communication: unsatisfactory

1 Point, Inexperienced Level
Work is poor with serious inaccuracies throughout.

Conceptual understanding: none apparent
Communication: very poor

0 Points
No relevant responses given.

Student-Designed Experiments Analytic Rubric

4 Points
Problem or hypothesis is clearly defined in a manner that can be tested.

Results are reproducible; controls are complete.

Observations are detailed, precise, and organized for easy interpretation.

Inferences are appropriate and strongly supported.

Communication is clear and well organized with correct use of the language.

3 Points
Problem or hypothesis is somewhat confused, but still allows for testing.

Results may not be totally reproducible; steps may be missing; controls are inadequate.

Observations are basic and adequate; organization may be lacking.

Inferences are mildly supported; some minor points may be missing; student may recognize inconsistencies.

Communication is satisfactory with some lapses in organization or clarity.

2 Points
Problem or hypothesis may be implied; experimental design is unrelated.

Results may not be reproducible due to confusing explanations and inadequate controls.

Observations are difficult to interpret; irrelevant ideas mixed with relevant.

Inferences are weak and partially supported; some supporting evidence is irrelevant; no acknowledgment of inconsistencies.

Communication is lacking with many errors in language and organization.

1 Point
Problem or hypothesis is neither defined nor implied; experimental design is lacking.

No variables controlled.

Observations are inadequate and generally unrelated; data is disorganized.

Inference is not supported or is supported by irrelevant information.

Communication is confused and disorganized.

Informal Assessment Direct Observations Checklist

Check all that apply.

- ❏ The student is able to work with little or no guidance.
- ❏ The student is able to stay on task for an extended period of time.
- ❏ The student recognizes the problem.
- ❏ The student is able to formulate an effective hypothesis.
- ❏ The student develops an effective experimental design.
- ❏ The student adapts materials in creative ways.
- ❏ The student makes unique observations.
- ❏ The student uses measuring devices correctly.
- ❏ The student makes complete observations.
- ❏ The student provides an in-depth analysis of data.
- ❏ The student's conclusions are well supported by the data.
- ❏ The student applies themes in a competent manner.
- ❏ The student acknowledges assumptions or inconsistencies in data.
- ❏ The student needs little prompting to respond.
- ❏ The student communicates orally.
- ❏ The student shows curiosity in performing the task.
- ❏ The student accepts different ideas from peers.
- ❏ The student chooses to test another's hypothesis.
- ❏ The student creatively uses information gained from others.
- ❏ The student anticipates and solves problems.
- ❏ The student uses mathematics when needed without prompting.
- ❏ The student acts in a responsible and safe manner.

Breaking new ground

to support

effective teaching

strategies:

comprehensive,

efficient, what

you need, where

you need it

"We are concerned as much with how science is taught as with the substance of what is taught . . ."

Fulfilling the Promise—Biology Education in the Nation's Schools

Instructional Planning

The *Biology: Visualizing Life* Annotated Teacher's Edition is organized along the lines of how you plan yearly, weekly, and daily activities. Your Annotated Teacher's Edition consists of two parts:

Weekly Planning

Teacher's Interleaf pages precede each chapter in the ATE.

Daily Planning

Teacher's Wrap surrounds the reduced pupil pages in the ATE.

Teacher's Interleaf— Weekly Planning

Teacher's Interleaf provides the background needed for the weekly planning of your course; most chapters cover about a week of classroom time. You'll find the following in the Interleaf pages.

Planning Guide
is your master plan for integrating all the materials available for the program into your teaching plan for each chapter.

Teacher's Wrap— Daily Planning Using the Lesson Cycle

Determining Prior Knowledge
This section can be used to determine how much students know about the subject to be studied and what misconceptions they could have that you will need to address in your teaching.

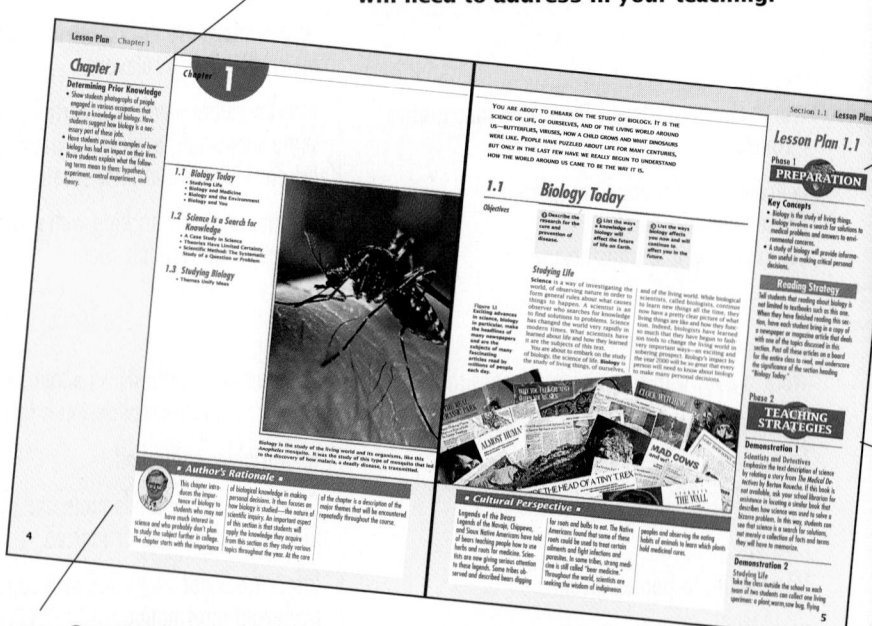

Chapter Opener
Author's Rationale — Each chapter begins with a rationale from the author as to the organization and intent of the chapter. Read this material first.

34T

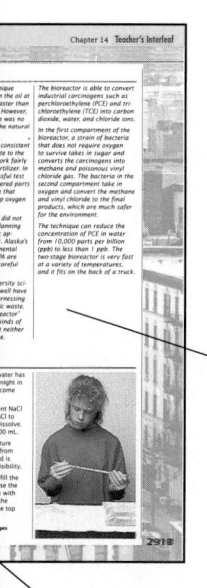

Meeting Individual Needs
This section contains a variety of strategies and activities tied to various learning modalities. You'll find ways to adapt traditional teaching methods in the Teacher's Wrap to suit the individual needs of your students. There are a number of ideas you can use to help all students have a successful experience with biology.

Portfolio Assessment
This section contains the many ideas you need to implement a comprehensive portfolio assessment program. Each section is divided into two parts based on students' selections and teacher selections.

Research Notes
provide you with a professional article on current research related to the material in the chapter.

Investigation Notes
give you the planning information that you will need for the Investigation found at the end of the chapter. Be sure to check this section as you plan for the lab.

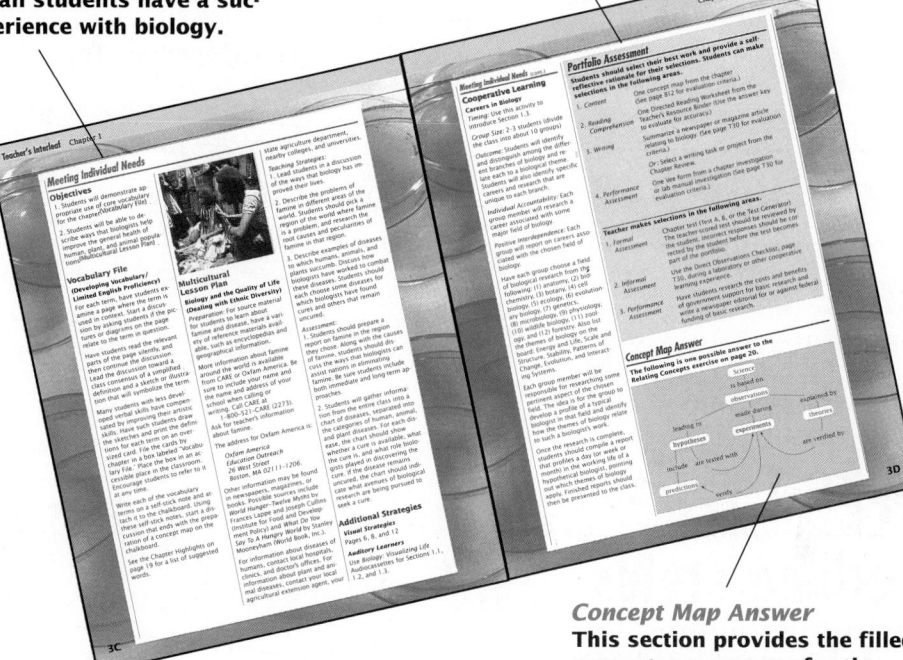

Concept Map Answer
This section provides the filled-in concept map answer for the concept-mapping activity in the Chapter Review.

Lesson Plans
Each lesson contains the following organized into a lesson-cycle format:

Phase 1
Preparation
Key Concepts emphasize the conceptual nature of the presentation.

Reading Strategy helps students with comprehension and study skills.

Visual Strategies are the core of the program. Here you find ideas to make effective use of the outstanding instructional illustrations throughout the program.

Phase 2
Teaching Strategies
Connections highlight interrelationships. You'll find Connections for other subjects like mathematics and social studies, Connections to other chapters,

and Connections to the six biological themes presented in Chapter 1.

Demonstrations provide concrete modeling for abstract ideas.

Historical Perspectives are historical backgrounds related to the text.

Cultural Perspectives provide information concerning people of various cultures associated with ideas presented in the text, or the influence of culture on a biological issue.

Matter of Fact provides a useful bit of information related to the ideas in the text.

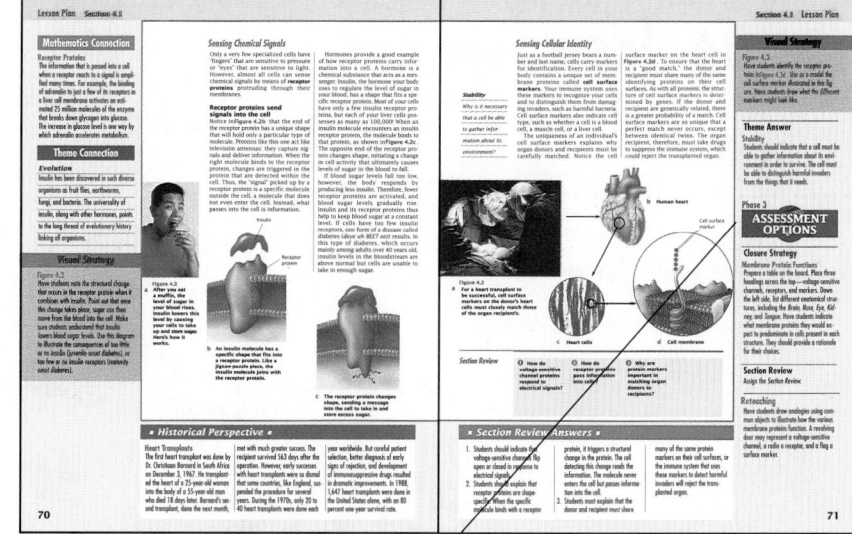

Phase 3
Assessment Options
Closure is a strategy for helping students summarize key concepts from the lesson.

Section Review assesses mastery of the section objectives.

Reteaching provides a strategy for reinstruction in a major concept of the section.

Portfolio Assessment prompts you to review the portfolio option listed in the interleaves for each chapter.

35T

Biology: Visualizing Life

brings you a range of supplements that support various teaching styles and learning modalities to ensure classroom success.

Laboratory Experiments

The student laboratory manual provides an introductory Focus Activity for every chapter that provides concrete conceptual models that students use in building their understanding of the chapter. The 46 Investigations in the manual provide solid hands-on equipment-based experiences. The Vee Form Report Sheets used in the manual provide the link between text concepts and laboratory Investigations. The Laboratory Experiments Teacher's Edition provides a thorough guide to all activities, safety guidelines, materials lists, and the use of the Vee technique.

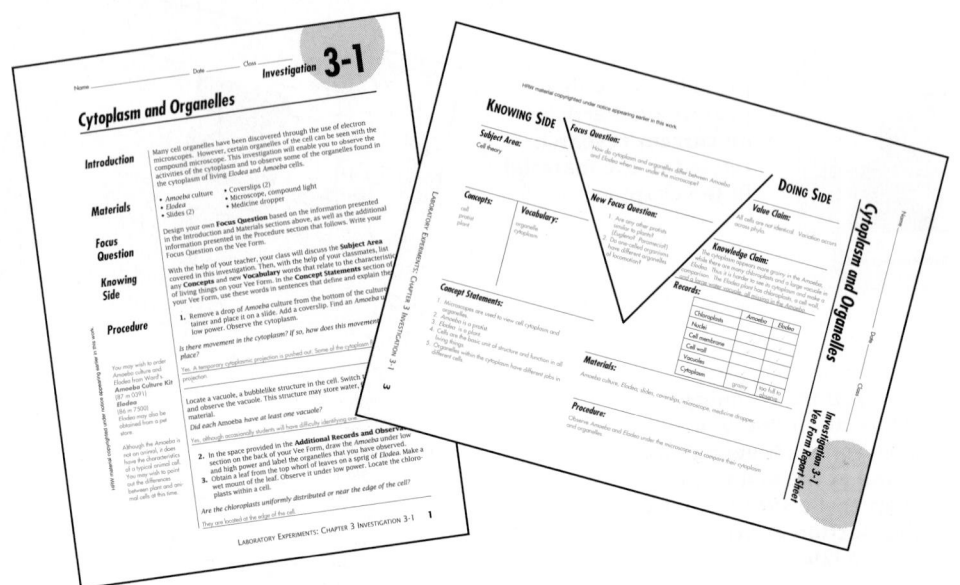

Teaching Transparencies

The overhead transparency package provides more than 160 of the text images for daily use in the classroom. A copymaster worksheet is included with each transparency to guide students in developing an understanding of each illustration.

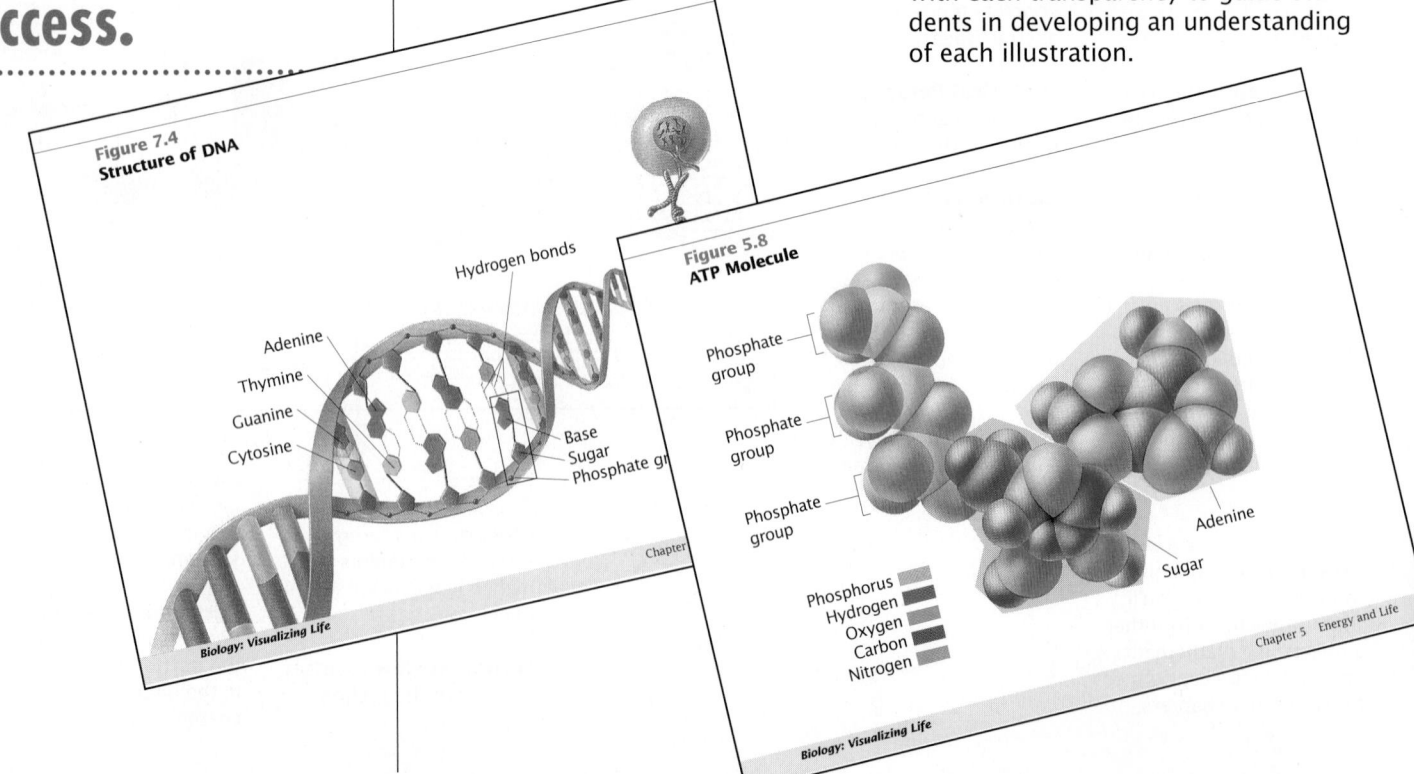

Teacher's Resource Binder

The binder contains loose-sheet copy-masters of:

- Laboratory Experiments manual
- Review Guide
- Chapter Tests

Review Guide

This student workbook focuses on reading comprehension, vocabulary development, reteaching, process skill development, and strategies for helping students extend concepts in the text.

Computer Test Generators

Using the test generators, you can select from a bank of more than 2,000 items to construct your own customized tests containing graphics. Test generators are available for Macintosh® and IBM® compatible PCs.

English/Spanish Audiocassettes

Each unit tape gives students an audio walk-through of the chapter in English and Spanish, focusing on key concepts. Students are prompted to pay particular attention to developing a complete understanding of text illustrations.

Spanish Resources

The Spanish Resources package helps students with reading abilities in Spanish, but limited English skills, to learn biology concepts. The translations cover key parts of *Biology: Visualizing Life*. The translation package contains:

- 160 color overhead transparencies and accompanying worksheets
- Chapter Highlights
- Complete text glossary

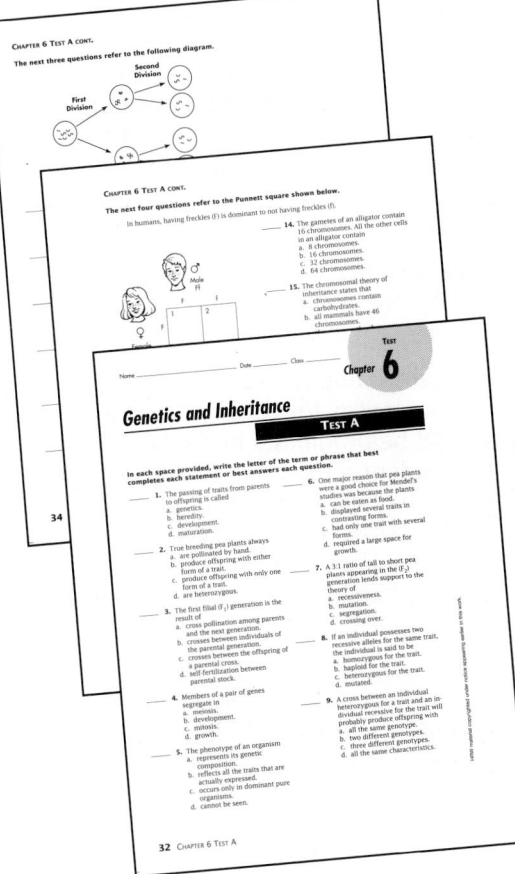

Biology Videodisc

Still images and video needed to support your classroom instruction are available on laser videodiscs designed to accompany *Biology: Visualizing Life*. Text images come to life, using a bar-code scanner or remote control, making the presentation of abstract ideas more concrete. The videodisc package includes a teacher's edition with bar codes and lesson plans for the program.

Breaking new ground

to help you

run a safe

and efficient

laboratory

program

General Safety Guidelines

1. Post laboratory rules in a conspicuous place in the laboratory.

2. Before the class begins an experiment, review safety rules and demonstrate proper procedures.

3. Never permit students to work in your laboratory without your supervision. No unauthorized investigations should ever be conducted, nor should unauthorized materials be in the laboratory.

4. Lock your laboratory (and storeroom) when you are not present.

5. Mark locations of eyewash stations, safety shower, fire extinguishers (ABC tri-class), chemical spill kit, first-aid kit, and fire blanket in the laboratory and storeroom. Check this safety equipment prior to conducting each investigation.

6. Post an evacuation diagram and procedures by every entrance to the laboratory.

7. Provide for labeled disposal containers for glass, sharps, and waste chemical reagents.

8. Allow no food or beverages in the laboratory. Caution students to keep their hands away from their face and to wash their hands with soap and water before leaving the laboratory.

9. Know the location for the master shut-off for laboratory circuits. Be sure that all outlets have correct polarity and have ground-fault interception. Polarity can be tested with an inexpensive (about $5) continuity tester available from most electronic hobby shops. All electrical equipment should have 3-prong plugs and 3-wire cords.

10. Follow prescribed procedures for any safety incident, including full documentation. Remind students that any safety incident, no matter how trivial, must be reported directly to you.

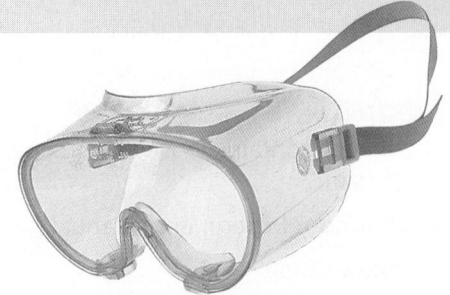

Personal Protective Equipment

Chemical goggles: (Meeting ANSI Standard Z87.1) These should be worn when working with any chemical or chemical solution other than water, when heating substances, using any mechanical device, or observing physical processes that could eject an object.

Face shield: (Meeting ANSI Standard Z87.1) Use in combination with eye goggles when working with corrosives.

Contact lenses: The wearing of contact lenses for cosmetic reasons should be prohibited in the laboratory. If a student must wear contact lenses prescribed by a physician, that student should be instructed to wear eye-cup safety goggles meeting ANSI Standard Z87.1 (similar to swimmer's cup goggles).

Eye-wash station: The device must be capable of delivering a copious, gentle flow of water to both eyes for at least 15 minutes. Portable liquid supply devices are not satisfactory and should not be used. A plumbed-in fixture or a perforated spray head on the end of a hose attached to a plumbed-in outlet is suitable if it is designed for use as an eye-wash fountain and meets ANSI Standard Z358.1. It must be within a 30-second walking distance from any spot in the room.

Safety Shower: (Meeting ANSI Standard Z358.1) Location should be within a 30-second walking distance from any spot in the room. Students should be instructed in the use of the safety shower in the event of a fire or chemical splash on their body that cannot be simply washed off.

Gloves: Polyethylene, neoprene rubber, or disposable plastic may be used. Nitrile or butyl rubber gloves are recommended when handling corrosives.

Apron: Rubber-coated cloth or vinyl (nylon-coated) halter is recommended.

Emergency Preparedness

What would you do if a student dropped a liter bottle of concentrated sulfuric acid? RIGHT NOW? Plan now how to effectively react BEFORE you need to.

1. Post the phone numbers of your regional Poison Control Center, Fire Department, Police, Ambulance, and Hospital on your telephone.

2. Practice fire and evacuation drills. Also have drills on what students MUST do if they are on fire or have chemical contact.

3. Assure that all personal and other safety equipment is available and tested frequently.

4. Compile an MSDS file for all chemicals. This reference resource should be readily accessible in case of a spill or other incident.

5. Provide for spill control procedures. Handle only those incidents that you feel comfortable in handling. Situations of greater severity should be handled by trained hazardous material professionals.

6. Students should never fight fires or handle spills.

7. Be trained in first aid and basic life support (CPR) procedures. Have first-aid kits and spill kits readily available.

8. Fully document ANY INCIDENT that occurs.

Safety With Animals

It is recommended that teachers follow the "Guidelines for the Use of Live Animals" established by the National Association of Biology Teachers, which is reproduced in the Teacher's Edition of *Laboratory Experiments.*

Safety in Handling Preserved Materials

The following practices are recommended when handling or dissecting any preserved specimen:

1. NEVER dissect road-kills or nonpreserved slaughterhouse material. Doing so increases the risk of infection.

2. Wear protective gloves and splash-proof safety goggles at all times when handling preserving fluids, preserved specimens, and during dissection.

3. Wear lab aprons. Use of an old shirt or smock is recommended.

4. Conduct dissection activities in a well-ventilated area.

5. Do not allow preservation or body-cavity fluids to contact skin. Fixatives do not distinguish between living or dead tissue.

Biological supply firms use formalin-based fixatives, of varying concentration to initially fix zoological and botanical specimens. WARD'S Natural Science Establishment provides specimens that are freeze-dried, and rehydrated in a 10% isopropyl alcohol solution. In these specimens, no other hazardous chemical is present.

Many suppliers provide fixed botanical materials in 50% glycerin.

Reduction of Free Formaldehyde

Currently, federal regulations mandate a permissible exposure level of 0.75 ppm for formaldehyde. Contact your supplier for an MSDS that details the amount of formaldehyde present as well as gas-emitting characteristics for individual specimens.

Pre-washing specimens (in a loosely covered container) in running tap water for 1–4 hours will dilute the fixative. Formaldehyde may also be chemically bound (thereby greatly reducing danger) by immersing washed specimens in a 0.5–1.0% potassium bisulfate solution overnight or by placing them in 1% phenoxyethanol holding solutions.

Safety With Microbes

Pathogenic (disease-causing) microorganisms are not appropriate investigational tools in the high school laboratory and should never be used.

Safety With Chemicals

Label student reagent containers with the substance's name and hazard class(es) (flammable, reactive, etc.).

Dispose of hazardous waste chemicals according to federal, state, and local regulations. Refer to the Material Safety Data Sheet for recommended disposal procedures.

Remove all sources of flames, sparks, and heat from the laboratory when any flammable material is being used.

Material Safety Data Sheets

The purpose of a Material Safety Data Sheet (MSDS) is to provide readily accessible information on chemical substances commonly used in the science laboratory or in industry.

The MSDS should be kept on file and referred to BEFORE handling ANY chemical. The MSDS can also be used to instruct students on chemical hazards, to evaluate spill and disposal procedures, and to warn of incompatibility with other chemicals or mixtures.

Resources

American Chemical Society Health and Safety Service

This service will refer inquiries to appropriate resources for finding answers to questions about health and safety.

American Chemical Society (ACS)
1155 Sixteenth Street, N.W.
Washington, D.C. 20036
(202) 872-4511

Hazardous Materials Information Exchange (HMIX)

Sponsored by the Federal Emergency Management Agency and the U.S. Department of Transportation, HMIX can be accessed through an electronic bulletin board, and it provides information regarding instructional material and literature listings, hazardous materials, emergency procedures, and applicable laws and regulations.

HMIX can be accessed by a personal computer having a modem (300, 1200, or 2400 baud) with communication parameters set to no parity, 8 data bits,

and 1 bit stop. Dial (312) 972-3275. The service is available free of charge. You pay only for the telephone call.

Safety Information References

Gessner, G.H., ed. *Hawley's Condensed Chemical Dictionary*. 11th Ed. Van Nostrand Reinhold, 1987. (Revised by N. Irving Sax).

A Guide to Information Sources Related to the Safety and Management of Laboratory Wastes from Secondary Schools. New York State Environmental Facilities Corp., 1985.

Lefevre, M.J. *The First Aid Manual for Chemical Accidents*. Dowdwen, 1989. (Revised by Shirley A. Conibeau).

Pipitone, D., ed., *Safe Storage of Laboratory Chemicals*. John Wiley, 1984.

Prudent Practices for Disposal of Chemicals from Laboratories. Committee on Hazardous Substances in the Laboratory, National Research Council, National Academy Press, 1983.

Prudent Practices for Handling Hazardous Chemicals in Laboratories, Committee on Hazardous Substances in the Laboratory, National Research Council, National Academy Press, 1981.

Strauss, H. and M. Kaufman, ed. *Handbook for Chemical Technicians*. McGraw-Hill, 1981.

WARD'S MSDS User's Guide. WARD'S, 1989.

Storing Chemicals

Never store chemicals alphabetically, as this greatly increases the risk of promoting a violent reaction.

Storage Suggestions:

1. Always lock the storeroom and all its cabinets when not in use.

2. Students should not be allowed in the storeroom and preparation areas.

3. Avoid storing chemicals on the floor of the storeroom.

4. Do not store chemicals above eye level or on the top shelf in the storeroom.

5. Be sure shelf assemblies are firmly secured to walls.

6. Provide for anti-roll lips on all shelves.

7. Shelving should be constructed out of wood. Metal cabinets and shelves are easily corroded.

8. Avoid metal, adjustable shelf supports and clips. They can corrode, causing shelves to collapse.

9. Acids, flammables, poisons, and oxidizers should each be stored in their own locking storage cabinet.

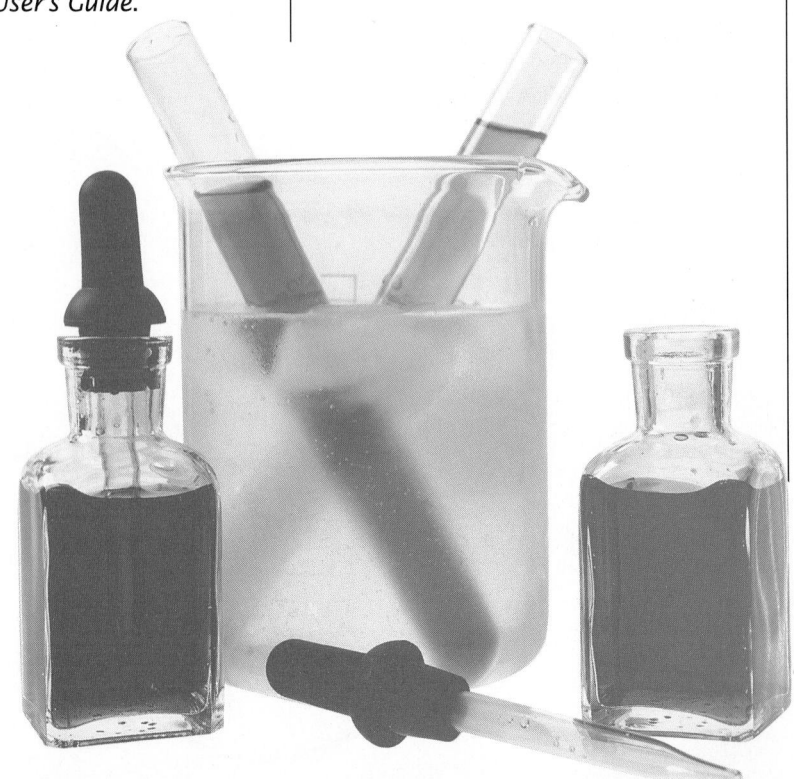

Master Material List

This list indicates all supplies needed to perform the in-text investigations as written (usually) for a class of 30 working in pairs.

WARD'S, the exclusive supplier for *Biology: Visualizing Life*, has created kits for each investigation, shown below. Each kit contains all supplies for a class of 32. Equipment and supplies may also be ordered separately using the charts that follow. Catalog numbers for WARD'S and the quantities to order are also provided.

WARD'S Natural Science Establishment, Inc.
5100 W. Henrietta
P.O. Box 92912
Rochester, New York 14692-9012

Phone: 1-800-962-2660
FAX: 1-800-635-8439

Investigation Kits

Investigation Title	WARD'S No.
Complete Set (Investigations 1–34)	36 M 0185•
1. How Do Scientists Seek Answers to Questions?	36 M 0151
2. What Properties Can Be Observed in Living Things?	36 M 0152
3. How Do Plant and Animal Cells Differ?	36 M 0153
4. Can All Molecules Diffuse Through a Membrane?	36 M 0154
5. How Does Temperature Affect the Rate of Photosynthesis?	36 M 0155
6. What Is Your Genetic Profile?	36 M 0156
7. How Does DNA Send Messages to the Cytoplasm?	36 M 0157
8. How Are DNA Fingerprints Interpreted?	36 M 0158
9. How Does Natural Selection Work?	36 M 0159
10. Can Life Arise From Nonlife?	36 M 0160
11. Comparing Structures of Animal Jaws	36 M 0161
12. Exploring a Soil Community	36 M 0162•
13. How Does Competition Affect Plant Growth?	36 M 0163
14. How Does Pollution Affect Organisms?	36 M 0164
15. How Do You Use a Classification Key?	36 M 0165
16. How Are Models of Viruses Constructed?	36 M 0166
17. What Are Slime Molds Like?	36 M 0167
18. Does Soil Type Affect the Germination of Seeds?	36 M 0168
19. How Is Flower Structure Related to Function?	36 M 0169•
20. Can Plants Be Grown Without Soil?	36 M 0170
21. How Does Segmentation Help an Earthworm Move?	36 M 0171
22. How Are Anole Lizards Well Adapted to Their Environment?	36 M 0172
23. How Do You Make a Key of the Major Animal Phyla?	36 M 0173
24. Can Pill Bugs Detect Differences in Moisture and pH?	36 M 0174
25. How Do Goldfish Respond to Light?	36 M 0175
26. How Do Down and Contour Feathers Differ?	36 M 0176
27. How Do Muscles and Bones Work Together?	36 M 0177•
28. How Does the Brain Interpret Information From the Eyes?	36 M 0178•
29. How Does the Endocrine System Work?	36 M 0179•
30. How Does Smoking Affect the Air Passages?	36 M 0180•
31. How Does Exercise Affect Pulse Rate?	36 M 0181•
32. How Do Antibody-Antigen Reactions Work?	36 M 0182
33. How Do You Test Foods for Nutrients?	36 M 0183•
34. How Are Sperm and Eggs Different?	36 M 0184

• Although the kits contain all the consumables needed to perform the investigation, they do not contain items such as brown paper, chicken wings, cigarettes, flowers, food substances, matches, and unlined paper (see "Miscellaneous"). These items should be purchased separately at a grocery store. In addition, Investigations 12 and 27 require access to a refrigerator and freezer.

Biological Supplies (includes living organisms)

Item	WARD'S No.	Qty.	Inv.
Amoeba Culture, vital stained	87 M 0380	1	14
Animal Phyla Survey Collection	62 M 0065	8	23
Anoles, 3	87 M 8135	3	22
Arthropod Classification Set	62 M 0032	11	15
Bean, Broad Windsor, seeds	86 M 8003	1	20
Bean, Lima, seeds	86 M 8008	2	13, 18
Contour Feather, microscope slide	92 M 3805	15	26
Contour/Down Feather Set, 10	69 M 2269	2	26
Daphnia Culture	87 M 5200	1	14
Down Feather, microscope slide	92 M 3806	15	26
Earthworm, live, 10	87 M 4660	2	21
Elodea, 10	86 M 7500	2	3, 5
Goldfish, 12	87 M 8100	1	25
Human Cheek Cell, microscope slide	93 M 6003	15	3
Pill Bug Culture	87 M 5520	2	24
Pond-Water Dry Mix	87 M 9055	1	2
Sea Star Development, microscope slide	92 M 8255	15	34
Sea Star Egg, microscope slide	92 M 8241	15	34
Sea Star Sperm, microscope slide	92 M 8238	15	34
Simulated Blood Typing Kit, class size	36 M 0022	1	32
Slime Mold Culture	85 M 4750	1	17
Tenebrio Larvae, 100	87 M 6250	1	22

Chemicals and Media

Some items in this category have limited shelf lives.

Item	WARD'S No.	Qty.	Inv.
Albumin, 100 g	39 M 0197	1	33
Amylase, 20 g	39 M 0058	1	4
Benedict's Solution, 500 mL	37 M 0698	1	4, 33
Biuret Reagent, 120 mL	37 M 0790	1	33
Clay, powdered, 500 g	46 M 0995	4	18
Dechlorinator	21 M 2292	1	25
Detain™	37 M 7950	8	2
Glucose, 500 g	39 M 1455	1	33
Lugol's Iodine Solution, 500 mL	39 M 1685	1	4, 33
Nutrient Agar, 1 lb. bottle	88 M 1500	1	17
Nutrient Agar Plates, 6	88 M 0905	2	12
Nutrient Broth (dehydrated)	88 M 0003	4	10
Salt Solution, 10% (makes 1 L)	37 M 9549	1	14
Sodium Bicarbonate, 500 g	37 M 5464	1	5
Sodium Hydroxide, dilute (makes 1 L)	37 M 9546	1	24
Starch, soluble, 100 g	39 M 3275	2	4, 33
Water, distilled, 3.8 L	88 M 7005	1	throughout

Laboratory Equipment

Item	WARD'S No.	Qty.	Inv.
Aspirator Pump	18 M 1570	8	30
Balance, triple beam	15 M 6057	15	12, 13
Beaker, 1000 mL	17 M 4080	8	22, 25
Beaker, 600 mL	17 M 4060	15	5, 22
Beaker, 400 mL	17 M 4050	15	4
Beaker, 250 mL	17 M 4040	8	4, 20
Beaker, 50 mL	17 M 4010	15	14
Burette Clamp	15 M 3917	8	30
Calculator	27 M 3055	15	31
Cold Pack, 12	15 M 8797	2	4, 14
Corks, 100	15 M 8350	1	16
Coverslips, 100	14 M 3555	1	2, 3, 14, 19, 32
Dialysis Tubing, 35 mm	14 M 4517	1	4
Dissection Pan Set	14 M 7011	15	12, 21, 24, 27
DNA and Molecular Model Kit	36 M 5523	15	7
Filter Paper, 9 cm, package	15 M 2835	1	24
Flashlight, pen	15 M 3264	8	25
Flask, Erlenmeyer, 250 mL	17 M 2982	36	10
Forceps	14 M 0999	15	2, 3, 12, 19
Funnel	17 M 0220	15	5
Glass Rods, 10	17 M 6010	2	17
Glass Tubing Smoking Model, 8	15 M 0508	1	30

Item	WARD'S No.	Qty.	Inv.
Gloves, disposable, 100	15 M 1072	1	throughout
Graduated Cylinder	18 M 1730	15	13
Hand Lens	25 M 1350	15	12, 19
Hot Plate	15 M 8055	8	4, 33
Lamp, heat	36 M 4168	8	22
Marker	726 M 0008	15	7
Meter Stick	15 M 4065	8	9
Microscope, compound light	24 M 2310	15	throughout
Microscope Slides, 72	14 M 3500	1	3, 19, 32
Microscope Slides, 12, concave	14 M 3510	2	2, 14
Mirror	15 M 9860	15	6, 11
Needle	14 M 0656	15	17
Net (fish)	21 M 3702	1	25
Personal Lab Safety Set	15 M 3044	30	throughout
Petri Dishes, 20	18 M 7101	1	17, 22
pH Indicator Strips, 200	15 M 2505	1	24
Pipette Filler	18 M 2340	15	14
Pipette, glass	14 M 3417	15	14
Pot, plastic, 3 in., 10	20 M 2130	5	13, 18, 20
Probe, blunt	14 M 0950	15	27
Razor Blades, single-edged, 100	14 M 4172	1	17
Refrigerator/Freezer	•	1	12, 27
Ring Stand	15 M 0719	8	30
Rubber Stoppers, one-hole, size 6	15 M 8486	2	10
Rubber Stoppers, solid size 6	15 M 8466	2	10
Ruler	14 M 0810	15	throughout
Scalpel Blade, refill	14 M 0709	15	27
Scalpel with Blade	14 M 0705	15	27
Scissors	14 M 0988	15	7, 25, 27
Stereomicroscope	24 M 4602	15	12, 14
Stopwatch	15 M 0512	15	1, 5, 25, 31
Swabs, sterile, 100	14 M 5502	1	12
Terrarium	21 M 2100	8	22
Test Tubes	17 M 0620	72	4, 5, 12, 33
Test Tube Rack	18 M 0010	8	4, 12, 33
Thermometer	15 M 1462	15	5, 14
Tray	18 M 3650	8	12
Trowel	20 M 7015	8	12
Watch Glasses, Syracuse	17 M 0530	15	22

• Access to a refrigerator and freezer is required for Investigations 12 and 27.

Miscellaneous

Item	WARD'S No.	Qty.	Inv.
Aluminum Foil	15 M 1009	1	10
Balloons, long, 72	15 M 9011	1	21
Bead Set	15 M 0504	1	8, 9
Cellophane, red, roll	14 M 8200	1	25
Cellophane Tape, roll	15 M 1957	8	25, 28
Chicken Wings	•	15	27
Cigarettes	•	15	30
Clay	36 M 4147	5	14, 16
Construction Paper, 50 sheets	15 M 9841	3	16, 22, 25
Cooking Oil, 500 mL	37 M 9539	1	33
Fertilizer, 1 lb. package	20 M 6020	1	20
Flowers	•	15	19
Food Substances (for testing)	•	8	33
Glue, white, 4 oz.	15 M 9806	15	19
Graph Paper, 100	15 M 3835	2	13, 21
Gravel, 25 lb.	21 M 1802	1	18
Index Cards, 1000	15 M 9807	1	28
Lens Paper, 50	15 M 8250	15	3
Light Bulb for Heat Lamp (200 W)	36 M 4173	8	22
Masking Tape, roll, 3	15 M 9828	3	9, 12, 16
Matches	•	30	30
Medicine Dropper, 12	17 M 0230	2	throughout
Oatmeal, 1 oz.	38 M 0590	1	17
Paper Towels, roll	15 M 9844	5	throughout
Paper, brown packing, roll	•	1	33
Paper, unlined	•	120	27, 28, 29,31
Pencils, colored, 12	15 M 2576	15	27, 29
Pencils (No. 2), 12	15 M 9816	3	28, 31
Pipe Cleaners, 12	82 M 1116	15	16
Plastic Bags, zipper type, 10	18 M 6924	5	12, 14
Potting Soil, 8 lb. bag	20 M 8306	5	13, 18, 20
Push Pin Set	15 M 0505	1	7
Sand, 5 kg	20 M 7425	2	18, 20
Straws, 500	15 M 9842	1	7
String (or Thread), 400 ft.	15 M 9863	2	4, 8
Styrofoam Balls, 144	82 M 1011	2	16
Toothpicks, 800	15 M 9840	1	2, 16
Vinegar, 1 pt.	39 M 0138	2	17, 24
Wax Pencil	15 M 1155	15	4, 10, 17, 33
Wicks, 12	23 M 1303	2	20
Wire, floral, roll	15 M 9838	7	16
Wooden Dowels, 25	15 M 9849	3	16

• Brown paper, chicken wings, cigarettes, flowers, food substances, matches, and unlined paper are not available through WARD'S. They should be purchased separately at a grocery store.

Enrichment

Additional related materials available from WARD'S.

Item	WARD'S No.	Qty.	Chap.
Canine Skull	65 M 5310	1	11
Cow Skull	65 M 5125	1	11
Crustaceans, slide set	175 M 0511	1	24
Dialysis Tubing Closures, 10	14 M 4526	3	4
Dicot Flower Model	81 M 1130	1	19
Discovery Scope Naturalist Kit	25 M 5502	15	2, 12
Extraction of Bacterial DNA Kit	88 M 8112	1	7
Five Kingdom Classification Kit	32 M 2208	1	15
Genetics Made Easy Demonstration Kit	36 M 1495	1	6
Germination Model	81 M 6142	1	18
Giant Brain Model	81 M 1080	1	28
Gorilla Skull Replica	80 M 1755	1	11
pH, Osmosis & Diffusion, slide set	170 M 9112	1	4
Photosynthesis Made Easy Kit	36 M 5737	1	5
Physarum polycephalum, microscope slide	91 M 2015	15	17
Plastic Human Skull	82 M 3031	1	11
Scientific Method Problem-Solving Lab Activity	36 M 5770	1	1
Slime Molds, slide set	173 M 0627	1	17
Soil—We Can't Grow Without It, slide set	175 M 0470	1	12
Soils & Associated Vegetation, slide set	173 M 0633	1	12
Working Muscular Arm Model	82 M 3823	1	27

Ancillary Materials and Media From Holt, Rinehart and Winston for *Biology: Visualizing Life*

Teacher's Resource Binder	**H76413-0**
Review Guide	
Laboratory Experiments, Pupil's Edition	
Chapter Tests	
Answer Keys	
Laboratory Experiments, Pupil's Edition	**H76414-9**
Laboratory Experiments, Teacher's Edition	**H76416-5**
Review Guide	**H76418-1**
Teaching Transparencies	**H76417-3**
Spanish Resources	**H76421-1**
English/Spanish Audiocassettes	**H76419-X**
Test Generator (IBM® PC and Compatibles)	**H98029-1**
Test Generator (Macintosh®)	**H98027-5**

Computerized Laboratories (Apple® II)	
Photosynthesis	**H12703-3**
Cellular Respiration	**H12387-9**
Genetics	**H12389-5**

Science in Action Video Series	
Laboratory Safety, Equipment, and Technique/ The Process of Science	**H53047-4**
The Atoms and Molecules of Life/DNA and Protein Synthesis	**H53043-1**
Cell Structure and Function/Cell Division	**H53042-3**
Photosynthesis and Cellular Respiration	**H53044-X**
Alternatives to Dissection: The Earthworm and the Frog	**H53039-3**

Holt Videodisc for *Biology: Visualizing Life*	**(available Fall '93)**

Media Resources from WARD'S

Item	WARD'S No.	Chap.
Books/Charts		
Adaptations Chart	33 M 0591	22
Animal and Plant Cell Chart	33 M 0634	3
Botany Chart Set	33 M 0593	20
Endocrine System Chart	33 M 5409	29
Germination of Seeds Chart	33 M 0574	18
Hydroponics Book	32 M 0626	20
Soils of the World Chart	33 M 5001	12
Films/Filmstrips		
Adaptations Filmstrip	70 M 1503	22
Diffusion & Osmosis Filmstrip Set	78 M 0340	4
DNA, set of transparencies	75 M 0150	7
Introducing Genetics Filmstrips	70 M 3400	6
Natural Selection Filmstrip	70 M 0989	9
Photosynthesis Slide Set	171 M 9883	5
Software		
Bio▪Cell, Apple II	74 M 0113	3
Bio▪Cell, IBM	74 M 0112	3
Biochemistry of Viruses/Viruses & Cancer, Apple II	74 M 2106	16
Classification System, Apple II	74 M 1644	15
Dichotomous Key to Pond Microlife, Apple II	74 M 1858	2
Dichotomous Key to Pond Microlife, Mac	74 M 0041	2
DNA—The Master Molecule, Apple II	74 M 1662	7
DNA—The Master Molecule, IBM	74 M 1665	7
DNA—The Master Molecule, Mac	74 M 1656	7
Endocrine System, Apple II+	74 M 1634	29
Genetics Problem Shop, Apple II	74 M 1686	6
Inner Body Works, IBM	74 M 0018	27–31
Inner Body Works, Mac	74 M 0017	27–31
Introductory Genetics, Apple II	74 M 1675	6
Natural Selection Program, Apple II	74 M 1670	9
Osmosis & Diffusion, Apple II	74 M 1664	4
Photosynthesis Advanced Unit, Apple II	74 M 1837	5
Photosynthesis Basic Unit, Apple II	74 M 1836	5
Photosynthesis & Transport, Apple II	74 M 1637	5
Plant Competition Program, Apple II	74 M 1605	13
Plant Growth Simulator Program, Apple II	74 M 1992	13
Plant Reproduction Program, Apple II	74 M 1638	19
Senses of the Body, Apple II	74 M 0162	28
The Plant Growth Package, Apple II	74 M 1986	18
The Scientific Method, Apple II	74 M 1876	1
The Study of Cells—Clip Art, Mac	74 M 3650	3

Video/Laserdisc		
Animal Kingdom, Video (VHS)	193 M 6440	23
Biological Classification and the Five Kingdoms, Video (VHS)	193 M 6435	15
Cell Biology Laserdisc	196 M 0006	3
Cell Biology Laserdisc Software	196 M 0008	3
Cell Biology Resource Series Videos	193 M 6456	3
Evolution: Inquiries into Biology (Laserdisc)	196 M 0022	11
Evolution, Video	193 M 0264	10
Exploring Hidden Worlds, Video	193 M 5500	2, 17
Genetic Engineering, Video	193 M 1503	6
"Home Sweet Hole," Adaptations Video	193 M 0260	22
Intriguing Methods of Pollination	193 M 0201	19
Lung Action & Function Video	193 M 2096	30
Photosynthesis Laserdisc	196 M 0020	5
Photosynthesis Video & Guide (VHS)	193 M 0314	5
Protist Ecology, Video	193 M 2002	2
Pumping Life (Circulatory System Video)	193 M 0009	32
Scientific Methods & Values, Video	193 M 0082	1
Skeletal & Muscle Action Video	193 M 2099	27
The Body Against Disease, (VHS)	193 M 0901	29
The Cell: Unit of Life Video	193 M 6442	3
Viruses Video (VHS)	193 M 2011	15

Credits

List of Abbreviations
All photographs attributed to SP/Foca were taken by Sergio Purtell/Foca Co., NY, NY.
AA = Animals Animals; AT = Alexander & Turner; CBMI = Claire Booth Medical Illustration; HS = Highlight Studios; KBI = Karapelou BioMedical Illustrations; KC= Kip Carter, M.S. certified medical illustrator; LM = Les Mintz/Incandescent Ink, Inc.; LPI = Lynne Prentice Illustration; MA = rep. Mattleson Associates, Ltd.; MC = MediChrome; MMA = Margulies Medical Art; MT = rep. Melissa Turk; PA = Peter Arnold, Inc.; RMF = rep. Rita Marie & Friends; RWS = rep. Richard W. Salzman S.F., CA; SS = Studio Sloth; TM = Teri J. McDermott, M.A., certified medical illustrator; TMMK = Teri J. McDermott, M.A., certified medical illustrator/Michael Kress-Russick, M.A., M.S.; VU = Visuals Unlimited; all photos of George Johnson, courtesy of George Johnson; globe on all wrap pages courtesy of NASA
t = top; b = bottom; c = center; l = left; r = right; bkg = background

15T SP/Foca; 16T–17T SP/Foca; 23T SP/Foca; 24T SP/Foca; 25T SP/Foca; 28T SP/Foca; 32T SP/Foca; 37T SP/Foca; 38T SP/Foca; 39T SP/Foca; 40T SP/Foca; 43T SP/Foca; 3A–D (bkg) SP/Foca; 3B (cl) Karen Kluglein/MA; (br) SP/Foca; 3C (tc) ©John Moss/MediChrome; 23A–D (bkg) ©Visuals Unlimited; 23B Edmond Alexander/AT; 23C SP/Foca; 43A–D (bkg) ©Visuals Unlimited; 43B (cl) SP/Foca; (br) Robert Margulies/MMA; 43C Claire Booth/CBMI; 67A–D (bkg) SP/Foca; 67B Claire Booth/CBMI; 67C SP/Foca; 87A–D (bkg) SP/Foca; 87B (cl) ©Walter H. Hodge/Peter Arnold, Inc.; (br) SP/Foca; 87C Claire Booth/CBMI; 115A–D (bkg) Robert Margulies/MMA; 115B SP/Foca; 135A–D (bkg) Claire Booth/CBMI; 135B (cl) Claire Booth/CBMI; (br) SP/Foca; 135C Robert Margulies/MMA; 155A–D (bkg) Edmond Alexander/AT; 155B (tc) John W. Karapelou/KBI; (bc) Courtesy WARD's Natural Science Establishment; 175A–D (bkg) SP/Foca; 175B (br) SP/Foca; (tc) ©Tom Ulrich/Visuals Unlimited; 197A–D (bkg) Laurie O'Keefe/HS; 197B (br) SP/Foca; (cl) Claire Booth/CBMI; 197C SP/Foca; 219A–D (bkg) Todd Buck/HS; 219B Randall Zwingler/LM; 219C SP/Foca; 245A–B (bkg) SP/Foca; 245B (cl) ©John Moss/MC; (br) SP/Foca; 245C Biruta Akerbergs Hansen/NSI; 265A–D (bkg) SP/Foca; 265B SP/Foca; 265C SP/Foca; 291A–D (bkg) SP/Foca; 291B SP/Foca; 317A–D (bkg) © Jeff Rotman/Peter Arnold, Inc.; 317B (cl) ©A.J. Copley/Visuals Unlimited; (br) SP/Foca; 317C SP/Foca; 337A–D (bkg) ©David Scharf/Peter Arnold, Inc.; 337B Edmond Alexander/AT; 337C (tr) Edmond Alexander/AT; (bl) ©David M. Phillips/Visuals Unlimited; 365A–D (bkg) ©T.E. Adams/Visuals Unlimited; 365B (bl) ©Cabisco/Visuals Unlimited; (cr) Lynn Prentice/LPI; 383A–D (bkg) SP/Foca; 383B SP/Foca; 401A–D (bkg) SP/Foca; 401B SP/Foca; 401C SP/Foca; 425A–D (bkg) SP/Foca; 425B (tc) ©Cabisco/Visuals Unlimited; (br) Deborah Daugherty/HS; 425C SP/Foca; 453A–D (bkg) SP/Foca; 453B (c) Claire Booth/CBMI; (br) SP/Foca; 453C SP/Foca; 479A–D (bkg) SP/Foca; 479B (tc) ©A. Kerstitch/Visuals Unlimited; (br) SP/Foca; 479C SP/Foca; 501A–D (bkg) SP/Foca; 501B SP/Foca; 501C (cl) ©Donald Specker/Animals Animals; (b) SP/Foca; 523A–D (bkg) SP/Foca; 523B SP/Foca; 523C SP/Foca; 547A–D (bkg) SP/Foca; 547B SP/Foca; 547C SP/Foca; 571A–D (bkg) SP/Foca; 571B (br) Pat Ortega; (bl) SP/Foca; 571C SP/Foca; 609A–D (bkg) SP/Foca; 609B (tc) NASA; (bc) SP/Foca; 609C Robert Margulies/MMA; 637A–D (bkg) Kevin A. Somerville/CBMI; 637B (tl) Kevin A. Somerville/CBMI; (br) SP/Foca; 637C Kip Carter/KC; 667A–D (bkg) Robert Margulies/MMA; 667B ©Werner H. Mueller/Peter Arnold, Inc.; 667C Peg Gerrity/HS; 689A–D (bkg) SP/Foca; 689B Peg Gerrity/HS; 689C SP/Foca; 709A–D (bkg) Teri J. McDermott/Michael Kress-Russick/TMMK; 709B (tc) NASA; (br) SP/Foca; 709C SP/Foca; 737A–D (bkg) John W. Karapelou/KBI; 737B (c) Tomo Narashima/SS; (br) SP/Foca; 737C Claire Booth/CBMI; 761A–D (bkg) Robert Margulies/MMA; 761B SP/Foca; 761C Teri J. McDermott/TM; 781A–D (bkg) ©Fred Hossler/Visuals Unlimited; 781B (cc) Robert Margulies/Virginia Ferrante/MMA; (br) SP/Foca; 781C Edmond Alexander/AT

45T

Unit 1

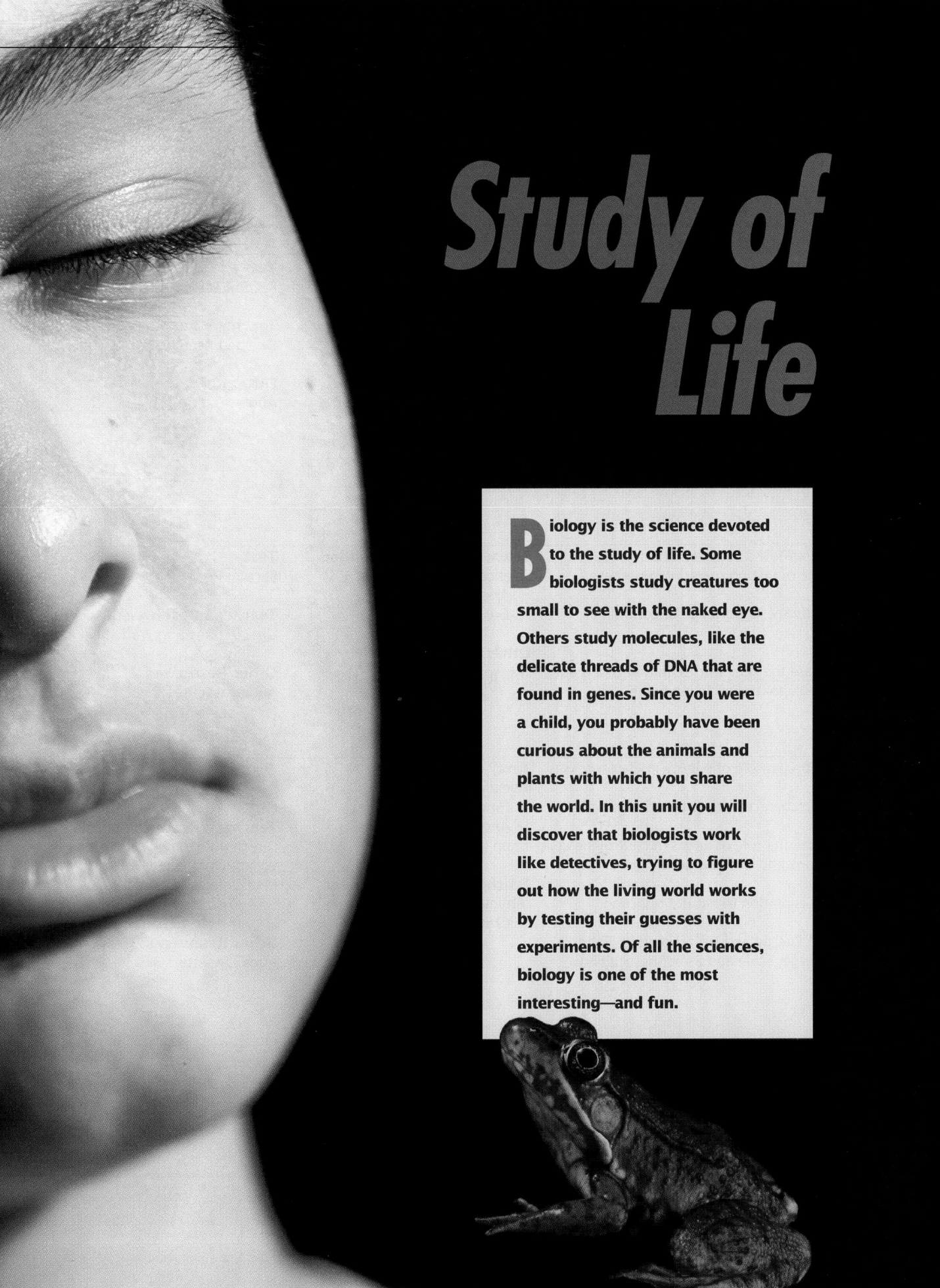

Study of Life

Biology is the science devoted to the study of life. Some biologists study creatures too small to see with the naked eye. Others study molecules, like the delicate threads of DNA that are found in genes. Since you were a child, you probably have been curious about the animals and plants with which you share the world. In this unit you will discover that biologists work like detectives, trying to figure out how the living world works by testing their guesses with experiments. Of all the sciences, biology is one of the most interesting—and fun.

Chapter 1 — The Science of Biology

Planning Guide

	Objectives/Themes	Classwork Resources	Homework Resources

1.1	**1.** Describe the research for the cure and prevention of disease.
	
	2. List the ways a knowledge of biology will affect the future of life on Earth.
	
	3. List the ways biology affects you now and will continue to affect you in the future.
	
	Theme: Interacting Systems
		Teacher's Resource Binder
		Focus Activity 1
		  Using Your Imagination:
		  Problem Solving
		Lab Investigation 1.1
		  Compound Microscope
		
		Other Resources
		Science in Action Video
		  Laboratory Safety, Equipment,
		  and Technique/The Process of
		  Science★
			Text
			Section Review, p. 10
			
			Teacher's Resource Binder
			Directed Reading Worksheet 1.1
			
			Other Resources
			Audiocassette 1.1★

1.2	**1.** Describe how a hypothesis is formed and the role of testing to verify hypotheses.
	
	2. Describe the importance of controls in testing hypotheses.
	
	3. Compare the scientific definition of a theory to the use of theory in language.
		Teacher's Resource Binder
		Science Skills Worksheet
		  Analyzing Scientific
		  Experiments
		
		Other Resources
		Transparencies 1–2
			Text
			Section Review, p. 15
			
			Teacher's Resource Binder
			Directed Reading Worksheet 1.2
			
			Other Resources
			Audiocassette 1.2★

1.3	**1.** Describe the role of energy in life processes.
	
	2. List the characteristic properties of cells.
	
	3. Describe the role of genes in controlling cell functions and development.
	
	4. Relate evolution to natural selection.
	
	Themes: Energy and Life, Evolution, Patterns of Change, Scale and Structure, Stability
		Text
		Investigation
		  How Do Scientists Seek
		  Answers to Questions?
		  pp. 22–23
		
		Teacher's Resource Binder
		Lab Investigation 1.2
		  Crime Lab
			Text
			Section Review, p. 18
			
			Teacher's Resource Binder
			Directed Reading Worksheet 1.3
			Vocabulary Review Worksheet★
			Reteaching Worksheet★
			  A Case Study in Biology
			
			Other Resources
			Audiocassette 1.3★

★Reteaching Options

Demonstrations	Assessment
1.1: pp. 5, 9 **1.2:** p. 13 **1.3:** pp. 16, 17, 18	Chapter Review pp. 20–21 Portfolio Assessment p. 3D Chapter Test—Teacher's Resource Binder Test Generator

Research Notes

Connection to History: Pseudoscience

Throughout the time after World War II, the United States and the Soviet Union were rivals in fields from rocketry to nuclear chemistry. But the republics of the former Soviet Union have always been, and still are, much weaker in the biological sciences, especially genetics. The reasons can be traced to one man's "bad" science or "pseudoscience."

In the late 1920s, a young scientist named T. D. Lysenko performed an experiment planting peas in the autumn for harvest in the winter. The plants grew well, and the experiment was described in the newspapers as a

great success, since it could help the new Soviet agricultural system increase production.

However, the data were never published for other scientists to examine and retest. Also, that winter was unusually mild. Lysenko had not controlled his experiment to be sure his conclusions were valid.

He went on to other experiments, which he claimed changed winter wheat into spring wheat, and wheat into rye, again using questionable experimental techniques. Lysenko publicized his ideas, and the Soviet government was eager to apply them and reap the benefits.

Even though none of his innovations worked as well as he promised, Lysenko became the most trusted science advisor of Soviet leader Joseph Stalin. Stalin favored Lysenko partly because Lysenko had been born into a peasant family, which fit well with Stalin's ideas about who should lead the Soviet Union. Lysenko's ideas were also what Stalin wanted to hear: quick, easy solutions to the Soviet Union's problems. But these criteria are not appropriate for judging the validity of scientific work.

In the meantime, instead of genetics and heredity, Lysenko embraced a theory that plants can pass on acquired characteristics to their offspring. His theory

was completely opposed to the science of genetics, but it fit very well with the beliefs of some Soviet leaders.

Instead of carefully testing this theory or examining arguments against it, Lysenko used his influence to silence those who disagreed. Eventually 3,000 researchers in genetics were dismissed from work. One by one, his most vocal opponents were arrested. Many were killed or died in prison. Soon, few researchers would work on experiments that might involve genetics. Work in biochemistry and many related fields came to a standstill.

Even after Stalin and Lysenko died and the government officially repudiated Lysenko's ideas, Soviet biologists found themselves starting from scratch, since an entire generation of researchers had been wiped out and there had been little contact with biologists from other nations.

The Soviet Union did not have a monopoly on pseudoscience. Even today, politicians and others make wild claims about things that are supposedly "scientifically proven," and several famous researchers are under fire because of allegations that they have not performed their experiments according to the scientific method.

Investigation Notes

How Do Scientists Seek Answers to Questions?

pages 22–23

Purpose: Students have an opportunity to practice problem-solving through experimentation. This investigation may also offer an opportunity for learning through discrepant events, since most students will erroneously hypothesize that heart rate will increase as a result of holding their breath.

Prelab Preparation

1. A clock with a second hand should be available for students who do not have a watch.

2. Although the investigation should not unduly stress the body, caution any student with a health problem to avoid potentially dangerous activity.

Answers will be found on pages 22–23.

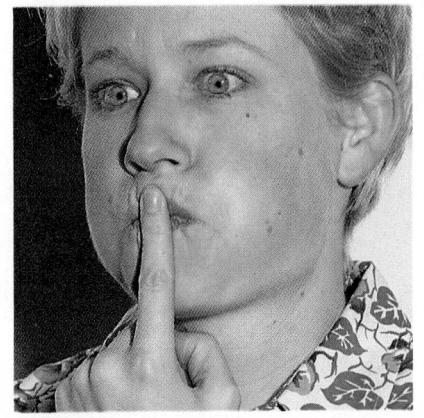

Meeting Individual Needs

Objectives

1. Students will demonstrate appropriate use of core vocabulary for the chapter (Vocabulary File).

2. Students will be able to describe ways that biologists help improve the general health of human, plant, and animal populations (Multicultural Lesson Plan).

Vocabulary File
(Developing Vocabulary/ Limited English Proficiency)

For each term, have students examine a page where the term is used in context. Start a discussion by asking students if the pictures or diagrams on the page relate to the term in question.

Have students read the relevant parts of the page silently, and then continue the discussion. Lead the discussion toward a class consensus of a simplified definition and a sketch or illustration that will symbolize the term.

Many students with less developed verbal skills have compensated by improving their artistic skills. Have such students draw the sketches and print the definitions for each term on an oversized card. File the cards by chapter in a box labeled "Vocabulary File." Place the box in an accessible place in the classroom. Encourage students to refer to it at any time.

Write each of the vocabulary terms on a self-stick note and attach it to the chalkboard. Using these self-stick notes, start a discussion that ends with the preparation of a concept map on the chalkboard.

See the Chapter Highlights on page 19 for a list of suggested words.

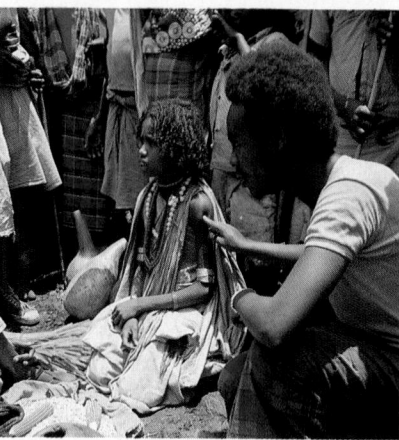

Multicultural Lesson Plan
Biology and the Quality of Life (Dealing with Ethnic Diversity)

Preparation: For source material for students to learn about famine and disease, have a variety of reference materials available, such as encyclopedias and geographical information.

More information about famine around the world is available from CARE or Oxfam America. Be sure to include your name and the name and address of your school when calling or writing. Call CARE at
    1–800–521–CARE (2273).
Ask for teacher's information about famine.

The address for Oxfam America is:

*Oxfam America
Education Outreach
26 West Street
Boston, MA 02111–1206.*

Other information may be found in newspapers, magazines, or books. Possible sources include *World Hunger–Twelve Myths* by Frances Lappe and Joseph Cullins (Institute for Food and Development Policy) and *What Do You Say To A Hungry World* by Stanley Mooneyham (World Book, Inc.).

For information about diseases of humans, contact local hospitals, clinics, and doctor's offices. For information about plant and animal diseases, contact your local agricultural extension agent, your state agriculture department, nearby colleges, and universities.

Teaching Strategies:
1. Lead students in a discussion of the ways that biology has improved their lives.

2. Describe the problems of famine in different areas of the world. Students should pick a region of the world where famine is a problem, and research the root causes and peculiarities of famine in that region.

3. Describe examples of diseases to which humans, animals, and plants succumb. Discuss how biologists have worked to combat these diseases. Students should each choose some diseases for which biologists have found cures and others that remain uncured.

Assessment:
1. Students should prepare a report on famine in the region they chose. Along with the causes of famine, students should discuss the ways that biologists can assist nations in eliminating famine. Be sure students include both immediate and long-term approaches.

2. Students will gather information from the entire class into a chart of diseases, separated into the categories of human, animal, and plant diseases. For each disease, the chart should show whether a cure is available, what the cure is, and what role biologists played in discovering the cure. If the disease remains uncured, the chart should indicate what avenues of biological research are being pursued to seek a cure.

Additional Strategies
Visual Strategies
Pages 6, 8, and 12

Auditory Learners
Use *Biology: Visualizing Life* Audiocassettes for Sections 1.1, 1.2, and 1.3.

Meeting Individual Needs (cont.)

Cooperative Learning
Careers in Biology

Timing: Use this activity to introduce Section 1.3.

Group Size: 2–3 students (divide the class into about 10 groups)

Outcome: Students will identify and distinguish among the different branches of biology and relate each to a biological theme. Students will also identify specific careers and research that are unique to each branch.

Individual Accountability: Each group member will research a career associated with some major field of biology.

Positive Interdependence: Each group will report on careers associated with the chosen field of biology.

Have each group choose a field of biological research from the following: (1) anatomy, (2) biochemistry, (3) botany, (4) cell biology, (5) ecology, (6) evolutionary biology, (7) genetics, (8) microbiology, (9) physiology, (10) wildlife biology, (11) zoology, and (12) forestry. Also list the themes of biology on the board: Energy and Life, Scale and Structure, Stability, Patterns of Change, Evolution, and Interacting Systems.

Each group member will be responsible for researching some pertinent aspect of the chosen field. The idea is for the group to develop a profile of a typical biologist in that field and identify how the themes of biology relate to such a biologist's work.

Once the research is complete, students should compile a report that profiles a day (or week or month) in the working life of a hypothetical biologist, pointing out which themes of biology apply. Finished reports should then be presented to the class.

Portfolio Assessment

Students should select their best work and provide a self-reflective rationale for their selections. Students can make selections in the following areas.

1. *Content* — One concept map from the chapter (See page 812 for evaluation criteria.)

2. *Reading Comprehension* — One Directed Reading Worksheet from the Teacher's Resource Binder (Use the answer key to evaluate for accuracy.)

3. *Writing* — Using the Vee Form, summarize a newspaper or magazine article relating to biology. (See page 22T for evaluation criteria.)

   Or: Select a writing task or project from the Chapter Review.

4. *Performance Assessment* — One Vee form from a chapter investigation or lab manual investigation (See page 22T for evaluation criteria.)

Teacher makes selections in the following areas.

1. *Formal Assessment* — Chapter test (Test A, B, or the Test Generator) The teacher-scored test should be reviewed by the student. Incorrect responses should be corrected by the student before the test becomes part of the portfolio.

2. *Informal Assessment* — Use the Direct Observations Checklist, page 33T, during a laboratory or other cooperative learning experience.

3. *Performance Assessment* — Have students research the costs and benefits of government support for basic research and write a newspaper editorial for or against federal funding of basic research.

Concept Map Answer

The following is one possible answer to the Relating Concepts exercise on page 20.

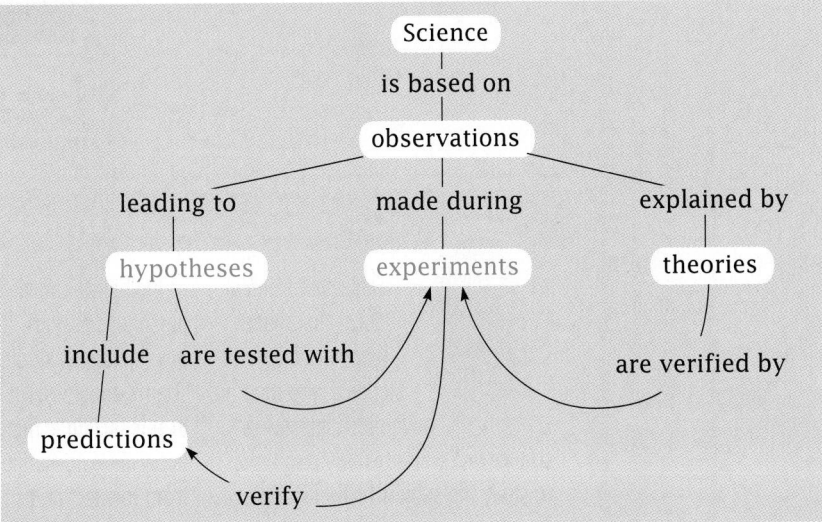

Chapter 1

Determining Prior Knowledge

- Show students photographs of people engaged in various occupations that require a knowledge of biology. Have students suggest how biology is a necessary part of these jobs.
- Have students provide examples of how biology has had an impact on their lives.
- Have students explain what the following terms mean to them: hypothesis, experiment, control experiment, and theory.

Chapter 1

The Science of Biology

1.1   Biology Today

- **Studying Life**
- **Biology and Medicine**
- **Biology and the Environment**
- **Biology and You**

1.2   Science Is a Search for Knowledge

- **A Case Study in Science**
- **Theories Have Limited Certainty**
- **Scientific Method: The Systematic Study of a Question or Problem**

1.3   Studying Biology

- **Themes Unify Ideas**

Biology is the study of the living world and its organisms, like this *Anopheles* **mosquito. It was the study of this type of mosquito that to the discovery of how malaria, a deadly disease, is transmitted.**

■ *Author's Rationale* ■

This chapter introduces the importance of biology to students who may not have much interest in science and who probably don't plan to study the subject further in college. The chapter starts with the importance of biological knowledge in making personal decisions. It then focuses on how biology is studied—the nature of scientific inquiry. An important aspect of this section is that students will apply the knowledge they acquire from this section as they study various topics throughout the year. At the core of the chapter is a description of the major themes that will be encountered repeatedly throughout the course.

4

YOU ARE ABOUT TO EMBARK ON THE STUDY OF BIOLOGY. IT IS THE SCIENCE OF LIFE, OF OURSELVES, AND OF THE LIVING WORLD AROUND US—BUTTERFLIES, VIRUSES, HOW A CHILD GROWS AND WHAT DINOSAURS WERE LIKE. PEOPLE HAVE PUZZLED ABOUT LIFE FOR MANY CENTURIES, BUT ONLY IN THE LAST FEW HAVE WE REALLY BEGUN TO UNDERSTAND HOW THE WORLD AROUND US CAME TO BE THE WAY IT IS.

1.1   Biology Today

Objectives

❶ Describe the research for the cure and prevention of disease.

❷ List the ways a knowledge of biology will affect the future of life on Earth.

❸ List the ways biology affects you now and will continue to affect you in the future.

Studying Life

Science is a way of investigating the world, of observing nature in order to form general rules about what causes things to happen. A scientist is an observer who searches for knowledge to find solutions to problems. Science has changed the world very rapidly in modern times. What scientists have learned about life and how they learned it are the subjects of this text.

You are about to embark on the study of biology, the science of life. **Biology** is the study of living things, of ourselves, and of the living world. While biological scientists, called biologists, continue to learn new things all the time, they now have a pretty clear picture of what living things are like and how they function. Indeed, biologists have learned so much that they have begun to fashion tools to change the living world in very important ways—an exciting and sobering prospect. Biology's impact by the year 2000 will be so great that every person will need to know about biology to make many personal decisions.

Figure 1.1
Exciting advances in science, biology in particular, make the headlines of many newspapers and are the subjects of many fascinating articles read by millions of people each day.

▪ Cultural Perspective ▪

Legends of the Bears

Legends of the Navajo, Chippewa, and Sioux Native Americans have told of bears teaching people how to use herbs and roots for medicine. Scientists are now giving serious attention to these legends. Some tribes observed and described bears digging for roots and bulbs to eat. The Native Americans found that some of these roots could be used to treat certain ailments and fight infections and parasites. In some tribes, strong medicine is still called "bear medicine." Throughout the world, scientists are seeking the wisdom of indigenous peoples and observing the eating habits of animals to learn which plants hold medicinal cures.

Key Concepts
- Biology is the study of living things.
- Biology involves a search for solutions to medical problems and answers to environmental concerns.
- A study of biology will provide information useful in making critical personal decisions.

Reading Strategy

Tell students that reading about biology is not limited to textbooks such as this one. When they have finished reading this section, have each student bring in a copy of a newspaper or magazine article that deals with one of the topics discussed in this section. Post all these articles on a board for the entire class to read, and underscore the significance of the section heading "Biology Today."

Phase 2
TEACHING STRATEGIES

Demonstration 1
Scientists and Detectives
Emphasize the text description of science by relating a story from *The Medical Detectives* by Berton Roueche. If this book is not available, ask your school librarian for assistance in locating a similar book that describes how science was used to solve a bizarre problem. In this way, students can see that science is a search for solutions, not merely a collection of facts and terms they will have to memorize.

Demonstration 2
Studying Life
Take the class outside the school so each team of two students can collect one living specimen: a plant, worm, sow bug, flying

insect, or any other small organism. Have the students examine their specimens in class, making as many observations as possible. Each team should think of a question or problem that stems from their observations. In this way, students will recognize two important attributes of a scientist: the ability to make keen observations, and the aspiration to be a problem solver.

Health Connection

Biology and Medicine
Invite a doctor or research scientist to speak to the class about the progress being made in detecting, treating, or curing a particular disease, perhaps using this opportunity to address the issue of AIDS.

Visual Strategy

Figure 1.2
Use this figure to address several major points about AIDS, especially the fact that anyone can get the disease. AIDS victims have been reported in both males and females, in every age group, in every ethnic group, and in every state of the United States.

Health Connection

The Top Three
Have students discuss what they think the top three causes of death in the United States are today (heart disease, cancer, and stroke), and what they think the top three causes are in the world as a whole.

Biology and Medicine

"What do I need to know about AIDS?"

Stop for a moment and think about the headlines you read in the newspaper this week, the news stories you heard on the radio or TV. Mixed in with bad news and politics is news that modern science is exploring new ways to cure genetically inherited disorders like cystic fibrosis and muscular dystrophy. The fight against infectious diseases like AIDS and malaria, and ways to protect the body from cancer and heart disease, affects everyone, as shown in **Figure 1.2**. The most direct impact of biology on our lives is in medicine, where scientific advances are improving health and health care every day.

Biological knowledge is used in stopping infectious diseases
At the beginning of this century, the infectious diseases flu, tuberculosis, and pneumonia were the top three causes of death in the United States. The flu epidemic of 1918 killed 22 million people worldwide in 18 months! Now, thanks to intense biological research, deaths from these three diseases are far less common. Because of the discovery of antibiotics like penicillin and the development of vaccines that prevent infection, these diseases will probably not be major causes of death in the United States in coming years.

Figure 1.2
AIDS is one of the most devastating diseases of the 1990s. A knowledge of biology helps us to understand what causes AIDS and how the virus that causes it is spread.

a The virus that causes AIDS is small but deadly. Because it may take years for the symptoms of infection to appear, people can spread the virus unknowingly.

b Anyone can get AIDS. Knowing how the virus is transmitted can be the first step in controlling its spread.

■ Matter of Fact ■
More people perished in the flu epidemic of 1918 than were killed in World War I.

The battle against disease is not over. More than 1 million people are likely to die of malaria this year alone. Spread by mosquitoes, this disease is prevalent in tropic regions—more than 250 million people suffer from malaria at any one time. Almost every child under the age of five who contracts malaria dies. The organism that causes malaria has a very complex life cycle, and is therefore difficult to eliminate. Using modern genetic engineering techniques, scientists are trying to design a vaccine that will attack at a critical stage of the organism's life cycle. Also, new ways to control the large mosquito population are being researched.

New strains of bacteria causing tuberculosis, the fatal lung disease, have recently arisen that are resistant to today's antibiotics.

Other scientists are working to find a cure for AIDS, a fatal disease caused by a virus that destroys the body's ability to defend itself from infections. Because the virus changes so quickly, normal vaccines don't work. No one yet knows what the solutions to these problems will be, but many approaches are being explored.

Biological knowledge is used in curing genetic disorders

Among the most disheartening medical problems are fatal disorders that result from inherited defective genes. Two common fatal human genetic disorders are cystic fibrosis and muscular dystrophy. Cystic fibrosis kills 1 in 1,800 Caucasian children. Their lungs are clogged with mucus because of a defect in a single gene. Muscular dystrophy kills about 1 in 10,000 humans. Their muscles waste away because of a defect in another gene.

Until recently, such genetic disorders were almost always fatal, and no cure was known. In 1990, biologists trying a new approach used genetic engineering techniques to transfer copies of a normally functioning gene from a healthy individual into a patient with defective copies of that gene. Gene transfer therapy offers the first hope that cures for previously fatal genetic disorders will be found. Progress with the new therapy is reported almost daily. A major effort is now underway to catalog every gene in the human body, a program that may soon open the avenue to curing many other genetic disorders.

This child has AIDS, which she acquired before her birth. Scientists predict that 40 million people world-wide will be infected with the virus that causes AIDS by the year 2000.

d **Politicians must also understand the biology of AIDS if they are to make informed decisions regarding testing, the allocation of research funds, and protecting the rights of AIDS victims.**

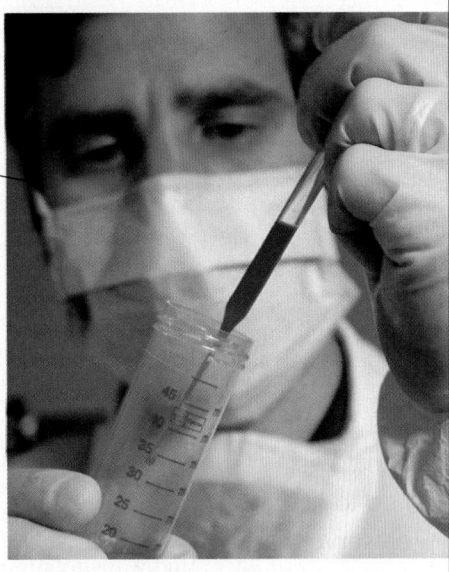

e **Researchers, like this one in the Clinical Immunology Lab at East Tennessee State University, use their knowledge of viruses to understand the biology of the AIDS virus.**

■ *Cultural Perspective* ■

Shaibasaburo Kitasato (Japanese)
Shaibasaburo Kitasato is credited with discovering the bacteria that cause bubonic plague, several weeks before Alexander Yersin of France published his discovery. Kitasato obtained the first pure tetanus in 1889. A year later, he discovered that inoculating animals with serum containing inactive tetanus toxin would prevent tetanus. In 1898, Kitasato isolated the bacteria that cause dysentery.

Visual Strategy

Figure 1.3

Review how scientists shown in this figure are searching for solutions concerning some major environmental problems facing the world.

Figure 1.3

Scientists throughout the world work to create solutions to global problems like toxic waste, destruction of the rain forests, and hunger.

a **Cleaning toxic spills is one way to preserve the planet for future generations. These researchers are testing contaminated soil in New Jersey.**

b **Destruction of the Amazonian rain forest must be avoided if we are to preserve its precious resources. This scientist is studying the bird life in the Cuyabeno Nature Reserve in the Amazon Basin.**

Biology and the Environment

"What can I do to help preserve the environment?"

When you were born, the world held somewhat more than 4 billion people. This year the world's population will pass 5.5 billion, and by the year 2000 it is estimated that it will exceed 6 billion. This exploding population is placing great stress on the planet, using a lot more energy, consuming more resources, and producing more waste than ever before in the history of the planet. One of the greatest challenges we face entering the next century is to find ways to support so many people without harming the Earth. The scientists in **Figure 1.3** use their knowledge and skill to seek solutions for our world and its future health.

Biological knowledge is used in feeding a hungry world

One of the most immediate challenges facing today's world is to produce enough food to feed its expanding population. Only three kinds of plants (rice, wheat, and corn) provide half of all human energy requirements worldwide. Researchers are presently working to increase the amount of food that can be obtained from a farm without demanding heavy use of fertilizers, pesticides, and equipment that consumes large amounts of energy. New experimental plants are being developed that are more resistant to disease and more tolerant of poor growing conditions.

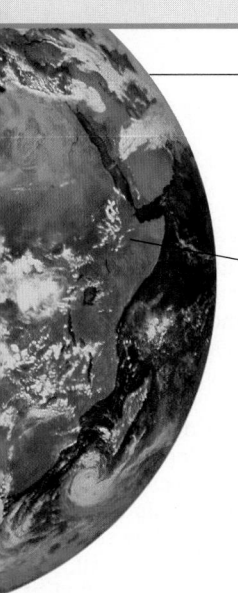

c   **The famine that causes misery for these Ethiopians may one day be reduced by biologists conducting agricultural research in the United States. New methods of growing food may help these people cultivate their arid East African land.**

Demonstration 3
Rain Forests
Display pictures of tropical rain forests. Discuss their significance in terms of the diversity of species that live in such areas.

Genetic engineers are also trying to improve the productivity of existing crops by adding genes to them that increase their resistance to pests, or increase growth rates.

Our impact on the environment is often negative

Think for a moment about the materials you consume and discard in a day—the plastic, paper, glass, and metal. We are beginning to exceed the capacity of the Earth to absorb the waste we generate. Dumping pollutants into the atmosphere, for example, produces "acid rain" that kills forests and poisons lakes. Other industrial wastes are destroying the ozone in the atmosphere that shields us from the sun's harmful rays. Indeed, the sheer volume of carbon dioxide released into the atmosphere by the burning of gasoline, oil, and coal is causing the Earth's temperature to rise alarmingly because carbon dioxide traps the sun's heat.

Preserving a world for your children

An old saying, attributed to many different sources from Native Americans to Amish farmers, states that "We do not inherit the world from our parents, we borrow it from our children." To ensure that our children have the resources they need tomorrow, we must stop wasting precious resources that cannot be replaced, like topsoil and groundwater.

Most important, we must not destroy the world's biological diversity. Unfortunately, we are doing just that, and on a very large scale. The world's tropical rain forests are being destroyed at a breathtaking rate, along with much of the world's biological richness. About 1.3 acres of rain forest per second are being cut for lumber or burned to create pastures for grazing cattle. If this practice continues at such an alarming rate, little or no rain forest will remain in 30 years, and the creatures that lived there, fully one fifth of the world's species, will be gone forever. Because of this incredible destruction, more animals and plants are expected to become extinct in your lifetime than at any time during the last several hundred million years—more even than during the great extinction 65 million years ago when the dinosaurs disappeared from the planet. Biologists and others are working to save as much as possible by creating preserves, and by educating the public about the need to save some of the world's biological richness for future generations.

9

Phase 3

ASSESSMENT OPTIONS

Closure Strategy
The Meaning of Science
The word science comes from the Latin *scientia,* which means knowledge. Have students check a dictionary for other words that come from this root. Have them explain how the Latin root meaning relates to each of these words.

Section Review
Assign the *Section Review.*

Reteaching
Have students select a career in which some knowledge of biology is necessary. Encourage them to go beyond the obvious (like a doctor, dentist, or nurse). Have each student write a brief report that summarizes how biology is used in that career.

Biology and You

Much of what you will learn in this course will give you information you will need in making critical personal decisions like the ones contemplated in **Figure 1.4**.

Should I smoke? Biologists now know how cigarette smoking causes lung cancer—chemicals in the smoke enter the cells of your lungs and damage them, and cancer results. Most people who die of lung cancer today are smokers. Smoking cigarettes is really a form of suicide. You'll learn a great deal about the dangers of smoking in Chapter 30.

How can I avoid developing heart disease? Biologists have learned that diet and exercise have a major influence over whether you are likely to have a heart attack or stroke. You will learn about what you can do to prevent heart disease in Chapter 31.

What are the risks in taking drugs? Most mind-altering drugs are addictive and thus very harmful. This includes the nicotine in cigarettes and the alcohol in wine, beer, and spirits. Biologists have learned much about the physical basis of drug dependency, and about the far-reaching effects drugs can have. You will learn about drug abuse and safe drug use in Chapter 30.

How does our species reproduce? The biology of reproduction is now well known, and because you may someday plan to raise a family of your own, you need to gain a clear understanding of these critical processes. You will learn about human reproduction in Chapter 34.

Figure 1.4
Asking questions is the first step to finding answers. A knowledge of biology will help answer many of the questions that these young adults have.

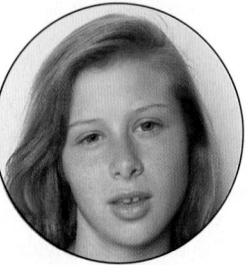

"Many of my friends smoke and they say smoking is all right as long as you quit when you're older. I wonder if that is true?"

"My uncle died three years ago at 45 of a heart attack. Will I have heart problems too?"

"I see other kids using drugs. They say they can stop whenever they want to. I wonder if that's true?"

"My sister is going to have a baby. What's going to happen to her?"

Section Review

❶ **List three decisions that you will be likely to make in the next 10 years that are related to biology.**

❷ **Why is research in the field of genetic engineering considered so important?**

❸ **How does a knowledge of biology help you in playing a role to preserve the Earth?**

▪ *Section Review Answers* ▪

1. Answers will vary.
2. Genetic engineering techniques could be used to replace defective copies of genes with healthy copies. This type of therapy could promise cures for individuals with genetic disorders.
3. Answers may vary. Students should suggest that biological knowledge can help create solutions to current problems such as global pollution, world hunger, and the loss of the world's biological diversity.

SCIENCE IS A WAY OF INVESTIGATING THE WORLD. A SCIENTIST IS AN OBSERVER WHO IS DRIVEN BY THE SEARCH FOR NEW KNOWLEDGE USING METHODS THAT DIFFER FROM THOSE USED BY WRITERS OR PHILOSOPHERS. PHILOSOPHERS MAY MAKE HYPOTHESES AFTER THINKING THROUGH A PROBLEM. HOWEVER, SCIENTISTS MAKE HYPOTHESES AFTER OBSERVING A NUMBER OF SPECIFIC CASES.

1.2 Science Is a Search for Knowledge

Objectives

❶ **Describe how a hypothesis is formed and the role of testing to verify hypotheses.**

❷ **Describe the importance of controls in testing hypotheses.**

❸ **Compare the scientific definition of a theory to the use of theory in language.**

A Case Study in Science

Perhaps the best way to see how science works is to look at a real case where science has been used to solve a problem and improve human health. We will study malaria, a disease that kills more humans than any other. The man you see in **Figure 1.5** contracted malaria and is now being treated with life-saving drugs. In 1941, more than 4,000 Americans died of malaria. In the year 2000, by contrast, fewer than five people are likely to die in the United States of malaria! This disease has been virtually eliminated from the United States as the result of one man's work. His story is a very real example of how science works. His investigations were simple. They involved careful observation and the formulation of clear questions. Finding the cause of malaria is one of the greatest medical advances of all time.

Observations suggest questions to investigate

In the summer of 1897 an English physician, Ronald Ross, worked in a remote field hospital in Secunderabad, India. Ross set out to find the cause of malaria. Of all tropical diseases, malaria was the greatest killer, taking more than a million lives a year in India alone. No one knew what caused the disease—most doctors thought it was brought about by poisonous mists or vapors. Working alone, Ross discovered the pattern by which the disease spread.

Ross observed that patients in the field hospital who did not have malaria were more likely to develop the deadly disease in the open wards (those without screens or netting) than in wards with closed windows or screens. Ross wondered why people in open wards were much more likely to get malaria than those in closed wards.

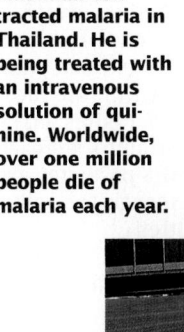

Figure 1.5
This man contracted malaria in Thailand. He is being treated with an intravenous solution of quinine. Worldwide, over one million people die of malaria each year.

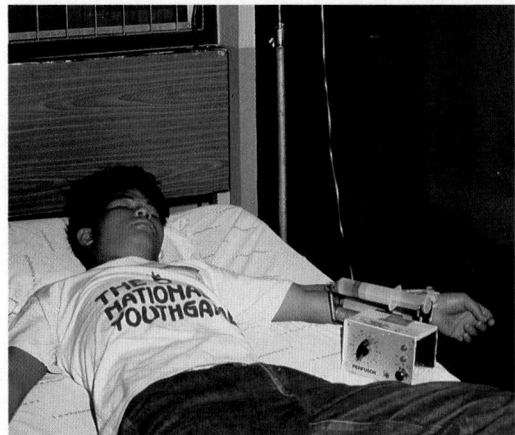

■ Cultural Perspective ■

Cinchona Bark
The native people of South America called the cinchona tree the "bark of barks" because it holds cures for human illnesses. With the help of natives, scientists discovered that bark of the cinchona tree contains the drug quinine, the cure for malaria.

Peruvian Native Americans also knew of the properties of the cinchona tree before A.D. 1500. This was documented when Native Americans offered the "fever tree" to Juan Lopez, a Jesuit missionary. Three centuries later it was recommended to King Charles III of Spain that

apothecaries be allowed to sell the effective antidote made from bark.

Lesson Plan 1.2

Phase 1

PREPARATION

Key Concepts
- Biologists begin their search for a solution to a particular problem or question by forming a hypothesis, a testable possible explanation for their observations.
- Biologists then make a prediction based on the hypothesis.
- Biologists may next test their prediction with the help of a controlled experiment.
- Biologists will reject any hypothesis not supported by the results of their experiments.
- Biologists may formulate a theory if the hypothesis is tested in many ways and supported by the results of their experiments.

Reading Strategy

Have students prepare a table, listing each of the steps of the scientific method as shown in Figure 1.6. As they read this section about the work of Ronald Ross, have students fill in the appropriate information for each step.

Phase 2

TEACHING STRATEGIES

Language Connection

The word malaria comes from the Italian words meaning "bad air," based on the commonly held idea that evil spirits and the night air caused diseases. Not until the 1870s was the germ theory of disease widely accepted.

Scientific Method

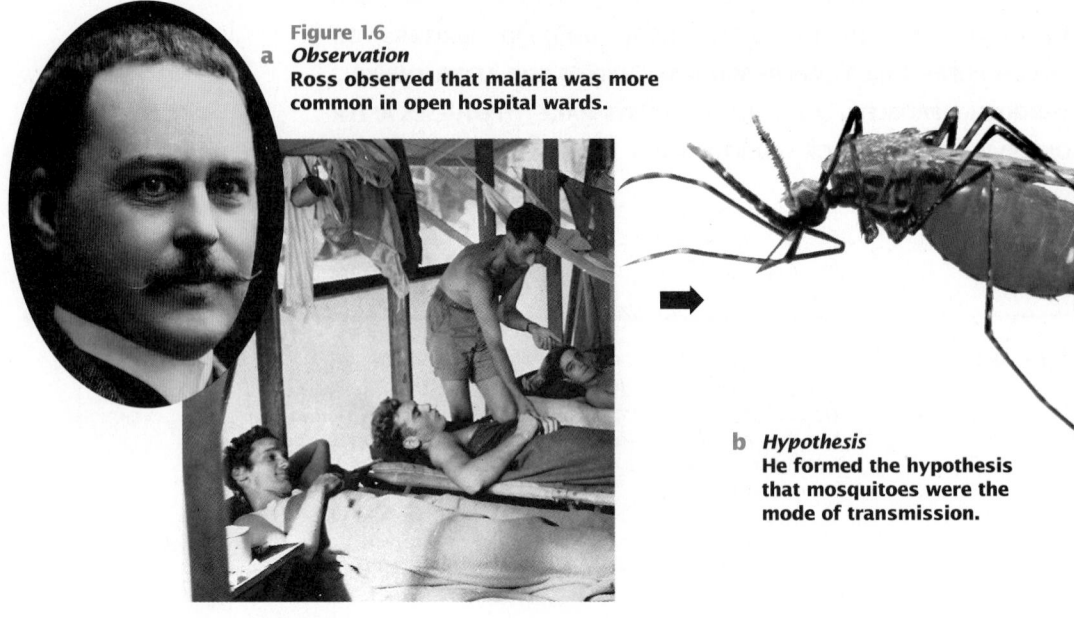

Figure 1.6
a *Observation*
Ross observed that malaria was more common in open hospital wards.

b *Hypothesis*
He formed the hypothesis that mosquitoes were the mode of transmission.

Hypothesis: The basis of further investigation

Observing this pattern, Ross suggested an explanation for why people in open wards were much more likely to get malaria, as shown in **Figure 1.6**. We call such an explanation a hypothesis. A **hypothesis** is a testable explanation for an observation. Ross proposed that mosquitoes in the open wards might be spreading the disease from patients with malaria to patients who did not have the disease. By observing the mosquitoes closely, Ross noted they were *Anopheles*. Using this fact, Ross formulated a hypothesis. Ross's hypothesis was that *Anopheles* mosquitoes were spreading the disease from one patient to another.

Predictions: The framework for testing hypotheses

If Ross's hypothesis was correct, then several consequences could reasonably be expected. We call these expected consequences predictions. A **prediction** is what you expect to happen if a hypothesis is accurate. Ross predicted that *if* the *Anopheles* mosquitoes were spreading malaria (hypothesis), *then* mosquitoes that had bitten malaria patients and sucked up some of their blood should have picked up the parasite *Plasmodium* (prediction), which is always present in the blood of malaria victims. Ross also predicted that parasites should be alive within the mosquito. Somehow the parasites make their way from the mosquito's stomach to its saliva so that the parasites are transferred with the mosquito's saliva to the next person bitten.

Testing under controlled conditions can verify predictions

The controlled test of a hypothesis is called an experiment. Ross did two types of experiments. He looked for living malaria parasites in *Anopheles* mosquitoes that had bitten malaria patients. He carefully dissected the mosquito's stomach and found the live parasites. He then located the mosquito's salivary gland and by careful dissection showed that the parasite spreads throughout an infected mosquito's body and was indeed present in the salivary gland.

c **Predictions**
He predicted that only mosquitoes that had bitten malaria patients would carry the parasite.

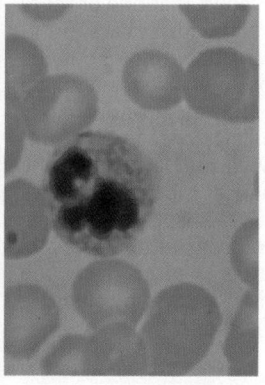

e **Theory**
After his theory was proposed, other scientists set out to verify his results to support or refute his theory.

d **Control experiments**
As a control experiment, Ross checked for the parasite in mosquitoes that had never been near malaria patients.

If a person is bitten by a malaria-carrying mosquito, that person will receive a dose of the parasite in the saliva left behind by the mosquito. To test this prediction, Ross carried out a control experiment. A **control experiment** is one in which the condition suspected to cause the effect is compared to the same situation without the suspected condition (a control group). Nothing else is changed or altered in any way. In Ross's experiment, the suspected condition was mosquitoes feeding on malaria victims. As a control, Ross checked mosquitoes that had not bitten someone with the disease to see if they also contained the parasites. If they did, then malaria patients could not possibly be the source of the parasites in the mosquitoes, and Ross's prediction must be wrong. Gathering newly hatched mosquitoes which had not yet fed, he allowed them to feed on malaria-free blood, and then he examined them as adults. Their stomachs and salivary glands lacked the parasite. The control group of mosquitoes did not contain malaria parasites.

Theories: Explanations for observations

A collection of related hypotheses that have been tested and supported is called a theory. A **theory** is a unifying explanation for a broad range of observations. Theories can have a major impact on science when they tie many accepted and proven hypotheses together into a unified concept. Ross's theory that malaria is transmitted by *Anopheles* mosquitoes carrying it from one person to another was an important milestone in medicine. The idea that malarial epidemics could be prevented by combating mosquitoes was first put forth in a letter written by Ross to the government of India in 1901. Before the end of that year, American army doctors had eliminated almost all malaria from Havana, Cuba, where malaria had reached an epidemic stage. The success of eliminating the deadly disease in these areas was brought about by the reduction of the mosquito population. There have been few advances in the history of medicine more dramatic than the discovery of the cause of malaria.

13

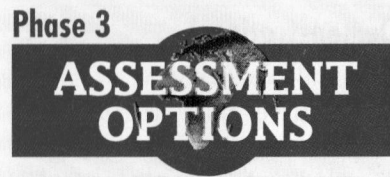
Closure Strategy

The Search for an Answer

Invite a scientist to speak about the use of the scientific method in his or her research work. Be sure that the speaker emphasizes the importance of insight and imagination. Encourage the speaker to cite examples of serendipitous discoveries, such as penicillin.

Science requires continued verification of hypotheses

The essence of science is to reject any hypothesis not supported by observations and the results of control experiments. A new hypothesis is examined very closely to see what it predicts, and the predictions are then rigidly tested. If the predictions are not supported, the hypothesis is rejected. If they are confirmed, the hypothesis is subjected to further verification. One very critical aspect of science is that a scientist's work is held up for review by other scientists. The validity of one's hypothesis is questioned by others until similar results are obtained from similar control experiments. This system of checking and rechecking hypotheses ensures that most, if not all, scientific information is factual information.

Hypotheses that do not explain observations are rejected

A scientist works by systematically showing that certain hypotheses are invalid, that is, they are not consistent with the results of experiments. The results of all experiments are used to evaluate alternative hypotheses. An experiment is successful when it shows that one or more of the alternative hypotheses are inconsistent with observations. By conducting experiments, Ross was able to eliminate the hypothesis that mosquitoes *could* transmit the malaria parasite without biting malaria victims. He retained the alternative hypothesis that if mosquitoes did bite malaria victims, then the mosquitoes could not transmit the parasite. Scientific progress is often made the same way a marble statue is, by chipping away the unwanted bits.

Theories Have Limited Certainty

Theories are the solid foundation of science, that of which we are most certain. There is no absolute certainty, however, no scientific "truth." The possibility always remains that future evidence will cause a theory to be revised or discarded. A scientist's acceptance of a theory is always provisional. See **Figure 1.7**.

The word "theory" is used differently by scientists and the general public. To a scientist, a theory represents that of which he or she is most certain; to the general public the word "theory" implies a lack of knowledge, a guess. How often have you heard someone say "It's only a theory," to imply lack of certainty? As you can imagine, confusion often results. In this text, the word theory will always be used in its scientific sense, as a generally accepted scientific principle.

Some theories, like the theory illustrated in **Figure 1.8**, are so strongly supported that the likelihood of their being rejected in the future is very small. Most of us would be willing to bet that the sun will rise in the east tomorrow, or that if an apple dropped, it would fall. In physics, the theory of the atom is universally accepted, although until recently no one had ever seen one. In biology, the theory of evolution by natural selection is so broadly supported by evidence that biologists accept it with as much certainty as they do the theory of gravity. We will examine the theory of evolution in Chapter 9. It is a very important theory to biologists because the theory of evolution provides the framework that unifies biology as a science.

Figure 1.7
Newton could have interpreted the falling apple from a different perspective.

...y of Continental Drift

Alfred Wegener thought the continents on Earth were once one giant continent. His idea was ridiculed by other scientists who did not think that the continents could move. After many years and much research, enough evidence was gathered to show his theory partially correct. New data showed the sea floor was spreading along with the continents. Continental drift theory was replaced by the theory of plate tectonics.

The continents were once part of a larger continent called Pangaea.

Heat and pressure beneath the Earth's mantle caused the continents to drift apart.

Continental shapes, and fossil and rock evidence indicate that Pangaea existed.

The theory of plate tectonics resolved some of the controversy with the theory of continental drift. As you can see in this illustration, all the continents rest on giant plates, which are indicated by the different shades of color. The arrows indicate their direction of movement. During a human lifetime, the plates do not seem to move a great deal. On a geologic time scale, however, the plates are very mobile.

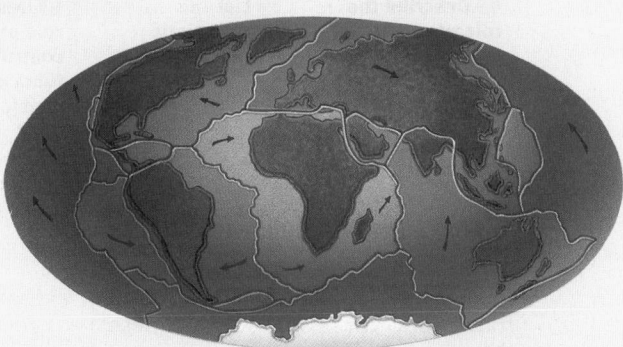

Figure 1.8
The theory of continental drift was tested many times before it became widely accepted.

Scientific Method: The Systematic Study of a Question or Problem

It was once fashionable to claim that scientific progress was the result of applying a series of steps called the "scientific method." In this view, science is a sequence of logical "either/or" steps, each step rejecting one of two incompatible alternatives. Trial-and-error testing could inevitably lead one through a maze of uncertainty. If this view were indeed true, a computer could be programmed to be a good scientist—but science is not done this way. If you ask successful scientists how they do their work, you will find that without exception they design experiments with a good idea of how their experiments are going to come out. Not just any hypothesis is tested, but rather a hunch or educated guess based on all the scientist knows and that allows his or her imagination full play. Because insight and imagination are so important in scientific progress, some scientists are better than others.

Section Review

❶ Under what conditions is a hypothesis supported?

❷ In the Ross case study, what was the relationship between his experiments and his hypothesis?

❸ How does the scientist's use of the word "theory" differ from the way the term is used by the general public?

▪ Section Review Answers ▪

1. Students should suggest the use of the scientific method to conduct a control experiment to test the hypothesis. A hypothesis is supported by observations and the results of experiments.

2. They were both based on his observations.

3. To a scientist, a theory represents that with which he or she is most certain; to the general public the word "theory" implies a lack of knowledge, a guess.

Section Review
Assign the Section Review.

Reteaching
Assign students to cooperative work groups of four. Have each group develop a concept map for this section.

15

Lesson Plan 1.3

Phase 1
PREPARATION

Key Concepts

Six major themes will form the framework for this text:

- Energy and Life
- Scale and Structure
- Stability
- Evolution
- Patterns of Change
- Interacting Systems

Reading Strategy

Be sure that students read this section with an awareness that these themes will be repeatedly encountered throughout this text. Consequently, they should memorize them and look for them as they continue reading the text.

Phase 2
TEACHING STRATEGIES

Demonstration 1

Energy and Life

Place some pond water containing hydra in a small dish and use an overhead projector to show students how they feed on *Daphnia* to obtain energy for their life processes. Emphasize that all living things require some form of energy.

SCIENTISTS HAVE BEEN STUDYING LIVING THINGS FOR SEVERAL HUNDRED YEARS. FROM THE MOUNTAIN OF INFORMATION COLLECTED, GENERAL PRINCIPLES OF PARTICULAR IMPORTANCE HAVE EMERGED. YOU WILL ENCOUNTER THESE PRINCIPLES REPEATEDLY AS YOU EXPLORE THE IDEAS IN THIS TEXT—THEY ARE THE FRAMEWORK OF BIOLOGY, THE SKELETON THAT SUPPORTS ALL YOU WILL LEARN.

1.3 Studying Biology

Objectives

1 Describe the role of energy in life processes.

2 List the characteristic properties of cells.

3 Describe the role of genes in controlling cell functions and development.

4 Relate evolution to natural selection.

Themes Unify Ideas

In this text, we have selected six unifying principles or themes to use as a framework for your study. These themes are introduced here, and you will see them repeatedly throughout the text. Your goal should be to understand how the many topics you study in biology are examples of themes.

Theme 1: Energy and Life

All organisms require energy to carry out life processes. Energy is used to grow and do work. Without it, life soon stops. Almost all the energy that drives life on Earth is obtained from the sun. Plants capture the energy of sunlight and use it to make complex molecules in a process called photosynthesis. These molecules then serve as the source of fuel for animals that eat them. The flow of energy among organisms, including the teenager shown in **Figure 1.9**, helps determine how organisms interact within their environment, which is another important concept in biology.

Figure 1.9
The teenager to the right is using the *energy* that he gets from the food he eats and, indirectly, from the sun.

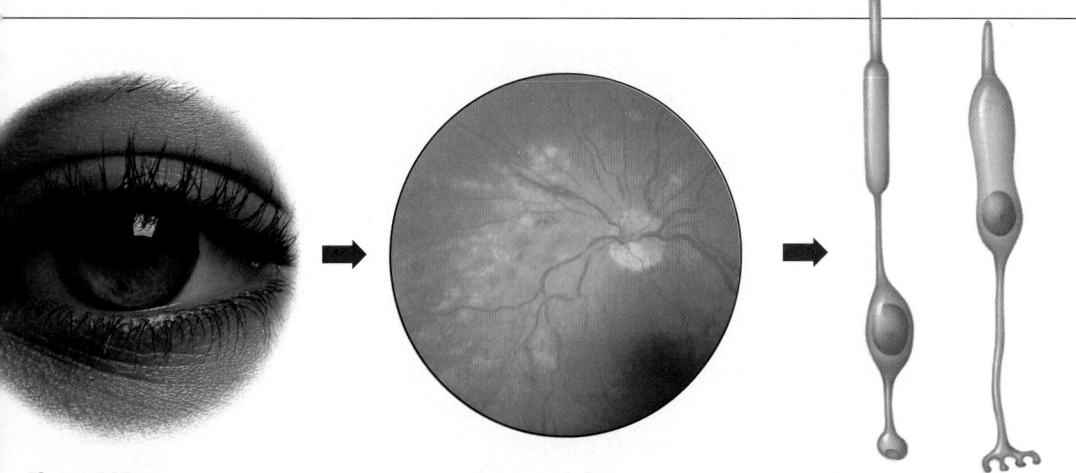

Figure 1.10

a The teenager's body has many levels of *structure*. The eye is a complex tool for sight.

b On a microscopic *scale*, the retina in the eye contains blood vessels. Light enters the eye and lands on the retina.

c On a more detailed *scale*, the retina is made of rods and cones, which send information to the brain.

Theme 2: Scale and Structure

Living things are made of the same materials as the rest of the universe, of atoms assembled into molecules. All the physical principles that apply to stars and home computers also apply to you and to every other living creature. Living things differ from non-living ones only in their degree of organization.

All organisms are composed of cells, tiny compartments surrounded by membranes. Your body has trillions of cells. Cells form the structures of your eye shown in **Figure 1.10**. The complex chemical processes that occur within cells are much the same in all organisms, and all cells have the same basic structure: a covering called a membrane that surrounds the cell and controls what information and materials enter and leave it; an internal fluid and skeleton that gives shape to the cell and supports the other things within it; and a central zone or nucleus which contains the cell's genes, the hereditary instructions coded within long complex molecules called DNA. Many cells have specialized structures within their cell walls called organelles that carry out some of the cell's activities. You will learn much more about cell processes in Chapters 3, 4, and 5. Cell activities are often influenced by outside molecules that attach to special proteins in the cell's membrane. Much of a cell's biology is determined by the nature of its membrane.

Theme 3: Stability

Control is an essential aspect of life. *Humans and other organisms must maintain a constant internal environment in order to function properly;* for example, your body temperature must not vary by more than a few degrees. Your body has many mechanisms for maintaining **homeostasis**, a word that means keeping things the same.

Theme 4: Evolution

Biologists have long suspected that life on Earth is the result of evolution. Life forms are slowly changing and have apparently been changing since Earth formed. Charles Darwin proposed the hypothesis that this change is the result of a long process called natural selection. Natural selection proposes that those organisms with more favorable mutations will be more likely to survive and reproduce. These

Demonstration 5

Patterns of Change

Have students bring in photographs of their immediate family. Have them describe their family resemblances to the rest of the class. Emphasize that all living things pass on instructions for development to their offspring.

Demonstration 6

Interacting Systems

Take students on a field trip to observe both plant and animal specimens. Have students explain how these organisms depend on their physical environment for survival. Emphasize that all living things interact with their physical environment.

Phase 3

ASSESSMENT OPTIONS

Closure Strategy

Your Role in a Biological World

Assign students to cooperative work groups of four. Have each group list five reasons why a knowledge of biology will be important to them. Have each group read their list to the rest of the class.

Section Review

Assign the *Section Review*.

Reteaching

Point out that biology is a large, complex, and growing science with many smaller fields such as biochemistry, botany, cell biology, ecology, embryology, genetics, microbiology, and physiology. Have each student select one of these specialties, do some library research, and write a short report. Be sure that students include the major themes they feel would be included as part of the field study.

---

favorable mutations better enable an organism to overcome the many challenges presented by its environment. Darwin's theory provides biology with a basis for understanding the diversity of life on Earth. *Evolution results from a long history of organisms adjusting to a diverse and changing environment.* Evolution provides the vast diversity of species that exist on Earth. A **species** is a group of organisms that look similar and can produce fertile offspring in their environment. *Natural selection leads to changes in species over time.*

Theme 5: Patterns of Change

In organisms composed of many cells, such as you, most of the cells are specialized, each performing distinct functions. For example, gland cells secrete hormones, muscle cells contract, and nerve cells conduct electrical signals. *All of the many different kinds of specialized cells, however, are descended from the same single fertilized egg cell—as the cells grow and divide, their genes manage an orderly process of change called development.* The process is controlled by genes, and is the same in all humans.

Children resemble their parents because instructions for development are passed from parents to offspring. These instructions are in the form of **genes**, which are segments of long DNA molecules. Sometimes damage to the DNA occurs. These changes, called mutations, are usually harmful, but sometimes mutations help an organism to better survive.

Theme 6: Interacting Systems

Living things interact with each other and with their environment in complex ways, like the hoverfly in **Figure 1.11**. Ecology is the study of complex communities of organisms in relation to their environment. *A living community is highly structured and interdependent. This interdependence is the result of a long process of evolution in which selection has favored cooperation.* This complex web of interactions is easily disrupted when the environment is polluted and individual species become extinct, as is happening in much of the world today.

Figure 1.11
This hoverfly is gathering nectar from the flower. Without the flower, the fly would not have food. Without the fly, the flower would not be able to pollinate and reproduce.

Section Review

❶ **Why would energy be considered a theme in biology?**

❷ **What properties do all cells share?**

❸ **What is the role of a gene in the development of an organism?**

❹ **How is evolution influenced by natural selection?**

■ *Section Review Answers* ■

1. All organisms require energy to carry out life processes.
2. Students should explain that all cells have the same basic structure (a membrane, internal fluid, a central zone or a nucleus, and DNA) and explain the function of each structure.
3. The development process is controlled by genes.
4. Evolution results from a long history of organisms adjusting to a diverse and changing environment. Natural selection leads to changes in species, which show characteristic patterns over time.

Chapter **1**

Highlights

Life is plentiful. In this rain forest alone, approximately 30 million different species of plants, animals and insects can be found.

	Key Terms	**Summary**
1.1 Biology Today 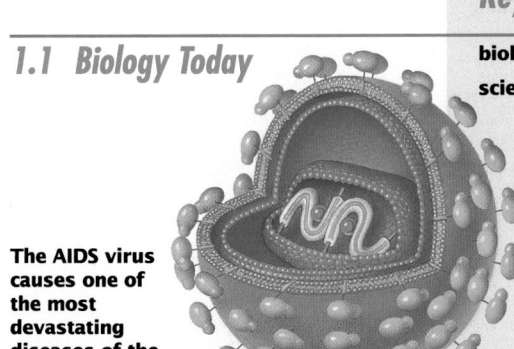 The AIDS virus causes one of the most devastating diseases of the 1990s.	biology (p. 5) science (p. 5)	• Biologists are combating infectious diseases. Smallpox, tuberculosis, and pneumonia have been largely conquered. AIDS and malaria are still being studied. • Genetic disorders may be eliminated by gene therapy involving the transfer of normally functioning genes to affected individuals. • As the population increases, Earth's ability to sustain the human population is being strained.
1.2 Science Is a Search for Knowledge 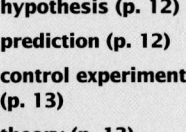 Using the scientific method, mosquitoes were discovered to be the carriers of the malarial parasite.	hypothesis (p. 12) prediction (p. 12) control experiment (p. 13) theory (p. 13)	• Scientists, by asking questions and making observations, determine scientific principles. • Scientific progress is made by posing hypotheses and testing predictions. Verification of hypotheses is required before they are widely accepted. • Controlled experiments are important in testing hypotheses. Control groups ensure that only one variable in an experiment has been changed. • A theory links well-supported hypotheses together as one concept.
1.3 Studying Biology 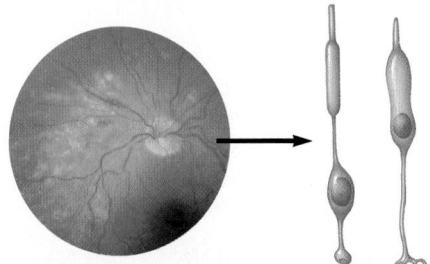 Scale and structure, as shown by this retina and the rods and cones, is one of the themes of biology.	development (p. 17) homeostasis (p. 17) genes (p. 17) species (p. 18)	• Energy in biological systems plays a role in organization—from cells to ecosystems. • Structure dictates function. Genes are the foundation of this concept. • Turning genes "on" and "off" produces different kinds of cells. • Mutations produce new versions of genes. • Individuals with favorable mutations are more likely to survive and reproduce. • Darwin proposed that evolution was the result of natural selection.

Chapter Review Answers

Understanding Vocabulary

1. **a.** Ecology is a branch of biology that involves the study of complex biological communities, while biology is the science of the living world.
   b. An idea or a guess about the relationship between variables is a hypothesis. When a hypothesis is supported by much evidence it is called a theory.
   c. Homeostasis means maintaining stability, as when the human body maintains a relatively constant body temperature. Development is the process by which genes control the orderly growth and division of cells.

Relating Concepts

2. Map answer is shown on page 3D.

Understanding Concepts

Multiple Choice

3. a	8. c
4. b	9. a
5. b	10. b
6. c	11. d
7. b	12. b

Completion

13. Ronald Ross, hypothesized
14. 5.5 billion, 6 billion
15. hypothesis, theory
16. genes
17. DNA

Short Answer

18. Scientists are developing new varieties of plants that are disease resistant and better able to tolerate poor growing conditions. Using genetic engineering, they are also making plants that grow faster and are more resistant to pests.

19. The tropical rain forests are the world's most diverse ecosystem. By destroying the world's rain forests, plants and animals that may be useful for medicines or food will be made extinct.

20. Some fatal diseases of the past are no longer threats to life since the discovery of antibiotics such as penicillin; the occurrence of malaria has been

---

Chapter **1** Review

Understanding Vocabulary

1. For each pair of terms, explain the differences in their meanings.
   a. ecology, biology
   b. hypothesis, theory
   c. homeostasis, development

Relating Concepts

2. Copy the unfinished concept map below onto a sheet of paper. Then complete the concept map by writing the correct word or phrase in each oval containing a question mark.

Understanding Concepts

Multiple Choice

3. Science is mainly concerned with
   a. asking questions and seeking answers.
   b. test tubes and beakers.
   c. making life easier for people.
   d. maintaining tropical rain forests.

4. Your knowledge of the relationships between such things as cigarette smoking and lung cancer and between diet and heart disease is important to you because
   a. biologists are learning more about lung cancer and heart disease.
   b. this knowledge is likely to affect personal decisions you make in the future.
   c. it will prepare you to be a health care professional in the next century.
   d. lung cancer and heart disease are typically associated with the use of drugs.

5. Gene transfer therapy
   a. involves transferring a defective copy of a gene into a healthy person.
   b. will enable biologists to cure some genetic disorders.
   c. can begin when every gene in the human body is cataloged.
   d. has produced a cure for AIDS.

6. Scientific principles are generated from
   a. tests.
   b. predictions.
   c. observations.
   d. variables.

7. In Ross's experiment, the control group was
   a. yellow fever mosquitoes.
   b. mosquitoes that had not bitten a person with malaria.
   c. stomachs and salivary glands of blood sucking mosquitoes.
   d. people living in places other than India.

8. The scientific method is
   a. a series of logical steps that leads to a final solution.
   b. trial-and-error tests of scientific problems.
   c. a process of investigation influenced by insight and imagination.
   d. a hunch or guess tested with the aid of computers.

9. Theories
   a. are always subject to revision.
   b. reflect scientific truths.
   c. always result in major scientific breakthroughs.
   d. are tested but never rejected.

10. The energy used by living things on Earth comes from
    a. burning fossil fuels.
    b. the sun.
    c. waste products of photosynthesis.
    d. DNA in plant and animal cells.

11. As proposed by Charles Darwin, life on Earth evolves by means of
    a. speciation.
    b. mutation.
    c. homeostasis.
    d. natural selection.

12. What biological theme is highlighted when studying the cell, its membrane and its organelles?
    a. energy and life
    b. scale and structure
    c. stability
    d. all of the above

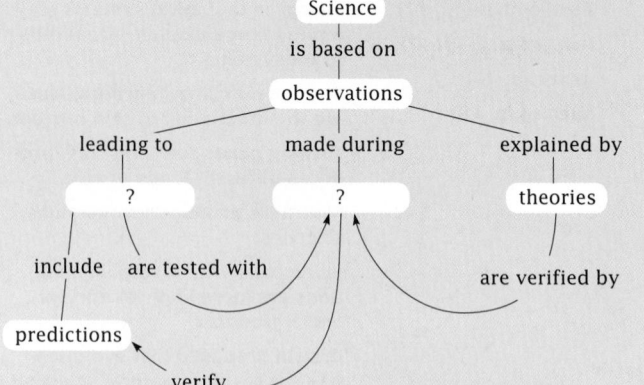

---

greatly reduced in some areas of the world because of altered health practices; and biological research continues in search of the cure for AIDS.

21. Ross's hypothesis that the *Anopheles* mosquitoes were spreading malaria led him to several predictions about what to expect if the hypothesis

were true. One of his predictions was that mosquitoes that have sucked up the blood of a malaria patient should contain the parasite *Plasmodium*.

Interpreting Graphics

22. A possible hypothesis is that the fish are dying from chemicals released into the river by the oil

refinery. From the hypothesis it may be predicted that if the chemicals released by the refinery are killing the fish, then the chemicals released by the refinery into the river will be found in high quantities in the bodies of the dead fish.

Completion

13. Mainly due to the work of _____ , malaria has been virtually eliminated from the United States. He _____ that malaria is spread from one person to another by the mosquitoes.

14. The world's population today is more than _____ people. By the year 2000, experts predict that the world's population will be about _____ .

15. When a _____ explains a set of observations it is called a _____ .

16. Segments of DNA molecules are called _____ .

17. Mutations are caused by changes to _____ .

Short Answer

18. What are scientists doing to help farmers produce enough food for the world's expanding population?

19. Why is it important that the world's tropical rain forests not be destroyed?

20. It has been said that medicine is the area of modern biology that has most affected our lives. What evidence supports this statement?

21. Describe the relationship between hypothesis and prediction in Ross's experiment.

Interpreting Graphics

22. Look at the photograph of a scientist at work. Based on your examination of the photo, write a hypothesis about what you think the scientist could be investigating. Then, write a prediction that you expect to occur if the hypothesis is true.

Reviewing Themes

23. *Interacting Systems*
How might the extinction of one species in an environment affect other organisms with which it interacts?

24. *Stability*
You perspire when you get too hot, and shiver when you get too cold. How do these actions help you maintain a constant body temperature?

25. *Evolution*
A politician who opposes the teaching of biological evolution in schools said her reason is "It's only a theory." Why are scientists likely to be upset by the politician's statement about biological evolution?

Thinking Critically

26. *Inferring Conclusions*
Scientific information is always changing, yet you are embarking on a yearlong study of scientific information. Why should you learn it?

27. *Compare and Contrast*
How is scientific progress a lot like constructing a stone statue by chipping away at the unwanted bits?

Cross-Discipline Connections

28. *Biology and History*
Edward Jenner conducted a daring experiment, the results of which led to the eradication of smallpox in many parts of the world. What was his hypothesis? What experience led him to formulate this hypothesis?

Discovering Through Reading

29. Read the article "The War Among the Greens," in *Newsweek*, May 4, 1992, page 78. What is the mission of the Sierra Club? How is the willingness of the Sierra Club's leadership to compromise on issues like clear-cutting in national forests likely to affect the future of life on Earth?

Thinking Critically

26. Science is a process of knowing, and scientific information is the product of this process. The information only changes when the outcomes of investigations indicate that changes are needed. Information you learn in this course will help you understand the changes that will come about in the future.

27. Scientific progress may be measured by hypotheses that are retained and rejected. Rejecting hypotheses that are inconsistent with new observations results in scientific progress. In much the same way, chipping away the bits of stone results in the construction of a statue.

Cross-Discipline Connection

28. The hypothesis was that scratching a person's skin with a bit of pus from a cowpox scab would prevent the person from contracting smallpox. Jenner's treatment of a woman with cowpox led him to believe that cowpox is a lot like smallpox but not as dangerous.

Discovering Through Reading

29. The Sierra Club's mission is to promote the preservation and proper management of unspoiled ecosystems. The leadership's willingness to compromise on clear-cutting may result in the destruction of vast timberlands. Without the trees, the soil will erode, destroying the habitats of many terrestrial organisms. The unanchored soil will also wash into streams, muddying waters and killing aquatic life.

Reviewing Themes

23. The extinction of one organism may lead to the extinction of others that depend on it for food.

24. Perspiration absorbs extra body heat, and as perspiration evaporates from your skin your body is cooled. When you shiver, blood vessels near the body's surface contract, helping to keep warm blood away from the skin. This action helps keep your body warm.

25. The politician's statement implies that the scientists are uncertain about the evolution of organisms on the Earth. Her use of the word theory is inconsistent with that of scientists, for whom theory represents that which is more certain.

Procedural Note

Remind students that they must hold their breath long enough to cause a potential reaction. They should not begin to record their pulse rate for at least 45 seconds after starting to hold their breath.

Procedure Answers

2. The error encountered is multiplied by the same factor used to arrive at the final answer.

3. Students should indicate that averaging will give an intermediate value that represents a mean between two extremes.

5. Answers will vary but each student should describe a logical control experiment that tests the stated hypothesis.

6. Answers will vary according to student's experimental design.

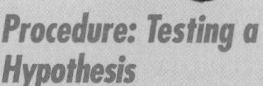

Chapter 1

Investigation

How Do Scientists Seek Answers to Questions?

Objectives

In this Investigation, you will:
- *formulate* a hypothesis
- *design* an experiment
- *test* a hypothesis
- *average* data
- *compare* experimental results
- *display* experimental results in graph form

Materials

- watch with second hand
- paper

Prelab Preparation

1. Form a cooperative group of four students. Work with one member of your group to complete steps 3–10.

2. In this investigation, you will explore the scientific methods used to answer a question. The question that you will attempt to answer is:
   What effect will holding your breath have on your pulse rate?

Procedure: Testing a Hypothesis

1. To find the pulse rate, have your teammate sit quietly at your desk. Find your partner's pulse in either wrist using your middle finger. Count the number of beats during a 60-second period. This number is the pulse rate. For example, a person might have a pulse rate of 70 beats per minute.

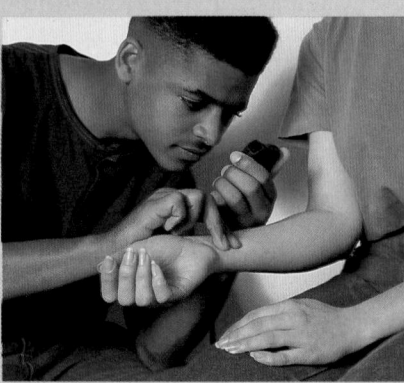

2. *Why would counting for a fraction of a minute and multiplying the result increase the error in your answer?*

3. To find the average pulse rate repeat step 1 twice. Average your three answers and record your result. *Why is it better to work with the average pulse rate than one single measurement?*

4. Hold your breath for 45 seconds. Record your pulse rate immediately after holding your breath.

5. Compare your average pulse rate with the rate after holding your breath. *Do you think your data matches those of your classmates?* Formulate a hypothesis that answers this question. Use Ross's hypothesis on page 12 as a guide.

6. Design an experiment to test your hypothesis. In designing your

experiment, one factor is varied and the response of another factor is measured. The factor varied is the independent variable. The factor that responds to the independent variable is called the dependent variable. *Identify the variables in your experimental design. What variables will you keep constant?*

7. Write out your experimental plan and have your teacher approve your experimental plan. Run the experiment and record your data in a table.

8. Compare your results with the other team in your group and with the other teams in your class. Record their data in your table.

9. Make a graph like the one below and show the data collected from each group in the class.

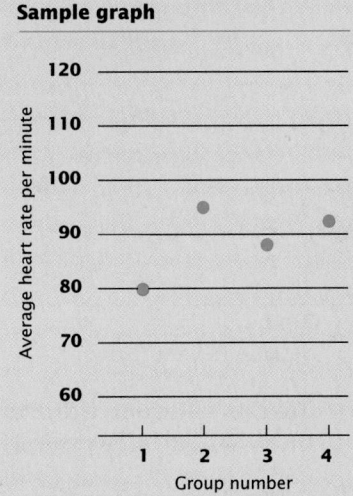

Sample graph

(y-axis: Average heart rate per minute — 120, 110, 100, 90, 80, 70, 60)

(x-axis: Group number — 1, 2, 3, 4)

10. Clean up your materials and wash your hands before leaving the lab.

Analysis

1. *Analyzing Data*
   What conclusions can be drawn from your data? Explain how the data support or refute your hypothesis.

2. *Analyzing Methods*
   Why must you know the average pulse rate during normal breathing?

3. *Evaluating Methods*
   What are some possible sources of error in this experiment?

4. *Evaluating Methods*
   What is the value of comparing results?

5. *Applying Concepts*
   Experiments are used to collect data from two or more groups of subjects or from the same subjects at two different times. The group exposed to changes in the independent variable is called the experimental group. The other group is not exposed to changes in the independent variable. This group is called the control group. Identify the experimental group and control group in your experiment.

Thinking Critically

How would the consumption of different kinds of foods affect your pulse rate? How would you form a hypothesis and test it? How do sugars act in the body? Would proteins or fats affect your pulse rate the same way? Is there an accurate way to predict the effect other than by direct experimentation?

Analysis Answers

1. Answers will vary but should relate to the stated hypothesis and should be supported by the data.
2. A control group is necessary to make the comparison between the pulse rate with the breath being held and not being held.
3. Answers will vary but should be logical and relate to the design of the experiment. Possible sources of error might include inaccuracies relating to timing, counting, recording data, and holding the breath long enough to create a response.
4. Comparison of data with other teams allows the investigator to check the reliability of the data.
5. The control group is represented by the data collected when the breath is not held. The experimental group is represented by the data collected when the breath is held.

Thinking Critically Answer

Students should describe an experiment that compares the effects of various foods on the pulse rate. Students should also suggest that sugar has the most noticeable effect on the pulse rate. Sugar levels in the blood can vary and cause changes in the nervous system (hyperglycemia or hypoglycemia), possibly speeding up the pulse rate. Students may have observed this themselves after eating a large amount of sweets or fasting.

23

Chapter 2
Discovering Life

Planning Guide

	Objectives/Themes	Classwork Resources	Homework Resources

2.1	**1.** Explain the difficulty in defining life using visually observable properties.
	
	2. Describe five properties shared by all living organisms.
	
	3. Relate the properties of life to the biological themes in Chapter 1.
	
	Themes: Energy and Life, Evolution, Patterns of Change, Scale and Structure, Interacting Systems, Stability
		Text
		Science in Action
		  Microscopy: The Invisible World
		  pp. 28–29
		
		Teacher's Resource Binder
		Focus Activity 2
		  Comparing Living and
		  Nonliving Things
		Lab Investigation 2.1
		  Designing Control Experiments
		Extension Worksheet
		  Are Viruses Alive?
		
		Other Resources
		Science in Action Video
		  The Atoms and Molecules of
		  *Life/DNA and Protein Synthesis**
		Transparency 3
			Text
			Section Review, p. 27
			
			Teacher's Resource Binder
			Directed Reading Worksheet 2.1
			
			Other Resources
			Audiocassette 2.1*

2.2	**1.** Relate atoms, elements, ions, and molecules to each other. **2.** Describe the structural features of an atom. **3.** Distinguish between covalent and ionic bonds and how they are formed. **Themes:** Interacting Systems, Scale and Structure	**Other Resources** Transparencies 4–6	**Text** Section Review, p. 32 **Teacher's Resource Binder** Directed Reading Worksheet 2.2 **Other Resources** Audiocassette 2.2*

2.3	**1.** Define sugar and describe the process that occurs in the formation of polysaccharides.
	
	2. Describe the solubility and energy storage properties of lipids.
	
	3. Explain the factors that affect the three-dimensional structure of proteins.
	
	4. Define nucleic acids and describe their functions.
	
	Themes: Scale and Structure, Energy and Life, Interacting Systems, Stability
		Text
		Investigation
		  What Properties Can Be Observed in Living Things?
		  pp. 42–43
		
		Teacher's Resource Binder
		Lab Investigation 2.2
		  Diversity of Life
		
		Other Resources
		Transparency 7
			Text
			Section Review, p. 38
			
			Teacher's Resource Binder
			Directed Reading Worksheet 2.3
			Vocabulary Review Worksheet*
			Reteaching Worksheet*
			  Macromolecular Chemistry
			
			Other Resources
			Audiocassette 2.3*

*Reteaching Options

Demonstrations
2.1: p. 26
2.2: pp. 31, 32
2.3: pp. 34, 35, 36, 37

Assessment
Chapter Review pp. 40–41
Portfolio Assessment p. 23D
Chapter Test—Teacher's Resource Binder
Test Generator

Research Notes

Connection to Biochemistry: Protein Structure

The abilities of a protein depend entirely upon its structure. Early experiments on polypeptides demonstrated that the secondary and tertiary structures of a protein (coiling and folding) depend mostly on the primary structure, the sequence of amino acids in the protein.

Researchers had reached this conclusion after growing polypeptides in vitro. All of the polypeptides began to snap into complicated coiled and folded patterns even before they were fully formed. Afterward, the polypeptides always had the same properties.

Chemists believe that each peptide bond is able to rotate until it reaches a certain position that is more stable. The net effect is that a characteristic coiling and folding pattern emerges after each bond has rotated to its stable position.

Since it worked so well in a test tube, for some time scientists assumed that the process worked the same way in living things. Recently, scientists have had to rethink this belief, as biochemists discover more about a family of proteins that they have dubbed chaperones. These proteins, which are found in both eukaryotes and prokaryotes, seem to regulate protein coiling and folding.

There seem to be two types of chaperone proteins. One type of chaperone binds to the protein very quickly, perhaps even as it is being synthesized by a ribosome. This chaperone prevents the protein from coiling or folding very much, possibly by making rotation easier, so that no single orientation for a peptide bond is favored.

Bound together, the new uncoiled protein and its chaperone make their way through the cell to their target, often an organelle. A second type of chaperone within the organelle helps the protein cross the membrane to the inside. It is easier for the protein to cross the membrane as a long strand of amino acids than as a globular coiled polypeptide.

Within the organelle, the second type of chaperone binds to the protein, but allows limited coiling and folding to take place. Then, when the protein is nearly ready, it is released to finish its coiling and folding and to begin performing its role as an enzyme or structural protein.

Biochemists first identified these chaperone proteins as gene products that were produced when cells were shocked with heat. This could be because heat tends to uncoil and unfold proteins so they no longer have the same shape, and no longer work properly. Production of these chaperones could help recoil and refold the proteins so that they regain their function in the cell's life. There is even some evidence that chaperones can detect and refold proteins that misfold during synthesis.

Investigation Notes

What Properties Can Be Observed in Living Things?
pages 42–43

Purpose: Students make simple observations of living organisms in a drop of pond water. The investigation should be approached as a get-acquainted experience rather than an in-depth study of the characteristics of life.

Prelab Preparation

1. Pond water cultures can be obtained from WARD'S or by collecting samples from a local pond. A net with a very fine weave can be used to concentrate samples that are collected locally, making classroom study easier and more interesting.

2. The materials list calls for "Detain," which is a nontoxic, protist-slowing agent available from WARD'S (37 M 7950) that will allow a student to see the organisms without having them rapidly move out of view.

3. You may want to prepare a short surprise quiz to give students before beginning the procedure, to be certain that the students have reviewed the use of the light microscope.

Answers will be found on pages 42–43.

Meeting Individual Needs

Objectives

1. Students will demonstrate appropriate use of core vocabulary for the chapter (Vocabulary File).

2. Students will distinguish between living and nonliving objects (Demonstration A).

Vocabulary File

If you are not already using the Vocabulary File refer to Chapter 1 for its preparation. See the Chapter Highlights on page 19 for a list of suggested words.

Demonstration A
Comparing Living and Nonliving Things
for page 25
(Developing Classification Skills/Kinesthetic Learners)

Materials and Preparation:
Display living and nonliving things at various numbered stations around the room. As much as possible, use objects that the students can pick up and touch while examining them. Be sure to clearly label any objects students should not try to pick up and touch (adjustment on microscope, live fish, etc.). Number the objects. A minimum of 10 objects is suggested.

The following list is intended only for guidance. The resources of your school will make your list different, but try to achieve a balance between living and nonliving things.

- animal cell slide ready for viewing under a microscope

- live fish in fishbowl or aquarium

- slide with salt crystals ready for viewing under a microscope

- carrot or celery stick

- commercial vitamin

- fossil

- potted plant

- mounted butterfly

- soft drink

- beaker filled with water (labeled "water")

Procedure: To introduce this demonstration, have students brainstorm to come up with evidence that they are alive. Usually, the ability to move is among the first things students will suggest. Have them turn to page 25 and scan the section on Movement.

Ask students to provide additional examples of living things that don't move and moving things that aren't alive. Have students sit still without moving. Then, brainstorm for further evidence that they are alive, even if they aren't moving. Guide the discussion into body functions such as heart beating, blood flowing, lungs expanding and contracting, etc.

Have students write the numbers from 1 to 10 on a sheet of paper, leaving room for additional writing next to each number. Then, have them go from one lab station to the other and examine the objects, deciding which are living and which are nonliving. They should record their decision for each station on the paper, next to the number of each station. Students should also include an explanation of their reasoning for classifying an object as living or nonliving.

Additional Strategies
Visual Strategies
Pages 25, 27, 30, 31, 33, 34, 35, and 38

Auditory Learners
Use *Biology: Visualizing Life* Audiocassettes for Sections 2.1, 2.2, and 2.3.

Meeting Individual Needs (cont.)

Cooperative Learning
Macromolecules in Living Things

Timing: Use this activity to introduce Section 2.3.

Group Size: 4 students

Outcome: Students will be able to describe examples of each of the types of macromolecules found in living things with their subunits, functions, and examples.

Individual Accountability: Each group member will research a single type of macromolecule.

Positive Interdependence: Each group will be able to relate the functions, examples, and subunits of different macromolecules.

Assign one of the following macromolecules to each member of each group: (1) carbohydrates, (2) lipids, (3) proteins, and (4) nucleic acids.

Have each group member research their macromolecule and prepare a report on an index card. The card should include the macromolecule type and the name of an example of such a macromolecule.

The cards may also include information such as the subunit(s) that the macromolecule is made of, the elements contained in the macromolecule, the shape or structure of the macromolecule, the function of the macromolecule, and the location of the macromolecule within the body or within cells.

Then have each group member show and explain his or her report to the other members of the group, so that they can illustrate the macromolecule on their own additional index cards. At the end of the activity, each group member should have index cards for each type of macromolecule.

Portfolio Assessment

Students should select their best work and provide a self-reflective rationale for their selections. Students can make selections in the following areas.

1. *Content* — One concept map from the chapter (See page 812 for evaluation criteria.)

2. *Reading Comprehension* — One Directed Reading Worksheet from the Teachers Resource Binder (Use the answer key to evaluate for accuracy.)

3. *Writing* — Using the Vee Form, summarize an article relating to the importance of chemicals to life. (See page 22T for evaluation criteria.)

   Or: Select a writing task or project from the Chapter Review.

4. *Performance Assessment* — One Vee form from a chapter investigation or lab manual investigation (See page 22T for evaluation criteria.)

Teacher makes selections in the following areas.

1. *Formal Assessment* — Chapter test (Test A, B, or the Test Generator) The teacher-scored test should be reviewed by the student. Incorrect responses should be corrected by the student before the test becomes part of the portfolio.

2. *Informal Assessment* — Use the Direct Observations Checklist, page 33T, during a laboratory or other cooperative learning experience.

3. *Performance Assessment* — Have students imagine they are part of a team designing an automated space probe. They should write a report describing methods and instruments to use for detecting life (or its absence) on other planets.

Concept Map Answer

The following is one possible answer to the Relating Concepts exercise on page 40.

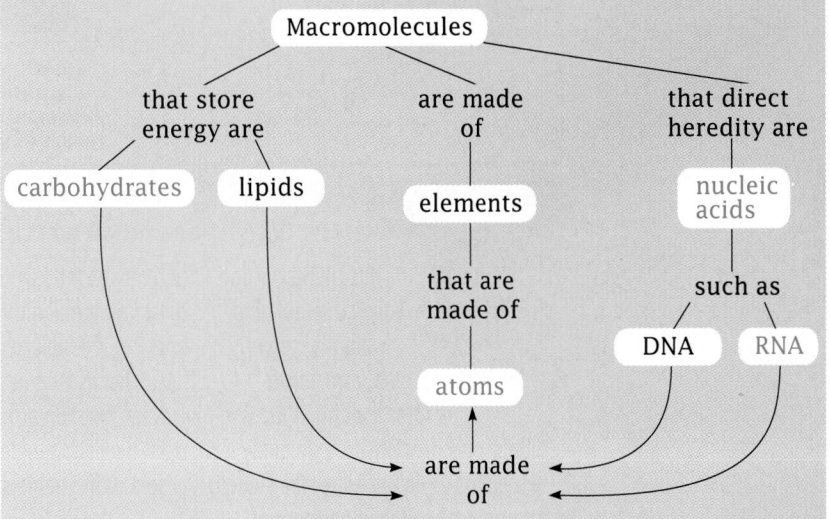

Chapter 2

Determining Prior Knowledge

- Ask students what criteria they would use to distinguish between living and nonliving things.
- Have each student draw and label the structure of an atom.
- Have students describe how any two atoms they have drawn would combine or bond.
- Draw the structural formula for water on the board. Ask students what information this formula provides.
- Ask students what they know already about carbohydrates, lipids, proteins, and nucleic acids. What do they have in common? How do they differ?

Chapter 2

Discovering Life

Review

- **biological themes (Section 1.3)**
- **the term *atom* (Glossary)**
- **the term *cell* (Glossary)**

2.1 What Is Life?

- **First Guesses at Defining Life**
- **Life Has Five Characteristic Properties**

2.2 Basic Chemistry

- **Atoms: The Basic Structural Units of Matter**
- **Formation of Bonds Stabilizes Atoms**

2.3 Molecules of Life

- **Organic Compounds Are Derived From Carbon**
- **Carbohydrates Are Energy Sources**
- **Lipids Store Energy**
- **Proteins Provide Structure and Increase Reaction Rate**
- **Nucleic Acids Contain Genetic Information**
- **Macromolecule Summary**

Are all the things you see when you walk along the seashore alive? How do you know? In order to decide, you will have to examine them more closely. In this chapter you will find out what to look for to define life.

■ Author's Rationale ■

This chapter is devoted to defining what we mean by the term "alive."

After looking at the limitations of characteristics commonly thought of as defining life, the student is introduced to the importance of heredity as the key criterion. Since most life-defining characteristics are noted at the molecular level, some basic chemistry is introduced here. This information provides the framework needed to understand cellular processes in Chapters 3–5.

IMAGINE THAT YOU ARE WALKING IN THE WOODS AND YOU UNEXPECTEDLY ENCOUNTER A LARGE FORMLESS BLOB LYING STILL ON THE FOREST FLOOR. IS THE BLOB ALIVE? HOW WOULD YOU DECIDE? MAKE A LIST OF THE THINGS YOU MIGHT DO TO DETERMINE WHETHER THE BLOB IS ALIVE. IN THIS SECTION YOU'LL FIND OUT IF YOUR LIST HAS THE SAME THINGS THAT A SCIENTIST WOULD DO.

2.1 *What Is Life?*

Objectives

❶ **Explain the difficulty in defining life using visually observable properties.**

❷ **Describe the five properties shared by all living organisms.**

❸ **Relate the properties of life to the biological themes in Chapter 1.**

First Guesses at Defining Life

Figure 2.1
If you found this blob on the ground, would you think it was a living thing? How would you be able to tell? What are the characteristics of living things?

In making your list about the blob shown in **Figure 2.1**, the first thing you would probably look for would be movement.

Movement Almost all animals move around. Squirrels dash along tree branches, sharks knife through the water, humans ride bicycles and run.

Everywhere you look, the world is alive with movement. However, movement from one place to another is not in itself a sure sign of life. A tree does not move about, for example, but it is alive. A cloud does move about, but it is not alive. Even if you could see the blob move on the ground in front of you, that would not be enough to tell you that the blob is alive.

Sensitivity So what should you do next? One thing you might do is poke the blob to see whether it responds. If you did that, you would be checking its **sensitivity**, its ability to respond to the stimulus of being poked.

Almost all organisms respond to stimuli. Deer flee from sounds they sense as dangerous, and plants grow towards light. Air movement, sound, light, and temperature are all stimuli. But not every stimulus produces a response. Can you imagine getting a response from kicking a redwood tree? If the blob just sits there after you poke it, and the tree doesn't move after you kick it, that doesn't mean they are not alive. Sensitivity, while a better criterion than movement, is still not a good single characteristic to define life.

Demonstration 1

Characteristics of Living Things

Help students *observe* that all living things are made of cells. Set up microscopes in the classroom with slides of unicellular organisms and sections of larger organisms that show cellular structure. Include examples from all five kingdoms of classification. Have students *compare* the appearance of the different cells. Explain that the cells of all living things are made of the same raw materials. Then ask students what they think makes cells different from one another. Encourage them to *infer* that cells are different because they organize the same materials differently.

Development   The next thing you might do is watch the blob to see if it changes. You would be looking for some signs of development. Development is an orderly progression that leads to greater specialization.

Most organisms exhibit development. The processes that change an acorn into an oak tree are development. You too are a product of the processes of development. You started life as a single cell too small to see. As you developed within your mother, some of your cells specialized to become nerve, muscle, and skin cells. But not all living things exhibit development. Bacteria do not develop. They simply grow and reproduce by splitting in half. Nor should all things be considered alive that undergo orderly, progressive change. The lines you see in rocks reflect the progressive, orderly laying down of material over a period of time, but the rock was never alive. The blob may or may not change as you watch it, but either way you cannot be sure if it is alive.

Complexity   At this point, you might walk up to the blob, pick it up, and examine it more closely, to see how complex it is. Study **Figure 2.2**.

All living things are complex. Even the simplest bacterium contains a bewildering array of molecules assembled into many intricate structures. Complex organization is essential to life. However, complexity is not solely a characteristic of life. A computer is also complex, and so is a television, but they are not alive. So even though complexity is a necessary condition of life, the fact that the blob may seem complex when you examine it does not tell you that the blob is alive.

Death   It might occur to you that the blob is the remains of a creature that was once alive but is now dead.

All living things die, while inanimate objects do not. Death is not the same thing as ceasing to function. A car that breaks down does not die. You may say, "The car died," but the now-broken car was never alive. Death is simply the termination of life. Unless one can detect life, death is a meaningless concept.

Figure 2.2
If you examined the blob more closely, you would be able to see details of its complex structure.

Life Has Five Characteristic Properties

You can now see that it is not so easy to determine whether something is alive. If you are going to determine whether the blob shown in **Figure 2.2** is alive, you will have to learn a great deal more about it. You will need to examine it more carefully to see how the blob resembles living things. All organisms that we know about share certain general properties. These characteristics have been passed down from the very first organisms to evolve on Earth. It is by these properties that we recognize other living things—they define what we mean by life. If the blob displays these properties, it is alive.

Sensitivity *Many living things are sensitive to external stimulation.* Slugs move toward light.

Complexity *Many living things are complex.* When amoebas exhaust the local supply of bacteria, they aggregate into a motile colony called a slug, which migrates into another area.

Development *Many living things grow and change.* When a slug reaches a new place, the colony differentiates into a basal portion, a stalk, and a swollen head.

Homeostasis *All living things maintain homeostasis.* Within the ameobal cell, levels of foods are kept high; if they fall the ameoba begins to seek other amoebas to form a slug.

Death *All living things die.* When nutrients are unavailable or the environment is unfavorable, the slug dies.

Reproduction *All living things reproduce.* The head of the slug bursts, releasing spores that drift on air currents to new areas.

DNA *All living things contain genetic information in DNA.* Each spore contains all the instructions for assembling a new amoeba.

Movement *Many living things move.* Individual slime mold cells move through moisture on the soil surface.

Cellular Organization *All living things are made of cells.* When spores from a slug encounter moisture, they develop into single-celled structures called amoebas.

Metabolism *All living things use energy.* Slime molds obtain their energy by ingesting bacteria in the soil.

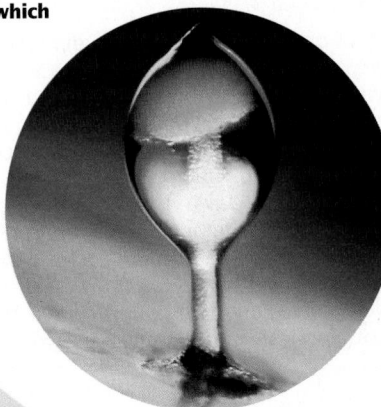

Figure 2.3
To be classified as a living thing, the blob must show the five characteristics listed in red type. The blob meets all the requirements for life. The figure also shows a summary of the first guesses about the blob in black type.

All living things are characterized by cellular organization, metabolism, reproduction, homeostasis, and heredity. Look at **Figure 2.3**. It shows how these characteristics apply to the blob. Living things use energy to grow and move in a process called **metabolism**. All living things must also maintain relatively stable internal conditions through a process called **homeostasis**. All living things contain genetic information. During reproduction, genetic information is passed on to offspring, a process called **heredity**.

The blob is alive because it possesses all five of these characteristics. The blob is a living organism called a cellular slime mold. These five properties define the core of your study of biology. Notice that the characteristics are closely related to the biological themes presented to you in Chapter 1. For example, the stability theme is directly related to the homeostasis characteristic. Stability is achieved through homeostasis. Patterns of change occur through heredity. Keep these relationships in mind as you study biology.

Section Review

❶ What is wrong with using movement as a single quality to define life?

❷ What are the five properties that define life?

❸ How is the fact that living things are composed of cells related to the theme of scale and structure?

▪ Section Review Answers ▪

1. Movement from one place to another is not in itself a sure sign of life. Some living things, such as plants, do not move, whereas nonliving things, such as clouds and water, do move.

2. All living things are characterized by cellular organization, metabolism, reproduction, homeostasis, and heredity.

3. Cells are a hierarchical component of living things.

Using the Feature

- Display pictures of the various kinds of microscopes. Ask students to briefly explain how each microscope magnifies objects.
- Display micrographs of an organism as it is viewed through each of the above microscopes. Ask students to describe the differences they see in the micrographs of the same organism.
- Show students several different micrographs. Ask them to observe each micrograph and infer which microscope produced the image.

Discussion

Guide a discussion by posing the following questions:

1. Why was the development of microscopes so important to science? *Microscopes made previously invisible organisms visible, allowing scientists to explain many misunderstood phenomena.*

2. What are some of the advantages and disadvantages of the electron microscope compared to other microscopes? *Advantages: very high magnification, specimen can be viewed in three dimensions. Disadvantages: specimens must be elaborately prepared before viewing and cannot be viewed while they are alive, microscopes are very expensive.*

Microscopy: *The Invisible World*

Why Are Microscopes Used?

If you hold a leaf out at arm's length, you can see the pattern of the veins and perhaps a few marks made by insects. As you bring the leaf closer you see more details of the structures, maybe a few cracks. But just when you are beginning to see a lot of detail, everything starts to blur. How can you get a closer look?

Scientists in the mid-1400s realized that they needed more than the human eye to study objects. As microscopy evolved, scientists have learned more than they could ever have imagined about plant and animal life.

Two important concepts relating to microscopes are magnification and resolution. Magnification is the ability of a microscope to make an image appear larger. Resolution is the ability to distinguish small, close objects. These concepts are equally important. If the details of a large image are unclear, the viewer would see only a fuzzy blur.

Kinds of Microscopes

Each type of microscope has its own strengths and limitations. Scientists have learned which microscopes can give the most information about whatever they are trying to see.

A compound microscope uses two lenses

Microscopes that use two lenses are called compound microscopes. A typical compound microscope, such as the one you use in biology class, has a light bulb or mirror in the base that shines light upward through the specimen. Light rays pass through the objective lens and then through the lens in the eyepiece. The image you see is magnified by both lenses. Total magnification is determined by multiplying the magnifications of the two lenses. If your microscope has a 10X eyepiece lens, and the 40X objective lens is in place, the object you are looking at appears 400 times larger than it actually is.

A biologist can use a compound microscope to study living cells. Cells appear to be essentially transparent, although there are small variations in thickness and density. As a result, the cell and some structures inside are visible, but the image is not very distinct. More details of the structures inside cells can be seen by thinly slicing cells and dyeing them with stains. Looking at a cell this way has obvious disadvantages—only one thin slice of cell is seen, and, of course, the cell is dead. However, sectioning and staining cells enables biologists to see many structures not visible in living cells.

These human cheek cells have been magnified 225X using a light microscope. They are seen using an ordinary bright field.

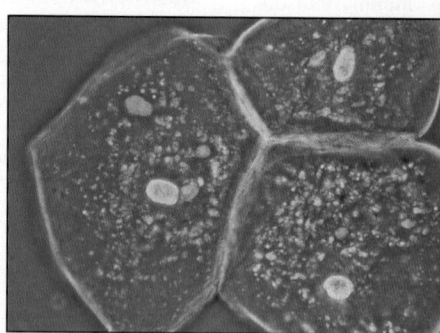

These human cheek cells have also been magnified 225X using a light microscope. The cells are unstained and are seen using phase contrast, which alters light waves so that details can be seen more easily.

n Action ▪ *Science in Action* ▪ *Science in Action* ▪ *Science*

This *Paramecium* has been magnified 1,000X using a scanning electron microscope (SEM). Note the numerous cilia it uses for locomotion.

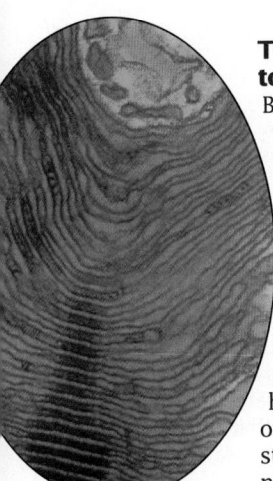

A transmission electron microscope (TEM) enables biologists to see details of the structure of cell organelles such as this Golgi body, magnified 13,000X.

TEMs and SEMs cause electrons to magnify objects

Because light has wavelike characteristics, there is a limit to the size of an object that can be viewed as a sharp, focused image. Practically speaking, bacteria with a diameter of 0.5 μm are about the smallest living things that can be distinguished using a good mass-produced light microscope. In the early part of this century, physicists showed that an accelerated stream of electrons also had wavelike properties similar to those of light. Microscopes using electrons instead of light to form images can magnify images at least 100 times as much as the light microscope.

In a transmission electron microscope (TEM), a stream of electrons passes through the specimen and strikes a fluorescent screen. By replacing the fluorescent screen with a piece of photographic film, a photograph called a transmission electron micrograph can be made. Sections of specimens viewed with a TEM are sliced much more thinly than sections prepared for the light microscope. These sections are treated with stains that block electrons, causing details to appear dark.

The scanning electron microscope (SEM) enables biologists to see detailed three-dimensional images of the surfaces of cells. Specimens are not sliced, but are placed on a small metal cylinder and coated with a very thin layer of metal. Like the picture on a television set, the image is formed one line at a time as the beam of electrons scans the specimen from side to side. The electrons that bounce off the specimen form an image that can be viewed on a video screen, or a scanning electron micrograph can be made. Because electrons would bounce off of the gas molecules in air, the stream of electrons and the specimen to be viewed must be placed in a vacuum chamber. Therefore, living cells cannot be viewed with an electron microscope.

The micrographs made with electron microscopes are always black-and-white, never colored. This is because the stream of electrons is only one wavelength. The fluorescent screen, the video screen, and the photographic film simply detect the presence or absence of electrons—light or dark. However, electron micrographs often have color added in the darkroom to make certain structures stand out in the micrograph.

New ways to look at cells

New video and computer techniques are extending the resolution and level of detail that can be detected by microscopes. The scanning tunneling electron microscope (STM) uses a needle-like probe to measure differences in voltage due to electrons on the surface of an object. A computer tracks the movement of the probe across the object, creating an image of the cell surface. The STM is used to view living cells.

This representation of DNA has been captured using a scanning tunneling electron microscope (STM).

Lesson Plan 2.2

Phase 1
PREPARATION

Key Concepts
- An atom consists of a core called a nucleus, which is surrounded by electrons.
- The nucleus contains two kinds of particles: protons and neutrons.
- Atoms combine with one another by forming bonds.
- An ionic bond forms when one atom loses one or more electrons and another atom gains one or more electrons.
- A covalent bond forms when two atoms share electrons.
- Both ionic and covalent bonds stabilize atoms in forming compounds.

Reading Strategy

Students often wonder why they must study chemistry in a biology course. Emphasize that a basic knowledge of chemistry is important in order to understand chemical reactions that are vital to life. Consequently, they should be aware of this chapter's importance to the material that will be covered in later chapters.

Phase 2
TEACHING STRATEGIES

Visual Strategy

Figure 2.4
Have students locate the protons, neutrons, and electrons in this figure. Emphasize that electrons do not have distinct orbits like those of the planets circling the sun. Rather, electrons travel in a mathematically defined region of probability called an electron cloud.

YOUR BODY IS A CHEMICAL MACHINE. EVERYTHING IT DOES, FROM THE SMASHING STROKE OF A TENNIS PLAYER TO THE DEEPEST THOUGHT OF A SCIENTIST OR POET, CAN BE UNDERSTOOD AS A CHEMICAL PROCESS. BECAUSE CHEMISTRY AND BIOLOGY ARE CLOSELY RELATED, A BRIEF INTRODUCTION TO CHEMISTRY WILL HELP YOU BETTER UNDERSTAND HOW LIVING ORGANISMS FUNCTION.

2.2 Basic Chemistry

Objectives

❶ **Relate atoms, elements, ions, and molecules to each other.**

❷ **Describe the structural features of an atom.**

❸ **Distinguish between covalent and ionic bonds and how they are formed.**

Figure 2.4
An atom consists of a nucleus surrounded by electrons that move about the nucleus at high speeds. This model shows regions of space outside the nucleus called energy levels.

Atoms: The Basic Structural Units of Matter

All matter in the universe is composed of tiny particles called atoms. An **atom** is the smallest particle of matter that can retain its chemical properties. For example, the smallest particle of carbon that still has all the chemical properties of carbon is a carbon atom.

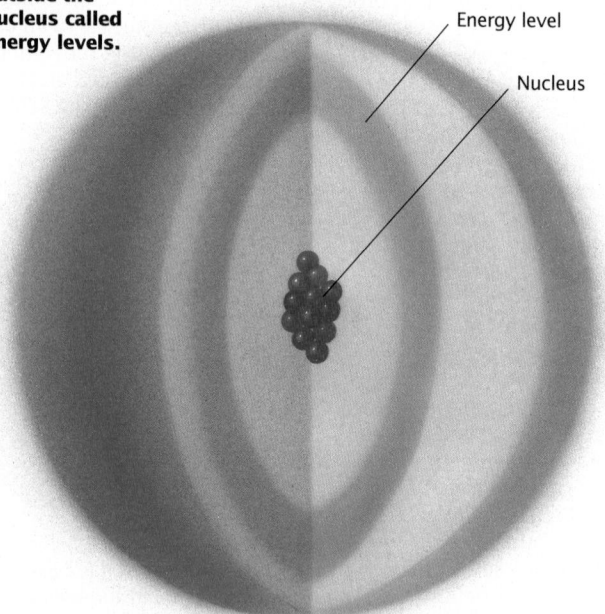

Energy level

Nucleus

An **element** is a substance composed of only one type of atom. There are 92 kinds of elements found in nature, but only 11 elements are common in living things. Over 99 percent of the atoms in your body are either nitrogen (N), oxygen (O), carbon (C), or hydrogen (H).

Atoms are composed of electrons, protons, and neutrons
An atom consists of a dense core called a nucleus surrounded by tiny moving particles called **electrons**. Electrons move about the nucleus in various energy levels. The nucleus contains two other kinds of particles, protons and neutrons. Most of the interior of an atom is empty space. If the nucleus of an atom were the size of an apple, the nearest electron could be more than a mile away. **Figure 2.4** shows a model for the structure of an atom.

Each proton in the nucleus has a positive charge, while the neutrons have no charge. Each electron has a negative charge. Atoms contain equal numbers of electrons and protons, therefore they have no charge.

▪ *Cultural Perspective* ▪

Dr. James Harris
(African-American)
Dr. James Harris, a nuclear chemist, co-discovered Elements 104 and 105 of the Periodic Table of Elements. Rutherfordium, Element 104, was discovered in 1969, and Hahnium, Element 105, in 1970.

Atoms can gain or lose energy

You may wonder why electrons are not pulled into the nucleus. It takes energy to overcome attractive forces and keep electrons moving about the nucleus. The energy needed to do this is similar to the energy it takes to hold an apple in your hand when gravity is pulling it toward the ground. The apple in your hand and electrons moving around the nucleus both possess energy. If you were to release the apple, it would fall to the ground. Electrons have energy of position, or potential energy, just like the apple. Electrons also have energy due to their motion, or kinetic energy.

Electrons move about the nucleus in different energy levels. The farther an electron is from the nucleus, the more energetic it is. An atom may have several energy levels stacked one on top of another, like the skin of an onion.

The sun's energy can be transferred to electrons by light photons. If a photon of light crashes into an atom, the transfer of energy jolts electrons to an energy level that is farther from the nucleus of the atom. The electron is able to maintain its new high energy level because it stores energy obtained from the photon.

Energy and Life

All interactions

among atoms

involve energy.

What atomic

particle carries

energy through

living systems?

Formation of Bonds Stabilizes Atoms

When an atom reacts, it gains, loses, or shares electrons. An atom that gains or loses electrons is called an **ion**. If the ion has more protons than electrons, it is said to be positively charged. If the ion has more electrons than protons, it is said to be negatively charged. Ions with positive charges are electrically attracted to ions with negative charges.

The force holding two atoms or ions together is called a **chemical bond**. All chemical bonds involve interactions between electrons in the energy levels farthest from the nucleus. These outer electrons determine an atom's chemical behavior. In other words, the atom's outermost electrons determine its ability to react with other atoms. Atoms bond to fill their outer energy levels with electrons. A full outer energy level makes an atom chemically stable. Carbon, nitrogen, and oxygen require eight electrons to fill their outer energy levels. Since these atoms all have fewer than eight electrons in their outer energy levels, they form bonds to gain, lose, or share electrons.

An **ionic bond** is the force of attraction between oppositely charged ions. To form an ionic bond, an atom loses one or more electrons so that it has a full outer energy level. The other atom involved in bonding gains one or more electrons so that it too has a full outer energy level. **Figure 2.5** shows a model for the formation of salt.

Figure 2.5

a **Sodium (Na) and chlorine (Cl) atoms are reactive. The sodium atom has one extra electron in its outer energy level. The chlorine atom is missing one electron in its outer energy level.**

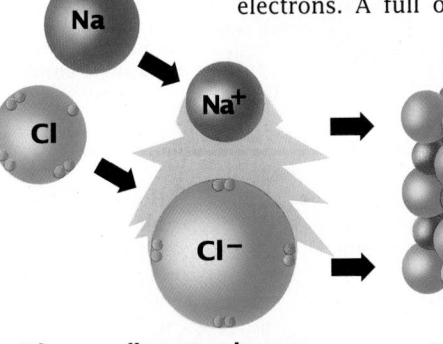

b **When a sodium atom loses an electron to a chlorine atom, two ions form. The force of attraction between the ions is an ionic bond.**

c **Oppositely charged sodium and chloride ions cluster to form salt crystals.**

31

Demonstration 3

Covalent Bonds

Again provide students with paper, scissors, and marble chips. Have students model the formation of covalent bonds as illustrated in Figures 2.6 and 2.7. Emphasize that in this case atoms come together not because of their opposite charges, but because the bonding electrons are shared by both atoms in the bond.

Phase 3

ASSESSMENT OPTIONS

Closure Strategy

Ionic or Covalent?

Borrow a conductivity apparatus from a chemistry teacher. Prepare both a salt and a sugar solution, labeling one A and the other B. Use the apparatus to test each for conductivity. Have students identify which is the salt solution and which is the sugar. Have them justify their choices based on their knowledge of the formation of ions versus molecules.

Section Review

Assign the *Section Review.*

Reteaching

Assign students to cooperative groups of four. Have each group develop a concept map for this section.

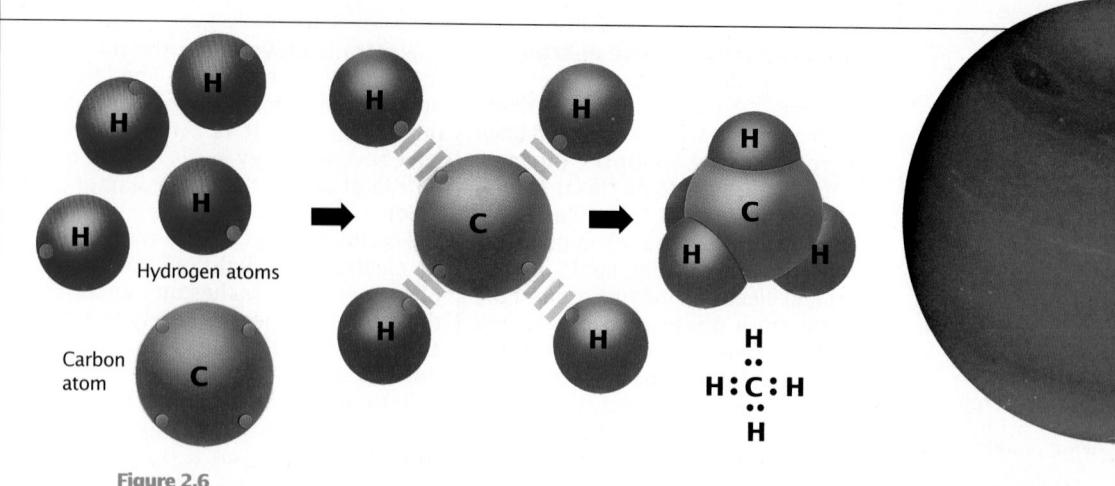

Figure 2.6

a **Hydrogen (H) and carbon (C) atoms are unstable.**

b **This carbon atom is electrically attracted to four hydrogen atoms.**

c **Methane forms when the carbon and hydrogen atoms share electrons.**

d **Methane gives Neptune's atmosphere its bluish-green appearance.**

Molecules result from the formation of covalent bonds

A second type of chemical bond results when two atoms share one or more electrons. A **covalent bond** is different from an ionic bond because electrons are *shared*, rather than *lost* or *gained*. **Figure 2.6** shows a model for covalent bonding between an atom of carbon and four atoms of hydrogen. The result is a molecule of methane.

A **molecule** is a group of atoms held together by covalent bonds. The force that holds these atoms together comes from sharing electrons. Most molecules in your body are made of more than two atoms, because most atoms must share electrons with more than one other atom to fill their outer energy levels. Oxygen atoms are able to form covalent bonds with two other atoms. Nitrogen atoms can form covalent bonds with up to three other atoms. Carbon atoms can form covalent bonds with as many as four different atoms.

Hydrogen requires just two electrons to fill its outer energy level. Each hydrogen atom is composed of a proton and a single electron. When two hydrogen atoms are close enough to each other, the two electrons can be shared equally by the two nuclei. The result is the diatomic (two atom) molecule of hydrogen gas (H_2), shown in **Figure 2.7**.

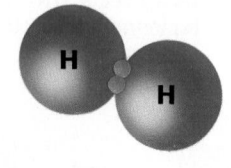

Figure 2.7
Two hydrogen atoms can form a covalent bond.

Section Review

❶ **How do atoms differ from ions?**

❷ **How are elements related to molecules?**

❸ **Describe the structure of an atom.**

❹ **How do covalent bonds differ from ionic bonds?**

▪ *Section Review Answers* ▪

1. An atom is the smallest particle of matter that can retain its chemical properties. An ion is an atom that has gained or lost electrons and as such possesses an electric charge.

2. A molecule is composed of atoms of one or more elements.

3. Atoms are composed of electrons, protons, and neutrons. Electrons have a negative charge and move about the nucleus in various energy levels. Protons have a positive charge and are found in the nucleus. Neutrons have no charge and are found in the nucleus.

4. A covalent bond is different from an ionic bond because electrons are shared, rather than lost or gained.

YOU HAVE JUST LEARNED HOW ATOMS COMBINE TO FORM BONDS. MOLECULES ALSO REACT TO FORM MACROMOLECULES. THERE ARE FOUR TYPES OF MACROMOLECULES IN YOUR BODY. THEY ARE COMPOSED ALMOST ENTIRELY OF CARBON, HYDROGEN, AND OXYGEN. IN THIS SECTION, YOU WILL LEARN ABOUT THE BASIC STRUCTURE AND FUNCTION OF EACH MACROMOLECULE.

2.3 *Molecules of Life*

Objectives

❶ Define sugar and describe the process that occurs in the formation of polysaccharides.

❷ Describe the solubility and energy storage properties of lipids.

❸ Explain the factors that affect the three-dimensional structure of proteins.

❹ Define nucleic acids and describe their functions.

Organic Compounds Are Derived From Carbon

Just as atoms can be joined to form molecules, molecules can be joined to build **macromolecules**. All organisms are composed of four major classes of macromolecules: proteins, lipids, carbohydrates, and nucleic acids. Each kind of macromolecule has different subunits. The primary component of all macromolecules is carbon.

The properties of carbon are important to biological systems, including those living in the coral reef shown in **Figure 2.8**. In the last section you read that carbon has four electrons in its outer energy level, and that a carbon atom seeks to fill

that energy level by sharing its electrons with other atoms. Carbon atoms form long chains that are the backbone of many different kinds of molecules. Molecules with carbon-carbon bonds are called **organic compounds**.

Figure 2.8
Diamonds are pure carbon. Carbon compounds are the primary components of living things, including plants and animals that live in this coral reef.

Demonstration 1
Carbohydrates
Display various foods that are high in carbohydrates, including potato, pasta, bread, corn, and cake.

Sports Connection

Why Carbohydrates?
Ask students why athletes who compete in long-distance or marathon events often eat foods high in carbohydrates the night before competition.

Connection: Chapter 5

Glucose
Tell students that glucose will be discussed in more detail when respiration and photosynthesis are covered in Chapter 5.

Visual Strategy

Figure 2.9
Tell students that each line in the structural formulas shown in this figure represents a covalent bond. Have them count the number of carbon, hydrogen, and oxygen atoms in the structural formula shown for glucose. Ask how many covalent bonds each carbon atom forms. Use this figure to show students how simple sugars are connected to form the more complex carbohydrates. The ways in which simple sugars can be connected vary. Compare the schematic illustrations for glycogen and starch as an example.

Language Connection

Prefixes
Understanding the terminology of organic compounds is easier when students know what certain prefixes mean. Have them look up the meanings of *macro-*, *poly-*, *di-*, and *mono-*. Have students use these prefixes to write a paragraph about carbohydrates.

Carbohydrates Are Energy Sources

A **carbohydrate** is composed of carbon, hydrogen, and oxygen in a ratio of one carbon atom to two hydrogen atoms to one oxygen atom. Some carbohydrates, such as table sugar, are simple, small molecules. Other carbohydrates, like the starch in potatoes, exist as chains hundreds of subunits long. Carbohydrates contain many carbon-hydrogen bonds. They are well suited to be energy sources because their bonds store considerable energy.

Carbohydrates can be either simple or complex molecules

Among the simplest carbohydrates are sugars, small molecules that taste sweet. While sugars may have as few as three carbon atoms, the sugars involved in energy storage, like glucose, have six. These sugars have the formula $C_6H_{12}O_6$. Carbohydrates are made by linking individual sugars together to form long chains called polysaccharides. Polysaccharides are insoluble in water. They can be deposited in specific storage areas in a cell. This ability to store energy in the form of polysaccharides lets organisms build up energy reserves.

Starch and glycogen are both complex carbohydrates

Starch is a polysaccharide composed of glucose subunits. Amylose is the simplest kind of starch. It exists as a long, unbranched chain of glucose molecules. Baking or boiling starchy plants such as potatoes and corn breaks these chains into shorter fragments that are soluble and can be used by the cell.

Humans consume a great deal of carbohydrates; the seeds of rice, wheat, and corn supply about two-thirds of all the calories used by people. Animals store glucose in the form of long, branched chains, called glycogen. **Figure 2.9** shows the structural differences of starch and glycogen.

Cellulose provides structural support

Many organisms use polysaccharides as structural molecules as well as for energy storage. Plants manufacture a polysaccharide called cellulose. Cellulose consists of glucose subunits linked in a way that most animals cannot break down. Cellulose forms in the cell walls of plants. When you eat plants containing cellulose, it passes through your body undigested. This undigested cellulose is called dietary fiber and is an important component of your diet. In contrast, cows and horses are able to graze on grass. These animals have bacteria that digest or break down cellulose in their intestines. You lack these bacteria and would starve on a diet of grass.

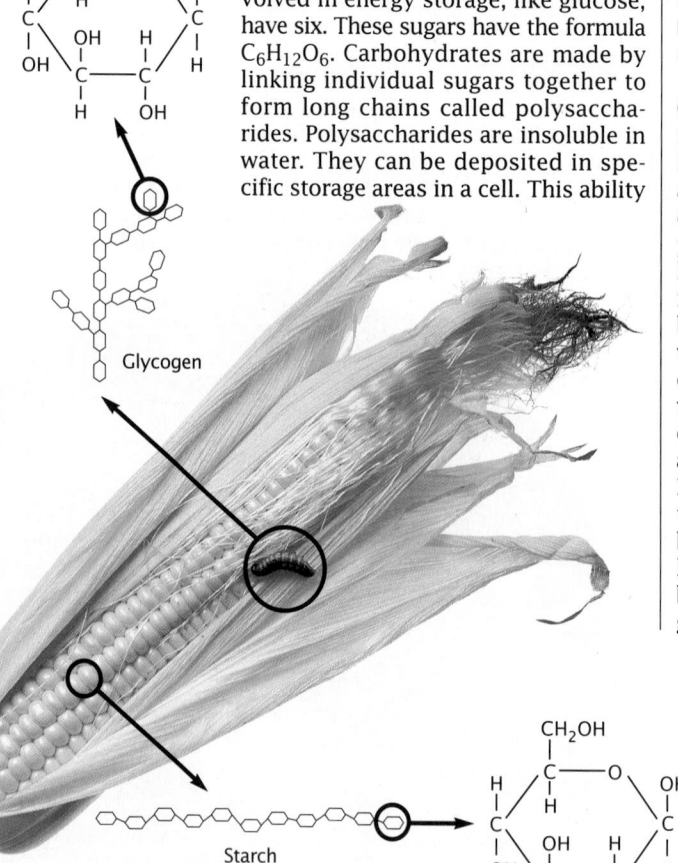

Figure 2.9
Starch, the carbohydrate found in corn, is chopped into glucose subunits and converted to glycogen when it is eaten by this worm. The glucose subunits are linked in different ways.

Table 2.1 Types of Fats

Type	Found in
Saturated Fat 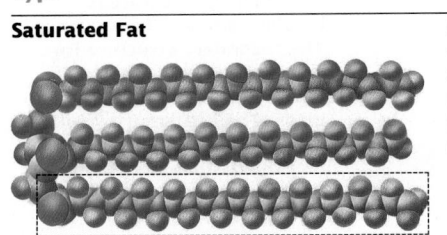 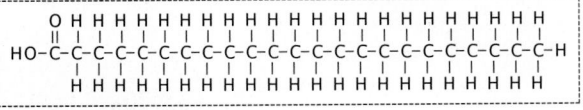 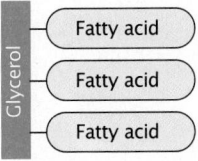	Butter Cheese Chocolate Beef Palm oil Coconut oil
Unsaturated Fat 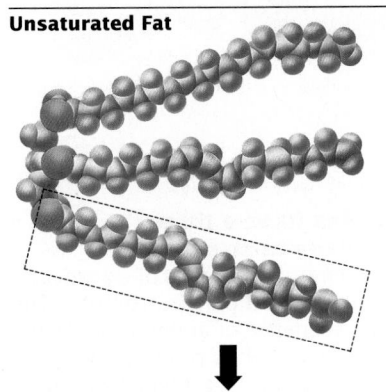 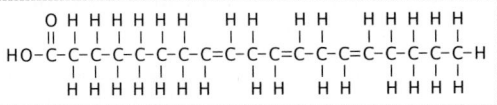 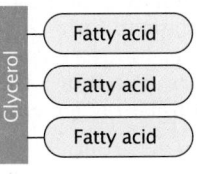	Avocado Olives Olive oil Peanuts Peanut oil Almonds Corn oil Fish Mayonnaise Safflower oil Sunflower oil

Lipids Store Energy

A **lipid** is not soluble in water, but is soluble in oil. The most important kind of lipid is fat, an energy storage molecule. Fats have more carbon-hydrogen bonds than carbohydrates and, therefore, can store more energy. They have three very long chains of CH_2 units that join at one end like the letter "E." The backbone of the structure is a glycerol molecule, and the branches are fatty acids made from CH_2 units.

There are two major categories of fats: saturated fats and unsaturated fats. In addition to fats, there are two other types of lipids: steroids and waxes. Steroids include the sex hormones, cholesterol, chlorophyll (an important pigment found in plants), and retinol (the vision pigment of your eyes). Earwax and beeswax are examples of lipids that are waxes.

Fats can be saturated or unsaturated

When the three chains of fatty acid molecules line up neatly side by side, the fat is said to be saturated. As you can see in **Table 2.1**, most carbon atoms in a saturated fat are bonded to two hydrogen atoms. The carbon atoms are saturated with hydrogen atoms. If some units of the three fatty acid chains are linked by double bonds, kinks occur in the chain. This type of fat is called an unsaturated fat. The carbon atoms could bond to additional hydrogen atoms if the double bonds between adjacent carbon atoms were broken. Unsaturated fats usually exist as liquids called oils at room temperature. In contrast, most saturated fats are solid at room temperature. Two exceptions are palm oil and coconut oil. Although both are liquid at room temperature, they are classified as saturated fats because they lack double bonds.

It is possible to make an oil into a solid fat by adding hydrogen. The double bonds become single bonds and the chains can then line up. Peanut butter is usually hydrogenated. The peanut fats are converted to saturated fats, so they don't separate as oils while the jar sits on the store shelf.

▪ *Cultural Perspective* ▪

Olive Oil

According to the *Journal of the American Medical Association*, olive oil, an unsaturated fat used mostly in southern Italy, helps lower cholesterol and may also lower blood pressure and control blood sugar. After studying 5,000 people living in northern, central, and southern Italy, scientists found that southern Italians, who use olive oil more than other oils and butter, had the lowest blood pressure and the healthiest levels of blood sugar.

Demonstration 4

Proteins

Display various foods that are high in proteins, including meat products, beans, and egg whites.

Demonstration 5

Protein Structure

Give each student a large paper clip and have them unbend it to form a straight wire. Then have them push four gumdrops onto the wire and space them equally apart. Tell them to compare their model to the primary structure of a protein shown in Figure 2.10. Next, students should either bend or coil the wire around a pencil to form the secondary structure. Have them fold the wire for the tertiary structure. Finally, two students should join their wires to see how the quaternary structure forms. In each case, students should compare their model to the appropriate figure in Figure 2.10.

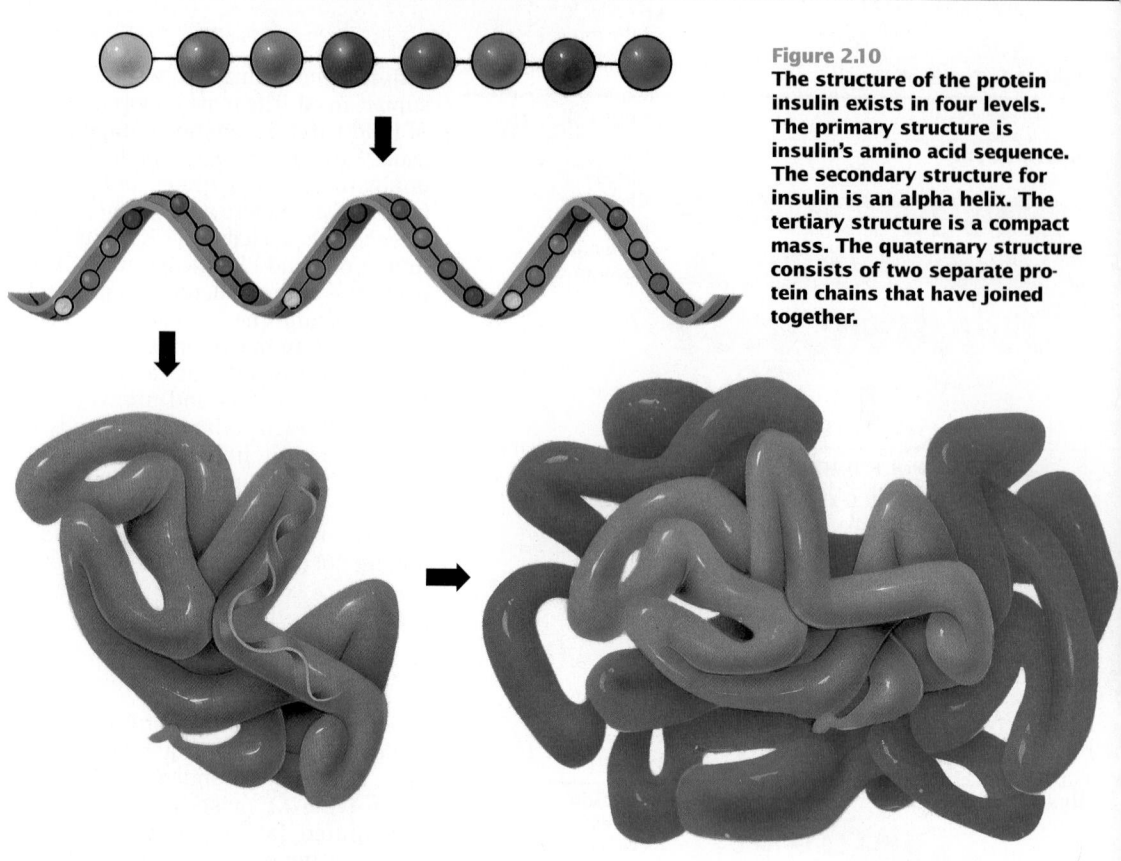

Figure 2.10
The structure of the protein insulin exists in four levels. The primary structure is insulin's amino acid sequence. The secondary structure for insulin is an alpha helix. The tertiary structure is a compact mass. The quaternary structure consists of two separate protein chains that have joined together.

Proteins Provide Structure and Increase Reaction Rate

A **protein** is composed of long chains of subunits called amino acids. There are 20 different kinds of amino acids that humans use. In a particular type of protein the positions of the component amino acids can vary along the chain. Thus, there is an almost endless variety of possible proteins. Think of a type of protein as a paragraph in a book made from 20 letters in the English alphabet. Any letter may be present at any position as the words of the paragraph are composed. A typical protein has approximately 100 amino acids linked together in its chain. There are many millions (actually 20^{100}) of different possible amino acid sequences for such a protein.

Proteins have a three-dimensional structure

The chemical properties of a particular protein depend on its structure. The actual sequence of amino acids in the protein is called its primary structure. Because amino acids interact with their neighbors, parts of the chain coil or bend. This coiling and bending determines the protein's secondary structure. In most proteins, the entire chain folds into a compact mass called its tertiary structure. When two or more folded proteins combine to form clusters, the mix of proteins forms a quaternary structure. **Figure 2.10** shows the four levels of structural arrangements for a protein.

In the next few chapters, you will study various types of proteins. Pay particular attention to the shapes of these macromolecules, because shape determines a protein's biological function.

Proteins function as structural molecules and enzymes

Proteins often play structural roles in organisms. Cartilage and tendons are made of a protein called collagen, as is the matrix of your skin and bone. A protein called keratin forms the horns of a rhinoceros and the feathers of a bird, as well as your own hair.

Proteins play a second very important role in organisms: they act as enzymes. Enzymes increase the rate at which chemical reactions occur such as those that take place during metabolism. Most chemical reactions necessary for growth, movement, and other body activities would not take place without enzymes.

Nucleic Acids Contain Genetic Information

The fourth major class of macromolecules is called **nucleic acids**. There are two types of nucleic acids: DNA (deoxyribonucleic acid) and RNA (ribonucleic acid). Subunits of DNA and RNA are called **nucleotides**. These nucleotides are grouped into units called genes, which encode information concerning how a given organism will grow and develop. RNA is involved in making working copies of genes. These copies are used in assembling amino acids to make proteins.

DNA is a double helix

A DNA molecule consists of two interlocking coil-shaped strands that resemble a spiral staircase. Chemists refer to this coiling structure as a double helix. Figure 2.11 shows the structure of a DNA double helix. DNA encodes the sequences of all the cell's proteins, as well as information that determines when each protein is to be produced. In simple cells like bacteria, the DNA exists as a long molecule. But in complex cells like those of your body, the DNA exists in numerous different segments. These segments, along with proteins, form compact bodies called **chromosomes**.

RNA helps in the synthesis of proteins

RNA molecules have a variety of shapes, depending on their function in the cell. Specific RNA molecules serve as scaffolds for the assembly of all the different proteins in the cell. Others exist in the cell as long, single-stranded threads that carry DNA's message from one part of the cell to another. You will learn more about how DNA and RNA work together to build proteins in Chapter 7.

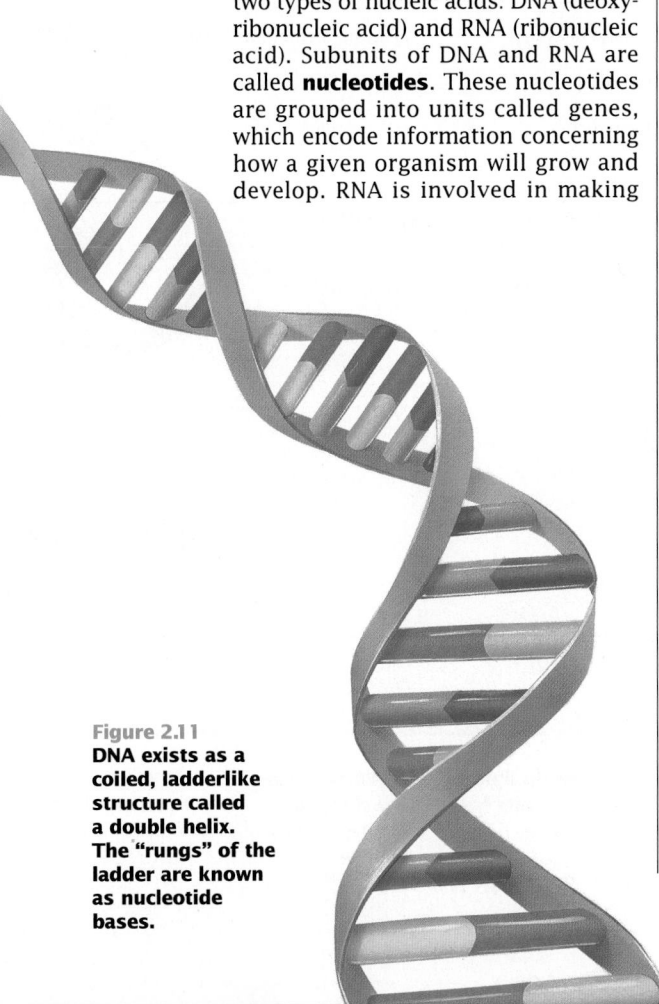

Figure 2.11
DNA exists as a coiled, ladderlike structure called a double helix. The "rungs" of the ladder are known as nucleotide bases.

▪ Cultural Perspective ▪

Dr. Herman Branson
(African-American)
Dr. Herman Branson researches protein structure and has assisted Dr. Linus Pauling, a Nobel Prize-winning chemist, in identifying the alpha and gamma helical structures of proteins.

Visual Strategy

Table 2.2
Tell students that the information summarized in this table will be most useful in helping them with the table they were instructed to complete in the Reading Strategy.

Phase 3

ASSESSMENT OPTIONS

Closure Strategy

What Is It?
Have students decide whether each of the following is a carbohydrate or lipid:
- $C_{12}H_{24}O_2$
- $C_{12}H_{24}O_{11}$
- $C_{57}H_{110}O_6$

Have them defend their choice.

Section Review

Assign the *Section Review.*

Reteaching

Prepare a table on the board with the headings given in the Reading Strategy. Fill in the information, asking for input from as many students as possible. Have students add information to their own tables as needed.

Macromolecule Summary

The structures of the four basic macromolecules (carbohydrates, lipids, proteins, and nucleic acids) have the same building blocks. All four are composed of long chains of similar subunits. Using **Table 2.2** you can compare and contrast the structures and functions of macromolecules.

Table 2.2 Classes of Macromolecules

Class	Subunit	Function	Example
Carbohydrates	Sugar	Stores energy	Starch, glycogen
		Structural component	Cellulose, chitin
Lipids	Fatty acid	Stores energy	Body fat
		Membrane bilayer	Plasma membrane
		Steroid hormones	Testosterone
		Pigments	Chlorophyll
Proteins	Amino acid	Catalysis by means of enzymes	Lactase
		Structural component	Hair, cartilage
		Peptide hormone	Insulin
Nucleic Acids	Nucleotide	Stores genetic information	DNA
		Makes proteins	RNA

Starch granules

Human fat cells

Hair

Chromosomes

Section Review

1. How is a sugar related to polysaccharides and starch?

2. How do lipids react in water?

3. What causes a protein to have a tertiary structure?

4. What is the function of DNA in cells?

▪ Section Review Answers ▪

1. They are all carbohydrates, composed of carbon, hydrogen, and oxygen. Polysaccharides and starch are long chains of carbohydrate macromolecules or complex carbohydrates.

2. Lipids are not soluble in water.

3. The sequence of amino acids causes a protein chain to fold into a compact mass.

4. DNA combined with proteins forms chromosomes that contain all of an organism's genetic information.

Chapter **2**

Highlights

These seeds are the start of new life, but is a seed itself alive? Use the five properties of life to decide.

	Key Terms	Summary
2.1 What Is Life? 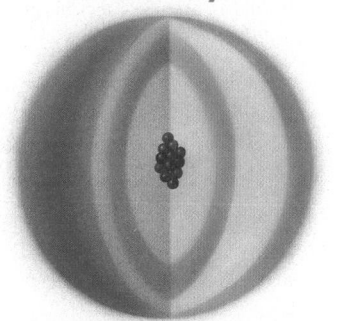 Is this blob alive? How would you go about answering this question?	sensitivity (p. 25) metabolism (p. 27) homeostasis (p. 27) heredity (p. 27)	• Some of the most obvious properties of life cannot be used alone to decide whether something is alive. • The characteristic properties of living things can also appear in nonliving things: batteries use energy, clouds move, oceans maintain constant temperatures.
2.2 Basic Chemistry Atoms are the basic structural units of matter.	atom (p. 30) element (p. 30) electron (p. 30) ion (p. 31) chemical bond (p. 31) ionic bond (p. 31) covalent bond (p.32) molecule (p. 32)	• Each of the 92 elements found on Earth is made of atoms. Nitrogen, oxygen, carbon, and hydrogen make up more than 99 percent of the atoms in your body. • When electrons absorb energy, they move to higher energy levels farther from the nucleus. When they release energy, they fall to lower energy levels. • Bonds form when atoms lose, gain, or share electrons. Molecules are groups of atoms held together by covalent bonds. Some compounds are held together by the force of attraction among ions.
2.3 Molecules of Life  Carbon is the primary chemical component of the macromolecules found in living things, including the plants and animals shown in this coral reef.	macromolecule (p. 33) organic compound (p. 33) carbohydrate (p. 34) lipid (p. 35) protein (p. 36) nucleic acid (p. 37) nucleotide (p. 37) chromosome (p. 37)	• Carbon is the most abundant element in living things. Molecules with carbon-carbon bonds are called organic compounds. • Organisms use carbohydrates to store energy and provide structural support. • Lipids are not soluble in water. Fats store energy more efficiently than carbohydrates because they have more carbon-hydrogen bonds. • The sequence of amino acids in a particular protein determines its shape and chemical properties. The shape of a particular protein determines its biological activity or function. • Nucleic acids contain genetic information and direct protein production.

Chapter Review Answers

Understanding Vocabulary

1. **a.** Movement does not fit the pattern because it is not one of the five characteristic properties of life.
   b. Compound does not fit the pattern since it is not a kind of atomic particle.
   c. Element does not fit the pattern because it is not one of the four classes of biological macromolecules.
   d. Tin does not fit the pattern because it is not an element common in living organisms.

Relating Concepts

2. Map answer is shown on page 23D.

Understanding Concepts

Multiple Choice

3. c	8. c
4. a	9. c
5. a	10. b
6. d	11. b
7. c	

Completion

12. ionic bonds, covalent bonds
13. atom, ion
14. chemical bond

Understanding Vocabulary

1. From each group of terms, select the one that does not fit the pattern and explain why it does not fit.
   a. homeostasis, movement, reproduction, heredity
   b. electrons, protons, compound, neutron
   c. protein, lipid, element, nucleic acid
   d. nitrogen, carbon, hydrogen, tin

Relating Concepts

2. Copy the unfinished concept map below onto a sheet of paper. Then complete the concept map by writing the correct word or phrase in each oval containing a question mark.

Understanding Concepts

Multiple Choice

3. Living things are made mostly of compounds that contain
   a. carbohydrates.   c. carbon.
   b. blood.   d. scandium.

4. Sugar, starch, and cellulose are examples of
   a. carbohydrates.   c. lipids.
   b. monosaccharides.   d. proteins.

5. A protein's secondary structure is caused by
   a. neighboring amino acids interacting.
   b. the amino acid sequence.
   c. folding of the total chain of amino acids.
   d. its clustering with other proteins.

6. Polysaccharides are formed
   a. from amino acids and nucleotides.
   b. when sugars and proteins react.
   c. when animal fats are hydrogenated.
   d. by linking sugars together.

7. Which class of macromolecules stores more energy than carbohydrates?
   a. proteins   c. lipids
   b. polysaccharides   d. nucleic acids

8. Molecules are formed
   a. from ionic bonds.
   b. by atoms carrying electrons.
   c. when electrons are shared.
   d. when protons gain or lose energy.

9. An electrically charged atom is a(n)
   a. element.   c. ion.
   b. organic.   d. molecule.
   compound.

10. Which of the following is not a characteristic property of life?
    a. maintains homeostasis
    b. uses energy
    c. undergoes development
    d. made of cells

11. Chemical bonds help atoms achieve chemical stability because bonding
    a. forms ions.
    b. fills an atom's outer energy level.
    c. prevents electrons from moving about the atom's nucleus.
    d. causes DNA to release energy.

Completion

12. Chemical bonds that result from the attraction of oppositely charged ions are called _____ , while those that result from atoms sharing electrons are called _____ .

13. The smallest particle of matter is a(n) _____ ; when it is electrically charged it is called a(n) _____ .

14. The force that holds two atoms together is called a(n) _____ .

Short Answer

15. Describe how the properties of carbon enable it to serve as the "backbone" of so many different kinds of molecules.

16. "Nucleic acids are biologically important molecules." Do you agree or disagree with this statement? Explain.

17. Name the four classes of biologically important macromolecules and the subunit of each.

Short Answer

15. Carbon needs eight electrons in its outer energy level to be stable, but it has only four. Therefore, carbon can form up to four covalent bonds, and in so doing forms long chains to which other molecules can attach.

16. DNA and RNA are nucleic acids. DNA encodes the sequence of all the cell's proteins and information regarding when the proteins are produced. RNA transfers the code from DNA that is used to build proteins from amino acid units.

17. carbohydrates and sugar; lipid and fatty acid; protein and amino acid; nucleic acid and nucleotide

Interpreting Graphics

18. Label the protons, neutrons, and electrons of the carbon atom shown above. Recall that a carbon atom requires four electrons to fill its outer energy level and become stable. How many electrons does a bonded carbon atom have moving around its nucleus?

19. Use a reference on food groups to help you answer the following. Name the carbohydrates you see in the picture. Name the proteins you see in the picture. Name the lipids you see in the picture.

Reviewing Themes

20. *Patterns of Change*
A pattern exists in the way electrons behave when covalent bonds are formed. Describe the pattern. How does this pattern differ from the pattern of electron behavior in ionic bonds?

21. *Energy and Life*
All living things store energy but they often do it in different ways. What are two forms of energy storage found in humans?

Thinking Critically

22. *Inferring Conclusions*
Why do football players and other athletes often consume a complex carbohydrate diet while training?

23. *Inferring Conclusions*
Butter and margarine are similar in appearance, but are chemically different. Margarine is actually an oil, sometimes called an unsaturated fat. During the manufacturing process, what is done to margarine to convert it to a hard fat?

24. *Analyzing Concepts*
Contrast the functions of proteins and carbohydrates in the human body.

Cross-Discipline Connections

25. *Biology and Health*
Dietary fiber is a carbohydrate that cannot be digested by the human body. Why do nutrition experts recommend that high fiber foods like whole-grain cereals and raw fruit be eaten regularly?

Discovering Through Reading

26. Read pages 115–119 of the article "Eloquent Remains," in *Scientific American*, May 1992. What is paleo-DNA? Of what value is paleo-DNA to molecular archaeologists?

usually in excess of 10 percent of a person's daily caloric intake, is considered a high protein diet.)

23. The oil is hydrogenated to make it look a lot like butter.

24. Proteins function in structural roles such as forming cartilage, tendons, skin, bones, and hair; they also catalyze chemical reactions. In contrast, carbohydrates serve as the source of energy needed to carry out bodily functions.

Cross-Discipline Connection

25. While supplying no nutrients, fiber helps to strengthen the intestine and ensure the speedy passage of food through the digestive tract. (In other words, a high-fiber diet keeps you regular!)

Discovering Through Reading

26. Paleo-DNA is the DNA of organisms that lived long ago. By analyzing paleo-DNA using the same procedures that are used to examine DNA from living organisms, molecular archaeologists are beginning to answer questions about the histories of man and the Earth. One study is investigating the possibility that all living humans descended from one woman. Note: This study is still largely discredited.

Interpreting Graphics

18. 10 electrons
19. Possible answers:
carbohydrates—bread
protein—cheese, milk, ham, student's hair and skin
lipids—cheese, milk, and mayonnaise

Reviewing Themes

20. When covalent bonds are formed, electrons are shared among atoms. In ionic bonds, electrons are not shared but transferred.

21. Humans store energy as carbohydrates and as fats.

Thinking Critically

22. Athletes eat a diet high in protein because heavy training requires protein. The protein is needed for tissue growth and maintenance. It is especially important since the human body cannot store protein as it does carbohydrates. (A diet that contains a high amount of protein,

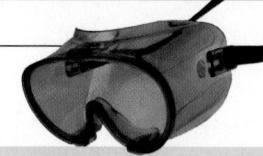

Prelab Preparation Answers

1. • Answers will vary but should list the five general characteristics of living things presented in the text.
   • Characteristics that might be easily observed could include: evidence of organization, cellular structure, and response to stimuli.
   • Characteristics that take a long period of time, e.g. growth and development.

2. • Characteristics such as growth and development, life span, reproduction, and the use of energy can require longer periods to become apparent.

3. • The first observations must be made using the low-power objective.
   • This is necessary to find the specimen on the slide.
   • The coarse adjustment should never be used while viewing under high power because the slide and the objective could be damaged.

Procedure Answers

4. Tilting the microscope at a sharp angle can cause the water of a wet-mount preparation to drain from beneath the coverslip.

9. A comparison of results allows scientists to evaluate the reliability of their work.

Chapter **2**

Investigation

What Properties Can Be Observed in Living Things?

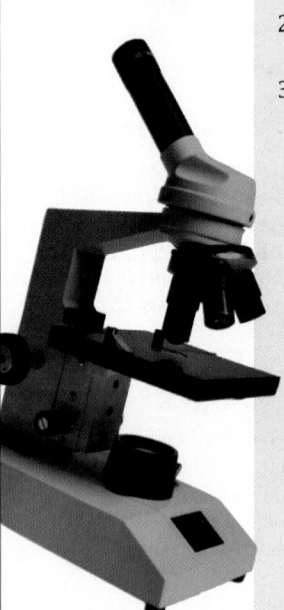

Objectives

In this investigation you will:
• *observe* and *identify* the characteristics of living things

Materials

• *Detain* slowing agent
• forceps
• microscope slide (with concave impression)
• coverslip
• compound light microscope
• medicine dropper
• pond water
• toothpick

Prelab Preparation

1. Review what you have learned about the characteristics of life by answering the following:
   • What are the characteristics of living things?
   • What characteristics might be easily observed in microscopic organisms?
   • What characteristics might be difficult to observe?

2. Explain why some characteristics are difficult to observe.

3. Read the information about the use and care of the compound light microscope in the Appendix.
   • What power objective must be used to make your first observations?
   • State the reasons why this objective must be used first.
   • Name the adjustment control that should never be used when you focus with the high power objective.

Procedure: Observing Characteristics of Living Things

1. Form a cooperative group of four students. Work with one member of your group to complete steps 2–8.

2. Using the medicine dropper, place several drops of pond water and a drop of *Detain* inside the depression on the slide. Stir with a toothpick.

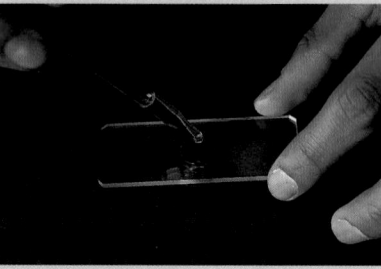

3. Carefully place the coverslip on top of the pond water.

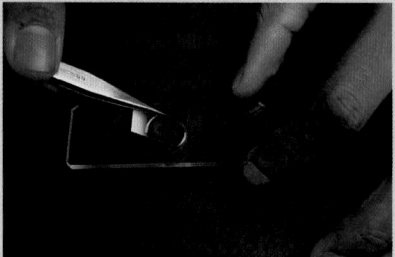

4. Make sure the microscope is level. Then, place the prepared slide on the microscope stage. *What might happen if the microscope were tilted at a sharp angle?*

5. Use low power to observe the pond water. Move the slide, as necessary, to inspect the entire drop of pond water.

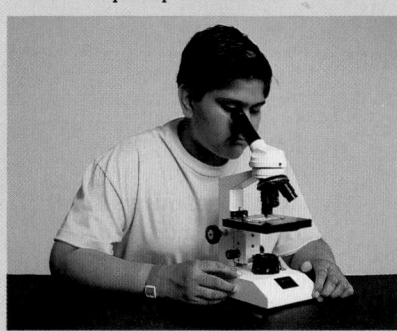

6. Make a drawing of the first thing you observe that appears to be alive. List the characteristics shown by this thing.

7. Repeat step 6 for each object you think might be classified as living.

8. Clean up your work area. Wash your hands before leaving the lab.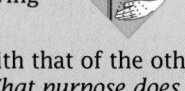

9. Combine your work with that of the other team in your group. *What purpose does a comparison of results serve in a scientific investigation?*

10. Pool your group's data with that of the rest of the class. Prepare a composite list of the characteristics you observed in the pond water.

Analysis

1. *Identifying Relationships*
   What characteristics can be used to distinguish between microscopic living and nonliving things?

2. *Analyzing Observations*
   List the properties of life that could not be observed in the pond water organisms. Explain why these characteristics cannot be seen with a microscope.

3. *Making Predictions*
   If you were to observe a drop of pond water for an extended period of time, what changes might occur?

4. *Making Inferences*
   Is movement an indication that an organism is alive? Why could movement be a misleading indicator for defining life?

Thinking Critically

1. What limitations does microscopic observation present in determining whether something is alive?

2. You see an object that you think is alive in the pond water. Devise a plan to test your prediction.

Analysis Answers

1. Answers will vary but are likely to suggest that cellular structure, organization, and response to stimuli are characteristics of life that can be seen using the compound light microscope.

2. Answers will vary but are likely to suggest that growth and development, reproduction, use of energy, and life span were not evident in the organisms observed. Observations of these characteristics require an extended length of time.

3. Observation for an extended length of time might reveal evidence of organisms adapting to changes in the environment, reproduction, or death.

4. Movement requires energy, which is a basic characteristic of life. Movement can be misleading because it may not be a result of internal activity; a nonliving object may be caught in a current or may be moved by the effects of gravity.

Thinking Critically Answers

1. Observing living things under the microscope is limited by what can be observed in a certain period of time. Observations of characteristics like cellular structure, organization, and response to stimuli can be done in a limited amount of time, whereas observations of characteristics such as growth and development, reproduction, use of energy, and life span require an extended length of time.

2. Answers will vary but should reflect that the study of biology has given them an understanding of the basic characteristics of life.

43

Chapter 3 Cells

Planning Guide

	Objectives/Themes	Classwork Resources	Homework Resources
3.1	**1.** Explain why cells must be small if they are to perform efficiently. **2.** Contrast the behavior of polar and nonpolar molecules. **3.** Describe how the structure of water shapes the membrane of the cell. **Themes:** Interacting Systems, Scale and Structure	**Teacher's Resource Binder** Focus Activity 3 *Modeling Cells: Surface Area-to-Volume* Lab Investigation 3.2 *Structure and Function of Cells* **Other Resources** Science in Action Video *Cell Structure and Function/ Cell Division***	**Text** Section Review, p. 49 **Teacher's Resource Binder** Directed Reading Worksheet 3.1 **Other Resources** Audiocassette 3.1*
3.2	**1.** Describe how phospholipids are organized to form a fluid cell membrane. **2.** Describe the functions of proteins in the cell membrane. **3.** Explain how cell membrane proteins interact with the lipid bilayer. **Themes:** Interacting Systems, Scale and Structure	**Text** Science in Action *Jack Wang: Protein Chemist* pp. 54–55 **Other Resources** Transparency 8	**Text** Section Review, p. 53 **Teacher's Resource Binder** Directed Reading Worksheet 3.2 **Other Resources** Audiocassette 3.2*
3.3	**1.** Name two characteristics that distinguish eukaryotes from prokaryotes. **2.** Relate each organelle to a task essential to the life of the cell. **3.** Contrast the structure of an animal cell with that of a plant cell. **4.** Describe how eukaryotes evolved from prokaryotes. **Themes:** Evolution, Scale and Structure, Stability	**Text** Investigation *How Do Plant and Animal Cells Differ?* pp. 64–65 Science, Technology, and Society *Do Your Cells Belong to You?* pp. 66–67 **Teacher's Resource Binder** Lab Investigation 3.1 *Cytoplasm and Organelles* Extension Worksheet *More Cell Structures* **Other Resources** Transparencies 9–11	**Text** Section Review, p. 60 **Teacher's Resource Binder** Directed Reading Worksheet 3.3 Vocabulary Review Worksheet* Reteaching Worksheet* *Mapping Cell Part Functions* **Other Resources** Audiocassette 3.3*

***Reteaching Options**

Demonstrations	Assessment
3.1: pp. 47, 48, 49 **3.2:** pp. 51, 52 **3.3:** p. 57	Chapter Review pp. 62–63 Portfolio Assessment p. 43D Chapter Test—Teacher's Resource Binder Test Generator

Research Notes

Connection to Physics: Magnetic Cells

There are still many structures within eukaryotic cells whose roles are not yet fully understood, and some that are still being discovered. Recently, Joseph Kirschvink of California Institute of Technology announced that he and his team of researchers had found crystals of the magnetic mineral magnetite (Fe_3O_4) in human brain cells.

Many types of bacteria also make magnetic particles of iron oxide and iron sulfide, which they apparently use as microscopic compass needles. Different kinds of bacteria produce different shapes of crystals, from rectangular to six-sided and eight-sided.

The particles are encased in an organic membrane within the cell to form a "magnetosome."

The magnetosomes in a bacterium spread out in a line, along the Earth's magnetic field. In this way, as the cell uses its flagella to swim, it will travel along a north-south axis. It is also possible that the bacteria oxidize the iron as a supply of energy.

Magnetic particles have also been found in tissue from migrating animals such as birds, bees, and fish, and seem to have a role in these animals' unerring sense of direction as they return to the same area over and over again.

The magnetic particles found in human brain tissue are similar to those found in bacteria, which indicates that they are formed by a biological process, instead of being a contaminant in the protoplasm.

As of yet, the role that these magnetic particles play remains uncertain. Since there is a much lower concentration of magnetite in humans than in the migrating animals, its use to provide a sense of direction seems unlikely.

Further study of the particles in humans could shed light on the controversial question of whether weak electromagnetic fields (EMFs) from electrical lines and appliances can cause adverse biological effects, such as cancer and other diseases. Until now, scientific research on the effects of EMFs had not definitively settled the question.

One of the main stumbling blocks in the research of EMF effects had been explaining how such fields could interact with cells, since researchers had assumed that the human body did not contain magnetic material. Although this assumption may now be invalid, it could be that the usual background electrical and magnetic "noise" of the cell, is stronger than any weak effect on the magnetic particles caused by the EMFs and the magnetic particles.

Investigation Notes

How Do Plant and Animal Cells Differ?

pages 64–65

Purpose: Students will learn to recognize the structural differences between plant and animal cells.

Prelab Preparation

1. Prepared slides of human cheek cells (WARD'S 93 M 6003) are used in response to the newly enacted OSHA blood-borne pathogen standard. Because saliva may contain traces of pathogens, using prepared slides provides protection measures against pathogens (hepatitis viruses, HIV) transmitted via body fluids.

2. The day before the lab, students should review the Appendix information on the care and use of the microscope and making wet mount slides.

3. Cleaning spills and identifying air bubbles consumes lab time. The use of prepared slides for both the plant and animal portions of the lab can save time and provide more opportunity for learning, if desired. It is suggested that you use slides from an onion tip and a whitefish if you choose to use only prepared slides. These examples clearly show the difference between plant and animal cells.

Answers will be found on pages 64–65.

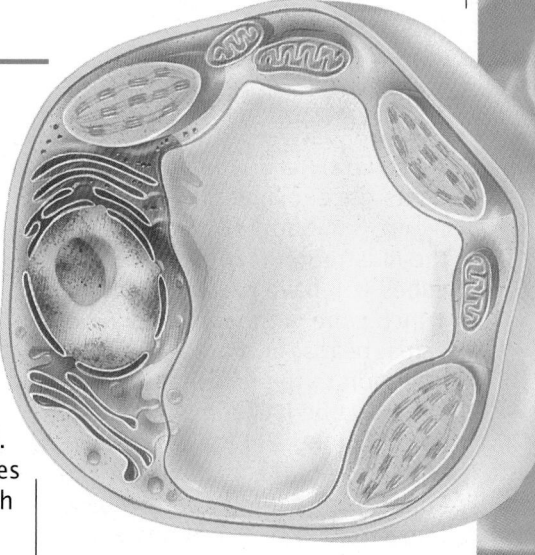

Meeting Individual Needs

Objectives

1. Students will demonstrate appropriate use of core vocabulary for the chapter (Vocabulary File).

2. Students will distinguish between measurements in two and three dimensions and understand the ratio of surface area to volume (Demonstration A).

3. Students will observe an example of water's polar nature and relate it to effects on cells (Demonstration B).

4. Students will use a model to demonstrate the differences that exist between prokaryotes and eukaryotes (Demonstration C).

Vocabulary File

If you are not already using the Vocabulary File, refer to Chapter 1 for its preparation. See the Chapter Highlights on page 61 for a list of suggested words.

Demonstration A

Surface Area-to-Volume Ratio
for page 47
(Developing Modeling Skills/Tactile Learners)

Materials: graph paper, tape, scissors, glue, small foam packing chips (or macaroni, beans, etc.)

Preparation: Prepare two cubes before class, using Figure 3.3 as a model to form a 1-inch and 2-inch cube.

Procedure: Read aloud to students the explanation under Figure 3.3. Show students how the illustration relates to the cubes you have prepared. Fill the 1-inch cube with foam packing chips, beans, or macaroni. Call attention to the fact that the second cube has edges that are twice as long. Fill the 2-inch cube with more of the same material. Pour the contents of both cubes into separate piles on the table. Have students count the items

and compare the relative amounts. These amounts are a rough measure of volume. Read aloud to students the paragraphs on page 51.

Demonstration B

Water's Properties
for page 48
(Developing Observational Skills/Tactile Learners)

Materials: Water, atomizer sprayer, and wax paper

Procedure: Give each student a small piece of wax paper. Squirt drops of water onto the wax paper. Let students play with the water, pushing it around and pushing drops together. Discuss how the water "felt." Relate the discussion to water's property of cohesion and the polarity of its molecules.

Demonstration C

Prokaryote vs. Eukaryote
for page 56
(Developing Modeling Skills/ Visual Learners)

Materials: two corrugated cardboard boxes (one plain, one with compartments), construction paper, large letter stencils, scissors, glue

Preparation: Use plastic wrap to enclose one of the compartments. Label it "*nucleus.*" Cut out construction paper letters for "*prokaryote*" and "*eukaryote.*" Glue "*prokaryote*" on the simple box and "*eukaryote*" on the box with compartments. Cover the labels with construction paper so students cannot see them. When the demonstration is finished, place the boxes on a shelf to be used later during a review, or with later chapters.

Procedure: Read aloud to students the contents of the chart on page 60. Display the boxes so that students can see the inside of each box. Ask students which box represents a prokaryote. Have students explain their choice. Expose the labels. Use the idea of compartments to lead into a discussion of organelles.

Additional Strategies

Visual Strategies

Pages 45, 46, 47, 48, 49, 50, 51, 52, 53, 56, 57, 58, 59, and 60

Auditory Learners

Use *Biology: Visualizing Life* Audiocassettes for Sections 3.1, 3.2, and 3.3.

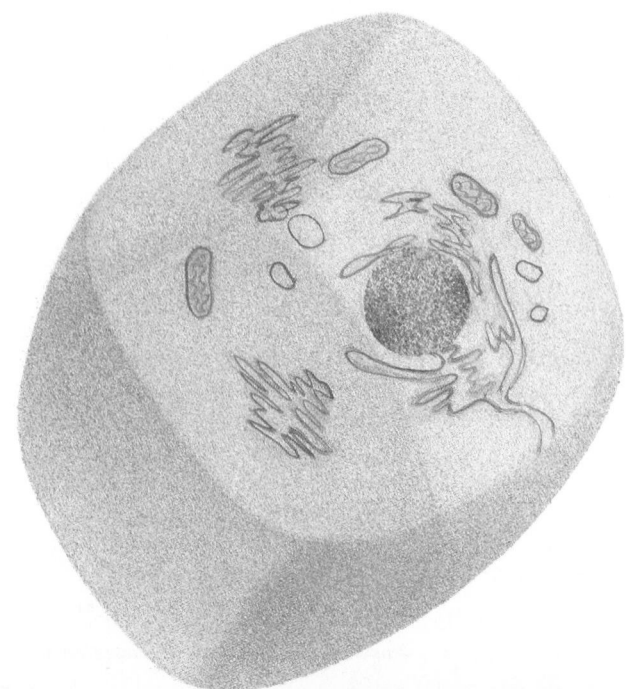

Meeting Individual Needs (cont.)

Cooperative Learning
Discovering Cells

Timing: Use this activity to introduce Section 3.1.

Group Size: 3–4 students

Outcome: Students will be able to recognize cell diversity and understand the historical context of the cell theory.

Individual Accountability: Each student is responsible for taking the role of an early scientist.

Positive Interdependence: Students will share observations and make comparisons with each other.

Have each student pick one of the following scientists and research his work.

> Robert Hooke
> Anton van Leeuwenhoek
> Matthias Schleiden
> Theodore Schwann
> Rudolph Virchow

Have each student take his or her scientist's role, and describe to the rest of the group what the scientist observed.

Ask students to discuss how scientists make reproducible observations and how their curiosity leads to the observation of more specimens. Point out that most of the figures in the text show far greater detail than the instruments used by these scientists.

Portfolio Assessment

Students should select their best work and provide a self-reflective rationale for their selections. Students can make selections in the following areas.

1. *Content* — One concept map from the chapter (See page 812 for evaluation criteria.)

2. *Reading Comprehension* — One Directed Reading Worksheet from the Teacher's Resource Binder (Use the answer key to evaluate for accuracy.)

3. *Writing* — Using the Vee Form, summarize a newspaper or magazine article relating to cells. (See page 22T for evaluation criteria.)

   Or: Select a writing task or project from the Chapter Review.

4. *Performance Assessment* — One Vee form from a chapter investigation or lab manual investigation. (See page 22T for evaluation criteria.)

Teacher makes selections in the following areas.

1. *Formal Assessment* — Chapter test (Test A, B, or the Test Generator) The teacher-scored test should be reviewed by the student. Incorrect responses should be corrected by student before the test becomes part of the portfolio.

2. *Informal Assessment* — Use the Direct Observations Checklist, page 33T, during a laboratory or other cooperative learning experience.

3. *Performance Assessment* — Have students design and create a model of a plant or animal cell, using different materials for each of the organelles. Models should be scientifically accurate.

Concept Map Answer

The following is one possible answer to the Relating Concepts exercise on page 62.

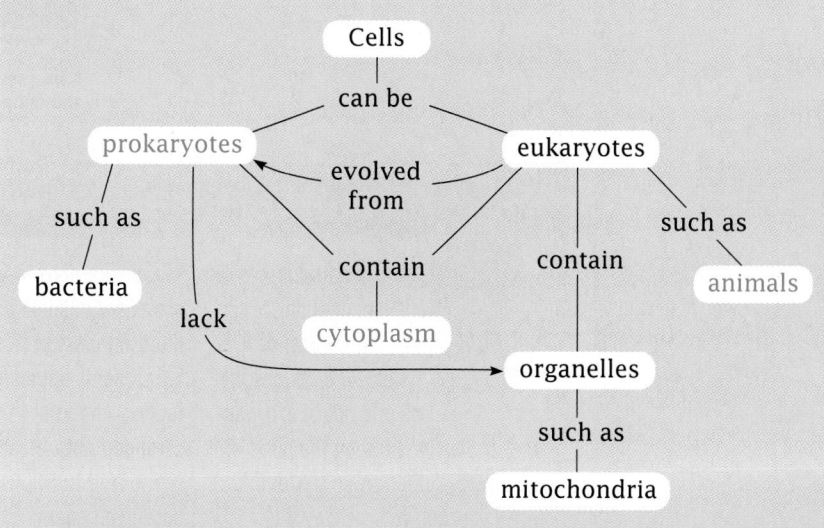

Chapter 3

Determining Prior Knowledge

- Have students draw a cell and label any parts they may remember from previous courses.
- Inflate a balloon several times with varying amounts of air. Have students identify the surface area and volume. See if they can determine which one increases more rapidly as the balloon gets larger.
- Pour some oil and water into a beaker. Ask students to explain their observations.

Chapter 3

Cells

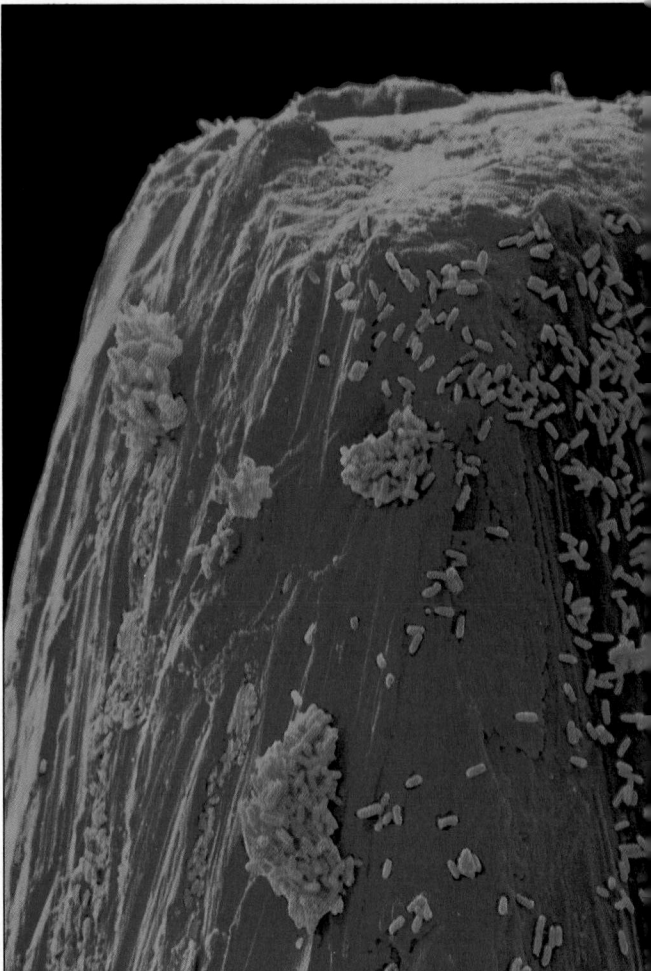

This color-enhanced scanning electron micrograph shows bacteria, one-celled organisms, on the point of a pin.

▪ Author's Rationale ▪

This chapter is critical in understanding the rest of the text. It describes the cell and how it is organized. The focus is not on the parts of the cell but rather on understanding why it is organized the way it is.

Surface area-to-volume ratio is presented in some detail because of the usefulness of this concept in helping students understand cell size limitations, and later in understanding the arrangements of cells and tissues for specialized functions, i.e., villi in the digestive system.

The plasma membrane is featured as the key element of cell structure, and how cells employ internal membranes to compartmentalize activities. The organizing role of water is presented here in the context of its most important function, rather than in isolation with a discussion of chemistry.

YOUR BODY IS COMPOSED OF MORE THAN 100 TRILLION CELLS. AS LONG AS CELLS CARRY OUT THEIR FUNCTIONS NORMALLY, YOU ARE GENERALLY UNAWARE OF THEM. BUT IN SOME DISEASES, SUCH AS CANCER, CELLS BEHAVE ABNORMALLY. IN THE SEARCH FOR THE CAUSES AND POSSIBLE CURES FOR CANCER, SCIENTISTS FIRST EXAMINE THE WORLD OF THE CELL TO SEE HOW CELLS WORK NORMALLY.

3.1 *World of the Cell*

Objectives

① Explain why cells must be small if they are to function efficiently.

② Contrast the behavior of polar and nonpolar molecules.

③ Describe how the structure of water shapes the membrane of the cell.

At the Edge of the Cell

Plants and animals are made up of a maze of tiny compartments. Each compartment is called a **cell**. Since the mid-1800s we have known that all living things are composed of cells. Most microscopic creatures like the one in **Figure 3.1** are single cells. A cell is the smallest unit that can carry on all of the activities of life.

In some ways, a cell is like a submarine. A submarine has a tough outer surface, called a hull, which wraps around the complex machinery that makes the submarine function. A cell is also filled with complex machinery and has a tough outer surface, which is called the **cell membrane**. The cell membrane serves the same purpose as the hull of a submarine. It separates what is inside from what is outside. Nothing gets into or out of the sub except through the hatches in the hull, and nothing gets into or out of a cell except through "gates" in its cell membrane.

Imagine that the submarine's hull suddenly disappears. The machinery would be scattered all over the sea floor. Without the cell membrane to hold together the machinery and the substances the cell needs for life, the cell would die. However, a cell membrane is much more than a simple container. The complex machinery that is inside the cell cannot function unless raw materials continuously enter into the cell and destructive waste products promptly leave the cell. By regulating what goes into and out of a cell, the cell membrane helps to maintain the internal environment of the cell.

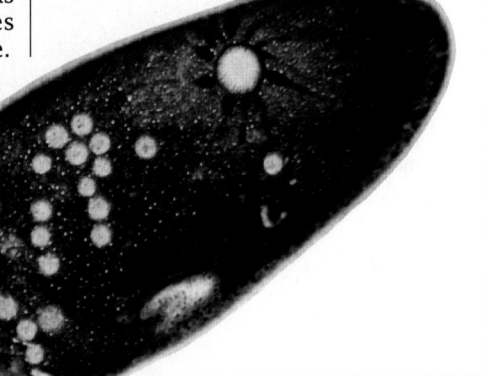

Figure 3.1
This microscopic protozoan, *Paramecium* **(133X), is a single cell.**

Lesson Plan 3.1

Phase 1

PREPARATION

Key Concepts
- A cell is the smallest unit of life.
- The cell membrane regulates what enters and exits a cell.
- A cell is limited in size so that all its interior structures are near the cell membrane.
- As a cell grows, its volume increases more rapidly than does its surface area.
- As water molecules cluster together, they force the nonpolar lipid molecules of the cell membrane into a layer.

Reading Strategy

Have students identify sentences that provide information about the section objectives.

In 1665, a group of scientists met in London. Robert Hooke demonstrated the use of a compound microscope he had just finished building to answer a simple question: Why does cork float? Hooke cut a thin slice of cork and had the scientists examine it with his microscope. The walls around the air spaces reminded Hooke of the small rooms in a monastery called cells. Referring to each roomlike space as a cell, Hooke introduced the term into the vocabulary of science.

Phase 2

TEACHING STRATEGIES

Visual Strategy

Figure 3.1
Have students locate the cell membrane on *Paramecium caudatum*. Have students name substances that might enter or exit the cell through its cell membrane.

What Limits the Size of a Cell?

What is the most obvious thing that can be said about cells? It is that they are very small. Your body has about 100 trillion cells. If these cells were each the size of a shoe box and were lined up end to end, they would stretch in a line about 30 billion km (18.6 billion mi.)— to the sun and back 100 times! So cells must be pretty small to cram that many into your body. Study **Figure 3.2**. What factors limit the size of a cell?

Cell parts cannot be too far from the cell membrane

Every bit of food and information needed by the cell must enter through the cell membrane. When cells are small, no part of their complex machinery lies too far from the area outside the cell. If a cell were larger, fewer of its interior structures could be near the cell membrane. That is bad for just the same reason that long supply lines are bad for an army—too many things could go wrong and responses to information would be too slow. Thus, small cells work more efficiently because their supply lines are short.

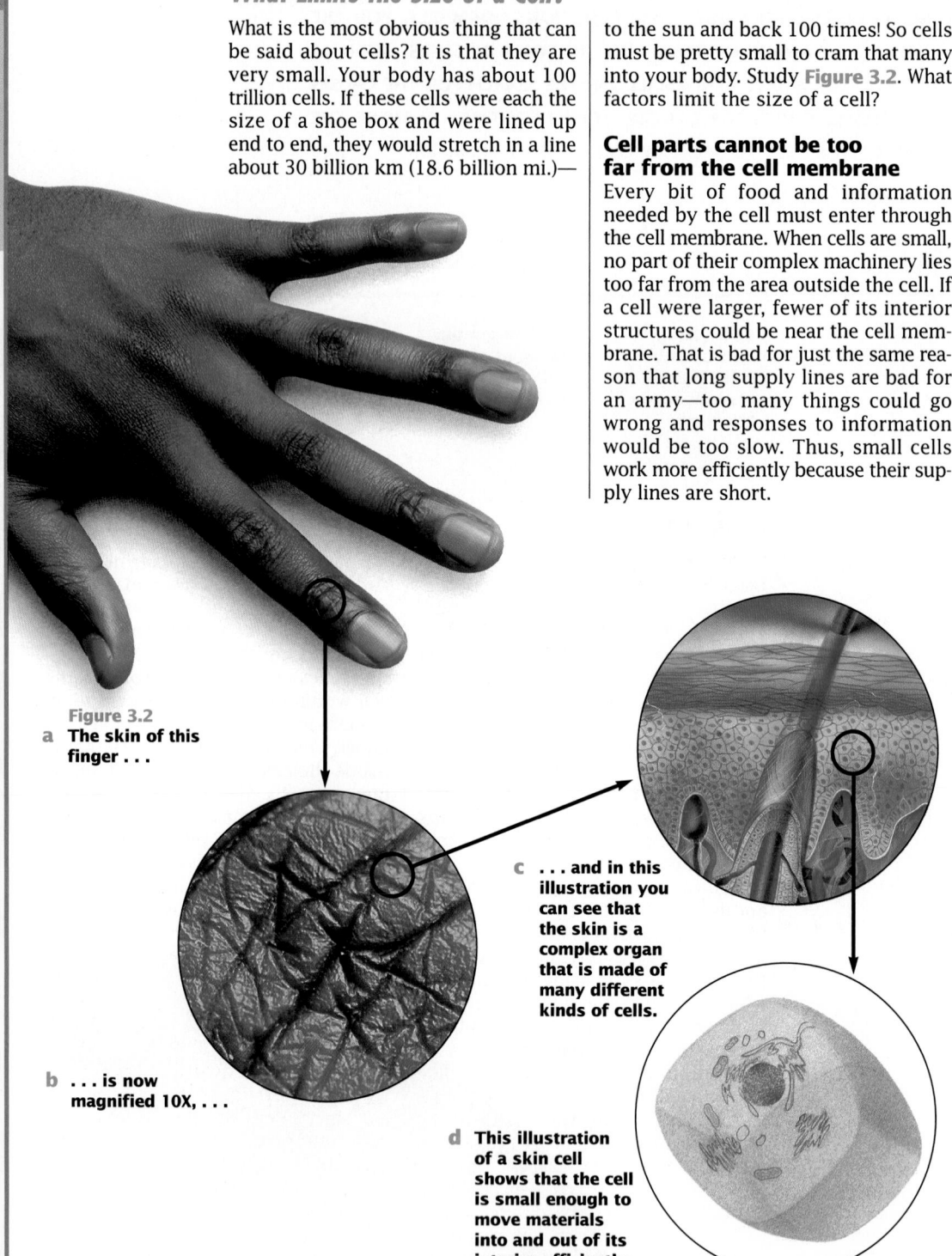

Figure 3.2

a **The skin of this finger . . .**

b **. . . is now magnified 10X, . . .**

c **. . . and in this illustration you can see that the skin is a complex organ that is made of many different kinds of cells.**

d **This illustration of a skin cell shows that the cell is small enough to move materials into and out of its interior efficiently.**

The Ratio of Surface Area to Volume in Cells

The surface area of a cell is a measurement of the exterior of the cell. The volume is a measurement of the internal contents of the cell.

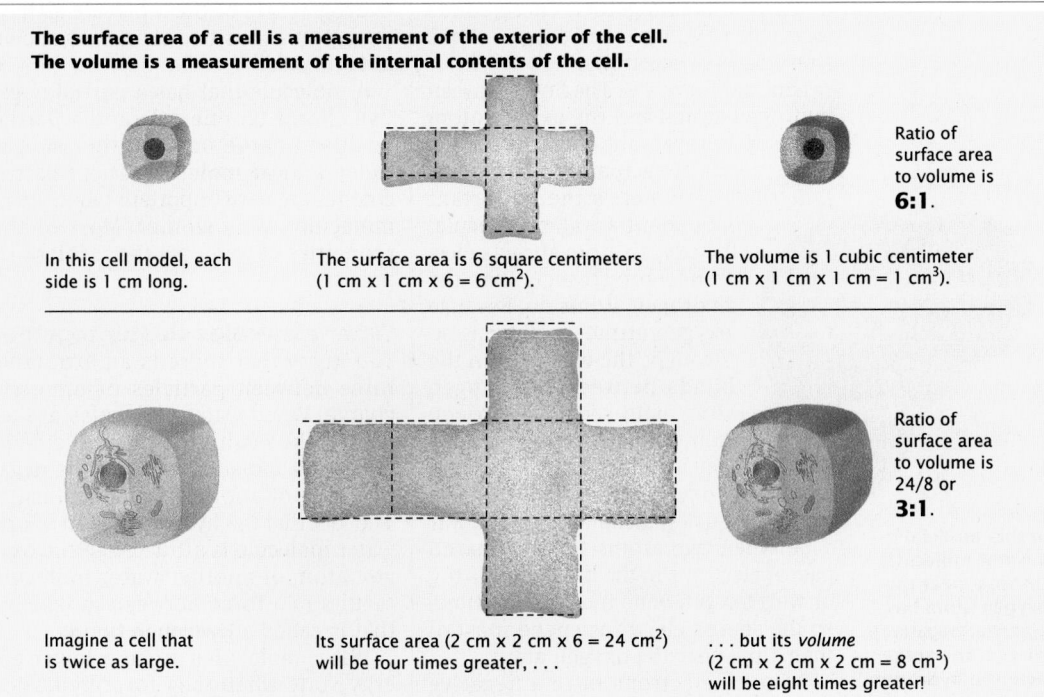

In this cell model, each side is 1 cm long.

The surface area is 6 square centimeters (1 cm x 1 cm x 6 = 6 cm²).

The volume is 1 cubic centimeter (1 cm x 1 cm x 1 cm = 1 cm³).

Ratio of surface area to volume is **6:1**.

Imagine a cell that is twice as large.

Its surface area (2 cm x 2 cm x 6 = 24 cm²) will be four times greater, . . .

. . . but its *volume* (2 cm x 2 cm x 2 cm = 8 cm³) will be eight times greater!

Ratio of surface area to volume is 24/8 or **3:1**.

Figure 3.3
As a cell gets larger, its volume increases at a faster rate than its surface area. A cell's surface area must be large enough to meet the needs of its volume.

A cell's volume increases faster than its surface area

As a cell grows, it takes in more food and creates more wastes. Since these substances must pass into and out of the "gates" in the cell membrane, the membrane must be large enough to service the cell's needs. As the cell grows, so does its membrane. But cells cannot grow indefinitely. So what limits cell size?

One factor is the relationship between the surface area and the volume of the cell. As a cell grows, its volume increases at a much faster rate than its surface area. A small cell, such as the one shown in the top row of **Figure 3.3**, has enough surface area to meet its needs. But a cell as large as the one shown in the bottom row might not. The ratio of a cell's surface area to its volume ultimately limits how large that cell can become. Cells cannot grow so large that their surface areas become too small to take in enough food and to remove enough wastes.

Water and the Cell

All cells are surrounded by water. Single-celled creatures swim in small ponds and also in vast oceans. Even the cells of your body, such as blood cells or skin cells, are surrounded by a thin film of water. Water is present inside the cell too. All the complex machinery inside the cell performs its functions in water. The cell membrane is shaped by the water found inside and outside of the cell. To understand how water can shape a cell membrane, you must first look closely at the structure of a water molecule.

Water is a polar molecule

The chemical formula for water is H_2O. A water molecule is made of two

Demonstration 1
Short vs. Long Distance
To illustrate the advantage of shorter distances for materials and information to travel, form two rows of students on opposite sides of the class. Have one row consist of twice as many students as the other. Have the first students in each row read a short passage you have selected. These students then relate the story to the next student in line. Each student in turn tells the next. Have the last student in each row write down what they heard and compare both versions to the original passage.

Visual Strategy

Figure 3.3
Many of your students will have a difficult time understanding the relationship between surface area and volume as a cell gets larger. Walk students through the calculations, explaining that surface area is equal to length × width × number of sides and that volume is equal to length × width × height.

Mathematics Connection

Determining the Surface Area-to-Volume Ratio
Have students determine the surface area-to-volume ratio for a cell that measures 3 cm on each side.

Theme Connection

Scale and Structure
The importance of structures having an extensive surface area will be reinforced in subsequent chapters. Alert students to look for this when covering such topics as the human respiratory and digestive systems, leaf structure, and adaptations in animals for living on land and in the air.

Connection: Chapter 2

Covalent Bonding

Review covalent bonding when introducing the structure of water. Point out that in some covalent bonds the electrons are shared equally between the atoms. Provide examples of such nonpolar molecules— H–H and O–O. This will set the stage for what happens when there is unequal sharing—polar molecules such as water are formed.

Visual Strategy

Figure 3.4

Guide students in using a molecular model kit to construct a water molecule. Have them compare their models to Figure 3.4. Ask what each stick in their model or each line in Figure 3.4 represents.

Visual Strategy

Figure 3.5

Ask students what the small purple and blue molecules represent in Figure 3.5c. Have students use the molecular model kits to demonstrate how a hydrogen bond forms.

Demonstration 2

Surface Tension of Water

Gently place a needle or razor blade on the surface of the water in a glass or beaker. The metal, although denser, will float. Have students explain how the clustering of water molecules might account for this phenomenon, known as surface tension. Ask if they have ever seen insects walk on water. Point out that the only liquid with a surface tension greater than that of water is mercury.

hydrogen atoms and one oxygen atom that are bonded together. These bonds form when hydrogen and oxygen atoms share pairs of electrons.

Look at **Figure 3.4**. The lines between hydrogen atoms and the oxygen atom are used to represent covalent bonds. These bonds are actually pairs of electrons. Now here's the important thing about a water molecule: the oxygen atom attracts electrons more strongly than the hydrogen atoms do. Because oxygen attracts electrons so strongly, the electrons in the bonds between the oxygen atom and each hydrogen atom are not shared equally. They are more likely to be near the oxygen atom. Think of this unequal sharing as a tug of war between two atoms for the shared pair of electrons in the bond. In this tug of war, oxygen wins most of the time, so the shared electrons spend most of their time near the oxygen atom.

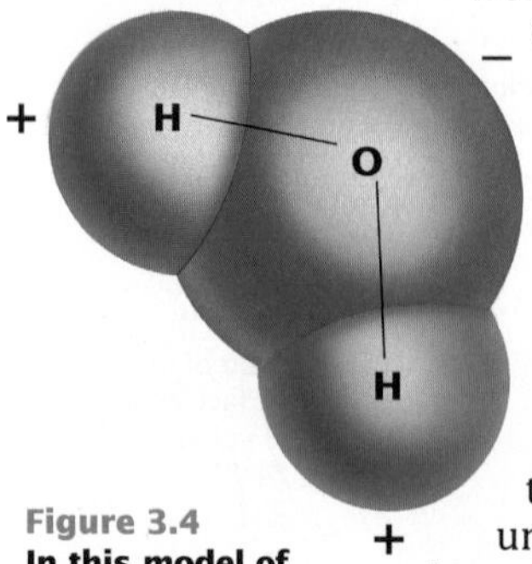

Because electrons have a negative charge, the oxygen part of the water molecule is slightly negative. The hydrogen atoms in the water molecule have slightly positive charges because electrons rarely spend their time near the positive hydrogen nuclei. We think of the oxygen side of the molecule as

Figure 3.4
In this model of a water molecule, the area near the oxygen atom has a partial negative charge; the areas near the hydrogen atoms have partial positive charges.

having a partial negative charge and the hydrogen side of the molecule as having a partial positive charge, as shown in **Figure 3.4**.

A molecule that has a partial negative charge on one side and a partial positive charge on the other side is called a **polar molecule**. These partial charges are very important when water molecules are together. Most of the properties of water are the result of its polarity.

Water molecules cluster together

You know that there is an attractive force between particles of opposite charge. When water molecules are together, the positively charged side of one water molecule attracts the negative side of another water molecule. The fact that the hydrogen atom of one water molecule is attracted to the oxygen atom of another water molecule results in a force between molecules that is called a **hydrogen bond**.

Water molecules are at a lower energy state when they form hydrogen bonds with each other. Since all things tend toward lower energy, there is a natural tendency for water molecules to form hydrogen bonds. When forming hydrogen bonds, the water molecules will cluster together as illustrated in **Figure 3.5**.

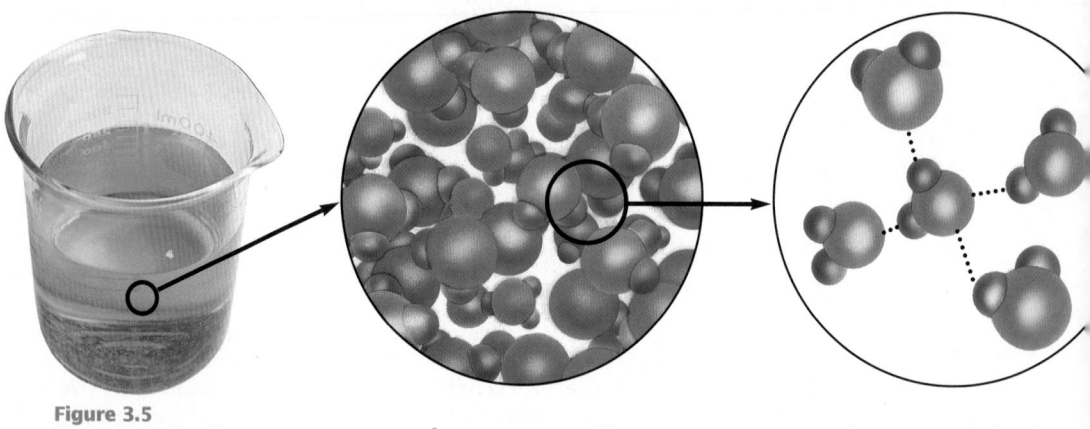

Figure 3.5
a **Water in its liquid state is composed of . . .**
b **. . . many rapidly moving water molecules.**
c **These water molecules form clusters due to the formation of hydrogen bonds.**

Water and the Cell Membrane

Now that you know something about the nature of water molecules, you can look more closely at how water shapes the cell membrane. The basic plan of the cell membrane begins with a sheet of lipids. You learned in Chapter 2 that lipids are molecules such as fats and oils. The interaction between water and lipids is what shapes the cell membrane.

Figure 3.6 shows what happens when oil is poured into a beaker of water and thoroughly mixed. Soon after mixing, small beads of oil form. Eventually the water and oil separate into two distinct layers. The water and oil won't stay mixed. Why? Water molecules start to cluster together because they form hydrogen bonds with each other. Unlike water, which is polar, lipids are **nonpolar molecules** because they have no negative and positive poles. Polar and nonpolar substances like oil and water will separate after being mixed. The oil, which is a nonpolar lipid, is not attracted to the water. Because the water molecules attract one another, the oil is pushed away. In this way, lipids only interact with each other. You can see the lipid and water layers in **Figure 3.7**.

Figure 3.6

a **If you pour oil into a beaker of water . . .**

b **. . . and stir it thoroughly, . . .**

c **. . . the oil and water will separate because water is polar and oil is nonpolar.**

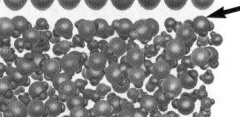

Figure 3.7
The force of water molecules pushing a sheet of lipids out of the way is the force that shapes the cell membrane.

Section Review

❶ What are the advantages of being composed of many small cells instead of a few large ones?

❷ Describe how a cell might be affected if it doubled in size.

❸ How does a polar molecule differ from a nonpolar molecule?

❹ Compare the interaction of water molecules with each other to that of water molecules with lipid molecules.

■ Section Review Answers ■

1. Answers may vary, but should explain that smaller cells are more efficient.
2. Answers may vary, but should explain that the cell's surface area/volume ratio would get smaller.
3. Answers should suggest that polar molecules maintain receptive charges while nonpolar molecules maintain a state of balance.
4. Answers should explain that water molecules attract each other but will not interact with lipid molecules because they are nonpolar.

Visual Strategy

Figures 3.6 and 3.7
If you have not done it as a way of tapping students prior knowledge, perform the oil and water demonstration in Figures 3.6 and 3.7. Have students refer to the appropriate step in the figures as you carry out the demonstration.

Demonstration 3
Solubility
The ability of substances to dissolve in water is an important chemical principle students will encounter in their study of subsequent topics. Enzymes must operate in an aqueous medium. Substances that are transported by osmosis must first dissolve in water. Food materials must first be digested into soluble compounds before they can be absorbed. Demonstrate the solubility of different substances in water, including butter, sugar, flour, and salt.

Phase 3

ASSESSMENT OPTIONS

Closure Strategy
Water Drops on a Penny
Provide each student with a penny, a dropper, and a small beaker of water. See who can place the most drops of water on the surface of the penny. Ask students to explain how it's possible for so many water drops to come together in such a small area. Have them try to repeat their results, this time using water to which a few drops of liquid soap have been added. Ask them to hypothesize why the same number of drops cannot be placed on the penny.

Section Review
Assign the *Section Review.*

Reteaching
Assign students to cooperative groups to make concept maps for this section.

Lesson Plan 3.2

Phase 1

PREPARATION

Key Concepts

- The cell membrane consists of a phospholipid bilayer containing proteins and other molecules.
- Each lipid layer consists of a polar head attracted to water and a nonpolar tail pushed away by water.
- The nonpolar tails block the passage of polar molecules through the cell membrane.
- The protein molecules serve as channels, receptors, and markers.
- The proteins contain nonpolar amino acids that anchor the molecules to the nonpolar lipid tail region.
- The proteins also contain polar amino acids that extend into the water inside and outside the cell.

Reading Strategy

Have students identify the section headings and explain their importance when reading a text.

Phase 2

TEACHING STRATEGIES

Visual Strategy

Figure 3.8c
Students will have difficulty in visualizing how the lipid bilayer shown in Figure 3.8c extends around the cell. Have them draw a complete circle of the bilayer, using a small circle for the polar head and two straight lines for the nonpolar tails.

YOU CANNOT JUDGE A BOOK BY ITS COVER, BUT A CELL IS SHAPED LARGELY BY THE KINDS OF PROTEINS IN ITS CELL MEMBRANE. SCIENTISTS STUDYING THE CELLS OF CYSTIC FIBROSIS PATIENTS HAVE FOUND A SINGLE DEFECTIVE PROTEIN IN THEIR CELL MEMBRANES THAT IS RESPONSIBLE FOR THE DISEASE. IF SCIENTISTS CAN CORRECT THE PROTEIN, THEY CAN CURE THE DISEASE.

3.2   Membrane Architecture

Objectives

❶ Describe how phospholipids are organized to form a fluid cell membrane.

❷ Describe the functions of proteins in the cell membrane.

❸ Explain how cell membrane proteins interact with the lipid bilayer.

Structure of the Lipid Bilayer

A cell membrane's framework consists of molecules called **phospholipids**. A phospholipid, as shown in **Figure 3.8a**, is a lipid in which the long chains or "tails" are joined to a short polar "head" that contains phosphorus. The long tails are nonpolar, so water molecules will tend to push them away. The heads are attracted to water because they form hydrogen bonds with the water molecules.

Water can interact with the polar heads and repel the nonpolar lipid tails best if the phospholipids are aligned into two layers. The polar heads of the phospholipids point toward the water that is inside and outside of the cell, and the tails point inward toward each other, as illustrated in **Figure 3.8b**. This double layer of phospholipids forms a tough yet flexible **lipid bilayer**, which is shown in **Figure 3.8c**.

Figure 3.8

a   **This phospholipid molecule . . .**

b   **. . . is part of a lipid bilayer.**

c   **The lipid bilayer forms the framework of the cell membrane.**

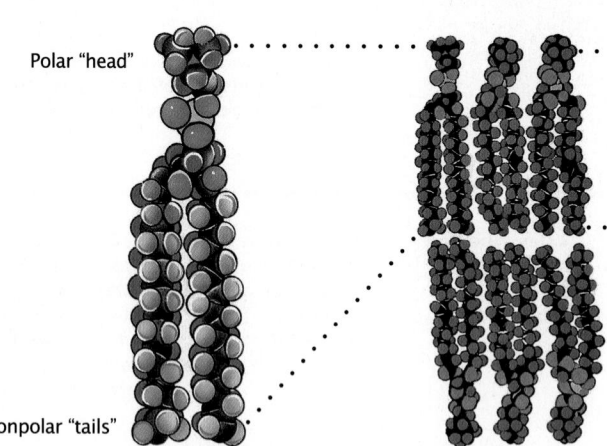

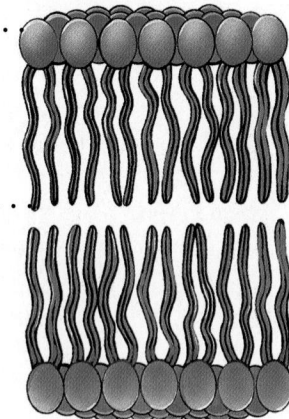

Polar "head"

Nonpolar "tails"

Characteristics and Functions of the Lipid Bilayer

Lipid bilayers stop polar molecules

The lipid bilayers that are found in cell membranes have two important characteristics. One important characteristic of the lipid bilayer is that most polar molecules cannot go across it. Polar molecules are attracted to the water inside the cell or to the water outside the cell. However, they cannot interact with the nonpolar tails of the phospholipids within the lipid bilayer. The result is that the interior part of the cell membrane forms a nonpolar zone that acts as a barrier to polar molecules. But most food molecules and other substances needed by the cell are polar. So the cell must have a way of allowing these molecules to cross the barrier. If the cell membrane were made only of a lipid bilayer, there would be no way for food and other things to pass in and out of the cell.

The solution is to have passageways through the barrier. The cell membrane has passageways made of proteins, such as those shown in **Figure 3.8d**. By making "gates" that can open and shut in the lipid bilayer, proteins enable the passageways to regulate precisely the substances that go into and out of cells. You will see later that cell membrane proteins have other roles, too.

Lipid bilayers are fluid

A second important characteristic of lipid bilayers is that their phospholipid and protein molecules are not rigidly fixed in place. These molecules move about like rubber life preservers floating on the surface of a swimming pool. The lipid bilayer is fluid.

Because they are not fixed in place, the phospholipid and protein molecules that make up the cell membrane can shift from one region of the cell membrane to another. This is important because cell membranes can be structured to fit the needs of different cell types.

Evolution

All cell membranes have a lipid bilayer. What does this suggest about evolutionary relationships among living things?

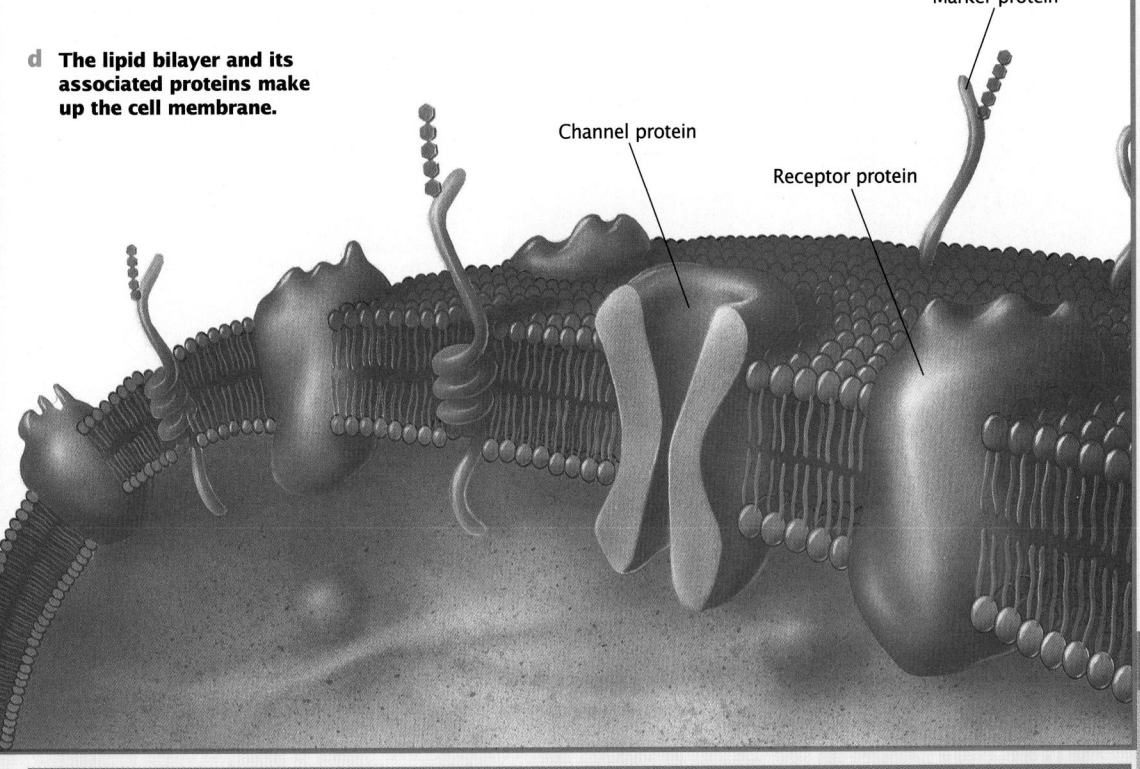

d **The lipid bilayer and its associated proteins make up the cell membrane.**

Marker protein

Channel protein

Receptor protein

■ *Historical Perspective* ■

The 1991 Nobel Prize in Medicine and Physiology was awarded to two German scientists for their work on channel proteins. Erwin Neher and Bert Sakmann were cited for discovering how these proteins control the passage of ions in and out of the cell. Their work may lead to the treatment of certain diseases with drugs that are specifically designed to enter cells through these channels.

Demonstration 1
Photo Analogies for the Cell Membrane
Display photographs of a toll gate, a stadium gate, an airport security gate, etc. that can serve as models of how the cell membrane functions. Have students explain the similarities to the cell membrane.

Demonstration 2
Model of a Cell Membrane
Have students float a thin layer of oil on water in a beaker. Have them sprinkle ground pepper on top of the oil. Ask how this model mimics the cell membrane.

Art Connection

Fluid Mosaic Model
Display a picture of artwork consisting of mosaic tiles. Ask students to explain why the structure of the cell membrane is referred to as the fluid mosaic model.

Theme Answer
Evolution
All cell membranes, including those surrounding the various organelles, have the same basic structure, consisting of 40 percent lipid and 60 percent protein. The types of lipids and proteins, though, vary not only among cells from different organisms but also within the cells of the same organism. A common structural arrangement points to an evolutionary relationship among all organisms. The variations point to adaptations by different organisms, and even by cells within an individual organism, to their particular environments.

Visual Strategy

Figure 3.8d
Point out that the cell membrane is not static but dynamic. Ask students what physical changes the proteins and lipids might undergo in the membrane shown in this figure.

51

Demonstration 3
Model of a Cell Membrane
Use toothpicks to connect a row of marshmallows together. Tell students that this represents the lipid bilayer. Poke several straws of different diameters through the marshmallows to represent protein channels. Embed some jellybeans into the marshmallows to represent receptor proteins. Use tape and toothpicks to make small flags and stick these into the marshmallows to represent marker proteins.

Language Connection

Words and Proteins
To enable students to understand the section heading (Proteins: A Limitless Variety), point out that the letters of the alphabet can be arranged to form the some 800,000 words (not counting plural forms of nouns or tense forms of verbs) that constitute the English language. There are fewer amino acids (20) than letters (26). But with proteins containing many more amino acids than words contain letters, the possibilities are much greater.

Visual Strategy

Figures 3.9, 3.10, 3.11
Have students compare these figures to the marshmallow model in order to reinforce the roles of the different types of proteins. Point out that the defective protein in cystic fibrosis mentioned in the section opener is a channel protein that helps pump chloride ions across the cell membrane. Without this protein, an imbalance of ion and fluid transport results, leading to the thick mucus that collects on the surfaces of respiratory cells.

Roles of Cell Membrane Proteins

If you were able to peer at the surface of a cell, it would look like a smooth sea of phospholipids interrupted by proteins sticking out from the surface, some like boulders, others like tall trees. Proteins that protrude from the cell membrane may serve as channels, receptors, or markers.

Channels allow some molecules to pass through the membrane
Proteins arranged into a shape like a doughnut form channels through the cell membrane. See the channel model in **Figure 3.9**. Many molecules and ions that are needed by the cell cross the membrane through these passageways. However, these channels are like locked doors—only people with a key can enter. In the same way, each channel will admit only certain molecules.

Figure 3.9
Channels
These proteins act like passageways through which only certain molecules pass.

Figure 3.10
Receptors
These proteins transmit information into the cell by reacting to certain other molecules.

Receptors transfer information across the membrane
Receptor proteins in the cell membrane are shaped like boulders, as shown in **Figure 3.10**. Receptors convey information from the world outside the cell to the inside of the cell. The end of the receptor that sticks out from the cell surface has a special shape that holds only a particular type of molecule. When a molecule of the right shape fits into the receptor, its presence causes a change at the other end of the receptor. This change, in turn, triggers responses inside the cell.

Markers aid cell identification
Cell surface markers are elongated proteins that often have carbohydrates on their surface, as shown in **Figure 3.11**. These protein markers are the "name tags" of the cell. Every cell of your body has markers on its surface saying that it is part of you and not of some other individual. These markers organized the first tissues of your body before you were born. As new tissues and organs formed, cell surface markers told your cells where to go and with which other cells to join. You will learn more in Chapter 4 about how cell surface markers help the cell to interact with its environment.

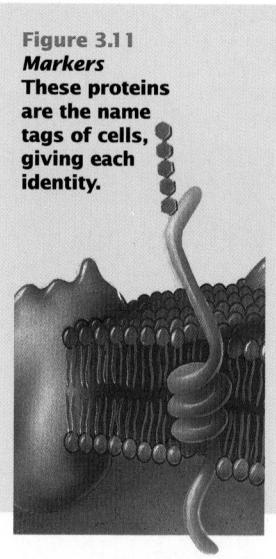

Figure 3.11
Markers
These proteins are the name tags of cells, giving each identity.

Proteins: A Limitless Variety

You remember from Chapter 2 that proteins are molecules made of subunits called amino acids. Proteins in the cell membrane can serve as channels, receptors, or markers. How can proteins play such different roles? And how do proteins, which can be polar, fit into the inner nonpolar region of the membrane bilayer? To answer these questions, take a closer look at the structure of protein.

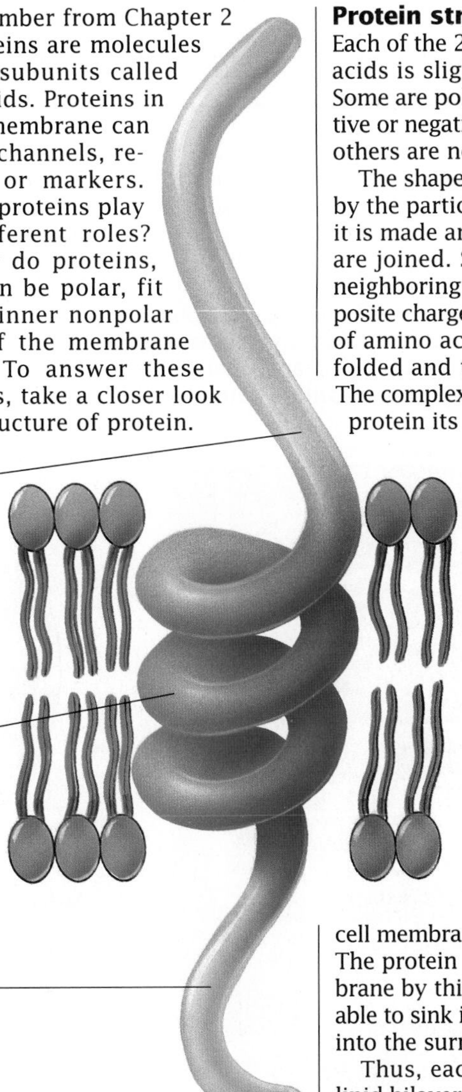

Figure 3.12

a The amino acids in this region of the protein are mostly polar. Therefore, this polar region is compatible with the water outside the cell.

b The amino acids in this region of the protein are mostly nonpolar. Therefore, this region is compatible with the nonpolar area in the center of the lipid bilayer.

c This region is also polar and is compatible with the water inside of the cell.

Protein structure is variable

Each of the 20 different kinds of amino acids is slightly different chemically. Some are polar; that is, they have positive or negative electrical charges, while others are nonpolar.

The shape of a protein is determined by the particular amino acids of which it is made and the order in which they are joined. Some amino acids attract neighboring amino acids that have opposite charges. As a result, the long line of amino acids in a protein becomes folded and twisted as in **Figure 3.12**. The complexity of the shape gives each protein its unique function.

Some proteins fit into a cell membrane

A protein that fits into the cell membrane has three sections. The two end sections, **Figure 3.12a** and **Figure 3.12c**, contain many polar amino acids that form hydrogen bonds with water. However, the middle section, **Figure 3.12b**, is made of many nonpolar amino acids. This nonpolar coil fits into the nonpolar interior of the lipid bilayer, allowing the protein to float in the cell membrane like a ship on the ocean. The protein is anchored into the membrane by this nonpolar region. It is unable to sink inside the cell or to float off into the surrounding water.

Thus, each cell is a prisoner of its lipid bilayer. A cell communicates with the outside world by means of the proteins embedded in its lipid shell.

Connection: Chapter 2

Review the structure of an amino acid. Remind students that a protein consists of a long chain of amino acids that are connected to one another with covalent bonds.

Visual Strategy

Figures 3.12a, 3.12b, 3.12c
Have students identify the nonpolar tails and polar heads of the lipid bilayer. Ask students which type of membrane protein (channel, receptor, or marker) is shown in this illustration.

Phase 3

ASSESSMENT OPTIONS

Closure Strategy
Making Analogies
Make a list of the following cells on the board and ask students to hypothesize which type of protein (channel, receptor, or marker) would be the most common in each cell membrane: a red blood cell, a brain cell, a kidney cell, and a tongue cell. Have students provide a reason for their choice.

Section Review
Assign the *Section Review*.

Reteaching
Have students select an object visible from the exterior of their house that would serve as a model for either the lipid bilayer or a protein of the cell membrane. For example, they may select a wall as the lipid bilayer, a door as a channel protein, a TV antenna as a receptor protein, or a name plate as a marker protein. Students should explain their analogies.

Section Review

1. How do phospholipids interact with water to form lipid bilayers?
2. Why are lipid bilayers said to be fluid?
3. Explain three roles proteins play in cell membranes.
4. Explain how the structure of a protein enables it to fit in the cell membrane.

■ Section Review Answers ■

1. The long "tails" of a phospholipid are nonpolar so water molecules push them away. The "heads" are polar and attracted to water.
2. The phospholipids are protein molecules and are not rigidly fixed in place.
3. Protein molecules in the cell membrane can serve as channels, receptors, or markers.
4. Proteins in the cell membrane have polar regions attracted to water, and a nonpolar region, which anchors the protein in the center of the lipid bilayer.

Using The Feature

- Ask students to focus on the aspects of Jack's upbringing that helped him to succeed in an academic career.
- This feature can also be used to highlight the process skills of observing and analyzing. Jack observes characteristics of proteins by analyzing their behavior in a particular matrix.

Discussion

Guide a discussion by posing the following questions:

1. What kind of difficulties do you suppose Jack and his family faced when they emigrated from China to the United States?
   Students should discuss any related cultural differences such as language.

2. What factor(s) contributed to Jack's positive attitude toward school and learning?
   Jack's parents taught him to respect well educated people.

3. What personality traits have enabled Jack to become a successful protein chemist?
   patience, perseverance, detail-oriented, inquisitiveness

Science in Action ▪ Science in Action ▪ Science in Action

Jack Wang: Protein Chemist

Why I Became a Protein Chemist

"I came to the United States from China when I was six years old. Although I don't recall much about moving to this country, I do remember that my parents expected me to excel in school. My parents were not scientists, but they were well educated and raised me to respect people who are scholars. As a result, I grew up with a natural desire for learning.

"In my sophomore year of high school, I took biology and learned about genetics. The teacher talked about the inheritance of traits. I enjoyed learning how features such as hair and eye color are passed on from generation to generation. When the time came for college, I already knew that I wanted to learn more about biology and how nature worked."

Jack spends his spare time with his wife and children. His two sons are in Little League. "We love baseball. During the Little League season, I coach the teams my sons play on. We also closely follow our favorite major league team— the New York Mets."

Name:	**Jack Wang**
Home:	**Lexington, Massachusetts**
Employer:	**Genetics Institute, Cambridge, MA**
Personal Traits:	• **Patience**
	• **Honesty**
	• **Perseverance**
	• **Detail Oriented**
	• **Friendliness**
	• **Inquisitiveness**

Action ▪ *Science in Action* ▪ *Science in Action* ▪ *Science*

The Fascination of Protein Chemistry

Career Path
....

High School:
• Biology
• Chemistry
• Physics
College:
• Cell Biology
• Organic Chemistry
• Genetics
Graduate School:
• Biochemistry

"I purify and characterize proteins. Purification involves passing a sample through a set of columns. This begins separating proteins found within the sample. Every protein has its own characteristics—different masses, shapes, and charges. By putting a protein mixture through a column with a particular matrix that selects for certain properties, you can separate the proteins. Some of the proteins will bind to the column more than others. You actually use a wide variety of columns, each with its own ability to select proteins. By running a mixture through several columns in a row, you end up eventually purifying the single protein you're after."

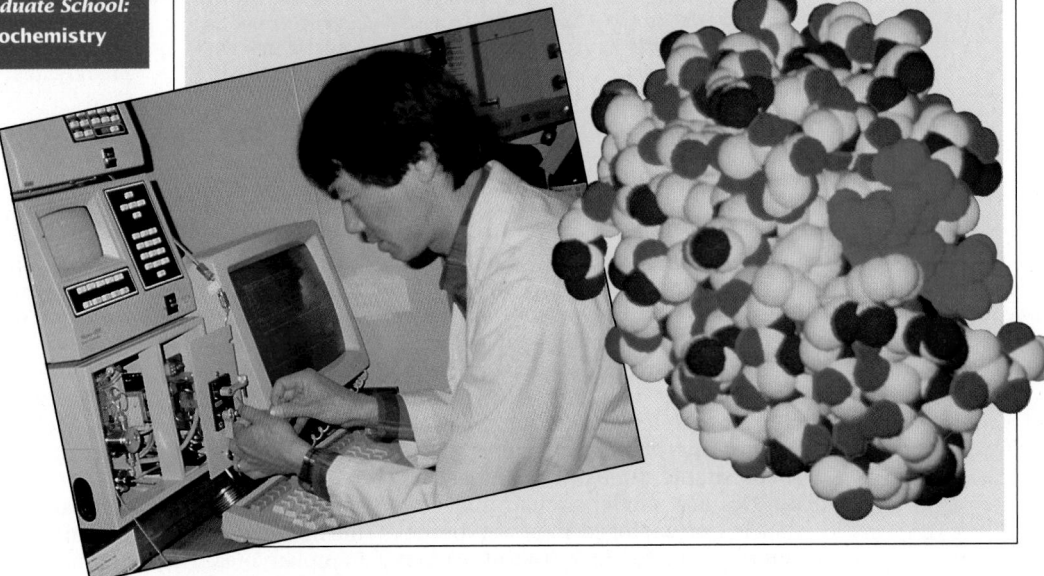

Research Focus

Dr. Wang works in the Protein Chemistry Department at the Genetics Institute in Cambridge, Massachusetts. He has worked on two projects. The first project developed a synthetic form of a protein called factor VIII. Factor VIII is found in the blood and is required for normal blood clotting. When this protein is missing from a person's blood, a severe bleeding disorder called hemophilia A results. People with this type of hemophilia often can receive the missing protein through concentrated blood products. Unfortunately, for years the blood supply was relatively impure and some hemophiliacs became infected with forms of hepatitis and the human immunodeficiency virus. Now, with the synthetic form of factor VIII that Dr. Wang helped develop, many hemophiliacs are safe from such dangers.

Dr. Wang's second project involves researching a human growth factor that stimulates the production of bone. "This protein is actually quite revolutionary," he says. "It's especially useful for someone who has a badly broken bone. The growth factor is undergoing clinical trials and probably won't be on the market for another three or four years."

Lesson Plan 3.3

Phase 1
PREPARATION

Key Concepts

- A eukaryotic cell is large, complex, and contains a nucleus.
- A prokaryotic cell is small, simple, and lacks a nucleus.
- All cells have a cell membrane, cytoplasm, and ribosomes.
- Only eukaryotic cells are divided into compartments that contain organelles.
- Such organelles include the nucleus, mitochondria, chloroplasts, endoplasmic reticulum, and the Golgi body.
- These organelles help the cell perform the basic life functions of reproduction, energy use, and homeostasis.
- Plant cells, unlike animal cells, have cell walls, vacuoles, and chloroplasts.
- Evidence supports the hypothesis that eukaryotes evolved from prokaryotes.

Phase 2
TEACHING STRATEGIES

Visual Strategy

Table 3.1

Have students identify the cell membrane, cytoplasm, and nucleus in the *Chilodonella*. Have students identify similarities and differences between the *Chilodonella* and bacterium. Review the information listed in the table so that students have a clear understanding of the differences between prokaryotes and eukaryotes.

WHEN YOU HAVE A CASE OF STREP THROAT, YOUR CELLS ARE UNDER ATTACK BY BACTERIA. BACTERIA ARE SINGLE-CELLED ORGANISMS. BACTERIAL CELLS ARE MUCH LESS COMPLICATED THAN YOUR CELLS. THEY ARE ALSO THE ANCESTORS OF YOUR CELLS. OVER THE COURSE OF TIME, BACTERIAL CELLS EVOLVED. THE RESULT WAS THE KINDS OF CELLS THAT MAKE UP YOUR BODY.

3.3 Inside the Cell

Objectives

❶ Name two characteristics that distinguish eukaryotes from prokaryotes.

❷ Relate each organelle to a task essential to the life of the cell.

❸ Contrast the structure of an animal cell with that of a plant cell.

❹ Describe how eukaryotes evolved from prokaryotes.

Two Types of Cells

All cells can be divided into two large categories. A eukaryotic cell, or **eukaryote** (*yoo KAR ee oht*), is a large, complex cell that contains a membrane-bound compartment called a **nucleus**. The nucleus contains DNA within chromosomes. A prokaryotic cell, or **prokaryote** (*pro KAR ee oht*), is a very small, simple cell that lacks a nucleus. Its DNA is a single, circular molecule that is not enclosed in a membrane-bound compartment. Fossils of the first cells on Earth, which existed about 3.5 billion years ago, reveal that these ancient cells were prokaryotes. Today, the only living prokaryotes are the bacteria. Compare the bacterium to the eukaryote in **Table 3.1**.

Eukaryotes and prokaryotes share several characteristics. For example, both kinds of cells have a cell membrane. They also contain **cytoplasm** (*SYT uh plaz uhm*). Cytoplasm is everything inside the cell membrane except the nucleus in eukaryotes, and the DNA in prokaryotes. These cells also contain **ribosomes** (*RY buh sohmz*), structures on which proteins are made.

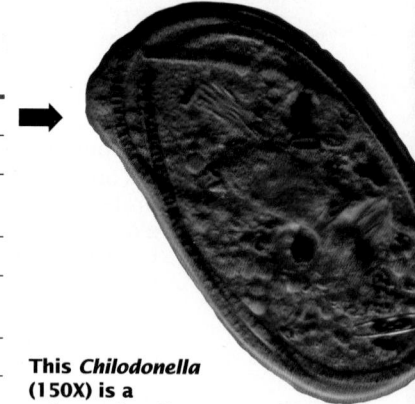

Table 3.1

Prokaryote	Eukaryote
No nucleus	Nucleus
No membrane-bound organelles	Many organelles
1–10 µm in size	2–1,000 µm in size
Evolved 3.5 billion years ago	Evolved 1.5 billion years ago
Only bacteria	All other cells

This bacterium (13,600X) is a prokaryotic cell.

This *Chilodonella* (150X) is a eukaryotic cell.

▪ Matter of Fact ▪

Use the following comparison to give students some idea of the small size of bacterial cells. Some bacteria climb on top of one another to a height of several millimeters. If the same number of humans climbed on top of one another, they would be over a mile high, or nearly four times higher than the world's tallest building.

Eukaryotic Cells Have Compartments

The nucleus is not the only compartment inside eukaryotic cells. The cytoplasm of eukaryotes contains many specialized parts called **organelles**, as shown in **Figure 3.13**. Each organelle is a specialized compartment that carries out a specific function. Examples of organelles include the nucleus and ribosomes. Many organelles, such as the nucleus, are membrane-bound.

Organelles isolate cell activities

A eukaryotic cell is like a building with many rooms. Imagine that you had to attend a class in a building where the gymnasium, the cafeteria, the library, and the boiler room were all located within the same room. It sounds confusing. Having separate compartments allows special activities to be restricted to particular places. This type of organization has many advantages over that of prokaryotic cells.

Figure 3.13
This is a diagram of a cell from a human liver.

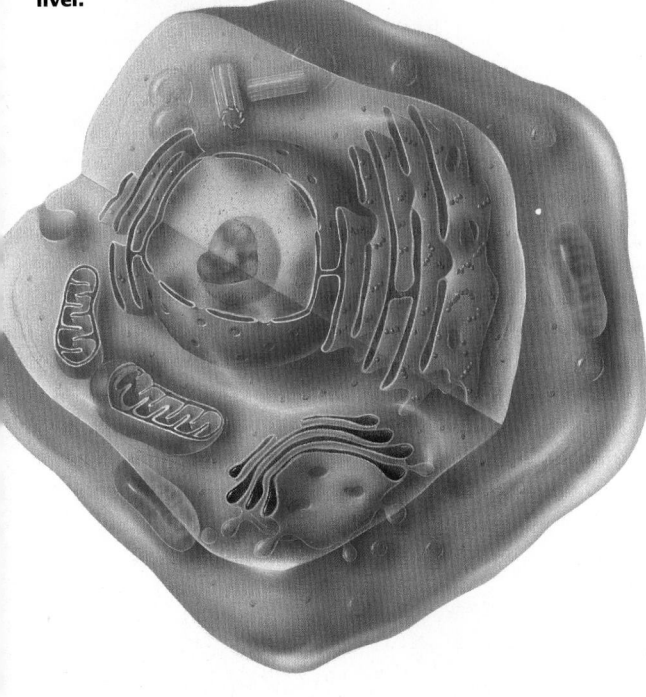

Organelles contribute to the specialization of eukaryotic cells

The organelles of a living eukaryotic cell are constantly growing, moving, reproducing, appearing, and disappearing. Multiple arrangements of these active, minute structures enable eukaryotic cells to specialize—that is, to perform special functions.

For example, a muscle cell in your leg, which must contract quickly, contains a greater number of energy-producing organelles than a bone cell does. Likewise, the striped, pulsating cells of heart muscle are far different from the boxlike cells that make up the growing layers of the skin on your hand. And skin cells differ greatly from the nerve cells that send electrical signals from your brain throughout your body. Other examples of specialized eukaryotic cells include gland cells that secrete hormones, and sperm cells that are capable of movement.

Keep in mind that all the cells of plants and animals originate from a single fertilized egg cell. As these cells divide and form new cells, they change in slightly different ways. At maturity, they become specialized. Single-celled eukaryotes evolved into complex, multicellular organisms because they had the ability to become specialized. These specialized cells make eukaryotes capable of survival in complex environments.

"Any living cell carries with it the experience of a billion years of experimentation by its ancestors."

Max Delbrück
Cell Biologist
Nobel Prize Winner

■ Historical Perspective ■

Almost 200 years passed before anyone recognized the importance of the cells first introduced to the scientific world by Hooke in 1665. In the 1830s, Theodor Schwann and Matthias Schleiden suggested that all organisms are composed of one or more cells. In the 1850s, Rudolf Virchow stated that cells can arise only from preexisting cells. Together, these observations led to the development of the cell theory, one of the great unifying themes in biology.

Demonstration 1
Colloidal Nature of Cytoplasm
Point out that the cytoplasm is a colloid and not a solution. Demonstrate the difference between the two by swirling both a beaker of water and a beaker containing some Jello. Ask students what problem would result if the cytoplasm were a solution and not a colloid.

Theme Connection

Stability
As stated in Chapter 1, control is an essential aspect of life. With their separate compartments, eukaryotic cells have greater control over their internal activities than do prokaryotic cells that lack internal organization. Have students imagine what control, if any, would be possible in a class like the one described in the text.

Visual Strategy

Figure 3.13
Although students at this point have read only about the nucleus and cell membranes, have them locate and describe the other organelles shown in this figure. Some of these organelles are discussed later in this section.

Theme Connection

Interacting Systems
Point out that the specialized cells discussed in the text must work in a coordinated fashion for the human body to function normally.

Demonstration 2
Variations Among Cells
Show students slides or transparencies of different cells. Ask them to describe similarities and differences among the various cell types. Stress that there is no such thing as a typical cell, although cells share certain characteristics.

57

Theme Connection

Energy and Life

Point out that the energy to carry out life processes comes from respiration in mitochondria and photosynthesis in chloroplasts.

Visual Strategy

Figure 3.14

Use this diagram to reinforce the characteristics of a eukaryotic cell. Have students select the items listed in the table on page 56 that can be used to identify the human liver cell illustrated in this diagram as a eukaryotic cell. Explain that each organelle described in this section is shown in greater magnification in the accompanying figures. Make sure students understand that each figure includes both an electron micrograph and an artist's rendition.

Visual Strategy

Figure 3.15

Point out the membrane that surrounds the nucleus. Ask students what they notice about the nuclear membrane. Have them suggest a possible reason for the presence of these pores in the membrane.

Visual Strategy

Figure 3.16

Use this diagram to demonstrate the importance of an extensive surface area resulting from the foldings of the internal membrane of a mitochondrion. Ask students how this might help the mitochondrion perform its function.

Theme Answer

Scale and Structure

The compartments in a beehive enable many separate and unrelated activities to be housed in one structure.

Cells Perform Basic Functions of Life

As you learned in Chapter 2, one of the five characteristics of living things is that they are composed of one or more cells. Like all living things, cells like the human liver cell in **Figure 3.14** display other properties shared by all organisms. Cells use energy. Cells maintain homeostasis. And cells reproduce.

You are already familiar with some of the ways in which cells exhibit the last two properties listed above. For example, you learned earlier that the cell membrane helps to maintain a constant internal environment. You also learned that DNA occurs in prokaryotic cells, and in eukaryotic cells where it is found in the nucleus.

Like the cell's nucleus, the other organelles in a eukaryotic cell are membrane-bound, specialized compartments whose contents are separated from the cytoplasm. Isolated from the hustle and bustle of the rest of the cell, each organelle of the cell stands ready to perform a specific task. The coordinated activities of the different organelles make it possible for cells to go about the business of living.

Cells reproduce

When a growing cell approaches a particular size, certain organelles, such as the nucleus shown in **Figure 3.15**, help it prepare to divide. Cell division enables single-celled organisms to reproduce by dividing in two. It also enables multicellular organisms to grow. In Chapter 4, you will learn how cells divide.

Scale and Structure

Activities in a eukaryotic cell are compartmentalized. What is the similar advantage of the compartments in a beehive?

Cells manufacture and release energy

Two kinds of organelles act as cellular powerhouses, playing essential roles in food manufacture and energy release. **Mitochondria** (*myt uh KAHN dree uh*) are found in all eukaryotic cells where they release the energy stored in food. A mitochondrion is shown in **Figure 3.16**. Your cells carry from 10 to several hundred of these mitochondria, each a tiny chemical factory breaking down food molecules to produce energy.

Chloroplasts, on the other hand, are organelles that have the amazing ability to make food in the form of sugars using air, water, and the energy from sunlight. This process is called photosynthesis. You will learn more about photosynthesis in Chapter 5. Chloroplasts are found only in algae, such as seaweed, and in green plants.

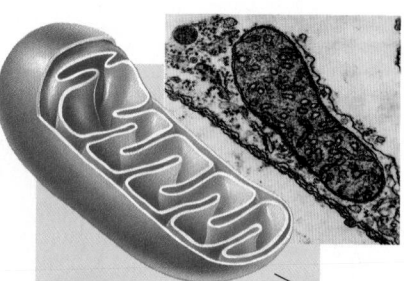

Figure 3.16
Mitochondrion
Without mitochondria your cells would be unable to produce the energy on which all of your activities depend.

Figure 3.15
Nucleus
The DNA inside the nucleus directs the cell's growth and development.

Cells maintain homeostasis

As you learned in Chapter 2, every living thing, no matter how big or how small, must be able to maintain a constant internal environment. Whether it is a one-celled organism or part of a larger living thing, a cell must be able to monitor and adjust internal conditions. You learned earlier in this chapter how the cell membrane helps to maintain homeostasis by controlling what can enter or exit a cell.

Some organelles help cells to maintain homeostasis by moving supplies from one part of a eukaryotic cell to another. In the small cell of a bacterium, a molecule can go from one place to another fairly quickly. But in large eukaryotic cells with intricately arranged membranes, molecular traffic needs to be directed more precisely. To accomplish this, eukaryotic cells have a system of membranes called the **endoplasmic reticulum** (*ehn duh PLAZ mihk rih TIHK yuh luhm*), or ER, as shown in **Figure 3.17**. In addition to transporting proteins, other parts of the ER help make lipids the cell needs.

The proteins and the lipids made by the ER are then transported to another organelle called the **Golgi** (*GOHL jee*) **body**. The Golgi body puts the finishing touches on these molecules and then releases them in membrane-wrapped bubbles called vesicles. Substances packaged in this way can be sent to particular places in the cell. Some types of vesicles fuse with the cell membrane, releasing their contents outside the cell. An illustration and photograph of a Golgi body are shown in **Figure 3.18**.

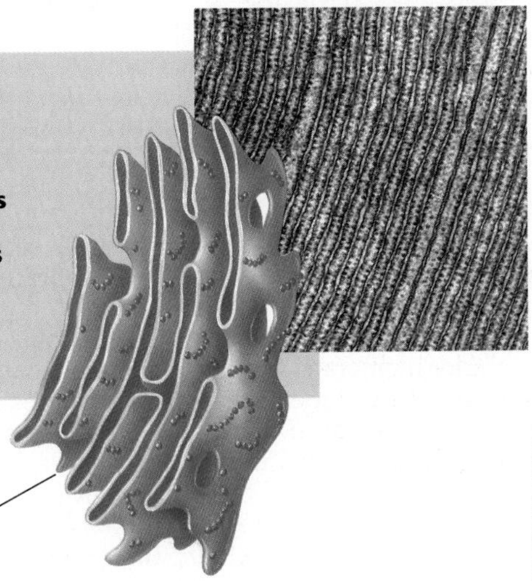

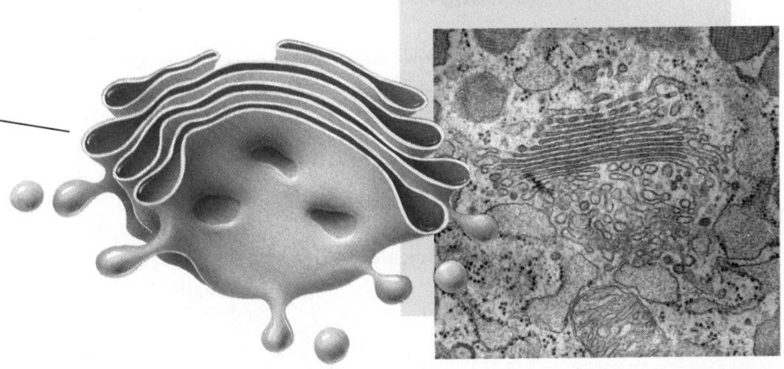

Theme Connection

Stability

Make sure students understand how the endoplasmic reticulum and Golgi body help the cell maintain a constant internal environment. Ask students to explain why cells that produce many secretions, including enzymes and hormones, have an extensive endoplasmic reticulum and Golgi body.

Visual Strategy

Figures 3.17 and 3.18
Have students identify the channels of the endoplasmic reticulum and Golgi body through which materials pass.

Figure 3.17
Endoplasmic reticulum
The ER is sometimes called "the highway of the cell." Many proteins made by the cell are threaded into the ER as they are made and are then moved to other parts of the cell.

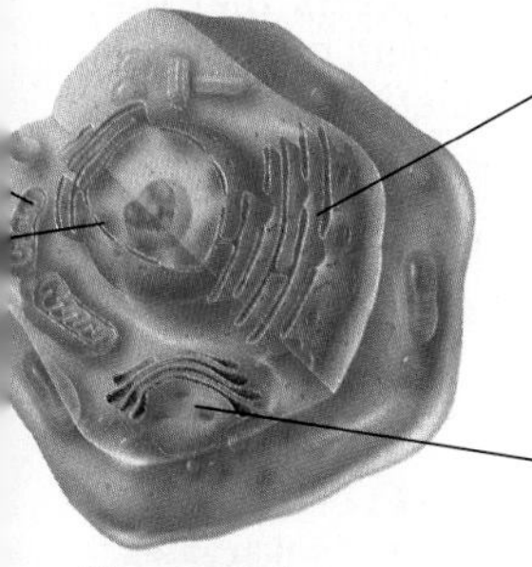

Figure 3.14
Human liver cell
This cell is one type of eukaryotic cell. It is made up of a membrane, cytoplasm, and different types of organelles.

Figure 3.18
Golgi body
Like a post office, the Golgi body labels molecules made in the ER with tags that specify their destinations.

Figure 3.19
Have students attempt to locate the nucleus in this plant cell. They should realize that the central vacuole can be so large that it squeezes the nucleus and other organelles against the cell wall.

Phase 3

ASSESSMENT OPTIONS

Closure Strategy

A Collage of the Cell

Have students assemble a collage of photographs that represent the various cell organelles. Examples might include a computer for the nucleus, highways for the endoplasmic reticulum, post office for the Golgi body, etc. Have them explain the reasons for each of their choices.

Section Review

Assign the *Section Review*.

Reteaching

Prepare a table for students to complete. Across the top, place the headings *Prokaryotic cell, Plant cell, Animal cell,* and *Function of organelle.* Down the left hand side, write the names of the organelles covered in this section. Have students indicate in which cell(s) these organelles are present and describe their function.

Kinds of Eukaryotic Cells

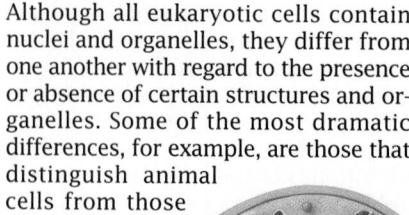

Figure 3.19
This illustration of a plant cell from timothy grass contains a cell membrane, a large central vacuole, chloroplasts that produce sugar, and a strong cell wall.

Although all eukaryotic cells contain nuclei and organelles, they differ from one another with regard to the presence or absence of certain structures and organelles. Some of the most dramatic differences, for example, are those that distinguish animal cells from those of plants.

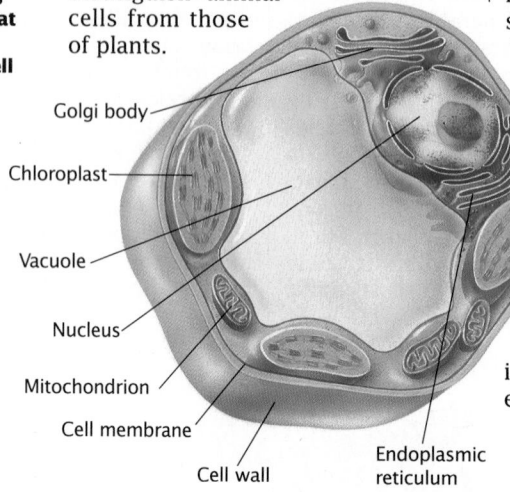

Golgi body

Chloroplast

Vacuole

Nucleus

Mitochondrion

Cell membrane

Cell wall

Endoplasmic reticulum

Plant cells have cell walls, vacuoles, and chloroplasts

Plant cells possess a **cell wall** in addition to a cell membrane. As **Figure 3.19** shows, the cell wall lies outside the cell membrane. Plant cell walls contain cellulose, a polysaccharide that gives strength and rigidity to the plant cell. The cells of algae, fungi, and some bacteria also have cell walls. Plant cells, unlike animal cells, store waste products, nutrients, and water in large, membrane-bound spaces called **vacuoles**. The pressure exerted by the stored water enables the plant to stand upright. When its vacuoles lack water, a plant will become limp.

Chloroplasts are also found in plant cells but not in animal cells. During photosynthesis, chloroplasts use energy of light to make sugars.

How Did Eukaryotes Evolve?

As biologists studied eukaryotes they were surprised to see that many organelles resembled bacteria. Some organelles have a double set of membranes. The interior membrane of such an organelle is like the cell membrane of a bacterium. It looks as if a bacterium has been engulfed by a much larger cell.

Scientists hypothesize that bacterial "trespassers" remained inside cells, gradually losing their ability to live independently. These invading bacteria became organelles, and eukaryotic cells

were the result. The first eukaryotic cells appeared about 1.5 billion years ago.

Additional evidence supports the hypothesis that eukaryotes evolved from prokaryotes. Some eukaryotic organelles such as mitochondria have their own DNA, which suggests that they once lived as independent cells. Eukaryotic organelles sometimes also have their own ribosomes, which are very similar to the ribosomes of bacteria. Also, some organelles divide in a manner similar to that of bacteria.

Section Review

❶ **What characteristics would help you distinguish a bacterium from a *Chilodonella*?**

❷ **Name five organelles found in cells and describe how each enables the cell to display the properties of life.**

❸ **In what ways does a plant cell differ from an animal cell?**

❹ **Do you think that eukaryotes could have evolved without prokaryotes? Explain why or why not.**

■ Section Review Answers ■

1. Answers may vary. Students should be able to relate characteristics of bacteria to those of the eukaryotes, and characteristics of *Chilodonella* with those of eukaryotes, with the obvious characteristic being the presence of or lack of a nucleus.

2. Answers will vary. Students should consider such factors as source of energy, reproduction, and the cells ability to maintain stability.

3. Answers will vary but should give the obvious characteristic that plant cells have a cell wall and animal cells have a cell membrane.

4. Answers will vary.

Chapter 3 *Highlights*

This child, who is 19 months old, has grown from a single cell to nearly 100 trillion cells.

Key Terms	Summary

3.1 World of the Cell

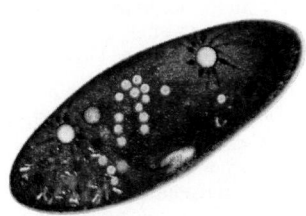

The cell, like this *Paramecium,* is the fundamental unit of life.

Key Terms:
cell (p. 45)

cell membrane (p. 45)

polar molecule (p. 48)

hydrogen bond (p. 48)

nonpolar molecule (p. 49)

Summary:
- A cell is the smallest unit of life.
- Small cells can take in food and communicate with each other more efficiently than large cells.
- Water molecules attract other water molecules and shape the cell membrane by repelling lipid molecules.

3.2 Membrane Architecture

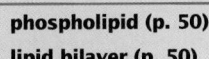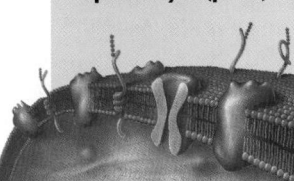

A cell membrane is made of a lipid bilayer, its associated proteins, and other molecules. The membrane is the interface between a cell and its environment.

Key Terms:
phospholipid (p. 50)

lipid bilayer (p. 50)

Summary:
- The cell membrane is a tough, flexible lipid bilayer made of phospholipids, protein and other molecules.
- Proteins in the cell membrane permit certain substances to go into and out of cells.

3.3 Inside the Cell

The eukaryotic cell is composed of cytoplasm and compartments called organelles. Compartments isolate cell activities.

Key Terms:
eukaryote (p. 56)

nucleus (p. 56)

prokaryote (p. 56)

cytoplasm (p. 56)

ribosome (p. 56)

organelle (p. 57)

mitochondrion (p. 58)

chloroplast (p. 58)

endoplasmic reticulum (p. 59)

Golgi body (p. 59)

cell wall (p. 60)

vacuole (p. 60)

Summary:
- All cells have DNA and cytoplasm.
- A eukaryotic cell has a nucleus and other membrane-bound organelles. A prokaryotic cell does not.
- Organelles are compartments that enable a cell to function by making and releasing energy, helping the cell to maintain homeostasis, and enabling the cell to reproduce.
- Unlike animal cells, plant cells have cell walls, vacuoles, and chloroplasts.

Chapter Review Answers

Understanding Vocabulary

1. **a.** The cell wall grows on the outside of the cell membrane in plants and fungi and functions to protect and provide support for the cell. The cell membrane is a selectively permeable lipid bilayer that serves to separate the inside of a cell from what's outside.

   b. A polar molecule has charged ends, sort of like a magnet, while a nonpolar molecule has no charged sites. Water is a polar molecule and lipids are nonpolar molecules.

   c. Eukaryotes have membrane-bound nuclei and partitioned cytoplasm. Prokaryotes lack these features and are thought to have given rise to eukaryotes.

   d. The chloroplast, found in some eukaryotic cells, is the cell organelle on which sugars are made. A mitochondrion is found only in eukaryotic cells and produces energy by breaking down food.

   e. Cytoplasm is the gel-like substance contained in all cells; organelles are specialized structures located in the cytoplasm. Ribosomes and mitochondria are cell organelles.

Relating Concepts
2. Map answer is shown on page 43D.

Understanding Concepts
Multiple Choice

3. a	8. d
4. a	9. a
5. d	10. a
6. c	11. d
7. b	

Completion
12. polar, nonpolar
13. cell membrane
14. organelles
15. cell wall

Short Answer
16. The polar water molecules shape the cell membrane by repelling the nonpolar lipid molecules.

Understanding Vocabulary

1. For each pair of terms, explain the differences in their meanings.
   a. cell wall, cell membrane
   b. polar molecule, nonpolar molecule
   c. prokaryote, eukaryote
   d. chloroplast, mitochondrion
   e. cytoplasm, organelle

Relating Concepts

2. Copy the unfinished concept map below onto a sheet of paper. Then complete the concept map by writing the correct word or phrase in each oval containing a question mark.

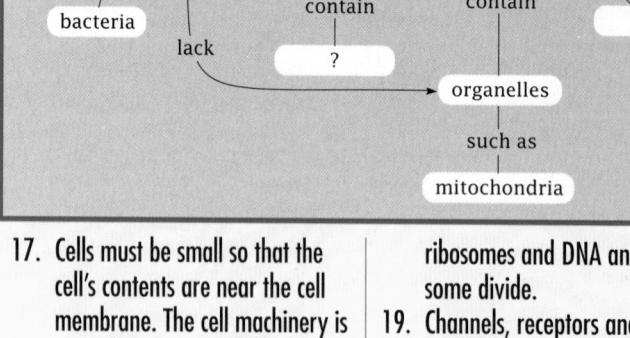

Understanding Concepts

Multiple Choice

3. The flexible lipid outer surface of a cell is called the
   a. cell membrane.   c. hull.
   b. phospholipid.   d. cytoplasm.

4. The growth of cells is limited by the ratio of
   a. surface area to volume.
   b. organelles to surface area.
   c. organelles to cytoplasm.
   d. nucleus to cytoplasm.

5. A cell membrane is composed of
   a. phospholipids.
   b. water molecules.
   c. proteins.
   d. phospholipids and proteins.

6. Which of the following does not enable the cell to communicate with its environment?
   a. channels   c. receptors
   b. phospholipids   d. markers

7. Besides the cell membrane, two components of all cells are
   a. cytoplasm, cell wall.
   b. DNA, cytoplasm.
   c. organelles, nucleus.
   d. DNA, mitochondria.

8. A part of the cell that functions to maintain homeostasis is the
   a. cytoplasm.   c. nucleus.
   b. cell wall.   d. cell membrane.

9. Single-celled organisms that do not have nuclei are called
   a. prokaryotes.   c. organelles.
   b. eukaryotes.   d. mitochondria.

10. Which of the following releases energy from nutrients?
    a. mitochondrion   c. Golgi body
    b. endoplasmic reticulum   d. protein channel

11. Lipids and proteins are transported through the cell by the
    a. nucleus.
    b. chloroplast.
    c. mitochondrion.
    d. endoplasmic reticulum.

Completion

12. Water molecules attract each other because they are _____ . Lipids are not attracted to water because they are _____ .

13. The double layer of phospholipids surrounding the cytoplasm is called a(n) _____ _____ .

14. Structures that perform specialized functions within the cell are called _____ .

15. Plant cells differ from animal cells in that they possess a(n) _____ , which is a thick layer of cellulose surrounding the cell membrane.

Short Answer

16. Explain how water shapes the cell membrane.

17. What advantage do small cells have over large cells?

18. List two kinds of evidence that suggest eukaryotes evolved from prokaryotes.

19. What are the names and functions of three cell membrane proteins?

---

17. Cells must be small so that the cell's contents are near the cell membrane. The cell machinery is able to obtain nourishment and to communicate with what lies beyond the cell by being near the cell membrane.

18. Eukaryotic organelles appear very similar to prokaryotic cells in that they (1) contain their own ribosomes and DNA and (2) some divide.

19. Channels, receptors and markers. Channels, shaped like donuts, permit some molecules to pass through the cell membrane. Receptors appear boulder-like and transfer information across the cell membrane. Markers look like long tails and help the cell identify other cells.

Interpreting Graphics

20. (1) The polar heads are pointed into the cytoplasm and away from the cell; the nonpolar tails are between the polar heads and are pointing toward each other. (2) Two layers, each composed

Interpreting Graphics

20. Look at the figure below (right) to answer the following questions. Identify the polar heads and the nonpolar tails in the lipid bilayer. How many layers make up the cell membrane? Why have the layers aligned themselves in the manner shown? Describe the pathway that molecules and ions would take to get inside the cell.

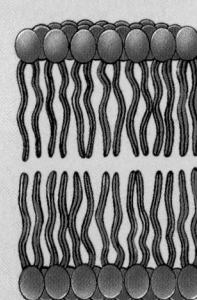

21. Compare the structure of each protein in the figure below to its function. Explain how the structure of each protein enables it to accomplish its specific task.

22. Look at the images of the three cells shown below. For each cell, place an X under each term that appropriately describes it.

Prokaryote Eukaryote Animal   Plant

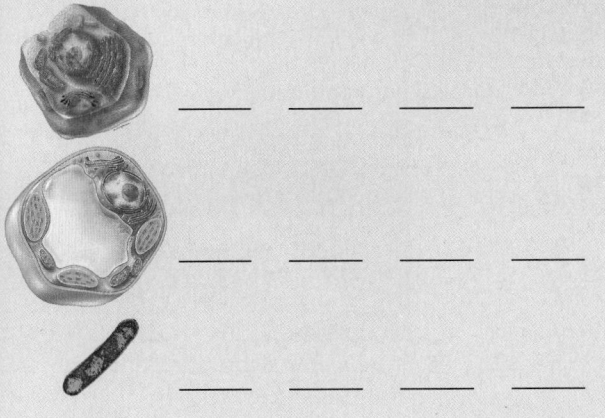

Reviewing Themes

23. *Stability*
    How does a cell membrane help a cell maintain homeostasis?

24. *Energy*
    How do chloroplasts and mitochondria function in the capture and release of energy within the cell?

Thinking Critically

25. *Inferring Conclusions*
    How would a cell be affected if the cell membrane were completely solid and watertight?

26. *Comparing and Contrasting*
    How does a bacterium swimming around in a pond differ from a cell in your body? How is it similar?

27. *Inferring Conclusions*
    To solve a hit-and-run case, police want to know if a microscopic sample of material scraped from a car bumper is animal or plant matter. What evidence should the police look for in the sample in order to eliminate the driver of the car as a suspect?

28. *Building on What You Have Learned*
    You learned in Chapter 2 that movement is a characteristic exhibited by most animals. What features of animal cells make movement possible?

Cross-Discipline Connection

29. *Biology and Art*
    Before photographs were commonly used in biology, drawings of cells and organisms were the most accurate way to share information. Describe the benefits and disadvantages of using art instead of a photograph.

Discovering Through Reading

30. Read the article "Cell Videos Catch Asbestos in the Act," in *Science News*, September 21, 1991, page 180. How do asbestos fibers get into cells? How do asbestos fibers cause cancer?

Reviewing Themes

23. A cell membrane helps a cell maintain homeostasis by regulating the substances that enter and exit the cell.

24. Chloroplasts use energy captured from sunlight to make food. Mitochondria break down food and transfer the chemical energy from the food to a form that can be used by the cell.

Thinking Critically

25. Substances could not enter and exit the cell. If living, the cell would soon die as nutrients are used and waste accumulates.

26. The two are different in that the bacterium is a prokaryote and is functioning as a single unit; the body cell is a eukaryote and functions as part of a whole. Both the bacterium and the body cell contain and are surrounded by water. In addition, both have a cell membrane, DNA and ribosomes.

27. To identify the sample as plant and not animal (or human matter), the scientists would look for the absence of lysosomes and the presence of cell walls, vacuoles, and chloroplasts.

28. Movement is possible due to the flexibility of the cell membrane, mitochondria that release the energy stored in food, and surface proteins that aid in molecular movement, information transfer, and cell identification.

Cross-Discipline Connection

29. In using art, particular features may be highlighted to make a point or shown from a perspective that could not be captured by photography. A composite picture may be drawn that reflects the best depiction of numerous observations. A major disadvantage of using art is that drawings show the artist's interpretation of what was observed.

Discovering Through Reading

30. Cells encapsulate asbestos fibers in membrane sacs. Fibers can interfere with chromosome division and activate an enzyme that signals cell proliferation.

---

of polar heads and nonpolar tails, make up the cell membrane. (3) The alignment is due to the polarity of the structures; the polar heads are attracted by water molecules, while the nonpolar tails are repelled by water. (4) The lipid bilayer contains proteins, through which many molecules and ions can pass.

Some small molecules may pass directly through the lipid bilayer.

21. The protein shown on the left is called a channel. It is circular with an opening through its middle. The opening enables select molecules and ions to enter and exit the cell. The protein shown on the right is called a receptor and it has no opening. When it comes in contact with certain molecules outside of the cell, a change occurs to the end of the receptor inside the cell. This change results in things happening inside the cell.

22. First cell: eukaryotic, animal; Second cell: eukaryotic, plant; Third cell: prokaryotic

Prelab Preparation Answers

1. • A cell wall is a thick, boxlike structure found outside plant cells. Cell membranes are very thin and not actually seen when using the light microscope.
   • Plant cells contain chloroplasts, organelles that can convert energy from sunlight into sugars.
   • Ribosomes, endoplasmic reticulum, Golgi bodies, mitochondria, cytoplasm, nucleus

Procedure Answers

3. Students should be able to see the cell wall, the nucleus, and chloroplasts.
4. Students will have a better view of chloroplasts and will possibly see a nucleolus and some cytoplasmic streaming.

Chapter **3**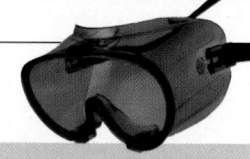

Investigation

How Do Plant and Animal Cells Differ?

Objectives

In this investigation you will:
- *observe* cell structures
- *compare* and *contrast* animal and plant cells
- *relate* the structure of a cell to its function

Materials

- lens paper
- compound light microscope
- *Elodea* plant
- forceps
- glass slides
- water
- coverslips
- medicine dropper
- prepared slide of human cheek cells
- paper towels

Prelab Preparation

1. Review what you have learned about cells by answering the following questions:
   - How does a cell wall differ from a cell membrane?
   - How is plant cell structure related to the ability of plants to make food from sunlight?
   - What organelles are found in both plant and animal cells?

2. Review the procedures for proper use of the microscope in the Appendix. Summarize the procedure for observing a specimen first at low power and then at high power.

3. Review the procedures in the Appendix for making a wet mount slide. Summarize the steps.

Procedure: Comparing Cells of Plants and Animals

1. Use lens paper to clean the microscope lenses.

2. Make a wet mount slide of a young *Elodea* leaf taken from the growing tip of the plant.

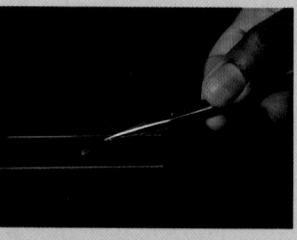

3. Observe your specimen under low power. *What structures do you recognize?*

4. Observe the specimen using high power. *What structures do you see that were not visible under low power?*

5. Draw your specimen as it appears at each magnification. Label the cell structures and note the magnification represented by each drawing.

6. Write a brief description of your observations next to each drawing.

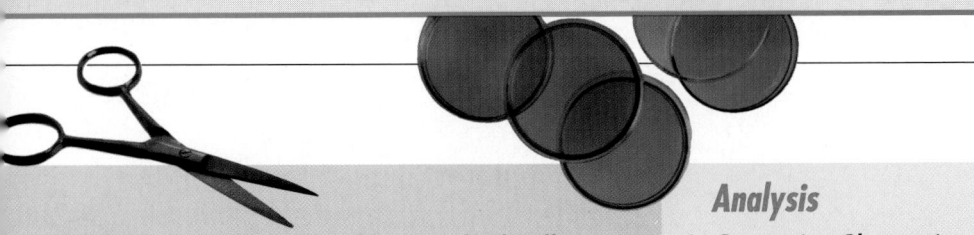

7. Obtain a prepared slide of human cheek cells.

8. Repeat steps 3 through 6.

9. Make a table similar to the one shown.

10. Use the table to record evidence suggesting that the cell structures listed in the table are present in your specimens. Use the line labeled "Other" to record observations of structures not listed in the table.

	Elodea leaf	Cheek cells
Cell membrane		
Cell wall		
Cytoplasm		
Nucleus		
Chloroplast		
Other		

11. Use the information in your table to write a summary that describes the differences between plant cells and animal cells.

12. Dispose of your *Elodea* specimen as directed by your teacher. Carefully clean and dry the slide and coverslip.

13. Return your laboratory materials to their proper places and wash your hands thoroughly before leaving the laboratory.

Analysis

1. *Comparing Observations*
Compare the sizes and shapes of animal and plant cells.

2. *Comparing Observations*
Compare the structures observed in animal and plant cells.

3. *Making Generalizations*
Based on your observations, describe a generalized plant cell and a generalized animal cell.

4. *Inferring Relationships*
What is the relationship between plant cell structure and the ability of plants to stand upright?

5. *Evaluating Methods*
Describe some problems that you had when preparing slides for viewing. How would you attempt to solve these problems the next time you make a slide?

Thinking Critically

1. Compare the organelles found in the plant and animal cells shown in **Figures 3.14** and **3.19**. What can you infer about the evolution of some of the organelles that are found both in plant cells and in animal cells?

2. For each of the organelles listed in your table, describe what function the cell would lose if the organelle were damaged or suddenly lost.

Analysis Answers

1. Answers will vary depending on the specimens observed. Students should indicate that, in general, animal and plant cells are similar in size although plant cells tend to be more box like in shape.

2. Answers may vary, but should indicate that animal and plant cells are similar in structure although plant cells have a cell wall and, possibly, chloroplasts which are not observed in animals.

3. Answers may vary, but should suggest that a typical plant cell will be box like in shape and possess a cell wall. It may contain chloroplasts and possibly a central vacuole. Animal cells are more irregular in shape and lack cell walls and chloroplasts.

4. The cell wall. It consists of layers of cellulose to provide strength and support for both plants and plant cells.

5. Answers will vary. Frequent problems include specimens that are too dark or too light, or in other ways difficult to see clearly. Specimens that are too dark may need to be cut thinner, given more illumination, or a light stain. Too-light specimens may need to be replaced with thicker specimens or stained more heavily.

Thinking Critically Answers

1. Student may describe the endosymbiotic theory, discussed on page 60.

2. Without a cell wall, the cell would collapse. Without a nucleus, a cell wouldn't have a control center. Without chloroplasts, plant cells would be unable to photosynthesize.

Using the Feature

This feature can be used to review a fascinating but little-known issue that poses a multitude of difficult ethical questions. Few students will be familiar with this issue, but most will have opinions about it once they have read the feature. You may want to use this feature to lead a discussion about the special ethical questions posed by the use or misuse of human tissues that otherwise would be discarded.

Discussion

Guide a discussion by posing the following questions:

1. What restrictions, if any, should be placed on the sale of human organs and tissues?
   Answers will vary. You may want to remind students that blood is one example of a tissue commonly and legally sold.

2. In what way does the sale of human tissue differ from the sale of animal or plant tissue?
   Answers will vary.

3. If a person's tissue is used by a company to make a profit, should that person be entitled to a share of the profits?
   Answers will vary.

Thinking Critically Answers

1. Students should discuss the Supreme Court of California's verdict that a patient forfeits ownership of his or her cells once a doctor removes them from the body.

2. Students should imagine a possible reaction to knowing this information before going to the doctor.

3. Students should provide convincing arguments for the fact that a person's cells are his or her own even if they are removed from the body.

4. Students should discuss the information that doctors could have provided John Moore before and after the surgery.

5. Students should address the issue taking into account the parent's action and the baby's rights.

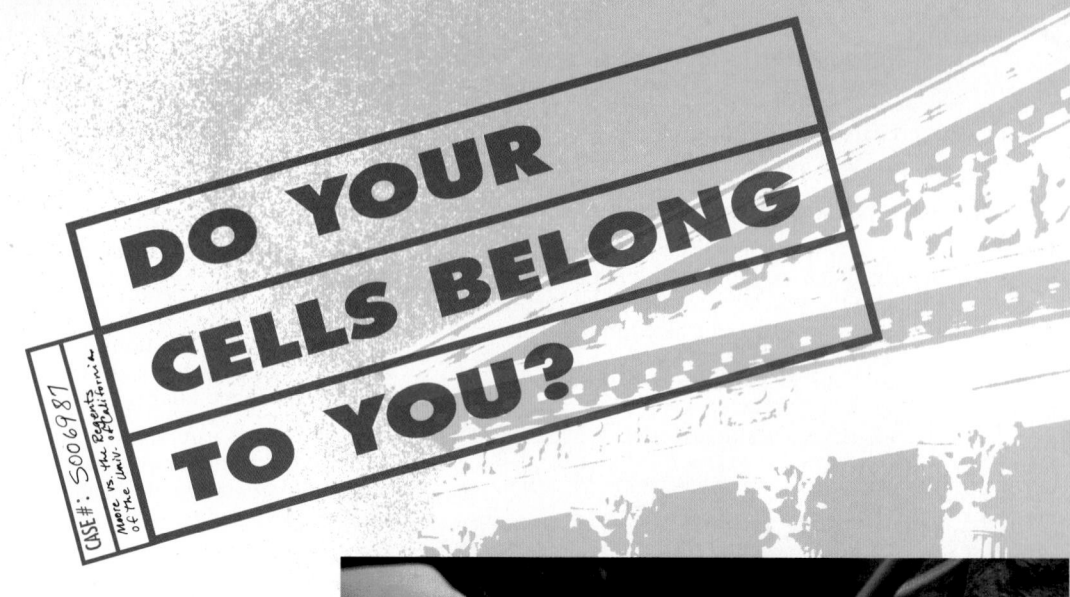

DO YOUR CELLS BELONG TO YOU?

CASE #: S006987
Moore vs. the Regents of the Univ. of California

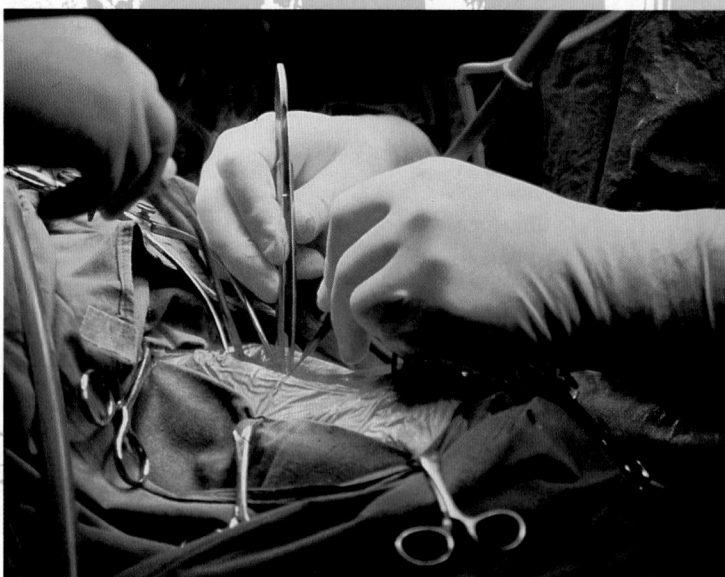

These doctors are performing spleen surgery.

Do you own your body? Are the cells of your arm, your liver, or your heart really yours? The answer might seem obvious: Yes! Of course! But that might not be so. Consider the highly controversial case of Moore v. the Regents of the University of California. *The court's decision might startle you.*

John Moore was thrilled with his cure. His leukemia had not come back since his spleen was removed. But now the doctors at the University of California wanted him to come back for some additional tests. Was something wrong?

Moore gave blood and skin samples to his doctor. The doctor said he should not be worried. Moore did not have leukemia; these were only follow-up tests.

But Moore soon found out that when the doctors who treated him realized that his cells were of a rare type, they applied for and were issued a patent on the cells they had removed from Moore. A patent allows the holder to own the product. The doctors now owned Moore's cells!

After the doctors had removed his spleen, they took the cells, worked with them, and sold their DNA and the DNA's products to a

Science, Technology, and Society ▪ Science, Technology,

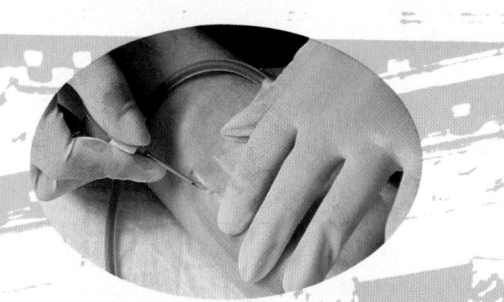

A nurse is taking a sample of blood.

biotechnology company. The company was going to market the cells and their products and make a tremendous profit. Didn't Moore have any right to the profits or at least a right to say, "Don't use my cells—I own them"?

The Case Goes to Court

John Moore took his case to court, and *Moore v. the Regents of the University of California* became one of the most publicized cases in medical history.

Moore's cells were special because their DNA produced a type of protein called lymphokine. Lymphokines are rare, and have been used to treat patients with cancer and blood diseases. One example of a lymphokine is interferon. Interferon is a chemical that blocks certain viruses from attacking cells within the body. If interferon is given to patients with cancer, the drug sometimes helps stop the spread of the

disease. Moore's lymphokine was much rarer than interferon and had the possibility of use as a treatment for other rare leukemias, which are forms of cancer. Because cells can be kept alive, growing, and also reproducing for years, doctors wanted to use Moore's cells for research and making money.

The Verdict Is In

After an extended hearing, the Supreme Court of California ruled in favor of the doctors. They said that once cells are removed from your body they are not considered yours anymore. They can be used by the doctors any way they like. The court said John Moore never expected to have his cells back when he gave permission to remove his spleen.

The court was afraid that scientific experimentation, a very necessary thing, would become more difficult if they didn't allow doctors to work with human cells freely. The work with human cells is extremely important because testing the effects of drugs and other treatments on living human cells must occur before they are used on human beings. If the scientists had to check into the ownership of the cells they use, it would take a great deal of time and study would be slowed.

Although the court did not allow John Moore to recover any money, they said he could go back to court and sue the doctors because they did not inform him that his cells were being used for this purpose. The court said Moore had a right to consent to this use.

Thinking Critically

❶ Do you agree with the court's decision not to allow John Moore the ownership of his cells? Why or why not?

❷ What do you think John Moore might have done if he had known beforehand that the doctors were going to experiment with and sell his cells' products?

❸ What arguments could John Moore's attorney have made to prove Moore's ownership of the cells?

❹ How much do you think the doctor should have told John Moore before his spleen was removed? How much after?

❺ Recently, a couple in California decided to have a baby so that the baby could donate bone marrow cells to their older child who suffered from leukemia. According to the Moore case, did those cells belong to the baby? Why or why not?

Acting on the Issue

❶ Find out what legislation may be under consideration in your state related to the sale of tissue.

❷ Find out if biotechnology companies now require consent forms before purchasing human tissue.

Chapter

4

The Living Cell

Planning Guide

	Objectives/Themes	Classwork Resources	Homework Resources

4.1	**1.** Explain how electrical currents affect voltage-sensitive channels.
	
	2. Describe how receptor proteins enable cells to sense chemical signals.
	
	3. List three different tasks performed by cell surface markers.
	
	Themes: Energy and Life, Interacting Systems, Patterns of Change
		Text
		Science in Action
		    Cystic Fibrosis: A Deadly Gene
		    pp. 72–73
		
		Teacher's Resource Binder
		Focus Activity 4
		    Demonstrating Diffusion
		Lab Investigation 4.1
		    Diffusion and Cell Membranes
		
		Other Resources
		Transparencies 12–13
		Science in Action Video
		    Cell Structure and
		    Function/Cell Division★
			Text
			Section Review, p. 71
			
			Teacher's Resource Binder
			Directed Reading Worksheet 4.1
			
			Other Resources
			Audiocassette 4.1★

4.2	**1.** Describe the difference between diffusion and osmosis. **2.** Distinguish facilitated diffusion from active transport. **3.** Explain how the sodium-potassium pump and the proton pump transport ions. **4.** Describe how large substances can enter and exit cells. **Themes:** Interacting Systems, Patterns of Change, Scale and Structure	**Teacher's Resource Binder** Extension Worksheet *Cellular Transport* **Other Resources** Transparencies 14–15	**Text** Section Review, p. 78 **Teacher's Resource Binder** Directed Reading Worksheet 4.2 **Other Resources** Audiocassette 4.2★

4.3	**1.** Compare and contrast cell division in bacteria and eukaryotes.
	
	2. Specify the number of chromosomes in a cell before and after cell division.
	
	3. List the events of mitosis and cell division in the correct sequence.
	
	4. Compare and contrast cell division in animals and plants.
	
	Themes: Evolution, Interacting Systems, Patterns of Change, Stability
		Text
		Investigation
		    Can All Molecules Diffuse Through a Membrane?
		    pp. 86–87
		
		Other Resources
		Transparency 16
			Text
			Section Review, p. 82
			
			Teacher's Resource Binder
			Directed Reading Worksheet 4.3
			Vocabulary Review Worksheet★
			Reaching Worksheet★
			    The Life of a Cell
			
			Other Resources
			Audiocassette 4.3★

★**Reteaching Options**

Demonstrations	**Assessment**
4.2: pp. 75, 76, 78 **4.3:** p. 80	Chapter Review pp. 84–85 Portfolio Assessment p. 67D Chapter Test—Teacher's Resource Binder Test Generator

Research Notes

Connection to Medicine: Glucose Transport

Along with the coupled-channel transport mechanism described in Section 4.1, glucose enters cells through a family of channel proteins. Since 1985, scientists have investigated a possible link between diabetes and problems in the pathway used by glucose to enter cells from the bloodstream.

The most prevalent form of diabetes is non-insulin dependent diabetes mellitus (NIDDM). In NIDDM, the body makes enough insulin, but the muscle, fat, and liver cells do not respond to the insulin by absorbing glucose.

As a result, too much glucose remains in the bloodstream (instead of entering the cells), and its osmotic pressure causes the blood to draw water from the tissues. Then, the kidneys

excrete this water and more salts than usual. Dehydration and salt loss can result, eventually causing blindness, strokes, kidney failure, and other disorders.

Some researchers have focused on a glucose channel protein called GluT4 in this form of diabetes. This protein has the unusual ability to move between the cell membrane and reservoirs within the cell.

In 1991, J.W. Slot of the University of Utrecht in the Netherlands led a team of researchers that monitored the GluT4 channel proteins. Before insulin was present, 1 percent of the cell's GluT4 was on the surface, with the remainder within the cell. After insulin was present, the proportion of GluT4 on the cell surface rose to 40 percent.

Scientists have theorized that the GluT4 proteins stay within the cell on the surface of small membrane bubbles called "vesicles."

When insulin is present, the vesicle fuses with the external cell membrane in a process similar to exocytosis, so that the channel proteins are now a part of the cell wall. Glucose is pumped into the cell. Eventually, another vesicle is formed from the GluT4 in a process similar to endocytosis.

Investigation Notes

Can All Molecules Diffuse Through a Membrane?
pages 86–87

Purpose: Students will use previously learned concepts and techniques to investigate new problems. Note that knowledge of enzymes is not necessary for success in this investigation.

Solution Preparation

1. The teacher and the students must all wear eye goggles and a lab apron during this Investigation and during preparation and cleanup. The use of disposable gloves is also necessary, since the solutions used are irritants.

2. Add 5 g of soluble starch to a small amount of cold water. Dilute to 100 mL with boiling water and stir until dissolved. This should be enough for 8–10 lab groups. Starch solution may develop mold if stored. Cool before using in the lab to avoid denaturing the enzymes.

3. Add 5 g of alpha amylase WARD'S (39 M 0058) to 100 mL of water. This should be enough for several classes.
NOTE: This solution does not have a long shelf life.

4. Benedict's solution is available ready-made from WARD'S (37 M 0698). To make it yourself, dissolve 17.3 g of sodium citrate and 10 g of sodium carbonate (Na_2CO_3) in 80 mL of distilled water in a beaker. In a smaller beaker, dissolve 1.7 g of copper sulfate ($CuSO_4 \bullet 5H_2O$) in about 10 mL of distilled water. While stirring the first solution, slowly pour the copper sulfate solution into it. Pour the combined solutions into a 100-mL graduated cylinder and dilute to make 100 mL.

This should provide enough for several classes. It can be stored for long periods of time.

5. Lugol's iodine (or "iodine-iodide solution") is available ready-made from WARD'S (39 M 1685).

NOTE: Iodine reacts with metal, skin, and many other substances.

To make Lugol's iodine, dissolve 1.0 g of potassium iodide (KI) in about 15 mL of distilled water. Using a porcelain spatula, place 0.7 g of iodine (I_2) in a 100-mL volumetric flask. Dilute to the 100-mL mark with water. Stopper and shake the flask. Additional KI may help dissolve the I_2.

This should provide enough for several classes. It should be stored in a glass stoppered bottle in a cool, dark place. Strong light can cause the iodine to react.

Prelab Preparation

While students conduct steps 2–4, soak dialysis tubing in warm water to make it pliable. Have students rub the tubing between their fingers to open.

Answers will be found on pages 86–87.

Meeting Individual Needs

Objectives

1. Students will demonstrate appropriate use of core vocabulary for the chapter (Vocabulary File).

2. Students will identify the process of diffusion (Demonstration A).

3. Students will interpret a model of the process of selective transport, facilitated diffusion, and active transport (Demonstration B).

Vocabulary File
(Developing Vocabulary/ Limited English Proficiency)

If you are not already using the Vocabulary File, refer to Chapter 1 for its preparation. See the Chapter Highlights on page 83 for a list of suggested words.

Demonstration A
Diffusion
for page 74
(Developing Observational Skills/Visual Learners)

Materials: ink or dye, medicine dropper, beaker of water (any size)

Preparation: Have ink or dye already in medicine dropper near beaker of water.

Procedure: Demonstrate ink diffusing in water as shown in Figure 4.4 and described in the visual strategy for that page. Point out that if salt or another

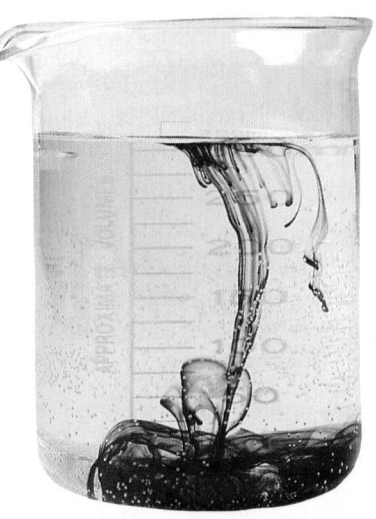

spice is added to a soup, its flavor soon spreads to all of the soup, not just the part where it was added. Ask students to identify other examples of diffusion. What substances are diffusing in these examples? Have the students describe the regions of high concentration and low concentration for their examples.

Demonstration B
Model for Selective Transport
for page 77
(Developing Modeling Skills/Tactile Learners)

Materials: shipping tube (the type used for mailing posters), 10 small wooden or plastic toy blocks, 10 small plastic foam balls (should be smaller than the blocks), utility knife, masking tape, glue, poster board

Preparation: Cut mailing tube in half. Cut five square openings on one side of the tube large enough for balls and blocks to easily pass through. Attach half of the poster board to one end of the mailing tube, to serve as a base for the model. Cut 5 square flaps out of poster board, 2 cm larger than holes in the tube. In the center of each flap cut a circle large enough to allow the balls to pass through, but not the blocks. Tape the flaps over holes cut in tube.

Procedure: Position the model upright on a table, resting on its poster board base. Explain that the tube is a model of the cell, with the flaps and holes representing channel proteins and the blocks and balls symbolizing molecules.

Be sure the square openings are covered with the flaps. Ask students to predict whether the round molecules will be able to pass through the outside. Have a student place eight balls through the top opening in the tube to test the predictions. Next ask students to predict if the square molecules can pass to the outside. Have another student

place eight blocks inside the tube. (NOTE: Balls should be able to fit through the holes in the flaps. Blocks should be unable to pass.) Have students give the name for the process that has taken place (Selective Transport).

Then, have a student empty the tube and place six to eight balls inside the tube while the flaps are down. Have students count aloud the number of round molecules inside and outside the tube. Ask which location has the greater concentration of molecules and which has the lowest concentration. Ask students which direction the molecules will move due to diffusion.

Transfer the molecules back and forth to achieve a balance. Point out that the round molecules were able to pass through the outside assisted by the cell model's holes. Have students name the mode of transport that achieved the balance of concentrations (Facilitated Diffusion).

After removing the round molecules, have a student lower the flaps and drop the blocks into the tube. Point out that no blocks are able to pass from inside the cell to outside. Lift the flaps and tilt the model to transfer some blocks outside. Lower the flaps. Ask students what was needed for the channels to change shape and allow the molecules to leave (Energy). Have students name this mode of transport (Active Transport).

Additional Strategies
Visual Strategies
Pages 69, 70, 71, 74, 75, 76, 77, 79, and 81

Auditory Learners
Use *Biology: Visualizing Life* Audiocassettes for Sections 4.1, 4.2, and 4.3.

Meeting Individual Needs (cont.)

Cooperative Learning
Methods of Cellular Transport

Timing: Use this activity to introduce Section 4.2.

Group Size: 5 students

Outcome: Students will be able to diagram and describe methods of cellular transport.

Individual Accountability: Each student will diagram one method of cellular transport and describe it to the group.

Positive Interdependence: Each group will be able to diagram and describe all five methods of cellular transport.

Assign one of the following methods of cellular transport to each group member: (1) endocytosis, (2) exocytosis, (3) diffusion, (4) osmosis, and (5) facilitated diffusion.

Have each group member draw a diagram to illustrate the method. Then have each group member use their diagram to describe and explain the method to the group.

Have groups make sure that all members can describe each method if they are asked to. Randomly call on a member from each group to describe one of the methods of cellular transport.

Portfolio Assessment

Students should select their best work and provide self-reflective rationale for their selections. Students can make selections in the following areas.

1. *Content* — One concept map from the chapter (See page 812 for evaluation criteria.)

2. *Reading Comprehension* — One Directed Reading Worksheet from the Teacher's Resource Binder (Use the answer key to evaluate for accuracy.)

3. *Writing* — Using the Vee Form, summarize a newspaper or magazine article relating to cellular transport. (See page 22T for evaluation criteria.)

   Or: Select a writing task or project from the Chapter Review.

4. *Performance Assessment* — One Vee form from a chapter investigation or lab manual investigation (See page 22T for evaluation criteria.)

Teacher makes selections in the following areas.

1. *Formal Assessment* — Chapter test (Test A, B, or the Test Generator) The teacher-scored test should be reviewed. Incorrect responses should be corrected by the student.

2. *Informal Assessment* — Use the Direct Observations Checklist, page 33T, in a lab or other cooperative learning experience.

3. *Performance Assessment* — Have students design an experiment that uses techniques from the investigation to measure the effects of insulin concentration on glucose intake in cells.

Concept Map Answer

The following is one possible answer to the Relating Concepts exercise on page 84.

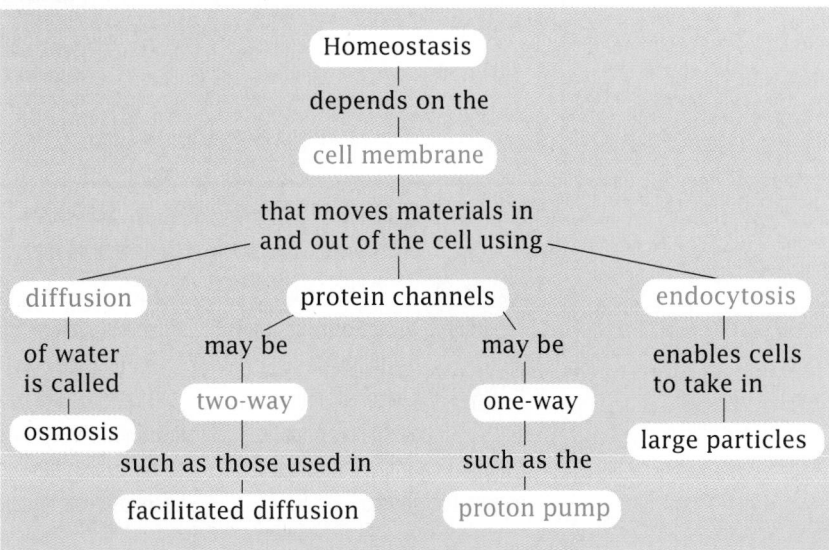

Chapter 4

Determining Prior Knowledge

- Ask students to name and describe the functions of the three major types of cell membrane proteins that were covered in Chapter 3.
- Open a bottle of perfume. Have students explain what happens to the molecules of perfume.
- Remind students that their bodies each contain 100 trillion cells. All these came from one original cell, the fertilized egg. Have them explain how one cell produced so many.
- Have students identify the cells in their body that are not constantly being replaced. Have them name the cells that are not replaced when damaged or destroyed.

Chapter 4

The Living Cell

Review

- **structure of cell membrane (Section 3.2)**
- **characteristics of prokaryotes and eukaryotes (Section 3.3)**

4.1 How Cells Receive Information

- **Sensing Electrical Signals**
- **Sensing Chemical Signals**
- **Sensing Cellular Identity**

4.2 Moving Into and Out of Cells

- **Diffusion and Osmosis**
- **Selective Transport of Substances**
- **How Large Particles Get Into and Out of Cells**

4.3 How Cells Divide

- **How Bacteria Divide**
- **How Eukaryotic Cells Divide**
- **Life Span of Cells**

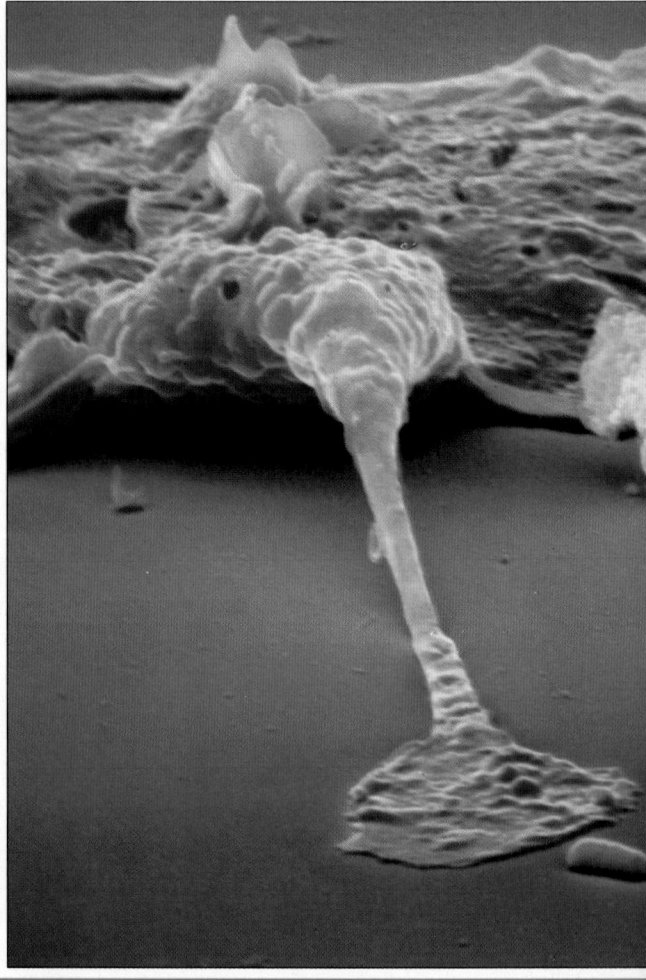

Cells are very much alive. A macrophage (top), a cell that attacks foreign intruders in the bloodstream, is about to engulf a harmful bacterium (lower right).

▪ Author's Rationale ▪

Of the many cellular activities that might be considered, four are particularly important:

- transporting materials across the plasma membrane
- obtaining energy
- learning about the environment
- reproducing

The focus will be on explaining how these processes work, rather than on describing the component parts. Chemiosmosis and proton pumps are introduced because of the key roles they play in the flow of energy, and their importance to future chapters.

LIKE YOU, CELLS NEED INFORMATION ABOUT THEIR ENVIRONMENTS. IN THE SAME WAY THAT YOUR SENSES—SUCH AS SIGHT, HEARING, AND TOUCH—ENABLE YOU TO GATHER INFORMATION, PROTEINS IN THE CELL MEMBRANE CONNECT THE CELL TO ITS ENVIRONMENT. THOSE PROTEINS ARE THE CELL'S AVENUES OF COMMUNICATION WITH THE WORLD AROUND IT.

4.1 How Cells Receive Information

Objectives

❶ **Explain how electrical currents affect voltage-sensitive channels.**

❷ **Describe how receptor proteins enable cells to sense chemical signals.**

❸ **List three different tasks performed by cell surface markers.**

Sensing Electrical Signals

Figure 4.1
Actions of the human body are controlled by a network of nerves.

a Nerves run throughout the body.

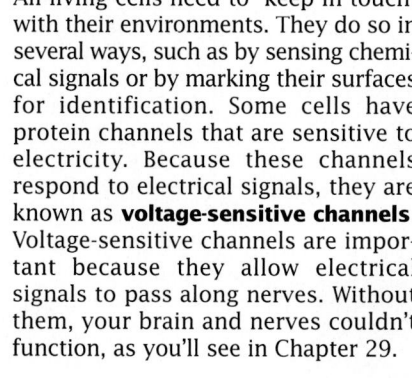

All living cells need to "keep in touch" with their environments. They do so in several ways, such as by sensing chemical signals or by marking their surfaces for identification. Some cells have protein channels that are sensitive to electricity. Because these channels respond to electrical signals, they are known as **voltage-sensitive channels**. Voltage-sensitive channels are important because they allow electrical signals to pass along nerves. Without them, your brain and nerves couldn't function, as you'll see in Chapter 29.

It is not difficult to understand how these channels work. They are like little doors with magnets on them. The doors usually remain closed. But they flip open or shut in response to electrical signals, as shown in **Figure 4.1c–d**. The center of each channel is occupied by a protein containing many charged amino acids. When an electrical signal reaches the channel, the door flips out of the way, and the channel opens to allow ions to enter the cell. You will see a voltage-sensitive channel in greater detail in Section 4.2.

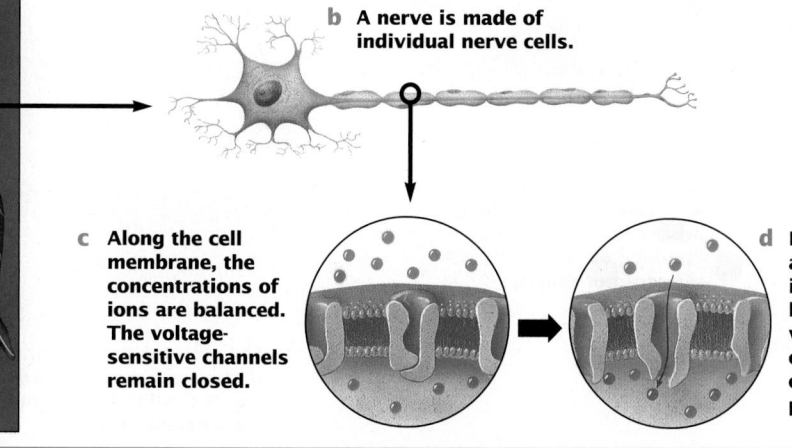

b A nerve is made of individual nerve cells.

c Along the cell membrane, the concentrations of ions are balanced. The voltage-sensitive channels remain closed.

d But when additional ions upset the balance, the voltage-sensitive channels spring open and ions pass through.

Lesson Plan 4.1

Phase 1

PREPARATION

Key Concepts
- Proteins form voltage-sensitive channels that control the passage of ions through the cell membrane.
- Proteins known as receptors enable the cell membrane to sense chemical signals.
- Proteins known as cell surface markers establish a cell's identity.

Reading Strategy

Have students take notes as they read this section. Point out that taking notes helps them learn as they read and gives them a record of the reading that they can use later for review.

Phase 2

TEACHING STRATEGIES

Visual Strategy

Figure 4.1
Use Figure 4.1b to show students the correlation between structure and function in a nerve cell. Also have students count the number of ions on each side of the membranes shown in Figures 4.1c and 4.1d and relate this to the closing and opening of the channels. Have students identify the lipid bilayer in these figures.

Mathematics Connection

Receptor Proteins

The information that is passed into a cell when a receptor reacts to a signal is amplified many times. For example, the binding of adrenalin to just a few of its receptors in a liver cell membrane activates an estimated 25 million molecules of the enzyme that breaks down glycogen into glucose. The increase in glucose level is one way by which adrenalin accelerates metabolism.

Theme Connection

Evolution

Insulin has been discovered in such diverse organisms as fruit flies, earthworms, fungi, and bacteria. The universality of insulin, along with other hormones, points to the long thread of evolutionary history linking all organisms.

Visual Strategy

Figure 4.2

Have students note the structural change that occurs in the receptor protein when it combines with insulin. Point out that once this change takes place, sugar can then move from the blood into the cell. Make sure students understand that insulin *lowers* blood sugar levels. Use this diagram to illustrate the consequences of too little or no insulin (juvenile-onset diabetes), or too few or no insulin receptors (maturity-onset diabetes).

Sensing Chemical Signals

Only a very few specialized cells have "fingers" that are sensitive to pressure or "eyes" that are sensitive to light. However, almost all cells can sense chemical signals by means of **receptor proteins** protruding through their membranes.

Receptor proteins send signals into the cell

Notice in **Figure 4.2b** that the end of the receptor protein has a unique shape that will hold only a particular type of molecule. Proteins like this one act like television antennas: they capture signals and deliver information. When the right molecule binds to the receptor protein, changes are triggered in the protein that are detected within the cell. Thus, the "signal" picked up by a receptor protein is a specific molecule outside the cell, a molecule that does not even enter the cell. Instead, what passes into the cell is information.

Hormones provide a good example of how receptor proteins carry information into a cell. A hormone is a chemical substance that acts as a messenger. Insulin, the hormone your body uses to regulate the level of sugar in your blood, has a shape that fits a specific receptor protein. Most of your cells have only a few insulin receptor proteins, but each of your liver cells possesses as many as 100,000! When an insulin molecule encounters an insulin receptor protein, the molecule binds to that protein, as shown in **Figure 4.2c**. The opposite end of the receptor protein changes shape, initiating a change in cell activity that ultimately causes levels of sugar in the blood to fall.

If blood sugar levels fall too low, however, the body responds by producing less insulin. Therefore, fewer receptor proteins are activated, and blood sugar levels gradually rise. Insulin and its receptor proteins thus help to keep blood sugar at a constant level. If cells have too few insulin receptors, one form of a disease called diabetes (*deye uh BEET eez*) results. In this type of diabetes, which occurs mainly among adults over 40 years old, insulin levels in the bloodstream are above normal but cells are unable to take in enough sugar.

Figure 4.2

a **After you eat a muffin, the level of sugar in your blood rises. Insulin lowers this level by causing your cells to take up and store sugar. Here's how it works.**

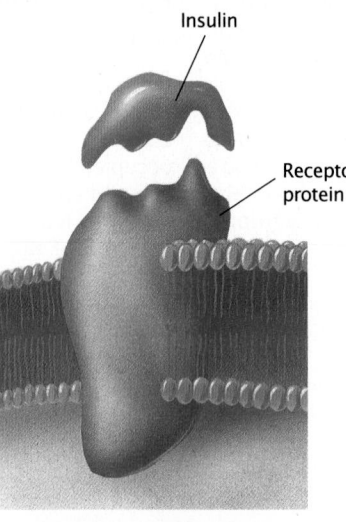

Insulin

Receptor protein

b **An insulin molecule has a specific shape that fits into a receptor protein. Like a jigsaw-puzzle piece, the insulin molecule joins with the receptor protein.**

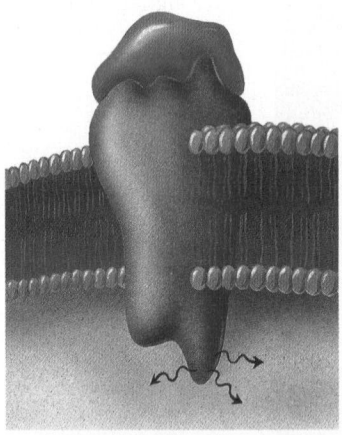

c **The receptor protein changes shape, sending a message into the cell to take in and store excess sugar.**

■ Historical Perspective ■

Heart Transplants

The first heart transplant was done by Dr. Christiaan Barnard in South Africa on December 3, 1967. He transplanted the heart of a 25-year-old woman into the body of a 55-year-old man who died 18 days later. Barnard's second transplant, done the next month, met with much greater success. The recipient survived 563 days after the operation. However, early successes with heart transplants were so dismal that some countries, like England, suspended the procedure for several years. During the 1970s, only 20 to 40 heart transplants were done each year worldwide. But careful patient selection, better diagnosis of early signs of rejection, and development of immunosuppressive drugs resulted in dramatic improvements. In 1988, 1,647 heart transplants were done in the United States alone, with an 80 percent one-year survival rate.

Sensing Cellular Identity

Just as a football jersey bears a number and last name, cells carry markers for identification. Every cell in your body contains a unique set of membrane proteins called **cell surface markers**. Your immune system uses these markers to recognize your cells and to distinguish them from damaging invaders, such as harmful bacteria. Cell surface markers also indicate cell type, such as whether a cell is a blood cell, a muscle cell, or a liver cell.

The uniqueness of an individual's cell surface markers explains why organ donors and recipients must be carefully matched. Notice the cell surface marker on the heart cell in **Figure 4.3d**. To ensure that the heart is a "good match," the donor and recipient must share many of the same identifying proteins on their cell surfaces. As with all proteins, the structure of cell surface markers is determined by genes. If the donor and recipient are genetically related, there is a greater probability of a match. Cell surface markers are so unique that a perfect match never occurs, except between identical twins. The organ recipient, therefore, must take drugs to suppress the immune system, which could reject the transplanted organ.

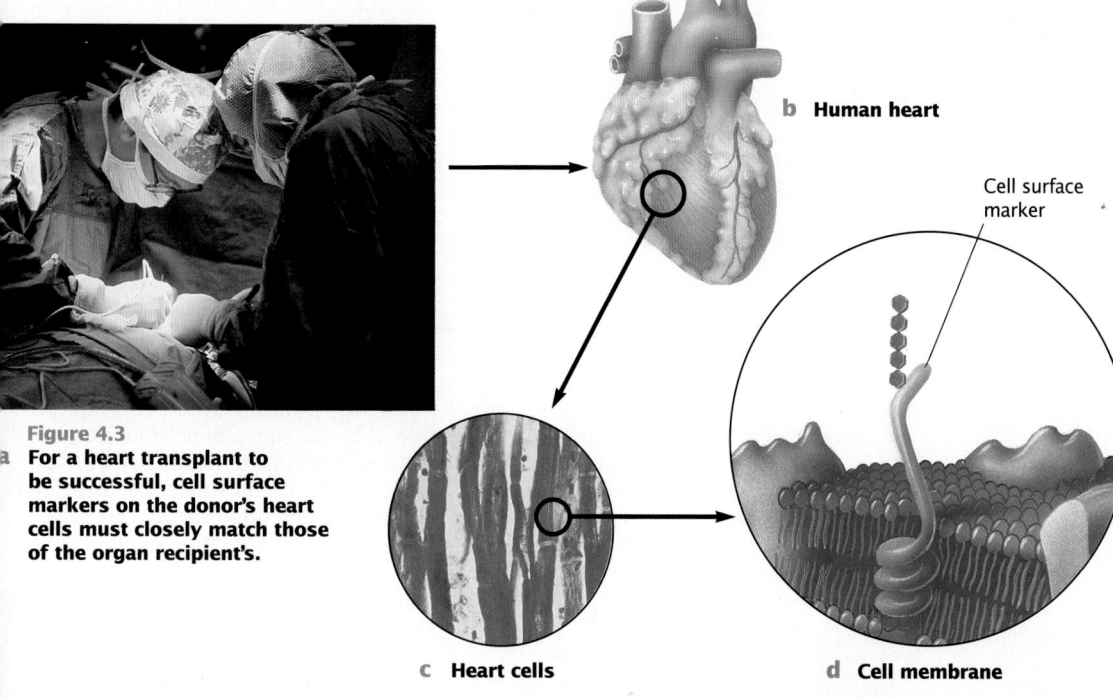

Figure 4.3

a **For a heart transplant to be successful, cell surface markers on the donor's heart cells must closely match those of the organ recipient's.**

b **Human heart**

c **Heart cells**

Cell surface marker

d **Cell membrane**

Section Review

❶ **How do voltage-sensitive channel proteins respond to electrical signals?**

❷ **How do receptor proteins pass information into cells?**

❸ **Why are protein markers important in matching organ donors to recipients?**

■ *Section Review Answers* ■

1. Students should indicate that voltage-sensitive channels flip open or closed in response to electrical signals.
2. Students should explain that receptor proteins are shape-specific. When the specific molecule binds with a receptor protein, it triggers a structural change in the protein. The cell detecting this change reads the information. The molecule never enters the cell but passes information into the cell.
3. Students must explain that the donor and recipient must share many of the same protein markers on their cell surfaces, or the immune system that uses these markers to detect harmful invaders will reject the transplanted organ.

Using the Feature

- This feature can be used to review the structure and function of protein channels in the cell membrane.
- Ask students to read the book *Alex: Life of a Child* by Frank Deford. This book accounts his daughter's life with cystic fibrosis.

Discussion

Guide a discussion by posing the following questions:

1. One of the major symptoms of cystic fibrosis is the build up of thick mucus in the lungs. What causes this to happen?
   High levels of salt in the cells of the lungs draw water out of the mucus, making it thick.

2. What causes the salt imbalance in these cells? What is the genetic basis for cystic fibrosis?
   Researchers have found a gene called the cystic fibrosis transmembrane conductance gene. The gene causes the construction of an irregularly shaped protein channel, which impairs proper ion transport.

3. Discuss treatments and a possible cure for cystic fibrosis.
   Students should discuss researchers' hopes to develop a spray that will deliver a normal CFTR gene into the lungs of CF patients, a medication that would force the defective protein to work normally, and a spray that would deliver the healthy protein directly into the lungs.

Cystic Fibrosis: A Deadly Gene

A Rare but Deadly Disease

The cystic fibrosis patient below is breathing into a vitalograph, a device that measures lung function.

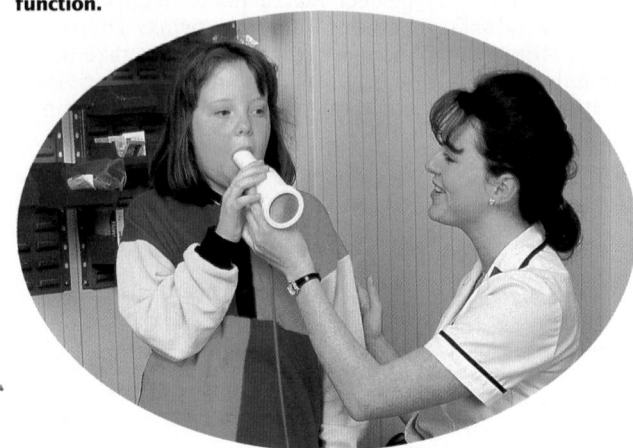

For a child with cystic fibrosis, life is mostly a series of respiratory infections, doctors' visits, and medications. Cystic fibrosis causes thick mucus to build up in the lungs, making breathing difficult.

The thick mucus also coats the hairlike projections lining air passages, weakening the body's immune system. Excess mucus interferes with the functioning of other organs too. In the liver and pancreas, mucus blocks the flow of digestive enzymes in the intestine, so food is not digested properly. Worn down by repeated bouts of illness, a cystic fibrosis patient rarely lives beyond his or her twenties.

Cystic fibrosis is the most common inherited disorder among Caucasian people. The disease, which begins in infancy, afflicts more than 25,000 Americans and causes 500 deaths every year. Forty years ago, the average life span of a cystic fibrosis patient was five years. Today, improved medical therapies and nutrient-rich diets have enabled cystic fibrosis sufferers to survive into adulthood.

Salt Imbalance Is a Clue

Research into the cause of cystic fibrosis reads like a detective story. One clue is that cystic fibrosis patients have excess amounts of sodium and chloride in their sweat, making it very salty. At the University of North Carolina, researchers found that salt imbalance causes thick mucus to accumulate in the patients' lungs. High levels of salt in lung cells draw water out of the mucus, causing it to thicken.

The level of salt in a cell is determined by the movement of ions across the cell membrane. Ions are carried across cell membranes by protein channels embedded in the cell membrane. Since the structure and functions of proteins are determined by genes, problems with ion transport can be assumed to have a genetic basis.

The X ray at left shows the chest of a healthy person. The white area in the center is the shadow of the heart.

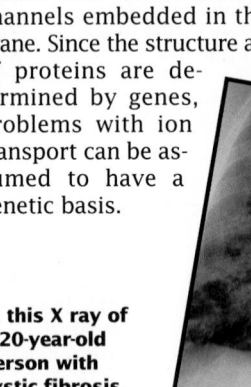

In this X ray of a 20-year-old person with cystic fibrosis, the airways are partially blocked with a thick buildup of mucus.

CFTR: The Defective Gene

The CFTR gene is located near the center of the long arm of chromosome 7.

In 1989, an American-Canadian research team found the defect in the gene that causes cystic fibrosis. The gene, called the cystic fibrosis transmembrane conductance regulator (CFTR) gene, produces a protein that usually helps maintain normal levels of chloride. In about 60 percent of cystic fibrosis patients, the protein made by the CFTR gene is missing an amino acid called phenylalanine. When this amino acid is missing, the protein doesn't fold into its correct shape and loses its function.

Researchers are not certain whether the defective protein is actually a transport molecule. Some research data indicate that the protein might instead be involved in a regulatory network that controls ion transport. Although most cystic fibrosis patients have the mutation causing the defective protein described above, another 61 mutations in the CFTR gene have been identified.

Cystic fibrosis (CFTR) gene

Treatment and a Possible Cure

Locating the cystic fibrosis gene gives researchers an opportunity to develop more effective treatments and possibly a cure for the disorder. In one experiment, viruses that cause the common cold were used to insert corrected CFTR genes into the lung cells of live laboratory rats. The cold viruses were first disabled, so they could not cause infection. Then they were modified to carry healthy CFTR genes. In tests on hundreds of rats, the altered viruses inserted the healthy CFTR gene into the cells lining the animals' lungs. Once inside the cells, the genes began to produce normal protein. Researchers hope to develop a spray that will deliver a normal CFTR gene into the lungs of cystic fibrosis patients and possibly cure the disease. But since the cells that line the lungs are replaced every few weeks, the gene therapy would have to be repeated on a regular basis.

Another possible therapy involves the defective protein that the CFTR gene produces. Using information from the CFTR gene, scientists have figured out the structure of the protein. With this knowledge, the scientists might be able to develop medications that could force the protein to work normally.

Another possibility is the use of live animals to produce large quantities of normal protein that could be incorporated into a lung spray for cystic fibrosis patients. Healthy CFTR genes would be inserted into the embryos of mice or goats. As adults, the females would produce the protein in their milk. The protein could then be harvested and sprayed directly into patients' lungs.

In 1991, Jeff Pinard, who suffers from a relatively mild form of cystic fibrosis, began to conduct genetic research at the University of Michigan. He hopes to isolate some of the DNA mutations that cause cystic fibrosis.

Lesson Plan 4.2

Phase 1

PREPARATION

Key Concepts

- Diffusion is the movement of molecules from a region of high concentration to a region of lower concentration until they are equally distributed.
- Osmosis is the diffusion of water through a cell membrane.
- Water inside a cell creates osmotic pressure by pushing against the cell membrane and cell wall.
- Transport of molecules and ions through a cell membrane channel is selective.
- Some channel proteins assist in selective transport by a process called facilitated diffusion.
- Some channel proteins assist in selective transport by a process called active transport.
- Substances too large to pass through membrane channels enter the cell by endocytosis and exit the cell by exocytosis.

Reading Strategy

As students read this section, have them list the various processes by which substances enter and exit cells. Next to each process, have them write the definition and indicate whether energy is required.

Phase 2

TEACHING STRATEGIES

Visual Strategy

Figure 4.4 a-c
Perform the demonstration shown in this figure. Make sure students recognize that random motion of molecules continues even when the ink particles are evenly dispersed. Have students give other examples of diffusion in a liquid.

ON A HOT DAY OR AFTER A HARD WORKOUT, YOU MIGHT REACH FOR A COLD SPORTS DRINK. THE SODIUM IONS IN THE DRINK REPLENISH THE ONES YOUR BODY LOST THROUGH PERSPIRATION. THESE IONS ENTER YOUR CELLS THROUGH PROTEIN CHANNELS IN THE CELL MEMBRANE. IN THIS SECTION YOU WILL SEE HOW CELLS TAKE IN NEEDED SUBSTANCES, SUCH AS SODIUM, AND DISCARD WASTES.

4.2 Moving Into and Out of Cells

Objectives

❶ Describe the difference between diffusion and osmosis.

❷ Distinguish facilitated diffusion from active transport.

❸ Explain how the sodium-potassium pump and the proton pump transport ions.

❹ Describe how large substances can enter and exit cells.

Diffusion and Osmosis

In addition to receiving information, cells need a way to move water molecules, food particles, and other ions through their membranes. Some molecules, such as water, pass through freely. Others must be carried through channels.

Diffusion mixes different molecules

Look at **Figure 4.4**, in which a student places ink droplets into a beaker of water. The ink slowly spreads throughout the water. What's happening? As the particles of ink dissolve in the water, their random motion soon carries them farther and farther out into the water until they are evenly distributed.

The mixing of two substances by the random motion of molecules is called **diffusion** (dih FYOO zhuhn). As shown in **Figure 4.4**, molecules diffuse from a region where their concentration is high (the ink) to a region where their concentration is lower (the water), until they are evenly dispersed.

Figure 4.4
a **When you place a few droplets of ink into a beaker of water, . . .**

b **. . . the ink will settle to the bottom . . .**

c **. . . and diffuse randomly through the water until the ink particles are evenly dispersed.**

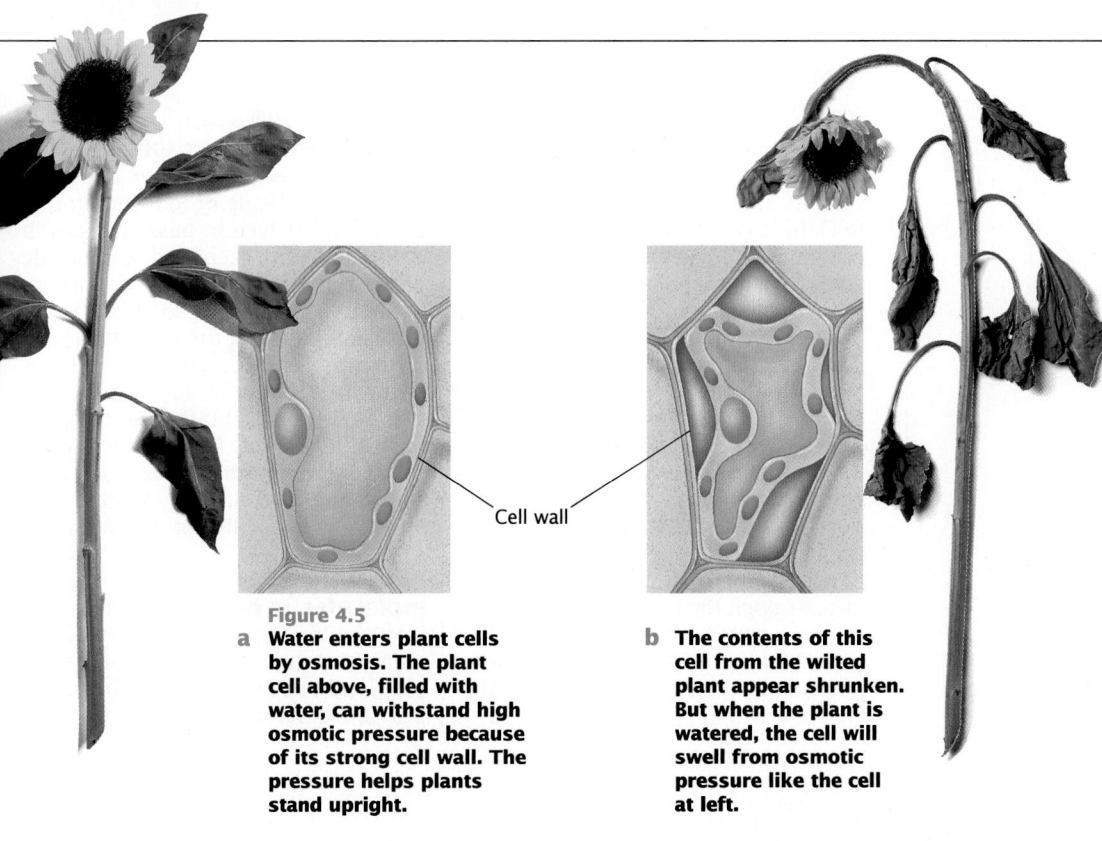

Figure 4.5
Have students identify the cell wall, cell membrane, cytoplasm, nucleus, chloroplasts, and vacuole as a way of reviewing some of the organelles covered in Chapter 3. Point out how the cell membrane has moved away from the cell wall in Figure 4.5b.

Cell wall

Figure 4.5

a **Water enters plant cells by osmosis. The plant cell above, filled with water, can withstand high osmotic pressure because of its strong cell wall. The pressure helps plants stand upright.**

b **The contents of this cell from the wilted plant appear shrunken. But when the plant is watered, the cell will swell from osmotic pressure like the cell at left.**

Demonstration 1

Volume vs. Concentration
Students have a difficult time understanding the difference between concentration and volume. To demonstrate the difference between the two, pour 500 mL of water into a liter beaker and 1,000 mL into another liter beaker. Students will obviously discern the difference in volumes. Place one drop of food coloring in each beaker and stir. Have students observe that the beaker with the lesser volume is more concentrated. Then add another drop to the larger volume to show that unequal volumes can have the same concentrations.

Water travels through membranes by osmosis

Looking at **Figure 4.4**, you might think that only the ink diffuses through the water. But water molecules can also diffuse. When water molecules diffuse through a cell membrane the process is called **osmosis** (*ahz MOH sihs*). Since water molecules are small, they can slip freely through the gaps between the phospholipids in the cell membrane. As a result, water molecules constantly move back and forth through the cell membrane. This free movement of water has a very important function: It enables cells to absorb water.

Unlike the water molecules randomly bustling about outside the cell, many of the water molecules inside the cell are busy interacting with sugars, proteins, and other polar molecules. As a result, more water molecules flow into the cell by osmosis than flow out of the cell. Therefore, there is often a net movement of water into the cell.

Water in a cell creates osmotic pressure

When water enters a cell, it creates pressure. This pressure is called **osmotic** (*ahz MAH tihk*) **pressure**. If osmotic pressure is very high, it can cause a cell to burst. Most cells cannot withstand high osmotic pressure unless their membranes are braced to resist the swelling. Many kinds of organisms, such as the plants in **Figure 4.5**, have cell walls to protect and support their cells.

What about organisms without cell walls? Many single-celled organisms have specialized organelles to pump water out. In multicellular animals, cells do not burst because a balance exists between the concentration of fluids inside and outside the cells. In your body, for example, the concentrations of salts, sugars, and other ions are about the same inside cells as in the fluids surrounding them. Thus, water enters the cells of the body at the same rate that it leaves. The cells do not burst.

Demonstration 2

Osmosis
Dissolve some salt, sugar, and protein in tap water. Pour a measured volume of the solution into a piece of dialysis tubing or a small, clear plastic sandwich bag. Seal the contents and submerge it in a beaker containing distilled water. The next day, remove the tubing or bag. Collect the contents and measure the volume. Have students explain why the volume increased. Relate the results to the text description of why there is often a net movement of water into a cell.

Patterns of Change

How does osmosis enable a cell to respond to changes within an organism?

Theme Answer

Patterns of Change
Students should indicate that osmosis allows water to enter and remain in the cell to help it carry on various life functions. The ability of the cell to allow water to enter is necessary for its survival.

Demonstration 3

Turgor Pressure in *Elodea*

Use a microprojector to show students *Elodea* cells. Slowly draw a 5-percent salt solution under the coverslip with a small piece of lens paper. Ask students to explain their observations.

Demonstration 4

Turgor Pressure in Lettuce

Show students some wilted lettuce. Have them discuss how osmosis has affected its cells. Place the lettuce in cold water and ask students the next day to explain their observations.

Sports Connection

Thirst Quenchers

Point out that thirst quenchers not only replace the water lost after a tough work-out or athletic contest but also replenish the salt ions lost through perspiration.

Visual Strategy

Figure 4.6

Have students count the number of particles on each side of the membrane. They should recognize that diffusion would result in particles moving through the membrane from top to bottom until they are evenly dispersed. But since protein channels accelerate this process, this is known as facilitated diffusion. Emphasize that facilitated diffusion always moves substances from a region where they are in higher concentration to one where they are in lower concentration.

Theme Connection

Stability

Glucose is an important source of energy for cells and usually slowly diffuses into cells. However, when a cell's demand for energy is high, not enough glucose enters a cell simply by diffusion. In such cases, a cell uses facilitated diffusion to obtain the needed glucose more quickly.

Selective Transport of Substances

Unlike water molecules, many substances cannot easily pass through the cell membrane. For example, molecules such as sugars and proteins are often too large to slip through the gaps in the cell membrane. In addition, these molecules are often polar and cannot pass through the nonpolar region of the lipid bilayer. Polar molecules and ions, therefore, use channels made of proteins to move in or out of the cell. A protein channel is shown in **Figure 4.6b**.

The transport of substances through membrane protein channels is called **selective transport**. This form of transport is said to be "selective" because each kind of channel will allow only a particular type of molecule or ion to pass through the membrane. Thus, the cell can control the substances that enter and leave. There are two modes of selective transport: facilitated diffusion and active transport.

Facilitated diffusion works in two directions

Some channels are like open doors. As long as a molecule or ion fits into the channel, it is free to pass through in either direction. Each kind of molecule or ion diffuses toward the side where it is least concentrated. Eventually, diffusion balances the concentrations of that molecule or ion on both sides of the membrane. Because some channels assist, or facilitate, the transport of substances, this form of diffusion is called **facilitated diffusion**. Sugar molecules and ions such as chloride ions enter cells by facilitated diffusion.

Active transport moves molecules in one direction

If all channels worked like open doors, many molecules and ions would simply flood out by facilitated diffusion. But some channels work only in one

Figure 4.6

a   *Ahhh!* There's nothing like the taste of a good thirst quencher after a tough workout. Have you ever wondered how the cells of your body replenish nutrients?

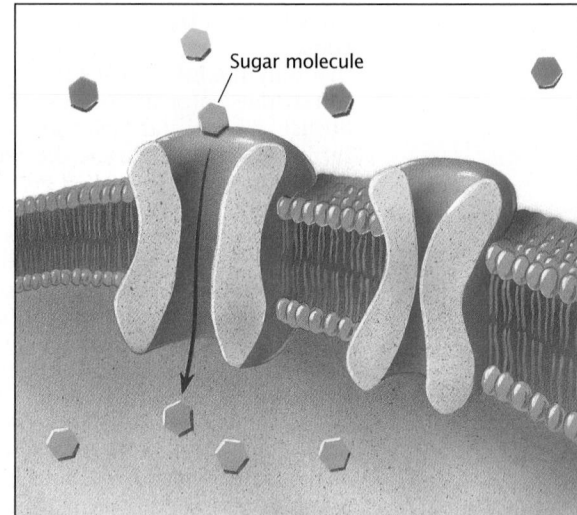

Sugar molecule

b   Your bloodstream carries minerals and sugars to your body's cells, which pull them in through protein channels.

direction. Like turnstiles at a subway station, these channels let certain molecules and ions into the cell, but do not let them out. Active transport enables a cell to stockpile certain substances in far greater concentrations than they occur outside the cell. Follow the steps in **Figure 4.7** to see how a channel can be one-way.

The cell must use some of its energy to change the shape of the channel protein and to move substances across the membrane to a region of higher concentration. The operation of these one-way channels, therefore, is called **active transport**. Active transport plays a critical role in acquiring sugars and other molecules, even when more of these molecules already exist inside the cell. Surprisingly, almost all the active transport in cells is carried out by only two kinds of channels: the sodium-potassium pump and the proton pump.

The sodium-potassium pump transports ions

The **sodium-potassium pump** is an active transport system that enables the cell to admit ions needed for important biological processes, such as the

conduction of nerve impulses through the body. Uniquely shaped receptor sites on protein channels enable the ions to move in one direction only.

As you can see in **Figure 4.7**, this mechanism works by actively pumping sodium ions out of cells and potassium ions into cells. In just one second, each channel can move more than 300 sodium ions out of the cell. More than one-third of all the energy expended by your body's cells is spent driving the sodium-potassium pump. When there are very few sodium ions in the cell, facilitated diffusion channels enable sodium ions to rush back into the cell.

There is one catch, however: The channels must be opened. Some are opened by electrical currents. Others are opened only when sodium ions are paired to partner molecules such as sugar or amino acids. Because so many sodium ions rush back in through these channels, large numbers of partner molecules are pulled through as well, even if they are already present in high concentrations within the cell. The protein channels that admit sodium ions and their partners are called coupled channels, as shown in **Figure 4.7d**.

Connection: Chapter 2

Protons
Review what a proton is.

Theme Connection

Energy and Life
The proton pump is an integral part of both photosynthesis and respiration, where it is used to synthesize adenosine triphosphate (ATP). ATP is the molecule that cells rely upon for their energy needs.

Visual Strategy

Figure 4.7
Students will have difficulty understanding how the sodium-potassium pump operates if they just read the text. Consequently, lead students step by step through each illustration. In Figure 4.7a, point out that the sodium ion is inside the cell. In Figure 4.7b, have students note that sodium is being actively pumped against the higher concentration outside the cell. In Figure 4.7c, the channel protein has changed its shape so that it can now pick up a potassium ion on the outside of the cell and actively pump it inside to a higher concentration. In Figure 4.7d, a sodium ion with an attached sugar molecule moves back into the cell by facilitated diffusion.

Figure 4.7
The sodium-potassium pump, an active transport system, actively pumps specific ions across the cell membrane in one direction.

a Sodium ions within the cell fit precisely into receptor sites on the channel protein.

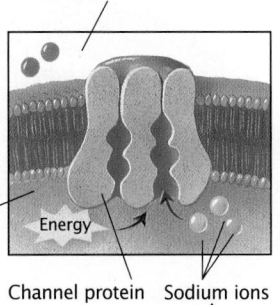

Outside the cell

Inside the cell

Energy

Channel protein   Sodium ions

Potassium ions

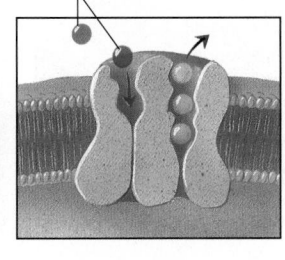

b The channel changes shape, pumping the sodium ions across the membrane. Potassium ions outside the cell move into receptor sites.

c The sodium ions are released and cannot reenter through this channel. At the same time, potassium ions are pumped across the channel into the cell.

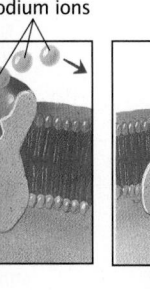

Potassium ions

Sodium ions

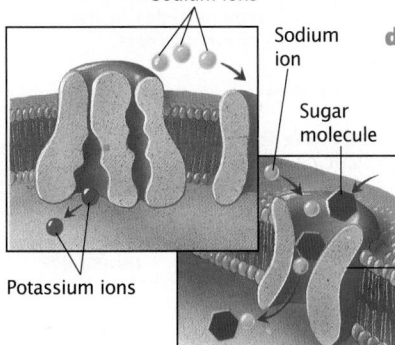

Sodium ion

Sugar molecule

Coupled channel

d Potassium ions enter the cell. Sodium ions outside the cell, along with sugar molecules, reenter the cell through a coupled channel.

Word Clues

Have students look up the meanings of endo-, exo-, and cyto-. Alert them to the fact that many biological structures and processes have names that provide clues to their meanings.

Demonstration 6

Endocytosis

Use a microprojector to show students how a *Didinium* can ingest a *Paramecium* by endocytosis.

Theme Answer

Energy and Life

Students should suggest that the cell could not survive. Without food, the cell would starve. Without a way for materials to exit, the cell would explode.

Phase 3

ASSESSMENT OPTIONS

Closure Strategy

Osmosis and Diffusion in Winter

During the winter, icy roads are sometimes salted. The plants lining such roads often do not survive. Ask students to explain why these plants die.

Section Review

Assign the *Section Review*.

Reteaching

Assign students to cooperative groups to make concept maps for this section.

The proton pump is another active transport mechanism

Although cells rely on the sodium-potassium pump to take in ions, they use a **proton pump** to pump protons across membranes. To function, this active transport system depends upon chemical or light energy. Just as the sodium-potassium pump causes a build-up of sodium ions outside the cell, the proton pump expels protons until large numbers of protons build up outside a membrane. Although the protons diffuse back in again, only certain channels are open to them.

The proton pump is the key to cell metabolism, as you will soon learn in Chapter 5. In green plants, proton pumps enable cells to convert light energy into sugars. Proton pumps also enable your cells to transform energy obtained from the food you eat into energy you can use. The mechanism of energy release by the action of proton pumps is called **chemiosmosis** (*kehm ee ahz MOH sihs*).

How Large Particles Get Into and Out of Cells

Energy and Life

What would happen to a cell if molecules or ions had no way to enter or exit?

Sometimes cells need to take in substances that are too big to pass through the protein channels in the cell membrane. For example, some eukaryotic cells take in food particles by extending their cell membranes out toward the particle and surrounding it. When the edges of the membrane meet and fuse together, the particle is captured within a sac inside the cell. The process of bringing particles into the cell by capturing them within a sac is called **endocytosis** (*ehn doh seye TOH sihs*), as shown in **Figure 4.8a**. The reverse process, ridding the cell of material by discharging it from sacs at the cell surface, is called **exocytosis** (*ehk soh seye TOH sihs*), as shown in **Figure 4.8b**.

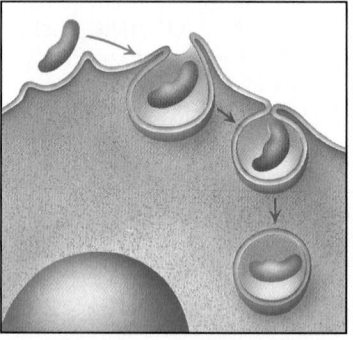

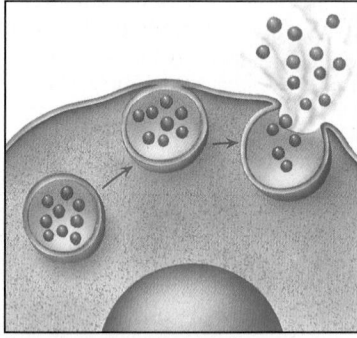

Figure 4.8

a **The cell membrane extends to engulf large particles in the process of endocytosis.**

b **Cellular wastes are discharged from sacs at the cell's surface in the process of exocytosis.**

Section Review

❶ How do diffusion and osmosis differ?

❷ What distinguishes facilitated diffusion from active transport?

❸ Name two active transport mechanisms that enable cells to store certain substances.

❹ Explain how a cell can take in substances too big to pass through the membrane.

▪ *Section Review Answers* ▪

1. Students should explain that in diffusion molecules move through the cell from an area of higher concentration to an area of lower concentration. In osmosis, water molecules diffuse across a cell membrane.

2. Facilitated diffusion always moves substances from an area of high concentration to an area of lower concentration. In active transport, protein channels accelerate the diffusion of substances through the cell membrane into areas of higher concentration.

3. The sodium potassium pump and the proton pump.

4. Students should describe the process of endocytosis.

AS CELLS BUSILY CARRY OUT THE FUNCTIONS OF LIFE, THEY GROW AND DEVELOP. YOU LEARNED IN CHAPTER 3 THAT CELLS OPERATE MOST EFFICIENTLY WHEN THEY HAVE A HIGH SURFACE AREA-TO-VOLUME RATIO. WHEN MOST CELLS REACH A CERTAIN SIZE, THEY EITHER STOP GROWING OR DIVIDE INTO TWO CELLS. CELL DIVISION IS ESSENTIAL FOR THE GROWTH AND DEVELOPMENT OF AN ORGANISM.

4.3   *How Cells Divide*

Objectives

❶ **Compare and contrast cell division in bacteria and eukaryotes.**

❷ **Specify the number of chromosomes in a cell before and after cell division.**

❸ **List the events of mitosis and cell division in the correct sequence.**

❹ **Compare and contrast cell division in animals and plants.**

How Bacteria Divide

When a healthy, living cell reaches a certain size, it will either stop growing or it will divide. This section will explain two types of cell division. First, you will read about how bacteria divide. Then you will learn about the more complex division of eukaryotic cells.

In bacteria, the process of cell division is fairly simple. Their hereditary information is encoded in a single circle of DNA, which is attached at one point to the inner surface of the cell membrane like a ring stuck to the inside of a balloon. Remember from Chapter 2 that DNA contains the instructions for making proteins and for carrying out the day-to-day activities of the cell. Each cell must have its own set of DNA instructions. Otherwise, it will have no way to make the proteins that are necessary for life. So, before a cell divides, it must first make a complete copy of its DNA so that each new cell has the same set of instructions.

In bacteria, cell division takes place in two stages. First the DNA is copied, and then the cell splits, as shown in **Figure 4.9**

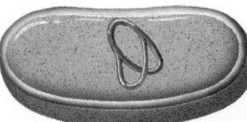

Figure 4.9
a **Starting at one point of its circle of DNA, this bacterium makes a copy of the molecule.**

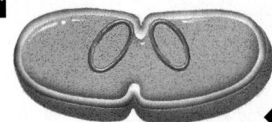

b **The two copies of hereditary information are attached to the cell membrane side-by-side.**

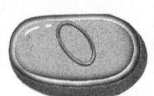

c **A new membrane and cell wall form between the DNA copies, gradually pinching inward.**

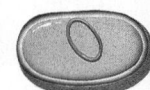

d **Eventually the cell is split into two cells. Each cell contains a circle of DNA, and is a distinct living bacterium.**

Lesson Plan 4.3

Phase 1
PREPARATION

Key Concepts
- In cell division in prokaryotes, the DNA is first copied. A complete set of DNA is passed into each of the two daughter cells that are produced when the original cell divides.
- In cell division in eukaryotes, the chromosomes are first copied. A complete set of chromosomes is passed to each of the two daughter cells that are produced when the original cell divides.
- Cells appear to be programmed to undergo only so many cell divisions.
- Cells that continue to divide uncontrollably lead to cancer.

Reading Strategy

Illustrations in texts are often paid little attention or even ignored. Have students recognize their importance, especially in science texts, by reading the descriptive material in all the figures before reading the text.

Phase 2
TEACHING STRATEGIES

Visual Strategy

Figure 4.9
Use this figure to emphasize that the two daughter cells contain the same genetic information as the parent cell. Have students compare the circle of DNA in each of the two daughter cells in Figure 4.9d to the single circle of DNA in the parent cell in Figure 4.9a. Ask whether a daughter cell or the parent cell has the higher surface area-to-volume ratio.

Prokaryotes vs. Eukaryotes
Students should quickly review the differences between prokaryotic and eukaryotic cells.

Demonstration 1

Copying the DNA Instructions
To illustrate the importance of making a copy of the entire DNA before the cell divides, show students a blueprint. Have students identify what information it provides. Then rip the blueprint in half. Have students explain how this would affect the builder. Have them compare this to a cell that would only get half the DNA if it did not make a complete copy prior to cell division.

Theme Connection

Patterns of Change

Students should recognize that cell division (mitosis) in prokaryotes and eukaryotes, as discussed in this section, does not promote change in the genetic instructions that get passed on to daughter cells.

Demonstration 2

Chromosomes
Project either a slide or overhead transparency of a cell whose chromosomes are clearly visible. Have students explain why the process of mitosis derives its name from *mitos*, the Greek word for thread.

Theme Answer

Evolution
Eukaryotes have chromosomes and therefore need a more complex mechanism of cell division to ensure that each cell has the same genetic information.

Evolution

Why do you think a more complex type of cell division evolved among eukaryotes?

How Eukaryotic Cells Divide

Eukaryotic cells carry far more DNA than bacteria. During cell division, the DNA in eukaryotic cells is packaged into tightly wound structures that are called **chromosomes**, as shown in **Figure 4.10**. Cell division plays a major role in the development of eukaryotic organisms. Just by looking at your hand, you can imagine how many millions of cells must have divided to form each crease on every finger. The magnified hand of the human fetus shown in **Figure 4.11b** highlights the importance of eukaryotic cell division.

A typical human cell contains 46 chromosomes. Instead of being attached to the cell membrane as with bacterial DNA, eukaryotic chromosomes are contained within the nucleus. Because eukaryotic cells have more DNA than do prokaryotic cells, and because the DNA is confined within a nucleus, eukaryotic cell division is more complex than bacterial cell division. First, each chromosome must be copied exactly. Next, the chromosomes must be sorted out precisely so that each new cell gets a complete set. Finally, the cell itself divides in half.

Two processes have evolved that enable eukaryotic cells to divide successfully: mitosis and cell division. **Mitosis** (*meye TOH sihs*) is the process by which the nucleus of a eukaryotic cell divides into two nuclei, each containing a complete set of the cell's chromosomes. Mitosis can be broken down into four distinct phases: prophase, metaphase, anaphase, and telophase. These phases are summarized in **Figure 4.12**.

In many cells, mitosis is followed by cell division, also called **cytokinesis** (*syt oh kuh NEE sihs*). During cell division, the cell divides into two cells, each with its own nucleus. Plant cells—which have strong cell walls that cannot be pinched like the cell membranes of animal cells—form a new cell wall in the center of the cell. This new cell wall divides the cell in half, like a partition dividing a room.

The end result of mitosis and cell division is two cells with the same genetic information, where only one cell existed before. To follow the events of eukaryotic cell division in an animal cell, see the steps shown in **Figure 4.12**.

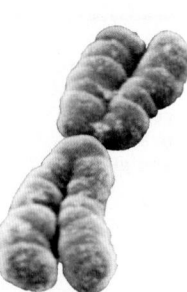

Figure 4.10
A chromosome and its copy may look something like an "X" because they generally are connected near the middle.

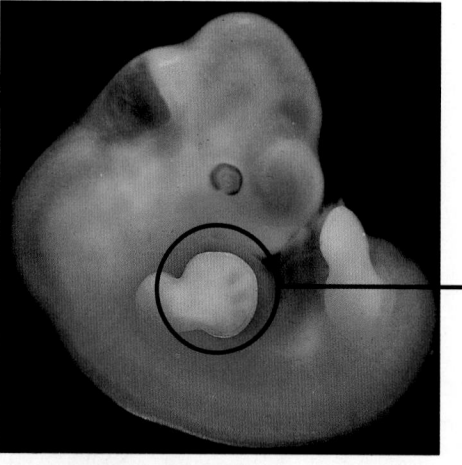
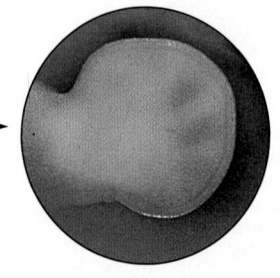
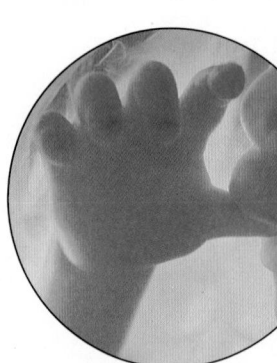

Figure 4.11
a   **Cell division plays an essential role in the growth of an organism.**

b   **For example, the hand of this human fetus is made of millions of cells. Some cells die, while others continue to divide until . . .**

c   **. . . the hand is fully formed. After birth, skin cells will keep dividing to replace those that wear out.**

■ *Cultural Perspective* ■

**Dr. Ernest Just
(African-American)**
Dr. Ernest Just was a pioneer in the study of embryology. He was a zoologist, a cell biologist, and one of the early researchers of egg fertilization, artificial parthenogenesis, and cell division. His major interests were fertilization and development of marine animal eggs. He was among the first to reveal the secrets of how cells reproduce. Dr. Just's writings on "The Biology of the Cell Surface" were published in 1939.

A Generalized Picture of Mitosis and Cell Division

Figure 4.12

a   *DNA replication*
This animal cell is ready to divide. Its chromosomes are not yet visible because they are extended and uncoiled. The DNA of each chromosome is copied. Each chromosome consists of two identical strands.

Nucleus

Cell membrane

b   *Prophase*
Mitosis begins. The chromosomes coil into short, fat rods. The nuclear envelope breaks up. A network of protein cables, called spindle fibers, assembles across the cell.

f   *Cytokinesis*
The cytoplasm is pinched in half to form two new cells. Each new cell contains identical DNA. After growth and DNA replication, these cells may divide again.

Nuclear envelope

Spindle fibers

Chromosomes

e   *Telophase*
Each side of the cell now has a complete set of chromosomes. A nuclear envelope surrounds each new set of chromosomes. The chromosomes uncoil so that proteins can be made. The spindle fibers disappear.

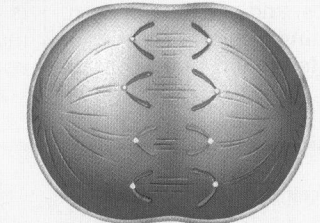

c   *Metaphase*
Each chromosome, attached to spindle fibers, lines up in the center of the cell.

d   *Anaphase*
Each chromosome separates from its identical copy. Chromosomes are reeled to opposite sides of the cell. The spindle fibers start to break down.

▪ *Matter of Fact* ▪

Life Spans
Unlike some cancer cells, which under the right conditions appear capable of living forever, no animal lives forever. Some life expectancies are mouse, 2–3 years; rabbit, 12; lion, 25; horse, 30; ostrich, 50; elephant, 60; human, 74; and tortoise, 100.

Art Connection

Continuity of Mitosis
By focusing on the individual stages, students often lose sight of the continuous nature of mitotic activity. Play a videotape of a movie, periodically pausing the action to isolate a single frame. Point out that each stage in mitosis is like a single frame in the movie, but that the whole process occurs continuously, just like a movie.

Visual Strategy

Figure 4.12
Project a slide of cells undergoing mitosis in an onion root tip. Have students use this figure to identify a cell in each of the six stages shown in the figure. Have them recognize that all the stages are not of equal length, as evidenced by the different number of cells in each stage. Ask them to identify which stage lasts the longest (contains the most cells). Also use the slide to reinforce the differences between mitosis in animal cells (as shown in Figure 4.12) with that in plant cells (as seen on the slide).

Theme Connection

Stability
Most cells are specialized, each performing a distinct function. Not all these specialized cells have the same ability to undergo mitosis. For example, human skin cells continue to divide throughout life to replace those lost by wear and tear. On the other hand, nerve cells stop dividing after the first few months of a baby's life. Although nerve cells are the most stable and have the longest life span, they cannot be replaced if they are damaged or destroyed. Other life spans include red blood cells, 120 days; stomach cells, 2 days; and sperm and egg cells, 2 days.

Health Connection

Cancer

Cancers can be either benign or malignant. Explain that benign tumors do not invade other body tissues. Malignant tumors do. Benign tumors that are not surgically removed may continue to grow. Large cancer growths can kill a person by interfering with an organ's normal function.

Phase 3

ASSESSMENT OPTIONS

Closure Strategy

Mistakes in Cell Division

Ask students to infer what might happen to a daughter cell that does not receive a complete copy of DNA or a full set of chromosomes.

Section Review

Assign the *Section Review.*

Reteaching

Assign students to cooperative groups of four. Provide each group with colored "pipe cleaners." Ask them to demonstrate what happens during mitosis in a cell containing five chromosomes, each represented by a different color.

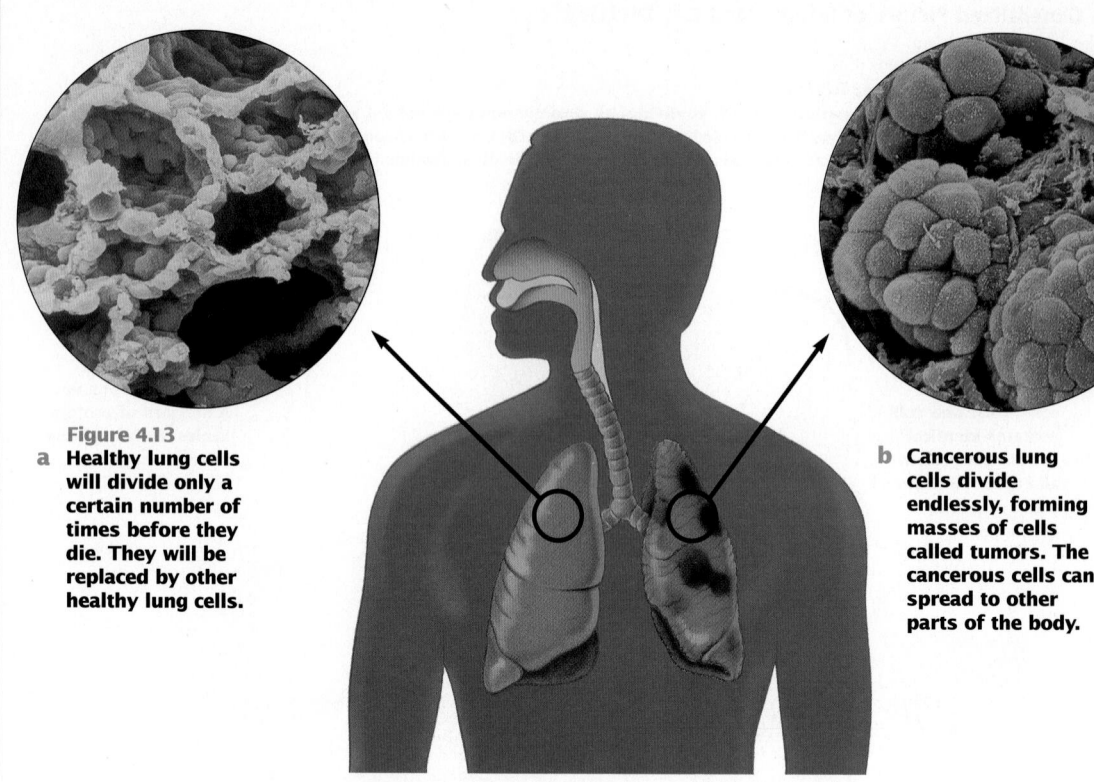

Figure 4.13

a **Healthy lung cells will divide only a certain number of times before they die. They will be replaced by other healthy lung cells.**

b **Cancerous lung cells divide endlessly, forming masses of cells called tumors. The cancerous cells can spread to other parts of the body.**

Life Span of Cells

No cell lives forever. Human cells, for example, appear to be programmed to undergo only so many cell divisions and then die, as if following a plan written into the genes. When human cells are grown in the laboratory, they divide about 50 times and then die. Even cells that are frozen for years under laboratory conditions, and are then thawed, die after reaching a certain number of divisions. Cells in your body contain hidden "hourglasses" whose grains of sand are cell divisions. When the "sand" runs out, the cells die. On rare occasions, some cells, such as those in **Figure 4.13b**, appear to disobey these instructions. **Cancer** is a disease in which cells grow and divide at an abnormally high rate. If their growth is stopped using various therapies, many cancers can be prevented from spreading.

Section Review

① **How does cell division differ between bacteria and eukaryotes?**

② **How does the number of chromosomes in a cell differ before and after cell division?**

③ **Summarize the events of mitosis and cell division in eukaryotes.**

④ **How does plant cell division differ from animal cell division?**

▪ *Section Review Answers* ▪

1. In bacteria, first the DNA is copied, then the cell divides. In eukaryotes, cell division is more complex. After the chromosomes are copied they must undergo mitosis before cytokinesis can occur.

2. Two cells with the same number of chromosomes exist where only one cell existed before.

3. See Figure 4.12 on page 81.

4. Plant cells cannot pinch the cytoplasm in half as animal cells do. Therefore, they form a new wall during cytokinesis.

Chapter **4**

Highlights

How do you smell perfume? Perfume molecules diffuse through the air, just like ink diffuses through water (p. 74).

	Key Terms	*Summary*
4.1 How Cells Receive Information 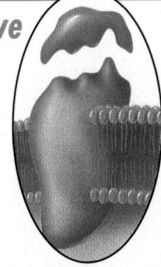 Insulin joins with a receptor protein to send chemical messages into the cell.	**voltage-sensitive channel** (p. 69) **receptor protein** (p. 70) **cell surface marker** (p. 71)	• Many proteins embedded in the cell membrane transmit information into cells. • Some proteins are cell surface markers that identify cells. Markers are unique to each individual.
4.2 Moving Into and Out of Cells 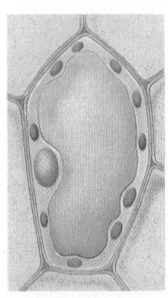 A strong cell wall prevents this plant cell from being burst by osmotic pressure.	**diffusion** (p. 74) **osmosis** (p. 75) **osmotic pressure** (p. 75) **selective transport** (p. 76) **facilitated diffusion** (p. 76) **active transport** (p. 77) **sodium-potassium pump** (p. 77) **proton pump** (p. 78) **chemiosmosis** (p. 78) **endocytosis** (p. 78) **exocytosis** (p. 78)	• Homeostasis is maintained in cells by the cell membrane, which controls movement of substances into and out of cells. • Some substances pass through the cell membrane by diffusion, the movement of molecules from a higher to lower concentration. • Water enters or leaves a cell by osmosis, the diffusion of water across a cell membrane. • Selective transport and facilitated diffusion do not require energy to move substances into cells. Active transport requires energy, and can move substances to a region of higher concentration. • Larger particles are moved into or out of the cell by endocytosis and exocytosis.
4.3 How Cells Divide 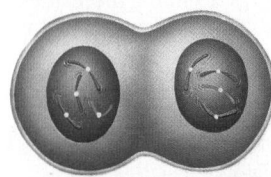 When cells reach a certain size, they either stop growing or divide. This cell is in the final stage of mitosis and will soon divide into two cells, each with identical DNA.	**chromosome** (p. 80) **mitosis** (p. 80) **cytokinesis** (p. 80) **cancer** (p. 82)	• Eukaryotic cell division consists of two processes: mitosis, or division of the nucleus, and cytokinesis, division of the cytoplasm. • In mitosis, the nucleus of a cell is divided into two nuclei, each with the same number of chromosomes as the parent cell. • In cytokinesis, the cytoplasm divides to form two distinct cells. • Most cells undergo a certain number of divisions and then die. Cancer cells, however, continuously grow and divide at an abnormally high rate.

Chapter Review Answers

Understanding Vocabulary

1. **a.** Diffusion is the random movement of molecules from an area of greater concentration to an area of lesser concentration; osmosis is the diffusion of water across the cell membrane.
   b. The process by which a eukaryotic cell's nucleus divides is mitosis; cell division follows mitosis and involves the splitting of the cell containing the two newly formed nuclei into two cells.
   c. Facilitated diffusion involves the use of specialized protein channels to transport substances across the cell membrane without the use of energy from the cell; active transport requires energy to transport molecules inside the cell, where these molecules are in higher concentration.
   d. Endocytosis is the process of transporting material into a cell by using a vesicle; exocytosis is the reverse process, whereby material exits the cell.

Relating Concepts

2. Map answer is shown on page 67D.

Understanding Concepts
Multiple Choice

3.	a	8.	a
4.	b	9.	c
5.	b	10.	b
6.	b	11.	b
7.	c	12.	a

Completion
13. plants
14. osmotic pressure
15. proton pump

Short Answer
16. Surface markers enable the body's immune system to distinguish body cells from harmful invaders and direct cells in organizing to form tissues and organs.
17. The fluids surrounding animal cells have the same concentration of dissolved substances as the fluids inside the cells.

Chapter 4 Review

Understanding Vocabulary

1. For each pair of terms, explain the differences in their meanings.
   a. diffusion, osmosis
   b. cell division, mitosis
   c. facilitated diffusion, active transport
   d. endocytosis, exocytosis

Relating Concepts

2. Copy the unfinished concept map below onto a sheet of paper. Then complete the concept map by writing the correct word or phrase in each oval containing a question mark.

Understanding Concepts
Multiple Choice

3. Voltage-sensitive channels help cells communicate by
   a. electrical signals.
   b. chemical stimuli.
   c. binding to molecules.
   d. insulin molecules.

4. Cells sense chemical signals by using
   a. hormones.
   b. receptor proteins.
   c. signaling messengers.
   d. surface markers.

5. The process by which water moves into and out of the cell is
   a. facilitated diffusion.
   b. osmosis.
   c. active transport.
   d. diffusion.

6. A cell uses some of its energy to move molecules by
   a. osmotic pressure.
   b. active transport.
   c. diffusion.
   d. osmosis.

7. The sodium-potassium pump
   a. requires no energy.
   b. moves potassium out of the cell.
   c. enables sugars to enter cells.
   d. works independently of channels.

8. Which is an example of active transport?
   a. sodium-potassium pump
   b. electron pump
   c. endocytosis
   d. facilitated diffusion

9. Particles too large to pass through protein channels in the cell membrane may enter the cell by
   a. exocytosis.
   b. selective transport.
   c. endocytosis.
   d. osmotic pressure.

10. During mitosis
    a. chromosomes are copied.
    b. chromosomes move to opposite sides of the cell.
    c. cytoplasm divides in half.
    d. a new cell wall forms in the center of the cell.

11. If a cell has 8 chromosomes before cell division, how many chromosomes will each of the two new cells have at the end of cell division?
    a. 16
    b. 8
    c. 4
    d. 32

12. During cell division in bacteria
    a. a circle of DNA is copied.
    b. chromosomes coil and move.
    c. the cell splits into three parts.
    d. two new nuclei are formed.

Completion
13. After mitosis, the cytoplasm of cells is pinched in half. In cells of _____ , cell walls are formed.

14. If placed in water, a cell would swell and possibly burst due to _____ .

15. A type of active transport system that uses light or chemical energy to move molecules that are required for cell metabolism is called the _____ .

Short Answer
16. What two helpful jobs are performed by the surface markers on cells?

17. Explain why animal cells do not burst due to osmotic pressure.

18. Describe the events that result in the formation of two complete nuclei in a eukaryotic cell.

Concept map:

Homeostasis
depends on the
?
that moves materials in and out of the cell using
? — of water is called — osmosis — such as those used in — facilitated diffusion
protein channels — may be — ? — such as the — ?
? — may be — one-way
? — enables cells to take in — large particles

---

18. Chromosomes are copied and then coiled into short, fat rods. Next, the nuclear envelope breaks and a network of spindle fibers is assembled. Chromosomes are separated from their partners and pulled to opposite ends of the cell by the spindle fibers. Upon arrival at opposite ends of the cell, each set of chromosomes uncoils and is surrounded by a nuclear membrane. Cell division soon follows.

Interpreting Graphics
19. c, e, b, a, d
20. Water molecules slip through gaps in the lipid bilayer. Potassium ions move across the membrane via the sodium-potassium pump. Sodium ions and sugar molecules enter the cell through coupled channels.

Interpreting Graphics

19. Look at the drawings of the five cells at different stages of cell division. Write the letters identifying each cell in the order that indicates the correct sequence of the events of cell division.

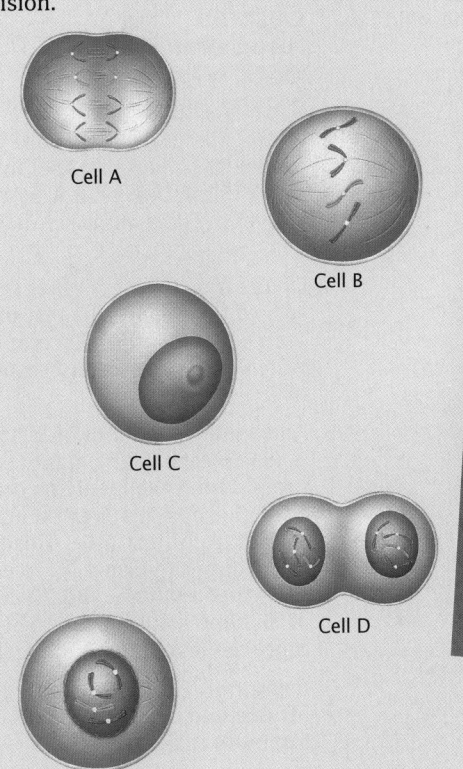

Cell A

Cell B

Cell C

Cell D

Cell E

20. Looking at the drawing below, explain how each molecule passed through the cell membrane.

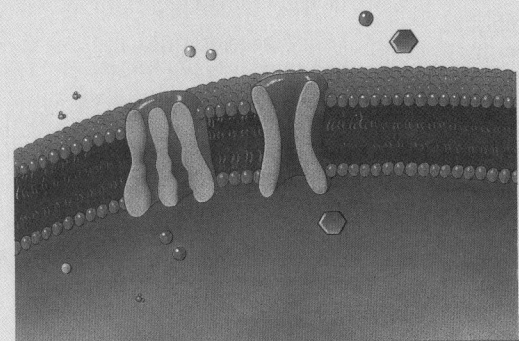

Reviewing Themes

21. *Interacting Systems*
How do cell surface markers affect the success of organ transplants? What can be done to increase the success of organ transplants?

22. *Patterns of Change*
How does mitosis promote genetic consistency?

23. *Evolution*
Bacteria simply split, but eukaryotic cells reproduce by mitosis. How has the evolution of mitosis in eukaryotic cells aided their reproduction?

24. *Energy and Life*
How is the cell's energy used in an active transport system like the sodium-potassium pump?

Thinking Critically

25. *Inferring Conclusions*
On the basis of your understanding of osmosis, explain what would happen to a marine clam placed in a freshwater aquarium.

26. *Building on What You Have Learned*
How does the structure of water enable it to interact with sugars, proteins, and other polar molecules within the cell?

Cross-Discipline Connection

27. *Biology and Drama*
William Shakespeare said that we are all actors and the world is our stage. With two students, write a skit showing the events of mitosis and cell division. Then, act out your play and judge its accuracy and drama.

Discovering Through Reading

28. Read "A Herpes Key," in *Discover*, November 1990, page 22. How does the herpes virus infect cells? What effect does covering cells with fibroblast growth-factor have on the rate of infection?

Thinking Critically

25. Water would diffuse into the cells of the clam because the concentration of water is greater on the outside of the cells. The cells of the clam would swell and eventually burst.

26. Because of their polarity, water molecules have the ability to attract and form strong bonds with polar molecules, such as sugars, proteins, and other molecules in the cell.

Cross-Discipline Connection

27. The play should clearly illustrate the chromosomes being copied and coiling into short, fat rods; the breaking apart of the nuclear membrane; chromosomes aligning themselves in the center of the cell and being pulled to opposite sides by spindle fibers; the assembling of nuclear membranes around the two sets of chromosomes; and cytoplasmic division with attention paid to differences between plant and animal cells.

Discovering Through Reading

28. The virus binds with a growth-factor receptor protruding from the cell membrane. Covering the cells with the protein reduced the infection rate by about 70 percent.

Reviewing Themes

21. Surface markers help the body's immune system distinguish the cells of one individual from those of another. Surface marker differences cause organs from a donor to be rejected by the recipient. The likelihood of a successful transplant may be increased by using drugs to suppress the immune system.

22. The two new cells resulting from mitosis have chromosomes that are identical to one another and identical to the original cell.

23. Mitosis increases the likelihood that both new cells will get complete sets of chromosomes.

24. Energy is used to pump potassium ions into the cell and sodium ions out.

Procedural Note

Review the procedures used to test for the presence of starch and sugar. A positive Benedict's test should show a brick-red precipitate. (In dilute solutions, an orange precipitate may be observed.) A positive Lugol's test will show a blue-black color. Caution the students to use only clean glasswear to avoid erroneous results.

Prelab Preparation Answers

1. Starch is a chain of monosaccharides, making it a much larger molecule than a simple sugar.
2. Answers may vary but should discuss lipid permeability, molecular size, and the use of carrier molecules as factors involved in selective permeability.

Procedure Answer

8. Rinsing the outside of each bag cleans off any debris that may react with the indicator and lead to false results.

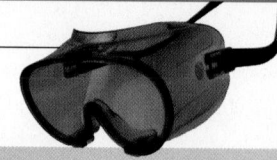

Chapter **4** Investigation

Can All Molecules Diffuse Through a Membrane?

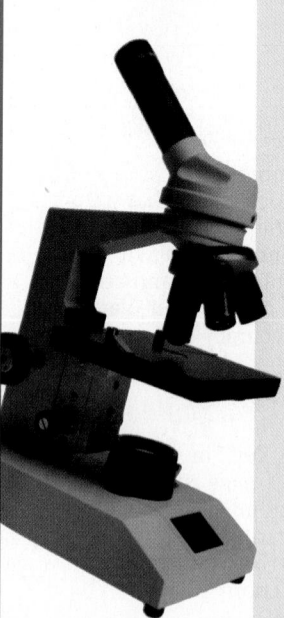

Objectives

In this investigation you will:
- *demonstrate* the movement of molecules through a selectively permeable membrane
- *perform* chemical tests for the presence of starches and sugars
- *collect* and *interpret* data

Materials

- lab apron
- safety goggles
- disposable gloves
- wax pencil
- two 250-mL beakers
- 8 test tubes
- water
- medicine dropper
- starch solution
- enzyme solution
- test-tube rack
- two 15-cm pieces of dialysis tubing
- heavy-duty thread
- rubber gloves
- Lugol's iodine solution
- Benedict's solution
- hot plate
- 400-mL beaker water bath

Prelab Preparation

Review what you have learned about sugars, starches, and diffusion by answering the following questions:
- How does starch differ from a simple sugar?
- How does a cell membrane control the passage of molecules into and out of a cell?

Procedure: Investigating Diffusion

1. Form a cooperative group of four students. Work with a member of your team to complete steps 2–16.

2. **CAUTION: Put on a lab apron, safety goggles, and disposable gloves.**

3. With a wax pencil, label one 250-mL beaker "B-1: Starch." Label the second 250-mL beaker "B-2: Starch and Enzyme." Fill each beaker half way with water.

4. Divide the eight test tubes into two sets of four. Label one test tube in each set as "Starch," "Enzyme," "B-1: Water," and "B-2: Water," respectively.

5. Add water to a depth of 2 cm (1 in.) to each of the eight test tubes. Using the medicine dropper, add 20 drops of starch solution to each "Starch" test tube. Rinse out the dropper thoroughly. To each "Enzyme" test tube add 20 drops of enzyme solution. Set the test tubes aside in a test-tube rack.

6. Using heavy-duty thread, tightly tie off one end of each piece of dialysis tubing to make two "bags."

7. Pour starch solution into the first bag until it is two-thirds full. Tightly tie the bag's opening with thread. Rinse the outside of the bag and place it in the "B-1: Starch" beaker.

8. Pour starch solution into the second bag until it is two-thirds full. Using a clean medicine dropper, add 20 drops of enzyme solution into the bag. Tie it tightly with thread. Rinse the outside of the bag and place it in the "B-2: Starch and Enzyme" beaker. *What is the purpose of rinsing the outside of each bag before continuing?*

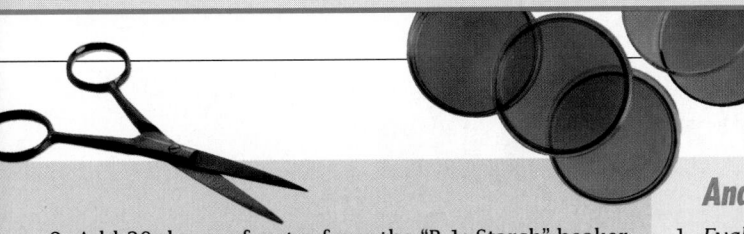

9. Add 20 drops of water from the "B-1: Starch" beaker to each test tube labeled "B-1: Water." Similarly, add 20 drops of water from the "B-2: Starch and Enzyme" beaker to each test tube labeled "B-2: Water."

10. Set the beakers aside for 15 minutes. Meanwhile, copy the table below and proceed to step 11.

		Presence of starch	Presence of sugar
Initial test (steps 4–13)	Starch solution		
	Enzyme solution		
	B-1 Water		
	B-2 Water		
Final test (step 15)	B-1 Water		
	B-2 Water		

11. **CAUTION: Lugol's iodine solution is poisonous; avoid skin/eye contact.**
Using one set of four test tubes, add two drops of Lugol's iodine solution to each test tube. If the liquid turns dark blue, starch is present. Record the results in your table.

12. **CAUTION: Use care when working with the hot plate and hot liquids.**
Fill a 400-mL beaker half way with water. Place it on a hot plate and heat until boiling. Heat the second set of test tubes in the boiling water bath for 5 minutes.

13. **CAUTION: Benedict's solution is an irritant; avoid skin/eye contact.**
Add 10 drops of Benedict's solution to each test tube. If the liquid turns orange, sugar is present. Record your results.

14. Empty and clean all the test tubes before continuing with the investigation.

15. After 15 minutes, use two test tubes to separately test the water in beaker B-1 for the presence of starch and sugar. Use another two test tubes to test the water in beaker B-2 for the presence of starch and sugar. Record your results.

16. Clean up your materials and wash your hands before leaving the lab.

Analysis

1. *Evaluating Methods*
Why was it necessary to make initial tests of the starch solution, the enzyme solution, and the water in each beaker?

2. *Analyzing Data*
Describe the contents of each dialysis bag, and discuss the evidence that supports your description.

3. *Analyzing Data*
Was starch or sugar present in the water of either beaker immediately after each bag was placed in the water? What evidence supports your answer?

4. *Analyzing Data*
What do your results suggest about the contents of the water in beakers B-1 and B-2 at the end of the investigation?

5. *Making Inferences*
Explain how the dialysis tubing demonstrates selective permeability.

6. *Making Inferences*
Why are some molecules unable to pass through a membrane while others move through freely?

Thinking Critically

Why would leaving out any of the procedural steps in this investigation affect your ability to draw valid conclusions from your results?

Analysis Answers

1. The initial tests were made to confirm that sugar was not present in the samples and that starch was present only in the starch solution.

2. Answers should reflect that the test results indicate only the presence of starch in both bags at the beginning of the experiment. At the conclusion of the experiment, a test of the water in the beaker with the bag containing starch and enzyme reveals the presence of sugar.

3. Answers should reflect that the test results indicate the absence of both starch and sugar in both samples.

4. Answers may vary but should reflect that the test results indicate the presence of sugar only in the B-1 sample and the absence of starch in both samples.

5. Answers should relate the selective permeability of the membrane to the passage of sugar but not starch.

6. Answers may vary but should suggest that the size of pores or spaces in the membrane could be one factor that contributes to selective permeability.

Thinking Critically Answer

Any alteration in the procedural steps may lead to erroneous test results.

Chapter 5 — Energy and Life

Planning Guide

	Objectives/Themes	Classwork Resources	Homework Resources

5.1	1. Distinguish between energy-storing and energy-releasing chemical reactions.
	2. Define activation energy and describe its role in chemical reactions.
	3. Describe the interaction between an enzyme and its substrate.
	4. Describe the role of enzymes in cells.
	Themes: Energy and Life, Interacting Systems, Scale and Structure
		Teacher's Resource Binder
		Focus Activity 5
		  Interpreting Labels: Stored Food Energy
		Lab Investigation 5.2
		  Release of Energy
		Other Resources
		Science in Action Video
		  Photosynthesis and Cellular Respiration*
			Text
			Section Review, p. 93
			Teacher's Resource Binder
			Directed Reading Worksheet 5.1
			Other Resources
			Audiocassette 5.1*

5.2	1. Explain how cells use energy. 2. Describe what is meant by the term biochemical energy. 3. Describe the role played by ATP in the cells. 4. Recognize the path of energy between plants and animals in the living world. **Themes:** Energy and Life, Interacting Systems	**Other Resources** Transparency 17	**Text** Section Review, p. 96 **Teacher's Resource Binder** Directed Reading Worksheet 5.2 **Other Resources** Audiocassette 5.2*

5.3	1. Summarize the evolution of photosynthesis.
	2. Describe how green plants and algae capture energy from sunlight.
	3. Explain how a plant cell uses light to make ATP and NADPH.
	4. Describe how photosynthesis provides the energy needed by all living things.
	Themes: Energy and Life, Interacting Systems, Evolution, Patterns of Change
		Teacher's Resource Binder
		Extension Worksheet
		  The Absorption of Light by Photosynthetic Pigments
		Lab Investigation 5.1
		  Plant/Animal Interrelationships
		Other Resources
		Transparencies 18–20
			Text
			Section Review, p. 101
			Teacher's Resource Binder
			Directed Reading Worksheet 5.3
			Other Resources
			Audio Review Tape 5.3*

5.4	1. Explain the importance of cellular respiration in living things.
	2. Summarize the process of glycolysis.
	3. Contrast fermentation with oxidative respiration.
	4. State the role of oxygen in oxidative respiration.
	Themes: Patterns of Change, Energy and Life
		Text
		Investigation
		  How Does Temperature Affect the Rate of Photosynthesis? pp. 110–111
		Discoveries in Science
		  What Is Life? Searching for Answers pp. 112–113
		Other Resources
		Transparency 21
			Text
			Section Review, p. 106
			Teacher's Resource Binder
			Directed Reading Worksheet 5.4
			Vocabulary Review Worksheet*
			Reteaching Worksheet*
			  Photosynthesis and Respiration
			Other Resources
			Audiocassette 5.4*

*Reteaching Options

Demonstrations
5.1: pp. 89, 90, 92, 93
5.2: pp. 94, 95
5.3: pp. 98, 99, 101
5.4: pp. 103, 105

Assessment
Chapter Review pp. 108–109
Portfolio Assessment p. 87D
Chapter Test—Teacher's Resource Binder
Test Generator

Research Notes

Connection to Computer Science: Designing Catalysts

For many years, scientists have been using enzymes from living things (and other catalysts) to speed up chemical reactions, thereby improving industrial production and purity. However, enzymes are often delicate chemicals that require a very narrow range of conditions to work. As a result, the disadvantages of using them to increase efficiency can often outweigh their advantages.

Since the job that enzymes can do is determined by their shape, much work has focused on under-

standing the shapes of existing enzymes. For examples, many drugs are designed to bind to specific enzymes so that they won't catalyze reactions in cells. These drugs are designed to have shapes complementary to the enzyme.

Some researchers are adopting a new approach, wherein they concentrate on the shape of the substrate they want to work with in order to try to design a sturdy enzyme-like catalyst that can speed reactions and still be able to withstand a broad range of conditions.

Before beginning the design phase, researchers at Sandia National Laboratories loaded some principles describing how different chemical groups affect the bending, stretching, and rigidity of a set of catalysts called metalloporphyrins. Metalloporphyrins are found in most living things, where they are catalysts in the metabolic processes that produce energy from food.

The scientists designed new metalloporphyrins with different

chemical groups attached to the molecules' frame, and the computer used the principles to predict the shape of the new compounds. By adjusting the subgroups attached to the metalloporphyrin molecule and their locations, they eventually created a catalyst whose predicted shape perfectly "fit" a carbon dioxide molecule.

Scientists made samples of the new catalyst molecules. X-ray crystallography evidence showed a close match between the actual structure of the catalyst and the computer's predictions. Tests continue on measuring how well the new catalyst works.

More catalysts are being designed. The metalloporphyrin that fits carbon dioxide could be useful in trying to control the large amounts of carbon dioxide in the atmosphere that cause global warming. A test run of another "designer" catalyst showed that it could break down poisonous contaminants in water about 100 times faster than a more traditional decontamination system.

Investigation Notes

How Does Temperature Affect the Rate of Photosynthesis?

pages 110–111

Purpose: Students formulate and test a hypothesis in this investigation.

Technique: Have the students identify the variables that must be controlled in this investigation. Such factors as the distance between the light source and the plant, the amount of water in the tube, and the concentration of carbon dioxide in the water might be mentioned.

Next, the students consider the variables that are controlled by using the same plant for both tests. Their discussion should include the number of leaflets on

the plant, the plant's length, its mass, its volume, and its general health.

Finally, have the students discuss the types of variables that might change as a result of using the same plant for both tests. These might include changes in the plant as a result of the initial test and changes in the concentration of carbon dioxide as the water sits in the beaker.

Prelab Preparation

Elodea can be purchased at most pet stores or obtained from WARD'S (86 M 7500). It can be kept alive indefinitely in an aquarium if given adequate illumination.

Answers will be found on pages 110–111.

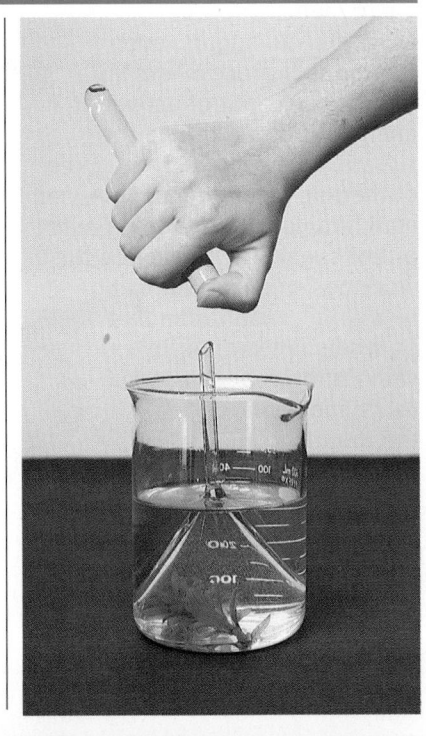

Meeting Individual Needs

Objectives

1. Students will demonstrate appropriate use of core vocabulary for the chapter (Vocabulary File).

2. Students will recognize an example of an enzymatic reaction (Demonstration A).

3. Students will model the structure and chemical properties of ATP (Demonstration B).

Vocabulary File
(Developing Vocabulary/ Limited English Proficiency)

If you are not already using the Vocabulary File, refer to Chapter 1 for its preparation. See the Chapter Highlights on page 107 for suggested words.

Demonstration A
Cells and Chemistry
for pages 89–91
(Developing Observational Skills/Multisensory Learners)

Materials: Irish potatoes (raw), knife, and paper towels

Procedure: Have students follow the arrows in Figure 5.1b that show an exothermic reaction similar to the one in which starch molecules release energy when they are changed into sugars, even though this process initially requires an input of activation energy. Contrast this to the situation in Figure 5.2b in an endothermic reaction such as when starch molecules are being made out of sugar. Then describe the workings of enzymes, using Figure 5.3. Emphasize that enzymes are present in saliva and produce chemical changes on food in the mouth.

Peel and cut raw Irish potatoes into chunks. Distribute and instruct students to chew slowly until their mouths become full of chewed potato and saliva. Caution students not to swallow. After a few minutes a sweet taste should be detected. Have students discard chewed potatoes into paper towel and dispose of them properly. Brainstorm with students about the chemical reactions that might have been happening.

Demonstration B
Energy and Life
for pages 94–95
(Developing Modeling Skills/Kinesthetic Learners)

Materials: three students

Preparation: Have three students stand in front of the class with arms to their sides.

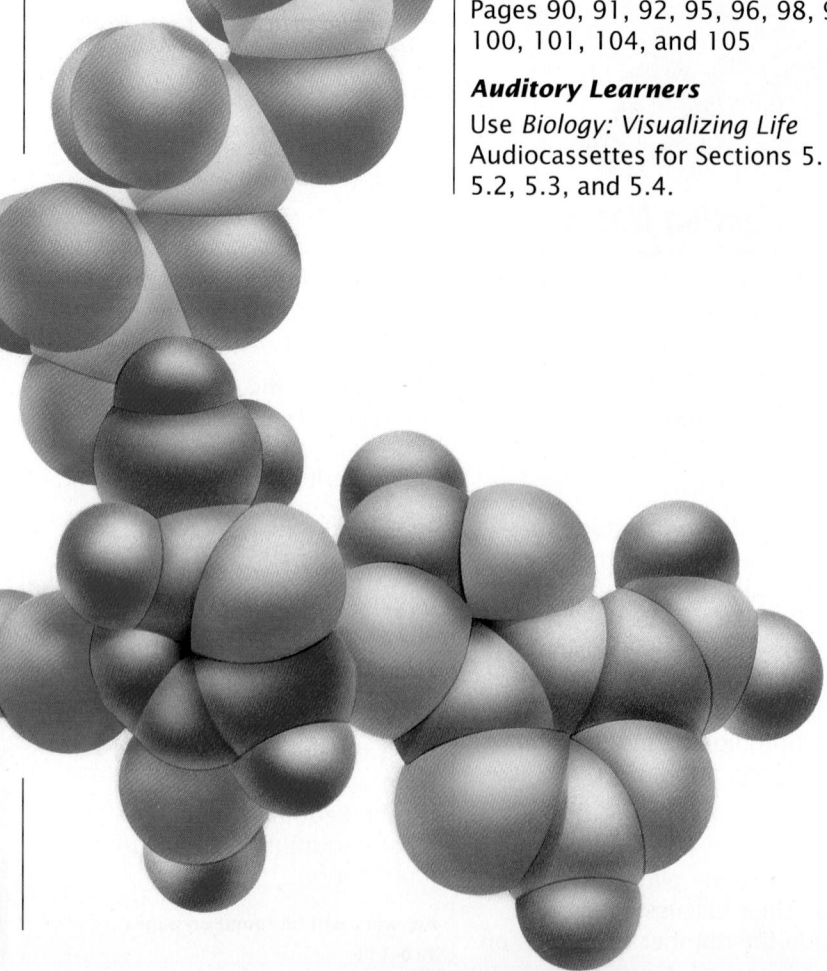

Procedure: Remind the class that for an organism to live, energy must be released by the cells. Read aloud "Cells Release Energy in a Series of Small Steps" on pages 94–95. Point out the importance of ATP as a form of "stored" energy.

Tell the class that the three students will represent the three phosphate groups bonded together in ATP. Have students link arms, with the third student also holding onto a chair or desk representing the adenosine part of the molecule. Have the first student jump as he/she breaks away from the others. Explain that the jump represents energy released. The new molecule has two phosphate groups and a new name, ADP. Write the energy equation on the chalkboard.

Additional Strategies
Visual Strategies
Pages 90, 91, 92, 95, 96, 98, 99, 100, 101, 104, and 105

Auditory Learners
Use *Biology: Visualizing Life* Audiocassettes for Sections 5.1, 5.2, 5.3, and 5.4.

Meeting Individual Needs (cont.)

Cooperative Learning

Modeling Cellular Respiration With and Without Oxygen

Timing: Use this activity to introduce Section 5.4.

Group Size: 4 students

Outcome: Students will use models to visualize the reactants and overall products of cellular respiration.

Individual Accountability: Each student will build a molecular model.

Positive Interdependence: Each group will model the overall process of cellular respiration with or without oxygen.

Assign each group a process — respiration with oxygen or cellular respiration without oxygen. Provide each group with materials to build molecular models. Gumdrops and toothpicks will do. Have each group build a glucose ($C_6H_{12}O_6$) molecule. In addition, the "with oxygen" group should build six oxygen (O_2) molecules. (You may want to put all of these chemical formulas, or more detailed structures, on the board for students to refer to.)

Then, have the "with oxygen" group build carbon dioxide (CO_2) and water (H_2O) molecules with parts from their models. The "without oxygen" group should build carbon dioxide and ethyl alcohol molecules (C_2H_5OH) with parts from their glucose molecules.

Each group should reach a consensus about how many of each type of molecule was needed for the products and the reactants of their reaction. Then, a "with oxygen" and a "without oxygen" group should explain their findings to each other.

Portfolio Assessment

Students should select their best work and provide a self-reflective rationale for their selections. Students can make selections in the following areas.

1. *Content*
One concept map from the chapter (See page 812 for evaluation criteria.)

2. *Reading Comprehension*
One Directed Reading Worksheet from the Teacher's Resource Binder (Use the answer key to evaluate for accuracy.)

3. *Writing*
Using the Vee Form, summarize a newspaper or magazine article relating to photosynthesis or cellular respiration. (See page 22T for evaluation criteria.)

Or: Select a writing task or project from the Chapter Review.

4. *Performance Assessment*
One Vee form from a chapter investigation or lab manual investigation (See page 22T for evaluation criteria.)

Teacher makes selections in the following areas.

1. *Formal Assessment*
Chapter test (Test A, B, or the Test Generator) The teacher-scored test should be reviewed by the student. Incorrect responses should be corrected by the student before the test becomes part of the portfolio.

2. *Informal Assessment*
Use the Direct Observations Checklist, page 33T, during a laboratory or other cooperative learning experience.

3. *Performance Assessment*
Have students design an investigation testing different variables affecting photosynthesis rates in *Elodea*, such as light intensity, light color, acidity, and salinity.

Concept Map Answer

The following is one possible answer to the Relating Concepts exercise on page 108.

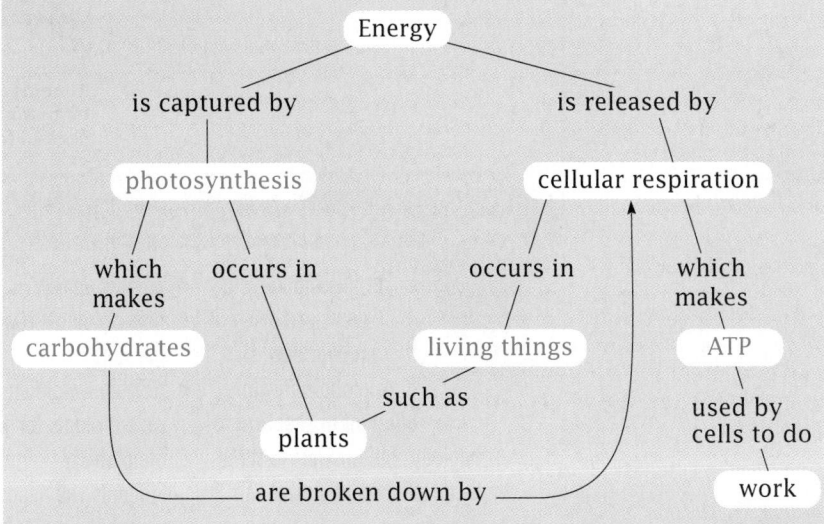

87D

Chapter 5

Determining Prior Knowledge

- Pour some 3-percent hydrogen peroxide into two test tubes. Add some sand to one and a small piece of fresh liver or potato to the other. Discuss the results with students, asking questions about chemical reactions and the role of enzymes.
- Give each student a cracker, cookie, candy, or something else to eat. As they chew, ask them how foods supply the energy they need. Have them describe life processes that are fueled by this energy.
- Use a plant as the focal point for determining what students know about photosynthesis. Ask questions concerning the role of light energy, what the reactants are, and what products are made.
- Have students exercise until they start becoming slightly winded. Ask them why their muscles start to tire and why they breathe faster as they exercise.

Chapter **5**

Energy and Life

Review

- carbohydrates (Section 2.2)
- cell structure (Section 3.3)
- the term *energy* (Glossary)

5.1 Cells and Chemistry
- Chemical Reactions in Living Things
- Actions of Biological Catalysts

5.2 Cells and Energy
- How Cells Use Energy
- Energy Flow in the Living World

5.3 Photosynthesis
- Harnessing the Sun's Energy
- Stage 1: Capturing Light Energy
- Stage 2: Using Light Energy to Make ATP and NADPH
- Stage 3: Building Carbohydrates

5.4 Cellular Respiration
- Releasing Energy From Organic Molecules
- Regulating Cellular Respiration

Running, jumping, and leaping over hurdles require lots of energy. In this chapter, you will learn that the energy needed for life comes ultimately from the sun.

▪ Author's Rationale ▪

In this chapter, students are given an overview of metabolism without a lot of chemical terminology. The answers to three simple questions summarize this chapter.

- What is energy?
- How do plants capture energy from the sun?
- How do we obtain this energy when we eat plants (or the animals that ate plants)?

Students will need to master the idea that energy-boosted electrons moving from one molecule to another are being used to drive proton pumps and make ATP.

JUST AS A RACE CAR DRIVER LEARNS HOW A CAR WORKS BY STUDYING ITS ENGINE, A SCIENTIST UNDERSTANDS HOW LIVING THINGS WORK BY STUDYING CELL CHEMISTRY. LIKE CARS, CELLS ARE COMPLEX MACHINES, FULL OF DELICATE DETAIL AND POWERED BY CHEMICAL ENERGY. TO UNDERSTAND HOW YOUR BODY WORKS, YOU MUST LOOK "UNDER THE HOOD" AT THE CHEMICAL MACHINERY IN YOUR CELLS.

5.1 Cells and Chemistry

Objectives

❶ Distinguish between energy-storing and energy-releasing chemical reactions.

❷ Define activation energy and describe its role in chemical reactions.

❸ Describe the interaction between an enzyme and its substrate.

❹ Describe the role of enzymes in cells.

Chemical Reactions in Living Things

Chemical reactions do not take place only in fizzing test tubes. Within cells, all the activities associated with life also are driven by chemical reactions.

A chemical reaction is the process of making or breaking the chemical bonds that link atoms. In plants, chemical reactions use light energy to form the chemical bonds of sugars such as glucose. The energy in these chemical bonds can then be used by cells to power their lives. A potato like the one in **Figure 5.1** is an excellent food source because it is crammed with glucose, linked in a thicket of long chains called starch. When the bonds of the glucose are broken, the stored energy is released.

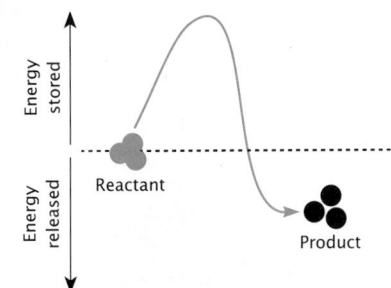

Figure 5.1

a When you eat a baked potato, starch is broken down into sugar in your mouth and small intestine. The energy in these sugar molecules will be used by your body to do work.

b Notice that the product contains less energy than the reactant. Reactions that release energy are called **exergonic reactions.**

Lesson Plan 5.1

Phase 1

PREPARATION

Key Concepts

- A chemical reaction involves the making and breaking of bonds between atoms.
- Chemical reactions either store or release energy.
- All chemical reactions that occur in an organism are referred to as metabolism.
- Enzymes provide the activation energy required to start chemical reactions in an organism.
- Enzymes are protein molecules with specific sites where substrates bind.
- Enzyme activity is affected by changes in either the temperature, pH level, or substrate concentration.

Reading Strategy

Point out that science textbooks have several reading aids. One such aid is the glossary. Have students practice using the glossary in this section by telling them to look up the definition of any word they find troublesome.

Phase 2

TEACHING STRATEGIES

Demonstration 1

Changing Starch Into Sugar
Provide each student with a cracker or piece of bread. Have them chew it until they notice the sweet taste resulting from the digestion of starch into sugars.

Demonstration 2

A Chemical Reaction in Humans

Demonstrate the connection between living things and chemical reactions by adding a digestive enzyme to a starch solution. Spray some laundry starch into a beaker, add some tap water and several drops of Lugol's iodine, and stir until the entire solution is blue-black. Remind students that starch turns blue-black in the presence of Lugol's iodine. Then have a student volunteer to place some of his or her saliva into a beaker and dilute it with water. Slowly add the diluted saliva to the starch solution while stirring. Have students observe what happens to the blue-black color. Discuss the result in terms of how a chemical reaction must have taken place—the starch must have been chemically changed by the saliva.

Connection: Chapter 2

Review the basic chemistry of bond formation, focusing on covalent bonds. Emphasize that most chemical reactions in living things involve either the making or breaking of covalent bonds.

Theme Connection

Energy and Life

The energy required to carry out all life processes comes from the chemical reactions that constitute metabolism.

Visual Strategy

Figure 5.2

Use this figure to show students that synthesis is an energy-storing process. Remind students that Chapter 2 discussed how simple sugars are bonded to form more complex carbohydrates, including starch.

Demonstration 3

Activation Energy

Use a picture of the *Hindenburg* disaster to underscore the effect of activation energy. Filled with explosive hydrogen gas, the

---

Any chemical reaction that results in a net release of energy is **exergonic**. Your cells put this released energy to work making proteins and other molecules of which your body is built. Building these molecules uses a lot of energy because many new chemical bonds must be made. Extra energy must be supplied to cause any chemical reaction in which the chemical bonds of the products have more energy than the bonds of the reactants. Any chemical reaction that requires a net input of energy is called **endergonic**. In the potato plant in **Figure 5.2**, the formation of glucose using the energy in light is an example of an endergonic reaction.

Thousands of chemical reactions are going on at any moment in each cell of a living plant and in every cell of your body, a symphony of living chemistry. All of these chemical reactions, taken together, are called **metabolism** (*muh TAB uh liz uhm*).

Figure 5.2

a **This potato plant uses light energy to form the chemical bonds of glucose.**

Chemical reactions need energy to get started

The heat from a flame ignites the twigs in a campfire. The spark from a spark plug causes gasoline in an engine to ignite. In both cases, an input of energy is used to start the chemical reaction. The amount of energy needed to cause a chemical reaction to start is called **activation energy**. Think of a boulder you must move up and over the top of a hill. To get it rolling downhill, you must first push it to the top. Activation energy is simply a chemical push that gets a reaction going.

Even if the product contains less energy than the reactants, activation energy must still be supplied before that reaction can occur. For example, the combustion of gasoline provides the energy needed to power an automobile. But only after the key is turned in the ignition will a spark from each spark plug ignite the gasoline in the engine's cylinders. The sparks provide the activation energy needed to trigger the burning of gasoline.

Reactions in cells must occur quickly

The burning of gasoline fuel to power an engine requires a spark or high temperature to get a reaction going. Cells also "burn" fuels. Like an engine, most cellular reactions would require very

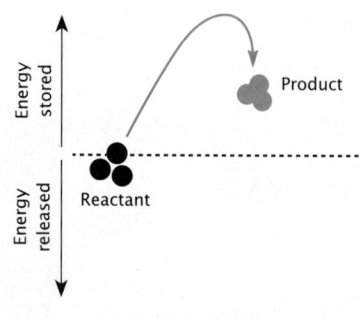

b **Reactions that store energy are endergonic reactions.**

■ *Historical Perspective* ■

Vitalism

In the 1800s, certain scientists, known as vitalists, argued that all living things contained some "vital spirit" that enabled them to carry out all life activities, including chemical reactions. How else, they argued, could such reactions, which needed high temperatures to get started, occur inside cells? Then in 1898, German chemists Eduard and Hans Buchner demonstrated that a substance isolated from yeast cells could cause fermentation to happen outside living cells and, more importantly, without the need for high temperatures. They named this substance *enzyme*, from the Greek word *zyme* meaning yeast or ferment. Vitalism was dead.

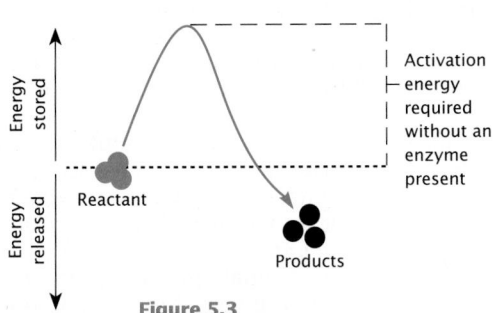

Figure 5.3

a **Activation energy must be supplied for most chemical reactions to occur.**

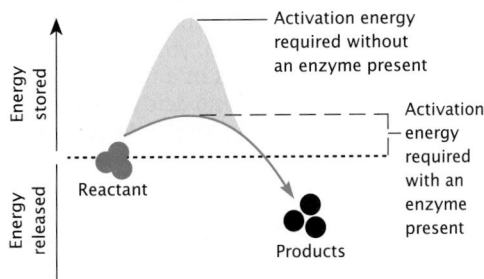

b **Enzymes lower the amount of activation energy required to start a reaction.**

high temperatures to proceed quickly enough to keep the cell functioning. But high temperatures would kill the cell. Fortunately, cellular reactions can occur quickly at relatively low temperatures through the action of enzymes. You read in Chapter 2 that enzymes are proteins that can hasten a chemical reaction.

How does an enzyme increase the speed of a reaction? A reaction could proceed faster if less energy were needed to get each molecule started. Think about the activation energy needed to move a heavy boulder over a hill. One way to reduce the amount of

energy necessary would be to reduce the hill's size. Digging away some of the ground in front of the boulder would reduce the amount of energy needed to send the rock rolling over the hill. Enzymes cause reactions to occur at a lower activation energy, as shown in **Figure 5.3**. Using an enzyme in a reaction is like lowering the top of the hill.

Enzymes are biological catalysts. **Catalysts** make a reaction proceed faster without themselves being used up during the reaction. An enzyme-catalyzed reaction is faster because it has a lower activation energy than does an uncatalyzed reaction.

Actions of Biological Catalysts

To understand the importance of enzymes, consider how red blood cells pick up carbon dioxide and deliver it to the lungs to be exhaled. Carbon dioxide is a cellular waste product that will poison the body if not

continuously removed. When carbon dioxide is converted to carbonic acid, it can be easily carried within the bloodstream. However, the chemical reaction that converts carbon dioxide to carbonic acid is very slow. Only 200 molecules of carbonic acid form in an hour. Fortunately, an enzyme present in the blood called carbonic anhydrase increases the rate of this reaction to 600,000 molecules of carbonic acid formed every second. The enzyme, shown in **Figure 5.4**, accelerates the reaction rate about 10 million times! Without carbonic anhydrase, your blood would quickly become poisoned with carbon dioxide.

Figure 5.4
This molecule is a computerized, three-dimensional model of carbonic anhydrase, an enzyme that enables your blood to remove carbon dioxide quickly. The colors represent different amino acids that make up the enzyme.

■ *Matter of Fact* ■

Slow Reactions
Chemical reactions in humans would occur slowly without enzymes. A single meal would take at least 50 years to be digested in the absence of digestive enzymes.

Hindenburg blew up in 1937 when, in some still-unknown manner, something provided the activation energy that sparked the explosion. Perhaps the activation energy came from a spark produced by an atmospheric discharge. Some suspect the work of saboteurs.

Visual Strategy

Figure 5.3
Use this figure to reinforce a point made in Demonstration 4—the amount of energy released is the same, with or without the enzyme. Have students use a ruler to measure the distance from the dotted line (reactant starting point) to a point midway through the products. This distance, the same in both cases, represents the energy released in this reaction.

Mathematics Connection

How Much Faster?
Work through the mathematics to show students that carbonic anhydrase accelerates the reaction rate slightly less than 11 million times:

600,000 molecules per second × 60 seconds × 60 minutes = 2,160,000,000 molecules per hour with enzyme

200 molecules per hour without enzyme = 10,800,000

Have students calculate the impact of a hypothetical enzyme that forms 500,000 molecules of a product every second. Without the enzyme, only 25 molecules per hour are formed.

Visual Strategy

Figure 5.5

Have students locate the reactant in Figure 5.5a and the products in Figure 5.5c. In so doing, students will realize that reactants and substrates are synonymous. Have students study the illustrations for the enzyme in each step. Point out the specificity of enzymes as shown in Figure 5.5a—only an enzyme with a particular shape (the proper active sites) will catalyze this reaction. Draw various substrates on the board and have students identify which ones the enzyme shown in this figure could bind to. Have them compare the enzymes shown in Figure 5.5a and 5.5c to see that the enzyme remains unchanged and can be reused.

Demonstration 4

Models for Enzyme Action

Show students a lock and several keys, only one of which will open the lock. Tell them that Emil Fischer suggested in 1894 that enzymes and substrate work like a lock and key. Try to open the lock with the various keys. Only the key that fits exactly works. This lock-and-key model was accepted for many years. Show students several gloves, only one of which fits your hand. Tell students that today, scientists think that enzymes and substrates work more like a glove and hand. This model is considerably more flexible than the lock and key. In this model, the enzyme does not have a rigid shape. Instead, the enzyme changes shape slightly as the substrate enters the active site. As the enzyme changes its shape, it fits more snugly around the substrate.

Theme Answer

By controlling the kinds and the amounts of enzymes present in a cell, the body can control the chemical reactions that take place within it.

Stability

A conductor controls the music of an orchestra by dictating when each instrument plays. How is this similar to the way your body controls chemical reactions?

Enzymes speed reactions by binding with specific molecules

How does an enzyme work? Typically, an enzyme is a large protein that consists of a folded chain of hundreds of amino acids. Each enzyme binds to a specific molecule and stresses the bonds of that molecule in such a way as to make a reaction more likely to occur. The molecule on which the enzyme acts is called a **substrate** (*SUHB strayt*).

The key to an enzyme's activity is its shape. Each enzyme has one or more deep folds on its surface. These crevices form pockets called **active sites**. As shown in **Figure 5.5**, an enzyme's substrate will fit into an active site just as your feet fit comfortably into your favorite pair of sneakers.

When the substrate molecule binds to the active site, the enzyme holds the substrate molecule close to a certain part of the enzyme surface. The enzyme may put a strain on a particular chemical bond in the substrate molecule and thus make the bond more likely to break. Or the enzyme may encourage the formation of a bond between two substrates by holding them near each other. Either way, the enzyme lowers the activation energy needed, which makes the reaction more likely to occur. When the reaction is complete, the products of the reaction are released, and the enzyme is ready to combine with an identical substrate molecule.

Because an enzyme must have a precise shape to work correctly, a cell can control an enzyme's activity by altering the enzyme's shape. The shapes of many enzymes can be altered by binding "signal" molecules to their surfaces. The new shape produced by binding the signal molecule acts to turn the enzyme "on" or "off."

Cells have thousands of different enzymes

A cell contains thousands of different kinds of enzymes, each promoting a different chemical reaction. Enzymes active at any one time in a cell determine what happens in that cell, just as traffic lights control the flow of traffic in a city. Not all cells contain the same enzymes. As you read this page, the chemical reactions going on in a nerve cell in your eye are very different from the reactions in one of your red blood cells, because the two kinds of cells contain different enzymes.

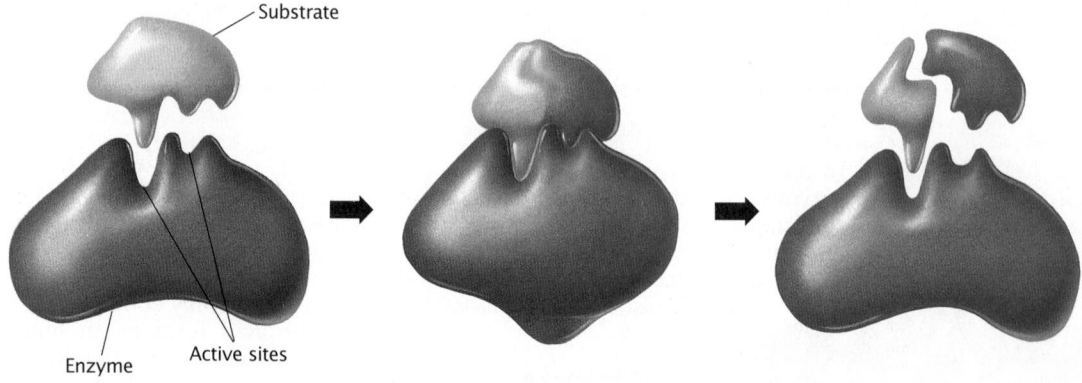

Figure 5.5

a **Enzymes act to lower the activation energy of a reaction. Here an enzyme is about to bind with its substrate.**

b **The enzyme binds to the substrate, stressing a particular bond and lowering the activation energy needed to break the bond.**

c **The products of the reaction are released. The enzyme returns to its original shape and is ready to combine with another substrate molecule.**

Figure 5.6
In some animals, enzymes that control pigmentation are active only at certain temperatures.

a The darker parts of the Siamese cat are also the cooler regions of its body. This is because the enzyme catalyzing the synthesis of the dark pigment is more active in cooler temperatures.

b Similarly, the fur of the adult northern seal is dark because the enzyme controlling pigmentation is more active in cold arctic temperatures. The newborn seal is white because the enzyme was less active in the warmer temperatures inside its mother's body.

Heat, acidity, and enzyme concentration affect enzymes

Each enzyme functions best within a certain temperature range. When temperatures become too low or too high, reaction rates decrease sharply. For instance, many enzymes in your body shut down when you have a high fever. If the internal body temperature of a human being were to reach 44°C (112°F), many enzymes would be destroyed and the individual would probably die. For additional examples of how enzymes are affected by temperature, look at **Figure 5.6**. How does temperature affect the pigmentation of the Siamese cat and the northern seal?

Another factor influencing enzyme activity is acidity, the concentration of acids in the body. When an organism's acidity is too high or too low, most enzymes cease to function properly. One exception, however, is the enzyme pepsin, which functions in the stomach's highly acidic environment.

Finally, the rate of an enzyme-catalyzed reaction is affected by the concentration of the enzyme and the substrate. The rate of a chemical reaction in the body, for example, can be accelerated by increasing the concentration of the enzyme necessary to catalyze that reaction. This is how your body controls its development.

Section Review

1 How can a chemical reaction store or release energy?

2 What is activation energy and how does it relate to cell metabolism?

3 Explain how an enzyme can increase the rate of a reaction.

4 Why do you think it is advantageous for the human body to have many different enzymes?

■ *Section Review Answers* ■

1. When a product contains more energy than the reactants, this is an energy-storing reaction. When a product contains less energy than the reactant, this is an energy-releasing reaction.

2. The amount of energy needed to cause a chemical reaction is called activation energy. Activation energy is necessary to initiate the reactions that comprise cell metabolism.

3. Enzymes speed reactions by binding with specific molecules called substrates and lowering the activation energy necessary for the reaction to occur.

4. Many different enzymes are needed for the many different reactions that occur in the human body.

Demonstration 5
Affecting Enzyme Action
Repeat Demonstration 2, but this time use a stopwatch and record how long it takes for the reaction to be completed (the blue-black color disappears). Repeat the process three more times, varying one factor in each case:
- Use hot water to study the effect of temperature on enzyme activity.
- Add hydrochloric acid to both the starch solution and saliva to study the effect of pH on enzyme action.
- Spray more starch into the beaker before adding the water to study the effect of increased substrate concentration on enzyme action.
- Place the times on the board and discuss the results with the class.

Phase 3

ASSESSMENT OPTIONS

Closure Strategy
Designing an Experiment
Amylase is an enzyme that digests starch. Assign students to cooperative work groups of four. Have each group design an experiment to test the effects of either different temperatures or various pH values on amylase's activity.

Section Review
Assign the *Section Review.*

Reteaching
Students often get the impression that with so many different enzymes, if one were missing, the impact would be insignificant. But such is not the case. In 1908, A.E. Garrod discovered that certain infants had a most unusual problem—their diapers turned black when wet with urine. Have students research the disease alkaptonuria, and write a report that discusses how Garrod was the first to discover the connection between genes, enzymes, and metabolism.

Lesson Plan 5.2

Phase 1

PREPARATION

Key Concepts

- Cells use energy to perform all their life activities.
- Cells release energy in small, gradual steps.
- Cells use the energy stored in ATP to drive many chemical reactions.
- Photosynthesis and respiration are responsible for the flow of energy in the living world.

Reading Strategy

Students often read science texts too quickly. Have students record how long it takes them to read the first page in this section. Then have them write down all they remember from their reading. Instruct them to double the amount of time spent reading the next page and again write down what they recall. Repeat the process for the third and final page in this section, again doubling the time spent on the second page. Have students check to see if there is a connection between time spent reading and amount of information retained.

Phase 2

TEACHING STRATEGIES

Demonstration 1

Uses for Energy

Assign students to cooperative work groups of four. Have each group spend five minutes listing as many life activities as possible that require energy. Then collate each group's list on the board. Remind students that nearly one-third of an organism's energy is used for driving the sodium-potassium pump of active transport (Chapter 4). Point out that the amount of energy used daily by an average adult for

---

IF YOU STOPPED EATING, YOU WOULD EVENTUALLY DIE. WHY? BECAUSE THE FOOD YOU EAT PROVIDES THE ENERGY NEEDED FOR LIFE. ALL THE PROPERTIES BY WHICH WE DEFINE LIFE—GROWTH, MOVEMENT, SENSITIVITY, AND REPRODUCTION—USE ENERGY. JUST AS LOGS MUST BE CONTINUALLY SUPPLIED TO KEEP A CAMPFIRE BURNING, ENERGY MUST BE CONTINUALLY SUPPLIED TO KEEP LIFE GOING.

5.2 Cells and Energy

Objectives

1. **Explain how cells use energy.**
2. **Describe what is meant by the term biochemical pathway.**
3. **Describe the role played by ATP in cells.**
4. **Recognize the path of energy between plants and animals in the living world.**

How Cells Use Energy

Figure 5.7
The photograph below is a colorized TEM of a bacterium (7,080X) that uses energy to spin its flagella as it moves through fluids.

Cells use energy to do all those things that require work. One of the most obvious of these is movement. Some bacteria swim about, propelling themselves through fluids by rapidly spinning long flagella, as shown in **Figure 5.7**. Cells also use energy to change their shape. White blood cells, for example, must extend and retract their cell membranes when they engulf invading bacteria in the bloodstream.

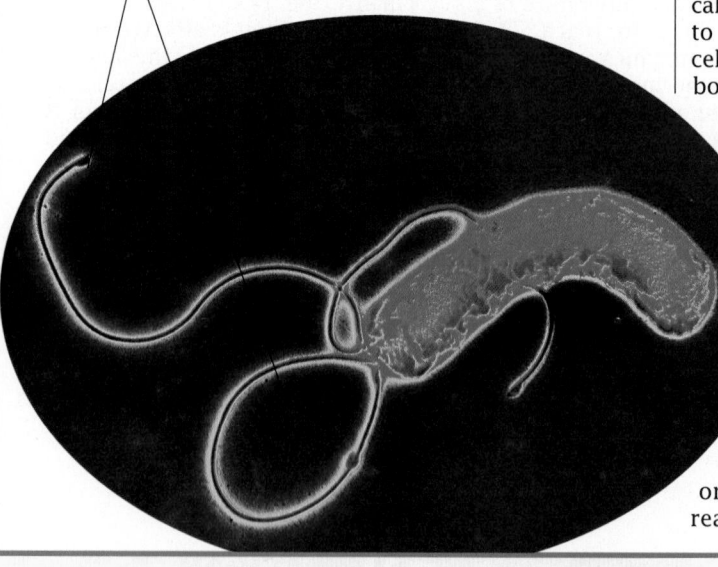

Flagella

Other cellular tasks, such as manufacturing new cellular components or maintaining and repairing cellular structures, require considerable energy. Cells also need energy to transport food into the cell and to expel wastes.

Cells release energy in a series of small steps

Perhaps you have heard that jogging or some other exercise "burns" a lot of calories. The word "burn" is often used to describe what happens when your cells release the energy from chemical bonds in food molecules.

Obviously the burning of food molecules in living cells differs from the release of energy that occurs in, say, the burning of logs in a campfire. When logs are burned, the energy contained in the wood is released all at once in the form of heat and light. But this is not what takes place in cells.

Instead, the energy stored in food molecules in living cells is gradually released in a series of chemical reactions. The product of one chemical reaction serves as a reactant in the next reaction. Such a series of linked chemical reactions is called a **biochemical pathway**. Each step of a biochemical pathway is catalyzed by a particular enzyme. The reactions that release energy from food molecules are just one of the many thousands of biochemical pathways that take place in living things.

Cells package energy in a molecule called ATP

When living cells break down molecules, part of the energy contained in the molecules is released as heat. But some of the energy is stored temporarily in a molecule called adenosine triphosphate, or **ATP**. Notice in **Figure 5.8** that ATP contains phosphate groups. The term "triphosphate" means the molecule contains three phosphate groups.

When the bond between the outermost two phosphate groups of ATP is broken, ATP becomes ADP (adenosine diphosphate). The term "diphosphate" means two phosphate groups. The reaction that forms ADP from ATP releases a sizable amount of energy. This equation summarizes that reaction:

$$ATP \xrightarrow{H_2O} ADP + P + energy$$

Almost all energy-requiring reactions in cells require less activation energy than is released by the splitting off of a phosphate from ATP. Therefore, ATP is able to power most of the cell's activities.

Cells use ATP to drive many chemical reactions

A steady supply of ATP is necessary to ensure that a cell can perform all the tasks essential for life. Making molecules, for example, uses energy. The reactions that build new molecules cannot begin until energy is supplied to the reaction from somewhere. That somewhere, as you might guess, is the breakdown of the cell's supply of ATP.

The breakdown of ATP to form ADP releases a great deal of energy. This energy can supply energy for other reactions. When an energy-requiring reaction in a cell is driven by the reaction of ATP to form ADP, it is called a coupled reaction. Coupled reactions occur often during cell metabolism.

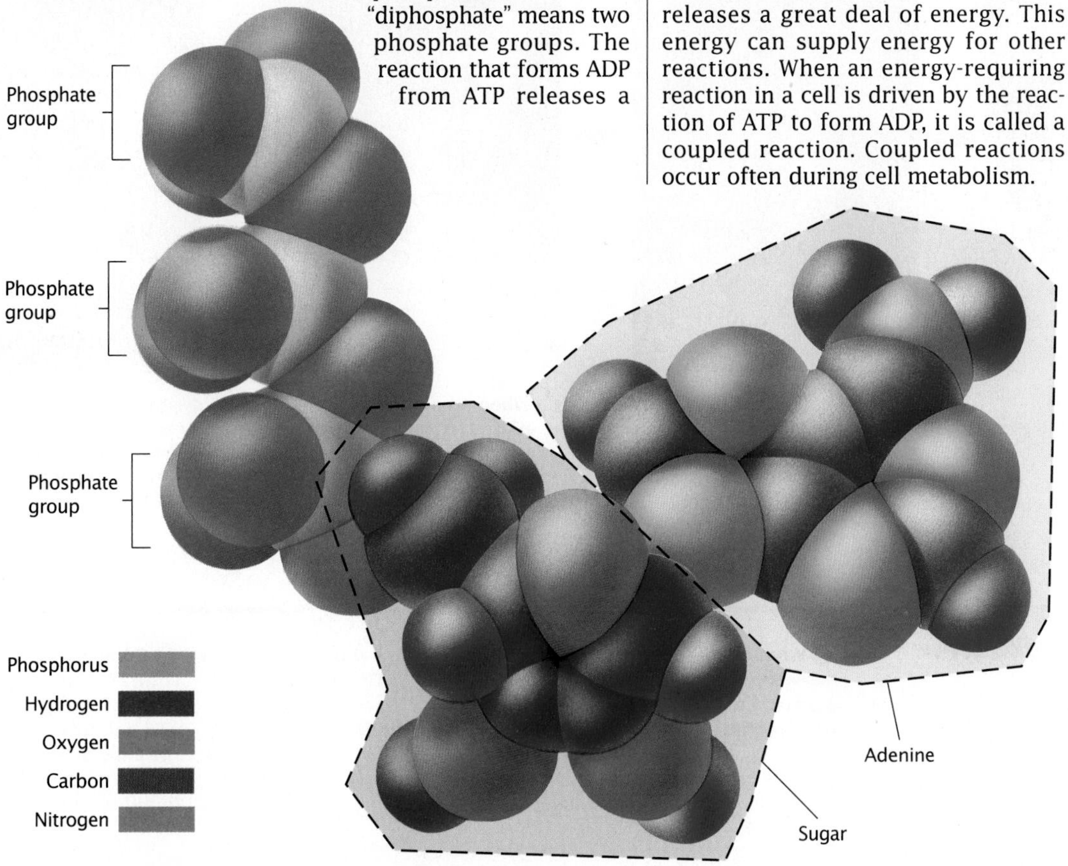

Figure 5.8
An ATP molecule is made of a sugar, an adenine molecule, and a chain made of three phosphate groups. When the bond between the outermost two phosphate groups is broken, ADP and phosphate are produced and energy is released.

Phosphate group

Phosphate group

Phosphate group

Phosphorus
Hydrogen
Oxygen
Carbon
Nitrogen

Adenine

Sugar

all his or her life activities is equivalent to the energy stored in approximately 3 kg (5 lbs.) of dynamite.

Demonstration 2
Small Steps
Use a staircase to demonstrate the advantage of releasing energy in small steps rather than in a single large one. Climb several steps and then jump to the floor. Exaggerate the impact on landing. Climb to the same level, but this time, walk down one step at a time. Have students compare this to what would happen to a cell if it released all the energy stored in molecules in one step.

Visual Strategy

Figure 5.8
Have students explain where they have previously encountered adenine in their reading. Have them locate where the bond is broken in ATP to produce ADP. Have students refer back to Figure 5.2b, where they should recall that the product has more stored energy than the reactants. Energy to drive such a reaction comes from the breakdown of ATP.

Demonstration 3
Recharging ATP
Use a rechargeable battery setup to illustrate how ATP can be compared to rechargeable batteries. The electrical energy represents the chemical energy in foods. The batteries represent ATP being continuously used and remade by cells.

95

Energy Flow in the Living World

Almost all the energy needed for life comes ultimately from the sun, which shines continuously on Earth. Using a process called **photosynthesis** (*foh toh SIHN thuh sihs*), green plants capture sunlight and use the energy to convert carbon dioxide and water into energy-storing carbohydrates. Oxygen is produced as a waste product. Photosynthesis is also performed by algae and some bacteria. Organisms that eat plants—and those that eat plant-eaters—use the energy in carbohydrates to fuel their own life processes. All living things use a process called **cellular respiration** to obtain energy from the bonds of food molecules. For humans, food molecules are the foods we eat, which include carbohydrates. For plants, the food molecules are carbohydrates made by photosynthesis.

As you can see in **Figure 5.9**, energy flows in connected pathways throughout the living world. The energy-requiring reactions of photosynthesis transform sunlight energy into chemical energy. And the energy-releasing reactions of cellular respiration enable living things to use chemical energy to do work. You will learn more about the processes of photosynthesis and cellular respiration in Sections 5.3 and 5.4.

Figure 5.9
a Light energy streaming down from the sun . . .

b . . . is converted by green plants into carbohydrates.

c When an animal eats plants, it uses energy within carbohydrates in the plants.

d And when that animal is eaten by another animal, energy is transferred.

Section Review

❶ What life activities performed by cells require energy?

❷ What are biochemical pathways? How do enzymes interact with biochemical pathways?

❸ How does ATP supply energy for the cell?

❹ Compare and contrast the roles of plants and animals in the flow of energy in the living world.

▪ Section Review Answers ▪

1. Cells use energy to do all those things that require work. Examples include movement, change of shape, manufacturing new cellular components, maintaining and repairing cellular structures, transporting food into the cell, and expelling waste.

2. Biochemical pathways are series of linked chemical reactions. The product of one chemical reaction serves as a reactant in the next reaction. Enzymes catalyze many biochemical reactions.

3. When the bond between the second and third phosphate groups of ATP is broken, ATP becomes ADP. The reaction that forms ADP from ATP releases a sizable amount of energy.

4. See Figure 5.9b–d.

WHAT DID YOU HAVE FOR LUNCH YESTERDAY? BEEF ENCHILADAS? THE BEEF CAME FROM A COW THAT ATE GREEN GRASS. NO MATTER WHAT YOU ATE, IF YOU TRACE THE FOOD BACK TO ITS ORIGIN YOU END UP WITH A GREEN PLANT. CLEARLY, YOU DEPEND ON GREEN PLANTS FOR ENERGY. PLANTS DEPEND ON SUNLIGHT TO MAKE FOOD, BECAUSE LIFE ITSELF IS POWERED BY THE SUN.

5.3   Photosynthesis

Objectives

① **Summarize the evolution of photosynthesis.**

② **Describe how green plants and algae capture energy from sunlight.**

③ **Explain how a plant cell uses light to make ATP and NADPH.**

④ **Describe how photosynthesis provides the energy needed by all living things.**

Harnessing the Sun's Energy

As you read in Section 5.2, the energy used by most living things comes ultimately from the sun. Each day, light energy that reaches Earth equals the energy of about 1 million atomic bombs. About 1 percent of that light energy is captured by photosynthesis and used to make energy-rich carbohydrates. Almost all living things, such as the cow in **Figure 5.10**, depend on the products of photosynthesis to survive.

Figure 5.10
The green plants that this cow is eating provide it with the energy it needs to carry out its life activities.

How did photosynthesis evolve?

Photosynthesis evolved billions of years ago among bacteria. Early forms of bacteria captured the electrons they needed for photosynthesis by stripping hydrogen atoms from hydrogen sulfide (H_2S), generating sulfur as a byproduct. For about 1 billion years, this early type of photosynthesis was the only kind of photosynthesis that occurred. In all that time, no oxygen gas existed in Earth's atmosphere. Approximately 3 billion years ago, a second method of photosynthesis evolved. In this new type of photosynthesis, electrons were obtained by removing hydrogen atoms from water (H_2O). Oxygen gas (O_2) was released into the atmosphere as a waste product. As a result of this type of photosynthesis, Earth's atmosphere is now rich in oxygen gas.

Photosynthesis takes place in three stages. In the first stage, energy is captured from light. In the second stage, the energy is used to make ATP and a high-energy compound called NADPH. During the third stage, the ATP and NADPH are used to power the manufacture of energy-rich carbohydrates from CO_2 in the air.

Lesson Plan 5.3

Phase 1

PREPARATION

Key Concepts

- The first photosynthetic organisms—bacteria that used hydrogen sulfide, releasing sulfur as a byproduct— evolved nearly 4 billion years ago.
- A billion years later, photosynthetic organisms evolved that used water, releasing oxygen as a byproduct.
- Photosynthesis involves using light energy to convert water and carbon dioxide into sugars and oxygen.
- In stage 1 of photosynthesis, light energy from the sun energizes electrons in chlorophyll.
- In stage 2, the energy in these excited electrons is used to make ATP and NADPH. In addition, water is split to provide electrons to replace those removed from chlorophyll. Oxygen is given off as a byproduct.
- In stage 3, ATP and NADPH are used to convert carbon dioxide into carbohydrates.

Reading Strategy

Emphasize to students that in addition to reading the text, they must also read the captions and pay particular attention to Figures 5.13 and 5.14 in order to understand what occurs in the three stages of photosynthesis.

Phase 2

TEACHING STRATEGIES

Connection: Chapter 3

Evolution of Life
Review the evolution of prokaryotes and eukaryotes as described at the end of Chapter 3.

Visual Strategy

Figure 5.11
Review each band shown in this figure. Point out that in addition to radio signals, the lowest energy portion of the spectrum also includes police car radios, shortwave sets, and television signals. Radar signals used by police are in the same band as microwaves. Ultraviolet light causes sunburn and helps in the formation of vitamin D. Have students explain why children who live in tropical climates do not need to drink as much milk. Show an X ray to demonstrate how powerful these waves are, since they can penetrate many substances, including bones. Gamma rays are produced in nuclear explosions and can penetrate through several centimeters of lead.

Demonstration 1

Chlorophyll
Boil some leaves in ethanol to extract the chlorophyll. Have students discuss what colors are absorbed by this pigment and what color is reflected. Show students a fluorescent light designed to enhance plant growth and ask why it emits a purplish color.

Demonstration 2

Other Pigments
Display photographs or show live specimens of plants with variegated leaves so that students recognize that they contain pigments other than chlorophyll. Have them explain why most leaves, even if they contain different pigments, look green. Have students hypothesize what happens to the various pigments in the fall.

Connection: Chapter 2

Electron Energy Levels
Review the structure of an atom, focusing on the energy levels occupied by electrons. Remind students that the higher the energy level (the farther away from the nucleus of the atom), the more energy an electron possesses.

Stage 1: Capturing Light Energy

If plants use light energy for photosynthesis, our first question might be: Where is the energy in light? Twentieth-Century physics has taught us that light actually consists of tiny packets of energy called **photons** (*FOH tahnz*). When light shines down on you, your body is being bombarded by a stream of photons smashing onto its surface. Some of these photons, such as X rays and ultraviolet light, carry a greater amount of energy, while others, such as radio waves, carry very little energy. Our eyes absorb only photons carrying intermediate amounts of energy, which is why we see only "visible" light, as shown in **Figure 5.11**. Plants are even choosier, absorbing mostly blue and red light. They reflect back what is left of the visible light, which is why most plants appear green.

Molecules called pigments absorb light
How can the molecules of a leaf or a human eye "choose" which photons to absorb? Molecules that absorb light are called **pigments**. In your eyes, the pigment is called retinal. The eyes of insects contain pigments that absorb photons of a higher energy level than those absorbed by retinal, so insects can see violet light that humans cannot.

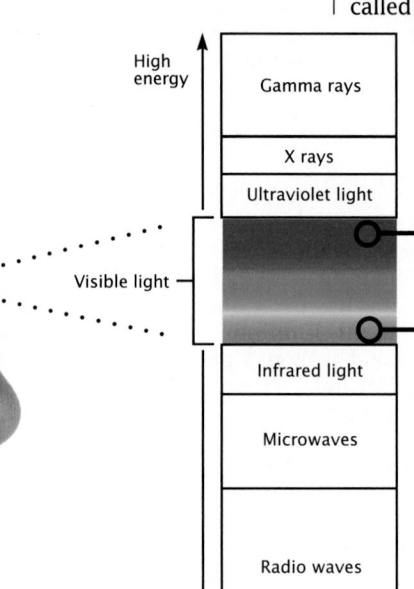

Figure 5.11
Visible light is that portion of the spectrum that humans can see. The spectrum consists of a range of waves from low-energy radio waves to high-energy gamma rays. These waves also range in wavelength. Green plants absorb red and blue wavelengths and reflect green.

But the insect pigment does not absorb low energy photons as well as retinal does, so insects cannot see the red light that humans can.

Plants capture light energy in chlorophyll
The major light-absorbing pigment in plants is chlorophyll. While it absorbs fewer kinds of photons than retinal, chlorophyll is much more efficient at capturing these photons. Where is chlorophyll located? Look at the structure of the leaf in **Figure 5.12**. In green plants, chlorophyll is found in chloroplasts within plant cells.

When atoms in a pigment absorb light, electrons are boosted to higher energy levels. The energy in the photons is transferred to the electrons, causing the move. Boosting an electron requires an exact amount of energy, just as when climbing a ladder you must

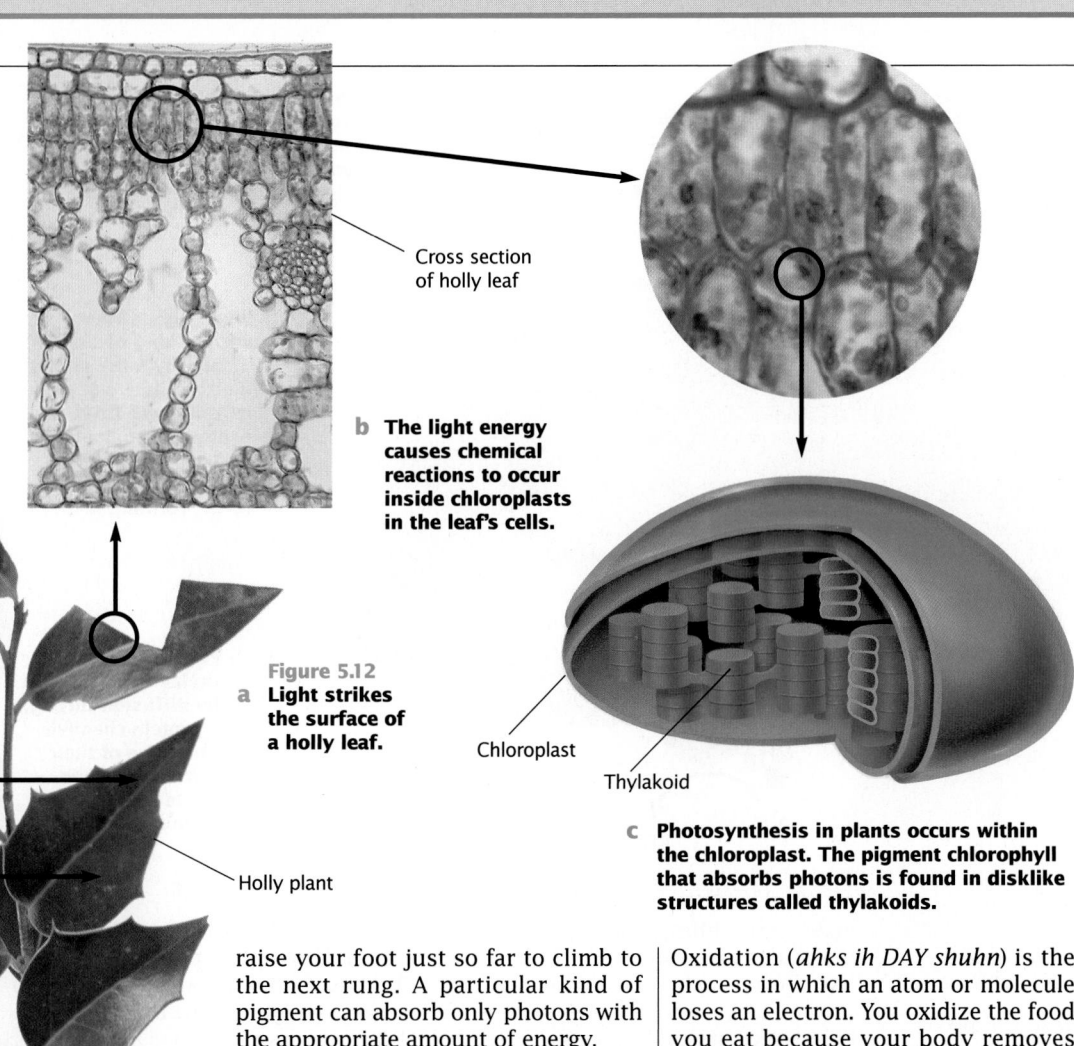

Cross section
of holly leaf

b **The light energy causes chemical reactions to occur inside chloroplasts in the leaf's cells.**

Figure 5.12
a **Light strikes the surface of a holly leaf.**

Chloroplast

Thylakoid

Holly plant

c **Photosynthesis in plants occurs within the chloroplast. The pigment chlorophyll that absorbs photons is found in disklike structures called thylakoids.**

raise your foot just so far to climb to the next rung. A particular kind of pigment can absorb only photons with the appropriate amount of energy.

The molecule building reactions of photosynthesis involve ferrying energetic (boosted) electrons from one molecule to another. Just as the action of a football game depends on moving the ball down the field, the energy flow in life depends on passing high energy electrons from one molecule to another. Passing an electron from one molecule to another transfers the energy contained in that electron from molecule to molecule as well.

Chemical reactions that involve the transfer of electrons from one atom or molecule to another are called **oxidation-reduction reactions**. Many of the chemical reactions of photosynthesis and cellular respiration are examples of oxidation-reduction reactions.

Oxidation (*ahks ih DAY shuhn*) is the process in which an atom or molecule loses an electron. You oxidize the food you eat because your body removes electrons from the food molecules during cellular respiration. Reduction is the process in which an atom or molecule gains an electron. Carbohydrate molecules produced by plants during photosynthesis contain many energy-rich electrons, and are said to be highly reduced.

Every oxidation involves a reduction: for every electron lost by one atom or molecule (oxidation), an electron is gained by another atom or molecule (reduction). In cells, electrons do not travel alone, but in the company of a proton. Recall that a proton and an electron together make a hydrogen atom. Thus, oxidation-reduction reactions usually involve the loss of hydrogen atoms from one molecule and the gain of hydrogen atoms by another molecule.

Figure 5.13

The sequential nature of stage 2 will be much clearer if each step in this figure is studied as students read the appropriate text passage. Emphasize that the process would come to a grinding halt if the electrons removed from chlorophyll were not replaced. Have students identify the source of these replacement electrons. Point out that the oxygen gas, although a byproduct of photosynthesis, is crucial to respiration, as they will see in the next section. Have students recognize that the light energy from the sun, temporarily captured by the excited electrons, has now been transferred to the chemical energy present in ATP and NADPH.

Theme Answer

Scale and Structure

When the hydrogen atom with an attached light-excited electron arrives at the proton-pump, the pump returns the excited electron back to its original energy level, releasing energy in the process. The released energy powers the pumping of a proton across the thylakoid membrane into the interior of the thylakoid. As this process repeats, protons are pumped inward. The pump is filled beyond capacity and protons are diffused through a protein channel. The force of their exit adds a phosphate to ADP, forming ATP.

Figure 5.13

a Inside a chloroplast, photosynthesis begins when light strikes a chlorophyll molecule in the membrane of a thylakoid (shown below). The light excites an electron that is joined to a proton donated by water.

b The energy carried by the electron powers a proton pump that transports a proton across the membrane into the thylakoid.

c When light strikes a second kind of chlorophyll, the excited electron does not use its energy to drive a proton pump. Instead, it is combined with a proton and joined to NADP, forming NADPH.

d Protons inside the thylakoid are driven by diffusion through a protein channel. The force of their exit adds a phosphate to ADP, forming ATP.

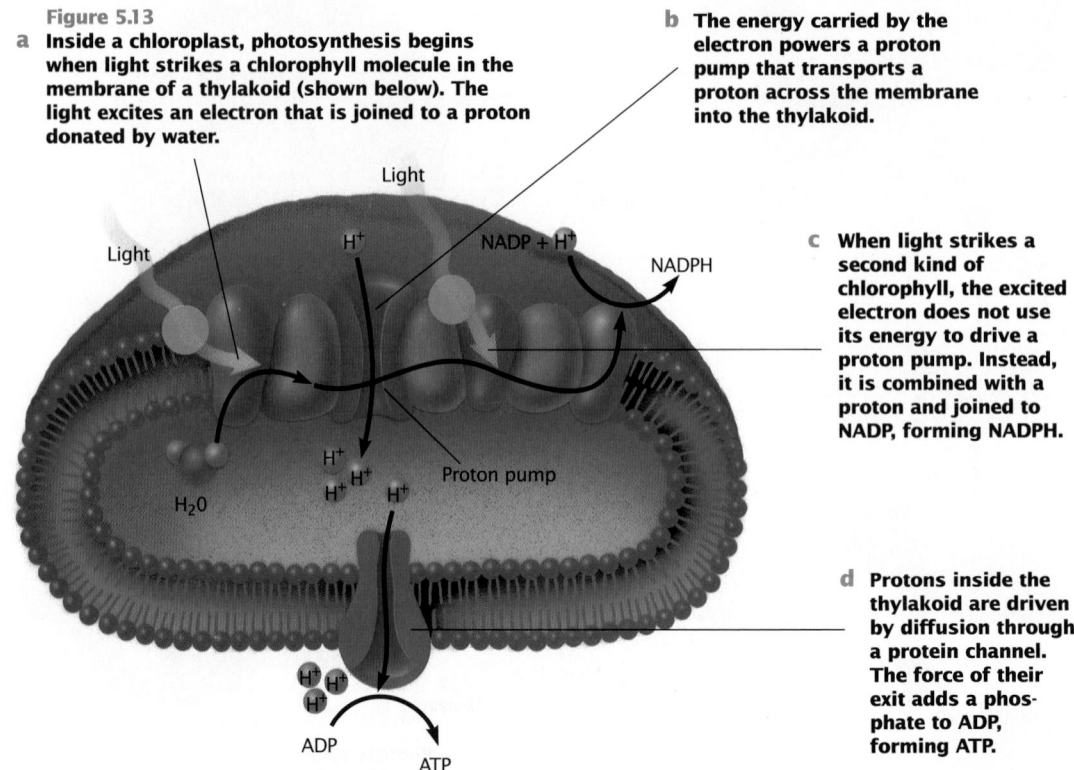

Stage 2: Using Light Energy to Make ATP and NADPH

Scale and Structure

How does the structure of the thylakoid enable green plants to capture light energy and transform it into chemical energy?

Inside the chloroplasts, chlorophyll molecules are contained within disk-like structures called thylakoids, as shown in **Figure 5.13**. When photons of light strike the chloroplasts, electrons are boosted within chlorophyll molecules to higher energy levels, as shown in **Figure 5.13a**. Each excited electron, traveling as part of a hydrogen atom, leaves the chlorophyll and jumps to a nearby protein in the membrane of the thylakoid. The electron is then passed from protein to protein, like a ball being passed down a line of people.

Soon the hydrogen atom carrying the light-excited electron arrives at its destination, a proton pump. The proton pump knocks the excited electron down to its original energy level, releasing energy in the process. This energy powers the pumping of a proton across the thylakoid membrane into the interior of the thylakoid. These events are shown in **Figure 5.13b**.

ATP is made when protons are forced through a protein channel

The payoff lies in what happens to protons pumped into the thylakoid. More protons are pumped inward, until the interior of the thylakoid is bursting at the seams. Straining to escape, the protons are driven by diffusion through the only exit available, a protein channel. As shown in **Figure 5.13d**, these protein channels use the force of the exiting protons to add a phosphate group to a molecule of ADP, making ATP.

A second kind of chlorophyll absorbs photons of higher energy than those absorbed by the ATP-making chlorophyll. The light-excited electrons of this second chlorophyll are carried by a hydrogen atom and attached to an electron carrier called NADP, forming NADPH. These steps are shown in **Figure 5.13c**. The ATP and NADPH will be used to help power the last stage of photosynthesis, the building of new carbohydrates.

Stage 3: Building Carbohydrates

The ultimate goal of photosynthesis is to capture carbon atoms from carbon dioxide in the air and use them to make carbohydrates that store energy. In a series of reactions, plants produce a number of carbon-containing molecules. From these molecules, plants can then assemble more complex carbohydrates such as glucose and other compounds needed for energy and growth. This series of reactions is called the Calvin cycle, for its discoverer Melvin Calvin of the University of California. The energy to fuel the Calvin cycle comes from ATP made during the second stage of photosynthesis. Follow the Calvin cycle in **Figure 5.14**.

The overall process of photosynthesis can be summarized as:

$$6CO_2 + 6H_2O \xrightarrow{\text{(light)}} C_6H_{12}O_6 + 6O_2$$

This equation indicates that carbon dioxide and water, in the presence of sunlight, will react to form sugars and oxygen gas.

Many plants store some of the sugars they produce by linking molecules to form complex carbohydrates such as starch. Starches can be stored either in the cells that formed them or in plant storage areas. For example, sugar made in the leaves of a potato plant is stored as starch in the potato tuber. The plant may later break down these starches to make the ATP needed by the cell, as you will see in Section 5.4.

Figure 5.14

a Using ATP generated from the second stage of photosynthesis, . . .

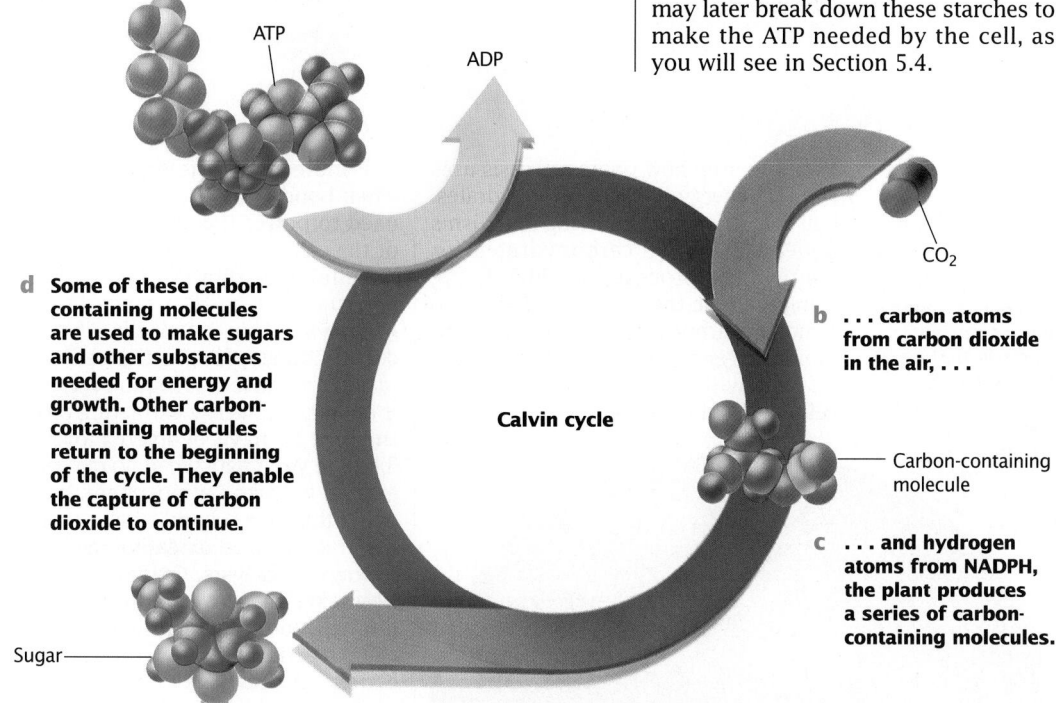

ATP

ADP

CO_2

Calvin cycle

b . . . carbon atoms from carbon dioxide in the air, . . .

Carbon-containing molecule

c . . . and hydrogen atoms from NADPH, the plant produces a series of carbon-containing molecules.

d Some of these carbon-containing molecules are used to make sugars and other substances needed for energy and growth. Other carbon-containing molecules return to the beginning of the cycle. They enable the capture of carbon dioxide to continue.

Sugar

Section Review

1 How did the appearance of a new kind of photosynthesis about 3 billion years ago change Earth's atmosphere?

2 How do green plants and algae capture energy from sunlight?

3 What happens to the ATP and NADPH made during the second stage of photosynthesis?

4 List the products of photosynthesis.

▪ Section Review Answers ▪

1. The appearance of a form of photosynthesis that used water as an electron donor meant that oxygen was released as a byproduct. Earth's atmosphere is now rich in oxygen gas.

2. Green plants and algae capture energy from sunlight using the pigment chlorophyll.

3. ATP is used as energy to fuel the Calvin cycle. NADPH provides energy and hydrogens. See Figure 5.14a–d.

4. Photosynthesis produces carbohydrates and oxygen.

Visual Strategy

Figure 5.14
Have students explain why stage 3 of photosynthesis is known as a cycle. Point out how some of the three-carbon molecules are used to keep the cycle going, as shown in Figure 5.14d. Remind students that the formula for glucose is $C_6H_{12}O_6$. Ask how many turns of the cycle are needed to make one glucose molecule from carbon dioxide. Use Figure 5.14 to point out where the ATP and NADPH made in stage 2 are used in stage 3.

Demonstration 4
Going Round in Circles
Show students a photo of a merry-go-round. Draw an analogy between it and the Calvin cycle. As long as people get on and others get off, the merry-go-round can keep operating. The same is true of the Calvin cycle. Have students explain what enters and what leaves the cycle to keep it going.

Phase 3

ASSESSMENT OPTIONS

Closure Strategy
Factors Affecting Photosynthesis
Have students list three factors that would increase the rate of photosynthesis. Have them describe what stage of photosynthesis each factor would affect.

Section Review
Assign the *Section Review*.

Reteaching
Write the overall equation for photosynthesis on the board. Take each of the molecules one at a time and have students discuss the details of its involvement in photosynthesis.

Lesson Plan 5.4

Phase 1
PREPARATION

Key Concepts

- Cellular respiration is the process by which living things release the energy stored in organic molecules.
- Glycolysis is the process by which glucose is broken down into two pyruvate molecules. In the process, two ATPs are gained.
- Fermentation is the process by which pyruvate is converted, in the absence of oxygen, into either alcohol and carbon dioxide or lactic acid.
- Oxidative respiration is the process by which pyruvate is broken down in the presence of oxygen into a two-carbon molecule. This molecule then enters the Krebs cycle, where NADH is produced to supply the electron transport system.
- As a result of the breakdown of glucose in the presence of oxygen, as many as 36 additional ATPs are produced.
- Oxygen serves as the final electron acceptor.

Reading Strategy

Before reading this section, have each student prepare a table with the title *Photosynthesis and Respiration*. Beneath the title, have them place two headings—*Similarities* and *Differences*. As they read this section, have them fill in their table.

Phase 2
TEACHING STRATEGIES

Connection: Chapter 3

Mitochondria
Remind students that mitochondria constitute some of the compartments that help make eukaryotic cells more efficient than prokaryotic cells.

102

ALTHOUGH ONLY PLANT CELLS WITH CHLOROPLASTS PRODUCE CARBOHYDRATES, ALL CELLS OF A PLANT USE THESE CARBOHYDRATES FOR ENERGY. IN BOTH PLANTS AND ANIMALS—INDEED, IN ALMOST ALL ORGANISMS—THE ENERGY FOR LIVING IS OBTAINED BY RECYCLING THE SUGARS PRODUCED BY PHOTOSYNTHESIS. THE ENERGY IN THESE MOLECULES IS RELEASED IN CELLULAR RESPIRATION.

5.4   *Cellular Respiration*

Objectives

❶ Explain the importance of cellular respiration to living things.

❷ Summarize the process of glycolysis.

❸ Contrast fermentation with oxidative respiration.

❹ State the role of oxygen in oxidative respiration.

Releasing Energy From Organic Molecules

You have seen how photosynthesis uses sunlight energy to make carbohydrates. You also know that all living organisms depend on these carbohydrates for energy. The process by which living things release the energy stored in the bonds of carbohydrates and other food molecules is called cellular respiration. As you will see, the first result of cellular respiration is the formation of ATP molecules. The energy released when bonds in ATP are broken is then used to power the chemical reactions of the cell.

Cellular respiration takes place in two stages. The first stage is called **glycolysis** (*gly KAHL uh sihs*). Glycolysis takes place in the cell's cytoplasm and does not require oxygen. It is an ancient energy-extracting process thought to have evolved more than 3 billion years ago, when no oxygen gas existed in Earth's atmosphere. In most living things, a second stage of cellular respiration called **oxidative respiration** follows glycolysis. Oxidative respiration takes place within mitochondria. It is far more effective than glycolysis at recovering energy from food molecules. Oxidative respiration is the method by which plant and animal cells get the majority of their energy.

Glycolysis breaks down glucose into two pyruvate molecules
Glycolysis is one of the most ancient biological processes we know. It evolved among bacteria, the first life forms on Earth. Bacteria, like those in **Figure 5.15**, relied upon glycolysis to make the ATP needed to drive

Figure 5.15
Glycolysis evolved among ancient bacteria that were similar to the photosynthetic bacteria (160X), below.

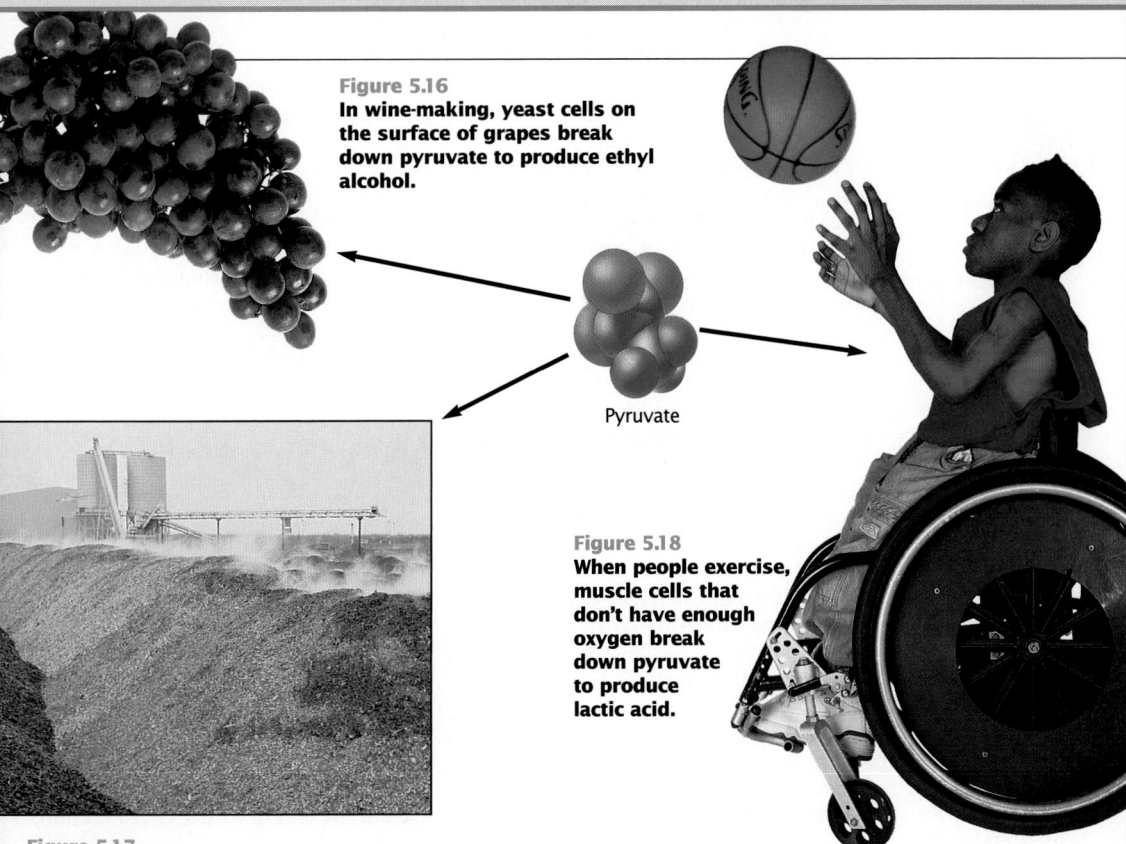

Figure 5.16
In wine-making, yeast cells on the surface of grapes break down pyruvate to produce ethyl alcohol.

Pyruvate

Figure 5.18
When people exercise, muscle cells that don't have enough oxygen break down pyruvate to produce lactic acid.

Figure 5.17
Bacteria act upon pyruvate to convert organic plant and animal wastes into "biogas"—a methane fuel used for heating, cooking, and, in some countries, for transportation. The biogas power-plant pictured above is in California.

chemical reactions within their cells. Today virtually every living organism, including you, still uses glycolysis.

The word "glycolysis" means "the splitting of glucose." In a series of 10 reactions, a molecule of glucose is split into two identical smaller molecules, each called **pyruvate** (*py ROO vayt*). Although the cell must use some ATP to begin glycolysis, the overall process produces more ATP than was used to initiate it. For each molecule of glucose that enters glycolysis, the cell harvests two molecules of ATP.

During glycolysis, an electron carried on a hydrogen atom is stripped from glucose. This electron is donated to an electron carrier molecule called NAD, forming NADH. For glycolysis to keep going, however, the electrons stripped away from the glucose molecules must be donated to some other organic molecule. This frees the NAD to go back and accept more hydrogens from glycolysis. Thus, glycolysis is followed by either fermentation or oxidative respiration.

Fermentation takes place in the absence of oxygen

Glycolysis evolved before the Earth's atmosphere contained oxygen. Consequently, the earliest energy-harvesting pathway did not require oxygen. The breakdown of organic compounds such as glucose in the absence of oxygen is called **fermentation**. During fermentation the hydrogen from NADH is attached to the pyruvate, forming lactic acid or ethyl alcohol (ethanol), the alcohol in beer and wine.

In the conversion to alcohol, pyruvate loses a molecule of carbon dioxide as it accepts an electron from NADH. This process regenerates NAD, which enables glycolysis to continue. Many microorganisms that live in the absence of oxygen use fermentation to produce small amounts of ATP. Some pathways of fermentation are shown in **Figures 5.16–5.18**.

■ *Matter of Fact* ■

Enzymatic Processes
Because of the economic importance of the wine industry, fermentation was the first enzymatic process to be thoroughly studied, primarily by Louis Pasteur. He was the first to recognize that yeast provides the enzymes to convert pyruvate into alcohol.

Theme Connection

Evolution
The fact that virtually every living organism uses glycolysis is another part of the molecular record supporting evolution. Other examples of this record are discussed in Chapter 9.

Demonstration 1
Glycolysis
Draw six connected circles on the board and inform students that each circle represents a carbon atom in glucose. Write glucose under the six circles. Then draw an arrow that leads to three connected circles. Place a 2 in front of these three circles. Have students explain what the three-carbon molecule is (*pyruvate*) and write its name under the circles. Also, have them explain what biochemical process should be written beneath the arrow. Include the NADH that is produced by placing an arrow connected to, but smaller than, the one used to denote glycolysis.

Demonstration 2
Fermentation
Draw two arrows leading away from the circles representing pyruvate. Have one arrow lead to the word alcohol, the other leading to the words lactic acid. In both cases, show that NADH is regenerated and can be used to keep glycolysis going. Also note that two ATPs are made, as a result of glucose being converted to either alcohol and carbon dioxide or to lactic acid.

Demonstration 3
Practical Applications
Arrange for a field trip to visit a winery, brewery, bakery, or cheese-making plant so that students can appreciate a practical application of fermentation.

Figure 5.19
Glucose is broken down into pyruvate during glycolysis. Two molecules of ATP are released. Without oxygen present, fermentation can follow, resulting in the formation of lactic acid or ethyl alcohol and carbon dioxide gas. When oxygen is present, oxidative respiration can occur. This process yields as many as 36 ATP molecules.

Fermentation occurs in your muscle cells when they are forced to operate without enough oxygen. Electrons freed by glycolysis are donated from NADH to pyruvate without the release of carbon dioxide, forming lactic acid. This process allows glycolysis to continue in your muscles as long as the supply of glucose holds out. Blood circulation removes excess lactic acid from muscles—but when lactic acid cannot be removed as fast as it is produced, your muscles cease to work well. Try raising and lowering your arms rapidly 100 times, for example. Because of this limit in removing lactic acid, the world record for running one mile is slightly under four minutes and not less.

Oxidative respiration occurs in the presence of oxygen

When Earth's atmosphere became rich in oxygen, an alternative to fermentation became possible. Instead of using the hydrogen atoms freed by glycolysis to form ethanol or lactic acid, the hydrogen atoms could now be attached to oxygen atoms, forming water. This

pathway is a wonderful alternative because the attachment process can be coupled to a proton pump and used to make more ATP. Not only that, the end product of glycolysis undergoes steps not possible in the absence of oxygen. Pyruvate is used to make even more ATP than is made during glycolysis and fermentation. **Figure 5.19** summarizes these processes.

The equation for the breakdown of glucose by oxidative respiration is:

$$C_6H_{12}O_6 + 6O_2 + ADP + P \rightarrow$$
$$6CO_2 + 6H_2O + ATP$$

This equation indicates that glucose and oxygen react to form carbon dioxide, water, and ATP molecules. Inside mitochondria, oxidative respiration picks up where glycolysis left off. Each of the two pyruvate molecules produced by glycolysis is oxidized, freeing a high-energy electron and a carbon, which is released as CO_2. The electron is donated to NAD, forming NADH, which will be used at the end of oxidative respiration.

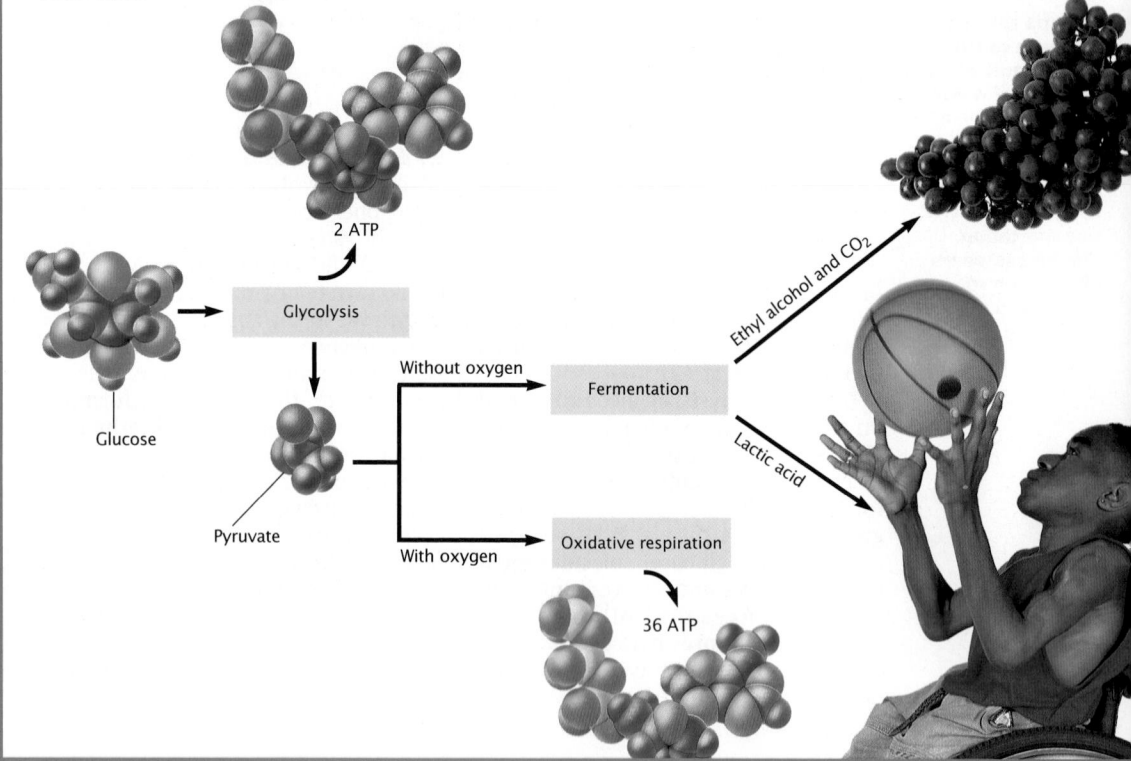

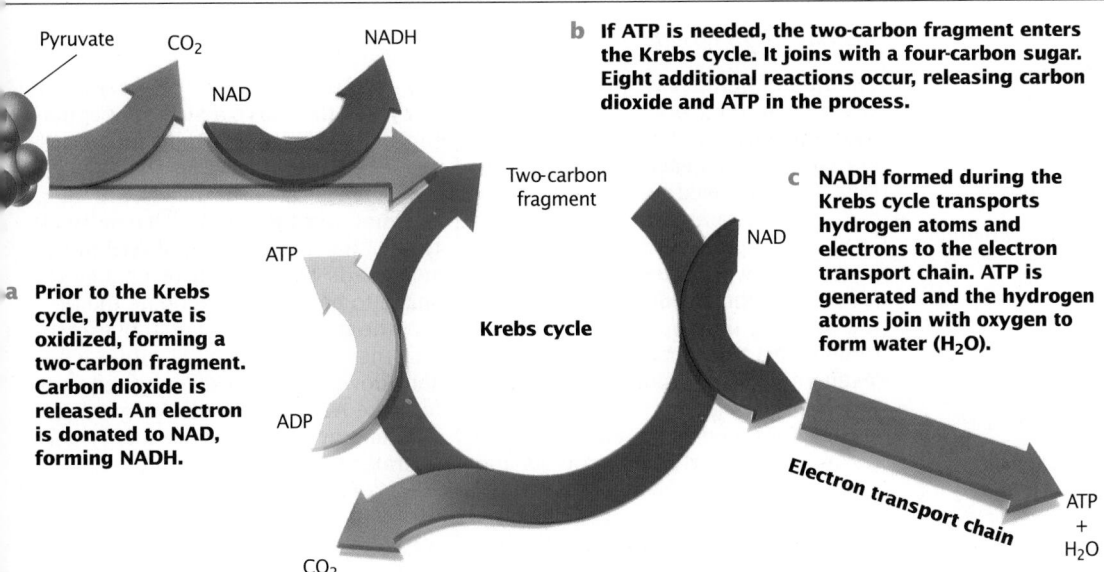

a Prior to the Krebs cycle, pyruvate is oxidized, forming a two-carbon fragment. Carbon dioxide is released. An electron is donated to NAD, forming NADH.

b If ATP is needed, the two-carbon fragment enters the Krebs cycle. It joins with a four-carbon sugar. Eight additional reactions occur, releasing carbon dioxide and ATP in the process.

c NADH formed during the Krebs cycle transports hydrogen atoms and electrons to the electron transport chain. ATP is generated and the hydrogen atoms join with oxygen to form water (H_2O).

Figure 5.20
Oxidative respiration occurs in two major steps: the Krebs cycle and the electron transport chain.

After each pyruvate is oxidized, only a two-carbon fragment remains. If the cell has enough ATP, the two-carbon fragment is funneled into fat synthesis and its energetic electrons are stored. If the cell needs ATP now, the two-carbon fragment will continue on to the next steps of oxidative respiration.

The Krebs cycle yields ATP and carbon dioxide

The two-carbon fragment, left over after the oxidation of pyruvate, joins with a four-carbon sugar. Then, in rapid-fire order, eight additional reactions occur. When the two-carbon fragment is used up, its two carbon atoms are expelled as two molecules of carbon dioxide. In addition, one ATP molecule has been made and four more energetic electrons have been harvested. All that remains is the original four-carbon sugar, now free to join with another two-carbon fragment. This cycle of nine reactions is known as the Krebs cycle, after Sir Hans Krebs, the biochemist whose work in the 1930s revealed how these reactions work. **Figure 5.20** summarizes the major events of the Krebs cycle.

The electron transport chain makes more ATP

The energetic electrons in the molecules of NADH that formed during the Krebs cycle are used to make ATP in a series of reactions known as the electron transport chain. The membranes of a mitochondrion contain proteins that serve as proton pumps. Using these proton pumps, a mitochondrion pumps protons outward. Driven by diffusion, the protons then pass back into the interior of the mitochondrion. The energy of the reentering protons is used by the mitochondrion to attach a phosphate group onto ADP, making new molecules of ATP. A living cell is never without a supply of ATP.

What happens to the electrons after the proton pumps have used their energy? The hydrogen atoms carrying them are joined to oxygen gas to form water. Because the electrons stripped from pyruvate need to find a final home, cellular respiration requires oxygen. The energy cannot be extracted from pyruvate without oxygen to siphon off the spent electrons. Otherwise, the proton pumps and other electron-ferrying components of the mitochondrion would soon become clogged with used electrons.

Oxidative respiration is very efficient. The breakdown of a molecule of glucose to pyruvate during glycolysis yields a net of only two ATP molecules, but oxidative respiration can yield as many as 36 additional molecules of ATP!

■ *Matter of Fact* ■

ATP
Each cell in the human body is estimated to use between 1 billion and 2 billion ATPs each minute. If you wish students to work with some large numbers, ask how many ATPs are used each minute by all 100 trillion cells in the body. See if they can calculate how many are used each hour. If they do, they'll certainly recognize the importance of oxidative respiration.

Visual Strategy

Figure 5.20
Have students follow pyruvate's entrance and subsequent fate in the Krebs cycle. Ask students what three molecules come out of this cycle. Tell them it's important that they understand carbon dioxide is a byproduct of chemical conversions formed during cellular respiration.

Theme Connection

Interacting Systems
Have students describe what happens to the carbon dioxide that is produced as waste during cellular respiration.

Demonstration 4

What Happens to Oxygen?
Have students imagine cars continuously entering a one-way, dead-end street. Have them describe what will happen. Tell students the Krebs cycle resembles cars on a one-way street. It keeps sending hydrogens via NADH down the electron transport system. Without oxygen, it would also be a dead end. But oxygen prevents a traffic jam of hydrogens by keeping the street open. By combining with hydrogen to form water, oxygen allows hydrogen to keep moving down the electron transport system.

Mathematics Connection

Efficiency of Aerobic Respiration
Tell students one molecule of glucose contains 686 Cal (kilocalories) of energy. Each ATP stores 7 Cal. Remind them that a total of 38 ATPs are made when one molecule of glucose undergoes oxidative respiration. Thus the efficiency is:

38 x 7 = 266 = almost 40% efficiency

Point out that an automobile engine is only about 25 percent efficient in converting the chemical energy in gasoline into mechanical energy to move the car.

Theme Connection

Stability

Use Figure 5.21 to show students that bio-chemical processes are regulated so that an organism's internal environment is maintained in a steady state of homeo-stasis. Have students explain what will happen when the level of ATP falls.

Theme Answer

Patterns of Change

When the body has produced enough ATP, feedback inhibition slows or stops an early reaction in the biochemical pathway. This occurs when a molecule binds to an en-zyme's regulatory site.

Phase 3

ASSESSMENT OPTIONS

Closure Strategy

Respiration and Growth

Have students discuss whether yeast cells would grow more quickly if they were carrying out cellular respiration with oxygen or without oxygen. Have them support their choice.

Section Review

Assign the *Section Review.*

Reteaching

Prepare a table on the board similar to the one described in the Reading Strategy on page 102. Review this section by filling in the information so that students see both the similarities and differences between photosynthesis and respiration.

Patterns of Change

How does the regulation of cellular respiration resemble a thermostat controlling the temperature in your home?

Regulating Cellular Respiration

The rate of cellular respiration slows down when your body's cells already have enough ATP. This is very sensible, but what signals each mitochondrion that it is supposed to slow down and cease ATP production? The control works through a system of feedback inhibition in which excess product shuts off the reaction. **Feedback inhibition** is the slowing or stopping of an early reaction in a biochemical pathway when levels of the end product of the path-way become high.

How does feedback inhibition work? Key reactions early in glycolysis and the Krebs cycle are catalyzed by enzymes that have a second "regula-tory" site. This site is the same shape as ATP, as shown in **Figure 5.21.**

When ATP levels in the cell are high, ATP molecules will likely become stuck to the site. The binding of ATP to the site goads the protein to change its shape to better accommodate the fit—and the new shape is not active as an enzyme. High levels of ATP thus act to shut down the processes the cell uses to make ATP. Like a well-designed car, your energy-producing machinery only operates when you step on the gas.

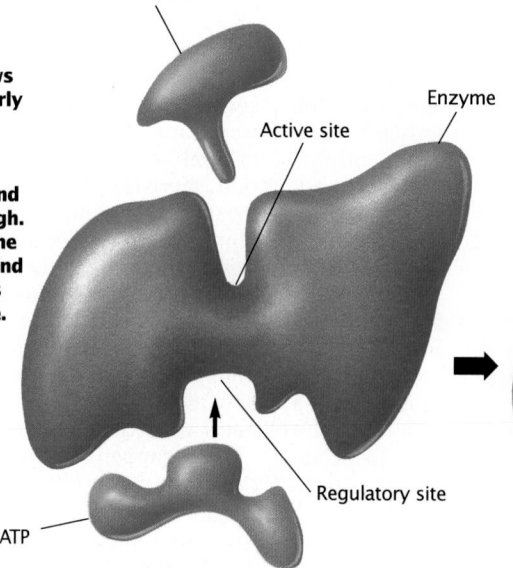
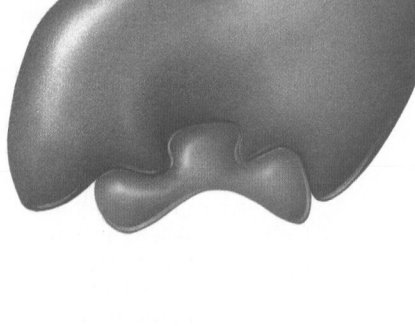

Figure 5.21 Feedback inhibition slows or stops an early reaction in a biochemical pathway when levels of the end product are high. Molecules of the end product bind to an enzyme's regulatory site.

Substrate

Enzyme

Active site

Regulatory site

ATP

a In this reaction, ATP binds with the regulatory site.

b The enzyme then changes shape. The active site is no longer able to catalyze its usual reaction, and no more ATP is made.

Section Review

1. **How is cellular respiration important for living cells?**

2. **Why do you think glycolysis might be particularly advantageous for certain organisms?**

3. **How do fermentation and oxidative respiration differ?**

4. **Explain how feedback inhibition enables cellular respiration to slow down when supplies of ATP are sufficient.**

▪ Section Review Answers ▪

1. ATP is formed during cellular respiration. The energy released when bonds in ATP are broken is used to power the chemical reactions of the cell.

2. The process does not require oxygen for cellular respiration.

3. Both provide living organisms with various amounts of energy in the form of ATP, but oxidative respira-tion occurs in the presence of oxygen and fermentation occurs in the absence of oxygen.

4. Students should suggest the control works through a system of feed-back inhibition in which excess product shuts off the reaction.

Chapter **5** *Highlights*

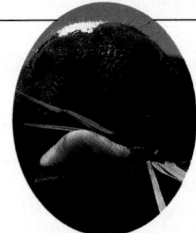

Green grass is the mainstay of this cow's diet. Using what you've learned, can you explain the link between sunlight and dairy products?

	Key Terms	*Summary*
5.1 Cells and Chemistry Enzymes, such as carbonic anhydrase, hasten chemical reactions. 	exergonic (p. 90) endergonic (p. 90) metabolism (p. 90) activation energy (p. 90) catalyst (p. 91) substrate (p. 92) active site (p. 92)	• Chemical reactions store or release energy needed for life. • Enzymes make reactions occur rapidly enough to sustain life. • Enzymes act on specific substrates, and are affected by heat, acidity, and substrate concentration.
5.2 Cells and Energy  Living things depend on the energy stored in green plants.	biochemical pathway (p. 95) ATP (p. 95) photosynthesis (p. 96) cellular respiration (p. 96)	• In a biochemical pathway, energy is released by the breakdown of chemical compounds. • The energy released is stored temporarily in a molecule called adenosine triphosphate, or ATP. • Photosynthesis captures the light energy available for life. • Cellular respiration releases energy stored in food molecules.
5.3 Photosynthesis Photosynthesis begins when light strikes a chloroplast. 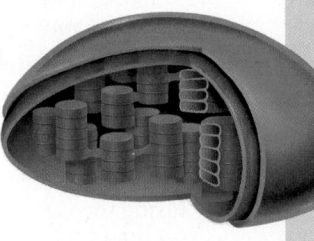	photon (p. 98) pigment (p. 98) oxidation-reduction reaction (p. 99)	• Photosynthesis evolved billions of years ago among bacteria. • Green plants and algae contain the pigment chlorophyll. • Light striking chlorophyll boosts electrons to higher energy levels. As electrons pass through a series of oxidation-reduction reactions, their energy is used to produce carbohydrates.
5.4 Cellular Respiration  Cellular respiration releases the energy necessary to carry out life activities.	glycolysis (p. 102) oxidative respiration (p. 102) pyruvate (p. 103) fermentation (p. 103) feedback inhibition (p. 106)	• Living things obtain energy from food through cellular respiration. • In glycolysis, glucose splits into two pyruvate molecules. • In the absence of oxygen, fermentation converts pyruvate to lactic acid or ethyl alcohol. With oxygen present, the Krebs cycle and the electron transport chain produce most of the ATP needed for life. • Feedback inhibition controls reactions of respiration.

Chapter Review Answers

Understanding Vocabulary

1. a. unchanged
   b. aerobic
   c. gain an electron
   d. releases energy

Relating Concepts

2. Map answer is shown on page 87D.

Understanding Concepts

Multiple Choice

3. b   8. b
4. b   9. c
5. c   10. d
6. a   11. b
7. a

Completion

12. energy
13. reduced, oxidized
14. cytoplasm, mitochondria
15. photosynthesis, respiration
16. food

Short Answer

17. Yes. In addition to producing carbohydrates by means of photosynthesis, plants also utilize carbohydrates as their source of energy. Like all living organisms, plants break down carbohydrates to release the stored energy through cellular respiration.

18. The plant would turn yellow and eventually die. This would occur because the plant cannot obtain the light energy needed to make ATP, the energy source needed to build glucose molecules needed by the plant.

19. carbon dioxide

20. photosynthesis

Interpreting Graphics

21. • glucose and oxygen
    • glucose and oxygen
    • The products of photosynthesis are the reactants for respiration.

22. a

Reviewing Themes

23. All animals would also die since animals either eat plants or eat other animals who eat plants.

Chapter 5 Review

Understanding Vocabulary

1. For each set of terms, complete the analogy.
   a. reactants:change ::enzyme:_____
   b. fermentation: anaerobic:: oxidative respiration:_____
   c. oxidation:lose an electron:: reduction:_____
   d. photosynthesis: stores energy:: cellular respiration:_____

Relating Concepts

2. Copy the unfinished concept map below onto a sheet of paper. Then complete the concept map by writing the correct word or phrase in each oval containing a question mark.

Understanding Concepts

Multiple Choice

3. Enzymes in the human body
   a. are changed during the reaction they catalyze.
   b. decrease the activation energy of a chemical reaction.
   c. are completely burned in the reactions.
   d. increase the speed of proton pumps.

4. The sum of all chemical reactions in the body is called
   a. meiosis.
   b. metabolism.
   c. respiration.
   d. activation energy.

5. Chemical reactions require energy to get started. What is this energy called?
   a. metabolism
   b. chemical energy
   c. activation energy
   d. potential pressure

6. Which is an example of a biochemical pathway?
   a. glycolysis   c. ATP
   b. pyruvate   d. catalyst

7. The formation of ATP involves the
   a. addition of a phosphate to ADP.
   b. reaction of chemical stimuli.
   c. bonding of carbon molecules.
   d. release of energy and ADP.

8. The energy supplied by the oxidation of glucose is used to
   a. break down ATP.
   b. make ATP from ADP.
   c. produce starch.
   d. change carbon dioxide to oxygen.

9. Products of glycolysis include
   a. chemical pathways.
   b. ATP and alcohol.
   c. pyruvate and ATP.
   d. glucose and energy.

10. When oxygen is unavailable, muscle tissue converts pyruvate to
    a. alcohol.   c. enzymes.
    b. ATP.   d. lactic acid.

11. The ultimate goal of photosynthesis is to
    a. make ATP from carbon dioxide.
    b. construct carbon-containing molecules that serve as an energy source.
    c. convert ADP to ATP by using energy from the sun.
    d. use enzymes to speed up chemical reactions.

Completion

12. Cells need _____ in order to move, transport food, and get rid of waste.

13. In oxidation-reduction reactions, the molecule that gains the electron is _____ , while the molecule that loses the electron is _____ .

14. Inside the cell, glycolysis occurs in the _____ , but oxidative respiration occurs in the _____ .

15. The sun's energy is used to produce food during _____ , while energy stored in food is released during _____ .

16. The energy needed by the human body to carry out the processes of life comes from _____ .

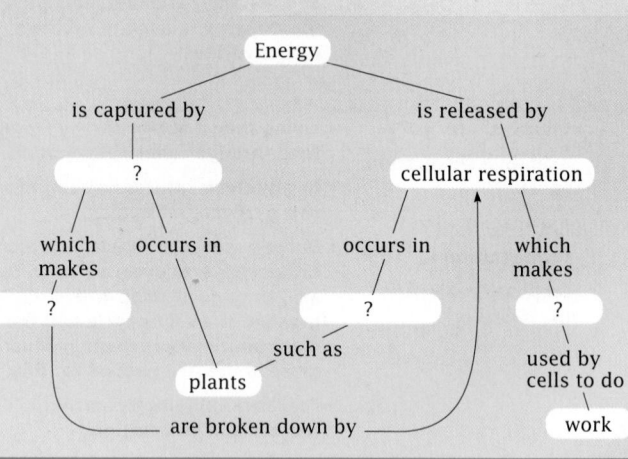

Short Answer

17. Do plants perform cellular respiration? Support your answer.

18. If you tried to grow a plant in a dark closet, what would happen? Explain your answer.

19. What is a green plant's source of carbon for photosynthesis?

20. Where does the glucose come from that the cell oxidizes or breaks down during respiration?

Interpreting Graphics

21. Look at the processes below.

Photosynthesis:

$$6CO_2 + 6H_2O \xrightarrow{\text{(light)}} C_6H_{12}O_6 + 6O_2$$

Oxidative respiration:

$$C_6H_{12}O_6 + 6O_2 + ADP + P \rightarrow$$
$$6CO_2 + 6H_2O + ATP$$

- What are the products of photosynthesis?

- What are the reactants for oxidative respiration?

- How are photosynthesis and respiration related?

22. Match the photograph below with one of these three pathways of cellular respiration:

a. Fermentation → lactic acid
b. Fermentation → ethyl alcohol and CO_2
c. Glycolysis

Reviewing Themes

23. *Interacting Systems*
If all the plants on Earth died, what would happen to the animals?

24. *Interacting Systems*
How does the feedback system of cellular respiration operate?

25. *Evolution*
What evidence suggests that glycolysis is a more ancient process than oxidative respiration?

Thinking Critically

26. *Inferring Conclusions*
Temperature affects the rate of enzyme reactions. How does freezing food affect the enzyme activity of organisms that cause decay?

27. *Comparing and Contrasting*
How is the interaction of a lock and key analogous to the interaction of an enzyme and its substrate?

28. *Inferring Conclusions*
What happens to the carbon dioxide gas produced by yeast cells when bread is baked?

29. *Building on What You Have Learned*
In Chapter 3 you read about the proton pump. How does this pump function during photosynthesis?

Cross-Discipline Connection

30. *Biology and Social Studies*
Yogurt is a fermented dairy product with an unusual history. Use library resources to find out how fermentation has been used in the production of yogurt, cheese, soybean products, wine, and beer.

Discovering Through Reading

31. Read the article "Oxygen, the Great Destroyer," in *Natural History*, August, 1992, pages 46–53. How do plants protect themselves from oxidation?

24. When adequate supplies of ATP are achieved, ATP molecules bind to sites on enzymes that catalyze critical reactions in glycolysis and the Krebs cycle. The binding of the ATP molecules to the enzymes stops the enzymes from catalyzing the reactions, thus reducing the amount of ATP produced.

25. All organisms use glycolysis, either alone or as a prelude to oxidative respiration. Also, the earth's early atmosphere is believed to have had no oxygen; glycolysis could have taken place in this early atmosphere, since it requires no oxygen.

Thinking Critically

26. Freezing stops the action of enzymes in organisms that cause decay, thus preserving food.

27. Only certain keys will fit in a lock; likewise, only one or very few substrates will bind with each enzyme.

28. The carbon dioxide gas is released, causing the bread to rise.

29. The proton pump moves protons into the chloroplast where they collect. The protons then escape the chloroplast through a protein channel. It is here that the energy from the excited proton is used to make ATP by adding a phosphate group to ADP.

Cross-Discipline Connection

30. Answers will vary. Students should find that fermentation plays many commercial roles.

Discovering Through Reading

31. Oxygen in its free state is a powerful oxidant. Plants have the capacity to produce chemical elements that control oxidative damage.

Procedural Note

Discuss the answers to the preparation questions. Ask the students to read the procedures to be used in the investigation before beginning their work. Review the purpose of a hypothesis in an experiment and clarify the independent and dependent variables involved in this investigation.

Prelab Preparation Answers

1. • A hypothesis is a testable explanation for an observation.
   • A control experiment is one in which the condition suspected to cause the effect is compared to the same situation without the suspected condition (a control group).
2. • During photosynthesis, green plants and algae capture sunlight and use the light energy to convert carbon dioxide and water into energy-storing carbohydrates. The reactants are carbon dioxide and water. The products are carbohydrates and oxygen.
   • The rate of photosynthesis refers to the rate at which the reaction takes place.
   • Answers may vary. Oxygen production would be easier to measure than carbohydrate production because carbohydrates are retained in the cell.
   • Light intensity, the wavelength or color of light, and temperature are some of the physical factors that could be suggested.

Procedure Answers

2. Answers will vary but should provide a rationale for a logical hypothesis.
3. Temperature is the variable being tested; the rate of photosynthesis is the variable being measured.

Investigation

How Does Temperature Affect the Rate of Photo-synthesis?

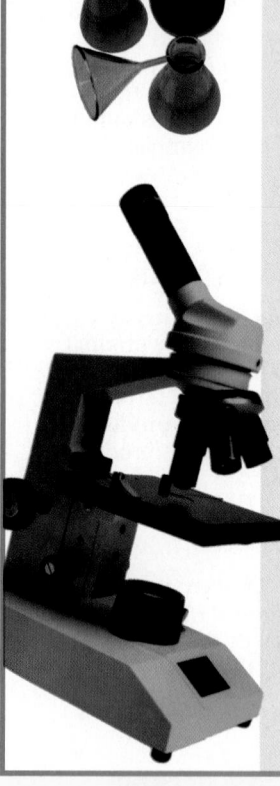

Objectives

In this investigation you will:
• *measure* the rate of photosynthesis
• *evaluate* the effect of temperature on the rate of photosynthesis

Materials

• water
• 600-mL beaker
• sodium bicarbonate ($NaHCO_3$)
• *Elodea* or other aquatic plant
• glass funnel
• test tube
• thermometer
• watch or clock with second hand

Prelab Preparation

1. Review what you have learned about a scientific method by answering the following questions:
   • What is a hypothesis?
   • What is a control experiment?

2. Review what you have learned about photosynthesis by answering the following questions:
   • Summarize the process of photosynthesis. List the reactants and the products.
   • What is meant by rate of photosynthesis?
   • Which product of photosynthesis would be easiest to measure in an aquatic plant? Explain your reasoning.
   • List the physical factors that might affect the rate of photosynthesis.

Procedure: Testing a Hypothesis

1. Form a cooperative team with a classmate to complete steps 2–10.

2. You and your partner are to design an experiment that demonstrates the effects of temperature on the rate of photosynthesis. With your partner, construct a hypothesis that you will test in your experiment. *Explain the reasoning for your hypothesis.*

3. Design a control experiment that tests your hypothesis. *What variable is being tested in your experiment? What variable is being measured in your experiment?* Be sure to include a table to contain your observations and data. Proceed with your experiment only after your experimental design and your table have been approved by your teacher.

4. Pour water into the 500-mL beaker until it is half full. Dissolve 3 g of sodium bicarbonate in the water.

5. Place a 6–10 cm (2–4 in.) length of *Elodea* in the bottom of the beaker. Place the glass funnel over the plant.

6. Fill a test tube with water. Placing your thumb tightly over the mouth of the test tube, invert the test tube, and place it over the end of the funnel.

7. Use a thermometer to find the temperature of the water in the beaker.

8. Set the beaker in direct sunlight or under a bright light. After two minutes observe the funnel. Record your observations.

9. Count the bubbles as they rise in the funnel's tube for a period of 210 seconds, recording the total number of bubbles at 30-second intervals.

10. Measure the temperature of the water again. If the temperature has changed by more than three degrees, find the average of the two temperatures. Record this average in your table as the water's temperature.

11. Pour out the water from the beaker and replace it with water differing in temperature by five to eight degrees. Add 3 g of sodium bicarbonate to the water and repeat steps 5–10.

12. On a separate sheet of paper, draw a graph like the one shown below. Graph your data. Use a separate line for data collected at each temperature.

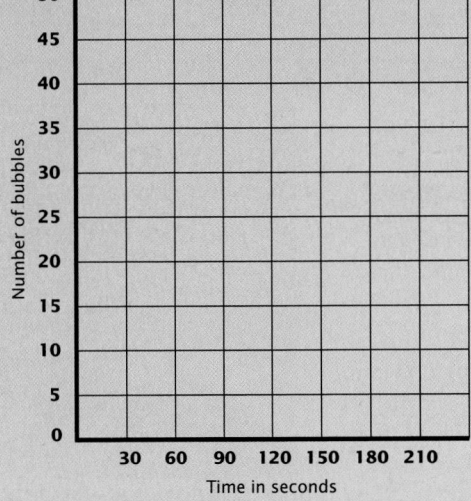

13. Clean up your materials and wash your hands before leaving the lab.

Analysis

1. *Summarizing Data*
   Summarize your results by describing the data illustrated in your graph.

2. *Analyzing Data*
   State a conclusion that relates to your hypothesis. Explain how your data support your conclusion.

3. *Making Inferences*
   Explain why the data give an indication of the rate of photosynthesis.

4. *Analyzing Relationships*
   Can you assume that raising the temperature of the water to 50°C (122°F) would continue the trend indicated by your data? Explain your answer.

Thinking Critically

1. Suggest a reason for adding sodium bicarbonate to the water in the beaker. Describe a control experiment that would determine the effects of sodium bicarbonate on the rate of photosynthesis.

2. What other physical factors might be affecting your experimental results?

Analysis Answers

1. Answers may vary but should adequately summarize the data collected during the experiment.
2. Answers may vary but should relate to the hypothesis.
3. It is expected that the data will show that the rate of photosynthesis is directly related to the temperature. The bubbles are assumed to be oxygen that is produced during photosynthesis.
4. No, because such high temperatures are likely to destroy enzymes, slowing or stopping photosynthesis.

Thinking Critically Answers

1. Adding sodium bicarbonate to the water provides a source of carbon dioxide for the *Elodea*. To determine the effects of sodium bicarbonate on the rate of photosynthesis, a control experiment would hold all other variables constant while comparing photosynthetic rates between tubes with and without sodium bicarbonate.
2. Answers will vary. Factors such as the distance between the light source and the plant, the amount of water in the tube, the health of the *Elodea*, and the concentration of carbon dioxide in the water might be mentioned.

Using the Feature

- Present the theory of spontaneous generation and show how it was eventually discredited through the experiments of Francesco Redi, Lazzaro Spallanzani, and Louis Pasteur. Have students research the great debate about the origin of life. Guide students to develop an appreciation for the complex nature of scientific inquiry, and to gain historical perspective on such a great scientific debate by simulating the debate. Divide students into groups. Discuss the pros and cons of spontaneous generation based on the evidence that could be gathered within that time frame. Discuss the implications of the theory of biogenesis. If living things come only from other living things, how did life on Earth begin?

- Guide students to develop an appreciation for the complex nature of scientific inquiry as you explain the cell theory. Tie it with Chapter 2 as you discuss cellular structure as one of the five characteristic properties of life. Name the contributors to the cell theory and what each contributed to its development.

- Introduce exobiology. Explain that exobiology is not just the study of extraterrestrial life, but that it includes the study of the origin of life on Earth. Familiarize students with developments that scientists infer must have occurred for life to begin on Earth. Build upon the concepts of basic chemistry introduced in Chapter 2, the cell processes described in Chapters 3, 4, and 5, and the introduction of biochemistry in Chapter 5. Have students research modern theories on the origin of life and report their findings to the class.

Discoveries in Science ▪ Discoveries in Science ▪ Disco

WHAT IS LIFE? SEARCHING FOR ANSWERS

1500

1500 Observations of fishes in dried ponds, and snakes and frogs emerging from the mud of riverbanks lead many people to believe that life **generates spontaneously** from non-living matter.

1600 Belgian physician **Jan Baptist van Helmont** places wheat grains in a sweaty shirt and returns 21 days later to find mice. He concludes that human sweat changes wheat grains into mice.

Anton van Leeuwenhoek

1668 The first to observe animalcules (bacteria) through a lens, **Anton van Leeuwenhoek**, a Dutch amateur scientist, reports his findings to scientists of The Royal Society of London. They conclude that larger animals may not be able to arise from nonliving matter, but microscopic organisms can.

Redi's experiment

1668 Observing maggots and flies developing on decaying meat, Italian physician **Francesco Redi** questions spontaneous generation. He concludes that if flies are kept away from rotten meat, maggots do not develop.

1920

1924 Soviet biochemist **A.I. Oparin** and British geneticist **J.B.S. Haldane** suggest that life may have come from nonliving matter on primitive Earth, when the atmosphere consisted of methane, ammonia, water vapor, and hydrogen. Energy supplied by electrical storms and ultraviolet light may have broken down atmospheric gases into elements, which reacted to form amino acids, the building blocks of proteins.

1953 American scientists **Stanley L. Miller** and **Harold C. Urey** of the University of Chicago test Oparin's hypothesis. Their experiment generates four amino acids, among other molecules.

1960 Exobiology, a term coined by American geneticist Joshua Lederberg, is the study of extraterrestrial life origins on Earth, or the existence of life outside Earth or its atmosphere.

Carl Sagan

1974 American astronomer **Carl Sagan** and other scientists from Cornell University beam a signal containing coded messages into outer space, with hopes of hearing from extraterrestrial life forms.

cience ▪ Discoveries in Science ▪ Discoveries in Science

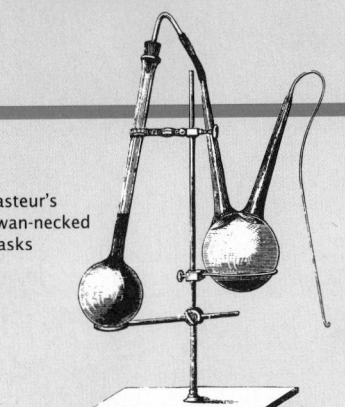
Pasteur's swan-necked flasks

1860

1750 The English Catholic priest **John Needham**, in support of his belief in spontaneous generation, reenacts Redi's experiments but concludes that life generates spontaneously.

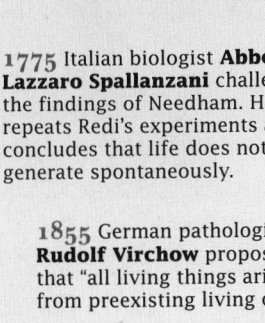

1775 Italian biologist **Abbe Lazzaro Spallanzani** challenges the findings of Needham. He repeats Redi's experiments and concludes that life does not generate spontaneously.

1855 German pathologist **Rudolf Virchow** proposes that "all living things arise from preexisting living cells."

1860 French chemist **Louis Pasteur** ends the spirited controversy over spontaneous generation. In a series of experiments, Pasteur proves that microorganisms appear only as contaminants from the air and not "spontaneously."

Louis Pasteur

2004

1976 Despite data collected on **Viking Missions 1 and 2**, scientists still question whether life exists on Mars.

1992 Program Sentinel (a radio telescope with 131,072 channels)—a project of **SETI** (Search for Extraterrestrial Intelligence)—sifts through a shower of radio waves falling on Earth for signals indicating extraterrestrial intelligence.

1985 Alexander Graham Cairns-Smith proposes that naturally occurring clay crystals may have served as the original templates for assembling amino acids into protein chains, providing the fundamental materials from which life originated.

2000 An **antenna** is proposed to be set up on the far side of the Moon to evaluate radio signals for possible intelligent extraterrestrial life.

Human expedition to Mars

2004 The focus of **future NASA missions**: settling the issue of life on Mars, past or present.

Moon

Discussion

Guide the discussion by posing the following questions.

1. State the theory of spontaneous generation, and tell how people used the theory to explain what they observed as the origin of life.
   Students should suggest that the theory of spontaneous generation states that living organisms can arise from nonliving matter. Accept any of the timeline examples starting on page 112.

2. What would have happened if Pasteur had tipped one of his flasks so that the broth in the flask came into contact with the inside of the curved neck?
   Students should explain that the curved-necked flask was used to allow air to enter and reach the sterile broth, but not dust or other particles. Had the flask tipped, the sterile broth would have become contaminated by organisms that were found inside the flask's opening but were unable to reach the broth, supporting Pasteur's conclusions.

3. Is a chemical reaction involved in Oparin's theory concerning the origin of life? Explain your reasoning.
   Students should respond with a yes, suggesting that a chemical reaction is the process of breaking chemical bonds to form new bonds. Also, they should explain that the suggested energy supplied by electrical storms and ultraviolet light was responsible for breaking down atmospheric gases into elements, which reacted to form amino acids.

4. Why is it necessary for the cell to take on different shapes?
   Explain how this is an advantage to organisms. Students should suggest that if only one cell shape was available, some structures would be inefficient and some life forms would not exist. A variety of cell shapes allows an organism to build structures with varied functions that have a greater degree of efficiency.

5. What gives a protein its unique characteristics?
   Students should suggest the sequence of amino acids and the resulting shape of the protein molecule.

113

Unit 2

Continuity of Life

Of all life's mysteries, one of the most puzzling has been heredity. Why do we resemble our parents? What mechanism told the tiny embryo that became you how to develop? We are now able to answer that question, for biologists have largely solved the mystery of heredity. In this unit, you will discover that the mechanism consists of chromosomes, which pass instructions from parent to child about how to grow and develop. Learning how heredity works has been one of the greatest triumphs of science.

Chapter 6

Genetics and Inheritance

Planning Guide

	Objectives/Themes	Classwork Resources	Homework Resources

6.1	1. Explain how Mendel discovered the laws of heredity using the information available to him.
	
	2. Outline the garden pea experiments of Gregor Mendel.
	
	3. Define the terms heterozygous, homozygous, dominant, and recessive.
	
	4. Compare and contrast each of Mendel's laws of heredity.
	
	Themes: Patterns of Change, Stability, Interacting Systems
		Teacher's Resource Binder
		Focus Activity 6
		  Interpreting Information in a Pedigree
		Extension Worksheet
		  Incomplete Dominance
		
		Other Resources
		Science in Action Video
		  *Cell Structure and Function/Cell Division**
		Transparencies 22–24
			Text
			Section Review, p. 122
			
			Teacher's Resource Binder
			Directed Reading Worksheet 6.1
			
			Other Resources
			Audiocassette 6.1*

6.2	1. Explain the relationship between genes and chromosomes. 2. Summarize the events that occur during meiosis. 3. Define crossing over and explain its role in evolution. 4. Describe how sex chromosomes determine the sex of humans. **Themes:** Evolution, Interacting Systems, Patterns of Change, Scale and Structure, Stability	**Teacher's Resource Binder** Lab Investigation 6.1 *The Case of the Long-Lost Son* **Other Resources** Transparency 25	**Text** Section Review, p. 126 **Teacher's Resource Binder** Directed Reading Worksheet 6.2 **Other Resources** Audiocassette 6.2*

6.3	1. Recognize the relationship between mutation and human genetic disorders.
	
	2. Explain the patterns of inheritance of cystic fibrosis and sickle cell anemia.
	
	3. Explain the purpose of genetic counseling.
	
	4. Describe some of the benefits that gene technology might offer.
	
	Themes: Evolution, Interacting Systems, Patterns of Change, Scale and Structure
		Text
		Investigation
		  What Is Your Genetic Profile?
		  pp. 134–135
		
		Teacher's Resource Binder
		Lab Investigation 6.2
		  A Human Pedigree
			Text
			Section Review, p. 130
			
			Teacher's Resource Binder
			Directed Reading Worksheet 6.3
			Vocabulary Review Worksheet*
			Reteaching Worksheet*
			  Genetics and Inheritance
			
			Other Resources
			Audiocassette 6.3*

*Reteaching Options

Demonstrations
6.1: pp. 118, 119, 120, 121, 122
6.2: p. 123
6.3: p. 128

Assessment
Chapter Review pp. 132–133
Portfolio Assessment p. 115D
Chapter Test—Teacher's Resource Binder
Test Generator

Research Notes

Connection to Medicine: Natural Selection and Diabetes

An epidemic of type II diabetes on the island of Nauru in the South Pacific may provide some evidence of natural selection in humans. The tiny island nation, which has a population of about 7,000, has a 60 percent rate of diabetes among older adults. In the United States, the rate is only 8 percent among a comparable part of the population.

Type II diabetes is also known as non-insulin dependent diabetes mellitus (NIDDM), because the body of an individual with this disease still produces insulin, but the cells do not respond to it appropriately.

Type II diabetes has both genetic and environmental risk factors. Among the environmental risk factors are low physical activity, high calorie intake, and obesity. These risk factors, which are characteristics of the lifestyle of many Western nations, are found in many parts of the developing world among people who experience sudden affluence.

The Nauruans had a vigorous lifestyle of fishing and farming until colonization by the British and Australians. The discovery of valuable phosphate mines on the island made them suddenly one of the richest peoples in the world. Their patterns of living changed, to include a sedentary life and an imported diet of high-calorie foods. As a result of these changes in environmental risk factors, a diabetes epidemic grew. The Nauruans had one of the world's shortest average lifespans.

In the past ten years, the incidence of diabetes has declined in the population, dropping to less than half of earlier levels. There has been little change in the environmental risk factors that lead to the disease, so the change could reflect changes in the genetics of the population. Perhaps those Nauruans genetically disposed to the disease have developed it already, and died, leaving behind a population with a higher proportion of genetically resistant individuals. Also, diabetic women in Nauru have less than half as many live births as other women do. Thus, susceptibility to diabetes is slowly becoming less frequent.

The NIDDM genotype must have been advantageous at one time for it to have spread so widely throughout the island's population. One hypothesis is that a fluctuating supply of food favored individuals who could binge in times of plenty and convert most of those calories into fat for times of scarcity.

The Nauruans' ancestors first immigrated to the island on long canoe voyages, and the island had a history of droughts and crop failures. During World War II, more than one-quarter of the islanders died of starvation when most of the population was deported by the Japanese. Such repeated pressures would have favored those who could store more food energy as fat.

Investigation Notes

What Is Your Genetic Profile?

pages 134–135

Purpose: Most students will find the exercise of developing a hereditary profile of themselves fun and interesting. Many students associate dominance with prevalence, and the five-finger trait is used to dispel this fallacy.

Technique: Have the students pool their data on the chalkboard in order to calculate the frequency of each trait in the class.

Several of the traits in this investigation do not follow the rules of Mendelian inheritance. However, students need not be aware of this to meet the objectives of this Investigation. (For more details, see *Mendelian Inheritance In Man* by McKusick, Johns Hopkins Press.)

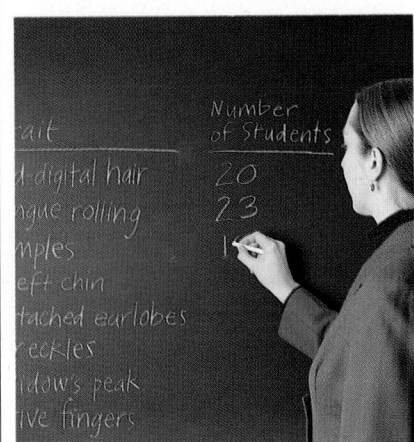

Prelab Preparation

A mirror can be used, if you prefer that the students work alone.

Answers will be found on pages 134–135.

Meeting Individual Needs

Objectives

1. Students will demonstrate appropriate use of core vocabulary for the chapter (Vocabulary File).

2. Students will develop an understanding of genetic patterns by modeling a pure breeding cross with cards (Demonstration A).

Vocabulary File
(Developing Vocabulary/ Limited English Proficiency)

If you are not already using the Vocabulary File, refer to Chapter 1 for its preparation. See the Chapter Highlights on page 131 for a list of suggested words.

Demonstration A
The Puzzle of Heredity
for pages 120–122
(Developing Organizational Skills/Kinesthetic Learners)

Materials: 18 5-by-8 inch non-ruled index cards, 2 black markers, poster boards, tape

Preparation: Print "Heredity Puzzle" on one side of the cards in large, bold letters. Turn cards over and prepare as follows:

Set 1—WW, ww;
Set 2—W, W, w, w;
Set 3—Ww, Ww;
Set 4—Ww, Ww;
Set 5—W, W, w, w;
Set 6—WW, Ww, Ww, ww.

Prepare the poster boards as shown in the diagrams. Display posters on the chalkboard tray or some other accessible spot with high visibility.

Poster Board Diagram #1

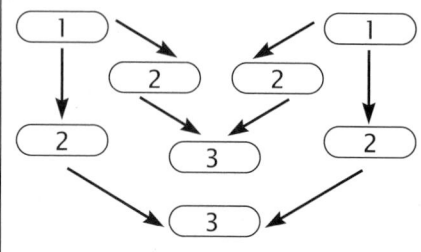

Poster Board Diagram #2
Possible Combinations that could occur in the F$_2$ Generation

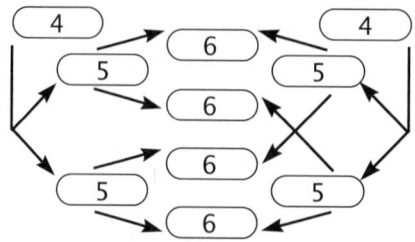

Procedure: Relate the difficulty of understanding genetics by observing changes in appearances to a mystery, in which we piece together clues that we can observe to understand things that we can't observe. Tell students that they will now re-create the solving of the genetic puzzle. Place Set 1 cards in your hand so that the side viewed by the students reads "Heredity Puzzle." Have a student take the cards from your hand and show them to the rest of the class. Brainstorm until students recognize that these represent the genotypes of pure-breeding parents. Redefine "homozygous."

Then, the student who drew the cards from Set 1 should tape them to the first chart (the one that matches Poster Board Diagram #1) in the places numbered "1." To the right of the diagram, have the student print "Pure Breeding Parental Generation—Homozygous." Have the class tell the student which set should be labeled as "recessive" and which should be described as "dominant."

Ask what happens when sperm and eggs (gametes) are formed from these parents. Guide answers to point out that the factors separate. Have another student take the Set 2 cards. Brainstorm until students recognize that these are the possible gametes that can be formed from the breeding pair (Set 1). Have the student tape the cards to the chart in places numbered "2," with the class checking the placement for accuracy. To the right

of the diagram, have the student print "Factors Separate When Gametes Are Formed."

Ask what is likely to be the outcome of the combination of the gametes. Have another student take the Set 3 cards. Brainstorm until students recognize that the cards in Set 3 show the possible combinations of the gametes in Set 2, which came from the breeding pair in Set 1. Have the student tape the cards to the diagram. Ask the student to print "An F$_1$ generation is produced." Mention "homozygous" and redefine "heterozygous." Ask the class which one the student should write underneath each of the offspring.

Tell students that if individuals in the newly formed F$_1$ generation are crossed with others in the F$_1$, the F$_2$ generation is produced.

Have another student take the cards in Set 4. Brainstorm until students understand that they represent members of the F$_1$ generation. Tell the students that the poster board which matches Poster Board Diagram #2 will be used to investigate the cross of two members of the F$_1$ generation. Have the student tape Set 4 cards on the appropriate spaces. Have the student print "F$_1$ Breeding Parental Generation." Ask the class whether the student should add "heterozygous" or "homozygous" to this description.

Follow steps similar to those for Sets 5 and 6 as were used for Sets 2 and 3. Have students indicate which of the following labels should be applied to the F$_2$ offspring: "heterozygous," "homozygous recessive," or "homozygous dominant."

Reteach the Punnett square model by showing how the results of the two crosses shown in the completed sheets would be recorded in a Punnett square.

Meeting Individual Needs (cont.)

Additional Strategies

Visual Strategies

Pages 117, 118, 120, 121, 124, 125, 126, 127, 128, 129, and 130

Auditory Learners

Use *Biology: Visualizing Life* Audiocassettes for Sections 6.1, 6.2, and 6.3.

Cooperative Learning

Probabilities

Timing: Use this activity to introduce Section 6.1.

Group Size: 4 students

Outcome: Students will predict the probability of each of the possible results of tossing two different coins, and they will test their predictions.

Individual Accountability: Each group member will record the results of tossing two different coins 25 times.

Positive Interdependence: Each group will reach a consensus on predictions and construct a table showing the results of tossing two different coins 100 times.

Explain the product rule for finding the probability of two or more independent events occurring together. Have each group member calculate the probability of all the possible results of tossing two different coins and compare their calculations with the rest of the group. Each group should reach a consensus on a set of group predictions.

Then, have group members take turns tossing two different coins and recording the results in a table until each group has made 25 tosses, for a total of 100 tosses for the group. Have groups compare predicted results with actual results and explain what they find and why to the class. Ask students to explain how flipping two coins can be compared to producing a generation of offspring. In what ways is this analogy imprecise?

Portfolio Assessment

Students should select their best work and provide a self-reflective rationale for their selections. Students can make selections in the following areas.

1. *Content* — One concept map from the chapter (See page 812 for evaluation criteria.)

2. *Reading Comprehension* — One Directed Reading Worksheet from the Teacher's Resource Binder (Use the answer key to evaluate for accuracy.)

3. *Writing* — Using the Vee Form, summarize a newspaper or magazine article relating to heredity. (See page 22T for evaluation criteria.)

   Or: Select a writing task or project from the Chapter Review.

4. *Performance Assessment* — One Vee form from a chapter investigation or lab manual investigation (See page 22T for evaluation criteria.)

Teacher makes selections in the following areas.

1. *Formal Assessment* — Chapter test (Test A, B, or the Test Generator) The teacher-scored test should be reviewed by the student. Incorrect responses should be corrected by the student before the test becomes part of the portfolio.

2. *Informal Assessment* — Use the Direct Observations Checklist, page 33T, during a laboratory or other cooperative learning experience.

3. *Performance Assessment* — Have students examine several traits among their family and extended family, such as eye color, hair color, height, etc. For each trait, have students indicate on a pedigree chart the patterns of heredity involved.

Concept Map Answer

The following is one possible answer to the Relating Concepts exercise on page 132.

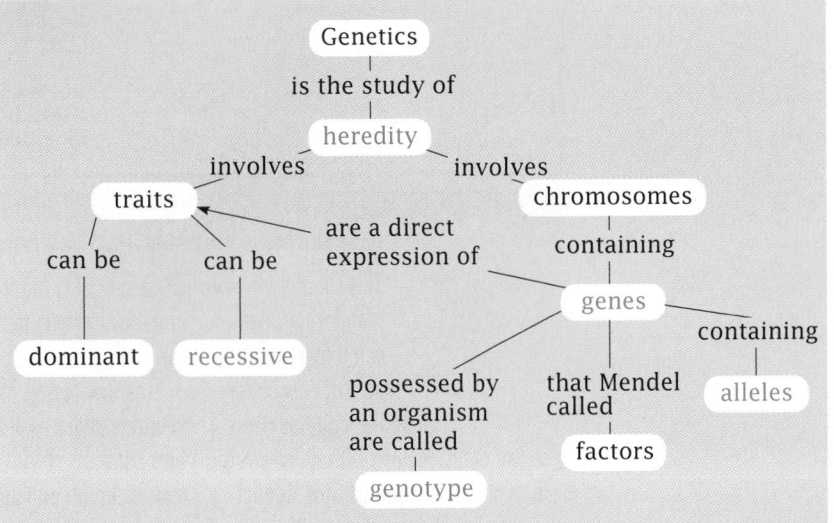

Chapter 6

Determining Prior Knowledge

- Have students give examples of traits that are inherited. Seek examples from a variety of organisms.
- Have students explain how these traits are passed from one generation to the next.
- Review chromosome activity during mitosis. Have students predict what the consequences would be if gametes were formed by mitosis.
- Display a chart used to test for color-blindness. Check to see if anyone is colorblind. Have students discuss why someone is colorblind.

Chapter 6

Review

- heredity (Section 1.3)
- chromosome (Section 4.3)
- mitosis (Section 4.4)
- the term *gene* (Glossary)

Genetics and Inheritance

6.1 The Puzzle of Heredity

- Gregor Mendel and the Garden Pea
- Visualizing Mendel's Model
- Mendel's Laws of Heredity

6.2 Chromosomes

- Genes and Chromosomes
- How Gametes Form: Meiosis

6.3 Human Genetic Disorders

- Mutations Are Changes in Genes
- Four Genetic Disorders
- Genetic Counseling and Technology

No two people are exactly alike. What you look like depends on what combination of genes you have inherited from your parents.

■ Author's Rationale ■

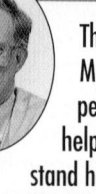

This chapter reviews Mendel's garden pea experiments to help students understand heredity as gene segregation. Meiosis is then introduced to explain the patterns that Mendel found. The chapter closes with a discussion of mutations, human genetic disorders and genetic counseling.

The key lesson of this chapter is that heredity—the most important criterion of life—is not mysterious. It can be explained by simple mechanisms and should be clearly understood for students to be biologically literate.

THE TRANSMISSION OF TRAITS FROM PARENTS TO THEIR OFFSPRING IS CALLED HEREDITY. FOR HUNDREDS OF YEARS, PEOPLE WERE PUZZLED BY THE PATTERNS OF HEREDITY. ONLY IN THIS CENTURY HAVE SCIENTISTS BEGUN TO UNDERSTAND THE MOLECULAR BASIS OF HEREDITY, A DISCOVERY THAT HAS ITS ROOTS IN THE ELEGANT EXPERIMENTS OF GREGOR MENDEL.

6.1 The Puzzle of Heredity

Objectives

❶ **Explain how Mendel discovered the laws of heredity using the information available to him.**

❷ **Outline the garden pea experiments of Gregor Mendel.**

❸ **Define the terms heterozygous, homozygous, dominant, and recessive.**

❹ **Compare and contrast each of Mendel's laws of heredity.**

Gregor Mendel and the Garden Pea

Figure 6.1
Mendel studied different traits of the garden pea, *Pisum sativum*, in an attempt to understand heredity.

The scientific study of heredity is called **genetics** (*juh NEHT ihks*). The basis for this science began when scientists crossed, or bred, varieties of animals and plants in an attempt to understand heredity. For example, in the 1790s the British farmer T.A. Knight crossed a type of garden pea plant that had purple flowers with one that had white flowers. What color flowers do you think the offspring had? People at that time thought traits blended the way paint colors mix. If heredity is a blending process, wouldn't all the flowers be lavender? But all of the offspring of Knight's cross had purple flowers! If two of these purple-flowering offspring were crossed, however, some of their offspring had purple flowers and some had white. Why was this so? Knight's studies puzzled people for many years.

Some of the most important pieces in the complex puzzle of heredity were assembled in the garden of an Austrian monastery in the 1860s. There lived Gregor Mendel, shown in **Figure 6.1**, a monk who was interested in both science and mathematics. Curious about the patterns of heredity, Mendel repeated the garden pea plant crosses done by Knight and others. The difference between Mendel's experiments and those done by earlier researchers was that Mendel counted the pea plants resulting from each cross. Mendel worked diligently to express the results of his experiments in terms of numbers. This quantitative approach to science, which counted and measured data, was just becoming fashionable in Europe. Mendel hoped that this detailed and numerical procedure would give some hint of how hereditary processes work. The results of his experiments changed the course of biology.

▪ Matter of Fact ▪

Approaches to Science Differ
As Mendel formulated the basis of heredity in Austria, Charles Darwin explained the basis of evolution in England. A major problem with Darwin's theory of natural selection was the inability to explain variations in a population. Mendel, whose work was not recognized in his lifetime, knew of Darwin's work. As a footnote in the history of science, a book was found in Darwin's library. One page, a scientific paper of no significance, is covered with Darwin's notes. On the opposite page bearing Mendel's results, Darwin made not a single note.

Perhaps, Darwin was unable to recognize the importance of a mathematical analysis of a biological principle.

Lesson Plan 6.1

Phase 1
PREPARATION

Key Concepts
- Gregor Mendel discovered the basic principles of heredity from his analysis of crosses using garden peas.
- Mendel's experiments were carefully designed, meticulously executed, and mathematically analyzed.
- Mendel reasoned that each of the seven traits he studied was controlled by two *factors*, known today as the two alleles of a gene.
- Mendel developed two laws of heredity: the *Law of Segregation* and the *Law of Independent Assortment*.

Phase 2
TEACHING STRATEGIES

Theme Connection

Patterns of Change
Genetics is the study of patterns involved in passing instructions for development from parents to their offspring. Emphasize that Mendel's two early laws represent the simplest pattern of heredity. Few traits in humans are inherited simply on the basis of his laws.

Visual Strategy

Figure 6.1
Use this Figure to point out that the flower is the pea plant's reproductive organ, and contains both male and female structures. Distinguish between *self-* and *cross-pollination*.

Demonstration 1

Reproductive Organs of Flowers
Show students a flowering pea plant. If not available, substitute another flowering plant so that students can see the male and female reproductive structures. If possible, display two plants, each with distinctly contrasting forms of a trait. Students should recognize that Mendel had no difficulty determining what forms of a trait each offspring had inherited.

Visual Strategy

Table 6.1

Walk the students through each of the three steps. Make sure they understand that the arrows in Steps 1 and 3 represent self-pollination and those in Step 2 denote cross-pollination. Assuming their parents represent a *P* generation, ask students who constitutes the F_1 generation, and who will make up the F_2 generation. In Step 3, have students note that three times as many plants with purple flowers were obtained as those with white flowers.

Mendel conducted his experiments in three steps

Mendel chose pea plants for his experiments because they are small, easy to grow, produce large numbers of offspring, and mature quickly. Mendel also took advantage of the reproductive characteristics of pea plants. He could either allow the plants to self-fertilize, or he could cross-fertilize them. Self-fertilization occurs when the sperm (located inside a pollen grain) of one flower fertilizes the egg (located in the base of the flower) of the same flower. Cross-fertilization occurs when pollen is transferred by hand from the flower of one plant to a flower of another.

Also, Mendel took advantage of the fact that many varieties of garden pea are available and can be easily grown. He selected seven pea plant traits for study. These seven traits were: flower color, seed color, seed shape, pod color, pod shape, flower position on the stem, and plant height. Each of these traits had two contrasting forms. For example, flower color could be purple or it could be white. In this text, we will follow only Mendel's experiments with flower color, although he studied the inheritance of all seven traits. The three steps that Mendel followed when he designed his experiments are shown in **Table 6.1**.

Table 6.1 How Mendel Designed His Experiments

	What Mendel Did	What Mendel Found
Step 1 ***Parental Generation*** First Mendel produced a parental (P) generation.	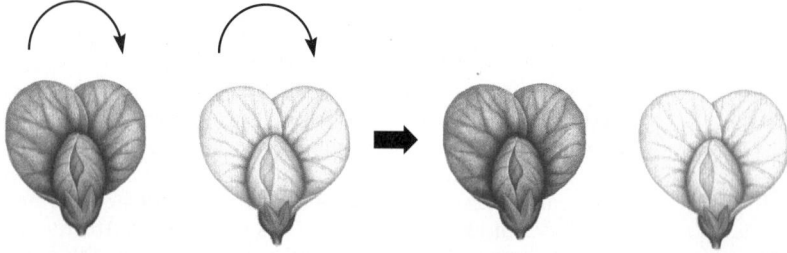 Mendel allowed the pea plants to self-fertilize for many generations. He then collected and grew the seeds from these plants. This would ensure that the plants he used for his research were "true-breeding," meaning that their offspring would produce only one form of a particular trait.	Mendel called the offspring the parental generation, or P generation for short. These plants were true-breeding, that is the white-flowering variety would only produce white flowers and purple-flowering variety would produce only purple flowers.
Step 2 ***F_1 Generation*** Next Mendel produced the F_1 generation.	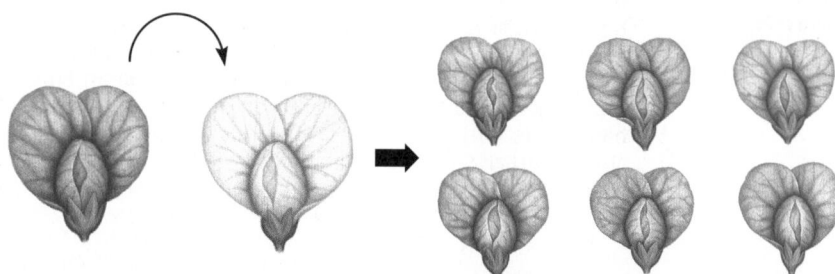 Mendel cross-fertilized the two different parental (P) varieties. He took pollen from a plant with white flowers and placed it on a flower that was purple. He then collected and grew the resulting seeds.	Mendel called the offspring the F_1 generation (F_1 for "first filial" generation, from the Latin word for son or daughter). Mendel found that all the F_1 plants had only purple flowers. No F_1 plants had white flowers. *(See Step 3 on the next page.)*

What Mendel Did	What Mendel Found

Step 3
F₂ Generation
Then Mendel produced the F₂ generation.

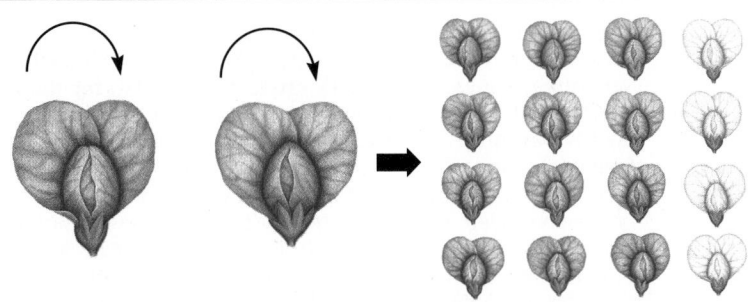

Mendel allowed the F₁ generation plants to self-fertilize. As before, he grew the seeds from these plants.

Mendel called the offspring the F₂ (second filial) generation. White flowers reappeared among the offspring. All together, Mendel grew 929 F₂ individuals. There were 705 plants with purple flowers and 224 plants with white flowers.

Patterns of Change

Explain how a form of a trait that appears in both parents could be absent from their children.

Mendel developed a model to explain his results

Mendel realized the ratio of purple-flowering plants to white-flowering plants in the F₂ generation was just about 3:1. Mendel obtained this same result for each cross. Surely this ratio must mean something important. To explain his results, Mendel came up with a simple model, a set of rules that could be used to accurately predict patterns of heredity. These rules summarize Mendel's ideas about inheritance.

1. Parents transmit information about traits to their offspring. Mendel called this information "factors."

2. Each individual has two factors for each trait, one from each parent. The two factors may or may not have the same information. If the two factors do have the same information (for example, if both have information for purple flowers), the individual is said to be **homozygous** (*hoh muh ZY guhs*) for the trait. If the two factors have different information (for example, one factor codes for purple flowers and the other for white flowers), the individual is said to be **heterozygous** (*heht uh roh ZY guhs*).

3. The alternative forms of a factor are called **alleles** (*uh LEELS*). The many alleles that an organism possesses make up its **genotype** (*JEE nuh typ*). An organism's physical appearance, which is determined by its alleles, is called its **phenotype** (*FEE nuh typ*).

4. An individual possesses two alleles for each trait. One allele is contributed by the female parent, and the other is contributed by the male parent. The two alleles in each pair are not affected by each other. They are passed on when an individual matures and produces gametes (eggs and sperm). During the formation of gametes, the paired alleles segregate (separate) randomly so that a gamete receives a copy of one allele or the other.

5. The presence of an allele does not ensure that the trait will be expressed in the individual that carries it. In heterozygous individuals, only the **dominant** allele achieves expression. The **recessive** allele is present but remains unexpressed. In Mendel's F₁ generation, purple flower color was caused by a dominant allele. For every pair of contrasting forms of a trait—tall versus short, or green seeds versus yellow seeds—the allele for one form of the trait was always dominant and the allele for the other form of the trait was always recessive.

Demonstration 3

Assigning Genotypes

Have students indicate what genotypes (*WW, Ww, or ww*) should be placed under each pea flower in Table 6.1.

Visual Strategy

Figure 6.2

Use this diagram to explain how a Punnett square is useful in predicting the expected results of a cross. Guide the students through the diagram. Tell them that the Step 2 referred to in the caption is not describing Rule 2 in the summary of Mendel's model on page 119. It refers to Table 6.1 on page 118. Also use this figure to review all the terms defined so far in this chapter.

Mathematics Connection

Punnett Square

A Punnett square is a mathematical model used to predict the results of a cross. Although mathematics had long been recognized as vital to the field of physical sciences, Mendel was among the first to apply it to the life sciences. Point out to students that Reginald Punnett, an English geneticist, developed his grid method some 30 years after Mendel reported his results.

Visualizing Mendel's Model

Mendel's model can be understood easily by diagraming Mendel's crosses. For example, consider Mendel's cross of parental (P) purple-flowering pea plants with parental white-flowering pea plants. For the sake of convenience, letters are used to represent alleles. The recessive allele is represented by the lower case letter "w" (for "white"), and the dominant allele is represented by the corresponding upper case letter "W." A plant that is true-breeding for the recessive white flower would be designated "ww," and a plant that is true-breeding for the dominant purple flower would be designated "WW." A heterozygote would be designated "Ww."

A simple diagram called a Punnett square can help you visualize crosses. Named for the British geneticist Reginald Punnett, a Punnett square is also a handy device for predicting the results of a cross.

In a Punnett square, the symbols for all the possible alleles carried by male gametes (sperm) are arranged along the top of the square, while all the possible alleles carried by the female gametes (eggs) are shown along the left side. By combining the symbol for an allele carried by a male gamete with the symbol for the allele carried by the female gamete, all the possible gamete combinations can be predicted, as shown in **Figure 6.2**.

Figure 6.2
This Punnett square illustrates a cross between two true-breeding varieties of garden pea plants. This was Step 2 of Mendel's experiment.

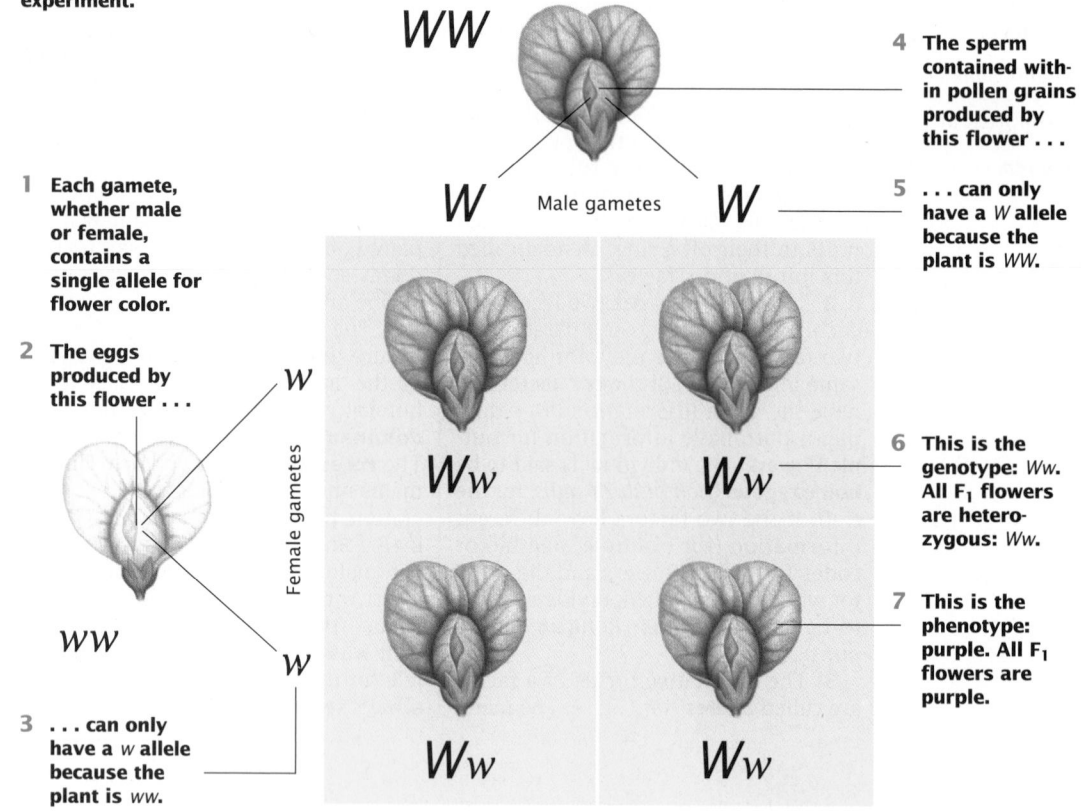

1 Each gamete, whether male or female, contains a single allele for flower color.

2 The eggs produced by this flower . . .

3 . . . can only have a *w* allele because the plant is *ww*.

4 The sperm contained within pollen grains produced by this flower . . .

5 . . . can only have a *W* allele because the plant is *WW*.

6 This is the genotype: *Ww*. All F$_1$ flowers are heterozygous: *Ww*.

7 This is the phenotype: purple. All F$_1$ flowers are purple.

Male gametes

Female gametes

Crossing two heterozygous plants

When heterozygous (Ww) F_1 individuals with purple flowers are allowed to self-fertilize, what will the resulting F_2 individuals be like? Let's predict what we would expect to find using Mendel's rules. To make a prediction, consider what scientists call probability. Probability is simply the likelihood that something will happen. For example, when you toss a coin, the probability that the coin will land with "heads" up is 50 percent, or one-half, because the coin is just as likely to fall showing "tails." Similarly, because of the segregation of alleles, the probability that an egg or sperm cell in one of Mendel's heterozygous F_1 pea plants (Ww) will contain the allele W is 50 percent, or one-half, just as in a coin toss. Obviously, the reverse is also true: the probability that a gamete will contain the allele w is also 50 percent or one-half.

Study the Punnett square for a cross involving two purple-flowered heterozygous plants (Ww) shown in **Figure 6.3**. In this particular cross, one-half the male and female gametes carry the allele for purple flowers (W). The other one-half of the gametes carry the allele for white flowers (w). You can see that Mendel's model clearly predicts that 75 percent of the F_2 generation will have purple flowers and 25 percent will have white flowers, a 3:1 ratio. The Punnett square also shows the genotypes found in the F_2 generation. Twenty-five percent of the F_2 is homozygous (ww) with white flowers, 50 percent heterozygous (Ww) with purple flowers, and the remaining 25 percent is homozygous (WW) with purple flowers. The 3:1 ratio Mendel had repeatedly observed is the expression of an underlying 1:2:1 ratio of genotypes in which the heterozygotes look like one of the homozygotes.

Figure 6.3
This Punnett square illustrates a cross between two F_1 individuals. This was Step 3 of Mendel's experiment.

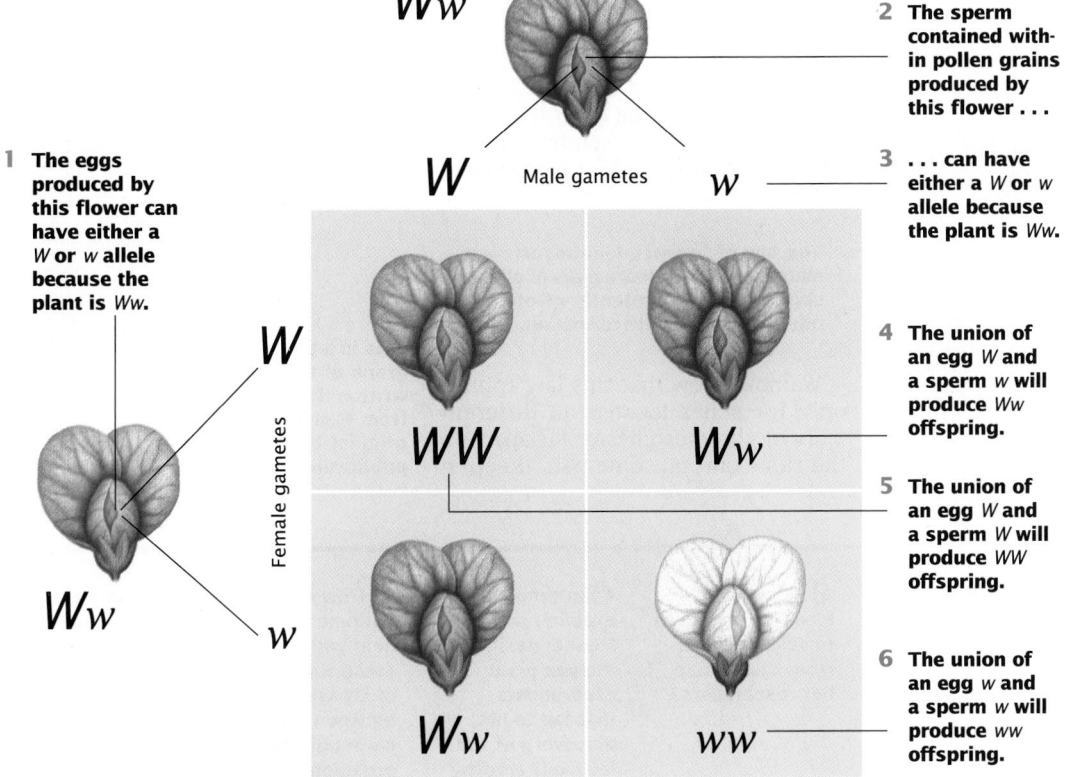

1 The eggs produced by this flower can have either a *W* or *w* allele because the plant is *Ww*.

Female gametes

Male gametes

Ww

W

w

W

w

WW

Ww

Ww

ww

2 The sperm contained within pollen grains produced by this flower . . .

3 . . . can have either a *W* or *w* allele because the plant is *Ww*.

4 The union of an egg *W* and a sperm *w* will produce *Ww* offspring.

5 The union of an egg *W* and a sperm *W* will produce *WW* offspring.

6 The union of an egg *w* and a sperm *w* will produce *ww* offspring.

Visual Strategy

Figures 6.2 and 6.3
Have students identify where the law of segregation is shown in these figures.

Theme Answer

Evolution
Because alleles segregate independently from one another, gametes will end up with different combinations of genes, providing the source of variation.

Phase 3

ASSESSMENT OPTIONS

Closure Strategy

Green vs. Yellow Pods
Inform students that Mendel discovered that green pods are dominant to yellow pods. Have them determine the phenotypes and genotypes expected in the offspring from a cross between a plant heterozygous for green pods and one that produces yellow pods. *Fifty percent would be heterozygous for green and 50 percent would be homozygous yellow.*

Section Review

Assign the *Section Review.*

Reteaching

Have students determine if they can roll their tongue. Inform them that this characteristic is inherited as a dominant trait. Have them check which members of their immediate family can roll their tongues and report their findings. In each case, have them identify the phenotype and genotype for each family member. Point out that tongue rollers may be homozygous dominant (*TT*) or heterozygous (*Tt*). See if they can determine whether the tongue rollers they identify are *TT* or *Tt*.

Mendel's Laws of Heredity

As you have just seen, Mendel's model analyzes the 3:1 ratio Mendel saw in the F$_2$ generation in a neat and very satisfying way. Similar patterns of heredity have been observed in countless other organisms. Mendel summarized his results by formulating two laws.

To describe how traits can disappear and reappear in a certain pattern from generation to generation, Mendel proposed what is now called the law or principle of segregation.

> The law of segregation states that the members of each pair of alleles separate when gametes are formed. A gamete will receive one allele or the other.

Evolution

How does the law of independent assortment provide a source of variation from one generation to the next?

You saw this law at work in the Punnett squares shown on pages 120 and 121.

Mendel went on to study the inheritance of two or more pairs of traits. He crossed pea plants with contrasting forms of two traits, such as flower color and plant height. He found that the inheritance of one trait did not influence the inheritance of the other trait. This idea is the basis for Mendel's second law, the law or principle of independent assortment.

> The law of independent assortment states that two or more pairs of alleles segregate independently of one another during gamete formation.

We now know that this law applies only for genes located on different pairs of chromosomes or far apart on the same chromosome pair. (Keep in mind that chromosomes had not yet been discovered in Mendel's time.)

Mendel's laws were not discovered for more than thirty years

Mendel was a member of a local science society. Each member tackled a scientific investigation that focused on an area of personal interest. The research would later be discussed at meetings. Mendel's paper describing his results was published in the science society journal in 1866. Unfortunately, his paper failed to arouse much interest, and the importance of his work was not recognized. In 1900, sixteen years after Mendel's death, several investigators independently rediscovered Mendel's pioneering paper. Scientists soon realized that chromosomes, discovered not long before Mendel's death, are the carriers of heredity.

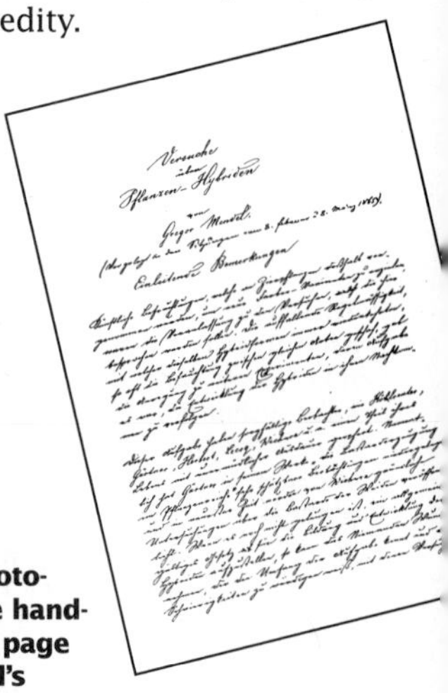

Figure 6.4
This is a photograph of the handwritten title page from Mendel's original 1866 publication.

Section Review

❶ How did Mendel's approach to science differ from that of earlier researchers?

❷ In three steps, explain how Mendel designed the pea plant experiments that led to his discovery of the laws of heredity.

❸ How could a Punnett square help you understand the results of crosses between pea plants or other organisms?

❹ How does the law of segregation differ from the law of independent assortment?

■ Section Review Answers ■

1. Students should indicate Mendel's mathematical approach to biology.
2. See **Table 6.1** on page 118 for an explanation.
3. A Punnett square helps to visualize the genotypes and phenotypes of two organisms and predict the results of a cross.
4. The law of segregation states that the members of each pair of alleles separate when gametes are formed so that a gamete will receive only one of the two alleles. The law of independent assortment states that two or more pairs of alleles segregate independently of one another during gamete formation.

MENDEL'S THEORY DISPELLED THE MYSTERY OF WHY TRAITS SEEMED TO APPEAR AND DISAPPEAR MAGICALLY FROM ONE GENERATION TO THE NEXT. STRIPPED OF RATIOS AND SYMBOLS, MENDEL'S WORK SHOWS US THAT PATTERNS OF HEREDITY REFLECT THE TRANSMISSION OF ENCODED INFORMATION FROM PARENTS TO OFFSPRING. THIS INFORMATION IS LOCATED ON CHROMOSOMES.

6.2 Chromosomes

Objectives

❶ **Explain the relationship between genes and chromosomes.**

❷ **Summarize the events that occur during meiosis.**

❸ **Define crossing over and explain its role in evolution.**

❹ **Describe how sex chromosomes determine the sex of humans.**

Genes and Chromosomes

Mendel concluded that factors containing information about traits are transmitted from parents to offspring. But what exactly are these factors? This question dominated biology for more than half a century after Mendel's work was rediscovered in 1900.

In 1909, a Danish biologist named Wilhelm Johannsen first used the word "gene" to describe the physical units of heredity that Mendel had called factors. A **gene** is a segment of the DNA molecule that carries the instructions for producing a specific trait. Recall that the alternative forms of a factor, or gene, are called alleles.

In Mendel's time, no one knew of chromosomes or genes. Chromosomes were first observed in 1879 by the German scientist Walther Flemming. In 1902, the American biologist Walter Sutton gave a convincing argument supporting an earlier notion that genes are located on chromosomes. Ten years later, Thomas Hunt Morgan, shown in **Figure 6.5**, verified Sutton's idea. Morgan presented clear-cut evidence that the presence of white eye color in fruit flies, which usually have red eyes, is associated with a particular gene on a particular chromosome. Biologists around the world soon accepted the chromosomal theory of inheritance—the genes on chromosomes are the units of inheritance. In Chapter 7 you will learn how the information in a gene results in the formation of individual traits.

Figure 6.5
The American biologist Thomas Hunt Morgan showed that the genes for specific traits of a fruit fly (inset) are located on the fruit fly's chromosomes.

Visual Strategy

Figure 6.6 a–c
Have students point to the tip of a chromosome in Figure 6.6c. Inform them that this region is known as a telomere, which is believed to play an important role in protecting chromosomes against damage and in directing them to take their proper positions when a cell divides.

Connection: Chapter 4

Chromosomes and Life Spans
Telomeres are also thought to play a time-keeping role by serving as a way of telling the cell how old it is. Scientists have determined that telomeres are shaved down to a slight extent each time a cell divides. The telomeres in a 70-year-old are much shorter than those of a child. Some scientists suggest that short telomere lengths may be the signal to a cell that it can no longer divide.

Health Connection

Chromosomes and Cancer
Telomeres are also thought to play a role in cancer. Recent experiments suggest that at a very advanced and highly aggressive stage of tumor development, telomere shrinkage may cease and even reverse itself. As the telomeres get longer, the cells begin to divide even more rapidly.

Theme Answer

Stability
Meiosis keeps the number of chromosomes in gametes constant so that each individual in a species will have the same number of chromosomes.

Theme Connection

Stability
Meiosis was predicted before it was observed. Once scientists understood how chromosomes behaved in mitosis, they realized that there must be another process to produce mature gametes. Ask students why there had to be another process. *Answers will vary. Accept any logical answer that deals with the reduction in chromosome number.*

Humans have forty-six chromosomes in most cells

If you could look at almost any cell in your body, you would find 46 chromosomes that are similar to the one shown in **Figure 6.6**. If you sorted the chromosomes by size and shape, you would find that they exist in 23 pairs. The members of almost every pair look similar, and contain genes that affect the same characteristics. Cells in which chromosomes occur in pairs are said to be **diploid** (2n) cells.

The chromosomes in your cells are about 40 percent DNA and 60 percent protein. The protein serves as a scaffold around which the slender thread of DNA is tightly wound. If the single DNA strand were stretched out in a straight line, it would be about 50 mm (2 in.) long.

The typical chromosome in your body contains thousands of genes. The information contained in each gene on every chromosome was used by your body to enable you to grow and develop, and now it helps run your body every day. If you have children someday, you will pass some of this information to them.

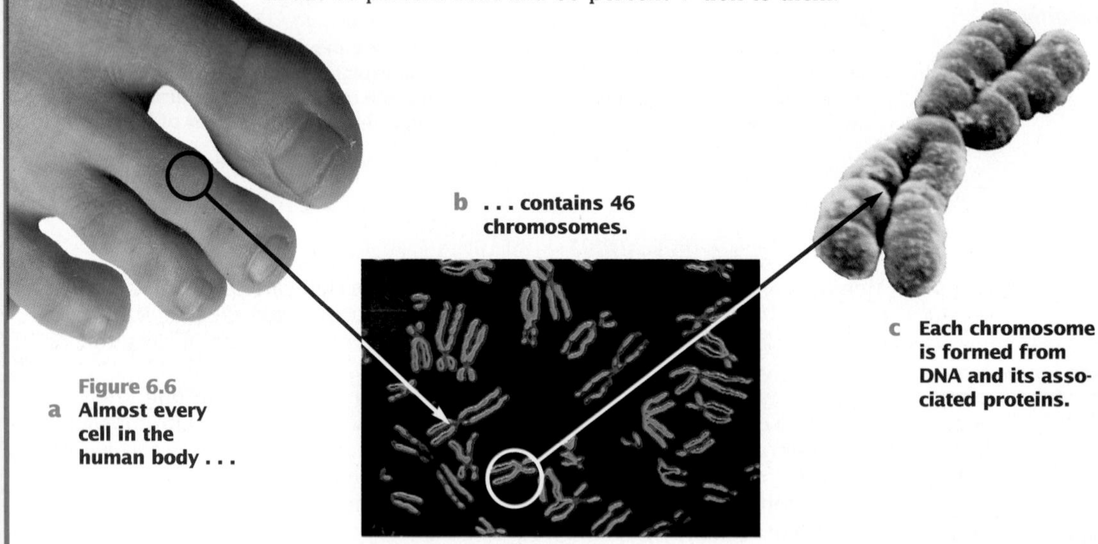

b ... contains 46 chromosomes.

Figure 6.6
a Almost every cell in the human body . . .

c Each chromosome is formed from DNA and its associated proteins.

How Gametes Form: Meiosis

Stability

Why is meiosis essential for the survival of a species?

In the previous section, you learned that almost all your cells have 46 chromosomes. Which cells are the exceptions? Your gametes—egg or sperm cells—have only 23 chromosomes. Cells such as gametes that contain only one chromosome of each pair are called **haploid** (n) cells.

Why must gametes have only 23 chromosomes? Consider this: If gametes had the same number of chromosomes as other body cells, when egg and sperm unite during fertilization the new individual would have twice as many chromosomes as its parents. Imagine how this number would soon become impossibly large with each new generation. However, the chromosome number does not double with each generation because eggs and sperm are formed by a special form of nuclear division called meiosis. **Meiosis** (*my OH sihs*) is a type of nuclear division in which the chromosome number is halved. Like mitosis, meiosis is followed by cell division. In humans, specialized reproductive cells with 46 chromosomes undergo meiosis and cell division to give rise to egg or sperm cells that have only 23 chromosomes each. Study this process in **Table 6.2** located on the next page.

A Generalized Picture of Meiosis

Table 6.2

Event	Sperm Formation	Egg Formation
Each cell nucleus that undergoes meiosis is diploid (2n). The cell nuclei shown here each have four chromosomes. One member of each chromosome pair is from one parent; the other chromosome is from the other parent.		
First, the amount of DNA doubles. Then, unlike mitosis, the members of every pair of similar chromosomes pair with one another. Occasionally paired chromosomes will exchange segments in a process called crossing over.		
The first meiotic division (Meiosis I) has separated each pair of chromosomes.		
The second meiotic division (Meiosis II) separates the two copies of each chromosome.		
During sperm formation, the four cells containing these nuclei develop heads and tails. During egg formation, only one of the cells containing these nuclei becomes fully mature. All cells are haploid (n).		

Visual Strategy

Table 6.2
Walk students through each step.

- First row —Remind them that these gametes are immature since they still contain the diploid number of chromosomes as a result of mitosis.
- Second row —Have students compare these two cells with the one shown in Figure 4.12c on page 81 to see the difference in chromosome behavior between mitosis (independent alignment) and meiosis (synapses).
- Third row —Point out that Meiosis I reduces the chromosome number from diploid to haploid. Also compare cytoplasmic division between sperm and egg.
- Fourth row —Indicate that Meiosis II is basically similar to mitosis since the two strands simply separate to produce individual chromosomes. Again compare cytoplasmic division between sperm and egg.
- Fifth row —Relate the production of four mature sperm cells versus only one mature egg cell as two different ways of boosting the chances of a successful fertilization. For a sperm, the chances are improved by mere numbers; for an egg, having one larger cell with more stored cytoplasmic materials gives it a greater chance of surviving. Ask students what might happen if four mature egg cells, each capable of being fertilized, were produced each time one egg underwent meiosis. *Less stored cytoplasmic material, less chance of surviving.*

▪ *Matter of Fact* ▪

How Many Offspring?

Organisms vary in the number of offspring they produce. Generally the number they produce depends on the likelihood of surviving. The elephant, with few enemies and a long life span, bears only one newborn every two years. At the other extreme is the ocean sunfish, which is subject to harsh environmental conditions and a large number of predators. The sunfish produces 300 million eggs in a single spawning. Other examples of prolific producers include mussels (25 million eggs), frogs (thousands of eggs), and queen termites (8,000).

Figure 6.7
During meiosis, parts of adjacent chromosomes may exchange DNA by means of crossing over.

Crossing over during meiosis recombines genetic material

During the first division of meiosis, the two members of each chromosome pair line up with each other side by side. While paired together, they may exchange segments of DNA, as the two chromosomes are doing in **Figure 6.7**. This reciprocal exchange of corresponding segments of DNA is called **crossing over**.

From the perspective of evolution, crossing over has an enormous impact. Exchanging segments of DNA between the members of a pair of chromosomes results in new combinations of genes in particular gametes, just as shuffling a deck of playing cards generates new combinations of cards dealt in a hand. These new combinations of genes act as one source of variation within a species. Variation is necessary for natural selection to occur.

Sex chromosomes determine what sex a child will be

One of your 23 pairs of chromosomes carries the genes that determine whether you are male or female. If you are female, this pair of chromosomes consists of two chromosomes designated as X chromosomes. If you are male, you have only one X chromosome. The other chromosome in the pair is a much smaller chromosome called a Y chromosome. A female can only produce eggs with an X chromosome. A male can produce sperm with either an X or Y chromosome. Thus individuals receiving an X chromosome from their father become females because they will be XX. Individuals that receive a Y chromosome from their father become males because they will be XY. The X and Y chromosomes are called the **sex chromosomes**. Study the process in **Figure 6.8**.

Figure 6.8
The sex of a child is determined by whether the father's sperm contains an X chromosome or a Y chromosome.

Section Review

❶ **Where are the instructions for specific traits located in cells?**

❷ **What happens to chromosomes in the first division of meiosis? What happens to them in the second division of meiosis?**

❸ **What are two biologically important outcomes of meiosis?**

❹ **Explain how the sperm and not the egg determines the sex of a child.**

▪ *Section Review Answers* ▪

1. Instructions for specific traits are located in segments of the DNA molecule called genes.
2. See Table 6.2 on page 125 for an explanation.

3. Meiosis maintains the number of chromosomes in the cells of new individuals of a species. It also results in new gene combinations, which contribute to variation within a species.

4. Only the sperm cell can produce either X or Y chromosomes.

MANY GENES PRESENT IN HUMAN POPULATIONS CAN CODE FOR
CHARACTERISTICS FAR MORE IMPORTANT THAN HAIR COLOR OR EYE
COLOR. SOME OF THE MOST DEVASTATING HUMAN DISORDERS RESULT
FROM ALLELES THAT LEAD TO MALFUNCTIONING OF IMPORTANT
PROCESSES IN THE BODY. WHILE ALLELES ARE PASSED FROM PARENT TO
CHILD, THE EXPRESSION OF A DEFECT FOLLOWS MENDEL'S SIMPLE IDEAS.

6.3   *Human Genetic Disorders*

Objectives

1. Recognize the relationship between mutation and human genetic disorders.

2. Explain the patterns of inheritance of cystic fibrosis and sickle cell anemia.

3. Explain the purpose of genetic counseling.

4. Describe some of the benefits that gene technology might offer.

Mutations Are Changes in Genes

For you to develop and function properly, your genes must code for certain proteins. Unfortunately, genes are sometimes damaged or copied incorrectly. A change in the gene is called a **mutation** (*myoo TAY shuhn*).

Mutations occur only rarely, but there are so many genes in our chromosomes that each of us carries dozens of mutations. The effects of a mutation can be helpful, harmful, or neutral, such as the one shown in **Figure 6.9**. Since mutations change genes at random, the chance that a mutation in a gene will improve an organism is very slim. For instance, imagine that you randomly changed a part of a blueprint for construction of an airplane. Some changes would not matter, but others would. Occasionally however, a mutation may have a beneficial effect. Mutations act as a source of the variation that is needed for a species to adapt to changing conditions or a new environment, and, thus, evolve over time.

Figure 6.9
A genetic mutation causes some humans to have six toes. In this case, the mutation neither helps nor harms the person.

Mutations in humans can cause genetic disorders

Most mutations are rare in human populations. Almost all mutations occur in recessive alleles and, therefore, are not expressed in heterozygous individuals. Because mutations are rare, it is unlikely that a person carrying a copy of the recessive allele caused by a mutation will marry someone who carries the same mutation. Instead, he or she will most likely marry a person who is homozygous dominant for that allele, a person who does not have the mutation. Therefore, their children will not be homozygous for the mutant allele and will not have the genetic disorder.

Lesson Plan 6.3

Phase 1

PREPARATION

Key Concepts

- A mutation is any change in the genetic information of an organism.
- Mutations are rare and can have a helpful, harmful, or neutral effect on an organism.
- Genetic disorders in humans originally brought about by a mutation include cystic fibrosis, sickle cell anemia, Down syndrome, and hemophilia.
- Genetic counseling is a process that helps identify the risk of having a child with a genetic disorder.
- Genetic technology includes techniques that may help correct a genetic disorder.

Reading Strategy

Have students outline a major paragraph in this section. Have them first decide what the main idea is and then list the important details that relate to this main idea.

Phase 2

TEACHING STRATEGIES

Visual Strategy

Figure 6.9
Use this figure to point out two important features of mutations: 1) although phenotypically visible here, not all mutations result in a phenotypic expression, and 2) mutations need not be harmful. Have students provide an example in which a mutation might be beneficial to an organism.

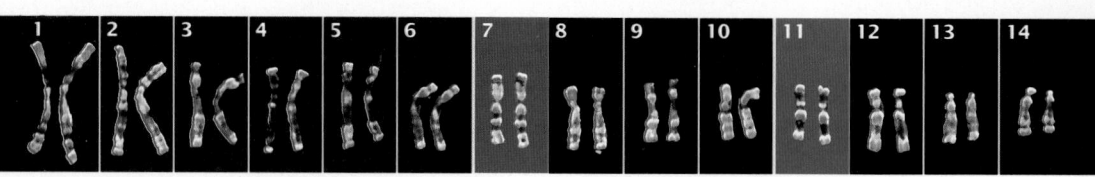

Demonstration 1

Punnett Square and Mutation

Have students explain how they would use a Punnett square to demonstrate that the couple described in the text (one partner carries the mutation, the other is homozygous normal) could not have children homozygous for the mutant allele. Ask them to predict what might happen if the person carrying a copy of the recessive allele married someone who also carried the same mutation. *The mutation will phenotypically express itself in the offspring.*

Connection: Chapter 3

Protein Channels

Question students to see how much they recall of protein channels and their role in cystic fibrosis.

Visual Strategy

Figure 6.10

Have students locate the four chromosomes involved. Point out that there are at least 3,000 human disorders caused by defects in the chromosomes. They range from such inconsequential conditions as certain forms of colorblindness and baldness to disorders that may be life-threatening.

Demonstration 2

Sickle Cell Anemia

Project a slide or transparency of normal and sickle red blood cells so that students can compare the two types. Inform students that the sickle cells may clog small blood vessels, often resulting in painful blood clots and depriving organs of their full supply of oxygen. People with sickle cell anemia have a shortened life expectancy.

Visual Strategy

Figure 6.12

Ask students what the phenotypes are for each offspring. *One homozygous normal, two heterozygous normal, one sickle cell anemic.*

Figure 6.10

The chromosomes highlighted in blue play a role in four serious genetic disorders. Each of the four disorders is discussed at length on pages 128 and 129.

Four Genetic Disorders

In some cases, particular mutant alleles have become common in human populations. The harmful effects that they produce are called genetic disorders. Four genetic disorders are described below. Not all genetic disorders are alike, however. Cystic fibrosis and sickle cell anemia are caused by mutations that result in harmful recessive genes. Down syndrome is caused when chromosomes fail to separate properly during meiosis. The individual is born with 47 chromosomes, as in **Figure 6.10**. Hemophilia is caused by a recessive gene, but the defective gene is on the X chromosome.

In the past, people with genetic disorders could do little to alleviate the harmful effects of their mutations. However, modern research into the causes and cures of such disorders is offering new encouragement to afflicted individuals. New techniques for identifying harmful genes can help couples determine any risks their children might face. New forms of gene therapy that are now being developed offer hope to sufferers of genetic abnormalities. As more information is learned about genes, scientists can begin to apply their new knowledge toward developing cures for genetic disorders.

	F	f
F	FF	Ff
f	Ff	ff

Figure 6.11

If both parents carry a defective copy of the CF gene (f), their child has a one-in-four chance of developing the disease.

Cystic fibrosis is caused by a recessive gene on chromosome 7

Cystic fibrosis (CF) is the most common fatal genetic disorder among people of European ancestry. The disease is carried by a recessive allele on chromosome 7. See **Figure 6.11**. In the United States, cystic fibrosis strikes one child in every 2,000. In these individuals, membrane channels that normally transport chloride ions into cells do not function. As a result, mucus accumulates in the lungs and pancreas, clogging important ducts in these organs. Cystic fibrosis patients have difficulty breathing and cannot properly digest their food. Most of the current treatments for CF patients can help relieve the symptoms of the disease, although they do not cure it.

Researchers have recently isolated the defective gene that causes cystic fibrosis. The discovery has enabled scientists to devise tests to identify people who carry the gene and who run the risk of having children with the disease. Isolation of the gene has led to the possibility of gene transfer therapy, in which a healthy copy of the CF gene is transferred to the lungs of CF patients to cure the disease.

Sickle cell anemia is caused by a recessive gene on chromosome 11

Sickle cell anemia is a recessive genetic disorder of a gene located on chromosome 11. See **Figure 6.12**. It is particularly common in African populations. People with sickle cell anemia have defective hemoglobin proteins that cause their red blood cells to be irregularly shaped. As a result, the red blood cells have difficulty moving through small blood vessels and cannot properly transport oxygen to tissues. The sickle cell mutation in the hemoglobin gene apparently first arose in Central Africa

Figure 6.12

If each parent carries a copy of the sickle cell gene, there is a one-in-four chance that a child's red blood cells will sickle like the cell on the left.

	S	s
S	SS	Ss
s	Ss	ss

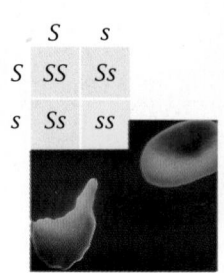

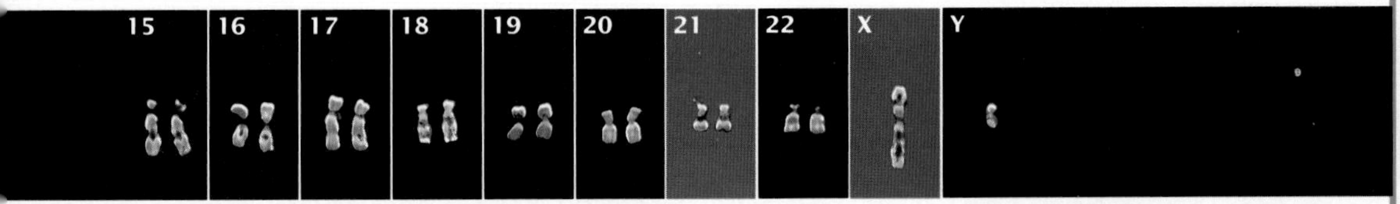

15
16
17
18
19
20
21
22
X
Y

centuries ago. There, up to 45 percent of the population is heterozygous for the sickle cell gene. Evolution has favored the sickle cell allele in Central Africa because heterozygous people are more resistant to malaria, a leading cause of illness and death in the tropics.

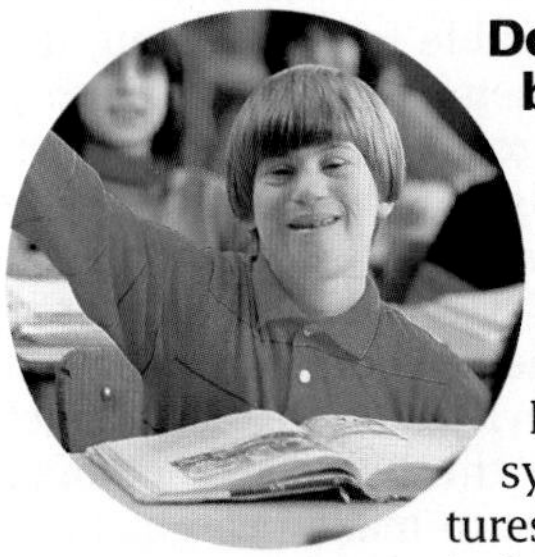

Down syndrome is caused by an extra copy of chromosome 21

The developmental features produced by having an extra copy of chromosome 21 were first described in 1866 by J. Langdon Down and are now known as Down syndrome. See **Figure 6.13**. The features that characterize Down syndrome include extra folds in the upper eyelids, a broad and somewhat flattened nose, short stature, and, most importantly, varying degrees of mental retardation. Although people with Down syndrome are physically challenged, they are able to lead active lives and make positive contributions to society.

Down syndrome occurs in about one out of every 1,000 children. This syndrome is much more common among children born to older mothers. In mothers older than 45 the risk is as high as one in sixteen births. Pregnant women over the age of 35 are usually advised to have a medical procedure called amniocentesis performed by their physician. Amniocentesis is a technique in which a small amount of the fluid surrounding the fetus is removed and analyzed. Fetal cells in this fluid are then examined for defects and abnormalities in chromosome number or structure. Results from amniocentesis can reveal whether or not the fetus has a genetic disorder. Along with Down syndrome, amniocentesis can detect other types of disorders such as hemophilia, sickle cell anemia and cystic fibrosis.

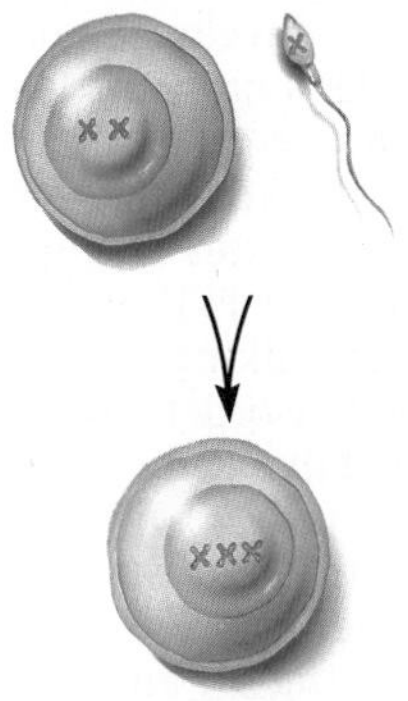

Egg
Sperm
XX
XXX
Zygote

Figure 6.13
When Chromosome 21 does not separate properly during meiosis in one parent, a child may receive three copies of the chromosome.

Hemophilia is the result of a recessive gene on the X chromosome

Hemophilia is a recessive genetic disorder in which the blood is slow to clot or does not clot at all. When you cut yourself, the blood in the immediate area of the cut solidifies into a clot that seals the wound. The blood clot is formed by several kinds of protein fibers that circulate in the blood. A mutation causing one of these proteins to be defective leads to hemophilia. See **Figure 6.14**. In afflicted individuals, small cuts are difficult to heal and internal bleeding can be fatal. However, treatments are available, such as injections of genetically engineered clotting factors that are lacking in the blood.

A dozen genes encode proteins involved in blood clotting, and mutations can occur in any of them. Two of the genes are located on the X chromosome. Any male who inherits a mutant copy of these alleles will develop hemophilia because his other sex chromosome is a Y chromosome, which lacks an allele of the gene. This pattern of heredity is called sex linkage. Traits that are determined by genes located on the X chromosome are said to be **sex-linked traits**.

	X	Y
X	XX	XY
X^H	X^HX	X^HY

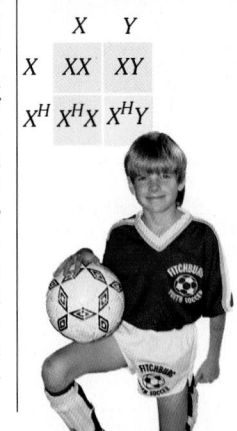

Figure 6.14
A female who is heterozygous for hemophilia can pass the gene for hemophilia to sons and daughters, but only the sons can have the disorder. With proper treatment, hemophiliacs, like this boy, can lead active lives.

Historical Perspective

Royal Families of Europe
Members of some European royal families often married among themselves as a practice of forming alliances. This increased the chances for producing homozygous recessive individuals with diseases such as hemophilia. The earliest cases of hemophilia were among descendants of Queen Victoria of England—members of the royal families of England, Spain, and Russia. Perhaps hemophilia's greatest impact is seen in the case of Nicholas II, the last czar of Russia, whose only son Alexis was hemophilic.

Theme Connection

Evolution
Evolution often favors the heterozygous individual. In the case of sickle cell anemia, homozygous dominant individuals are susceptible to malaria; homozygous recessive individuals have the disease. Heterozygous individuals do not contract malaria and do not have sickle cell anemia.

Social Studies Connection

Farming
The superiority of heterozygous individuals was recognized by plant geneticists in the 1930s. Because of their work, hybrid (heterozygous) corn caused a revolutionary improvement in crop yields owing to the plants' increased size and hardiness.

Visual Strategy

Figure 6.13
Point out that the zygote should contain only two copies of chromosome 21, but in fact has three. Have students review Table 6.2 to see how chromosome pairs separate normally in the first meiotic division. Sometimes, however, they don't. If it is chromosome pair 21, then Down syndrome results. Point out that other chromosome pairs, including the sex chromosomes, can also fail to separate properly.

Visual Strategy

Figure 6.14
Most students will have difficulty understanding the reason for the superscript *H*. Inform them that the female carries one allele for hemophilia and is shown having the genotype X^HX. The male is normal and is *XY*. Review the phenotypes for each of the children. Ask students what the odds are of this couple having a son with hemophilia.

Visual Strategy

Figure 6.15

Explain to students how to construct a pedigree. Inform students what the half-filled circle and the filled-in square represent. Point out the genotypes of the mother and father illustrated in Figure 6.15.

Phase 3

ASSESSMENT OPTIONS

Closure Strategy

Huntington's chorea, a severe neuromotor disease, is caused by a dominant allele. People so afflicted develop muscular shakiness, with symptoms similar to intoxication. Paralysis soon ensues. Once the symptoms appear, death usually occurs in about 15 years. Interestingly, Huntington's chorea does not manifest itself until some time in adulthood. Have students use a Punnett square to show what is expected if a woman who carries the dominant allele marries a normal male.

Section Review

Assign the *Section Review*.

Reteaching

Prepare a table on the board. Down the left-hand side, list the genetic disorders covered in this chapter. Across the top, place the headings *Dominant allele, Recessive allele, Nondisjunction,* and *Sex-linked.* Have students check the appropriate boxes. Tell students to check the library for information regarding other human genetic disorders and have them add their findings to this table. Compile a summary on the board and discuss the results with the class.

Genetic Counseling and Technology

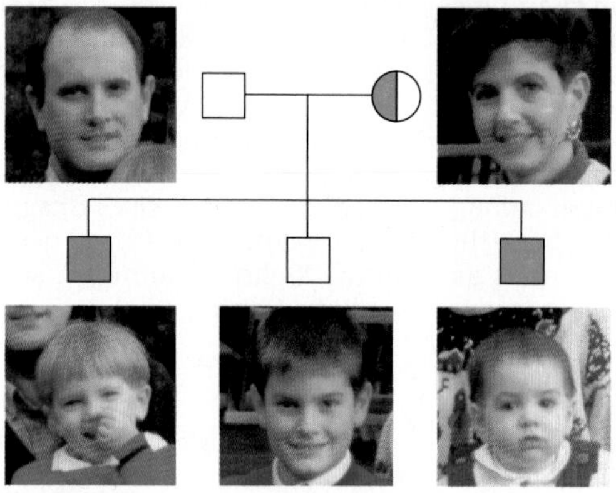

Figure 6.15

a Two boys in this family have hemophilia, but no one else does. How can this happen?

b This pedigree shows that the mother carries the gene (shaded) for hemophilia, and that two of her sons have received this gene.

Most genetic disorders cannot be cured. However, researchers are learning much about them and making progress towards successful therapies. In the absence of a cure, sometimes the only thing to do is avoid having children who could possibly be affected by a genetic disorder. Genetic counseling is a process that helps identify parents at risk for having children with genetic defects. With such information, parents can decide whether or not to have children.

You might ask, "If most genetic disorders are caused by mutations that affect recessive alleles, how do parents know if they are heterozygous for a damaging trait?" Usually it is difficult to be sure, but if one family member is affected by a genetic disorder such as cystic fibrosis or hemophilia, there is a possibility that other family members could be heterozygous for the disorder. These heterozygous individuals are said to be carriers of the trait. In such cases, genetic counselors can prepare a family **pedigree** *(PEHD uh gree)*, a record that shows how a trait is inherited over several generations, as shown in **Figure 6.15.** They will look for the presence of any relatives who have a genetic disorder, and assess the likelihood that a particular individual could be a carrier.

In some cases, therapy is available for genetic disorders if they are diagnosed early enough. For example, phenylketonuria (PKU) is a disorder in which affected individuals lack an enzyme responsible for converting the amino acid phenylalanine into another amino acid, tyrosine. As a result, phenylalanine builds up in the body. People with this disorder suffer severe mental retardation. If PKU is diagnosed shortly after birth, the newborn can be placed on a low phenylalanine diet. Such a diet ensures that the baby gets enough phenylalanine to make proteins, but not enough to do any damage. The child maintains the low phenylalanine diet until six years of age. At this age, the child's brain is fully developed and PKU is usually no longer a problem. Because this disease can be easily diagnosed after birth by inexpensive laboratory tests, many states require testing of all newborns for PKU.

Gene technology is making it possible to correct genetic disorders by replacing copies of defective genes with copies of healthy ones. In 1990, this approach was tried for the first time on human patients. In one case, a healthy copy of a gene encoding an enzyme was successfully transferred. In another case, the transferred gene was a potent cancer-fighting gene.

Section Review

❶ Describe how a mutation could result in a genetic disorder.

❷ Compare the pattern of inheritance of a dominant genetic disorder with that of a recessive genetic disorder.

❸ How is a pedigree used to trace the inheritance of a genetic disorder?

❹ Describe how gene technology could be used to cure a genetic disorder.

■ Section Review Answers ■

1. Students should explain that a mutation is a change in a gene that can have helpful, neutral, or harmful effects (as in the case of a genetic disorder).

2. See Figures 6.12, 6.13, and 6.14 beginning on page 128 for an explanation.

3. A pedigree visually shows the inheritance of traits over several generations. It can show the likelihood of passing on a genetic disorder to a future generation.

4. Students should indicate that gene technology has made it possible to replace some defective genes with copies of healthy genes.

Chapter **6** *Highlights*

These children are brothers. Even though they have the same parents, they both look very different.

Alternative Assessment
Assign students to cooperative work groups. Have each group develop a genetics problem based on the information in this chapter. Collect the problems, write them down on the board, and have each group work out the solutions.

	Key Terms	Summary
6.1 The Puzzle of Heredity  Mendel used pea plants in his attempts to understand patterns of heredity.	genetics (p. 117) homozygous (p. 119) heterozygous (p. 119) allele (p. 119) genotype (p. 119) phenotype (p. 119) dominant (p. 119) recessive (p. 119)	• Mendel's experiments with pea plants marked the beginning of genetics, the scientific study of heredity. • Mendel noted that factors now called genes transmit information about traits from parents to offspring. The different forms of genes are called alleles. • Organisms that have two identical alleles for a particular trait are said to be homozygous for that trait. Organisms that have two different alleles for that same trait are said to be heterozygous.
6.2 Chromosomes  This chromosome formed from DNA and its associated proteins.	gene (p. 123) diploid (p. 124) haploid (p. 124) meiosis (p. 124) crossing over (p. 126) sex chromosome (p. 126)	• Chromosomes contain DNA and proteins. They carry genetic information from one generation to the next. • Meiosis is a type of nuclear division that results in the formation of haploid gametes. • A pair of chromosomes known as sex chromosomes carries the genes that determine whether an individual is male or female.
6.3 Human Genetic Disorders 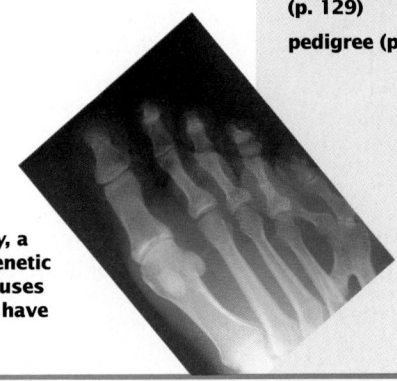 Occasionally, a harmless genetic mutation causes a person to have six toes.	mutation (p. 127) sex-linked trait (p. 129) pedigree (p. 130)	• A mutation is a change in a gene. • Cystic fibrosis, sickle cell anemia, Down syndrome, and hemophilia are four genetic disorders. • Genetic counseling can help identify parents at risk for having children with genetic defects. • Therapy is available for some genetic disorders if they are diagnosed early enough. • Gene technology offers hope for correcting genetic disorders by replacing copies of defective genes with copies of healthy ones.

Chapter Review Answers

Understanding Vocabulary

1. a. recessive
   b. Bb
   c. diploid
   d. male

Relating Concepts

2. Map answer is shown on page 115D.

Understanding Concepts

Multiple Choice

3.	a	8.	b
4.	d	9.	a
5.	a	10.	c
6.	a	11.	d
7.	c		

Completion

12. heterozygous, homozygous
13. phenotype
14. alleles
15. male
16. sex-linked

Short Answer

17. Mendel read about the work done by Knight using garden pea plants, followed a quantitative approach (i.e., statistical) in his experiments, and found encouragement for his work among members of his "science club."

18. Beginning with mixed-breeding (i.e., homozygous and/or heterozygous), individuals for a trait would have resulted in F_1 generations with different genotypes for every cross. Therefore, identification of the dominant and recessive forms of the trait would have been very difficult, if not impossible.

19. Rather than returning to the diploid condition, the cell would contain twice the number of chromosomes.

Interpreting Graphics

20. a. Cc
    b. cleft chinned
    c. cc
    d. 3:1

Understanding Vocabulary

1. For each set of terms, complete the verbal analogy by filling in the blank.
   a. seen in F_1 generation:dominant:: not seen in F_1 generation: _____
   b. homozygous: BB::heterozygous: _____
   c. 23 chromosomes: haploid::46 chromosomes: _____
   d. XX chromosomes: female::XY chromosomes: _____

Relating Concepts

2. Copy the unfinished concept map below onto a sheet of paper. Then complete the concept map by writing the correct word or phrase in each oval containing a question mark.

Understanding Concepts

Multiple Choice

3. The scientific study of the inheritance of traits is called
   a. genetics.
   b. genotype.
   c. heredity.
   d. osmosis.

4. Mendel studied contrasting traits of _____ pea plants in his experiments.
   a. dominant
   b. mutant
   c. white-flowering
   d. true-breeding

5. The crossing of a white-flowering plant with a purple-flowering plant (ww x WW) results in plants with the genotype
   a. Ww.
   b. purple.
   c. WW or ww.
   d. purple and pink.

6. The formation of an equal number of W and w gametes from a Ww individual demonstrates
   a. independent assortment.
   b. segregation.
   c. dominance.
   d. recessiveness.

7. The phenotypic ratio resulting from the cross Ww x Ww is
   a. 1:1.
   b. 1:2:1.
   c. 3:1.
   d. 2:1.

8. A gene is
   a. a complete molecule of DNA.
   b. a short segment of DNA.
   c. made of many chromosomes.
   d. a tiny particle first seen by Mendel in the 1860s.

9. Crossing over
   a. produces variation in the chromosomes.
   b. enables the second division to occur.
   c. causes the number of chromosomes in a cell to be reduced by half.
   d. produces XY but not XX chromosomes.

10. Cell division that halves the number of chromosomes is called
    a. mitosis.
    b. endocytosis.
    c. meiosis.
    d. zygote.

11. The purpose of genetic counseling is to
    a. repair defective genes.
    b. cure genetic defects that occur in unborn children.
    c. tell parents that they cannot have children.
    d. help parents understand the nature and risk of genetic disorders.

Completion

12. An organism showing the dominant trait may be homozygous or _____ , but an organism showing the recessive trait must be _____ .

13. An organism's _____ is the physical expression of its genotype.

14. Different forms of the same gene are called _____ .

15. A human baby with the sex chromosomes XY is a _____ .

16. A trait is _____ if a male has only one allele of the gene for the trait, as in the case of hemophilia.

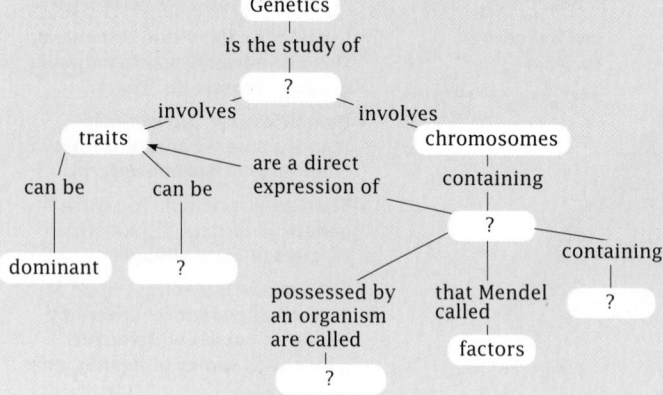

Short Answer

17. Mendel was a very meticulous and diligent experimenter. What else is known about Mendel that might help explain the success of his experiments?

18. To begin his experiments, Mendel used true-breeding pea plants. How might the results of his experiments have been different if he had started with pea plants that were not true-breeding?

19. During meiosis, the number of chromosomes is cut in half. What would happen when a sperm fertilizes an egg if the number of chromosomes was not reduced by one-half during meiosis?

Interpreting Graphics

20. A Punnett square is a diagram used to show the results of a cross. The Punnett square below shows the results of a cross between parents with cleft chins. Study the Punnett square carefully.

$$C$$

	(a)	(b)
C	CC	Cc
	(c)	(d)
c	Cc	

What is the genotype of the father?
What is the phenotype of the offspring in block a?
What is the genotype of the offspring in block d?
What is the phenotypic ratio for the offspring produced by the cross?

Reviewing Themes

21. *Stability*
Two tall pea plants are crossed. How does knowing that both parent plants are heterozygous (Tt) for this trait and that only the dominant trait is expressed help you predict the genotypic and phenotypic ratios for the offspring?

22. *Evolution*
How does crossing-over contribute to genetic diversity?

Thinking Critically

23. *Inferring Conclusions*
When Mendel crossed true-breeding purple flowers (WW x ww), he found that only purple flowers appeared in the F_1 generation. A ratio of 3 purple flowers to 1 white flower appeared in the F_2 generation. Suppose that Mendel found only pink flowers in the F_1 generation and a ratio of 1 purple flower to 2 pink flowers to 1 white flower in the F_2 generation. How would you explain these results?

24. *Inferring Conclusions*
A husband and wife have three children, two girls and a boy. The couple is expecting a fourth child. What is the likelihood that the child will be a boy?

Cross-Discipline Connection

25. *Biology and History*
Identify the contributions of these people to today's understanding of genetics.

1900	Hugo deVries
1901–1903	W. S. Sutton
1909	W. Johannsen
1910–1916	T. H. Morgan, A. H. Sturtevant, H. J. Muller

Discovering Through Reading

26. Read the article "Hand-Me-Down Genes," in *Newsweek*, January 27, 1992, page 53. What are the benefits of knowing your family's medical history?

Reviewing Themes

21. This information can be used in setting up a Punnett square and carrying out the possible crosses. Following the law of independent assortment, one would predict that the genotypic ratio is 1:2:1 and the phenotypic ratio is 3:1.

22. Crossing over permits the exchange of DNA segments by breakage and reunion during the first division of meiosis. This exchange results in new combinations of genes in an organism's gametes.

Thinking Critically

23. Flower color is an expression of incomplete rather than complete dominance.

24. The likelihood that the child will be a boy (having XY chromosomes rather than XX chromosomes) is 50 percent.

Cross-Discipline Connection

25.

1900	H. de Vries replicated the work of Mendel and rekindled interest in his conclusions.
1901–1903	W.S. Sutton showed the connections between cell biology and Mendel's findings.
1909	W. Johannsen introduced the terms gene, genotype, and phenotype.
1910–1916	T.H. Morgan and H.J. Muller discovered that genes are located on chromosomes.

Discovering Through Reading

26. As scientists learn more about the heredity links for diseases, knowledge of your family's medical history may suggest treatments that could lengthen your life.

133

Procedural Note

Review the answers to the preparation. The *Focus Activity* provides the necessary foundation for using pedigrees for the analysis of these traits.

Prelab Preparation Answers

- *Phenotype:* the expressed physical traits determined by alleles
  Genotype: the type of alleles that an organism possesses
  Dominant: alleles that achieve expression
  Recessive: alleles that are present but remain unexpressed
- *Pedigree:* a record that shows how a trait is inherited over several generations. It can show the likelihood of passing on a genetic disorder to a future generation.
- Circles represent females and squares represent males in a pedigree. Parents are joined by a marriage line and are connected to their offspring by another line.

Procedure Answers

3. Answers will vary for Exercises a–h.
5. When both parents with the same trait produce an offspring with a different trait, the offspring's trait must be recessive. In such a case parents are heterozygous.
6. Answers may vary but should explain that tongue rolling is a dominant trait and having five fingers is a recessive trait.
7. Answers will vary.

Chapter  **6**

Investigation

What Is Your Genetic Profile?

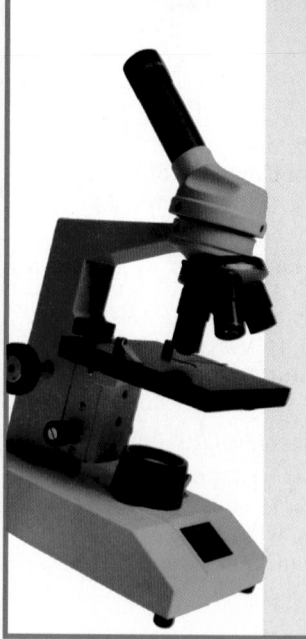

Objectives

In this investigation you will:
- *observe* a variety of human traits
- *collect* and *record* data about your phenotype
- *infer* possible genotypes from your data
- *calculate* percentages

Materials

- mirror

Prelab Preparation

Review what you have learned about genetics by answering the following questions:
- Define the terms "phenotype," "genotype," "dominant," and "recessive."
- What is a pedigree? Why are pedigrees often necessary when studying human hereditary traits?
- When reading a pedigree, how do you distinguish males from females and children from parents?

Procedure: Making Your Genetic Profile

1. Form a cooperative group of two students.
2. Make a table similar to the one shown at the top of the next page for recording your observations.
3. Working with a partner, check for each trait described below. Record your phenotype in the appropriate column in your table.
   a. **Mid-digital hair**
      Each segment of a finger is called a digit. *Is hair present on the middle digit of any of your fingers?*
   b. **Tongue rolling**
      Look in the mirror. *When you stick out your tongue, can you roll up the edges on each side?*
   c. **Dimples**
      When you smile, are there small indentations in your cheeks?
   d. **Cleft chin**
      Do you have an indentation in the middle of your chin?
   e. **Attached earlobes**
      Do the tips of your earlobes hang partially free, or are they completely attached to the side of your head?
   f. **Freckles**
      Do you have small reddish-brown spots on your skin?
   g. **Widow's peak**
      Pull your hair back over your forehead. *Does the hairline come down to a short point in the middle of your forehead or does it go straight across?*
   h. **Five fingers**
      Were you born with five fingers on each hand?

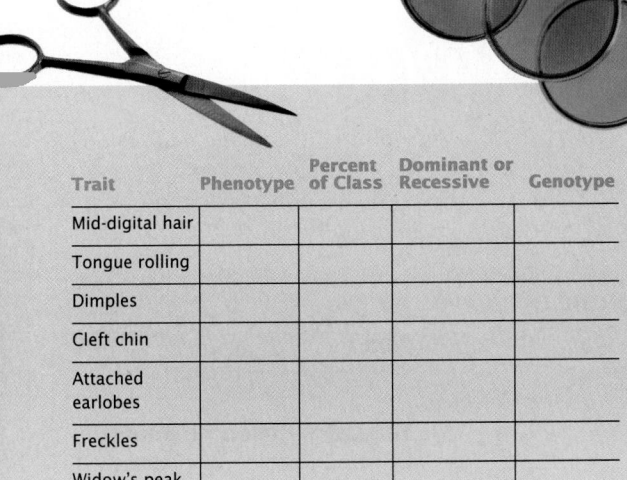

Trait	Phenotype	Percent of Class	Dominant or Recessive	Genotype
Mid-digital hair				
Tongue rolling				
Dimples				
Cleft chin				
Attached earlobes				
Freckles				
Widow's peak				
Five fingers				

4. Compare your data with that of your classmates. Calculate the percentage of the class that exhibits each trait.

5. Look at the pedigree below. Colored figures represent individuals that possess the trait. Explain why individual IV-8 and her parents provide the evidence that attached earlobes is a recessive trait.

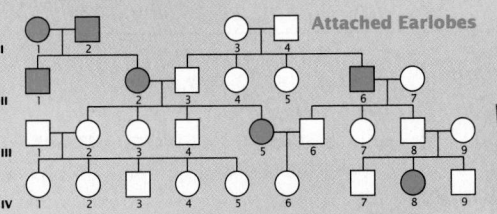

Attached Earlobes

6. Analyze the pedigrees for tongue rolling and five fingers. Explain whether each trait is dominant or recessive. Each of the other traits listed in your table is dominant.

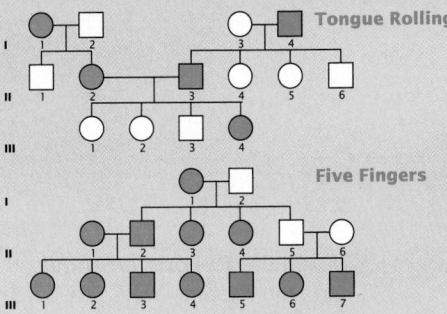

Tongue Rolling

Five Fingers

7. Based on the available information, use appropriate symbols for each trait to record your possible genotypes in your table.

Analysis

1. *Analyzing Data*
Does the information you collected and studied during this investigation indicate that dominant traits are the most common? Explain.

2. *Evaluating Methods*
Look at the pedigree for five fingers. Explain why individuals II-5, II-6, and their children are the most important for analyzing whether this trait is dominant or recessive.

Thinking Critically

What would happen to your percentages if you were to perform this investigation with five other classes and were to record their data?

Analysis Answers

1. Answers should use the data relating to the five-finger trait to explain that dominance is unrelated to the degree of commonality of a trait.

2. These individuals show that two parents with the same trait have at least one child with a different trait. This could only be possible when both parents are carriers of a recessive allele. These recessive alleles pair in the child to produce the recessive trait.

Thinking Critically Answer

Answers will vary, but very little change should be expected. The percentages for the classes will eventually approach the percentages found in the general population.

Chapter 7 How Genes Work

Planning Guide

	Objectives/Themes	Classwork Resources	Homework Resources
7.1	**1.** Explain how researchers concluded that DNA is the genetic material. **2.** Summarize how scientists determined the structure of DNA. **3.** Describe the structure of the DNA molecule. **4.** Summarize the process of DNA replication. **Themes:** Stability, Patterns of Change, Interacting Systems, Scale and Structure	**Text** Discoveries in Science *Identifying the Genetic Molecule* pp. 242–243 **Teacher's Resource Binder** Focus Activity 7 *Making Models* **Other Resources** Science in Action Video *The Atoms and Molecules of Life/DNA and Protein Synthesis*[*] Transparency 26	**Text** Section Review, p. 140 **Teacher's Resource Binder** Directed Reading Worksheet 7.1 **Other Resources** Audiocassette 7.1[*]
7.2	**1.** Identify and describe the two stages of gene expression. **2.** Explain why the genetic code is said to be universal. **3.** Compare and contrast the roles of the three types of RNA. **4.** Explain the relationships among codons, anticodons, and amino acids. **Themes:** Scale and Structure, Interacting Systems, Stability	**Text** Science in Action *Diana Punales-Morejon: Genetic Counselor* pp. 146–147 **Teacher's Resource Binder** Lab Investigation 7.1 *Protein Synthesis Drama* Extension Worksheet *Making a Protein* **Other Resources** Transparencies 27–30	**Text** Section Review, p. 145 **Teacher's Resource Binder** Directed Reading Worksheet 7.2 **Other Resources** Audiocassette 7.2[*]
7.3	**1.** Explain why cells must regulate gene expression. **2.** Summarize how a gene can be switched off and on. **3.** Distinguish between exons and introns. **4.** Define transposons and explain how they affect gene expression. **Themes:** Stability, Patterns of Change, Evolution, Scale and Structure	**Text** Investigation *How Does DNA Send Messages to the Cytoplasm?* pp. 154–155	**Text** Section Review, p. 150 **Teacher's Resource Binder** Directed Reading Worksheet 7.3 Vocabulary Review Worksheet[*] Reteaching Worksheet[*] *The Structure of DNA* **Other Resources** Audiocassette 7.3[*]

[*]**Reteaching Options**

Demonstrations	**Assessment**
7.1: pp. 138, 139 **7.2:** pp. 142, 144 **7.3:** pp. 149, 150	Chapter Review pp. 152–153 Portfolio Assessment p. 135D Chapter Test—Teacher's Resource Binder Test Generator

Research Notes

Connection to Chemistry: Unusual Amino Acids

Along with the 20 amino acids normally found in living things, there are several similar "nonstandard" amino acids, such as hydroxyproline or iodotyrosine, that are found in some living things.

These nonstandard amino acids usually have additional atoms or groups of atoms substituted onto the normal amino acid framework. They are favored by protein engineers who substitute a nonstandard amino acid in a

protein where an analogous standard one usually is, then see how the protein's shape and activity is affected.

Chemists and biochemists at the University of California's Irvine and Berkeley campuses have come up with two approaches for "tricking" ribosomes in test tubes into making the proteins with the substitutions.

In one, the mRNA coding for the protein is altered, and one of the "STOP" codons is put where the substitution should take place. In a cell, these codons usually signal the end of translation, but in this technique, special tRNA molecules with anticodons that match the "STOP" codon are added, with the nonstandard amino acid iodotyrosine attached. As a result, protein translation often continues past the "STOP," with iodotyrosine put in about 73 percent of the time.

Another approach involves actually expanding the genetic code. The usual nucleotides can make 64 different codons. Scientists have created a "65th" codon that uses a new set of base-pairing nucleotides, "iso-cytosine" and

"iso-guanine," not found in living things. When one of the new nucleotides is used as part of a codon, and scientists add tRNA with the matching anticodon and iodotyrosine attached, the substitution occurs up to 94 percent of the time.

Work continues on adapting this process for microbes, since the test-tube approach produces only very small quantities of protein. Researchers are now working on using plasmids to incorporate nonstandard base-pairs into DNA, and checking on whether those base-pairs can be copied and transcribed faithfully. Another hurdle involves engineering new enzymes within cells that will make tRNA with the new anticodons and the correct nonstandard amino acid attached.

Genetically engineered proteins containing these nonstandard amino acids can help scientists learn more about the mechanisms of these proteins, and may help in designing improved proteins for a variety of uses, from industrial catalysts to drugs.

Investigation Notes

How Does DNA Send Messages to the Cytoplasm?
pages 154–155

Purpose: The students are asked to build on the skills developed in the modeling activity in this chapter to construct a model that illustrates the formation of mRNA from DNA.

Technique: Discuss the structure of DNA before beginning the investigation. Clarify its role in protein synthesis and identify the locations within the cell where these activities take place. (The investigation may be scheduled before reading or discussing the material in the chapter relating to transcription.)

Prelab Preparation

1. The materials for this investigation are similar to those used in the Focus Activity. However, it is best not to re-use the straws from the previous activity.

2. Be sure not to use push pins with pins longer than 0.75 cm (0.25 in.). Such pins may be accidentally pushed completely through the straw, pricking students' fingers. A special DNA and molecular model kit that will work well with this investigation is available from WARD'S (36 M 5523).

Answers will be found on pages 154–155.

Meeting Individual Needs

Objectives

1. Students will demonstrate appropriate use of core vocabulary for the chapter (Vocabulary File).

2. Students will create and interpret an outline describing DNA and RNA (Teaching Strategy A).

Vocabulary File
(Developing Vocabulary/ Limited English Proficiency)

If you are not already using the Vocabulary File, refer to Chapter 1 for its preparation. See the Chapter Highlights on page 151 for a list of suggested words.

Teaching Strategy A
Relating DNA and RNA
for pages 139–145
(Developing Organizational Skills/Visual Learners)

Materials: blank transparency, overhead projector, and projector markers (preferably at least 6 different colors)

Preparation: Place the following points on the transparency as a framework for an outline:

   I. DNA is made of subunits containing three parts.

   II. DNA has a shape of a double helix.

   III. Replication copies the genetic information from DNA into more DNA.

   IV. Transcription transfers genetic information from DNA to RNA.

   V. RNA is made of subunits containing three parts.

Procedure: Have students scan pages 139–145. Tell them to make notes on points that might fit into this outline describing DNA and its function.

Brainstorm with students about DNA and RNA's components, structure, replication, and transcription. Use questions that lead to the completion of the outline. As students respond,

fill in corresponding parts of the outline on the transparency.

As you work through Part I of the outline, ask students to name the three parts of a DNA subunit, as well as the name given to the subunit itself. Ask students to name the four bases and indicate which ones are complementary.

Alongside the specific points in the outline, include diagrams for each process. Instead of being exact chemical structures, these should just be colored shapes that indicate which bases are complementary. Students should include these diagrams in their outlines, to assist them in studying. You may want to color-code the bases as in the DNA diagram shown on this page, and Figure 7.4 on page 139:

Adenine: red
Thymine: orange (or yellow)
Cytosine: light blue (or green)
Guanine: dark blue

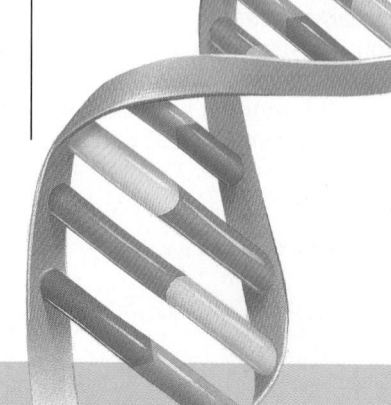

Be sure to refer to Figure 7.4, and compare the roles of the base pairs and the sugar-phosphate backbone to the sides and rungs of a "twisted" ladder that form a double helix.

Before you work through the sections about replication, remind students of cell division as discussed in Chapter 4 (pages 79–81). Remind them that when a cell divides, the chromosomes and DNA must be duplicated. Describe duplication as a three-step process:

1. The two strands separate, and the bonds between base pairs are broken.
2. DNA polymerase moves along the exposed strand, adding complementary nucleotides to each exposed nucleotide.
3. Each strand, with its complementary nucleotides, now forms two new DNA molecules.

Then, compare and contrast replication and transcription. What are the products of each?

Brainstorm with students about the names of the cell structures studied in Chapter 3. Point out that after DNA transcription takes place in the nucleus, the ribosome uses mRNA as a template for building a protein.

Be sure students recognize that in RNA, uracil is used in place of thymine.

Additional Strategies
Visual Strategies

Pages 137, 138, 139, 141, 142, 143, and 144

Auditory Learners

Use *Biology: Visualizing Life* Audiocassettes for Sections 7.1, 7.2, and 7.3.

Meeting Individual Needs (cont.)

Cooperative Learning
Practicing Codes

Timing: Use this activity to introduce Section 7.2.

Group Size: 4 students

Outcome: Students will construct and solve genetic codes.

Individual Accountability: Each student will create codes using the English alphabet and the genetic code.

Positive Interdependence: The group members will decode each other's coded messages.

Before starting Section 7.2, demonstrate a simple code by writing two single-column lists of the alphabet on separate sheets of paper or acetate transparencies. Place the lists side by side, matching the letters. Then, displace one list by one or two letters. (For example, A=Z, B=A, C=B, and so on.) Have each student use the code to write a short message and give it to another group member for decoding.

After Section 7.2 has been introduced, have each student write a 30-nucleotide-long DNA code. Have the group decode both the complementary mRNA code and the translated amino acid chain.

Portfolio Assessment

Students should select their best work and provide a self-reflective rationale for their selections. Students can make selections in the following areas.

1. *Content*   One concept map from the chapter (See page 812 for evaluation criteria.)

2. *Reading Comprehension*   One Directed Reading Worksheet from the Teacher's Resource Binder (Use the answer key to evaluate for accuracy.)

3. *Writing*   Using the Vee Form, summarize a newspaper or magazine article relating to DNA or gene expression. (See page 22T for evaluation criteria.)

   Or: Select a writing task or project from the Chapter Review.

4. *Performance Assessment*   One Vee form from a chapter investigation or lab manual investigation (See page 22T for evaluation criteria.)

Teacher makes selections in the following areas.

1. *Formal Assessment*   Chapter test (Test A, B, or the Test Generator) The teacher-scored test should be reviewed by the student. Incorrect responses should be corrected by the student before the test becomes part of the portfolio.

2. *Informal Assessment*   Use the Direct Observations Checklist, page 33T, during a laboratory or other cooperative learning experience.

3. *Performance Assessment*   Have students re-create the mRNA and DNA sequences that correspond to one of the following polypeptides: angiotensin, cholecystokinin, somatostatin, or vasopressin. (Amino acid sequences are in *The Merck Index.*)

Concept Map Answer

The following is one possible answer to the Relating Concepts exercise on page 152.

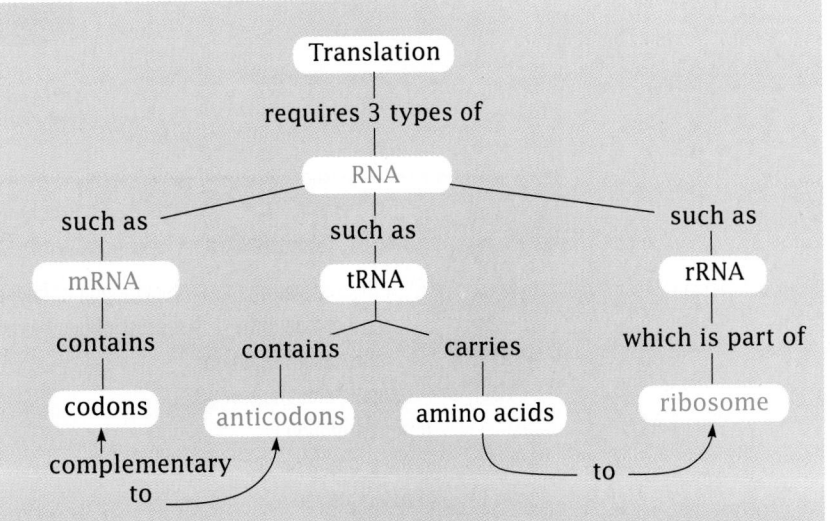

Chapter 7

Determining Prior Knowledge

- Use a microprojector to display the chromosomes in an onion root tip cell. Have students discuss the connection between chromosomes, genes, and DNA.
- Check to see how much students recall about the structure of proteins and DNA, as described in Chapter 2.
- Have students explain what is meant by the statement, "DNA is the genetic material."

Chapter 7

Review

How Genes Work

The genes that determine physical traits and regulate body functions are passed from parent to child.

▪ Author's Rationale ▪

This chapter describes how DNA encodes information about proteins, how that information is passed to ribosomes, and how ribosomes use it to assemble proteins. The key element of this chapter is the explanation of how cells control the process of gene expression.

ONE OF THE MOST IMPORTANT ADVANCES IN HUMAN KNOWLEDGE WAS THE REALIZATION THAT OUR UNIQUE CHARACTERISTICS ARE ENCODED WITHIN MOLECULES OF DNA. THE APPLICATIONS OF THIS KNOWLEDGE HAVE PROFOUNDLY INFLUENCED THE WAY PEOPLE THINK ABOUT THEMSELVES, THEREBY MAKING THE BIOLOGICAL NATURE OF HUMAN BEINGS SEEM APPROACHABLE.

7.1 *Understanding DNA*

Objectives

❶ Explain how researchers concluded that DNA is the genetic material.

❷ Summarize how scientists determined the structure of DNA.

❸ Describe the structure of the DNA molecule.

❹ Summarize the process of DNA replication.

How Scientists Discovered That DNA Is the Genetic Material

As you learned in Chapter 6, chromosomes are made of DNA and protein. Because chromosomes and heredity are linked, biologists hypothesized that either DNA or protein was the genetic material. But which molecule was it?

Griffith discovers the process of transformation

In 1928 the British microbiologist Frederick Griffith was using mice to study the bacterium that causes pneumonia. One type, or strain, of bacteria was enclosed in coats made of complex sugar molecules. When Griffith infected mice with this strain, the mice died. In the other strain of bacteria, the coat was absent. Surprisingly, this strain did not kill mice. Griffith found that if he first killed the coated bacteria by boiling them, they were then harmless to mice. When Griffith mixed the dead coated bacteria (harmless) with live uncoated bacteria (harmless), the mixture killed mice! He examined the blood of the dead mice and found that the live uncoated bacteria had grown coats. Somehow the live uncoated bacteria had acquired the ability to make coats from the dead coated strain. Griffith called this process **transformation.** His experiment is summarized in **Figure 7.1**.

Figure 7.1
Griffith's experiment showed that live uncoated bacteria acquired the ability to make coats from dead coated bacteria.

a When Griffith infected mice with live coated bacteria, the mice died.

b When Griffith infected mice with live uncoated bacteria, they lived.

c When Griffith infected mice with dead coated bacteria, the mice lived.

d When Griffith infected mice with a mixture of dead coated and live uncoated bacteria, they died.

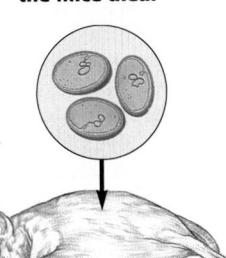

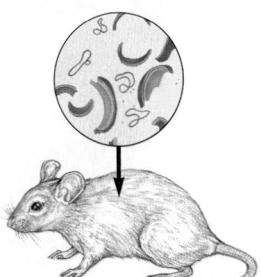

■ *Historical Perspective* ■

The Discovery of Nucleic Acid
DNA was discovered in the same decade that Darwin was formulating his theory of natural selection and Mendel was laying the foundation of genetics. In 1869, Friedrich Miescher, a German chemist, extracted a substance from the nuclei of white blood cells. His analysis revealed that the substance was white, contained phosphorus, and was slightly acidic. He called it nucleic acid.

Lesson Plan 7.1

Phase 1
PREPARATION

Key Concepts
- A series of experiments, first starting with the work of Frederick Griffith in 1928, has shown that DNA is the genetic material.
- In the 1950s, James Watson and Francis Crick determined that the structure of the DNA molecule consists of a double helix.
- The DNA molecule can make a copy of itself in a process called replication.

Reading Strategy

This section illustrates how the discoveries made by one scientist often serve as the foundation on which other scientists build. Students should understand this connection as they read this section, starting with Griffith's work in the 1920s and concluding with Watson's and Crick's work in the 1950s.

Phase 2
TEACHING STRATEGIES

Visual Strategy

Figure 7.1
This figure will help students better understand Griffith's experiments as described in the text. Lead them through each step. Ask why Griffith was puzzled by the results of the experiment he performed (shown in Figure 7.1d). Griffith thought he had made some error in his experimental procedure, so he repeated this experiment many times over. Each time, he got the same result. The only conclusion Griffith could arrive at was that some substance from the dead, coated bacteria must have changed or transformed the live, uncoated bacteria into live, coated bacteria.

137

Demonstration 1

Protein or DNA?

Review the structure of both proteins and DNA so that students understand that proteins have a more complex structure. Use photographs or overhead transparencies to show that proteins are highly folded structures. Point out that proteins are made up of 20 kinds of amino acids, and have students contrast their structure with that of DNA and its four bases. After comparing the two, have students explain why scientists were reluctant to abandon the idea that protein, not DNA, was the genetic material.

Demonstration 2

A Most Useful Scientific Tool

Show students an ordinary blender; the simpler, the better. Tell students that after infecting the bacteria with the viruses, Hershey and Chase had to separate any viral material remaining outside, from the infected bacteria. They used a very simple blender to do this.

Demonstration 3

Virus Infection

Show students an electron micrograph of viruses infecting bacterial cells. Have them identify the viral protein coats.

Visual Strategy

Figure 7.2

Use this figure to reinforce the text description of Hershey's and Chase's experiment. Make sure students understand how Hershey and Chase arrived at their conclusion: Since only DNA entered the bacteria where the new viruses were produced, DNA must carry the genetic instructions needed to make these new viruses. Point out that Hershey and Chase were lucky, to some extent. Certain viruses, but not the ones Hershey and Chase used, also inject some or all of their protein into the cell. Ask students how Hershey's and Chase's experiments would have been affected if the virus they used had been the type that injected its protein, in addition to its DNA, into the bacteria.

Avery shows that DNA is the transforming principle

What was the material that transformed the bacteria from harmless (without coats) to deadly (with coats)? Was this "transforming principle" made of DNA or of protein? In 1944 biologist Oswald Avery and two colleagues at Rockefeller University in New York showed that

Figure 7.2
Hershey and Chase used radioactive labeling to identify the DNA and the protein coats of viruses. Their experiment showed that DNA, not protein, is the hereditary material of viruses.

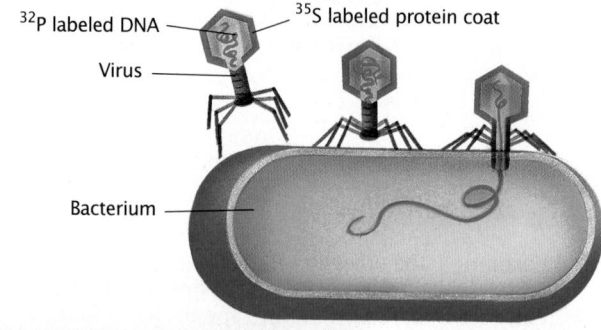

^{32}P labeled DNA ^{35}S labeled protein coat

Virus

Bacterium

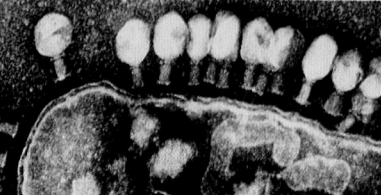

a   The viral DNA (red) is injected into the bacterium, where it will direct the production of new viruses. The protein coat (green) remains outside the bacterium.

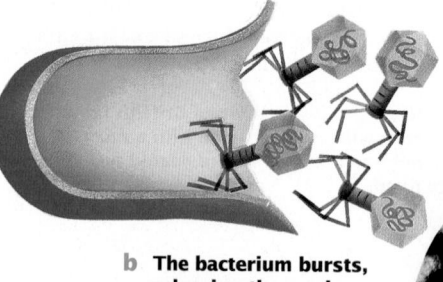

b   The bacterium bursts, releasing the newly made viruses. The new viruses contain DNA labeled with ^{32}P.

DNA is the transforming principle. He extracted DNA from bacteria with coats and added it to a population of bacteria without coats. Some of the bacteria that grew from this mixture now had coats.

In a separate experiment, Avery found that when he added protein-destroying enzymes to bacteria, transformation still occurred. However, when DNA-destroying enzymes were added to bacteria, transformation did not occur. Avery's work provided clear evidence that DNA was the genetic material in these bacteria.

Hershey and Chase confirm that DNA is the genetic material

At first, Avery's results were not widely appreciated. Many biologists were reluctant to give up the idea that protein was the genetic material. In 1952, however, Alfred Hershey and Martha Chase, two scientists at Cold Spring Harbor Laboratory on Long Island, New York, performed an experiment using viruses that infect bacteria. The viruses attach to the surfaces of bacteria and inject their hereditary information into the cells like tiny hypodermic needles. Once inside the bacteria, this hereditary information directs the production of hundreds of new viruses. When the new viruses are mature, they burst out of the infected bacteria and attack new cells. These bacteria-infecting viruses have a very simple structure: a core of DNA surrounded by a protein coat.

To identify the hereditary material, Hershey and Chase used radioactive phosphorus (^{32}P) to label the DNA and radioactive sulfur (^{35}S) to label the protein coats in the viruses. The Hershey-Chase experiment is summarized in **Figure 7.2**. When Hershey and Chase examined the new viruses that burst out of the bacteria, they found that the viruses contained the radioactive phosphorus (^{32}P) label, but not the radioactive sulfur (^{35}S) label. The conclusion was undeniable—DNA is the hereditary material.

How Scientists Determined the Structure of DNA

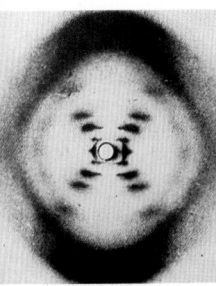

Figure 7.3
This photograph produced by Rosalind Franklin suggested that DNA was helical— the form indicated by the pattern of X-ray reflections in the photograph.

After the experiments of Hershey and Chase, scientists began to study DNA intensely. Scientists already knew that DNA was composed of subunits called **nucleotides** (*NOO klee oh tydz*). Every nucleotide has three parts: a sugar, a phosphate group, and a base. The sugar and phosphate group are the same in every nucleotide, but there are four different bases. The two larger bases, adenine and guanine, are called **purines** (*PYUR eenz*). The two smaller bases, cytosine and thymine, are called **pyrimidines** (*py RIHM uh deenz*).

In 1949 Erwin Chargaff, a biochemist working at Columbia University in New York City, made a key discovery about the chemical structure of DNA by studying the DNA from different organisms. Chargaff found that the amount of adenine in a DNA molecule always equals the amount of thymine (A=T). Likewise, the amount of guanine always equals the amount of cytosine (G=C). These observations, now known as Chargaff's rules, suggested that DNA has a regular structure.

The value of Chargaff's rules became clear when Rosalind Franklin, a chemist working at King's College in London, began studying the structure of DNA using X-ray diffraction. Franklin's X-ray diffraction images, one of which is shown in **Figure 7.3**, suggested that the DNA molecule resembled a tightly coiled spring, a shape called a helix.

Watson and Crick build a model showing DNA's structure

In the early 1950s a young American scientist, James Watson, went to Cambridge, England, on a research fellowship. At Cavendish Laboratories he met Francis Crick, a British physicist interested in DNA. Together, Watson and Crick attempted to construct a model of DNA. They applied the clues provided by Chargaff's rules and Franklin's X-ray diffraction studies. Using tin and wire models of the bases, sugars, and phosphate groups, Watson and Crick deduced that the structure of the DNA molecule is a **double helix**, a spiral staircase of two strands of nucleotides whose bases face each other. The double helix is held together by weak hydrogen bonds between the bases. Adenine can only form hydrogen bonds with thymine, and guanine can form hydrogen bonds only with cytosine. **Figure 7.4** shows the structure of DNA. Francis Crick and James Watson were awarded the Nobel Prize in 1962 for their work in formulating this brilliant model.

Figure 7.4

a **Inside the nucleus of a cell are chromosomes, which contain long strands of DNA.**

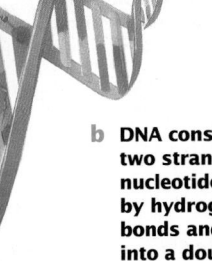

b **DNA consists of two strands of nucleotides, joined by hydrogen bonds and twisted into a double helix.**

c **Every DNA nucleotide contains a sugar, a phosphate group, and a base.**

Adenine
Thymine
Guanine
Cytosine
Hydrogen bonds
Base
Sugar
Phosphate group

■ *Matter of Fact* ■

Rosalind Franklin
Although Franklin's X-ray diffraction photograph was crucial to solving the puzzle of the structure of DNA, she never received credit. She died from cancer before her accomplishments were fully recognized.

ASSESSMENT OPTIONS

Closure Strategy

How Mutations Affect Replication
To study the effects of a mutation, have students work with the materials they used to model DNA replication. Instruct students to change, insert, or delete a base and determine how the replication process is affected.

Section Review

Assign the *Section Review.*

Reteaching

Provide students with the percentage of any one of the bases in an organism's DNA. For example, inform them that chemical analysis revealed that 32 percent of the bases in a particular DNA consisted of adenine. Have them calculate the percentages for the other bases.

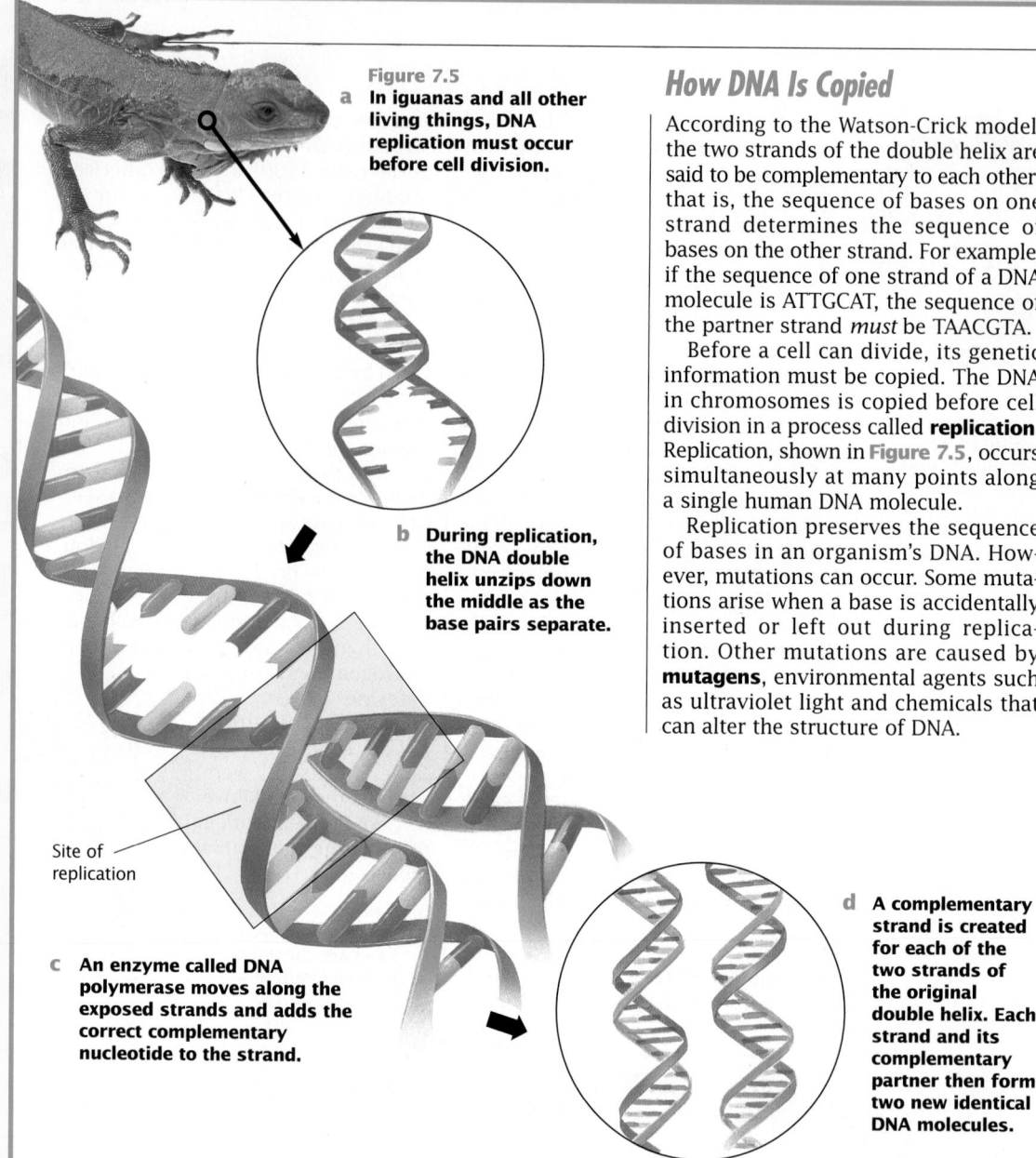

Figure 7.5

a In iguanas and all other living things, DNA replication must occur before cell division.

b During replication, the DNA double helix unzips down the middle as the base pairs separate.

Site of replication

c An enzyme called DNA polymerase moves along the exposed strands and adds the correct complementary nucleotide to the strand.

d A complementary strand is created for each of the two strands of the original double helix. Each strand and its complementary partner then form two new identical DNA molecules.

How DNA Is Copied

According to the Watson-Crick model, the two strands of the double helix are said to be complementary to each other; that is, the sequence of bases on one strand determines the sequence of bases on the other strand. For example, if the sequence of one strand of a DNA molecule is ATTGCAT, the sequence of the partner strand *must* be TAACGTA.

Before a cell can divide, its genetic information must be copied. The DNA in chromosomes is copied before cell division in a process called **replication**. Replication, shown in **Figure 7.5**, occurs simultaneously at many points along a single human DNA molecule.

Replication preserves the sequence of bases in an organism's DNA. However, mutations can occur. Some mutations arise when a base is accidentally inserted or left out during replication. Other mutations are caused by **mutagens**, environmental agents such as ultraviolet light and chemicals that can alter the structure of DNA.

Section Review

① What did Griffith, Avery, and Hershey and Chase each contribute to identifying the genetic material?

② What evidence did Watson and Crick use to deduce the structure of DNA?

③ What are the restrictions on the four different nucleotide bases when pairing in the double helix?

④ Describe the process by which DNA is copied.

▪ Section Review Answers ▪

1. Students should credit Griffith with the process of transformation; Avery with naming DNA as the transforming principle in Griffith's experiments; and Hershey and Chase with confirming DNA as the hereditary material.

2. Students should explain that Watson and Crick used Chargaff's rules and Franklin's X-ray diffraction studies to determine the structure of DNA.

3. Students should indicate that adenine can only form hydrogen bonds with thymine, and guanine can only form hydrogen bonds with cytosine.

4. See Figure 7.5 on page 140.

THE DISCOVERY THAT DNA IS THE GENETIC MATERIAL LEFT STILL MORE QUESTIONS UNANSWERED. HOW IS THE INFORMATION IN DNA USED? SCIENTISTS NOW KNOW THAT DNA DIRECTS THE CONSTRUCTION OF PROTEINS. PROTEINS DETERMINE THE SHAPES OF CELLS AND SPEED THE RATES OF CHEMICAL REACTIONS SUCH AS THOSE THAT OCCUR DURING METABOLISM AND PHOTOSYNTHESIS.

7.2 *How Proteins Are Made*

Objectives

❶ **Identify and describe the two stages of gene expression.**

❷ **Explain why the genetic code is said to be universal.**

❸ **Compare and contrast the roles of the three types of RNA.**

❹ **Explain the relationships among codons, anticodons, and amino acids.**

The Transfer of Genetic Information

Figure 7.6
During gene expression, the information in DNA is used to assemble proteins.

Once scientists understood the structure of DNA, they were able to learn how a specific protein is built from information found in the DNA of one gene. Scientists now know that DNA is used as a blueprint to make a similar molecule called **ribonucleic acid**, or RNA for short. This RNA then directs the formation of proteins. The use of genetic information in DNA to make proteins is called **gene expression**. Gene expression takes place in two stages. The first stage is called **transcription**. During transcription, an RNA copy of a gene is made. During **translation**, the second stage of gene expression, three different kinds of RNA work together to assemble amino acids into a protein molecule. Gene expression is summarized in **Figure 7.6**.

Through gene expression, the messages encoded in DNA direct all cellular activities. For example, when you eat carbohydrates such as those found in a bowl of cereal or a slice of bread, certain genes direct the production of a protein called insulin. Insulin helps your body maintain its blood sugar level. Other genes direct the production of hundreds of other structural proteins and enzymes.

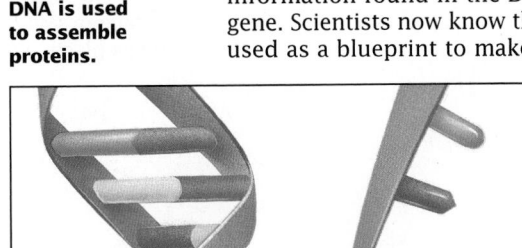
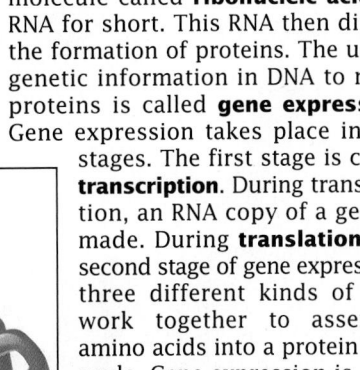

DNA → Transcription → RNA → Translation → Protein

Lesson Plan 7.2

Phase 1

PREPARATION

Key Concepts
- DNA directs the synthesis of RNA in a process called transcription.
- RNA directs the synthesis of proteins in a process called translation.
- Three types of RNA are involved in protein synthesis: messenger RNA (mRNA), transfer RNA (tRNA), and ribosomal RNA (rRNA).
- The genetic information in DNA is transcribed into mRNA as a triplet combination of bases known as a codon.
- An mRNA codon is then matched by a triplet combination in tRNA known as an anticodon.
- This codon-anticodon match determines which amino acid is inserted into the protein being synthesized on the ribosomes.
- The genetic code in DNA is universal, since a particular codon specifies the same amino acid in all organisms.

Reading Strategy

Most students will have difficulty understanding how DNA directs the synthesis of proteins. Visualization will help. Instruct students to visualize the material as they read this section. Tell them to construct a mental image for each main idea they read. Studying the accompanying figures will help them in this task.

Phase 2
TEACHING STRATEGIES

Visual Strategy

Figure 7.6
Use this figure to distinguish between transcription and translation.

Proteins

Review how amino acids are joined to form protein molecules. Remind students that proteins, as enzymes, play a vital role in cell metabolism, since most biochemical reactions would not occur without them.

Demonstration 1

Transcription

Have each student write down a random sequence of 18 bases that might be found in a segment of a DNA molecule. Then instruct them to transcribe this sequence into RNA. Check to make sure that each A in DNA was correctly transcribed into the complementary U. Have students keep their sequences for later use, when they will examine how translation operates.

Visual Strategy

Figure 7.7

Have students compare Figures 7.5 and 7.7. They must understand that in both replication and transcription, the DNA double helix unwinds. Point out one major difference: cells contain little DNA polymerase but much RNA polymerase. Consequently, while few portions of a DNA molecule can undergo replication simultaneously, many RNAs can be transcribed simultaneously from the same DNA segment.

Demonstration 2

Site of Transcription

Show students either an overhead transparency, or a slide, of a eukaryotic cell. Remind them that transcription occurs in the nucleus. The RNA that is synthesized must then travel to the cytoplasm.

How DNA Makes RNA

Transcription is the process by which genetic information encoded in DNA is transferred to an RNA molecule. Genetic information must be copied because DNA cannot leave the nucleus. Just as an architect protects building plans from loss or damage by keeping them in a central place, your cells protect genetic information by keeping the DNA safe within the nucleus. Instead of sending out the DNA, copies of genes are sent out into the cell to direct the assembly of proteins. These working copies of genes are made of a single strand of RNA.

RNA is chemically similar to DNA except that its sugars have an additional oxygen atom, and the base thymine (T) is replaced by a structurally similar base called uracil (U). RNA occurs in three different forms. Figure 7.7 shows transcription of one form of RNA, messenger RNA (mRNA). Just as monks once copied manuscripts by faithfully transcribing each letter, so enzymes in your cells' nuclei make mRNA copies of your genes by copying their nucleotides.

During transcription, each gene is copied from a fixed starting position called a promoter site. Here, an enzyme called RNA polymerase binds to one strand of the DNA double helix and moves along the DNA strand like a train on a track. As it passes over each nucleotide, the RNA polymerase pairs the base in that nucleotide with its complementary RNA base. Cytosine is paired with guanine. But because RNA contains uracil instead of thymine, adenine is now paired with uracil. In this way, a complementary strand of mRNA is gradually built. Nucleotide sequences located at the end of the genes tell the RNA polymerase where to stop. In eukaryotes, after transcription of a gene is finished, the mRNA passes out of the nucleus through pores in the nuclear membrane and into the cytoplasm. There the second stage of gene expression, translation, takes place.

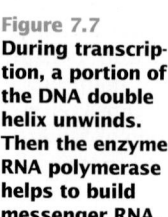

Figure 7.7
During transcription, a portion of the DNA double helix unwinds. Then the enzyme RNA polymerase helps to build messenger RNA.

mRNA

RNA polymerase

DNA

Visual Strategy

Figure 7.8
To make sure that students know how to read the chart, provide examples of codons and ask for the amino acids that are coded. Have students identify the mRNA codons that stop the transcription process.

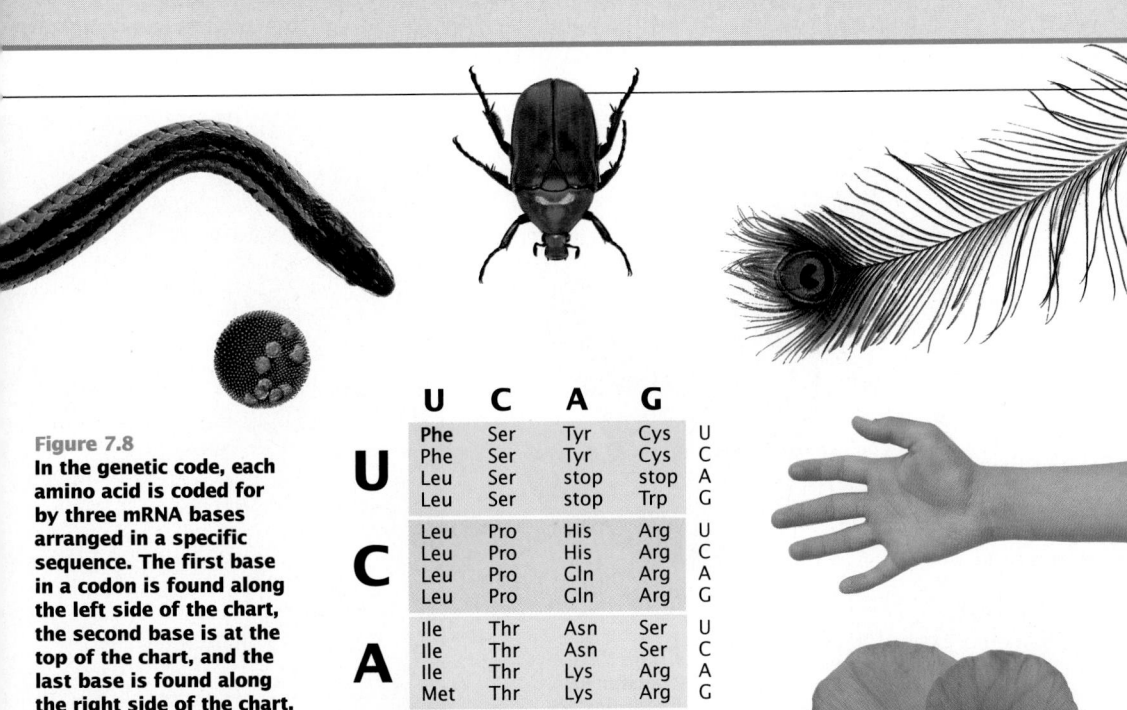

Figure 7.8
In the genetic code, each amino acid is coded for by three mRNA bases arranged in a specific sequence. The first base in a codon is found along the left side of the chart, the second base is at the top of the chart, and the last base is found along the right side of the chart. For example, the amino acid alanine (Ala) is coded for by the triplets GCU, GCC, GCA, and GCG.

	U	C	A	G	
U	Phe	Ser	Tyr	Cys	U
	Phe	Ser	Tyr	Cys	C
	Leu	Ser	stop	stop	A
	Leu	Ser	stop	Trp	G
C	Leu	Pro	His	Arg	U
	Leu	Pro	His	Arg	C
	Leu	Pro	Gln	Arg	A
	Leu	Pro	Gln	Arg	G
A	Ile	Thr	Asn	Ser	U
	Ile	Thr	Asn	Ser	C
	Ile	Thr	Lys	Arg	A
	Met	Thr	Lys	Arg	G
G	Val	Ala	Asp	Gly	U
	Val	Ala	Asp	Gly	C
	Val	Ala	Glu	Gly	A
	Val	Ala	Glu	Gly	G

Theme Answer
Evolution
Students should explain that the relationship between DNA and different organisms is universal because DNA is found in all living things, and the genetic code is the same in nearly all organisms.

Mathematics Connection

Why a Triplet Codon?
Have students explain why either a one- or two-base codon system could not work. Tell them to keep in mind how many different amino acids are found in proteins.

The Genetic Code

How is mRNA translated into the sequence of amino acids that make up proteins? How can the four nucleotide bases found in mRNA carry instructions to build the thousands of proteins your body needs? Every three nucleotides in mRNA specify a particular amino acid. Each nucleotide triplet in mRNA is called a **codon** *(KOH dahn)*. The order of bases in a codon determines which amino acid will be added to a growing protein chain. In turn, the order of amino acids will determine the structure and function of a protein.

Evolution

How is the DNA

of all organisms

similar?

To learn more about how mRNA directs amino acids to join in a specific order, biologists performed laboratory experiments using artificial mRNA to direct protein production. An mRNA that contained only the nucleotide uracil (U), for example, made a protein that consisted entirely of the amino acid phenylalanine (Phe). This information told scientists that the codon "UUU" codes for the amino acid phenylalanine. These experiments ultimately revealed the genetic code, which is shown in **Figure 7.8**. The names of amino acids have been abbreviated in the chart. The **genetic code** is the correspondence between nucleotide triplets in DNA and the amino acids in proteins. Any of the four bases (U, C, A, G) found in mRNA can occur at any of the three positions of a codon. Thus, there are 64 different possible three-letter codons $(4 \times 4 \times 4 = 64)$ in the genetic code. Since there are 64 possible codons, but only 20 different amino acids occur in proteins, more than one codon may specify a single amino acid.

The genetic code is the same in nearly all organisms, so it is said to be universal. For example, the code for phenylalanine is the same in bacteria and humans.

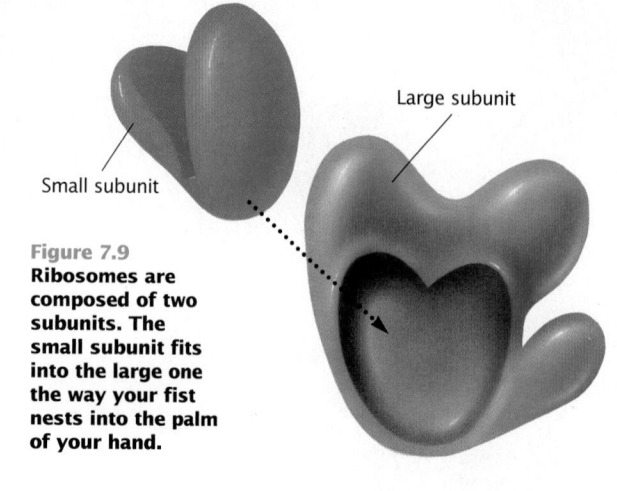

Small subunit

Large subunit

Figure 7.9
Ribosomes are composed of two subunits. The small subunit fits into the large one the way your fist nests into the palm of your hand.

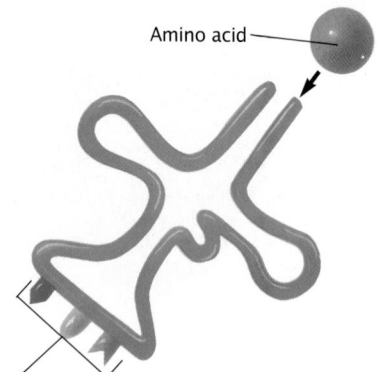

Amino acid

Anticodon

Figure 7.10
Transfer RNA molecules are chains about 80 nucleotides long, folded into a compact shape. The anticodon is a three-nucleotide sequence at one end of the tRNA. An amino acid is attached at the opposite end.

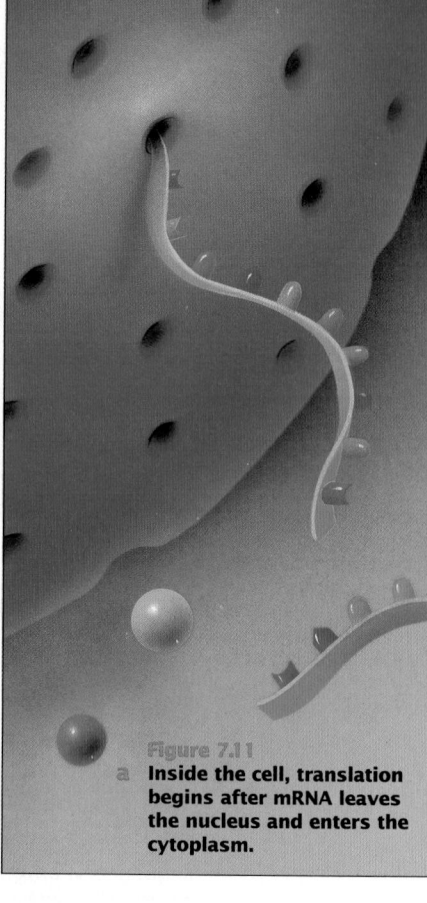

Figure 7.11
a **Inside the cell, translation begins after mRNA leaves the nucleus and enters the cytoplasm.**

How RNA Makes Proteins

After transcription in eukaryotes, the mRNA strand leaves the cell's nucleus and travels into the cytoplasm. During translation, the mRNA works with two other types of RNA to build proteins by joining amino acids. Translation occurs on ribosomes, complex organelles that contain a special kind of RNA called ribosomal RNA (rRNA). **Figure 7.9** shows that each ribosome consists of two subunits. The smaller ribosomal subunit contains a short rRNA sequence that is complementary to the mRNA codon that signals "start" in all genes. Translation begins when the mRNA "start" codon binds to the small ribosomal subunit. The large ribosomal subunit attaches to form a complete ribosome with a strand of mRNA running through it. Just as factories use blueprints to direct the assembly of cars, so do ribosomes use mRNA to direct the assembly of proteins.

The pocket, or dent, in the small ribosomal subunit has just the right shape to bind a third kind of RNA molecule, transfer RNA (tRNA). Transfer RNA carries amino acids to the ribosome. The three-nucleotide sequence shown on one end of the transfer RNA molecule in **Figure 7.10** is called an **anticodon**. Anticodons are complementary to mRNA codons.

Like the address on an envelope, the anticodon ensures that an amino acid is delivered to its proper "address" on the mRNA as a protein is being assembled. And just as a mail carrier checks the address before delivering a letter, so the small ribosomal subunit checks

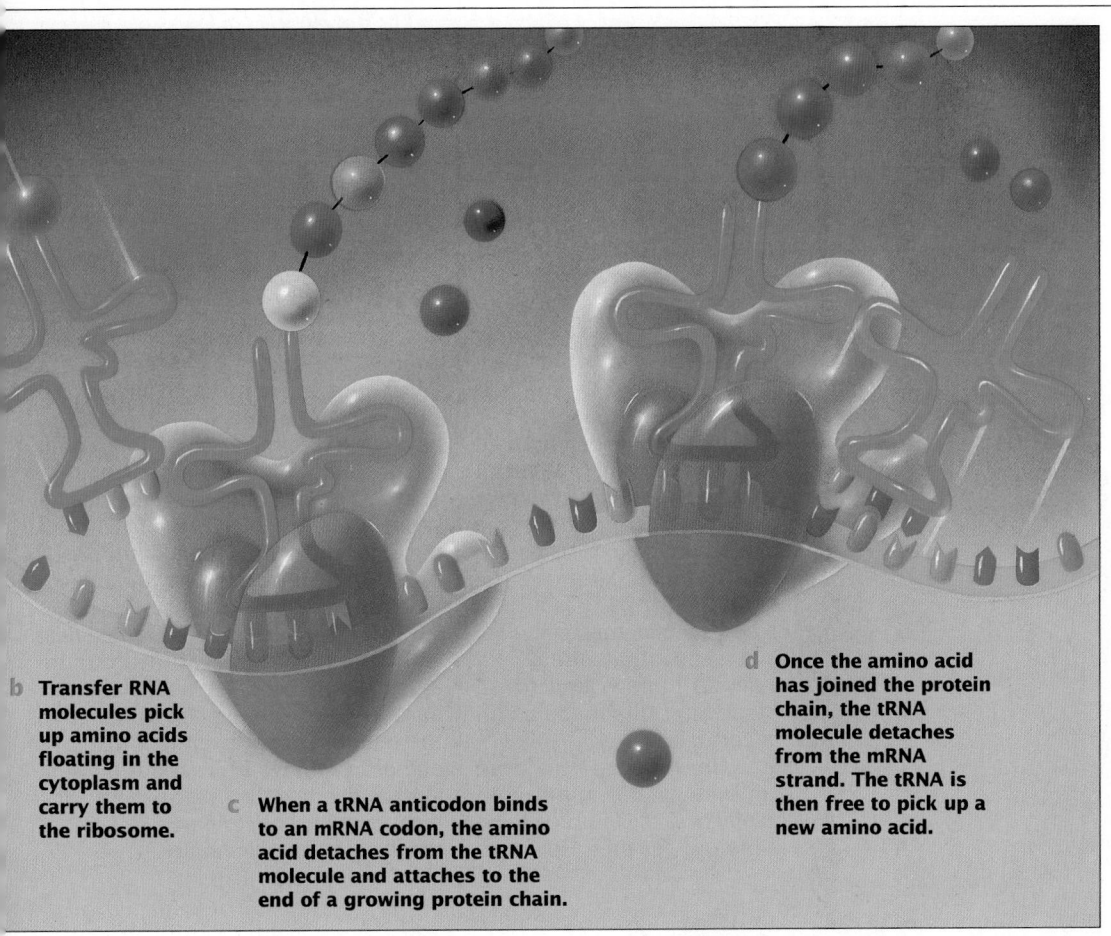

b Transfer RNA molecules pick up amino acids floating in the cytoplasm and carry them to the ribosome.

c When a tRNA anticodon binds to an mRNA codon, the amino acid detaches from the tRNA molecule and attaches to the end of a growing protein chain.

d Once the amino acid has joined the protein chain, the tRNA molecule detaches from the mRNA strand. The tRNA is then free to pick up a new amino acid.

Phase 3

ASSESSMENT OPTIONS

Closure Strategy

Working Backward

Using the abbreviations found in Figure 7.8, provide students with a sequence of three amino acids. Have students write an mRNA sequence that could code for these amino acids, then write the appropriate DNA sequence. Compare several responses to show students that more than one sequence of DNA bases can code for the same sequence of amino acids.

Section Review

Assign the *Section Review.*

Reteaching

Write down a nine-base DNA sequence. Have students transcribe the mRNA, then translate the amino acid sequence. Next, change any one of the nine DNA bases and have students check to see how both transcription and translation would be affected.

the tRNA anticodon to see that it is complementary to the mRNA codon.

As the mRNA passes through the ribosome, one tRNA after another is selected to match the sequence of mRNA codons. Amino acids are added to the end of the growing protein chain until the end of the mRNA sequence is reached. At this point, a "stop" codon is encountered for which there is no anticodon on any tRNA molecule. With nothing to fit into the tRNA site, the ribosome complex falls apart and the newly assembled protein is released into the cell. **Figure 7.11** shows the steps of translation.

Section Review

1 Outline the path genetic information travels when proteins are made.

2 Explain the process by which DNA is copied into RNA.

3 What is the genetic code? Why is it said to be universal?

4 Explain the process by which proteins are made from RNA.

▪ Section Review Answers ▪

1. See Figure 7.6 on page 141.
2. See Figure 7.7 on page 142.
3. The genetic code is the correspondence between nucleotide triplets in DNA and the amino acids in proteins. The genetic code is the same in nearly all organisms, so it is said to be universal.

4. See Figures 7.9–7.11 starting on page 144.

Using the Feature

- Use this feature as a springboard for a class discussion of the challenges that Diana Punales-Morejon might have faced upon her arrival in this country as a five-year-old. Students with similar personal histories might be willing to share some of their personal experiences with the class.

- This feature can also be used to focus on the process skills of designing and evaluating. In her work, Diana designs surveys to analyze patient's perceptions toward prenatal care and reproductive functions.

- Emphasize that other subjects (such as psychology) can be studied in conjunction with biology to provide a foundation for an exciting career in a people-oriented field.

Discussion

Guide a discussion by posing the following questions.

1. What difficulties did Diana face when she moved to the United States from Cuba as a young girl?
   Students should discuss the language barrier Diana encountered.

2. What personal traits does Diana possess that make her a successful genetic counselor?
   Diana is caring, sociable, and has good communication skills.

3. Who played an important role in Diana's life as a high school student? What kind of influence did this person have?
   Diana fondly remembers her high school biology teacher whose enthusiasm for science made her very eager to learn.

4. In what other disciplines does Diana have a strong interest?
   teaching and psychology

Diana Punales-Morejon:
Genetic Counselor

How I Became Interested in Genetic Counseling

"When I was five years old, my family moved to the United States from Cuba. We didn't speak any English, so I had to learn the language in school. It was very difficult at first, but soon I picked it up and began enjoying the time I spent at school, talking to my new friends. My favorite teacher was my high school biology teacher, a fantastic woman. She took the class on field trips to laboratories and science museums. We could sense that she really enjoyed biology and loved teaching; this made us all eager to learn.

"In college, I majored in biology and minored in psychology because I wanted to be a biology teacher. However, after graduation, I ended up working in a biomedical lab, doing research. I enjoyed the work, but most of all I missed the personal contact with people. I began looking for a career that could include both biology and psychology. That's when I came across genetic counseling. Genetics has always fascinated me and the counseling component satisfied my interest in psychology, so it sounded perfect.

"Today, as a genetic counselor, I deal with issues that can be difficult to understand and handle. It usually takes a lot of teaching to explain genetics to the people I'm counseling. But that's one of the things I like about my job. Most of the time they can't believe what technology is able to detect. It's mind-boggling for many people."

Diana enjoys hiking around town with her 10-month-old daughter, Amanda Isabel.

Name:	**Diana Punales-Morejon**
Home:	**Union City, New Jersey**
Employer:	**Beth Israel Medical Center, New York, NY**
Personal Traits:	• **Caring**
	• **Dedicated**
	• **Detail-oriented**
	• **Good communication skills**
	• **Sociable**

Action ▪ Science in Action ▪ Science in Action ▪ Science

The Satisfaction of Genetic Counseling

"I was hired in 1987 as a genetic counselor. Within a year, I was promoted to Coordinator of the Genetic Counseling Program. I teach medical, nursing, and graduate students. I also see patients for genetic counseling—women and couples that are at risk for having children with birth defects and genetic disease.

"It is the most incredible feeling to be able to tell someone that their child, who was at risk, will be normal. That feeling lasts forever. The worst part, the part that never gets easy for me, is when a couple has prenatal testing and the test shows the baby will be affected. That is the worst thing I will ever do in my life—

especially when people want and plan for the pregnancy.

"Fortunately, abnormalities occur only 5 percent of the time. The good far outweighs the bad, but it never gets any easier. A family having an abnormal pregnancy is faced with difficult and painful choices."

Research Focus

Diana coordinates all research projects for the genetic counseling department of her hospital. She is now studying social, psychological, and cultural issues by analyzing patients' perceptions and attitudes towards prenatal care and reproductive functions. She has developed a survey that is administered to patients who are referred for prenatal diagnosis and genetic counseling. The survey asks patients about their knowledge of genetics and prenatal tests as well as their feelings about terminating abnormal pregnancies.

Diana also looks at how demographic factors like race, religion, education, income, and gender impact patients' views about testing and reproductive options. She wants to see if patients of different races and cultures use testing differently. She is also interested in identifying patients' major concerns. Some individuals are more anxious about the pain involved; others are more worried about the length of time they must wait for results. It is important to give people the kind of information they're most anxious about first, before discussing their medical history.

Lesson Plan 7.3

Key Concepts

- In prokaryotes, repressor proteins attach to DNA and switch genes off by preventing transcription.
- An inducer combines with the repressor so that it can no longer attach to the DNA and inhibit transcription.
- Activator proteins facilitate transcription by helping the DNA unwind.
- Enhancer proteins facilitate transcription by exposing the RNA polymerase binding sites.
- In eukaryotes, the DNA contains intron and exon regions. Exons are the regions that code for proteins.
- Transposons are genes that can move from one location in a chromosome to another.

Reading Strategy

As students read this section, instruct them to draw an analogy to show where control mechanisms or regulations are necessary. For example, without signal lights, traffic would be chaotic. Similarly, without regulation, genes would produce proteins in a chaotic fashion.

Theme Answer

Stability

Students should explain that eukaryotic cells regulate gene expression within themselves by switching transcription of genes on or off. This ensures that they will have the proteins necessary to function properly.

THE TRANSLATION OF A GENE INTO A PROTEIN IS ONLY PART OF GENE EXPRESSION. EVERY CELL MUST ALSO BE ABLE TO REGULATE THE USE OF PARTICULAR GENES. JUST AS A CONDUCTOR CONTROLS HOW LOUD AND HOW FAST THE DIFFERENT INSTRUMENTS IN AN ORCHESTRA PLAY, A CELL DETERMINES WHEN PARTICULAR GENES ARE TRANSCRIBED BY CONTROLLING WHEN GENES ARE SWITCHED "ON" AND "OFF."

7.3 Regulating Gene Expression

Objectives

1 **Explain why cells must regulate gene expression.**

2 **Summarize how a gene can be switched off and on.**

3 **Distinguish between exons and introns.**

4 **Define transposons and explain how they affect gene expression.**

Switching Genes On and Off

Stability

What mechanisms provide a eukaryotic cell with the proteins it needs to function properly?

Cells control the expression of their genes by determining when individual genes are to be transcribed. Each gene possesses special regulatory sites, which act as points of control. Specific regulatory proteins within the cell bind to these sites, switching transcription of the gene on or off.

The best understood regulatory mechanisms are those used by prokaryotes. Some genes in prokaryotes are expressed nearly all the time, while others are rarely used. Genes that are expressed only occasionally are said to be switched off. They are transcribed only when the proteins are needed. In these genes, transcription cannot occur because a large molecule called a **repressor protein** is bound to the DNA in front of the genes, as shown in **Figure 7.12a**. The repressor protein blocks transcription by preventing the RNA polymerase from moving along the gene. If someone placed a brick wall between your chair and desk, you could not begin your work until the wall was removed. In the same way, transcription cannot begin until the repressor protein is removed.

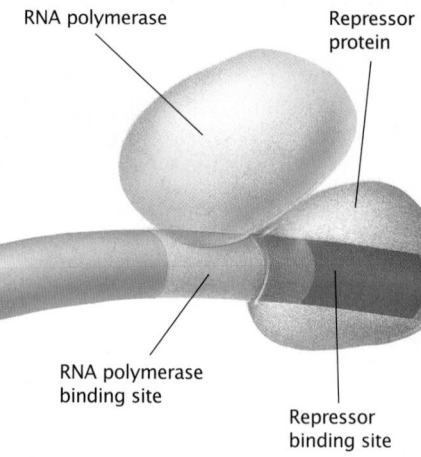

RNA polymerase

Repressor protein

RNA polymerase binding site

Repressor binding site

Figure 7.12
a **A repressor protein sits on the DNA and blocks RNA polymerase from attaching to its binding site.**

Transcription begins when an inducer is present

For transcription to begin, molecules called **inducers** must bind to the repressor protein. The binding causes the repressor protein to change its shape so that it no longer fits the DNA. As a result, the repressor protein falls off, removing the barrier to transcription. When this happens, the gene is switched on.

In bacteria, gene expression controls the digestion of lactose, the sugar found in milk. When a bacterium encounters lactose, the lactose acts as an inducer by binding to the repressor protein and altering its shape. The altered repressor protein then detaches from the DNA molecule, which allows the RNA polymerase to transcribe the genes needed to produce the enzyme responsible for digesting lactose.

Activators help DNA unwind

Because RNA polymerase binds to only one strand of the DNA double helix, it is necessary that the DNA molecule partially unwind to expose the bases at the promoter site. In many genes, this unwinding cannot take place without the help of a regulatory protein called an activator. An activator binds to the DNA in this region and helps it unwind.

Cells are able to switch genes "off" by binding signal molecules to the activator protein. This prevents the activator from binding to the DNA molecules. As a result, the double helix cannot unwind and the genes cannot be transcribed.

Activators enable a cell to carry out a second level of control. When a bacterium already has plenty of energy, the level of another signal molecule—a special "I-need-sugar" signal molecule —decreases. Without being prodded by this signal molecule, the activator protein cannot bind to the DNA molecule's unwinding site. As a result, the genes are not transcribed, even though the repressor protein does not block the RNA polymerase.

Enhancers expose binding sites in eukaryotic DNA

A third level of control is exercised by regulating access to the gene. For example, because human chromosomes are large and complex, it is not easy for RNA polymerase to find its way to the beginning of a particular gene. To make themselves more accessible to RNA polymerase when switched on, many genes in eukaryotes possess special sequences called enhancers. When activated by specific protein molecules, enhancers aid in exposing the RNA polymerase binding site. Unlike promoters and activator proteins, which are found at the beginning of a gene, enhancers are usually located far from the start of the gene, often as far as several thousand nucleotides away.

Genes for digesting lactose

RNA polymerase

Repressor protein

Inducer

b **When the inducer is present, it binds with the repressor protein and changes its shape. The altered repressor protein detaches from the DNA strand, enabling the polymerase to attach and begin transcribing the gene.**

Demonstration 1
Repressors and Inducers

Wrap some clay around a meter stick. Inform students that the clay represents a repressor that, when bound to DNA (meter stick), prevents RNA polymerase from attaching. Use a different color clay to represent the inducer. Attach it to the repressor and have the repressor fall off. Tell students that RNA polymerase can now attach to the DNA. Have students predict what will happen when the inducer is no longer present. Relate this simulation of gene control to the illustrations in Figure 7.12.

Demonstration 2
Activators

Place two meter sticks close together. Tell students that they represent the two strands of DNA. As long as the two strands are connected, RNA polymerase cannot bind and begin transcription. Demonstrate what happens when an activator protein binds to the DNA by moving the two sticks apart.

■ *Historical Perspective* ■

Jumping Genes

The Nobel Prize that Barbara McClintock received in 1983 was in recognition of the studies of corn she conducted nearly 40 years earlier. This long delay reflected the unwillingness of the scientific community to accept the fact that genes can jump from one location to another. The concept of genes moving about in the genome did not fit into the framework scientists had developed about genetic material. Only in the last 10 years, with the discovery of transposons in both prokaryotes and eukaryotes, have McClintock's studies been recognized as the pioneering efforts in this area.

Demonstration 3

Introns and Exons

Write the following sentence on the board: "Both boys and girls love school and enjoy spending time with their friends and family." Before showing it to students, add extra letters between each word so the entire sequence looks like nonsense. Ask students to remove the introns so the exons make sense. Determine how many different ways the words (exons) can be juggled to make a complete sentence (a functional protein).

Phase 3

ASSESSMENT OPTIONS

Closure Strategy

Almost Infinite Possibilities

Each human is capable of making millions of different antibody proteins. Yet no one has millions of genes to code for all these proteins. Ask students how transposons make it possible to produce so many different antibodies.

Section Review

Assign the *Section Review.*

Reteaching

Review how mitosis produces daughter cells with the same number and kind of chromosomes as the parent cell. Since the 100 trillion cells in the human body are the products of mitotic divisions, each must contain a complete set of genetic instructions. Discuss how genes must be turned on and off in different ways in the various types of cells to produce a normal adult.

Architecture of the Gene

While it is tempting to think of a gene as a single uninterrupted stretch of DNA that codes for proteins, this simple arrangement occurs only in bacteria. In eukaryotes, a gene contains a series of sequences called exons and introns. **Exons** are the portions of a gene that actually get translated into proteins. They are interrupted by noncoding portions of the DNA called **introns**. In most genes, the introns far outweigh the protein-encoding exons. In fact, in most cases, less than 10 percent of a human gene consists of exons. Like cars on a rural highway, exons are scattered here and there within genes. Introns are separated from exons during transcription.

Figure 7.13
Barbara McClintock received a Nobel Prize in 1983 for her discovery of transposons in Indian corn.

Exons play a role in evolution

How could such a complicated system have survived the process of natural selection? The answer is that it adds evolutionary flexibility. Each exon encodes a different part of a protein. One exon may influence which molecules an enzyme is able to recognize, while another may determine whether a protein will respond to particular signal molecules. By possessing introns and exons, cells can shuffle exons between genes and create new combinations. Natural selection probably favored the intron-exon system of organization because cells are able to manufacture many different proteins by juggling exons between genes. The many thousands of proteins that occur in human cells appear to have arisen from only a few thousand exons!

Genes can jump to new locations

A few genes in chromosomes have the ability to move from one location in the chromosome to another. These genes are known as **transposons** (*tranz POH zahnz*). Once every few thousand cell divisions, a transposon jumps to a new location in the same chromosome or in a different chromosome. Transposons often inactivate the genes they jump into, creating mutations. Barbara McClintock, a geneticist working in the Cold Spring Harbor Laboratory on Long Island, New York, discovered transposons in the course of her studies of corn. The spotted and streaked patterns seen in the Indian corn shown in **Figure 7.13** result from the interactions of transposons that control its kernel pigments.

Section Review

❶ How do repressor proteins and inducer molecules affect transcription?

❷ How do activator proteins help RNA polymerase bind to a strand of DNA?

❸ How do the arrangements of genes differ in eukaryotes and prokaryotes?

❹ What effect do transposons have on other genes?

▪ *Section Review Answers* ▪

1. A repressor protein blocks RNA polymerase from moving along a gene. An inducer molecule binds to the repressor protein causing the repressor protein to change its shape so that it no longer fits the DNA, removing the barrier to transcription.

2. Activator proteins bind to the DNA and help it unwind.

3. In prokaryotes, genes occur in a single stretch of DNA. In eukaryotes, genes exist in a series of sequences called exons (coding) and introns (noncoding).

4. Transposons have the ability to inactivate genes, creating mutations.

Chapter **7** *Highlights*

Watson and Crick built this model of DNA in 1953.

Alternative Assessment
Assign four students to a cooperative work group. Remind everyone that DNA contains the genetic code. Have each group write a short message based on a code system they have developed. Place each group's message on the board and have the other groups try to decipher the code.

	Key Terms	Summary
7.1 Understanding DNA 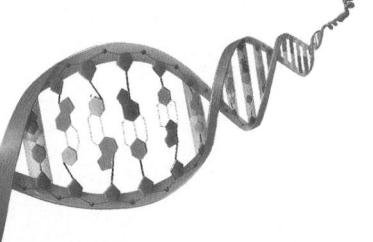 It took scientists many years to realize that DNA is the genetic material and to determine its structure.	transformation (p. 137) nucleotide (p. 139) purine (p. 139) pyrimidine (p. 139) double helix (p. 139) replication (p. 140) mutagen (p. 140)	• Griffith, Avery, and Hershey and Chase performed experiments that helped show that DNA is the hereditary material. • DNA is composed of subunits called nucleotides. Each nucleotide contains a sugar, a phosphate group, and one of four bases. • Watson and Crick showed that the DNA molecule is a double helix. • Before cell division, DNA copies itself in a process called replication. The DNA separates into two strands, and new complementary bases attach to the exposed base.
7.2 How Proteins Are Made 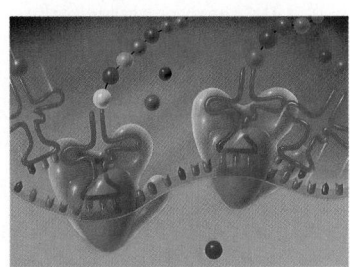 Proteins are made when amino acids are assembled by mRNA and tRNA molecules at the ribosome.	ribonucleic acid (p. 141) gene expression (p. 141) transcription (p. 141) translation (p. 141) codon (p. 143) genetic code (p. 143) anticodon (p. 144)	• During transcription, the genetic message from DNA is transferred to RNA. • During translation, RNA directs the production of specific proteins encoded by genes. • Each group of three nucleotides in mRNA is called a codon. Codons specify amino acids. • Each sequence of three nucleotides in a tRNA molecule is called an anticodon. Anticodons complement the nucleotide sequence in mRNA codons.
7.3 Regulating Gene Expression 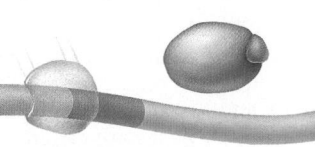 Cells can control gene activity. When an inducer binds to the repressor molecule, it falls off of DNA, allowing transcription to begin.	repressor protein (p. 148) inducer (p. 149) exon (p. 150) intron (p. 150) transposon (p. 150)	• In prokaryotes, transcription is regulated by repressor proteins. A repressor protein blocks RNA polymerase from transcribing a gene. • In eukaryotes, genes are fragmented. Exons are the portions of a gene that are translated into proteins. Introns are noncoding regions of DNA. • Transposons are genes that can jump to new locations on chromosomes.

Chapter Review Answers

Understanding Vocabulary

1. **a.** A nucleotide is a subunit of DNA and RNA that is composed of a nitrogen base, a sugar, and a phosphate group. A gene is a segment of DNA within a chromosome that codes for a particular protein.
b. Replication is the process of copying DNA. Transcription is the building of an RNA complement to a strand of DNA.
c. A codon is a nucleotide triplet in mRNA (or DNA). An anticodon is a nucleotide triplet at the end of a tRNA molecule that matches up with a codon in mRNA.
d. Exons are the portions of the DNA that are first transcribed into RNA and then translated into proteins. Introns are portions of DNA that are not translated into proteins.

Relating Concepts

2. Map answer is shown on page 135D.

Understanding Concepts

Multiple Choice

3.	b	8.	c
4.	c	9.	c
5.	c	10.	d
6.	c	11.	a
7.	a		

Completion

12. James Watson, Francis Crick
13. adenine, cytosine
14. nucleotides
15. RNA polymerase
16. transposons

Chapter 7 Review

Understanding Vocabulary

1. For each set of terms, explain the differences in their meanings.
   a. nucleotide, gene
   b. replication, transcription
   c. codon, anticodon
   d. exon, intron

Relating Concepts

2. Copy the unfinished concept map below onto a sheet of paper. Then complete the concept map by writing the correct word or phrase in each oval containing a question mark.

Understanding Concepts

Multiple Choice

3. Hershey and Chase's experiment with viruses that infect bacteria showed that
   a. protein gets into the bacterial cells.
   b. DNA is the genetic material.
   c. DNA contains radioactive sulfur.
   d. viruses undergo transformation.

4. The technology used by Franklin to investigate the structure of DNA is called
   a. translation.
   b. transcription.
   c. X-ray diffraction.
   d. photography.

5. Which represents the correct base pairing for DNA?
   a. G-T, A-C   c. T-A, G-C
   b. A-G, A-T   d. G-C, C-A

6. DNA and RNA are similar in that both have
   a. thymine as a nitrogen base.
   b. a single-stranded helix shape.
   c. nucleotides containing sugars, nitrogen bases, and phosphates.
   d. the same sequence of nucleotides for the amino acid phenylalanine.

7. If a segment of a DNA strand is CGTAGC, the complementary RNA strand is
   a. GCAUCG.   c. ATGCAT.
   b. CGUAGC.   d. AUGCAU.

8. The genetic code contains directions for
   a. copying DNA.
   b. constructing a double helix.
   c. ordering amino acids in proteins.
   d. removing repressor proteins.

9. The site where proteins are built from amino acids is the
   a. nucleus.
   b. endoplasmic reticulum.
   c. ribosome.
   d. gene.

10. To bring amino acids to the ribosome is the function of
   a. DNA.   c. rRNA.
   b. mRNA.   d. tRNA.

11. Which substance remains in the nucleus during translation?
   a. DNA   c. tRNA
   b. mRNA   d. rRNA

Completion

12. The first double helix model of DNA was constructed by _____ and _____ .

13. According to Chargaff's rule, in DNA the amount of _____ equals the amount of thymine, and the amount of _____ equals the amount of guanine.

14. The subunits that make up DNA are called _____ . They contain a sugar, a phosphate group, and a nitrogenous base.

15. The enzyme responsible for building a complementary strand of mRNA from DNA is called _____ .

16. Genes that move from one location on a chromosome to another and often inactivate other genes are called _____ , or jumping genes.

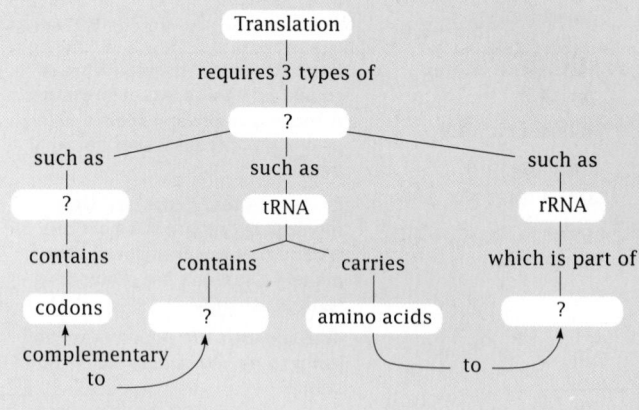

Short Answer

17. Griffith's work unveiled the process of transformation in which live uncoated bacteria had acquired the ability to make coats from the dead coated strain.

18. In the process of replication DNA polymerase adds complementary nucleotides to an exposed DNA strand.

19. The genetic code is the correspondence between nucleotide triplets in DNA and the amino acids in proteins. It is said to be universal because it is the same in nearly all organisms.

20. Messenger RNA: carries sequence of nucleotides from nucleus to ribosome. Ribosomal RNA: present in ribosomes. Transfer RNA: carries amino acids to ribosomes.

21. Sometimes they jump in the middle of genes.

Short Answer

17. What was the importance of Griffith's work with strains of the bacterium *Streptococcus pneumoniae*?

18. Explain the process by which DNA is copied.

19. What is the genetic code? Why is it said to be universal?

20. What are the roles of the three types of RNA?

21. How do transposons inactivate genes?

Interpreting Graphics

22. The figure below shows the events of translation. Recall that translation follows transcription and is the process by which a protein molecule is assembled according to the mRNA code.

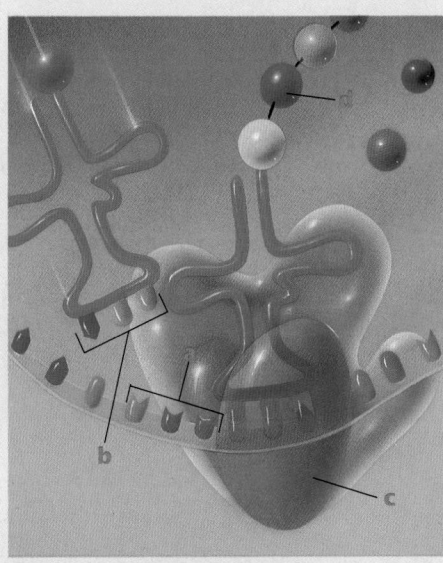

- What does **a** represent in the model?

- Part **c** represents the organelle inside the cell where translation occurs. What is the name of this organelle?

- What is the relationship between **a** and **b** during translation?

- What structure is represented by **d**?

Reviewing Themes

23. *Scale and Structure*
    How is the design of the ribosome's structure appropriate for carrying out protein synthesis?

24. *Evolution*
    What evidence is provided by the genetic code that all organisms have a common ancestry?

Thinking Critically

25. *Inferring Conclusions*
    How would the operation of RNA polymerase be affected if the repressor protein were not bound to the proper site on a gene?

26. *Inferring Conclusions*
    The codon "UUU" codes for the amino acid phenylalanine. What is the complementary DNA nucleotide to "UUU"?

27. *Building on What You Have Learned*
    In Chapter 4 you learned about mitosis and cell division. Why is it important that DNA replication occur before mitosis and cell division?

Cross-Discipline Connection

28. *Biology and Language Arts*
    As you learned in this chapter, proteins are compounds that consist of one or more chains of amino acids. Do library research to find the origin of the word "protein" and why it is used to describe amino acid chains.

Discovering Through Reading

29. Read pages 64–67 of the article "Life's Off-Switch" in *Discover*, July, 1991. How did Hwang's creative thought and persistence lead him to discover the protein that stops DNA replication? What was Kornberg's contribution to the discovery?

Cross-Discipline Connection

28. Webster's Dictionary states that the word protein is Greek in origin, derived from the word *proteios,* meaning first. The name protein is well chosen for chemical compounds consisting of one or more chains of amino acids because they rank first among the substances of life, according to chemists Robert Morrison and Robert Byrd. Proteins are found in all living cells; they constitute a major portion of the animal body. Proteins are the most abundant material in blood, skin, muscles, nerves, enzymes, and other substances necessary for life.

Discovering Through Reading

29. Rather than tracking a function and finding the enzyme that caused DNA replication to stop, as Kornberg and others had done, Hwang considered the problem from a different perspective. He decided to search for a protein to fit the spot on DNA that bulged before the two strands separated, to determine if it would function as an off-switch. He searched bacterial proteins for more than eight months before finding one that stops DNA replication. By knowing what Kornberg and others working in the lab had tried before, Hwang knew where not to focus his efforts.

Interpreting Graphics

22. - A is a codon on mRNA.
    - C is the ribosome.
    - They are complementary.
    - D is an amino acid.

Reviewing Themes

23. Transfer RNA can easily bind to the pocket in the small ribosomal subunit.

24. It is universal for almost all organisms, from bacteria to humans.

Thinking Critically

25. RNA polymerase would not be blocked from contacting the gene and transcription could not be stopped.

26. AAA

27. Replication must occur first to ensure that the DNA copies can be distributed to the two daughter cells that are formed.

Prelab Preparation Answers

- DNA functions like a computer program that directs the formation of proteins in the cell.
- It is located in the nucleus of the cell.
- A nucleotide is a subunit of a nucleic acid; it is composed of a sugar, a phosphoric acid, and a nitrogenous base.
- Students' answers should describe a double helix of nucleotides held together by complementary pairing between the base pairs, with adenine pairing with thymine and guanine pairing with cytosine.
- Proteins are made at the ribosomes, which are located in the cytoplasm of the cell.

Procedure Answers

5. red, blue, red, blue, green, blue, blue, red, yellow, blue, green, green, green, red, blue, red, yellow
   The straw pieces represent a phosphate group and the sugar deoxyribose.

9. The correct sequence of strand II is red, white, red, white, green, white, white, red, yellow, white, green, green, green, red, white, red, yellow.

How Does DNA Send Messages to the Cytoplasm?

Investigation

Objectives

In this investigation you will:
- *compare* and *contrast* the molecular structures of DNA and messenger RNA
- *design* models of DNA and messenger RNA to show how DNA sends messages to the cytoplasm

Materials

- soda straws (cut into 46 3-cm pieces)
- 13 red pushpins
- 9 blue pushpins
- 8 yellow pushpins
- 10 green pushpins
- 5 white pushpins
- 45 paper clips
- metric ruler
- a pair of scissors
- black felt-tip marker

Prelab Preparation

Review what you have learned about DNA by answering the following questions.
- What is the function of DNA?
- Where is DNA located in the eukaryotic cell?
- What is a nucleotide?
- Describe the basic structure of a DNA molecule.
- Where are proteins made in the cell?

Procedure: Making and Sending Messages From DNA

1. Form a cooperative team with another student to complete steps 2–11.

2. **CAUTION: Pointed objects can cause injury.** Cut soda straws into 3-cm segments to make 46 segments.

3. Insert a pushpin midway along the length of each segment of soda straw. Push a paper clip into one end of each 3-cm segment of straw until it touches the pushpin.

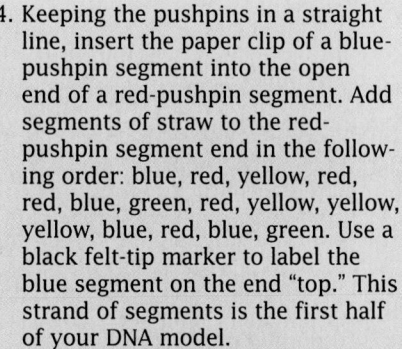

4. Keeping the pushpins in a straight line, insert the paper clip of a blue-pushpin segment into the open end of a red-pushpin segment. Add segments of straw to the red-pushpin segment end in the following order: blue, red, yellow, red, red, blue, green, red, yellow, yellow, yellow, blue, red, blue, green. Use a black felt-tip marker to label the blue segment on the end "top." This strand of segments is the first half of your DNA model.

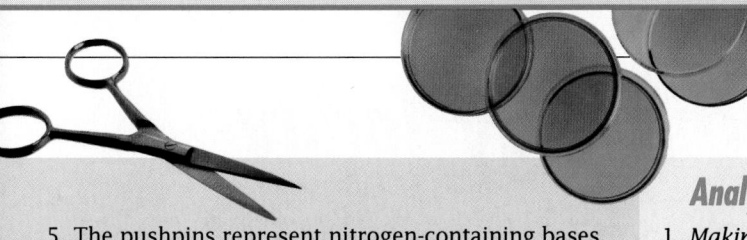

5. The pushpins represent nitrogen-containing bases, with red representing adenine (A), blue representing thymine (T), yellow representing guanine (G), and green representing cytosine (C). *What is the correct sequence of colors for a complementary strand for this model?* Construct this complementary strand using the remaining soda straw segments. *What do the pieces of soda straw represent?*

6. Place the first half of your DNA model (strand I) and its complementary strand (strand II) parallel to each other on the table. Make sure that the pushpins are in the correct order in both strands.

7. Make a sketch of your model and label the nucleotides A, T, G, or C.

8. Use a black marker to color a band around the remaining 15 pieces of soda straw. Each piece represents a phosphate group and the sugar ribose. These pieces are needed to make RNA nucleotides.

9. Separate strand I of the DNA molecule from strand II. Now use the ribose-containing nucleotides to form a strand of messenger RNA that pairs with strand I of the DNA molecule. Recall that RNA-DNA pairing is similar to DNA-DNA pairing except that uracil replaces thymine in the RNA molecule. Use white pushpins to represent uracil molecules. *What is the correct sequence of colors for an mRNA strand complementary to strand I of the DNA model?*

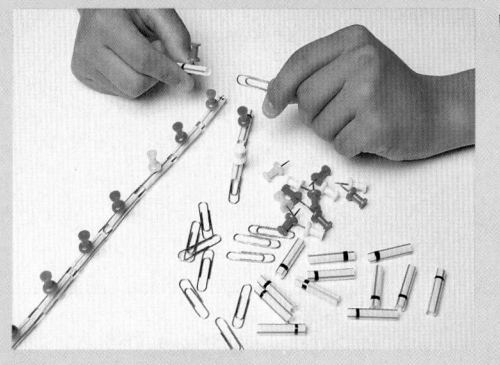

10. Make a sketch showing the pairing of the nitrogen-containing bases in the DNA nucleotides with those in the messenger RNA nucleotides. Label the bases A, T, G, C, or U.

11. Move the messenger RNA to the side and bring the two DNA strands back together.

Analysis

1. *Making Comparisons*
   How is messenger RNA similar to DNA? How is it different?

2. *Making Comparisons*
   Compare the sequence of nucleotides in strand I of the DNA molecule to the sequence of nucleotides in the messenger RNA.

3. *Making Comparisons*
   Compare the sequence of nucleotides in strand II of the DNA molecule with the sequence of nucleotides in the messenger RNA.

4. *Identifying Relationships*
   Why can messenger RNA be thought of as a "message" from DNA?

5. *Identifying Relationships*
   Where in the cell is messenger RNA made?

6. *Making Inferences*
   Where will the messenger RNA go after it is made?

Thinking Critically

1. What are the advantages of having DNA remain in the nucleus rather than allowing it to move about the cell?

2. According to your models, how might a mutation arise?

Analysis Answers

1. Students should indicate that mRNA is composed of only one strand, rather than two, and that it contains uracil but no thymine.
2. Students should describe the complementary relation of the mRNA strand to strand I of the DNA molecule.
3. The mRNA strand has the same sequence of bases as strand II of the DNA molecule, except that uracil in mRNA replaces thymine in DNA.
4. Students should point out that the sequence of mRNA nucleotides is formed according to the sequence of bases in strand I of the DNA molecule.
5. Messenger RNA is formed in the nucleus of the cell.
6. After being formed, mRNA moves to the cytoplasm of the cell.

Thinking Critically Answers

1. Students might suggest that the DNA is less likely to be damaged by remaining in the nucleus. Also, messages can be transcribed and sent to many locations in the cell while the DNA remains in a central location.
2. Possible answers may include sequencing the wrong color for a complementary strand, or damaging a straw segment.

Chapter 8 — Gene Technology Today

Planning Guide

	Objectives/Themes	Classwork Resources	Homework Resources

8.1	1. Outline the genetic engineering experiment performed by Cohen and Boyer.
	
	2. Explain the four stages involved in a gene transfer experiment.
	
	3. Describe how scientists use restriction enzymes in genetic engineering experiments.
	
	4. Discuss some of the safety concerns associated with genetic engineering.
	
	Themes: Patterns of Change, Interacting Systems
		Teacher's Resource Binder
		Focus Activity 8
		    Making a Genetic Engineering Model
		Extension Worksheet
		    Genetic Engineering
		
		Other Resources
		Transparencies 31–33
			Text
			Section Review, p. 160
			
			Teacher's Resource Binder
			Directed Reading Worksheet 8.1
			
			Other Resources
			Audiocassette 8.1*

8.2	1. Explain the role of the Ti plasmid in agricultural research.
	
	2. Describe how herbicide-resistant genes in crop plants can benefit the environment.
	
	3. Explain how genetic engineering techniques have been used to improve crop yields.
	
	4. Describe how gene transfers are being used to make livestock more productive.
	
	Themes: Patterns of Change, Interacting Systems
		Teacher's Resource Binder
		Lab Investigation 8.1
		    Investigating Human Karyotypes
		
		Lab Investigation 8.2
		    Transforming Genetic Information
			Text
			Section Review, p. 164
			
			Teacher's Resource Binder
			Directed Reading Worksheet 8.2
			
			Other Resources
			Audiocassette 8.2*

8.3	1. Describe how transferring human genes into bacteria can benefit human health.
	
	2. Explain how gene transfers could be used to combat genetic disorders.
	
	3. Summarize the goals of the Human Genome Project and the ethical issues it raises.
	
	4. Describe three uses of DNA profiling.
	
	Themes: Patterns of Change, Interacting Systems
		Text
		Investigation
		    How Are DNA Fingerprints Interpreted? pp. 172–173
		Science, Technology, and Society
		    Genetic Screening: Helpful or Harmful? pp. 174–175
		
		Other Resources
		Transparency 34
			Text
			Section Review, p. 168
			
			Teacher's Resource Binder
			Directed Reading Worksheet 8.3
			Vocabulary Review Worksheet*
			Reteaching Worksheet*
			    Molecular Cutting and Pasting
			
			Other Resources
			Audiocassette 8.3*

*Reteaching Options

Demonstrations
8.1: pp. 158, 159
8.2: pp. 161, 162
8.3: p. 166

Assessment
Chapter Review pp. 172–173
Portfolio Assessment p. 155D
Chapter Test—Teacher's Resource Binder
Test Generator

Research Notes

Connection to Medicine: Gene Therapy

In the early 1990s, the first tests of gene therapy were approved by the National Institutes of Health. Instead of treating symptoms, gene therapy tries to cure diseases by aiming at their genetic causes. The following techniques and diseases were among the first to be tried.

• *ADA:* Patients with a rare defect in which white blood cells did not produce enough of adenosine deaminase enzyme were injected with their own white blood cells, which had been genetically engineered to include a functioning gene for that enzyme. Injections of the cells every two to three months worked far better than daily injections of the enzyme itself.

• *Retroviruses:* Retroviruses are stripped of their harmful genes, and human genes that are lacking in the patient are added,

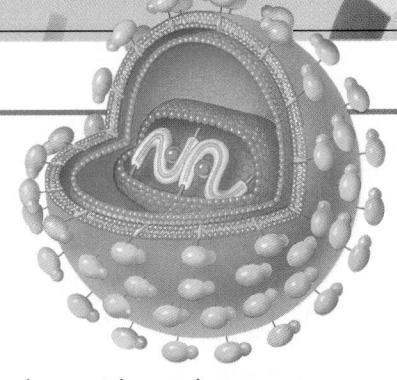

along with regulatory sequences. The viruses then infect the body of the patient, changing the DNA in all cells they infect to now include the missing gene.

• *Suicide Gene:* To protect against possible unknown side effects to the retrovirus treatment method, a "suicide" gene can be inserted that will cause the altered cells to die in response to a chemical signal. If the altered cells need to be stopped, the patient receives a drug that provides that chemical signal.

• *Common Cold Viruses:* Airborne viruses, such as those that cause colds, will be genetically engineered to deliver genes to the lung cells of cystic fibrosis patients. The gene will provide working copies of a cell protein that helps break down mucus.

• *Direct Injection:* DNA injected directly into muscle tissue is often picked up by cells. This technique could be useful in making vaccines for AIDS and influenza.

• *Fibroblasts:* Since cells of the connective tissue cling together, genetically altered cells can be transplanted under the skin. There, they will grow into a tiny mass, providing hormones or proteins for the whole body. If something goes wrong, they can easily be removed surgically.

• *Bone Marrow:* Most blood cells are derived from the bone marrow. By taking bone marrow samples, modifying the cells genetically, and reinjecting them, scientists can ensure that at least some of the body's new blood cells will have the genes needed.

As scientists learn more about genetics and about the genes that are involved in diseases, even more possibilities for gene therapy will emerge.

Investigation Notes

How Are DNA Fingerprints Interpreted?
pages 172–173

Purpose: This investigation gives the students the opportunity to analyze data similar to that from DNA samples.

A good summary of the use of DNA fingerprinting can be found in "DNA Fingerprints—Witness For The Prosecution," *Discover,* June 1988, pages 44–52. A discussion of some of the difficulties and complications involved can be found in "When Science Takes the Witness Stand," *Scientific American,* May 1990, pages 46–53.

Prelab Preparation

1. Each team should be supplied with a container containing beads of different sizes.

2. WARD'S has developed a special "kit" of beads the appropriate sizes specifically for this investigation (15 M 0504).

3. Be sure to set aside a container of beads for the "crime scene." You may determine the results: one team's container may have a distribution of beads that is more similar to the "crime scene" set, or more than one set that match can be found, or else you can arrange for none of the sets to match.

Answers will be found on pages 172–173.

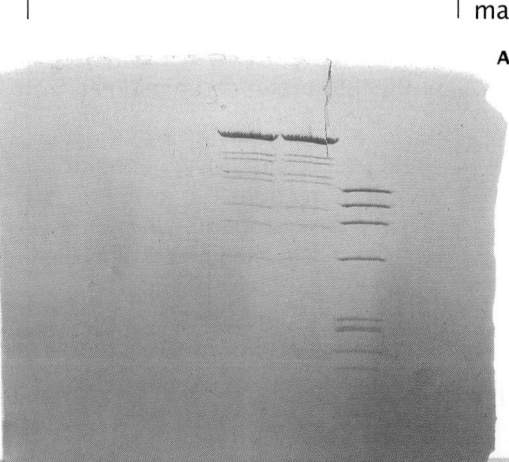

Meeting Individual Needs

Objectives

1. Students will demonstrate appropriate use of core vocabulary for the chapter (Vocabulary File).

2. Students will use models to visualize the stages involved in gene transfer (Demonstration A).

Vocabulary File
(Developing Vocabulary/ Limited English Proficiency)

If you are not already using the Vocabulary File, refer to Chapter 1 for its preparation. See the Chapter Highlights on page 169 for a list of suggested words.

Demonstration A
Genetic Engineering Model
for pages 157–160 and 165
(Developing Modeling Skills/ Visual Learners)

Materials: modeling clay, scissors, orange and white construction paper, pins, 2 plastic foam wreaths or rings (available from craft stores), orange ribbon, tape, 4 red oblong balloons, transparency, overhead projector, projector pens

Preparation: Write the following points on the transparency, leaving room after each one:

> I. Insulin gene isolated on the chromosome
>
> II. Insulin gene inserted into a plasmid
>
> III. Bacterial cell infected with plasmid
>
> IV. Bacteria containing plasmid are cloned
>
> V. Insulin is produced and collected for human usage.

Use the modeling clay to fashion a chromosome. Pin a small circle from the orange construction paper to one arm of the chromosome.

Wrap one of the wreaths with orange ribbon, pinning the ribbon securely onto the wreath. Cut both the orange wreath and the unwrapped wreath in the same places, so that a segment of each can be removed.

Inflate and tie off the red balloons. Cut four small flat rings out of the white construction paper. Color a small segment of each ring orange.

Procedure: Brainstorm with students about genetic engineering and what genetically engineered products they are aware of. Some students may know about insulin. Ask a student with diabetes or a student who is a friend or relative of someone with diabetes to explain what they know of the disease and why insulin is needed to control the illness.

Tell students that together you and they will work through the basic steps of the genetic engineering process using color-coded models to help them keep track of the sources of all genetic material involved. (Remind students that the items in question are not really colored.)

Show the clay chromosome model. Ask students to explain what the orange circle represents. (The gene for insulin.) Remind them that in genetic engineering, the insulin gene must come from some organism. This should help students realize that the first step is locating a working gene for insulin in a human.

At this point, reveal the first point of the outline. Near the first point of the outline, draw a sketch of the chromosome and the insulin gene. Be sure students realize that this chromosome is only one of the 46 human chromosomes.

Ask students what the white wreath represents. They should be able to recognize it as a plasmid. Ask students what must happen to the isolated insulin gene so it can be inserted successfully into a bacterial cell. Reveal the second point of the outline.

Remove the small segment from the white wreath. Replace it with the orange-wrapped segment. Near the second point of the outline, draw a sketch of the plasmid with the insulin gene.

Take one of the white and orange construction paper rings, and tape it to a red balloons. Students should recognize that the ring represents the insulin-gene carrying plasmid from the previous point of the outline. Ask students what types of organisms take up plasmids.

When students realize the red balloon represents a bacterial cell, reveal the third point of the outline. Draw a sketch of the bacterial cell containing the plasmid with the insulin gene.

Tell students to look at Figure 8.11c, on page 165. Then, show them the three other red ballons, all of which have white and orange construction paper rings taped to them. Reveal the fourth point of the outline, drawing a sketch of several cells with plasmids.

Remind students of their study of DNA and RNA in Chapter 7. Be certain students realize that once within a cell, the expression of the insulin gene involves the same natural steps, whether it is within a human cell or a bacterial cell. Emphasize that the product will be the same kind of insulin, with the same amino acids in the same order. Reveal the last point on the outline. Include a sketch of a portion of the DNA for insulin (colored orange) being transcribed, and the insulin produced being stored in vials.

Additional Strategies
Visual Strategies

Pages 157, 158, 159, 160, 161, 163, 164, 165, and 166

Auditory Learners

Use *Biology: Visualizing Life* Audiocassettes for Sections 8.1, 8.2, and 8.3.

Meeting Individual Needs (cont.)

Cooperative Learning

Applying Genetic Technology

Timing: Use this activity to introduce Section 8.1.

Group Size: 4 students

Outcome: Students will be able to list the possible benefits and hazards of releasing genetically altered plants and animals into the environment, such as in agriculture.

Individual Accountability: Each group member is responsible for listing advantages or disadvantages of using genetically altered organisms.

Positive Interdependence: Each group will debate the advantages and disadvantages of using genetically altered organisms.

Divide each group into two pairs. Have one pair list all the possible advantages of using genetically altered plants and animals and have the other pair list the disadvantages. Have each pair present their list to the other pair in the group.

Then, have the pairs switch positions for a debate discussion. The pair that listed advantages should now argue that introducing genetically altered organisms would be disadvantageous, and the pair that listed disadvantages should argue that it would be advantageous.

Upon conclusion of the debates, each group should combine their lists of advantages and disadvantages to try to come up with a consensus statement regarding releasing genetically engineered organisms into the environment. Challenge students to suggest precautions or regulations that could resolve some of the disadvantages.

Portfolio Assessment

Students should select their best work and provide a self-reflective rationale for their selections. Students can make selections in the following areas.

1. *Content* — One concept map from the chapter (See page 812 for evaluation criteria.)

2. *Reading Comprehension* — One Directed Reading Worksheet from the Teacher's Resource Binder (Use the answer key to evaluate for accuracy.)

3. *Writing* — Using the Vee Form, summarize a newspaper or magazine article relating to gene technology. (See page 22T for evaluation criteria.)

    Or: Select a writing task or project from the Chapter Review.

4. *Performance Assessment* — One Vee form from a chapter investigation or lab manual investigation (See page 22T for evaluation criteria.)

Teacher makes selections in the following areas.

1. *Formal Assessment* — Chapter test (Test A, B, or the Test Generator) The teacher-scored test should be reviewed by the student. Incorrect responses should be corrected by the student before the test becomes part of the portfolio.

2. *Informal Assessment* — Use the Direct Observations Checklist, page 33T, during a laboratory or other cooperative learning experience.

3. *Performance Assessment* — Have students imagine they are members of Congress who are reconsidering funding for the Human Genome Project during a budget crisis. Have each one prepare and give a short speech on his or her position. Hold a mock vote after all speeches.

Concept Map Answer

The following is one possible answer to the Relating Concepts exercise on page 170.

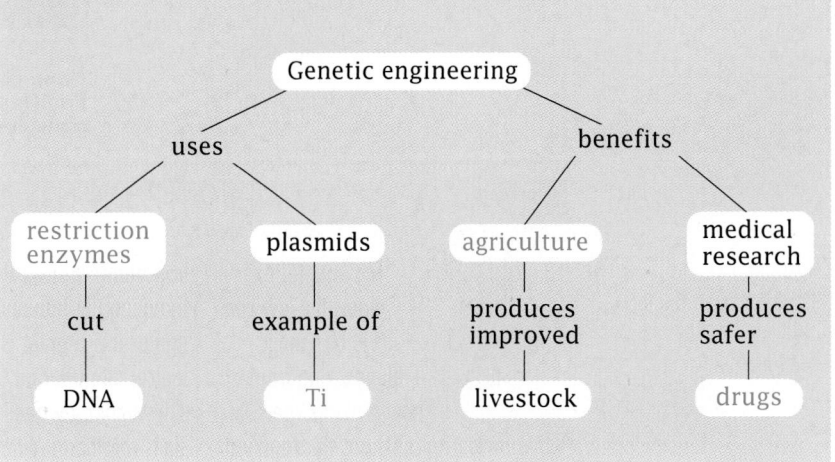

Chapter 8

Determining Prior Knowledge

- Genetic engineering has received much attention from the media. Have students discuss reports they may have read in the news or seen on TV.
- Certain precautions and restrictions have been taken when conducting genetic engineering experiments. Ask why such concerns have been raised.
- Ask students what vaccinations they have received. Have them explain how vaccines are prepared.
- Show the class a photograph of a DNA fingerprint. Use it as the basis for determining how much students know about the human genome project or the use of DNA fingerprints by criminologists.

Chapter 8

Review

- DNA structure (Section 7.1)
- replication (Section 7.1)
- protein synthesis (Section 7.2)

Gene Technology Today

8.1 Genetic Engineering

- What Is Genetic Engineering?
- How to Move a Gene From One Organism to Another

8.2 Transforming Agriculture

- The Ti Plasmid
- Resistance to Plant-killing Chemicals
- Nitrogen Fixation
- Resistance to Insects
- Genetic Engineering in Livestock

8.3 Advances in Medicine

- Making Miracle Drugs
- Making Vaccines
- Human Gene Therapy
- Gene Sequencing

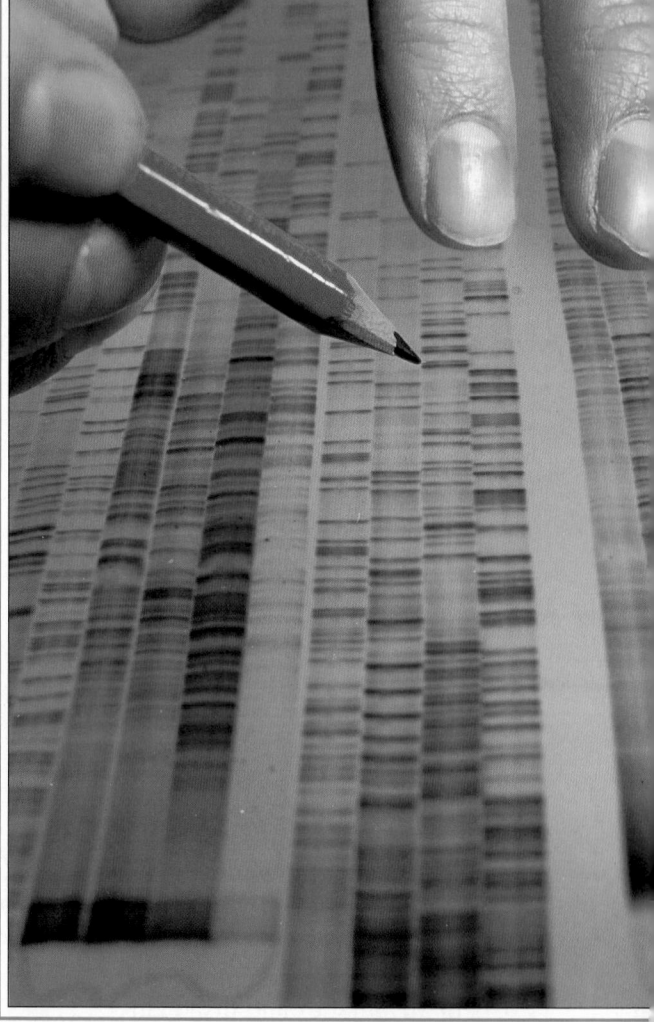

Using genetic engineering techniques, such as this DNA profile, scientists can isolate individual genes and transfer them from one species to another.

▪ Author's Rationale ▪

This chapter explains the key concept of genetic engineering from a "how-it-works" perspective. These concepts are then applied in looking at three areas where genetic engineering has been particularly important: agriculture, medicine, and mapping the human genome. Coverage of this chapter is essential for biological literacy, and to give students skills they may need in making decisions related to genetic technologies.

IN MYTHOLOGY, A CHIMERA IS A CREATURE WITH THE HEAD OF A LION, THE BODY OF A GOAT, AND THE TAIL OF A SERPENT. ALTHOUGH CHIMERAS DO NOT EXIST IN NATURE, SCIENTISTS CAN CREATE THEM IN LABORATORIES BY ALTERING THE GENES OF LIVING ORGANISMS. AS YOU WILL LEARN IN THIS CHAPTER, SCIENTISTS ARE USING THESE TECHNIQUES TO TRANSFORM AGRICULTURE AND REVOLUTIONIZE MEDICINE.

8.1   *Genetic Engineering*

Objectives

❶ Outline the genetic engineering experiment performed by Cohen and Boyer.

❷ Explain the four stages involved in a gene transfer experiment.

❸ Describe how scientists use restriction enzymes in genetic engineering experiments.

❹ Discuss some of the safety concerns associated with genetic engineering.

What Is Genetic Engineering?

The first chimera was created in 1973 by Stanley Cohen and Herbert Boyer, two geneticists at the University of California at San Francisco. Cohen and Boyer set out to insert a gene from an African clawed frog into a bacterium. They were hoping that the bacteria could then be used to make the protein encoded by the frog gene.

First, Cohen and Boyer isolated the gene that coded for frog ribosomal RNA. Then, they added that gene to bacterial DNA. Their experiment created the first living cells that had DNA from a foreign organism added to their own DNA. **Recombinant DNA** is a molecule formed when fragments of DNA from two or more different

organisms are spliced together in a laboratory. Did the bacteria with the recombinant DNA use the frog gene? Yes! The genetically engineered bacteria cells produced frog rRNA, just as Cohen and Boyer had hoped. **Figure 8.1** outlines their experiment.

This experiment ushered in a new age in biology, one that explored the possibilities of moving genes from one organism to another. In approximately 20 years, the growing ability of researchers to transfer DNA from one organism into another has revolutionized biology. Moving genes from the chromosomes of one organism to those of another is called **genetic engineering**. Today specific genes from human chromosomes can be routinely transferred into bacteria.

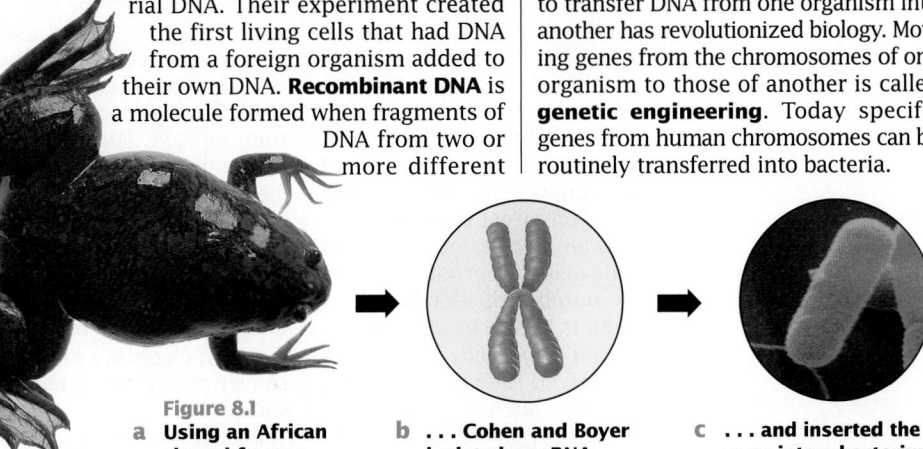

Figure 8.1

a   Using an African clawed frog as their experimental organism, . . .

b   . . . Cohen and Boyer isolated an rRNA gene from one of its chromosomes . . .

c   . . . and inserted the gene into a bacterium. This bacterium then produced frog rRNA.

Lesson Plan 8.1

Phase 1
PREPARATION

Key Concepts
- Genetic engineering is the transfer of genes between different organisms, producing recombinant DNA molecules.
- Genetic engineering is conducted in four stages:
  1. Restriction enzymes are used to cleave the DNA of two different organisms.
  2. The DNAs are combined to form the recombinant DNA.
  3. The recombinant DNA is inserted into target cells that are cloned.
  4. Target cells are examined to see if the particular gene of interest is functioning.

Reading Strategy

One problem students often encounter in biology is learning vocabulary. This chapter provides examples of how the meanings of words can sometimes be deciphered from context clues. As students read this chapter, have them closely examine the words, phrases, and sentences surrounding the bold faced terms. In this section, have them look for a restatement where the context actually includes a definition of the bold faced term.

Phase 2
TEACHING STRATEGIES

Visual Strategy

Figure 8.1
Use this figure to point out that the first genetic engineering experiment broke down a natural barrier between eukaryotes and prokaryotes established by evolution billions of years ago.

157

Visual Strategy

Figure 8.2 a–c
Have students follow this figure as you illustrate it in a slightly different way. Show the DNA double helix as a series of letters representing the nitrogenous bases. Draw a short segment of one strand on the board as:

   G A A T T C

Next draw the complementary strand:

   C T T A A G

Then show where *Eco*RI cleaves the DNA molecule (as indicated by an •) to produce sticky ends:

   G • A A T T C
   C T T A A • G

The breaks (•) allow the two strands to separate, producing the *sticky* ends:

   G   A A T T C
   C T T A A   G

Explain that the ends are sticky because they could match up with one another again or with a different DNA segment that has a complementary sequence. Students will not be able to see from these board illustrations that several such fragments are produced. Consequently, refer them to Figure 8.2c and emphasize that *Eco*RI breaks apart the frog's DNA to produce many fragments. Each fragment is double-stranded except at opposite ends where there is a sticky single-stranded piece. Have students locate these sticky ends and compare them to the illustrations on the board.

Demonstration 1

Cleaving DNA
Cut some Velcro® into strips. Stick two strips together to represent the two strands in the frog DNA molecule as illustrated in Figure 8.2b. To demonstrate Figure 8.2c, use a pair of scissors (*Eco*RI) to cut the Velcro® (DNA) into smaller strips. As you cut, simulate how the *Eco*RI cleaves DNA; don't cut through the two Velcro® strips at the same spot. Cut one strip at some point, and then cut the other strip a short distance from this spot. Thus, sticky ends, similar to those illustrated in Figure 8.2c will be produced. Save strips to use with Demonstration 2 on page 159.

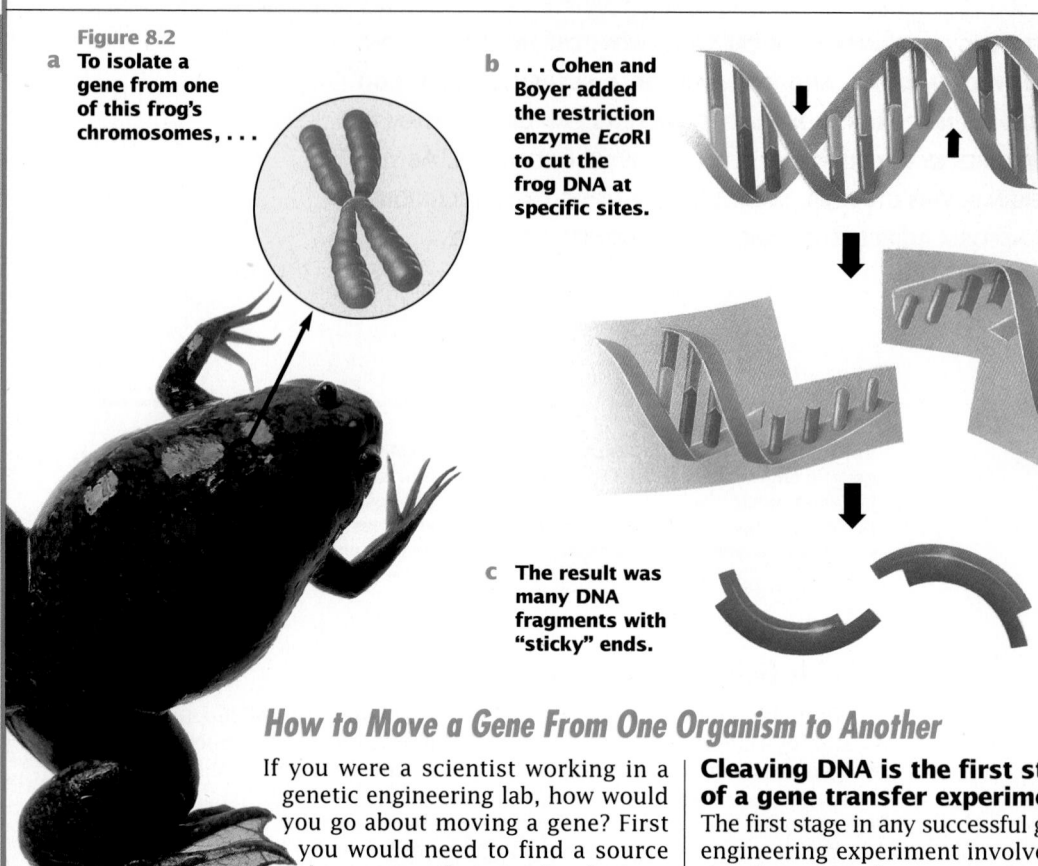

Figure 8.2

a **To isolate a gene from one of this frog's chromosomes, . . .**

b **. . . Cohen and Boyer added the restriction enzyme *Eco*RI to cut the frog DNA at specific sites.**

c **The result was many DNA fragments with "sticky" ends.**

How to Move a Gene From One Organism to Another

If you were a scientist working in a genetic engineering lab, how would you go about moving a gene? First you would need to find a source chromosome that contains the gene you wish to isolate. Next you would need to select a target cell into which the gene could be moved. And finally, you would need to figure out a way to transfer the isolated gene into the target cell. Every gene transfer experiment is performed in four distinct stages:

1. *Cleaving DNA* The source chromosome is cut into fragments of DNA.

2. *Producing recombinant DNA* The DNA fragments containing the desired gene are inserted into viral or bacterial DNA. The recombinant DNA is then allowed to infect target cells.

3. *Cloning target cells* Infected target cells are allowed to reproduce. Growing a large number of identical cells from one cell is known as **cloning**.

4. *Screening target cells* Target cells that have received the particular gene of interest are isolated.

Each of these stages will be explained in greater detail using the Cohen-Boyer experiment as an example.

Cleaving DNA is the first step of a gene transfer experiment

The first stage in any successful genetic engineering experiment involves cutting the source chromosome into fragments to obtain copies of the gene you wish to transfer. In their experiment, Cohen and Boyer used chromosomal DNA from *Xenopus laevis*, the African clawed frog.

In order to cut the frog DNA, Cohen and Boyer used special enzymes called **restriction enzymes**. Restriction enzymes recognize and bind to specific short sequences of DNA, and then cut the DNA at a specific site within that sequence. Cohen and Boyer used a restriction enzyme called *Eco*RI, which cuts DNA whenever it encounters the sequence CTTAAG. The sequence of the opposite strand is GAATTC, the same sequence written backward. **Figure 8.2b** shows the restriction enzyme *Eco*RI cutting a DNA sequence found within a frog chromosome.

Restriction enzymes do not make a straight cut through both strands of DNA. Instead, the cut is offset a few

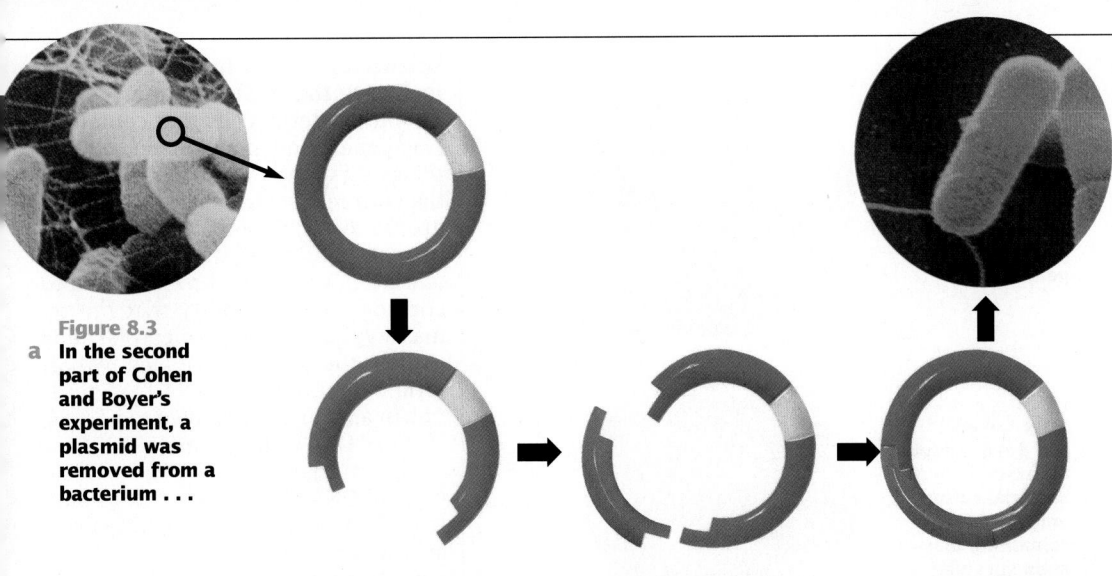

Figure 8.3

a In the second part of Cohen and Boyer's experiment, a plasmid was removed from a bacterium . . .

b . . . and treated with *Eco*RI, leaving sticky ends.

c Fragmented frog DNA was then inserted into the bacterial plasmid.

d The new DNA was used as a vector to infect a bacterial cell.

Evolution

What evidence revealed by gene technology suggests that all organisms are related?

bases. For example, in the sequence CTTAAG, *Eco*RI cuts each strand between the A and G, a site four bases apart on both strands, creating DNA fragments with single-stranded "sticky ends."

Because the two single-stranded sticky ends have complementary sequences, they can pair back up and heal the break (which is why they are called sticky), *or they can pair with any other DNA fragment cut by the same enzyme.* DNA cut by the restriction enzyme *Eco*RI can be joined to DNA from any other organism that has also been cut by *Eco*RI, because they have the same sticky ends. One of the fragments Boyer and Cohen cleaved from frog DNA contained the rRNA gene they sought to transfer.

Restriction enzymes are among the basic tools of genetic engineering. More than 200 different restriction enzymes have been isolated and identified. Most restriction enzymes attack only one specific sequence. By trial and error, biologists can often find a restriction enzyme that cuts out the gene that they seek.

Restriction enzymes are used to produce recombinant DNA
Cohen and Boyer also used *Eco*RI to cut circular forms of bacterial DNA called **plasmids**. Because each plasmid contained only one *Eco*RI site, it was cut open in only one place. By mixing the frog DNA fragments with the plasmids cut open by *Eco*RI, Cohen and Boyer produced recombinant DNA.

It was easy for the ends of the opened plasmids to stick to the ends of the frog DNA fragments because they had complementary sticky ends. An enzyme that can join the ends was added to bond the plasmids and fragments together, resulting in new DNA. Once formed, the new DNA served as delivery agents, or **vectors**, which would carry the frog ribosomal RNA gene into the target bacterial cells.

Figure 8.3 shows how the frog DNA was inserted into a plasmid. The vector used by Cohen and Boyer contained two critical genes: one replicated the plasmid DNA, and the other made the cell resistant to an antibiotic called tetracycline. This second gene is important during the screening stage.

Connection: Chapter 4

Mitosis in Prokaryotes
Ask students why clones would be produced when the bacterial cells divide.

Visual Strategy

Figure 8.4
Have students identify the plasmids that survived in Figure 8.4b and explain why they were not killed by tetracycline. They had the gene for resistance to tetracycline.

Health Connection

Dangerous Microbes
When genetic engineering became a reality in the 1970s, biologists were concerned about the potential dangers of producing new strains of pathogenic and lethal microorganisms. They gathered in 1975 at Asilomar, California, where strict protocols for developing recombinant DNAs were formulated. Even today, scientists using genetic engineering must first obtain approval from several agencies.

Phase 3

ASSESSMENT OPTIONS

Closure Strategy

"Shotgun" Experiments
The first genetic engineering experiments became known as "shotgun" procedures since the restriction enzymes could cleave a chromosome at numerous sites. Ask students to suggest why the term "shotgun" was an appropriate description.

Section Review

Assign the *Section Review*.

Reteaching

One definition of engineering is the application of scientific principles to design, construct, and operate systems. Ask students why the term "genetic engineering" is both appropriate and descriptive.

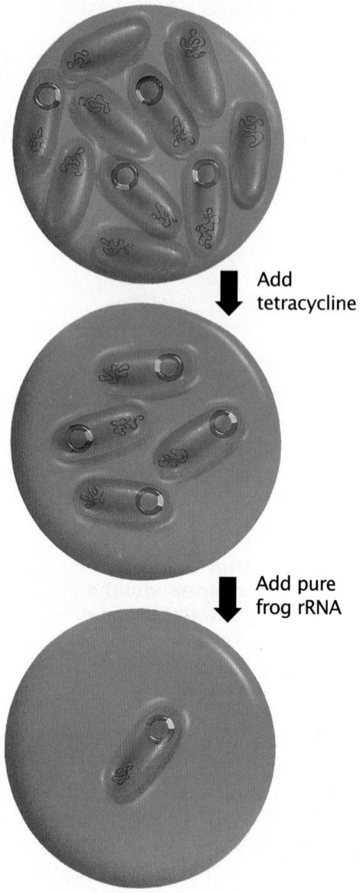

Figure 8.4
a Cohen and Boyer screened a population of bacteria to locate cells containing the frog rRNA gene.

Add tetracycline

b After Cohen and Boyer added tetracycline to the bacterial culture, only bacteria containing plasmids survived.

Add pure frog rRNA

c The pure frog rRNA that Cohen and Boyer added to the culture stuck to bacteria containing the one plasmid with the frog rRNA gene.

Target cells are cloned
To produce clones, Cohen and Boyer added recombinant DNA to a culture of bacteria. They placed the bacterial culture under conditions that encouraged the bacteria to take the recombinant DNA into their cells. Each bacterial cell then reproduced, forming clones. Some of the clones contained recombinant DNA, while others did not.

Scientists screen target cells to locate the desired gene
To find bacterial cells that contained frog genes, Cohen and Boyer did two things. First, they added tetracycline to the culture. Bacteria that did not take up the recombinant DNA were killed because they lacked the gene for resistance to tetracycline. Now they had to consider the possibility that the remaining bacteria contained plasmids without the frog gene. To identify bacteria containing the rRNA frog gene, Cohen and Boyer attempted to pair the bacterial DNA with pure frog rRNA. Only a cell carrying the frog rRNA gene would contain DNA that would stick to pure frog rRNA. Patient searching revealed a colony with cells containing DNA that stuck to the pure frog rRNA. These were the cells they sought—bacterial cells that carried the frog gene. These cells could be grown to produce large quantities of frog DNA and its rRNA gene product. Figure 8.4 shows the steps involved in screening bacteria.

Precautions ensure that genetic engineering is safe
Because a genetic engineer can, in principle, move *any* gene from one organism to another, it is important to be careful not to introduce potentially dangerous genes into organisms that might escape from the laboratory. For this reason, particular care is taken when cancer cells or disease-causing organisms are used as gene sources in gene transfer experiments. Scientists select target cells that cannot survive outside the laboratory, and potentially dangerous experiments are forbidden. In more than a decade, no dangerous accident has ever occurred.

Section Review

❶ Describe the first experiment that successfully transferred a gene from a frog to a bacterium.

❷ Define the term "genetic engineering." List the four distinct stages involved in a gene transfer.

❸ How are restriction enzymes used in gene transfer experiments?

❹ How can scientists ensure that genetic engineering experiments are safe?

■ Section Review Answers ■

1. See Figure 8.1 on page 157 for an explanation.
2. Genetic engineering is defined as moving genes from the chromosome of one organism to those of another. The four stages are: (1) cleaving DNA, (2) producing recombinant DNA, (3) cloning target cells, and (4) screening target cells.
3. Restriction enzymes recognize and bind to specific short sequences of DNA. They cut the DNA at a specific site within the sequence leaving fragments with sticky ends.
4. Scientists select target cells that cannot survive outside the laboratory. Potentially dangerous experiments are forbidden.

ONE OF THE GREATEST SUCCESSES OF GENETIC ENGINEERING HAS BEEN
THE MANIPULATION OF GENES IN CROP PLANTS AND LIVESTOCK. GENE
TRANSFERS HAVE MADE CROP PLANTS MORE RESISTANT TO DISEASE,
HERBICIDES, AND INSECTS. GENETIC ENGINEERING HAS ALSO BEEN USED
TO INCREASE THE MILK PRODUCTION AND GROWTH RATE OF LIVESTOCK.

8.2 *Transforming Agriculture*

Objectives

❶ **Explain the role of the Ti plasmid in agricultural research.**

❷ **Describe how herbicide-resistant genes in crop plants can benefit the environment.**

❸ **Explain how genetic engineering techniques have been used to improve crop yields.**

❹ **Describe how gene transfers are being used to make livestock more productive.**

The Ti Plasmid

Figure 8.5

a The tumor-causing gene in the Ti plasmid . . .

b . . . is removed and replaced with the desired gene.

c The plasmid can then be inserted into bacteria, which can transfer the gene to plant cells.

For years, genetic engineering in plants was difficult because scientists lacked an appropriate vector to deliver desirable genes. Unlike bacteria, plants normally contain few viruses or plasmids that can act as delivery agents. Recently, however, scientists discovered the Ti plasmid, an unusual bacterial plasmid shown in **Figure 8.5**. The Ti plasmid causes the development of large tumors in plants. **Figure 8.6** shows a tree that has been infected by a crown gall tumor, which is caused by the bacteria containing the Ti plasmid. When the Ti plasmid infects a plant cell, it inserts itself into the plant cell's chromosomes. To transform the Ti plasmid into an effective genetic engineering vehicle, scientists first remove the tumor-causing gene from the Ti plasmid. The vacant space in the now harmless plasmid can be filled by the desired gene. Unfortunately, bacteria carrying the Ti plasmid cannot be used to insert DNA into plants that produce cereal grains such as corn, rice, and wheat. Researchers are presently developing powerful new techniques for introducing useful genes into these plants, such as shooting the cells with "gene guns."

Figure 8.6
This tree has a crown gall tumor, an abnormal growth caused by the Ti plasmid.

Lesson Plan 8.2

Phase 1
PREPARATION

Key Concepts

- Scientists have used the Ti plasmid to insert specific genes into plants.
- Genetic engineering experiments in plants have been carried out to develop crops that are resistant to weedkillers and insects and crops that can convert nitrogen into a usable form.
- Genetic engineering experiments in animals have been carried out to develop larger and more productive livestock.

Reading Strategy

In this section have students infer the meanings of any words they don't know. Tell them to look for information about the unknown words so that they can draw some reasonable conclusion, or infer its meaning. Then have them check the meaning of the word in the dictionary.

Phase 2
TEACHING STRATEGIES

Visual Strategy

Figure 8.5
Have students locate the "desired gene" that is being inserted into the Ti plasmid shown in Figure 8.5b. Ask students for examples of a "desired gene" that scientists might use in such an experiment.

Demonstration 1

Crown Gall Disease
Bring in a plant or part of a tree that has a crown gall tumor if one is available. Ask students if they have ever seen anything growing on trees like the mass shown in Figure 8.6.

Demonstration 2
Beneficial Bacteria
Show students a plant with nitrogen-fixing nodules on its roots. Have students examine the nodules with a stereomicroscope.

Interacting Systems
Use Figure 8.8 to show that some relationships between two organisms are beneficial. The plants benefit by obtaining the nitrates produced by the bacteria. The bacteria benefit by obtaining the products of photosynthesis that are carried out by the plant.

Farming and Oil Production
The beneficial effects of leguminous (nitrogen-fixing) plants have long been recognized by farmers. Crop rotation has been practiced for hundreds of years. This practice involves alternating plantings of a nonleguminous crop, such as corn, with a leguminous one, such as soybeans. When supplemental nitrogen is needed, farmers must turn to nitrogen-rich fertilizers. Several million barrels of oil are used each day to produce these fertilizers.

Resistance to Plant-killing Chemicals

A recent improvement in agriculture has been the development of crop plants that are resistant to the chemical glyphosate, a powerful weed killer that also kills most actively growing plants. Glyphosate kills plants by destroying an enzyme that plants need to make certain amino acids. Genetic engineers found a kind of bacterium in which this enzyme is resistant to glyphosate. They then isolated the gene that codes for the enzyme. Shooting the gene in like a bullet, they successfully transferred the glyphosate-resistant gene into crop plants such as the wheat plant shown in **Figure 8.7**.

This advance is of great interest to farmers. The farmer simply treats a field with glyphosate, and all growing plants die except the crop, which is resistant to glyphosate. After it is applied, glyphosate is quickly broken down in the environment. Glyphosate is not harmful to humans because they do not make the amino acids it affects. These qualities make it a great improvement over most commercial weed-killing chemicals, which can be highly toxic. In addition, much of the tragic erosion of fertile topsoil could be prevented if cropland did not have to be intensively cultivated to remove weeds.

Figure 8.7
Scientists have made crop plants, like this wheat plant, resistant to plant-killing chemicals.

Nitrogen Fixation

Nitrogen is an element that plants must have in order to make proteins and DNA. The most abundant source of nitrogen in the environment is the atmospheric gas N_2. However, plants cannot obtain nitrogen from the air. All of the nitrogen that plants need must be obtained from the soil. Bacteria living in the roots of plants such as soybeans, peanuts, and clover provide plants with nitrogen by converting N_2 gas from the atmosphere into a form that plants can use. **Figure 8.8** shows these bacteria. The process of converting nitrogen into a form that plants can use is called **nitrogen fixation**.

Because crops use nitrogen rapidly, most farmers add high-nitrogen fertilizers to the soil. Worldwide, farmers applied more than 65 million metric tons of nitrogen fertilizers in 1990. Since high-nitrogen fertilizers are costly, using them adds a considerable expense to a farmer's budget. Farming would be much cheaper if major crops such as wheat, rice, and corn could be genetically engineered to carry out nitrogen fixation. The bacterial genes for nitrogen fixation have been successfully inserted into plants. However, these genes do not seem to function properly in their new hosts. Many experiments are being performed to find a way to overcome this difficulty.

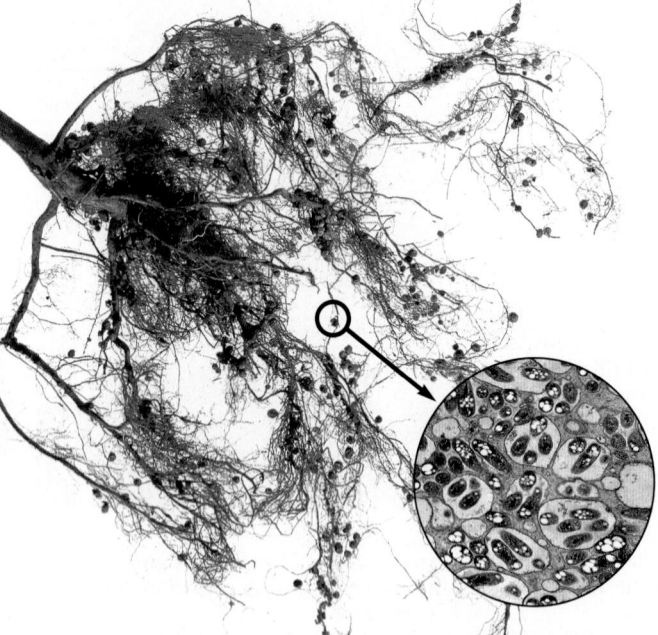

Figure 8.8
Nodules on the roots of this soybean plant contain populations of bacteria that convert nitrogen to a form the plant can use.

Resistance to Insects

Genetic engineering techniques can also be used to make crops resistant to destructive insects, such as the locust and other pests. Resistant crops would not need to be sprayed with insecticides, which are expensive and can be harmful to other organisms in the environment. Today more than 40 percent of all chemical insecticides are used to kill insects that eat cotton plants. Biologists are now trying to produce cotton plants that are resistant to these pests, so that insecticides would not be needed.

One successful approach to making plants resistant to insects uses bacteria that produce enzymes responsible for killing destructive caterpillars. When the genes coding for these enzymes are inserted into tomato plants, for example, the new enzymes make the plants highly toxic to insects called tomato hornworms. Hornworms that eat any of the tomato plants quickly die.

Many pests attack the roots of important plants. To combat these pests, genetic engineers are introducing the same insect-killing enzyme into a bacterium that colonizes the roots of crop plants. If an insect eats these roots, it will consume the bacteria and be killed by the enzyme. **Figure 8.9** shows how genetic engineering could improve a cotton plant.

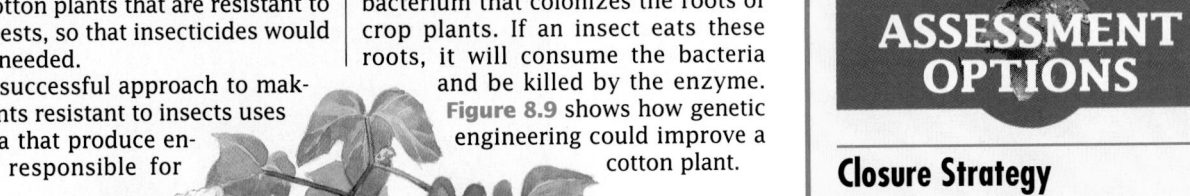

Figure 8.9
Plants, such as this cotton plant, could be genetically engineered:

a to produce an enzyme that kills pests, so that highly toxic pesticides would not be needed,

b to resist weed-killing chemicals such as glyphosate, so that highly toxic herbicides would not be needed,

c and to carry out nitrogen fixation, so that farmers would not need to apply fertilizers.

Theme Answer

Patterns of Change

There is a medical advantage where the potential is to aid in preventing and curing illness. For example, growth hormones are a potential treatment for dwarfism, a disorder in which the pituitary gland fails to make enough growth hormone.

Figure 8.10
When genetically–engineered bovine growth hormone is injected into dairy cattle such as these, each cow's milk production increases by about 10 percent. The United States government is presently reviewing this experimental procedure to determine if it is acceptable for widespread use by American farmers.

Genetic Engineering in Livestock

Patterns of Change

What are some advantages of genetically engineered proteins such as human growth hormone?

Genetic engineering techniques can also be used to produce bigger, more productive livestock. For example, injecting growth hormone into dairy cows, such as the ones shown in **Figure 8.10**, increases their milk production. The milk is no different from milk produced by other cows. Growth hormone is the natural signal that cows use to increase their milk production. In this case, the hormone is used in higher amounts than occur naturally.

Instead of extracting growth hormone from the pituitary glands of dead cows, the hormone is obtained in the laboratory using genetic engineering.

The gene containing the instructions for producing growth hormone has been introduced into bacteria. These bacteria produce growth hormone so inexpensively that it can be added as a supplement to a cow's diet.

Biologists have introduced extra copies of the genes that code for growth hormones into the chromosomes of cows and hogs. These attempts are likely to create new, leaner, fast-growing cattle and hogs. Human growth hormone is now being tested in humans as a potential treatment for dwarfism, a disorder in which the pituitary gland fails to make enough growth hormone.

Section Review

❶ How is the Ti plasmid used to insert genes into plant cells? What are its shortcomings as a genetic engineering tool?

❷ Explain how genetic engineering techniques can make crop plants resistant to weed killers such as glyphosate.

❸ How can genetic engineering reduce the amount of insecticides used in agriculture?

❹ What effects can genetic engineering have on livestock?

▪ Section Review Answers ▪

1. See Figure 8.5 for an explanation. Bacteria carrying the Ti plasmid cannot be used as a tool for transferring DNA into plants that produce grains.
2. Students should explain that genetic engineers have inserted weedkiller-resistant genes from bacteria into certain kinds of crop plants.
3. Students should explain that genetic engineers have inserted bacterial genes that produce insect-killing enzymes into plants.
4. Students should discuss the role of genetic engineering in the production of new varieties of leaner, faster growing, larger, and more productive livestock.

MUCH OF THE EXCITEMENT SURROUNDING GENETIC ENGINEERING HAS FOCUSED ON ITS POTENTIAL TO AID IN PREVENTING AND CURING ILLNESS. MAJOR ADVANCES HAVE BEEN MADE IN THE PRODUCTION OF PROTEINS THAT CAN TREAT ILLNESS, IN THE DEVELOPMENT OF NEW VACCINES TO COMBAT DISEASE, AND IN THE REPLACEMENT OF DEFECTIVE GENES WITH HEALTHY ONES.

8.3 Advances in Medicine

Objectives

❶ Describe how transferring human genes into bacteria can benefit human health.

❷ Explain how gene transfers could be used to combat genetic disorders.

❸ Summarize the goals of the Human Genome Project and the ethical issues it raises.

❹ Describe three uses of DNA profiling.

Making Miracle Drugs

Figure 8.11

a To make large amounts of insulin, scientists first obtain the gene that produces insulin.

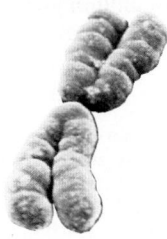

Many human illnesses occur because our bodies fail to make critical proteins. One example of such an illness is diabetes. In one form of diabetes a person cannot make a protein called insulin. As a result, they cannot regulate the levels of sugar in their blood. Diabetes can be treated if the body is supplied with insulin.

Until recently, it was not practical to use proteins as drugs because they were difficult to obtain. The body contains only small amounts of proteins like insulin, making them difficult and expensive to obtain. In most cases, proteins must be obtained from cow or pig tissue. Now genetic engineering can be used to provide large quantities of proteins in a short period of time. Genes that produce medically important proteins can be inserted into bacteria. Because bacteria can be easily grown in bulk, large amounts of the desired protein can be inexpensively prepared. **Figure 8.11** shows how genetic engineering is used to produce insulin.

Some proteins now being produced by genetically engineered bacteria dissolve blood clots, helping to prevent heart attacks and strokes. Genetically engineered bacteria are also used to produce proteins that prevent high blood pressure, and proteins that help regulate kidney function.

b The insulin gene is then inserted into a bacterial plasmid.

c The bacteria containing the plasmid are cloned and insulin is produced.

d The insulin can be collected and injected into diabetics, such as this young man.

 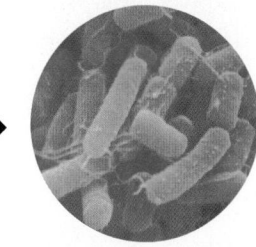  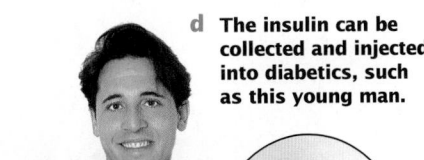

▪ Matter of Fact ▪

Amino Acid Sequence
Insulin was the first protein in which the amino acid sequence was deciphered. A relatively small protein, insulin contains 51 amino acids. In 1943, the British scientist Frederick Sanger set out to determine its amino acid sequence. Relying primarily upon enzymes to cleave the molecule and chromatography to analyze the fragments, Sanger finally obtained the solution. But it took him 10 years! Today, the amino acid sequence for a protein of approximately the same size would take about a week to determine, thanks to modern technology.

Lesson Plan 8.3

Phase 1

PREPARATION

Key Concepts
- Genetic engineering experiments have produced bacteria capable of synthesizing insulin and other proteins of medical value to humans.
- Genetic engineering experiments have produced vaccines designed to protect against certain viral diseases, including herpes and hepatitis.
- Genetic engineering experiments have introduced certain genes into humans with the eventual goal of curing inherited disorders and treating cancer.
- Genetic engineering experiments are underway to determine the nucleotide base sequence for every human gene.

Reading Strategy

In this section have students use comparison to determine the meaning of new words. For example, the term *tumor necrosis factor* is used. Students should be able to decipher the meaning of necrosis by reading the next sentence, "TNF attacks and kills cancer cells."

Phase 2
TEACHING STRATEGIES

Visual Strategy

Figure 8.11
Use this figure to point out that isolation of the insulin gene, as with many other human genes, is not performed as described earlier. Students should realize isolating a specific gene is much more complicated when dealing with human chromosomes. In fact, the locations of most genes are not known. Consequently, scientists must often use alternative means to obtain a desired gene.

Demonstration 1

Viruses

Project overhead transparencies or display electron micrographs of viruses so that students can see that they come in a variety of shapes and sizes. Tell students that viruses basically consist of a protein coat surrounding a nucleic acid core of either DNA or RNA. In order to reproduce, a virus must infect a cell, then take control of the cell's "machinery" to direct the synthesis of new viral particles. Point out that most vaccinations are aimed at preventing viral diseases. Mention that the major problem in developing a vaccine against the AIDS virus is that it mutates so frequently and consequently changes its antigenic properties.

Connection: Chapter 2

Viruses—Living or Nonliving?

List several characteristics of viruses on the board, including the facts that they contain nucleic acids capable of mutating and that they must infect a cell to reproduce. Have students identify any characteristics described in Chapter 2 to support the position that viruses are "alive." Ask students which characteristics cannot be applied to viruses. Ask students what, if any, conclusion can be made.

Visual Strategy

Figure 8.12

Point out that the surface protein coded by the DNA fragment taken from a virus serves as a marker. Remind students that protein markers establish a cell's, or in this case a virus's, identity. This marker protein must be recognized as foreign by an infected person, who will then respond by making antibodies. Ask students what will happen if the DNA fragment inserted into the harmless virus does not produce a surface protein.

Making Vaccines

A **vaccine** is a harmless version of a disease-causing microbe (bacterium or virus) injected into animals or people so that their immune systems will develop defenses against the disease. The injected version serves as a model for the body, which responds by making defensive proteins called **antibodies**. If a vaccinated animal is exposed to the disease-causing microbe, it will immediately begin large-scale production of microbe-attacking antibodies. The antibodies will stop the growth of the microbe before the disease can develop.

Traditionally, vaccines have been prepared either by killing the microbe or rendering it unable to grow. This ensures that injecting a vaccine into your body will not make you sick. The problem with this approach is that any failure in the killing or weakening of the disease-causing microbe can result in introducing the disease into the very patients in need of protection. While the majority of such vaccines are extremely safe, a tiny percentage of vaccinated individuals contract the disease from the vaccine. This small, but real, danger is one reason why rabies vaccines are administered only when a person has actually been bitten by an animal suspected of carrying rabies.

Harmless microbes can be made into piggyback vaccines

Using genetic engineering techniques, genes from disease-causing microbes can be inserted into harmless bacteria or viruses. These harmless microbes can be used to stimulate your body to make disease-attacking antibodies. For example, a vaccine against the genital herpes virus, which produces small blisters on the genitals, has been made using genetic engineering techniques as shown in **Figure 8.12**. Genes that encode the surface proteins of the herpes virus are inserted into a harmless virus. This viral vector carries the genes "piggyback" into the human body, stimulating the person's immune system to make protective antibodies against the virus. Individuals who are injected with the vaccine are able to fight infections from the Herpesvirus II.

Genetically engineered vaccines also offer hope for other diseases. A vaccine against the hepatitis B virus, which causes a sometimes fatal inflammation of the liver, is also available. Scientists are also working to produce a vaccine that will protect people against malaria. Transmitted by mosquitoes, malaria affected more than 250 million people in 1992, and over 1 million of them died.

Figure 8.12

 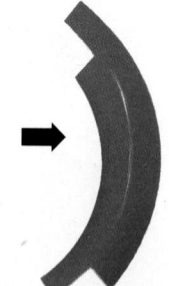 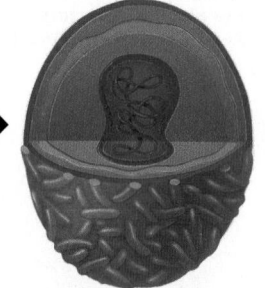 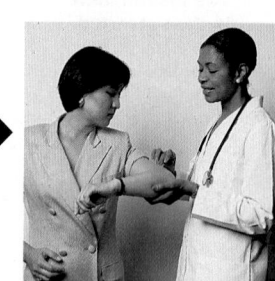

a To produce a vaccine against genital herpes, the virus that causes genital herpes is obtained from the cells of a person with the disease.

b A DNA fragment containing the gene that codes for the herpes surface protein . . .

c . . . is inserted into a harmless cowpox virus. The fragment causes the virus to make herpes surface proteins.

d A person injected with the engineered virus containing the desired gene can make antibodies against the genital herpes virus.

▪ Historical Perspective ▪

Louis Pasteur's development in 1884 of a vaccine against rabies provided the basis for most of our modern vaccinations. He used the term vaccination in recognition of the work performed in 1798 by the English country doctor, Edward Jenner. Jenner observed that milkmaids who contracted cowpox, a mild disease in farm animals, never contracted smallpox. Smallpox was a feared disease since it afflicted 95 percent of the population, killing nearly 10 percent of those who got the disease. Moreover, half of those who survived were left disfigured with scars, and many survivors were blinded.

Jenner inoculated a young farm boy with fluid taken from a sore from a milkmaid with cowpox. He subsequently inoculated the boy with fluid from a smallpox sore. The boy survived and never developed smallpox. Jenner called the process vaccination, from the Latin word *vacca* for cow.

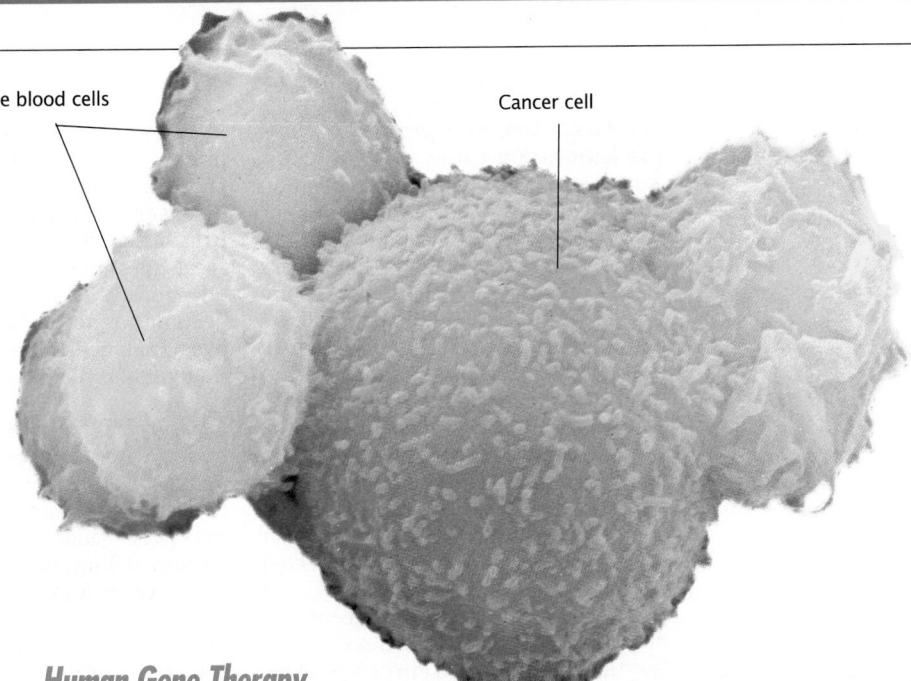

White blood cells

Cancer cell

Figure 8.13
White blood cells containing tumor necrosis factor are attacking this cancer cell.

Human Gene Therapy

Patterns of Change

If a woman with genetically altered bone marrow cells has children, would the genetic change be passed on to them?

Most human genetic diseases are due to an individual's lack of a normally functioning copy of a particular gene. The problem usually arises when both of the individual's parents contribute defective copies of a gene. One obvious way to cure such disorders is to give the person a working copy of the defective gene. Until recently, this approach was not practical for three reasons. First, the defective gene was difficult to identify and isolate. Second, it was hard to transfer a "healthy" copy of such a gene into the cells of body tissues that use it. Finally, it was necessary to find a way to keep the altered cells or their offspring alive in the body for a long time. With genetic engineering, it is now possible to overcome these difficulties. Gene transfers are being attempted as a way of combating a variety of genetic disorders, including cystic fibrosis and muscular dystrophy.

Human gene therapy has proven successful

Among the first successful attempts at human gene therapy was the transfer of an enzyme-encoding gene into a girl suffering from a rare blood disorder that was caused by the lack of this enzyme. Using genetic engineering, the gene was isolated, cloned, and inserted into cells taken from the girl's blood-cell-producing bone marrow. These cells were returned to the girl's bones where they began to produce the enzyme her body lacked. Because this kind of bone marrow cell actively divides within the bone marrow, researchers hope that offspring of the introduced cells will continue to secrete the enzyme into her blood for a long time.

Human gene transfers may also help in the battle against cancer

All humans possess white blood cells that secrete a protein called TNF (tumor necrosis factor) into the blood. TNF attacks and kills cancer cells. Unfortunately, TNF cannot work unless it encounters a cancer cell. This does not often happen. Recently, genetic engineers developed a method of adding the gene-encoding TNF protein to a kind of white blood cell that is very effective at locating cancer cells, but not very effective at harming them. Armed with this new TNF weapon, however, these white blood cells will become more like cruise missiles with a deadly payload homing in on cancer cells. **Figure 8.13** shows three white blood cells attacking a cancerous tumor.

Connection: Chapter 6

Patterns of Heredity
Use a Punnett square to review how a child may be born with a disease caused by a recessive gene if both parents carry one copy of this gene.

Theme Connection

Interacting Systems
Both human gene therapy and the human genome project have received, and will continue to receive, much attention from the media. Students should be made aware of the importance of these issues in their future. Have students research these two issues, focusing on the arguments offered both for and against these projects. Conduct a class discussion on the relative merits of both positions so students will have an opportunity to engage in bioethical decision-making.

Theme Answer

Patterns of Change
No, because she passes on her gametes, not her bone marrow.

■ *Matter of Fact* ■

Gene Therapy
In 1990, the case of a four year-old girl was brought to the attention of doctors. Missing an enzyme known as ADA, the girl has a genetic disorder called severe combined immune deficiency syndrome (SCIDS). Without ADA, a person's immune system cannot function, much like someone with AIDS. Emphasize that someone with SCIDS, unlike AIDS, has an inherited condition and therefore is a candidate for gene therapy. Preliminary tests have shown that the girl has started to produce small amounts of ADA. Since this pioneering attempt, others with various hereditary defects have undergone gene therapy.

Mathematics Connection

The Human Genome

Scientists estimate that human chromosomes contain between 50,000 and 100,000 genes. Of the 2,000 genes mapped so far on particular chromosomes, some 400 are disease-related. Another 3,000 genes have been partially deciphered to date. Scientists hope to decipher the remaining genes by 2005. But their task is formidable. Since a single gene may be as large as 30,000 bases, the human genome contains some 3 billion bases. Remind students that each base can be represented by one of four letters, A, T, G, and C. Inform students that their biology text contains approximately 291,500 words. Have them assume the average word contains 6 letters. Ask them to calculate how many books, equivalent in size to their biology text, 3 billion letters would fill.

Phase 3

ASSESSMENT OPTIONS

Closure Strategy

The Human Genome to Decipher

If someone were to be selected to have his or her genome deciphered, ask students who that person should be. Have students defend their choice. Point out that no one individual will serve as the source, but that the final sequence will be a composite of information from hundreds of individuals. Have students suggest why this method will be followed in the Human Genome Project.

Section Review

Assign the *Section Review*.

Reteaching

Assign students to cooperative work groups. Have each group develop a concept map for this section.

Gene Sequencing

Scientists are mapping the human genome

Genetic engineers are presently attempting to catalog, locate, and sequence every gene on the 46 human chromosomes. Gene sequencing is the process of determining the order of nucleotide bases within a gene. Biologists call the entire collection of genes within human cells the **human genome**. The effort to determine the nucleotide base sequence of every human gene is called the **Human Genome Project**. Since the human genome contains approximately 3 billion nucleotides, the project is not a small one. This United States project was launched in 1988 and is expected to cost several billion dollars.

Our expanding knowledge of the human genome and our increasing ability to manipulate it present some difficult ethical questions. Who should have access to a person's genetic profile? Who decides what kinds of human gene transfer experiments should and should not be done? Which diseases should we attempt to cure first? There are no simple answers to these questions. These and many other questions will arise as gene technology becomes a more essential part of our lives. Each question will require careful thought and planning.

DNA profiling can be used to identify unknown DNA

In 1984, Alec Jeffreys, a British geneticist, devised a way to visually identify the DNA in genes. This technique, called **DNA profiling**, can identify the base sequences in a sample of DNA. The end result, shown in **Figure 8.14**, is a photograph with a pattern of dark bands that reflect the composition of a DNA molecule. Because these images can be used to establish the identity of a person, they are often called **DNA fingerprints**.

DNA profiling is based on the theory that it is extremely unlikely for two people to have identical DNA. Therefore, when the process is used to analyze DNA in blood, semen, bone, or hair, DNA profiling can help identify criminals. Criminal investigation is just one use for this new technology.

Scientists can isolate small fragments of DNA and locate genes that cause disorders such as neurofibromatosis, a painful condition causing benign tumors. In addition, doctors can use DNA profiling to reveal hereditary relationships and to measure the success of bone marrow transplants in leukemia patients.

Figure 8.14
This scientist is examining a DNA profile, or fingerprint, to determine the nucleotide base sequence of a portion of one DNA molecule.

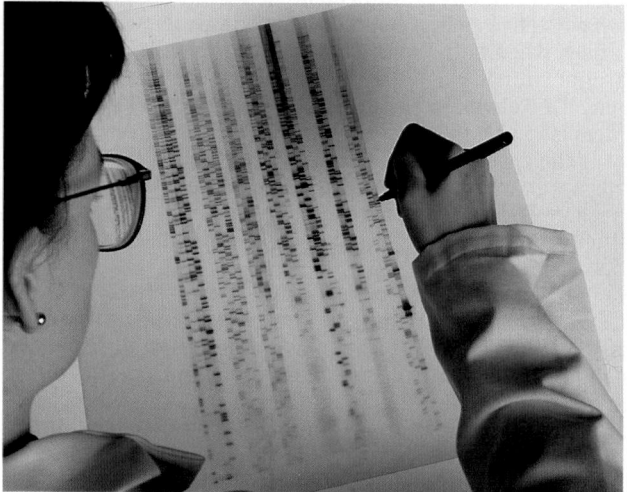

Section Review

1 Explain how genetic engineering can be useful in the treatment of human illnesses such as diabetes.

2 How could human genetic therapy cure genetic disorders such as cystic fibrosis?

3 What are the goals of the Human Genome Project? What ethical issues does it raise?

4 Describe how DNA profiling could be used to help determine a suspected criminal's innocence or guilt.

■ Section Review Answers ■

1. Students should explain that genes that produce medically important proteins such as insulin can be inserted into bacteria. Since bacteria can be easily grown in bulk, large amounts of the desired protein can be inexpensively prepared.

2. Students should explain that some human genetic disorders are being treated and corrected by inserting copies of healthy genes into individuals lacking them.

3. Students should indicate that the goals of the Human Genome Project are to catalog, locate, and sequence every human gene. The ethical issues are those of determining access, priorities, and types of experiments and cures.

4. Students should indicate that DNA profiling reveals DNA patterns that vary distinctly from one individual to the next.

Chapter **8** *Highlights*

Gene therapy offers hope for patients with genetic disorders such as this child with cystic fibrosis.

Alternative Assessment

Assign students to cooperative work groups. Give each group the assignment of designing a board game to be known as the Human Genome Project. Tell them that they are free to choose the format of the game and the way in which a winner is determined. However, any questions they include as part of the game must come from the information covered in this chapter.

	Key Terms	Summary
8.1 Genetic Engineering  Bacterial plasmids make genetic engineering possible.	recombinant DNA (p. 157) genetic engineering (p. 157) cloning (p. 158) restriction enzyme (p. 158) plasmid (p. 159) vector (p. 159)	• Over the last 20 years, genetic engineers have learned how to move genes from one organism to another. • Every gene transfer starts by cleaving DNA into small fragments using restriction enzymes. • Fragments containing the desired gene are then transferred to a target cell using a vector. • After removing uninfected cells, scientists search for cells that have taken up the desired gene.
8.2 Transforming Agriculture  Cows injected with genetically engineered growth hormone will produce more milk.	nitrogen fixation (p. 162)	• Genetic engineers have manipulated the genes of certain kinds of crop plants to make these plants resistant to weed killers and destructive pests. • Genetic engineers are looking for a way to transfer genes for nitrogen fixation into crop plants. • Adding genetically engineered growth hormone to the diet of livestock increases milk production in dairy cows and increases the weight of cattle and hogs.
8.3 Advances in Medicine 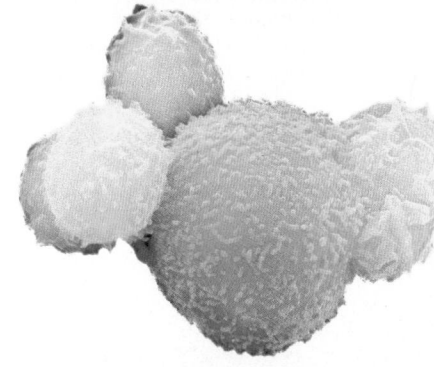 Scientists hope that they will soon be able to use genetic engineering to treat diseases such as cancer.	vaccine (p. 166) antibody (p. 166) human genome (p. 168) Human Genome Project (p. 168) DNA profiling (p. 168) DNA fingerprint (p. 168)	• Genetic engineering techniques can be used to manufacture proteins such as insulin and vaccines. • Some human genetic disorders are being treated and "corrected" by inserting copies of healthy genes into individuals lacking them. • The human genome is the entire collection of genes within human cells. The Human Genome Project seeks to locate, catalog, and read the base sequence of every human gene. • Scientists use DNA profiling to reveal DNA patterns that vary distinctively from one individual to the next.

Chapter Review Answers

Understanding Vocabulary

1. **a.** Recombinant DNA is created when pieces of DNA from different species are joined together, while a restriction enzyme is a protein that cuts DNA into small pieces.
b. The procedure used to produce recombinant DNA is called genetic engineering, while DNA profiling is a technique used to visually identify the base sequences of DNA in genes.
c. A vector is an infective agent, often a bacterium, that functions as a vehicle to transport a gene into a host cell. A plasmid is a ring-shaped piece of bacterial DNA that can be replicated.

Relating Concepts

2. Map answer is shown on page 155D.

Understanding Concepts

Multiple Choice

3.	b	8.	d
4.	a	9.	c
5.	a	10.	a
6.	a	11.	b
7.	d		

Completion

12. sticky ends
13. disease-causing, harmless
14. human genome, 3 billion

Short Answer

15. They are *sticky* because the restriction enzymes make offset cuts through the strands of the DNA molecule rather than making straight cuts.

16. Gene transfers have made certain plants more resistant to disease, herbicides, and insects. Also, gene transfers have been used to increase the production and growth rates in both crops and livestock.

Chapter 8 Review

Understanding Vocabulary

1. For each pair of terms, explain the differences in their meanings:
   a. recombinant DNA, restriction enzyme
   b. genetic engineering, DNA profiling
   c. vector, plasmid

Relating Concepts

2. Copy the unfinished concept map below onto a sheet of paper. Then complete the concept map by writing the correct word or phrase in each oval containing a question mark.

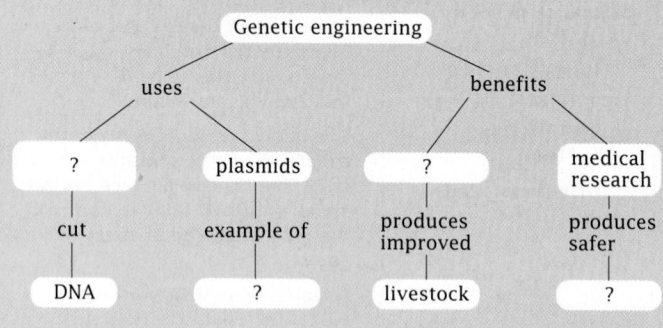

Understanding Concepts

Multiple Choice

3. The genetic engineering experiment of Cohen and Boyer produced
   a. frog clones.
   b. recombinant DNA.
   c. synthetic insulin.
   d. prokaryotic bacteria.

4. The screening stage of the Cohen and Boyer experiment involved
   a. locating cells that received the gene for frog rRNA.
   b. searching cells for plasmids not resistant to tetracycline.
   c. cutting frog DNA into fragments and inserting them into plasmids.
   d. marking DNA segments resistant to tetracycline.

5. Which is *not* one of the stages of gene transfer?
   a. DNA profiling
   b. producing recombinant DNA
   c. screening
   d. cleaving DNA

6. Scientists are able to transfer segments of DNA from one organism to another by using
   a. bacterial plasmids.
   b. frog chromosomes.
   c. cotton plants.
   d. piggyback vaccines.

7. Growing a large number of identical cells from one cell is known as
   a. nitrogen fixation.
   b genetic engineering.
   c. DNA profiling.
   d. cloning.

8. The role the Ti plasmid plays in agricultural research is that of a
   a. restriction enzyme.
   b. weed killer.
   c. nitrogen fixer.
   d. vector.

9. Genetically engineered plants that are able to fix nitrogen will
   a. make crop rotation a necessity.
   b. kill soil bacteria.
   c. reduce the use of fertilizers.
   d. be resistant to chemicals that kill insects.

10. An advantage of a genetically engineered human protein such as insulin is
   a. it can be produced in large amounts at low cost.
   b. less is needed to produce the same effect.
   c. it is safer than naturally produced insulin.
   d. it dissolves blood clots in addition to treating diabetes.

11. Cystic fibrosis, muscular dystrophy, and other genetic disorders
   a. are vectors used in genetic engineering.
   b. may be treated by replacing defective genes.
   c. can be cured in babies by genetic analysis.
   d. are too deadly to be considered by genetic engineers.

Completion

12. Restriction enzymes are used to cut DNA segments. The cut leaves _____ _____ .

13. To make "piggyback" vaccines, genetic engineers insert genes from a(n) _____ virus into the chromosomes of a(n) _____ virus, which is then put into the human body.

14. The complete collection of genes within human cells is called the _____ ; it contains about _____ nucleotides.

Short Answer

15. What makes the DNA fragments cut by restriction enzymes "sticky"?

16. In what ways can genetic engineering affect agriculture?

Interpreting Graphics

17. The figures below show four distinct stages of a gene transfer experiment.

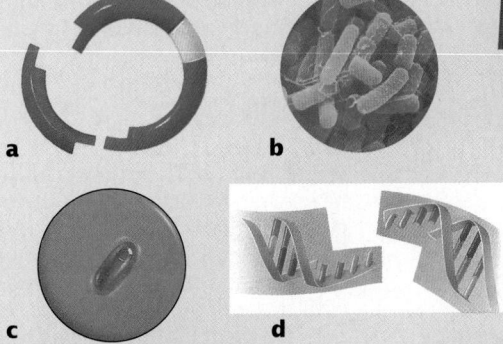

a

b

c

d

- How should the figures be ordered to show the proper sequence of events?

- What is occurring in **a**? In **c**?

Reviewing Themes

18. *Interacting Systems*
Why would farmers be interested in genetically engineered crop plants that are resistant to the weed-killing chemical glyphosate?

19. *Patterns of Change*
How has genetic engineering changed the way some vaccines are prepared?

Thinking Critically

20. *Inferring Conclusions*
Why are restriction enzymes said to be the basic tools of genetic engineering?

21. *Comparing and Contrasting*
How are organisms produced by genetic engineering different from organisms produced by sexual reproduction?

22. *Building on What You Have Learned*
In Chapter 7 you learned about the double helix model of DNA first constructed by Watson and Crick. How is Watson and Crick's model of DNA related to genetic engineering?

Cross-Discipline Connection

23. *Biology and Agriculture*
Several Texas farmers were recently asked to plant a genetically engineered variety of cotton that produces tan, green, and red cotton. What are the possible advantages of genetically engineering new varieties of different colored cotton?

Discovering Through Reading

24. Read the article "Barnyard Bioengineers" in *Newsweek*, September 9, 1991. How are animals being used to produce drugs to improve human health? Why are animal-rights activists interested in the work of "barnyard bioengineers"?

Thinking Critically

20. There are more than 200 identified restriction enzymes that can be used to cut many different gene segments for insertion into bacteria for transfer to another organism.

21. Two organisms from the same species produce an offspring of the same species by sexual reproduction, but in genetic engineering DNA from two different species is combined. The product of genetic engineering may be a new species.

22. Watson's and Crick's model predicted the cleaving of DNA as shown by the use of restriction enzymes.

Cross-Discipline Connection

23. Farmers enjoy the higher prices they can earn from planting colored cotton. Garments made from colored cotton will not need to be dyed and are reported to be less likely to fade when washed.

Discovering Through Reading

24. Human genes are spliced into the DNA of an animal fetus, and the implanted genes cause the production of proteins needed by humans. Unfortunately, only a small number of embryos in which the genes are planted develop to produce the proteins. Animal-rights activists wish to protect the animals involved in the research from harm.

Interpreting Graphics

17. • d,a,b,c
  • inserting a gene into a plasmid, screening target cells

Reviewing Themes

18. Farmers could use glyphosate to kill weeds in their fields without fear of crop damage. In addition, fields sprayed with glyphosate would not need to be intensively cultivated to remove weeds, and the erosion of topsoil typically associated with intense cultivation would be lessened.

19. Vaccines were prepared by killing or weakening the disease-causing bacteria and then injecting them into the human body. Their presence resulted in the production of antibodies. Much less dangerous are the genetically engineered vaccines; they are modified bacteria that display the harmful bacteria's surface proteins. These surface proteins trigger the production of antibodies.

Procedural Notes

1. Clarify that, at the present time, DNA fingerprinting does not result in a direct reading of the complete code of the DNA molecule. Discuss the processes involved in developing DNA fingerprints: explain how restriction enzymes cut at specific sequences of nucleotides. Describe how gel electrophoresis "sorts" the molecules by size (and polarity).

2. Challenge students' critical thinking skills by having them discuss the strengths and weaknesses of the model of the DNA fingerprinting process used in this Investigation.

3. Some students may also be aware of the controversy surrounding the use of DNA fingerprinting in criminal cases. An additional activity might involve a class debate on the use of this technology. (Possible sources of information include the articles referred to in the Investigation Notes in the interleaf preceding this chapter.)

Prelab Preparation Answers

- Answers may vary, but should describe the double helical structure of the DNA molecule and the complementary base pairing found in that structure.

- A photograph with a pattern of dark bands that reflects the composition of a DNA molecule. DNA fingerprinting can be used to establish the identity of a person.

Procedure Answers

3. Answers will vary.
6. B
7. Answers will vary.

Chapter 8

Investigation

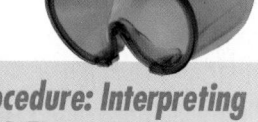

How Are DNA Fingerprints Interpreted?

Objectives

In this investigation you will:
- *observe* and *interpret* similarities and differences among various representations of DNA fingerprints

Materials

- container
- colored beads (of various shapes and lengths)
- heavy-duty thread

Prelab Preparation

Review what you have learned about DNA by answering the following questions:
- Describe the structure of DNA.
- What are DNA fingerprints? What can they be used for?

Procedure: Interpreting DNA Fingerprints

1. The process of determining a DNA fingerprint involves three steps. First, DNA is extracted from cells and cut into small fragments by restriction enzymes. The containers on the desks in your classroom contain beads that represent such fragments of DNA. Each container represents a different source. Remove the beads from the container at your desk and sort them into groups based on size and shape. No two containers hold the exact same assortment of beads. However, one container holds an assortment of beads similar to those in the container at your teacher's desk. These beads represent fragments of DNA taken from cells found at the scene of a crime.

2. In the second step of determining a DNA fingerprint, the DNA fragments are separated by size. Arrange your collection of beads in a line from the shortest to the longest.

3. Using a piece of heavy-duty thread, string your beads in the order you have arranged them. Tie the two ends together in a knot so that the beads will not fall off. Compare the numbers and sizes of your beads with those of your classmates. *How does the combination of beads on your string compare to other combinations of beads in the class?*

4. In the third step of determining a DNA fingerprint, the DNA fragments are treated with a radioactive substance so they can be seen. The substance sticks to specific parts of the fragments and produces a pattern of stripes when exposed to X-ray film. The resulting pattern is unique for every individual. Study

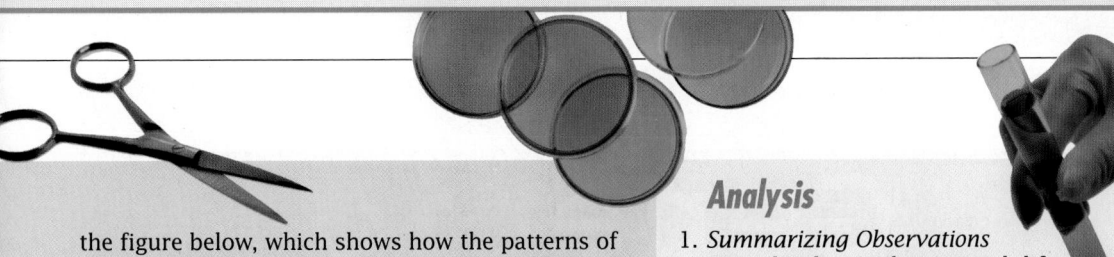

the figure below, which shows how the patterns of stripes might appear in a photograph. These illustrations represent DNA fingerprints from four separate sources.

Crime Scene   **Suspects**

A   B   C

5. Imagine that a crime has been committed and that a sample of cells belonging to the criminal is collected at the crime scene. The DNA fingerprint of these cells is represented by the illustration labeled "Crime Scene." Study the pattern of stripes in this DNA fingerprint.

6. Compare the DNA fingerprint of the cells obtained at the crime scene with the DNA fingerprints of the samples collected from the three suspects labeled A, B, and C. *Which suspect's DNA fingerprint most closely matches the DNA fingerprint of the cells obtained at the crime scene?*

7. Under direction from your teacher, compare the arrangement of beads on your string with the arrangement of beads on the string in the container in the front of the room. *How do your beads compare with the evidence?*

Analysis

1. *Summarizing Observations*
   Describe the similarities and differences between the DNA fingerprint from the crime scene sample and the DNA fingerprints of the suspects shown in the figure to the left.

2. *Analyzing Data*
   What evidence is there that the cells left at the crime scene belong to one of the suspects?

3. *Drawing Conclusions*
   Why is it unlikely that two individuals would have identical DNA?

4. *Evaluating Methods*
   Why is the technique of labeling and photographing DNA fragments called DNA fingerprinting?

5. *Inferring Relationships*
   Why do similarities exist among all the samples in the figure to the left?

Thinking Critically

1. Imagine you are on a jury and DNA fingerprinting evidence is introduced. Explain how you would regard such evidence.

2. Explain how the DNA fingerprinting procedure could be used by scientists who study fossils.

Analysis Answers

1. Students should discuss the similarities and differences in number, location, pattern, and size of bands between a crime scene sample and samples from suspects.

2. The samples from the crime scene and suspect B are identical in their fragment bands.

3. Students should discuss the differences in genetic makeup of individuals.

4. Students should indicate that it is extremely unlikely for two people to have identical DNA making DNA as unique as an individual's fingerprints.

5. Most of the genes in the human genome are common to most individuals.

Thinking Critically Answers

1. Answers will vary, but should discuss the validity of the test.

2. DNA fingerprinting can be used on samples of bone to reveal hereditary patterns in fossils.

Using the Feature

- This feature can be used to discuss human genetic disorders other than those covered in the text. Possible topics include Tay-Sachs disease and phenylketonuria.
- This feature can also be used to discuss some of the techniques of genetic counseling such as pedigrees, karyotypes, and amniocentesis.

Discussion

Guide a discussion by posing the following questions:

1. Who should be entitled to accessing information revealed during genetic screening procedures?
   Students should point out that this information could be used in a discriminatory way. For example, denying someone insurance because of their genetic make up.

2. Does society have the right to lay down rules for people at risk of producing offspring with a genetic disorder?
   Students should indicate that this practice conflicts with ideas of personal freedom.

3. Suppose that genetic screening uncovers some serious defect. Who, if anyone, is to inform relatives who might be similarly affected, or who might be carriers?
   Ideally, family members should feel some sense of responsibility toward relatives.

4. Would it be desirable to have a national repository of information containing what is known about each person's genotype?
   It might be useful in identifying some genetic disorders, but there is potential for abuse.

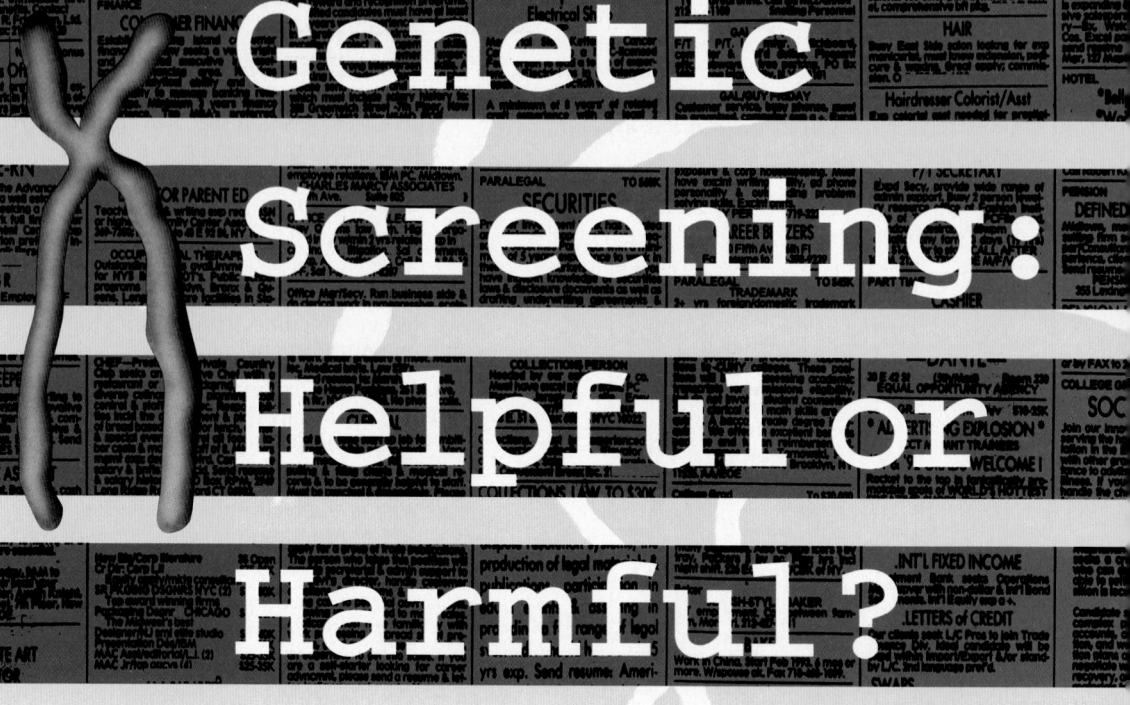

Genetic Screening: Helpful or Harmful?

Have you ever wondered about your future? Do you wonder if you will be rich or famous? Do you wonder if you will be healthy? The time is rapidly approaching when scientists will help foretell the future. The Human Genome Project could provide the tools geneticists need for diagnosing and treating many disorders. But many people are against this project. Read on to find out why.

It is the year 2020 and Bill Islet has applied for a job with a company in his area. During the interview, Bill is told that as a condition of employment he will have to undergo some DNA testing. This testing will determine if Bill is a carrier of any genetic diseases. Because some genetic diseases are very serious and require extensive medical treatment, the company cannot obtain health insurance coverage if it employs people likely to have these diseases.

"But I am not sick," says Bill. "Why would you test me for a disease that I don't have?"

Bill's situation could very well occur as a spinoff from the Human Genome Project. The Human Genome Project is an effort by many scientists to map human chromosomes. These maps will tell scientists what genes you have and where they are located.

The technology to screen for many genetic diseases or conditions could be available from the gene mapping project. Specific policies for how this information will be used are yet to be developed. There is the very real possibility that this technology could be misused. Because of these concerns about misuse of genetic screening information, there are many opponents of the Human Genome Project.

Some people would like to know if they are carrying a gene that could result in a disease, such as Huntington's disease. Symptoms of Huntington's disease do not appear until middle age. The disease begins with lapses of memory and with irritability. As it progresses, the victim loses muscle control, experiencing spasms and extreme mental illness. The

Science, Technology, and Society ▪ Science, Technology,

disease eventually causes death. The unfortunate aspect of Huntington's disease is that by the time symptoms appear, genes for the disease have already been passed on to offspring.

Those against testing contend that because the disease is so devastating and is incurable, no one would want to know if they had it. Forcing someone to be tested could become difficult, because such tests are seen as an invasion of privacy. If a person has the test, the next issue becomes who should be given the results of such testing? Consider Bill's case. Should the insurance company or his prospective employer have access to his genetic profile? What about Bill's family? If Bill is married, his wife might want to know his genetic profile. Bill's offspring might inherit a genetic disease from him. Should he be legally bound to share this information with his wife?

Under most circumstances, courts have ruled that medical records are private and cannot be accessed without permission. A patient must sign a release if he or she wants an employer, insurance company, or relatives to see medical test results. However, insurance companies can require certain tests before agreeing to provide insurance. Employers can also make the passing of a medical

exam a condition of employment. Such practices place people in situations where their records become known. Insurance companies that provide health care and life insurance want healthy clients. Insurance companies do not want to pay for long, expensive treatments for those who have the potential to become seriously ill. For example, if Bill was shown to carry a cystic fibrosis gene (a recessive trait), he would not be affected by the disease. But if his wife also carried the gene, their child might be affected by the disease. Health care for a child with cystic fibrosis (CF) is quite expensive. Therefore, if Bill were a carrier of CF, he could have trouble getting health insurance for himself, his wife, and their child.

Some people see the value in mapping the human genome, but feel that the project should stop there. Spinoff technologies, like genetic screening tests, should not be pursued. Those supporting the project, however, see reasons to explore the positive uses of this technology. For example, gene therapy could be used to cure fatal or painful diseases.

Should the development of a technology be limited when we know it can help some, yet harm others?

Thinking Critically Answers

1. Students' responses will vary but should be logical and well-argued.
2. Students' responses will vary but should discuss the advantages and disadvantages of the project.
3. Students' responses will vary but could indicate that many diseases may be eliminated.
4. Students' responses will vary but could include cases of mistaken identity, invasion of privacy, and the expense of developing a genetic screening program that focuses on "bad genes."

Thinking Critically

❶ If you were in Bill's situation, would you agree to be tested?

❷ The cost of the Human Genome Project will be about $88 million per year for 15 years. Should the government spend money on this project? Why or why not?

❸ Some people say that knowing about our own personal genome could change the human population physiologically. How do you think the human population could change? What benefits could result from these changes?

❹ List three disadvantages of creating an extensive bank of criminal genomes.

Acting on the Issue

❶ Find out what diseases, if any, are screened for by law in your state.

❷ Find a genetic counselor at your local hospital or clinic. Ask the counselor to come and speak to your class about his or her role in

advising prospective parents. How does the counselor expect his or her role to change in the future?

❸ Propose possible legislation that could result from the Human Genome Project.

❹ Write for information on the Human Genome Project:
Human Genome Project
Cold Spring Harbor, NY
10098

Chapter 9
Evolution and Natural Selection

Planning Guide

Objectives/Themes	Classwork Resources	Homework Resources

9.1 **1.** Summarize Darwin's beliefs about the origin of species before he sailed around the world.
 	
2. Identify two observations from Darwin's voyage that led him to question his beliefs.
 	
3. Describe two major ideas Darwin put forth in *The Origin of Species*.
 	
Themes: Evolution, Patterns of Change
	Teacher's Resource Binder
	Focus Activity 9
	  Comparing Observations of
	  Body Parts
	Lab Investigation 9.1
	  Peppered Moth Survey
	
	Other Resources
	Transparency 35
		Text
		Section Review, p. 179
		
		Teacher's Resource Binder
		Directed Reading Worksheet 9.1
		
		Other Resources
		Audiocassette 9.1*

9.2 **1.** Describe the conditions necessary for fossils to form. **2.** List one example in the fossil record indicating that evolution has occurred. **3.** Explain how comparisons of organisms can reveal evidence of evolution. **4.** Describe the important evidence for evolution found in proteins and DNA. **Themes:** Evolution, Interacting Systems, Patterns of Change, Scale and Structure	**Text** Science in Action *Sexual Selection* pp. 184–185 **Other Resources** Transparencies 36–37	**Text** Section Review, p. 183 **Teacher's Resource Binder** Directed Reading Worksheet 9.2 **Other Resources** Audiocassette 9.2*

9.3 **1.** Describe how natural selection occurs.
 	
2. Summarize the effects of natural selection on the peppered moth and on the sickle cell allele.
 	
3. Describe how natural selection can lead to the formation of new species.
 	
4. Contrast the hypotheses of punctuated equilibrium and gradualism.
 	
Themes: Evolution, Patterns of Change, Stability
	Text
	Investigation
	  How Does Natural Selection
	  Work? pp. 196–197
	
	Teacher's Resource Binder
	Extension Worksheet
	  Two Important Bones
	
	Other Resources
	Transparencies 38–39
		Text
		Section Review, p. 192
		
		Teacher's Resource Binder
		Directed Reading Worksheet 9.3
		Vocabulary Review Worksheet*
		Reteaching Worksheet*
		  Natural Selection in Action
		
		Other Resources
		Audiocassette 9.3*

*Reteaching Options

Demonstrations
9.2: pp. 180, 181
9.3: pp. 187, 189, 190, 191, 192

Assessment
Chapter Review pp. 194–195
Portfolio Assessment p. 175D
Chapter Test—Teacher's Resource Binder
Test Generator

Research Notes

Connection to Behavior: Kin Selection

In the 1960s, the British evolutionist William Hamilton proposed a theory of "kin selection." According to the theory, members of a species may evolve behaviors that aid close relatives, helping genes identical to their own spread through a population. Until recently, evidence for the theory was hard to find, because differences in behavior toward relatives could often be ascribed to other causes.

Experiments by David Pfennig of Cornell University have provided what may be evidence for the theory. In a specialized form of spadefoot toads, the tadpoles become cannibalistic if they are fed a diet of whole prey. Other individuals not fed on such a diet become omnivores. The cannibals prefer to associate with nonrelatives, which they often devour. Cannibals do not eat relatives.

Pfennig says the fact that omnivorous tadpole relatives tend to group together and cannibalistic tadpoles avoid their relatives is evidence of kin selection. He notes that since it is possible to make omnivores and cannibals that are genetically similar, the differences seen in the two are unlikely to be due to other causes such as heredity.

In addition, several studies of bird behavior indicate that there may be times when evolutionary pressures can cause birds to favor certain relatives over others.

Biologists from Cornell University analyzed the behavior of the white-fronted bee-eater of Kenya. In this species, Merops bullock-oides, the birds live in clans. The older birds often disrupt the mating rituals of younger ones. In fact, the most common case of disruption involves fathers harassing their sons during the sons' mating attempts.

Often, the harassed birds end up helping feed the older birds' offspring. A statistical analysis indicated that survival rates for birds with similar genetic makeup were about the same whether young birds helped their parents raise siblings or mated and raised their own offspring.

Other researchers in the United States videotaped the feeding behavior of the eastern bluebird, Sialia sialis, and found that the fathers consistently fed their daughters more often than they fed their sons. No difference was found for mothers. A similar pattern was detected in parakeets.

It is theorized that the sons may eventually compete with the father birds for food and nesting sites, since they tend to settle where they were raised. Daughters, on the other hand, tend to move away when they mate, and would not threaten the fathers' territory.

Some biologists believe an alternative theory: the daughters are being trained to have high expectations, so they will mate with males that are able to feed them as well as their fathers. Such good providers are likely to be better adapted for their environment.

Investigation Notes

How Does Natural Selection Work?

pages 196–197

Purpose: This investigation gives the students the opportunity to model the process of natural selection. Beads of different sizes are used to represent prey while the students play the role of the predator. Students will test the relationship between bead size and the likelihood of being captured.

Prelab Preparation

1. WARD'S has developed a special "kit" of beads the appropriate sizes specifically for this investigation (15 M 0504). Plastic beads of different sizes are readily available at reasonable cost from a craft store. Select five different sizes, with the largest being 15–20 mm in diameter and the smallest about 1 mm. Have each size be a different color to facilitate reference to the different beads during a discussion.

2. Discuss natural selection and the example of industrial melanism before scheduling this investigation.

Answers will be found on pages 196–197.

Meeting Individual Needs

Objectives

1. Students will demonstrate appropriate use of core vocabulary for the chapter (Vocabulary File).

2. Students will be able to explain the basic aspects of natural selection (Teaching Strategy).

3. Students will be able to report on sickle cell anemia and the sickle cell trait, and describe how it relates to their family history (Multicultural Lesson Plan).

Vocabulary File
**(Developing Vocabulary/
Limited English Proficiency)**

If you are not already using the Vocabulary File, refer to Chapter 1 for its preparation. See the Chapter Highlights on page 193 for a list of suggested words.

Teaching Strategy
Adaptation and the Changing of Species
for pages 179–183
**(Building Organizational Skills/
Visual Learners)**

Materials: fossils from your area

Procedure: Point out that students themselves are living things. Ask them if they look the same as they did when they were six months of age. Brainstorm briefly about the changes in the students between now and then. Ask students to consider whether such changes represent evolution as you read aloud to them the section of the text entitled "Darwin's Mechanism for Evolution," page 179.

Continue the discussion so students become aware that a change in species (not just in an individual) over time is known as evolution. Ask students how this takes place. Be sure students' answers include that natural selection is a process in which organisms with traits well-suited to the environment survive and

reproduce at a greater rate than organisms less-suited to the environment.

If you live in a geographic region that has fossil beds, display fossils common to your area. Brainstorm about the fossils, challenging students to discuss clues about whether they are fossils of a marine, aquatic, or terrestrial environment. Compare the fossils to shells, etc., of animals living in the area today. Use students' input to direct the discussion into natural selection and changes of the environment.

Recall activities of Chapter 6 (Genetics and Inheritance) so that students realize that there is variation in each species.

Multicultural Lesson Plan
Sickle Cell Anemia

(Dealing with Ethnic Diversity)

Preparation: Invite a local sickle cell organization to send literature and/or a speaker to the class. Write to either of the organizations below for more details.

*The National Association
for Sickle Cell Disease
111 West 57th Street
Suite #1108
New York, NY 10019*

*The National Sickle Cell
Disease Research Foundation
820 Fifth Avenue
New York, NY 10036*

Teaching Strategies
1. Introduce the sickle cell trait and sickle cell anemia, either through literature from a sickle cell organization or a speaker.

2. Be sure to point out that sickle cell anemia is not limited to Africans and African-Americans. Since travel throughout the world has become more common, individuals with the trait have participated in interracial marriages and cohabitation, bringing the trait from its original geographic location to new sites and new groups of people.

3. Review the history of anti-malarial campaigns in the United States and other countries. Have students name the countries where the sickle-cell trait is still advantageous.

4. Be sure students are reminded that individuals must be aware of their families' medical history to know if they are at risk, since the risk pattern depends upon both sets of inherited genes.

Assessment: Students should take notes and do further research on sickle cell anemia or some other genetic trait. Each student should ask family members if relatives have the trait or anemia. Students should create a family medical tree or a family medical history as it relates to sickle cell anemia. (Be sure to provide guidelines or examples for these documents.)

Where appropriate, allow students to perform research on other genetic illnesses that may be present in their families or communities. (For example, Tay-Sachs disease, a disorder caused by a deficiency of the enzyme hexosaminidase A, is common among families of Eastern European origin.)

Additional Strategies
Visual Strategies
Pages 178, 181, 182, 183, 187, 188, 189, and 190

Auditory Learners
Use *Biology: Visualizing Life* Audiocassettes for Sections 9.1, 9.2, and 9.3.

Meeting Individual Needs (cont.)

Cooperative Learning
Fossil Types

Timing: Use this activity with Section 9.2.

Group Size: 4 students

Outcome: Students will compare the formation of different kinds of fossils.

Individual Accountability: Each student will illustrate the origin of one type of fossil.

Positive Interdependence: The group members will compare the relative scientific value of each kind of fossil.

Assign one of the following fossil types to each group member:

> (1) petrifications (hard parts replaced by minerals)
>
> (2) molds and casts
>
> (3) imprints
>
> (4) footprints and trackways

Ask students to do research in the library on their type of fossil, and prepare an outline or diagram illustrating the formation of this type of fossil.

Each group member should present his or her findings to the rest of the group. The group should discuss the kinds of information that can be obtained from each fossil type, and which types of information have been lost.

Portfolio Assessment

Students should select their best work and provide a self-reflective rationale for their selections. Students can make selections in the following areas.

1. *Content* — One concept map from the chapter (See page 812 for evaluation criteria.)

2. *Reading Comprehension* — One Directed Reading Worksheet from the Teacher's Resource Binder (Use the answer key to evaluate for accuracy.)

3. *Writing* — Using the Vee Form, summarize a newspaper or magazine article relating to evolution or natural selection. (See page 22T for evaluation criteria.)

   Or: Select a writing task or project from the Chapter Review.

4. *Performance Assessment* — One Vee form from a chapter investigation or lab manual investigation (See page 22T for evaluation criteria.)

Teacher makes selections in the following areas.

1. *Formal Assessment* — Chapter test (Test A, B, or the Test Generator) The teacher-scored test should be reviewed by the student. Incorrect responses should be corrected by the student before the test becomes part of the portfolio.

2. *Informal Assessment* — Use the Direct Observations Checklist, page 33T, during a laboratory or other cooperative learning experience.

3. *Performance Assessment* — As a model for natural selection, have students record how TV shows compete over a season with other shows for air time, viewers, ratings, and commercial support. Challenge students to infer which characteristics enable a TV show to survive cancellation.

Concept Map Answer

The following is one possible answer to the Relating Concepts exercise on page 194.

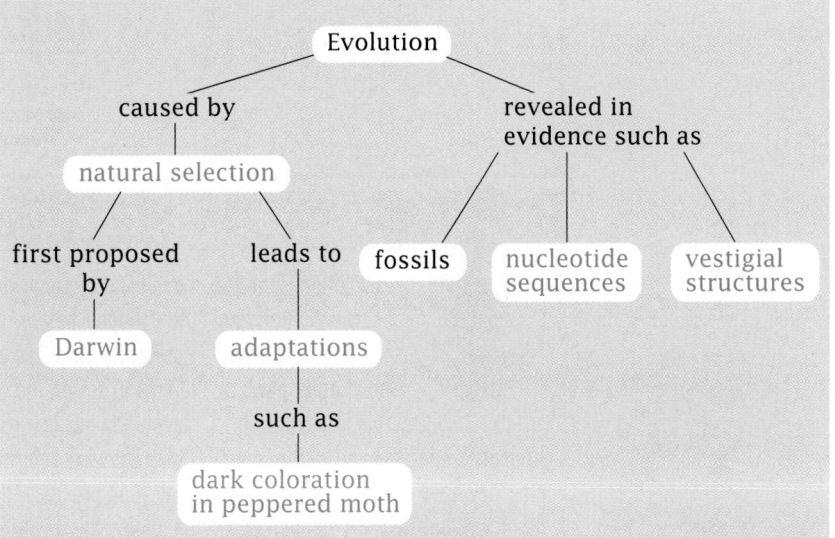

Chapter 9

Determining Prior Knowledge

- Have students discuss what evolution means to them. Elicit responses to questions concerning the types of organisms subject to evolution, the time required for evolutionary change, and the evidence scientists rely upon to explain the evolutionary process.
- Show students a fossil. Ask how scientists use fossils to learn about evolution.
- Display photographs of animals in their natural habitat. Ask students to describe any adaptations they can detect. Have students hypothesize what would happen to the animals if their environment changed.
- Show photographs of various breeds of dogs. Ask students if they are members of the same or different species. Have them justify their answers.

Chapter 9

Evolution and Natural Selection

Review

- evolution (Section 1.3)
- sickle cell anemia (Section 6.3)
- DNA structure (Section 7.1)
- the term *allele* (Glossary)

9.1 Charles Darwin

- **Voyage of the *Beagle***
- **Darwin's Finches**
- **Darwin's Mechanism for Evolution**

9.2 The Evidence for Evolution

- **Understanding the Fossil Record**
- **How Fossils Are Dated**
- **Comparing Organisms**

9.3 Natural Selection

- **How Natural Selection Causes Evolution**
- **The Peppered Moth: Natural Selection in Action**
- **The Puzzle of Sickle Cell Anemia**
- **How Species Form**
- **Does Evolution Occur in Spurts?**

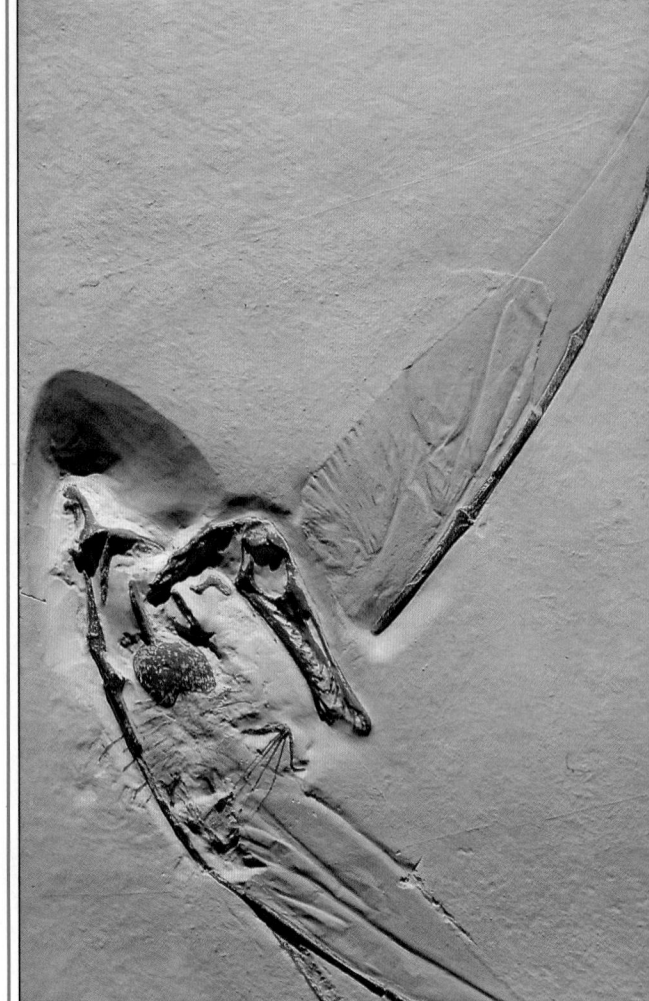

Fossils, such as this 150 million-year-old pterosaur, a flying reptile, are one line of evidence for the occurrence of evolution.

▪ Author's Rationale ▪

Evolution is the fundamental unifying theme of biology. This chapter plays a key role in advancing students' understanding of the principles they have studied thus far. Assuming that students will perceive evolution as somewhat controversial, the chapter begins with an objective presentation of the evidence. The text focuses on the observation that evolution has occurred, then follows with Darwin's reasoning that natural selection provides the mechanism that explains why evolution happens.

OUR EARTH IS RICH IN LIFE. ABOUT 1.4 MILLION SPECIES HAVE BEEN NAMED AND MILLIONS MORE ARE THOUGHT TO EXIST. WHY ARE SOME SPECIES MORE ALIKE THAN OTHERS? WERE THEY ALL CREATED AT ONCE, OR HAVE SPECIES EVOLVED? THESE QUESTIONS HAVE BEEN ASKED FOR THOUSANDS OF YEARS, BUT IT WAS NOT UNTIL 1859 THAT THE WORK OF CHARLES DARWIN OFFERED ANSWERS.

9.1 Charles Darwin

Objectives

❶ **Summarize Darwin's beliefs about the origin of species before he sailed around the world.**

❷ **Identify two observations from Darwin's voyage that led him to question his beliefs.**

❸ **Describe the two major ideas Darwin put forth in *The Origin of Species*.**

Voyage of the Beagle

Figure 9.1
Below you can see the course of the *Beagle*, the ship in which Darwin sailed around the world. On this voyage, Darwin collected thousands of specimens of plants, animals, and fossils.

On December 27, 1831, H.M.S. *Beagle* sailed from England to survey the coast of South America. On board as the ship's unpaid naturalist was Charles Darwin, a 22-year-old who had graduated from Cambridge University. Darwin's observations during his five years at sea would eventually change the way we think of ourselves and our world.

The son of a wealthy doctor, Darwin was not an attentive student, spending more time outdoors than in school. As a medical student in Edinburgh, he was horrified by operations, which were performed without anesthetic. For two years Darwin skipped lectures to spend time collecting biological specimens. In desperation, his father sent him to Cambridge University to train to be a minister. After graduating from Cambridge in 1831, Darwin was recommended by one of his professors for the position on the *Beagle*. Follow the voyage of the *Beagle* in **Figure 9.1**.

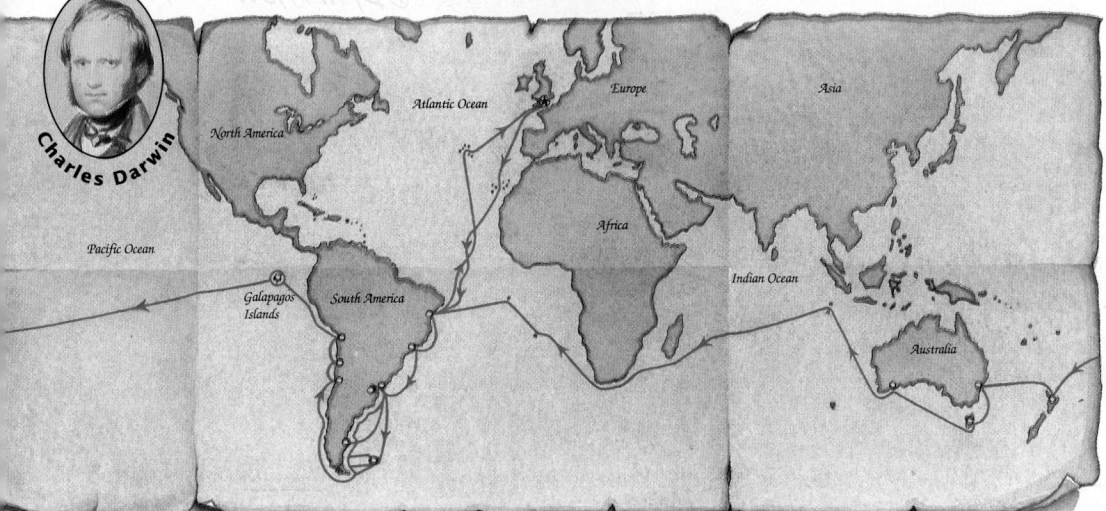

Charles Darwin

North America
Atlantic Ocean
Europe
Asia
Pacific Ocean
Africa
Indian Ocean
Galapagos Islands
South America
Australia

Lesson Plan 9.1

Phase 1

PREPARATION

Key Concepts

- Evolution is the change in species over time.
- In 1859, Charles Darwin proposed a mechanism to explain how evolution occurs.
- Darwin's mechanism, known as natural selection, was derived from his observations of nature conducted during a five-year voyage on the *Beagle*.
- Natural selection is the process in which organisms well suited to their environment survive and reproduce to a greater extent than organisms less suited to the environment.
- Darwin used the finches of the Galapagos Islands as an example of how natural selection operates.

Reading Strategy

Have students recognize that reading in science does not always involve learning the meaning of terms or understanding how certain processes operate. In some cases, as in this section, the reading may focus on the history behind the development of a major scientific theory. Since this passage contains only one boldfaced term, caution students not to skim through the reading of this section. Tell them to read this section much as they would a good novel.

177

Phase 2

TEACHING STRATEGIES

Visual Strategy

Figure 9.2
Have students describe how each finch's bill matches its diet.

WRONG

When the *Beagle* sailed from England, Darwin, like most people of his time, believed in creationism. Creationism is the idea that God was responsible for the creation of new species, or kinds, of organisms. According to creationists, God designed each kind of animal and plant to match its particular habitat. Thus, places with similar environments have similar kinds of plants and animals. Moreover, creationists believe that species are unchanging.

During his journey Darwin often left the ship to collect specimens of animals, plants, and fossils. His observations led him to doubt creationism and its assumption of unchanging species.

Darwin's Finches

Darwin repeatedly saw patterns in how kinds of animals and plants differed, patterns suggesting that species changed over time and gave rise to new species. On the Galapagos Islands, 1,000 km (600 mi.) from the coast of Ecuador, Darwin collected several species of finches. All these species were similar, but each was specialized to catch food in a different way, as shown by the different shapes of the birds' bills in **Figure 9.2.** Some species had heavy bills for cracking open tough seeds. Others had slender bills for catching insects. One finch species even used a twig to probe for insects.

All the species of finches closely resembled one species of South American finch. In fact, all the plants and animals of the Galapagos Islands were very similar to those of the nearby coast of South America. If each one of these plants and animals had been created to match the habitat on the Galapagos Islands, why did they not resemble the plants and animals of islands with similar environments that lie off the coast of Africa? Why did they instead resemble those of the adjacent South American continent? Darwin felt that the simplest explanation was that a few organisms from South America must have migrated to the Galapagos Islands in the past. These few kinds of animals and plants then changed during the years they lived in their new home, giving rise to many new species. Change in species over time is known as evolution.

DEFINITION

Figure 9.2
The blue-black grassquit (inset), native to the Pacific coast from Mexico to Chile, is thought to be the ancestor of the Galapagos finches (below). Darwin attributed the differences in bill size and feeding habits among these finches to evolution that occurred after the birds migrated to the Galapagos Islands.

a   **The woodpecker finch captures insects with its grasping bill.**

b   **The crushing bill of the large ground finch enables it to feed on seeds.**

c   **The cactus finch uses its probing bill to feed on cactuses.**

▪ *Historical Perspective* ▪

A Meeting of the Minds

Like Darwin, Wallace traveled extensively and collected many specimens. Unable to sleep because of a fever one night, Wallace had a sudden thought: What if only those organisms adapted to their environment could survive and reproduce? In that case, he reasoned, "the fittest alone would continue the race." In two days, Wallace wrote a 20-page paper in which he described his theory and gave supporting evidence. When Darwin received Wallace's paper, he was at a loss as to what to do. If Darwin published his own ideas, he felt people would say he stole them from Wallace. Somewhat anxious about the course events had taken, he turned to his friends for advice. A month later, two of his closest friends presented the theory of Darwin and Wallace at a scientific meeting in England.

Alfred Russel Wallace

1859

Figure 9.3
In 1859 Darwin published his famous book, *The Origin of Species*. He accomplished much of his work in his study at Down House in Kent, England (above, right). Darwin is shown at age 73 (above).

Darwin's Mechanism for Evolution

Darwin returned to England in 1836 and for 20 years gathered evidence supporting his ideas about evolution—but he did not publish them. He accomplished much of his work in Down House in Kent, England, shown in **Figure 9.3**. Then in 1858 another biologist, Alfred Russel Wallace, sent Darwin an essay putting forth these same ideas. This drove Darwin to finally publish his work.

When Darwin's book, *On the Origin of Species by Means of Natural Selection*, appeared in November of 1859, it stirred up great controversy. Darwin's conclusion that species changed over time and gave rise to new species contradicted the prevailing beliefs that God created all living things and that living things did not change. Furthermore, the implication that apes were close relatives of humans was unacceptable to many people.

In *The Origin of Species*, as the book is commonly known, Darwin not only presented much evidence that evolution occurred, but he also proposed that natural selection was its mechanism. **Natural selection** is a process by which organisms with traits well suited to an environment survive and reproduce at a greater rate than organisms less suited to that environment.

Because Darwin presented a mechanism as well as evidence for evolution, his arguments were compelling. His views were soon accepted by biologists around the world. Since the rediscovery of Mendel's ideas about genetics in the early 1900s, genetic principles have been added to Darwin's ideas, forming the modern theory of evolution.

Section Review

❶ **Identify two major beliefs of creationism.**

❷ **Describe two observations Darwin made on his voyage that led him to doubt creationism.**

❸ **Explain the two major ideas Darwin presented in *The Origin of Species*.**

Connection: Chapter 1

Evolution—Fact or Theory?
Emphasize that evolution is considered a scientific fact, as documented by several lines of evidence that will be discussed in Section 9.2. Natural selection is a theory Darwin advanced to explain *how* evolution occurs. Have students review how a hypothesis develops into a theory. Evidence that supports Darwin's theory will be covered in Section 9.3.

Phase 3

ASSESSMENT OPTIONS

Closure Strategy
Organisms and Their Environment
Have students explain what would happen to a population of organisms if conditions in the environment were to gradually change, so that those organisms that had been less suited now became more suited.

Section Review
Assign the *Section Review*.

Reteaching
Place photographs of various breeds of dogs on the board. Describe several different environments. Have students select the breeds that would be best adapted to each environment you have described. Have them provide reasons for their choices.

■ *Section Review Answers* ■

1. Creationists believe that their God designed each kind of plant and animal to match its particular habitat, and that species are unchanging.

2. Students should discuss the role of the finches as they explain the patterns Darwin saw in how kinds of plants and animals differed, patterns that suggested species changed over a period of time to give rise to new species.

3. The two major ideas are evolution and natural selection. Students' answers should define evolution and natural selection and indicate that natural selection is the mechanism for evolution.

Lesson Plan 9.2

Phase 1

PREPARATION

Key Concepts
- Evidence for evolution comes from the fossil record, including examples of gradual changes over time.
- Fossils can be dated based on the rate at which radioactive elements decay.
- Evidence for evolution comes from a comparison of homologous structures in different organisms.
- Evidence for evolution comes from vestigial structures.
- Evidence for evolution comes from a comparison of the embryological development of different organisms.
- Evidence for evolution comes from a comparison of molecules, especially DNA, from different organisms.

Reading Strategy

Have students assume the roles of jurors as they read this section. Tell them to write down all the evidence they can gather from their reading to support the statement that evolution has occurred.

Phase 2
TEACHING STRATEGIES

Demonstration 1
Fossils
Show students examples of fossils such as casts and molds.

Theme Answer
Stability
The evolution of a species represents many changes.

MUCH OF WHAT YOU HAVE READ SO FAR IN THIS BOOK WAS UNKNOWN TO DARWIN WHEN HE WROTE *THE ORIGIN OF SPECIES*. THE NUCLEIC ACIDS DNA AND RNA HAD NOT BEEN DISCOVERED. THE STRUCTURE OF PROTEINS WAS UNKNOWN. MENDEL'S GENETIC EXPERIMENTS WERE UNPUBLISHED. THESE AND OTHER DISCOVERIES HAVE SINCE CONTRIBUTED ADDITIONAL EVIDENCE FOR EVOLUTION.

9.2   The Evidence for Evolution

Objectives

❶ Describe the conditions necessary for fossils to form.

❷ List one example in the fossil record indicating that evolution has occurred.

❸ Explain how comparisons of organisms can reveal evidence of evolution.

❹ Describe the important evidence for evolution found in proteins and DNA.

Understanding the Fossil Record

Stability

Explain why the evolution of one species from another can be considered a case of instability.

More than a century has passed since Darwin's death in 1882. During this period, a great deal of new evidence has accumulated supporting the theory of evolution, much of it far stronger than that available to Darwin and his contemporaries. This evidence has come from a variety of sources, including studies of fossils, comparisons of the structures of organisms, and the rapidly expanding knowledge about DNA and proteins.

Figure 9.4
This photograph shows the fossil of a fish that formed in limestone in Brazil 110 million years ago.

Fossils are any traces of dead organisms

What are fossils? Most people think of fossils as shells or old bones. Actually, fossils are any traces of dead organisms. Tracks of dinosaurs, footprints of human ancestors, insects trapped in sticky tree sap, impressions of leaves or skin, and animals buried in tar are fossils. A photograph of a fossil is shown in **Figure 9.4**. For fossils to form, very special conditions are necessary. If a skeleton or shell is to fossilize, the dead animal must be buried by sediment. Burial usually only occurs on the ocean floor, in swamps, in mud, or in tar pits. Calcium in the bone or in the shell is slowly replaced by other harder minerals. Unless the sediment is very fine and no oxygen is present to promote decay, soft tissues such as those found in skin or muscle do not fossilize.

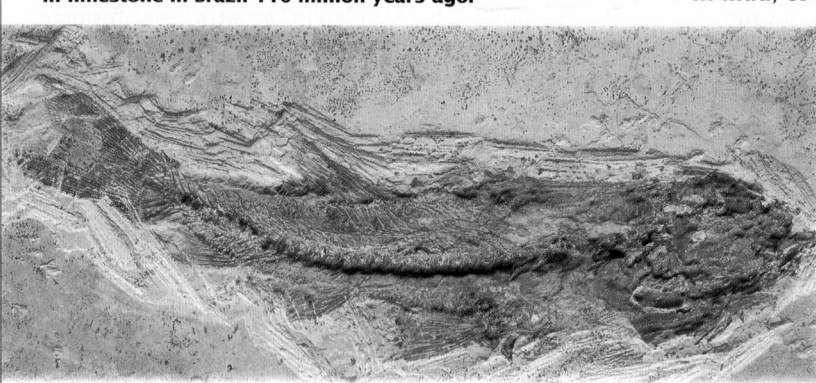

How Fossils Are Dated

[handwritten: 4.5 BILLION YRS OLDS]

Since the late 1940s scientists have been able to determine the ages of rocks and fossils by measuring the amount of radioactive decay, or breakdown, of radioactive atoms in the rock. A radioactive atom contains an unstable combination of protons and neutrons. Since it is unstable, a radioactive atom will eventually change into a more stable atom of another element. For example, carbon-14, a rare form of carbon found in tiny amounts in living things, decays into nitrogen. The term **half-life** describes how long it takes for one-half of the radioactive atoms in a sample to decay. For example, the half-life of carbon-14 is 5,730 years. Thus, a sample that initially contained 12 g of carbon-14 will have 6 g of carbon-14 left after 5,730 years and 3 g of carbon-14 left after 11,460 years. Since carbon-14 decays relatively rapidly, other isotopes with longer half-lives are more often used to date fossils.

Because the rate of decay of a radioactive element is constant, scientists can use the amount of radioactive element remaining in a rock or fossil to determine its age. This technique is called **radioactive dating**.

Evolution is a very slow process; the transformation of one species into another requires at least thousands of years. Using radioactive dating, scientists have determined that the Earth is about 4.5 billion years old, ancient enough for all species to have been formed through evolution.

Transitional forms link new species to old

Because new species form from existing species, Darwin predicted that transitional forms, intermediate stages between older and newer species, would be found in the fossil record. When *The Origin of Species* was published, no intermediates had been found. Darwin recognized that this was a weakness in his theory. But there are now many good examples of evolutionary transitions. For instance, modern horses are descendants of dog-sized animals with four toes on each front foot and three toes on each back foot—modern horses have only one toe per foot. As you can see in **Figure 9.5**, fossil intermediates between this 60 million-year-old ancestor of the horse and the modern horse reveal a history of slow transformation. Over time, the length and size of the limbs increased, and the number of toes on each foot decreased until only one toe was left. A detailed picture of horse evolution would be somewhat more complicated than that shown in **Figure 9.5**, but there is no doubt that a general trend toward the modern one-toed horse did occur.

Figure 9.5
The evolution of the modern horse began with *Hyracotherium* (below) about 60 million years ago. Notice the change from four toes to one toe on each front foot.

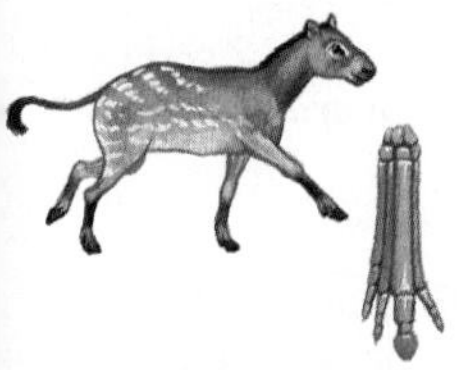

a *Hyracotherium*
60 million years ago

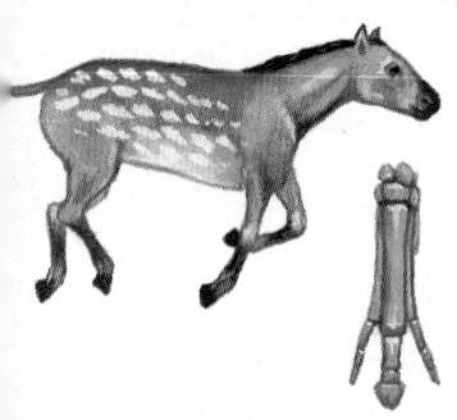

b *Mesohippus*
About 30 million years ago

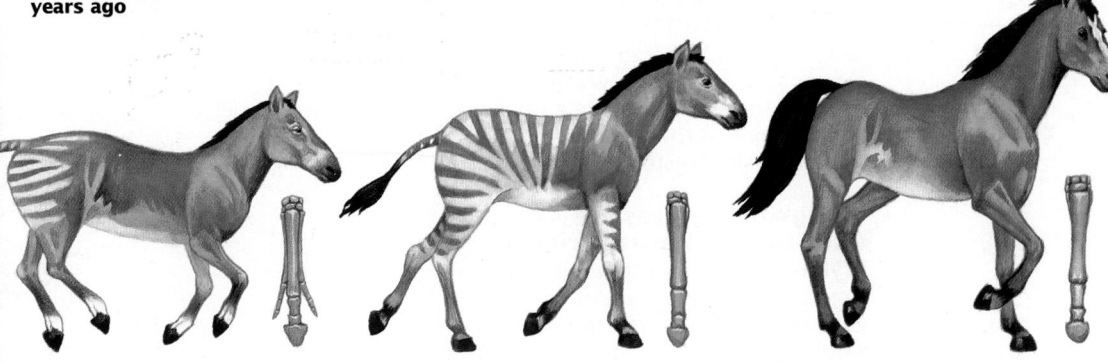

c *Merychippus*
About 20 million years ago

d *Pliohippus*
14 million to 7 million years ago

e *Equus* (modern horse)
10,000 years ago

■ *Matter of Fact* ■

Examples of radioactive isotopes with longer half-lives than carbon-14 include uranium-235 (700 million years), potassium-40 (1.25 billion years), uranium-238 (4.5 billion years), and tin-124 (more than a hundred thousand trillion years).

[handwritten: NB →]

Connection: Chapter 2

Basic Chemistry

Review the structure of an atom so that students have a better understanding of how a radioactive element decays. Point out that as carbon-14 decays into nitrogen-14, it emits particles and energy in the form of radioactivity.

Demonstration 2

Circles and Half-Life

Use colored chalk to draw 40 circles on the board. Above the circles, indicate that each circle represents a carbon-14 atom starting at time 0. Beneath these circles, draw 20 circles using the same color as above, and 20 circles of a different color. Above these circles indicate that 20 circles represent carbon-14 atoms and the other 20 represent nitrogen-14 atoms that would be present after 5,730 years, or one half-life. Continue for another half-life to show 10 circles representing carbon-14 and 30 circles representing nitrogen-14 after 11,460 years have passed.

Demonstration 3

Pennies and Half-Life

Place 100 pennies in a large, flat, covered container. Shake the pennies several times, open the cover, and remove either the *heads* or *tails*. Record the number of pennies removed and those remaining. Repeat the process until no pennies are left in the tray. If *heads* were removed the first time, then *heads* must be removed every time. Have students graph the results, plotting pennies remaining in the container on the x-axis, and the number of times the container was shaken on the y-axis. Relate the results to the half-life of a radioactive element.

Visual Strategy

Figure 9.5

Ask students to describe differences between the modern horse and *Hyracotherium*, besides those mentioned in the text. Point out that in the modern horse, the single toe is covered by a hoof that aids in running.

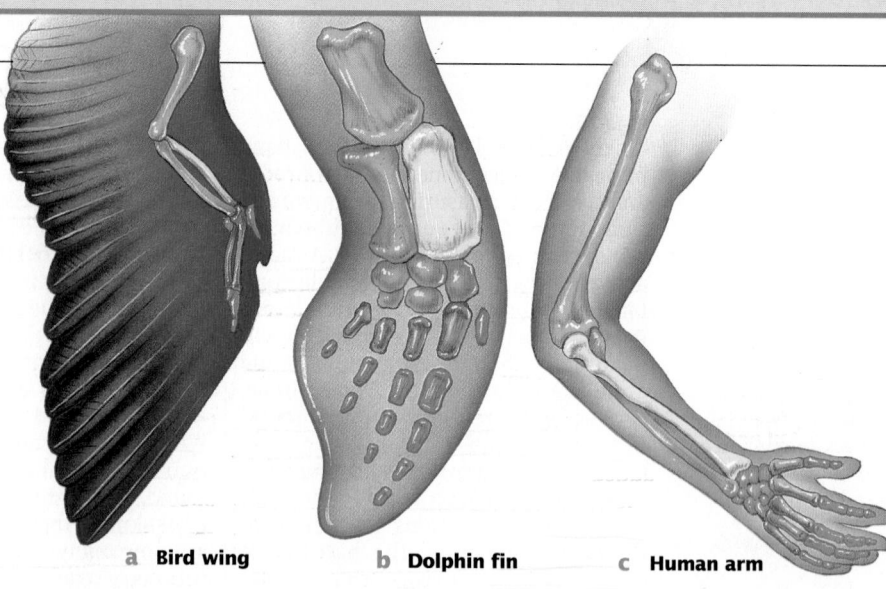

Figure 9.6
The bones in the front limbs of the bird, the dolphin, and the human are homologous structures. Homologous bones are shown in the same color on each diagram.

a   **Bird wing**   b   **Dolphin fin**   c   **Human arm**

Comparing Organisms

Comparing the way organisms are put together provides important evidence for evolution. Your arm appears quite different from the wing of a bird or the front fin of a dolphin. Yet if you examine **Figure 9.6**, you can see that the placement and order of bones in these limbs are very similar. Biologists say that these three limbs are homologous. **Homologous structures** are structures that share a common ancestry. Homologous structures are similar because they are modified versions of structures that occurred in a common ancestor. Although suited for flying, swimming, and grasping, the limbs of the animals above are modified versions of the front fins of their common fish ancestor.

Vestigial structures are clues to evolutionary origins

If you were designing a submarine, would you include a set of wheels in your design? Of course not. Wheels would serve no purpose on a submarine. However, structures without function are found in living things. A whale propels itself with its powerful tail and has no need for hind limbs or

Scale and Structure

Name two structures in your cells that are homologous to structures found in bacterial cells.

the pelvis to which they attach. Nevertheless, whales still have a reduced pelvis that serves no apparent purpose, as shown in **Figure 9.7**. Structures with no purpose are known as **vestigial structures**. Vestigial structures are remnants of an organism's evolutionary past. The whale's pelvis is evidence of its evolution from four-legged, land-dwelling mammals.

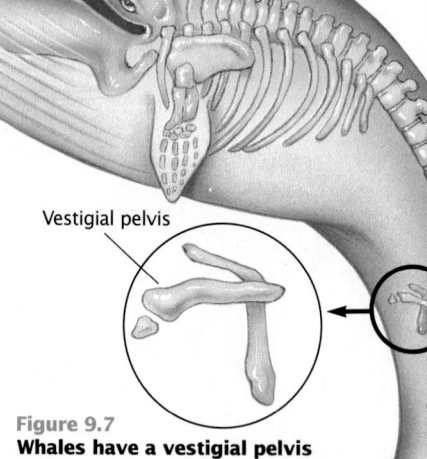

Vestigial pelvis

Figure 9.7
Whales have a vestigial pelvis which serves no apparent purpose. This pelvis is a remnant of the whale's evolutionary past.

Developmental patterns show evolutionary relationships

Much of our evolutionary history can be seen in the way human embryos develop. Early in development, human embryos and embryos of all other vertebrates are strikingly similar, as shown in **Figure 9.8**. In later stages of development, a human embryo develops a coat of fine fur. The similarity of these early developmental forms strongly suggests that the process of development has evolved. New instructions on how to grow have been added to old instructions inherited from ancestors.

Figure 9.8
The five-week-old human embryo (a) and the four-day-old chicken embryo (b) each have a bony tail and gill pouches similar to those of fishes.

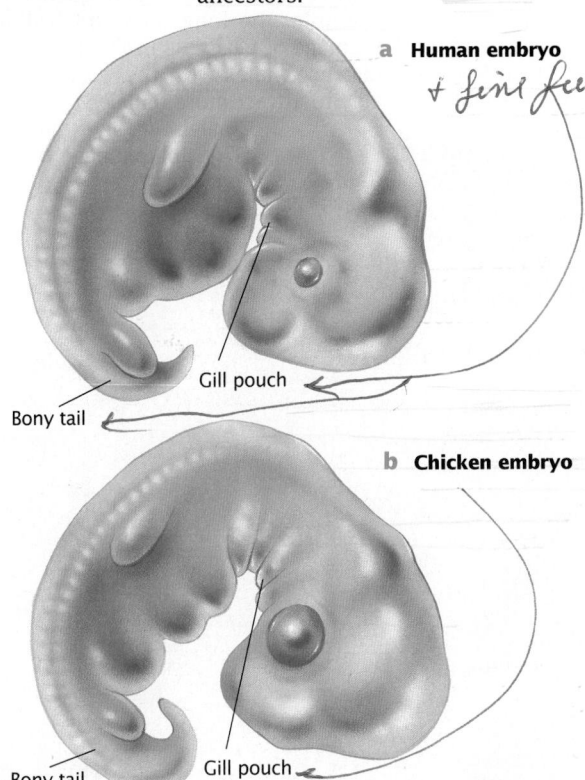

a **Human embryo**

+ fine fur

Gill pouch

Bony tail

b **Chicken embryo**

Gill pouch

Bony tail

DNA and proteins contain evidence of evolution

Although complete fossil histories for living organisms are rare, an organism's history is written in the sequence of nucleotides making up its DNA. If species have changed over time, their genes also should have changed. The theory of evolution predicts that genes will accumulate more alterations in their nucleotide sequences over time. Thus, if we compare the genes of several species, closely related species will show more similarities in nucleotide sequences than will distantly related species. Closely related species also will show more similarities in the amino acid sequences in their proteins. This is because the amino acid sequence in a protein reflects the nucleotide sequence of the gene coding for that protein.

For example, to see how closely related chimpanzees, dogs, and rattlesnakes are to humans, scientists examined the sequence of amino acids in the protein cytochrome *c*, an essential participant in cellular respiration. They found that human cytochrome *c* and chimp cytochrome *c* are identical in all 104 amino acids. This high degree of similarity indicates our very close kinship to chimpanzees. A dog's cytochrome *c* differs from human cytochrome *c* in 13 amino acids, indicating that dogs are fairly distant relatives. But dogs are more closely related to us than are rattlesnakes, whose cytochrome *c* differs from ours in 20 amino acids. In most cases, the evolutionary relationships indicated by DNA or protein sequences confirm those suggested by comparative anatomy and by developmental patterns.

Section Review

1 Why is it unlikely that you will be fossilized?

2 Explain why transitional species, such as the sequence of ancestral horses, are crucial evidence for evolution.

3 How does the whale's vestigial pelvis provide evidence in support of evolution?

4 Explain how sequences of amino acids in proteins can be used to reveal relationships among organisms.

■ *Section Review Answers* ■

1. To be fossilized the dead organism must be buried by sediment in slow-moving bodies of water. Humans are normally placed in coffins and buried in dry areas.

2. Students should explain that the sequence of ancestral forms of horses shows a transitional form to link the old species with the new species.

3. Students should indicate that the whale's vestigial pelvis is evidence of its evolution from four-legged, land-dwelling mammals.

4. Students should explain that closely related organisms have more similar DNA nucleotide sequences and more similar amino acid sequences in their proteins than do distantly related organisms.

Sexual Selection

Using the Feature

Sexual selection is another mechanism by which evolution occurs. Emphasize that sexual selection is not a rival or replacement for natural selection as an explanation of evolution. In fact, both types of selection operate through greater reproduction by individuals with favored traits. In sexual selection, however, traits are advantageous if they promote mating success, even if their effects on survival are neutral or negative.

Discussion

Guide the discussion by posing the following questions.

1. List some traits not mentioned in the feature that might have evolved by sexual selection.
   Possible answers include bright coloration in some songbirds, gaudy coloration in fishes, and swords of swordtails.

2. Are there differences between males and females in these traits?
   Because it usually favors different traits in each sex, sexual selection often leads to sexual dimorphism, phenotypic differences between males and females in structures other than the sex organs. Sexual dimorphism is extreme in many species: a male gorilla weighs nearly twice as much as a female.

3. Identify some examples of sexual dimorphism in humans.
   On average, men are 8 percent taller, 20 percent heavier, and have greater muscle mass. Men also have facial hair and more hair on the body.

4. Why is the beard in men a good candidate for a sexually selected character?
   Like the peacock's tail, the beard seems to be purely ornamental; its insulating value must be small, since women (and beardless men) survive cold climates without one.

5. Describe a control group for Malte Andersson's experiment on widowbirds.
   The control group would have normally sized tails.

The Mating Game

Why does the male peacock (below, right) have brilliant plumage, while the female (below) is brown? Darwin proposed that male peacocks evolved beautiful feathers because males with long tails and bright plumage attracted more mates than males with dull plumage. This evolutionary mechanism is called sexual selection.

The plumage of a male peacock is brilliant green and blue. The peacock's brightly colored, elaborate tail can measure five times its body length. In contrast, the female, or peahen, is drab brown and has a small tail. How have these differences between the sexes evolved? A long tail and bright plumage cannot be essential for survival, since the peahen survives without them. In fact, the male's showiness hampers his survival by making him more visible to predators. It seems that the gaudy plumage of

the peacock should never have been permitted by natural selection.

Charles Darwin proposed the mechanism of sexual selection to account for the evolution of traits like the peacock's tail. In sexual selection, traits that enhance the ability of individuals to acquire mates increase in frequency. For example, the long tails and brilliant plumage of peacocks probably evolved because males with these traits were able to attract more females than were males with short tails and dull plumage.

Action ■ Science in Action ■ Science in Action ■ Science

Sexual Selection at Work

competition for MATES

What kinds of traits are advantageous for acquiring mates? Darwin noticed that in most species males compete for opportunities to mate with females. This competition may involve direct interactions between males, such as threatening displays or combat. Therefore, traits that make males more intimidating or better at combat will be favored by sexual selection. Examples of these traits include large body size, antlers of deer, manes of lions, horns of bighorn sheep, and the oversized jaws of the beetle shown at the left.

Male–male competition can also take subtler forms. A male can gain a reproductive advantage over other males by interfering with their reproduction. In some species of worms, butterflies, and snakes, for instance, the male seals the female's reproductive tract after mating so that other males cannot mate with her.

Darwin also recognized that in many species females choose their mates. Consequently, males with characteristics preferred by females are chosen more often and so have more offspring than males with less attractive traits. Therefore, traits attractive to females become more common.

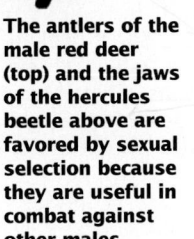

The antlers of the male red deer (top) and the jaws of the hercules beetle above are favored by sexual selection because they are useful in combat against other males.

Female Choice in Selecting a Mate

Darwin's suggestion that female choice could influence evolution was initially rejected by most scientists. However, recent research has supported Darwin's proposal. For instance, zoologist Malte Andersson of the University of Gothenburg, in Sweden, studied female preference in the long-tailed widowbird of Kenya. Like the peacock, the male widowbird has an extremely long tail, which is about 0.5 m (19 in.) in length. The female's tail, in contrast, is only about 7 cm (less than 3 in.) long. Andersson shortened the tails of one group of males by clipping their tail feathers. He glued the tail pieces removed from these males onto the tails of another group of males. After releasing the birds, Andersson found that four times as many females settled in the territories of the males with lengthened tails. Since females usually mate with the male that controls the territory they settle in, this experiment indirectly reveals a preference for long tails among female widowbirds.

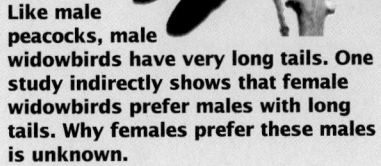

Like male peacocks, male widowbirds have very long tails. One study indirectly shows that female widowbirds prefer males with long tails. Why females prefer these males is unknown.

6. Andersson's study did include a control group of birds: half had their tail feathers clipped and then reattached, the other half were not altered. The clipped and unclipped birds were equally successful at attracting females. What was the purpose of cutting and reattaching the tails of some males?

   It was possible that clipping the tail feathers affected the birds' behavior in some way, perhaps reducing or enhancing their attractiveness. A comparison of clipped and unclipped males in the control group shows that the treatment had no such effect.

7. Andersson's study clearly shows that female widowbirds prefer males with elongated tails. Form a hypothesis explaining why widowbirds do not have longer tails.

   The most likely hypothesis is that males with longer tails suffer higher mortality—perhaps they are more apparent to predators or less proficient at flying.

8. How could this hypothesis be tested?

   Artificially lengthen the tails of some males, then compare the death rate of these males to that of males with normal tails.

185

Lesson Plan 9.3

Phase 1

PREPARATION

Key Concepts

- Natural selection is the process by which organisms adapted to their environment survive and reproduce at a greater rate than those organisms that have not adapted.
- Natural selection involves variation within a population, overproduction of individuals, a struggle for existence, survival, and reproduction of those organisms over time.
- Natural selection can explain the change that has occurred in the coloration of peppered moths in Britain.
- Whereas Darwin suggested natural selection occurs at a very slow, constant rate, recently scientists have proposed that evolution can occur at an irregular rate, in a process known as punctuated equilibrium.

Reading Strategy

Most students read without taking any notes. Tell students that of the total time they spend reading through this section, about 80 percent of the time should be devoted to actually reading, while the remaining 20 percent should be spent taking notes on what they are reading.

HAS NATURAL SELECTION AFFECTED YOUR LIFE DIRECTLY? YES, BECAUSE YOUR BODY HAS BEEN SHAPED BY NATURAL SELECTION. FOR EXAMPLE, THE ABILITY OF YOUR EYES TO FOCUS, THE WAY YOUR HANDS GRIP OBJECTS, YOUR UPRIGHT POSTURE, YOUR LARGE BRAIN, THE COLOR OF YOUR SKIN, AND NUMEROUS OTHER CHARACTERISTICS ARE ALL RESULTS OF EVOLUTION BY NATURAL SELECTION.

9.3   *Natural Selection*

Objectives

❶ Describe how natural selection occurs.

❷ Summarize the effects of natural selection on the peppered moth and on the sickle cell allele.

❸ Describe how natural selection can lead to the formation of new species.

❹ Contrast the hypotheses of punctuated equilibrium and gradualism.

How Natural Selection Causes Evolution

Darwin not only stated that evolution has occurred, he also proposed its mechanism—natural selection. The key factor in natural selection is the environment. The environment presents challenges that only individuals with particular traits can meet. Thus, the environment "selects" which organisms will survive and reproduce. Traits possessed by organisms successful at survival and reproduction are more likely to be transmitted to the next generation. These traits, therefore, will become more common. Compare the modern giraffe in **Figure 9.9** with its short-necked ancestors in **Table 9.1**. This table explains in detail how natural selection could drive the evolution of long-necked giraffes from short-necked ancestors.

Figure 9.9
How did giraffes evolve long necks? Natural selection is the mechanism by which their long necks may have evolved.

Table 9.1 The Process of Natural Selection

Steps	Explanation	Example
Variation is the raw material for natural selection.	Every species contains genetic variation: individuals differ because they carry different alleles for a certain trait. As you learned in Chapter 6, mutation is the source for new variation. In addition, sexual reproduction and crossing over produce individuals with unique combinations of alleles.	Giraffes were born with alleles for varying neck lengths. Some had long necks, some had short necks.
Living things face a constant struggle for existence.	Organisms produce more offspring than can survive. These offspring emerge into a hostile world where they must evade predators and compete with other individuals for limited supplies of food and living space.	Giraffes with longer necks could reach the leaves in tall trees. Those with shorter necks could not.
Only some individuals survive and reproduce.	Some individuals survive the challenges of life better than others. Perhaps a particular allele makes them more drought tolerant, or more efficient, or more resistant to disease. These individuals are more likely to survive and produce offspring.	The giraffes with longer necks were better at getting food than were short-necked giraffes. Consequently, long-necked giraffes produced more offspring than did giraffes with short necks.
Natural selection causes genetic change.	Each generation consists of offspring of individuals that successfully reproduced. Thus, it contains an increased proportion of individuals with traits that promote survival and reproduction than did the previous generation. The same is true for subsequent generations. Over time, the alleles for successful traits will increase in frequency, while alleles for traits that reduce survival and reproduction will decline in frequency.	Since more long-necked than short-necked giraffes were being born, over time, long-necked giraffes became common, and short-necked giraffes became rare. Eventually, long-necked giraffes replaced short-necked giraffes.
Species adapt to their environment.	Selection tends to make a population better suited to its environment. The environment determines the direction of genetic change. An allele favored in one environment may not be favored in another. *TEST*	Long-necked giraffes are well adapted for browsing on the foliage of tall trees, which is out of the reach of most other animals.

TEACHING STRATEGIES

Visual Strategy

Table 9.1

Lead students through each step of the natural selection process as described in this table. Also use this table to review some terms previously covered: mutation, allele, phenotype, and genotype. Emphasize that natural selection acts on the phenotype, in this case long necks in giraffes. Stress that in the struggle for existence, organisms often do not physically compete with one another, a misconception held by most students. Rather, those with certain traits are more likely to survive and reproduce, passing the genes for those traits on to their offspring.

Demonstration 1

The Process of Natural Selection

Display a photograph of a group of animals alongside a photograph of an environment where such an animal is not likely to be found. For example, juxtapose photographs of camels and tundra. Have students identify any traits that would be advantageous for the animals to have in such an environment. Examples might include camels with thicker wool coats to protect against the cold.

Mathematics Connection

Overproduction

Not long after returning to England, Darwin read a paper by Thomas Malthus. The paper, which first appeared in 1798, warned that the human population was increasing so rapidly that it would soon be impossible to feed everyone. Darwin realized that Malthus' conclusion was true for all species, not just humans. In fact, Darwin did some mathematical calculations and determined that a single pair of elephants could be the start of a population of 19 million elephants in 750 years, if all offspring survived.

Indiv do NOT adapt
Species adopt

Visual Strategy

Figure 9.10

Have students identify which moth is adapted to the environment shown in this figure. Ask to what type of environment would the dark moth be adapted. Consequently, make students aware that any variation may be an adaptation—it all depends on the time and place.

Theme Connection

Interacting Systems

The study of ecology uncovers adaptations organisms possess to survive and reproduce in their environment. To illustrate the interaction between organisms and their environment, draw two circles on the board. Label one *variations*, and the other *environment*. Next, draw the two circles so that they overlap. Place the label *adaptations* in the area of overlap. Have students describe how the relative size of this overlap would change in populations that vary in the degree to which they are adapted to their environments. Have them explain what the small area of overlap may indicate about a species' future.

Visual Strategy

Table 9.2

Lead students through the information in this table so they see that the environment determines whether or not a particular trait is an adaptation. Recently, England has instituted pollution control measures. Forests near cities are once again becoming covered with lichens. Have students hypothesize what Kettlewell would find today if he repeated his experiments near Birmingham.

The Peppered Moth: Natural Selection in Action

Over many generations natural selection gradually changes a species in response to the demands of its environment. **Adaptation** is the process by which a species becomes better suited to its environment. The word "adaptation" can also refer to any change in a trait that increases the likelihood an organism will survive or reproduce. For another example of natural selection in action, look closely at the light and dark peppered moths in **Figure 9.10**. Until the 1850s, dark gray peppered moths were rare, and were treasured by British butterfly and moth collectors. Almost all peppered moths were cream-colored. Around 1850, however, dark peppered moths started to become more common, usually in heavily industrialized areas. By 1950, peppered moth populations near industrial centers consisted almost entirely of dark individuals.

Why did the frequency of dark peppered moths increase? Darwin's theory of evolution by natural selection suggests a hypothesis. The color change coincided with a great increase in the number of factories in England. Previously white tree trunks were blackened by heavy pollution from these factories. Perhaps dark moths sitting on soot-darkened bark escaped being eaten by birds because it was hard for the birds to see the dark moths against the dark background. Light-colored moths, on the other hand, would have stood out against a dark background and would have been easy prey for hungry birds. H.B.D. Kettlewell, a British biologist, tested this hypothesis in the late 1950s. **Table 9.2** describes Kettlewell's experiments.

Figure 9.10
Studies of light and dark peppered moths provide an example of adaptation.

Table 9.2 How Kettlewell Demonstrated Natural Selection in Peppered Moths

	What Kettlewell Did	What Kettlewell Found
Step 1	Kettlewell knew that coloration was an inherited trait. He raised large numbers of both light and dark moths in the laboratory. He released equal numbers of light and dark moths into a forest near Birmingham, England, where trees were blackened by soot. For later recognition, each of these moths was marked with a dot on the underside of its wings.	Kettlewell set out rings of traps to recapture moths that survived. He found that two-thirds of the recaptured moths were dark. More moths that matched the dark tree trunks had survived.

	What Kettlewell Did	What Kettlewell Found
Step 2	Again using marked moths, Kettlewell released equal numbers of light and dark moths into an unpolluted forest in Dorset, England. Trees here were white, not black.	When Kettlewell set out traps, two-thirds of the moths that were recaptured were light. Again, more moths that matched the color of the tree trunks survived.

	What Kettlewell Did	What Kettlewell Found
Step 3	Kettlewell set up hidden cameras in both forests to record the capture of moths by birds.	Films showed that birds were more likely to capture light moths on the dark trunks near Birmingham. In Dorset, birds were more likely to eat the dark moths, which were obvious against the light trunks.

The Puzzle of Sickle Cell Anemia

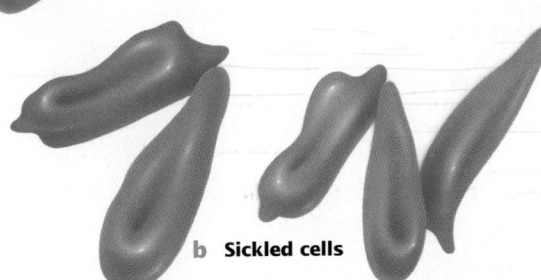

a Oxygenated red blood cells

Increasing level of oxygen in the bloodstream

b Sickled cells

Figure 9.11

When oxygen levels in the blood are high, the red blood cells of people homozygous for the sickle cell allele appear normal (a). But when oxygen levels are low, some of their cells collapse into a sickle shape (b) and cannot function normally. In people heterozygous for the sickle cell allele, only a few red blood cells sickle at low oxygen levels. Yet these individuals are malaria resistant.

Sickle cell anemia, as you learned in Chapter 6, is a hereditary disease that affects hemoglobin molecules, the proteins in human blood that carry oxygen. This trait is apparent in the cells in **Figure 9.11b**. Persons homozygous for the defective sickle cell allele have sickle cell anemia and often die at an early age. Heterozygous individuals, who have both a defective and a normal allele, are healthy. The disease is now known to have originated in central Africa, where 1 in 100 people is homozygous for the defective allele and thus has the disease. In the United States, however, only 1 African-American out of every 500 has sickle cell anemia, and the disease is almost unknown in other races.

Why has natural selection not acted against the sickle cell allele in Africa and reduced its frequency? Why is this potentially fatal allele very common?

The sickle cell allele confers an unexpected advantage in Africa

The defective allele is common in central Africa because people who are heterozygous for the sickle cell allele are much less susceptible to malaria. Over one million people die of malaria each year, most of them in central Africa and tropical Asia.

In central Africa, the number of deaths from sickle cell anemia is far lower than would occur from malaria if heterozygous individuals were not malaria-resistant. One in five individuals is heterozygous for the sickle cell allele and survives malaria. But only 1 person in 100 is homozygous for the allele and dies of anemia. Stated simply, natural selection has favored the sickle cell allele in central Africa because the payoff in survival of heterozygotes is greater than the price in death of homozygotes.

You can see that natural selection is acting on the sickle cell allele in opposite directions. On one hand, selection tends to eliminate the sickle cell allele because of its lethal effects on homozygotes. On the other hand, selection tends to favor the sickle cell allele because it protects heterozygotes from malaria. Biologists use the term **balancing selection** to refer to the situation in which two opposing selective forces affect the frequency of an allele in a population.

Selection does not favor the sickle cell allele in the United States

Why is the frequency of individuals who are homozygous for sickle cell anemia so much lower among African-Americans? After all, most African-Americans descended from Africans who arrived in this country hundreds of years ago. The answer lies in the distribution of malaria. Malaria is a tropical disease that does not occur in North America today. Therefore, the protection against malaria that is associated with being heterozygous for the sickle cell allele offers no real advantage in the United States. As a result, the allele for sickle cell anemia has become far less common in the United States than it is in Central Africa. Biologists call unopposed selection **directional selection**. Directional selection moves the frequency of a particular allele (and that of the trait the allele produces) in one direction.

Connection: Chapter 6

Inheritance of Sickle Cell Anemia
Use a Punnett square to review what offspring are expected from two parents heterozygous for sickle cell anemia.

Connection: Chapter 7

DNA and Protein Synthesis
Review how DNA directs the synthesis of proteins. Point out that hemoglobin contains 574 amino acids. The entire structural difference between normal hemoglobin and the sickle cell molecule consists of two amino acids. In two of the four chains that make up the molecule, glutamic acid has replaced valine in the sixth position along the chain.

Demonstration 2

Balancing Selection
Draw the three possible genotypes for the sickle cell trait as SS, Ss, and ss. Ask students what is likely to happen to people in Africa with each of these genotypes. Place an *X* over the SS genotype to indicate those who die from malaria and an *X* over the ss genotype to indicate those who die from sickle cell anemia. Point out that natural selection tends to eliminate the sickle cell allele in one situation (ss) but tends to favor it in another situation (Ss), resulting in balancing selection both for and against the s allele.

Visual Strategy

Figure 9.11
Have students compare the shapes of normal and sickled red blood cells. Discuss what is likely to happen when sickled cells attempt to pass through narrow capillaries.

▪ Cultural Perspective ▪

Is All Variation Adaptive?
Based on his research on population genetics and molecular evolution, Motoo Kimura has proposed the neutral theory of evolution. Kimura argues that much of the variation in natural populations is selectively neutral, that is, the different forms of traits neither help nor hinder survival and reproduction. According to Kimura, changes in the frequency of these traits are due to chance, not to natural selection.

189

RACES

Demonstration 3

How Many Species

Review the definition of a species. Show students photographs of viceroy and monarch butterflies. Ask whether they are members of the same or different species to demonstrate that structural similarities are not the criteria used to define a species. Do the same with photographs of a horse and a donkey, pointing out that their offspring (mules) are sterile.

Visual Strategy

Figure 9.12

Have students identify the genus and species names for these seaside sparrows. Ask why three, rather than the usual two, Latin names are given in this case.

How Species Form

Because natural selection favors changes that increase an organism's chances of surviving and reproducing, it will continuously shape a species to improve the fit between the species and its environment. Recall from Chapter 1 that a species is a group of individuals that can interbreed and produce fertile offspring, but that cannot breed with any other such group. When populations of a species are found in several different kinds of environments, selection will act to make each population suit its particular environment.

Populations in different places thus become increasingly different, as each becomes better suited to the particular challenges of living where it does.

Separate populations of a species can eventually become quite distinct, if their environments differ enough. These populations form what biologists call ecological races, as shown by the example of the sparrows in **Figure 9.12**. **Ecological races** are populations of the same species that differ genetically because they have adapted to different living conditions.

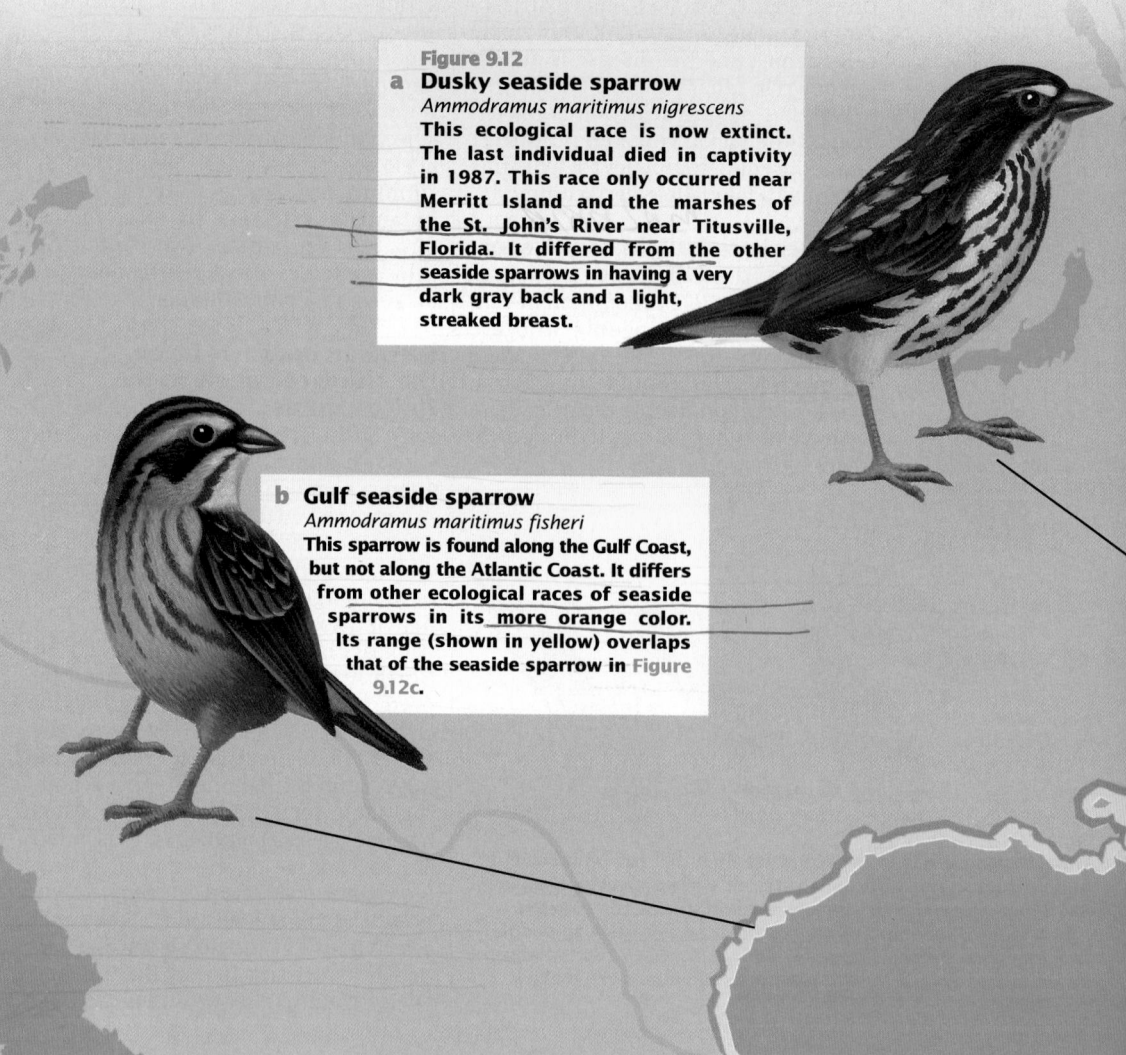

Figure 9.12

a Dusky seaside sparrow
Ammodramus maritimus nigrescens
This ecological race is now extinct. The last individual died in captivity in 1987. This race only occurred near Merritt Island and the marshes of the St. John's River near Titusville, Florida. It differed from the other seaside sparrows in having a very dark gray back and a light, streaked breast.

b Gulf seaside sparrow
Ammodramus maritimus fisheri
This sparrow is found along the Gulf Coast, but not along the Atlantic Coast. It differs from other ecological races of seaside sparrows in its more orange color. Its range (shown in yellow) overlaps that of the seaside sparrow in **Figure 9.12c**.

PUNCTUATED EQUILIBRIUM :: FOSSIL RECORD : TEST
VS GRADUALISM DISCONTINUOUS

Members of different ecological races are not yet different enough to belong to different species, but they have taken the first step. The differences among human races, while they may seem large to some of us, are actually very small in an evolutionary sense. Notice that each sparrow in **Figure 9.12** has a three-word scientific name, which indicates a subspecies, a distinct group within a species.

Ecological races form new species

Ecological races often become increasingly different. The accumulation of differences between species or populations is called **divergence**. Divergence occurs because natural selection favors different survival strategies in different environments. Eventually, races can accumulate so many differences that biologists consider them separate species.

TEST

Demonstration 4

Divergent Evolution

Use a map of the Galapagos Islands to discuss how divergent evolution occurs. The 13 species of finches described earlier are believed to have evolved from a small ancestral group that came from the South American mainland. Once they arrived, the original population spread out to the various islands, where varying selection pressures would have favored divergent evolution. Ask students what would have happened if all the islands had the same type of environment.

c Seaside sparrow
Ammodramus maritimus maritimus
This sparrow is widely distributed along the Atlantic and Gulf coasts from Massachusetts to southern Texas. It is a common bird, noted for making "crash landings" into marsh grasses. This is by far the most common and widespread of the four ecological races of seaside sparrow. Its range is shown in pink.

d Cape Sable seaside sparrow
Ammodramus maritimus mirabilis
This olive-green seaside sparrow is found only in Cape Sable, on the southwestern tip of Florida. It is a very local ecological race. No other type of seaside sparrow lives in the area.

191

Demonstration 5
Gradualism vs. Punctuated Equilibrium
Display the box scores for two football games that end with the same total points scored by both teams. However, one score should show points being scored in each quarter, while the other should show points scored in the first and last quarter. Have students select which game could represent gradualism, and which could represent punctuated equilibrium.

Phase 3

ASSESSMENT OPTIONS

Closure Strategy
Natural Selection in Insects
When DDT was first introduced, it was highly effective as an insecticide. Over time, however, DDT became less and less effective. In fact, many populations of insects are now resistant to DDT. Have students explain how natural selection acted in this case.

Section Review
Assign the *Section Review.*

Reteaching
Sharks and alligators are considered living fossils since they have undergone little change over millions of years. Have students discuss why these organisms have not evolved to any appreciable extent.

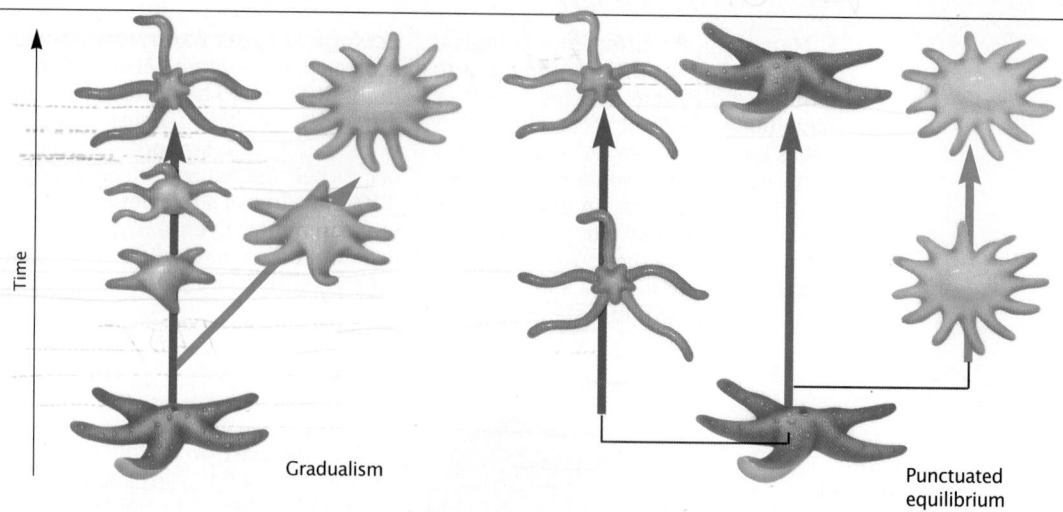

Gradualism

Punctuated equilibrium

Figure 9.13

a According to the hypothesis of gradualism, the starfish evolved at a slow, constant rate.

b According to the hypothesis of punctuated equilibrium, starfish would have evolved at an irregular rate, forming new species in short bursts, followed by long periods of constancy.

Does Evolution Occur in Spurts?

Biologists are now engaged in a debate about the rate at which evolution proceeds. Following Darwin's lead, most biologists have assumed that species formation is a slow, gradual process that goes on all the time. The hypothesis that evolution occurs at a slow, constant rate is known as **gradualism**.

Recently, some biologists have challenged gradualism. These scientists argue that major environmental upheavals (for example, a comet striking the Earth or volcanoes erupting) have had a major impact on species formation. Such upheavals caused large-scale extinctions that left many habitats vacant and open for colonization by other species. These drastic changes did not occur often and were separated by quiet periods lasting tens of millions of years. Consequently, species formation occurred rapidly after major upheavals. (Keep in mind that a rapid occurrence in geological time lasts many thousands of years.) Short periods of rapid species formation were followed by long periods during which little evolution occurred. The hypothesis that evolution occurs at an irregular rate through time is known as **punctuated equilibrium**.

Punctuated equilibrium predicts that the fossil record should be very discontinuous and that fossils should exhibit little evidence of change over long periods of time. Transitional forms should be rare because new species evolve so rapidly. Is this prediction supported by fossil evidence? There is considerable disagreement among biologists on this point. Some groups of organisms appear suddenly in the fossil record, as if they had evolved very rapidly, and then remain almost unchanged for millions of years. Other groups show gradual change, as predicted by gradualism. Compare these two hypotheses in **Figure 9.13**.

Section Review

❶ Describe the steps of natural selection.

❷ The British have instituted pollution controls on factories. How do you think this will affect the evolution of the peppered moth?

❸ Describe how natural selection can lead to the formation of new species.

❹ Contrast the hypotheses of punctuated equilibrium and gradualism.

■ Section Review Answers ■

1. Students should summarize the steps described in Table 9.1 on page 187.

2. Tree trunks will likely be returning to their natural white color. Dark peppered moths should therefore become less common.

3. Students should explain divergence and how ecological races can form new species.

4. Gradualism is the hypothesis that evolution occurs at a constant rate. Punctuated equilibrium is the hypothesis that evolution occurs at an irregular pace.

Chapter **9** *Highlights*

Using radioactive dating, scientists found that this fossil of a duck-billed dinosaur skull dates back to the late Cretaceous period, about 125 million years ago.

	Key Terms	Summary
9.1 Charles Darwin  The observations of a young Charles Darwin changed our ideas about the world.	natural selection (p. 179)	• Darwin's observations led him to doubt the beliefs of creationism. • Darwin proposed that new species formed by the slow transformation of existing species. • Darwin proposed that natural selection caused evolution. • Natural selection occurs because organisms with traits that enable them to survive are more likely to leave offspring.
9.2 The Evidence for Evolution  Fossils show that modern horses evolved from dog-sized animals with four toes on each front foot.	half-life (p. 181) radioactive dating (p. 181) homologous structure (p. 182) vestigial structure (p. 182)	• Fossils are preserved traces of dead organisms. Fossils form only under specific conditions. • Radioactive dating enables us to determine the ages of fossils. • The fossil record provides evidence that older species gave rise to more recent species. • Organisms with homologous structures share common ancestry. • Closely related organisms have more similarities in their DNA than do distantly related organisms.
9.3 Natural Selection  Dark peppered moths became increasingly common when pollution darkened forest trees in England.	adaptation (p. 188) balancing selection (p. 189) directional selection (p. 189) ecological race (p. 190) divergence (p. 191) gradualism (p. 192) punctuated equilibrium (p. 192)	• Adaptation is the process by which organisms become well suited to their environments. • In response to the darkening of tree trunks by pollution, some peppered moth populations evolved from cream-colored to dark gray. • In Africa, the sickle cell anemia allele is favored because it produces resistance to malaria in heterozygotes. • Gradualism is the hypothesis that evolution occurs at a constant rate. Punctuated equilibrium is the hypothesis that evolution occurs at an irregular rate.

Chapter Review Answers

Understanding Vocabulary

1. **a.** Adaptation is a process whereby a population becomes better suited to its environment. Natural selection is the process in which organisms with traits well suited to the environment survive and reproduce at a greater rate than organisms less suited to the environment.

   b. Homologous structures are those that share a common ancestry, while vestigial structures are structures with no apparent purpose but that are considered evidence of an organism's evolutionary past.

   c. Balancing selection involves equilibrium between selective forces in a population, as in the case of the sickle cell allele, whereas unopposed selection is called directional selection.

   d. Gradualism is characterized by the slow accumulation of small changes over long periods of time, while punctuated equilibrium involves long periods with few evolutionary events followed by a period of rapid change.

Relating Concepts

2. Map answer is shown on page 175D.

Understanding Concepts

Multiple Choice

3.	a	8.	b
4.	c	9.	d
5.	a	10.	b
6.	a	11.	a
7.	a		

Completion

12. natural selection, The Origin of Species
13. Galapagos, finches
14. divine creation
15. fossils
16. half-life
17. vestigial structures
18. ecological races

Short Answer

19. The dead organism is buried by sediment. The calcium that composes the organism's skeleton or shell is

194

Understanding Vocabulary

1. For each pair of terms, explain the differences in their meanings.
   a. adaptation, natural selection
   b. homologous structures, vestigial structures
   c. balancing selection, directional selection
   d. gradualism, punctuated equilibrium

Relating Concepts

2. Copy the unfinished concept map below onto a sheet of paper. Then complete the concept map by writing the correct word or phrase in each oval containing a question mark.

Understanding Concepts

Multiple Choice

3. Darwin was prompted to publish his ideas by
   a. the arrival of Wallace's essay.
   b. his declining health.
   c. permission from the Queen.
   d. encouragement from his family.

4. Darwin concluded that new species arose from
   a. divine creation.
   b. nonliving matter.
   c. existing species.
   d. extinct species.

5. Which of the following is not an example of directional selection?
   a. sickle cell anemia in Africa
   b. color change in peppered moths
   c. sickle cell anemia in the United States
   d. evolution of the modern horse

6. Evolutionary relationships among species can be deduced from
   a. nucleotide sequences in DNA.
   b. scientists using small clocks.
   c. radioactive atoms.
   d. ecological races.

7. Which species' cytochrome *c* is most similar to our own?
   a. chimpanzee   b. dog
   c. rattlesnake   d. poppy

8. The mechanism proposed by Darwin to explain how evolution occurs is
   a. speciation.   c. adaptation.
   b. natural selection. d. punctuated equilibrium.

9. Using radioactive dating, scientists have determined that
   a. 90 percent of the earth was once covered by water.
   b. the earth is about 10 billion years old.
   c. carbon-14 has a half-life of 500 years.
   d. the Earth is old enough for evolution to have produced all species.

10. Which of the following is *not* a factor in natural selection?
    a. genetic variation
    b. evolution of the individual
    c. survival
    d. struggle for existence

11. Studies of the peppered moth in England illustrate the operation of
    a. adaptation.   c. speciation.
    b. gradualism.   d. punctuated equilibrium.

Completion

12. Darwin proposed the mechanism of _____ _____ to explain evolution. This mechanism is described in his book entitled _____ .

13. On the _____ Islands, Darwin witnessed an example of evolution in the beaks of small birds called _____ .

14. Before his voyage on the *Beagle*, Darwin believed that _____ best explained the origin of life on Earth.

15. Naturally preserved remains or traces of dead organisms are called _____ .

16. The time required for one-half of the radioactive atoms in a sample to decay is its _____ _____ .

---

replaced by hard minerals. Typically, only organisms that lived in the ocean or in swamps or bogs are preserved as fossils.

20. that whales descended from organisms with hind limbs

21. Gradualism is the hypothesis that evolution occurs at a constant rate. Punctuated equilibrium is

the hypothesis that evolution occurs at an irregular pace.

Interpreting Graphics

22. • *Mesohippus, Merychippus, Equus*
    • There would be no transitional form to connect the old species with the new.

Reviewing Themes

23. The kind of food eaten by each species of finch was dictated by its bill. For example, birds with heavy bills cracked open tough seeds, while insects were eaten by birds with slender bills.

24. The organisms share a common ancestry.

17. Human structures with no apparent purpose include the tail bone and wisdom teeth. These structures are called _____ _____ .

18. Populations of the same species that differ genetically as a result of their different living conditions are called _____ _____ .

Short Answer

19. What conditions are necessary for an animal fossil to form?

20. What does the whale's vestigial pelvis reveal about the evolution of whales?

21. Explain how the hypotheses of gradualism and punctuated equilibrium differ.

Interpreting Graphics

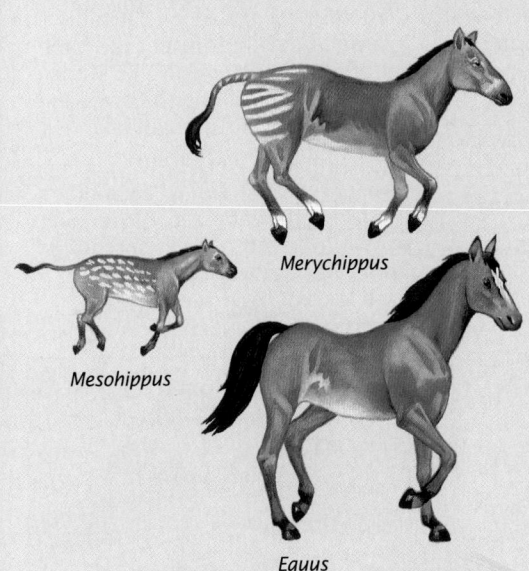

Merychippus

Mesohippus

Equus

22. These sketches show three stages in the evolution of the horse.

- List the three stages in their correct evolutionary order.

- Suppose that radioactive dating showed that *Equus* was older than *Mesohippus*. How would this discovery weaken the evidence for evolution provided by this example?

Reviewing Themes

23. *Scale and Structure*
Darwin noted that the Galapagos finches had specialized bills. What relationship did Darwin observe between the birds' bill types and the ways in which they caught food?

24. *Scale and Structure*
The bones in the human arm are very similar to the bones in a bird's wing and a dolphin's flipper. What does this similarity in structure tell about the ancestry of these animals?

25. *Patterns of Change*
What evidence supports the idea of evolution as an ongoing process?

Thinking Critically

26. *Inferring Conclusions*
Punctuated equilibrium supports the notion that transitional forms of life are rare in the fossil record because evolution occurred so rapidly. What else could explain the absence of transitional forms?

27. *Building on What You Have Learned*
In Chapter 7 you learned that DNA is a molecule made up of nucleotides that code for amino acids. What can biologists learn about two species by comparing their DNA sequences?

Cross-Discipline Connection

28. *Biology and Art*
Darwin observed how different finches adapted to a specific environment. Considering Darwin's observations, identify an environment and then draw a picture showing the human adaptations needed to match that environment.

Discovering Through Reading

29. Read "In the Beginning," in *Discover*, October, 1990, pages 98–102. What conclusions about the origin of life on Earth did Dr. J. William Schopf draw from his examination of the Precambrian rocks?

Discovering Through Reading

29. They found a filamentous fossil that appears to be an ancestral form of a modern cyanobacteria. The discovery is important because the filamentous fossil is likely an ancestor of all life on earth. He concluded that life on Earth began about 4 billion years ago and microorganisms were the only form of life for about 2 billion years. These microorganisms were oxygen-producing photosynthetic forms. The oxygen they released enabled non-bacterial life forms to survive.

25. The rise in frequency of dark peppered moths and the decline in frequency of the sickle cell allele in the United States are two evolutionary events that have occurred in the last few centuries.

Thinking Critically

26. Transitional forms may be rare because of the incompleteness of the fossil record.

27. DNA sequences can be used to deduce the relationships between species.

Cross-Discipline Connection

28. Students' drawings should show an adaptation that enables the human to survive in a special niche.

Prelab Preparation Answers

- Natural selection is a process in which organisms with traits well suited to the environment survive and reproduce at a greater rate than organisms less suited to the environment.
- Explain that variation is the raw material on which natural selection acts.
- Natural selection favors the changes that increase an organism's chances of surviving and reproducing. It continuously shapes a species to improve the fit between the species and its environment.

Procedure Answer

5. It ensures that beads below a certain size will not be "captured." Thus, there is selection on bead size.

Chapter 9

Investigation

How Does Natural Selection Work?

Objectives

In this investigation you will:
- *model* the process of natural selection

Materials

- beads of five different sizes and shapes
- masking tape
- meter stick

Prelab Preparation

Review what you have learned about natural selection by answering the following questions:
- What is natural selection?
- What is the role of variation in natural selection?
- How does natural selection lead to adaptation?

Procedure: Modeling Natural Selection

1. Make a data table like the one shown below.

Prey Type	Survivors of each generation			
	First	Second	Third	Fourth
1				
2				
3				
4				
5				

2. Form a cooperative group of four students. Work with one member of your group to complete steps 3–13.

3. With your partner, use tape to mark a 50 × 50 cm square on the table. The square will represent the habitat of one predator and five kinds of prey. Beads will be the prey, and you and your partner will take turns being the predator.

4. Select 20 beads, four of each of the five different sizes. Close your eyes. Have your partner spread the 20 beads throughout the habitat. Each kind of bead represents a different prey type.

5. To hunt, have your partner position your hand so that it is approximately 5–10 mm above the table surface at one edge of the square. Slowly move your hand across the square, trying to keep it at the same height above the surface. *Why is it important to keep your hand at a constant height?*

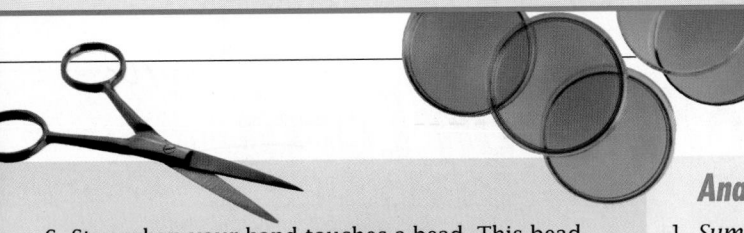

6. Stop when your hand touches a bead. This bead has been captured. Your partner should remove it.

7. Continue moving your hand across the square, stopping to capture each bead you touch. If you reach the edge of the square, have your partner reposition your hand at a new spot along the edge of the square, then start hunting again.

8. Continue hunting until 50 percent of the prey have been captured.

9. Count the survivors of each kind of prey. Record these data in the column labeled "First" in your data table.

10. The prey now reproduce. Assume each survivor has one offspring. Add the appropriate number of beads to the habitat.

11. Conduct another round of hunting, again stopping after 50 percent of the prey have been "eaten." Count the survivors and record their numbers in your data table in the column labeled "Second."

12. Now allow the prey to reproduce. Again, each survivor produces one offspring. Distribute the offspring in the habitat.

13. Exchange roles with your partner and repeat steps 4–12, recording the number of survivors in the third and fourth generations.

14. Pool the data for both teams in your group. Plot the number of survivors of each kind of prey over the four generations in a graph.

Analysis

1. *Summarizing Data*
   Describe what happened to the number of each type of prey over the four generations.

2. *Analyzing Data*
   What evidence suggests that certain types of prey are better adapted to the habitat than others?

3. *Making Inferences*
   What adaptation of the prey contributes most to their chance of survival in this habitat?

4. *Making Inferences*
   Describe an example in nature in which larger size might be a disadvantage.

5. *Making Inferences*
   Why might the shapes of similarly sized beads affect your ability to capture them?

Thinking Critically

1. Suppose that you slid your hand along the surface of the table, instead of moving it slightly above the surface. How might this different way of hunting change your results?

2. In what way does this experiment differ from events that might occur in a natural habitat?

Analysis Answers

1. Small beads should increase in frequency, and large beads should decrease in frequency.

2. Beads larger than 5-10 mm are likely to be captured, and so decline in frequency. Beads smaller than 5-10 mm are passed over, and so increase in frequency.

3. Small size is favored.

4. Answers may vary, but should be logical. For example, large grasshoppers may have a greater risk of predation from birds than do smaller grasshoppers.

5. Answers may vary, but might suggest that the shape of an object might affect your ability to feel it.

Thinking Critically Answers

1. There is no selection in this case, because all bead sizes are equally likely to be captured. Any increases or decreases in frequency of different bead sizes are due to chance.

2. Students should point out some of the unrealistic features of this simulation. For instance, hunting continues until the arbitrary value of 50 percent of the prey have been captured.

Chapter 10 History of Life on Earth

Planning Guide

	Objectives/Themes	Classwork Resources	Homework Resources
10.1	**1.** Contrast the three explanations for the origin of life. **2.** Describe the importance of the Miller-Urey experiment. **3.** Summarize the reasons scientists think RNA, not DNA, was the first genetic material. **4.** Describe how the first cells might have evolved. **Themes:** Energy and Life, Evolution	**Teacher's Resource Binder** Focus Activity 10 *Analyzing Adaptations: Living on Land* **Other Resources** Transparencies 40–41	**Text** Section Review, p. 202 **Teacher's Resource Binder** Directed Reading Worksheet 10.1 **Other Resources** Audiocassette 10.1*
10.2	**1.** Recognize the great age of the Earth. **2.** Compare and contrast the two major groups of bacteria. **3.** Identify the major change in the early atmosphere caused by bacteria. **4.** Describe the evolutionary relationships between prokaryotes and eukaryotes. **Themes:** Evolution, Interacting Systems, Patterns of Change, Scale and Structure	**Text** Journeys *Journey Into the Past* pp. 208–209 **Other Resources** Transparency 42	**Text** Section Review, p. 205 **Teacher's Resource Binder** Directed Reading Worksheet 10.2 **Other Resources** Audiocassette 10.2*
10.3	**1.** Explain how ozone was critical to the development of life on Earth's surface. **2.** Recognize how the relationship between plants and fungi enabled both to invade the land. **3.** Identify the importance of flight in the evolution of insects. **4.** Describe the role insects and plants played in each other's evolutionary success. **Themes:** Evolution, Interacting Systems, Patterns of Change	**Text** Journeys *Tour of a Dinosaur* p. 213	**Text** Section Review, p. 210 **Teacher's Resource Binder** Directed Reading Worksheet 10.3 **Other Resources** Audiocassette 10.3*
10.4	**1.** Identify the defining characteristic of vertebrates. **2.** List three adaptations that enabled amphibians to colonize land. **3.** Compare the ability of a reptile to survive on land with that of an amphibian. **4.** Identify three differences between mammals and reptiles. **Themes:** Evolution, Patterns of Change, Scale and Structure	**Text** Investigation *Can Life Arise From Nonlife?* pp. 218–219 **Teacher's Resource Binder** Lab Investigation 10.1 *Animals of the Future* Extension Worksheet *The Evolution of the Horse*	**Text** Section Review, p. 214 **Teacher's Resource Binder** Directed Reading Worksheet 10.4 Vocabulary Review Worksheet* Reteaching Worksheet* *History of Life on Earth* **Other Resources** Audiocassette 10.4*

*Reteaching Options

Demonstrations
10.1: p. 202
10.2: p. 205
10.3: p. 207
10.4: p. 212

Assessment
Chapter Review pp. 216–217
Portfolio Assessment p. 197D
Chapter Test—Teacher's Resource Binder
Test Generator

Research Notes

Connection to Chemistry: Self-Replicating Molecules

Although scientists like Cech (mentioned on page 201) have made great strides in demonstrating that RNA molecules are capable of self-replication, many others believe that life arose from even simpler molecules that could self-replicate.

In labs across the world, scientists are trying to design self-replicating chemical systems that are even simpler than RNA. The goal is to show that life possibly could have arisen first from some other self-replicating precursor. As discussed in the chapter, the actual mechanism involved in the development of life cannot be measured scientifically.

Research is focusing on molecules that can act as catalysts for the formation of similar molecules. Chemists have identified several sets of molecules that have this property. In this case, two different molecules, A and B, bond to form a third molecule, C.

The shape of C is such that it has weak attractions (such as hydrogen bonds) for additional A and B molecules. When they are attracted to the C molecule, they are held together so that they bond, forming another C molecule. Then, the attractions are broken, and the old and new C molecules are ready to assist in making more Cs out of A and B.

In 1992, an important breakthrough was announced. Scientists at the Massachusetts Institute of Technology had already designed two sets of precursor molecules for two different self-replicating molecules. But when they "recombined" these two systems by mixing one precursor from one set with another precursor from the other set, the compound formed was a faster self-replicator than either of their original self-replicating molecules. When they mixed the other precursors, they found that the other "recombinant" molecule was unable to self-replicate.

Earlier work had shown that individual self-replicating molecules could be changed or "mutated" by exposure to ultraviolet light, and that some of these molecules were faster self-replicators than the original molecules. These results indicate that some sort of natural selection could sort out the early replicators. Successful self-replicators would make more and more copies of themselves, and rapidly outpace less successful molecules.

Such studies can help us to understand what features of life on Earth are essential for any life, and which are merely peculiarities brought about through the conditions on Earth or some other means.

Related work is underway in computer science, in which different self-replicating "programs" compete for memory space in a computer. More insights on the different types of competition involved in the chemical and computer systems may help scientists understand the balances in the ecosystems of nature.

Investigation Notes

Can Life Arise From Nonlife?

pages 218–219

Purpose: This investigation gives the students an opportunity to collect, analyze, and discuss experimental data related to the origin of life under current conditions on Earth. The students recreate the essentials of Pasteur's experiments on spontaneous generation.

Prelab Preparation

1. A large quantity of glassware is required for this investigation. If necessary, it can be conducted as a class demonstration with data being collected from one set of flasks.

2. Be sure to have a lubricant such as glycerin or petroleum jelly for use when inserting tubes through stoppers. Wear heavy gloves to avoid puncture injury.

3. The flasks should be sterilized in an autoclave or a pressure cooker at 15 pounds (psi) of pressure for 15 minutes. If neither an autoclave nor a pressure cooker is available, boiling may produce adequate results, but expect all the flasks to eventually show contamination.

Answers will be found on pages 218–219.

Meeting Individual Needs

Objectives

1. Students will demonstrate appropriate use of core vocabulary for the chapter (Vocabulary File).

2. Students will observe the chemical properties of lipids and aqueous solutions and relate them to the role of the lipid membrane in living cells (Demonstration A).

3. Students will describe the pattern of increasing complexity in the history of life on Earth through visual activities. (Teaching Strategy A).

Vocabulary File
(Developing Vocabulary/ Limited English Proficiency)

If you are not already using the Vocabulary File, refer to Chapter 1 for its preparation. See the Chapter Highlights on page 215 for a list of suggested words.

Demonstration A
Making Membrane Models
for pages 202–203
(Developing Modeling Skills/ Visual Learners)

Materials: stoppered flask filled with water, dropper, and oil (such as sesame seed oil, olive oil, or cooking oil)

Procedure: Ask students what separates the cell's contents from its surroundings. Brainstorm until students mention "cell membrane" as an answer. Reteach the role of lipids in the cell membrane, as discussed in Section 3.2.

Unstopper the flask. Drop a few drops of oil into the water. Re-stopper the flask. Shake. Watch oil and water separate, with oil forming small globules.

Some students may relate this to preparing an Italian salad dressing when oil is mixed with vinegar.

Read aloud the first two paragraphs of "Origin of First Cells," page 202. Use this demonstration and discussion to introduce Section 10.2, "Early Life in the Sea."

Teaching Strategy A
Complexity in the Parade of Life
for pages 203–214
(Developing Classification Skills/Tactile Learners)

Materials: models of prokaryotic and eukaryotic cells prepared for Chapter 3, bulletin board with illustrations for each of the five classification kingdoms

These may be either sketches or photographs. If you have a student artist, have him/her draw some illustrations for you. Frequently, teachers' catalogs and promotional material have very colorful illustrations, including representatives from the five kingdoms.

If budget allows, keep several preserved specimens that students may touch and examine that closely follow the parade of vertebrates on page 211, such as in the following list:

    lamprey eel
    hogfish
    small shark
    perch
    frog
    pigeon

Procedure: Refer to the box models of prokaryotic and eukaryotic cells, and recall the material from Chapter 3 (page 60). Place these models in an area visible to all. Have students recall that the boxes are models of two types of cells. Ask students which type of cell is the oldest and why. Lead brainstorm discussion about prokaryotic and eukaryotic characteristics.

Be sure students realize that the formation of additional membranes and a nucleus in the eukaryotes signify a more highly developed cell. Read aloud to the students the first two paragraphs of the section entitled "Dawn of Eukaryotes." Use this to introduce the five classification kingdoms.

Refer to the bulletin board with the names and illustrations of the five kingdoms. Be sure students realize that as living organisms evolved, each step brought about a more detailed development. Have students scan "The Journey Through Time," pages 208–209.

Until this time, the specimens should be kept stored in a stock-room out of sight. Keep them covered with damp paper towels between viewing times. Using the text as a reference, bring each specimen out at the appropriate time in the lesson. Some students have only seen pictures of these animals. For many it will be a bit of a shock, a dose of reality that they find awesome, and that will generate interest. Students love it, and it neatly complements the parade described in the book.

Additional Strategies
Visual Strategies
Pages 200, 201, and 211

Auditory Learners
Use *Biology: Visualizing Life* Audiocassettes for Sections 10.1, 10.2, 10.3, and 10.4.

Meeting Individual Needs (cont.)

Cooperative Learning
Timeline of Life

Timing: Use this activity to introduce Section 10.1.

Group Size: 4 students

Outcome: Students will construct a timeline of the history of life.

Individual Accountability: Each group member is responsible for listing events in the history of life, and the time they occurred, on index cards.

Positive Interdependence: Each group will construct their own timeline of the history of life on Earth.

Assign each group member two of the following events:

    (1) Formation of the Earth

    (2) Earliest Fossils of Bacteria

    (3) Earliest Known Eukaryotes

    (4) Multicellularity Dominant in Cambrian Period

    (5) Plants and Animals Invade Land

    (6) Dinosaurs and Early Mammals Arise

    (7) Evolution of Birds

    (8) Extinction of Dinosaurs.

Have group members scan the chapter for the time periods for their event. They should make an index card with the event and the time it occurred. Then, give each group a 2-m-long piece of string and 8 paper clips. Have groups attach the index cards to the string using the paper clips at distances in proportion to the time elapsed since each event began.

Compare timelines prepared by different groups and discuss any differences found. Point out that humans and apes are believed to have diverged about 5 million years ago. Have students point out where along their timeline this point would occur.

Portfolio Assessment

Students should select their best work and provide a self-reflective rationale for their selections. Students can make selections in the following areas.

1. *Content* — One concept map from the chapter (See page T30 for evaluation criteria.)

2. *Reading Comprehension* — One Directed Reading Worksheet from the Teacher's Resource Binder (Use the answer key to evaluate for accuracy.)

3. *Writing* — Using the Vee Form, summarize a newspaper or magazine article relating to early forms of life. (See page 22T for evaluation criteria.)

    Or: Select a writing task or project from the Chapter Review.

4. *Performance Assessment* — One Vee form from a chapter investigation or lab manual investigation (See page 22T for evaluation criteria.)

Teacher makes selections in the following areas.

1. *Formal Assessment* — Chapter test (Test A, B, or the Test Generator) The teacher-scored test should be reviewed by the student. Incorrect responses should be corrected by the student before the test becomes part of the portfolio.

2. *Informal Assessment* — Use the Direct Observations Checklist, page 33T, during a laboratory or other cooperative learning experience.

3. *Performance Assessment* — Have students research the natural history of a group of species and create an evolutionary tree/timeline indicating how the species are related to one another and how long ago they diverged from one another.

Concept Map Answer

The following is one possible answer to the Relating Concepts exercise on page 216.

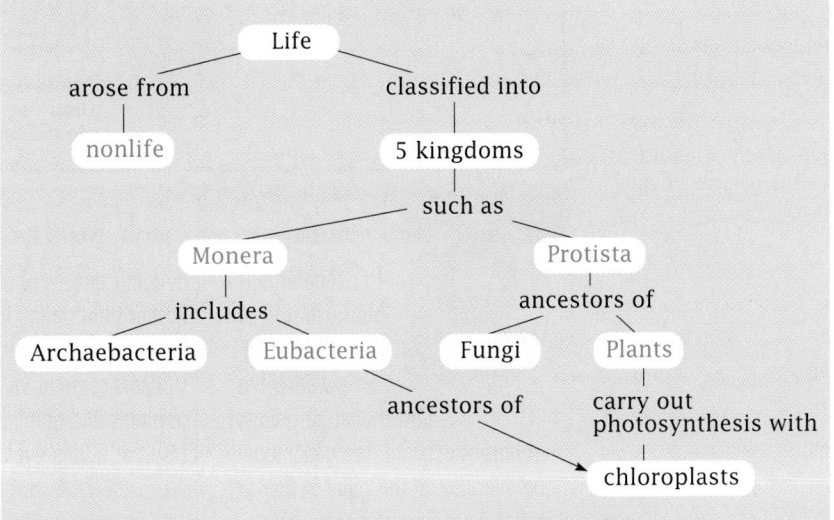

Chapter 10

Determining Prior Knowledge

- Have students suggest various ways in which the first life form may have originated on Earth.
- Assign students to cooperative work groups of four. Have each group construct a model, draw a picture, or assemble a collage that depicts their impressions of the first life form on Earth. Have each group discuss their representation with the rest of the class.
- Show students photographs of a plant, an invertebrate, a fish, an amphibian, a reptile, a bird, and a mammal. Have students arrange these organisms in the order in which they think they evolved.

Chapter 10

History of Life on Earth

Review

- prokaryotes and eukaryotes (Section 3.3)
- structure of RNA (Section 7.2)
- natural selection (Section 9.3)

10.1 Origin of Life
- How Did Life Begin?
- Origin of Life's Chemicals
- Origin of the First Cells

10.2 Early Life in the Sea
- Earliest Life: Bacteria
- Dawn of the Eukaryotes
- Life Blooms in the Ancient Seas

10.3 Invasions of the Land
- The Importance of Ozone
- Plants and Fungi Colonize Land
- Invasion of the Arthropods
- Insects Were the First Flying Animals

10.4 Parade of Vertebrates
- Animals With Backbones

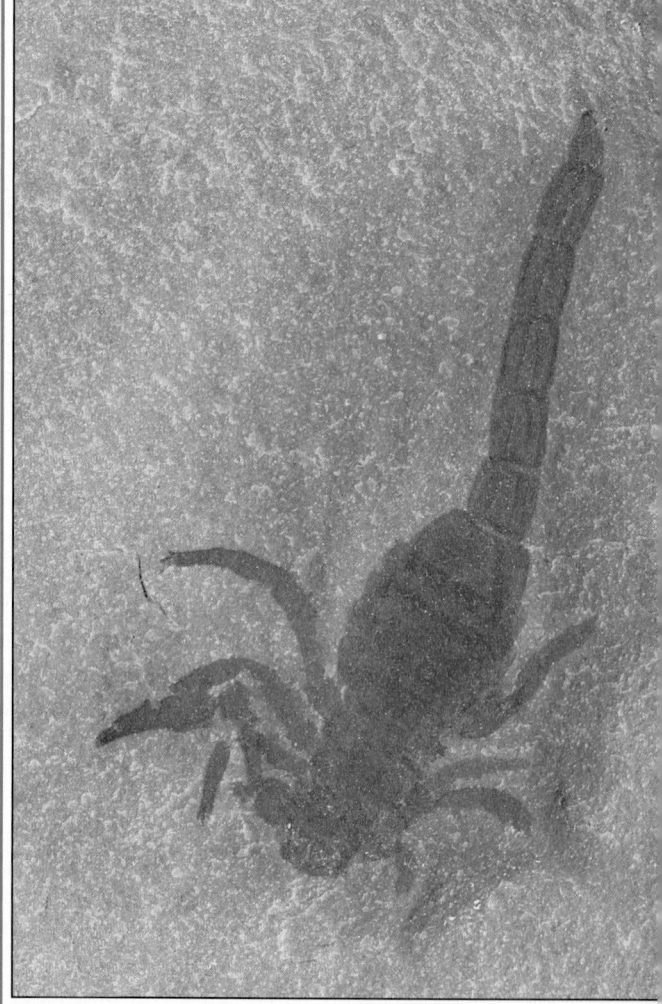

All living things, including this scorpion that lived nearly 410 million years ago, evolved from the same simple life forms that were found on the young planet Earth.

■ Author's Rationale ■

This chapter tells a historical story of life on Earth. An overview describes how the major groups of organisms evolved. The advantage of the overview at this point in the text is that it provides students with a dynamic picture of life's diversity. Rather than seeing the Earth as a large zoo with many types of animals, students come to view the diversity of plants and animals on Earth as the result of a long evolutionary process.

WITH LIFE ALL AROUND US, IT IS DIFFICULT TO IMAGINE A TIME WHEN THERE WAS NO LIFE ON EARTH, WHEN NO GRASSES GREW AND NO FISHES SWAM IN THE SEA. THE EARTH IS MUCH OLDER THAN LIFE, HOWEVER. THE STUDY OF RADIOACTIVE DECAY IN ROCKS REVEALS THAT THE EARTH IS SOME 4.5 BILLION YEARS OLD, 1 BILLION YEARS OLDER THAN THE OLDEST FOSSILS. WHERE DID LIFE COME FROM?

10.1   Origin of Life

Objectives

❶ Contrast the three explanations for the origin of life.

❷ Describe the importance of the Miller-Urey experiment.

❸ Summarize the reasons scientists think RNA, not DNA, was the first genetic material.

❹ Describe how the first cells might have evolved.

How Did Life Begin?

There were no witnesses to the origin of life, and you cannot go back in time to see for yourself. But if you could, what would the possibilities be? In principle, there are at least three ways life could have begun:

1. *Extraterrestrial origin* Some scientists hypothesize that life originated on another planet outside our solar system. Life was then carried here on a meteorite or an asteroid and colonized the Earth. How life arose on other planets, if it did, is a question we cannot hope to answer soon.

2. *Creation* Many people believe that life was put on Earth by divine forces. In this view, common to many of the world's religions, the forces leading to life cannot be explained by science.

3. *Origin from nonliving matter* Most scientists think that life arose on Earth from inanimate matter, after the newly formed Earth had cooled, as shown in **Figure 10.1**. First, random events produced stable molecules that could reproduce themselves. Then, natural selection favored changes in these molecules that increased their rate of reproduction, leading eventually to the first cell.

This chapter will examine the third option. The first two possibilities are not considered because they are not testable and thus fall outside the realm of science. However, the demonstration that life could have originated from nonliving matter does not rule out the other two possibilities. Since these possibilities cannot be tested, either can be accepted as a matter of faith.

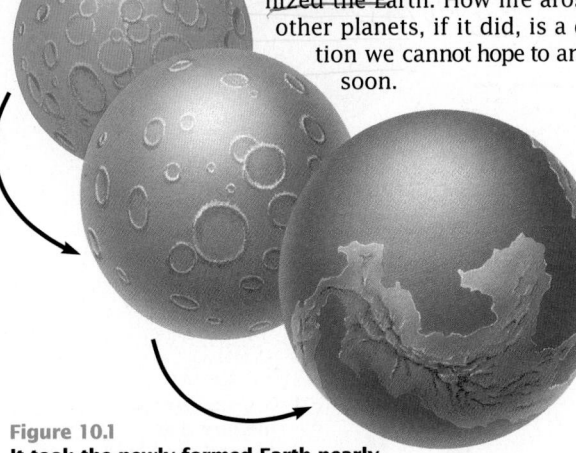

Figure 10.1
It took the newly formed Earth nearly 1 billion years to cool enough for the first life to appear.

Lesson Plan 10.1

Phase 1
PREPARATION

Key Concepts
- The Earth was formed some 4.5 billion years ago.
- Scientists think that life arose on Earth from nonliving matter approximately 3.5 billion years ago.
- Stanley Miller and Harold Urey demonstrated that organic compounds can form from inorganic chemicals under conditions thought to have been present on the primitive Earth.
- RNA was probably the first genetic molecule that could replicate and catalyze chemical reactions.
- The first cells probably arose when membranes formed around aggregates of organic compounds.

Reading Strategy
Show students how to outline their reading. Instruct them to first read this section and then decide what the main ideas are. Have them write down the main ideas in their own words, using Roman numerals for each one. Have them next outline both supporting details and subdetails for each main idea, using the following form:
I. Main Idea
  A. Supporting Detail
    1. Subdetail
  B. Supporting Detail
    1. Subdetail
    2. Subdetail

Phase 2
TEACHING STRATEGIES

Connection: Chapter 1

A Scientific Hypothesis
Review the prerequisite that any valid hypothesis must be testable, thus forming the basis for further investigation. Conse-

quently, both extraterrestrial origin and creation are excluded from further consideration since neither can be tested.

Art Connection

The Primitive Earth

Have students create a drawing, collage, painting, or any other art form that represents their interpretation of what the primitive Earth may have looked like some 4 billion years ago.

Visual Strategy

Figure 10.2

Have students identify which part of this apparatus represents the primitive ocean and which part stands for the primitive atmosphere. Emphasize that this experiment did not prove that these organic compounds were formed under such conditions some 3.5 billion years ago, only that they could have formed.

Connection: Chapter 2

Organic Compounds

All the molecules Miller and Urey used as their building blocks, with the exception of methane, are inorganic compounds. Point out that all the molecules that have been synthesized in their apparatus are organic compounds.

Origin of Life's Chemicals

GAS from volcano

You learned in Chapter 2 that all organisms are composed of the same chemicals, just as all cars are assembled from the same materials. Cars are made from steel, glass, and rubber; cells are made from proteins, nucleic acids, lipids, and carbohydrates. The source of these molecules, the raw materials of life, must be discovered to understand how life originated from nonliving matter.

Life's building blocks can form spontaneously

To find the source of the molecules of life, take a mental journey back 4.5 billion years to the origin of the Earth. The Earth and the rest of the solar system had just condensed from a cloud of interstellar matter. Life was not possible then, because the Earth's surface, which is solid rock today, was so hot it was liquid. By a little less than 4 billion years ago, the surface had cooled enough for a solid crust to form. Water condensed in the atmosphere and fell as rain. At this time, there was no oxygen in the air. Instead, the atmosphere was rich in hydrogen and many other gases that spewed out from the many volcanoes. Energy in the form of lightning and ultraviolet light from the sun stimulated many chemical reactions.

What chemical reactions could occur in this atmosphere? To answer this question, in 1953 Stanley Miller and Harold Urey re-created the conditions of the early Earth inside two connected flasks, as illustrated in **Figure 10.2**. The atmosphere Miller and Urey introduced into the flasks contained simple molecules that were probably found in the early Earth's atmosphere: hydrogen, carbon dioxide, methane, water vapor, nitrogen, ammonia, and carbon monoxide. They heated the mixture and zapped it with electrical sparks to simulate lightning. Within days, a dark, smelly mixture developed. When this mixture was analyzed, Miller and Urey found that many complex molecules had formed, including some amino acids, the building blocks of proteins. Using slightly different combinations of starting molecules, Miller and other scientists have been able to generate many amino acids, nucleotides found in DNA and RNA, as well as lipids, carbohydrates, and ATP.

NO   O₂

carbo   DNA   AA   RNA   ATP

Figure 10.2
This apparatus was used by Stanley Miller, when he was a graduate student studying under Harold Urey, to simulate the conditions present on a young Earth.

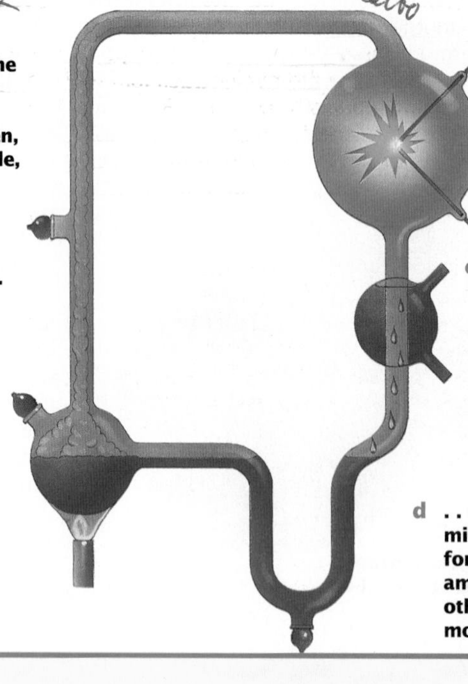

CO_2
CO
N_2
NH_3
CH_4
H_2

a **Water was heated and the water vapor was mixed with hydrogen, carbon dioxide, carbon monoxide, nitrogen, ammonia, and methane.**

b **The mixture of gases was sparked with electricity to simulate lightning.**

c **The gases were cooled using a glass tube filled with circulating cold water, . . .**

d **. . . and the dark mixture that formed contained amino acids and other complex molecules.**

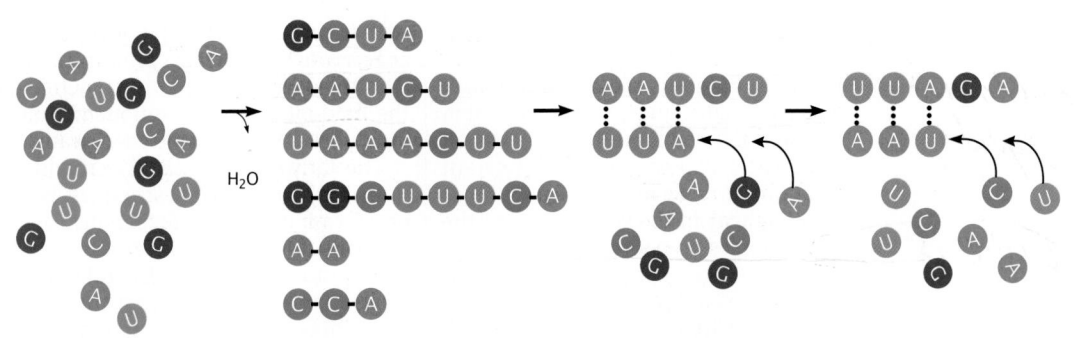

a **Guanine, cytosine, uracil, and adenine may have existed in a random mixture in the ancient seas.**

b **When the water was removed from the environment, perhaps by evaporation, chains of nucleotides formed.**

c **Some of the newly formed nucleotide chains were able to catalyze other chemical reactions, . . .**

d **. . . such as reactions in which nucleotide chains make copies of themselves.**

Connection: Chapter 5

Enzymes
Review the role of enzymes as biological catalysts.

Visual Strategy

Figure 10.3
Have students synthesize a nucleotide chain different from any of those shown in Figure 10.3b. Next have them assemble the complementary strand as illustrated in Figure 10.3c.

Figure 10.3
Under conditions present on the young Earth, nucleotides produced from simple gas molecules may have spontaneously assembled into chains of RNA molecules.

RNA was probably the first genetic molecule

The basic building blocks of life can assemble spontaneously, requiring no force more mysterious than simple chemistry. But there was still a long way to go before life could arise. How did amino acids link like beads in a necklace to form the chains we call proteins? And how did nucleotides join to form long chains of DNA and RNA?

As you learned in Chapter 7, cells link amino acids together into proteins, but only according to instructions written in DNA and carried in RNA. Similarly, cells synthesize DNA and RNA, but only with the aid of enzymes, which are proteins. So how could proteins form without DNA and RNA, and vice versa?

Although nucleic acids are only produced by living things today, small chains of nucleotides will form spontaneously under conditions that were probably found on the early Earth. For instance, short RNA molecules are able to condense from individual nucleotides without the assistance of enzymes.

More important, the chemist Thomas Cech and co-workers at the University of Colorado have recently shown that RNA molecules, like enzymes, can catalyze chemical reactions. In your cells, for example, RNA molecules catalyze the synthesis of other RNA molecules. Cech received the 1989 Nobel Prize in chemistry for this discovery. In addition, other scientists have discovered RNA molecules that are able to produce complementary copies of part of their own nucleotide sequence.

The discovery that RNA can catalyze its own synthesis suggests the following hypothesis. First, RNA nucleotides formed from simple gas molecules in much the same way as in experiments similar to those done by Miller and Urey. Nucleotides then assembled spontaneously into small chains, as shown in **Figure 10.3**. These small chains were able to catalyze chemical reactions, such as protein synthesis, and were able to make copies of themselves. Once replicating molecules like these appear, natural selection and evolution are possible. Molecules that can replicate faster or more efficiently will become more common than slower-replicating or less-efficient molecules.

COACERVATES ✓

Demonstration 1

Coacervates

In a test tube, mix 5 mL of a 1 percent gelatin solution with 3 mL of a 1 percent gum Arabic solution. Add 0.1 M HCl in drops until the solution turns cloudy. Have teams of two students each prepare a wet mount of the solution and examine their slide with a microscope under high power. Have students locate a coacervate and compare it to the one shown in Figure 10.4. Tell them to look for signs of movement, growth, and separating or joining with another coacervate.

Connection: Chapter 7

From DNA to RNA

Most students cannot see how RNA could possibly develop before DNA, especially after learning that genetic information is transferred from DNA to RNA in the process of transcription. Tell students that certain viruses, including HIV (which causes AIDS), contain RNA as their genetic material. This viral RNA is released into a cell where it is translated by an enzyme into DNA.

Phase 3

ASSESSMENT OPTIONS

Closure Strategy

The First Cell

Have students consider whether the first cells were prokaryotes or eukaryotes. Have them support their choice.

Section Review

Assign the *Section Review*.

Reteaching

Have students describe how the work of Miller and Urey exemplifies the scientific method, as illustrated in Figure 1.7 on page 14.

Origin of the First Cells

cell membrane
lipids

LIPIDS

How did the first cells form? The crucial feature that separates the cell from its environment is the cell membrane, which contains lipids, as you learned in Chapter 3. If you mix a lipid such as oil with water you can see an important lipid property. Small spherical bubbles of oil appear in the water because the oil molecules do not mix with the water molecules. Scientists think that similar tiny spheres of lipid may have been the first stage in the origin of the cell. When mixed with water, certain lipids will form a bubble that is called a **coacervate** (*koh AS uhr vayt*), which has a double-layered membrane much like the lipid bilayer of the cell membrane. Figure 10.4 shows some coacervates.

The early oceans probably contained numerous small lipid coacervates, each one forming and then dispersing. Over millions of years, coacervates that could survive longer by taking in molecules and energy from their surroundings would have become more common than the here-today-gone-tomorrow kind. When a means arose to transfer this ability to "offspring" coacervates, probably through self-replicating RNA, life had begun.

If RNA was the first genetic material, when did DNA evolve? Most scientists think DNA evolved after simple cells had arisen. The advantage of DNA over RNA as a genetic material may have been that DNA ensured the safety of the hereditary information by storing it in a central location.

Our vision of the origin of life is incomplete. No scientist has been able to create life from nonlife in the laboratory. Yet many of the important steps that might have led to living cells have been worked out. Also remember that scientists have been performing experiments regarding the origin of life for fewer than 50 years. Simple molecules were combining in the ancient seas for hundreds of millions of years before life resulted.

Figure 10.4
These spheres are coacervates, small spheres that form when lipids are mixed with water. Coacervates have a double-layered membrane similar to the cell membrane.

Section Review

❶ **Explain why the hypothesis that life arose from nonliving matter is a scientific hypothesis.**

❷ **What were the results of the Miller-Urey experiment?**

❸ **Describe the evidence that suggests RNA was the first genetic molecule.**

❹ **Describe one hypothesis that explains how the cell membrane might have evolved.**

■ *Section Review Answers* ■

1. Students should explain that the origin of life from nonliving matter is a scientific hypothesis because it is testable, such as in experiments like those of Miller and Urey.

2. Many complex molecules formed, including some amino acids, the building blocks of proteins.

3. RNA molecules, like enzymes, can catalyze chemical reactions and are able to produce complementary copies of part of their own nucleotide sequence.

4. It is believed that the cell membrane formed when lipids mixed with water and formed a bubble called a coacervate, a double-layered membrane much like the lipid bilayer of the cell membrane.

**SOON AFTER THE EARTH'S SURFACE COOLED, LIFE AROSE IN THE
ANCIENT SEAS. THE FIRST ORGANISMS TO APPEAR ON THE PLANET
WERE BACTERIA, WHICH ARE SINGLE-CELLED PROKARYOTES. THESE
EARLY BACTERIA ARE THE ANCESTORS OF MODERN BACTERIA AND OF
ALL THE MANY DIFFERENT KINDS OF ORGANISMS LIVING TODAY,
INCLUDING YOU.**

10.2   Early Life in the Sea

Objectives

❶ Recognize the great age of the Earth.

❷ Compare and contrast the two major groups of bacteria.

❸ Identify the major change in the early atmosphere caused by bacteria.

❹ Describe the evolutionary relationships between prokaryotes and eukaryotes.

Earliest Life: Bacteria

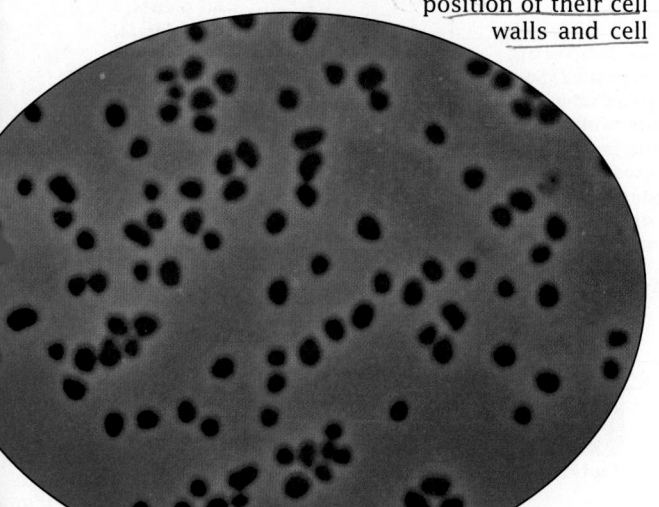

**Figure 10.5
These archae-
bacteria produce
methane, other-
wise known as
"swamp gas."**

Several independent studies estimate the Earth's age as 4.5 billion years. So far, no fossils have been found in the oldest rocks, which are about 3.8 billion years old. The oldest fossils that have been discovered occur in 3.5 billion-year-old rocks that were once sediments on the ocean floor. The tiny fossils that were found in these ancient rocks were bacteria.

Biologists separate the bacteria into two groups that differ in the composition of their cell walls and cell membranes, and in the structure of some of their proteins. Because the differences between these two groups are so great, it is likely that they diverged early in the history of life. One group is the **eubacteria** (*YOO bak TIHR ee ah*) or "true bacteria," the most common bacteria today. Most living bacteria, including those that cause disease and decay, are eubacteria. Eukaryotic cells contain mitochondria and chloroplasts, which are probably descendants of ancient eubacteria.

Representatives of the second group of bacteria, the **archaebacteria** (*AHR kee bak TIHR ee ah*) or "ancient bacteria," are shown in **Figure 10.5**. Archaebacteria are rare today. They are found mainly in hostile environments where conditions resemble those of the early Earth (such as hot springs, very salty lakes, and swamps). The cell wall, cell membrane, RNA polymerase (the enzyme that copies DNA into messenger RNA), and ribosomes of archaebacteria are much more like those of eukaryotes than are those of eubacteria. Most biologists now think that archaebacteria are direct ancestors of eukaryotes.

Key Concepts

- Bacteria are divided into two groups: eubacteria (the majority of today's bacteria and the ancestors of mitochondria and chloroplasts) and archaebacteria (the ancestors of eukaryotes).
- About 3 billion years ago, a group of photosynthetic eubacteria known as cyanobacteria evolved and released oxygen into the oceans and atmosphere.
- About 1.5 billion years ago, the first eukaryotes, known as protists, evolved from bacteria. In turn, protists gave rise to more complex life forms, including plants, animals, and fungi.
- From about 600 million to 500 million years ago, a time known as the Cambrian period, almost all the major groups of animals evolved.

Reading Strategy

Much of what is introduced in this section will be discussed in more detail in later chapters. Consequently, this section can help students practice using the index when reading. Have them identify three topics introduced in this section and use the index to find additional information on these subjects.

Phase 2

TEACHING STRATEGIES

Connection: Chapter 3

Evolution of Eukaryotes
Remind students that scientists think certain bacterial *trespassers* became trapped inside other cells, gradually losing their independence. Some 1.5 billion years ago, these invading bacteria evolved into organelles such as the mitochondria and chloroplasts.

[Handwritten top margin: O₂ from oceans. ↑ ATMOSPHERE (SATURATED)]

Connection: Chapter 5

Photosynthesis

Cyanobacteria carry out photosynthesis much like plants. However, they lack the chloroplasts of plant cells. Instead, their photosynthetic pigments are located on folded membranes in the cytoplasm. These pigments include green chlorophyll and blue phycocyanin, giving these organisms their blue-green color.

Theme Connection

Evolution

Use Figure 10.6 to emphasize that the classification of organisms into five kingdoms is based primarily on evolutionary relationships.

[Handwritten left margin:]
1 MONERANS — are prokaryotic
2 PROTISTS: algae, FUNGI
3 Plants
4 Animal — 9-10 phylum
5 Fungi

[Handwritten: 21% O₂   O₂ from BACTERIA]

Though small, bacteria changed the atmosphere

The modern atmosphere is about 21-percent oxygen. Where did this oxygen come from, since the early Earth's atmosphere lacked oxygen? It was produced by bacteria. About 3 billion years ago, a group of photosynthetic eubacteria known as **cyanobacteria** (*sy uh noh bak TIHR ee ah*) evolved. As they carried out photosynthesis, cyanobacteria released oxygen gas into the oceans. After hundreds of millions of years, when the waters of the ancient oceans had soaked up all the oxygen they could hold, the oxygen produced by photosynthesis began to bubble out of the oceans into the air. Over the billions of years that followed, more and more oxygen was added to the air until the modern composition of the atmosphere was achieved.

Dawn of the Eukaryotes

For about 2 billion years, bacteria were the only living things on Earth. About 1.5 billion years ago, the first eukaryotic cells evolved from bacteria. Fossils of early eukaryotes show that they were much larger than bacteria, and that they had internal membranes.

Based on evolutionary relationships, biologists classify living things into five great kingdoms: monerans, protists, plants, animals, and fungi. Study the representatives of these five groups in **Figure 10.6.** Monerans are prokaryotic. Of the four eukaryotic kingdoms, the most diverse by far is the kingdom that first arose from bacteria, the protists. Protists are important to the history of life because they are the ancestors of the three other eukaryotic kingdoms. Members of the other three eukaryotic kingdoms—fungi, plants, and animals—are made of many cells. Having more than one cell is known as multicellularity. **Multicellularity** is a relatively recent evolutionary event. The first known fossils of multicellular organisms are found in 630 million-year-old rocks, which are nearly 1 billion years younger than the first eukaryotes. These earliest multicellular organisms were animals that did not have hard shells or bones and, as a result, were not well preserved as fossils. Many appeared to have been very flat and thin, like pancakes, and they probably floated on the surface of the ancient seas.

Figure 10.6
Living things are classified into five separate kingdoms based on their evolutionary relationships.

[Handwritten: PROTISTS ← 3 eukaryotic KINGDOMS]

Members of the most ancient kingdom, Monera, are prokaryotic. Members of the other four kingdoms are eukaryotic.

Monera (Bacteria)

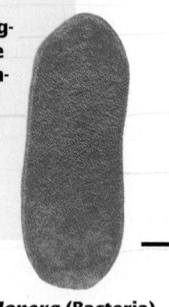

Many protists are single-celled. Some, such as seaweed, are made of many cells. Some protists are photosynthetic, while others hunt bacteria or other protists for their food.

Protista (Protists)

All plants, all animals, and most fungi are multicellular.

Plantae (

Animalia (

Fungi (Fung

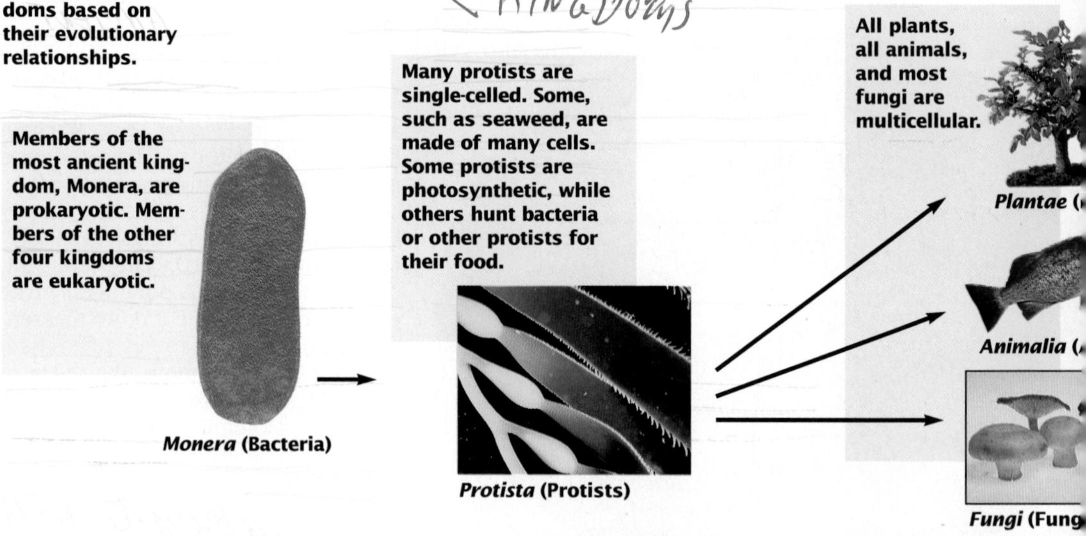

500 - 600 million

Life Blooms in the Ancient Seas

Patterns of Change

What do fossils from the Cambrian period reveal about the rate at which evolution occurs?

The appearance of multicellular animals led to a great blossoming of life in the Earth's oceans. Almost all the major groups of multicellular organisms that survive today, except plants, originated during less than 100 million years, from a little under 600 million years ago to about 500 million years ago, a time period known as the Cambrian period.

Biologists classify the bacteria, protists, and animals into major groups called **phyla** (singular, phylum). Among the major animal phyla are mollusks (snails are examples of mollusks), sponges, and arthropods (insects and crabs are examples of arthropods). Most of the major animal phyla alive today evolved during the Cambrian period.

Animals unlike any living today also originated during the Cambrian. A rich collection of Cambrian fossils has been discovered on a rocky mountain slope in eastern British Columbia, Canada, in a rock formation called the Burgess Shale. There you can see the remains of many bizarre "oddball" animals that are members of extinct phyla, such as those illustrated in Figure 10.7 below. For unknown reasons, many of the phyla represented in the Burgess Shale became extinct before the end of the Cambrian period.

Five mass extinctions have occurred in Earth's history

Five times during the history of life, a large percentage of existing species have become extinct within a short period of time. These large extinction events are called **mass extinctions**. The first mass extinction occurred about 440 million years ago, some 60 million years after the end of the Cambrian period. Scientists are not sure what caused these mass extinctions.

Figure 10.7
This artist's rendition shows some of the unusual Burgess Shale animals. Since these animals are known only from fossils, their true coloration is unknown.

Section Review

❶ A human lifetime is about 75 years. Calculate the number of human lifetimes in 4.5 billion years.

❷ Identify two differences between eubacteria and archaebacteria.

❸ Describe how cyanobacteria changed the composition of the atmosphere.

❹ Explain why this statement is true: A human cell is descended from both eubacteria and archaebacteria.

Demonstration 1
Geologic Time
Show the class an overhead transparency of a geologic time scale. Point out that the last 600 million years is divided into three eras—the Paleozoic, Mesozoic, and Cenozoic. Each era is characterized by the kinds of organisms that appeared at that time. The end of each era was marked by mass extinction of organisms. For example, the Paleozoic era ended with the mass extinction of more than 90 percent of all ocean-dwelling species.

Theme Answer
Patterns of Change
The fact that most of the major animal phyla originated within this short period of time indicates that the rate of evolution is not constant.

Phase 3
ASSESSMENT OPTIONS

Closure Strategy
The Seas
Have students explain why scientists think that early life forms flourished in the seas but not on land.

Section Review
Assign the *Section Review*.

Reteaching
Have students write a short essay to show how this section illustrates the six major themes introduced in Chapter 1.

■ *Section Review Answers* ■

1. Sixty million lifetimes.
2. Students should suggest that eubacteria and archaebacteria differ in the composition of their cell walls and cell membranes, and in the structure of some of their proteins.
3. As cyanobacteria evolved, they carried out photosynthesis, releasing oxygen gas into the oceans. After the oceans had soaked up all the oxygen they could hold, oxygen began to bubble out of the oceans into the air. As more and more oxygen was added to the air, the modern composition of the atmosphere was achieved.
4. Eukaryotic cells are thought to have evolved from archaebacteria, but mitochondria are thought to be the descendants of eubacteria.

Lesson Plan 10.3

Phase 1

PREPARATION

Key Concepts
- About 400 million years ago, the ozone layer developed, permitting life to move onto land.
- Mycorrhizae, a partnership between plants and fungi, enabled the first living things to make the transition from water to land.
- Arthropods, especially insects, were the first animals to thrive on land.

Reading Strategy

Several techniques can help students better remember what they read. One such technique is visualization. Have students practice visualization by forming a picture or image in their mind of what they read. This section is suitable for practicing this technique since it graphically describes how life made the successful transition from ocean to land.

Phase 2

TEACHING STRATEGIES

Connection: Chapter 5

UV Light
Have students refer to Figure 5.11 on page 98 to see where ultraviolet light waves are located in the electromagnetic spectrum. Discuss how the energy of UV light waves compares to that of visible light.

WHAT WAS THE EARTH LIKE 500 MILLION YEARS AGO? YOU WOULD FIND THAT IT WAS VERY DIFFERENT FROM TODAY'S EARTH. EVEN THOUGH THE SEAS TEEMED WITH LIFE, THE DRY LAND WAS UNINHABITED, AS IT HAD BEEN FOR THE EARTH'S ENTIRE HISTORY. THE GREAT VARIETY OF TERRESTRIAL LIFE HAS EVOLVED ONLY IN THE LAST 400 MILLION YEARS.

10.3 Invasions of the Land

Objectives

❶ **Explain how ozone was critical to the development of life on Earth's surface.**

❷ **Recognize how the relationship between plants and fungi enabled both to invade the land.**

❸ **Identify the importance of flight in the evolution of insects.**

❹ **Describe the role insects and plants played in each other's evolutionary success.**

UV light damage DNA

$O_2 \xrightarrow{UV} O_3$

The Importance of Ozone

Figure 10.8
a Ozone, composed of three oxygen atoms, protects Earth's surface by absorbing ultraviolet radiation in the upper atmosphere.

Until just over 400 million years ago, there was no life on the dry rocky surface of the land because high levels of ultraviolet light from the sun bombarded the Earth. Ultraviolet light damages DNA. So much ultraviolet light was reaching Earth that life could not survive out of water, which absorbs ultraviolet rays.

Life was only able to move onto land because of a change in the atmosphere. Recall that photosynthesis carried out by cyanobacteria was adding oxygen gas to the atmosphere. As large amounts of oxygen began to diffuse into the upper atmosphere, ultraviolet rays broke apart some of the oxygen molecules, which then recombined to form **ozone**. Ozone (O_3), shown in **Figure 10.8**, is a gas that has the remarkable and very fortunate property of absorbing ultraviolet radiation. In the upper atmosphere, ozone acted like a great shield, blocking out ultraviolet radiation. By about 400 million years ago, enough ozone had formed in the atmosphere to make the Earth's surface a safe place to live.

In this context, you can understand why so many people are worried about the ongoing destruction of Earth's ozone shield by industrial chemicals. You will learn more about the ozone shield in Chapter 14.

O

O

O

b The protective layer of ozone is being destroyed by industrial chemicals. The dark center of this satellite image indicates an area of very low ozone concentration over Antarctica.

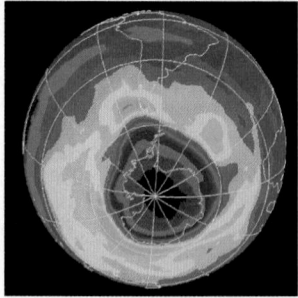

Plants and Fungi Colonize Land

The first living things to populate the surface of the land were plants and fungi. Together plants and fungi solved a particularly difficult challenge: how to survive on bare rock. Each brought to this task a unique ability. Plants, which evolved from green algae, were able to carry out photosynthesis. However, they could not extract minerals from the soilless, rocky surface of ancient Earth. Fungi, which evolved from protists, could not make sugars from sunlight but were adept at absorbing minerals. *FUNGI ABSORB*

The solution to the challenge of living on dry land was a unique biological partnership between plants and fungi called **mycorrhizae** (*MY koh REYE zee*). Mycorrhizae are close associations between the roots of plants and fungi. Fungi actually grow on or into the plant root and then branch out into rock or soil, as shown in **Figure 10.9**. The fungi provide minerals absorbed from rock or soil to the plant. The plant provides food to the fungi. This kind of "you-help-me-and-I-help-you" partnership is called **mutualism**. Fossils show that the earliest plants, which lived about 410 million years ago, had mycorrhizae. The partnership between plants and fungi continues today. Indeed, 80 percent of plant species have mycorrhizae associated with their roots.

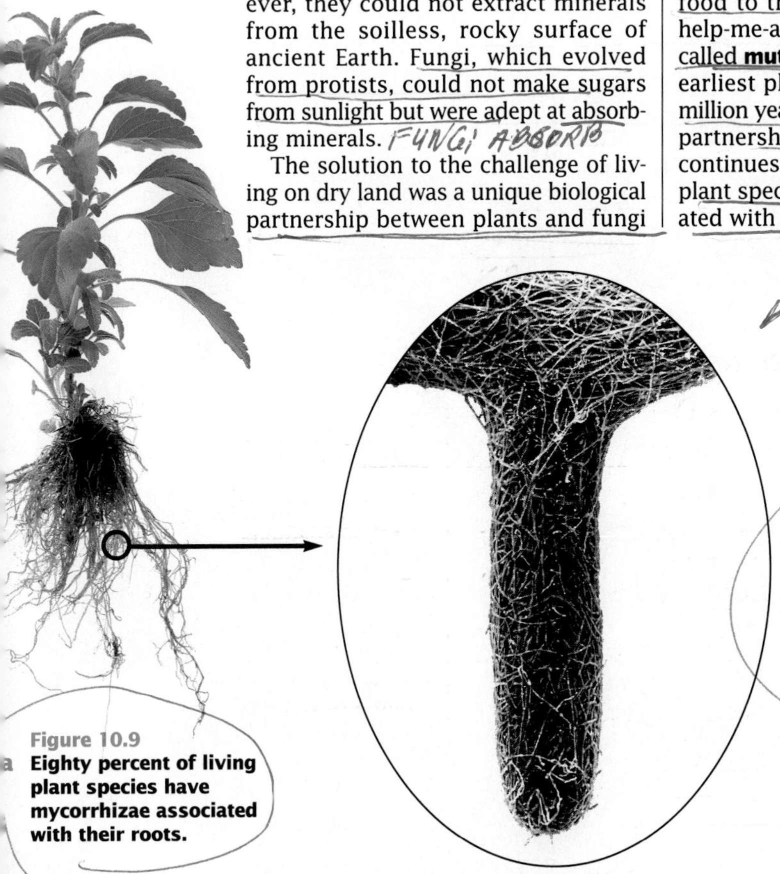

b This is a single root hair. The hairlike strands on the root hair are fungi, which help the plant absorb minerals in return for food from the plant.

Figure 10.9
a **Eighty percent of living plant species have mycorrhizae associated with their roots.**

Invasion of the Arthropods — *exoskeleton*

The fossil record reveals that plants covered the surface of Earth within 80 million years of their initial invasion. Animals soon followed plants onto land. The first animals to leave the water were the arthropods, a kind of animal with a hard body covering and jointed legs. Crabs and lobsters are examples of existing aquatic arthropods. The first arthropods to live on land were scorpions, carnivorous relatives of spiders with two large pincers on their front legs and a venomous stinger at the end of their tails. The arthropod invasion of the land and other major events in Earth's history are illustrated in the *Journey Into the Past* on pages 208–209.

Demonstration 1
Mycorrhizae

Have students use microscopes to examine fungal strands growing on plant roots. Emphasize that mycorrhizae are most common in plants that grow in nutrient-poor soil. Have students explain this observation.

Demonstration 2
Arthropods

Display either specimens or photographs of various arthropods. Have students consider what features enabled these organisms to invade the land with such great success. Their small size, rapid and prolific reproductive ability, exoskeleton, and specialized body parts have made arthropods the most successful animal group on Earth. Point out that they constitute nearly 80 percent of the known animal species.

Using the Feature

- It may be difficult for students to get a feel for the lengths of time described in this feature. This analogy may provide some perspective. Imagine the Earth's 4.5-billion-year history as a single day, with the Earth forming at 12:01 A.M. Life, in the form of tiny bacteria, first appeared at just after 5:00 A.M. No eukaryotes appeared until 4:00 P.M. There was no life on land until 10:00 P.M. The demise of the dinosaurs came at 20 minutes before midnight. The first humans did not originate until less than one minute before midnight.
- Have students select one period and research its organisms. Have students present reports to the class.
- Suggest that students who want additional information read *Life on Earth* by David Attenborough.

Discussion

Guide the discussion by posing the following questions.

1. Explain the rarity of fossils older than about 600 million years.
   Before this time, most organisms were single-celled bacteria and protists. These organisms were unlikely to form fossils because they lacked hard parts. Also, their fossils are small and hard to find.

2. Why was the appearance of photosynthetic bacteria that released oxygen such an important event?
   All euykaryotes and many bacteria require oxygen for oxidative respiration. These organisms could not have evolved in the oxygen-poor atmosphere that existed prior to the advent of these photosynthetic bacteria. Release of oxygen also made possible the ozone layer, which is essential for life on land.

3. What are mass extinctions?
   Short periods during which the extinction rate is very high.

Journey Into the Past

The history of human existence is only a tiny fraction of the story of life on Earth. From the fragmentary evidence provided by the fossil record, scientists have reconstructed much of life's history.

Sabertooth cat

Giant ground sloth

American lion

Mammoth

Mastodon

The Page Museum and the lake pit at the La Brea Tar Pits in Los Angeles, California

Scientists have excavated millions of fossils from the oozing tars of the La Brea Tar Pits in Los Angeles, California. These excavations have uncovered the remains of sabertooth cats, mastodons, and other mammals that became extinct about 10,000 years ago. Some of these animals were encountered by the first human settlers of North America.

■ *Journeys* ■ *Journeys* ■ *Journeys* ■ *Journeys* ■ *Journeys*

The Geological Time Scale

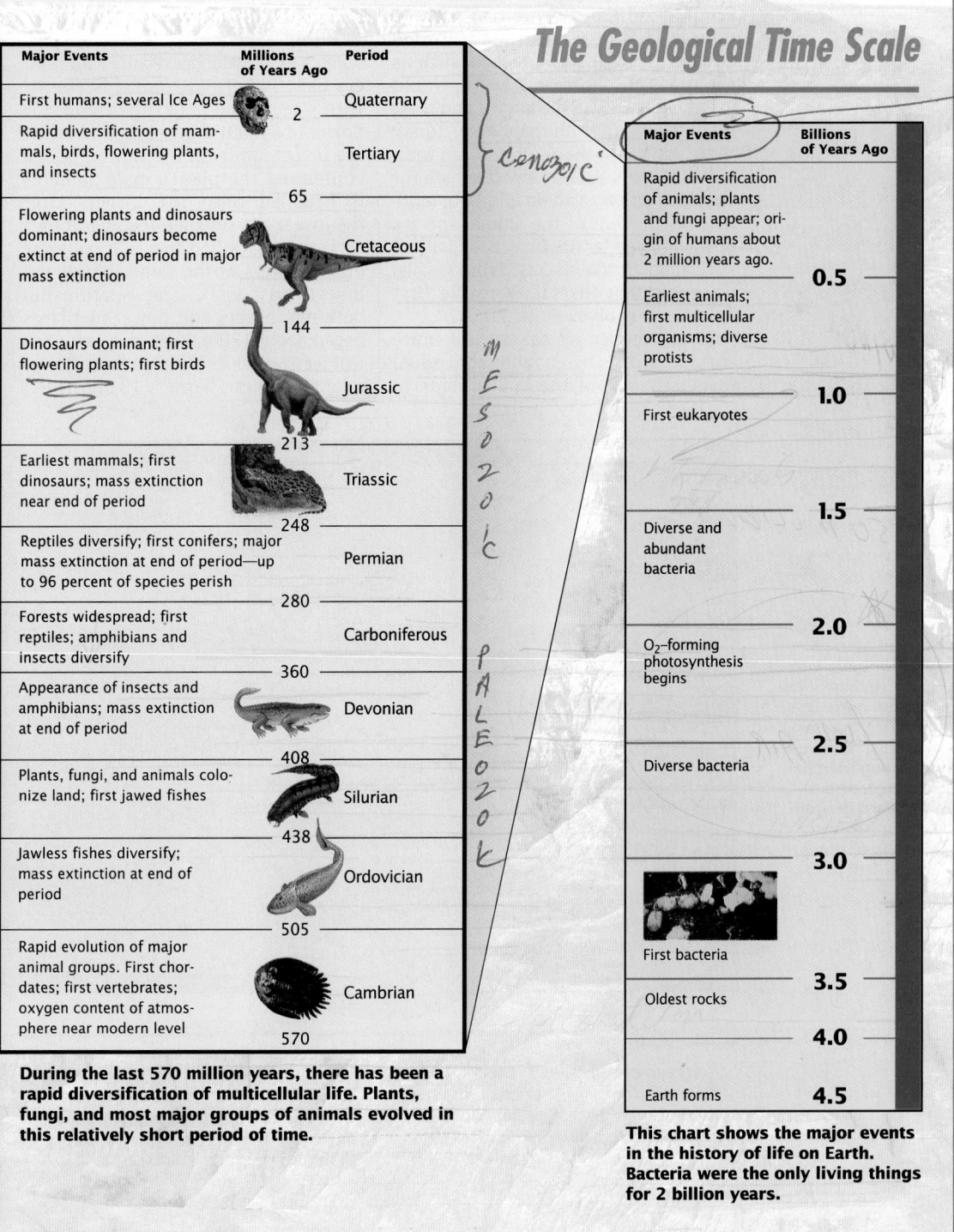

Major Events	Millions of Years Ago	Period
First humans; several Ice Ages	2	Quaternary
Rapid diversification of mammals, birds, flowering plants, and insects		Tertiary
	65	
Flowering plants and dinosaurs dominant; dinosaurs become extinct at end of period in major mass extinction		Cretaceous
	144	
Dinosaurs dominant; first flowering plants; first birds		Jurassic
	213	
Earliest mammals; first dinosaurs; mass extinction near end of period		Triassic
	248	
Reptiles diversify; first conifers; major mass extinction at end of period—up to 96 percent of species perish		Permian
	280	
Forests widespread; first reptiles; amphibians and insects diversify		Carboniferous
	360	
Appearance of insects and amphibians; mass extinction at end of period		Devonian
	408	
Plants, fungi, and animals colonize land; first jawed fishes		Silurian
	438	
Jawless fishes diversify; mass extinction at end of period		Ordovician
	505	
Rapid evolution of major animal groups. First chordates; first vertebrates; oxygen content of atmosphere near modern level		Cambrian
	570	

During the last 570 million years, there has been a rapid diversification of multicellular life. Plants, fungi, and most major groups of animals evolved in this relatively short period of time.

Major Events	Billions of Years Ago
Rapid diversification of animals; plants and fungi appear; origin of humans about 2 million years ago.	
Earliest animals; first multicellular organisms; diverse protists	0.5
First eukaryotes	1.0
Diverse and abundant bacteria	1.5
O_2–forming photosynthesis begins	2.0
Diverse bacteria	2.5
	3.0
First bacteria	
Oldest rocks	3.5
	4.0
Earth forms	4.5

This chart shows the major events in the history of life on Earth. Bacteria were the only living things for 2 billion years.

Theme Connection

Evolution

Equipped with wings, insects were able to exploit an environment almost totally devoid of any other animal life form. As a consequence, they evolved to become the dominant terrestrial organisms on the Earth. As proof of their success, it is estimated that more than a billion insects are alive at any one time.

Theme Answer

Interacting Systems

Yes. Mutualism is a symbiotic relationship in which all participating species benefit. Certain species of insects are attracted by certain species of flowering plants offering sweet nectars or other foods. Insects that land on the flowers are brushed with pollen (the male gametes). After feeding, the insects move to other plants, carrying pollen that will pollinate flowers of the same species.

Phase 3

ASSESSMENT OPTIONS

Section Review

Assign the *Section Review*.

Reteaching

Assign students to cooperative work groups of four. Have each group design a hypothetical animal that would be well adapted in making the transition from ocean to land some 400 million years ago. Have each group describe their organism and its adaptations to the rest of the class.

Interacting Systems

Is the relationship between insects and flowering plants a case of mutualism? Why or why not?

ı₅ᵀ WINGS

Insects Were the First Flying Animals

From the initial scorpion invaders, a unique kind of terrestrial arthropod soon evolved: insects. Insects such as the dragonfly in **Figure 10.10**, would eventually become the largest and most diverse group of animals ever. Today there are more than 200 million individual insects alive at any one time for each person on earth. In addition, more than 70 percent of the animal species discovered so far are insects. What was special about the insects? From fossils, we know that insects were the first animals to evolve wings.

Flying opened up the world's entire surface to insects, enabling individual insects to patrol the countryside in search of food, mates, or nesting sites. A flying insect can also transport objects long distances. Plants with flowers have benefited from this ability. With sweet nectars or other food, the flower of a plant attracts insects. When the insect lands on the flower, pollen containing the plant's male gametes is brushed onto the insect. After the insect has finished feeding, it flies off, carrying pollen that will pollinate other flowers of the same species the insect may visit. The relationship between insects and flowering plants began about 120 million years ago. You will learn more about the insect-flower relationship in Chapters 13 and 18.

Figure 10.10
About 370 million years ago, forests might have looked like the one illustrated here. Dragonflies that lived at this time, although similar in structure to modern dragonflies, had wingspans of more than 1 m (3 ft.).

Section Review

❶ Why is ozone important for life on dry land?

❷ Explain how plants and fungi were able to move onto land together.

❸ What advantages does a flying insect have over a non-flying insect?

❹ Describe the relationship between insects and flowering plants.

▪ Section Review Answers ▪

1. Students should explain that ultraviolet light destroys DNA and that life on land would not have been possible without ozone to absorb ultraviolet radiation.
2. Students should suggest plants and mycorrhizae fungi were able to move onto land together because

of their mutualistic relationship. The roots of plants provided food for the fungi. The fungi grew on or in the roots of the plants and branched out to provide absorbed minerals from rock or soil.

3. Flying insects can seek out distant territories for food, mates, or nesting sites.
4. Students should suggest that insects are the pollinators for flowering plants, receiving food in return.

THINK OF A COMMON ANIMAL. WAS THE ANIMAL YOU THOUGHT OF
A HORSE? AN ELEPHANT? A GUPPY? A CAT? THE ANIMAL YOU
THOUGHT OF WAS PROBABLY A VERTEBRATE. VERTEBRATES ARE THE
ANIMALS MOST FAMILIAR TO US, NOT ONLY BECAUSE WE ARE
VERTEBRATES BUT BECAUSE ALL LAND ANIMALS BIGGER THAN YOUR
FIST ARE VERTEBRATES.

10.4 *Parade of Vertebrates*

Objectives

❶ **Identify the defining characteristic of vertebrates.**

❷ **List three adaptations that enabled amphibians to colonize land.**

❸ **Compare the ability of a reptile to survive on land with that of an amphibian.**

❹ **Identify three differences between mammals and reptiles.**

Animals With Backbones

worms

Along with the organisms you saw in **Figure 10.7**, the Burgess Shale also contains fossils of a 5-cm (2-in.) wormlike animal that is the earliest known chordate. Members of this phylum have a flexible rod of cartilage known as the **notochord** that extends along the back. In most chordates, the notochord only exists for a short time during early embryonic development and is then replaced by the vertebral column, or backbone. Chordates with a vertebral column are called **vertebrates**. BONE

CARTILAGE

The first vertebrates lacked jaws

The earliest vertebrates were jawless fishes with bony skeletons. These small fishes appear to have fed in a head-down position, their fins helping to keep them upright while they sucked up organic particles from the bottom. For 100 million years jawless fishes were the only vertebrates. Today, only two kinds of jawless fishes remain. These are the eel-like, parasitic lampreys and the scavenging hagfishes. Examples of a sea lamprey and a pre-historic jawless fish are found in **Figure 10.11**a and **b**, respectively.

Jaws evolved about 400 million years ago

Scientists know that the first fishes with jaws evolved approximately 400 million years ago. These jawed fishes rapidly replaced the jawless fishes in the oceans. Jaws enabled fishes to bite instead of suck and thus become efficient predators. Early jawed fishes dominated the seas for 50 million years before being replaced by swifter swimmers, the sharks and bony fishes.

Figure 10.11
Both the lamprey (above) and the extinct fish (right) are jawless. Lampreys still thrive in the ocean and in fresh waters.

b The 13 cm (5 in.) *Hemicyclaspis* lived during the early Devonian period.

Lesson Plan 10.4

Phase 1

PREPARATION

Key Concepts
- Vertebrates are organisms that possess a vertebral column or backbone.
- The first vertebrates to evolve were jawless fishes.
- Other vertebrate groups that evolved include jawed fishes, sharks, bony fishes, amphibians, reptiles, mammals, and birds.

Reading Strategy

Have students prepare a table. Across the top, have them write: *Evolutionary history, Structural features, Special adaptations,* and *Examples*. Down the left, have them write: *Jawless fish, Jawed fish, Sharks, Bony fishes, Amphibians, Reptiles, Mammals,* and *Birds*. Have students fill in the information as they read this section.

Phase 2

TEACHING STRATEGIES

Visual Strategy

Figure 10.11
Point out that lampreys are the only parasitic vertebrates. These organisms obtain their food by attaching their round, suckerlike mouths to the bodies of other fishes and using their teeth or tongues to gnaw a hole. Lampreys then suck the blood and other body fluids from their victims.

211

3 SIDED ♡

Demonstration 2
Skeletons
Pass pieces of cartilage and bone around the class so that students can note the differences between the two.

Demonstration 3
Amphibians
Display a variety of live or preserved amphibians, including frogs, toads, and salamanders. Have students describe the major differences between these animals and fishes.

Demonstration 4
Reptiles
Display a variety of live or preserved reptiles, including turtles, lizards, and snakes. Have students describe the major differences between these animals and amphibians.

Theme Connection

Evolution

Throughout the Mesozoic era, for more than 150 million years, dinosaurs ruled the Earth. Then they suddenly disappeared. The reasons for their extinction have been debated by scientists for years. Some argue that a sudden climatic change—perhaps the result of the impact of a large asteroid—was the cause; others point to competition from mammals. Perhaps their disappearance was the result of several factors. In any case, they did leave something—a line of descendants leading to the modern birds.

Shark skeletons are not made of bone
CARTILAGE

From the early jawed fishes evolved a very efficient predator, the shark. Shark skeletons are made of cartilage rather than bone, making sharks lighter and more buoyant than the early jawed fishes. Sharks also have large, strong, mobile fins, which allow them to swim fast and to adjust their motion quickly through the water.

Bony fishes are versatile and abundant
The sharks were largely replaced by particularly versatile newcomers, the bony fishes. Trout, catfish, perch, bass, guppies, carp, and most other familiar fishes are bony fishes. As their name indicates, bony fishes have a skeleton of bone instead of cartilage. Over half of all living vertebrate species are bony fishes.

Amphibians, though adapted to life on land, reproduce in moist environments
350 MILLION

The first vertebrates on land were amphibians, which evolved from bony fishes about 350 million years ago.

O₂ from AIR

Three crucial adaptations enabled amphibians to become successful on land. First, amphibians absorb oxygen from the air with lungs. Second, in amphibians, blood flows from the heart to the lungs, where it picks up oxygen. The blood then returns to the heart to be repumped throughout the body. Thus, oxygen-rich blood flows more rapidly to the muscles and organs than it does in a fish. Third, amphibians walk on four sturdy limbs, which evolved from the fins of their fish ancestors.

Although terrestrial, amphibians constantly lose water through their moist skins, through which they also absorb oxygen. Hence they generally must remain in moist places. Amphibians also must lay their eggs in water or in moist environments. Modern amphibians include toads, salamanders, and frogs, such as the tree frog in **Figure 10.12**.

3 chamber ♡

Reptiles reproduce out of water
Unlike amphibians, reptiles thrive in dry climates because both their skin and their eggs are largely watertight. Examples of living reptiles include snakes, lizards, turtles, and crocodiles.

Reptiles evolved from amphibians about 300 million years ago. During the following 50 million years, these earliest reptiles gave rise to a variety of species that gradually replaced the amphibians as the dominant land animals. Reptiles enjoyed a special advantage over amphibians during this period, because it was a time of widespread drought. About 240 million years ago, at the end of the Permian period, another mass extinction occurred. Scientists estimate that about 96 percent of all species became extinct.

Those reptiles lucky enough to survive found themselves in a garden of opportunity—a world full of food that no one else was eating and space that was no longer occupied by others. A great burst of evolution occurred among the reptiles. The most famous and most impressive reptiles, the dinosaurs, arose about 220 million years ago, soon after the Permian mass extinction. You can read much more about the dinosaurs in the *Tour of a Dinosaur* on page 213.

Figure 10.12
The tropical red–eyed tree frog has suction pads on its toes, enabling it to cling to vertical surfaces.

Tour of a Dinosaur

Dinosaurs evolved at the same time as mammals, 220 million years ago, and for over 150 million years dominated life on Earth. Fossils of Brachiosaurus were first discovered in 1907 in East Africa. It lived 160 million years ago. Its fossils have also been found in the western United States.

Brachiosaurus had its nostrils on top of its head! No one knows why.

All dinosaurs had a hole in the skull, in front of the eye socket. So do birds and crocodiles.

The simple teeth suggest that *Brachiosaurus* fed on plants.

Flat Top

Brachiosaurus is the largest dinosaur of which a complete skeleton has been found. From nose to tail it is about 23 m (75 ft.) long.

The sauropods were the largest dinosaurs ever. This one, **Brachiosaurus,** is estimated to have weighed 73,000 kg (80 tons)—as much as 1,000 people!

A long flexible neck was characteristic of sauropods. Each of the 15 neck vertebrae of *Brachiosaurus* was as much as 1 m (3 ft.) long.

Dinosaurs are classified into two orders: "Lizard-hipped" dinosaurs (includes sauropods such as *Brachiosaurus*) and "bird-hipped" dinosaurs.

Brachiosaurus was as tall as a four-story building. Standing next to one, you would not reach its knee.

Sauropod front legs are much longer than hind legs.

Since *Brachiosaurus* fossil tracks show no marks of a dragging tail, they probably held their tails off the ground.

Brachiosaurus tracks show a narrow stance, telling us its legs were positioned beneath the body, unlike the legs of modern reptiles, and directly supported its immense weight.

All sauropods have five toes, while most other dinosaurs have three. All dinosaurs have hinged ankles.

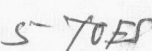 *3 vs 5 TOES*

Using the Feature

- In popular perception, dinosaurs were huge, stupid, sluggish beasts that were doomed to extinction once the "superior" mammals came on the scene. Popular movies such as *Fantasia* perpetuate this perception, which, however, is increasingly at odds with modern research. Emphasize to students that dinosaurs were not ill-adapted monstrosities. For one thing, mammals and dinosaurs appeared almost simultaneously. But it was the dinosaurs that diversified rapidly, while the mammals remained small and mainly nocturnal until the extinction of the dinosaurs.
- Have students investigate the new perceptions of dinosaurs reflected in books such as *Digging Dinosaurs* by John Horner and James Gorman.

Discussion

Guide the discussion by posing the following questions.

1. What evidence suggests that *Brachiosaurus* ate plants? *It had simple teeth.*
2. Show a picture of the stance of a lizard (found on page 483). How does the stance of the lizard compare to that of a mammal such as a dog? *The lizard's limbs jut out from the sides of the body, while the dog's are directly below the body.*
3. What evidence suggests that *Brachiosaurus's* limbs were positioned beneath its body, like a dog's? *Fossilized tracks show a narrow stance.*

scale → feathers
4 chambers

Theme Connection

Stability

Have students describe how amphibians and reptiles (ectotherms) would differ from mammals and birds (endotherms) in their ability to maintain homeostasis.

Theme Answer

Evolution

Although mammals and dinosaurs coexisted for 150 million years, the diversity of mammals was low during this time, probably due to competition from dinosaurs. When the dinosaurs died out, mammals diverged rapidly and exploited the resources formerly used by the dinosaurs.

Phase 3

ASSESSMENT OPTIONS

Closure Strategy

Animal Research

Laboratory animals are often used in many types of research work, from testing the effectiveness of drugs to checking the toxicity of cosmetic products. In many cases, the animals used for such purposes are vertebrates. Both opponents and proponents for using vertebrates have been vocal and active in seeking support for their positions. Assign students to cooperative work groups of four. Instruct each group to choose either the opposing or supporting side and prepare a position paper supporting their viewpoint.

Section Review

Assign the *Section Review*.

Reteaching

On the board, prepare a table similar to the one described in the Reading Strategy. Fill in the information by asking students what they wrote in their individual tables.

Evolution

What effect did the extinction of the dinosaurs have on the evolution of mammals?

"Same Temp"

Mammals have hair and produce milk

As dinosaurs were reaching massive size, mouse-sized mammals were also evolving. Mammals, such as the zebras in **Figure 10.13**, are animals with hair that produce milk to feed their young.

Another feature of mammals is the four-chambered heart. This type of heart is more efficient than the three-chambered heart found in most living reptiles. Mammals are **endotherms**. Endotherms are able to regulate their body temperature through internal mechanisms. On the other hand, **ectotherms** cannot regulate their temperature internally and must absorb heat from their surroundings. Reptiles and amphibians are ectotherms.

For more than 150 million years, mammals and dinosaurs coexisted. During this time, mammals were small and did not evolve into many species.

Birds are the descendants of dinosaurs

Although dinosaurs are extinct, descendants of small, insect-eating dinosaurs are still with us today. These descendents are the birds, which evolved about 150 million years ago. Bird feathers evolved from the same scales that protected the dinosaurs so well. Feathers are one of the features that enable birds to fly. Like mammals, birds are endotherms with four-chambered hearts. Two examples of birds are shown in **Figure 10.14**.

Birds, mammals, and dinosaurs coexisted until the sudden extinction of the dinosaurs 65 million years ago. Like the world after the Permian extinction, the post-dinosaur world offered many opportunities for evolution. In response, mammals and birds diverged rapidly and filled the nearly empty world.

Figure 10.13
All mammals, including these zebras, produce milk to feed their young.

Figure 10.14
a **Hummingbirds, such as this bee hummingbird, are among the smallest birds. Some are only 5 cm (2 in.) long.**

b **The ostrich is the largest bird, standing almost 1.8 m (6 ft.) tall. Although the ostrich cannot fly, it has feathers, like all other birds.**

Section Review

1 **Explain why you are considered to be a vertebrate.**

2 **Describe three adaptations that enable amphibians to live on land.**

3 **Compare the reptilian adaptations for living on land with the adaptations of amphibians.**

4 **Describe two differences between mammals and reptiles.**

▪ Section Review Answers ▪

1. Humans have a backbone.
2. Students should explain that amphibians evolved lungs, circulation to the lungs, and limbs for movement on land.
3. Students should explain that unlike amphibians, reptiles are better adapted to living in dry climates because their skins and their eggs are watertight.
4. Possible answers may include that mammals are endothermic, have hair, produce milk, and give birth to live young. Reptiles are ectothermic, have scales, and lay eggs.

Chapter **10** *Highlights*

There are many stars similar to the sun in our galaxy. Could life exist on another planet?

Key Terms	Summary
10.1 Origin of Life  Lipids may have formed protective coatings around the earliest molecules.	coacervate (p. 202)	• Life arose from nonliving matter present on the early Earth. • Miller and Urey showed that simple molecules could react to form some of life's building blocks. • RNA can catalyze its own synthesis and other chemical reactions. • RNA was the first self-replicating molecule. DNA evolved later.
10.2 Early Life in the Sea 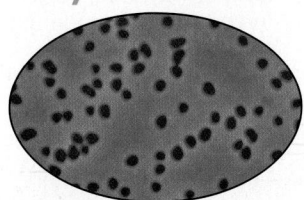 Archaebacteria, though rare today, may be the direct ancestors of eukaryotes.	eubacteria (p. 203) archaebacteria (p. 203) cyanobacteria (p. 204) multicellularity (p. 204) phylum (p. 205) mass extinction (p. 205)	• The Earth is 4.5 billion years old. • The oldest fossils are 3.5 billion-year-old bacteria. Eukaryotes evolved from prokaryotes about 1.5 billion years ago. • Multicellular organisms arose about 630 million years ago. • There have been five mass extinctions that have wiped out many of the Earth's inhabitants.
10.3 Invasions of the Land Ozone made it possible for life to come onto land. Presently, the thinning of the ozone layer is threatening all life. 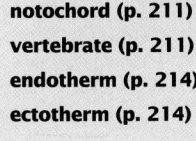	ozone (p. 206) mycorrhizae (p. 207) mutualism (p. 207)	• By 400 million years ago, enough ozone had formed to make life on land possible. • Plants and fungi invaded the land about 400 million years ago. • Arthropods, including insects, followed the plants and fungi onto land.
10.4 Parade of Vertebrates 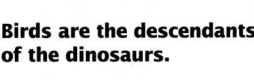 Birds are the descendants of the dinosaurs.	notochord (p. 211) vertebrate (p. 211) endotherm (p. 214) ectotherm (p. 214)	• Chordates have a rod of cartilage called the notochord that runs along the back. • In vertebrates, the vertebral column replaces the notochord. • Vertebrates include sharks, bony fishes, amphibians, reptiles, birds, and mammals. • The first vertebrates were fishes without jaws. • The dinosaurs died out about 65 million years ago.

Alternative Assessment

Have each student prepare a blank bingo card with 25 squares, labeling the center one *FREE*. Write 24 terms, names, or short phrases from this chapter on the board. Have the students randomly place each term in one of the squares on their bingo card. Next, have students cross out the appropriate box each time a definition is read or a clue is provided. Have students yell *bingo!* when either a horizontal or vertical line has been completed.

Chapter Review Answers

Understanding Vocabulary
1. **a.** birds
   b. eubacteria
   c. arthropods

Relating Concepts
2. Map answer is shown on page 210D.

Understanding Concepts
Multiple Choice

3. c		8.	c
4. a		9.	a
5. b		10.	a
6. c		11.	a
7. b		12.	d

Completion
13. RNA
14. bacteria
15. ultraviolet radiation
16. mutualism
17. vertebrates, notochord

Short Answer
18. The ability to carry out photosynthesis that released oxygen gas enabled cyanobacteria to change Earth's atmosphere.
19. The cell wall, cell membrane, RNA polymerase, and ribosomes of archaebacteria resemble those of eukaryotes more than those of eubacteria.
20. The discovery that RNA is able to catalyze its own synthesis suggests that it was the first self-replicating molecule.

Interpreting Graphics
21. The purpose of the discharged electricity was to simulate lightning. Amino acids would be found in the bottom portion of the U-shaped trap, at point d.

Chapter **10** Review

Understanding Vocabulary

1. For each set of terms, complete the verbal analogy by filling in the blank.
   a. evolved from fishes: amphibians::evolved from dinosaurs: _____
   b. ancestors of eukaryotes: archaebacteria::ancestors of mitochondria and chloroplasts: _____
   c. first life on land: plants and fungi::first animals on land: _____

Relating Concepts

2. Copy the concept map below onto a sheet of paper. Complete the concept map by writing the correct word or phrase in each oval containing a question mark.

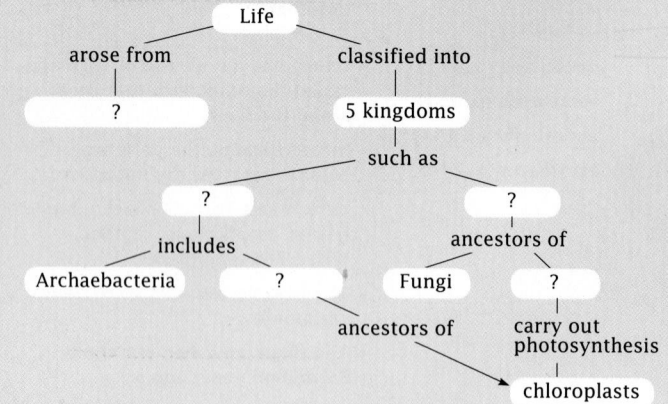

Mastering Concepts

Multiple Choice

3. The only testable hypothesis for the origin of life on Earth is
   a. divine creation.
   b. natural selection.
   c. origin from non-living matter.
   d. extraterrestrial origin.

4. Among the products of Miller and Urey's experiment were
   a. amino acids.
   b. prokaryotes.
   c. bacterial cells.
   d. ammonia and hydrogen.

5. Insects were the first animals to
   a. live on land.
   b. fly.
   c. live in the sea.
   d. have four-chambered hearts.

6. Protists are *not* the ancestors of
   a. animals.
   b. plants.
   c. archaebacteria.
   d. fungi.

7. Scientists estimate that the age of the Earth is about
   a. 10 million years.
   b. 4.5 billion years.
   c. 2 quadrillion years.
   d. 2,000 years.

8. The oldest fossils of eukaryotes are approximately
   a. 4.5 billion years old.
   b. 3.8 billion years old.
   c. 3.5 billion years old.
   d. 1.5 billion years old.

9. In a mycorrhizal partnership, the fungus gets nutrients from the
   a. soil.   c. air.
   b. plant.   d. water.

10. Two unique characteristics of mammals are hair and
    a. ability to produce milk.
    b. scales.
    c. four-chambered heart.
    d. feathers.

11. In what way are lampreys different from other living fishes?
    a. They lack jaws.
    b. They live in water as juveniles and adults.
    c. They use gills to breathe.
    d. They have no brain.

12. The animal that is not a reptile is
    a. snake.   c. dinosaur.
    b. crocodile.   d. shark.

Completion

13. The first genetic material is thought to have been _____ .

14. Ancient _____ are the ancestors of all life on Earth.

15. Ozone in the Earth's atmosphere absorbs most of the _____ _____ from the sun.

16. An association between different kinds of organisms in which both benefit is called _____ .

17. Animals with a backbone are called _____ . These same animals have a(n) _____ made of flexible cartilage during embryonic development.

Short Answer

18. What ability of cyanobacteria enabled them to produce oxygen and change the Earth's atmosphere?

19. Why do scientists think that archaebacteria are the ancestors of eukaryotes?

20. What is the importance of Thomas Cech's work for understanding the origin of life on Earth?

Interpreting Graphics

21. The diagram below shows the apparatus used by Miller and Urey in their 1953 experiment.

    • What is the purpose of discharging electricity into the flask labeled "c"?
    • Where would you expect to find amino acids in the apparatus? Explain your answer.

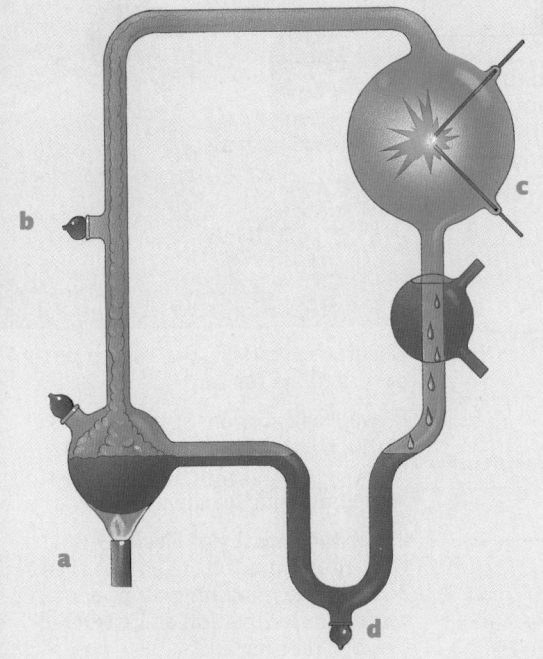

Reviewing Themes

22. *Scale and Structure*
    Why are amphibians sometimes called transitional land-dwellers?

23. *Patterns of Change*
    What evidence shows a cause-and-effect relationship between the evolution of life on Earth and changes to the Earth's physical environment?

Thinking Critically

24. *Inferring Conclusions*
    How might the structure of plant cells be different if eubacteria had not evolved?

25. *Inferring Conclusions*
    If insects had not evolved wings, how would it have affected their invasion of land?

26. *Building on What You Have Learned*
    As you learned in Chapter 2, all living things make proteins from the same 20 kinds of amino acids. Explain how this fact supports the idea that all life shares a common ancestor.

Cross-Discipline Connection

27. *Biology and Language*
    "*Saurus*" is the Greek word for lizard. Find out what the names of these dinosaurs mean: *Brontosaurus*, *Tyrannosaurus*, *Stegosaurus*, *Gorgosaurus*. How does each dinosaur's name reflect its habits and structure?

Discovering Through Reading

28. Read the article "Skinning the Dinosaur," in *Discover*, March, 1989, pages 39–43. Why are many paleontologists skeptical of Stephen Czerkas's hypotheses about dinosaurs? What questions does he hope to answer about *Carnotaurus* and other dinosaurs by careful study of fossilized skin?

Reviewing Themes

22. Even though amphibians spend time on land, they are closely tied to the water for survival. Their skin must remain moist, and they lay their eggs in water or moist places.

23. An anaerobic environment led to the evolution of life from nonliving matter; living organisms able to carry out photosynthesis produced an aerobic environment; and atmospheric oxygen led to the formulation of the ozone layer that protects life forms from harmful radiation.

Thinking Critically

24. Plant cells would not have chloroplasts, since the ancestors of modern eubacteria became chloroplasts. Thus, plant cells might contain a different organelle to convert energy from the sun (or another energy source) into a form that can be used by the plant.

25. They would not have been as successful as they were.

26. If life had multiple origins, it is unlikely that all living things would use the same 20 amino acids.

Cross-Discipline Connection

27. *Brontosaurus* means "thunder lizard." *Tyrannosaurus* means "tyrant lizard." *Stegosaurus* means "roof or thatch lizard." *Gorgosaurus* is a combination of "*Gorgon*"and "*saurus*." *Brontosaurus* was named for its great size—up to 27,000 kg (30 tons). *Tyrannosaurus* was a large, fierce predator. *Stegosaurus'* name refers to the plates on its back. *Gorgosaurus* is named for the Gorgons, three hideously ugly monsters of mythology.

Discovering Through Reading

28. Czerkas is an artist with no formal training in paleontology. Questions of interest to Czerkas include: Were dinosaurs multicolored creatures? Were they coldblooded or warmblooded? Did dinosaurs have skin glands that secreted odors or sweat?

Procedural Note

Compare the conditions that existed on earth when life first formed to those of today. Tell the students that this investigation will show how evidence might be collected to support the hypothesis that life could not arise from nonliving materials today.

Prelab Preparation Answers

1. The early Earth's atmosphere lacked oxygen but was rich in hydrogen. In addition, since the Earth had no protective ozone layer, more ultraviolet light from the sun reached the surface than does today.

2. The modern atmosphere is 21 percent oxygen and 78 percent nitrogen. Ultraviolet light is absorbed by ozone in the upper atmosphere.

3. The reactions that produce nucleotides and amino acids from simple molecules do not occur in the oxygen-rich modern atmosphere.

Chapter 10 Investigation

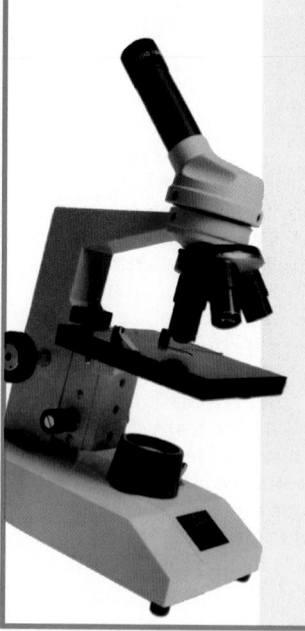

Can Life Arise From Nonlife?

Objectives

In this investigation you will:
- *test* whether life can arise from nonlife today
- *infer* conclusions about the origin of life on Earth

Materials

- one wax pencil
- six 250-mL Erlenmeyer flasks
- nutrient broth medium
- two solid rubber stoppers
- one rubber stopper with a straight plastic tube through it
- one rubber stopper with a curved plastic tube through it
- aluminum foil

Prelab Preparation

Review what you have learned about the origin of life on Earth by answering the following questions:
- What conditions existed on the early Earth?
- How were conditions on the early Earth different from conditions today?
- Why doesn't life arise from nonlife today?

Procedure: Investigating the Origin of Life

1. Form a cooperative group of six students. Each member of the group should be responsible for preparing one flask. Make a table like the one shown on the following page.

2. Collect six 250-mL Erlenmeyer flasks, two solid rubber stoppers, one stopper with a straight plastic tube through it, and a stopper with a curved plastic tube through it.

3. Label the flasks 1–6.

4. **Caution. Put on safety goggles and a lab apron. Leave them on while preparing the flasks.** Add 100 mL of nutrient broth medium to each flask.

5. Observe each flask and record your observations for Day 0.

6. Leave Flask 1 open. Seal Flask 2 with a solid rubber stopper. Take Flasks 1 and 2 to the location assigned by your teacher.

7. Cover the mouths of Flasks 3–6 with aluminum foil. Then put Flasks 3–6 and the remaining stoppers aside for sterilization as instructed by your teacher.

8. After sterilization, remove the foil from all the flasks. Leave Flask 3 open. Seal Flask 4 with a solid rubber stopper. Carefully push the stopper with the straight plastic tube into the mouth of Flask 5. Similarly, use the stopper with the curved plastic tube to close Flask 6.

9. Place Flasks 3–6 with Flasks 1 and 2.

10. Observe the flasks after 1, 2, 3, 7, and 14 days. Record your observations for each day in your table.

Flask	#1	#2	#3	#4	#5	#6
Day 0						
1						
2						
3						
7						
14						

11. Clean up your materials and wash your hands before leaving the lab.

Analysis

1. *Summarizing Data*
Describe what happened in each flask.

2. *Analyzing Data*
Cloudiness is an indication of bacterial contamination of the broth. Which flask showed the first signs of contamination? Why do you think contamination occurred in that flask first?

3. *Comparing Observations*
Did Flask 2 or Flask 4 turn cloudy first? Would this result have been different if you had not completely sealed Flask 4? Explain.

4. *Evaluating Methods*
What hypothesis is being tested by Flasks 3 and 4?

5. *Comparing Observations*
Did Flask 5 or Flask 6 show contamination first? Explain why this flask was contaminated faster.

6. *Making Inferences*
What evidence supports the conclusion that, under modern conditions, life does not arise from nonliving materials?

Thinking Critically

A classmate says that the experiment you have just completed shows that life could not have arisen from nonlife at any time in Earth's history. How could you refute this statement?

Analysis Answers

1. All the flasks are expected to be initially clear. Flask 1 should become cloudy first, followed by Flasks 2 and 3. Flask 5 should become cloudy within the first week and Flasks 4 and 6 should remain uncontaminated.

2. Flask 1 should show the first sign of contamination, since it was not sterilized and was left exposed to the air.

3. Flask 2 turned cloudy before Flask 4 because neither the medium nor the flask had been sterilized. Yes. Contamination would have taken longer but exposure to airborne microorganisms would have eventually led to contamination.

4. Flasks 3 and 4 test the hypothesis that a solid stopper prevents contamination.

5. Flask 5. The straight tube allowed airborne microorganisms a small but direct pathway to the medium. The curved tube trapped entering microorganisms.

6. Answers may vary but should logically explain how the data collected from the flasks support the hypothesis that life cannot arise from nonliving materials. For example, for Flasks 4 and 6, in which living things were killed and were prohibited from entering, no evidence of contamination was observed.

Thinking Critically Answer

This experiment demonstrates only that life cannot arise from nonlife under current conditions.

Chapter **11** *Human Evolution*

Planning Guide

Objectives/Themes	Classwork Resources	Homework Resources

11.1 **1.** List two distinctive features of primates.
 	
2. Describe one adaptation of modern prosimians to nighttime activity.
 	
3. Identify two differences between monkeys and prosimians.
 	
4. Recognize the close evolutionary relationship between humans and apes.
 	
Themes: Evolution, Patterns of Change, Scale and Structure
	Teacher's Resource Binder
	Focus Activity 11
	  Comparing Primate Features
	Lab Investigation 11.1
	  Human Evolution
	
	Other Resources
	Transparency 43
		Text
		Section Review, p. 225
		
		Teacher's Resource Binder
		Directed Reading Worksheet 11.1
		Reteaching Worksheet*
		  Distinctive Features of Primates
		
		Other Resources
		Audiocassette 11.1*

11.2 **1.** Contrast the characteristics of chimpanzees with those of humans. **2.** List three reasons why hominid fossils are rare. **3.** Describe evidence that human ancestors walked upright before their brains enlarged. **4.** Compare and contrast australopithecines and modern humans. **Themes:** Evolution, Patterns of Change	**Other Resources** Transparencies 44–45	**Text** Section Review, p. 230 **Teacher's Resource Binder** Directed Reading Worksheet 11.2 **Other Resources** Audiocassette 11.2*

11.3 **1.** List two key characteristics of *Homo habilis.*
 	
2. Contrast the skeletal features of *Homo habilis* with those of *Homo erectus.*
 	
3. Summarize the evidence that *Homo sapiens* evolved in Africa.
 	
4. Describe evidence that Neanderthals had a complex culture.
 	
Themes: Evolution, Patterns of Change, Scale and Structure
	Text
	Investigation
	  Comparing Structure of Animal Jaws pp. 240–241
	
	Teacher's Resource Binder
	Science Skills Worksheet
	  Studying Physical Evidence
	
	Other Resources
	Transparency 46
		Text
		Section Review, p. 236
		
		Teacher's Resource Binder
		Directed Reading Worksheet 11.3
		Vocabulary Review Worksheet*
		
		Other Resources
		Audiocassette 11.3*

*Reteaching Options

Demonstrations	**Assessment**
11.1: pp. 222, 223, 224 **11.2:** p. 227	Chapter Review pp. 238–239 Portfolio Assessment p. 219D Chapter Test—Teacher's Resource Binder Test Generator

219A

Research Notes

Connection to Paleontology: African Eve Remains Unproven

For many years, some paleontologists used fossil evidence to argue that modern humans gradually developed in several different continents at once, over a period of nearly a million years. Another group of paleontologists believe the evidence indicates that modern humans developed in Africa, and then spread to other continents.

In 1987, molecular biologists led by Allan Wilson of the University of California at Berkeley produced a family tree of human origins based on analyses of the small amounts of DNA in human mitochondria. On the basis of this evidence, they proposed that all humans alive today descended from a single female ancestor who lived in Africa 200,000 years ago.

The news that there may have been an "African Eve," made headlines throughout the world. However, other analyses of the same data led to different results, calling some of Wilson and his colleagues' techniques into question.

Wilson's group used mitochondrial DNA because it is passed on from mother to offspring in the egg cells. The sperm of the father never contributes mitochondria to a newly conceived fetus. Since the 37 genes of human mitochondrial DNA do not recombine, they can only change by random mutation. The more differences between two samples that share a common ancestor, the longer ago they must have diverged.

The group used the polymerase chain reaction (PCR) to create enough DNA for analysis from a data set that eventually included 241 individuals of ethnic groups from all continents. Using a computer to analyze the differences among the groups, they came up with a genealogical tree based in Africa made up of 182 types of mitochondrial DNA.

But some paleontologists still disagreed, arguing that in certain areas, some characteristics seen both in modern humans and in ancient hominids would have to evolve twice: once in the hominids, and then again in the descendants of modern human immigrants from Africa. Skulls were discovered in China that had a combination of ancient and modern characteristics, supporting the hypothesis of independent evolution throughout the Old World.

The biochemical evidence of the theory was challenged in early 1992 when several other geneticists noted that the group had failed to enter the data in a random order. The results obtained by a computer program can be affected by the order of entry.

In addition, many runs must be performed to achieve consensus between the hundreds of possible trees each run can produce. Wilson's group had only done one run.

Other workers using these refinements generated many other equally likely trees, some of which did not start in Africa. Substantial doubts have arisen over the timing of 200,000 years suggested by the group as well. In February of 1992, Wilson's group withdrew their support of the theory of an "African Eve," saying that the evidence was inconclusive.

Investigation Notes

Comparing Structure of Animal Jaws
pages 240–241

Purpose: This activity gives the students the opportunity to infer relationships about tooth structure and diet. Emphasize that inferences should be based on observations and logical assumptions.

Prelab Preparation

1. As examples of skull variations, WARD'S has the following skulls available: human (82 M 3031), canine (65 M 5310), cow (65 M 5125), and gorilla (80 M 1755). As an alternative, the lower jaw of a cow's skull could be obtained from a local meat packing plant.

2. The use of a mirror is recommended because some people are self-conscious about others looking in their mouth.

Answers will be found on pages 240–241.

Meeting Individual Needs

Objectives
1. Students will demonstrate appropriate use of core vocabulary for the chapter (Vocabulary File).

2. Students will identify several human traits that demonstrate adaptations to fit the human environment (Teaching Strategy A).

Vocabulary File
(Developing Vocabulary/ Limited English Proficiency)

If you are not already using the Vocabulary File, refer to Chapter 1 for its preparation. See the Chapter Highlights on page 237 for a list of suggested words.

Teaching Strategy A
Adaptations of Humans and Other Primates
for pages 221–223
(Developing Observational Skills/Multisensory Learners)

Materials: pencils and paper for each student

Procedure: Remind students of the common children's activity of making tracings of the outline of a hand. Explain that in order to help evaluate how well students are adapted to the human environment, they will do this again. Tell students to observe what both of their hands do as they repeat this procedure.

Students should lay their hand down on a sheet of paper. They should trace around it, making an outline in the shape of their hand. Be sure they trace closely to the fingers so that the joints can be seen.

Instruct students to silently read Section 11.1, "Evolution of Primates" on page 221. When students have finished, remind them that flexible fingers and opposable thumbs are two distinctive features of primates. Ask students how these features are advantageous, and brainstorm so students are aware that certain tasks for living can be performed because of these hand features.

During the hand tracing exercise, were students' fingers flexible and able to bend independently? Could their thumbs bend towards their fingers while their fingers are bent? What task were they able to do as a result? (Hold a pencil and write) Be sure that students realize that the characteristics of flexible fingers and opposable thumbs are shared by primates.

Remind students of the importance of vision in primates, as described in the reading selection.

The students should place their hands over their left eyes, and determine if they can see with their right eyes alone. Then they should reverse the procedure. Ask if they can detect a difference in the field of vision for each eye. Students should be able to detect that the angle of view was slightly different for the different eyes. Explain that this is a result of binocular vision.

With a hand covering one of their eyes, the students should try to pick up a piece of paper on the desk in front of them. Although they will be able to do this, students should perceive that it is more difficult using one eye than both. Ask students what advantage comes with binocular vision. (The ability to perceive depths and distances)

Explain to the students that with grasping hands and binocular vision, a large brain is needed to control hand-eye coordination. Since humans share these primate characteristics, humans can be considered primates.

Additional Strategies
Visual Strategies
Pages 221, 225, 226, 227, 228, 229, 231, 232, 234, 235, and 236

Auditory Learners
Use *Biology: Visualizing Life* Audiocassettes for Sections 11.1, 11.2, and 11.3.

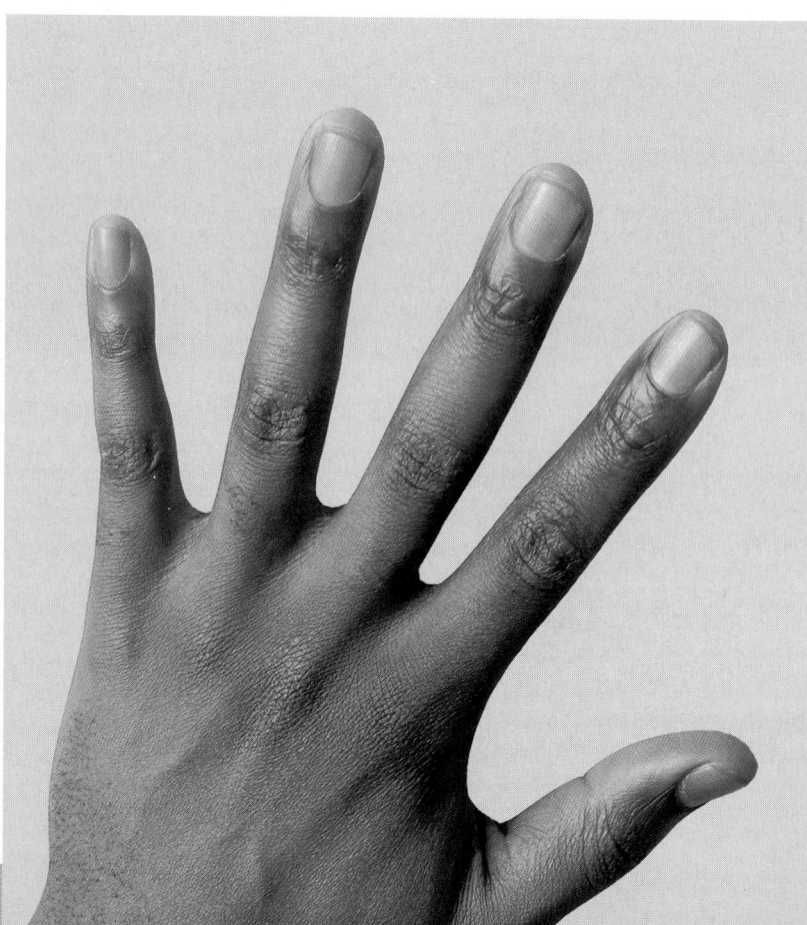

Meeting Individual Needs (cont.)

Cooperative Learning
Primate Characteristics

Timing: Use this activity to introduce Section 11.1.

Group Size: 4 students

Outcome: Students will compare the characteristics of various primates.

Individual Accountability: Each group member will research two kinds of primates.

Positive Interdependence: The group will cooperatively prepare an outline of primate characteristics.

Assign two of the following kinds of primates to each group member: (1) tree shrew, (2) lemur, (3) tarsier, (4) monkey, (5) gibbon, (6) orangutan, (7) gorilla, and (8) chimpanzee. Have each student make an outline of the primate's traits and habitat. Then the group should compare outlines and discuss the major similarities and differences among the primates.

Finally, have each group prepare a table summarizing common primate traits and major differences. Students should observe the facial characteristics of the primates and describe the transition in facial characteristics from the tree shrew to humans.

Portfolio Assessment

Students should select their best work and provide a self-reflective rationale for their selections. Students can make selections in the following areas.

1. *Content*	One concept map from the chapter (See page 812 for evaluation criteria.)
2. *Reading Comprehension*	One Directed Reading Worksheet from the Teacher's Resource Binder (Use the answer key to evaluate for accuracy.)
3. *Writing*	Using the Vee Form, summarize a newspaper or magazine article relating to the evolution of humans. (See page 22T for evaluation criteria.)
	Or: Select a writing task or project from the Chapter Review.
4. *Performance Assessment*	One Vee form from a chapter investigation or lab manual investigation (See page 22T for evaluation criteria.)

Teacher makes selections in the following areas.

1. *Formal Assessment*	Chapter test (Test A, B, or the Test Generator) The teacher-scored test should be reviewed by the student. Incorrect responses should be corrected by the student before the test becomes part of the portfolio.
2. *Informal Assessment*	Use the Direct Observations Checklist, page 33T, during a laboratory or other cooperative learning experience.
3. *Performance Assessment*	Have students research ancient hominid lifestyles and environment in order to write a scientifically accurate story about a day in the life of an ancient hominid.

Concept Map Answer

The following is one possible answer to the Relating Concepts exercise on page 238.

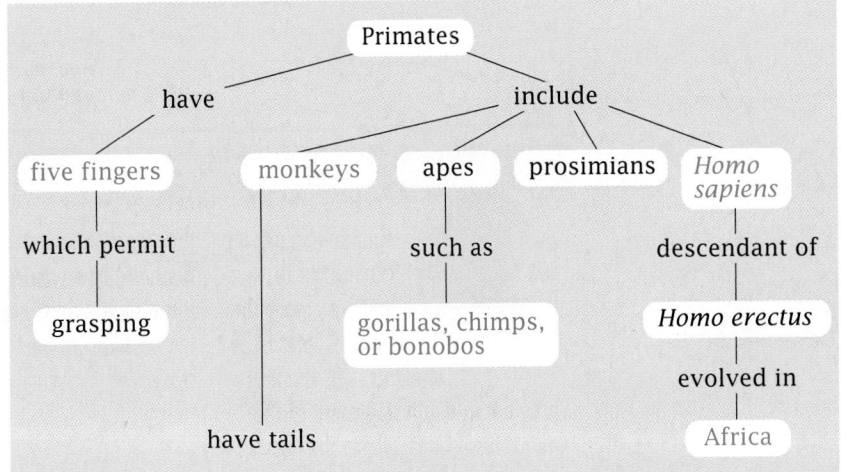

Chapter 11

Determining Prior Knowledge

- Display photographs of primates, including lemurs, monkeys, apes, and humans. Have students name the traits shared by these primates.
- Have students draw a diagram that depicts their concept of the evolutionary relationship between monkeys, apes, and humans.
- Show students a photograph depicting an early hominid, such as a Neanderthal. Elicit responses to questions concerning how human evolution occurred, including where the first modern humans originated and what characteristics they possessed.

Chapter 11

Human Evolution

Review

- **DNA structure (Section 7.1)**
- **use of DNA and proteins to study evolution (Section 9.2)**
- **natural selection (Section 9.3)**

11.1 Primates

- Evolution of Primates
- Anthropoids Are Day-Active Primates
- Apes

11.2 Evolutionary Origins of Humans

- Searching for the Fossil Relatives of Humans
- Characteristics of the Earliest Hominids
- Branches of the Hominid Evolutionary Tree

11.3 The First Humans

- Evolution of the Genus *Homo*
- Our African Origins
- The Origin of *Homo sapiens*

The scene above was painted in a cave in Algeria about 4,000 years ago. By this time, humans had domesticated cattle (shown above) and other animals, and grew crops for food.

▪ Author's Rationale ▪

Human evolution is the part of the evolutionary story that is of most interest to students. This chapter provides in-depth treatment of the major fossil finds, where they occurred, who found them, and how they were interpreted. Students see firsthand how science works. Although it is a highly controversial subject, it is the best known part of the evolution story.

IN 1871 DARWIN PUBLISHED *THE DESCENT OF MAN.* IN THIS BOOK, HE SUGGESTED THAT HUMANS EVOLVED AND WERE MOST CLOSELY RELATED TO THE AFRICAN APES—THE GORILLA AND THE CHIMPANZEE. ALTHOUGH LITTLE FOSSIL EVIDENCE EXISTED IN 1871 TO SUPPORT DARWIN'S CASE, NUMEROUS FOSSIL DISCOVERIES MADE SINCE DARWIN'S DEATH STRONGLY SUPPORT HIS HYPOTHESIS.

11.1   Primates

Objectives

1 List two distinctive features of primates.

2 Describe one adaptation of modern prosimians to nighttime activity.

3 Identify two differences between monkeys and prosimians.

4 Recognize the close evolutionary relationship between humans and apes.

Evolution of Primates

5 flex. fingers   rodents → primates

Look closely at your hand. You have five flexible fingers. Animals with five flexible fingers are called **primates**. Monkeys, apes, and humans, are examples of primates. All primates are mammals, animals that have hair and suckle their offspring on milk. Primates most likely evolved from small, insect-eating, rodentlike mammals that lived about 60 million years ago.

Like nearly all of today's primates, the earliest primates dwelt in trees. Natural selection favored traits enabling these early primates to capture insects while scampering through the trees. Unlike rodents, which have clawed feet, primates have grasping hands and feet. Primates can grip limbs, hang from branches, and seize food. Some primates can even use tools. Compare your hand to the hands of the other primates in **Figure 11.1**. The hand of a primate can grasp an object because it has an **opposable thumb**. An opposable thumb stands out at an angle from the other fingers and can be bent toward them to grip an object. All primates except for humans also have opposable big toes.

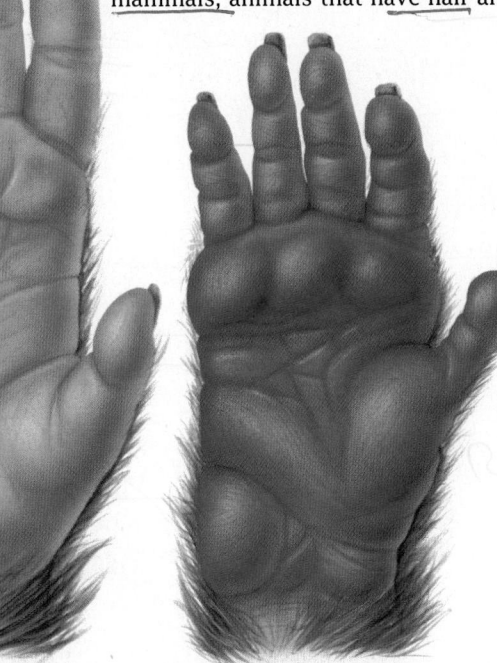

Figure 11.1
Primates have five flexible fingers.

a Chimpanzee hand   **b** Baboon hand   **c** Lemur hand

Lesson Plan 11.1

Phase 1
PREPARATION

Key Concepts

- Primates are mammals that have opposable thumbs, binocular vision, and large brains.
- Prosimians, the first primates to evolve, are small, furry, nocturnal animals that live in trees.
- Anthropoids, which include monkeys, apes, and humans, are primates that evolved later. Anthropoids are diurnal animals possessing color vision and large brains.
- Humans and chimpanzees shared a common ancestor as recently as 5 million years ago.

Reading Strategy

Students rarely use the index found at the end of most textbooks. As they read, have them list as many items as possible that are discussed in this section and that can also be found listed in the index.

Phase 2
TEACHING STRATEGIES

Visual Strategy

Figure 11.1
Have students measure the angle formed between the thumb and the rest of the hand for each of the organisms shown in this figure. Then have each student place one hand down flat on a piece of paper, place their four fingers together, and spread their thumb as far apart as possible. Tell them to trace their hand with a pencil. Have them remove their hand, and measure the angle formed between their thumb and the rest of their hand. Ask which organism is capable of extending its thumb the most.

30-40 MILLION YR PROSIMIANS
PRE SIMIANS are 1ST PRIMATES

Demonstration 1
Opposable Thumbs

Have students attempt to carry out ordinary tasks without the use of their thumbs. Have them prepare three lists: impossible, difficult, and easy. Have them place each task attempted under the appropriate heading.

Demonstration 2
Binocular Vision

Assign students to work in cooperative groups of two to test their binocular vision. They are to take turns holding two pencils, each at a different distance, in front of their partner. Tell them to hold the two pencils so their partner can see both. With both eyes open, each student should attempt to grab the pencil that is farther away with just their thumb and forefinger. Repeat this process several times, changing the pencil's distances with each trial. Then have students test their ability to perform the same task, keeping one eye closed.

Social Studies Connection

Last Refuges for Prosimians

In addition to the lemurs, modern prosimians include tarsiers, which are found only in the Philippines, Indonesia, and Malaysia. On a world map, have students locate the geographic areas where prosimians still exist.

Figure 11.2
The large eyes of this loris reveal that it is a nighttime hunter. Native to India, Sri Lanka, and Southeast Asia, lorises creep slowly through the trees, hunting insects and small animals. They grow to about 25 cm (10 in.) in length.

Another adaptation to living in trees is the position of the eyes in the skull. The eyes of rats and squirrels are in the sides of their heads. The fields of vision of their eyes do not overlap. The eyes of primates, in contrast, are located in the front of the face, like the eyes of the loris in **Figure 11.2**. Each eye of a primate sees a slightly different view of the same scene. The brain merges the two views in perceiving the distances to objects. This type of vision is called **binocular vision**. The ability to judge distance is advantageous when jumping from one branch to another or when stalking prey.

Other mammals, such as cats, have binocular vision, but only primates have both binocular vision and grasping hands. These features require an enlarged brain to process information from the eyes and to coordinate hand movements. Large brains and increased intelligence thus became hallmarks of the primates.

Prosimians were the first primates

The first primates were **prosimians** (meaning "before monkeys"). Fossils 30 million to 40 million years old show that prosimians were common in North America, Europe, Asia, and Africa. Only a few species of prosimians, such as the loris shown in **Figure 11.2**, survive today. Many of these surviving species are nighttime hunters. You can tell by the disproportionately large size of their eyes. Large eyes are necessary to capture what little light is available at night or in dark forests. All 24 surviving species of lemurs, cat-sized prosimians with long tails for balancing, live on Madagascar, an island about 180 km (300 mi.) off the east coast of Africa. Today the native vegetation of Madagascar is being rapidly destroyed by an expanding human population. As the lemur's forest home disappears, some of our oldest living relatives are becoming extinct in the wild.

Anthropoids Are Day-Active Primates

About 35 million to 40 million years ago, a revolutionary change occurred in how primates lived: they became active during daytime. How do we know this change occurred? Fossil skulls of primates that lived at this time have much smaller eye sockets than do prosimian skulls, suggesting that they were active during the day. The new day-active primates were **anthropoids.** Monkeys, apes, and humans are the existing anthropoids.

Since daytime activity places different demands on the eye, many changes in eye design probably evolved at this time. One of these changes was the development of color vision. Of course, we can't actually examine 35 million-year-old anthropoid eyes to see if they could see color. Soft tissues such as those of the eye are rarely preserved as fossils. Instead, we infer that early anthropoids had color vision because all living anthropoids see color. Therefore, color vision probably arose early in anthropoid history.

The brains of anthropoids are larger than prosimian brains. Larger brain size seems to be associated with the more complicated behavior patterns of anthropoids. Anthropoids replaced prosimians rather rapidly. In the fossil record, prosimian fossils become rare as anthropoid fossils become common.

Scale and Structure

House cats lack color vision. From this information, what can you conclude about the time of day during which house cats probably are most active?

Figure 11.3
Monkeys, such as the adult baboon (below, left), take care of their young for a longer time than do most other mammals.

SMALL
EYE
SOCKET

Figure 11.4
This spider monkey, like many other New World monkeys, has a long tail that functions as a limb for grasping branches.

Scientists don't know exactly why prosimians lost the evolutionary competition, but they suspect that the larger brains and color vision of anthropoids better adapted them to living in trees.

Monkeys have complex social interactions

Monkeys are anthropoids with tails. Two groups of monkeys occur today. Old World monkeys, such as the baboons in **Figure 11.3**, live in Asia and Africa. New World monkeys, such as the spider monkey in **Figure 11.4**, inhabit Central America and South America. Monkeys feed mainly on fruits and leaves rather than on insects. They live in groups in which complex social interactions occur. Monkeys tend to care for their young for a longer time than most other mammals, except for humans and apes. This long period of dependency seems to be necessary for the development of the large brains of monkeys, apes, and humans.

Demonstration 4

Molecular Clocks

Assign students to cooperative work groups and have each one select a name for a hypothetical organism. Draw a long sequence of DNA bases on the board and inform students that this represents the genetic information for an organism known as Blorg. Have each group copy the sequence on paper. Inform students that spontaneous mutations occur at a constant rate. Tell students that every 15 seconds, each group is to randomly change one base in the DNA sequence. But instruct students that each group will not start their *molecular clocks* at the same time. Instead, have each group start at different times. After an appropriate amount of time, instruct all the groups to stop their clocks. Then have each group compare the similarities and differences in their DNA with that of the Blorg. Have students construct a diagram showing the evolutionary history of Blorg and their hypothetical organisms, based on the comparisons of the DNAs.

Apes

Unlike monkeys, apes lack tails. Apes also have larger brains than monkeys. The existing apes are the gibbons, the orangutan, the gorilla, the bonobo (*BAHN uh boh*), and the chimpanzee. With their large brains, apes are capable of learning a greater variety of behaviors than any mammal except human beings. Once common, apes are rare today. Their natural forest habitat has largely disappeared as the growing human population has cleared woodland to make way for farms. Modern apes are confined to relatively small areas in Africa and Asia. No apes ever existed in the wild in North or South America.

DNA sequences reveal our close kinship to chimpanzees

Modern apes are not our direct ancestors. Humans share a common ancestor with the living apes, but the apes from which humans descended are extinct. Nevertheless, much can be learned about our own evolution by studying the modern apes, such as the chimpanzees in **Figure 11.5** below. For example, studies of DNA and protein sequences have revealed a great deal about the relationships among the living apes and about their relationship to humans. You learned in Chapter 9 how differences in nucleotide sequences in DNA, or differences in amino acid sequences in proteins, can be used to determine evolutionary relationships. If the average rates of change in these sequences are known, it is possible to calculate approximately when evolutionary divergences occurred. When DNA from apes and humans is compared, gibbon DNA shows the greatest number of differences in nucleotide sequence. Scientists think the species that gave rise to gibbons diverged from the common ancestor of humans and the other apes about 10 million years ago, as shown in **Figure 11.6**. The next ape to evolve, the ancestor of the orangutan, split off approximately 8 million years ago.

Humans shared a common ancestor with the gorilla, the chimpanzee, and the bonobo until about 7 million years ago. At this time, the gorilla's ancestor diverged from the common ancestor of the chimpanzee, the bonobo, and humans. The ancestor of humans did not begin to diverge from the ancestor of the chimpanzee and bonobo until about 5 million years ago. Because this divergence began so recently, the genes of humans and chimpanzees have not accumulated many differences. Overall, the nucleotide sequences of human and chimpanzee genes differ by only 1.6 percent. As a result, most of the proteins encoded by your genes are very similar or even identical to the corresponding proteins in a chimpanzee.

Figure 11.5
The physical structure, DNA sequences, and protein sequences of humans are more similar to those of bonobos and chimpanzees, such as this pair from Zaire, than to those of any other living species.

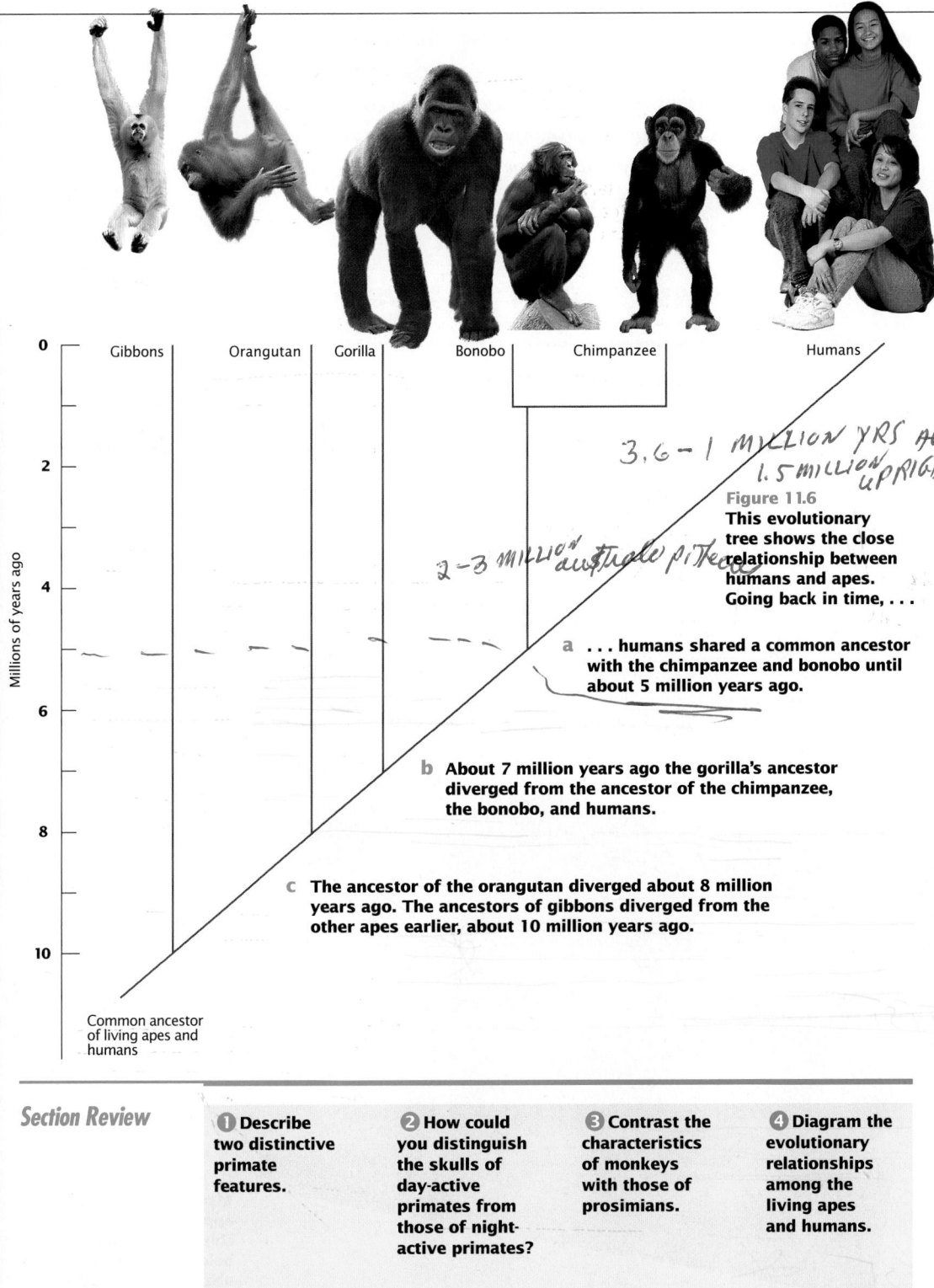

Figure 11.6
This evolutionary tree shows the close relationship between humans and apes. Going back in time, . . .

a **. . . humans shared a common ancestor with the chimpanzee and bonobo until about 5 million years ago.**

b **About 7 million years ago the gorilla's ancestor diverged from the ancestor of the chimpanzee, the bonobo, and humans.**

c **The ancestor of the orangutan diverged about 8 million years ago. The ancestors of gibbons diverged from the other apes earlier, about 10 million years ago.**

(handwritten notes on figure:)
3.6 – 1 MILLION YRS ago
1.5 MILLION UPRIGHT
2 – 3 MILLION australo pithecus

Visual Strategy

Figure 11.6
Use this figure to review the evolutionary relationships between humans and apes.

Phase 3

ASSESSMENT OPTIONS

Closure Strategy
Concept Maps
Have students draw concept maps to show the evolutionary relationships among the various primates.

Section Review
Assign the *Section Review*.

Reteaching
Show students pictures of various primates. Have them identify any traits that could be used to classify each organism as a primate.

Section Review

❶ **Describe two distinctive primate features.**

❷ **How could you distinguish the skulls of day-active primates from those of night-active primates?**

❸ **Contrast the characteristics of monkeys with those of prosimians.**

❹ **Diagram the evolutionary relationships among the living apes and humans.**

■ *Section Review Answers* ■

1. Possible answers include a highly developed brain, opposable thumbs, grasping hands, and binocular vision.

2. The eye sockets of night-active primates are much larger than those of the day-active primates.

3. Prosimians lack color vision and have smaller brains than monkeys.
4. See Figure 11.6 on page 225.

Section 11.2

Phase 1
PREPARATION

Key Concepts
- Humans and their ancestors are known as hominids.
- Hominid fossils are rare because the ancestors of humans were few in number until the development of agriculture about 10,000 years ago. Also, few individuals were fossilized, those that were are embedded in rock formations deep in the earth.
- Compared with the other primates, hominids displayed bipedal locomotion, possessed larger brains, and had smaller jaws and teeth.
- The first hominids, which appeared some 3.5 million years ago, are the australopithecines.

Reading Strategy

A good way of organizing ideas and information contained in a reading assignment is concept mapping. Students should read the information about concept mapping in the Appendix, and use it build a concept map of this section.

Phase 2
TEACHING STRATEGIES

Visual Strategy

Figure 11.7
Use this figure to illustrate that despite their 98.4 percent genetic similarity, humans and chimpanzees are obviously very different phenotypically. Students should recognize that changing less than 2 percent of the genetic information has a great impact on phenotypic expression. As an example, remind them that a single base alteration in the DNA is the cause of sickle cell anemia.

226

WRITING, SCULPTING, AND DIALING A TELEPHONE: THESE ARE JUST A FEW OF THE MANY TASKS YOU CAN DO WITH YOUR HANDS. YOUR HANDS ARE FREE TO DO THESE TASKS BECAUSE YOU WALK ON TWO LEGS INSTEAD OF ON FOUR. UPRIGHT WALKING WAS ONE OF THE FIRST ADAPTATIONS TO EVOLVE AMONG OUR ANCESTORS AFTER THE HUMAN AND CHIMPANZEE LINES DIVERGED ABOUT 5 MILLION YEARS AGO.

11.2   Evolutionary Origins of Humans

Objectives

1. **Contrast the characteristics of chimpanzees with those of humans.**
2. **List three reasons why hominid fossils are rare.**
3. **Describe evidence that human ancestors walked upright before their brains enlarged.**
4. **Compare and contrast australopithecines and modern humans.**

Searching for the Fossil Relatives of Humans

No one could mistake the girl in **Figure 11.7** for a chimpanzee, despite their 98.4-percent genetic similarity. That 1.6-percent genetic difference—accumulated in the 5 million years since the lines that gave rise to humans and chimps began to separate—results in substantial differences in physical appearance. Not only is the chimp covered with long hair, but its skull, teeth, limbs, and feet differ from those of a human. Moreover, chimps walk on four legs, while humans walk on two.

Studies of DNA nucleotide sequences reveal the close relationship of humans to the apes. However, current technology cannot show how the differences in form that are so evident to us today came about. Instead, we must dig into the fossil record to find out what our closest relatives were like. Humans and their closest fossil relatives are known as **hominids** (HAHM uh nihdz).

Figure 11.7
Many differences are apparent between a chimp and a human being, despite their nearly identical genes.

Hominid fossils are rare and hard to find

As you learned in Chapter 9, fossils do not form very often. Conditions have to be right, or a dead body will simply decay, leaving no evidence of what it looked like. Not surprisingly, hominid fossil hunters face difficulties in finding fossils. Until the development of agriculture about 10,000 years ago, hominids seem to have been rare. For example, early hominids left few footprints relative to other animals. Add to this difficulty the fact that only individuals that died on the shores of a lake or swamp or fell into a mud pit would be covered by sediment, a necessary step in fossilization. Furthermore, fossils that formed 3 million to 5 million years ago are embedded within rock formations deep in the ground. These fossils can only be found if the rock formation is exposed by wind, water, or earth movements. Once exposed, however, fossils are quickly destroyed by erosion. Since most hominid fossils are only fragments, scientists must reassemble them like pieces of a jigsaw puzzle. A fossil hunter is shown in **Figure 11.8** searching for fossils embedded in rock.

How can the age of a fossil be determined? Because the fossil is too precious, it cannot be ground up for radioactive dating. And we cannot simply date the rock in which the fossil is found, because fossil-bearing rock is composed of little bits of rock broken off from other rocks. Therefore, radioactive dating would yield the age of the original rock that broke into bits, not the rock that formed much later around the fossil. So how do scientists find the age of a fossil? First, they know that volcanic rock can be dated with great accuracy. If a nearby volcano erupted within a few thousand years of the time when the fossil was deposited, scientists can date the rock that formed as a result of the eruption. A fossil embedded nearby can be said to have a similar age. *TEST*

How are hominid fossils recognized?

How can you tell whether a fossil is from a hominid or some other kind of animal? After all, fossils don't come with identification tags. Identification is possible because different kinds of animals exhibit different details of structure. For instance, hominid jaws are quite different from those of chimpanzees, as shown in **Table 11.1**. Other structural differences are apparent in the skull, spine, pelvis, and other bones, as you will soon see. Only by painstaking comparisons can scientists distinguish hominid fossils from those of apes or other animals.

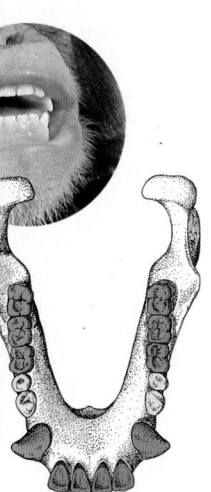

Figure 11.8
In Africa, fossil hunter Louis Leakey searches for fossils of human ancestors.

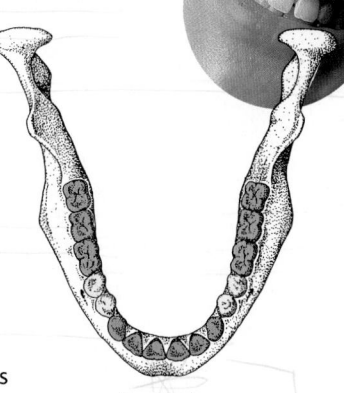

Table 11.1 Comparison of Chimpanzee and Human Jaws

Chimpanzee	Human
U-shaped jaw with molars arranged in two parallel rows	Arc-shaped jaw with molars not parallel
Space between incisors and canines	No space between incisors and canines
Relatively long canines	Short canines

■ Molars ■ Premolars ■ Canines ■ Incisors

Chimpanzee jaw

Human jaw

RM

Visual Strategy

Table 11.2

Review the characteristics that distinguish gorillas from australopithecines. Have students trace the gorilla's spine to see the C-shaped curve and the hominid's spine to see the S-shaped curvature. Point out the difference in the location where the spinal cord enters the skull. Its location in hominids allowed the head to be balanced on top of the spinal cord. In fossils, the location of this opening reveals whether the primate walked bipedally. Inform students that, even when standing upright, the first australopithecines were only about 1 m (3.5 ft.) to 1.5 m (5 ft.) tall and weighed about 25 kg (55 lbs.).

Connection: Chapter 3

Surface Area and Volume

Review the difference between surface area and volume. Show students a model or picture of the human brain. Tell them a modern human's brain is characterized by not only a larger volume, but more importantly, by a much greater surface area. Point out the folds of the cerebral hemisphere. Discuss how these folds greatly increase the total surface area without affecting the volume that this portion of the brain occupies.

Table 11.2 Comparison of Gorilla and Australopithecine Skeletons

Gorilla	Australopithecine
Skull atop C-shaped spine	Skull atop S-shaped spine
Spinal cord exits near rear of skull	Spinal cord exits at bottom of skull
Arms longer than legs; arms and legs used for walking	Arms shorter than legs; only legs used for walking
Tall and narrow pelvis	Bowl-shaped pelvis, centering the body weight over the legs
Femurs (thigh bones) angled away from pelvis when walking upright	Femurs angled inward so legs are directly below body to carry its weight

■ Skull   ▢ Spine   ▢ Arms

■ Pelvis   ▢ Femurs

Gorilla

Australopithecine

Characteristics of the Earliest Hominids

If you were looking for early hominid fossils, where would you search? In 1871 Charles Darwin considered this question and decided that Africa was the most likely place to find these fossils. Darwin concluded that because the closest relatives of humans, the chimpanzees and gorillas, lived in Africa, humans probably evolved there too. As Darwin predicted, fossils of the earliest hominids were uncovered in eastern and southern Africa. Dating from about 3.6 million to 1 million years ago, these fossils represent species that evolved after the human line split off from the line leading to chimpanzees and bonobos.

At least five species of these early hominids have been discovered, all assigned to the genus *Australopithecus*. Members of this genus are called australopithecines (*aw stray loh PIHTH uh seenz*). Even the earliest australopithecines show clear differences from apes.

Australopithecines walked upright

Australopithecines were **bipedal**—that is, they walked upright on two legs. In fact, upright walking is a characteristic of all hominids. Although apes can walk upright for short distances, they do so with an awkward, waddling gait. How do we know that australopithecines that died 3.5 million years ago walked upright? Comparisons of australopithecine skeletons to those of apes such as the gorilla have revealed how they walked, as shown in **Table 11.2**.

Australopithecines also had large brains. A chimp's brain occupies a volume of about 400 cm^3 (24 cu. in.), about the size of an orange. Australopithecine brains were slightly larger, ranging from 400 to 550 cm^3 (24–34 cu. in.). Though larger than an ape's brain, australopithecine brains were still much smaller than the brains of modern humans, which average 1,350 cm^3 (83 cu. in.), about the size of a small melon.

Branches of the Hominid Evolutionary Tree

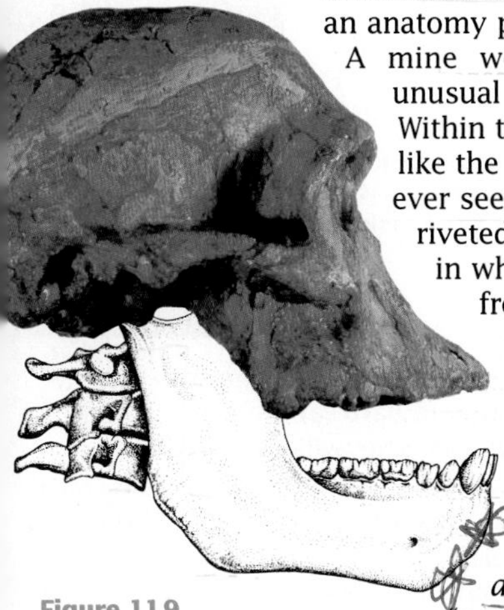

The first australopithecine fossil was discovered in 1924 by Raymond Dart, an anatomy professor in South Africa. A mine worker brought Dart an unusual rock-hard chunk of earth. Within this chunk was a skull unlike the skull of any ape Dart had ever seen. One detail of the fossil riveted Dart's attention: the rock in which it was imbedded was from a geological formation thought to be millions of years old. At that time, the oldest reported hominid fossils were only 500,000 years old.

Dart named the new species *Australopithecus africanus*, which means "southern ape from Africa" (this name is unfortunate, since we no longer think *Australopithecus* was an ape). Dart argued that *A. africanus* was the direct ancestor of humans, the long-sought "missing link" between apes and humans. Most scientists at the time dismissed Dart's discovery as the skull of a young ape. Yet many *A. africanus* fossils have been discovered since 1924, and we now know that this species was a hominid, not an ape. **Figure 11.9** shows a skull of *A. africanus*. This species had jaws that are more rounded than those of an ape and teeth that resembled those of a human. The brain of *A. africanus* was larger than the brain of an ape, about 440 cm^3 (27 cu. in.). *Australopithecus africanus* was bipedal and existed from 2.5 million to 3 million years ago. Whether *A. africanus* was a direct ancestor of humans is still under debate.

Figure 11.9
The spinal cord of *A. africanus* exited from the bottom of its skull, indicating that this hominid walked upright. This species lived about 2.5 million to 3 million years ago. (The lower jaw and spinal cord have been added by an illustrator.)

Another important find was the footprints of a group of bipedal animals that walked across wet volcanic ash about 3.5 million years ago in what is now northern Tanzania. The footprints, shown in **Figure 11.10**, were preserved when the ash hardened. They reveal small but very humanlike feet, lacking the ape's opposable toe. Our ancestors or very close relatives walked upright only 1.5 million years after diverging from the chimpanzee line.

Other australopithecines have been discovered

A stockier kind of australopithecine was unearthed in South Africa in 1938. Called *A. robustus*, this species had massive teeth and jaws. In 1959 the fossil hunter Mary Leakey discovered the even more solidly built *A. boisei*, shown in **Figure 11.11**, in Tanzania. Nicknamed "nutcracker man" because of its massive jaws, *A. boisei* had a large, bony ridge running along the crest of the head to anchor powerful jaw muscles. Excavations in 1985 turned up yet another australopithecine, a massively boned species resembling *A. boisei*.

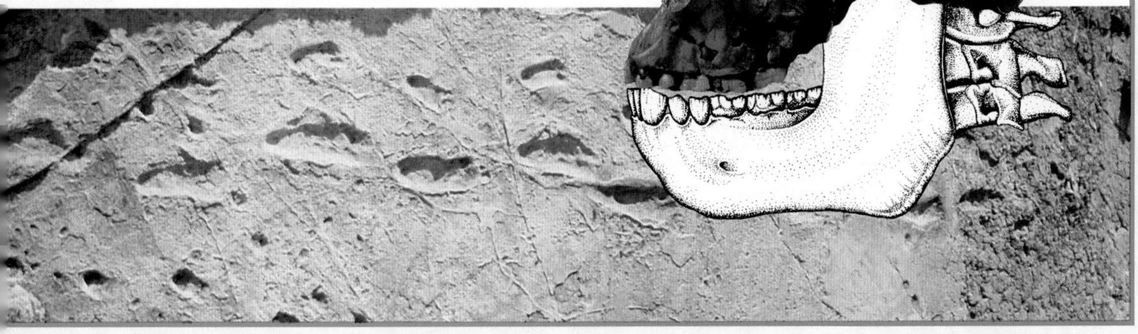

Figure 11.11
This 1.8 million-year-old *A. boisei* skull has massive jaws (illustrated below) and a bony ridge along the crest of its head.

Figure 11.10
The footprints below show that bipedal hominids had evolved by about 3.5 million years ago.

Visual Strategy

Figure 11.9
Inform students that the skull Dart studied was that of a young child, estimated to be five or six years old. Consequently, scientists could not determine how adults of the species looked until 12 years later when they discovered fossils of adult australopithecines.

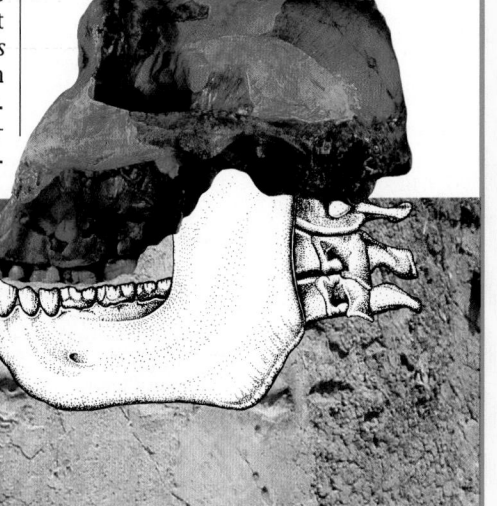

Phase 3

ASSESSMENT OPTIONS

Closure Strategy

Missing Link

In what is now known as the Piltdown forgery, the skull of a modern human was carefully joined with the jaw of an orangutan and planted as a fossil. The fossil was discovered in 1911 in Piltdown, England, and was claimed to be the remains of a common ancestor of apes and modern humans. However, chemical tests conducted in 1953 revealed the hoax. Have students describe what the Piltdown skull must have looked like.

Section Review

Assign the *Section Review*.

Reteaching

Have students prepare a list of items to show how australopithecines were intermediate in form between apes and humans.

Theme Answer

Patterns of Change

It is assumed that the largely hairless condition of modern humans had not evolved yet.

Patterns of Change

Like an ape, Lucy is covered with hair in the reconstruction on this page. What assumptions about evolution are made by showing Lucy in this way?

Figure 11.12
Because major portions of Lucy's limb bones and pelvis were preserved, scientists were able to determine that she walked upright.

Lucy is the oldest hominid

In 1974 Donald Johanson went to the Afar Desert of Ethiopia in search of early hominid fossils. There he found the most complete, best preserved skeleton of a prehuman hominid ever discovered. Nicknamed "Lucy," the skeleton was over 40-percent complete, as shown in **Figure 11.12**. Lucy was found to be nearly 3 million years old. Johanson assigned the skeleton the scientific name *Australopithecus afarensis*. Since the discovery of Lucy, many other fossils of *A. afarensis*, some over 3.5 million years old, have been unearthed in Ethiopia and in Tanzania. Since some of these fossils were found near the site of the bipedal footprints shown on the previous page, it is possible that *A. afarensis* individuals left these footprints.

What do we know about Lucy's species? A reconstruction of how a living *A. afarensis* individual might have looked is shown in **Figure 11.13**. Lucy herself was small by modern standards, just over 1 m (3 ft. 6 in.) tall. Other *A. afarensis* specimens range up to 1.5 m (5 ft.) tall, with males much larger than females. The shape of Lucy's

Figure 11.13
This reconstruction illustrates how Lucy might have looked 3 million years ago.

pelvis indicates that she was female, and the inward angle of her femur shows that she walked upright.

Despite being bipedal, *A. afarensis* was quite apelike in many respects. With a protruding face and fairly large jaws, an *A. afarensis* skull resembles a chimpanzee skull more than a human skull. In size, *A. afarensis* teeth are intermediate between most ape and human teeth. For instance, the canines of *A. afarensis* were shorter than those of apes, but still longer than human canines. Most important, the brain of *A. afarensis* was only about 400 cm³ (24 cu. in.), the size of a chimpanzee brain. Johanson's discovery demonstrated that hominids walked upright before they evolved brains larger than ape brains.

Scientists now know that humans are closely related to the australopithecines. But which australopithecines are our ancestors? Did humans evolve from the large-boned *A. boisei* or from the smaller *A. africanus*? Scientists disagree on the relationship among the australopithecines and on how this genus is related to our genus, *Homo*. The evolution of *Homo* is the subject of the next section.

Section Review

| ❶ Identify four structural differences that distinguish humans from apes. | ❷ Describe three difficulties you might face in trying to find hominid fossils. | ❸ Evaluate: "During human evolution, upright posture and large brain size evolved at the same time." | ❹ Describe three differences between *A. afarensis* and modern humans. |

■ *Section Review Answers* ■

1. See Table 11.2 on page 228.
2. Hominid fossils are rare, often fragmentary, and usually buried deep within rock formations.

3. This statement is false, *Australopithecus afarensis* walked upright but had an ape-sized brain.

4. *A. afarensis* features were more apelike than human. *A. afarensis* had a protruding face with fairly large jaws and teeth sized between those of apes and humans. The skull of *A. afarensis* resembles the chimpanzee's more than the human's.

BECAUSE OF YOUR LARGE COMPLEX BRAIN, YOU ARE ABLE TO READ THIS BOOK, WHILE A CHIMPANZEE IS NOT. YOUR BRAIN IS ABOUT THREE TIMES LARGER THAN A CHIMP'S BRAIN AND MORE THAN TWICE THE SIZE OF THE LARGEST *AUSTRALOPITHECUS* BRAIN. LARGE BRAINS ARE A CHARACTERISTIC OF OUR GENUS, *HOMO*, AND ARE RESPONSIBLE FOR OUR COMPLEX CULTURE.

human 3 X > chimp Brain skull

11.3 The First Humans

Objectives

❶ List the two key characteristics of *Homo habilis.*

❷ Contrast the skeletal features of *Homo habilis* with those of *Homo erectus.*

❸ Summarize the evidence that *Homo sapiens* evolved in Africa.

❹ Describe evidence that Neanderthals had a complex culture.

Evolution of the Genus Homo

TEST

Humans belong to *Homo sapiens*, the third and only surviving species of the genus *Homo*. All members of this genus are called humans. About 2 million years ago, the first member of our genus, *Homo habilis*, evolved from australopithecine ancestors. By about 1.5 million years ago, *Homo habilis* was replaced by its larger-brained descendent, *Homo erectus*. In turn, *H. erectus* gave way to its even larger-brained descendent, *H. sapiens*.

There are many hypotheses describing which australopithecines gave rise to humans. **Figure 11.14** shows two of the more popular hypotheses. Most researchers think that *Australopithecus afarensis*, the oldest known hominid, is the ancestor of all other hominids, including ourselves. *TEST*

Figure 11.14
These evolutionary trees represent two hypotheses of how *Homo sapiens* evolved from australopithecines.

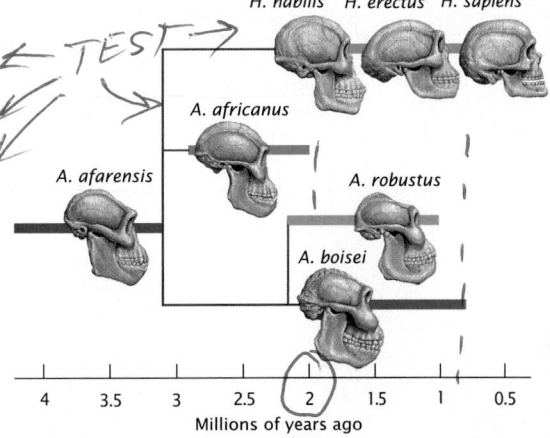

a Hypothesis #1: *Australopithecus afarensis* gave rise to three lines of hominids: two lines of australopithecines and one line that led to the genus *Homo*.

b Hypothesis #2: *Australopithecus afarensis* gave rise to three lines of hominids. About 2 million years ago, *A. boisei* diverged from its common ancestor with *A. robustus*.

← TEST →

■ **Matter of Fact** ■

What's in a Name
Lucy received her name after the Beatles' song, "Lucy in the Sky With Diamonds," which Johanson and his colleagues had listened to the night before unearthing her fossil remains.

Lesson Plan 11.3

Phase 1
PREPARATION

Key Concepts
- One of the species of australopithecines is the ancestor of the genus *Homo*, which consists of three species.
- Modern humans belong to *Homo sapiens*, a descendant of *Homo erectus*, which in turn evolved from *Homo habilis*.
- *Homo habilis* had a large brain and used crude stone tools.
- *Homo erectus* was the first hominid to use specialized tools, to hunt, to build shelters, and to form social groups.

Reading Strategy

One way to improve students' ability to remember what they read involves repetition. Have students repeat orally any written information they consider important. Suggest that they repeat their words aloud several times as a way of reinforcing their reading.

Phase 2
TEACHING STRATEGIES

Visual Strategy

Figure 11.14
Have students compare and contrast the two hypotheses illustrated in this figure. Use this figure to impress upon students that the lineage of *Homo sapiens* is unclear, primarily because so little fossil evidence exists.

Our African Origins

Since the publication of Darwin's *Descent of Man* in 1871, scientists have spread over the globe in search of fossil remains that tell the story of the origin of humans. Scientists have unearthed fossils, tools, and other remains belonging to *Homo habilis* and *H. erectus*, the ancestors of our species. Follow the map in **Figure 11.15** to see where many of these finds were discovered.

Homo habilis evolved in Africa
The earliest fossils of *H. habilis* have been discovered in East Africa. These fossils are over 2 million years old. *Homo habilis* is known for its large brain; the brain of one specimen is about 775 cm³ (47 cu. in.), nearly twice the size of Lucy's brain. *Homo habilis* fossils have been found with crude stone tools. Most scientists think that *H. habilis* fashioned and used these tools for hunting and scraping. *Homo habilis* existed from about 2 million to 1.5 million years ago, when it was replaced by *H. erectus*.

Homo erectus migrated out of Africa
Homo erectus was a larger species than *H. habilis*—about 1.5 m (5 ft.) tall. It also had a large brain (about 1,000 cm³, or 61 cu. in.) and walked erect.

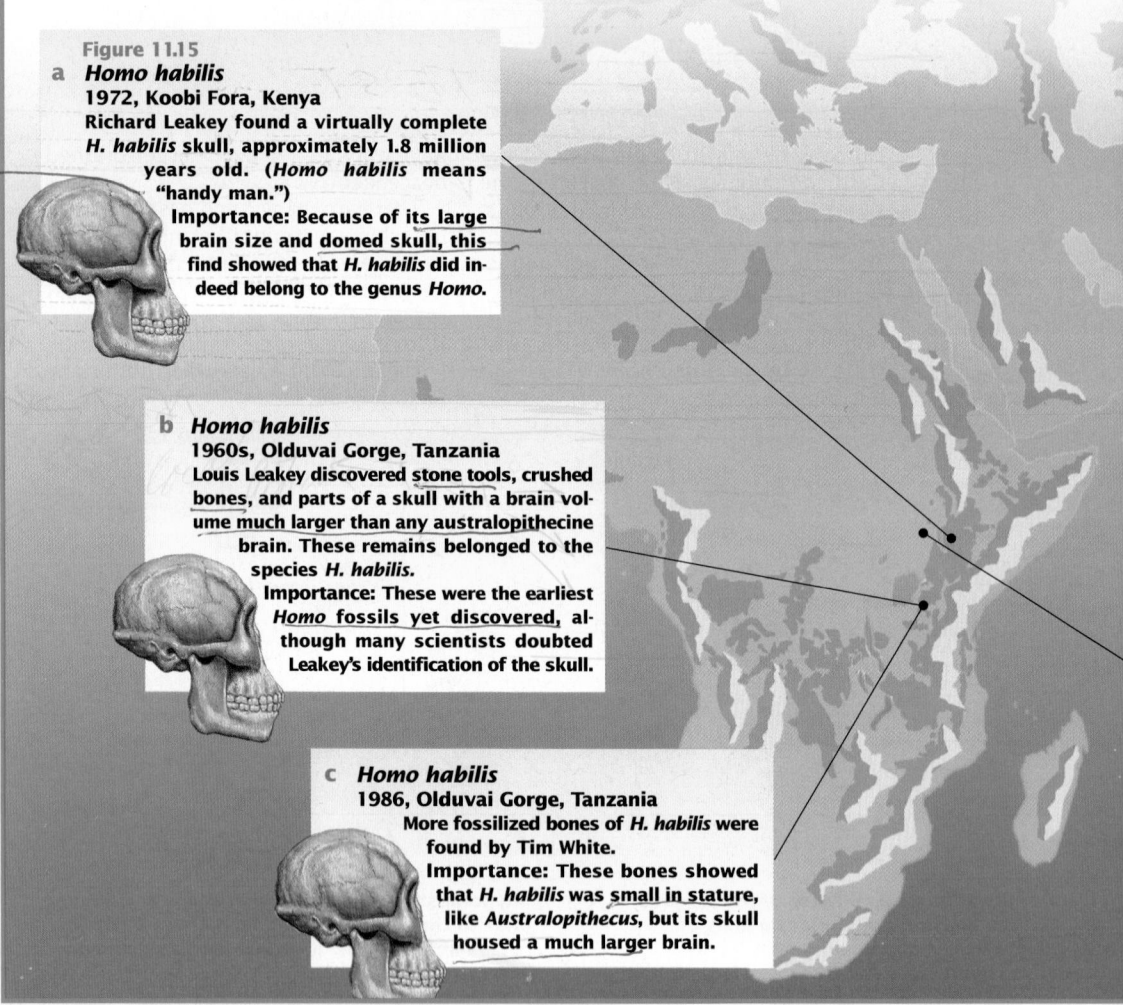

Figure 11.15

a *Homo habilis*
1972, Koobi Fora, Kenya
Richard Leakey found a virtually complete *H. habilis* skull, approximately 1.8 million years old. (*Homo habilis* means "handy man.")
Importance: Because of its large brain size and domed skull, this find showed that *H. habilis* did indeed belong to the genus *Homo*.

b *Homo habilis*
1960s, Olduvai Gorge, Tanzania
Louis Leakey discovered stone tools, crushed bones, and parts of a skull with a brain volume much larger than any australopithecine brain. These remains belonged to the species *H. habilis*.
Importance: These were the earliest *Homo* fossils yet discovered, although many scientists doubted Leakey's identification of the skull.

c *Homo habilis*
1986, Olduvai Gorge, Tanzania
More fossilized bones of *H. habilis* were found by Tim White.
Importance: These bones showed that *H. habilis* was small in stature, like *Australopithecus*, but its skull housed a much larger brain.

Where did *H. erectus* originate? Since its immediate ancestor, *H. habilis,* lived in Africa, it should come as no surprise that the earliest *H. erectus* skulls have been found in Africa. *Homo erectus* first evolved in Africa about 1.5 million years ago. Its appearance marked the beginning of an expansion of human populations across the globe. Far more successful than *H. habilis, H. erectus* quickly became widespread and abundant in Africa and migrated into Asia and Europe about 1 million years ago.

Homo erectus probably lived in small groups of 20 to 50 individuals. We know how these early humans lived because scientists have found the remains of their living places. Some *Homo erectus* groups lived in caves, and there is evidence that they also built crude wooden shelters. They successfully hunted large animals, butchered them using flint and bone tools, and cooked them over fires. *Homo erectus* living sites in China contain the remains of horses, bears, elephants, deer, and rhinoceroses. *Homo erectus* survived for more than 1 million years, longer than any other species of human. These very adaptable humans disappeared in Africa and Europe only about 500,000 years ago, while modern humans were evolving. Interestingly, *H. erectus* survived much longer in Asia, until about 250,000 years ago. *Homo erectus* was without serious doubt our immediate ancestor. From the neck down, we are almost identical to this early human.

d *Homo erectus*
1927–1938, Peking (now Beijing), China
Scientists unearthed a *Homo erectus* skull that closely resembled Java Man's skull (found in 1890). It was nicknamed "Peking Man." Fourteen other well-preserved skulls, together with lower jaws and other bones, crude stone tools, and the ashes of campfires were also discovered.
Importance: These finds showed that both Java Man and Peking Man belonged to the same species, *H. erectus.*

e *Homo erectus*
1890, Java (now Indonesia)
Physician Eugene Dubois discovered the top of a skull that could house a brain much larger than any ape's brain. It belonged to *Homo erectus* and was nicknamed "Java Man."
Importance: Java Man's skull proved to be older than any other hominid fossils discovered up to that time.

f *Homo erectus*
1984, Lake Turkana, Kenya
Scientists discovered a nearly complete skeleton of *Homo erectus*, dated at about 1.5 million years old (1 million years older than the skulls found in Java and Peking).
Importance: The age of this fossil suggested that *H. erectus* first evolved in Africa.

  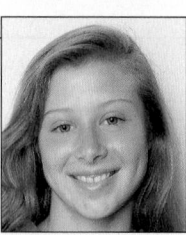 

The Origin of Homo sapiens

Figure 11.16
Scientists disagree about how and when different racial groups, some of which are represented by the people above, evolved.

Scale and Structure

How do the size and shape of the skull of Homo habilis *show that this species had a larger brain than* Australopithecus?

Where *Homo sapiens* evolved has been a much-debated issue among scientists. Some scientists argue that our species evolved simultaneously from different groups of *H. erectus* that lived in different areas of the Old World. According to these scientists, the differences among racial groups, such as among the young people in **Figure 11.16**, reflect the descent of each racial group from a different population of *H. erectus*. In contrast, other scientists contend that our species evolved in one place, in Africa, and then spread to the rest of the world. Therefore, the unique characteristics of each race represent adaptations to local conditions that arose after our species migrated out of Africa.

Recently, scientists studying DNA sequences in mitochondria have uncovered evidence supporting the hypothesis that *H. sapiens* evolved in Africa and then spread to the rest of the world, as shown in **Figure 11.17**. Mitochondria carry their own DNA and reproduce independently of the cell's chromosomes. Moreover, mitochondrial DNA is inherited only from the mother. The egg contains mitochondria, which give rise to all the mitochondria in the body of the child; sperm do not contribute any mitochondria to the offspring. Unlike DNA in the nucleus, mitochondrial DNA is not shuffled by meiosis or crossing over. Any changes in mitochondrial DNA are caused by mutations. Scientists have determined the sequence of nucleotides in mitochondrial DNA samples from people on all continents. Since DNA accumulates mutations over time, the oldest DNA should show the largest number of mutations. It turns out that the greatest number of different mitochondrial DNA sequences occurs among modern Africans. Thus, we can conclude that humans have been living in Africa longer than on any other continent. This evidence suggests that our species evolved in Africa, and that the differences among racial groups emerged after *H. sapiens* descended from *H. erectus*.

Figure 11.17
This map shows one hypothesis suggesting that *Homo sapiens* evolved first in Africa and then migrated to other continents.

Visual Strategy

Figure 11.18
Use this figure to point out the massive skulls Neanderthals possessed. Explain that the Neanderthal's larger brain does not mean they were more intelligent than modern humans. Although smaller, the brain of a modern human has a much more developed cerebrum.

Theme Connection

Evolution

What happened to the Neanderthals is not known with certainty. Perhaps they were exterminated in warfare by modern *Homo sapiens* migrating out of Africa. No evidence, however, exists to support this possibility. Perhaps Neanderthals and modern humans interbred, although again, no evidence has been found to support the idea. Another possibility is that Neanderthals were eliminated by natural selection. The better-equipped *Homo sapiens*, with their more advanced tools, complex social organizations, and perhaps a language capability, were better adapted to the environment than were Neanderthals. Neanderthals may have simply been unable to compete with the newcomers for food and shelter. Consequently, they became extinct.

Figure 11.18
Fossil evidence shows that Neanderthals lived complex lifestyles. They hunted, built shelters for their families, buried their dead, and protected themselves from cold temperatures.

Neanderthals were the first *Homo sapiens* in Europe

Homo sapiens migrating from Africa first appeared in Europe about 130,000 years ago. The first fossil of one of these individuals was found in 1856 in the Neander Valley of Germany. Hence, these early humans were called Neanderthals (*nee AN dur THALZ;* the word "thal" means valley in Old German). By 70,000 years ago, Neanderthals were common in Europe and in parts of the Middle East and North Africa. They made diverse stone tools, including scrapers, spearheads, and hand axes. Neanderthals were more powerfully built, shorter, and stockier than humans today. Their skulls were massive, with protruding faces and bony brow ridges. Their brains were even larger than those of modern humans. They lived in huts, caves, or open air sites, as shown in **Figure 11.18**. Neanderthals took care of their injured and sick and buried their dead, often placing food, weapons, and even flowers with the dead bodies. Such attention to the dead suggests that they may have believed in a life after death. This is the first evidence of the symbolic thought processes characteristic of modern humans.

Modern *Homo sapiens* evolved in Africa and replaced the Neanderthals

A new version of *Homo sapiens* essentially identical to modern humans appeared in Africa as early as 100,000 years ago. By about 40,000 years ago, these modern humans had migrated out of Africa and had replaced the Neanderthals of southwest Asia. The modern humans then spread across Europe, coexisting and possibly interbreeding with Neanderthals.

By about 34,000 years ago, the Neanderthals had disappeared. The modern humans that replaced them had skulls identical to those of present-day humans. In fact, one of these early modern humans would look just like you or me (groomed and wearing modern clothes, of course).

Visual Strategy

Figure 11.20

Tell students that the meaning of these cave paintings has long been a matter of debate. They may have had some spiritual significance. Some of the animals are marked with darts or wounds, suggesting that these people believed they could control what happened to animals by directing the appropriate action against its image. In addition, many of the animals appear to be pregnant, suggesting they may have symbolized fertility.

Social Studies Connection

The Human Cultural Evolution

Cave paintings came to an abrupt end about 8,000 to 10,000 years ago. In fact, cave dwelling appears to have ceased as a way of life at that time. Instead, as the ice sheets began to retreat, *Homo sapiens* began a new way of life—agriculture. As the glaciers retreated, once-cold grasslands gave way to forests and land that could be farmed. The most important event in the cultural evolution of *Homo sapiens* was the change from hunting to agriculture.

Phase 3

ASSESSMENT OPTIONS

Closure Strategy

Is it Homo sapiens?

Assume that a nearly complete fossil skeleton has just been found. Ask students what evidence they would need to justify the claim that the fossil was an early member of *Homo sapiens,* and not a member of a different species.

Section Review

Assign the *Section Review.*

Reteaching

Ask students why scientists disagree on the evolutionary history of modern humans.

236

Figure 11.19
Modern *Homo sapiens* used stone tools, such as these, about 10,000 years ago.

Figure 11.20
The painting below was discovered in a cave in Lascaux, France, about 50 years ago. It was painted by our *Homo sapiens* ancestors 17,000 to 20,000 years ago.

These modern humans used tools made of bones, horn, and stone, as shown in **Figure 11.19**. Their preserved villages and burial grounds show that they had complex social organization. And the shape of certain regions of the skull suggests that modern *H. sapiens* had full language capabilities.

At the time anatomically modern *H. sapiens* moved into Europe, the world was cooler than it is now. A huge sheet of ice extended from the Arctic into northern Europe; southern Europe was covered with grasslands inhabited by large herds of grazing animals. Modern *H. sapiens* painted pictures of these animals deep within caves throughout Europe, as shown in **Figure 11.20**.

Modern *H. sapiens* eventually spread throughout Asia and Australia and entered North America, which they reached at least 13,000 years ago. At that time, the ice sheets had begun to

retreat and a land bridge connected Siberia and Alaska. By 10,000 years ago, about 10 million people lived throughout the entire world.

Like all other living things, humans are the product of evolution. Our evolution has been marked by a progressive increase in brain size. As a result, humans are the only animal able to make complex tools, a capability that has enabled us to change the world around us. Although not the only animal capable of conceptual thought, we have extended this ability until it has become the hallmark of our species. We use symbolic language and can shape concepts out of experience with words. This ability has led to the accumulation of knowledge that can be transmitted from one generation to the next. Thus, we have what no other animal has ever had, cultural evolution. Through culture, we have found ways to mold our environment to our needs. We control our biological future in a way never before possible—an exciting potential with weighty responsibility.

Section Review

❶ **Identify two characteristics of *Homo habilis*.**

❷ **Contrast two features of the skeletons of *Homo habilis* and *Homo erectus*.**

❸ **How does mitochondrial DNA analysis support the hypothesis that *Homo sapiens* evolved in Africa?**

❹ **Describe some of the cultural traits of Neanderthals.**

■ *Section Review Answers* ■

1. *Homo habilis* had a large brain (approximately 775 cm^3) and a dome-shaped skull. They were the first to make stone tools.
2. *Homo erectus* had a larger brain and was taller than *Homo habilis*.
3. Scientists have determined the sequence of nucleotides in mitochondrial DNA from people on all continents. The oldest DNA should show the largest number of mutations. The greatest number of different mitochondrial DNA sequences occurs among modern Africans, thus supporting the hypothesis that *Homo sapiens* evolved in Africa.
4. Answers might include: use of stone tools, living in huts, caves, or open air sites; caring for the injured and sick; burial of the dead.

Chapter 11 *Highlights*

"I don't know what it means but I like the look of it."

Assessment Alternative
Have students imagine that they are gathered around a campfire while on a fossil dig in Africa. Suddenly, hominids begin to be drawn to the fire, including *Australopithecus africanus, Australopithecus robustus, Australopithecus boisei, Australopithecus afarensis, Homo habilis, Homo erectus,* and *Homo sapiens.* Have them describe how they would distinguish one from the other.

Key Terms	Summary

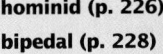

11.1 Primates

Many New World monkeys, such as this spider monkey, have long tails that can be used as an extra limb.

Key Terms:
primate (p. 221)
opposable thumb (p. 221)
binocular vision (p. 222)
prosimian (p. 222)
anthropoid (p. 223)

Summary:
- Prosimians, monkeys, apes, and humans are primates. They have five fingers and opposable thumbs. Primates also have binocular vision.
- The first primates were prosimians.
- Anthropoids evolved about 35 million to 40 million years ago. Apes, monkeys, and humans are anthropoids.
- Humans and chimpanzees share 98.4 percent of their DNA nucleotide sequences.

11.2 Evolutionary Origins of Humans

Because of its massive jaws, *Australopithecus boisei* was nicknamed "nutcracker man."

Key Terms:
hominid (p. 226)
bipedal (p. 228)

Summary:
- Humans and their closest fossil relatives are known as hominids.
- The earliest hominids belong to the genus *Australopithecus.*
- Fossils show that upright walking evolved soon after the evolutionary line leading to modern humans diverged from the line leading to chimpanzees.
- *Australopithecus afarensis* walked upright but had an ape-sized brain.

11.3 The First Humans

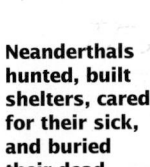

Neanderthals hunted, built shelters, cared for their sick, and buried their dead.

Summary:
- The genus *Homo* evolved about 2 million years ago in Africa. The first member of our genus was *Homo habilis.*
- *Homo habilis* was later replaced by *Homo erectus. Homo erectus* migrated out of Africa to Europe and Asia.
- Mitochondrial DNA analysis reveals that *Homo sapiens* evolved in Africa.
- *Homo sapiens* that lived in Europe, North Africa, and the Middle East, from 130,000 to 40,000 years ago are known as Neanderthals.
- Modern *Homo sapiens* evolved in Africa about 100,000 years ago and migrated to Europe about 40,000 years ago.

Chapter Review Answers

Understanding Vocabulary

Choose the term or phrase that does not fit the pattern and explain why it does not fit.

1. **a.** *Baboon* does not fit because it is not an ape.
   b. *Long canines* does not fit the pattern because they are a characteristic of chimps; the others are characteristics of modern humans.
   c. *Brain about 400 cm³* does not fit because it is not an attribute associated with *Homo erectus*; the others are attributes of *Homo erectus*.
   d. *Lemur* does not fit the pattern because it is not an anthropoid.

Relating Concepts

2. Map answer is shown on page 219D.

Understanding Concepts

Multiple Choice

3.	a	9.	b
4.	b	10.	d
5.	a	11.	d
6.	c	12.	a
7.	c	13.	d
8.	d	14.	c

Completion

15. *Australopithecus africanus*
16. *Australopithecus*; 3 million
17. 1.5 million; Africa

Short Answer

18. Binocular vision allows the visual fields of both eyes to overlap, enabling the animal to judge distance accurately.
19. Their evolutionary divergence occurred quite recently, about 5 million years ago.
20. Hominid fossils are rare because hominids were uncommon until recently; locations where hominids lived have generally been unsuitable for fossil formation; and fossils are difficult to recover because they are buried deep in the ground and are easily destroyed when exposed to the weather.

Chapter 11 Review

Understanding Vocabulary

1. For each set of terms or phrases, choose the term or phrase that does not fit the pattern and explain why it does not fit.
   a. baboon, chimpanzee, gorilla, gibbon
   b. bipedal, arms shorter than legs, long canines, brain about 1,350 cm³ (83 cu. in.)
   c. migrated to Europe and Asia, walked erect, controlled fire, brain about 400 cm³ (29 cu. in.)
   d. anthropoid, *Homo sapiens*, gorilla, lemur

Relating Concepts

2. Copy the concept map below onto a sheet of paper. Then complete the map by writing the correct word or phrase in each oval containing a question mark.

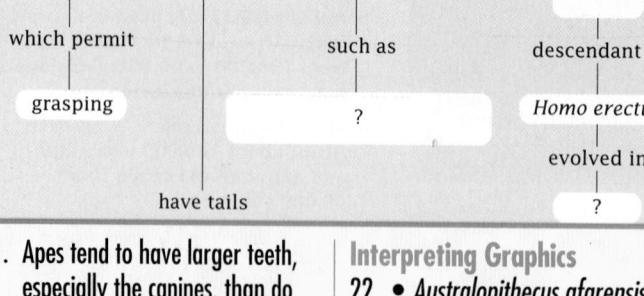

Understanding Concepts

Multiple Choice

3. All primates have grasping fingers and
   a. binocular vision.   c. tails.
   b. color vision.   d. long canine teeth.

4. The first animals with an opposable thumb were
   a. chimpanzees.   c. cats.
   b. prosimians.   d. monkeys.

5. The DNA nucleotide sequences of this species are least like those of humans.
   a. gibbon   c. orangutan
   b. bonobo   d. gorilla

6. Which characteristic does not indicate that australopithecines walked upright?
   a. angle formed by femur and pelvis
   b. shape of spinal column
   c. size of teeth
   d. position of hole through which spinal cord leaves skull

7. The oldest hominid fossils are those of
   a. *Homo erectus*.
   b. *Australopithecus boisei*.
   c. *Australopithecus afarensis*.
   d. *Homo habilis*.

8. *Australopithecus afarensis* is like modern humans in
   a. length of canine teeth.
   b. brain size.
   c. height.
   d. bipedal posture.

9. The skeleton nicknamed "Lucy" provided evidence to suggest that
   a. *A. afarensis* was more humanlike than previously thought.
   b. upright walking evolved before large brains.
   c. Lucy was really a chimpanzee.
   d. humans coexisted with dinosaurs.

10. The common ancestor of both *Homo sapiens* and *Australopithecus robustus* is
    a. *Homo habilis*.
    b. *Homo erectus*.
    c. *Australopithecus boisei*.
    d. *Australopithecus afarensis*.

11. Which is the accepted evolutionary sequence for the genus *Homo*?
    a. *H. sapiens > H. erectus > H. habilis*
    b. *H. erectus > H. sapiens > H. habilis*
    c. *H. habilis > H. sapiens > H. erectus*
    d. *H. habilis > H. erectus > H. sapiens*

12. This species migrated out of Africa.
    a. *Homo sapiens*
    b. *Australopithecus afarensis*
    c. *Homo habilis*
    d. *Australopithecus boisei*

13. Neanderthals did *not*
    a. bury their dead.
    b. have larger brains than modern humans.
    c. make and use tools.
    d. live 10,000 years ago.

14. Neanderthals are included in the species
    a. *Australopithecus boisei*.
    b. *Homo erectus*.
    c. *Homo sapiens*.
    d. *Australopithecus africanus*.

21. Apes tend to have larger teeth, especially the canines, than do hominids. Also, the jaws of hominids tend to be smaller and more rounded than ape jaws.

Interpreting Graphics

22. • *Australopithecus afarensis*
    • *A. africanus*
    • *Homo sapiens*

Reviewing Themes

23. Madagascar is the home of all the species of lemurs that survive today. The destruction of Madagascar's natural vegetation will likely lead to the extinction of these species.

Completion

15. Raymond Dart discovered the first specimen of _____ .

16. The most complete skeleton from the genus _____ is called Lucy. She lived in Africa about _____ years ago.

17. *Homo erectus* is thought to have evolved about _____ years ago on the continent of _____ .

Short Answer

18. What is the advantage of binocular vision?

19. The nucleotide sequences of chimpanzee and human genes differ by only 1.6 percent. What does this reveal about the evolutionary relationship between these two primates?

20. If Donald Johanson invited you on his next expedition to Ethiopia, what problems would you face when trying to find hominid fossils?

21. On this expedition, you discover what might be a hominid jaw. How can you tell if this jaw is a hominid jaw and not an ape jaw?

Interpreting Graphics

22. Look at the evolutionary tree below.

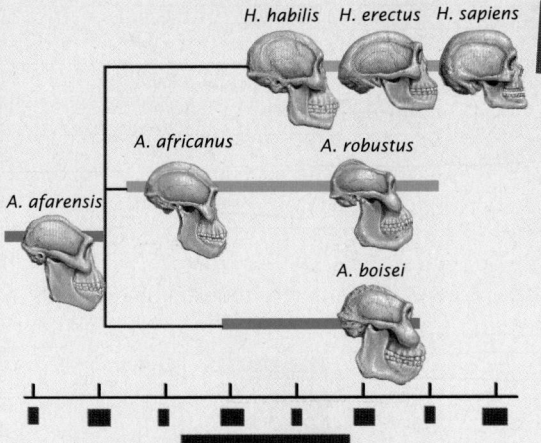

- Which species is the common ancestor of all the other species?
- Which species is the direct ancestor of *A. robustus*?
- Which species exists today?

Reviewing Themes

23. *Interacting Systems*
Why is there concern about the destruction of Madagascar's forests?

24. *Scale and Structure*
How are monkeys and prosimians different?

25. *Evolution*
Why is it inaccurate to say that humans evolved from chimpanzees?

26. *Interacting Systems*
Fossil evidence suggests that *Homo erectus* was nomadic. Where did their travels take them?

Thinking Critically

27. *Inferring Conclusions*
Based on your understanding of fossils and the history of human evolution, explain how error could have been introduced into this history.

28. *Inferring Conclusions*
Why do you think much controversy surrounds the study of human evolution?

Cross-Discipline Connection

29. *Biology and Physical Education*
The opposable thumb enables primates to perform tasks that other animals cannot. Test the truth of this statement by taping your thumb to your forefinger and then attempting some common tasks such as writing your name or opening a door. How has your performance changed?

Discovering Through Reading

30. Read "Machiavellian Monkeys," in *Discover*, June 1991, pages 69–73. To what do Byrne and Whitten attribute our superior intelligence?

28. Some people claim that human evolution is at odds with religious teachings, while others contend that scientists disagree about the evidence for human evolution provided by fossils and DNA.

Cross-Discipline Connection

29. It is harder to write your name, and difficult to open a door, with the thumb taped.

Discovering Through Reading

30. Consistent with Humphrey's hypothesis, Byrne and Whitten claimed that the evolution of primate intelligence is a result of social interaction, primarily deceptive social interaction. The study of monkeys paints a picture of the social life of early hominids, thus providing a glimpse of how cognitive skills may have been honed over time.

---

24. Monkeys have smaller eye sockets than prosimians, and they have color vision. Monkeys are typically active during the day and eat fruits and leaves, while prosimians are nocturnal and hunt insects.

25. Humans and chimpanzees evolved from a common ancestor. To say that humans evolved from chimpanzees is inaccurate.

26. Their travels took them to Asia and Europe, far from their native Africa.

Thinking Critically

27. The fossil record is incomplete and most interpretations of lineage have been based on small fossils that could easily be misidentified or misinterpreted.

Procedural Note

1. Discuss the difference between an inference, an observation, and a hypothesis.
2. Although most students know the names and locations of the various types of teeth, a review of this information might be helpful.

Prelab Preparation Answers

- apes in general
- incisors
- molars

Procedure Answers

5. Students will likely have two fewer molars. The wisdom teeth do not usually erupt until early adulthood.
6. molars: grinding; premolars: grinding; canines: biting and gripping; incisors: cutting

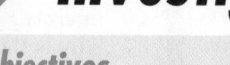

Chapter 11

Investigation

Comparing Structure of Animal Jaws

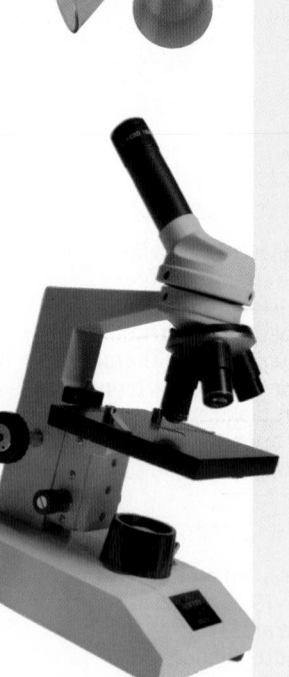

Objectives

In this investigation you will:
- *observe* the jaws and teeth of four kinds of animals
- *infer* the diet of all four animals from the shapes of their teeth and jaws

Materials

- mirror

Prelab Preparation

Review what you have learned about human evolution by answering the following questions:
- Which animals are the closest relatives of humans?
- Which teeth do you use for biting off a chunk of food?
- Which teeth do you use to chew?

Procedure: Comparing Animal Jaws

1. Form a cooperative team with another student to complete steps 2–7.
2. Make a table similar to the one shown below for recording your observations.

	Human	Chimpanzee	Horse	Coyote
Molars				
Number				
Location				
Size and shape				
Premolars				
Number				
Location				
Size and shape				
Canines				
Number				
Location				
Size and shape				
Incisors				
Number				
Location				
Size and shape				

3. Use a mirror to examine the teeth in your lower jaw.

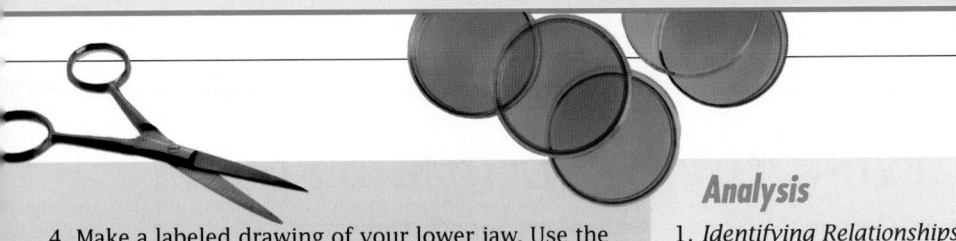

4. Make a labeled drawing of your lower jaw. Use the figure shown below as a guide to identify the types of teeth present in your lower jaw.

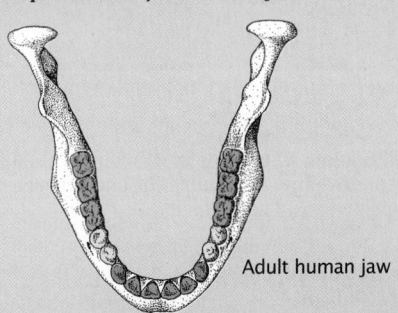

Adult human jaw

5. Look at your molars. Note their relative size and the shape of their biting surfaces. *How many molars do you have in your lower jaw? Does it have fewer molars than does the jaw shown above? If so, why?* Record your observations in your table.

6. Repeat step 5 for your premolar, canine, and incisor teeth. *What function does each kind of tooth serve?*

7. Repeat steps 5 and 6 using the jaws of the chimpanzee, horse, and coyote, below.

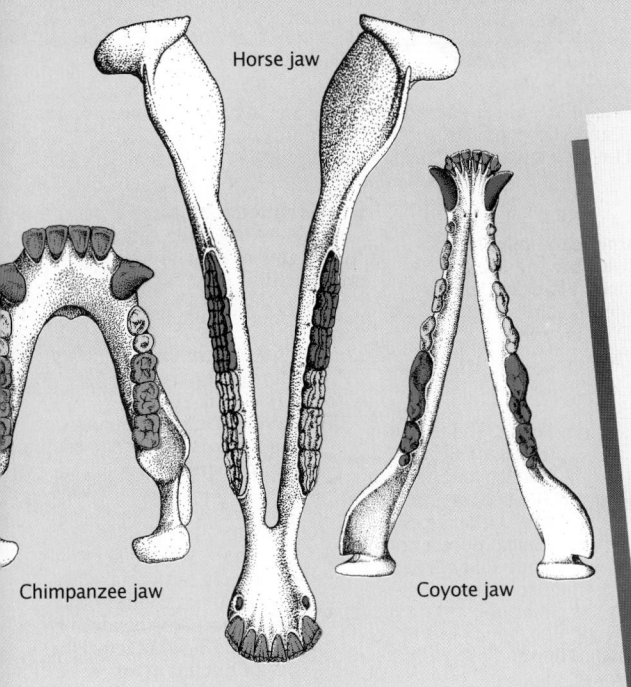

Horse jaw

Chimpanzee jaw

Coyote jaw

Analysis

1. *Identifying Relationships*
   On what does the coyote feed? How is each kind of tooth adapted for this diet?

2. *Identifying Relationships*
   Describe how the horse's teeth are adapted for a diet of plants.

3. *Comparing Structures*
   Describe three similarities between your teeth and the teeth of the chimpanzee. Now describe two differences.

4. *Inferring Conclusions*
   On what do chimpanzees feed? Explain your answer.

5. *Inferring Conclusions*
   Suppose you were shown only the canine teeth of these animals. Would you still think the chimpanzee was closely related to humans? Explain your answer.

6. *Evaluating Methods*
   How might studying living animals help biologists better understand the extinct ancestors of these animals?

Thinking Critically

Gorillas live in small groups composed of a few females, some offspring, and a few males. Each group is controlled and defended by a large male. One way males ward off rival males is by exposing their large canines in a threatening display. Explain how you would test this hypothesis: Gorillas have larger canines than humans because male gorillas use their canines in defense.

Analysis Answers

1. Students should indicate that the coyote is a carnivore. Its canine teeth are for holding and tearing prey, the row of sharp incisors are for cutting meat free, and the premolars and molars are for slicing flesh.

2. Broad, flattened molars are for grinding and chewing, and sharp incisors are for cutting plant material free.

3. Similarities: Chimpanzees and humans have the same number of each kind of tooth; the same total number of teeth; and similarly shaped incisors, premolars, and molars. Differences: Chimpanzees have a space between incisors, canines, and molars; canines are longer; and the chimpanzee's incisors jut out.

4. Chimpanzees are largely herbivores, eating various types of leaves, barks, fruits, and buds. Although their large canines suggest that chimpanzees eat meat, they do so rarely.

5. Canines of the chimpanzee are much larger than the human's. Based on canines alone, humans would not appear to be closely related to the chimpanzee.

6. We can infer the habits of extinct organisms by studying the habits of their living relatives.

Thinking Critically Answer

If large canines are an adaptation for signaling in males, then we would not expect females to possess such large canines. Thus, we could test this hypothesis by comparing the size of canine teeth in males and females. It turns out that canines of males are much larger than canines of females, even when differences in body size are factored out.

Using the Feature

- Explain how Walter Sutton combined his observation of chromosome behavior with his previous knowledge of inheritance to infer that chromosomes carry the "factors" that Mendel had referred to years earlier. Have students research Walter Sutton's observations of chromosome behavior.

- Have students research Thomas Hunt Morgan's work with *Drosophila*, and write a brief report about how Morgan's research led to scientists' understanding of inheritance patterns introduced in Chapter 7.

- Tell students that new techniques are surfacing that will enable scientists to construct entire chromosomes, rather than individual genes. Scientists have already constructed artificial yeast cell chromosomes by isolating those regions that govern the essential activities of the chromosome and splicing those regions into a bacterial plasmid. The plasmid was then cut, creating a linear chromosome, and reinserted into the yeast cell. Researchers can learn more about the mechanisms of chromosomal segregation and replication by studying these artificial chromosomes. For example, scientists have determined that chromosome length affects the rate of segregation error. Encourage students to think about how such knowledge might be used to determine the cause of Down syndrome and other chromosomal abnormalities.

- Emphasize that genetic engineering allows humans to almost instantly alter life forms while evolution may take thousands of years to bring about change in a species.

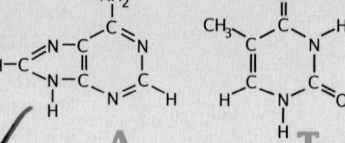

Discoveries in Science ▪ Discoveries in Science ▪ Disco

IDENTIFYING THE GENETIC MOLECULE

1850

1871 Friedrich Miescher, a German scientist, isolates the nucleic acids DNA and RNA from cell nuclei.

A   T   G   C

Lubber grasshopper

1903 American biochemist **Phoebus A. Levene** shows that DNA contains four nitrogenous bases: adenine (A), thymine (T), guanine (G), and cytosine (C), which are shown in the diagram above.

1910–1915 Thomas Hunt Morgan, an American geneticist, confirms that the genes of fruit flies occur in linear sequences on chromosomes.

Thomas Hunt Morgan

1902 Walter S. Sutton, an American cytologist studying grasshopper genetics, proposes that genes are found on chromosomes. Based on the work of Gregor Mendel, scientists know that genes control an organism's traits.

Chromosomes stained with fuchsin

1914 Robert Fuelgen, a German scientist, discovers that nucleic acids can be stained with the red dye fuchsin, helping confirm that nucleic acids are found in chromosomes.

1950

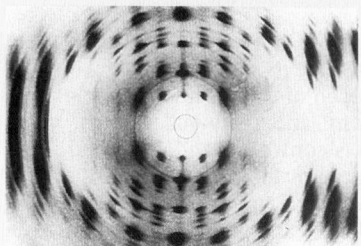

X-ray diffraction photograph of DNA

1952 American scientists **Alfred Hershey** and **Martha Chase** purify DNA and show that DNA, rather than protein, is the genetic material. The race to discover the structure of DNA quickens.

1953 American biologist **James Watson** and British biophysicist **Francis Crick** determine that the structure of DNA is indeed a double helix.

1961 Francis Crick confirms the hypothesis that three sequential nucleotides in DNA code for one amino acid.

1961 American scientists **Marshall Nirenberg**, **J.H. Matthei**, and **Severo Ochoa** demonstrate the three-base genetic code for DNA and mRNA.

1951–1952 British scientists **Maurice Wilkins** and **Rosalind Franklin** use X-ray diffraction to produce images of DNA. In 1952, Franklin produces a DNA image suggesting that DNA is a spiral molecule called a helix.

1954 The use of Chargaff's ratios brings scientists to the realization that the sequence of DNA bases represents a code that carries hereditary information.

1971 Biochemist **Marie Maynard Daly**, an African American, researches the atomic bonding characteristics of nucleic acids.

Rosalind Franklin

cience ▪ **Discoveries in Science** ▪ **Discoveries in Science**

Edward Tatum

1941 American scientists **George Beadle** and **Edward Tatum** show that each gene mutation in bread mold leads to a change in one of the mold's enzymes. This links changes in genes to changes in proteins.

George Beadle

1950

1947 Erwin Chargaff shows that for every adenine there is a thymine, and for every guanine there is a cytosine. No one can explain why these proportions of A to T and G to C are found in DNA.

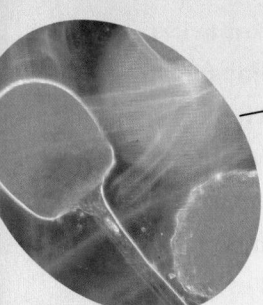
Reproductive structures of black bread mold

1944 American scientists **Oswald Avery**, **Maclyn McCarty**, and **Colin MacLeod** provide evidence that DNA is the genetic material, an idea greeted with skepticism. Others believe that proteins are the genetic material.

1992

1975 Scientist **F. Agnes Stroud-Lee**, a Santa Clara Pueblo (Tewa) Native American, studies abnormalities in chromosomes resulting from birth defects, chemicals, and radiation.

F. Agnes Stroud-Lee

1972–1973 American molecular biologists **Paul Berg**, **Stanley Cohen**, and **Herbert Boyer** cut genes into pieces using enzymes. These pieces are then inserted into another organism's genes. This technique, similar to combining two strips of film, is called gene splicing.

1992 American geneticist **Mark Dubnick**, who is visually challenged, writes scientific computer applications for genetic engineering, such as DNA sequencing, cloning, and mapping human genomes.

Susumu Tonegawa

1988 James Watson becomes the director of the Human Genome Project (HUGO). Scientists working on this project plan to map the sequence of the 3 billion nucleotides that make up the human genome.

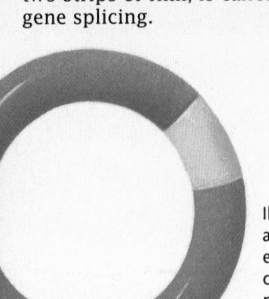
Illustration of a genetically engineered circular DNA molecule

1987 Geneticist **Susumu Tonegawa**, a Japanese-American, is awarded the Nobel Prize for discovering how genes change to create antibodies. His findings open the door to new experimental ideas regarding cells of the immune system.

Discussion

Guide the discussion by posing the following questions.

1. What were the inferences Sutton made on which he based his chromosome theory?

   Sutton based his chromosome theory on the following inferences: chromosomes, like Mendel's factors, occur in pairs; and the sperm and the egg are the only possible carriers of hereditary factors.

2. What precautions should be taken before releasing genetically altered life forms into the environment?

   Answers will vary but students should suggest that regulations should exist that consider the type of life forms developed, and how they will affect the environment.

243

Unit 3

One of the primary concerns of modern biology is the threat to the environment posed by today's high-tech society. Global warming, acid rain, ozone holes, disappearing rain forests—our world is under great stress, and great care will be needed to prevent further damage. In this unit you will discover how living communities like this pine forest function, how human activities are damaging many of the world's natural communities, and what is being done today to protect the fragile Earth. Few areas of biology are as important to your future, and to that of your children.

The Environment

Chapter 12 Ecosystems

Planning Guide

Objectives/Themes	Classwork Resources	Homework Resources

12.1 1. Identify the components of an ecosystem.
 	
2. Discuss the flow of energy through ecosystems.
 	
3. Identify the different trophic levels in an ecosystem.
 	
4. Explain why ecosystems can only contain a few trophic levels.
 	
Themes: Energy and Life, Interacting Systems, Scale and Structure, Stability
	Teacher's Resource Binder
	Focus Activity 12
	  Making a Food Web
	Lab Investigation 12.1
	  Mapping an Environmental Site
	
	Other Resources
	Transparencies 47–49
		Text
		Section Review, p. 252
		
		Teacher's Resource Binder
		Directed Reading Worksheet 12.1
		
		Other Resources
		Audiocassette 12.1*

12.2 1. Recognize that nutrients, water, and carbon cycle within ecosystems. 2. Describe how nitrogen and water are recycled within ecosystems. 3. Understand the role of plants in the cycling of materials within ecosystems. 4. Trace the pathway of carbon between living organisms and the environment. **Themes:** Energy and Life, Interacting Systems, Stability	**Teacher's Resource Binder** Extension Worksheet *The Phosphorus Cycle* **Other Resources** Transparencies 50–52	**Text** Section Review, p. 256 **Teacher's Resource Binder** Directed Reading Worksheet 12.2 **Other Resources** Audiocassette 12.2*

12.3 1. Identify the characteristics of freshwater ecosystems.
 	
2. Identify the factors that determine the ecosystem type found in a particular area.
 	
3. Contrast the seven major terrestrial ecosystems.
 	
4. Identify the major ocean ecosystems.
 	
Themes: Interacting Systems, Stability
	Text
	Investigation
	  Exploring a Soil Community
	  pp. 264–265
		Text
		Section Review, p. 260
		
		Teacher's Resource Binder
		Directed Reading Worksheet 12.3
		Vocabulary Review Worksheet*
		Reteaching Worksheet*
		  Establishing Ecosystems
		
		Other Resources
		Audiocassette 12.3*

*Reteaching Options

Demonstrations
12.1: pp. 248, 249, 250, 251
12.2: p. 256
12.3: pp. 257, 260

Assessment
Chapter Review pp. 262–263
Portfolio Assessment p. 245D
Chapter Test—Teacher's Resource Binder
Test Generator

Research Notes

Connection to Agriculture: Insects as Farmers' Friends

In 1986, the Indonesian government faced a terrible problem. Since 1984, rice production had grown so that the nation could be self-sufficient in food production, but an infestation of the brown planthopper insect was eating enough rice each year to feed 3 million people. The infestation, along with a rapidly growing population, threatened the Indonesians' self-sufficiency.

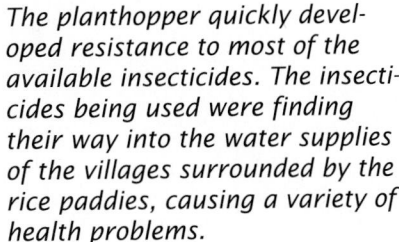

The planthopper quickly developed resistance to most of the available insecticides. The insecticides being used were finding their way into the water supplies of the villages surrounded by the rice paddies, causing a variety of health problems.

Rice farmers cried out for more government subsidies to buy new insecticides. However, in a bold step, the government outlawed all but a handful of the insecticides. Money that had been spent on subsidies of the insecticides was redirected to Indonesia's formerly neglected integrated pest management (IPM) program.

The IPM program, run by the United Nations' Food and Agriculture Organization, focuses on using natural means and knowledge of ecology, rather than insecticides, to control pests.

In each village, farmers meet regularly in an IPM field school to observe rice plants and insects (both pests and beneficial ones) in their ecosystems. Since insecticides kill off beneficial insects, farmers are taught techniques for irrigation management,

weather forecasting, and fertilizer use. Farmers also share their experiences to arrive at a consensus on how best to protect the crop and maximize their yield.

After their training is complete, the farmers spread their knowledge to the rest of the village by putting on a play in the style of traditional folk theater about growing rice.

Not only is insecticide use in Indonesia decreasing sharply, but rice yields increased 10 percent between 1986 and 1992. A test of the techniques came in 1990 when an infestation of the white rice stemborer threatened the rice crop.

Although farmers pleaded for subsidized insecticides, government officials at all levels, from the capital to the villages, insisted on IPM techniques. Nearly all rice paddies were spared a return of the pest. Most of the reinfested paddies were ones in which one of the few remaining legal insecticides was used.

The U.N. hopes to repeat this success story in similar situations elsewhere in Asia.

Investigation Notes

Exploring a Soil Community
pages 264–265

Purpose: This investigation gives the students an opportunity to explore a soil community. In doing so, they will gather evidence that should reveal the existence of different trophic levels and an ecological pyramid.

Technique: The investigation should be approached as an exploration rather than a detailed field study of a community.

Prelab Preparation

1. Select a variety of habitat types and locations from which the students will collect samples. If necessary, the samples can be collected during one class period and analyzed the next day.

2. The use of chemical anesthetics to treat the litter and soil samples is discouraged because of some chemicals' dangerous properties. If access to a refrigerator is unavailable, the use of an anesthetic such as ethyl acetate may be necessary.

Answers will be found on pages 264–265.

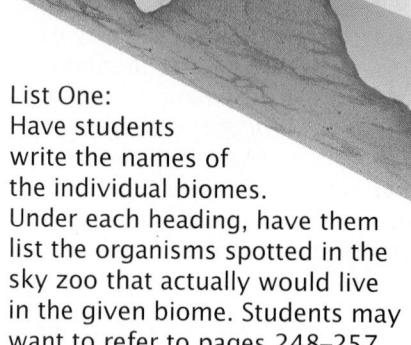

Meeting Individual Needs

Objectives

1. Students will demonstrate appropriate use of core vocabulary for the chapter (Vocabulary File).

2. Students will explore and interpret a classroom simulation modeling the diversity of an ecosystem (Demonstration A).

3. Students will relate a model to the different cycles within an ecosystem (Demonstration B).

Vocabulary File
(Developing Vocabulary/ Limited English Proficiency)

If you are not already using the Vocabulary File, refer to Chapter 1 for its preparation. See the Chapter Highlights on page 261 for a list of suggested words.

Demonstration A
Examples of Ecosystems
for pages 250–252
(Developing Classification Skills/Tactile Learners)

Materials and Preparation
Prepare at least four to six toys, cut-out pictures, or sketches for each of the biomes discussed in the chapter. Closely follow the list of organisms found in the text on pages 258–259. Place a name tag on each organism. Prior to the second day of this demonstration, hang each one from the ceiling of the room with yarn, being sure to mix up the members of different biomes to make a kind of mixed "air" zoo. Vary lengths of yarn so the representations of the organisms will hang at different levels—all fairly high.

Procedure: Have students scan the chapter quickly as an overview. Brainstorm with students about how each of the key terms would relate to the different biomes described in the chapter (refer to pages 257–259).

List One:
Have students write the names of the individual biomes. Under each heading, have them list the organisms spotted in the sky zoo that actually would live in the given biome. Students may want to refer to pages 248–257.

List Two: For each biome, have students prepare a possible food chain identifying producers, consumers, etc. Students may want to refer to pages 250–252.

Discuss possible trophic levels for each ecosystem. Be sure students understand the importance that diversity among living things has for supporting organisms in higher trophic levels.

Demonstration B
Cycles Within Ecosystems
for pages 253–256
(Developing Modeling Skills/Tactile Learners)

Materials: construction paper, scissors, markers, blank white press-on labels, glue

Preparation: Cut strips of construction paper. Glue strips together to form loops that are linked into chains such as those elementary students make for Christmas decorations. On labels, use the marker to write the names of factors involved in the carbon, water, and nitrogen cycles. Be sure that the factors of each cycle are glued into the same chain. For example, all factors of the water cycle should be glued together. Then, make the chain into an extended loop. Do this for each cycle. Then glue all the cycles together. Make enough sets to drape around the room for all to view as you move around in the brainstorm session.

Procedure: Read aloud to students Section 12.2 "Cycles Within Ecosystems," pages 253–256.

Read aloud to students the names of the factors on the chains, and have students identify the cycle from the names on the chains.

Brainstorm with students about the cycles of carbon, nitrogen and water. In what way do these cycles depend on each other? How does life in each ecosystem depend on these cycles? What role do plants play in keeping these cycles in balance? What role do humans play in keeping these cycles in balance? Describe ways in which the balances of each cycle are threatened.

Additional Strategies
Visual Strategies
Pages 250, 251, 252, 253, 255, 256, 258, and 260

Auditory Learners
Use *Biology: Visualizing Life* Audiocassettes for Sections 12.1, 12.2, and 12.3.

Meeting Individual Needs (cont.)

Cooperative Learning
Biomes

Timing: Use this activity to introduce Section 12.4.

Group Size: 4 or 5 students (Divide the class into 7 groups.)

Outcome: Students will be able to distinguish seven different biomes.

Individual Accountability: Each group member will be responsible for performing part of the research on one biome.

Positive Interdependence: Groups will be able to describe one biome to the class.

Assign one of the following biomes to each group: (1) tropical rain forests, (2) savannas, (3) deserts, (4) temperate grasslands, (5) deciduous forests, (6) coniferous forests, (7) tundra.

Have each group member complete one or more of the following tasks: color in the range of the biome on a world map, find an illustration of the biome, list the common plants found in the biome, list the common animals found in the biome, and describe the climate and amount of precipitation found in the biome. Have each group report the information to the class.

Portfolio Assessment

Students should select their best work and provide a self-reflective rationale for their selections. Students can make selections in the following areas.

1. *Content* — One concept map from the chapter (see page 812 for evaluation criteria.)

2. *Reading Comprehension* — One Directed Reading Worksheet from the Teacher's Resource Binder (Use the answer key to evaluate for accuracy.)

3. *Writing* — Using the Vee Form, summarize a newspaper or magazine article relating to ecosystems. (See page 22T for evaluation criteria.)

   Or: Select a writing task or project from the Chapter Review.

4. *Performance Assessment* — One Vee form from a chapter investigation or lab manual investigation (See page 22T for evaluation criteria.)

Teacher makes selections in the following areas.

1. *Formal Assessment* — Chapter test (Test A, B, or the Test Generator) The teacher-scored test should be reviewed by the student. Incorrect responses should be corrected by the student before the test becomes part of the portfolio.

2. *Informal Assessment* — Use the Direct Observations Checklist, page 33T, during a laboratory or other cooperative learning experience.

3. *Performance Assessment* — Have students create a history, through words, diagrams, or pictures, of a single water molecule or carbon or nitrogen atom as it passes through the cycles in the biosphere.

Concept Map Answer

The following is one possible answer to the Relating Concepts exercise on page 262.

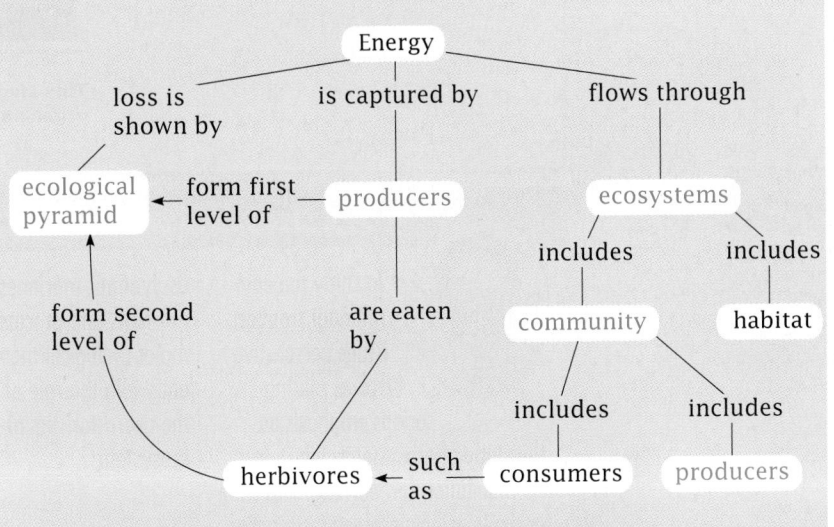

Chapter 12

Determining Prior Knowledge

- Have students provide examples of environmental problems of global concern.
- Display a diagram of a simple food web. Have students explain what it represents. Have them identify the producers, herbivores, carnivores, and omnivores.
- Show students a picture of a greenhouse. Have them explain how the air inside is warmed. Ask, "What is the greenhouse effect?" and, "How did it get its name?"
- Inform students that one of the major ecosystem types is desert. Have them name as many other types of ecosystems as possible. Have students describe the major features of each one.

Chapter **12**

Ecosystems

Review

- **photosynthesis**
  (Section 5.3)

- **cellular respiration**
  (Section 5.4)

- **metabolism**
  (Section 5.4)

12.1 What Is an Ecosystem?

- **State of Our World**
- **Ecology and Ecosystems**
- **Why Study Ecology?**
- **Energy in Ecosystems**
- **How Many Trophic Levels Can an Ecosystem Contain?**

12.2 Cycles Within Ecosystems

- **Nutrient Cycles**
- **Water Cycle**
- **Carbon Cycle**

12.3 Kinds of Ecosystems

- **Freshwater Ecosystems**
- **Terrestrial Ecosystems**
- **Ocean Ecosystems**

This slug, mushroom, and spider are just three of the many organisms found in a deciduous forest ecosystem.

■ Author's Rationale ■

To study the environment from an issues perspective involves placing serious emphasis on the study of ecosystems, which lays the foundation for learning. This chapter explains how ecosystems function as dynamic machines that cycle nutrients and process energy—two major biological themes. The chapter ends with the role of climate in shaping the characteristics of each type of ecosystem.

YOU CANNOT PICK UP A NEWSPAPER TODAY WITHOUT SEEING NEWS ABOUT THE ENVIRONMENT. ENVIRONMENTAL ISSUES ARE IMPORTANT TO EVERYONE BECAUSE WE ALL HAVE TO LIVE IN A WORLD WE SEEM TO BE DESTROYING. WE NEED DETAILED KNOWLEDGE OF HOW THE WORLD WORKS SO THAT WE CAN PREVENT FURTHER ABUSE TO OUR PLANET AND PERHAPS BEGIN TO REPAIR THE DAMAGE WE ALREADY HAVE DONE.

12.1   What Is an Ecosystem?

Objectives

1 Identify the components of an ecosystem.

2 Discuss the flow of energy through ecosystems.

3 Identify the different trophic levels in an ecosystem.

4 Explain why ecosystems can only contain a few trophic levels.

State of Our World

More than 5 billion humans live on Earth. Scientists estimate that between 10 million and 80 million other species share the world with us. Yet we seem to be rapidly destroying our planet's ability to support us and its other inhabitants. For example, here are a few changes humans have made to the Earth *in just the last year*. About 17 million hectares (40 million acres) of forest have been burned down or cut, like the forest shown in **Figure 12.1**. That is an area almost as large as the state of Washington. As many as 50,000 species of animals and plants have become extinct, disappearing forever. The world's human population has grown by 92 million people, mostly in the world's poorer nations. These 92 million people were born into a world in which more than 1 billion people are not adequately nourished.

For centuries people have believed that the environment was there for them to use as they saw fit. But, today's environmental problems teach us that the environment is not a passive stage on which we can act as we please. Rather, we share the environment with other organisms, resulting in a complex network of interactions on which we all depend. Changes made to the environment can have serious consequences, not all of which are predictable.

Figure 12.1
This section of forest in the Quinleute Reservation in Washington has been logged. Although the trees will provide many useful products, cutting the forest has a devastating effect on the organisms living there.

■ Matter of Fact ■

Population Statistics
Nearly two-thirds of the people alive today live in Asia, with most of them in India and China. If all the people in China were to stand on top of one another, they would extend into space more than three times the distance from the Earth to the Moon. During the 1970s, the world population increased annually at a rate of nearly 2 percent. If this trend continues, the world's human population in the year 2000 will exceed 6 billion.

Lesson Plan 12.1

Phase 1
PREPARATION

Key Concepts
- Ecology is the study of the relationship between organisms and their environment.
- An ecosystem is a self-sustaining collection of organisms and their physical environment.
- Both food chains and food webs describe the flow of energy through an ecosystem.
- An ecological pyramid shows the amount of energy present at each level within a food chain.

Reading Strategy

Use this section to have students expand upon their reading. Instruct students to bring in newspaper or magazine articles that deal with current environmental problems. Have each student write a summary of each article and make a brief report to the class.

Phase 2
TEACHING STRATEGIES

Theme Connection

Interacting Systems
Form cooperative work groups of four students. Have each group develop a plan that addresses one of the environmental problems brought to the class's attention as a result of the reading strategy. Each group may have to do some library research.

Interacting Systems

Use Figure 12.2 to point out that an ecosystem involves the interaction between organisms and their physical environment. Have students point out examples of interactions that are illustrated in this figure.

Demonstration 1

An Ecosystem

Assign students to cooperative work groups of four. Tell each group to select an ecosystem they would like to simulate, perhaps a pond, grassland, marsh, or open field. Have each group design a mini-ecosystem that includes at least three types of plants and three types of small animals such as pill bugs, snails, worms, ants, or other invertebrates. Have them set up their mini-ecosystem in a large glass jar covered with cheesecloth. The jars should be exposed to sunlight and should be watered every few days. Have the groups make periodic observations and report to the rest of the class.

Demonstration 2

Tropical Rain Forest

Show either a photograph or slide of a tropical rain forest. Ask students to describe what communities and habitats they see. Do the same with a desert or tundra ecosystem to impress students with the diversity present in a rain forest.

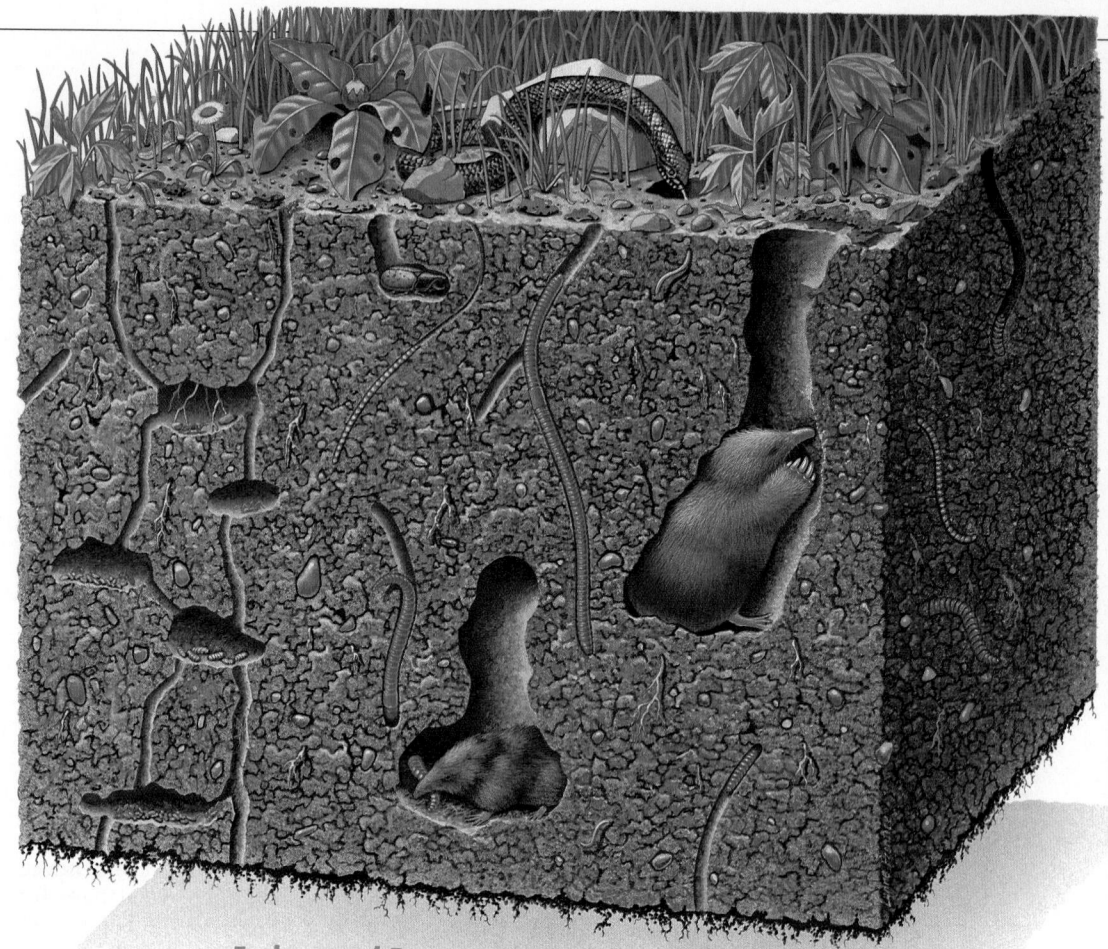

Ecology and Ecosystems

Figure 12.2
The inhabitants of this soil ecosystem include earthworms, insects, snakes, moles, bacteria, and a variety of plants and fungi.

In 1866, the German biologist Ernst Haeckel gave a name to the study of how organisms fit into their environment. He called it **ecology**, from the Greek words *oikos* (house, place where one lives) and *logos* (study of). Ecology, then, is the study of the "house" in which we live. Most of our environmental problems could be avoided if we treated the world in which we live the same way we treat our own homes.

Ecology is the study of the interactions of organisms with one another and with their physical environment. The organisms that live in a particular place, such as a forest, are known as a **community**. Ecologists, the scientists who specialize in ecology, call the physical location of a community its **habitat**. You can think of a habitat as a neighborhood and of a community as the residents of the neighborhood. The sum of the community and habitat is called an ecological system or ecosystem. An **ecosystem** is a self-sustaining collection of organisms and their physical environment.

Imagine that you could collect every organism living in an ecosystem. Figure 12.2 shows just some of the inhabitants of a soil ecosystem. If you visited a tropical rain forest, you could collect many more species than occur in this soil ecosystem. As many as 100 species of trees can be found in 1 hectare (2.5 acres) of South American rain forest. The **diversity** of an ecosystem is a measure of the number of species living there and how common each species is. Tropical rain forests are the most diverse terrestrial ecosystems.

Why Study Ecology?

You study ecology because you need to know about the place in which you live. If you are going to prevent pollution, conserve resources, and save the world for your children to live in, then you need to know how your world works—just as you need to study how any complex machine works in order to keep it running properly. Do you think a car would run for long if its owner had no idea of the need for water, oil, and gasoline? Remember that there is a fundamental difference between a car and an ecosystem, however. A car that receives no attention from its owner will eventually break down. Ecosystems, on the other hand, have been functioning for billions of years without human tending. Only now, when we have caused them great damage, do some ecosystems require our help to continue.

Ecosystems are very complex

It is very difficult to understand how an ecosystem works, because it can contain hundreds or even thousands of interacting species. Nevertheless, you can gain a basic understanding of how an ecosystem works by asking two questions. From where does the energy needed by particular animals and plants come? How do organisms in ecosystems maintain adequate amounts of the minerals and other inorganic substances they need?

Answering these questions will give you a pretty good idea of how an ecosystem normally works. You will then be in a position to ask how an ecosystem might be expected to respond to a disturbance. To make such predictions, ecologists build a model, a simplified version of the ecosystem. An ecosystem model consists of a series of hypotheses that describe how the ecosystem functions: how energy moves through the ecosystem, how species interact, and so on. In some cases, ecologists express their models as mathematical equations and then solve these equations for various situations. **Figure 12.3** is a schematic representation of an ecological model.

Because ecosystems are so complex, no ecosystem model can consider all the factors affecting that ecosystem. Nevertheless, seeing what happens to the ecosystem model when a variable is changed helps ecologists predict what might happen if some component of the real ecosystem is altered. If, for example, one species in an ecosystem becomes extinct, or the amount of available energy declines, an ecological model can help scientists predict the possible consequences. Ecological models allow you to look into the future, but your vision can be only as accurate as the hypotheses used to build the model.

Figure 12.3
Ecologists use models to predict the possible outcomes of a disturbance in an ecosystem such as the soil ecosystem shown below.

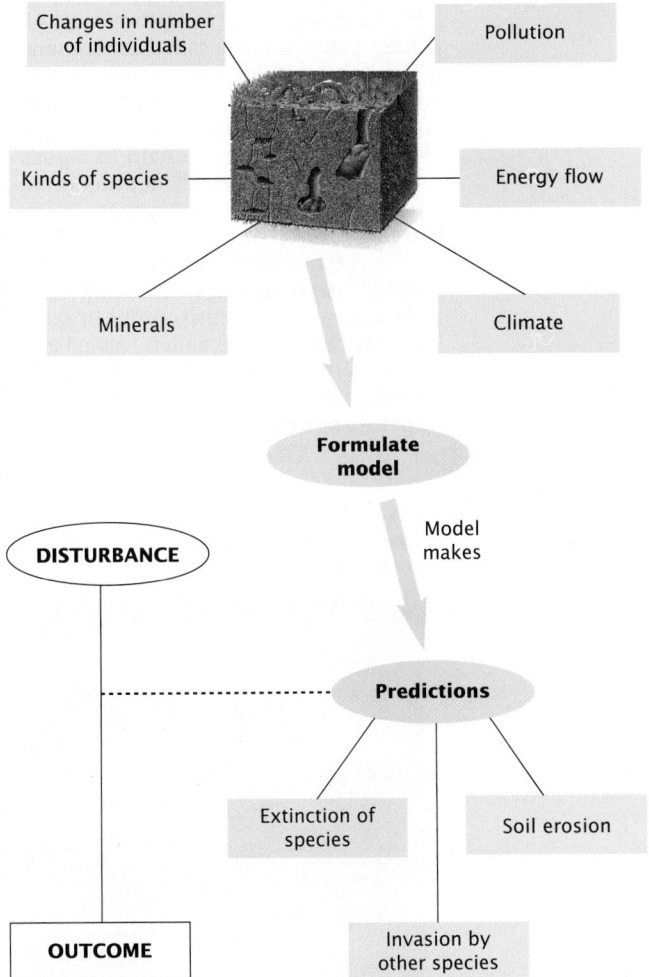

Demonstration 3

An Ecological Model

Have students construct an ecological model for the mini-ecosystem their group established. Be sure the model consists of a series of hypotheses that describe how energy flows through their ecosystem, how both plant and animal species interact, and how a particular disturbance might affect their ecosystem.

Theme Connection

Energy and Life

Have students use their ecological model to see how energy flows from one organism to another.

Connection: Chapter 5

Photosynthesis
Review the major features of photosynthesis so students can appreciate the role of producers in a food chain.

Visual Strategy

Figure 12.4
Have students identify the producers and consumers in this food chain. Have them hypothesize what would happen if all the cod were removed from this particular ecosystem by fishermen.

Demonstration 4

Food Chains
Assign students to cooperative work groups of four. Have each group give an example of a food chain with at least four levels, starting with a specific producer.

Visual Strategy

Figure 12.5
Remind students that mushrooms are heterotrophic fungi. Point out that by breaking down dead animals, plants, and organic material decomposers return substances to the environment to be reused by other members of the ecosystem.

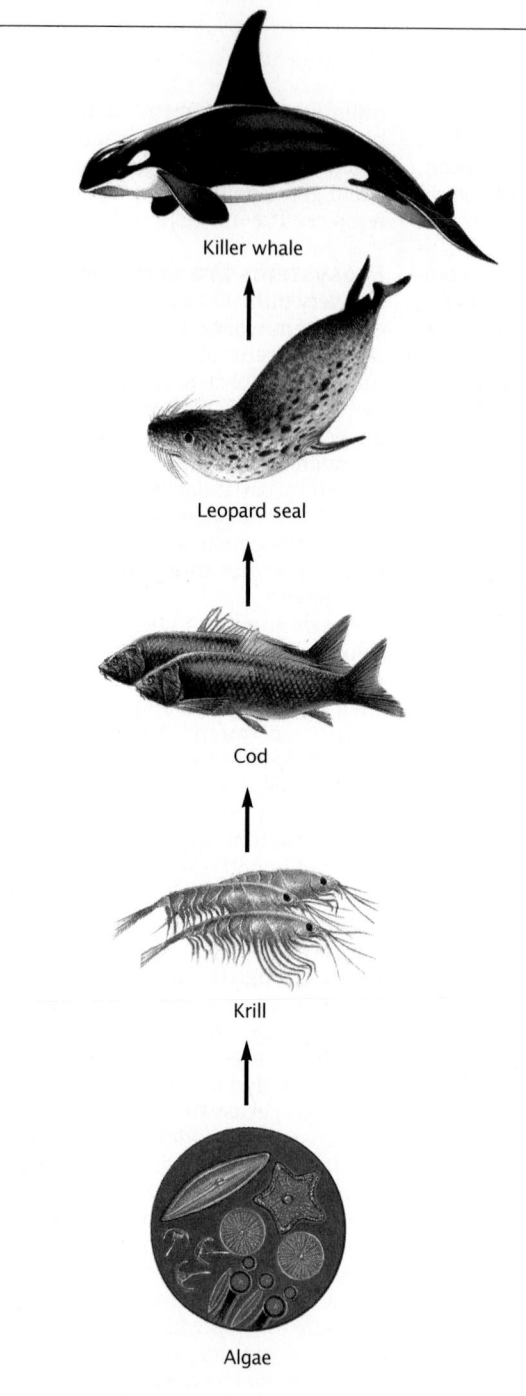

Killer whale

Leopard seal

Cod

Krill

Algae

Figure 12.4
This food chain shows the path of energy transfer in an ocean ecosystem. Algae, which are photosynthetic, are the producers in this ecosystem.

Energy in Ecosystems

An ecosystem uses energy because its living members use energy. A flower blooming, a squirrel running along a tree branch, a worm burrowing through the soil—every action of even the tiniest creatures requires energy. The most important factor determining how many and what kinds of organisms live in an ecosystem is the amount of energy available.

From where do the organisms in an ecosystem obtain energy? Life exists because organisms called **producers** take in energy from their surroundings and store it in complex molecules. Except for a few kinds of bacteria, producers capture energy from the sun through photosynthesis. Plants, some bacteria, and algae are producers. All other organisms are called **consumers**. They obtain their energy by consuming other organisms. Animals, most protists and bacteria, and all fungi are consumers. The food chain in **Figure 12.4** shows how energy is transferred from producers to consumers in an ocean ecosystem.

Each ecosystem contains consumers called **decomposers**. Decomposers obtain energy by consuming organic wastes (feces, urine, fallen leaves) and dead bodies. Fungi, such as the mushrooms shown in **Figure 12.5**, and some species of bacteria, are decomposers.

Figure 12.5
These scarlet waxy caps are decomposers.

Interacting Systems

If pollution drastically reduced the number of algae in the food web shown below, would killer whales be affected? Explain.

Energy flows from producers to consumers

To follow the movement of energy through an ecosystem, ecologists assign each organism to a **trophic** (feeding) **level**. A trophic level is a group of organisms whose energy sources are the same number of steps away from the sun. Producers such as plants are in the first trophic level. Plants and other organisms that make their own food are **autotrophs**. Animals that eat plants are in the second trophic level. Animals that feed on plant-eaters are in the third trophic level.

Creatures in the second trophic level are **herbivores** (plant eaters). Cows, caterpillars, elephants, and ducks are herbivores. All organisms at the third trophic level or above are **carnivores**

(flesh eaters). Carnivores at the third level feed on herbivores, while carnivores above the third trophic level feed on other carnivores. Tigers, hawks, weasels, pelicans, and killer whales are carnivores. Some animals cannot be classified as either carnivores or herbivores. **Omnivores** such as bears and humans eat both plants and animals. Because they cannot make their own foods, organisms living in trophic levels above the first trophic level are **heterotrophs**. Most animals feed at more than one trophic level and eat several different species at each trophic level. As shown in **Figure 12.6**, energy moves through an ecosystem in a complex network of feeding relationships called a food web. Notice that the food chain in **Figure 12.4** is just one part of this food web.

Figure 12.6

This food web shows how energy flows through an ocean ecosystem as one organism is eaten by another.

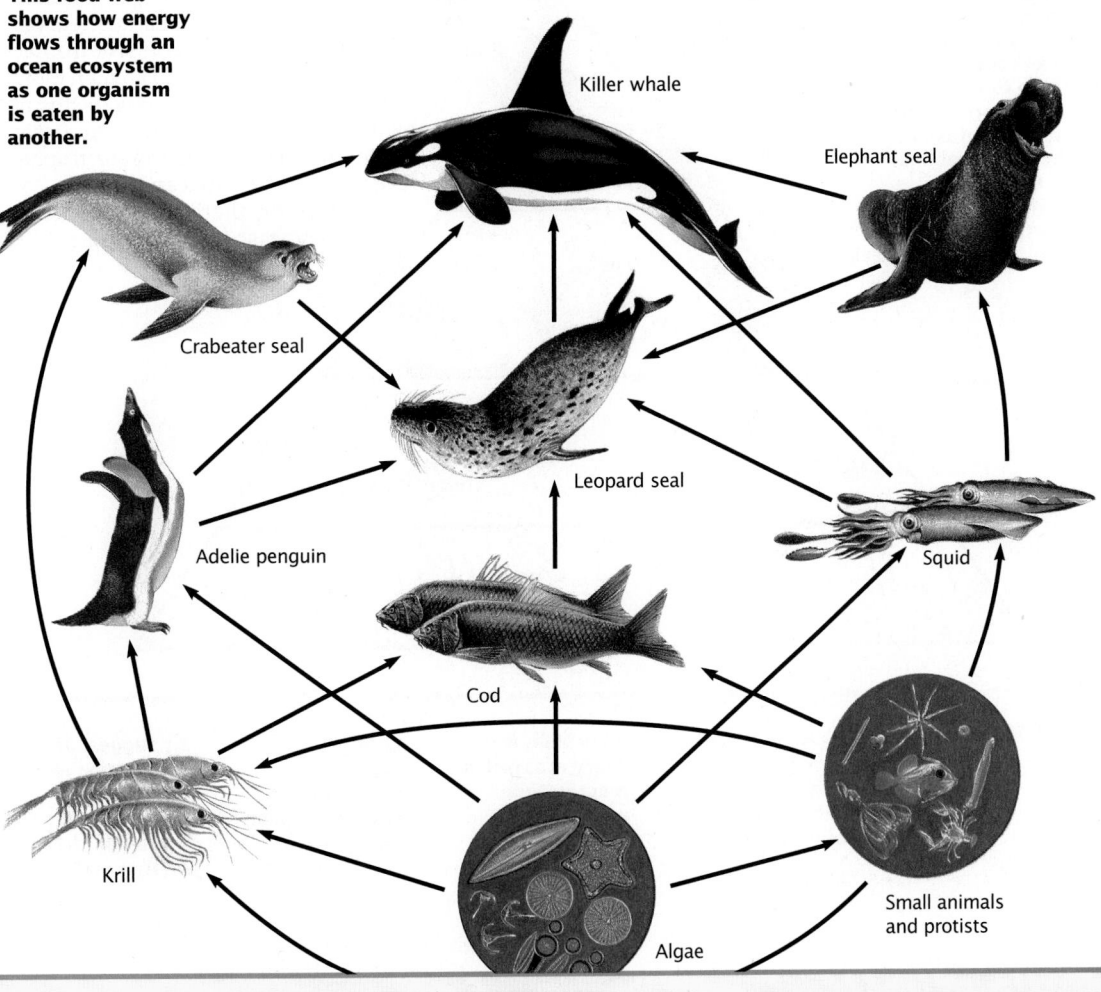

251

Visual Strategy

Figure 12.7
Have students identify the ecological role played by each organism in this figure.

Mathematics Connection

How Much Gets to the Top?

Have students calculate the percentage of energy originally present in the grass that eventually gets to the hawk.

Theme Answer

Energy and Life

Because 90 percent of the energy at any trophic level is not transferred to the next, much more energy is required to raise animals for meat than to raise an equivalent amount of grain or vegetables. For instance, feeding 1,000 kg (2,200 lbs.) of grain to cattle only results in 100 kg (220 lbs.) of meat—90 percent less food than if the grain served as human food.

Phase 3

ASSESSMENT OPTIONS

Closure Strategy

A Lopsided Pyramid

Draw an ecological pyramid on the board. Include a very small base instead of the large one normally shown for the producer level I. Have students explain why such a pyramid can accurately reflect a community in either a tropical rain forest or temperate deciduous forest.

Section Review

Assign the *Section Review.*

Reteaching

Have students bring in pictures of various organisms. Ask the class to see what food webs can be constructed. Remove a picture of a different organism from each food web and have students predict what would happen.

252

Energy

More food would

be available for

the growing

human population

if humans did

not eat meat.

Explain why.

Figure 12.7
This ecological pyramid shows the amount of energy at each of four trophic levels in an ecosystem. There is 1,000 times more energy stored in grass at the first level than in Swainson's Hawks at the fourth trophic level.

How Many Trophic Levels Can an Ecosystem Contain?

A plant absorbs energy from the sun and uses it to make carbohydrates such as cellulose, the major component of cell walls. Only about one-half of the energy captured by a plant becomes part of the plant body, however. Part of the remaining energy is used to make ATP during cellular respiration. The remaining energy escapes as heat. Similar losses of energy occur at each trophic level of an ecosystem.

In the 1950s, ecologist Howard Odum determined how much energy was present at each trophic level in a Florida stream ecosystem. He captured animals and plants, measured their energy content, and built a model of how energy passed through the ecosystem. Odum found that when a herbivore eats a plant, only about 10 percent of the energy present in the plant's molecules ends up in the herbivore's molecules. The other 90 percent of the energy is lost, some as the cost of doing work (breathing, walking, chewing) and much more as heat. Likewise, when a carnivore eats the herbivore, only 10 percent of the energy in the herbivore goes toward making carnivore molecules. At each trophic level, the energy stored in the organisms is about one-tenth that of the level below it.

You can see the loss of energy from one trophic level to the next by looking at the ecological pyramid shown in **Figure 12.7**. Because energy diminishes at each successive trophic level, few ecosystems can contain more than five trophic levels. Organisms at higher trophic levels tend to be fewer in number than those at lower trophic levels. On the African plains, for instance, there are about 1,000 zebras, gazelles, and wildebeest for each lion.

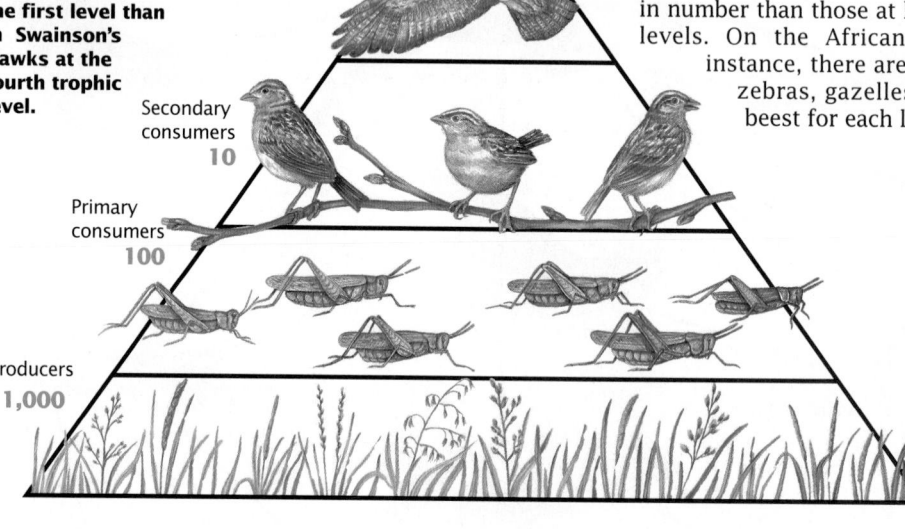

Tertiary consumers
1

Secondary consumers
10

Primary consumers
100

Producers
1,000

Section Review

① **What is the difference between a habitat and a community?**

② **Explain why every ecosystem must include producers.**

③ **At which trophic level would you place humans? Explain your answer.**

④ **Suggest an explanation for why there are fewer lions than zebras on the African plains.**

■ Section Review Answers ■

1. A community is a group of organisms that live in a particular place. Habitat is the physical location of a community.

2. Producers are the ultimate energy source for all the organisms in an ecosystem.

3. Humans are omnivores and so do not neatly fit into any trophic level. They would probably fit best at the highest level, since they have no regular predators.

4. Because there is less energy available at higher trophic levels, organisms at these trophic levels, such as lions, tend to be fewer in number than organisms at lower trophic levels, such as zebras.

MANY OF THE OBJECTS YOU USE EVERY DAY ARE DESIGNED TO BE THROWN AWAY AFTER YOU ARE DONE WITH THEM. IN CONTRAST, THE PHYSICAL COMPONENTS OF AN ECOSYSTEM ARE USED AGAIN AND AGAIN. ECOLOGISTS CALL THIS CONTINUAL REUSE "CYCLING." MATERIALS THAT CYCLE WITHIN ECOSYSTEMS INCLUDE NITROGEN, WATER, AND CARBON.

12.2   Cycles Within Ecosystems

Objectives

❶ Describe the results of Bormann and Likens's experiments.

❷ Describe how nitrogen and water are recycled within ecosystems.

❸ Explain the role of plants in the cycling of materials within ecosystems.

❹ Trace the pathway of carbon between living organisms and the environment.

Nutrient Cycles

Unlike energy, which flows through an ecosystem, nutrients such as calcium and nitrogen circulate within an ecosystem. To study how nutrients cycle, a team of scientists led by ecologists Herbert Bormann and Gene Likens carried out studies at Hubbard Brook in New Hampshire beginning in the 1960s. **Figure 12.8** shows the experimental site at Hubbard Brook. Bormann and Likens wanted to determine if rainwater removed nutrients from ecosystems. They built small dams so that they could measure how much water left the ecosystem in the stream at the base of the valley. They found that water leaving the ecosystem contained few nutrients. They concluded that the trees very efficiently prevented nutrients from leaving the ecosystem.

Having built an ecological model, Bormann and Likens were able to make predictions about what would happen if the ecosystem were disturbed. For example, they knew that nutrients such as calcium were held by the trees of the forest. Their ecological model predicted that much more calcium would be lost if the trees were cut down.

Bormann and Likens test their model of nutrient cycling

To test their ecological model, Bormann and Likens cut the trees and vegetation from one portion of the forest. For several years they monitored calcium and other nutrients in the water runoff into Hubbard Brook. The ecosystem lost the ability to retain nutrients. The runoff of calcium, for example, was six times greater than it had been before the trees were cut. Other studies have confirmed that minerals and other nutrients pass from organisms to habitat and back again in delicate cycles that are easily disturbed.

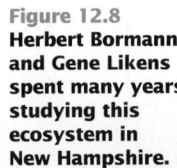

Figure 12.8
Herbert Bormann and Gene Likens spent many years studying this ecosystem in New Hampshire.

Phase 1
PREPARATION

Key Concepts
- Materials, including nitrogen, water, and carbon, are recycled through ecosystems.
- The burning of fossil fuels is upsetting the carbon cycle and warming the atmosphere in a phenomenon known as the greenhouse effect.

Reading Strategy
As students read this section, have them individually prepare several questions that can be used as the basis of a class discussion to review the material. Tell students that not all their questions should be based on factual recall. At least one should involve critical thinking, application, or problem solving.

Phase 2
TEACHING STRATEGIES

Theme Connection

Interacting Systems
Discuss with students the similarity between nutrient cycling by nature and waste recycling by humans. Arrange for the class to visit a recycling center.

Visual Strategy

Figure 12.8
Have students identify living and non-living components of the Hubbard Brook ecosystem.

Connection: Chapter 8

Gene Technology

Review how genetic engineering is being used to develop major crops that can carry out nitrogen fixation without the help of symbiotic bacteria.

Theme Connection

Energy and Life

When discussing processes that provide the energy of life, photosynthesis and respiration come immediately to mind. However, without the nitrogen fixation that bacteria carry out, there would be no life. Point out that the electrical energy from lightning results in nitrogen fixation. However, the impact on the nitrogen cycle is negligible.

Theme Answer

Patterns of Change

Students should suggest that plant growth in ecosystems is often severely limited by the availability of nitrogen in the soil. Therefore, the addition of nitrogen-containing fertilizer should enhance plant population growth, stimulating population growth among consumers.

Patterns of Change

How would adding nitrogen-containing fertilizer affect the ecosystem shown below?

Figure 12.9
The white-tailed deer and speckled alder play important roles in the nitrogen cycle in an ecosystem found in the northeastern United States.

Bacteria play a key role in the nitrogen cycle

Organisms must have nitrogen to make proteins and nucleic acids. **Figure 12.9** shows how nitrogen cycles through an ecosystem in the northeastern United States. As you learned in Chapter 8, most living things cannot use the nitrogen gas in the air. The two nitrogen atoms that make a molecule of nitrogen gas are held together by a strong chemical bond that is difficult to break. The variety of life found on earth is possible only because a few kinds of bacteria have enzymes that can break this strong bond. Nitrogen atoms are then free to bond with hydrogen atoms to form ammonia molecules. Conversion of nitrogen gas to ammonia is called **nitrogen fixation**. Ammonia is a form of nitrogen that plants can absorb and use to make proteins. Since animals cannot absorb nitrogen from the soil, they must obtain nitrogen by eating plants or other animals.

Nitrogen-fixing bacteria live in the soil or within the roots of plants such as peas, clover, alfalfa, beans, and alder trees. The growth of plants in ecosystems is often severely limited by the availability of nitrogen in the soil. When an organism dies, the nitrogen in its body is released by decomposers. Animal wastes, such as dung and urine, as well as plant materials like leaves and bark, also contain nitrogen. These materials are also broken down by decomposers. Thus, decomposers play a vital role in ecosystems by returning nitrogen to the soil.

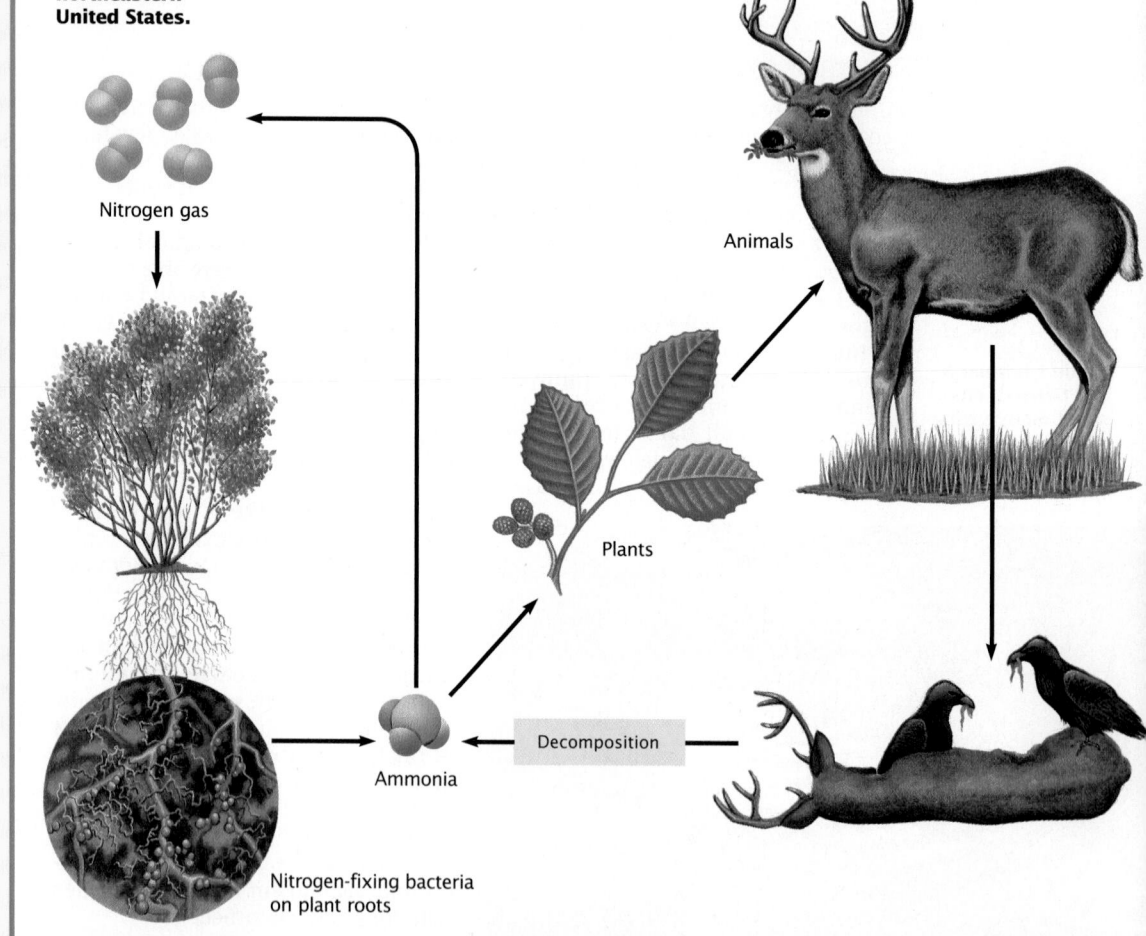

Nitrogen gas

Plants

Animals

Decomposition

Ammonia

Nitrogen-fixing bacteria on plant roots

■ Matter of Fact ■

Ecological Impact

The greatest impact of cutting down trees in the Hubbard Brook ecosystem was on the nitrogen cycle. In the undisturbed forest areas, the ecosystem accumulated nitrogen at the rate of about 2 kg (5.5 lbs.) per hectare (about 2.5 acres) per year. Where the trees had been cut, nitrogen was lost from the soil at the rate of about 120 kg per hectare (about 650 lbs. per acre) per year.

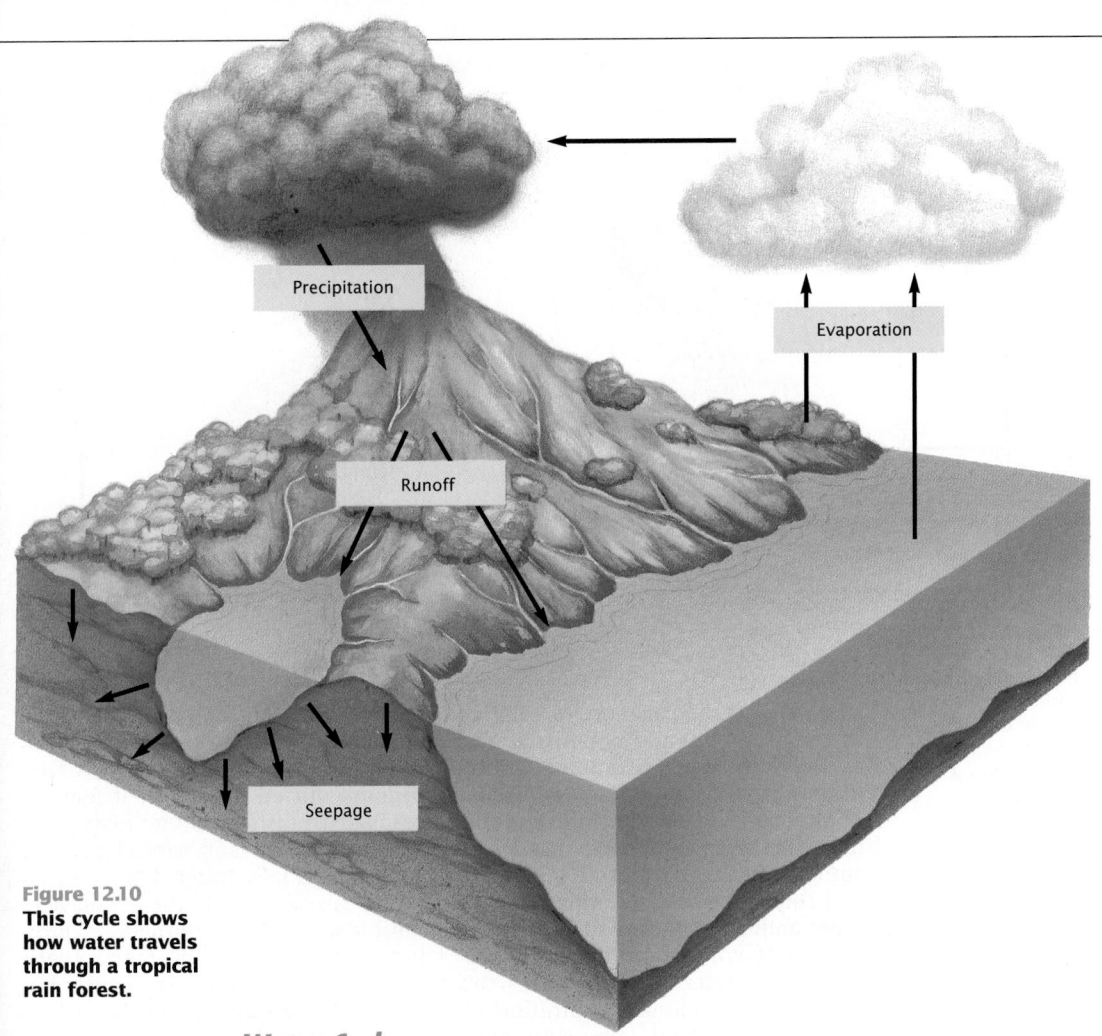

Precipitation

Evaporation

Runoff

Seepage

Figure 12.10
This cycle shows how water travels through a tropical rain forest.

Visual Strategy

Figure 12.10
Emphasize that the driving force behind the water cycle is the sun. Most of the water returned to the atmosphere by evaporation comes from the oceans.

Water Cycle

Water is perhaps the most important nonliving component of an ecosystem. To a large degree, availability of water determines the diversity of organisms in an ecosystem. Few species live in the desert where there is little water, while many species live in the tropical rain forest where water is plentiful. Water is constantly moving within ecosystems in the cycle illustrated in **Figure 12.10**.

Plants play an important role in the water cycle

In tropical rain forests, where there are dense concentrations of trees and other plants, more than 90 percent of the moisture that enters the ecosystem passes through plants and evaporates from their leaves. In a very real sense, these plants create their own rain.

When forests are cut down, the water cycle is broken. Moisture cannot be returned to the atmosphere by plants. Instead, water drains into streams and rivers and eventually flows into the ocean. Moreover, without protection by the roots of trees and other plants, the soil is easily carried away by runoff. As a result, nutrient cycles also are broken. Because both water and nutrients can no longer cycle in a forest ecosystem after the trees are cut down, extensive cutting can convert lush forests into deserts. Tragically, such a transformation is presently occurring in many tropical rain forests.

Connection: Chapter 5

Life Processes

Review the overall equations for photosynthesis and respiration so that students can see how the carbon is recycled by these two processes.

Visual Strategy

Figure 12.11

Use this figure to emphasize once again that plants carry out respiration. Point out that although respiration is shown three times and photosynthesis only once, the two processes are actually in balance, or at least they were until combustion started to add more carbon dioxide to the atmosphere than plants could handle.

Demonstration 1

The Greenhouse Effect

Arrange for a class trip to a local nursery, private home, or other site where students can explore a greenhouse. Discuss how certain variables, including the number of plants, the thickness of the glass, and the amount of sunlight, might affect the temperature.

Phase 3

ASSESSMENT OPTIONS

Closure Strategy

The Oxygen Cycle

Have students diagram how oxygen is recycled. Instruct them to use the same components that are included in the illustration for the carbon cycle in Figure 12.11.

Section Review

Assign the *Section Review*.

Reteaching

Have students develop a concept map for each of the cycles described in this section.

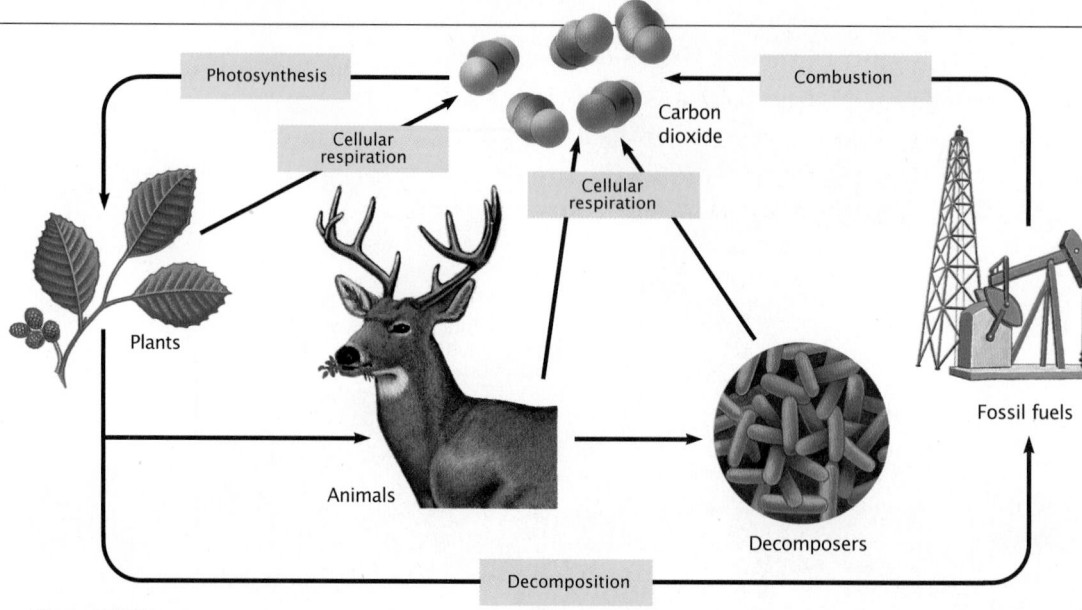

Figure 12.11
This cycle shows how carbon moves within an ecosystem in an industrialized nation such as the United States.

Carbon Cycle

Like water, carbon also cycles between the environment and organisms. The Earth's atmosphere contains carbon in the form of carbon dioxide. Plants use carbon dioxide to build organic molecules during photosynthesis. Consumers obtain energy-rich molecules that contain carbon by eating plants or other animals. As these molecules are broken down, carbon dioxide is produced and released into the Earth's atmosphere. Cellular respiration by decomposers and photosynthetic organisms also returns carbon dioxide to the atmosphere. **Figure 12.11** shows how carbon cycles within an ecosystem.

Humans are overloading the carbon cycle

Large amounts of carbon are tied up in wood and may stay trapped there for hundreds of years, only returning to the atmosphere when the wood is burned. Over millions of years, plants that become buried in sediment may be gradually transformed into fossil fuels such as coal, oil, and natural gas. The carbon originally trapped by plants is not released back into the atmosphere until fossil fuels are burned. By burning large amounts of fossil fuels, humans are increasing the concentration of carbon dioxide in the atmosphere. Carbon dioxide traps heat from the sun within the atmosphere, much like glass panes trap the sun's heat in a greenhouse. The ability of gases such as carbon dioxide to retain the sun's heat, and in so doing to warm the atmosphere, is called the **greenhouse effect**. You will learn more about the greenhouse effect in Chapter 14.

Section Review

❶ Explain the significance of Bormann and Likens's experiments at Hubbard Brook.

❷ Explain the role of nitrogen-fixing bacteria in the nitrogen cycle.

❸ How do you think deforestation affects the water cycle?

❹ Describe how human interference in the carbon cycle may be causing an increase in global temperatures.

■ Section Review Answers ■

1. Bormann and Likens showed that ecosystems effectively recycle most of their nutrients. Vegetation is responsible for this recycling.

2. With the aid of bacteria, these plants transform atmospheric nitrogen into ammonia, a form of nitrogen plants can use.

3. Plants draw water from the soil and evaporate it from their leaves. Without plants, runoff increases, often carrying away soil and nutrients.

4. By burning fossil fuels, humans are increasing the levels of carbon dioxide in the atmosphere. Since carbon dioxide traps the sun's heat, increased temperatures may result.

ALL ECOSYSTEMS ARE CONNECTED. THE DESTRUCTION OF THE RAIN FORESTS IN BRAZIL, FOR EXAMPLE, WILL PERSONALLY AFFECT YOU AND EVERYONE ELSE IN THE WORLD. IF HUMANS ARE GOING TO TAKE ACTION TO PRESERVE THE WORLD FOR FUTURE GENERATIONS, IT IS IMPORTANT TO UNDERSTAND WHAT THE MAJOR ECOSYSTEMS OF THE WORLD ARE LIKE AND HOW THEY FUNCTION.

12.3   Kinds of Ecosystems

Objectives

❶ Identify the importance of plankton in freshwater ecosystems.

❷ Identify the factors that determine the ecosystem type found in a particular area.

❸ Contrast the seven major terrestrial ecosystems.

❹ Identify the major ocean ecosystems.

Freshwater Ecosystems

Freshwater ecosystems include lakes, ponds, and rivers. These ecosystems are very limited in area. Inland lakes cover 1.8 percent of the Earth's surface, and rivers and streams cover about 0.3 percent. Although small in total area, freshwater ecosystems support a rich array of life, including fishes, amphibians, insects, turtles, crocodiles, and many plants. A diverse biological community of microscopic organisms called **plankton** lives near the surface of lakes and ponds. Plankton contain photosynthetic organisms that are the base of aquatic food webs. All freshwater habitats are strongly connected to land ecosystems. Nutrients flow from terrestrial ecosystems into freshwater ecosystems. In addition, many land animals come to the water to feed or reproduce.

Figure 12.12 shows a freshwater ecosystem. Ponds and lakes usually have three zones in which organisms occur: a shallow "edge" zone, an open-water surface zone, and, in deep lakes and ponds, a deep-water zone to which little light can penetrate.

Figure 12.12
Freshwater ecosystems, such as this pond, support a wide variety of plants and animals.

Lesson Plan 12.3

Phase 1

PREPARATION

Key Concepts
- Freshwater ecosystems include lakes, ponds, and rivers.
- The main types of terrestrial ecosystems are tropical rain forest, savanna, desert, temperate grassland, deciduous forest, coniferous forest, and tundra.
- Ocean ecosystems include shallow ocean waters, open ocean surfaces, and deep ocean waters.

Reading Strategy

Emphasize the importance of reading and studying Figure 12.13 and Table 12.1, since they contain most of the information concerning the various ecosystems.

Phase 2

TEACHING STRATEGIES

Demonstration 1

A Freshwater Ecosystem
Conduct a field trip to study a freshwater ecosystem. Have students collect specimens, including plankton, for further examination in class.

Theme Connection

Interacting Systems
Freshwater ecosystems also include swamps and marshes, often called wetlands. Most students do not recognize the ecological importance of these ecosystems. Tell students that wetlands are homes for many different kinds of plants and animals. Point out that these natural homes are threatened in many coastal areas by the construction of human homes

257

and developments. Have students check for a wetland preservation organization in their area and report on its work.

Visual Strategy

Table 12.1

Lead students through the descriptive information on the various biomes. Use the table to dispel some misconceptions most students have:

- Deserts are not the only biomes that have low yearly precipitation. What marks a desert as unique is the high rate at which the little water that does fall evaporates.
- Larger animals are not primarily found in the warmer biomes. True, elephants are large. But as a rule, warm-blooded animals tend to be much larger in colder climates. Larger animals have a smaller surface area-to-volume ratio, thus reducing their heat loss. Remind students about the relationship between surface area and volume as a cell gets larger (Chapter 3).
- The tundra is not barren as far as life is concerned. Although the vegetation supports a limited number of animal species, during the warm season migrating visitors, including flies and mosquitoes, make their appearance.

Terrestrial Ecosystems

Major ecosystems that occur over wide areas on land are called **biomes**. The seven biomes are: (1) tropical rain forests; (2) savannas; (3) deserts; (4) temperate grasslands; (5) deciduous forests; (6) coniferous forests; and (7) tundra. These biomes differ remarkably from one another because they evolved in different geographic locations. The kinds of animals and plants that live in an ecosystem depend on the physical nature of the habitat: the soils, the terrain, and the climate. **Table 12.1** compares the Earth's seven biomes.

a Tropical rain forest, Australia

b Savanna, Kenya

Table 12.1 Biomes

	Biome	Yearly Precipitation	Characteristics
a	**Tropical rain forests**	250 cm (100 in.)	Little temperature variation; abundant moisture
b	**Savannas**	90–150 cm (36–60 in.)	Open; widely spaced trees; seasonal rainfall
c	**Deserts**	20 cm (8 in.)	Dry; sparse vegetation; scattered grasses
d	**Temperate grasslands**	10–60 cm (4–24 in.)	Rich soil; tall, dense grasses
e	**Deciduous forests**	75–250 cm (30–100 in.)	Warm summers; cool winters
f	**Coniferous forests**	20–60 cm (8–24 in.)	Short growing season; cold winters
g	**Tundra**	25 cm (10 in.)	Open; wind-swept; dry; ground always frozen

c Desert, Australia

d **Temperate grassland, Oklahoma**

Inhabitants	Location	Comments
More species than any other biome	Tropical regions of South America, Central America, Asia, Africa, Australia	Forests and their species being destroyed at an alarming rate
Lions, rhinoceroses, elephants, giraffes, gazelles	Parts of Africa, South America, Australia	Conversion to agriculture threatens inhabitants
Kangaroo rats, camels, cactuses	Parts of Africa, Asia, Australia, North America	Inhabitants must conserve water
Buffalo, prairie grasses	Central North America, central Asia	American prairies were once the home of huge numbers of bison
Raccoons, deer, maples, oaks, hickories	Europe, north-eastern United States, eastern Canada	Deciduous trees lose their leaves every year
Elk, moose, needle-leaved evergreens	Northern Asia, northern North America	Primary source of the world's lumber
Caribou, lemmings, wolves, mosses, lichens	Far northern Asia, northern North America	Covers one-fifth of the Earth's land surface

g **Tundra, Alaska**

e **Deciduous forest, New Hampshire**

f **Coniferous forest, Montana**

■ *Cultural Perspective* ■

The Neem Tree

In India, the neem tree provides so many useful products that it has been called "the village pharmacy." Its leaves are used to make medicinal teas and to repel insects. Juice from the tree is used to treat skin problems, and the twigs are used to clean the teeth.

Visual Strategy

Figure 12.13
Lead students through the descriptive information on the ocean ecosystems.

Demonstration 2

Creatures of the Deep
Show students pictures of some of the bizarre organisms that live in deep ocean waters. Have them look for adaptations each organism has for surviving in such an environment.

Phase 3

ASSESSMENT OPTIONS

Closure Strategy

Biodiversity
Ask, "What characteristics of a tropical rain forest make it suitable to be the home for so many different species of plants and animals?"

Section Review

Assign the *Section Review.*

Reteaching

Show students either photographs or slides of various biomes. Ask them to identify each one and discuss its characteristics.

Ocean Ecosystems

Nearly three-quarters of the Earth's surface is covered by ocean. Three types of ocean ecosystems are shown in **Figure 12.13**. Shallow ocean waters are small in area but contain most of the ocean's diversity. Many fishes swim in the open ocean surface, feeding on plankton. Photosynthetic plankton account for about 40 percent of all photosynthesis on earth. There is increasing evidence that pollution is harming photosynthetic plankton. If significant numbers of plankton are destroyed, the oxygen you breathe will be slowly depleted from the Earth's atmosphere. The deep ocean waters are cold and dark. Among the few residents of the deep ocean are some of the most bizarre organisms found on earth. Many organisms in the deep ocean have light-producing body parts that they use to attract mates or lure prey. Because photosynthesis cannot occur in the deep ocean, most of these organisms prey on other deep sea residents or scavenge the dead bodies of organisms that have fallen from above.

Figure 12.13
Each type of ocean ecosystem supports different kinds of living organisms. This representation is not drawn to scale.

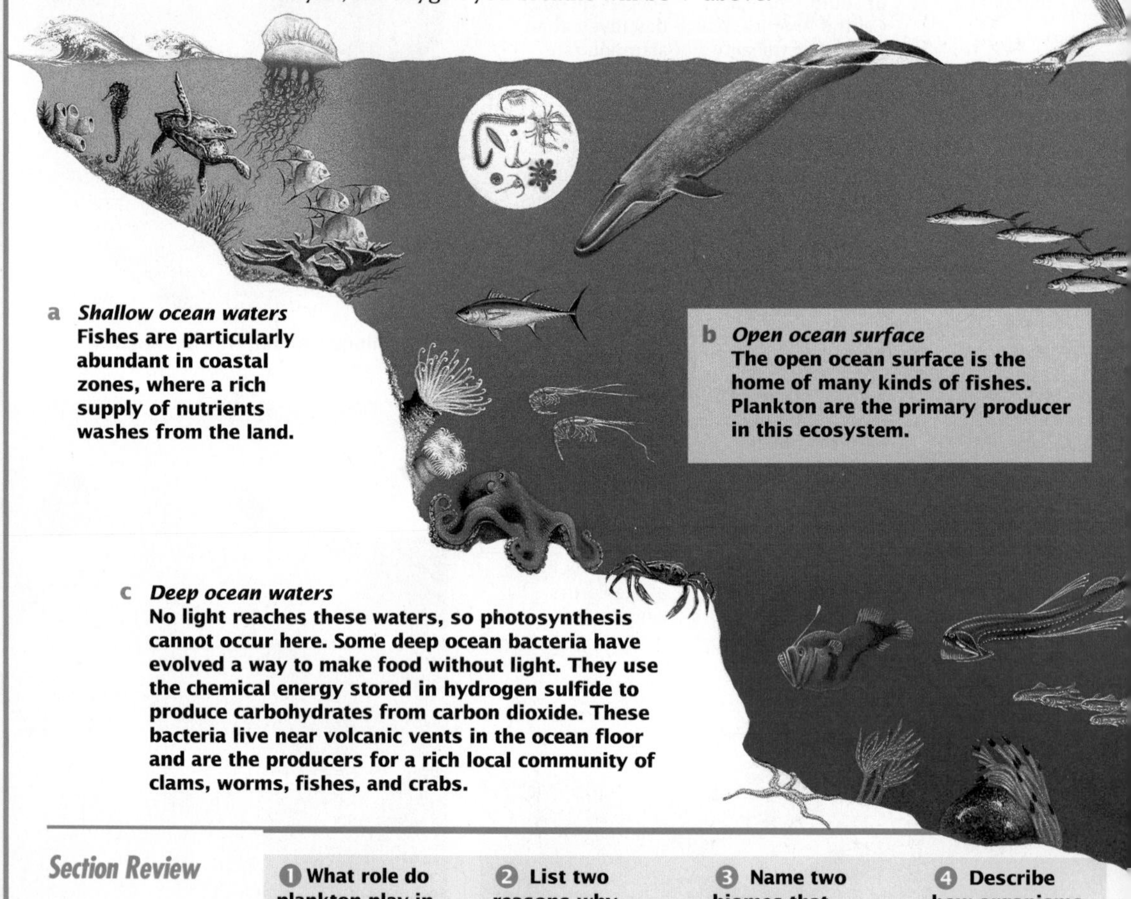

a **Shallow ocean waters**
Fishes are particularly abundant in coastal zones, where a rich supply of nutrients washes from the land.

b **Open ocean surface**
The open ocean surface is the home of many kinds of fishes. Plankton are the primary producer in this ecosystem.

c **Deep ocean waters**
No light reaches these waters, so photosynthesis cannot occur here. Some deep ocean bacteria have evolved a way to make food without light. They use the chemical energy stored in hydrogen sulfide to produce carbohydrates from carbon dioxide. These bacteria live near volcanic vents in the ocean floor and are the producers for a rich local community of clams, worms, fishes, and crabs.

Section Review

❶ **What role do plankton play in a freshwater ecosystem?**

❷ **List two reasons why tropical rain forests do not occur in the United States.**

❸ **Name two biomes that occur where there is very little precipitation.**

❹ **Describe how organisms living deep in the ocean obtain their food.**

▪ Section Review Answers ▪

1. Plankton contains photosynthetic organisms that are the base of aquatic food webs.

2. Variations in temperature and smaller amounts of precipitation
3. Possible answers are deserts, temperate grasslands, tundra, and coniferous forests.

4. They obtain their food through predation, scavenging, and from bacterial producers.

Chapter **12** *Highlights*

If the tropical rain forests disappear, so will this butterfly, from Trinidad.

Assessment Alternative
Have each student select both a plant and an animal and describe how they are adapted to their particular biome.

	Key Terms	*Summary*
12.1 What Is an Ecosystem?  **Decomposers, such as these mushrooms, are an important part of every ecosystem.**	ecology (p. 248) community (p. 248) habitat (p. 248) ecosystem (p. 248) diversity (p. 248) producer (p. 250) consumer (p. 250) decomposer (p. 250) trophic level (p. 251) autotroph (p. 251) herbivore (p. 251) carnivore (p. 251) omnivore (p. 251) heterotroph (p. 251)	• Ecology is the study of how living things fit into their environment. • An ecosystem is a group of interacting organisms and their physical environment. • Producers capture energy and store it in complex molecules. Consumers obtain energy by feeding on producers or other consumers. Decomposers eat dead organisms, animal wastes, fallen leaves, twigs, and other debris. • Autotrophs are organisms such as plants that make their own food. Heterotrophs cannot make their own food and must obtain it from other organisms.
12.2 Cycles Within Ecosystems  **Bormann and Likens studied how nutrients cycle within ecosystems at this site in New Hampshire.**	nitrogen fixation (p. 254) greenhouse effect (p. 256)	• Materials such as water, nitrogen, and carbon move through ecosystems in cycles. • Some bacteria absorb nitrogen gas and convert it to ammonia. The process of transforming nitrogen gas into ammonia is known as nitrogen fixation. • The burning of fossil fuels releases large amounts of carbon dioxide into the atmosphere. Carbon dioxide in the atmosphere retains heat from the sun, a phenomenon known as the greenhouse effect.
12.3 Kinds of Ecosystems  **The tundra is one of the seven biomes.**	plankton (p. 257) biome (p. 258)	• Photosynthetic plankton are the basis of the food web in aquatic ecosystems. • On land, there are seven major types of ecosystems, which are called biomes. • There are three major types of ecosystems found in the ocean.

Chapter Review Answers

Understanding Vocabulary

1. **a.** *Producers* does not fit the pattern, because it is not a type of consumer.
   b. *Energy* does not fit the pattern, because it is not a material that cycles within an ecosystem.
   c. *Photosynthesis* does not fit the pattern, since it does not add carbon dioxide to the atmosphere.

Relating Concepts

2. Map answer is shown on page 245D.

Understanding Concepts
Multiple Choice

3. b   8. c
4. c   9. b
5. a   10. a
6. c   11. a
7. d

Completion

12. ecology
13. nitrogen fixation
14. herbivores, omnivores
15. biome, community

Short Answer

16. Decomposers break down complex molecules, thereby recycling mineral nutrients.
17. Nitrogen fixation converts atmospheric nitrogen to a form usable by plants.
18. An ecological model is a set of hypotheses representing how an ecosystem operates; they may be expressed as mathematical equations. Ecologists use ecological models to test predictions about the effect of changes to an ecosystem.
19. Diversity is measured by the number of species in an ecosystem.

Chapter 12 Review

Understanding Vocabulary

1. For each set of terms, identify the term that does not fit and explain why.
   a. producers, decomposers, herbivores, omnivores
   b. water, energy, carbon, nitrogen
   c. photosynthesis, cellular respiration, decomposition, combustion

Relating Concepts

2. Copy the unfinished concept map below onto a sheet of paper. Then complete the concept map by writing the correct word or phrase in each oval containing a question mark.

Understanding Concepts
Multiple Choice

3. Which of the following organisms would *not* be part of the same trophic level?
   a. photosynthetic bacteria
   b. fungi
   c. grass
   d. trees

4. Bormann and Likens found that cutting trees from part of the Hubbard Brook ecosystem
   a. reduced calcium loss.
   b. decreased rainfall.
   c. increased calcium loss.
   d. increased rainfall.

5. The form of nitrogen usable by plants is
   a. ammonia.
   b. nitrous oxide.
   c. nitrogen gas.
   d. nucleic acids.

6. Which does *not* add carbon dioxide to the atmosphere?
   a. cellular respiration
   b. gasoline-burning cars
   c. photosynthesis
   d. forest fires

7. The biome with the greatest diversity is
   a. grassland.
   b. temperate deciduous forest.
   c. coniferous forest.
   d. tropical rain forest.

8. Tundra and desert are both
   a. cold.
   b. very diverse.
   c. dry.
   d. near the equator.

9. Which biome is *not* found in the United States?
   a. desert
   b. tropical rain forest
   c. deciduous forest
   d. temperate grasslands

10. The first trophic level in the open ocean contains
    a. plankton.
    b. herbivorous clams.
    c. carnivorous sharks.
    d. bacteria and fungi.

11. Which organisms would *not* be in a deep ocean ecosystem?
    a. photosynthetic bacteria
    b. clams
    c. crabs
    d. carnivorous fishes

Completion

12. Ernst Haeckel coined the term _____ to describe the study of how organisms interact with each other and with their environment.

13. Nitrogen gas is converted to ammonia during the process of _____ .

14. Because they eat only plants, cows are called _____ . Humans who eat plants and animals are called _____ .

15. The African savanna is the _____ inhabited by lions, giraffes, and cheetahs. Lions, giraffes, cheetahs, and other organisms that live on the African savanna make up a(n) _____ .

Short Answer

16. What is the ecological role of decomposers such as bacteria and fungi?

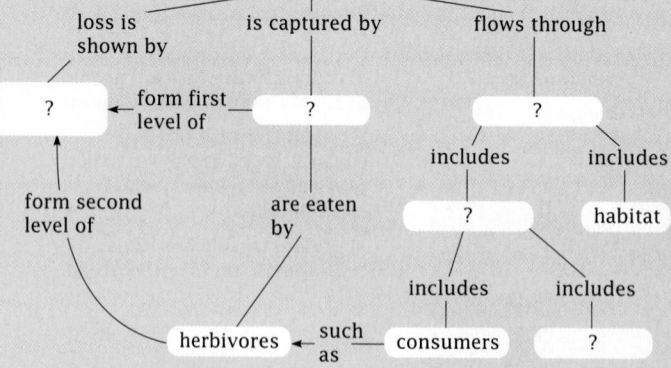

Interpreting Graphics

20. • algae
    • cod, leopard seal, and killer whale
    • krill
    • Killer whale; organisms at higher trophic levels tend to be fewer in number than those at lower trophic levels.

Reviewing Themes

21. As rain forests are destroyed, the earth's climate will change and many plant and animal species will become extinct. Climatic changes are due to less oxygen and water vapor in the air and the increase in carbon dioxide as the rain forests are burned. The lost plant and animal species could be sources of medicine and food.

17. What are the benefits of nitrogen fixation to plants?

18. What is an ecological model? How are ecological models used by ecologists?

19. How is diversity of an ecosystem measured?

Interpreting Graphics

20. Look at the food chain of a marine ecosystem shown below.

    • What is the producer in this ecosystem?

    • Which organisms are carnivores?

    • Which organisms are herbivores?

    • Which organisms should be least common? Explain.

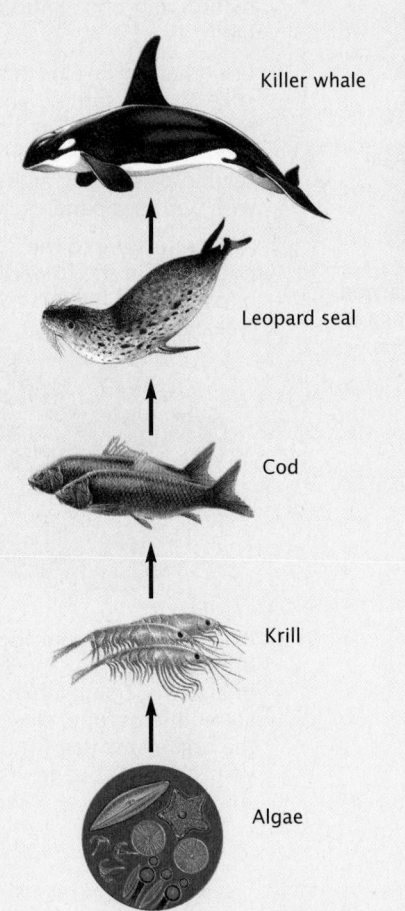

Killer whale

Leopard seal

Cod

Krill

Algae

Reviewing Themes

21. *Interacting Systems*
    If destruction of tropical rain forests continues at its present rate, most of the world's rain forests will be gone within 40 years. What are two likely effects of the destruction of tropical rain forests?

Thinking Critically

22. *Inferring Conclusions*
    Clear-cutting—cutting all trees in a certain area—was a practice used extensively by loggers in the early part of this century. It had a devastating effect on the water cycle. What can be done to restore the water cycle in clear-cut areas?

23. *Comparing and Contrasting*
    Describe the characteristics of the biome in which you live. What plants and animals live there? Which of the biomes described in this chapter is most like the one in which you live?

24. *Building on What You Have Learned*
    Based on what you learned about photosynthesis and cellular respiration in Chapter 5, why can plants act as producers and consumers, while animals can only act as consumers?

Cross-Discipline Connection

25. *Biology and History*
    By watching Native Americans, European settlers learned that placing a dead fish and corn seeds in the same hole produced a greater yield. How did this affect soil nutrients? How is this practice continued today?

Discovering Through Reading

26. Read the article "Why American Songbirds Are Vanishing," in *Scientific American*, May 1992, pages 98–104. What types of ecological damage explain the vanishing of many bird species? What can be done to increase the number of songbirds in North America?

Cross-Discipline Connection

25. Bacteria and fungi in the soil decompose the fish, releasing nutrients such as nitrogen into the soil. The soil nutrients can then be used by the corn plant. Today's farmers enrich the soil through crop rotation and use of artificial fertilizers.

Discovering Through Reading

26. Growth of suburbs has drawn in more nest parasites, such as bluejays, raccoons, and opossums. Deforestation is eliminating the birds' wintering grounds. Consolidating and expanding the largest tracts of forest, preserving wetlands, and slowing tropical deforestation can increase the number of songbirds in North America.

Thinking Critically

22. Planting trees in clear-cut areas might eventually lead to the restoration of the water cycle.

23. Responses will vary depending on location, but they should include information about precipitation, climate, and plant and animal life.

24. Animals cannot photosynthesize.

Procedural Note

Review the answers to the preparation questions and discuss the methods and locations to be used during the investigation. If possible, have each member of a cooperative group investigate a different type of soil community.

Prelab Preparation Answers

- Energy moves from producers, which capture solar energy, to herbivores, carnivores, and omnivores.
- A pyramid of energy is used to visualize the amount of energy found at each trophic level. A pyramid of numbers represents the number of individual organisms at each level.
- Decomposers feed on dead organisms, fallen leaves, and animal wastes. They are important in returning nutrients to the soil.

Procedure Answers

5. Freezing immobilizes the animals, making them easier to observe.
11. The organisms growing on the petri dish are bacteria and fungi, which are decomposers.

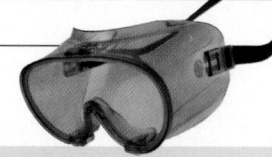

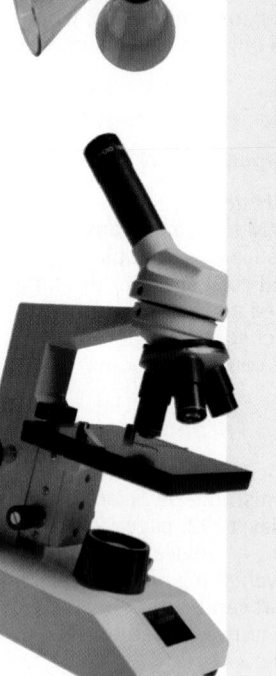

Chapter 12

Investigation

Exploring a Soil Community

Objectives

In this investigation you will:
- *observe* a soil community
- *measure* the total mass present at each trophic level in the community

Materials

- metric ruler
- plastic bags
- rubber bands
- hand trowel
- freezer
- pan
- forceps
- hand lens or stereomicroscope
- triple-beam balance
- test tube
- distilled water
- test tube rack
- sterile swab
- petri dish of nutrient agar
- tape

Prelab Preparation

Review what you have learned about communities by answering the following questions:
- How does energy move through a community?
- What is an ecological pyramid?
- What role do decomposers play in a community?

Procedure: Exploring a Soil Community

1. Form a cooperative group of four students. Work with one member of your group to complete steps 2–13. Make a table like the one shown on the opposite page. Steps 2–4 should be carried out in a nearby field or forest. The less disturbed your chosen area is, the better your results will be.

2. Collect the leaf litter from a 25 x 25 cm (10 x 10 in.) area and place it in a plastic bag. Use a rubber band to seal the bag.

3. Observe the surface of the soil and record your observations in your table.

4. Use a hand trowel to remove a cube-shaped sample of soil that measures about 15 cm (6 in.) on each side. Place the soil sample in a second plastic bag and seal the bag with a rubber band.

5. After returning to the lab, place both bags in the freezer for 5 hours or overnight. *What is the purpose of freezing the bags?*

6. Pour the contents of the bag of leaf litter into the pan and search for organisms. Using a hand lens or stereomicroscope, closely observe the organisms you find. Record your observations in your table.

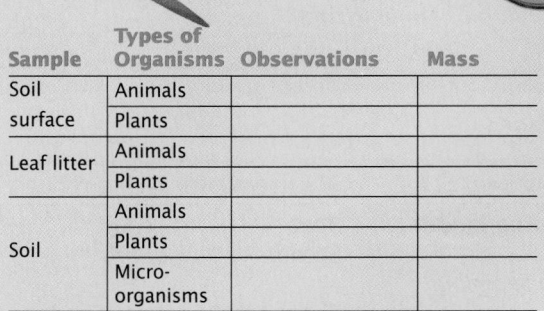

Sample	Types of Organisms	Observations	Mass
Soil surface	Animals		
	Plants		
Leaf litter	Animals		
	Plants		
Soil	Animals		
	Plants		
	Micro-organisms		

7. Using forceps, separate the animal and plant material into two piles. Use the balance to find the mass of each pile. Return the leaf litter to the plastic bag.

8. Open the plastic bag containing your soil sample. Place a sample of soil about the size of a pea in a test tube. Add enough distilled water to the test tube so that the soil sample is covered in water. Set the tube aside.

9. Now perform steps 6 and 7 for the remainder of the soil sample.

10. Remove a sterile swab from its package. Dip the swab into the test tube containing the soil-water mixture. Remove excess moisture by rotating the swab against the side of the tube. Open the cover of the petri dish and lightly rub the moistened swab over the surface of the agar in a zigzag pattern. Close the lid of the dish and seal it with tape, as your teacher directs.

11. Label the petri dish and incubate it at room temperature for one to two days. Then record your observations. **Do not open the petri dish**. After making your observations, dispose of the dish as directed by your teacher.
What kinds of organisms are growing on the petri dish?

12. Share your observations with the other members of your group.

13. Clean up your materials and wash your hands before leaving the lab.

Analysis

1. *Summarizing Observations*
Compare the leaf litter to the soil in terms of the mass and variety of animals.

2. *Analyzing Observations*
What evidence suggests that an ecological pyramid exists in this community?

3. *Analyzing Observations*
What evidence suggests that different trophic levels exist in this community?

4. *Making References*
On what do the soil microorganisms feed in this community?

5. *Comparing Observations*
How does the community you investigated compare to the community studied by the other team in your group?

Thinking Critically

1. Is there evidence that predators occur in this community? Explain your answer.

2. Where does a soil ecosystem start and end? What are the boundaries of an ecosystem?

3. Suppose that a fire destroyed the leaf litter of this community. How would this disturbance affect the community?

Analysis Answers

1. Answers will vary according to the data collected but should reflect the observation of a variety of organisms.

2. Answers should indicate that a pyramid of biomass exists.

3. Observations of organisms with mouthparts for feeding on different foods—plants, other animals—would be evidence.

4. They feed on decaying material.

5. Answers will vary.

Thinking Critically Answers

1. Observation of arthropods with biting mouthparts—centipedes, for instance—would be evidence for the presence of predators in this community.

2. Answers may vary. Students should recognize that the boundaries of an ecosystem are not always well defined. Although the soil ecosystem "ends" at the soil surface, its members can cross into the leaf litter above. Moreover, most of the nutrients in the soil ecosystem stem from rotting material from the ecosystem above the surface.

3. The community would eventually cease to exist unless the leaves that were destroyed by fire were replaced by other leaves or another energy source.

Chapter 13 How Ecosystems Change

Planning Guide

	Objectives/Themes	Classwork Resources	Homework Resources

13.1	**1.** Recognize the role of coevolution in shaping the structure of ecosystems.
	
	2. Relate the characteristics of flowers to their coevolution with insects.
	
	3. Describe how plants and their herbivores have coevolved.
	
	4. Contrast parasitism, mutualism, and commensalism.
	
	Themes: Evolution, Interacting Systems, Scale and Structure, Stability
		Teacher's Resource Binder
		Focus Activity 13
		    Using Random Sampling
			Text
			Section Review, p. 271
			
			Teacher's Resource Binder
			Directed Reading Worksheet 13.1
			Reteaching Worksheet*
			    Complex Interactions in Ecosystems
			
			Other Resources
			Audiocassette 13.1*

13.2	**1.** Describe the components of an organism's niche.
	
	2. Describe the process of succession.
	
	3. Relate the stability of an ecosystem to its diversity.
	
	4. Recognize the roles of ecosystem size and latitude in determining diversity.
	
	Themes: Interacting Systems, Patterns of Change, Stability
		Text
		Science in Action
		    Carmen R. Cid: Ecologist
		    pp. 280–281
		Journeys
		    History of an Ecosystem
		    pp. 276–277
		
		Teacher's Resource Binder
		Lab Investigation 13.1
		    Detergents as Pollutants
		
		Other Resources
		Transparencies 53–56
			Text
			Section Review, p. 279
			
			Teacher's Resource Binder
			Directed Reading Worksheet 13.2
			
			Other Resources
			Audiocassette 13.2*

13.3	**1.** Explain how changes in natural habitats can have a drastic impact on ecosystems.
	
	2. Recognize the devastating effects that exotic species have on native organisms.
	
	3. List three ways to reduce human impact on ecosystems.
	
	Themes: Interacting Systems, Patterns of Change, Stability
		Text
		Investigation
		    How Does Competition Affect Plant Growth? pp. 288–289
		Science, Technology, and Society
		    Deforestation: Are We Losing Only the Trees? pp. 290–291
		
		Teacher's Resource Binder
		Extension Worksheet
		    Population Growth Curves
			Text
			Section Review, p. 284
			
			Teacher's Resource Binder
			Directed Reading Worksheet 13.3
			Vocabulary Review Worksheet*
			
			Other Resources
			Audiocassette 13.3*

*Reteaching Options

Demonstrations
13.1: pp. 267, 268, 270, 271
13.2: pp. 272, 274, 278
13.3: pp. 282, 283, 284

Assessment
Chapter Review pp. 286–287
Portfolio Assessment p. 265D
Chapter Test—Teacher's Resource Binder
Test Generator

Research Notes

Connection to Ecology: Ecosystem Invaders

The delicate balance of an ecosystem can be disrupted by humans who introduce exotic species and don't even realize what they have done. An example is found in the Great Lakes, which are North America's largest freshwater reservoirs.

The Great Lakes carry more commercial shipping than any other freshwater system in the world. Many ships carry ballast, in the form of water, rocks, or sand, in special tanks to help balance the ship when it is not carrying cargo. When they are ready to take on cargo, that ballast is flushed out into the surrounding water.

What many shippers don't realize is that frequently their ballast contains "stowaways," plants and animals that can cause trouble for the new ecosystem. Ecologists believe that the sudden appearance of several exotic species in the Great Lakes is due to such "stowaways." Each one threatens to upset the ecosystem, to the point of possibly causing extinctions of some species.

The ruffe is a small perchlike fish that was formerly found in Europe. It has adapted so well to conditions there that it can withstand even arctic temperatures. It also has evolved to mature early, and to spawn more frequently than other fish. The ruffe was first detected in the Great Lakes in 1987. Because it competes so aggressively, scientists have already noted declines in most other fish populations in the Great Lakes.

The spiny water flea is about 3 mm long, and it dines on Daphnia, a microscopic algae-eating crustacean. Ordinarily found in waters near Leningrad, the spiny water flea was first seen in the Great Lakes in 1984. Since its arrival, two of three common species of Daphnia have almost disappeared from the Great Lakes. Since Daphnia is also a popular food for many small fish, ecologists are concerned that these fish could be in danger.

Starting in November 1992, ships were required to exchange their ballast for seawater before entering the Great Lakes. Presumably any freshwater "stowaways" would not do well in the salty ocean, and any saltwater organisms in the new ballast would not survive long in the Great Lakes. To protect other freshwater ports, Canada, Australia, and the United States are calling for nations to voluntarily adopt such safety measures for shipping throughout the world.

Investigation Notes

How Does Competition Affect Plant Growth?
pages 288–289

Purpose: Students design a controlled experiment that tests a hypothesis in this Investigation. Excellent results should be achieved fairly quickly. Students will conduct the experiment in pairs and pool their data with that of their group for final analysis. The emphasis of this investigation is on problem solving rather than on specific concepts related to plant growth.

Technique: This investigation can also be used to develop a better understanding of data analysis. Most students' use of statistics is limited to calculating an arithmetic mean, or average. While the mean is useful, it can be misleading when it is the only parameter used.

It might be helpful to have the students find the median and mode in addition to the mean. The median is the middle value of the measurements, and the mode is the most frequent value. In a normally distributed sample, the mean, median, and mode should have the same value.

This experiment has no set end point and can run for weeks. It is important that the plants are watered regularly. A standard amount of water—enough to moisten the soil without causing them to become waterlogged—should be applied to the pots about every other day. The exact amount of water needed will vary with the size and shape of pots used.

Prelab Preparation

Broad bean seeds and small pots are available from garden supply stores or WARD'S. If necessary, small paper cups could be substituted for pots.

Answers will be found on pages 288–289.

Meeting Individual Needs

Objectives

1. Students will demonstrate appropriate use of core vocabulary for the chapter (Vocabulary File).

2. Students will create models of organisms reacting to each other in an ecosystem (Demonstration A).

3. Students will recognize applications of the concepts of niches and change using a simple ecosystem (Demonstration B).

Vocabulary File
(Developing Vocabulary/ Limited English Proficiency)

If you are not already using the Vocabulary File, refer to Chapter 1 for its preparation. See the Chapter Highlights on page 285 for a list of suggested words.

Demonstration A
Interaction Within Ecosystems
for pages 270–271
(Developing Communication Skills/Kinesthetic Learners)

Materials: construction paper (at least 10 colors: red, blue, yellow, green, purple, brown, pink, orange, gray, and black), scissors, two envelopes, large sheet of white paper

Preparation: Cut small rectangles from construction paper. Cut three for each color listed. Place two of each color in an envelope. Staple or glue the third one to a sheet of white paper. By each color rectangle glued to this sheet, write the following color code for identification of names of organisms to be represented:

red—a blue cornflower and a bee

blue—a yellow sunflower and a bee

yellow—a hawkmoth and a scarlet lobelia flower

green—a cactus and a jackrabbit

purple—a cabbage butterfly and a mustard plant

brown—an oak tree and mistletoe

pink—fungus and green alga

orange—*E. coli* bacteria and a human intestine

gray—barnacles and a gray whale

black—a person and poison ivy.

Procedure: Introduce the term "coevolution" by reading aloud to students "Avoiding Being Eaten: Plants and Herbivores," on page 269. Have students independently scan the section entitled "Three Types of Close Species Interaction." Brainstorm with students to come up with additional examples of each type of interaction, in order to make sure students understand these concepts.

Tape the color-code sheet to the bulletin board, and have students draw a single rectangle out of the envelope. After drawing the rectangles, students match colors to find a partner. If partners are not compatible, they may make one partner trade—only one.

Then partners use the color-code sheet to identify the organisms they are to represent. Partners have 20 minutes to prepare a verse, skit, rap, or song that explains their interaction. After this preparation time, partners make their presentation to the class. The other students in the class should identify the type of interaction being portrayed.

Demonstration B
Ecosystem Niches and Change
for pages 272–274
(Building Observational Skills/Visual Learners)

Materials: several sprigs of *Elodea*, several live snails, battery jar, water

Preparation: Fill a battery jar with water. Place two sprigs of *Elodea* and two or three snails inside the jar. Set the jar on a window ledge that has sunlight each day. Leave the jar open to the air.

Procedure: Have students observe the jar during class and discuss what they observe. They should be able to see bubbles rising to the surface of the water during the day when the water becomes warm. Ask them if they can identify what is in the bubbles and why the bubbles are formed. If students don't realize this shows that photosynthesis is occurring, refer back to Chapter 5 and the discussion of photosynthesis. Ask whether the plants are producers or consumers at this time.

Discuss the small watery environment. Have students brainstorm about the way it is similar to and different from an ecosystem. Have students identify organisms and describe the niches they occupy. Be sure students notice the stability of this small environment. No organisms from outside are added to it, and no organisms are taken out of it. The jar will simply sit on the window ledge in the presence of sunlight.

Check the jar every third day and write down the date and observations. Note changes that may be taking place. Ask questions such as: Is the *Elodea* growing? Is the *Elodea* dying? Is competition present? Relate observations to the text.

Additional Strategies
Visual Strategies
Pages 270, 273, 274, 278, and 279

Auditory Learners
Use *Biology: Visualizing Life* Audiocassettes for Sections 13.1, 13.2, and 13.3.

Meeting Individual Needs (cont.)

Cooperative Learning
Symbiosis

Timing: Use this activity to introduce Section 13.1.

Group Size: 3 students

Outcome: Students will compare the three kinds of symbiosis discussed in the chapter: parasitism, commensalism, and mutualism.

Individual Accountability: Each group member will describe one kind of symbiosis and give examples.

Positive Interdependence: Groups will make a table comparing different kinds of symbiosis.

Assign one of the following types of symbiosis to each group member: (1) parasitism, (2) commensalism, and (3) mutualism. Ask each group member to write a brief report about their type of symbiosis. The report should include two sets of examples and a discussion of how the symbiosis is advantageous to one or both partners.

After completing their individual reports, group members should explain their results to the other students in the group. The entire group will work together to prepare a table comparing the types of symbiosis.

Portfolio Assessment

Students should select their best work and provide a self-reflective rationale for their selections. Students can make selections in the following areas.

1. *Content* One concept map from the chapter (See page 812 for evaluation criteria.)

2. *Reading Comprehension* One Directed Reading Worksheet from the Teacher's Resource Binder (Use the answer key to evaluate for accuracy.)

3. *Writing* Using the Vee Form, summarize a magazine or newspaper article relating to symbiosis or ecosystem change. (See page 22T for evaluation criteria.)

   Or: Select a writing task or project from the Chapter Review.

4. *Performance Assessment* One Vee form from a chapter investigation or lab manual investigation (See page 22T for evaluation criteria.)

Teacher makes selections in the following areas.

1. *Formal Assessment* Chapter test (Test A, B, or the Test Generator) The teacher-scored test should be reviewed by the student. Incorrect responses should be corrected by the student before the test becomes part of the portfolio.

2. *Informal Assessment* Use the Direct Observations Checklist, page 33T, during a laboratory or other cooperative learning experience.

3. *Performance Assessment* Have students write a newspaper editorial for or against development in their town, based on consideration of the environmental impact it will have on the ecosystem.

Concept Map Answer

The following is one possible answer to the Relating Concepts exercise on page 286.

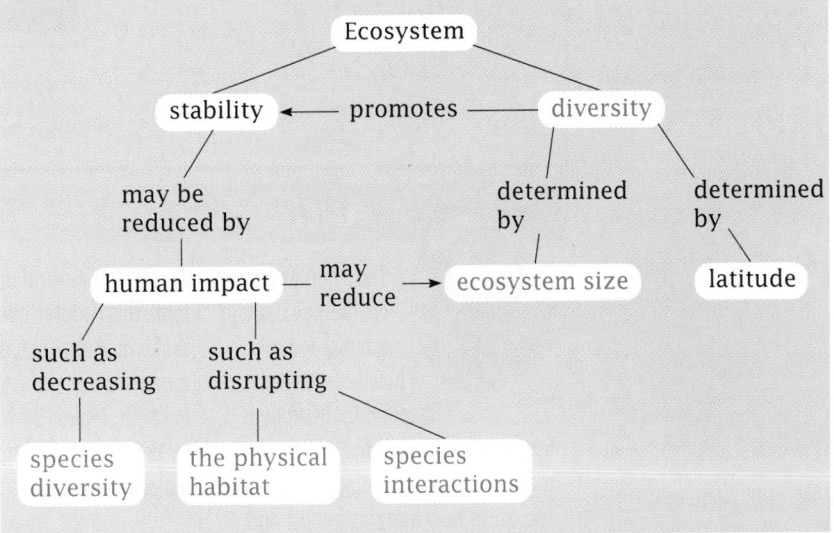

Chapter 13

Determining Prior Knowledge

- Show students a plant in bloom and a cactus that is not in bloom. Have them explain how the flowering plant has adapted to promote reproduction and how the cactus has adapted to favor its survival.
- Show the class a videotape of a volcanic eruption. Have students describe what will happen to the local ecosystem over a period of time.
- Have students describe one specific example of how humans have had a negative impact on the environment.

Chapter 13

How Ecosystems Change

Review

- **natural selection** (Section 9.3)
- **structure of ecosystems** (Section 12.1)
- **biomes** (Section 12.3)

13.1 Interactions Within Ecosystems

- **Evolution and Ecosystems**
- **Coevolution Shapes Species Interactions**
- **Avoiding Being Eaten: Plants and Herbivores**
- **Three Types of Close Species Interactions**

13.2 Ecosystem Development and Change

- **Ecosystem Lifestyles**
- **Competing Organisms Coevolve**
- **Competition and Ecosystem Development**
- **Ecosystem Stability**
- **What Makes Some Ecosystems More Diverse Than Others?**

13.3 How Humans Disrupt Ecosystems

- **Human Impact on Ecosystems**
- **Modifying the Environment**

Ecosystems are constantly changing. Here in Yellowstone National Park, naturally occurring forest fires enable new trees to grow.

■ Author's Rationale ■

This chapter focuses on the key issue in ecology today—how ecosystems respond to disruptions. Students learn what factors tend to make an ecosystem more stable. They recognize how humans impact and disrupt the natural order of an ecosystem. This chapter outlines principles that will shape a student's thinking on environmental issues for years to come, and as such it is a highly relevant part of his or her study of biology.

EVERY ECOSYSTEM ON EARTH, WHETHER IT IS FROZEN TUNDRA OR TROPICAL RAIN FOREST, IS A COMPLEX NETWORK OF INTERACTING SPECIES. TO PRESERVE THE EARTH'S FAST-DISAPPEARING ECOSYSTEMS, IT IS ESSENTIAL TO UNDERSTAND THE NATURE OF THESE INTERACTIONS AND HOW THEY ARE SHAPED BY NATURAL SELECTION AND THE PHYSICAL ENVIRONMENT.

13.1   *Interactions Within Ecosystems*

Objectives

❶ Recognize the role of coevolution in shaping the structure of ecosystems.

❷ Relate the characteristics of flowers to their coevolution with insects.

❸ Describe how plants and their herbivores have coevolved.

❹ Contrast parasitism, mutualism, and commensalism.

Evolution and Ecosystems

Figure 13.1
In this photograph, you can see that the protective coloration of this spider conceals it from both predators and prey.

Species evolve in response to the challenges posed by their environments. As a result, animals, plants, and other creatures in an ecosystem possess characteristics that fine-tune them for living where they do. For example, many plants in desert ecosystems have waxy

coatings on their leaves that help them retain water. For the same reason, desert animals often hide underground during the hottest part of the day; an animal's behavior is just as much an adaptation to its environment as are its physical characteristics.

An organism's survival and reproduction also depend on interactions with other living members of its ecosystem, as seen in the photograph in **Figure 13.1**. These interactions influence the evolution of a species just as do its interactions with the physical environment. For instance, many plant species have evolved tough leaves that protect against being eaten by herbivores. Of course herbivores evolve too, and many have evolved flatter, larger teeth that are better suited to grinding the very tough leaves they eat. Cows and horses have teeth such as these. **Coevolution** occurs when two or more species evolve in response to each other.

Lesson Plan 13.1

Phase 1
PREPARATION

Key Concepts
- Coevolution is the process by which two or more species evolve in response to one another.
- Examples of coevolution include the interactions between flowering plants and their animal pollinators, and between plants and herbivores.
- Relationships between two or more species include parasitism, mutualism, and commensalism.

Reading Strategy

An understanding of how natural selection operates is crucial to understanding the interactions within ecosystems as described in this section. Consequently, have students reread Table 9.2 on page 188 to review the process of natural selection before they begin reading this section.

Phase 2
TEACHING STRATEGIES

Demonstration 1

Coevolution

Some of the most obvious examples of coevolution are seen in predator-prey relationships. As natural selection improves the predator's skills, the prey also may evolve in response. The prey may develop new ways to avoid its predator. In turn, new mechanisms may evolve in the predator to help it detect its prey. Show students a photograph of a predator and its prey (fox and rabbit or hawk and mouse) and have them describe how each might evolve in response to the other.

Demonstration 2

Plants and Pollinators

Show students photographs or specimens of orchids, snapdragons, and irises. Point out their deep nectar-secreting structures and unusual platforms on which only certain pollinators can land. These flowering plants attract only certain pollinators that are efficient and precise in their job.

Demonstration 3

Would It Attract You?

Contact a florist to borrow a *Rafflesia* plant in blossom. Have students sniff its pungent aroma. If one is not available, point out to students that its flowers are among the world's largest, measuring nearly 1 m (3 ft.) in diameter.

Demonstration 4

Wind-Pollinated Flowers

Show students photographs or specimens of wind-pollinated flowers (wild oat, bog cotton, and the grasses, including wheat, rye, and corn). Have them note the differences between wind-pollinated flowers and flowers that are pollinated by insects or other animals. Wind-pollinated ones lack color and do not produce odors. Point out how these plants are adapted to their environment, producing copious amounts of pollen and growing in dense groupings, two traits that enhance the effect of wind.

Coevolution Shapes Species Interactions

Coevolution shapes the many ways animals and plants interact within ecosystems, creating a complex web of interactions among the organisms that live there. For example, producers, herbivores, and carnivores in a mature forest ecosystem have adjusted to one another during millions of years of evolution.

One of the most dramatic examples of coevolution is the side-by-side evolution of flowering plants with insects. Plants are immobile and cannot actively search for mates. Instead of moving to seek a mate, many species of flowering plants rely upon animals to transport their male gametes within grains of pollen. Insects are attracted to a plant's flowers by bright colors, pleasant odors, and food, which is usually sweet nectar. As the insect feeds, sticky pollen attaches to its body. After feeding at the flowers of one individual, the pollen-covered insect flies to flowers of another individual. At this next feeding stop, some of the insect's load of pollen is rubbed onto the female reproductive structures of the flowers. Fertilization is now able to take place. Animals that carry pollen from flower to flower are known as pollinators.

Within each species of flowering plant, those individuals that are better at attracting pollinators will leave more offspring. Thus, the attractive features of each species of flowering plant have evolved in concert with the preferences of its pollinators. For instance, bees cannot see the color red, and plant species pollinated by bees rarely have red flowers. Bee-pollinated flowers are usually yellow or blue. In comparison, the *Rafflesia* flower in **Figure 13.2** is pollinated by flies that feed on dead organisms. To attract its pollinators, this flower releases a powerful, nauseous odor like the stench of rotting flesh.

Pollinators, in turn, have evolved traits that enable certain species to specialize on particular species of flowers when seeking nectar. Scarlet lobelia flowers, for example, produce nectar at the bottom of a very long, tube-shaped flower. The hawk moth is the only pollinator with a tongue long enough to reach the bottom of the flower.

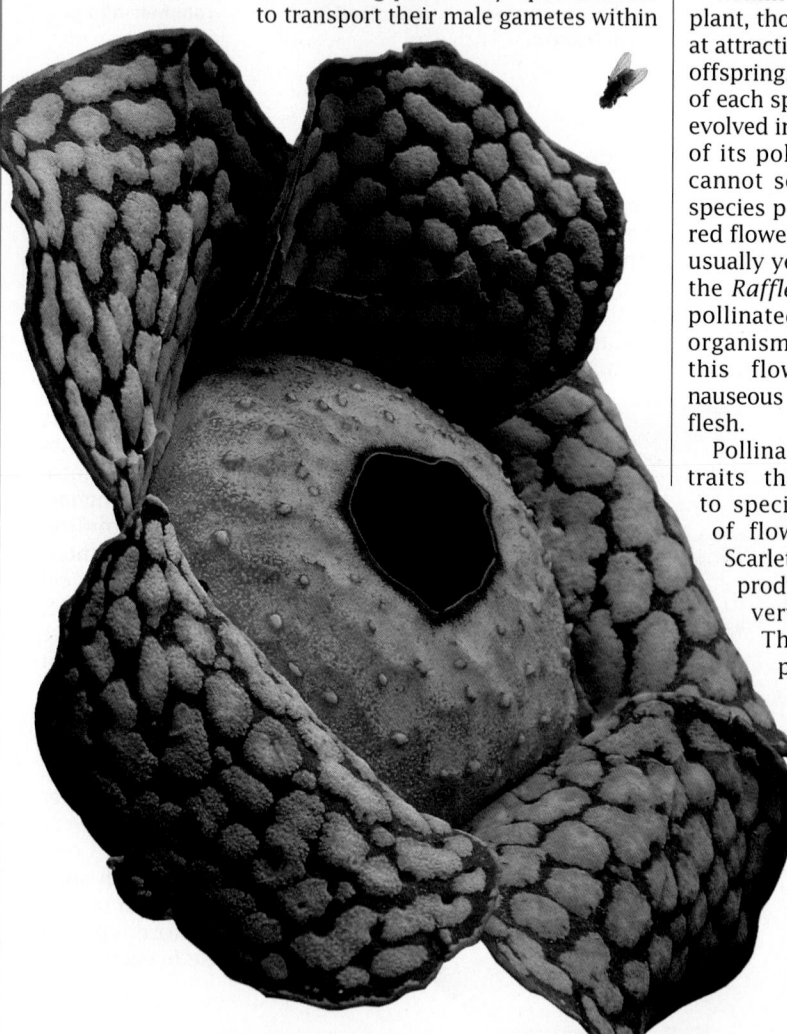

Figure 13.2
The pungent aroma of rotting meat attracts flies to this giant *Rafflesia* flower. The flies will pollinate it, and then transfer pollen from this flower to another *Rafflesia* flower.

Avoiding Being Eaten: Plants and Herbivores

Figure 13.3
The spines of this golden barrel cactus protect it from herbivores.

Being eaten is not beneficial for plants. Herbivores can kill plants by feeding on them, just as carnivores kill their prey. Therefore, characteristics that enable plants to protect themselves from herbivores are favored by natural selection. The biological structures of many of the Earth's ecosystems have been determined largely by the ways plants avoid being eaten and the ways by which herbivores succeed in eating them.

Plants defend themselves from herbivores

Touch the stem of a rose bush and you might experience a familiar plant defense against herbivores. Many species of plants, such as roses and the cactus in **Figure 13.3**, employ physical defenses such as thorns, prickles, sticky hairs, and tough leaves. The most crucial plant defenses, however, are chemical. Virtually all plant species produce chemicals that protect them against herbivores. Some of these defensive chemicals are poisons that kill the animal that eats the plant, while other chemicals simply make the plant taste bad. Most herbivores learn to avoid plants that possess defensive chemicals. Anyone who has broken out in a rash from contact with poison ivy, shown in **Figure 13.4**, has been the victim of plant chemical defenses. Poison ivy produces a gummy oil called urushiol that causes severe blistering in many people.

As a rule, each group of closely related plant species has a unique battery of chemical defenses. Poison ivy, poison oak, and poison sumac produce urushiol. The plants of the mustard family produce a group of defensive chemicals called mustard oils. Mustard oils are the source of the pungent aromas and tastes characteristic of such plants as mustard, cabbage, radish, capers, and horseradish. The same tastes that we enjoy signal the presence of chemicals that are toxic to many groups of insects.

Many herbivores overcome plant defenses

Over time, herbivores have evolved ways to overcome the chemical defenses of plants. Because different plant species produce different chemicals, coevolution has resulted in a very specialized pattern of feeding; certain kinds of herbivores feed exclusively on particular kinds of plants.

Cabbage butterflies provide a good example of how counteractive measures have evolved in response to the chemical defenses of plants. Although most insects avoid plants of the mustard family, the caterpillars of cabbage butterflies eat these plants voraciously. These caterpillars are able to eat plants of the mustard family because they have evolved the ability to break down mustard oils into harmless chemicals. As a result of this evolutionary breakthrough, cabbage butterflies have been able to use a new food resource—plants of the mustard family—without competition from other insect herbivores.

Figure 13.4
Touch the leaves of poison ivy and the chances are good that you'll get a rash from the toxins on their surfaces.

Health Connection

Plant Chemicals
Some of the chemicals that plants produce as a defensive mechanism include nicotine, caffeine, curare, mescaline, cocaine, opium, and tetrahydrocannabinol (the active ingredient of marijuana). Though toxic in large doses, curare is administered in small doses to surgical patients to relax their muscles.

Connection: Chapter 1

Malaria
Review how Ronald Ross discovered that mosquitoes were responsible for transmitting malaria.

Theme Answer

Evolution
The rabbits varied in their ability to resist the myxoma virus. Those with high resistance survived and passed this ability on to their offspring. The frequency of resistant rabbits rapidly increased. Consequently, this effort to control the rabbit population failed.

Demonstration 5

Parasitism
Show students a photograph of someone with athlete's foot. The disease is caused by a fungus that enters the foot through a break in the skin. The moisture in locker rooms and showers is ideal for the growth and spread of the fungus.

Demonstration 6

Mutualism
Pass around a sealed jar containing termites and pieces of wood. Point out that protozoans in the digestive tracts of the termites digest the cellulose in the wood, providing the termites with nutrients. In turn, the termites provide the protozoans with food and a place to live.

Visual Strategy

Figure 13.5
Have students suggest why this lichen is known as the British soldier.

Three Types of Close Species Interactions

Evolution

Variation is essential for natural selection.

What role did variation play in the long-term survival of rabbits in Australia?

Three types of species interactions involve particularly close relationships among the participants. **Symbiosis** is a close, long-term association between two or more species. The three types of symbiotic relationships are parasitism, mutualism, and commensalism.

Parasites and their hosts coevolve

Worldwide, between 200 million and 300 million people suffer from malaria. People who have malaria play host to a single-celled parasite that was injected into their blood by the bite of a mosquito. Parasites obtain nutrition by feeding on their host. How is a parasite different from a predator? A parasite usually does not kill its host, and it is usually smaller than the organism on which it feeds. The relationship between a parasite and its host is called **parasitism**.

Both host and parasite coevolve in response to each other. In 1859, 12 rabbits were introduced to Australia. No rabbits naturally occur on the Australian continent. By the 1940s there were millions of rabbits swarming the countryside. To control the rabbit population, the Australian government introduced the viral disease myxomatosis (*mihk suh muh TOH suhs*) from South America. At first, myxomatosis was very deadly to the rabbits. More than 99 percent of the Australian rabbit population was killed by the initial introduction of the virus. The few surviving rabbits continued to breed, so the virus was reintroduced. This time, 90 percent of the rabbits died. By the third introduction of the virus, only 50 percent of the rabbits were killed.

Why did a smaller percentage of the rabbit population die each time the virus was introduced? Tests on the rabbits and the virus showed that both were evolving. Because rabbits resistant to myxomatosis were more likely to survive and reproduce, the rabbit population contained a greater proportion of resistant individuals after each introduction. Also, instead of becoming more deadly to overcome the rabbit's resistance, the virus had actually become less virulent, or deadly. If a virus kills its host rabbit too quickly, that rabbit cannot spread the virus to other rabbits. A virus that allows its hosts to live longer is able to infect more hosts.

All parties benefit in mutualism

Not all coevolution involves antagonistic relationships such as those between plants and herbivores or between parasites and their hosts. **Mutualism** is a symbiotic relationship in which all participating species benefit. For example, a lichen, such as the British soldier lichen in **Figure 13.5**, is a mutualistic partnership between a fungus and a green alga. The fungus absorbs nutrients for both partners from the surface on which the lichen is growing. The alga carries out photosynthesis to provide food for itself and its fungal partner. Mycorrhizae, which you read about in Chapter 10, are mutualistic associations between plants and fungi. In Chapter 12, you

Figure 13.5
The British soldier lichen is made of a mutualistic relationship between a fungus and an alga.

Alga
Fungus

▪ *Matter of Fact* ▪

A Hare Brained Scheme
In 1859, rabbits were introduced to Australia by an Englishman, who imported them from Europe to grace his estate. Six years later, he had killed nearly 20,000 rabbits on his property and estimated that 10,000 were still alive and devouring his vegetation.

Less than 30 years after they were introduced, 20 million rabbits were killed by Australians in the state of New South Wales alone. This is a striking example of what happens when an organism is introduced into an ecosystem in which there is no natural predator for that organism.

Figure 13.6
Stinging anemones live on the claws of this female boxing crab. She guards herself and the red eggs she is carrying under her abdomen by using the anemones like boxing gloves, jabbing them at attacking predators. This is an example of a commensal relationship.

learned about the relationship between nitrogen-fixing bacteria and the roots of plants such as peas and beans. These bacteria provide the plants with a source of nitrogen. In exchange, the nitrogen-fixing bacteria receive a place to live (swellings on the plant roots) and sugars produced by the plant.

Mutualism also has played an important role in the coevolution of humans and the microbes that live in our intestines. Within the human large intestine live immense colonies of the bacterium *Escherichia coli*. These bacteria have ready access to food while providing us with vitamin K, which is necessary for blood to clot. Animals that lack these bacteria, such as birds, must consume food that contains the necessary amounts of vitamin K.

Figure 13.7
Barnacles on this gray whale have found a safe way to obtain food by riding the whale's back. The whale does not seem to be bothered by the presence of barnacles.

Commensalism is taking without harming

Commensalism is an ecological relationship in which one species benefits and the other is not obviously affected. A very intriguing example of commensalism is the boxing crab, which is described in **Figure 13.6**. The crab benefits from the protection anemones provide, and the anemones apparently are not harmed or helped.

The crusty growths seen on the back of the gray whale in **Figure 13.7** are actually small animals called barnacles. Barnacles hitch a ride on the whale. In doing so, they gain protection from predators and transportation to new sources of food (tiny animals they filter from the water). Apparently, the whale neither benefits nor is harmed by the presence of barnacles.

Demonstration 7
Show students a photograph of epiphytes growing on trees in a tropical rain forest. These epiphytes benefit from living high in the air, where they receive more sunlight and rain than if they lived on the deeply shaded forest floor. The trees are neither harmed nor benefited. Point out that many species of orchids are epiphytes. Use this opportunity to reiterate the ecological importance of tropical rain forests.

Phase 3

ASSESSMENT OPTIONS

Closure Strategy
A Life Long Relationship
Have students explain why parasites usually do not kill their hosts.

Section Review
Assign the *Section Review*.

Reteaching
Have students imagine they are on an expedition in a tropical rain forest when suddenly they come upon a very strange-looking organism. Upon closer examination, they discover that there are actually two organisms attached to one another. Have students describe how they would determine what kind of symbiotic relationship these organisms share.

Section Review

❶ Describe one example of coevolution.

❷ Do you think that the drab flowers of some grasses and trees are pollinated by insects? Explain.

❸ Describe some changes that have occurred in caterpillars and their food plants as a result of coevolution.

❹ What kind of ecological relationship do your cells have with their mitochondria? Explain.

■ *Section Review Answers* ■

1. Answers will vary. Students should suggest species that have evolved in response to other living members of their ecosystems.

2. No. These plants are pollinated by wind. They do not need to produce attractive flowers.

3. Plants evolved poisonous chemicals that deter caterpillars. In response, caterpillars have evolved the ability to detoxify the defensive chemicals of their food plants.

4. Mutualistic. Cells provide a place to live. Mitochondria produce energy used by the cells.

Lesson Plan 13.2

Phase 1

PREPARATION

Key Concepts
- A niche is the role an organism plays in its ecosystem.
- Different species sharing resources tend to evolve reduced competition.
- Ecosystems develop by succession, whereby a progression of species replacements takes place.
- The diversity of species within an ecosystem is related to the size of the ecosystem and its latitude.

Reading Strategy

Emphasize the importance of reviewing by briefly going over what students have just learned. One way is for students to talk to each other about what they just read or explain it to someone else.

Phase 2
TEACHING STRATEGIES

Demonstration 1
Habitat vs. Niche
Show students a picture of a house and tell them that this represents a habitat, a place where its inhabitants live. Tell them that a family of four occupies the house—a father, mother, a teenage girl, and a young baby boy. Have them describe what possible *niche (role)* each might play. The teenage girl, for example, might be a high school senior who works part time, takes care of her baby brother when her mother works, likes Italian food, and enjoys playing Monopoly with her family.

OUR EFFECT ON THE ENVIRONMENT NOW EXTENDS TO ALL PARTS OF THE GLOBE, TO EVERY LIVING THING IN THE WORLD. WHETHER THE WORLD'S ECOSYSTEMS CAN SURVIVE DEPENDS ON HOW THEY REACT TO HUMAN INTERFERENCE. HOW AN ECOSYSTEM RESPONDS TO CHANGES DEPENDS CRITICALLY ON THE RELATIONSHIPS AMONG ORGANISMS IN THAT ECOSYSTEM.

13.2 Ecosystem Development and Change

Objectives

1. Describe the components of an organism's niche.

2. Describe the process of succession.

3. Relate the stability of an ecosystem to its diversity.

4. Recognize the roles of ecosystem size and latitude in determining diversity.

Ecosystem Lifestyles

Changes are a natural part of the history of any ecosystem. However, to save ecosystems threatened by human interference, it is essential to first understand that every organism in an ecosystem plays a role in the ecosystem. An organism might be prey for one species and predator of another. The sum of an organism's interactions with its physical environment and with other organisms is its **niche**.

A niche describes how an organism lives, the "job" it performs in the ecosystem. A niche can be described as the position a species occupies in the movement of energy through the ecosystem. The niche of grass growing in a meadow, for example, is that of producer, while the niche of deer that eat the grass is that of herbivore. The niche of an organism also includes the climate it prefers, the time of day it feeds, the time of year it reproduces, what it likes to eat, and where it finds its food. Each organism in an ecosystem, like the earthworm that is shown in **Figure 13.8**, has a unique niche. The total niche that an organism could potentially use within an ecosystem is that organism's **fundamental niche**.

Niche

Brings minerals to the soil's surface

Helps air and water to penetrate soil

Carries dead plant matter underground

Figure 13.8
Every organism has a niche, or job, in the ecosystem. An earthworm's niche includes many activities that enhance the soil.

Figure 13.9

a The barnacle *Chthamalus stellatus* can live in both shallow and deep water on a rocky coast. These areas are its fundamental niche.

b The barnacle *Balanus balanoides* prefers to live in deep water, which is its fundamental niche.

c When the two barnacles live together, *Chthamalus* is restricted to shallow water, its realized niche. What is the realized niche of *Balanus*?

Competing Organisms Coevolve

Sometimes organisms are unable to occupy their entire fundamental niche because another species already occupies part of it. The barnacle *Chthamalus*, for example, is able to live on rocks in both shallow and deep ocean waters. It is not usually found in deep water, because another species of barnacle, *Balanus*, prefers to live at these depths. In a series of experiments performed in the 1960s, ecologist Joseph Connell of the University of California removed all the barnacles from rocks and then added the two kinds of barnacles in equal numbers. As **Figure 13.9** shows, Connell found that *Chthamalus* won out only in shallow water where *Balanus* could not live. In the deeper water where both could live, *Balanus* eliminated *Chthamalus*.

Situations in which two or more organisms attempt to use the same resource are called **competition**. The two kinds of barnacles in Connell's experiment competed for living space. Competition often prevents an organism from occupying all of its fundamental niche. That part of a fundamental niche that an organism actually occupies as a direct result of competition is called its **realized niche**. The fundamental niche of the barnacle *Chthamalus* extended down to deep water. Its realized niche was restricted by competition to shallow water.

Competition can cause changes in an ecosystem

When two species compete intensely for the same resource, one species usually wins. The losing species may be driven to extinction within the ecosystem. The process in which one species is out-competed and dies out within an ecosystem is called **competitive exclusion**. Competitive exclusion is rare in most ecosystems, however, because selection tends to favor evolutionary changes that decrease competition. As a result, competing species reduce their use of common resources, and their niches become less similar.

In the 1950s the Princeton ecologist Robert MacArthur showed how potential competitors often compete very little because of subtle differences in their niches. MacArthur studied five species of warblers (small insect-eating songbirds), all of which fed in the same spruce trees. Each species spent most of its time feeding in a different part of the tree. As a result of these different feeding habits, the five species of warblers were not really in direct competition with one another. In effect, these warbler species share the wealth available to them in their ecosystem. The result of this resource sharing is that the ecosystem supports many more kinds of lifestyles, which makes the ecosystem more complex.

■ *Matter of Fact* ■

The Competitive Exclusion Principle

The competitive exclusion principle was developed by the Russian biologist, G.F. Gause. His classic experiment involved growing two species of *Paramecium*. When the two were grown in the same culture and thus competed for the same food source, one species reproduced faster than the other, which soon died out. By the way, the surviving species is significantly smaller than the one it overtook, proving that a larger size is not always an advantage.

Demonstration 2
The Pioneers
Because of their ability to survive in harsh conditions, lichens and mosses have earned the name pioneers. Have students use a stereomicroscope to examine specimens of these pioneers, which can survive in rocky terrains where there are few nutrients.

Visual Strategy

Figure 13.10
Use this figure to distinguish between primary and secondary succession. Since there was growth before Mount St. Helens erupted, secondary succession occurred in this case.

Competition and Ecosystem Development

Competition plays an important role in how ecosystems develop. The role of competition in the development of ecosystems is most easily seen when a serious disruption creates new habitat. New habitat is formed, for example, when a volcano forms a new island, or a glacier recedes and exposes bare soil, or a fire burns all the vegetation in an area. In every case scientists have been able to study, the empty habitat is quickly occupied. **Figure 13.10** shows some of the events that followed the eruption of Mount St. Helens in Washington. The first organisms to move into new habitat are small, fast-growing plants—you would probably call them "weeds." These early "settlers" are specialized for life under harsh conditions, such as on bare rock, and are able to eke out a living where few others could.

Competition drives change in a developing ecosystem
The initial weedy colonists do not remain in the ecosystem very long, however, because their pioneering efforts soon make the ground more hospitable. As a result of this improvement, later plant arrivals soon out-compete and replace the original inhabitants. This second wave of immigrants is replaced in turn by still other species better able to compete in the new environment. As the ecosystem matures, niches become more and more finely subdivided, and species become more and more interdependent. Hence, the diversity of the ecosystem usually increases as succession proceeds.

The regular progression of species replacement in a developing ecosystem is called **succession**. When succession takes place on land where nothing has ever grown before, it is called **primary succession**. When it occurs in areas where there has been previous growth, as in abandoned fields or forest clearings, it is called **secondary succession**. Succession does not continue indefinitely. Eventually, if the ecosystem is undisturbed for a long time, a community that is resistant to change results. Although no two episodes of succession are exactly alike, the progression of species replacement tends to result in similar communities in similar physical conditions. That is why biomes such as tropical rain forests are so similar wherever they occur.

Figure 13.10
a **A coniferous forest covered the slopes of Mount St. Helens before its eruption on May 18, 1980.**

b **The eruption leveled about 18,000 hectares (44,000 acres) of forest.**

c **Now, more than 10 years later, a new ecosystem has become established in the devastated areas.**

Ecosystem Stability

Why does succession stop? The last stage in a series of successional stages is able to absorb disruption without major change better than could its predecessors. Early successional stages, for instance, are easily changed by the invasion of new competing species. In contrast, the final stage of succession is able to resist invasion by potential competitors.

The ability of an ecosystem to resist change in the face of disturbance is known as **stability**. During succession, early stages show low stability, later stages are more stable, and the final community is the most stable.

What factors promote stability?

What makes some ecosystems more stable than others? Most ecologists now agree that more diverse ecosystems will be more stable than less diverse ecosystems. A more diverse ecosystem contains a more complex web of interactions among species than does a less diverse ecosystem. Alternate links in the web of species interactions are more likely to be available to compensate for the disruptions such as the loss of a species.

Even ecosystems that are very diverse contain points of vulnerability, however. A species whose niche affects many others in the ecosystem and that cannot be readily replaced if lost is called a **keystone species**. Because keystone species are the focus of many biological interactions, these species represent points where the web of species interactions can come unraveled.

In the 1960s, ecologist Robert Paine of the University of Washington discovered an excellent example of a keystone species. Paine worked on a 15-species ecosystem along the Washington coast. As shown in **Figure 13.11**, when he removed all the sea stars from this ecosystem, one species of mussel that the sea star ate began to thrive and out-competed many other species in the ecosystem. The number of species in this ecosystem fell from 15 to 8. The sea star in Paine's ecosystem was a keystone species.

To preserve natural ecosystems, it is essential to promote their biological diversity. It is very important to realize that diverse ecosystems can be damaged if key species are lost.

Figure 13.11

a **The sea star in the above photograph is prying open a mussel. It is the primary predator in a 15-species ecosystem that was studied by Robert Paine. His experiments showed the importance of a keystone species.**

b **When the sea stars were removed, the mussels thrived. They out-competed other organisms in the ecosystem, thereby reducing the total number of species in the ecosystem from 15 to 8.**

Theme Connection

Stability

The replacement of species by others during succession leads to a final community that is most stable. Point out that stability is a characteristic of all levels in nature—from chemical bonds that form when atoms combine with one another, to the final community in an ecosystem undergoing succession.

Using the Feature

- Use the feature as an illustration of the process of succession described on page 274.

Discussion

Guide the discussion by posing the following questions.

1. Would the newly exposed soil of Step 2 be suitable for agriculture? Explain.
   No. Its nutrients have been leached away by meltwater.

2. What part do nitrogen-fixing bacteria have in the process of succession?
   The early colonists depend on nitrogen-fixing bacteria for their supply of nitrogen.

3. Describe the plants that colonize the newly exposed soil.
   These colonists are small, fast-growing, and hardy.

4. Why is the appearance of pine trees dependent on the presence of alders?
   Alders have nitrogen-fixing bacteria in their roots. Their surplus nitrogen enters the soil and is available for pine trees, which lack nitrogen-fixing bacteria.

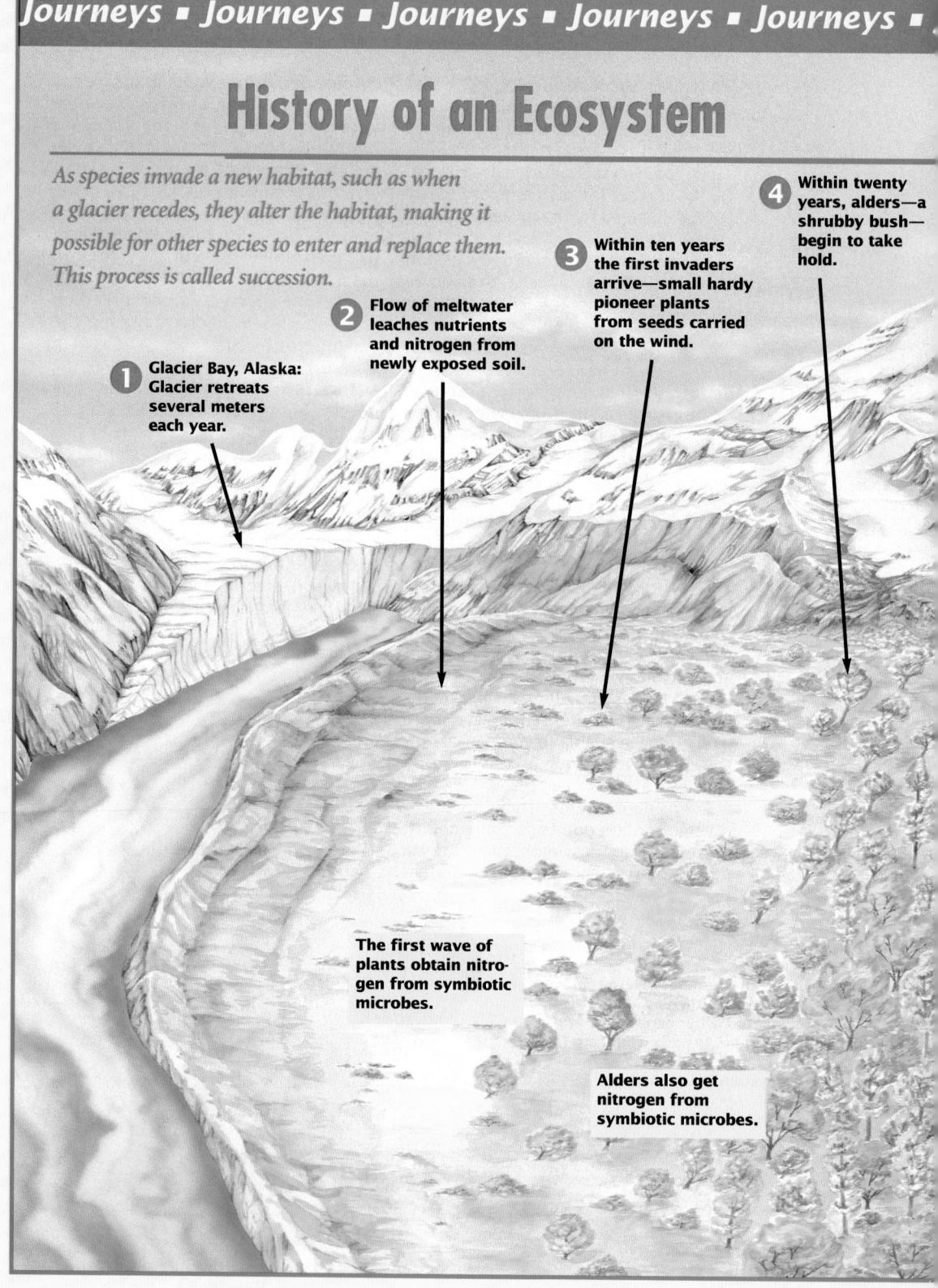

Journeys ■ *Journeys* ■ *Journeys* ■ *Journeys* ■ *Journeys* ■

History of an Ecosystem

As species invade a new habitat, such as when a glacier recedes, they alter the habitat, making it possible for other species to enter and replace them. This process is called succession.

1 Glacier Bay, Alaska: Glacier retreats several meters each year.

2 Flow of meltwater leaches nutrients and nitrogen from newly exposed soil.

3 Within ten years the first invaders arrive—small hardy pioneer plants from seeds carried on the wind.

4 Within twenty years, alders—a shrubby bush—begin to take hold.

The first wave of plants obtain nitrogen from symbiotic microbes.

Alders also get nitrogen from symbiotic microbes.

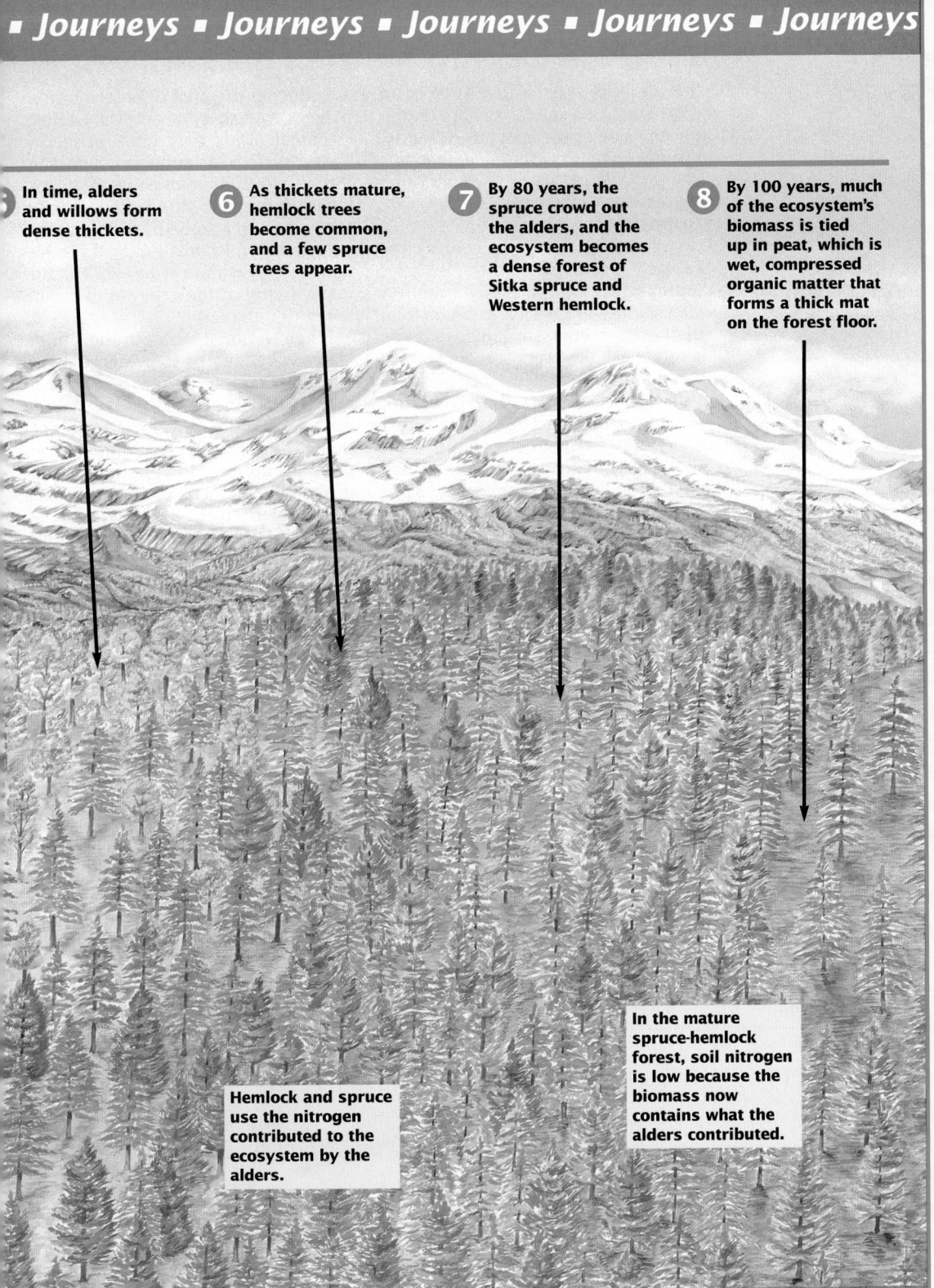

In time, alders and willows form dense thickets.

6 As thickets mature, hemlock trees become common, and a few spruce trees appear.

7 By 80 years, the spruce crowd out the alders, and the ecosystem becomes a dense forest of Sitka spruce and Western hemlock.

8 By 100 years, much of the ecosystem's biomass is tied up in peat, which is wet, compressed organic matter that forms a thick mat on the forest floor.

Hemlock and spruce use the nitrogen contributed to the ecosystem by the alders.

In the mature spruce-hemlock forest, soil nitrogen is low because the biomass now contains what the alders contributed.

Visual Strategy

Table 13.1
Use this table to illustrate that when the size of an ecosystem is reduced, the number of different species in that ecosystem is also reduced.

Demonstration 3

Stability in an Ecosystem
Have students refer back to Figure 12.2 on page 248 and Figure 12.6 on page 251. Remind them that each figure represents an ecosystem. Ask them which would be the more stable. Have them support their choice.

Why Are Some Ecosystems More Diverse Than Others?

What determines the diversity of an ecosystem? Two key factors are important: the size of the ecosystem and its latitude (distance from the equator).

Larger ecosystems support more species

Ecosystems that occupy large areas and are not subdivided into isolated communities will usually contain a wider variety of physical habitats than small ecosystems. Large ecosystems therefore usually support more species than small ecosystems. Geography often acts to restrict the size of an ecosystem, however. An island forest can be only as big as the island. Even on the mainland, terrain or human activity can limit the size of an ecosystem. A river can cut a forest into two ecosystems, and so can a road or a fence. These divisions have little effect on organisms that can readily cross them (for example, eagles that fly over fences), but they serve to isolate organisms that cannot (deer that are unable to jump high fences).

Reducing the area of an ecosystem reduces the variety of physical habitats it contains. Thus, the number of species the ecosystem can support declines. In today's world, human activity is causing ecosystems to shrink. Reduction in ecosystem area sometimes produces a situation in which many animal species, like the gray wolf in **Figure 13.12**, formerly living in the ecosystem are no longer found there.

When National Parks in the United States were created to preserve diversity, they contained more species of mammals than they do today. As you can see in **Table 13.1**, isolating a portion of a large ecosystem into a small park did not provide enough habitat diversity to maintain the many species that formerly lived in the larger area. Different species have died out in different parks, so no species has become entirely extinct, but the pattern of species loss is clear. If we want to maintain animal diversity, the National Parks will have to be managed carefully and lost species reintroduced.

Table 13.1   The Effect of Area on Ecosystem Diversity

National Park	Area (km²)	Percentage of Large Mammal Species Lost Since Founding
Bryce Canyon	144	36
Lassen Volcano	426	43
Crater Lake	641	31
Rocky Mountain	1,049	31
Yosemite	2,083	25
Sequoia-Kings Canyon	3,389	23
Grand Canyon	4,931	4
Yellowstone	10,328	4

Figure 13.12
The gray wolf needs a large range to survive. As shown in the table, the smaller a natural preserve, the greater the chance that native organisms will be lost.

Table 13.2 The Effects of Latitude on Diversity

Latitude	Number of Bird Species	Location
70°N	26	Northern Alaska
30°N	153	Southwest Texas
10°N	600	Central America

Visual Strategy

Table 13.2
Have students identify the latitudes where the most and least species diversity would exist. Have students hypothesize what would happen to the diversity of species at increasing altitudes within a particular latitude.

Phase 3

ASSESSMENT OPTIONS

Closure Strategy
Backyard Succession
Have students explain what would happen if a lawn were not mowed for an extended period of time.

Section Review
Assign the *Section Review*.

Reteaching
Have students research the fires that occurred in Yellowstone National Park in 1988. Instruct them to focus on the role fires play in the Yellowstone ecosystem.

Latitude affects diversity
Latitude has a great influence on ecosystems because both moisture and temperature vary with distance from the equator. The tropics, which are closest to the equator, have the highest species diversity for two reasons. First, latitude helps determine the length of the growing season. The greater the amount of food produced by plants and other producers, the more consumers an ecosystem can support. In the tropics, with ample sunlight, warm temperatures, and generous rainfall throughout the year, the growing season never stops.

Second, latitude plays a major role in determining climatic stability. The tropical climate does not vary much from season to season or from year to year. These unchanging physical conditions in the tropics have provided a long evolutionary window for specialized relationships to coevolve. In temperate or arctic regions, by contrast, weather can vary a great deal from one year to the next, and does vary from season to season. Conditions do not usually persist long enough for coevolution to foster as many new relationships as in the tropics. The effects of latitude on diversity are shown in **Table 13.2.**

Section Review

① **Describe the niche of humans.**

② **What features of weeds make them well suited for their role in early successional stages?**

③ **Explain what makes high-diversity ecosystems more stable than low-diversity ecosystems.**

④ **List two factors that make tropical ecosystems more diverse than temperate-zone ecosystems.**

■ Section Review Answers ■

1. Students should recognize that humans affect all other species on the globe, eating some, competing with some, altering the habits of most.
2. Weeds are small, fast-growing plants that are specialized for life under harsh conditions and are able to survive where others may not.
3. Highly diverse ecosystems contain a more complex web of interactions among species. Chances are better that alternate links in the web of species interactions will be available to compensate for the effect of a disruption.
4. Students should discuss how latitude affects the length of the growing season, and climatic stability, and how these factors affect diversity.

Using the Feature

- Use this feature to discuss the potential impact women and minorities can have in all areas of biology. Ask students to brainstorm about ways to encourage them to enter careers in the sciences.
- This feature can also be used to discuss the difficulties Carmen may have faced when she moved to Brooklyn, New York from Havana, Cuba. Ask students who have been in a similar situation to share their experiences.

Discussion

Guide a discussion by posing the following questions:

1. Discuss two factors in Carmen's background that made studying ecology so attractive to her.
   Carmen had grown up in a large crowded city and was overwhelmed by the woods and wetlands she visited while in college. She also had an enthusiastic ecology professor.

2. What personal traits does Carmen possess that make her a successful ecologist?
   She is curious, ambitious, and energetic. She also obviously loves spending time outdoors.

3. Besides research, what else does Carmen do professionally?
   She is an associate professor of biology and teaches ecology.

Science in Action ▪ Science in Action ▪ Science in Action

Carmen R. Cid: Ecologist

How I Became Interested in Ecology

"Ever since I was a little girl, my father, a well-educated man, fostered my interest in biology. When I was 12, my family moved from Havana, Cuba, to Brooklyn, New York. In high school, my curiosity about biology deepened thanks to my biology teacher. She was a great role model who shared her love of biology while demanding respect from a tough audience of inner-city teenagers.

"I entered college with the intention of becoming a biology teacher. However, in my junior year, I took a plant ecology course and fell in love with the outdoors. The professor, an enthusiastic naturalist, would take us out of New York City into the nearby woods and wetlands. I had always lived in big cities and this was my first opportunity to visit the lovely, unspoiled environment of a forest. Until then, I had thought of trees and other plants only as decorations along sidewalks or in parks. I never thought of plants as being very much alive. When you grow up with no connection to plants, and then someone gives names to things you thought were lifeless, it's like making a whole bunch of new friends. My professor taught me a new way to look at nature, and I soon saw plants as living, growing organisms, interacting with each other."

Carmen loves to spend her spare time traveling to places with waterfalls and lush, green mountains.

Name:	**Carmen R. Cid**
Home:	**Willimantic, Connecticut**
Employer:	**Eastern Connecticut State University**
Personal Traits:	• Curiosity
	• Ambition
	• Patience
	• Energy
	• Attention to detail

Research Focus

Dr. Cid is currently examining the effect of ground leaf litter (the layer of dead leaves) that accumulates in the fall on seedling growth and competition in the spring. She is also investigating the factors that affect diversity and regeneration of plant species in wetlands. The results of her research can be used to restore and manage previously disturbed forest and wetland communities.

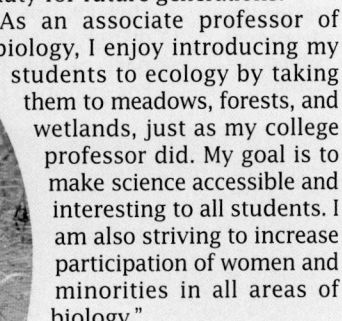

Career Path:

High School:
• Biology

College:
• Biology
• Ecology

Graduate School:
• Botany
• Ecology

The Excitement of Ecology

"Ecology is a challenging science that seeks to discover how a community of living organisms, such as a forest, is organized.

"This calls for exciting detective work. As an ecologist, I design experiments that enable me to piece together the puzzle of forest species interactions. One of my projects researched short-lived annual plants living in forests where mostly long-lived species such as trees are found. Why weren't these annuals found in surrounding fields where more light and nutrients were available? Why did these annual plants grow in patches throughout the forest? Once I answered these and many other questions, I had the information to solve the puzzle of species interactions in the forest. Ultimately, the data I collect from all my experiments can show me how to manage, restore, and maintain species diversity in forests and wetlands. My work gives me the satisfaction of knowing that I am helping to preserve a place of great natural beauty for future generations.

"As an associate professor of biology, I enjoy introducing my students to ecology by taking them to meadows, forests, and wetlands, just as my college professor did. My goal is to make science accessible and interesting to all students. I am also striving to increase participation of women and minorities in all areas of biology."

Lesson Plan 13.3

PREPARATION

Key Concepts

• Humans have a negative impact on ecosystems by disrupting physical habitats, decreasing species diversity, and destroying interactions among species.

• Such human impact cannot be avoided but must be minimized.

Reading Strategy

Use this section as an opportunity for students to do some *outside* reading. Have each student read and summarize a newspaper or magazine article that relates to the information discussed in this section.

Phase 2
TEACHING STRATEGIES

Demonstration 1

Destroying Physical Habitats
Show pictures of cities and point out that urbanization destroys or endangers numerous ecosystems.

HUMAN ACTIVITIES ARE DISRUPTING THE DELICATE BALANCE OF THE WORLD'S ECOSYSTEMS. THE FATE OF EVERY ECOSYSTEM ON EARTH DURING THE NEXT CENTURY MAY BE INFLUENCED MORE BY ITS ABILITY TO SURVIVE HUMAN DISRUPTION THAN BY ANY OTHER FACTOR. WE MUST UNDERSTAND HOW WE ARE DISRUPTING ECOSYSTEMS IN ORDER TO REDUCE THE MOST DAMAGING ASPECTS OF THAT INTERFERENCE.

13.3   How Humans Disrupt Ecosystems

Objectives

❶ Explain how changes in natural habitats can have a drastic impact on ecosystems.

❷ Recognize the devastating effects that exotic species have on native organisms.

❸ List three ways to reduce human impact on ecosystems.

Human Impact on Ecosystems

Climate, coevolution, competition—these driving forces of evolutionary change have molded ecosystems over long periods of time to create the world seen today. The most important single influence on natural ecosystems today, however, is human activity. To get some sense of how ecosystems are responding to the massive impact of our species, it is important to realize that disruption and disturbance are a natural part of the life of any ecosystem. For example, fires ignited by lightning normally burn forest ecosystems, and usually the forest quickly returns. Today, however, humans are disturbing ecosystems on a greater scale and to a greater degree than ever before. By making large-scale changes, like the development of ocean-front property shown in **Figure 13.13**, humans tear apart the webs of species interactions that stabilize ecosystems and enable them to bounce back from disturbances.

Figure 13.13
Construction of this housing development in the Florida Keys greatly altered the habitat of the island.

Disrupting physical habitat

Ever since humans learned to grow crops about 10,000 years ago, natural habitats have been altered and replaced with habitats of human construction. Forests are felled, swamps are drained and filled, and rivers are diverted, all to make room for buildings, parking lots, roads, and farms. Now humans are changing not only local habitats but the whole globe. Burning of fossil fuels may be changing the world's climate, and industrial chemicals are destroying the Earth's ozone shield. These changes to natural habitats reduce their ability to support living things.

■ Cultural Perspective ■

Tagua Nuts
The ivory-like tagua nut grows on palm trees in the Ecuadorian rain forest. An American clothing company has made a deal with the Tagua Initiative, a marketing development program, to purchase 5 million buttons made from tagua nuts. This purchase helps give economic value to standing forests, a boon to their conservation. Tagua nuts also are a natural substitute for ivory, for which large numbers of elephants have been slaughtered.

Stability

How can the introduction of an exotic species reduce the stability of an ecosystem?

Decreasing species diversity

Altering habitats in drastic ways often exterminates the native organisms that have evolved to fit natural habitats, not the artificial habitats humans create. For instance, conversion of forest to farmland or pasture reduces the number of species from hundreds or thousands to only a few. Similarly, massive logging of virgin forests followed by the planting of a single species of tree as a future lumber "crop" diminishes diversity of the forest ecosystem.

Destroying interactions among species

Intentionally or unintentionally, the kinds of species interactions that promote diversity often are eliminated. Removal of predators, whether wolves or insects, often reduces diversity, as you saw in the earlier keystone species example. The intentional or accidental introduction of exotic species from other parts of the world, such as the kudzu plant shown in **Figure 13.14**, can disrupt competitive balances that have coevolved among native species. Freed from the controls imposed by their natural predators and diseases, exotic species often easily out-compete native species, displacing many or all of them. The result is a simplified ecosystem in which far less competition occurs. By removing or introducing organisms without regard to their effects on native species, humans often wreak havoc on the complex web of biological relationships within ecosystems.

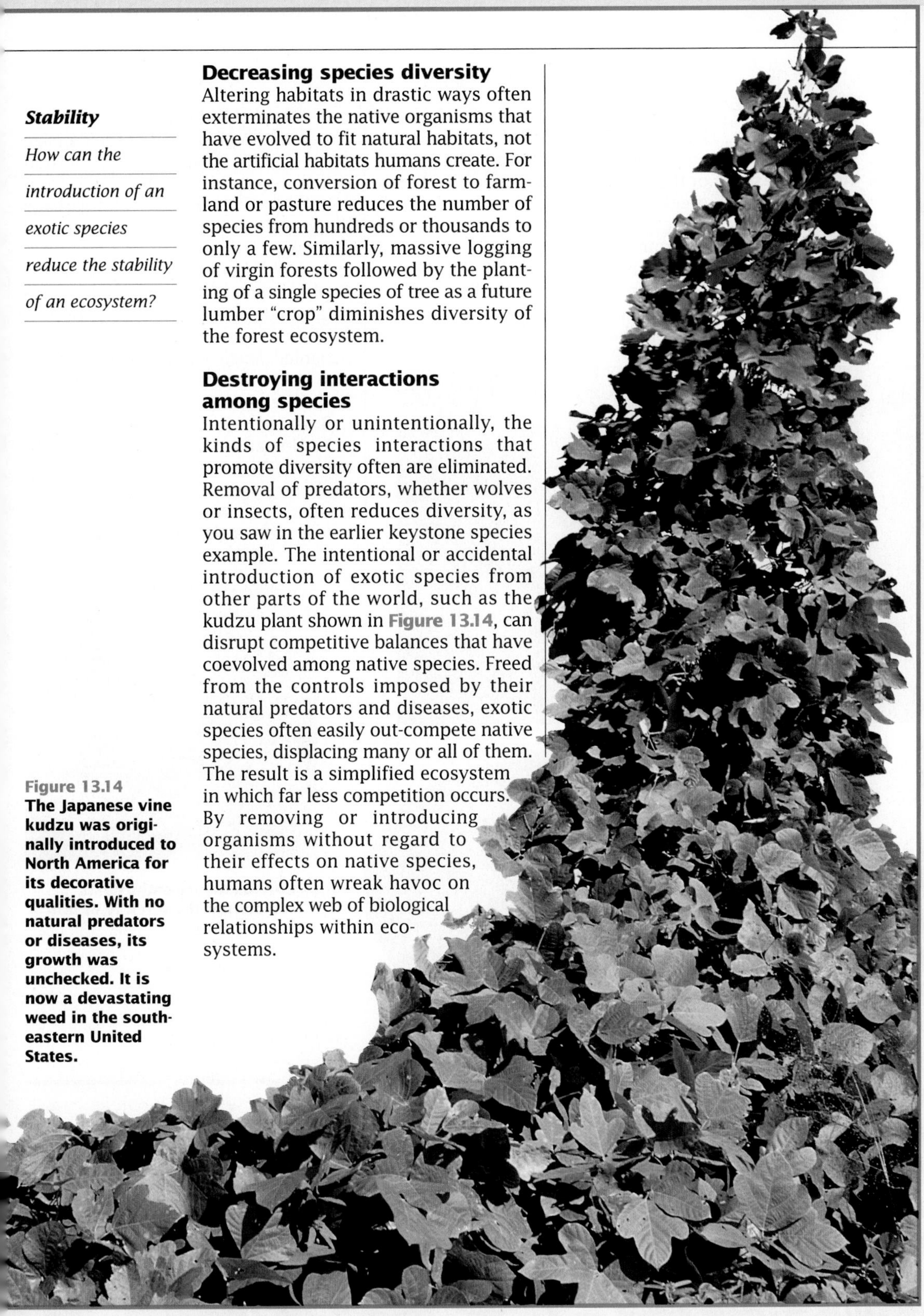

Figure 13.14
The Japanese vine kudzu was originally introduced to North America for its decorative qualities. With no natural predators or diseases, its growth was unchecked. It is now a devastating weed in the southeastern United States.

Demonstration 2

Decreasing Species Diversity

Show side-by-side comparisons of a natural forest and a forest planted as a future lumber crop. The picture of the natural forest should show several different species of trees, whereas that of the planted forest should depict a single species.

Demonstration 3

Destroying Species Interactions

Show photographs of oak and maple trees ravaged by gypsy moths, Dutch elm trees destroyed by Japanese beetles, or crops stripped bare by starlings. Point out that each of these destructive species was brought into the United States, either accidentally or intentionally, and that they have proliferated, since they have no natural predators in this country.

Theme Answer

Stability

The introduction of an exotic species can reduce the stability of an ecosystem through overpopulation, since the introduced species often has no natural predators or disease to keep its growth in check.

Demonstration 4

Minimizing the Human Impact

Contact a local organization to arrange for the class to become involved in some community effort that is aimed at minimizing human impact on the environment.

Phase 3

ASSESSMENT OPTIONS

Closure Strategy

Soil Conservation

Point out that sediment from soil erosion is the largest pollutant of water. Each year, almost 4 billion tons of eroded soil enters streams and rivers in the United States. This soil makes the water cloudy, decreasing the amount of light available for photosynthesis. Several techniques have been employed to reduce soil erosion, from crop rotation to contour plowing. Invite a farmer or agricultural research scientist to speak to the class about these methods.

Section Review

Assign the *Section Review.*

Reteaching

Assign students to cooperative work groups of four. Have each group prepare a list of specific examples that reflect the negative impact humans have had on their local environment.

Modifying the Environment

Figure 13.15

a **These cypress trees are located in the Okefenokee swamp in Georgia. The Okefenokee is a protected wetland.**

b **This black rhino in the Nairobi National Park in Kenya is protected by law. Because the horns of the rhino are so valuable, poachers have killed off large numbers of these animals.**

It is unrealistic to expect that humankind will be able to avoid disrupting the environment in the future. The growth of the human population is simply too great, and the appetite of industrial society is too voracious. But the environment must be modified in such a way that the fabric of the Earth's ecosystems is disturbed as little as possible. **Figure 13.15** shows some examples of what should be done:

1. *Disrupt the physical habitat as little as possible* A great deal of the damage done today is unnecessary. Tidal wetlands could be protected from conversion to home sites and garbage dumps, without impairing our standard of living.

2. *Avoid decreasing species diversity* Replacement of forests need not be with stands of only one species of tree, as often occurs when logged forests are replanted. Single-species stands are created simply because they make logging in the area easier.

3. *Avoid disrupting species interactions* Humans should not intrude into natural ecosystems unless compelled to do so to preserve them or by economic necessity. Introductions of exotic species should be avoided.

c **Curious nature enthusiasts can disrupt the serenity of bird nesting sites.**

Section Review

❶ How can a change in a natural habitat reduce the number of species occurring there?

❷ How are exotic species such as kudzu able to spread rapidly when introduced to a new environment?

❸ List three steps that can be taken to minimize human impact on the environment.

▪ *Section Review Answers* ▪

1. The native organisms are often exterminated, because they are unable to live in the altered habitat.

2. Exotic species introduced into new environments rapidly spread because they have no natural predators or diseases to control their growth.

3. Disrupt the physical habitat as little as possible; avoid decreasing species diversity; and avoid disrupting species interactions.

Chapter **13**   *Highlights*

Naturally occurring forest fires are part of the process of succession.

	Key Terms	Summary
13.1 Interactions Within Ecosystems This boxing crab has a commensal relationship with the anemones she is wearing. 	coevolution (p. 267) symbiosis (p. 270) parasitism (p. 270) mutualism (p. 270) commensalism (p. 271)	• Species evolve in response to other living members of their ecosystems. This process, called coevolution, shapes the species interactions in an ecosystem. • Insects and flowers coevolve as do plants and the herbivores that attempt to eat them. • A symbiosis is a close, long-term relationship between species. • In parasitism, one species (the parasite) lives on or in another species (the host). Mutualism is a symbiotic relationship in which all parties benefit. Commensalism is a relationship in which one species benefits and the other is neither helped nor harmed.
13.2 Ecosystem Development and Change  The sea star in the 15-species ecosystem studied by Robert Paine is an example of a keystone predator.	niche (p. 272) fundamental niche (p. 272) competition (p. 273) realized niche (p.273) competitive exclusion (p. 273) succession (p. 274) primary succession (p. 274) secondary succession (p. 274) stability (p. 275) keystone species (p. 275)	• An organism's niche is the sum of all its interactions in its environment, including interactions with other organisms. • Competition occurs when organisms attempt to use the same resource. • Succession, the regular progression of species replacements in a developing ecosystem, is driven by competition. • Stability is the ability of an ecosystem to resist change. More diverse ecosystems are usually more stable than less diverse ecosystems. • An ecosystem's diversity is partly determined by its latitude and its size.
13.3 How Humans Disrupt Ecosystems The introduction of exotic species, such as kudzu, can damage ecosystems. 		• Human activities disrupt ecosystems in three main ways: by altering natural habitats, by reducing species diversity, and by destroying interactions among species.

Chapter Review Answers

Understanding Vocabulary

1. **a.** Parasitism is a type of symbiotic relationship in which one organism in the relationship benefits and the other is harmed. Mutualism is also a type of symbiotic relationship, but here both parties benefit.

   b. An organism's fundamental niche is the total niche that it is potentially able to use, while an organism's realized niche is the one the organism actually occupies.

   c. Primary succession occurs on land where nothing has grown before, while secondary succession occurs on land where there has been previous growth.

   d. Competition occurs when two or more organisms try to use the same resources. Competitive exclusion is the process in which one species is outcompeted and dies out in the ecosystem.

Relating Concepts

2. Map answer is shown on page 274D.

Understanding Concepts

Multiple Choice

3. c   9. c
4. d   10. a
5. c   11. c
6. c   12. b
7. b   13. b
8. c

Completion

14. coevolution
15. symbiosis, commensalism
16. primary, secondary
17. niche

Short Answer

18. Caterpillars of the cabbage butterfly have evolved the ability to break down mustard oils, which enables them to eat the leaves of plants of the mustard family.

19. Mutualism, because both benefit.

Understanding Vocabulary

1. For each pair of terms, explain the differences in their meanings.
   a. parasitism, mutualism
   b. fundamental niche, realized niche
   c. primary succession, secondary succession
   d. competition, competitive exclusion

Relating Concepts

2. Copy the unfinished concept map below onto a sheet of paper. Then complete the concept map by writing the correct word or phrase in each oval containing a question mark.

Understanding Concepts

Multiple Choice

3. Plants that have thick, waxy coatings on their leaves to reduce water loss are likely to be found in
   a. tropical rain forests.
   b. conifer forests.
   c. deserts.
   d. tundra.

4. Thorns and prickles enable plants to
   a. attract pollinators.
   b. absorb water.
   c. capture additional carbon dioxide.
   d. discourage herbivores.

5. When the niches of two species overlap and one species dies out, it is an example of
   a. speciation.
   b. ecosystem stability.
   c. competitive exclusion.
   d. coevolution.

6. Robert MacArthur found that five species of warblers could share the same spruce forest because
   a. they are active at different times of the day.
   b. there was only one bird per tree.
   c. they fed in different parts of the tree.
   d. some species ate seeds, while others ate insects.

7. The regular progression of species replacement in a developing ecosystem is called
   a. mutualism.
   b. succession.
   c. competition.
   d. commensalism.

8. The growth of weeds on rock exposed by a recent landslide is an example of
   a. secondary succession.
   b. competitive exclusion.
   c. primary succession.
   d. ecosystem stability.

9. When a keystone species is removed from an ecosystem, the number of species in the ecosystem
   a. remains the same.
   b. increase in number.
   c. decrease in number.
   d. become extinct.

10. The most stable ecosystems are those that
    a. are most diverse.
    b. form when disruption creates a new habitat.
    c. contain exotic species.
    d. have the fewest species.

11. Ecosystem diversity increases when
    a. humans change natural ecosystems.
    b. latitude increases.
    c. ecosystem size increases.
    d. species become extinct.

12. The final stage of succession in an ecosystem
    a. lacks keystone species.
    b. is able to resist invasion by new species.
    c. is not a stable community.
    d. has few species.

13. Which of the following reduces human impact on ecosystems?
    a. removing predators
    b. preserving natural habitats
    c. disrupting species interactions
    d. introducing exotic species

Ecosystem → stability ← promotes → ?
stability — may be reduced by — human impact — may reduce →
? — determined by — ? — determined by — latitude
human impact — such as decreasing — ? ; such as disrupting — ? — ?

---

20. Latitude, or distance from the equator, determines the length of the growing season and climatic stability. A longer growing season leads to greater food production and to more consumers. Climate stability provides more time for coevolution.

21. Humans disrupt physical habitats by diverting rivers, clearing forests, and draining swamps.

Interpreting Graphics

22. The *Rafflesia* flower (center) is pollinated by the fly. The moth pollinates the flower at top. The bee pollinates the flower at bottom. To attract its fly pollinators,

the center flower has the color and odor of decaying flesh. The flower at top has its nectar within a long tube, so only the long tongue of the hawk moth can reach to the bottom of the tube. The bottom flower is a color that is visible to bees.

Completion

14. The evolution of two or more species in response to each other is called _____ .

15. A long-term association between two species is called a(n) _____ . When this association enables one species to benefit while the other is not affected, it is called _____ .

16. Succession on a newly formed island is called _____ succession. Succession in an abandoned field is called _____ succession.

17. An organism's _____ describes how it lives within an ecosystem.

Short Answer

18. Describe how the caterpillar of the cabbage butterfly has overcome the mustard plant's defenses.

19. Crocodile birds eat leeches and food particles from the mouths of crocodiles. What type of symbiosis is exhibited by this relationship? Explain.

20. Explain how and why an ecosystem's size and latitude affect its diversity.

21. Identify three ways in which humans disrupt physical habitats.

Interpreting Graphics

22. Match each flower with its pollinator. Describe how you think each flower and its pollinator were probably shaped by coevolution.

Reviewing Themes

23. *Patterns of Change*
    What patterns do you think might be seen when succession occurs on bare soil exposed by a glacier's retreat?

24. *Evolution*
    Why was the disease myxomatosis introduced into the Australian rabbit population? What evidence suggests that the virus that causes myxomatosis and the rabbits coevolved?

Thinking Critically

25. *Inferring Conclusions*
    What adaptations in flowering plants ensure that they will be pollinated?

26. *Compare and Contrast*
    Why would competition between two animals of the same species be greater than competition between two animals of different species?

27. *Building on What You Have Learned*
    What actions would you advise the people who inhabit the world's tropical rain forests to take to avoid lowering diversity?

Cross-Discipline Connection

28. *Biology and Art*
    Draw a plant that shows one or more defenses against herbivores. Challenge a classmate to draw a herbivore with adaptations that enable it to eat the plant. When both drawings are completed, ask other classmates to judge whether the plant is likely to be eaten by the herbivore.

Discovering Through Reading

29. Read the article "Plight of the Plover," in *Science News*, December 7, 1991, pages 382–383. How has the presence of humans on beaches in the eastern United States affected the natural habitat of the piping plovers? What actions have been taken to ensure that the plovers have a peaceful nesting habitat? What can the average citizen do to help save the birds from extinction?

26. Animals of the same species have more similar niches than do animals of different species. Hence, competition is ususally more intense between members of the same species.

27. Advise the inhabitants to leave as much of the land as possible undisturbed. Encourage them to use portions of land that are not likely to adversely affect land left undisturbed. For example, runoff from land cleared for farming could muddy waters that flow into undisturbed areas and adversely affect animals and plants that need clean water. Another suggestion is to plant crops that do not remove all the soil nutrients and, in some cases, allow the native vegetation to reclaim the land cleared for farming.

Cross-Discipline Connection

28. Plant drawings should show some evidence of defense such as thorns or thick leaves. Herbivore drawings should show adaptations that can overcome the plant's defenses.

Discovering Through Reading

29. Human presence disturbs the birds, keeping them from foraging for food. Government-owned beaches were closed during plover breeding season. The average citizen can stay away from the birds when on the beach and keep the beach clean, so as not to attract unwanted predators.

Reviewing Themes

23. Primary succession occurs. Barren land is colonized by small, fast-growing plants that make the ground suitable for larger plants that outcompete the original inhabitants. The second generation of inhabitants is later replaced by other species that are better able to compete in the new environment. The diversity of the ecosystem increases as succession progresses.

24. The disease was introduced to control the rabbit population. The evidence for coevolution is that over time, fewer and fewer rabbits were killed, while the virus responsible for the disease was becoming less deadly.

Thinking Critically

25. Plants attract animals to carry pollen to other plants of the same species. Some of the features of flowers that attract pollinators include attractive colors, scents, and food.

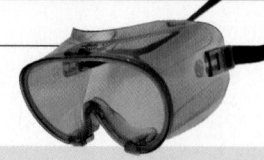

Procedural Note

Discuss the preparation questions and review the elements of good experimental design.

Prelab Preparation Answers

- Competition occurs when two or more organisms attempt to use the same resource. Competition occurs between individuals of the same species (intraspecific competition) and can occur between individuals of different species (interspecific competition).
- water, nutrients, light, and space.
- A species may evolve reduced use of resources for which it must compete with another species.

Procedure Answers

4. Answers should be logical.
6. Students should describe how their experiment meets the criteria of a control experiment.
7. Answers should include amount of light, soil, and water.

Chapter 13 Investigation

How Does Competition Affect Plant Growth?

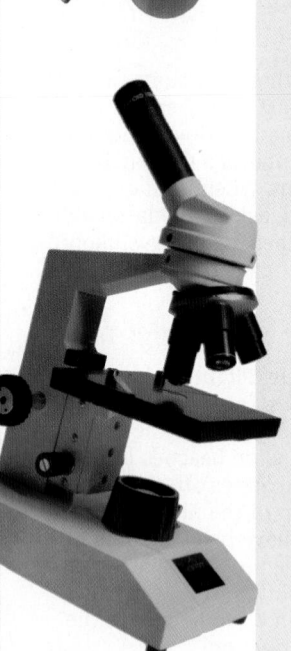

Objectives

In this investigation you will:
- *design* a control experiment that enables you to *observe* the effects of competition on plant growth

Materials

- three 8-cm (3-in.) pots
- potting soil
- 20 lima bean seeds
- centimeter ruler
- triple-beam balance
- beaker or graduated cylinder
- graph paper

Prelab Preparation

Review what you have learned about competition by answering the following questions:
- What is competition?
- What resources would you expect plant seedlings to compete for?
- How can competition affect the evolution of a species?

Procedure: Demonstrating the Effects of Competition

1. Form a cooperative team with a classmate to complete steps 2–10.
2. With your partner, discuss the objectives of the investigation.
3. State a hypothesis that addresses the following question: *Does competition affect the growth of plants?*
4. Design a control experiment to test your hypothesis that uses the available materials. Be sure to include a table to contain your observations and data. *Explain why you chose your hypothesis.*
5. Proceed with your experiment only after your experimental design and table have been approved by your teacher.
6. Describe the design of your experiment. *Explain why your design represents a control experiment.* You are now ready to plant and water your seeds.
7. Fill three pots with soil up to 2.5 cm (1 in.) from the rim. Each seed should be planted about 1 cm (approximately 0.5 in.) to 2 cm (approximately 0.75 in.) below the surface. Tamp down the soil covering the beans, but do not pack the soil or else the seedlings will not be able to break through the soil. *What factors will you hold constant for all pots?*

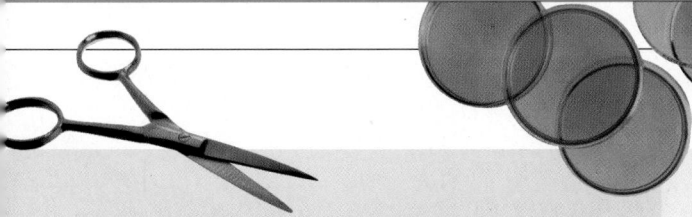

8. Allow the plants to grow for one week after germination before you begin to record measurements.

9. You must decide which structures of the seedling you will measure to indicated growth. Explain the reasons for your choice. Collect measurements for those plant structures that you chose. Calculate the average measurement for each structure, and use these averages to answer the analysis question. You might also want to note the condition of the plants as an indicator of their health.

10. For each structure that you measure, prepare a graph that shows the average growth of the structure in each pot over time.

11. Clean up your materials and wash your hands before leaving the lab.

Analysis

1. *Summarizing Data*
Summarize the data collected throughout your experiment.

2. *Evaluating Methods*
Why is a graph of the data more useful than a table?

3. *Analyzing Data*
Do the measurements for each of the different structures support your hypothesis? Explain.

4. *Inferring Conclusions*
What evidence indicates that competition affects the growth of your seedlings?

5. *Applying Concepts*
When might a homeowner apply the principles illustrated in this investigation?

Thinking Critically

1. How is the competition seen in this investigation different from the competition between the barnacles *Balanus* and *Chthamalus* described on page 273?

2. Suppose your experimental design were as follows: you planted 1 bean in the first pot, 9 beans in the second pot, and 10 beans in the third pot. Describe two flaws in this experimental design.

Analysis Answers

1. Answers should fully describe the students' results.
2. The graph is more useful for visualizing and interpreting data.
3. Answers will vary depending on individual's hypothesis.
4. Answers should indicate that plants in crowded pots grew more slowly.
5. Answers will vary. One possible answer might be when considering landscaping.

Thinking Critically Answers

1. In the barnacle example, individuals of two different species are competing for space on rocks. In this investigation, competition occurs between individuals of one species. Since the example involves individuals from different species, intraspecific competition is more likely to occur than interspecific competition. Perceptive students might notice that intraspecific competition is one of the key points in Darwin's formulation of natural selection.

2. One flaw is that this design does not allow average values to be collected for all three pots. Also, the design only tests two conditions: very low and very high crowding. It ignores the effects of moderate crowding.

Using the Feature

- Have students locate on a map these countries, which contain large tracts of rain forest: Brazil, Peru, Colombia, Papua New Guinea, Indonesia, Malaysia, and Zaire. Lest students think that deforestation is only a problem of tropical countries, point out that essentially all of Europe's original forests have been cut down; only tracts of intensively managed stands remain. Just 15 percent of the original forest remains in the United States, most of it in Alaska.
- Have students research the economic, political, and social conditions of the countries mentioned above. Have students explain why these conditions encourage rain forest destruction.

Discussion

Guide the discussion by posing the following questions.

1. What is the connection between deforestation and rising carbon dioxide levels?
   Deforestation removes vegetation, which soaks up carbon dioxide. Also, forests are often burned, and combustion releases carbon dioxide.

2. How does rain forest destruction alter local climates?
   Trees draw water from the soil, and the water evaporates from their leaves. The evaporating water becomes rain. When trees are removed, the water runs off and does not reach the atmosphere.

3. Why is agricultural land created by felling rain forest not productive over the long term?
   Tropical soils are poor in nutrients, most of which are contained in the vegetation. Crops rapidly exhaust the nutrients in the soil.

Deforestation: Are We Losing Only the Trees?

This logging truck is removing trees that were once part of the tropical rain forest.

How many products made of paper or wood do you use in a day? We are highly dependent on trees as a natural resource, yet the amount of forested land is dwindling worldwide.

After trees of the tropical rain forest are cut down, the land often erodes.

Worldwide, the rate of deforestation—or forest loss—has reached record levels. Each year, about 17 million hectares (40 million acres) are deforested—usually by cutting or burning. Deforestation is most severe in the tropical regions of Africa, Asia, Central America, and South America. More than 50 percent of the rain forests have already been cut for firewood and timber or burned to open up land for agriculture or livestock. About 2 percent of the remaining area of rain forest is destroyed each year. Two percent may sound like a small amount, but it means that an area the size of Florida is cleared each year. At this rate, no rain forest will remain by the middle of the next century.

The countries with large stands of tropical rain forest, such as Brazil, Indonesia, and Zaire, are poor and have rapidly increasing populations. Timber from the rain forest provides these countries with essential income. Clearing the forest makes land available for the expanding populations.

Deforestation has disastrous consequences on both a local and global scale. Cropland or pasture created by clearing rain forest is productive for only a short time—sometimes as little as two years. Although it supports luxuriant growth, tropical soil is usually poor in nutrients. Most of the nutrients are held within the tissues of plants. Burning or cutting the vegetation breaks the nutrient cycles that sustain the forest. Unless supplemented by artificial fertilizers, the soil is quickly exhausted

Science, Technology, and Society ▪ Science, Technology,

by agriculture or ranching. Without the protection that vegetation gives, topsoil on cleared patches is easily carried away by wind and water. Trees also affect the local climate. Water that evaporates from their leaves falls back on the forest as rain. Deforested areas often rapidly turn into virtual deserts.

Tropical rain forest also affects the levels of carbon dioxide in the atmosphere, thereby influencing global climate. In the tropics, the growing season is year-round. Photosynthesis by the lush vegetation withdraws large amounts of carbon dioxide from the atmosphere.

Reduced forest area means lessened absorption of carbon dioxide. Even worse, burning of rain forests releases carbon dioxide; about 20 percent of the carbon dioxide added to the atmosphere each year comes from burning rain forest. Thus, deforestation is a major contributor to the rising carbon dioxide levels that are projected to escalate global warming.

Another serious consequence of deforestation is extinction of species. Tropical rain forests are the most diverse terrestrial ecosystems, home to more than one-half of the world's species. The diversity in the rain forests is so great that most of the species are unknown to science. As the rain forests are destroyed, species are becoming extinct much faster

Pacific yew tree

than they can be identified and studied. Many of these species may be beneficial. For instance, in 1978 a species of grass closely related to corn was discovered in Mexico. The newly discovered species is much more disease-resistant than corn and is a perennial (it lives longer than one growing season, unlike corn). If genes from this grass could be transferred to corn, a strain of disease-resistant, perennial corn could be created to increase food production. But this potentially important new species was discovered in the nick of time—just one week before its habitat was scheduled to be burned and cleared.

Valuable species are threatened by forest clearing throughout the world. One of these is the Pacific yew tree, which grows only in old growth forests of the Pacific Northwest. Old-growth forests are ancient ecosystems: some of their trees may be more than 1,000 years old. More than 90 percent of the old-growth forest in the lower 48 states has been logged and replanted. Yew trees were not replanted because they were considered useless by the lumber industry. The bark of the "useless" Pacific yew yields taxol, an effective treatment for a variety of cancers. No one knows how many other medicines, foods, and other products are disappearing with the world's forests.

<table>
<tr><td>***Thinking Critically***</td><td>❶ Think of three ways that the rain forest might be more valuable when left standing than when cut.</td><td>❷ List two ways that wood could still be harvested without destroying the forest.</td><td>❸ Give three reasons why scientists (not environmentalists) would want to preserve the rain forest.</td></tr>
<tr><td>***Acting on the Issue***</td><td>❶ Look for rain forest products in your supermarket. How does buying these products benefit the rain forest?</td><td>❷ Write a letter to a logging company in the Northwest to find out how the company manages its forest land.</td><td>❸ Research 25 species endangered by the destruction of the rain forest. What efforts are being made to save these species?</td></tr>
</table>

Thinking Critically Answers

1. Possible answers include value of rain forest products, reduction of carbon dioxide levels, role in local climates.
2. Cutting only selected individual trees; cutting small patches of forest.
3. Possible answers include aesthetic value, effects on local and global climate, value as a source of new products, genes contained within its inhabitants.

291

Chapter **14** *The Fragile Earth*

Planning Guide

Objectives/Themes	Classwork Resources	Homework Resources
14.1 **1.** List some ways human health is affected by pollution. **2.** Explain the cause and the effects of acid rain. **3.** Identify the causes and effects of ozone depletion. **4.** Describe why global temperatures are rising. **Themes:** Patterns of Change, Interacting Systems	**Teacher's Resource Binder** Focus Activity 14 *Determining the Amount of Refuse* **Other Resources** Transparencies 57–58	**Text** Section Review, p. 297 **Teacher's Resource Binder** Directed Reading Worksheet 14.1 **Other Resources** Audiocassette 14.1*
14.2 **1.** Explain two ways pollution has been reduced in the United States. **2.** Identify alternatives to fossil fuels, and describe some benefits of each. **3.** Contrast nonrenewable resources with renewable resources. **4.** Recognize the relationship of human population growth to environmental problems. **Themes:** Interacting Systems, Stability, Patterns of Change,	**Text** Science in Action *How to Help Save the Environment* pp. 304–305 **Teacher's Resource Binder** Lab Investigation 14.1 *Acid Rain Effects* Science Skills Worksheet *Using Maps* **Other Resources** Transparency 59	**Text** Section Review, p. 303 **Teacher's Resource Binder** Directed Reading Worksheet 14.2 **Other Resources** Audiocassette 14.2*
14.3 **1.** List some examples of successful solutions to environmental problems. **2.** List the five basic elements necessary to solve any environmental problem. **3.** Recognize your role in solving environmental problems. **Themes:** Interacting Systems, Patterns of Change, Stability	**Text** Investigation *How Does Pollution Affect Organisms?* pp. 312–313 Discoveries in Science *Agriculture and Technology: Feeding the World's Population* pp. 314–315 **Teacher's Resource Binder** Lab Investigation 14.2 *Oil-Degrading Microbes*	**Text** Section Review, p. 308 **Teacher's Resource Binder** Directed Reading Worksheet 14.3 Vocabulary Review Worksheet* Reteaching Worksheet* *Two Critical Issues* **Other Resources** Audiocassette 14.3*

*Reteaching Options

Demonstrations
14.1: pp. 295, 296
14.2: pp. 299, 300

Assessment
Chapter Review pp. 310–311
Portfolio Assessment p. 291D
Chapter Test—Teacher's Resource Binder
Test Generator

Research Notes

Connection to Ecology: Microbes Eating Toxic Waste

Scientists are beginning to "domesticate" microorganisms for specific uses. Some microbes may be able to slurp up oil spills and solve other environmental problems. So far, the results have been mixed.

The disastrous 1989 oil spill of the Exxon Valdez in Alaska's Prince William Sound (Figure 14.3) provided scientists an opportunity to test such techniques. The method tested used fertilizer solutions that stimulate the natural activity of bacteria already in the environment to break down oil.

Results of field studies by the Environmental Protection Agency were ambiguous. The best results were achieved when workers first used steaming hot water to spread out the oil on beaches into a thin layer and then sprayed a fertilizer containing urea and oleic acid.

At some sites, this technique appeared to break down the oil at rates about two years faster than traditional approaches. However, on other test sites, there was no net improvement over the natural process.

The difficulty in finding consistent improvement could relate to the fact that the bacteria work fairly well even without the fertilizer. In addition, the less successful test sites were in more sheltered parts of the Sound. It could be that wave action helps stir up oxygen for the microbes.

The crisis of the oil spill did not allow time for careful planning and a rigorous scientific approach to the treatment. Alaska's Department of Environmental Conservation and the EPA are planning several more careful studies in the future.

A group of Cornell University scientists led by William Jewell have expanded the idea of harnessing bacteria to clean up toxic waste. They have made a "bioreactor" that uses two different kinds of bacteria to do a job that neither could do effectively alone.

The bioreactor is able to convert industrial carcinogens such as perchloroethylene (PCE) and trichloroethylene (TCE) into carbon dioxide, water, and chloride ions.

In the first compartment of the bioreactor, a strain of bacteria that does not require oxygen to survive takes in sugar and converts the carcinogens into methane and poisonous vinyl chloride gas. The bacteria in the second compartment take in oxygen and convert the methane and vinyl chloride to the final products, which are much safer for the environment.

The technique can reduce the concentration of PCE in water from 10,000 parts per billion (ppb) to less than 1 ppb. The two-stage bioreactor is very fast at a variety of temperatures, and it fits on the back of a truck.

Investigation Notes

How Does Pollution Affect Organisms?

pages 312–313

Purpose: This investigation gives the student an opportunity to explore the effects of heat and NaCl on simple organisms. As a result of their work, they should realize that, in addition to lethal effects, pollutants can adversely affect a habitat by causing motile organisms to be displaced, changing the community structure.

Prelab Preparation

1. As always, only alcohol thermometers should be used, since mercury thermometers pose a safety hazard if broken.

2. Be sure the distilled water has been allowed to sit overnight in an open container to become oxygenated.

3. To prepare a 10 percent NaCl solution, add 10 g of NaCl to about 50 mL of water. Dissolve. Add water to bring to 100 mL.

4. A special Amoeba culture (Protocolor) is available from WARD'S (87 M 0380), and is prestained to enhance visibility.

5. Use a pipette filler to fill the pipette with culture. Close the narrow tip of the pipette with modeling clay. Remove the pipette filler and plug the top with clay.

Answers will be found on pages 312–313.

Meeting Individual Needs

Objectives

1. Students will demonstrate appropriate use of core vocabulary for the chapter (Vocabulary File).

2. Students will relate the destruction of the rain forest to endangering the traditions and the health of its human inhabitants (Multicultural Lesson Plan).

3. Demonstrate the greenhouse effect on a small scale (Demonstration).

Vocabulary File

(Developing Vocabulary/ Limited English Proficiency)

If you are not already using the Vocabulary File, refer to Chapter 1 for its preparation. See the Chapter Highlights on page 309 for a list of suggested words.

Multicultural Lesson Plan

Rain Forest Destruction (Dealing with Ethnic Diversity)

Preparation: For source material, gather encyclopedias, and information from your local water utility about water's mineral content.

Teaching Strategies

1. The teacher will ask students to recall from Chapter 12 the types of animals and plants found in the rain-forest ecosystem. Students also describe the characteristics of the rain forest.

2. Students will brainstorm about reasons why the rain forests are being destroyed. Be sure students relate the conflicts involving rain forests to the value different cultures place on the rain forest. For many Western cultures, the rain forest is a place to drill for oil and harvest cheap timber. For some governments of nations with rain forests, the land under the rain forests is valuable for farming, to boost the economy and help the nation reach agricultural self-sufficiency. For many Native Americans and tribal people on other continents, the rain forest is their home.

3. Students will discuss the possible effects the destruction of the rain forest will have on the lives of those who call it home. Be sure students discuss the political, social, economic, health-related, and psychological problems that arise when native people are displaced from their homes and traditional ways of life.

As a concrete example of a health-related problem associated with deforestation, the teacher should point out that oil drilling and extensive burning of the rain forest can pollute the water of streams and rivers in the rain forest. The Cofan Indians of Ecuador, who formerly relied upon these sources for drinking water, now depend on rainwater.

Assessment: Students will research the topic of drinking water to determine which minerals are found in tap water and bottled water in the United States, which minerals are found in stream water and rainwater, and what problems can result from an inadequate supply of these minerals. Students will determine which water is healthier for people of the rain forest (rain or river/stream).

Demonstration

The Greenhouse Effect: Global Warming
for page 297
(Developing Modeling Skills/ Visual Learners)

Materials: Large jar with lid, tea bags, battery jar, petri dish, gravel

Procedure: Read aloud to students page 297. Have students brainstorm about the effects that global warming could have on their hometown. Would there be flooding, drought, or other changes? Encourage students to explore other less direct effects of global warming. For example, without cold winters, hibernating mammals' seasonal habits may be disrupted.

The greenhouse effect already affects the conditions on Earth, making them favorable for life. Without some aspects of the greenhouse effect, the Earth would be more like Mars, which experiences a wide range of temperatures, from –123°C (–190°F) to 23°C (80°F). Because the pollution of the Industrial Age has added to the greenhouse gases, the atmospheric processes that regulate temperature may no longer be balanced.

Several options can used to demonstrate the greenhouse effect.

- Make some sun tea. Explain the greenhouse effect beforehand, so that as the tea is brewing, students can relate the principles of the atmosphere trapping heat to the jar of water trapping heat.

- Place a petri dish lid partially filled with water on the bottom of the inside of a battery jar. Surround this with small gravel so that the temperature can easily be read. Cover. Either place the battery jar on a window ledge which receives direct sunlight, or else place the jar on a lab counter with a lab lamp to simulate sunlight. Have students make regular observations of the temperature inside the jar.

- If the weather is cool, shut off all heating to the room. Raise the blinds to the top of the windows for full sun exposure. This will accomplish some passive solar heating of the room. This can be related to the greenhouse effect.

Follow up with brainstorm sessions about sources adding to the CO_2 in the atmosphere. Which ones have remained fairly constant over centuries? Which ones have increased dramatically in the past 200 years?

Meeting Individual Needs (cont.)

Additional Strategies

Visual Strategies

Pages 294, 295, 297, 298, 301, 302, 303, and 307

Auditory Learners

Use *Biology: Visualizing Life* Audiocassettes for Sections 14.1, 14.2, and 14.3.

Cooperative Learning

The Ozone Layer

Timing: Use this activity to introduce Section 14.1.

Group Size: 3 students

Outcome: Students will be able to describe conditions and problems associated with the depletion of the ozone layer.

Individual Accountability: Each group member will be responsible for completing part of the research on the ozone layer.

Positive Interdependence: Groups will prepare a report about the ozone layer.

Have each group member use references to research information about one of the following topics: (1) conditions that contribute to the depletion of the ozone layer, (2) long-term effects of damage to the ozone layer, and (3) methods to prevent depletion of the ozone layer.

Have group members combine their information into a single report on the ozone layer.

Portfolio Assessment

Students should select their best work and provide a self-reflective rationale for their selections. Students can make selections in the following areas.

1. *Content*	One concept map from the chapter (See page 812 for evaluation criteria.)
2. *Reading Comprehension*	One Directed Reading Worksheet from the Teacher's Resource Binder (Use the answer key to evaluate for accuracy.)
3. *Writing*	Using the Vee Form, summarize a magazine or newspaper article relating to human impact on the planet. (See page 22T for evaluation criteria.)
	Or: Select a writing task or project from the Chapter Review.
4. *Performance Assessment*	One Vee form from a chapter investigation or lab manual investigation (See page 22T for evaluation criteria.)

Teacher makes selections in the following areas.

1. *Formal Assessment*	Chapter test (Test A, B, or the Test Generator) The teacher-scored test should be reviewed by the student. Incorrect responses should be corrected by the student before the test becomes part of the portfolio.
2. *Informal Assessment*	Use the Direct Observations Checklist, page 33T, during a laboratory or other cooperative learning experience.
3. *Performance Assessment*	Have students create a plan to promote and increase recycling in their school or town. If paper and aluminum are already being recycled, consider plastic and other materials.

Concept Map Answer

The following is one possible answer to the Relating Concepts exercise on page 310.

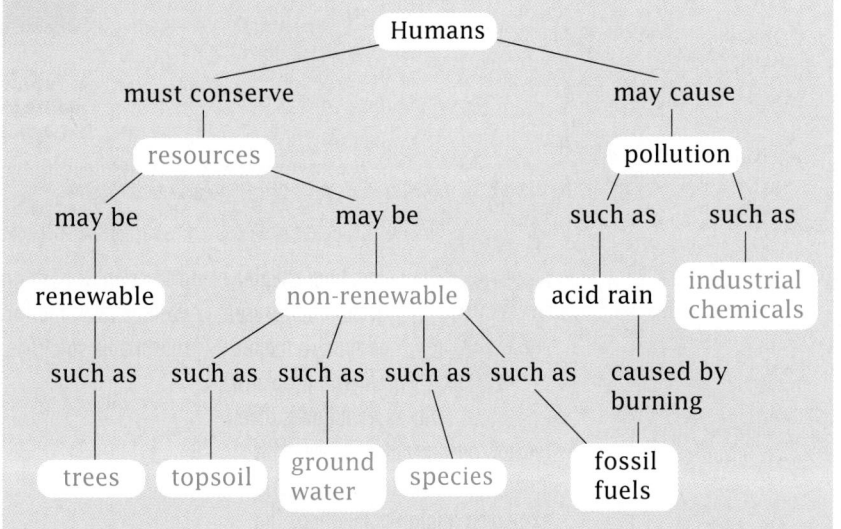

Chapter 14

Determining Prior Knowledge

- Have students provide examples of pollution in their local community and in areas they have visited, especially cities.
- Show students a piece of coal. Have them explain how it was formed, how it is used, and how it is contributing to atmospheric pollution.
- Show students a log. Have them discuss the benefits wood offers as an energy source. Ask students to name concerns that have been raised about cutting forests for lumber and firewood.
- Have students discuss examples of how an individual can contribute to solving an environmental problem.

Chapter 14

The Fragile Earth

Review

- ozone (Section 10.2)
- carbon cycle (Section 12.2)
- habitat destruction (Section 13.4)

14.1 Planet Under Stress

- **Humans Have Damaged the Environment**
- **Pollution's Toll**
- **Destroying the Ozone Layer**
- **Global Warming**

14.2 Meeting the Challenge

- **Reducing Pollution**
- **Finding Enough Energy**
- **Conserving Nonrenewable Resources**
- **The Deeper Problem: Population Growth**

14.3 Solving Environmental Problems

- **Environmental Problems Can Be Solved**
- **Steps Toward Saving the Environment**
- **What You Can Contribute**

In recent years people have begun to realize that human activities threaten the existence of life on our planet.

■ Author's Rationale ■

This final chapter in the unit focuses on today's major ecological problems. An explanation is given for how each problem arose, which inevitably leads back to the world's exploding human population. The discussion of each problem is kept positive, with the focus on developing reasonable solutions.

IMAGINE FOR A MOMENT EVERY PLASTIC WRAPPER, CUP, AND CONTAINER YOU HAVE EVER THROWN AWAY PILED UP BESIDE YOU. ALL OF THIS PLASTIC IS STILL IN THE ENVIRONMENT. PLASTICS ARE JUST ONE OF THE MANY KINDS OF CHEMICALS THAT HUMANS ARE ADDING TO ECOSYSTEMS, SOMETIMES WITH DAMAGING EFFECTS TO OUR HEALTH AND THE HEALTH OF ECOSYSTEMS.

14.1 *Planet Under Stress*

Objectives

1. **List some ways human health is affected by pollution.**

2. **Explain the cause and the effects of acid rain.**

3. **Identify the causes and effects of ozone depletion.**

4. **Describe why global temperatures are rising.**

Humans Have Damaged the Environment

Our world is a patchwork of interconnected ecosystems. Because of this interdependence, damage done to any one ecosystem can have ill effects on others. Burning sulfur-rich coal in Missouri kills trees in Canada; dumping refrigerator coolants in California destroys atmospheric ozone over Antarctica and leads to higher rates of skin cancer in Paris. Biologists call such widespread effects on our world global change.

How much trash do you think you throw away? On average, each American tosses out almost 2 kg (4 lbs.) of unwanted paper, metals, glass, plastics, food, and other items *every day*. We cannot banish our wastes from existence—garbage does not just disappear when you haul it to the curb or drop it into a dumpster. We release our wastes into the environment: the water, the air, and the soil, as shown in **Figure 14.1**.

Pollution is anything potentially harmful that humans add to an ecosystem. Automobile exhaust is an example of pollution, as are poisonous chemicals dumped into rivers and heat from nuclear power plant cooling towers. The substance added to the environment is called a pollutant. Automobile exhaust, for example, contains a variety of pollutants. Until recently, people felt that the environment could absorb and dilute pollution without suffering ill effects. It is now apparent that this was a mistaken attitude.

Pollution can drastically damage the health of ecosystems as well as the health of human beings. Some pollutants are poisonous. Others are carcinogens. A **carcinogen** is a substance that causes cancer. Examples of carcinogens include industrial chemicals such as dioxin and benzene, and asbestos (a fireproof covering).

Figure 14.1
The chemicals that smokestacks pour into the air are released into the environment.

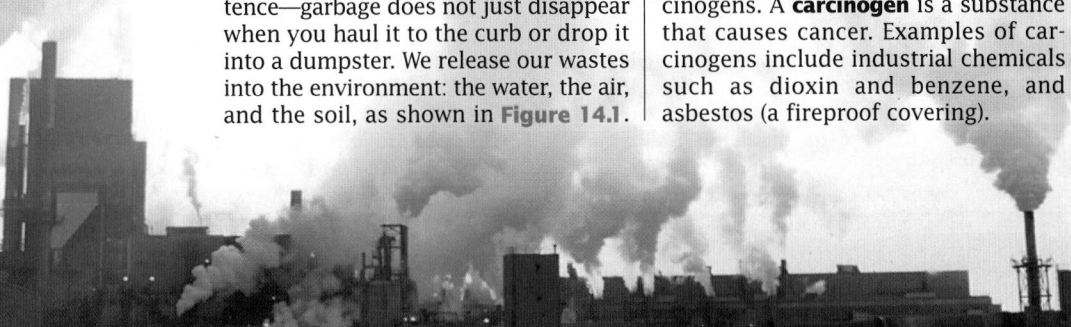

Section 14.1

Phase 1
PREPARATION

Key Concepts
- Pollution is the process by which substances potentially harmful to ecosystems are added to the environment.
- Pollution has endangered our water supply, promoted acid precipitation in the atmosphere, and depleted the ozone layer.
- Increasing levels of carbon dioxide in the atmosphere may lead to global warming in the future.

Reading Strategy

This is another section where students can realize that outside reading is an extension of their textbook. Schedule the class to meet with the librarian to be familiarized with the resources available in your school or town library, including periodicals, journals, scientific encyclopedias, data bases, and microfiches. Give each student an assignment to use one of these resources to locate an article dealing with the material in this section.

Phase 2
TEACHING STRATEGIES

Theme Connection

Interacting Systems
Demonstrate the occurrence of interactions between ecosystems by underscoring how damage done to one ecosystem reverberates throughout others.

Visual Strategy

Figure 14.3

So that students may appreciate the size of these supertankers, point out that their capacity is equivalent to the tanks of nearly 10 million average-sized automobiles. The *Exxon Valdez* spilled more than 10 million gallons of oil within the first five hours, eventually pouring about 240,000 barrels into Prince William Sound. Normally, an oil spill results in patches that cover 20 to 30 percent of the affected area. But the volume spilled by the *Exxon Valdez* was so great that the oil covered the entire surface of the water, killing all the plankton. Ask students how such a spill affected the food chains in Prince William Sound.

Connection: Chapter 12

Freshwater Ecosystems

When rainfall washes chemicals, including nitrogen-containing fertilizers, into bodies of water, a freshwater ecosystem can be seriously disrupted. For example, when large amounts of nitrates and phosphates are washed into a lake or pond, they stimulate an explosive growth of algae. Only the topmost algal layer receives enough light for photosynthesis; the lower layers die and are decomposed by bacteria. These bacteria use up large amounts of the oxygen dissolved in the water, resulting in the death of organisms that require a high level of oxygen to survive.

Connection: Chapter 9

Lichens and Pollution

Lichens are a sensitive indicator of air pollution. Because they rapidly absorb substances from rainwater, lichens are particularly susceptible to airborne pollutants. Remind students how pollution from factories darkened the lichens on the trees in forests near industrialized cities in England.

Pollution's Toll

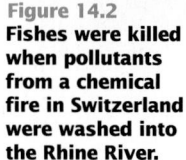

Figure 14.2
Fishes were killed when pollutants from a chemical fire in Switzerland were washed into the Rhine River.

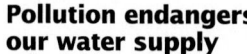

Figure 14.3
Oil spilled from the *Exxon Valdez* damaged miles of coastline and killed thousands of marine organisms.

Ecosystems cannot absorb this sort of chemical punishment indefinitely. Too much pollution disrupts the delicate web of relationships that binds the world's ecosystems together. In eastern Europe, a century of unrestrained pollution has destroyed forests and rendered lakes nearly lifeless. Eighty percent of Poland's deep wells are polluted, and one-fourth of its soil is far too contaminated for safe farming. One-fourth of Czechoslovakia's rivers cannot support fish. One-third of Bulgaria's forests are damaged or dying. Intensive efforts are underway to reverse this damage, but no one knows if they will succeed.

Pollution endangers our water supply

Humans need water for drinking, irrigation, and industry, yet we have a very casual attitude toward water pollution. Every day, wastes are poured down the sink, flushed down the toilet, or dumped into rivers and lakes without considering where they will end up. For instance, while putting out a fire in a chemical warehouse in Basel,

Switzerland, in November 1986, firefighters washed 27 metric tons (30 tons) of mercury and pesticides into the Rhine River. A deadly wall of poison pollutants flowed down the Rhine, killing everything as it moved through Germany and Holland to the sea. **Figure 14.2** shows some of the fishes killed by the chemical pollutants.

The poisoning of the Rhine is just one example of a worldwide problem. In 1989 an oil tanker, the *Exxon Valdez*, ran aground off the coast of Alaska. If the *Exxon Valdez* had been loaded no higher than the waterline, little oil would have spilled—but it was loaded much higher than that. The weight of the above-waterline oil forced thousands of tons of oil out of a rip in the ship's hull. The spilled oil polluted many miles of coastland and killed fishes, marine mammals, and birds. **Figure 14.3** shows the *Exxon Valdez*.

The high productivity of modern agriculture is based on the widespread use of insecticides to kill insect pests, herbicides to control weeds, and fertilizers to enrich the soil. Rainfall washes these chemicals, some of which are carcinogenic or toxic, into rivers, lakes, and the ocean. It is no coincidence that one of the highest rates of cancer in the United States is seen in the Mississippi Delta region. Here the river draining our country's agricultural heartland empties into the sea, carrying with it herbicides, insecticides, and industrial chemicals.

Acid rain threatens forests and lakes

Many coal-burning power plants use high-sulfur coal because it is cheap and plentiful. The smoke produced when high-sulfur coal is burned smells bad (like rotten eggs), blackens buildings, and kills local trees. Smokestacks more than 65 m (210 ft.) tall were introduced as a way to burn high-sulfur coal without public outcry. The intent of those who designed plants with tall stacks was to release the sulfur-rich smoke high in the atmosphere, where the winds would disperse and dilute it.

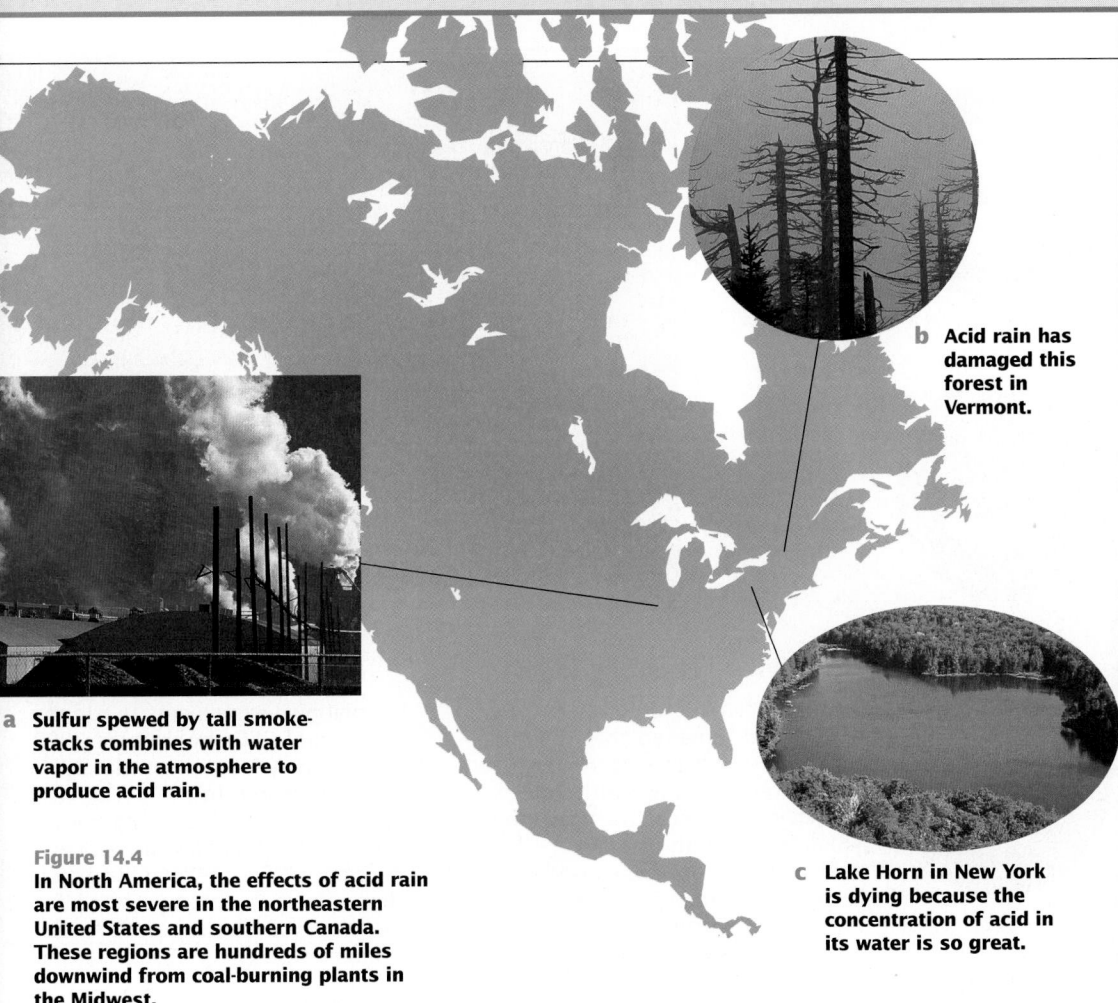

a Sulfur spewed by tall smoke-stacks combines with water vapor in the atmosphere to produce acid rain.

b Acid rain has damaged this forest in Vermont.

c Lake Horn in New York is dying because the concentration of acid in its water is so great.

Figure 14.4
In North America, the effects of acid rain are most severe in the northeastern United States and southern Canada. These regions are hundreds of miles downwind from coal-burning plants in the Midwest.

In the 1970s, it became clear to ecologists that tall stacks were not eliminating the problems of sulfur-rich coal, just exporting the ill effects elsewhere. In the upper atmosphere, sulfur released by smokestacks combines with water vapor to produce sulfuric acid. When the water vapor later condenses and falls back to earth as rain or snow, it carries the sulfuric acid with it. Since moisture travels great distances high in the atmosphere, the acid is far from its source when it falls. Beginning in the 1970s, ecologists reported that the lakes of Sweden were beginning to die. These lakes could no longer support life. Also dying were the trees of the great Black Forest of Germany. **Figure 14.4** shows some areas of North America damaged by acid rain.

What was causing lakes and forests to die? Scientists discovered that the rain in these areas was unusually acidic, a phenomenon they called **acid rain** (although it is more correctly called acid precipitation, since snow can be acidic as well). When an **acid** is dissolved in water, the resulting solution has a higher concentration of hydrogen ions (H^+) than does pure water. Scientists describe the acidity of a solution using a logarithmic value called **pH**. A solution with a low pH has a high concentration of hydrogen ions. Rainwater normally has a pH value of 5.6. Pure water has a pH of 7.0. Rainfall in some areas of the northeastern United States, however, has a pH value of 3.8, almost 100 times more acidic than the typical value for the rest of the country.

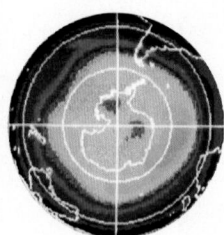

1982

Ozone hole

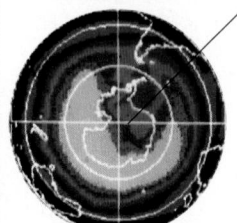

1987

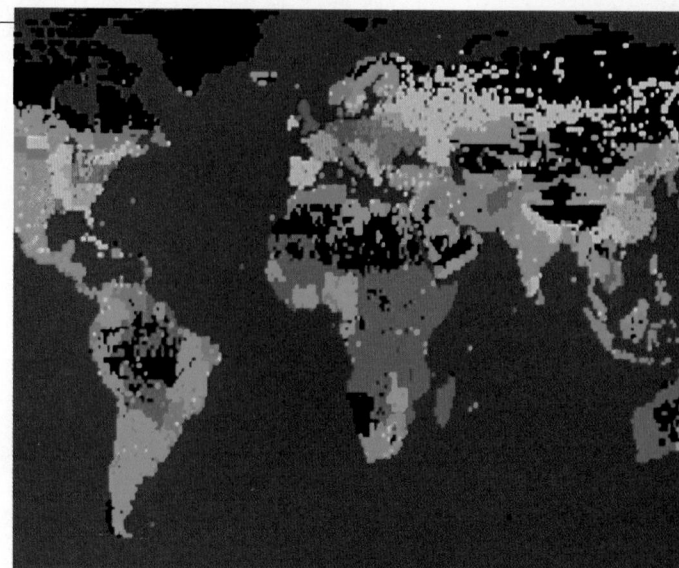

1989

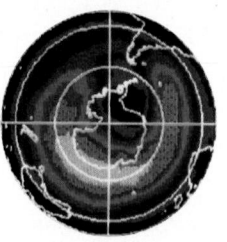

1990

Figure 14.5
Although the level of ozone in the upper atmosphere varies, these satellite photos taken above Antarctica show that a hole in the ozone layer may be increasing in size.

Destroying the Ozone Layer

In 1985, British researchers in the Antarctic discovered that the ozone concentration had become unexpectedly lower over the South Pole than elsewhere in the Earth's atmosphere. This decrease in the level of the atmospheric ozone has caused a "hole" in the Earth's atmosphere, as shown in **Figure 14.5**. Currently, the ozone hole is about the size of the United States and grows wider and deeper every year.

What is destroying the ozone? The culprit is a class of chemicals that scientists once thought was harmless: **chlorofluorocarbons** (CFCs). Since their invention in the 1920s, CFCs have been manufactured to be used as the coolant in refrigerators and air conditioners, as the propellant in aerosol cans, and as the foaming agent during production of plastic foam containers.

Eventually CFCs escaped into the atmosphere and accumulated there. High above the South and North poles, CFCs stuck to frozen water vapor and began to attack ozone molecules. Just as an enzyme carries out a reaction in your cells without being changed itself, so CFCs catalyze the conversion of ozone (O_3) into oxygen (O_2). The drop in ozone worldwide is now more than 3 percent, and each year the rate of ozone loss increases.

International agreements to ban the production of CFCs by the year 2000 have been signed. But no one knows if the ban has come in time. The vast majority of CFCs produced have not yet reached the Earth's upper atmosphere. Furthermore, CFCs are long-lived molecules; they can destroy ozone for more than 100 years before finally breaking down.

Ozone depletion leads to health problems

Recall from Chapter 10 that atmospheric ozone blocks ultraviolet (UV) radiation. Ozone depletion is frightening because exposure to high levels of UV radiation can cause severe health problems. In humans, exposure to UV radiation causes skin cancer and cataracts, an eye disorder that can lead to blindness if not treated. Experts estimate that each 1-percent drop in atmospheric ozone leads to a 6-percent increase in the incidence of skin cancer. Thus the drop of just over 3 percent in ozone concentration that has already occurred is estimated to have caused an increase of as much as 20 percent in skin cancer. In addition, food crops like wheat and corn are damaged by increased exposure to UV radiation.

▪ *Matter of Fact* ▪

Global Warming

For more than 200 years, our industrial society has grown on a diet of cheap energy. Much of this energy has been obtained by the burning of fossil fuels. **Figure 14.6a** shows that most of the carbon dioxide released by burning fossil fuels comes from industrialized nations. When something is burned, its molecules are broken apart. They combine with oxygen, freeing energy as a result. Coal, oil, and natural gas are the remains of ancient organisms, and are rich in carbon. When fossil fuels are burned, this carbon combines with oxygen to produce carbon dioxide. Centuries of fossil-fuel burning have released large quantities of carbon dioxide into the atmosphere.

Unfortunately, too much carbon dioxide is harmful. As you learned in Chapter 12, carbon dioxide absorbs solar energy, trapping heat in the atmosphere. That is why the Earth is warm, and the moon (which has no atmosphere) is very cold. This ability to retain the sun's heat and in so doing to warm the atmosphere is called the greenhouse effect. Most scientists think that the increased levels of carbon dioxide are causing increases in global temperatures, or global warming. **Figure 14.6b** compares the release of carbon dioxide from all sources with the rate of global temperature increase.

Global warming may have many serious consequences

Although predicting worldwide climatic change is complicated, scientists can make projections using mathematical models of the atmosphere. By the year 2050, the world's average annual temperature is expected by most scientists studying the problem to have risen between 1.5°C and 4.5°C (3°–8°F). Based on this, the city of Dallas, Texas, would have 78 days each year in which the temperature would be greater than 38° C (100° F), compared with only 19 such days currently.

That 1.5°C–4.5°C change is approximately equal to the temperature change from the last ice age 15,000 years ago until now. However, that change in temperature is now occurring over a period of less than 100 years. Obviously, that kind of climatic change will vastly change patterns of rainfall and temperature, which will have a major and unpredictable impact worldwide. The midwestern United States, which produces much of the world's food, would become hotter and drier. Agricultural production there would probably fall significantly.

Figure 14.6a
Carbon dioxide released from fossil fuels is highest in areas shaded red and lowest in areas shaded blue.

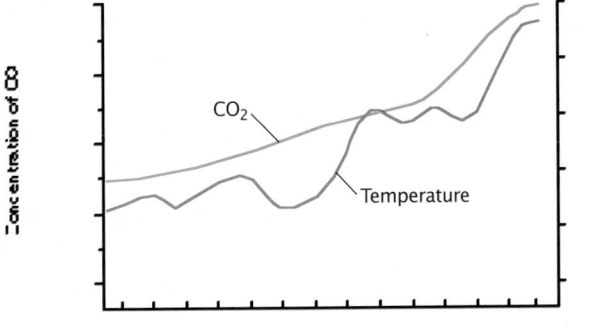

Amount of Carbon Dioxide in the Atmosphere

Concentration of CO₂

CO₂

Temperature

temperature change (°C)

Figure 14.6b
The concentration of carbon dioxide (blue) in the atmosphere has risen since 1860. Temperature (red), based upon the 1951–1980 average, has risen in the last two decades.

Section Review

① **Define carcinogen and give one example.**

② **Describe the cause of acid rain.**

③ **Explain two health effects of ozone depletion.**

④ **How does tropical deforestation contribute to global warming?**

Visual Strategy

Figure 14.6a–b
Have students hypothesize why the concentration of carbon dioxide began to increase significantly starting around 1850. Point out the two major contributing factors—the Industrial Revolution, and the accelerated clearing of forests.

Social Studies Connection

Lost Islands
Scientists estimate that rising temperatures might cause some low-lying islands like the Marshall Islands and the Maldives to be completely flooded, forcing their inhabitants to move to nearby continents. Have students locate these islands on a map and indicate where their inhabitants may be forced to relocate.

Phase 3

ASSESSMENT OPTIONS

Closure Strategy

The Year 2000
Have students hypothesize which environmental concern covered in this section will have the greatest impact on their community in the year 2000. Have them provide reasons for their choice.

Section Review
Assign the *Section Review.*

Reteaching
Have the class select a local ecosystem and investigate what, if any, form of pollution has affected it.

▪ *Section Review Answers* ▪

1. A carcinogen is a substance that causes cancer. Examples may vary. Some possible answers are benzene, asbestos, and dioxin.

2. Sulfur released in smoke combines with water vapor in the upper atmosphere to form sulfuric acid. As water vapor condenses and forms rain or snow, sulfuric acid is included with it.

3. Students should explain the effects of human exposure to UV radiation and that they are subject to skin cancer, eye cataracts, etc.

4. Deforestation destroys tropical vegetation, which absorbs large amounts of carbon dioxide. In addition, tropical rain forests are often burned, releasing carbon dioxide.

Section 14.2

Phase 1

PREPARATION

Key Concepts

- Special laws and "pollution permits" have been enacted to reduce pollution in the United States.
- Nonrenewable energy resources include oil, coal, and natural gas.
- Renewable energy resources include solar energy and wind power.
- Three nonrenewable resources warrant immediate attention: fertile soil, groundwater supply, and existing species.
- The growth of the human population is the major factor contributing to environmental pollution.

Reading Strategy

Emphasize that students cannot expect to remember every single point made in their reading assignments. What is important to remember are the main ideas. Be sure students recognize that the objectives listed at the beginning and the headings highlighted throughout the section are important guides to focusing on the main ideas.

Phase 2

TEACHING STRATEGIES

Visual Strategy

Figure 14.7
Use this figure to emphasize that the responsibility to reduce pollution encompasses a wide spectrum—from individuals to international organizations. Have students discuss how an individual can make a contribution toward this effort.

IN THE UNITED STATES TODAY, PRACTICALLY EVERYONE FROM INDIVIDUAL CITIZENS TO LARGE CORPORATIONS PARTICIPATES IN RECYCLING IN SOME WAY. SEVENTY PERCENT OF THE GLASS IN EVERY GLASS BOTTLE WAS PRESENT IN SOME PREVIOUS GLASS CONTAINER. AS YOU WILL SEE IN THIS SECTION, RECYCLING IS JUST ONE WAY WE ARE MEETING THE CHALLENGES POSED BY OUR ENVIRONMENTAL PROBLEMS.

14.2   Meeting the Challenge

Objectives

1 Explain two ways pollution has been reduced in the United States.

2 Identify alternatives to fossil fuels, and describe some benefits of each.

3 Contrast nonrenewable resources with renewable resources.

4 Recognize the relationship of human population growth to environmental problems.

Reducing Pollution

The pattern of global change overwhelming our world is very disturbing. Human activities are placing severe stress on the global ecosystem, and we must quickly find ways to reduce the harmful impact. There are four areas in which it is particularly important to find solutions: reducing pollution, finding enough energy, preserving irreplaceable resources, and curbing human population growth. Governments all around the world are now making serious efforts to reduce pollution. In some cases, such as CFC production, the efforts involve international agreements to reduce or stop production of the pollutant. In most cases, however, pollution problems are national or local and require action by individual governments. Individuals can also have an impact. The man shown protesting in **Figure 14.7** hopes to improve the quality of the water he and his family drink. Two approaches, both effective, have been taken to curb pollution in this country. The first approach is to pass laws limiting how much pollution can be released. In the last 20 years, laws have begun to significantly curb the spread of pollution by setting stiff standards for what can be released into the environment. All new cars must have effective catalytic converters to reduce the amount of pollution they release. Similarly, the Clean Air Act of 1990 requires that power plants install scrubbers on their smokestacks to restrict sulfur emissions.

A second approach to curbing pollution has been to increase the consumer costs directly by placing a "tax" on pollution. These taxes, often imposed as "pollution permits," are becoming an increasingly important part of laws that regulate pollution. They are a key element of the 1990 Clean Air Act.

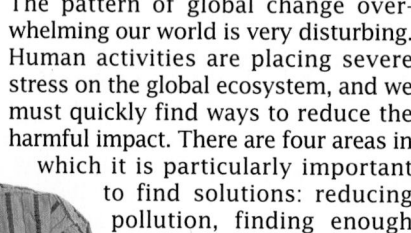

Figure 14.7
To solve our environmental problems, action is needed from both governments and individuals.

▪ Matter of Fact ▪

CFC Treaty
In September of 1987, representatives from 46 nations agreed to the Montreal Protocol on Substances that Deplete the Ozone Layer. This treaty mandated a 50 percent world cut in CFC production by the year 2000. The Environmental Protection Agency estimated that these reductions would reduce the number of skin cancer cases by 137 million worldwide. In 1990, the Montreal protocol was significantly strengthened. Ninety-three nations, including the United States, agreed to stop using CFCs by the year 2000.

Finding Enough Energy

Our dependence on fossil fuels has led to serious pollution problems and has resulted in rapid draining of fuel supplies. The known reserves of oil and natural gas will be nearly depleted by the middle of the next century. Fossil fuels are nonrenewable energy sources. **Nonrenewable resources** do not replenish themselves naturally, while **renewable resources** do. Trees are a renewable resource. New trees can be grown to replace those cut down.

Alternatives to fossil fuels do exist

Nuclear power—capturing the energy released when radioactive atoms break apart—is an important alternative to fossil fuels. More than 70 percent of France's electricity is produced by nuclear power plants. For all its promise of plentiful energy, nuclear power presents three areas of concern that must be addressed if it is to provide energy for our future. These areas are safe operation, waste disposal, and security.

New nuclear power plant designs are much safer than those of the past. The best of these designs virtually eliminates the possibility of loss-of-coolant explosions, such as the one that occurred in 1986 at Chernobyl in Ukraine. Spent nuclear fuel is very radioactive, and the power plants themselves eventually wear out. Spent fuel and components of power plants will remain radioactive for thousands of years. For this reason, wastes must be disposed of in a safe manner, usually by burying them in a remote location. Finally, security is an issue because spent nuclear fuel can be used to recover plutonium, which can be used to make atomic weapons.

Until nuclear power's problems are solved, it is important to develop other alternatives to burning oil, coal, and gas. Solar energy and wind power are two promising sources of energy.

Energy conservation can reduce reliance on fossil fuels

The most cost-effective way to meet our future energy needs is conservation, using energy more carefully. As much as 75 percent of the electricity used in the United States and Canada is wasted through the use of inefficient appliances, according to scientists at the Lawrence Berkeley Laboratory in California. The use of highly efficient motors, lights, heaters, air conditioners, refrigerators, and light bulbs, like the one shown in **Figure 14.8**, could lead to large energy savings.

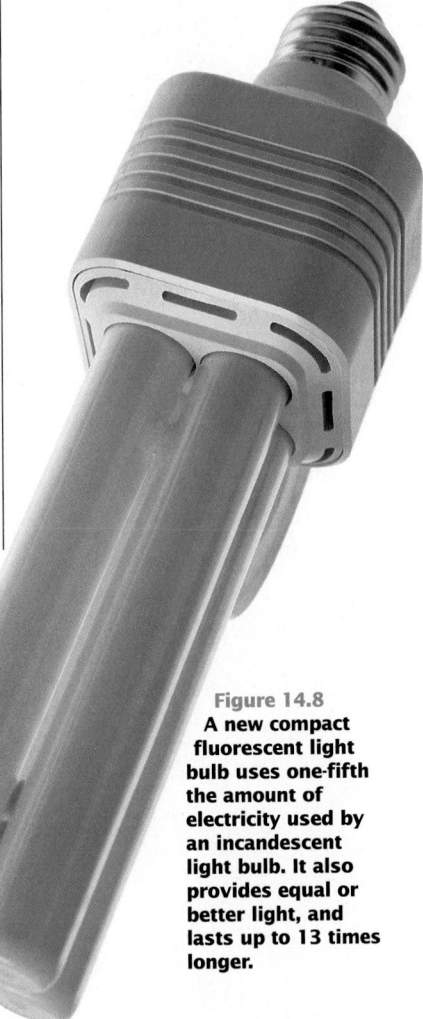

Figure 14.8
A new compact fluorescent light bulb uses one-fifth the amount of electricity used by an incandescent light bulb. It also provides equal or better light, and lasts up to 13 times longer.

Energy and Life

Living things other than humans also require energy resources. Give examples of renewable and nonrenewable resources in nature.

▪ *Matter of Fact* ▪

Nuclear Power
The first nuclear power plant in the United States went into operation in 1957. More than 25 years later, only 4 percent of the nation's energy comes from nuclear power. On a global scale, nuclear power accounts for only 1 percent of the world's ener-gy supply, less than that supplied by hydroelectricity (2 percent).

Demonstration 2
Soil Conservation
Most students are not familiar with the various techniques used in soil conservation. Point out in Figure 14.9c that the mounds formed by the plow and the crops prevent water from running straight down the slope, and taking the topsoil with it. Show students photographs of other farming methods utilized to conserve soil—strip cropping, terracing, crop rotation, and use of windbreaks and cover crops. Discuss how each method helps to reduce soil erosion.

Theme Answer
Interacting Systems
Groundwater supplies water for many agricultural areas. Should plants die from polluted ground water or lack of ground water, the topsoil in which plants grow will dry up and become exposed to wind and rain, which causes erosion.

Conserving Nonrenewable Resources

While a polluted stream can be cleaned up, no one can restore an extinct species. Worldwide, three sorts of nonrenewable resources merit particular attention.

Topsoil is the basis for agriculture
One of the great strengths of the United States over the last two centuries has been its strong agriculture. The United States has been able to grow extraordinary amounts of crops because of its particularly fertile soils. These soils have accumulated slowly during hundreds of thousands of years. Although it takes hundreds of years for an inch of topsoil to form, we are allowing topsoil to be lost at a rate of inches each decade. By intensively cultivating crops—repeatedly turning the soil over to eliminate weeds—we permit rain and wind to carry topsoil away. Our country has lost one-fourth of its topsoil since 1950. Loss of topsoil is a worldwide problem as well. Each year, nearly 22 billion metric tons (24 billion tons) of topsoil are blown or washed off the world's farmland. Figure 14.9 shows how some farmers have implemented contour farming in an effort to prevent the erosion of topsoil.

Ground water supplies are being depleted
A second resource that we are depleting is ground water. Ground water is water trapped beneath the soil, largely in porous rock. This water seeped into its underground reservoirs very slowly during the last 12,000 years. We use water in thousands of ways, especially for irrigating crops and in our homes. But we should not waste it, for it accumulates very slowly.

Today, there is very little control over the use of ground water, and much of what is used is wasted to water lawns, wash cars, and run fountains. A great deal more ground water is inadvertently polluted by poor disposal of chemical wastes. Once pollution enters the ground water, there is no effective means of removing it.

Interacting Systems

How might the loss or pollution of ground water affect the loss of topsoil?

Figure 14.9

a When farmers cut down trees and vegetation from land that will become a field, erosion can occur.

b In an effort to save land that has begun to erode, farmers can plant rows on different levels or "contours."

c In the long run, contour farming limits erosion.

▪ *Matter of Fact* ▪

Water
Although the Earth has an abundant supply of water, more than 97 percent is salt water. In addition, nearly 75 percent of the Earth's fresh water is bound as ice or snow in the icecaps at the poles. With the average American using more than 300 L (80 gal.) of water per day, the ground-water supply is being seriously affected. Have students hypothesize what will happen to water consumption in the year 2000 as a result of global warming.

a The humpback chub inhabits rivers in Arizona, Colorado, Utah, and Wyoming.

b The number of black-footed ferrets in Wyoming increased after scientists bred them in captivity.

Figure 14.10
Among the many endangered species found in the United States are the humpback chub, the black-footed ferret, and the whooping crane.

c The whooping crane's range extends from the Canadian Rocky Mountains to Texas.

Species are disappearing

During the last 20 years, about half of the world's tropical rain forests have been destroyed, burned to make pasture land or cut for timber. Each year the rate of loss increases as the human population of the tropics grows; nearly 2,000 hectares (4,800 acres) of forest are cut each minute, an area larger than Indiana each year. At this pace, all the world's rain forests will be gone in about 30 years. It is estimated that one-fifth or more of the world's species of animals and plants will become extinct, more than a million species lost, if the rain forests are destroyed.

Why is the loss of species important? As you learned in Chapter 13, removing even one species from an ecosystem can seriously disrupt the workings of that ecosystem. Moreover, as species disappear, so do our chances to learn about them and of their potential benefits. Like burning a library without reading the books, we do not know what we are wasting. All we can be sure of is that we cannot retrieve it.

Many important and useful plants are discovered each year. For instance, the periwinkle, a garden plant native to Madagascar, has been used to develop two drugs used to treat leukemia, a type of cancer that affects white blood cells. A child with Hodgkin's disease (a form of leukemia) has a 90-percent chance of survival if treated with these drugs. Without the drugs, the child would have only a 20-percent chance of living.

The loss of species means that we lose organisms we might need. With the advent of genetic engineering, scientists now have the ability to screen plants and animals for desirable genes, ones that confer resistance to pests or spur more rapid growth. In the coming century, genetic engineering could lead to major improvements in agriculture—but only if the vast library of genes contained in the world's species is there to be searched. **Figure 14.10** shows three endangered species native to the United States.

Visual Strategy

Figure 14.11

Use this figure to point out that three major spurts occurred in the history of the human population. The first, occurring some 1 million years ago and illustrated in Figure 14.11a, was due to the development of efficient tools for hunting and gathering. The second, occurring some 10,000 years ago and illustrated in Figure 14.11b, resulted from agriculture. The third, beginning in the mid 1600s but becoming more significant in the mid 1800s, was attributed to improved medical care and the conquest of many diseases.

Mathematics Connection

Zero Population Growth

To stabilize population growth in their countries, many governments are promoting zero population growth in their countries, primarily by encouraging families to have no more than two children. Also a country's immigration must equal its emigration. However, many parents continue to have more than two children, especially in agricultural societies. Tell students that the population of the United States is currently about 255 million. Ask them what would happen to this country's population if each family had four children.

The Deeper Problem: Population Growth

If we were to solve the many problems mentioned in this chapter, we would only buy time to address the fundamental problem: there are too many of us. Figure 14.11 shows the increase in the world's human population over time.

Ten thousand years ago, when agriculture first developed, there were about 10 million people on Earth, distributed over all the continents except Antarctica. As new, more dependable sources of food became available as a result of agriculture, the human population began to grow more rapidly. The world population passed 5 billion people in early 1987, and will hit 6 billion before the year 2000.

What factors triggered the human population explosion?

Starting in the seventeenth century, the human population began to increase sharply. A population will grow when the birth rate (number of births per thousand people per year) exceeds the death rate (number of deaths per thousand people per year). From the time agriculture was introduced approximately 10,000 years ago until about 1650, the birth rate was only slightly higher than the death rate. The result of this small difference was fairly slow population growth. With the spread of better sanitation and improved medical care that began around 1650, the death rate plunged while the birth rate remained relatively constant. As a result, the human population began to grow more rapidly.

Currently, the death rate (as a world average) is about 10 deaths per thousand people per year, and the birth rate is about 28 births per thousand people per year. The difference between these two figures yields an annual population growth rate of approximately 1.8 percent. This number may seem small, but it leads to a doubling of the world's population in only 39 years. Nearly 92 million more people are born each year than die. More than 250,000 people are added to the world population each day, more than 175 each minute. Our world cannot continue to support such growth.

The human population continues to grow

Many countries are devoting considerable attention to slowing the growth rates of their populations,

Figure 14.11
In the past, the human population grew very slowly. Technology has increased life span and decreased infant mortality to the point where the human population can now double in just 39 years.

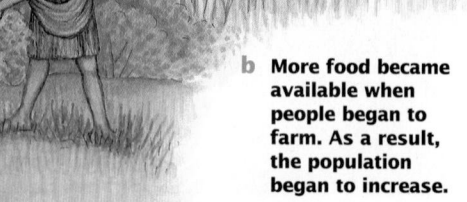

a In pre-agricultural times, humans lived by hunting and gathering. During this period, the population size remained relatively stable.

b More food became available when people began to farm. As a result, the population began to increase.

▪ Cultural Perspective ▪

Population control in China

China's population of 1.1 billion is the largest of any country. China has a very strict population control program, which rewards those who have small families and punishes those who have large families. Couples who agree to have only one child receive economic benefits, including higher salaries, more food, and better housing. One parent of families with two children must be sterilized. The program features an incentive educational effort and free access to contraceptives. Between 1972 and 1985, China reduced its birth rate by nearly 50 percent.

and there are genuine signs of progress. If these efforts continue, the United Nations estimates that the world's population may stabilize by the close of the next century at about 13 to 15 billion people, three times the number living today. No one knows whether the world can support 13 to 15 billion people indefinitely. Finding a way to do so is the greatest task we will face in the coming years. The quality of life available for your children in the next century will depend to a large extent on our success.

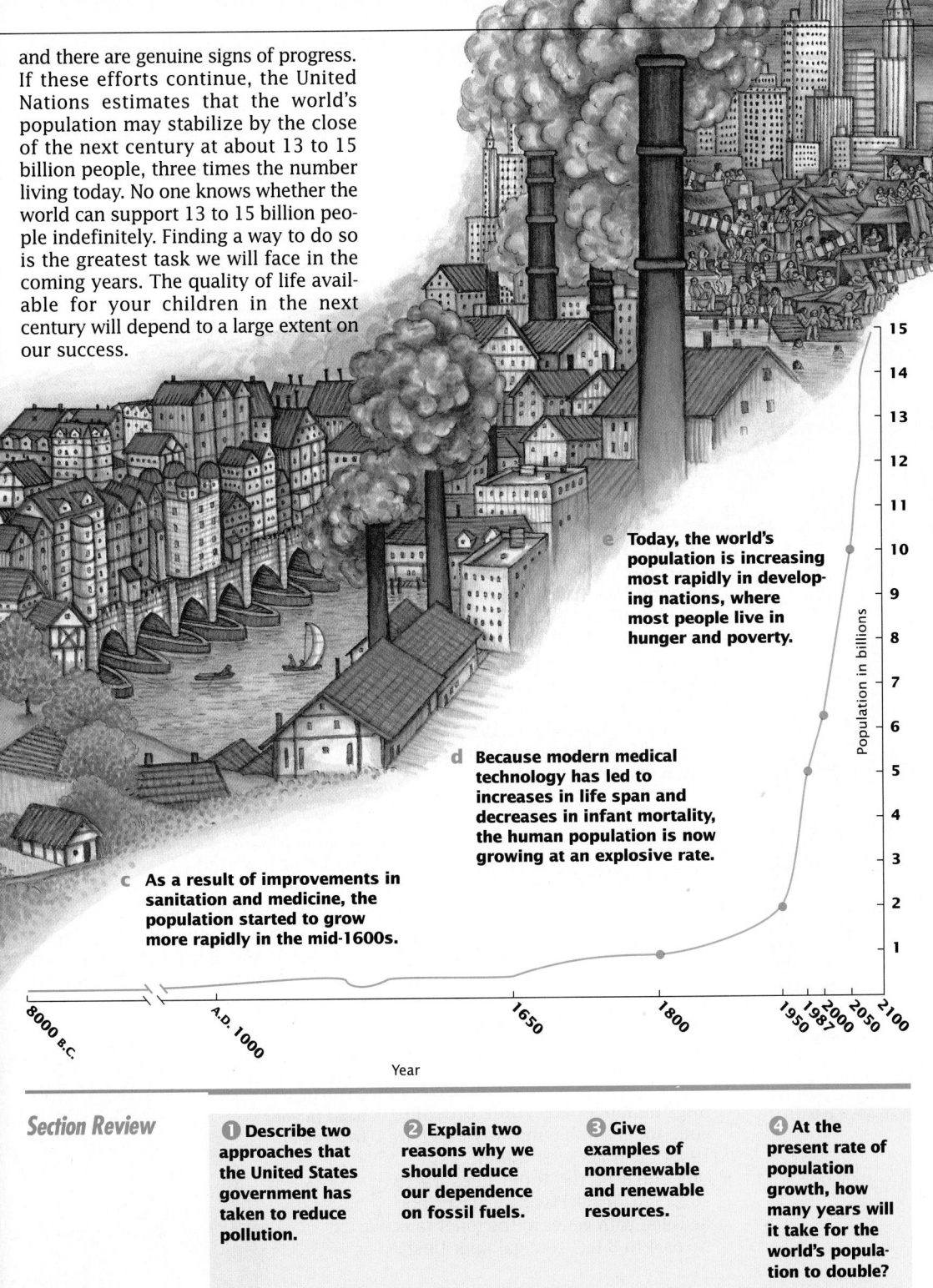

e Today, the world's population is increasing most rapidly in developing nations, where most people live in hunger and poverty.

d Because modern medical technology has led to increases in life span and decreases in infant mortality, the human population is now growing at an explosive rate.

c As a result of improvements in sanitation and medicine, the population started to grow more rapidly in the mid-1600s.

Population in billions

15
14
13
12
11
10
9
8
7
6
5
4
3
2
1

8000 B.C.   A.D. 1000   1650   1800   1950 1987 2000 2050 2100

Year

Section Review

1 Describe two approaches that the United States government has taken to reduce pollution.

2 Explain two reasons why we should reduce our dependence on fossil fuels.

3 Give examples of nonrenewable and renewable resources.

4 At the present rate of population growth, how many years will it take for the world's population to double?

■ *Section Review Answers* ■

1. laws that limit release of pollution, taxes on pollution
2. Fossil fuel supplies are limited and can be exhausted. Burning fossil fuels releases large amounts of pollution, including carbon dioxide.
3. Renewable resources include trees, wind power, and solar energy. Nonrenewable resources include topsoil, ground water, fossil fuels, and living species.
4. 39 years

Using the Feature

Use this feature to show students that their personal decisions can influence the quality of the environment. Also emphasize that many of these decisions are inexpensive and do not require wholesale lifestyle changes. Point out that this is not an exhaustive list of actions. There is a plethora of books on the market that make additional suggestions. Some of these are *Living In the Environment* by G. Tyler Miller; *Fifty Ways You Can Help Save the Planet* by Tony Campolo and Gordon Aeschliman; *50 Simple Things You Can Do To Save the Earth*, *50 More Things You Can Do To Save the Earth*, and *The Recycler's Handbook*, all by the EarthWorks Group of Berkeley, California; and *One Earth, One Future: Our Changing Global Environment* by the National Academy of Sciences.

Discussion

Guide the discussion by posing the following questions.

1. List three ways to reduce energy and save money.
   Turn off lights and appliances when not in use; walk, ride a bike, or take a bus or train to school; lower water heater setting; proper automobile maintenance; use compact fluorescent light bulbs.

2. How could riding a bike to school affect the global climate?
   Bicycles are one form of transportation that does not produce the greenhouse gas carbon dioxide.

3. Identify five products that can be recycled.
   newspapers; aluminum cans, plates, and foil; glass containers; office and school paper; some kinds of plastic containers.

How to Help Save the Environment

Can you help save the environment? Although it may seem that all important environmental actions and decisions are the responsibility of the courts or the government, there are many simple and inexpensive ways that you can contribute. Some ways are listed below. You can find out more by writing to the organizations listed on the next page.

Conserving Energy

The United States is an energy hog. With less than 5 percent of the world's population, we consume 25 percent of the world's energy. Nearly 90 percent of our energy is derived from nonrenewable fossil fuels—coal, oil, and natural gas. Extracting fossil fuels from the earth causes significant environmental damage. Burning fossil fuels for energy creates pollution; an automobile emits its own weight in pollutants each year. Here are some ways you can reduce your consumption of energy.

- Turn off all lights and appliances when you leave a room.

- Set the water heater temperature to 54°C (130°F).
- Use compact fluorescent light bulbs. Although more expensive than incandescent bulbs, compact fluorescent bulbs use one-fourth as much electricity and last 10 to 13 times longer.
- Reduce your use of automobiles. Walk, ride a bicycle, or take public transportation (buses or trains) to work or school.
- If you drive a car, properly inflated tires increase fuel efficiency by 5 percent. A tuneup increases fuel efficiency by 9 percent.

You can help conserve energy by riding your bicycle more often for transportation.

Conserving Water

The United States is also the world's largest consumer of water. Water in many western states is already in short supply. By the year 2000, most of the United States west of the Mississippi is projected to face water shortages. Here are some ways you can reduce your water use.

- Turn off faucets when not in use. Letting the faucet run while brushing your teeth or shaving wastes 11 to 19 L (3 to 5 gal.) per minute.
- One flush of the toilet uses 19 to 26 L (5 to 7 gal.). Placing a plastic bottle filled with water in the toilet tank saves 4 to 8 L (1 to 2 gal.) per flush.

- Showers account for as much as 32 percent of home water use. Inexpensive, low-flow shower heads are easy to install, and they reduce shower water use by 50 percent.

You can conserve water by installing a low-flow shower head in your bathroom.

Reducing Pollution and Waste

Each year, Americans generate 138 billion kg (153 million tons) of garbage. This mountain of trash is dumped into landfills or is burned, creating

By recycling newspapers, you can help reduce the amount of material added to your local landfill.

pollution. Moreover, much of this waste could be reused or recycled. Here are some ways to help reduce waste and pollution.

- *Recycle* Aluminum products (cans, foil, and pie plates), glass containers, office and school paper, newspaper, and many plastic containers can be recycled. Many communities, school groups, and private organizations collect recyclable materials.

- *Properly dispose of toxic and hazardous materials* These materials include many household products, (insecticides, oven cleaner, furniture cleaners, oil-based paint), used automobile batteries, and used motor oil. No toxic or hazardous substance should be poured down the drain or thrown into the trash; contamination of ground water can result. Many communities will collect and properly dispose of your hazardous and toxic materials.

Contacting Environmental Agencies

For more information on these and other actions you can take to help the environment, write to these organizations:

Natural Resources Defense Council
40 West 20th Street
New York, NY 10011

Environmental Defense Fund
1616 P Street NW, Suite 150
Washington, D.C. 20036

Citizens for a Better Environment
33 East Congress, Suite 523
Chicago, IL 60605

Rocky Mountain Institute
1739 Snowmass Creek Road
Snowmass, CO 81654

World Resources Institute
1735 New York Avenue NW
Washington, D.C. 20037

World Wildlife Fund
1250 24th Street NW
Washington, D.C. 20037

Section 14.3

Phase 1

PREPARATION

Key Concepts

- Some efforts to solve environmental problems have been successful.
- The five basic steps to solving an environmental problem are assessment, risk analysis, public education, political action, and follow-through.
- An individual can make a difference in solving an environmental problem.

Reading Strategy

Use the information in Table 14.1 as the basis for a class discussion. After students have read the various steps, have them provide specific examples of what they have done to save the environment. Moreover, use the table as a springboard for ideas of what the class might undertake as a project aimed at solving some local environmental problem.

Phase 2

TEACHING STRATEGIES

Chemistry Connection

Modifying CFCs

Industrial chemists are attempting to modify CFCs by replacing one chlorine or one fluorine atom with one hydrogen atom. Such a modified CFC would break down before it reached the ozone layer. However, these modified CFCs have different physical properties, requiring a redesign of air conditioners and refrigerators. In addition, these modified CFCs are not efficient. Automobile air conditioners, for example, would have to be bulkier and heavier, cutting fuel consumption. Point out to students that saving the environment often involves a trade-off.

WHEN EUROPEAN SETTLERS ARRIVED IN NORTH AMERICA, 60 MILLION TO 125 MILLION BISON LIVED HERE. BY 1889, ONLY 85 BISON REMAINED. CONSERVATION EFFORTS HAVE ENABLED THE BISON POPULATION TO REBOUND TO ABOUT 65,000. SAVING THE BISON FROM EXTINCTION IS ONE EXAMPLE OF AN ENVIRONMENTAL PROBLEM IN WHICH ACTION LED TO A SOLUTION.

14.3 Solving Environmental Problems

Objectives

❶ **List some examples of successful solutions to environmental problems.**

❷ **List the five basic elements necessary to solve any environmental problem.**

❸ **Recognize your role in solving environmental problems.**

Environmental Problems Can Be Solved

The most important fact to remember about the environmental crisis is that each of its many problems is solvable. A polluted lake can be cleaned up, a dirty smokestack altered to remove toxic gases, and the waste of key resources stopped. Each requires a clear understanding of the problem and a commitment to doing something about it. The extent to which American families recycle aluminum cans and newspapers shows that people want to become part of the solution, rather than remaining part of the problem. Progress is being made. Newspapers carry success stories daily. In the United States since 1970 eight national parks have been established. As a result, 80 million acres of land have been protected as wilderness and 34 million acres of farmland particularly vulnerable to soil erosion have been withdrawn from production. Many previously endangered species are better off than they were in 1970, including the pronghorn antelope, the wild turkey, the bald eagle, and the peregrine falcon, which is shown in **Figure 14.12**.

Pollution control efforts have been particularly successful. Emissions of sulfur dioxide, carbon monoxide, and soot, which were 200 million tons in 1970, have been reduced by more than 30 percent. The release of toxic chemicals into the environment (notably the insecticide DDT, and the carcinogens asbestos and dioxin) has been banned outright. The Environmental Protection Agency estimates that private firms and public agencies are spending about $100 billion per year on pollution control, double the figure of 10 years ago and five times the figure of 1970. In the same period the population of the United States increased by about 40 million people and the number of cars grew by more than 60 million vehicles. Had this progress not been made, environmental quality in the United States would almost certainly have declined dramatically.

Figure 14.12
Peregrine falcons are found throughout the continental United States. They can often be seen roosting on buildings or bridges in small towns as well as large cities.

Steps Toward Saving the Environment

There are five steps to solving an environmental problem: assessment, risk analysis, public education, political action, and follow-through. Table 14.1 outlines these steps using as an example a lake damaged by chemicals.

Table 14.1 Steps Toward Solving an Environmental Problem

Steps	Plan of action	What you can do
Assessment	Data must be collected and experiments performed in order for scientists to construct an ecological model of an ecosystem. The model predicts how the environment will respond to changes.	You can volunteer to test the water of a nearby lake to determine the degree of damage done by chemical pollution. You also can record changes in the numbers of animals and plants found in and around the lake.
Risk Analysis	Using the ecological model, scientists can predict the effects of environmental intervention. Scientists will evaluate the potential for solving the environmental problem as well as the potential for any adverse effects of the proposed solution.	A local college or university can test the water samples that you collect. Scientists there can suggest ways to decrease the amounts of chemicals entering the lake.
Public Education	When a clear choice can be made, the public must be informed. This involves explaining the problem in terms people can understand. Costs and expected results of each alternative should be presented and explained.	The people of your town or city will have to decide whether they are willing to pay for the ditches necessary to divert agricultural runoff from the lake. You could go to a city council or town meeting and emphasize the importance of saving the lake for recreation.
Political Action	The public, through its elected officials, selects a course of action and implements the plan. Individuals can have a major impact by exercising their right to vote. Many voters do not understand the magnitude of what they can achieve by writing letters and supporting special interest groups.	It is important to start learning about environmental issues now, so that you can make good decisions when you are old enough to vote. Voters in your town or city will need to ask themselves if the cost of preventing water pollution is worth the potential benefits.
Follow-through	The results of any action taken should be monitored carefully to see if the environmental problem is being solved and to evaluate and improve the initial assessment and modeling of the problem. We learn by doing.	If the town does agree to channel runoff water away from the lake, you can volunteer to continue collecting water samples. From these samples, scientists can determine how well the lake is recovering.

Visual Strategy

Table 14.1
See the Reading Strategy for suggestions on how to use this table as the basis for a class discussion.

Figure 14.13

a We can all do our part to solve the Earth's environmental problems. Members of a community transformed a vacant lot in New York City into a community garden.

b The garden was once similar to this vacant lot across the street.

What You Can Contribute

Interacting Systems

How can your decision to walk or ride a bike to school affect global temperatures?

Each of us can significantly contribute to solving today's environmental problems. The energy and materials we consume every day place a great and growing strain on the environment. Americans consume hundreds of times more resources than people in less developed countries. The simplest way to help is for each of us to cut down on needless resource consumption. Turn off the lights when you leave a room, purchase energy-saving light bulbs, share a ride to school. You also can recycle paper, glass, and metal cans. The cumulative efforts of large numbers of people can result in significant reductions in resource consumption.

The difference between the two photographs in **Figure 14.13** shows the impact that students like you can make by joining a community effort to improve your neighborhood.

The biggest impact you can have is to make a very serious effort to *understand* the environment. You cannot preserve what you do not understand. Although solving the world's environmental problems will take the efforts of many different kinds of people—politicians, economists, engineers—the issues are largely biological. Your knowledge of biology is an essential tool you will need to contribute to the effort.

Section Review

❶ Describe two examples of reductions in pollution that have occurred in the United States since 1970.

❷ List the five essential steps to solving environmental problems.

❸ Describe two things you can do to help save the environment.

■ *Section Review Answers* ■

Chapter **14** *Highlights*

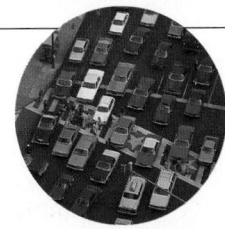

Automobiles are a major source of the carbon dioxide that causes global warming.

Assessments Alternative

Have the class organize a display entitled *The Fragile Earth.* Tell students that the objective of the display is to inform the rest of the school about the severity of the environmental issues raised in this chapter. Consequently, the display should cover a wide range of issues, provide specific information, and show how an individual can make a difference. Make arrangements to have the display exhibited in a common area for the entire student body to see.

	Key Terms	Summary
14.1 Planet Under Stress Water pollution and acid rain threaten our water supply and the inhabitants of rivers, lakes, and streams. 	pollution (p. 293) carcinogen (p. 293) acid rain (p. 295) acid (p. 295) pH (p. 295) chlorofluorocarbon (p. 296)	• Human activities are causing changes to the entire planet. • Pollution is anything potentially harmful that humans add to ecosystems. • Sulfur-containing pollution from coal-burning plants mixes with water in the atmosphere, causing acid rain. Acid rain is killing forests and lakes. • The protective ozone shield that surrounds our planet is being destroyed by chlorofluorocarbons. • The world's climate is warming as large amounts of carbon dioxide are released into the Earth's atmosphere.
14.2 Meeting the Challenge  This black-footed ferret is just one of the growing number of plants and animals classified as endangered species.	nonrenewable resource (p. 299) renewable resource (p. 299)	• Humans need to reduce pollution, find enough energy, and conserve nonrenewable resources. • Nuclear power is one alternative to fossil fuels, although significant problems associated with its use remain to be solved. • We must act now to save non-renewable resources such as topsoil, ground water, and species diversity. • The exploding human population is the single greatest threat to the world's future.
14.3 Solving Environmental Problems  You can take action to improve the environment.		• Almost all environmental problems can be solved. • Every citizen can help by reducing needless consumption of resources and by learning about ecology.

Chapter Review Answers

Understanding Vocabulary

1. a. carbon dioxide
   b. nonrenewable
   c. global warming

Relating Concepts

2. Map answer is shown on page 287d.

Understanding Concepts

Multiple Choice

3. c   8. b
4. a   9. b
5. a   10. a
6. d   11. b
7. c   12. c

Completion

13. pH, pH, 14, 5.6
14. conservation
15. ground water

Short Answer

16. The water used by the people of the Mississippi Delta comes mainly from the Mississippi River. This water supply is contaminated with toxic and carcinogenic chemicals applied to farmland and dumped by industrial plants up river.

17. Laws have been passed to limit pollution and taxes have been levied to make pollution costly.

18. They are all nonrenewable resources that are being consumed or destroyed at a rapid rate.

19. Since 1650 the human population has grown rapidly; some call it a population explosion. Reasons for the rapid growth include improvements in agriculture, medicine, and sanitation.

20. Follow-through is essential to determine whether the environmental problem is responding to the correction measures. Alternative measures may be required.

Chapter 14 Review

Understanding Vocabulary

1. For each set of terms, complete the analogy.
   a. acid precipitation: sulfuric acid::greenhouse effect: _____
   b. tree:renewable:: gasoline: _____
   c. too little atmospheric ozone:skin cancer::too much carbon dioxide: _____

Relating Concepts

2. Copy the unfinished concept map below onto a sheet of paper. Then complete the concept map by writing the correct word or phrase in each oval containing a question mark.

Understanding Concepts

Multiple Choice

3. Carcinogens are substances that
   a. destroy the ozone layer.
   b. spoil the air.
   c. cause cancer.
   d. are used as coolants.

4. The major cause of acid rain is
   a. burning high-sulfur coal.
   b. automobile exhaust.
   c. nuclear power plants.
   d. increased global temperatures.

5. Which is a major cause of increased atmospheric carbon dioxide?
   a. burning fossil fuels
   b. acid precipitation
   c. reforestation
   d. agriculture

6. The ozone layer is being destroyed by
   a. acid rain.
   b. nuclear waste.
   c. ultraviolet radiation.
   d. chlorofluorocarbons.

7. A possible consequence of global warming is
   a. increased rain fall.
   b. higher rates of cancer.
   c. decreased crop yields in the midwestern United States.
   d. cooler temperatures in the United States.

8. Which is an example of a renewable resource?
   a. gasoline   c. wood
   b. coal   d. topsoil

9. As much as 75 percent of the electricity used in the U.S. and Canada is wasted by
   a. tall smokestacks.
   b. inefficient appliances.
   c. industrial plants that employ too few people.
   d. burning high-sulfur coal.

10. Rain-forest species are becoming extinct primarily due to
    a. loss of habitat.   c. pollution.
    b. hunting.   d. disease.

11. Population growth may be determined by
    a. adding the yearly birth rate and death rate.
    b. subtracting the yearly death rate from the yearly birth rate.
    c. subtracting the yearly birth rate from the population total.
    d. multiplying the yearly birth rate by the yearly death rate.

12. What is the first step in solving an environmental problem?
    a. risk analysis
    b. political action
    c. assessment
    d. public education

Completion

13. Scientists measure acidity using the _____ scale. The _____ for normal rainwater is about _____ , while the _____ for pure water is about _____ .

14. The best and least expensive way to meet our immediate energy needs is _____ , which means using less energy more wisely.

15. Water held in underground reservoirs of porous rock is called _____ _____ .

Interpreting Graphics

21. • slow rate of growth; rapid rate of growth; no growth
    • At point z the population growth is zero. Such a growth rate is most likely the result of limited supplies of some required resource, perhaps food or places to live. The curve in Figure 14.11 resembles the first part of this curve, but human population continues to grow explosively.

Reviewing Themes

22. acid rain caused the trees to die; reducing the amount of high sulfur coal that is burned by factories and power plants in the midwestern United States, or requiring that the factories and power plants clean the sulfur from their emissions before they are released into the air

23. turn off lights, stereo, and television when leaving an unoccupied room; ride the school bus or a bicycle to school rather than driving a car; recycle newspapers and soft-drink cans

Short Answer

16. Suggest a likely cause of the high rate of cancer among people who live in the Mississippi Delta region of the United States.

17. Describe two government approaches that have been taken to curb pollution in the United States.

18. What do rain-forest species, topsoil, and ground water have in common?

19. Describe the changes in human population size since 1650. What are the reasons for the changes?

20. Summarize the importance of follow-through in solving an environmental problem.

Interpreting Graphics

21. Biologists have studied the population growth of many kinds of organisms. The figure below shows a growth curve for a typical population.

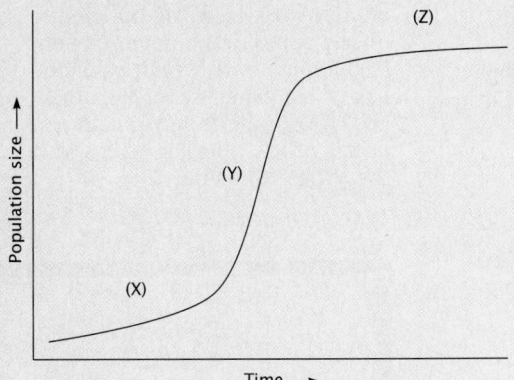

- Describe the rate of population growth in the portions of the curve marked X, Y, and Z.

- Compare this graph with **Figure 14.11** on pages 302–303. What labeled region of the growth curve shown above is absent from **Figure 14.11**? Suggest an explanation for the differences you observe between the growth curve shown above and the one in **Figure 14.11**.

Reviewing Themes

22. *Interacting Systems*
    What has caused many forest trees in the northeastern U.S. and southeastern Canada to die? What can be done to help the forests survive?

23. *Patterns of Change*
    To preserve the Earth for future generations, humans must give up wasteful habits and adopt new ones. What three daily routines could you adopt to help solve environmental problems?

Thinking Critically

24. *Inferring Conclusions*
    Many fast-food restaurants have chosen to package their food in paper, rather than plastic foam containers. How might this decision affect your health?

25. *Building on What You Have Learned*
    You learned about genetic engineering in Chapter 8. What is the relationship between genetic engineering and efforts to preserve species in tropical rain forests and other ecosystems?

Cross-Discipline Connection

26. *Biology and Geography*
    Find out which plant and animal species are endangered or threatened in your state. On a state map, locate the counties containing these species. Research the effects of habitat loss on species in your county and in surrounding counties.

Discovering Through Reading

27. Read the article "The Bottom Line," in *Parenting*, May 1992, pages 67–68. Why are disposable diapers at the center of an environmental controversy? How does the article show that clear choices are not always possible when considering environmental issues?

Cross-Discipline Connection

26. Suggest that students contact state agencies for listings and locations of endangered species. Or, contact environmental groups such as the Sierra Club, the Nature Conservancy, the Audubon Society, Bat Conservation International, Earth First!, etc. Visit with an ecologist or naturalist if there is a university in your city. Students should relate habitat loss with decrease or extinction of a species.

Discovering Through Reading

27. Disposable diapers are not reusable like cloth diapers, and because so many parents are now using disposable diapers, they are filling and polluting landfills. As is true for cloth diapers, the use of disposable diapers has both benefits and disadvantages. The benefits and disadvantages associated with both types of diapers make a clear choice impossible.

Thinking Critically

24. The production and use of fewer styrofoam containers will result in less chlorofluorocarbons being released into the atmosphere. Less CFCs mean less ozone loss, reducing the likelihood of exposure to high levels of ultraviolet radiation.

25. The genes contributed by a number of species have led to genetically engineered medical and agricultural products. The contributions of species that disappear as the rain forests or other ecosystems are destroyed will never be known. Preserving the species is the best way to not limit the potential benefits of genetic engineering.

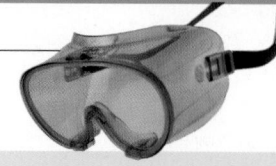

Procedural Note

1. Discuss the nature of pollution and pollutants with students before beginning the investigation.
2. Review the procedure for making wet-mount preparations, if necessary.

Prelab Preparation Answers

1. • Pollution is some undesirable change in the physical, chemical, or biological characteristics of an ecosystem.
   • Industrial chemicals, heat, and radioactivity are some common pollutants.

Procedure Answers

7. The *Amoeba* tends to move away from the salt solution.
9. Most of the *Daphnia* collect near the middle of the tube where the temperature is moderate.

Chapter 14

Investigation

How Does Pollution Affect Organisms?

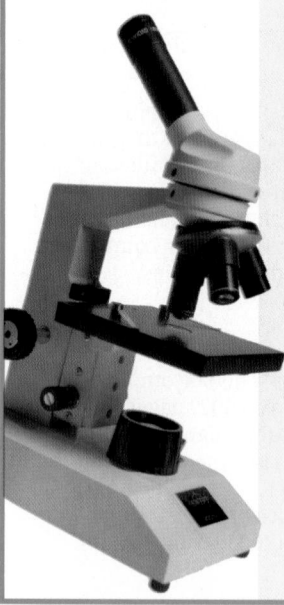

Objectives

In this investigation you will:
• *observe* the effects of two forms of pollution on microorganisms
• *relate* environmental conditions to an organism's ability to survive

Materials

• cold pack or ice in self-sealing plastic bag
• hot pack or hot water in self-sealing plastic bag
• thermometer
• distilled water
• *Daphnia* culture
• 50-mL beaker
• 5-mL pipette
• modeling clay
• paper towels
• *Amoeba* culture
• medicine droppers
• depression slide
• coverslip
• compound light microscope
• 10-percent table salt (NaCl) solution
• stereomicroscope

Prelab Preparation

1. Review what you have learned about pollution by answering the following questions:
   • What is pollution?
   • What are some examples of pollutants found in the environment?
2. Review the procedures for proper use of the microscope in the Appendix.
3. Review the procedures for making a wet-mount slide in the Appendix.

Procedure: Investigating the Effects of Pollution

1. Form a cooperative group with another student to complete steps 2–10.

2. Obtain a hot pack or fill a plastic bag with hot tap water having a temperature of at least 60° C. Obtain a cold pack or fill a plastic bag with ice cubes. Record the actual temperature of each pack or bag.

3. Mix approximately 3 mL of distilled water and approximately 3 mL of *Daphnia* culture in a 50-mL beaker. Swirl the mixture gently. Fill the pipette with the mixture from the 50-mL beaker. Use modeling clay to close the narrow tip of the pipette. Use modeling clay to close the top of the pipette.

4. Place a paper towel on the surface of your work area. Lay the pipette on the paper towel. Cover one end of the pipette with a hot pack or bag of hot water. Cover the other end of the pipette with a cold pack or bag of ice. Note the time and set the apparatus aside.

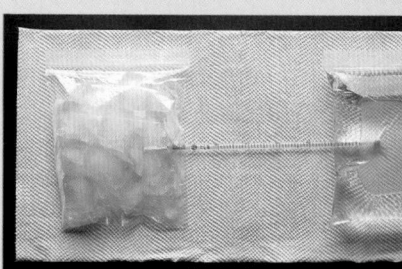

5. Make a wet-mount slide with a drop of *Amoeba* culture. Observe the slide under low power. Make a

drawing of an *Amoeba* and record your observations.

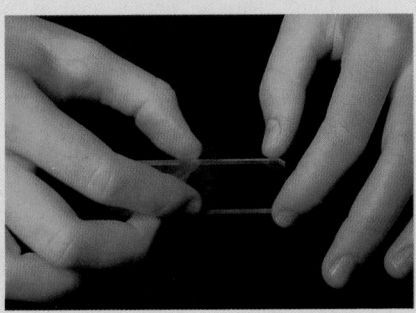

6. Use a clean medicine dropper to place a drop of 10-percent NaCl solution at the edge of the coverslip. Draw the solution under the coverslip by placing a small piece of paper towel against the opposite edge of the coverslip.

7. Examine the slide under high power. *How does an* Amoeba *react to the salt solution?* Record your observations.

8. Pick up the pipette after 15 minutes. Feel the pipette's surface at both ends and in the middle. Record your observations.

9. Use a stereomicroscope to examine the culture in each part of the pipette. *Where have most of the* Daphnia *gathered?* Record your observations.

10. Dispose of the *Daphnia* mixture as directed by your teacher. Clean up your materials and wash your hands before leaving the lab.

Analysis

1. *Comparing Observations*
Compare the response of *Daphnia* to heat with the response of *Amoeba* to salt.

2. *Making Inferences*
Heat and salt are necessary for life. Explain why they could be considered pollutants.

3. *Making Inferences*
How might excessive heat affect an organism that cannot move?

4. *Making Predictions*
How might heat pollution affect the fish populations of a river?

5. *Identifying Relationships*
What would happen to an *Amoeba* if it could not move away from the salt solution?

6. *Evaluating Methods*
Suppose you neglected to completely fill the pipette before corking it with modeling clay. How would this error affect your results?

Thinking Critically

1. Why do you think it is advantageous for an organism to avoid temperature extremes?

2. Near volcanic vents on the ocean floor, bacteria live in water with a temperature of 250°C (480°F). What adaptations would you expect these bacteria to have that enable them to live at such high temperatures?

Analysis Answers

1. *Amoeba* and *Daphnia* moved away from the environmental extremes—high or low temperature, high salt concentrations.

2. When ordinarily beneficial substances are present in abnormal concentrations, they may adversely affect organisms. In such cases, these substances can be considered pollutants.

3. Answers may vary but could indicate that excessive heat can often be fatal for organisms.

4. Thermal pollution may cause certain fish populations to migrate to new areas or it may kill others that are unable to move.

5. Higher concentrations of salt can cause water to leave the cell through the process of osmosis. The *Amoeba* would probably die.

6. If the pipette is not filled before sealing, an air bubble will remain. When the pipette is placed on the paper towel, this air bubble may prevent the organisms from moving to the middle of the pipette.

Thinking Critically Answers

1. High temperatures can be harmful or lethal for organisms because proteins lose their shape when heated. Cells of organisms are unable to function properly when their enzymes have degraded because of heat. Similarly, very low temperatures greatly slow chemical reactions and reduce enzyme function.

2. One adaptation that would enable these bacteria to live in such hostile conditions would be proteins that remain stable at much higher temperatures than proteins found in most other organisms.

313

Using the Feature

- Provide students with some perspective on the interactions of plants and humans. Remind students that plants, as autotrophs, are essential to heterotrophic nutrition.
- Use the time line to emphasize the environmental impact of plant crops. Discuss the fact that in selecting plants for nutritious qualities people have turned many natural areas into cultivated fields. Also, explain that when over cultivation of these areas depletes nutrients in the soil or drought occurs, crops fail and famine may result.
- The principal technique discussed in this feature is agricultural engineering. Use the time line to emphasize that farmers and scientists, from the time of the ancients, have always looked toward creating new varieties of plants that could withstand environmental variability. Explain that technology has always been used to increase crop yields, to develop food plants more resistant to bacterial and fungal infections and insects, and to create plants with high nutritional value.
- Point out that in developing food crops, people may have initially overlooked some of the beneficial traits of the "wild" variety of a plant. In restoring such traits to crop plants, today's scientists can sometimes correct "mistakes" of the past.

Discussion

Guide the discussion by posing the following questions.

1. What group of plants can be considered the world's main source of food? Explain your answer.

   The grains of grasses, or cereals, are considered the world's main source of food. Students should explain that "wild" varieties of these plants have always existed in abundance. Early farmers domesticated these plants and developed many uses for them. Today's researchers have continued to improve the work of early farmers so that today's grass and cereal plant crops are more hardy, more productive, and have a wider use.

Discoveries in Science ▪ Discoveries in Science ▪ Disco

AGRICULTURE AND TECHNOLOGY: FEEDING

8000 B.C.

8000 B.C. The **Natufians of Palestine** are considered the first farmers.

6000 B.C. Domesticated cattle, cultivated crops, massive pit silos, and granaries are found in parts of **ancient Greece**. Nitrogen-fixation and crop rotation maintain soil fertility. Irrigation is used in drier areas.

Egyptian farmers

1750 B.C. Tradition gives Chinese **Emperor Shen Nung** the title "inventor of agriculture." It is said that he carved a piece of wood into a plowshare, bent another piece to make a handle, and taught the world the advantages of plowing and weeding.

1300 B.C. Ancient Egyptians prosper from agricultural development and an integrated social system. Wider uses of plants advance the field of medicine. Farming techniques like plowing, raking, and manuring are practiced in Egypt. Animal breeding for special purposes is used.

Emperor Shen Nung

1896

1896 George Washington Carver, an African-American agricultural chemist, revolutionizes Southern agriculture. He discovers 325 different products that can be made from peanuts, 118 products from the sweet potato, 75 from the pecan, and hundreds of products from cotton and corn stalks.

1962 Naturalist author **Rachel Carson** warns of the severe environmental impact of DDT and the danger it poses to human food sources. DDT sprayed on crops destroys insect pests and endangers other animal life.

George Washington Carver

1975 Richard E. French, a Yakima Native American and forester, protects cultural foods in forest areas. He monitors fire management, fire and timber trespass activities, disease control, and insect control for the Bureau of Indian Affairs in Portland, Oregon.

Rachel Carson

*ience ▪ **Discoveries in Science** ▪ **Discoveries in Science***

THE WORLD'S POPULATION

Pliny the Elder

1450 Native Americans develop advanced systems of agriculture. Native American farmers, in various parts of the Americas, are producing crops such as corn, peanuts, peppers, cocoa beans, squash, rubber trees, tobacco, and tomatoes. Europeans first learn of these crops from Native Americans.

1793

Native American farmers

A.D. 43 Roman historian **Pliny the Elder** writes about revolutionary changes in Roman agriculture, including the introduction of complex plows fitted with wheeled forecarriages pulled by oxen or donkeys.

1793 Catherine Green develops plans for the cotton gin. Fearful of public scorn, she directs her employee **Eli Whitney** to build the machine.

Cotton gin

1991

Deforestation of rain forest by gold prospectors

1991 Ben Villalón is a Hispanic-American virologist and geneticist at Texas A&M University's Agricultural Experiment Station in Weslaco, Texas. He breeds disease- and insect-resistant varieties of chilies. He and his group have developed 15 new varieties, including the world's first mild jalapeño.

1989 Discoveries of **Dr. J.E. Henry** lead to the commercialization of protozoans as bacterial insecticides. Dr. Henry is a Chippewa Native American and a professor at Montana State University.

1980's Ecological consequences of **tropical deforestation** begin to receive widespread notice—land is stripped of nutrients; extensive erosion occurs; watersheds are destroyed.

1990 University of Georgia agronomist **Dr. David E. Radcliffe**, physically challenged, studies the physical behavior of soil when subjected to a variety of disturbances. He provides valuable agricultural information, determining which types of soil are more subject to runoff, or erosion, and their tendency towards crusting or compaction.

David Radcliffe

Chili peppers

2. Give examples of ways in which the potential for famine in the world can be decreased.
Answers will vary. The potential for famine can be decreased if farmers use plants and cultivation technologies that are compatible with their region. For example, planting cassava in the desert.

3. How do people act as selecting agents in the evolution of food plants?
People act as selecting agents when they propagate only those plants that have characteristics they find valuable, such as nutritional value, robustness, etc.

315

Unit 4

Diversity of Life

The greatest hallmark of life on Earth is its incredible diversity. Living things occupy every available nook and cranny, from the boiling waters of hot springs to the tangled growth of rain forests. Even the perpetual darkness at the bottom of the oceans, too deep for light to penetrate, is teeming with life, bizarre forms unlike any that live near the sea's surface. In this unit you will discover some of the richness of life's tapestry, exploring the many living things that are NOT animals like you.

Chapter 15 Classifying Living Things

Planning Guide

	Objectives/Themes	Classwork Resources	Homework Resources

15.1	**1.** Explain why scientists use scientific names instead of common names.
	
	2. Describe the scientific system of naming organisms.
	
	3. Explain why scientific names are in Latin.
	
	4. Recognize the role of Linnaeus in creating the modern system of naming organisms.
	
	Theme: Scale and Structure
		Teacher's Resource Binder
		Focus Activity 15
		   Grouping Things You Use Daily
		Lab Investigation 15.1
		   Classification
		
		Other Resources
		Transparency 60
			Text
			Section Review, p. 322
			
			Teacher's Resource Binder
			Directed Reading Worksheet
			   15.1
			
			Other Resources
			Audiocassette 15.1*

15.2	**1.** Describe the system scientists use to classify organisms.
	
	2. Describe how the classification of living things reflects their evolutionary history.
	
	3. Summarize the methods of classification used by taxonomists.
	
	4. Define the term "species."
	
	Themes: Evolution, Scale and Structure, Stability
		Teacher's Resource Binder
		Extension Worksheet
		   Classification Keys
		
		Other Resources
		Transparencies 61–64
			Text
			Section Review, p. 329
			
			Teacher's Resource Binder
			Directed Reading Worksheet
			   15.2
			
			Other Resources
			Audiocassette 15.2*

15.3	**1.** Describe the weaknesses of Linnaeus' two-kingdom classification system.
	
	2. List the five kingdoms of organisms.
	
	3. Identify two characteristics of members in each of the five kingdoms.
	
	Themes: Evolution, Scale and Structure
		Text
		Investigation
		   How Do You Use a
		   Classification Key?
		   pp. 336–337
			Text
			Section Review, p. 332
			
			Teacher's Resource Binder
			Directed Reading Worksheet
			   15.3
			Vocabulary Review Worksheet*
			Reaching Worksheet*
			   Classifying Living Things
			
			Other Resources
			Audiocassette 15.3*

*Reaching Options

Demonstrations
15.1: pp. 319, 321
15.2: pp. 324, 326, 327, 329
15.3: pp. 330, 331, 332

Assessment
Chapter Review pp. 334–335
Portfolio Assessment p. 317D
Chapter Test—Teacher's Resource Binder
Test Generator

Research Notes

Connection to Paleontology: Classifying the Burgess Shale Creatures

The Burgess Shale fossils of the Cambrian period pose several challenges to paleontologists. Merely describing and categorizing the variety of living creatures represented by the fossils is difficult.

Also challenging is the explanation for how and why so many living things appeared at this time, when they are not to be found in earlier fossil records.

The Burgess Shale fossils are difficult to describe and categorize since they do not fit tidily into all of the categories that taxonomists have used for living creatures. Initially, at least eight specific fossils were found that did not seem to fit into already defined phyla. Often, they were close matches, but had too many of a certain body part to belong to a certain phyla.

It is also difficult to determine from a fossil what role different body parts play. What may look like a head may actually be a tail, and so on.

However, over time, some of the Burgess Shale creatures have been classified into possible phyla by paleontologists. For example, one organism, Amiskwia, may actually be an example of a marine worm belonging to the little-known phyla Chaetognatha.

Closer looks at other organisms have revealed clues that help classify them. Another organism, Hallucigenia, which was thought to have two rows of spiny legs and long tentacles on top might originally have been viewed upside

down. The long tentacles might actually be the legs, and the two rows could be spiny armor on the organism's top.

Several theories have been proposed to account for the diversity of life found in these Cambrian fossils. One theory maintains that recent evolutionary advances opened up a series of ecological niches for organisms, which evolved rapidly as a result.

According to another theory, chance, rather than genetic fitness for the environment, was the factor driving organisms to evolve into such varied forms.

On the other hand, maybe the forms aren't so varied after all. Just because these organisms do not fit into categories made by taxonomists does not imply that they are extremely diverse. It could be that the categories are not yet accurate.

Understanding how these creatures relate to today's organisms helps scientists to understand the pathways of evolution and improve their systems of categorization for living things.

Investigation Notes

How Do You Use a Classification Key?

pages 336–337

Purpose: This activity gives students the opportunity to observe organisms and to use a biological key. Keys are invaluable tools to biologists, naturalists, and those who wish to identify specimens.

Prelab Preparation

1. Display a wide variety of arthropods. It is best to provide the specimens to students in sealed plastic vials. Students

should leave the vials closed during their observation. In this way, exposure and/or contact with preservatives is eliminated.

2. If desired, this investigation can be performed with two cooperative groups working together, dividing the specimens among each group. Have each team check the others' work.

3. Place a card next to each specimen with its name or an identification number for reference during comparisons.

Answers will be found on pages 336–337.

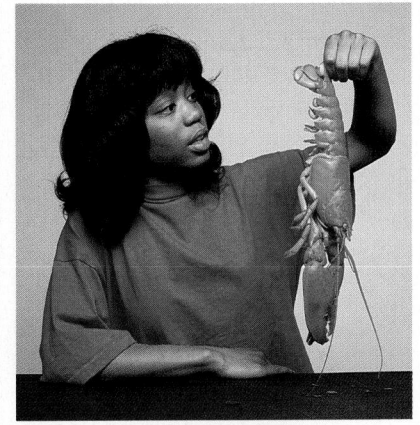

Meeting Individual Needs

Objectives

1. Students will demonstrate appropriate use of core vocabulary for the chapter (Vocabulary File).

2. Students will be able to explain the importance of using scientific names for animals and plants, and relate them to their common names in various languages (Multicultural Lesson Plan).

Vocabulary File

(Developing Vocabulary/ Limited English Proficiency)

If you are not already using the Vocabulary File, refer to Chapter 1 for its preparation. See the Chapter Highlights on page 333 for a list of suggested words.

Multicultural Lesson Plan

Exploring Differences in Languages

(Dealing with Ethnic Diversity)

Preparation: If a foreign exchange student is available, arrange for him/her to visit the biology class in order to help you and the students compare names of animals in different languages. As an alternative, bilingual students, a foreign-language teacher, or students taking foreign-language classes can assume this role. Be sure to have several foreign language-English dictionaries available, especially a Latin-English one. Field guides that list scientific names may also be helpful.

Teaching Strategies:

1. Students will make a chart using common names of living things. Students will list the common names of each living thing in various foreign languages, with the help of the exchange student, or other foreign language-speakers or references. Choose the languages that best fit your classroom situation and resources. (Recommended languages include Japanese, Latin, Spanish, Hindi, German, Chinese, and Swahili).

Be sure to go beyond the examples of languages given in Figure 15.3. Include such terms as man, woman, dog, cat, potato, pine tree, horse, yam, cow, etc. Then, have students determine the scientific name for the living thing named and add it to the chart.

2. Be sure to point out that in Romance languages (such as French, Spanish, Italian, and Portuguese) the common names are often very similar to the Latin name on which the scientific name is based. There are even a few instances in which the English language retains words based on the Latin. Examples include "canine" for dog, "pork" (from the Latin "porcine") for pig, and "herb" for annual plants used as spices and medicine (from the Latin "herba" for grass).

3. Discuss the naming of newly discovered species for which there are no Latin words. For example, the Romans did not know about microorganisms or dinosaurs. Be sure students are aware of the factors that can influence the scientific name given to an organism, such as the organism's appearance, its lifestyle, and the circumstances surrounding its discovery.

Assessment: Students will provide common and scientific names for "new" forms of life. Be sure students have access to a Latin-English dictionary.

One alternative is to have each student draw or sketch a fictitious organism they imagine might inhabit another planet. Then, students can exchange drawings, and create names for their newly discovered creatures.

Another possibility which may take less time is to challenge students to create common and scientific names for fictitious creatures in books. Children's books are a good source for such creatures. Standards likely to be found in most school library systems include Maurice Sendak's *Where the Wild Things Are,* and many of the works of Dr. Seuss.

Additional Strategies

Visual Strategies

Pages 319, 320, 325, 326, 327, 328, 329, and 331

Auditory Learners

Use *Biology: Visualizing Life* Audiocassettes for Sections 15.1, 15.2, and 15.3.

Meeting Individual Needs (cont.)

Cooperative Learning
Diversity and Classification

Timing: Use this activity to introduce Section 15.1.

Group Size: 2 students

Outcome: Students will be able to describe different schemes for classifying objects.

Individual Accountability: Each group member is responsible for creating a scheme for classifying objects and deducing the scheme that his or her partner used for classifying objects.

Positive Interdependence: Each group will discuss the benefits and disadvantages of their classification systems, for comparison with the rest of the class.

Give each pair of students a set of common objects. Possibilities could include a set of classroom objects, a set of kitchen tools, or a set of nuts and bolts and other hardware. One member of the pair should sort the objects into groups according to some scheme or rule. The other member of the pair should look at the way the objects were separated into groups and guess what rule was used.

Then, have the students switch roles, with the other member of the pair using a different rule to sort the objects. The first member of the pair should guess what rule is now being used. The students should then discuss possible benefits and disadvantages for their classification systems.

Close the activity with a class discussion about the problems encountered classifying objects. Be sure students understand that classification systems should be comprehensive, but not too complicated.

Portfolio Assessment

Students should select their best work and provide a self-reflective rationale for their selections. Students can make selections in the following areas.

1. *Content*	One concept map from the chapter (See page 812 for evaluation criteria.)
2. *Reading Comprehension*	One Directed Reading Worksheet from the Teacher's Resource Binder (Use the answer key to evaluate for accuracy.)
3. *Writing*	Using the Vee Form, summarize a magazine or newspaper article relating to taxonomy or bio-diversity. (See page 22T for evaluation criteria.)
	Or: Select a writing task or project from the Chapter Review.
4. *Performance Assessment*	One Vee form from a chapter investigation or lab manual investigation (See page 22T for evaluation criteria.)

Teacher makes selections in the following areas.

1. *Formal Assessment*	Chapter test (Test A, B, or the Test Generator) The teacher-scored test should be reviewed by the student. Incorrect responses should be corrected by the student before the test becomes part of the portfolio.
2. *Informal Assessment*	Use the Direct Observations Checklist, page 33T, during a laboratory or other cooperative learning experience.
3. *Performance Assessment*	Have students create a classification system for all of the students in the school, describing the criteria for dividing students into groups, and giving examples of specific students.

Concept Map Answer

The following is one possible answer to the Relating Concepts exercise on page 334.

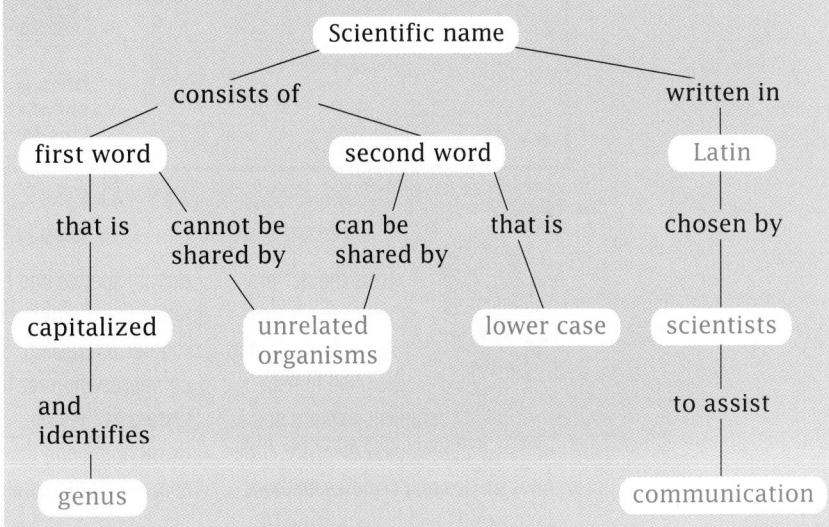

Chapter 15

Determining Prior Knowledge

- Write *Homo sapiens* on the board. Have students explain the significance of scientific names.
- Have students explain why a dog and cat belong to different species.
- Have the class assemble a wide assortment of objects in the classroom—chalk, a pencil, an eraser, tape, a ruler, a beaker, a flask, a microscope, overhead projector, etc. Have students develop some sort of classification system to place the objects into various categories. Tell students they must have a rationale for their system.

Chapter 15

Classifying Living Things

Earth is teeming with life. To make order out of the chaos, scientists use a five-kingdom classification system.

▪ Author's Rationale ▪

This chapter provides the framework for principles designed to help students explore and understand biological diversity. It defines species and discloses methods taxonomists and biologists use to classify species into hierarchical systems. The use of analogies keeps the chapter interesting. Students develop an appreciation for the five-kingdom system as they observe difficulties in classifying living things into two kingdoms—Plantae and Animalia.

DO THE WORDS "CANE," "CHIEN," AND "HUND" MEAN ANYTHING TO YOU? THESE ARE ITALIAN, FRENCH, AND GERMAN WORDS FOR "DOG." COULD YOU CONDUCT A CONVERSATION ABOUT DOGS WITH SOMEONE WHO SPOKE ONLY ONE OF THESE LANGUAGES? SCIENTISTS—REGARDLESS OF THEIR NATIVE LANGUAGES—CAN COMMUNICATE ABOUT DOGS BECAUSE THEY HAVE THE SAME NAME FOR THE DOG: *CANIS FAMILIARIS.*

15.1   *The Need for Naming*

Objectives

① Explain why scientists use scientific names instead of common names.

② Describe the scientific system of naming organisms.

③ Explain why scientific names are in Latin.

④ Recognize the role of Linnaeus in creating the modern system of naming organisms.

The Importance of Scientific Names

What do you and your classmates call the "bug" shown in **Figure 15.1**? You will probably have many common names for this animal, including sow bug, pill bug, wood louse, roly-poly, and potato bug. If some of your classmates are from other countries, you could collect an even longer list of names.

Yet if you asked a biologist to name this creature, you might receive only one answer: *Porcellio scaber.* Each kind of organism on Earth is assigned a unique two-word **scientific name**.

Porcellio scaber is the scientific name of the animal shown below. *Homo sapiens*, which means "wise man," is our species' scientific name. All biologists, regardless of their native languages, use scientific names when speaking or writing about organisms.

Most organisms also have common names. Why don't scientists use common names? Although adequate for everyday use, common names are too ambiguous for scientific communication. For one thing, as you have seen, an organism can have more than one common name. In addition, science is an international endeavor, and an organism rarely has the same name in different languages. Finally, one common name often refers to more than one kind of organism. The plant we know as corn in North America is called maize in Great Britain. To a resident of Britain, corn is the plant we call wheat. When a biologist writes a scientific paper on *Zea mays*, however, other scientists know the subject of the paper is the American "corn" plant. The use of scientific names enables all scientists to exchange information about an organism and to be certain that they are referring to the same organism.

Figure 15.1
Below are some of the common names for this animal (240X). Scientists have assigned it a single scientific name: *Porcellio scaber.*

Sow bug

Wood louse

Potato bug

Roly-poly

Pill bug

Visual Strategy

Table 15.1
Have students use the information in this figure to give reasons why these two oak trees belong to different species.

What's in a Scientific Name?

The first word in a scientific name describes the organism in a general way. The second word identifies the exact kind of living thing. We use a similar naming system in our everyday speech. For instance, your last name identifies your family, while your first name specifies exactly who you are.

The first word of a scientific name is the name of the **genus** (*JEE nuhs*) to which the organism belongs. (The plural of genus is "genera.") A genus is a group of organisms that share major characteristics. For example, all oak trees produce acorns. Therefore, all oak trees are assigned to the genus *Quercus*, which means "oak" in Latin.

There are dozens of different kinds of oak trees. One kind is tall, while another is low and spreading. Leaves and acorns of different oak trees can vary in size and shape, as shown in **Table 15.1**. The second word in a scientific name identifies one particular kind of organism within the genus. For example, *Quercus rubra* is the red oak. *Quercus phellos* is the willow oak. Scientists call each different kind of organism a **species** (*SPEE sheez*). (The plural of species is "species.") The correct name for an organism must include *both* parts of its scientific name. The red oak is properly called *Quercus rubra*, not just *rubra*.

Table 15.1 Comparison of Red Oak and Willow Oak

	Red oak	Willow oak
Genus name	*Quercus*	*Quercus*
Scientific name	*Quercus rubra*	*Quercus phellos*
Traits	Acorns about 25 mm (1 in.) long	Acorns about 15 mm (0.5 in.) long
	Common in open Northeastern forests; tolerant of city soot and cold temperatures	Popular shade tree found in the South; grows well in rich, moist soil
	Lobed leaves	Unlobed, narrow leaves

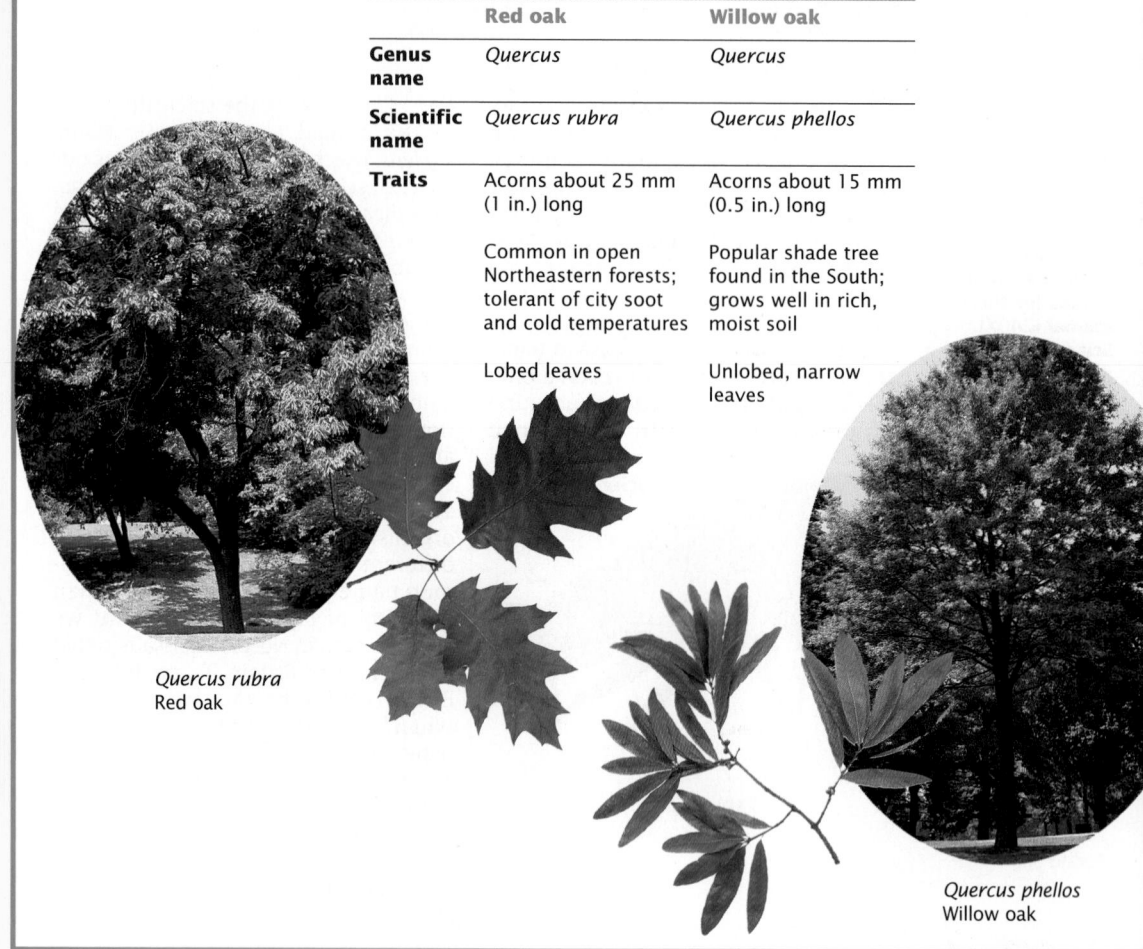

Quercus rubra
Red oak

Quercus phellos
Willow oak

Figure 15.2

a   The second word of a scientific name can come from various sources. The green anole lizard *Anolis carolinensis* and the chickadee *Parus carolinensis* are both found in North Carolina and South Carolina.

b   *Tyrannosaurus rex*, which means "tyrant-lizard king," was named for its enormous teeth and tremendous size. This dinosaur measured about 15 m (49 ft.) in length.

c   The frog *Rhinoderma darwinii* was named to honor Charles Darwin.

Scientific names must conform to a set of rules

The name given to a newly discovered species must conform to rigorous rules formulated by an international commission of scientists. All scientific names must be made of Latin words or of terms constructed according to the rules of Latin grammar. Two different organisms cannot be assigned the same scientific name. Since all members of a genus will share their genus name, the second word in the name of each member of that genus must be different. Only one member of the genus *Homo* can be given the name *sapiens*. Organisms in different genera cannot have the same genus name but can share the second word of their scientific names.

For example, the green anole lizard *Anolis carolinensis* and the chickadee *Parus carolinensis*, shown in **Figure 15.2a**, share the name *carolinensis* because they both occur in North Carolina and South Carolina.

When choosing a name for a species, biologists often pick a name that describes the appearance or lifestyle of an organism. One of the most vivid names belongs to the fierce-looking dinosaur *Tyrannosaurus rex*, shown in **Figure 15.2b**. *Tyrannosaurus rex* means "tyrant-lizard king," a fitting name for a carnivore with teeth 15 cm (6 in.) long. Sometimes scientific names are a tribute to the discoverer of a species or to an admired colleague or teacher of the discoverer. The frog *Rhinoderma darwinii* shown in **Figure 15.2c**, the lizard *Liolaemus darwinii*, and the bird *Rhea darwinii* are three of the many species named for Charles Darwin.

Demonstration 2

What's in a Name?

Place the following scientific names on the board. Have students suggest how these organisms received their species names: *Carnegiea gigantea* (a giant saguaro cactus); *Nymphaea ordorata* (a fragrant water lily); *Canis familiaris* (a domestic dog); *Viola tricolor* (a pansy with three-colored flowers); and *Peromyscus californicus* (a mouse common in California). Emphasize that the rules require both genus and species names be included when referring to an organism. The species name alone is not sufficient. For example, *Drosophila melanogaster* is a fruit fly, whereas *Thamnophis melanogaster* is a garter snake.

Mathematics Connection

$(a + b)^2$

In biology, the two-part naming system is called binomial nomenclature. In mathematics, $(a + b)^2$ represents a binomial expansion. Have students explain the meaning of both the biological and mathematical terms.

Phase 3

ASSESSMENT OPTIONS

Closure Strategy

Scientific Names

Have students identify what is incorrect about the following "scientific names": spotted leopard; *Tabanus b.*; *Acer Rubrum*; and *buono animale*.

Section Review.

Assign the *Section Review*.

Reteaching

Assign students to cooperative groups of four and have each group develop a concept map for this section.

Why are scientific names in Latin?

Why are scientific names written in a language that is no longer spoken? Why aren't they in English, or Russian, or Chinese? In the Middle Ages, when scientists began to name organisms, Latin was used in academic circles. Hence, scientists and other scholars found it easier to communicate with each other in Latin. They wrote books in Latin, wrote letters to each other in Latin, spoke Latin when they met, and named organisms in Latin.

Although scientists no longer communicate with each other in Latin, it is easier and more logical to retain the Latin names for living things than to rename all 1.5 million living organisms in a more modern language. **Figure 15.3** shows how useful a Latin name can be when scientists from around the world communicate with each other.

Figure 15.3
If you asked scientists from around the world what they might call the animal in Figure 15.1, they would all agree on one name: *Porcellio scaber*.

"תחבית (*tah hah VEET*) or *Porcellio scaber*"
—Yehoshua Anikster, Botanist
Israel

"Cloporte (*klo PORT*) or *Porcellio scaber*"
—Paul Melançon, Chemist and biochemist
Canada

"Mangy sowbug or *Porcellio scaber*"
—Maria Alma Solis, Entomologist
United States

"Cucaracha (*koo kah RAH chah*) or *Porcellio scaber*"
—Ernest H. Williams, Jr., Marine biologist
—Lucy Bunkley-Williams, Microbiologist
Puerto Rico

"КОЛОРАДСКИЙ ЖУК (*koh loh RAD skee JOOK*) or *Porcellio scaber*"
—Andrei Lapenis, Earth systems scientist
Russia

Linnaeus devised the two-name system

The modern system of naming organisms was the brainchild of Swedish botanist Carl von Linné (1707–1778). As was the fashion in his time, Von Linné gave himself a Latin name: Carolus Linnaeus. In Linnaeus' day, organisms were given very long Latin names (sometimes more than 15 words), which were often changed according to the whims of particular authors. Linnaeus assigned a standard, two-word Latin name to each organism known in his time.

Writing a scientific name is simple

When you write a scientific name, always capitalize the genus name. Begin the second word with a lowercase letter. Both parts of a scientific name are underlined or written in italics. The scientific name for humans can be written either Homo sapiens or *Homo sapiens*. After the first use of the full scientific name, the genus name can be abbreviated as a single letter if the meaning is clear. For example, since *Homo sapiens* has just been mentioned, *H. sapiens* is acceptable.

Section Review

1 List the problems associated with the use of common names to identify living things.

2 Why is it necessary to use both words of a scientific name to correctly identify an organism?

3 Explain the advantages of using Latin for scientific names.

4 Evaluate the accuracy of this statement: "Linnaeus was the first scientist to give species Latin names."

■ *Section Review Answers* ■

1. Different organisms can have the same common name. An organism may have many common names.
2. Different organisms can share a genus name—for example *Homo sapiens* and *Homo erectus*—or a species name—for example *Drosophila melanogaster* and *Thamnophis melanogaster*. However, each organism will have a unique combination of genus and species names.
3. We still use Latin names because changing the Latin names of the 1.5 million living organisms into a more modern language would be difficult and illogical.
4. Linnaeus was the first to assign a standard two-word Latin name to every organism, but previous scientists had assigned Latin names to organisms.

BECAUSE THE PRODUCTS IN A STORE ARE ORGANIZED, YOU CAN GO STRAIGHT TO THE AISLE WHERE SCHOOL AND OFFICE SUPPLIES ARE KEPT TO FIND A PEN, INSTEAD OF SEARCHING THE ENTIRE STORE. LIVING THINGS HAVE BEEN SIMILARLY ORGANIZED BY SCIENTISTS, BUT NOT ONLY FOR CONVENIENCE. LIVING THINGS ARE ORGANIZED INTO GROUPS TO REFLECT EVOLUTIONARY RELATIONSHIPS.

15.2 *Classification: Organizing Life*

Objectives

❶ Describe the system scientists use to classify organisms.

❷ Describe how the classification of living things reflects their evolutionary history.

❸ Summarize the methods of classification used by taxonomists.

❹ Define the term "species."

Classification of Living Things

Figure 15.4
This page from a manual called a herbal was published in Turkey in the tenth century. The Greek text classifies the plant illustrated below according to its medicinal uses.

Humans have been classifying organisms for thousands of years. The Greek philosopher Aristotle (384–322 B.C.) grouped animals according to their physical similarities. Aristotle's classification included some of the groups recognized today, such as the mammals. In the Middle Ages, herbalists used plants to treat disease and needed to know how to identify which plants were poisonous and which had healing powers. Herbalists produced manuals of plant types, known as "herbals," like the one shown in **Figure 15.4**. In herbals, plants were organized by their medicinal uses.

Today biologists classify organisms not by their usefulness but by their physical, chemical, and behavioral similarities. These similarities reveal evolutionary relationships. The science of classifying living things is called **taxonomy**. Taxonomists are scientists who practice taxonomy.

Like the system for naming organisms, the system of classification was derived by Linnaeus. Linnaeus wrote a huge encyclopedia of life, the *Systema Naturae* ("system of nature"). This work described all organisms then known. It classified living things into a hierarchy in which individuals are assigned to groups, groups are collected into larger groups, and these larger groups are part of still larger groups. A similar system is used in the U.S. Army. Each soldier belongs to a squad containing about nine soldiers. Four squads are organized into a platoon. Each platoon belongs to a company of about 150 people, and so on.

Lesson Plan 15.2

Phase 1
PREPARATION

Key Concepts
- Taxonomy is the science of classifying organisms.
- Organisms are classified in a hierarchical system that includes kingdom, phylum, class, order, family, genus, and species.
- Classification of organisms reflects their evolutionary relationships.
- In classifying organisms, taxonomists use a variety of information including evolutionary history, structural features, behavioral patterns, methods of reproduction, life cycles, patterns of development, and similarities in DNA.
- A species is a group of organisms that are able to interbreed with each other to produce fertile offspring.

Reading Strategy

Again emphasize that taking notes is important when reading. A student's notes will be useful when reviewing the material. Have students suggest some timesaving steps for taking notes. Tell them one way is to use symbols and abbreviations whenever they can. Have students develop symbols and abbreviations for words in this section.

Phase 2

TEACHING STRATEGIES

Social Studies Connection

Classification and Geography

List the seven different classification groups on the board. Next to *kingdom*, write *continent*; next to *phylum*, write *country*. Have students provide the appropriate geographic descriptions (state, county, city, etc.) for the remaining classification groups. Students should see that characteristics become more specific in going from kingdom to species.

Theme Answer

Evolution

In a biological heirarchy of classification, organisms are assigned to a group because they share distinctive characteristics with other members of that group. Organisms share distinctive characteristics because they are descended from a common ancestor. The more similarities two organisms share, the more recently they shared a common ancestor.

Demonstration 1

Classifying Buttons

Assign students to cooperative work groups. Give each group 12 different buttons to arrange in a hierarchical system. Tell students that the 12 buttons belong to one large group that can be divided into two smaller ones—one with holes, the other with no holes. Further divide each group on the basis of color, size, shape, etc. Explain that the more groups two buttons share, the more features they have in common.

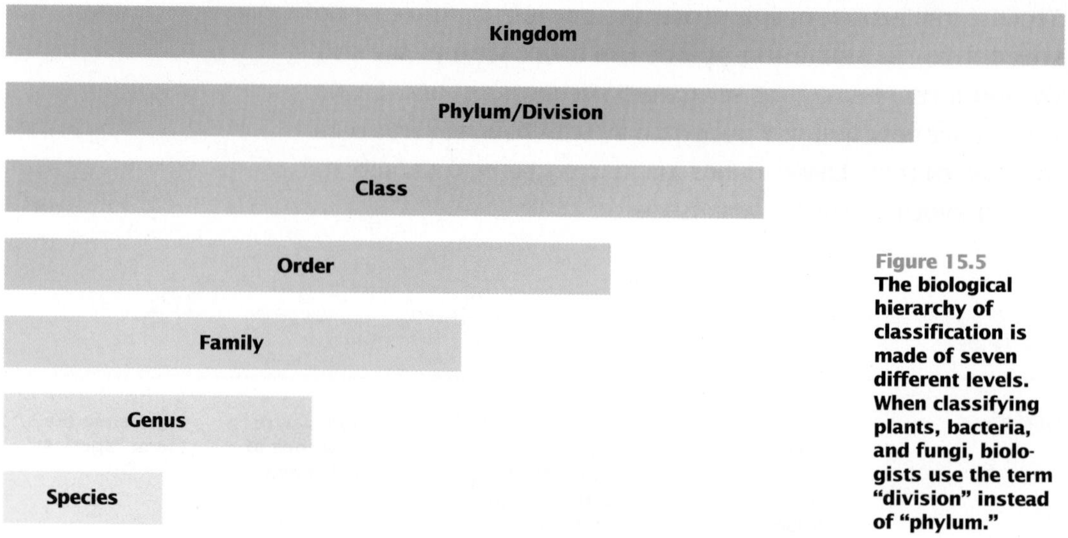

Kingdom

Phylum/Division

Class

Order

Family

Genus

Species

Figure 15.5
The biological hierarchy of classification is made of seven different levels. When classifying plants, bacteria, and fungi, biologists use the term "division" instead of "phylum."

Organisms are classified by similarity

Evolution

How is the hierarchy of classification based upon evolutionary relationships among organisms?

In the army, soldiers are assigned to a group without regard to their appearance; there are no companies of only tall soldiers or of only blond soldiers, for instance. In a biological classification, however, organisms are assigned to a group because they share distinctive characteristics with the other members of that group. The biological hierarchy of classification has seven different levels: kingdom, phylum, class, order, family, genus, and species, as shown in **Figure 15.5**.

The smallest group in biological classification is the species. Similar species are collected into a genus. Similar genera are united into a **family**. Similar families are combined into an **order**. Similar orders are collected into a **class**. Similar classes are united into a **phylum** (*FEYE luhm*). Finally, similar phyla are collected into a **kingdom**. The term **division** is substituted for phylum in classifications of plants, bacteria, and fungi. **Table 15.2**, at right, illustrates the classification of our species, *Homo sapiens*.

The more classification categories two species share, the more traits they have in common. For instance, a house cat and a guppy belong to only two of the same categories: the kingdom Animalia and the phylum Chordata. The guppy and the cat both have bony skeletons, notochords, and nerve cords. But only the cat breathes with lungs, is covered with fur, walks on four legs, and nurses its young with milk. The guppy breathes with gills, has numerous fins and scales, and does not nurse its young. The house cat, *Felis cattus*, and the mountain lion, *Felis concolor*, belong to six of the same classification categories: kingdom Animalia, phylum Chordata, class Mammalia, order Carnivora, family Felidae, and genus *Felis*.

If more specific groups are needed, the seven classification categories can be subdivided into smaller units. For example, the phylum Chordata includes all animals that develop notochords. The great majority of the members of the Chordata also have backbones, but a few members do not. To reflect this difference, animals with backbones are placed into their own subphylum: Vertebrata. Two other subphyla contain the chordates that do not have vertebrae.

■ *Matter of Fact* ■

Hierarchical Groups

Before Linnaeus, scientists assigned organisms into three hierarchical groups: species, genus, and kingdom. Linnaeus and subsequent taxonomists added the other categories.

Table 15.2   The Classification of Modern Humans

	Homo sapiens	Homo erectus	Australo-pithecus	Gorilla	Elephant	Fish	Snake Sea star	Earthworm Snail	

Kingdom
Animalia

Chordates, sea stars, earthworms, snails, jellyfish, sponges, clams, and insects, are members of the kingdom Animalia.

Phylum
Chordata

Mammals, fishes, reptiles, birds, and amphibians are chordates, members of the phylum Chordata.

Class
Mammalia

Primates and elephants, along with cats, dogs, horses, kangaroos, whales, bats, seals, dolphins, and many others, belong to the class Mammalia.

Order
Primates

Members of the family Hominidae, along with pro-simians, monkeys, and apes such as the gorilla, make up the order Primates.

Family
Hominidae

The family Hominidae includes the genus *Homo* and the extinct genus *Australopithecus.*

Genus
Homo

Homo sapiens belongs to the genus *Homo,* along with the extinct species *Homo habilis* and *Homo erectus* (shown here).

Species
Homo sapiens

Modern humans belong to the species *Homo sapiens.*

Visual Strategy

Table 15.2
Use this table to show students that *kingdom* is the broadest classification group, while *species* is the narrowest. Have students identify the organisms that drop out at each successively lower classification level.

■ Cultural Perspective ■

Different Cultures, Same Taxonomy

Are species natural units that are recognized by different cultures? Or are they artificial distinctions imposed on nature by Western scientists? It appears that the former is true. In 1928 Ernst Mayr, then a young ornithologist, traveled to New Guinea to collect bird specimens. Mayr found that the local Arfak people recognized 136 kinds of birds, just one less than he did. Moreover, Mayr and the Arfak recognized the same species except for two very similar species that the Arfak did not distinguish.

Connection: Chapter 13

Coevolution

Distinguish between coevolution and convergent evolution. Coevolution occurs when two or more species evolve in different ways in response to one another. Convergent evolution occurs when two or more species evolve in a similar way in response to their physical environment.

Visual Strategy

Figure 15.6

Have students locate analogous structures that have evolved in the shark and dolphin in response to living in water. Have them describe the differences between the two to show that they outnumber the similarities.

Demonstration 2

Convergent Evolution

Show the class a picture of a seal and a penguin. Have students identify analogous structures (streamlined, fishlike bodies; webbed appendages; layers of insulating fat). Emphasize that their differences far outweigh their similarities; as a result, the seal and penguin belong to different classes. The seal is classified as a mammal, the penguin as a bird. Have students explain what phylum and kingdom they share.

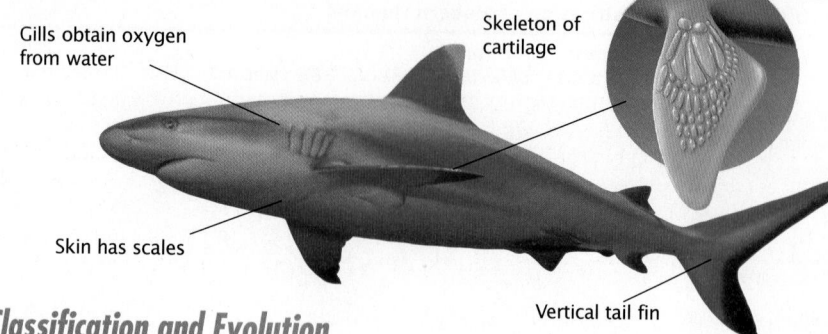

Figure 15.6
Although the shark and the dolphin look similar, they belong to different classes. This is an example of convergent evolution. Study the differences in the diagrams at the top and bottom of this page.

a **This gray reef shark belongs to the class Chondrichthyes (*kahn DRIHK thees*), which means "cartilage fish."**

b **The Atlantic spotted dolphins below belong to the class Mammalia.**

Gills obtain oxygen from water

Skeleton of cartilage

Skin has scales

Vertical tail fin

Classification and Evolution

Linnaeus' classification is based on the fact that different degrees of similarity exist among organisms. For instance, mountain lions more closely resemble house cats than do guppies. For Darwin, classification provided strong evidence supporting evolution. Organisms are similar because they descended from a common ancestor. The more similarities two organisms share, the more recently they shared a common ancestor. For instance, the ancestor of the house cat diverged from that of the mountain lion just a few million years ago. In contrast, the ancestor of guppies and the far-distant ancestor of cats diverged about 390 million years ago. Thus, the more classification categories two organisms share, the more closely related they are.

Similarity does not guarantee close relationship

Compare the two ocean-dwelling animals in **Figure 15.6**. Both have streamlined bodies, paddlelike fins, and flattened tails. Would you say these organisms are closely related? Many people classified both as "fish." Although both belong to the subphylum Vertebrata, they are now placed in different classes.

The shark and dolphin are an example of **convergent evolution**. In convergent evolution, organisms evolve similar features independently, often because they live in similar habitats. Similar features that evolved through convergent evolution are known as **analogous characters**. Homologous characters, which you read about in Chapter 9, are similar because of common ancestry. Analogous characters are similar because of similar selection of two or more species.

Convergent evolution creates problems for taxonomists because it means that similar appearance does not guarantee common ancestry. Taxonomists, therefore, view a classification as a hypothesis of relationships between organisms. Like any hypothesis, a classification can be tested against the available evidence and disproved or supported. Because the number of differences between sharks and dolphins far exceeds the number of similarities, it is easy to reject the hypothesis that these animals are close relatives.

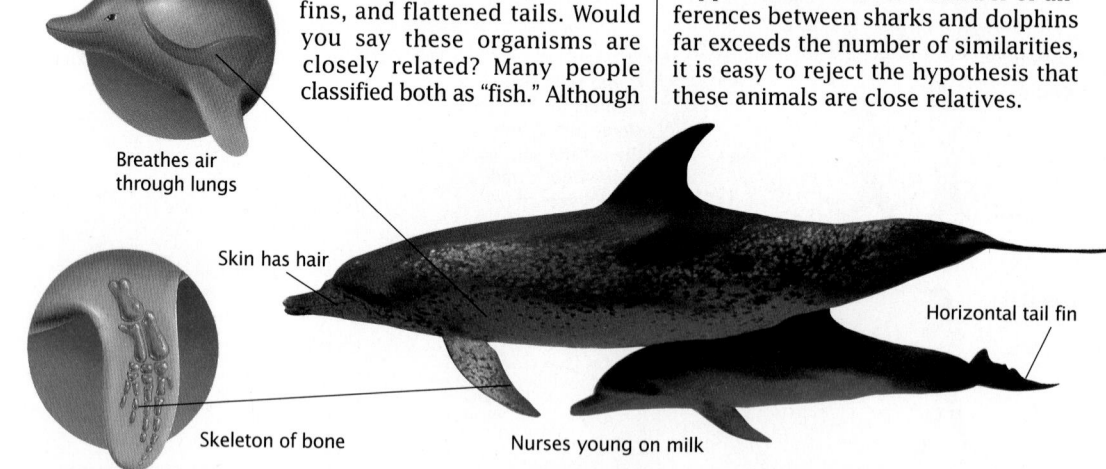

Breathes air through lungs

Skin has hair

Skeleton of bone

Nurses young on milk

Horizontal tail fin

Methods of Taxonomy

The example of the shark and dolphin illustrates the difficulty in determining which similarities will be useful when classifying an organism. There are two alternative methods of choosing which similarities are useful. The first method is cladistics (*kluh DIHS tihks*), from the Greek word *klados*, meaning "branch." The scientists who use cladistics seek to determine the order in which evolutionary lines diverged, or branched. To do so, they consider only a restricted set of characters of the organisms they want to classify. Organisms are assigned to a group because they share unique characters not found in any other organisms. These unique characters are called **derived characters**. For example, all species of mammals share the derived characters of hair and the ability to produce milk. Using patterns of shared derived characters, scientists construct branching diagrams called **cladograms,** which show the evolutionary relationships among groups of organisms. A cladogram of the major groups of plants is shown in **Figure 15.7**.

The second method for classifying organisms is phenetics (*fuh NEHT ihks*). Scientists using phenetics consider as many characters of organisms as possible and classify organisms into groups based on overall degree of similarity. Phenetics does not attempt to reconstruct the relationships among organisms, just to produce groups that can be named.

Both cladistics and phenetics have drawbacks. Phenetic classifications do not always reveal the relationships among organisms. And while cladistics is able to reconstruct the sequence of evolutionary divergence, it does not reveal the amount of difference between groups.

The taxonomy used in this book and by many biologists reflects both overall similarity and the sequence of evolutionary divergence—a compromise between phenetics and cladistics.

Energy and Life

How could methods of obtaining energy be used as derived characters in a cladogram of the five kingdoms?

Figure 15.7
This cladogram shows the evolutionary relationships among the major divisions of plants. A derived character distinctive to each group is marked on the cladogram. These derived characters indicate the divergence of one group from the common ancestor shared with the others.

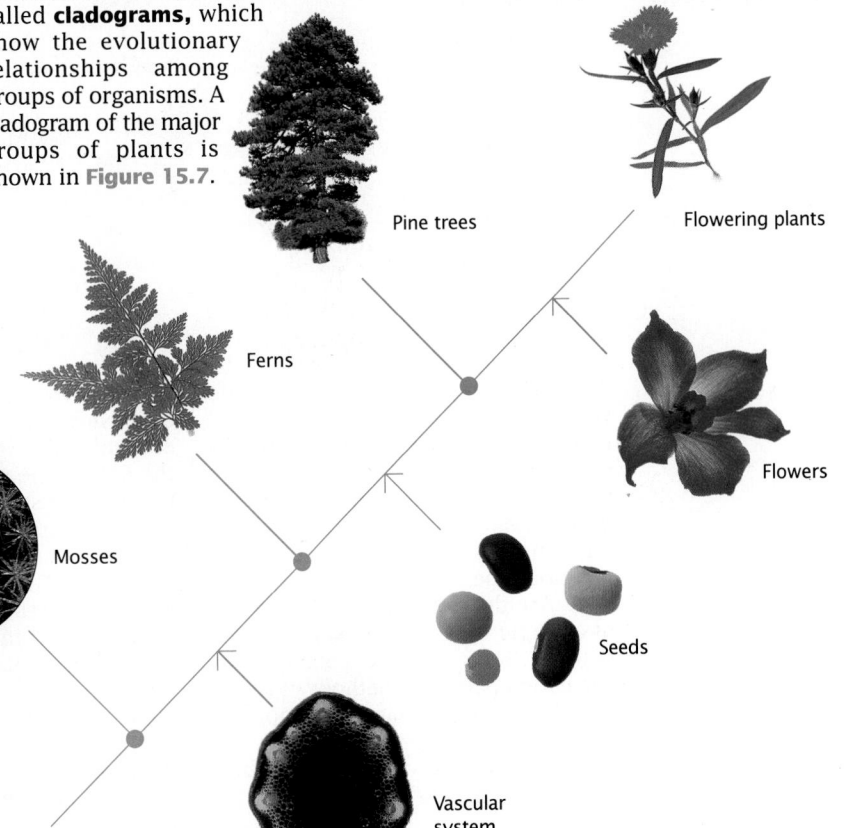

Pine trees

Flowering plants

Ferns

Flowers

Mosses

Seeds

Vascular system

Demonstration 3
Cladistics

Show the class photographs of a bird, a crocodile, a turtle, a lizard, and a snake. Point out that, in a traditional classification, the last four are grouped in the class *Reptilia*, while the bird belongs to the class *Aves*. Have students discuss what observable characteristics the four have in common to belong to the class *Reptilia*. Cladists focus on the order in which evolutionary lines diverged, or branched, as revealed by derived characters. Consequently, a cladist would group crocodiles with birds since they share several derived characters. Traditional taxonomists argue that putting crocodiles and birds in the same group, with their obvious physical differences, makes no sense.

Visual Strategy

Figure 15.7

Have students examine this cladogram to identify what derived character caused the fern, pine tree, and flowering plant each to branch at a particular point.

Theme Answer
Energy and Life

Heterotrophy and autotrophy are not confined to a single kingdom—Protista and Monera contain both heterotrophs and autotrophs. Thus, these features cannot be considered derived characters with which to make a cladogram of the five kingdoms.

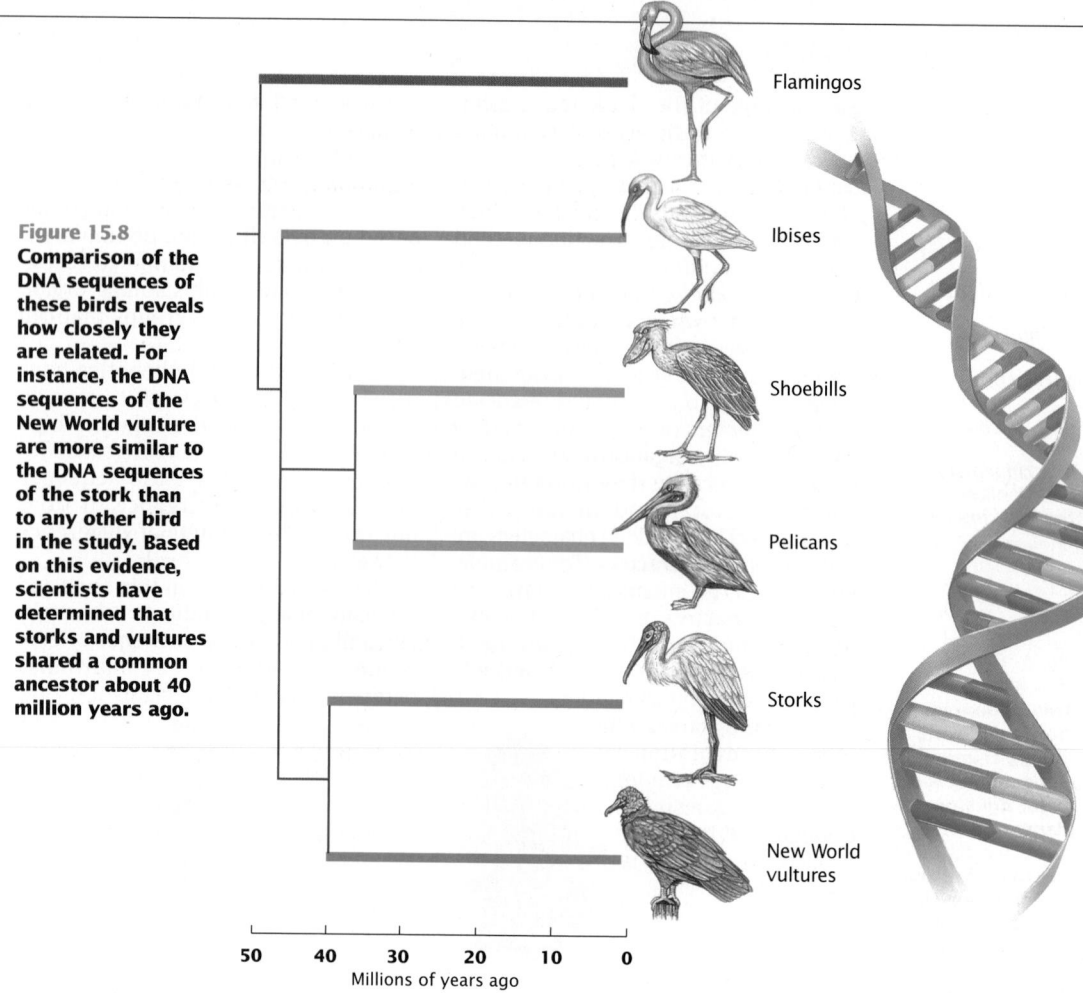

Figure 15.8
Comparison of the DNA sequences of these birds reveals how closely they are related. For instance, the DNA sequences of the New World vulture are more similar to the DNA sequences of the stork than to any other bird in the study. Based on this evidence, scientists have determined that storks and vultures shared a common ancestor about 40 million years ago.

Flamingos

Ibises

Shoebills

Pelicans

Storks

New World vultures

50   40   30   20   10   0
Millions of years ago

Taxonomy and Technology

What characteristics of organisms do you think might be useful for determining evolutionary relationships? As you might expect, biologists have traditionally compared the appearances of organisms in order to discover the relationships among them. For example, does the animal have four or six legs? Are the flowers of a plant made of a single tube or of many separate petals? Biologists also consider the behavioral patterns, methods of reproduction, life cycles, and development from fertilization to adulthood.

Technological advances have enabled biologists to study the genes that produce the traits used to classify organisms. Taxonomists use techniques of molecular biology to compare the DNA nucleotide sequences of different organisms. Comparisons of DNA sequences are especially important for the taxonomist because mutations—changes in DNA—are random events. As time passes, more mutations tend to accumulate in the DNA of a particular species. Thus, DNA acts as a "molecular clock." As shown in **Figure 15.8**, the more similar the DNA sequences of two species, the more recently their common ancestor must have lived, and the more closely they are related.

What Is a Species?

What does it mean to say that house cats belong to one species while mountain lions belong to another? What is a species? In one sense, a species is just a level in the classification system to which scientists assign very similar organisms. But in a more profound sense, a species is the basic unit of evolution. Over time, species change and give rise to new species in a process known as speciation. Indirectly, speciation gives rise to new genera, new families—all of the so-called higher classification categories. A new genus, for example, is not produced by the transformation of all members of an existing genus. A new genus is formed when one species in the "old" genus accumulates enough changes to be considered not only a new species but a member of a new genus.

Biologists have traditionally defined a species as a group of organisms that are able to interbreed with each other to produce fertile offspring and that usually do not reproduce with members of any other groups. This definition works well for most animals and many plants. For example, the horse and the zebra belong to different species. Although they can mate, the resulting offspring, the "zebroid" shown in **Figure 15.9**, is sterile.

Reproductive barriers between sexually reproducing species are not always perfect. **Hybrids** are offspring that result from interbreeding by individuals of different species. Coyotes, dogs, and wolves are all separate species in the genus *Canis*. Interbreeding between dogs and coyotes, and between dogs and wolves produces fertile hybrids.

A species is a unique kind of organism

Despite these occasional complications, the classification of organisms into species has worked very well since Linnaeus. The word "species" simply means "kind" in Latin. Therefore, a species is basically a unique kind of organism. Members of a species share at least one inherited characteristic not found in other similar organisms. The characteristic that sets a species apart might be the shape of a leaf, beak, flower, or tooth. It could be an unusual mating signal or a unique DNA nucleotide sequence. In sexually reproducing species, this distinctive characteristic is maintained from generation to generation because members of different species do not usually interbreed.

Figure 15.9
The zebra and horse belong to the same genus (*Equus*) but are members of different species. When they mate, they produce sterile offspring, known as a "zebroid," like the one below.

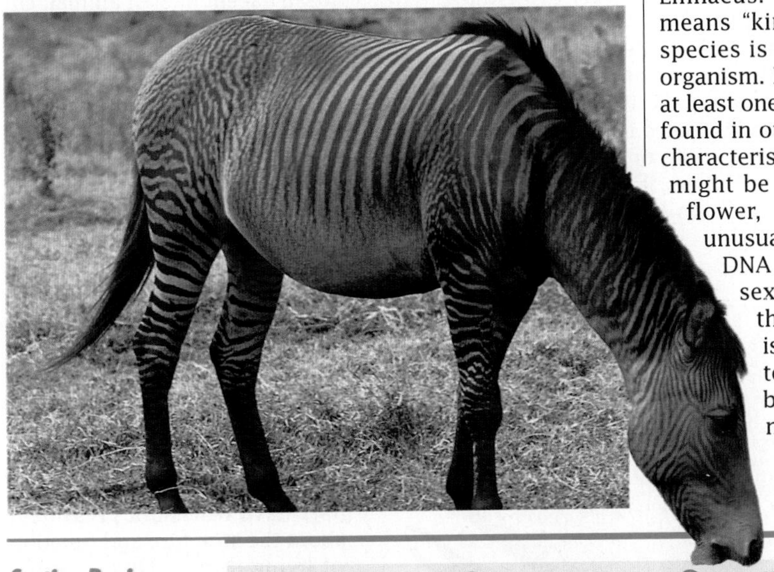

Section Review

1. **Explain why two species in the same genus must also belong to the same family.**
2. **Explain how convergent evolution can make classification difficult.**
3. **Describe how cladistics classifies organisms.**
4. **Why aren't the breeds of domestic dogs considered separate species?**

Section Review Answers

1. Because the classification system is hierarchical, two organisms that belong to a classification category must also belong to the same categories higher in the hierarchy.
2. Through convergent evolution organisms can have similar characteristics without being closely related. Such similarities complicate the task of defining groups of related organisms.
3. Cladistics identifies related organisms by shared, uniquely-evolved, or derived, characters. A cladistic classification is based on the derived evolutionary relationships.
4. Although they differ anatomically, the breeds of dogs can interbreed.

Visual Strategy

Figure 15.9
Tell students that when a horse and donkey mate and produce offspring, the result is a sterile mule. Have students compare a mule with the *zebroid* shown in this figure. Ask them if they would classify the horse and donkey as members of the same species, and have them explain the reason for their answer.

Demonstration 5

What Is a Species?
Show students photographs of plants, bacteria, and *Amoebas*. Point out that many plants can reproduce asexually, and that others can form fertile hybrids with other plants; bacteria reproduce asexually; and amoebae reproduce by binary fission. Have students discuss how these organisms differ from the text definition of a species. Emphasize that what appears to be a simple problem in biology (definition of a species) has no simple solution.

Phase 3 ASSESSMENT OPTIONS

Closure Strategy

Classification
Have students provide examples from everyday situations where some sort of classification system makes life easier. In each case, have them explain the rationale for the classification system.

Section Review

Assign the *Section Review*.

Reteaching

Assign students to cooperative work groups of four. Take the class to the school parking lot to make a list of cars, including make, model, and color. Have each group develop a classification system for all the cars.

Lesson Plan 15.3

Phase 1

PREPARATION

Key Concept
• All organisms have been grouped into five kingdoms: Monera, Protista, Fungi, Plantae, and Animalia.

Reading Strategy

Have students prepare a table. Across the top, instruct them to place the names of the five kingdoms. Along the left-hand side, have them write down *Distinguishing features, Examples,* and *Additional comments.* Have the students complete their tables as they read this section.

Phase 2

TEACHING STRATEGIES

Demonstration 1

Animal or Plant?

Use a microprojector to show the students living *Euglena.* Have them decide whether they would place it into the kingdom Plantae (since it has green chlorophyll) or kingdom Animalia (since it moves).

THE ABILITY TO MOVE IS PROBABLY ONE TRAIT THAT YOU ASSOCIATE WITH ANIMALS, WHILE THE ABILITY TO PHOTOSYNTHESIZE IS CHARACTERISTIC OF PLANTS. *EUGLENA* IS A SINGLE-CELLED ORGANISM THAT IS ABLE TO MOVE AND TO PERFORM PHOTOSYNTHESIS. IS *EUGLENA* A PLANT OR AN ANIMAL? THERE ARE A NUMBER OF ORGANISMS THAT CANNOT BE CATEGORIZED AS ANIMALS OR PLANTS.

15.3 Five Kingdoms

Objectives

❶ **Describe the weaknesses of Linnaeus' two-kingdom classification system.**

❷ **List the five kingdoms of organisms.**

❸ **Identify two characteristics of members in each of the five kingdoms.**

Five Kingdom System

Figure 15.10
Like plants, sea anemones cannot move from place to place. However, they are classified as animals. They are heterotrophs, using tentacles to paralyze and capture small fishes.

Following Linnaeus' lead, biologists of his time classified every living thing into either kingdom Plantae or kingdom Animalia. This seemed logical because most familiar organisms are either plants or animals. If it is green, has leaves, and is rooted in soil, it is a plant. If it is furry, slimy, or scaly, and it runs, crawls, or slithers, it is an animal. However, numerous living things do not quite fit either description.

For example, where would a mushroom fit? It doesn't seem like an animal. On the other hand, is it a plant if it doesn't have leaves and isn't green?

If you had to fit a mushroom into one of the two kingdoms—Plantae or Animalia—you would most likely place it in kingdom Plantae. That is what Linnaeus did. After all, a mushroom does look a little like a plant. It grows from the soil and doesn't move from one place to another.

Does this mean that any organism that doesn't move from place to place is a plant? Sea anemones, such as the one in **Figure 15.10**, are organisms that are firmly attached to one spot and look like flowers. They eat other organisms, however, and are animals. Since Linnaeus' time, biologists have learned a great deal about the structure and function of living things. This information has enabled them to make increasingly precise distinctions among the major groups of organisms. Most biologists now use a five kingdom system of classification. The five kingdoms are Monera, Protista, Fungi, Plantae, and Animalia.

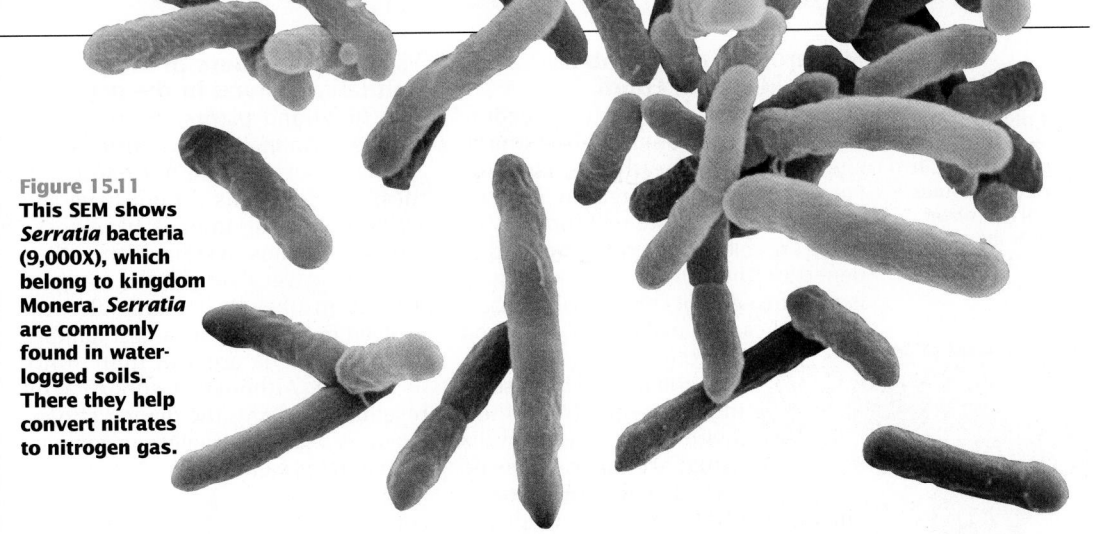

Figure 15.11
This SEM shows *Serratia* bacteria (9,000X), which belong to kingdom Monera. *Serratia* are commonly found in water-logged soils. There they help convert nitrates to nitrogen gas.

Scale and Structure

Which traits of bacteria are shared by all living things?

Figure 15.12
This colony of *Volvox* (50X) belongs to the kingdom Protista. *Volvox* are made of hollow balls, each containing 500 to 500,000 photosynthetic cells. They are found in fresh-water habitats.

All prokaryotes are in kingdom Monera

Kingdom Monera contains all prokaryotes, which are also called bacteria. Prokaryotes have adapted to almost every environment on Earth, including environments with extremes of acidity, temperature, and salt content. In addition, bacteria have evolved more ways to obtain food than have all eukaryotic organisms combined. Monera represent the most ancient kingdom. The first cells, which appeared about 3.5 billion years ago, were bacteria. Bacteria gave rise to the first eukaryotes, members of kingdom Protista. **Figure 15.11** illustrates some typical bacteria.

During the last decade, studies of DNA have shown that there are two fundamentally different kinds of bacteria, the archaebacteria (*ahr kee bak TIHR ee uh*) and the eubacteria (*yoo bak TIHR ee uh*). Many biologists think the differences between these two groups of bacteria are significant enough to justify placing the archaebacteria into a sixth kingdom.

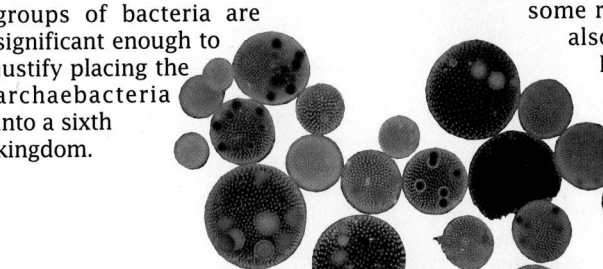

Members of the kingdom Protista were the first eukaryotes

As you learned in Chapter 10, protists evolved from bacteria. Multicellularity then evolved many times among the protists. The three most successful multicellular groups—plants, animals, and fungi—are assigned to their own kingdoms. Other multicellular eukaryotes and nearly all single-celled eukaryotes (except for a few one-celled fungi such as yeasts) are assigned to kingdom Protista. Thus, in a way, this kingdom is a catchall kingdom. It contains organisms that do not fit in one of the other three eukaryotic kingdoms—Plantae, Animalia, and Fungi. Plants, animals, and fungi most likely descended from three different ancient protists.

Protists include protozoa such as *Amoeba* and *Paramecium*, and algae such as seaweeds and kelps. Slime molds and water molds, which bear some resemblance to fungi, also belong to kingdom Protista. **Figure 15.12** shows a colony of *Volvox*, protists that capture light energy during the process of photosynthesis.

Connection: Chapter 2

What Is Life?
Review the five characteristics that all living things have in common.

Connection: Chapter 3

Prokaryotes
Review the distinguishing characteristics of the two major categories of cells: prokaryotes and eukaryotes.

Visual Strategy

Figure 15.11
Use this figure to emphasize that monerans thrive in environments that support no other form of life. Monerans exist in the icy regions of Antarctica, the dark depths of the ocean, and the boiling waters of hot springs. Some can even live without oxygen.

Visual Strategy

Figure 15.12
Use this figure to emphasize that protists are so different from each other that some scientists suggest the kingdom should be broken up into nearly 20 separate kingdoms.

Theme Answer

Scale and Structure
All living things use energy, reproduce, maintain homeostasis, contain genetic material, and are made of one or more cells.

Demonstration 2

Protists
Have students use their microscopes to examine various representatives of this kingdom, including *Amoebas*, *Paramecia*, *Volvox*, diatoms, and algae. Have students prepare a table, listing similarities and differences among the organisms they observe.

Demonstration 3

Fungi

Have students use stereomicroscopes to examine various specimens of fungi, including mushrooms and molds.

Connection: Chapter 5

Photosynthesis

Review the major features of photosynthesis, the life process carried out by all members of the kingdom Plantae.

Demonstration 4

Animals

Show students photographs of some unusual and interesting animal specimens, such as sponges, corals, sand dollars, tunicates, and water bears. Have each student bring in a photograph of the most unusual animal they can find. Point out that all animals must take in food materials and are made of cells lacking cell walls.

Phase 3

ASSESSMENT OPTIONS

Closure Strategy

Name That Kingdom

Show the class photographs of various organisms. Have students assign each one to a kingdom and provide reasons for their choice.

Section Review

Assign the *Section Review*.

Reteaching

Use the chalkboard or overhead projector to prepare a table like the one suggested in the Reading Strategy. Have student volunteers fill in the appropriate information from their own tables.

Figure 15.13
This mushroom belongs to the kingdom Fungi. It secretes enzymes that break down organic matter, which it then absorbs.

Mushrooms are members of the kingdom Fungi

Difficult to classify in a two-kingdom system, a mushroom like the one in **Figure 15.13** fits neatly into kingdom Fungi in the five-kingdom system. Instead of roots, stems, and leaves, fungi are made of thin filaments that penetrate the soil or decaying organisms. Fungi do not contain chloroplasts and cannot make their own food by photosynthesis. Instead, they obtain food by absorbing nutrients from their environment. The cell walls of fungi are made from the polysaccharide chitin (*KEYE tihn*). The many differences between plants and fungi suggest that they evolved from different protist ancestors.

Kingdom Plantae contains multicellular photosynthetic organisms

Kingdom Plantae includes only organisms that most people would recognize as "plants," such as mosses, flowers, and trees. All members of the plant kingdom are multicellular and terrestrial. Almost all plants obtain energy by photosynthesis, which is carried out in chloroplasts within their cells. Plant cells have cell walls made of the polysaccharide cellulose. Leaves of a fern are shown in **Figure 15.14**.

Figure 15.14
The leaves of this Dallas fern contain chloroplasts, which carry out photosynthesis.

The first members of kingdom Animalia evolved in the ocean

Like fungi and plants, organisms in kingdom Animalia are multicellular. Animals, however, do not photosynthesize. Their cells do not have cell walls. Nearly all animals have some kind of nervous system, although it might be a very simple one. As you learned in Chapter 10, animals first evolved in the sea. The largest number of animal phyla are still found only in the sea. Although it superficially resembles a plant, the demon stinger in **Figure 15.15** is actually a venomous, predatory sea animal.

Figure 15.15
This demon stinger, which measures about 8 cm (4 in.) across, looks like a plant but belongs to kingdom Animalia. It hides in the sand in the South Pacific Ocean with only its mouth and eyes exposed. There it lies waiting for crabs and small fish to cross its path. Its spines can cause painful wounds to anyone unfortunate enough to step on it.

Section Review

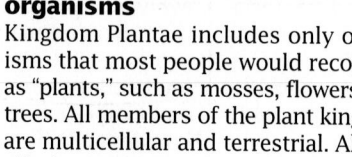

❶ **Explain why it would be difficult to classify a mushroom using the two-kingdom classification system.**

❷ **What are the names of the five kingdoms of organisms?**

❸ **List two characteristics of protists.**

❹ **What are two differences between plants and fungi? What are two characteristics both share?**

■ *Section Review Answers* ■

1. Students should explain that mushrooms do not fit the basic characteristics of either the plant or animal kingdom.

2. The five kingdoms are Monera, Protista, Fungi, Plantae, and Animalia.

3. Prokaryotes have no true nucleus, membrane-bound organelles, or chromosomes.

4. Fungi do not contain chloroplasts and cannot make their own food by photosynthesis, whereas plants do; the cell walls of fungi are made from the polysaccharide chitin, whereas the cell walls of plants are made from the polysaccharide cellulose. Plant and fungal cells have nuclei and membrane-bound organelles.

Chapter 15 *Highlights*

Interbreeding among members of the genus *Canis* produces fertile hybrids, such as this hybrid of a dog and a wolf.

Key Terms	Summary
15.1 The Need for Naming	
scientific name (p. 319)	• To ensure accurate communication of information, biologists assign a unique two-word scientific name to each organism.
genus (p. 320)	• The first word of a scientific name is the name of the genus to which the organism belongs. The second word identifies the kind of organism within the genus.
species (p. 320)	• The system of scientific names used today was developed by Linnaeus in the eighteenth century.

15.1 The Need for Naming

The scientific name for the red oak is *Quercus rubra*.

15.2 Classification: Organizing Life

In the Middle Ages, herbalists classified plants according to their medicinal uses.

Key Terms	Summary
taxonomy (p. 323)	• Scientists classify organisms into a hierarchical system of groups within groups.
family (p. 324)	
order (p. 324)	• Linnaeus classified organisms into groups according to their similarities. Since Darwin, classifications have reflected the evolutionary relationships among organisms.
class (p. 324)	
phylum (p. 324)	
kingdom (p. 324)	
division (p. 324)	
convergent evolution (p. 326)	• Analogous characters are similar but do not share a common ancestry. Convergent evolution is the evolution of similar characteristics in unrelated organisms.
analogous character (p. 326)	
derived character (p. 327)	• A species is a group of organisms possessing unique characteristics.
cladogram (p. 327)	
hybrid (p. 329)	

15.3 Five Kingdoms

The sea anemone is one example of an organism that looks like a plant but is classified as an animal.

• Linnaeus classified all living things into either the plant kingdom or the animal kingdom. Today biologists use a five-kingdom classification system.

• The five kingdoms are Monera, Protista, Fungi, Plantae, and Animalia.

• Many biologists think the kingdom Monera should be separated into two kingdoms, the archaebacteria and the eubacteria.

Assessment Alternative

Use the names of the five kingdoms as the categories for a Jeopardy game. Assign students to cooperative work groups of four. Have each group develop acceptable questions and answers for all five categories. Arrange competitions by having two cooperative groups use the information developed by a third group.

Chapter Review Answers

Understanding Vocabulary

1. **a.** Phylum is the second highest classification category for animals and protists. Division is the category at the same level used for plants, fungi, and bacteria.

   b. Protista is a kingdom of mainly unicellular eukaryotes. Monera is a kingdom of prokaryotes.

   c. Cladistics is a method of reconstructing the relationships among organisms by using uniquely-evolved (derived) characters. Phenetics is a method of classifying organisms based on overall similarity.

   d. A species is a group of organisms that can interbreed and bear fertile offspring. A hybrid is an offspring resulting from interbreeding by members of different species.

Relating Concepts

2. Concept map answers are on page 317D.

Understanding Concepts

Multiple Chioice

3. c	8. d
4. d	9. d
5. a	10. d
6. c	11. b
7. b	12. b

Completion

13. taxonomy
14. genus
15. common ancestor
16. Linnaeus; genus
17. Plantae

Short Answer

18. Other scientists often would be unsure what organisms were being discussed, since there is not always a 1:1 correspondence between a common name and an organism.

Understanding Vocabulary

1. For each pair of terms, explain the differences in their meanings.
   a. phylum, division
   b. Protista, Monera
   c. cladistics, phenetics
   d. species, hybrid

Relating Concepts

2. Copy the unfinished concept map below onto a sheet of paper. Then complete the concept map by writing the correct word or phrase in each oval containing a question mark.

Understanding Concepts

Multiple Choice

3. The scientific name for humans is correctly written as
   a. Homo sapiens.
   b. Homo Sapiens.
   c. *Homo sapiens.*
   d. *homo sapiens.*

4. Red oak and willow oak
   a. are in different families.
   b. cannot be distinguished from each other.
   c. are the same species.
   d. are members of the same genus.

5. In classifications of plants, bacteria, and fungi, the term "division" is substituted for
   a. phylum.   c. family.
   b. class.   d. order.

6. For grasshoppers and locusts to be in the same family, they must also be in the same
   a. order.   c. genus.
   b. group.   d. species.

7. Organisms in different genera can
   a. share the first word of their scientific names.
   b. be in different families.
   c. have the same scientific name.
   d. be members of the same species.

8. Two members of the same _____ would be the most closely related.
   a. order   c. family
   b. class   d. genus

9. Which living organism is *not* correctly matched with its kingdom?
   a. mouse: Animalia
   b. *Euglena:* Protista
   c. mushroom: Fungi
   d. bacteria: Protista

10. The first cells were members of the kingdom
    a. Plantae.   c. Fungi.
    b. Protista.   d. Monera.

11. Both plants and fungi
    a. secrete enzymes that break down organic matter.
    b. have cell walls.
    c. are photosynthetic.
    d. have roots.

12. Almost all single-celled eukaryotes are classified in the kingdom
    a. Fungi.   c. Plantae.
    b. Protista.   d. Monera.

Completion

13. The science of classifying living things is called _____ .

14. The scientific name for corn is *Zea mays. Zea* is the _____ name.

15. The more similarities two organisms share, the more recently they shared a(n) _____ .

16. The system of naming organisms used today was devised by _____ . According to this system, every organism is given a two-word Latin name that includes its capitalized _____ name.

17. Members of the kingdom _____ are multicellular photosynthetic organisms that live on land.

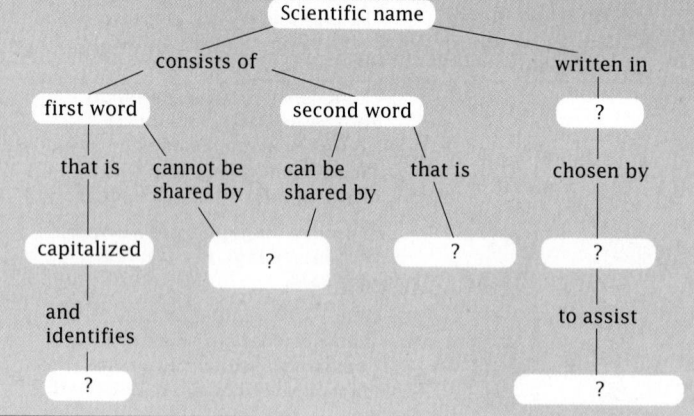

Short Answer

18. What problems might scientists encounter if they used common names instead of scientific names when communicating their findings?

19. What is the relationship between evolution and taxonomy?

20. What is a cladogram? How does it help scientists classify living organisms?

21. Why is a mushroom classified as a fungus rather than as a plant?

Interpreting Graphics

22. Examine the cladogram shown below.

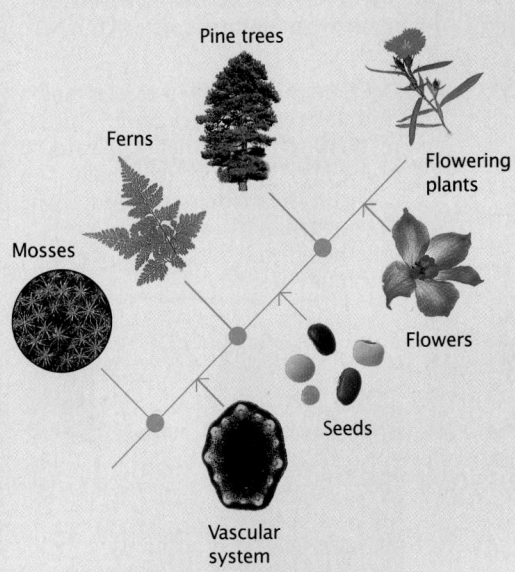

Pine trees

Ferns

Flowering plants

Flowers

Mosses

Seeds

Vascular system

- What is the derived character possessed by ferns?
- What derived characters are shared by pine trees and flowering plants?
- Which type of plant is most ancient?
- Which type of plant evolved most recently?
- Which type of plant lacks a vascular system?

Reviewing Themes

23. *Scale and Structure*
What features common to both the shark and the dolphin suggest that convergent evolution has occurred?

24. *Evolution*
What characteristics besides physical appearances are used by biologists to determine evolutionary relationships among organisms?

Thinking Critically

25. *Comparing Ideas*
Compare analogous characters and homologous characters.

26. *Building on What You Have Learned*
Based on what you learned in Chapter 11, which pair of organisms—humans and chimpanzees or chimpanzees and prosimians—are more closely related?

Cross-Discipline Connection

27. *Biology and Language Arts*
A mnemonic is a code used to aid memory. In the mnemonic "King Philip came over from Geneva, Switzerland," the first letter of each word is the same as the first letter of each of the seven levels of classification. Design your own mnemonics to help you remember the five kingdoms and two characteristics of each.

Discovering Through Reading

28. Read the article "A Biologist Whose Heresy Redraws Earth's Tree of Life," in *Smithsonian*, August 1989, pages 71–81. What did Margulis' work reveal about evolutionary relationships among the five kingdoms?

---

19. Taxonomy is an attempt to create groups of organisms that reflect evolutionary relationships.

20. A cladogram is a branching diagram, based on patterns of shared, derived characters, that shows the evolutionary relationships of a set of organisms. Groups identified on a cladogram can be categories in classification.

21. A mushroom cannot photosynthesize, has cell walls of chitin, not cellulose, and is heterotrophic.

Interpreting Graphics
22. • vascular system
    • seeds
    • mosses
    • flowering plants
    • mosses

Reviewing Themes
23. streamlined bodies, fins, flattened tail
24. Scientists can use nucleotide sequences in DNA or RNA, amino acid sequences in proteins, behavior patterns.

Thinking Critically
25. Analogous characters are similar but have a different evolutionary origin. The similarity of homologous characters is due to their common evolutionary heritage.
26. humans and chimpanzees

Cross-Discipline Connection
27. Answers will vary.

Discovering Through Reading
28. Margulis proposed that mitochondria and chloroplasts were descended from bacteria that lived within early eukaryotic cells. She has also proposed that flagella are descended from spiral bacteria.

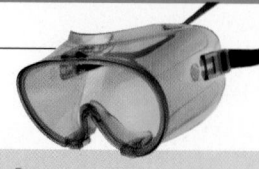

Procedural Note

Review the answers to the prelab questions and discuss the value of classification systems in everyday life.

Procedural Note

Review the answers to the prelab questions and discuss the value of classification systems in everyday life.

Prelab Preparation Answers

1. Most grocery stores classify things according to type of product. For example, fresh meats, fresh produce, canned vegetables, and paper goods are some of the classification categories of a grocery store.
2. Living things are classified into a hierarchy based on their similarities in anatomy, behavior, and biochemistry.
3. Classifications are attempts to create groups that reflect the evolutionary relationships among organisms.

Chapter 15

Investigation

How Do You Use a Classification Key?

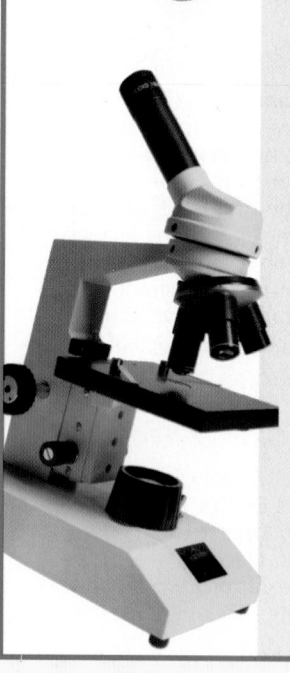

Objectives

In this investigation, you will:
• *use* a classification key to identify organisms

Materials

• 10 preserved specimens, photographs, or drawings of arthropods

Prelab Preparation

Review what you have learned about classification by answering the following questions:
• How are items classified or grouped in a grocery store?
• How are living things classified?
• What is the relationship of classification to evolution?

Procedure: Using a Classification Key

1. Make a table similar to the one on the next page.

2. Form a cooperative group with three other students in your class. Work with a member of your group to complete steps 3–6.

3. By examining 10 members of the phylum Arthropoda, the arthropods, you will determine to which class each arthropod belongs. You will look at members of five classes: Arachnida (spiders and ticks), Insecta (insects), Crustacea (crabs and lobsters), Chilopoda (centipedes), and Diplopoda (millipedes). Select one specimen and record its identity in the first row of your table.

4. Examine the specimen closely and find its breathing structure. *Gills* are feathery structures near the top of the legs. *Spiracles* are small holes on each side of the abdomen. In your table, record the type of breathing structure that your specimen has.

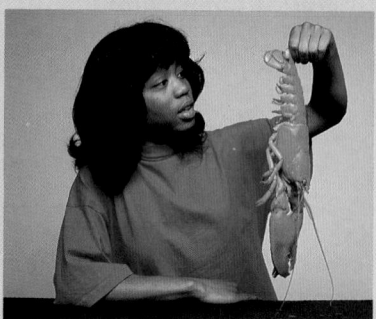

5. Count the number of legs, and the number of legs on each segment of the body. Record these values in the appropriate columns in your table.

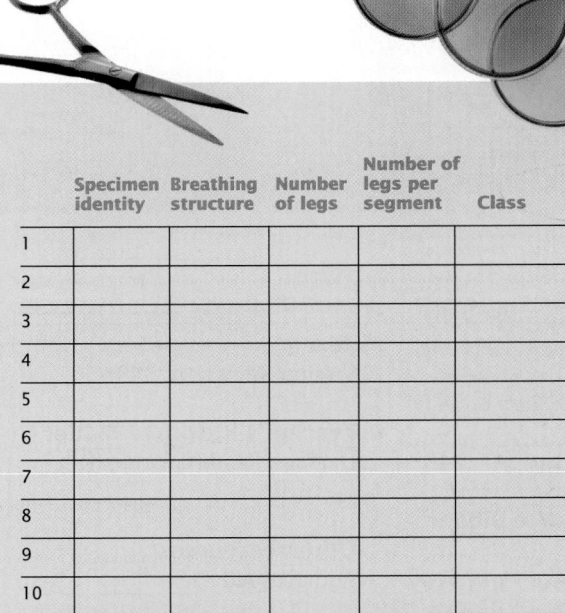

	Specimen identity	Breathing structure	Number of legs	Number of legs per segment	Class
1					
2					
3					
4					
5					
6					
7					
8					
9					
10					

6. Using the data from your specimen, follow the paths of the classification key below until you find the class to which the specimen belongs. Choose the path that best matches the features of the organism. Record your choice in the column labeled "Class" in your table.

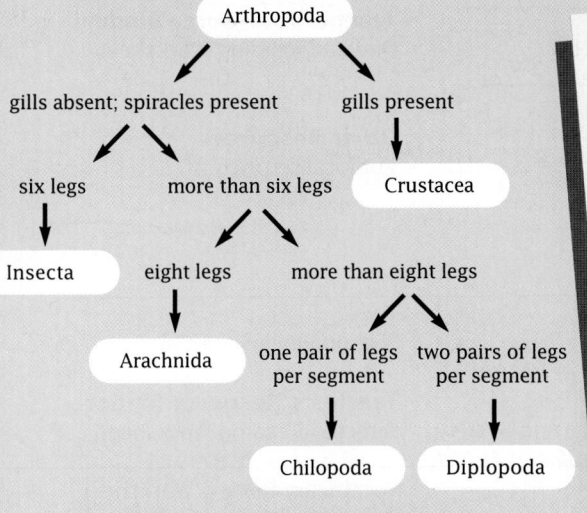

7. Repeat steps 3–6 for the remaining specimens.

8. Exchange data tables with the other team in your group and check their classifications against yours.

Analysis

1. *Evaluating Methods*
Why is a key based on physical characteristics easier to use and more accurate than one based on an organism's habitat or behavior?

2. *Making Inferences*
How might a a person use a classification key when going on a nature walk through a forest?

3. *Evaluating Methods*
How could you modify this key to classify different insects?

4. *Inferring Relationships*
What can you conclude about the evolutionary relationships among your organisms?

Thinking Critically

Turn back to **Figure 11.6** on page 225, which shows the relationships of humans to apes. Make your own classification key that would enable you to classify these different primates. You should be able to explain why you chose the characteristics that separate these species.

Analysis Answers

1. Physical characteristics can be readily observed. A key should not require vast amounts of prior knowledge. Habitat and behavior patterns are difficult to deduce from preserved specimens.
2. Answers may vary, but might suggest such things as identifying edible or poisonous plants.
3. Answers may vary but should suggest that additional branches would be added using characteristics that would distinguish between insects.
4. Students can reasonably conclude that the members of each class are more closely related to each other than to the members of the other classes. From the information given, however, students cannot determine which classes are most closely related.

Thinking Critically Answer

Answers will vary but the characters chosen should lead to dichotomous choices.

Chapter 16 Bacteria and Viruses

Planning Guide

	Objectives/Themes	Classwork Resources	Homework Resources

16.1	**1.** Describe the structure of a bacterial cell.
	
	2. Contrast a bacterial cell with a eukaryotic cell.
	
	3. Explain how bacteria reproduce.
	
	4. Recognize the diverse ways bacteria obtain nutrition.
	
	Themes: Energy and Life, Evolution, Patterns of Change, Scale and Structure, Stability
		Text
		Journeys
		    Tour of a Bacterium p. 341
		Science in Action
		    Lyme Disease pp. 344–345
		
		Teacher's Resource Binder
		Focus Activity 16
		    Using Bacteria to Make Food
		
		Other Resources
		Transparencies 65–66
			Text
			Section Review, p. 343
			
			Teacher's Resource Binder
			Directed Reading Worksheet
			    16.1
			
			Other Resources
			Audiocassette 16.1*

16.2	**1.** Describe three beneficial effects of bacteria.
	
	2. List five human diseases caused by bacteria.
	
	3. Summarize three ways to prevent bacterial disease.
	
	4. Recognize the importance of antibiotics in fighting bacterial diseases.
	
	Theme: Interacting Systems
		Teacher's Resource Binder
		Lab Investigation 16.1
		    Antibiotics and Zones of Inhibition
		Lab Investigation 16.2
		    Bactericidal Effect of Soap
		
		Other Resources
		Transparency 67
			Text
			Section Review, p. 352
			
			Teacher's Resource Binder
			Directed Reading Worksheet
			    16.2
			
			Other Resources
			Audiocassette 16.2*

16.3	**1.** Describe the structure of a virus.
	
	2. Explain why viruses are not living organisms.
	
	3. Describe how a virus reproduces.
	
	4. List four diseases that are caused by viruses.
	
	Themes: Energy and Life, Patterns of Change, Interacting Systems, Scale and Structure
		Text
		Investigation
		    How Are Models of Viruses Constructed? pp. 362–363
		Science, Technology, and Society
		    Testing for HIV: Who Is At Risk? pp. 364–365
		Discoveries in Science
		    Humans and Their Viruses pp. 450–451
		
		Teacher's Resource Binder
		Extension Worksheet
		    Viral Invasion and the Body's Response
		
		Other Resources
		Transparencies 68–70
			Text
			Section Review, p. 358
			
			Teacher's Resource Binder
			Directed Reading Worksheet
			    16.3
			Vocabulary Review Worksheet*
			Reaching Worksheet*
			    Bacteria or Virus?
			
			Other Resources
			Audiocassette 16.3*

*Reteaching Options

Demonstrations
16.2: pp. 347, 349, 350

Assessment
Chapter Review pp. 360–361
Portfolio Assessment p. 337D
Chapter Test—Teacher's Resource Binder
Test Generator

Research Notes

Connection to Medicine: Can Doctors Transmit HIV?

Many patients and some doctors are quite alarmed about the possibility that they may become infected with the HIV by each other. Two studies published in May 1992 by the Centers for Disease Control in Atlanta indicate that although doctors can infect patients in certain circumstances, it is extremely unlikely.

In one study of 32 HIV-infected health-care providers, only 84 patients out of 15,795 had HIV, and none of those cases could be confirmed as being due to infection by the health-care providers. In another study, DNA "fingerprint" analysis of HIV by researchers at the Centers for Disease Control indicated that doctor-to-patient transmission had occurred in a Florida dentist's practice. The dentist had continued his practice two years after he had been diagnosed with AIDS.

Most of the evidence already suggested that transmission from the dentist was the likeliest cause of AIDS in some of the patients. Of the seven who had tested positive for the HIV virus, five said they had no other risk factors for the disease. All of the patients had invasive procedures such as root canals or tooth extractions. However, this evidence did not establish conclusive proof that they contracted the virus from the dentist.

The CDC study used the polymerase chain reaction to rapidly create many copies of the HIV DNA for the protein coat of the virus. This portion of the viral DNA changes and mutates very rapidly, producing different "strains" of the HIV virus. Scientists sequenced the DNA from this area and found that for the five patients without risk factors, the strains of patient HIV were very similar to the dentist's HIV strains.

These strains were also found to be different than "control" strains from 35 unrelated HIV-infected people in the same area. For the

other two patients, who had other risk factors, the strains were as closely related to these control strains as they were to the dentist's strains, indicating that they probably were not infected by the dentist.

Some scientists criticized the CDC study, saying that the technique did not provide the sort of certainty that would be needed in a courtroom. The CDC's results are likely to be employed as evidence in a $15 million lawsuit by one of the patients against the dentist's insurance company. There are also some who criticize the CDC's procedures, their statistical analyses, and their interpretation of the results.

CDC researchers admit that they cannot explain how the dentist infected the patients. They have concluded that it was neither through sexual contact nor deliberate transmission. It could be that bodily fluids from a wound on the dentist may have come in contact with the patients' blood during the procedures.

Investigation Notes

How Are Models of Viruses Constructed?
pages 362–363

Purpose: This investigation gives students an opportunity to express themselves creatively, while learning about virus structure. The models provide concrete examples that help students better understand spatial relationships.

Technique: This investigation is not recommended as a home activity. Some students may need additional help with the scaling aspects of the investigation and those less confident in their creative abilities are likely to avoid the experience or turn to others for help. For these students, making the attempt is more important than the artistic quality of the final results.

Prelab Preparation

1. Materials for model-building are available through craft and art supply stores. Clay, plastic foam pieces, and pipe cleaners tend to be the most popular items when students perform this investigation.

2. For teachers wanting a set of models illustrating the complete life cycle of a bacteriophage, WARD'S has an eight-model set available (81 M 4741).

Answers will be found on pages 362–363.

Meeting Individual Needs

RON

Objectives

1. Students will demonstrate appropriate use of core vocabulary for the chapter (Vocabulary File).

2. Students will describe bacteria and viruses and relate them to their effects upon people (Teaching Strategy A).

Vocabulary File
(Developing Vocabulary/ Limited English Proficiency)

If you are not already using the Vocabulary File, refer to Chapter 1 for its preparation. See the Chapter Highlights on page 359 for a list of suggested words.

Teaching Strategy A
Bacteria and Viruses
for pages 339–358
(Developing Classification Skills/Verbal Learners)

Preparation: Gather trivia facts from the chapter about bacteria and viruses. These will be used for a contest. Be sure you have the questions and answers in an easy-to-read format.

Suggested questions and answers include:

Name a shape that bacteria can have. (Rod, sphere, or spiral)

Name the type of bacteria that stain purple. (Gram-positive)

Name the type of bacteria that stain pink. (Gram-negative)

Why do different bacteria stain different colors? (Some have 1-layer cell walls, others have 2 layers)

What type of bacteria are found in pea and bean roots? (Nitrogen-fixing)

What type of bacteria recycles nutrients? (Decomposer)

Name a food product that requires bacteria. (Sauerkraut or pickles)

Name diseases caused by bacteria. (Tetanus, typhoid, diphtheria)

How is water purified of bacteria? (Chlorine treatment)

How is food kept bacteria-free? (Pasteurization and cooling)

What drugs can treat bacterial diseases? (Antibiotics)

What do viruses need to reproduce? (Living cells)

What is in the virus core? (Genetic material)

What protects the core? (Protein coat)

Name diseases caused by viruses. (Measles, mumps, chickenpox, AIDS, herpes)

Are antibiotics effective against viruses? (No)

What procedure can defend against viruses? (Vaccination)

Procedure: Have students read the chapter for their homework before this Teaching Strategy is used. Tell the students they will compete for some prize you have selected. Divide the class into two or three teams. Have each team stand in a line. For a round, only the people at the front of the line may answer the question, and only if they signal that they know the answer before the other team. If the answer given is incorrect, the other teams' designated answerers may try to answer it.

After the question round is over, have those at the front of the line return to the back of the line, so a new set of answerers is ready. Keep score, adding one point for each correct answer, and subtracting $\frac{1}{2}$ point for incorrect ones. Play as long as needed. (Going through the questions more than once may provide the drill some students need.)

Additional Strategies
Visual Strategies

Pages 339, 340, 341, 342, 343, 346, 348, 350, 352, 355, 357, and 358

Auditory Learners

Use *Biology: Visualizing Life* Audiocassettes for Sections 16.1, 16.2, and 16.3.

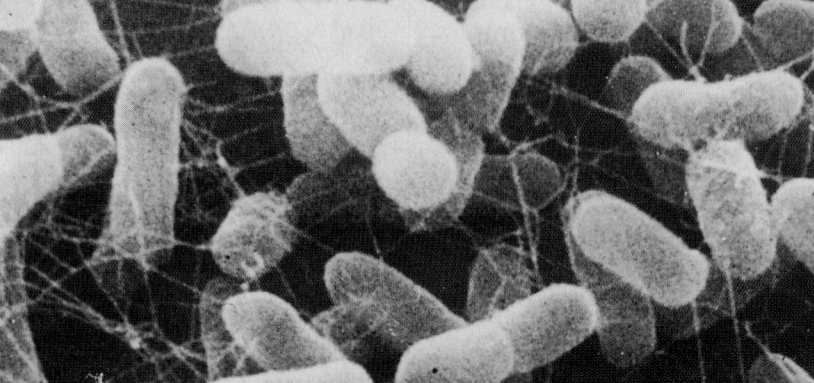

Meeting Individual Needs (cont.)

Cooperative Learning
Fighting Bacterial Diseases

Timing: Use this activity to introduce Section 16.2.

Group Size: 3 or 4 students

Outcome: Students will be able to describe characteristics of several bacterial diseases.

Individual Accountability: Each group member is responsible for researching one bacterial disease.

Positive Interdependence: Each group will compare the reports and make an expanded table of characteristics for the bacterial diseases.

Assign each group member one or more of the bacterial diseases from Table 16.1 on page 348. Students should research each disease they are assigned in order to determine the name of the bacteria that cause the disease, the bacteria's shape and size, the type of body systems the bacteria disrupts, geographic regions most affected by the disease, measures that can help prevent the spread of the disease, and methods involved in the cure (if any).

Then, members of each group should work together to compile their findings on a chart, concept map, or similar organizational device. The group should come up with a summary sentence for the chart that describes the three most important steps to take in order to avoid the bacterial diseases they studied.

Portfolio Assessment

Students should select their best work and provide a self-reflective rationale for their selections. Students can make selections in the following areas.

1. *Content* — One concept map from the chapter (See page 812 for evaluation criteria.)

2. *Reading Comprehension* — One Directed Reading Worksheet from the Teacher's Resource Binder (Use the answer key to evaluate for accuracy.)

3. *Writing* — Using the Vee Form, summarize a magazine or newspaper article relating to bacteria or viruses. (See page 22T for evaluation criteria.)

   Or: Select a writing task or project from the Chapter Review.

4. *Performance Assessment* — One Vee form from a chapter investigation or lab manual investigation (See page 22T for evaluation criteria.)

Teacher makes selections in the following areas.

1. *Formal Assessment* — Chapter test (Test A, B, or the Test Generator) The teacher-scored test should be reviewed by the student. Incorrect responses should be corrected by the student before the test becomes part of the portfolio.

2. *Informal Assessment* — Use of the Direct Observations Checklist, page 33T, during a laboratory or other cooperative learning experience.

3. *Performance Assessment* — Have students prepare a detailed personal health history with the help of their family doctor. Students should determine which illnesses were caused by viruses or bacteria.

Concept Map Answer

The following is one possible answer to the Relating Concepts exercise on page 360.

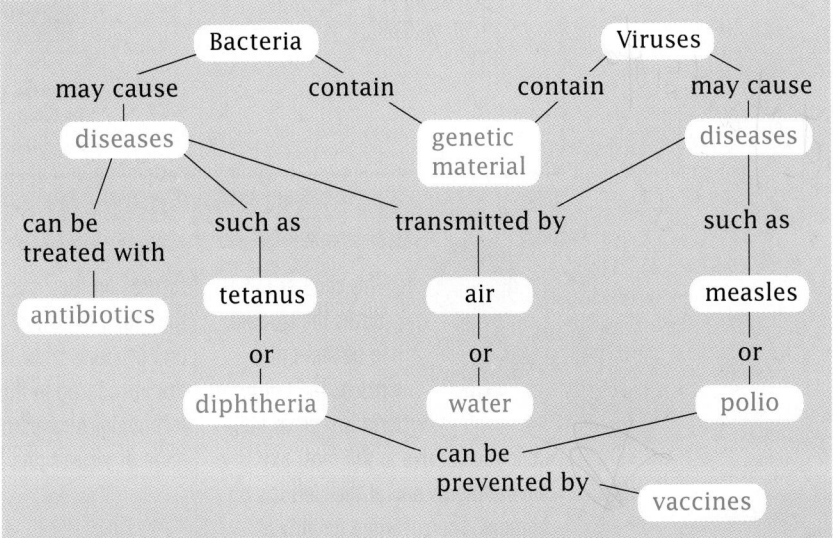

Chapter 16

Determining Prior Knowledge

- Have students explain why bacteria are considered prokaryotes.
- Show the class a container of yogurt and ask what its connection is to bacteria. Have students give other examples of foods that are prepared with the help of bacteria.
- Determine how many students have had strep throat. Have them describe their experiences, especially any medical treatment they received.
- Check students' knowledge about viruses and the diseases they cause. Ask questions about AIDS: What causes AIDS? How does the HIV virus affect the body? How is the virus transmitted between people? Why are scientists having a difficult time developing a vaccine against HIV?

Chapter 16

Review

- prokaryotes (Section 3.3)
- cell division (Section 4.3)
- five kingdoms (Section 10.2)

Bacteria and Viruses

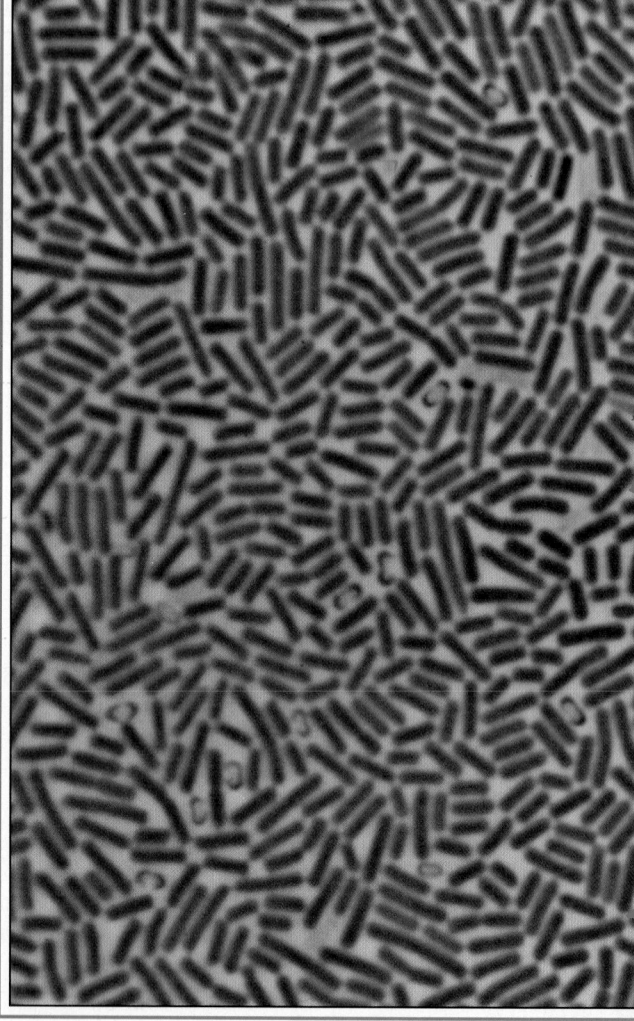

A large group of bacteria can look like abstract art, but it is really life at its least complex level.

■ Author's Rationale ■

This chapter introduces the student to bacteria and viruses. The nature of bacteria is seen in their ability to survive in the most extreme environments and in their effects on humans. The coverage on viruses focuses on their structure and on how they take over host cell machinery and produce other viruses. Vaccines are introduced in this chapter as a technological application of our knowledge of viruses and bacteria.

ALTHOUGH YOU DO NOT SEE THEM, BACTERIA AFFECT YOUR LIFE IN NUMEROUS WAYS. IN YOUR LARGE INTESTINE, BACTERIA SYNTHESIZE VITAMIN K, WHICH IS ESSENTIAL FOR YOUR NUTRITION. BACTERIA THRIVE IN YOUR MOUTH AND MAY CAUSE CAVITIES IN YOUR TEETH. BACTERIA ALSO CAUSE MANY DISEASES, INCLUDING CHOLERA, TYPHUS, PNEUMONIA, AND TUBERCULOSIS.

16.1 Bacteria

Objectives

❶ Describe the structure of a bacterial cell.

❷ Contrast a bacterial cell with a eukaryotic cell.

❸ Explain how bacteria reproduce.

❹ Recognize the diverse ways bacteria obtain nutrition.

Bacteria Are Small, Simple, and Successful

Figure 16.1
A human red blood cell (250X) is approximately 100 times larger than the bacteria shown below and to the right.

In many ways, bacteria are the most successful organisms on earth. For one thing, bacteria are the oldest group of organisms. The earliest known fossils are 3.5 billion-year-old bacteria. Today, bacteria can be found living almost everywhere on the globe, even in some very hostile habitats. Certain bacteria live beneath more than 400 m (1,200 ft.) of ice in Antarctica, and others live near deep sea volcanic vents where temperatures reach 360° C. Bacteria also occur in great abundance. One gram ($\frac{1}{28}$th of an ounce) of rich soil can contain around 2.5 billion bacteria.

It is obvious that bacteria are very small,

otherwise you could see them without a microscope. To give you an idea of how small bacteria are, look at **Figure 16.1.** A human red blood cell dwarfs the *Escherichia coli* bacterium that is normally found in the human intestine. If you made a chain of *E. coli* bacteria laid end to end, the chain would have to be more than 250 bacteria long just to be visible to the unaided eye.

Most species of bacteria are one of three different shapes: spherical, spiral, or rod-shaped. **Figure 16.1** clearly illustrates these shapes. *Streptococcus* bacteria that cause strep throat are spherical, for example. Spherical bacteria often link to form long chains of cells.

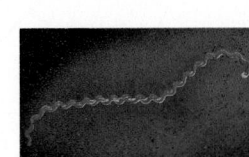

a **These rodshaped bacteria, *Bacillus subtilis*, produce antibiotics.**

b **These spherical bacteria, *Staphylococcus aureus*, cause skin infections.**

c **This spiral bacterium, *Leptospira*, sometimes causes liver and kidney damage.**

▪ Matter of Fact ▪

The Kingdom Monera
Have students explain to which kingdom bacteria belong. Point out that the kingdom name *Monera* is derived from the Greek word *moneres*, meaning single. Although many bacteria are single-celled, both the rod-shaped and spherical-shaped bacteria form

clumps or chains. Only the spiral-shaped bacteria exist exclusively as single cells.

Theme Answer

Scale and Structure

Many antibiotics interfere with cell wall synthesis. Gram-positive bacteria are susceptible to these antibiotics, but Gram-negative bacteria are not.

Connection: Chapter 3

The Cell Membrane

Review the structure and function of the cell membrane.

Connection: Chapter 2

Organic Compounds

Review the major characteristics of lipids and polysaccharides.

Visual Strategy

Figure 16.2

Have students identify the cell membrane and cell wall in Figure 16.2. Point out that penicillin inhibits the formation of a cell wall in dividing bacteria. Without a cell wall, bacterial cells easily burst and die since their cell membranes cannot withstand the pressure exerted by the water in the cytoplasm.

ARE PROKARYOTES

The structure of a bacterial cell is simple

Scale and Structure

How does the structure of the bacterial cell wall affect its susceptibility to antibiotics?

All bacteria are prokaryotes, members of the kingdom Monera. Recall from Chapter 3 that prokaryotes lack cell nuclei, chromosomes, and membrane-bound organelles such as mitochondria, Golgi bodies, and chloroplasts. You can see the internal and external structure of a bacterial cell in the *Tour of a Bacterium* on the next page.

Bacteria have one of two kinds of cell walls

Like a cell from your body, a bacterial cell is enveloped by a cell membrane made of a double layer of lipids. Unlike your cells, bacterial cells have an outer cell wall composed of polysaccharides. In some bacteria, this cell wall is surrounded by yet another layer, which is made up of polysaccharides bound to molecules of lipid.

The chemical difference between these two types of bacterial cell walls can be revealed using a special staining procedure known as **Gram staining**. In this procedure, dyes are added to a sample of bacteria as a microscope slide is being prepared. As shown in **Figure 16.2**, bacteria are either Gram-negative or Gram-positive. The difference between Gram-negative and Gram-positive bacteria is important in diagnosing and treating diseases caused by bacteria. Gram-negative bacteria are unaffected by many antibiotics, the drugs used to treat bacterial infections. These antibiotics cannot penetrate the additional layer outside the cell wall.

As you learned in Chapter 4, bacteria reproduce by splitting in two. Except for occasional mutations, each new cell is exactly like the parent cell. Genetic material in some cases is transferred from one bacterium to another.

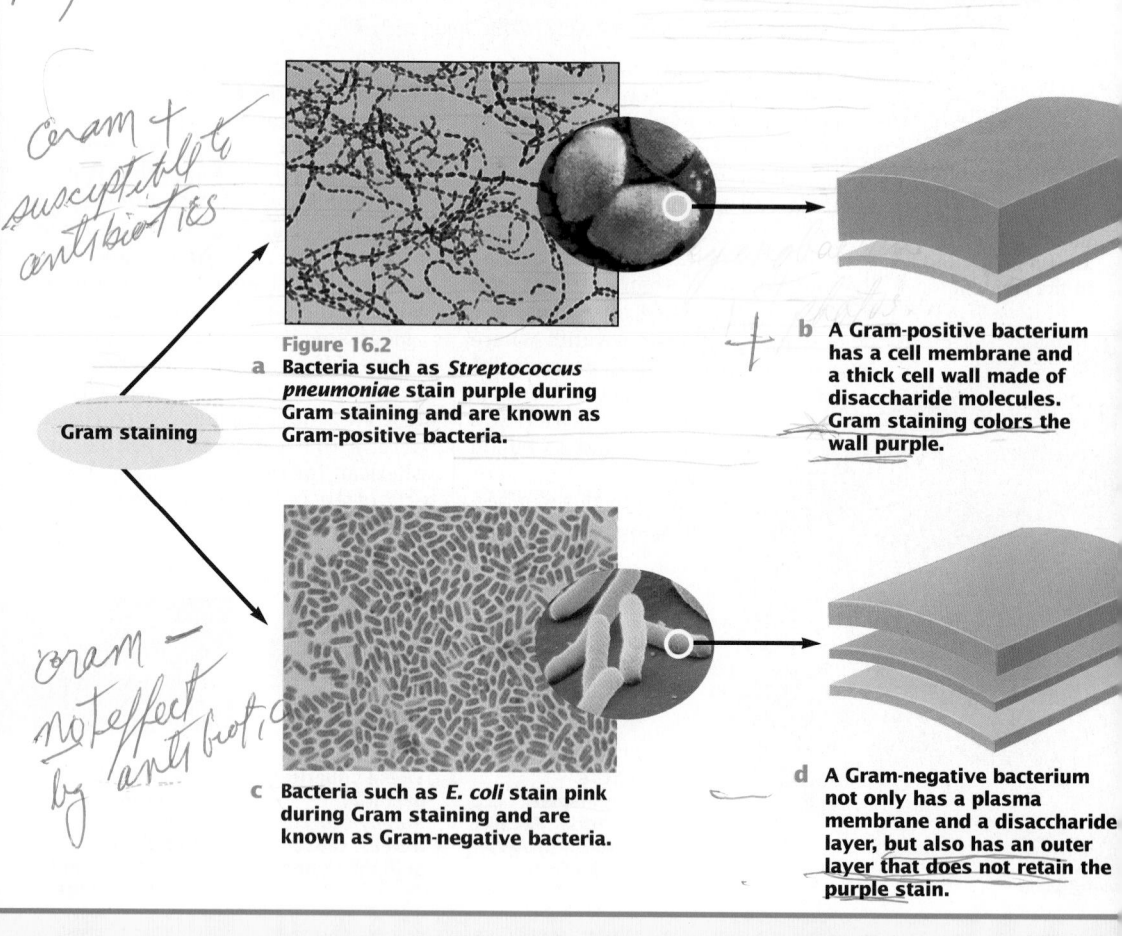

Figure 16.2

Gram staining

a **Bacteria such as *Streptococcus pneumoniae* stain purple during Gram staining and are known as Gram-positive bacteria.**

b **A Gram-positive bacterium has a cell membrane and a thick cell wall made of disaccharide molecules. Gram staining colors the wall purple.**

c **Bacteria such as *E. coli* stain pink during Gram staining and are known as Gram-negative bacteria.**

d **A Gram-negative bacterium not only has a plasma membrane and a disaccharide layer, but also has an outer layer that does not retain the purple stain.**

Tour of a Bacterium

Bacteria are single cells too small to see without a microscope. This one, Escherichia coli, is perhaps the most thoroughly studied organism in the world. It is used in genetic engineering and in studies of protein synthesis and the control of gene expression.

Billions of *E. coli* live in the human intestine.

Sticking out from the outer surface are pili, by which cells attach to surfaces or other bacteria.

Proteins are made on ribosomes that float free in the cytoplasm.

All of the genes are located on a single molecule of DNA. There is no nucleus.

The cytoplasm is not divided into separate compartments.

E. coli is Gram-negative. It has an additional layer of lipid and polysaccharide outside of the cell wall.

Flagella rotate like propellers to drive the cell through its environment.

The strong cell wall is a network of polysaccharides cross-linked by short chains of amino acids.

Bacteria reproduce by dividing in half as frequently as every 20 minutes.

Using the Feature

Use this feature to introduce students to the structure of a typical bacterium, *E. coli*, which is found in great abundance in the human intestine. Call attention to the differences between the structure of *E. coli* and that of eukaryotic cells. Also emphasize that the relationship between humans and their intestinal bacteria is mutualistic: the bacteria synthesize vitamin K—which is essential for blood clotting—in exchange for food and a place to live.

Discussion

Guide the discussion by posing the following questions.

1. How is the cell wall of *E. coli* different from the cell wall of a Gram-positive bacterium?
   The cell wall of E. coli *is covered by an additional layer of lipeds and polysaccharides.*

2. Identify two differences between *E. coli* and a eukaryotic cell.
   E. coli *has no nucleus, lacks membrane-bound organelles, and is much smaller.*

3. Would penicillin, an antibiotic that obstructs cell wall synthesis, kill *E. coli*? Explain.
   No. E. coli *is a gram-negative bacterium; its cell wall is protected from antibiotics by its covering of lipids and polysaccharides.*

Theme Answer

Energy and Life

Bacteria are quite versatile in the ways they obtain nutrients. They include autotrophs that carry out photosynthesis or chemosynthesis, and heterotrophs that ingest organic materials. This versatility in obtaining energy is a major reason for their success in life.

Connection: Chapter 5

Photosynthesis

Review the major features of photosynthesis as the process that converts light energy into chemical energy.

Visual Strategy

Figure 16.3

Have students explain why cyanobacteria are also known as blue-green bacteria. Tell students the steps of photosynthesis in cyanobacteria are much the same as in eukaryotes. But as prokaryotes, cyanobacteria lack chloroplasts. Their chlorophyll is found folded on membranes in the cytoplasm. In addition, cyanobacteria contain phycobilins, some of which give these organisms their blue-green color. Phycobilins also allow cyanobacteria to use different wavelengths of light for photosynthesis.

Social Studies Connection

The Color of Seas

Both the Red Sea and the Black Sea owe their names to photosynthetic bacteria that thrive in these waters. Occasional blooms of red-colored and black-colored bacteria occur in these seas, giving each its characteristic color.

Bacteria reproduce rapidly

Bacteria, unlike most other organisms, do not undergo mitosis or meiosis. Instead, a bacterium first duplicates its DNA so that there is enough DNA for two cells. Then the bacterium splits into two identical cells. Each cell receives one molecule of DNA and some cytoplasm, a process you saw in Chapter 4. Some kinds of bacteria are able to divide as much as five times in an hour.

Bacterial reproduction does not include meiosis, so it does not allow for the recombination that crossing over and sexual reproduction provide.

Reproduction occurs very rapidly in bacteria. If you were to place a single bacterium into a culture dish containing an abundant supply of food, you could find more than 600,000 bacteria in the dish after only four hours. After six hours, the bacterial population of the dish could reach 476 million.

Energy and Life

Bacteria use many different energy sources. How does this enable them to live in a wide variety of places?

How Bacteria Obtain Nutrition

A glass of milk left out of the refrigerator provides a wealth of food for bacteria. Within hours, bacteria colonize the milk and break down its supply of sugar, causing the milk to curdle. Other species of bacteria feed on organic material in sewage. Still other species consume industrial products such as nylon, and pesticides. Some bacteria are able to metabolize petroleum and may be used to clean oil spills. A major reason for the success of bacteria is the wide variety of foods they can use.

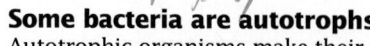

Some bacteria are autotrophs

Autotrophic organisms make their own food by using simple molecules. All autotrophic eukaryotes are photosynthetic: they capture solar energy and use it to make food. Many autotrophic bacteria, such as the freshwater cyanobacteria shown in **Figure 16.3**, are also photosynthetic. As you learned in Chapter 10, cyanobacteria were probably the first photosynthetic organisms to produce oxygen, starting about 3 billion years ago.

Figure 16.3

a **Microscopic examination of the water in freshwater ponds such as this one in Kansas reveals . . .**

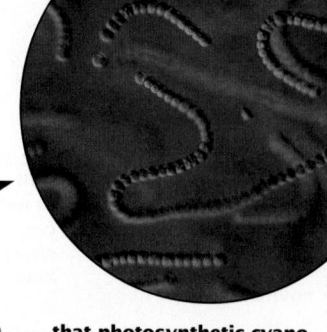

b **. . . that photosynthetic cyanobacteria such as *Nostoc* are a major component of the pond's ecosystem.**

Parasitic

MOST BACT are Heterotrophs ← Saprositic

Not all autotrophic bacteria are photosynthetic, however. In Chapter 12 you studied the communities that surround volcanic vents on the ocean floor. The organisms in these very deep ocean communities cannot perform photosynthesis because no light reaches these depths. Instead, bacteria such as those shown in **Figure 16.4** are able to use the energy stored in the inorganic compound hydrogen sulfide. Similarly, some kinds of soil-dwelling bacteria obtain energy from ammonia. Other bacteria that live in swamps use methane. These bacteria use a process called **chemosynthesis** to make complex organic molecules from the energy in inorganic molecules. All chemosynthetic organisms are prokaryotes.

Heterotrophic bacteria are consumers

Most bacteria cannot make their own food and are therefore heterotrophs. Many feed on dead animals and animal wastes; dead plants; and fallen leaves, branches, and fruit. Other types of heterotrophic bacteria are parasites. Parasitic bacteria cause many diseases, as you will see later in the chapter.

Figure 16.4
The canals in Venice, Italy are one of the world's great tourist attractions, but an examination of the water reveals . . .

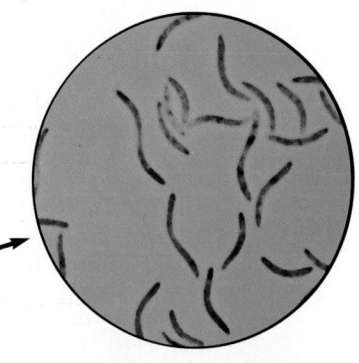

b . . . the presence of large numbers of sulfur bacteria much like *Desulfovibrio gigas.* These bacteria metabolize sulfur, releasing an odor that is similar to rotten eggs. The smell detracts from the beauty of the canals.

Section Review

❶ **How does Gram-staining help a doctor prescribe treatment for a bacterial infection?**

❷ **List three structures found in a eukaryotic cell but not in a bacterial cell.**

❸ **How do bacteria reproduce?**

❹ **Explain the two ways in which autotrophic bacteria obtain energy.**

■ *Section Review Answers* ■

1. The doctor has to know whether the bacteria causing the infection are Gram-negative or Gram-positive before prescribing an antibiotic. Gram-negative bacteria are covered with an additional protein layer and are unaffected by many antibiotics.

2. Answers may vary. Possible answers are a cell nucleus, chromosomes, and membrane-bound organelles.

3. The DNA is first copied and then the cell divides. A complete set of DNA is passed into each of the two daughter cells that are produced when the cell divides.

4. Students should discuss photosynthesis and chemosynthesis.

Using the Feature

Arthropods are important vectors of disease throughout the world. Mosquitoes carry malaria, yellow fever, encephalitis, and dengue fever. Lice transmit typhus; fleas carry bubonic plague. Lyme disease is only one of the diseases spread by ticks. Some others are Rocky Mountain spotted fever, Texas cattle fever, Q fever, and relapsing fever.

Discussion

Guide the discussion by posing the following questions.

1. What are the early symptoms of Lyme disease?
   a red rash that begins at the site of the tick bite, fever and chills, nausea, fatigue, and headache

2. What are some of the possible consequences of untreated Lyme disease?
   meningitis, arthritis, severe pain, and organ damage

3. Describe three ways to prevent Lyme disease.
   Use tick repellents and wear long sleeves, a hat, and long pants tucked into the shoes when in brushy or wooded areas. Inspect your clothes and body for ticks after being in tick-infested areas. Wear light-colored clothing, on which ticks are more easily discerned. Promptly remove any attached ticks.

4. White-tailed deer carry *Borrelia* but do not get Lyme disease. Explain why this is evidence of a longer coevolutionary relationship between deer and *Borrelia* than between humans and *Borrelia*.
   Parasites and their hosts coevolve, with the host evolving greater resistance and the parasite evolving reduced virulence. Thus, a parasite will have largely benign effects on a host it has been associated with for a long time. Its effects on a recently acquired host, however, will be more harmful.

Lyme Disease

Biology and the Latest in Outdoor Fashion

"Tuck your pant legs into your socks. Roll down your shirt sleeves. Those clothes are too dark." These are not the comments of a friend who doesn't like the fashion statement you make. They are advice from the Centers for Disease Control suggesting ways you can reduce your chances of contracting an ailment called Lyme disease.

Lyme disease is caused by a spiral-shaped bacterium called *Borrelia burgdorferi*. This bacterium lives inside several species of ticks, the most common being members of the genus *Ixodes*. When a tick carrying these bacteria bites a human to feed on blood, the bacteria can be transferred to a human, who may develop the disease.

Lyme disease was first identified in the town of Lyme, Connecticut, in 1975. The disease, however, is found in almost every state. It is common among pets, especially dogs. Ticks carrying the Lyme disease bacteria are spread by deer, rodents, other mammals, and birds.

The first symptom of Lyme disease is usually a red rash that begins at the site of the tick bite and spreads outward. Other early symptoms include fever and chills, nausea, fatigue, and headache. If untreated, later symptoms include severe pain, arthritis, meningitis, and internal organ damage, including brain damage.

Lyme disease is treated with various antibiotics. With early treatment, a patient can be cured before any severe damage occurs. However, diagnosis is difficult to confirm because no reliable

The center of the characteristic "bulls-eye" rash of Lyme disease is the tick bite.

Lyme disease is caused by this bacterium, *Borrelia burgdorferi*. This bacterium is carried mainly by ticks of the genus *Ixodes*. When young, these ticks are as small as poppy seeds. Adults that have recently fed can be as large as jelly beans.

White-tailed deer carry *Borrelia*, but are not affected by it.

Action ▪ Science in Action ▪ Science in Action ▪ Science

test for the presence of this bacterium exists. The best "cure" is prevention. Cover up when you journey into wooded, grassy, or bushy areas. Long sleeves and pants, tucked-in pant legs, and hats help prevent ticks from coming into contact with your skin. Wear light-colored clothing so ticks on your clothes can be seen easily. You can also use tick repellents sparingly on your clothing and skin.

After you have been in tick-infested areas, inspect your body for ticks. Ticks often move to warm, hair-covered areas of the body. Get help inspecting your scalp. Carefully remove attached ticks with tweezers, wrap them in tissue, and flush them down the toilet. Do not smash or burn them as you may release the bacteria they contain. Do not attempt to remove ticks using nail polish, ointments, or a hot match head, as doing so may force the tick's gut contents—which may include the disease-causing bacteria—into your body.

Inspect pets that have been in wooded or grassy areas. Comb out loose ticks and remove any attached ticks with tweezers. Tick repellents for pets also can be used. Have dogs vaccinated against Lyme disease.

By taking a few simple steps and using some common sense, you can enjoy the outdoors, free of big worries about tiny pests. If you think you may have Lyme disease, see your doctor immediately. Treatment is more successful when initiated early.

A Tick's Life

The deer tick, *Ixodes dammini,* has three life stages that differ from each other in size: nymph (about the size of a poppy seed), larva, and adult (when swollen with blood, it can be the size of a jelly bean). During the adult stage, deer ticks lay large numbers of eggs.

Deer ticks usually feed on field mice and white-tailed deer and only need three blood meals during their two-year life cycle. *Borrelia*, one of several kinds of internal parasites carried by ticks, can be ingested during one of these blood meals and transferred to a host during a subsequent blood meal.

Since ticks are unable to fly and can only crawl slowly, they hide in trees or tall grass, waiting for a possible host to wander into their jumping range. They have poor vision and rely on their keen sense of smell. They are especially sensitive to carbon dioxide and butyric acid, a rancid-smelling chemical occurring on the skin of many animals, including humans. The scents immediately activate the hungry tick, sending it leaping in the direction of the odor.

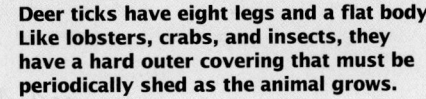

Deer ticks have eight legs and a flat body. Like lobsters, crabs, and insects, they have a hard outer covering that must be periodically shed as the animal grows.

Lesson Plan 16.2

Phase 1

PREPARATION

Key Concepts

- Some bacteria are helpful, since they cycle nutrients, fix nitrogen, and are used in the preparation of certain foods and drugs.
- Some bacteria are harmful, since they cause certain diseases and are responsible for food poisoning.
- Harmful bacteria can be controlled by proper sanitation, good hygiene, temperature treatments, antibiotics, and vaccinations.

Reading Strategy

Another pattern used in science textbooks is cause and effect. In this pattern, a direct relationship is shown between two things—one (the cause) is responsible for a certain result (the effect). Have students list cause-and-effect relationships as they read this section.

Phase 2

TEACHING STRATEGIES

Visual Strategy

Figure 16.5
Use this figure to introduce the ways in which bacteria are both helpful and harmful. Emphasize that only a small percentage of bacterial species are harmful. Many more are not only beneficial but also necessary, such as the decomposers and nitrogen-fixing bacteria.

THE DISEASE TUBERCULOSIS IS CAUSED BY THE BACTERIUM *MYCOBACTERIUM TUBERCULOSIS*. TUBERCULOSIS PATIENTS ARE OFTEN TREATED WITH STREPTOMYCIN, AN ANTIBACTERIAL DRUG THAT IS PRODUCED BY BACTERIA OF THE GENUS *STREPTOMYCES*. THIS EXAMPLE SHOWS HOW BACTERIA ARE BOTH HARMFUL AND BENEFICIAL TO HUMANS.

16.2   How Bacteria Affect Humans

Objectives

1. **Describe three beneficial effects of bacteria.**
2. **List five human diseases caused by bacteria.**
3. **Summarize three ways to prevent bacterial diseases.**
4. **Recognize the importance of antibiotics in fighting bacterial diseases.**

Beneficial Bacteria

Although you probably know that some bacteria can cause harm—such as when they destroy food or make you ill—you might not be fully aware of the tremendous benefits bacteria provide. For instance, bacteria maintain crucial links in nutrient cycles that make essential elements such as nitrogen and sulfur available to plants and, indirectly, to humans. Bacteria are also very important in the manufacture of food and life-saving drugs. **Figure 16.5** summarizes the ways bacteria affect humans in their everyday lives.

Figure 16.5
Bacteria in a soybean field may have beneficial effects, whereas other bacteria can have damaging effects on human beings.

Decomposers are nutrient recyclers

Recall from Chapter 12 that decomposers are organisms that return nutrients to the environment by breaking down organic matter. When a plant or animal dies, bacteria from the air and soil settle on the dead organism. The bacteria begin to grow, releasing carbon dioxide, water, nitrogen, phosphorus, and sulfur—nutrients that plants need to grow. Without decomposers, most of these nutrients would be locked away in the bodies of dead organisms.

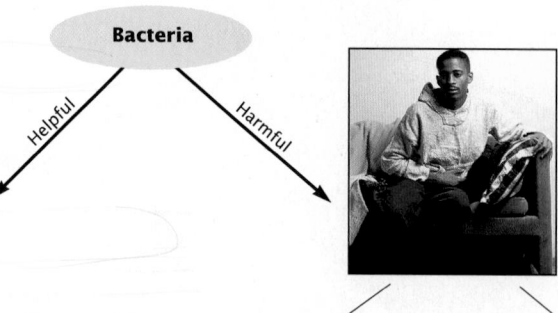

Bacteria

Helpful

Harmful

| Nutrient cycling | Nitrogen fixation | May cause diseases | May cause food poisoning |

−NH₂

N−H H

ATMOSPHERE N₂ ——— BACTERIA NH₃ SOIL

Section 16.2

Nitrogen-fixing bacteria enrich the soil

Interacting Systems

How would an ecosystem be affected if its nitrogen-fixing bacteria were destroyed?

Because nitrogen is a component of proteins, plants cannot photosynthesize, grow, or reproduce without it. Much of the nitrogen available for plants is produced by nitrogen-fixing bacteria. Nitrogen-fixing bacteria transform atmospheric nitrogen, which cannot be absorbed by plants, into ammonia, a nitrogen compound that plants can absorb. No other organisms have this ability.

Nitrogen-fixing bacteria are found in the soil, in aquatic ecosystems, and within the roots of some plant species. The legumes, plants including the peas and beans, contain nitrogen-fixing bacteria in swellings on their roots. These bacteria enable legumes to grow in nitrogen-poor soils where other plants cannot.

Bacteria are used to manufacture food and drugs

Like the young woman and man in **Figure 16.6**, when you eat yogurt or olives, you are eating foods that are the product of bacterial decomposition. Humans have learned that decomposition is occasionally beneficial because it adds flavor to food. For example, bacteria convert cabbage and cucumbers into tangy sauerkraut and pickles.

Modern technology is taking advantage of the enormous genetic diversity among the bacteria. Using genetic engineering technology, biologists can now "reprogram" bacteria to manufacture any protein for which a gene has been isolated. For example, bacteria now produce most of the insulin needed by diabetics in the United States. Before genetic engineering, insulin had to be isolated from the pancreases of animals killed in slaughterhouses. Other drugs and products produced by genetically engineered bacteria are described in Chapter 8.

Figure 16.6
The young woman below is eating yogurt, while the young man on the right is eating olives. Yogurt and olives are made when the bacteria indicated are added during the food manufacturing process.

Theme Answer
Interacting Systems
Nitrogen-fixing bacteria provide nitrogen to plants and to the soil. Without these bacteria, the amount of available nitrogen would fall. Plant growth would be reduced, and the amount of energy available to higher trophic levels would decline.

Connection: Chapter 8
Genetic Engineering
Remind students that the bacterial genes for nitrogen fixation have been successfully inserted into plants. However, scientists have not yet been able to get these genes to function in their new hosts.

Demonstration 1
Benefits of Bacteria
Display some of the food products manufactured with the help of bacteria, including buttermilk, ice cream, yogurt, sour cream, olives, cheese, pickles, sauerkraut, and vinegar.

Streptococcus thermophilus and *Lactobacillus bulgaricus*

Leuconostoc mesenteroide and *Lactobacillus plantarum*

Yogurt

Pathogenic bacteria release TOXINS

Bacteria and Disease

Occasionally your body is a temporary home for parasitic bacteria with serious effects—they cause disease or infection. A disease-causing agent is called a **pathogen**. Pathogenic bacteria are harmful because they damage their host's tissues. This damage results from either direct attacks on the host's cells or from poisonous substances called toxins that many bacteria release.

causes DISEASE

How are bacterial diseases actually transmitted? Each species of pathogenic bacteria has a characteristic way of being carried to new hosts: in water, in the air, in food, by insects, or by direct human contact. **Table 16.1** lists some bacterial diseases and their modes of transmission.

Water can carry pathogens

Cholera is a serious disease transmitted through polluted water. Cholera bacteria produce a strong toxin that causes acute diarrhea and vomiting, which can lead to rapid dehydration. If untreated, this severe loss of water can be fatal within 24 hours. Cholera bacteria are spread in drinking water that has been contaminated by the feces of infected individuals. Any fish that are caught in contaminated water can also carry cholera bacteria. In 1991 a large cholera epidemic struck Peru and rapidly spread to the rest of South America and Central America. At least 1,700 people died in the first year of this devastating outbreak.

Table 16.1 Common Bacterial Diseases

Disease	Mode of Transmission	Symptoms
Tuberculosis	Airborne water droplets	Fatigue, persistent cough, bleeding in lungs; can be fatal
Diphtheria	Airborne water droplets	Fever, sore throat, fatigue
Scarlet fever	Airborne water droplets	Rash, fever, sore throat
Bubonic plague	Fleas	Swollen glands, bleeding under skin; often fatal
Typhus	Lice	Rash, chills, fever; often fatal
Tetanus	Dirty wounds	Severe, prolonged muscle spasms
Cholera	Contaminated water	Severe diarrhea, vomiting; often fatal
Typhoid	Contaminated water and food	Headaches, fever, diarrhea, rash; often fatal
Leprosy	Personal contact	Nerve damage, skin lesions, tissue degeneration
Lyme disease	Ticks	Rash, pain, swelling in joints

SALMONELLA

Figure 16.7
Just in case the chicken you are preparing is contaminated with *Salmonella*, you should use hot, soapy water to wash the surfaces that have been touched by the raw chicken.

BOTULISM

Figure 16.8
A sneeze sends thousands of bacteria into the air. Some of these bacteria may be pathogens.

Food can be contaminated by bacteria

Although certain bacteria can produce yogurt and cheese, most bacteria in food are not helpful. Pathogenic bacteria can contaminate foods and cause food poisoning.

One of the most dangerous kinds of food poisoning is botulism, which is caused by a toxin released by the bacterium *Clostridium botulinum*. Consumption of less than one-millionth of a gram of this toxin causes paralysis and death. *C. botulinum* normally lives in the soil, but it can grow in canned foods that have not been properly sterilized. Because oxygen kills botulism bacteria, they cannot grow in fresh and frozen foods, which contain oxygen.

Another type of food poisoning is caused by *Salmonella* bacteria found in pork, eggs, poultry, and other foods. The symptoms of *Salmonella* food poisoning are diarrhea, vomiting, and abdominal cramps. Although the symptoms usually pass quickly, a serious infection can cause dehydration, a drastic loss of water from the body. In the very young or in the elderly, severe dehydration can be fatal. Hot, soapy water will destroy *Salmonella* bacteria encountered during food preparation, as shown in **Figure 16.7**.

Bacteria travel through the air

A sneeze or cough sprays out a shower of tiny droplets, as you can see in **Figure 16.8**. Visible in the circular photograph are some of the 10,000 to 100,000 bacteria that are carried in the water droplets produced by a single sneeze. The bacteria that cause diphtheria, scarlet fever, whooping cough, and tuberculosis drift through the air in droplets such as these. Most airborne diseases affect the respiratory systems of their victims. Tuberculosis bacteria, for example, invade the lungs. The scars they leave are often visible as very large shadows on X rays of the lungs.

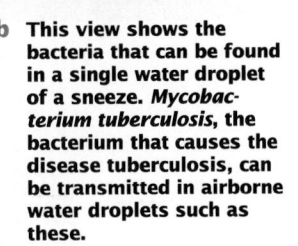

b This view shows the bacteria that can be found in a single water droplet of a sneeze. *Mycobacterium tuberculosis*, the bacterium that causes the disease tuberculosis, can be transmitted in airborne water droplets such as these.

Consumer Connection

Food Poisoning
Invite the school's human biology teacher or a local dietitian to speak to the class about food poisoning.

Demonstration 2
Clean Chickens
Wipe a sterile cotton swab across a piece of chicken, streak it on a sterile petri dish containing nutrient agar, and incubate at 37°C for 24–48 hours. Repeat this procedure, but this time wipe a piece of chicken that has been thoroughly rinsed with water. Compare the bacterial growth on the two cultures. Be sure to sterilize the bacterial cultures before disposing of them. Use the demonstration to discuss the dangers of *Salmonella* poisoning and the precautions to guard against it.

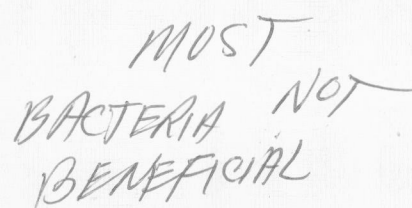
MOST BACTERIA NOT BENEFICIAL

Visual Strategy

Figure 16.9

Use this figure to point out that nearly 200 years elapsed between the discovery of bacteria and the awareness that they could cause a variety of diseases. Bacteria and other microorganisms were first seen and described by Anton van Leeuwenhoek, who worked with a simple microscope in the late 1600s. Not until the mid1800s, however, was the germ theory of disease accepted, as a result of the efforts of Louis Pasteur in France, Robert Koch in Germany, Joseph Lister in England, and Ignaz Semmelweis in Austria.

Demonstration 3

Effects of Temperature on Bacterial Growth

Have students design an experiment to test the effects of temperature on bacterial growth.

Controlling Bacterial Diseases

Today most Americans have little experience with serious bacterial diseases. Advances in medicine and sanitation have brought freedom from bacterial disease to the industrialized countries of the world only in the last 100 years. Much of the world's population is still subject to these life-threatening but preventable bacterial diseases.

Sanitation and hygiene prevent bacterial diseases

In the industrialized countries, drinking water is filtered and then purified with chlorine, a chemical that kills bacteria. Similarly, sewage is collected and treated to remove pathogens before it is discharged into rivers or the ocean. Thus, cholera, typhoid, and other diseases of contaminated water are almost unknown in the industrialized countries.

The clean conditions found in many of the industrialized countries are not characteristic of some of the less-developed countries. Because of poverty, many of these countries cannot provide clean drinking water for all of their citizens. In the poorest areas, as shown in Figure 16.9, the same river that provides water for drinking and bathing might also serve as a sewer. Each year about 25 million people die from typhoid, cholera, and other diseases of contaminated water in the less-developed countries.

Figure 16.9

a Water sanitation is of critical importance in controlling bacterial diseases. In poor, rural areas, such as this area of Peru, some of the poorest people are forced to cook, drink, and bathe in the same water.

b Modern methods of water purification, such as those used in these treatment plants, can almost eliminate the possibility of bacterial contamination of water.

VACCINE

Energy and Life

How do viruses reproduce if they cannot use energy?

Figure 16.10
The preparation of vaccines involves inactivation of the disease-causing organism and its subsequent introduction into the body.

Heat and cold protect food from bacterial contamination

Bacterial contamination of food can be prevented with either heat or cold. **Pasteurization** involves heating food so that the threat of bacterial contamination is removed. Pasteurization is doubly effective because it kills most bacteria and destroys their toxins. Pasteurization eliminates the possibility of contracting diseases such as botulism from canned food, or brucellosis or tuberculosis from milk.

Cooling food to a few degrees above freezing also prevents bacterial contamination. Although bacteria are not killed by the cold, their rate of growth is greatly slowed, and their population size remains small. Most salmonella infections are caused by food that has been left out of the refrigerator too long, allowing bacteria to grow.

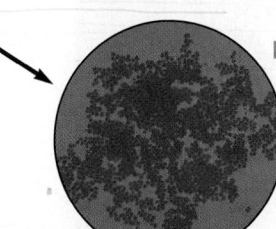

a Whooping cough vaccine is prepared by first culturing the bacteria responsible for whooping cough.

b The bacteria, *Bordetella pertussis,* are then killed using heat or chemicals.

c The dead bacteria are purified and cleansed, and the vaccine is packaged.

Vaccination stimulates the body's defenses

Preventing the contamination of food and drinking water are only two ways to protect yourself from infection. Vaccinating against disease is another way. Before you started school, you were probably vaccinated against bacterial diseases such as diphtheria, whooping cough, and tetanus. A **vaccine** is a solution containing pathogens or their toxins that have been made harmless, usually by treatment with heat, chemicals, or by genetic engineering. How does a vaccine protect you against pathogenic bacteria? As shown in **Figure 16.10**, a vaccine against whooping cough, for example, is a solution of whooping cough bacteria that have been killed with heat or chemicals. These dead bacteria cannot cause disease, but once inside your body they stimulate your immune system to make defenses against them. Once you have been vaccinated against whooping cough, your immune system has defenses ready to destroy any live whooping cough bacteria before they have a chance to make you ill.

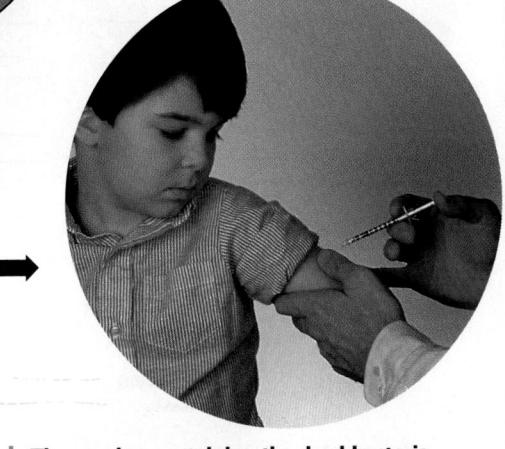

d The vaccine containing the dead bacteria is injected. This child's body will now produce antibodies that fight *Bordetella pertussis.*

■ *Cultural Perspective* ■

Medicinal Models
The Egyptians and ancient Chinese used mold from bread and cereal to treat wounds and infections. Thus, it is evident that the Egyptians and Chinese were aware of the healing properties of some molds long before penicillin was discovered.

351

PENICILLIN MOLD STOP BACTERIAL GROWTH

Figure 16.11
Use this figure to point out that Fleming made another serendipitous find. Suffering from a bad head cold, Fleming was working in his lab when a droplet fell from his nose onto a culture dish containing bacteria. He noticed that bacteria near the area where the droplet had fallen were slowly destroyed. Subsequent work revealed that the droplet contained an enzyme called lysozyme. This enzyme, found also in tears and saliva, breaks down the cell walls of many types of bacteria. Have students discuss what benefit is derived from having lysozyme in these bodily fluids.

Phase 3

ASSESSMENT OPTIONS

Closure Strategy

Treating a Bacterial Infection
Have students assume that they are suffering from a painful sore throat. Have them describe how they would determine which antibiotic to use to treat their sore throat. What experimental procedures would they employ? Would they start by experimenting on humans?

Section Review

Assign the *Section Review*.

Reteaching

Have students hypothesize what would happen if all the bacteria in the world were destroyed.

Figure 16.11
Alexander Fleming's discovery of penicillin is one of the most important medical milestones of the century. Millions of lives have been saved because of his observations. The fungus that produces penicillin is shown below.

Antibiotics are used to treat bacterial diseases

Before the 1940s, doctors had few treatments for bacterial diseases. Whether a patient recovered or died often depended more on the type of disease and the strength of the patient than it did on the efforts of the doctor. This grim situation changed because of a chance event in 1928. In that year, Alexander Fleming, the British physician shown in **Figure 16.11**, found a blue-green mold growing on one of his bacterial samples. At first Fleming was angry because his experiment had been contaminated. Before throwing out the sample, however, he noticed that bacteria did not grow near the mold. The mold was apparently releasing a chemical that was poisonous to bacteria. Fleming isolated this substance and named it **penicillin**, after the *Penicillium* mold that produced it.

In the early 1940s, scientists following up on Fleming's discovery showed that penicillin was a very effective treatment for many bacterial diseases. Antibacterial drugs such as penicillin are called **antibiotics**. There have been many other antibiotics discovered since penicillin; tetracycline and streptomycin are two common examples. Some antibiotics prevent bacteria from making new cell walls.

a **The genus *Penicillium* is a common fungus. It is the common blue mold often seen on an old orange.**

b **The organism is microscopic and reproduces by forming spores on the ends of fungal branches. It is easy to culture, and modern manufacturing methods can produce large quantities of penicillin.**

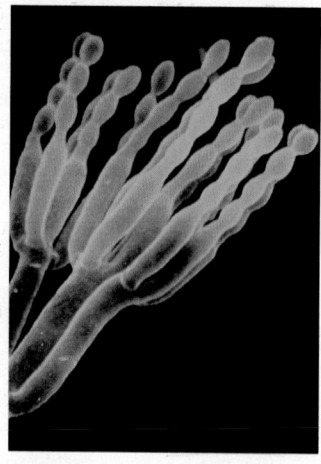

c **Penicillin is still one of the most common medicines used today to combat bacterial infections.**

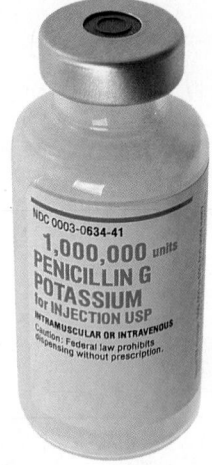

Section Review

❶ Describe one way in which bacteria are used to treat a disease.

❷ Name four diseases caused by bacteria.

❸ Explain why refrigeration and pasteurization are effective in preventing bacterial contamination of food.

❹ Suggest a hypothesis to explain why antibiotics will kill bacteria without destroying cells in your body.

▪ *Section Review Answers* ▪

1. Answers may vary. One possible answer might be that through genetic engineering, bacteria can be reprogrammed to manufacture any protein for which a gene has been isolated, as in the case of bacteria used to produce insulin for diabetics.

2. Answers may vary. See Table 16.1 on page 348.
3. Cooling food to a few degrees above freezing inhibits bacterial growth. Pasteurization is effective because it kills most bacteria and destroys their toxins.

4. Answers will vary. One possible hypothesis is that your cells lack cell walls, the target of many antibiotics.

THE WORLD HEALTH ORGANIZATION ESTIMATES THAT MORE THAN 10 MILLION PEOPLE THROUGHOUT THE WORLD ARE INFECTED BY **HIV**, THE VIRUS THAT CAUSES **AIDS**. THE NUMBER OF INFECTED PEOPLE GROWS RAPIDLY EACH YEAR. TO FIND A CURE FOR **AIDS** OR A VACCINE AGAINST **HIV** WILL REQUIRE A CLEAR UNDERSTANDING OF WHAT VIRUSES ARE, HOW THEY REPRODUCE, AND HOW THEY AFFECT THEIR HOSTS.

16.3 Viruses

NOT A CELL — genetic material + protein coat

Objectives

❶ **Describe the structure of a virus.**

❷ **Explain why viruses are not living organisms.**

❸ **Describe how a virus reproduces.**

❹ **List four diseases that are caused by viruses.**

What Is a Virus?

Think back to the last time you had a cold or the flu. Did you ask the doctor for antibiotics to kill the flu or cold "bacteria" that were making you so miserable? If you did, the doctor would have explained that colds and flu are not caused by bacteria, so antibiotics would have no effect. Colds and flu are caused by **viruses**. Viruses are microscopic particles that invade the cells of plants, animals, fungi, and bacteria. Viruses often destroy the cells they invade.

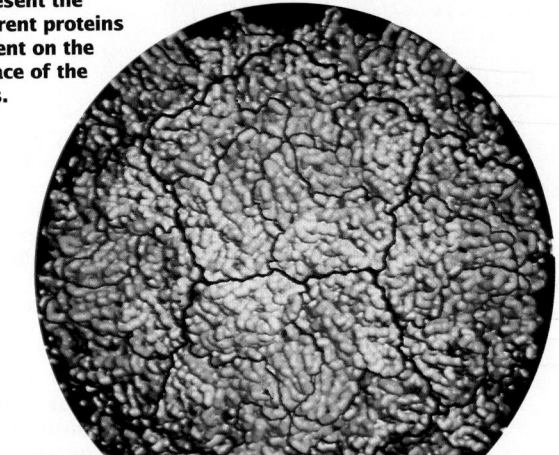

Figure 16.12
The computer-generated colors on this image of the polio virus represent the different proteins present on the surface of the virus.

Viruses are small, simple particles

If you could open a virus, what would you find inside? Would you find cytoplasm, ribosomes, and mitochondria, like you would find in one of your own cells? No, because a virus is not a cell. A typical virus, such as the polio virus shown in **Figure 16.12**, is composed of a core of genetic material surrounded by a protein "coat." The protein coat protects the genetic material and enables the virus to invade its host cell. The *Tour of a Virus* on the next page shows the structure of the Human Immunodeficiency Virus (**HIV**). This virus causes Acquired Immune Deficiency Syndrome (**AIDS**).

In many viruses, DNA is the genetic material. A few viruses have RNA instead and are known as RNA viruses. The viruses that cause AIDS, polio, and the flu are RNA viruses. Viruses are parasitic and can only reproduce inside the cells of their hosts. Because viruses are so small, viral genetic material has room for only a few genes, usually only genes coding for the protein coat and for enzymes that enable the virus to take over its host cell.

■ Matter of Fact ■

Viruses
Viruses were first detected 100 years ago, when scientists were searching for something smaller than bacteria that could cause a disease in tobacco plants. Known as tobacco mosaic disease, the disease caused the leaves to develop a patchwork of

light- and dark-green areas. In the late 1800s, this disease was destroying the tobacco crop in Russia. In 1892, a Russian biologist, Dimitri Iwanowski, discovered that whatever caused this disease was smaller than a bacterium and was contagious. In 1935, an American biologist, Wendell

Stanley, isolated the tobacco mosaic virus.

Lesson Plan 16.3

Phase 1
PREPARATION

Key Concepts

- Viruses are small particles, consisting of a nucleic acid core surrounded by a protein coat which, in some viruses, are covered by a layer of lipids.
- Although viruses have genetic material and can evolve, they are not considered living since they are not made up of cells, cannot use energy, and can reproduce only when inside living cells.
- Viruses reproduce by taking control of a host cell, where new virus particles are assembled.
- Viruses cause many diseases, including AIDS, smallpox, polio, flu, and colds.
- Vaccinations can protect an individual against certain viral diseases.

Reading Strategy

A third pattern often seen in science textbooks is the problem-solving pattern, which gives a description of how a particular problem or question was solved through experimentation. As students read about Jenner's work in this section, have them answer the following questions, which help to identify a problem-solving pattern.

- What was the problem Jenner was attempting to solve?
- How did Jenner conduct his experiments?
- How did his results lead to a solution of the problem?

Phase 2
TEACHING STRATEGIES

Connection: Chapter 7

Viruses
Review the experiments of Hershey and Chase.

Using the Feature

Human Immunodeficiency Virus (HIV) is the cause of the disease AIDS. The World Health Organization estimates that 40 million people will be infected with HIV by the year 2000. HIV is a retrovirus. Retroviruses have RNA as their genetic material and carry an enzyme, reverse transcriptase, that makes a DNA copy of their RNA when they infect a cell. Not all RNA viruses are retroviruses (the influenza virus, for example). Use this feature to highlight the differences between viruses and bacteria. Point out the lack of cell structure and metabolism in HIV. Also point out that HIV, a fairly large virus, is about one-tenth the size of *E. coli.*

Discussion

Guide the discussion by posing the following questions.

1. How is the genome of HIV different from the genome of a cell?
   HIV's genome is composed of RNA and has very few genes.

2. Could a particle of HIV lacking reverse transcriptase replicate itself? Explain.
   No. Reverse transcriptase makes a DNA copy of HIV's genome. It is this copy that takes over an infected cell and directs the production of new viruses.

3. Penicillin blocks cell wall synthesis. Why can't penicillin be used to cure AIDS?
   Like all viruses, HIV lacks a cell wall and so is unaffected by penicillin (and all other antibiotics).

4. Identify two functions of HIV's proteins.
   Reverse transcriptase copies the virus's genome; surface proteins enable the virus to recognize and enter a host cell; the protein core protects the genome.

Tour of a Virus

The disease AIDS is caused by a human virus called HIV, human immunodeficiency virus.

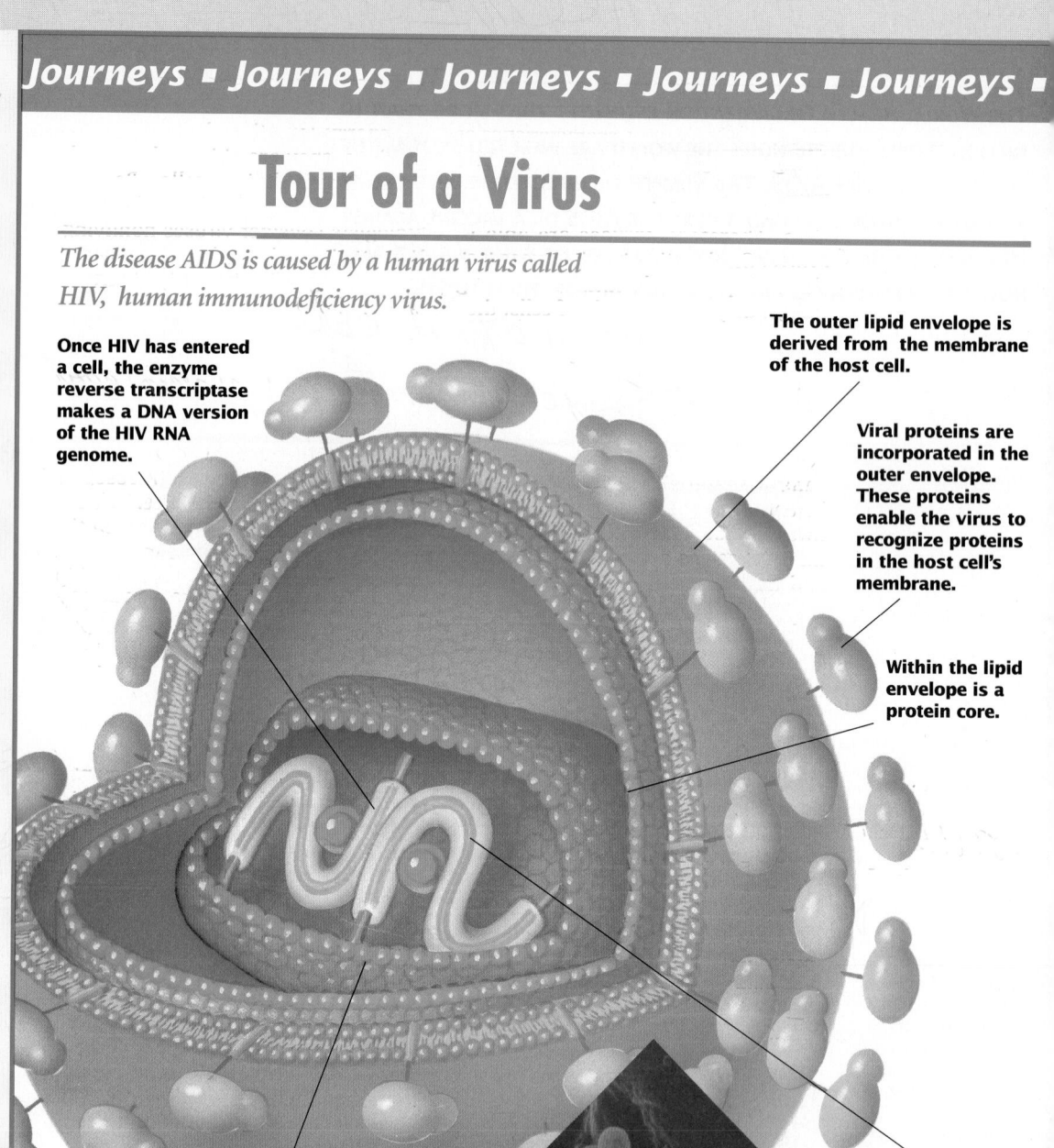

Once HIV has entered a cell, the enzyme reverse transcriptase makes a DNA version of the HIV RNA genome.

The outer lipid envelope is derived from the membrane of the host cell.

Viral proteins are incorporated in the outer envelope. These proteins enable the virus to recognize proteins in the host cell's membrane.

Within the lipid envelope is a protein core.

The HIV genome is a strand of RNA with three major genes and five smaller genes. Occasionally, a particle of HIV contains two copies of the RNA.

A protein coat surrounds the RNA, which is the genetic material of HIV.

The DNA copy of the HIV genome takes over the host cell and directs the production of thousands of new HIV particles, which bud out of the cell or are released when the cell bursts.

NON LIVING

Are viruses alive?

Viruses do have some characteristics of living things. They have genetic material that is transmitted to future generations and that can change over time. Therefore, viruses are able to evolve. On the other hand, viruses lack three very critical features of living things: they are not made of cells, they cannot make proteins, and they cannot use energy. Even though viruses can reproduce, they are able to do so only when inside living cells. Because viruses lack these essential features, biologists consider viruses nonliving. Nevertheless, since viruses are active inside living cells, the study of viruses is part of biology.

How Viruses Reproduce

To reproduce, a virus must insert its genetic material, which contains the instructions for making new viruses, into a host cell. This viral genetic material seizes control of its host cell and transforms it into a virus factory. All viruses, whether they attack your cells or bacteria, reproduce by taking over the reproductive machinery of a cell. Follow how HIV reproduces in the example below.

HIV seizes control of host cells

HIV cannot infect a cell it cannot enter. HIV is able to enter only those cells that have a particular receptor protein in their cell membranes. HIV recognizes these cells because the virus contains a protein that will bind to the receptor protein in the cell membrane. You can think of the relationship between the viral protein and the cell membrane protein as being like the relationship between a key and lock. If the key (the viral protein) fits the lock (the cell protein), then the cell is opened to invasion. Otherwise, the virus is locked out.

The main host cell for HIV is a type of white blood cell known as the helper T cell. Helper T cells occur in the blood and in the lymphatic system. They play a very crucial role in the body's ability to fight infection. Follow the events that occur when HIV infects a helper T cell in **Figure 16.13**.

Figure 16.13
HIV enters a human helper T cell and then takes over the cell's machinery. The cell is directed to make HIV proteins and RNA. Newly made viruses then leave the cell and continue the cycle of infection.

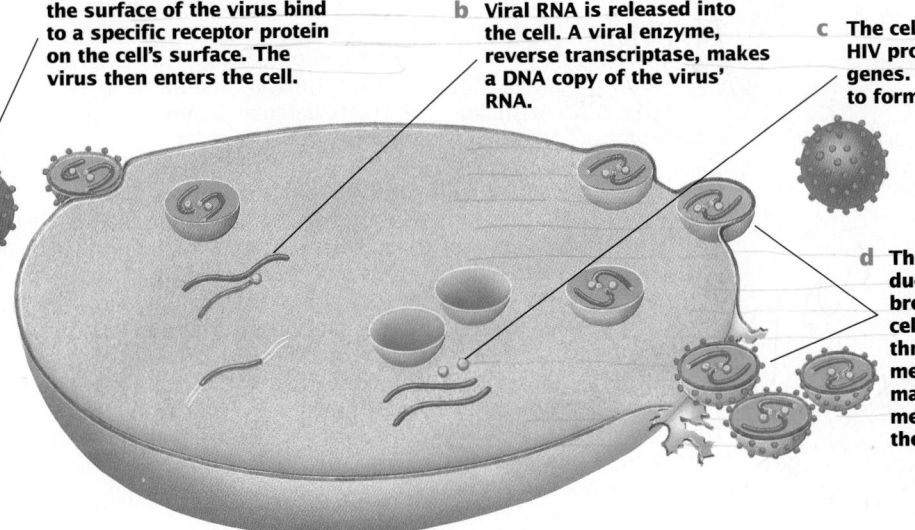

a The knob-shaped proteins on the surface of the virus bind to a specific receptor protein on the cell's surface. The virus then enters the cell.

b Viral RNA is released into the cell. A viral enzyme, reverse transcriptase, makes a DNA copy of the virus' RNA.

c The cell then produces HIV proteins and genes. These assemble to form new viruses.

d The newly produced viruses may break out of the cell by budding through the cell membrane. Or they may burst the cell membrane, killing the cell.

Connection: Chapter 2

Characteristics of Life
Review the five characteristic properties of all living things. Be sure that students understand that since viruses are not considered living, they are not assigned to any of the five kingdoms.

Language Connection

Virus
Virus is derived from the Latin word *virus*, which means poison. The ability of a virus to cause disease is known as virulence. Have students explain why both virus and virulence are appropriate terms.

Visual Strategy

Figure 16.13
Use this figure to review how virus infection and reproduction occur. In Figure 16.13a, point out the connection to Chapter 3, where the roles of the various cell membrane proteins were discussed.

Mathematics Connection

Viral Reproduction
A single virus that infects a bacterial cell can produce about 100 virus particles in 20 minutes. Have students calculate how many viruses there would be four hours after a single bacterial cell infected with a single virus was placed in a bacterial culture.

Health Connection

AIDS

Invite the school's health teacher or nurse to discuss HIV and AIDS with the class. Have students follow up this presentation by arranging a visit to the library to search for current information in available scientific resources.

[handwritten margin notes: MEASLES →; INTERFERE ① cell processes / protein production OR cell wall production; antibiotics WORK BY; VIRUSES NOT ALIVE "cannot be KILLED"; AZT]

AIDS is fatal

HIV causes AIDS by killing helper T cells, which coordinate the body's attack on pathogens. However, HIV may not begin to destroy large numbers of helper T cells for a period lasting from a few months to over 10 years after infection. During this period, an infected person may experience only mild or flu-like symptoms, or no symptoms at all. An infected person can transmit HIV to others during this period. Scientists now think that practically everyone infected with HIV will develop AIDS. Eventually, the virus begins to destroy a large percentage of an infected person's helper T cells. When the number of helper T cells has fallen to very low levels, the person is said to have AIDS. AIDS patients usually die of cancer or diseases a healthy immune system would defeat. You will read more about HIV transmission and AIDS prevention in Chapters 32 and 34.

Figure 16.14
The tell-tale blotchy rash of measles has become relatively uncommon since the development of the measles vaccine.

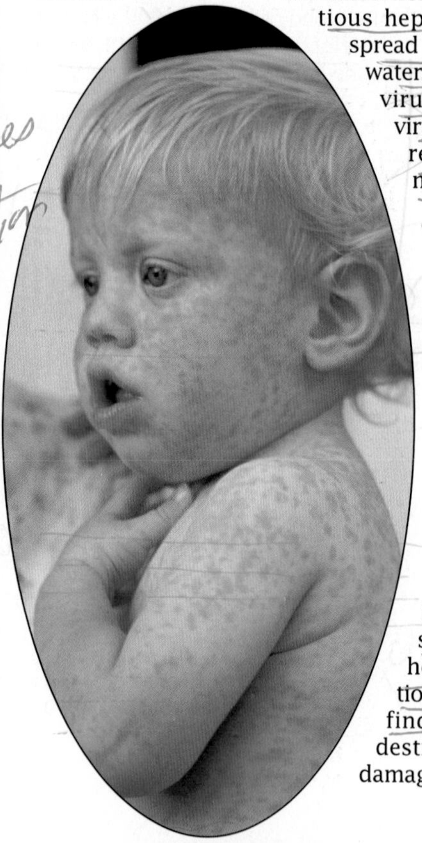

Diseases Caused by Viruses

Viruses cause many serious human diseases in addition to AIDS, such as measles, shown in **Figure 16.14**, and smallpox. Like pathogenic bacteria, pathogenic viruses are transmitted from host to host in characteristic ways. Most of the viral diseases listed in **Table 16.2** are airborne. A few, such as infectious hepatitis and polio, can spread through contaminated water. Insects also transport viruses. The yellow fever virus, common in tropical regions, is carried by mosquitoes.

There are defenses against viruses

Why don't physicians treat viral diseases such as colds, flu, and AIDS with antibiotics? Antibiotics work by interfering with cellular processes such as protein production or cell-wall synthesis, which do not occur in viruses. Moreover, since a virus uses its host cell for reproduction, it is very difficult to find any drugs that will destroy the virus without damaging the host.

A drug called azidothymidine (AZT) blocks an enzyme essential for DNA replication. Many AIDS patients are now being treated with AZT, which can prolong the lives of many of these patients. Unfortunately AZT cannot cure AIDS; it only slows the course of the disease. Furthermore, AZT is very toxic and has many side effects.

Vaccination also protects against viral diseases

Vaccination is the only effective defense against most viral diseases. Recall that vaccines against bacterial disease are composed of dead bacteria. But viruses are not alive and cannot be killed. Instead, vaccines against viral diseases contain viruses made harmless by treatment with chemicals or by genetic engineering. These harmless viruses stimulate the immune system to create defenses against the harmful form of the virus.

Smallpox is no longer a killer

Eliminating smallpox from the world is vaccination's greatest triumph. Smallpox virus produced tiny pustules or sores (small "pox") on its victim's skin. These pustules developed scabs and often turned into permanent, disfiguring scars. Far worse, smallpox killed half of those who contracted it.

Because smallpox was so deadly, cures and ways to prevent smallpox had been sought for centuries.

▪ Matter of Fact ▪

Vaccines and Vaccinations

The first vaccination, as pointed out in the text discussion of smallpox, was administered by Edward Jenner in 1798. In 1894, Louis Pasteur developed the first rabies vaccine. In 1954, Jonas Salk developed the first polio vaccine. He prepared the vaccine by treating the virus with formaldehyde. In the early 1960s, Albert Sabin developed an oral vaccination to protect against polio. Today, scientists are attempting to develop a vaccine against HIV. Tell students that the best natural defense against viral infections is the body's immune system.

In 1798, an English doctor named Edward Jenner, shown in **Figure 16.15**, discovered a way to immunize people against smallpox. He noticed that milkmaids, who were continually exposed to cattle, did not develop smallpox. He formed the hypothesis that milkmaids developed immunity to smallpox because they had been exposed to cowpox, a disease like smallpox that infects cattle but produces only very mild symptoms in humans. Jenner tested his hypothesis by injecting seepage from cowpox sores into a boy. The boy did not get smallpox, even when Jenner deliberately exposed him to smallpox virus.

Vaccination against smallpox became commonplace in the industrialized countries. As a result, smallpox rapidly disappeared from these countries. The last case in the United States occurred in 1949. Because the disease persisted in some of the poorest nations, the World Health Organization launched a worldwide vaccination campaign against smallpox. In 1967, the year this campaign was launched, 10 million to 15 million cases of the disease occurred. Just 11 years later, smallpox was eliminated. The last known person to contract smallpox is shown in **Figure 16.16**.

Why does injection with cowpox virus provoke immunity to smallpox? Jenner couldn't explain how his vaccine worked, but scientists now can. The cowpox virus and the smallpox virus have very similar protein coats. After vaccination, the immune system creates defenses that recognize the shape of the cowpox protein coat. These defenses cannot differentiate a cowpox virus from a smallpox virus, so they destroy either virus whenever it is encountered.

Visual Strategy

Figure 16.16

Use this figure to point out that vaccination has eliminated smallpox. As late as 1950, there had been as many as 2 million new cases of smallpox every year. In 1977, the World Health Organization declared smallpox extinct and recommended that laboratories destroy their supplies of any remaining viruses. However, the story of the conquest of smallpox had a tragic ending. The following year, a medical photographer in England contracted smallpox. She had been working at a university that was in the process of destroying its supply of smallpox virus stored in a laboratory. However, samples of the virus escaped through the ventilating system. Not having been vaccinated, the photographer contracted the disease. Feeling responsible for what had happened, the head of the laboratory committed suicide. The photographer died five days later.

shape of Protein coat

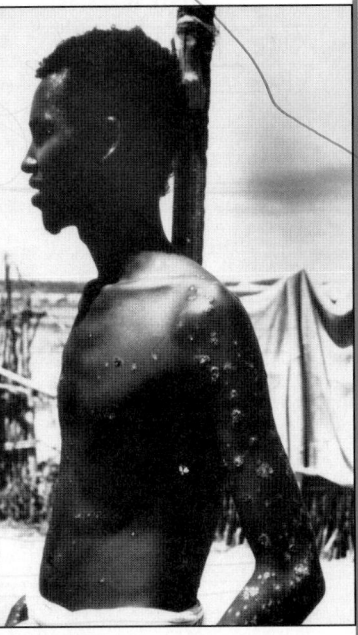

Figure 16.16
Ali Maow Maalin of Merka, Somalia, contracted smallpox in 1977 at the age of 23. His was the last known case of smallpox in the world.

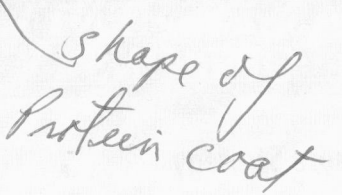

Figure 16.15
Edward Jenner was able to decrease the threat of smallpox by vaccinating people with the cowpox virus.

Connection: Chapter 6

Mutations

Remind students that a change in a gene is called a mutation. Review how a change in DNA will result in a change in RNA, and, in turn, a change in the protein that is synthesized.

Visual Strategy

Table 16.2

Have students identify the viral diseases for which they have been vaccinated. Point out that viruses have also been implicated in the production of certain cancers in humans. Viruses have long been known to cause cancer in animals, including chickens, mice, and cats. The viral genes responsible for promoting cancer are called oncogenes.

Phase 3

ASSESSMENT OPTIONS

Closure Strategy

A Viral Epidemic

Have students assume that a viral epidemic has been declared in their community. Discuss what steps should be taken to control the epidemic and eventually eliminate it.

Section Review

Assign the *Section Review*.

Reteaching

Invite a physician to speak to the class about viruses and viral diseases. Have the class review the information in this section as a way of preparing questions to ask the physician.

Evolution

Why is it difficult for researchers to develop effective vaccines against those viruses that change the shapes of their surface proteins?

Vaccines do not protect against all viral diseases

Why has it been so difficult to develop a vaccine against HIV? And why do millions of us suffer from colds and flu each year? Unfortunately, vaccination is effective only against viruses with nonvarying proteins on their surfaces, proteins that remain unchanged generation after generation. Smallpox virus, measles virus, and polio virus have surface proteins that remain the same from one viral generation to the next. HIV, cold viruses, and flu viruses, however, have surface proteins whose genes mutate often. As a result, the shape of these surface proteins often changes over just a few generations. Such a change produces a form of virus not recognized by the immune system. Although you can be vaccinated against influenza, each kind of influenza vaccine produces immunity against just one form, or strain, of the virus. That is why a person vaccinated against the flu can get the flu many times in a lifetime, each time from a different strain. Of all the changeable viruses, HIV changes most rapidly. **Table 16.2** lists some common viral diseases.

Table 16.2 Common Viral Diseases

Disease	Transmitted by	Symptoms
Chickenpox	Air currents	Rash, fever
Measles	Air currents	Blotchy rash, high fever, congestion in nose and throat
Rubella (German measles)	Air currents	Rash, swollen glands
Mumps	Air currents	Swollen salivary glands
Influenza (flu)	Air currents	Headache, muscle aches, sore throat, cough; historically, one of the great "killer" diseases
Smallpox	Air currents	High fever, pustules on skin; often fatal; now extinct
Infectious hepatitis	Contaminated food or water	Fever, chills, nausea, swollen liver, jaundice, pain in the joints
Polio (Polio virus)	Contaminated food or water	Headache, stiff neck, possible paralysis
Yellow fever	Mosquitoes	Nausea, fever, aches, liver cell destruction; can be fatal
AIDS (HIV)	Sexual contact, contaminated blood products, contaminated hypodermic needles and syringes	Immune system failure; fatal

Section Review

❶ Draw a virus and label its parts.

❷ Describe three ways in which viruses differ from living cells.

❸ How does a virus reproduce?

❹ Explain why a vaccination against influenza will not give you lifelong protection.

■ *Section Review Answers* ■

1. See *Tour of a Virus* on page 354.
2. Viruses are not made up of cells; they cannot make proteins; they cannot use energy.
3. A virus reproduces by taking over the reproductive machinery of its host cell, where new viruses are assembled.
4. Influenza viruses mutate rapidly. Hence, the proteins on their outer surface change swiftly. The body does not recognize altered viruses as pathogens it has been vaccinated against. These viruses can therefore cause disease.

Chapter **16** *Highlights*

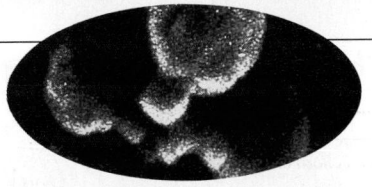

These are human immunodeficiency viruses seen through a scanning electron microscope.

Alternative Assessment

Scientists have recently discovered two types of particles, viroids and prions, that are even smaller than viruses. Viroids have been identified as the cause of several diseases in plants. Have students use library resources to obtain information about these two particles in preparing a brief oral report to share with the class.

	Key Terms	Summary
16.1 Bacteria Unlike eukaryotic cells, bacterial cells do not have a nucleus. 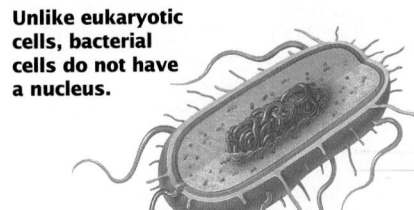	Gram staining (p. 340) flagellum (p. 341) chemosynthesis (p. 343)	• Bacterial cells, the oldest group of organisms on Earth, are much smaller than eukaryotic cells. • Most bacteria are spherical, spiral, or rod-shaped, and they have one of two types of cell walls. • Bacterial cells lack membrane-bound organelles and chromosomes. They have ribosomes, cytoplasm, and one molecule of DNA. • The kingdom Monera contains autotrophs and heterotrophs.
16.2 How Bacteria Affect Humans The discovery of penicillin by Alexander Fleming is one of the great scientific and medical achievements of this century. 	pathogen (p. 348) pasteurization (p. 351) vaccine (p. 351) penicillin (p. 352) antibiotic (p. 352)	• Beneficial bacteria include decomposers and nitrogen-fixing bacteria. Bacteria also make certain foods and drugs. • Many human diseases are caused by bacteria. • Vaccines help the body's immune system resist infection, thereby reducing the incidence of many bacterial diseases. • Antibiotics are antibacterial drugs. Alexander Fleming discovered the first antibiotic, penicillin, in 1928.
16.3 Viruses Viruses are responsible for many diseases, including polio, herpes, AIDS, and the common cold. 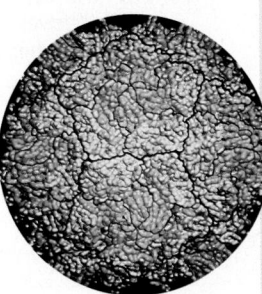	virus (p. 353) HIV (p. 353) AIDS (p. 353)	• Viruses are not living organisms. They are small particles that invade cells. • Viruses consist of RNA or DNA surrounded by a coat of protein. • A virus can only reproduce by controlling a cell. It may kill the cell or it may stay dormant. • Viruses cause many diseases, including AIDS. • Smallpox was eliminated by an intensive vaccination program. • Some viruses change rapidly, and it is hard to make a vaccine to combat the many forms.

Chapter Review Answers

Understanding Vocabulary
1. a. nonliving
   b. heat
   c. inorganic molecules
   d. harmless
   e. disease

Relating Concepts
2. Concept map answers are on page 337D.

Understanding Concepts
Multiple Choice

3. d	8. a
4. b	9. b
5. a	10. c
6. c	11. b
7. a	12. b

Completion
13. rod, spiral, sphere
14. Gram-staining; prescribe treatments
15. medicine and sanitation
16. immune system

Short Answer
17. Gram-negative bacteria have an additional layer that covers the cell wall. Some antibiotics, such as penicillin, interfere with cell wall synthesis. Because they cannot penetrate the additional layer around the cell membrane, these antibiotics do not affect Gram-negative bacteria.
18. See Table 16.1, page 348.
19. AZT does not cure AIDS, it only slows the replication of HIV.
20. Antibiotics interfere with cellular processes, such as cell wall synthesis and protein synthesis, that do not occur in viruses.
21. Ethically, Jenner's action would not be tolerated today, because the boy could have been infected with smallpox.

Chapter 16 Review

Understanding Vocabulary

1. For each set of terms complete the analogy.
   a. bacteria: living::viruses: _____
   b. refrigeration: cold::pasteurization: _____
   c. photosynthesis: light::chemosynthesis: _____
   d. pathogen:harmful::vaccine: _____
   e. HIV:cause::AIDS: _____

Relating Concepts

2. Copy the unfinished concept map below onto a sheet of paper. Then complete the concept map by writing the correct word or phrase in each oval containing a question mark.

Understanding Concepts

Multiple Choice

3. Which cell organelle is present in both bacteria and eukaryotic cells?
   a. nucleus   c. Golgi body
   b. chloroplast   d. ribosome

4. Which is *not* a source of energy used by bacteria?
   a. hydrogen sulfide
   b. viral proteins
   c. sunlight
   d. dead plants

5. Bacteria that cause disease are called
   a. pathogens.   c. antibiotics.
   b. vaccines.   d. Gram-positive.

6. Which of the following diseases is caused by a bacterium?
   a. hepatitis   c. cholera
   b. polio   d. yellow fever

7. Which bacterium causes food poisoning?
   a. *Salmonella*   c. *Streptococcus*
   b. *Lactobacillus* d. *Penicillium*

8. The bacterium that causes botulism
   a. produces a deadly toxin.
   b. causes fever and diarrhea.
   c. causes fatal lung infections.
   d. is transmitted by lice.

9. Penicillin was isolated and named by
   a. Louis Pasteur.
   b. Edward Jenner.
   c. Alexander Fleming.
   d. Robert Hooke.

10. Pathogenic bacteria are not in the drinking water of industrialized countries because
   a. these bacteria have been eliminated by vaccination.
   b. it has been boiled.
   c. it is filtered and purified with chlorine.
   d. it is pasteurized.

11. A vaccine can protect you against a disease because it
   a. destroys bacterial toxins before they can make you sick.
   b. stimulates your immune system against a pathogen.
   c. kills any pathogenic bacteria in your body.
   d. changes pathogenic bacteria into harmless bacteria.

12. Viruses are considered nonliving because they
   a. can evolve.
   b. cannot make proteins or use energy.
   c. are smaller than bacteria.
   d. lack genetic material.

Completion

13. Species of bacteria have one of three different shapes. The three shapes are _____ , _____ , and _____ .

14. The procedure used to distinguish between Gram-positive and Gram-negative bacteria is called _____ . The results of this procedure can help physicians to _____ _____ .

15. In industrialized countries, many deadly bacterial diseases have been all but eliminated in the last 100 years by advances in _____ and _____ .

16. AIDS is caused by a virus that attacks the _____ _____ .

Bacteria — may cause — ? — contain — ? — contain — ? — may cause — Viruses

can be treated with — ? — such as — tetanus — transmitted by — air — such as — measles

? — or — ? — or — ? — or — ?

can be prevented by — ?

Interpreting Graphics
22. viral RNA injected into host cell—b
   viruses are released—d
   new viruses are made—c
   a virus attaches to the host cell's surface—a

Reviewing Themes
23. Helper T cells coordinate the immune system's attack on pathogens. Reduction in the number of helper T cells impairs the body's ability to fight off disease.

24. Flu viruses mutate very rapidly. Hence, the shape of their surface proteins changes rapidly. A flu shot will provide protection against one form of flu virus, but not against one with slightly different surface proteins.

Short Answer

17. How do the cell walls of Gram-positive bacteria differ from those of Gram-negative bacteria? How does this difference affect the effectiveness of antibiotics used to treat bacterial infections?

18. Identify a bacterial disease that is transmitted through the air, another bacterial disease that is transmitted by an insect, and another transmitted by contaminated water. What are the symptoms of each disease?

19. The synthetic drug AZT is used by many people who have AIDS. Does AZT cure AIDS? Explain your answer.

20. Explain why antibiotics work against bacterial diseases but not against viral diseases.

21. When testing his smallpox vaccine, Jenner deliberately exposed a boy to the smallpox virus. Was Jenner's action justified? Explain your answer.

Interpreting Graphics

22. Examine the figure shown below.

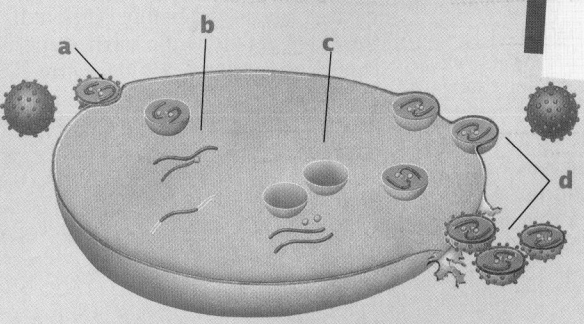

Match each lettered event in the figure with the sentence that best describes it.
  a. Viral RNA is injected into the host cell.
  b. Viruses are released.
  c. New viruses are made.
  d. A virus attaches to the host cell surface.

Arrange the events in the correct sequence. Using your imagination, describe some ways in which a drug or vaccine could affect events shown in the figures above.

Reviewing Themes

23. *Interacting Systems*
HIV attacks and destroys helper T cells. How does having fewer helper T cells affect the human immune system?

24. *Evolution*
Hundreds of thousands of people get flu shots each year. Yet many of them still get the flu. Why are flu shots often ineffective?

25. *Scale and Structure*
How does the protein on the surface of a virus affect the kind of cells it can attack?

Thinking Critically

26. *Inferring Conclusions*
What could happen if all heterotrophic bacteria were abruptly removed from Earth?

27. *Comparing and Contrasting*
Using AIDS as an example, distinguish between infection and disease.

28. *Building on What You Have Learned*
Which of the five properties of life described in Chapter 2 would have to be changed before viruses could be considered alive.

Cross-Discipline Connection

29. *Biology and History*
Robert Koch and Joseph Lister are two individuals associated with significant advances in the study of bacteria. Do library research and write short descriptions of the work of each individual.

Discovering Through Reading

30. Read the article "Control of Rabies in Wildlife," in *Scientific American*, June 1992, pages 86–92. What is the goal of the research discussed by Winkler and Bögel? What recent advances bring this goal closer to reality?

28. cellular organization, homeostasis, metabolism

Cross-Discipline Connection
29. Descriptions should have correct grammar, punctuation, and usage.

Discovery Through Reading
30. The goal of the research is to control rabies in wildlife through immunization. The development of new baits in which the vaccine can be concealed, new vaccines that can immunize the animal when eaten, and genetically engineered vaccines that cannot cause rabies are the important advances.

25. Before a virus can infect a cell, one of its surface proteins must bind to a protein on the cell's surface. This binding can only occur if the shape of the viral protein is complementary to the shape of the cellular protein. Thus, a virus can only infect a cell with matching proteins.

Thinking Critically
26. Heterotrophic bacteria are essential players in the nutrient cycles: they fix nitrogen and release nutrients by decomposing dead organisms. Without heterotrophic bacteria, far lower amounts of nutrients would be available in ecosystems, which would be able to support fewer organisms.

27. Disease is the ill effects caused by a pathogen. In AIDS, these effects include a depressed immune system, which leaves the sufferer open to attack by other pathogens. Infection occurs when a pathogen enters the body. A person can be infected with HIV for 10 years or longer before the disease AIDS develops.

Procedural Note
Review the answers to the preparation questions.

Prelab Preparation Answers

1. While viruses tend to be very small, there is a wide range of sizes, ranging from 0.01 μm to about 1 μm. Prokaryotic cells are about 1 μm in diameter, and eukaryotic cells are about 10 μm in diameter. Viruses can be small because they lack most of the structures found in living cells: organelles, chromosomes, and cytoplasm.

2. A typical virus has an inner core of nucleic acid and a protective outer coat of protein. In some viruses, the protein coat is surrounded by a layer of lipids and proteins called the envelope. The protein coat protects the genetic material and, in unenveloped viruses, enables the virus to enter its host cell. The nucleic acid carries instructions for making new viruses. A virus cannot carry out any cellular functions, has no organelles, and can only reproduce when inside a cell.

Procedure Answer

7. Answers may vary but should be logical.

Chapter 16 Investigation

How Are Models of Viruses Constructed?

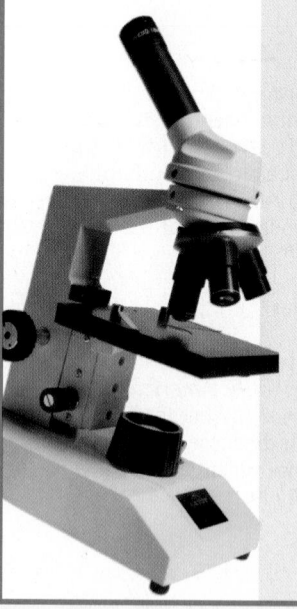

Objectives
In this investigation, you will:
- *construct* models of typical viruses
- *recognize* size relationships between different viruses

Materials
- toothpicks
- clay
- plastic foam in different shapes
- construction paper
- pipe cleaners
- wooden dowels
- wire hangers
- insulated electrical wire
- tape
- cork

Prelab Preparation
Review what you have learned about viruses by answering the following questions:
- Compare the sizes of viruses with the sizes of bacterial and eukaryotic cells. Why are viruses so much smaller than most cells?
- Draw the structure of a typical virus. Label the parts and describe the function of each part. How does a virus differ from a cell?

Procedure: Building Models of Viruses

1. Make a table similar to the one below. Use this table to record information about your models.

Virus	Actual size	Scaling factor	Scale size
Mumps	0.2 μm		
Potato X	0.01 x 0.5 μm		
Tobacco mosaic	0.018 x 0.3 μm		
Polio	0.028 μm		
HIV	0.1 μm		

2. Form a cooperative team of five students. Work as a team to complete steps 3–9. Each member of the team will take responsibility for the construction of one virus model.

3. Look at the sketches of the five viruses presented below. Although the sketches show the basic structure of the viruses, they are not drawn to scale. These viruses vary greatly in size, as shown by the data in the table.

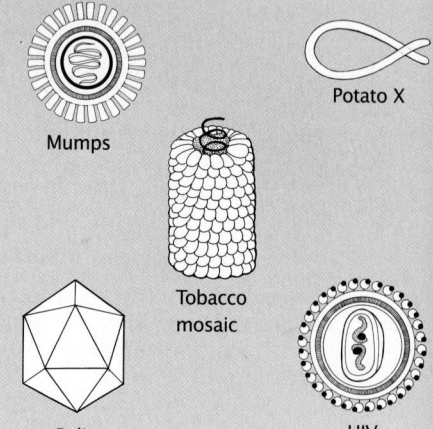

Mumps

Potato X

Tobacco mosaic

Polio

HIV

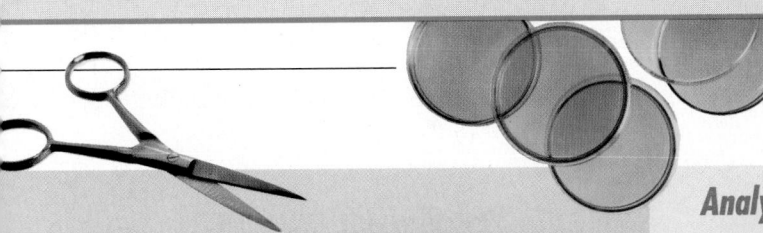

4. Use the information found in the sketches and the table to build three-dimensional models of each virus. Be as original as you wish in your choice of materials, *but all five models must be made correctly to scale.*

5. Begin by finding the range of sizes in these viruses. The smallest value is 0.01 μm, the diameter of the potato virus. The length of the potato virus is the largest value, 0.5 μm. By dividing the largest value by the smallest, you find that the largest model dimension is 50 times greater than the smallest model dimension. In other words, the *range factor* for these viruses is 50.

6. Next, decide on a practical scaling factor. The *scaling factor* is similar to the magnification power of a microscope. It indicates how many times larger than actual size you plan to make your model. To make this decision, you should consider the materials you plan to use and the range factor.

7. Before making a final decision on the scaling factor, however, you should also consider that the range factor is 50. So choose your scaling factor carefully, taking into account the amount of construction materials available. Record a scaling factor for your models. *Justify your choice of this scaling factor.*

8. Multiply the dimensions of each virus by your scaling factor and record that information in your table. Be sure to convert your units correctly.

9. Use your imagination and the data in your table to build your models. When finished, you and your classmates will display your work.

Analysis

1. *Analyzing Data*
   How many times larger is the mumps virus than HIV?

2. *Evaluating Methods*
   How are these models more useful than a drawing or a photograph?

3. *Applying Methods*
   A hypothetical virus has a diameter of 0.005 μm and a length of 0.6 μm. Based on your scaling factor, to what dimensions would you build a model of this virus?

4. *Recognizing Relationships*
   If your virus models were real, what kinds of biological molecules would make up their outer shells? What kinds of molecules would they have inside?

Thinking Critically

1. In Chapter 3 you learned about the limitations placed on cell size by surface area-to-volume ratios. Explain why the size of a virus is not limited by the ratio of its surface area to its volume.

2. Why is it advantageous for viruses to be small?

Analysis Answers

1. The mumps virus is twice as large as HIV.

2. The models allow you to visualize things in three dimensions. You can more easily see how the sizes of different viruses relate to one another. For example, you get a better idea of the relationship between diameter and length.

3. Answers will vary according to the scaling factor.

4. The outer layer of some viruses is made of protein, while other viruses are surrounded by a layer of lipids and proteins. Inside the virus is genetic material, either DNA or RNA.

Thinking Critically Answers

1. Viruses, unlike cells, do not absorb or excrete materials across a membrane. Therefore, their sizes are not limited by the amount of membrane surface area through which substances can pass.

2. A virus that enters its host cell must be smaller than the cell it enters. Small size also increases the likelihood of a virus being transmitted to a new host, especially for viruses transmitted through the air.

Using the Feature

Here is a summary of the Kimberly Bergalis case. In 1990, seven patients of a Florida dentist who had AIDS were found to be HIV positive. Further investigation revealed that five of these patients, including Kimberly Bergalis, had not engaged in behaviors that put them at risk for contracting HIV. In 1992, scientists from the Centers for Disease Control (CDC) in Atlanta showed that these five patients were probably infected by their dentist (for more information, see *Science* 256:1165-1171). The scientists could not pinpoint how the virus was transmitted, however. The dentist had used a mask and gloves and had always sterilized his instruments. Despite an exhaustive search, the CDC has found no other cases of transmission from health-care worker to patient.

The most commonly used test for HIV is the ELISA, which detects the presence of antibodies to HIV in the blood. If an ELISA test yields a positive result, a further, more conclusive test is often given. This test, the Western blot, detects the presence of viral proteins in the blood.

Discussion

Guide the discussion by posing the following questions.

1. Why was Kimberly Bergalis's case unusual?
She was infected through dental procedures, not through the usual routes of HIV transmission.

2. Do you think Kimberly Bergalis's estate is entitled to compensation from the dentist who infected her?
Students should give logical answers.

3. The ELISA test for HIV infection produces some false positives. That is, some people who test positive turn out not to be infected. Should the occurrence of false positives affect who has access to HIV test results?
False positives increase the likelihood of abuses of test results. For instance, an uninfected person could be denied health insurance based on the positive test result. False positives support the case for confidentiality of test results.

Testing for HIV:

Who Is at Risk?

If you have been to the doctor or dentist recently, you have probably noticed changes in their procedures. Rubber gloves and masks are precautions against the spread of HIV. Health-care workers now treat every patient as if he or she is infected with HIV. How should the patient assess the risk of contracting the virus from health-care workers?

Kimberly Bergalis's dentist was ill from AIDS but could still work. Eventually, however, he died from complications due to AIDS. Kimberly Bergalis later tested positive for HIV, the virus that causes AIDS. She was not an intravenous drug user, she did not have unprotected sex, and she had not had any blood transfusions. After some investigation, it was determined that Kimberly was probably infected by her dentist. There had never been a case of this sort, so doctors and scientists were skeptical of this finding. But when five more patients of this same dentist tested positive for HIV infection, the experts began to take Kimberly's case more seriously. As a result of her situation, Kimberly Bergalis became an advocate of mandatory HIV testing for health-care workers. She died of AIDS in 1991.

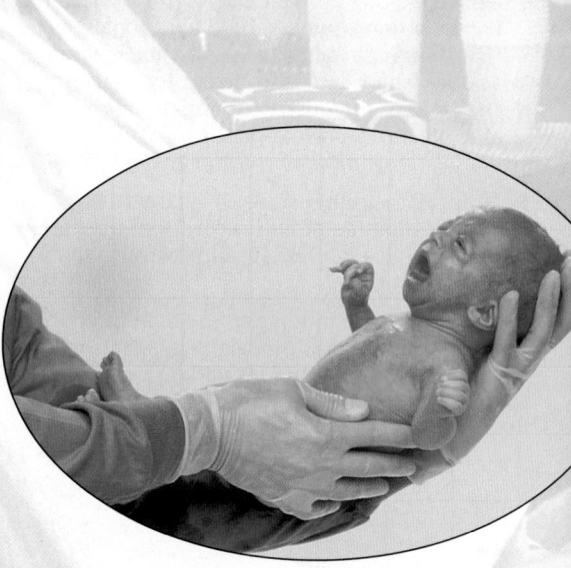

Belinda Mason was infected in the same way as Kimberly but voiced the opposite view on mandatory HIV testing. She feels that testing health-care workers will create the feeling that those infected with HIV are a threat to others.

■ *Science, Technology, and Society* ■ *Science, Technology,*

If Kimberly Bergalis could be infected by her dentist, then can dentists, doctors, and other health-care workers be infected by their patients? Some doctors have already stopped treating AIDS patients because they feel there is a threat to their careers and their lives.

Doctors who perform what are called "invasive procedures" are at a higher risk for contracting HIV from their patients. Invasive procedures include surgery, delivering babies, and extracting teeth. Procedures such as injections and blood transfusions put the physician or health-care worker at some risk because of possible needle sticks and an increased likelihood of exposure to blood.

One of the main questions raised by cases like Kimberly's is "Who should have to be tested?" Even though the number of patients infected with the HIV from health-care workers has been small, some people feel the potential risk is enough to require mandatory testing of doctors, nurses, and technicians. It is also felt that the results of these tests should be made known to patients with whom these health care workers have contact.

Doctors argue that if they are forced to be tested, they want their patients to be tested as well. Today, by law, a person cannot be tested for HIV without his or her consent. The actual testing isn't the only issue to be resolved. Of great concern is the fact that a doctor might refuse to treat a patient if he or she knows the patient is HIV positive. Recently, a man who was HIV positive tried to find a dentist. He told dentists of his condition, and of the 25 dentists he visited, none were willing to treat him.

The Centers for Disease Control (CDC) in Atlanta has issued guidelines for all health-care workers, including doctors, nurses, and dentists. The CDC requires wearing rubber gloves and careful handling of blood products. Health-care workers with open wounds should not perform any invasive procedures. These guidelines leave it up to the doctor, dentist, nurse, or technician as to whether they are tested for HIV or whether they disclose the result of a test.

Because of concerns that the guidelines were not strict enough, the Senate passed a bill that imposes stiff penalties on health-care workers who know that they have HIV and still perform invasive procedures. The bill specifically lists what procedures qualify for these restrictions. Legislation is being considered at the state level that would require doctors to inform patients of their HIV test status.

In a *Newsweek* poll taken in June 1991, 90 percent of those surveyed said they would like to know the HIV status of their doctor or dentist. Sixty percent thought that doctors who were HIV positive should not practice.

Thinking Critically

1. Students should give a logical answer.
2. Students should give a logical answer.
3. Students should give a logical answer.
4. Students should support their choice. Mandatory testing would lead to early identification of infected individuals, which would allow them to avoid infecting others and to receive treatment. It would also provide public health officials with better estimates of the numbers of infected individuals, so that resources for treatment and prevention can be better allocated. Mandatory testing could be abused to infringe on civil rights. Some doctors have refused to treat HIV-infected patients. Insurance companies may deny them coverage. Employers may refuse to hire infected individuals. Being HIV positive is also seen as a social stigma.
5. Students should give logical answers.
6. Both issues involve biological knowledge that could be misused.

Thinking Critically

❶ If you were a patient of Kimberly Bergalis's dentist, would you be tested? Why or why not?

❷ In your opinion, should doctors be forced to be tested for HIV?

❸ If your dentist tested positive for HIV, would you continue to be treated by this dentist? Why or why not?

❹ If everyone were tested for HIV when taking their driver's test, who should have access to the results? List three advantages and three disadvantages of this idea.

❺ Should doctors be allowed to pick their patients?

❻ What similarities are there between the HIV testing issue and the genetic screening testing issue on page 174?

Acting on the Issue

❶ Find out what legislation related to HIV testing may be under consideration in your state.

❷ Contact a local AIDS education organization and see if they can send a speaker to your class to discuss the testing issue.

❸ Find the AIDS hotline for your community. Call to find out where one can go for information and testing.

Chapter 17 *Protists*

Planning Guide

	Objectives/Themes	Classwork Resources	Homework Resources

17.1	**1.** Identify the characteristics shared by all members of the kingdom Protista.
	
	2. Describe the evolutionary relationship between bacteria and protists.
	
	3. Summarize the relationships of protists to the other three eukaryotic kingdoms.
	
	Themes: Evolution, Scale and Structure
		Teacher's Resource Binder
		Focus Activity 17
		  Observing Protists
		Lab Investigation 17.1
		  Observing Pond Water
		
		Other Resources
		Transparency 71
			Text
			Section Review, p. 369
			
			Teacher's Resource Binder
			Directed Reading Worksheet 17.1
			
			Other Resources
			Audiocassette 17.1*

17.2	**1.** Recognize the variation in shape, size, and method of obtaining energy among protists.
	
	2. Explain the role autotrophic protists play in ecosystems.
	
	3. Recognize the evolutionary connection between green algae and plants.
	
	4. Identify three kinds of heterotrophic protists.
	
	Themes: Energy and Life, Evolution, Interacting Systems
		Text
		Journeys
		  Tour of a Protist p. 371
		
		Teacher's Resource Binder
		Lab Investigation 17.2
		  Prokaryotic and Eukaryotic Algae
		Science Skills Worksheet
		  Biochemistry of Algae
		
		Other Resources
		Transparency 72
			Text
			Section Review, p. 375
			
			Teacher's Resource Binder
			Directed Reading Worksheet 17.2
			Reteaching Worksheet*
			  Some Common Protists
			
			Other Resources
			Audiocassette 17.2*

17.3	**1.** Explain how malaria is transmitted. **2.** Describe the methods used to control malaria. **3.** Explain why malaria-control methods have not been entirely successful. **4.** Name three human diseases caused by protists. **Theme:** Interacting Systems	**Text** Investigation *What Are Slime Molds Like?* pp. 382–383 **Other Resources** Transparencies 73–74	**Text** Section Review, p. 378 **Teacher's Resource Binder** Directed Reading Worksheet 17.3 Vocabulary Review Worksheet* **Other Resources** Audiocassette 17.3*

*Reteaching Options

Demonstrations
17.1: p. 368
17.2: pp. 370, 372, 374

Assessment
Chapter Review pp. 380–381
Portfolio Assessment p. 365D
Chapter Test—Teacher's Resource Binder
Test Generator

Research Notes

Connection to Marine Biology: What Causes Fish Kills?

Fish kills can be a frightening phenomenon. Suddenly, all of the fishes in a lake or river or part of the ocean float to the top of the water, dead and stinking. Scientists from North Carolina State University investigating fish kills in rivers along the southern part of the United States' Atlantic coast believe that the culprit is a protist, a phytoplankton that produces a fish-killing toxin so that it can feed on bits of dead fish flesh.

For years, scientists have known that similar "blooms" of dinoflagellates, diatoms, and other types of microalgae are responsible for the "red tides" that happen in different parts of the world. In this phenomenon, so many of the microorganisms are present that the water appears to turn red, yellow, or green. The fish-kill phytoplankton was more difficult to detect because it can cause massive fish kills even when present in small amounts.

Initially, the phytoplankton is in the form of inactive cysts in the sediments on the bottom of a riverbed. On the approach of live fishes, the cysts become photosynthetic dinoflagellates. These cells produce a neurotoxin that causes disorientation, lethargy, suffocation, and then death in both freshwater and saltwater fishes. Even shellfish, such as blue crabs and scallops, are often killed.

The dinoflagellates swim toward bits of dead fish flesh and attach themselves, digesting the debris. Some of the cells form gametes for reproduction. Others enter an amoeboid stage whose purpose is still unknown.

Immediately thereafter, the cells re-form cysts and sink to the bottom, ready to repeat the cycle. Because the cells attack almost any kind of fish, scientists believe that they may be stimulated by a common substance found in fish excrement. The reproduction of these cells is also enhanced by the presence of phosphates found in many fertilizers. The increase in fish kills could be due to agricultural pollution.

Although the dinoflagellate has only been conclusively linked to fish kills in the southeastern United States, the researchers from North Carolina believe that it is a worldwide phenomenon.

Researchers hypothesize that similar toxic blooms could also be responsible for the deaths of whales and porpoises. In 1991, pelicans in Monterey Bay, California, died from feeding on anchovy that had eaten diatoms containing another toxin. In the late 1980s, 26 people in Guatemala died and hundreds more became ill after eating shellfish that had absorbed a similar poison.

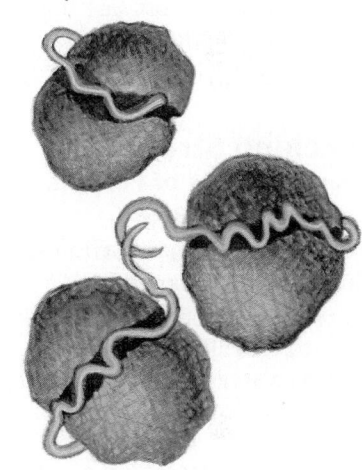

Investigation Notes

What Are Slime Molds Like?

pages 382–383

Purpose: This investigation gives students the opportunity to observe a living plasmodial slime mold and collect evidence of its diverse nature.

Prelab Preparation

1. If your compound microscopes are not shaped to allow use with a petri dish, a stereomicroscope can be substituted. However, cytoplasmic streaming may be difficult to observe.

2. If you have enough dissecting needles they can be used instead of sewing needles.

3. Plasmodial slime molds can be obtained from WARD'S (85 M 4750). Schedule the material to arrive about one week before use. Cytoplasmic streaming can best be observed on petri dish cultures of nutrient agar (WARD'S 88 M 1500). Make one dish for each team. The cultures can be reused over the course of the day.

4. Prepare the nutrient agar by melting in a water bath. Pour a 2–4mm layer of sterile nutrient agar into the bottom of the dish and allow to cool. Transfer a small portion of plasmodium to the agar and incubate at room temperature for 24 hours.

5. Since vinegar is an irritant, be certain students wear lab goggles.

Answers will be found on pages 382–383.

Meeting Individual Needs

RM

Objectives

1. Students will demonstrate appropriate use of core vocabulary for the chapter (Vocabulary File).

2. Students will identify characteristics of protists (Teaching Strategy A).

3. Students will name different protists that demonstrate the diversity of this kingdom (Teaching Strategy B).

Vocabulary File
(Developing Vocabulary/ Limited English Proficiency)

If you are not already using the Vocabulary File, refer to Chapter 1 for its preparation. See the Chapter Highlights on page 379 for a list of suggested words.

Teaching Strategy A
Characteristics of Protists
for pages 367–369
(Developing Organizational Skills/Visual Learners)

Preparation: About three weeks prior to teaching this chapter, set a jar of water on a window ledge. Green algae should form in that length of time. If an aquarium is set up in the classroom or lab, algae may already be forming in it.

Procedure: Use an overhead projector to fill in the following outline as Section 17.1 is read aloud. Students should copy the outline and keep it to study.

Characteristics of Protists:
Single-celled or Multicellular
Autotrophic or Heterotrophic
Movement
Response to Environment
Reproduction

Evolutionary Relationships:
  Prokaryotes
  ↓
Archaebacteria and eubacteria
  ↓   ↓
          Mitochondria and
          chloroplasts
  ↓   ↓
    Protists such as *Euglena* *HAS*

Have students observe the algae. Then, have a class discussion about the ways in which algae demonstrates the characteristics of protists.

auto Trophic *PROTISTS* *PLANTS*

Teaching Strategy B
The Protist March Game
for pages 370–375
(Developing Communication Skills/Kinesthetic Learners)

Materials: masking tape, 20 manila envelopes, paper, several sets of dice

Preparation: Attach the masking tape to the floor to form 20 squares large enough for students to stand in. Number the squares 1 through 20. Label square 7 "Go Forward Four Spaces." Label square 13 "Go Back Four Spaces."

Number each of the manila envelopes 1 through 20. For each numbered envelope (except #7 and #13), write five specific questions that relate to protists mentioned in the chapter and place the questions inside the envelopes. For example: "What protist is multicellular and can grow to 100 m in length?" or "Name a single-celled heterotrophic protist that ingests food through a gullet."

Keep a sheet for yourself that has the questions and the answers. (The answer for the first example question is "kelp," and for the second it is "*Paramecium*.") Place each envelope of questions within the labeled square whose number corresponds with the number on the outside of the envelope.

Procedure: Students should read Section 17.2 prior to playing the game. To play the game, first have students determine a playing order in some random way (such as drawing numbered slips of paper out of a box). The first player rolls the dice and advances to the square that matches the number rolled. The player has one minute to scan the text for the answer to the first question in the envelope that matches that square. If the player gives the correct answer before the minute is up, he or she stands in that square until it is his or her turn again. If the answer is incorrect, the player goes to the end of the player line and waits for another turn. A round of play ends when each player has rolled the dice and tried to answer a question. (To speed up play, have each player roll two dice instead of one.)

The next round of play should begin with the player who is furthest along the "march of the protists" rolling the dice to reach a new square. Play for a preset time. The winner is the player who makes the most progress towards the twentieth square. Suggested prizes could be a free period, a free soft drink from the cafeteria, etc.

Additional Strategies
Visual Strategies

Pages 367, 369, 370, 372, and 377

Auditory Learners

Use *Biology: Visualizing Life* Audiocassettes for Sections 17.1, 17.2, and 17.3.

Meeting Individual Needs (cont.)

Cooperative Learning
Diversity Among Protists

Timing: Use this activity to introduce Section 17.2.

Group Size: 4 students

Outcome: Students will be able to describe four characteristic types of protists.

Individual Accountability: Each group member is responsible for creating a scheme for describing one characteristic type of protist.

Positive Interdependence: Each group will construct a table describing the four categories of protists.

Assign one of the following categories of protists to each group member: (1) unicellular autotrophs, (2) multicellular autotrophs, (3) slime and water molds, and (4) protozoa. Have each group member read Section 17.2 and take notes on the method of movement, life cycle, method of reproduction, and response to stimuli of organisms in their category. Students should also identify several organisms that fit in their category.

Have each group member share the information from his or her notes with the other members of the group. Then, the group should complete a table with the following column headings: "Category of Protist," "Method of Movement," "Life Cycle," "Method of Reproduction," "Response to Stimuli," and "Examples of Category."

Portfolio Assessment

Students should select their best work and provide a self-reflective rationale for their selections. Students can make selections in the following areas.

1. *Content* — One concept map from the chapter (See page 812 for evaluation criteria.)

2. *Reading Comprehension* — One Directed Reading Worksheet from the Teacher's Resource Binder (Use the answer key to evaluate for accuracy.)

3. *Writing* — Using the Vee Form, summarize a magazine or newspaper article relating to protists, algae, or diseases caused by protists. (See page 22T for evaluation criteria.)

   Or: Select a writing task or project from the Chapter Review.

4. *Performance Assessment* — One Vee form from a chapter investigation or lab manual investigation (See page 22T for evaluation criteria.)

Teacher makes selections in the following areas.

1. *Formal Assessment* — Chapter test (Test A, B, or the Test Generator) The teacher-scored test should be reviewed by the student. Incorrect responses should be corrected by the student before the test becomes part of the portfolio.

2. *Informal Assessment* — Use the Direct Observations Checklist, page 33T, during a laboratory or other cooperative learning experience.

3. *Performance Assessment* — Have students research protists to create an evolutionary tree showing how the different types are related to one another.

Concept Map Answer

The following is one possible answer to the Relating Concepts exercise on page 380.

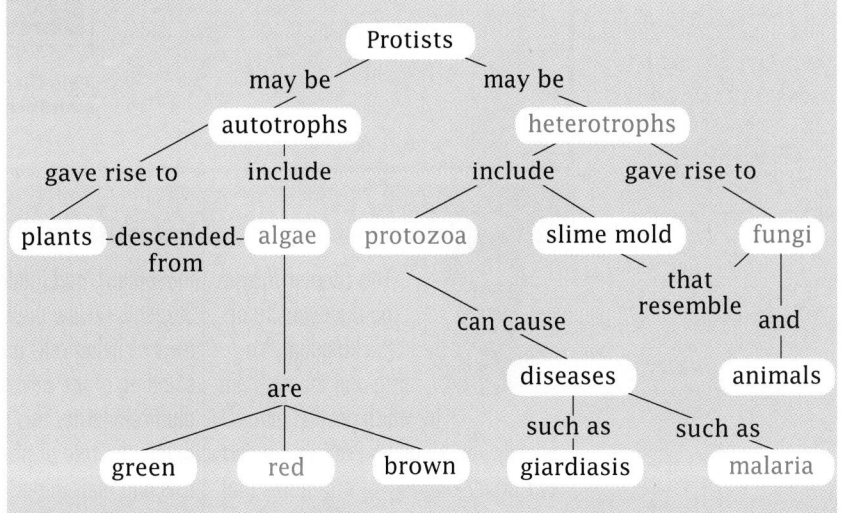

Chapter 17

Determining Prior Knowledge

- Have students describe the major features of the kingdom Protista and distinguish between autotrophs and heterotrophs.
- Show students a photograph, slide, or overhead transparency of a protozoan. Have them identify as many organelles as possible.
- Fill a small drawer with a wide variety of assorted and nonrelated items. Ask if anyone has a similar "junk" drawer at home. Have students explain why the kingdom Protista can be compared to a "junk" drawer.

Chapter 17

Protists

Review

- **evolution of protists (Section 10.2)**
- **autotrophs and heterotrophs (Section 12.1)**
- **the terms *endocytosis* and *exocytosis* (Glossary)**

17.1 What Is a Protist?

- **Characteristics of Protists**
- **Evolutionary Relationships Among Protists**

17.2 Protist Diversity

- **Classification of the Protists**
- **Autotrophic Protists**
- **Heterotrophic Protists**

17.3 Diseases Caused by Protists

- **Protists and Disease**

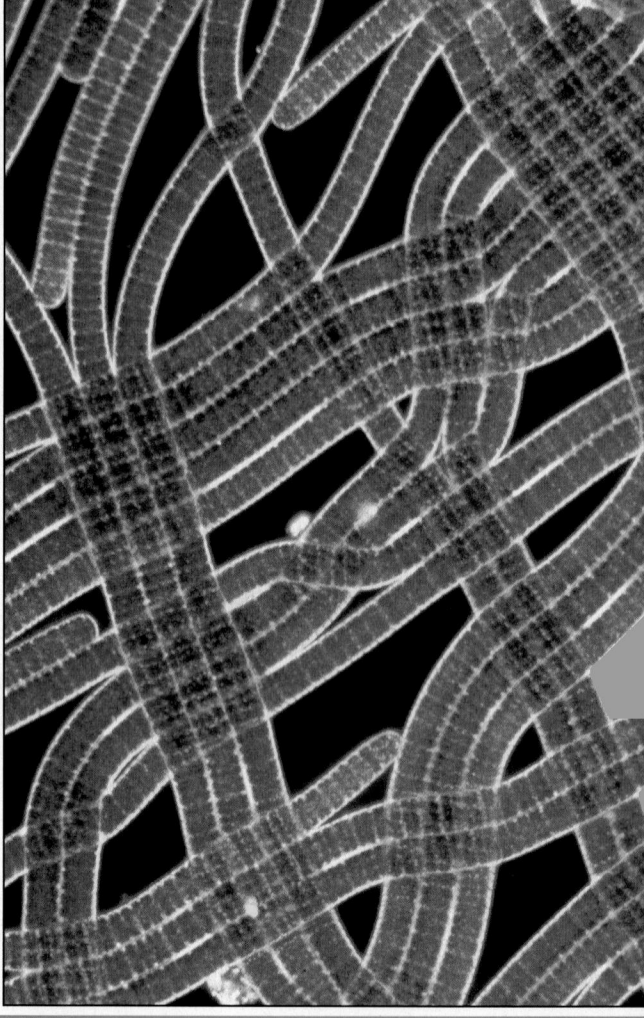

This filamentous alga can photosynthesize, but it is not a plant. It a multicellular protist. Like almost all protists, it is aquatic.

■ Author's Rationale ■

This chapter focuses on the diversity of the kingdom Protista and the methods by which protists obtain energy. The kingdom Protista is a sort of catchall kingdom of organisms that vary in size, structure, manner of movement, and mode of reproduction. Students review the basic characteristics of living cells and gain an understanding of the advantages of multicellularity. This chapter also covers the role of photosynthetic protists and their importance in an ecosystem. The relevance of this chapter becomes apparent as students explore the benefits of protists, as well as the diseases they cause.

TEN MILLION PEOPLE ARE INFECTED WITH THE **AIDS** VIRUS WORLDWIDE, AND THAT NUMBER IS INCREASING. YET DID YOU KNOW THAT MALARIA, A DISEASE CAUSED BY A PARASITIC PROTIST, INFECTS 20 TO 30 TIMES MORE PEOPLE THAN THE **AIDS** VIRUS? MALARIA KILLS MORE THAN 1 MILLION PEOPLE EACH YEAR. LIKE BACTERIA AND VIRUSES, PROTISTS EXERT MAJOR INFLUENCES ON HUMANS.

all single cell eukaryotes + algae

17.1   What Is a Protist?

Objectives

❶ Identify the characteristics shared by all members of the kingdom Protista.

❷ Describe the evolutionary relationship between bacteria and protists.

❸ Summarize the relationships of protists to the other three eukaryotic kingdoms.

Characteristics of Protists

If you examine a drop of pond water under a microscope, you might see the organism shown in **Figure 17.1** swimming on the slide. This tiny one-celled creature is called *Euglena*. To which kingdom does *Euglena* belong? It cannot be a member of the kingdom Monera because it has organelles and a nucleus. Its lack of a cell wall and its ability to move suggest that *Euglena* might be an animal—but it is single-celled, and all animals are multicellular. Also, if you look carefully you can see that this *Euglena* has chloroplasts; it is photosynthetic. Scientists classify organisms such as *Euglena* into the kingdom Protista. The kingdom Protista contains nearly all the single-celled eukaryotes (except the yeasts) as well as multicellular algae.

Protists have complex cells

Like all protists, *Euglena* is eukaryotic. Its nucleus, which contains chromosomes, is separated from the cytoplasm by a membrane. Also, membrane-bound organelles such as chloroplasts and mitochondria are found in the cell's cytoplasm. Chloroplasts occur in autotrophic protists such as *Euglena*, but they are lacking in heterotrophic protists. *Euglena* has the unique ability to shift between being autotrophic and heterotrophic. In some conditions, such as abundant light, it produces its own food. When light is scarce it feeds on bacteria and protists. Most protists, however, are always one or the other, either autotrophs or heterotrophs.

Figure 17.1
This scanning electron micrograph of *Euglena* shows that it is a single cell. Is *Euglena* a bacterium, a plant, a fungus, an animal, or a protist?

NOT Monera because nuclear membrane + organells

No cell wall But chloroplast NOT an animal

Lesson Plan 17.1

Phase 1
PREPARATION

Key Concepts
- The kingdom Protista includes a wide diversity of organisms, including autotrophs and heterotrophs.
- Protists are eukaryotic cells that are capable of movement, respond to stimuli, and undergo mitosis and meiosis.
- Plants evolved from autotrophic protists, while fungi and animals evolved from different heterotrophic protists.

Reading Strategy

Have students practice the three-step method of reading:
- survey
- read and take notes
- review
Tell students to focus on the main ideas during all three steps.

autotrophic heterotrophic

Phase 2
TEACHING STRATEGIES

Visual Strategy

Figure 17.1
Have students locate the flagellum. Point out that *Euglena* has a red-pigmented eyespot that is sensitive to light. Discuss what purpose this eyespot might serve.

Demonstration 1

Locomotion in Protists

Use a microprojector to show how *Euglena*, *Paramecium*, and *Amoeba* move. Have students compare their observations of living organisms with those shown in Figures 17.2, 17.3, and 17.4.

Theme Connection

Scale and Structure

Point out that humans have cells that utilize the three principal methods of locomotion shown in Figures 17.2, 17.3, and 17.4. Sperm cells swim by means of a flagellum. Cells lining the respiratory passages are covered with cilia to control the movement of bronchial fluids. White blood cells move by pseudopods. Have students discuss the significance of each of these locomotive features in human cells.

Demonstration 2

Responses in Protists

Assign students to cooperative work groups of four. Have each group design an experiment to test the responses of *Euglena* and *Paramecium* to various environmental stimuli including light, gravity, and heat.

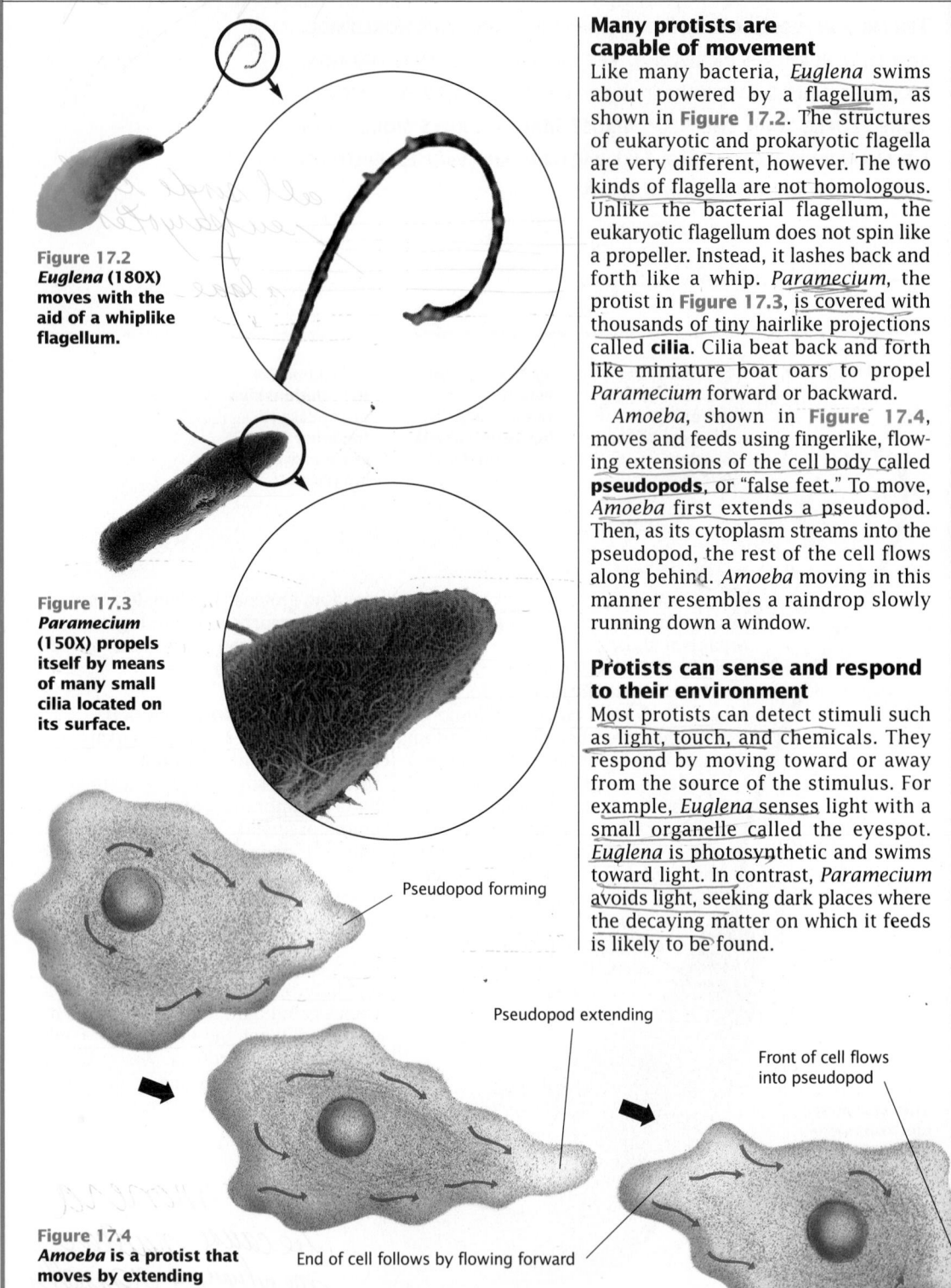

Figure 17.2
***Euglena* (180X) moves with the aid of a whiplike flagellum.**

Figure 17.3
***Paramecium* (150X) propels itself by means of many small cilia located on its surface.**

Pseudopod forming

Pseudopod extending

Front of cell flows into pseudopod

End of cell follows by flowing forward

Figure 17.4
***Amoeba* is a protist that moves by extending pseudopods.**

Many protists are capable of movement

Like many bacteria, *Euglena* swims about powered by a flagellum, as shown in **Figure 17.2**. The structures of eukaryotic and prokaryotic flagella are very different, however. The two kinds of flagella are not homologous. Unlike the bacterial flagellum, the eukaryotic flagellum does not spin like a propeller. Instead, it lashes back and forth like a whip. *Paramecium*, the protist in **Figure 17.3**, is covered with thousands of tiny hairlike projections called **cilia**. Cilia beat back and forth like miniature boat oars to propel *Paramecium* forward or backward.

Amoeba, shown in **Figure 17.4**, moves and feeds using fingerlike, flowing extensions of the cell body called **pseudopods**, or "false feet." To move, *Amoeba* first extends a pseudopod. Then, as its cytoplasm streams into the pseudopod, the rest of the cell flows along behind. *Amoeba* moving in this manner resembles a raindrop slowly running down a window.

Protists can sense and respond to their environment

Most protists can detect stimuli such as light, touch, and chemicals. They respond by moving toward or away from the source of the stimulus. For example, *Euglena* senses light with a small organelle called the eyespot. *Euglena* is photosynthetic and swims toward light. In contrast, *Paramecium* avoids light, seeking dark places where the decaying matter on which it feeds is likely to be found.

▪ Matter of Fact ▪

The fossil record indicates that the first protists evolved at least 1.5 billion years ago.

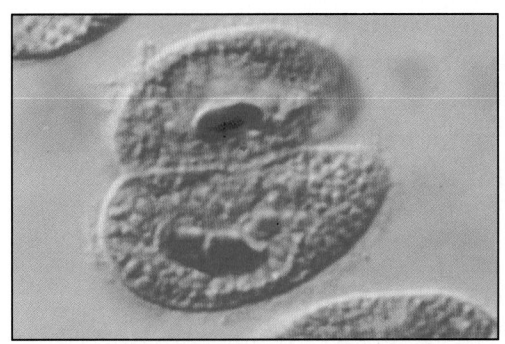

a   **The two paramecia that result from mitosis are genetically identical.**

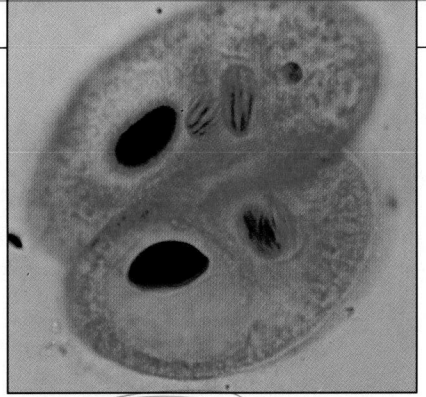

b   **After conjugation, the two paramecia are not genetically identical.**

Figure 17.5
Paramecium usually divides by mitosis. However, meiosis followed by conjugation is necessary for the individual to survive.

Both meiosis and mitosis occur among protists

As you read in Chapter 16, bacterial reproduction is a very simple process: bacteria simply split in two. Since eukaryotes typically have numerous chromosomes, they need a more complex mechanism to separate chromosomes evenly. The two mechanisms for accomplishing this sorting, meiosis and mitosis, evolved first in the protists. Many protists, including *Euglena*, reproduce by the process of mitosis alone. *Paramecium* and others can carry out both meiosis and mitosis. After meiosis, two paramecia join together and exchange part of their DNA. This process, called conjugation, is analogous to mating in animals and pollination in plants. The two types of reproduction are shown in **Figure 17.5**. An occasional exchange of DNA is required. If a *Paramecium* is isolated and unable to conjugate, it will die.

Evolutionary Relationships Among Protists

Eukaryotes are descended from prokaryotes, but in two different ways. First, the eukaryotic cell itself arose through modification and elaboration of an ancestral prokaryotic cell. Second, other prokaryotes gave rise to two of the organelles that exist within the eukaryotic cell. Mitochondria evolved from heterotrophic bacteria, while chloroplasts evolved from autotrophic bacteria. Thus a protist such as *Euglena*, which has both mitochondria and chloroplasts, is a union of three evolutionary lines.

Most protists live in a water environment. As you learned in Chapter 15, protists are the ancestors of the three major terrestrial eukaryotic kingdoms. For instance, plants evolved from autotrophic protists, while animals and fungi descended from different heterotrophic protists.

Section Review

① **Describe the structural differences between a bacterial cell and a protist cell.**

② **Contrast bacterial reproduction with reproduction in protists.**

③ **Summarize the ways in which protists are related to bacteria.**

④ **Draw an evolutionary tree showing the relationship of plants, animals, and fungi to protists.**

■ *Section Review Answers* ■

1. A protist cell has a membrane-bound nucleus with chromosomes, and membrane-bound organelles such as chloroplasts and mitochondria. A bacterial cell does not have a nucleus, membrane-bound organelles, or chromosomes.

2. In bacterial reproduction, the DNA is copied and then the cell divides. Many protists reproduce only by mitosis. Others carry out both mitosis and meiosis.

3. Eukaryotes are descended from bacteria in two ways. First, eukaryotic cells arose through modification and elaboration of an ancestral prokaryotic cell. Second, prokaryotes gave rise to mitochondria and chloroplasts.

4. See Figure 10.6 on page 204.

Visual Strategy

Figure 17.5
Have students explain why the two paramecia shown in Figure 17.5a are genetically identical. Discuss why those in Figure 17.5b are not genetically identical.

Connection: Chapter 4

Mitosis
Review mitosis as the process by which eukaryotic cells divide.

Connection: Chapter 6

Meiosis
Review meiosis as the process by which gametes are formed.

Phase 3

ASSESSMENT OPTIONS

Closure Strategy

Sporozoans
Sporozoans are animal-like protists that do not move by means of either flagella, cilia, or pseudopods. Have students suggest what features these protists share.

Section Review
Assign the *Section Review*.

Reteaching

Have students prepare a table that summarizes the similarities and differences between monerans and protists.

Lesson Plan 17.2

PREPARATION

Key Concepts

- Protists can be autotrophic or heterotrophic, microscopic or very large. They may be composed of a single cell or millions of cells.
- Autotrophic protists, commonly known as algae, include green, brown, and red algae.
- Heterotrophic protists include protozoa and fungus-like organisms (slime molds and water molds).
- Evidence suggests that modern plants evolved from green algae and that animals evolved from protozoa.

Reading Strategy

Have students prepare a table with the headings *Autotrophic protists* and *Heterotrophic protists*. Down the left side, have them write *Examples*, *Major features*, *Common uses*, and *Additional comments*. Instruct students to complete the table as they read this section.

Phase 2

TEACHING STRATEGIES

Demonstration 1

Protist Diversity

Display photographs of various protists so that students can appreciate their great diversity. Such organisms might include *Amoeba*, diatoms, *Volvox*, sea lettuce, kelp, slime molds, and fire algae.

Visual Strategy

Figure 17.6

Point out that *Fucus* is a member of the group of brown algae that makes up most of the seaweeds found in both temperate and polar regions.

370

LOOK FOR CARRAGEENAN IN THE LIST OF INGREDIENTS OF YOGURT, ICE CREAM, AND OTHER DAIRY PRODUCTS. IT IS ADDED TO THESE FOODS AND TO PAINTS AND COSMETICS TO CREATE A SMOOTH TEXTURE. CARRAGEENAN, A JELLYLIKE MATERIAL EXTRACTED FROM THE CELL WALLS OF A RED ALGA KNOWN AS IRISH MOSS, IS ONE OF MANY PRODUCTS WE OBTAIN FROM PROTISTS.

17.2   Protist Diversity

Objectives

1. **Recognize** the variation in shape, size, and method of obtaining energy among protists.
2. **Explain the** role autotrophic protists play in ecosystems.
3. **Recognize** the evolutionary connection between green algae and plants.
4. **Identify** three kinds of heterotrophic protists.

Classification of the Protists

Figure 17.6
Although these two organisms look very different, they are both members of the kingdom Protista.

Look at the two dissimilar organisms in **Figure 17.6**. Figures **17.6a** and **17.6b** show a brown alga, or kelp, which is a type of seaweed. Kelp is multicellular and autotrophic. Some kelps can grow to 100 m (328 ft.) in length. The creature in **Figure 17.6c** is *Didinium*. It is single-celled and also heterotrophic. *Didinium* is so small that it cannot be seen without a microscope. Both of these organisms are classified as protists. The members of the kingdom Protista exhibit a greater range of sizes and a greater variety of structure than do the members of any other kingdom.

Great diversity is also found in protist metabolism. Among the eukaryotic kingdoms, the kingdom Protista is the only one that includes both heterotrophs and autotrophs. Autotrophic protists, including both unicellular and multicellular forms, are photosynthetic, as are plants. Heterotrophic protists, such as *Paramecium* and *Amoeba*, obtain their energy by consuming other organisms or by parasitism. You can see the structure of a heterotrophic protist in the *Tour of a Protist* on the next page.

Separating the protists into autotrophs and heterotrophs is an artificial classification that does not show the complexity of their evolutionary relationships. Because of its diversity, biologists disagree on how many phyla the kingdom Protista should contain.

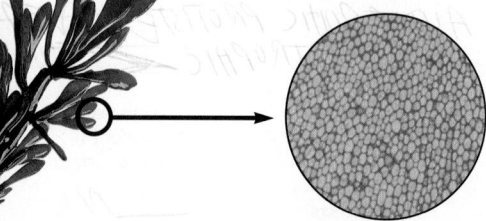
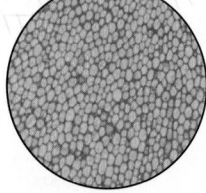
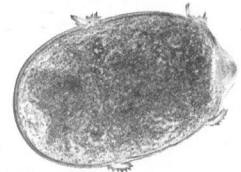

a   A kelp, such as this *Fucus*, is a protist. It is a large organism that is autotrophic . . .

b   . . . and is composed of millions of cells.

c   *Didinium* (150X), also a protist, is unicellular and heterotrophic.

Tour of a Protist

Paramecium *is a kind of protist called a ciliate. It is covered with fields of flagella called cilia that move it through the water.*

Paramecium spontaneously absorbs water. Excess water is squeezed back out by contractile vacuoles.

Paramecium has two nuclei. The macronucleus contains fragmented chromosomes used in routine cellular functions, and divides by pinching in two.

The micronucleus contains the cell's chromosomes, and divides by mitosis.

Paramecium ejects wastes by exocytosis.

Food is ingested through a cilia-lined gullet and taken into the cell by endocytosis as food vacuoles.

A *Paramecium* recognizes members of the opposite sex by special protein molecules that ring the gullet like a necklace.

Paramecium is found in streams and ponds.

Paramecium is unicellular but so complex that some biologists prefer to call them "organisms without cell boundaries" rather than single cells.

Using the Feature

This feature shows some of the features of *Paramecium*, a ciliate common in ponds and streams. Ciliates have the most complex cells of any organism. The organelles of *Paramecium* are analogous to the organs of an animal; like a kidney, the contractile vacuol removes excess water; like a stomach, the food vacuole digests food.

Discussion

Guide the discussion by posing the following questions.
1. What causes *Paramecium* to absorb water?
   Since Paramecium *lives in fresh water, its internal fluids have a higher salt concentration than its environment. It gains water through osmosis.*
2. What is the function of the macronucleus? *It contains pieces of chromosomes that control routine cellular functions.*
3. How does *Paramecium* recognize a member of the opposite sex? *By a ring of proteins that surrounds the gullet.*

Connection: Chapter 12

Food Webs

Have students refer back to Figure 12.6 on page 251 to review how algae (photosynthetic plankton) form the base of this ocean ecosystem.

Visual Strategy

Figure 17.7

Have students examine the diatom shown in Figure 17.7a. Point out that the prefix *dia* means *through* or *apart* while the root is derived from the Greek word *tomos*, which means *to cut*. Have students explain why the term diatom is therefore an appropriate name.

Have students identify the wavelengths of light absorbed by the algae shown in Figure 17.7b. Emphasize that red algae contain a pigment that absorbs blue light, the only wavelength that can penetrate deep water. Consequently, red algae are able to live more than 100 m (320 ft.) below the surface.

Demonstration 2

Staying Afloat

Use a microprojector to show students *Spirogyra*, a green alga that is part of the phytoplankton of ponds and streams. During the day, the oxygen produced by photosynthesis gets trapped by *Spirogyra's* long filaments. At night, the oxygen slowly dissolves into the water. Have students explain how this affects where *Spirogyra* will be found during the day and at night.

algae = autotrophic protists

Autotrophic Protists

Do you think anything you did today involved autotrophic protists? Before you answer "no" without hesitation, consider these activities. Did you drive or ride in a car? Did you brush your teeth? Did you eat ice cream, cheese, or yogurt? Did you eat fish or other seafood? Autotrophic protists or their products are involved in all these activities. Autotrophic protists are often called algae (*AL jee*). Despite being grouped under one name, not all kinds of algae are closely related.

Although they can be found in moist soil, on damp rocks, and on the shady sides of trees, most species of algae live in fresh water or salt water, as shown in **Figure 17.7**. Algae serve the same ecological function in aquatic ecosystems as plants serve in terrestrial ecosystems. They form the base of the aquatic food web, directly or indirectly providing food for all aquatic consumers. Thus, fishes, whales, and humans who eat seafood depend on algae. Algae also produce large amounts of oxygen. About one-third of the oxygen in the atmosphere is produced by algae.

algae → O₂

Most kinds of algae are unicellular

Unicellular algae are one part of a community of aquatic organisms called plankton. The organisms found in plankton float near the surface in fresh water or salt water. Heterotrophic bacteria, fish larvae, crustaceans, and heterotrophic protists are also considered to be part of the plankton. Photosynthetic plankton are called **phytoplankton** (*fy tuh PLANK tuhn*), which means "plant plankton," because they are photosynthetic.

Phytoplankton require light for photosynthesis, and so they usually are found near the surface. A wide variety of adaptations help them stay afloat. Some beat their flagella and "tread water." Others have fins and spines that act as water wings. Others store extra food as oil, which buoys them up near the surface. When algae full of oil die and sink to the bottom, they can be buried under mud and sand. Over millions of years, heat and pressure within the earth transform the oil from the algae into crude-oil deposits. The

Figure 17.7
Autotrophic protists are found in the open ocean, along the shore, and in freshwater habitats. They range from unicellular to multicellular, yet all are photosynthetic.

a **Microscopic algae like this diatom serve as food for many marine animals.**

b **Red algae such as this *Porphyra* often are attached to rocks at the edge of the sea.**

c **Brown algae such as *Fucus* can be many meters in length.**

ALGAE — Predecessor PLANTS (handwritten)

diatoms are aquatic algae (handwritten)

gasoline used today was formed partly from the remains of algae that died millions of years ago.

Look at the object in **Figure 17.7a**. These objects may look like hubcaps or jewelry, but they are the remains of aquatic algae called diatoms. Diatoms form intricately patterned shells made of silica, a glasslike substance. When diatoms die, their shells settle to the ocean floor. Humans harvest fossilized deposits of these shells to make abrasives for toothpaste and silver polish.

If you visit the beach, you might experience a colorful but dangerous effect of marine algae. Occasionally, the ocean turns reddish and thousands of dead fishes, seabirds, turtles, and dolphins wash up onto the beach. These "red tides" are caused by phytoplankton known as dinoflagellates (*deye noh FLAJ uh lihtz*). Population explosions of certain species of dinoflagellates occur irregularly in coastal areas. The dinoflagellates release toxins that are poisonous to vertebrates. Because fishes and shellfish concentrate large amounts of these toxic secretions in their internal organs, health authorities often do not allow fishing during red tides.

Many green, brown, and red algae are multicellular

There are three groups of multicellular algae—the green, brown, and red algae. The brown algae and the red algae are commonly called "seaweeds."

You might have seen algae like the ones in **Figure 17.7b** and **17.7c** attached to rocks at low tide or washed up on a beach.

Green algae, like the pond scum in **Figure 17.7d**, are distinguished from other algae by their kinds of photosynthetic pigments, by their use of starch to store food, and by their unique form of cell division. Plants share these characteristics with the green algae. Therefore, scientists think that green algae are the ancestors of the plant kingdom.

Kelp, which you saw in **Figure 17.6a**, is a brown alga that often grows in dense stands or thick floating mats. Kelp "forests" occur off the coast of California. The Sargasso Sea, an area of the Atlantic Ocean east of Florida, is so thick with floating kelp that early European explorers feared their sailing ships could be trapped there (an unreasonable fear, it turns out).

Red algae tend to be smaller and less spectacular than the brown algae. Distinctive photosynthetic pigments give red algae their color. These same pigments enable some species to photosynthesize in waters more than 200 m (650 ft.) deep, where there is little light.

Red and green algae are harvested for many uses. *Porphyra*, one of the red algae, is particularly popular in Far Eastern countries such as Japan and China, where it is known as *nori*. Nori is cultivated as a crop on coastal algae "farms."

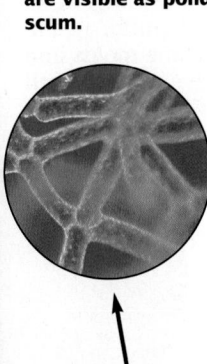

d Many kinds of green algae live in freshwater habitats. These filamentous green algae, called *Hydrodictyon*, are microscopic. But large numbers of green algae often are visible as pond scum.

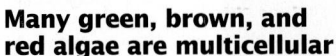

Demonstration 4

Slime Molds—Which Kingdom?

Show the class photographs of different slime molds. Although they are classified as protists, have students discuss what other kingdom these organisms might belong to. Point out that slime molds and water molds have sometimes been placed in the kingdom Fungi because of similarities in appearance and method of nutrition. However, several major features, especially ones involving reproduction, distinguish them as protists. Help students distinguish between the "amoeba" portion of the slime mold life cycle and the protistan genus *Amoeba*.

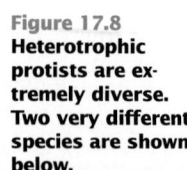

herbivores
carnivores
PROTISTS

Heterotrophic Protists

The word "carnivore" brings to mind fierce, sharp-toothed animals like lions, tigers, wolves, and sharks. The word "herbivore," on the other hand, calls forth images of dull, plodding creatures like cows and sheep. You don't normally think of tiny one-celled protists as carnivores and herbivores, but most heterotrophic protists play one of these two ecological roles. In the plankton, for example, some heterotrophic protists "graze" on phytoplankton while others prey on these grazers. Other important heterotrophic protists are parasites or decomposers.

Slime molds and water molds superficially resemble fungi

If you looked carefully through moist, decaying leaves on a forest floor, you might find a slime mold. A slime mold spends much of its life as a single-celled mass called an amoeba that is found in damp environments and feeds on bacteria and decaying matter. If food or even moisture become scarce, the individual amoebas come together, forming a mass resembling the slime mold shown in **Figure 17.8a**. The mass produces reproductive structures, also shown in **Figure 17.8a** (inset).

If your ancestors arrived in the United States from Ireland in the 1840s, they were probably fleeing the effects of a heterotrophic protist. In Ireland, from 1845 to 1847, a famine killed more than 1 million people and forced another 3 million to emigrate, mainly to the United States. This famine occurred because the potato crop (the staple of the Irish diet) was almost wiped out by an outbreak of late blight. Late blight is a disease caused by *Phytophthora infestans*, a protist that is a water mold, shown in **Figure 17.8b**.

Despite some similarities, slime molds and water molds are not close relatives of fungi. For one thing, fungi have cell walls made of chitin, a complex polysaccharide. Slime molds and water molds either have cellulose cell walls or have no cell walls at all.

Figure 17.8
Heterotrophic protists are extremely diverse. Two very different species are shown below.

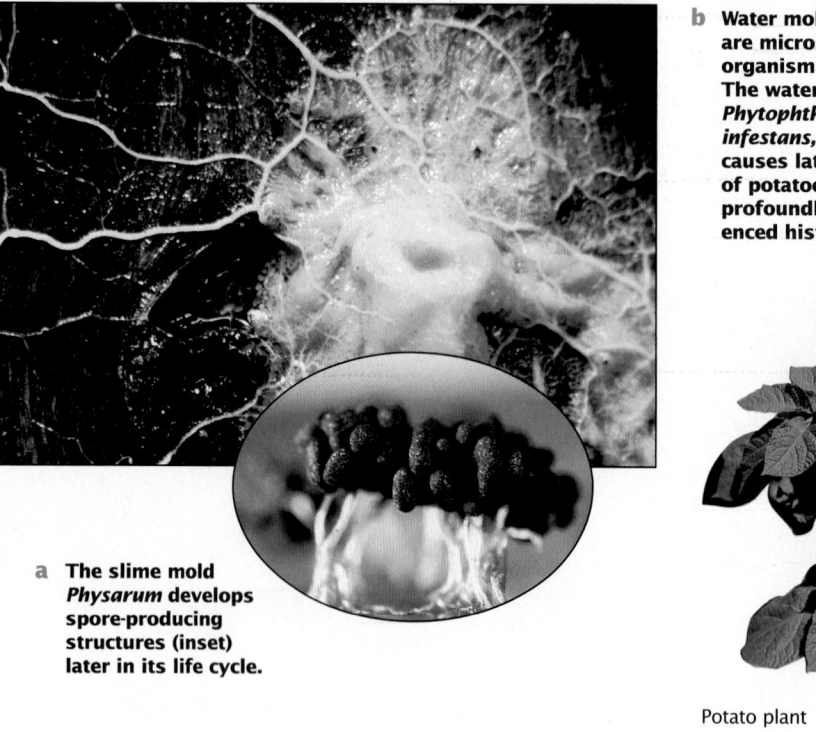

a **The slime mold *Physarum* develops spore-producing structures (inset) later in its life cycle.**

b **Water molds are microscopic organisms. The water mold *Phytophthora infestans*, which causes late blight of potatoes, has profoundly influenced history.**

Potato plant

Protozoa include the ancestors of animals

About 300 years ago, scientists using the first microscopes saw what appeared to be tiny animals darting about in drops of water. They called these organisms protozoa, which means "first animals." Today scientists do not classify protozoa into the animal kingdom, but animals are thought to have descended from extinct protozoa.

Figure 17.9
***Stentor* is a ciliate that lives anchored to a solid surface and gathers food by the whirling action of its cilia.**

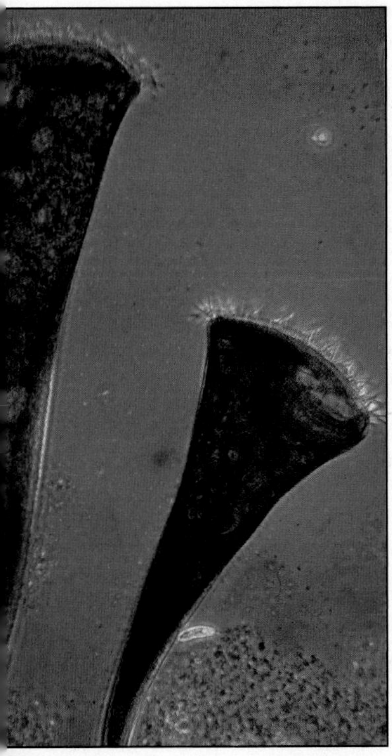

Paramecium, Didinium, and the *Stentor* in **Figure 17.9** are members of a group of protozoa called ciliates. Ciliates are covered with numerous cilia, which function in locomotion and feeding.

Not all groups of protozoa are closely related. Indeed, based on recent molecular analysis of DNA, some biologists argue that ciliates should be placed in a separate kingdom of their own.

The ancestor of animals probably belonged to a group of protozoa known as the zoomastigotes (*ZOH mast ih gohtz*). Zoomastigotes are one-celled, heterotrophic protists that have at least one flagellum. Some zoomastigotes, such as the *Trichonympha* illustrated in **Figure 17.10**, are covered with flagella. If your house has ever been attacked by termites, you can blame *Trichonympha.* Without these protists living in their intestines, termites could not digest wood.

The tiny, ornate shells shown in **Figure 17.11** are made by marine protozoa called foraminiferans (*fawr uh MIHN ih fur ihnz*). When foraminiferans die, their calcium shells sink to the ocean floor. Over millions of years, these shells have accumulated, forming huge limestone deposits like the White Cliffs of Dover in southeastern England.

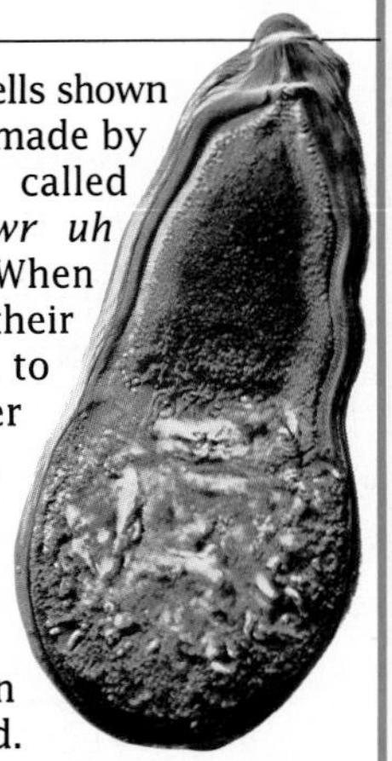

Figure 17.10
***Trichonympha* is a zoomastigote that lives in a termite's intestines.**

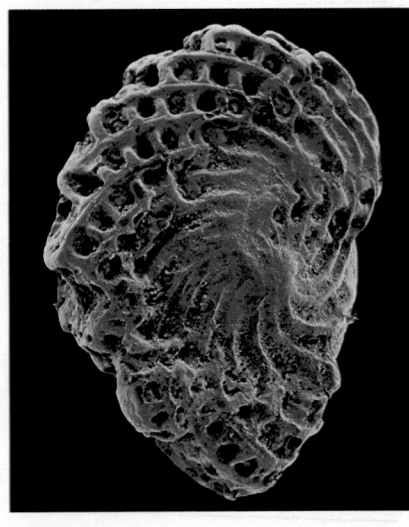

Figure 17.11
Foraminiferans secrete tiny, ornate shells that accumulate on the bottom of the sea.

Section Review

1. **Explain why the statement, "All protists are microscopic" is untrue.**
2. **Summarize the importance of algae in aquatic ecosystems.**
3. **Describe two pieces of evidence indicating that plants evolved from green algae.**
4. **Name two kinds of heterotrophic protists.**

Social Studies Connection

African Sleeping Sickness

Another zoomastigote is *Trypanosoma gambiense,* which causes African sleeping sickness in humans. This parasite reproduces in the blood, where it releases toxins that cause weakness, fever, and sleepiness. If left untreated, the victim will eventually die. The protist lives in the blood of both wild and domesticated animals in Africa and is transmitted by the bite of the tsetse fly.

Phase 3
ASSESSMENT OPTIONS

Closure Strategy

Looking for Protists

Have students assume that they are assigned to collect a variety of protists in a pond. Ask where in the pond they would look for the following organisms: *Amoeba, Paramecium,* unicellular algae, and *Stentor.* Have students provide reasons to support each location they chose.

Section Review

Assign the *Section Review.*

Reteaching

On the board, prepare a table similar to the one described in the Reading Strategy. Fill in the information, having students volunteer suggestions based on what they wrote in their own tables.

Section Review Answers

1. Many red, brown, and green algae are macroscopic—some brown algae reach 100 m (320 ft.).
2. Algae form the base of the aquatic food web, directly or indirectly supplying food for all aquatic consumers. Algae also produce large amounts of oxygen.
3. Green algae and plants have the same photosynthetic pigments in their chloroplasts, starch to store food, and have a unique form of cell division.
4. Possible answers include slime molds, water molds, *Amoeba,* and *Paramecia.*

Lesson Plan 17.3

Phase 1

PREPARATION

Key Concepts
- Protozoans transmitted by mosquitoes cause malaria.
- Malaria can be treated by taking drugs that inhibit growth of the protozoa, and it can be prevented by reducing the size of mosquito populations.
- Protists are responsible for causing many diseases in humans.

Reading Strategy

Students often read chapters in science texts as isolated fragments of information and rarely, if ever, see "threads" that weave throughout the book. This section offers an opportunity to show students how the readings in different chapters are connected. In Chapter 1, malaria was discussed in terms of scientific experimentation. In Chapter 9, malaria was again brought up to show how natural selection operates in the case of sickle cell anemia. In this chapter, malaria is covered in the context of a disease caused by a protist. Make students aware of these connections as they read this section.

Phase 2

TEACHING STRATEGIES

Connection: Chapter 1

Malaria
Review how Ronald Ross discovered that mosquitoes transmitted malaria.

IN CENTRAL AFRICA, THE DISEASE SLEEPING SICKNESS IS CAUSED BY A PROTIST TRANSMITTED BY THE BITE OF A LARGE FLY, THE TSETSE FLY. THIS DISEASE IS OFTEN FATAL TO HUMANS AND DOMESTIC ANIMALS SUCH AS CATTLE, SHEEP, AND PIGS. LIVESTOCK CANNOT BE RAISED IN MUCH OF CENTRAL AFRICA BECAUSE OF THE PRESENCE OF SLEEPING SICKNESS.

17.3 Diseases Caused by Protists

Objectives

1. **Explain how malaria is transmitted.**
2. **Describe the methods used to control malaria.**
3. **Explain why malaria-control methods have not been entirely successful.**
4. **Name three human diseases caused by protists.**

Protists and Disease

Although many protists have no effect on the health of human beings, some parasitic protozoa are a major cause of disease in the world. Millions of people are infected by, and die from, various protist-caused diseases each year. Disease-causing protists are transmitted mainly by insects and contaminated water.

Figure 17.12
When it pierces the skin, an infected mosquito of the genus *Anopheles* (like the one on the right) can transmit the parasite that causes malaria.

Malaria is the great killer
Currently, from 200 million to 300 million people are infected with malaria. This disease kills more than 1 million people each year, making it one of the most serious human diseases. Most deaths caused by malaria occur among children under the age of five. It is a leading cause of death among children in tropical countries. This disease has affected the evolution of our species. As you learned in Chapter 9, the allele for sickle-cell hemoglobin, a mutant form of hemoglobin, is prevalent in malaria-prone areas of Africa, Asia, and the Middle East. This mutant hemoglobin produces resistance to the parasite that causes malaria.

A parasitic protozoan causes malaria
Malaria is caused by protozoa of the genus *Plasmodium*, which has a very complex life cycle. *Plasmodium* is carried between human hosts by mosquitoes, particularly mosquitoes of the genus *Anopheles*, like the one shown in **Figure 17.12**. When a female mosquito bites and feeds, she injects saliva into the wound to prevent her victim's blood from clotting. The mosquito's saliva transfers *Plasmodium* into the human bloodstream. (Male mosquitoes do not spread the disease because they feed

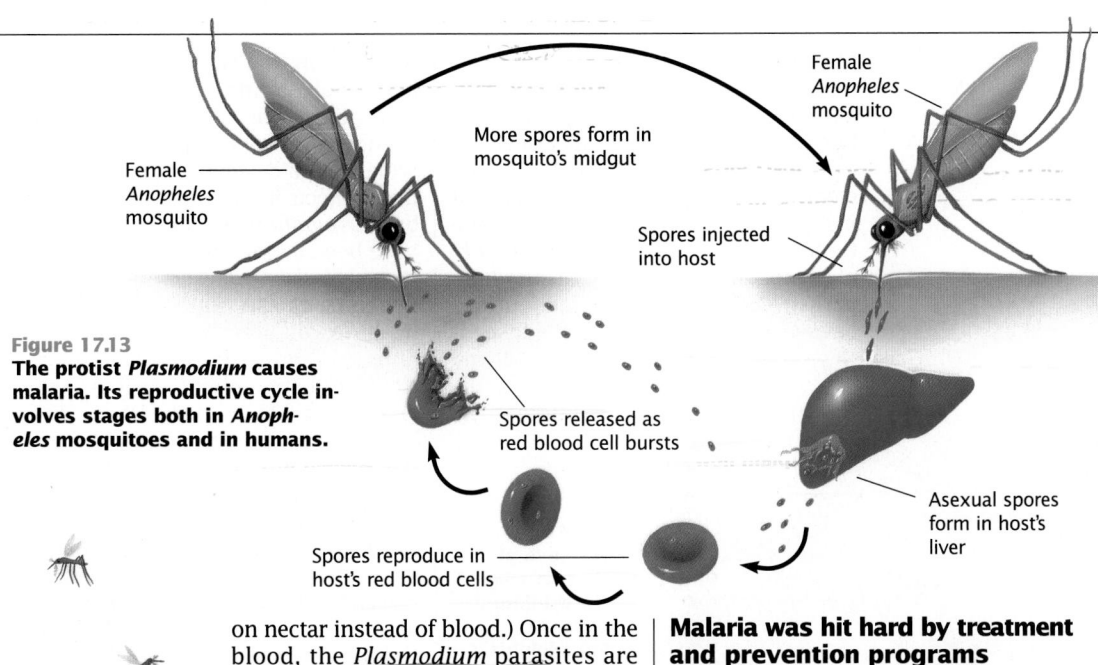

Figure 17.13
The protist *Plasmodium* causes malaria. Its reproductive cycle involves stages both in *Anopheles* mosquitoes and in humans.

Female *Anopheles* mosquito

Female *Anopheles* mosquito

More spores form in mosquito's midgut

Spores injected into host

Spores released as red blood cell bursts

Spores reproduce in host's red blood cells

Asexual spores form in host's liver

on nectar instead of blood.) Once in the blood, the *Plasmodium* parasites are carried to the liver, where they reproduce. They then reenter the bloodstream, penetrate the red blood cells, and reproduce again. Every 48 or 72 hours, depending on the species of *Plasmodium*, a new generation of parasites bursts out of infected red blood cells, as shown in **Figure 17.13**, and then invades other red blood cells. Each of these outbreaks can destroy up to 40 percent of the host's red blood cells.

Female mosquitoes acquire *Plasmodium* when they bite infected humans. *Plasmodium* does not make the mosquito ill, so she is able to fly off and transmit the parasite to other humans. Since mosquitoes are more abundant in the tropics, nearly all of malaria's victims live in Africa, Asia, and South America.

The symptoms of malaria follow a cycle that corresponds to the reproduction of *Plasmodium* in the blood. When the parasites emerge from the blood cells, the host experiences high fever, delirium, and sweating. Severe chills follow the fever and can last until the next outbreak of *Plasmodium* parasites. Malaria sometimes weakens its victims so much that they die from other infections. Also, the drastic loss of blood cells causes anemia and can result in fatal brain or kidney damage.

Malaria was hit hard by treatment and prevention programs

Chemical treatments for malaria are hundreds of years old. In the 1600s, quinine, a bitter chemical derived from the bark of a tropical tree, was found to reduce the symptoms of malaria, although it did not cure the disease. Derivatives of quinine, such as chloroquine and primaquine, are used today to treat infected individuals and to prevent malaria in healthy individuals. If you travel to the tropics, you will probably have to take regular doses of one of these drugs to prevent growth of malaria parasites in your body.

One way to reduce the number of cases of malaria is to reduce the size of mosquito populations. To control *Anopheles* mosquitoes, swamps and ditches where mosquitoes breed are drained. Powerful insecticides such as DDT are also applied to kill mosquitoes. Through these two measures, malaria has been eradicated from the southern United States and southern Europe, where the disease was common until the 1940s. The World Health Organization introduced similar mosquito population control measures into parts of Africa and South America. Initially, these programs greatly reduced the worldwide incidence of malaria. Some optimists predicted the elimination of the deadly disease.

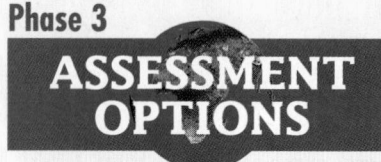

Connection: Chapter 16

Vaccinations

Remind students that vaccinations stimulate the body's defenses against a variety of diseases, including those caused by bacteria and viruses. Have students suggest a reason why a vaccine against malaria has been so difficult to develop.

Phase 3

ASSESSMENT OPTIONS

Closure Strategy

Another Protist-caused Disease
The disease primary amoebic meningencephalitis (PAM), is caused by an amoeba that lives in the bottom of swimming holes and lakes. The amoeba can enter the body with water that may be inhaled through the nose and then swallowed. Once inside the body, the amoeba travels to the brain. Almost immediately, headaches and fever develop. In 24 hours, the infected person enters a coma, followed by death in 48 hours. Fortunately, this disease is rare. The best ways to prevent infection are to wear a nose clip and to avoid the bottom of freshwater swimming areas.

Section Review

Assign the *Section Review*.

Reteaching

Assign students to cooperative groups of four. Have each group develop a concept map for this section.

Malaria struck back

As you have seen, malaria has not been eliminated. To the contrary, the number of cases of malaria has been increasing since the mid-1970s. Human efforts to reduce the incidence of malaria have been thwarted by evolutionary advances in both the *Anopheles* mosquito and the *Plasmodium* parasite. In many areas, mosquitoes have evolved immunity to the insecticides used to control them. Thus, mosquito populations are increasing.

Additionally, the *Plasmodium* parasites have evolved immunity to drugs like chloroquine that have traditionally been used to control them. Efforts are underway to develop a malaria vaccine that protects individuals from infection by *Plasmodium*. So far, these efforts have been largely unsuccessful.

Protists cause many diseases

Imagine you are hiking in the mountains on a hot day. You come upon a clear, cold stream and are tempted to take a drink. Many people have given in to this temptation and later regretted it. Even the cleanest-looking streams often have parasitic protozoa called *Giardia*. Most people who swallow *Giardia* show no ill effects. But a few individuals suffer diarrhea, cramps, nausea, and vomiting, symptoms of the disease giardiasis (*jee ahr DEYE uh sihs*). *Giardia* enters streams and lakes in the feces of infected humans and animals. In rare cases, *Giardia* has reached urban water systems as shown in **Figure 17.14**. Philadelphia's water system was contaminated for several days in 1984, forcing thousands of residents to boil their drinking water to kill *Giardia* protozoa. Other examples of diseases caused by protists are listed in **Table 17.1**.

Figure 17.14

a For several days in 1984, the drinking water in Philadelphia had to be boiled. The water was contaminated by . . .

b . . . the protist *Giardia*, which causes severe intestinal pain.

Table 17.1 Diseases Caused by Protists

Disease	Host	Organism
Amoebic dysentery	Humans	*Entamoeba*
Malaria	Humans	*Plasmodium*
Toxoplasmosis	Humans, cats	*Toxoplasma*
Giardiasis	Humans	*Giardia*
Sleeping sickness	Humans, tsetse flies	*Trypanosoma*
Leishmaniasis	Humans, sand flies	*Leishmania*
Late blight	Potatoes	*Phytophthora*

Section Review

❶ Why might the use of mosquito netting in malaria-prone areas decrease the number of malaria cases?

❷ Describe two methods by which the incidence of malaria was reduced.

❸ Explain why malaria has not been eliminated, despite intense efforts to do so.

❹ What are three human diseases caused by protists?

■ *Section Review Answers* ■

1. Students should suggest that the netting is to protect humans from mosquitoes, which carry *Plasmodium*, the protozoan that causes malaria.

2. To control *Anopheles* mosquitoes, insecticides are applied, or the breeding grounds are destroyed by draining many ditches and swamps. Drugs such as chloquine are taken to prevent infection.

3. Because mosquitoes have evolved an immunity to many insecticides, malaria has not been eliminated and mosquito populations are increasing. *Plasmodium* has also evolved an immunity to drugs that have been used to control it.

4. Answers may vary. See Table 17.1 on page 378.

Chapter **17**   *Highlights*

Amoeba and other protists are remarkably sensitive to their environments.

	Key Terms	Summary
17.1 What Is a Protist? 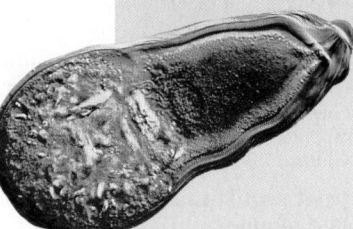 *Euglena* is a unicellular protist that is photosynthetic.	cilia (p. 368) pseudopod (p. 368)	• The kingdom Protista contains the eukaryotes that are not plants, animals, or fungi. • Both mitosis and meiosis occur among protists. • Protists descended from bacteria. Plants, animals, and fungi are the descendants of different protist ancestors.
17.2 Protist Diversity *Trichonympha* is a single-celled protist that lives in the guts of termites and digests wood.	phytoplankton (p. 372)	• The kingdom Protista is the most diverse eukaryotic kingdom. • Both autotrophs and heterotrophs are included among the protists. • Most protists are unicellular but some are multicellular. • Green algae are probably the ancestors of plants. • Zoomastigotes are probably the ancestors of animals.
17.3 Diseases Caused by Protists 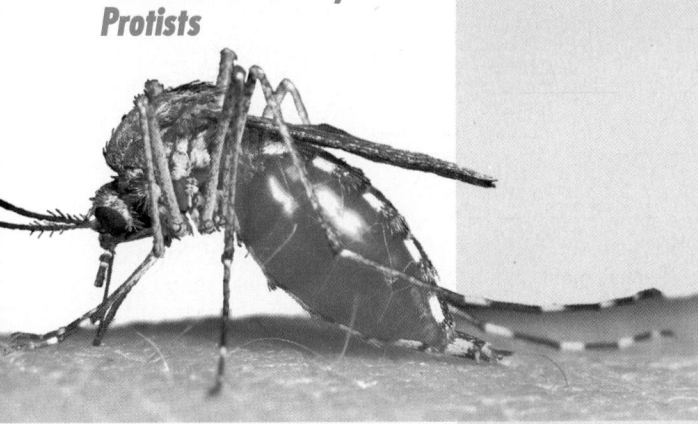 The female *Anopheles* mosquito carries a protist called *Plasmodium* that causes malaria.		• Malaria is a disease caused by a parasitic protozoan in the genus *Plasmodium*. About 200 million to 300 million people are currently infected with the malaria parasite. • *Plasmodium* is transmitted by mosquitoes, especially species in the genus *Anopheles*. • Intense efforts to eliminate malaria initially showed promise. These efforts have failed to eliminate malaria because of the evolution of drug-resistance in the *Plasmodium* parasite and pesticide resistance in mosquitoes. • Other protists cause amoebic dysentery, giardiasis, and sleeping sickness.

PROTISTA

BACTERIA

Chapter Review Answers

Understanding Vocabulary

1. **a.** Algae are not structures with which protists move.
   b. Salmonella is not a protist.
   c. Red algae are not heterotrophic protists.
   d. Rabies is not a protist-caused disease.

Relating Concepts

2. Map answer is on page 365D.

Understanding Concepts

Multiple Choice

3.	a	8.	b
4.	d	9.	a
5.	b	10.	c
6.	c	11.	a
7.	d		

Completion

12. animals, plants
13. autotrophic; phytoplankton
14. tsetse flies; mosquitoes
15. malaria; five

Short Answer

16. Their cells have a nucleus, chromosomes, and membrane-bound organelles.
17. raging fever, sweating, delirium followed by chills; in the tropics
18. making toothpaste and silver polish

Chapter 17 Review

Understanding Vocabulary

1. Identify the word or phrase that does not fit the pattern and explain why.
   a. pseudopods, flagella, algae, cilia
   b. kelp, *Amoeba*, *Salmonella*, slime mold
   c. red algae, slime molds, protozoa, foraminiferans
   d. African sleeping sickness, rabies, malaria, giardiasis

Relating Concepts

2. Copy the unfinished concept map below onto a sheet of paper. Then complete the concept map by writing the correct word or phrase in each oval containing a question mark.

Understanding Concepts

Multiple Choice

3. Members of the kingdoms Protista and Monera are different in that protists
   a. have cell organelles.
   b. have cell walls.
   c. lack nuclei.
   d. are much smaller than monerans.

4. *Euglena* move with the aid of structures called
   a. cilia.   c. feet.
   b. pseudopods.   d. flagella.

5. What organisms are responsible for a "red tide"?
   a. green algae   c. kelp
   b. dinoflagellates   d. water molds

6. What organisms form the base of an aquatic food chain?
   a. plants   c. algae
   b. heterotrophic   d. zooplankton
      protists

7. Brown algae are
   a. unicellular.
   b. animal-like.
   c. heterotrophic.
   d. multicellular.

8. What type of protist caused the Irish potato famine?
   a. carrageenan   c. slime mold
   b. water mold   d. brown alga

9. The spread of the disease giardiasis may be stopped by
   a. boiling contaminated drinking water.
   b. draining swamps.
   c. spraying DDT.
   d. vaccinating domestic animals.

10. The seaweeds that wash up on beaches are
    a. unicellular fungi and protists.
    b. autotrophic bacteria.
    c. multicellular algae.
    d. water molds.

11. Efforts used to control malaria include
    a. spraying insecticides to kill mosquitoes.
    b. vaccinating travelers with quinine.
    c. draining wetlands where *Plasmodium* parasites breed.
    d. treating drinking water with chlorine.

Completion

12. Fungi and _____ evolved from heterotrophic protists, while _____ evolved from autotrophic protists.

13. Algae are _____ protists. Single-celled varieties of algae that float near the surface in fresh water or salt water are called _____ .

14. The protist that causes sleeping sickness is transmitted by _____ , while the protist responsible for malaria is transmitted by _____ .

15. The protist-caused disease _____ kills more people each year than AIDS. Most of its victims are less than _____ years old.

Short Answer

16. Name two characteristics that animals and protists have in common.

Protists
may be — autotrophs   may be — ?
gave rise to   include   include   gave rise to
plants -descended- ?   ?   slime mold   ?
        from   that resemble   and
                        can cause   diseases   animals
                                    such as   such as
        are
green   ?   brown   giardiasis   ?

17. What are the symptoms of malaria? Where do most of the victims of malaria live?

18. What are two uses for the shells of diatoms?

Interpreting Graphics

19. The photograph to the right shows kelp, a brown alga.
    • How is this kelp similar to a green plant? How is it different?
    • The round structures contain trapped air. How might these structures be advantageous?

20. Look at the photographs of the four protists shown below:

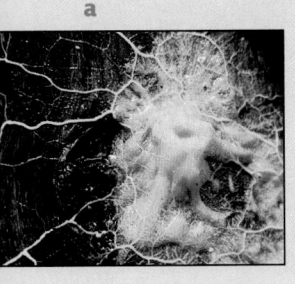

a

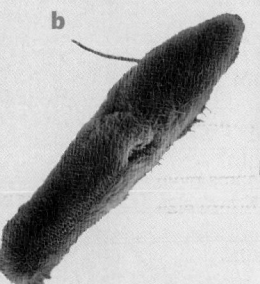

b

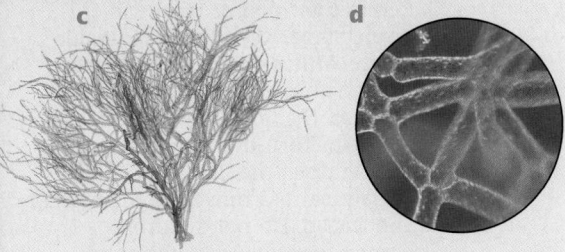

c

d

• Which is a green alga?
• Which is a funguslike protist?
• Which is a red alga?
• Which is a protozoan?

Reviewing Themes

21. *Evolution*
    What evidence suggests that modern algae and plants share a common ancestor?

22. *Energy and Life*
    Describe the connection between algae that lived millions of years ago and gasoline.

Thinking Critically

23. *Inferring Conclusions*
    Explain why some biologists have described Protista as the "catch all" kingdom.

24. *Compare and Contrast*
    Slime molds and water molds were once thought to be fungi. Yet they are classified as protists. Why?

25. *Building on What You Have Learned*
    In Chapter 16, you learned that AIDS is caused by HIV. Both HIV and *Plasmodium* parasites have thwarted scientists' efforts to control the diseases they cause. Although the two are very different, what common factor enables HIV and *Plasmodium* to continue to cause disease?

Cross-Discipline Connection

26. *Biology and Home Economics*
    Do library research to find out about algae that are used as vegetables in Asian-American cooking.

Discovering Through Reading

27. Read the article, "Algae help to clean up contaminated water," in *New Scientist*, May 25, 1991, page 24. How can algae be used to extract heavy metal ions?

Interpreting Graphics

19. • It is multicellular and photosynthetic; it has different photosynthetic pigments in its chloroplasts.
    • These structures serve as floats to keep the kelp near the surface

20. • d
    • a
    • c
    • b

Reviewing Themes

21. They have the same photosynthetic pigments, method of cell division, and food storage molecule (starch).

22. The algae died and sank to the ocean bottom. Over millions of years, heat and pressure transformed the oil in their cells into crude oil, which is refined to produce gasoline.

Thinking Critically

23. Organisms that don't fit neatly into one of the other eukaryotic kingdoms are grouped in the kingdom Protista. Thus, representatives of this kingdom exhibit great diversity in cellular organization, size, and metabolism.

24. Unlike fungi, slime molds and water molds do not have cell walls made of chitin.

25. HIV and *Plasmodium* evolve rapidly, thereby thwarting efforts to control them. *Plasmodium* has evolved resistance to anti-malarial drugs. HIV's surface proteins evolve quickly, complicating the task of developing a vaccine.

Cross-Discipline Connection

26. Answer will vary.

Discovering Through Reading

27. Algae are killed and processed into a matrix of silica gel. The cell walls of the algae absorb heavy metal ions.

Procedural Note

1. Discuss the reasons the five-kingdom system of classification was developed.
2. Review the use of the microscope and demonstrate how to observe the slime mold in the petri dish culture.

Prelab Preparation Answers

1. • Slime molds are members of the kingdom Protista.
   • Although resembling fungi in their feeding habits and reproductive structures, slime molds lack the chitinous cell walls characteristic of fungi.

Procedure Answers

7. The separated piece will rejoin the remainder of the slime mold.
8. The slime mold recoils from the vinegar.
9. The slime mold moved toward and onto the oatmeal flake. While the slime mold does move, it does so very slowly. Marking its position provides a reference point for comparison to its later position.

Chapter 17

Investigation

What Are Slime Molds Like?

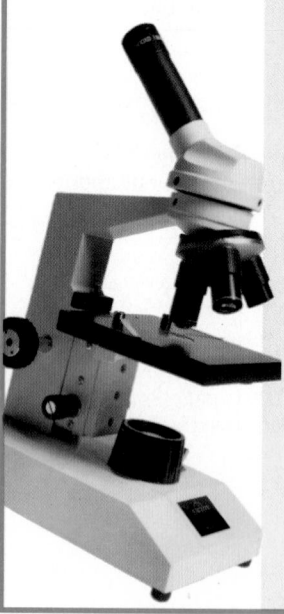

Objectives

In this investigation you will:
• *observe* slime molds
• *contrast* slime molds with animals and plants

Materials

• slime mold culture
• compound microscope
• wax pencil
• oatmeal flakes
• sewing needle
• single-edged razor blade
• glass rod
• beaker containing a small amount of vinegar

Prelab Preparation

1. Review what you have learned about slime molds by answering the following questions:
   • To which kingdom do slime molds belong?
   • Slime molds were once classified in the kingdom Fungi. Why are they no longer considered to be members of this kingdom?
2. Review the procedures for using the compound microscope in the Appendix.

Procedure: Observing Slime Mold

1. Form a cooperative group with another student. Work with your partner to complete steps 2–10 below. Obtain a petri dish containing a slime mold culture.

2. Remove the cover from the petri dish and observe the slime mold. Record your observations.

3. Make a sketch of the petri dish and its contents.

4. With the lid of the petri dish closed, turn the dish upside down and place it on the stage of the microscope. Observe one branch of the slime mold under low power for two minutes. Record your observations. Make a sketch of the slime mold as it appears under low power.

5. Remove the dish from the microscope stage. Keep the dish upside down. With a wax pencil, trace the outline of one branch of the slime mold on the bottom surface of the dish. Turn the dish right side up and open the lid. Place a flake of oatmeal 1–2 mm from the branch you traced. Do not disturb this branch for a few minutes as you conduct the next three procedures.

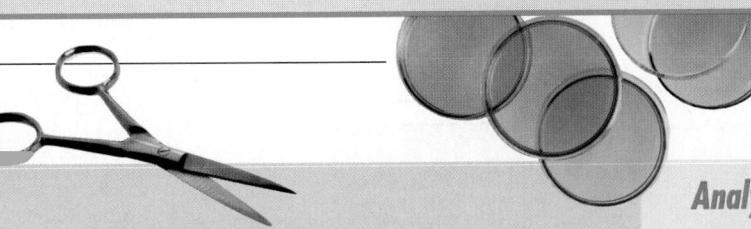

6. Use a clean, fine-tipped sewing needle to puncture the surface of one branch of the slime mold. Record your observations.

7. **CAUTION: Razor blades are sharp and can cause injury.**
Use a razor blade to cut a small piece of slime mold from one branch. Place the piece next to the body of the slime mold. Observe the slime mold for a few minutes. *What happened to the separated piece?* Record your observations.

8. Dip a clean glass rod into a beaker of vinegar. While observing a branch of the slime mold, touch it with the glass rod. *How does the slime mold respond to vinegar?* Record your observations.

9. Now observe the flake of oatmeal. *How did the slime mold react to the oatmeal? How did tracing the outline of the slime mold help you to observe its behavior?*

10. Your teacher has prepared a petri dish containing a slime mold that has been left in the light for 24 hours. Examine the slime mold in this dish and make a drawing of your observations.

11. Clean up your materials and wash your hands before leaving the lab.

Analysis

1. *Analyzing Observations*
What evidence suggests that the slime mold is alive?

2. *Making Comparisons*
Describe two differences between a plant and a slime mold.

3. *Making Comparisons*
In what ways do slime molds resemble animals?

Thinking Critically

Slime molds live among decaying material on the forest floor. Why is it advantageous for the slime mold to be able to form fruiting bodies when exposed to light?

Analysis Answers

1. Answers may vary but should discuss evidence of organization, response to stimuli, movement, and repair.
2. Slime molds lack chloroplasts. They are heterotrophic, as evidenced by their movement toward the oatmeal.
3. Answers may vary but should discuss evidence of movement and the slime mold's response to stimuli.

Thinking Critically Answer

Fruiting bodies produce spores, which function as escape pods. Spores enable the slime mold to endure extremely harsh conditions, such as drying that is associated with exposure to light.

383

Chapter 18 Fungi and Plants

Planning Guide

Objectives/Themes	Classwork Resources	Homework Resources

18.1 1. List the features shared by most fungi.
 	
2. Explain how fungi obtain nutrients.
 	
3. Describe the main differences among the four divisions of fungi.
 	
4. Summarize the ecologic and economic roles of fungi.
 	
Theme: Energy and Life, Evolution, Interacting Systems, Scale and Structure
	Teacher's Resource Binder
	Focus Activity 18
	    Comparing Plant Adaptations
	Science Skills Worksheet
	    Fungal Antibiotics
	
	Other Resources
	Transparencies 75–76
	Science in Action Video
	    *Photosynthesis and Cellular Respiration***
		Text
		Section Review, p. 389
		
		Teacher's Resource Binder
		Directed Reading Worksheet 18.1
		Reteaching Worksheet*
		    Fungi and Plants
		
		Other Resources
		Audiocassette 18.1*

18.2 1. Describe the challenges that faced early land plants. 2. Describe the adaptations that enable plants to survive on land. 3. Define "alternation of generations." 4. Compare the characteristics of nonvascular plants with those of vascular plants. **Themes:** Evolution, Interacting Systems, Patterns of Change, Scale and Structure, Stability	**Teacher's Resource Binder** Lab Investigation 18.1 *Plant Growth and Soil Environment* **Other Resources** Transparency 77	**Text** Section Review, p. 393 **Teacher's Resource Binder** Directed Reading Worksheet 18.2 **Other Resources** Audiocassette 18.2*

18.3 1. Explain how the development of seeds changed today's landscape.
 	
2. Summarize the events involved in reproduction for gymnosperms.
 	
3. Explain the role flowers play in the life cycle of angiosperms.
 	
4. List four uses angiosperms have in your daily life.
 	
Themes: Energy and Life, Evolution, Interacting Systems, Patterns of Change, Scale and Structure, Stability
	Text
	Investigation
	    How Does Soil Type Affect the Germination of Seeds?
	    pp. 400–401
		Text
		Section Review, p. 396
		
		Teacher's Resource Binder
		Directed Reading Worksheet 18.3
		Vocabulary Review Worksheet*
		
		Other Resources
		Audiocassette 18.3*

*Reteaching Options

Demonstrations
18.1: pp. 386, 388
18.2: pp. 390, 391

Assessment
Chapter Review pp. 398–399
Portfolio Assessment p. 383D
Chapter Test—Teacher's Resource Binder
Test Generator

Research Notes

Connection to Natural History: Fantastic Fungus

What is the biggest living thing? What is the oldest living thing? In the past, people considered the giant redwood trees to be among the oldest and the largest living things on the planet. However, starting in 1992, scientists in Canada and Michigan demonstrated that several fungi rivaled the largest and oldest known plants and animals.

Most people think of fungi as isolated individuals, like the mushrooms that dot a forest floor. However, the visible part of a mushroom is actually only the reproductive structure. The thallus, the vegetative structure, is often unseen. Instead of moving, like spores from the reproductive phase do, the thallus seeks out new food sources by growing toward them vegetatively.

In one case, the scientists examined a basidiomycete, Armillaria bulbosa, which has an underground thallus in the form of a network of hyphae called rhizomorphs, which absorb nutrients from tree roots. Using genetic techniques, the scientists demonstrated that samples taken from a continuous area larger than 15 hectares (37.5 acres) were all from the same genetic individual.

They hypothesized that these samples came from a single individual, and not from reproduction, since there is no evidence of crossing over or other DNA rearrangements. If this is true, since the area covered by the fungus is about 635 m (2080 ft.) across and the fungus has an optimum growth rate of about 0.2 m (0.7 ft.) per year, the fungus would be almost 1,500 years old. However, it is unlikely that the fungus could have grown this fast for its entire lifetime, so it could be even older than that. The fungus definitely predates a 300-year-old forest that existed on the site until a 1928 fire.

Estimates of the total mass of this individual fungus range from 100,000 kg to 500,000 kg (220,000 lbs. to 11 million lbs.). By comparison, an adult blue whale, considered one of the largest animals, can have a mass of 200,000 kg (440,000 lbs.). Some giant redwoods have total masses of a million kg (2.2 million lbs.), but they are made up mostly of wood composed of dead plant cells.

The scientists point out that since fungi receive far less attention than animals and plants, it is likely that there are even larger fungi individuals not yet discovered.

Along with searching for these other fungi, scientists are also trying to explore the secret of this fungus's success to determine whether it is genetically more fit for the environment, or if its preeminence is merely a matter of chance. Understanding more about this individual fungus could provide evidence of the specific mechanisms that cause evolution.

Investigation Notes

Does Soil Type Affect the Germination of Seeds?

pages 400–401

Purpose: This investigation provides the student with the opportunity to practice scientific problem solving by designing an experiment. Keep in mind that in this exercise, the results of the experiment are not as important as the methods used to obtain them.

Prelab Preparation

Have sufficient quantities of materials available for students to work in teams of four. Bean seeds, potting soil, sand, and gravel can also be purchased at local garden supply stores. Powdered clay may be available through your art department or can be purchased at a craft store.

Answers will be found on pages 400-401.

Meeting Individual Needs

Objectives

1. Students will demonstrate appropriate use of core vocabulary for the chapter (Vocabulary File).

2. Students will research and explore the use of fungi as food, or in food preparation, in different cultures (Multicultural Lesson Plan).

3. Students will identify differences in plants and their seed structures (Demonstration A).

Vocabulary File
(Developing Vocabulary/ Limited English Proficiency)

If you are not already using the Vocabulary File, refer to Chapter 1 for its preparation. See the Chapter Highlights on page 397 for a list of suggested words.

Multicultural Lesson Plan
Fungi as Foods (Dealing with Ethnic Diversity)

Preparation: Assign students to bring in a recipe from their own culture or from another culture in which yeast is used.

Teaching Strategies:
1. Point out to students that yeast is a fungus. Describe the role of yeast in bread recipes. Yeast can cause dough to "rise" as it consumes some of the sugar present in the dough, making bubbles of carbon dioxide. Contrast breads made with yeast (such as rolls and loaves) with unleavened breads (such as Mexican corn tortillas and Jewish matzo).

2. Before yeast was sold in stores, Americans made their own yeast. Have students make their own yeast in the class or lab using this recipe:

> *Recipe for yeast:*
> One cup flour
> One cup potato water
> (put a potato in water and let sit for a day)
> Salt
> Sugar

Mix and let stand for several hours. (Sometimes more than one form of yeast is made by this reaction. As a result, it does not always produce the same amount of "rising" that packaged yeast does.)

3. Discuss each recipe brought in by the students and the use of yeast in it. With the help of the home economics teacher, have student groups prepare traditional breads from various cultures using recipes brought in by students or found in recipe books. As an alternative, have students volunteer to prepare the breads at home and bring them to school. Be sure each student has an opportunity to sample each type of bread. Choose recipes to also provide examples of bread made without yeast. Have students guess which ones have yeast. Suggested examples include: beaten biscuits—African-American tradition, pizza dough—Italian, St. Lucy buns—traditional Swedish bread, pita bread—Middle Eastern, Matzo—Jewish, corn tortilla—Mexican.

Assessment: Students will research different types of mushrooms and other fungi used for human consumption and food preparation by people of different cultures throughout the world and prepare a chart of fungi, their scientific names, and their uses.

Demonstration A
Basic Differences in Plants
for pages 250–252
(Developing Classification Skills/Multisensory Learners)

Materials: moss, weeds or other herbaceous plants, (preferably from the campus), Boston fern plant, scalpels, hand magnifiers, pine cones, grapes (preferably not seedless), peanuts (salted or unsalted)

Preparation: Locate an area on campus that maintains enough dampness for mosses to grow (probably under a corner eave of the roof). Early in the school year, transplant some to a plant tray. Keep it constantly damp. During a few weeks' time, one cycle of the alternation of generations will have taken place. Students can observe the moss frequently over several weeks. If growing moss inside is unsuccessful, adapt the procedure by taking students outside for observations.

Procedure: Have students recall their observations about mosses. Give each student a frond from the Boston fern. Ask them to observe it. Also hand out samples of herbaceous plants. Have students cross-section stems from the fern and the other plants, and leaflets from the moss, and use hand magnifiers to observe them. Start a discussion with the following questions: What differences did you find? What similarities did you find?

If necessary, read aloud Section 18.2 to students and/or brainstorm general characteristics of plants.

Distribute pine cones, grapes, and peanuts for observation. Read aloud the sections of text titled "Gymnosperms: Plants With Naked Seeds" and "Angiosperms: Flowering Plants." Discuss the differences and similarities observed in the seeds. Have students classify each one, based on the text. Be sure to break open the peanuts to observe cotyledon. Eat the grapes and peanuts.

Additional Strategies
Visual Strategies
Pages 385, 386, 387, 393, and 394

Auditory Learners
Use *Biology: Visualizing Life* Audiocassettes for Sections 18.1, 18.2, and 18.3.

Meeting Individual Needs (cont.)

Cooperative Learning
Four Divisions of Fungi

Timing: Use this activity to introduce Section 18.1.

Group Size: 4 students

Outcome: Students will be able to describe characteristics and methods of reproduction of the four divisions of fungi.

Individual Accountability: Each group member is responsible for describing characteristics, methods of reproduction, and examples for one of the divisions of fungi.

Positive Interdependence: Each group will be able to describe characteristics, methods of reproduction, and examples for each of the divisions of fungi.

Assign one of the following divisions of fungi to each group member: (1) zygomycetes, (2) ascomycetes, (3) basidiomycetes, (4) deuteromycetes. Have each group member read the chapter and take notes on the characteristics and methods of reproduction of the type of fungus assigned. Encourage students to draw illustrations and diagrams to use when sharing information with the other members of the group.

Have each group be certain that each member knows the information about all of the divisions, so that they will be prepared to discuss them. Then, randomly call on a member from each group to discuss characteristics, methods of reproduction, and examples of the four divisions of fungi.

Portfolio Assessment

Students should select their best work and provide a self-reflective rationale for their selections. Students can make selections in the following areas.

1. *Content* — One concept map from the chapter (See page 812 for evaluation criteria.)

2. *Reading Comprehension* — One Directed Reading Worksheet from the Teacher's Resource Binder (Use the answer key to evaluate for accuracy.)

3. *Writing* — Using the Vee Form, summarize a magazine or newspaper article relating to fungi or plant structure. (See page 22T for evaluation criteria.)

   Or: Select a writing task or project from the Chapter Review.

4. *Performance Assessment* — One Vee form from a chapter investigation or lab manual investigation (See page 22T for evaluation criteria.)

Teacher makes selections in the following areas.

1. *Formal Assessment* — Chapter test (Test A, B, or the Test Generator) The teacher-scored test should be reviewed by the student. Incorrect responses should be corrected by the student before the test becomes part of the portfolio.

2. *Informal Assessment* — Use the Direct Observations Checklist, page 33T, during a laboratory or other cooperative learning experience.

3. *Performance Assessment* — Have students illustrate and describe an imaginary plant that features adaptations allowing it to live and reproduce floating in the air instead of on land or in the water.

Concept Map Answer

The following is one possible answer to the Relating Concepts exercise on page 398.

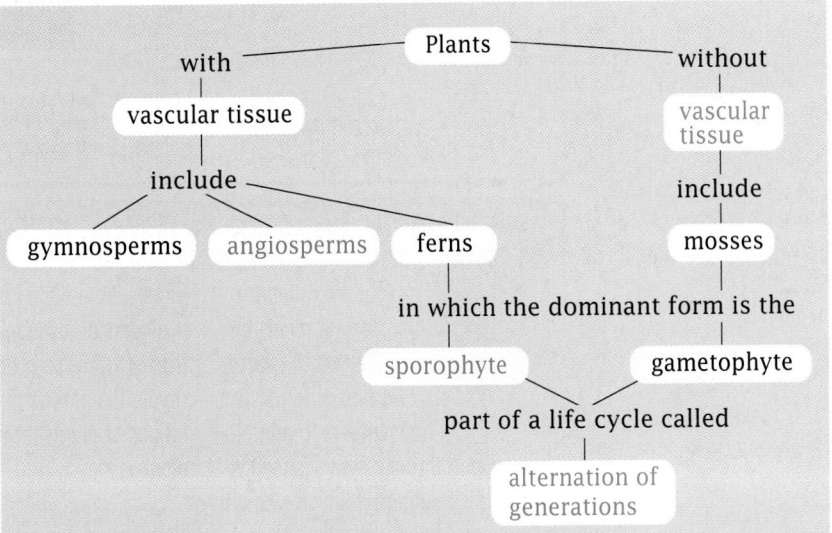

Chapter 18

Determining Prior Knowledge

- Have students describe what happens to oranges, bread, and cheese when left in the refrigerator too long.
- Have students explain why mushrooms are not considered plants.
- Show students examples of the following: fungi, moss, algae, an evergreen, and an angiosperm. Have them explain their similarities and differences.
- Show students a piece of fruit and a pine cone. Have them compare and discuss the structure and function of both.
- Have students explain what they already know about fungi.
- Have students draw a storyboard on seed plant development.

Chapter 18

Fungi and Plants

Review

- ratio of surface area to volume
  (Section 3.1)
- invasions of the land
  (Section 10.3)
- five kingdoms
  (Section 15.3)

18.1 Fungi

- **The Kingdom Fungi**
- **Kinds of Fungi**
- **Fungi in Nature**
- **Fungi and Human Life**

18.2 Early Land Plants

- **Challenges of Life on Land**
- **Adaptations to Life on Land**
- **Nonvascular Plants**
- **Evolution of Vascular Plants**
- **Vascular Plants Without Seeds**

18.3 Seed Plants

- **Vascular Plants With Seeds**
- **Gymnosperms: Plants With Naked Seeds**
- **Angiosperms: Flowering Plants**

Mushrooms, club mosses, and seed plants can all be found growing in the leaf litter that carpets the forest floor.

■ *Author's Rationale* ■

The evolutionary story of plants begins with the introduction of fungi. The first plants were able to invade land through symbiosis with fungi. This relationship involved plants as the providers of food through photosynthesis and fungi as the providers of inorganic nutrients extracted from the ground. The chapter covers the parade of successful adaptations of plants through flowering plants.

FUNGI AND PLANTS WERE THE FIRST EUKARYOTIC ORGANISMS TO INVADE LAND FROM THE SEA. TOGETHER WITH BACTERIA, FUNGI DECOMPOSE ORGANIC MATTER IN THE ENVIRONMENT. THEIR ACTIVITIES ARE AS NECESSARY TO THE CONTINUED EXISTENCE OF THE WORLD AS ARE THOSE OF PLANTS. AS YOU WILL SEE, FUNGI ALSO HAVE A DIRECT IMPACT ON HUMAN LIFE, PROVIDING FOOD AND MEDICINE.

18.1   Fungi

Objectives

❶ **List the features shared by most fungi.**

❷ **Explain how fungi obtain nutrients.**

❸ **Describe the main differences among the four divisions of fungi.**

❹ **Summarize the ecologic and economic roles of fungi.**

The Kingdom Fungi

Fungi (*FUHN jeye*) are a group of eukaryotic organisms, most of which are multicellular. Although little is known about their origins, fungi are at least 400 million years old and probably older. Like plants, fungi are terrestrial. Fungi make up a diverse kingdom that includes the mold and mushrooms shown in **Figure 18.1**. The body of a fungus is made up of many slender filaments called **hyphae** (*HY fee*), which are barely visible to the naked eye. Each hypha of a fungus has cell walls that contain chitin (*KYT uhn*). Chitin is a polysaccharide that is also found in the outer skeletons of insects.

Figure 18.1
Fungi come in many forms, . . .

a   **. . . from mold growing on a melon, . . .**

b   **. . . to this poisonous fly agaric mushroom . . .**

Fungi digest organic matter
You may think of fungi as mold growing on a loaf of bread or on an orange. Just as you use these items for food, fungi can also use them for nutrients. Unlike plants, fungi lack chloroplasts and cannot carry out photosynthesis. Instead, like animals, they are heterotrophs and obtain energy and nutrients from other organisms. Whereas animals digest their food with enzymes in their stomachs, fungi secrete enzymes that digest food outside their bodies. They then absorb the nutrients. Thus, mold growing on bread is actually digesting it a little at a time, and absorbing the nutrients.

c   **. . . and this edible shiitake mushroom.**

Lesson Plan 18.1

Phase 1

PREPARATION

Key Concepts
- Fungi are eukaryotes; most are multi-cellular and feed by decomposing organic matter.
- The body of a fungus is made up of hyphae, and the cell wall of each hypha contains chitin.
- Fungi reproduce asexually and sexually, and are characterized by their method of reproduction.
- There are four divisions of fungi: zygomycetes, ascomycetes, basidio-mycetes, and deuteromycetes.
- Fungi have adapted to almost every environment where organic material and moisture are available.
- Although some fungi are beneficial, many cause disease in plants and animals.

Reading Strategy

Have students read the bold subheadings to grasp the main ideas of the lesson and prepare a summary before reading the section for clarity.

Phase 2
TEACHING STRATEGIES

Visual Strategy

Figure 18.1
If you have not yet done it as a way of determining student knowledge, have students examine these three examples of fungi to determine what they may have in common and how they differ.

385

Visual Strategy

Figure 18.2

In examining the life cycle of mushrooms, emphasize that the ability of fungi to reproduce both ways allows them to survive in almost any environment. Have students predict ways in which fungi such as molds might find their way to bread stored in refrigerators.

Mathematical Connection

Asexual Yeasts

Point out that in the 1950s scientists discovered that yeasts live as diploid and haploid organisms. Emphasize that when conditions are favorable, they live as diploid organisms that reproduce asexually by budding. Scientists first used yeasts to research recessive mutations. Diploid yeasts express a recessive trait in one-fourth of their offspring. However, two of the haploid offspring of diploid yeasts will express a recessive trait. Yeast cells reproduce once every two hours, allowing many mutations to be studied over a small period of time. As an example, have students determine how many yeast cells in a 24-hour period will result from the budding of one yeast cell.

Demonstration 1

Spore Prints

Demonstrate the use of a field guide to mushrooms. Make spore prints of several mushroom species. Have students examine them with a video-microscope or hand lens. Have them observe the differences and use the field guide to identify the species of mushrooms used in this demonstration. Emphasize that not all mushrooms are edible.

Theme Answer

Interacting Systems

Fungi and bacteria are the decomposers of the biosphere.

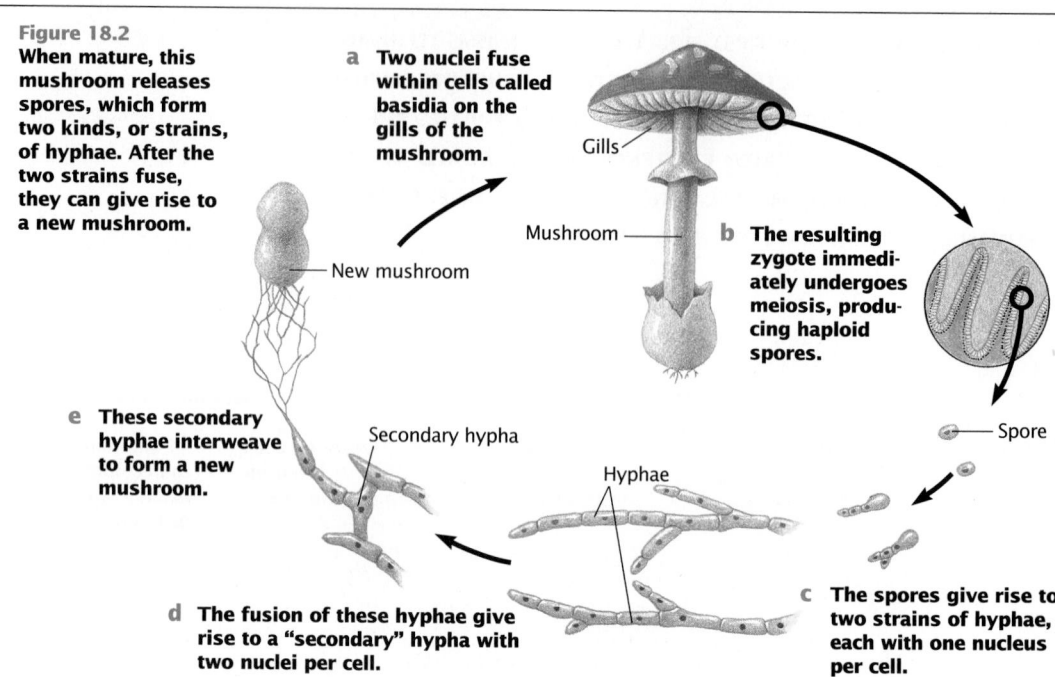

Figure 18.2
When mature, this mushroom releases spores, which form two kinds, or strains, of hyphae. After the two strains fuse, they can give rise to a new mushroom.

a Two nuclei fuse within cells called basidia on the gills of the mushroom.

— Gills

— New mushroom

Mushroom —

b The resulting zygote immediately undergoes meiosis, producing haploid spores.

— Spore

e These secondary hyphae interweave to form a new mushroom.

Secondary hypha —

Hyphae —

d The fusion of these hyphae give rise to a "secondary" hypha with two nuclei per cell.

c The spores give rise to two strains of hyphae, each with one nucleus per cell.

Interacting Systems

What important role in nature do fungi share with bacteria?

In breaking down organic matter, some fungi attack living plants and animals, whereas other fungi attack dead organisms. Both provide rich sources of organic molecules. For example, some fungi grow on healthy elm trees and others grow on animal carcasses. The enzymes that fungi secrete break down the organic matter, which the fungi absorb through the cell walls of their hyphae. A hypha releases chemicals that cause other hyphae to grow nearby. The hyphae spread throughout the food source. As a result, fungi have a very high ratio of surface area to volume. As decomposers, fungi help clear dead plants, animals, and other organic wastes from the environment. In the process, they return to the soil nitrogen, phosphorus, and other nutrients from the bodies of plants and animals. Imagine what the world would look like if fungi did not exist!

Fungi often reproduce by spores

Undoubtedly you have seen fruits and vegetables covered with green patches of mold, or a shower curtain speckled with mildew. If you use a microscope to examine these fungi, you will see that many of their hyphae are tipped with knobs or short strings of beadlike spheres that contain reproductive spores. A fungal **spore** is a reproductive cell that is capable of developing into a new organism. Fungal spores are light and flow through the air, often to new sites. Most spores are well adapted to life on land because they can withstand dry conditions. When a spore lands on a food source and enough moisture is present, it grows into a threadlike hypha. The hypha grows rapidly, branching to form a tangled mass.

Among the molds, much reproduction is asexual. Sexual reproduction occurs in many varieties of fungi. Spores form during both asexual and sexual reproduction. Fungi that reproduce sexually exchange genetic material at some point during their life cycle.

The parts of a fungus you normally see are reproductive structures that produce spores. A mushroom, as shown in **Figure 18.2**, is the reproductive part of a fungus that is growing underground. The fungal hyphae grow rapidly, eventually lifting the cap of the mushroom above ground. Underneath the cap are the spores, which are shed when the cap opens like an umbrella.

■ *Matter of Fact* ■

Mating Strains

Molds reproduce sexually when contact occurs between hyphae from two genetically different molds, called *mating strains*. The mating strains are called plus (+) and minus (−) rather than male and female, because they have identical shapes and functions.

Kinds of Fungi

Like plants, fungi are classified into divisions rather than phyla. Three divisions separate fungi according to the structures they possess for sexual reproduction. Fungi in which sexual reproduction is rare, or has not been recognized, are placed into a fourth division. These divisions are summarized in **Table 18.1**.

Bread molds form structures called zygospores

You may have had some unpleasant experiences with the smallest division of fungi, the zygomycetes (*zy goh MEYE seets*). This division includes the molds, such as *Rhizopus,* that frequently grow on bread. Other zygomycetes live on decaying plant or animal matter in the soil. During sexual reproduction, zygomycetes form a reproductive structure called a zygospore, a thick-walled zygote that can remain dormant for months. When conditions are favorable, a zygospore produces erect stalks tipped with spore-producing structures.

When the spores reach maturity, they are often picked up by the wind and blown about.

Sac fungi form sacs of spores

Without the division of fungi called the ascomycetes (*as coh MEYE seets*), you wouldn't be able to enjoy bread or rolls. On the other hand, without these fungi, grapes and other fruits would not be attacked by powdery mildews and would remain edible longer. The ascomycetes include the yeasts, mildews, morels, and truffles. The feature that unites the ascomycetes is the saclike reproductive cell that forms after hyphae fuse together (the prefix "asco" means sac). Spores are produced inside these sacs, which eventually burst open. In some ascomycetes, spores are thrown as far as 30 cm (12 in.). As a whole, ascomycetes are not completely understood, and thousands of additional species—some undoubtedly of significant economic importance—await scientific description.

Table 18.1 Divisions of Fungi

Division	Sexual Reproduction	Examples	
Zygomycetes	Zygospores form, later producing hyphae that produce spores	Some of the black molds on bread and other foods	*Rhizopus* on bread
Ascomycetes	Spores are produced in saclike reproductive structures	Yeast, mildews, morels, and truffles; fungi that cause Dutch elm disease and chestnut blight	Yeast
Basidiomycetes	Spores form at tips of clublike structures	Toadstools, puffballs, mushrooms, rusts, and smuts	*Amanita muscaria* mushroom
Deuteromycetes	Unusual, rare, or not observed	Many common molds; molds that add flavor to Roquefort cheese	Roquefort cheese mold

Visual Strategy

Table 18.1
Have students visit a library and research the divisions of fungi. Have them extend the table to include the habitat, the number of species, the diseases they cause, and the economic use of each division of fungi.

Theme Connection

Evolution

Point out that it is not known how fungi evolved, but the popular belief is that the first true fungi were unicellular and gave rise to multicellular coenocytic fungi. Coenocytic fungi have filaments in place of the internal cross walls exhibited by club fungi and sac fungi. Coenocytic fungi are believed to have given rise to club fungi, which in turn gave rise to sac fungi.

Connection: Chapter 6

Yeast-Modeling Genes
Tell students that since scientists synthesized a functional artificial chromosome in *Saccharomyces cerevisiae,* yeasts have become a preferred organism in genetic research. Explain that while yeasts are simple eukaryotes with true cellular organelles, their biochemistry closely resembles that of plants and animals, offering great insight into plant and animal genetics. Emphasize that yeasts are used as models for identifying human genes that influence cancer, Down syndrome, and other genetic disorders.

■ *Matter of Fact* ■

Predatory Fungus
Pleurotus ostreatus is a predatory fungus that captures prey for food. It secretes a substance that makes its prey, usually worms, sluggish. The fungal cells surround the prey, penetrate its body, and absorb its contents.

Demonstration 2

Budding Yeast

Pour a packet of dry yeast into a petri dish. Have students observe it through a hand lens. Pour another packet of yeast into a container of warm water, and another packet into warm water in which 2 to 3 teaspoons of sugar have been dissolved. Have students observe and compare the events taking place in both containers of water. Have them make an inference based on their observations. Bubbles of carbon dioxide in the sugar water indicate the growth of yeast cells and lead to the inference that yeast needs sugar in order to grow.

Theme Connection

Energy and Life

Point out that fungi, through decomposition, return carbon dioxide to the atmosphere, where green plants reuse it to make food.

Connection: Chapter 14

Environmental Indicator

Point out that lichens are one of the first groups of organisms to suffer from acid rain, making them good indicators of the pH of rain.

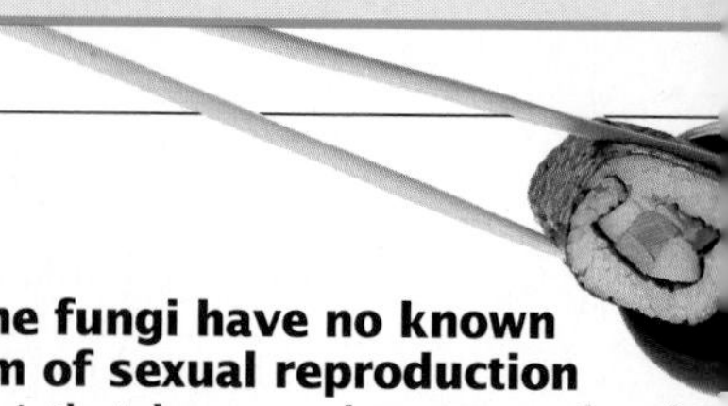

Figure 18.3
Soy sauce, used to flavor sushi (right) and other foods, is made by the fermentation of soybeans by certain fungi.

Many fungi form clublike structures

Most of the organisms you call mushrooms belong to a third division of fungi, the basidiomycetes (*buh sid ee oh MY seets*). This division also includes toadstools, puffballs, rusts, and smuts. These fungi form reproductive structures shaped like wooden clubs, and thus are called club fungi (the prefix "basidio" means "little club"). The clublike structures are found under the caps of mushrooms and toadstools or under the "shelves" of bracket fungi. Spores form and mature on the tips of the clubs and are released.

Some fungi have no known form of sexual reproduction

Fungi that have no known mode of sexual reproduction make up a fourth division called the deuteromycetes (*doot uh roh MEYE seets*), or the Fungi Imperfecti. Some of these fungi have particular economic importance. The flavors of gourmet cheeses such as Roquefort and Camembert are produced by particular strains of molds in this group. Another species is used to ferment soy sauce, shown in **Figure 18.3**. In addition, most of the fungi that cause human skin diseases, such as athlete's foot, are also deuteromycetes.

Fungi in Nature

Figure 18.4
Lichens are resistant to extremes in temperature and moisture. The lichen *Caloplica elegans* is shown here growing on a rock overlooking the St. Lawrence River in upstate New York.

There are very few places on land that at least one species of fungi does not call home. Fungal spores are found in almost any environment. That is why many molds and mushrooms seem to spring up in any location with the right amount of moisture and food. As you read in Chapter 10, many fungi grow in or on the roots of certain plants, in symbiotic associations called mycorrhizae. When mycorrhizal fungi are present, they help transfer nutrients from the soil to the roots and enable plants to thrive. The symbiosis of fungi and plants in mycorrhizae played a critical role in the successful invasion of land by life from the sea. Many of the earliest known plants had mycorrhizae. Fungi provided mineral nutrients for the plants, and plants provided photosynthetically produced energy for the fungi. Recently, researchers have found that the destruction of mycorrhizae by acid rain is playing a key role in the deaths of many forests. Without the mycorrhizae, forest trees are unable to absorb minerals from soil.

If you have ever been in the woods, you may have noticed rocks or logs spotted with orange or green patches called lichens, as shown in **Figure 18.4**. A **lichen** (*LY kuhn*) is an organism that consists of a fungus and an alga living in a symbiotic relationship. The fungus absorbs minerals and other nutrients from the rock and retains water the alga needs for photosynthesis. In turn, the alga produces carbohydrates that the fungus absorbs as food. Lichens can live in very harsh environments. For instance, they are found high on the slopes of Mt. Everest in the Himalaya Mountains of Asia.

Because they absorb water and nutrients directly from the air, lichens are extremely susceptible to environmental pollution. For example, lichens are generally absent in and around cities because automobile traffic and industrial activity produce pollutants that destroy chlorophyll. As a result, the alga cannot make food for itself or for the fungus. Because of this sensitivity, scientists can use lichens as indicators of air quality.

▪ *Cultural Perspective* ▪

Mold Can Heal?

Ancient Egyptians and Chinese used bread molds to treat wounds. Not until contemporary scientists discovered that penicillin is extracted from mold were the Egyptians and Chinese given credit for their awareness of the healing properties of mold.

Fungi and Human Life

Figure 18.5
In 1928, Alexander Fleming discovered that *Penicillium notatum* killed bacteria. Ten years later, the drug penicillin was purified. Penicillin is effective in curing several bacterial diseases including pneumonia, scarlet fever, diphtheria, and many others. Today, penicillin is the world's most widely used antibiotic.

When most of us think of fungi, we think of decay—mold spoiling bread, mildew speckling a shower curtain, or athlete's foot between our toes. However, there are many ways in which fungi work for our benefit.

Take the single-celled ascomycete yeast, for example. How would you be able to enjoy freshly baked bread without *Saccharomyces cerevisiae*, better known as common baker's yeast? *Saccharomyces* is a fungus that is able to turn sugar into carbon dioxide or alcohol. These byproducts enable us to make bread and alcoholic beverages.

Fungi provide medically valuable compounds

If you forget about that orange in the kitchen, it may become covered with a bluish-white fungus called *Penicillium*. In 1928 Alexander Fleming noticed that bacteria he was growing in petri dishes had been killed by the fungus *Penicillium notatum*.

Fleming's accidental discovery, shown in **Figure 18.5**, led to the development of the world's most widely used antibiotic, penicillin.

Other fungi have also been used to revolutionize the field of medicine. In 1972 the Swiss immunologist Jean Borel found a type of fungus that produces a substance capable of suppressing the immune system's response to transplanted organs. This substance, called cyclosporine, opened up a new frontier in medicine. Before it became available in 1979, fewer than half of all kidney transplants were successful. Now, with the widespread use of cyclosporine, the survival rate has risen to 90 percent.

Yeast is now a hot topic in many scientific laboratories. Because yeast cells are eukaryotes, they make better genetic engineering subjects for proteins than do bacteria. Research with yeast may lead to innovative treatments for diseases such as cancer and AIDS.

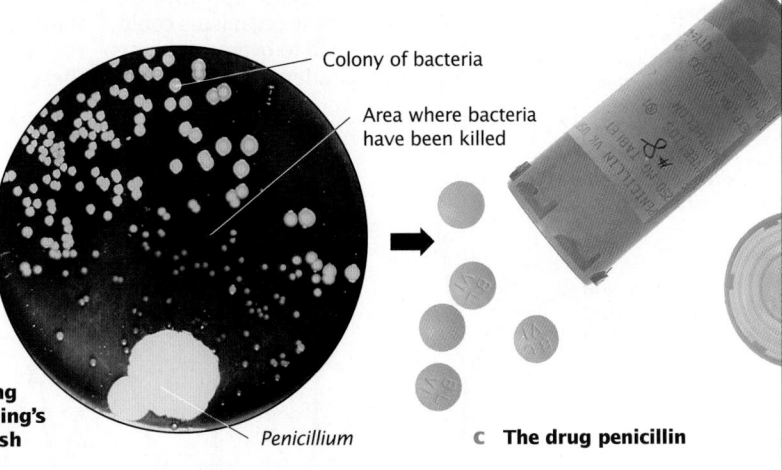

a **Penicillium with spores**

b **Penicillium killing bacteria in Fleming's original petri dish**

Colony of bacteria

Area where bacteria have been killed

Penicillium

c **The drug penicillin**

Section Review

❶ Describe some reasons why biologists have classified fungi into a separate kingdom.

❷ How do fungi obtain nutrients?

❸ Summarize the methods of reproduction for the four divisions of the kingdom Fungi.

❹ How have fungi proven valuable to human life?

■ *Section Review Answers* ■

1. Biologists have classified fungi into a separate kingdom because fungi possess characteristics not found in plants and animals.

2. Fungi secrete enzymes that break down organic matter, which they absorb through the cell walls of their hyphae.

3. See Table 18.1 on page 387.

4. Answers will vary but should discuss both commercial and medical benefits of fungi.

Lesson Plan 18.2

Phase 1
PREPARATION

Key Concepts
- Green plants and fungi first invaded land in a mutual symbiosis called mycorrhizae.
- The first successful land plants—mosses, liverworts, and hornworts—had no vascular tissues or seeds. They evolved a cuticle to prevent water loss and transported material by osmosis and diffusion.
- Club mosses, horsetails, and ferns were among the first plants to evolve vascular tissue for moving water and nutrients throughout the plant; they do not produce seeds.
- Land plants alternate generations from a haploid gametophyte to a diploid sporophyte.

Reading Strategy

Have students incorporate all the subheadings in dark print into a concept map. Explain that they will need to read the text before determining how to use the terms in dark print.

Phase 2
TEACHING STRATEGIES

Demonstration 1
Plant Characteristics
Show students specimens of green algae such as *Ulva* or *Spirogyra*, mosses, and liverworts. Give them hand lenses and dissecting scopes to examine the specimens. Have them develop a list of characteristics common to each plant.

WE TAKE FOR GRANTED THE PRESENCE OF GREEN PLANTS FOUND ALMOST EVERYWHERE ON EARTH. FOR A LONG TIME, HOWEVER, THE LANDSCAPE WAS BARREN. THE EARLIEST LAND PLANTS PROBABLY EVOLVED FROM AN ANCIENT FORM OF GREEN ALGAE THAT GREW AT THE EDGES OF OCEANS AND LAKES. THEIR SUCCESSFUL INVASION OF THE LAND INVOLVED COOPERATION WITH FUNGI.

18.2   Early Land Plants

Objectives

1. **Describe** the challenges that faced early land plants.
2. **Describe the** adaptations that enable plants to survive on land.
3. **Define** "alternation of generations."
4. **Compare the** characteristics of nonvascular plants with those of vascular plants.

Challenges of Life on Land

Figure 18.6
The early land plant *Rhynia* evolved certain features to overcome the environmental challenges of life on land. This illustrated reconstruction of *Rhynia* is based on fossil evidence.

The green algae that were probably the ancestors of today's land plants first invaded the land about 400 million years ago. Before these organisms could live on land, they had to overcome three environmental challenges. First, they had to find a means of conserving water. Second, they had to develop a way to reproduce on land. And third, they had to absorb minerals from the rocky surface. **Figure 18.6** presents these challenges and their solutions.

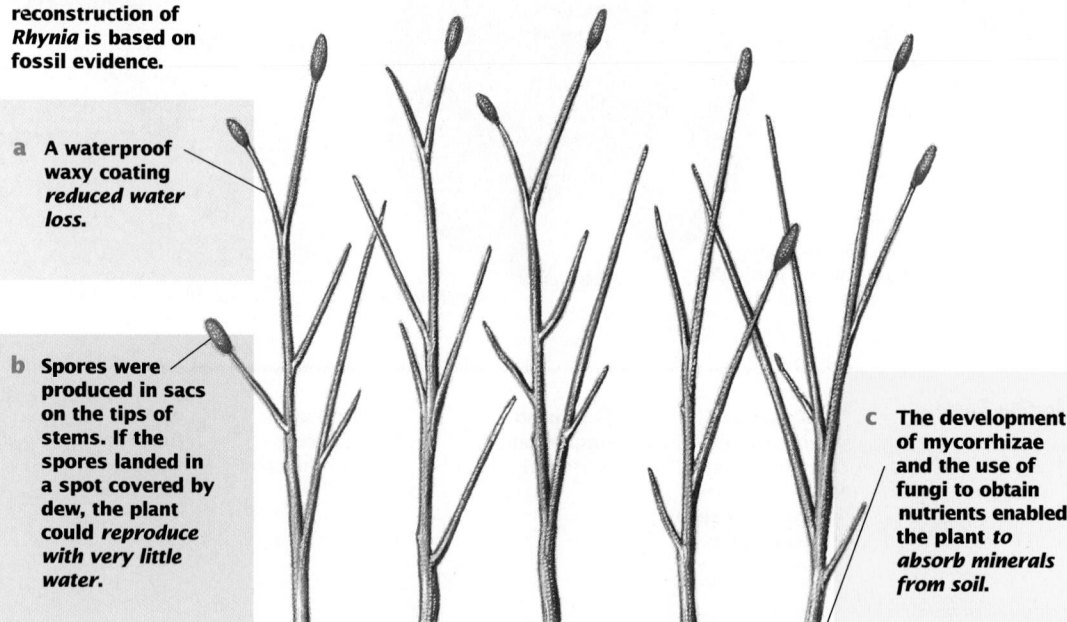

a **A waterproof waxy coating *reduced water loss*.**

b **Spores were produced in sacs on the tips of stems. If the spores landed in a spot covered by dew, the plant could *reproduce with very little water*.**

c **The development of mycorrhizae and the use of fungi to obtain nutrients enabled the plant *to absorb minerals from soil*.**

Adaptations to Life on Land

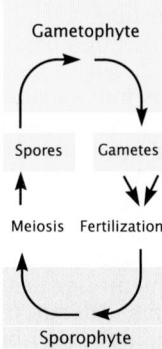

Figure 18.7
In a plant's life cycle, a haploid gametophyte alternates with a diploid sporophyte.

Not all plants solved the problems associated with life on land in the same way. Some plants developed an internal system of interconnected tubes and vessels called **vascular tissues**. Vascular tissues transport water and nutrients throughout the plant. Most of the plants you are familiar with—ferns, oak trees, roses, grasses, house plants—are vascular plants. The surface of a vascular plant is covered by a **cuticle**, a waxy, waterproof layer that helps reduce the amount of water that evaporates from the plants. Other plants lack vascular tissues. Nonvascular plants transport water and nutrients by osmosis and diffusion, much as algae do.

In humans, body cells are diploid, and meiosis produces gametes that are haploid. Plants have a life cycle in which a haploid form alternates with a diploid form. The haploid form is called the **gametophyte** (*guh MEET uh feyet*) because it produces gametes. When two gametes fuse they grow into a diploid form, the **sporophyte** (*SPOHR uh feyet*). This is the "plant" that you usually see. Some of the sporophyte's cells undergo meiosis, producing haploid spores. Each spore then develops into a haploid gametophyte. Thus, in a plant's life cycle a haploid stage is followed by a diploid stage. This cycle is referred to as the **alternation of generations**, shown in **Figure 18.7**.

Nonvascular Plants

The first successful land plants had no vascular tissues or seeds. Nonvascular plants transport materials by osmosis and diffusion and therefore need a large supply of water to survive. Because many of these plants produce sperm that must swim to the egg to fertilize it, they also need a film of water for sexual reproduction. For these reasons, most nonvascular plants, such as mosses, hornworts, and the liverworts shown in **Figure 18.8**, grow close to the ground in moist and shady environments found in forests and swampy shorelines.

Nonvascular plants are relatively small—most are less than 20 cm (8 in.) tall and many are even less than 2.5 cm (1 in.) tall. Many are anchored by rootlike structures called **rhizoids** (*REYE zoydz*). The green, leaflike structures where photosynthesis occurs are only one or two cells thick and are usually protected by a waxy cuticle only on the upper surface. Water enters through pores on the lower surface, which is not covered by a cuticle.

Figure 18.8
Liverworts live in damp places. *Marchantia* produces spores in 2-cm high structures that resemble open umbrellas.

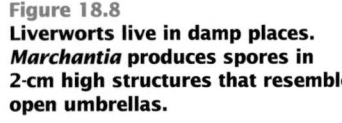

■ Matter of Fact ■

If It's Called "Moss," Is It Moss?
A variety of plants are called mosses, but are they really mosses? No, only members of the class Muscopsida are "true mosses." Other plants referred to as "moss" are not. For example: Irish moss is red algae; Spanish moss

is a flowering vascular plant; and reindeer moss is a lichen.

391

Sporophyte

Gametophyte

Mathematics Connection

Determining the Mass of Dry *Sphagnum*

Have students determine the mass of dry sphagnum moss (available at garden stores). Have them place the moss in a beaker of water for 10–15 minutes. Have them make an inference as to what the mass will be. Afterwards have them remove the moss and determine its mass. Have students compare their findings with their inferences.

Connection: Chapter 14

Pioneer Mosses

Tell students that almost all mosses are small and insignificant in appearance. Yet they play a major role in the establishment of new plants. Tell them mosses are the first plants to appear after a heavy rainfall, flooding, or landslide. Have students brainstorm some inhospitable places where mosses might grow. Ask them how mosses might benefit from being pioneer plants. Explain that pioneer mosses do not face competition from other plants.

Figure 18.9
Mosses are commonly found in moist, shady environments. The leafy haploid gametophyte (lower portion of inset) produces gametes through mitosis. The emerging stalks of the sporophyte are also shown (above, right).

Mosses are nonvascular plants

You have probably seen green, cushiony patches of moss growing in sidewalk cracks, in a dense carpet on a forest floor, or on decaying wood. The greatest number of these nonvascular plants grow best in areas of high humidity, although some mosses are extremely drought-resistant.

In mosses the gametophyte is the dominant form

A clump of moss is actually many individual plants growing close together. The soft, green, leaflike moss plant with which you are most familiar is the haploid gametophyte, which produces gametes through mitosis. A gametophyte is shown in Figure 18.9. Mosses need a film of water to reproduce because the sperm have flagella and must swim through water to reach the egg. Therefore, fertilization can only take place when the moss gametophytes are covered with moisture.

The sporophyte generation begins with the fertilized egg, called a zygote. As the zygote develops into a mature sporophyte, it depends on the gametophyte for water and nutrients. The sporophyte looks like a long stalk with a capsule on top, as shown in Figure 18.9. Inside, cells are undergoing meiosis, producing haploid spores. When the spores are mature, the capsule opens and the spores fall out. If a spore lands in a warm, moist environment, it will grow into a long filament that can carry out photosynthesis. As this filament grows, it periodically produces budlike structures that develop into the familiar green moss gametophytes.

Mosses are pioneer plants

Mosses are among the first plants to grow in otherwise barren areas. By anchoring soil, mosses help prevent erosion. After mosses start to grow, other plants are soon established.

The most important moss to humans is *Sphagnum*. *Sphagnum* is the main component in peat, an organic fuel used in Ireland, Canada, Siberia, and in other northern regions. Peat is cut from bogs that are mainly composed of decomposing mosses and other plants.

Evolution of Vascular Plants

The more than 250,000 species of grasses, trees, ferns, shrubs, and flowers that cover the Earth are vascular plants. Vascular plants have vascular tissues that carry water, nutrients, and the products of photosynthesis throughout the plant. Vascular tissues can transport fluids over long distances, from roots buried deep in the soil to treetops more than 50 m (163 ft.) above the ground. For this reason, vascular plants can grow larger and more complex than nonvascular plants. These adaptations are major evolutionary innovations, giving vascular plants tremendous advantages over nonvascular plants and the ability to expand their geographic range.

▪ Historical Perspective ▪

Peat Bogs

Peat bogs that formed during the Carboniferous period were the sites of the first step in the formation of coal. The plants that lived in swampy areas died and fell into the boggy waters where there was little oxygen and few decomposing bacteria. They were later subjected to pressure that caused them to harden and form coal. The major producer of a peat bog is a moss that belongs to the genus *Sphagnum.*

Vascular Plants Without Seeds

The first vascular plants to evolve were those without seeds, such as club mosses, horsetails, and the ferns shown in **Figure 18.10.** You are probably familiar with the fern's arched, ruffled leaves called fronds. Fronds are supported by short, thick underground stems with strong roots.

About 325 million years ago, during the Carboniferous period, dense forests with tall tree ferns covered much of the Earth. Today, most ferns prefer the moist soil of the tropics. They are also abundant in woods in temperate regions. Yet ferns inhabiting the areas north of the Arctic Circle show that ferns have adapted to other, harsher climates.

In ferns the sporophyte is the dominant form

Figure 18.10 outlines the reproductive life cycle of a fern. The sporophyte in vascular plants, in this case the fern frond, is larger and more complex than the gametophyte. Spores form on the underside of fronds in tiny capsules resembling insect eggs. When spores mature, they are released and dispersed by the wind. If a spore lands in a suitable environment, it develops into a heart-shaped gametophyte, which anchors itself to the soil by rhizoids. The gametophyte produces gametes, which can fuse together in a thin film of water to form a zygote. As the sporophyte matures, it produces a tightly coiled leaf called a fiddlehead. The fiddlehead unrolls into a new frond.

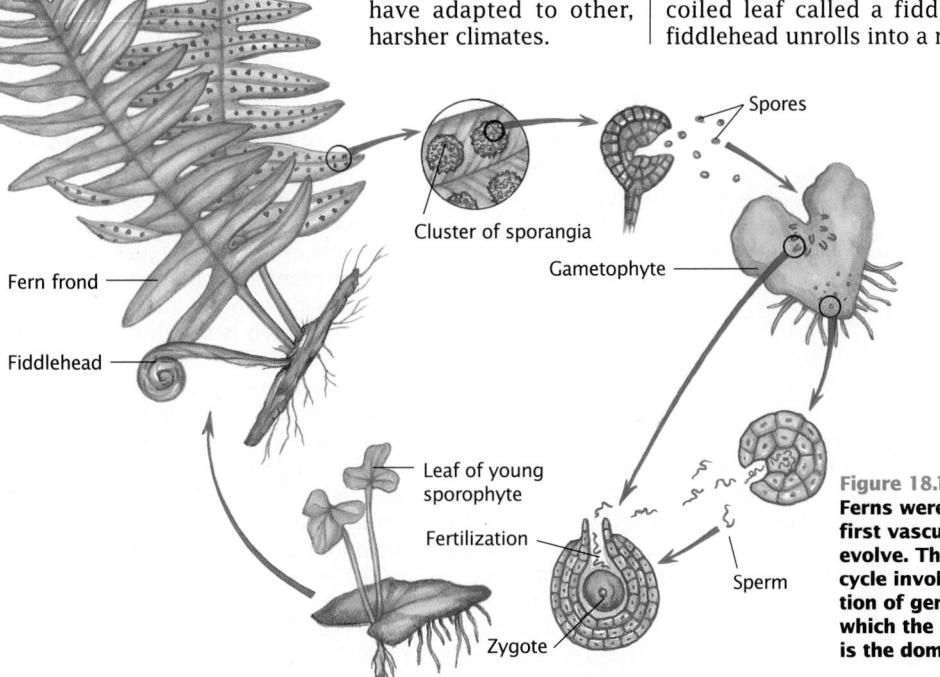

Spores

Cluster of sporangia

Gametophyte

Fern frond

Fiddlehead

Leaf of young sporophyte

Fertilization

Sperm

Zygote

Figure 18.10
Ferns were among the first vascular plants to evolve. The fern life cycle involves alternation of generations in which the sporophyte is the dominant phase.

Section Review

❶ **Describe how the sperm traveled to the egg among early land plants.**

❷ **What adaptations enabled plants to survive on land?**

❸ **How do sporophytes differ from gametophytes?**

❹ **Compare and contrast the characteristics of a moss with those of a fern.**

■ Section Review Answers ■

1. The sperm would swim in thin sheets of water to the egg.
2. Mycorrhizae, a waxy cuticle, and seed development enabled plants to survive on land.
3. Sporophytes are diploid, whereas gametophytes are haploid.
4. Mosses are nonvascular plants. Their life cycles involve alternation of generation in which the gametophyte is the dominant phase. Ferns are vascular plants whose life cycles also involve alternation of generation, but the sporophyte is the dominant phase.

Lesson Plan 18.3

PREPARATION

Key Concepts

- Seeds enable vascular plants to survive in unfavorable environments by storing energy and providing protection for the embryonic plant.
- Seed plant embryos have leaflike structures called cotyledons that store food or help absorb food stored elsewhere in the seed.
- Seed plants exhibit various means of seed dispersal.
- The two groups of seed plants are the gymnosperms and angiosperms. The gymnosperms typically produce naked seeds in cones, while the angiosperms surround the seed with a fruit.

Reading Strategy

Have students relate what they read in this section to the major differences seen in early land plants and fungi.

Phase 2

TEACHING STRATEGIES

Visual Strategy

Figure 18.11

Explain that seeds are well adapted to survival in a wide variety of climates. Emphasize that they can survive in a dormant state if conditions are unfavorable for growth. Have students explain how new seedlings appear after a forest has been destroyed by fire. Explain that some seeds are encased in seed coats so hard that fire cannot destroy the embryo.

THE EVOLUTION OF THE SEED WAS A CRITICAL STEP IN THE DOMINATION OF LAND BY PLANTS. THE SEED PROTECTS THE EMBRYONIC PLANT FROM DRYING OUT. IT ALSO STORES FOOD THAT THE RAPIDLY GROWING YOUNG PLANT USES AS A READY SOURCE OF ENERGY. SOME SEEDS ARE PROTECTED BY FRUITS THAT HELP DISTRIBUTE THEM IN A VARIETY OF WAYS.

18.3   Seed Plants

Objectives

❶ **Explain how the development of seeds changed today's landscape.**

❷ **Summarize the events involved in reproduction for gymnosperms.**

❸ **Explain the role flowers play in the life cycle of angiosperms.**

❹ **List four uses angiosperms have in your daily life.**

Vascular Plants With Seeds

Although early land plants such as ferns had vascular tissue, they were still confined to moist environments because their sexual reproduction depended on a film of water. In contrast, land plants that evolved later were fully adapted to life on land and were able to live in dry places. One of their evolutionary achievements is woody tissue, which lends strength to plants so they can grow tall and compete for sunlight. Another development—more complex vascular tissue—enables water and nutrients to be carried to these new heights. These land plants also developed two features that freed them from requiring a film of water for reproduction—seeds and flowers.

Leaves

Seed coat

Cotyledons

Stem

Figure 18.11
A seed is a storage container that houses a plant embryo. Food inside keeps the embryo alive. A tough outer seed coat protects the emerging young seedling from drying out.

Roots

Seeds enable vascular plants to survive in unfavorable environments

A **seed** is a specialized structure that develops from the fertilized egg of certain plants. Each seed contains a partially developed plant called an embryo, which is capable of growing into a mature plant, as shown in **Figure 18.11**. A plant embryo has leaflike structures called **cotyledons** (*kaht uh LEED uhnz*) that store food or help absorb food stored elsewhere in the seed. A hard outer seed coat covers the embryo and cotyledons, protecting them from physical injury and drought.

The seed is a remarkable evolutionary development. Seeds protect plant embryos from harsh conditions. As a result, an embryo can lie dormant for years and still grow into a healthy plant when conditions improve. Some seeds have stickers, burrs, or feathery wings that enable them to travel long distances on animals or in the wind. The seeds of flowering plants, in turn, helped vascular plants spread to new areas.

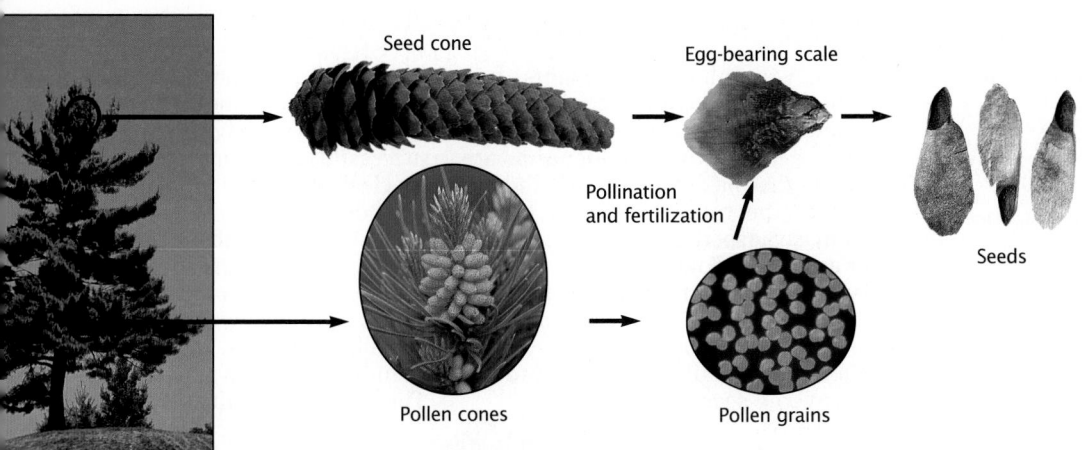

Seed cone

Egg-bearing scale

Pollination and fertilization

Seeds

Pollen cones

Pollen grains

Pine tree

Figure 18.12
A pine tree is a gymnosperm that produces both pollen cones and seed cones. After pollination, new seeds are formed and fall to the ground. There, they germinate to form a new tree.

Gymnosperms: Plants With Naked Seeds

The first land plants to evolve seeds were gymnosperms. **Gymnosperms** (*JIHM nuh spermz*) are plants with naked ovules (that is, not surrounded by an ovary). Many gymnosperms, such as pine, spruce, and fir trees, produce their seeds in cones. They are better adapted to life on land than earlier land plants because they are dispersed by seeds rather than spores. The name gymnosperm combines the Greek words for "naked" and "seed." The seeds of most of these plants develop uncovered on scales within cones. Gymnosperms include redwoods in California, bald cypresses standing knee-deep in Southern swamps, palm-like cycads found in warmer climates, and *Ginkgo* trees found on city streets. One gymnosperm, a bristlecone pine in Nevada, may be the oldest tree in the world—about 5,000 years old. The world's biggest tree is also a gymnosperm, a giant sequoia in California nicknamed after General Sherman of the Civil War. This giant sequoia stands more than 80 m (260 ft.) tall and measures 20 m (60 ft.) around its base.

Many gymnosperms produce gametes within cones

In seedless vascular plants, water is required for the sperm to reach and fertilize the egg. Gymnosperms possess an evolutionary development that enables them to reproduce without a film of water: tiny male gametophytes called **pollen grains**. The tree forms two kinds of cones, in each of which meiosis produces two kinds of spores. In seed cones, one kind of spore develops into the female gametophyte containing egg cells. In pollen cones, the other spores develop into pollen grains. Pollen grains are much smaller and lighter than the gametophytes of earlier plants. Each pollen grain has a pair of air sacs that help carry it in the wind. Pollen grains are shed from the cones in huge quantities, often appearing as a sticky, yellow layer on the surface of ponds, lakes, and even on windshields.

Pollen from the pollen cones is carried to the seed cones in a process called **pollination**. Grains of pollen float in the air and settle down in the scales of the female cones. Here the pollen grain completes its development, growing a slender tube that will deliver the sperm cell to the female gametophyte. Fertilization occurs when the sperm cell fuses with the egg, forming a zygote. The zygote develops into a seed. This process is shown in **Figure 18.12**.

Most species of pine have seeds with thin, flat wings attached. These wings help catch air currents, which carry the seed to new areas. Once a seed has fallen to the ground, it may lie dormant for years. When conditions are favorable, however, the seed will germinate and will begin to grow into a young tree.

Theme Answer

Evolution

The fact that gymnosperm and angiosperm plants produce seeds, have similar life cycles, and carry on pollination, suggests a close evolutionary relationship.

Theme Connection

Energy and Life

Tell students that when they eat real maple syrup, they are actually ingesting boiled sap that once flowed through the veins of sugar maple trees. Have students explain how the sap develops within the trees. Students should understand that sap contains sugar that is formed during photosynthesis, which is stored in the tree for energy.

Phase 3

ASSESSMENT OPTIONS

Closure Strategy

Assign students to cooperative groups of four. Assign individual sections of this chapter to each group. Have students create a concept map for their section. Have them place their maps on an overlay to be presented to the class on the overhead projector. Compare the two groups' mappings of the same section. Have students presenting Section 18.2 make a linkage to the Section 18.1 presentation before they begin. Have students presenting Section 18.3 make a linkage to Section 18.2.

Section Review

Assign the *Section Review*.

Reteaching

Have students design questions and procedures for a game of Jeopardy on this chapter.

Angiosperms: Flowering Plants

Evolution

What traits shared by both angiosperms and gymnosperms suggest a close evolutionary relationship?

When you think about plants, you usually think of the seed plants that evolved most recently—angiosperms (*AN jee uh sperms*). **Angiosperms** are plants that have flowers, which help ensure the transfer of gametes. In many angiosperms, the colors, shapes, or aromas of flowers attract a particular insect, bird, or other animal, which carries pollen from one plant to another. Angiosperms produce seeds enclosed in fruits, in contrast to the uncovered seeds of the gymnosperms. Fruits protect seeds and help disperse them in various ways.

For example, an animal that eats fruit may later deposit the seeds elsewhere.

Angiosperms have been extremely successful. Of the more than 250,000 species of vascular plants, about 235,000 species are angiosperms. These include shrubs, herbs, grasses, vegetables, and many trees—nearly all the plants you encounter daily. Some examples are shown in Figure 18.13. Virtually all our food is derived, directly or indirectly, from flowering plants. They are also valuable sources of timber, textiles, and medicines, with many more products awaiting discovery.

Figure 18.13

a **The fruit of the angiosperm *Vitis* (grape) adds flavor to an open-faced sandwich.**

b **The nutty taste of peanut butter comes from the angiosperm *Arachis hypogaea* (peanuts).**

c **The bread for this sandwich came from the kernels of the angiosperm *Triticum aestivum* (wheat).**

d **T-shirts are produced from the angiosperm *Gossypium* (cotton).**

Section Review

❶ **How do seeds increase a plant's ability to survive on land?**

❷ **How do male gametes reach female gametes in gymnosperms?**

❸ **What function do flowers serve?**

❹ **What are four common products manufactured from angiosperms?**

■ *Section Review Answers* ■

1. Seeds protect plant embryos from harsh conditions. The embryo inside can lie dormant and later grow into a healthy plant when conditions become more favorable.

2. After pollination, the male gametophyte grows a slender tube to deliver the sperm cell to the female gametophyte, so that fertilization can occur.

3. Flowers help ensure the transfer of gametes. Their colors and odors often attract carriers that carry pollen from one plant to another.

4. Angiosperms have been used to produce medicines, textiles, timber, and most of our foods.

Chapter **18** *Highlights*

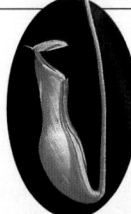

This monkey cup, a vascular plant, traps insects within its long tube. The plant has enzymes that break down an insect's tissues to absorb its nutrients.

Alternative Assessment

Have students make an individual or group concept map, using the key terms on this page that show the interrelationship between fungi and plants.

	Key Terms	Summary
18.1 Fungi  Fungi include molds, morels, and mushrooms, such as this shiitake mushroom.	fungus (p. 385) hypha (p. 385) spore (p. 386) lichen (p. 388)	• Fungi are eukaryotic organisms made up of slender filaments called hyphae. Their cells have cell walls containing chitin. • Fungi secrete enzymes that break down organic material, which they then absorb for food. • Fungi are classified into three divisions based on their sexual reproductive structures. Some fungi display no evidence of sexual reproduction and are placed into a fourth division. • While many fungi spoil food and cause disease, some are helpful to humans.
18.2 Early Land Plants  Lichens, which consist of fungi and algae living in a symbiotic relationship, are shown here growing on a rock.	vascular tissue (p. 391) cuticle (p. 391) gametophyte (p. 391) sporophyte (p. 391) alternation of generations (p. 391) rhizoid (p. 391)	• Land plants have certain adaptations that enable them to live in a dry environment. These adaptations reduce water loss, move water and nutrients throughout the plant, and enable plants to reproduce in dry environments. • Land plants have a complex life cycle that involves an alternation of generations between a haploid gametophyte and a diploid sporophyte. • Mosses and ferns are living examples of early land plants that still require the presence of a film of water for reproduction.
18.3 Seed Plants 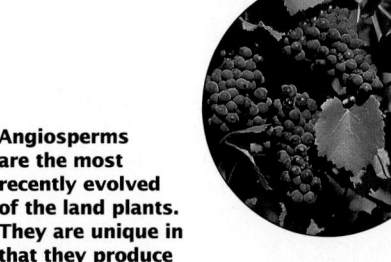 Angiosperms are the most recently evolved of the land plants. They are unique in that they produce flowers and fruits.	seed (p. 394) cotyledon (p. 394) gymnosperm (p. 395) pollen grain (p. 395) pollination (p. 395) angiosperm (p. 396)	• Seed plants are able to reproduce without the presence of a film of water. Seeds can survive long periods of time in unfavorable environments. • Gymnosperms are seed plants with uncovered ovules. Seeds are usually in cones. • Angiosperms are seed plants that have flowers and fruits.

Chapter Review Answers

Understanding Vocabulary

1. **a.** A spore is a haploid, reproductive cell that can be carried by the wind or water and develops into a new organism. A seed is a diploid structure that develops from a fertilized plant egg. Each seed contains a partially developed embryo that can grow into a mature plant.
b. Mycorrhizae is a symbiotic relationship between fungi and plants, whereas lichens are a fungus and an alga living in a symbiotic relationship.
c. Nonvascular tissue uses osmosis and diffusion to transport materials, and vascular tissue uses an internal system of interconnected tubes and vessels to transport materials throughout the plant.
d. The gametophyte is the haploid form of a fungus or plant; the sporophyte is a diploid form of a fungus or plant that is formed by the fusion of two gametes.
e. A plant that produces its seeds in cones is called a gymnosperm, while a plant that has flowers and whose seeds develop inside a fruit is called an angiosperm.

Relating Concepts

2. Map answer is shown on page 383D.

Understanding Concepts
Multiple Choice

3. c	7. c
4. a	8. a
5. c	9. d
6. d	10. c

Completion
11. lichens
12. baker's yeast
13. algae, water, water
14. cotyledons, seed coat

Short Answer
15. They form spores at the tips of clublike structures.

Chapter 18 Review

Understanding Vocabulary

1. For each pair of terms, explain the difference in their meanings.
   a. spore, seed
   b. mycorrhizae, lichen
   c. nonvascular, vascular
   d. gametophyte, sporophyte
   e. gymnosperm, angiosperm

Relating Concepts

2. Copy the unfinished concept map below onto a sheet of paper. Then complete the concept map by writing the correct word or phrase in each oval containing a question mark.

Understanding Concepts

Multiple Choice

3. Fungi are classified according to
   a. the way they get nutrients.
   b. the diseases they cause.
   c. their structures for sexual reproduction.
   d. the environment in which they grow.

4. Human ailments caused by fungi include
   a. athlete's foot.   c. rusts.
   b. truffles.   d. penicillin.

5. Which structure enabled early land plants to live on land?
   a. seed   c. waxy cuticle
   b. flower   d. fruit

6. Which plant lacks vascular tissue?
   a. rose   c. ginkgo
   b. fern   d. moss

7. Which structure is part of the fern sporophyte?
   a. cone   c. frond
   b. capsule   d. flower

8. Which is needed for fertilization in ferns but not needed for fertilization in gymnosperms?
   a. wind   c. a film of water
   b. insects   d. pollen tube

9. Both angiosperms and gymnosperms
   a. develop flowers.
   b. bear fruit.
   c. are pollinated by insects.
   d. produce seeds.

10. Structures produced by angiosperms but not produced by gymnosperms include
    a. seeds and pollen.
    b. pollen and egg-bearing cones.
    c. flowers and fruit.
    d. gametophytes and sporophytes.

Completion
11. Because of their sensitivity to air pollution, _____ are used as indicators of air quality.

12. The yeast *Saccharomyces cerevisiae*, commonly called _____ , plays an important role in bread-making.

13. The ancestors of today's land plants were green _____ . To be successful, early land plants had to conserve _____ , absorb minerals, and reproduce with very little _____ .

14. A seed contains an embryo and one or more leaflike _____ that are protected by the outer _____ _____ .

Short Answer
15. Basidiomycetes include rusts, toadstools, and puffballs. What do these fungi have in common?

16. Give two examples of how fungi are helpful to humans.

17. Why are mosses called pioneer plants?

18. How do the two types of spores in pine trees differ in the gametes they produce?

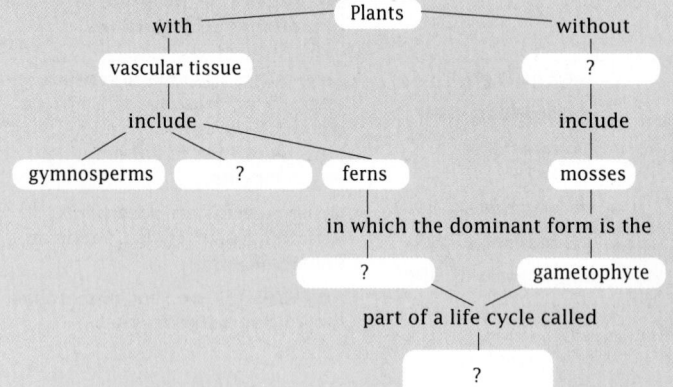

---

16. Cyclosporine is the product of a fungus; it suppresses the human immune system's response to transplanted organs. Yeast cells are used as subjects in genetic engineering experiments. These experiments may lead to treatments for AIDS and cancer.

17. Mosses are typically the first plants to grow in an uninhabited area. They provide the nutrients and soil conditions needed by other plants that soon follow.

18. The female spore develops into a gametophyte with egg cells, while the male spore develops into immature gametophytes called pollen. Pollen is small and transported by wind to the female gametophyte. The female gametophyte remains in a cone until after its union with the male gamete. This union results in a zygote that develops into a seed.

Interpreting Graphics

19. Examine the three photographs below. Identify each plant or plant part as either a moss, fern, gymnosperm, or angiosperm. Indicate whether the plant is a vascular or nonvascular plant and if it is the gametophyte or sporophyte generation of the plant.

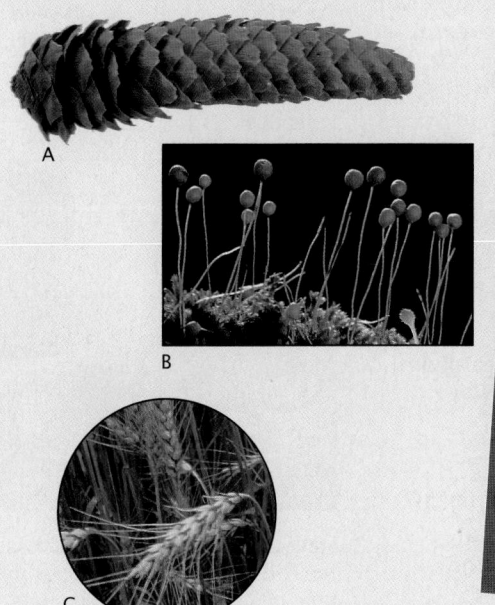

A

B

C

20. Study the diagram below and answer the following questions.

- Which event is shown at A?
- What structures are produced at B?
- What structures are produced at C?
- Which event is shown at D?
- Which half of the life cycle is haploid?
- Which half of the life cycle is diploid?

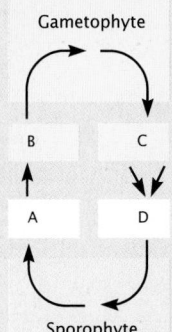

Gametophyte

B   C

A   D

Sporophyte

Reviewing Themes

21. *Energy and Life*
What energy source is used by the embryo in a seed before photosynthesis can begin in the new plant?

22. *Patterns of Change*
Plants have a life cycle that involves gametophyte and sporophyte generations. How are the gametophyte and sporophyte generations of a plant linked through fertilization and meiosis?

Thinking Critically

23. *Comparing and Contrasting*
Fungi were once considered plants. What features distinguish fungi from plants?

24. *Inferring Conclusions*
Explain the observation that moss tends to grow on the north side of a tree. Justify your answer.

25. *Inferring Conclusions*
How might the landscape of Earth differ if seeds had not evolved?

26. *Comparing and Contrasting*
Describe the ways in which gymnosperms and angiosperms distribute pollen. Which method of pollen delivery is more efficient? Explain.

Cross-Discipline Connection

27. *Biology and History*
In the 1700s, the long-leaf pine forests of North Carolina were valued for their tar, pitch, rosin, and turpentine. Many of these products were shipped to England. Why were these products called "naval stores"? Why did the British need them?

Discovering Through Reading

28. Read the article "Jekyll-Hyde Mushrooms," in *Natural History*, March 1992, pages 46–52. How do fungi such as *Arthrobotrys* obtain nutrients?

Thinking Critically

23. Fungi are heterotrophs and plants are autotrophs. Plant bodies are greatly differentiated, while fungi lack differentiation except among the reproductive cells. The cell walls of most fungi contain chitin, but the cell walls of plants contain cellulose. Fungi lack chlorophyll and do not have alternation of generations.

24. The south side of trees may receive direct sunlight during most of the day, while the east and west sides may receive direct sunlight during parts of the day. Only the north side of trees is shaded from the sun all day, remaining moist enough for moss to grow.

25. Land plants would be found only in areas where water is plentiful—next to rivers, lakes, oceans, or where the humidity is very high. Dry areas would be totally void of plants and other organisms that depend on plants for food.

26. In gymnosperms, pollen is moved by wind to the female cone; animals are the primary movers of pollen in angiosperms. The movement of pollen in angiosperms is more efficient. More pollen is deposited on the pistil, and less is wasted.

Cross-Discipline Connection

27. Tar, pitch, rosin, and turpentine were called naval stores because they were used in building and maintaining wooden ships. Naval stores were needed by the British to build ships and to maintain their large number of war and merchant vessels.

Discovering Through Reading

28. *Arthrobotrys* obtains nutrients by sending out an extensive system of filaments through soil or rotting logs to capture nematodes. Strands of hyphae adhere to the nematode, releasing paralyzing toxins. Then, the fungus digests the worm from inside its body.

Interpreting Graphics
19. A gymnosperm, vascular, sporophyte
B moss, nonvascular, sporophyte
C angiosperm, vascular, sporophyte
20. • meiosis
• spores
• gametes
• fertilization
• The second half of the cycle, or the gametophyte generation, is haploid.
• The first half of the cycle, or the sporophyte generation, is diploid.

Reviewing Themes
21. In many plants, the energy required is stored as food in cotyledons.
22. The diploid sporophyte produces haploid spores by meiosis. The spores then develop into gametophytes that produce gametes. The gametes join during fertilization to form the diploid zygote, which develops into a new sporophyte.

399

Prelab Preparation Answers

1. • A *hypothesis* is a testable possible explanation for an observation.
   • A *control experiment* is one in which the condition suspected to cause the effect is compared to the same situation without the suspected condition (a control group).
2. Answers will vary but should suggest the type of soil, particle size, soil pH, texture, drainage properties, composition, amount of humus, etc.

Procedure Answers

3. Answers will vary but should relate to germination and the properties of soil.
4. Answers will vary but should clearly relate to the problem.
6. Answers will vary but should describe an experiment in which four groups of seeds are treated identically except for the type of soil used. Control experiments allow data to be compared from groups that differ only in the variable to be tested.

Chapter 18

Does Soil Type Affect the Germination of Seeds?

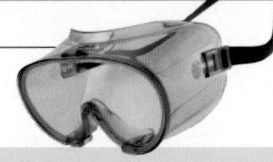

Investigation

Objectives

In this investigation you will:
• *design* an experiment to test a hypothesis
• *compare* the germination of seeds to the conditions of different soil types

Materials

• lima bean seeds
• paper cups (or small flowerpots)
• potting soil
• sand
• gravel
• powdered clay
• water

Prelab Preparation

1. Review what you have learned about scientific methods by answering the following questions:
   • What is a hypothesis?
   • What is a control experiment?
2. A gardener wonders how the conditions of the soil in the backyard might affect the germination of lima bean seeds. What properties of the soil might affect the rate of germination of the seeds?

Procedure: Determining If Soil Type Affects Germination

1. Form a cooperative group of four students. Work with a member of your team to complete steps 2–11.
2. You and your partner are to design an experiment that demonstrates the effects of soil type on seed germination. With your partner, discuss how you might complete your task, using the materials on your desk.

3. State a hypothesis that addresses the following question: *Does the type of soil affect the germination of a seed?*
4. Using the materials on your table, design a control experiment that tests your hypothesis. Keep in mind that lima bean seeds will usually germinate when they are planted about 2 cm (1 in.) below the surface of the soil and are adequately watered. *What steps will you use in your experiment?*

5. For each step specify the materials you will use, the amounts that will be used, how measurements will be made, and the conditions under which the procedures will take place.

6. Working with your partner, proceed with your experiment only after having your experimental design approved by your teacher. *What makes your experiment a control experiment?*

7. Determine a period of time to observe your experiment. Make a table similar to the one shown below for recording your data.

Observation of Seeds Grown in Different Types of Soils

Soil type	Day 1	Day 2	Day 3	Day 4	Day 5
Potting soil					
Sand					
Gravel					
Clay					

8. Record your observations in your table. After the period of observation, combine your data with that of the other team in your group.

9. Make a bar graph showing the number of seeds that germinated in each type of soil.

10. Explain how the information does or does not support your hypothesis. If your data show the need to change your hypothesis, state a new one. Explain why you changed your original hypothesis.

11. Clean up your materials and wash your hands before leaving the lab.

Analysis

1. *Summarizing Data*
   Summarize the data collected throughout your experiment.

2. *Analyzing Information*
   Do the results of your observations support your hypothesis? Explain.

3. *Analyzing Methods*
   What is the importance of keeping careful records when doing an experiment?

4. *Communicating Information*
   Of what value is sharing observations with other scientists?

5. *Making Inferences*
   What aspect of soil is most important to the germination of seeds?

6. *Applying Concepts*
   How might a gardener apply the principles illustrated in this investigation?

Thinking Critically

1. What relationship do you observe between rates of growth of the lima bean seeds and the type of soil?

2. Is a soil best suited for germinating seeds likely to be the best soil for growing plants? Explain your reasoning.

Analysis Answers

1. Students' responses should reflect the data collected in their charts.

2. A clearly stated hypothesis usually states the variable being tested and measured in the experiment.

3. Records tell what has been done and the results of these actions. This information is necessary to draw valid conclusions and plan future investigations.

4. Sharing observations makes it possible to check the validity of one's work and to identify sources of error.

5. Answers may vary but might suggest that the aspects of soil that affect drainage and evaporation also affect the rate of germination.

6. Answers may vary but might suggest that for best results, soil types and properties should be a major concern when planting seeds.

Thinking Critically Answers

1. Answers may vary. Lima beans grow best in rich soil that does not have too much nitrogen but retains moisture.

2. Soil that is best suited for germination may not be best suited for growing plants, because the needs of the seed and the needs of the plant are not identical.

Chapter 19 Plant Form and Function

Planning Guide

	Objectives/Themes	Classwork Resources	Homework Resources

19.1	**1.** Describe the functions of roots and shoots.
	
	2. Compare and contrast vascular tissues, ground tissue, and dermal tissue.
	
	3. Describe the role of meristems in plant growth.
	
	4. Distinguish between monocots and dicots.
	
	Themes: Patterns of Change, Scale and Structure
		Text
		Journeys
		  Tour of a Leaf p. 405
		Science in Action
		  Eloy Rodriguez: Biochemist pp. 408–409
		
		Teacher's Resource Binder
		Focus Activity 19
		  Inferring Function From Structure
		
		Other Resources
		Transparency 78
		Science in Action Video
		  *Photosynthesis and Cellular Respiration***
			Text
			Section Review, p. 407
			
			Teacher's Resource Binder
			Directed Reading Worksheet 19.1
			
			Other Resources
			Audiocassette 19.1*

19.2	**1.** Summarize how water is transported through plants.
	
	2. Explain the manner in which carbohydrates are moved through plants.
	
	3. State the functions of three kinds of plant hormones.
	
	4. Explain how plant growth responds to changes in seasons.
	
	Themes: Energy and Life, Interacting Systems, Patterns of Change, Stability
		Teacher's Resource Binder
		Science Skills Worksheet
		  Understanding Processes and Graphs
		Lab Investigation 19.1
		  Stomata and Transpiration Rates
		
		Other Resources
		Transparency 79
			Text
			Section Review, p. 414
			
			Teacher's Resource Binder
			Directed Reading Worksheet 19.2
			
			Other Resources
			Audiocassette 19.2*

19.3	**1.** Describe the structure of a flower.
	
	2. Summarize the processes of pollination and fertilization in typical flowering plants.
	
	3. Explain the survival value of double fertilization.
	
	4. Describe some adaptations of flowers that help ensure pollination.
	
	Themes: Interacting Systems, Scale and Structure, Stability
		Text
		Investigation
		  How Is Flower Structure Related to Function? pp. 424–425
		
		Other Resources
		Transparencies 80–81
			Text
			Section Review, p. 420
			
			Teacher's Resource Binder
			Directed Reading Worksheet 19.3
			Vocabulary Review Worksheet*
			Reteaching Worksheet*
			  Plant Form and Function
			
			Other Resources
			Audiocassette 19.3*

*Reteaching Options

Demonstrations
19.1: pp. 404, 406
19.2: p. 410
19.3: pp. 417, 418, 419

Assessment
Chapter Review pp. 422–423
Portfolio Assessment p. 401D
Chapter Test—Teacher's Resource Binder
Test Generator

Research Notes

Connection to Chemistry: Vitamins for Plants

In a few years, people may tend drooping and wilting plants with solutions of common vitamins such as vitamin C and vitamin E. Research by Dale Norris and his team at the University of Wisconsin shows that such vitamins seem to improve plant growth and health, apparently by stimulating the plant's own stress-defense systems.

Norris, who is an entomologist, began to work with plants when he noticed that some cell walls in plants have stress-sensitive proteins similar to those in insect nerve cells. These proteins apparently respond to any type of stress by producing chemical messengers that stimulate the plant's genes to produce stress-defense hormones. This system is called into action whenever the plant faces stress, whether from exposure to dangerous chemicals, a fungal infection, or an insect attack.

However, these stress-sensitive trigger proteins can easily undergo chemical change and be oxidized and disabled by reactive chemicals. When this happens, the plant's main line of defense against a broad range of threats is weakened.

Vitamins C and E both are strong antioxidizing agents and can prevent this damage to the proteins. Norris has found that the vitamins work best in extremely dilute solutions (just a few parts per million). Overdoses of vitamins can cause more harm than good in plants.

A correct dosage of vitamins can provide beneficial effects when applied in a variety of ways, such as being sprayed in the soil, painted on stalks and trunks, or applied to leaves. Apparently stimulating the stress protein in some of the cells triggers a response in the entire plant. Researchers have found similar responses in plants as diverse as snap beans, sweet corn, elm trees, and soybeans.

It also appears that common garden pests such as caterpillars will avoid plants that have been treated with vitamins when they are given a choice. In addition, University of Wisconsin researchers have shown that the vitamins can help plants resist microbial infections, herbicides, bruising, and drought. Other investigators have confirmed some of Norris's findings, and some companies are beginning to pursue commercial applications. Current research is focusing on determining the most effective doses.

Investigation Notes

How Is Flower Structure Related to Function?
pages 424–425

Purpose: In this investigation, students examine the structure of a flower, because flowers represent such an important step in the development of land plants.

Prelab Preparation

1. Try to obtain complete flowers with easily identifiable parts. Flowers such as gladioluses, snapdragons, azaleas, and any member of the lily family are good choices.

2. Florists will often donate flowers that cannot be sold. Flowers can also be brought by students from a home garden. Be sure students first seek permission to pick flowers. Remind them that they should never pick flowers in public parks.

3. Additional materials that may help clarify the investigation include WARD'S dicot flower model (81 M 1130) and a microscope slide with mixed pollen (91 M 7001).

Answers will be found on pages 424–425.

Meeting Individual Needs

Objectives

1. Students will develop core vocabulary for the chapter (Vocabulary File).

2. Students will gather information about and explain different examples of dyes used in various cultures (Multicultural Lesson Plan).

Vocabulary File
(Limited English Proficiency/ Learning Disabled)

If you are not already using the Vocabulary File, refer to Chapter 1 for its preparation. See the Chapter Highlights on page 421 for a list of suggested words.

Multicultural Lesson Plan
Dyes
(Dealing with Ethnic Diversity)

Materials and Preparation: Gather marigolds, walnuts, fresh red grapes, red cabbage, several T-shirts, and several rubber bands. Be sure students have access to references about plants and dyes, such as *Vegetable Dyeing—151 Color Recipes for Dyeing Yarns and Fabrics with Natural Materials,* by Alma Lesch; *Dye from Plants,* by Seonaid Robertson; and *Dye Plants and Dyeing,* by Ethel J. Schekty. Before class or during it, boil either the flowers of the marigold or the walnuts in water to make a yellow dye. Boil the red grapes or the red cabbage (or each one) for a red dye.

Teaching Strategies:

1. Describe the importance that dyeing and colored fabrics had in other cultures. Students will discuss examples of the importance of color that remain in today's culture: for example, colors for schools or sports teams. There was a time in American culture when making and using dyes was such an important component of culture that it was a part of everyday life. Today, some Africans, Mexicans, and aboriginal people still use plants to make dyes.

2. Bring out the dyes, and explain to the students how each one was made. Encourage them to bring in more natural substances to use in making dyes. (Nuts, leaves, and fruits are likely choices.) Try boiling them in water to see if a dye is formed from them.

3. With all of the dyes available, students may tie-dye T-shirts. Knot and tangle the tee-shirt, using the rubber band to hold it together. Dunk the shirt into one type of dye. Then remove it, and adjust the knots and rubber bands before dunking into another color dye. After repeating this process several times, the T-shirt will be covered with colorful and intricate patterns. Caution students to launder the T-shirts in cold water (not warm or hot) to keep the dye in them as long as possible.

Assessment:

1. Students should research the way in which dye production has been shifted in Western cultures from individuals to industries. Have students research the circumstances of the discovery of the first synthetic dye, "mauve." Students are to find the names of companies that specialize in selling dyes, paints, crayons, and other materials that we now use to color materials, houses, clothing, etc. Where possible, they should identify the sources of the dyes—are they natural or synthetic?

2. Students will research the uses of the dye called henna. They will be able to tell the scientific classification of its source, describe where it can be found in various continents, discuss its ancient use by the Egyptians and other cultures, and describe the way it is used in the United States today. Students will bring to class or list products that contain henna.

Additional Strategies
Visual Strategies

Pages 403, 404, 406, 411, 412, 415, 416, 417, 418, and 420

Auditory Learners

Use *Biology: Visualizing Life* Audiocassettes for Sections 19.1, 19.2, and 19.3.

Meeting Individual Needs (cont.)

Cooperative Learning
Pollination of Flowers

Timing: Use this activity to introduce Section 19.3.

Group Size: 2 students

Outcome: Students will be able to depict an example of the relationship between flower structure and the pollinating process.

Individual Accountability: Each group member will be responsible for creating a model of a flower or an animal well suited for pollinating it.

Positive Interdependence: Each group will be able to explain the relationships between a flower and an animal capable of pollinating it.

Give each pair of students a variety of materials, such as pipe cleaners, construction paper, paper rolls, tape, glue, and scissors. Students in each pair should work together to construct models of a flower and an animal, such as an insect, bird, or mammal, that would be well suited to pollinating that particular flower. Then have each pair demonstrate and explain the pollination process, using their models.

Portfolio Assessment

Students should select their best work and provide a self-reflective rationale for their selections. Students can make selections in the following areas.

1. *Content* — One concept map from the chapter (See page 812 for evaluation criteria.)

2. *Reading Comprehension* — One Directed Reading Worksheet from the Teacher's Resource Binder (Use the answer key to evaluate for accuracy.)

3. *Writing* — Using the Vee Form, summarize a magazine or newspaper article relating to plant parts and their roles. (See page 22T for evaluation criteria.)

   Or: Select a writing task or project from the Chapter Review.

4. *Performance Assessment* — One Vee form from a chapter investigation or lab manual investigation (See page 22T for evaluation criteria.)

Teacher makes selections in the following areas.

1. *Formal Assessment* — Chapter test (Test A, B, or the Test Generator) The teacher-scored test should be reviewed by the student. Incorrect responses should be corrected by the student before the test becomes part of the portfolio.

2. *Informal Assessment* — Use the Direct Observations Checklist, page 33T, during a laboratory or other cooperative learning experience.

3. *Performance Assessment* — Have students create scientifically accurate models of the reproductive structures of flowering plants.

Concept Map Answer

The following is one possible answer to the Relating Concepts exercise on page 422.

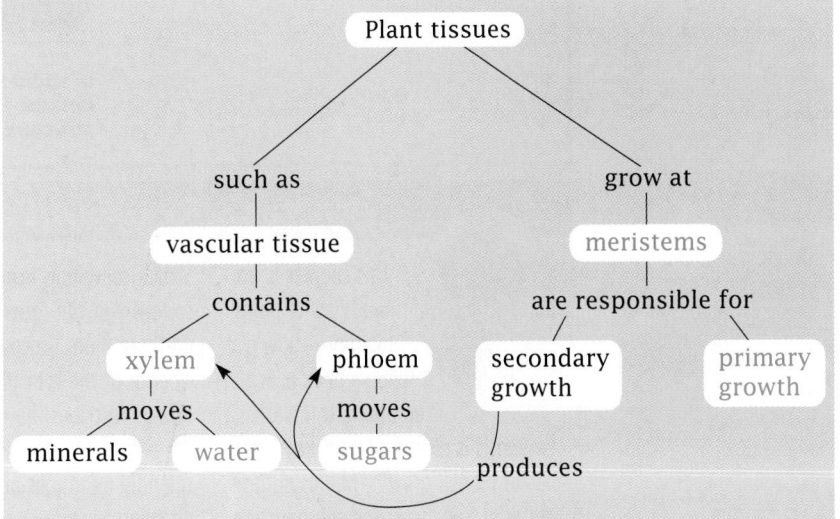

Chapter 19

Determining Prior Knowledge

- If this was not done in Chapter 18, have students draw a storyboard of the life cycle of a seed plant.
- Display pictures of a variety of fruits and vegetables. Have students locate the seeds.
- Have students explain where fruit develops on a flowering plant. Have them explain the purpose of fruit.
- Have students explain the difference between fertilization and pollination.
- Have students explain the functions of the following plant organs: roots, shoots, leaves, and flowers.

Chapter 19

Plant Form and Function

Review

- ratio of surface area to volume (section 3.1)
- plant cells (Section 3.3)
- photosynthesis (Section 5.3)
- the terms *osmosis* and *cotyledon* (Glossary)

19.1 The Plant Body
- Roots
- Shoots
- Plant Tissues

19.2 How Plants Function
- How Water Moves Through Plants
- Regulating Plant Growth: Plant Hormones
- Other Factors Affecting Plant Growth

19.3 Reproduction in Flowering Plants
- Architecture of a Flower
- Pollination and Fertilization
- How Seeds Are Dispersed
- Plant Cell Growth and Differentiation

In addition to their beauty, plants provide us with oxygen to breathe and food to eat. In this chapter you will learn how the structure of a plant helps it carry out its daily activities.

■ Author's Rationale ■

This chapter is an overview of plant biology—how a plant's body is organized and how it works. Emphasis is placed on four key activities of plants: growth, photosynthesis, material transport, and reproduction.

Structural details are de-emphasized so that you can focus your instruction on these plant processes. The coverage of each of the four plant activities, in effect, describes the organization of the plant body. These activities are splendid examples of structure dictating function.

THE STRUCTURE OF A PLANT ENABLES IT TO CAPTURE ENERGY AND BUILD MOLECULES DURING PHOTOSYNTHESIS. BELOW THE GROUND, ROOTS ANCHOR THE PLANT AND ABSORB WATER AND MINERALS. ABOVE THE GROUND, SHOOTS HOLD THE LEAVES HIGH TO ABSORB SUNLIGHT AND TAKE IN CARBON DIOXIDE FOR PHOTOSYNTHESIS. NO ANIMAL MAKES ITS OWN FOOD IN THIS WAY.

19.1 The Plant Body

Objectives

❶ **Describe the functions of roots and shoots.**

❷ **Compare and contrast vascular tissues, ground tissue, and dermal tissue.**

❸ **Describe the role of meristems in plant growth.**

❹ **Distinguish between mono-cots and dicots.**

Roots

Although orchids, cactuses, and pine trees do not look alike, these and all plants are similar in the way they grow. Throughout their lives, vascular plants develop by adding new cells at the ends of roots and shoots. The roots are the part of the plant below the ground. They anchor the plant in the soil and absorb water and minerals from the soil. **Figure 19.1** shows the roots of a radish. Roots make up about one-third of the total

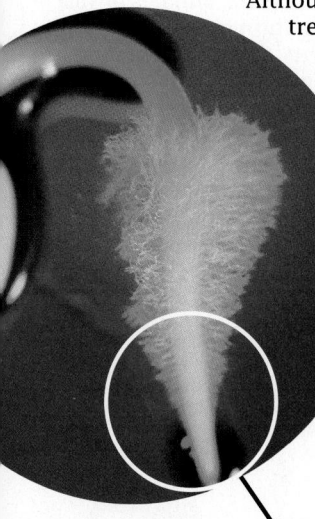

Figure 19.1
The roots of plants absorb water and minerals necessary for growth. Absorption occurs through the root hairs. Cell division occurs inside the root cap.

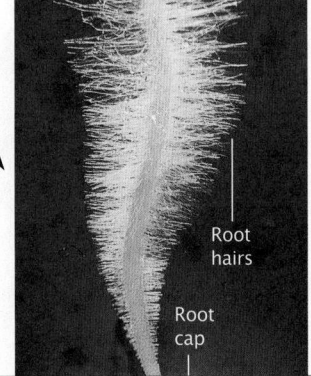

Root hairs

Root cap

dry weight of a plant. The roots of most plants do not usually extend into the earth beyond a depth of 3 m to 5 m (9–15 ft.). However, the roots of mesquite trees can grow to be more than 50 m (about 164 ft.) long. The roots of some plants have specialized functions such as food storage or water storage. For example, when you eat a carrot or sweet potato, you are eating a root that stores large amounts of carbohydrates.

Root hairs increase the surface area of a root

When a seed germinates, the first part to start growing is the root of the new plant. The end of the root is covered by the root cap, a thimble-like cluster of cells that protects the tip of the root. The root elongates in the region just above its root cap. As the root grows longer, the cells on the outside of the root cap are sloughed off and replaced by the new cells underneath. Above this region are slender projections called root hairs. Each individual root hair is a single cell that penetrates the space between soil particles. Virtually all absorption of water takes place through root hairs, which greatly increase the surface area of the root.

▪ Matter of Fact ▪

Root Systems
A plant can be anchored by either one main root called a "taproot," or branching "fibrous roots" that grow over a wide area. Generally, taproot systems penetrate deeper into the soil than fibrous root systems. Fibrous root systems cling tightly to soil particles, which make them well suited as erosion preventing ground cover.

Key Concepts
- Roots anchor the plant in soil and absorb water and minerals from the soil.
- Shoots consist of stems and leaves. Stems support growth above ground and transport water and food between roots and leaves.
- Plants consist of three types of tissue: vascular tissue, ground tissue, and dermal tissue.
- Xylem tissue transports raw materials toward photosynthetic cells and phloem tissue transports sugars toward non-photosynthetic cells.
- Plants grow in regions of active cell division called meristems.
- There are two types of flowering plants—monocots and dicots.

Reading Strategy

Tell students that the subheadings represent the main ideas of the lesson. Have them use all the subheadings to develop a summary of Section 19.1. Have students read the paragraphs to give the subheads more meaning.

Phase 2

TEACHING STRATEGIES

Visual Strategy

Figure 19.1
Review the needs of a vascular plant, such as support, anchor, absorption, photosynthesis, and reproduction. Have students explain how root hairs and the main root accomplish these tasks. Explain that root hairs are thin enough to squeeze between soil particles and absorb water and minerals, while the thicker root anchors the plant.

Connection: Chapter 12

Evolution of Plant Adaptations

Generally, the functional adaptation of different stem and leaf types evolves within the constraints of the local population of other plants and animals, as well as in relation with other ecological factors such as climate, soil, etc. Depending on the ecosystem, many plants have adaptions that perform special functions, such as "tendrils" that allow plants to climb, modified stems that function as storage units, and shoots called "runners" that grow over the ground to carry out asexual reproduction.

Demonstration 1

Comparing Stems and Stem Structures

Show students cuttings from woody and herbaceous stems. Include some modified supporting stem structures, such as tendrils from clematis, morning glories, cucumbers, pole beans, wild grapes, etc. Lead them to identify the specific adaptations and functions of various types. For example: runners of strawberries for asexual reproduction.

Theme Connection

Energy and Life

Every day, humans consume a variety of roots, stems, and leaves. Of the thousands of species of edible plants, only a few dozen are widely cultivated and sold as vegetables.

Visual Strategy

Figure 19.2

Walk through the major adaptations of these very different stems. Have students explain how the height and surface area of these three plant leaf types vary with the adaptations of their stems.

Shoots

The shoot is the portion of the plant that consists of stems and leaves and in many cases, flowers and fruits. The stem supports the leaves and enables them to receive sunlight. Leaves are the major site of photosynthesis. Compare the shoots of the rose plant, cactus, and strawberry plant shown in **Figure 19.2**.

Stems connect roots to leaves

Stems support leaves, flowers, and fruits, usually holding them off the ground. They vary greatly in shape and size from one plant species to another, as is evident when comparing a redwood trunk with a grass stalk. Stems contain vascular tissues that transport substances between roots and leaves. Some stems are modified for storage. For example, cactuses are able to store a considerable amount of water inside their stems. White potatoes are underground stems swollen with stored food, usually in the form of starch.

Figure 19.2
To survive in different environments, plants have evolved different types of leaves and stems.

a The stem of this rose plant has prickles, sharp outgrowths that serve as a type of protection.

b This cactus has modified leaves called spines that protect it from predators and extreme temperatures.

c The horizontal stems of this strawberry plant spread over the ground and ultimately may root to establish a new plant.

Leaves are the main sites of photosynthesis

In Chapter 5 you learned that photosynthesis enables plants to capture sunlight energy and use it to make carbohydrates. Leaves are specialized to carry out photosynthesis. They have many cells packed with chloroplasts, the chlorophyll-containing organelles essential to photosynthesis.

Regardless of their size, most leaves are relatively thin and flat. A thin, flat shape maximizes the ratio of surface area to volume. This helps a plant efficiently capture the sunlight and carbon dioxide needed for photosynthesis. A waxy outer layer called a cuticle prevents the leaf from losing too much water and drying out. Carbon dioxide enters the leaf, and water and oxygen exit through tiny pores called **stomata** (*stoh MAH tuh*). Inside the leaf are layers of photosynthetic cells containing chloroplasts. Bundles of vascular tissue run through the leaf, moving water with its dissolved minerals to photosynthetic cells and carrying the products of photosynthesis away from them. These bundles are the veins you see when you hold a leaf up to the light. You can read more about leaves in the *Tour of a Leaf* on page 405.

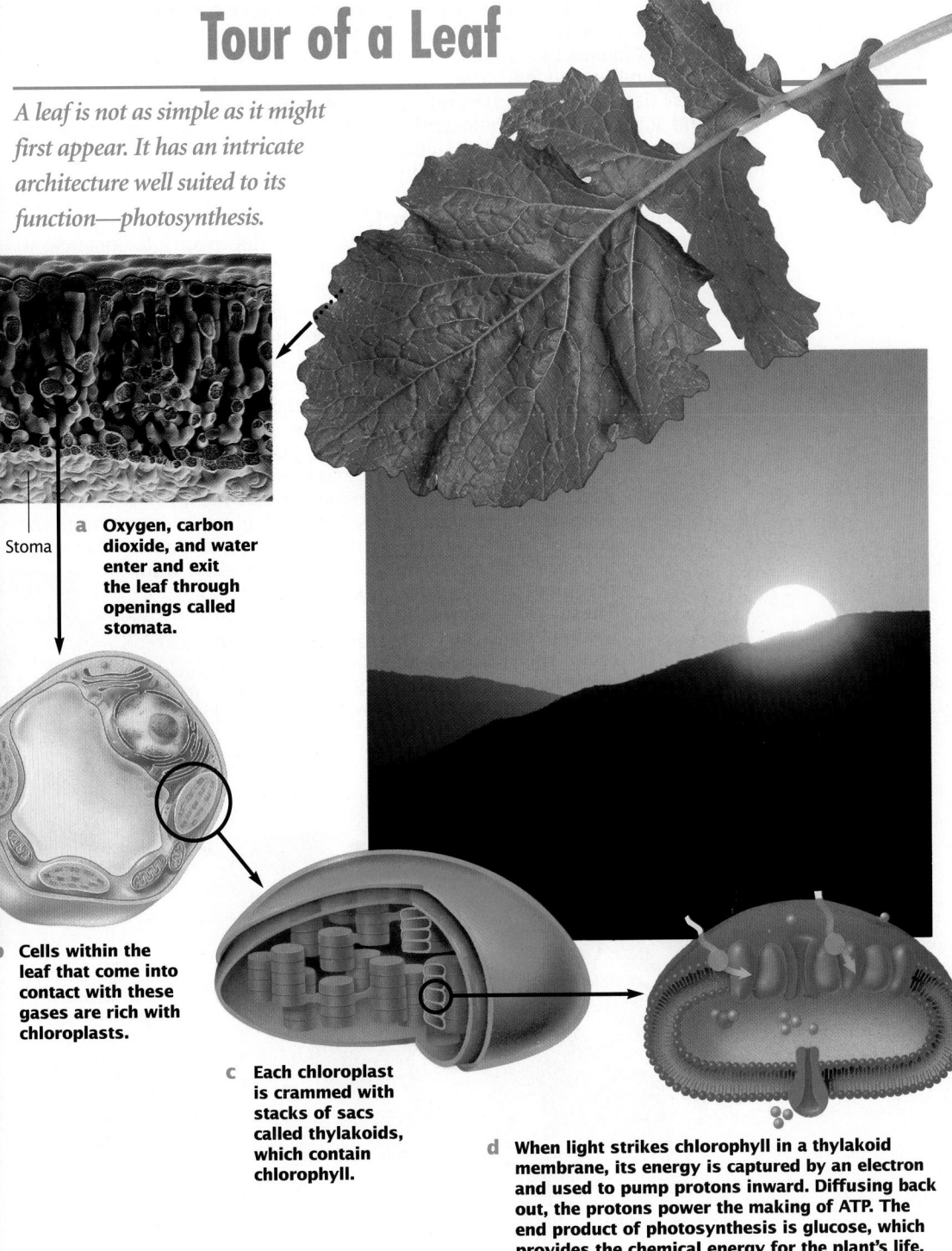

Tour of a Leaf

A leaf is not as simple as it might first appear. It has an intricate architecture well suited to its function—photosynthesis.

Stoma

a **Oxygen, carbon dioxide, and water enter and exit the leaf through openings called stomata.**

b **Cells within the leaf that come into contact with these gases are rich with chloroplasts.**

c **Each chloroplast is crammed with stacks of sacs called thylakoids, which contain chlorophyll.**

d **When light strikes chlorophyll in a thylakoid membrane, its energy is captured by an electron and used to pump protons inward. Diffusing back out, the protons power the making of ATP. The end product of photosynthesis is glucose, which provides the chemical energy for the plant's life.**

Using the Feature

- This feature can be used to review the material on photosynthesis in Chapter 5.
- Bring a variety of leaves to class, such as pine needles, maple, oak, or ash. Or have students collect a variety of leaves from your local area. Make field guides available so that students can classify the leaves and identify their structures.
- Have students use references to research CAM (crassulacean acid metabolism) photosynthesis. CAM plants, which are adapted to hot, dry climates, include a variety of succulents and cactuses. Photosynthesis in these plants is modified so that the light reactions occur during the day with stomata closed. At night, the stomata are open, carbon dioxide is taken in, and the dark reactions proceed. Plants with CAM photosynthesis thus avoid excessive water loss during the heat of the day.

Discussion

Guide the discussion by posing the following questions:

1. How might the shape of the leaf affect its ability to capture sunlight?
   The greater the surface area exposed to the sun, the greater the amount of light that will be captured for photosynthesis.

2. Why is it an advantage for deciduous trees to lose all their leaves during the winter?
   By losing their leaves, the tree has less surface area from which to lose water and less bulk to maintain during the winter.

3. Would leaves of rain forest plants have more or fewer stomata per square centimeter of surface than leaves of cold climate plants?
   Tropical rain forest plants have leaves with many stomata. Leaves of plants in colder, drier regions have fewer stomata.

Demonstration 2

Viewing Xylem Function

Slice a stalk of celery in half. Place the two halves in two beakers, each containing a different color of water (red and blue work best). Students should be able to see the veins in the leaves turn color and readily view the xylem in cross section through the stem. Keep the celery in colored water, to be used again with Figure 19.4 on page 410.

Visual Strategy

Figure 19.3

Walk students through Figures 19.3 a and b. Use Demonstration 2, *Viewing Xylem Function*, to point out how phloem and xylem tissue differ.

Connection : Chapter 18

Comparison of Early and Late Land Plants

Have students explain how mosses, hornworts, liverworts, and ferns differ from vascular plants.

Theme Connection

Patterns of Change

A tree's age and annual growth are permanently recorded inside the tree in growth rings. The amount of secondary xylem produced during one growing season shows up as a ring. Scientists use tree-ring dating to piece together the history of weather on Earth. In years when rainfall is plentiful, rapid growth is shown by broad rings. In dry years, the rings are narrow, showing that little growth took place. By comparing rings from many trees in an area, it can be determined which years had more rainfall.

Plant Tissues

The leaves, roots, and stems of plants are made of different tissues. As you learned in Chapter 18, vascular tissue is the "plumbing system" of a plant. Vascular tissue enables water, minerals, and sugars made by photosynthesis to move through the roots, stems, and leaves of a plant. Water and minerals are transported through vascular tissue called **xylem** (*ZY luhm*), shown in **Figure 19.3a**. Sugars are transported through another vascular tissue called **phloem** (*FLOH ehm*).

The bulk of the plant body is made of **ground tissue**, which supports the vascular tissue. Some cell types found in ground tissue are designed for storage, while other cell types have thickened walls that lend support to the plant. A layer of tightly packed, flattened cells makes up the plant's **epidermis**. These cells secrete a waxy substance that protects the plant from water loss.

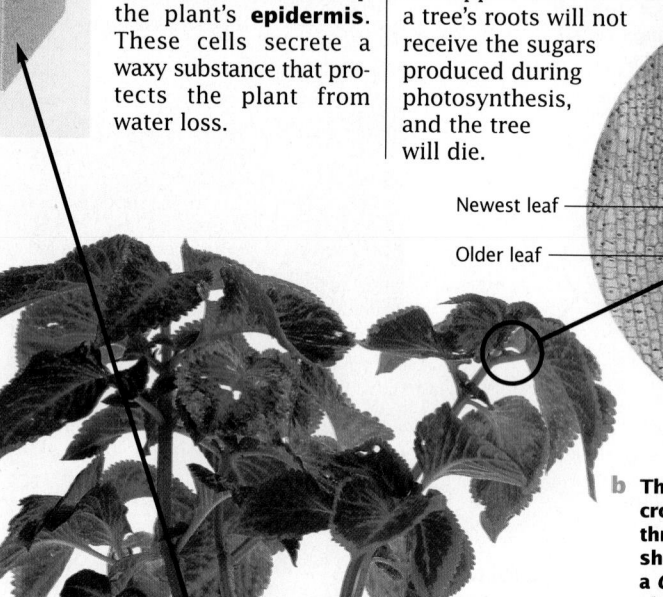

Figure 19.3

a **The xylem of this *Coleus* plant is made of elongated cells that connect end to end and transport water throughout the plant. The phloem contains cells that transport sugars. The epidermis covers and protects the plant body. The ground tissue supports the vascular tissue.**

Ground tissue

Epidermis   Xylem   Phloem

Coleus plant

Meristems are regions of active cell division

Plants grow in regions of active cell division called **meristems** (*MEHR uh stemz*), shown in **Figure 19.3b**. Every time a cell in the meristem divides, one cell remains in the meristem and the other cell becomes more specialized as it matures.

Meristems at the tips of roots and shoots enable plants to increase in length. The lengthening of roots and shoots is called primary growth. Annuals—plants that die after one season of growth—may have only primary growth. Woody plants—trees and shrubs—show secondary growth at meristems that run like cylinders through stems. Secondary growth causes plant bodies to thicken by producing new xylem and new phloem. Wood consists mainly of accumulated xylem. Secondary phloem forms the inner part of bark. If a ring of bark is stripped from the trunk a tree's roots will not receive the sugars produced during photosynthesis, and the tree will die.

Newest leaf

Older leaf

Bud

b **This vertical cross section through the shoot tip of a *Coleus* plant shows the meristem that produces new plant cells.**

Flowering plants are either monocots or dicots

As you learned in Chapter 18, cotyledons are leaflike structures that store or absorb food in a seed. The number of cotyledons is used to classify flowering plants into two large groups: the monocots (about 65,000 species) and the dicots (about 170,000 species). Monocots include grasses, palms, lilies, and orchids. The crop plants wheat, corn, rice, rye, and barley are also monocots. Dicots include many of the familiar flowering plants, shrubs, trees, and cactuses. Monocots and dicots are similar in structure and function, but they differ in distinctive ways. Evidence suggests that monocots may have evolved from primitive dicots. **Table 19.1** shows some of the characteristic differences between most monocots and dicots.

Table 19.1 Summary of Differences Between Monocots and Dicots

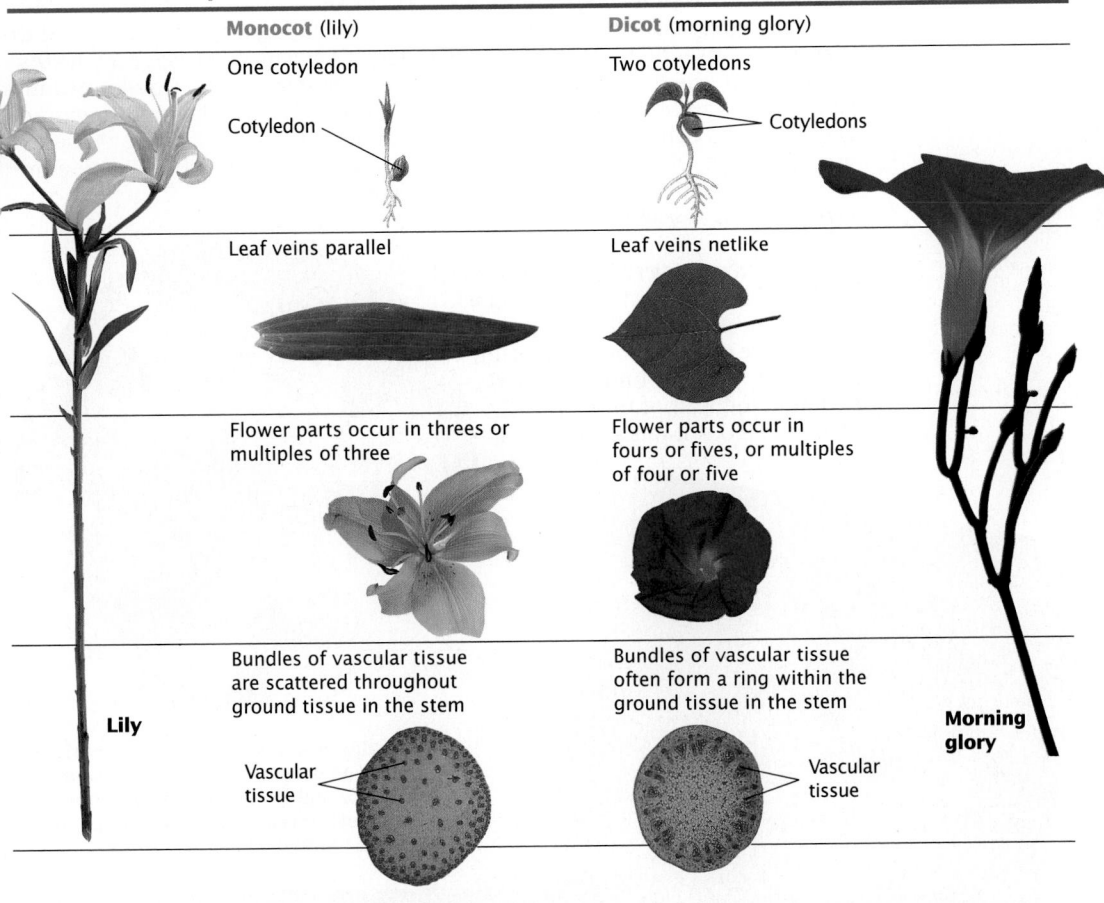

Monocot (lily)

One cotyledon
Cotyledon

Leaf veins parallel

Flower parts occur in threes or multiples of three

Bundles of vascular tissue are scattered throughout ground tissue in the stem

Vascular tissue

Lily

Dicot (morning glory)

Two cotyledons
Cotyledons

Leaf veins netlike

Flower parts occur in fours or fives, or multiples of four or five

Bundles of vascular tissue often form a ring within the ground tissue in the stem

Vascular tissue

Morning glory

Section Review

① **What are the functions of roots and shoots?**

② **Explain the functions of vascular tissue, ground tissue, and epidermis.**

③ **How do meristems enable a plant to grow?**

④ **What do the similarities between monocots and dicots suggest?**

Connection: Chapter 15

Display a variety of leaves or pictures of leaves. Include gymnosperms and both monocot and dicot angiosperms. Obtain as great a diversity of leaves as possible. If possible, provide students with illustrated taxonomic keys for plants of your region and illustrated guides to houseplants. Have students identify as many of the specimens as possible.

Phase 3

ASSESSMENT OPTIONS

Closure Strategy

Plant Structure and Function

Summarize this section by setting up a table on the chalkboard or overhead transparency. Across the top write the headings: *Structure* and *Function*. Down the left write the headings: *Root, Root hairs, Stems, Stoma, Leaf, Xylem, Phloem, Ground tissue, Epidermis, Innerstem.* Have students complete the table by summarizing the structure and function of each plant part.

Section Review

Assign the *Section Review.*

Reteaching

Have students consider the specific requirements of plants carried on a spaceship whose mission may take months. Tell them the astronauts would depend on these plants for some oxygen/carbon dioxide exchange and production of fruits. Some consideration might be given to the amount of space, and the plants' growth of roots, stems, and leaves. Have students design plants best suited for such a task.

■ *Section Review Answers* ■

1. Roots anchor the plant in soil, and absorb water and minerals from the soil. Shoots consist of stems and leaves. Stems support the leaves and enable them to receive sunlight.

2. Vascular tissue moves materials throughout the plant. Ground tissue stores and supports, and the epidermis protects the plant from water loss.

3. Plants grow at meristems. Meristems at the tips of roots and shoots allow the plant to increase in length.

4. Their similarities suggest they are closely related.

Using the Feature

- Eloy mentions that his family, namely his grandfather, helped develop his interest in biology. You can use this feature to get students to discuss important people in their lives and how they have influenced them.
- Eloy descibes himself as being creative. Ask students if they think creativity is a valuable trait to posess as a scientist.
- Have students research articles that describe recent findings of animals using plants as medicines.

Discussion

1. What personal traits have enabled Eloy to become a successful biochemist?
   He describes himself as creative, hard-working, fair, compassionate, and demanding.
2. What else does Eloy do in addition to his research?
   Eloy also runs several programs aimed at attracting minority and female students to the sciences.
3. Eloy was involved in creating a new scientific discipline called zoopharmacognosy. What four disciplines make up this new field of study?
   Zoopharmacognosy combines anthropology, behavioral ecology, organic chemistry, and botany.

Eloy Rodriguez: Biochemist

How I Became Interested in Biology

"I grew up in the county with the lowest average income per person in the United States—Hidalgo, Texas. I lived close to an open field where my grandfather would take me on nature walks. He would stop and point out different plants and insects, telling me all he knew about them. We would often gather seeds and bring them home to germinate and cultivate. The plants we grew brought life to our otherwise dreary-looking house. Since then, I have always been fascinated by the life around me.

"My family was very close. I was one of 67 cousins living within a five-block radius. The entire family stressed the importance of education. As a result, 64 of us earned undergraduate degrees, and four of us eventually received Ph.D.s. It was a good thing that I paid no attention to a high school counselor who advised me to go to a technical school, which was standard advice for many minority students at the time.

As an undergraduate, I was hired to clean a lab. I met a researcher who disliked laboratory work. He accepted my offer of help and I learned to extract compounds. I loved doing those experiments. My work in the laboratory changed my life."

In addition to his research, Dr. Rodriguez also runs several programs that help minority and female students to become engaged in science and the process of critical thinking.

Name:	**Eloy Rodriguez**
Home:	**Irvine, California**
Employer:	**University of California–Irvine**
Personal Traits:	• **Creative**
	• **Hard-working**
	• **Fair**
	• **Demanding**
	• **Compassionate**

Action ▪ Science in Action ▪ Science in Action ▪ Science

*Research
Focus*

Dr. Rodriguez and his colleague Dr. Richard Wrangham of Harvard have developed a new scientific discipline, which they call zoopharmacognosy—the chemistry of plants used medicinally by animals. It combines anthropology, behavioral ecology, organic chemistry, and botany. Dr. Rodriguez's research has led to the discovery of new drugs that are useful against cancer and viruses.

In his research, Dr. Rodriguez isolates natural drugs, determines their structure, and tries to understand how they work. How do they stop cancer cells? How do they kill parasites? By combining medicine, chemistry, the study of animal behavior, and botany, Dr. Rodriguez and his colleagues have come up with a multidisciplinary approach to answering these questions.

One plant Dr. Rodriguez has studied is used by wild chimpanzees to remove parasites. By learning about the plant's structure, he discovered a drug that is effective against human intestinal parasites, fungi that cause skin infections, and possibly cancer.

The Wild World of Biochemistry

To determine how a plant can cure a disease or ailment, Dr. Rodriguez studies chemicals extracted from plant cells, such as this one.

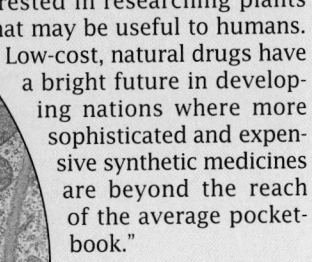

"I have a passion for what I do. I search the world for natural medicines. In the past 10 years, I've visited the rain forests deep in South America and Africa. The indigenous people taught me about their medicines and the plants from which they are derived. To find out more about these natural medicines, I brought many plants back to my laboratory. By extracting chemicals out of the plant cells and running various experiments on them, I've been able to study how these chemicals affect human cells that are infected by bacteria or other damaging organisms. Essentially, I try to figure out how these plant compounds act as medicines.

"Having investigated the biochemistry of many plants from all over the world, I believe that thousands of species of plants possess medicinal properties and other undiscovered uses. Many are potential sources of oils, food, pesticides, and other products. However, these applications may go undiscovered unless we start conserving our limited resources, many of which are already endangered. I hope that many other people will become interested in researching plants that may be useful to humans. Low-cost, natural drugs have a bright future in developing nations where more sophisticated and expensive synthetic medicines are beyond the reach of the average pocketbook."

409

Lesson Plan 19.2

PREPARATION

Key Concepts

- Water movement in plants results from capillary action and the pull created by transpiration.
- The transport of sugars throughout a plant is known as translocation.
- Auxins effect growth in plant stems, gibberellin stimulates stem growth and causes some kinds of plants to sprout and flower, and ethylene causes fruits to ripen.
- Photoperiodism is the plants' response to light and dark periods.

Reading Strategy

Form students into cooperative groups of four. Have each group construct a crossword puzzle on this section. Have them use the bold or italicized words, and the main concepts, to complete this task.

Phase 2

TEACHING STRATEGIES

Demonstration 1

Capillary Action
Review Demonstration 2 on page 406. Explain that the change in coloration in the leaves and stalk of celery are brought about by capillary action.

Theme Answer

Scale and Structure
The long narrow tubes contained in xylem tissue are designed for capillarity—the tendency of liquids to move into tiny, passages whenever a liquid comes in contact with air. Phloem tissue is porous and conducts sugars from the leaves to the root of the plant.

ALTHOUGH A PLANT MAY NOT APPEAR AS COMPLEX AS AN ANIMAL, ITS INTERNAL STRUCTURE IS MORE COMPLEX THAN YOU MIGHT THINK. ITS VASCULAR SYSTEM SENDS WATER, CARBOHYDRATES, AND MINERALS FROM ONE PLANT PART TO ANOTHER. LIKE YOUR BODY, A PLANT REGULATES ITS GROWTH WITH HORMONES, CHEMICALS THAT ACT AS MESSENGERS TO COORDINATE ACTIVITIES.

19.2 How Plants Function

Objectives

❶ **Summarize** how water is transported through plants.

❷ **Explain** the manner in which carbohydrates are moved through plants.

❸ **State** the functions of three kinds of plant hormones.

❹ **Explain** how plant growth responds to changes in seasons.

How Water Moves Through Plants

Scale and Structure

Explain how the structures of the vascular tissues make them an efficient means of transportation for water, minerals, and sugars.

In northern California, giant redwood trees can be up to 117 m (348 ft.) tall. How do trees move water from roots deep in the soil up to branches and leaves at such heights? Water enters the vascular system by osmosis. But it takes more than just osmosis to move water up to the top of a tall tree!

Several forces cause the movement of water in plants. Water enters root hairs and moves into the xylem tissue. Xylem cells form long narrow tubes. The water in these tubes is attracted to the xylem cell walls. This attraction tends to pull water up the tubes. The force of attraction that causes water to move up narrow tubes such as those in the xylem is called **capillary action**. **Figure 19.4a** shows the xylem of a zucchini plant's stem.

The major force behind the movement of water in plants, however, is the fact that when water evaporates from leaves, the upward movement of one molecule tugs on the molecules below. Thus the loss of water from the leaves is responsible for water flow through the plant. More than 90

Figure 19.4

a **Capillary action is the name for one of the mechanisms that moves water from the roots to the leaves.**

Xylem   Phloem

Zucchini plant

▪ Historical Perspective ▪

Weeds
Many weeds have been successful throughout history and are thought to have traveled here with new settlers in the 1500s. They have successfully invaded new habitats because of their abilities to produce adventitious buds capable of growing new stems. They

also have roots that contract and pull the apical meristem to safety under the soil. Most weeds grow leaves at ground level and only put up a tall reproductive structure when the time is right for seed dispersal. In a new environment, the invaders did not have natural predators to keep their

populations in check. They also were able to compete with the native plants for resources.

b Transpiration is the process by which water is lost to the environment through the above-ground portion of the plant.

c Stomata, most frequently found on the underside of a leaf, open and close to regulate water loss.

Phloem

Xylem

d Translocation is the process by which carbohydrates made during photosynthesis are carried to the rest of the plant via phloem.

percent of the water taken up by the roots of a plant is lost to the atmosphere as water vapor. Water passes primarily through the stomata in the leaves, as shown in **Figure 19.4b**. The loss of water by the leaves and the stem of the plant is called **transpiration**. Water that evaporates from leaves is continually replaced with water entering the roots.

Stomata regulate water loss

The only way that plants can control water loss is to close their stomata. Stomata are numerous—an average-sized sunflower leaf has about 2 million stomata. Each stoma is formed by two pickle-shaped cells called guard cells. The walls of guard cells are flexible and thicker on the inner surface than on the outer surface. When guard cells are swollen with water, they curve outward and the stoma opens. When guard cells lose water, they collapse and the stoma closes. The stomata of the zucchini plant, shown in **Figure 19.4c**, are open during the day and closed at night. Therefore, very little water is lost from leaves at night.

How carbohydrates are transported

Most of the carbohydrates made in the photosynthetic green parts of plants are moved through the phloem tissue to other parts of the plant. **Figure 19.4d** shows the phloem tissue of a zucchini plant. The transport of sugars made by photosynthesis from the leaves to the rest of the plant is known as **translocation**. Translocation is responsible for moving carbohydrate building blocks to actively growing regions of the plant. In some plants, such as potatoes or sugar beets, carbohydrates are concentrated in storage structures. When the plants need these carbohydrates, they can be broken down into smaller molecules and moved through the phloem. **Figure 19.4** shows the processes involved in the transport of water, minerals, and carbohydrates through the tissues of a zucchini plant.

411

Interacting Systems

A plant grows as a result of interactions between the plant and its environment. Light, moisture, and temperature influence plant growth. Sunlight provides energy for photosynthesis, and seasonal changes in the length of the light period can determine when a plant flowers. Plants require water for structural support, and to perform metabolic functions. Plant functions are limited to certain ranges of temperature and other environmental conditions. In many climates extreme conditions cannot be avoided, so plants have to make adaptations in order to survive. Plant hormones regulate the growth and development of some of these adaptations. Different hormones have distinctly different effects, allowing interaction between the plant and its environment.

Visual Strategy

Figure 19.5
The term *auxin* comes from the Greek word meaning "to increase." Have students explain the relationship between the word *auxin* and the explanation given in this figure.

Regulating Plant Growth: Plant Hormones

Plants continue to grow throughout their lives. With plenty of water, air, and sunlight, leaves and branches will form and develop, flowers will blossom, and fruit will ripen and drop. Where and how these events occur is controlled by chemical messengers called hormones. As you will see, hormones control many aspects of plant life.

Auxins stimulate the elongation of plant cells

Have you ever seen a houseplant bending toward the light? This growth pattern is caused by plant hormones called auxins (*AWK sihns*). Auxins cause cell walls to become more flexible. As a result, a cell grows longer. Experiments have shown that auxins migrate to dark sides of plants in response to light, causing the cells there to elongate.

The presence of auxins is what makes a plant bend toward the light as it grows. As shown in **Figure 19.5**, when auxins are present, cells in that portion of the plant elongate.

Gardeners often use synthetic auxin in a commercial powdered product to dust the cut ends of leaf cuttings. The powder contains auxins that speed the formation of roots for the new plant. Some types of auxins have been used as herbicides that cause weeds to literally grow themselves to death.

Auxins can also inhibit plant growth. For example, the presence of auxins at the tip of the shoot blocks the active growth of the branches along the sides of a plant. Gardeners often make plants grow more side branches by pinching off the shoot tip at the top of the plant where auxins are produced.

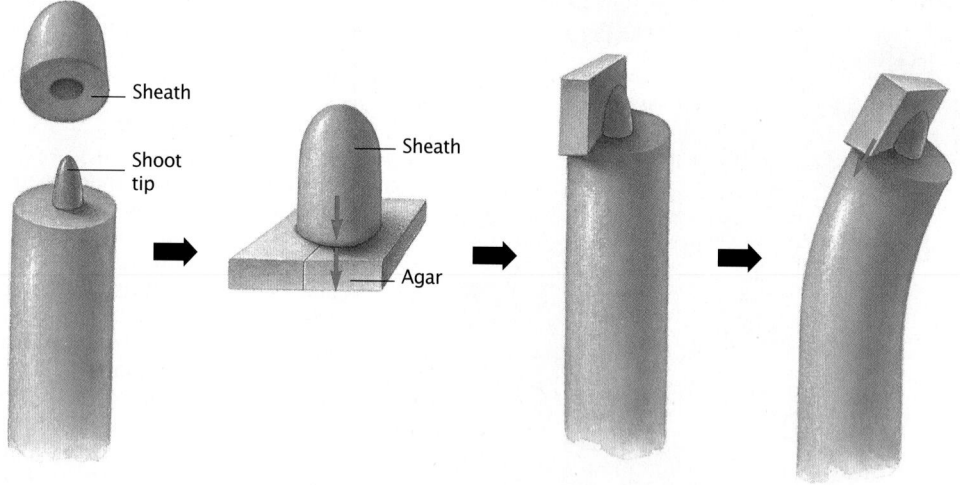

Figure 19.5

a To better understand how auxins work, scientists cut off the protective sheath surrounding the tip of a young oat seedling . . .

b . . . and placed it on a block of a gelatinlike substance called agar. Auxins (red arrow) diffused into the agar.

c When a piece of the block was put on one side of another seedling's tip, . . .

d . . . the plant began to grow in the opposite direction. Cells in contact with auxins in the agar elongated.

Figure 19.6

a This California poppy was grown under normal conditions.

b This California poppy was treated with gibberellin so that it would grow more quickly, and be larger than a normal poppy.

Gibberellin stimulates rapid growth

In 1926 Japanese scientists reported that rice seedlings infected with the fungus *Gibberella* had stems that grew abnormally long. They found that extracts of the fungus brought about the same effects in uninfected plants. Nine years later, the substance was identified and called gibberellin, after the fungus the scientists studied. Gibberellins not only dramatically increase stem growth, but also cause some kinds of plants to germinate and to flower. Their most dramatic effects can be seen in dwarf or miniature strains of plants. These strains are often genetic mutants that do not produce gibberellins. When treated with gibberellins, they grow as tall as normal varieties. Gibberellins are used to increase the size of plants, such as the California poppy shown in **Figure 19.6b**. They are also used to delay the ripening of citrus fruits and to speed the flowering of strawberries.

Ethylene controls the ripening of some fruits

In ancient China, farmers ripened fruits in rooms that contained burning incense. More recently, citrus growers ripened their fruits in rooms that contained stoves burning kerosene. In both instances, ethylene gas brought about the results. In 1934 scientists discovered that ethylene is produced naturally by fruits, such as the bananas shown in **Figure 19.7**, as well as by flowers, seeds, leaves, and even roots. Ethylene also appears to be the main factor in the formation of specialized cells that form before leaves drop off plants.

Today ethylene is used commercially to ripen bananas, honeydew melons, and mangoes. These fruits are harvested when still green. Some growers still use natural ethylene to ripen pears and peaches by wrapping each fruit individually in tissue paper. The paper keeps ethylene from escaping from the fruit and hastens ripening.

Figure 19.7
Ethylene is the hormone that determines when bananas and other fruits ripen.

Connection: Chapter 4

Cytokinins

Cytokinins are hormones that stimulate cell division in plants. They are mostly concentrated in endosperm and young fruits. Cytokinins cause plant cells to divide many times without differentiating or specializing. When a balance of auxins and cytokinins is reached, together the hormones stimulate normal plant growth.

Home Economics Connection

Ripe Fruits

To demonstrate the effect of ethylene on the ripening of fruits, place a green banana in a plastic bag by itself and another green banana in a plastic bag with a ripe apple. The banana in the bag with the apple will start to turn yellow within a few hours. The green banana alone will take a day or two. Have students form hypotheses about what they observe. For example: The green banana with the apple turns yellow faster because the ripe apple has molecules that affect fruit ripening.

▪ *Matter of Fact* ▪

Other Plant Hormones

Plants produce several other kinds of hormones. *Abscissic acid* causes dormancy in seeds and buds by inhibiting cellular activity. *Maleic hydrazine* works with abscissic acid to maintain dormancy. *Cytokinin* stimulates cell division in plants, causing plant cells to divide many times without change.

Stability

The growth of many plants varies with the seasons. Many plants have adapted to cold climates by losing their leaves each winter. The process of shedding leaves is called "abscission." Brought on by seasonal changes, abscission is the plant's way of reducing water loss due to transpiration. With no leaves, the tree has less surface area from which to lose water and less bulk to maintain during the winter.

Phase 3

ASSESSMENT OPTIONS

Closure Strategy

Moving Plants

In what way is phototropism analogous to an animal's movements? How can these phototrophic responses increase the plant's ability to survive in a given environment?

Section Review

Assign the *Section Review*.

Reteaching

Assign students to cooperative work groups of four. Have each group develop a concept map for this section. Use the chalkboard or overhead projector to present finished maps. Have each group elect a member to talk through their map with the class.

Other Factors Affecting Plant Growth

The growth of many plants varies with the seasons. Have you ever wondered why? The growth of plants such as strawberry plants and apple trees is affected by the length of daylight. These plants contain a pigment that is sensitive to the amount of daylight. As light gives way to darkness, the pigment initiates changes that alter growth.

Some plants respond to changes in day length by producing flowers. Long-day plants, such as the irises shown in **Figure 19.8**, produce flowers during the longer days of summer, when the nights are shorter. Short-day plants, such as goldenrods, begin to produce flowers during the short days of spring or fall. The response of plants to periods of light and dark is called **photoperiodism**.

Plants have other mechanisms that enable them to grow in response to changes in their environment. Occasionally, some parts of plants grow faster than other parts, causing the plant to bend. You have already read that uneven light can cause auxins to stimulate growth of cells on the shady sides of stems. Uneven growth causes the stem to bend toward the light. Roots respond differently; they grow downward, whereas stems grow upward. These movements are clearly beneficial. Stems that grow upward can receive more light than those that do not. Roots that grow downward will reach water and minerals. Differences in concentrations of auxins are responsible for this growth with or against gravity.

Length of Darkness Determines When Plants Bloom

Figure 19.8

a Irises are long-day plants. They bloom in the summer when the days are longer.

b Goldenrods, which are short-day plants, bloom in the fall.

Early summer

Day length is greatest in the summer months.

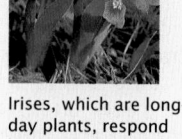

Irises, which are long-day plants, respond by producing flowers.

Goldenrods, which are short-day plants, do not bloom.

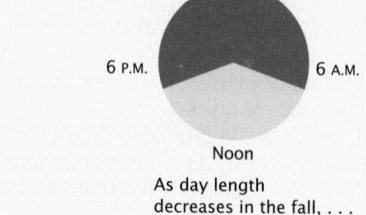

Late fall

As day length decreases in the fall, . . .

. . . irises stop flowering.

Goldenrods produce flowers during this time.

Section Review

① Trace the path of water through a plant.

② How are carbohydrates moved through a plant?

③ List three plant hormones and explain their effects on plant development.

④ How can a change in season affect plant growth?

■ Section Review Answers ■

1. Water enters the cells of the root by osmosis and travels to the xylem tissue. Capillary action and transpiration pull water up through a plant.
2. Carbohydrates move through a plant by a process called translocation.

3. Auxins break the chemical bonds in cellulose and cause the cell wall to become more flexible, allowing plant growth and causing the plant to bend toward light. Gibberellin stimulates rapid growth and can delay the ripening of some fruits and speed the flowering of some plants. Ethylene stimulates fruits to ripen.
4. The length of darkness determines when plants bloom.

FLOWERING PLANTS ARE THE DOMINANT FORM OF PLANT LIFE FOUND ON EARTH TODAY. WHILE YOU MAY APPRECIATE FLOWERS FOR THEIR DECORATIVE ROLE, THEY ARE ACTUALLY THE VISIBLE EVIDENCE OF A FLOWERING PLANT'S ABILITY TO REPRODUCE. EACH FLOWER HELPS ENSURE THAT A PARTICULAR PLANT SPECIES WILL PRODUCE MORE OF ITS OWN KIND.

19.3 *Reproduction in Flowering Plants*

Objectives

① Describe the structure of a flower.

② Summarize the processes of pollination and fertilization in typical flowering plants.

③ Explain the survival value of double fertilization.

④ Describe some adaptations of flowers that help ensure pollination.

Architecture of a Flower

Figure 19.9 shows the sexual reproductive structure of angiosperms, the flower. Most flowers share certain basic features. They consist of four whorls, or circles, of parts. The two outer whorls of a flower, the sepals and petals, protect the flower and attract insects and animal pollinators. The inner whorls of a flower contain the male and female gametophytes. One inner whorl of the flower consists of **stamens** (*STAY mehnz*), the male parts of the flower that produce pollen grains. Pollen is produced in the tip of the stamen, an area called the **anther**. The innermost whorl of the flower consists of the pistil. The base of the **pistil** is the ovary.

Inside an ovary, egg cells are produced within **ovules**. The tip of the pistil is the stigma, the site of pollination.

Petal   Sepal

Figure 19.9
If you slice the flower of a lily in half, you can see all of its reproductive structures.

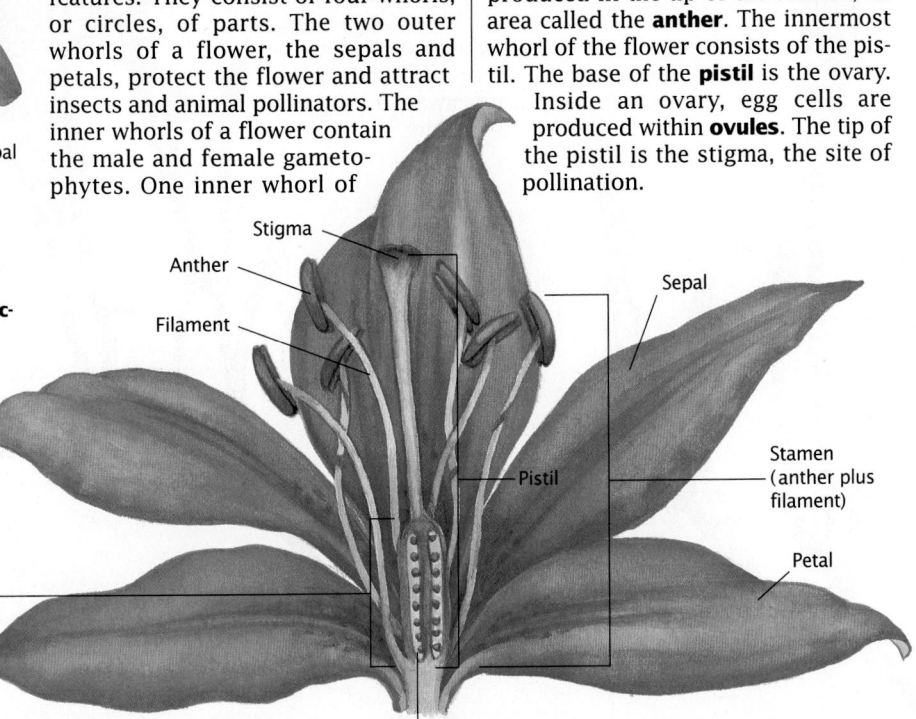

Stigma
Anther
Filament
Sepal
Pistil
Stamen (anther plus filament)
Petal
Ovary
Ovule

Kinds of Flowers
Flowers differ in the number and kinds of parts they possess. *Perfect, complete flowers,* like violets and roses, contain all the essential and nonessential parts. *Incomplete flowers,* such as grasses, lack one or more of the essential or nonessential parts.

Imperfect flowers, such as spinach, contain the reproductive structure of only one sex.

Lesson Plan 19.3

Phase 1
PREPARATION

Key Concepts
• Flowers are the reproductive organ of angiosperms.
• Flowers typically have the following parts: stamens, pistil, anther, ovules.
• Pollination occurs when pollen is transferred to a pistil. Fertilization occurs when a sperm cell fuses with an egg cell, producing a diploid zygote. Pollination is necessary in order for fertilization to occur.
• In double fertilization, one sperm cell fuses with an egg cell to form the embryo and a second sperm cell fuses with two nuclei to form nutritive tissue for the embryo.
• After fertilization, the flower's ovary becomes the plant's fruit and the ovule becomes the seed.

Reading Strategy

Have students develop a concept map incorporating the objectives for this section.

Phase 2
TEACHING STRATEGIES

Visual Strategy

Figure 19.9
Tell students that the parts of the flower that produce the gametes and carry out sexual reproduction are called *essential flower parts*. The delicate essential flower parts are protected and adorned by *nonessential flower parts*. Have students use this figure to name the essential and nonessential flower parts.

Pollination and Fertilization

For sexual reproduction to occur, pollen containing the male gametes must reach the stigma. Since flowering plants cannot move about to seek their mates, they must rely on other methods to move pollen. Pollination is the transfer of pollen grains from the anther to the stigma, the top of the pistil. In some plants, pollen falls from the anthers onto the stigma of the same flower, which may result in **self-pollination**. Wind can transfer pollen from the flowers of one plant to the flowers of another. However, most flowering plants, from magnolia trees to orchids to the morning glory in **Figure 19.10a**, depend on animals to transport their pollen. The flowers of the plants must attract pollinators. In Chapter 13 you learned how flowers and their pollinators coevolved, or evolved in response to changes in each other. Some flowers, such as the morning glory shown in **Figure 19.10b**, have attention-catching "advertisements" such as brightly colored petals or scents. Others secrete a sugary liquid called nectar, which pollinators use as food. As an animal explores these flowers, pollen sticks on its body. When the animal wanders away, it transfers the pollen to the stigma of another flower. By attracting animals, flowering plants can ensure **cross-pollination**. Cross-pollination is the transfer of pollen to another plant of the same species. This ensures genetic recombination, and thus produces wider genetic variety than does self-pollination.

Fertilization in flowers occurs in two stages

When the pollen grain of a flowering plant reaches the stigma, a pollen tube may begin growing through the tissue of the pistil and into the ovule, which contains egg cells. Each pollen grain contains two sperm, which travel down

Figure 19.10
a **Flowers, the reproductive structures of angiosperms, begin as buds that grow out of a plant's shoots.**

b **Pollinators are attracted to the bright colors or scents of many flowers. As they fly from one flower to another, they may transfer pollen.**

▪ Cultural Perspective ▪

Artificial Pollination

Artificial pollination was practiced as early as 3500 B.C. by farmers in ancient Mesopotamia. Mesopotamian farmers noted that the date palm trees were either male or female, and that if they brought clusters of staminate flowers in contact with pistillate flowers, fruit production was controlled. This ancient method of artificial pollination is still being used today.

the pollen tube. One sperm fuses with the egg to form a zygote, as shown in **Figure 19.10c**. The other sperm fuses with two nuclei inside the ovule, forming a tissue that will become a source of nutrition for the plant developing inside the seed. This event, in which one sperm fuses with an egg and a second sperm fuses with two nuclei, is called **double fertilization**. Double fertilization has great survival value because each new generation carries its own initial source of nutrition.

Remember that pollination is not the equivalent of fertilization. Fertilization involves the union of egg and sperm and may not occur until weeks or months after pollination has taken place. Sometimes it may not follow pollination at all.

Seeds and fruits form after fertilization

Fertilization causes rapid changes to occur in the flower. The ovule develops into a seed, often with a tough coat protecting the developing plant and its food supply. As you read in Chapter 18,

development of the seed was a major factor in the success of flowering plants on land. Seeds enable plant species to survive in unfavorable environments. As the seed develops, the ovary grows larger and develops into a **fruit**. A fruit is an enlarged ovary of a flowering plant that contains seeds. You can see the fruit of a morning glory plant in **Figure 19.10d**. Many "vegetables," including tomatoes, string beans, cucumbers, and squash, are technically fruits. Whether a flower's ovary will turn into a fruit depends on whether it is fertilized. If the egg cells go unfertilized, the flower will normally wither and drop.

Some fruits lack seeds

Have you ever wondered why navel oranges and pineapples have no seeds? Scientists describe the ability of plants to develop fruits without the fertilization of eggs as parthenocarpy (*pahr thuh noh KAHR pee*). Seedless varieties of grapes and citrus fruits occur naturally because they contain high levels of the plant hormone auxin. Researchers can induce parthenocarpy in fruits such as tomatoes and watermelons by applying auxin to plants at certain stages in their development.

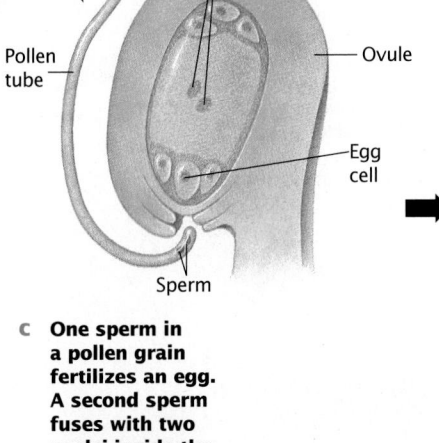

Pollen tube

Pistil

Nuclei

Pollen tube

Ovule

Egg cell

Sperm

c One sperm in a pollen grain fertilizes an egg. A second sperm fuses with two nuclei inside the ovule to form nutritive tissue. The petals, anthers, and sepals wither as the fruit begins to develop.

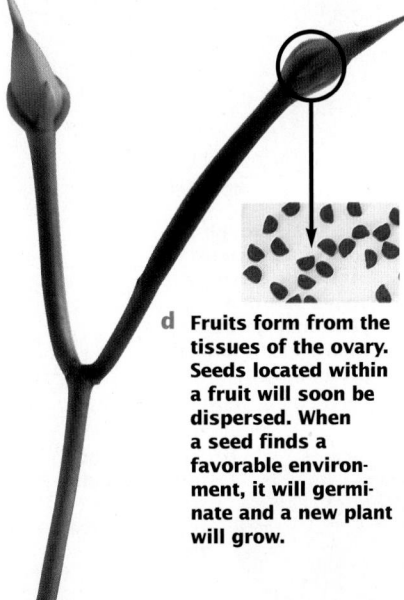

d Fruits form from the tissues of the ovary. Seeds located within a fruit will soon be dispersed. When a seed finds a favorable environment, it will germinate and a new plant will grow.

Visual Strategy

Figure 19.10
Refer students to Figure 19.10c. Have them explain why the pollen tube is an essential part of fertilization. Explain that without the pollen tube, the sperm cells would have no way of traveling down the stigma and reaching the ovule.

Theme Connection

Patterns of Change
Tell students that a fruit is a ripened ovary containing one or more seeds. As soon as the ovary changes color, it becomes either fleshy or dry. All fruits can be classified into one of three groups. Fruits such as peaches, tomatoes, and beans that form a single ovary are *simple fruits*. Fruits, such as berries, that form from flowers with many pistils on the same flower are *aggregate fruits*. Pineapples and figs are *multiple fruits*, many single fruits that have fused together to form one single structure. Have students brainstorm examples of fruits for each fruit type.

Demonstration 1

True Fruits?
To help students distinguish between the different types of fruit, dissect some of the following fruits and have students observe their internal structures: plums, grapes, tomatoes, oranges, olives, apples, lemons, melons, (simple fruits); raspberries, strawberries (aggregate fruits); and pineapples and figs (multiple fruits).

▪ *Cultural Perspective* ▪

Kola Nuts
Kola nuts are cultivated in the West Indies, West Africa, and South Africa. They are the fruits of several types of evergreen trees. People who live in African countries call the nuts *guru* or *goora* nuts and chew them like gum. Kola nuts are also used to make cola soft drinks and medicines. The nuts contain caffeine and theobromine. Caffeine has a stimulating effect and assists in combating fatigue.

Demonstration 2

Seed Dispersal

Tell students that fruits are typically eaten and dispersed to new environments by animals, and that most dry fruits possess structural adaptations that enhance wind or water dispersal. Show students as many of the following as possible: bean, pea, dandelion, sunflower, spinach, corn, walnut, apple, acorn, coconut, marigold, cucumber, and apple. Have students speculate about the dispersal method of each.

Visual Strategy

Figure 19.11

Explain that tiny seeds may be an adaptation to fit more seeds in the fruit, whereas large seeds may assure enough food for a growing embryo, and burrs aid in dispersal by animals. Bring a variety of seeds or fruits to class, including tiny seeds with burrs, wind-borne dandelion seeds, or maple seed helicopters. Have students speculate about the advantages of the different adaptations.

How Seeds Are Dispersed

Figure 19.11
Plant seeds are dispersed by a variety of mechanisms. Several examples are shown below.

Once mature, seeds are ready to be dispersed. **Figure 19.11** shows how various types of seeds are dispersed. Some fruits, such as the plump, fleshy peaches, tomatoes, and watermelons you buy at the grocery store, are eaten by animals. When fleshy fruits ripen, their sugar content increases and the fruit becomes soft and juicy. Their colors often change from green to bright red, yellow, or orange. These changes, signs that the seeds are ripe and ready for dispersal, aid in catching the attention of hungry animals. When fruits are eaten by birds or mammals, the seeds are spread by passing unharmed through the digestive tract. The success of these sweet, colorful fruits as a method of seed dispersal is important in the coevolution of animals and flowering plants. Plants such as maple trees, tumbleweeds, and dandelions have extremely light fruits and seeds that can be carried by the wind. The fruits and seeds of many plants growing in or near water, like coconuts, are adapted for floating. Some fruits or seeds, such as burdocks, have hooks that catch the feathers, fur, or clothing of passing animals, giving the seed a free ride to a new home.

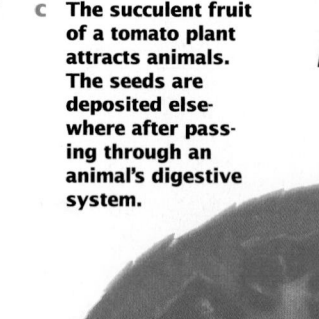

d **Humans can disperse seeds too. Have you ever spit watermelon seeds on the ground? One of those seeds may have germinated and grown into a new watermelon plant.**

c **The succulent fruit of a tomato plant attracts animals. The seeds are deposited elsewhere after passing through an animal's digestive system.**

a **The sharp, tough coat of a peach pit cannot be eaten by animals. A new plant may germinate wherever the animal drops the pit.**

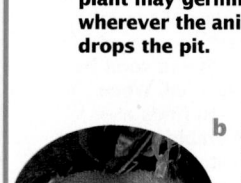

b **Squirrels and other animals carry seeds from one area to another.**

▪ Matter of Fact ▪

Self-Dispersal of Seeds

Some plants use an explosive or propelling action to disperse their seeds. Seeds of the Scotch gorse, an evergreen shrub, become warped and dry in the summer. When daytime temperatures are very hot, the pods will suddenly explode with a force that scatters the seeds in all directions. The dwarf mistletoe fruit absorbs water until the pressure within its tissues causes the fruit to burst open, throwing seeds as far as 14.5 m (48 ft.). The seeds of some grasses produce long bristles that coil and uncoil in response to air moisture, causing the seeds to creep along the ground. Similar structures in some wall ivies help bury the ivy seeds in small crevices in rocks or bricks.

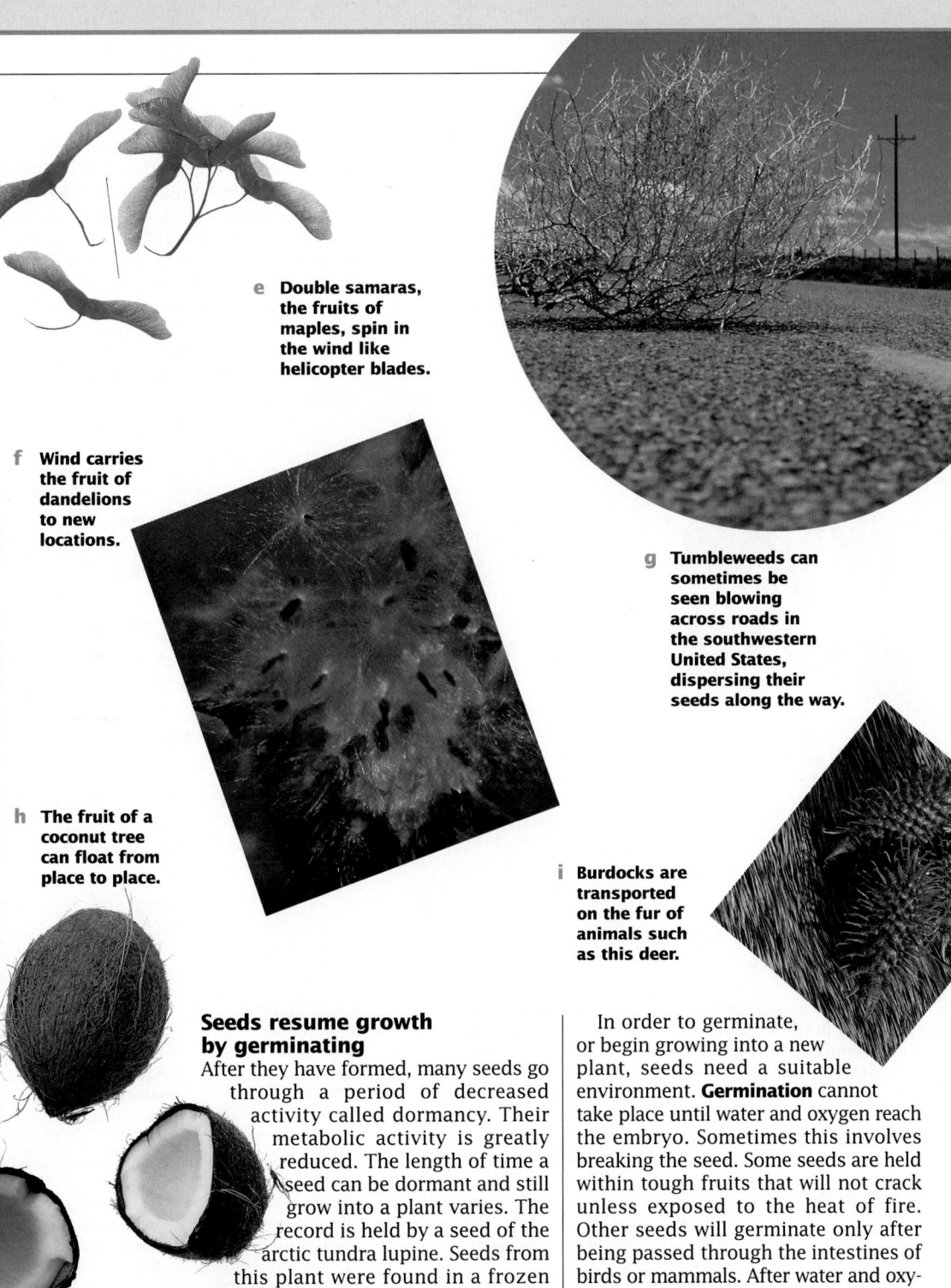

e **Double samaras, the fruits of maples, spin in the wind like helicopter blades.**

f **Wind carries the fruit of dandelions to new locations.**

g **Tumbleweeds can sometimes be seen blowing across roads in the southwestern United States, dispersing their seeds along the way.**

h **The fruit of a coconut tree can float from place to place.**

i **Burdocks are transported on the fur of animals such as this deer.**

Seeds resume growth by germinating

After they have formed, many seeds go through a period of decreased activity called dormancy. Their metabolic activity is greatly reduced. The length of time a seed can be dormant and still grow into a plant varies. The record is held by a seed of the arctic tundra lupine. Seeds from this plant were found in a frozen animal burrow in the Canadian Yukon and were estimated to be 10,000 years old. When these seeds were planted, they sprouted in 48 hours.

In order to germinate, or begin growing into a new plant, seeds need a suitable environment. **Germination** cannot take place until water and oxygen reach the embryo. Sometimes this involves breaking the seed. Some seeds are held within tough fruits that will not crack unless exposed to the heat of fire. Other seeds will germinate only after being passed through the intestines of birds or mammals. After water and oxygen enter the seed, the embryo swells, grows, and breaks through the seed coat. The young plant is called a seedling.

419

Visual Strategy

Figure 19.12
Review meristems and the advantages of double fertilization. Students will be aware that the potato "eyes" have something to do with reproduction. Introduce them to the possible cellular mechanism of vegetative propagation by asking them what tissue in the "eye" produces primary growth. Have students explain the function of the tuber.

Phase 3

ASSESSMENT OPTIONS

Closure Strategy

Floral Adaptations
Tell students to design a plant's flower placement and aromatic qualities if it lives near riverbanks where all the insects feed on carrion and lay their eggs in animal dung. The wild ginger plant is such an example. Its flower is brown, lies low on the ground, and smells like rotting meat.

Section Review

Assign the *Section Review*.

Reteaching

Have students grow new plants from sweet potato eyes suspended by toothpicks over a jar of water. Have them explain why part of the potato must remain with the eye. If time is a factor, have students work with rapidly-growing plants, such as *Brassica*, a member of the mustard family.

420

Figure 19.12
Although large-scale production of vegetables by tissue culture is not economically feasible, the procedure is used on an experimental basis to grow potatoes with specific genetic characteristics.

Plant Cell Growth and Differentiation

When a new plant starts to develop, all its cells are identical. They all have the same genes because they all come from the same fertilized egg. With time, hormones act on some cells but not others. The hormones turn some genes on and others off. Cells differentiate, becoming different from one another. Some become vascular tissue, others become ground tissue or epidermis.

Even after they have differentiated, many kinds of plant cells retain the potential to form other types of cells. As illustrated in **Figure 19.12**, scientists can grow entire potato plants just from pieces of stem tissue. Each cell in the tissue contains all the genetic information needed to produce a new plant. Many other plant cells also have the same ability. A single plant can be cut into many pieces and the pieces grown into many identical plants. Plants such as bananas and chrysanthemums are also produced in this way.

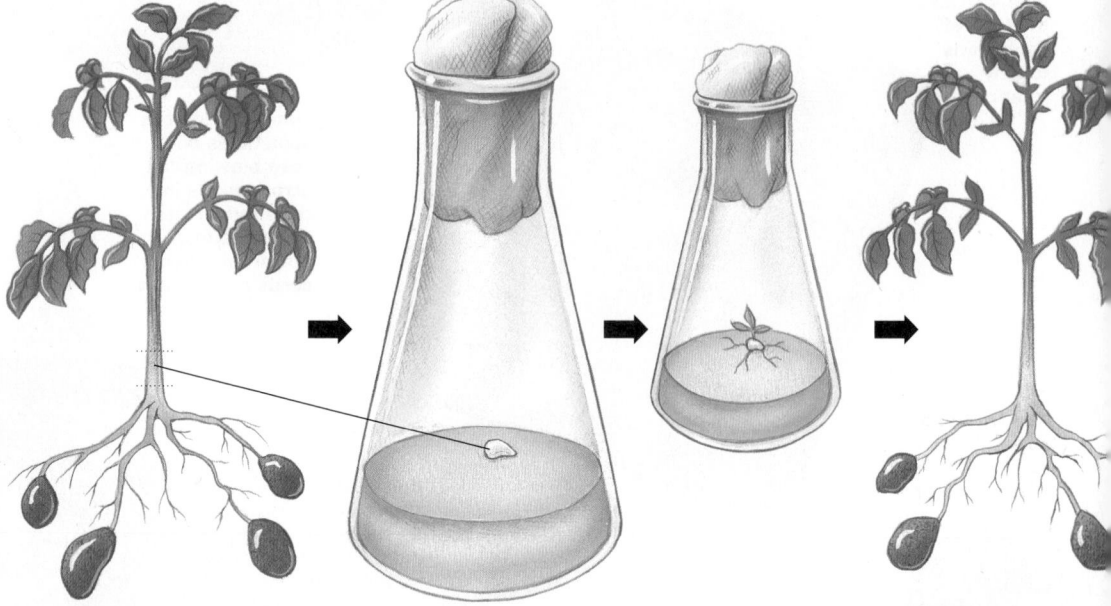

a To grow a new potato plant in tissue culture, scientists first isolate tissue from the center of the stem.

b The stem tissue is then placed in a flask containing nutrients and hormones.

c A new plant then begins to grow in the flask culture.

d When the plant is large enough, it is planted in soil and allowed to grow to maturity.

Section Review

❶ **Explain the role each part of a flower plays in reproduction.**

❷ **Summarize how a flower becomes a fruit.**

❸ **How has double fertilization contributed to the success of flowering plants?**

❹ **Explain some ways in which seeds can be dispersed.**

▪ Section Review Answers ▪

1. The sepals and petals protect the flower and attract pollinating insects. Stamens are the male parts of the flower that produce pollen grains that develop in the anther. The ovary is the female part of the plant. Chambers inside the ovary called ovules produce egg cells.

2. See Figure 19.10 beginning on page 416.

3. For plants that utilize double fertilization, each new generation carries its own temporary food supply.

4. See Figure 19.11 on pages 418 and 419 for examples of seed dispersal mechanisms.

Chapter **19** *Highlights*

Rose hips, the fruit of the rose plant, are sometimes used to make herbal tea.

	Key Terms	Summary
19.1 The Plant Body This rose plant has roots and shoots that absorb and transport all the molecules the plant needs. 	stoma (p. 404) xylem (p. 406) phloem (p. 406) ground tissue (p. 406) epidermis (p. 406) meristem (p. 406)	• Roots absorb nutrients. Shoots consist of stems and leaves. • Leaves are the main site of photosynthesis. Carbon dioxide enters the leaf, and water and oxygen exit through the stomata. • Vascular tissue contains xylem and phloem. Xylem transports water and minerals. Phloem transports sugars. • Vascular tissue is found within ground tissue, which makes up most of the plant. The epidermis covers the outside of the plant. • Regions of cell division are called meristems. Cell division in meristems increases the length and girth of a plant.
19.2 How Plants Function Osmosis, transpiration, and translocation regulate the essential molecules taken in and discarded by plants. 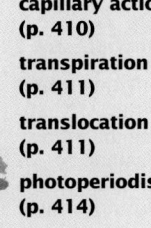	capillary action (p. 410) transpiration (p. 411) translocation (p. 411) photoperiodism (p. 414)	• The loss of water by transpiration pulls water into the plant at the roots. • Stomata in leaves open and close to regulate the amount of water lost. • Plant growth and development is regulated by hormones. • Many plants differ in the amount of daylight needed to bloom.
19.3 Reproduction in Flowering Plants  Flowers are the reproductive structures that have helped make angiosperms the most widespread type of plant on earth.	stamen (p. 415) anther (p. 415) pistil (p. 415) ovule (p. 415) self-pollination (p. 416) cross-pollination (p. 416) double fertilization (p. 417) fruit (p. 417) germination (p. 419)	• Flowers usually consist of sepals, petals, stamens, and ovaries containing ovules. • During fertilization in flowering plants, one sperm fuses with an egg to form an embryo, and another fuses with two nuclei to form nutritive tissue. • After fertilization, an ovary contains seeds and will become a fruit. • Germination occurs when seeds find favorable conditions.

Chapter Review Answers

Understanding Vocabulary
For each set of terms, complete the verbal analogy.
1. a. phloem
   b. ovaries
   c. cross-pollination
   d. shoots

Relating Concepts
2. Map answer is shown on page 401D.

Understanding Concepts
Multiple Choice

3. c	8. c
4. a	9. c
5. b	10. b
6. d	11. a
7. c	

Completion
12. monocots, dicots
13. gibberellins
14. epidermis, vascular

Short Answer
15. Primary growth occurs at the tips of roots and shoots and results in the lengthening of these parts. Secondary growth occurs at meristem regions that encircle stems and shoots and results in thickening of the plant.
16. Pollination is not equivalent to fertilization. Fertilization involves the union of egg and sperm. Pollination involves the transfer of pollen from the anther to the stigma. Both are male parts. Pollination can occur weeks or months before fertilization.
17. It is called double fertilization because two sperm are involved. One sperm fuses with the egg to form a zygote and the second sperm fuses with two nuclei inside the ovule to form a tissue (the endosperm) used by the developing plant.
18. String beans, cucumbers, and squash are fruits because they are enlarged ovaries of flowering plants that contain seeds.

Chapter 19 Review

Understanding Vocabulary

1. For each set of terms, complete the analogy by filling in the blank.
   a. water transport: xylem::sugar transport: _____
   b. third whorl: stamens::inner-most whorl: _____
   c. same flower:self-pollination:: different flower: _____
   d. absorbs water: roots::supports leaves: _____

Relating Concepts

2. Copy the unfinished concept map below onto a sheet of paper. Then complete the concept map by writing the correct word or phrase in each oval containing a question mark.

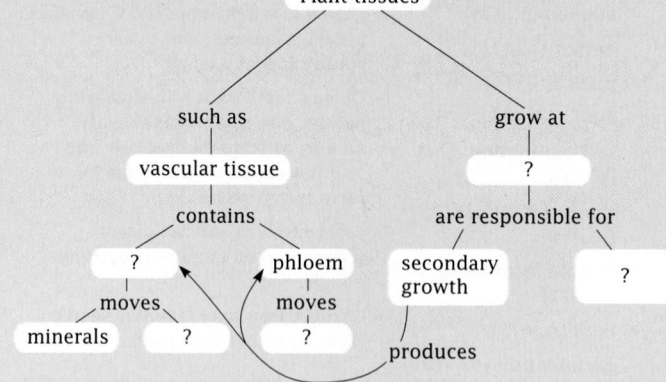

Understanding Concepts

Multiple Choice

3. Stomata are tiny pores in leaves that enable
   a. plants to grow at the tips of roots and shoots.
   b. plants to bend toward light.
   c. carbon dioxide to enter and water and oxygen to exit.
   d. pollen tubes to grow through the pistil of a flower.

4. Growth in plants occurs in regions of active cell division called
   a. meristems.   c. capillaries.
   b. root cap   d. ovules.
      and leaves.

5. Plant structures that anchor the plant and absorb minerals and water are
   a. stems.   c. shoots.
   b. roots.   d. meristems.

6. Leaves are structures specialized to
   a. transport water.
   b. protect seeds.
   c. absorb minerals.
   d. carry out photosynthesis.

7. Auxins are plant hormones that enable plants
   a. to flower during the winter.
   b. to attract pollinators.
   c. to bend toward light.
   d. to transport sugars through shoots.

8. In vascular plants, translocation is a process that enables
   a. water molecules to move up a plant.
   b. fruits to develop without seeds.
   c. sugars to move through a plant.
   d. flowers to develop during the winter.

9. Fruits tend to ripen when treated with
   a. water.   c. ethylene.
   b. auxins.   d. nectar.

10. Pollen is produced in the male parts of a flower called
    a. ovules.   c. petals.
    b. stamens.   d. stigma.

11. The ability of a plant to develop fruit without fertilization is called
    a. parthenocarpy.
    b. germination.
    c. transpiration.
    d. double fertilization.

Completion

12. Crop plants such as corn and wheat are called _____ because they have one cotyledon, while most shrubs and trees are called _____ because they have two cotyledons.

13. The action of plant hormones called _____ can be seen in their effects on dwarf or miniature strains of plants.

14. Plant cells that protect the plant body make up the _____ tissue, while cells that transport water, minerals, and sugars make up the _____ tissue.

Short Answer

15. How do primary growth and secondary growth differ?

16. Distinguish between pollination and fertilization in angiosperms.

17. Explain why fertilization in angiosperms is called double fertilization.

18. String beans, cucumbers, and squash are typically called vegetables. Why are they really fruits?

Interpreting Graphics

19. Look at the three pictures presented below. Explain the mechanism by which the seed in each picture is dispersed.

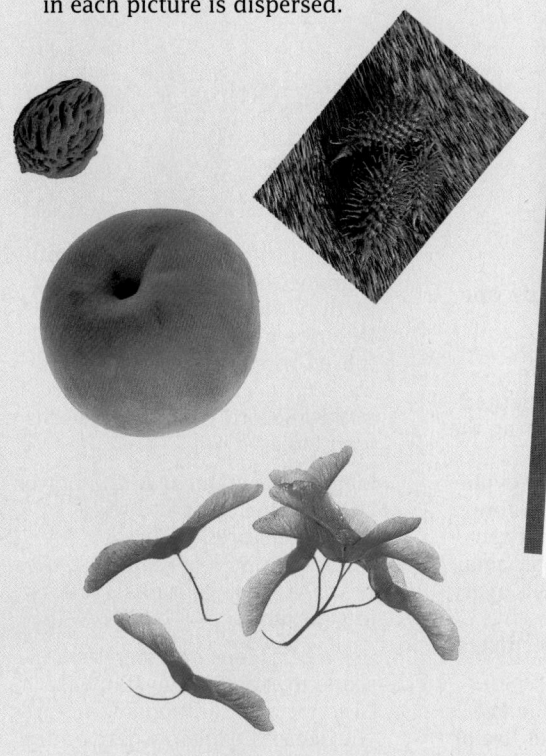

20. Study the figure below. Describe the role each labeled area plays in photosynthesis.

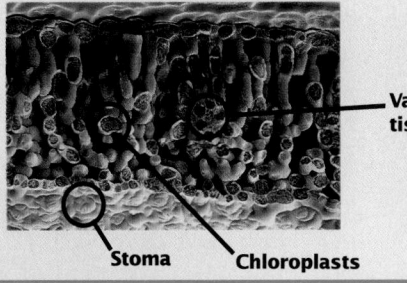

Vascular tissue

Stoma   Chloroplasts

Reviewing Themes

21. *Patterns of Change*
    When rings are counted to determine the age of a tree, is primary or secondary growth being measured? Explain your answer.

22. *Evolution*
    What evidence suggests that animals and flowering plants have coevolved?

Thinking Critically

23. *Inferring Conclusions*
    Would a 3 m (about 10 ft.) tree with no leaves but healthy roots be able to move water to its top? Explain your answer.

24. *Comparing and Contrasting*
    Distinguish between pollination and fertilization.

25. *Building on What You Have Learned*
    In Chapters 10 and 18 you read about the symbiotic association between fungi and plants that is called mycorrhizae. Why do you think plants with mycorrhizae develop fewer root hairs than plants without mycorrhizae?

Cross-Discipline Connection

26. *Biology and Mathematics*
    Locate a tall tree near your school. How can you determine the height of the tree without leaving the ground?

Discovering Through Reading

27. Read the article "A Tree Grows in Baja," in *Natural History*, June 1991, pages 54–58. Where do boojum trees commonly thrive? What combination of features enables the boojum to reach up to 26 m (86 ft.) and survive hundreds of years in an arid climate?

Thinking Critically

23. No. Water would not flow through the plant because transpiration could not occur without leaves.

24. Pollination is the transfer of pollen from anther to stigma, while fertilization is the joining of egg and sperm. Pollination may precede fertilization by weeks or months.

25. Mycorrhizae reduce a plant's need for root hairs because they assist in transferring nutrients from the soil to the roots.

Cross-Discipline Connection

26. The height of the tree can be determined using simple geometry and an astrolabe. An astrolabe can be constructed by attaching one end of a 30-cm piece of thread to a large paper clip and then attaching the other end of the thread to the center of the straight edge of a protractor. Align the top of the tree with the straight edge of the protractor. At this point, the height of the tree is equal to the distance the person using the astrolabe is away from the tree, plus his/her height.

Discovering Through Reading

27. The boojum tree grows in Baja, California and Sonora, Mexico. The trees store large amounts of water in soft, succulent tissues during the rainy season.

Interpreting Graphics

19. See Figure 19.11a, e, and i for explanation.

20. During photosynthesis, carbon dioxide enters the leaf, while water and oxygen exit, by way of the stomata. Vascular bundles move water to photosynthetic cells and carry products away from them. Chloroplasts house chlorophylls that absorb energy for the plant.

Reviewing Themes

21. Secondary growth is being measured. Secondary growth produces new xylem and phloem and increases the width of a tree.

22. The bright colors of ripe fruits attract animals who eat the fruit and the seeds inside. The fruits provide nourishment for the animals, and the seeds are dispersed as they pass through an animal's digestive tract.

Procedural Note

1. Be sure students realize angiosperms are the most successful and varied group of plants. In large part, this is due to the development of the flower. The diversity of flowers is an indication that this basic structure can be adapted to a great variety of conditions.

2. Throughout this investigation, emphasize the need for students to produce accurate and detailed descriptions of what they are observing.

3. Techniques to help students maintain thoroughness include emphasizing correct terminology, comparing the parts of the flower to each other, and trying to create a logical sequence for the description.

Preparation Answers

1. • A flower is a mechanism that ensures the transfer of gametes.
   • A *sepal* is one of the leaflike structures that grow out from the base of a flower and enclose the bud before it blooms.
   • A *petal* is a modified leaf that grows between the sepals and the part of the flower that produces gametes; it carries out sexual reproduction.
   • A *stamen* is the male part of a flower.
   • *Anther* is the tip of the stamen, where pollen is produced.
   • The *ovary* is the female part of a plant.
   • An *ovule* is that portion of the ovary in which the eggs are produced.
   • *Pollen* is the male gametophyte of seed plants.
   • A pollen tube grows through the pistil to the ovule. One sperm fuses with an egg to form an embryo, and another fuses with two nuclei to form nutritive tissues.

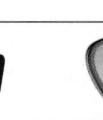

Chapter 19 **Investigation**

How Is Flower Structure Related to Function?

Objectives

In this investigation you will:
• *identify* the parts of a flower
• *relate* the structure of each part of a flower to its function
• analyze a flower to identify patterns in nature

Materials

• a complete flower
• hand lens or stereomicroscope
• forceps
• white paper
• glue
• medicine dropper
• glass slide
• coverslip
• compound light microscope

Prelab Preparation

1. Review what you have learned about flowers by answering the following questions.
   • State the function and evolutionary advantage of a flower.
   • Explain the function of each of the following: sepal, petal, stamen, pollen, anther, ovary, and ovule.
   • Explain how pollen fertilizes an ovule in an ovary.

2. Review the procedures in the Appendix for the proper use of the microscope.

Procedure: Determining the Relationship Between Flower Structure and Function

1. Form a cooperative group of two students and study the figure below. Make a table similar to the one shown at the right.

2. Working with your partner, carefully examine the flower at your desk. Identify the sepals and the petals. Describe the color and shape of these two structures. *Where are the sepals located in relation to the petals?* Record your observations in your table.

3. Identify the stamens in the flower. Describe the color and shape of these structures. *Where are the stamens located in relation to the petals?* Record your observations in your table.

4. Use a hand lens or a stereomicroscope to examine the stamens. Describe any additional structures you see and explain where on the stamen they are located. Record your observations in your table.

5. Identify the ovary of your flower. In your table, describe the color and shape of the ovary. *Where is the ovary located in relation to the other parts of the flower you studied?* Locate the stigma. Record your observations in your table.

6. Use forceps to carefully remove the sepals. Arrange these structures on a piece of white paper. *What is the function of the sepals?*

Procedure Answers

2. The sepals are located on the first outer whorl of the flower, below the petals located on the second outer whorl.

3. The stamen is located in the third or inner whorl of the flower, whereas the petals are located on the second outer whorl.

5. The ovary is located on the fourth and innermost whorl of the flower.

6. Sepals protect the flower.

7. Petals protect the flower and attract insects and other animals.

8. The anthers produce pollen grains on the tips of the stamen.

9. The ovary produces the eggs.

12. pollen grains
    Pollen grains are the male gametes of seed plants.

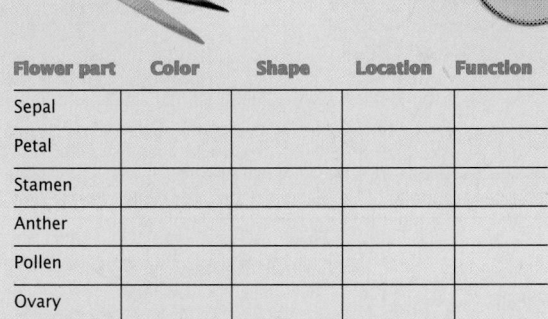

Flower part	Color	Shape	Location	Function
Sepal				
Petal				
Stamen				
Anther				
Pollen				
Ovary				

7. Carefully remove the petals of your flower. If the petals are fused together, remove them in one piece. Arrange the petals above the sepals in a row on the paper. *What is the function of the petals?*

8. Count and remove the stamens. *What is the function of the stamens and the anthers?*

9. Examine the ovary of your flower. *What is the function of the ovary?*

10. Except for one stamen, glue each flower part to the piece of paper. Label each part.

11. Place a drop of water on a glass slide. While holding the top of a stamen over the water, strike it gently with a fingernail. Place a coverslip on the drop of water.

12. Using low power of a compound light microscope, observe the specimen. *Identify what structure you are observing. What is the function of the specimen?* Switch to high power and sketch what you see.

13. Glue the remaining stamen in the appropriate row on the piece of paper.

Analysis

1. *Identifying Relationships* How does the location of each flower part provide evidence of its function?

2. *Identifying Relationships* How does the shape of each flower part in your table provide evidence of its function?

3. *Evaluating Methods* What is the importance of accurate descriptions in scientific inquiry?

4. *Recognizing Relationships* What is the relationship between the number of sepals, petals, and stamens in your flower?

Thinking Critically

How do you think the flower you examined might be pollinated? Explain how the structure of the flower is well suited for this type of pollination.

Analysis Answers

1. Students should indicate that as you advance from the outer to the inner parts of the flower, each section becomes more important.

2. The sepals are short, sturdy, and form an armor-like housing that protects the ovary. The petals are full and broad, protecting the ovary and the stamen. Petals also attract flying insects looking for a place to light. The stamen and anthers are very delicate, tall, and flexible, for transferring pollen. The ovary is very well protected in the innermost part of the flower and has a sturdy outer covering protecting all of the plant's eggs.

3. Accurate descriptions in scientific inquiry are important in making identifications and classifications, for comparing observations, and for determining reliability.

4. Answers will depend on the kind of flower studied.

Thinking Critically Answer

Answers will depend on the kind of flower studied but should include a description of petal shape and color and any noticeable fragrance.

Chapter 20 Plants in Our Lives

Planning Guide

	Objectives/Themes	Classwork Resources	Homework Resources
20.1	**1.** Summarize the uses of grains and legumes. **2.** Compare fruits with vegetables. **3.** List examples of vegetables that are stems, roots, or leaves. **4.** Identify two sources of sugar. **Themes:** Energy and Life, Scale and Structure	**Text** Journeys *History of Corn* pp. 430–431 Science in Action *Fermentation* pp. 434–435 **Teacher's Resource Binder** Focus Activity 20 *Relating Root Structure to Function* Lab Investigation 20.1 *Vitamin C From Foods* **Other Resources** Transparency 82	**Text** Section Review, p. 433 **Teacher's Resource Binder** Directed Reading Worksheet 20.1 **Other Resources** Audiocassette 20.1*
20.2	**1.** List three important wood products. **2.** Recognize the role of plants as sources for drugs. **3.** Describe how turpentine and rubber are made. **4.** Identify the source of two kinds of natural fibers. **Themes:** Interacting Systems, Patterns of Change, Scale and Structure	**Teacher's Resource Binder** Lab Investigation 20.2 *Regulatory Chemicals of Plants* **Other Resources** Transparency 83	**Text** Section Review, p. 440 **Teacher's Resource Binder** Directed Reading Worksheet 20.2 **Other Resources** Audiocassette 20.2*
20.3	**1.** Explain how crop rotation increases crop yield. **2.** Contrast the benefits and limits of the Green Revolution. **3.** Describe the benefits offered by genetic engineering and the discovery of new crop plants. **4.** Predict some ways in which hydroponics might be used in the future. **Themes:** Evolution, Interacting Systems, Patterns of Change, Stability	**Text** Investigation *Can Plants Be Grown Without Soil?* pp. 448–449 **Teacher's Resource Binder** Extension Worksheet *Polyploidy*	**Text** Section Review, p. 444 **Teacher's Resource Binder** Directed Reading Worksheet 20.3 Vocabulary Review Worksheet* Reteaching Worksheet* *Plants in Our Lives* **Other Resources** Audiocassette 20.3*

*Reteaching Options

Demonstrations
20.1: pp. 429, 432
20.2: p. 436
20.3: p. 444

Assessment
Chapter Review pp. 446–447
Portfolio Assessment p. 425D
Chapter Test—Teacher's Resource Binder
Test Generator

Research Notes

Connection to Medicine: Hairy Root Chemical Factories

The medicines found in many plants are difficult to manufacture using standard chemical techniques or genetically altered bacteria. Scientists and pharmacological companies are continually searching for methods to boost the production of such useful chemicals. One technique applies principles of genetic engineering and hydroponics to grow roots in culture, instead of growing the whole plant.

In this technique, plant cells are infected with the bacteria Agrobacterium rhizogenes, *which transfer genes to the plant cells, causing them to undergo continuous growth. The cells repeatedly branch, forming rootlets that are covered with fine hairs. The hairy roots are then placed in an antibiotic culture, which kills the bacteria.*

Since the genetic material from the bacteria is still in the plant cells, they continue to grow rapidly. When placed in a liquid medium with the necessary nutrients, these hairy roots can grow several thousand times bigger in a month's time.

These cultures of root cells will produce the same array of chemicals that an entire plant would. Since the roots grow quickly, they can be harvested repeatedly. Often, such systems can be stimulated to produce more of the desired chemicals than ordinary plants will.

There are several advantages to hairy roots. For example, their cells mutate less frequently and go through fewer developmental changes than cells in whole plants, so they continue to make the same chemicals more consistently. However, growing

hairy roots in culture also involves some challenges. For example, the roots frequently weave together so tightly that it can be difficult for the nutrient-carrying growth medium to reach all of the root hairs. Similarly, the roots often grow so quickly that they clog openings and tubes carrying the nutrient media.

Chemical engineer Jacqueline V. Shanks and her colleagues at Rice University in Houston are using hairy roots of the rosy periwinkle of Madagascar to produce a precursor to vincristine, the drug used to treat some forms of juvenile leukemia.

Other researchers are working with Chinese cucumber hairy roots to produce ribosome-inactivating proteins that can inhibit replication of HIV. And marigold hairy roots can be used to produce chemicals that will combat crop infections by fungi and nematodes.

Investigation Notes

Can Plants Be Grown Without Soil?
pages 448–449

Purpose: In this investigation, students design their own controlled experiment to learn about hydroponics.

Prelab Preparation

To reduce the chances of bacterial and fungal infections, soak the seeds for 15 minutes in a solution of 1 part bleach to 4 parts water.

Answers will be found on pages 424–425.

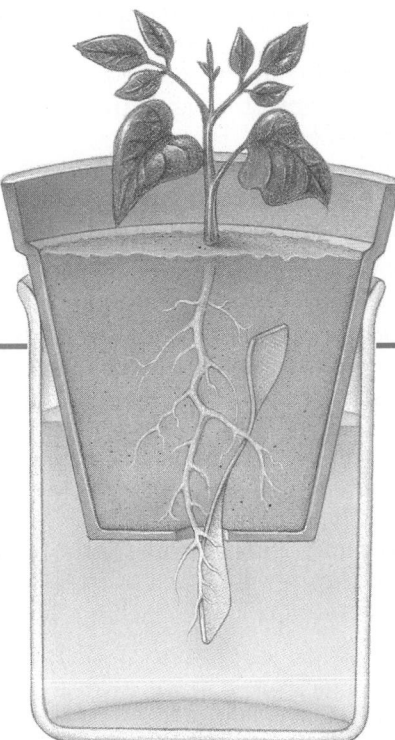

425B

Meeting Individual Needs

Objectives

1. Students will demonstrate appropriate use of core vocabulary for the chapter (Vocabulary File).

2. Students will provide examples relating common products to plants (Teaching Strategy A).

3. Students will gather information about and explain different examples of roots used as foods in various cultures (Multicultural Lesson Plan).

Vocabulary File
(Limited English Proficiency/ Learning Disabled)

If you are not already using the Vocabulary File, refer to Chapter 1 for its preparation. See the Chapter Highlights on page 445 for a list of suggested words.

Teaching Strategy A
Uses of Plants
for pages 427–440
(Developing Classification Skills/Kinesthetic Learners)

Materials: bulletin board, construction paper

Procedure: To reinforce the relevance of this chapter, involve students in a contest. Divide the class into teams. Have each team choose a plant name: the Yams, the Wheaties, etc. Have them make name labels for each team out of construction paper. Attach these to the bulletin board.

For three days have students collect labels from cans, bottles, or boxes indicating names of plants or plant parts that have been used for food, clothing, medicine, etc. Staple these labels under each team's name. The winning team is the one with the most labels that are different but accurate. Provide some sort of prize or reward, such as free library time.

Multicultural Lesson Plan
Root Foods
(Dealing with Ethnic Diversity)

Preparation: Gather materials for students to make a children's book: construction paper, glue, cardboard, scissors, tape, etc. Be sure students have access to reference works that will include information about root foods. Possible sources include books such as *Roots: Their Place in Life and Legend*, by Vernon Quinn (Stokes Publishing Co.), and *Edible Wild Plants of Eastern North America*, by Merritt Lyndon Fernalk and Alfred Charles Kinsey (Harper and Row, Inc.).

Teaching Strategies: Start a discussion about which parts of plants students use for food. Most students will mention leaves and fruit, but roots and tubers are not usually among the first they consider.

Have students speak with grandparents or elderly members of the community and record the various ways they use root foods. In addition, students may interview people at open-air markets, or managers of produce departments at grocery or specialty stores.

Assessment: Students will share their research and compile it into a children's alphabet book. For each letter of the alphabet, they should find a root food with a name beginning with that letter. Students will locate or draw a picture of the root, or food made from it, and place the picture on the page along with a large capital letter and a brief explanation or picture indicating which cultural group eats the food.

> *Example:*
> F is for fufu. Fufu is made from the root of the cassava plant and is enjoyed by the people of West Africa.

Additional Strategies
Visual Strategies
Pages 427, 428, 429, 432, 436, 438, 441, 442, and 443

Auditory Learners
Use *Biology: Visualizing Life* Audiocassettes for Sections 20.1, 20.2, and 20.3.

Meeting Individual Needs (cont.)

Cooperative Learning
Pollination of Flowers

Timing: Use this activity to introduce Section 20.3.

Group Size: 4 students

Outcome: Students will evaluate the benefits and disadvantages of the Green Revolution

Individual Accountability: Each group member will research techniques used in the Green Revolution.

Positive Interdependence: Pairs of students will debate one another and compare their results.

Within each group, all students will research the topic, and one pair will debate whether the Green Revolution was a success in increasing food production in a reasonable way. The other pair of students will offer rebuttals to the first students.

Be sure students address issues such as economic fairness, environmental impact, and positive and negative effects on biodiversity.

Instruct students to listen carefully, offer constructive criticism, and allow time for questions. When the debates are over, have the groups report their conclusions to the class. Discuss ways in which lessons from the Green Revolution could be applied without harmful effects.

Portfolio Assessment

Students should select their best work and provide a self-reflective rationale for their selections. Students can make selections in the following areas.

1. *Content*   One concept map from the chapter (See page 812 for evaluation criteria.)

2. *Reading Comprehension*   One Directed Reading Worksheet from the Teacher's Resource Binder (Use the answer key to evaluate for accuracy.)

3. *Writing*   Using the Vee Form, summarize a magazine or newspaper article relating to uses for plants. (See page 22T for evaluation criteria.)

    Or: Select a writing task or project from the Chapter Review.

4. *Performance Assessment*   One Vee form from a chapter investigation or lab manual investigation (See page 22T for evaluation criteria.)

Teacher makes selections in the following areas.

1. *Formal Assessment*   Chapter test (Test A, B, or the Test Generator) The teacher-scored test should be reviewed by the student. Incorrect responses should be corrected by the student before the test becomes part of the portfolio.

2. *Informal Assessment*   Use the Direct Observations Checklist, page 33T, during a laboratory or other cooperative learning experience.

3. *Performance Assessment*   Have students plan and create a set of flower beds and flower boxes for the school using plants that are adapted to local conditions.

Concept Map Answer

The following is one possible answer to the Relating Concepts exercise on page 446.

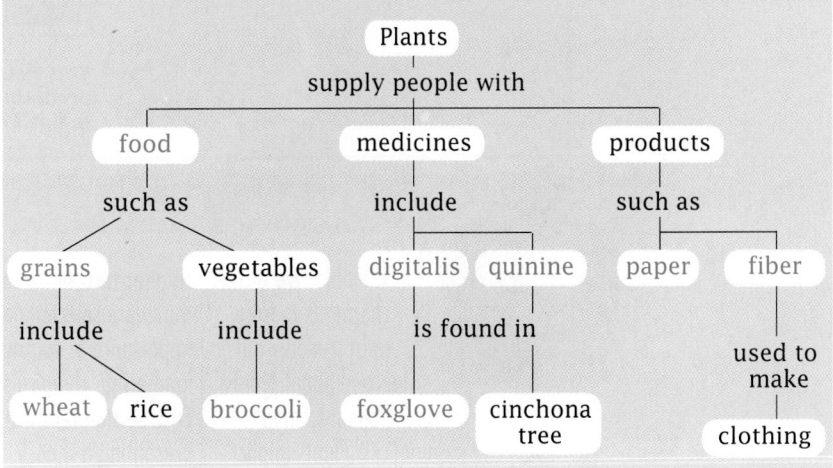

Chapter 20

Determining Prior Knowledge

- Have students explain where sugar comes from.
- Have students identify objects in the room or on their bodies that are made from plant products. If it does not occur to them, have students explain where the oxygen they breathe comes from.
- Have students explain what the immediate or ultimate source is of all food and most of the shelter used by wildlife.
- Have students identify the parts of the plant that are used for food and give an example of each.
- The role of plants in supplying food for wildlife is of primary importance. Have students discuss their value as a source of water.
- Discuss with students which agricultural crops have the greatest appeal to wildlife. Have them explain which crops are intended for man's own consumption or for his domestic livestock.

Chapter **20**

Plants in Our Lives

Review

- **photosynthesis** (Section 5.4)
- **nitrogen fixation** (Section 8.2)
- **seeds and fruits** (Section 19.3)

20.1 *Plants as Food*
- **Important Grains**
- **Food From Other Plant Parts**
- **Sources of Sugar**

20.2 *Other Uses for Plants*
- **What Is Wood?**
- **Drugs From Plants**
- **Other Plant Products**
- **Landscaping and Gardening**

20.3 *Plant Use in the Future*
- **Improvement in Food Crops**
- **Growing Plants Without Soil**

This field in the western United States is producing wheat, one of the most important grains grown in the world today.

▪ *Author's Rationale* ▪

The objective of this chapter is to take what students have learned about plants and relate it to their everyday experiences. Plants impact each of us in countless ways. The chapter focuses on four applications that were selected on the basis of their importance to society: agriculture, lumbering, chemicals, and ecology. Each section tells a story rather than presenting a theory, to help students recognize the issues that affect them and that will require decisions to be made, now and later in life.

PLANTS ARE AN IMPORTANT PART OF OUR LIVES. YOU ALREADY KNOW THAT GREEN PLANTS PRODUCE THE OXYGEN WE BREATHE. THEY ALSO PROVIDE US WITH ALL OUR FOOD; EVERYTHING WE EAT COMES FROM A PLANT OR AN ANIMAL THAT ATE A PLANT. FOSSILS SUGGEST THAT HUMANS BEGAN GROWING PLANTS FOR FOOD BETWEEN 7,000 AND 9,000 YEARS AGO.

20.1   *Plants as Food*

Objectives

❶ Summarize the uses of grains and legumes.

❷ Compare fruits to vegetables.

❸ List examples of vegetables that are stems, roots, or leaves.

❹ Identify two sources of sugar.

Important Grains

Did you know that most of the food people eat comes from fruits? As you read in Chapter 19, fruits are the parts of a flowering plant that contain the plant's seeds. When you think of fruits, you probably think of oranges, peaches, and apples. However, the fruits that provide the most food to humans are wheat, rice, corn, and oats. The fruits of these cereal plants are called **grains**. Grains are rich in carbohydrates,

protein, vitamins, and fiber. About 85 percent of human food comes from wheat, rice, oats, and corn. These grain crops occupy more than 70 percent of the world's farmland, as shown in **Figure 20.1**. Usually grains are first separated from stalks. Then they are ground into flour or meal and cooked. Some grains are eaten without being cooked. Corn, oats, and barley are also fed to livestock.

Figure 20.1
Wheat, rice, oats, and corn are grown all over the world.

Wheat
Rice
Oats
Corn

Lesson Plan 20.1

Phase 1
PREPARATION

Key Concepts
- Plants form the base of the food chain on which human survival depends.
- Wheat, rice, corn, and oats are the fruits of cereal plants.
- Fruits are the reproductive parts of plants, whereas the vegetative parts of plants are the roots, stems, and leaves, all of which are edible in some plants.
- Sugar is stored in the stem or root of a plant.

Reading Strategy

Have students read for the main ideas of this lesson by identifying any sentences that provide specific information about the objectives for this section.

Phase 2
TEACHING STRATEGIES

Visual Strategy

Figure 20.1
Discuss how the first plants to be cultivated were cereals, about 11,000 years ago in the Fertile Crescent region of the Middle East. Tell students that people propagated plant crops based on individual plant's characteristics that they found most valuable, and selectively altered the evolution of food plants.

Visual Strategy

Figure 20.2
Most people in the world get their protein primarily from plants. Have students explain why. They could suggest economic reasons; or in many countries, the eating of some kinds of meat is prohibited for religious, ethical, or cultural reasons.

Connection: Chapter 14

Changing Plant Resources

Since the colonization of this country, drastic changes have been made in our rich plant resources. Removal of magnificent virgin stands of forests and the destruction of the prairie sod have rendered extensive areas uninhabitable for such native wildlife as the wild turkey, ruffled grouse, heath hen, bison, antelope, and moose. The draining of marshes and swamps for agricultural purposes has eliminated much of the vegetation that furnished a favorable habitat for waterfowl, muskrats, and other important wildlife. Silting due to farmland erosion has destroyed much of the aquatic plant growth in streams, ponds, and lakes that formerly supported large numbers of fish and waterfowl. But in spite of what we have done to our original native flora, we still have rich plant resources. All in all, agriculture has tremendously increased the supply of choice plant foods relished by many kinds of wildlife.

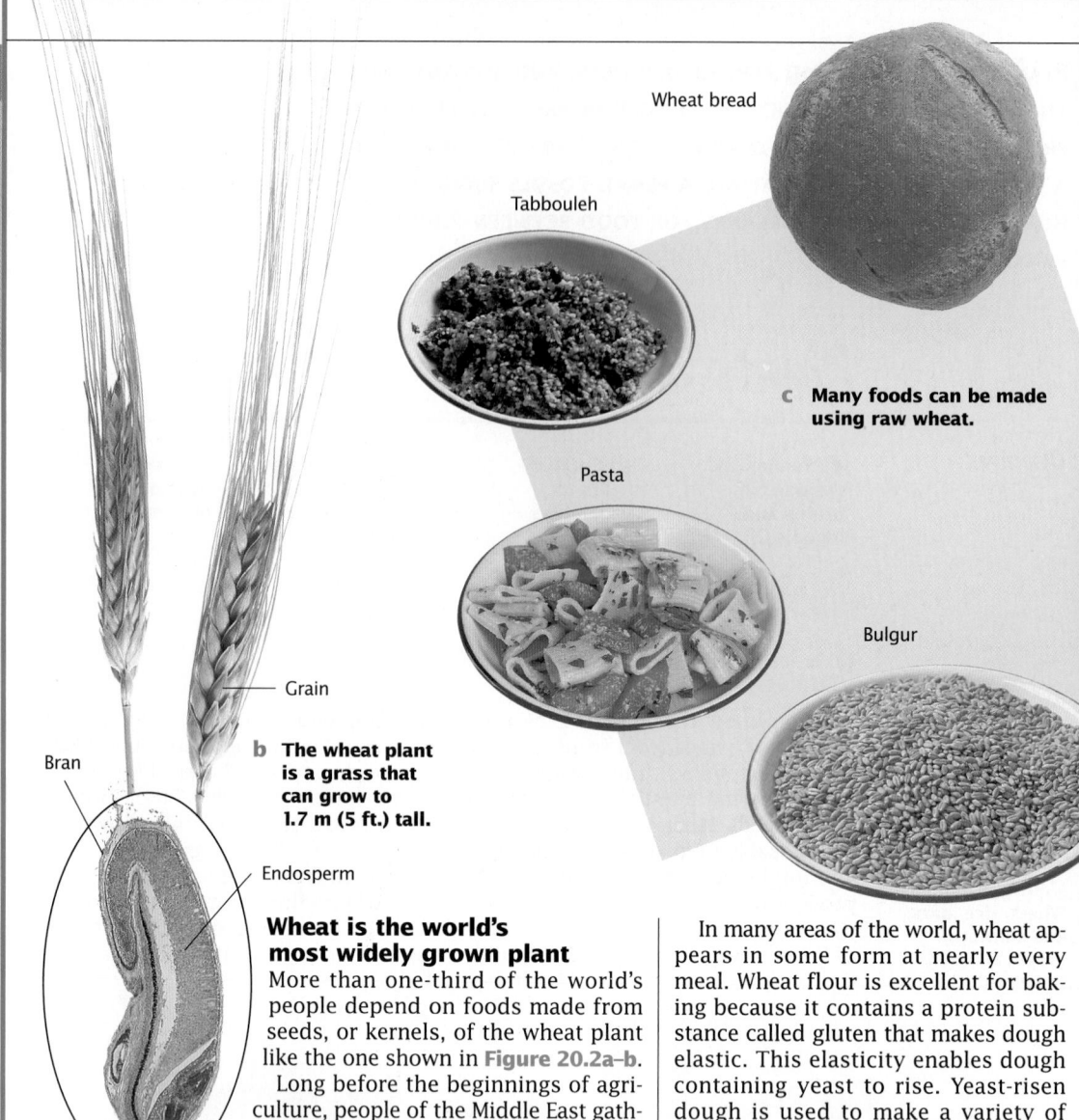

Wheat bread

Tabbouleh

c   Many foods can be made using raw wheat.

Pasta

Bulgur

Grain

b   The wheat plant is a grass that can grow to 1.7 m (5 ft.) tall.

Bran

Endosperm

Figure 20.2
a   This section of a wheat kernel shows the starchy endosperm used in making white flour, and the outer bran layer, included in whole wheat flour.

Wheat is the world's most widely grown plant

More than one-third of the world's people depend on foods made from seeds, or kernels, of the wheat plant like the one shown in **Figure 20.2a–b.**

Long before the beginnings of agriculture, people of the Middle East gathered and chewed the kernels of wild wheat. Later, early farmers selected kernels from their best wheat plants to use as seeds for planting. Over time, this practice resulted in the evolution of improved kinds of wheat. During the 1900s scientists used selective breeding to develop new varieties of wheat that produce large amounts of grain and tolerate cold temperatures, disease, insects, and other threats. The wheat that is grown today bears little resemblance to its wild ancestor.

In many areas of the world, wheat appears in some form at nearly every meal. Wheat flour is excellent for baking because it contains a protein substance called gluten that makes dough elastic. This elasticity enables dough containing yeast to rise. Yeast-risen dough is used to make a variety of breads, and other baked goods. Pasta is made by adding water to ground wheat to form a thick paste, which is then pressed into spaghetti and other shapes. In the Middle East, wheat kernels are often boiled or soaked and then pounded until they crack. The cracked particles, called bulgur (*BUHL guhr*), are then dried and used to prepare dishes such as tabbouleh, pilavi, or kisir. Pita, or pocket breads, made from wheat flour, may accompany these meals. **Figure 20.2c** shows a variety of foods made from wheat.

Figure 20.3

a **The rice plant is a grass that grows 80 to 180 cm (32 to 72 in.) tall. When the rice is ripe, the plant turns golden yellow.**

b **Because rice requires a great deal of water to grow, it is grown in fields of standing water called paddies.**

Rice is one of the world's most important food crops

More than half the people in the world eat rice as the main part of their meal. Rice is grown in moist environments, especially near rivers and flood plains, as shown in **Figure 20.3**.

No one knows exactly when or where rice originated, but it probably grew wild and was gathered and eaten by people in Southeast Asia thousands of years ago. Archaeologists have found evidence that people cultivated rice for food as early as 5000 B.C. in southern China. From there, rice spread north to Japan and west to India.

Although low in protein, rice is an excellent source of carbohydrates. Rice is usually processed into white rice, which contains few vitamins. In societies where people live mainly on rice, vitamin-rich sauces such as soy sauce are added to cooked rice to make meals more nutritious. White rice has had its hull and bran layers removed during processing. Brown rice, however, still contains the nutrient-rich bran layers on the outside of the kernel. **Table 20.1** shows the nutritional contents of rice and wheat. Rice is included in many processed foods, including breakfast cereals, soup, baby food, and flour. Breweries use broken rice kernels to make mash, an important ingredient in the making of beer. In Japan, rice kernels are used to make an alcoholic drink called sake (*SAHK ee*), or rice wine. But rice isn't just for eating and drinking. In industry, the outer coat of the rice kernel is used as an ingredient in products such as insulation and cement. Many people in Asia use the dried stalks of rice plants to thatch roofs and to make sandals, hats, and baskets.

Table 20.1 Protein and Dietary Fiber

Grain (1 oz.)	Protein (g)	Fiber (g)
Brown rice	2.3	1.0
White rice	2.0	0.3
Whole wheat flour	3.9	3.6
White wheat flour	2.9	0.8

■ *Cultural Perspective* ■

Food for Celebrations
The food we eat is also used to celebrate various cultural holidays around the globe. The festival of Boy's Day in Japan is an example. Traditionally, boys are bathed in water in which sword-shaped leaves of the iris plant have soaked. These leaves are symbols of strength. To ensure a life of strength and good luck, the boys eat rice wrapped in leaves of iris, bamboo, and oak. Boy's Day has recently been renamed Children's Day, to honor both boys and girls.

429

Using the Feature

- This feature can be used in conjunction with a world map to highlight the areas of the world where corn was originally cultivated, where it was introduced, and where it thrives today. Small groups of students can be assigned to research major corn-producing countries such as China, Argentina, Brazil, France, India, Mexico, Romania, the former Soviet Union, and Yugoslavia.

- You can also use this feature to encourage students to share corn-based recipes that were staples of their ancestors. Students can be asked to prepare a dish at home to bring in for other students to sample.

Discussion

Guide the discussion by posing the following questions:

1. Where and by whom was corn first used for food?
   Native American farmers such as the Aztecs, Mayas, and Incas first cultivated corn.

2. What kinds of food are made with corn?
   Students' responses will vary but will include tortillas, corn bread, hominy, grits, corn syrup, magarine, salad dressing, and even popcorn.

3. About three-fourths of the United States' corn crop comes from the Corn Belt. What states make up this region?
   Answers should include Iowa, Illinois, Indiana, Ohio, Minnesota, and Nebraska.

4. Why is the variety *Zea diploperennis*, which was discovered in Mexico in 1978, considered a valuable resource?
   Zea diploperennis is a disease-resistant variety of perennial corn. It can be crossed with Zea mays to produce fertile hybrids that are resistant to many viral diseases.

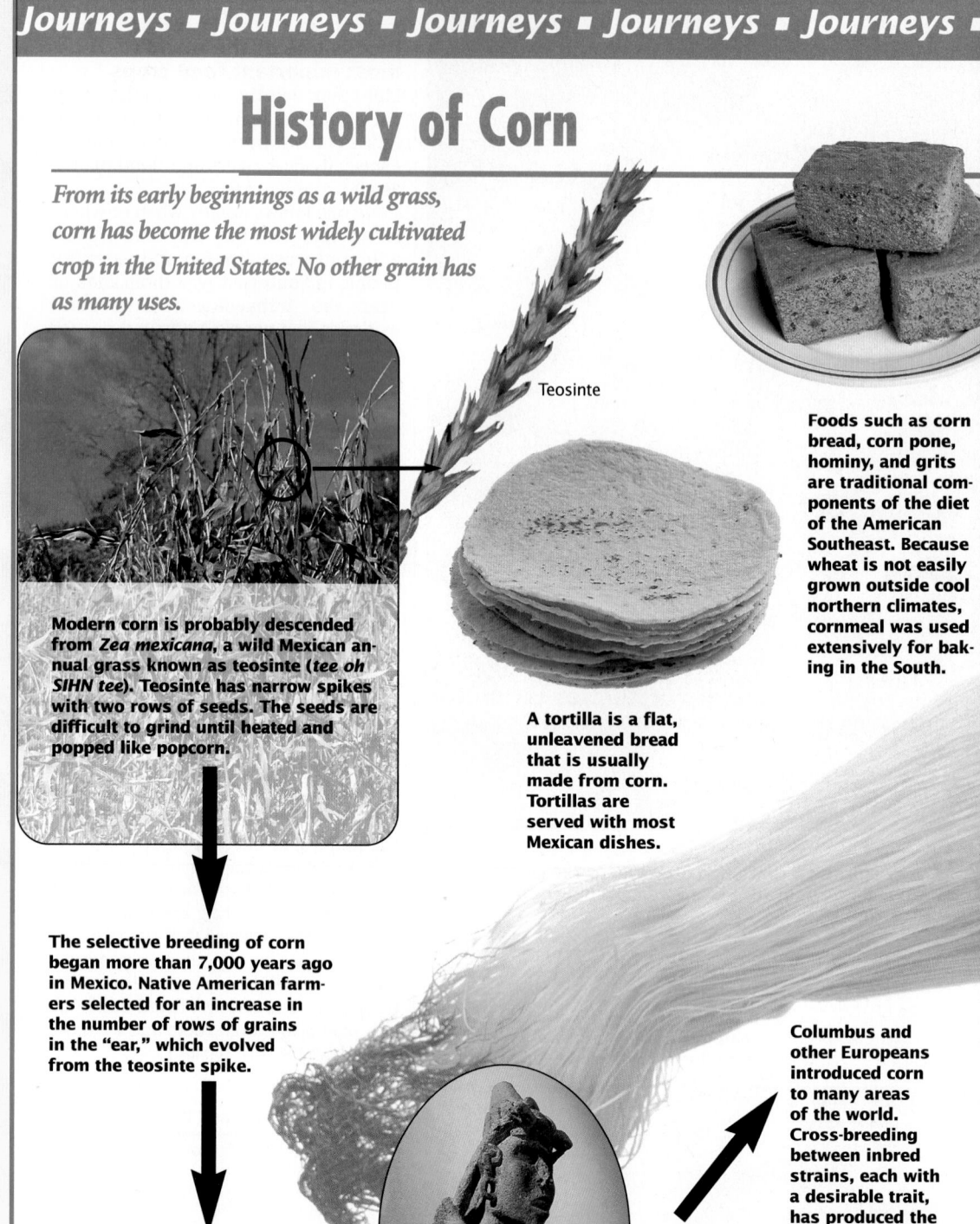

History of Corn

From its early beginnings as a wild grass, corn has become the most widely cultivated crop in the United States. No other grain has as many uses.

Teosinte

Modern corn is probably descended from *Zea mexicana*, a wild Mexican annual grass known as teosinte (*tee oh SIHN tee*). Teosinte has narrow spikes with two rows of seeds. The seeds are difficult to grind until heated and popped like popcorn.

A tortilla is a flat, unleavened bread that is usually made from corn. Tortillas are served with most Mexican dishes.

Foods such as corn bread, corn pone, hominy, and grits are traditional components of the diet of the American Southeast. Because wheat is not easily grown outside cool northern climates, cornmeal was used extensively for baking in the South.

The selective breeding of corn began more than 7,000 years ago in Mexico. Native American farmers selected for an increase in the number of rows of grains in the "ear," which evolved from the teosinte spike.

Corn was cultivated by the Aztecs of central Mexico, the Mayas of southern Mexico and northern Central America, and the Incas of western South America.

Quetzalcoatl, Aztec Lord of the Winds, carries corn for a successful harvest

Columbus and other Europeans introduced corn to many areas of the world. Cross-breeding between inbred strains, each with a desirable trait, has produced the high-yield hybrid corn varieties grown today.

■ *Journeys* ■ *Journeys* ■ *Journeys* ■ *Journeys* ■ *Journeys*

Native Americans are still growing corn. Here a Hopi man is shown tending to corn in the Oraibi area in northern Arizona. American colonists in the 1600s and 1700s learned how to grow corn from the Native Americans.

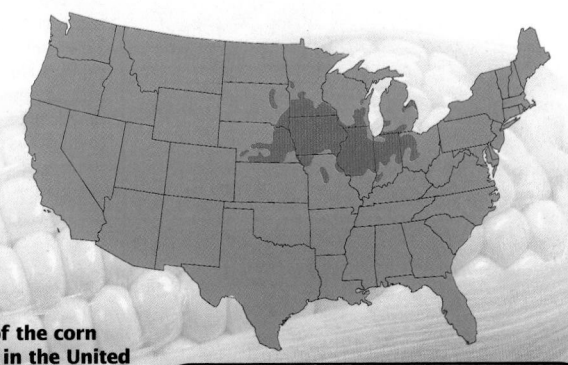

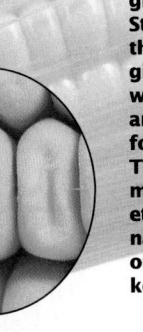

Most of the corn grown in the United States comes from the Corn Belt, a region of the Midwest where the climate and soil are ideal for growing corn. The most commonly grown variety is dent corn, so named for the dent on the top of each kernel.

Zea diploperennis

Today, corn is one of the world's chief foods for domestic animals. Seventy percent of the U.S. corn crop is consumed by animals. Corn is also used to produce corn syrup, margarine, cornstarch, salad dressing, and many industrial products.

In 1978, a disease-resistant variety of perennial corn was discovered in a small clearing in Jalisco, a west-central state of Mexico. The variety was named *Zea diploperennis*. *Z. diploperennis* can be crossed with *Z. mays* to produce fertile hybrids that are resistant to many viral diseases. Newly discovered species such as *Z. diploperennis* are valuable resources because they provide new sources of genetic variability.

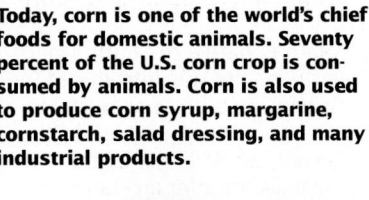

Demonstration 2

Food Products

Supply students with examples of root crops such as carrots, rutabagas, and yams; and fruits such as squash, tomatoes, chili peppers, pineapples, and walnuts. Also, bring in some leaves from spinach, lettuce, and herbs (sage, basil, oregano, etc.), and some stems of asparagus and broccoli. Have students categorize the foods as fruits, cereals, or legumes; or roots, stems, or leaves, and list at least one nutritional benefit derived from each food.

Theme Connection

Interacting Systems

George Washington Carver, an African-American scientist, studied botany, geometry, bacteriology, zoology, and entomology. His agricultural research revolutionized the economy of the southern United States by encouraging farmers not to depend on a single plant, cotton. He urged farmers to instead grow peanuts, sweet potatoes, soybeans, and pecans. Dr. Carver discovered more than 300 industrial products that could be made from the peanut. He also found that more than 100 industrial products could be made from the sweet potato and that 75 industrial products could be made from the pecan.

Figure 20.4

a   Soybeans are very high in protein. They are used to make tofu, soy milk, cheese, and many types of meat substitutes.

Dahl

Lentils

b   Lentils, another type of legume, are also high in protein. They can be made into dahl, a spicy lentil dish, shown above.

Legumes are high in protein

Legumes (*LEHG yooms*) are plants that are members of the pea or bean family. As you read in Chapter 8, legumes have nitrogen-fixing bacteria living in their roots. These bacteria provide legumes with extra nitrogen, which is needed to make proteins.

Legumes contain more protein than most plants. This makes them important additions to a grain diet. The soybean, which is shown in **Figure 20.4a**, has been a nutritious staple in the Chinese diet for several thousand years. Soybeans are cooked down into a custard-like consistency, pressed into small square cakes, and transformed into a highly nutritious food called tofu—also known as bean curd or soybean cakes. Discovered by Chinese royalty in 164 B.C., tofu has been the protein staple of the East Asian diet for 2,000 years and is an important food for more than 1 billion people. Other legumes such as lentils are popular in Egypt and India, and can be ground into flour or made into dahl, as indicated in **Figure 20.4b**.

Foods From Other Plant Parts

Figure 20.5 Here, New Guinean men are stacking yams in preparation for a feast.

The stems, roots, and leaves of plants are vegetative, or nonreproductive plant parts. Vegetables like spinach, lettuce, and cabbage are leaves, while carrots, radishes, and turnips are roots. Celery is a leaf stalk. Broccoli and cauliflower are immature flowers.

The white potato is an important food staple in many regions of the world. The edible parts of the potato plant are modified stems called **tubers**, which grow underground and store starch.

The potato is native to South America. When the potato was introduced to Europe, it became a great success, especially in Ireland. It grew well in the country's cool, moist climate and provided enough calories for an entire family. The potato blight of 1845, caused by the protist *Phytophthora infestans*, virtually destroyed the Irish potato crop. As a result, at least 1 million Irish people died of starvation.

Yams are also tubers. In many parts of the world, yams are an essential food crop and play an important cultural role as well, shown in **Figure 20.5**.

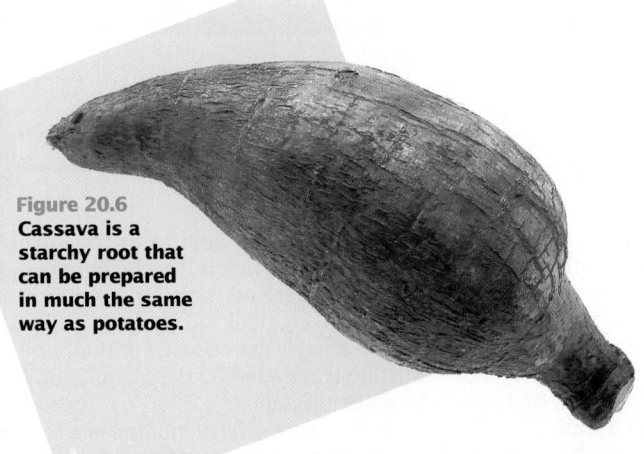

Figure 20.6
Cassava is a starchy root that can be prepared in much the same way as potatoes.

Interacting Systems

How do humans benefit from the ability of plants to store the products of photosynthesis?

Roots are also a source of food

Some plant roots are food sources. These plants store food in swollen parts of their root systems during their first year of growth, then use the food to produce flowers the next year. Some roots, such as the carrot, have been grown as food in Europe for at least 2,000 years. Other important root crops include turnips, beets, sweet potatoes, and cassava, also called manioc. Cassava, shown in Figure 20.6, is the staple food of more than 500 million people around the world. It contributes more than one-third of the calories consumed in Africa.

Humans eat the leaves of many plants

Besides carrying out photosynthesis, leaves also provide us with vitamins and minerals. Humans eat the leaves of cabbage, parsley, spinach, chard, rhubarb, and lettuce. Some forms of lettuce may have been cultivated as early as 4500 B.C., and lettuce is even depicted on Egyptian tombs.

Sources of Sugar

Sugar is by far the most important sweetener in the world today. You probably don't think about the fact that it comes from the stem or root of a plant. Most of the sugar that you eat comes from sugar cane or sugar beets. These plants produce a sugar called sucrose, the sugar kept in sugar bowls. Sugar cane is a tall grass believed to have been first used by peoples of the Far East. Sugar cane grows tall, sturdy stalks that contain a large amount of juice from which sugar and syrup are made. The process of producing sugar from sugar cane has been practiced in India since 3000 B.C.

Sugar beets have large, fleshy roots like carrots, in which they store sucrose. The Romans ate sugar beets, but it was not until the eighteenth century that the Germans discovered that the roots contained considerable amounts of sugar.

Figure 20.7
Sugar cane, such as this crop growing near Clewiston, Florida, is a major source of the world's sugar.

Section Review

1 Name some legumes that are part of your diet.

2 Describe the difference between a fruit and a vegetable.

3 What two vegetables are tubers?

4 Where does most sugar come from?

■ *Section Review Answers* ■

1. Answers should list members of the bean or pea family.

2. Fruits are the seed-bearing, reproductive portions of the plant. Vegetables are the vegetative, nonreproductive parts of plants such as the root, stem, or leaves of plants.

3. Answers should include the white potato and the yam.

4. Most of the sugar we eat comes from sugar cane and sugar beets.

Using the Feature

- This feature can be used to review the material on cellular respiration in Chapter 5.
- You should use this feature as an opportunity for students to research the use of ethanol-containing gasolines in cities with air pollution.

Discussion

1. How do biomass fuels differ from gasoline?
   Biomass fuels are alternative fuels made from fermented plant matter.

2. What are three advantages of using biomass fuels?
   Biomass fuels are renewable, they burn cleaner than petroleum-based fuels, and because they don't release carbon dioxide, their use may help reduce global warming.

3. What difficulties does large scale fermentation present to industries?
   Different organisms metabolize only certain sugars, so industry must find the right combinations of organisms to ferment the various sugars found in plant matter.

4. How are scientists working to solve these problems?
   Some are experimenting with a technique in which plant matter is treated with acid and enzymes and then rapidly fermented by yeast. Others are using genetic engineering techniques to create organisms that can efficiently ferment many kinds of sugars. Still other scientists are seeking the right combination of microbes and enzymes to produce ethanol by fermentation.

Fermentation

Fuel From Plants

Imagine pulling up to a gas station and filling up your car—not with gasoline, but with corn fuel. Corn, as well as wheat, wood chips, grass clippings, and other plants, can be fermented to form so-called biomass fuels such as ethanol and methanol. Some biomass fuels are already being used on a limited scale. Ethanol, for example, can be added to gasoline to improve combustion. Methanol might someday become the fuel of choice in California, which is struggling to cut down on air pollution from millions of motor vehicles.

Concerns about global warming and decreasing supplies of petroleum have served as an incentive to increase research on alternative fuels. Biomass fuels, say advocates, have a lot of attractive characteristics. They are renewable, since the starting materials are plants or plant wastes such as corn plants left in the field after the corn is harvested. They burn cleaner than petroleum-derived fuels. And they can help reduce global warming, since the carbon dioxide gas released during combustion is recycled by the plants grown to make the fuel.

At first glance, fermentation is a relatively simple process. Certain bacteria and yeasts can break down sugars for energy even when oxygen is limited or completely absent. The simple compounds left at the end of the process—alcohols and fatty acids—still contain energy. These chemicals can be collected and burned as fuel.

Fermentation on an industrial scale, however, is not quite that simple. Different organisms metabolize particular sugars under specific conditions. For instance, some organisms can ferment only five-carbon sugars such as

Fermentation vats

Since ancient times, people have used fermentation to convert sugar to alcohol to make beverages such as wine.

The process of winemaking is not very different from the process that produces ethanol. After the grapes are crushed, yeast is added in an oxygen-free environment so that fermentation can occur. One-way gas valves enable the carbon dioxide that is produced to escape without allowing oxygen to enter. In this environment, alcohol is produced.

arabinose and xylose. Other microbes can handle only six-carbon sugars such as glucose and mannose. Given the mix of sugars in various plant materials, coming up with the right combination of microbes can be quite a challenge.

One method of biomass fuel production that is currently under study is a technique called simultaneous saccharification-fermentation (SSF). In SSF, plant materials like corncobs are treated with dilute sulfuric acid. This releases cellulose and lignin, polysaccharides that give plants their rigid cell structure. Then the mixture is put into a container with the enzyme cellulase and yeast. The enzyme breaks the polysaccharides down into glucose, which the yeast rapidly ferments to ethanol.

Scientists working on fermentation technology are also looking at genetically engineered organisms to perform fermentation. For example, one researcher has used the bacterium *Zymomonas mobilis* as a source of genes for two enzymes that produce ethanol during fermentation. Splicing the *Zymomonas* genes into another bacterium, *Escherichia coli*, has resulted in a strain of microbe that ferments all the sugars found in different plants. The genetically engineered microbe has shown promise in laboratory experiments.

Ethanol From Wood

Fermentation of Wood Chips

1. *Wood is treated with dilute sulfuric acid:* cellulose and xylan are released.

2. *Two rounds of enzyme treatments are applied:* large compounds are broken down into five- and six-carbon sugars.

3. *Bacteria and yeast ferment the sugars to ethanol.*

Not everyone believes it is necessary to rely on genetically engineered organisms to produce ethanol by fermentation. Some scientists think that the right combination of naturally occurring microbes and enzymes can do the trick. One system, still in the experimental stage, uses a

process to make fuel from hardwood wastes left over by the forestry industry. Wood chips are treated with dilute sulfuric acid to free cellulose and xylan, another polysaccharide. Additional enzyme treatments break these large compounds down into five- and six-carbon sugars. Finally, an assortment of bacteria and yeasts are added to vats to ferment the sugars to ethanol.

As researchers find microbes that perform fermentation more quickly and that efficiently handle the types of plant matter available, the use of biomass fuels will increase. As much as 50 percent of the nation's liquid fuel needs could eventually be met with fuels generated from biomass.

Wood chips (above) will be treated with dilute sulfuric acid and enzymes. In fermentation vats (left), yeast (right) ferments the sugars in the wood chips to ethanol.

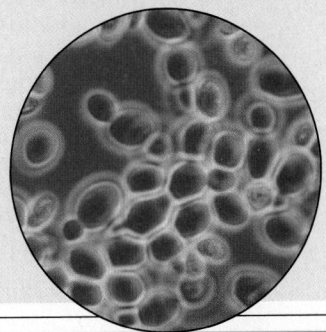

435

Lesson Plan 20.2

Phase 1

PREPARATION

Key Concepts
- Plants have a wide variety of uses and are essential in the production of many products, such as medicines, clothing, and building materials.
- Plants are used artistically for aesthetic purposes.

Reading Strategy

As they read, have students keep a running list of the uses for plants described in this section. When they have finished their reading, have them make a table using the items in their list.

Phase 2
TEACHING STRATEGIES

Visual Strategy

Figure 20.7a

Have students identify a product that could be substituted for each of the following forest products: pencils, writing paper, paper containers, cardboard, lumber, turpentine, particleboard, plywood, wood mulch chips, and firewood.

Demonstration 1

Making Paper

Shred and soak newspaper in a bleach solution overnight. Have students press the pieces together on top of a screen. Smooth it out by rolling over it with a rolling pin and let it dry. The next day you should have a fairly good surface upon which to write.

436

PLANTS HAVE MANY VALUABLE USES IN ADDITION TO PROVIDING FOOD FOR ANIMALS AND PEOPLE. TREES PROVIDE LUMBER AND COOKING FUEL FOR HEAT, AND RAW MATERIALS FOR MANUFACTURED PRODUCTS SUCH AS PAPER, PLASTICS, AND RAYON FABRIC. IN ADDITION TO TREES, OTHER PLANTS ARE USED TO MAKE FIBERS, MEDICINES, AND FOOD SEASONINGS.

20.2   Other Uses for Plants

Objectives

① **List three important wood products.**

② **Recognize the role of plants as sources for drugs.**

③ **Describe how turpentine and rubber are made.**

④ **Identify the source of two kinds of natural fibers.**

What Is Wood?

After food and oxygen, the most valuable resource that plants produce for people is wood. In Chapter 19 you learned that plants grow when the cells in their meristems divide. When plants that live more than one growing season get thicker, their meristems are producing secondary xylem and secondary phloem. **Wood,** which is shown in **Figure 20.7a**, is secondary xylem. Thousands of products are made from wood, including paper goods of all kinds.

Trees are harvested for lumber

Wood from trees that have been cut down and sawed into boards and planks is called lumber. Nearly three-quarters of the lumber in the United States is used for construction. The rest goes to factories that make products such as boxes, crates, toys, railroad cars, boats, and items shown in **Figure 20.7b–d**. Wood chips and sawdust can be treated with chemicals to make wood pulp. Manufacturers use wood pulp to make paper, rayon, and other products.

Bark
Sapwood
Heartwood

Figure 20.7

a **This cross section of a cedar tree shows the nonliving heartwood, the water-conducting sapwood, and the bark. Many products are made from wood. One of the most important is paper, which forms the graphic backdrop of these two pages.**

▪ *Cultural Perspective* ▪

Trees

Many cultures understand the importance of the tree. In the United States and Canada, Arbor Day is celebrated. Arbor is another word for tree. The idea for Arbor Day originated with Sterling Morton, a newspaperman who understood that trees enrich the soil

and help it hold water. On the first Arbor Day, the state of Nebraska offered prizes to groups and people who planted the most trees. One million trees were planted in Nebraska on that day. After Morton's death, Arbor Day was celebrated on April 22, Morton's birthday. In California, Arbor

Day is celebrated on Luther Burbank's birthday, March 7, because Burbank developed many new varieties of trees and plants. Israel's equivalent to Arbor Day is called Tu B'Shebat (*Too buh shuh vat*). The fifteenth day of Shebat (a Hebrew month) is called the New Year for Trees.

c **Many musical instruments, like this violin, are made from wood.**

b **Wood has been used in art for centuries. This sculpture was created by a Native American who lived in the Florida Keys before Columbus arrived in the New World.**

d **Furniture is one of the many things that are made from wood. This chair is called an Eames chair and it is made of laminated plywood.**

Some wood is used as fuel

Wood has been used as a fuel since prehistoric times. It is the main fuel for most people in developing countries and for more than half the people in the world. A family that uses wood for heating and cooking burns about a ton of wood each year. As a result, wood is being cut for fuel faster than trees can regrow. In some African countries, wood has become so scarce that the average family spends one-third of its income on wood.

Paper is one of our most important plant products

Papermaking begins when wood chips are ground and chemically treated to produce wood pulp. Wood pulp contains cellulose fibers, which are found in all plant cell walls. When a mixture of cellulose fiber and water is filtered through a fine screen, the fibers tangle in a mass. This mass is pressed between huge rollers and allowed to dry, forming paper. Papermaking fibers come from many different plants, including bamboo, cotton, sugar cane, wheat, and rice.

Paper can be recycled and the fibers used to make new paper. But because it is expensive to remove inks and dyes, the cost of recycling paper often exceeds the cost of making paper from wood.

▪ *Cultural Perspective* ▪

Celebrations and Trees

Long ago, the people of Palestine celebrated Tu B'Shebat by planting a tree to honor the birth of a child born during that year. A cedar tree was planted if the child was a boy and a cypress tree was planted for a girl.

History Connection

Willow Bark

At one time, members of the Crow tribe chewed willow bark to relieve headaches. Scientists have discovered that the bark of the weeping willow tree contains a chemical that turns into salicylic acid in the human body. Today we call this acid aspirin.

Visual Strategy

Figure 20.8

Have students write a report on plants that are used to make medicines and other useful substances. There are a number of plants from which students may choose. For example, *Atropa belladonna, Salix alba, Flex vomitoria, Papaver somniferum, Rhus radicans,* and *Nicotiana tabacum.*

Connection: Chapter 19

Opium

The opium poppy, which differs from the Oriental poppy, is a source of commercial opium. It is collected from the milky juice found in unripe fruits. Opium juice is generally used as a sedative. Also, it is the source of morphine and codeine. Misuse of opium and its derivatives has resulted in addiction in humans, and has prompted control legislation at state, federal, and international levels. This plant has probably done much to shape world history, particularly in the Orient.

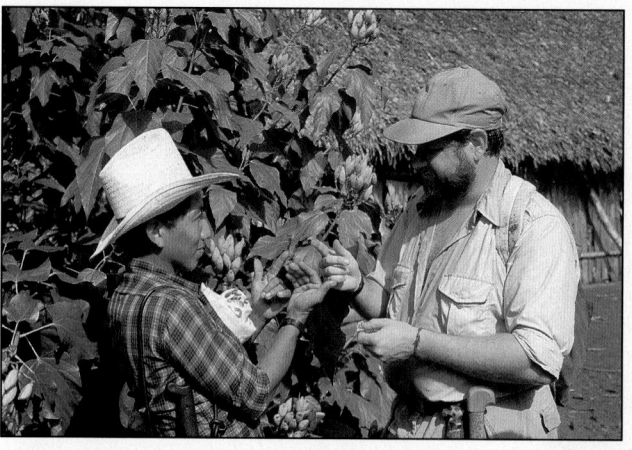

Figure 20.8
The Native American doctor (left) is teaching the Western doctor (right) how to use plants native to the Amazonian rain forest to treat some illnesses.

Drugs From Plants

People have always used plants to relieve pain and cure ailments or disease. Some people still use plants to cure ailments, and many doctors, such as the one in Figure 20.8, are studying the cures these plants provide. Today plants produce many substances used in medicinal drugs. For instance, sweet potatoes provide an extract used in producing steroid hormones for birth control pills and cortisone. Figure 20.9 shows foxglove, a plant that produces digitalis, used to treat heart problems. High blood pressure is controlled by reserpine (*REHS ur pihn*), a drug obtained from the shrub *Rauwolfia.* The May apple was used by Cherokees to kill parasitic worms. Research on the May apple has led to drugs that kill viruses and treat cancer of the testis, and to a spray that protects crops from beetles.

Quinine, aspirin, and ephedrine were first derived from plants

Throughout recorded history, no disease has ever caused more human deaths than malaria. The first medicine used to treat malaria successfully came from the cinchona (*sihn KOH nuh*) tree in South America. In the middle of the seventeenth century, Jesuit missionaries found that Native Americans had a remedy for malaria made by boiling cinchona bark in water. The Native Americans called the tree bark "quina." The medicine isolated from its bark was called quinine, (*KWEYE neyen*). Today quinine is made synthetically. It is still widely used to prevent malaria.

Solutions made by soaking leaves of the white willow were often placed on aching areas of the body. The ingredient in willow that reduced the pain was isolated in 1827 and called salicin. A derivative called acetylsalicylic acid can be swallowed to relieve all types of pain. The original makers of this drug called it aspirin. Aspirin is the most widely used drug in the world today.

Ephedrine (*eh FEH drihn*) can be obtained by soaking the dried stems of the gymnosperm *Ephedra sinica.* These stems have been prescribed in China for centuries as a stimulant, and for the treatment of high blood pressure, hay fever, and asthma. Ephedrine is used today as an ingredient in decongestants.

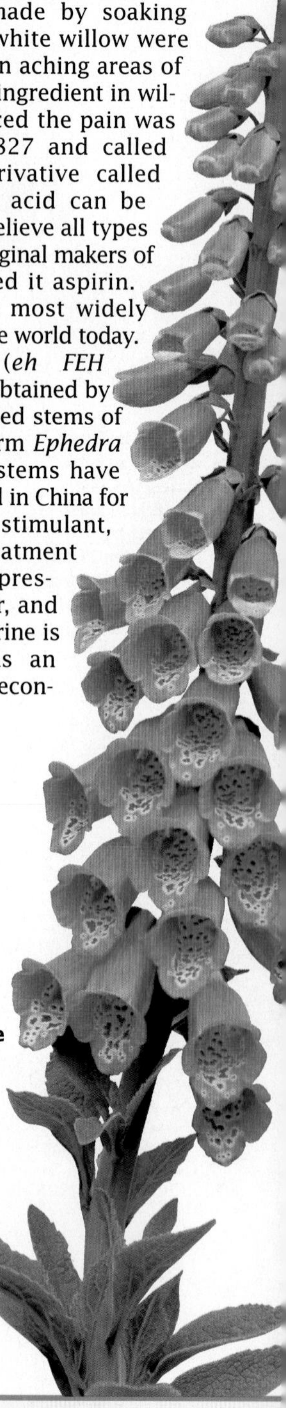

Figure 20.9
Foxglove is an extremely poisonous European plant. The leaves of the plant produce digitalis, which is effective in stabilizing the heart's action.

Figure 20.10
The rosy periwinkle, found in Madagascar, is a plant that produces the chemical vincristine, which is used to treat childhood leukemia.

The rosy periwinkle produces a leukemia-fighting drug

One of the most recently discovered drug-producing plants is the rosy periwinkle, *Catharanthus roseus*, shown in **Figure 20.10**. It originally came from Madagascar, an island near the southeastern coast of Africa, and is often planted in flower beds. The plant contains a chemical called vincristine (*vihn KRIHS teen*), which is used to treat certain forms of leukemia. This disease once killed nearly every child who had it, but thanks to vincristine, the lives of many patients now are saved.

Figure 20.11
An ethical question has arisen concerning the yew tree. Should the yew trees be destroyed to benefit those few individuals who need cancer treatment? Or should the people go untreated to preserve the yew tree?

A chemical from yew trees may help combat cancer

Scientists have recently discovered that a chemical isolated from the yew tree, *Taxus*, appears to have cancer-fighting properties. The chemical, called taxol, is found in the bark of yew trees. Clinical tests have shown that taxol reduces the sizes of cancerous tumors in some patients. An obstacle to further testing, however, is the fact that only very small amounts of taxol are made by a single yew tree. Approximately 8,000 pounds of yew bark are needed to produce one pound of taxol. Unfortunately, scientists have not yet found a way to make taxol in the laboratory, and yew trees remain the only source. Most of the taxol used for clinical testing comes from the bark of the Pacific yew, shown in **Figure 20.11**. It is a tree that has been widely eliminated by clearcutting. Biologists are now searching for yew varieties that contain high levels of taxol. If these varieties could be grown in cultivation, the availablity of taxol would greatly increase.

Like the rosy periwinkle and the yew, many other plant species might also have valuable medicinal properties. The cancer-fighting ability of taxol is yet another example of why species preservation may be important for our own survival.

Social Studies Connection

Periwinkle Plants
Researchers have discovered that the periwinkle plant, a native of the rain forest in Madagascar, contains a chemical that is helpful in the treatment of Hodgkin's disease. It has been suggested that, if the people of Madagascar received even a small profit from the plant, it would be the country's largest source of income and could raise the standard of living for its population. However, drug companies searching for and finding valuable botanicals throughout developing countries refuse to pay the people for their knowledge or products. Recently, Prince Charles of England called for a system of payment to indigenous peoples who share their encyclopedic knowledge of the rain forest resources.

■ *Cultural Perspective* ■

Garlic
Tell students that many Italians believe that garlic contains healing properties, and it is often used in their diet. Recently researchers have found that garlic does indeed have beneficial properties. Garlic contains a compound that activates enzymes in the liver to destroy aflatoxin, a carcinogen that is the leading cause of liver cancer. Also, garlic may serve to protect humans against carcinogens found in charcoal-broiled meats, polluted air, and cigarette smoke. Thus, researchers are now suggesting that people eat garlic, but no more than two cloves a day, because garlic also has chemicals that can irritate the intestines.

Phase 3

ASSESSMENT OPTIONS

Closure Strategy

Plant Uses

Review this section by asking students the following questions.

- What are some of the important products of the forest industry? *fuel, building lumber, paper, cleaning compounds, insecticides, cosmetics, medicines, soil mulches, soil conditioners, turpentine, and rosin*
- Why are natural fibers still often preferred over synthetic ones? *They are preferred for their strength and resilience.*
- What are some medicines that are derived from plants? *digitalis, morphine, codeine, and quinine*

Section Review

Assign the *Section Review*.

Reteaching

Students may not realize the importance of plants in their daily lives. To help them understand the significance of plants, have each student make a timetable of a typical day, listing the plant products he or she uses routinely. A timetable might begin with the following entries:

7:00 A.M.—wake up, put on robe (fabric made of cotton fibers); blow stuffy nose (facial tissue made of wood pulp); make bed (sheets made of cotton, bed frame made of wood)

7:15 A.M.—shower and shampoo (soap and shampoo made of fats, waxes, and fragrances from plants, towel made of cotton)

7:25 A.M.—eat breakfast (orange juice and corn flakes made from plants, eggs obtained from grain-fed chickens); read back of cereal box (made from wood pulp)

440

Other Plant Products

Another important plant product is the fiber used in clothing. Cloth made from cotton has been worn for centuries. Cotton thread is spun from the strong, fine fibers attached to cotton seeds, shown in **Figure 20.12**. Cotton is still the world's most important plant fiber. The stems of flax plants yield a softer, more durable fiber that is used to make a cloth known as linen. Although synthetic fibers now make up more than 30 percent of the world's fibers, natural fibers are still prized for their durability, comfort, and strength.

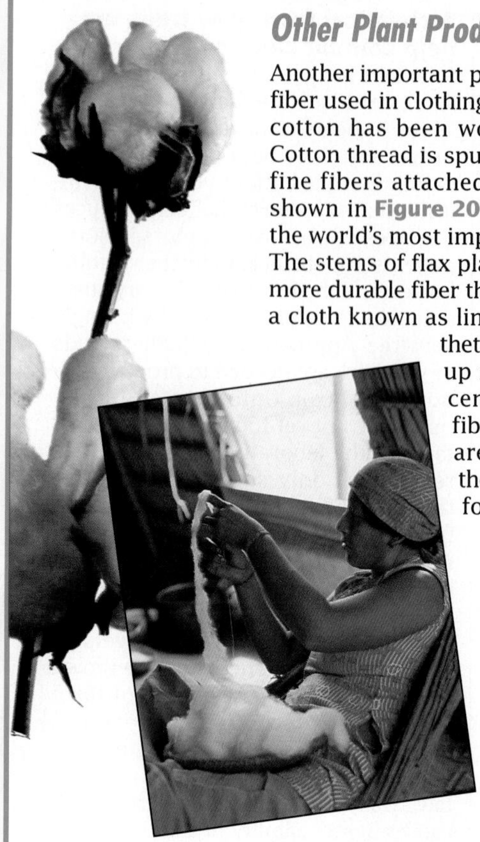

Figure 20.12
The fibers that grow in the seed pod, or boll, of the cotton plant are being spun into thread by this Brazilian woman.

Most turpentine is made from pine trees

Turpentine is a colorless liquid used to remove paint. This highly inflammable substance is made from pine trees that grow in the southeastern United States. Turpentine is made when the pine trees are converted to wood pulp. A vapor containing turpentine forms during the pulping process. The cooled vapor is a liquid that contains turpentine. Turpentine is also used to make disinfectants, insecticides, medicines, and perfumes.

Rubber plants produce latex

Rubber is one of the most important raw materials obtained from plants. Latex, a milky white liquid, is extracted from rubber trees. Native Americans of Central and South America made rubber balls and waterproof shoes from latex. Today, natural rubber comes from trees planted on plantations in Asia. Most of today's rubber, however, is manufactured synthetically from petroleum products.

Landscaping and Gardening

Many people cultivate plants. Lawn grasses are grown by many people in the United States, but millions also grow trees, shrubs, flowering plants, vegetables, and houseplants. Most houseplants are native to tropical forests, areas with dim light like that found in most buildings. Buildings and roads are usually landscaped by planting trees, lawns, and shrubs that make them more attractive. In addition, the skilled use of trees and shrubs in landscape design can prevent erosion and control runoff. All this plant care has given rise to huge industries employing millions of people worldwide in occupations that supply gardeners with machinery, fertilizer, pesticides, and plants.

Section Review

1. How is paper made from wood?
2. Name three drugs that are derived from plants. Identify the plant that is the source for each.
3. How are cotton and linen made?
4. Describe three additional uses for plants.

■ Section Review Answers ■

1. Students should explain how wood pulp is used to make paper. See Figure 20.7 on pages 436 and 437.

2. Possible answers might include the drug digitoxin produced by the foxglove plant, the drug reserpine produced by *Rauwolfia*, and the drug quinine from the cinchona tree.

3. Cotton thread is spun from fine, strong cotton fibers attached to cotton seeds. Linen is made from the stems of flax plants.

4. Possible answers include decoration, paint, and lumber.

SCIENTISTS APPLY KNOWLEDGE ABOUT PLANT GROWTH, DEVELOPMENT, REPRODUCTION, AND GENETICS TO MAKE AGRICULTURE AS EFFICIENT AS IT CAN POSSIBLY BE. GREATER CROP YIELDS WILL BE OBTAINED WHEN PLANTS ARE DEVELOPED THAT ARE MORE RESISTANT TO DISEASE, AND WHEN GROWING TOLERANCES CAN BE CHANGED TO SUIT LESS THAN IDEAL ENVIRONMENTS.

20.3 *Plant Use in the Future*

Objectives

❶ Explain how crop rotation increases crop yield.

❷ Contrast the benefits and limits of the Green Revolution.

❸ Describe the benefits offered by genetic engineering and the discovery of new crop plants.

❹ Predict some ways in which hydroponics might be used in the future.

Improvements in Food Crops

As growing populations convert more and more farmland into places to live, farmers and agricultural scientists strive to make the remaining farmland produce more food more efficiently. Efforts to increase the amount and quality of food have focused on developing new varieties of crops, as well as on improving the nutritional value of crops.

As you read in Chapter 8, genetic engineers are developing plants that are more resistant to diseases and pests.

Farmers use various methods to increase crop yield

Many food crops use up the available nutrients in the soil in just a few growing seasons. Farmers improve the productivity of their land by using fertilizers or by plowing under crop remains to enrich the soil, as shown in **Figure 20.13**. **Crop rotation** is the alternating cultivation of two or more crops in the same field. After a season of growing corn, for example, a farmer may plant soybeans in the field. Soybeans and other legumes contain nitrogen-fixing bacteria that convert the nitrogen in the atmosphere into a form that plants can use. The nitrogen compounds that remain in the soil provide nutrients for the next grain crop.

Figure 20.13
Plowing mixes the plant material left after harvesting into the soil. Incorporating the plant material helps to recycle the organic matter and plant nutrients such as nitrogen, phosphorus, potassium and sulfur.

The Green Revolution introduced improved crops to many parts of the world

Scientists estimate that 6.2 billion people will populate the Earth by the year 2000. Small gains can still be made in creating more farmland, but the most promising approach to feeding more people seems to lie in improving existing crops. One of the most comprehensive efforts to introduce high-yield crops into poor agricultural regions is an intensive plant breeding program called the **Green Revolution**. The Green Revolution began in Mexico and India in the 1950s. At that time, wheat was difficult to grow in tropical areas because heavy downpours often knocked down the tall, slender stalks. Once the wheat was matted on the ground, it quickly rotted. In 1953 Dr. Norman Borlaug, shown in **Figure 20.14**, started the Green Revolution by crossing varieties of wheat sent to him from around the world, and selecting for plants better able to grow in tropical Mexico. By 1963 Borlaug had developed a dwarf wheat variety that grew quickly, had high yields, and was resistant to disease. These wheat varieties prospered in tropical countries around the world. For his work, Dr. Borlaug received a Nobel Peace Prize in 1970.

The long-term success of the Green Revolution has been limited, however. New crop varieties often required expensive fertilizers, irrigation, and pesticides. Many poor farmers could not afford to buy the materials and machinery needed to cultivate the grains properly.

Figure 20.14
Dr. Norman Borlaug started the Green Revolution and won a Nobel Prize in 1970.

Genetic engineering could increase the nutritional value of crop plants

Although food crops are rich in carbohydrates, many are relatively poor sources of protein. The most common grains, legumes, and root crops do not provide all the amino acids humans need. To improve the protein content of crops, genetic engineers are attempting to transfer genes for "essential amino acids" (the amino acids humans cannot produce themselves) into grains. This research will be important in countries where people depend on a plant-based diet. Some cotton plants that have been genetically engineered are shown in **Figure 20.15**.

Figure 20.15
a **The cotton plants, above, have been genetically engineered to produce larger cotton bolls.**

b **As you can see, the genetically engineered boll on the left is much larger than the boll on the right.**

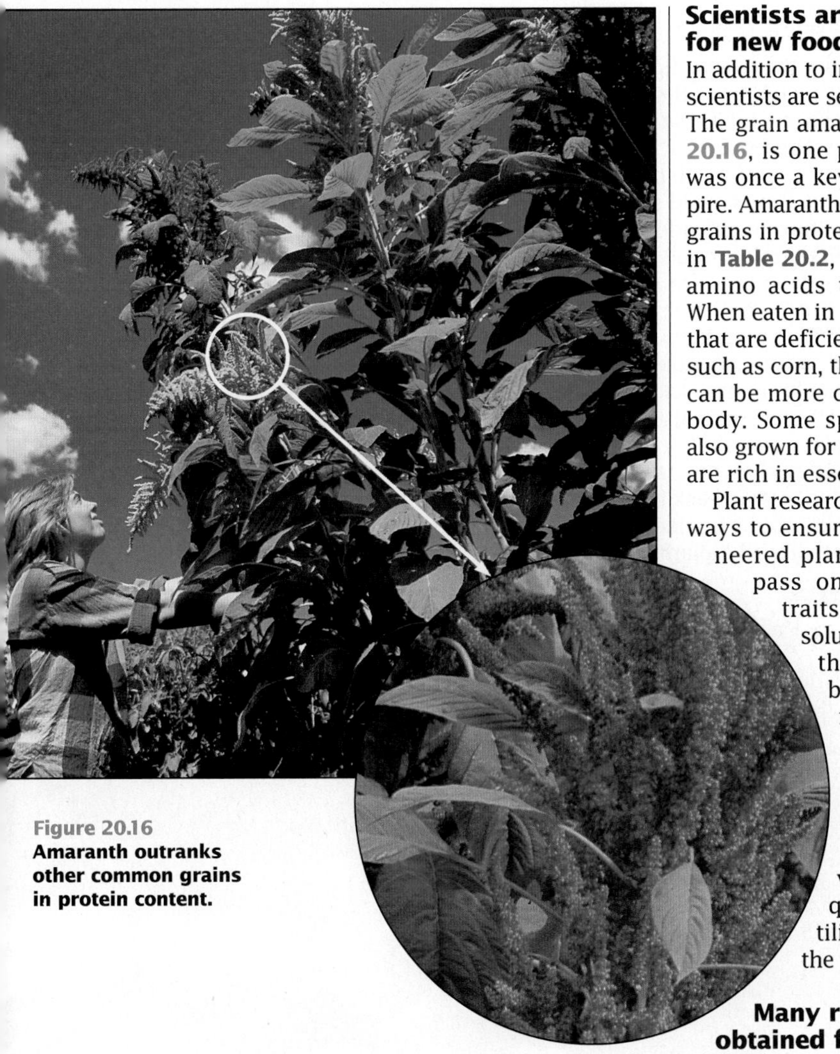

Figure 20.16
Amaranth outranks other common grains in protein content.

Scientists are also searching for new food sources

In addition to improving existing crops, scientists are seeking new food sources. The grain amaranth, shown in **Figure 20.16**, is one promising discovery. It was once a key food of the Aztec empire. Amaranth outranks other common grains in protein content, as indicated in **Table 20.2**, and contains important amino acids that most grains lack. When eaten in combination with grains that are deficient in some amino acids, such as corn, the protein in both grains can be more completely used by the body. Some species of amaranth are also grown for their outer leaves, which are rich in essential nutrients.

Plant researchers are also looking for ways to ensure that genetically engineered plant varieties are able to pass on their new, improved traits to their offspring. One solution is artificial seeds. In this technique, plant embryos are encased in artificial seed coats. These "engineered seeds" produce plants of very high quality. Gene technologists are also experimenting with new crop varieties that do not require special types of fertilizers or that will grow in the desert without irrigation.

Table 20.2 Protein, Carbohydrate and Fat Content

Source (1 oz.)	Protein (g)	Carbohydrate (g)	Total Fat (g)
Grain amaranth	4.1	18.87	1.8
Corn flour	2.0	21.8	1.1
Oat flour	3.5	21.5	0.5
Barley	3.5	20.8	0.7
Whole wheat flour	3.9	20.6	0.5

Many resources can be obtained from desert plants

Substances produced by certain desert plants can be used as substitutes for scarce natural resources. Often these plants are easier to obtain than the natural resources they replace. For example, the sperm whale, now an endangered species, once supplied a very high-grade oil used in cosmetics, lubricating oils, and floor wax. Scientists have found, however, that oil from the desert shrub jojoba (*hoh HOH buh*) has many of the same properties possessed by sperm whale oil. Guayule (*gwah YOO lee*), another desert plant, produces a substitute for natural rubber.

Visual Strategy

Figure 20.16
Explain that amaranth provides abundant food and shelter for winter birds. Young growth can be cooked like spinach and mixed with strong-flavored greens for human consumption. The small, shining fruits were parched or ground and used by Indians as food. The young plants can be eaten raw as a salad. Have students research amaranth and its uses, both old and new.

▪ Cultural Perspective ▪

The Quighao Plant
The Chinese have been using the Quighao plant, *Artemisia annua*, in traditional medicine for more than 2,000 years to treat chills and fever associated with malaria. In 1972, Chinese scientists isolated artemisinin, the active ingredient that fights the malaria parasite, from the Quighao plant. As a result, 2 million doses were administered in Chinese clinics confirming its effectiveness. The World Health Organization Program for Research and Training in Tropical Diseases has been working with Chinese scientists, as well as with the Walter Reed Army Institute of Research, to complete the preclinical trials. The first clinical trials for the drug artemether are scheduled to start in the near future in Kenya, Nigeria, Malawi, Papua New Guinea, Thailand, and Vietnam.

Demonstration 1
Hydroponics vs. Soil-Grown Fruits and Vegetables
Bring in a number of pieces of the same fruits and vegetables that were grown in both media. Have students compare external characteristics, as well as taste and internal structures.

History Connection

Plant Use
Research the historical importance and use of jojoba (*Simmondsia chinensis*), tepary bean (*Phaseolus acutifolius*), and amaranth (*Amaranthus hypochondriacus*).

Phase 3
ASSESSMENT OPTIONS

Closure Strategy
Space Ecosystems
Have students design their own space-based, self-contained ecosystem for the mission or planet of their choice. Students' designs should be realistic and take into account all pertinent environmental factors. Are the designs dependent on technological advancements yet to come, or are they feasible with today's technology? Are the designs indefinitely self-supporting, or are there problems that would limit their life span?

Section Review
Assign the *Section Review*.

Reteaching
Have students draw pictures of at least one of the plant uses and techniques presented in this section.

Growing Plants Without Soil

Figure 20.17
These lettuce plants are shown growing in a chamber of a working model of the "salad machine," which contains nutrient-rich water. Many garden variety plants can be grown this way. NASA hopes that a machine like this one will be able to supply a crew of four with three salads per week.

During the twenty-first century, people may spend months and even years in space. Technology must provide ways to supply air, water, food, and warmth, and to dispose of or recycle wastes. Plants could serve several functions during extended space missions. For example, much of the space and energy required for storing and carrying foods might be saved by raising food crops while in space. Also, fresh food would no doubt be a welcome change from prepackaged, freeze-dried fare. Plants could also purify the spaceship's atmosphere, taking up carbon dioxide exhaled by astronauts and releasing oxygen back into the air. Human wastes could be used to fertilize the plants. Finally, plants could provide a psychological benefit by surrounding people with living reminders of Earth. Plants could create a pleasant, natural oasis in the high-technology environment of a space station.

To solve the problem of growing plants in such confined areas and in very arid regions of the world, scientists are using a technology called **hydroponics**. Hydroponics is the growing of plants in nutrient solutions instead of soil. Plants are grown in containers filled with water or with coarse sand, gravel, or other materials to which nutrients have been added, as shown in **Figure 20.17**. Although it is expensive, hydroponics can be an effective way to raise certain crops in places where good farmland is scarce.

Section Review

① Describe the benefits of planting soybeans for one season in a field where corn is usually grown.

② Explain how the Green Revolution increased crop yields.

③ How could genetic engineering improve crop plants?

④ In what areas do you think hydroponics might be used?

■ *Section Review Answers* ■

1. Soybeans, and other legumes, contain nitrogen-fixing bacteria that convert nitrogen in the atmosphere into a form that plants can use. The nitrogen that remains in the soil provide nutrients for the next grain crop.

2. Dr. Borlaug conducted extensive plant breeding programs, from which was developed a dwarf wheat variety that grew quickly, had high yields, and was resistant to disease.

3. Genetic engineering could improve the protein content of some crops.

4. Students should suggest arid regions or places where good farmland is scarce.

Chapter **20** *Highlights*

Wheat, rice, and corn are eaten all over the world by many different peoples.

Alternative Assessment

Have students show the interrelationships between the three sections in this chapter by doing a brief concept map showing only those relationships.

	Key Terms	*Summary*
20.1 Plants as Food	grain (p. 427) legume (p. 432) tuber (p. 432)	• The most important foods are fruits called grains. • Legumes, which are high in protein, provide a valuable supplement to grains. • Other plant parts such as stems, roots, and leaves are also used as food. • Sugar is produced from sugar beets and sugar cane.
 Eighty-five percent of human food comes from three grains: wheat (above), rice, and corn.		
20.2 Other Uses for Plants	wood (p. 436)	• Wood is a valuable resource that is used for construction and fuel and for the manufacture of paper, rayon, and other items. • Many drugs have been isolated from plants, including birth control pills, cortisone, digitalis, and reserpine. The drugs vincristine and taxol are also produced from plants. • Other important plant products include turpentine, natural rubber, and fibers. Ornamental plants are used for homes and gardens.
This violin is one of the many products made from wood.		
20.3 Plant Use in the Future	crop rotation (p. 441) **Green Revolution** (p. 442) hydroponics (p. 444)	• Genetic engineers are striving to improve the nutritional value of crops. • New foods will result from the rediscovery of ancient ones and from the development of new varieties. • The Green Revolution was an intensive plant-breeding program that introduced high-yield crops into agriculturally poor regions. • Some desert plants produce substitutes for scarce natural products. • Hydroponics makes it possible to grow plants in areas where farmland is unavailable.
Growing plants on a space station may have to be done without soil. 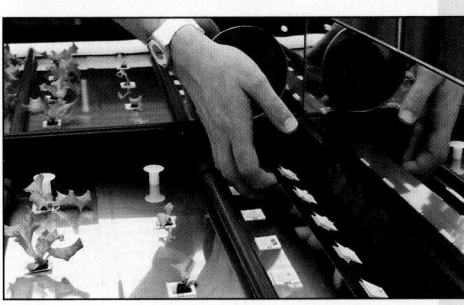		

Chapter Review Answers

Understanding Vocabulary

1. **a.** Sugar cane does not fit the pattern, since it is not a grain.
   b. Morphine is not a wood product and thus does not fit the pattern.
   c. Turpentine is not a medicine obtained from a plant.
   d. A peach is a true fruit and would not be considered a vegetable.

Relating Concepts

2. Map answer is shown on page 425D.

Understanding Concepts

Multiple Choice

3.	c	8.	b
4.	b	9.	c
5.	a	10.	a
6.	c	11.	d
7.	c	12.	d

Completion

13. sugar cane, sugar beets
14. taxol
15. lumber
16. flax

Short Answer

17. Legumes are the seeds of the bean and pea family. Legumes are important to the diets of many people because they are high in protein.
18. The sauces are added to flavor the rice and because they contain essential vitamins absent from rice.
19. A tangled fiber mixture is then pressed and dried.
20. Disinfectants, insecticides, medicines, and perfumes are made from turpentine.

Interpreting Graphics

21. *a* is a kernel of corn, a fruit.
    b is a cassava, a root.
    c is spinach, leaves
    d shows yams, which are roots

Chapter 20 Review

Understanding Vocabulary

1. For each set of terms, choose the term that does not fit the pattern and explain why it does not fit.
   a. corn, wheat, sugar cane, rice
   b. lumber, morphine, fuel wood, paper
   c. quinine, ephedrine, aspirin, turpentine
   d. peach, turnip, spinach, celery

Relating Concepts

2. Copy the unfinished concept map below onto a sheet of paper. Then complete the concept map by writing the correct word or phrase in each oval containing a question mark.

Understanding Concepts

Multiple Choice

3. Pasta, breads, and tabbouleh are made from
   a. rice.
   b. tofu.
   c. wheat.
   d. tubers.

4. Which grain has the highest protein content?
   a. barley
   b. grain amaranth
   c. corn
   d. wheat

5. The starchy portion of the wheat kernel is the
   a. endosperm.
   b. husk.
   c. bran.
   d. germ.

6. A root crop that provides more than one-third of the calories consumed by people in Africa is the
   a. potato.
   b. carrot.
   c. cassava.
   d. turnip.

7. The active ingredient in aspirin was first obtained from
   a. sugar beets.
   b. yew trees.
   c. white willow leaves.
   d. cotton plant roots.

8. The drug vincristine is used to treat
   a. malaria.
   b. childhood leukemia.
   c. cancer.
   d. AIDS.

9. What is the world's most important plant fiber?
   a. silk
   b. polyester
   c. cotton
   d. rubber

10. Most of the natural rubber harvested today is grown in
    a. Asia.
    b. South and Central America.
    c. North America.
    d. Africa.

11. Fertilizing and crop rotation are ways to
    a. improve the nutritional value of crops.
    b. stop fungal diseases.
    c. eliminate irrigation.
    d. increase crop yields.

12. What is the accomplishment for which Dr. Borlaug received a Nobel Peace Prize in 1970?
    a. finding new uses for jojoba
    b. using genetic engineering to develop new crop varieties
    c. developing hearty varieties of corn
    d. starting the Green Revolution

Completion

13. Sources of table sugar include _____ and _____ .

14. The drug _____ is often obtained from the bark of yew trees.

15. Wood that has been sawed into boards for construction is called _____ .

16. A soft fiber known as linen comes from the stems of _____ plants.

Short Answer

17. What are legumes? Why are legumes important to the diets of many of the world's people?

18. What are the nutritional advantages of adding soy sauce and fish sauce to meals consisting of rice?

Reviewing Themes

22. In early times, kernels from the best plants were used as seed for the next planting. Selective breeding has been used recently, and techniques of genetic engineering are currently being tried to improve wheat.

23. Farmers and agricultural scientists are focused on producing more food for the world's growing population. Farmers are using fertilizers, rotating crops, and planting improved crop varieties. Agricultural scientists are improving the protein content of food through genetic engineering and breeding programs. Also, they are working with artificial seeds and experimenting with hydroponics.

19. Explain how paper is made.
20. What products are made from turpentine?

Interpreting Graphics

21. Look at the pictures presented below. Each picture shows an edible plant part. Identify the plant part as a fruit, stem, root, or leaves and give its common name.

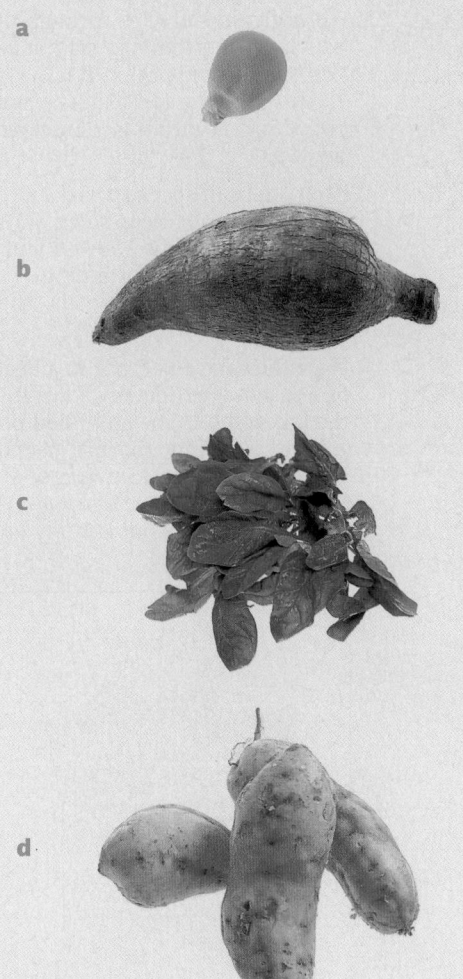

a

b

c

d

Reviewing Themes

22. *Patterns of Change*
    Modern wheat looks very different from its wild ancestor. What methods have been used to develop the robust varieties of wheat grown today?

23. *Stability*
    Describe several ways that farmers and agricultural scientists are striving to maintain adequate food supplies for growing populations.

Thinking Critically

24. *Inferring Conclusions*
    Why is tofu a popular food among people who are vegetarians?

25. *Comparing and Contrasting*
    What is hydroponics? What are two advantages of hydroponics?

26. *Building on What You Have Learned*
    In Chapter 14 you learned about several environmental problems. What environmental problems are associated with the burning of wood for cooking and heating?

Cross-Discipline Connection

27. Biology and Health
    Do library research to discover the difference between white flour and whole wheat flour. Why are products made with whole wheat flour more healthful than those made with white flour?

Discovering Through Reading

28. Read the article "Death and Taxus," in *Natural History*, September 1992, pages 20–23. Why did Nicholson's expedition to Mexico collect stems and leaves from different yew trees?

Cross-Discipline Connection
27. Students should suggest that whole wheat flour is more healthful because it requires less processing than white flour.

Discovering Through Reading
28. They need to know which species of yew held the most taxol, when it produced the substance, and whether or not the plant could be manipulated to increase its yield of the drug.

Thinking Critically
24. The diets of vegetarians lack the proteins provided by meats. Tofu, a product made from soybeans supplies the proteins needed by vegetarians.

25. Hydroponics is a method of growing plants without soil. Advantages are that plants can be grown in places where soil is not available, and the needs of plants can be studied because the nutrients provided can be carefully regulated.

26. The cutting of large numbers of trees leaves the ground barren and susceptible to erosion. Also, the burning of wood produces ash and gases such as carbon dioxide that pollute the atmosphere.

Procedural Note

1. Review the answers to the prelab questions to emphasize the role of soil in plant growth.
2. In designing their own experiments, many students mistakenly believe that the goal is to confirm their initial hypothesis. Be sure students realize that their goal should be valid testing and a correct conclusion regarding the hypothesis, and not necessarily a correct hypothesis to begin with.

Prelab Preparation Answers

- Students should suggest a source of sunlight, water, air, and warmth.
- Soil is the source of water, minerals, and other nutrient compounds necessary for plant growth. Soil also provides a physical medium that holds the plant erect. Plant roots obtain oxygen from air pockets in the soil. Fertilizers provide nutrients.
- Hydroponics is a technique in which plants are grown in a nutrient solution instead of soil.

Procedure Answer

5. Light intensity, temperature, and moisture should be held constant for both pots during the experiment. The physical medium in which the plant grows is the independent variable being tested.

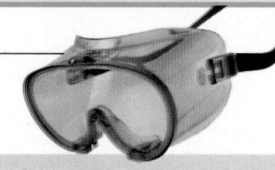

Chapter 20 Investigation

Can Plants Be Grown Without Soil?

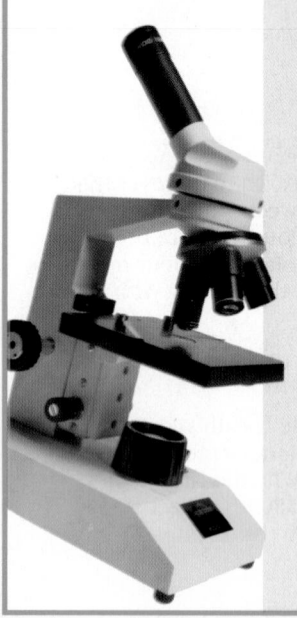

Objectives

In this investigation you will:
- *design* a control experiment
- *compare* the growth of plants cultivated using hydroponic methods with those grown in soil

Materials

- bean seeds
- two small clay pots
- potting soil
- glass-wool wick
- sand
- beaker
- commercial water-soluble fertilizer
- centimeter ruler

Prelab Preparation

Review what you have learned about plants by answering the following questions:
- What do plants need to grow?
- How do soil and fertilizer help a plant grow?
- How does growing plants by hydroponics differ from growing plants in soil?

Procedure: Growing Plants Without Soil

1. Form a cooperative group of four students. Work with a member of your group to complete steps 2–7.

2. After discussing with your partner the problem that is the topic of this investigation, state your hypothesis. Explain why you chose this hypothesis.

3. Pull a glass-wool wick through the hole in the bottom of a clay pot. Add sand to the pot until it is 2.5 cm from the top. Fill a second pot with soil until it is 2.5 cm from the top.

4. Plant six bean seeds in each pot. Water both pots. Keep the planting medium moist (but not wet) and allow the seeds to germinate. Record the number of seeds that germinate in each pot.

5. Place the sand-filled pot in a beaker of nutrient solution as shown in the illustration. The soil-filled pot will be treated in a normal fashion. *What factors should be held constant for both pots during the experiment? What is being tested in this experiment?*

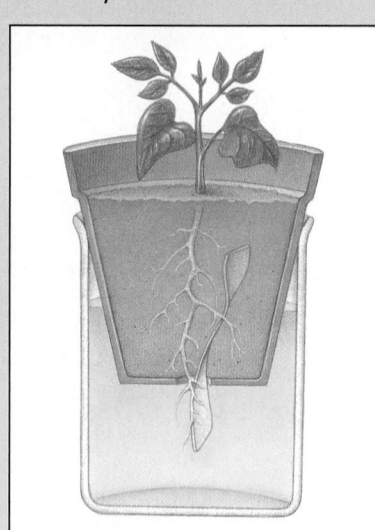

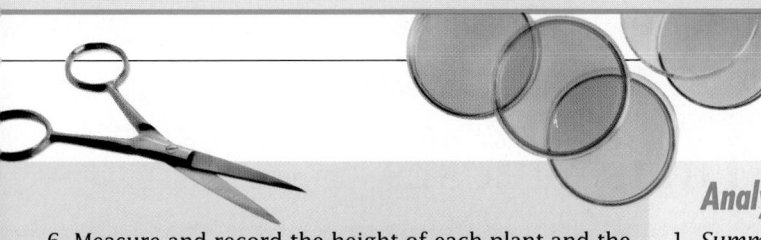

6. Measure and record the height of each plant and the number of leaves of each plant at weekly intervals as directed by your teacher. Record this information each week in a table similar to the one shown below.

Pot 1:

Soil	Week#	1	2	3	4
Average Height					
Average Number of Leaves					
Average Leaf Width					

Pot 2:

Sand	Week#	1	2	3	4
Average Height					
Average Number of Leaves					
Average Leaf Width					

7. Calculate the average height, average number of leaves, and average leaf width for the plants from each pot on the same days that height and leaf number are measured. Record the differences between the seedlings in the two pots. Record your data in your table.

8. At the end of the observation period, make a bar graph that compares the germination rates observed in the two pots. Make line graphs that compare average height, average number of leaves, and average leaf width of the seedlings in each pot at weekly intervals.

Analysis

1. *Summarizing Data*
   Summarize the data collected during the experiment.

2. *Analyzing Data*
   Do you observe a similar pattern between the two pots for height, leaf number, and leaf width?

3. *Analyzing Data*
   Do the data support your hypothesis? Explain your answer.

4. *Making Inferences*
   State the relationships you observed between rates of growth and the method of cultivation of each of the plants in the pot.

Thinking Critically

What are some advantages of growing plants using hydroponics? What are some possible disadvantages?

Analysis Answers

1. Summaries will vary depending on data collected.
2. Answers will vary but should be consistent with data collected.
3. Answers will vary depending on the data collected. Seedlings grown under hydroponic conditions should grow as well or better than those grown in soil.
4. Students should suggest that plants are likely to grow best under conditions that provide an optional amount of nutritional substances.

Thinking Critically Answer

Possible advantages might include: plants can be grown on spaceships, in arid regions and areas where good soil is not available. One possible disadvantage might be the expense.

Using the Feature

- Have students research and write a report on present-day scientists, for example, scientists at the Centers for Disease Control in Atlanta, who are searching for preventions or cures for viral diseases, and the precautions they take to ensure that they do not become contaminated with the viruses they work with.

- Have students choose a scientist from the timeline, or a viral disease not discussed in the timeline, to research. Have them write a report that discusses the process their scientist followed in identifying and isolating the viral agent with which he or she is associated. Also, have students discuss any advancements being made toward the development or testing of a vaccine for a cure of the viral diseases.

- Have students research and write a report on some viral disease such as polio, that is now preventable. In their report have students discuss the process scientists followed in identifying the cause of the disease, in isolating the virus, in formulating a vaccine, and in testing the vaccine.

Discussion

Guide the discussion by posing the following questions.

1. Did viruses evolve before or after cells? Explain your reasoning.
   They evolved after cells. Students should realize that viruses are parasites that cannot exist without cells.
   Note: *You might suggest that scientists believe that viruses probably formed spontaneously from nonliving organic material or evolved from existing cells.*

2. Explain how AZT might help victims of AIDS.
   AIDS is caused by the retrovirus HIV, which must use reverse transcriptase to reproduce. By blocking the reverse transcriptase, the drug stops the virus from reproducing.

HUMANS AND THEIR VIRUSES

900

Thomas Thatcher

900 Rhazes, a Persian physician, writes the first accurate descriptions of infectious diseases: plague, measles, consumption, tuberculosis, smallpox, and rabies.

1678 British physician **Thomas Thatcher** writes the first medical treatise published in America on smallpox and measles.

1721 Onesimus, an African slave, describes to his American owner the process of inoculation used in Africa for the treatment of smallpox.

1790s English physician **Edward Jenner** observes that milkmaids who have had cowpox rarely contract smallpox. He develops a vaccination against smallpox.

1930

Wendell Stanley

1930s American biochemist **Wendell Stanley** proves that viruses are not living organisms but are chemical matter.

1937 Max Theiler, a South African microbiologist, develops the vaccine for yellow fever.

Max Theiler

1943 Infantile paralysis epidemic kills almost 1,200 in U.S. and cripples thousands more.

1949 African-American physician **Dr. Jane C. Wright** continues the work of her father, **Dr. Louis Tompkins Wright**. They believe cancer stems from viral agents, and they study the effect of various drugs for the treatment of cancer.

1954 American physician **Dr. Jonas Salk** develops poliomyelitis vaccine.

Polio virus

*ience ■ **Discoveries in Science** ■ **Discoveries in Science***

Yellow fever epidemic, Memphis.

1930

Yellow fever virus

1878 Memphis, Tennessee, reels from an outbreak of **yellow fever**. Fifty-two hundred of the 19,600 residents die. Memphis loses its city charter due to the decrease in population.

1881 Cuban physician **Carlos Finlay** suggests that yellow fever is transmitted by the bite of the common household mosquito.

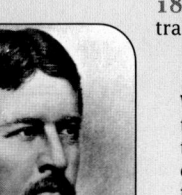

1900 U.S. Army physician **Walter Reed** establishes that the bite of certain mosquitoes transmits yellow fever. The experiments also show how the fever might be controlled.

1901 Cuban physician **Juan Guiteras** verifies the cause of yellow fever independently of Walter Reed.

1927 The **yellow fever virus** is isolated.

Walter Reed

1966 American **Francis Peyton Rous** receives the Nobel Prize in medicine and physiology for discovery of a cancer virus.

PRESENT

1969 Americans **M. Delbrück, A. D. Hershey**, and **S. E. Luria** win Nobel Prizes for discovering the genetic structure of viruses.

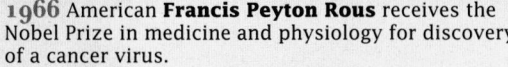
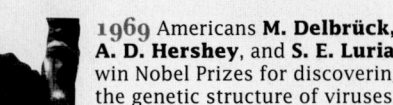

1978 Smallpox virus is now extinct in the wild. Scientists debate whether the only smallpox virus in existence (in two high-security laboratories) should be eliminated. The world's last known case of naturally occurring smallpox was reported in Ali Maow Maalin of Merka, Somalia, in 1977.

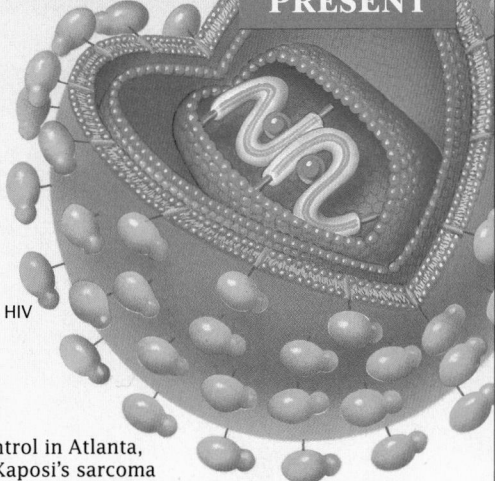

HIV

1981 The Centers for Disease Control in Atlanta, Georgia, discovers in its study of Kaposi's sarcoma more than 500 cases of a mysterious disease that knocks out the immune system.

1983 HIV, the virus that causes **AIDS**, is identified.

Ali Maow Maalin

1980s - present Antiviral drug AZT slows the reproduction of HIV.

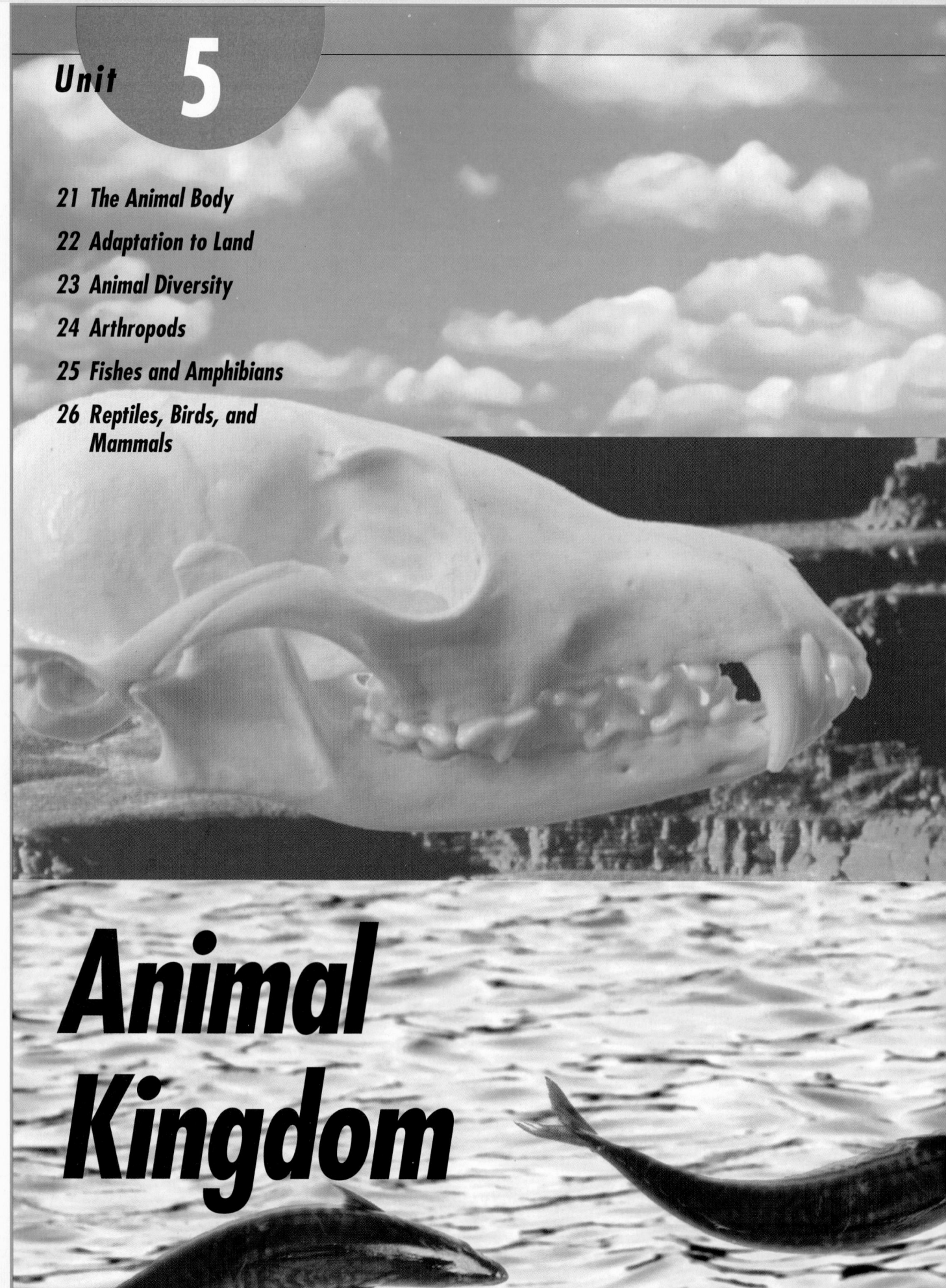

Unit 5

Animal Kingdom

You are an animal, and share a common heritage with earthworms and dinosaurs, butterflies and sea stars. It is no accident that the fingers of your hand have bones like those in a bird's wing. In this unit you will discover how the animal body has been shaped by its long evolutionary journey, from the simplest sponge to worms, insects, and vertebrates. Evolution has molded animals to suit many ways of living, often altering their design to take advantage of new opportunities in the environment.

Chapter 21 The Animal Body

Planning Guide

Objectives/Themes	Classwork Resources	Homework Resources

21.1 1. List three characteristics of animals.
 	
2. Recognize the advantage of multicellularity.
 	
3. Identify the difference between tissues and specialized cells.
 	
4. List three features found in cnidarians but not in sponges.
 	
Themes: Evolution, Interacting Systems, Scale and Structure
	Text
	Science in Action
	   Alan Shipley: Research Associate pp. 460–461
	
	Teacher's Resource Binder
	Focus Activity 21
	   Recognizing Patterns of Symmetry
	
	Other Resources
	Transparencies 84–87
		Text
		Section Review, p. 459
		
		Teacher's Resource Binder
		Directed Reading Worksheet 21.1
		
		Other Resources
		Audiocassette 21.1*

21.2 1. Contrast the body plans of bilaterally and radially symmetrical animals.
 	
2. List three structures derived from mesoderm.
 	
3. Describe the advantage of a one-way gut.
 	
4. Contrast the three kinds of body cavities.
 	
Themes: Evolution, Patterns of Change, Scale and Structure
	Teacher's Resource Binder
	Science Skills Worksheet
	   The Animal Body
	Lab Investigation 21.1
	   Roundworms and Earthworms
	
	Other Resources
	Transparencies 88–92
	Science in Action Video
	   *Alternatives to Dissection: The Earthworm and the Frog**
		Text
		Section Review, p. 467
		
		Teacher's Resource Binder
		Directed Reading Worksheet 21.2
		
		Other Resources
		Audiocassette 21.2*

21.3 1. Contrast the body plans of segmented and nonsegmented animals.
 	
2. Compare the exoskeleton of arthropods with the endoskeleton of vertebrates.
 	
3. Summarize the differences between protostomes and deuterostomes.
 	
4. List two traits that reveal the relationship between chordates and echinoderms.
 	
Themes: Evolution, Patterns of Change, Scale and Structure
	Text
	Investigation
	   How Does Segmentation Help an Earthworm Move? pp. 478–479
	
	Other Resources
	Transparencies 93–97
		Text
		Section Review, p. 474
		
		Teacher's Resource Binder
		Directed Reading Worksheet 21.3
		Vocabulary Review Worksheet*
		Reaching Worksheet*
		   Defining Taxonomic Traits
		
		Other Resources
		Audiocassette 21.3*

*Reteaching Options

Demonstrations
21.1: p. 457
21.2: pp. 462, 466
21.3: pp. 469, 470, 472, 473, 474

Assessment
Chapter Review pp. 476–477
Portfolio Assessment p. 453D
Chapter Test—Teacher's Resource Binder
Test Generator

Research Notes

Connections to Developmental Biology: Clues From a Roundworm

How does a single fertilized egg multiply to form a complex organism with a specialized body plan? Scientists investigating this question are using the nematode Caenorhabditis elegans *(see "Exploration of a Roundworm," page 465)* to better understand the development of this creature's body plan. These studies may also provide insights into the history and purpose of different structures and the advantages of a given body plan.

A fully grown C. elegans *contains 1,090 cells. Scientists at the Massachusetts Institute of Technology have completed a "fate map" for the organism that is something like a "family tree" for these cells, showing where each one came from and how it is related to the others.*

With this map, scientists are able to observe which cells are disrupted by genetic mutations, and what abnormalities in behavior and body plan are caused by these changes. For example, a certain mutation in the ced-9 gene can cause many cells to die before the animal reaches maturity. As a result, the animal is missing several different motor neurons, causing body motion to be very uncoordinated. Such mutants are also frequently infertile, due to incomplete development of eggs and the egg-laying mechanism.

Scientists frequently devote great efforts to understanding a lot about very simple things, in order to apply what they learn to more complicated systems and situations. C. elegans *is a good example. Studies like this one could give researchers some of the tools they need to tackle human diseases that have genetic and developmental bases.*

In addition, as part of the Human Genome Project, researchers will attempt to map all of the genes found in this animal. This project, which involves only six pairs of chromosomes and about 10,000 genes, will serve as a "test case" for the handling of the huge volumes of data to be generated in the mapping of the much larger human genome of 23 pairs of chromosomes and more than 60,000 genes.

Investigation Notes

How Does Segmentation Help an Earthworm Move?

pages 478–479

Purpose: Segmentation is a significant advance in the evolution of animals. This investigation gives the students an opportunity to observe earthworm locomotion and to relate it to the earthworm's segmented structure.

Prelab Preparation

1. Live earthworms can be obtained from a local bait store as well as from WARD'S (87 M 4660).

2. After use, earthworms may be released into any area with loosened soil.

3. For the simulation of earthworm movement, use long, thin balloons. Add a small amount of water to each balloon (balloons should be less than half full), and then tie the ends in a knot. Tie two balloons together to form a short chain. This simple model works better with water than with air.

Answers will be found on pages 478–479.

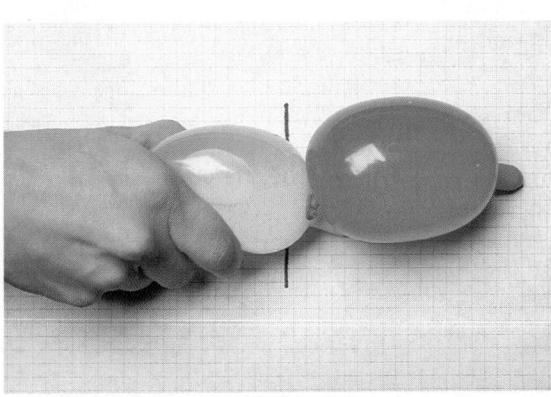

Meeting Individual Needs

Objectives

1. Students will demonstrate appropriate use of core vocabulary for the chapter (Vocabulary File).

2. Students will relate the development of body cavities using an outline and visual diagram (Teaching Strategy A).

3. Students will describe the body structure of a specific animal and relate it to the animal's lifestyle (Teaching Strategy B).

Vocabulary File

(Developing Vocabulary/ Limited English Proficiency)

If you are not already using the Vocabulary File, refer to Chapter 1 for its preparation. See the Chapter Highlights on page 475 for a list of suggested words

Teaching Strategy A

Body Cavities
for pages 455–469
(Developing Organizational Skills/Visual Learners)

Materials: overhead projector, transparency, transparency pens

Procedure: Introduce the unit by reading aloud to students the statement on page 455. Emphasize that the body plan of an animal is its overall structure—the way its parts fit together. Have students read pages 459–463 in order to gather information for an outline.

Using Figure 21.7 on page 463 as a model, sketch the developing flatworm embryo on the right side of the overhead. Have students copy these. Color cells of ectoderm blue, mesoderm red, and endoderm yellow.

Begin a discussion about development and gastrulation, asking students to contribute information for the outline. To the left of the sketches write the outline as topics are mentioned. Use the diagrams to explain each topic. Students should copy the outline

from the overhead projector, and keep it for future reference.

The outline should include at least the following main points with appropriate details:

I. Gastrulation
    A. Ectoderm
    B. Endoderm
    C. Mesoderm

II. Body Symmetry
    A. None
    B. Radial
    C. Other

III. Body Cavity Classifications
    A. Acoelomate
    B. Pseudocoelomate
    C. Coelomate

Teaching Strategy B

Body Plan and Lifestyle
for pages 468–474
(Developing Classification Skills/Visual Learners)

Materials: Computer printout banner, animal pictures for mobiles, index cards (one for every two students), reference works such as a wildlife encyclopedia or field guides for the types of animals studied

Preparation: Using a computer, print out a banner that reads "Welcome to the Animal Kingdom." Tape this over the doorway or entrance to the classroom.

Cut out pictures of animals using biological supply house ads, catalogs, and posters. Mount them on poster board. Hang them as mobiles from the ceiling of the room. Include the following animals: sponge, hydra, liver fluke, planarian, ascarid, snail, earthworm, wasp, starfish, and lancelet. This will be the start

of a "sky zoo" of animal pictures that you will be able to use and refer to throughout this unit.

Write the following instructions on each index card:

(1) Describe my lifestyle.
(2) Name my type of body cavity.
(3) Name my type of symmetry.
(4) Determine whether I am segmented.
(5) Explain how my body plan relates to my lifestyle.

Procedure: Assign students to work in groups of two. Assign an animal to each group. Give each group an index card. Allow one day for research using text and other references.

On the second day, each group should take the picture of their animal off the mobile and present the facts about their animal to the class.

Additional Strategies

Visual Strategies

Pages 455, 457, 459, 462, 463, 464, 465, 466, 467, 468, 469, 470, 472, and 474

Auditory Learners

Use *Biology: Visualizing Life* Audiocassettes for Sections 21.1, 21.2, and 21.3.

Meeting Individual Needs (cont.)

Cooperative Learning
Body Plans

Timing: Use this activity to introduce Section 21.1.

Group Size: 5 students

Outcome: Students will be able to describe the advantages of different developments in animal body plans.

Individual Accountability: Each group member is responsible for information on one stage of development of animal body plans.

Positive Interdependence: Each group will complete a timeline listing each stage of development with its adaptations.

Assign one of the following innovations in body plan to each group member: (1) Body Cavity; (2) Internal Organs, Bilateral Symmetry, and Cephalization; (3) Coelom; (4) Multicellularity; and (5) Radial Symmetry, Extracellular Digestion, and Specialized Tissues. (Note: these developments are purposely described out of chronological order.) Each group member should scan the chapter for information about the benefits and implications of these innovations in body plan of development and report the information to the other members of the group.

The group should then construct a timeline that puts the stages in order and summarizes their main points.

Portfolio Assessment

Students should select their best work and provide a self-reflective rationale for their selections. Students can make selections in the following areas.

1. *Content* — One concept map from the chapter (See page 812 for evaluation criteria.)

2. *Reading Comprehension* — One Directed Reading Worksheet from the Teacher's Resource Binder (Use the answer key to evaluate for accuracy.)

3. *Writing* — Using the Vee Form, summarize a magazine or newspaper article relating to body plans or to the animals discussed, especially sponges, or worms.

   Or: Select a writing project from the Chapter Review.

4. *Performance Assessment* — One Vee form from a chapter investigation or lab manual investigation (See page 22T for evaluation criteria.)

Teacher makes selections in the following areas.

1. *Formal Assessment* — Chapter test (Test A, B, or the Test Generator) The teacher-scored test should be reviewed by the student. Incorrect responses should be corrected by the student before the test becomes part of the portfolio.

2. *Informal Assessment* — Use the Direct Observations Checklist, page 33T, during a laboratory or other cooperative learning experience.

3. *Performance Assessment* — Have students use a computer graphics program or other means of animation to create a short "movie" of the development of a simple animal.

Concept Map Answer

The following is one possible answer to the Relating Concepts exercise on page 476.

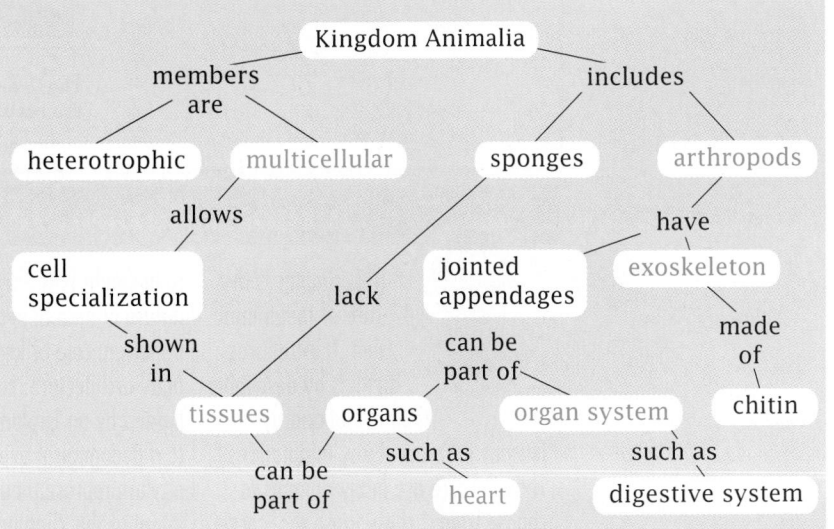

Chapter 21

Determining Prior Knowledge

- Have students give examples of the various methods animals use to get food.
- Display a natural sponge and a piece of coral. Have students explain what they know about these two materials, where they come from, and how they are formed. Ask students what a sponge is. Ask them to note similarities between the sponge and the coral, such as the irregular form, and the softness and hardness.
- Display photographs of a wide variety of the following animals, both invertebrates and vertebrates. Show animals that vary considerably in appearance. Have students identify characteristics these organisms may have in common.
- Display several of each kind of preserved specimens: flatworms, roundworms, segmented worms, mollusks, echinoderms, and primitive chordates. Have students group those with similar characteristics.
- Ask students what comes to mind when they think of worms. Have them list their responses on the chalkboard. Lead them to identify bilateral symmetry and cephalization.

Chapter 21

The Animal Body

Review

- **natural selection** (Section 9.3)
- **animal evolution** (Sections 10.3 and 10.4)
- **classifying living things** (Section 15.2)

21.1 The Advent of Tissues

- **Animal Body Plans**
- **Many Cells Are Better Than One: Sponges**
- **Tissues Enable Greater Cell Specialization: Cnidarians**
- **Regularly Arranged Animals**

21.2 Origin of Body Cavities

- **Heading Toward Complexity: Flatworms**
- **A One-Way Gut and a Body Cavity: Roundworms**
- **A Better Body Cavity: Mollusks**

21.3 Four Innovations in Body Plan

- **Segmented Worms: Annelids**
- **Limbs and Skeletons: Arthropods**
- **An Embryonic Revolution**
- **Echinoderms**
- **The Most Successful Deuterostomes: Chordates**

These arctic fox cubs are vertebrates. Their body plan is the result of more than 3 billion years of evolution.

▪ Author's Rationale ▪

This chapter is the core of the animal unit. It introduces animals by describing their evolutionary journey in terms of key adaptations. Though there are many animals to choose from, I chose those that serve to highlight both the progressive nature of animal evolution and the important role of key elements in body architecture. Each stage is highlighted by an Exploration and is linked to a pedagogical overview of the phylum representing the stage. The intent of this chapter is to minimize detail and to emphasize the evolutionary processes that have produced the diversity we see today.

From microscopic worms to blue whales, animals occur in a great variety of sizes and shapes. This chapter describes the evolutionary journey leading to today's great diversity of animals, in terms of a series of key adaptations in body architecture. These adaptations reveal both the progressive nature of evolution and the importance of body design.

21.1 The Advent of Tissues

Objectives

① **List three characteristics of animals.**

② **Recognize the advantage of multicellularity.**

③ **Identify the difference between tissues and specialized cells.**

④ **List three features found in cnidarians but not in sponges.**

Animal Body Plans

Figure 21.1
These animals show just a part of the diversity of the kingdom Animalia.

What do the animals in **Figure 21.1** have in common? These animals, and all the other members of the kingdom Animalia, share three characteristics. First, all animals are heterotrophs that ingest their food, digesting it within the body. Second, all animals are multicellular. Third, as you learned in Chapter 3, animal cells lack cell walls.

Despite these similarities, it is obvious that the animals below differ in shape and structure. The **body plan** of an animal is its overall structure, the way its parts fit together. You learned in Chapter 15 that animals evolved from heterotrophic protists. The first animals probably evolved from colonies of protist-like cells. How could

Lesson Plan 21.1

Phase 1
PREPARATION

Key Concepts
- The kingdom Animalia shares three characteristics. All animals are heterotrophic, multicellular, and lack cell walls.
- The most important advantage to multicellularity is the specialization of cells with a division of labor.
- Tissues are the first great innovation in the animal body plan.
- Most animal species exhibit some type of symmetry.

Reading Strategy

Have students use the most important concepts from reading this section and make three concept maps entitled: *Animal Body Plans; Sponges;* and *Regularly Arranged Animals.*

Phase 2
TEACHING STRATEGIES

Visual Strategy

Figure 21.1
After students have finished reading and concept-mapping this section, display once again the photographs you displayed during the Determining Prior Knowledge section. Have students group the animals based on body plan.

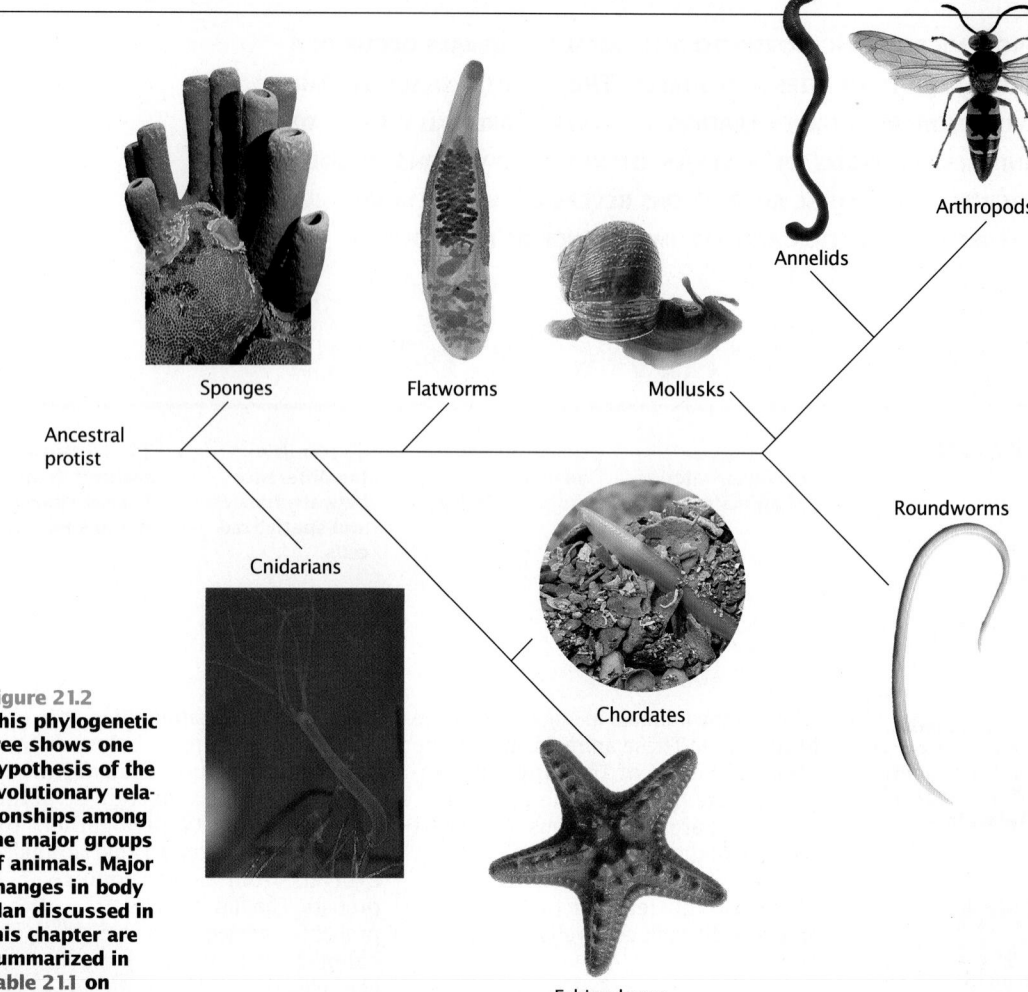

Figure 21.2
This phylogenetic tree shows one hypothesis of the evolutionary relationships among the major groups of animals. Major changes in body plan discussed in this chapter are summarized in Table 21.1 on page 474.

such simple organisms give rise to as complex an animal as you, with a brain, eyes, and 10 toes? The complex body plan of humans did not spring directly from the simple body plans of the earliest animals. Instead, our body plan is the sum of many additions and alterations to the body plans of earlier animals. The driving force for this process has been natural selection, which favors changes in body plan that increase the likelihood of survival and reproduction.

This chapter traces the major evolutionary changes in animal body plan. The phylogenetic tree in **Figure 21.2** illustrates the evolutionary stages discussed in this chapter. Each stage, representing a unique body plan, is profiled in an **Exploration**. Keep in mind that the body plan of an animal determines the lifestyle it leads, the way it functions in its environment. Changes in body plan often result in different body functions and entirely new ways of making a living. Each animal body plan is successful for a certain lifestyle. You might hear simple animals such as sponges described as "lower" or "primitive," implying that they are somehow inferior to "higher" animals such as humans. This is not the case. Although sponges cannot fly or run, they are well adapted for clinging to rocks and filtering food particles from the water. Sponges have been successfully leading this lifestyle for more than 600 million years.

Many Cells Are Better Than One: Sponges

As you learned in Chapter 10, animals belong to one of the three multicellular kingdoms of life. The most important advantage to multicellularity is the specialization of cells. Individual cells of an animal can specialize on a single task such as digestion or reproduction. Cell specialization enables division of labor among cells in an animal's body. Division of labor is advantageous because a specialized cell can carry out its task more effectively than can a cell that must carry out many tasks.

Sponges are simple, multicellular animals

Sponges, like the one shown in **Figure 21.3**, are members of the phylum Porifera. They demonstrate the advantages of division of labor. A sponge extracts food and oxygen from the water flowing through its body. The **Exploration** below shows the simple body plan of a sponge. The outer layer of cells protects the sponge. Cells called choanocytes (*koh AN oh seyts*) line the internal chamber and produce water currents by beating their flagella. Choanocytes engulf and digest small animals and small organic particles that are drawn into the sponge.

Although each of the sponge's different cell types is specialized to carry out the different functions needed for survival, there is very little coordination among cells. The flagella of choanocytes beat independently, for instance. Some sponges can pass through fine silk mesh, each cell slipping through a different opening and then slowly reassembling on the other side. If isolated, any one of a sponge's cells can eventually grow into a new sponge.

Figure 21.3
Sponges are found in fresh water and salt water. All sponges, such as these tube sponges, have a simple body plan composed of specialized cells.

EXPLORATION OF A SPONGE

Stage 1: Multicellularity

The body of a sponge is not symmetrical and is not composed of any organized tissues.

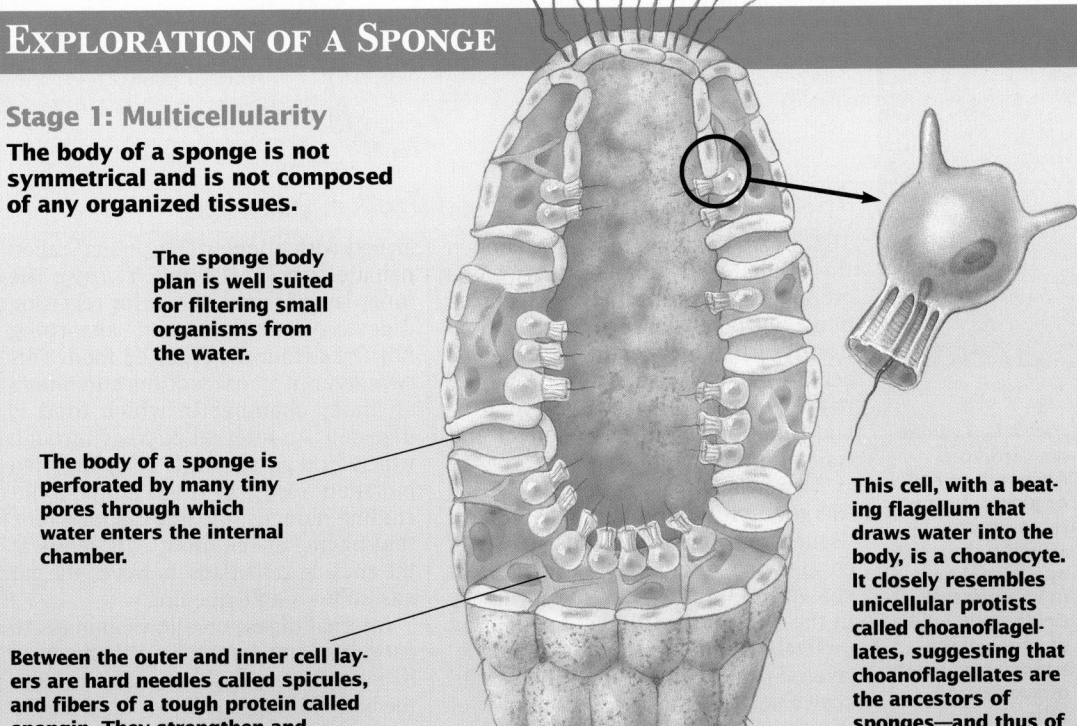

The sponge body plan is well suited for filtering small organisms from the water.

The body of a sponge is perforated by many tiny pores through which water enters the internal chamber.

Between the outer and inner cell layers are hard needles called spicules, and fibers of a tough protein called spongin. They strengthen and protect the sponge. You encounter spongin as bath sponges.

This cell, with a beating flagellum that draws water into the body, is a choanocyte. It closely resembles unicellular protists called choanoflagellates, suggesting that choanoflagellates are the ancestors of sponges—and thus of all animals.

EXPLORATION OF A CNIDARIAN

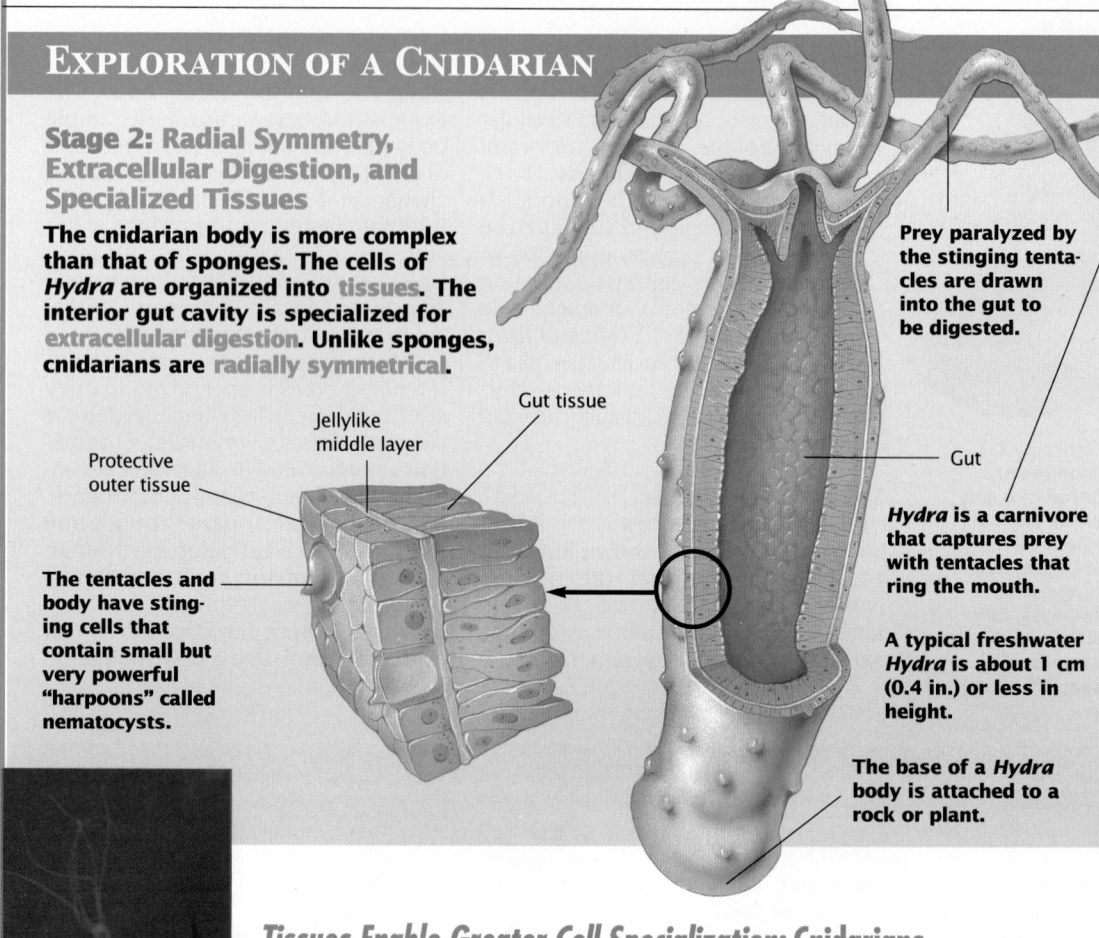

Stage 2: Radial Symmetry, Extracellular Digestion, and Specialized Tissues

The cnidarian body is more complex than that of sponges. The cells of *Hydra* are organized into tissues. The interior gut cavity is specialized for extracellular digestion. Unlike sponges, cnidarians are radially symmetrical.

Jellylike middle layer

Gut tissue

Protective outer tissue

The tentacles and body have stinging cells that contain small but very powerful "harpoons" called nematocysts.

Prey paralyzed by the stinging tentacles are drawn into the gut to be digested.

Gut

Hydra is a carnivore that captures prey with tentacles that ring the mouth.

A typical freshwater *Hydra* is about 1 cm (0.4 in.) or less in height.

The base of a *Hydra* body is attached to a rock or plant.

Figure 21.4
Hydra is a cnidarian carnivore. Anchored to rocks or plants, it captures its prey by stinging them with specialized cells in its tentacles. Cnidarians, unlike sponges, have tissues composed of specialized cells.

Tissues Enable Greater Cell Specialization: Cnidarians

Tissues are the first great innovation in the animal body plan. A **tissue** is a group of similar cells that are organized into a functional unit. Your body is composed of many types of tissues. The activities of cells in a tissue are coordinated. For example, cells in nervous tissue function together to collect and transmit information.

The phylum Cnidaria (*nye DAIR ee uh*) was the first group of animals to have tissues. This phylum includes jellyfish, corals, sea anemones, and hydras, shown in **Figure 21.4**. As you can see in the **Exploration** above, a cnidarian (*ny DAIR ee uhn*) is basically a double-layered bag of cells with a jellylike substance between the layers. Each layer of cells is a tissue. The outer cell layer forms a protective covering for the cnidarian. This outer layer contains cells armed with stinging "harpoons" called nematocysts (*NEHM uh toh sihsts*). The inner layer is specialized for releasing digestive enzymes and absorbing nutrients from the digested food. This two-layered construction surrounds an inner chamber in which food is digested. An internal passage through which food passes while being digested is called a **gut**. In most animals, including humans, the gut is a long tube that begins at the mouth and ends at the anus. In cnidarians, however, the gut has only a single opening.

Internal digestion allows animals to eat organisms larger than themselves by taking pieces of those organisms into the body for digestion. A sponge has no gut and is able to consume only organisms small enough to be absorbed by the cells lining its internal cavity.

Tissue layers form early in the development of the embryo

Figure 21.5 shows how the two-layered construction of a cnidarian is formed during the development of the embryo. This process of forming layers of cells is called **gastrulation** (*gas troo LAY shuhn*). Gastrulation takes place in the developing embryos of all animals except sponges. In humans, gastrulation occurs when the embryo is a little over one week old. The inner layer of cells formed by gastrulation is **endoderm** ("inner skin"), and the outer layer of cells is **ectoderm** ("outer skin").

The layers that result from gastrulation will produce all the tissues of the adult body. These layers develop into the same body parts in all animals. Endoderm gives rise to the lining of the gut, and ectoderm becomes the outer layer of skin and the nervous system.

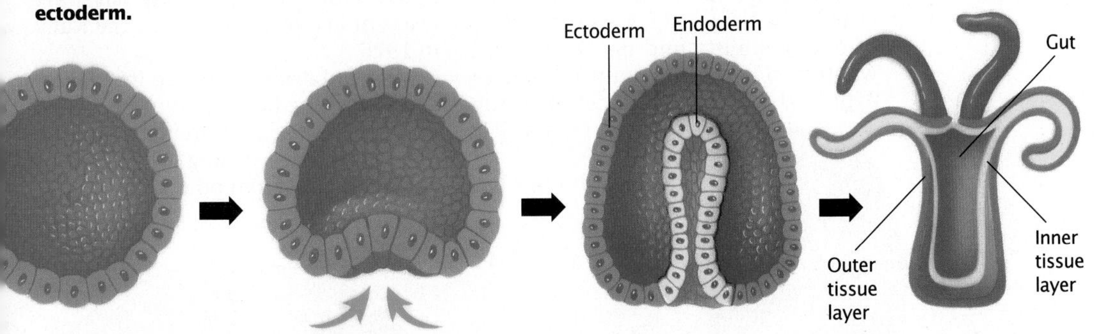

Figure 21.5
If you were to push on a tennis ball with your thumb, it would simulate what happens during the formation of the endoderm and ectoderm.

a A ball of cells forms from continuous divisions of a single fertilized egg.

b Gastrulation begins when cell divisions force one surface inward.

c The inner layer of cells is the endoderm; the outer layer is the ectoderm.

d In *Hydra*, ectoderm forms the outer tissue and nervous system, and endoderm forms the tissue that lines the gut.

Ectoderm Endoderm Gut

Outer tissue layer Inner tissue layer

Regularly Arranged Animals

Scale and Structure

Explain why sponges are unable to live on land.

Compare the shape of the cnidarian shown in the Exploration with the shape of the sponge you saw earlier. Most sponges are asymmetrical; they lack a regular arrangement of body parts. In contrast, the body parts of cnidarians are regularly arranged around the center of the body. A hydra's arms, for instance, seem to radiate from the body axis (an imaginary line through the center of the body) like spokes radiate from the hub of a wheel.

A test for symmetry is to imagine slicing the body along its axis. If such a slice produces approximate mirror-image body halves, then the animal is symmetrical. **Radial symmetry** is the wheel-like symmetry of hydras and sea anemones. You will soon see that radial symmetry is very different from the type of symmetry shown by most other animals, including humans.

Section Review

① Explain why a protist such as *Amoeba* is not considered an animal, but a sponge is.

② Why can a multicellular organism be more efficient than a unicellular one?

③ Why are the choanocytes of a sponge not considered a tissue?

④ List three differences between a sponge and a hydra.

Visual Strategy

Figure 21.5
Use a tennis ball to demonstrate this figure.

Theme Answer

Scale and Structure
A sponge requires water to obtain food and to support its body.

Phase 3

ASSESSMENT OPTIONS

Closure Strategy

Sponge Adaptations
Have students write a "performance review" of sponge adaptations. Suggest that they focus on the advantages and drawbacks of being sessile as well as on the sponge's ability to reproduce sexually and asexually. For example, tell students that being sessile allows the sponge to obtain food carried on currents, but prevents it from hunting. Provide students with copies of book reviews and reviews of plays from which they may draw their ideas.

Section Review

Assign the *Section Review*.

Reteaching

Have students imagine cnidarians without stinging cells or tentacles. Have them explain how the lives of these animals would be different.

▪ Section Review Answers ▪

1. Unlike a sponge, an *Amoeba* is not a multicellular organism.

2. The cells of a multicellular organism can specialize for a single task.

3. Choanocytes function independently, not in coordination, as do the cells of a tissue.

4. A hydra exhibits radial symmetry, extracellular digestion, and specialized tissues, while a sponge does not.

Using the Feature

- Alan Shipley's career path does not emphasize formal education. He earned much of his experience in the military. Use this feature as an example of alternate routes that can be taken to leading productive, fulfilling, and lucrative lives. Ask students to describe training options other than formal education and the military. Some students may want to discuss their plans for the future.

- Alan Shipley states that he learns by doing. You can use this feature to get students thinking about the ways in which they learn best. Engage them in a discussion about their learning styles by asking them about the classes and activities they enjoy and why.

Discussion

Guide the discussion by posing the following questions.

1. In addition to the personal traits listed in the feature, what are some other qualities that have enabled Alan to become a successful research associate?
Alan has an inherent interest in science and is marked by his curiosity and desire to learn.

2. Alan's ambitions do not stop at being a research associate. What does he do in addition to working at the Marine Biological Laboratory?
He also has a side business in electronics, and on weekends, he enjoys fishing with friends.

3. Alan's work at the Marine Biological Laboratory involves communicating with people. Describe some of these interactions.
He assists visiting scientists and scholars by teaching them about techniques in biological research.

Science in Action ▪ Science in Action ▪ Science in Action

Alan Shipley: Research Associate

How I Became Interested in Science

"I left school at age 15, then traveled around the country and joined the Navy when I was 17. That was in the '60s—the Vietnam War era. Both my parents were in the Navy so I felt a strong sense of responsibility and patriotism and wanted to serve my country, and to explore life. I had to experience the real world. I got a GED and a high school diploma through the Navy at a high school in Hawaii.

"In the Navy, I was a sonar technician. That's where I learned electronics, physics, and oceanography. After six years of active service, I left the Navy in 1976.

"As a kid I was interested in science. I've always had an inherent curiosity. Though I left school prematurely, I've never stopped learning. I continually read and learn from others by doing."

In his free time, Alan Shipley can occasionally be found on a boat, fishing in the waters off the coast of Massachusetts with his good friend Patrick O'Malley, a fellow engineer at the Marine Biological Laboratory. He also has a side business in electronics that he does on weekends.

Name:	**Alan Shipley**
Home:	**Sandwich, Massachusetts**
Employer:	**Marine Biological Laboratory, Woods Hole, MA**
Personal Traits:	**• Honest**
	• Resourceful
	• Friendly
	• Helpful
	• Cooperative

Action ▪ Science in Action ▪ Science in Action ▪ Science

Research Focus

Mr. Shipley works for Dr. Lionel Jaffe at the National Vibrating Probe Facility located at the Marine Biological Laboratory in Woods Hole, Massachusetts. They are funded by the National Institute of Health, Division of Research Resources, to design, develop, build, and teach visitors the use of new and innovative techniques in biological research. Thus, the laboratory provides an invaluable research environment for visiting investigators from around the world. The lab's main focus is the investigation of ionic currents and gradients as they relate to living organisms. Dr. Jaffe has perfected a technique of measuring electrical fields around living organisms. It is a non-invasive technique and is applicable to many different disciplines of scientific research. Mr. Shipley is also involved in trying to introduce an inexpensive way to study environmental pollution at minute levels in order to avoid damage before it becomes apparent by other means.

The Satisfaction of Working With Scientists

Career Path

U.S. Navy:
• Electronics
• Physics
• Oceanography

Junior College:
• Business
• Electronics
• Labor Relations

"A jack-of-all-trades like me was ideal to fit into this environment. I had to be able to do a wide range of things in order to effectively assist the visiting investigators. Working for Dr. Jaffe for ten years, I've learned by doing. I've worked on a few hundred projects that have involved different aspects of biology and physical science. We're on the cutting edge of technology. Everything we do is brand new. The education I receive here every day is irreplaceable.

"Recently I got a call from a student who was trying to find a good Ph.D. thesis. I suggested that he work here. After less than a week, he had accurate data with an interesting story. I have had two other Ph.D. students in the past do their work in our lab. I've been proud to be able to work with and assist them. It's been very satisfying for me. As an NIH facility, the Marine Biological Laboratory provides a marvelous opportunity for scientists to do a good project, publish and try to get some more funding."

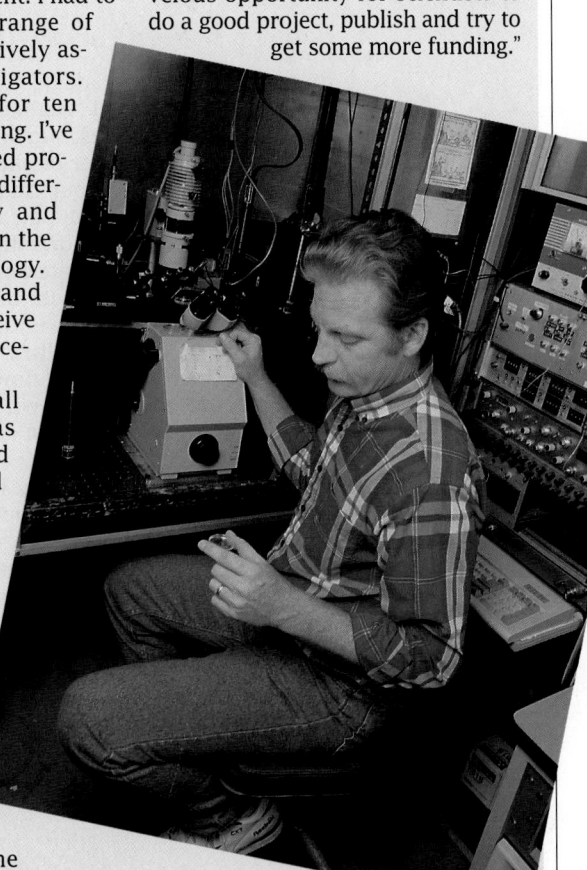

Lesson Plan 21.2

Phase 1

PREPARATION

Key Concepts

- Flatworms are bilaterally symmetrical, cephalized, and have organs.
- An acoelomate is an organism without a body cavity.
- A pseudocoelomate is an organism with a body cavity that develops between the mesoderm and endoderm.
- Coelomates are organisms with a body cavity that lies within the mesoderm.

Reading Strategy

Have students read for understanding by listing all the important ideas in this section and organizing them in outline form.

Phase 2

TEACHING STRATEGIES

Visual Strategy

Figure 21.6

Review with students the characteristics of flatworms. Emphasize the evolutionary advantages of bilateral symmetry and cephalization, as well as the evolutionary limitations of an incomplete digestive system and the lack of a coelom. Point out that flatworms are the least specialized phylum of those having a mesoderm, bilateral symmetry, and cephalization.

Demonstration 1

Worms

Show students preserved specimens, living specimens, or photographs of flatworms—such as flukes and tapeworms—and other organisms such as roundworms, segmented worms, and insect larvae. Have them discuss the different characteristics observed in flatworms and the other organisms.

462

ALTHOUGH CNIDARIANS SHARE A FEW CHARACTERISTICS WITH FAMILIAR ANIMALS, THEY PROBABLY STILL SEEM QUITE UNUSUAL TO YOU. HOW MANY RADIALLY SYMMETRICAL ANIMALS HAVE YOU SEEN TODAY, FOR INSTANCE? CHANGES IN SYMMETRY AND OTHER ASPECTS OF THE BODY PLAN EVENTUALLY RESULTED IN ANIMALS THAT ARE MORE FAMILIAR-LOOKING AND MORE COMPLICATED.

21.2   Origin of Body Cavities

Objectives

❶ **Contrast the body plans of bilaterally and radially symmetrical animals.**

❷ **List three structures derived from mesoderm.**

❸ **Describe the advantage of a one-way gut.**

❹ **Contrast the three kinds of body cavities.**

Heading Toward Complexity: Flatworms

Your body is neither radially symmetrical nor asymmetrical. You have two arms, two legs, two eyes, two kidneys, two lungs—one member of each pair on the right side of your body and the other member of the pair on the left. The human body shows **bilateral symmetry**. (The word bilateral means "two sides.") Any animal with bilateral symmetry can be separated into nearly mirror-image halves by drawing an imaginary line lengthwise down the middle of the body, as shown in the flatworm in **Figure 21.6**. Most animals are bilaterally symmetrical.

Look again at the hydra in the Exploration on page 458. Are you able to locate its head?

Most bilaterally symmetrical animals have definite head and tail ends, but radially symmetrical animals such as cnidarians do not. **Cephalization** (*sehf uh lih ZAY shuhn*) is the evolution of a definite head end. Animals that have heads are often active and mobile, moving through their environment headfirst. It is advantageous for sensory organs to be concentrated in the head so that an animal can test for food, danger, hiding places, and mates as it enters new surroundings.

Figure 21.6
All flatworms are bilaterally symmetrical. If you were to cut this flatworm along the dotted line, the result would be two halves that are almost mirror images.

Brain area

Head (cephalized) region

Light-detecting eyespot

Branched intestine

Mouth

Interpreting and evaluating the information obtained by sensory organs requires a complex structure. Cephalization is usually accompanied by the evolution of a collection of nervous tissue at the front end of the animal: a brain or brainlike structure. A brain located near the sensory organs is advantageous because information can rapidly cross the short distance between brain and receptor. This enables an animal to respond quickly to stimulation.

Flatworms are bilaterally symmetrical and cephalized

Flatworms, members of the phylum Platyhelminthes (*plat ih hehl MIHN theez*), were perhaps the first group of animals to be bilaterally symmetrical and cephalized. The dark spots on the head of the flatworm in **Figure 21.6** are eyespots that can detect light, but they cannot focus an image like your eyes can.

Flatworms have organs

Flatworms are also one of the first major groups of animals to evolve organs. **Organs** are collections of different kinds of tissue that are dedicated to one function. For instance, the heart of a vertebrate is made up of muscle tissue, connective tissue, and nervous tissue, all of which function together to pump blood. Organs are usually found as units in larger systems known as organ systems. An **organ system** is a group of interrelated organs that carries out one essential body function. A digestive system, for instance, breaks down food and absorbs nutrients, each organ playing a role in these processes.

Where do organs come from during development? During gastrulation, endoderm gives rise to the gut lining and ectoderm gives rise to nervous tissue and skin. A flatworm's body is more than these two simple kinds of tissue. In flatworms and all other major groups of animals that evolved after cnidarians, a third tissue layer forms between the ectoderm and endoderm during gastrulation, as illustrated in **Figure 21.7**. This third layer is known as **mesoderm**, which means "middle skin." Mesoderm develops into muscle, reproductive organs, and circulatory vessels.

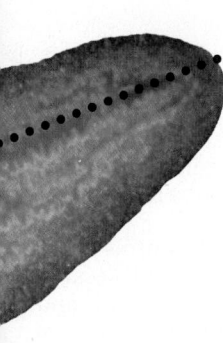

Figure 21.7
In flatworms, the process of gastrulation gives rise to three types of tissue layers.

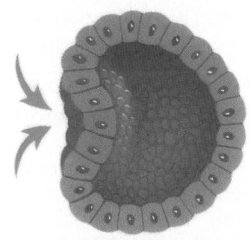

a After gastrulation begins, . . .

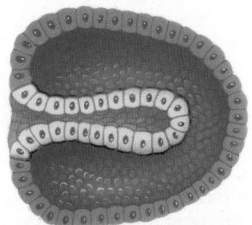

b . . . the endoderm and ectoderm form. Endoderm (yellow) gives rise to gut tissue. Ectoderm (blue) gives rise to nervous tissue and skin.

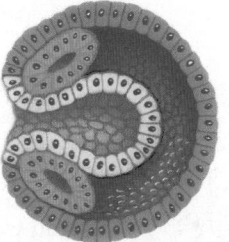

c The mesoderm forms in the space between the endoderm and the ectoderm. Mesoderm (red) gives rise to muscle, reproductive organs, and circulatory vessels.

d The embryo eventually develops into an adult flatworm.

EXPLORATION OF A FLATWORM

Stage 3: Internal Organs, Bilateral Symmetry, and Cephalization

The evolution of the mesoderm allowed the formation of organs, which first appeared in flatworms such as the liver fluke. Flatworms are bilaterally symmetrical and have a distinct head. The body plan of the liver fluke is composed of solid layers of tissues surrounding a central gut.

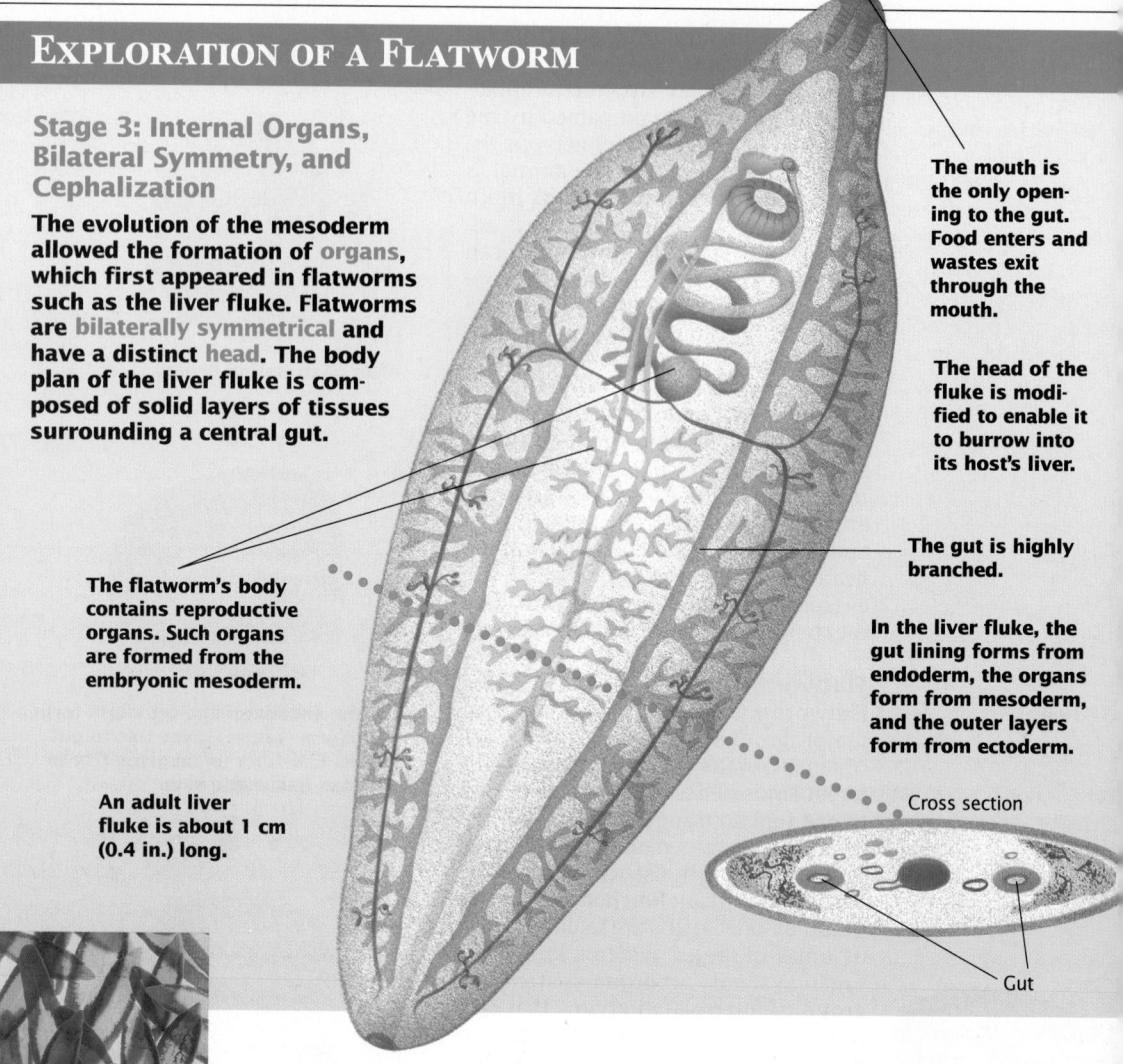

The mouth is the only opening to the gut. Food enters and wastes exit through the mouth.

The head of the fluke is modified to enable it to burrow into its host's liver.

The gut is highly branched.

In the liver fluke, the gut lining forms from endoderm, the organs form from mesoderm, and the outer layers form from ectoderm.

Cross section

Gut

The flatworm's body contains reproductive organs. Such organs are formed from the embryonic mesoderm.

An adult liver fluke is about 1 cm (0.4 in.) long.

Figure 21.8
Liver flukes are parasitic flatworms. Their acoelomate body plans require that they be thin to allow substances to pass easily to all organs.

Flatworms are solid worms

If you were to cut a flatworm in half across its body, as shown in the **Exploration** above, you would see that the gut is completely surrounded by tissue and organs. This solid body construction is termed **acoelomate** (*ay SEEL oh mayt*), meaning without a body cavity.

Flatworms are thin because of their acoelomate body construction. Dissolved substances such as carbon dioxide, oxygen, and nutrients cannot diffuse rapidly through the solid bodies of flatworms, such as the liver fluke in **Figure 21.8**. Flatworms are small or thin (or both), which shortens the distance these substances must move. In addition, the gut of a flatworm is highly branched so that it runs close to most of the tissues.

The guts of all flatworms have only one opening, the mouth. Because these animals consume food and eliminate wastes through the same opening, two-way movement of material occurs within the gut. The two-way gut is less efficient at extracting nutrients than a one-way gut such as yours. If animals with an acoelomate body plan eat when food is already in the gut, newly consumed food can mix with partially digested food and wastes.

A One-Way Gut and a Body Cavity: Roundworms

Figure 21.9
Roundworms, such as this nematode, have a body cavity, the pseudocoelom. The presence of the pseudocoelom means that roundworms are not packed with solid tissues, as are flatworms. The pseudocoelom allows for more efficient diffusion of nutrients to body organs.

Roundworms (phylum Nematoda), or nematodes, such as *Caenorhabditis elegans* in **Figure 21.9**, have a cavity within the body and a one-way gut with two openings. Food is taken in through the mouth, and wastes are eliminated through an opening at the other end of the gut, the **anus**. This arrangement allows a one-way movement of food through the gut. Also, different regions of the digestive tube can be specialized for different digestive activities. The front part of the gut is adapted for ingesting food. The middle region breaks down food and absorbs nutrients. The last region expels waste products.

Specialization of different regions of the gut brings with it a potential problem, however. Since the absorption of nutrients occurs in only one part of the digestive tract, there must be some effective way of distributing those absorbed nutrients to other parts of the body. Roundworms have a fluid-filled cavity between the gut and the body wall, as illustrated in the **Exploration**. This body cavity is called a **pseudocoelom** (*SOO doh see luhm*), which means "false body cavity." This term is somewhat misleading. The cavity is real, but it differs from the body cavities of most other animals.

The pseudocoelom permits rapid diffusion of nutrients and other dissolved substances over the short distances between tissues and organs. Diffusion is enhanced by body movements, which cause the fluid within the body cavity to move. Nevertheless, diffusion is a slow process. Pseudocoelomate animals must either be very small—most are less than 2.5 mm (0.1 in.) in length—or have body shapes that maintain short distances between organs and the body surface. For this reason, nematodes are usually thin and threadlike.

Visual Strategy

Figure 21.9
Show students a spadeful of quality garden soil. Have them guess how many individual nematodes it might contain. Suggest as many as a million. Explain that nematodes are tiny and are important in aerating the soil and distributing organic and mineral material. Tell students there are about 50 species that are parasitic and cause damage to crops, livestock, and humans.

EXPLORATION OF A ROUNDWORM

Stage 4: Body Cavity

The major body plan innovation in roundworms, such as this nematode, is the presence of a body cavity between the gut and the body wall. This cavity is the pseudocoelom. After its evolution, animals were not constrained by a solid body, as were flatworms. Nematode organs could now form away from the gut because nutrients could diffuse through the new body cavity.

Nematode adults consist of very few cells. The nematode *C. elegans* has only 1,000 cells and is the only animal whose complete cellular anatomy is known.

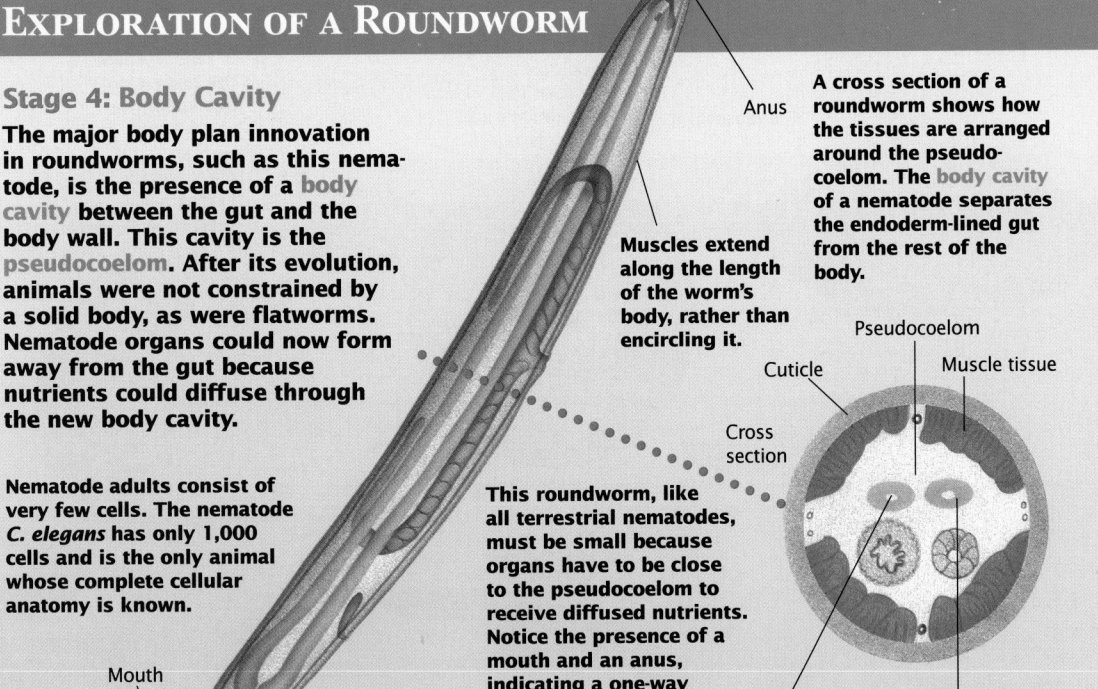

Anus

Mouth

Muscles extend along the length of the worm's body, rather than encircling it.

This roundworm, like all terrestrial nematodes, must be small because organs have to be close to the pseudocoelom to receive diffused nutrients. Notice the presence of a mouth and an anus, indicating a one-way digestive tract.

A cross section of a roundworm shows how the tissues are arranged around the pseudocoelom. The body cavity of a nematode separates the endoderm-lined gut from the rest of the body.

Pseudocoelom

Muscle tissue

Cuticle

Cross section

Gut

Digestive tissue

A Better Body Cavity: Mollusks

Most animals with a body cavity have a coelom, or true body cavity. A **coelom** (*SEE luhm*) is a fluid-filled body cavity that lies completely within the mesoderm. The coelom separates the muscles of the body wall from the muscles that surround the gut.

A major advantage of the coelom is that it allows interactions between mesoderm and endoderm to occur during development. This interaction is necessary in order for local regions of the digestive tract to become highly specialized. For example, your stomach is a locally specialized portion of the gut that developed from both endoderm and mesoderm. In a pseudocoelomate, by contrast, mesoderm and endoderm are separated by the fluid-filled pseudocoelom. Interactions between these two layers are limited, and a high degree of digestive specialization is not possible. The fluid-filled coelom also provides a body cavity in which organs can develop and against which muscles can operate. The body constructions of acoelomates, pseudocoelomates, and coelomates are compared in **Figure 21.10**.

In coelomates, as in flatworms, the gut tube is surrounded by solid tissue that is a barrier to rapid diffusion. But most coelomates have a **circulatory system**, a network of blood-carrying vessels. The circulatory system brings nutrients and oxygen to the tissues and removes wastes and carbon dioxide. Blood is usually propelled through the circulatory system by contractions of one or more muscular hearts.

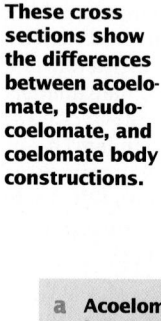

These cross sections show the differences between acoelomate, pseudocoelomate, and coelomate body constructions.

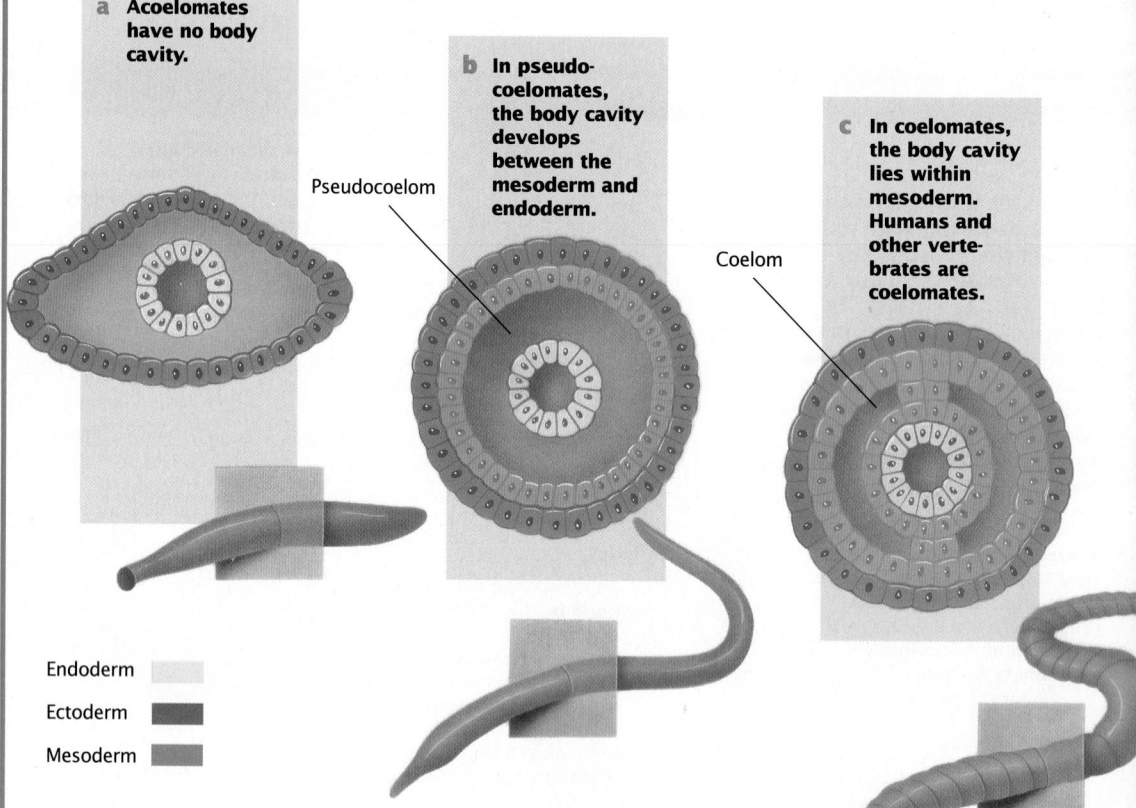

a Acoelomates have no body cavity.

b In pseudocoelomates, the body cavity develops between the mesoderm and endoderm.

Pseudocoelom

c In coelomates, the body cavity lies within mesoderm. Humans and other vertebrates are coelomates.

Coelom

Endoderm
Ectoderm
Mesoderm

EXPLORATION OF A MOLLUSK

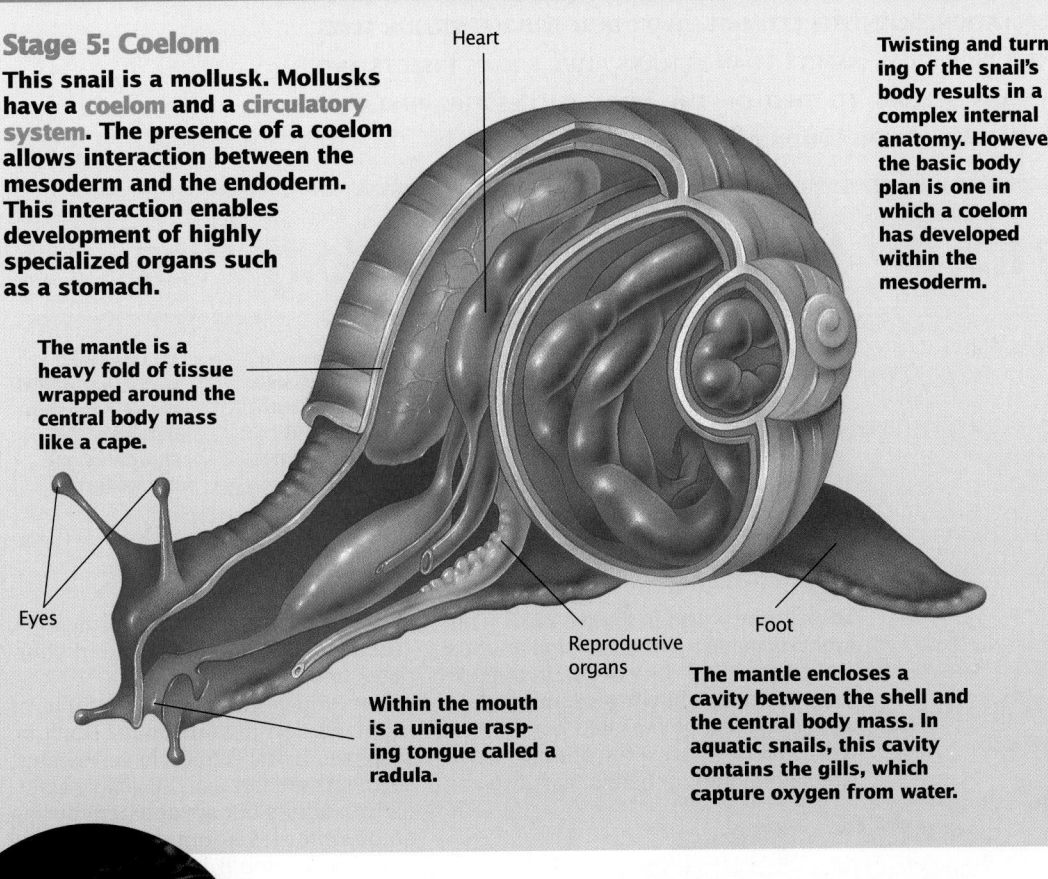

Stage 5: Coelom

This snail is a mollusk. Mollusks have a coelom and a circulatory system. The presence of a coelom allows interaction between the mesoderm and the endoderm. This interaction enables development of highly specialized organs such as a stomach.

The mantle is a heavy fold of tissue wrapped around the central body mass like a cape.

Heart

Twisting and turning of the snail's body results in a complex internal anatomy. However, the basic body plan is one in which a coelom has developed within the mesoderm.

Eyes

Within the mouth is a unique rasping tongue called a radula.

Reproductive organs

Foot

The mantle encloses a cavity between the shell and the central body mass. In aquatic snails, this cavity contains the gills, which capture oxygen from water.

Figure 21.11
Mollusks, such as this snail, are more complex than nematodes.

Mollusks are coelomates with a circulatory system

A coelom and a circulatory system first evolved in the mollusks, phylum Mollusca. Snails (such as the one in **Figure 21.11**), clams, squids, and mussels belong to this phylum. The mollusk body plan is illustrated in the **Exploration** above.

Section Review

1 Explain the difference between bilateral symmetry and radial symmetry.

2 What structures in your body are derived from mesoderm?

3 Summarize the advantage of a one-way gut.

4 Diagram the two types of body cavities.

■ *Section Review Answers* ■

1. Bilateral symmetry is the arrangement of an organism's body parts so that one-half of the body is an approximate mirror-image of the other half, whereas radial symmetry is the arrangement of body parts around a central point.

2. Mesoderm tissue gives rise to muscle, reproductive organs, and circulatory vessels.

3. Students should suggest that one way is more efficient at extracting nutrients.

4. See Figure 21.10 on page 466.

Visual Strategy

Figure 21.11
Use preserved specimens, fresh specimens, or photographs of the internal organs of the following: a clam, an oyster, a mussel, and a scallop. Point out the heart and the circulatory system in this group of animals. Discuss with students the potential problem a snail might have because its anus is above its head. Have students form a hypothesis about the way snails keep waste products out of their mouths and gills.

Mathematics Connection

Snail Movement
Many snails move at a speed of less than 8 cm (3 in.) per minute. This means that if a snail did not stop to rest or eat, it could travel 4.8 m (0.003 mi.) per hour.

Phase 3
ASSESSMENT OPTIONS

Closure Strategy

Body Structure
Assign students to cooperative work groups of three. Have each student in each group write down the major body plans, the type of symmetry, the structures derived from the mesoderm, and the different kinds of body cavities. When students have finished their lists, have them exchange their papers within their groups and discuss each other's work. Tell them to add any characteristics that might have been omitted.

Section Review

Assign the Section Review.

Reteaching

Have students develop a table that displays major characteristics of body plans, symmetry, body cavities, and additional systems to solve problems created by greater sophistication.

467

Lesson Plan 21.3

Phase 1
PREPARATION

Key Concepts
- Annelids were the first animals to have segmented bodies.
- Arthropods and chordates are segmented.
- All chordates share three basic characteristics at some stage in their lives: a notochord, a dorsal nerve cord, and pharyngeal slits.
- Protostomes are organisms whose development is based on molecules that act as developmental signals in different parts of the egg.
- Deuterostomes are organisms whose development is controlled by genes within cells.

Reading Strategy

Have students make the following concept maps after reading this section and identifying the major concepts: Segmented Worms; Arthropod Limbs/Skeletons; Echinoderm Characteristics; and Chordates.

Phase 2
TEACHING STRATEGIES

Visual Strategy

Figure 21.12
Provide students with live or preserved earthworms and hand lenses. Have them identify the worms' anterior ends. Have them count the segments and compare their results.

THERE ARE MORE THAN 5 BILLION PEOPLE ON EARTH. THIS FIGURE MAY SEEM LARGE, BUT IT IS TINY COMPARED WITH THE INSECT POPULATION. SCIENTISTS ESTIMATE THAT THERE ARE 200 MILLION TIMES MORE INDIVIDUAL INSECTS THAN HUMANS ALIVE TODAY. INSECTS AND HUMANS BELONG TO TWO OF THE MOST SUCCESSFUL PHYLA OF ORGANISMS—ARTHROPODA AND CHORDATA.

21.3   Four Innovations in Body Plan

Objectives

1 Contrast the body plans of segmented and nonsegmented animals.

2 Compare the exoskeleton of arthropods with the endoskeleton of vertebrates.

3 Summarize the differences between protostomes and deuterostomes.

4 List two traits that reveal the relationship between chordates and echinoderms.

Segmented Worms: Annelids

Look at the worm in **Figure 21.12**. This worm is made up of many similar units linked together, like beads in a necklace. Animals showing **segmentation** are composed of repeated body units. Three very successful animal phyla are segmented: annelids (earthworms and their relatives), arthropods (insects, crustaceans, and spiders), and chordates (mostly vertebrates).

Can you point out an example of segmentation in your body? Don't be surprised if you cannot. In vertebrates, segments are not usually visible externally in adults but are apparent during embryonic development. Vertebrate muscles develop from repeated blocks of tissue that occur in the embryo. Another example of segmentation is the vertebral column, which is a stack of very similar vertebrae.

What great advantage does segmentation provide? Its main advantage is the evolutionary flexibility it offers. A small change in an existing segment can produce a new kind of segment with a different function. As illustrated in the **Exploration** shown on the next page, some segments of the earthworm are modified for reproduction, some for feeding, and some for eliminating wastes.

Figure 21.12
The bristle worm is an example of an annelid. Like the earthworm on the next page, its body is segmented. Specialized segments perform specific tasks.

EXPLORATION OF AN ANNELID

Stage 6: Segmentation

Annelids, such as this earthworm, were the first organisms to evolve a body plan that consisted of segments. Most segments are separated by partitions that cross the coelom. In each segment, parts of the excretory, circulatory, and nervous systems are repeated.

Gut

A cross section of the earthworm shows the gut and coelom.

Coelom

A pair of excretory organs and a nerve center are located in each segment.

A small brain coordinates the activities of the segments.

Gut tube

Brain

Mouth

Nerve cord

Blood vessels

Bristles called setae occur on each segment. Earthworms crawl by anchoring setae to the ground and pulling against them.

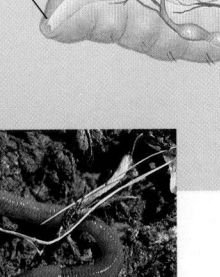

**Figure 21.13
The earthworm is an annelid. Members of this phylum have segmented bodies.**

Annelids have segments that are specialized

Annelids (phylum Annelida), such as the earthworm in **Figure 21.13**, were the first segmented animals to evolve. The earthworm body plan is shown in the **Exploration** above. The basic body plan of an annelid is a tube within a tube. The gut tube, which extends from mouth to anus, is suspended within the larger tube of the coelom. Note that the body is partitioned internally between segments. This partitioning limits diffusion of materials from segment to segment. A circulatory system overcomes this limitation by transporting materials between segments.

The front segments of annelids are modified to house a small brain and sense organs. Each segment along the body is controlled by an individual nerve center. A nerve cord running along the underside of the worm connects these nerve centers with the brain so that all of the body's activities can be coordinated.

Visual Strategy

Figure 21.13
Discuss whether the setae appear all along the body, and how many there are per segment. Have students determine if all their earthworms have the same number of segments in front of the clitellum (reproductive region).

Demonstration 1

Display of Specimens
Prepare leech and earthworm specimens for students to observe. Have them compare and identify analogous structures in each. Discuss the similarities and differences between earthworms and leeches. Have students identify the segments and the anterior and posterior suckers.

Theme Connection

Scale and Structure

Have students discuss how the fluid-filled coelomic chamber provides support, and how the circular and longitudinal muscles coordinate to produce an accordion-like movement.

Visual Strategy

Figure 21.14
Insects have successfully adapted to many environments and have proliferated in number and variety. Ask students to identify one way insects have been very successful in getting away from predators. The evolution of wings for flight in insects is probably one of the most successful adaptations in the group.

Demonstration 2

Arthropod Diversity
Display many specimens or photographs of different arthropods. Have students note the major differences in body shape, appendages, etc. How are the arthropods alike?

EXPLORATION OF AN ARTHROPOD

Stage 7: Jointed Appendages, Exoskeleton, and Wings

Arthropods have a coelom, segmented bodies, and jointed appendages. The three body regions of an insect, such as this wasp, are the head, thorax, and abdomen. Each region is actually composed of a number of segments that fuse during development. The presence of a strong exoskeleton made of chitin, a complex muscular system, and wings permit this wasp to move quickly from place to place.

Like most insects, wasps have two pairs of wings attached to the thorax.

A wasp is an insect. Like all arthropods, it has a segmented body and jointed appendages. This body plan has helped insects become one of the most successful animal groups.

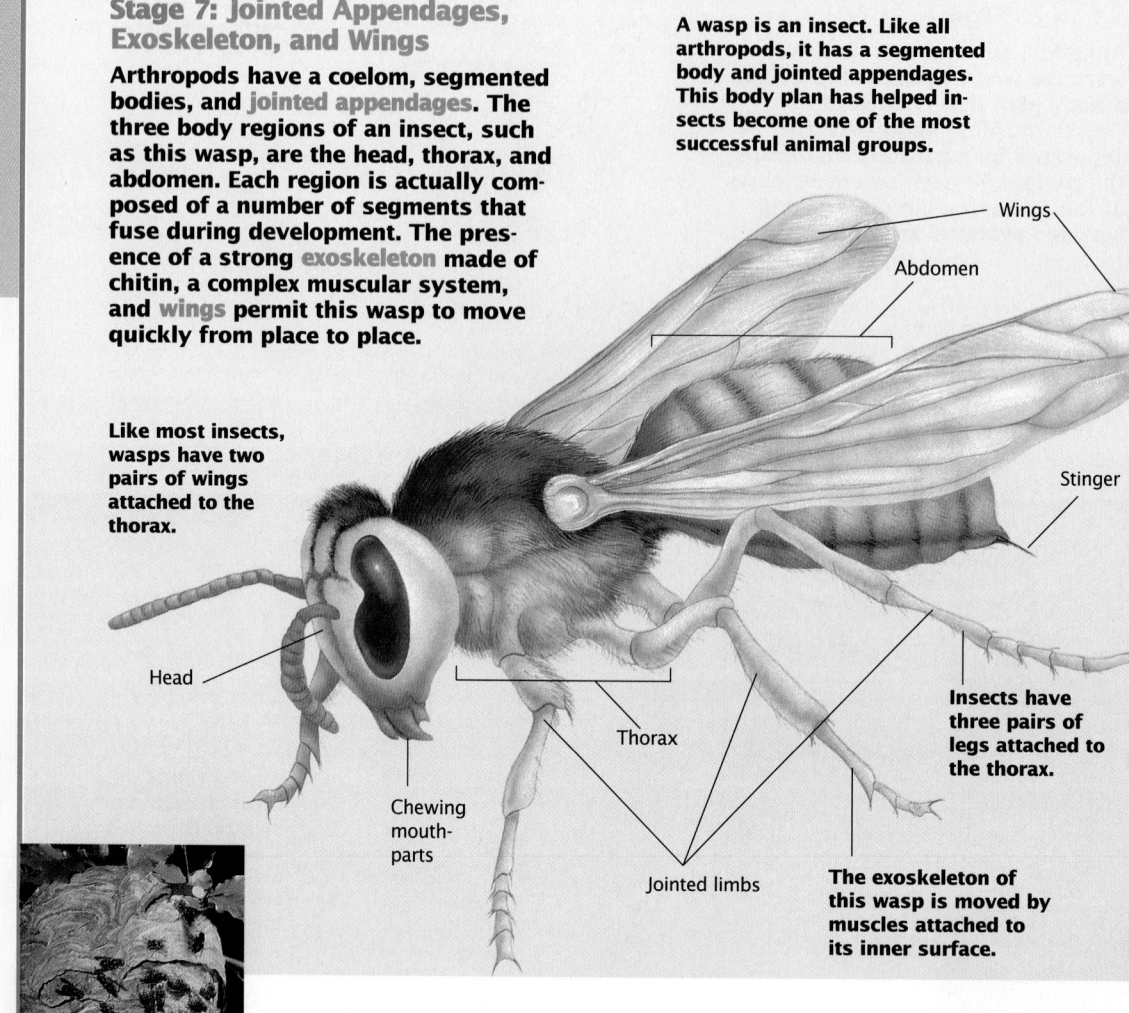

Wings

Abdomen

Stinger

Head

Chewing mouthparts

Thorax

Jointed limbs

Insects have three pairs of legs attached to the thorax.

The exoskeleton of this wasp is moved by muscles attached to its inner surface.

Figure 21.14
Wasps are highly social insects that build a nest where the colony lives and breeds. This nest is made of a paperlike material and may house hundreds of wasps.

Limbs and Skeletons: Arthropods

The name arthropod comes from the Greek words *arthros,* meaning jointed, and *podes,* meaning feet. The great success of arthropods, such as the wasps in **Figure 21.14,** is due largely to their jointed appendages. These appendages have evolved to perform a variety of tasks. Some appendages serve as limbs for walking or grasping. Others function as antennae for sensing the environment or as mouthparts for chewing, sucking, or poisoning prey. For example, a scorpion seizes and tears apart its prey with appendages that have been modified into large pincers capable of grasping and holding tightly.

Like annelids, the basic body plan of arthropods is segmented, as shown in the **Exploration** above. Individual segments of an arthropod often exist only during early development, however, and fuse into functional groups in adults.

For example, caterpillars have many segments, while butterflies have only three main body units—head, thorax, and abdomen—each composed of several fused segments. Arthropods have an external skeleton or **exoskeleton**. The arthropod exoskeleton is made of chitin (*KYT uhn*), a tough polysaccharide. The muscles lie within the skeleton and attach to its inner surface. As it grows, an arthropod periodically sheds its rigid exoskeleton. Many arthropods change their body form as they develop. A butterfly begins life as a wormlike caterpillar, later transforming into the flying adult. A silverfish, on the other hand, does not change as it develops; it merely grows larger.

An Embryonic Revolution

Humans and the other chordates are more closely related to the echinoderms—the phylum that includes the sea star—than to the arthropods. There seems to be little resemblance between humans and sea stars (often called "starfish"). For one thing, sea stars are radially symmetrical, as you can see in the Exploration on the next page.

In animals with tissues, except for echinoderms and chordates, the first opening that forms in the embryo during gastrulation eventually becomes the mouth. Animals that develop in this fashion are known as **protostomes**. The word protostome means "first mouth," and refers to the fact that the initial depression that starts gastrulation becomes the mouth of the adult organism. In protostomes, the developmental fate of each cell of the early embryo is usually determined when that cell first appears.

Echinoderms and chordates are **deuterostomes**. The word deuterostome means "second mouth," and refers to the fact that the first opening in the embryo does not form the mouth—it becomes the anus of the adult animal. The mouth develops from an opening that appears later in development. In these animals, all the cells of the early embryo are identical. This means that any isolated cell can develop into a complete organism.   shows some protostomes and deuterostomes.

Figure 21.15

Figure 21.15
Below are examples of protostomes and deuterostomes. Recent studies using DNA nucleotide sequences support the division of the coelomates into these two groups.

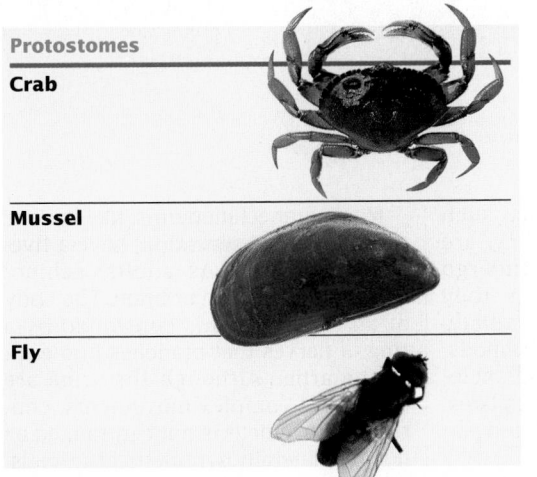

Protostomes
Crab
Mussel
Fly

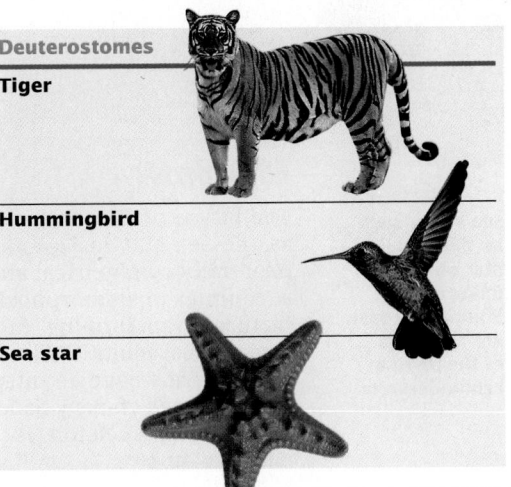

Deuterostomes
Tiger
Hummingbird
Sea star

Visual Strategy

Figure 21.16
Have students examine prepared slides of echinoderm larvae and explain what makes them bilaterally symmetrical. If no slides are available, use the drawing on page 472 to prompt a discussion.

Connection: Chapter 10

Evolution
The fossil record of echinoderms dates back to the Cambrian period, more than 500 million years ago.

Demonstration 3

Sea Star Symmetry
Demonstrate the concept of symmetry by using a flat-sided mirror and drawings or photographs showing the front view of a human and the upper surface of a sea star. Have students observe both images while a volunteer places the flat side of the mirror against the human and turns it in various directions—horizontal, vertical, and diagonal. Have another volunteer place the mirror against the sea star and turn it in various directions. Have students compare their observations. If you would like to, use the larval and adult sea star in the text.

EXPLORATION OF AN ECHINODERM

Stage 8: Deuterostome Development and Endoskeleton

Echinoderms, such as this sea star, have coeloms and are deuterostomes. Sea stars are bilaterally symmetrical as larvae. As adults, they have five-part, radial symmetry and an endoskeleton. Sea stars move by means of a system of water-filled canals known as the water vascular system. The water vascular system causes tube feet on the bottom of the sea star to extend and retract.

Sea stars have a delicate skin stretched over a calcium-rich endoskeleton of spiny plates.

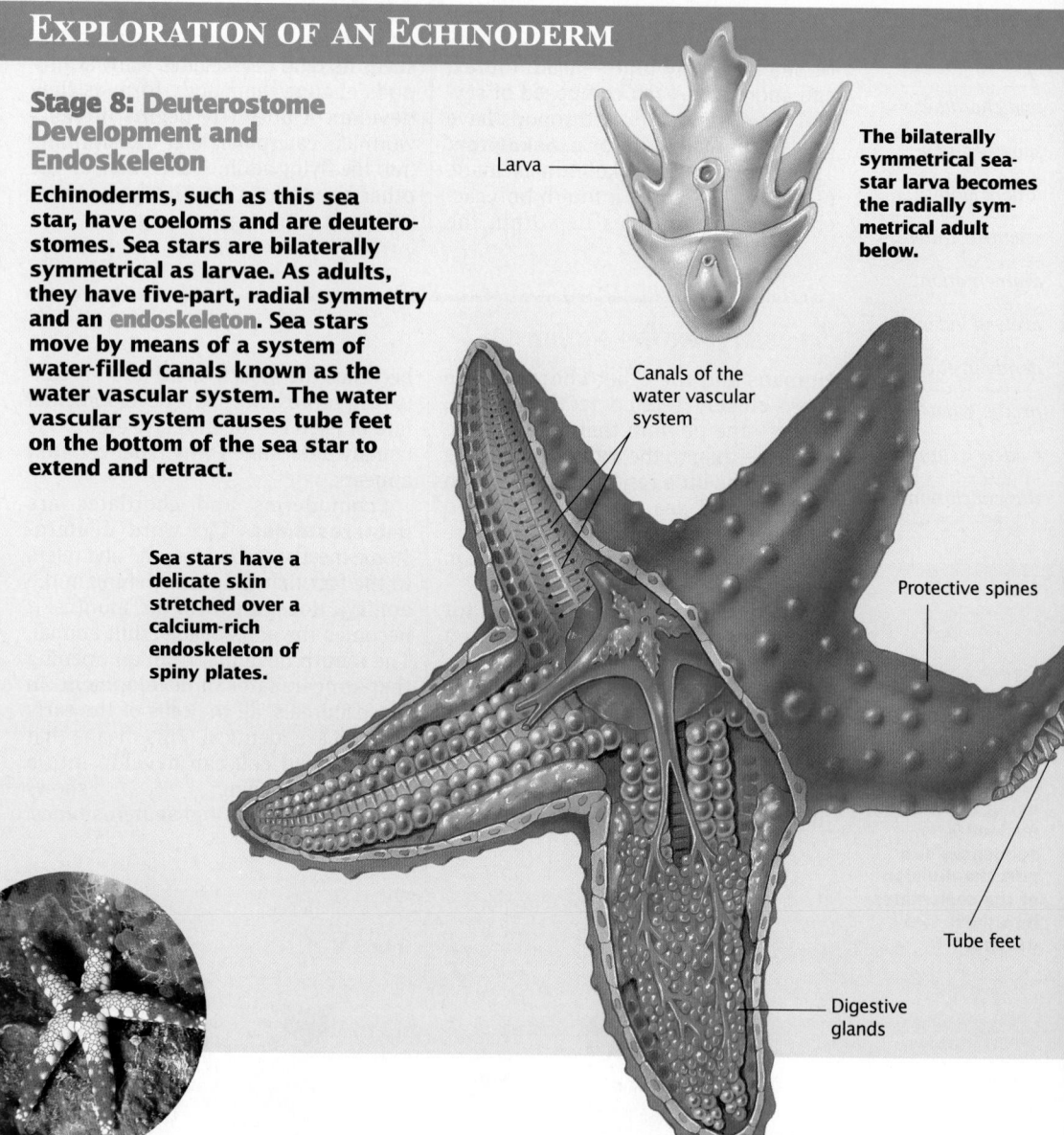

Larva

The bilaterally symmetrical sea-star larva becomes the radially symmetrical adult below.

Canals of the water vascular system

Protective spines

Tube feet

Digestive glands

**Figure 21.16
Sea stars, such as this small sea star on a deep ledge off the Philippine coast, are members of the phylum Echinodermata.**

Echinoderms

The larvae of all echinoderms, such as the sea star in **Figure 21.16**, are bilaterally symmetrical and undergo a complex metamorphosis as they mature, transforming into radially symmetrical adults. Unlike arthropods, echinoderms have an internal skeleton composed of many bonelike plates. An internal skeleton is called an **endoskeleton**.

Most adult echinoderms, like the one shown in the **Exploration**, have a five-part body plan. As adults, echinoderms have no head or brain. The body of an echinoderm is controlled by a ring of nerves that branches into each of the arms. Although the arms are capable of complex movements, control of movement is not centralized as it is in bilaterally symmetrical animals.

Figure 21.17
The lancelet is a chordate, a member of the same phylum as human beings. Although it does not resemble a human, the lancelet has characteristics that it shares with humans and all other chordates.

The Most Successful Deuterostomes: Chordates

Fishes, amphibians, reptiles, birds, and mammals are chordates, as is the lancelet shown in **Figure 21.17**. Lancelets do not resemble familiar chordates such as birds, dogs, or humans. Nonetheless, lancelets and all chordates share three key features:

1. dorsal (along the back) hollow nerve cord,

2. dorsal supportive rod called the notochord,

3. slits in the pharynx (the region of the digestive tract just behind the mouth) that connect the pharynx to the outside.

This unique combination of characteristics can be seen in the lancelet shown in the **Exploration** below. All chordates have all three of these characteristics at some time during their lives. For example, humans have pharyngeal slits, a nerve cord, and a notochord as embryos. As adults, humans retain only the nerve cord and one pair of pharyngeal slits, which are the eustachian tubes that connect the throat to the middle ear.

In addition to these three principal characteristics, chordates have a number of other distinguishing features. The body plan is segmented. All chordates have a tail that extends beyond the anus, at least during embryonic development.

Demonstration 4
Chordate and/or Vertebrate
Show students an unlabeled photo of a human embryo. Have them determine whether the embryo is a chordate. Then have them determine if it is also a vertebrate. Have students explain the basis for their answer. Then identify the embryo as human.

Theme Connection

Scale and Structure
Tails that form in human embryos reach their greatest length during the second month of embryonic life, then usually disappear. Tell students that in a few cases infants are born with short tails, which are surgically removed soon after birth.

EXPLORATION OF A LANCELET

Stage 9: Notochord

The lancelet *Branchiostoma* is a chordate. It is a coelomate animal that has a notochord, pharyngeal slits, and a dorsal nerve cord. The notochord persists throughout the life of *Branchiostoma*. In most other chordates, such as birds and humans, the notochord is replaced during embryonic development by the vertebral column.

The lancelet is modified for life as an ocean-dwelling bottom feeder. It takes in water through its mouth, passes it over its pharyngeal slits, and filters out small prey.

The body is pointed at both ends and has no head.

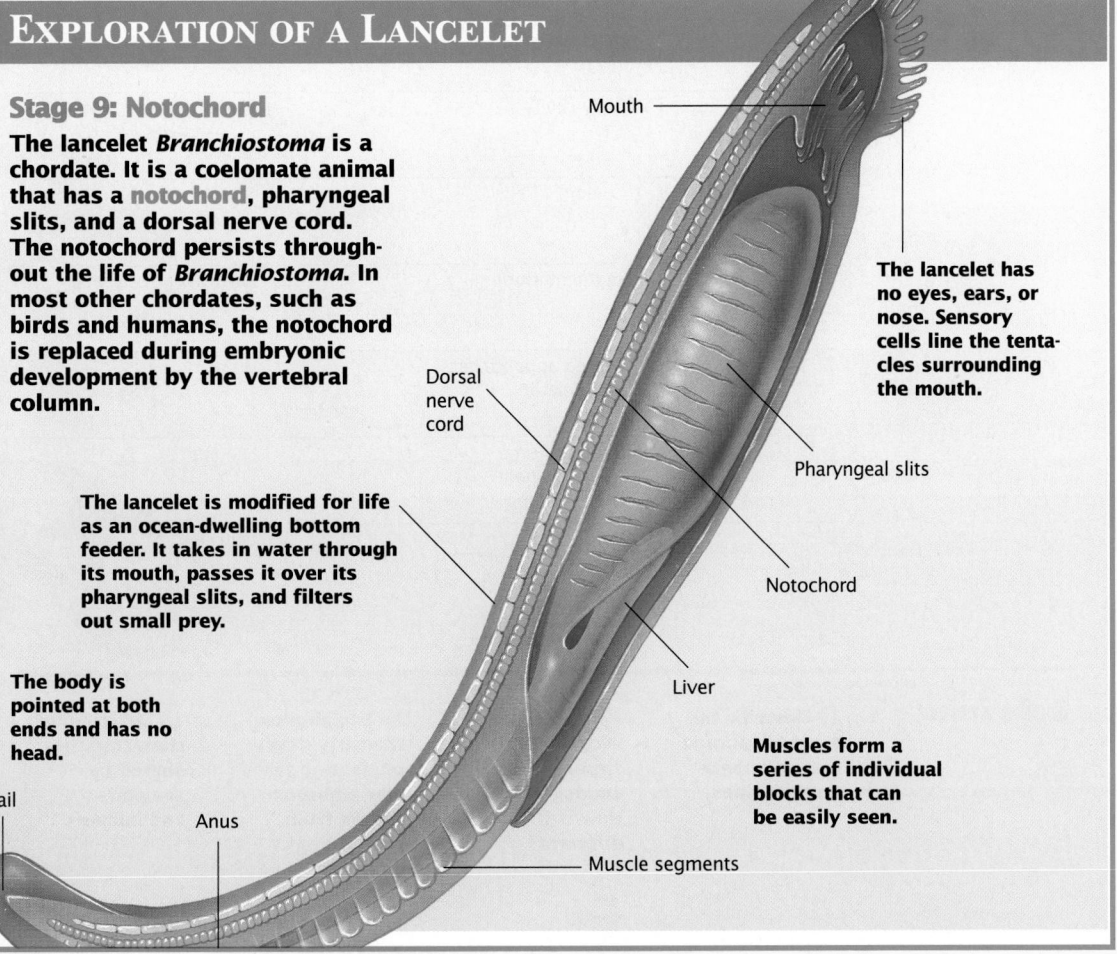

Mouth

The lancelet has no eyes, ears, or nose. Sensory cells line the tentacles surrounding the mouth.

Dorsal nerve cord

Pharyngeal slits

Notochord

Liver

Muscles form a series of individual blocks that can be easily seen.

Tail

Anus

Muscle segments

Visual Strategy

Table 21.1
Have students qualify this statement: Every stage on this table has the characteristics of the previous stages.

Demonstration 5

Endoskeleton
Display a human skeleton or a drawing of a human skeleton. Encourage students to point out what replaced the notochord.

Phase 3

ASSESSMENT OPTIONS

Closure Strategy

Common Ancestors
Have students explain why biologists have concluded that the phyla Mollusca and Annelida evolved from a common ancestor. Emphasize that they should give specific examples of evidence that led to this conclusion, especially specific structures and their functions.

Section Review

Assign the *Section Review.*

Reteaching

Have students explain why the word "worm" is not a precise biological term. Have students design flashcards to test their knowledge of major ideas in this section.

Vertebrates have a sturdy endoskeleton

Most chordates belong to the subphylum Vertebrata, the vertebrates. Fishes, amphibians, reptiles, birds, and mammals are vertebrates, but the lancelet is not. In most vertebrates, the notochord is replaced by the vertebral column during development.

Vertebrates also have an endoskeleton of bone or cartilage against which the muscles work. The vertebrate endoskeleton is lightweight and provides effective support.

Table 21.1 below summarizes the many changes in the animal body that you have read about throughout this chapter.

Table 21.1   The Animal Body: An Evolutionary Journey

Stage	Milestone	Typical organism	
1	Multicellularity	Sponge	
2	Radial symmetry Extracellular digestion Specialized tissues	Hydra	
3	Internal organs Bilateral symmetry Cephalization	Liver fluke	
4	Body cavity	Nematode	
5	Coelom	Snail	
6	Segmentation	Earthworm	
7	Jointed appendages Exoskeleton Wings	Wasp	
8	Deuterostome development Endoskeleton	Sea star	
9	Notochord	Lancelet	

Section Review

❶ Describe the great advantage that segmentation provides.

❷ How is an exoskeleton similar to an endoskeleton? How is it different?

❸ Explain how the early development of sea star embryos differs from that of insect embryos.

❹ Describe two characteristics shared by sea stars and humans.

■ *Section Review Answers* ■

1. Its main advantage is the evolutionary flexibility it offers. A small change in an existing segment can produce a new kind of segment with a different function.

2. They are similar in that they both provide support. They differ in their location: an endoskeleton is internal, while an exoskeleton is external.

3. In an insect, the fate of each cell in the embryo is largely controlled by molecules within the egg. In a sea star, each cell's fate is largely controlled by its genes.

4. Both are deuterostomes and both have an endoskeleton.

Chapter 21 *Highlights*

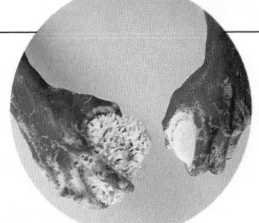

A natural sponge is the skeleton of a marine sponge.

21.1 The Advent of Tissues	*Key Terms*	*Summary*
Sponges have the simplest body plans among animals. They have no symmetry and no tissues. 	body plan (p. 455) tissue (p. 458) gut (p. 458) gastrulation (p. 459) endoderm (p. 459) ectoderm (p. 459) radial symmetry (p. 459)	• Multicellularity enables individual cells to specialize on one life task. • Sponges are multicellular animals without tissues. Sponges filter small food particles from the water. • Cnidarians have two tissue layers and show radial symmetry.
21.2 Origin of Body Cavities 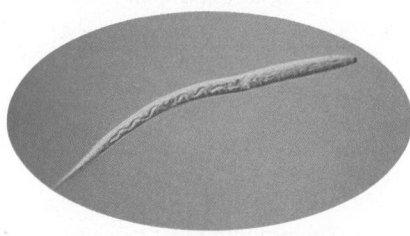 Nematodes have two openings to the gut, allowing for specialization of digestive activities.	bilateral symmetry (p. 462) cephalization (p. 462) organ (p. 463) organ system (p. 463) mesoderm (p. 463) acoelomate (p. 464) anus (p. 465) pseudocoelom (p. 465) coelom (p. 466) circulatory system (p. 466)	• Flatworms are bilaterally symmetrical and cephalized. • Nematodes have a one-way gut and a pseudocoelom, a cavity lying between the gut and the muscles of the body wall. • A coelom is a fluid-filled body cavity that lies within the mesoderm. Mollusks have a coelom and a circulatory system that distributes nutrients and oxygen to the tissues.
21.3 Four Innovations in Body Plan Arthropods have jointed appendages, a segmented body, and an exoskeleton. 	segmentation (p. 468) exoskeleton (p. 471) protostome (p. 471) deuterostome (p. 471) endoskeleton (p. 472)	• Annelids were the first animals to have a segmented body. Chordates and arthropods are also segmented. • An exoskeleton of chitin, segmentation, and jointed appendages are characteristics of the arthropods. • The embryos of echinoderms and chordates develop similarly. These two groups of animals are closely related. • At some time in their lives, all chordates have a notochord, a dorsal nerve cord, and pharyngeal slits.

Alternative Assessment

Assign students to cooperative work groups of four. Have each group build a concept map summarizing the chapter and including the major ideas presented on page 475. Concept maps should include all key terms and the concepts presented in each summary section. This will be a rather large undertaking. Suggest that students write the major concepts on small pieces of paper that can be moved around on a lab table.

Chapter Review Answers

Understanding Vocabulary

1. **a.** Both are produced during gastrulation; the inner layer is called the endoderm, while the outer layer is called the ectoderm.

   b. A tissue is composed of similar cells that are organized into a functional unit, while an organ is a collection of different tissues that perform a single function.

   c. Radial symmetry refers to the body configuration that appears wheel-like, while organisms that have bilateral symmetry have two sides that are mirror images.

   d. In protosomes, cell fate is determined by molecules within the egg. Genes within the cells control development in deuterostomes.

Relating Concepts

2. Map answer is shown on page 451D.

Understanding Concepts

Multiple Choice

3. b	7. b
4. a	8. a
5. c	9. c
6. c	10. a

Completion

11. Porifera
12. gastrulation
13. exoskeleton, endoskeleton
14. Echinodermata
15. mouth, anus

Short Answer

16. Division of labor means that cells are specialized and can perform different tasks. Division of labor is advantageous because each cell does not have to perform the same function but can benefit from the functions performed by other cells.

Understanding Vocabulary

1. For each pair of terms, explain the difference in their meanings.
   a. endoderm, ectoderm
   b. tissue, organ
   c. radial symmetry, bilateral symmetry
   d. protostomes, deuterostomes

Relating Concepts

2. Copy the unfinished concept map below onto a sheet of paper. Then complete the concept map by writing the correct word or phrase in each oval containing a question mark.

Understanding Concepts

Multiple Choice

3. Which characteristic is shared by all animals?
   a. radial symmetry
   b. multicellularity
   c. segmentation
   d. pseudocoelom

4. Which body structures do *not* develop from mesoderm?
   a. brain and nerve
   b. muscles
   c. arteries and veins
   d. reproductive organs

5. Which animal is bilaterally symmetrical?
   a. sponge   c. snail
   b. sea star   d. hydra

6. Internal digestion, symmetry, and tissues are features found in
   a. annelids but not in cnidarians.
   b. both nematodes and sponges.
   c. cnidarians but not in sponges.
   d. both protists and sponges.

7. Which of these animals is cephalized?
   a. sponge   c. jellyfish
   b. liver fluke   d. sea star

8. Which of the following is *not* a characteristic of mollusks?
   a. pseudocoelom
   b. circulatory system
   c. radula
   d. mantle

9. Which is *not* an example of a segmented animal?
   a. wasp
   b. earthworm
   c. roundworm
   d. spider

10. What do chordates and echinoderms have in common?
    a. gene-controlled development
    b. a head and a brain
    c. radially symmetrical larvae
    d. hollow dorsal nerve cord

Completion

11. Members of the phylum _____ are multicellular, but their cells are not organized into tissues or organs.

12. The process that leads to the formation of tissue layers during development is called _____ .

13. The skeleton of arthropods is called a(n) _____ because the muscles lie inside the skeleton. However, the skeleton of vertebrates is covered by muscles and is called a(n) _____ .

14. The phylum most closely related to the chordates is the _____ .

15. In protostomes, the first opening in the embryo becomes the _____ and the second opening becomes the _____ .

Short Answer

16. What is meant by the phrase "division of labor"? Why is it advantageous for the cells of an animal to exhibit division of labor?

17. The brain and the sense organs in most bilaterally symmetrical animals are located in a head region. Why is this arrangement advantageous?

Concept map: Kingdom Animalia — members are heterotrophic, ? ; includes sponges, ? ; allows cell specialization; lack jointed appendages; have ? ; cell specialization shown in ? ; organs can be part of ? ; jointed appendages can be part of ? ; ? made of chitin; organs such as ? ; such as digestive system.

17. Proximity of the brain to the sense organs enables rapid information transfer from the sense organs to the brain, thus allowing a quick response by the organism.

18. In a nematode, food is ingested in the front of the gut, broken down and absorbed in the middle region, and expelled in the last region. In contrast, food is taken in and expelled from the same opening in the flatworm gut.

19. The features are: a dorsal hollow nerve cord, a notochord, and slits in the pharynx.

18. How does the movement of food in a nematode compare with the movement of food in a flatworm?

19. What three features do all chordates have at some time during their lives?

Interpreting Graphics

20. Look at the diagrams below.

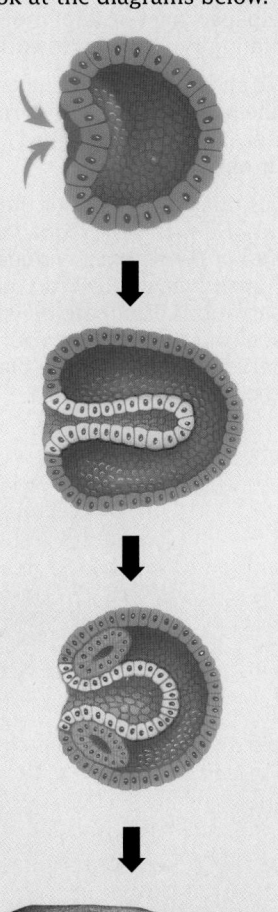

- What process is illustrated in these diagrams?
- Identify the mesoderm, endoderm, and ectoderm.

Reviewing Themes

21. *Patterns of Change*
    What evidence suggests that humans and sea stars are more closely related than humans and crabs?

22. *Evolution*
    How does segmentation provide for evolutionary flexibility?

23. *Structure and Function*
    List three functions performed by the jointed appendages of arthropods.

Thinking Critically

24. *Inferring Conclusions*
    Explain why animals that have a pseudocoelom are usually very small.

25. *Comparing and Contrasting*
    How is the gut of a jellyfish different from the gut of a roundworm?

26. *Building on What You Have Learned*
    In Chapter 15 you learned about how organisms are classified. Combine your knowledge of classification systems with your knowledge of body cavity types to explain why earthworms and sand worms are grouped in the phylum Annelida but flatworms are not.

Cross-Discipline Connection

27. *Biology and Health*
    Hookworms are parasitic roundworms that infect humans. Look for information in your library about the symptoms associated with hookworm infection. What can humans do to prevent becoming infected?

Discovering Through Reading

28. Read the article "Invasion of the Zebra Mussels," in *Discover*, January 1991, page 44. What problems are caused by zebra mussels in the Great Lakes? How are mussels being controlled?

Thinking Critically

24. Diffusion is the mechanism by which nutrients move from the gut to other parts of the body in the pseudocoelom. Because diffusion is such a slow process, the time it takes for nutrients to reach the cells of the animal necessitates that the animal be small.

25. The gut of a jellyfish has only one opening; food enters and waste is excreted through the same opening. The gut of a roundworm has two openings; food enters through the mouth and waste is excreted through the anus.

26. Earthworms and sandworms are quite different from flatworms in terms of body cavity. Earthworms and sandworms are segmented and have a coelom and a circulatory system. Flatworms, on the other hand, are acoelomates. They have a two-way gut, and lack a circulatory system.

Cross-Discipline Connection

27. The hookworm lives in the intestine of animals and humans, and its eggs exit the body in the feces. In locations where the climate is warm and sewage disposal and treatment are poor, the eggs hatch into larvae in the soil. People often come into contact with hookworms while walking barefoot on contaminated soil. Hookworm infection may be prevented by wearing shoes.

Discovering Through Reading

28. The mussel population is growing rapidly, clogging pipes, and causing shortages of phytoplankton needed by other animals in the Great Lakes. Controls tried include chlorine treatments and jet-spray washing.

Interpreting Graphics

20. • gastrulation
    • The mesoderm (red) forms in the space between the endoderm (yellow) and the ectoderm (blue).

Reviewing Themes

21. Humans and sea stars share the deuterostome pattern of development.

22. Segmentation enables greater differentiation, because changes to segments can result in segments with very different functions.

23. The appendages may be used for walking, sensing the environment, and as mouthparts.

Procedural Note

Review the structure of an earthworm, Pay particular attention to the role of the circular and longitudinal muscles of the body wall. Remind students to keep the earthworms on moist paper to avoid injuring the worms.

Prelab Preparation Answers

- Segmentation is the division of the body into generally similar, repeated parts.
- Vertebrate muscle is segmented, and so is the vertebral column.
- Annelids, arthropods, and chordates are segmented.

Procedure Answers

4. Each segment initially elongates as the circular muscles contract. It will then shorten as the longitudinal muscles contract. Students should observe an alternating pattern of elongation and shortening.

8. The rear balloon lengthens, forcing the front balloon forward. When released, however, the balloons return to their previous positions. In the earthworm, extension of the segment is followed by contraction and thickening. The segments do not slide back because they are anchored by setae.

9. Balloons do not slide back when the tip of the front balloon is anchored.

Chapter 21 Investigation

How Does Segmentation Help an Earthworm Move?

Objectives

In this investigation you will:
- *observe* a moving earthworm
- *relate* its pattern of locomotion to its segmented structure
- simulate an earthworm's movement

Materials

- pan
- paper towels
- water
- live earthworm
- long balloons
- graph paper

Prelab Preparation

- Review what you have learned about segmentation by answering the following questions:
- Define the word segmentation.
- Identify one example of segmentation in your body.
- What groups of animals are segmented?

Procedure: Observing Earthworm Movement

1. Form a cooperative team with another student to complete steps 2–9.

2. Line the bottom of a pan with six layers of paper towels. Thoroughly moisten the paper with tap water. The paper must be kept moist to avoid injury to the earthworm.

3. Place an earthworm at one end of the pan. Observe the worm closely as it moves. Record your observations.

4. Focus on one segment of the earthworm. *How does the diameter and length of the segment change as the worm moves?* Record your observations. Compare the shape changes in the segment you observe with those in the diagram of the earthworm below.

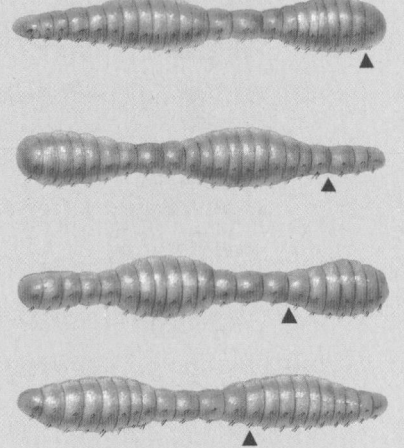

5. Carefully touch the underside of the earthworm. The bristles you feel are called setae. Setae occur on each segment of the earthworm. As the earthworm moves forward, setae anchor it to the soil so that it does not slip back and so that the segments have something to pull

against. Carefully return the earthworm to its container.

6. You will now simulate earthworm movement using two balloons containing water. From your teacher, get two balloons that have been tied together. Be careful with the balloons since rough treatment can easily break them. Each water-filled balloon represents one segment of an earthworm. Place the balloons in a straight line on a sheet of graph paper. Mark the position of each balloon on the graph paper.

7. Hold the rear balloon as shown in the figure below and gently squeeze it along its length. While squeezing the balloon, have your partner mark the position of the ends of each balloon on the graph paper. Your fingers simulate the action of muscles in the earthworm's body wall.

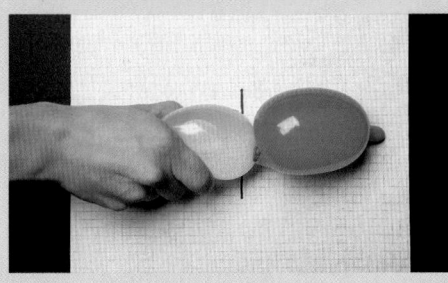

8. Now release the balloon. *What happens to the position of both balloons? How is this change different from what you saw in the earthworm's segments?*

9. Squeeze the rear balloon again. Hold the tip of the front balloon down on the graph paper. Now stop squeezing the balloon. *What happens to the position of both balloons?* Holding the tip of the balloon down simulates the effects of the earthworm's setae.

10. Clean up your materials and wash your hands before leaving the lab.

Analysis

1. *Evaluating Methods*
Like its exterior, the earthworm's coelom is segmented. Each segment of the coelom is separated from the others by membranes called septa. How is this pattern of internal division reflected in the balloon model of the earthworm?

2. *Inferring Relationships*
Relate the movement of fluid within a segment of the earthworm to the changes in shape you observed in that segment.

3. *Making Inferences*
Suppose the earthworm's coelom was not partitioned internally. Would the earthworm still be able to move in the same way? Explain.

4. *Making Inferences*
Explain why segments must be able to both elongate and shorten in order for the earthworm to move.

Thinking Critically

Explain why an earthworm's method of locomotion is not an effective way to move through water.

Analysis Answers

1. Each balloon represents a segment of the earthworm. As in the earthworm's segments, fluid in the balloons cannot flow from one balloon to the other.

2. Contraction of the circular muscles compresses the fluid, causing the segment to elongate.

3. Partitions between segments are critical to the earthworm's ability to move. If fluid could move between segments, contraction of one segment's circular muscles would propel fluid into the next segments. The extension of the segment that normally results from contraction of these muscles would not occur.

4. Elongation extends the body forward, while shortening pulls the back of the worm along behind.

Thinking Critically Answer

The earthworm can propel itself forward only if setae are able to attach to the substrate. In water, this attachment could not occur.

Chapter 22 Adaptation to Land

Planning Guide

Objectives/Themes	Classwork Resources	Homework Resources

22.1 **1.** List some animal groups that successfully made the transition to life on land.
 	
2. Explain the importance of the skeleton in arthropods and vertebrates on land.
 	
3. Describe the evolution of amphibian limbs from the fins of lobe-finned fishes.
 	
4. Relate the structure of the vertebrate ear to the differences between air and water.
 	
Themes: Evolution, Scale and Structure
	Teacher's Resource Binder
	Focus Activity 22
	  Comparing Animal Eggs
	
	Other Resources
	Transparency 98
	Science in Action Video
	  *Alternatives to Dissection:
	  The Earthworm and the Frog*★
		Text
		Section Review, p. 484
		
		Teacher's Resource Binder
		Directed Reading Worksheet
		  22.1
		
		Other Resources
		Audiocassette 22.1★

22.2 **1.** Recognize the role of the exoskeleton in preventing water loss.
 	
2. Contrast the skins of amphibians and reptiles.
 	
3. Summarize the evolution of lungs.
 	
4. Describe two ways animals rid themselves of wastes while conserving water.
 	
Themes: Evolution, Patterns of Change, Scale and Structure, Stability, Interacting Systems
	Text
	Journeys
	  Evolution of the Lung
	  pp. 490–491
	
	Other Resources
	Transparencies 99–100
		Text
		Section Review, p. 493
		
		Teacher's Resource Binder
		Directed Reading Worksheet
		  22.2
		
		Other Resources
		Audiocassette 22.2★

22.3 **1.** Recognize the advantages of internal fertilization for land animals.
 	
2. Compare a reptilian egg with an amphibian egg.
 	
3. Summarize two advantages of a placental mammal's development.
 	
Themes: Evolution, Patterns of Change, Scale and Structure, Stability
	Text
	Investigation
	  *How Are Anole Lizards Well
	  Adapted to Their Environment?*
	  pp. 500–501
	
	Teacher's Resource Binder
	Lab Investigation 22.1
	  *Insect Behavior and
	  Bioluminescence*
	Extension Worksheet
	  Osmotic Balance in Fish
		Text
		Section Review, p. 496
		
		Teacher's Resource Binder
		Directed Reading Worksheet
		  22.3
		Vocabulary Review Worksheet★
		Reteaching Worksheet★
		  *How Terrestrial Arthropods
		  Breathe*
		
		Other Resources
		Audiocassette 22.3★

★Reteaching Options

Demonstrations
22.1: pp. 481, 482, 483, 484
22.3: pp. 495

Assessment
Chapter Review pp. 398–499
Portfolio Assessment p. 479D
Chapter Test—Teacher's Resource Binder
Test Generator

Research Notes

Connection to Paleontology: The First Land Animals?

The movement of animals from the sea to the land was one of the most important steps in evolutionary history. For many years, paleontologists found no fossils of land animals from times before the early Devonian period, about 398 million years ago. Because this was millions of years after plants had already colonized the land, scientists theorized that animals did not adapt to land until plants had been there for some time. However, fossils found in 1990 point to a far more ancient origin for land animals.

The rocks were found in Shropshire, England, and were analyzed by a team led by Andrew Jeram of the Ulster Museum in Belfast, Northern Ireland, Paul A. Selden of the University of Manchester, and Dianne Edwards of the University of Wales.

To analyze the fossils, the researchers dissolved the rocks around the fossils in hydrofluoric acid. They were able to recover various pieces of animal exoskeletons, such as legs and trunk segments from centipedes. The scientists also found the body of a spiderlike arthropod that was 1.3 mm long, about the size of a flea.

The researchers had already determined that the rocks containing the fossils dated from the Silurian period, about 414 million years ago. During this time, the most complex plants on the continents were only a few millimeters high, and covered parts of the land like a carpet.

Since both of these arthropods are believed to have been predatory animals and not herbivores, scientists have begun to search for other early animals that may have served as prey for these creatures. The fossils' discoverers hypothesize that the prey were other arthropods that were so small they are hard to spot in fossils. They believe these smaller animals fed mostly on decayed plant matter.

This community of animals seems to have an anatomy that is fully adapted to land, with features such as strong legs. The researchers believe this indicates that the transitional species that first crawled out of the oceans might be even older than these land animals. If they are correct, it may signal a need to modify the traditional theories. Instead of millions of years between the time when the plants colonized the land and the time when they were followed by the first land animals, the two events could have happened much more closely in time.

Investigation Notes

How Are Anole Lizards Well Adapted to Their Environment?

pages 500–501

Purpose: This investigation gives students an opportunity to observe some of the advanced characteristics of reptiles. The anole, also known as the American chameleon, exhibits a number of easily observable adaptations, one of which is the ability to change its coloration. Although background color may affect the animal's coloration, variables such as temperature, emotional state, and light intensity play far more significant roles.

Prelab Preparation

1. Anole lizards should be kept in a large terrarium containing twigs and branches for climbing. Since these animals will not drink from a dish, spray the plants each day with water and place some cotton balls soaked with water in the aquarium. They can be fed *Tenebrio* larvae (mealworms) or crickets. The animals do best in a temperature range of 18°C to 26°C.

2. Do not feed the lizards for two days before they are to be used. Cover one-half of the floor of a dry aquarium with green construction paper, the other half with orange construction paper. Place the aquarium under a fluorescent light. At least one hour before class, put the lizards in the aquarium.

Answers will be found on pages 500–501.

Meeting Individual Needs

Objectives

1. Students will demonstrate appropriate use of core vocabulary for the chapter (Vocabulary File).

2. Students will identify and describe features that enable animals to live on land. (Demonstration A).

3. Students will identify and describe advantages of internal fertilization (Demonstration B).

Vocabulary File
(Developing Vocabulary/ Limited English Proficiency)

If you are not already using the Vocabulary File, refer to Chapter 1 for its preparation. See the Chapter Highlights on page 497 for a list of suggested words.

Demonstration A
Staying Moist in a Dry World
for pages 481–493
(Developing Observational Skills/Visual Learners)

Materials: hand magnifiers, gills extracted from a preserved crayfish and a preserved perch, lungs extracted from a preserved frog, live crayfish and live perch, dissecting trays for gills and lungs, and pictures of crayfish and perch suitable for hanging as mobiles

Preparation: Leave the banner "Welcome to the Animal Kingdom" from Chapter 21 above the doorway for the rest of the chapters about animal studies. Add the crayfish and perch mobiles to those of animals previously studied. Two or three days

before teaching this chapter, set up a tank of live crayfish and perch in the classroom. Often, students know of a "crawdad" hole and are happy to bring some to school. Oatmeal is a clean food that can be eaten by the crayfish during the few days they will be in the room.

Procedure: Read aloud to students "Supporting the Body," pages 482–483. Reteach the concepts of segmentation, endoskeleton, and exoskeleton from Chapter 21. Ask students to describe the types of skeletons found in the crayfish and the perch. Take two small live crayfish from the tank. Place them on a large glass plate on the top of a desk. Caution students not to poke the crayfish with their fingers.

Have students brainstorm about adaptations shown by the crayfish that enable it to live on land for short periods of time. Be sure students notice the use of legs on land, and the exoskeleton that holds water long enough for the crayfish to be able to live a little while on land. As long as there is water between the carapace and the gills, the crayfish will suffer no damage. This can be for several minutes if the room is not too warm.

Be sure to place crayfish back into the tank of water before they dry out too much. Read aloud to

students "Watertight Skin" and "Gas Exchange," on pages 485–487. Brainstorm about the differences of gills in the two live specimens. Be sure students realize the reasons why the fish must stay in water.

Students should closely examine the two sets of gills in the dissecting trays, comparing them to each other and to the lungs of the frog.

Demonstration B
Internal Fertilization
for pages 494–496
(Developing Observational Skills/Visual Learners)

Materials: several three-day incubated chicken eggs from a feed or a poultry store, petri dishes and color pencils

Procedure: Have students scan Section 22.3. Then, lead a discussion about internal fertilization. Be sure students discuss the characteristics required for eggs that are internally fertilized.

Next, break an incubated egg into a petri dish. The egg should be at the stage of development that clearly shows all the parts illustrated in Figure 22.16, page 495. Students should draw the egg that is in the petri dish, labeling the drawing using the text as a guide. Have them color the embryo brown and the yolk yellow, with red blood vessels.

Relate the structures in the students' drawings to those of eggs without shells, as described on page 496.

Additional Strategies
Visual Strategies
Pages 486, 487, 488, 489, 493, 494, and 496

Auditory Learners
Use *Biology: Visualizing Life* Audiocassettes for Sections 22.1, 22.2, and 22.3.

Meeting Individual Needs (cont.)

Cooperative Learning

Adapting to Land

Timing: Use this activity to conclude Section 22.3.

Group Size: 5 students

Outcome: Students will be able to apply examples of adaptations that allow animals to leave the sea.

Individual Accountability: Each group member is responsible for information on one adaptation or aspect of life on dry land.

Positive Interdependence: Each group will complete a description and a sketch of a hypothetical animal capable of making the transition from sea to land.

Assign one of the following challenges to life to each group member: (1) dehydration; (2) body support; (3) gas exchange; (4) waste removal; and (5) internal fertilization.

Each group member should scan the chapter for information about the adaptations needed to meet these challenges of life on dry land, and report the information to the other members of the group.

The group should then work together to describe and sketch a hypothetical animal that has adaptations to meet these challenges. Emphasize that this is a chance for students to use creativity. The goal is not a single "correct" creature, but rather an imaginary creature that could survive dry conditions.

Portfolio Assessment

Students should select their best work and provide a self-reflective rationale for their selections. Students can make selections in the following areas.

1. *Content* — One concept map from the chapter (See page 812 for evaluation criteria.)

2. *Reading Comprehension* — One Directed Reading Worksheet from the Teacher's Resource Binder (Use the answer key to evaluate for accuracy.)

3. *Writing* — Using the Vee Form, summarize a magazine or newspaper article relating to land animals, especially to gas exchange and internal fertilization. (See page 22T for evaluation criteria.)

   Or: Select a writing task or project from the Chapter Review.

4. *Performance Assessment* — One Vee form from a chapter investigation or lab manual investigation (See page 22T for evaluation criteria.)

Teacher makes selections in the following areas.

1. *Formal Assessment* — Chapter test (Test A, B, or the Test Generator) The teacher-scored test should be reviewed by the student. Incorrect responses should be corrected by the student before the test becomes part of the portfolio.

2. *Informal Assessment* — Use the Direct Observations Checklist, page 33T, during a laboratory or other cooperative learning experience.

3. *Performance Assessment* — Have students create a chart comparing and contrasting the different components of eggs for arthropods, fish, amphibians, reptiles, birds, and mammals, and detailing adaptations to life on land.

Concept Map Answer

The following is one possible answer to the Relating Concepts exercise on page 498.

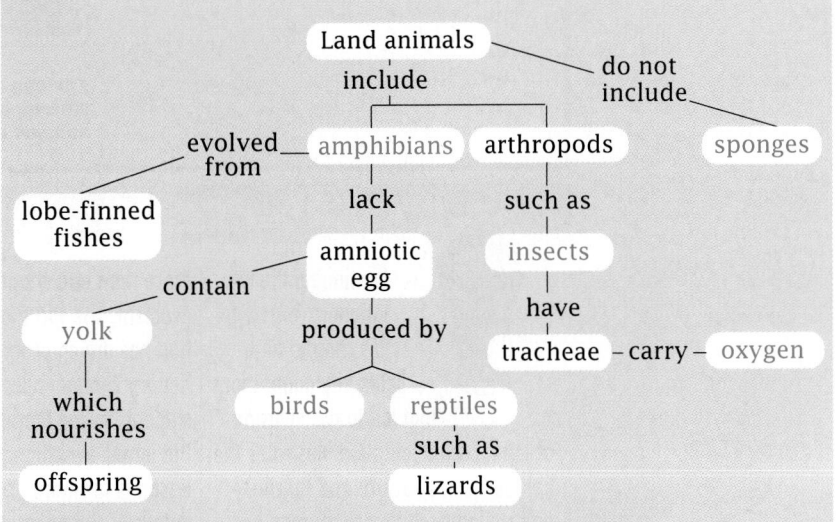

Chapter 22

Determining Prior Knowledge

- Have students imagine and draw how the first land animal might have looked.
- Have students explain what problems animals would have to solve when moving from an aquatic to a terrestrial environment.
- Have students explain why an amniotic egg that resists drying is so important for land animals.

Chapter 22

Review

- diffusion (Section 4.2)
- innovations in body plan (Section 21.3)
- the terms *endoskeleton* and *exoskeleton* (Glossary)

Adaptation to Land

The egg from which this young snapping turtle is emerging is one of the many adaptations that land animals evolved in their move from the sea onto land.

■ Author's Rationale ■

Continuing the evolutionary journey, this chapter highlights key adaptations that led to the invasion of land. Each adaptation illustrates the link between structure and function, and students see that there may be more than one adaptive solution to a problem. The focus is on how adaptation has driven progressive evolutionary change, a lesson that permits students to see themselves as part of the larger evolutionary picture. This lesson is far more important than the details of the adaptations, which are treated as stories rather than taxonomic details.

LIFE AROSE IN THE SEA AND STAYED THERE MORE THAN 3 BILLION YEARS UNTIL A PROTECTIVE SHIELD OF OZONE FORMED. WITHOUT THIS SHIELD, AQUATIC ANIMALS COULD NOT LIVE ON LAND. THE REQUIREMENTS FOR LIVING ON LAND ARE VERY DIFFERENT FROM THOSE FOR LIVING IN THE SEA. IN THIS CHAPTER, YOU WILL EXAMINE SOME KEY PROBLEMS ANIMALS FACED IN MOVING ONTO LAND.

22.1 Leaving the Sea

Objectives

① List some animal groups that successfully made the transition to life on land.

② Explain the importance of the skeleton in arthropods and vertebrates on land.

③ Describe the evolution of amphibian limbs from the fins of lobe-finned fishes.

④ Relate the structure of the vertebrate ear to the differences between air and water.

Which Animals Live on Land?

The animal body evolved over many millions of years in the sea. All of the major changes in body plan you read about in Chapter 21 took place in the sea. But the evolutionary journey of animals did not end there.

From the sea, animals invaded the land. Of the major animal phyla, only sponges, cnidarians, and echinoderms—animals that pump sea water through their bodies—were left behind. Members of every other major phylum can be found on land. Flatworms live in damp leaf litter. Nematodes and annelids burrow through the soil, and snails creep over damp ground at night. These animals, however, are found mainly in moist habitats. Only two groups of animals have fully adapted to life on dry land: arthropods and vertebrates such as the mudskipper shown in **Figure 22.1**. How these two groups evolved ways to survive the many challenges of living out of water is one of biology's most fascinating stories.

Figure 22.1
This mudskipper is an animal that has adapted to life in water and on land. Underwater, it breathes through gills. On land, it can absorb oxygen through the lining of its mouth. Mudskippers use their enlarged front fins to scurry across mud flats.

▪ *Matter of Fact* ▪

Walking Catfish
Do walking catfish walk? No. The walking catfish makes its way from one pond to another using its side fins and tail to help it crawl over the ground.

Lesson Plan 22.1

Phase 1
PREPARATION

Key Concepts
- Arthropods and vertebrates are the most successful land animals.
- Land animals have strong flexible limbs for moving on land.
- Many land animals have sturdy skeletons for support.
- Sound travels much faster in water than in air.

Reading Strategy

After reading the material in this section, have students work in cooperative groups of three to develop three concept maps: *Animals on Land, Body Support,* and *Hearing.*

Phase 2
TEACHING STRATEGIES

Demonstration 1
What's the Difference?
Provide the class with specimens of a fish and a terrestrial mammal. (A goldfish and a hamster are good choices.) If live animals are not available, provide pictures of any fish and any terrestrial mammal from nature books or magazines. Have students suggest how each is adapted to living either in water or on land. Have students explain what characteristics would have to change in order for fishes to live on land and for the mammal to live in water. Remind students that cetaceans—whales and dolphins—have readapted to life in water. Have students explain ways dolphins are unlike fishes. Students should suggest lungs, viviparity, and endothermy.

Demonstration 2

Appendicular Comparisons

Have students examine specimens or photographs of skeletons of fishes, snakes, turtles, frogs, mammals, etc. Have students look for the similarities and differences in the appendicular skeletons and speculate about reasons for the differences. Have them discuss the behavior of the animals in Figure 22.2c and 22.2d that would require the differences in the appendages.

Figure 22.2
Several changes in limb structure took place as terrestrial vertebrates evolved from aquatic vertebrates.

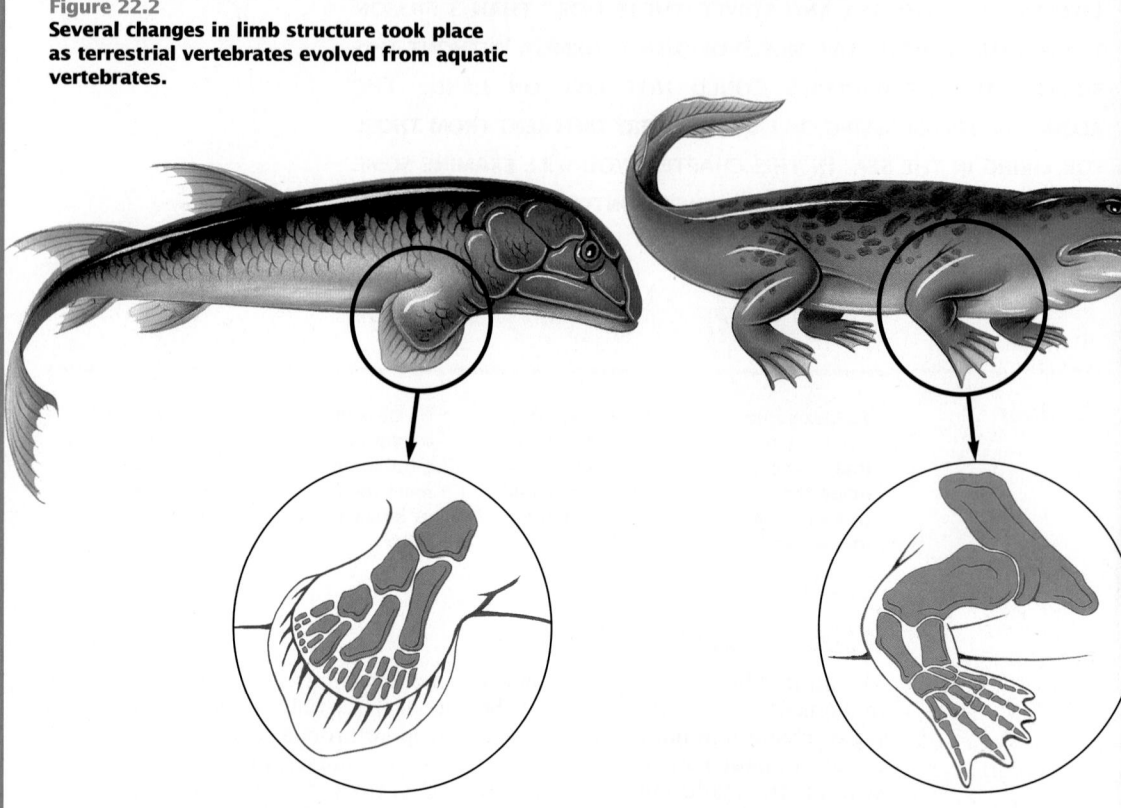

a Scientists think that a lobe-finned fish was the ancestor of the earliest land vertebrates. The drawing above shows the bones in the fin of an extinct lobe-finned fish.

b The earliest known land vertebrate is the amphibian *Ichthyostega*, which lived about 350 million years ago. It had four sturdy legs. Like its fish ancestors, it had a tail fin and some bony scales.

Supporting the Body

Water is about 1,000 times denser than air. Because of water's density, you can float on your back in a pool. For some aquatic animals, water provides much of the support necessary to keep their bodies from collapsing under the pull of gravity. That is why a jellyfish stranded on the beach cannot maintain its shape.

A variety of adaptations enabled animals that left the sea to overcome the loss of physical support. In nematodes and earthworms, the body cavity is filled with fluid under pressure that helps stiffen the body. In arthropods and vertebrates, support of the body is largely taken over by the skeleton. A land animal's skeleton holds up its body against gravity, much like beams and girders hold up a skyscraper.

Limbs play an important role in supporting vertebrates on land. When a terrestrial animal is standing, its legs bear the entire weight of its body. In this way, legs function like the pillars that hold up the roof of a building. Unlike pillars, however, animal legs have flexible joints where movement occurs.

In both arthropods and vertebrates, legs evolved from limbs adapted for movement in water. **Figure 22.2a–b** illustrates the evolution of amphibian limbs from the limbs of fishes. Recall

Demonstration 3

Comparing Locomotion in Vertebrates

Have students observe a salamander swimming in an aquarium. Suggest that they observe its movements from above and then from the side as it swims. Have them describe their observations. Place the salamander in a terrarium and have students observe the animal both from above and from the side. Ask students if the salamander had difficulty walking on land. What impeded its movement?

Next, place a lizard in the terrarium. Ask how the lizard's movement differs from that of the salamander? Place a snake in water and observe it from above as it swims. Remove it from the water and place it in the terrarium. Ask students if the snake uses its muscles on land in the same way it does in water. Explain that it is difficult to observe the function of the snake's scales in locomotion. The scales keep the snake from sliding backward. Have students observe their own legs and compare them with the legs of other vertebrates.

Front view of amphibian posture

Front view of reptilian posture

c The limbs of a modern amphibian, such as the salamander above, are positioned at the sides of its body. When it walks on land, its belly drags on the ground.

d The limbs of a reptile, such as the iguana above, are located partway beneath its body. These limbs carry greater body weight than amphibian limbs do, and help reptiles run faster.

from Chapter 10 that amphibians, the first vertebrates on land, evolved from fishes. The limbs of amphibians are homologous to fins of fishes. But most fishes have thin, paddlelike fins that would be useless for supporting and moving an animal on land. As you can see in **Figure 22.2a–b**, the arrangement and structure of bones in the limb of an early amphibian are similar to those in the fin of a lobe-finned fish. Because of these similarities, scientists think the first amphibians were descendants of the lobe-finned fishes, a group whose modern members include the coelacanth and the lungfishes.

Compare the posture of the amphibian in **Figure 22.2c** to that of the reptile in **Figure 22.2d**. Amphibian limbs are short and join the body horizontally; an amphibian's belly drags on the ground as it walks. Reptilian limbs are positioned slightly beneath the body. Limbs positioned in this way support more body weight than do the limbs of amphibians. Mammalian limbs are positioned directly beneath the body, raising the belly well off the ground. Compared with reptilian and amphibian limbs, limbs located directly beneath the body support the animal's weight more effectively. They also enable higher running speeds and give the animal a greater field of vision, helping it locate food and be alert to danger.

Demonstration 4

"Hearing" Through the Jaw

Assign students to groups of two. Have one member of each group rest his or her chin on the surface of a desk (making sure that no other part of the body touches the desk). The other group member should tap on the desk with a pen or pencil. Then have the first student plug his or her ears while the other taps on the desk. Have students switch roles and repeat the demonstration. Students should contrast the quality and volume of sound with ears unplugged (when sound is transmitted both through air and jaw) with that when the ears were plugged (when transmission mainly occurs through the jaw, as in snakes.)

Phase 3

ASSESSMENT OPTIONS

Closure Strategy

Have students speculate on future evolutionary trends in animals that may cause changes in their physical appearance. Have them explain their reasoning. Accept any well-reasoned suggestion. Remind them again about the readaptation of cetaceans to water.

Section Review

Assign the *Section Review*.

Reteaching

Provide students with descriptions of specific insect habitats. Have students create hypothetical insects and patterns of development best suited for survival in the habitat described.

Hearing Airborne Sounds

Most of the senses work as well in air as in water, but hearing is an exception. In the ocean, whales are able to communicate with low-pitched "songs" that travel as far as 200 km (125 mi.) because water is much denser than air. The denser the medium through which sound travels, the faster and farther the sound can travel.

Water's density makes detecting sound in water easier than doing so on land. Most fishes can hear. In addition, fishes are able to sense physical disturbances in the water, such as those caused by the swimming movements of nearby fishes, with their **lateral line** system. The lateral line is a row of pressure-sensitive cells that lie within a fluid-filled canal. One network of canals runs down each side of the fish. The lateral line is shown in **Figure 22.3**.

In air, the pressure changes caused by sound are much smaller because of air's lower density. Terrestrial animals have evolved structures that amplify airborne sounds so that they are detectable.

The ear is the sound amplifying system of land vertebrates. In mammals,

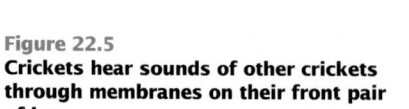

Figure 22.3
Fishes can sense sounds or the movements of other fishes with a row of pressure-sensitive cells called the lateral line.

Lateral line

Figure 22.4
Many mammals, such as this kit fox, can point their ears in the direction of a sound to pinpoint where the sound is coming from.

such as the kit fox in **Figure 22.4**, the eardrum and three small bones magnify sound waves and transmit them to the inner ear, where sounds are detected. In salamanders and snakes, sound is conducted to the inner ear entirely through bones of the jaw and skull. These animals rely more on ground vibrations (which transmit low-frequency sounds even better than does water) than on airborne sounds. You will learn more about the workings of the ear in Chapter 28.

Insects can hear well, but they evolved different solutions to the challenges of hearing in air. Some insects detect sounds with sensitive hairs on their legs, body, or head. Others have eardrum-like structures on the body or legs. Crickets, for instance, hear the songs of other crickets through membranes on their front pair of legs, as you can see in **Figure 22.5**.

Figure 22.5
Crickets hear sounds of other crickets through membranes on their front pair of legs.

Hearing mechanism

Section Review

1. **Which groups of animals are not found on land?**

2. **Describe the role of the skeleton in the life of a terrestrial animal.**

3. **What evidence suggests that amphibian limbs evolved from the fins of lobe-finned fishes?**

4. **Compare sound transmission in air to sound transmission in water.**

▪ *Section Review Answers* ▪

1. sponges, cnidarians, and echinoderms
2. The skeleton provides support by holding up the body against gravity.

3. The arrangement and structure of bones in the limbs of early amphibians are similar to that of the fin of the lobe-finned fish. See Figure 22.2 starting on page 482.

4. Water is much denser than air, so sounds travel much farther and faster in water.

CONSERVING BODY WATER IS ONE OF THE GREATEST CHALLENGES OF LIVING ON LAND. WHEN ANIMALS FIRST MOVED ONTO LAND, THEY FACED THE RISK OF LOSING EXCESSIVE AMOUNTS OF WATER THROUGH EVAPORATION. IN RESPONSE, MANY TERRESTRIAL ARTHROPODS AND VERTEBRATES EVOLVED TRAITS ENABLING THEM TO CONSERVE BODY WATER.

22.2   *Staying Moist in a Dry World*

Objectives

❶ **Recognize the role of the exoskeleton in preventing water loss.**

❷ **Contrast the skins of amphibians and reptiles.**

❸ **Summarize the evolution of lungs.**

❹ **Describe two ways animals rid themselves of wastes while conserving water.**

Watertight Skin

If you leave a wet towel in the sun, what happens to it? The towel dries by evaporation. Since an animal's body is about 70 percent water, an animal on land is like a wet towel. Its cells are full of water and, like the fibers of the wet towel, are surrounded by water. Like the towel, an animal on land is in danger of drying out by evaporation. Losing even small amounts of body water can be dangerous. Losing 15 percent to 20 percent of the water in your body would be fatal. How do terrestrial animals prevent large amounts of water loss? In this section, you will examine some of the adaptations that enable animals like the desert lizard in **Figure 22.6** to exist on dry land.

Figure 22.6
Many desert lizards, such as this African spiny-tailed lizard, have specialized scales on the head that can trap dew and funnel it to their mouths.

Arthropods and vertebrates have waterproof coatings

What would happen if you sealed a wet towel in a plastic bag? Because the bag is waterproof, water could no longer escape from the towel into the atmosphere. For the same reason, sandwiches stay moist when they are sealed in sandwich bags. The bodies of arthropods and vertebrates are wrapped in a watertight coating. In arthropods, this coating is the exoskeleton. In vertebrates, it is the skin.

Recall from Chapter 21 that the arthropod exoskeleton is a stiff outer coat composed of chitin. In addition to providing protection and physical support, the exoskeleton stops water from leaving the terrestrial arthropod's body. In spiders and insects the outer layer of the exoskeleton contains waxes that enhance the waterproofing effect of the exoskeleton.

Phase 2

TEACHING STRATEGIES

Visual Strategy

Figure 22.8
Have students explain why gills would not be efficient mechanisms of gas exchange on land if not kept moist. Students should suggest that gills would collapse without the support of water and would quickly dry out in air.

Figure 22.7
Reptiles have dry, scaly skin that is watertight. The skin of a banded rock rattlesnake is shown above.

Vertebrate skin can be moist and thin, or dry and thick

Have you ever touched a frog? What did its skin feel like? Amphibians secrete a slippery mucus that is responsible for their slimy texture. Like the wax on an insect, this mucus coating helps limit evaporation. Despite this mucus coating, the skins of amphibians are not watertight. As you will see in the next section, amphibians absorb oxygen and release carbon dioxide directly through their skin. To do this, their skin must be moist and thin. Therefore, to keep from drying out, amphibians must remain in water or in moist environments.

A reptile is not slimy like an amphibian. Reptilian skin is dry and covered with tough scales, as shown in **Figure 22.7**. A watertight skin of scales was a significant evolutionary step, completely freeing reptiles from the necessity of living in a wet environment. Mammals and birds, which evolved from reptiles, also have skin that is dry and relatively watertight.

Gas Exchange

You breathe about 12 times each minute. Have you ever considered why you breathe? From the perspective of water conservation, breathing is very costly. About 300 mL (9 oz.) of water evaporates from your lungs each day. You breathe, despite the water loss, because breathing is essential for gas exchange. Gas exchange is the process of absorbing oxygen from the environment and ridding the body of carbon dioxide. This "swap" of molecules is necessary for all animals, since they require oxygen to carry out cellular respiration, and they release carbon dioxide as a waste product.

Gas exchange occurs by diffusion, which you learned about in Chapter 4. Oxygen diffuses across the cell membrane into the cell, and carbon dioxide diffuses out. The cells that make carbon dioxide and use oxygen must be close to the environment, because diffusion works well only over short distances. Recall from Chapter 21 that flatworms and nematodes cannot be thick, since diffusion alone carries oxygen to their innermost tissues. The chordates, arthropods, mollusks, echinoderms, and some annelids can be large because they carry out gas exchange with specialized structures.

Gills are structures for aquatic gas exchange. Aquatic arthropods, fishes, and amphibian larvae have gills. Amphibian larvae have external gills. The gills of crabs and many other aquatic arthropods are located in chambers behind the head, as shown in **Figure 22.8**. The gills of fishes also lie directly behind the head, as you can see in **Figure 22.9**.

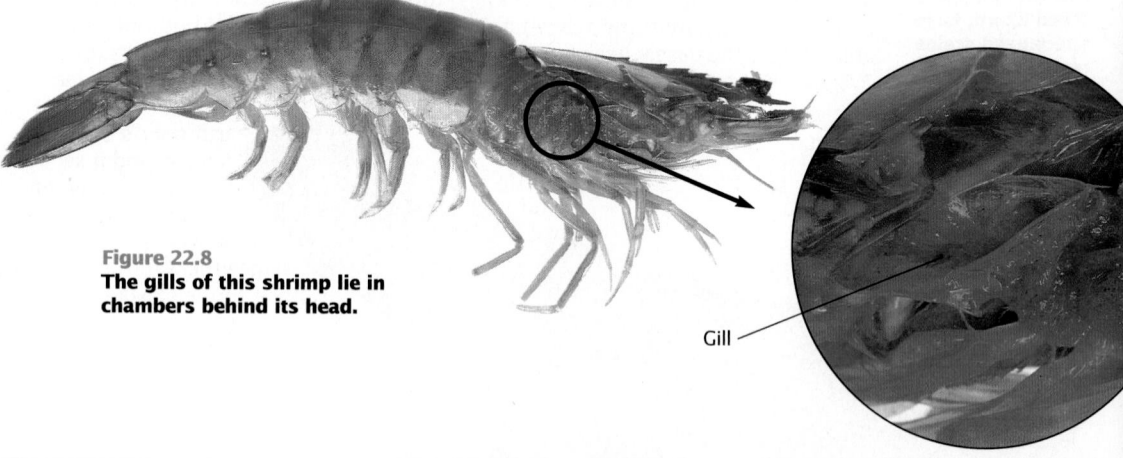

Figure 22.8
The gills of this shrimp lie in chambers behind its head.

Gill

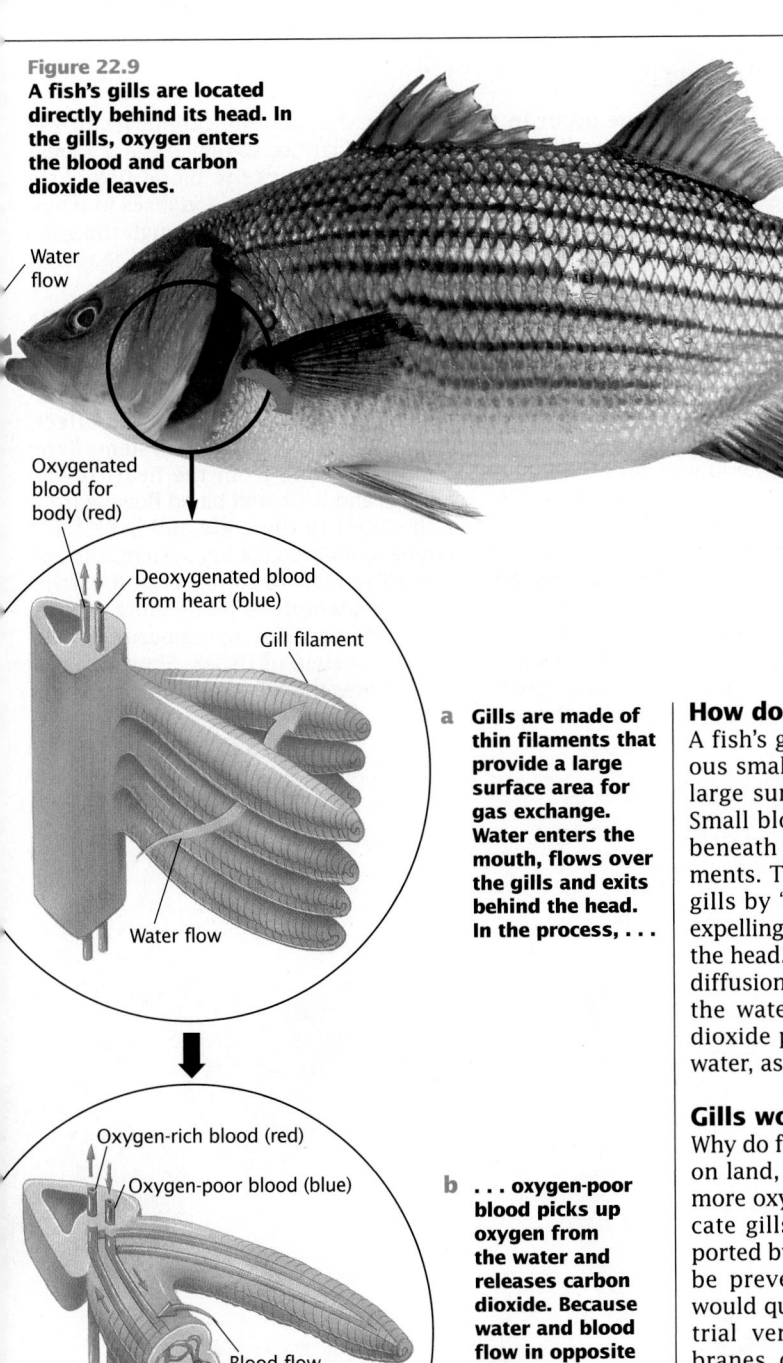

Figure 22.9
A fish's gills are located directly behind its head. In the gills, oxygen enters the blood and carbon dioxide leaves.

Water flow

Oxygenated blood for body (red)

Deoxygenated blood from heart (blue)

Gill filament

Water flow

a Gills are made of thin filaments that provide a large surface area for gas exchange. Water enters the mouth, flows over the gills and exits behind the head. In the process, . . .

Oxygen-rich blood (red)

Oxygen-poor blood (blue)

Blood flow

Blood vessels

Water flow

b . . . oxygen-poor blood picks up oxygen from the water and releases carbon dioxide. Because water and blood flow in opposite directions, diffusion is particularly efficient.

How do gills work?

A fish's gills are composed of numerous small filaments, which provide a large surface area for gas exchange. Small blood vessels carry blood just beneath the surface of the gill filaments. The fish pumps water over its gills by "swallowing" water and then expelling it through an opening behind the head. As water flows over the gills, diffusion occurs. Oxygen passes from the water to the blood, and carbon dioxide passes from the blood to the water, as shown in **Figure 22.9**.

Gills would collapse on land

Why do fishes suffocate when stranded on land, even though air contains far more oxygen than water? A fish's delicate gills would collapse if not supported by water. Even if the gills could be prevented from collapsing, they would quickly dry out in air. In terrestrial vertebrates, respiratory membranes deep within the body have replaced gills. These respiratory membranes line the interior surfaces of sacs called lungs.

487

Theme Answer

Stability

Sea water has a higher concentration of salts than vertebrate tissues. Thus, ocean-dwelling vertebrates lose water by osmosis.

Visual Strategy

Figure 22.10

Point out that oxygenated and deoxygenated blood mix in the single atrium of the amphibian heart. This mixing does not occur in the four-chambered hearts of birds, mammals, and crocodiles.

Vertebrate Lungs

Stability

Vertebrates living in the ocean also lose water to their environment. Explain why this occurs.

How does gas exchange occur in the lungs? Vertebrate lungs are lined with moist, thin tissues through which gas exchange occurs. A network of fine blood vessels called capillaries runs close to this lining. Oxygen-rich air is brought into the lungs by inhalation. Oxygen diffuses across the lining of the lungs into the capillaries. At the same time, carbon dioxide moves from the capillaries into the air in the lungs. Exhalation expels the "used" air from the lungs. You can follow the evolution of vertebrate lungs in the *Evolution of the Lung* feature on pages 490–491.

Terrestrial vertebrates have a double-loop circulatory system

In fishes, the heart pumps blood from the heart to the gills, where the blood picks up oxygen before flowing to the rest of the body. This creates the "single-loop" system, shown in

Figure 22.10a. The capillaries of the gills are narrow, so they present a great deal of resistance to blood flow. As a result, the flow of blood loses much of its force as it passes through the gills. Circulation from the gills to the rest of the body is sluggish.

The fine capillaries of an amphibian's lung also slow down blood flow. In amphibians, however, the blood returns to the heart for repumping before circulating to the body's tissues. In effect, there are two circulatory systems here: blood flowing from the heart to the lungs and back, and blood flowing from the heart to the body and back. This type of dual circulatory system is found in all terrestrial vertebrates. It carries oxygenated blood much more rapidly through the body than does the circulatory system of fishes. **Figure 22.10b** illustrates the "double-loop" circulation of an amphibian.

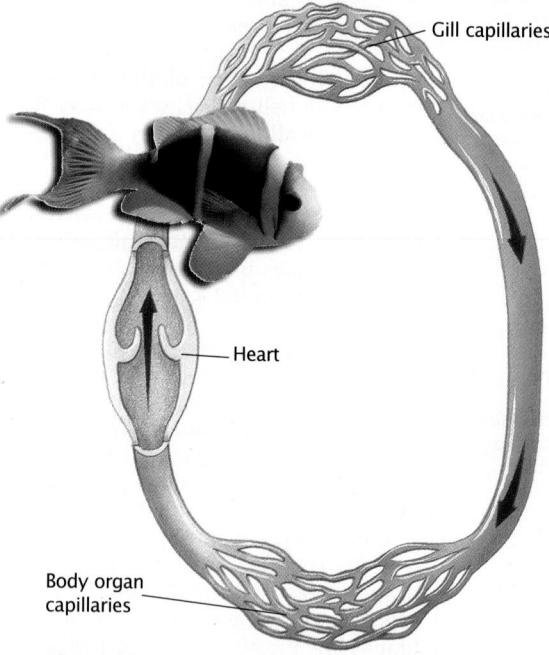

Gill capillaries

Heart

Body organ capillaries

Figure 22.10

a In fishes, the heart pumps blood to the gills, where it picks up oxygen. The oxygenated blood then flows to the rest of the body. This creates a "single-loop" system.

Lung capillaries

Heart

Body organ capillaries

b In amphibians and other land vertebrates, the heart first pumps blood to the lungs, where it picks up oxygen. The oxygenated blood returns to the heart to be pumped to the rest of the body, creating a "double loop

■ *Matter of Fact* ■

Avian Air Sacs

Since flight is energetically demanding, a bird requires much oxygen. The air sacs in birds are adaptations that maintain a supply of oxygen-rich air in the lungs while adding buoyancy.

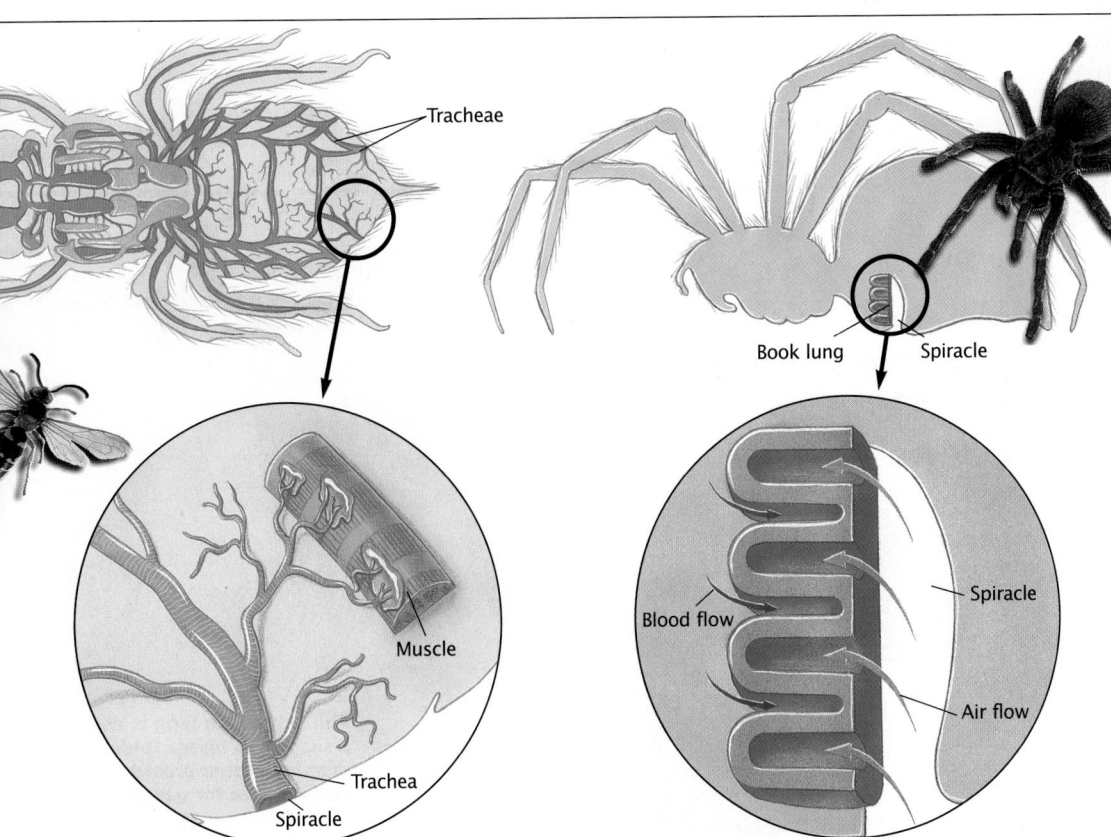

Visual Strategy

Figure 22.11
Have students explain how the tracheal system places a limit on insect size.

Figure 22.11
Tracheae leading into an insect's body from pores on the surface deliver oxygen directly to the cells and transport carbon dioxide outside the body.

Figure 22.12
In spiders, oxygen is absorbed through book lungs. The many folds of the book lungs provide a large surface area through which diffusion can occur.

How Terrestrial Arthropods Breathe

Land-dwelling arthropods, which are much smaller than most vertebrates, evolved different gas exchange systems. Insects breathe through **tracheae** (*TRAY kee ee*). Tracheae (singular, trachea) are tubes that lead into an insect's body cavity from pores called spiracles in the surface of the exoskeleton, as shown in **Figure 22.11**. Each trachea branches and rebranches, penetrating deep into the body. The tips of the tiny branches are open and carry oxygen directly to cells where it is needed. They also allow carbon dioxide to exit. This system is a very direct and efficient way of delivering oxygen and picking up carbon dioxide, and so is a good design to support the very active lifestyles of

most insects. Notice that oxygen passes directly from the tracheae to the tissues. In your body, oxygen is first exchanged in the lungs, and then is carried to the tissues by blood. Insect blood does not transport oxygen.

Most spiders breathe through **book lungs**. As shown in **Figure 22.12**, book lungs are highly folded sacs, which are located inside the body, that open to the outside through a spiracle. Air moves into the sacs, where it is moistened so the oxygen in the air can dissolve and pass into the blood. Having respiratory membranes inside the body allows the surface where gas exchange occurs to remain moist without losing large amounts of water to evaporation.

489

Using the Feature

Although they are a hallmark of terrestrial vertebrates, lungs first evolved in lobe-finned fishes, the group that includes today's coelacanth and lungfishes. The earliest fishes with lungs probably lived in shallow, oxygen-poor tropical waters. The advantages of air breathing in such environments are obvious. Lungs probably began as small outpocketings from the pharynx and served as auxiliary respiratory structures. Students may have seen aquarium fishes (which do not have lungs) gulping air at the surface. The early fishes with lungs may have obtained air in a similar way.

Only seven extant fish species—the lungfishes of Africa, South America, and Australia—have lungs. The swim bladder characteristic of most bony fishes is homologous to the lung, however.

Discussion

Guide the discussion by posing the following questions.

1. Lungs are absent in salamanders of the family Plethodontidae. How do you think these animals carry out gas exchange?
    through the skin and lining of the mouth

2. Why can't reptiles conduct gas exchange in the same way as the salamanders of Question 1?
    The scaly, watertight reptilian skin is not permeable to oxygen and carbon dioxide.

3. How does the reptilian lung compensate for the lack of skin respiration?
    It contains many alveoli, giving it a much larger surface area than an amphibian lung.

4. Identify the advantage of the bird's two-cycle respiratory system.
    Fully oxygenated air is always in the lungs.

Evolution of the Lung

Land vertebrates evolved respiratory membranes deep within their bodies, providing a large surface area where gas exchange can occur in a moist environment.

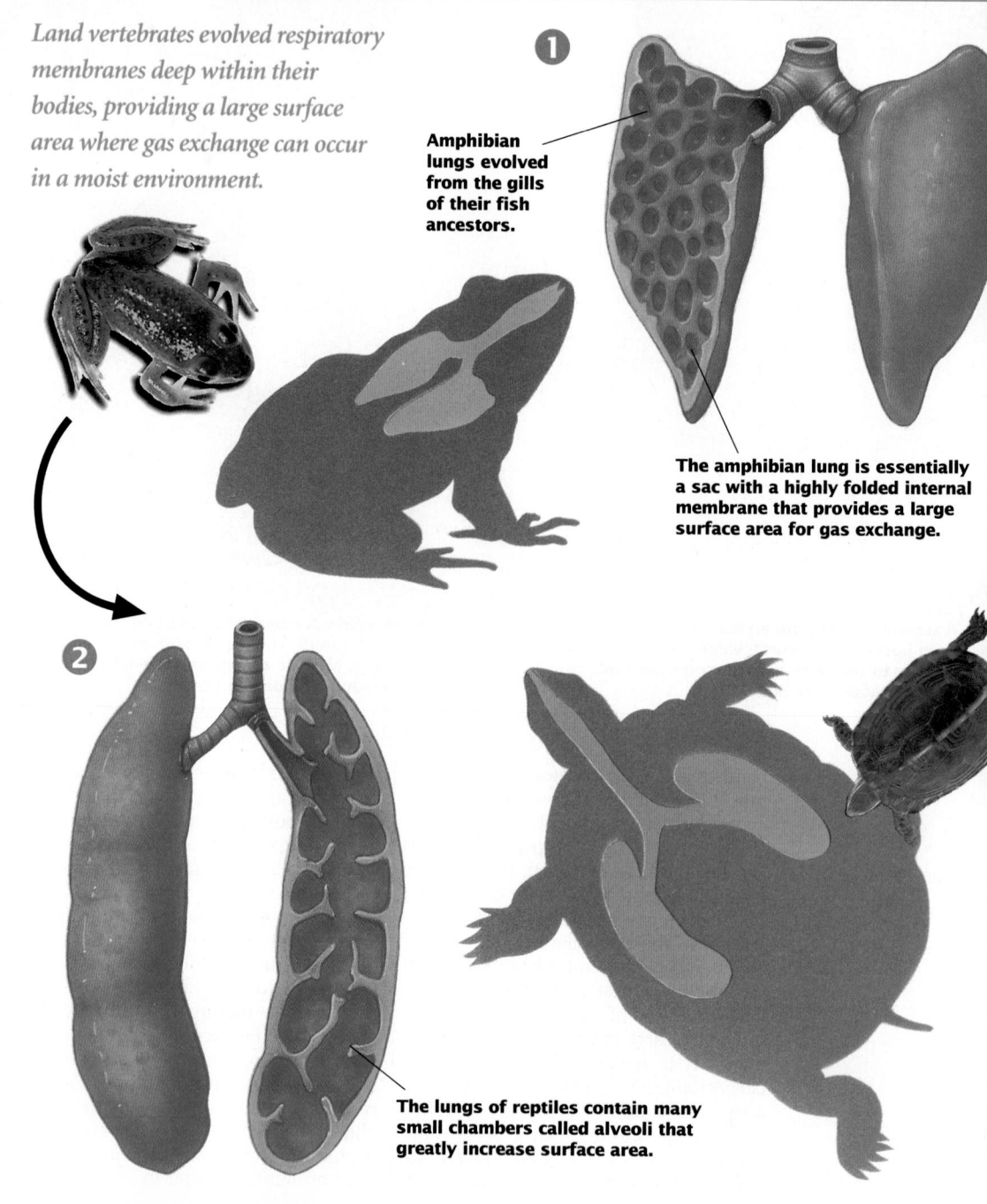

❶ Amphibian lungs evolved from the gills of their fish ancestors.

The amphibian lung is essentially a sac with a highly folded internal membrane that provides a large surface area for gas exchange.

❷

The lungs of reptiles contain many small chambers called alveoli that greatly increase surface area.

490

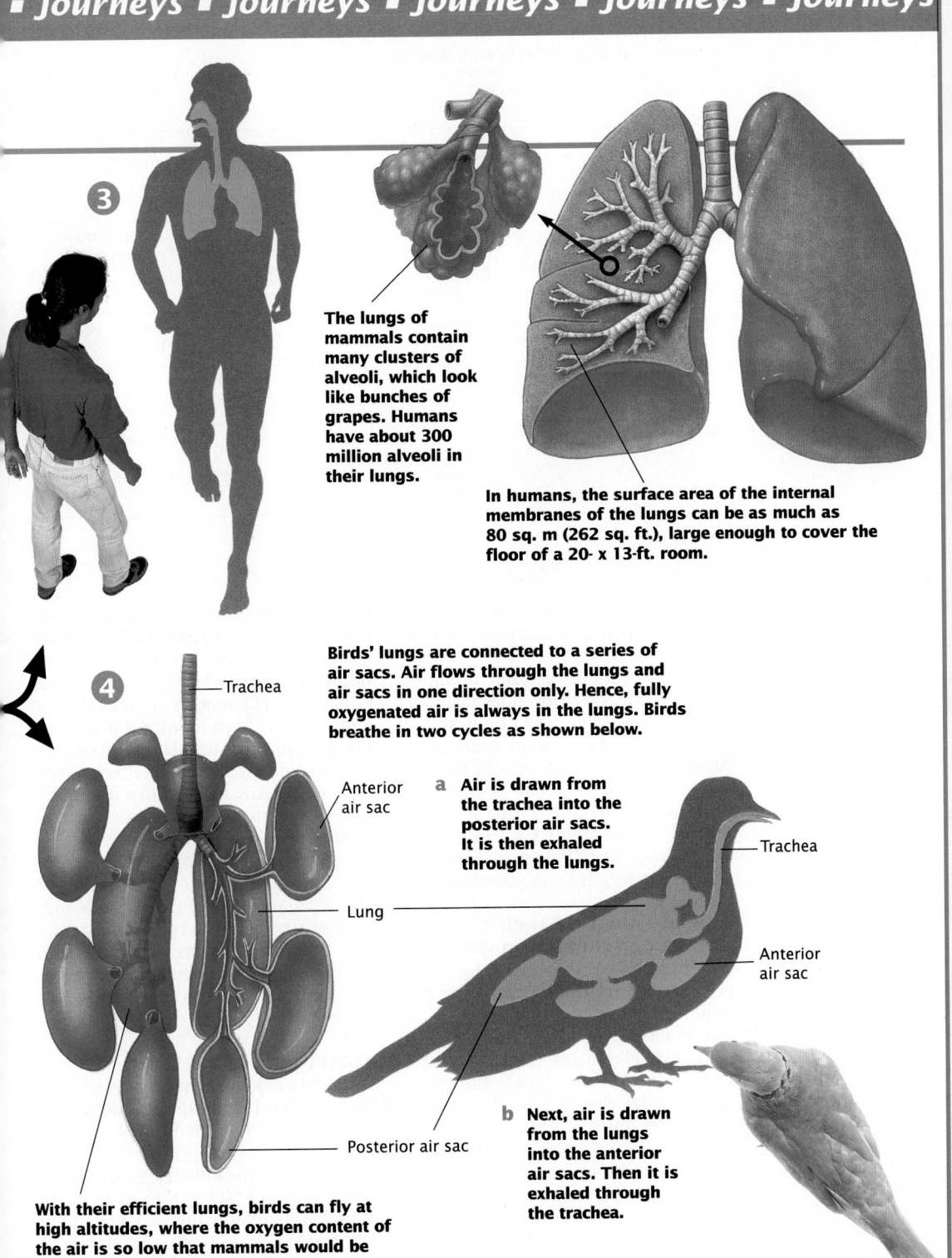

③

The lungs of mammals contain many clusters of alveoli, which look like bunches of grapes. Humans have about 300 million alveoli in their lungs.

In humans, the surface area of the internal membranes of the lungs can be as much as 80 sq. m (262 sq. ft.), large enough to cover the floor of a 20- x 13-ft. room.

④

Trachea

Birds' lungs are connected to a series of air sacs. Air flows through the lungs and air sacs in one direction only. Hence, fully oxygenated air is always in the lungs. Birds breathe in two cycles as shown below.

Anterior air sac

a Air is drawn from the trachea into the posterior air sacs. It is then exhaled through the lungs.

Trachea

Lung

Anterior air sac

Posterior air sac

b Next, air is drawn from the lungs into the anterior air sacs. Then it is exhaled through the trachea.

With their efficient lungs, birds can fly at high altitudes, where the oxygen content of the air is so low that mammals would be unable to get enough oxygen to move.

491

Theme Answer

Stability

Kidneys help land animals maintain homeostasis by removing nitrogen-containing wastes from the blood by regulating the amount of water and salt in the body.

Table 22.1 Nitrogen-Containing Wastes

Type	Structure (nitrogen in yellow)	Solubility in water	Animal type
Ammonia		Soluble	Most aquatic animals
Urea		Soluble	Amphibians, mammals
Uric acid		Insoluble	Insects, reptiles, birds
Guanine		Insoluble	Spiders

Getting Rid of Wastes While Conserving Water

Stability

What role do the kidneys play in homeostasis?

Despite its covering of skin, your body is not completely watertight. Every day you lose about 2.5 L (2.4 qt.) of water—about 2 percent of the total amount of water in your body. How does that much water escape each day? About 300 mL (9 oz.) evaporates from your lungs. Another 500 mL (15 oz.) evaporates from your skin as sweat, cooling you in the process. The majority of the water you lose each day, about 1.5 L (45 oz.), is excreted as urine. Urine carries nitrogen-containing wastes from your body.

Animals must rid their bodies of nitrogen-containing wastes

When animals break down the amino acids found in foods containing protein, nitrogen is released as ammonia. Ammonia is toxic to animals and must be quickly eliminated from the body in very dilute form. Since the water required for dilution is plentiful for freshwater animals, they excrete their nitrogen directly as highly dilute ammonia. Land animals, however, cannot afford to lose so much body water. Instead, they spend some energy converting the ammonia to less toxic forms that need less water for dilution.

In vertebrates, ammonia is converted into one of two different compounds in the liver. Most mammals excrete nitrogen as **urea**, a less toxic compound of nitrogen. Elimination of urea requires some water, because urea is toxic if highly concentrated. Many birds and reptiles (and insects) excrete their waste nitrogen as **uric acid**. Unlike urea, uric acid is a nontoxic solid, and so can be eliminated with minimal water loss. (The white "deposits" birds leave on cars and statues is a combination of uric acid and feces.) Once produced, urea or uric acid are released by cells into the blood and are removed from the blood by the kidneys. The nitrogen-containing wastes are illustrated in **Table 22.1**.

Kidneys are blood filters

Vertebrate kidneys remove wastes from the blood and regulate the amounts of water and salts in the body. Blood circulates in the vertebrate body under pressure, and the kidneys act as filters, much as a car's oil filter does. As blood flows through the kidneys, its pressure forces water and dissolved substances such as sodium, potassium, chloride,

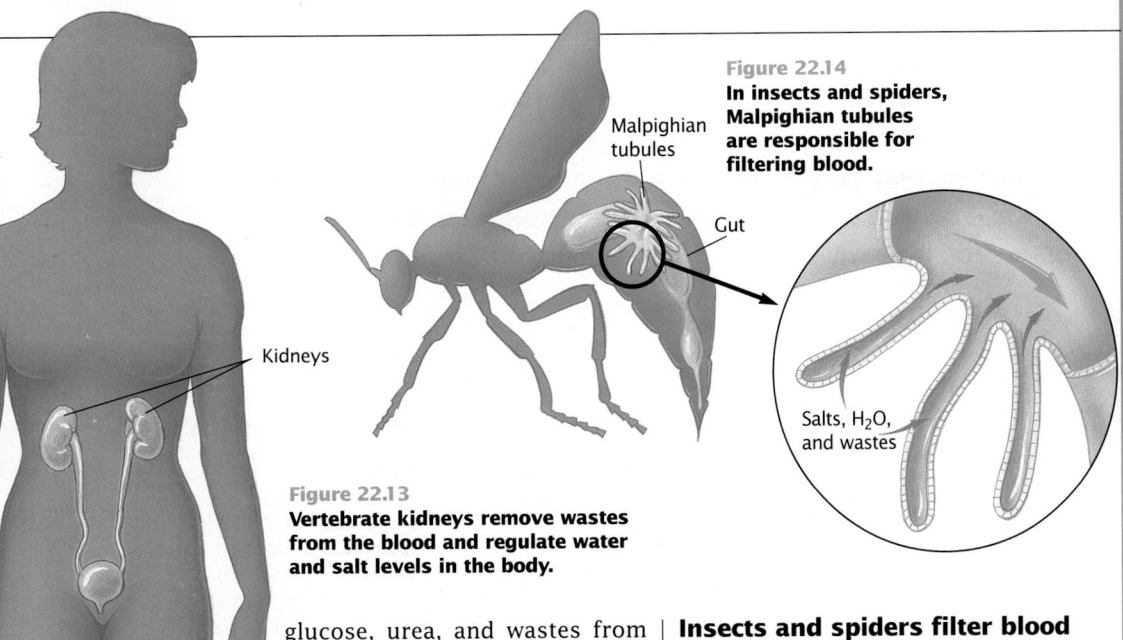

Figure 22.14
In insects and spiders, Malpighian tubules are responsible for filtering blood.

Malpighian tubules

Gut

Salts, H₂O, and wastes

Kidneys

Figure 22.13
Vertebrate kidneys remove wastes from the blood and regulate water and salt levels in the body.

glucose, urea, and wastes from blood vessels into a series of coiled tubes. Thus the kidneys are pressure filters, as is the oil filter in a car. Vertebrate kidneys are shown in **Figure 22.13**. Proteins and blood cells remain behind in the blood.

Most of the filtered substances are useful to the body, and it would be wasteful to discard them. Instead they are reabsorbed back into the bloodstream. Reabsorption is very selective. Urea, wastes, and some water are not reabsorbed, and pass out of the body as urine. The body regulates how much water is reabsorbed and therefore can control how much water leaves the body as urine. For instance, if you drink a large amount of water, your brain senses an increase in blood pressure and releases hormones that cause the kidneys to reabsorb less water. As a result, you produce more urine.

Insects and spiders filter blood through Malpighian tubules

Since blood does not circulate under pressure in insects and spiders, these terrestrial animals do not have pressure-filter kidneys. The organs responsible for filtering blood in insects and spiders are the **Malpighian** (*mal PIHG ee uhn*), **tubules** shown in **Figure 22.14**. Salts and wastes are actively transported from the blood into the Malpighian tubules. The high concentration of salts in the Malpighian tubules *draws* water from the blood. The contents of the Malpighian tubules are released into the gut, where useful salts such as potassium are reabsorbed. Water follows the salts by osmosis. Wastes such as uric acid and guanine are not reabsorbed and remain in the gut as a nearly dry paste that is excreted with the feces. Malpighian tubules are very efficient water conservation organs.

Section Review

① **How would the exoskeleton of a land arthropod be different from that of an aquatic arthropod? Explain.**

② **Contrast the amount of evaporation from amphibian skins with that from reptilian skins.**

③ **Describe the differences between amphibian lungs and mammalian lungs.**

④ **List two adaptations that enable land vertebrates to get rid of wastes while still conserving water.**

Visual Strategy

Figures 22.13 and 14
The excretory systems of fishes, birds, mammals, and arthropods differ in their structure and their function. As students saw in Table 22.1, the tasks for removing waste products are different based on the type of waste product. Have students explain how these differences are related to living on land.

Phase 3

ASSESSMENT OPTIONS

Closure Strategy
Systems Adaptations
Have students work in small cooperative groups to prepare poster-size tables illustrating how the skeletal, the integumentary, the circulatory, the respiratory, and the excretory systems of each group of animals are adapted to water, land, or both. Along with the text, the tables can contain graphics.

Section Review
Assign the *Section Review*.

Reteaching
For each of the following systems, have students write a description of one adaptation to a terrestrial environment: skeletal, integumentary, circulatory, respiratory, and excretory systems. Sample adaptations include adult amphibian limbs, reptile scales, and amniote eggs. Have students briefly explain and support their answers.

▪ *Section Review Answers* ▪

1. The exoskeleton of land arthropods is covered by a waxy coating for conserving water. Aquatic arthropods lack this coating.

2. Amphibians exchange gases through their skin, thus constantly losing moisture. To keep from drying out, amphibians must remain in water or moist environments. Reptilian skin is dry and covered with tough scales, preventing water loss.

3. See "Evolution of the Lung" on pages 490–491.

4. kidneys, uric acid as a nitrogenous waste

Lesson Plan 22.3

Phase 1

PREPARATION

Key Concepts
- Amphibians and fishes produce eggs without shells that are fertilized externally.
- Most land animals produce watertight eggs that are fertilized internally.
- Development in placental mammals is completed within the mother's body.

Reading Strategy

Have students make two concept maps from this section. One on internal fertilization; the other on eggs without shells. Within the scope of their map, external fertilization should also be compared.

Phase 2

TEACHING STRATEGIES

Visual Strategy

Figure 22.15
Have students begin at the left and work their way to the right of the picture considering how each animal mates and lays eggs; the whereabouts of the mating and the laying; and whether the parent exhibits any parental care during incubation or after hatching.

REPRODUCING ON LAND IS VERY DIFFERENT FROM REPRODUCING IN WATER. THE ADAPTATIONS TO LAND THAT YOU HAVE SEEN SO FAR WOULD NOT HAVE EVOLVED IF ANIMALS HAD NOT ALSO EVOLVED ADAPTATIONS THAT ENABLE THEM TO REPRODUCE ON LAND. IN THIS SECTION, YOU WILL SEE SOME OF THE ADAPTATIONS THAT MAKE IT POSSIBLE FOR TERRESTRIAL ANIMALS TO REPRODUCE FAR FROM WATER.

22.3 Reproducing on Land

Objectives

❶ **Recognize the advantages of internal fertilization for land animals.**

❷ **Compare a reptilian egg to an amphibian egg.**

❸ **Summarize two advantages of a placental mammal's development.**

Internal Fertilization

Figure 22.15
Each animal shown below lays a different type of egg. The toads, for instance, lay strings of eggs in a gelatinous coating that is permeable to water.

The toads you see in **Figure 22.15** are amphibians that have come from miles around to breed in the pond. The behavior of these toads reflects the strong ties between amphibians and water. Amphibians must reproduce in water or in damp environments. As you will see in this section, insects, mammals, birds, and reptiles have broken the reproductive tie to water.

When egg and sperm unite, they must do so in a moist environment. For many aquatic animals, including most fishes, finding a moist environment is easy: they release their gametes into the surrounding water. Fertilization takes place outside the body of either parent. This type of fertilization is called **external fertilization**. External fertilization is less common on land, because both sperm and egg run the risk of drying out and dying.

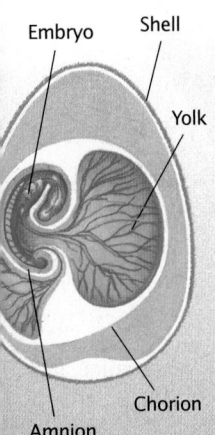

Figure 22.16
Bird and reptilian eggs, such as the turtle egg below, have a watertight protective coating called the chorion and an internal membrane called the amnion, which encloses the embryo in a watery environment.

Embryo   Shell

Yolk

Chorion

Amnion

Most terrestrial animals reproduce by internal fertilization

How did terrestrial animals overcome the limitations of external fertilization? Recall that each animal carries its own supply of water around inside its body. Thus, if gametes can be transferred directly between individuals, the risk that gametes will dry out disappears. In almost all land animals, including flatworms, roundworms, annelids, mollusks, arthropods, mammals, reptiles, and birds, the male deposits his sperm directly inside the female. Fertilization occurs inside the female's body. This type of fertilization is called **internal fertilization**. Most amphibians lack internal fertilization. They reproduce by external fertilization, as did their fish ancestors. Now you can see one reason why amphibians such as toads must return to water to reproduce.

If you go to a pond a few days after toads breed, you will find that the water contains strings of small, round eggs, like those shown in **Figure 22.15** (far left). These eggs are the second reason that amphibians must reproduce in water or moist places. The embryonic toad develops within the egg's jellylike coating. This coating is freely permeable to carbon dioxide, oxygen, and water.

Amphibian eggs removed from moisture soon dry out and die.

The eggs of reptiles and birds are watertight

Unlike amphibian embryos, the embryos of reptiles are surrounded by a watertight protective membrane called the **chorion** (*KAWR ee ahn*). The chorion is impermeable to water, but it does permit oxygen to enter the egg and carbon dioxide to leave. Lying within the chorion is another membrane, the **amnion** (*AM nee uhn*). The amnion encloses the embryo within a watery environment, as shown in **Figure 22.16**. This kind of egg is called an **amniotic egg**. Amniotic eggs evolved first in reptiles and are also found in birds and mammals. In the eggs of reptiles and birds, a rich yolk supplies nutrition to the developing embryo. A tough shell surrounds and protects the eggs of birds, reptiles, and three species of egg-laying mammals.

Like the eggs of birds and reptiles, the eggs of most terrestrial arthropods are watertight and very yolky, providing nutrients for the young during development. Arthropod eggs are usually either thick-walled and resistant to drying or are encased in some sort of waxy protective covering.

Demonstration 1
Observing a Chicken Egg
Soak a chicken egg overnight in vinegar to dissolve the shell so that students can see inside of it. Although it is not likely that a store-bought egg will be fertile, it can still be used to illustrate the placement of the yolk and the amnion membrane. If possible, use a fertile egg to demonstrate exactly what is seen in this figure. Realize, of course, that doing this will kill the embryo.

▪ *Matter of Fact* ▪

External vs. Internal Eggs
Soft, externally fertilized eggs are vulnerable to predators and to changes in the environment. That is why externally fertilizing animals produce more eggs than internally fertilizing animals. There is an inverse relationship between the number of offspring produced and the amount of parental care invested on each.

Visual Strategy

Figure 22.17
If possible, borrow a latex model of a placenta with umbilical cord from a community college's anatomy and physiology department. Also, if possible, locate a specimen preserved in formalin.

Phase 3

ASSESSMENT OPTIONS

Closure Strategy
Animal Development
Review what students know about the birth and parental care of puppies and kittens, the laying and hatching of frog eggs, and the laying, incubating, and hatching of fowl. Ask them to describe the advantages of each strategy.

Section Review
Assign the *Section Review.*

Reteaching
Provide students with descriptions of fictional animals. Include their habitat and their way of life. Have students infer the best type of fertilization and development for their imaginary animals and have them defend their choices.

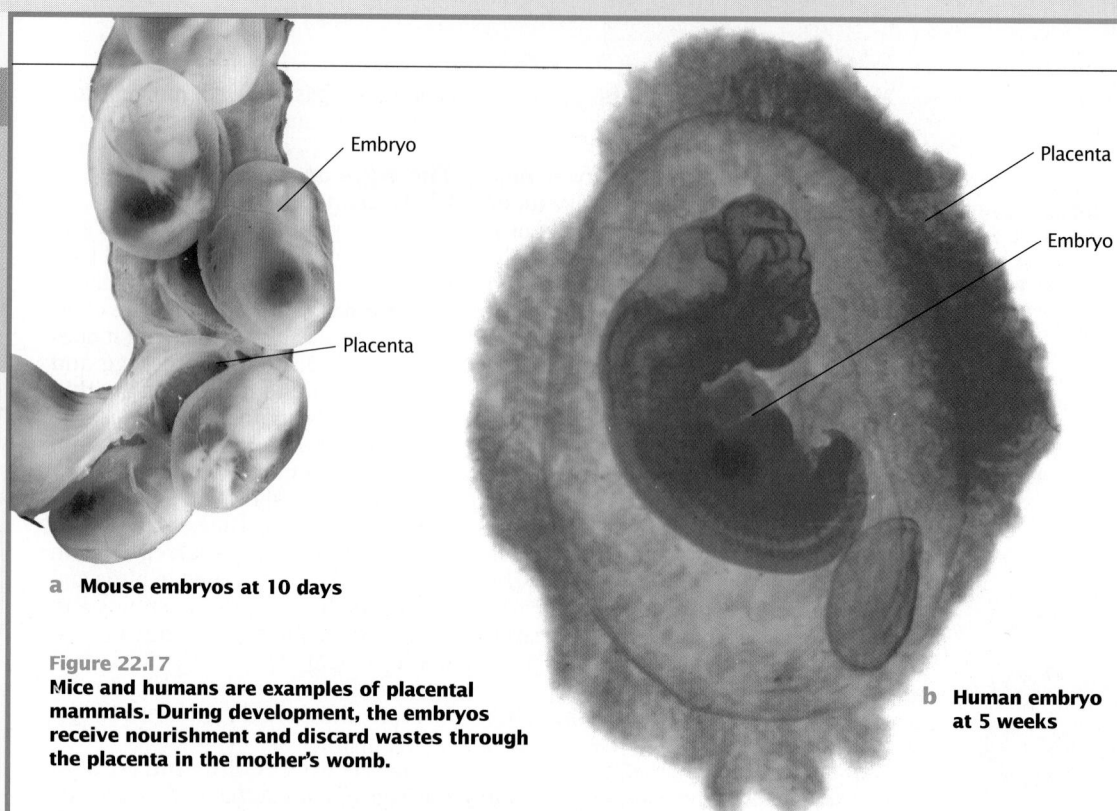

a   **Mouse embryos at 10 days**

Figure 22.17
Mice and humans are examples of placental mammals. During development, the embryos receive nourishment and discard wastes through the placenta in the mother's womb.

b   **Human embryo at 5 weeks**

Eggs Without Shells

Once the egg of a reptile or bird has been laid, the parent cannot provide further nourishment to the offspring until hatching. Furthermore, the egg is exposed to environmental hazards such as predators, overheating, or freezing.

Except for platypuses and echidnas (*ee KIHD nuhz*), mammals do not lay eggs. Instead, their eggs develop until birth inside the mother's body. No shell forms around the egg. Among marsupial mammals, such as kangaroos and opossums, birth occurs very early, and the offspring matures in the mother's pouch, nourished by suckling her milk.

Among placental mammals (mice and humans, for instance), birth occurs much later, when development is essentially complete. A human embryo develops for nine months inside its mother. During its development, the embryo of a placental mammal is nourished by the mother through a unique structure called the **placenta** (*pluh SEHN tuh*), as shown in **Figure 22.17**. The placenta is made of embryonic and maternal membranes. Nutrients and oxygen move from mother to embryo through the placenta, while carbon dioxide and other wastes leave the embryo.

Section Review

❶ Why isn't external fertilization an effective way to reproduce on dry land?

❷ Describe the advantages of the shelled egg over the amphibian egg.

❸ Describe two hazards faced by reptilian eggs but not by eggs of placental mammals.

▪ Section Review Answers ▪

1. Students should suggest that both the sperm and the egg could dry out.
2. Students should suggest that amphibian eggs removed from moisture would soon dry out, whereas shelled eggs are watertight.
3. Answers may vary. Students should suggest that reptilian parents cannot provide further nourishment to the unhatched offspring, and that the egg is exposed to environmental hazards, such as predators and temperature extremes.

Chapter 22 *Highlights*

"Gee, evolution is slow."

Alternative Assessment
Have students make a summary concept map based on the information found on the highlights page. Encourage students to include transition linkages between the section information of their maps.

	Key Terms	Summary
22.1 Leaving the Sea  **Reptiles have limbs partway beneath their bodies to support their weight more effectively.**	lateral line (p. 484)	• Life on land places different demands on animals than life in water. • The most successful land groups are the arthropods and the vertebrates. Both of these groups have sturdy skeletons that support their bodies out of water. They also have strong flexible limbs for moving on land. • The high density of water makes sound easy to detect. Land animals hear well only with an amplifying system, such as the ear.
22.2 Staying Moist in a Dry World **Amphibian blood is pumped through a "double-loop" system.** 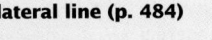	gill (p. 486) trachea (p. 489) book lung (p. 489) urea (p. 492) uric acid (p. 492) Malpighian tubule (p. 493)	• The exoskeleton of arthropods serves as a barrier to evaporation. Birds, reptiles, and mammals minimize water loss by means of their watertight skins. • Animals obtain oxygen from the air and release carbon dioxide. In land vertebrates, this gas exchange occurs in the lungs. In insects, tracheae carry oxygen. • Breakdown of amino acids produces the toxic byproduct ammonia. Aquatic animals excrete dilute ammonia. Terrestrial animals transform ammonia into urea or uric acid, which is eliminated with less water loss.
22.3 Reproducing on Land 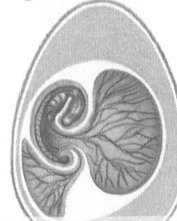 **Reptilian eggs have a protective shell and a watery internal environment.**	external fertilization (p. 494) internal fertilization (p. 495) chorion (p. 495) amnion (p. 495) amniotic egg (p. 495) placenta (p. 496)	• Most amphibians reproduce in water by external fertilization. Most land animals reproduce by internal fertilization. • The eggs of reptiles, birds, and mammals are surrounded by watertight membranes. In most mammals, development is completed within the mother's body.

Chapter Review Answers

Understanding Vocabulary

1. **a.** Kidneys do not fit the pattern because they are not gas exchange structures.
   b. Turtle does not fit the pattern because its eggs are not fertilized externally and are not laid in water.
   c. Blood is not a nitrogen-containing waste that is excreted by animals.
   d. Malpighian tubules are not a component of the amniotic egg.

Relating Concepts

2. Map answer is shown on page 479D.

Understanding Concepts

Multiple Choice

3. c	8. a
4. c	9. d
5. d	10. b
6. a	11. b
7. b	12. c

Completion

13. lateral line
14. lungs, book lungs
15. urine
16. kidneys, Malpighian tubules

Short Answer

17. You could tell by touching it. An amphibian feels slimy, but a reptile is dry and covered with scales.
18. As the density of a medium increases, the speed and distance traveled by sound also increase.
19. The land animals would become dehydrated and would die, due to the large amounts of water needed to dilute ammonia.
20. The kidney filters waste and useful substances from the blood. The useful substances such as potassium, chloride, and glucose are reabsorbed back into the bloodstream, while the waste—urea or uric acid—is removed.

Chapter 22 Review

Understanding Vocabulary

1. For each list of terms, identify the one that does not fit the pattern and describe why it does not fit.
   a. kidneys, lungs, tracheae, gills
   b. turtle, fish, toad, frog
   c. blood, ammonia, urea, uric acid
   d. chorion, amnion, shell, Malpighian tubules

Relating Concepts

2. Copy the unfinished concept map below onto a sheet of paper. Then complete the concept map by writing the correct word or phrase in each oval containing a question mark.

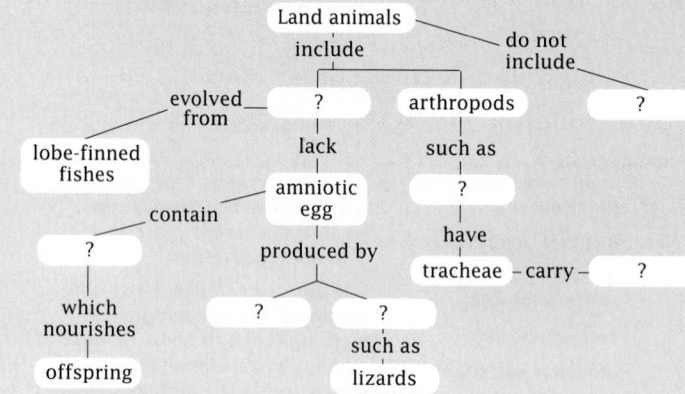

Understanding Concepts

Multiple Choice

3. Which animal group lives on land?
   a. sponges
   b. cnidarians
   c. roundworms
   d. echinoderms

4. Water serves to counteract the force of gravity for aquatic animals. What body structure takes the place of water for terrestrial vertebrates and arthropods?
   a. blood plasma   c. skeleton
   b. skin   d. lungs

5. Amphibian limbs are thought to be modified
   a. fish scales.   c. reptile limbs.
   b. lateral lines.   d. fish fins.

6. The waterproof coating that covers vertebrates and helps prevent water loss is called
   a. skin.   c. body armor.
   b. endoskeleton.   d. exoskeleton.

7. Which animal has the most efficient respiratory system?
   a. cow   c. lizard
   b. bird   d. crab

8. Compared with the single-loop circulation of a fish, the double-loop circulation of an amphibian has
   a. increased rate of blood flow.
   b. decreased ammonia removal.
   c. decreased rate of blood flow.
   d. increased oxygen removal from tissues.

9. In which group of land vertebrates does fertilization usually occur outside the female's body?
   a. mammals   c. birds
   b. reptiles   d. amphibians

10. Which of the following is a placental mammal?
    a. opossum   c. kangaroo
    b. mouse   d. toad

11. Gas exchange occurs in the gills of crabs and fishes by
    a. osmosis.
    b. diffusion.
    c. active transport.
    d. proton pumps.

12. The nitrogen-containing waste excreted by birds and reptiles is
    a. ammonia.   c. uric acid.
    b. urine.   d. urea.

Completion

13. Fishes detect disturbances in water with their _____ .
14. In mammals, gas exchange occurs in the _____ , while in spiders it occurs in the _____ .
15. Most of the water lost from the human body is eliminated as _____ .
16. In mammals, wastes are removed from the blood by the _____ . In insects, the same function is performed by the _____ .

21. Water provides the support and moisture required for gills to function. Out of water, gill membranes collapse and dry out.

Interpreting Graphics

22. • fish
    • a
    • b

Reviewing Themes

23. The similarities in limb bones between the earliest land vertebrate and lobe-finned fishes led scientists to believe that the two forms were related.

Short Answer

17. Your friend caught an animal in her backyard. How could you tell if it is a reptile or an amphibian without harming the animal?

18. Explain the relationship between the density of a medium and the speed and distance that sound travels through it.

19. Fish eliminate nitrogen-containing wastes as ammonia. What complications would be associated with land animals also eliminating waste in this form?

20. Describe how kidneys enable a mammal to remove wastes and regulate how much water is retained by the body.

21. Explain why gills don't function out of water.

Interpreting Graphics

22. Look at the figure shown below.

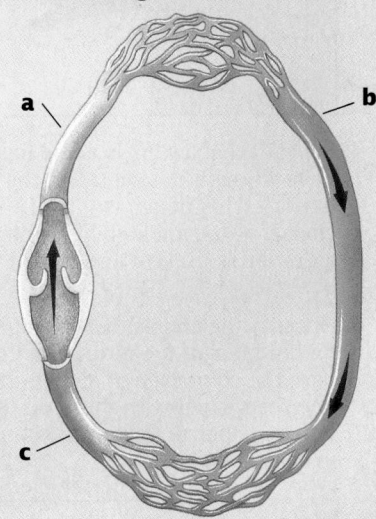

- In what kind of animal does this type of circulatory system occur?
- At which point would you expect to find oxygenated blood?
- At which point would you expect to find the highest blood pressure?

Reviewing Themes

23. *Evolution*
What evidence has led scientists to conclude that the earliest land vertebrate evolved from lobe-finned fishes?

24. *Energy and Life*
Land animals expend energy to convert ammonia into urea or uric acid. What do these animals gain from this conversion?

Thinking Critically

25. *Inferring Conclusions*
The world-record time for running the mile is just under four minutes. How might this record be different if humans had air sacs like birds do?

26. *Comparing and Contrasting*
What makes the lungs of mammals more efficient than the lungs of amphibians?

27. *Building on What You Have Learned*
In Chapter 4 you learned about diffusion. What role does diffusion play in the exchange of gases in the gills of fish?

Cross-Discipline Connection

28. *Biology and Art*
Draw a picture of a fictional animal that is adapted for survival in one of the seven terrestrial ecosystems studied in Chapter 12. Then write a description of how the animal reproduces, breathes, and eliminates wastes.

Discovering Through Reading

29. Read the article "Grasshoppers Change Coats to Beat the Heat," in *Science News*, August 24, 1991, page 119. What differences did scientists find between the lipid coats of southern and northern grasshoppers?

Cross-Discipline Connection

28. The fictional animal and its offspring may take any form. The form must be able to survive in the chosen terrestrial ecosystem. The written description should tell how the animal reproduces, breathes, and gets rid of waste. The description should be consistent with the drawing.

Discovering Through Reading

29. The lipid coats of northern grasshoppers tend to melt at lower temperatures.

---

24. The conversion enables animals to remove nitrogen-containing wastes with minimal water loss.

Thinking Critically

25. The world record time would be less because the respiratory system would function to provide more oxygen needed by the runner.

26. Mammalian lungs have much greater surface area than do amphibian lungs. Mammalian lungs can absorb more oxygen.

27. Diffusion is the mechanism by which oxygen moves from water to blood and carbon dioxide moves from blood to water.

Procedural Note

1. Review the answers to the preparation.
2. Students are to avoid handling the animals.
3. Caution students not to grasp the animals' tails.

Prelab Preparation Answers

- Lizards and other reptiles have a watertight skin covered by overlapping scales. Reptiles (other than snakes and a few lizards) also have four limbs. Reptiles also lay shelled eggs in which the embryo develop within two membranes: the chorion and the amnion.
- Reptile limbs are partway beneath the body. Mammal limbs are directly underneath the body, while amphibian limbs jut out from the sides of the body.

Procedure Answers

2. No, it is dry and scaly.
3. Human limbs are directly underneath the body to more effectively support its weight. The lizard's limbs are partway beneath its body.
5. The lizard will darken when on the orange square, but it will not become orange.
6. *Tenebric* larvae
7. The anoles are drawn to the light.
8. When exposed to cold, the anole darkens.

Chapter 22 Investigation

How Are Anole Lizards Well Adapted to Their Environment?

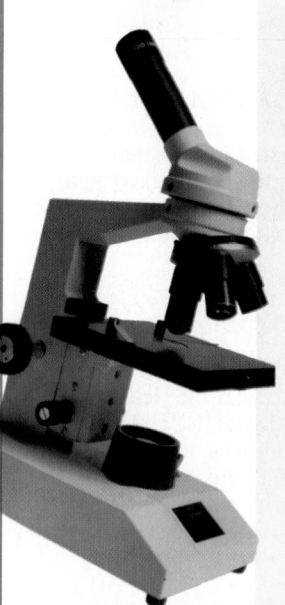

Objectives

In this investigation you will:
- *observe* structural and behavioral characteristics of an anole lizard
- *relate* these characteristics to the ability of the lizard to survive on land

Materials

- terrarium
- live anole lizards
- one piece of orange construction paper large enough to cover one-half of the terrarium floor
- one piece of green construction paper the same size as the orange paper
- live *Tenebrio* larvae or crickets
- small pieces of apple
- small dish
- desk lamp or heat lamp
- 1,000-mL beaker or large glass jar
- 600-mL beaker
- crushed ice

Prelab Preparation

Review what you have learned about reptiles by answering the following questions.
- What features of reptiles enable them to live on land?
- How are the limbs of a reptile different from the limbs of an amphibian and a mammal?

Procedure: Observing Anole Lizards

1. Form a cooperative team with another student to complete steps 2–9.

2. Observe one lizard in the terrarium. Record its coloration. Describe the texture of its skin. *Does its skin appear to be moist?*

3. Closely observe the lizard's limbs and posture. *How does the orientation of the lizard's hind limbs compare with the orientation of your legs?*

4. Beneath the jaw, male anole lizards have a pouch of skin called the dewlap. This pouch is pink. Look closely at the anole to determine if a dewlap is present and record your observations.

5. Observe the anoles that are sitting on the floor of the terrarium. Compare the coloration of anoles sitting on green paper with those on the orange paper. *Is there any difference in coloration?*

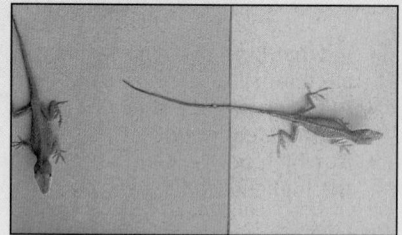

6. Watch the anoles as your teacher places a dish with live *Tenebrio* larvae (or crickets) and bits of apple into the terrarium. *Which food do the anoles prefer?*

7. Place a lamp near the glass at one corner of the terrarium. *What is the response of the anoles to the lamp?* Record your observations.

8. Observe as your teacher removes one anole and places it in a 600-mL beaker filled with crushed ice. *What is the anole's response?* Be sure to note any color change.

9. Your teacher will now introduce a new anole into the terrarium. Closely observe the behavior of the residents and the introduced anole. Record your observations.

Analysis

1. *Analyzing Observations* How do the limbs and skin of the lizard differ from those of an amphibian?

2. *Making Inferences* How are a lizard's limbs and skin better adapted for life on land than those of an amphibian?

3. *Analyzing Observations* What causes color change in anoles? Explain your answer.

Thinking Critically

1. Lizards are ectotherms that absorb heat from their surroundings. Yet lizards are able to maintain a fairly constant body temperature. Explain how lizards are able to control their body temperature.

2. What is the advantage of the anole's ability to change color, if its primary stimulus is not a change in background coloration? What evidence from your observations supports your answer?

3. What function does the male anole's dewlap serve? Why is it advantageous for the male to be able to retract the dewlap?

Analysis Answers

1. Answers should describe the dry, scaly body covering, claws, and the larger, stronger limbs of the lizard.

2. Answers should suggest that the scaly body covering makes the animal more waterproof, which also reduces drying. The limbs are better adapted for crawling and climbing than those of most amphibians.

3. The fact that the animal may have changed color when handled (even while on the green paper), and that it changed color as a result of lower temperature, indicates that changes in coloration can be induced by factors other than background color.

Thinking Critically Answers

1. Lizards maintain a nearly constant body temperature throughout the day by moving into and out of the sun, and by changing their orientation to the sun's rays.

2. Answers may vary but should suggest that the presence of predators might induce a color change, making the lizard more difficult to see by predators. The lizard darkens when cold to enable it to absorb more heat from the sun.

3. The dewlap is a signaling structure. Males advertise themselves to females and warn off other males by extending and retracting the dewlap in a specific rhythm. The ability to retract the brightly colored dewlap when not signaling is an adaptation for camouflage.

Chapter 23 Animal Diversity

Planning Guide

	Objectives/Themes	Classwork Resources	Homework Resources

23.1	**1.** Contrast the lifestyles of sponges and cnidarians.
	
	2. Compare the polyp and medusa stages of the cnidarian life cycle.
	
	3. Describe the life cycle of the beef tapeworm.
	
	4. List two parasitic nematodes that can live in humans.
	
	Themes: Evolution, Interacting Systems, Patterns of Change
		Text
		Journeys
		  Tour of a Jellyfish p. 506
		
		Teacher's Resource Binder
		Focus Activity 23
		  Observing Some Major
		  Animal Groups
		Lab Investigation 23.1
		  Snails
		
		Other Resources
		Transparencies 101–104
			Text
			Section Review, p. 510
			
			Teacher's Resource Binder
			Directed Reading Worksheet
			  23.1
			
			Other Resources
			Audiocassette 23.1*

23.2	**1.** Identify the three main classes of mollusks.
	
	2. Summarize the evolutionary advantages of segmentation.
	
	3. Describe two ways annelids affect humans.
	
	4. List five kinds of arthropods.
	
	Themes: Evolution, Interacting Systems, Scale and Structure
		Text
		Journeys
		  Tour of a Mollusk p. 513
			Text
			Section Review, p. 516
			
			Teacher's Resource Binder
			Directed Reading Worksheet
			  23.2
			
			Other Resources
			Audiocassette 23.2*

23.3	**1.** Identify three kinds of echinoderms.
	
	2. List two characteristics of the phylum Echinodermata.
	
	3. Contrast the three subphyla of chordates.
	
	Themes: Evolution, Scale and Structure
		Text
		Investigation
		  How Do You Make a Key of
		  the Major Animal Phyla?
		  pp. 522–523
		
		Teacher's Resource Binder
		Extension Worksheet
		  The Water-Vascular System
			Text
			Section Review, p. 518
			
			Teacher's Resource Binder
			Directed Reading Worksheet
			  23.3
			Vocabulary Review Worksheet*
			Reteaching Worksheet*
			  Animal Diversity
			
			Other Resources
			Audiocassette 23.3*

*Reteaching Options

Demonstrations
23.1: pp. 504, 505, 507, 509
23.2: p. 512

Assessment
Chapter Review pp. 520–521
Portfolio Assessment p. 501D
Chapter Test—Teacher's Resource Binder
Test Generator

Research Notes

Connections to Chemistry: Sticky Mussels

Many products of modern life, from tooth fillings to cars, depend on adhesive technology. Fewer items are put together using screws, nails, and welding. However, most glues do not work well in water.

Water's unique chemical properties attack adhesive materials and can also prevent an adequate bond from forming in the first place. The powerful forces unleashed by freezing and thawing water can pull glued surfaces apart. To meet these challenges, some scientists are turning to nature's champion at underwater gluing, the mussel.

Mussels have evolved the ability to make a strong adhesive because their survival depends on being able to cling to rocks in a turbulent ocean. They use a material called byssus, which is woven into threads that help them cling to any hard surface.

The mussels use their muscular foot to probe a surface, clean it of dust, and then create a water-tight vacuum. Glands pour byssus into the foot, forming first a foamy gel at the point of contact, and then adhesive threads, attaching the mussel to the surface. Byssus works better than any other adhesive discovered so far.

Scientists such as J. Herbert Waite of the University of Delaware are working to unravel the complicated biochemical pathways that make byssus. It takes a lot of work to get the protein out of the mussels and purify it. So far, they have already found several precursor proteins for byssus. A biotechnology company, Genex Corp. of Gaithersburg, MD, has found and cloned the gene coding for one of the precursors.

They have discovered that byssus is composed of polymer fibers mostly made of keratin, the same protein found in fingernails, and a resin protein whose amino acids all have hydroxyl groups. However, they still have not discovered exactly how these two components are mixed together to form such a strong adhesive. There is some evidence that catechol oxidase, an enzyme which seems to assist in byssus formation, may also be an ingredient.

There are many applications for a successful adhesive that can withstand water. It could be used instead of stitches to help close surgical wounds, to reattach severed nerves, and for dental work. Long-lasting barnacle-repelling coatings and adhesives also could have wide applications for the shipping industry.

Investigation Notes

How Do You Make a Key of the Major Animal Phyla?

pages 522–523

Purpose: Students observe representative specimens of each of the major animal phyla discussed in the chapter and use these observations as a basis for the construction of a classification key. This investigation gives the students some concrete experiences with each phylum.

Prelab Preparation

Have each lab station act as a learning center for one phylum. Include one or more examples of organisms in that phylum along with any supplementary resources you have available.

Answers will be found on pages 522–523.

Meeting Individual Needs

Objectives

1. Students will demonstrate appropriate use of core vocabulary for the chapter (Vocabulary File).

2. Students will distinguish between different organisms' body plans and lifestyles (Teaching Strategy).

Vocabulary File
(Developing Vocabulary/ Limited English Proficiency)

If you are not already using the Vocabulary File, refer to Chapter 1 for its preparation. See the Chapter Highlights on page 519 for a list of suggested words.

Teaching Strategy
Body Plans and Lifestyle
for pages 503–518
(Developing Inferential Skills/Visual Learners)

Materials: scalpel; preserved specimen of *Grantia*; dropper bottle filled with water; slides and microscopes; pictures of the sponge (*Grantia*), jellyfish, tapeworm, earthworm, clam, squid, leech, butterfly, tarantula, and sand dollar (if pictures are not available of all of these, have students sketch and color some); poster board; and yarn for mobiles

Preparation: Cut the *Grantia* in half and place the pieces on slides. Do not drop water over the *Grantia* until immediately before students view them. Mount and cut out the pictures of the animals for mobiles. Have these mobiles ready for hanging in the sky zoo, but do not hang them before the class session.

Procedure: Recall from the environmental unit that people are not participants in the food chain. Ask students for examples of the "diversity" of animals available for food that could be found in a supermarket. These different animals are important food sources for humans. Lead a discussion about how their body forms and lifestyles make them useful as food sources.

Ask students for examples of items that could be purchased in a supermarket for the purpose of "control" of pesky animals. Lead a discussion about insect repellents and other pest control items. Ask students to describe why the animals targeted are considered pests. Make sure students realize that diversity of body form and lifestyle of animals is directly related to their effects on humans.

For example, ants pursue sweet-tasting things for their sugar content, since these will provide them with energy. This means ants are often found where food is stored.

Scan the chapter one section at a time. At the end of each section, ask the class if there is an example of the animal discussed in the sky zoo. If not, have students hang a mobile that represents that type of animal from the set prepared for this chapter. Have students use the pictures to suggest a description of the body plan and unique features of that animal.

Reread aloud the section on sponges, page 503. Now is the time to drop water over the *Grantia.* The water makes the spicules more visible. Explain to students that the spicules will look like slivers of glass. After students have observed the spicules, start a discussion, relating this to the diversity of endoskeletons and exoskeletons from Chapter 22.

Additional Strategies
Visual Strategies
Pages 503, 504, 508, 509, 510, 512, 515, 516, 517, and 518

Auditory Learners
Use *Biology: Visualizing Life* Audiocassettes for Sections 23.1, 23.2, and 23.3.

Meeting Individual Needs (cont.)

Cooperative Learning
Animal Phyla

Timing: Use this activity to introduce Section 23.1.

Group Size: 7 students

Outcome: Students will be able to describe characteristics and examples of several animal phyla.

Individual Accountability: Each group member will describe one phyla.

Positive Interdependence: Each group will be able to describe several animal phyla.

Assign one of the following phyla to each group member: (1) Porifera, (2) Cnidaria, (3) Platyhelminthes, (4) Nematoda, (5) Mollusca, (6) Annelida, (7) Echinodermata.

Each group member should read the chapter and take notes on the assigned phylum. Students should share the information in their notes with other members of the group.

Then the group should prepare a table comparing and contrasting these animal phyla.

The table should have the following column headings: phylum, symmetry, body structure, habitat, reproduction, feeding and digestion, and specific examples. Tell students that you will randomly call on a member from each group to discuss characteristics of one of these animal phyla.

Portfolio Assessment

Students should select their best work and provide a self-reflective rationale for their selections. Students can make selections in the following areas.

1. *Content*   One concept map from the chapter (See page 812 for evaluation criteria.)

2. *Reading Comprehension*   One Directed Reading Worksheet from the Teacher's Resource Binder (Use the answer key to evaluate for accuracy.)

3. *Writing*   Using the Vee Form, summarize a magazine or newspaper article relating to animal physiology or the animals studied in this chapter. (See page 22T for evaluation criteria.)

    Or: Select a writing task or project from the Chapter Review.

4. *Performance Assessment*   One Vee form from a chapter investigation or lab manual investigation (See page 22T for evaluation criteria.)

Teacher makes selections in the following areas.

1. *Formal Assessment*   Chapter test (Test A, B, or the Test Generator) The teacher-scored test should be reviewed by the student. Incorrect responses should be corrected by the student before the test becomes part of the portfolio.

2. *Informal Assessment*   Use the Direct Observations Checklist, page 33T, during a laboratory or other cooperative learning experience.

3. *Performance Assessment*   Have students create a diagram of the life cycle of a parasitic nematode or flatworm that is not depicted in the text.

Concept Map Answer

The following is one possible answer to the Relating Concepts exercise on page 520.

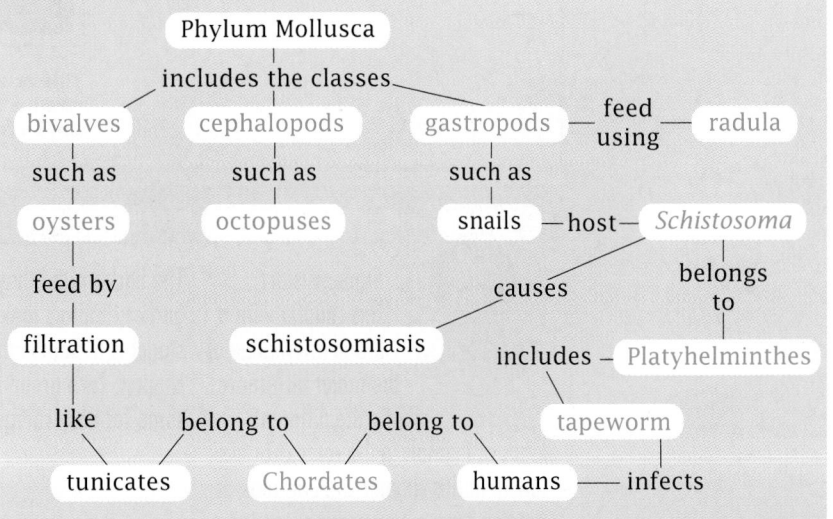

Chapter 23

Determining Prior Knowledge

- Ask students if a starfish (sea star) is really a fish.
- Display a variety of invertebrates represented in these three sections. Have students identify each specimen and name one or two ways in which each affects humans. Students should gain an appreciation of the fact that invertebrates exhibit many forms, and that as lowly as these animals may seem, they are of great importance to humans.
- Show students a dried sponge and a dried sea star. Discuss the skeletons of these animals. Ask students about the existence of skeletons in other invertebrate groups.
- Display pictures of coral and sponges in a reef. Then show some dried specimens of coral and sponges. Have them describe the specimens and compare them—live versus dead.

Chapter 23

Animal Diversity

Review

- animal tissues (Section 21.1)
- origin of body cavities (Section 21.2)
- innovations in body plan (Section 21.3)

This ocellated nudibranch is a marine mollusk. Unlike its relative the snail, it lacks a shell. This species of nudibranch can be found in both the Pacific and Indian Oceans.

▪ Author's Rationale ▪

Students start this chapter with a broad knowledge of the major differences among the different kinds of animals in the context of evolution. In this chapter students look at animals from a taxonomic perspective.

The background they received from earlier chapters makes it easier for students to handle the details of this chapter. Each group is treated the same for easy comparison.

CHAPTER 21 TOLD THE STORY OF THE EVOLUTION OF ANIMAL BODY ARCHITECTURE. IN THIS CHAPTER, YOU WILL RETURN TO THE MAJOR PHYLA OF ANIMALS FOR A LONGER VISIT. YOU WILL DISCOVER WHERE MANY OF THESE ANIMALS LIVE, HOW THEY FEED, AND HOW THEY REPRODUCE. YOU WILL ALSO LEARN HOW SOME OF THESE ANIMALS AFFECT HUMANS, BOTH POSITIVELY AND NEGATIVELY.

23.1 Sponges, Cnidarians, and Simple Worms

Objectives

❶ Contrast the lifestyles of sponges and cnidarians.

❷ Compare the polyp and medusa stages of the cnidarian life cycle.

❸ Describe the life cycle of the beef tapeworm.

❹ List two parasitic nematodes that can live in humans.

Sponges

Sponges are probably most familiar to you as absorbent pads used for wiping up spills or cleaning dishes. The sponges you buy in the store are usually manufactured, not derived from the animals known as sponges. There are more than 9,000 species of sponges (phylum Porifera), most of which are marine. About 150 species live in fresh water. An adult sponge, like the one in  **Figure 23.1**, spends its life attached to a hard surface. For this reason, it was not until 1765 that zoologists realized that sponges are animals, not plants. Most sponges are asymmetrical and grow to conform to the surface on which they live.

A sponge's body is perforated by holes (Porifera means "pore-bearer") that lead to an inner water chamber. Sponges pump water through these pores and expel it through a large opening at the top of the chamber. While water is passing through the body, nutrients are engulfed, oxygen is absorbed, and wastes are eliminated.

As you learned in Chapter 21, between the inner and outer layer of many sponges is a jellylike layer. Embedded within this layer is a network of needle-like **spicules**. Spicules form the skeleton of the sponge. The middle layer of some large sponges contains a mesh of tough protein called spongin. These kinds of sponges have been harvested for centuries because of the ability of their spongin skeletons to soak up water and release it when squeezed. Today, most of the sponges you can buy are copies of this spongin mesh manufactured from plastic or cellulose.

Figure 23.1
Sponges vary widely in size, shape, and color. This purple tube sponge lives in the ocean surrounding Bonaire, an island about 95 km (60 mi.) north of Venezuela.

■ Matter of Fact ■

A Japanese Wedding
Tell students that in Japan, up to a century ago, it was customary to give to a newlywed couple a Hexactinellid, or glass sponge, with a pair of shrimp parents living inside. The shrimp entered the osculum of the sponge when they were smaller and courting to mate. They gained protection from the sponge and filtered the water brought in by the sponge, for food. Their young remained with them until they were developed enough to make it on their own, then they left through the osculum by which their parents entered. The parents by this time were too big to leave and so remained within the sponge. Suggest that the significance was "till death do us part." This custom has dropped out of favor, due perhaps to the difficulty of trawling for these sponges and their inhabitants.

Lesson Plan 23.1

Phase 1

PREPARATION

Key Concepts
- Sponges are nonmobile; they attach themselves to rocks and filter water for food and oxygen.
- Spicules form the skeleton of sponges.
- Sponges reproduce both sexually and asexually.
- Jellyfish, hydras, corals, and sea anemones are cnidarians.
- Cnidarians have a gut that opens to the outside and tentacles with nematocysts.
- Cnidarians have two distinct life stages—the polyp and the medusa.
- Most species of flatworms, like flukes and tapeworms, are internal parasites that carry out their life cycles in more than one host.
- Flatworms can reproduce asexually by regeneration.
- Most roundworms or nematodes are parasitic and reproduce sexually.

Reading Strategy

Have students make storyboards about the life cycles of sponges, flatworms, cnidarians, and roundworms from the reading of this section.

Phase 2

TEACHING STRATEGIES

Visual Strategy

Figure 23.1
Have students examine this figure and explain why sponges might have been classified as plants by early taxonomists.

503

Visual Strategy

Figure 23.2

Discuss the characteristics of sponge reproduction as students view this diagram. Emphasize that sponges have the ability to reproduce both sexually and asexually. Have students infer how it might be advantageous to the sponge to reproduce both sexually and asexually.

Mathematics Connection

Calculating the Volume of Water Circulated in Sponges

Ask students how much water an average sponge circulates during a day. Tell them that one type of sponge, 10 cm (4 in.) tall and 1 cm (0.4 in.) in diameter, pumps about 23 qts. of water through its body in a day. Have students calculate the metric volume of water. They should suggest 22.5 L.

Demonstration 1

Natural Sponges

Have students examine natural sponges and record the structures they observe. After students have completed their observations, compile a list on the chalkboard of the structures that students identified. Then have students discuss the possible functions of each of the structures they identified.

Art Connection

Sponge Models

Provide students with three colors of modeling clay with which to make a model of a sponge or a cnidarian. Each color of clay should represent a cell layer. Students should include as many structural features as possible in their models, including incurrent pores, osculum, collar cells, spicules, and the internal cavity of the sponge. The cnidarian model should represent either the polyp or medusa body shape. Spicules, nematocysts, and collar cells can be represented by adding pins or paper clips to the models.

Sponges reproduce sexually and asexually

Most sponges are able to reproduce asexually. In some sponges, new individuals bud from the parent; in others, the parent sponge breaks into many fragments, and each fragment grows into a new sponge. Sponges can also reproduce sexually, as illustrated in **Figure 23.2.**

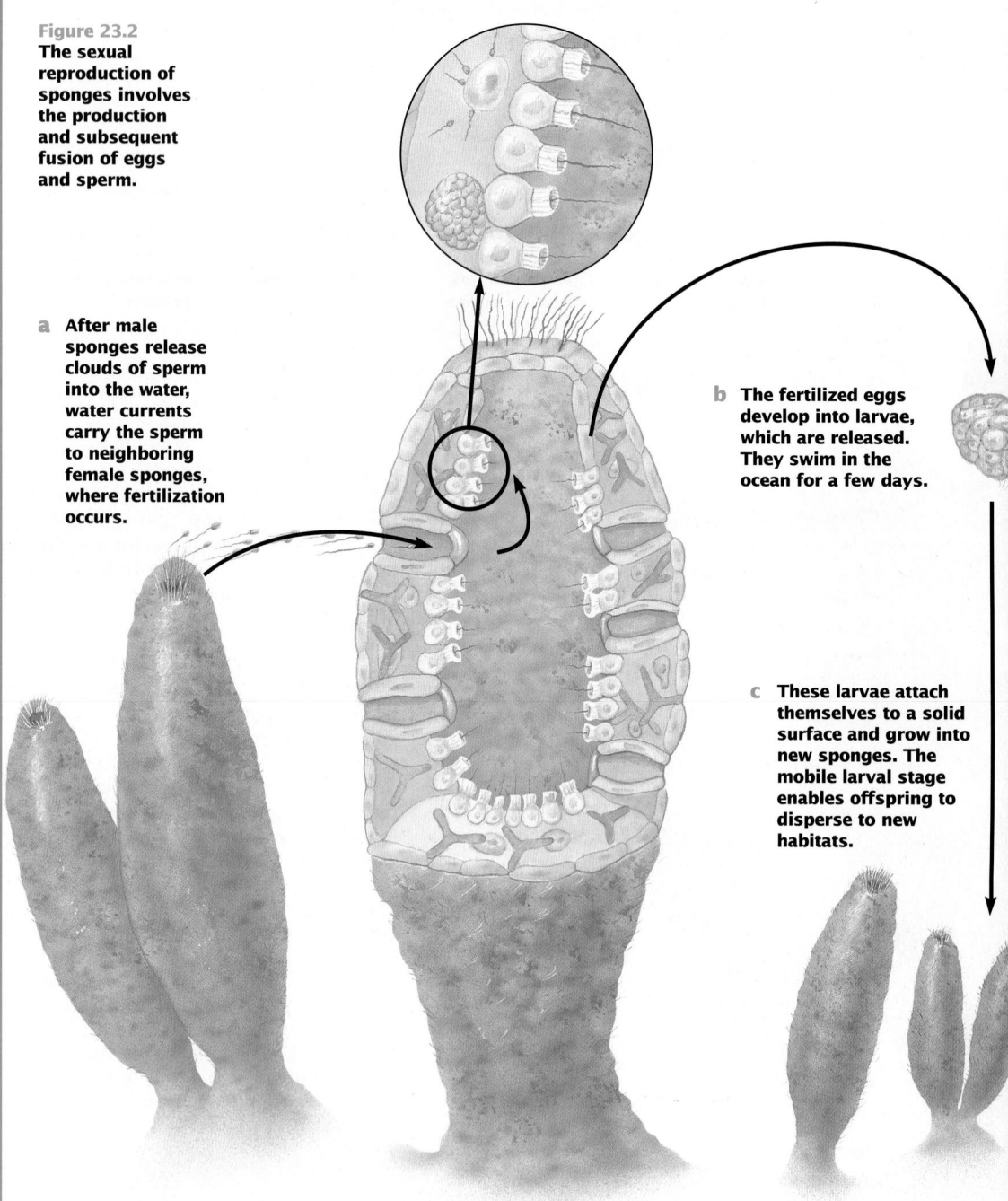

Figure 23.2
The sexual reproduction of sponges involves the production and subsequent fusion of eggs and sperm.

a **After male sponges release clouds of sperm into the water, water currents carry the sperm to neighboring female sponges, where fertilization occurs.**

b **The fertilized eggs develop into larvae, which are released. They swim in the ocean for a few days.**

c **These larvae attach themselves to a solid surface and grow into new sponges. The mobile larval stage enables offspring to disperse to new habitats.**

Cnidarians: Jellyfish and Relatives

Corals, jellyfish, sea anemones, and hydras are members of the phylum Cnidaria. The name "cnidaria" comes from the Greek word meaning "nettle" and refers to the stinging structures that are characteristic of these animals.

Cnidarians have a gut with only one opening to the outside. This opening is surrounded by a ring of tentacles used to capture food and defend against predators. Cells in the tentacles and outer body surface are armed with stinging, harpoon-like structures called **nematocysts** (*NEHM uh toh sihsts*). The projectile fired by a nematocyst contains toxins that can cause paralysis.

Most cnidarians are harmless to people. If you touch the tentacles of a seashore anemone, you will feel a sticky sensation as its short projectiles barely pierce your skin. A few jellyfish and corals, however, have very potent toxins that cause a painful, burning rash. The stings of tropical sea wasps can be fatal. To protect against their stings, Australian surfers often wear pantyhose.

Many cnidarians have two distinct life stages

In Chapter 18 you learned about alternation of generations in plants. Like plants, many cnidarians have two different body forms during their life cycle, as illustrated in **Figure 23.3**. The two forms are called the **polyp** stage and the **medusa** (plural, medusae), or jellyfish, stage. Polyps generally live attached to a hard surface. Sea anemones, hydras, and corals are polyps.

Not all cnidarians go through both polyp and medusa stages. Some remain as either a polyp or medusa throughout their lives. You can read more about cnidarians in the *Tour of a Jellyfish* on page 506.

Figure 23.3

a **Polyps bud to produce more polyps and, in some cnidarians, to produce the medusa stage of the life cycle.**

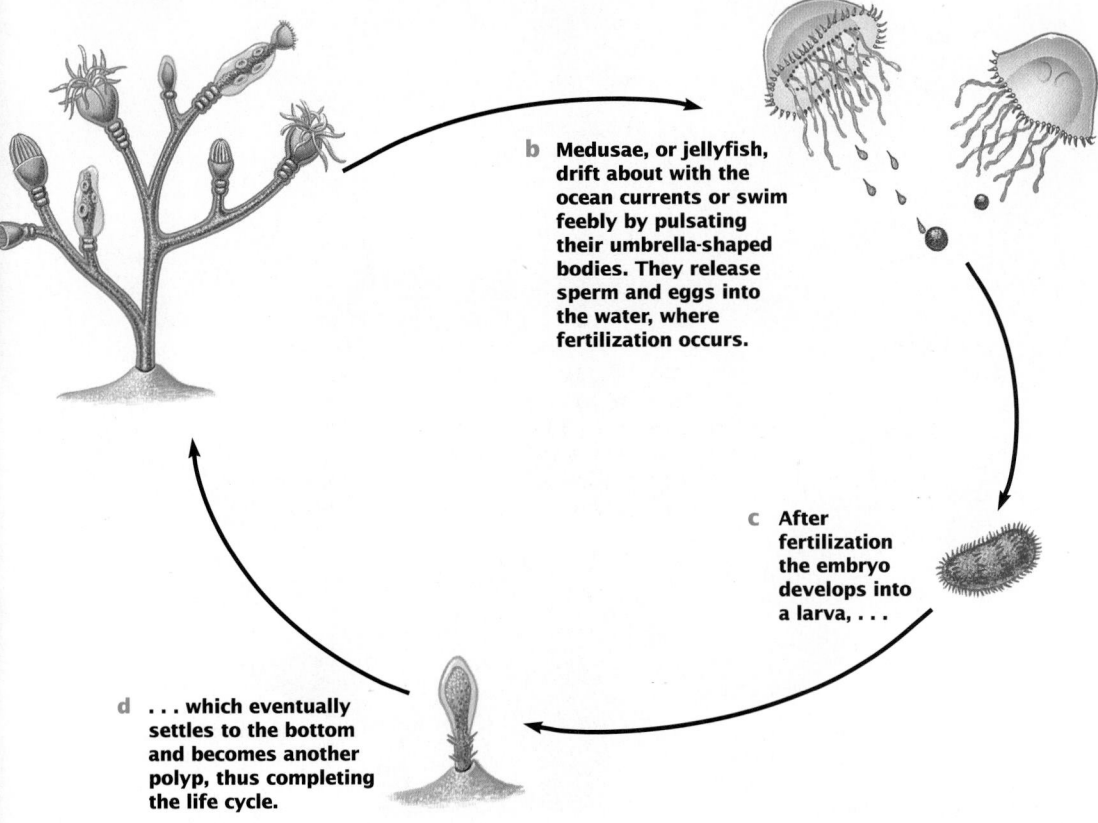

b **Medusae, or jellyfish, drift about with the ocean currents or swim feebly by pulsating their umbrella-shaped bodies. They release sperm and eggs into the water, where fertilization occurs.**

c **After fertilization the embryo develops into a larva, . . .**

d **. . . which eventually settles to the bottom and becomes another polyp, thus completing the life cycle.**

Using the Feature

Review the structure of cnidarians, as discussed on pages 458-459. Point out the two tissue layers, radial symmetry, and the gut with extracellular digestion. Unlike sponges, cnidarians have a simple nervous system, a nerve net, which enables them to respond to their environment. They do not have a brain, however.

Discussion

Guide the discussion by posing the following questions.

1. How does a jellyfish capture prey?
   It paralyzes prey with stinging cells in its tentacles.
2. What is the function of nematocysts?
   Nematocysts are stinging structures that release toxins.
3. During which stage of a cnidarian's life does sexual reproduction occur?
   the medusa stage

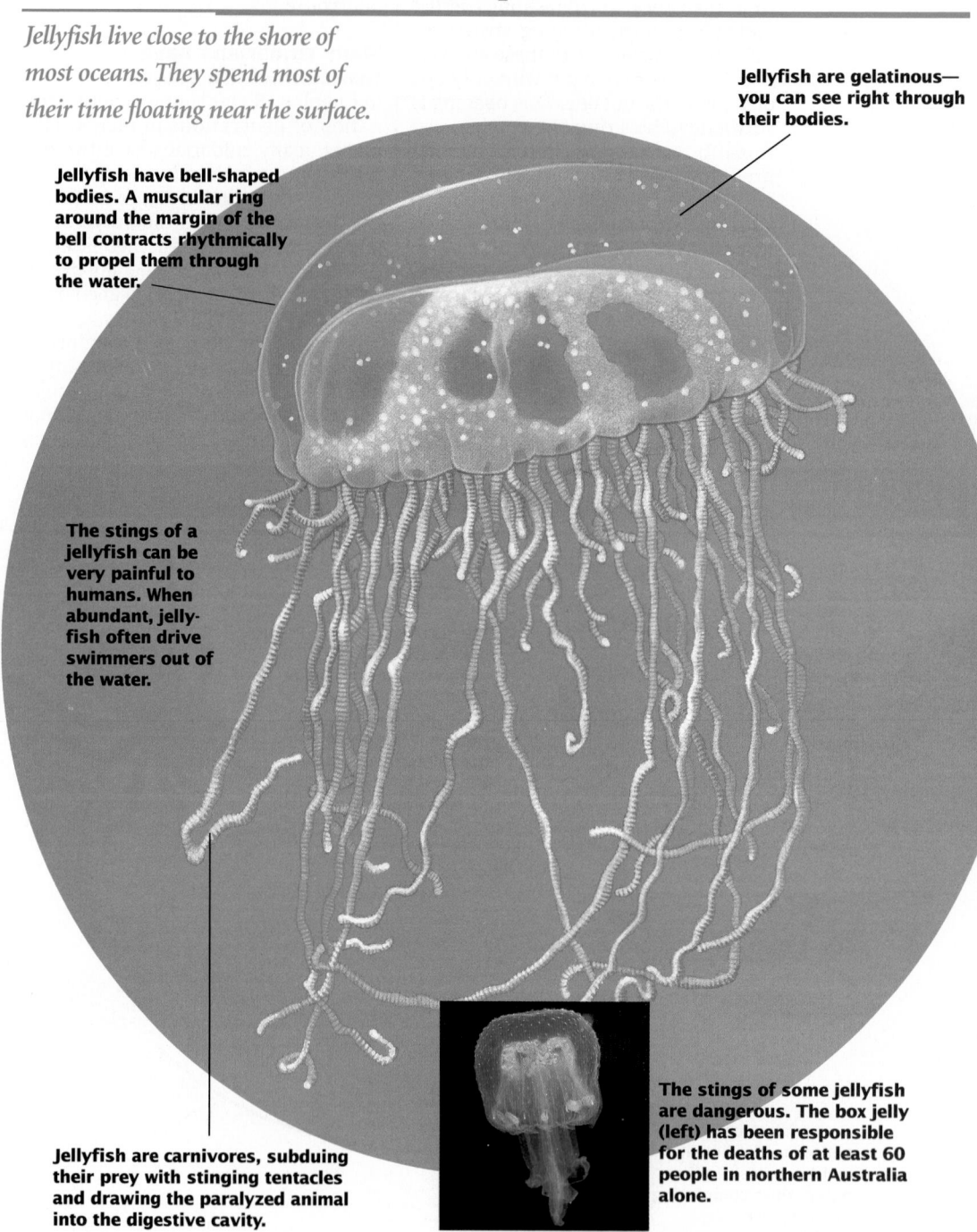

Journeys ■ *Journeys* ■ *Journeys* ■ *Journeys* ■ *Journeys* ■

Tour of a Jellyfish

Jellyfish live close to the shore of most oceans. They spend most of their time floating near the surface.

Jellyfish are gelatinous—you can see right through their bodies.

Jellyfish have bell-shaped bodies. A muscular ring around the margin of the bell contracts rhythmically to propel them through the water.

The stings of a jellyfish can be very painful to humans. When abundant, jellyfish often drive swimmers out of the water.

The stings of some jellyfish are dangerous. The box jelly (left) has been responsible for the deaths of at least 60 people in northern Australia alone.

Jellyfish are carnivores, subduing their prey with stinging tentacles and drawing the paralyzed animal into the digestive cavity.

Flatworms

There are about 20,000 species of flatworms (phylum Platyhelminthes). Most free-living flatworms, such as planarians, are aquatic. They are common in shallow oceans, usually in protected habitats. Freshwater flatworms often reside in gravel or under sunken objects. A few species live on land but always in very damp areas. The flatworm body is soft and unprotected against predators and dehydration.

Flatworms can reproduce asexually by regeneration

If you cut a planarian in half, either lengthwise or across the body, each half will regrow into a complete worm. This ability to regrow lost parts is called regeneration. Many free-living flatworms can reproduce asexually through regeneration. Scientists are very interested in understanding how flatworms can regenerate, in hopes of applying this information to humans. People can regenerate some damaged parts, such as healing new skin after a cut or scrape. Human abilities to regenerate are far less dramatic than those of flatworms, however.

Flatworms also reproduce sexually. Most species are **hermaphrodites**. Hermaphrodites (*huhr MAHF roh deyets*) contain both male and female reproductive systems. During mating, two flatworms exchange sperm so the eggs of both flatworms are fertilized. A flatworm usually does not fertilize itself.

Most flatworms are parasites

Most species of flatworms are parasitic, infecting a variety of hosts, including humans. Nearly 300 million people are afflicted with the disease schistosomiasis (*shihs tuh soh MEYE uh sihs*). Each year schistosomiasis kills 800,000 people in Asia, Africa, Latin America, and the Middle East. Microscopic flatworms in the genus *Schistosoma* cause this deadly disease. **Figure 23.4** shows the life cycle of *Schistosoma*.

Figure 23.4

a After hatching, *Schistosoma* larvae invade freshwater snails.

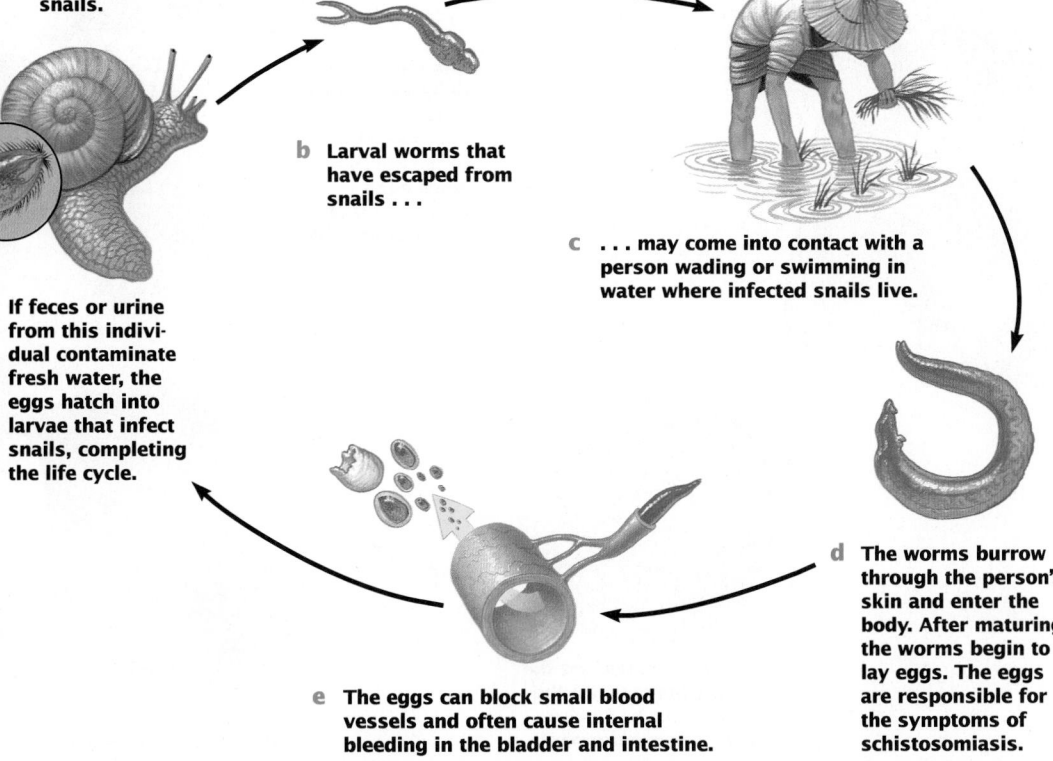

b Larval worms that have escaped from snails . . .

c . . . may come into contact with a person wading or swimming in water where infected snails live.

f If feces or urine from this individual contaminate fresh water, the eggs hatch into larvae that infect snails, completing the life cycle.

d The worms burrow through the person's skin and enter the body. After maturing, the worms begin to lay eggs. The eggs are responsible for the symptoms of schistosomiasis.

e The eggs can block small blood vessels and often cause internal bleeding in the bladder and intestine.

Theme Answer

Interacting Systems

Answers will vary but should suggest the spread of schistosomiasis by infested feces or soil washed into lakes by floods or water runoff.

Demonstration 4

Planaria

Have students observe live planaria specimens through a stereomicroscope or hand lens. Encourage them to observe how they move about and respond to touch, light, and changes in temperature. Have students make drawings of planaria.

Connection: Chapter 9

Evolution

Explain that bilateral symmetry and cephalization, along with a mesoderm germ layer, are benchmarks in the evolutionary history of animals.

Visual Strategy

Figure 23.6

Explain that tapeworms are found outside their hosts only when they are eggs. Sheep liver flukes can be found outside their hosts in egg, cyst, and larval stages.

Figure 23.5
The hooks on the end of the beef tapeworm can be thrust out and buried in the wall of its host's intestine.

In addition to *Schistosoma*, parasitic flatworms include flukes, which attack the liver or lungs, and tapeworms. The head of the beef tapeworm, which commonly infects humans, is shown in **Figure 23.5**. These parasites can grow to more than 10 m (30 ft.) in length. Follow the tapeworm's life cycle in **Figure 23.6**.

Tapeworms are so specialized for the parasitic lifestyle that they do not have a digestive system. Nutrients from the host's digested food are absorbed directly through the tapeworm's skin. Infection by the beef tapeworm often causes no symptoms, but it can cause pain, discomfort, nausea, and abdominal swelling.

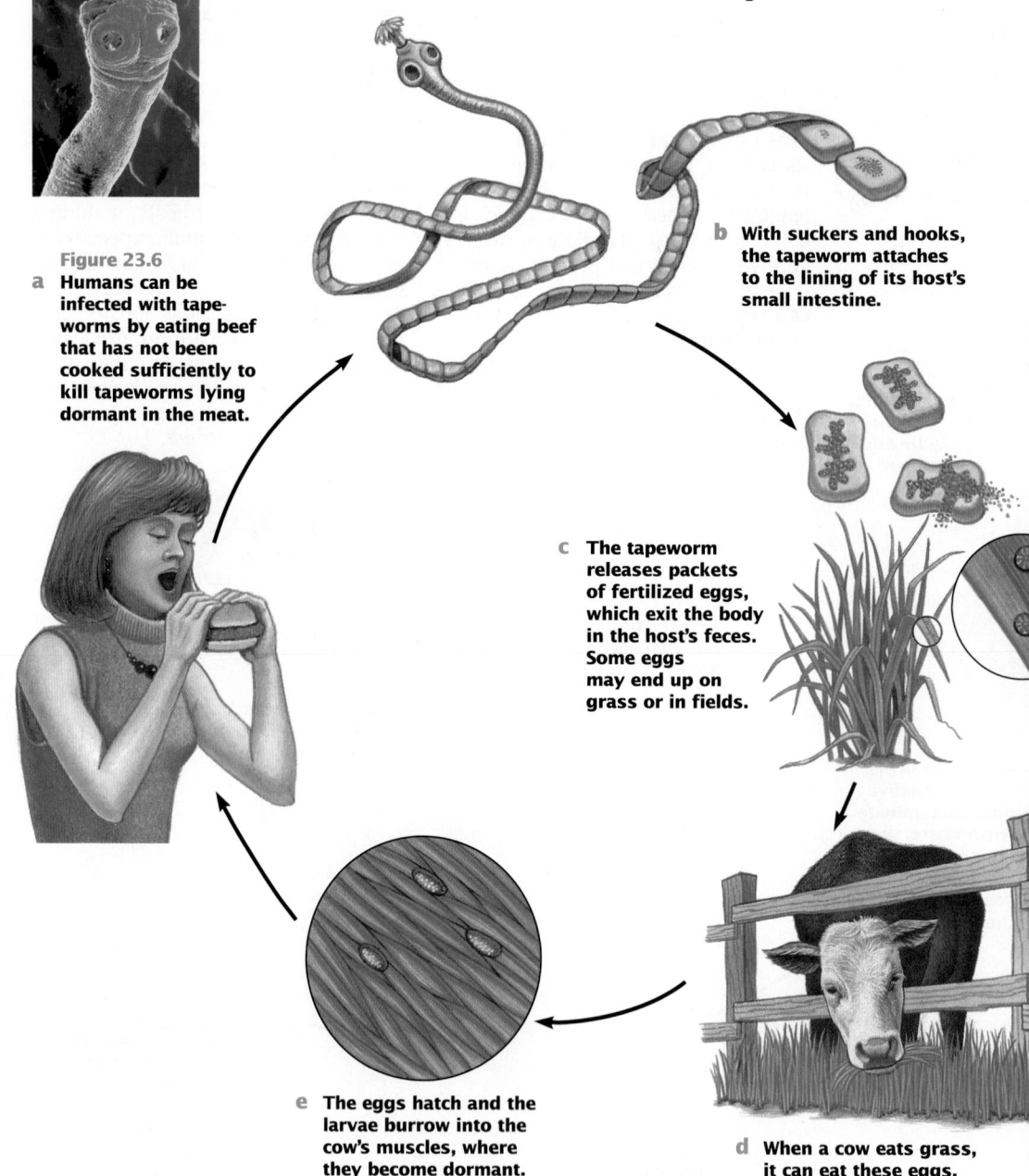

Figure 23.6

a **Humans can be infected with tapeworms by eating beef that has not been cooked sufficiently to kill tapeworms lying dormant in the meat.**

b **With suckers and hooks, the tapeworm attaches to the lining of its host's small intestine.**

c **The tapeworm releases packets of fertilized eggs, which exit the body in the host's feces. Some eggs may end up on grass or in fields.**

d **When a cow eats grass, it can eat these eggs.**

e **The eggs hatch and the larvae burrow into the cow's muscles, where they become dormant.**

Roundworms

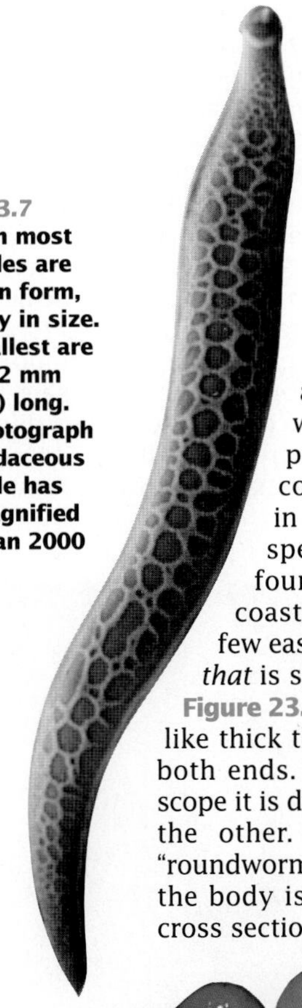

Figure 23.7
Although most nematodes are similar in form, they vary in size. The smallest are about 0.2 mm (.008 in.) long. This photograph of a predaceous nematode has been magnified more than 2000 times.

Roundworms, or nematodes (phylum Nematoda), are extremely abundant; a single spadeful of soil can contain more than 1 million individual nematodes. Nematodes also occur in all aquatic environments and in the bodies of plants and animals as parasites. The great abundance of nematodes was illustrated by one very patient zoologist who counted 90,000 nematodes in a single rotting apple. One specimen of nematode is found only in the damp felt coasters under beer mugs in a few eastern European towns. Now *that* is specialization. As shown in Figure 23.7, roundworms are shaped like thick threads that are tapered at both ends. In fact, without a microscope it is difficult to tell one end from the other. The common name of "roundworm" comes from the fact that the body is circular when viewed in cross section.

The sexes are separate in nearly all roundworms, and the males are often smaller than the females. After mating, the female secretes a tough case around each fertilized egg and deposits the eggs in the environment, where development takes place. Many parasitic roundworms have complex life cycles, regularly passing from one host to another.

Nematodes play an important role in research on genetics and development. Scientists have learned the complete cellular structure for the nematode *Caenorhabditis elegans*, which has only about 1,000 cells. In addition, scientists are close to identifying the locations of all 10,000 genes of this nematode.

Some nematodes cause disease

Nematodes are known to infect virtually all kinds of animals and plants, and some cause millions of dollars of damage each year to livestock and crops. For instance, puppies and adult dogs often must be "wormed," treated with drugs to kill parasitic nematodes such as the intestinal roundworm *Ascaris*. Heavy infestations of *Ascaris* can be fatal to puppies. *Ascaris* also infects livestock and humans. Figure 23.8 shows *Ascaris suum* in the intestine of a pig.

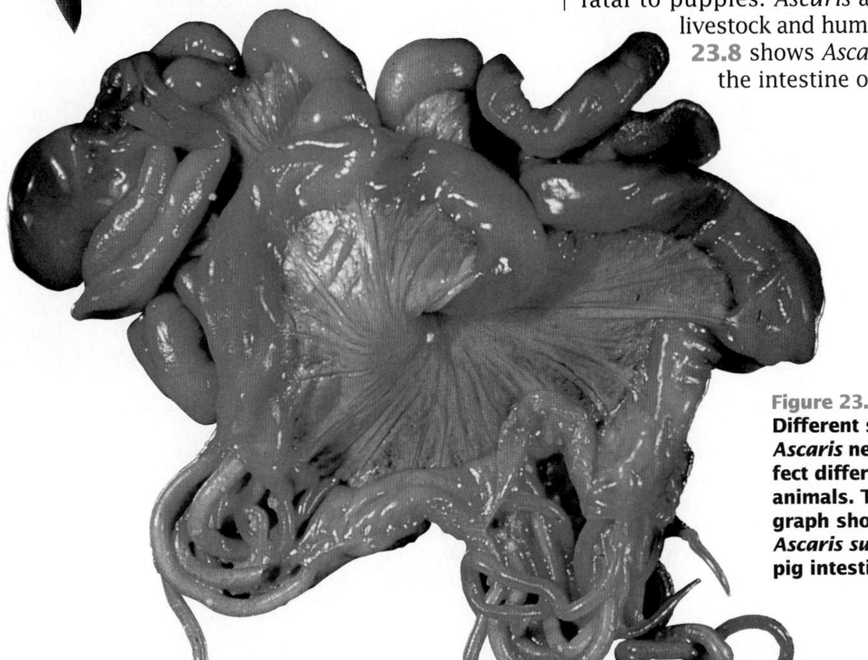

Figure 23.8
Different species of *Ascaris* nematodes infect different types of animals. This photograph shows adult *Ascaris suum* within a pig intestine.

■ Matter of Fact ■

Roundworms

There are more than 12,500 species of roundworms. The transmission of human-specific roundworms is the reason why human feces should not be used as fertilizer, as is done in many parts of the developing world.

Visual Strategy

Figure 23.7

Obtain a preserved heart with heartworms from a veterinarian for students to observe. A photograph from an invertebrate zoology text may be substituted. Tell students that dogs are often host to parasitic worms and can be given pills during mosquito season to prevent heartworm infestation. Have students speculate on the role mosquitoes play in heartworm infestation. Explain that dogs are not the only animals vulnerable to worms. Humans are vulnerable to approximately 50 roundworm species, with only about 10 of these species being harmful to humans.

Demonstration 5

Preventative Posters

Have students work in cooperative groups of three to develop and produce a poster that warns of the dangers of worm infestation and tells ways to prevent it. Encourage all members of the group to agree on the main ideas to be communicated. Have one group member write the copy for the poster. Have another member develop illustrations for the poster, and another design the poster using the copy and illustrations.

Visual Strategy

Figure 23.8

Ascaris is common where human wastes are not treated and where food is not properly cooked. The adult worms live in the intestines of humans. Other vertebrates, such as pigs, are also infected, but usually with species other than *lumbricoides*. A female *Ascaris* may lay 200,000 eggs a day. Have students identify the parasitic worm that usually infects pigs.

Visual Strategy

Figure 23.9
Tell students that *Ascaris* males are smaller than the females, 23 to 39 cm (6 to 10 in.); the diameter of the female is about 1 cm (0.39 in.) smaller than the male. Living specimens are white, yellow, or pink. Have students compare the sexes of all the animals illustrated in this section.

Phase 3

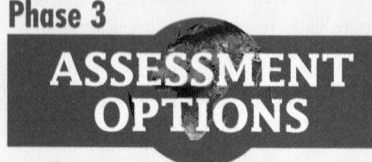

ASSESSMENT OPTIONS

Closure Strategy

Roundworms
Have students summarize, in writing, the major structural and behavioral features of roundworms, and compare these features to those of flatworms, which were discussed earlier in the section. Have students identify those features that cause roundworms to be considered more advanced than flatworms.

Section Review

Assign the *Section Review*.

Reteaching

On the chalkboard, create the structure for the following table. Down the left side write the names of the phyla: *Platyhelminthes, Nematoda,* etc. Across the top, write the headings: *Body shape/plan, Habitat, Nervous control, Digestion, Symbiotic relationship, Reproduction,* and *Examples.* Have students supply the information necessary to complete the table.

About 50 species of nematodes parasitize humans. Almost 1 billion people are infected by the nematode *Ascaris lumbricoides,* shown in **Figure 23.9**. This nematode spends its adult life inside the intestine of its host. Large numbers of adult nematodes in the intestine can cause pain and intestinal blockage. Because juvenile worms bore through the tissues of the lungs, infection with *A. lumbricoides* can also lead to pneumonia.

Each day a female *A. lumbricoides* lays as many as 200,000 eggs, which pass out of her host in feces. These eggs are very durable and can survive in the soil for 10 years. To be infected, a person must consume the nematode's eggs. This occurs when uncooked vegetables are eaten, or when dirty hands are not properly washed before meals. *A. lumbricoides* infections occur primarily in areas where sanitary facilities are poor or where human feces are used for fertilizer.

In tropical regions of Africa and Asia the nematode *Wuchereria bancrofti* causes the disease known as filariasis *(fihl uh REYE uh sihs)*. These nematodes are parasites of the circulatory system. When present in large numbers, they can clog lymphatic vessels, causing fluid accumulation, skin thickening, and extreme swelling. The condition elephantiasis, shown in **Figure 23.10**, results.

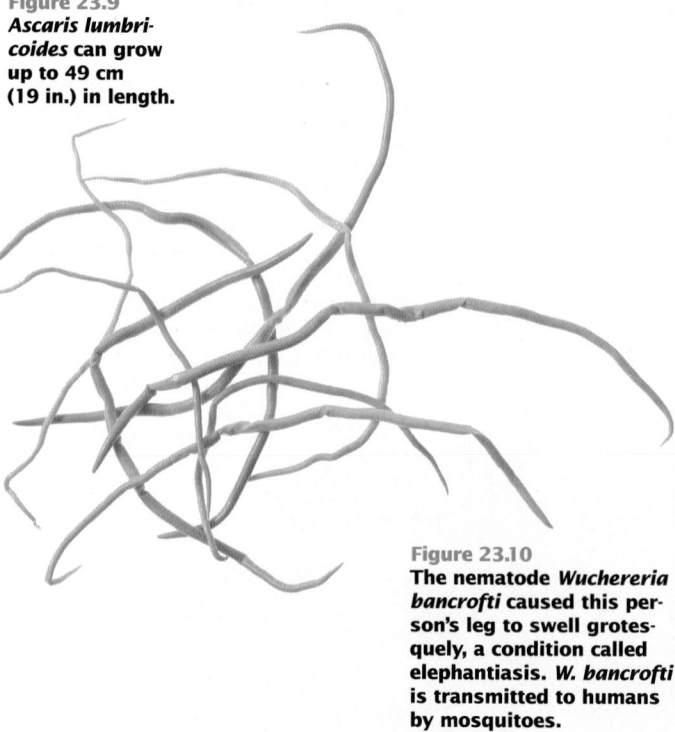

Figure 23.9
Ascaris lumbricoides can grow up to 49 cm (19 in.) in length.

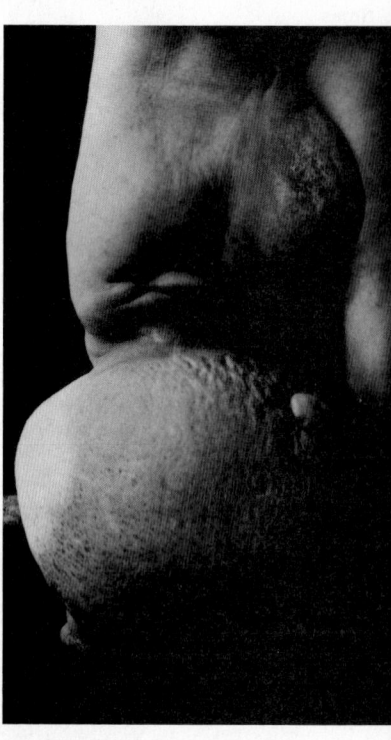

Figure 23.10
The nematode *Wuchereria bancrofti* caused this person's leg to swell grotesquely, a condition called elephantiasis. *W. bancrofti* is transmitted to humans by mosquitoes.

Section Review

① Distinguish between the ways sponges and cnidarians capture food.

② Diagram the life cycle of a cnidarian that goes through polyp and medusa stages.

③ Explain two ways infection by the beef tapeworm could be prevented.

④ Name two examples of parasitic nematodes that can infect humans.

■ *Section Review Answers* ■

1. Sponges absorb nutrients from water passing through their bodies. Cnidarians capture their food with tentacles armed with nematocysts.

2. See Figure 23.3 on page 505.

3. Students should suggest eating beef that has been sufficiently cooked or insuring proper treatment and disposal of human wastes.

4. Students should discuss *Ascaris lumbricoides* and *Wuchereria bancrofti*.

CHANCES ARE YOU HAVE NEVER SEEN A NEMATODE OR FLATWORM. IN THIS SECTION, YOU WILL LEARN ABOUT MOLLUSKS, ANNELIDS, AND ARTHROPODS, THREE GROUPS OF ANIMALS THAT ARE PROBABLY MORE FAMILIAR TO YOU. YOU CAN FIND MEMBERS OF THESE GROUPS CRAWLING IN A FIELD OR GARDEN, BURROWING IN THE SOIL, FLYING OVERHEAD, OR AS PART OF YOUR EVENING MEAL.

23.2 *Mollusks, Annelids, and Arthropods*

Objectives

❶ **Identify the three main classes of mollusks.**

❷ **Summarize the evolutionary advantages of segmentation.**

❸ **Describe two ways annelids affect humans.**

❹ **List five kinds of arthropods.**

Mollusks

Figure 23.11
Scallops (below) and razor clams (below right) are two examples of aquatic mollusks.

If you visit a shop where fresh seafood is sold, look for some of these mollusks: clams, scallops, mussels, squids, octopuses, and oysters. With more than 100,000 living species, the phylum Mollusca is the second largest animal phylum on Earth, exceeded in size only by the phylum Arthropoda, which includes insects, spiders, and crustaceans. **Figure 23.11** shows two types of mollusks that are often eaten by humans. A shell is a feature of many mollusks. The shell (or shells) provides protection for the body and a solid structure for muscle attachment. The rest of the body is soft; in fact, the name "mollusk" comes from the Latin word for "soft." All mollusk shells are lined with and secreted by a fleshy fold of tissue called the mantle. Another characteristic of mollusks is a muscular foot. Among mollusks, the foot has evolved to perform several functions, including locomotion, burrowing, and prey capture.

Mollusks are one of the earliest groups of animals to evolve efficient excretory organs, the nephridia. **Nephridia** (*nee FRIHD ee uh*) are small tubules that collect wastes from the body fluids and discharge them to the outside.

Key Concepts

• There are three main classes of mollusks: class Bivalvia, the bivalves, which include clams, scallops, and mussels; class Gastropoda, the univalves, which include snails and slugs; and class Cephalopoda, the head-foot mollusks, which include squids, octopuses, and nautiluses.

• Mollusks were the first animals to evolve nephridia for collecting and discharging waste.

• Mollusks have three basic and distinct body parts: the head, the foot, and the visceral mass. Their soft bodies are covered by the mantle, which secretes the shell.

• Annelids are segmented worms, which include earthworms, leeches, and a variety of marine worms, such as tube worms, feather dusters, and clam worms.

• Arthropods are characterized by jointed appendages, a segmented body, and an exoskeleton of chitin.

• Phylum Arthropoda includes spiders, insects, centipedes, scorpions, shrimps, crabs, and lobsters.

Reading Strategy

Based on what they can recall from previous science classes or life experiences, have students list the basic characteristics of mollusks, annelids, and arthropods. Have them compare their list with what they read in this section.

TEACHING STRATEGIES

Demonstration 1

Snails

Explore simple snail behavior. Test the snail's response to touch, moisture, gravity, ammonium hydroxide, and light. Note: Land snails can be found in many gardens and wooded areas, or they can be obtained from a biological supply house. Store them in a cool, moist terrarium. Supply bits of lettuce on which they may feed.

Visual Strategy

Figure 23.12

Explain that members of class Gastropoda have radula. Have students describe how gastropods feed.

Theme Connection

Interacting Systems

Also called sea arrows, squids are the basic food of sperm whales and of many other marine animals. Harvested by the ton as bait for codfish and other marine species, they are also used as fertilizer.

Mathematics Connection

Dried Squid

Some types of squid are dried and shipped as food. Three tons of wet squid produce 1 ton of dried squid. Have students calculate the percentage of water loss during the drying process.
66.66 percent

Slugs and snails are gastropods

Slugs and snails, which you often see in a garden or aquarium, belong to the class Gastropoda. Gastropods are the most diverse group of mollusks. They are found in the ocean, in fresh water, and on land. Snails and slugs move by creeping slowly on a muscular foot. The shell, when present, is often brightly colored and is usually coiled.

Just inside the mouth of gastropods and some other mollusks is a unique feeding organ called the **radula** (*RAJ oo lah*). The radula shown in **Figure 23.12** is a flexible structure that is covered by rows of teeth, like a carpenter's file. The radula is used to scrape free small particles of food. In some predatory mollusks, the radula is a sharp fang that is used to stab prey and inject venom.

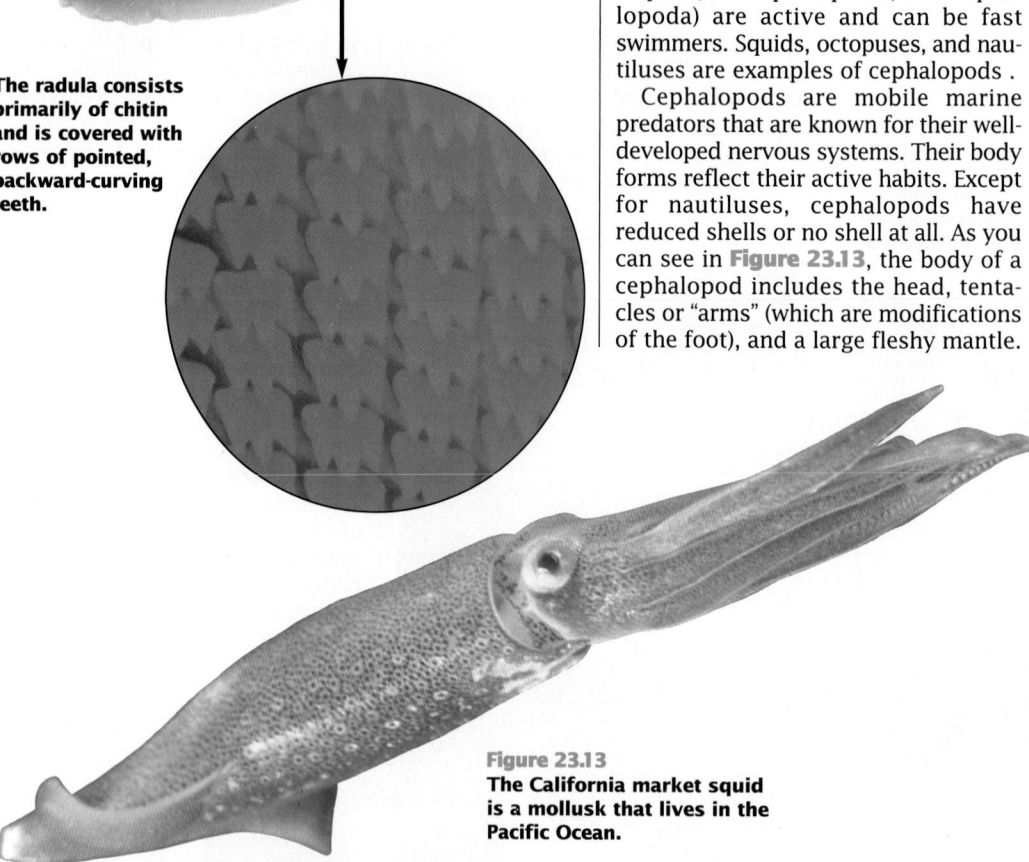

Figure 23.12

a **A snail's shell is made of layers of calcium carbonate. The shell forms a protective barrier against predators.**

b **The radula consists primarily of chitin and is covered with rows of pointed, backward-curving teeth.**

Oysters, mussels, clams, and scallops are bivalves

If you are a fan of oysters, mussels, scallops, or clam chowder, you have enjoyed the delicate taste of bivalves (class Bivalvia). Some of these animals are of great value, supporting commercial fishing industries around the world.

Bivalves have two shells. Unlike other mollusks, bivalves feed by filtering small particles from the water. They lack a radula. Most bivalves are sedentary, and some, such as oysters and mussels, permanently fix themselves to hard surfaces as adults. The foot is wedge-shaped and is used for digging in sand or mud or for secreting tough attachment threads.

Squids, octopuses, and nautiluses are cephalopods

In contrast to the bivalves and gastropods, the cephalopods (class Cephalopoda) are active and can be fast swimmers. Squids, octopuses, and nautiluses are examples of cephalopods .

Cephalopods are mobile marine predators that are known for their well-developed nervous systems. Their body forms reflect their active habits. Except for nautiluses, cephalopods have reduced shells or no shell at all. As you can see in **Figure 23.13**, the body of a cephalopod includes the head, tentacles or "arms" (which are modifications of the foot), and a large fleshy mantle.

Figure 23.13
The California market squid is a mollusk that lives in the Pacific Ocean.

▪ *Cultural Perspective* ▪

Dr. Godfrey Bourne (African American)

Dr. Godfrey Bourne, a professor at Florida Atlantic University, studied small hawk like birds known as snail kites while working on his Ph.D. at the University of Michigan. Dr. Bourne discovered that the kites feed primarily on a genus of snail that is an agricultural pest. Dr. Bourne was born in Guyana.

■ Journeys ■ Journeys ■ Journeys ■ Journeys ■ Journeys

Tour of a Mollusk

Octopuses have the largest brains of any invertebrate. They are eight-armed mollusks that are usually bottom dwellers. They have no shell.

Except for having horizontal pupils, the eyes of the octopus look remarkably like vertebrate eyes.

An ink sac releases a cloud of dark ink out of the rectum to confuse enemies.

The eight arms of an octopus bear sucking discs for seizing prey, such as crabs or other mollusks.

Octopuses swim by forcefully expelling water from the mantle cavity—a sort of jet propulsion.

Octopuses can change color by contracting tiny muscle cells that surround their pigment-containing cells.

Using the Feature

Review the structure of a mollusk shown on page 467. Unlike the snail, the octopus has no shell. In addition to a radula, an octopus has a sharp beak with which it tears apart prey. Some octopuses produce poison in their salivary glands. Octopuses are the most intelligent invertebrates. They can be trained to perform many tasks, including distinguishing between objects and opening jars to get food.

Discussion

Guide the discussion by posing the following questions.

1. To which class of mollusks do octopuses belong?
   Cephalopoda
2. How is the octopus's eye different from yours?
   The octopus's eye has a horizontal pupil.
3. What is the function of the ink sac?
   It releases a dark cloud to cover the octopus's escape.

Earthworms and Mythology

Tell students that earthworms were once thought to "rain down," as in "worm weather." In reality, they are really forced out of their burrows when the burrows become filled with water. Also, earthworms are sensitive to vibrations in the soil and may be brought to the surface by vibrations stemming from a variety of sources.

Social Studies Connection

Earthworm Farms

Have students explain what an earthworm farm is. One characteristic of a productive earthworm farm is the presence of hundreds of thousands of earthworms per acre. Aeration of soil by earthworms is extremely beneficial to growing crops. Worms are raised in backyards, garages, and cellars. A large worm farm may have five acres of outdoor pits. A small one may be a simple trench or a box in a backyard. The most common earthworm farmed is called a red wiggler, or manure worm, because it is at home in the waste droppings of farm animals. Buyers of worms include farmers, gardeners, and anglers.

The mantle is thick and muscular. It covers the internal organs and a large mantle cavity. A tubelike siphon derived from the foot serves as an outlet for water leaving the mantle cavity. When the mantle muscles contract, water is forced rapidly out the siphon, allowing these animals to move by what is essentially jet propulsion. This is the usual manner of movement in squids. Octopuses use this method to escape predators.

Annelids: Segmented Worms

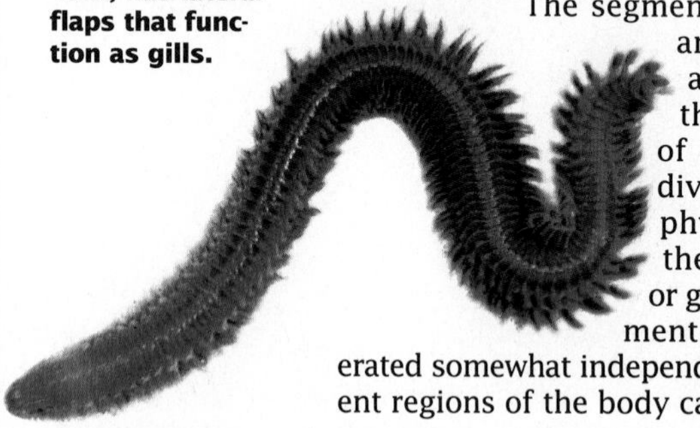

Figure 23.14
Each body segment of this clam worm, found on Long Island, New York, has lateral flaps that function as gills.

The 15,000 or so species of segmented worms (phylum Annelida) include earthworms, leeches, and a host of marine species such as tube worms, feather dusters, and the clam worm shown in **Figure 23.14**.

The segmented body of annelids has allowed for the evolution of tremendous diversity in the phylum. Since the segments or groups of segments can be operated somewhat independently, different regions of the body can specialize for different functions, as shown in **Figure 23.15**. Such specializations include the suckers of leeches, localized reproductive organs in earthworms, and tube-secreting segments in certain marine worms.

Earthworms feed by sucking in soil and decaying matter. Organic material is digested, and wastes are eliminated as castings, smooth blobs of soil you can see at the entrances to earthworm burrows. Charles Darwin calculated that a single earthworm could eat its own weight in soil every day. The action of earthworms is beneficial to plants because it breaks up and aerates the soil, and because castings are rich in nutrients.

Figure 23.15
Segmentation is a distinctive feature of annelids. The body of an earthworm can have more than 100 segments.

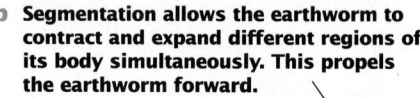

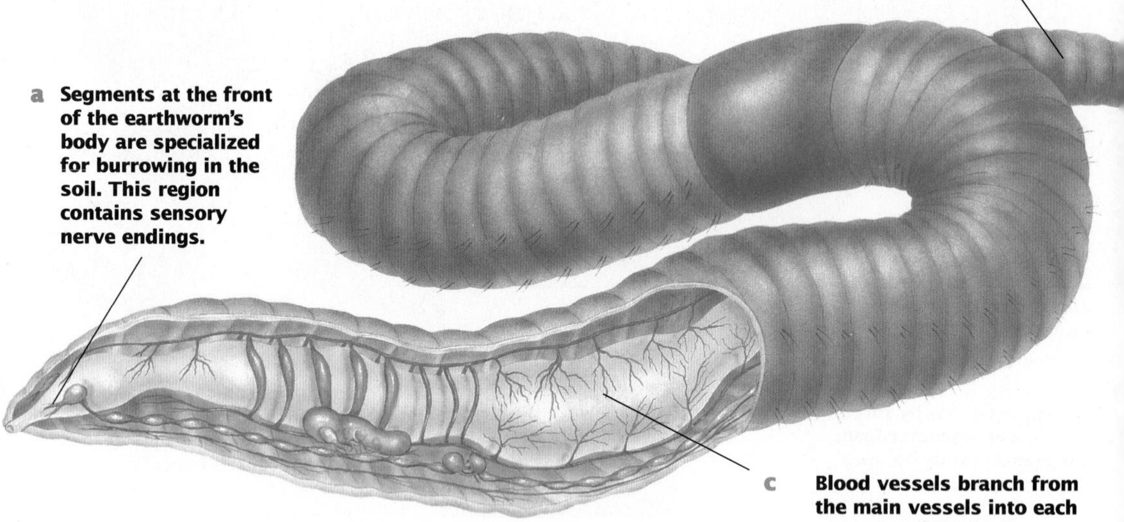

b **Segmentation allows the earthworm to contract and expand different regions of its body simultaneously. This propels the earthworm forward.**

a **Segments at the front of the earthworm's body are specialized for burrowing in the soil. This region contains sensory nerve endings.**

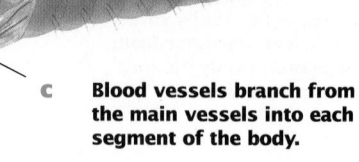

c **Blood vessels branch from the main vessels into each segment of the body.**

■ *Matter of Fact* ■

Earthworms

Darwin contended that 1 hectare (2.5 acres) of soil might contain 155,000 earthworms, which in a year might bring 18 tons of soil to the surface, and in 20 years might build a new layer 11 cm. (3 in.) thick.

Leeches are parasitic annelids

Externally, leeches are distinct from the other annelids. The body is somewhat flattened and not quite as "wormlike" as an earthworm's body, as you can see in Figure 23.16. Some leeches are scavengers or predators in fresh water or salt water. Many leech species are at least part-time parasites. They feed on the bodies of other animals, usually vertebrates. Suckers on the front and back of the leech attach to its host.

Parasitic leeches use enzymes or three sharp, bladelike teeth to make an opening in the host's skin through which blood is sucked. Some leeches release proteins into the host that prevent blood from clotting and stopping their meal. Scientists have isolated one of the proteins produced by leeches. This protein is being used in research on blood clotting. Another leech protein has been effective in preventing the spread of lung cancer. Some leeches secrete a local anesthetic, so their host is not even aware that it is providing a meal for the leech.

For centuries, leeches were used to "bleed" sick patients because people believed that excess blood caused disease. This practice fell out of favor when scientists showed that viruses, bacteria, and other microorganisms cause disease. Recently leeches have made a medical comeback. The large blood-sucking leech *Hirudo medicinalis* is used by modern physicians to liquify and remove blood from bruises and severely damaged tissue.

Figure 23.16
The body of this water leech is segmented, just like the body of an earthworm.

Arthropods: The Most Abundant Animals

Arthropods (phylum Arthropoda) dominate virtually all habitats, both in numbers of individuals and in numbers of species. Nearly 1 million species of arthropods have been described. Some specialists believe that as many as 30 million species still remain to be discovered.

As you learned in Chapter 21, arthropods have a segmented body that is covered by an exoskeleton of chitin. Most important, they have jointed appendages. The phylum Arthropoda includes a variety of familiar animals, including those shown in Figure 23.17. Spiders, insects, centipedes, scorpions, shrimp, crabs, and lobsters all belong

Figure 23.17
a Both the praying mantis (above) and honeybees (right) are helpful insects and are common throughout the United States.

b This tarantula is found in Asia.

■ *Cultural Perspective* ■

Insects as Food

Entomologists recognize over 300 species of insects that are eaten as foods in many countries around the world, especially in Asia, Africa, South America, and Mexico. Asians consume giant water bugs. Roasted termites are enjoyed in South America. In Columbia, winged leaf-cutter ants can be purchased as a snack in movie theaters. In Japan, insect farmers harvest wasps, which are cooked, canned, and sold in stores. Cakes are made from the eggs of water boatmen. Fried caterpillars and chocolate-covered ants and bees are sold in some stores in the United States.

Visual Strategy

Figure 23.18
Have students examine the photographs in this figure. Have them identify as many characteristics of arthropods as they can.

Phase 3

ASSESSMENT OPTIONS

Closure Strategy

Classify This Fossil
Have students imagine that they have found an invertebrate fossil. Have them identify characteristics that would aid them in classifying this fossil as a mollusk, an annelid, or an arthropod.

Section Review

Assign the *Section Review.*

Reteaching

Have students explain how mollusks, annelids, and arthropods differ. Have them add this information to the table prepared for the Reteaching on page 510.

Figure 23.18
Despite the diversity of their body shapes, insects, ticks, scorpions, centipedes, and butterflies, are all arthropods.

to the phylum Arthropoda. Stop for a moment to consider the many ways in which arthropods affect our lives. Lobsters, crabs, and shrimp are sources of food. Some insects are serious pests of food crops and cause millions of dollars in damage each year. Other insects are extremely beneficial as pollinators of crop plants. Still others are parasites.

Malaria and sleeping sickness, two of the world's most serious diseases, are transmitted by biting insects.

Look at the arthropods shown in **Figure 23.18**. They give you some idea of the variety within this phylum. You will read much more about arthropods in Chapter 24.

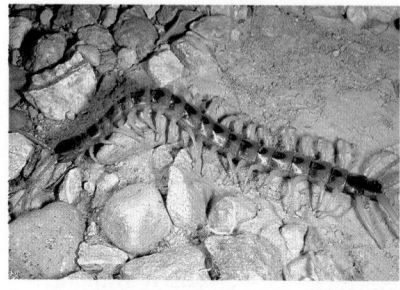

a The ladybird beetle kills pests that attack crop plants and flowers such as this buttercup.

b The giant centipede is found in Texas.

c The scorpion *Pandinus imperoitor* is found in Africa.

d The tick that transmits Lyme disease is found throughout the United States.

e Most monarch butterflies live in the United States during the summer, and migrate to Mexico or South America in the winter.

Section Review

❶ Name the major classes of mollusks. Give an example of each.

❷ What are the advantages of a segmented body?

❸ Why would the fertility of a farmer's field decline if all earthworms were removed from the soil?

❹ Name five different kinds of arthropods.

■ *Section Review Answers* ■

1. • class Gastropoda: slugs, snails
   • class Bivalvia: oysters, mussels, clams, and scallops
   • class Cephalopoda: squids, octopuses, and nautiluses

2. Since each segment or group of segments can be operated independently, different regions of the body can be specialized for different functions.

3. The action of earthworms breaks and aerates the soil. Their castings are rich in nutrients.

4. insects, spiders, crustaceans, centipedes, and millipedes

ECHINODERMS AND CHORDATES BELONG TO A FUNDAMENTALLY DIFFERENT EVOLUTIONARY LINE FROM THE ANNELIDS, MOLLUSKS, AND ARTHROPODS. ECHINODERMS AND CHORDATES ARE DEUTEROSTOMES, WHICH ARE DISTINGUISHED FROM THE GROUPS YOU STUDIED IN THE OTHER SECTIONS OF THIS CHAPTER BY A UNIQUE FORM OF EMBRYONIC DEVELOPMENT.

23.3 Echinoderms and Chordates

Objectives

1. **Identify three kinds of echinoderms.**

2. **List two characteristics of the phylum Echinodermata.**

3. **Contrast the three subphyla of chordates.**

Echinoderms: Sea Stars and Their Relatives

Sea stars, sea cucumbers, sand dollars, and sea urchins are echinoderms (phylum Echinodermata). As adults, nearly all echinoderms show a five-part radial symmetry: parts of the body radiate from its center like the arms of a five-pointed star, as shown in **Figure 23.19**. The sand dollar shown in **Figure 23.20** also has five-part body symmetry, but it is less obvious.

In addition to radial symmetry, echinoderms have a **water vascular system**. As you learned in Chapter 21, the water vascular system is a complex arrangement of fluid-filled tubes that operate the tube feet on the animal's lower surface. The water vascular system functions in locomotion, feeding, and gas exchange.

Echinoderms have no excretory organs, and thus have no efficient means of regulating water balance. That is why echinoderms are restricted to marine habitats. Most of these animals obtain oxygen by diffusion through thin parts of the body surface or through the walls of the tube feet. Internal distribution of oxygen and nutrients occurs mainly by circulation of fluid within the coelom.

Figure 23.19
Many sea stars are brightly colored. This stubby-armed sea star lives in New Guinea's Mandang Harbor.

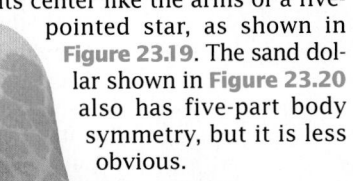

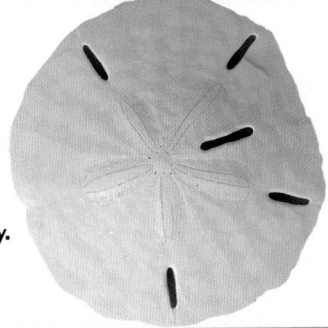

Figure 23.20
All echinoderms, including this sand dollar, display five-part radial symmetry.

■ *Matter of Fact* ■

Sea Cucumbers
Sea cucumbers are more sensitive than sea stars. Their tentacles show greater ability to select suitable food, which includes plants and animals. When disturbed, a sea cucumber may eviscerate most of its internal organs. These can be completely regenerated.

Lesson Plan 23.3

Phase 1
PREPARATION

Key Concepts
- Echinoderms and chordates are deuterostomes that are characterized by a unique form of embryonic development.
- Echinoderms exhibit a five-part radial symmetry and a water vascular system.
- Chordates are characterized by a notochord, a dorsal nerve cord and pharyngeal slits.

Reading Strategy

From their reading, have students make a double concept map with echinoderms on one map and chordates on another, and a linkage connecting the two maps.

Phase 2
TEACHING STRATEGIES

Visual Strategy

Figure 23.19
Sea stars have a deep groove along each arm that is lacking in brittle stars. Have students examine a sea star for the groove and predict how these animals move about.

Theme Connection

Interacting Systems
Point out that sea stars are slow-moving organisms that are often easy to see. Have students identify the characteristics that help sea stars survive predation. Be sure to relate the meanings of *echino* (spiny) and *derm* (skin).

Visual Strategy

Figure 23. 22
Have students describe the characteristics of this semitransparent, fishlike, pointed-tailed (lancelike) animal and have them determine the root of its name.

Social Studies Connection

Lancelets
Normally lancelets are not considered economically important, but near Amoy, China, fishermen harvest large quantities of lancelets from August to April, which are used as food.

Phase 3

ASSESSMENT OPTIONS

Closure Strategy

Vertebrate Characteristics
Have each student write three characteristics of vertebrates. Have students exchange papers and comment on each other's work. Then have students state the characteristics as you list them on the chalkboard. Reintroduce any characteristics students may have overlooked.

Section Review

Assign the *Section Review.*

Reteaching

Based on what they know about the characteristics of echinoderms, have students draw and label an imaginary echinoderm exhibiting all the proper characteristics.

Chordates

Like echinoderms, chordates (members of the phylum Chordata) are deuterostomes. As you learned in Chapter 21, all chordates share three features: a notochord, a dorsal nerve cord, and pharyngeal slits. Three subphyla make up the phylum Chordata.

Tunicates lose their chordate characteristics as adults

Adult tunicates, or sea squirts (subphylum Urochordata), live attached to rocks or the sea floor, as shown in **Figure 23.21**. Like sponges, tunicates obtain food by filtering particles of food from the water. They secrete a leathery or jellylike covering composed of a substance similar to cellulose. This outer body layer is known as a tunic, hence the name of the group. During development, urochordates go through a free-swimming larval stage. During this stage the three chordate features are apparent. Adult tunicates retain only the pharyngeal slits.

Figure 23.21
Unlike their larval form, adult tunicates, such as sea squirts, are not free-swimming.

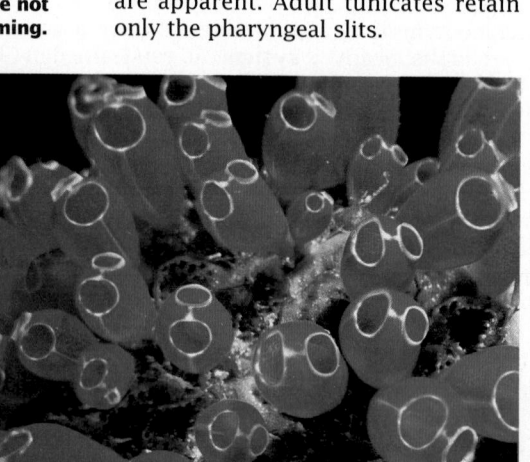

Figure 23.22
Unlike tunicates, lancelets retain all of their chordate characteristics throughout their lives.

Lancelets retain their chordate features throughout their lives

Lancelets (subphylum Cephalochordata) are about 5–8 cm (2–3 in.) long. **Figure 23.22** shows the lancelet *Branchiostoma*. Lancelets bury their tails in sand or mud and keep their heads exposed in the overlying water. As water enters the mouth, large particles are sorted out by the tentacles. The water passes into the pharynx, where food is trapped in the mucus-covered lining, and exits through gill slits.

Vertebrates have backbones

With 42,500 species, the vertebrates (subphylum Vertebrata) are the largest and most successful group of chordates. Much of this success is due to the increased complexity of many organ systems plus the addition of an internal skeleton. The distinguishing feature of the vertebrates is the vertebral column, which gives the group its name. Fishes, amphibians, reptiles, birds, and mammals (including humans) are members of this very diverse subphylum. You will learn more about vertebrates in Chapters 25 and 26.

Section Review

① **Name three kinds of echinoderms.**

② **List two echinoderm characteristics.**

③ **Which feature distinguishes the vertebrates from the other subphyla of chordates?**

■ *Section Review Answers* ■

1. sea stars, sand dollars, and sea cucumbers

2. Echinoderms have water vascular systems and exhibit a five-part radial symmetry.

3. the vertebral column

Chapter 23 *Highlights*

Many bivalves are important sources of human food.

Alternative Assessment
Have students divide into three groups that represent each section of this chapter. Then further divide each group into pairs of students. Have each pair of students make up questions for the different levels of difficulty in the game Jeopardy. The next day, play the game as a review using the students' questions.

	Key Terms	Summary
23.1 Sponges, Cnidarians, and Simple Worms  When the rostellum of this tapeworm is extended, it can dig into human tissue.	spicule (p. 503) nematocyst (p. 505) polyp (p. 505) medusa (p. 505) hermaphrodite (p. 507)	• Sponges live attached to rocks. They filter food from the water. • Jellyfish, hydras, corals, and sea anemones are cnidarians. Cnidarians are characterized by stinging nematocysts. • Most flatworms are parasites. Parasitic flatworms of humans include the beef tapeworm and *Schistosoma*, which causes the disease schistosomiasis. • Roundworms, or nematodes, are among the most abundant organisms. Some nematodes are parasites.
23.2 Mollusks, Annelids, and Arthropods  Snails belong to the class Gastropoda, one of the three major groups of mollusks.	nephridium (p. 511) radula (p. 512)	• Clams, mussels, snails, slugs, octopuses, and squids are mollusks. • Most mollusks have a protective shell and feed by using a rasping structure called a radula. • The three main classes of mollusks are Gastropoda, Bivalvia, and Cephalopoda. • Earthworms, leeches, and certain marine worms belong to the phylum Annelida, the segmented worms.
23.3 Echinoderms and Chordates 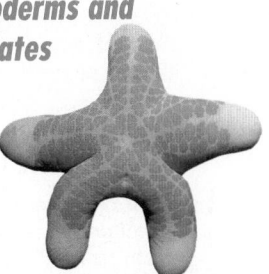 This sea star is an echinoderm. Like you, sea stars are deuterostomes.	water vascular system (p. 517)	• Echinoderms and chordates are deuterostomes. • Echinoderms have a five-part radial symmetry. They also have a water vascular system that operates their tube feet. • The three main groups of chordates are the tunicates, lancelets, and vertebrates. You belong to the vertebrates.

Chapter Review Answers

Understanding Vocabulary

1. **a.** attached
   b. *Wuchereria bancrofti*
   c. segmented
   d. cephalopod
   e. vertebrate

Relating Concepts

2. Map answer is shown on page 501D.

Understanding Concepts

Multiple Choice

3. a	7. d
4. b	8. c
5. c	9. d
6. a	10. b

Completion

11. Cephalopoda, Gastropoda, Mollusca
12. Echinodermata
13. regenerate

Short Answer

14. The adult form retains pharyngeal slits, only one of the three characteristics found in chordates. A notochord and nerve cord are missing.
15. They have complex organ systems and an internal bony skeleton.
16. Snails are mobile and feed as they move using an organ called a radula. The radula is used to scrape small particles of food from surfaces. In contrast, mussels are attached and feed by filtering small particles of food from the water.
17. All filter particles of food from the water.
18. They are segmented.

Understanding Vocabulary

1. For each set of terms, complete the analogy.
   a. medusa:free-floating::polyp: _____
   b. schistosomiasis: *Schistosoma*::filariasis: _____
   c. flatworms:unsegmented::annelids: _____
   d. snail:gastropod::squid: _____
   e. sea squirt:Urochordata::human: _____

Relating Concepts

2. Copy the unfinished concept map below onto a sheet of paper. Then complete the concept map by writing the correct word or phrase in each oval containing a question mark.

Understanding Concepts

Multiple Choice

3. The skeleton of the sponge is composed of tiny needle-like structures called
   a. spicules.
   b. nematocysts.
   c. bone.
   d. pharyngeal slits.

4. The actions of earthworms are beneficial to farmers because they help
   a. kill weeds.
   b. restore soil fertility.
   c. fix nitrogen.
   d. destroy plant parasites.

5. A phylum of animals characterized by radial symmetry and a water vascular system is
   a. Chordata.
   b. Arthropoda.
   c. Echinodermata.
   d. Cnidaria.

6. Tapeworms live in the human body as
   a. parasites.
   b. free-living organisms.
   c. prey.
   d. mutualists.

7. Animals, such as the earthworm, that contain both male and female reproductive systems are called
   a. generatives.
   b. parasites.
   c. reproducers.
   d. hermaphrodites.

8. Which animal is *not* a representative of the class Bivalvia?
   a. clam
   b. mussel
   c. nautilus
   d. oyster

9. Spiders, centipedes, and crabs are members of the phylum
   a. Mollusca.
   b. Annelida.
   c. Echinodermata.
   d. Arthropoda.

10. Animals that have a notochord, a dorsal nerve cord, and pharyngeal slits are members of the phylum
    a. Echinodermata.
    b. Chordata.
    c. Annelida.
    d. Cnidaria.

Completion

11. Octopuses are members of the class _____ , while slugs are members of the class _____ . Both octopuses and slugs are members of the phylum _____ .

12. The phylum _____ includes starfish, sand dollars, and sea urchins.

13. If a planarian is cut in half, each piece is able to _____ its missing half.

Short Answer

14. If only adult forms were considered, why would the sea squirt not be classified as a chordate?

15. How are members of subphylum Vertebrata different from other chordates?

16. How does the feeding mechanism of mussels differ from that of snails?

17. Describe the methods by which tunicates, clams, and sponges obtain food.

18. What one characteristic of annelids helps distinguish them from other kinds of worms?

Phylum Mollusca — includes the classes — ? / ? / ? — feed using — ?
such as — ? ; such as — ? ; such as — snails — host — ?
feed by — filtration ; causes ; belongs to
like — tunicates — belong to — ? — belong to — humans — infects ; schistosomiasis ; includes — ? ; ?

Interpreting Graphics

19. Look at the diagram of the *Schistosoma* life cycle.

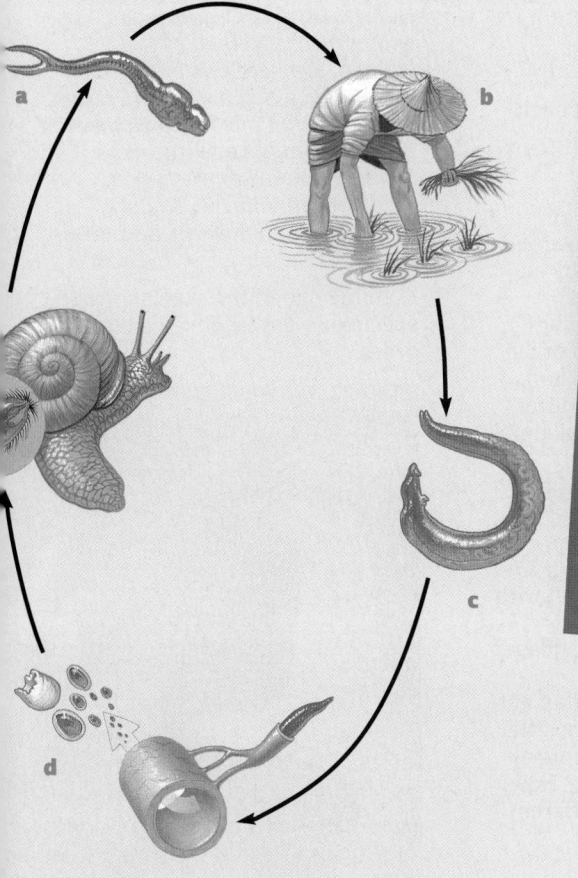

- Suggest two points at which the life cycle of *Schistosoma* could be interrupted, thus preventing human infections. Describe the actions required to interrupt the life cycle at each point.
- Explain why *Schistosoma* infections are very rare in the United States.

Reviewing Themes

20. *Interacting Systems*
    When a swimmer is stung by a jellyfish, what causes the pain felt by the swimmer?

21. *Scale and Structure*
    In which biome would free-living flatworms most likely be abundant? Explain your answer.

Thinking Critically

22. *Inferring Conclusions*
    Having learned about the life cycle of the beef tapeworm, would you order a very rare steak in a restaurant? Explain your answer.

23. *Comparing and Contrasting*
    How are the adult stages of sponges and jellyfish the same? How are they different?

24. *Building on What You Have Learned*
    In Chapter 18 you learned about alternation of generations in plants. Alternation of generations is also a characteristic found among many members of the animal kingdom. Name one animal that exhibits alternation of generations and describe the body form for each stage of the life cycle.

Cross-Discipline Connection

25. *Biology and Health*
    Do library research on the disease trichinosis. What causes this disease? How can trichinosis infections be prevented?

Discovering Through Reading

26. Read the article "Leeches: when bleeding is exactly what you want," in *RN*, September 1991, pages 31–33. According to Nurse Strangio, why was Maria Vargas in the hospital? How were leeches used to improve Ms. Vargas's condition?

Thinking Critically

22. No. Beef that has not been sufficiently cooked may contain tapeworm eggs that could hatch and become attached to the lining of small intestine.

23. The adult forms of both sponges and jellyfish are sexually active and appear very different than their juvenile forms. They are different in that the adult sponge is attached and the adult jellyfish is free-swimming.

24. Cnidarians exhibit alternation of generations. See Figure 23.3 on page 505.

Cross-Discipline Connection

25. Students should explain that trichinosis is a disease caused by eating undercooked pork contaminated with trichina cysts. They should suggest that the cysts can be killed by freezing the meat and suitably cooking it.

Discovering Through Reading

26. Ms. Vargas was in the hospital because two of the fingers on her right hand were severed in a work-related accident. Leeches were attached to her reattached finger. The leeches removed the blood that accumulated in the finger due to poor venous circulation.

Interpreting Graphics

19. • Destroy the hosting *Schistosoma* larva. People can wear protective gear such as wading boots or avoid swimming in infected areas.
    • Students might suggest that proper sewage treatment and climate are factors that make Schistosoma infections rare in the United States.

Reviewing Themes

20. When the swimmer's skin contacts the tentacles of a jellyfish, nematocysts in cells of the tentacles are discharged. Like miniature harpoons, the nematocysts pierce the swimmer's skin and release a toxin. It is the toxin that causes the painful, burning rash.

21. Students might suggest that free-living flatworms would thrive in shallow waters, usually in protected habitats.

Procedural Note

Review the techniques introduced in Chapter 15 for making and using a classification key. Be sure students understand that the key will based upon observations of specimens and other resources including the textbook.

Prelab Preparation Answers

- Tissue is a group of similar cells that carry out a common function, whereas an organ is several types of tissue that together perform a bodily function.
- Asymmetrical animals lack regular arrangement, whereas symmetrical animals express some form of regular arrangement or balance. Bilateral symmetry is the arrangement of an organism's body parts, so that one-half of the body is a mirrored image of the other half. Radial symmetry is an arrangement of body parts around a central point.
- Gills are structures for aquatic gas exchange. Lungs are structures for terrestrial gas exchange. The tracheae deliver oxygen from pores on the surface directly to the cells and transport carbon dioxide out. Book lungs have folds that provide a large surface area through which diffusion can occur.
- Exoskeleton is the exterior skeleton, whereas the endoskeleton is the interior skeleton.
- Jointed appendages are any movable extensions of the body that meet at certain points and do not contain vital organs.
- Segmentation is the repeating of body units.
- Animals with body cavities are provided structural support, whereas animals without body cavities lack structural support. In coelomates, the body cavity lies within the mesoderm allowing for more specialization of body parts, whereas in pseudocoelomates the body cavity develops between the mesoderm and the endoderm.

Procedure Answer

4. Answers will vary depending upon specimens observed.

Chapter 23 Investigation

How Do You Make a Key of the Major Animal Phyla?

Objectives

In this investigation you will:
- *compare* and *contrast* the distinguishing features of the major animal phyla
- *construct* a key for identification of specimens

Materials

- specimens from each of the nine major animal phyla

Prelab Preparation

Review what you have learned about the animal body by answering these questions:
- What is the difference between a tissue and an organ?
- How do asymmetrical and symmetrical animals differ? Distinguish between bilateral and radial symmetry.
- What is the function of gills? Of lungs? How are book lungs different from tracheae?
- Contrast an exoskeleton with an endoskeleton.
- What is a jointed appendage?
- Define segmentation.
- How does the body of an animal with a body cavity differ from the body of an animal without a body cavity? How is a pseudocoelom different from a coelom?

Procedure: Making a Key

1. Form a cooperative group of four students. Work with one member of your group to complete steps 2-7.
2. Make a table with eight columns and nine rows. Label the columns *Highest Level of Organization*, *Type of Symmetry*, *Method of Gas Exchange*, *Type of Skeleton*, *Type of Appendages*, *Segmentation Present*, *Type of Body Cavity*, and *Number of Body Openings*. Label the rows *Sponge*, *Cnidarian*, *Flatworm*, *Roundworm*, *Annelid*, *Mollusk*, *Arthropod*, *Echinoderm*, and *Chordate*.
3. Examine one of the specimens. The specimens can be observed in any order.

4. Answer the following questions about the specimen. Use your textbook and other available resources to find the information not readily observed in the specimen. Record your answers in your table.
   - What is the highest level of organization shown by the specimen? Does it have tissues or organs, or is it organized only at the cellular level?
   - Is it symmetrical? If so, what type of symmetry does it show (radial or bilateral)?
   - Does it have specialized gas exchange structures such as gills, tracheae, book lungs, or lungs?
   - If it has a skeleton, what type of skeleton does it have (endoskeleton or exoskeleton)?
   - Does it have jointed appendages?
   - Is it segmented?
   - Does it have a body cavity? If so, what kind (coelom or pseudocoelom)?
   - Does it have one body opening (mouth), or does it have two (mouth and anus)?

5. Repeat Steps 3 and 4 for each specimen.

6. You are now ready to make your key. Use your table to choose features that enable you to distinguish between the phyla. When you are finished, all the phyla should be accounted for in your key. If your teacher gives you a new specimen of animal, you should be able to determine the phylum to which it belongs.

7. Exchange keys with the other team in your group. Use the other team's key to determine the phylum to which each specimen belongs. Share any suggestions for improving the key with the other team.

Analysis

1. *Analyzing Observations*
   Which phylum of animals is least like the others? Explain.

2. *Evaluating Methods*
   Is your key identical to the key of the other team in your group? Why can two keys be different but equally correct?

3. *Evaluating Methods*
   Why isn't the color of the specimens a good distinguishing characteristic for constructing a key?

4. *Evaluating Methods*
   Do you think your key would be useful for identifying these animals in the wild? Explain.

Thinking Critically

Explain the difference between making a classification key and classifying organisms based on their evolutionary relationships.

Analysis Answers

1. Sponges are least like the other groups of animals. The lack of tissue suggests these organisms are extremely primitive.

2. It is likely that the keys of different teams will differ. The keys may be equally well constructed, while choosing different traits in different sequences to use.

3. Convergent evolution could cause very different organisms to have a similar color or appearance without sharing a common ancestor.

4. Answers will vary but should be logical. In general, the biggest weakness of any classification key is that it only works for the organisms it was designed to identify. The key may not work for new or unusual specimens.

Thinking Critically Answer

In a classification key, organisms are assigned to a group because they share distinctive characteristics with the other members of that group. The biological hierarchy of classification is made up of seven different levels: kingdom, phylum, class, order, family, genus, and species. The more classification categories two species share, the more traits they have in common. In evolutionary relationships organisms can share similar traits, but not share a common ancestor.

Chapter 24 Arthropods

Planning Guide

	Objectives/Themes	Classwork Resources	Homework Resources
24.1	**1.** Identify three kinds of arachnids. **2.** Contrast the functions of chelicerae and pedipalps in spiders. **3.** Describe three ways spiders use silk. **4.** List two arachnids that directly affect people. **Theme:** Evolution, Interacting Systems, Scale and Structure	**Teacher's Resource Binder** Focus Activity 24 *Observing Insect Behavior*	**Text** Section Review, p. 529 **Teacher's Resource Binder** Directed Reading Worksheet 24.1 **Other Resources** Audiocassette 24.1*
24.2	**1.** Contrast the anatomy of spiders with that of insects. **2.** Compare incomplete and complete metamorphosis. **3.** Describe five ways that insects affect your life. **4.** Identify two differences between millipedes and centipedes. **Themes:** Evolution, Interacting Systems, Patterns of Change, Scale and Structure	**Text** Science in Action *Genaro Lopez: Entomologist* pp. 530–531 **Teacher's Resource Binder** Lab Investigation 24.1 *Live Crickets* Extension Worksheet *Spider Webs* **Other Resources** Transparencies 105–106	**Text** Section Review, p. 538 **Teacher's Resource Binder** Directed Reading Worksheet 24.2 **Other Resources** Audiocassette 24.2*
24.3	**1.** Identify three kinds of crustaceans. **2.** List three differences between crustaceans and insects. **3.** Describe the importance of crustaceans for the ecology of the sea. **Themes:** Evolution, Scale and Structure, Stability, Interacting Systems	**Text** Investigation *Can Pill Bugs Detect Differences in Moisture and pH?* pp. 546–547 **Teacher's Resource Binder** Lab Investigation 24.2 *Comparison of Five Arthropod Classes*	**Text** Section Review, p. 542 **Teacher's Resource Binder** Directed Reading Worksheet 24.3 Vocabulary Review Worksheet* Reteaching Worksheet* *Insects* **Other Resources** Audiocassette 24.3*

*Reteaching Options

Demonstrations
24.2: pp. 533, 534, 537

Assessment
Chapter Review pp. 544–545
Portfolio Assessment p. 523D
Chapter Test—Teacher's Resource Binder
Test Generator

Research Notes

Connections to Behavioral Sciences and Chemistry: Bees and Chemical Passwords

Some of the most fascinating of the insects are those that live in colonies, such as bees and ants. A puzzling aspect of these insect communities is the methods that individuals belonging to the communities use to differentiate community members from non-members.

Michael Breed, of the University of Colorado, and Glennis Julian, formerly a student at Pomona College in Claremont, California, have determined that honeybees rely on their sense of smell for cues about which insects belong to their hive.

In a control experiment, they used two organic compounds, hexadecane and methyl docosonoate, to test whether honeybees could detect invaders. They removed young bees from a hive and raised them in separate groups of 10 bees each. Some bees were exposed to one or the other of the compounds, some were exposed to both, and others were exposed to neither compound.

When bees treated with one chemical were placed with bees treated with the other, they were bit and stung repeatedly, sometimes even until they were dead. However, when both chemicals were present, the hexadecane cue seemed to have a much stronger effect.

Bees used to both chemicals accepted bees smelling only of hexadecane, but attacked those smelling only of methyl docosonoate. Similarly, bees expecting only methyl docosonoate would attack those smelling of both compounds.

The predominance of one cue over the other suggests that bees have simple priorities to help them make sense out of the many chemicals and smells likely to be found within a hive. Scientists are still working to isolate the actual compounds bees use naturally within the hive for chemical cues.

Often, commercial beekeepers must place a new queen in a hive. Sometimes the hive rejects the new queen. Beekeepers can apply the new research and treat both the hive and the new queen with the same compound or combination, helping to make the bees in the hive more receptive to the new queen.

Investigation Notes

Can Pill Bugs Detect Differences in Moisture and pH?

pages 546–547

Purpose: In this investigation students observe some simple isopod behaviors. An isopod's response to moisture is an example of a kinesis.

An animal exhibiting kinesis moves about randomly until it encounters a favorable environmental condition. Then, it stops moving. This type of behavior differs from a taxis, which is the direct movement of an animal toward or away from a stimulus.

Point out to the students that the design of an experiment might prevent the recording of a crucial observation and lead to an erroneous conclusion. For example, if the experimental design does not take the possibility of kinesis into account, one might incorrectly conclude that pill bug movement is due to taxis.

Solution Preparation

Make a dilute solution by dissolving about 4 g of sodium hydroxide in 1 L of water.

Prelab Preparation

1. Pill bugs can be obtained from WARD'S (87 M 5520) and are easily maintained in a plastic container containing crumbled moist paper and a piece of potato. Distribute the pill bugs to the students in small paper cups.

2. Pill bugs may be released into the environment if appropriate. Place them near or under fallen logs or stones in shaded or wooded areas.

Answers will be found on pages 546–547.

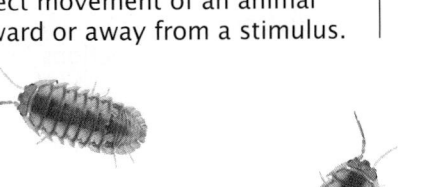

Meeting Individual Needs

Objectives

1. Students will demonstrate appropriate use of core vocabulary for the chapter (Vocabulary File).

2. Students will identify examples of arthropod diversity (Teaching Strategy A).

3. Students will explain how to help scientists track butterflies (Teaching Strategy B).

Vocabulary File
(Developing Vocabulary/ Limited English Proficiency)

If you are not already using the Vocabulary File, refer to Chapter 1 for its preparation. See the Chapter Highlights on page 543 for a list of suggested words.

Teaching Strategy A
Diversity of Arthropods' Body Plans
for pages 524–542
(Developing Organizational Skills/Visual Learners)

Materials: overhead projector, transparency, projector pens, photographs of arthropods, spider silk

Preparation: Photos of the following animals should be hung before class in the "sky zoo": spider, scorpion, horseshoe crab, grasshopper, monarch butterfly, bee, fly, water bug, ant, water flea, copepod, and lobster. Have one or two strands of silk from a spider web available to be passed around so students can feel the strength. (Spider webs should be easy to find.)

Procedure: Open by asking students to identify new zoo members. Scan the entire chapter with students. Have them suggest what the main ideas are so that the result is an outline written on the overhead projector. Recall and relate the new animals in the sky zoo to the outline as it is developed. Students are to copy the outline and use it to study with. An outline should cover spiders, insects, and crustaceans. For each group, describe the characteristic body regions, the appendages, and any features unique to that group.

During the discussion of the outline dealing with spiders, pass threads taken from a spider's web. Be sure to note that among the spider's unique features are the web-producing organs. During the insect discussion, also describe complete and incomplete metamorphosis.

Close by recalling special features observed earlier in animal studies —for example, the live crayfish used in Demonstration A for Chapter 22.

Teaching Strategy B
Tracking Migrations of Monarch Butterflies
for pages 534–537
(Developing Communication Skills/Visual and Verbal Learners)

Preparation: Have the librarian obtain a copy of *The Travels of Monarch X* , by Ross E. Hutchins, illustrated by Jerome P. Connally. This is a Weekly Reader Children's Book Club selection, published by Rand McNally & Company in 1966.

Procedure: Lead a discussion comparing and contrasting complete and incomplete metamorphosis. Read the entire book to the students. Tell students the date of the copyright and point out that butterflies are still being tagged today. The tagging continues because keeping records of butterfly flights is one way to understand patterns of change in the environment.

For example, a cold and rainy spring in 1992 decreased the number of migrating butterflies. Efforts to track tagged monarchs intensified that year. Every year there are articles in local papers asking people to be on the lookout for the tagged butterflies.

If you see one, net it, copy down the information, and let it go to continue the flight. If the butterfly seems weak, put it in a jar with some sugar water for a day or two and let it eat and rest before releasing it. The information obtained from the tag should be taken to the local chapter of the Audubon Society who can see that it gets to the researchers.

Even if monarchs may not be common where you live, flights over cities and through storms often are confusing and some butterflies get far off the regular path of migration.

Close by discussing how the differences in lifestyles of arthropods affect the general food chain of the world.

Additional Strategies
Visual Strategies
Pages 525, 526, 527, 528, 529, 532, 533, 534, 535, 536, 538, 539, 541, and 542

Auditory Learners
Use *Biology: Visualizing Life* Audiocassettes for Sections 24.1, 24.2, and 24.3.

Meeting Individual Needs (cont.)

Cooperative Learning
Types of Arthropods

Timing: Use this activity to introduce Section 24.1.

Group Size: 3 students

Outcome: Students will be able to describe three of the types of arthropods and explain why they are such successful animals.

Individual Accountability: Each group member will describe one type of arthropod.

Positive Interdependence: Each group will be able to list explanations for the success of arthropods in adapting to a wide variety of environments.

Assign one of the following types of arthropods to each member of each group: (1) arachnids, (2) insects, and (3) crustaceans. Have group members read the chapter and take notes on the types of arthropods assigned. Students should note the number of body segments, number of legs, modifications of appendages, habitat, method of reproduction, and provide examples of organisms of this type.

Students should share this information with other group members. Using this information, the group should discuss reasons why arthropods are among the most successful animals on Earth.

Portfolio Assessment

Students should select their best work and provide a self-reflective rationale for their selections. Students can make selections in the following areas.

1. *Content* — One concept map from the chapter (See page 812 for evaluation criteria.)

2. *Reading Comprehension* — One Directed Reading Worksheet from the Teacher's Resource Binder (Use the answer key to evaluate for accuracy.)

3. *Writing* — Using the Vee Form, summarize a magazine or newspaper article relating to specific arthropods and their effects on the environment. (See page 22T for evaluation criteria.)

   Or: Select a writing task or project from the Chapter Review.

4. *Performance Assessment* — One Vee form from a chapter investigation or lab manual investigation (See page 22T for evaluation criteria.)

Teacher makes selections in the following areas.

1. *Formal Assessment* — Chapter test (Test A, B, or the Test Generator) The teacher-scored test should be reviewed by the student. Incorrect responses should be corrected by the student before the test becomes part of the portfolio.

2. *Informal Assessment* — Use the Direct Observations Checklist, page 33T, during a laboratory or other cooperative learning experience.

3. *Performance Assessment* — Have students report on insects and arthropods found in the school building and their homes, and the insecticides (if any) used to control them.

Concept Map Answer

The following is one possible answer to the Relating Concepts exercise on page 544.

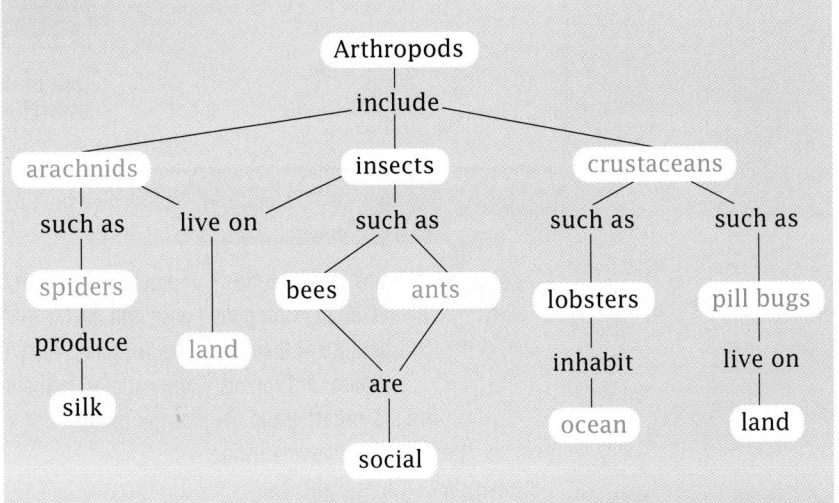

Chapter 24

Determining Prior Knowledge

- Have students explain the difference between an insect and an arachnid.
- Have students draw a spider. Ask them how many legs are on their representation.
- Ask students if spiders have eyes. If so, how many?
- Have students brainstorm why spiders make webs.
- Ask students if spiders hunt in groups, as seen in some horror movies.

Chapter 24

Arthropods

Review

- **innovations in body plan** (Section 21.3)
- **gas exchange (Section 22.2)**
- **the terms** *exoskeleton, gill, trachea, book lungs,* **and** *Malpighian tubules* **(Glossary)**

24.1 Spiders and Their Relatives

- **Characteristics of Arachnids**
- **Spiders**
- **Other Arachnids**

24.2 Insects, Millipedes, and Centipedes

- **Insects**
- **Millipedes and Centipedes**

24.3 Crustaceans

- **Crustaceans Are Successful Aquatic Arthropods**
- **Crustacean Diversity**

One of the most prominent features of this green-headed horsefly is its compound eyes.

▪ Author's Rationale ▪

This chapter is devoted to arthropods because of their importance. Arthropods are the largest group of animals. They display some interesting characteristics, such as flight. Also, arthropods have a major impact on humans in terms of the spread of disease and the destruction of crops. As the arthropod story unfolds, students are exposed to the incredible diversity of this group.

THERE ARE ABOUT 1 MILLION KNOWN SPECIES OF ARTHROPODS, FAR MORE SPECIES THAN BELONG TO ANY OTHER PHYLUM. ARTHROPODS HAVE A GREAT IMPACT ON HUMANS AND ON WORLD ECOLOGY. WHILE SOME ARTHROPODS TRANSMIT DISEASES AND DESTROY BILLIONS OF DOLLARS WORTH OF CROPS EACH YEAR, OTHERS POLLINATE IMPORTANT CROPS OR SERVE AS HUMAN FOOD.

24.1 *Spiders and Their Relatives*

Objectives

❶ **Identify three kinds of arachnids.**

❷ **Contrast the functions of chelicerae and pedipalps in spiders.**

❸ **Describe three ways spiders use silk.**

❹ **List two arachnids that directly affect people.**

Characteristics of Arachnids

What do you see when you look closely at a spider? One of the spider's key features is the pair of appendages called **chelicerae** (*kuh LIHS uh ree*). Chelicerae, which are located at the front of a spider's body, are the hallmark of spiders and their relatives—scorpions, mites, ticks, and horseshoe crabs. Chelicerae are positioned far forward on the body and are modified for feeding. The chelicerae of spiders, such as the tarantula shown in **Figure 24.1**, are poison-delivering fangs. A second pair of appendages, known as **pedipalps**, lies just behind the chelicerae. Pedipalps are used to capture and manipulate prey, and for courtship displays and mating. Spiders, scorpions, mites, and ticks are **arachnids** (*uh RAK nihdz*), members of the class Arachnida.

Remember from Chapter 21 that segmentation is a characteristic of the arthropods. In arachnids, body segments are fused to form two body regions: the cephalothorax (*sehf uh luh THAWR aks*) and the abdomen. Arachnids have four pairs of legs, which are attached to the cephalothorax.

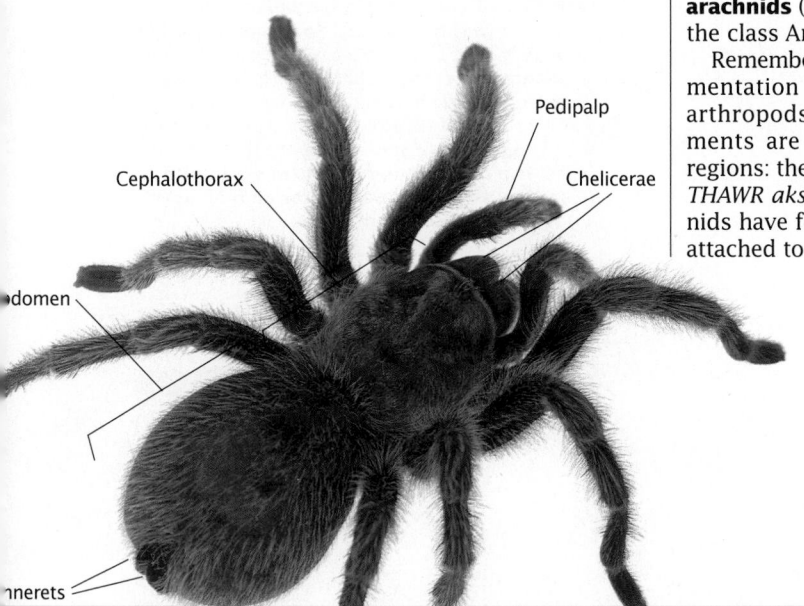

Pedipalp

Cephalothorax

Chelicerae

Abdomen

nnerets

Figure 24.1
The tarantula is one of the world's largest spiders. Tarantulas and all other arachnids have chelicerae.

525

Connection: Chapter 12

No Web, No Nest

Some spiders do not build webs or nests. Daddy longlegs, or harvestman spiders, are such arachnids. For protection, they have noxious stink glands. They may even shed legs to escape an enemy. On the other hand, some spiders, like the social spiders of the Amazon rain forest, build group webs and communally feed on birds they ensnare.

Visual Strategy

Figure 24.2

Explain that spiders spin different types of threads. Dry threads are used to form the radial framework of the web. Sticky threads that trap victims are arranged in a spiral formation around the radial threads. Spiders walk only on the nonsticky threads. Ask students if spiders stick to their own webs. Have them explain their answers.

Spiders

There are about 35,000 kinds of spiders. While a few live in fresh water, the majority are land dwellers. Indeed, spiders are thought to have been among the first arthropods to successfully invade the land, soon after their close relatives the scorpions. Spiders are predators, feeding on nearly any other animal of manageable size.

In addition to the characteristics that make arthropods so successful, spiders have several unique features that make them particularly well adapted to their lifestyles. First, they have efficient organs of excretion and water balance, so they can rid their bodies of wastes while still conserving precious water. Second, spiders are able to breathe air by using book lungs. Recall from Chapter 22 that book lungs allow air to come close to the spider's blood in a humid chamber where gas exchange occurs. Third, nearly all spiders are poisonous. The chelicerae of spiders are fangs equipped with poison glands. Fourth, and perhaps most important, spiders have the ability to produce silken threads, which they use in a great variety of ways.

Silk and fangs are a deadly combination

You may have watched a spider building its web. At the end of a spider's abdomen are small nozzle-like structures called **spinnerets**. Spinnerets direct the flow of silk from silk-producing glands in the abdomen. Spiders can produce different kinds of silk threads in various diameters.

Spider silk is composed of a complex structural protein that is produced in the glands as a liquid. As the liquid silk flows out through holes in the spinnerets, it solidifies into the threads that make up the web. As illustrated in **Figure 24.2**, there are as many different types of webs as there are types of spiders.

Figure 24.2

Each kind of spider has a specialized web that enables it to catch its prey. Some spiders build webs in high tree branches, while others build horizontal sheet webs that catch insects that drop from above or are carried by the wind. Some spiders build silk-lined burrows beneath the ground. These spiders extend threads that act as trip lines around the burrow entrance. Then they sit inside, waiting for an unsuspecting insect to become entangled.

Spider silk is elastic and extremely strong, about as strong as a nylon thread of the same diameter. Spiders use silk for many purposes. Threads are used as safety lines when spiders dangle from branches or drop to the ground. Many spiders also use silk to line their nests or burrows, to wrap prey, or to fashion cocoons for their young. Some newly hatched spiderlings spin long, thin threads on which they ride the winds over great distances.

A spider's web is a sticky trap to ensnare prey, such as the bee caught in the web shown in **Figure 24.3**. Constructing a web is a complex architectural feat that requires several steps. When an insect strikes the web and struggles, the spider is alerted by the threads' vibrations. The spider then rushes to the victim, bites it, and injects poison and digestive enzymes. The ability to stun or kill the prey quickly with poison from the chelicerae is a great advantage. It saves the spider from having to wrestle with large or potentially dangerous prey, such as bees and wasps.

Although nearly all spiders produce poison, only about 20 kinds are considered dangerous to people. Two such species are found in the United States: the American black widow spider and the brown recluse spider, both shown in **Figure 24.4**. Most spiders are more beneficial than harmful to humans. They are the world's champion pest controllers, consuming millions of insects in their daily activities. Many of these insects are agricultural pests.

Spiders have courtship rituals

Spiders are solitary animals. When they do seek a mate, it is important that males and females of the same species are able to recognize each other. Males are usually smaller than the females and could easily be mistaken for a meal if some signal did not identify the male to his prospective partner. Spiders have evolved complex behaviors that ensure

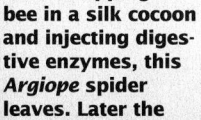

Figure 24.3
After wrapping a bee in a silk cocoon and injecting digestive enzymes, this *Argiope* spider leaves. Later the spider returns to its web and sucks out the now-liquid prey, and discards the empty carcass.

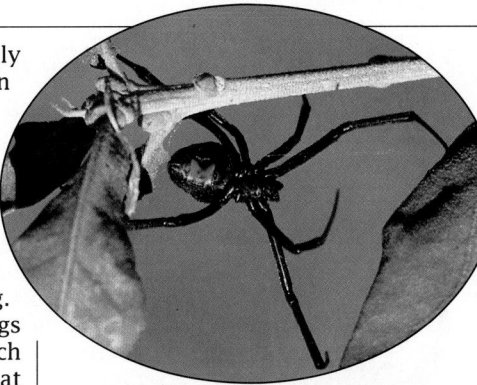

Figure 24.4
a **The American black widow spider is most common in the southern United States. The bite of the female injects a powerful venom that acts on the nervous system.**

b **The brown recluse spider can be recognized by the violin shape on its upper surface. They live in dark places such as basements and woodpiles. Their venom causes tissue damage in the area of the bite.**

successful mating and fertilization of the female's eggs.

Mating behavior in the black widow spider provides an example of the dangers a male spider faces. As the male approaches the web of a female, he drums a specific rhythm on the strands of the web to alert the female that he is a potential mate and not a meal. If he vibrates the female's web inappropriately, she will kill and eat him. After mating, the female often kills the male anyway, wrapping and saving him for her offspring's first dinner.

Pedipalps

Chelicerae

Cephalo-
thorax

Abdomen

Stinger

Figure 24.5
This African scorpion is about 18 cm (7 in.) long. Scorpions found in the United States range from 5–13 cm (2–5 in.) in length.

Other Arachnids

More than 400 million years ago, scorpions were the first arthropods to invade land. They share many of the adaptations that enable spiders to survive on land, including book lungs and efficient excretory and water conservation organs. Unlike the poison-injecting chelicerae of spiders, scorpions' chelicerae are ripping claws. Their pedipalps have evolved into enlarged pincers used to capture prey. Scorpions feed mostly on insects. As you can see in **Figure 24.5**, the scorpion's stinger is found at the tip of its tail. While the scorpion holds its victim in its pincers, it brings its tail forward to jab into the prey, and then tears the prey into pieces.

Scorpions live in the warmer regions of the world. About 20 scorpion species are found in the United States, mostly in the southwest. Scorpions hide during the day and hunt at night. Although it is painful, a scorpion's sting is rarely lethal to humans. Like spiders, scorpions perform elaborate mating rituals.

Most mites and ticks are parasites

If dust makes you sneeze, you are reacting not to the dust itself, but to the tiny mites found in the dust. Over 30,000 species of mites and ticks have been described. A number of these species are free-living on land or in water, but many are parasites on the bodies of other animals. The free-living forms are mostly scavengers or predators on other tiny creatures. If you have done much camping or hiking in brushy terrain, chances are you have spent some time removing ticks from your body. Most ticks are blood-suckers that feed on vertebrates. Their sharp chelicerae are specialized for slicing skin. Although normally quite small, some ticks can swell

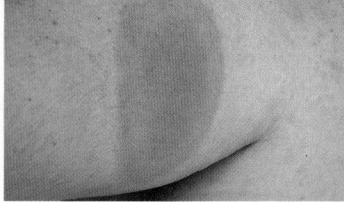

Figure 24.6
About one-half of all the people bitten by the tick that causes Lyme disease develop a "bulls-eye rash" within a few days.

to 3 cm (just over 1 in.) in length after a full meal of blood. Recall from Chapter 16 that ticks transmit Lyme disease and Rocky Mountain spotted fever to humans, and Texas cattle fever to livestock. **Figure 24.6** shows the rash that usually occurs a few hours after a person has been bitten by the tick that causes Lyme disease.

Mites, such as the red water mite in **Figure 24.7**, are much more diverse than ticks. They parasitize virtually all groups of animals and many plants. The incredible specialization of mites is shown by two species, *Demodex folliculorum* and *Demodex brevis*. One

Figure 24.7
The red water mite is an active swimmer. The larval stages of water mites may be parasitic on insects or fishes.

species lives only in hair follicles, the other in oil glands, respectively, of the human forehead. Mites have serious direct and indirect effects on humans. Many, such as chiggers, burrow just beneath the skin, where they cause severe itching. Mites also carry viruses that plague food crops, such as mosaic viruses of rye and wheat. The red-legged mite and the winter grain mite both feed on stored crops, destroying many tons of grain each year. Other mites cause feather loss in birds, decreased wool production in sheep, and mange in dogs. If you find yourself itching a bit after reading this section, don't be alarmed; we are all host to some of these parasites.

Horseshoe crabs are ancient arthropods

The most ancient arthropod with chelicerae is not an arachnid, but the horseshoe crab shown in **Figure 24.8**. There are only five species of horseshoe crabs alive today. They live in shallow oceans, where they plow through sandy bottoms and use their clawlike legs to grasp small animals on which they feed. They also scavenge on dead animals.

Horseshoe crabs are closely related to the extinct giant "water scorpions," or eurypterids, such as the one shown in **Figure 24.9**. The eurypterids flourished about 300 million years ago. Some eurypterids reached lengths of 3 m (10 ft.).

Figure 24.8
This horseshoe crab is not really a crab at all, but a close relative of the scorpion.

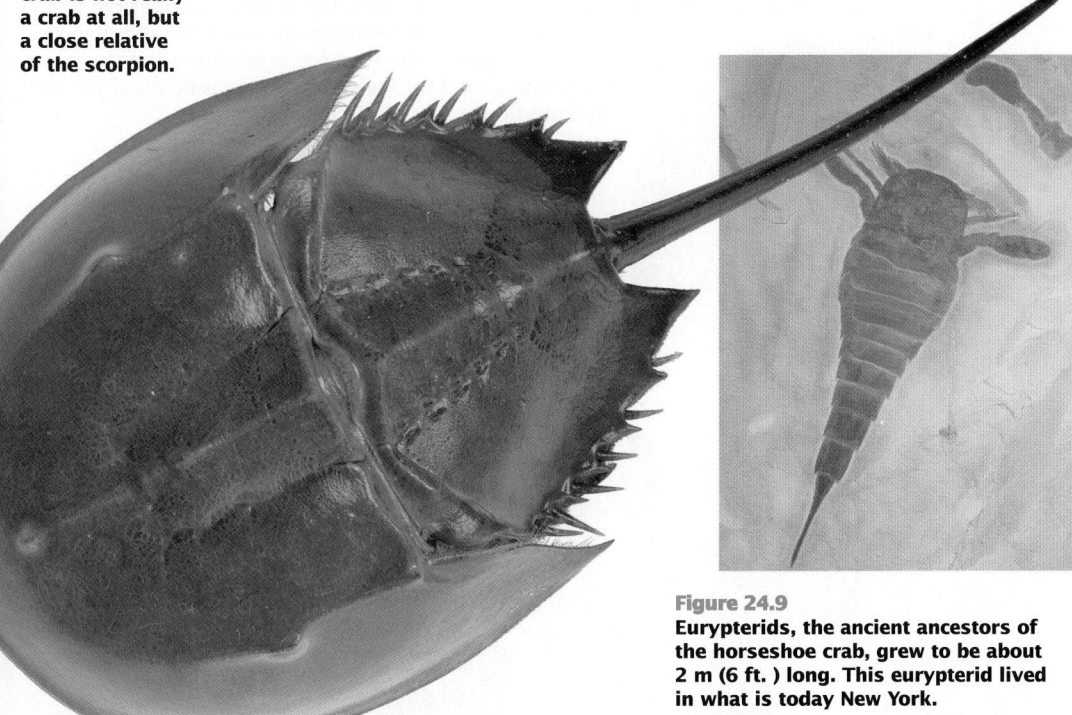

Figure 24.9
Eurypterids, the ancient ancestors of the horseshoe crab, grew to be about 2 m (6 ft.) long. This eurypterid lived in what is today New York.

Section Review

1. **Name three arachnids that are not spiders.**
2. **Contrast the functions of pedipalps and chelicerae in spiders.**
3. **Describe two ways spiders use silk other than for building webs.**
4. **How is your life affected by arachnids?**

Section Review Answers

1. scorpions, mites, and ticks
2. The chelicerae are poison-delivering fangs, whereas pedipalps are used to capture or manipulate prey, and for courtship displays and mating.
3. Possible answers might include that the silk may be used as a safety line, to line the nest or burrows, to wrap prey, to fashion cocoons for their young, to aid in riding the winds over long distances, and to ensnare prey.
4. Students may discuss arachnids as a potential source of disease or as annoying pests.

Visual Strategy

Figure 24.8

Tell students that the horseshoe crab is related to the scorpion. It can grow to 78 cm (20 in.) long. Explain that the horseshoe crab is protected by its shell and hiding ability. It is not aggressive and feeds mostly at night. Tell students that even though horseshoe crabs live in the ocean, they breathe air. Suggest that horseshoe crabs were once commonly fed to hogs and chickens and used as fertilizer. Some people consider them good food, but generally they are of little economic importance, in spite of their large size. Have students speculate about why the horseshoe crab and scorpion are related.

Phase 3

ASSESSMENT OPTIONS

Closure Strategy

Characteristics of Arachnids

Organize students into cooperative work groups of three. Have each student in each group write the characteristics of spiders, scorpions, ticks, and mites. Then have them share their information with other group members.

Section Review

Assign the *Section Review.*

Reteaching

Show students preserved specimens of spiders, scorpions, ticks, and mites. Have them identify each of the organisms and state their reasons for each identification. Have them make a table with the name of the organisms down the left side. Across the top, use the following headings: *Size, Habitat, Behavior, Feeding,* and *Structure.*

Using the Feature

Genaro Lopez grew up in Texas and spoke only Spanish until elementary school. Today he often acts as an interpreter in Spanish-speaking neighborhoods. You can use this feature to show the benefits of being bilingual. Many of your students may be fluent in more than one language. Find out what other languages are spoken in the class. Ask students to discuss their thoughts on being bilingual.

Genaro's career also offers him opportunities to travel and explore other countries. You can use this feature to engage students in a discussion about careers that enable a person to travel.

Discussion

Guide the discussion by posing the following questions:

1. What personal traits have enabled Genaro to become a successful entomologist?
   He is curious, hard-working, adventurous, ambitious, and fun-loving. He is also bilingual, which has enabled him to act as an interpreter for a pest control company in Spanish-speaking communities and to appear on a Spanish television show with a collection of live roaches.

2. Genaro's enthusiasm is not limited to his career. What hobby does he invest time in?
   He's a bicycle racer.

3. To which American cities and foreign countries has Genaro traveled?
   Genaro has traveled to the Amazon River in Colombia, South America, to Mexico, and to Miami, Los Angeles, and New York City. (You may want to have students identify these countries on a world map.)

Science in Action ▪ *Science in Action* ▪ *Science in Action*

Genaro Lopez: Entomologist

How I Became an Entomologist

"I grew up in Brownsville, Texas, where I spoke only Spanish until the first grade. I have always loved the outdoors. As a child, I went camping, hiking, and fishing whenever I could. In high school, my science classes were my favorite.

"Although neither of my parents finished high school, they encouraged me to go to college. A bunch of my friends talked about going, and I thought if they could do it, I certainly could. I started working for Robert Baker, a professor who was only five years older than I was. It's amazing how a good teacher can influence a person. Although I was only an undergraduate, he had me doing stuff that only graduate students did—like traveling to the Amazon River in Colombia, South America, to collect bats. On spring break, we'd go to Mexico to collect animals and plants. I acted as an interpreter and scientist. I'd never seen such peculiar life forms before.

"I loved working in the tropics—just being outside and discovering all the different trees, animals, and insects. I like to say science and I see the world in the same way—a little bit of fact and a whole lot of mystery."

Dr. Lopez is also a bicycle racer. "When you're going along at 25–28 mph and the guy in front of you is only three inches away, you know you're just a split second away from crashing. You have to make quick decisions. It's exciting. You have to be finely tuned."

Name:	**Genaro Lopez**
Home:	**Brownsville, Texas**
Employer:	**University of Texas at Brownsville, Brownsville, TX**
Personal Traits:	• Adventurous • Curious • Hard-working • Ambitious • Fun-loving

Action ▪ Science in Action ▪ Science in Action ▪ Science

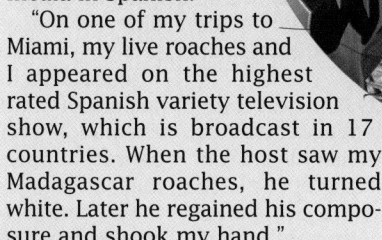

Career Path

High School:
• Science
• Literature

College:
• Zoology
• Chemistry

Graduate School:
• Entomology
• Ecology

Applying Science to Life

"During the academic year, I teach general biology, human anatomy, and physiology to college students. In the summer, I work as a consultant urban entomologist for a pest control company. I act as their Hispanic spokesperson. Every summer they fumigate entire apartment complexes in poor areas of New York City, Miami, and Los Angeles. Several hours after spraying, we go back into the buildings and look at all the dead roaches to see what species are more prevalent in different areas. I explain what's happening to the residents and media in Spanish.

"On one of my trips to Miami, my live roaches and I appeared on the highest rated Spanish variety television show, which is broadcast in 17 countries. When the host saw my Madagascar roaches, he turned white. Later he regained his composure and shook my hand."

Research Focus

Dr. Lopez has an ongoing project studying the tiger beetles along the edge of salt marshes in the lower Laguna Madre of Texas. He has found three species of beetles there. He samples their populations to learn how they coexist. It's an ecological question: How can three similar species share the same resources? He's trying to learn how cooperation, rather than competition, works in nature.

His hypothesis is that one species is most abundant in the spring, a second is most abundant in the summer, and a third is most abundant in the fall. That way they divide the available food equally.

Through his research, Dr. Lopez hopes to show that living organisms such as the beetle he is studying often separate their competition, so that they can successfully coexist.

531

Lesson Plan 24.2

Phase 1

PREPARATION

Key Concepts

- Insects have three main body parts: the head, the thorax, and the abdomen.
- Insects are classified by the number and type of wings.
- Insect development involves complete or incomplete metamorphosis.
- Ants, bees, wasps, and termites are highly social and live in organized societies of related individuals, with a division of labor among society members.
- Millipedes and centipedes are closely related to insects because they share the same head structure.
- Millipedes have two pairs of legs per segment. Centipedes have one pair of legs per segment.

Reading Strategy

Have students describe the beneficial and harmful effects of insect interaction with humans. Encourage students to expand the information that they will read in this chapter with information they have read or gained as a result of personal experience.

Phase 2
TEACHING STRATEGIES

Visual Strategy

Figure 24.10
Explain that insects live in almost every freshwater and terrestrial habitat. Insects are successful because of the characteristics they have in common with other arthropods: their small size, their adaptation to a wide variety of environmental conditions, and their ability to fly. Emphasize that insects are classified into more than thirty different orders based primarily on the type of wings.

532

AT THIS MOMENT, THERE ARE OVER 200 MILLION TIMES MORE INSECTS THAN HUMANS LIVING ON EARTH. THE NUMBER OF INSECT SPECIES IS FAR GREATER THAN THE NUMBER IN OTHER ANIMAL GROUPS. IN FACT, THERE ARE MORE SPECIES OF BEETLES THAN SPECIES OF ALL NON-INSECT ANIMALS COMBINED. INSECTS AND THEIR CLOSE RELATIVES, THE MILLIPEDES AND CENTIPEDES, HAVE MANDIBLES INSTEAD OF CHELICERAE.

24.2   Insects, Millipedes, and Centipedes

Objectives

❶ Contrast the anatomy of spiders with that of insects.

❷ Compare incomplete and complete metamorphosis.

❸ Describe five ways that insects affect your life.

❹ Identify two differences between millipedes and centipedes.

Insects

Figure 24.10
The bodies of all insects, including this grasshopper, consist of three distinct regions: head, thorax, and abdomen. Three pairs of legs are attached to the thorax. Most insects have wings, which are also attached to the thorax. The evolution of wings was a major reason for the diversification of insects.

Like other arthropods, insects are segmented. In insects, these segments have fused into three body regions: the head, the thorax, and the abdomen, as shown in **Figure 24.10**.

A hallmark of insects is the diversity of the specialized appendages attached to their heads. Unlike arachnids, insects have antennae, appendages specialized for sensing the environment. Insect mouthparts are also extremely varied, enabling insects to feed on a greater variety of foods than other arthropods. Among the insects are predators and parasites, bloodsuckers and plant-sapsuckers, scavengers and wood-eaters. The chelicerae and pedipalps of spiders are simple compared with the highly specialized mouthparts of insects. Instead of chelicerae, insects have jaws, or **mandibles**, and other mouthparts that have evolved into different shapes for grinding, scraping, piercing, and sucking.

Insect legs have also evolved into a variety of shapes. The legs of beetles are modified for walking, while the legs of grasshoppers are adapted for jumping. Ant legs are suited for burrowing.

Like other terrestrial arthropods, insects have evolved mechanisms for gas exchange and water conservation. Insect tracheae are extremely efficient in delivering oxygen directly to the body's cells. Insects have a thick, waxy exoskeleton that prevents water loss. Their Malpighian tubules enable them to excrete nitrogenous wastes with little loss of water. **Table 24.1** shows some of the major orders of insects.

Forewing

Thorax

Antennae

Hind wing

Head

Jumping legs

Abdomen

Table 24.1 Major Orders of Insects

Order	Approximate Number of Species	Main Characteristics	Examples	
Coleoptera "Shield-winged"	350,000	Two pairs of wings (front pair covers transparent hind pair); heavy, armored exoskeleton; biting and chewing mouthparts; complete metamorphosis	Beetles, weevils	
Diptera "Two-winged"	120,000	Transparent front wings; hind wings reduced to knobby balancing organs; sucking, piercing, lapping mouthparts; complete metamorphosis	Flies, mosquitoes	
Lepidoptera "Scale-winged"	120,000	Two pairs of broad, scaly wings; hairy bodies; tubelike sucking mouthparts; complete metamorphosis	Butterflies, moths	
Hymenoptera "Membrane-winged"	100,000	Two pairs of transparent wings; mobile head; well-developed eyes; chewing and sucking mouthparts; stinging; many species social; complete metamorphosis	Ants, bees, wasps	
Hemiptera "Half-winged"	60,000	Two pairs of wings or wingless; piercing, sucking mouthparts; incomplete metamorphosis	Giant water bug, bedbug, chinch bug	
Orthoptera "Straight-winged"	20,000	Two pairs of wings or wingless; biting and chewing mouthparts in adults; incomplete metamorphosis	Grasshoppers, crickets, cockroaches, mantids	
Odonata "Toothed"	5,000	Two pairs of transparent wings; chewing mouthparts; incomplete metamorphosis	Dragonflies, damselflies	
Isoptera "Equal-winged"	2,000	Two pairs of wings, but some stages are wingless; chewing mouthparts; social insects with division of labor; incomplete metamorphosis	Termites	
Siphonaptera "Tube-wingless"	1,200	Small, wingless, flattened body; piercing and sucking mouthparts; jumping legs; complete metamorphosis	Fleas	

534

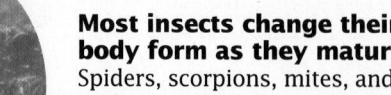

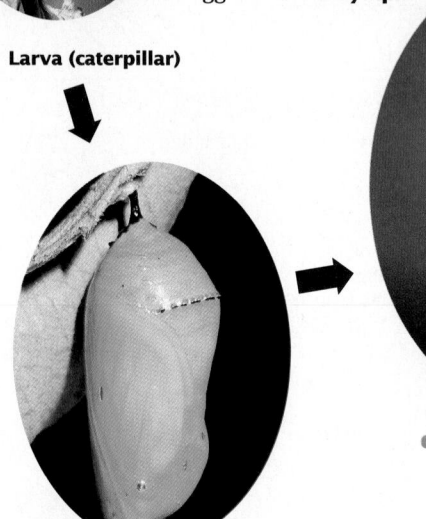

a **Egg**

b **Larva (caterpillar)**

c **Chrysalis**

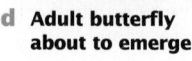

d **Adult butterfly about to emerge**

e **Adult butterfly**

Figure 24.11
The life cycle of the monarch butterfly illustrates complete metamorphosis.

Most insects change their body form as they mature
Spiders, scorpions, mites, and ticks all undergo **direct development**. In this kind of development, a miniature copy of the adult form hatches from the egg. Some species of insects also undergo direct development.

Most insects, however, hatch as a form that is not identical to the adult. As it grows, the young insect changes to become more and more like its parents. Each of these changes is called a **metamorphosis** (*meht uh MAWR fuh sihs*). If the metamorphosis into the adult form involves a series of gradual changes, it is called **incomplete metamorphosis**. Insects that undergo incomplete metamorphosis include grasshoppers, dragonflies, mayflies, and cockroaches. In this type of development, the insect that emerges from the egg is called a **nymph** (*NIHMF*). The nymph is a smaller version of the adult insect species, similar in structure but without wings or mature reproductive organs.

In **complete metamorphosis**, the juvenile form usually does not resemble the adult. When the egg of an insect hatches, an immature form called a larva emerges. The larva looks nothing like the adult. Larvae, like nymphs, cannot yet fly. When the larval stage is complete, the insect enters a stage called the pupa (*PYOO puh*) or chrysalis (*KRIHS uh lihs*), depending on the species. Many pupae form a covering around themselves. This covering is called a cocoon. During this stage, most insects remain immobile while larval tissues and organs are replaced with new tissues and organs. At the end of the pupal stage, a fully formed mature adult emerges, as shown in Figure 24.11e. About 90 percent of insect species undergo complete

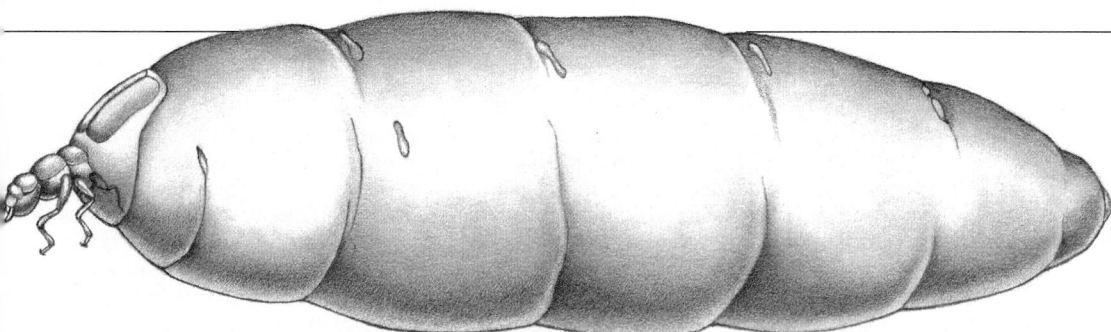

Visual Strategy

Figure 24.12
Encourage students to analyze the benefits of social organization. For example, have them consider the advantages of division of labor in other social insects—in bees and in ants.

Theme Connection

Energy and Life
Tell students that termites feed mostly on wood and other plant material. Because they are destructive to wooden objects, great care must be taken to control the establishment of colonies. Termites can eat the interior out of a wooden table, leaving only the thinnest shell—later causing it to suddenly collapse. Have students explain why it is necessary to spray some homes for termites.

Figure 24.12
A termite nest usually has a single queen who lays all the eggs. A queen is about 6 cm (2.5 in.) long. A nest of termites may contain as many as 5 million individuals.

metamorphosis, including flies, beetles, ants, bees, wasps, butterflies, and moths.

What is the evolutionary advantage of the complicated insect life cycle? During the life cycle, the various stages are not only physically different from each other, they are ecologically different as well. As a result, at different times in its life a single animal can exploit different habitats and different food resources. For example, the leaf-chewing caterpillar will eventually transform into a nectar-drinking butterfly. This ecological separation of young from adults eliminates competition between the two life-cycle phases.

Ants, bees, wasps, and termites live in complex societies

Ants, termites, and certain kinds of bees and wasps are highly social. These insects live in highly organized "societies" of related individuals, with division of labor among the society members. Different individuals serve the group by performing different functions. In a termite colony, workers gather food, tend offspring, and excavate the tunnels and chambers of the colony. Soldiers, as the name implies, defend the colony against attack. Enlarged jaws and heads enable them to carry out this role. Both soldiers and workers are sterile. Only the king and queen of the colony reproduce. The role played by an individual in the colony is its caste. **Figure 24.12** shows representatives of the four castes of a termite colony. All four caste members are shown to scale and are pictured slightly more than three times their actual size. Insects have no choice about their caste in the society. Their genes determine the caste to which they will belong.

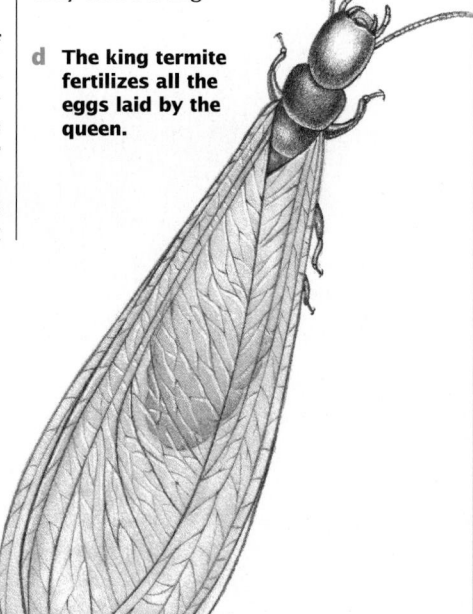

Worker termites search for food, which they supply to the queen, king, and soldiers.

c **Soldier termites guard the passages in the nest and the covered trails that radiate out from the nest.**

d **The king termite fertilizes all the eggs laid by the queen.**

Visual Strategy

Figure 24.13

Many beekeepers have threatened to close their hives upon the arrival of the "killer bees" in the United States. Beekeepers do not wish to deal with these aggressive bees. Have students discuss what effects, other than raising the cost of honey and beeswax, the closing of the hives might have. You might suggest that there would be fewer pollinators, which would affect the production of fruit and flowers.

Insects affect people in many ways

Imagine a world without insects. You might at first imagine a better world without all the pests that plague us. While some insects are merely nuisances, others cause billions of dollars of crop losses each year. These insects either eat the plants directly or transmit diseases to plants. **Figure 24.13** shows several insects that affect humans. The use of pesticides to control insect pests is only partially effective, since many harmful insects have evolved resistance to pesticides. Also, many pesticides are toxic or carcinogenic to humans and other organisms.

Insects destroy our homes and transmit diseases to humans, livestock, and wildlife. In Africa, biting flies transmit African river blindness and sleeping sickness. Fleas carry bubonic plague and were responsible for transmitting this disease throughout Europe and Asia in the fourteenth century. Fleas also serve as an intermediate host for dog and cat tapeworms. You might want to look back at the tables of human diseases in Chapters 16 and 17 and count how many of these diseases are insect-borne.

In spite of these distasteful and often fatal consequences of sharing the planet with insects, humans benefit from insects in many ways. Important crops such as apples, cherries, pears, carrots, cotton, and oranges depend on insects for pollination. The food supply for humans would be much lower were it not for insect pollinators. Humans also harvest a variety of insect products. For

Figure 24.13
Although many of the millions of insects found on Earth are harmful to humans, many others are helpful.

a **Bees pollinate many plants, such as this dandelion, and produce honey.**

b **Boll weevils are responsible for billions of dollars worth of damage to cotton plants each year.**

c **Clothes moths eat clothing such as wool sweaters.**

■ *Matter of Fact* ■

Solitary Bees
Over 85 percent of all bees are solitary; they do not live in hives. Many crops, such as alfalfa, are pollinated entirely by solitary bees.

example, silk is spun by the caterpillars of silk moths.

Perhaps most important, insects provide vital links in the food webs of nearly every ecosystem on Earth. Many freshwater fishes—trout, for example—feed largely on insects and insect larvae. Many birds, bats, amphibians, small reptiles, and small mammals eat insects.

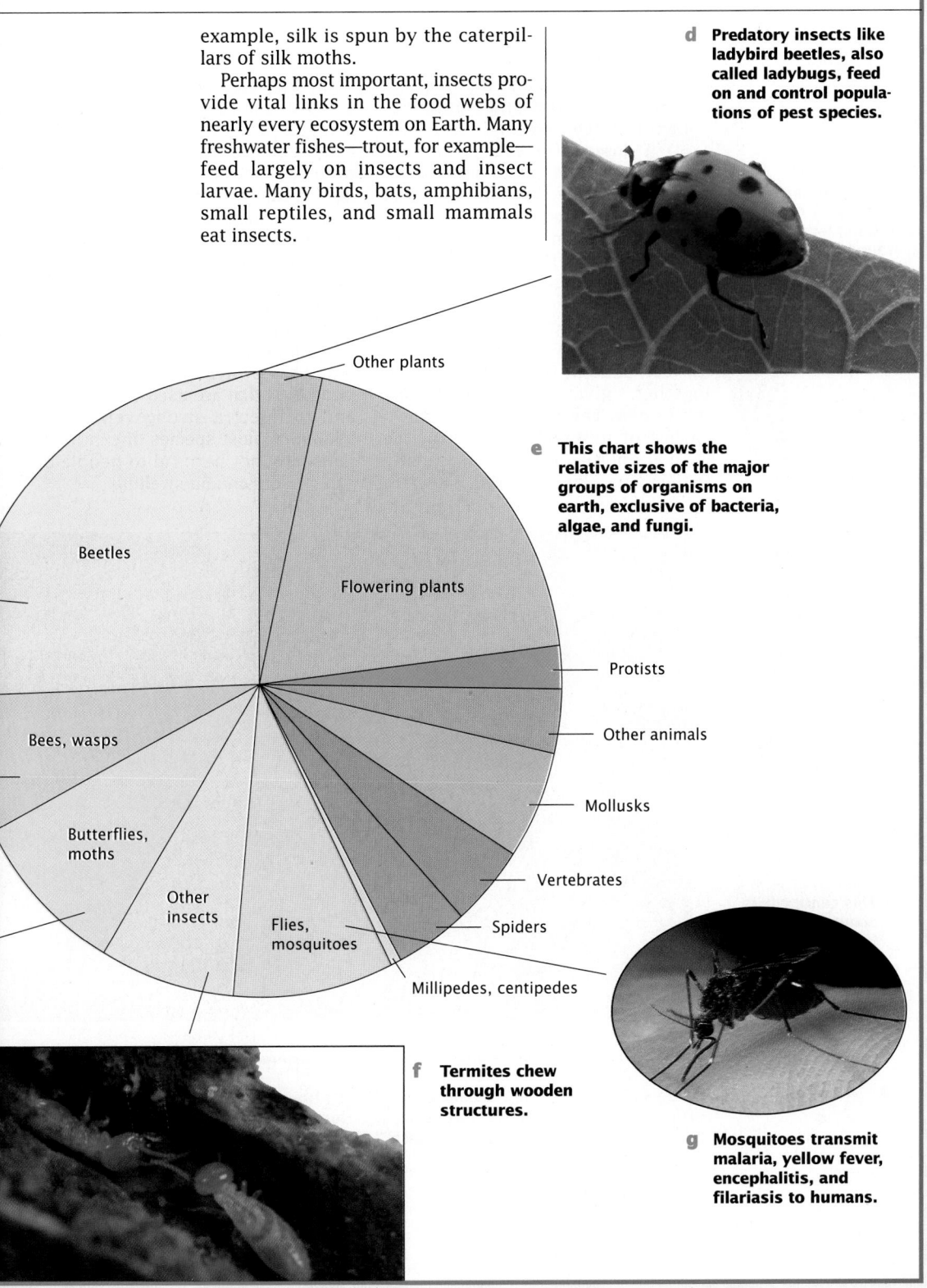

d Predatory insects like ladybird beetles, also called ladybugs, feed on and control populations of pest species.

e This chart shows the relative sizes of the major groups of organisms on earth, exclusive of bacteria, algae, and fungi.

Other plants

Beetles

Flowering plants

Bees, wasps

Protists

Other animals

Mollusks

Butterflies, moths

Vertebrates

Other insects

Flies, mosquitoes

Spiders

Millipedes, centipedes

f Termites chew through wooden structures.

g Mosquitoes transmit malaria, yellow fever, encephalitis, and filariasis to humans.

Demonstration 3

Insect Collections

Retrieve old insect collections that may be gathering dust in your prep or supply room, or in some drawer in the classroom. Ask students how they would begin the task of producing a collection.

Visual Strategy

Figure 24.14

Explain that millipedes, like the one in Figure 24.14, are common in dark moist places, such as meadows and gardens, and in decaying food materials. Millipedes have more than 30 body segments, stink glands, and use the legs on their seventh segment for mating. They protect themselves by curling so the hard plates of their back cover the leg-bearing surface.

Connection: Chapter 12

Centipedes

Explain that centipedes favor damp places such as cellars. They can have as many as 173 pairs of legs or as few as 15. They feed on insects that are captured and killed by poison forced through openings in special poison-bearing claws. They will kill many small animals but are usually not harmful to humans, though their bite may be painful.

Phase 3

ASSESSMENT OPTIONS

Closure Strategy

The Grasshopper
- Provide students with an unlabeled drawing of a grasshopper. Have them label the external structures.
- Have students draw a concept map that distinguishes between incomplete and complete metamorphosis.

Section Review

Assign the *Section Review.*

Reteaching

Have students relate the social behaviors and division of labor in insects to their own societies at home, at school, etc. and speculate about the difficulties one might face if he or she lived in complete isolation.

Millipedes and Centipedes

Elongated and multilegged, millipedes and centipedes scarcely resemble six-legged insects. However, centipedes, millipedes, and insects all share the same fundamental head structure.

The name "millipede" means "thousand feet," although no millipede really has that many feet. The segments of the long, wormlike millipede in Figure 24.14 appear to bear two pairs of legs each. In fact, pairs of segments are fused together, giving this misleading impression. The 10,000 or so known species of millipedes found on the Earth are herbivores, feeding mostly on decaying plant material. They are common inhabitants of forest floors and are important ecologically as decomposers.

Centipedes (the name means "hundred feet") have one pair of walking legs on each body segment. Worldwide, there are about 2,500 known species of centipedes. Unlike the millipedes, centipedes are predators. The centipede shown in Figure 24.15 feeds on worms and other arthropods. A few large tropical centipedes kill and eat frogs and lizards. The first pair of limbs has evolved into large clawlike structures that are used to grasp prey and to inject a strong venom. The venom of most species of centipedes is usually not harmful to people but can cause pain and swelling.

Figure 24.14
This giant millipede is 25 cm (10 in.) long. It is found in Malaysia.

Figure 24.15
This centipede is about 12 cm (5 in.) long. It is found in deserts in the southwestern United States.

Section Review

❶ **Describe two differences between a spider and an insect.**

❷ **Compare the metamorphosis of grasshoppers with that of butterflies.**

❸ **How do some insects benefit crop plants?**

❹ **Compare the diets of millipedes and centipedes.**

▪ Section Review Answers ▪

1. Insects have antennae specialized for sensing the environment and arachnids do not. Instead of chelicerae, insects have mandibles for grinding, scraping, piercing, or sucking, depending on the insect and the shape of the mandible.

2. Butterflies undergo complete metamorphosis, whereas grasshoppers undergo incomplete metamorphosis. Metamorphosis involves a series of changes in which the insect becomes more and more like its parents. Incomplete metamorphosis involves a nymph—a smaller version of the parent insect.

3. Insects pollinate various crops.

4. Millipedes are herbivores feeding on decaying plant material, whereas centipedes are predators.

ALTHOUGH INSECTS ARE VERY ABUNDANT ON LAND, RELATIVELY FEW SPECIES OF INSECTS HAVE MADE THE TRANSITION TO MARINE LIFE. OF THE ARTHROPODS, CRUSTACEANS SUCH AS CRABS, LOBSTERS, AND SHRIMP ARE COMMON OCEAN INHABITANTS. CRUSTACEANS ARE ALSO IMPORTANT IN FRESHWATER ECOSYSTEMS. ONLY A FEW SPECIES OF CRUSTACEANS LIVE ON LAND.

24.3 Crustaceans

Objectives

❶ Identify three kinds of crustaceans.

❷ List three differences between crustaceans and insects.

❸ Describe the importance of crustaceans for the ecology of the sea.

Crustaceans Are Successful Aquatic Arthropods

Figure 24.16
Crustaceans have thick calcium-rich exoskeletons that are generally much heavier than those of insects and spiders. Without the buoyancy of water, many large crustaceans cannot support their weight, and, therefore cannot move around on dry land.

Lobsters, crabs, shrimp, prawns, barnacles, and pill bugs are crustaceans, members of the subphylum Crustacea. The roughly 30,000 species of crustaceans are even more varied in body form than are the insects. They range in size from microscopic forms that drift in the sea to huge lobsters.

Crustaceans have appendages for many kinds of locomotion—swimming, crawling, burrowing, even jumping. They also have specialized appendages for sensing the environment, mating, brooding eggs and embryos, feeding, gas exchange, attracting mates, and defense. Like insects, crustaceans have mandibles. **Figure 24.16** shows some of the specialized appendages of the crayfish.

With very few exceptions (like pill bugs, which you saw in Chapter 15), crustaceans have not been able to successfully invade the land. Crustaceans never evolved several of the key adaptations that are crucial to survival on land. The exoskeleton of crustaceans is heavier, and less watertight than that of insects, so they face problems of water loss. Recall also that insects have very efficient excretory and water balance organs, called Malpighian tubules, that help them get rid of wastes while conserving water. Crustacean excretory organs are far less efficient in accomplishing these functions. Perhaps most important, crustaceans have gills. Although gills are very efficient under water, they collapse in air.

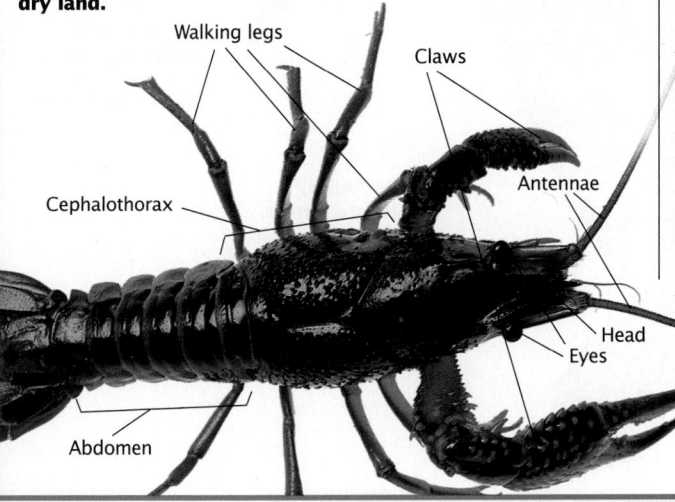

Walking legs
Claws
Cephalothorax
Antennae
Head
Eyes
Abdomen

Lesson Plan 24.3

Phase 1
PREPARATION

Key Concepts
- Class Crustacea includes crayfish, lobsters, crabs, shrimps, and pill bugs.
- Crustaceans typically have bodies divided into two major parts, the cephalothorax and the abdomen. Even so, crustaceans remain a diverse group of organisms in appearance.
- Lobsters, shrimps, crabs, and crayfish are decapods with five pairs of feet.
- Pill bugs are isopods with seven pairs of similar walking legs.
- Copepods like *Daphnia* are the smallest crustaceans and are a vital link between the producers and the rest of the food web.

Reading Strategy

Tell students that there are approximately 30,000 species that make up the class Crustacea. Have students identify, from their reading, members of the class Crustacea and their characteristics.

Phase 2
TEACHING STRATEGIES

Visual Strategy

Figure 24.16
Have students examine the external structure of the crayfish illustrated in this figure and the lobster in the Tour of the Crustacean on page 540. Have them explain how these illustrations display adaptation in these creatures. For example, antennae facilitate hunting and movement in low light.

539

Using the Feature

Have the students review the characteristics of arthropods on pages 470-471. Note some of the differences between the wasp on page 470 and the lobster on page 540. For instance, the wasp has six limbs, the lobster has ten; the wasp has only one pair of antennae, the lobster has two pairs. Also point out some of their similarities: an exoskeleton of chitin, segmentation, and jointed appendages.

Discussion

Guide the discussion by posing the following questions.

1. Why is it necessary for the lobster to shed its shell in order to grow?
   The shell is stiff and cannot stretch to accomodate a larger animal.

2. List two reasons crustaceans have not been successful on land.
   They have gills, which collapse out of water, and lack a light, watertight exoskeleton and efficient excretory organs.

3. In some crabs, which are also decapods, one of the male's claws is much larger than the other. The female's claws are equal in size. Suggest a function for the male's enlarged claw.
   It may be used in contests with other males, or to attract females.

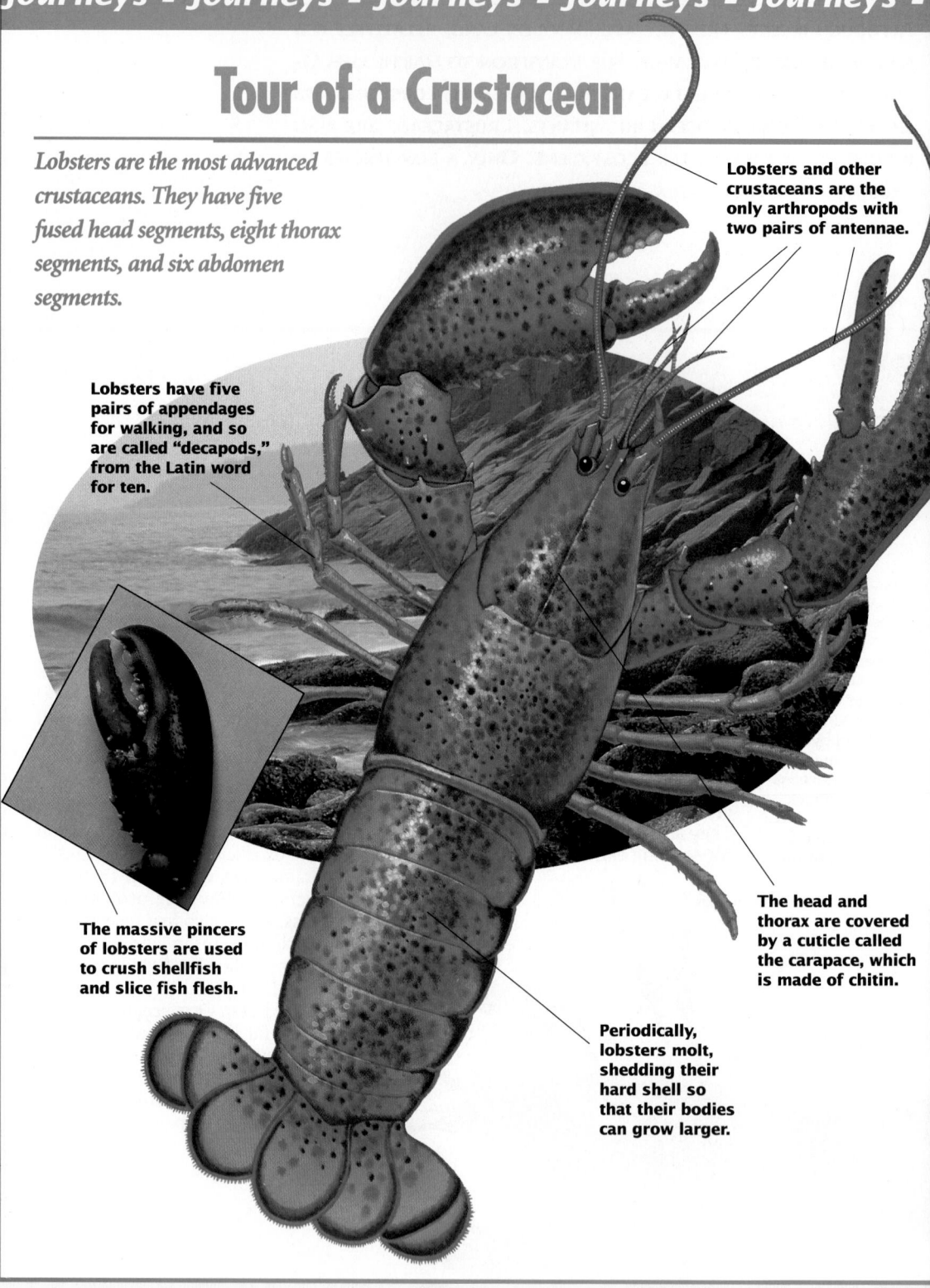

Tour of a Crustacean

Lobsters are the most advanced crustaceans. They have five fused head segments, eight thorax segments, and six abdomen segments.

Lobsters and other crustaceans are the only arthropods with two pairs of antennae.

Lobsters have five pairs of appendages for walking, and so are called "decapods," from the Latin word for ten.

The massive pincers of lobsters are used to crush shellfish and slice fish flesh.

The head and thorax are covered by a cuticle called the carapace, which is made of chitin.

Periodically, lobsters molt, shedding their hard shell so that their bodies can grow larger.

Crustacean Diversity

Lobsters, shrimp, crabs, and crayfish are all **decapods**. They have five pairs of thoracic legs (the name "decapod" means "ten feet"). The first pair of legs is often an enlarged set of claws used for food gathering and for defense, as you can see in **Figure 24.17**. Lobsters, shrimp, and crayfish all resemble one another in that the abdomen is large and muscular.

Barnacles don't look very much like other crustaceans, or even other arthropods, for that matter. However, as larvae they clearly resemble their crustacean relatives. As adults, barnacles live inside the walls of their calcium shells, which are attached to solid objects like rocks, pilings, ships' hulls, and even whales. If you look along almost any rocky seashore, you will probably find two different kinds of barnacles. Acorn, or volcano, barnacles have their shells cemented directly to the rock on which they live. The second kind, the goose barnacles, sit atop a fleshy stalk, which attaches to the rock.

Pill bugs and their relatives are called isopods, which means "same feet." They have seven pairs of similar walking legs on the thorax. They are closely related to another group of crustaceans called the amphipods. The name "amphipod" refers to having two different kinds of feet. If you turn over a stone along the beach or lake shore, you are sure to find some examples of amphipods. Beach hoppers are very common on sandy coastlines. Often, you can find these animals by looking for their burrows or by turning over piles of seaweed lying on the sand.

Figure 24.17
Crustaceans exhibit a wide variety of body forms. They have many different kinds of appendages.

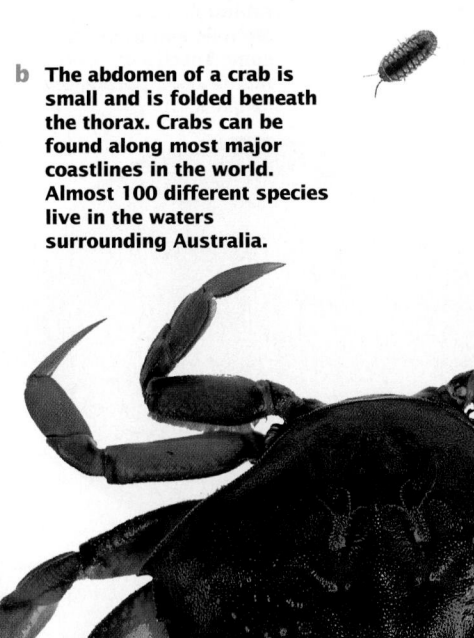

a Shrimp use their legs and long "feelers" to bury themselves in the sand when a predator approaches.

b The abdomen of a crab is small and is folded beneath the thorax. Crabs can be found along most major coastlines in the world. Almost 100 different species live in the waters surrounding Australia.

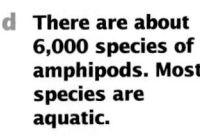

c Pill bugs are often found under rotting logs in forests throughout the continental United States.

d There are about 6,000 species of amphipods. Most species are aquatic.

e These common acorn barnacles feed by extending their feathery legs into the water to filter small food particles.

■ *Matter of Fact* ■

The Robber Crab

The "robber crab," *Bingus latro*, lives on South Pacific islands and climbs coconut palms in search of food. It uses its pincers to cut the nut from the tree and to open the husk.

Visual Strategy

Figure 24.19
Explain that copepods, as small as they are, are an important link in the basic food chain for some whales. Some relatives of copepods are found in the ocean, sometimes in such great abundance that they color the water red for miles around. Emphasize that they are not to be confused with the red tide caused by dinoflagellates.

Phase 3

ASSESSMENT OPTIONS

Closure Strategy

Catching Crayfish
Have students consider the body structure of the crayfish and its defense mechanisms. Have them speculate on why the best way to catch crayfish is with a net.

Section Review

Assign the *Section Review*.

Reteaching

Ask students if they have ever eaten crustaceans. Have them speculate about ways these animals might affect humans. They may need to visit the library, or have a supply of library books available to them in your room. Ask them which crustaceans and crustacean parts are edible. Have them explain ways humans come in contact with crustaceans. For example: some crabs destroy rice fields; isopods bore holes into wood and concrete harbor structures; and copepods carry parasitic worms.

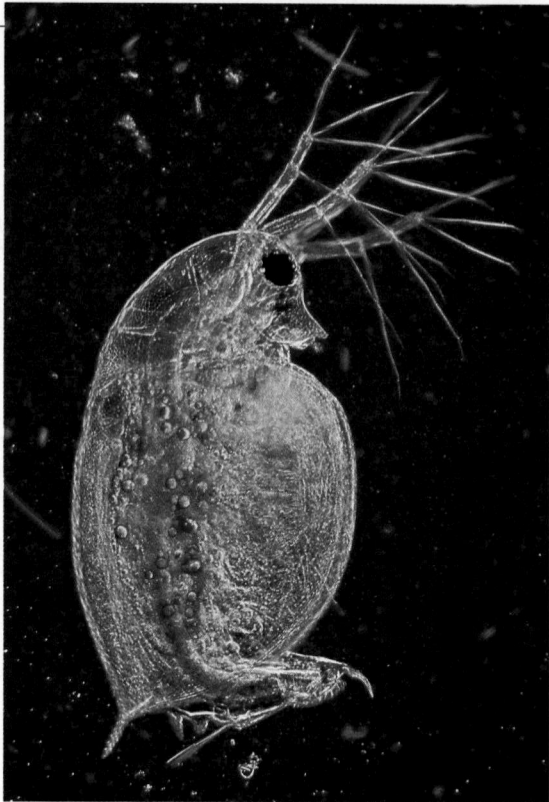

Figure 24.18
Large numbers of *Daphnia* can be found in almost any sample of water taken from a nutrient-rich lake or pond. There are nearly 400 species of *Daphnia* in freshwater environments. They are between 0.2 and 3 mm (.008 and 0.1 in.) in length.

Among the most important crustaceans in freshwater environments are the so-called water fleas or cladocerans (*kluh DAHS uhr uhns*), like *Daphnia* shown in **Figure 24.18**. *Daphnia* feed on phytoplankton. In high concentrations, water fleas consume huge quantities of photosynthetic organisms and in turn are fed upon by small predators, such as young fishes. In many freshwater habitats, these little crustaceans are a vital link between the producers and the rest of the food web.

Look at the photograph in **Figure 24.19**. This tiny crustacean is perhaps the most important animal on Earth. It is a **copepod** (*KOH puh pahd*). While some kinds of copepods live on the sea bottom, and a few are parasitic, the vast majority are part of the zooplankton, the heterotrophic organisms that feed on phytoplankton. Copepods occur both in the sea and in fresh water, often in incredibly high concentrations. Their abundance follows the seasonal changes in concentrations of phytoplankton. Like the water fleas in lakes, the copepods link the ocean's photosynthetic life to the rest of the ocean's food web. Copepods are consumed by a variety of small predators, which are eaten by larger predators, and so on. Virtually all animal life in the open sea depends on the copepods, either directly or indirectly. Although humans do not eat copepods directly, our sources of food from the ocean would disappear without the copepods.

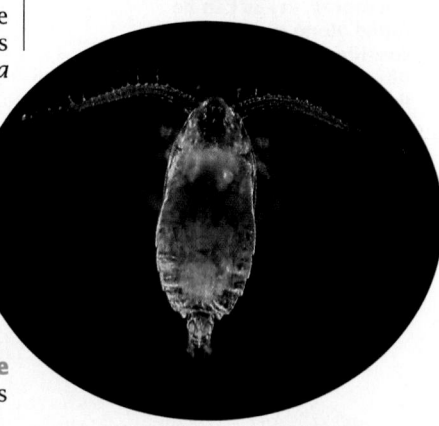

Figure 24.19
Although most copepods are pale and transparent, some species are brilliant red, orange, purple, blue, or black.

Section Review

❶ Name three kinds of crustaceans.

❷ Describe three differences between insects and crustaceans.

❸ How would the ecology of the sea and the human food supply be affected if all crustaceans died?

■ *Section Review Answers* ■

1. Lobsters, barnacles, and pill bugs are three possible answers.

2. The exoskeletons of crustaceans are heavier and less watertight than those of insects. Insects have Malpighian tubules to help them get rid of waste and conserve water. The crustacean excretory system is less efficient. Crustaceans have gills, insects do not.

3. Students should suggest that crustaceans such as copepods link the ocean's photosynthetic life to many species of the ocean's food web. Also, they should suggest that our source of food from the ocean would disappear without copepods.

Chapter **24** *Highlights*

"Ernie! Look what you're doing —take those shoes off!"

Alternative Assessment

Assign students to cooperative work groups of three. Have one member from each group summarize one subsection of this section. Then have them share their summaries with other group members.

	Key Terms	Summary
24.1 Spiders and Their Relatives  Spiders and their relatives are characterized by chelicerae.	chelicera (p. 525) arachnid (p. 525) pedipalp (p. 525) spinneret (p. 526)	• Spiders, scorpions, mites, ticks, and horseshoe crabs have chelicerae, a specialized pair of appendages at the front of the body. They have two main body units: the cephalothorax and the abdomen. Spiders, scorpions, mites, and ticks are arachnids. • The chelicerae of spiders are poison-delivering fangs. Spider silk is used as a safety line, to line burrows, to protect the young, and to trap food. • Scorpions' chelicerae are adapted for tearing prey. The pedipalps of scorpions are pincers for grasping prey. • Many ticks and mites are parasites. Some transmit diseases to humans and domestic animals.
24.2 Insects, Millipedes, and Centipedes  There are more species of insects on Earth than any other group of animals.	mandible (p. 532) direct development (p. 534) metamorphosis (p. 534) incomplete metamorphosis (p. 534) nymph (p. 534) complete metamorphosis (p. 534)	• There are more species of insects than any other group of animals. Insects have three main body units: the head (with three sets of mouthparts), the thorax (with three sets of legs), and the abdomen. • Insect development involves either complete or incomplete metamorphosis. • Millipedes have two pairs of legs per segment and feed on decaying matter. Centipedes have only one pair of legs per segment and are predators.
24.3 Crustaceans  Most crustaceans live in or close to the ocean.	decapod (p. 541) copepod (p. 542)	• Crabs, lobsters, shrimp, pill bugs, and barnacles are crustaceans, subphylum Crustacea. Most crustaceans are aquatic. • The crustaceans known as copepods are extremely important links in marine food webs.

Chapter Review Answers

Understanding Vocabulary
1. a. complete metamorphosis
   b. pedipalps
   c. gills
   d. fleas

Relating Concepts
2. Map answer is shown on page 523D.

Understanding Concepts
Multiple Choice

3.	c	8.	d
4.	d	9.	c
5.	a	10.	a
6.	d	11.	a
7.	b		

Completion
12. incomplete, complete
13. caste
14. barnacles
15. centipedes, millipedes
16. spinnerets

Short Answer
17. The fangs of a spider inject digestive enzymes in addition to poison. The enzymes liquefy the tissues of the victim, which can then be sucked from the carcass.
18. The juvenile and adult forms do not compete for food or habitat. The farmer has to contend with both the juvenile and adult who, because of their ecological differences, will attack different crops or different parts of a single plant.
19. Insects pollinate many crops that humans use as food, including apples and oranges. Humans also harvest silk that is produced by silk moth caterpillars for use in clothing.
20. Crustaceans breathe using gills and are unable to support their heavy skeleton without the help of water. In addition, their exoskeletons are not waterproof and their excretory organs are not very efficient.

Chapter 24 Review

Understanding Vocabulary

1. For each set of terms, complete the analogy.
   a. grasshopper: incomplete metamorphosis::ant: _____
   b. first appendages:chelicerae:: second appendages: _____
   c. insects: tracheae:: crustaceans: _____
   d. Lyme disease: ticks::bubonic plague: _____

Relating Concepts

2. Copy the unfinished concept map below onto a sheet of paper. Then complete the concept map by writing the correct word or phrase in each oval containing a question mark.

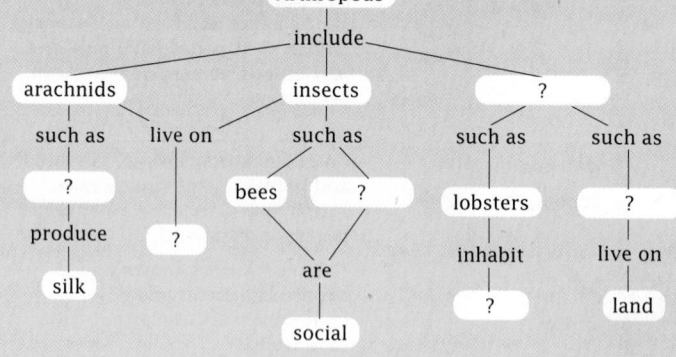

Understanding Concepts

Multiple Choice

3. In spiders, the chelicerae are modified
   a. spinnerets.
   b. antennae.
   c. fangs.
   d. reproductive organs.

4. Where is the greatest variability in the insect body?
   a. number of segments
   b. digestive tract
   c. wing structure
   d. mouthparts

5. Which two arthropods have chelicerae?
   a. scorpion and spider
   b. moth and mite
   c. flea and horseshoe crab
   d. crayfish and tick

6. Which group of arthropods has body segments fused to form the cephalothorax and the abdomen?
   a. insects
   b. centipedes
   c. millipedes
   d. spiders

7. Rocky Mountain spotted fever and Lyme disease are carried by
   a. scorpions.
   b. ticks.
   c. flies.
   d. fleas.

8. How are spiders different from insects?
   a. spiders have antennae and pedipalps; insects do not
   b. insects have four pairs of legs; spiders have six pairs of legs
   c. spiders undergo metamorphosis; insects do not
   d. insects breathe using tracheae; spiders breathe using book lungs

9. Which is *not* a crustacean?
   a. an isopod
   b. a crab
   c. a mite
   d. a barnacle

10. Pill bugs are unusual crustaceans because they
    a. live on land.
    b. have thoracic legs.
    c. show no body segmentation.
    d. feed on decaying animals and plants.

11. Copepods are said to be the most important animals on earth because they are
    a. a critical link in several marine food chains.
    b. found in both the ocean and fresh water.
    c. affected by seasonal changes.
    d. easier to collect and study than other arthropods.

Completion

12. In _____ metamorphosis, an egg hatches into a juvenile that looks like its parents. In _____ metamorphosis, however, the young that hatches from an egg does not look like its parents.

13. In a termite colony, each individual fills a role known as its _____ .

14. Crustaceans known as _____ secrete calcium shells in which they live.

15. _____ are predators and have one pair of legs per segment, while _____ are herbivores and have two pairs of legs per segment.

---

21. One is the black widow. Actually, the female is dangerous but the male is not. The female black widow can be identified by the red hourglass-like pattern on its abdomen.

Interpreting Graphics

22. • Students should suggest the exoskeleton, the number of legs, the well-developed pedipalps and chelicerae, and the segmented body.

• Its exoskeleton and water-conserving organs enable it to conserve water. The well-developed pedipalps, the chelicerae, and the stinger allow it to capture prey and defend itself.

16. Spider body structures that direct the flow of silk from the silk-producing glands are called _____ .

Short Answer

17. Spiders are predators, but have no teeth or jaws. How do they obtain nutrients from their prey?

18. An insect that undergoes complete metamorphosis is both anatomically and ecologically different during the various stages of its life cycle. How is this beneficial for the insect? How does this pose a problem for the farmer on whose crops the insect feeds?

19. What are two ways that insects benefit humans?

20. Unlike insects, crustaceans have not been able to successfully invade the land. Offer an explanation for why this is so.

21. Name one spider found in the United States that is dangerous to humans. How can it be identified?

Interpreting Graphics

22. Look at the photograph of the arthropod below.

- What characteristics enable you to recognize this animal as an arthropod?
- Explain the adaptations that enable this arthropod to conserve water, capture prey, and defend itself.

Reviewing Themes

23. *Evolution*
The promise of pesticides to eliminate all insect pests without harming humans has not been realized. Why not?

24. *Scale and Structure*
Spiders are terrestrial predators. What body features make them well adapted to this lifestyle?

Thinking Critically

25. *Inferring Conclusions*
Rather than spraying with chemicals, a farmer releases thousands of ladybird beetles in his fields. Why would the farmer do this?

26. *Making Predictions*
How would a spider be affected if it could not make silk?

27. *Comparing and Contrasting*
How are pupa and nymphs alike? How are they different?

28. *Building on What You Have Learned*
In Chapter 22, you learned that terrestrial animals have many adaptations that enable them to live on land. How are the eggs of many arthropods well-suited for survival on land?

Cross-Discipline Connection

29. *Biology and Health*
Head lice are sometimes a problem in schools. Do library research to discover the symptoms of lice infestation and to learn what can be done to stop the spread of head lice.

Discovering Through Reading

30. Read the article "Spin Doctor," in *Discover*, May 1992, pages 32–36. Why are Dr. Lewis and the Offices of Naval and Army Research interested in the silk spun by spiders? What actions are being considered by Dr. Lewis to produce large quantities of silk?

27. Pupa and nymphs are both juvenile forms of insects. Nymphs develop into adults via incomplete metamorphosis; they resemble the adult form during the series of gradual changes. On the other hand, pupa are typically the final stage of complete metamorphosis; they appear very different from the adult.

28. Students should suggest that the eggs of arthropods are usually microscopic but encased in some watertight covering.

Cross-Discipline Connection

29. The symptoms are the itching of the scalp and a rash on the neck. The spread of head lice can be stopped by not sharing combs, brushes, or clothing with other people. If lice are discovered, wash hair with a medicated shampoo.

Discovering Through Reading

30. It is believed that silk, because of its strength and flexibility, will be the fiber of the future. It has applications in clothing, parachute cord, sutures, and ligament and tendon replacement. Lewis' research team has isolated the genes from the gland where silk is produced. Their next step will be to engineer the genes into organisms, like yeast or bacteria, so that silk can be mass produced.

Reviewing Themes

23. Insects have evolved resistance to pesticides, requiring the development of new pesticides to eliminate the mutant varieties. These new and improved pesticides are often harmful to humans and other animals as well as insects.

24. They breathe air using book lungs, they are poisonous, and they produce silk. In addition, they are efficient at getting rid of waste and conserving water.

Thinking Critically

25. Lady bird beetles are predaceous insects and are used to control insect pests without pesticides.

26. The spider would soon die because its movement would be restricted and it would be unable to capture prey.

Procedural Note

1. Review the answers to Prelab Preparation questions.
2. Caution the students about the use of acidic and alkaline solutions in the laboratory.
3. Have students wear safety goggles and rubber gloves when handling the solutions.

Prelab Preparation Answers

1. • Crustacea
   • Pill bugs are generally found in moist, dark places, for example, under rocks, rotting logs, leaves, etc.
2. • The acidity of a solution.
   • An acidic solution has a pH less than 7.
     A basic solution has a pH greater than 7.
     A neutral solution has a pH equal to 7.

Procedure Answers

4. No. They tend to move randomly.
5. Yes. They prefer moist areas and tend to collect near the water.

Chapter 24 Investigation

Can Pill Bugs Detect Differences in Moisture and pH?

Objectives

In this investigation you will:
- *observe* a terrestrial crustacean
- *identify* a pill bug's responses to pH and moisture differences

Materials

- ruler
- paper towels
- pan
- 10 pill bugs
- 3 medicine droppers
- distilled water
- dilute sodium hydroxide solution
- vinegar
- 3 filter paper disks
- safety goggles
- lab apron
- pH paper

Prelab Preparation

1. Review what you have learned about crustaceans by answering the following questions.
   - To which class of arthropods do pill bugs belong?
   - In what habitats are pill bugs found?

2. Review what you have learned about acids in Chapter 14 by answering the following questions.
   - What does a solution's pH indicate?
   - What range of pH values indicates an acidic solution? A basic solution? A neutral solution?

Procedure: Observing Pill Bugs

1. Form a cooperative group of four students. Work with one member of your group to complete steps 2–10. Make three tables similar to the one shown below. Mark one table "Control," one "Water," and one "Acidic/Basic."

30-second intervals	Trial 1			Trial 2			Trial 3		
	Number of pill bugs in quadrant								
	1	2 3 4		1 2 3 4			1 2 3 4		
1									
2									
3									
4									
5									
6									
7									

2. Use a ruler and a pencil to divide a paper towel into four equal quadrants. Then, in the center of the paper towel, draw a circle about 8 cm (3 in.) in diameter. Place the marked paper towel on top of five unmarked sheets of paper towel and place the stack in a clean pan.

3. Place 10 pill bugs in the center of the circle. Observe their responses over a period of four minutes.

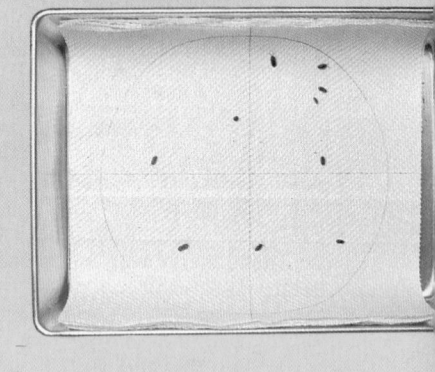

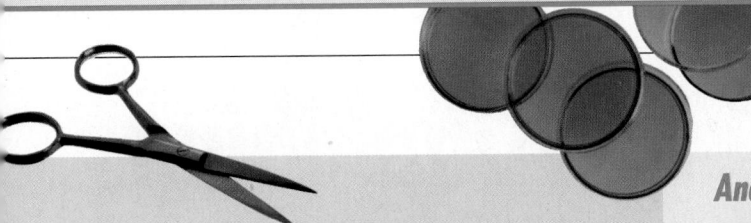

4. In the table marked "Control," record the number of pill bugs present in each quadrant after each 30-second interval. Repeat the test for two more trials. Calculate the average number of pill bugs in each quadrant for each 30-second interval. *Do the pill bugs show a preference for one of the quadrants?*

5. Using one medicine dropper, moisten a disk of filter paper with distilled water. Place the disk in the middle of one quadrant. Repeat step 3, but record the results in the table marked "Water." *Do the pill bugs now show a preference for a particular quadrant? Why do they prefer this quadrant?*

6. **CAUTION: Wear safety goggles and a lab apron while working with the sodium hydroxide and vinegar solutions, as they can injure the skin and eyes. Thoroughly wash any area that has been contaminated by these solutions.**

7. Use small strips of pH paper to find the pH of the distilled water, vinegar, and sodium hydroxide solutions. Record your results.

8. Using a clean medicine dropper, soak a fresh disk of filter paper with vinegar. Using a different medicine dropper, soak another disk with dilute sodium hydroxide. Place each disk in a separate quadrant of the paper towel.

9. Repeat step 3, recording the results in the table marked "Acidic/Basic."

10. Return the pill bugs to their storage container. Clean up your materials and wash your hands before leaving the lab.

Analysis

1. *Summarizing Observations* Describe the behavior of the pill bugs when placed on clean, dry paper.

2. *Summarizing Observations* How did pill bug behavior change when the water-soaked disk was added to the paper towel?

3. *Analyzing Observations* Are pill bugs attracted directly to moisture, like a moth to light, or do they move randomly until moisture is encountered? Explain how your observations support your statement.

4. *Evaluating Methods* Why was it necessary to first observe the animals on a dry paper towel?

5. *Analyzing Observations* How do pill bugs respond to differences in pH?

6. *Making Inferences* How does the pill bug's response to moisture and pH increase its chance of survival?

7. *Making Predictions* How do you think a pill bug would respond to light? Explain the reasons for your prediction.

Thinking Critically

How does the fact that pill bugs are crustaceans help explain why they are attracted to moist conditions?

Analysis Answers

1. The pill bugs move randomly on the clean, dry paper towel.

2. The pill bugs tend to collect on or near the water-soaked disk after it has been added to the paper towel.

3. Pill bugs move randomly until moisture is encountered. Students should explain that the pill bugs were observed to move randomly until they encountered the moist paper, at which time they stopped moving.

4. Observation of the pill bugs on a dry paper towel provided a control for the experimental condition, in which moisture was introduced.

5. Answers will vary but should indicate that the pill bugs did not collect on the acidic or alkaline surfaces.

6. A pill bug's response to moisture and pH tends to keep it in a favorable environment, thus increasing its chances of survival.

7. Answers will vary but might suggest that a pill bug is likely to move randomly until it is away from the light. This behavior, similar to its response to moisture and pH, would keep a pill bug in a favorable environment.

Thinking Critically Answer

Students should indicate that all crustaceans have gills. Gills collapse when they are not kept moist.

Chapter 25 Fishes and Amphibians

Planning Guide

Objectives/Themes	Classwork Resources	Homework Resources

25.1 **1.** List three characteristics of agnathans.
 	
2. Describe how a lamprey feeds.
 	
3. Describe how jaws are thought to have evolved.
 	
4. Contrast sharks and rays with agnathans.
 	
Themes: Energy and Life, Evolution, Scale and Structure
	Text
	Science in Action
	  Phyllis Stout: Microscope Slide Technician pp. 554–555
	
	Teacher's Resource Binder
	Focus Activity 25
	  Observing a Frog
	Lab Investigation 25.1
	  Fish Morphology and Behavior
	
	Other Resources
	Transparency 107
		Text
		Section Review, p. 553
		
		Teacher's Resource Binder
		Directed Reading Worksheet 25.1
		
		Other Resources
		Audiocassette 25.1*

25.2 **1.** Identify three differences between bony fishes and sharks.
 	
2. Describe the importance of the swim bladder for bony fishes.
 	
3. Summarize the importance of lobe-finned fishes in the evolution of land vertebrates.
 	
Themes: Evolution, Scale and Structure, Stability
	Text
	Journeys
	  Tour of a Fish p. 558
	Investigation
	  How Do Goldfish Respond to Light? pp. 568–569
	
	Other Resources
	Transparency 108
		Text
		Section Review, p. 560
		
		Teacher's Resource Binder
		Directed Reading Worksheet 25.2
		
		Other Resources
		Audiocassette 25.2*

25.3 **1.** Identify two characteristics that enabled amphibians to invade the land.
 	
2. Describe the life cycle of a frog.
 	
3. Identify the three orders of living amphibians.
 	
Themes: Evolution, Patterns of Change, Scale and Structure
	Text
	Journeys
	  Tour of a Frog p. 563
	Science, Technology, and Society
	  Endangered Species: How Far Do We Go to Save Them? pp. 570–571
	
	Teacher's Resource Binder
	Science Skills Worksheet
	  Amphibian Diversity
	
	Other Resources
	Transparencies 109–110
	Science in Action Video
	  *Alternatives to Dissection: The Earthworm and the Frog**
		Text
		Section Review, p. 564
		
		Teacher's Resource Binder
		Directed Reading Worksheet 25.3
		Vocabulary Review Worksheet*
		Reteaching Worksheet*
		  Comparing Organisms
		
		Other Resources
		Audiocassette 25.3*

*Reteaching Options

Demonstrations	Assessment
25.1: pp. 550, 552 **25.2:** p. 557 **25.3:** p. 561	Chapter Review pp. 566–567 Portfolio Assessment p. 547D Chapter Test—Teacher's Resource Binder Test Generator

Research Notes

Connection to Ecology: Amphibians as Environmental Monitors

Since 1990, ecologists have warned that global populations of amphibians were declining dramatically. Many scientists believe that amphibians are ecological "indicators," and that their decline might be a sign of some harmful global environmental changes that could eventually affect all life. Other researchers believe that more traditional explanations can explain the disappearance of the amphibians.

The golden toad of Costa Rica, which is found only in that country's Monteverde Cloud Forest Preserve, is either extinct, or may be soon.

But habitat destruction is not adequate to explain the near extinction of the golden toad, according to Alan Pounds, a researcher at the preserve. Ever since the golden toad's discovery, the area it was found in has remained undisturbed. Some researchers are testing a hypothesis that unusually warm temperatures in Costa Rica over the past few years may have affected the toad's population.

Duke University's Joseph H. K. Pechmann led an extensive study of a few species of salamanders and one type of frog in a single pond in South Carolina. The results appear to suggest that changes in the populations of the species found in the pond depended solely on the dryness of the weather, and not on habitat change.

During times of drought, the populations plummeted, but when moist weather returned, the populations swelled, apparently because breeding was more successful when there was more time before the pond dried up. In fact, all of the trends in population observed corresponded directly to trends in the amount of rainfall.

This study indicates that often natural pressures that are not directly related to human intervention or pollution can cause wide fluctuations in animal populations, and can possibly even cause extinction. This provides support for some scientists who doubt that the rate of extinction of amphibians today is any greater than in the past. They point out that there is not enough historical data available about amphibians to determine whether the rate of extinction has increased recently.

Both sides of the debate agree that habitat conservation is a good idea, but more extensive studies on biological diversity are necessary to make better judgments about when a habitat and its inhabitants are in danger.

Investigation Notes

How Do Goldfish Respond to Light?

pages 568–569

Purpose: This investigation gives students an opportunity to test a hypothesis about the response of a goldfish to red light. The emphasis should be on the problem-solving process.

Prelab Preparation

1. Goldfish can also be obtained from pet stores. Have a sufficient supply to ensure that individual specimens are not overused during the course of the day. Only use aquarium or nonchlorinated water.

2. Goldfish should not be released into the environment. Arrange to have unwanted fish taken to a pet store or given to students.

3. If necessary, one set of cardboard boxes can be used throughout the day.

Answers will be found on pages 568–569.

Meeting Individual Needs

Objectives

1. Students will demonstrate appropriate use of core vocabulary for the chapter (Vocabulary File).

2. Students will describe characteristics of bony fishes and their functions (Teaching Strategy A).

3. Students will observe the metamorphosis from frog eggs to frogs and explain how the characteristics of each stage relate to the environment (Teaching Strategy B).

Vocabulary File
(Developing Vocabulary/ Limited English Proficiency)

If you are not already using the Vocabulary File, refer to Chapter 1 for its preparation. See the Chapter Highlights on page 565 for a list of suggested words.

Teaching Strategy A
Characteristics of Fish
for pages 549–560
(Developing Classification Skills/Visual Learners)

Materials: picture of lamprey, picture of shark, picture of perch (or any other bony fish), posterboard, scissors, unlined index cards

Preparation: Prepare fish pictures for hanging as mobiles in the sky zoo. Draw a large outline of a bony fish on the posterboard. Cut slits into the outline in which "scales" can be inserted. Don't forget to cut slots in the paired fins of the outline of the bony fish.

With scissors, round off one end of an index card to resemble a fish scale. Cut the edges off the other end and taper it to fit into the slit made in the outline of the fish. On the back of each scale print a characteristic with a text page-reference for that given trait.

Procedure: Open the unit by asking students about fishing experiences—what is commonly being caught in local lakes, etc.

Hold up the lamprey picture. Have students silently read page 550 as a reference for the lamprey. Then ask if a lamprey has been caught recently in local water. Discuss the lamprey's habitat and characteristics. Upon completion of the discussion, hang the lamprey in the sky zoo. Repeat the procedure for the shark. Refer to text pages 552 and 553.

Display the bony fish outline. Have students pick a scale from the fish and research the chapter to find a description of that characteristic and its functions, if appropriate. After students' reports on bony fish characteristics, close with a brainstorm/discussion of structural differences between lampreys, sharks, and bony fishes.

Teaching Strategy B
Metamorphosis of Amphibians
for pages 561–564
(Developing Observation Skills/Visual Learners)

Materials: cut-out mounted picture of a frog for sky zoo, posterboard, frog eggs, battery jar, stone aerator with pump

Preparation: Order frog eggs from WARD'S (87 M 8205) or have one of your student fishers bring them in from a local lake. Place them in the battery jar with water and aeration. Be sure there is a stone or rock large enough for a young frog to crawl onto. Tape a posterboard chart on the back of the battery jar and record changes of metamorphosis of the frogs on the chart.

Procedure: Direct students' attention to the sky-zoo frog and the eggs. Ask students what happens as the eggs turn into frogs. Hang the frog in the sky zoo. Discuss Figure 25.21 to prepare students for recording observation data.

Take a poll to determine how many students have actually observed the metamorphosis of a frog.

On the chart, record dates and the changes observed. This is a long-term strategy, which will carry over into time devoted to other units, but the interest it generates among the students makes it well worth the effort.

As the observations are recorded over a period of time, be sure to stop and relate the observations to the basic characteristics a frog needs to survive on land. Offer a prize for the student who most closely estimates the day and time the frog will climb onto the rock. By this time, the unit on the human body should be in progress, and this could also be a time for relating characteristics of various vertebrate bodies (such as lungs).

Additional Strategies
Visual Strategies
Pages 549, 550, 551, 552, 553, 559, and 564

Auditory Learners
Use *Biology: Visualizing Life* Audiocassettes for Sections 25.1, 25.2, and 25.3.

Meeting Individual Needs (cont.)

Cooperative Learning

Amphibians

Timing: Use this activity to introduce Section 25.3.

Group Size: 3 students

Outcome: Students will be able to describe adaptations that amphibians have developed for living on land.

Individual Accountability: Each group member is responsible for information on one order of amphibians.

Positive Interdependence: Each group will complete a table listing each amphibian order with its adaptations.

Assign one of the following orders of amphibians to each group member: Anura, Urodela, and Apoda. Have each group member find out the adaptations this order has developed for living on land and report the information to the other members of the group.

The group should then construct a table that lists each order of amphibians and the adaptations they have made for living on land.

Each group should interpret the information in their table to determine which order has the most successful adaptations for survival on land. Each group should record the name of this order and provide a rationale for their decision below the table.

Portfolio Assessment

Students should select their best work and provide a self-reflective rationale for their selections. Students can make selections in the following areas.

1. *Content* — One concept map from the chapter (See page 812 for evaluation criteria.)

2. *Reading Comprehension* — One Directed Reading Worksheet from the Teacher's Resource Binder (Use the answer key to evaluate for accuracy.)

3. *Writing* — Using the Vee Form, summarize a magazine or newspaper article relating to fish or amphibians. (See page 22T for evaluation criteria.)

   Or: Select a writing task or project from the Chapter Review.

4. *Performance Assessment* — One Vee form from a chapter investigation or lab manual investigation (See page 22T for evaluation criteria.)

Teacher makes selections in the following areas.

1. *Formal Assessment* — Chapter test (Test A, B, or the Test Generator) The teacher-scored test should be reviewed by the student. Incorrect responses should be corrected by the student before the test becomes part of the portfolio.

2. *Informal Assessment* — Use the Direct Observations Checklist, page 33T, during a laboratory or other cooperative learning experience.

3. *Performance Assessment* — Have students design an experiment that will test what sorts of noises goldfish are able to detect. Experiments should test noises of different pitch, loudness, and location.

Concept Map Answer

The following is one possible answer to the Relating Concepts exercise on page 566.

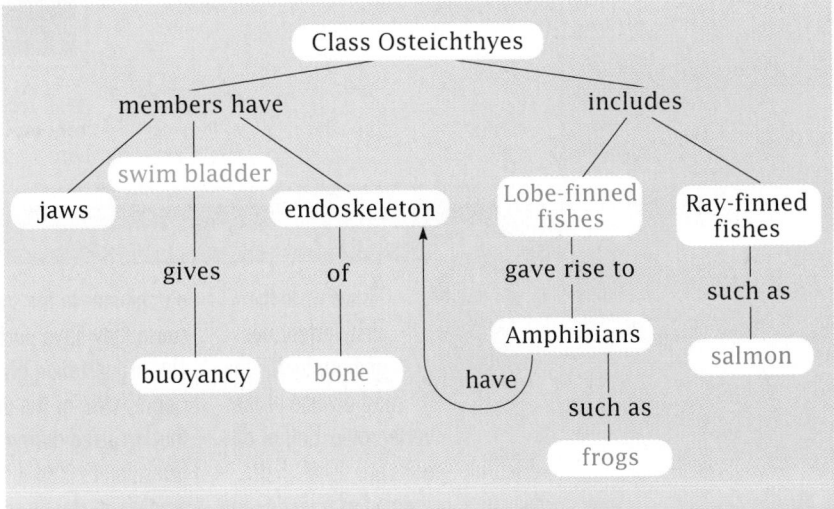

Chapter 25

Determining Prior Knowledge

- Have students draw a storyboard of the life cycle of a frog or fishes.
- Ask students why dolphins and whales are not fishes.
- Have students think of as many different fishes as possible that represent extremes of size, shape, color, and way of life.
- Have students explain the difference between a chordate vertebrate and a nonchordate vertebrate.
- Have students explain how sharks and bony fishes are different, yet alike.
- Have students explain why, long ago, people thought that tadpoles and adult frogs/toads were two different animal species.
- Have students explain the differences between frogs and toads.

Chapter 25

Fishes and Amphibians

Review

- characteristics of chordates (Section 21.3)
- characteristics of vertebrates (Section 21.3)
- evolution of the amphibian heart (Section 22.2)

25.1 Early Fishes
- The First Fishes: Class Agnatha
- Evolution of Jaws
- Sharks and Rays: Class Chondrichthyes

25.2 Bony Fishes
- Structure of a Bony Fish
- Major Groups of Bony Fishes

25.3 Amphibians
- The First Land Vertebrates
- The Tie to Water
- Kinds of Amphibians

This frog in Botswana lays its eggs in water. Amphibians rely on the presence of water for fertilization and survival.

■ *Author's Rationale* ■

Fishes were the first vertebrates and are by far the most diverse of the vertebrates. Half of all vertebrate species are fishes. Fish diversity is presented as a fascinating evolutionary story. We present amphibians in the same chapter because they have parallels to fishes. This organization gives students a clearer view of the many changes that occurred during the evolutionary transition to land. In some ways an amphibian closely resembles a fish with feet: it is still adapted to a water environment. But in other ways it has been radically transformed.

You and most of the animals you commonly see are vertebrates. As you learned in Chapter 10, vertebrates are chordates that have a vertebral column, or spine. There are approximately 40,000 species of vertebrates today and over half of them are fishes. The first vertebrates evolved more than 500 million years ago.

25.1 Early Fishes

Objectives

❶ **List three characteristics of agnathans.**

❷ **Describe how a lamprey feeds.**

❸ **Describe how jaws are thought to have evolved.**

❹ **Contrast sharks and rays with agnathans.**

The First Fishes: Class Agnatha

The first vertebrates evolved about 510 million years ago. They were jawless fishes belonging to the class Agnatha (Agnatha means "without jaws"). Thick bony plates covered the bodies of these ancient agnathans (*AG na thuns*), as you can see in **Figure 25.1**. Bone and cartilage first evolved in this group of fishes. Agnathans were an abundant and diverse group for about 150 million years, until they were largely replaced by jawed fishes about 360 million years ago. Only 63 species of agnathans exist today.

The living agnathans are the lampreys and hagfishes, which are shown on the next page. These eel-like creatures have scaleless, slimy skin and lack paired fins. The skeleton is mostly composed of cartilage, and there is no well-developed vertebral column. The notochord, which is replaced by vertebrae in most vertebrates, functions as the major support structure in adult lampreys and hagfishes. The gills of agnathans lie within pouches that branch from the pharynx. Water exits these pouches through several openings behind the head.

Figure 25.1
Drepanaspis was an agnathan that fed on the ocean bottom. Its gaping mouth did not have jaws. This species became extinct about 360 million years ago.

Lesson Plan 25.1

Phase 1
PREPARATION

Key Concepts

- The first vertebrates to evolve were jawless fishes of class Agnatha, which have cartilaginous skeletons, a notochord, eel-like bodies, and gill pouches.
- Jaws are believed to have evolved from one or more gill arches present in jawless fishes.
- The first vertebrates with jaws were members of class Acanthodii. They were followed by members of the class Placodermi. Acanthodians had bony internal skeletons. Placoderms had heavy, bony platelike skin. Both classes had paired fins and strong jaws.
- Members of class Chondrichthyes are fishes with skeletons of cartilage, jaws, gill slits, teeth that move forward to replace lost teeth, and sandpaper-like skin. Sharks, rays, and skates are the modern members of class Chondrichthyes.

Reading Strategy

For future reference, have students make a table comparing invertebrates with vertebrates, using information acquired from reading this chapter and Chapter 24.

Phase 2
TEACHING STRATEGIES

Visual Strategy

Figure 25.1
Have students note the features on the fish's ventral surface. Have them discuss whether this fish was a predator, scavenger, or parasite; the advantages of such a body; and why they think *Drepanaspis* became extinct.

Demonstration 1
Hagfish and Lampreys
As you discuss their characteristics, pass around photographs or specimens of lampreys and hagfish. Emphasize the characteristics of lampreys. Point out the hagfish's gill openings, slitlike mouth, and poorly developed eyes. Discuss their feeding habits. Usually students are fascinated by the way hagfish bore into their prey and eat the entrails, leaving the skin intact. Explain that hagfish often attack fish caught in fishing nets and are of particular concern to commercial fishermen.

Visual Strategy

Figure 25.2
Tell students that sometimes fish are caught that have circular scars caused by feeding lampreys. Have students explain why lampreys are called parasites. Explain that lamprey attacks do not necessarily kill prey outright, but can weaken them, often to the point of death.

Connection: Chapter 13

Agnathan Survivors
Tell students that for more than 100 million years, agnathans were the only vertebrates. Except for the lampreys and hagfish, all the major groups of agnathans are now extinct. Suggest that one reason why they have survived may be because their eating habits are specific. Explain that as scavengers and parasites, they may have faced few competitors.

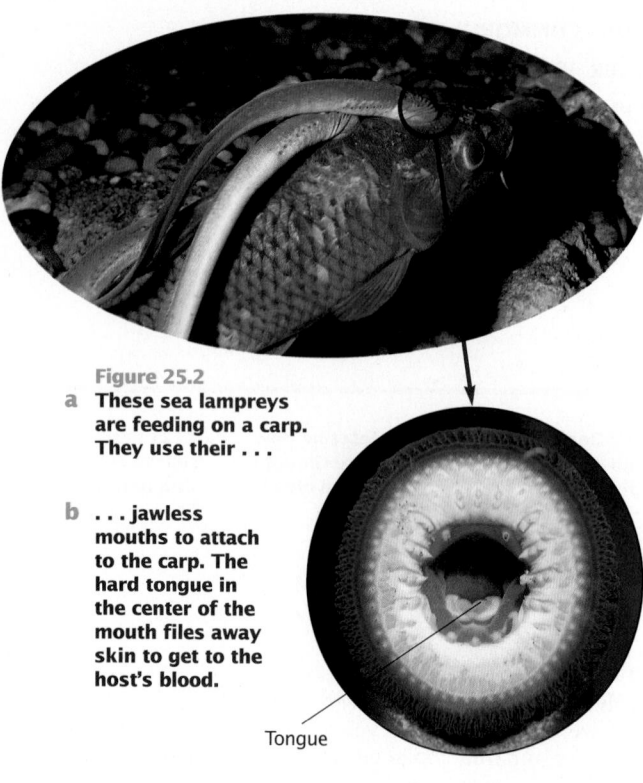

Figure 25.2

a These sea lampreys are feeding on a carp. They use their . . .

b . . . jawless mouths to attach to the carp. The hard tongue in the center of the mouth files away skin to get to the host's blood.

Tongue

Figure 25.3
Hagfishes are jawless scavengers. They invade the bodies of dying or dead animals and feed on the internal organs.

Lampreys are parasites
Most kinds of lampreys spend their entire lives in fresh water. A few species live in the sea as adults but migrate into fresh water to breed. In a way, an adult lamprey's lifestyle is similar to that of a leech. As shown in **Figure 25.2**, lampreys are external parasites that feed on other fishes. A lamprey's mouth is recessed within a funnel-like structure. Sharp teeth in the funnel help the lamprey hook onto its host. The rim of the funnel functions as a suction cup, which fastens the lamprey to its host. A rough tongue scrapes off small particles of the host's skin and flesh. The lamprey sucks in these particles along with the host's blood. Like some leeches, the lamprey secretes an anticoagulant that prevents its host's blood from clotting.

After a lamprey has fed, it drops off of its host. Damage to the host can be severe since the wound left by the lamprey may become infected or cause the host to bleed to death. For this reason, large lamprey populations can cause great damage to populations of other fishes. For instance, when a canal that allowed ships to bypass Niagara Falls was deepened in the early 1900s, the ocean lamprey was able to move from the Atlantic Ocean into the Great Lakes. By the 1940s and 1950s, this lamprey was abundant enough to cause a serious decline in the commercial and sport fishing industries of the Great Lakes. Lamprey populations were eventually reduced by treating the Great Lakes with poisons toxic to lamprey larvae.

Hagfishes are scavengers
Hagfishes, such as the one in **Figure 25.3**, are scavengers that generally feed on dead or dying animals, such as large invertebrates or other fishes. When a hagfish locates a potential meal, it enters the body of the other animal by squirming in through the gill openings, the mouth, or the anus. Once inside, it feeds on the internal organs of the animal, biting with jawlike folds of muscle that close side to side.

■ **Matter of Fact** ■

Hagfish
Hagfish are bottom dwellers that live only in cold ocean water. They are sometimes called "slime eels" because they have mucus-secreting glands all over their bodies. They produce so much mucus that, when agitated, one hagfish can turn a bucket of water into a jellylike substance within minutes. Hagfish have both male and female sex organs. They may produce eggs one season and sperm the next. Eggs are fertilized externally. Young hagfish hatch from the egg as miniature versions of the adult.

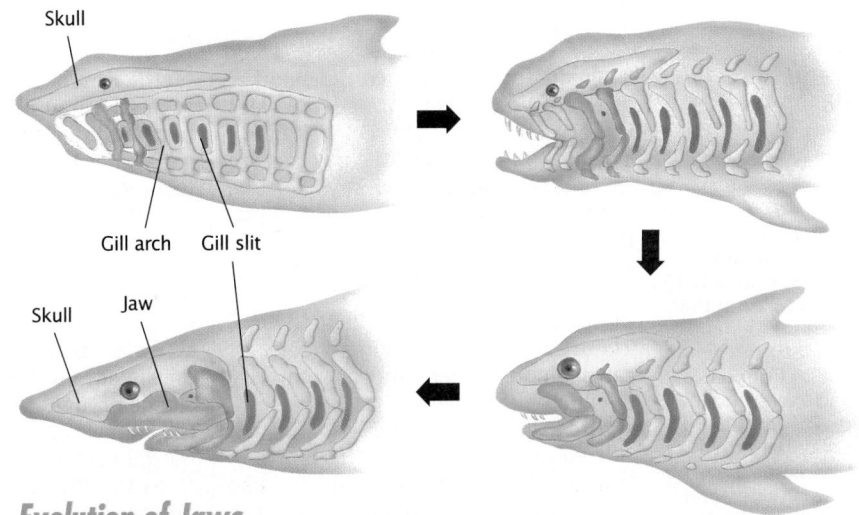

Figure 25.4
In the process of evolution, the gill arches present in jawless fishes probably moved forward to form a a jaw.

Skull

Gill arch   Gill slit

Skull   Jaw

Evolution of Jaws

Because they lack jaws, agnathans can only eat food that can be sucked into their mouths. Apart from the agnathans, all vertebrates have movable jaws. Animals with jaws can exploit a much wider range of foods than can jawless animals. Scientists think that jaws evolved from one or more of the gill arches that support the pharynx in agnathans, possibly in the way shown in **Figure 25.4**.

The earliest jawed fishes were acanthodians (*uh KAN thoh dee uhns*), members of the class Acanthodii. Although superficially resembling sharks, the acanthodians or "spiny fishes" had a bony internal skeleton and were definitely not sharks. Acanthodians evolved about 435 million years ago. Another group of early jawed fishes was the placoderms (*PLAK uh durms*), class Placodermi, which evolved about 400 million years ago. Placoderms (meaning "plate skin")

were armored with heavy, bony plates and had strong jaws, as in the predator *Dunkleosteus* illustrated in **Figure 25.5**. Both placoderms and acanthodians show a characteristic found in most other vertebrates: paired appendages, in this case paired fins. Placoderms and acanthodians diversified rapidly. Perhaps because of jaws and paired fins, these fishes largely replaced the agnathans. Lampreys and hagfishes may have survived because of their specialized feeding habits. Placoderms became extinct about 345 million years ago, and the acanthodians died out about 270 million years ago. Scientists do not know why these once-diverse groups became extinct.

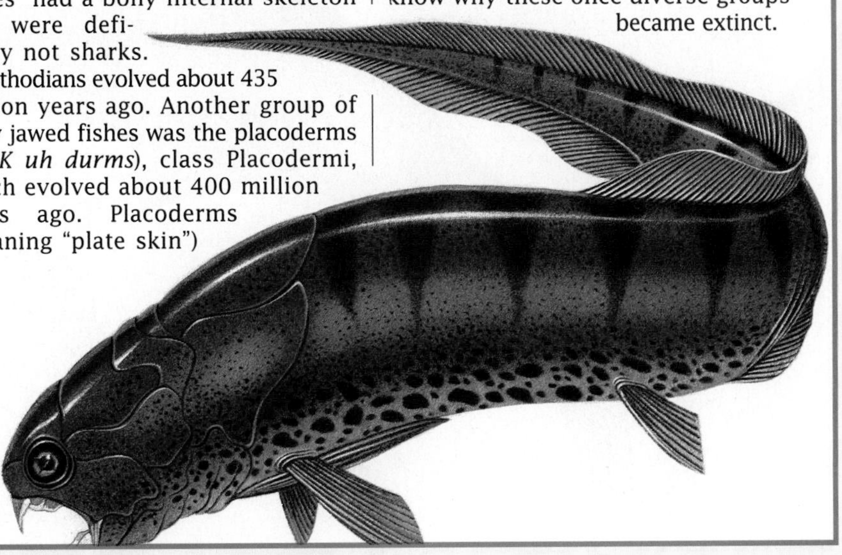

Figure 25.5
***Dunkleosteus*, a placoderm that was a predator in the ancient seas, might have looked like this drawing. Its skull and jaws were over 65 cm (2 ft.) long.**

Visual Strategy

Figure 25.4
Explain that the jaws of all vertebrates evolved through modifications in one or more gill arches (areas between gill slits). Explain that the gill arches moved forward in relation to the animal's body and changed form over a period of time. Progressive modification of the arches behind the mouth produced modern-day jaws. Have students describe how jaws evolved.

Visual Strategy

Figure 25.5
Have students give examples of how the features pictured here adapted *Dunkleosteus* to its ecological niche.

Connection: Chapter 10

Devonian Period
The Devonian period began about 410 million years ago and lasted approximately 50 million years. Explain that this period of time was considered "the age of the fishes," a period during which fishes were rapidly becoming abundant and diverse. Emphasize that by the end of the Devonian period, most of the jawless fishes were extinct. The placoderms survived into the next period, but eventually became extinct. Have students research the Devonian period and report their findings to the class.

Sharks and Rays: Class Chondrichthyes

Soon after placoderms evolved, another group of jawed fishes arose, the class Chondrichthyes (*kahn DRIHK thees*). The modern members of this class are the sharks, skates, and rays. Unlike the bony skeletons of acanthodians, the skeletons of sharks and rays are composed of a flexible substance called cartilage (*KAHRT'l ihj*). Chondrichthyes means "cartilage fishes." There are about 850 living species of cartilaginous fishes, the vast majority of which live in salt water. **Figure 25.6** shows a shark and a ray.

If you were to touch a shark or ray, you would notice rough, sandpaper-like skin. This texture results from the many small scales that are embedded in the skin. As shown in **Figure 25.7**, shark scales are very similar to shark teeth, which probably evolved from scales. The mouth of a cartilaginous fish includes upper and lower jaws, generally armed with rows of hard teeth. As the outer teeth are lost or broken, they are replaced by others moving up from behind.

The gills of sharks and rays open to the outside through a series of slits, as shown in **Figure 25.8**. One gill slit, the spiracle, opens directly to the outside on the side or top of the head. Water can be brought to the pharynx through the spiracle and then expelled through the gill slits. This arrangement not only frees the mouth for feeding but is a great advantage for sharks and rays that lie on the bottom of the ocean, where drawing water in through the mouth would also bring in sediment.

Figure 25.6
Sharks, skates, and rays have two sets of paired fins: pectoral fins and pelvic fins. The dorsal fin (on the back) and the caudal fin (the tail) are unpaired. In the skates and rays, the pectoral fins are greatly enlarged into a pair of large, winglike fins.

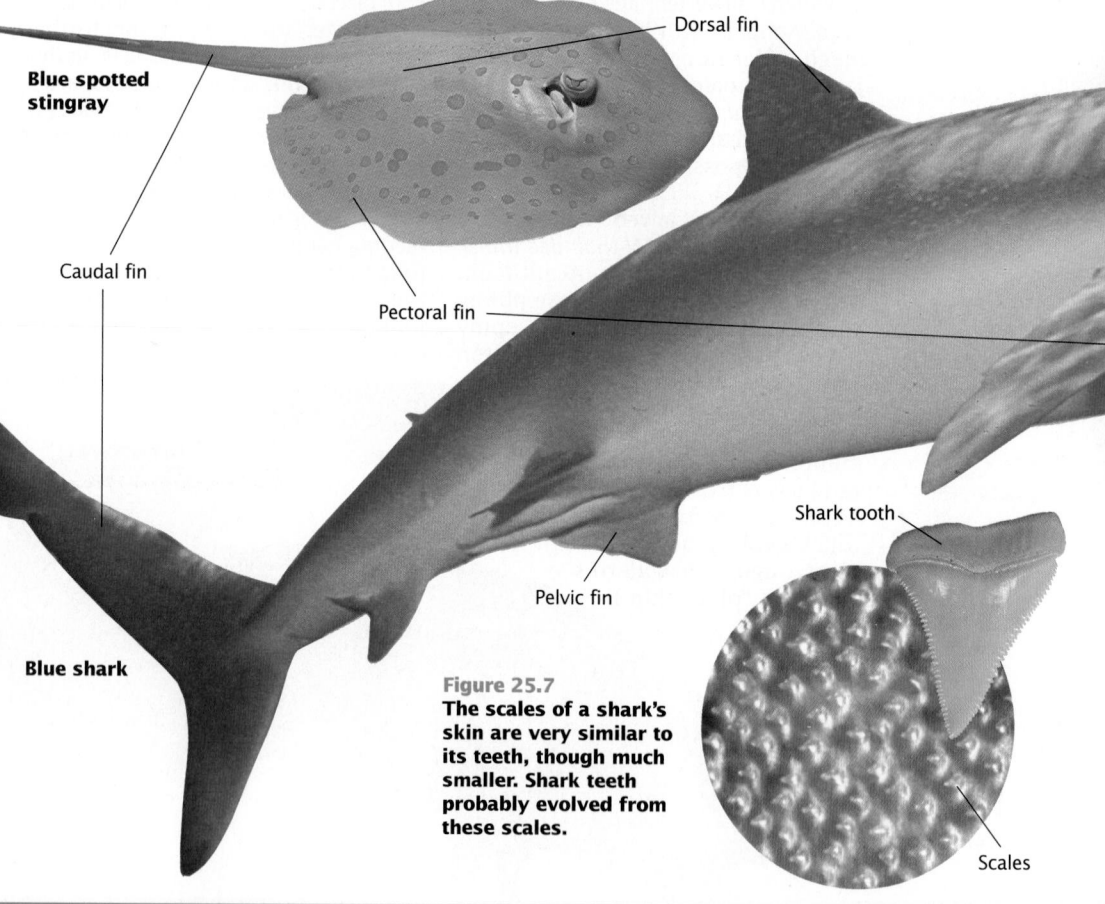

Dorsal fin

Blue spotted stingray

Caudal fin

Pectoral fin

Pelvic fin

Blue shark

Shark tooth

Figure 25.7
The scales of a shark's skin are very similar to its teeth, though much smaller. Shark teeth probably evolved from these scales.

Scales

■ *Matter of Fact* ■

Man-Eaters?
Some people think that all species of sharks are "man-eaters." Of the 370 known species of sharks, only 27 have attacked humans. One expert estimates that the chances of being attacked by a shark are equal to the chances of being struck by lightning.

How sharks detect and capture prey

Most sharks and rays are carnivores. Although sharks have the reputation of being man-eaters, only a few species of sharks are dangerous to humans. Three well-developed senses enable sharks and rays to detect their prey. First, they have an acute sense of smell. Second, they have the ability to sense electric currents in water. This sense is particularly useful in detecting the small electric currents generated by the muscle movements of animals. The electric ray *Torpedo* is also able to produce a powerful electric current, which it uses to deter predators and to stun prey. Third, sharks and rays have a lateral line system. As you learned in Chapter 22, the lateral line system is a series of pressure-sensitive cells that lie within canals along the sides of a fish. Changes in pressure caused by a fish or other animal swimming nearby can be detected by the cells in the lateral line.

Given the variety of shapes and sizes of cartilaginous fishes, it is not surprising that they exhibit a variety of feeding methods. Like the largest whales, the largest shark (the whale shark) and the largest ray (the manta ray) feed on plankton. The whale shark may exceed 13 m (45 ft.) in length, and the manta ray can have a "fin span" of nearly 6 m (20 ft.). Neither the whale shark nor the manta ray has teeth. Instead, their mouths are filled with a bony mesh that traps tiny organisms. Both of these large animals cruise the oceans and filter small crustaceans and protists from the water.

Sharks and rays have internal fertilization

The pelvic fins of male cartilaginous fishes are modified into a pair of claspers, which are used to transfer sperm to the female during mating. Thus, fertilization in these animals is internal. Skates, rays, and some sharks lay eggs. The extremely yolky fertilized eggs are usually housed in an elaborate leathery case. Many species of sharks do not lay eggs. Instead, the female keeps the yolky eggs inside her body until they hatch, and the young sharks are born alive. In some sharks, the female provides nutrients to her developing young through a membranous sac somewhat like the placenta.

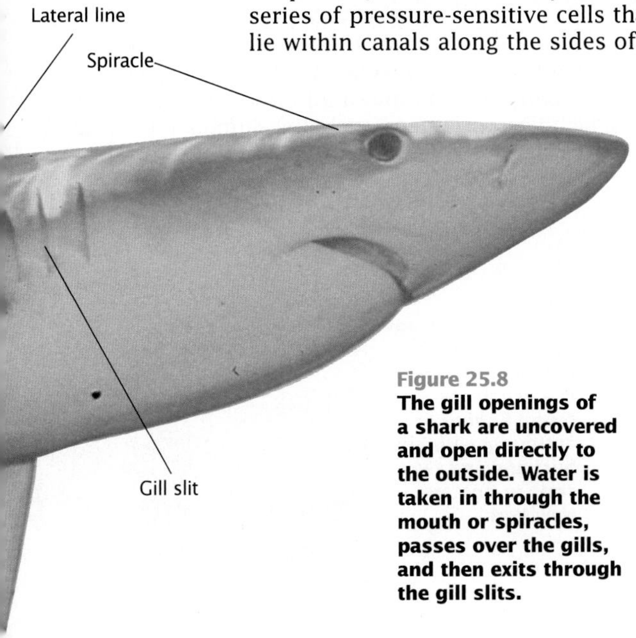

Scale and Structure

Why is it advantageous for a shark to be able to replace lost teeth throughout its life?

Lateral line

Spiracle

Gill slit

Figure 25.8
The gill openings of a shark are uncovered and open directly to the outside. Water is taken in through the mouth or spiracles, passes over the gills, and then exits through the gill slits.

Section Review

1. **Identify two characteristics of agnathans.**

2. **Explain two adaptations that enable a lamprey to be an external parasite of fishes.**

3. **Diagram the stages in the evolution of jaws.**

4. **List two differences between sharks and lampreys.**

■ *Section Review Answers* ■

1. Possible answers include: jawless, eel-like creatures with slimy skin that lack paired fins; skeleton is mostly composed of cartilage; no well-developed vertebral column; gills that lie within pouches that branch from the pharynx.

2. Students should suggest the funnel-like mouth, sharp teeth, and the file-like tongue.

3. See Figure 25.4 on page 551.

4. Possible answers include: sharks have upper and lower jaws, rough sandpaper-like skin, two sets of paired fins and gill slits; whereas lampreys are jawless, have slimy skin, and lack paired fins and gill slits.

Theme Answer

Scale and Structure

A shark's teeth are adapted to its feeding habits. Struggles that ensue while trying to capture prey can cause the loss of teeth. If sharks could not grow new teeth, they would starve. A single shark may use more than 20,000 teeth in a lifetime.

Visual Strategy

Figure 25.8
Tell students that sharks and rays can be distinguished by the placement of their gill slits. The gill slits of a ray are found beneath its wide fins. Have students describe where the gill slits of the shark are found.

Phase 3

ASSESSMENT OPTIONS

Closure Strategy

Comparing Fishes

Form cooperative groups of four students. Have one member of the group draw a table with the following headings across the top: *Skeleton*, *Body shape*, and *Reproduction*. Down the left side write *Class Agnatha* and *Class Chondrichthyes*. Through group discussions, have students complete the chart. Have them copy the finished chart and save it for use with Section 25.2.

Section Review

Assign the *Section Review*.

Reteaching

Have students make a table comparing characteristics of each class of fishes. Have them leave room for the classes yet to be discussed in Section 25.2. Across the top, write *Lampreys, Hagfish, Sharks,* and *Rays*. Down the left side write: *Habitat, Fin number and location, Gill type, and location, Sense organs,* and *Feeding habits*. Have a column for any other general characteristics as well; for example, *Body shape*.

Using the Feature

Phyllis has accomplished much without a college degree. You can use this feature to discuss the career possibilities available to people who do not wish to earn a college degree. Ask students what the alternatives are. Refer them back to the *Science in Action* in Chapter 21, featuring Alan Shipley, who also chose an alternate route.

Discussion

1. How was Phyllis able to pursue a career in science without a college degree?

Phyllis started out with a clerical job at WARD'S Biological Supply Company and seized opportunities for advancement. She trained in the Osteology Department and learned how to process, assemble, and repair skeletons. She was also trained to prepare slides, which is what she does now.

2. What is Phyllis' philosophy for reaching goals?

Phyllis feels that if someone has the motivation and desire to do something, nothing can stop him from accomplishing it.

3. What personal traits have enabled Phyllis to become a successful microscope slide technician?

She is dedicated, enthusiastic, patient, outgoing, and detail oriented.

Science in Action ▪ Science in Action ▪ Science in Action

Phyllis Stout: Microscope Slide Technician

How I Became Interested In Science

"I can't remember a time in my life that I wasn't involved with science. I loved my high school biology classes, especially when we did lab work. My sister once needed a skull for a science project. My father knew about a biological supply company through his work and was able to find it for her there. When I graduated from high school I wasn't sure if I could really pursue a career in science without a college degree, but I wanted to try.

"I started out as a Microscope Slide Order Filler at WARD'S, a biological supply company. It was more of a clerical job than science-related, but it was a great foot in the door. As soon as an opportunity for advancement came along, I took it— I wanted hands-on experience! For the next six months, I trained in the Osteology Department. I learned how to process, assemble and repair human and animal skeletons. My specialty became disarticulating and reassembling skulls using the Beauchene mounting technique, which lets students see and study all the individual parts of the skull. As far as I know, WARD'S has the only Osteology Lab of its kind in the country. Back around the turn of the century, WARD'S prepared and mounted all kinds of unusual specimens, including Jumbo, P.T. Barnum's circus elephant!"

When Phyllis is not peering through a microscope, she can usually be found fishing or antiquing with her husband and two sons in Upstate New York.

Name:	**Phyllis Stout**
Home:	**North Chili, New York**
Employer:	**WARD'S Natural Science, NY**
Personal Traits:	**• Sense of Humor**
	• Dedicated
	• Patient
	• Detail Oriented
	• Enthusiastic
	• Outgoing

"When my two children came along, I was able to work part-time. It would have been almost impossible to try and juggle family, job and college at that time. When an opportunity to work in the Microscope Slides Department came along, I decided to go back to work full-time. It was a complete switch from working with large skeletons to dissecting tiny specimens under a microscope! I've now been trained to prepare slides from the beginning to the final steps. I still often work with bones, only now I slice them paper thin for students to study individual cells. Last year, our Micro Slide team prepared over 200,000 slides that were distributed to students all over the world.

"I personally feel that if an individual has the motivation and desire to do something, no barriers can stop them. I am proud of my accomplishments even without a college degree. Now the results of my on-the-job education are in turn used to teach others."

The Craft of Preparing Micro Slides

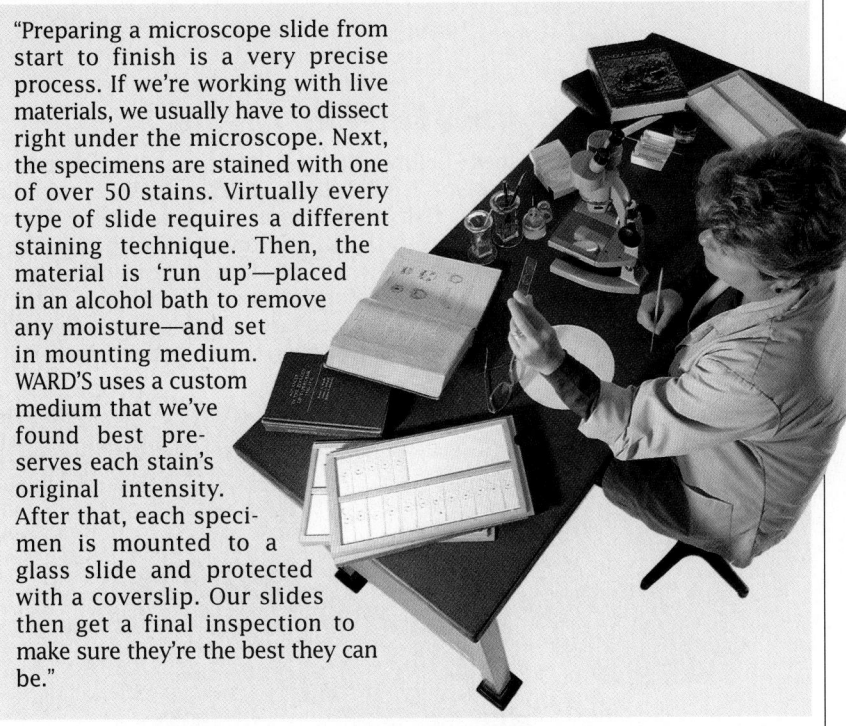

"Preparing a microscope slide from start to finish is a very precise process. If we're working with live materials, we usually have to dissect right under the microscope. Next, the specimens are stained with one of over 50 stains. Virtually every type of slide requires a different staining technique. Then, the material is 'run up'—placed in an alcohol bath to remove any moisture—and set in mounting medium. WARD'S uses a custom medium that we've found best preserves each stain's original intensity. After that, each specimen is mounted to a glass slide and protected with a coverslip. Our slides then get a final inspection to make sure they're the best they can be."

*Research
Focus*

Phyllis Stout has participated in a number of interesting projects at WARD'S Natural Science in Rochester, New York. One of her most memorable was in the Osteology Lab, preparing a huge bison skeleton for a college in Ontario, Canada. After processing and assembling the skeleton in a six-week time frame, she hand-delivered the 6' X 10' bison to the college, where it then had to be transported to its home in a tiny "museum" room on an upper floor of the building! Most recently, Phyllis worked as part of an effort to re-outfit the university system of Kuwait after the Gulf War.

Lesson Plan 25.2

Phase 1

PREPARATION

Key Concepts

- Bony fishes belong to the class Osteichthyes, which means "bone fish."
- Bony fishes have skeletons composed largely of bone instead of cartilage, and they are covered by scales formed of bone.
- The gills of bony fishes are housed in a common chamber on each side of the head and are covered by the operculum.
- Bony fishes usually have a swim bladder.
- In almost all bony fishes, fertilization and development occurs outside the body.
- There are two major groups of bony fishes: lobe-finned fishes and ray-finned fishes.
- Lobe-finned bony fishes may be closely related to amphibians.

Reading Strategy

After reading this section have students complete the table from the Closure Strategy in Section 25.1, comparing the features of class Osteichthyes with Agnatha and Chondrichthyes. In their comparisons, have them focus on body shape, skeleton, sense organs, and reproduction.

OF THE NEARLY 20,000 LIVING SPECIES OF FISHES, 18,000 SPECIES ARE BONY FISHES. BONY FISHES EVOLVED ABOUT 400 MILLION YEARS AGO AND ARE ONE OF EVOLUTION'S GREATEST SUCCESS STORIES. THE BONY FISHES ALIVE TODAY INCLUDE FAMILIAR ANIMALS SUCH AS TROUT, BASS, PERCH, TUNA, SWORDFISH, GUPPIES, CATFISH, BLUEFISH, MACKEREL, AND COD.

25.2   Bony Fishes

Objectives

❶ **Identify three differences between bony fishes and sharks.**

❷ **Describe the importance of the swim bladder for bony fishes.**

❸ **Summarize the importance of lobe-finned fishes in the evolution of land vertebrates.**

Structure of a Bony Fish

Figure 25.9
a **The scales of bony fishes are usually made of bone. They are not tooth-like as are the scales of a shark.**

Bony fishes belong to the class Osteichthyes (*ahs tee IHK thees*), which means "bone fish." What are the differences between sharks and bony fishes? As shown in **Figure 25.9**, most bony fishes have a skeleton composed largely of bone and are covered by scales formed of bone. However, these scales are very different from the tooth-like scales of sharks and rays, as you can see by comparing **Figure 25.9a** to **Figure 25.7**. The gills of bony fishes are housed in a common chamber on each side of the head. Each chamber is covered by a hard plate called the operculum (*oh PUR kyoo luhm*). The

b **The bones of this bony fish are clearly evident in this 40 million year old fossil.**

c **The gills of bony fishes are covered by a protective plate, the operculum. Gills provide a large surface area for the rapid uptake of oxygen and the release of carbon dioxide.**

Operculum
Gills

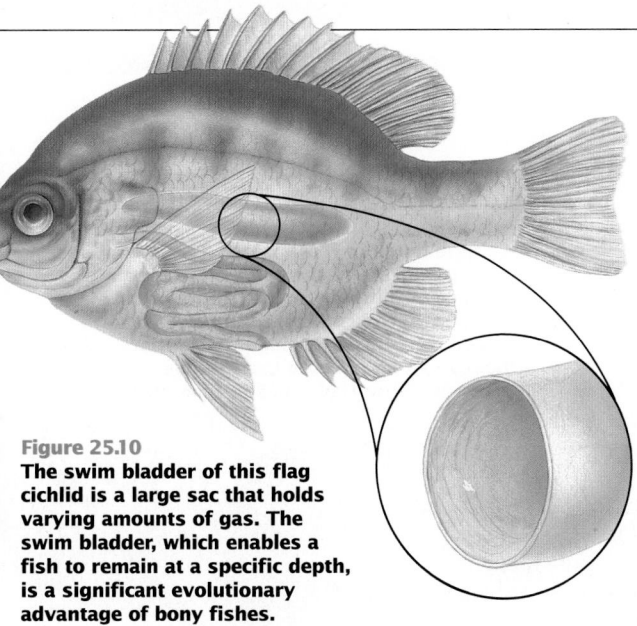

Figure 25.10
The swim bladder of this flag cichlid is a large sac that holds varying amounts of gas. The swim bladder, which enables a fish to remain at a specific depth, is a significant evolutionary advantage of bony fishes.

upper tail lobe is larger than the lower lobe in sharks, but the lobes are more symmetrical in bony fishes.

A critical difference between bony fishes and sharks is the presence of a **swim bladder** in most bony fishes. This gas-filled sac, shown in the cutaway view in **Figure 25.10**, gives the fish buoyancy. By regulating the amount of gas in the swim bladder, a fish can adjust its buoyant density. Thus, the fish can remain at a particular depth without expending energy through swimming. Although its oil-rich liver gives it some buoyancy, a shark is denser than water and will sink if it stops swimming.

Like sharks, bony fishes sense pressure changes in water with a lateral line system. Vision is also an important sense for most bony fishes. Most bony fishes have color vision. The bright colors of many fishes such as those in **Figure 25.11** serve as signals to potential mates and rivals. Fishes living

Figure 25.11
The coloration of these glassy sweepers allows members of the species to easily recognize one another, which is especially important for schooling and mating.

in different environments are sensitive to different colors. For instance, deep-sea fishes are more sensitive to blue light than to red light; red light is absorbed by the ocean, so little red light penetrates to the depths inhabited by these fishes.

Many species of bony fishes can detect electric fields. For example, the knife fishes of South America produce their own weak electric fields. Knife fishes detect disturbances in their fields, which can be caused by rocks or other objects, by prey or predators, or by other fishes of the same species. Knife fishes live in silty streams where vision is limited. They use their electric fields to navigate, to find prey, to avoid predators, and to signal to other members of their species.

Most bony fishes are egg-layers
The eggs and sperm of bony fishes are typically shed into the water or into a nest fashioned by the parents. In almost all bony fishes, fertilization and development occur outside the female's body. In livebearers, such as guppies, mollies, and swordtails, the male transfers his sperm directly to the female via a modified fin near his tail. Fertilization is internal, and the young develop inside the female. You can read more about fishes in the *Tour of a Fish* on the next page.

557

Using the Feature

With more than 18,000 species, the class Osteichthyes, the bony fishes, is the largest class of vertebrates. From the locations of the earliest fossils of bony fishes, it appears that this group originated in fresh water, only later invading the oceans. Salmon are one of the few fish species that live in both fresh water and salt water. Adult salmon live from three to five years in the ocean. They then return to the stream in which they hatched to spawn. During spawning, female salmon lay from 2,000 to 10,000 eggs. In some species, the adults die after spawning; in others, the adults return to the sea. The young salmon hatch and make their way to the sea. Most of them perish during this journey—a Canadian study reported that less than one percent of hatchlings returned to spawn.

Discussion

Guide the discussion by posing the following questions.

1. A salmon's tissues have a higher salt content than fresh water, but a lower salt content than salt water. What happens to a salmon's salt and water balance in fresh water and in salt water?

In fresh water, it tends to lose salt and gain water by osmosis. In salt water, it tends to lose water and gain salt.

2. How fast can a 20-cm long salmon swim?

approximately 2 m per second

3. What is the function of the salmon's slimy coating?

to reduce friction, thus lessening the amount of energy used for swimming

Tour of a Fish

There are more than 18,000 species of bony fishes. This fish is a salmon, one of the few fishes that live in both salt water and fresh water.

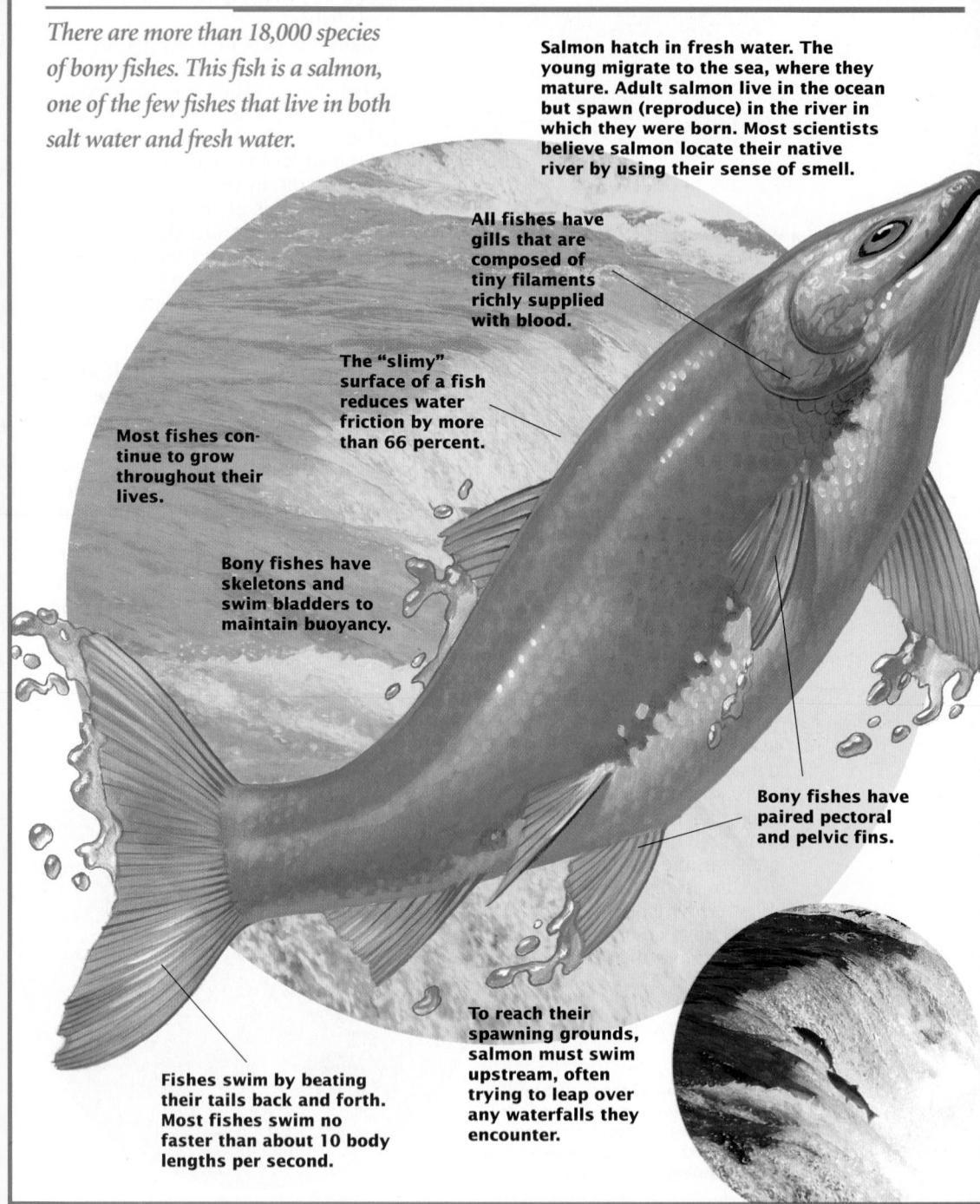

Salmon hatch in fresh water. The young migrate to the sea, where they mature. Adult salmon live in the ocean but spawn (reproduce) in the river in which they were born. Most scientists believe salmon locate their native river by using their sense of smell.

All fishes have gills that are composed of tiny filaments richly supplied with blood.

The "slimy" surface of a fish reduces water friction by more than 66 percent.

Most fishes continue to grow throughout their lives.

Bony fishes have skeletons and swim bladders to maintain buoyancy.

Bony fishes have paired pectoral and pelvic fins.

Fishes swim by beating their tails back and forth. Most fishes swim no faster than about 10 body lengths per second.

To reach their spawning grounds, salmon must swim upstream, often trying to leap over any waterfalls they encounter.

Major Groups of Bony Fishes

Patterns of Change

The living species of coelacanth is very similar to the extinct coelacanth species. What does this similarity indicate about the rate of evolution in the coelacanths?

Look carefully at the fins of the salmon shown in the *Tour of a Fish* on page 558. Salmon belong to the group of bony fishes known as the ray-finned fishes. All but seven species of bony fishes are **ray-finned fishes**. In these fishes, the fins are fan-shaped and are supported by thin bony rays. Ray-finned fishes are the most successful and diverse group of vertebrates.

The other group of bony fishes is the **lobe-finned fishes**. In this group, the fins are fleshy and are supported by central bones. The existing lobe-finned fishes consist of six species of lungfishes and

one species of **coelacanth** (*SEE luh kanth*). **Table 25.1** lists the major groups of fishes.

As their name suggests, lungfishes have functional lungs. These fishes live in Africa, South America, and Australia. A lungfish from Africa is shown in **Figure 25.12**. Some lungfishes inhabit

Figure 25.12
Lungfishes, like this African lungfish, have fleshy, lobed fins. Lungs enable these fishes to live in water that is low in oxygen.

Table 25.1 Major Groups of Fishes

Class	Approximate Number of Species	Major Characteristics	Examples	
Agnatha	63	No jaws; no paired appendages	Lampreys, hagfishes	
Acanthodii	Extinct	Jaws; spiny paired fins	Acanthodians	
Placodermi	Extinct	Jaws; bony armor	Placoderms	
Chondrichthyes	850	Skeleton of cartilage; no swim bladder; spiracle; internal fertilization	Sharks, skates, rays	
Osteichthyes	18,000	Paired fins supported by bony rays; bony skeleton; most have swim bladder	Ray-finned fishes	
	7	Paired lobed fins; bony skeleton; extinct forms are ancestors of amphibians	Lobe-finned fishes	

Theme Answer
Patterns of Change
The rate of evolution in coelacanths is very slow. They have survived virtually unchanged for more than 80 million years.

Visual Strategy

Figure 25.12
Evolution has provided ways for lungfishes to be borderline amphibious; for example, lungs double as swim bladders, a very efficient use of one structure for two functions. Ask students which organisms these lungfishes might be closely related to. Suggest amphibians.

Theme Connection

Evolution
Ray-finned fishes evolved about 400 million years ago. Emphasize that most fishes belong to this group. Scientists think that ray-finned fishes evolved in freshwater habitats and that some later migrated to the oceans. They soon replaced cartilaginous fishes as the dominant kind of fish on earth. Bony fishes are thought to have evolved in fresh water, because of the type of rock formation in which their fossils are found.

Visual Strategy

Table 25.1
Have students use *Major Characteristics of Groups of Fishes* as the title of a concept map of the table information. Have them consider *jaws, appendages, spiracles, swim bladder, skeleton, bone, cartilage, ray fins, lobed fins,* and *fertilization* as the main ideas.

Figure 25.13
Latimeria chalumnae, the coelacanth, was named to honor Marjorie Courtney-Latimer, a museum curator in South Africa. She recognized and preserved the first coelacanth specimen in 1938.

stagnant water that is low in oxygen. Other lungfishes live in ponds that dry up annually. These fishes survive these conditions by burying themselves in mud at the bottom of the pond. Lungs enable these fishes to supplement their oxygen intake by breathing air.

Coelacanths, in contrast, inhabit deep ocean environments. Fossils of coelacanths from the time of the dinosaurs are common, but no coelacanth fossils younger than 80 million years old have been found. Coelacanths were thought to be extinct until a specimen was captured off the eastern coast of South Africa in 1938. About 200 specimens of coelacanths have since been captured.

Figure 25.13 shows an example of a preserved coelacanth.

Lobe-finned fishes are the ancestors of amphibians

The role of lobe-finned fishes in evolution is important. Zoologists are convinced that land vertebrates evolved from lobe-finned fishes that are now extinct. Until recently, scientists thought that coelacanths, which have sturdier fins than lungfishes, were the closest living relatives of the amphibians. However, comparisons of mitochondrial DNA reported in 1990 suggest that lungfishes, not coelacanths, are the closest living relatives of modern amphibians.

Section Review

❶ How could you tell whether a fish was a shark or a bony fish?

❷ Explain how a swim bladder enables a fish to conserve energy.

❸ What have studies of mitochondrial DNA revealed about the relationships between amphibians and lobe-finned fishes?

■ *Section Review Answers* ■

1. Refer to Table 25.1 on page 559 for characteristics of Chondrichthyes and Osteichthyes.

2. Students should suggest that the swim bladder enables fishes to remain at various depths without expending energy through swimming.

3. Students should explain that the amphibians' closest living relatives are the lungfishes, and not the coelacanth.

THE NAME "AMPHIBIAN" IS DERIVED FROM THE GREEK WORDS MEANING "DOUBLE" AND "LIFE." THIS NAME REFLECTS THE FACT THAT ALTHOUGH MOST AMPHIBIANS ARE TERRESTRIAL, THEY MUST REPRODUCE IN WATER OR IN MOIST ENVIRONMENTS. AMPHIBIANS WERE THE FIRST VERTEBRATES TO WALK ON LAND. THEY EVOLVED FROM LOBE-FINNED FISHES ABOUT 350 MILLION YEARS AGO.

25.3   Amphibians

Objectives

❶ Identify two characteristics that enabled amphibians to invade the land.

❷ Describe the life cycle of a frog.

❸ Identify the three orders of living amphibians.

The First Land Vertebrates

Approximately 4,200 species of amphibians exist today. Amphibians include frogs, toads, salamanders, and the wormlike, legless caecilians (*see SIHL ee uhns*). Two examples of amphibians are shown in **Figure 25.14** and **Figure 25.15**. As adults, amphibians are clearly different from fishes, showing many adaptations for land life. Except for caecilians, all amphibians have legs. Most amphibians have lungs for gas exchange, although some salamanders lack lungs. Gas exchange also occurs across the thin, moist skin of amphibians, as you learned in Chapter 22. The lack of a skin that is resistant to drying is one feature that limits amphibians to moist environments.

Another difference between fishes and amphibians is evident in the amphibian circulatory system. Amphibians achieve more efficient circulation than fishes because of their double-loop circulatory system. As you saw in **Figure 22.10b** on page 488, a fish's heart pumps only deoxygenated

Figure 25.14
The Asiatic fire-bellied toad lives near slow-flowing rivers in the low-lands of central and eastern Europe. Although adapted to land, amphibians such as this toad must have access to a watery environment to reproduce.

Figure 25.15
This spotted newt is an amphibian. Limbs enable the newt to move on land.

■ *Matter of Fact* ■

Lungs or Gills
Not all amphibians have lungs or gills. Some salamanders have neither. They respire almost entirely through the skin, which contains many capillary networks. Salamanders supplement this breathing by "mouth breathing."

Lesson Plan 25.3

Phase 1
PREPARATION

Key Concepts
- Amphibians, which include frogs, toads, salamanders, and caecilians, were the first terrestrial vertebrates.
- Except for caecilians, all amphibians have legs, webbed feet, and toes that lack claws.
- Amphibians use their gills, lungs, skin, and mouth in respiration. They have a well-developed circulatory system and lack watertight skin, which limits them to moist environments.
- Amphibians undergo metamorphosis to change from an aquatic larval stage to a terrestrial adult form.
- There are three orders of amphibians: Anura, Urodela, and Gymnophiona.

Reading Strategy

Interesting writing makes reading a pleasurable experience. Have students read a selection on salamanders and newts from E.B. White's collection of humorous poems about animals.

Phase 2
TEACHING STRATEGIES

Demonstration 1
Display a variety of live or preserved amphibians, including frogs, toads, salamanders, and, if possible, a caecilian. Have students explain how these animals are different from the fishes they studied in the last section.

Language Connection

Amphibian
The word *amphibian* means "double life." Have students explain the connection between "double life" and *amphibian*.

Connection: Chapter 19

Adapting to Land
Review the adaptations of plants to life on land. Have students compare the obstacles that amphibians faced in their movement to land with those faced by plants.

Theme Connection

Interacting Systems
Toads usually live in arid areas and are protected by thick, bumpy skin, whereas salamanders live in cool or tropical areas of the world. Caecilians live in tropical forests and have long, slender, wormlike bodies with no limbs. Discuss the features that distinguish caecilians from toads and salamanders. Encourage students to infer why caecilians are classified as amphibians.

Theme Connection

Evolution
Explain that during the Mississippian and Pennsylvanian periods—from 360 million to 290 million years ago—amphibians were the most important vertebrates on land. Modern-day amphibians did not appear until the Mesozoic era—from 240 million to 63 million years ago. By that time, most of the other amphibians had died out.

Connection: Chapter 9

Salamanders
Tell students that scientists believe the salamander—of all modern amphibians—most closely resembles ancestral amphibians. Have students make inferences as to how ancestral amphibians might have looked.

blood, sending it to the gills where it picks up oxygen. When the blood reaches the gills, it slows down while passing through narrow capillaries. The amphibian heart, in contrast, pumps both deoxygenated and oxygenated blood. Deoxygenated blood returning from the body is sent back out by the heart to the lungs to absorb oxygen. This oxygen-rich blood, which also slows as it passes through the narrow capillaries of the lungs, returns to the heart to be pumped to the rest of the body. Thus, amphibians are able to pump oxygenated blood at higher pressures and faster rates of flow than are fishes. The amphibian heart is partly divided, and some mixing of oxygenated and deoxygenated blood occurs as the blood flows through the heart.

The Tie to Water

As you learned in Chapter 22, amphibians do not lay watertight eggs and so must reproduce in water or in moist environments. Many species of frogs and toads, for example, congregate in ponds during the spring in preparation for mating. The males generally arrive first and begin noisy mating calls that attract nearby females. You have probably heard the raucous symphony of male frogs on warm spring evenings. Once mating begins, the male grasps a female and holds her. While she releases eggs, he simultaneously releases sperm. Fertilization takes place externally. Follow the development of a frog in **Figure 25.16**. You can read more about frogs in the *Tour of a Frog* on the next page.

Salamander reproduction follows a slightly different pattern than that of frogs and toads. Salamanders that live on land return to the water to breed. The males deposit packets of sperm in the water and then perform complex courtship behaviors to attract the females. The females draw the sperm packets into their reproductive openings, and fertilization occurs internally. Eggs are then deposited in the water where they continue developing. Salamander larvae are not as different from the adults as tadpoles are from adult frogs. For example, salamander larvae are carnivorous, like their parents, and they retain external gills until the time of metamorphosis.

Figure 25.16

a The life cycle of a frog involves large-scale changes in body form. First, a mass of eggs is laid in a wet or moist environment.

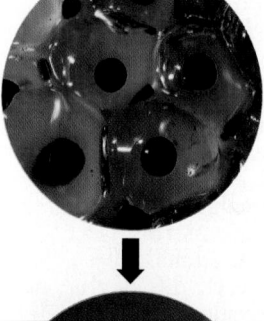

b The young tadpole emerges from the egg with external gills, which are later replaced by internal gills. After feeding and growing, the tadpole begins to transform into an adult frog.

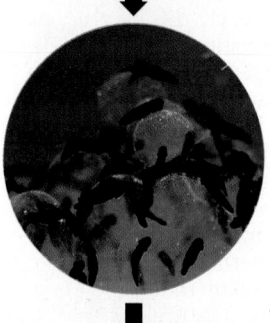

c Dramatic changes occur in the tadpole. The tail and gills recede. Lungs and front and hind limbs grow. Feeding habits may also change. Herbivorous tadpoles change into carnivorous adults.

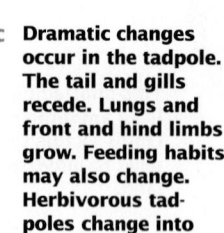

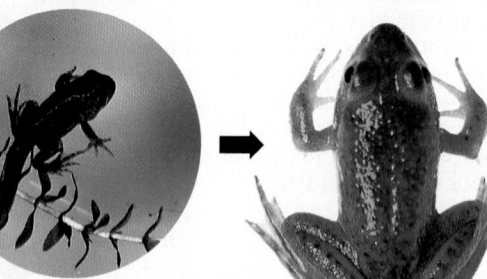

d The adult frog has completely lost its tail and gills. Its lungs enable it to breathe air. The changes that transform a tadpole into an adult frog are called metamorphosis.

Tour of a Frog

There are about 2,000 species of frogs and toads, all of them carnivores. All frogs, including this red-eyed tree frog, go through complete metamorphosis, changing from water-living tadpoles with gills to land-dwelling adults with lungs.

During winter, most frogs hibernate in the soft mud at the bottom of pools and streams. Tree frogs hibernate in the decaying material of the forest floor.

Toads are a kind of frog with short legs, stout bodies, and wart-covered skin.

Northern frogs can survive prolonged freezes; they prepare for winter by adding a chemical to their blood that serves as anti-freeze.

Most frogs are solitary except during the breeding season, when males call noisily to attract females.

Frog fossils have been found that are 150 million years old.

Adult frogs have long, powerful hind legs and no tails; that is why they are called anurans, from the Greek for "without a tail."

Frogs have moist, hairless, scaleless skin, through which they carry out much of their gas exchange. All frogs produce skin poison. The skin poison of South American *Dendrobates* frogs is more toxic than that of any spider or snake.

The world's largest frog is *Conraua goliath*, which is more than 30.5 cm (1 ft.) long from mouth to anus and weighs over 3.5 kg (7 lb.). This species lives in west Africa and eats animals as large as rats and ducks.

Using the Feature

Frogs and toads are the most familiar and widespread amphibians. They live on all continents except Antarctica, and are even found north of the Arctic Circle in Scandinavia. Many species of frogs lay their eggs in bodies of water, but many others have evolved ways to protect eggs laid out of water from dessication. For instance, in several frog species (known as marsupial frogs), the females carry the eggs in a pouch on the back. South African tree frogs construct an arboreal foam nest in which the eggs develop and hatch.

- Have students research some of the anuran adaptations for protecting their eggs from dessication and predation.
- According to some reports, population sizes of frogs and other amphibians have fallen steeply. Have students research the apparent decline in amphibian abundance. A good starting point is "Silence of the Frogs," in *The New York Times Magazine*, December 13, 1992, pages 36-39, 64, 66, and 76.

Discussion

Guide the discussion by posing the following questions.

1. Identify two differences between frogs and salamanders.
   Salamanders have a tail and internal fertilization.
2. How do frogs in temperate climates survive winter?
   They hibernate in the mud on pond bottoms or in decaying material on the forest floor. Some add an antifreeze to their blood.

Visual Strategy

Table 25.2
Ask students why more species of frogs and toads exist compared to other amphibians. Explain that caecilians are restricted to tropical regions, and salamanders are restricted to moist habitats not far from bodies of water. Toads and frogs are found anywhere there is a body of water for breeding.

Theme Connection

Energy and Life

Most amphibian larvae eat algae and plant material, but salamander larvae feed on small water animals. Most adult amphibians use their tongues to capture insects and other kinds of small animals. Other large amphibians, such as bullfrogs, prey on snakes, small mammals, and birds. One group of South American frogs feeds mainly on other frogs.

Phase 3
ASSESSMENT OPTIONS

Closure Strategy

Fishes and Amphibians

Have students explain how these animals differ from the fishes studied in Sections 25.1 and 25.2.

Section Review

Assign the *Section Review*.

Reteaching

Have students describe the characteristics and adaptations of a particular type of amphibian. Each student should read their description aloud and have the other members of the class try to guess which amphibian is being discussed.

564

Kinds of Amphibians

Figure 25.17 shows a tropical burrowing amphibian called a caecilian. Caecilians are members of the order Gymnophiona (from the ancient Greek words for "naked" and "snakelike"). Caecilians burrow through the soil and feed on earthworms and other small animals.

Frogs and toads are **anurans**. They belong to the order Anura (meaning "without a tail" in Greek). This order is the largest amphibian order, containing over 3,600 species. As adults, frogs and toads are insect-eaters. They have large mouths, often with long tongues, and hind legs specialized for jumping. The body form of frogs and toads is distinct: the head and trunk are fused and there is no tail. Frogs and toads are found in a variety of habitats. Some species are completely aquatic, while others spend some time on land. Toads, which you might see in a garden or park, are mostly terrestrial, only returning to water to breed. The skin of a toad is dry and warty and is more resistant to evaporation than the skin of other amphibians.

Salamanders belong to the order Urodela ("visible tail"). The body shape of a salamander is more like that of reptiles. A salamander has a distinct head, trunk, and tail. The limbs are set at right angles to the body. Like frogs, toads, and caecilians, salamanders are carnivores. **Table 25.2** lists three orders of amphibians and their main characteristics.

Figure 25.17
Caecilians are limbless, one of their many adaptations for burrowing.

Table 25.2 Orders of Amphibians

Order	Approximate Number of Species	Major Characteristics	Examples
Gymnophiona	160	Wormlike body with no limbs; tail short or absent; restricted to tropics	Caecilians 
Anura	3,680	Head and trunk fused, no tail; lungs; limbs specialized for jumping	Frogs, toads 
Urodela	360	Body has distinct head, trunk, and tail; limbs set at right angles to body	Salamanders, newts 

Section Review

1. Compare an amphibian to a fish. In what important ways are they different?

2. Describe the stages in the life cycle of a frog.

3. Name one representative of each amphibian order.

■ Section Review Answers ■

1. Most amphibians have legs and lungs, and they have a double-loop circulatory system with a heart that pumps both deoxygenated and oxygenated blood; fishes do not have legs and lungs, and their heart pumps only deoxygenated blood, sending it to the gills, where it picks up oxygen.

2. See Figure 25.16 on page 562.

3. See Table 25.2 on page 564.

Chapter **25** Highlights

A male Darwin's frog carries developing offspring in his vocal sacs. Here, a young frog has just been released.

Alternative Assessment

Using all the summary statements on this page, have students construct a concept map comparing early fishes, bony fishes, and amphibians. Be sure to encourage them to make linkages from one group to another.

	Key Terms	**Summary**
25.1 Early Fishes 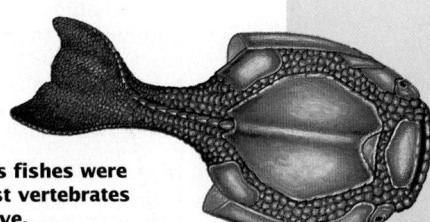 Jawless fishes were the first vertebrates to evolve.		• Jawless fishes were the first vertebrates to evolve. • The jawless fishes include lampreys, which are external parasites on other fishes, and hagfishes, which are scavengers. • Acanthodians were the first vertebrates with jaws. • Sharks and rays have skeletons of cartilage.
25.2 Bony Fishes  Most bony fishes are ray-finned fishes.	swim bladder (p. 557) ray-finned fishes (p. 559) lobe-finned fishes (p. 559) coelacanth (p. 559)	• Unlike sharks and rays, most bony fishes have skeletons of bone. • Bony fishes have a swim bladder, a gas-filled sac that helps them to maintain position in the water. • Most bony fishes are ray-finned fishes. • A group of bony fishes known as the lobe-finned fishes are thought to be closely related to the amphibians. Studies of mitochondrial DNA suggest that lungfishes are the closest relatives of amphibians.
25.3 Amphibians Most amphibians, including this frog, return to water to reproduce. 	anuran (p. 564)	• Amphibians were the first terrestrial vertebrates. • Frogs, toads, salamanders, and caecilians are amphibians. • Because their skins are not watertight, most amphibians live in water or in damp environments. • Amphibian eggs are not watertight and will dry out if not kept wet or moist. • Caecilians are legless, tropical amphibians. They burrow in moist soils and eat worms and other small animals. • Frogs and toads, the anurans, are adapted for jumping. • Salamanders have four limbs and a tail. They are carnivorous and have internal fertilization.

Chapter Review Answers

Understanding Vocabulary
1. **a.** Osteichthyes
   b. caecilians
   c. lamprey

Relating Concepts
2. Map answer shown on page 447D.

Understanding Concepts
Multiple Choice

3.	a	7.	a
4.	b	8.	d
5.	c	9.	a
6.	c	10.	d

Completion
11. Acanthodii, gill arches
12. Chondrichthyes, cartilage
13. lobe-finned, lungs, skin
14. Anura, Urodela

Short Answer
15. spiracle
16. They are able to adjust their density by regulating the amount of air in the swim bladder.
17. The young and the adults are not competing for the same food.
18. Lungs, double-loop circulatory system, and sturdy limbs enable amphibians to live on land. Thin, moist skin, and the need to reproduce in water keep amphibians in or near water.

Chapter 25 Review

Understanding Vocabulary

1. Complete each analogy by providing the missing word or phrase.
   a. skeleton of cartilage:chondrichthyes::skeleton of bone: ____
   b. adapted for jumping:frogs and toads::adapted for burrowing: ____
   c. scavenger:hagfish::parasite: ____

Relating Concepts

2. Copy the unfinished concept map below onto a sheet of paper. Then complete the concept map by writing the correct word or phrase in each oval containing a question mark.

Understanding Concepts

Multiple Choice

3. Fishes that have a notochord and pharyngeal gill slits but lack paired fins are members of the class
   a. Agnatha.
   b. Placodermi.
   c. Chondrichthyes.
   d. Osteichthyes.

4. Lampreys and leeches are alike in that they both
   a. have jaws.
   b. are external parasites.
   c. prevent their hosts from bleeding.
   d. live in fresh water but not sea water.

5. Sharks and hagfishes are different in that
   a. sharks live in the ocean, hagfishes do not.
   b. hagfishes are scavengers, sharks are parasites.
   c. sharks have paired fins, hagfishes do not.
   d. sharks have a skeleton made of cartilage, hagfishes have a bony skeleton.

6. What is the function of the swim bladder in bony fishes?
   a. an oxygen reservoir
   b. digestion
   c. buoyancy
   d. a source of energy

7. As a result of DNA studies, scientists think that amphibians are most closely related to which group?
   a. lung fishes
   b. ray-finned fishes
   c. coelacanths
   d. agnathans

8. What characteristics enable amphibians to live on land?
   a. scaly skin and lungs
   b. bony skeleton and lobed fins
   c. appendages and moist skin
   d. lungs and double-loop circulation

9. How do tadpoles differ from frogs?
   a. tadpoles have gills and a tail; frogs do not
   b. tadpoles are carnivorous; frogs are herbivorous
   c. frogs show body segmentation; tadpoles do not
   d. frogs inhabit decaying animals and plants; tadpoles live in water

10. Caecilians are members of the order
    a. Urodela.
    b. Amphibia.
    c. Osteichthyes.
    d. Gymnophiona.

Completion

11. Members of the class ____ are the earliest known jawed fishes. Scientists believe that the jaws evolved from ____ .

12. Skates and rays are members of the class ____ . Their skeletons are made of ____ .

13. Amphibians are thought to have evolved from ____ fishes. Their ____ and thin, moist ____ enable them to conduct gas exchange on land.

14. Toads and frogs are members of the order ____ , while salamanders are members of the order ____ .

Interpreting Graphics
19. • 1991
    • 1989
    • More rain increases the size of the pond, resulting in a greater breeding area and thus more frogs.
    • more frogs

Reviewing Themes
20. The early stage that hatches from the egg has external gills, which are replaced by internal gills that are covered. Later on, the tail and gills recede and limbs and lungs develop. The diet also changes from plants to insects as the tadpole becomes a frog.

21. Lobe-finned fishes are the evolutionary ancestors of amphibians.
22. lungs, which enable lungfishes to extract oxygen from the air

Short Answer

15. Once it was thought that sharks would die if they stopped swimming. What structural adaptation enables sharks to rest on the bottom and still move clean water over their gills?

16. Sharks must swim to keep from sinking, but many bony fishes can remain stationary at a given depth without expending much energy. How do bony fishes do this?

17. Tadpoles are herbivorous, while frogs are carnivorous. What benefit is associated with this change in feeding habit?

18. What characteristics enable amphibians to live on land? What characteristics require that most live in or near the water throughout their lives?

Interpreting Graphics

19. An ecologist studying a population of frogs in a small pond recorded the amount of spring rainfall and the number of frogs seen each year for six years. Construct a graph of her data, which appear in the table below.

Rainfall Amounts and Numbers of Frogs

Year	Spring Rainfall (cm)	Number of Frogs Observed
1988	13.1 cm	53
1989	14.7 cm	75
1990	13.4 cm	61
1991	12.5 cm	34
1992	13.6 cm	65
1993	12.8 cm	38

- In which year were the fewest frogs observed?

- In which year was the amount of rainfall the greatest?

- What relationship exists between the amount of spring rainfall and the number of frogs in the pond? Form a hypothesis to explain this relationship.

- If 14.0 cm of rain falls in the spring of 1994 will the ecologist likely see more frogs than during 1993?

Reviewing Themes

20. *Patterns of Change*
Describe the changes that occur as a tadpole becomes an frog.

21. *Evolution*
Why would evolutionary biologists consider lobe-finned fishes important?

22. *Scale and Structure*
What structural adaptation enables lungfishes to live in slow-moving, stagnant water?

Thinking Critically

23. *Inferring Conclusions*
Adult female salmon smell and taste the water in order to return to their freshwater spawning grounds where they were hatched. What effect might polluted water have on their journey?

24. *Comparing and Contrasting*
How does fertilization among sharks that are born alive differ from fertilization among most bony fishes?

25. *Building on What You Have Learned*
What type of evidence did scientists use to show that lungfishes rather than coelacanths are the closest living relatives of amphibians?

Cross-Discipline Connection

26. *Biology and Mathematics*
Experiments have shown that the cruising speed of most fishes is about .4 kilometers per hour for each centimeter of length. If this is true, what is the cruising speed of a 20-cm-long barracuda? Of an 8-cm oscar?

Discovering Through Reading

27. Read the article "This Fish Don't Get No Respect," in *Field & Stream*, July 1992, pages 34–35. In what kind of waters can gar be found? Why is the flesh of the gar not considered good eating?

Cross-Discipline Connection

26. The barracuda's cruising speed is about 8 km (5 mi.) per hour and the oscar's is about 3.2 km (2 mi.) per hour. The burst speed for the two fishes is 12 and 4.8 km (8.7 and 3 mi.) per hour, respectively.

Discovering Through Reading

27. Gar can be found in waters where other fish cannot survive. Because they can breathe air through their swim bladder, gar can live in poorly oxygenated and polluted waters. Gar are not considered delicious table fare because they are ugly fish, eat almost anything, and often live in polluted waters.

Thinking Critically

23. The polluted water may cause the salmon to go up the wrong river and not reach their spawning grounds. The number of offspring will be fewer, since salmon will mate and lay eggs only in the spawning grounds where they were hatched.

24. In sharks, fertilization occurs inside the female's body, but fertilization is external in most bony fishes.

25. By comparing DNA sequences, scientists were able to determine the evolutionary relationships among the lungfishes, coelacanths, and amphibians. The differences between the DNA sequences of the two types of lobe-finned fishes and the amphibians indicated that it had been longer since coelacanths and amphibians had shared a common ancestor than it had been since lungfishes and amphibians shared a common ancestor.

Procedural Note

Before students perform the investigation, review the answers to the Prelab Preparation questions. Caution the students to handle the fish with care. The fish are likely to respond to white light by changing their orientation, but not to red light.

Prelab Preparation Answers

- Osteichthyes
- yes

Procedure Answer

4. The box isolates the fish and allows control of the amount and direction of light.

Chapter **25** Investigation

How Do Goldfish Respond To Light?

Objectives

In this investigation you will:
- *observe* a goldfish's behavior
- *test* the responses of goldfish to white light and red light

Materials

- cardboard box to fit over large glass jar or 1,000-mL beaker
- scissors
- cellophane tape
- black paper
- dechlorinated water
- large glass jar or 1,000-mL beaker
- aquarium fish net
- goldfish
- flashlight
- red cellophane
- wristwatch or clock with a second hand

Prelab Preparation

Review what you have learned about fishes by answering the following questions.
- To which class of fishes does a goldfish belong?
- Do fishes have color vision?

Procedure: Observing Goldfish Behavior

1. Form a cooperative team of four students. Work with your team to complete steps 2–10.

2. Cut a hole slightly smaller than the bulb end of the flashlight in the center of one side of the cardboard box. Make a similar hole in the bottom of the box. Tape one edge of a piece of black paper over each hole to form a flap.

3. **Caution: Use care when handling live animals.** Add 500 mL of dechlorinated water to a clean jar or 1,000-mL beaker. Use a net to transfer a goldfish to the jar. Allow the fish to become accustomed to its surroundings. Do not tap on the jar. Watch the fish for a few minutes. Record your observations.

4. Place the cardboard box over the jar so that the flap on the bottom of the box is directly above the jar. With the room lights down or off, shine the flashlight through the hole above the jar while watching the fish through the other hole. Record your observations. *Why is the box necessary?*

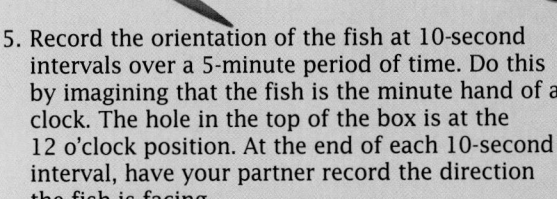

5. Record the orientation of the fish at 10-second intervals over a 5-minute period of time. Do this by imagining that the fish is the minute hand of a clock. The hole in the top of the box is at the 12 o'clock position. At the end of each 10-second interval, have your partner record the direction the fish is facing.

6. Repeat step 5 with the light shining through the hole in the side of the box while you watch through the hole in the top. Record your observations.

7. Tape a piece of red cellophane over the flashlight glass. Then, repeat steps 5 and 6.

8. Return the fish to the aquarium.

9. Combine your data with the other team in your group.

10. Make bar graphs of the combined data. Have each bar represent the number of times the fish was observed in each orientation around the imaginary face of the clock. Make a separate graph for each lighting situation.

11. Clean up your materials and wash your hands before leaving the lab.

Analysis

1. *Summarizing Data*
   Describe the response of the goldfish to white light.

2. *Summarizing Data*
   Describe the response of the goldfish to red light.

3. *Making Predictions*
   Based on your observations, how might the fish respond to white light coming from below the beaker?

Thinking Critically

What is the adaptive advantage of the goldfish's response to white light?

Analysis Answers

1. Answers should indicate that the fish will initially tilt toward the light, but will usually reorient itself so that it is parallel to the bottom of the beaker.

2. Answers should indicate no apparent response to red light.

3. Answers may vary but might predict that the fish might tilt farther than usual but would not orient itself upside down.

Thinking Critically Answer

The direction of light is one cue by which the goldfish orients itself. The cue can be overridden by the fish's sense of direction and gravity.

Using the Feature

Most biologists agree that humans are causing the sixth mass extinction in Earth's history. The number of species disappearing each year is unknown, as is the total number of species on Earth. Edward O. Wilson, Curator in Entomology at Harvard's Museum of Comparative Zoology, totals the number of named species at slightly more than 1.4 million. According to Wilson, adding the numerous undiscovered species would raise the total to somewhere between 10 million and 100 million. If Wilson's numbers are correct, then the majority of species are still uncollected and undescribed. Most of these undescribed species are small and unspectacular: insects and other invertebrates, bacteria, fungi, and protists. However, large organisms—vertebrates included—are occasionally discovered. A new species of whale was named in 1991, for instance.

- Have students investigate different efforts to conserve species. Include efforts aimed at particular species—such as captive breeding, gene and seed banks, and projects of zoos and botanical gardens—and those directed at whole ecosytems—such as extractive reserves and preserves.
- Have students research the CITES (Convention on International Trade in Endangered Species) treaty and evaluate its effectiveness.

Endangered Species: How Far Do We Go to Save Them?

To what lengths should we go to save an endangered species? Should efforts to preserve the species take precedence over economic development? Or are the goals of species conservation and development compatible?

The Endangered Species Act, passed by Congress in 1973, is the cornerstone of efforts to preserve endangered species. This act requires the federal government to compile and maintain a list of endangered species (those in immediate danger of extinction) and threatened species (those declining in abundance and likely to be endangered soon). Once added to this list, a species is protected under federal law. It is illegal to kill, injure, capture, import, or export a listed species. Disrupting the habitat of a listed species is also prohibited. Furthermore, the government is forbidden from constructing, funding, or authorizing projects that would threaten listed species.

Under the protection of the Endangered Species Act, the American alligator, the bald eagle, the peregrine falcon, and the brown pelican recovered from near extinction. When efforts to save species have clashed directly with development, however, bitter controversies have resulted.

For instance, in 1973 a small (about 8 cm, or 3 in.), brownish fish was captured in the Little Tennessee River in eastern Tennessee. This fish belonged to an undescribed species; it was named the snail darter, after its primary food. In 1975 the snail darter was declared endangered. The snail darter's future looked bleak because its only known habitat was to be flooded after completion of the Tellico dam on the Little Tennessee River. A lawsuit was filed to stop construction of the dam. Tellico dam supporters argued that the economic gains from the dam, such as increased electric

Science, Technology, and Society ▪ Science, Technology,

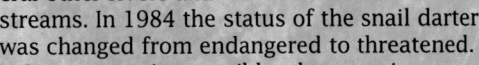

power generation and enhanced recreational opportunities, far exceeded the value of a small fish. Opponents countered that in addition to eliminating the snail darter, the dam would destroy thousands of acres of rich farmland and would flood important historical sites, such as an early Cherokee village. In 1978 the Supreme Court ruled against the dam, and construction was stopped.

Congress responded to this decision by amending the Endangered Species Act to create a committee that could grant exceptions to the act's provisions. This committee, nicknamed the "God Committee" because it held the power of life and death over species, also decided against the dam. After Congress voted to exempt the dam from all federal laws—including the Endangered Species Act—the project was completed. Economic considerations had won, and Congress had decided to sacrifice the snail darter.

Although the snail darter seemed to teeter on the verge of extinction, it survived all legal and bureaucratic battles over its preservation. Several hundred snail darters that had been transplanted to a nearby river thrived in their new habitat.

Additional natural populations of snail darters were discovered in several other rivers and streams. In 1984 the status of the snail darter was changed from endangered to threatened.

Is compromise possible when species preservation and development clash? In California's Coachella Valley, about 160 km (100 mi.) east of Los Angeles, such a compromise appears to have been reached. By the mid-1980s, much of the land in the valley had been converted from desert to golf courses, subdivisions, and hotels. Development was rapidly consuming the habitat of the Coachella Valley fringe-toed lizard, which lives only in this valley. This lizard, about 23 cm (9 in.) from nose to tip of tail, is adapted for life on fine, wind-blown sand. Instead of prohibiting development, bulldozing the last remnants of the lizard's habitat, or fighting a long and costly court battle, environmentalists and developers agreed to a compromise. Using $25 million in federal, state, and private funds, 5,300 hectares (13,000 acres) of land were purchased and set aside as a preserve. Development was permitted on the remaining land in the valley.

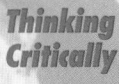

Thinking Critically

❶ If scientists find a plant with important medical uses, should they be allowed to remove the plant from its natural habitat? Give three reasons why or why not.

❷ If an underdeveloped country were to destroy a species in order to build up technologically, should the United States government interfere? Why or why not?

❸ During the Gulf War, Iraq intentionally spilled crude oil into the Persian Gulf. How should world powers deal with a country that maliciously damages the environment?

❹ You are the attorney for the environmentalists in the snail-darter case. Give three arguments in support of your case to stop construction of the dam.

❺ How do you think biology education has affected the public's reaction to endangered species?

Acting on the Issue

❶ Obtain a book from your local library on endangered species, and list those species that are endangered in your area.

❷ Find a copy of the Endangered Species Act and summarize it for your class.

❸ Write to your representatives in Congress to find out what legislation is pending that is related to the environment.

❹ Write to your representatives in Congress in support of legislation that you feel has a positive impact on the environment.

Discussion

Guide the discussion by posing the following questions.

1. Can a monetary value be assigned to a species?

   This question has no simple answer. Students should suggest that while it may be easy to appraise the products of a species (by the market value of its skin, flesh, fruits, or other products), it is difficult to fix a value on the ecological role a species plays, or on its aesthetic contribution.

2. Do humans, as the most intelligent species, have a moral obligation to be the steward of other species? Explain.

   Again, this question is not easily answered. Students should be able to defend their position.

Thinking Critically

1. Three reasons to remove the plant are its medical uses, its value for its discoverers, and its value for the country in which it grows. Three reasons to leave it are that it may have a crucial ecological function, that removing it may lead to its extinction, and that native peoples may depend on it.

2. Students should suggest that the effects of a species may transcend national boundaries. Thus, destruction of a species in another country may affect the United States and may require action from our government.

3. Students may suggest a variety of punishments for such acts.

4. The dam would eliminate the snail darter, flood thousands of acres of rich farmland, and obliterate important historical sites.

5. Hopefully, biology education has increased the public's appreciation of endangered species.

571

Chapter 26 Reptiles, Birds, and Mammals

Planning Guide

Objectives/Themes	Classwork Resources	Homework Resources
26.1 **1.** Identify three adaptations that make reptiles well suited to terrestrial life. **2.** Contrast ectothermy and endothermy. **3.** Describe one hypothesis that explains the disappearance of the dinosaurs. **4.** List the four orders of living reptiles. **Themes:** Energy and Life, Evolution, Interacting Systems, Scale and Structure, Stability	**Text** Journeys *Tour of a Lizard* p. 576 **Teacher's Resource Binder** Focus Activity 26 *Vertebrate Skeletons* Lab Investigation 26.1 *Field Identification of Flying Birds* **Other Resources** Transparencies 111–114	**Text** Section Review, p. 583 **Teacher's Resource Binder** Directed Reading Worksheet 26.1 **Other Resources** Audiocassette 26.1*
26.2 **1.** List two similarities between birds and reptiles. **2.** Identify two differences between birds and reptiles. **3.** Identify two functions of feathers. **4.** Describe two bird adaptations, other than feathers, for flight. **Themes:** Evolution, Scale and Structure, Stability	**Text** Journeys *Tour of a Bird* p. 587 **Other Resources** Transparencies 115–116	**Text** Section Review, p. 589 **Teacher's Resource Binder** Directed Reading Worksheet 26.2 **Other Resources** Audiocassette 26.2*
26.3 **1.** List the unique characteristics of mammals. **2.** Identify two features of monotremes. **3.** Contrast the manner of development of marsupials and placentals. **Themes:** Evolution, Patterns of Change, Scale and Structure	**Teacher's Resource Binder** Extension Worksheet *Senses and Adaptations of Birds* **Other Resources** Transparencies 117–118	**Text** Section Review, p. 595 **Teacher's Resource Binder** Directed Reading Worksheet 26.3 **Other Resources** Audiocassette 26.3*
26.4 **1.** Identify three functions of hair. **2.** List two mammalian structures that contain keratin. **3.** Contrast the teeth of herbivores and carnivores. **4.** Describe how bats navigate in the dark. **Themes:** Evolution, Scale and Structure, Stability	**Text** Journeys *Tour of a Mammal* p. 599 Investigation *How Do Down and Contour Feathers Differ?* pp. 604–605 Discoveries in Science *Deciphering the Fossil Record* pp. 606–607	**Text** Section Review, p. 600 **Teacher's Resource Binder** Directed Reading Worksheet 26.4 Vocabulary Review Worksheet* Reteaching Worksheet* *Bird Origins and Characteristics* **Other Resources** Audiocassette 26.4*

*Reteaching Options

Demonstrations	Assessment
26.1: pp. 573, 575, 581	Chapter Review pp. 602–603
26.2: p. 585	Portfolio Assessment p. 571D
26.3: p. 591	Chapter Test—Teacher's Resource Binder
26.4: pp. 596, 597	Test Generator

Research Notes

Connections to Paleontology: Dinosaurs: Ectotherms or Endotherms?

For years, scientists believed that all dinosaurs were sluggish, ecto-thermic animals. Recent fossil finds and measurements using a new technique provide evidence that suggests otherwise. If some of the dinosaurs were endo-therms, it could change a lot of assumptions about how life evolved on the planet, and about why the dinosaurs died out.

Fossils have been found indicat-ing that some dinosaurs were living in climates too cold for ectotherms. Other fossils have skeletons that seem adapted for activities requiring a high rate of metabolism, such as running quickly for extended periods of time. Such activities would re-quire a faster metabolism than that of an ectotherm.

However, the best evidence for en-dothermic dinosaurs comes from an isotopic dating technique in-volving isotopes of oxygen. When a living thing incorporates oxygen into bone or shell, the ratio of oxygen-16 to oxygen-18 present depends on the temperature. Paleontologist Reese Barrick of the University of Southern California and oceanographer William Showers of North Carolina State University tried the technique out on living things and dinosaur fossils.

*Instead of trying to measure exact temperatures, they com-pared the temperature measure-ments in the extremities with those in the center of the body. Ectotherms have extremities that are colder than the core of their body, **but in endotherms, the** temperature remains constant throughout the body.*

When the researchers analyzed two young dinosaurs, a young Orodromeus and a member of the ceratops family, they found that they fit into the endothermic pattern of having only 3–4°C difference. Results were not as clear-cut with a large herbivore, Camarasaurus. Although no consistent variation was found from core to extremities, a few of the bones were more varied than those found in endotherms.

*Barrick hypothesizes that large dinosaurs may have been en-dotherms in their youth, but eventually lost temperature con-trol as they grew, **aged, and their** metabolism slowed. At any rate, only small or young dinosaurs seem to exhibit signs of being endotherms.*

Critics of Barrick and Showers's theories point out that the miner-als contained in fossils may not be the minerals contained in the living thing. The process of fossilization involves the slow replacement of organic material by minerals. As a further test, Barrick and Showers will test a fossil crocodile found in the same rocks as the dinosaur. The crocodile should show up as an ectotherm. If it does, that would indicate that the technique and its results are valid.

Investigation Notes

How Do Down and Contour Feathers Differ?
pages 604–605

Purpose: In this activity, observa-tion is emphasized rather than the terminology of feather structure. Students examine the central shaft and vane

macroscopically. Using the micro-scope, the students should detect the presence of barbs, barbules, and hooks. These observations form the basis for relating structure to function.

Prelab Preparation

Feathers collected in the field should be avoided because they can carry disease-causing organisms.

Answers will be found on pages 604–605.

Meeting Individual Needs

Objectives

1. Students will demonstrate appropriate use of core vocabulary for the chapter (Vocabulary File).

2. Students will distinguish characteristics of reptiles, birds, and mammals (Teaching Strategy).

Vocabulary File
(Developing Vocabulary/ Limited English Proficiency)

If you are not already using the Vocabulary File, refer to Chapter 1 for its preparation. See the Chapter Highlights on page 601 for a list of suggested words.

Teaching Strategy
Terrestrial Animals
for pages 573–599
(Developing Classification Skills/Visual Learners)
Materials: pictures of a lizard, a bird, and a mouse to hang in the "sky zoo" as mobiles; dice; index cards; and markers

Preparation: On one side of the unlined index cards print "REPTILE? BIRD? MAMMAL?" On the other side of the cards print characteristics of one type of animal, based on information from the text. For example, the opposite side of one "REPTILE? BIRD? MAMMAL?" card might read "dry, scaly skin," a characteristic of reptiles. About 20 such cards will be needed.

Procedure: Display the lizard for the sky zoo. Read aloud relevant sections of pages 573–583 with emphasis on adaptations that have made it possible for reptiles to live a terrestrial life.

Next, display the bird for the sky zoo. Again, read aloud relevant sections, using pages 584–589 for reference.

Repeat the same procedure for the mouse (mammal), using pages 590–600 as reference. Hang the reptile, bird, and mammal to complete the sky zoo.

(Once the unit on the animal kingdom is completed, the zoo may be taken down.)

On the overhead projector, list adaptations important for each of these three classes of animals. Have students copy the information for use studying.

Divide the class into 4–5 teams. Roll dice to determine the order of play. Stack the cards, with the side describing characteristics down. The first team is given the first card. After a minute's discussion, they should identify whether the characteristics are those of reptiles, birds, or mammals. For each correct answer, the team gains one point. For each incorrect answer, the team loses one point, and the next team has a chance to answer the question. Award some prize to the team with the most points.

Additional Strategies
Visual Strategies

Pages 574, 575, 580, 584, 588, 590, 591, 592, 593, 594, 596, 598, and 600

Auditory Learners

Use *Biology: Visualizing Life* Audiocassettes for Sections 26.1, 26.2, 26.3, and 26.4.

Cooperative Learning
Observing Bird Behavior

Timing: Use this activity to conclude Section 26.2.

Group Size: 2–4 students

Outcome: Students will hypothesize about how bird behavior enhances the bird's ability to survive.

Individual Accountability: Each group member is responsible for observing and recording bird behavior.

Positive Interdependence: Each group will compare observations and consult field guides to attempt to link behavior with adaptations for survival.

Have each group go to a nearby location to observe birds. Good locations include parks, vacant lots, cemeteries, beaches, fields, ponds, and similar sites. Provide each group with a field guide to birds of your area and suggest that each student bring a notebook and pencil.

Students should record observations of bird behavior, such as interaction with members of its own species, behavior directed at other animals and other bird species, feeding behavior, singing, and other solitary behavior. Be sure students realize that the field notes can include written notes, sketches, and shorthand methods for recording sounds.

After observation, the group should compare behavioral notes and identify the birds using the field guide. For each behavior observed, the group should develop a hypothesis explaining it. Emphasize that students should concentrate on making plausible hypotheses that fit the evidence, not on merely guessing what they think the "right" answer is.

Portfolio Assessment

Students should select their best work and provide a self-reflective rationale for their selections. Students can make selections in the following areas.

1. *Content* — One concept map from the chapter (See page 812 for evaluation criteria.)

2. *Reading Comprehension* — One Directed Reading Worksheet from the Teacher's Resource Binder (Use the answer key to evaluate for accuracy.)

3. *Writing* — Using the Vee Form, summarize a magazine or newspaper article relating to reptiles, birds, or mammals. (See page 22T for evaluation criteria.)

   Or: Select a writing task or project from the Chapter Review.

4. *Performance Assessment* — One Vee form from a chapter investigation or lab manual investigation (See page 22T for evaluation criteria.)

Teacher makes selections in the following areas.

1. *Formal Assessment* — Chapter test (Test A, B, or the Test Generator) The teacher-scored test should be reviewed by the student. Incorrect responses should be corrected by the student before the test becomes part of the portfolio.

2. *Informal Assessment* — Use the Direct Observations Checklist, page 33T, during a laboratory or other cooperative learning experience.

3. *Performance Assessment* — Have students write a newspaper editorial about animal control and pet issues in their neighborhood or assist a local animal shelter with a public education campaign.

Concept Map Answer

The following is one possible answer to the Relating Concepts exercise on page 602.

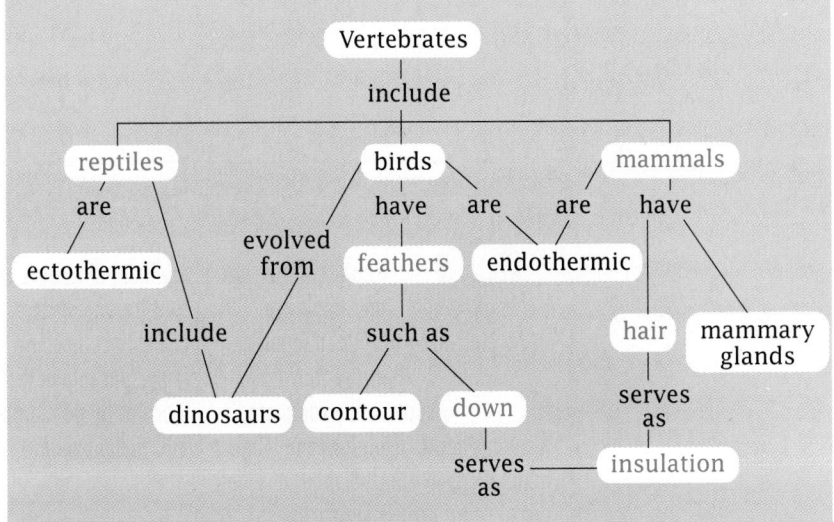

Chapter 26

Determining Prior Knowledge

- Randomly list characteristics of reptiles, birds, and mammals on the chalkboard or overhead projector. Have students identify the respective groups. Display photographs or preserved specimens of various reptiles, birds, and mammals. Include dried bird legs, skeletons, fur, horns, antlers, claws, talons, etc. Have students identify each specimen displayed. Ask if reptiles are the only animals with scales. Have them describe the skin found on the legs of birds.
- Have students identify which of the three types of animals lays eggs. Be prepared to discuss birds, egg-laying snakes, and egg-laying mammals.
- Have students discuss whether all birds can fly. Have them consider penguins (which seem to fly underwater but not in air). Also consider flightless cormorants, ostriches, rheas, and emus.
- Discuss with students the similarities between birds and dogs. For example, they both pant to regulate their temperature, since they do not sweat.
- Have students explain what it means when an animal is said to be ectothermic or endothermic.

Chapter 26

Review

- placenta
  (Section 22.3)

- amniotic egg
  (Section 22.3)

- the terms *ectotherm* and *endotherm* (Glossary)

Reptiles, Birds, and Mammals

26.1 Reptiles
- **Reptilian Adaptations to Terrestrial Life**
- **The Age of Reptiles**
- **The Survivors**

26.2 Birds
- **Birds Evolved From Reptiles**
- **How Birds Fly**
- **Major Orders of Birds**

26.3 Introduction to Mammals
- **Evolution of Mammals**
- **Mammalian Characteristics**
- **Egg-Laying Mammals: The Monotremes**
- **Pouched Mammals: The Marsupials**
- **True Placental Mammals**

26.4 Mammalian Adaptations
- **Hair Has Many Functions**
- **Claws, Hooves, Horns, and Antlers**
- **Food and Feeding**
- **Flying Mammals**

It looks as if this hawk plans to make a meal of this tortoise, but he is really looking for a place to perch. Birds and mammals evolved from reptiles, which evolved from amphibians.

■ Author's Rationale ■

Reptiles, birds, and mammals are grouped together because they have successfully completed the transition to living on land. Their adaptations to the challenges of living out of water are somewhat similar. In addition, dinosaurs are given significant treatment here, even though they are extinct, because they played an important role in the evolution of vertebrates by dominating the land for 200 million years.

REPTILES (CLASS REPTILIA) WERE THE FIRST FULLY TERRESTRIAL VERTEBRATES. THEIR FREEDOM FROM AQUATIC ENVIRONMENTS WAS MADE POSSIBLE BY SEVERAL ADAPTATIONS THAT MADE THEM AND THEIR EGGS ESSENTIALLY WATERTIGHT. MODERN REPTILES INCLUDE SNAKES, LIZARDS, TURTLES, CROCODILES, AND ALLIGATORS. THEY REPRESENT THE LIVING MEMBERS OF A GROUP THAT ONCE DOMINATED THE LAND.

26.1 Reptiles

Objectives

❶ Identify three adaptations that make reptiles well suited to terrestrial life.

❷ Contrast ectothermy and endothermy.

❸ Describe one hypothesis that explains the disappearance of the dinosaurs.

❹ List the four orders of living reptiles.

Reptilian Adaptations to Terrestrial Life

Hold a snake or lizard and you will discover that the skin of a reptile, contrary to what many people think, is not wet and slimy. Instead, reptilian skin is dry and covered with tough, hard, platelike scales, as shown in **Figure 26.1** . This dry skin forms a barrier to water loss in land environments. Reptilian skin is resistant to water loss because it contains large amounts of lipids and the protein keratin. Keratin is the tough, wear-resistant material that composes your hair and fingernails. In some reptiles, thick bony plates develop beneath the scales, such as those that form the shells of turtles.

Figure 26.1
This Phillippine sail-finned lizard has dry, scaly skin that protects its body from drying out and from being cut or scratched while moving over the ground.

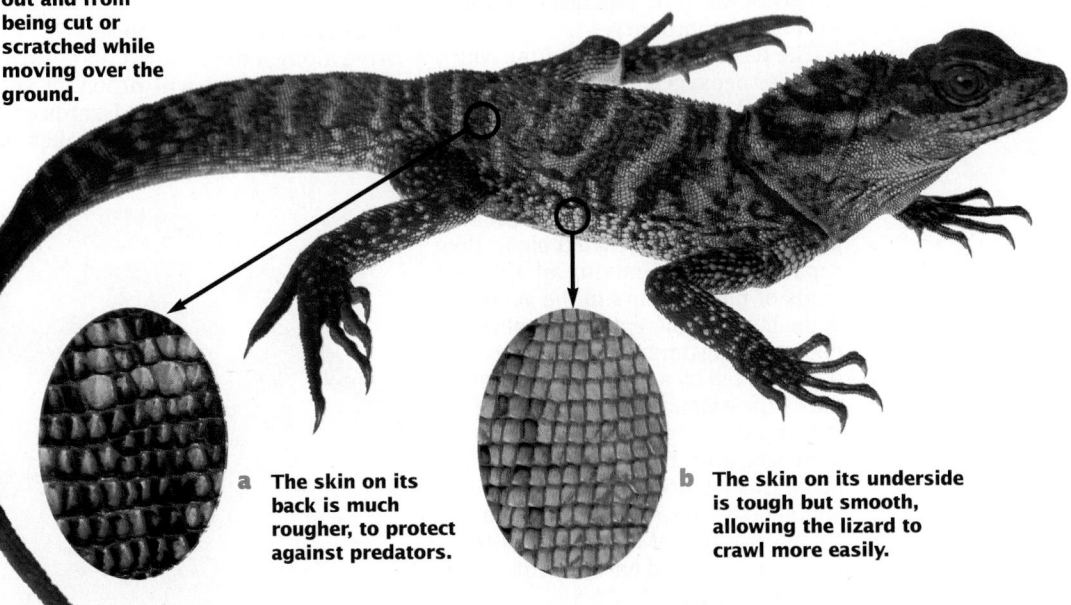

a The skin on its back is much rougher, to protect against predators.

b The skin on its underside is tough but smooth, allowing the lizard to crawl more easily.

Lesson Plan 26.1

Phase 1
PREPARATION

Key Concepts
- Reptiles evolved before birds and mammals.
- Reptiles have dry skin and scales that prevent dehydration. They are ectothermic creatures whose amniotic eggs are watertight. Also, they conserve water by excreting a pasty waste.
- Adaptations in early reptiles led to the evolution of all higher vertebrate forms.
- Dinosaur extinction may have been related to climatic change.
- Four orders of reptiles exist today: Squamata, Testudines, Crocodylia, and Rhynchocephalia. Some species are endangered.

Reading Strategy

Have students read for the main ideas of this lesson and identify any sentences that provide specific information about this section's objectives.

Phase 2
TEACHING STRATEGIES

Demonstration 1
Observing a Lizard
Have students examine the dry, scaly skin on the dorsal surface and sides of an anole lizard. Also, have them note the smooth underbelly.

Visual Strategy

Figure 26.2

Use an overhead projector to show transparencies of the human heart. Point out differences between a human heart and the other hearts in Figure 26.2. Point out the name and location of each chamber. Pay particular attention to the movement of oxygen-rich and oxygen-poor blood through each heart. Have students compare the heart's evolution among the different reptile groups. Explain that unlike the human heart, the frog heart has a single ventricle, as can be seen in Figure 26.2a. Tell students that most reptiles have begun to develop a separation of the ventricles as shown in Figure 26.2b. The hearts of crocodiles, birds and mammals ensure that oxygen-rich and oxygen-poor blood do not mix. In humans, the division of the ventricles is complete. Have students explain why the division of the ventricles is said to be complete in humans.

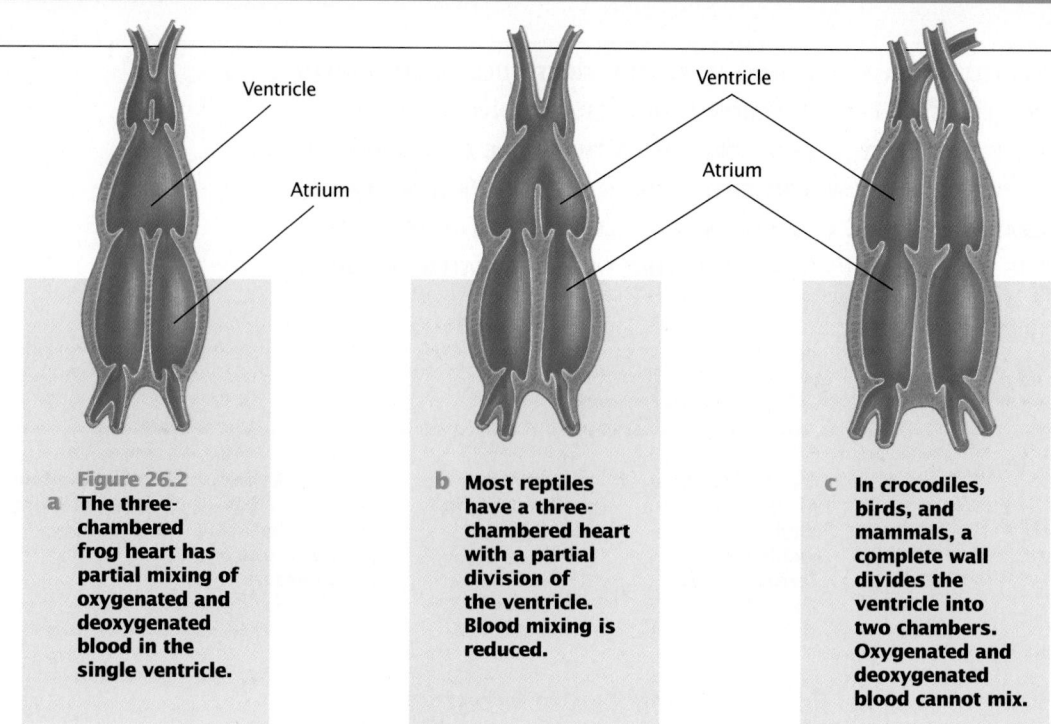

Figure 26.2

a The three-chambered frog heart has partial mixing of oxygenated and deoxygenated blood in the single ventricle.

b Most reptiles have a three-chambered heart with a partial division of the ventricle. Blood mixing is reduced.

c In crocodiles, birds, and mammals, a complete wall divides the ventricle into two chambers. Oxygenated and deoxygenated blood cannot mix.

An important adaptation of reptiles to life on dry land is the amniotic egg, which is resistant to water loss. As you learned in Chapter 22, the reptilian egg protects the vulnerable embryo and provides for its needs as it develops. Since the amniotic egg contains its own supply of water, reptiles need not travel to water to reproduce.

The fish heart has one atrium, which receives deoxygenated blood from the body, and one ventricle, which pumps blood to the gills. In the amphibian heart, such as the frog heart shown in Figure 26.2a, a wall separates the atrium into two chambers, so that oxygenated and deoxygenated blood flow separately. Some mixing of the two kinds of blood occurs in the ventricle. In the hearts of all reptiles except crocodiles and alligators, the ventricle is partially divided by a wall of tissue that reduces the amount of blood mixing, as shown in Figure 26.2b. In crocodiles, alligators, birds, and mammals, this partition is complete, as you can see in Figure 26.2c. The circulatory system of reptiles helps them meet the increased energy demands of an active terrestrial lifestyle.

You read about another reptilian adaptation for water conservation in Chapter 22. Reptiles excrete nitrogenous waste as uric acid, a form that requires very little water for dilution. Reptile urine contains so little water that it is paste rather than liquid.

Are reptiles "cold-blooded"?

Drive along a desert road on a warm morning and you will probably see lizards basking in the sun. Lizards and other reptiles, such as the snakes in

Figure 26.3
These red-sided garter snakes are emerging from their den, where they have spent the winter. By clustering together, they are able to conserve heat.

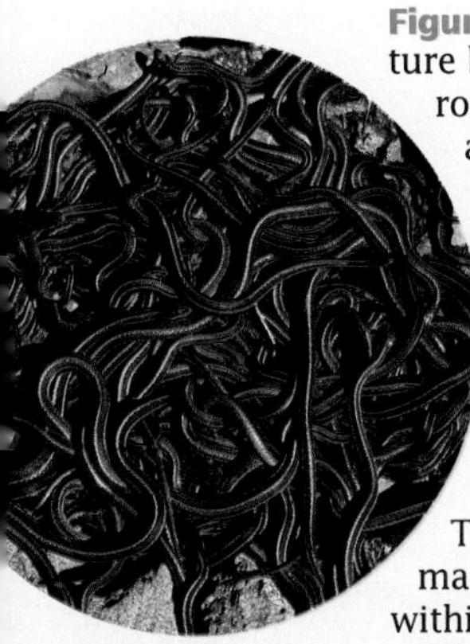

Figure 26.3, raise their body temperature by absorbing heat from their surroundings. Reptiles, like fishes and amphibians, are ectotherms. Ectotherms cannot regulate their body temperature through metabolism. Consequently, their body temperature changes with the temperature of their surroundings. Birds and mammals, in contrast, produce their body heat internally through metabolism. They are endotherms. Endotherms maintain their body temperature within narrow limits.

Throughout the day, the body temperature of an ectotherm often follows the temperature of its surroundings. Body temperature falls at night, when the air is cool. In the morning, many reptiles seek sunny places, letting the sun's rays warm their bodies. Once warmed, many reptiles maintain a relatively constant body temperature by moving into and out of the sunshine, their bodies warming and cooling, as described in Figure 26.4. You might hear ectotherms called "cold-blooded," but this description is inaccurate. On warm days, some desert lizards have body temperatures higher than yours.

Endotherms such as birds and mammals can sustain activity for longer periods of time than can reptiles and amphibians. Birds and mammals can also be active on colder days and live in colder climates than can reptiles and amphibians. There is a high cost to endothermy, however. It requires large amounts of food. A mouse must eat about 10 times as much food as a similarly sized lizard.

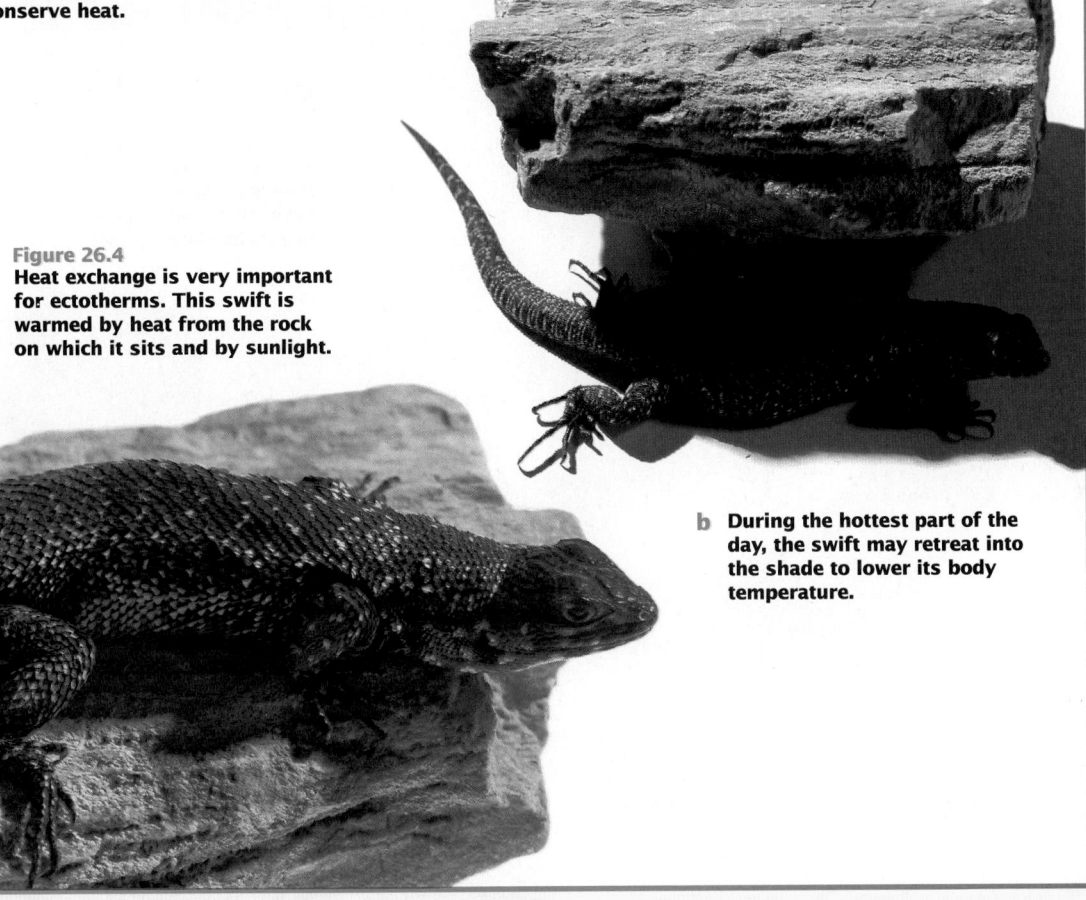

Figure 26.4
a **Heat exchange is very important for ectotherms. This swift is warmed by heat from the rock on which it sits and by sunlight.**

b **During the hottest part of the day, the swift may retreat into the shade to lower its body temperature.**

Visual Strategy

Figure 26.3
Have students discuss other animals that group themselves together for one reason or another. Remind students that grouping behaviors evolved to solve certain problems. Ask students to describe how forming groups can be advantageous to other animals.

Demonstration 2

Temperature Regulation
Set up a lamp 1 to 1.5 m (3 to 4 ft.) away from a piece of white paper and a piece of black paper. Expose papers to the light source for approximately 5 minutes. Have students touch the papers and compare the temperature. Or at 5-minute intervals, use a thermometer strip on each paper to compare their temperatures. Also, set up a shaded area and record the temperature in the shade. Discuss with students how the color of a reptile is related to its body temperature regulation.

Using the Feature

- Lizards slightly exceed snakes in number of species (3,800 to 3,000). Lizards are found on all continents except Antarctica, but are most abundant in deserts and the tropics. The lizard shown here is a day gecko from Madagascar. Unlike this species, most geckos are nocturnal. Students who have visited the South or California may be familiar with Mediterranean geckos, which were unintentionally introduced to North America from southern Europe. These lizards are frequently seen on walls and near lights, where they hunt for insects.
- Have students research Madagascar's unusual flora and fauna, which evolved during this island's long isolation. Have them concentrate on the reptiles of the island.

Discussion

Guide the discussion by posing the following questions.

1. The glass snake is a legless European lizard. How can the glass snake be distinguished from a true snake?
   A snake has a "floating" jaw with five hinges.

2. Lizards are most closely related to which group of living reptiles?
   snakes

3. Imagine that you go away on vacation and accidentally leave your pet lizard and mouse behind. Which pet could live the longest without food? Explain.
   The ectothermic lizard needs about one-tenth the food of the mouse and so would live longer.

Tour of a Lizard

The largest group of reptiles today is the lizards. This gecko is a very fast runner. Most living lizards are small; few are bigger than a squirrel.

All reptiles have a tough, dry, scaly skin. The outer layer of scales is shed periodically.

Most lizards have external ears; snakes do not.

Most lizards have four legs, but some are legless. These lizards are not snakes because they lack the "floating jaws" that allow snakes to swallow large prey.

A gecko has adhesive toe pads that enable it to walk upside down and on vertical surfaces.

Many lizards drop their tails when seized by a predator; the tail regrows in about seven weeks.

Modern lizards evolved late in the age of dinosaurs.

Most lizards lay leathery, shelled eggs that do not dry out when exposed to air.

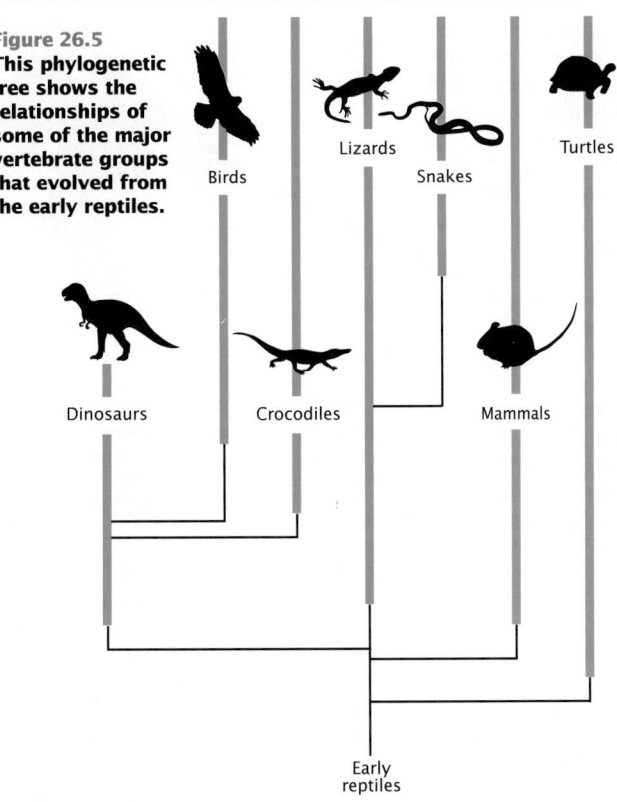

Figure 26.5
This phylogenetic tree shows the relationships of some of the major vertebrate groups that evolved from the early reptiles.

Birds

Lizards

Snakes

Turtles

Dinosaurs

Crocodiles

Mammals

Early reptiles

Figure 26.6
The 2-m (6-ft. 6-in.) *Ichthyosaurus* gave birth to live young, like modern day whales. The 2.3-m (7-ft. 6-in.) body of the *Plesiosaurus* was adapted for maneuverability to catch fish. Both existed during the Jurassic period.

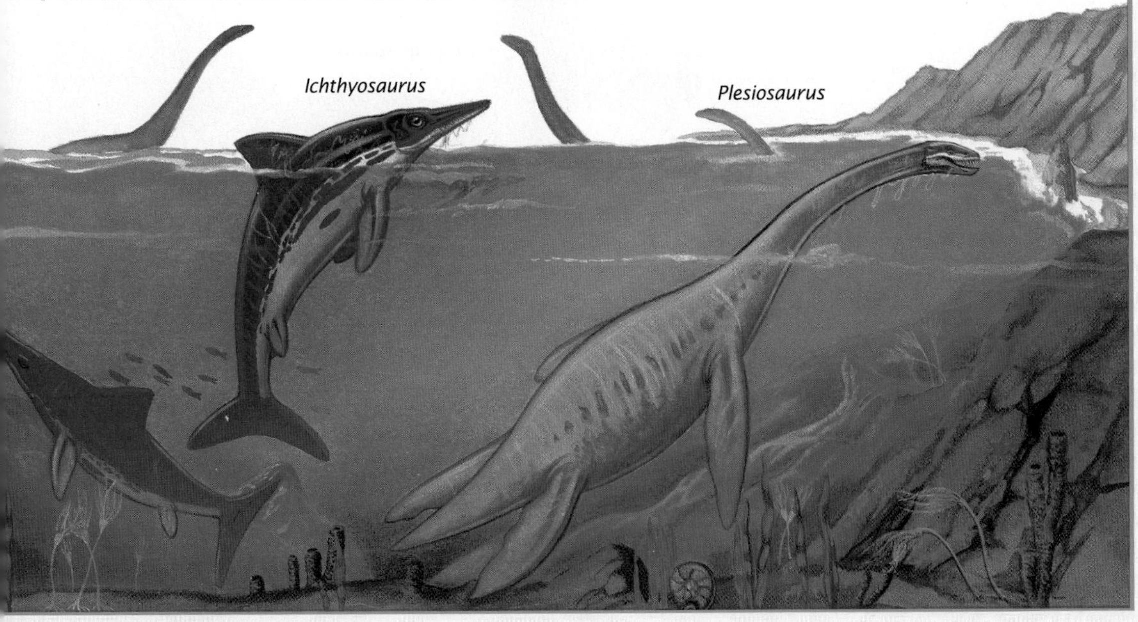

The Age of Reptiles

The first reptiles evolved about 320 million years ago, when the world was entering a dry period. Well suited to dry conditions, reptiles diversified rapidly after their initial appearance, giving rise not only to the ancestors of modern reptiles, but also to the wide variety of dinosaurs and other reptiles that were the dominant animals on Earth for over 170 million years. The period of reptile dominance, which lasted from 250 million to 65 million years ago, is called the Age of Reptiles. The evolutionary relationships of the reptiles and their descendants are shown in **Figure 26.5**.

Two groups of reptiles inhabited the oceans while the dinosaurs lived on land. Ichthyosaurs (*IHK thee oh sawrs*) were fully adapted for an aquatic existence. Ichthyosaurs ("fish lizards" in Greek) resembled dolphins, having a pointed snout, streamlined body, fins, and a flattened tail. The long-necked plesiosaurs (*PLEE see oh sawrs*) had barrel-shaped bodies with paddlelike fins. Plesiosaurs and ichthyosaurs probably fed on fish. These creatures are shown in **Figure 26.6**.

Dinosaurs, meaning "terrible lizards" in Greek, evolved from reptiles that were only 0.5 m to 1 m (2 ft. to 3 ft.) in

Ichthyosaurus

Plesiosaurus

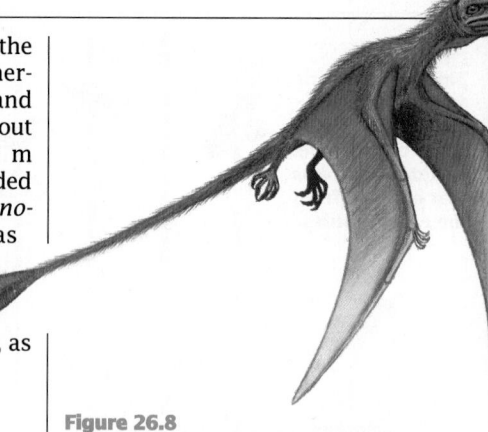

Dinosaur Diversity

The amount of diversity among different phylogenetic groups is related to their habitat and niche. Dinosaurs were no different in that respect. Have students speculate why the dinosaurs were such a diverse group.

length. Among the dinosaurs were the largest land animals, such as the herbivores *Apatosaurus*, *Diplodocus*, and *Brachiosaurus*. *Brachiosaurus* was about 23 m (75 ft.) long and about 17 m (56 ft.) tall. The dinosaurs also included the largest land carnivore, *Tyrannosaurus rex*. This great predator was over 5 m (16 ft.) tall when standing on its hind legs. Dinosaurs, known for their large size, were varied in size and habits, as you can see in **Figure 26.7**.

Pterosaurs were the first vertebrates to fly

No reptiles can fly today, but one group of flying reptiles existed alongside the dinosaurs: the pterosaurs (*TEHR oh sawrs*), such as *Rhamphorhynchus* shown in **Figure 26.8**. These reptiles were flying 75 million years before the first birds. During the 160 million years they existed, pterosaurs were a diverse

Figure 26.8

Pterosaurs such as *Rhamphorhynchus* were the first group of vertebrates to evolve the ability to fly. They evolved 75 million years before the first bird.

group. Some were as small as a sparrow. Others had wingspans of 11 m (35 ft.), greater than the wingspan of many small airplanes.

How did pterosaurs fly? Did they only glide? Or were they capable of flapping their wings? By comparing pterosaur fossils to birds and bats, scientists have been able to draw some

Figure 26.7

The popular conception of dinosaurs is of huge lumbering animals like *Stegosaurus*, which was 6 m (20 ft.) long and weighed up to 1,500 kg (2 tons). But many dinosaurs, such as *Struthiomimus*, were small and fast. Although the species shown here did not all exist at the same time, this illustration shows some of the great diversity among dinosaurs.

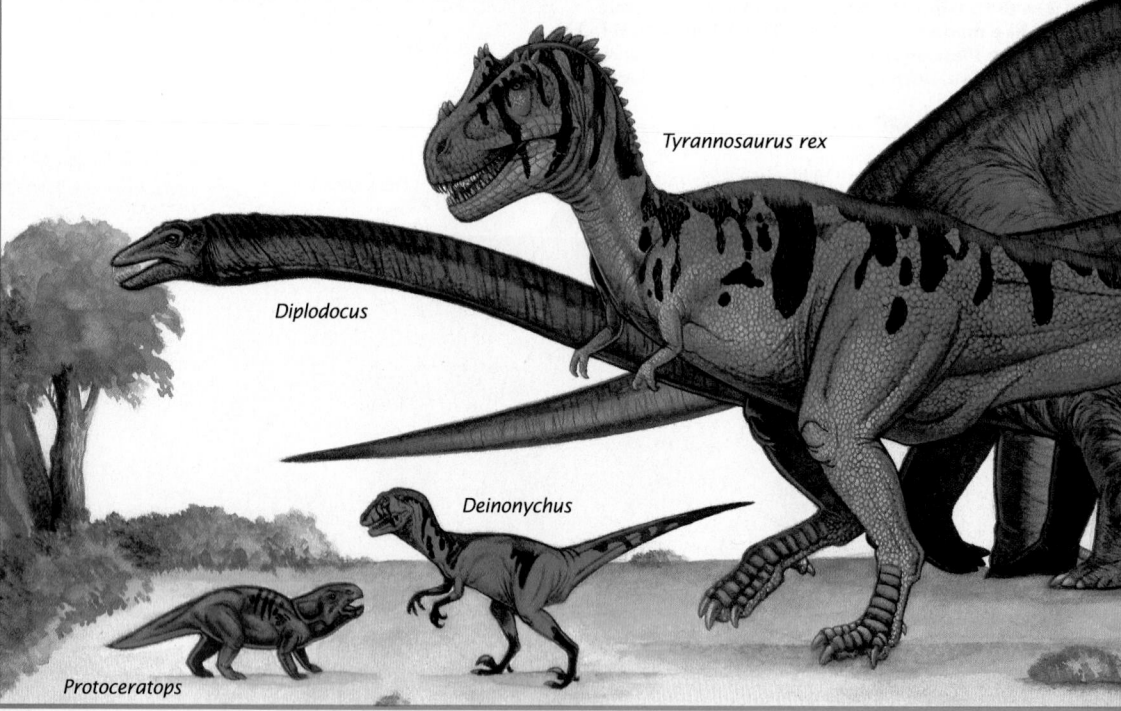

Tyrannosaurus rex

Diplodocus

Deinonychus

Protoceratops

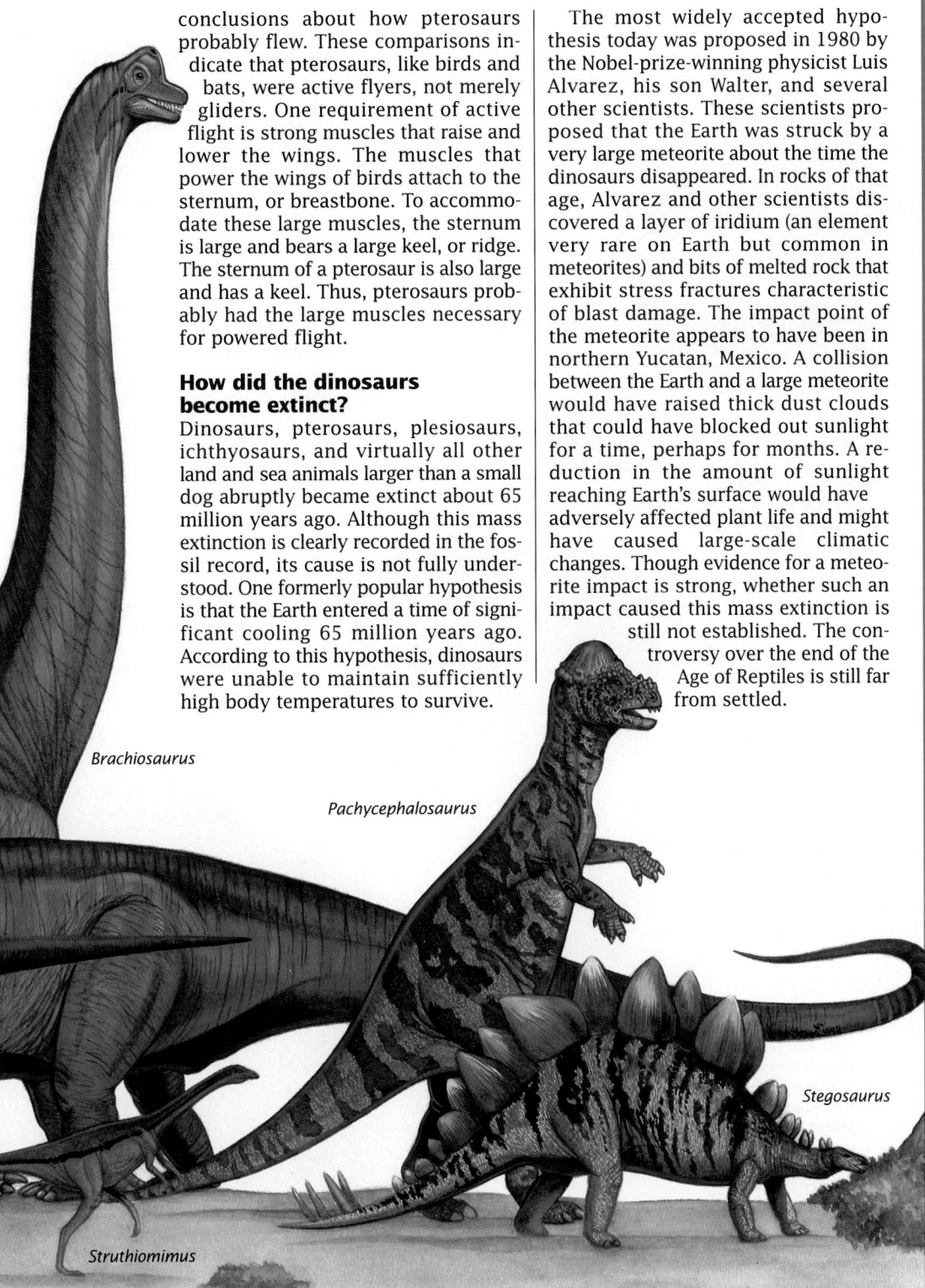

conclusions about how pterosaurs probably flew. These comparisons indicate that pterosaurs, like birds and bats, were active flyers, not merely gliders. One requirement of active flight is strong muscles that raise and lower the wings. The muscles that power the wings of birds attach to the sternum, or breastbone. To accommodate these large muscles, the sternum is large and bears a large keel, or ridge. The sternum of a pterosaur is also large and has a keel. Thus, pterosaurs probably had the large muscles necessary for powered flight.

How did the dinosaurs become extinct?

Dinosaurs, pterosaurs, plesiosaurs, ichthyosaurs, and virtually all other land and sea animals larger than a small dog abruptly became extinct about 65 million years ago. Although this mass extinction is clearly recorded in the fossil record, its cause is not fully understood. One formerly popular hypothesis is that the Earth entered a time of significant cooling 65 million years ago. According to this hypothesis, dinosaurs were unable to maintain sufficiently high body temperatures to survive.

The most widely accepted hypothesis today was proposed in 1980 by the Nobel-prize-winning physicist Luis Alvarez, his son Walter, and several other scientists. These scientists proposed that the Earth was struck by a very large meteorite about the time the dinosaurs disappeared. In rocks of that age, Alvarez and other scientists discovered a layer of iridium (an element very rare on Earth but common in meteorites) and bits of melted rock that exhibit stress fractures characteristic of blast damage. The impact point of the meteorite appears to have been in northern Yucatan, Mexico. A collision between the Earth and a large meteorite would have raised thick dust clouds that could have blocked out sunlight for a time, perhaps for months. A reduction in the amount of sunlight reaching Earth's surface would have adversely affected plant life and might have caused large-scale climatic changes. Though evidence for a meteorite impact is strong, whether such an impact caused this mass extinction is still not established. The controversy over the end of the Age of Reptiles is still far from settled.

Brachiosaurus

Pachycephalosaurus

Stegosaurus

Struthiomimus

Visual Strategy

Figures 26.9–12
Have students compare the four orders of reptiles and give examples of these orders. Have them infer reasons why modern reptiles are divided into these four orders.

The Survivors

The mass extinction 65 million years ago spared four groups of reptiles: lizards and snakes (order Squamata), turtles and tortoises (order Testudines), crocodiles and alligators (order Crocodylia), and the tuatara (order Rhynchocephalia).

Of these four groups, alligators, such as the one in **Figure 26.9**, and crocodiles are the most closely related to dinosaurs. Twenty-two species of crocodiles and alligators live in tropical and subtropical regions of the world. Two species, the American alligator and the American crocodile, occur in the United States. Crocodiles and alligators lead a largely aquatic life, feeding on aquatic animals such as fishes and turtles and on terrestrial animals that come to drink or feed in the water, occasionally including humans. Crocodiles and alligators have been hunted intensely for their hides, which are used to make handbags, shoes, and other leather products. Because of overhunting, three species of crocodiles are now endangered.

All turtles, such as the box turtle shown in **Figure 26.10a**, have a shell. A turtle's shell is composed of bony plates that are fused together. The vertebrae and ribs are fused to the interior of the shell, as shown in **Figure 26.10b**. Turtles lack teeth but have a sharp beak. Most turtles spend some of their

Figure 26.9
The American alligator and other crocodilians differ from other reptiles in having a four-chambered heart. All crocodilians are carnivores and all lay eggs.

b **Part of the skeleton of a turtle is the hard shell. The backbone runs along the underside of the shell.**

Vertebrae

Figure 26.10
a **This Chinese box turtle is just one of 250 species of turtles and tortoises. Although some turtles live in water, they must deposit their eggs on dry land.**

Figure 26.11
The tuatara is the sole survivor of the order Rhynchocephalia. The name tuatara means "spiny crest."

time in water. An extreme case is the sea turtles, which only leave the ocean to lay their eggs. The limbs of sea turtles are flattened and paddlelike for steering and propulsion in the ocean. Tortoises, on the other hand, are almost completely terrestrial and have elephant-like limbs for walking.

The tuatara, like the one shown in Figure 26.11, is the only surviving member of the order Rhynchocephalia. This lizardlike reptile closely resembles its relatives that were common 150 million years ago. Tuataras live only on a few islands near New Zealand.

Snakes and lizards belong to the largest order of reptiles, order Squamata. There are about 6,800 species of snakes and lizards, and they live on every continent except Antarctica. All snakes lack limbs, movable eyelids, and external ears. Snakes are carnivores,

feeding on animals much larger than themselves. A snake's jaw has five joints (your jaw has only one), which give the jaw great flexibility. This flexibility enables the snake to engulf prey several times its own diameter, as shown in Figure 26.12. Most species of snakes are nonpoisonous, but a few species can deliver a poisonous bite. Rattlesnakes, coral snakes, copperheads, and cobras are examples of poisonous snakes.

Figure 26.12
The structure of the snake's jaw and the flexible nature of its skin and skeleton allow the snake to eat animals larger in diameter than its own body. It may take this snake an hour or more to swallow its prey.

a **After suffocating its prey by constriction, . . .**

b **. . . the snake begins the long process of swallowing.**

Demonstration 3
Snake Feeding
Many students have pet snakes. Arrange for a student to bring in a hungry snake so the class can witness its feeding behavior. Many students may be afraid of snakes. Be aware of these feelings as you display the animal in the classroom.

Art Connection

Orders of Reptiles

Refer students to Table 26.1. Have them discuss what kinds of things should be included in a general display to educate other students about reptiles. Divide the tasks among students. Have them display their materials in a school display case.

Unlike snakes, most lizards have four limbs and external ears. Lizards are most abundant in the tropics and in deserts. Most species are carnivorous, feeding on insects, small mammals, or other lizards. A few species are herbivorous. Only two species of lizards are poisonous: the Gila monster, shown in **Figure 26.13**, of the southwestern United States and northern Mexico, and the Mexican beaded lizard of western Mexico. **Table 26.1** summarizes the major reptile orders.

Figure 26.13
The glands that produce the Gila monster's poison are located in the lower jaw. The lizards chew it into their victims.

Table 26.1 Orders of Reptiles

Order	Approximate Number of Living Species	Main Characteristics	Examples	
Ornithischia	Extinct	Mostly plant-eating dinosaurs with two pelvic bones facing backwards, as in a bird's pelvis; hole in the skull in front of eye socket; legs positioned beneath the body; Over 150 genera	*Triceratops, Stegosaurus, Iguanodon*	*Stegosaurus*
Saurischia	Extinct	Flesh-eating and plant-eating dinosaurs with one pelvic bone facing forward, the other backward, as in a lizard's pelvis; terrestrial with three or sometimes five toes; hole in skull in front of eye socket; legs positioned beneath body; over 200 genera	*Tyrannosaurus, Brontosaurus, Brachiosaurus*	*Tyrannosaurus rex*
Pterosauria	Extinct	Flying reptiles with wings of skin between fourth finger and body; wing span of early (Jurassic) *Rhamphorhynchus* was typically 60 cm (2 ft.), of later (Cretaceous) *Pteranodon* over 7.5 m (25 ft.)	*Pteranodon, Pterodactylus, Rhamphorynchus*	*Rhamphorynchus*
Plesiosauria	Extinct	Marine reptiles with very large paddle-shaped fins, a barrel shaped body, and long jaws with sharp teeth; some had a snakelike neck twice as long as the body; others had a short neck and elongated skull about 3.7 m (12 ft.) in length	*Plesiosaurus, Elasmosaurus, Kronosaurus*	*Plesiosaurus*

Ichthyosauria	Extinct	Marine reptiles with streamlined bodies up to 3 m (10 ft.) in length; the four legs modified into balancing fins; apparently fast swimmers, with many body similarities to modern fishes such as tuna or mackerel	*Ichthyosaurus*
Squamata suborder Sauria	3,800	Lizards; largely terrestrial with limbs set at right angles to body; dry skin of scales; teeth; anus is in transverse (sideways) slit	Anoles, geckos, horned lizards
Squamata suborder Serpentes	3,000	Snakes; largely terrestrial; no legs; scaly skin shed periodically; teeth	Rattlesnakes, garter snakes
Testudines	250	Body encased in shell of bony plates; sharp, horny jaw edges without teeth; vertebrae and ribs fused to shell	Turtles, tortoises, terrapins
Crocodylia	25	Four-chambered heart; extended jaw with socketed teeth; five digits on forelimbs, four digits on hind limbs; anus is a longitudinal (lengthwise) slit	Crocodiles, alligators
Rhynchocephalia	1	Sole survivor of a group that largely disappeared about 100 million years ago. Skull like those of early Permian reptiles; fused, wedgelike, socketless teeth; primitive eye under skin of forehead	Tuatara

Section Review

❶ Describe two adaptations to land shown by reptiles.

❷ What is one disadvantage of endothermy?

❸ What evidence indicates a meteorite collided with Earth about the time the dinosaurs became extinct?

❹ List a representative of each of the four living orders of reptiles.

Phase 3

ASSESSMENT OPTIONS

Closure Strategy

Reptiles

Reptiles have been portrayed in literature as evil. Have students discuss facts they have read that either support or disprove negative portrayals of reptiles.

Section Review

Assign the *Section Review*.

Reteaching

Divide students into cooperative groups of four. Have each group write the characteristics of one of the orders of living reptiles. List the characteristics on the chalkboard. Reintroduce any characteristics students may have overlooked. Focus on common characteristics.

▪ Section Review Answers ▪

1. Two adaptations that allowed reptiles to adapt to a terrestrial environment are the amniotic egg and dry, scaly skin containing keratin.

2. Students should indicate that endotherms require large quantities of food.

3. Students should discuss Luis Alvarez's discovery of iridium, a rare element not found on Earth but common in meteorites.

4. See Table 26.1 on pages 582–583.

Lesson Plan 26.2

Phase 1

PREPARATION

Key Concepts

- Birds evolved from reptiles but still retain many reptilian characteristics. Birds lay amniotic eggs and have scales on their legs and feet.
- Only birds have feathers.
- The adaptations that have enabled birds to fly include a lightweight skeleton, endothermy, and feathers.
- The feet and beaks of birds help to determine the order to which they belong.

Reading Strategy

Have students read for the main ideas of this lesson. Have them identify sentences that provide specific information about the objectives for this section.

Phase 2

TEACHING STRATEGIES

Visual Strategy

Figure 26.14
Bring in samples of any kind of fossil to make the photograph in the figure more imaginable. Have students relate what they see in the picture to a bird skeleton.

ALTHOUGH THE DINOSAURS DIED OUT 65 MILLION YEARS AGO AT THE END OF THE CRETACEOUS PERIOD, THEIR DESCENDANTS, THE BIRDS, SURVIVED. TODAY BIRDS ARE THE MOST DIVERSE GROUP OF LAND VERTEBRATES, WITH ABOUT 8,800 SPECIES. ALL BIRDS HAVE FEATHERS, AND ALMOST ALL BIRDS ARE CAPABLE OF POWERED, SUSTAINED FLIGHT.

26.2   Birds

Objectives

1 List two similarities between birds and reptiles.

2 Identify two differences between birds and reptiles.

3 Identify two functions of feathers.

4 Describe two bird adaptations, other than feathers, for flight.

Birds Evolved From Reptiles

Birds (class Aves) evolved about 150 million years ago. The oldest known bird fossils were found in fine limestone in Bavaria. These specimens were named *Archaeopteryx*, which means "ancient wing." *Archaeopteryx* shared many characteristics with the small dinosaurs from which it evolved, including teeth in sockets and a long, bony tail. But as you can easily see in **Figure 26.14**, *Archaeopteryx* had feathers, a characteristic unique to birds.

Modern birds lack teeth and have only a vestigial tail, but they still retain many reptilian characteristics. For instance, birds lay amniotic eggs, although the shell of bird eggs is hard rather than leathery. Scales are also present on the feet and lower legs of birds.

Feathers are unique to birds
Feathers have replaced scales as the body covering of birds. Feathers are flexible, strong structures that can be regrown and that make an excellent wing for flying. Feathers provide most of the surface area of a bird's wing. A bird can change the surface area of its wing and alter its flight patterns by spreading or collapsing its wing feathers and the wing itself.

Like wing feathers, a bird's tail feathers can be spread or collapsed, changing their effective surface area. Tail feathers can also be used for braking and steering during flight. Watch the action of the tail of a bird in flight and during landing to see some of the actions of these feathers.

Figure 26.14
Archaeopteryx lithographica was about the size of a pigeon. It had feathers.

▪ *Matter of Fact* ▪

Flight of the Birds!
The arctic tern makes a migration flight from pole to pole, a flight of almost 40,008 km (24,859 mi.). Some birds fly nonstop to their migratory lands, sometimes flying hundreds of miles. Other birds fly until they must stop to restock their fat reserves, and take off again in two to three days. Birds are incredible flying machines. It's obvious why their skeletons are so spongy—to reduce air freight!

As shown in **Figure 26.15**, birds have two major types of feathers. Most of a bird's feathers are **contour feathers**. These feathers cover the body of the bird and give the wings and tail their shape. Contour feathers also insulate against heat loss. Fine **down feathers** growing underneath or among the contour feathers are specialized for insulation. The down feathers of eider ducks are used in sleeping bags because they are a lightweight, effective insulation.

Bird skeletons are lightweight

The skeleton of a bird is adapted for flight. The bones are thin and hollow. Many are reinforced by internal struts, like the wings of an airplane. The sternum is large and has a keel, providing solid anchorage for some of the large flight muscles.

Bird wings are modified forelimbs. The bones of the forelimbs fully support and move the wings. The finger bones are very tiny, but the arm and hand bones are long, providing strength and enabling complex movements of the wings. As you learned in Chapter 9, the bones of a bird's wing are homologous to the bones of your arm and hand.

Birds are endothermic and active

Flight is an energy-demanding activity. Birds, like mammals, are endothermic. Endothermy enables birds to meet the energetic demands of flight. In addition, birds have a four-chambered heart with separate circulatory loops to the lungs and to the body. Therefore, oxygen-rich blood is rapidly delivered to tissues where it is needed, without mixing with deoxygenated blood. Bird respiration is very efficient because their system of air sacs permits air to flow in only one direction through the lungs, as explained in Chapter 22. You can read more about birds in the *Tour of a Bird* on page 587.

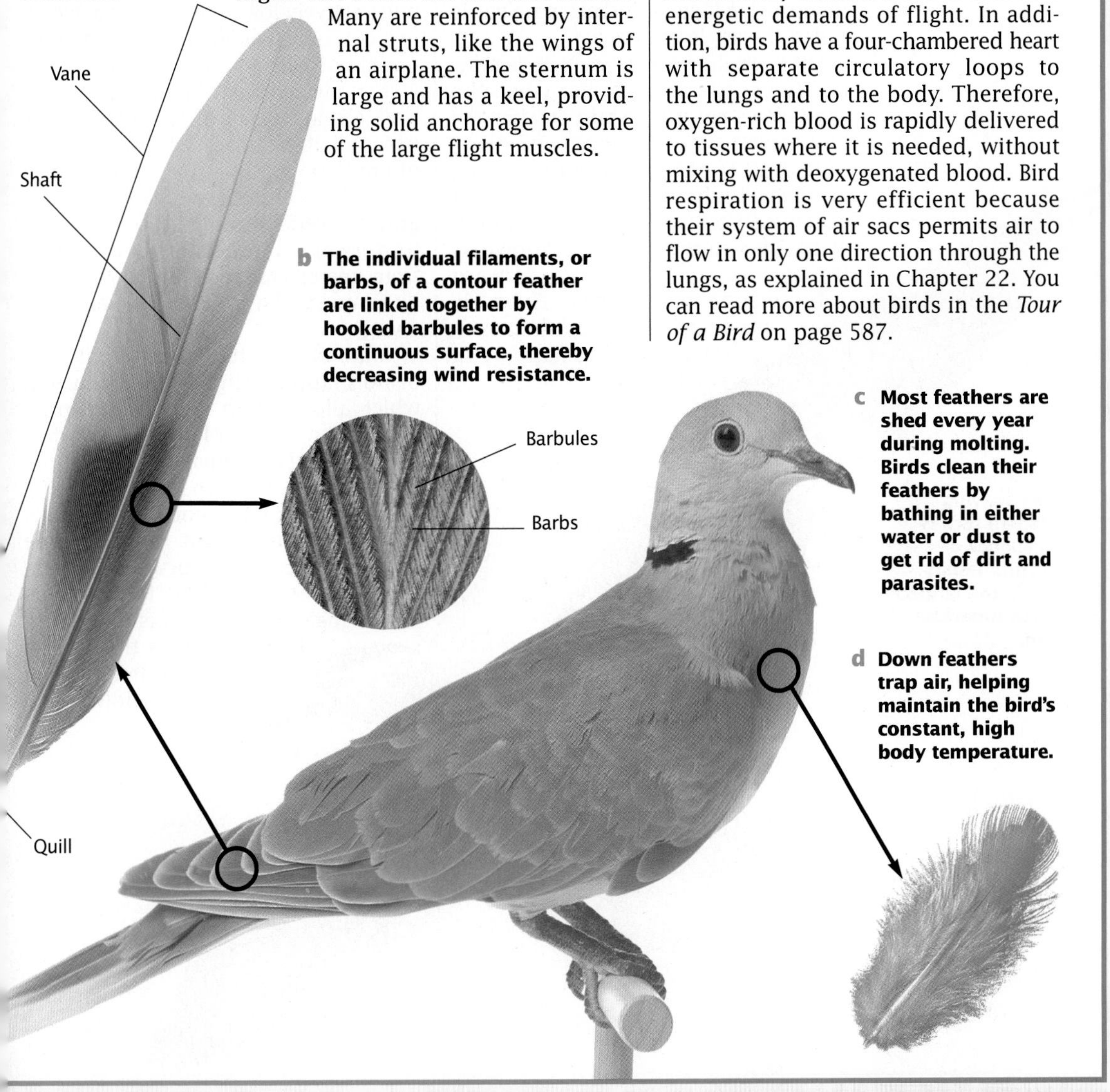

Figure 26.15
Contour feathers and down feathers have different functions. Down feathers serve primarily as insulation. Contour feathers provide insulation, steering, balance, and coloration.

Vane

Shaft

b The individual filaments, or barbs, of a contour feather are linked together by hooked barbules to form a continuous surface, thereby decreasing wind resistance.

Barbules

Barbs

Quill

c Most feathers are shed every year during molting. Birds clean their feathers by bathing in either water or dust to get rid of dirt and parasites.

d Down feathers trap air, helping maintain the bird's constant, high body temperature.

Demonstration 1

Feathers

Bring in enough feathers for your students to examine under the microscope and to handle in general. Have students use Figure 26.15b as a guide to distinguish the differences in the feathers' barbs. Have them discuss what would happen to the feather if the bird never groomed itself.

Demonstration 2

Comparing Bones

Bring in thoroughly cleaned leg or wing bones from a turkey or chicken. Guide the students in an examination of the bones. After they have carefully examined the exterior of the bones, have them break open the bones and observe the numerous air spaces inside. For contrast, show students a long bone from a cow or pig (available from a meat market). Discuss how the internal structure of a bird's bones helps it to fly. Discuss why pigs can't fly!

■ *Matter of Fact* ■

Feathers

Do all birds have the same number of feathers? No. Whistling swans often have more than 25,000 feathers, whereas ruby-throated hummingbirds have as few as 950 feathers.

Theme Connection

Interacting Systems

There is an interaction of inheritance and learning in bird migration. Birds have an innate directional preference and a nervous system "odometer" that tells them when they have reached their destination. This is evident in some species when the young birds leave earlier than their parents on the migratory trek. If adults and young are picked up along their migratory route and taken to a suitable area, the adults will fly back to the place in which they spent the last winter. The young, however, will continue to fly in the direction they were going when captured, for the amount of miles left on their "odometer."

How Birds Fly

Birds must overcome gravity in order to achieve and maintain flight. The wing is the feature that enables birds (and airplanes) to fly. An upward force known as lift is generated when air passes across the surface of a wing. Lift is produced when the pressure of the air passing over the top of the wing is less than that of the air passing under the wing. This pressure difference can be created in two ways. First, the wing can be arched, creating a pocket on the underside of the wing. Second, the front edge of the wing can be held higher than the back edge, increasing what is called the angle of attack.

Lift is produced only when air is moving across the wing, either when the wind blows toward the bird or when the bird travels forward. Since wind direction and speed are unpredictable, efficient flight demands that birds generate force to propel themselves forward. This forward force is known as thrust. Birds typically move forward by beating their wings. Thrust is produced largely by the downstroke during flapping of the wings.

Active flight requires strong muscles to raise and lower the wings. The muscles that power the wings of birds during flight attach to the sternum, or breastbone. The sternum is large and keeled, like the hull of a boat, to accommodate these large muscles.

There are different kinds of bird flight

Birds fly in a variety of ways. The most common method is called flapping flight. It is also the most energy-demanding form of flight. The bird must actively lower and raise the wings in a complex twisting pattern that produces both lift and forward thrust.

If you watch birds in the air, you will notice that not all of them flap their wings all of the time. The simplest form of nonflapping flight is gliding. A bird always loses altitude as it glides because gravity overcomes the lift provided by its forward descent. Birds dropping from treetops to the ground often do so by simple gliding; it requires very little energy.

Some birds exhibit a different kind of nonflapping flight called soaring, which allows them to overcome gravity and stay aloft for long periods while spending only minimal energy. Soaring birds, such as the tern in **Figure 26.16**, are generally large-bodied and have relatively small flight muscles. The large body results in forward momentum once the bird is flying, thus maintaining lift. It is this ability to maintain lift that makes soaring different from simple gliding.

Figure 26.16
This soaring Caspian tern is a member of the gull family and is the largest North American tern.

■ *Cultural Perspective* ■

Uses of Birds and Their Feathers

Through the ages, birds and their feathers have been a part of many cultures. The Indians of North and South America have used bird symbols in their rituals, and bird feathers, beaks, and talons in their costumes.

Costa Rica uses a bird on its currency. Birds have been important in mythology, as icons in certain religions, and as pets around the world.

■ *Journeys* ■ *Journeys* ■ *Journeys* ■ *Journeys* ■ *Journeys*

Tour of a Bird

Birds, such as this cedar waxwing, evolved from dinosaurs. The similarities between dinosaurs and birds prompted Thomas Huxley, a colleague of Darwin, to call birds "glorified reptiles."

Ranging from 40°C–42°C (104°F–108°F), bird body temperatures are higher than the lethal limit for mammals.

All special adaptations found in flying birds contribute to two ends: more power and less weight.

Birds are like airplanes in many ways. They have wings for lift, a tail for steering, and wing slots to avoid stalling at slow speeds.

A bird's neck is highly flexible, with elaborately interwoven and subdivided muscles.

All flying birds have a keeled sternum to which powerful flight muscles are attached.

Most birds molt at least once a year, a few feathers at a time. They survive winter cold by ruffling their feathers to provide insulation.

Feathers evolved from reptilian scales. A feather consists of a shaft and several hundred parallel barbs.

There are 8,800 species of living birds. All birds have forelimbs modified into wings, are covered with feathers, have horny beaks, and lay hard-shelled eggs.

The bones of birds are laced with air cavities, making them light and strong at the same time. The skeleton weighs less than the feathers.

Many birds migrate seasonally. The bobolink commutes 6,400 km (4,000 mi.) each year between North America's Great Lakes region and central South America.

Using the Feature

Birds are the descendents of dinosaurs. Although the traditional classification places birds and reptiles in different groups, a cladistic classification would place them in the same group.

Review avian adaptations for flight: feathers, wings, a lightweight skeleton, and a high metabolic rate.

Discussion

Guide the discussion by posing the following questions.

1. Why is a bird's body temperature higher than a mammal's?
   The bird's metabolism is higher because of the high energetic demands of flight.
2. What feature of a bird's skeleton contributes to its low weight?
   The bones are hollow.
3. Identify two similarities between birds and reptiles.
   Both have scales and lay amniotic eggs.

587

Visual Strategy

Table 26.2
Prepare eight different lab stations. Use examples of things that fit the eight of the sixteen bird orders shown in this figure. Place field guides to birds at each station. Assign students to cooperative groups of three or four, depending on the size of the class. Assign each group one lab station. Have them determine all they can about their order of birds. The following day, have each group report its findings to the class.

Theme Connection

Interacting Systems
In Africa there is a bird called the honeyguide. It has a mutualistic behavioral relationship with honey badgers. The bird's call alerts the badger to a beehive and actually leads the badger to the hive. The badger tears apart the hive, leaving a supply of food for the honeyguide. Humans now follow the honeyguide and perform the same service for the bird. Missionaries in the 1960s reported that honeyguides came into their tent-churches to steal the wax from the church candles.

Major Orders of Birds

You can often tell a great deal about the habits and food of a bird by examining its beak and feet. For instance, carnivorous birds such as eagles have curved talons for seizing prey and a sharp beak for tearing apart their meal. The beaks of seed-eating birds such as finches are short, thick, seed-crushers. What birds are common in your neighborhood? What sorts of adaptations do they show? Twenty-eight orders of living birds have been described. **Table 26.2** lists 16 of the most common orders of birds, showing examples. Pay particular attention to the feet and beaks of these birds.

Table 26.2 Major Orders of Birds

Order	Approximate Number of Living Species	Main Characteristics	Examples	
Passeriformes	5,276	Songbirds; perching feet; well-developed vocal organs; dependent young; largest bird order, containing 60 percent of all bird species	Sparrows, robins, warblers, crows, starlings, mockingbirds	
Apodiformes	428	Small bodies, short legs, rapid wing beat; hummingbirds are the smallest birds	Hummingbirds, swifts	
Piciformes	383	Sharp, chisel-like bills for pounding through wood in search of insects; grasping feet	Woodpeckers, toucans, honeyguides	
Psittaciformes	340	Well-developed vocal organs; large powerful bills for crushing seeds	Parrots, cockatoos	
Charadriiformes	331	Shorebirds; typically with long, slender, probing bills and long, stiltlike legs	Gulls, terns, plovers, auks, sandpipers,	
Columbiformes	303	Stout bodies; perching feet	Pigeons, doves	
Falconiformes	288	Birds of prey; day-active carnivores; sharp pointed beaks for tearing flesh; keen vision; strong fliers	Eagles, hawks, falcons, vultures	
Galliformes	268	Rounded bodies; often limited flying ability	Chickens, quail, grouse, pheasants	

■ Matter of Fact ■

Excess Baggage?
Songbirds are known to reduce their gonad size during long migratory flights only to have them increase again for mating, sometimes more than 200 times. Why carry excess baggage if you don't need it?

Gruiformes	209	Marsh dwellers; diverse body shapes; long, stiltlike legs	Rails, coots, bitterns, cranes	
Anseriformes	150	Waterfowl; webbed toes; broad bill with filtering ridges at margins	Swans, geese, ducks	
Ciconiiformes	114	Long-legged waders; often with large bodies	Storks, herons, ibises	
Strigiformes	146	Nocturnal birds of prey; large eyes; powerful beaks and feet	Owls	
Procellariiformes	104	Sea birds; tube-shaped bills; many can fly for long periods of time	Albatrosses, petrels	
Sphenisciformes	18	Marine; flightless; confined to Southern Hemisphere; thick coat of insulating feathers; wings modified as paddles for swift swimming	Penguins	
Dinornithiformes	2	Primitive; small and flightless; found only in New Zealand	Kiwis	
Struthioniformes	1	Large; flightless; only two toes; long, strong running legs	Ostrich	

Section Review

❶ What evidence indicates that birds evolved from reptiles?	❷ Name two differences between a lizard and a bird.	❸ Ostriches have feathers but cannot fly. What functions do feathers perform for an ostrich?	❹ List two features of the bird's skeleton that suit it for flight.

Phase 3

ASSESSMENT OPTIONS

Closure Strategy

Birdcalls

Obtain a recording of birdcalls and play the songs of local species for the class. Have students try to identify the calls. Point out that birds use calls for species recognition. Urge students to suggest reasons why species recognition is important to birds. List these reasons on the chalkboard.

Section Review

Assign the *Section Review*.

Reteaching

- Review the characteristics that distinguish birds from other vertebrates. Discuss the adaptations for flight in the skeleton and other anatomical and physiological mechanisms.
- To help students gain a better understanding of the complexity of bird behavior, have them design concept maps on the following areas: *Courtship, Nesting, Care of young, Migration,* and *Navigation.*
- Have students use their field guides to show the relationship between the feet of birds and their feeding habits and habitats.
- Have students create a concept map showing the features and systems of a bird and how they are adapted to flight.

▪ *Section Review Answers* ▪

1. Students should suggest that birds lay amniotic eggs, and that scales are also present on the feet and lower legs of birds.

2. Use Table 26.1 on pages 582–583 and Table 26.2 starting on page 588 for discerning differences between lizards and birds.

3. Students should explain that feathers act as insulation against heat loss.

4. Possible answers include: birds have thin, hollow bones with air pockets; the sternum is large and has a keel for solid anchorage of large flight muscles; and they have modified forelimbs that support and move the wings.

Lesson Plan 26.3

Phase 1

PREPARATION

Key Concepts

- True mammals evolved at the same time as dinosaurs.
- Mammals are distinguished by two basic characteristics: the presence of hair on the body, and mammary glands that secrete milk.
- Mammalian ability to regulate body temperature allowed mammals to invade colder habitats where reptiles could not live.
- Monotremes are egg-laying mammals.
- Marsupials are pouched mammals that give birth to young that complete development outside their mother's body, in the pouch.
- Placental mammals complete development inside their mothers before birth.

Reading Strategy

Have students make a concept map illustrating the important mammalian characteristics. They can be guided by the objectives on page 590 and by reading the accompanying information.

Phase 2

TEACHING STRATEGIES

Visual Strategy

Figure 26.17
Have students look at a geologic time scale to see when the therapsid reptiles lived.

YOU ARE A MAMMAL. SO ARE MOST OF THE ANIMALS THAT HUMANS EAT, USE AS WORK ANIMALS, AND KEEP AS PETS. ALL MAMMALS HAVE HAIR OR FUR ON THEIR BODIES, A CHARACTERISTIC NOT FOUND IN ANY OTHER KIND OF ANIMAL. MAMMALS ARE ALSO ENDOTHERMIC, LIKE BIRDS. FEMALE MAMMALS PRODUCE MILK WITH WHICH THEY NURSE THEIR OFFSPRING.

26.3   Introduction to Mammals

Objectives

① List the unique characteristics of mammals.

② Identify two features of monotremes.

③ Contrast the manner of development of marsupials and placentals.

Evolution of Mammals

Figure 26.17
The animals illustrated below are members of a group of reptiles called therapsids. At first glance, these therapsids seem to be mammals. The similarity is not surprising: mammals evolved from therapsids.

Mammals (class Mammalia) arose from early reptiles called therapsids (*thur AP sihdz*), some of which are illustrated in **Figure 26.17.** The fossil record provides a well-documented transition between these reptiles and mammals, with fossil forms ranging from reptiles with a few mammalian characteristics to true mammals. True mammals appeared about 230 million years ago, about the same time as the first dinosaurs. These early mammals were small, about the size of mice. For 165 million years—all of the time the dinosaurs flourished—mammals were a minor group that changed little. In the 65 million years since the dinosaurs became extinct, mammals have rapidly diversified to fill the ecological opportunities made available by the disappearance of the dinosaurs. There are now about 4,500 species of mammals.

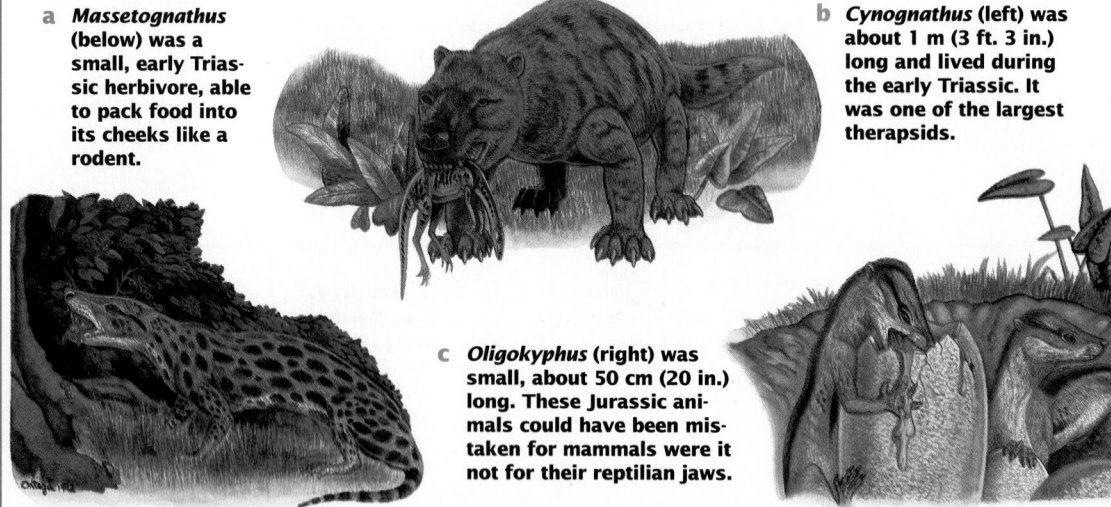

a *Massetognathus* (below) was a small, early Triassic herbivore, able to pack food into its cheeks like a rodent.

b *Cynognathus* (left) was about 1 m (3 ft. 3 in.) long and lived during the early Triassic. It was one of the largest therapsids.

c *Oligokyphus* (right) was small, about 50 cm (20 in.) long. These Jurassic animals could have been mistaken for mammals were it not for their reptilian jaws.

Figure 26.18
The grizzly bear, *Ursus horribilis*, is a large mammal, standing nearly 2.8 m (9 ft.) tall and weighing close to 800 kg (1,760 lbs.). Like all mammals, it has fur and nurses its young on milk.

a Grizzly bears are omnivores, eating mostly plants and fruits. But they do have sharp teeth for pulling apart flesh.

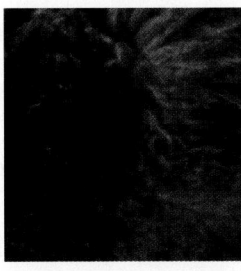

b The grizzly bear's name comes from the fact that its fur is silver tipped, giving it a grizzled appearance.

c One or two tiny and nearly naked cubs are born in midwinter, during hibernation. The mother is able to produce milk even though she does not eat or drink until spring.

d The four-chambered heart of the bear is very efficient, rapidly delivering oxygen to the body's muscles and organs.

Ventricles

Atria

Mammalian Characteristics

Mammals are distinguished by two characteristics: the presence of hair or fur on the body, and the ability to produce milk. Even the apparently naked whales and dolphins grow sensitive bristles on their snouts. Evolution of fur and the ability to regulate body temperature through metabolism enabled mammals to inhabit colder environments than could be tolerated by ectothermic reptiles and amphibians.

Like birds, mammals have a four-chambered heart. There is no mixing of deoxygenated and oxygenated blood in the heart.

Female mammals have **mammary glands** that secrete milk. Newborn mammals, which are born without teeth, suckle this rich milk until able to feed on their own. The grizzly bear shown in **Figure 26.18** shows all of these key mammalian characteristics.

Demonstration 1

Display pictures of many different animals, such as a whale, seal, bat, horse, cow, dog, giraffe, kangaroo, platypus, mole, and donkey. Include anatomical illustrations of these animals along with photographs. Discuss the similarities and differences among these animals. Have students point out characteristics and structural features common to each animal. Encourage students to speculate why all these animals belong to the same class.

Visual Strategy

Figure 26.18
Use Figure 26.18d to help students understand how mammalian organ systems differ from those of lower vertebrates. Review the similarities and differences between the organ systems of exothermic organisms and mammals. Encourage students to analyze the relationships between a high metabolic rate and the need for a constant food supply, efficient respiratory and circulatory systems, and a complex brain.

Demonstration 2

Mammals
Bring a live mammal such as a dog or cat to class. Have students point out the visible mammalian characteristics, such as body hair, outer ears and mammary glands if a female. If possible, ask a vet to pay a visit and describe other characteristics.

Egg-Laying Mammals: The Monotremes

Energy and Life

Explain why birds and mammals can live in Antarctica but reptiles cannot.

Imagine an egg-laying, furry animal with a large bill and webbed feet. Such an animal exists in Australia—the duckbill platypus. The platypus and five species of echidnas (*ee KIHD nuhs*), or spiny anteaters, are the only living **monotremes**. A platypus and an echidna are shown in Figure 26.19a and b. Although they are mammals, monotremes have some reptilian characteristics. They lay shelled eggs, and the structure of their pelvis is very similar to that of the early reptiles. Also like reptiles, monotremes have a single opening through which feces, urine, and reproductive products leave the body. Scientists think monotremes are more closely related to the early mammals than are any other living mammals.

In addition to their reptilian features, monotremes have both defining mammalian features: fur and functional mammary glands. Young monotremes drink their mother's milk after they hatch from eggs. Because of their strange mouths and the absence of

b This echidna can be found in New Guinea and eastern Australia. It uses its long nose to root for insects and other small animals.

well-developed nipples on the females, the young cannot suckle. Instead, the milk oozes onto the mother's fur and the young lap it off with their tongues.

The platypus is a good swimmer. It uses its bill much as a duck does, rooting in the mud for worms and other soft-bodied animals. Echidnas have very strong, sharp claws, which they use for burrowing and digging for insects. The echidna probes for insects, especially ants and termites, with its long beaklike snout.

Among living mammals, only monotremes lay eggs. In the rest of the mammals, the offspring are born after developing for some time inside the mother. **Marsupials** such as kangaroos, koalas, and opossums carry the offspring internally for only a short time. The young are born at an early developmental stage and finish developing in a pouch on the mother's abdomen. In contrast, the offspring of **placental mammals** remain inside their mother until development is essentially complete. They are nourished through a structure known as the placenta, which was described in Chapter 22. Dogs, cats, cattle, horses, humans, and most other mammals are placental mammals.

Figure 26.19

a The platypus, about half the size of a house cat, is a mammal that lays eggs. The young are nourished by milk from the mother's mammary glands.

Pouched Mammals: The Marsupials

The major difference between marsupials and placentals is their pattern of embryonic development. In marsupials, the fertilized egg is surrounded by a shell membrane, but no shell forms around the egg. During most of its early development the marsupial embryo is nourished by the abundant yolk within the egg. Shortly before birth, the shell membrane is lost and a short-lived placenta forms. Still early in its development, the baby marsupial is born.

Newborn marsupials emerge hairless and tiny, as you can see in **Figure 26.20**. Smaller than your thumbnail, they must crawl up to the pouch, where they can nurse.

Marsupials evolved not long before placental mammals, about 100 million years ago. Today, all but about 20 species of marsupials live in New Guinea and Australia, where placental mammals did not occur until introduced by humans in recent times. Marsupials in Australia and New Guinea have diversified to fill ecological positions occupied by placental mammals elsewhere in the world. For example, kangaroos are the Australian grazers, playing the ecological role that antelope, horses, and buffalo perform elsewhere. The only marsupial found in North America is the Virginia opossum.

Figure 26.20
This newborn opossum, North America's only marsupial species, crawls to the mother's nipples, which are in a pouch.

b **After several months, the young opossums are able to grasp the hair on their mother's back.**

True Placental Mammals

Mammals that produce a true placenta, through which the embryos are nourished for their entire development, are called placental mammals. The majority of mammal species, including humans, are placental mammals. Placental mammals are very diverse, ranging in size from 1.5-g (0.05-oz.) bats to 100,000-kg (110-ton) whales. They have invaded the air and returned to the sea. Fourteen of the 17 orders of living mammals are described in **Table 26.3**.

Social Studies Connection

Australia and Its Mammals
Most of Australia's mammals are marsupials or monotremes that became isolated 70 million years ago when Australia split off and drifted away from the continents of Antarctica and South America.

Visual Strategy

Figure 26.20
Have students describe what they know about pouched animals such as the kangaroo and the opossum. When students have finished reading, have them compare the birth and development of marsupials to that of other mammals.

Visual Strategy

Table 26.3

Assign students to cooperative work groups of two. Assign each group an order from the table. Have each group become experts on their order by researching a pertinent area of interest relating to the assigned order. For example: historical vs. current range; feeding habits; mode of living; migratory habits, if any; general physical and behavioral characteristics; number of known members of the order; extinct or endangered members of the order; representative animals of the order; and significance of the order to humans or the biosphere. Group members should compile their findings into a group report and present it to the class.

Mathematics Connection

Have students solve the following problem: A certain female field mouse always gives birth to a litter of eight, half of which are female. All surviving females achieve sexual maturity and, if mated, give birth at exactly three months of age and every three months after that, until death. Their average life span is two years. In order to survive, each field mouse needs 1 sq. m (3.3 sq. ft.) of grass.

A pair of adult field mice—a male and a pregnant female—arrive, floating on a log, at a grassy island. The island is football field size, 91 m × 49 m (299 ft. × 160 ft.). The day after they arrive, the female gives birth. There are no predators, and all the descendants survive until the island can no longer provide enough space.

How many field mice are there after three months? six months? nine months? When can the island no longer support the population of mice?

After 3 months, 50; after 6 months, 250; after 9 months, 1,550; The island has an area of 4,459 sq. m (14,665 sq. ft.) and will run out of space before the end of the first year, when 7,050 mice will be present.

Table 26.3 Major Orders of Mammals

Order	Approximate Number of Living Species	Main Characteristics	Examples
Monotremata	3	The only egg-laying mammals; once widespread, now found only in Australia and New Guinea	Platypus, echidnas
Marsupialia	280	Primitive mammals; have an abdominal pouch in which young are reared	Kangaroos, koalas, opossums
Rodentia	1,814	Small herbivores with chisel-like, incisor teeth that grow continuously	Squirrels, rats, mice, beavers, porcupines
Chiroptera	986	The only flying mammals; elongated fingers that support a thin wing membrane; mainly fruit or insect eaters; many fly at night, navigating by sonar	Bats
Insectivora	390	Small, chiefly night-active mammals; feed on insects; sharp-snouted; spend most of their time underground; the most primitive placental mammals	Moles, shrews, hedgehogs
Carnivora	240	Land-living predators; teeth adapted for seizing prey and shearing flesh; there are no native families in Australia	Dogs, bears, cats, wolves, otters, weasels
Primates	233	Largely tree dwellers; binocular vision and an opposable thumb; large brains; the end product of a line that branched off early from other mammals; retains many primitive characteristics	Prosimians, apes, monkeys, humans

■ *Matter of Fact* ■

Vampire Bats

Vampire bats live only in the tropics and feed on the blood of other mammals, especially cows and horses. The vampire bat uses its front teeth to make a small cut in the skin of the victim and then laps up the blood with its tubelike tongue.

Artiodactyla	211	Hoofed mammals with two or four toes; large herbivores; most are grass eaters	Sheep, pigs, cattle, deer, giraffes	
Cetacea	79	Aquatic, streamlined bodies; front limbs modified into broad flippers; no hind limbs; nostrils are blowholes on top of head; hairless except on muzzle	Whales, dolphins, porpoises	
Lagomorpha	69	Rodentlike mammals with four upper incisors, rather than the two seen in rodents; hind legs often longer than forelegs, an adaptation for jumping	Rabbits, hares, pikas	
Pinnipedia	34	Marine carnivores with limbs modified for swimming; feed mainly on fish	Seals, sea lions, walruses	
Edentata	30	Mostly insect eaters; many are toothless, but some have degenerate, peglike teeth	Sloths, anteaters, armadillos	
Perissodactyla	17	Hoofed mammals with one or three toes; herbivores with teeth adapted for chewing	Horses, zebras, rhinoceroses, tapirs	
Proboscidea	2	Enormous herbivores with long trunks; two upper incisors elongated as tusks; the largest living land animals	Elephants	

Section Review

❶ Why are monotremes classified as mammals even though they lay eggs?

❷ List two differences between monotremes and other mammals.

❸ Contrast the pattern of development of a marsupial with that of a placental mammal.

Theme Connection

Stability

The blunt front end of many whale and dolphin heads contains a huge organ of oil that contributes to buoyancy and focuses powerful sound waves for long-distance communication. The sperm whale can remain submerged for up to 85 minutes. It has been known to dive to depths of 2.25 km (1.4 mi.).

Phase 3

ASSESSMENT OPTIONS

Closure Strategy

Comparing Mammals

Have students develop a concept map that distinguishes among the three types of mammals, based on the development of their young.

Section Review

Assign the *Section Review*.

Reteaching

Show students pictures of different mammals. Have them classify these mammals according to the orders described in Table 26.3. Include pictures of as many different kinds of mammals as possible. If possible, show the mammals in their native habitats. Include some nonmammals—birds, fish, or lizards—to test students' understanding. Describe a habitat and feeding strategy and have students identify order(s) of placental mammals to which these may apply.

■ *Section Review Answers* ■

1. Students should indicate that monotremes have two defining features of mammals: hair and mammary glands.

2. Possible answers might include: they lay eggs, they have large bills, and they have webbed feet.

3. Students should indicate that marsupials carry their young internally for a short period of time before giving birth to offspring in early developmental stages, which continue to develop outside the mother's body, in a pouch on the mother's abdomen. In contrast, the placental offsprings' development is essentially complete. They are nourished through the placenta.

Lesson Plan 26.4

Phase 1

PREPARATION

Key Concepts

- Mammalian hair has many functions, including camouflage, sensory structures, defense weapons, and insulation.
- Claws, hooves, horns, and antlers are made of keratin.
- Mammals may be herbivores, omnivores, or carnivores.
- Among mammals, only bats are capable of powered flight.
- Bats use sonar to navigate.

Reading Strategy

Have students work in cooperative groups of four. Construct a concept map on mammalian adaptation by using the major subheadings and reading the accompanying information.

Phase 2

TEACHING STRATEGIES

Visual Strategy

Figure 26.21

If your students have pets such as gerbils, hamsters, rats, mice, cats, dogs, etc., have them bring in samples of hair and compare them under the microscope.

Demonstration 1

Insulation

Bring in a down pillow, quilt, or sleeping bag to illustrate the insulating qualities of the underfeathers, which serve the same function as underhair.

596

ALTHOUGH THERE ARE FEWER SPECIES OF MAMMALS THAN REPTILES, MAMMALS THRIVE IN A FAR GREATER VARIETY OF HABITATS AND LEAD A GREATER VARIETY OF LIFESTYLES THAN REPTILES. IN THIS SECTION, YOU WILL LEARN ABOUT SOME OF THE UNIQUE ADAPTATIONS OF MAMMALS THAT SUIT THEM TO THEIR WIDE RANGE OF HABITATS AND LIFESTYLES.

26.4 *Mammalian Adaptations*

Objectives

1 Identify three functions of hair.

2 List two mammalian structures that contain keratin.

3 Contrast the teeth of herbivores and carnivores.

4 Describe how bats navigate in the dark.

Hair Has Many Functions

All mammals have hair, including the seeming hairless whales and dolphins. A hair extends from a bulblike structure known as a hair follicle, which lies below the skin surface. Hair is composed mainly of air spaces and dead cells filled with the protein keratin.

Most mammals are covered by a coat of hair, shown in **Figure 26.21**. Two different types of hair make up this coat. **Guard hairs** are the long, thick outer hairs responsible for the coat's color. Between the guard hairs grows a dense coat of **underhair** consisting of thinner, shorter hairs. You can see the underhair of a dog or cat by brushing back the guard hairs with your hand.

What are the functions of hair? One of its functions is insulating against heat loss. Endothermic animals such as mammals often maintain body temperatures higher than the temperature of their surroundings and so tend to lose body heat. The dense underhair coat of many mammals reduces the amount of body heat that escapes, performing the same function as a layer of fiberglass insulation in the attic of a house.

Another function of hair is camouflage. The coloration and pattern of a mammal's coat usually match its background. Hairs also function as sensory structures. The whiskers of cats and dogs are stiff hairs that are sensitive to touch. Mammals that are active at night or that live underground often rely on their whiskers to locate prey or to avoid colliding with objects. Hair can also serve as a defense weapon. For instance, porcupines and hedgehogs protect themselves with long, sharp, stiff hairs.

Figure 26.21
All mammals have fur, which primarily functions as insulation. Both underhair and guard hair aid in this function.

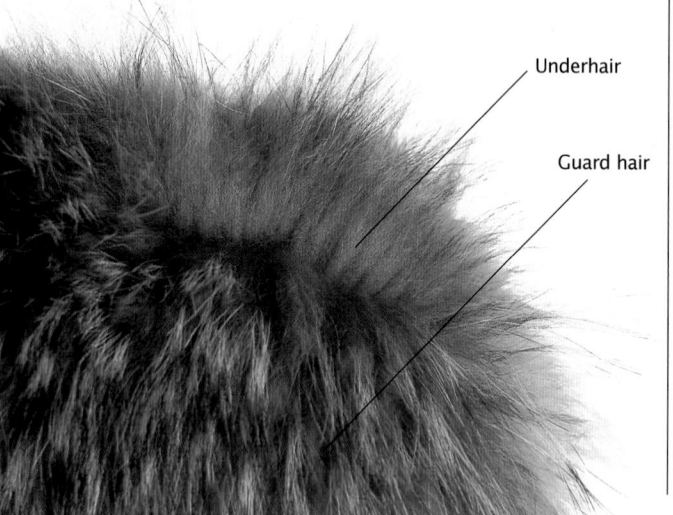

Underhair

Guard hair

Claws, Hooves, Horns, and Antlers

Keratin is a versatile protein that is a component of many mammalian structures. For example, keratin is found in claws, fingernails, and hooves. Hooves, such as the zebra's hoof shown in **Figure 26.22**, are specialized keratin pads on the toes of horses, cattle, sheep, antelopes, and other running mammals. The horns of cattle, sheep, and antelopes are composed of a core of bone surrounded by a sheath of keratin. The bony core is attached to the skull, and the horn is not shed. The horns of rhinoceroses are made of hairlike fibers of keratin that are compacted into a very hard structure.

Deer grow and shed a set of antlers each year. Deer antlers, which are grown only by males, are made of bone. While growing, antlers are covered by a thin layer of skin known as velvet. The velvet dies and is scraped off when

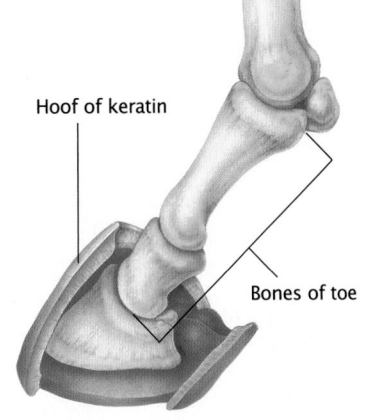

Hoof of keratin

Bones of toe

Figure 26.22
A zebra's foot is modified so that the zebra actually walks on a single toe that is covered with a hoof made of keratin.

the antlers are fully grown. The male deer uses his antlers to attract females and to combat other males. After the breeding season, the male sheds his antlers. Some examples of horns and antlers are shown in **Figure 26.23**.

Figure 26.23

a **The hollow horns of the South African springbok are permanent. They use these horns to fight with other males.**

b **The antlers of the mule deer are used for combat and to attract females. They drop off every winter and regrow during the spring and summer.**

■ *Cultural Perspective* ■

Animal Bones as Jewelry
Indigenous peoples in Brazil often wear jewelry that consists of animal bones, claws, teeth, and feathers. Jaguar claws, for example, are used in necklaces. The brightly colored feathers of the toucan are used in earrings.

Demonstration 2
Hoof Clippings and Other Keratin Stuff
Display keratin-containing structures for the class to examine. Contact local horse stables for hoof clippings from recently shod horses. A local pet store may sell cow hooves. Check your supply room for horns, or the grocery store for pigs feet with attached hooves.

Demonstration 3
Bony Structures
Display antlers or photographs of animals with antlers. Emphasize that antlers are not made of keratin, but bone. Antlers may be shed.

Food and Feeding

Mammals feed on a variety of foods. Horses, giraffes, antelopes, and elephants are herbivores. Lions, wolves, seals, and sperm whales are carnivores. The blue whale, the largest animal, filters small crustaceans from the ocean. It is usually possible to determine a mammal's diet by examining its teeth. For example, look at the skull of the coyote (a carnivore) and the deer (a herbivore) shown in **Figure 26.24**. The coyote's long canine teeth are suited for biting and holding prey. Its premolar and molar teeth are triangular and sharp for shearing off chunks of flesh. The deer's canines, in contrast, are small. It clips off mouthfuls of plants with its flat incisors. The deer's molars are large, and their surfaces are covered with ridges to form an effective grinding surface that can break up tough plant tissues.

Cellulose is the major component of plant cell walls, and thus is a major constituent of the plant body. Mammals do not have enzymes that can digest cellulose. Herbivorous mammals rely on a mutualistic partnership with bacteria that can produce cellulose-splitting enzymes. Mammals such as cows, buffaloes, antelopes, goats, deer, and giraffes have huge four-chambered stomachs that function as storage and fermentation vats. The first chamber is the largest and holds a large population of bacteria. When the animal swallows, chewed plant material passes into this chamber. Bacteria partly digest the material, which is then regurgitated and chewed again. A cow chewing its cud is rechewing this partly digested food. After another thorough grinding, the cud is swallowed and further digested in the stomach. It then passes from the stomach into the intestines.

Rodents, horses, elephants, and rabbits are herbivores but have relatively small stomachs lacking mutualistic bacteria. These animals do not chew a cud. Bacteria that aid in digestion live in a pouch that branches from the large intestine.

Even with these complex adaptations for breaking down cellulose, a mouthful of plants is less nutritious than a mouthful of flesh. Herbivores must consume large amounts of plant material to gain sufficient nutrition. An elephant eats 135–150 kg (300–400 lbs.) of food per day. You can read more about mammalian adaptations in the *Tour of a Mammal* on page 599.

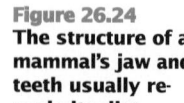

Figure 26.24
The structure of a mammal's jaw and teeth usually reveals its diet.

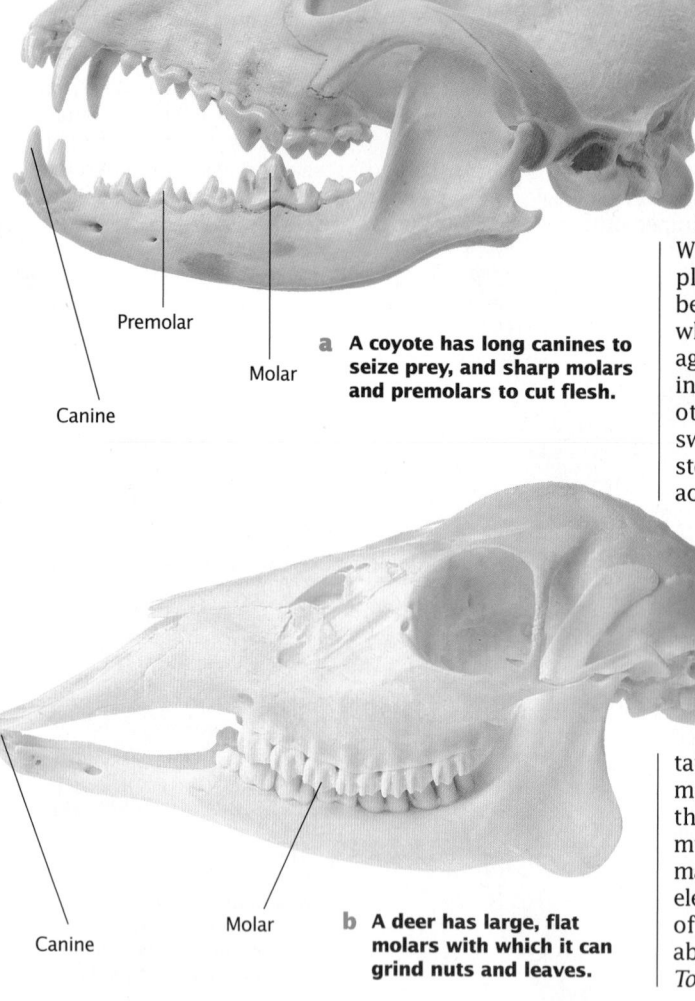

Premolar

Molar

Canine

a **A coyote has long canines to seize prey, and sharp molars and premolars to cut flesh.**

Canine

Molar

b **A deer has large, flat molars with which it can grind nuts and leaves.**

■ *Journeys* ■ *Journeys* ■ *Journeys* ■ *Journeys* ■ *Journeys*

Tour of a Mammal

Mammals, such as this field mouse, are endotherms and maintain a constant body temperature. Mammals have highly developed brains.

Rodents, the most numerous and widely distributed of all mammals, include mice, rats, squirrels, beavers, and porcupines.

Most rodents have cheek pouches for holding food.

Whiskers are special sensory hairs that provide a tactile sense.

Mammalian teeth are replaced only once, unlike reptilian teeth, which are replaced throughout the lifetime. The chisel-like incisors of rodents grow continually, an adaptation for gnawing.

Smaller mammals must eat more to remain active because they lose more heat due to their greater surface area-to-volume ratio. Per gram of body mass, a mouse consumes five times as much as a dog.

Except for primates, all mammals have definite mating seasons.

The fur coat of mammals has two layers: dense underhair for insulation, and coarse guard hair for protection and coloration. Most mammals shed their fur twice a year, in spring and in fall.

Many mammals in northern climates hibernate during winter months, reducing their body temperature to within a few degrees of their surroundings.

Reptilian scales occur on the tails of rodents.

All female mammals nourish their young with milk-secreting glands.

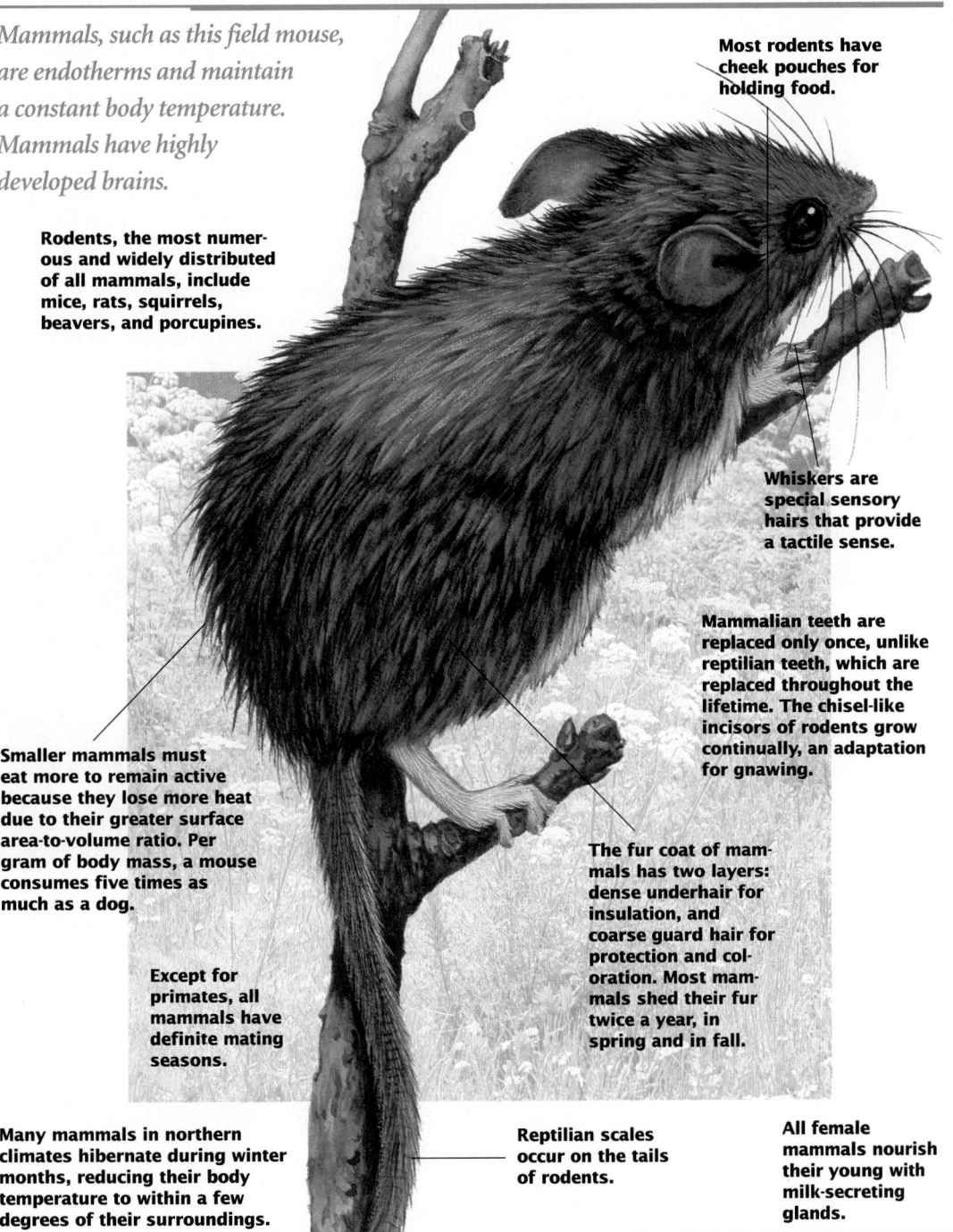

Using the Feature

Nearly all large terrestrial animals are mammals. Yet, the majority of mammals are small—44 percent of mammal species are rodents, like this field mouse. All mammals have hair or fur, and females have milk-producing glands. The mammalian jawbone is a single bone, whereas the jawbone of reptiles is composed of five bones.

Mammals are endotherms, metabolically regulating their body temperature. Endothermy allows mammals to live in cold climates, including the Arctic and Antarctic.

Discussion

Guide the discussion by posing the following questions.

1. Why must a mouse eat more food per gram of body mass than a dog?
   The mouse is smaller and therefore has a higher surface area-to-volume ratio than the dog. The mouse loses heat more rapidly than the dog and must compensate by having a higher metabolism.
2. What is the function of underhair?
   insulation
3. In what way are the incisors of rodents unusual?
   They continue to grow throughout life.

Visual Strategy

Figure 26.26
When bats fly they emit high-pitched squeaks that bounce back from a solid object. They are able to fly wingtip to wingtip with other bats through dark caves without hitting each other or the cave walls. Some bats hibernate and some migrate, but they are all very social mammals.

Language Connection

Catchy Animal Phrase
Encourage students to use library references to look up common phrases referring to characteristics of animals. Such phrases might include: happy as a lark, sings like a bird, blind as a bat, dog tired. Have students determine whether each phrase is based on fact. For example, are bats really blind? Have students compile a list from some of their collected phrases to share with the class.

Phase 3

ASSESSMENT OPTIONS

Closure Strategy

Design a Mammal
Have students design a mammal to fit a particular habitat. Tell them to draw this fictional mammal and describe any particular anatomical or physiological adaptations that would enable it to survive and reproduce successfully. Provide the students with information such as habitat restrictions, whether the mammal is nocturnal or diurnal, and perhaps habits of some of its predators.

Flying Mammals

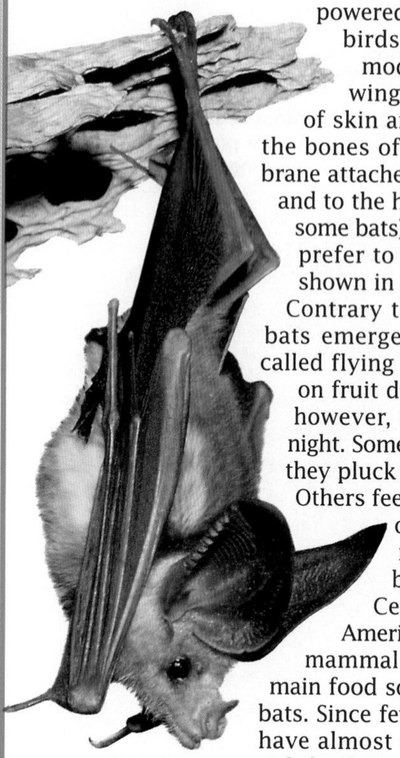

Bats are the only mammals capable of powered flight. Like the wings of birds, the wings of bats are modified forelimbs. The bat wing is a leathery membrane of skin and muscle supported by the bones of four fingers. The membrane attaches to the side of the body and to the hind leg (and to the tail in some bats). When resting, most bats prefer to hang from their legs, as shown in **Figure 26.25**.

Contrary to popular belief, not all bats emerge only at night. The so-called flying foxes, for instance, feed on fruit during the day. Most bats, however, are active at dusk or at night. Some eat fish and frogs, which they pluck from the water's surface. Others feed on nectar from flowers or prey on small mammals, including other bats. Vampire bats of Central America and South America drink blood from large mammals. Flying insects are the main food source for most nocturnal bats. Since few birds fly at night, bats have almost exclusive access to this rich food supply.

Figure 26.25
Bats, such as this California leaf-nosed bat, are social animals. They roost together in groups of thousands, and some species hunt cooperatively.

How do bats navigate in the dark?
How can bats fly in dark caves and hunt their food at dusk or even at night? Do bats have particularly sensitive vision? Or do they use another sense to find their way around? Late in the eighteenth century, the Italian scientist Lazzaro Spallanzani showed that a blinded bat could fly without crashing into things and could still capture flying insects. However, when Spallanzani plugged the ears of the bat, it was unable to navigate and collided with objects. Spallanzani concluded that bats "hear" their way through the world. It was not until the late 1930s that bats' ability to "fly blind" was explained.

Bats have evolved a sonar system that functions much like the sonar devices used by ships to locate underwater objects. As a bat flies, it emits extremely high-pitched sounds well above our range of hearing. These high-frequency pulses are emitted through the mouth or, in some cases, through the nose. The sound waves reflect off obstacles or flying insects, and the bat hears the echo. Through sophisticated processing of this echo within its brain, a bat can determine not only the direction of an object but also the distance to the object. This sonar system enables bats to navigate and to capture their prey at night, as shown in **Figure 26.26**.

Figure 26.26
"Blind as a bat" is an inaccurate phrase. This bat was able to use its sonar to catch this moth in midair in the dark.

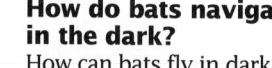

Section Review

❶ **List two functions of hair.**

❷ **Name two keratin-containing structures on your body.**

❸ **Contrast the teeth of a coyote with those of a deer.**

❹ **Explain why a blinded bat can catch prey but a deafened one cannot.**

▪ Section Review Answers ▪

1. Hair functions as insulation, sensory structures, defense weapons, and camouflage.
2. Students should suggest nails and hair.
3. Students should explain that coyotes have long canine teeth suited for biting and capturing prey. Their premolars and molars are triangular and sharp for tearing chunks of flesh. In contrast, deer have large flat molars, with surfaces covered with ridges for grinding and breaking plant tissue.
4. Students should indicate that bats navigate by hearing, not by sight.

Chapter 26 *Highlights*

"Well, of course
I did it in cold
blood, you idiot!
. . . I'm a reptile!"

Alternative Assessment
Arrange for a "Bring a Photo of Your Pet Day" for anyone having a pet reptile, bird, or mammal. Have each owner formulate questions about the characteristics and identity of his or her pet. Ask students questions and encourage discussion.

Key Terms	Summary
26.1 Reptiles  Reptiles control their body temperatures through their behavior.	• Reptiles have dry, largely watertight skin, lay watertight eggs, and are ectotherms. • Reptiles evolved about 320 million years ago. • Pterosaurs were the only reptiles that evolved the ability to fly. • Dinosaurs became extinct about 65 million years ago. • The surviving reptiles include crocodiles and alligators, turtles, the tuatara, lizards, and snakes.
26.2 Birds  Some birds may have more than 25,000 feathers on their body. contour feather (p. 585) down feather (p. 585)	• The first birds evolved about 150 million years ago. • All birds have feathers, are endothermic, and lay eggs. • Birds have down feathers and contour feathers. • The skeletons of birds are lightweight. The bones are hollow and thin.
26.3 Introduction to Mammals Mammals, such as these bears, receive nourishment from their mother's mammary glands while young.  mammary gland (p. 591) monotreme (p. 592) marsupial (p. 592) placental mammal (p. 592)	• Mammals evolved from reptiles about 230 million years ago. • Mammals have hair and are endothermic. Female mammals have mammary glands. • Monotremes are mammals that lay eggs. • Marsupial mammals are born early and complete their development in the mother's pouch. • Placental mammals nourish their young via the placenta throughout development.
26.4 Mammalian Adaptations The skull of a coyote has long canines and strong jaws.  guard hair (p. 596) underhair (p. 596)	• Hair serves as insulation, as camouflage, as a sensory structure, and as a defense. • Bats are the only flying mammals. Bats are able to navigate and capture prey in the dark by means of a sonar system.

Chapter Review Answers

Understanding Vocabulary

1. **a.** An endotherm is an animal whose body temperature does not fluctuate with the temperature of its surroundings. An ectotherm is an animal whose body temperature follows the temperature of its surroundings.

   b. Contour feathers give the wings and tail shape and are used in flight; they provide some insulation. Down feathers are specialized for insulation and grow between and underneath the contour feathers.

   c. Monotremes are mammals that have features similar to reptiles, including single openings for waste and reproduction, and young that hatch from eggs. Marsupials are also mammals but have few reptile-like features. They are very similar to placental mammals but differ, in that the young move to their mother's pouch early during development.

   d. Guard hairs are long and thick and give color to an animal's coat; underhair is thin and short and serves to keep body heat from escaping.

Relating Concepts

2. Map answer is shown on page 571D.

Understanding Concepts

Multiple Choice

3. c	8. a
4. d	9. b
5. a	10. c
6. a	11. b
7. b	

Completion

12. pterosaurs, reptiles
13. Squamata, Testudines
14. dinosaurs, 65 millions years ago
15. marsupials, pouch

Understanding Vocabulary

1. For each pair of terms, explain the differences in their meanings.
   a. endotherm, ectotherm
   b. contour feathers, down feathers
   c. monotremes, marsupials
   d. guard hair, underhair

Relating Concepts

2. Copy the unfinished concept map below onto a sheet of paper. Then complete the concept map by writing the correct word or phrase in each oval containing a question mark.

Understanding Concepts

Multiple Choice

3. The skin of reptiles forms a watertight barrier because it
   a. is sticky and covered with slime.
   b. is covered with scales and hair.
   c. contains fats and the protein keratin.
   d. is covered with feathers.

4. Which animal is not an ectotherm?
   a. lizard   c. toad
   b. alligator   d. chicken

5. What evidence suggests that reptiles and birds are closely related?
   a. both lay amniotic eggs
   b. scales and feathers are found on birds and reptiles
   c. both are endothermic
   d. birds and reptiles live where mammals cannot

6. Which of the following are not bird adaptations for flight?
   a. a cartilaginous skeleton
   b. an efficient respiratory system
   c. contour feathers
   d. the ability to regrow lost feathers

7. Mammals are different from any other vertebrates in that they
   a. lay eggs.
   b. produce milk.
   c. are endotherms.
   d. have structures that contain keratin.

8. In placental mammals,
   a. embryos are nourished through a placenta throughout development.
   b. embryos are nourished by yolk before attaching to the placenta.
   c. the young finish developing in a pouch.
   d. young develop in hard-shelled eggs.

9. Which is not a function of hair?
   a. insulation   c. camouflage
   b. navigation   d. sensory structures

10. Which structures do not contain keratin?
    a. fingernails   c. teeth
    b. hooves   d. hair

11. In Spallanzani's experiment to determine how bats navigate in the dark, what variable did he manipulate?
    a. the type of insect used as food for the bat
    b. sense organs: he blocked them one at a time
    c. the time of day the experiment was carried out
    d. bat feeding preferences

Completion

12. The first flying vertebrates were the _____ . They were not birds but _____ .

13. Snakes are members of the order _____ , while turtles belong to the order _____ .

14. Birds are most closely related to _____ , which became extinct _____ years ago.

15. Kangaroos are _____ . Newborn kangaroos crawl from the mother's birth canal to her _____ , where they finish developing.

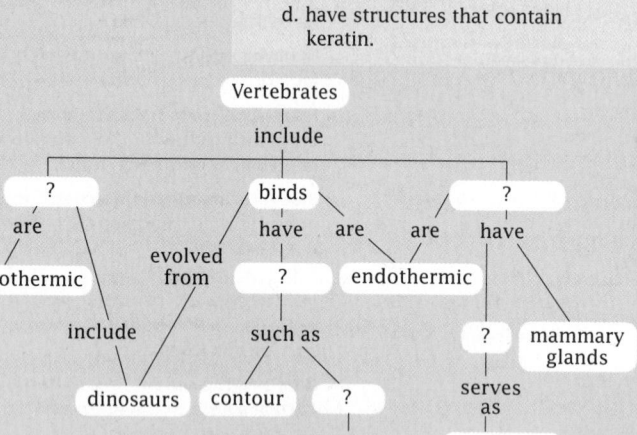

Short Answer

16. "Cold-blooded" may lead some people to believe that the blood of reptiles is "cold." On warm days, some lizards have body temperatures higher than ours. Ectothermic indicates that the animal's body temperature fluctuates with the temperature of its surroundings.

17. Alvarez proposed that a large meteorite struck the Earth and that the dust clouds kicked up by the meteorite blocked the Sun. The absence of sunlight caused the plants on which the dinosaurs fed to die, thus starving the dinosaurs into extinction.

18. A bat uses a type of sonar. It emits high-pitched sound waves that reflect off insects. The bat processes the echo in its brain to determine the exact location of the insect.

19. They lay shelled eggs, have a pelvis that is similar to that of early reptiles, and have a single opening through which feces, urine, and reproductive products leave the body.

Short Answer

16. How could use of the term "cold-blooded" lead to a misunderstanding about how reptiles regulate their body temperature? Why is it more appropriate to call them ectothermic?

17. Describe the hypothesis proposed by Luis Alvarez and his colleagues about the end of the Age of Reptiles.

18. Explain how a bat is able to locate a flying insect on a dark night.

19. Why do scientists think that monotremes are more closely related to reptiles than are either marsupials or placental mammals?

20. Explain why birds can live in colder climates than reptiles.

Interpreting Graphics

21. Examine the animals pictured below.

- Identify the class to which each animal belongs.
- Which animal is endothermic?
- In which animal is the ventricle of the heart incompletely divided?

Reviewing Themes

22. *Patterns of Change*
    If you were a marsupial, how would you have spent the early months of your life?

23. *Evolution*
    What evidence suggests that reptiles and birds have a common ancestor?

24. *Scale and Structure*
    How could you tell if a mammalian skull you found was that of a herbivore or a carnivore?

Thinking Critically

25. *Inferring Conclusions*
    How would losing 50 percent of its down feathers affect a bird?

26. *Compare and Contrast*
    How are underhair and down feathers alike? How are they different?

27. *Comparing and Contrasting*
    How are an echidna and a crocodile alike? How are they different?

28. *Building on What You Have Learned*
    In Chapter 16 you learned about bacteria. How are bacteria helpful to herbivorous mammals?

Cross-Discipline Connection

29. *Biology and English*
    Identify fears that people have about bats by surveying students in your school. Then learn if their fears are justified by writing a letter to Bat Conservation International, P.O. Box 162603, Austin, Texas 78716.

Discovering Through Reading

30. Read the article "Listening to the Mockingbird," in *National Wildlife*, June/July 1992, pages 12–16. How did the mockingbird get its name? According to scientists who have studied mockingbirds, why do male mockingbirds sing? What effect does the male's singing have on females?

24. You could examine the teeth. A herbivore would have flat incisors and molars that are flat with ridges. A carnivore would have long and sharp canine teeth and sharp, triangular-shaped molars.

Thinking Critically

25. Loss of down feathers would affect a bird's ability to insulate itself against the cold; it would not affect its ability to fly.

26. Both serve to insulate the body of endotherms. Underhair is found on mammals, while birds have down feathers.

27. They both are egg-layers, have similarly shaped pelvises, and have a single opening through which feces, urine, and reproductive products pass. Besides physical appearance, they are different in that only the echidna has mammary glands and hair.

28. Bacteria live within the animal's stomach. There the bacteria digest the cellulose in the plants eaten by the mammals.

Cross-Discipline Connection

29. Each letter should ask at least one question about bats and should be free of spelling and grammatical errors.

Discovering Through Reading

30. The name has several native American origins; all stem from the mockingbird's ability to imitate sounds that it hears. Male mockingbirds sing to attract females. It seems that females are attracted to males that have the greatest repertoire of songs. Upon hearing the male sing, the female's reproductive system is "reset," which culminates in mating.

---

20. Birds are endotherms, which means that they can regulate their body temperature. As ectotherms, reptiles are unable to regulate their body temperature. Their body temperature fluctuates with the temperature of their surroundings. Thus, reptiles are not well adapted for cold climates.

Interpreting Graphics

21. • Reptile, Aves, Mammalia
    • the bird and whale
    • the turtle

Reviewing Themes

22. After birth, I would have crawled to my mother's pouch. There I attached to my mother's nipple and stayed in her pouch until I was large enough to get about on my own.

23. Both lay amniotic eggs and have scales.

Procedural Note

1. Before beginning this activity, have the class discuss the possible functions performed by feathers .
2. Review the answers to the preparation questions.

Prelab Preparation Answers

- Down feathers act as insulation.
- Down feathers grow close to the body between the contour feathers.
- Contour feathers function as insulation, but also streamline the bird and make up the wings and tail.
- Contour feathers are found all over the body of a bird. They usually overlay down feathers.

Procedure Answers

4. The contour feather has a long, stiff shaft to which the barbs attach. The down feather has only a short shaft. Also, the barbs of the contour feather are less flexible than the wispy filaments of the down feather.
5. The vane is stiffer than the down feather (because of the interlocking barbules on the barbs).
7. No. Down feathers lack the interlocking hooklets and barbules.

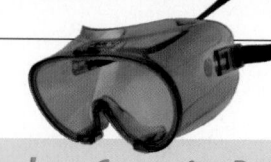

Chapter **26** # Investigation

How Do Down and Contour Feathers Differ?

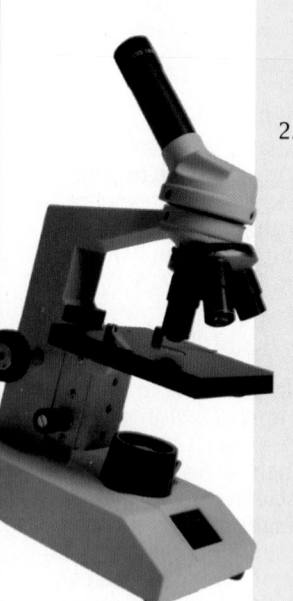

Objectives

In this investigation you will:
- *observe* down and contour feathers
- *contrast* the structure of down and contour feathers
- *relate* the structure of each kind of feather to its function

Materials

- contour feather
- down feather
- compound light microscope
- prepared slide of contour feather
- prepared slide of down feather

Prelab Preparation

1. Review what you have learned about feathers by answering the following questions:
   - What is the function of down feathers?
   - Where are down feathers found on a bird?
   - What functions do contour feathers perform?
   - Where would you find contour feathers on a bird?

2. Review the procedures in the Appendix for using the compound microscope.

Procedure: Comparing Down and Contour Feathers

1. Form a cooperative team with another student to complete steps 2–7.

2. Examine a down feather. The stiff base of the feather is known as the quill. The quill extends into the flesh to anchor the feather. Hold the feather by the quill and wave the feather in the air. Describe the texture of the feather. Record your observations.

3. Examine a contour feather. Using the photograph below as a guide, identify the quill, shaft, vane, and barbs.

4. Hold the feather by the quill and gently bend the tip with your other hand. Be careful not to break the feather. Next, hold the feather by the quill and wave the feather through the air. Record your observations. *What makes the contour feather stiffer than the down feather?*

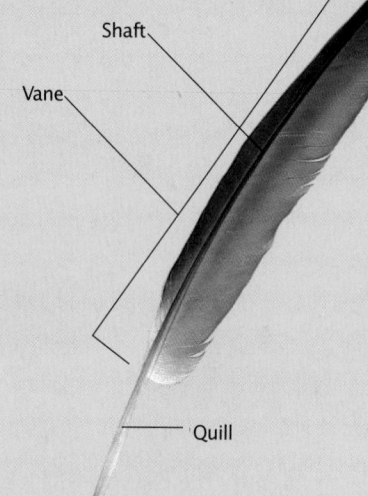

Shaft

Vane

Quill

5. Examine the vane of the contour feather. *How does the texture of the vane compare with the texture of the down feather?*

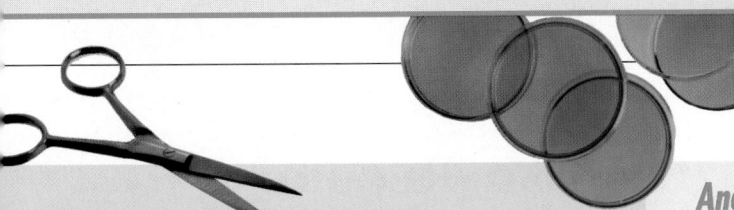

6. Observe the prepared slide of the down feather under low power. Now observe it under high power. Make a drawing of the feather as it appears under under high power.

7. Repeat step 6 for the contour feather. Using the photograph above as a guide, you should be able to identify the barbules that branch from each barb and the hooklets that interlock the barbules. *Were these structures visible on the down feather?*

8. Clean up your materials and wash your hands before leaving the lab.

Analysis

1. *Identifying Relationships* Which kind of feather has a longer quill? How does the difference in quill length relate to the function of each kind of feather?

2. *Identifying Relationships* The contour feathers of the tail and back edge of the wings are often called flight feathers. What features make contour feathers better than down feathers for flight?

3. *Inferring Conclusions* Birds spend much time preening their feathers. They rearrange and straighten their feathers and align the barbs on individual feathers. Explain why preening is crucial for the feathers to function effectively.

Thinking Critically

Unlike a bird's wing, a bat's wing is a thin membrane of skin stretched across elongated fingers. By contrasting the structure of feathers and hair, explain why a wing could not be made of hair.

Analysis Answers

1. The contour feather has a longer quill. This quill more firmly attaches it to the flesh. As the outer layer of feathers, contour feathers are subjected to greater stresses as the bird flies, especially the contour feathers on wings and tail.

2. With their stiffened shafts and interlocking vanes, contour feathers are a nearly solid unit. This reduces their wind resistance and is also responsible for the bird's wing shape. Downy feathers are too flimsy to be able to form a wing.

3. It is advantageous for the bird to be as streamlined as possible in order to reduce its wind resistance. Preening keeps the bird streamlined.

Thinking Critically Answer

A hair is a straight, thin shaft, but a feather has flat vanes, which increase its surface area. A wing of hairs would not have an essentially solid surface like a wing of feathers. Air would flow through the hair wing, instead of flowing over it.

Using the Feature

- Make the connection with Chapter 10 and Chapter 15, by telling students that before the 1800s, dinosaurs were unknown. Explain that fossil hunters would occasionally find a dinosaur tooth or bone without knowing what it was they had. In 1822, Mary Ann Mantell, the wife of an English physician named Gideon Mantell, found a large tooth partly buried in a rock. She showed it to her husband (a fossil collector), and he proposed that the tooth came from an iguana-like reptile. He therefore named the species Iguanodon, which means "iguana tooth." It was not until the remains of several of these large creatures had been discovered that they were identified as dinosaurs. In 1841, Sir Richard Owen classified this peculiar group of reptiles as Dinosauria and named its members "dinosaurs."

- Tell students that Baron Cuvier proposed the theory of "geological catastrophe" to explain why many fossil animals were no longer alive. His theory suggests that great volcanic upheavals and similar catastrophes destroyed many life forms. Have students research other theories and report their findings to the class.

- Explain that scientists believe that many forms of plants and animals once existed but left no fossil record of their existence. Inform students that the oldest fossils are microscopic traces of bacteria that lived about 3.5 billion years ago. The oldest animal fossils are the remains of invertebrates estimated to be about 700 million years old. The oldest fossils of vertebrates are fossil fish about 500 million years old.

- Tell students that fossils are very easy to find. They can be found in every state of the United States. To develop a sense of the importance of fossils, take students on a fossil hunt. As resources for determining fossil sites, contact the geology department of a local college or state university, the local land office, or the book "Roadside Geology" for your state or area. Also, to identify the type of fossils that you may find, check the local library for field books on fossils.

DECIPHERING THE FOSSIL RECORD

1800

Baron Georges Cuvier

1800s French naturalist **Baron Georges Cuvier** excels in comparative anatomy (comparing animal structures) and pioneers the founding of paleontology (the study of fossils). Cuvier reconstructed extinct animals by comparing their fossils to the skeletons of modern animals.

1842 Fossil finds of large, extinct reptiles lead **Sir Richard Owen**, an English scientist, to suggest that these reptiles belong to a group of reptiles that were unlike any living animals, the dinosaurs.

Sir Richard Owen

1855 Geologists **Sir John Dawson** (Canadian) and **Sir Charles Lyell** (British) discover bones of amphibians in layers of the earth associated with tree ferns and early gymnosperms.

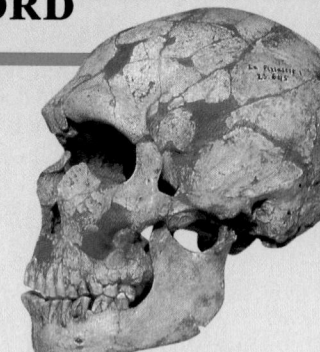

Neanderthal skull

1856 **Neanderthal skull** is found in Feldhofer Cave near Dusseldorf, Germany.

1861 *Archaeopteryx*, the earliest known bird, is discovered in a limestone quarry in Germany.

1933

1959 Anthropologist **Louis S. B. Leakey** and his wife, **Mary Leakey**, find the skull of *Australopithecus boisei* in Tanzania. It dates back to more than one million years.

Mary Douglas Leakey

1960 British anthropologist **Mary Leakey** and son **Jonathan** discover the first *Homo habilis* fossils at Olduvai Gorge in Tanzania. *Homo habilis* is the oldest member of our genus, and lived in Africa 2 million years ago.

1972 **Richard Leakey**, son of Louis and Mary Leakey, discovers a 1.75-million-year-old skull of *Homo habilis* in western Kenya.

Louis S. B. Leakey

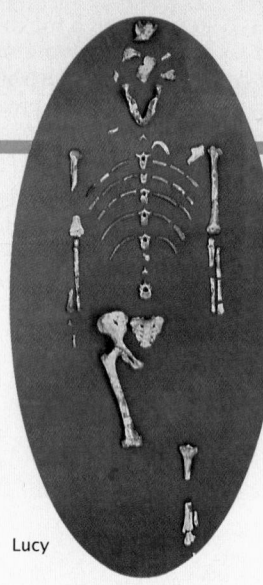

Lucy

1974 In Ethiopia, American anthropologist **Donald Johanson** discovers "Lucy," the 3 million-year-old skeleton of a bipedal female. Lucy is classified as *A. afarensis* and is the oldest known hominid.

*cience ▪ **Discoveries in Science** ▪ **Discoveries in Science***

1932

Dinosaur eggs

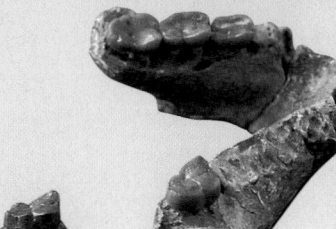

Ramapithecus jaw

1924 Raymond Dart, a South African anatomy professor, discovers the "Taung child," the skull of a four- to six-year-old *Australopithecus africanus* child who died approximately 1.5 million years ago.

Raymond Dart

1903 American explorer **Roy Chapman Andrews** leads expeditions into the Gobi Desert. His team finds dinosaur eggs and the remains of Earth's largest land animals, dinosaurs thought to be 95 million years old.

1927 Canadian anthropologist **Davidson Black** and French philosopher and paleontologist **Pierre Teilhard de Chardin** discover fossilized bones of *Homo erectus* near Beijing, China.

1932 American anthropologist **George E. Lewis** discovers jaw fragments and teeth of the ape *Ramapithecus,* which lived 8 million to 14 million years ago in northern India.

1990

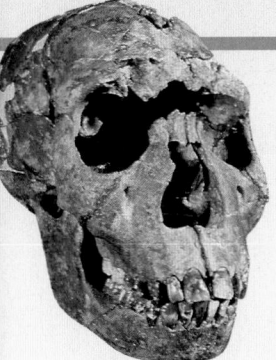

Homo erectus skull

1985 Kamoya Kimeu, Richard Leakey's Kenyan assistant, finds a nearly complete *Homo erectus* skeleton in western Kenya. The skeleton is that of a 12-year-old boy who died 1.6 million years ago.

1987 Wendy Slobada, a recent Canadian high school graduate, discovers a new dinosaur eggshell site near Milk River, Alberta, Canada; this site yields a fossil of an unhatched dinosaur.

Wendy Slobada

1988 In Argentina, University of Chicago paleontologist **Paul Sereno** (American) discovers what could be the oldest known dinosaur fossil, 230 million-year-old *Herrarasaurus.*

1992 Brian Anderson, an American graduate student at the University of California, Riverside, discovers rare fossilized impressions of dinosaur skin in the Books Cliffs of Utah.

- To make a connection with Chapter 15, explain that fossil species, like plant and animal species, are classified according to their biological features and how closely related they are to one another. Explain that generally features like the skull, teeth, limbs, shells, skeleton, etc., (the hard parts of the organism) are fossilized.

- Make a connection with Chapter 10 by explaining that most scientists are familiar with the geological time scale. Explain that usually when a fossil is found, it is found in the presence of other fossil species. If the paleontologist can position the other fossil species on the geological time scale, an approximate time frame can be assigned to the new find. Point out that the assigned date is not to be confused with the actual age of the organism. Rather, it describes the time period in which the organism lived. Explain that actual age is determined by measuring radioactive isotopes in the rocks that contain fossils. Have students research dating fossils and report their findings to the class.

Discussion

Guide the discussion by posing the following questions.

1. Why do scientists consider *Archaeopteryx* the skeletal link between reptiles and birds?
   Students should indicate that the link is based on the similarities of bone structure of dinosaurs and birds, and the presence of scales on the legs and feet of birds.

2. Would bone or fur be more likely to become fossilized?
   Students should suggest bone would be more likely to become fossilized. Fur would decay before being replaced by minerals.

3. Why would a hoax such as the Piltdown man be unsuccessful today?
   Students should suggest radioactive dating as the key to authenticating fossils.

607

Unit 6

Human Life

E very human life begins when
a single sperm cell penetrates
an egg cell and fertilization
occurs. As the fetus grows, the
organ systems needed to maintain
life develop. Those systems are
put to work daily in everything
we do. As we age, organ systems
slow down and cease to function
properly. Although every person,
like every living thing, must die,
the cycle of life continues. In this
unit, you will learn how life begins
and how the body's organ systems
work together.

Chapter 27 The Human Body

Planning Guide

	Objectives/Themes	Classwork Resources	Homework Resources

27.1	**1.** Explain the roles of the four kinds of tissues in the human body.
	
	2. Describe four types of connective tissues.
	
	3. List four organ systems of the human body.
	
	Themes: Interacting Systems, Scale and Structure, Stability
		Teacher's Resource Binder
		Focus Activity 27
		  Comparing Skeletal Joints
		
		Other Resources
		Transparency 119
		
		Text
		Science in Action
		  Videoscope Surgery
		  pp. 614–615
			Text
			Section Review, p. 613
			
			Teacher's Resource Binder
			Directed Reading Worksheet
			  27.1
			
			Other Resources
			Audiocassette 27.1*

27.2	**1.** Compare and contrast the dermis and the epidermis.
	
	2. Describe the functions of the different components of the dermis.
	
	3. Explain why acne is a common problem for adolescents.
	
	4. Name two ways to reduce the risk of skin cancer.
	
	Themes: Interacting Systems, Patterns of Change, Scale and Structure, Stability
		Teacher's Resource Binder
		Lab Investigation 27.1
		  Sweat Gland Activity
		
		Other Resources
		Transparency 120
			Text
			Section Review, p. 619
			
			Teacher's Resource Binder
			Directed Reading Worksheet
			  27.2
			
			Other Resources
			Audiocassette 27.2*

27.3	**1.** Draw a diagram of a typical long bone and label its parts.
	
	2. Differentiate a fracture from a sprain.
	
	3. Discuss the causes and effects of osteoporosis.
	
	4. Identify five types of joints in your body and explain the movement each permits.
	
	Themes: Interacting Systems, Patterns of Change, Scale and Structure
		Text
		Discoveries in Science
		  Uncovering the Secrets of the Body pp. 810–811
		
		Other Resources
		Transparencies 121–122
			Text
			Section Review, p. 625
			
			Teacher's Resource Binder
			Directed Reading Worksheet
			  27.3
			Reteaching Worksheet*
			  Types of Joints
			
			Other Resources
			Audiocassette 27.3*

27.4	**1.** Compare and contrast the three types of muscles.
	
	2. Explain how muscles work to move a bone.
	
	3. Describe how different forms of exercise affect muscles.
	
	4. Discuss the dangers of taking anabolic steroids.
	
	Themes: Scale and Structure, Stability, Interacting Systems
		Text
		Journeys
		  Tour of a Skeletal Muscle p. 629
		Investigation
		  How Do Muscles and Bones Work Together? pp. 636–637
		
		Teacher's Resource Binder
		Extension Worksheet
		  Strength of Bones and Muscles
		
		Other Resources
		Transparency 123
			Text
			Section Review, p. 632
			
			Teacher's Resource Binder
			Directed Reading Worksheet
			  27.4
			Vocabulary Review Worksheet*
			
			Other Resources
			Audiocassette 27.4*

*Reteaching Options

Demonstrations
27.1: p. 612
27.2: pp. 618, 619
27.3: pp. 621, 622, 623
27.4: pp. 626, 627, 630

Assessment
Chapter Review pp. 634–635
Portfolio Assessment p. 609D
Chapter Test—Teacher's Resource Binder
Test Generator

Research Notes

Connection to Space Sciences: NASA Bioreactor Grows Tissues

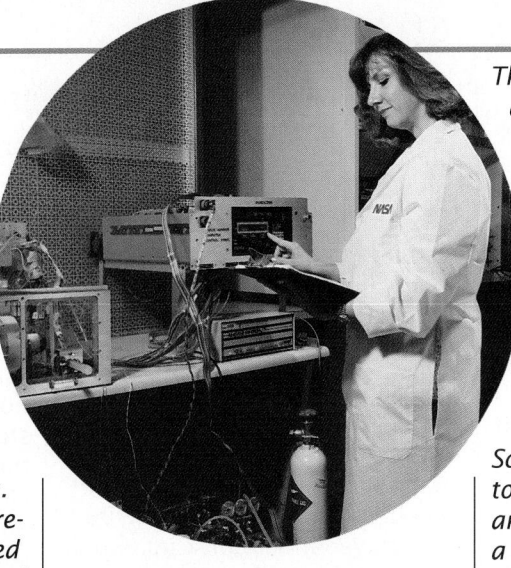

Human cells grown in culture are often very fragile and do not develop advanced stages of organization. As part of the space shuttle's life sciences program, NASA scientists created a bioreactor to grow cells in a liquid culture while protecting them from the forces of launching and landing. Researchers who tested the bioreactor on Earth got an unexpected surprise: cells grown in it reached an advanced stage with a three-dimensional structure resembling that of natural tissue.

NASA's bioreactor is called a "rotating wall bioreactor," and it consists of a sealed roller bottle with a cylinder in the middle. Cells are suspended in a culture medium between the two rollers, which are rotated together with a motor. As a result, the liquid culture rotates as if it were solid, and the rotating liquid is not sloshed back and forth by external forces.

The structures that the cells form are more complex than those grown in other cultures because the conditions inside the reactor allow different types of cells to grow closer to each other and interact chemically. Researchers working with lung and liver cells found evidence that tubular glands and blood vessels began to form in the cultured tissue. In addition, enzymes that formerly were made only by intact organs can now be made by lung and liver cells in this special type of culture.

The reactor is being used to grow artificial brain tumors and colon cancers to test antigens, antibodies, and white blood cell responses. Other researchers are growing tissues that can be infected with a common virus, unlike those in standard cultures. Another team is growing cartilage in pieces that are larger than those grown with other techniques.

Scientists at NASA look forward to the day when the bioreactor and its cell cultures are flown on a shuttle mission. It is possible that in microgravity the process of tissue formation might work even better than on Earth. Some are predicting that within 20 years, entire organs for transplants could be grown in space.

Investigation Notes

How Do Muscles and Bones Work Together?
pages 636–637

Purpose: In this activity, students observe the muscles, bones, tendons, and ligaments of the wing. Simple movements of the wing illustrate the interactions of joints, cartilage, and opposing muscle pairs.

Prelab Preparation

1. Use fresh, raw chicken wings that have been soaked in 70 percent ethanol and refrigerated overnight in order to kill *Salmonella* and act as a fixative. Cover chicken wings with plastic wrap to reduce vapors. Store wings in an explosion-proof refrigerator.

2. Be sure to mark a dotted line on the wing from the exposed end of the upper wing to the first joint, so students know where to make the first cut.

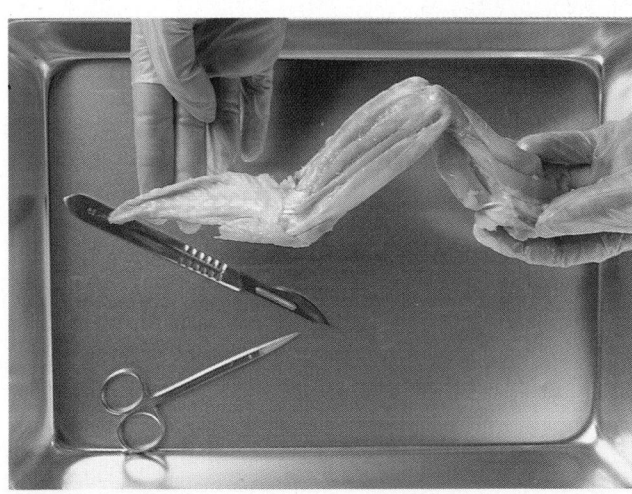

3. This laboratory will work better with pairs of students working together rather than with individuals working alone.

Answers will be found on pages 636–637.

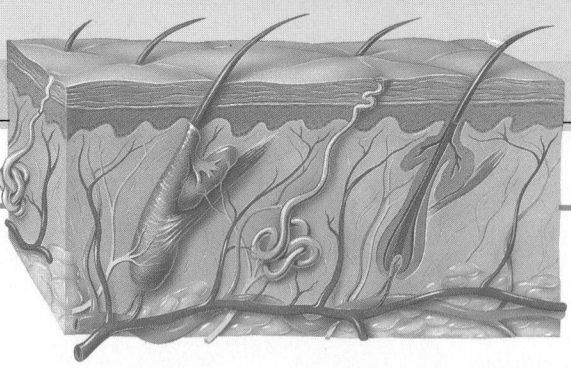

Meeting Individual Needs

Objectives

1. Students will demonstrate appropriate use of core vocabulary for the chapter (Vocabulary File).

2. Students will relate tissue types and organ systems to examples of their functions (Teaching Strategy).

3. Students will relate the prevention of osteoporosis to their own lifestyle (Multicultural Lesson Plan).

Vocabulary File
(Developing Vocabulary/ Limited English Proficiency)

If you are not already using the Vocabulary File, refer to Chapter 1 for its preparation. See the Chapter Highlights on page 633 for a list of suggested words.

Teaching Strategy
Functions of Body Systems
for pages 611–613
(Developing Organizational Skills/Visual Learners)

Materials: unlined index cards and markers

Preparation: On one side of the card, print "Relating Systems and Tissues." On the other side of the card, print a single phrase relating to a specific aspect or feature of a type of tissue or an organ system described in the text. The following is a suggestion for phrases on one group of cards:

    pinprick
    goose bumps
    carry nutrients
    regulate body temperature
    cool body with sweat
    keratin
    melanin
    acne
    malignant melanoma

All of these in some way relate to the skin or epithelial tissue. Make similar sets of cards for the circulatory system, the digestive system, the reproductive system, bones, muscles, and nerves. Lay these cards with the phrase side face down. Spread them out over a long lab table.

Procedure: Divide the class into two teams. Each team alternates taking turns, with different team members taking turns at playing. Each player should pick up two cards and turn them over so that the word/phrase sides of both cards are facing up. If the student thinks they relate, he or she should keep them. If the student does not think they match or relate to the same body system, he or she should turn the cards over again.

A player from the other team then has a turn. When all cards have been chosen, each student must show his or her pairs and explain how each pair is related. In the explanations, references may be made to the text. The winning team is the one with the most correct choices/ explanations.

Be prepared for students to combine items in unexpected ways. Some of them can relate to more than one tissue type or organ system. For example, "goose bumps" could relate to skin, muscle, or nerves.

Multicultural Lesson Plan
Osteoporosis
(Dealing with Ethnic Diversity)

Preparation: Help students find adults who they may interview about osteoporosis. Possible sources include family members, churches, retirement homes, nursing homes, and senior citizens' centers. Gather literature about osteoporosis from local doctors, clinics, or hospitals. If possible, arrange for a guest speaker to discuss the disease with students.

Teaching Strategies:
1. Use the literature on osteoporosis, or a group speaker, to help start a discussion of the causes, symptoms, and prevention of osteoporosis. Be certain students realize that although this disease mostly strikes the elderly, students' lifestyle choices now can affect their own health later.

2. Have students interview the elderly from their family, from local churches, in retirement and nursing homes, and at senior citizens' centers. They should gather information about the age, sex, and ethnic group of those who have suffered broken bones or have hunched backs.

Assessment:
1. Students will prepare a report detailing their findings and noting any patterns present. (In general, osteoporosis tends to strike white and Asian women, although its incidence in the African-American population is increasing.) The report should also include causes and preventions of osteoporosis, indicating changes that students can make now in their lifestyles to help prevent osteoporosis.

Additional Strategies
Visual Strategies
Pages 611, 612, 613, 616, 617, 620, 621, 622, 623, 624, 625, 627, and 630

Auditory Learners
Use *Biology: Visualizing Life* Audiocassettes for Sections 27.1, 27.2, 27.3, and 27.4.

Meeting Individual Needs (cont.)

Cooperative Learning
Muscles, Bones, and Skin

Timing: Use this activity to introduce Section 27.2.

Group Size: 3–6 students

Outcome: Students will describe major muscles, bones, and protective coverings of the body.

Individual Accountability: Each group member is responsible for one of three body systems.

Positive Interdependence: Each group will work together to prepare a presentation for the class about the muscles, bones, and protective coverings of a specific area of the human body.

Assign a different area of the human body to each group. Suggestions include: head, arm, leg, hand, or foot. Ask each group to divide up the task of investigating the muscular, skeletal, and epithelial tissue present in each body part, and the functions of the tissue. Research resources should include anatomy books, anatomical charts, and other library resources.

After initial research, ask group members to assemble their information into a presentation for the class. Each group member should know how many bones, muscles, and types of epithelial tissues are found in the body part they studied. The group should also be able to describe the functions of these components.

Portfolio Assessment

Students should select their best work and provide a self-reflective rationale for their selections. Students can make selections in the following areas.

1. *Content* — One concept map from the chapter (See page 812 for evaluation criteria.)

2. *Reading Comprehension* — One Directed Reading Worksheet from the Teacher's Resource Binder (Use the answer key to evaluate for accuracy.)

3. *Writing* — Using the Vee Form, summarize a magazine or newspaper article relating to body systems or tissues. (See page 22T for evaluation criteria.)

   Or: Select a writing task or project from the Chapter Review.

4. *Performance Assessment* — One Vee form from a chapter investigation or lab manual investigation (See page 22T for evaluation criteria.)

Teacher makes selections in the following areas.

1. *Formal Assessment* — Chapter test (Test A, B, or the Test Generator) The teacher-scored test should be reviewed by the student. Incorrect responses should be corrected by the student before the test becomes part of the portfolio.

2. *Informal Assessment* — Use the Direct Observations Checklist, page 33T, during a laboratory or other cooperative learning experience.

3. *Performance Assessment* — Have students research muscle pairs and create a working model of an arm, using pulleys, rods, and string.

Concept Map Answer

The following is one possible answer to the Relating Concepts exercise on page 634.

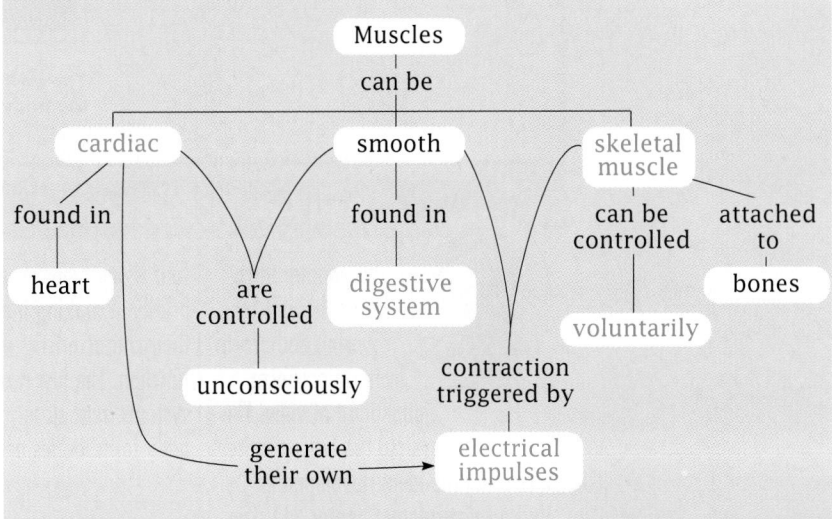

Chapter 27

Determining Prior Knowledge

- Show the class models or photographs of various human organs, such as the heart, lung, stomach, and brain. Have students identify to which body system each belongs and also have them name other organs belonging to the same system.
- Have students use a stereomicroscope to examine a piece of chicken skin. Have them discuss various functions of the skin.
- Ask if any student has had a sprain, fracture, or pulled muscle. Have each student describe what happened and how the injury was treated.
- Show students photographs of people engaged in various physical activities. Ask what each type of exercise does for muscles.

Chapter 27

The Human Body

Review

- cell division (Section 4.3)
- tissues (Section 21.1)
- innovations in body plan (Section 21.3)

The swift strokes of a crew team require the coordination of the body's skin, bones, muscles, and other tissues.

▪ Author's Rationale ▪

This chapter introduces the unit by covering body form from an organizational point of view. The first section lays out the basic architecture using the ideas developed in Unit 5 (particularly Chapter 21). The four tissue types should be covered briefly in making the point of how their characteristics match their function. The last two sections cover systems most closely associated with body form: bones and muscles.

YOUR BODY CONSISTS OF MANY SPECIALIZED TISSUES SUCH AS THOSE FOUND IN SKIN, BONES, MUSCLES, NERVES, AND BLOOD VESSELS. THESE TISSUES, EACH MADE OF SPECIALIZED CELLS, ARE ARRANGED DIFFERENTLY IN EACH PART OF YOUR BODY. THIS PROCESS OF SPECIALIZATION AND ARRANGEMENT BEGAN BEFORE YOU WERE BORN WHEN A SINGLE CELL BEGAN TO DIVIDE.

27.1 *Tour of the Human Body*

Objectives

① **Explain the roles of the four kinds of tissues in the human body.**

② **Describe four types of connective tissues.**

③ **List four organ systems of the human body.**

Similar Cells Form Tissues

Figure 27.1
The human body is made of four types of tissue, as shown below. All work together to make the body a functional unit capable of all life activities.

All of the trillions of cells in the human body arise from a single cell—the fertilized egg. The first few cells that form by cell division after fertilization look alike. Soon, however, the new cells become specialized to carry out particular tasks.

As cells specialize, they begin to group together to form tissues. A **tissue** is a group of similar cells that work together to perform a specific function, such as movement or protection. Your body contains four types of tissues—epithelial tissue, muscle tissue, nerve tissue, and connective tissue, as shown in **Figure 27.1**.

Epithelial tissue covers and protects

Epithelial (*ehp uh THEE lee uhl*) **tissue** is made of tightly connected cells that are arranged in flat sheets, often only a few cells thick. Just as canvas protects machinery that must be stored outdoors, epithelial tissue prevents damage to the cells that lie beneath it. Skin is an example of epithelial tissue. Epithelial tissue also lines spaces within the body and covers the inner and outer surfaces of your internal organs.

Some epithelial tissue has an entirely different function. This epithelial tissue contains exocrine glands. An **exocrine gland** is a cell or group of cells that produces and releases secretions onto a body surface. One type of exocrine gland in your skin secretes sweat, which helps cool your body. Exocrine glands in the epithelial tissue lining your digestive system produce enzymes that break down food.

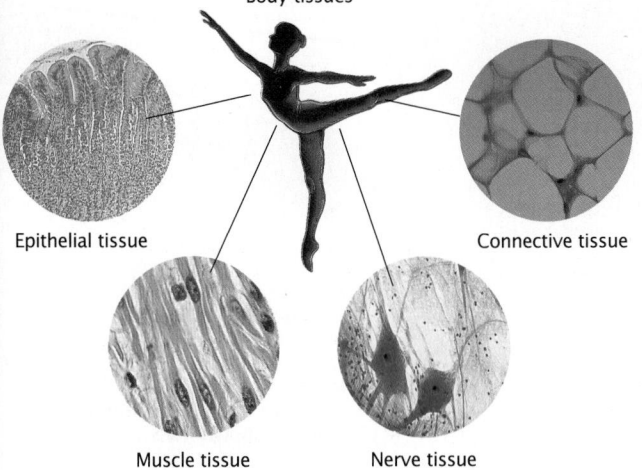

Body tissues

Epithelial tissue

Connective tissue

Muscle tissue

Nerve tissue

■ *Cultural Perspective* ■

Imhotep
(African-Egyptian)
Imhotep is the first physician, known by name, in history. He lived about 2650 B.C. in ancient Egypt. Many early Greek physicians studied under Imhotep. After his death he was elevated to the status of a god in Egypt,

and Greeks identified him with their own god of healing, Asclepius. Temples and statuettes were erected in Imhotep's honor. A statue of him stands in the Hall of Immortals in the International College of Surgeons in Chicago.

Demonstration 1

Epithelial Tissue

Use a microprojector to show a prepared slide of epithelial cells, so that students can visualize the arrangement of epithelial tissue into flat sheets. Point out that epithelial tissue covers surfaces and lines cavities.

Demonstration 2

Muscle Tissue

Show the class a poster or diagram illustrating the human muscular system. Point out that the body has more than 600 muscles, accounting for about 40 percent of the body's weight.

Demonstration 3

Nerve Tissue

Show the class a piece of telephone cable. Compare a neuron to one of the colored conducting cables that transmits signals over long distances. Point out that a nerve is actually a bundle of many neurons—usually hundreds or even thousands. Just like the individual wires in a telephone cable, each neuron is capable of transmitting a separate message.

Demonstration 4

Connective Tissue

Display a human skeleton or pass various bone specimens around the class so that students can examine one type of connective tissue. Students should recognize that bone, although strong and rigid, is amazingly light, accounting for less than 20 percent of body weight.

Visual Strategy

Figure 27.2

Use this figure to reinforce a major difference between connective tissue and the other three types; it contains relatively few cells and is formed mostly from cell products like fibers and fluid. Students should understand that connective tissue is the most widely distributed tissue in the body since it joins, supports, and protects the other types.

Muscle tissue moves the body

Have you ever awakened with sore muscles? If so, you probably became aware of how often you use those particular muscles. **Muscle tissue** moves the parts of your body. Even when you are sitting still, muscles are at work moving food through your digestive system, maintaining your posture, and moving blood through your heart.

Muscle tissue is made of cells that contract, or shorten, and then return to their normal length. Another characteristic of muscle tissue is that it responds to electrical stimulation. Electrical signals control when a muscle will contract.

Nerve tissue sends electrical signals through the body

As you walk up a flight of stairs, the electrical signals that cause the muscles in your legs to contract are produced by nerve tissue. **Nerve tissue** is found in your brain, nerves, and sense organs. It contains cells called neurons that are able to generate electrical impulses and transfer the impulses to other cells. Neurons have long extensions that can carry electrical impulses long distances. Nerve tissue also contains other cells that nourish and protect the neurons that carry the electrical signals.

Connective tissue joins, supports, and transports

The bones of your skeleton and the ligaments that join them are examples of **connective tissue**. Connective tissue actually contains few cells. Most of the tissue consists of fluid and fibers secreted by the cells in connective tissue. **Figure 27.2** shows the four basic types of connective tissue, all of which can be found in your knee. They are cartilage, bone, connective tissue proper, and blood.

Figure 27.2
The graceful leap of a dancer depends on many types of tissues including muscle, nerve, and epithelial tissues. These tissues and all four types of connective tissue can be seen in this longitudinal section of a human knee.

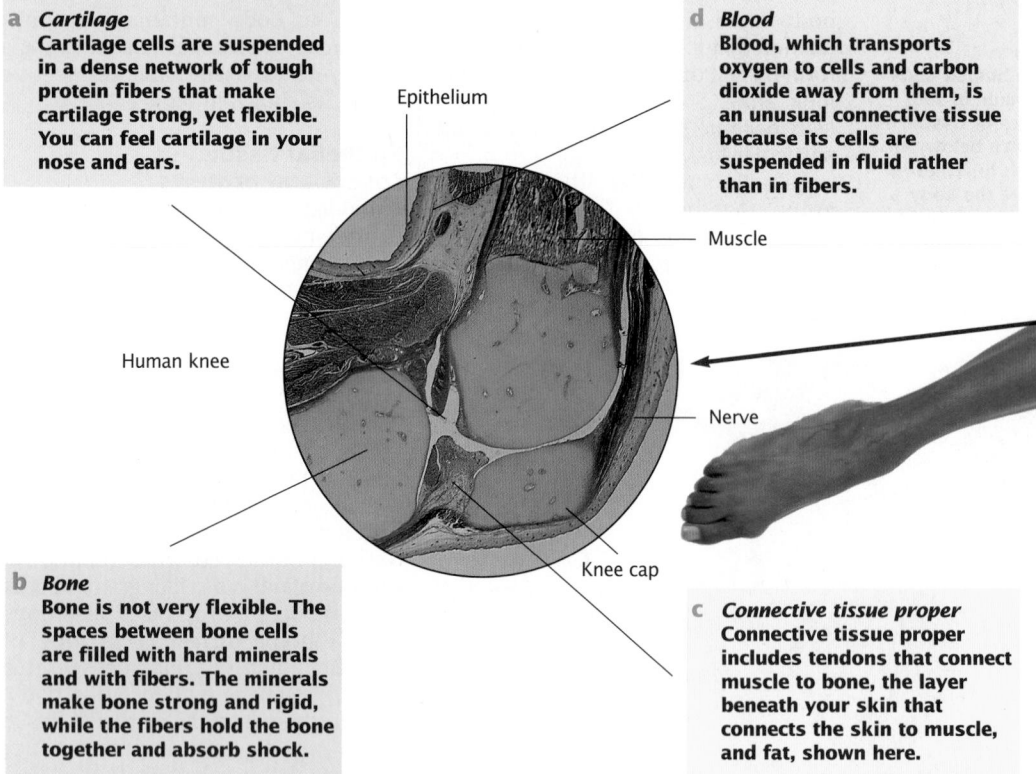

a **Cartilage**
Cartilage cells are suspended in a dense network of tough protein fibers that make cartilage strong, yet flexible. You can feel cartilage in your nose and ears.

b **Bone**
Bone is not very flexible. The spaces between bone cells are filled with hard minerals and with fibers. The minerals make bone strong and rigid, while the fibers hold the bone together and absorb shock.

d **Blood**
Blood, which transports oxygen to cells and carbon dioxide away from them, is an unusual connective tissue because its cells are suspended in fluid rather than in fibers.

c **Connective tissue proper**
Connective tissue proper includes tendons that connect muscle to bone, the layer beneath your skin that connects the skin to muscle, and fat, shown here.

Epithelium

Muscle

Nerve

Human knee

Knee cap

■ *Cultural Perspective* ■

Curtis L. Parker, Ph.D.
(African-American)
Dr. Curtis L. Parker, a zoologist, developed a method to make chick embryo cells differentiate in culture the way they do when they form muscle and cartilage in a living animal. Dr. Parker performed research in the areas of limb development, differentiation, and regeneration.

Tissues Form Organs

How can only four kinds of tissues make up all the different organs in your body? The answer is simple: different tissues are combined in different ways to form each organ. An **organ** is a structure composed of a number of tissues that work together to perform a specific job in the body.

For example, the largest organ of your body—your skin—contains all four types of tissues. Your skin is made of many watertight sheets of epithelial tissue covering a cushioning layer of connective tissue. Nerve tissue in your skin makes it sensitive to touch. Tiny muscles can make the fine hairs that cover most of your body stand on end when you are cold. All of these tissues cooperate to protect your body.

Organs that work together form an **organ system**. An example of an organ system is the circulatory system. The heart and the blood vessels are separate organs that work together to carry substances through the body. **Figure 27.3** shows four organ systems you will learn more about in this chapter and in chapters to come.

Figure 27.3
Each organ in this dancer's body performs a specific task related to other organs in its system. For instance, in the circulatory system, the heart pumps blood through vessels that deliver oxygen to the lungs and to other organs.

a Circulatory system

b Skeletal system

c Reproductive system

d Digestive system

Section Review

① How does each of the four tissue types function in the human body?

② How might a disease that slowly destroys nerve cells affect muscle tissue in the body?

③ Name four connective tissues found in the knee.

④ List four organ systems of the human body.

Visual Strategy

Figure 27.3
To review the major concepts of this section, have students identify an organ and tissue type found in each of the four systems illustrated in this figure.

Theme Connection

Interacting Systems and Stability
Be sure students understand that all the systems must work together in a coordinated manner in order for the body to function normally and maintain homeostasis.

Phase 3

ASSESSMENT OPTIONS

Closure Strategy

Organ Transplants
Have students list body parts that surgeons have transplanted in humans. Have them identify whether the part is a tissue or organ. Their list may include such organs as the heart, kidney, liver, lung, skin, and pancreas, and such tissues as blood, bone marrow, cornea, lens, and bone.

Section Review
Assign the *Section Review.*

Reteaching
Use a microprojector to show students prepared slides of the four tissue types. Have them identify each type and explain the reasons for their answer.

■ *Section Review Answers* ■

1. Epithelial tissue covers and protects the body and lines spaces within the body. Muscle tissue moves the body. Nerve tissue sends electrical signals through the body. Connective tissue joins, supports, and transports.

2. Students should suggest that muscles would not receive signals to contract. Over a period of time, muscle tissue would begin to atrophy.

3. See Figure 27.2 on page 612.
4. See Figure 27.3 on page 613.

Science in Action ▪ *Science in Action* ▪ *Science in Action*

Using the Feature

- You can use this feature to lead an informal discussion about students' encounters with surgery. Some students themselves may have undergone surgery; some may have family members or friends who have had an operation. To those students who volunteer information, ask what kind of surgery was performed, how long the person was in the hospital, how the person felt during recovery, etc. Would the person have benefited from videoscope surgery? Why or why not?

- Students might enjoy hearing some firsthand accounts of the activities in an operating room. Invite a surgeon, nurse, or other surgical assistant to talk about his or her experiences in the operating room. This person can explain their responsibilities during surgery and the training it took to acquire them.

Discussion

1. Videoscope surgery is said to be rapidly displacing conventional surgical techniques. Describe the instruments and techniques used in videoscope surgery. *Videoscope surgery uses slender fiber-optics tubes that are inserted deep inside the body through tiny (1-cm-long or less) incisions. Some of these tubes are for grasping, snipping, and stapling internal organs; others are for projecting images onto a videoscreen for the surgeon to study.*

Videoscope Surgery

Futuristic Surgery? The Time Is Now

In a popular futuristic science-fiction series, doctors perform surgery without spilling a drop of blood. Knives and scalpels are obsolete, and patients are up and around in no time. Now, thanks to fiber optics and video technology, bloodless surgery may not be too far in the future.

Increasingly, physicians are using a technique called videoscope surgery to diagnose and treat injured and diseased body organs. In videoscope surgery, a surgeon makes a small incision and inserts a long slender fiber-optic tube into the body. The tube has a tiny camera lens at the end that transmits images to a nearby television screen in the operating room. Surgical instruments for grasping, snipping, and stapling can be inserted through other small incisions. The surgeon controls the instruments from outside the body while watching the screen.

In 1987, videoscope surgery was used for the first time to remove a diseased gallbladder. In traditional gallbladder surgery, a 7-cm to 13-cm (3-in. to 6-in.) cut is made along the patient's abdomen. With the new surgery, four small incisions are made that heal faster and hurt less. Videoscope surgery is now used on about three-quarters of the American patients having their gallbladders removed.

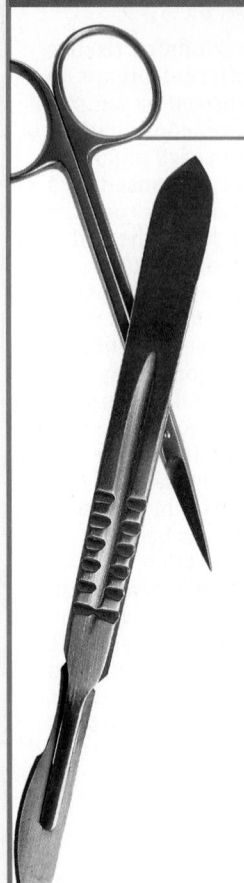

Will these surgical instruments soon be outdated? One study predicts that 40 percent of all surgery will be performed with videoscopes by the next century.

Widespread Applications

In addition to gallbladder surgery, the new technique is rapidly being adapted to other kinds of surgery. For instance, one surgeon has removed a diseased kidney this way, after using instruments to maneuver the organ into a small sac and chop it into smaller pieces. Another surgeon recently began using videoscope surgery to perform lung biopsies to identify cancerous tissue. In the past, biopsy surgery involved slicing through chest and rib muscles. Afterward, a patient would spend several days in intensive care and then have weeks of painful recovery. With the new technique, blunt-tipped instruments are pushed gently through the muscles, which stretch temporarily. When finished, the surgeon closes up the small incision with a few stitches and a bandage.

Although videoscope surgery is likely to have a great impact in surgery of the chest and abdomen, other operations have already benefited from fiber-optic instruments. For example, the torn cartilage associated with many joint injuries, especially in the knee, can be repaired using an instrument called an arthroscope. Arthroscopic surgery has become widely used on athletes who used to have to undergo extensive surgery to treat their knee injuries.

Gallbladder removal		
Surgery:	Traditional	Videoscope
Incision(s):	one 7-cm to 13-cm (3-in. to 6-in.) cut	four 0.5-cm (0.25-in.) punctures
Average cost:	$4,250	$3,500
Hospital stay:	5 to 8 days	overnight
Recovery time:	4 to 6 weeks	5 to 7 days
Complications:	1% to 9.4%	5.1%
Deaths:	0.2% to 0.6%	0.1%

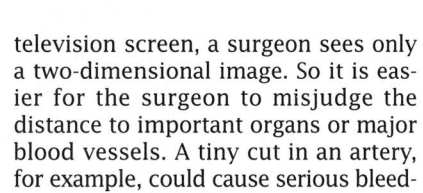

■ *Action* ■ *Science in Action* ■ *Science in Action* ■ *Science*

Taking Precautions

As videoscope surgery becomes more common, however, some experts have become concerned. Operating by video camera requires a lot of practice. Surgeons have to acquire a new set of skills. For example, when looking at a television screen, a surgeon sees only a two-dimensional image. So it is easier for the surgeon to misjudge the distance to important organs or major blood vessels. A tiny cut in an artery, for example, could cause serious bleeding that would be hard to stop.

In videoscopic gallbladder surgery, if the surgeon is relatively new at the technique, patients stand a greater risk of injury to the bile duct, the tube connecting the gallbladder to the small intestine. Consequently, some medical societies are drafting guidelines to ensure that surgeons are well trained in this new technique. One recommendation is to have doctors practice on animals first. Then, surgeons who are already experts in the technique will supervise the newly trained surgeons' first operations on humans.

In addition to better training and more practice, video surgeons also need improvements in their equipment. Stiff necks from straining to watch a television screen and stiff elbows from using awkwardly designed instruments are common problems. Nevertheless, the field is progressing quickly. Eventually, say some doctors, cutting a patient open for certain kinds of surgery might go the way of the dinosaurs.

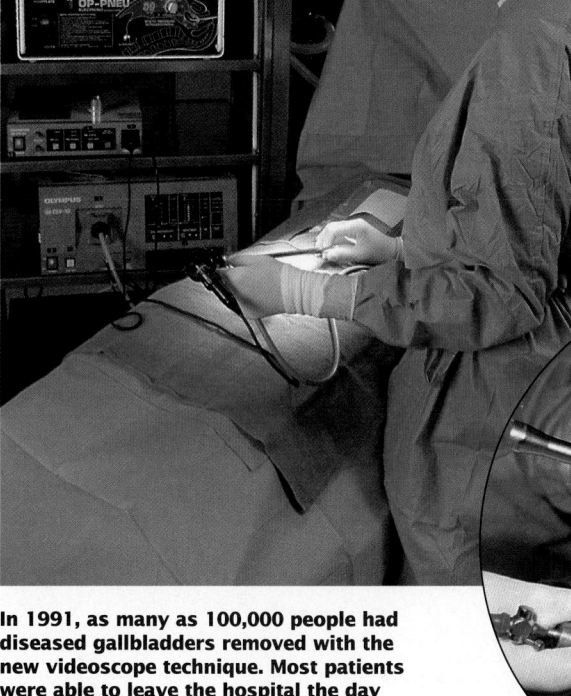

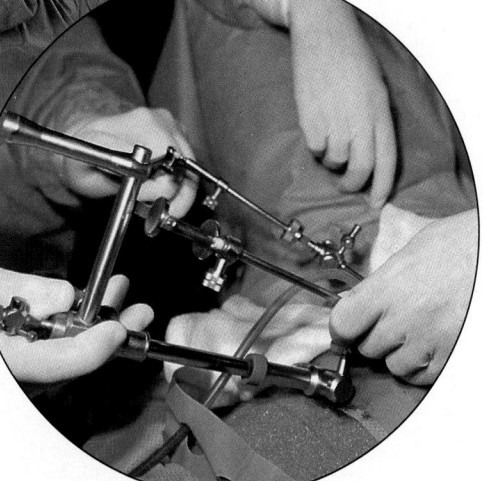

In 1991, as many as 100,000 people had diseased gallbladders removed with the new videoscope technique. Most patients were able to leave the hospital the day after surgery with only slight bruising and a bandage.

2. Why has videoscope surgery become so popular?
When it is correctly performed, videoscope surgery can dramatically reduce surgical trauma.

3. List three applications of videoscope surgery.
It is being used to remove gallbladders and kidneys, and to operate on joint injuries, especially in the knee.

4. Discuss a concern experts have about videoscope surgery.
Surgeons will have to acquire a new set of skills to safely manipulate the somewhat cumbersome equipment. Medical societies are drafting guidelines to ensure that surgeons are well trained in this new technique.

615

Lesson Plan 27.2

Phase 1

PREPARATION

Key Concepts

- The skin is composed of two layers: an outer epidermis and an underlying dermis.
- The dermis contains epithelial, muscle, nerve, and connective tissues.
- The epidermis is made entirely of epithelial tissue containing melanin and hair follicles.
- Skin disorders include acne and skin cancer.

Reading Strategy

Drawing inferences from data is a scientific process. An inference is drawing a conclusion or making a deduction from facts stemming from data or observations. Have students practice this scientific skill by writing three inferences based on their reading of this section. For example, they may infer that persons with fewer blood vessels in their dermis will bleed less when cut, or that people with more sweat glands will need to drink more water on hot summer days.

Phase 2

TEACHING STRATEGIES

Visual Strategy

Figure 27.4

Instruct students to identify examples of the four tissue types illustrated in this figure. Have them calculate the relative thickness of the dermis and epidermis by using a metric ruler to measure each of the layers shown in this figure.

616

SKIN IS A LOT MORE THAN JUST AN OUTER COVERING. IT IS THE LARGEST—AND ONE OF THE MOST IMPORTANT—ORGANS OF THE BODY. IT CUSHIONS AND PROTECTS INTERNAL ORGANS AND PREVENTS THE LOSS OF THE FLUIDS THAT BATHE ALL YOUR CELLS. SMALL CUTS, ABRASIONS, AND BURNS ARE REPAIRED WITH REMARKABLE SPEED. OLD SKIN CELLS ARE CONTINUALLY REPLACED BY NEW, HEALTHY CELLS.

27.2 Skin

Objectives

1. **Compare and contrast the dermis and the epidermis.**
2. **Describe the functions of the different components of the dermis.**
3. **Explain why acne is a common problem for adolescents.**
4. **Name two ways to reduce the risk of skin cancer.**

The Dermis

Figure 27.4
On the surface, skin looks uniform. But underneath it is a complex organ made of blood vessels, nerve fibers, glands, and muscles.

Your skin is composed of two main parts: an outer layer called the epidermis, and an underlying, thicker layer. This inner part of your skin is called the **dermis**. Shown in **Figure 27.4**, the dermis is composed mainly of connective tissue. Fibers in the dermis make your skin tough, flexible, and elastic. The dermis also contains nerves, muscles, blood vessels, and glands.

Nerves and muscles run throughout your skin

When a friend taps you on the shoulder or you accidentally prick yourself with a pin, you immediately react. Nerves in the dermis make it possible for you to respond appropriately with a "hello" or an "ouch." These nerves enable you to sense pressure, temperature, and, of course, pain.

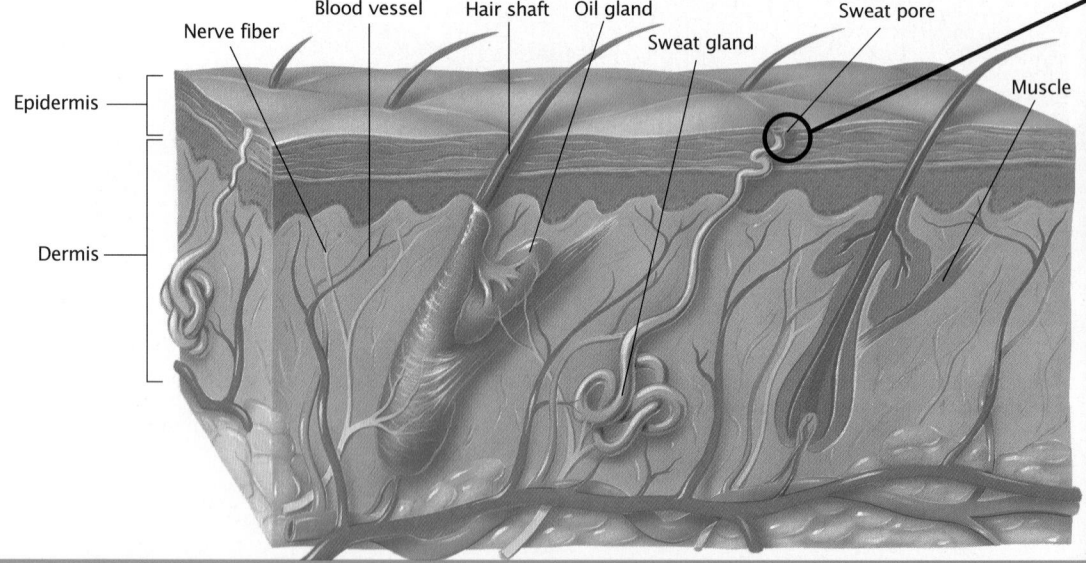

Nerve fiber · Blood vessel · Hair shaft · Oil gland · Sweat gland · Sweat pore · Muscle · Epidermis · Dermis

---

▪ Cultural Perspective ▪

Avicenna
(Persian)

Avicenna (Ibn Sina), a philosopher and physician from Persia (now called Iran), authored *The Canon of Medicine* around 1030. It stood as the Bible of medical knowledge for 500 years.

Stability

How does sweating

enable the human

body to maintain

homeostasis?

Your dermis also has tiny muscles that are attached to the hairs in your skin. When you are cold or afraid, the muscles contract, pulling the hairs upright. This same process happens in the skin of other mammals. The fur of a cat, for instance, will stand up when the cat is threatened by a dog, making the cat look larger and more dangerous. Or when the cat is cold, its fur fluffs up and traps more air near its body. Since trapped air is a good insulator, this helps the cat stay warm. The muscles in your skin behave just like those of a cat. However, since you don't have fur, you just get goose bumps—a leftover from our evolutionary past.

Blood vessels carry nourishment

Your skin, like all other living parts of your body, requires nourishment to live. This nourishment is supplied by blood that courses through tiny blood vessels in the dermis. In addition to carrying nutrients, the blood in these vessels carries away waste products and helps regulate body temperature. Blood radiates heat into the air as it passes near the surface of the skin. If your body becomes too hot, the blood vessels enlarge, allowing more blood to flow through the dermis near the body surface. This is why the skin of light-complected people often becomes reddish during strenuous exercise.

Sweat glands cool your body

Sweat is another way by which your body removes excess heat. Your skin contains about 100 sweat glands per square centimeter. Why is sweat important? The evaporation of sweat from the surface of your skin removes heat much more efficiently than simply radiating heat from the blood into the air. Without sweat, you would have great difficulty cooling your body on a hot day or after exercising like the person in **Figure 27.5**

Not all sweat is the same. Most sweat is about 99-percent water, mixed with small amounts of salts, acids, and waste products. However, the sweat from certain sweat glands, called apocrine glands, also contains proteins and fatty acids. These substances provide a rich source of food for bacteria that live on the skin. The waste products of these bacteria give sweat an unpleasant odor. Most apocrine glands are located in the armpits and groin area. You have probably noticed that sweat glands are activated by nervousness or stress, even in cool temperatures.

Figure 27.5
The evaporation of sweat from the surface of your skin helps cool your body during and after a tough workout.

Theme Answer
Stability
Homeostasis is an organism's ability to maintain a constant body temperature, regardless of outside temperature. The evaporation of sweat from the skin ensures that your body loses excess heat more efficiently than it would by radiating heat from the blood into the air.

Theme Connection

Stability
Be sure students recognize the skin's role in homeostasis: helping to eliminate wastes and regulate body temperature.

Visual Strategy

Figure 27.5
Use this figure to point out that a significant amount of heat can be lost through evaporation of sweat. To impress students with this homeostatic mechanism, relate the story of Dr. Charles Blagden, secretary of the Royal Society of London some 200 years ago. To demonstrate how mammals can regulate their body temperature, he, a few friends, a dog, and a steak spent 45 minutes in a room where the temperature measured 126°C (260°F). Upon emerging from the room, everyone and the dog were fine. The steak, though, was cooked!

Mitosis

Have students name the type of cell division responsible for producing new skin cells. Point out that a complete "surface armor" is generated every 27 days by mitotic divisions in the lower epidermis.

Demonstration 1

Epidermis

Have students use a stereomicroscope to examine the epidermis of their hands. Have them identify any skin marks, including moles, freckles, and birthmarks. Point out that most skin marks are formed from patches of pigment. Some birthmarks are the result of a collection of small blood vessels in an area. Tell students to examine their fingerprints. Point out that fingerprints are caused by ridges and indentations formed by the interlocking of the dermis and epidermis.

Demonstration 2

Albinism

Show the class a photograph of an albino person. Ask how this person's epidermis differs from that of most people. Point out that melanin is also responsible for eye color—a small amount produces blue eyes, more pigment results in green eyes, while a large amount is responsible for brown eyes.

Demonstration 3

Hairs

Use a microprojector to show a variety of hair samples carefully removed from student volunteers. Have the class compare the texture, thickness, and coloration of the hairs.

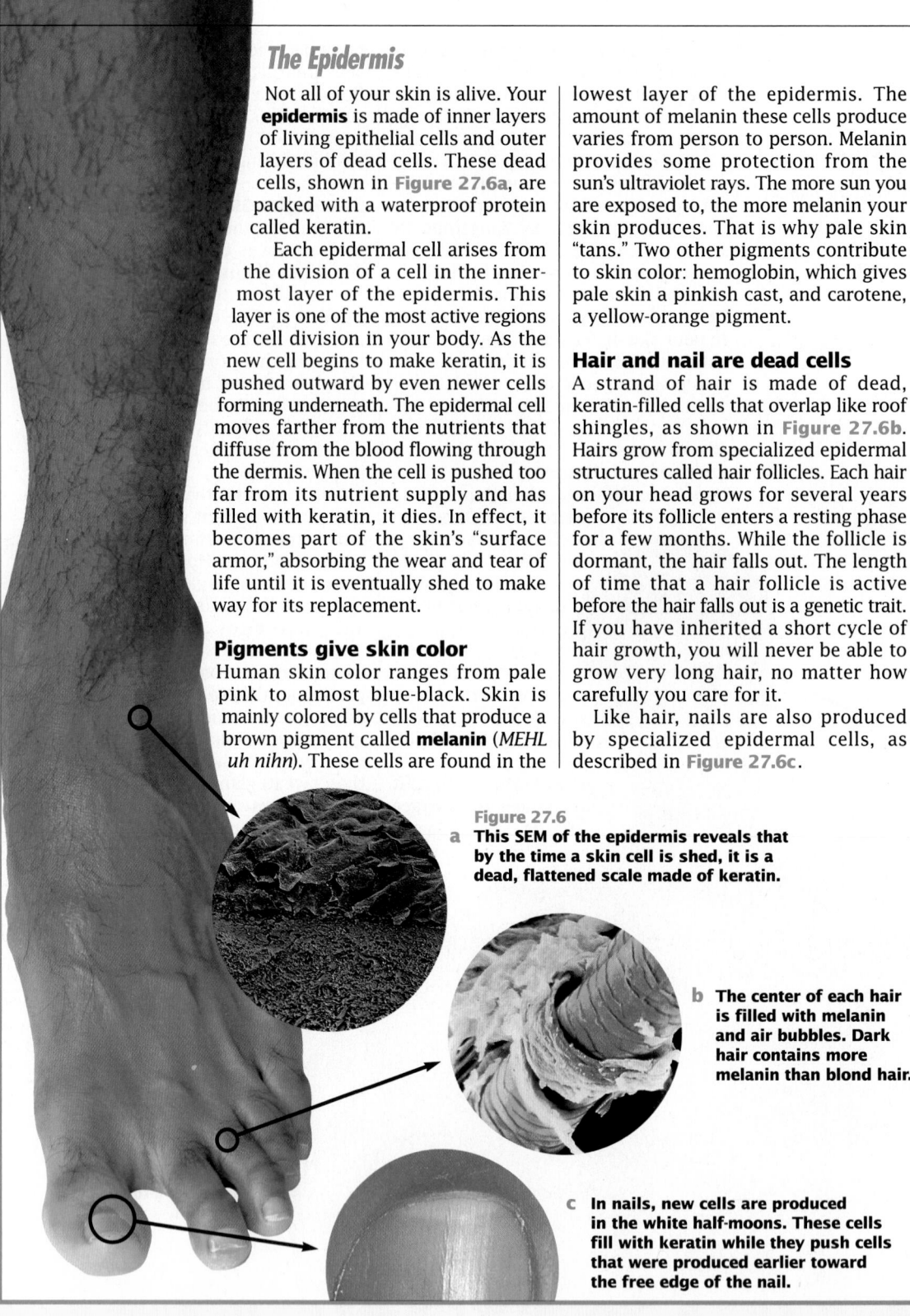

The Epidermis

Not all of your skin is alive. Your **epidermis** is made of inner layers of living epithelial cells and outer layers of dead cells. These dead cells, shown in **Figure 27.6a**, are packed with a waterproof protein called keratin.

Each epidermal cell arises from the division of a cell in the innermost layer of the epidermis. This layer is one of the most active regions of cell division in your body. As the new cell begins to make keratin, it is pushed outward by even newer cells forming underneath. The epidermal cell moves farther from the nutrients that diffuse from the blood flowing through the dermis. When the cell is pushed too far from its nutrient supply and has filled with keratin, it dies. In effect, it becomes part of the skin's "surface armor," absorbing the wear and tear of life until it is eventually shed to make way for its replacement.

Pigments give skin color

Human skin color ranges from pale pink to almost blue-black. Skin is mainly colored by cells that produce a brown pigment called **melanin** (*MEHL uh nihn*). These cells are found in the lowest layer of the epidermis. The amount of melanin these cells produce varies from person to person. Melanin provides some protection from the sun's ultraviolet rays. The more sun you are exposed to, the more melanin your skin produces. That is why pale skin "tans." Two other pigments contribute to skin color: hemoglobin, which gives pale skin a pinkish cast, and carotene, a yellow-orange pigment.

Hair and nail are dead cells

A strand of hair is made of dead, keratin-filled cells that overlap like roof shingles, as shown in **Figure 27.6b**. Hairs grow from specialized epidermal structures called hair follicles. Each hair on your head grows for several years before its follicle enters a resting phase for a few months. While the follicle is dormant, the hair falls out. The length of time that a hair follicle is active before the hair falls out is a genetic trait. If you have inherited a short cycle of hair growth, you will never be able to grow very long hair, no matter how carefully you care for it.

Like hair, nails are also produced by specialized epidermal cells, as described in **Figure 27.6c**.

Figure 27.6

a **This SEM of the epidermis reveals that by the time a skin cell is shed, it is a dead, flattened scale made of keratin.**

b **The center of each hair is filled with melanin and air bubbles. Dark hair contains more melanin than blond hair.**

c **In nails, new cells are produced in the white half-moons. These cells fill with keratin while they push cells that were produced earlier toward the free edge of the nail.**

Skin Disorders

Since your skin is the most exposed part of your body, it is often damaged. The damage may be minor—a blister, an insect bite, or a small cut. Other damage is more serious, as in the case of skin cancer. Your skin may also be affected by changes within your body.

Acne is caused by overactive oil glands

Acne is a common, undesirable side effect of adolescence. Normally, oil glands in your skin produce just enough oil to waterproof your hair and to seal moisture into the skin. However, high levels of sex hormones produced in the body during adolescence increase the oil production of the oil glands.

Excessive oil production can clog the oil glands, causing a buildup of oil. This accumulation of oil can appear as a whitehead. If this material is exposed to air, it darkens, forming a blackhead. If the oil builds up to the point that the oil gland actually bursts, the area around the gland becomes red and inflamed. Bacteria from the skin may also infect the damaged area, making the inflammation worse. The pimple that forms is evidence that the body is fighting the tissue damage and the invading bacteria. Although acne cannot be prevented, it can usually be controlled with proper skin care.

Skin cancer is caused by DNA mutations in skin cells

Cancers occur when mutations in DNA cause cells to lose the ability to stop dividing when they are supposed to. The sun's ultraviolet light is known to cause mutations. Years of exposure

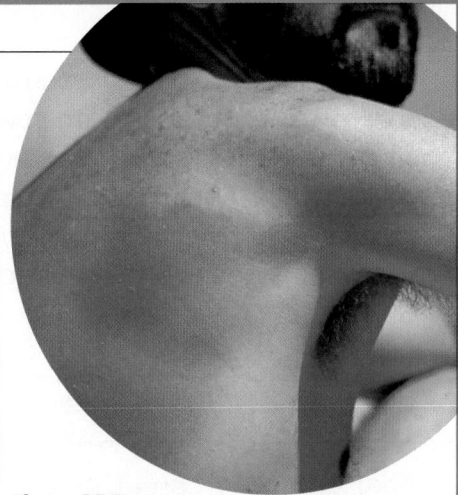

Figure 27.7
Sunburn can be painful and can lead to skin cancer and premature aging of the skin. To prevent sunburn, use a sunscreen and avoid direct sunlight.

to sunlight may result in the mutations that can cause cancer of the skin.

The danger of a particular skin cancer depends on the type of skin cell that becomes cancerous. The most common skin cancers have a very high cure rate when detected early. These include cancers that arise from the rapidly dividing cells of the lower layer of the epidermis, or the cells in the first stages of keratin production. About 1 percent of all skin cancers, however, result from mutations in the melanin-producing cells. These cancers, called malignant melanomas, are very dangerous and have a very low cure rate.

The most effective ways to reduce your risk of skin cancer are to minimize your exposure to sunlight and to use a sunscreen. Sunbathers, like the person in **Figure 27.7**, increase their chances of skin cancer. They also age their skin prematurely.

Section Review

① Which part of the skin has the simpler structure, the dermis or the epidermis? Explain your answer.

② What do nerves, blood vessels, glands, and connective tissue contribute to the skin?

③ How could keeping your skin very clean help minimize the effects of acne?

④ What are two effective ways to reduce your risk of skin cancer?

▪ Section Review Answers ▪

1. The epidermis has a simpler structure than the dermis, which contains most of the living structures in the skin.

2. Students should indicate that nerves sense pressure, temperature, and pain; blood vessels house the blood carrying nourishment and waste; sweat glands regulate body temperature; and connective tissue attaches the skin to muscle.

3. Students should suggest that cleansing the skin aids in the removal of excess oil and controls the amount of bacteria growing on the skin's surface.

4. Students should suggest minimizing exposure to sunlight and use of sunscreen.

Health Connection

Skin Disorders
Invite the health teacher or school nurse to address the class about proper skin care that can help control acne. See if a dermatologist would be willing to speak about skin cancer and other serious skin disorders.

Demonstration 4
Severe Sunburn
Show the class a photograph of a person with severe sunburn. Remind students that melanin offers some protection by absorbing ultraviolet rays. However, melanin cannot absorb all of them, thus allowing UV rays to penetrate into the skin where they damage enzymes, destroy blood vessels, and even kill cells. Sun poisoning occurs when toxic materials from dead skin cells enter the bloodstream. Discuss how sunscreen products with different numerical ratings provide varying degrees of protection.

Phase 3

ASSESSMENT OPTIONS

Closure Strategy
Artificial Skin
Scientists have developed artificial skin for use in cases of severe skin burns. This artificial skin is temporarily useful while allowing skin grafts time to grow. Have students suggest several features scientists most likely incorporated into the design of this artificial skin.

Section Review
Assign the *Section Review*.

Reteaching
Assign students to cooperative groups of four. Have each group develop a concept map for this section.

Lesson Plan 27.3

Phase 1

PREPARATION

Key Concepts

- Bones, found in all shapes and sizes, are dynamic structures that are continuously being broken down and repaired.
- As a human embryo develops, minerals harden most of the cartilage, which is gradually replaced by bone.
- Long bones, such as those found in the arms and legs, consist of a dense outer layer that surrounds an inner core of marrow.
- All the bones in the body form the skeleton, which supports the body, enables it to move, and protects internal organs.
- Bones meet at joints and are held together by ligaments, which can become sprained when stretched too far.

Reading Strategy

Have students practice how to survey a section before they actually start reading. Instruct them to take no more than five minutes to glance through this section, focusing on the objectives, section headings, and illustrations. Students should realize that surveying will help them identify and focus on the main ideas as they read.

Phase 2

TEACHING STRATEGIES

Visual Strategy

Figure 27.8
Point out that the adult body (206 bones) has fewer bones than a human at birth (about 270 bones). As in the case of the skull, many bones fuse together during growth.

YOU MAY THINK OF YOUR SKELETON AS JUST A RIGID FRAMEWORK OF BONE THAT ENABLES YOU TO SIT, STAND, OR RUN. BONE ITSELF MAY APPEAR AS LIFELESS AS ROCK. HOWEVER, BONE IS IN FACT A DYNAMIC CONNECTIVE TISSUE MADE OF LIVING CELLS. THROUGHOUT YOUR LIFE, THESE CELLS CONTINUE TO PRODUCE THE MANY FIBERS AND MINERALS THAT FILL THE SPACES BETWEEN THE CELLS.

27.3   Bones

Objectives

❶ Draw a diagram of a typical long bone and label its parts.

❷ Differentiate a fracture from a sprain.

❸ Discuss the causes and effects of osteoporosis.

❹ Identify five types of joints in your body and explain the movement each permits.

Figure 27.8
The skull of an infant has more separate bones than that of an adult. As the child grows, many of these bones will fuse together. The "channels" you see in the infant skull will disappear by the age of four.

Bone Structure and Growth

Throughout your childhood and adolescence, bone cells build more and more bone as your body grows. Compare the size and shape of the infant's skull with the adult's skull in **Figure 27.8.** Even in adults, specialized bone cells continue to break down and rebuild

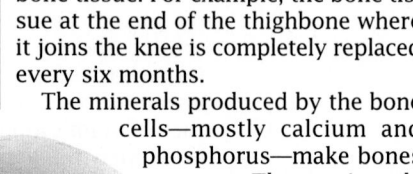

b   Infant skull

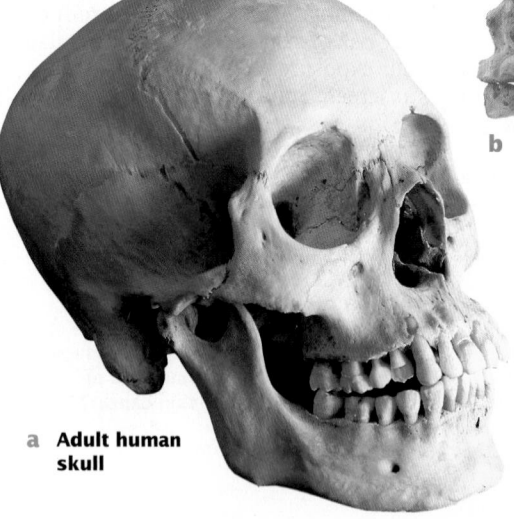

a   Adult human skull

bone tissue. For example, the bone tissue at the end of the thighbone where it joins the knee is completely replaced every six months.

The minerals produced by the bone cells—mostly calcium and phosphorus—make bones strong. These minerals also regulate an amazing variety of activities in your body. For example, nerves and muscles cannot work without the proper level of calcium. Bones act as a warehouse for storing calcium. If the level of calcium falls, bone cells release more calcium into the blood. The storage and release of calcium in your bones help maintain a precise level of calcium inside your body.

The human body contains bones of all shapes and sizes. The long bones of your arms and legs are shaped like cylinders. Curved, flat plates of bone form the part of the skull that protects your brain. Wrists and ankles contain many small bones that look like pebbles, and bones of the face and spine have unusual, irregular shapes.

Bone growth begins with cartilage

Bone growth begins long before birth. The basic shape of a long bone, such as an arm bone, is first formed in cartilage. Later, the cartilage cells begin to be replaced by cells that form bone. While you are growing, long bones still have a region of cartilage near each end that allows bones to grow longer. When you reach your adult size, these regions, too, are converted to bone.

Figure 27.9 shows a cross section of the femur in the thigh, a typical long bone. The outer layer of a long bone consists mainly of minerals and mature bone cells "trapped" by the minerals that they have deposited. To form this dense outer layer, bone cells deposit minerals in concentric rings, leaving a canal in the center of each group of rings. Each canal contains a small blood vessel that carries nutrients to the bone cells.

Inside the ends of a long bone are bone cells and minerals with large spaces in between, like a sponge. These spaces, and the entire center of the long middle part of the bone, are filled with **marrow**. The marrow inside long bones produces blood cells in newborns. As you get older, this marrow gradually changes its job to storing fat. In adults, most blood cells are produced in the marrow of flat bones like the sternum, or breastbone.

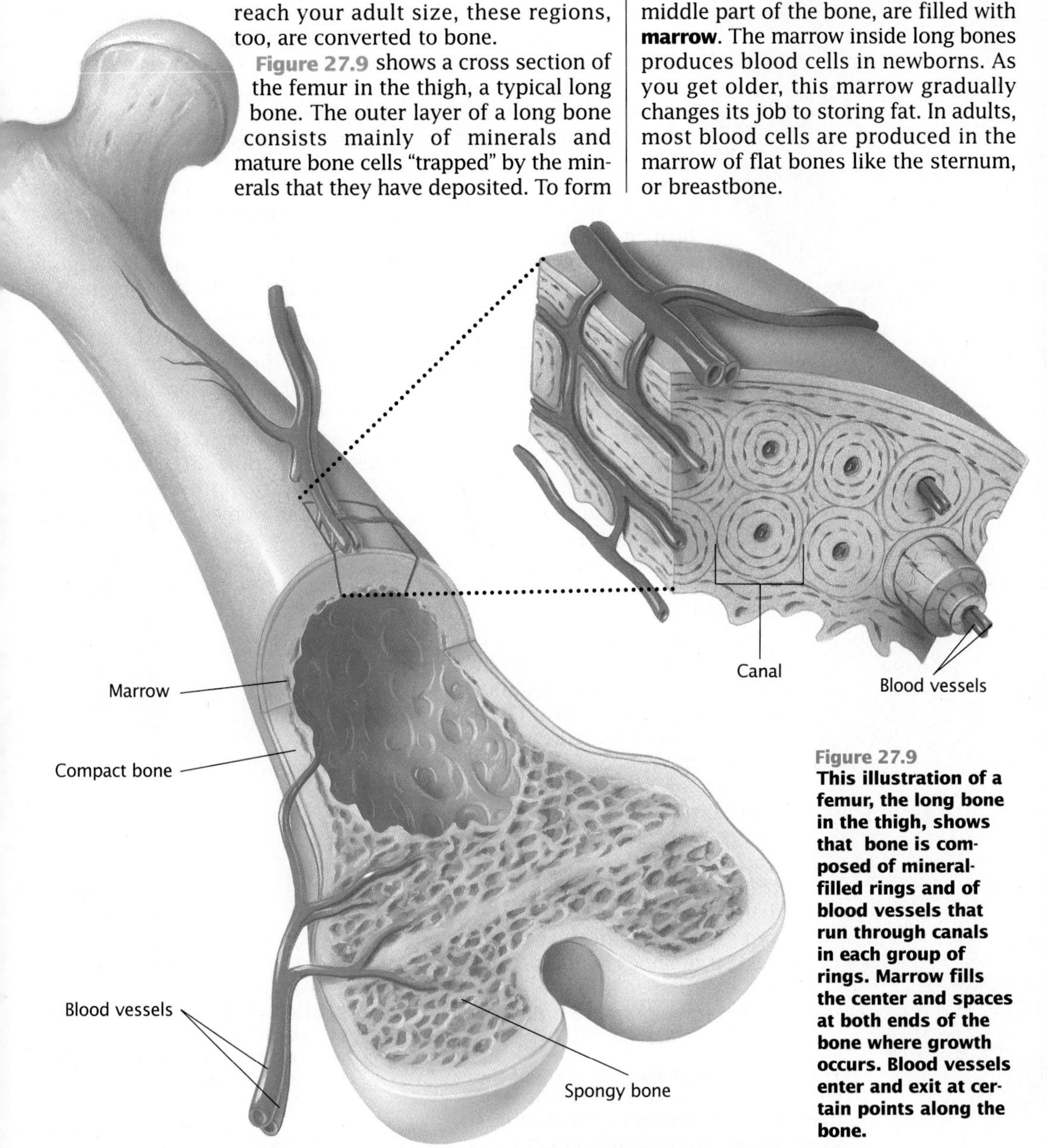

Marrow

Compact bone

Blood vessels

Spongy bone

Canal

Blood vessels

Figure 27.9
This illustration of a femur, the long bone in the thigh, shows that bone is composed of mineral-filled rings and of blood vessels that run through canals in each group of rings. Marrow fills the center and spaces at both ends of the bone where growth occurs. Blood vessels enter and exit at certain points along the bone.

Demonstration 1

The Human Skeleton
Show the class a model or photograph of the human skeleton. See how many different shapes students can identify among the bones that are found in a human skeleton.

Demonstration 2

A Long Bone
Have students crack open the long bone from a chicken leg and use a stereomicroscope to examine its structure. Other types of long bones may be obtained from the meat department at a supermarket.

Visual Strategy

Figure 27.9
Point out that in newborns, the marrow in long bones is red (from producing blood cells), but that in adults, most of the marrow is yellow (from storing fat). Have students relate this structural change to the functional changes that occur in long bones as a person ages.

▪ *Cultural Perspective* ▪

Dr. Louis Tompkins Wright
(African-American)
Dr. Louis Tompkins Wright (physician) was a pioneer in science. He originated a method of operation on fractures of the knee joint and invented a special brace for patients with head and spine injuries. Dr. Wright was one of the first African-American graduates of Harvard Medical School. He was the first African-American doctor appointed to a municipal hospital position in New York City.

Demonstration 3

Broken Bones

Contact a doctor's office or hospital to obtain X rays of various broken bones. Have students try to identify the bone involved and determine whether the fracture was simple or compound. If possible, invite an orthopedic surgeon to talk to the class, not only about treating fractures, but also about bone replacements, including artificial hips and knees.

Visual Strategy

Figure 27.10

Have students discuss what first must be done to this broken bone before the doctor sets it in a cast. Point out that a pin is sometimes necessary to hold together the two broken ends of a bone.

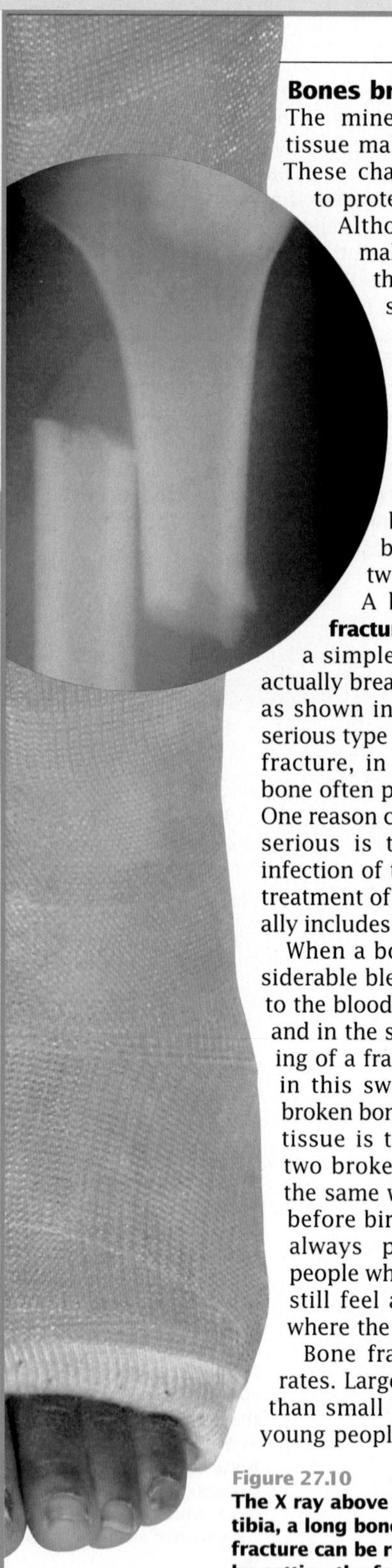

Bones break under stress

The minerals deposited in bone tissue make bones hard and rigid. These characteristics enable bone to protect and support the body. Although the fibers in bones make them much less brittle than a piece of chalk, severe stress placed on a bone may cause it to break. For example, certain types of falls while snow skiing can cause the bones in the lower leg to break at the top of the ski boot. In contact sports like football, a bone may break because a limb has been twisted.

A broken bone is called a **fracture**. A bone fracture may be a simple crack, or the bone may actually break into two or more pieces as shown in **Figure 27.10**. The most serious type of fracture is a compound fracture, in which pieces of broken bone often protrude through the skin. One reason compound fractures are so serious is that they may result in infection of the bone. Because of this, treatment of compound fractures usually includes large doses of antibiotics.

When a bone breaks, there is considerable bleeding caused by damage to the blood vessels in the bone itself and in the surrounding tissues. Healing of a fracture begins as the blood in this swollen region around the broken bone begins to clot. The bone tissue is then rebuilt between the two broken ends of bone in much the same way as bone tissue forms before birth. The rebuilding is not always perfect, however. Many people who have broken a bone can still feel a thicker region of bone where the fracture occurred.

Bone fractures heal at different rates. Large bones heal more slowly than small bones, and the bones of young people heal more quickly than those of older people. Holding the broken ends close to each other and keeping them completely still speeds healing of bones. This is why bone fractures are often treated by encasing the fractured limb in a cast.

Osteoporosis causes bones to become brittle

As bones grow longer, they also grow thicker and denser. In young adults, the density of bone usually remains relatively constant as bone tissue is broken down and replaced at a steady rate. During middle age, bone replacement gradually becomes less efficient, and bones become less dense. The loss of bone density that may eventually result is called **osteoporosis** (*ahs tee oh puh ROH sihs*). Compare the healthy bone with the bone that has undergone severe mineral loss after the onset of osteoporosis, shown in **Figure 27.11**.

Osteoporosis can cause bones to become light, brittle, and easily broken. In the United States, more than 600,000 bone fractures a year result from osteoporosis. Severe osteoporosis in the bones of the spine often changes the posture of very old people. Although both men and women lose bone as they

Figure 27.10

The X ray above shows a fracture of the tibia, a long bone in the lower leg. This fracture can be repaired without surgery by setting the fracture with a cast.

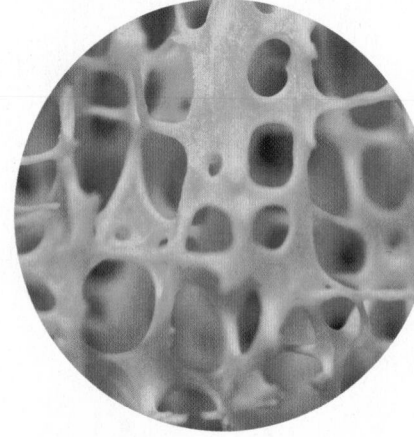

Figure 27.11

a In healthy bone tissue (magnified 10X in this photograph) minerals are continuously replaced, so the bone remains strong.

age, women are at a greater risk for osteoporosis for two reasons. First, women's bones are usually smaller and lighter than men's bones. Therefore, the loss of the same amount of bone in a man and a woman could result in the woman having thin, fragile bones, while the man's bones might still be quite strong. Second, the production of female sex hormones declines rapidly during menopause. Because sex hormones help to maintain bone density, this decline in hormone production increases the rate of bone loss.

Researchers have discovered that exercise can increase the amount of minerals deposited in bone. For people in their teens and twenties, regular exercise and a balanced diet that includes plenty of calcium can actually increase bone density. And for older people, regular exercise can slow the bone loss that can lead to osteoporosis. That's good news for the tennis player in **Figure 27.12**.

You may think that osteoporosis is something that you won't have to think about for a long time. However, bone density can be increased only during your teens and twenties. Regular exercise and a healthy diet will make you healthier now and will also pay off later. The stronger your bones are now, the less likely you are to be affected by osteoporosis later.

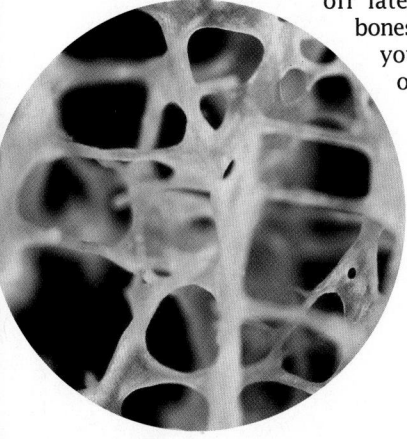

b After the onset of osteoporosis, minerals are not rapidly replaced. As a result, bones become brittle and can break easily.

Figure 27.12
This woman is over 60 years old and is still very active. Regular exercise at any age can help slow the bone loss that can lead to osteoporosis.

Visual Strategy

Figure 27.13

Point out that the vertebrae are separated by discs of cartilage that serve as cushions. A slipped disc occurs when a disc protrudes, pressing on nerves. A disc may even break. This condition, known as a ruptured disc, can cause severe pain and even lead to paralysis. Also point out that the bottom two pairs of ribs, unlike the others, are not attached to the sternum. These two pairs are known as floating ribs.

Connection: Chapter 11

Hominids

Remind students that upright walking is a characteristic shared only by hominids. Apes can also walk upright, but only for short distances, since their pelvic girdles do not permit true bipedalism.

Health Connection

Childbirth

Tell students that a human female has a broader pelvis than a male. Discuss how this structural difference is advantageous for giving birth.

The Skeleton

All of the bones in the body make up the skeleton. The skeleton supports the body's weight, enables it to move, and protects many of its internal organs.

A typical human skeleton is shown in **Figure 27.13**. The central part of the skeleton consists of the skull, spine, and ribs. The skull consists of many fused bones that protect the brain and form the shape of the face. The lower jaw, the mandible, is the only bone of the skull that moves easily. The spine is made up of vertebrae that support the trunk and allow flexibility. The vertebrae surround and protect the spinal cord. Attached to the spine are 12 sets of ribs. The ribs protect the heart, lungs, and other organs in the chest cavity. Some ribs are attached to the sternum by cartilage.

Two frameworks of bone, called girdles, connect the arms and legs to the central skeleton. The arms are attached to the pectoral girdle, and the legs are attached to the pelvic girdle. The pectoral girdle is connected to the spine only by connective tissue and muscles. This permits the arms and shoulders to move freely. By contrast, hipbones in the pelvic girdle attach directly to the lower part of the spine, so that the legs can bear the full weight of the body. This is why legs cannot move as freely as arms.

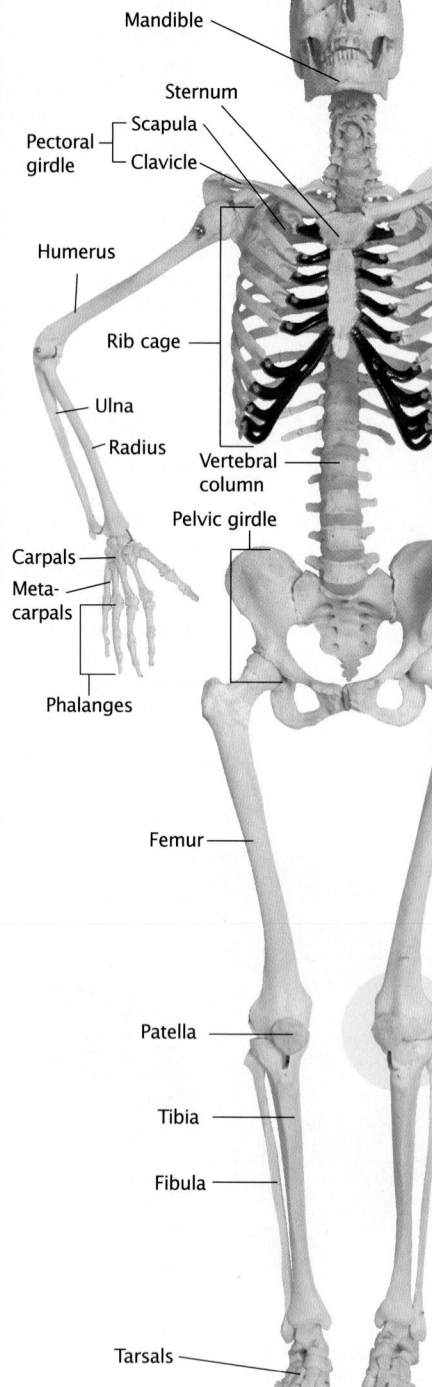

Figure 27.13
The adult human skeleton has 206 bones. Several major bones are identified in this skeleton.

▪ *Cultural Perspective* ▪

Dr. Augustus A. White III
(African-American)
Dr. Augustus A. White, an orthopedic surgeon and educator, specialized in the research of fracture healing and the spine. As a professor of Orthopedic Surgery at Yale Medical School, Dr. White assisted in the establishment of the first orthopedic biomechanic laboratory. He authored over 100 scientific publications, books, and articles. His most noted work is "The Clinical Biomechanics of the Spine."

Figure 27.14
A joint is where two bones connect. Some examples of joints are shown here, along with the kinds of movements they allow. The wires, screws, and bolts holding these bones together represent the functions of ligaments in the body.

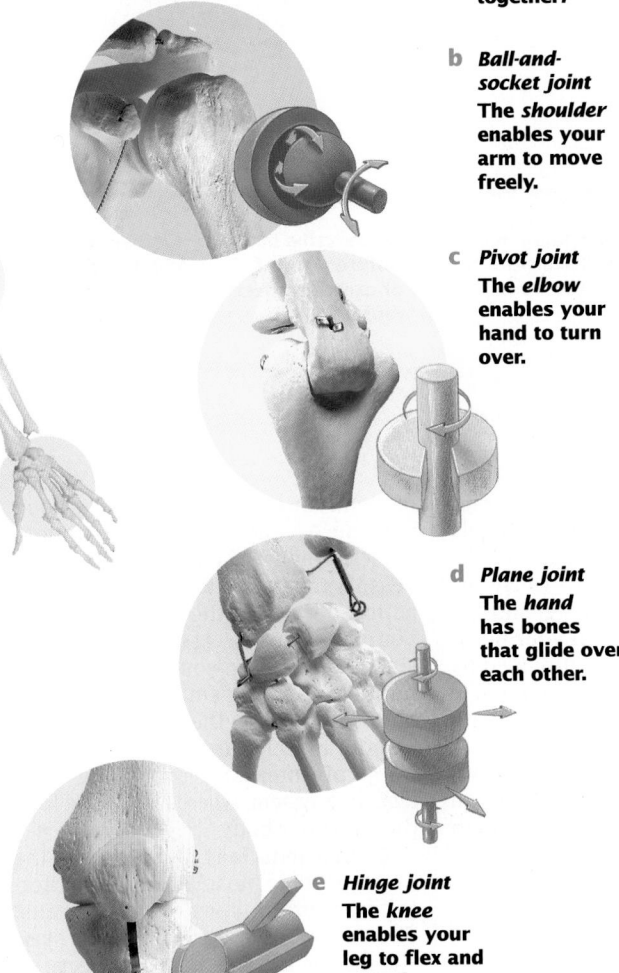

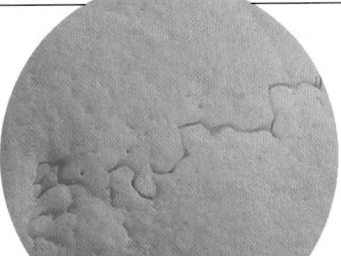

a *Suture joint*
The *skull* is immovable because its bones are fused together.

b *Ball-and-socket joint*
The *shoulder* enables your arm to move freely.

c *Pivot joint*
The *elbow* enables your hand to turn over.

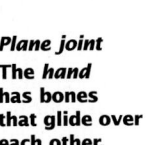

d *Plane joint*
The *hand* has bones that glide over each other.

e *Hinge joint*
The *knee* enables your leg to flex and extend.

Bones fit together with joints

The place where two or more bones connect is called a **joint**. Bones are joined to each other by strong elastic bands of connective tissue called **ligaments**. Joints vary greatly in their flexibility. For example, the wrist and thumb can be moved in many directions, but the joints in your pelvis can move only slightly. Some examples of joints are shown in **Figure 27.14**.

Although ligaments are very elastic, there is a limit to how far they can stretch. When ligaments are stretched too far, the injury that results is called a **sprain**. If the ligament is actually torn by overstretching, it will not heal and must be repaired surgically.

The knee is a particularly complex joint. Because of the upright posture of humans, the leg bones joined at our knees must carry the body's full weight. Many ligaments are needed to connect bones at the knee and make the joint as stable as possible. Knee injuries may result when these ligaments are damaged. Such injuries are particularly common in sports like football and basketball.

Section Review

1 Draw a diagram of a typical long bone and label dense bone, spongy bone, marrow, and blood vessels.

2 How do fractures and sprains differ?

3 What actions can you take now to prevent osteoporosis later in life?

4 List five different types of joints in your body and give an example of each.

■ *Section Review Answers* ■

1. See Figure 27.9 on page 621.
2. A fracture is a broken bone, whereas a sprain is the overstretching of a ligament.

3. Students should suggest regular exercise and a balanced diet that includes plenty of calcium.

4. See Figure 27.14 on page 625.

Visual Strategy

Figure 27.14
Use this figure to have students demonstrate with their own bodies how each type of joint permits a different kind of movement. Have students identify which types allow the greatest flexibility.

Sports Connection

Football Knees
Explain why many football players suffer from bad knees. Besides causing sprains and torn ligaments, getting tackled from the side may also damage the cartilage above the tibia that helps cushion the knee joint. When this cartilage is lost, the long bones begin to rub against each other, causing bone chips that accumulate in the knee joint.

Phase 3

ASSESSMENT OPTIONS

Closure Strategy

Bone vs. Cartilage
Inform students that some vertebrates, like sharks, have skeletons made entirely of cartilage. Have students list advantages and disadvantages of having a human skeleton made of cartilage rather than bone.

Section Review

Assign the *Section Review*.

Reteaching

Have students use their microscopes to examine prepared slides of cartilage and bone. Have students describe how their structure is related to the function each performs in the body.

Lesson Plan 27.4

Phase 1

PREPARATION

Key Concepts

- The three major types of muscle include skeletal muscle, smooth muscle, and cardiac muscle.
- Movement of bones requires two sets of muscles; one contracts while the other relaxes.
- Movement is also possible because of tendons that connect muscles to bones.
- Movement of muscles involves the sliding action of two threads of protein—actin and myosin.
- Skeletal muscle can be either slow-twitch (used in endurance performances) or fast-twitch (used in short-term events).
- Muscles can undergo either anaerobic, aerobic, or resistance exercise.
- Anabolic steroids taken to increase muscle size can cause serious side effects.
- Overuse of muscles can lead to a "pulled muscle," or can cause tendinitis.

Reading Strategy

Use this section to promote extension readings. Have students select one of the concepts presented in this section and check library resources to locate additional information. Have students present either a brief oral or written report of their outside readings.

Phase 2

TEACHING STRATEGIES

Demonstration 1

Muscle Types

Use a microprojector to show students prepared slides of the three types of muscle. Have students describe similarities and differences.

LARGE, POWERFUL MUSCLES IN THE LEGS CAN PROPEL AN ATHLETE HIGH ENOUGH TO SLAM-DUNK A BASKETBALL OR FAST ENOUGH TO RUN 100 M (300 FT.) IN LESS THAN 10 SECONDS. SMALLER MUSCLES ENABLE YOUR EYES TO READ THE WORDS ON THIS PAGE. A SMILE OR A FROWN WOULD NOT BE POSSIBLE WITHOUT THE ACTION OF MUSCLES BENEATH THE SKIN IN YOUR FACE.

27.4   Muscles

Objectives

❶ **Compare and contrast the three types of muscles.**

❷ **Explain how muscles work to move a bone.**

❸ **Describe how different forms of exercise affect muscles.**

❹ **Discuss the dangers of taking anabolic steroids.**

The Actions of Muscles

Your body has more than 600 muscles, each containing hundreds of muscle cells. A muscle can move a part of the body when electrical signals cause the muscle cells to contract, or get shorter. For most of the muscles in your body, these electrical signals are provided by nerves. After contracting, a muscle relaxes and returns to its original length until the next signal.

Figure 27.15 on the next page shows the three main types of muscle tissue and their functions in the human body. **Skeletal muscle** is attached to the bones of the skeleton. Some of these muscles move bones to perform actions like walking, grasping, or bending. Skeletal muscles are sometimes called voluntary muscles because you can control their actions. You also use skeletal muscles for activities that happen without your thinking about them, such as blinking or maintaining your posture. However, you can consciously control these actions at any time.

Unlike skeletal muscle, **smooth muscle** is usually not under conscious control. Most of the actions of smooth muscle occur without your being aware of them. For example, when you go outside on a cold, sunny day, smooth muscles reduce the diameter of blood vessels in your skin. This automatic action decreases the amount of heat your body loses. At the same time, smooth muscles in your eye cause your pupils to become smaller so you will not be blinded by the bright sunshine. Smooth muscles are also found in the walls of internal organs like your digestive system, where they move food through your body.

Cardiac muscle is found only in the heart. Cardiac muscle cells are different from other types of muscle cells. The contraction of a smooth or skeletal muscle is triggered by an electrical signal from a nerve. However, cardiac muscle cells are "self-starters"—they generate their own electrical signals that cause them to contract. Nerves to the heart can control how fast the heart beats, but cardiac muscle will contract even if these nerves are cut. Connections among the muscle cells coordinate the contractions of the individual cells. This causes the entire heart to beat as the cells contract together.

Making Your Skeleton Move

A skeletal muscle can move a bone like the one in your thigh by pulling on it. When the muscle contracts, it pulls on the bone to which it is attached. No muscle can push a bone; muscles can only pull bones.

Muscle pairs work together

Moving a bone requires the cooperation of two sets of muscles. Pairs of muscles run parallel on either side of a joint in the skeleton. Why are two sets of muscles needed to move a bone? When a muscle pulls on a bone, it moves the bone in one direction. Since the muscle cannot push on the bone, another muscle is needed to pull the bone in the opposite direction. Contracting one set of muscles while the other set relaxes can cause a particular joint to bend. Contracting the other set of muscles while the first set relaxes causes the joint to straighten. **Figure 27.15 d–e** shows how the main pair of muscles that control the human elbow work.

What would happen if both sets of muscles contracted at the same time? Under normal circumstances, the bone would not move because the muscles would be pulling in opposite directions at the same time. However, large muscles in your legs are so powerful that if all of the muscles contracted strongly at the same time, they could break a bone. This is prevented by a nerve message (coordinated by the brain) sent from one set of muscles to the other. When one set of muscles contracts, the other set is signaled to relax. The intricately coordinated sequence of muscle contractions alternately bends and straightens joints at the hip, knee, and ankle, allowing you to walk or run smoothly.

Figure 27.15
The human body contains three main types of muscle tissue that move different parts of the body.

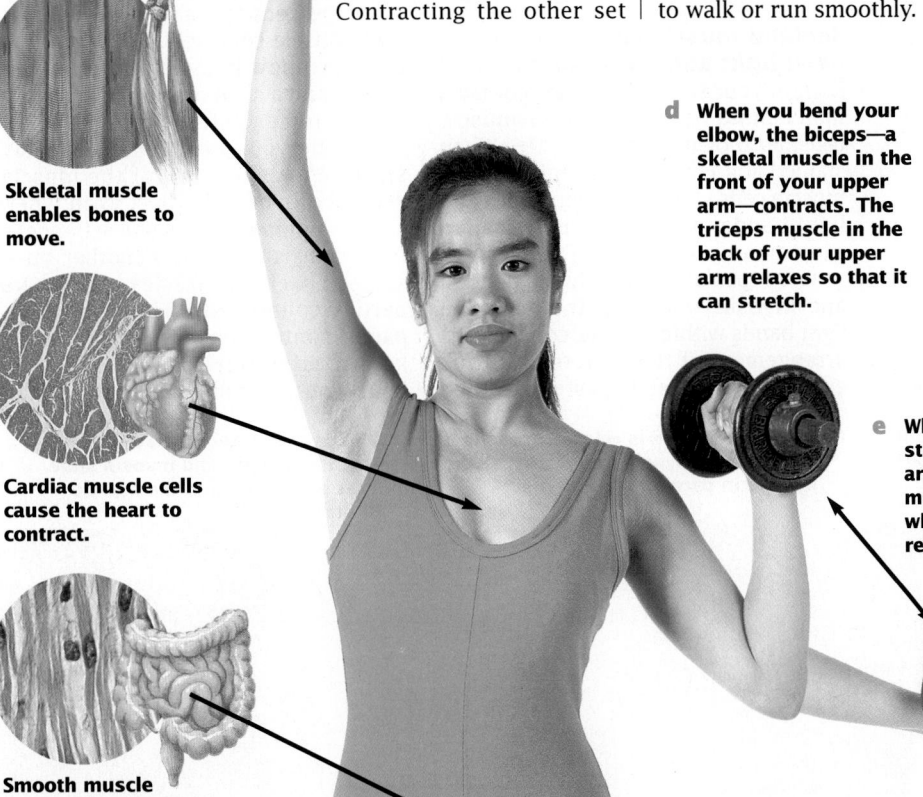

a **Skeletal muscle enables bones to move.**

b **Cardiac muscle cells cause the heart to contract.**

c **Smooth muscle moves food through digestive organs.**

d **When you bend your elbow, the biceps—a skeletal muscle in the front of your upper arm—contracts. The triceps muscle in the back of your upper arm relaxes so that it can stretch.**

e **When you straighten your arm, the triceps muscle contracts while the biceps relaxes.**

Tendons attach muscles to bones

Some muscles are attached directly to bone. Most muscles, however, are connected to bones by **tendons**. For example, the biceps muscle in your arm is attached by tendons to a bone in your shoulder and to another bone in your forearm. Tendons are made of the same tough, elastic connective tissue as ligaments. The difference between tendons and ligaments is that ligaments connect a bone to another bone, while tendons connect a muscle to a bone.

Sometimes the distance between a muscle and the bone to which it is attached is quite long. There are no muscles in your fingers—only long tendons that extend from muscles in your forearm and attach to the bones in your fingers. When you bend your fingers, you are actually contracting muscles in your forearm. Move your fingers and watch the movement of the muscles in your forearm.

Skeletal muscle cells have light and dark bands

Each of your skeletal muscles contains many bundles of long, thin muscle cells, as shown in **Figure 27.16**. When stained and viewed with a microscope, muscle cells show alternating dark and light bands in repeating units called sarcomeres (*SAR kuh mihrz*).

Two types of protein threads—actin and myosin—make up the dark and light bands within each sarcomere. The arrangement of these protein threads enables them to slide past each other.

When a muscle cell receives an electrical signal from the nervous system, the sets of actin and myosin threads slide past one another, and the entire muscle cell shortens. You can read more about actin and myosin in the *Tour of a Skeletal Muscle* on page 629.

Nerves send messages to muscles to generate force

If the muscles in your arms contracted with the same force when picking up a pencil as when picking up a bowling ball, the pencil would probably go flying up to the ceiling. How does a muscle create just enough force to perform a particular job?

The contraction of each muscle cell generates a certain amount of force that pulls on the bone. But not every muscle cell contracts with the rest of the muscle. Signals from the nervous system make a muscle generate more force by causing more muscle cells to contract. When you pick up a pencil, a small percentage of muscle cells in each muscle contracts. But when you pick up a bowling ball, more muscle cells contract in order to provide enough force to lift the ball.

Your nervous system continuously sends signals to some of the muscle cells in all your muscles, even when you are not moving. As one set of cells tires, signals are sent to another set— so your skeletal muscles are always partly contracted. This constant, partial contraction of your muscles is known as muscle tone. Muscle tone helps you maintain your posture.

Figure 27.16
The photos below show the relaxation and contraction of the frontalis muscles of the forehead.

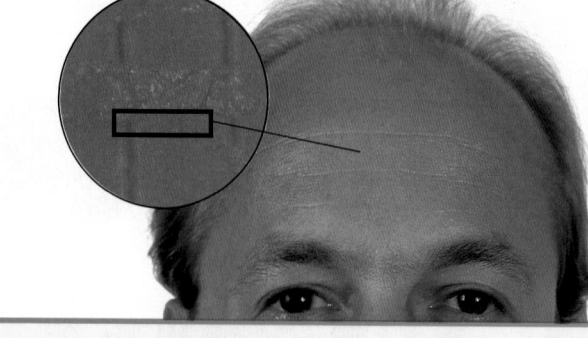

a   **When the muscle is relaxed, threads of actin and myosin are at rest, side by side in parallel stacks.**

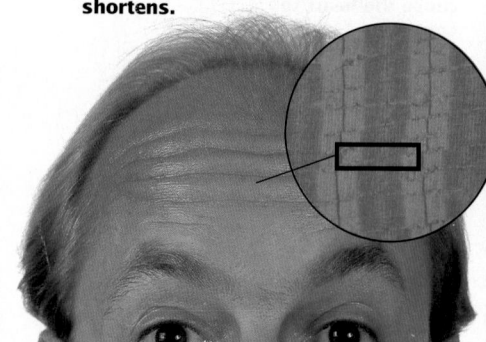

b   **When the muscle contracts, the threads of actin and myosin slide past each other and the muscle shortens.**

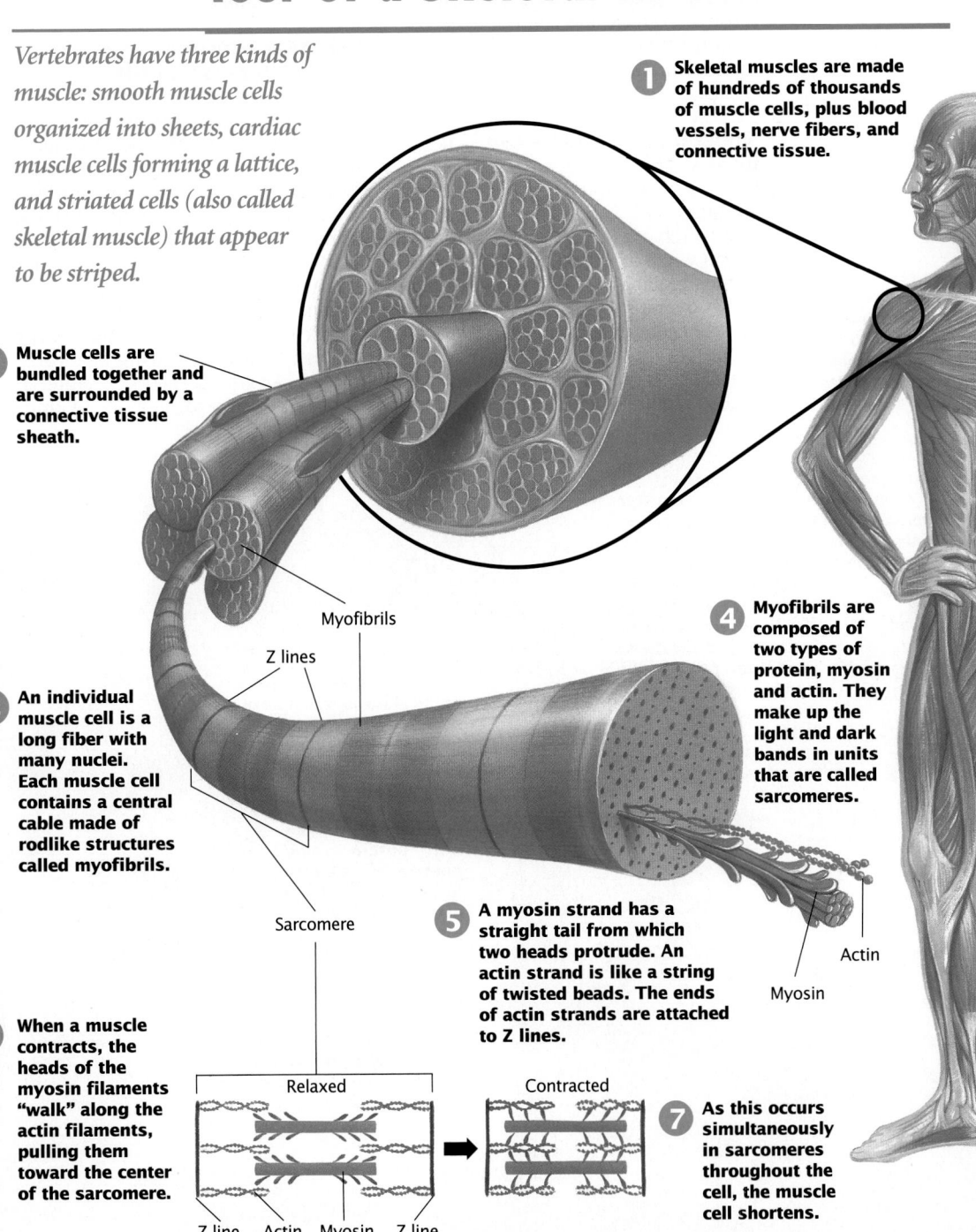

■ *Journeys* ■ *Journeys* ■ *Journeys* ■ *Journeys* ■ *Journeys*

Tour of a Skeletal Muscle

Vertebrates have three kinds of muscle: smooth muscle cells organized into sheets, cardiac muscle cells forming a lattice, and striated cells (also called skeletal muscle) that appear to be striped.

1 Skeletal muscles are made of hundreds of thousands of muscle cells, plus blood vessels, nerve fibers, and connective tissue.

2 Muscle cells are bundled together and are surrounded by a connective tissue sheath.

Myofibrils

Z lines

3 An individual muscle cell is a long fiber with many nuclei. Each muscle cell contains a central cable made of rodlike structures called myofibrils.

4 Myofibrils are composed of two types of protein, myosin and actin. They make up the light and dark bands in units that are called sarcomeres.

Sarcomere

5 A myosin strand has a straight tail from which two heads protrude. An actin strand is like a string of twisted beads. The ends of actin strands are attached to Z lines.

Actin

Myosin

6 When a muscle contracts, the heads of the myosin filaments "walk" along the actin filaments, pulling them toward the center of the sarcomere.

Relaxed

Contracted

7 As this occurs simultaneously in sarcomeres throughout the cell, the muscle cell shortens.

Z line Actin Myosin Z line

Using the Feature

- This feature can be used to review cellular respiration in Chapter 5. Begin this review by asking students how the cells in a skeletal muscle would get energy if they were working to lift a weight. Then ask students how the cells in a skeletal muscle would get their energy if they were working to swim across the English Channel.
- Take this oppportunity to discuss muscle disorders such as muscular dystrophy and myasthenia gravis.

Discussion

1. List the units that comprise a skeletal muscle, beginning with the protein molecules.
   Actin and myosin make up sarcomeres, which are the light and dark bands found in myofibrils. Myofibrils are bundled together and surrounded by a connective tissue sheath. Hundreds of thousands of these bundles, plus blood vessels and nerve fibers, make up a skeletal muscle.

2. How does the organization of smooth muscle tissue and cardiac muscle tissue differ from the striated appearance of skeletal muscle?
   Smooth muscle cells are organized into sheets. Cardiac muscle cells form a lattice. (You may want to show the class photos and slides of these two tissue types for clarification.)

3. Describe what happens to myosin and actin filaments when a muscle contracts.
   When a muscle contracts, the heads of the myosin filaments "walk" along the actin filaments, pulling them toward the center of the sarcomere.

Demonstration 3
White and Dark Meat
Show students pieces of white and dark chicken meat. Tell them that white meat consists mostly of fast-twitch muscles, while dark meat is mainly composed of slow-twitch muscles. In a chicken, which walks around some 14 hours a day, leg muscles (dark meat) are mostly slow-twitch, while wing muscles (white meat) are primarily fast-twitch. Have students hypothesize whether fast- or slow-twitch muscles would be found in each of the following: a rabbit's leg (*fast*), wing of a migratory bird (*slow*), and a frog's leg (*fast*).

Visual Strategy

Figure 27.17
Tell students that the average human has a 50-50 distribution of fast-twitch and slow-twitch muscles. However, muscles of long-distance runners and swimmers average 80 percent slow-twitch, while those of sprinters and weight lifters average about 75 percent fast-twitch. Have students discuss which type of muscle would be best for each of the following players on a football team: tackle, running back, and quarterback.

Connection: Chapter 5

Anaerobic Respiration
Review the major features of anaerobic respiration, emphasizing that only a fraction of the energy stored in foods is released during this process.

What Exercise Does for Muscles

Most people can increase the flexibility of their joints, build endurance, and increase their strength and agility through exercise and conditioning. Different kinds of exercise affect your muscles in different ways, as shown in **Figure 27.17**.

Skeletal muscle cells can be fast-twitch or slow-twitch
Every skeletal muscle in your body contains two types of muscle cells. Fast-twitch muscle cells are called into action when a person, such as a pole-vaulter, requires speed and quick movements over a short period. Slow-twitch muscle cells respond more slowly but do not tire as quickly as fast-twitch muscle cells. When an activity such as cycling requires endurance, slow-twitch

Figure 27.17
Each exercise, from cycling to sprinting to weight training, affects muscles in different ways.

muscle cells are working to provide it. You can read more about fast- and slow-twitch muscles in **Figure 27.17a**.

Anaerobic exercises use fast-twitch muscle cells
Some activities demand that your muscles work at high intensity for a very brief period. Such activities are called **anaerobic** (*an uh ROH bihk*) **exercise**. These short bursts of activity use fast-twitch muscle cells. Running up stairs, sprinting, and making a dash for home plate in a baseball game are examples of anaerobic exercise. **Figure 27.17b** explains how anaerobic exercise works.

Muscle cells rapidly use large amounts of energy during anaerobic exercise. As you learned in Chapter 5, cells must have oxygen to produce

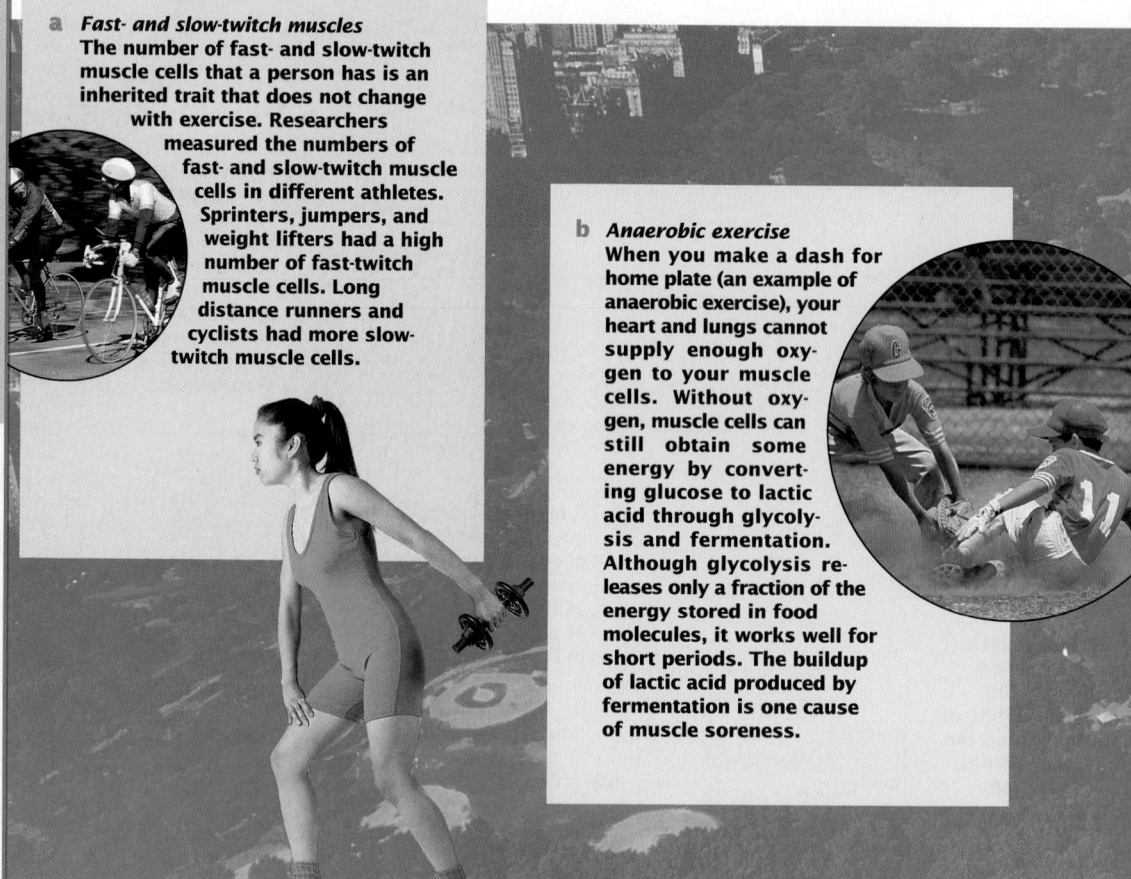

a *Fast- and slow-twitch muscles*
The number of fast- and slow-twitch muscle cells that a person has is an inherited trait that does not change with exercise. Researchers measured the numbers of fast- and slow-twitch muscle cells in different athletes. Sprinters, jumpers, and weight lifters had a high number of fast-twitch muscle cells. Long distance runners and cyclists had more slow-twitch muscle cells.

b *Anaerobic exercise*
When you make a dash for home plate (an example of anaerobic exercise), your heart and lungs cannot supply enough oxygen to your muscle cells. Without oxygen, muscle cells can still obtain some energy by converting glucose to lactic acid through glycolysis and fermentation. Although glycolysis releases only a fraction of the energy stored in food molecules, it works well for short periods. The buildup of lactic acid produced by fermentation is one cause of muscle soreness.

the maximum amount of energy from food molecules.

After you stop anaerobic exercise, your body needs extra oxygen to burn up the excess lactic acid and return your energy reserves to normal. This need for extra oxygen is called oxygen debt. Oxygen debt is the reason you must breathe rapidly and deeply for a few minutes after a hard run.

Aerobic exercises demand a continuous supply of oxygen

Steady, low-intensity exercise like jogging, in-line skating, or swimming laps is called **aerobic** (*ehr OH bihk*) **exercise**. The slow, steady pace of aerobic exercise ensures that your lungs and heart can deliver oxygen to your muscles at the same rate at which the muscle cells are using it. Your muscles can use this continuous supply of oxygen to extract

the maximum amount of energy from food molecules and continue the exercise for as long as 20 minutes or more. **Figure 27.17c** describes some of the benefits of aerobic exercise.

Resistance exercises increase muscle size and strength

Aerobic exercise will increase the size and strength of your muscles somewhat, but it won't make you look like a bodybuilder. To significantly increase the size of your muscles, resistance exercises are the most effective form of exercise. **Resistance exercises** are exercises that require exerting a great deal of force against a heavy or immovable object. Weight lifting and using exercise machines that provide resistance are forms of resistance exercises. You can read more about resistance exercise in **Figure 27.17d**.

Sports Connection

Aerobic Exercises

Invite a physical education teacher to talk to the class about aerobics. Have them discuss the difference between low- and high-impact aerobic exercises.

Sports Connection

Resistance Exercises

Invite a student or person who "works out" at the school or local gym to demonstrate and discuss some of the resistance exercises they perform to build up their muscles. To impress students with how resistance exercising can build up muscles, display photographs, or show a video, of body builders.

c   *Aerobic exercise*
The amount of oxygen that can be taken into the body and delivered to the muscles can be increased, in most people, through aerobic training. Training strengthens the chest muscles so that more air—and oxygen—enters the body with each breath. Aerobic training helps the heart pump blood more efficiently and increases the number of blood vessels in the muscles. This increased ability to supply oxygen to the muscles results in an increase in endurance.

d   *Resistance exercise*
Exercising will not increase the number of your muscle cells, but resistance exercises will increase muscle size because each muscle cell gets bigger. However, resistance exercises are much less effective than aerobic exercises in strengthening the heart and lungs, and in increasing the number of blood vessels in the muscle. Because resistance exercises do not significantly improve the ability of your body to deliver oxygen to your muscles, they do not increase your endurance.

Phase 3

ASSESSMENT OPTIONS

Closure Strategy

Muscular Dystrophy

Have students research and submit a brief oral or written report about this disease, which is characterized by the gradual wasting of skeletal muscle. Tell them to include information that scientists have recently uncovered.

Section Review

Assign the *Section Review*.

Reteaching

Assign students to cooperative groups of four. Have each group develop a concept map for this section.

Figure 27.18
Lyle Alzado, former defensive lineman for the Denver Broncos, Los Angeles Raiders, and Cleveland Browns died of a rare type of brain cancer. Alzado attributed the cause of his cancer to steroid use.

Anabolic steroids are dangerous

Some people are tempted to experiment with anabolic steroids to increase the size of their muscles. Anabolic steroids are powerful synthetic compounds that chemically resemble the male sex hormone, testosterone. The use of steroids may produce serious side effects—including severe acne, cancer, failure to reach full height, heart disease, and psychological disorders. Some males who use anabolic steroids develop female-like breasts and shriveled testes. Females who use these chemicals may develop facial hair, deepening of the voice, and male-pattern baldness. Many of these symptoms of steroid abuse are irreversible. The use of anabolic steroids may have caused death in some cases, as you can see in **Figure 27.18**

"If I win this battle, it will be the best one. . . . If I lose this battle, whatever I've done has been real wrong."
Lyle Alzado
1949–1992

Overuse causes muscle injuries

Overusing muscles by exercising too much or without proper conditioning can lead to muscle injury. A muscle strain, commonly called a "pulled muscle," is the overstretching or even tearing of a muscle. Muscle strain may occur when a muscle is overused or when strenuous exercise is done before doing warm-up exercises. **Tendinitis** is a painful inflammation of a tendon caused by too much friction or stress on the tendon.

Section Review

❶ Describe the three main types of muscle.

❷ How might your ability to move your arm be affected by an injury to the biceps muscle? To the triceps muscle?

❸ If you wanted to increase your endurance, which kind of exercise would be most effective?

❹ List three of the dangers associated with anabolic steroids.

▪ *Section Review Answers* ▪

1. See Figure 27.15 on page 627.
2. Students should suggest that injury to the biceps muscle would affect bending the arm. Injury to the triceps muscle would affect straightening or stretching the arm.
3. aerobic exercise

4. Possible answers include severe acne, cancer, stunted growth, heart disease, psychological disorders; female-like breasts and shriveled testes in males; and facial hair, deepening voice, and male-pattern baldness in females.

Chapter **27** *Highlights*

Aerobic exercise, such as cross-country skiing, increases oxygen flow to the muscles.

Alternative Assessment
Assign students to cooperative groups of four. Have each group develop a crossword puzzle for this chapter, including the clues for all the words in the puzzle. Then have the groups challenge one another to solve each other's puzzles.

	Key Terms	Summary
27.1 Tour of the Human Body 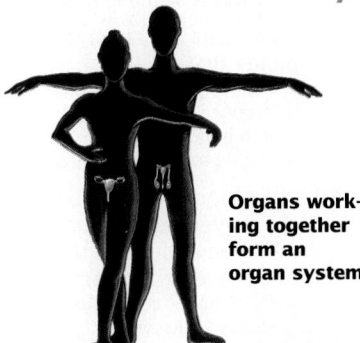 Organs working together form an organ system.	tissue (p. 611) epithelial tissue (p. 611) exocrine gland (p. 611) muscle tissue (p. 612) nerve tissue (p. 612) connective tissue (p. 612) organ (p. 613) organ system (p. 613)	• Specialized cells are organized into groups called tissues. • Four tissue types in the body are: epithelial (for protection), muscle (for movement), nerve (for communication), and connective (for support and transportation). • Tissues work together to form organs. Organs work together to form organ systems.
27.2 Skin  Dead skin cells flake off and are replaced.	dermis (p. 616) epidermis (p. 618) melanin (p. 618)	• Skin cushions and protects the body. The dermis is the inner layer, made of living cells. The epidermis is the outer layer, made mostly of dead cells.
27.3 Bones 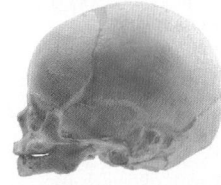 As you grow, bone cells build more bone tissue.	marrow (p. 621) fracture (p. 622) osteoporosis (p. 622) joint (p. 625) ligament (p. 625) sprain (p. 625)	• Bone is a living tissue made of cells that deposit minerals. • Marrow is important in blood cell production and fat storage. • Many of the skeleton's joints are held in place by ligaments.
27.4 Muscles  Exercise helps increase muscle strength, agility, and joint flexibility.	skeletal muscle (p. 626) smooth muscle (p. 626) cardiac muscle (p. 626) tendon (p. 628) anaerobic exercise (p. 630) aerobic exercise (p. 631) resistance exercise (p. 631) tendinitis (p. 632)	• The three basic types of muscle tissue are skeletal muscle, smooth muscle, and cardiac muscle. • Moving a bone requires two sets of muscles working in opposition. • Muscles may attach directly to bone or by means of tendons. • Muscle action depends on the sliding action of protein threads (actin and myosin). • Three types of exercise (anaerobic, aerobic, and resistance) make different demands on muscles.

Chapter Review Answers

Understanding Vocabulary
1. a. nerve tissue
   b. epidermis
   c. sprain
   d. skeletal muscles
   e. aerobic exercise

Relating Concepts
2. Map answer is shown on page 609D.

Understanding Concepts
Multiple Choice

3. b	9. c	
4. d	10. c	
5. d	11. c	
6. b	12. b	
7. b	13. c	
8. c		

Completion
14. organ
15. dermis, epidermis
16. sprain
17. possible answers: heart disease, psychological disorder, severe acne, cancer
18. smooth, cardiac

Short Answer
19. Most apocrine glands are located in the armpits and groin area. These glands contain protein and fatty acid on which bacteria feed. The waste products of bacteria produce an odor.
20. Clogged oil glands cause the formation of whiteheads and blackheads. The area around the gland becomes inflamed if the oil gland bursts and becomes infected by bacteria.
21. Bleeding occurs at the site of the break. Healing begins with clotting. Cartilage cells form and are replaced by bone cells.

Chapter 27 Review

Understanding Vocabulary

1. For each set of terms, complete the analogy.
   a. cartilage:connective tissue::sense organs: _____
   b. contains blood vessels:dermis:: contains melanin: _____
   c. broken bone:fracture::ligament injury: _____
   d. digestion:smooth muscles::walking: _____
   e. weight lifting:resistance exercise::jogging: _____

Relating Concepts

2. Copy the unfinished concept map below onto a sheet of paper. Then complete the concept map by writing the correct word or phrase in each oval containing a question mark.

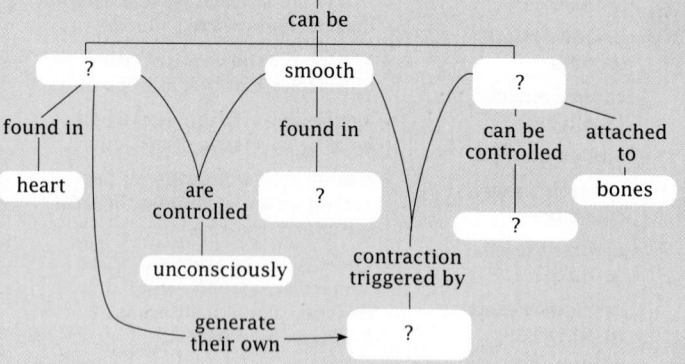

Understanding Concepts

Multiple Choice

3. The primary function of epithelial tissue is to
   a. join bones and muscles.
   b. cover and protect the body.
   c. move parts of the body.
   d. help the body maintain posture.

4. Which is *not* a type of connective tissue found in the human body?
   a. blood   c. skin
   b. ligament   d. muscle

5. Which sequence best describes organization of the body from the simplest to most complex?
   a. cell, tissue, organ system, organ
   b. organ system, organ, cell, tissue
   c. tissue, cell, organ, organ system
   d. cell, tissue, organ, organ system

6. Tiny nerve endings that enable you to sense pressure, pain, and cold are located in
   a. the epidermis.
   b. the dermis.
   c. keratin-filled cells.
   d. exocrine glands.

7. Melanin is a brown pigment that
   a. is produced in bone marrow.
   b. provides some protection from the sun's ultraviolet rays.
   c. is found in cardiac muscle.
   d. distinguishes fast-twitch muscles from slow twitch muscles.

8. Calcium and phosphorus are minerals that are
   a. present in sarcomeres in skeletal muscles.
   b. found in anabolic steroids.
   c. produced and stored in bones.
   d. make the skin waterproof.

9. Ligaments are strong elastic bands that
   a. connect muscle to bones.
   b. produce red blood cells.
   c. join bones together.
   d. enable muscle to contract.

10. What kind of muscles help in the movement of food in the intestine and cause the pupil of the eye to contract?
    a. those under conscious control
    b. skeletal muscles
    c. smooth muscles
    d. cardiac muscles

11. The dense material in bone is mainly composed of
    a. marrow.   c. calcium.
    b. cartilage.   d. connective tissue.

12. When you bend your elbow,
    a. the triceps muscle contracts and the biceps relaxes.
    b. the biceps muscle contracts and the triceps relaxes.
    c. only the triceps contracts.
    d. only the biceps contracts.

13. The dark and light bands in skeletal muscles are made of
    a. hemoglobin and carotene.
    b. fast-twitch and slow-twitch cells.
    c. actin and myosin.
    d. calcium and phosphorus.

Completion

14. The heart is composed of a number of tissues that work together. It is an example of a(n) _____ .

15. The _____ is the inner layer of skin that contains most of the living structures, while the _____ is the outer layer of the skin that contains both living and dead cells.

16. The injury resulting when ligaments are over-stretched is called a(n) _____ .

17. Dangers linked to the use of anabolic steroids include _____ , _____ , and _____ .

18. Two types of involuntary muscles include the _____ and _____ .

Short Answer

19. Where are apocrine glands located in your body? How do apocrine glands differ from other sweat glands?

20. Describe what causes the pimples associated with acne.

21. Describe what occurs when a broken bone begins to heal.

22. What is osteoporosis? Why are women more likely than men to suffer from osteoporosis?

23. Why are two sets of muscles required to move a bone?

24. What causes a muscle to exert different amounts of force?

Interpreting Graphics

25. Study the cross-section of the knee shown below. Answer the questions that follow.

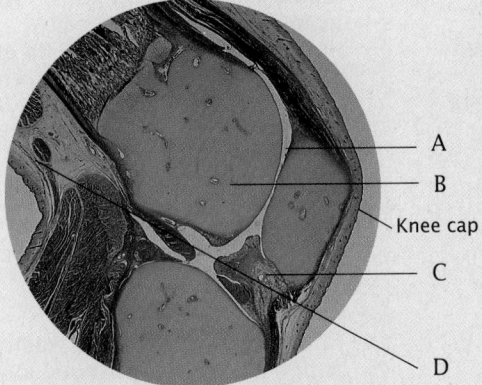

Knee cap

- What is the function of tissue A?
- What is tissue B made of?
- What might happen if tissue C were torn and damaged?
- What tissue is identified as D?

Reviewing Themes

26. *Patterns of Change*
When infrequently used muscles are worked hard, they feel sore the next day. What is the cause of muscle soreness?

27. *Evolution*
What do goose bumps reveal about our evolutionary past?

Thinking Critically

28. *Inferring Conclusions*
In some countries, potential athletes have been selected for particular sports at a very young age based on the percentages of fast-twitch and slow-twitch muscle cells they have. If an examination of leg muscle tissue reveals that athlete A has fewer fast-twitch muscle cells than athlete B, which athlete would have been chosen for training as a marathon runner? Explain your answer.

29. *Building on What You Have Learned*
As you read in Chapter 14, scientists have noted a reduction in the Earth's ozone layer. How is the possible destruction of the ozone layer related to an increase in the number of warnings to use sunscreens and to avoid the midday sun?

Cross-Discipline Connection

30. *Biology and History*
Luigi Galvani lived in Italy during the eighteenth century. What did Galvani discover about muscles when he touched the leg of a frog to his machine that produced an electric spark?

Discovering Through Reading

31. Read the article "I'm Sick and I'm Scared," in *Sports Illustrated*, July 8, 1991, pages 20–27. What does Alzado identify as the cause of his illness? What advice does he give to young athletes?

Reviewing Themes

26. The buildup of lactic acid, a product of fermentation, in the muscles. The buildup is due to the oxygen debt brought about by anaerobic exercise.

27. In our evolutionary ancestors, as in most mammals, the contraction of muscles in the skin pulls the hair upright. The hair in this position provides warmth. But, since humans have little or no hair over most of the body, the contraction of muscles in the skin causes goose bumps. Unfortunately, they provide little warmth.

Thinking Critically

28. Athlete A would have been chosen for training as a marathon runner because she has a greater percentage of slow-twitch muscles. Slow-twitch muscles are needed for the endurance associated with running a marathon.

29. The ozone protects the Earth from harmful ultraviolet rays. Without the protection provided by the ozone, more cases of skin cancer are very likely.

Cross-Discipline Connection

30. Luigi Galvani discovered that electrical currents could cause contractions in nerves and muscles.

Discovering Through Reading

31. Alzado said his cancer was the result of taking anabolic steroids and growth hormones. He urged young athletes not to take anabolic steroids or growth hormones.

22. Osteoporosis is a condition associated with loss of bone density. Women are more likely to suffer from osteoporosis because their bones are smaller and the production of female sex hormones declines during menopause.

23. Muscles cannot push on a bone. For movement to occur when a muscle pulls on a bone, another muscle is needed to pull the bone in the opposite direction.

24. Bundles of muscle cells contract, exerting the force needed to cause the muscle to contract.

Interpreting Graphics
25. • cushions the bones in a joint
• minerals and fibers
• tendinitis
• blood

Procedural Note

1. Have students wear lab aprons, goggles, and gloves.
2. Caution students to use the scalpels and scissors with care.

Prelab Preparation Answers

- Muscles move bones by pulling on them.
- Tendons attach muscles to bones.
- The difference between tendons and ligaments is that ligaments connect a bone to another bone, whereas tendons connect muscle to a bone.
- Cartilage cushions contact between bones.

Procedure Answers

4. Students should suggest that muscle tissue looks smooth.
5. elbow
6. Tendons will attach muscles to bones. Ligaments will attach bone to bone.
10. Tendons will attach muscles to bones. Ligaments will attach bone to bone.
11. Students should indicate that joints are more loose.
13. It is tough and very smooth.

Chapter **27** **Investigation**

How Do Muscles and Bones Work Together?

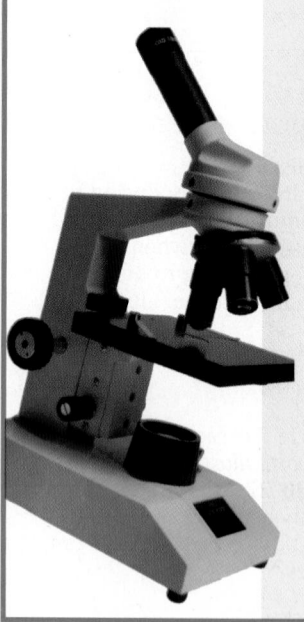

Objectives

In this investigation, you will:
- *observe* the movement of opposing pairs of muscles
- *relate* the structure of a bird's wing to its function
- *compare* and *contrast* the range of motion of a bird's wing to that of the human arm

Materials

- disposable gloves
- chicken wing
- dissecting pan
- scissors
- blunt probe
- scalpel
- paper towel
- unlined paper
- red, yellow, blue, and purple colored pencils

Prelab Preparation

Review what you have learned about muscles and bones by answering the following questions.
- How does a muscle move a bone?
- How are muscles attached to bones?
- What is the difference between a tendon and a ligament?
- Why is cartilage found between bones?

Procedure

1. Form a cooperative group of four students. Work with a member of your team to complete steps 2–14.
2. **CAUTION: Raw chicken may be contaminated by *Salmonella*. Keep your hands away from you face and mouth throughout this investigation.** Put on a pair of disposable gloves. Place a chicken wing in a dissecting pan.
3. Compare your chicken wing to the figure below.

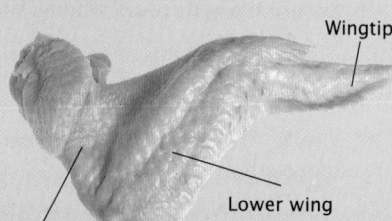

Identify the three main parts of the wing: the upper wing, the lower wing, and the wingtip.

4. **CAUTION: Handle dissecting instruments carefully to avoid injury.** At the cut end of the upper wing, slip the tip of the scissors between the skin and the muscles underneath. *What does the muscle tissue look like?*

5. Cut the skin lengthwise, to the joint between the upper wing and the lower wing. *Which joint in your body corresponds to this joint in the chicken wing?*

6. Using a scalpel and scissors, carefully remove the skin from the joint between the upper and lower wing. Be careful not to cut any tendons or ligaments. *How will you recognize tendons and ligaments?*

7. Remove the skin of the lower wing in the same way that you removed the skin from the upper wing. Leave the skin on the wingtip.

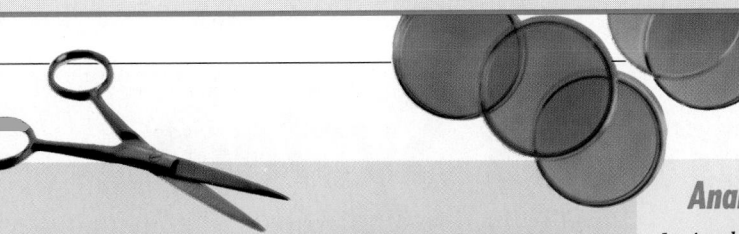

8. Using scissors, remove any tissues covering the muscle. Use a blunt probe to separate the individual muscles from each other without tearing them. On a sheet of unlined paper, draw your chicken wing, showing all of its separate muscles.

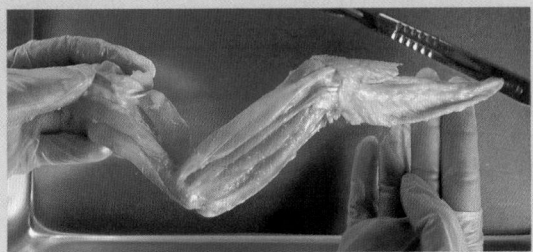

9. Straighten the chicken wing and hold it horizontally above the tray. Pull on each of the muscles and note the movement that each muscle causes. Turn the wing upside down and bend the joints. Again pull on each muscle and note how the bones move. On your drawing, color red each muscle that bends a joint. Color yellow each muscle that straightens a joint.

10. Find as many tendons as possible in your chicken wing. Determine where each tendon connects to a bone. *How will you decide if the structure that you have found is a tendon or a ligament?* With the blue pencil, add tendons to your drawing.

11. Cut crosswise through each muscle. Then follow each half of the muscle to the point it attaches to the bone and cut only the tendon. *How did removing the muscles and tendons from the wing affect the stability of this joint?*

12. Closely examine the joint between the upper wing and the lower wing and identify the ligaments. Using a purple pencil, add ligaments to your drawing.

13. Bend and straighten the joint and observe how the bones fit together. The shiny, white covering of the joint surfaces is made of cartilage. *What characteristics of the joint enable it to move smoothly?*

14. Clean up your materials and wash your hands before leaving the lab. **Follow your teacher's directions for proper disposal of the chicken wing.**

Analysis

1. *Analyzing Structure* How does tendon tissue differ from muscle tissue?

2. *Recognizing Relationships* Describe how bones and cartilage form joints.

3. *Relating Ideas* Explain how muscles, tendons, bones, ligaments, and joints combine to allow the back-and-forth movement of the lower chicken wing.

Thinking Critically

1. Compare the arrangement of bones in a chicken wing with the bones found in the human arm shown in **Figure 27.13**. Describe the similarities and differences between the limbs of these two species.

2. What can you infer about the evolution of limb structure in vertebrates?

3. The chicken is not capable of sustained flight. How do you think the structure of the chicken wing differs from the wing structure of a bird that flies?

Analysis Answers

1. Students should explain that tendons are made of white, tough, elastic connective tissue whereas muscle tissue is made of cells that contract or shorten and return to their normal length.

2. Two bones are linked together by cartilage, which prevents the ends of bones from grinding together.

3. Answers will vary but should include a discussion of the action of opposing pairs of muscles across a joint between a stationary and a movable bone. The discussion should also mention the function of the ligaments in binding bones together and the tendons that attach muscles to bones.

Thinking Critically Answers

1. Similarities and differences should be easily seen. Accept any reasonable answer.

2. Answers will vary but should infer a common ancestor based upon similar bone structure, even though the use of the appendage itself may be different.

3. Students might suggest that the wings of a bird that flies will have longer bones than those in a chicken's wings (for support in flight) and a larger muscle mass to provide strength for flying. Students might also suggest that the bones of such a bird will be lighter, thus making flight easier.

Chapter 28

The Nervous System

Planning Guide

Objectives/Themes	Classwork Resources	Homework Resources
28.1 1. Describe the structure of a neuron. 2. Explain how a nerve impulse travels along a myelinated neuron. 3. Explain how a nerve impulse is carried across a synapse. 4. Explain how nerves and muscles interact. **Themes:** Energy and Life, Interacting Systems, Patterns of Change, Scale and Structure	**Text** Journeys *Tour of a Neuron* p. 643 **Teacher's Resource Binder** Focus Activity 28 *Determining Experimental Bias* Lab Investigation 28.1 *Taste and Smell Sense Perception* **Other Resources** Transparencies 124–125	**Text** Section Review, p. 646 **Teacher's Resource Binder** Directed Reading Worksheet 28.1 **Other Resources** Audiocassette 28.1*
28.2 1. Explain the functions of each of the three divisions of the nervous system. 2. Compare and contrast the functions of the three main parts of the brain. 3. Compare sensory neurons, motor neurons, and the central nervous system. 4. Discuss two techniques scientists use to study the brain. **Themes:** Scale and Structure, Interacting Systems	**Text** Journeys *Evolution of the Brain* pp. 650–651 Science in Action *Jim Moreland: National Pharmaceutical Sales Manager* pp. 656–657 **Teacher's Resource Binder** Extension Worksheet *Sleep* **Other Resources** Transparency 126	**Text** Section Review, p. 655 **Teacher's Resource Binder** Directed Reading Worksheet 28.2 **Other Resources** Audiocassette 28.2*
28.3 1. List three stimuli that sense organs react to. 2. Explain how the inner ear helps you to maintain balance. 3. Explain how your ear detects sound. 4. Compare and contrast the functions of rods and cones. **Themes:** Energy and Life, Interacting Systems, Patterns of Change, Scale and Structure, Stability	**Text** Investigation *How Does the Brain Interpret Information From the Eyes?* pp. 666–667 **Other Resources** Transparencies 127–129	**Text** Section Review, p. 662 **Teacher's Resource Binder** Directed Reading Worksheet 28.3 Vocabulary Review Worksheet* Reteaching Worksheet* *Identifying Structures* **Other Resources** Audiocassette 28.3*

*Reteaching Options

Demonstrations
28.1: pp. 640, 641
28.2: pp. 648, 649, 653, 654, 655
28.3: pp. 659, 660, 661, 662

Assessment
Chapter Review pp. 664–665
Portfolio Assessment p. 637D
Chapter Test—Teacher's Resource Binder
Test Generator

Research Notes

Connection to Physiology: Brain Cells' Fountain of Youth?

For many years scientists believed that brain cells stopped growing and dividing shortly after birth. Many studies had failed to detect evidence of recent growth in adult brains. Researchers at the University of Calgary in Alberta, Canada, have found that this doesn't necessarily mean that the neurons can no longer divide.

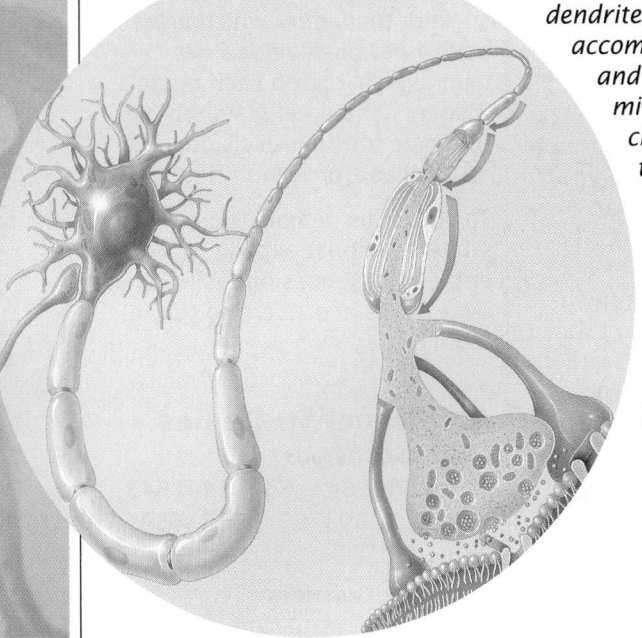

Samuel Weiss and Brent Reynolds placed 1,000 brain cells from adult mice in culture. Some of the cells were placed in a culture with a protein called EGF, which is known to have a role in developing skin and nervous tissue in embryos. In the non-EGF cultures, all of the cells died. Most of the cells in the EGF cultures died, but about 15 from every 1,000 lived. When placed into separate dishes with EGF, these cells divided to form a clump of new cells.

The new cells began to take the shape of neurons—complete with dendrites, axons, and their accompanying glial cells— and produced neurotransmitters and other chemicals that indicate they were nerve cells. These precursor cells, which form various types of nerve cells, seem analogous to the "stem" cells in bone marrow that form all of the different cells of the blood.

These results may have been overlooked earlier because there is little or no EGF in adult brains. Although human trials of EGF are a long way off, it could be that diseased or injured portions of the brain may be able to heal themselves with the help of this growth factor.

Unrelated research has already shown that scientists can use glial cells and collagen to prompt portions of rat spinal cords to regenerate after they are severed. Someday this approach, alone or with EGF, could help people who have sustained a head injury or have neurological disorders such as Parkinson's disease. However, researchers caution that the growth of new neurons in the brains of adults could cause adverse effects such as scrambling memories and disrupting brain patterns.

Investigation Notes

How Does the Brain Interpret Information From the Eyes?

pages 666–667

Purpose: This investigation provides students with an opportunity to learn about the functioning of the human eye by exploring their own vision. Students will see examples of how the brain combines information from both eyes, how color vision operates, and how information about an image persists after being initially viewed.

Scientists use all their senses to observe the world around them. Normally, we trust what we perceive, but sometimes our sensory perceptions can fool us. This exercise shows students that what we see might not be real.

Prelab Preparation

Students will need an unlined index card, plain white paper, and a pencil.

Answers will be found on pages 666–667.

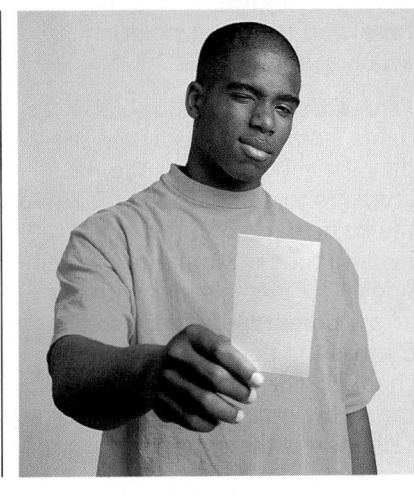

Meeting Individual Needs

Objectives

1. Students will demonstrate appropriate use of core vocabulary for the chapter (Vocabulary File).

2. Students will interpret the different components of a model of an active neuron (Demonstration A).

3. Students will interpret and develop a model of a nerve impulse traveling along an axon (Demonstration B).

Vocabulary File
(Developing Vocabulary/ Limited English Proficiency)

If you are not already using the Vocabulary File, refer to Chapter 1 for its preparation. See the Chapter Highlights on page 663 for a list of suggested words.

Demonstration A
Model of a Nerve Cell
for page 639
(Developing Modeling Skills/ Visual Learners)

Materials: glue, scissors, mailing tube, pipe cleaners, construction paper

Preparation: Cut a 15-cm (6-in.) length of mailing tube, then split it in half lengthwise. Attach two pipe cleaners to opposite edges of the tube. Cut two arrows out of construction paper and glue them onto the pipe cleaners, one pointing toward the mailing tube, and the other pointing away from it.

Procedure: Read aloud to students from page 639. Then display the model. Challenge the students to compare the model to the basic parts of a neuron. (The mailing tube is similar to the cell body. The pipe cleaner with the arrow pointing toward the cell body is the dendrite, the input channel. The pipe cleaner with the arrow pointing away from the cell body is the axon, the output channel.)

Demonstration B
Model of a Nerve Impulse
for pages 640–641
(Developing Modeling Skills/ Multisensory Learners)

Materials: glue, scissors, mailing tube, small round stickers of two different colors, construction paper

Preparation: Cut three 27-cm (10-in.) lengths of mailing tube, then split them in half lengthwise. Label one color of stickers "+" and the other "–." Cut six thick arrows out of one color of construction paper, and six thin arrows out of another color. The thick arrows should be labeled "Na+" and the thin arrows "K+."

The three lengths of mailing tube represent the phases of nerve impulse movement along an axon, as shown in Figure 28.2c on page 641. At the end of one model, attach the "K+" arrows to the sides of the mailing tube, pointing outward. A little farther from the end, attach the "Na+" arrows to the sides of the mailing tube, pointing inward.

On the second tube, attach the arrows in the same order, but closer to the middle. On the third tube, the arrows should be placed closer to the opposite end.

Procedure: Read aloud to students from pages 640–641. Then display the models. Explain that they are consecutive views of the same axon as a nerve impulse travels down it. Challenge the students to identify the different aspects of the models. (The thick, Na+ arrows mark the beginning of the nerve impulse as sodium ions rush into the cell. The thin, K+ arrows mark the end of the impulse, with potassium slowly leaking out of the cell to restore the resting potential.)

Challenge students to use the stickers to identify which areas have a net positive or a net negative charge on each model. Students should explain the rationale for their decisions, on the basis of the ion flow.

To close this demonstration, students should explain nerve impulse transmission, using their own words and referring to the model.

Additional Strategies
Visual Strategies
Pages 640, 641, 642, 644, 645, 646, 652, 653, 654, 658, 659, 660, and 661

Auditory Learners
Use *Biology: Visualizing Life* Audiocassettes for Sections 28.1, 28.2, and 28.3.

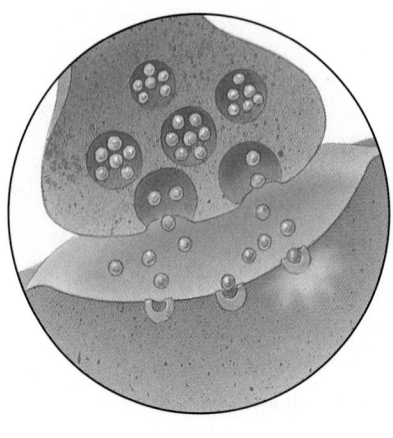

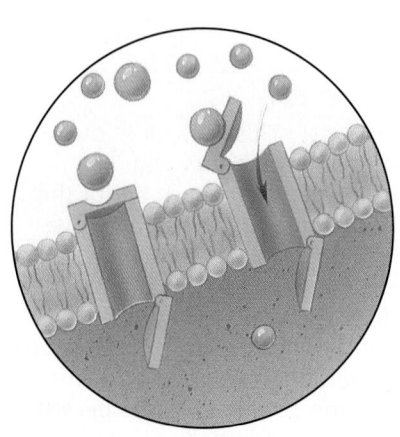

Meeting Individual Needs (cont.)

Cooperative Learning
Parts of the Nervous System

Timing: Use this activity to introduce Section 28.2.

Group Size: 3 students

Outcome: Students will relate different parts of the nervous system to the parts of the body they affect.

Individual Accountability: Each group member is responsible for one aspect of poster construction.

Positive Interdependence: Each group will work together to prepare a poster detailing the relationships between a part of the nervous system and the parts of the body.

Assign one of these parts of the nervous system to each group: (1) cerebellum and motor neurons, (2) brain stem, (3) cerebrum and sensory neurons. One member of each group should create an outline of the body and a drawing of their portion of the nervous system within the body. Another group member should draw in the specific body parts that are affected by the assigned portion of the nervous system. The third group member should label all parts of the chart and indicate additional nervous system functions that do not relate to specific body parts.

Portfolio Assessment

Students should select their best work and provide a self-reflective rationale for their selections. Students can make selections in the following areas.

1. *Content* — One concept map from the chapter (See page 812 for evaluation criteria.)

2. *Reading Comprehension* — One Directed Reading Worksheet from the Teacher's Resource Binder (Use the answer key to evaluate for accuracy.)

3. *Writing* — Using the Vee Form, summarize a magazine or newspaper article relating to the nervous system or sense organs. (See page 22T for evaluation criteria.)

   Or: Select a writing task or project from the Chapter Review.

4. *Performance Assessment* — One Vee form from a chapter investigation or lab manual investigation (See page 22T for evaluation criteria.)

Teacher makes selections in the following areas.

1. *Formal Assessment* — Chapter test (Test A, B, or the Test Generator) The teacher-scored test should be reviewed by the student. Incorrect responses should be corrected by the student before the test becomes part of the portfolio.

2. *Informal Assessment* — Use the Direct Observations Checklist, page 33T, during a laboratory or other cooperative learning experience.

3. *Performance Assessment* — Have students design an experiment to test the thresholds of their senses of hearing, vision, taste, and smell.

Concept Map Answer

The following is one possible answer to the Relating Concepts exercise on page 664.

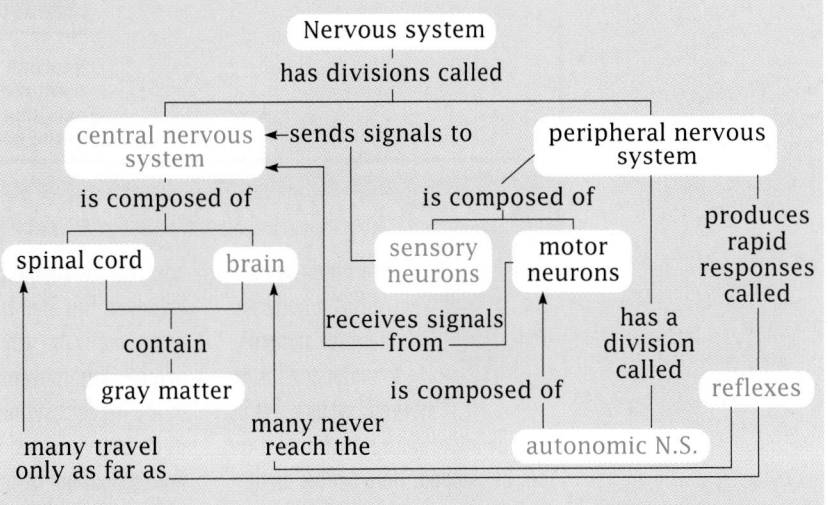

Chapter 28

Determining Prior Knowledge

- Form cooperative groups of two students. Darken the classroom and have students take turns observing what happens to their partner's eyes when the lights are turned on. Have students explain their observations.
- Show students a picture of the human brain and a computer. Have them provide examples of how the two are alike.
- Have several students volunteer to wear earplugs and a blindfold while the class discusses what they know about sense organs. Have the students remove the earplugs and blindfold after 20 minutes and then have them describe what was discussed while they were "disabled."

Chapter 28

The Nervous System

Review

- **sodium-potassium pump** (Section 4.1)
- **protein receptors** (Section 4.1)
- **nerve tissue** (Section 27.1)

28.1 How a Nerve Carries a Message

- The Neuron
- Nerve Impulses
- The Synapse
- Nerve-Muscle Junctions

28.2 The Nervous System

- The Central Nervous System
- Three Parts of the Brain
- The Peripheral Nervous System
- The Autonomic Nervous System
- How Scientists Study the Brain

28.3 The Sense Organs

- Sensing Internal Information
- Sensing Sound
- Sensing Light
- Touch, Smell, and Taste

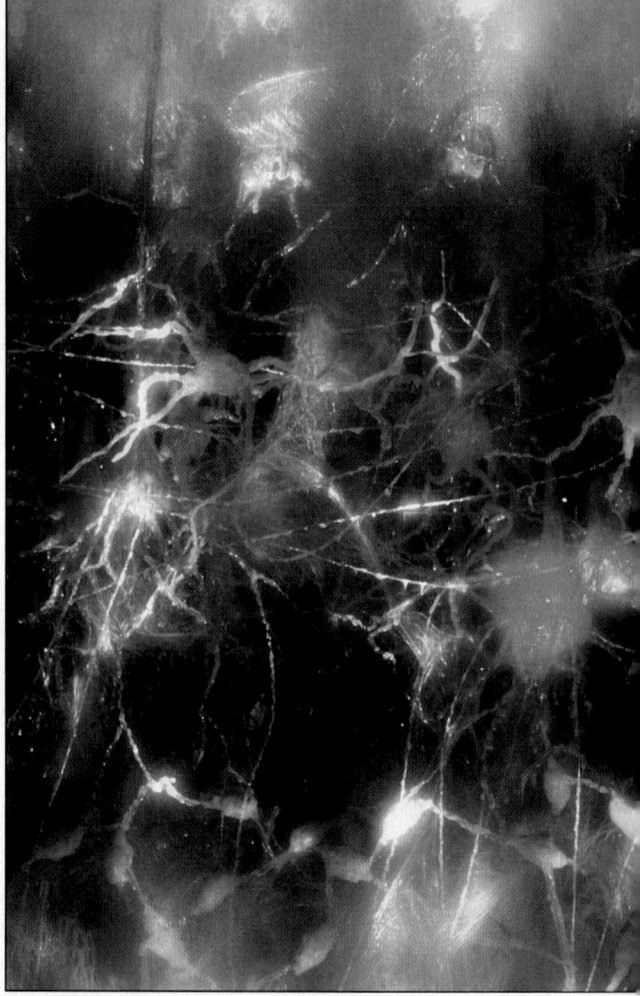

Neurons such as these brain cells are specialized to carry electrical messages. Their membranes are rich in gated ion channels that can open and close, routing messages along nerves in the body.

■ Author's Rationale ■

Perhaps the most difficult body system for students to understand is the nervous system. Yet its overall importance is such that it cannot be skipped. Here I focus on describing how it works by looking at how channel membranes create nerve impulses. This idea is not difficult to grasp, nor is the way ATP drives myofilament movement. Anatomy of the nervous system is presented as an overview, and instructional time should not be bogged down with anatomical details. If students understand how nerves and muscles function, then they have learned a great deal.

WHEN YOU TOUCH A HOT STOVE, YOUR HAND INSTANTLY JERKS BACK BECAUSE OF A MESSAGE FROM YOUR SPINAL CORD. WHEN YOU SWAT A FLY, THE MUSCLES OF YOUR ARM CONTRACT QUICKLY BECAUSE OF A MESSAGE FROM YOUR BRAIN. THESE MESSAGES TRAVEL ALONG NERVES, WHICH ARE BUNDLES OF CELLS THAT CARRY ELECTRICAL SIGNALS THROUGHOUT YOUR BODY.

Lesson Plan 28.1

Phase 1

PREPARATION

28.1   *How a Nerve Carries a Message*

Objectives

① **Describe the structure of a neuron.**

② **Explain how a nerve impulse travels along a myelinated neuron.**

③ **Explain how a nerve impulse is carried across a synapse.**

④ **Explain how nerves and muscles interact.**

The Neuron

Figure 28.1
Neurons such as these brain cells receive information and send it to other parts of the body.

The basic units of communication in the nervous system are cells called **neurons** (*NOO rahnz*). Neurons carry electric signals like the electric wires in a house. When you flip a light switch, an electric current travels through a wire and a light bulb comes on. When your finger touches an uncomfortably hot surface, a message travels along neurons to your spine. Your spine sends out orders along other neurons that cause your muscles to withdraw your hand.

Neurons are specialized to carry signals throughout the body

Although they differ greatly in the details of their structure, neurons with the same basic architecture occur in the brain, the spinal cord, and the nerves that travel throughout the body. As you can see in **Figure 28.1**, short, slender branches called **dendrites** (*DEHN dryts*) extend from one end of the neuron's body. Dendrites are input channels. They receive information from other neurons or from sensory cells. At the other end of the cell body there is a single, long, tubelike extension called an **axon**. Axons are output channels. It is along the axon that the neuron sends out messages to other neurons or directly to the muscles.

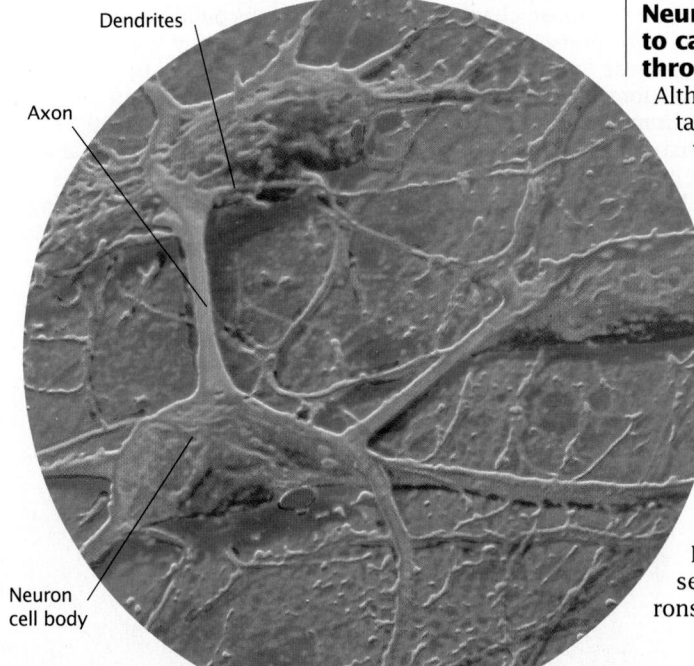

Dendrites

Axon

Neuron cell body

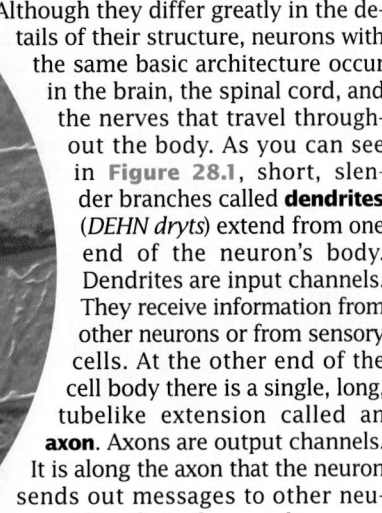

Key Concepts

- A nerve cell is called a neuron and consists of a cell body, dendrites, and an axon.
- A nerve impulse involves the movement of sodium and potassium ions across the cell membrane of a neuron.
- When not conducting an impulse, a neuron exhibits a resting potential in which the inside of the neuron is more negatively charged than the outside.
- When conducting an impulse, the neuron exhibits an action potential in which the inside of the neuron momentarily becomes more positively charged than the outside.
- Myelin sheaths surrounding neurons allow impulses to travel faster.
- Neurotransmitters carry impulses from one neuron to another across a gap known as a synapse.
- Many drugs, including opium and alcohol, interfere with the function of neurotransmitters.
- Some neurotransmitters travel from nerves to muscles, where they trigger muscle contraction.

Reading Strategy

Perhaps the most basic skill in understanding what one reads is knowing what every word means. This section is an excellent place for students to practice how to learn a word's meaning by analyzing the word itself for clues. For example students should recognize that a neurotransmitter must somehow be involved in "transmitting" signals through the nervous system, or that an action potential must refer to a neuron that is in "action" or in the process of conducting a nerve impulse.

Phase 2

TEACHING STRATEGIES

Demonstration 1

How Long Can an Axon Be?

Tell students that most of the cell bodies of neurons involved in controlling muscle movement are located in the spinal cord. Consequently, their axons must extend all the way from the spinal cord to the muscles they innervate. Demonstrate how long an axon can be by measuring the distance from a volunteer student's spine—at a point just below the neck—to the fingertips, and from the base of the spine to the toe.

Demonstration 2

Measuring Electrical Activity

Check with a physics teacher to arrange a simple demonstration to show the class how electrodes conduct electrical activity detectable by a voltmeter or an oscilloscope.

Connection: Chapter 2

Electrons and Ions

Remind students that electrons are the negatively charged particles that move around the atom and that ions are atoms with either a positive or negative charge.

Visual Strategy

Figure 28.2

Have students compare the sequence of events shown in this figure with the sequence they carried out in the previous demonstration. Have them identify where both the resting and action potentials would be found in each of the three phases that are shown. Be sure students realize that the resting potential is restored with the help of the sodium-potassium pump that reestablishes the original ion gradients.

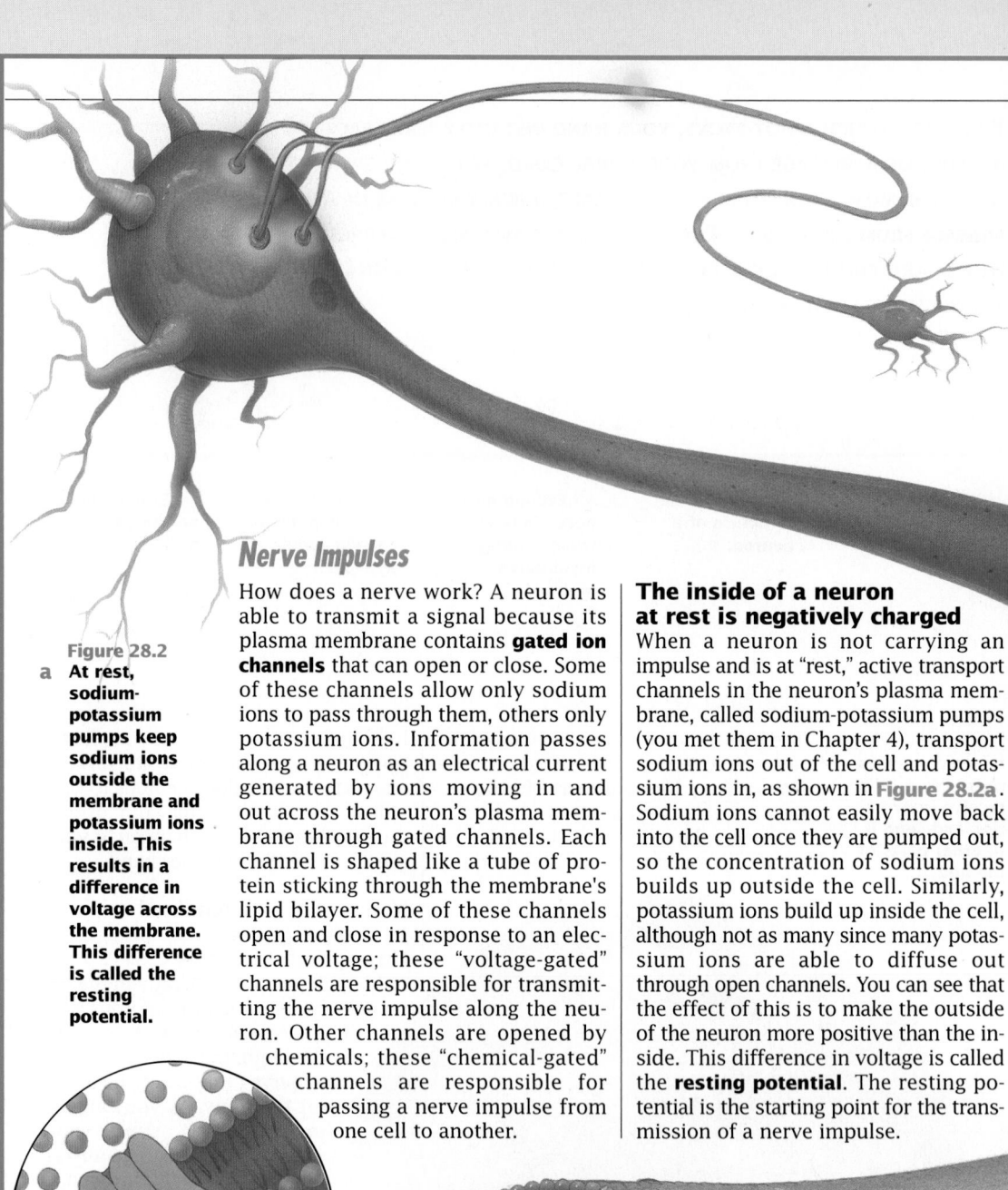

Figure 28.2

a **At rest, sodium-potassium pumps keep sodium ions outside the membrane and potassium ions inside. This results in a difference in voltage across the membrane. This difference is called the resting potential.**

Nerve Impulses

How does a nerve work? A neuron is able to transmit a signal because its plasma membrane contains **gated ion channels** that can open or close. Some of these channels allow only sodium ions to pass through them, others only potassium ions. Information passes along a neuron as an electrical current generated by ions moving in and out across the neuron's plasma membrane through gated channels. Each channel is shaped like a tube of protein sticking through the membrane's lipid bilayer. Some of these channels open and close in response to an electrical voltage; these "voltage-gated" channels are responsible for transmitting the nerve impulse along the neuron. Other channels are opened by chemicals; these "chemical-gated" channels are responsible for passing a nerve impulse from one cell to another.

The inside of a neuron at rest is negatively charged

When a neuron is not carrying an impulse and is at "rest," active transport channels in the neuron's plasma membrane, called sodium-potassium pumps (you met them in Chapter 4), transport sodium ions out of the cell and potassium ions in, as shown in **Figure 28.2a**. Sodium ions cannot easily move back into the cell once they are pumped out, so the concentration of sodium ions builds up outside the cell. Similarly, potassium ions build up inside the cell, although not as many since many potassium ions are able to diffuse out through open channels. You can see that the effect of this is to make the outside of the neuron more positive than the inside. This difference in voltage is called the **resting potential**. The resting potential is the starting point for the transmission of a nerve impulse.

▪ *Matter of Fact* ▪

Squid Study

The squid was the first organism used to measure the resting potential of a neuron. Because of its large, long axons that innervate muscles responsible for expelling water from the body cavity, the squid made an ideal specimen for studies in nerve physiology.

A nerve impulse disrupts the resting potential

A nerve impulse starts when pressure, or other sensory inputs, disturbs a neuron's plasma membrane, causing sodium channels on a dendrite to open. As a result, sodium ions flood into the neuron from outside, and for a brief moment the inside of the membrane becomes more positive in that immediate area of the cell. This sudden local reversal of electric voltage across the neuron membrane is called an **action potential**.

The sodium channels in the small patch of membrane with the action potential remain open for only about half a millisecond. However, the change in voltage causes nearby voltage-gated sodium ion channels to open, which starts the action potential moving down the neuron, as the opening of the gated channels causes nearby voltage-gated channels to open, like a chain of falling dominoes.

When the action potential has passed, the voltage-gated sodium channels snap closed again and the resting potential is restored. So a nerve impulse is actually the movement of an action potential along a neuron as a series of voltage-gated ion channels open and close. Examine the details of how nerve impulses move along an axon in **Figure 28.2b**

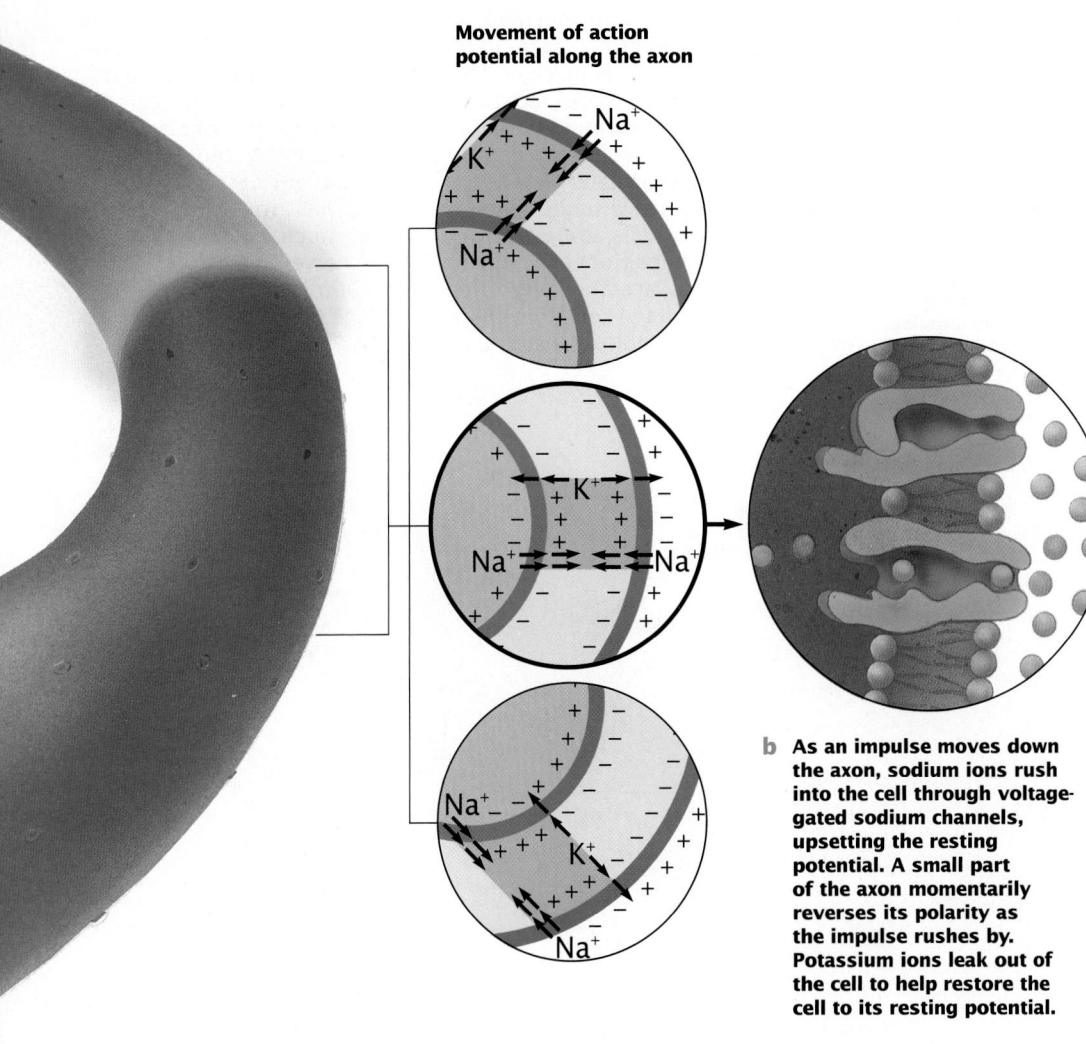

Movement of action potential along the axon

b As an impulse moves down the axon, sodium ions rush into the cell through voltage-gated sodium channels, upsetting the resting potential. A small part of the axon momentarily reverses its polarity as the impulse rushes by. Potassium ions leak out of the cell to help restore the cell to its resting potential.

Theme Answer

Evolution

They enable nerve impulses to be transmitted more rapidly.

Social Studies Connection

Multiple Sclerosis

Tell students that multiple sclerosis is most prevalent in northern Europe and North America, where the diet is high in animal fat. Have the class suggest an explanation for this correlation.

Visual Strategy

Figure 28.3

Use this figure to have students describe how the impulse "hops" along a myelinated fiber. Have them explain what happens to the movement of ions across portions of the cell membranes that are covered with myelin. Students should realize that these stretches do not undergo an action potential and consequently do not have to expend energy in actively pumping sodium and potassium ions to reestablish a resting potential. To impress students with how effective myelin is as a shortcut, point out that in some large, myelinated fibers, an impulse can travel 200 m (650 ft.) per second, while in small, unmyelinated fibers, it can travel no faster than a few millimeters (much less than 1 in.) per second.

Evolution

Why are myelin sheaths considered an evolutionary advancement?

Myelin sheaths speed signals through the nervous system

A neuron that reaches from the tip of your index finger to your spinal cord is very long. Some vertebrates have neurons that are even longer. Think of one stretching from a giraffe's foot to its spine—a distance of 5 m (16 ft.)! It takes time for a nerve impulse to travel down a long neuron. If the nerve impulse is too slow, you risk burning your finger on the hot stove. Similarly, a giraffe would be in serious trouble if the message from its brain to its legs were slower than a charging lion.

Humans, giraffes, and other vertebrates have evolved ways of speeding up nerve signals. Many of their neurons have axons encased in myelin (*MEY uh lihn*) sheaths. A **myelin sheath**, illustrated in **Figure 28.3**, is made of special cells that wrap their fatty cell membranes in layers around the axon. Between zones of myelin, the neuron is exposed for a short interval. The exposed gap is called a node. When a nerve impulse travels along a myelinated axon, it jumps from one node to the next. Nodes are the only sites with exposed voltage-sensitive channels that can respond to the arrival of

an electrical charge. Jumping from node to node like this is much faster than traveling along the full length of the bare axon.

Destruction of large patches of myelin characterize a disease called multiple sclerosis. In multiple sclerosis, small, hard plaques appear throughout the myelin. Normal nerve function is impaired, causing symptoms such as double vision, muscular weakness, loss of memory, and paralysis.

Some animals, like cockroaches, have neurons with giant axons that are extra thick and carry impulses faster than normal axons. These giant axons run from receptors in the roach's abdomen to its brain. If you swat at a cockroach, the giant axons carry the message that there is a current of air caused by your hand. When the brain gets the message, the cockroach scurries away, quickly evading your descending hand. You can read much more about nerves in the *Tour of a Neuron* on page 643.

Figure 28.3
Myelin sheaths wrap the axon membrane many times, forming an insulating sheath. The nerve impulse then "hops" from node to node, thereby speeding up the transmission of electrical signals.

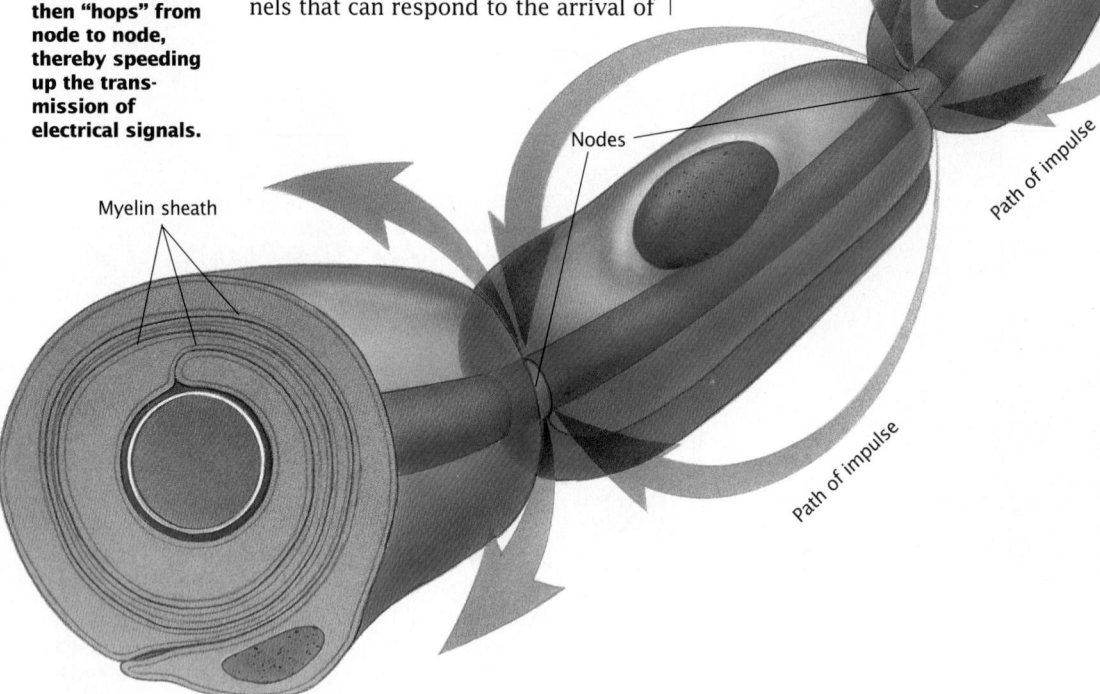

Myelin sheath

Nodes

Path of impulse

Path of impulse

■ *Journeys* ■ *Journeys* ■ *Journeys* ■ *Journeys* ■ *Journeys*

Tour of a Neuron

Nerves are made of cells called neurons. A motor neuron like this one is specialized to transmit messages rapidly to muscle cells.

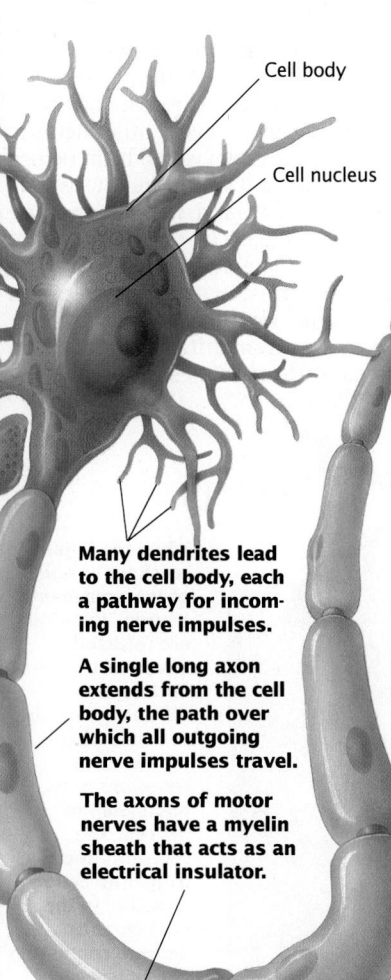

Cell body

Cell nucleus

Impulses travel along a neuron by successively opening ion channels at exposed gaps, like a chain of falling dominos.

Neurons connect to other neurons or to muscles at tiny gaps called synapses. Chemicals called neurotransmitters carry the nerve signal across the synapse. Muscle-nerve synapses use the neurotransmitter acetycholine.

Many dendrites lead to the cell body, each a pathway for incoming nerve impulses.

A single long axon extends from the cell body, the path over which all outgoing nerve impulses travel.

The axons of motor nerves have a myelin sheath that acts as an electrical insulator.

At the far side of the synapse, the neurotransmitter opens a gated sodium channel through the plasma membrane of the cell receiving the nerve impulse. This enables the nerve impulse to continue along the next cell.

The tips of many axons are branched.

Sodium

Sodium channel

Plasma membrane

Using the Feature

- You can use this tour to discuss axonal transport. The distribution of molecules in neurons can be a problem because axons can be quite long. A unique transport mechanism called axonal transport moves needed materials along the axons in two directions—toward the ends of axons and toward the cell body. After explaining this to students, tell them that certain viruses such as the herpes virus and the rabies virus use axonal transport as a vehicle to make their way to the nerve cell bodies where they multiply and cause their damage.

- You can also use this feature to inform students about disorders that affect the nervous system. Discuss the origins and symptoms of some common nervous system disorders such as poliomyelitis, cerebral palsy, Parkinson's disease, multiple sclerosis, epilepsy, Tay-Sachs disease, rabies, Reye's Syndrome, and even headaches. You may want to have students research these disorders at the library and present information to the class.

Discussion

1. How do dendrites and axons differ structurally and functionally?
   Dendrites are short branches that carry information to the cell body. Axons are long extensions that carry information away from the cell body. A neuron has many dendrites, but only a single axon.

2. How do myelin sheaths speed up nerve impulses?
   Myelin sheaths wrap around sections of an axon, leaving small gaps of the plasma membrane exposed. A nerve impulse travels along the neuron by successively opening ion channels at exposed gaps, instead of the entire axon.

3. How is a nerve impulse carried across a synapse?
   Chemicals called neurotransmitters travels across the synapse to a dendrite on an adjacent neuron, where it opens gated sodium channels. This enables the nerve impulse to continue along the next cell.

643

Visual Strategy

Figure 28.4

Use this figure to illustrate how an impulse is transmitted across a synapse. Have students identify the end of the axon with its neurotransmitter sacs and the dendrite with its cell membrane receptors. Point out that the amount of neurotransmitter released is related to the rate at which impulses reach the end of the axon—the stronger the stimulus, the faster and more often impulses arrive. Tell students that synapses control the direction an impulse can travel. Neurotransmitters are released only at the ends of axons, and travel to adjacent dendrites. Thus, an impulse can travel in only one direction across a synapse—from axon to dendrite.

Health Connection

Neurotransmitters

Some evidence points to a connection between a low level of neurotransmitters and depression. For example, some forms of depression have been successfully treated with drugs that promote the buildup of neurotransmitters across synapses. Have students hypothesize how one's health might be affected by an abnormally high level of neurotransmitters.

Mathematics Connection

How Many Neurons?

Each axon may have one or more synapses with as many as 1,000 other neurons. Have the class calculate how many neurons could be stimulated if each of these 1,000 neurons, in turn, forms synapses with another 1,000 neurons. Students should readily see how quickly and widely an impulse can travel. In reality, some of these impulses are excitatory, while others are inhibitory. Each neuron evaluates the total, averages the input, and responds accordingly. For example, if the overall total is excitatory, the neuron transmits the impulse to its neighbor.

The Synapse

Nerve impulses can travel only so far along a cell membrane. Eventually they reach the end of the axon, usually positioned very close to another neuron or to a muscle cell. This junction of a neuron with another cell is called a **synapse** (*SIHN ahps*). When a nerve impulse gets to the end of an axon, its message must cross the synapse if the message is to continue. The neuron carrying the message toward the synapse is called the presynaptic neuron. The neuron that transmits the message away from the synapse is called the postsynaptic neuron.

Messages do not "jump" across synapses. Instead, the messages are carried by chemical messengers called **neurotransmitters**. These chemicals are packaged in tiny sacs in the end of the axon, as you can see in **Figure 28.4a–b**. When a nerve impulse reaches the axon end, it causes the sacs to release the neurotransmitters into the synapse. The neurotransmitters diffuse across the synapse and bind to receptors in the membrane of the cell on the other side, passing the signal to that cell. Neurotransmitter molecules in the synapse are then quickly removed so that another message can cross the synapse. Depending on the type of neurotransmitter, the molecules may be absorbed back into the first neuron or they may be broken down by enzymes.

Synapses are the slowest part of the nervous system. This is because electricity travelling along a neuron moves faster than chemicals crossing a synapse. What is the evolutionary advantage of synapses, if impulses travel faster along axons? Imagine what it would be like if there were no synapses, if every neuron in your body were connected to every other neuron. The nervous system would be of no use at all, because it would be impossible to move your hand without moving every other part of the body at the same time. The advantage to having many neurons, with gaps between them, is that we can control and receive information from different parts of the body at different times.

Figure 28.4

a Nerve messages are transferred from one neuron to another using chemical messengers called neurotransmitters.

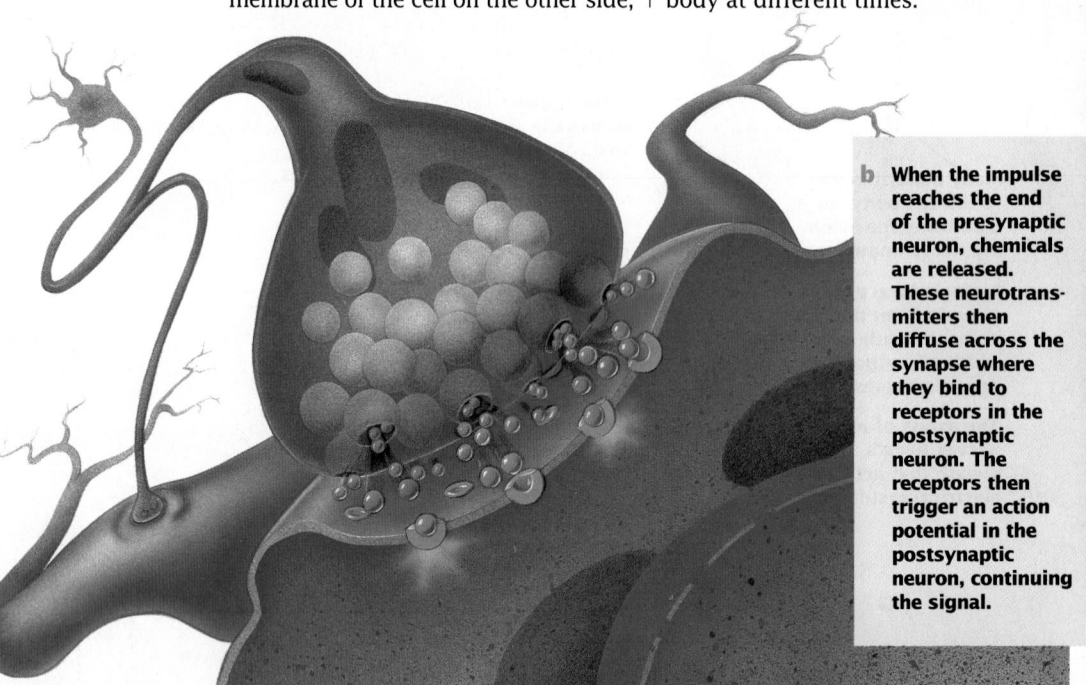

b When the impulse reaches the end of the presynaptic neuron, chemicals are released. These neurotransmitters then diffuse across the synapse where they bind to receptors in the postsynaptic neuron. The receptors then trigger an action potential in the postsynaptic neuron, continuing the signal.

■ Cultural Perspective ■

Dr. Emmeline Edwards (African-American)

Dr. Emmeline Edwards, who grew up in Haiti, studies the role that neurochemicals play in coping with stress. Information gained from her research can be applied toward the development of drugs for treating human depression. Dr. Edwards works at the department of psychiatry at the State University of New York in Stony Brook.

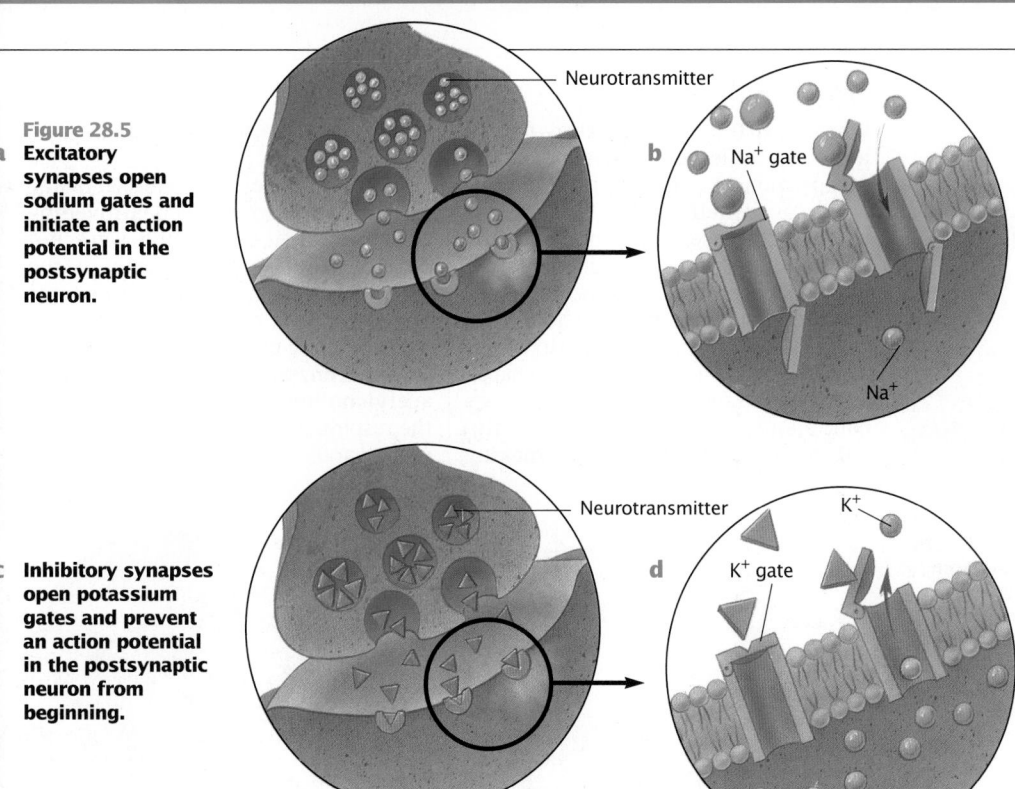

Figure 28.5
Excitatory synapses open sodium gates and initiate an action potential in the postsynaptic neuron.

Inhibitory synapses open potassium gates and prevent an action potential in the postsynaptic neuron from beginning.

Energy and Life

Why is the term

resting potential

misleading when

describing a

neuron's cell

membrane?

Different kinds of synapses use different ion gates

Vertebrate nervous systems use dozens of different kinds of neurotransmitters, each with specific receptors on postsynaptic membranes. They fall into two general classes, depending on whether they excite or inhibit passage across the synapse.

In an excitatory synapse, the receptor protein is a chemical-gated sodium channel that is closed at rest. On binding with a neurotransmitter that it recognizes, the sodium channel opens, allowing the sodium ions to flood inward, as shown in **Figure 28.5a-b**. If enough sodium ion channels are opened by neurotransmitters, an action potential begins in the postsynaptic membrane.

In an inhibitory synapse, the receptor protein is a chemical-gated potassium channel. Binding with its neurotransmitter opens the potassium channel, leading to the exit of positively-charged potassium ions and a more negative interior of the postsynaptic membrane, as shown in **Figure 28.5c-d**. This inhibits the start of an action potential, because the negative voltage change inside means that even more sodium ion channels must be opened to get the domino effect started among voltage-gated sodium channels and to start an action potential.

An individual nerve cell can possess both kinds of synaptic connections to other nerve cells. When signals from both excitatory and inhibitory synapses reach the body of the neuron, the excitatory effects (which cause less internal negative charge) and the inhibitory effects (which cause more internal negative charge) interact with one another. The result is a process of **integration** in which the various excitatory and inhibitory electrical effects tend to cancel or reinforce one another. Neurons often receive many inputs. A single motor neuron in the spinal cord may have as many as 50,000 synapses on it.

Visual Strategy

Figure 28.6
Use this figure to point out that the acetylcholine from a single axon may stimulate just a few muscle fibers, as in the case of those that move the eyeball, or more than 1,000 fibers, as in the case of the biceps. Have students hypothesize what might happen if only a few muscle fibers were stimulated in the biceps when someone tried to lift a heavy object.

Connection: Chapter 27

Muscle Contraction
Remind students how actin and myosin filaments slide past each other when muscles contract.

Phase 3

ASSESSMENT OPTIONS

Closure Strategy

The Knee-Jerk
Have students take turns lightly tapping each other's leg slightly below the knee. If possible, have them use a rubber-headed hammer borrowed from the school nurse or a local physician. Have students write a brief report summarizing their observations. Tell them that their report must include the following terms: dendrite, axon, neurotransmitter, synapse, resting potential, action potential, acetylcholine, and nerve-muscle junction.

Section Review

Assign the *Section Review.*

Reteaching

On the chalkboard, list the major themes that were presented in Chapter 1. Have the class review this section to identify examples that can be used to illustrate each of these themes.

Nerve-Muscle Junctions

When your finger moves away from a hot stove, a nerve impulse makes the biceps muscle contract, jerking your arm away from the stove. The message to contract crosses from a neuron to the muscle at a synapse. When the nerve impulse reaches the synapse, it releases the neurotransmitter acetylcholine (*uh seet uhl KOH leen*) from the tip of the axon. Acetylcholine crosses the synapse and binds to receptors on the muscle's cell membrane. This causes a signal to pass along the muscle, making it contract. Examine the diagram in **Figure 28.6.**

After acetylcholine is released by the muscle's receptors, it is destroyed by enzymes. If acetylcholine were not destroyed, it would remain in the synapse and the muscle would continue to twitch. This characteristic is used to make nerve gas, a type of poison that prevents enzymes from breaking down acetylcholine. As a result, muscles in the respiratory and nervous systems become paralyzed.

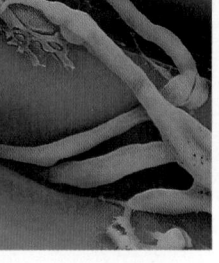

Figure 28.6
a As you can see in the scanning electron micrograph above, the motor neuron is directly attached to the muscle.

b When a nerve impulse reaches the nerve-muscle junction, acetylcholine is released into the synapse. It attaches to the muscle receptors, causing the muscle to contract.

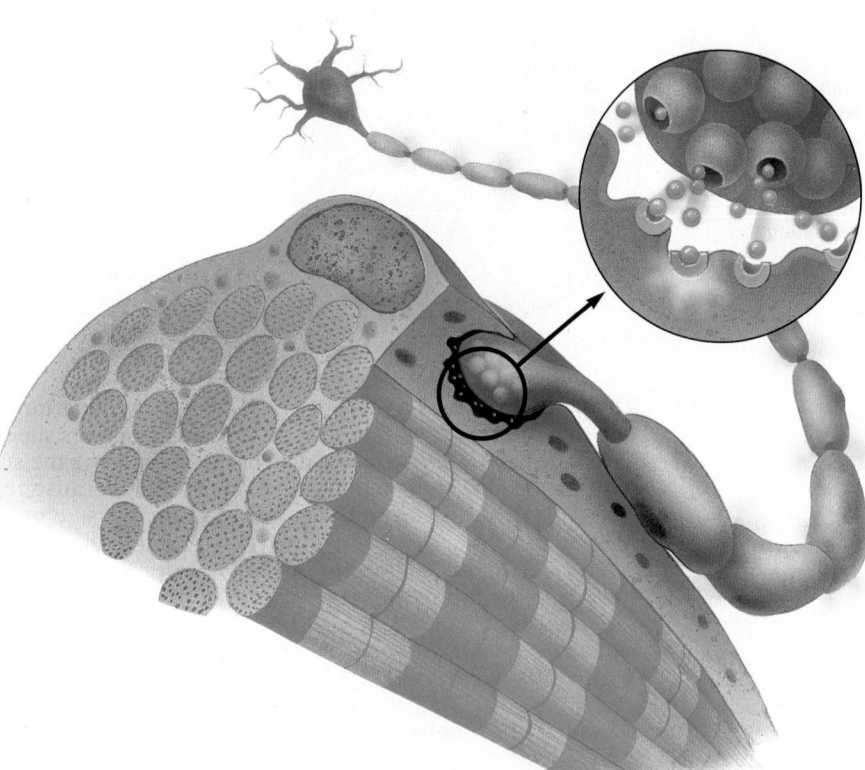

Section Review

❶ How is a neuron specialized for carrying electrical signals?

❷ How do myelin sheaths speed up nerve impulses?

❸ How do neurotransmitters carry a nerve impulse from one cell to another?

❹ What causes a muscle to contract?

■ Section Review Answers ■

1. Neurons are composed of dendrites, which serve as input channels, and axons that act as output channels.

2. When a nerve impulse travels along the myelinated axon, it jumps from one node to the next.

3. Neurotransmitters are chemical messengers that diffuse across the synapse and bind to receptors in the membrane of the cell on the other side, passing the signal to that cell.

4. See Figure 28.6 on page 646.

SIGNALS FROM NERVE CELLS ENABLE YOU TO PLAY A PIANO, THROW A BASEBALL, WRITE IN YOUR NOTEBOOK, OR JUST SIT AND THINK. A HIGHLY DEVELOPED BRAIN COORDINATES ACTIVITIES AND ENABLES YOU TO LEARN, IMAGINE, REMEMBER, AND REASON. THE COMPLEX SYSTEM THAT COORDINATES AND CONTROLS BODY FUNCTIONS IS THE NERVOUS SYSTEM.

28.2 *The Nervous System*

Objectives

❶ **Explain the functions of each of the three divisions of the nervous system.**

❷ **Compare and contrast the functions of the three main parts of the brain.**

❸ **Compare the central nervous system with the peripheral nervous system.**

❹ **Discuss two techniques scientists use to study the brain.**

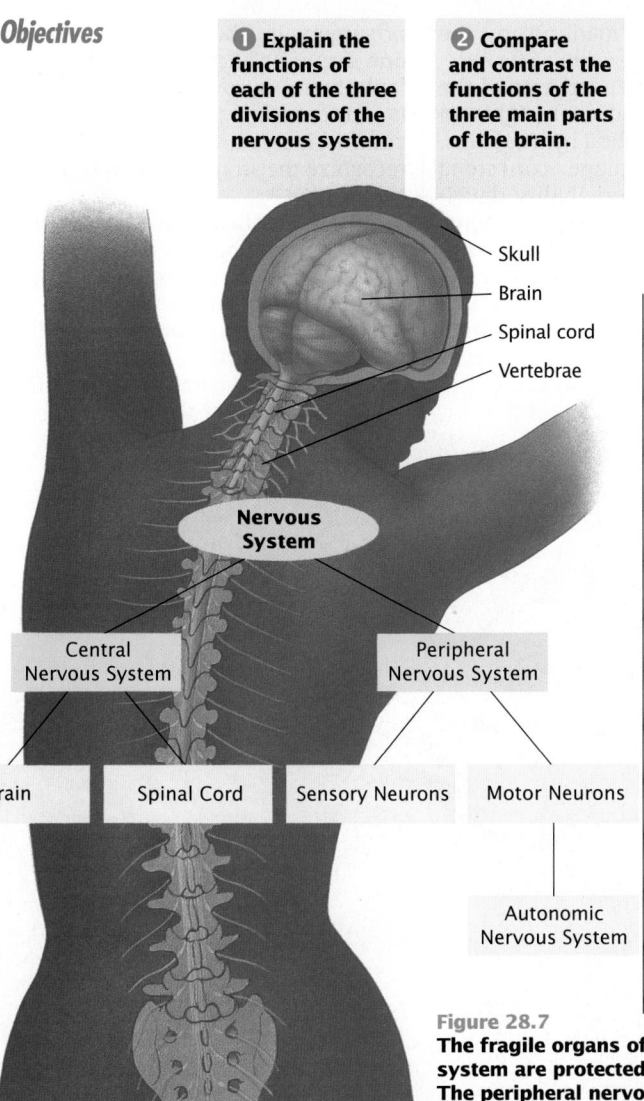

- Skull
- Brain
- Spinal cord
- Vertebrae

Nervous System

Central Nervous System — Peripheral Nervous System

Brain | Spinal Cord | Sensory Neurons | Motor Neurons

Autonomic Nervous System

Figure 28.7
The fragile organs of the human central nervous system are protected by the vertebrae and the skull. The peripheral nervous system is all the nerves with the exception of the brain and spinal cord.

The Central Nervous System

If you think of the nervous system as a business, the brain and spinal cord are the managers at the main office, directing and coordinating all major activities. They are the control center of the nervous system. For this reason, the brain and spinal cord are called the **central nervous system**.

The human brain, shown in **Figure 28.7**, is constantly receiving, analyzing, and storing information about conditions both inside and outside the body. It is also the source of thoughts, emotions, and moods.

In invertebrates such as worms and insects, the central nervous system is composed of a nerve cord with a collection of neurons at the front end. Vertebrates have a more complex brain made up of many parts. Apes, dolphins, and whales have exceptionally well-developed brains. The human brain is more complex and highly developed than the brain of any other animal. It consists of billions of neurons that enable people to use written language and technology.

Phase 2

TEACHING STRATEGIES

Demonstration 1

The Human Brain

Pass a model of the human brain around the class. Have students identify each of the three main parts by comparing the model to the brain illustrated in Figure 28.8. Point out that the brain is one of the most active organs, using nearly 20 percent of the body's energy supply.

Health Connection

Meningitis

Remind students that the brain is encased in the skull. In addition, the brain is covered by tough protective membranes known as meninges. Inflammation of these meninges, brought about by bacteria or viruses, is known as meningitis and can be fatal if not treated.

Connection: Chapter 11

Early Hominids

Dispel any misconception that modern *Homo sapiens* possesses the largest brain of any human in history. Remind students that Neanderthals had larger brains than those of some modern humans. Obviously, brain size is not related to intelligence. The brain of modern humans is much more convoluted than that of the Neanderthal, suggesting that the number of convolutions might be related to intelligence. The relative size of the brain in relation to the rest of the body may be a better indicator of intelligence. Whales and porpoises have brains larger than those of humans. When differences in body size are factored out, humans have the largest brains.

Demonstration 2

A Map of the Human Brain

Show students a diagram illustrating where the control of various functions is centered in the cerebrum.

Three Parts of the Brain

When a vertebrate embryo develops, the central nervous system starts out as a tube. The front part of the tube develops three bulges that form the brain. The rest of the tube forms the long, straight spinal cord. The brain has three main parts: the cerebrum, the cerebellum, and the brain stem, as shown in Figure 28.8.

The cerebrum is the control center of the brain

About 85 percent of the weight of the human brain is made up by the **cerebrum** (*suh REE bruhm*). The cerebrum is the large rounded area of the brain divided by a groove into right and left halves called cerebral hemispheres. It functions in language, conscious thought, memory, personality development, vision, and other sensations. The cerebrum, which looks like a wrinkled mushroom, is positioned over the rest of the brain. It contains thick layers of unmyelinated neurons, which look gray. We refer to this gray layer when we speak of our "gray matter."

The right and left cerebral hemispheres are linked by a bundle of neurons

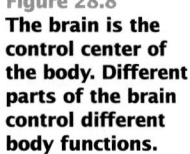

Figure 28.8
The brain is the control center of the body. Different parts of the brain control different body functions.

called a tract. This tract tells each half of the brain what the other half is doing. Surprisingly, each half of the brain controls muscles and glands on the opposite side of the body. In general, the left brain controls language and speech. The right side is important to performing mathematics and music.

Researchers have found that the two sides of the cerebrum can operate as two different brains. For instance, in some people the tract between the two hemispheres has been cut by accident or surgery. In laboratory experiments, one eye of an individual with such a "split brain" is covered and a stranger is introduced. If the other eye is then covered instead, the patient does not recognize the stranger who was just introduced.

Sometimes blood vessels in the brain are blocked by blood clots, causing a disorder called a stroke. During a stroke, circulation to an area in the brain is blocked and brain tissue dies. A severe stroke in one side of the cerebrum may cause paralysis of the other side of the body.

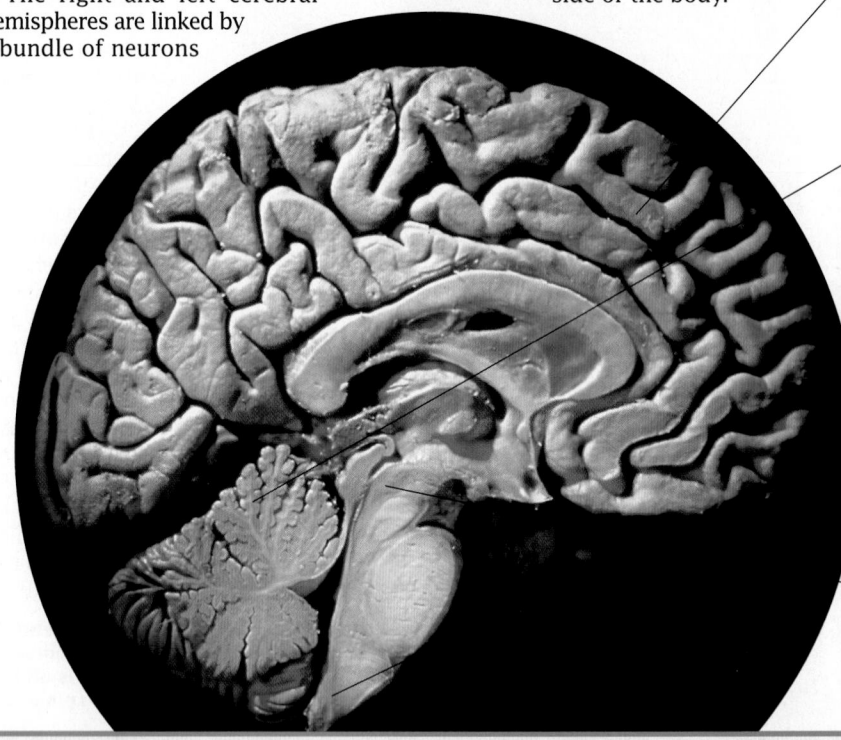

▪ *Cultural Perspective* ▪

William E. Thomas, Ph.D.
(African-American)
As a postdoctoral fellow at Harvard Medical School, William E. Thomas studied the properties of cells in the cerebral cortex of mammals. As a result, he developed a tissue culture system to facilitate study of the

neurons and other cells in the cortex of rats. Dr. Thomas believed that this was the first step in understanding the higher functions of human memory, learning, and emotions.

The cerebellum coordinates muscle movements

At the rear of the brain is a structure known as the **cerebellum** (*sehr uh BEHL uhm*). As shown in **Figure 28.8b**, the cerebellum controls balance, posture, and coordination. This small cauliflower-shaped structure, while well-developed in mammals, is even more developed in birds. Birds perform more complicated feats of balance than most mammals, because they move through the air as well as along the ground. Imagine the kind of balance and coordination needed for a bird to land on a branch at precisely the right moment. You can learn more about vertebrates in the *Evolution of the Brain* on pages 650-651.

The brain stem controls vital body processes

The cerebrum is connected to the spinal cord by the **brain stem**. This stalklike structure contains nerves that control your breathing, swallowing, digestive processes, and action of the heart and blood vessels, as shown in **Figure 28.8c**. Inside the brain stem, major sensory and motor pathways cross over between the body and the cerebrum. This is why each cerebral hemisphere controls the opposite side of the body.

A network of nerves called the reticular formation runs through the brain stem and connects to other parts of the brain. Their widespread connections make these nerves essential to consciousness, awareness, and sleep. One part of the reticular formation filters sensory input, enabling you to sleep through repetitive noises such as traffic, yet awaken instantly when a telephone rings.

The upper end of the brain stem contains the **hypothalamus**. The hypothalamus helps regulate blood pressure, heart rate, body temperature, hunger and thirst (indicated in **Figure 28.8d**), urination, sexual drives, and emotions.

The spinal cord carries information to and from the brain

The spinal cord is a cable of nerve tissue extending from the brain through the backbone. It is actually a bundle of neurons that, like the brain, contains gray matter. The gray nerve-cell bodies form a column in the center of the cord, which is coated with bundles of white nerve fibers. The backbone—a tunnel of bone formed by the rings of the vertebrae—surrounds the cord. Messages from the body and the brain run up and down the spinal cord.

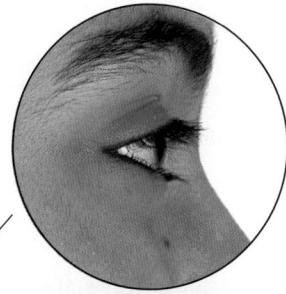

a *Cerebrum*
One of the many functions of the cerebrum is to translate what your eyes see into useful information.

b *Cerebellum*
The cerebellum is necessary for this gymnast to balance on the narrow beam of wood.

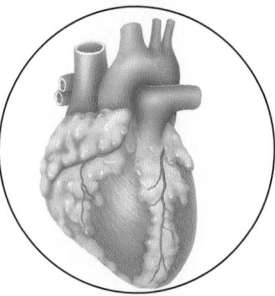

c *Brain stem*
The brain stem controls vital body processes that you do not consciously control, like your heartbeat.

d *Hypothalamus*
Certain parts of the hypothalamus control hunger and thirst.

Demonstration 3

The Cerebellum

Show the class a videotape, in slow motion if possible, of someone performing a complicated high dive or balance-beam routine. Have students relate this type of performance to the function of the cerebellum. Point out that alcohol has a significant effect on the synapses in the cerebellum. Have students discuss why too much drinking causes a person to stagger and stumble.

Health Connection

"Brain Dead"

When no electrical activity in the brain—especially in the region of the brain stem—can be registered, a person is declared "brain dead." In such cases, the vital body processes controlled unconsciously by the brain stem must be sustained by a life-support system.

Language Connection

Drawing Analogies

The reticular formation can be compared to a railroad switch tower that determines which of several incoming trains will be allowed to pass. Have students suggest other analogies that illustrate the role of the reticular system.

Connection: Chapter 27

Vertebrae

Remind students that the vertebrae support and protect the spinal cord. Discuss what keeps the vertebrae from rubbing against one another.

649

Using the Feature

- The human brain is not the largest in absolute size; whales and elephants have bigger brains. But per kilogram of body mass, humans do have the largest brains. Our large, complex brains exemplify three trends in the evolution of the brain. The first trend is toward increased size, especially among birds and mammals. A 1-kg (2.2-lb.) mammal will have a much larger brain than a 1-kg (2.2-lb.) fish or reptile. The second trend is toward increasing specialization of brain regions. The third trend is towards increased size and complexity of the cerebrum. This trend is most apparent in humans, in which the cerebrum accounts for 85 percent of the brain's mass.

- Differences in brain size obviously account for some of the differences in intelligence between species. Large-brained chimpanzees can use tools, while pea-brained mice cannot. But do brain size differences explain differences in intelligence between members of the same species? Did Einstein, Shakespeare, and Darwin have exceptionally large brains? Have students research the debate over the connection between brain size and intelligence.

Discussion

Guide the discussion by posing the following questions.

1. Identify the functions of the cerebrum and cerebellum.

   The cerebrum's functions are interpreting sensory data, learning, association, and, in humans, language, memory, and conscious thought. The cerebellum coordinates movement.

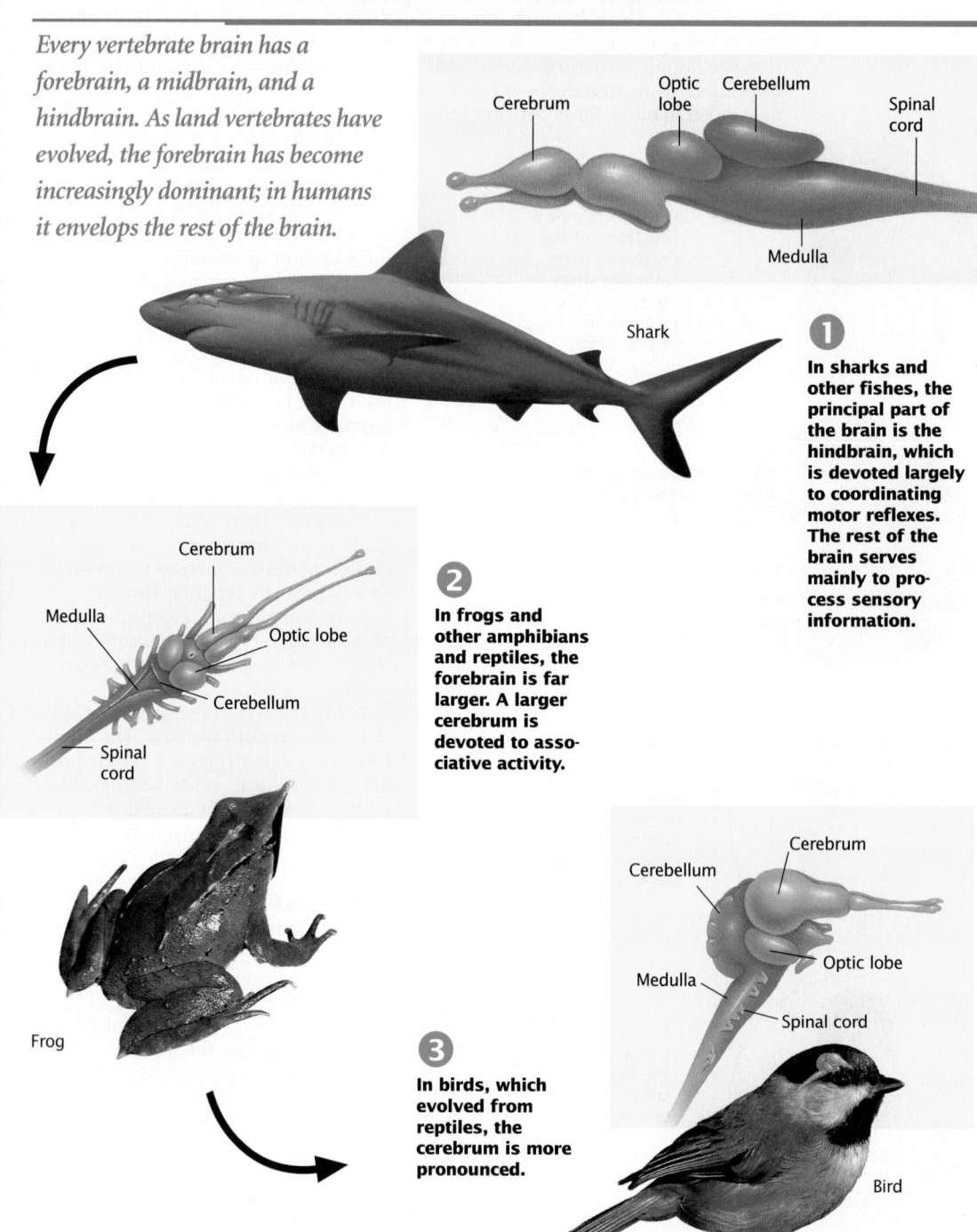

Journeys ▪ Journeys ▪ Journeys ▪ Journeys ▪ Journeys ▪

Evolution of the Brain

Every vertebrate brain has a forebrain, a midbrain, and a hindbrain. As land vertebrates have evolved, the forebrain has become increasingly dominant; in humans it envelops the rest of the brain.

Cerebrum • Optic lobe • Cerebellum • Spinal cord • Medulla

Shark

1 In sharks and other fishes, the principal part of the brain is the hindbrain, which is devoted largely to coordinating motor reflexes. The rest of the brain serves mainly to process sensory information.

Cerebrum • Medulla • Optic lobe • Cerebellum • Spinal cord

2 In frogs and other amphibians and reptiles, the forebrain is far larger. A larger cerebrum is devoted to associative activity.

Frog

Cerebrum • Cerebellum • Optic lobe • Medulla • Spinal cord

3 In birds, which evolved from reptiles, the cerebrum is more pronounced.

Bird

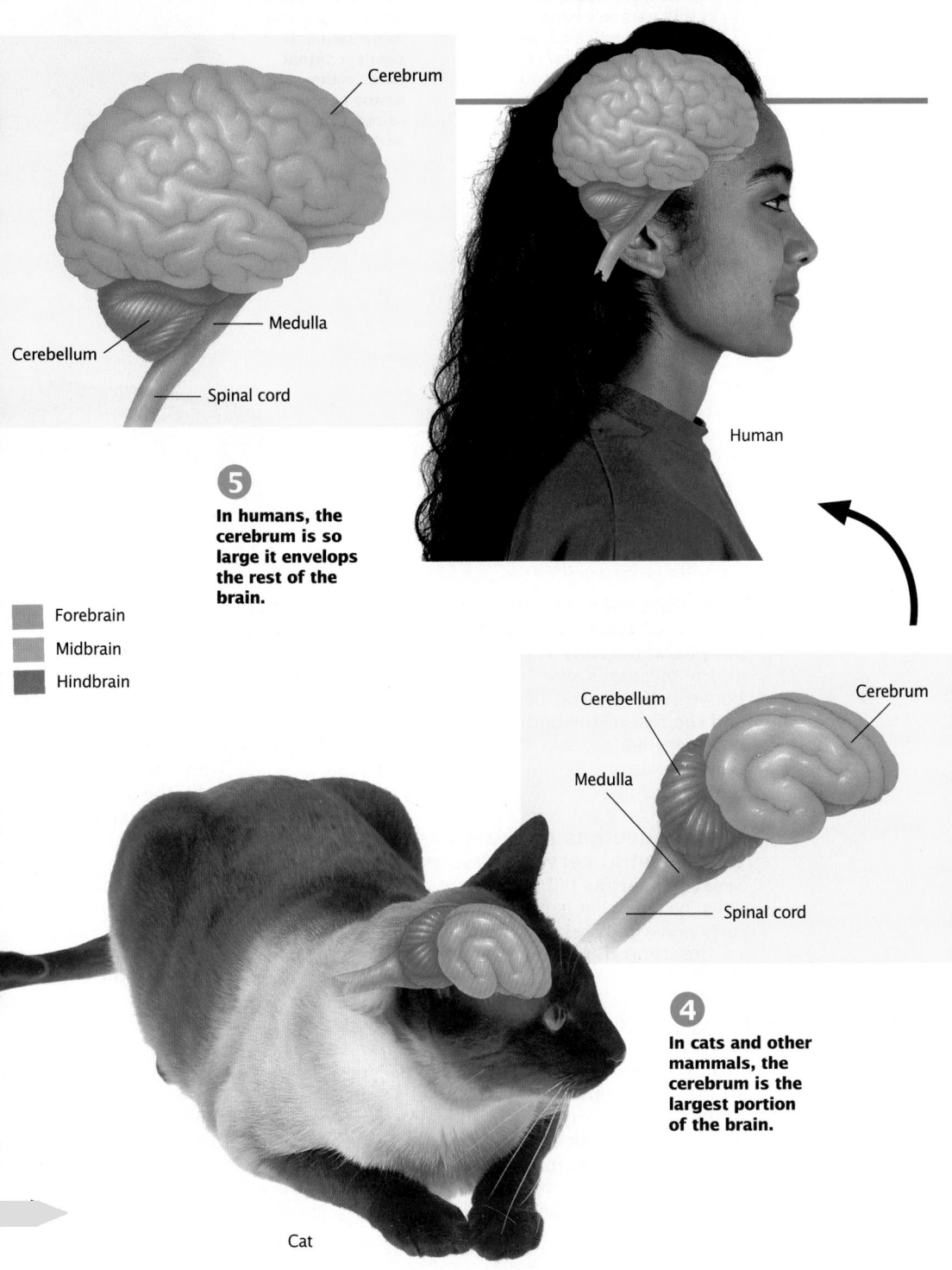

Cerebrum

Medulla

Cerebellum

Spinal cord

Human

5

In humans, the cerebrum is so large it envelops the rest of the brain.

Forebrain

Midbrain

Hindbrain

Cerebellum

Cerebrum

Medulla

Spinal cord

4

In cats and other mammals, the cerebrum is the largest portion of the brain.

Cat

2. In a shrew's brain, the olfactory region of the brain is very large—about the size of the rest of the brain. What does this indicate about the importance of the sense of smell to the shrew?
The size of a brain region devoted to a particular sense indicates the importance of that sense to the animal. For a shrew, smell is the most important sense.

3. How would the size of the olfactory region of the human brain compare to the size of the visual region?
For humans, sight is much more important than smell; thus, the visual region is much larger.

4. Birds have a large, complex cerebellum. Explain why.
The cerebellum must be complex to organize the complicated motions of flight.

Visual Strategy

Figure 28.9

Use this figure to have students trace the path an impulse takes along a sensory neuron to the spinal cord and out from the spinal cord to a motor neuron. Emphasize that most nerves, like the ones shown in this figure, contain both sensory and motor neurons running side by side. In this way, students will not develop the misconception that the two types of neurons must be structurally separate.

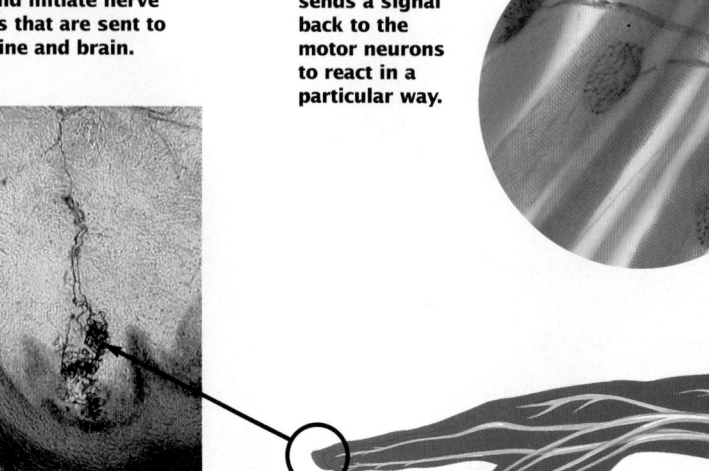

Figure 28.9

a The sensory neurons in this person's hand react to stimuli such as heat and initiate nerve signals that are sent to the spine and brain.

b After the brain has received the information, it sends a signal back to the motor neurons to react in a particular way.

The Peripheral Nervous System

All of the nervous system outside the spinal cord and brain is known as the **peripheral nervous system**. It carries all the messages sent back and forth between the central nervous system and the rest of the body. The peripheral nervous system has two main types of neurons: sensory neurons and motor neurons.

Sensory neurons relay signals to the central nervous system

Sensory neurons tell the central nervous system what is happening. They carry nerve impulses from sense organs to the central nervous system, as shown in **Figure 28.9a**.

Sense organs are organs that react to changes inside and outside the body. They detect many different things, including changes in blood pressure, strain on ligaments, and smells in the air. Sense organs include complex organs, such as eyes and ears. Your skin has many small structures called sensory receptors, which enable you to sense pressure and temperature.

Motor neurons deliver information to muscles and glands

Motor neurons are partners of sensory neurons. Motor neurons carry information from the central nervous system to a muscle or gland, as shown in **Figure 28.9b**. They act on the information delivered by the sensory neurons. If your eyes see a runaway truck speeding toward you, the central nervous system sends messages through motor neurons to glands that secrete adrenaline. The adrenaline increases your heartbeat and breathing rate. The central nervous system also sends messages through motor neurons to many muscles, which contract and get the body out of there—fast!

In each segment of the spine, sensory nerves go into the cord and motor nerves come out of it. The motor nerves control most of the muscles below the head. This is why injuries to the spinal cord often paralyze the lower part of the body. A muscle is paralyzed and cannot move if its motor neurons are damaged.

▪ Cultural Perspective ▪

Acupuncture

Acupuncture, a traditional way of treating pain and disease, originated in China but has been used by the Japanese, Koreans, and Vietnamese. In acupuncture, needles are inserted into various parts of the body. Chinese philosophers believe that an imbalance occurs between *yin* and *yang*, causing pain and disease. Some scientists believe that acupuncture works because needles inserted along certain lines of the body connect the body's organs in a certain way and influence organ functions. Some believe that acupuncture works by increasing the brain's production of natural pain-killers, endorphins. Others believe that acupuncture works through the nervous system by triggering signals that interrupt pain messages sent to the brain.

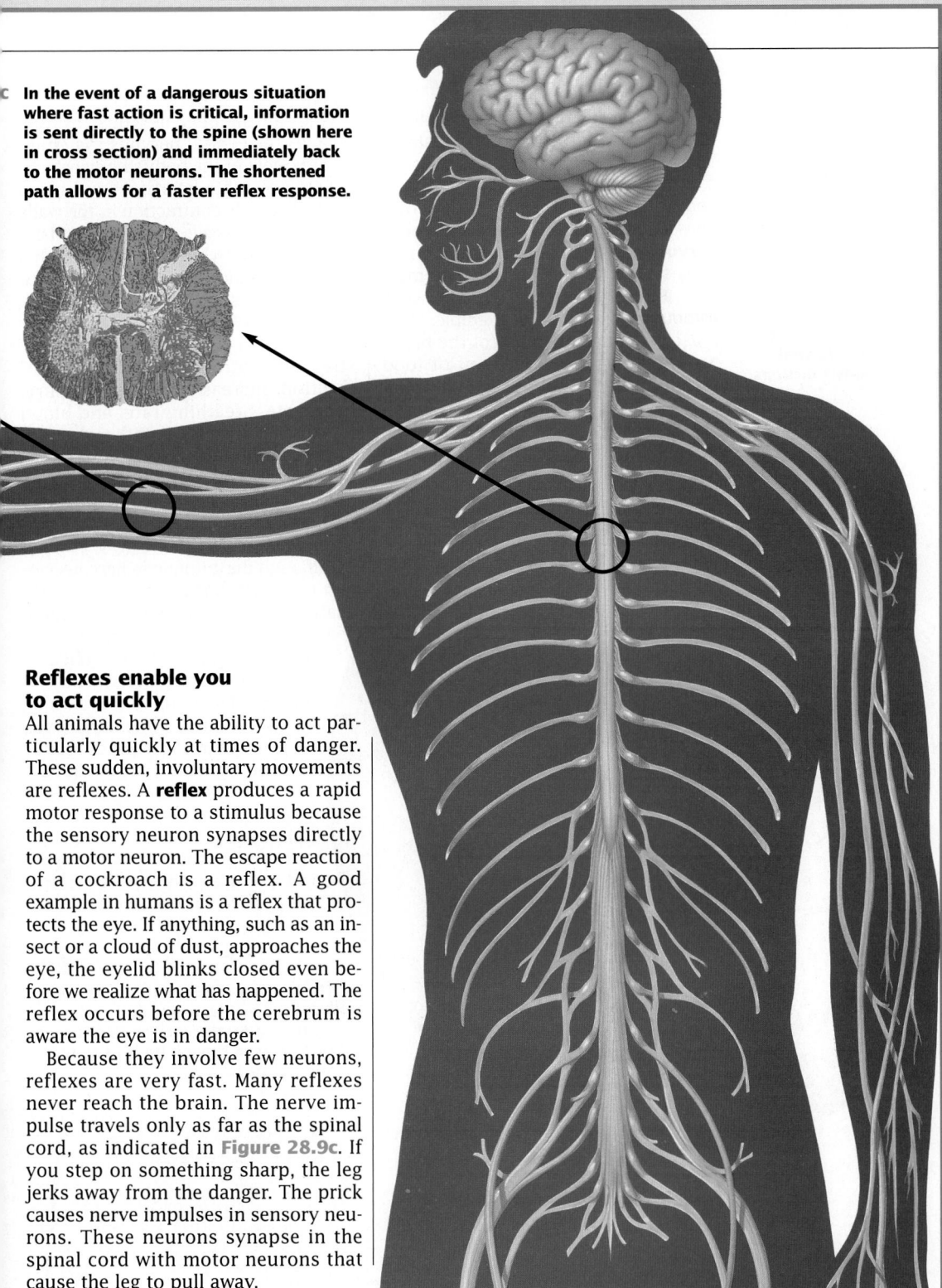

In the event of a dangerous situation where fast action is critical, information is sent directly to the spine (shown here in cross section) and immediately back to the motor neurons. The shortened path allows for a faster reflex response.

Reflexes enable you to act quickly

All animals have the ability to act particularly quickly at times of danger. These sudden, involuntary movements are reflexes. A **reflex** produces a rapid motor response to a stimulus because the sensory neuron synapses directly to a motor neuron. The escape reaction of a cockroach is a reflex. A good example in humans is a reflex that protects the eye. If anything, such as an insect or a cloud of dust, approaches the eye, the eyelid blinks closed even before we realize what has happened. The reflex occurs before the cerebrum is aware the eye is in danger.

Because they involve few neurons, reflexes are very fast. Many reflexes never reach the brain. The nerve impulse travels only as far as the spinal cord, as indicated in **Figure 28.9c**. If you step on something sharp, the leg jerks away from the danger. The prick causes nerve impulses in sensory neurons. These neurons synapse in the spinal cord with motor neurons that cause the leg to pull away.

Theme Connection

Stability

Emphasize how the autonomic nervous system is involved in homeostasis in numerous ways.

Demonstration 5

Opposite Effects

Show the class a photograph of an automobile. Discuss how the car accelerates and slows down. Have students compare these actions to the workings of the autonomic nervous system—one division acts as the accelerator, the other as the brake. Underscore the efficiency of such an arrangement by pointing out that a car can slow down without a brake, but the consequences might be disastrous.

Visual Strategy

Figure 28.10

Have students describe what the electrocardiogram shown in this figure might look like if the person were awake and active.

The Autonomic Nervous System

Some motor neurons are active all the time, even when the body is asleep. These neurons carry messages from the central nervous system that keep the body going even when it is not active. These are the neurons of the **autonomic nervous system**. The autonomic nervous system carries messages to muscles and glands that usually work without our noticing. For example, these muscles and glands control the blood pressure and the movement of food through the digestive system even during sleep, as shown in Figure 28.10.

The autonomic nervous system enables the central nervous system to govern most of the body's homeostasis. It helps regulate heartbeat and helps control muscle contraction in the walls of the blood vessels, digestive, urinary, and reproductive tracts. It also helps stimulate glands to secrete tears, mucus, and digestive enzymes.

One division of the autonomic nervous system dominates in times of stress. It controls the "fight-or-flight" reaction, increasing blood pressure, heart rate, breathing rate, and blood flow to the muscles. Another division of the autonomic nervous system has the opposite effect. It conserves energy by slowing the heartbeat and breathing rate, and by promoting digestion and elimination.

Although the autonomic nervous system can carry out its tasks automatically, it is not completely independent of voluntary control. For instance, breathing is controlled by the autonomic nervous system, but one can decide to stop breathing for a short time. However, any voluntary control of the autonomic nervous system that endangers life disturbs homeostasis of the brain tissue, causing unconsciousness. The autonomic nervous system then takes over again and restores normal functions. This is why you cannot hold your breath indefinitely.

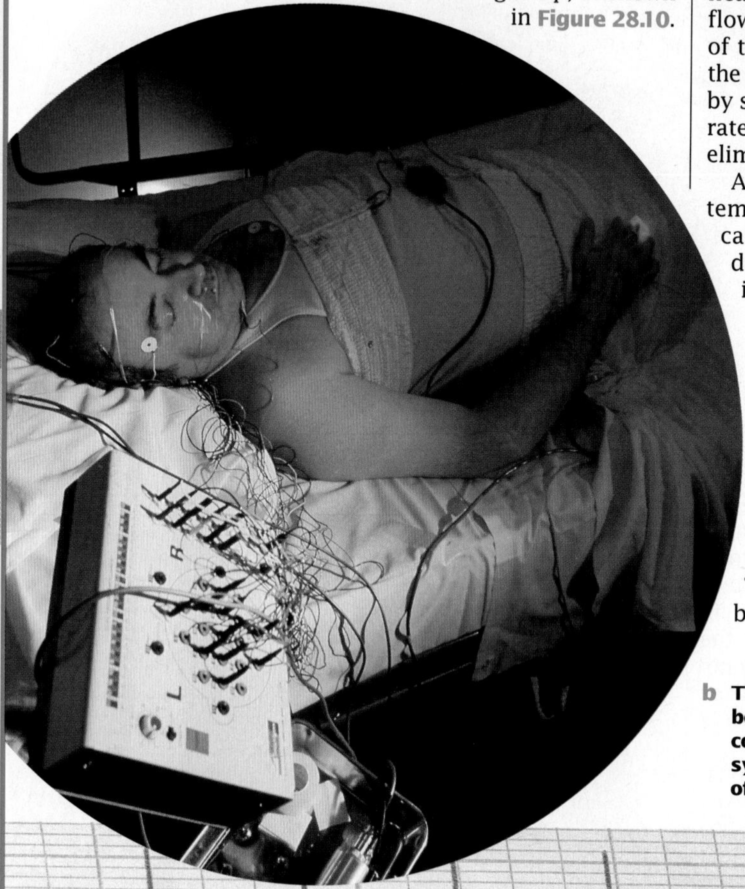

Figure 28.10

a **This sleeping man may appear to be inactive, but his brain is really quite active. Not only is he dreaming, but his brain is also maintaining his vital body functions.**

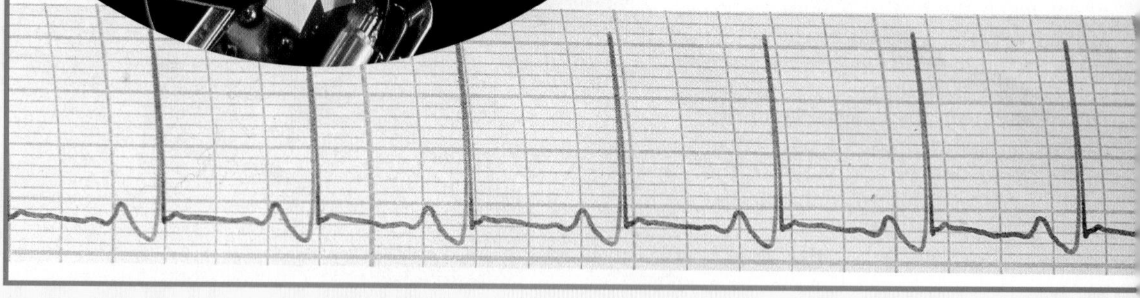

b **The beating of his heart, as shown below on an electrocardiogram, is controlled by the autonomic nervous system even though he is in a state of deep sleep.**

How Scientists Study the Brain

How do we know the functions of the different parts of the brain? The different regions of the brain have been investigated in several ways. Researchers can map the activities of each area by stimulating different parts of the brain with electricity. Or they may observe which functions of the body or thought processes are lost when parts of the brain are deliberately or accidentally destroyed. A newer method is to find out which parts of the brain use the most ATP while a person performs some act, such as recalling information, as shown in **Figure 28.11**. The neurons that use the most ATP are the most active.

Figure 28.11
Using new imaging techniques, scientists have been able to study the brain while it does certain jobs, like recalling information. In the images below, the colored areas are the most active.

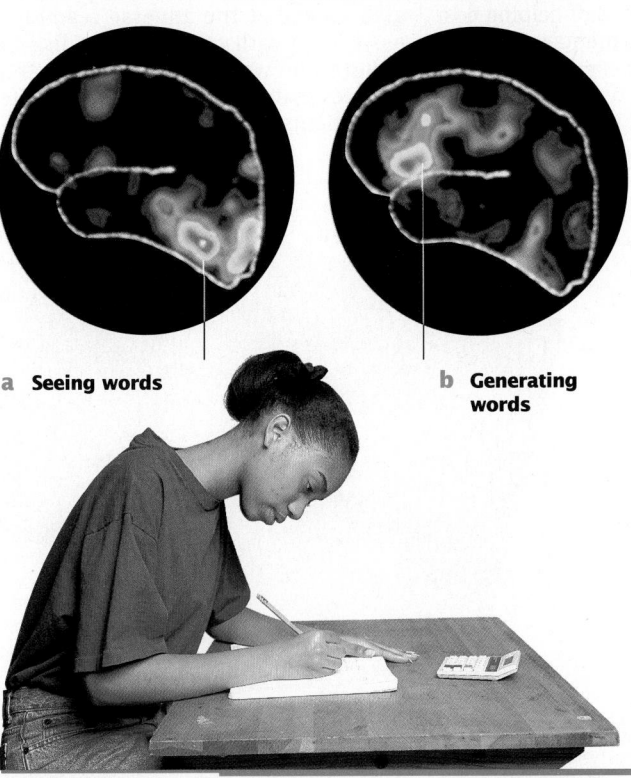

a **Seeing words**

b **Generating words**

Learning differs from memory

Learning and memory are brain functions that are especially well developed in humans but remain poorly understood by scientists. Learning and memory are not the same thing. As young children we learn many skills, such as talking, but we do not remember learning them.

Memory lasts for different lengths of time. When you take notes in class, you often remember what the teacher says only long enough to write it down. This is short-term memory, which can last for a few seconds or a few hours. We use it for things like remembering to do an errand or cramming for a test. If information is stored for any length of time, it is transferred to long-term memory. Here it may remain for life. Much long-term memory is stored subconsciously, meaning that we do not know we know it.

Scientists still do not fully understand sleep

Sleep is another mysterious function of the nervous system. Even after 50 years of research, scientists still have no solid ideas about why you must sleep each night, nor about why sleeping affects your temper, alertness, and emotions. And scientists do not understand why some people need more sleep than others. Although you may think of sleep as giving the brain a rest, the brain is active during sleep periods. It must maintain breathing and the rate of heartbeat. Scientists hypothesize that sleep permits the brain to restore biochemical functions depleted by the day's activities and to process and reorganize information taken in during the day.

Section Review

❶ **Summarize the functions of each division of the nervous system.**

❷ **Discuss the different functions of the cerebrum.**

❸ **Compare the functions of the central nervous system, sensory neurons, and motor neurons.**

❹ **What are two ways scientists can study the brain?**

▪ *Section Review Answers* ▪

1. The central nervous system is the control center. The peripheral nervous system carries messages back and forth between the central nervous system and the body.
2. The cerebrum is the control center of the brain and functions in language, conscious thought, memory, personality development, vision, and other sensations.
3. Sensory neurons deliver information to the CNS, and the motor neurons carry information away from the CNS.
4. Scientists can map the areas by stimulating different parts of the brain with electricity; they may observe which body functions or thought processes are lost when parts of the brain are destroyed; or they may find out which parts of the brain use more ATP while performing some act.

655

Using the Feature

- This feature can be used to show that a background in biology does not necessarily limit one's future to laboratories or classrooms. Here is an example of a career that is indirectly related to biology. In the pharmaceutical industry, people who have an interest in biology and who are outgoing and people-oriented are able to combine these traits by helping doctors, patients, and hospitals.

Discussion

Guide the discussion by posing the following questions:

1. Explain how James is involved in the lives of dialysis patients, AIDS patients, and people infected with HIV.
   James works for a pharmaceutical company that is responsible for manufacturing drugs used in treating the above conditions.
2. What personal traits enable James to be a successful pharmaceutical sales manager?
   He is people-oriented and likes to travel, two qualities necessary for being involved in sales.
3. What did James study in college?
   He majored in political science and history.
4. Where did he receive his training in science?
   The company he works for put him through an extensive training program in which he studied anatomy, pharmacology, chemistry, and biology.

James Moreland:
National Pharmaceutical Sales Manager

How I Became Interested in Pharmaceuticals

"Though I majored in political science and history both in high school and in college, science has always intrigued me. I took the required courses—biology, chemistry, and physics. When I got out of college, I started reading about life science. That's how I got into the industry. I liked the idea of helping hospitals, doctors, and patients with medications to relieve illness.

"My pharmaceutical company put me through a very extensive training program. I studied anatomy, pharmacology, chemistry, and biology. I learned how drugs are put together, how they interact in the body, what they're good for, and what they're not good for. We also looked at the adverse reactions associated with some drugs. I like to believe that we provide products and services that improve the quality of life for patients."

James is married and has two children. His son, Kareem, is 20 and his daughter, Jessica, is 10.

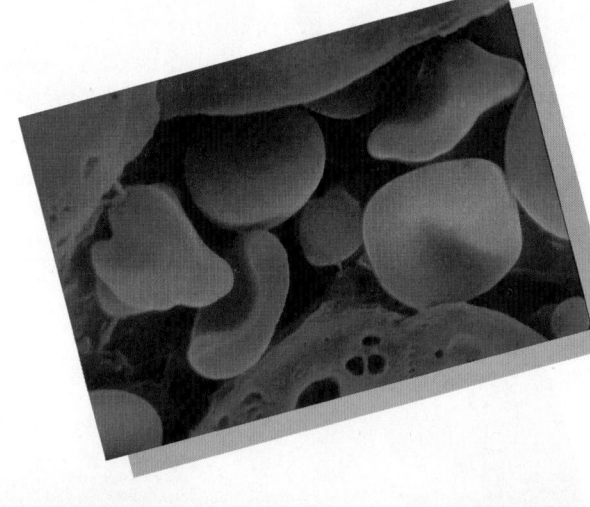

Name:	**James Moreland**
Home:	**Buckingham, Pennsylvania**
Employer:	**Ortho Biotech, division of Johnson & Johnson, Raritan, NJ**
Personal Traits:	• **People-oriented**
	• **Likes to travel**
	• **Articulate**
	• **Curious**

Action ▪ *Science in Action* ▪ *Science in Action* ▪ *Science*

The Excitement of the Pharmaceutical Industry

"The company I work for manufactures genetically-engineered drugs. For example, we make Orthoclone OKT-3. This drug is especially helpful to patients about to receive a kidney transplant. When the body receives a new organ, the immune system initially rejects the organ. The body knows the organ is foreign and builds antibodies for that organ. Orthoclone OKT-3 shuts down the immune system for two weeks, giving the new organ an opportunity to get adjusted to the new body. Thus when the immune system comes back on, it doesn't recognize the new organ as foreign. There is a much better chance of accepting that organ and not experiencing rejection, which could be complicated and even result in death. This same drug is also useful with liver and heart transplants.

"Another major product I am responsible for is a synthetically-produced version of erythyropoietin, a hormone produced by the kidneys that stimulates the production of red blood cells. Red blood cells carry oxygen to all parts of the body. All people need red blood cells in order for their heart, brain, and lungs to function. If your body does not produce enough red blood cells, you will become anemic. You begin to feel tired and cannot function normally.

"Our synthetically-produced version of erythyropoietin is helpful for people on dialysis, a process that purifies the blood during kidney failure. Dialysis patients can become very anemic. They can take synthetically-produced erythyropoietin to compensate for their inability to naturally produce erythyropoietin. As a result, the amount of blood and iron in the body is increased, thereby providing the patient with energy and enabling him or her to function at a near normal level.

"Of the approximately 120,000 patients on dialysis therapy, 90,000 are on this drug. It makes a tremendous difference in their treatment.

"Anemia also occurs in patients who are being treated for AIDS or HIV infection. As part of their therapy, they take an antiviral drug called AZT in order to arrest the virus. As a result of taking AZT, their bodies become highly immune-suppressed. AZT can affect the bone marrow, which is responsible for generating red blood cells. The patient will probably become anemic because the body is not producing enough red blood cells.

"Erythyropoietin therapy can increase the production of red blood cells, let the patient continue his or her AZT drugs, and even allow the doctor to increase the AZT dosage."

Career Path
....

High School:
- **Language Arts**
- **Math**
- **Biology**

College:
- **Political Science**
- **History**
- **Biology**

Lesson Plan 28.3

Phase 1

PREPARATION

Key Concepts

- Sensory neurons located in sense organs and other parts of the body detect stimuli and transmit impulses to the central nervous system.
- Sensory neurons in the inner ear enable a person to hear, maintain balance, and monitor orientation in space.
- Sensory neurons in the eye enable a person to see colors, fine details, and minute movements.
- Sensory neurons in the tongue enable a person to distinguish between sweet, sour, bitter, and salty tastes.

Reading Strategy

This section presents the opportunity to stress the importance of concentrating on the use of one sense (sight) while keeping the other senses (especially hearing) "turned off" when reading. Have students discuss how focusing on visual sensory input would help the cerebrum retain what is read.

Phase 2

TEACHING STRATEGIES

Theme Connection

Stability

Have students discuss how sensing internal information is important in maintaining homeostasis.

Visual Strategy

Figure 28.12
Have students discuss how an inner-ear infection would most likely affect the man shown in this figure.

VIVID REPORTS ABOUT YOUR ENVIRONMENT STREAM INTO YOUR BRAIN THROUGH YOUR EYES, EARS, SKIN, NOSE, AND MOUTH. THE FACT IS, THERE ARE MORE THAN A DOZEN DIFFERENT TYPES OF SENSORY CELLS THAT DETECT CHANGES OUTSIDE AND INSIDE YOUR BODY. TOGETHER, THEY HELP YOU INTERACT WITH THE WORLD THAT SURROUNDS YOU.

28.3   *The Sense Organs*

Objectives

❶ **List three stimuli to which sense organs react.**

❷ **Explain how the inner ear helps you maintain balance.**

❸ **Explain how your ear detects sound.**

❹ **Compare and contrast the functions of rods and cones.**

Sensing Internal Information

Many receptors that enable the body to receive information from the environment are located in highly specialized organs called **sense organs**. The most familiar sense organs are the eyes, ears, nose, and tongue. These organs have receptors that can respond to stimuli by producing nerve impulses in a sensory neuron. Your body's receptors detect stimuli such as light, heat, or pressure. The receptor converts the energy of a stimulus into electrical energy that can travel in the nervous system.

Receptors inside the body inform the central nervous system about the condition of the body. For instance, temperature receptors throughout the body detect changes in temperature. This information travels to the hypothalamus, which helps control body temperature.

Receptors in the joints, tendons, and muscles detect changes in the position of parts of the body. Receptors in the ear tell us the position of the head. Together, all these receptors inform the brain where the body is in three dimensions. This knowledge is essential to move freely and maintain your balance, like the man in **Figure 28.12**.

Receptors in the inner ear sense position in space
The ear is really two sense organs in one. It not only detects sound waves,

Figure 28.12
a **In-line skating requires balance and coordination. Without sense organs in this man's inner ear telling him how to balance, where his legs are, and where he is in space, he would be in big trouble.**

b The semicircular canals are three looped rings in the inner ear.

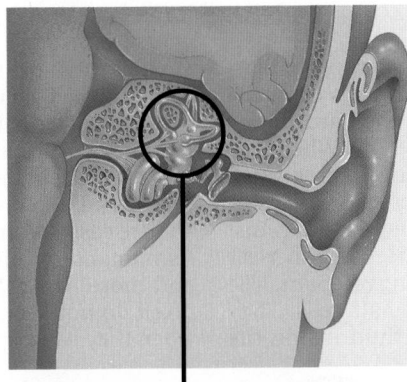

c Each canal has a swelling at one end, which is lined with receptor cells inside. Tiny hairs protrude from the receptor cells into a jellylike fluid.

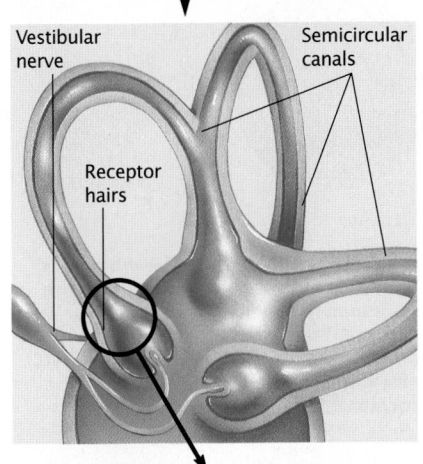

Vestibular nerve

Semicircular canals

Receptor hairs

d When your head is upright, the fluid is still and the hairs are upright.

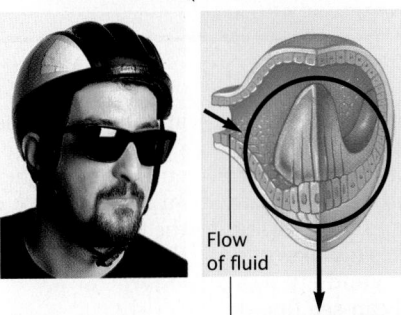

Flow of fluid

e Any movement causes the fluid to slide over the hairs and bend them in the opposite direction. The hair cells send messages to the brain about your position in space.

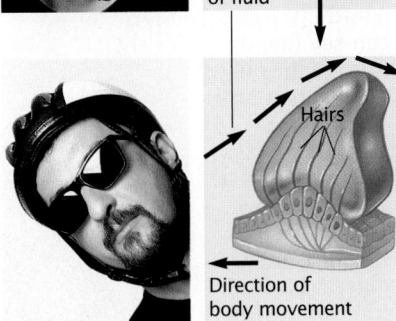

Hairs

Direction of body movement

it also senses the position of the head, whether it is still, moving in a straight line, or rotating.

The receptor cells for your sense of position are located in canals and chambers of the inner ear, shown in **Figure 28.12c**. Each receptor cell has a bundle of tiny hairs protruding from it. In the semicircular canals, the receptor cells are covered with a jellylike cap. Each canal is full of fluid and contains hair cells. When the fluid moves, it stimulates the hair cells, as shown in **Figure 28.12d-e**. The three canals lie in three different planes, so they can detect head movement in any direction.

The semicircular canals detect when the head changes speed or direction. They do not react if the head rotates at a constant speed or moves in a straight line. Traveling in a car or airplane at constant speed gives no sense of movement. The semicircular canals detect movement only if the car or airplane turns. This is because when the head is moving in a straight line, the fluid in the canals does not move. When the head changes direction, however, the fluid tends to keep going in its original direction, and so it bends the hair cells.

The brain interprets stimuli from sense organs

Messages from sense organs to the central nervous system are all in the form of nerve impulses. How does the brain know whether an incoming nerve impulse indicates a shining light, a melodious sound, or a familiar odor? This information is built into the "wiring" in the pathways of neurons that synapse with each other, and into the location in the brain where the information arrives. The brain "knows" it is responding to light when it gets a message from a sensory neuron that comes from light receptor cells. Neurons from the eye may send impulses for other reasons. For instance, when you press your fingertips gently against the corners of your eyes, you "see stars." This is because the brain treats any impulse from the eye as light, even though the eye received no light.

Demonstration 1

The Cochlea
Show the class a snail shell and inform them that the cochlea has a similar spiral shape.

Visual Strategy

Figure 28.12
Point out how the three semicircular canals lie at right angles to each other. Have students discuss how a person's sense of balance would be affected if the three canals were oriented in the same plane.

Demonstration 2

Seeing Stars
Darken the room. Have students close their eyes while gently tapping or pressing a fingertip against the corner of their eyes. Ask which part of their brain "saw stars."

■ *Cultural Perspective* ■

Dwayne D. Simmons, Ph.D.
(African-American)
An assistant professor at Pepperdine University, Dr. Dwayne D. Simmons researches the auditory nervous system. He studies the developmental changes in the neurons of the inner ear. Simmons's 1986 research led to

the discovery that interactions between sensory neurons and the receptors were much more complex than previously thought.

Demonstration 3

Loud Noises
Play a "hard rock" tape at a volume high enough to be disconcerting to an "average" listener. Tell students the loudness of sounds is measured in decibels, ranging from a sound of 0 decibels (db), which is just audible, to a sound of 140 db, which is painful. Rock music may be as loud as 130 db, and a jet taking off can produce a 150-db sound.

Demonstration 4

The Eye
Show the class either a model or photograph of the human eye and have them trace the path light takes from the cornea to the retina. Also display several optical illusions to point out how light waves can "fool" a person. Examples can include illusions that involve wavy lines, color combinations, or angles of intersection that the brain has difficulty interpreting correctly.

Sensing Sound

When we hear, we detect waves of pressure in the air. These waves enter the ear canal and strike the eardrum as shown in **Figure 28.13**. When the eardrum moves, it moves three small bones that stretch across a cavity inside the ear. The third bone presses against the snail-shaped cochlea, the inner ear that contains the hearing receptors. The hearing receptors are hair cells. They detect waves in the fluid inside the cochlea. When sound strikes the eardrum, the fluid in the cochlea moves. This stimulates hair cells, which stimulate the sensory nerve that carries information about the sound to the brain. The hair cells in the cochlea are delicate and are easily destroyed by loud noises. As humans get older, they lose more and more hair cells, so older people often do not hear as well as they did when they were younger.

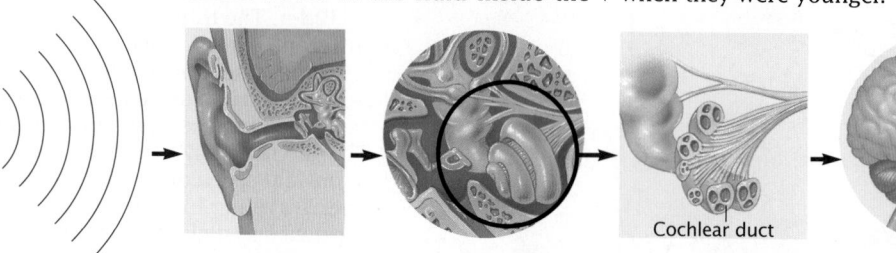

Figure 28.13

a A sound wave is a vibration in the air that enters the ear canal . . .

b . . . and strikes the eardrum. The sound causes the eardrum to vibrate. Behind the vibrating eardrum, in the middle ear, are . . .

c . . . three small bones that move in response to the eardrum. These bones transfer the vibrations to the cochlea.

d The vibrations travel through the cochlear duct toward the auditory nerve.

e Nerve impulses travel to the brain. The brain then translates them into a sound you can understand.

Sensing Light

Almost all animals have receptors that use pigments to absorb light energy. Changes in pigments trigger nerve impulses in sensory neurons. In humans, the light receptors are in the eyes. Humans and other primates (monkeys and apes) have extremely good eyesight. They see in color and can see fine details and movements. Birds are the only animals with better eyesight than primates.

As light enters the eye, it first passes through the cornea. The cornea is a transparent, protective covering over the eye. The light is focused by the lens, which changes shape when it is pulled by muscles around its edges. The pupil is the opening in the middle of the iris. The iris is a diaphragm that controls the amount of light entering the eye by altering the diameter of the pupil. The pupil gets larger in dim light and smaller in bright light. The retina is a layer at the back of the eye containing the light receptors and neurons. Follow the passage of light through the eye in **Figure 28.14**.

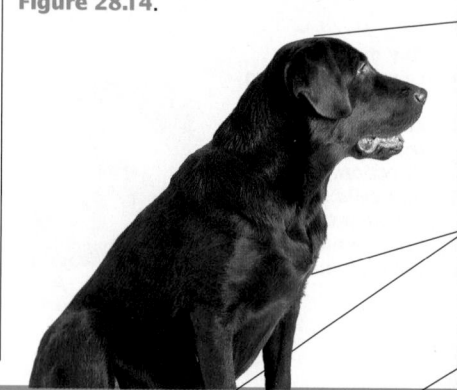

Rods and cones are receptors in the eye

When light reaches the retina, it stimulates about 125 million receptor cells called rods and cones. Rods and cones generate nerve impulses that travel to the brain along a nerve called the optic nerve. **Rods** are cells that are extremely sensitive to light and can detect various shades of gray even in dim light. However, they cannot distinguish colors, and they produce poorly defined images. **Cones** are cells that detect color, produce sharp images, and are important for seeing in bright light. The center of the retina contains a tiny pit densely packed with cones. This area produces the sharpest image. We tend to move our eyes so that the image of an object we want to see clearly falls on this area. No rods or cones exist at the point where the optic nerve enters the retina. Therefore, impulses cannot be transmitted, resulting in a "blind spot" in the field of vision.

The pigments in rods and cones are made from pigments called carotenoids. Animals cannot make carotenoids; they obtain the raw material for their visual pigments from plants. This is why eating carrots is said to be good for night vision. The orange color in carrots is due to the presence of carotenoids.

Other animals see things you cannot

Although humans have excellent eyesight, other animals can see things we cannot see. For instance, the human lens filters out ultraviolet light. Honeybees do not have ultraviolet filters in their eyes, so they can detect ultraviolet light. If you look through polarizing sunglasses, glare disappears and you can see lines of polarized light. Polarized light can be detected by many animals, such as squids, octopuses, amphibians, and sea turtles. These animals can see better in the glare of sunlight on water than humans can. They can also tell where the sun is, even on a cloudy day. This gives them information they use to find their way around.

Figure 28.14
▶ **The human eye is an organ specialized for sensing light. Light that reflects off this dog is received by receptors at the back of the eye. The light signal is transformed into electrical impulses that travel to the brain. The brain receives the signal and translates it into meaningful information.**

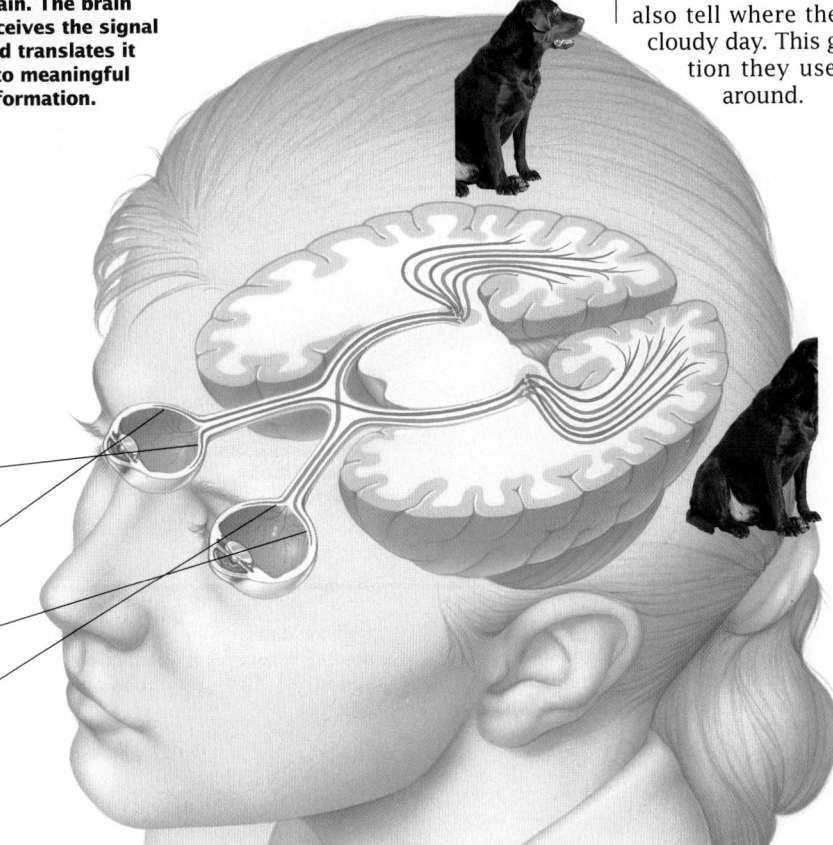

b **Each eye receives about three-quarters of the image. The brain combines this information so that you see the entire image.**

Demonstration 5
Colorblindness

Show the class a chart used to test a person for colorblindness. See if any students are colorblind. Point out that colorblindness is caused by a chemical disorder in the cones. Complete colorblindness is extremely rare. A deficiency in the red or green cones is most common, affecting about 5 percent of the American population.

Visual Strategy

Figure 28.14

Use this figure to point out that the image formed on the retina is actually upside down and backward. In the eye illustrated in this figure, have students locate where the inverted image forms. Then have them trace the path impulses follow to the cerebrum where the image is reversed so that the object appears right side up.

Demonstration 6

Show the class a picture of a hawk. Point out that the human eye has about 150,000 cones per sq. mm (2,300 cones per sq. in.), whereas the hawk has nearly 1 million per sq. mm (15,600 per sq. in.). Hawks, consequently, have a visual acuity about eight times greater than humans. Have students hypothesize the evolutionary advantage of this sharpened acuity.

Demonstration 7

Smell Versus Taste

Much of what we call flavor in foods is actually the result of smell rather than taste. To demonstrate this, have students work in pairs. Have them take turns closing their eyes and holding their nose while their partner places small pieces of various foods on their tongue. Ask if they can identify the food being placed on their tongue. Foods that can be tested include various fruits, an onion, and a potato.

Phase 3

ASSESSMENT OPTIONS

Closure Strategy

Sensory Deprivation

Sensory deprivation chambers have been developed to isolate the occupant from as much sensory input as possible. Have students describe how a person might react in such a situation. Have them research to find out how such chambers are constructed and operate.

Section Review

Assign the *Section Review*.

Reteaching

Assign students to cooperative groups of four. Have each group choose a sensory perception they would least like to lose. Instruct each group to defend their choice.

Touch, Smell, and Taste

Animals have dozens of types of receptors that respond to pressure when they come into contact with something. Human touch receptors are concentrated on the tongue, lips, fingertips, and face, reflecting the importance of the head and hands in our lives.

Inside our bodies we have pressure receptors that constantly monitor blood pressure in the large arteries. Other pressure receptors in the joints, tendons, and muscles detect movement and degree of stretch, and help control how you move.

Chemical receptors enable us to taste and smell

Although we are not as good at sensing chemicals as most mammals, we can taste our food and smell odors. Our external chemical receptors are involved in the sense of smell and of taste as shown in **Figure 28.15**. We have four types of taste receptors on the tongue: sweet, sour, bitter, and salty taste receptors. Smell is very important in telling us about our food. When you have a bad cold and your nose is stuffed up, food has little taste. The "hot" sensation of foods such as chili peppers is detected by pain receptors, not chemical receptors.

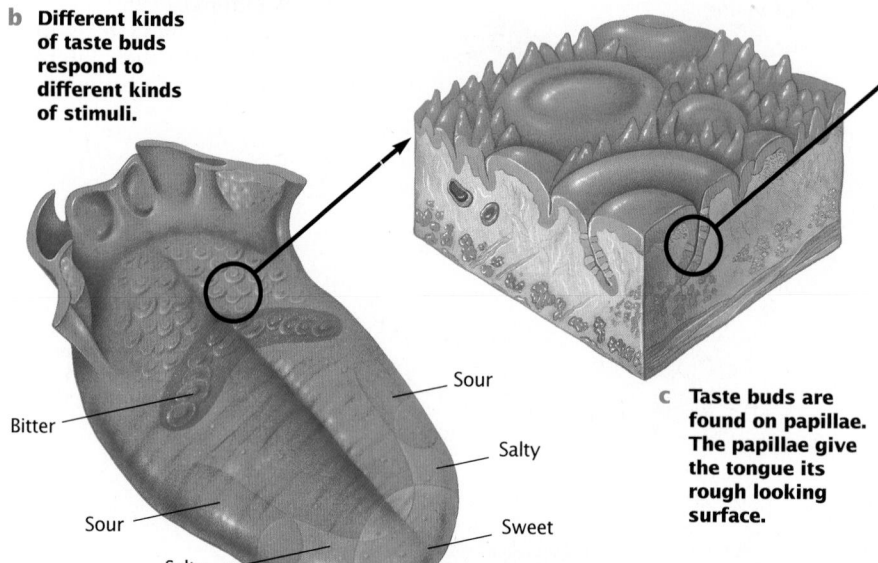

Figure 28.15

a **The taste buds on this man's tongue enable him to taste the sour lemon.**

b **Different kinds of taste buds respond to different kinds of stimuli.**

Bitter

Sour

Salty

Sour

Salty

Sweet

c **Taste buds are found on papillae. The papillae give the tongue its rough looking surface.**

d **Receptors in the taste bud react to chemicals in the food. This reaction causes a nerve impulse to be sent to the brain.**

Section Review

❶ **What are three types of stimuli that sense organs react to?**

❷ **Explain how the inner ear helps us sense our position in space.**

❸ **How does the ear detect sound?**

❹ **Contrast the functions of the rods and cones.**

▪ Section Review Answers ▪

1. Students should suggest that body receptors detect stimuli such as light, heat, or pressure.

2. The semicircular canals, which are located toward the center of the head from the eardrum, help the body monitor position in space.

3. See Figure 28.13 on page 660.

4. Rods are light-sensitive cells that detect various shades of gray in dim light, but cannot distinguish colors and produce poorly defined images. Cones are cells that detect color, produce sharp images, and are necessary for bright light vision.

Highlights

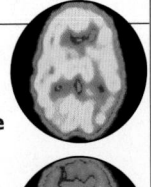

Positron Emission Tomography (PET) scans are useful for detecting activity in the brain. These scans compare deoxy-glucose levels in a normal (top) and a depressed (bottom) person.

Alternative Assessment
Have students formulate questions from the chapter for playing a Jeopardy game as a review technique.

	Key Terms	*Summary*
28.1 How a Nerve Carries a Message 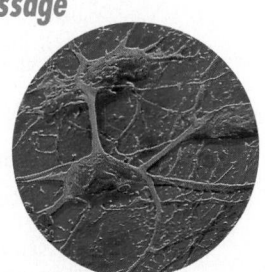 Neurons are the body's messengers. They enable you to interact with the world around you.	neuron (p. 639) dendrite (p. 639) axon (p. 639) gated ion channels (p. 640) resting potential (p. 640) action potential (p. 641) myelin sheath (p. 642) synapse (p. 644) neurotransmitter (p. 644) integration (p. 645)	• The nervous system consists of the brain, spinal cord, and nerves. • The basic unit of the nervous system is the nerve cell, or neuron. • Action potentials travel along an axon, briefly changing the balance of sodium and potassium on the two sides of the membrane. • Myelin sheaths speed up nerve impulses by enabling them to jump from node to node. • Nerves communicate with other cells by sending neurotransmitters across synapses.
28.2 The Nervous System 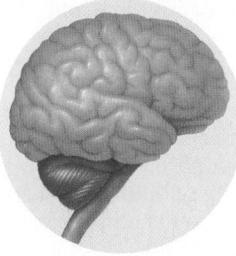 The brain translates data from your sensory neurons into useful information.	central nervous system (p. 647) cerebrum (p. 648) cerebellum (p. 649) brain stem (p. 649) hypothalamus (p. 649) peripheral nervous system (p. 652) sensory neuron (p. 652) motor neuron (p. 652) reflex (p. 653) autonomic nervous system (p. 654)	• The brain consists of the cerebrum, cerebellum, and brain stem. The brain and spinal cord make up the central nervous system. • Sensory neurons tell the central nervous system what is going on; motor neurons are used by the central nervous system to send messages to the muscles and glands. • Reflexes involve few neurons and require a short time to cause action. • The autonomic nervous system is always active, regulating involuntary activities.
28.3 Sense Organs 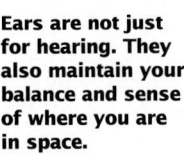  Ears are not just for hearing. They also maintain your balance and sense of where you are in space.	sense organ (p. 658) rod (p. 661) cone (p. 661)	• Sense organs detect internal stimuli, such as position, temperature, and balance, and external stimuli, such as sound, light, and chemicals. • Vision requires two types of receptor cells: rods and cones. Rods are useful in dim light. Cones are used in bright light for detail and color.

Chapter Review Answers

Understanding Vocabulary

1. **a.** Reflex does not fit, since it is not a part of the brain.
   b. Action potential does not fit the pattern because it is not a part of a neuron.
   c. Cochlea is not a part of the eye.
   d. Neuron is not a form of energy that sensory receptors react to.

Relating Concepts

2. Map answer is shown on page 637D.

Understanding Concepts

Multiple Choice

3.	a	8.	b
4.	a	9.	c
5.	d	10.	c
6.	c	11.	b
7.	a	12.	b

Completion

13. brain, spinal cord, peripheral
14. synapse
15. ion channels
16. cerebrum
17. reflex
18. sweet, sour

Short Answer

19. During an action potential, sodium channels on a dendrite open, allowing sodium ions to flood into the neuron. The change in voltage causes nearby voltage-gated sodium ion channels to open, which starts the action potential moving down the neuron. When the action potential has passed, the voltage-gated sodium channels snap closed again and the resting potential is restored.

20. Without the myelin, the impulse would travel much slower than it would with the myelin.

21. The brain stem controls heart rate and breathing. A blow to the back of the skull could damage the brain stem and interfere with these vital processes.

22. His or her ability to hear would be impaired.

Chapter 28 Review

Understanding Vocabulary

1. Identify the term that does not fit the pattern and explain why it does not fit the pattern.
   a. cerebellum, cerebrum, reflex, brain stem
   b. cell body, action potential, axon, dendrite
   c. pupil, retina, cochlea, iris
   d. chemical, sound, light, neuron

Relating Concepts

2. Copy the unfinished concept map below onto a sheet of paper. Then complete the concept map by writing the correct word or phrase in each oval containing a question mark.

Understanding Concepts

Multiple Choice

3. Chemical messengers used by neurons to communicate with other neurons across a synapse are called
   a. neurotransmitters.
   b. hormones.
   c. reflexes.
   d. enzymes.

4. The resting potential across a neuron's plasma membrane is achieved when
   a. the inside of a neuron is negatively charged.
   b. the inside of a neuron is positively charged.
   c. sodium ions are transported into the neuron.
   d. the charges on the inside and outside of the neuron are balanced.

5. The process of integration is the result of
   a. myelinated axons speeding up nerve impulses.
   b. light waves striking sensory cells in the retina.
   c. acetylcholine attaching to receptor proteins in muscle cells.
   d. signals from excitatory and inhibitory synapses.

6. In myelinated nerve fibers, the action potential
   a. responds to changes in the body cell.
   b. moves from dendrite to axon.
   c. jumps from node to node.
   d. causes muscle twitching.

7. The neurotransmitter acetylcholine functions in
   a. nerve to muscle communication.
   b. restoring resting potential.
   c. the senses of taste and smell.
   d. speeding up nerve impulses.

8. Motor neurons are cells that
   a. carry signals to the central nervous system.
   b. carry information to a muscle or a gland.
   c. are found in gray matter in the brain.
   d. contain rods and cones.

9. Damage to the cerebellum could result in
   a. increased levels of hunger and thirst.
   b. the inability to interpret light waves.
   c. difficulty in coordinating body movements.
   d. memory loss.

10. Muscle contraction in the walls of blood vessels and the digestive tract is controlled by
    a. the spinal cord.
    b. mineral crystals in the semicircular canals.
    c. the autonomic nervous sytem.
    d. the cerebellum.

11. What parts of the ear make you aware of your movement and help you keep your balance?
    a. the eardrum
    b. the semicircular canals
    c. the cochlea
    d. the auditory nerve

12. Light waves entering the eye stimulate
    a. rods and cones in the iris.
    b. rods and cones in the retina.
    c. rods and cones in the blind spot.
    d. hair cells in the retina.

Completion

13. The central nervous system is composed of the _____ and the _____ _____ . Sensory neurons and motor neurons make up the _____ nervous system.

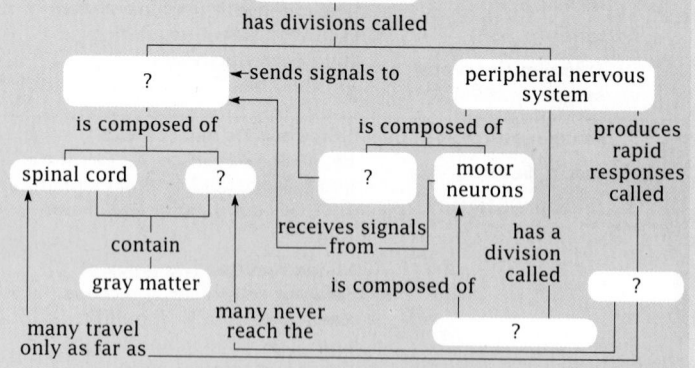

14. The space at the junction of a neuron and another cell is called a _____ .

15. A neuron carries a signal along its plasma membrane through gated _____ _____ , which can open and close.

16. The human ability to use language is a function of the part of the brain called the _____ .

17. A sudden, involuntary reflex that enables animals to react quickly to stimuli is called a _____ .

18. Four different taste sensations are _____ , _____ , bitter, and salty.

Short Answer

19. Explain what happens to the sodium channels in a neuron's plasma membrane during an action potential.

20. Suppose that the axons of a giraffe were not sheathed by myelin. How would an impulse traveling from its legs to its brain be affected?

21. Why is a blow to the back of the skull near the brain stem extremely dangerous?

22. How would a person be affected if an accident damaged the hairs inside the cochlea?

Interpreting Graphics

23. Study the figure to the right and answer the questions that follow.

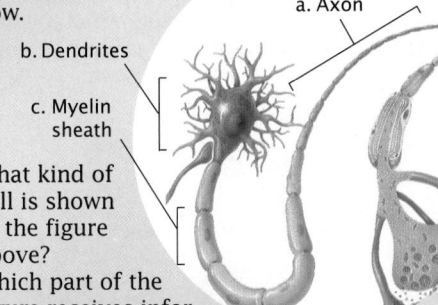

a. Axon
b. Dendrites
c. Myelin sheath

- What kind of cell is shown in the figure above?
- Which part of the figure receives information from other cells?
- Which part of the figure sends messages to other cells?
- Which part of the figure would be damaged by a disease called multiple sclerosis?

Reviewing Themes

24. *Stability*
A child may try to hold his breath until he gets his way. How does the autonomic nervous system prevent the child from harming his body?

Thinking Critically

25. *Inferring Conclusions*
Sufferers of vertigo feel dizzy and disoriented in certain situations, such as on a roller coaster. What is the relationship between vertigo and the functioning of the semicircular canals?

26. *Relating Information*
What accounts for the fact that although each eye has a blind spot, your field of vision appears whole?

27. *Building on What You Have Learned*
How do sodium-potassium pumps function in helping to achieve resting potential of a neuron's cell membrane?

Cross-Discipline Connection

28. *Biology and History*
A. L. Hodgkin and A. F. Huxley conducted research using the giant axons of squids. What did their research reveal that led to their winning a Nobel Prize?

Discovering Through Reading

29. Read the article "Left Brain May Serve as Language Director" in *Science News*, March 7, 1992, page 149. How is performance of one task affected by another task performed at the same time? What evidence was there that language is a task governed by the left hemisphere of the brain? What limits are there to Dr. Corina's study?

Interpreting Graphics

23. • nerve cell
   • b
   • a
   • c

24. Functioning of the autonomic nervous system will cause the child to pass out, restoring involuntary control of breathing.

Thinking Critically

25. The semicircular canals contain receptor cells that enable a person to sense the body's position in space. Damage to them can result in vertigo.

26. Each eye receives partial information about an image. The brain combines the information, filling in the blind spots, so that you see the entire image.

27. The sodium-potassium pumps move potassium ions into the cell and sodium ions out of the cell.

Cross-Discipline Connection

28. They found that an electrical potential difference across the cell membrane contributes to the functioning of a neuron.

Discovering Through Reading

29. The performance of one left-brain task is inevitably disrupted by the performance of another left-brain task. The discovery that tasks with linguistic meaning disrupt right-handed tapping indicates that the left hemisphere governs language. The study focused only on right-handed individuals.

Procedural Note

Discuss the preparation questions and review the structure of the human eye. The students should be familiar with how information is transmitted from the eyes to the brain.

Prelab Preparation Answers

1. • Students should describe the cornea, lens, pupil, iris, retina, rods, cones, and optic nerve and give their functions.
   • Rods and cones are stimulated by light to generate nerve impulses. Rods function in dim light but cannot perceive colors. Cones are responsible for the detection of color.
   • The optic nerves carry nerve impulses from each eye to the visual center in the the brain.

Procedure Answers

2. Students will notice that the X disappears when the card is a few centimeters from their eyes.
5. A "hole" appears in your palm.
7. The tiger appears to be inside the cage.
9. Students should see a red heart bordered in blue.

Chapter 28 Investigation

How Does the Brain Interpret Information From the Eyes?

Objectives

In this investigation you will:
• *observe* and *analyze* the functions of the retina
• *perform* experiments using your field of vision
• *interpret* and *evaluate* observations

Materials

• unlined 3"- x 5"- index card
• plain white paper
• tape
• pencil

Prelab Preparation

1. Review what you have learned about the human eye by answering the following:
   • Describe the structure of the human eye.
   • Explain the function of the rods and cones.
   • Describe the location and function of the optic nerves.

2. On an unlined index card, draw an **X** that is about the size of the one in this sentence. The **X** should be on an imaginary line halfway between the long edges and 1 in. from one short edge of the index card. On the same imaginary line about 3 in. from the **X**, draw an **O** that is approximately the same size.

Procedure: Interpreting Information From Your Eyes

1. Hold the marked index card in front of you at arm's length. Close your right eye and stare at the **O**. Move the card slowly toward you while continuing to stare at the **O**.

2. *What do you notice about the **X** as you bring the card slowly toward you?* Record your observations, noting the distance of the card from your eye when you notice a change in your field of vision.

3. Roll a sheet of plain white paper into a tube about 1 in. in diameter. Secure it with tape.

4. Look through the tube with your left eye. Place your right hand, palm facing you, beside the tube. While looking through the tube with your left eye, look at your open hand with your right eye.

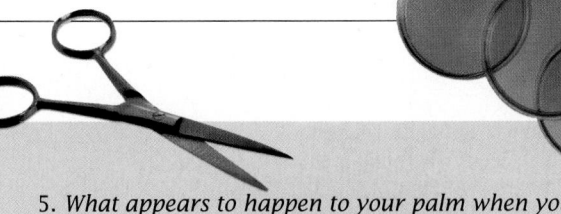

5. *What appears to happen to your palm when you look through the tube and at your hand?* Repeat the procedure, looking through the tube with your right eye and using your left hand.

6. Look at the picture of the tiger and the cage shown below.

Place the long edge of an index card between the tiger and the cage. Next, place the tip of your nose against the upper edge of the card and look at the cage with one eye and the tiger with the other.

7. *What appears to happen to the images when you look at the tiger and cage at the same time?*

8. Place a piece of plain white paper in front of you. Stare at the black dot on the drawing of the heart on this page for about 20 seconds.

Then, immediately stare at the white paper for about 10 seconds.

9. *What do you see when you look at the plain white paper after staring at the dot in the middle of the heart?*

Analysis

1. *Analyzing Relationships*
What is the relationship between the optic nerve and the disappearance of the **X** on the index card?

2. *Interpreting Ideas*
Why did a "hole" appear in your palm when you looked through the tube and at your hand?

3. *Analyzing Theory*
How does the theory of color vision help explain the afterimage observed with the heart illustration?

4. *Inferring Relationships*
How can the phenomena that you observed with the tiger and cage and the heart afterimage explain how you see motion in a movie?

Thinking Critically

1. Each of your eyes has a "blind spot," as you demonstrated in this investigation. Why don't you notice two "holes" in everything you look at?

2. When you close one eye, you still do not notice the "hole" in your field of vision. Why not?

Analysis Answers

1. The point where the optic nerve enters the retina is known as the optic disc or blind spot. This area does not contain any rods or cones and cannot perceive images. The X on the card seems to disappear when it is at the distance at which its image falls on the blind spot.

2. A "hole" appears in your palm because the brain combines the images perceived by both eyes.

3. Humans have three different types of cone cells, each sensitive to different wavelengths. Staring at the heart stimulates the mechanism that perceives green and yellow. A red heart bordered in blue is seen when the eyes look away, because the removal of the stimulus shuts down the green-yellow mechanism and triggers the red-blue mechanism.

4. Images perceived by the eye tend to persist for a brief second after the image is actually seen. Thus, the images of a motion picture do not appear to be a series of separate still frames, but are perceived as moving.

Thinking Critically Answers

1. The visual fields of each eye overlap. The information from each eye contributes to a complete image.

2. The brain "fills in" the image.

Hormones

Planning Guide

Objectives/Themes	Classwork Resources	Homework Resources
29.1 **1.** Identify two components of the endocrine system. **2.** Explain the role of the hypothalamus and pituitary glands in regulating body systems. **3.** Describe two examples of negative feedback. **Themes:** Interacting Systems, Scale and Structure, Stability	**Teacher's Resource Binder** Focus Activity 29 *Graphing Growth Rate Data* Lab Investigation 29.1 *Effects of Hormones on Circulation* **Other Resources** Transparency 130	**Text** Section Review, p. 673 **Teacher's Resource Binder** Directed Reading Worksheet 29.1 **Other Resources** Audiocassette 29.1*
29.2 **1.** Describe how a target cell responds to a hormone. **2.** Explain how a steroid hormone affects a target cell. **3.** Explain how a peptide hormone affects a target cell. **Themes:** Patterns of Change, Interacting Systems	**Teacher's Resource Binder** Extension Worksheet *Hormonal Reaction to Stress* **Other Resources** Transparencies 131–132	**Text** Section Review, p. 676 **Teacher's Resource Binder** Directed Reading Worksheet 29.2 **Other Resources** Audiocassette 29.2*
29.3 **1.** Name three endocrine glands and list the hormones each produces. **2.** Explain the effects of producing too much or too little thyroxine. **3.** Describe diabetes and explain how the condition can be treated. **4.** List two effects prostaglandins have on cells. **Themes:** Interacting Systems, Patterns of Change, Scale and Structure, Stability	**Text** Investigation *How Does the Endocrine System Work?* pp. 688–689 **Other Resources** Transparency 133	**Text** Section Review, p. 684 **Teacher's Resource Binder** Directed Reading Worksheet 29.3 Vocabulary Review Worksheet* Reaching Worksheet* *Identifying Structures* **Other Resources** Audiocassette 29.3*

*Reteaching Options

Demonstrations
29.1: pp. 670, 672
29.3: pp. 678, 679, 683

Assessment
Chapter Review pp. 686–687
Portfolio Assessment p. 667D
Chapter Test—Teacher's Resource Binder
Test Generator

Research Notes

Connection to Physiology: Hormones and Jet Lag

Scientists are investigating the activity of a hormone from the pineal gland called melatonin, which seems to play a role in synchronizing the body's natural rhythms with daytime and nighttime.

Other researchers are working on an enzyme in a photosynthetic bacterium, Rhodobacter capsulatus, which enhances the ability of photopigments to respond to light in dim conditions. Surprisingly, the two lines of research have converged, with the discovery that the enzyme that catalyzes melatonin synthesis in vertebrates is remarkably similar to the one in the bacteria.

This discovery, along with the fact that the amino acid sequences for the two enzymes are very different from all other known vertebrate proteins, indicates that both proteins have a common ancestor. Although the production of such an enzyme would be an advantageous adaptation for a photosynthetic organism, it is remarkable that such a mechanism for detecting darkness and light would survive for so long in living organisms that no longer rely on photosynthesis.

The interplay between these enzymes and the hormone could provide insight into a variety of modern ailments. For example, the phenomenon of "jet lag" occurs when people travel across time zones and the sun sets earlier or later than where they came from. The body does not have time to readjust the cycles of hormones to match the new daytime and nighttime, and fatigue ensues.

In 1988, an initial study was made with an experimental melatonin pill. Jet lag symptoms were reduced by as much as 50 percent. Endocrinologists believe such pills could help the body adjust to the new time zone faster, and help workers avoid fatigue in their new setting. Pills for travelers could be on the market within a few years.

Seasonal affective disorder (SAD) is another malady that seems to be related to the hormone-enzyme interplay of the light cycle. In the wintertime, people suffering from SAD find themselves depressed, sleepy, and craving carbohydrates. During winter's shorter days, which have less sunlight, people have higher average levels of melatonin, as they do during nighttime.

New treatments for SAD involve exposing patients to bright lights daily. This can reverse the symptoms of SAD in just two to four days. If the light therapy is stopped, or if patients are given melatonin, the symptoms return. Similar light therapy also can help patients suffering from severe "cluster headaches."

Investigation Notes

How Does the Endocrine System Work?
pages 688–689

Purpose: The investigation gives the students an opportunity to apply their knowledge of problem solving to endocrine glands. The students analyze experimental data about a hypothetical system to ascertain relationships among the parts of the system.

Prelab Preparation

Only paper and pencil are needed.

Answers will be found on pages 688–689.

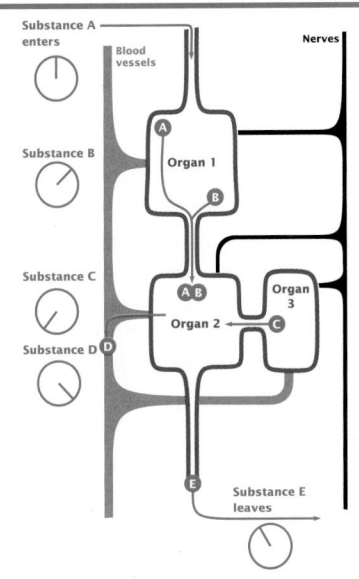

Meeting Individual Needs

Objectives

1. Students will demonstrate appropriate use of core vocabulary for the chapter (Vocabulary File).

2. Students will relate the different sources and targets of hormones (Demonstration A).

3. Students will relate hormonal imbalances to the diseases they can cause (Demonstration B).

Vocabulary File
(Developing Vocabulary/ Limited English Proficiency)

If you are not already using the Vocabulary File, refer to Chapter 1 for its preparation. See the Chapter Highlights on page 685 for a list of suggested words.

Demonstration A
What Hormones Do
for pages 669–673
(Developing Modeling Skills/ Visual Learners)

Materials: construction paper, pipe cleaners, glue, stick-on white rectangular labels, markers

Procedure: Have each student construct a model to summarize the endocrine system. Distribute pipe cleaners and construction paper to each student. Students should fashion a human body out of pipe cleaners, and then glue the pipe-cleaner person to the left side of the paper. Using the text on page 669 as an example, write the correct name for each endocrine gland on a label. On the edge of the right side of the paper, press down these labels. Draw lines from the label to the part of the pipe-cleaner body containing the gland it represents. Inside the label, also write the function of that gland, and some of the hormones it produces. Students may refer to the chart on page 683.

Demonstration B
Chemical Malfunctions
for pages 669–684
(Developing Organizational Skills/Verbal Learners)

Procedure: Lead a discussion about disease. Be sure students mention the various agents that can cause diseases, such as bacteria and viruses, that were studied. Point out that one's own body can be responsible for diseases, especially those involving hormonal imbalances.

Briefly discuss the three most common disorders of this type:

Graves' disease—the thyroid gland produces too much thyroxine. Signs are rapid, irregular heartbeat, nervousness, and loss of weight.

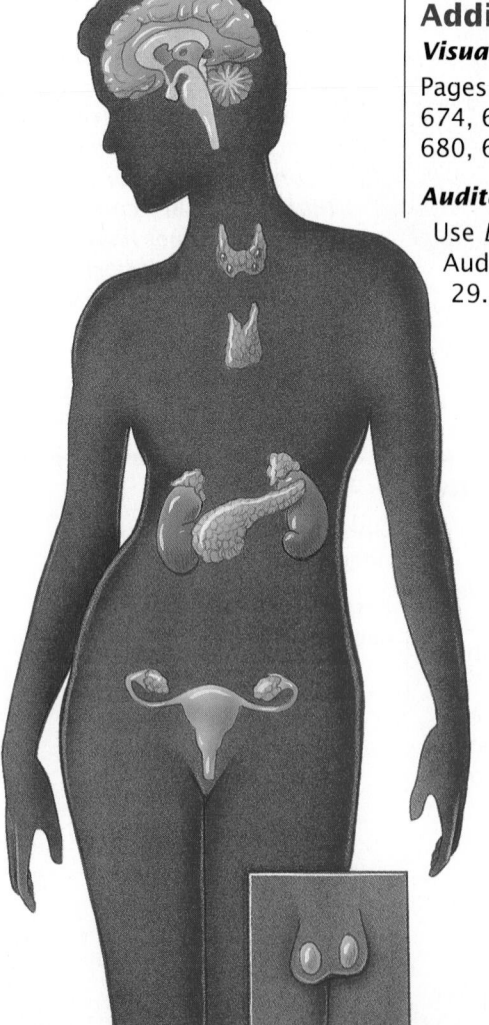

Goiter—the thyroid gland produces too little thyroxine as a result of insufficient iodine in the diet. The principal symptom is a swollen thyroid gland.

Diabetes—insulin and glycogen are secreted in unbalanced amounts by glands in the pancreas. Cells cannot take in sufficient amounts of glucose, which results in toxic byproducts that can disrupt heart rate, breathing, and brain activity.

Re-scan the chapter, discussing in depth how the body is kept in balance by the secretions of hormones. List these on the board. Be sure students realize that the endocrine and nervous systems are the two main control systems of the body, and that they are so interconnected that it can be difficult to separate them.

Additional Strategies
Visual Strategies

Pages 669, 670, 671, 672, 673, 674, 675, 676, 677, 678, 679, 680, 681, 682, 683, and 684

Auditory Learners

Use *Biology: Visualizing Life* Audiocassettes for Sections 29.1, 29.2, and 29.3.

Meeting Individual Needs (cont.)

Cooperative Learning
Steroids: Helpful or Harmful?

Timing: Use this activity to introduce Section 29.2.

Group Size: 4 students

Outcome: Students will be able to list the pros and cons of using steroids.

Individual Accountability: Each group member is responsible for listing helpful and harmful effects of using steroids.

Positive Interdependence: Each group will reach a consensus on whether steroid use is helpful or harmful.

Divide each group into two pairs. One pair should do research and develop a list of all possible helpful effects of using steroid drugs to improve athletic ability. The other pair should do research to develop a list of the harmful effects of using steroids.

After each pair presents their list to the other pair, the group should have a debate about the pros and cons. Each group should come to an agreement regarding whether steroids are helpful or harmful. If you have conflicting viewpoints among groups, encourage group-to-group debates. (It should be clear that the evidence indicates that the risks of using steroids far outweigh the benefits.)

Portfolio Assessment

Students should select their best work and provide a self-reflective rationale for their selections. Students can make selections in the following areas.

1. *Content* — One concept map from the chapter (See page 812 for evaluation criteria.)

2. *Reading Comprehension* — One Directed Reading Worksheet from the Teacher's Resource Binder (Use the answer key to evaluate for accuracy.)

3. *Writing* — Using the Vee Form, summarize a magazine or newspaper article relating to hormones or hormonal disorders. (See page 22T for evaluation criteria.)

   Or: Select a writing task or project from the Chapter Review.

4. *Performance Assessment* — One Vee form from a chapter investigation or lab manual investigation (See page 22T for evaluation criteria.)

Teacher makes selections in the following areas.

1. *Formal Assessment* — Chapter test (Test A, B, or the Test Generator) The teacher-scored test should be reviewed by the student. Incorrect responses should be corrected by the student before the test becomes part of the portfolio.

2. *Informal Assessment* — Use the Direct Observations Checklist, page 33T, during a laboratory or other cooperative learning experience.

3. *Performance Assessment* — Have students create a computer program that models the system and feedback inhibition described in the Investigation.

Concept Map Answer

The following is one possible answer to the Relating Concepts exercise on page 686.

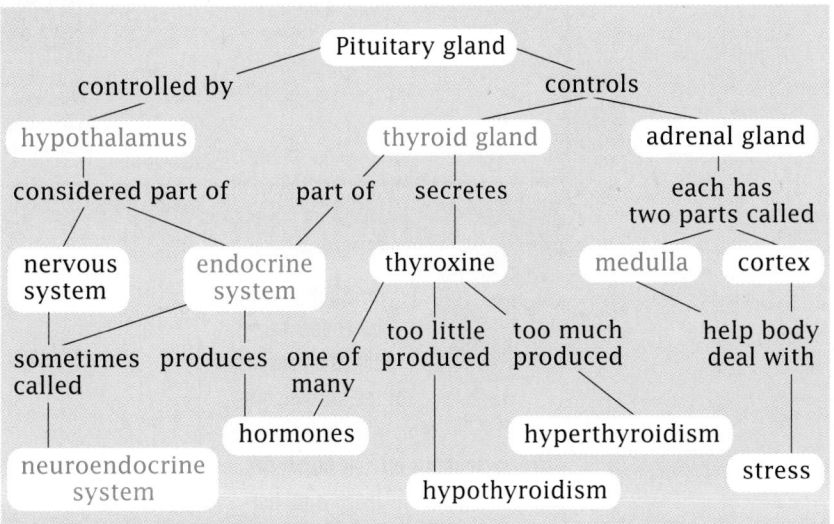

Chapter 29

Determining Prior Knowledge

- Once students have been seated at the start of class, tell them that you have decided to give a surprise quiz on what was covered in yesterday's session. After they have had a chance to absorb what you've said, discuss what physiological changes their bodies underwent as they experienced this "stressful" situation.
- Ask if anyone has—or knows of someone who has—diabetes. Discuss what they know about this disease.
- Show the class photographs of people who obviously have a permanent or temporary hormonal imbalance—for example, someone who is extremely tall or short in height or someone who is covered with excessive body hair.

Chapter 29

Review

- **cell membrane architecture (Section 3.1)**
- **protein structure (Section 3.2)**
- **chemical signals (Section 4.1)**
- **nervous system (Section 28.2)**

Hormones

29.1 What Hormones Do

- **Hormones: Chemical Signals**
- **The Hypothalamus-Pituitary Connection**
- **Regulating Hormone Release**

29.2 How Hormones Work

- **Hormone Receptor Proteins**
- **Steroid Hormones**
- **Peptide Hormones**

29.3 Glands and Their Functions

- **The Adrenal Glands**
- **The Thyroid Gland**
- **The Pancreas**
- **Other Glands and Hormones**

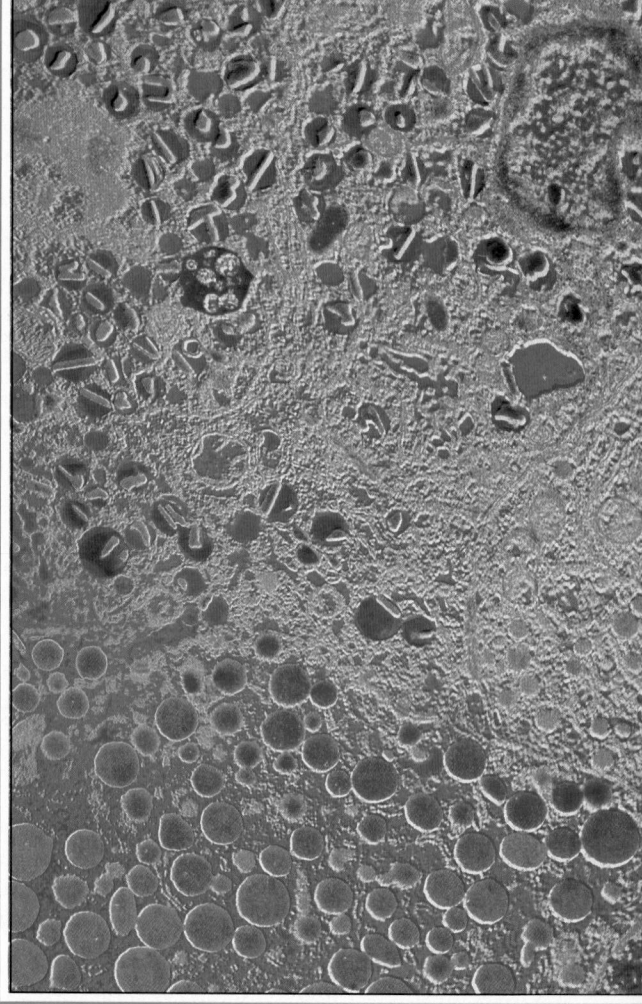

This colorized TEM shows hormone-secreting islet cells of the pancreas. The green, yellow, and brown cells on top secrete the hormone insulin. The red cells at the bottom secrete glucagon.

■ Author's Rationale ■

This chapter treats the endocrine system as an extension of the nervous system, while placing a strong emphasis on how hormones work. Students will learn what hormones do and how they do it.

AFTER FIGHTING A BLAZING FIRE FOR MANY HOURS, A FIREFIGHTER NEARS EXHAUSTION. SUDDENLY A WALL COLLAPSES! THE FIREFIGHTER RUSHES TO CONTROL THE FIERY DEBRIS. HOW? CHEMICAL SIGNALS CARRIED BY THE BLOODSTREAM ENSURE THAT THE FIREFIGHTER'S BODY GETS THE EXTRA OXYGEN AND ENERGY-SUPPLYING SUGARS NEEDED TO SUSTAIN PEAK PERFORMANCE UNTIL THE CRISIS PASSES.

29.1 *What Hormones Do*

Objectives

❶ Identify two components of the endocrine system.

❷ Explain the role of the hypothalamus and pituitary glands in regulating body systems.

❸ Describe two examples of negative feedback.

Hormones: Chemical Signals

In the emergency described above, the signals traveling through the firefighter's blood are small molecules called hormones. A **hormone** is a chemical signal, made in one place and delivered to another, that regulates the body's activities. Various organs scattered throughout the body are sources of hormones. Organs that produce most of the hormones in your body are called **endocrine** (EN duh krihn) **glands**. Endocrine glands make up the endocrine system, shown in **Figure 29.1**. Other organs, such as the brain and kidneys, also produce hormones.

Endocrine glands secrete hormones directly into the bloodstream. The hormones travel to a specific tissue or organ called a **target**. Once a hormone arrives at its target, the hormone will elicit a specific response. For example, hormones speed up the heart. They help digest food and enable the body to use the food as fuel. Without hormones, humans would not grow or mature. Like the brain and nerves, hormones are essential to maintaining homeostasis.

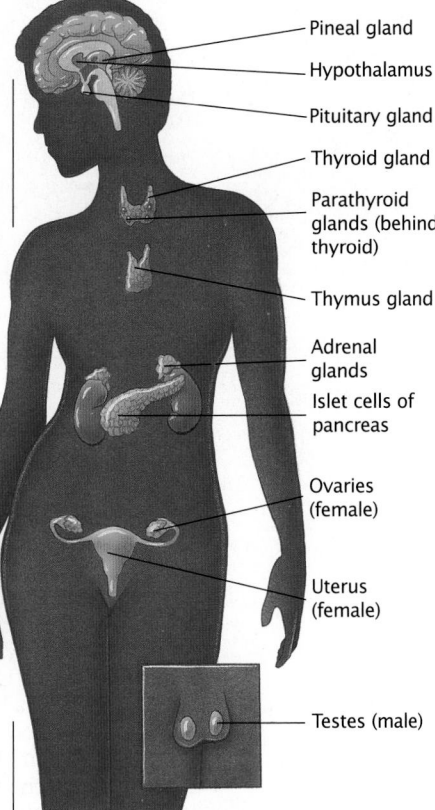

- Pineal gland
- Hypothalamus
- Pituitary gland
- Thyroid gland
- Parathyroid glands (behind thyroid)
- Thymus gland
- Adrenal glands
- Islet cells of pancreas
- Ovaries (female)
- Uterus (female)
- Testes (male)

Figure 29.1
The endocrine system regulates overall metabolism, growth, reproduction, and maintenance of homeostasis.

Visual Strategy

Figure 29.2

Have students identify what sensory organs Cary Grant might have used to sense the oncoming danger. To review the connection between the nervous and muscular systems, discuss what other effect the nervous system had, besides activating the adrenal gland. Point out that long after his muscles stopped contracting (since they are no longer being stimulated by nervous impulses), Grant's heart continued to beat faster (since adrenaline remained in the bloodstream long after the plane had disappeared).

Theme Connection

Stability

Emphasize the involvement of the nervous and endocrine systems in homeostasis.

Demonstration 1

The Hypothalamus-Pituitary Connection

Show the class a diagram or overhead transparency of the brain with the hypothalamus and pituitary gland clearly identified. Mention that the pituitary, lying just beneath the hypothalamus, is directly under its influence. The blood stream carries hormones released by the hypothalamus that either stimulate or inhibit the pituitary gland.

Figure 29.2
In an emergency, the nervous system sends a signal to the adrenal glands. The glands secrete hormones that signal a particular receptor or target cell. The hormone and the receptor bind, changing the shape of the receptor and effecting changes in cellular activities. For instance, the cell increases output of energy-supplying sugars.

Nervous system → Gland → Hormone → Target cell → Changes in cell activity

The effects of hormones last longer than nerve impulses

Imagine that you start to cross a street and suddenly a car speeds around the corner. Instantly, nerve impulses deliver the message—"Danger!"—from your eyes to your brain. In a matter of seconds, you sprint back to the safety of the curb. Nerve impulses prompt a nearly instantaneous response to a change in the environment.

Hormones, on the other hand, are released more slowly than nerve impulses, but their effects usually last longer. Imagine that the reckless driver crashes into another car and then speeds away. Within seconds, hormones from your adrenal glands cause your heart to beat faster, pumping extra oxygen through your bloodstream. Your brain becomes more alert; you might think quickly enough to look at and remember the license plate number of the hit-and-run driver. You race down the street to see if anyone is injured. After this is over, you may find yourself shaking a little. **Figure 29.2** summarizes the events involved in an emergency.

The effects of some hormones will probably last 10 or 20 minutes. But when the body has to deal with an emergency that lasts longer, other adrenal hormones are released. These hormones can maintain extra energy levels for several hours after an emergency occurs.

The Hypothalamus-Pituitary Connection

The endocrine system and the nervous system are the two main control systems of the body. They are so closely linked that they often are considered a single system—the **neuroendocrine system**. In Chapter 28 you learned that the hypothalamus is the part of your brain that regulates body temperature, breathing, hunger, and thirst. The hypothalamus can also be considered the master switchboard of the endocrine system. Your hypothalamus is continuously checking conditions inside your body. Are you too hot or too cold? Are you running out of fuel? How about your blood pressure? Is it too high or too low? If your internal environment starts to get out of balance, your hypothalamus has several ways to set things right again. For example, the hypothalamus can send a nerve signal to another part of the brain—the medulla—to speed up or slow down your heart rate. The hypothalamus also sends out commands in the form of hormones, thus acting like an endocrine gland.

The pituitary gland secretes and stores hormones

All of the hormones produced by the hypothalamus move through a slender thread of tissue to the **pituitary (puh TOO uh tehr ee) gland**. The pituitary gland produces at least six different hormones in response to the hormones released from the hypothalamus. As described in **Figure 29.3 a–b**, two of these hormones, growth hormone and prolactin, have direct effects on tissues in the body. Growth hormone, for instance, stimulates protein synthesis and cell division in target cells. It also profoundly influences the growth of cartilage and bone. Because these pituitary hormones are controlled by "releasing" hormones from the hypothalamus, the brain exercises direct control over the endocrine system.

The pituitary gland secretes four other hormones that control the activities of a number of other endocrine glands such as the thyroid gland and the adrenal glands. Some of these glands are discussed in more detail later in this chapter.

In addition to producing hormones, the pituitary gland stores hormones made in the hypothalamus. Two of these hormones, oxytocin (*ahk see TOHS ihn*) and vasopressin (*vay so PREHS ihn*), discussed in **Figure 29.3c–d**, are released from the pituitary gland when needed by the body.

Figure 29.3
The pituitary gland, located at the base of the hypothalamus, was formerly called the "master gland" because so many of its hormones regulate other endocrine functions.

Visual Strategy

Figure 29.3
Use this figure to impress students with the fact that the pituitary gland—about the size of a pea—is a powerful organ, producing hormones that have a far-reaching impact. At one time, the pituitary was referred to as the "master gland," since its hormones controlled the action of other endocrine glands. Have students identify some of the pituitary's target organs shown in this figure.

a *Growth hormone* stimulates general growth of the body, particularly of the skeleton. When too little growth hormone is produced, a condition called dwarfism can result. With too much growth hormone, a condition called gigantism can occur.

b *Prolactin* affects breast tissue in nursing mothers, causing glands in the breasts to produce milk.

c *Oxytocin* is the hormone that causes contractions of the uterus during labor.

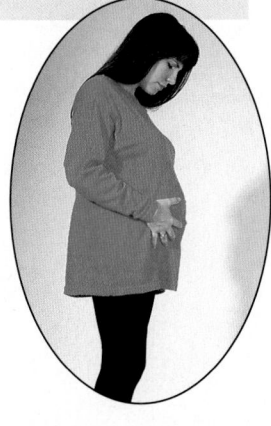

d *Vasopressin* causes the kidneys to form more concentrated urine, thereby conserving water in the body.

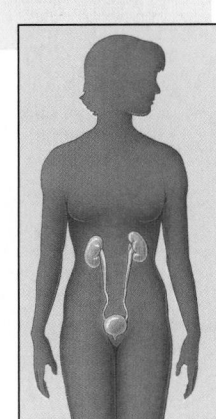

■ *Cultural Perspective* ■

Choh Hao Li
(Chinese-American)
Choh Hao Li, a Chinese-American endocrinologist, isolated and identified five hormones of the pituitary gland. He discovered that the pituitary gland's growth hormone contains a chain of 256 amino acids. In 1970, he discovered a method for synthesizing the hormone and set the record for creating the largest synthesized protein molecule.

Regulating Hormone Release

Because the body produces more than 30 hormones, it must be able to regulate the release of these hormones. For example, endocrine glands usually do not secrete their hormones at a constant rate. A nerve impulse may cause a gland to increase or decrease its rate of hormone production. In most cases, however, chemical signals, including other hormones, regulate the function of the endocrine system through **negative feedback**. Negative feedback is a process by which a change in an environment causes a response that returns conditions to their original state. For example, negative feedback enables a thermostat to keep your classroom at a constant temperature. Suppose the thermostat is set for 22°C (72°F). The temperature in the classroom is not always exactly that temperature. If it is a chilly winter day and the room temperature falls below 22°C, the furnace will come on and blow warm air into the room until the temperature rises. Then the furnace shuts off, and the temperature slowly begins to fall. Just about the time you start feeling a little chilly—at about 20°C or 21°C (68°F or 69°F)—the furnace switches on again.

Parathyroid glands regulate levels of calcium in the blood

The way a thermostat maintains the temperature in a room is similar to the way most endocrine glands help a body maintain homeostasis. For example, the parathyroid glands, four tiny oval glands embedded in the back of the thyroid, secrete a hormone that regulates the level of calcium in the bloodstream. Calcium is a mineral necessary for proper growth, healthy teeth and bones, nerve function, and muscle contraction. While some calcium is stored inside cells, most of it is stored in bones. When the level of calcium in the blood drops even slightly, the parathyroid glands are stimulated to secrete parathyroid hormone. This hormone causes the release of calcium from bone into the blood. When the level of calcium in the blood rises to a certain level, the parathyroid glands halt production of the hormone. The diagram in **Figure 29.4** summarizes these events.

Parathyroid glands

↓

Parathyroid hormone (PTH)

↓

Bone receives PTH

↓

Releases calcium into blood

↓

Parathyroid cells detect calcium levels

↓

Is the level of calcium adequate?

No → (loops back)   Yes ↓

PTH production ceases

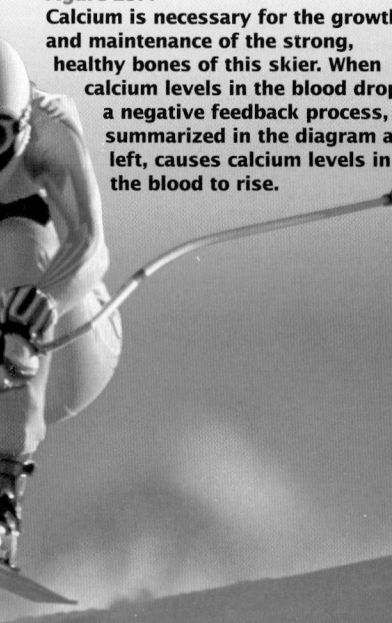

Figure 29.4
Calcium is necessary for the growth and maintenance of the strong, healthy bones of this skier. When calcium levels in the blood drop, a negative feedback process, summarized in the diagram at left, causes calcium levels in the blood to rise.

Figure 29.5
When you are cold, the hypothalamus sends signals to the pituitary gland, which signals the thyroid to produce thyroxine. Thyroxine speeds up metabolism until your body temperature rises slightly. Then the system shuts off.

Negative feedback also maintains body temperature

Imagine that you are waiting for the bus on a windy winter day, shivering in the cold. Even though you are not aware of it, your hypothalamus is monitoring your body temperature. As you get colder, your hypothalamus produces a hormone that stimulates your pituitary gland. Your pituitary gland then secretes a hormone that stimulates the thyroid

gland. The thyroid gland, in turn, secretes thyroxine, a hormone that speeds up the rate of metabolism in your body, keeping your body temperature at a normal level. When the thyroxine level reaches a certain point, the secretion of the thyroid-stimulating hormone by the pituitary is reduced, and the thyroid stops secreting thyroxine. This negative feedback process is summarized in **Figure 29.5**.

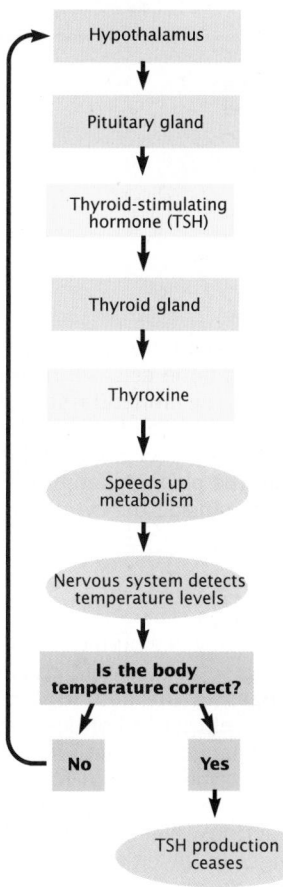

Hypothalamus → Pituitary gland → Thyroid-stimulating hormone (TSH) → Thyroid gland → Thyroxine → Speeds up metabolism → Nervous system detects temperature levels → Is the body temperature correct? → No / Yes → TSH production ceases

Section Review

1. **Define the term "hormone" and name four endocrine glands in the body.**

2. **Why are the nervous system and endocrine system often called the neuroendocrine system?**

3. **Explain how the parathyroid glands regulate levels of calcium in the blood.**

4. **Explain how the hypothalamus helps maintain a constant body temperature.**

▪ *Section Review Answers* ▪

1. A hormone is a chemical signal, made in one place and delivered to another, that regulates the body's activities. Students might suggest any of the following: pineal, pituitary, thyroid, parathyroid, thymus, adrenal glands.

2. Students should suggest that the nervous system and the endocrine system are closely linked and are the body's two main control systems.

3. See the flow chart in Figure 29.4 on page 672.

4. See the flow chart in Figure 29.5 on page 673.

Lesson Plan 29.2

Phase 1

PREPARATION

Key Concepts

- Steroid hormones bind with receptor proteins inside the cell cytoplasm, forming a complex that then attaches to the DNA, where it activates specific genes.
- Peptide hormones bind with appropriate receptor proteins on the cell membrane, activating the formation of cyclic AMP that acts as a second messenger inside the cell.

Reading Strategy

This section requires an understanding of material described earlier in the text. Tell students that reading a science text often requires looking back and rereading parts of previous sections. In this way, understanding the new material becomes easier, since it is based on a foundation of previous knowledge.

Phase 2

TEACHING STRATEGIES

Visual Strategy

Figure 29.6

See if students recall which other class of organic compounds illustrated in this figure, besides protein, makes up the cell membrane. Have them compare the shape of the receptor protein before the hormone binds with the shape after it binds.

HORMONES ARE CARRIED THROUGHOUT THE HUMAN BODY IN THE BLOODSTREAM. ONLY TARGET CELLS ACTUALLY RESPOND TO THE MESSAGE THAT HAS BEEN SENT. AFTER A HORMONE ARRIVES AT ITS TARGET, THE HORMONE CAUSES CERTAIN ACTIONS TO OCCUR. TWO TYPES OF HORMONES CAUSE CHANGES IN CELLS IN DIFFERENT WAYS, AS YOU WILL LEARN IN THIS SECTION.

29.2   How Hormones Work

Objectives

❶ Describe how a target cell responds to a hormone.

❷ Explain how a steroid hormone affects a target cell.

❸ Explain how a peptide hormone affects a target cell.

Hormone Receptor Proteins

As you read in Chapter 4, the plasma membrane of the cell contains receptor proteins. A receptor protein has a unique shape that will only hold a particular type of molecule. If a cell has a receptor protein that will hold a particular hormone molecule, then the cell will respond to that hormone. For example, bone cells have receptor proteins that will hold parathyroid hormone molecules. A parathyroid hormone molecule will fit into one of these receptor proteins like a key fitting into a lock, as shown in **Figure 29.6**. Fitting the hormone molecule into the receptor changes the receptor's shape, which causes the cell's activities to change. In this case, the bone cell will begin to break down the minerals stored in the bone so that calcium can be released. While many receptor proteins are located in the plasma membrane, others are found inside the cytoplasm of the cell.

Hormones affect enzymes in cells

How can the binding of a hormone molecule with a receptor protein cause drastic changes in the activities of a cell? The main effect of a hormone on a cell is to change the activity or amounts of enzymes present in that cell. Recall that enzymes speed up chemical reactions. When a hormone causes changes in the enzymes in a cell, it causes changes in the chemical reactions that are happening inside the cell.

Figure 29.6

a **When a hormone comes into contact with a plasma-membrane receptor protein that will hold it, the hormone binds to the receptor protein like a key fitting into a lock.**

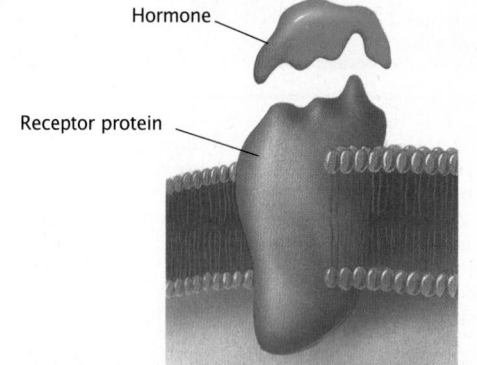

Hormone

Receptor protein

b **The receptor protein changes shape, causing changes within the cell.**

■ *Cultural Perspective* ■

Sandra A. Murray, Ph.D.
(African-American)
Dr. Murray is an associate professor of Neurobiology, Anatomy, and Cell Science at the University of Pittsburgh, where she studies the cellular mechanism of hormone action.

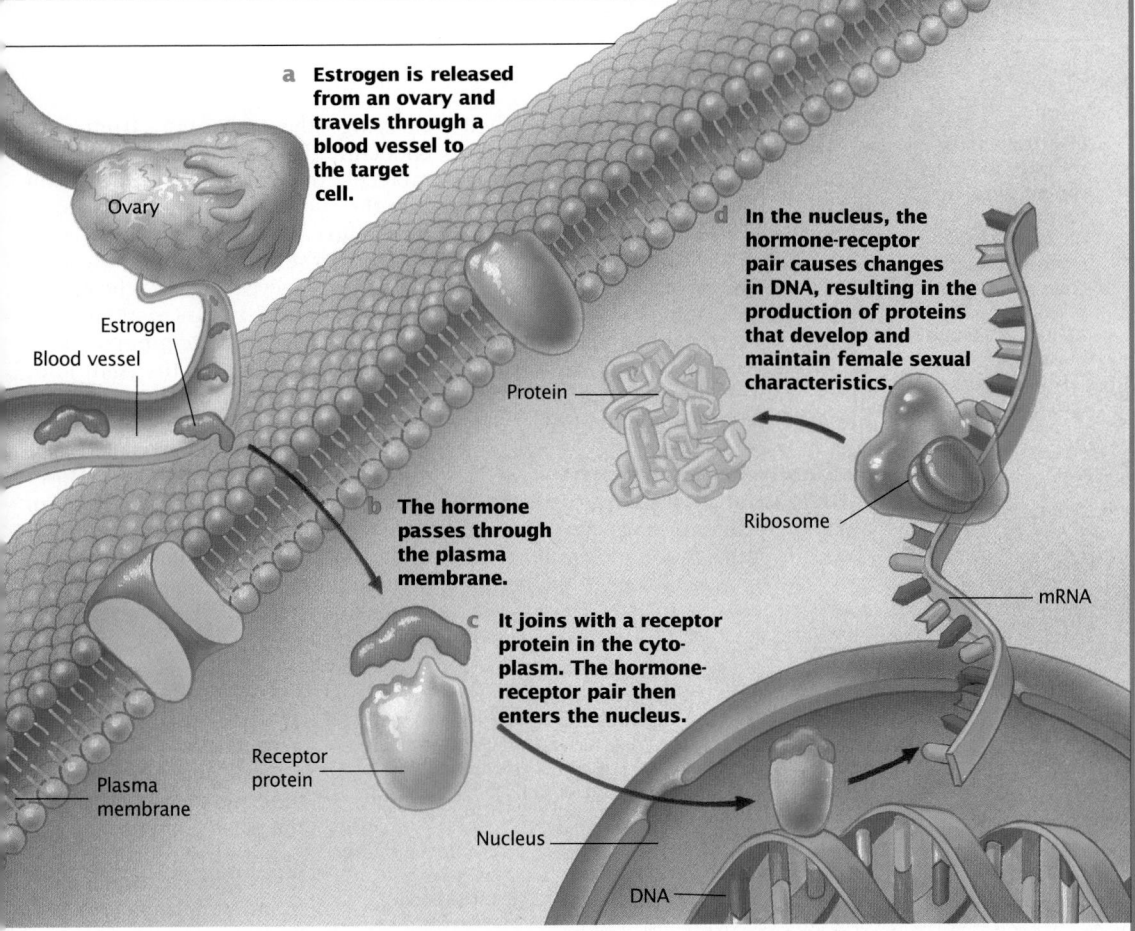

a Estrogen is released from an ovary and travels through a blood vessel to the target cell.

Ovary

Estrogen

Blood vessel

d In the nucleus, the hormone-receptor pair causes changes in DNA, resulting in the production of proteins that develop and maintain female sexual characteristics.

Protein

Ribosome

mRNA

b The hormone passes through the plasma membrane.

c It joins with a receptor protein in the cytoplasm. The hormone-receptor pair then enters the nucleus.

Receptor protein

Plasma membrane

Nucleus

DNA

Steroid Hormones

Figure 29.7
The steroid hormone estrogen influences development of female sexual traits. The way steroid hormones, such as estrogen, cause changes in cellular activity is described above.

Hormones assembled from cholesterol are called **steroid hormones**. Since cholesterol is a lipid, steroid hormones can pass through the lipid bilayer of the plasma membrane. Within the cytoplasm, a steroid hormone can bind to a receptor protein. The hormone-receptor complex can then pass into the nucleus, where it will trigger changes in the chromosomes. The male sex hormone testosterone and the female sex hormones estrogen and progesterone are examples of steroid hormones.

Steroid hormones activate genes

Recall that every cell in your body (except sperm or egg cells) contains all of the genes that you inherited from your parents. However, not all genes are actively being used to make proteins. Different sets of genes are switched on in different types of cells. For example, the genes that are active in liver cells are different from the genes that are active in skin cells.

One way to alter a cell's enzyme activity is to change which genes are switched on. Steroid hormones work by activating specific genes in the target cell, as shown in **Figure 29.7**. First, a steroid hormone passes through the plasma membrane and joins a receptor protein in the cytoplasm. Then the hormone-receptor unit passes into the nucleus and attaches itself to the DNA, as shown in **Figure 29.7d**. This attachment activates certain genes in that cell, causing particular proteins, including new protein enzymes, to be produced.

▪ Cultural Perspective ▪

Richard H. Pointer, Ph.D.
(African-American)
While earning his Ph.D., Dr. Pointer studied the regulation of glycogen metabolism in liver cells. Dr. Pointer is a biochemist at Howard University, where he researches how hormones regulate other aspects of metabolism.

Connection: Chapter 5

ATP

Have students refer back to Figure 5.8 on page 95 to review the structure of ATP. Have them identify the two phosphate groups that are removed to form AMP, whose ends are then joined to form cyclic AMP.

Visual Strategy

Figure 29.8

Lead students through each of the steps illustrated in this figure so that they understand how protein hormones bind to receptors on the cell membrane surface. Point out that some adults become diabetic because of a decrease in the number of receptors on target cell membranes. Although such individuals produce insulin, it cannot function as well when the number of insulin receptor proteins has decreased.

Phase 3

ASSESSMENT OPTIONS

Closure Strategy

Which Is Faster?

Have students hypothesize which acts faster—steroid or peptide hormones. After providing time for discussion, point out that peptide hormones tend to have immediate effects, since they quickly activate enzymes that serve to catalyze chemical reactions. Steroid hormones take longer to bring about their responses, but their effects last longer.

Section Review

Assign the *Section Review*.

Reteaching

Have students describe how protein hormones and neurotransmitters are similar in the ways they affect their target cells.

Peptide Hormones

Figure 29.8
Liver cells have receptor proteins for the peptide hormone glucagon, which is made by the pancreas. Glucagon uses the cyclic AMP as a second messenger to release glucose into the blood.

Hormones made of amino acids joined to form small peptides or larger proteins are called **peptide hormones**. Recall that amino acids dissolve easily in water. This is because amino acids and water molecules are both polar molecules. However, the positive and negative charges on peptide hormones prevent them from passing through the lipid bilayer of a plasma membrane. Thus, peptide hormones must send messages from outside the cell.

Second messengers carry information into the cytoplasm

Since peptide hormones cannot enter the cell, some other molecule must carry the message from the cell membrane into the cytoplasm. The peptide hormone acts as the first messenger, carrying the message from the endocrine gland to the cell surface. The molecule that carries the information into the cytoplasm is called a **second messenger**.

One of the most common second messengers is cyclic AMP. Cyclic AMP is made from ATP by an enzyme that removes two phosphate groups, forming AMP. The ends of the AMP join, forming a circle. Even though many hormones use cyclic AMP as a second messenger, these hormones cause many different effects in target cells. How? Each target cell has different enzymes in its cytoplasm that are activated by cyclic AMP. **Figure 29.8** shows how the peptide hormone glucagon (*GLOO kuh gahn*) signals liver cells to release glucose into the blood.

Second messengers are hormone amplifiers

A single hormone molecule binding to a receptor in the plasma membrane can result in the formation of many second messengers in the cytoplasm. Each second messenger, in turn, can activate many molecules of a certain enzyme. Sometimes this enzyme activates another enzyme, enabling each hormone molecule to have a tremendous effect inside a cell, even though it never actually enters the cell.

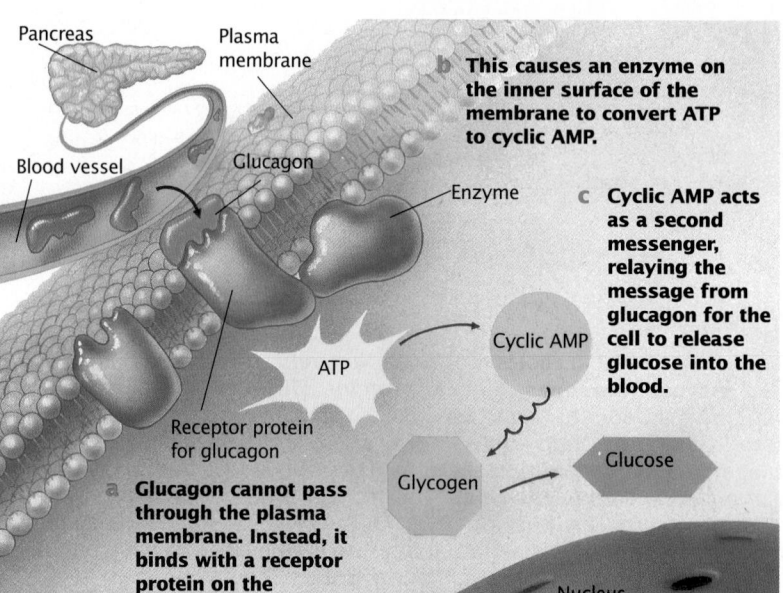

b This causes an enzyme on the inner surface of the membrane to convert ATP to cyclic AMP.

c Cyclic AMP acts as a second messenger, relaying the message from glucagon for the cell to release glucose into the blood.

a Glucagon cannot pass through the plasma membrane. Instead, it binds with a receptor protein on the cell surface.

Labels: Pancreas, Plasma membrane, Blood vessel, Glucagon, Enzyme, Cyclic AMP, ATP, Receptor protein for glucagon, Glycogen, Glucose, Nucleus

Section Review

❶ How does a receptor protein respond to a hormone molecule?

❷ Name the two ways hormones affect enzymes.

❸ How does the hormone estrogen deliver a message to a target cell?

❹ How does the hormone glucagon deliver a message to a target cell?

■ *Section Review Answers* ■

1. See Figure 29.6 on page 674.
2. Students should suggest that hormones change the activity or amount of enzymes involved in chemical reactions within the cell.
3. See Figure 29.7 on page 675.
4. See Figure 29.8 on page 676.

YOUR BODY FUNCTIONS DEPEND ON THE DELICATE INTERACTION BETWEEN YOUR ENDOCRINE SYSTEM AND YOUR BODY TISSUES. THE ENDOCRINE SYSTEM ORCHESTRATES SEVERAL PROCESSES NEEDED BY THE BODY TO COPE WITH STRESS, TO REGULATE METABOLISM, TO CONTROL BLOOD SUGAR LEVELS, TO CARRY ON RESPIRATION, AND EVEN TO REPRODUCE.

29.3 Glands and Their Functions

Objectives

❶ **Name** three endocrine glands and list the hormones each produces.

❷ **Explain** the effects of producing too much or too little thyroxine.

❸ **Describe** diabetes and explain how the condition can be treated.

❹ **List** two effects prostaglandins have on cells.

The Adrenal Glands

Eight different endocrine glands in the body produce dozens of different hormones. More than two dozen hormones are produced by the **adrenal glands**. The adrenal glands are two almond-sized glands located on top of the kidneys. Each adrenal gland is two endocrine glands in one. The inner part of each adrenal gland is called the adrenal medulla, and the outer part is called the adrenal cortex.

Figure 29.9
This is a famous movie scene. But if it were a real, life-threatening emergency, the adrenal medulla would secrete the hormone epinephrine. This hormone speeds up the heart rate and increases blood flow to the muscles.

The adrenal medulla helps the body react to a sudden crisis

The adrenal medulla is different from other glands because of the signal that activates it. Rather than being activated by hormones, the adrenal medulla is stimulated to release its hormones by nerves that run from the hypothalamus directly to the adrenal glands.

The adrenal medulla produces two hormones—epinephrine (*ep uh NEF rihn*), or adrenaline, and norepinephrine. These hormones produce what is called the "fight-or-flight" reaction. They are secreted in response to sudden stresses such as fear, anger, pain, or physical exertion. In a fraction of a second, the adrenal medulla can respond to any emergency that may arise, such as the scenario in **Figure 29.9**. The heart beats faster and blood flow increases to the heart and muscle cells. At the same time, air passages in the lungs relax and more oxygen is delivered throughout the body. For this reason, epinephrine is sometimes administered to people with asthma, a condition in which air passages in the lungs swell, making breathing difficult. During an acute asthma attack, epinephrine can help relax the swollen air passages.

Lesson Plan 29.3

Phase 1
PREPARATION

Key Concepts
- The adrenal glands produce hormones that help the body react to sudden crises and deal with long-term stress.
- The thyroid gland produces thyroxin, a hormone that regulates the body's metabolism.
- Too much thyroxine results in an increased metabolic rate, leading to a condition known as Graves' disease.
- Too little thyroxine results in a decreased metabolic rate, leading to a condition known as goiter.
- The pancreas produces two hormones, insulin and glucagon, which regulate blood glucose levels.
- Other hormones include the sex hormones produced by the gonads, melatonin produced by the pineal gland, and prostaglandins produced by all cells in the body.

Reading Strategy

Have students practice how to outline their reading. Outlining will be most helpful in this section, where many details are presented about how hormones affect body structure and function. Check the suggestions for outlining that were given in the Reading Strategy for Section 10.1 on page 199.

Phase 2
TEACHING STRATEGIES

Visual Strategy

Figure 29.9
Refer students to Figure 29.1 on page 669. Have students locate the adrenal glands in this figure.

Figure 29.10
Have the class discuss how this figure illustrates the close connection between the nervous and endocrine systems.

Demonstration 1

The Bearded Lady in the Circus
Show students a photograph of such a woman. Point out that the adrenal cortex is also the source of male sex hormones. A tumor in the adrenal cortex may result in the increased production of these sex hormones, enhancing the appearance of such male secondary sex characteristics as facial hair.

Theme Answer

Scale and Structure
Students should indicate that the adrenal gland has two hemispheres, the adrenal medulla and adrenal cortex, which secrete different hormones to perform various functions.

Figure 29.10
In 1989, hurricane Hugo ripped through the island of St. Croix, leaving many people without water. As these hurricane victims can tell you, standing in a long line for many hours can be stressful. The diagram to the right illustrates how the endocrine system releases hormones to help the body deal with such situations.

Scale and Structure

Why are the adrenal glands considered two separate glands?

The adrenal cortex helps the body deal with long-term stress

While the effects of epinephrine and norepinephrine wear off within a few minutes, hormones produced by the adrenal cortex enable the body to handle stress for hours or even days. Cortisol (KAWRT uh sawl) is one of the main hormones produced by the adrenal cortex. Cortisol increases the amount of energy available to the body by forming glucose from fats and proteins. Hormones released by the hypothalamus and the pituitary gland control the release of cortisol. For instance, standing in line for many hours can be stressful. The endocrine pathway in **Figure 29.10** shows how your body is supplied with energy during such circumstances. Even when you are not under stress, the level of cortisol fluctuates during the day. This is because you need different amounts of energy for different activities.

Cortisone is a compound that is similar to cortisol. Cortisone is produced synthetically and is used as a drug for the treatment of arthritis and a variety of other diseases. The prolonged use of cortisone, however, can be hazardous to your health. Possible side effects include the loss of resistance to infections, loss of bone and muscle protein, poor wound healing, and excess fat deposits, especially in the face. Because of these side effects, cortisol and other drugs chemically similar to cortisone are usually prescribed in the lowest effective dosages and are used only for short periods.

The Thyroid Gland

The thyroid gland is located in the neck, just below the Adam's apple. As you have read, the thyroid gland releases thyroxine, which regulates the body's metabolic rate. Thyroxine is also necessary for normal growth and development of the brain, bones, and muscles during childhood. In addition, thyroxine maintains a normal heart rate and affects reproductive functions.

Too much thyroxine causes Graves' disease

Some common disorders of the endocrine system are caused by problems with the thyroid. If too much thyroxine is produced, the condition that results is called **hyperthyroidism**, or Graves' disease. People with Graves' disease have a rapid, irregular heart rate, feel very nervous, and lose weight. Olympic gold medalist Gail Devers, pictured in **Figure 29.11**, began to experience the symptoms of Graves' disease just prior to the 1988 Summer Olympics. Concern about an irregular heartbeat also led to the diagnosis of Graves' disease in former President George Bush in 1991. Most people with Graves' disease can lead normal, productive lives once they are treated.

Graves' disease seems to be caused by a malfunction of the immune system. Recall that hormones work by binding to receptor proteins, which causes changes in the activity of the cell. The binding of a hormone to receptor proteins in thyroid cells causes the thyroid gland to release thyroxine. In a person with Graves' disease, however, antibodies made by the immune system fit into these receptor proteins instead. As a result, even though too much thyroxine already is present in the body, the thyroid gland still releases more.

Graves' disease is usually treated with a dose of radioactive iodine. The iodine collects in the thyroid gland and, over the course of several months, slowly destroys thyroid tissue. It is hard to control just how much of the gland is destroyed, however, so the patient often makes too little thyroxine after the treatment. Thyroxine levels are then brought up to normal by taking pills containing the hormone.

Figure 29.11
In 1988, American sprinter Gail Devers experienced severe weight loss, uncontrollable shaking fits, and loss of vision in her left eye. Nearly two years later she was diagnosed with Graves' disease, a condition in which too much thyroxine is produced by the thyroid gland. With proper treatment, however, Devers went on to win the gold medal in the women's 100-meter dash in the 1992 Summer Olympics.

Visual Strategy

Figure 29.11
Have students explain why a person with Graves' disease is not likely to be overweight but rather is likely to have a thin body.

Demonstration 2
Bulging Eyes
Show the class a photograph of a person with pronounced bulging eyes. Point out that hyperthyroidism also causes fluid to accumulate behind the eyes, making them protrude.

679

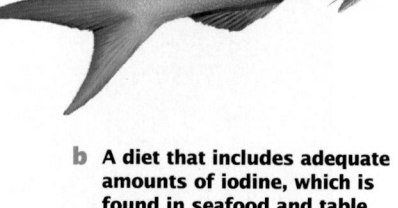

b   A diet that includes adequate
amounts of iodine, which is
found in seafood and table
salt, can prevent a goiter
from developing.

Figure 29.12
a   The lack of
iodine in this
woman's diet
has caused her
thyroid gland
to swell,
producing a
goiter.

Low levels of thyroxine cause hypothyroidism

The production of too little thyroxine
is called **hypothyroidism**. As thyroid
cells make thyroxine, they attach three
or four iodine molecules to it. If not
enough iodine is present in the diet, the
thyroid cells cannot do this. The
"unfinished" thyroid hormone builds
up and the thyroid gland begins to
swell. A swollen thyroid gland is called
a **goiter** (GOY tuhr), as shown in Figure
29.12a.

Iodine is found in sea water and
collects in seaweed, fish, and other
marine organisms. People who live near
the ocean usually include fish and other
seafood in their diets and rarely have
thyroid problems caused by a lack of
iodine. However, goiters used to be
quite common in inland regions. The
midwestern part of the United States
was once called the "goiter belt." Nowa-
days, small amounts of iodine are
added to ordinary table salt. By including the kinds of food shown in Figure
29.12b, goiters can be prevented.

Some people who get plenty of
iodine in their diet still suffer from low
thyroid levels. Like people with Graves'
disease, these people are suffering from
a misdirected attack by antibodies
made by their own immune systems.
In this case, however, the antibodies
damage the thyroid gland. The damaged thyroid gland cannot produce
adequate thyroxine.

Symptoms of hypothyroidism include
dry skin, low energy levels, and feeling
cold. Since the thyroid gland controls
metabolic rate, and people with too
much thyroid hormone lose weight,
being overweight is sometimes attributed to an "underactive thyroid." However, even very low thyroxine levels
cause a weight gain of only 5 lbs. to
10 lbs., so being seriously overweight
is not caused by low thyroxine levels.

The Pancreas

The pancreas contains small clusters of hormone-secreting cells that secrete two hormones—insulin and glucagon—that help regulate how much glucose is dissolved in the blood. In addition to secreting hormones, the pancreas also secretes digestive juices into the small intestine.

Insulin lowers blood glucose levels

After you eat, enzymes break down the carbohydrates in the food into glucose and other simple sugars. These sugars pass through the walls of the small intestine into your bloodstream, causing glucose levels in your blood to rise. **Insulin** is a hormone that enables the cells of certain tissues to take in glucose molecules. The glucose molecules are then transported into the cytoplasm of the cells, where they can be used to supply the cell's energy needs or stored for later use. Taking glucose into cells has the effect of lowering glucose levels in the blood. Therefore, insulin helps keep glucose levels steady. Insulin also increases protein production and fat production inside cells.

Unlike insulin, glucagon raises the level of glucose in the blood. Glucagon is a hormone that helps convert glycogen stored in the liver into glucose when the body needs more energy. The homeostatic control of insulin and glucagon is shown in **Figure 29.13**.

Figure 29.13
The hormones insulin and glucagon help maintain normal blood glucose levels, illustrated by the balanced "seesaws" in Steps 3 and 6 below. Levels of these hormones are controlled by negative feedback, as shown below.

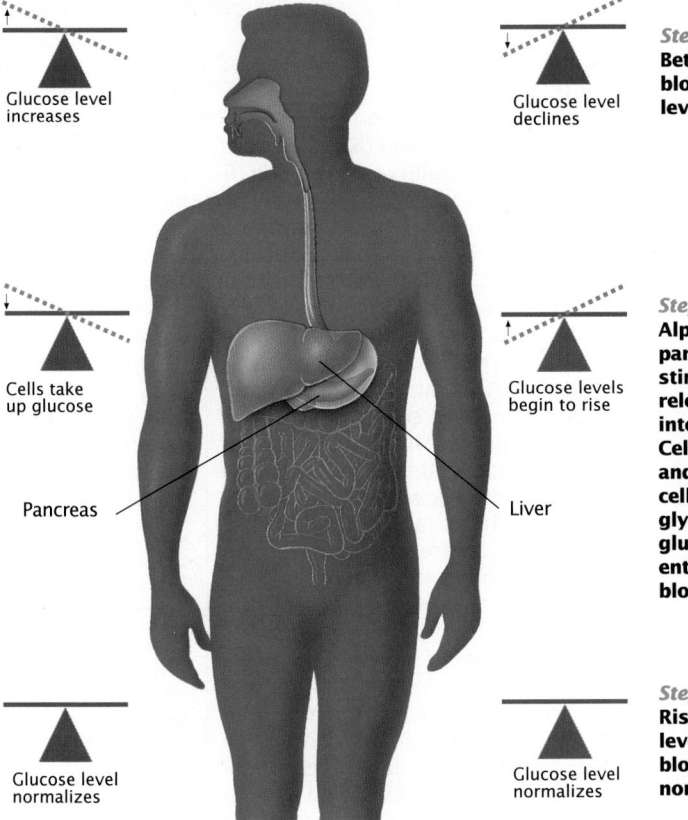

Step 1
After you eat a meal, glucose enters the bloodstream faster than cells can use it. Glucose levels rise.

Glucose level increases

Step 2
Beta cells of the pancreas are stimulated to release insulin into the blood. Cells take up glucose.

Cells take up glucose

Step 3
Cells of the liver and other body cells use glucose for energy or convert it to the polysaccharide glycogen. The level of glucose in the blood returns to normal.

Glucose level normalizes

Pancreas

Liver

Step 4
Between meals, blood glucose levels drop.

Glucose level declines

Step 5
Alpha cells of the pancreas are stimulated to release glucagon into the blood. Cells of the liver and other body cells convert glycogen to glucose, which enters the bloodstream.

Glucose levels begin to rise

Step 6
Rising glucose levels return blood sugar to its normal level.

Glucose level normalizes

Visual Strategy

Figure 29.14
Point out that injections or a pump must be used to administer insulin to a diabetic, since the hormone would be digested if taken by mouth. Scientists are currently working to develop a pill or nasal inhalant that would successfully deliver insulin into the bloodstream.

Visual Strategy

Table 29.1
Tell students that glucose serves as a cell's primary source of energy. Have them explain the connection between this fact and some of the symptoms of diabetes listed in this table.

Mathematics Connection

How Much Insulin?
Taking the correct dosage of insulin is vital to balance blood glucose levels. In normal individuals, the blood glucose level ranges between 80 and 90 mg per 100 mL of blood. Present a case in which a diabetic determines that her glucose level is 220 mg per 100 mL of blood. She knows that one "unit" of insulin will reduce her glucose level by 3 mg per 100 mL of blood. Have students calculate how many "units" of insulin she should take to bring her glucose to a normal level.

Connection: Chapter 8

Genetic Engineering
At one time, Type I diabetics depended on insulin extracted from the pancreas of cows, pigs, and sheep. Comparable—but not identical—to human insulin, this animal insulin was less than ideal, since diabetics often experienced unpleasant side effects. Today, diabetics depend on a synthetic human insulin produced by genetic engineering.

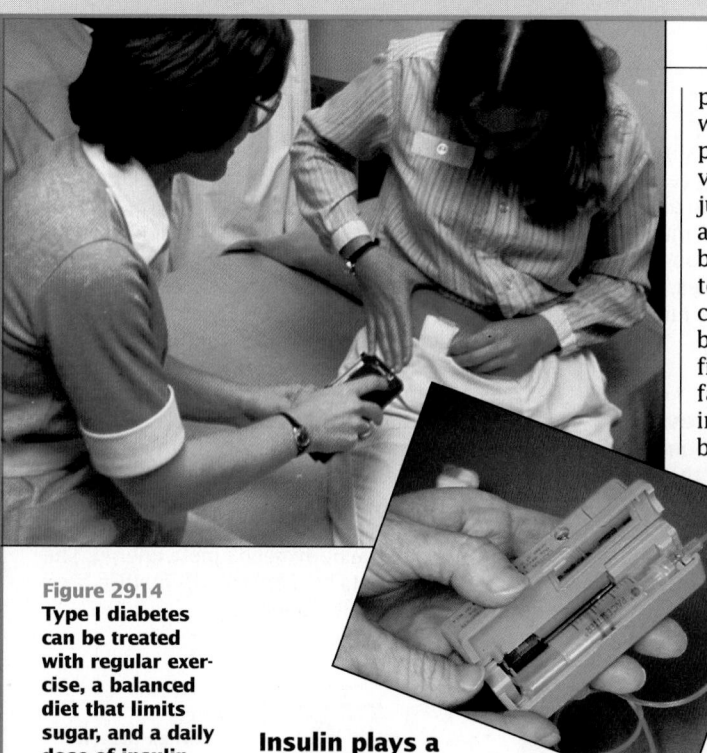

Figure 29.14
Type I diabetes can be treated with regular exercise, a balanced diet that limits sugar, and a daily dose of insulin that can be administered through an insulin pump (inset). The pump delivers a continuous, 24-hour supply of insulin through a tube that is inserted into the abdomen. The person can increase the insulin supply whenever necessary, such as before meals.

Insulin plays a role in diabetes
The inability of cells to take in glucose from blood and tissue fluids is called **diabetes mellitus** *(deye uh BEET eez muh LYT uhs)*, or simply diabetes. Even though the level of glucose in the blood is very high, some kinds of cells, such as muscle cells, will "starve" for lack of glucose. Lacking sugars, the starving cells break down proteins and fats for energy. The excessive breakdown of proteins can cause tissue damage. An even more serious situation is the buildup of dangerous levels of waste products that results from rapid fat breakdown. These toxic byproducts can disrupt heart rate, breathing, and brain activity. Severe, untreated diabetes can lead to coma and death.

There are two major types of diabetes. People who have Type I diabetes produce almost no insulin. Type I diabetes usually develops in childhood or early adolescence. Before the role of insulin was discovered in the 1920s, Type I diabetes was always fatal. Today, people with Type I diabetes can give themselves daily injections of insulin. Some diabetics use a device called an insulin pump to supply their bodies with the insulin they need every day. An insulin pump is shown in **Figure 29.14**. Even with these treatments, mimicking the precise control systems of the body is very difficult. Diabetics sometimes misjudge the amount of insulin they need and can suffer the effects of having blood sugar levels that are too high or too low. If too much insulin is given, cells take too much glucose out of the blood. The brain does not receive sufficient glucose, and disorientation and fainting may occur. Excessively high insulin levels can generally be treated by taking in sugar quickly, such as by drinking orange juice.

Type II diabetes usually occurs in overweight people who are over 40 years old. Diabetics with this form of the disease produce insulin, but cells do not take in enough glucose. This happens when there are not enough insulin receptors on the cells, when the receptors are somehow defective, or when the level of insulin production is simply not adequate.

Most people with Type II diabetes are obese, weighing at least 20 percent more than their ideal weight. Obesity can cause a decrease in insulin receptors. For this reason, weight control is an important part of the treatment for this form of diabetes. Since most Type II diabetics produce adequate insulin, treatment with insulin is rarely necessary. All diabetics, however, must carefully control their diets and get regular exercise to manage their disease. The symptoms of Type I and Type II diabetes are summarized in **Table 29.1**.

Table 29.1 Warning Signs of Diabetes

Type I	Type II
Frequent urination	Any of the Type I symptoms
Unusual thirst	
Extreme hunger	Frequent infections
Unusual weight loss	Blurred vision
Extreme fatigue	Cuts or bruises that are slow to heal
Irritability	Tingling or numbness in the hands or feet
	Recurring skin, gum, or bladder infections

Other Glands and Hormones

The body contains several other glands that produce important hormones. The hormones that are responsible for the development and maintenance of sexual characteristics are **sex hormones**. In males, the sex hormone testosterone *(tehs TAHS tuh rohn)* is produced by the testes. In females, the ovaries produce the sex hormones estrogen *(EHS truh jehn)* and progesterone *(proh JEHS tuh rohn)*. Sex hormones control the maturation of sperm or egg cells and affect sexual behavior.

The release of sex hormones is controlled by the hypothalamus and the pituitary gland. Males and females produce the same hormones in these glands, but the pituitary hormones cause the release of testosterone in males and estrogen and progesterone in females. You will learn more about the sex hormones in Chapter 34.

Anabolic steroids are a synthetic form of testosterone

Athletes who use anabolic steroids in an attempt to build bigger, stronger muscles are usually using some form of synthetic testosterone. As you learned in Chapter 27, steroid use poses serious health risks. Liver and kidney disorders and high blood pressure can occur. Women who use anabolic steroids may experience menstrual irregularities. Males may develop aggressive behavior, a decreased sperm count, and experience impotence.

A variety of other glands produce important hormones

The endocrine glands release many other hormones. The parathyroid glands, tiny glands located on the back of the thyroid gland, control calcium levels. The pineal *(py NEE uhl)* gland, a pea-sized gland near the base of the brain, secretes a hormone called melatonin. Melatonin appears to inhibit reproductive activities. The pineal gland stops functioning at about the time of puberty, an event that may be related to sexual maturation. The functions of these glands and others are summarized in **Table 29.2**.

Table 29.2 Endocrine Glands, Organs, and Their Functions

Source	Hormone	Main target(s)	Effect(s)
Pineal gland	Melatonin	Gonads (indirectly)	Inhibits reproductive activities; stops functioning at about onset of puberty
Parathyroid	Parathyroid hormone	Bone, kidney	Elevates calcium levels in blood
Thymus	Thymosin	Lymphocytes	Plays role in immune response
Ovary	Estrogen	Female reproductive structures	Maturation of reproductive organs; influences egg maturation and release
	Progesterone	Uterus, breasts	Prepares and maintains uterine lining for pregnancy; stimulates breast development
Testis	Testosterone	Male reproductive structures	Maturation of reproductive organs; influences sperm formation

Connection: Chapter 2

Lipids

Prostaglandins are fatty acids, components of lipids. Remind students that lipids are high in carbon and hydrogen and are insoluble in water. Point out that prostaglandins are among the most powerful of all known chemicals in the body, since they are able to bring about dramatic effects in very small quantities.

Visual Strategy

Figure 29.15

Point out that aspirin manufacturers were at first unwilling to label their products with a warning about Reye's syndrome. Their position, at the time, was that no scientific evidence existed to support the claim that a child who had the flu and who took aspirin might develop Reye's syndrome. Recent studies have reported that more than 1,000 children have died as a result of taking aspirin while suffering from flu.

Phase 3

ASSESSMENT OPTIONS

Closure Strategy

Hormones

Assign students to cooperative groups of four. Have each group select two hormones and specifically describe how too much or too little of each would affect body structure or function.

Section Review

Assign the *Section Review*.

Reteaching

Assign students to cooperative groups of four. Have each group develop a concept map for this section.

Figure 29.15
Teenagers are warned not to take aspirin. It has been associated with Reye's syndrome, a potentially fatal disease.

Prostaglandins may cause fever and pain

A group of lipids called **prostaglandins** *(pras tuh GLAN dihnz)* also function in the body as hormones. Prostaglandins are produced in small quantities by specialized cells in many parts of the body. They act locally rather than through blood transport.

There are dozens of different kinds of prostaglandins, which have many different effects on the cells in their immediate vicinity. Prostaglandins help blood to clot and can raise blood pressure by causing the smooth muscle surrounding blood vessels to constrict. They also cause fever and increase pain during inflammation. The uterus is sensitive to prostaglandins. A rise in prostaglandin levels is an important part of the events that lead to labor (the forceful contractions of the uterus during childbirth). High levels of prostaglandins in the uterus can also cause menstrual cramps.

Aspirin and ibuprofen interfere with the activities of prostaglandins, which is why these drugs are so effective in treating menstrual cramps, headaches, muscle pains, and inflammatory diseases such as arthritis. However, teenage girls should not take aspirin for menstrual cramps or any other reason. The use of aspirin by children and teenagers is associated with Reye's syndrome, a potentially fatal disease of the liver and central nervous system. Although many people are aware that young children should not take aspirin, they do not know that most deaths from Reye's syndrome are among teenagers. Warnings about Reye's syndrome appear on aspirin bottles, as shown in **Figure 29.15**.

Section Review

1. **Name three major endocrine glands in your body and the hormones they produce.**

2. **What occurs if the thyroid gland produces too much or too little thyroxine?**

3. **Distinguish between Type I and Type II diabetes.**

4. **Why are aspirin and ibuprofen effective in treating headaches?**

■ *Section Review Answers* ■

1. The adrenal gland produces epinephrine, norepinephrine, and cortisol; the thyroid gland produces thyroxine; the pancreas produces insulin and glucagon.

2. Too much thyroxine results in Graves' disease, whereas too little thyroxine results in a goiter, and slowly destroys the thyroid gland.

3. Type 1 diabetics produce almost no insulin, causing some cells to starve for glucose. In Type II diabetes, insulin is produced but the cells do not take in enough glucose, because insulin receptors on the cells are defective or not enough insulin is produced.

4. Aspirin and ibuprofen interfere with the activities of prostaglandins.

Chapter 29 *Highlights*

When the body produces too little growth hormone, dwarfism can result. When too much growth hormone is produced, gigantism can occur.

Alternative Assessment

Have each student prepare a blank Bingo card with 25 squares, labeling the center one "FREE." Write 24 terms, names, or short phrases from this chapter on the board. Then instruct each student to randomly place each term in one of the squares on their Bingo card. Instruct students to cross out the appropriate box each time a definition is read or a clue is provided. Have students yell "Bingo!" when either a horizontal or vertical line has been completed.

	Key Terms	Summary
29.1 What Hormones Do  In a negative feedback process, thyroxine speeds up metabolism, keeping body temperature constant.	hormone (p. 669) endocrine gland (p. 669) target (p. 669) neuroendocrine system (p. 670) pituitary gland (p. 671) negative feedback (p. 672)	• Endocrine glands secrete hormones into the bloodstream. • The effects of the endocrine system are similar to those of the nervous system, but these effects usually occur more slowly and last longer. • The brain is directly connected to the endocrine system by the hypothalamus, which helps regulate the body's internal environment through the pituitary gland. • A negative feedback system turns off the supply of hormones when they are not needed.
29.2 How Hormones Work 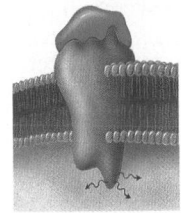 Hormones bind with receptors like keys fitting into locks.	steroid hormone (p. 675) peptide hormone (p. 676) second messenger (p. 676)	• Hormones travel throughout the body in the bloodstream. They affect only the appropriate target cells. • Steroid hormones pass through the cell membrane. Peptide hormones act from outside the cell by means of second messengers. • Hormones act by affecting kinds or amounts of enzymes produced.
29.3 Glands and Their Functions  If the thyroid gland produces too much thyroxine, a treatable condition called Graves' disease can result.	adrenal gland (p. 677) hyperthyroidism (p. 679) hypothyroidism (p. 680) goiter (p. 680) insulin (p. 681) diabetes mellitus (p. 682) sex hormone (p. 683) prostaglandin (p. 684)	• Eight endocrine glands produce dozens of hormones in the human body. • The adrenal medulla helps the body react quickly to crises. The adrenal cortex makes steroids that control long-term stress and regulate energy levels. • The thyroid gland regulates metabolism. The pancreas makes digestive enzymes but also regulates blood sugar by releasing insulin. • Other hormones include the sex hormones made in the testes and ovaries, and prostaglandins.

Chapter Review Answers

Understanding Vocabulary
1. a. steroid hormones
    b. hormones
    c. adrenal cortex
    d. diabetes

Relating Concepts
2. Map answer is shown on page 667D.

Understanding Concepts

Multiple Choice

3.	b	7.	c
4.	d	8.	b
5.	d	9.	b
6.	b		

Completion
10. bloodstream
11. neuroendocrine system
12. pituitary gland
13. testes, ovaries

Short Answer
14. They function in the cells in which they are produced.
15. The pituitary gland secretes hormones that affect many body functions directly and indirectly by controlling the activities of other glands.
16. Production of too much thyroxine by the thyroid gland causes the disease. Its symptoms include irregular heart rate, weight loss, and a nervous feeling. Treatment usually involves giving the patient a dose of radioactive iodine to destroy thyroid tissue, followed by daily doses of thyroxine, typically in pill form.
17. Oxytocin causes contractions of the uterus during labor, and vasopressin causes the kidneys to form a concentrated urine.

Understanding Vocabulary

1. For each set of terms, complete the analogy.
    a. polar molecules:peptide hormones::nonpolar molecules: _____
    b. short-term effects:nerve impulses::long-term effects: _____
    c. epinephrine:adrenal medulla:: cortisol: _____
    d. thyroxine:Graves' disease::insulin: _____

Relating Concepts

2. Copy the unfinished concept map onto a sheet of paper. Then complete the concept map by writing the correct word or phrase in each oval containing a question mark.

Understanding Concepts

Multiple Choice

3. A hormone is a chemical signal that
    a. binds to voltage-sensitive channels.
    b. regulates growth, metabolism, or reproduction.
    c. enables skeletal muscles to contract.
    d. carries electrical signals throughout the body.

4. Hormones travel through the bloodstream to an organ or tissue called a
    a. goiter.
    b. second messenger.
    c. prostaglandin.
    d. target.

5. Which part of the nervous system is also considered to be part of the endocrine system?
    a. pituitary gland
    b. adrenal medulla
    c. cerebrum
    d. hypothalamus

6. An example of a hormone that passes through the plasma membrane and affects cell activity is
    a. glucagon.
    b. estrogen.
    c. cyclic AMP.
    d. glycogen.

7. Insulin and glucagon are two hormones secreted by the pancreas that
    a. speed up heart rate.
    b. cause contractions of the uterus.
    c. regulate metabolism of glucose.
    d. stimulate growth of the body.

8. Which gland is responsible for the dramatic effects felt in response to sudden stresses such as fear or anger?
    a. adrenal medulla   c. kidneys
    b. adrenal cortex   d. thyroid

9. Testosterone and estrogen are two hormones that
    a. break down fats and proteins.
    b. are responsible for the development and maintenance of sexual characteristics.
    c. are secreted by the pineal gland.
    d. release calcium from bone.

Completion

10. Endocrine glands secrete hormones into the _____ , which carries the signal to the target organ or tissue.

11. Because the nervous system and endocrine system are so closely linked, they are often considered a single system called the _____ _____ .

12. All hormones produced by the hypothalamus move through a slender thread of tissue to the _____ _____ .

13. The _____ and the _____ are the glands that secrete sex hormones.

Short Answer

14. Why might prostaglandins be called "local" hormones?

15. Explain why the pituitary gland is sometimes called the "master gland."

16. What causes Graves' disease? What are its symptoms and how is it treated?

17. Name two hormones that are produced by the hypothalamus and stored in the pituitary gland. What are their functions?

Interpreting Graphics

18. Study the diagram below and answer the questions that follow.

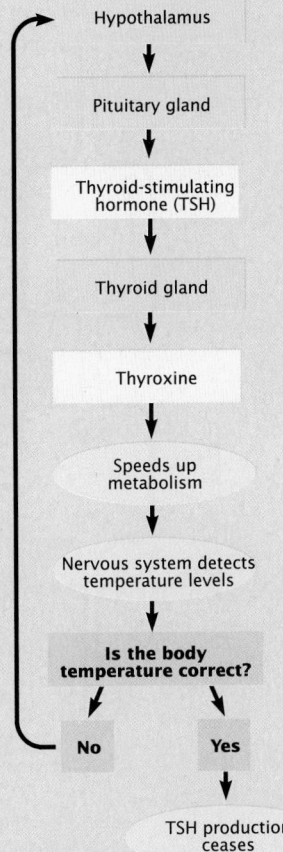

- What process controls thyroxine production?
- Which organ monitors body temperature?
- Which organ produces a hormone that stimulates the thyroid gland?
- What hormone speeds up metabolism?
- What happens when the body temperature is correct?

Reviewing Themes

19. *Patterns of Change*
Although goiters were common in the United States years ago, few people today suffer from them. Why?

20. *Stability*
What are two ways that the hypothalamus can help the body maintain homeostasis?

Thinking Critically

21. *Comparing and Contrasting*
How do polypeptide hormones and steroid hormones differ in the way they affect the operations of their target cells?

22. *Building on What You Have Learned*
In Chapter 28, you learned about neurotransmitters. How are the actions of hormones and neurotransmitters alike? How are they different?

Cross-Discipline Connection

23. *Biology and Language*
A form of Japanese poetry called haiku has 3 lines with 5 syllables in the first line, 7 syllables in the second line, and 5 syllables in the third line. Write a haiku using what you know about the endocrine system.

Discovering Through Reading

24. Read the article "The Long and Short of It," in *The Economist*, July 11, 1992, pages 79–81. What physical characteristic is the target of human growth hormone therapy? What physical characteristic is the target of sex hormone therapy? What are the risks involved with both these treatments?

Interpreting Graphics

18. • negative feedback
    • hypothalamus
    • pituitary gland
    • thyroxine
    • TSH ceases production

Reviewing Themes

19. The iodine needed for the production of thyroxine is added to table salt. Many people use table salt regularly.

20. It can send out commands in the form of nerve signals or hormones.

Thinking Critically

21. Polypeptide hormones bind to receptors on the cell membrane; they do not enter their target cells. The binding of the hormones to the receptors activates enzymes attached to the receptor site inside the cell membrane. The enzymes function as second messengers. In contrast, steroid hormones go through the cell membrane and combine with receptor molecules in the cytoplasm.

22. Both are released by a cell and change the behavior of another cell. They are different, in that hormones carry a message over great distances in the body while neurotransmitters carry a message from one cell to a second cell that nearly touches the first.

Cross-Discipline Connection

23. Any haiku that includes information about the endocrine system is acceptable.

Discovering Through Reading

24. height; to halt excessive tallness; possible answers might include tumors, rapid growth of organs affected by puberty, emotional and physical trauma.

Procedural Note

Discuss the preparation questions. Be sure that the students have an understanding of an endocrine gland, target organ, hormone, or negative feedback control. While this investigation can be completed individually, a team effort is likely to provide better results.

Prelab Preparation Answers

- Endocrine glands make most of the body's hormones.
- A hormone is a chemical signal—made in one place and delivered to another—that regulates the body's activities.
- A target is a specific tissue or organ from which a hormone will elicit a specific response.
- bloodstream and nerves

Procedure Answers

3. A, E
4. blood vessels, nerves
5. 50 minutes
6. Organ 1, Organ 3, Organ 2
7. Students should suggest that Substance B triggers the organ that makes Substance D.
8. from the bloodstream or the nerves
9. Students should indicate that the bloodstream is sending messages to produce Substance C.
10. Substance D, pinch off the bloodstream and leave the nerves open.

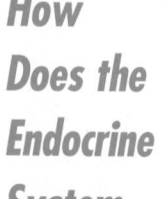

How Does the Endocrine System Work?

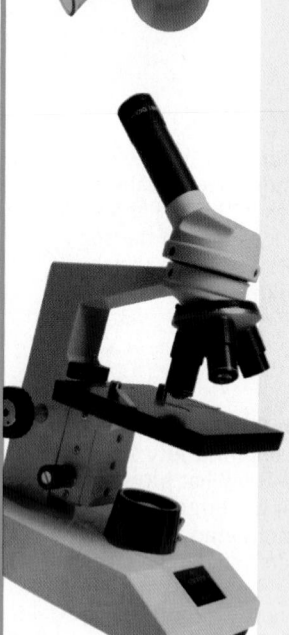

Objectives

In this investigation you will:
- *describe* how hormones are regulated
- *relate* endocrine glands and hormones to their target organs
- *interpret* a diagram detailing cause-and-effect processes

Materials

- unlined paper
- colored pencils

Prelab Preparation

Review what you have learned about the endocrine system by answering the following questions:
- What do endocrine glands do in the body?
- What is a hormone?
- What is a target organ?
- Describe two ways that messages can be carried from one tissue or organ to another in the body.

Procedure: Analyzing Endocrine Function

1. Form a cooperative group of 2–4 students to perform steps 2–10.
2. The figure below represents an organ system that produces Substance E from Substance A. Locate the three organs in the figure.

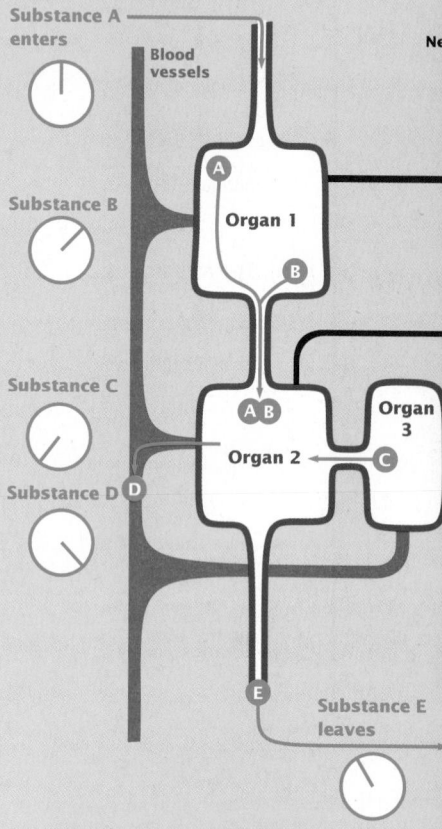

3. Each organ is capable of producing substance B, C, D, or E. *Which substance enters the organ system at the top of the figure? Which substance leaves the organ system at the bottom of the figure?*

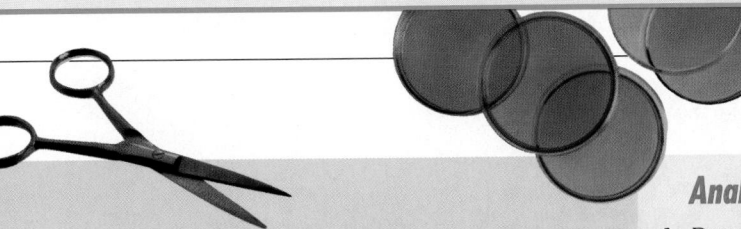

4. In order to function properly and maintain homeostasis, the organ system must be able to send and receive information. *Which part of the figure would enable an organ to send and receive chemical messages? Which part of the organ system would enable it to send and receive electrical messages?*

5. The clock faces throughout the figure show a minute hand that represents the approximate time each substance is present in the organ system. When Substance A enters the organ system, the minute hand is at twelve o'clock. *How many minutes later does Substance E leave the system?*

6. Study the figure. *Which organ produces Substance B? Which organ produces Substance C? Which organ produces Substance D?*

7. A scientist studying this organ system notes that when Substance A and Substance B are both present in Organ 2, the organ produces Substance D. To figure out whether Substance A or Substance B triggers the organ to produce Substance D, the scientist injects each substance separately into Organ 2. When Substance A is injected, Substance D is not detected in the organ system. When Substance B is injected, the amount of Substance D in the blood vessel leaving Organ 2 increases. *What conclusions can you draw from these observations?*

8. When Substance B is present in Organ 2, Organ 3 produces Substance C. Describe two possible ways that Organ 3 could receive the message directing it to produce Substance C.

9. The scientist studying the organ system decides to test the hypothesis that the nerve from Organ 2 sends the message to Organ 3 to produce Substance C. The scientist blocks the nerve and notes that Organ 3 continues to produce Substance C. *What conclusion can you draw from this observation?*

10. The scientist's second hypothesis is that Organ 3 receives its message through the bloodstream. *Which substance would be the messenger that directs Organ 3 to release Substance C? Describe an experiment the scientist could perform to test the hypothesis.*

Analysis

1. *Drawing Conclusions*
   Which substance, or substances, in the organ system is a hormone? Explain your answer.

2. *Summarizing Observations*
   Explain how the organ system described in this investigation provides an example of how endocrine glands function.

3. *Applying Concepts*
   Identify one of each of the following in this system: endocrine gland, target organ, and hormone.

Thinking Critically

1. What would happen if Organ 3 were malfunctioning and failed to produce Substance C?

2. Which substance acts as a switch to stop the production of Substance D and eventually Substance C?

Analysis Answers

1. Substance D, hormones are chemical messengers that travel through the bloodstream.
2. Students should explain negative feedback.
3. Organ 2, Organ 3, Substance D

Thinking Critically Answers

1. Without substance C, substance E would probably not be produced.
2. Substance B

Chapter 30

Drugs and the Nervous System

Planning Guide

	Objectives/Themes	Classwork Resources	Homework Resources
30.1	**1.** Discuss information that research has yielded about psychoactive drugs.	**Teacher's Resource Binder** Focus Activity 30 *Collecting Data Through a Survey* Lab Investigation 30.1 *How Drugs Affect the Heartbeat Rate*	**Text** Section Review, p. 694
	2. Summarize how drugs affect the nervous system.		**Teacher's Resource Binder** Directed Reading Worksheet 30.1
	3. Describe addiction in terms of physiology.		
	4. Explain the role of neuromodulators in nerve impulse transmission.	**Other Resources** Transparencies 134–135	**Other Resources** Audiocassette 30.1*
	Themes: Interacting Systems, Patterns of Change, Scale and Structure, Stability		
30.2	**1.** Explain how the body controls pain.	**Other Resources** Transparency 136	**Text** Section Review, p. 698
	2. Summarize why narcotics are addictive.		**Teacher's Resource Binder** Directed Reading Worksheet 30.2 Reaching Worksheet* *Psychoactive Drugs*
	3. Describe the action of cocaine at the synapse.		
	Themes: Interacting Systems, Patterns of Change, Stability		**Other Resources** Audiocassette 30.2*
30.3	**1.** Describe how nicotine affects the nervous system.	**Text** Investigation *How Does Smoking Affect the Air Passages?* pp. 708–709	**Text** Section Review, p. 704
	2. Describe the effects of alcohol on the nervous system.		**Teacher's Resource Binder** Directed Reading Worksheet 30.3 Vocabulary Review Worksheet*
	3. Describe two dangers associated with alcohol abuse.	**Teacher's Resource Binder** Extension Worksheet *Tobacco*	
	4. Summarize the effects of hallucinogens on the nervous system.	**Other Resources** Transparency 137	**Other Resources** Audiocassette 30.3*
	Themes: Interacting Systems, Patterns of Change		

*Reteaching Options

Demonstrations
30.1: pp. 692, 693
30.2: pp. 695
30.3: pp. 700, 701, 702, 703

Assessment
Chapter Review pp. 706–707
Portfolio Assessment p. 689D
Chapter Test—Teacher's Resource Binder
Test Generator

Research Notes

Connection to Genetics: Heredity and Alcoholism

Kenneth Blum of the University of Texas at San Antonio and Ernest P. Noble of the University of California at Los Angeles sparked a controversy in 1990 when they said their study indicated a statistically significant link between alcoholism and a single gene coding for dopamine receptors on the surface of brain cells. Dopamine is a neurotransmitter known for evoking pleasure and other sensations.

Blum and Noble's results indicate that one allele (A1) of the D2 gene is linked to alcoholism. Another study also indicates a link between the A1 allele and alcoholism, as well as other disorders such as Tourette's syndrome, attention-deficit disorder, and autism.

Blum and Noble believe that since people with the A1 allele tend to have fewer dopamine receptors on their brain cells, such people tend to overindulge to compensate for this deficiency in absorbing pleasure-inducing dopamine. Critics counter that the decreased numbers of receptors could be an effect of alcoholism, rather than a cause of it.

However, several other scientists have found no significant association between the A1 allele and alcoholism in their investigations, especially when comparing alcoholics to others within the same ethnic group. Critics of Blum and Noble's work complain that since they did not make similar comparisons, they haven't controlled for variables such as ethnic group.

These researchers say Blum and Noble's focus on severely ill alcoholics may mean that the study's results are actually indicators of illnesses that alcohol can intensify, rather than of alcoholism itself. Others point out that the A1 allele is far from the part of the D2 gene that actually codes for the proteins in the dopamine receptor. Thus far, there is no direct molecular evidence that the A1 allele affects the functioning of the dopamine receptor gene.

Proponents of Blum and Noble's theory reply that the critics did not keep alcoholics out of their control groups and neglected alcoholics with severe medical problems, so that the statistical significance of the genetic differences could no longer be detected. They also claim that unpublished studies with other markers closer to the essential part of the D2 gene still show the same relationship.

Investigation Notes

How Does Smoking Affect the Air Passages?
pages 708–709

Purpose: This investigation will concentrate on the tars in cigarette smoke. The investigation is designed to give students a clear example of how the lungs and bronchial tubes of a heavy smoker might actually look. While nicotine acts on the nervous system, it is the tars in cigarette smoke that cause genetic changes that can cause lung cancer.

Prelab Preparation

1. A pre-assembled apparatus is available from WARD'S (15 M 0508).

2. Before assembling the apparatus, be sure to lubricate both ends of the glass tubing before inserting it into the stopper. Avoid cutting your hands by using a towel to hold both the tubing and the stopper during insertion.

3. After inserting the glass tubing through the rubber stopper, cut off the closed end of the rubber top of the medicine dropper. Use a short piece of rubber tubing to connect this end of the medicine dropper top to the glass tubing.

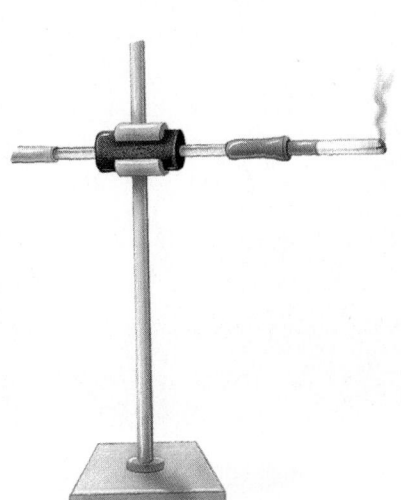

Put the cigarette into the other end of the rubber top of the medicine dropper.

4. Use another short piece of rubber tubing to connect the aspirator to the open end of the glass tubing.

5. Each team should use two or three cigarettes to perform the investigation. The investigation can also be approached as an experiment by comparing filtered and non-filtered cigarettes or low-tar and regular brands.

6. The equipment should be used either under a fume hood or near an open window to provide adequate ventilation.

7. Be certain that there are no flammable materials near the lab activity.

Answers will be found on pages 708–709.

Meeting Individual Needs

Objectives

1. Students will demonstrate appropriate use of core vocabulary for the chapter (Vocabulary File).

2. Students will compare the types of advertising targeted to different groups and distinguish patterns (Multicultural Lesson Plan).

Vocabulary File
(Developing Vocabulary/ Limited English Proficiency)

If you are not already using the Vocabulary File, refer to Chapter 1 for its preparation. See the Chapter Highlights on page 705 for a list of suggested words.

Multicultural Lesson Plan
Tobacco and Alcohol Ads (Dealing with Ethnic Diversity)
Preparation:

1. Prepare a collection of magazines targeted to specific groups. Examples could include *Woman's Day, Cosmopolitan, Life, Hispanics, Jet, Black Enterprise, Ebony,* etc.

2. If possible, arrange a bus field trip or walking tour to view and count billboards for tobacco and alcohol in different neighborhoods of your town.

Teaching Strategies

1. If possible, invite a local speaker from the education department of your local Alcoholics Anonymous chapter to speak to the students about the dangers of alcohol abuse. They should take notes for the test during this talk. Be sure this speaker emphasizes the role that ads and other media messages can play in fostering alcoholism.

2. Another possibility for a guest speaker might be a representative of the American Heart or Lung Association or a local doctor who can talk about the dangers of tobacco abuse. This speaker should also emphasize the role that ads and other media messages can play in fostering addiction.

3. If no guest speakers are available, be sure students understand the physiological basis for addiction and for the diseases associated with tobacco use and alcoholism (emphysema, cirrhosis, etc.). Be sure students realize that there are no TV or radio ads for "hard" liquor and tobacco.

Assessment

1. As the students go on their class field trip or examine the magazines, they should keep track of the number and types of ads for tobacco and alcohol. Students should analyze each ad on a variety of levels. Questions like these should be addressed: Who is the target of this ad? How and why should it appeal to the target person? What will this ad make them want to do? Why does the advertiser want them to do this?

2. Students should compile their findings into a report about such ads. The report should point out underlying patterns in the ads and their placement, so that the students are viewing them from the perspective of the advertiser as well as the audience.

Additional Strategies
Visual Strategies

Pages 692, 694, 695, 696, 697, 699, 700, 702, and 703

Auditory Learners

Use *Biology: Visualizing Life* Audiocassettes for Sections 30.1, 30.2, and 30.3.

Meeting Individual Needs (cont.)

Cooperative Learning
Dangers of Common Drugs

Timing: Use this activity to introduce Section 30.1.

Group Size: 2–3 students (the class should be divided into 11 groups)

Outcome: Students will be able to describe the effects and dangers of psychoactive drugs.

Individual Accountability: Each group member is responsible for researching the effects of a specific psychoactive group.

Positive Interdependence: Each group will prepare a report on the effects of a specific psychoactive drug.

Assign each group one of the following substances to research: (1) alcohol, (2) barbiturates, (3) cocaine, (4) methamphetamines, (5) inhalants, (6) heroin, (7) LSD, (8) tobacco, (9) marijuana, (10) PCP, and (11) "designer drugs."

Each group should research the following topics: immediate effects on the body, psychoactive effects, long-term effects, effects on unborn children, addictiveness, legal status, and areas of origin.

When students complete their research they should work with their group to pool their findings into a finished report to be presented to the class.

Portfolio Assessment

Students should select their best work and provide a self-reflective rationale for their selections. Students can make selections in the following areas.

1. *Content* — One concept map from the chapter (See page 812 for evaluation criteria.)

2. *Reading Comprehension* — One Directed Reading Worksheet from the Teacher's Resource Binder (Use the answer key to evaluate for accuracy.)

3. *Writing* — Using the Vee Form, summarize a magazine or newspaper article relating to drugs or drug abuse. (See page 22T for evaluation criteria.)

   Or: Select a writing task or project from the Chapter Review.

4. *Performance Assessment* — One Vee form from a chapter investigation or lab manual investigation (See page 22T for evaluation criteria.)

Teacher makes selections in the following areas.

1. *Formal Assessment* — Chapter test (Test A, B, or the Test Generator) The teacher-scored test should be reviewed by the student. Incorrect responses should be corrected by the student before the test becomes part of the portfolio.

2. *Informal Assessment* — Use the Direct Observations Checklist, page 33T, during a laboratory or other cooperative learning experience.

3. *Performance Assessment* — Have students write a newspaper editorial for or against the specific tactics used by the federal, state, or local government in its "war" on drugs.

Concept Map Answer

The following is one possible answer to the Relating Concepts exercise on page 702.

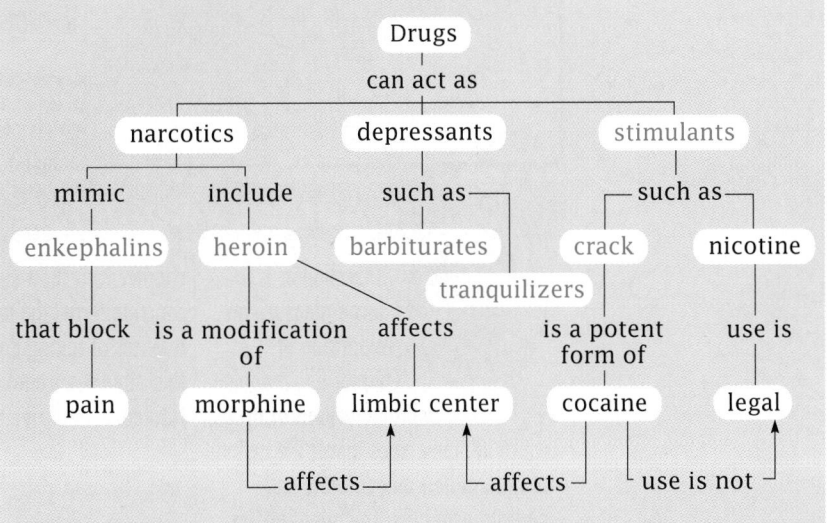

Chapter 30

Determining Prior Knowledge

- Have students list drugs, both legal and illegal, that are addictive. Have them explain how each affects the body.
- Have students explain the difference between drug use (taking a drug prescribed for a specific purpose and as directed), drug misuse (taking too much of a prescribed drug or for a purpose for which it is not intended), and drug abuse (taking a substance that is not intended to be taken into the body).
- Use a microprojector to show the class several *Daphnia*. Point out the heart and have students determine the number of beats per minute. Place a drop of alcohol on the slide, wait a minute or two, then have students determine the number of heartbeats per minute. Discuss how alcohol affects heartbeat rate. Ask how alcohol might affect the human circulatory system.

Chapter 30

Review

- protein receptors (Section 4.1)
- action potential (Section 28.1)
- neurotransmitters (Section 28.1)

Drugs and the Nervous System

30.1 Drugs and Addiction

- **What Are Psychoactive Drugs?**
- **How Psychoactive Drugs Affect Nerves**
- **The Nature of Addiction**

30.2 Narcotics and Cocaine

- **Narcotics**
- **How the Body Controls Pain**
- **Why Narcotics Are Addictive**
- **Action of Cocaine**

30.3 Dangerous Social Drugs

- **Nicotine**
- **Alcohol**
- **Barbiturates and Tranquilizers**
- **Amphetamines**
- **Hallucinogens and Marijuana**

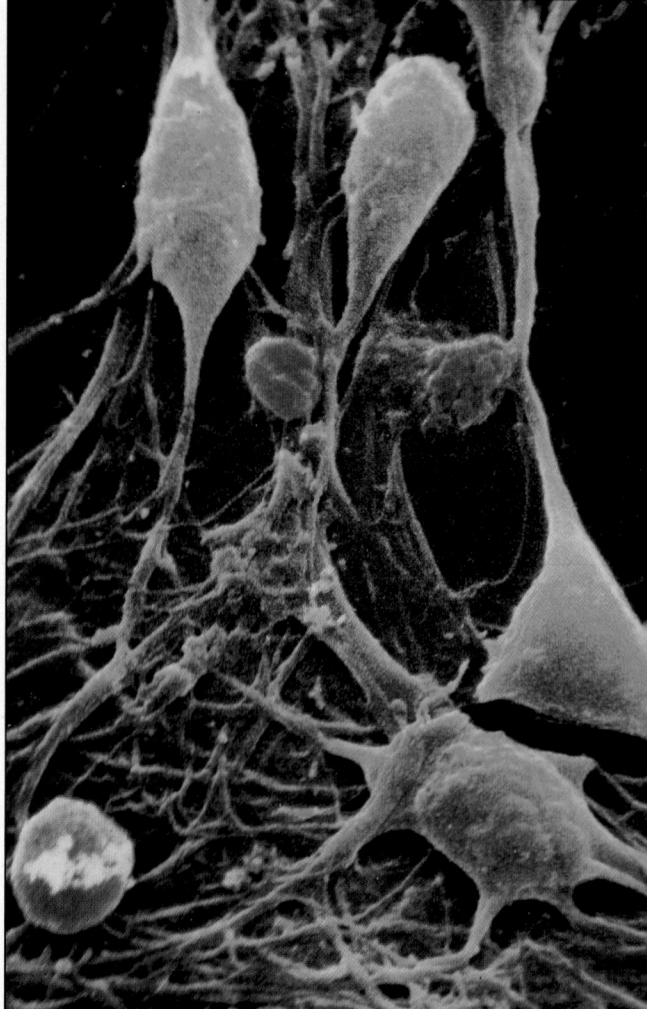

Nerve cells like these are the targets of psychoactive drugs. Drugs cause changes in the synapses between nerve cells, which can lead to addiction.

■ Author's Rationale ■

This chapter is in large measure an extension of Chapter 27. It deals with the actions of drugs, which in many cases mimic the action of natural neuromodulators and neurotransmitters. The intent of this chapter is to show students in very concrete terms the unavoidable consequences of taking dangerous drugs. In this context, alcohol and nicotine are treated as addictive drugs.

FEW SOCIAL PROBLEMS IN THIS COUNTRY HAVE A GREATER IMPACT ON PEOPLE'S LIVES THAN THE SPREADING ABUSE OF ADDICTIVE DRUGS. ADDICTION TO HEROIN, COCAINE, AND CRACK UTTERLY DESTROYS PEOPLE'S LIVES, WHILE ADDICTION TO DRUGS LIKE ALCOHOL AND NICOTINE CONDEMNS HUNDREDS OF THOUSANDS OF OTHER PEOPLE TO EARLY DEATHS.

30.1 Drugs and Addiction

Objectives

❶ **Discuss information that research has yielded about psychoactive drugs.**

❷ **Summarize how drugs affect the nervous system.**

❸ **Describe addiction in terms of physiology.**

❹ **Explain the role of neuromodulators in nerve impulse transmission.**

What Are Psychoactive Drugs?

Figure 30.1
This woman is addicted to crack cocaine. Drug addiction is a serious personal and social problem that can destroy lives and communities. Understanding the biology of drug addiction can be a first step in tackling drug problems.

Drugs are chemicals put into the body that are not normally found there. Heroin, cocaine, and alcohol are drugs, as are aspirin, penicillin, and caffeine. **Psychoactive drugs** such as alcohol, cocaine, heroin, and nicotine affect the nervous system. Psycho-active drugs are often addictive. The woman in **Figure 30.1** is addicted to crack, a form of cocaine. Widespread use of addictive psychoactive drugs has created major social problems, particularly in our nation's inner cities.

Research has shown how addictive drugs work

Scientists have recognized an important fact: addiction to psychoactive drugs is a physiological response, one that involves drug molecules and nerve cell membranes. Addiction is the human body's attempt to cope with chemical disruption of its signaling systems. Psychoactive drugs, taken to fool the nervous system into providing pleasure or taken to control pain, cause the nervous system to adapt physiologically to their presence, leading to addiction. Scientists are beginning to understand that the body's response is a straightforward chemical response. There is no way to prevent addiction with willpower, any more than willpower can stop a bullet when playing Russian roulette with a loaded gun. To understand the nature of addiction, you must focus your attention on how nerves communicate with one another, for it is in attempting to disrupt this process that addictive drugs cause their damage.

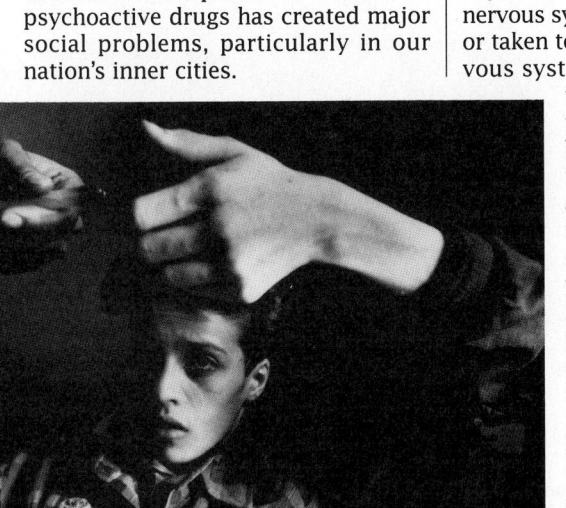

Lesson Plan 30.1

Phase 1

PREPARATION

Key Concepts
- Psychoactive drugs are chemical substances that affect the nervous system.
- Neurotransmitters carry impulses across a synapse, from one neuron to another.
- Neurotransmitters are usually reabsorbed or broken down by enzymes after transmitting the impulses.
- Neuromodulators regulate the amount of neurotransmitters present in a synapse.
- Addiction to psychoactive drugs is a physiological response.
- Some psychoactive drugs work like neuromodulators, slowing down or inhibiting the destruction of neurotransmitters. As a result, a neuron eventually will have fewer receptor proteins for the neurotransmitter.

Reading Strategy

Stress the importance of recalling what has been previously read. In this way, new material can be more easily integrated into an overall framework that connects information presented in different chapters. Have students review information they read about the structure and function of the nervous system.

Phase 2

TEACHING STRATEGIES

Theme Connection

Stability
The body's physiological processes are kept in balance through a variety of homeostatic mechanisms. Taking psychoactive drugs disrupts this stability, placing the body in a precarious position.

Visual Strategy

Figure 30.2
Use this figure to review how neurotransmitters carry an impulse across a synapse. Have students describe what would happen if the enzyme shown in Figure 30.2b were absent or not functioning properly.

Demonstration 1

Neurotransmitters

Use sugar cubes and small marbles to make several depressions in a piece of clay. Tell students that the sugar cubes and marbles represent neurotransmitters, the clay depicts the dendrite of a neuron, and the depressions symbolize two different receptor proteins. Show students how each "neurotransmitter" fits into its specific receptor to stimulate the neuron. Remove the sugar cubes and crush them to signify that they have been destroyed. Ask students how this affects the neuron. Have them contrast this to the situation in which the marbles are still in their receptors.

How Psychoactive Drugs Affect Nerves

Most addictive drugs act at the ends of nerve cells. In a house, all the wires are connected to each other by switches. A current passes from one wire to the next when a switch makes a physical connection between two wires. In your body, nerve cells are not connected at all. Instead they are separated from one another by tiny gaps called synapses that act as chemical switches. A current passes from one nerve cell to another when a synapse makes a chemical connection between the two nerve cells. Addictive drugs work by short-circuiting these chemical connections. In effect, they "hot-wire" certain nerve pathways in the brain.

A nerve impulse passes across a synapse carried by a chemical called a neurotransmitter, as shown in **Figure 30.2a**. Your body has dozens of different kinds of neurotransmitters. Each is present at the ends of particular kinds of nerves. The neurotransmitter does not actually carry an electric charge. Instead, it opens ion channels on the far side of the synapse, an action which has an electrical effect. Some neurotransmitters open sodium ion (Na^+) channels, letting sodium ions flood the receiving nerve cell. This action starts a nerve impulse that travels down that neuron. Other kinds of neurotransmitters open potassium ion (K^+) channels, letting potassium ions rush out of the receiving nerve cell and making it harder for the nerve to fire. This action is called inhibition. A nerve cell body can have many axons in close contact with it, some stimulating and others inhibiting. The summed effect is how your brain integrates signals.

Neurotransmitters work by binding to specific receptor proteins in the membrane on the far side of the synapse. Each kind of neurotransmitter fits into its own type of receptor, as shown in **Figure 30.2b**. The reason all the nerves in your brain don't fire all at once when a signal arrives is that different nerves have different combinations of neurotransmitters and receptors. A signal can only pass between two nerves if the sending nerve releases a neurotransmitter for which the receiving nerve has a receptor.

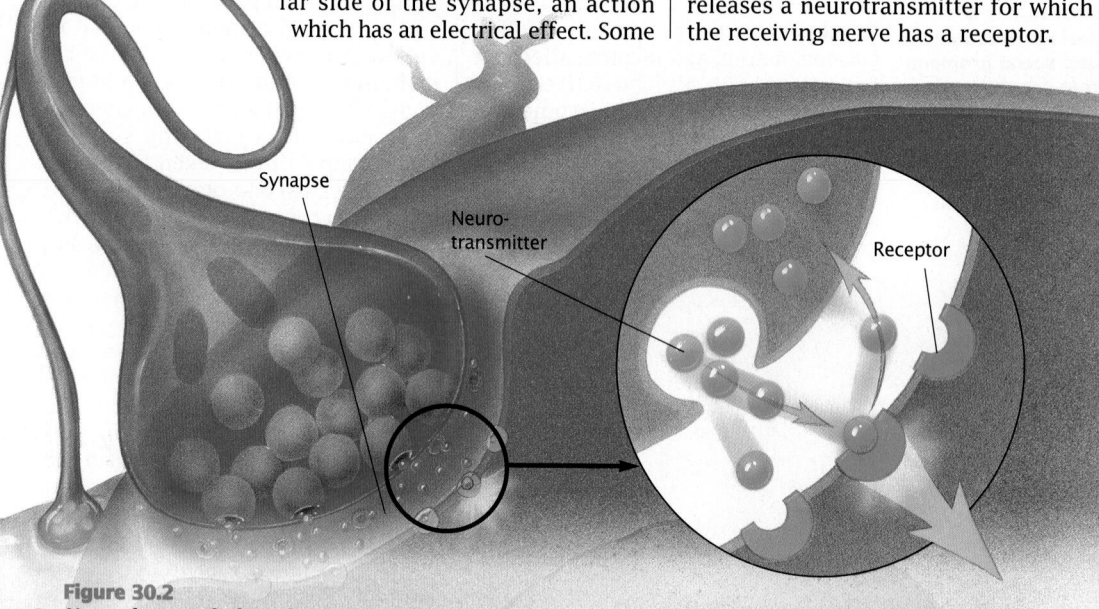

Synapse

Neuro-transmitter

Receptor

Figure 30.2

a *Normal transmission of an impulse across a synapse.*
When a nerve impulse reaches a synapse, the electrical signal is transferred across the gap to the next neuron by a chemical called a neurotransmitter.

b Neurotransmitter molecules bind to receptor proteins in the receiving neuron's membrane, initiating a new nerve impulse. Then the neurotransmitter molecules are reabsorbed or broken down.

When a neurotransmitter has passed a signal to the far side of a synapse, it is either destroyed or reabsorbed by the nerve cell that released it. If it were not destroyed or reabsorbed, it would just keep firing the nerve cell again and again. Nerve gas works by inhibiting an enzyme that normally destroys the neurotransmitter acetylcholine (*uh seet uhl KOH leen*) within synapses between nerve cells and muscle cells. When the acetylcholine cannot be broken down, muscle cells cannot cease contracting and death results.

Neuromodulators prolong the action of neurotransmitters

Your body often prolongs the transmission of a signal across a synapse by slowing the destruction of neurotransmitters. When this happens, the receptor proteins on the far side of the synapse continue to encounter the neurotransmitter for a longer period of time. Your body is able to do this naturally with certain chemicals called **neuromodulators**. Neuromodulators last longer, compared with neurotransmitters. Some of them aid in releasing a neurotransmitter into the synapse; others inhibit the reabsorption of a neurotransmitter. And still other neuromodulators delay the breakdown of neurotransmitters after their reabsorption. This leaves more of a neurotransmitter in the neuron ending to be released when the next signal arrives.

How drugs alter mood

Mood, pleasure, pain, and other mental states are determined by particular groups of nerves in the brain that use special sets of neurotransmitters and neuromodulators. Mood, for example, is strongly influenced by the neurotransmitter serotonin (*sihr uh TOH nuhn*). Many researchers think that depression results from a shortage of serotonin. It is difficult to treat depression directly with serotonin (it has too many other effects). However, depression can be treated successfully with drugs that act as serotonin neuromodulators, as shown in **Figure 30.3**. One class of serotonin modulators blocks a specific enzyme, which then slows the breakdown of serotonin after it is reabsorbed from synapses. Prozac®, the world's top-selling antidepressant, inhibits the reabsorption of serotonin. Antidepressants thus act to increase the amount of serotonin in the synapse by slowing its removal and destruction. This allows the nerve impulses necessary for transmission of proper mood.

Connection: Chapter 5

Enzymes

Remind students that an enzyme binds to a specific molecule (its substrate) in such a way as to make a reaction more likely to occur. Have students identify the substrate in Figure 30.3b. Show the complex interactions that can occur in a synapse. The antidepressant drug binds and inhibits the enzyme, which consequently can no longer bind and break down serotonin (its substrate).

Demonstration 2

Addiction

Refer students to Figure 30.4 on page 694. Use a small marble and some clay to simulate addiction. Flatten two pieces of clay. Use a marble to make depressions in both, but one should have significantly more depressions than the other. Show the class both pieces of clay and ask which would represent the surface of a neuron in a person addicted to a psychoactive drug. Students should recognize that the one with fewer depressions (receptors) represents a desensitized neuron from an addict. Emphasize that fewer receptors are needed since so many neurotransmitters are present for longer periods of time.

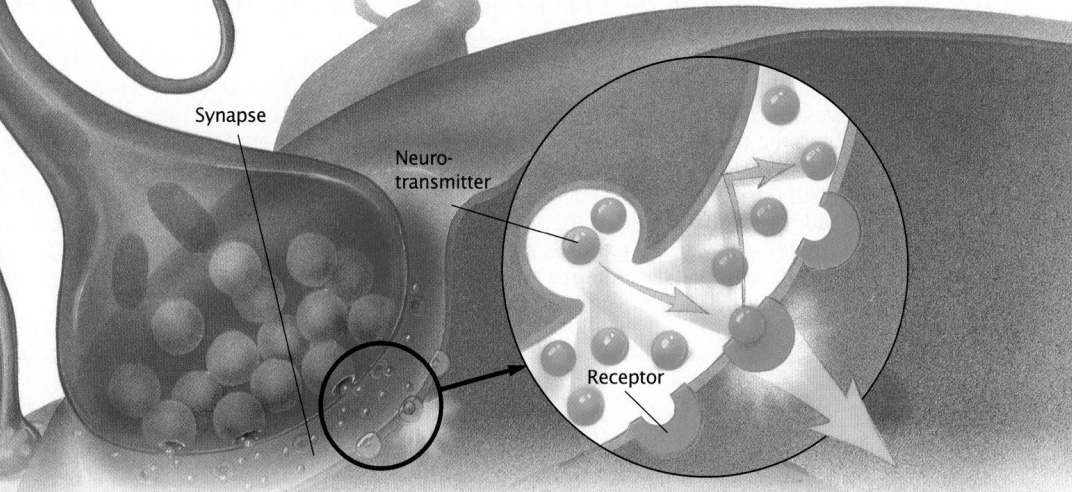

Figure 30.3

a *Drug-altered transmission of impulse across the synapse* Depression can result from a shortage of the neurotransmitter serotonin (shown as green spheres).

b The antidepressant drug Prozac® works by blocking reabsorption of serotonin from the synapse. This action leaves more serotonin in the synapse, making up for the shortage.

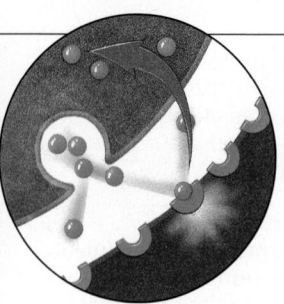
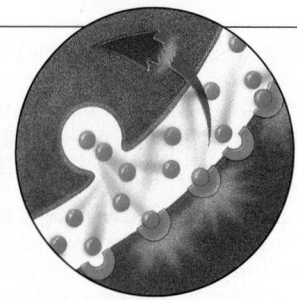
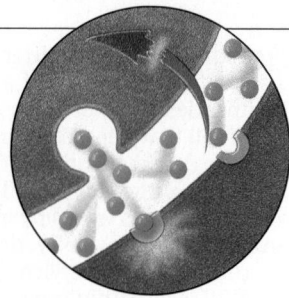

Figure 30.4
a In a normal synapse, neurotransmitters are rapidly reabsorbed.

b When a drug blocks removal of a neurotransmitter, receptors across the synapse are flooded with excess neurotransmitters.

c The receiving nerve responds to this overload by lowering the number of its receptors in the synapse.

d Now if the drug is withdrawn, excess neurotransmitters can again be removed, leaving too few in the synapse to fire the reduced number of receptors.

The Nature of Addiction

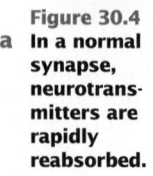

Stability

How is the relationship between neurotransmitters and receptor proteins altered by taking drugs?

When a cell is exposed to a chemical signal for a prolonged period, it tends to lose its ability to respond to the stimulus with its original intensity. You are familiar with this loss of sensitivity— when you sit in a chair, how long are you aware of the chair? Nerve cells are particularly affected by this loss of sensitivity. If receptor proteins within synapses are exposed to high levels of neurotransmitter molecules for prolonged periods, that nerve cell will often respond by inserting fewer receptor proteins into the membrane. This feedback process is a normal part of the functions of all neurons. It is a simple mechanism that has evolved to make the cell more efficient by adjusting the number of "tools" (receptor proteins) in the membrane "workshop" to suit the workload. For example, if someone takes a drug that is a neuromodulator, that drug causes abnormally large amounts of neurotransmitter to remain in the synapses for long

periods of time. Neurons will often adjust by producing fewer receptors as shown in **Figure 30.4**. The result of this action is **addiction**. The decreased number of receptors creates a less sensitive nerve pathway. Physiologically, the only way a person can then maintain normal functioning of the nerve pathway is to continue taking the drug. You can see that addiction is not simply a psychological state to be overcome by willpower. It is a physiological dependence caused by changes in neurons.

When the drug is stopped, the body must accept that it is not functioning "normally" until the number of receptors in the affected synapses have time to readjust. Sometimes in drug treatment programs the drug is withdrawn slowly, so that the number of receptor proteins gradually adjusts upward. However, there is no way to avoid the necessity of eliminating the drug—and no avoiding addiction if the drug is not eliminated.

Section Review

1. What information has research yielded about addictive drugs?

2. How do psychoactive drugs affect the nervous system?

3. Explain why addiction is defined as a physiological response.

4. How do neuromodulators differ from neurotransmitters?

A NATIONAL SURVEY OF HIGH SCHOOL STUDENTS CONDUCTED BY THE CENTERS FOR DISEASE CONTROL INDICATES THAT ABOUT 8 PERCENT OF TEENS IN THE UNITED STATES HAVE TRIED COCAINE, OR ABOUT 2.5 MILLION TEENS. WHAT IS MORE ALARMING IS THE POSSIBILITY THAT THESE TEENS ARE ON THE ROAD TO ADDICTION. YOU WILL LEARN THAT THE DECISION TO USE COCAINE IS THE DECISION TO BECOME AN ADDICT.

30.2 *Narcotics and Cocaine*

Objectives

❶ Explain how the body controls pain.

❷ Summarize why narcotics are addictive.

❸ Describe the action of cocaine at the synapse.

Narcotics

Narcotics are powerful drugs used to relieve pain and induce sleep. Many of the most potent narcotics are derived from chemicals extracted from *Papaver somniferum*, one of many species of the poppy plant. The term "narcotic" comes from the Greek word meaning "to make numb," which describes the drug's ability to soothe pain. Because narcotics are dangerously addictive, their use is strictly controlled. Therefore, the word "narcotic" also has a legal meaning: a chemical substance whose use is restricted by law. The sap that oozes from the cut seed pod shown in **Figure 30.5** forms a substance called opium, and derivatives of this drug are called opiates. There is evidence that opium was prepared and used for medicinal purposes as long as 6,000 years ago by the Sumerian and Assyrian civilizations. The major active ingredient in opium, morphine, is far more potent than opium and is one of the most effective pain relieving drugs known. A relatively simple chemical modification of morphine produces heroin, a highly addictive

drug that induces an intense feeling of well-being when injected into the bloodstream. Abuse of heroin is among the most serious drug problems in society today. To understand narcotics, you need to look at how the human body controls pain and pleasure.

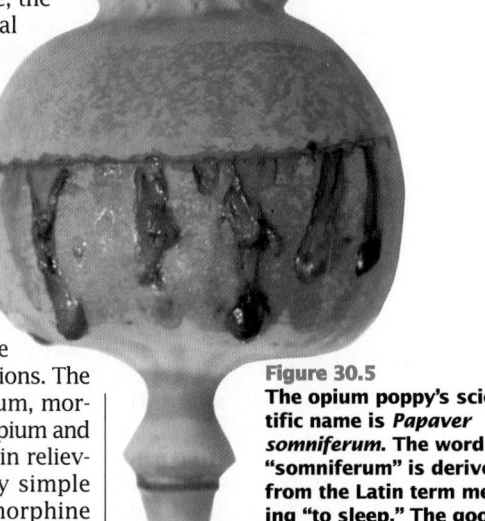

Figure 30.5
The opium poppy's scientific name is *Papaver somniferum*. The word "somniferum" is derived from the Latin term meaning "to sleep." The gooey substance contains opium.

■ Matter of Fact ■

The Coca Plant
The coca plant, *Erythroxylum coca*, is a native of South America. The plant has been cultivated since the Inca civilization. The plant is grown primaily in Peru, Bolivia, and Colombia. The natives of these countries have traditionally chewed the leaves, which are rich in calcium, phosphorus, and vitamins A and B_2. The leaves have the effect of an appetite suppressant and help workers work long hours on little food. Tea made from coca leaves is drunk as a preventive medicine for altitude sickness. It is used in hospitals as a local anesthetic, and in some Latin American hospitals, it is the only anesthetic available. In many countries the use of coca in the form of cocaine and crack threatens people's lives.

Lesson Plan 30.2

Phase 1

PREPARATION

Key Concepts
- Narcotics act by blocking pain signals from reaching the brain.
- Enkephalin, a type of endorphin, is a substance naturally produced by the body in response to pain.
- Heroin causes an increase in the number of opiate receptors in spinal neurons. In the absence of heroin, these neurons are stimulated to transmit pain signals.
- Methadone blocks these opiate receptors, allowing for the gradual withdrawal from heroin addiction.
- Cocaine, unlike a narcotic, is a stimulant that acts by blocking the reabsorption of the neurotransmitter, dopamine.

Reading Strategy

Have students bring in newspaper and magazine articles about drugs and drug abuse. Display the articles on a bulletin board. Encourage the class to read them as they go through the material in this section.

Phase 2

TEACHING STRATEGIES

Demonstration 1
Narcotics
Invite a narcotics officer to talk to the class. Ask the speaker to focus on encounters where opium, morphine, or heroin were involved.

Visual Strategy

Figure 30.5
Have students conduct library research to identify countries where opium is grown. Also have them determine what role opium plays in each nation's economy.

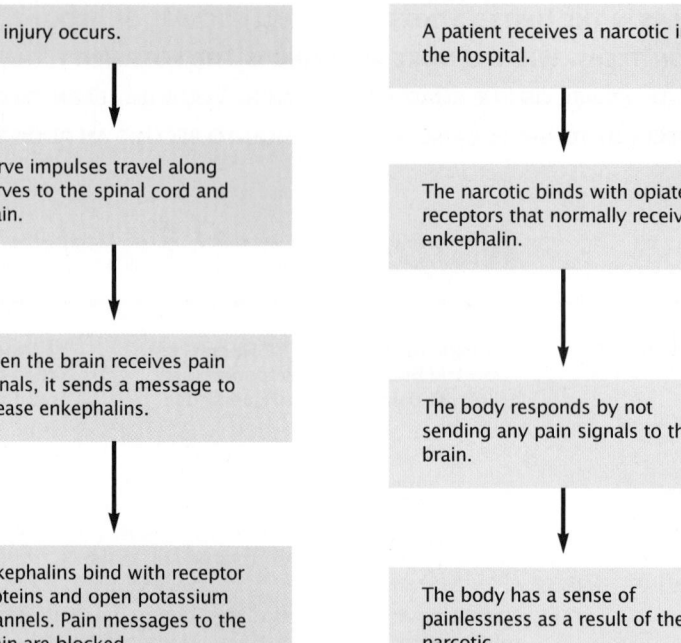

Figure 30.6
Narcotics such as heroin and morphine mimic the body's normal response to pain. The model on the left shows the sequence of responses after a painful injury. This process involves the production of enkephalins. On the right, you can see how narcotics work in a manner similar to enkephalins.

a Normal response to pain

An injury occurs.

Nerve impulses travel along nerves to the spinal cord and brain.

When the brain receives pain signals, it sends a message to release enkephalins.

Enkephalins bind with receptor proteins and open potassium channels. Pain messages to the brain are blocked.

b Action of narcotics

A patient receives a narcotic in the hospital.

The narcotic binds with opiate receptors that normally receive enkephalin.

The body responds by not sending any pain signals to the brain.

The body has a sense of painlessness as a result of the narcotic.

Scale and Structure

Why is a knowledge of cell membrane structure necessary to understand how drugs affect nerves?

How the Body Controls Pain

Pain serves a very important function in your body by warning the brain of injury. Without pain, the accidental grasping of a hot frying pan on the stove may seriously burn your hand. As shown in **Figure 30.6a**, pain begins as a signal at damaged nerve endings near the surface of the skin and travels up the spinal cord to the brain. Pain signals are transmitted from one neuron to the next by a neurotransmitter called "substance P." The release of substance P into the synapse requires the presence of a neuromodulator, in this case a prostaglandin.

After a pain signal has been received by the brain, further pain signals usually serve no useful purpose. To shut off a pain signal, the brain signals nerves leading to the spine to release a neuromodulator called an **enkephalin** (*ihn KEHF uh luhn*). Enkephalins open potassium channels in spinal neurons, which blocks pain messages from traveling to the brain. Enkephalins are one of several endorphins (*ehn DAWR fuhns*), natural pain relievers our bodies release in response to pain and stress. The presence of endorphins may help explain why under certain circumstances, such as on a battlefield, a person may receive a severe injury and not immediately feel pain.

Narcotics mimic enkephalins

Narcotics function by imitating enkephalins, as shown in **Figure 30.6b**. Their molecular structure is similar to that of enkephalins. Therefore, they easily bind to the same receptor proteins to which enkephalins bind. These enkephalin receptor proteins are called "opiate receptors," because the binding of opiates to them was discovered before enkephalins were discovered. Narcotics such as morphine and heroin are potent pain-blocking drugs because they do the job of enkephalins; they act as neuromodulators to block pain signals from traveling up the spine to the brain.

Why Narcotics Are Addictive

Figure 30.7
The limbic system of the brain delivers pleasurable sensations to the body. With narcotic use, the chemistry of the limbic system changes.

Limbic system

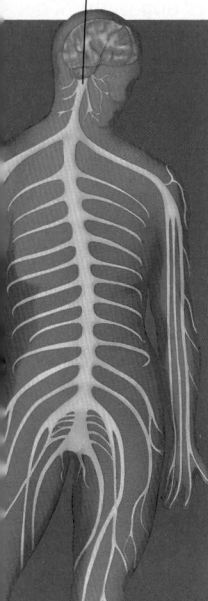

Narcotics have a second activity in the body. They interact with the brain's limbic system, the "pleasure center," producing an intense feeling of well-being. Heroin is very effective in stimulating the limbic system, shown in **Figure 30.7**. With continued use, however, a problem arises. In this case, the neurons become desensitized not by lowering the number of receptor proteins, but rather by adjusting their internal chemistry. Thus, the result of prolonged narcotic use is an increase in the number of opiate receptors in spinal neurons. Addiction is a result of the internal chemical changes that compensate for the stimulation of heroin by desensitizing the nerve cells. More of the drug must be taken in order to achieve the same level of pleasure.

What happens when the individual stops taking the narcotic? Removing the inhibitory action of the narcotic causes nerve cells to become hypersensitive. Unpleasant symptoms result, including anxiety, tremors, and heightened sensitivity to pain. Sensitivity to pain results because the changes in potassium ion levels that the body's natural enkephalins produce are no longer enough to inhibit the over-compensated spinal nerves. As a result, pain signals traveling up the spine are not blocked, and the body becomes very receptive to pain. The narcotic is required in significant amounts just to make the person's body feel normal. The addiction trap has now sprung shut; any withdrawal of the narcotic is very unpleasant, with symptoms ranging from fever and headache to severe pain.

Heroin addicts are often treated with methadone, a synthetic narcotic that also mimics enkephalin. Methadone fits opiate receptors but produces only a mild high. The amount of methadone taken is controlled by a physician or clinic, as shown in **Figure 30.8**. The dosage is decreased with time, allowing the body to adjust to lower levels of the drug. Such treatment, however, does not solve the basic problem—the person is still addicted to a drug. The only real cure for heroin addiction is to stop taking the drug and suffer through the painful withdrawal.

Figure 30.8
Methadone is the only approved treatment for heroin addiction. Methadone attaches to the receptors that cause the strong craving for heroin, thereby lessening the symptoms of withdrawal.

Health Connection

Cocaine and Crack

Cocaine can either be inhaled (snorted) or injected. When it is snorted, the drug is absorbed into the bloodstream through the mucous membranes of the nasal cavity. Cocaine causes blood vessels in the nasal area to constrict, reducing blood flow and causing the passage to become dry. Repeated use can produce holes in the septum that divides the two nasal passages. Crack is considered one of the most addictive drugs; evidence indicates that even a single use can result in addiction.

Phase 3

ASSESSMENT OPTIONS

Closure Strategy

Drug Interactions

Have students describe what would probably happen if a person took heroin (a depressant) and cocaine (a stimulant) at the same time.

Section Review

Assign the *Section Review*.

Reteaching

Write on the chalk board: *opium, heroin, methadone, cocaine,* and *crack.* Discuss how each drug affects the nervous system.

Action of Cocaine

Unlike narcotics, which are depressants, cocaine is a stimulant. Extracted from the leaves of coca plants that grow at high altitudes in the mountains of South America, cocaine acts directly on the limbic system to produce an intense euphoria. A German scientist discovered how to extract cocaine from coca leaves in the mid-1800s. Many physicians at first considered it a miracle drug, prescribing it for all sorts of physical and mental ailments; it even was added to soft drinks. Unfortunately, cocaine is highly addictive. Today United States law forbids the importation, manufacture, and use of cocaine for nonmedical purposes, and the medical use is extremely limited.

Figure 30.9
Cocaine binds with the dopamine transporter to block the reabsorption of dopamine. Dopamine is a neurotransmitter that helps send pleasure messages to the brain.

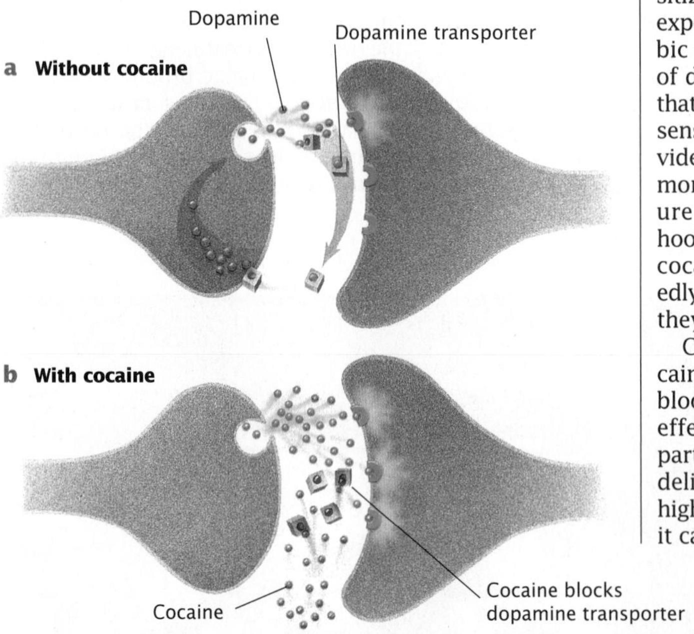

Dopamine

Dopamine transporter

a Without cocaine

b With cocaine

Cocaine

Cocaine blocks dopamine transporter

Cocaine blocks the reabsorption of dopamine

Despite being illegal and highly addictive, cocaine is still used by many people. The effect is to make any pleasurable sensation more intense. However, cocaine is highly addictive, not because of the pleasure it provides but because of the changes it creates in the brain. As shown in **Figure 30.9b**, cocaine works by preventing the reabsorption of the neurotransmitter dopamine. The dopamine that is trapped in the synapse keeps stimulating neurons, producing euphoria.

Prolonged use of cocaine produces addiction because the brain adjusts to its presence. The addiction is the direct result of changes in the number of receptor proteins, as nerve cells desensitize in response to the drug. Repeated exposure to cocaine may cause the limbic system to decrease its population of dopamine receptors, with the effect that the pleasure pathway becomes less sensitive without the amplification provided by the drug. Over time, more and more cocaine is required for the pleasure pathway to work at all. Monkeys hooked up intravenously to a source of cocaine will inject themselves repeatedly, rejecting food, sex, and sleep, until they die.

Crack is a highly potent form of cocaine that can be smoked, entering the bloodstream through the lungs (a very effective delivery route). It provides a particularly intense "high" because it is delivered to the limbic system in such high concentrations, and for this reason it can produce addiction very quickly.

Section Review

① **What role do enkephalins and endorphins play in the body?**

② **What two effects of narcotics make them so addictive?**

③ **How does cocaine addiction differ from heroin addiction?**

④ **Why is crack capable of quickly producing addiction?**

▪ *Section Review Answers* ▪

1. Enkephalins are neuromodulators that are signaled by the brain to shut off pain signals. Endorphins are natural pain relievers the body releases in response to pain and stress.

2. Students should explain the narcotic's ability to mimic enkephalins and interact with the brain's limbic system.

3. Students should suggest that heroin binds to opiate receptors, blocking pain signals to the brain, whereas cocaine blocks the reabsorption of dopamine.

4. Students should suggest that because crack can be smoked, it rapidly enters the bloodstream in high concentrations from the lungs. It is also very potent.

NOT ALL DANGEROUS DRUGS ARE ILLEGAL. IT IS INTERESTING TO NOTE THAT MORE DEATHS AND SUFFERING ARE CAUSED BY THE DRUGS NICOTINE AND ALCOHOL, WHICH ARE LEGAL, THAN BY NARCOTICS AND COCAINE. THE FACT THAT YOU ARE MORE LIKELY TO BECOME ADDICTED TO LEGAL DRUGS SHOULD CAUSE YOU TO CONSIDER CAREFULLY YOUR USE OF THEM.

30.3   *Dangerous Social Drugs*

Objectives

❶ Describe how nicotine affects the nervous system.

❷ Describe the effects of alcohol on the nervous system.

❸ Describe two dangers associated with alcohol abuse.

❹ Summarize the effects of hallucinogens on the nervous system.

Figure 30.10
Harmful chemicals in cigarette smoke can damage the lungs of smokers and the people around them. In addition, cigarette smoking causes many other hazards to individuals and society.

Nicotine

Drugs called stimulants stimulate the central nervous system and speed up body processes. You have already read about one stimulant, cocaine. Another drug, nicotine, is the most widely used stimulant in our society. Nicotine is the highly addictive drug found in tobacco. Cigarettes, shown in **Figure 30.10**, are among the most addictive and deadly substances available to the general public. The Surgeon General estimates that in a typical year in the 1980s, 130,000 cancer deaths, 170,000 heart disease deaths, and 50,000 chronic lung disease deaths were the direct result of smoking. The health problem arises because chemicals in cigarette smoke are toxic. Other substances, mostly tars, are highly mutagenic, causing changes in DNA that can lead to cancer.

Table 30.1   Cigarette Facts

Item	
Number of cigarettes smoked in USA each year	529 billion
Percentage of USA population that smokes	29%
Number of lung cancer deaths in USA each year	160,000
Percentage of lung cancer deaths who were smokers	96%
Rate of death from lung cancer among smokers each year	2 per 1,000

▪ *Cultural Perspective* ▪

Tobacco
The word *tobacco* is a Spanish adaptation of the term the natives of the Caribbean used for cigar. The majority of the natives in South America used tobacco before Columbus arrived in the new world. Tobacco was chewed or smoked for ceremonies surrounding war, peace, puberty, the harvest, or death. Tobacco was also burned as an incense, sprinkled in leaf form, or buried with the dead. Most uses suggested the notion of sacrificial offerings.

Secular use of tobacco was also popular. It was used to relieve stress from work or war. It was also used to cure disease and as an anesthetic to treat wounds. Tobacco contains the physiologically active alkaloid nicotine.

Visual Strategy

Figure 30.11

Have students observe how blood flow through this woman's hand changes as she smokes. Point out that in addition to constricting blood vessels, smoking also increases the level of carbon monoxide in the blood. Carbon monoxide reduces the level of oxygen carried by hemoglobin, thus depriving cells of an essential component of aerobic respiration. Have students explain how this would affect a smoker.

Demonstration 1

Healthful Cigarettes?

Have students examine the labels of various brands of cigarettes, including those advertised as being low in tar and nicotine. Prepare a table on the board, listing each brand and the amount of tar and nicotine it contains. Have the class examine the data and draw as many conclusions as possible.

Figure 30.11
Smoking affects all areas of the body. In this experiment, a woman placed her hand under a thermograph, which measures skin temperature. She then smoked a cigarette. The sequence of thermograms you see here shows that the hand gets colder and colder, a direct result of the cigarette's effect on blood flow.

Nicotine is addictive

When a smoker inhales a cigarette, the nicotine in the cigarette smoke reaches the brain within 10 seconds. The nicotine acts as a neuromodulator, binding to a variety of receptors in the brain, particularly those normally targeted by the neurotransmitter acetylcholine. The binding of these receptors produces a mild, short-lived pleasurable feeling and frequently has a calming effect. It also decreases blood flow, as shown in **Figure 30.11**. The addiction to nicotine is similar to that of cocaine. The brain adjusts to prolonged exposure to nicotine by making fewer receptor proteins to which nicotine can bind, and the body eventually requires nicotine to maintain a "normal" feeling. Because the pleasurable effect is short-lived, the drug is

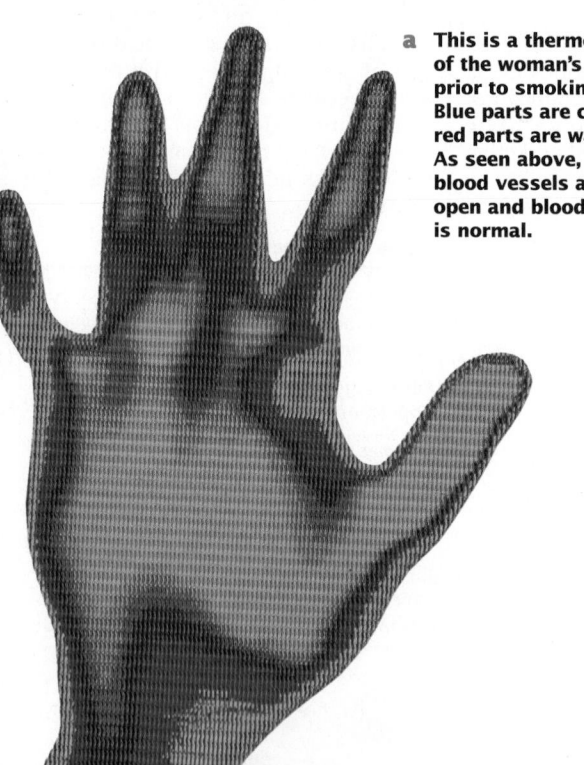

a This is a thermogram of the woman's hand prior to smoking. Blue parts are cool; red parts are warmer. As seen above, the blood vessels are open and blood flow is normal.

b After three puffs on the cigarette, blood vessels begin to constrict. Blood flow to the fingers decreases and the skin temperature cools.

▪ *Matter of Fact* ▪

A person who smokes is also likely to have other bad health habits. A 1988 study of more than 4,000 people revealed that those who smoked were 2.5 times more likely than nonsmokers to consume 8 or more drinks per week and 1.3 times more likely to take drugs.

Figure 30.12

Comparing the levels of lung cancer in many countries shows a relationship between the incidence of lung cancer in men between 35 and 44 years of age and cigarette consumption 20 years earlier. The risk of cancer increases with the number of cigarettes smoked daily.

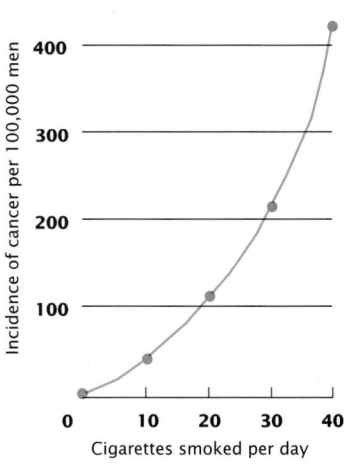

Smoking Causes Cancer

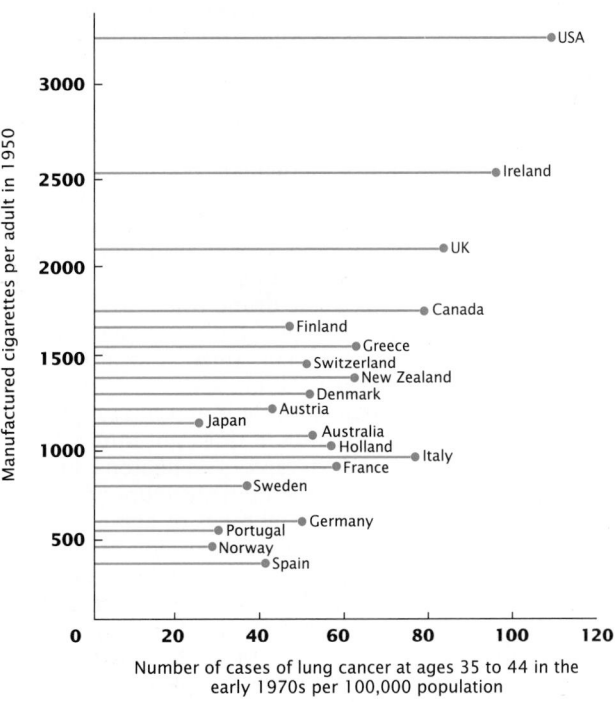

Smoking and Lung Cancer

c After more puffs, the hand continues to cool as blood vessels are constricted even further.

typically used frequently. A pack-a-day habit is 7,500 cigarettes a year with 75,000 inhaled puffs. Craving a cigarette is similar to the craving a cocaine user has for cocaine. In both cases, the craving results from brain cells adjusting the number of receptors in response to the loss of the drug. Addiction to nicotine can be overcome; half the men and nearly half the women who have ever smoked in this country have quit—more than 50 million people. But another 50 million Americans still smoke. Why don't they try to quit? Many do try, but it is very difficult to overcome the addiction. Most studies indicate a success rate—at least two years' abstinence—of about 20 percent.

The chemicals in cigarette smoke are dangerous, as indicated in **Figure 30.12,**

even to those who do not smoke. When nonsmokers inhale cigarette smoke, they, too, inhale tar and nicotine. In a study of smokers and nonsmokers confined in airplanes, blood samples taken from nonsmokers contained high levels of nicotine. And in a 1992 study of women in Italy who died accidentally, almost all those married to smokers had precancerous lesions in their lungs; those women married to nonsmokers did not. For these reasons, smoking is banned from many public places.

You do not have to smoke tobacco to get nicotine from it. Nicotine is also absorbed by the body from chewing tobacco and from snuff. These "smoke-less" forms of tobacco are just as addictive as cigarettes. The tars in chewing tobacco and snuff lead to increased risk of mouth and throat cancer. Chewing tobacco instead of smoking it simply trades one form of deadly cancer for another.

Theme Answer

Patterns of Change

Alcohol abuse can cause considerable damage to the body. It causes psychological as well as physiological adjustments within the brain. Continued use can cause cirrhosis, which can be fatal.

Connection: Chapter 5

Fermentation

Review how yeast and some other microorganisms, in the absence of oxygen, convert organic compounds such as glucose into ethyl alcohol. Ask students what glucose is converted to by human muscle cells in the absence of oxygen.

Demonstration 5

Alcoholic Beverages

Point out that beer, wine, and liquor differ in their ethanol concentrations. However, a 12-oz. can of beer, a 4-oz. glass of wine, and a 1-oz. shot of liquor all contain about the same amount of ethanol. Tell students that "proof" reflects the alcohol concentration in a beverage. Alcohol content is determined by dividing the proof number in half—80 proof liquor has a 40 percent alcohol content.

Connection: Chapter 3

The Lipid Bilayer

Remind students that a double layer of phospholipids form a tough, yet flexible, lipid bilayer in the cell membrane.

Visual Strategy

Figure 30.13

After students have read the quotation included in this figure, conduct a class discussion on reasons why people become alcohol abusers.

Alcohol

Patterns of Change

What changes can occur in the body as a result of prolonged alcohol use?

Of all psychoactive drugs, ethyl alcohol is the most popular. Consumed for centuries as wine from fermented grapes or as beer from fermented grain, ethyl alcohol acts quickly and powerfully to reduce inhibitions and produce a sense of well-being. Unfortunately, it also impairs judgment. Consuming alcohol can alter both judgment and reaction time, a condition know as being "drunk." Many high school students die each year in automobile accidents resulting from being drunk while driving, making alcohol a truly dangerous drug. Other students suffer from alcoholism, as discussed in **Figure 30.13**.

Ethyl alcohol is an unusual psychoactive drug. It does not interact with specific receptors on neurons. Instead, it alters the lipid structure within the plasma membranes of nerve cells, inhibiting the transmission of all nerve signals. This action inhibits pain (that's why alcohol used to be administered as an anesthetic) and also inhibits the nerves that repress the limbic system, causing feelings of pleasure and inhibition. The pleasure from drinking alcohol is a direct result of this release of the limbic system.

Alcohol abuse can cause disorders

Addiction to alcohol results from psychological as well as physiological adjustments within the brain. The addiction that results from excessive use is particularly powerful and long lasting. Alcoholism presents a major social problem in many countries around the world, and only wrenching psychological adjustments can free a habitual drinker from the addiction. People who drink large amounts of alcohol over long periods of time cause considerable damage to their bodies. In particular, chronic alcohol use forces the liver to use alcohol as an energy source instead of fats. Eventually liver cells, become unable to function properly, accumulating massive fat deposits and causing the liver to swell. That is why 75 percent of all alcoholics have enlarged livers. Eventually liver tissue is replaced by scar tissue, a condition called **cirrhosis**, which can be fatal.

Children born to women who drink during pregnancy may suffer from **fetal alcohol syndrome** (FAS), a condition that can cause physical and mental disabilities in newborn babies. Some of the effects of FAS are low birth weight, general weakness, and heart defects. The long-term effects include lack of coordination, slow growth, short attention span, hyperactivity, and learning difficulties due to mental retardation. Unfortunately, alcohol does the greatest damage during early pregnancy, when the woman often is unaware that she is pregnant. Given this information, the obvious conclusion is that pregnant women, or women likely to become pregnant, should not drink alcohol.

Figure 30.13
Because drinking is often seen as a normal activity, many of the terrible ravages of alcohol abuse go unnoticed until the effects become pronounced.

"I started drinking when I was 15 years old. To me, it was a very big obsession. It was always on my mind: when was I going to drink, where was I going to get it. It used up a lot of my energy, time, and life.

"I was able to replace that obsession with a spirituality. I relieved the self-doom, the negative feelings, the feeling sorry for myself. Quitting created a positive outlook in my life. I became physically more rested. Now I wake up feeling good about myself—as opposed to feeling miserable. When I was drinking, I expected things to get worse. Now I know things will get better. I also now know I am a person I do like after all."   Frank L.

■ Cultural Perspective ■

Alcoholic Beverages

Before the people of the Americas had contact with Europeans, many with highly developed agricultural societies made and used alcoholic beverages. In Mesoamerica, parts of the Caribbean, the Southwest, and the Southeast, alcoholic beverages were made from domesticated and wild plants. Mexicans made at least 40 varieties. There was corn beer, maguey wine, sotol wine, and balche, a drink made from fermented honey. Southwestern Indians made wine from cactuses, and persimmon wine was made by the Southeastern Indians. The intake of alcohol for the Papagos and Pimas Indians of the Southeast was used as part of the ceremony to bring rain. Among the Aztecs, intoxication served to induce medication and prophecy. Public drunkenness was frowned upon and in some instances punishable by death.

Barbiturates and Tranquilizers

In addition to alcohol, a variety of other psychoactive drugs act by depressing the central nervous system. Barbiturates and tranquilizers are two major examples of these drugs. Barbiturates are compounds that were first developed in 1862 in Germany. Because they depress the central nervous system and inhibit pain, barbiturates soon became popular as sleeping pills and as a treatment to reduce anxiety. In the 1950s, American scientists synthesized a new group of anxiety-reducing compounds that have come to be known as tranquilizers.

Barbiturates are often prescribed in small doses to induce sleep or relieve tension. Larger doses are sometimes used to treat sleep disorders and extreme anxiety. However, these drugs should be used only under a doctor's care, for they are addictive and their misuse can be fatal. Excessive doses of barbiturates, for example, can result in coma, a condition of deep unconsciousness. Taken in combination with alcohol, another depressant of the central nervous system, barbiturates can be deadly. The combined effects of the two drugs can stop respiration.

Amphetamines

As described in **Figure 30.14**, amphetamines are dangerous and addictive drugs. They are available legally only with a physician's prescription. Often prescribed as diet pills, amphetamines also acts to sharpen attention and alertness. Sometimes nicknamed "uppers," amphetamines act in the brain to increase levels of a variety of neurotransmitters called monoamine neurotransmitters. Monoamine neurotransmitters adjust body activities for danger, quickening heartbeat, increasing alertness, and shutting off hunger—like the "fight-or-flight" hormone adrenaline. Addiction occurs because the brain adjusts to the continued presence of the drug by lowering its sensitivity to such stimulation; it makes fewer monoamine receptors. When the drug is withdrawn, normal alertness is more difficult to achieve without amphetamines, so the body craves more.

Figure 30.14
Drug addiction can take a serious toll on a person's life. This former user recounts a life lost and regained.

"I started doing drugs when I was 14 or 15. I liked them better than alcohol. My drug use escalated to a point that when I remembered to leave the house with my coat, it would be something major. Since I quit six years ago, everything I've ever wanted from my life, the kind of person I am, has come true. Now I eat, I shower, I have friends that don't steal things from me, and I don't steal things from my friends. Now my life isn't about constantly getting high. Now I feel like a human being."

Lucy T.

■ Cultural Perspective ■

Peyote
Peyote, the fruit of the *Lophophora williamsii* cactus that grows in northern Mexico and along the Rio Grande Valley, is another hallucinogen used by Mexican tribes and the Apache for both sacred and secular purposes. The Mexicans and Apaches learned to cut off the rounded top of the plant, dry it, and brew it into a tea. Peyote was also used as an appetite and thirst suppressant.

Language Connection

Hallucinogen

"Hallucinogen" comes from a Latin word meaning "to dream." Have students explain why this is an appropriate derivation.

Health Connection

Marijuana

Marijuana contains not only THC, but more than 400 different chemicals. When marijuana is smoked, more than 2,000 chemicals are produced. Some of these are 50 to 100 percent more carcinogenic than the chemicals found in cigarette smoke.

Phase 3

ASSESSMENT OPTIONS

Closure Strategy

Alcoholism and Drugs

Have students explain why an alcoholic is more likely to be affected by any of the drugs discussed in this section. *Explain that since their liver function has been impaired, any drug taken will not be as quickly metabolized.*

Section Review

Assign the *Section Review.*

Reteaching

Tell students to prepare a table that summarizes what they have read. Down the left side of the paper, have them list each drug that was discussed in this section. Across the top, have them write the headings *Neurological effects* and *Psychological effects.* Instruct them to fill in the required information.

Hallucinogens and Marijuana

Figure 30.15 Hallucinogens make the user perceive a very different reality. The pictures below were produced by this artist while under the effects of LSD.

Drugs that distort the way the brain translates signals from the sensory organs are called **hallucinogens** (*huh LOO sih nuh jehnz*). LSD is a hallucinogen derived from a fungus that grows on rye in grainfields. Scientists believe that LSD inhibits the neurotransmitter serotonin, which usually acts in the brain to reduce free association. By inhibiting serotonin's action in the brain's association centers, LSD unleashes associations at random. The brain begins to tell itself that it is seeing three-foot mosquitoes, the insides of the body, or other unreal visions called hallucinations, as shown in **Figure 30.15.** Mescaline is a hallucinogen derived from peyote, a small cactus that grows in the southwestern United States and Mexico. When dried and ingested, mescaline alters the transmission of norepinephrine. Hallucinogens do not have any currently acceptable medical use and are illegal.

Cannabis sativa, more commonly known as marijuana, is a plant that grows throughout the world. While the cannibis plant is usually linked to marijuana today, for many years it was harvested as a commercial crop for its fiber, which was used in the production of rope. The dried leaves, flowers, or stems of the plant are the source of marijuana. Marijuana can be eaten, but it is more often smoked like tobacco. It has a wide range of effects, including many of those associated with alcohol and hallucinogens. Like ethanol, it acts to dampen all nerve activity; like LSD, it interacts with serotonin. The active ingredient is a chemical called tetrahydrocannabinol (THC). In 1990 researchers discovered that THC binds to a specific receptor in the brain. The natural function of this receptor is not yet understood. While marijuana is sometimes used in controlled doses to treat pain and disease, its use by the general public is illegal.

Section Review

❶ How does nicotine affect the nervous system?

❷ How does alcohol differ from other drugs in its effect on the nervous system?

❸ What are the dangers of abusing alcohol?

❹ How do hallucinogens affect the nervous system?

▪ Section Review Answers ▪

1. Nicotine acts as a neuromodulator, binding to a variety of receptors in the brain, especially the receptors targeted by the neurotransmitter, acetylcholine. It produces a mild, short-lived pleasurable and calming feeling.

2. Alcohol does not interact with specific receptors on neurons. Instead, it alters the lipid structure within the plasma membranes of nerve cells, inhibiting the transmission of all nerve signals.

3. Alcohol abuse can cause severe damage to the body. It causes psychological and physiological adjustments in the brain. It can cause cirrhosis of the liver, and fetal alcohol syndrome in babies born to alcoholic mothers.

4. Hallucinogens interact with the neurotransmitter, serotonin, and distort how the brain interprets sensory information.

Chapter 30 *Highlights*

Smokeless tobacco is just as addictive as cigarettes.

Alternative Assessment

Designer drugs are synthetic compounds. A single dose of a designer drug has resulted in permanent brain damage and even death. Assign students to cooperative groups of four and have each group research a designer drug. Have each group present a brief oral report to the class.

	Key Terms	Summary
30.1 Drugs and Addiction  **Nerve signals are passed from one neuron to another by neurotransmitters.**	psychoactive drug (p. 691) neuromodulator (p. 693) addiction (p. 694)	• Neurotransmitters and neuro-modulators interact to pass signals from one neuron to another across the synapse. • Addiction to drugs results be-cause neurons adjust to different levels of stimulation caused by the drugs.
30.2 Narcotics and Cocaine  **The narcotics morphine and heroin are made from the opium poppy.**	narcotic (p. 695) enkephalin (p. 696)	• Narcotics such as heroin and mor-phine affect the nervous system's control of pain perception by mimicking enkephalins, which are neuromodulators used to block pain signals. • Withdrawal from narcotics results in pain because neurons are more sensitive to pain, and pain signals are no longer blocked. • Cocaine, a stimulant, acts by blocking dopamine reabsorption at synapses. Since dopamine transmits "pleasure messages," its continued presence causes a constant pleasurable feeling. With-drawal from cocaine results in reduced pleasure sensations.
30.3 Dangerous Social Drugs **One cigarette contains more than 2,000 chemicals, of which 200 are known to cause cancer.** 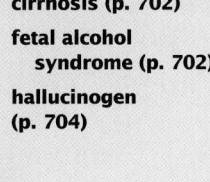	cirrhosis (p. 702) fetal alcohol syndrome (p. 702) hallucinogen (p. 704)	• Nicotine in tobacco acts by bind-ing to brain receptors for acetyl-choline; this causes mild pleasurable feelings. • Alcohol, unlike many other drugs, does not act on specific sites but acts on the entire cell membrane to inhibit all signals. • Barbiturates act on a part of the midbrain, causing sleep or decreasing alertness. • Amphetamines function like adrenaline, causing increased alertness. Hallucinogens, such as LSD and marijuana, interact with serotonin and distort how the brain interprets sensory informa-tion.

Chapter Review Answers

Understanding Vocabulary

1. **a.** Neurotransmitters are chemicals that transmit impulses from neuron to neuron, whereas neuromodulators prolong the stimulation between neurons.
b. In the brain, amphetamines increase the level of monoamine neurotransmitters responsible for alertness, whereas tranquilizers depress the central nervous system, reducing anxiety and inhibiting pain.
c. Enkephalin is a neuromodulator that combines with receptor proteins to shut off pain signals, whereas opium is a chemical that mimics enkephalin to bind with receptor proteins that shut off pain signals.
d. Cocaine and crack both bind with dopamine transmitters to block the reabsorption of dopamine. However, the reaction time to addiction is different. Cocaine addiction occurs over a period of time, whereas an addiction to crack occurs quickly.

Relating Concepts

2. Map answer is shown on page 689D.

Understanding Concepts
Multiple Choice

3.	a	8.	a
4.	c	9.	a
5.	a	10.	c
6.	c	11.	b
7.	d		

Completion

12. Possible answers are serotonin or dopamine.
13. amphetamines
14. limbic, pain perception
15. opium
16. neurotransmitter
17. lower, speed
18. cirrhosis
19. *Cannabis sativa*

Chapter 30 Review

Understanding Vocabulary

1. For each pair of terms, explain the differences in their meanings:
   a. neurotransmitter, neuromodulator
   b. amphetamine, tranquilizer
   c. enkephalin, opium
   d. cocaine, crack

Relating Concepts

2. Copy the unfinished concept map below onto a sheet of paper. Then complete the concept map by writing the correct word or phrase in each oval containing a question mark.

Understanding Concepts

Multiple Choice

3. Drugs that affect the nervous system are called
   a. psychoactive drugs.
   b. socially acceptable drugs.
   c. depressants.
   d. narcotics.

4. Which drug acts as a neuromodulator and binds to receptors targeted by the neurotransmitter acetylcholine?
   a. alcohol   c. nicotine
   b. morphine   d. cocaine

5. What chemicals are naturally present in the body and affect the levels of neurotransmitters in a synapse?
   a. neuromodulators   c. stimulants
   b. narcotics   d. enkephalins

6. Pain messages to the brain are naturally blocked by
   a. opium.
   b. substance P.
   c. enkephalins.
   d. prostaglandins.

7. In the body, heroin binds to the same receptor proteins for enkephalins and
   a. causes hallucinations.
   b. decreases blood flow.
   c. stimulates muscle cells.
   d. blocks pain.

8. Cocaine produces intense feelings of well-being because it
   a. blocks the reabsorption of dopamine.
   b. mimics the action of enkephalins.
   c. is both a stimulant and a depressant.
   d. desensitizes the brain.

9. The most widely used stimulant in our society is
   a. nicotine.   c. alcohol.
   b. caffeine.   d. marijuana.

10. Alcohol inhibits transmission of nerve signals by
    a. increasing levels of monoamine neurotransmitters.
    b. breaking down cyclic AMP.
    c. changing the lipid structure of plasma membranes of nerve cells.
    d. blocking the reabsorption of dopamine in the synapses.

11. Which drug inhibits the neurotransmitter serotonin?
    a. cocaine   c. alcohol
    b. LSD   d. nicotine

Completion

12. Mood is strongly influenced by the neurotransmitter _____ .

13. Often prescribed as diet pills, _____ also act to sharpen attention and alertness.

14. Heroin affects the brain's _____ system, or pleasure center, and stops _____ by mimicking enkephalins.

15. Morphine and heroin are refined forms of _____ .

16. Dopamine is a(n) _____ that sends pleasure messages to the brain.

17. Barbiturates _____ the rate of nervous system activity, while amphetamines _____ the rate of nervous system activity.

Drugs
can act as
narcotics   depressants   ?
mimic   include   such as   such as
?   ?   ?   ?   nicotine
?
that block   is a modification of   affects   is a potent form of   use is
pain   morphine   limbic center   cocaine   legal
affects   affects   use is not

Short Answer

20. Students should suggest the presence of endorphins.
21. Students should indicate that constant use of heroin increases the number of neurotransmitter receptors in a synapse.

22. Prolonged use of cocaine causes the brain to adjust to its presence. The addiction is the direct result of changes in the number of receptor proteins, as nerve cells desensitize in response to the drug.

23. Students should describe fetal alcohol syndrome, a condition that causes physical and mental disabilities in newborn babies.

18. Many alcoholics suffer from _____ of the liver, a condition in which healthy tissue is replaced by scar tissue.

19. The source of marijuana is the plant _____ _____ .

Short Answer

20. Why might a person who is injured in an accident feel no pain for some time after the accident?

21. Why does the body of a heroin user become extremely sensitive to pain when heroin is no longer taken?

22. How does a person develop a tolerance for the drug cocaine?

23. Describe a condition that could affect a newborn baby if the mother drank alcohol during pregnancy.

24. Describe two studies that show that cigarette smoke can be dangerous to non-smokers.

Interpreting Graphics

25. Complete the chart below.

Drug	Source	Drug type	Affect on neurons
	Opium poppy	Narcotic	Mimics enkephalins
Cocaine	Coca plant	Stimulant	
	Tobacco plant	Stimulant	Binds to acetylcholine receptors
Alcohol	Fermentation of grapes or grain	Depressant	
Mescaline	Peyote cactus		Alters transmission of norepinephrine

Reviewing Themes

26. *Patterns of Change*
When a drug is no longer taken, how are the number of protein receptors affected?

27. *Interacting Systems*
Coaches warn teenage athletes that cigarette smoking will adversely affect their physical performance. What is the basis for these warnings?

Thinking Critically

28. *Inferring Conclusions*
Early formulations of a cola soft drink contained a derivative of the coca plant. Why is the coca plant derivative no longer an ingredient in cola soft drinks?

29. *Applying Concepts*
An author spends at least six hours of every day writing. His friends who want him to join them on their vacation claim that he is addicted to his work. Would you describe the author's addiction as physiological or psychological? Explain.

30. *Building on What You Have Learned*
In Chapter 5 you learned about the process of fermentation. How is fermentation involved in the production of ethyl alcohol?

Cross-Discipline Connection

31. *Biology and Civics*
Smoking is now prohibited in many public buildings and areas in many cities. What scientific findings have led to the enactment of these laws?

Discovering Through Reading

32. Read the article "Drunk on the Job," in *Motor Trend*, June 1992, page 8. What was the nature of the experiment conducted by *Motor Trend* employees? What were the results of the experiment?

Thinking Critically

28. Students should suggest the addictive nature of the drug over a period of time.

29. Psychological, students should suggest that physiological addiction involves changes that take place within the cells of the body.

30. In cellular respiration, glucose is broken down by glycolysis to form pyruvic acid and 2 ATP molecules. Pyruvic acid is broken down further by fermentation to form ethyl alcohol.

Cross-Discipline Connection

31. Students should discuss the 1992 study of Italian women with precancerous lesions on their lungs, and the study of nonsmokers confined in airplanes.

Discovering Through Reading

32. Four drivers drove a short, tight road course. Three were given drinks and one remained sober. The results showed that a drinker who is driving cannot think clearly and is particularly poor at dividing his or her attention or dealing with surprises.

24. Students could suggest studies performed on nonsmokers and smokers on an airplane, or the 1992 study done on Italian women who were found to have precancerous lesions on their lungs.

Interpreting Graphics
25. • opium, morphine, heroin
   • blocks the reabsorption of dopamine
   • nicotine
   • alters the lipid structure within the plasma membranes of nerve cells
   • hallucinogen

Reviewing Themes
26. The number of protein receptors are few. Over a period of time, the body must accept the fact that it is not functioning properly and gradually adjust the number of receptor proteins upward.

27. Students should suggest studies done on smoking and lung cancer.

Procedural Note

1. Review the Prelab Preparation questions and discuss the construction of the apparatus with students.
2. Clean glass tubing is required for each investigation.
3. Determine if any student has a medical condition such as asthma, which might preclude participation.
4. Be sure the room is well ventilated and that smoke alarms are not set off by the experiment.

Prelab Preparation Answers

- tar, nicotine, carbon monoxide, carbon dioxide, water, ash particles
- tar, nicotine, and carbon dioxide

Procedure Answers

3. the rubber top of a medicine dropper
9. Brown-black substances collect on the inner surface of the glass tubing.
10. The tube's temperature increases as the cigarette is smoked.

Chapter **30** *Investigation*

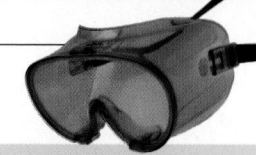

How Does Smoking Affect the Air Passages?

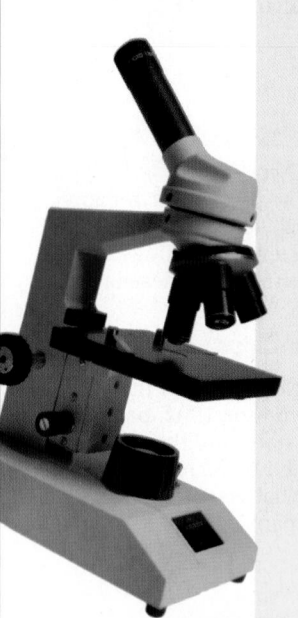

Objectives

In this investigation you will:
- *build* a model that represents an air passage in the human respiratory system
- *observe* tobacco smoke as it appears in the respiratory system
- *infer* the effect of tobacco smoke on cells

Materials

- preassembled glass tubing model
- rubber tubing
- aspirator
- burette clamp
- ring stand
- cigarettes
- matches

Prelab Preparation

Review what you have learned about tobacco by answering the following questions:
- What substances are found in tobacco smoke?
- Which of these substances might be harmful to body tissues?

Procedure: Examining the Effects of Smoking

1. Form a cooperative group of four students. Work as a team to complete steps 2–12.
2. Obtain a glass tubing model made from glass tubing, a rubber stopper, two segments of rubber tubing, and the rubber top of a medicine dropper.
3. Compare the model with the diagram shown below.

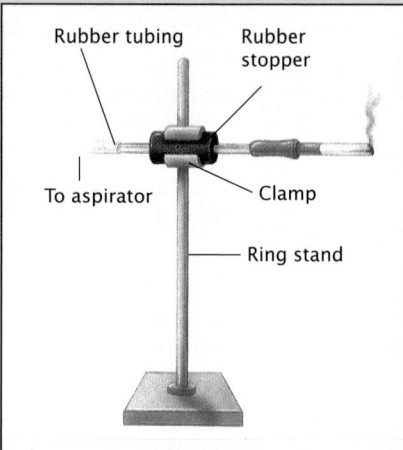

Identify all the labeled parts. *Which end of the model will be used as a cigarette holder?*

4. Connect a 20-cm piece of rubber tubing to the side arm of an aspirator. Connect another 20-cm piece of rubber tubing to the bottom end of the aspirator and place the end of the tubing in the sink near the drain.
5. Use a burette clamp to hold the model by the rubber stopper at an appropriate level on the ring stand, as shown in the figure.
6. Place a cigarette in the end of the rubber holder.

7. Turn the water on. The flow of water through the aspirator will create suction that will draw air through the model.

8. Ignite the cigarette by placing a lighted match at the tip of the cigarette. Continue to draw air through the model using the aspirator and the faucet to "smoke" the cigarette.

9. Use the apparatus to "smoke" two or three cigarettes. *What appears on the inside of the glass tubing as the cigarettes are smoked?*

10. Gently touch the glass tubing. *How has the temperature of the glass changed since beginning the investigation?*

11. Make sure that all cigarettes are extinguished and dispose of the remains. Disconnect the aspirator and cigarette holder from the glass tubing. Remove the glass tubing from the burette clamp and describe the odor of the material in the glass tubing.

12. Clean up your materials and wash your hands before leaving the lab.

Analysis

1. *Interpreting Models*
What part of the human respiratory system does the glass tubing represent?

2. *Analyzing Observations*
What evidence is there that substances in cigarette smoke can coat air passages?

3. *Analyzing Observations*
What substances are probably coating the inside of the glass tubing?

4. *Making Inferences*
What effect might these substances have on the surface of cells that line human respiratory passages?

Thinking Critically

1. Explain how a scientist could determine the chemicals present in cigarette smoke.

2. Explain how smoking might affect the fetus of a woman who smokes, even though the fetus never actually inhales any of the smoke.

Analysis Answers

1. The glass tubing represents the air passages in the human respiratory system.
2. The gases condense into a brown-black coating on the inside of the glass tube.
3. primarily tar
4. Answers will vary but should suggest that tar will inhibit the normal functioning of the cilia.

Thinking Critically Answers

1. The chemical nature of cigarette smoke can be determined by filtering the captured cigarette smoke and performing a chemical identification test on the filtrate and filter.
2. Students may suggest that the fetus derives its oxygen and nutrients from its mother's blood. Any chemicals from the cigarette will be exchanged through the placenta.

Chapter 31

Circulation and Respiration

Planning Guide

	Objectives/Themes	Classwork Resources	Homework Resources
31.1	**1.** Explain the three functions of circulation. **2.** Describe the components of blood. **3.** Name the three major types of blood vessels. **4.** Describe the role of the lymphatic system. **Themes:** Energy and Life, Scale and Structure, Stability, Interacting Systems	**Text** Journeys *Tour of the Blood Vessels* p. 715 **Teacher's Resource Binder** Focus Activity 31 *Determining Lung Capacity* **Other Resources** Science in Action Video *Photosynthesis and Cellular Respiration**</sup>	**Text** Section Review, p. 717 **Teacher's Resource Binder** Directed Reading Worksheet 31.1 **Other Resources** Audiocassette 31.1*
31.2	**1.** Describe the structure of the heart. **2.** Trace the two routes blood can take through the body. **3.** Define blood pressure and explain how it is measured. **4.** Describe the causes and symptoms of two cardiovascular diseases. **Themes:** Interacting Systems, Patterns of Change, Scale and Structure, Stability	**Text** Journeys *Evolution of the Heart* pp. 722–723 Science in Action *Cholesterol and Your Health* pp. 726–727 **Teacher's Resource Binder** Lab Investigation 31.1 *Comparing Normal and Sickled RBCs* Extension Worksheet *The Heartbeat* **Other Resources** Transparencies 138–143	**Text** Section Review, p. 725 **Teacher's Resource Binder** Directed Reading Worksheet 31.2 Reaching Worksheet* *The Circulatory System* **Other Resources** Audiocassette 31.2*
31.3	**1.** Explain how the diaphragm and rib muscles work to move air into and out of the lungs. **2.** Describe the pathway by which oxygen from air travels to a cell in the body. **3.** Explain how your body regulates your breathing rate. **4.** Describe three diseases of the respiratory system. **Themes:** Energy and Life, Interacting Systems, Patterns of Change, Scale and Structure, Stability	**Text** Investigation *How Does Exercise Affect Pulse Rate?* pp. 736–737 **Other Resources** Transparency 144	**Text** Section Review, p. 732 **Teacher's Resource Binder** Directed Reading Worksheet 31.3 Vocabulary Review Worksheet* **Other Resources** Audiocassette 31.3*

*Reteaching Options

Demonstrations	Assessment
31.1: pp. 713, 714 **31.2:** pp. 719, 720, 724 **31.3:** pp. 729, 730	Chapter Review pp. 734–735 Portfolio Assessment p. 709D Chapter Test—Teacher's Resource Binder Test Generator

Research Notes

Connection to Space Sciences: Low Gravity and Circulation

The Spacelab Life Sciences-1 mission of NASA's space shuttle has provided valuable evidence for biologists studying how the human body adjusts during spaceflight. Before, during, and after a shuttle flight in June 1991, researchers monitored several astronauts' cardiovascular systems.

Since some of the effects of weightlessness resemble prolonged bed rest, researchers hypothesized that the astronauts' circulations would show the same patterns seen in people confined to bed. At first, such people have an increase in cardiovascular output, the volume of blood pumped by the heart. Then there is a slow decrease to normal levels over the following days.

The researchers also expected blood pressure to fluctuate with the cardiovascular output. On the other hand, they expected a slow

increase in filling pressure in the heart as it became easier for blood from the lower body to reach the heart.

The results of the experiment showed that the hypotheses the researchers made before the mission were mostly wrong. The astronauts had an increase in cardiovascular output, but it remained constantly larger than normal instead of gradually decreasing back to normal levels.

The astronauts' blood pressures remained constant throughout

the mission, with the cardiovascular system filling organs with more blood than usual. The filling pressure of the heart decreased dramatically for the entire flight.

The researchers say that these physiological changes do not affect people's ability to work in space. However, they can cause adverse effects once astronauts return to Earth. They found that the ability of the astronauts to exercise was not fully recovered until seven days after their return to Earth.

The rejection of the scientists' hypotheses provides them with valuable insights into how to lessen the adverse effects of spaceflight. It also demonstrates that the circulatory system is not yet very well understood. Studies of the system on Earth have been plagued by misleading gravitational effects, which change results for persons standing and sitting. More work will be needed to better understand the circulatory system.

Investigation Notes

How Does Exercise Affect Pulse Rate?

pages 736–737

Purpose: Pulse rate is a measure of the rate at which the left ventricle contracts and forces blood into the arteries. This investigation examines the effect of exercise on pulse rate. While individuals will differ in absolute pulse rates, resting rates are expected to be lower than those taken just after exercising.

Prelab Preparation

1. The teacher must be certain that individuals doing the strenuous exercises in this lab have no physical restrictions. The school nurse should be consulted beforehand to check students planning to participate. Those

students with health problems or who are reluctant should not be required to perform the exercise test. Students not performing the exercise test can collect and record data, and calculate averages.

2. Students will need access to a watch, clock, or timer with a secondhand. Be sure that the stool, bench, or chair used for exercising is stable and sturdy. You may wish to use the bleachers in the gymnasium to complete this investigation.

3. All students should have their resting pulse rate recorded by their partner.

Answers will be found on pages 736–737.

Meeting Individual Needs

Objectives

1. Students will demonstrate appropriate use of core vocabulary for the chapter (Vocabulary File).

2. Students will explain the different blood types and be able to relate their own blood type to their ability to receive and donate blood (Multicultural Lesson Plan).

3. Students will relate the operation of the circulatory system and tests of the system to their health (Teaching Strategy).

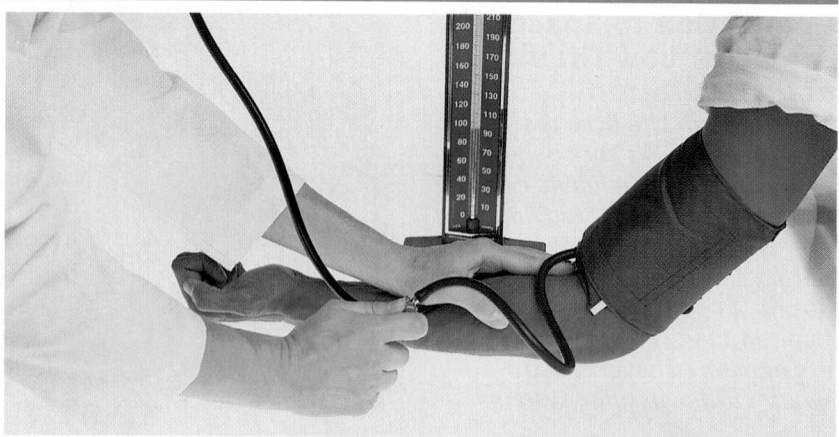

Vocabulary File
(Developing Vocabulary/ Limited English Proficiency)

If you are not already using the Vocabulary File, refer to Chapter 1 for its preparation. See the Chapter Highlights on page 733 for a list of suggested words.

Multicultural Lesson Plan
**Blood Donation
(Dealing with Ethnic Diversity)**

Preparation: Contact your local chapter of the American Red Cross to determine how students can assist in organizing a blood drive. Ask for a local education program coordinator to be a guest speaker for the class.

Teaching Strategy: Have the school nurse or a guest speaker from the Red Cross visit the class to discuss the history of blood typing and transfusions, with a special emphasis on the role of Dr. Charles Drew. Be sure students realize that there is still a need for donated blood.

Assessment:
1. Students should help plan and execute a blood drive at their school or elsewhere. They can help recruit and register donors, design and create posters and other publicity for the drive, and serve as workers during the drive.

2. If possible, students should learn the blood types of each person in their family (including themselves). After they bring the results to class, they should work with others in the class to determine who among their friends, classmates, and family could donate or receive blood from them if the need arose. They should present the results in a chart format.

Teaching Strategy
The Circulatory System
for pages 711–732
(Developing Organizational Skills/Multisensory Learners)

Preparation: This teaching strategy relies on the use of the school nurse as a guest speaker. Be sure to schedule a preclass meeting with the nurse so the material to be covered is understood by both you and the nurse.

Materials: The nurse may request help in obtaining some lab materials from the local hospital or a doctor's office.

Procedure: Discuss the flowing of blood as an introduction to the school nurse's demonstrations. Refer students to the text references listed before each explanation/demonstration from the nurse. Read aloud what is needed for students to focus on the biology behind the explanation to be given by the nurse.

• Pages 712–713: Nurse explains how blood counts are done in a medical lab.

• Page 716: Nurse explains how blood typing is done.

• Pages 718–719: Nurse listens to a student's heart and hears a "lubb-dup" sound, which she allows other students to hear.

• Page 724: Nurse demonstrates how blood pressure is taken using student volunteers. (This test can be recalled later and also used as an introduction to the pulse rate investigation that follows this chapter.)

• Page 732: Nurse explains how a hospital lab measures lung capacity. Be sure to relate this to students who may be using an inhaler for asthma.

• Pages 731–732: Nurse explains how blood gas testing is done. Be sure students are told that the blood is taken at the wrist for this test. Nurse should explain why, leading into a discussion of the differences between veins and arteries.

Lead a class discussion about the ways that the information from these tests can help students be better informed about their health.

Additional Strategies
Visual Strategies

Pages 712, 713, 714, 716, 717, 718, 719, 720, 721, 729, 730, and 731

Auditory Learners

Use *Biology: Visualizing Life* Audiocassettes for Sections 31.1, 31.2, and 31.3.

Meeting Individual Needs (cont.)

Cooperative Learning

Measuring Lung Capacity

Timing: Use this activity to introduce Section 31.3.

Group Size: 4 students

Outcome: Students will be able to calculate their lung capacity.

Individual Accountability: Each group member is responsible for calculating his or her lung capacity.

Positive Interdependence: Each group will share the materials and help each other measure lung capacity.

Mix bubble solution for the class using 4 L (1 gal.) of water, 1 cup of dishwashing liquid, and 50 drops of glycerin. Give each group 1 cup of bubble solution, and give each group member a straw. Have group members use the following procedure:

1. Pour bubble solution onto a flat surface.

2. Use your hand to wet an area about 40 cm (16 in.) in diameter.

3. Dip a straw into the bubble solution.

4. Place the straw just touching the soapy surface.

5. Take a breath and gently expel all the air through the straw to form a bubble.

6. After the bubble pops, use a meter stick to measure the inside diameter of the ring of soap suds left by the bubble.

7. Calculate lung capacity using the formula $V = \frac{4}{3}\pi r^3$ divided by 2 to determine the volume of half a sphere.

8. Repeat the procedure to blow a total of three bubbles.

9. Average data from the three trials.

Portfolio Assessment

Students should select their best work and provide a self-reflective rationale for their selections. Students can make selections in the following areas.

1.	*Content*	One concept map from the chapter (See page 812 for evaluation criteria.)
2.	*Reading Comprehension*	One Directed Reading Worksheet from the Teacher's Resource Binder (Use the answer key to evaluate for accuracy.)
3.	*Writing*	Using the Vee Form, summarize a magazine or newspaper article relating to the circulatory system or the respiratory system, or diseases affecting them. (See page 22T for evaluation criteria.)
		Or: Select a writing task or project from the Chapter Review.
4.	*Performance Assessment*	One Vee form from a chapter investigation or lab manual investigation (See page 22T for evaluation criteria.)

Teacher makes selections in the following areas.

1.	*Formal Assessment*	Chapter test (Test A, B, or the Test Generator)` The teacher-scored test should be reviewed by the student. Incorrect responses should be corrected by the student before the test becomes part of the portfolio.
2.	*Informal Assessment*	Use the Direct Observations Checklist, page 33T, during a laboratory or other cooperative learning experience.
3.	*Performance Assessment*	Have students design their own posters, similar to those on page 732, warning of the dangers of smoking and other threats to the respiratory system.

Concept Map Answer

The following is one possible answer to the Relating Concepts exercise on page 734.

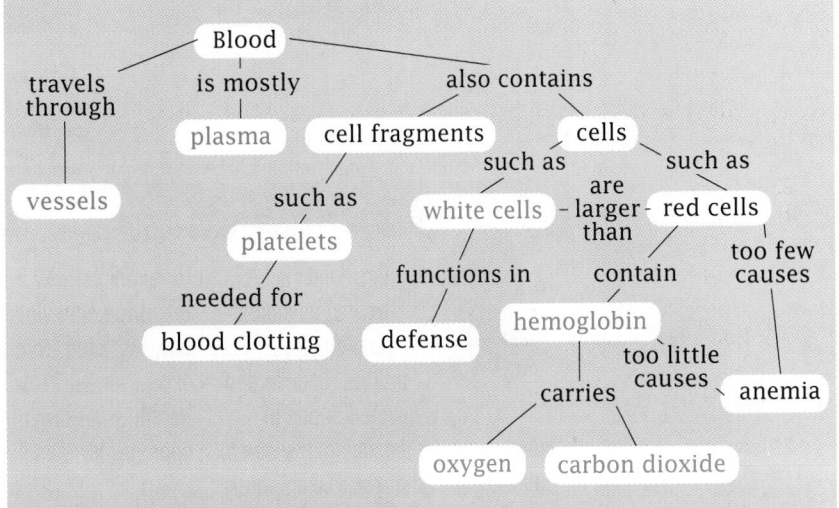

Chapter 31

Determining Prior Knowledge

- Use a microprojector to show the class an *Amoeba*, a *Paramecium*, or any other single-celled organism. Have students explain why such an organism does not need a circulatory system.
- Use food coloring to prepare a dark red solution. Tell students that the fluid is blood. Have them explain what they would find in this "blood" if they examined it under a microscope. Have them describe how blood travels through the body.
- As students are entering class, play a recent hit song that has the word "heart" in the title. When the song is finished, ask students what they know about the structure and function of the human heart.
- Have students take a deep breath and hold it for as long as possible. Have them explain what was happening to oxygen and carbon dioxide as they held their breath, and why they could no longer hold their breath.

Chapter **31**

Review

- gas exchange (Section 22.2)
- evolution of the lung (Section 22.2)
- the terms *heterozygous* and *homozygous* (Glossary)

Circulation and Respiration

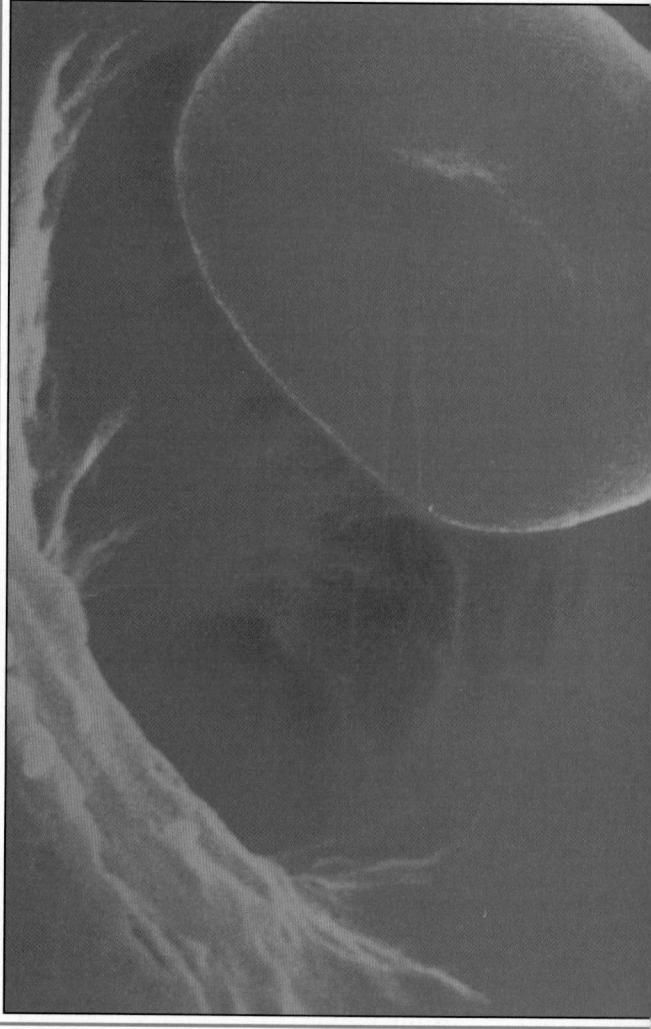

A red blood cell is shown here entering a blood vessel. The cell will travel on a 60,000-mile journey throughout the human body.

■ Author's Rationale ■

Circulation is treated as a basic physiological process. Each section is devoted to one of three elements: the piping, the pump, and the gas exchanger. The emphasis is on how structure leads to function; for example, why it is necessary for the atrium and ventricle to be electrically insulated from one another, except for a narrow bridge of tissue. Health issues are introduced where appropriate.

INSIDE YOUR BODY, BLOOD FLOWS ALONG A VAST INNER HIGHWAY ON A 60,000-MILE JOURNEY, FAR ENOUGH TO ENCIRCLE THE PLANET TWO AND ONE-HALF TIMES. BLOOD NOURISHES AND CLEANSES THE ENTIRE BODY, DELIVERING FOOD AND OXYGEN TO EVERY CELL AND PICKING UP WASTES. BLOOD TRAVELS THROUGH EVERY LIVING TISSUE, PROPELLED BY THE BEATING OF THE HEART.

31.1 *Circulation*

Objectives

① Explain the three functions of circulation.

② Describe the components of blood.

③ Name the three major types of blood vessels.

④ Describe the role of the lymphatic system.

Transporting Materials Through the Body

All living things must capture materials from their environment to use in carrying on their life processes. Many bacteria, single-celled protists, and simpler multicellular animals live within a liquid environment, enabling materials to diffuse directly into and out of the organism through plasma membranes. However, larger animals with many cells stacked in layers, such as earthworms, armadillos, and humans, cannot rely solely on diffusion to supply needed materials and carry away wastes. In these organisms, a circulatory system transports oxygen, carbon dioxide, food molecules, hormones, and other materials to and from the cells of the body. In addition, the circulatory systems of mammals and birds help maintain their constant body temperatures. The circulatory system also carries cells that help protect the body from disease. All of these functions help the body maintain homeostasis and are essential to survival.

Our modern understanding of the human circulatory system began with the work of the seventeenth-century English physician William Harvey. Harvey demonstrated that blood circulates in one direction in a closed circuit throughout the body. An illustration of his experiment is shown in **Figure 31.1**.

Figure 31.1
In 1628, William Harvey demonstrated that blood travels from the heart to the body's limbs and then back again. If an arm is bound above the elbow, swellings appear in the lower arm. Harvey used this as evidence that blood flow back to the heart was slowed.

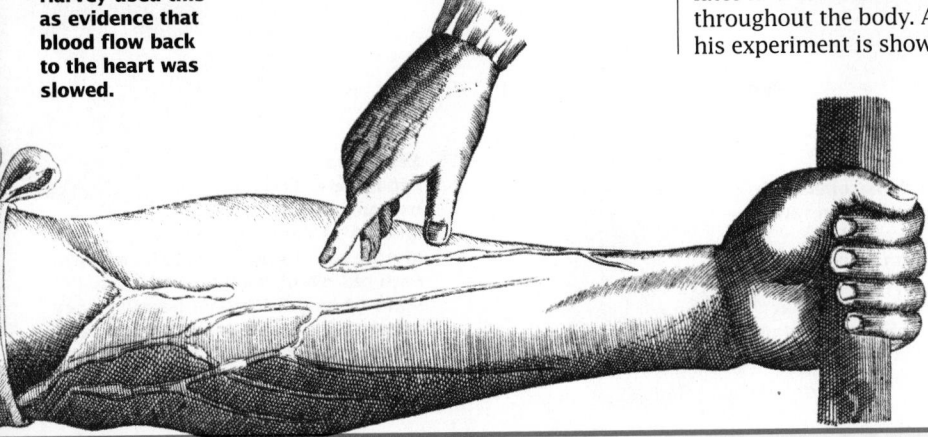

Lesson Plan 31.1

Phase 1
PREPARATION

Key Concepts

- The circulatory system transports materials, helps maintain a constant body temperature, and functions to protect the body from disease.
- The blood is composed of plasma, red blood cells, white blood cells, and platelets.
- Red blood cells contain hemoglobin that transports oxygen to all cells in the body.
- White blood cells help defend the body against disease.
- Platelets are essential for clotting, a process that is necessary to repair injured blood vessels.
- Arteries are thick, muscular, elastic vessels that transport blood away from the heart.
- Capillaries are vessels whose walls are only one cell thick, which allows substances to easily diffuse between the blood and cells.
- Marker proteins on the surface of red blood cells are responsible for the four major blood types: A, B, AB, and O.
- Marker proteins are also responsible for Rh factor.
- The lymphatic system consists of vessels that return fluid and proteins to the blood, away from the heart.

Reading Strategy

This section offers another opportunity for an "outside" reading assignment. Check with the school librarian to obtain a description of William Harvey's work and make it available for students to read. While reading about his experiments, students should appreciate Harvey's simple but brilliant piece of scientific work.

Phase 2

TEACHING STRATEGIES

Theme Connection

Stability
Have students identify the ways in which the circulatory system functions in homeostasis.

Connection: Chapter 27

Tissues
Remind students that tissue is a group of cells that work together to perform a specific function. What kind of tissue is blood?

Mathematics Connection

Blood Numbers
A human adult has about 5 L (1.25 gal.) of blood. The plasma is more than 90 percent water. The body contains some 30 trillion red blood cells and about 60 billion white blood cells. Every second, about 2 million new red blood cells are made in the bone marrow to replace those that die at the end of their 120-day life span. Have students calculate how many new red blood cells are made every hour.

Visual Strategy

Figure 31.2
Have students name substances (salts, enzymes, hormones, vitamins, minerals, nutrients, cellular waste products) that would be found in the plasma portion of the blood shown in this figure.

Connection: Chapter 8

Genetic Engineering
Ask students how genetic engineering could provide a cure for people suffering from sickle cell anemia. Remind them that DNA directs the synthesis of proteins, including hemoglobin.

Blood: A Liquid Tissue

Blood has been called the river of life. It is the tissue responsible for transporting nearly everything within the body. Among all the body's tissues, blood is the only liquid tissue. The material that makes blood a liquid is a protein-rich substance called **plasma** (*PLAZ muh*). Plasma makes up approximately 55 percent of blood. The other 45 percent is made mostly of cells—red blood cells, white blood cells, and platelets.

Red blood cells carry oxygen to other body tissues
Most of the cells that make up blood are red blood cells. There are approximately 5 million red blood cells per cubic millimeter of blood! The main function of a **red blood cell** is to carry oxygen from the lungs to all cells of the body. Red blood cells also carry carbon dioxide from cells back to the lungs to be exhaled. The structure of a red blood cell is well suited to this function.

As shown in **Figure 31.2a**, red blood cells are shaped like round cushions, squashed in the center. Red blood cells are filled with a protein called hemoglobin, an iron-containing molecule that gives blood its red color. Oxygen binds easily to the iron in hemoglobin, making red blood cells efficient oxygen carriers. A single red blood cell contains about 250 million hemoglobin molecules. Each red blood cell can carry about 1 billion molecules of oxygen.

A condition in which a person has too few red blood cells or not enough hemoglobin is called **anemia** (*uh NEE mee uh*). Anemia results in a shortage of oxygen in the cells of the body. Recall from Chapter 6 that people who have sickle cell anemia have an abnormal form of hemoglobin that causes cells to become sickle-shaped. The sickled red blood cells clog small blood vessels and interfere with oxygen delivery. Sickle cell anemia is a serious disease that cannot be cured and can be fatal.

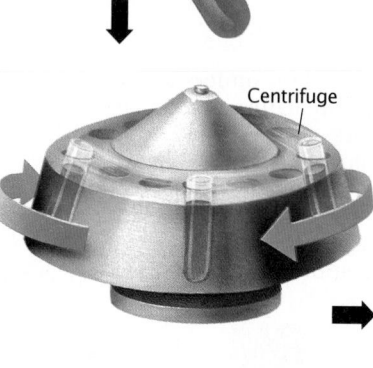

Vacuum syringe

Centrifuge

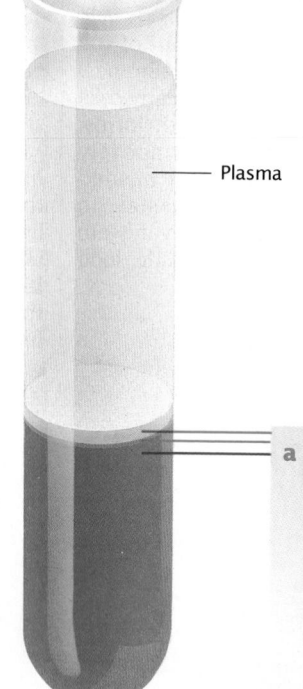

Plasma

Figure 31.2
After a blood sample is collected in a vacuum syringe, it is spun rapidly in a device called a centrifuge (above). This rotation forces the blood to settle in layers, with the heaviest cells (red blood cells) at the bottom of the tube, white blood cells and platelets in the middle, and plasma at the top.

a   Red blood cells
Red blood cells carry oxygen from the lungs to all cells of the body, and carry carbon dioxide back to the lungs to be exhaled. Each red blood cell has an average life span of only 120 days because they mature without a nucleus. About 2 million new red blood cells are produced every second in the bone marrow to replace those that are worn out.

■ Cultural Perspective ■

Donella J. Wilson, Ph.D.
(African-American)
Dr. Wilson studies normal and abnormal development of red blood cells at Meharry Medical College in Nashville, Tennessee.

White blood cells defend the body

White blood cells form a mobile army that protects the body against invading bacteria, viruses, or other foreign cells. White blood cells called macrophages (*MAK roh fayj iz*) act as a clean-up crew, scavenging worn-out or dead cells. While red blood cells are confined to the bloodstream, white blood cells are able to squeeze in and out of blood vessels. They move toward areas of tissue damage and infection by responding to chemicals released by damaged cells. By following the chemical trail, macrophages are able to migrate to infected areas in the body. There, they gather in large numbers to engulf and digest foreign cells.

As shown in **Figure 31.2b**, white blood cells are colorless and irregularly shaped. Most of them are manufactured and stored in bone marrow until they are needed by the body. Although they are larger than red blood cells, they are considerably less numerous; normally there are 4,000 to 11,000 white blood cells per cubic millimeter of blood. Whenever white blood cells are mobilized for action, the body speeds up their production so that twice the normal number may appear within a few hours. A white blood cell count over 11,000 generally indicates the presence of a bacterial or viral infection in the body.

Platelets help repair damaged blood vessels

Platelets are not really cells; they are fragments of cells formed from large white blood cells within bone marrow. Each cubic millimeter of blood contains between 250,000 and 500,000 tiny platelets.

Platelets are essential for the clotting process that occurs when blood vessels are injured, as shown in **Figure 31.2c**. When you accidentally cut your finger, platelets come into contact with the ends of the broken blood vessels. There, the platelets swell and stick to the rough surfaces created by the injury as well as to each other. Certain of these platelets rupture and release a protein, which triggers a series of reactions that forms a clot. The clot stops the bleeding and hardens into a patch over the injured area. In time, the injury is repaired by the growth of cells that replace those that were damaged.

Patterns of Change

What can cause the number of white blood cells in the body to increase?

b *White blood cells*
White blood cells help defend the body against invading bacteria and other intruders. Some white blood cells move against the flow of the bloodstream. Others squeeze through the walls of blood vessels. They arrive at the damaged area by following chemical signals given off by infected cells. The white blood cells then gather and attack foreign intruders.

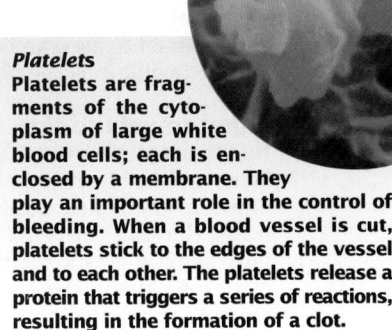

c *Platelets*
Platelets are fragments of the cytoplasm of large white blood cells; each is enclosed by a membrane. They play an important role in the control of bleeding. When a blood vessel is cut, platelets stick to the edges of the vessel and to each other. The platelets release a protein that triggers a series of reactions, resulting in the formation of a clot.

Theme Connection

Scale and Structure

Emphasize the relationship between structure and function of arteries, veins, and capillaries. Point out that arteries are thick, muscular, and elastic; veins have valves; and capillaries have walls that are only one cell thick. Ask how each of these structural features relates to the vessel's function.

Visual Strategy

Figure 31.3

Have students trace the path blood follows in this figure. Be sure they understand that blood coming from the heart circulates as follows: artery → capillary → vein → back to heart.

Demonstration 3

Muscles and Blood Flow

Show students a photograph of a person in a hospital bed. Ask why hospital workers try to get these people to walk, even if only for a few minutes each day. Students should recognize that muscle movements squeeze the walls of veins, thus preventing blood from accumulating in parts of the body like the legs.

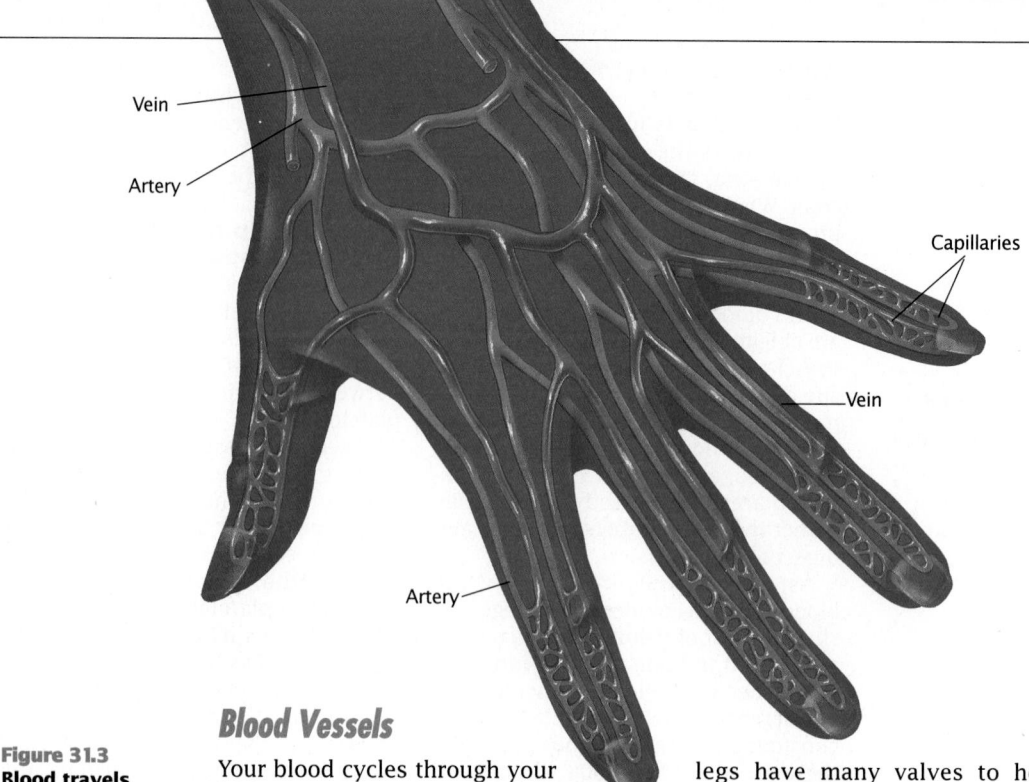

Figure 31.3
Blood travels through all parts of the body (such as the hand above) in blood vessels. Arteries (red) carry blood away from the heart to the body's organs and tissues. Veins (blue) return blood to the heart. Exchange of oxygen and cellular wastes occurs in tiny tubes called capillaries (purple). These tubes connect arteries to veins.

Blood Vessels

Your blood cycles through your body 1,440 times a day, traveling through tubes called **blood vessels**. Blood vessels carry blood from the heart to the organs in the body and back again to the heart. They have flexible walls so that they can change diameter when more blood is needed. The blood vessels of the hand are shown in **Figure 31.3**.

Blood travels away from the heart through arteries

Arteries are the blood vessels that carry blood away from the heart. In most instances, arteries carry oxygen-rich blood. Arteries have highly elastic, muscular walls, enabling them to expand and contract to accommodate changes in blood volume as the heart pumps.

Blood travels to the heart through veins

Veins are the blood vessels that return blood to the heart. Unlike the artery walls, the walls of veins are thin and only slightly elastic. Veins have flaps of tissue called valves that open and close to prevent blood from flowing backward. The veins in your arms and legs have many valves to help return blood to the heart against gravity. When valves do not work properly, blood tends to build up within the vein. The walls of the vein become stretched and twisted, causing a condition called varicose veins.

Materials are exchanged in the capillaries

Since the purpose of blood circulation is to transport substances, it is important that these substances be able to move from the blood into the surrounding cells. A network of tiny tubes called **capillaries** makes this transfer of substances possible. Their walls are only one cell thick, so substances are able to diffuse through them—oxygen and nutrients leave the blood and enter the fluid surrounding cells, and carbon dioxide and wastes from cells enter the blood in the capillary to be carried away. Networks of capillaries connect arteries to veins. Most capillaries are so narrow that blood cells must pass through them in single file. The structures of capillaries, arteries, and veins are shown in the *Tour of the Blood Vessels* on page 715.

■ *Journeys* ■ *Journeys* ■ *Journeys* ■ *Journeys* ■ *Journeys*

Tour of the Blood Vessels

Arteries, veins, and capillaries are the three main types of blood vessels in the human body. The walls of these vessels are well constructed for the passage of blood throughout the body.

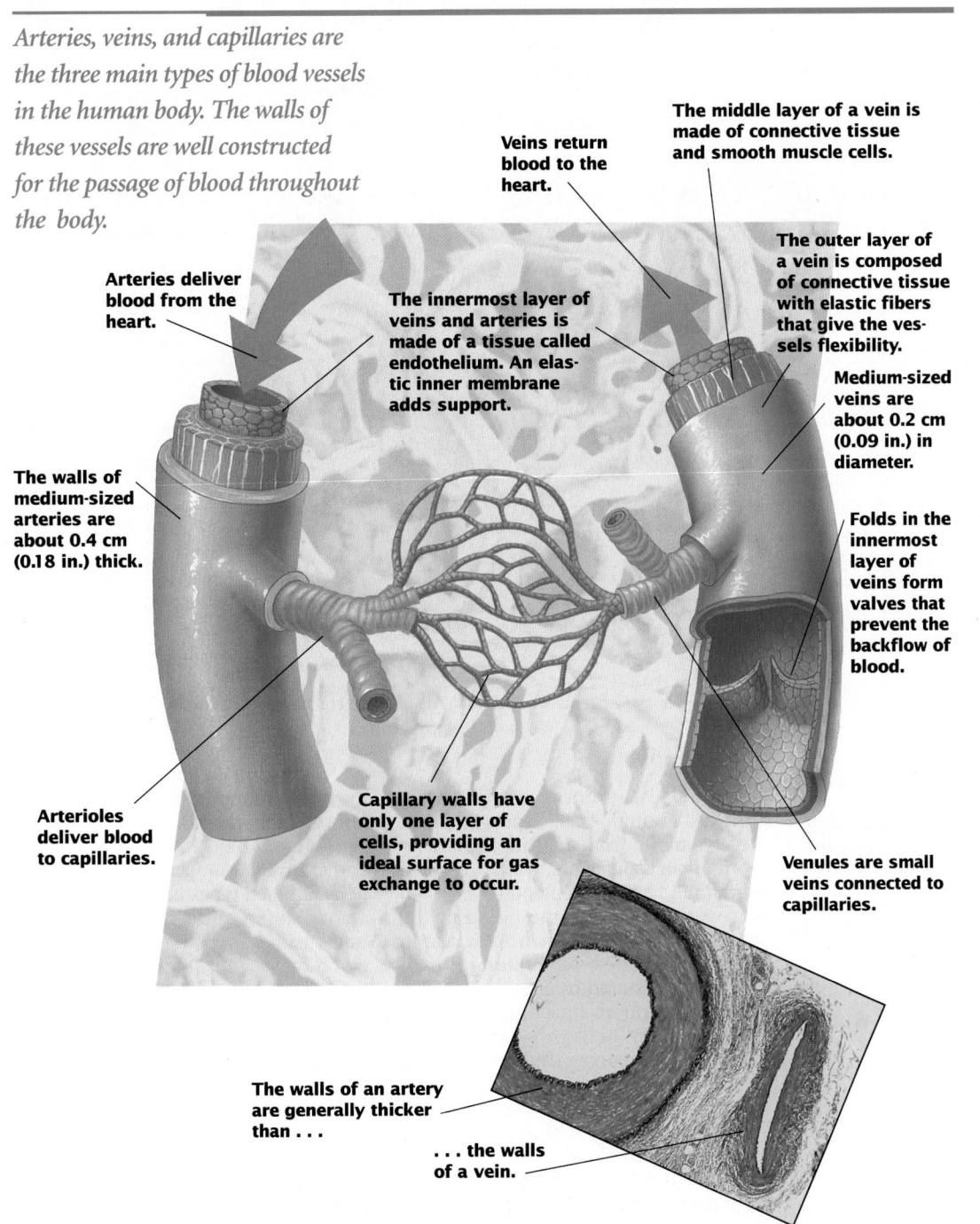

Arteries deliver blood from the heart.

Veins return blood to the heart.

The middle layer of a vein is made of connective tissue and smooth muscle cells.

The innermost layer of veins and arteries is made of a tissue called endothelium. An elastic inner membrane adds support.

The outer layer of a vein is composed of connective tissue with elastic fibers that give the vessels flexibility.

The walls of medium-sized arteries are about 0.4 cm (0.18 in.) thick.

Medium-sized veins are about 0.2 cm (0.09 in.) in diameter.

Folds in the innermost layer of veins form valves that prevent the backflow of blood.

Arterioles deliver blood to capillaries.

Capillary walls have only one layer of cells, providing an ideal surface for gas exchange to occur.

Venules are small veins connected to capillaries.

The walls of an artery are generally thicker than . . .

. . . the walls of a vein.

Using the Feature

- Blood vessels are the site where many cardiovascular disorders originate. Use this feature as a foundation for a discussion about atherosclerosis and coronary artery spasms, two principle underlying causes of coronary artery disease. Hypertension is the most common disorder affecting the heart and blood vessels and can be explained at this time as well. Ask students to research the causes for these disorders, the dangers they pose, and any ways in which they can be prevented, treated, or cured.

- Students may be interested in learning about some of the drugs and techniques used in treating cardiovascular diseases. Present information about vasodilators (nitroglycerin), fibrinolytic drugs (streptokinase), inotropics (digitoxin), antiarrhythmics (quinidine, lidocaine), cholesterol-lowering agents (clofibrate), antihypertensives (diuretic drugs and beta blockers), and anticoagulants. Some surgical techniques to discuss include cardiac catheterization and open-heart surgery.

Discussion

Guide the discussion by posing the following questions.

1. What are the three main types of blood vessels in the human body?
   arteries, veins, and capillaries

2. Capillaries have walls that are only one cell thick. How does this structure assist their function?
   It provides an ideal surface for gas exchange to occur.

3. How is the backflow of blood in veins prevented?
   Veins have folds of tissues called valves, which prevent blood from flowing in the wrong direction through veins.

715

Visual Strategy

Figure 31.4 and Table 31.1
Blood transfusions have been accepted as a routine procedure of medical practice for nearly 100 years. Prior to 1900, however, blood transfusions often led to serious complications and occasionally death. In the early 1900s, Karl Landsteiner discovered that sometimes the red blood cells taken from one member of his laboratory staff would clump when mixed with those taken from another lab worker. In other cases, two different blood samples would not clump. As a result, Landsteiner concluded that there were different categories of blood. Soon after, the four major blood types were determined.

Connection: Chapter 6

The Genetics of Rh Factor
Inform students that a person with Rh-positive blood can be either homozygous dominant (+/+) or heterozygous (+/−). An Rh-negative person must be homozygous recessive (−/−). Have students use a Punnett square to demonstrate why no concerns about Rh factor exist if both parents are Rh negative.

"I gave blood for the first time this year. It didn't hurt at all!"

"The nurse told me that all blood is screened for signs of infectious diseases before it's transfused."

"By donating blood, you could help save someone's life."

"Did you know that nearly 99 percent of the blood used for transfusion in the U.S. is drawn from volunteer blood donors?"

Table 31.1   Major Blood Groups: Their Frequencies and Transfusion Capabilities

Blood type	Frequency in the United States (%)				Blood that can be received
	African Americans	Asians	Caucasians	Native Americans	
O	46	38	45	92	O only (universal donor)
A	28	27	41	8	A and O
B	23	21	10	1	B and O
AB	13	4	4	0	A, B, AB, and O (universal recipient)

Figure 31.4
To be eligible to donate blood, like the volunteers above, a person must generally be 17 years of age, weigh at least 100 lbs., and be in good health.

Blood Types

Occasionally an injury or disorder is serious enough that a person must receive blood from another person. As with organ transplants, which you read about in Chapter 4, a blood transfusion can succeed only if the blood of the recipient and donor match. Each year in the United States, 12 million pints of blood are donated by 8 million volunteer blood donors, including the young volunteers in **Figure 31.4**. Among the factors to be considered in matching blood is **blood type**. Blood type is determined by the presence or absence of specific marker proteins found on the surfaces of red blood cells.

The most familiar blood group system uses the letters A, B, and O to label the different modifications of one kind of protein markers. Under this system, the primary blood types are A, B, AB, and O. **Table 31.1** shows the distributions of blood types in the United States among different racial groups and the transfusion capabilities of the different blood types. You will learn more about the basis of the different blood types in Chapter 32.

Another type of marker protein on the surface of red blood cells is the Rh factor, so named because it was originally identified in rhesus monkeys. People whose blood contains the Rh factor are said to be Rh positive (Rh+). People whose blood does not contain the Rh factor are said to be Rh negative (Rh−).

The Lymphatic System

All the cells in your body are bathed in a clear, watery fluid that helps move materials between the capillaries and body cells. This fluid is formed from parts of the blood that diffuse out of capillaries—water, proteins, and other nutrients. If this fluid had no way of returning to the circulatory system, body tissues would soon become flooded and would swell up. At the same time, the constant loss of fluid from the blood would eventually drain the circulatory system. These excess fluids and proteins are returned to the blood by a system of vessels called the **lymphatic system**. Lymph nodes are located at various places along the lymphatic vessels, as shown in **Figure 31.5**. Each lymph node contains many white blood cells that help defend the body against cancerous cells or disease-causing organisms. Lymph nodes also return water, proteins, and other nutrients to the blood.

Stability

Explain why swollen lymph nodes are often considered a symptom of illness.

Lymph node

Figure 31.5

a Lymphatic vessels and lymph nodes located throughout the body make up the lymphatic system.

b Lymph nodes filter foreign matter floating in the body's fluid and prevent cancerous cells, bacteria, and other disease-causing organisms from entering the bloodstream.

Section Review

❶ **List three functions of the circulatory system.**

❷ **Summarize the functions of the three components of blood.**

❸ **Distinguish between arteries and veins.**

❹ **How does the lymphatic system help the body fight infection?**

▪ *Section Review Answers* ▪

1. transports vital materials throughout the body; maintains homeostasis; and protects the body against disease

2. See Figure 31.2, starting on page 712.

3. Arteries are blood vessels that carry blood away from the heart, whereas veins are blood vessels that return blood to the heart.

4. See Figure 31.5 on page 717.

Theme Answer

Stability
Students should explain that swollen lymph nodes contain large numbers of bacteria or virus particles.

Visual Strategy

Figure 31.5
Have students examine the diagram of the human body in this figure to locate where lymph nodes are concentrated; for example, the abdomen, the armpits, and the groin. Point out that the tonsils and adenoids are made of lymphatic tissue. At one time, they were removed at the first sign of recurrent problems. With the recognition of their involvement in the lymphatic system, doctors do not surgically remove them unless necessary.

Phase 3

ASSESSMENT OPTIONS

Closure Strategy
Position Is Crucial
Have students explain why arteries are generally found far below the body's surface, why veins are located closer to the surface, and why capillaries are in close contact with every cell of the body.

Section Review
Assign the Section Review.

Reteaching
Have students describe what the blood sample in the test tube shown in Figure 31.2 might look like if it were taken from someone with either leukemia (cancer of the white blood cells) or an iron deficiency.

Lesson Plan 31.2

Phase 1

PREPARATION

Key Concepts

- The heart is separated into right and left halves and divided into four chambers: two atria and two ventricles.
- Blood travels through the body as follows: body cells → right atrium → right ventricle → lungs → left atrium → left ventricle → body cells.
- Specialized cells, known as the pacemaker, control the heart's beating.
- Pulmonary circulation involves transport of blood between the heart and lungs.
- Systemic circulation involves transport of blood between the heart and rest of the body.
- Blood pressure is the result of blood exerting force on the walls of blood vessels.
- Cardiovascular diseases include heart attacks and hypertension.

Reading Strategy

Have students bring in newspaper or magazine articles that deal with the heart, including any coronary disorders. Have each student write a summary of his or her article and make a brief report to the class at the conclusion of this lesson.

Phase 2

TEACHING STRATEGIES

Visual Strategy

Figure 31.6
Have students calculate how much blood their hearts would pump while riding a bicycle for 15 minutes. To dramatize the volume, fill a liter flask with water and ask how many flasks would be filled by their hearts in that period of time.

718

BLOOD COULD NOT MEET THE BODY'S NEEDS IF IT DID NOT FLOW. BLOOD FLOWS BECAUSE THE CIRCULATORY SYSTEM INCLUDES A PUMP— THE HEART—THAT FORCES BLOOD TO MOVE THROUGHOUT THE BODY. KEEPING FIT, NOT SMOKING, AND EATING FOODS LOW IN ANIMAL FATS WILL REDUCE YOUR RISK OF DISEASES AFFECTING THE HEART AND BLOOD VESSELS.

31.2   How Blood Flows

Objectives

❶ **Describe the structure of the heart.**

❷ **Trace the two routes blood can take through the body.**

❸ **Define blood pressure and explain how it is measured.**

❹ **Describe the causes and symptoms of two cardiovascular diseases.**

The Heart

The heart is a muscular organ that pumps blood throughout the body. When you are sitting still, your heart pumps about 5 L (5.3 qt.) of blood each minute. If you are riding a bike, like the person in **Figure 31.6**, your heart may have to pump up to seven times that amount per minute.

Your heart is divided into left and right halves. The right half pumps blood to the lungs to pick up oxygen and release carbon dioxide. The left half of the heart pumps the oxygen-rich blood to the rest of the body—your head, arms, legs, and all the tissues and organs in between.

As shown in **Figure 31.7**, each half of the heart has an upper and a lower chamber. The upper chamber, called the **atrium**, receives blood coming into the heart. The lower chamber, called the **ventricle**, pumps blood out of the heart. The upper and lower chambers are separated from each other by flaplike valves that control the direction of the blood flow inside the heart. Blood flows into both atria at the same time, and the atria contract together. Similarly, the ventricles contract together. All this activity causes the heartbeat, a sound that is usually described as "lubb dup."

Figure 31.6
When you are active, your heart pumps up to 35 L (37 qt.) of blood each minute. That's seven times the amount your heart pumps when you are at rest.

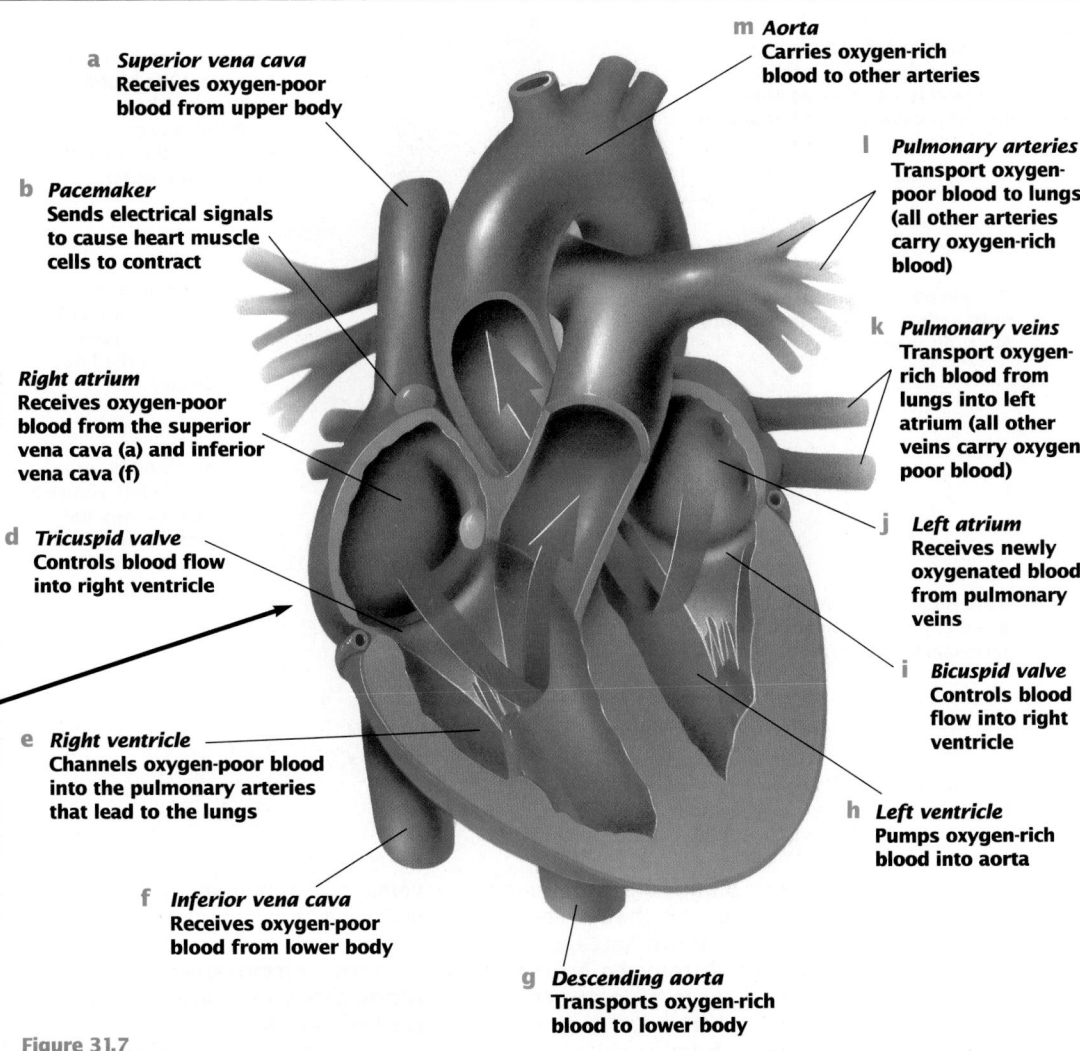

a *Superior vena cava*
Receives oxygen-poor blood from upper body

b *Pacemaker*
Sends electrical signals to cause heart muscle cells to contract

c *Right atrium*
Receives oxygen-poor blood from the superior vena cava (a) and inferior vena cava (f)

d *Tricuspid valve*
Controls blood flow into right ventricle

e *Right ventricle*
Channels oxygen-poor blood into the pulmonary arteries that lead to the lungs

f *Inferior vena cava*
Receives oxygen-poor blood from lower body

g *Descending aorta*
Transports oxygen-rich blood to lower body

m *Aorta*
Carries oxygen-rich blood to other arteries

l *Pulmonary arteries*
Transport oxygen-poor blood to lungs (all other arteries carry oxygen-rich blood)

k *Pulmonary veins*
Transport oxygen-rich blood from lungs into left atrium (all other veins carry oxygen-poor blood)

j *Left atrium*
Receives newly oxygenated blood from pulmonary veins

i *Bicuspid valve*
Controls blood flow into right ventricle

h *Left ventricle*
Pumps oxygen-rich blood into aorta

Figure 31.7
The human heart pumps oxygen-poor blood to the lungs and oxygen-rich blood to the rest of the body.

The "lubb" sound relates to the closing of the valves that lead from the atrium to the ventricle. The "dup" comes very shortly afterward and is related to the closing of the valves between the ventricles and the arteries that lead to the lungs and the rest of the body.

The four-chambered human heart, shown in **Figure 31.7**, is similar in design to the heart of birds. The *Evolution of the Heart* on pages 722–723 summarizes the evolution of the vertebrate heart.

What makes the heart beat?
The heartbeat originates in heart tissue. It begins to beat in the embryo, before any nerves connect it to the brain. It can continue to beat during transplant surgery, after all nerves have been cut. How is this possible?

Each heartbeat is started by the **pacemaker**, a small bundle of cells at the entrance to the right atrium. An electrical signal from the pacemaker, shown in **Figure 31.7b**, travels through the heart muscle cells in the right and left atria, causing them to tighten, or contract. When the signal reaches the right and left ventricles, they also contract. These contractions cause the chambers to squeeze the blood through the heart and push it to other parts of the body.

■ *Matter of Fact* ■

The Heart
During an average lifetime, the heart beats some 2.5 billion times and pumps enough blood to fill about 2,000 swimming pools.

Demonstration 1
The Heart
Obtain a sheep or cow heart from a local butcher. Ask the butcher to leave attached to the heart as many blood vessels as possible. Dissect the heart to show the students its division into right and left halves and the separation of the two atria from the two ventricles.

Visual Strategy
Figure 31.7
Show the class a model or photograph of the human heart. Have them compare it with the one shown in this figure. Point out how the right side of the heart is always shown on the left, as if the heart were in a person facing the viewer. Have students trace the path blood follows through the heart. Have them compare the thickness of the walls of the atria and ventricles and explain the reason for this difference.

Demonstration 2
How to Use a Stethoscope
Invite the school nurse to demonstrate how a stethoscope can be used to hear the sounds of a heartbeat. Have students identify which valves in Figure 31.8 make the "lubb" sound, and which are responsible for the "dup" sound. Inform students that heart murmurs are caused by blood leaking back through faulty heart valves. As the blood backs up, it produces a sloshy noise, heard as a heart "murmur."

Health Connection
Irregular Heartbeat
Malfunctioning of the pacemaker can result in a disorder known as fibrillation, a condition in which the heart contractions become irregular and rapid. In such cases, surgeons can implant an artificial pacemaker powered by batteries. This pacemaker operates by delivering electrical stimuli to the heart at regular intervals.

Visual Strategy

Figure 31.8
Have students locate the pulmonary arteries. Have them explain how these arteries differ from all the others in the body. Do the same with the pulmonary veins. Have students explain why the heart is considered a double pump, so that they see how two circular paths of blood are involved. Also have them identify the heart chamber having the thickest wall and explain how this structural feature is related to its function.

Demonstration 3

An Athlete's Heart
Display a photograph of an athlete engaged in some type of strenuous activity. Point out that during vigorous activity, an athlete's heart rate climbs rapidly, but so does the stroke volume—the amount of blood that passes through the heart with each beat. Although an unconditioned person's heart also beats faster during vigorous exercise, it does so with significantly less increase in stroke volume. Consequently, less blood is being pumped out to the body, causing him or her to tire more quickly.

Demonstration 4

An Electrocardiogram
Ask a local physician or Emergency Medical Technician (EMT) to show the class an electrocardiogram and explain what it represents. If possible, have him or her exhibit abnormal electrocardiograms and discuss what is malfunctioning in each case.

Figure 31.8
Pulmonary circulation carries blood from the heart to the lungs and back to the heart. Follow this pathway of blood flow beginning with oxygen-poor blood entering the right atrium.

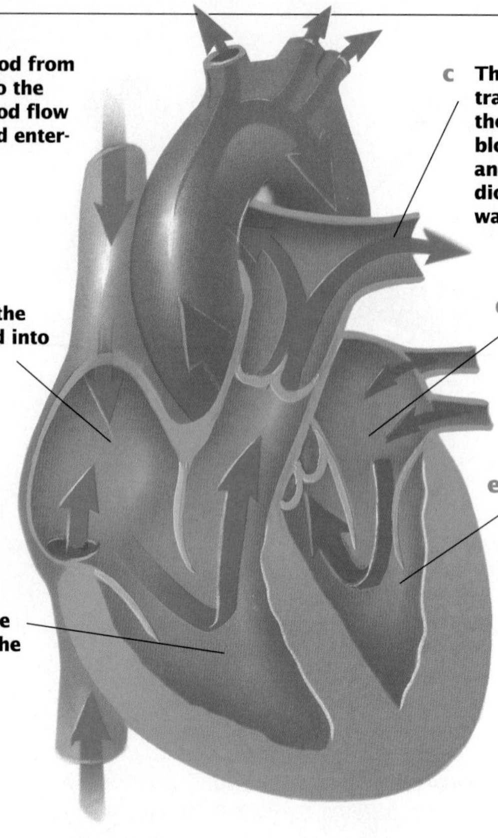

a   Oxygen-poor blood enters the right atrium and is pumped into the right ventricle.

b   From the right ventricle, the blood is pumped through the pulmonary arteries.

c   The pulmonary arteries transport the blood to the lungs. There the blood picks up oxygen and gets rid of carbon dioxide and other wastes.

d   The pulmonary veins transport the newly oxygenated blood to the left atrium.

e   The blood is pumped into the left ventricle, which pumps the blood through the aorta (top) to the rest of the body.

Circulatory Pathways

Your heart pumps blood through two major circulatory pathways. Pulmonary circulation, shown in **Figure 31.8**, carries blood from the heart to the lungs and back to the heart. The pathway of blood from the heart to other parts of the body is called systemic circulation.

Blood picks up oxygen and releases carbon dioxide in the lungs
Blood returning to the heart from the body is low in oxygen and high in carbon dioxide. This blood is pumped through the right side of the heart into arteries that lead to the lungs. These arteries, the pulmonary arteries, are the only arteries that carry oxygen-poor blood. As blood flows through the capillaries in the lungs, it picks up oxygen and gets rid of carbon dioxide. Leaving the lungs, the blood travels through veins, back to the heart. These veins, the pulmonary veins, are the only veins that carry oxygen-rich blood.

Systemic circulation carries blood from the heart to the rest of the body
Blood returning to the heart from the lungs is ready to deliver oxygen to the rest of the body. Notice in **Figure 31.8e** that oxygen-rich blood is pumped through the left side of the heart and out the aorta, the largest artery in the body. From the aorta, blood flows to all parts of the body through a system of increasingly smaller arteries. Systemic circulation has three branches of special importance: the branch that carries blood to the heart, the branch that carries blood to the digestive tract and liver, and the branch that carries blood to and from the kidneys. Several of the major veins and arteries of systemic circulation are shown in **Figure 31.9**.

■ Matter of Fact ■

Causes of Death
The number of deaths due to heart attacks, hypertension, and stroke is greater than the number due to cancer and accidents combined.

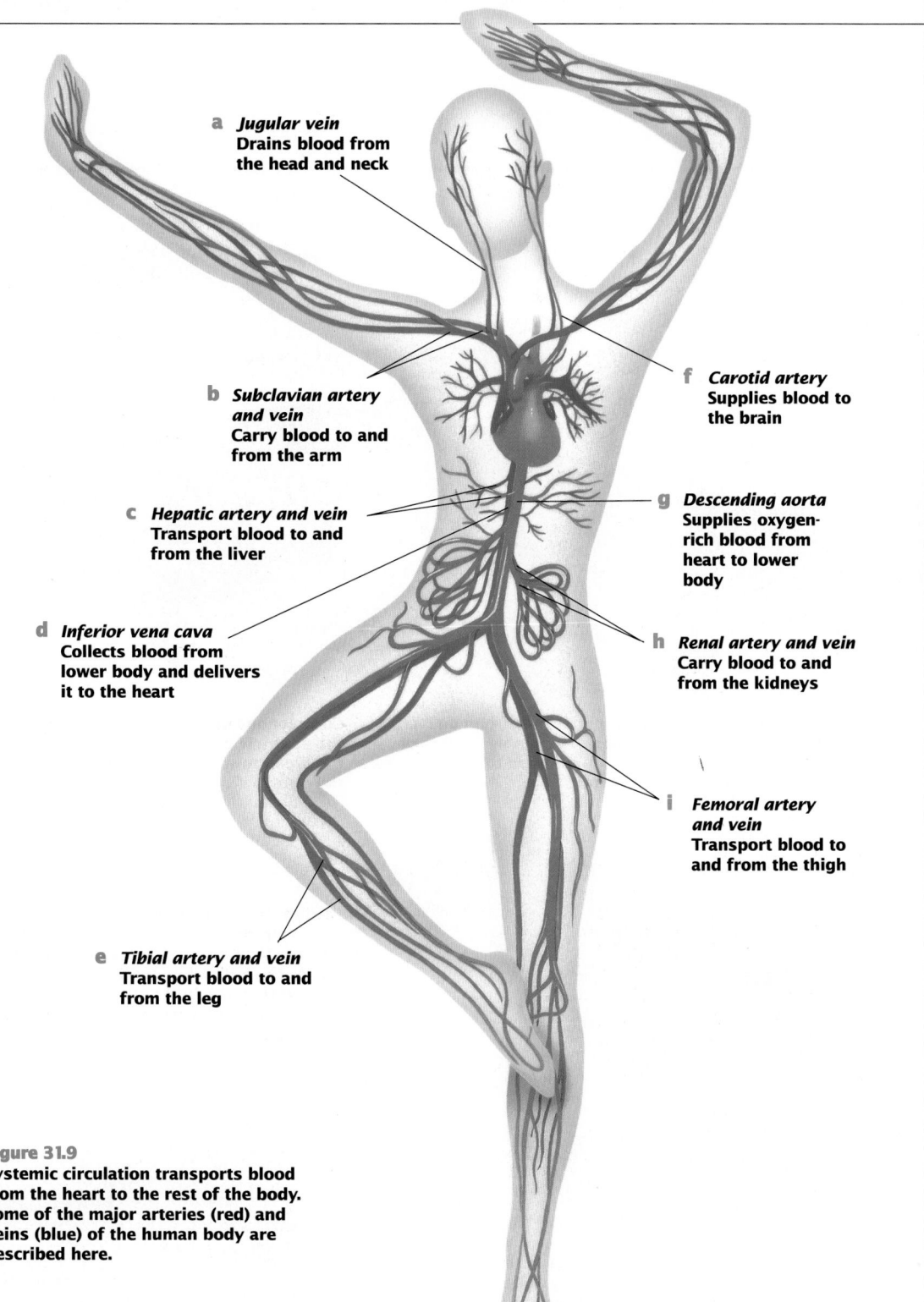

a **Jugular vein**
Drains blood from
the head and neck

b **Subclavian artery
and vein**
Carry blood to and
from the arm

c **Hepatic artery and vein**
Transport blood to and
from the liver

d **Inferior vena cava**
Collects blood from
lower body and delivers
it to the heart

e **Tibial artery and vein**
Transport blood to and
from the leg

f **Carotid artery**
Supplies blood to
the brain

g **Descending aorta**
Supplies oxygen-
rich blood from
heart to lower
body

h **Renal artery and vein**
Carry blood to and
from the kidneys

i **Femoral artery
and vein**
Transport blood to
and from the thigh

Figure 31.9
**Systemic circulation transports blood
from the heart to the rest of the body.
Some of the major arteries (red) and
veins (blue) of the human body are
described here.**

Visual Strategy

Figure 31.9
Instruct students to trace a drop of blood
that travels from the brain to the leg,
naming all the arteries and veins that are
involved. Be sure that they describe how
blood returning from the brain to the heart
must first travel to the lungs and then
back to the heart before beginning its
voyage to the leg.

Using the Feature

This feature traces the evolution of the heart, from the single-cycle heart of a fish to the two-cycle hearts of amphibians, reptiles, birds, and mammals. Point out that a fish's heart pumps only deoxygenated blood. Blood flows to the gills, where it picks up oxygen, but also where it slows substantially while flowing through the narrow gill capillaries. Blood pressure falls by about 50 percent in the gills. Consequently, delivery of oxygen to the tissues is relatively slow.

Adult amphibians have lungs, not gills, and have two, partially separated circulatory loops. Blood flows from heart to lungs and back through a new feature, pulmonary circulation. The two components of pulmonary circulation are the pulmonary artery, which carries deoxygenated blood from the ventricle to the lungs, and the pulmonary vein, which carries the blood back to the heart. Because blood is re-pumped by the heart after slowing in the capillaries of the lung, the blood pressure in the initial portion of the systemic circulation is about twice as high in an amphibian as in a fish. This arrangement creates a problem not found in the fish heart: mixing of deoxygenated and oxygenated blood in the ventricle.

In reptiles other than crocodiles, mixing is greatly reduced by a partial division of the ventricle. In crocodiles, birds, and mammals, this partition is complete.

Evolution of the Heart

Contractions of the vertebrate heart propel blood through the body. Follow the evolution of heart structure below.

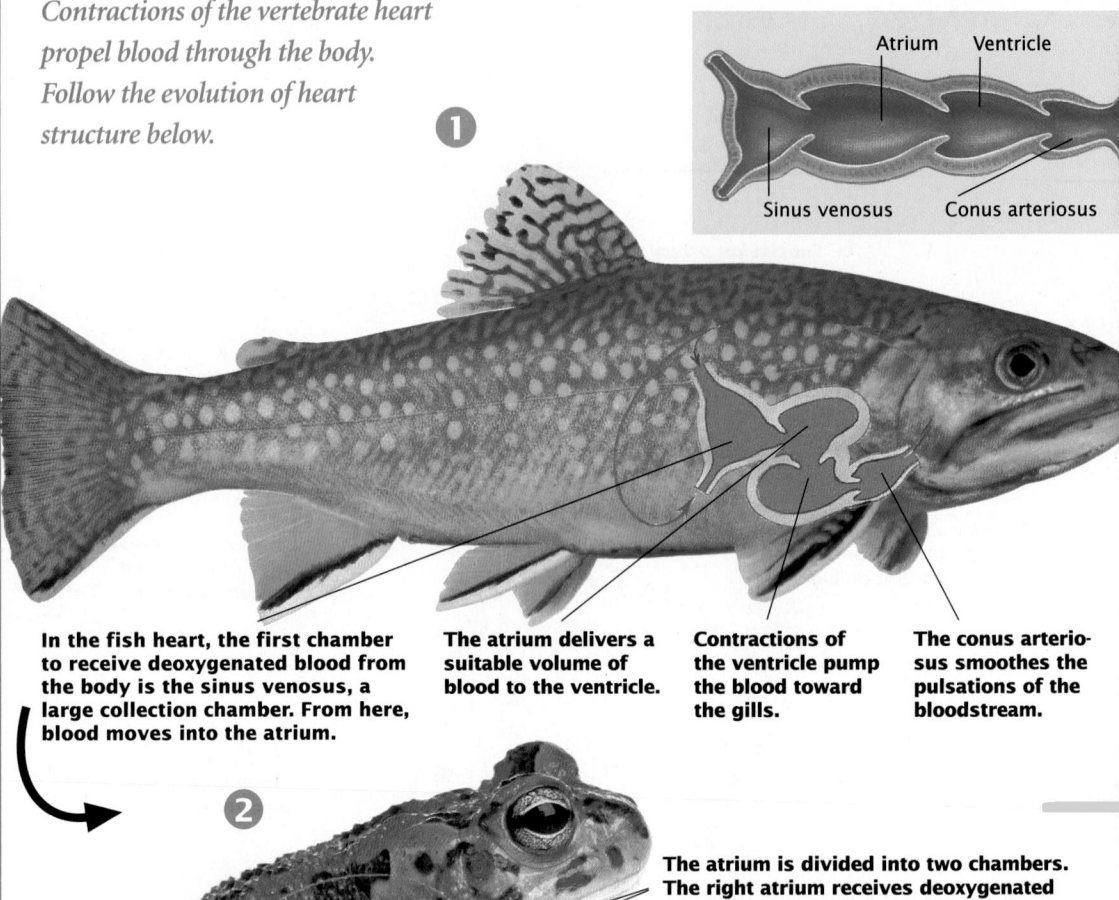

❶

Atrium   Ventricle
Sinus venosus   Conus arteriosus

In the fish heart, the first chamber to receive deoxygenated blood from the body is the sinus venosus, a large collection chamber. From here, blood moves into the atrium.

The atrium delivers a suitable volume of blood to the ventricle.

Contractions of the ventricle pump the blood toward the gills.

The conus arteriosus smoothes the pulsations of the bloodstream.

❷

The atrium is divided into two chambers. The right atrium receives deoxygenated blood from the body and pumps it into the ventricle. The left atrium receives oxygenated blood from the lungs, also sending it to the ventricle. Some mixing of oxygenated and deoxygenated blood occurs in the ventricle.

In the amphibian heart, the sinus venosus is reduced in size. The pulmonary vein carrying oxygenated blood from the lungs enters the left atrium.

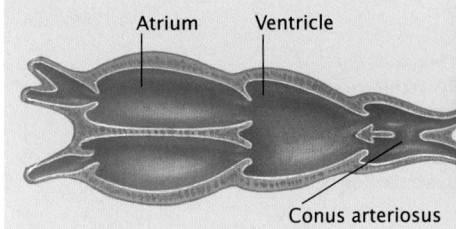

Atrium   Ventricle
Conus arteriosus

■ *Journeys* ■ *Journeys* ■ *Journeys* ■ *Journeys* ■ *Journeys*

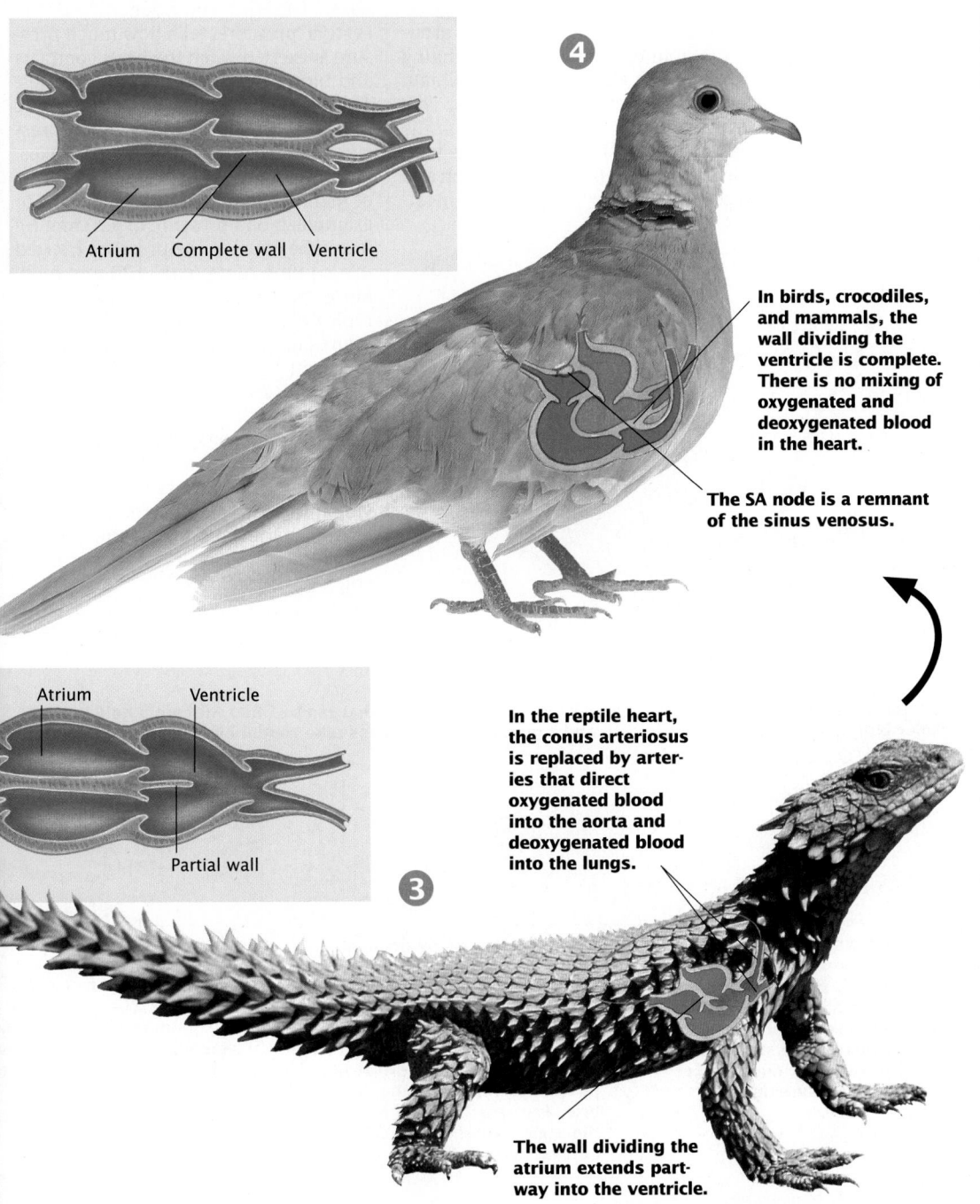

4

Atrium Complete wall Ventricle

In birds, crocodiles, and mammals, the wall dividing the ventricle is complete. There is no mixing of oxygenated and deoxygenated blood in the heart.

The SA node is a remnant of the sinus venosus.

Atrium Ventricle

Partial wall

3

In the reptile heart, the conus arteriosus is replaced by arteries that direct oxygenated blood into the aorta and deoxygenated blood into the lungs.

The wall dividing the atrium extends partway into the ventricle.

Discussion

Guide the discussion by posing the following questions.

1. In the mammalian heart, the left ventricle is much larger than the right. Explain this size disparity.

   The right ventricle only needs to generate enough pressure to send blood the short distance to the lungs and back. In contrast, the left ventricle must create enough pressure to circulate the blood throughout the body. The left ventricle must be stronger to generate the greater pressure, and so it is larger.

2. In embryonic mammals, the pulmonary artery is connected directly to the aorta through a vessel known as the ductus arteriosus. The ductus arteriosus closes and ceases to function after birth. What is the function of the ductus arteriosus before birth?

   Since the embryo does not breathe until born, blood needn't flow to the lungs. The ductus arteriosus diverts blood from the pulmonary artery into the aorta.

723

Demonstration 5

Blood Pressure

Invite the school nurse to demonstrate and explain how a sphygmomanometer is used to measure blood pressure. Correlate the demonstration with the steps illustrated in Figure 31.10. Then have students attempt to measure each other's blood pressure.

Sports Connection

Activity and Blood Pressure

Point out that blood pressure varies with activity, being higher during times of vigorous activity. Have students check this correlation by taking each other's blood pressure before and after exercising.

Blood Pressure

When ventricles contract, blood is forced into the arteries, which exerts pressure on the walls of the blood vessel. This force is called **blood pressure**. When ventricles relax, the pressure decreases. The muscular, elastic walls of the arteries are able to adjust to the changing pressure. The elasticity helps maintain the pressure between heartbeats. This way blood is kept flowing through the body continuously. When you take your pulse, you are feeling the expansion and relaxation of an artery with each heartbeat.

How is blood pressure measured?

As described in **Figure 31.10**, two numbers are used to register blood pressure. The first number, called the systolic pressure, tells how much pressure is exerted when the heart contracts and blood flows through the arteries. The second number is the diastolic pressure, which tells how much pressure is exerted when the heart relaxes. Blood pressure is expressed in terms of millimeters of mercury (mm Hg). For example, blood pressure of 120 mm Hg would be equal to the pressure exerted in a column of mercury 120 mm high. Blood pressure in a healthy adult is typically about 120/80 mm Hg. These figures indicate that the blood is pushing against the artery walls with a pressure of 120 mm Hg as the heart contracts and 80 mm Hg as the heart rests. Blood pressure steadily rises with increasing age as the arteries become less elastic. Blood pressure figures provide information about the conditions of the arteries and are useful in diagnosing high blood pressure and hardening of the arteries.

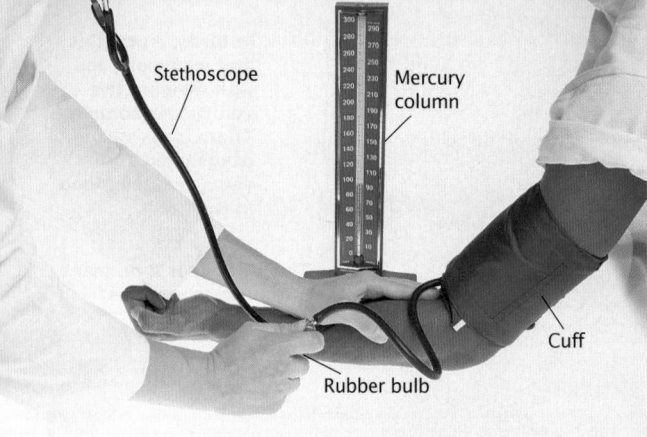

Stethoscope
Mercury column
Cuff
Rubber bulb

Figure 31.10
A kit for measuring blood pressure includes a stethoscope, a cuff that can be filled with air, a hollow rubber bulb that pumps air into the cuff, and a gauge or hollow glass tube containing a column of mercury.

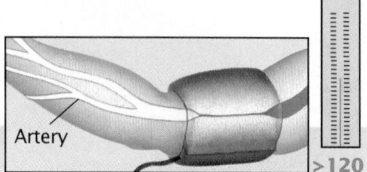

Artery
>120

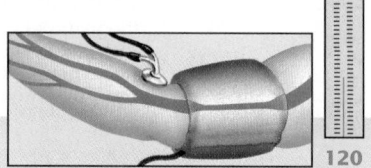

120

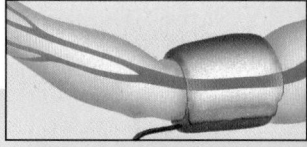

a The cuff is wrapped around the patient's upper arm. When the rubber bulb is pumped, air inflates the cuff. This squeezes an artery in the arm until no blood passes through. The stethoscope is used to listen for the sound of blood flow to ensure that the artery is closed.

b The cuff is gradually deflated until blood begins to flow into the arm. A sound of blood pulsing can be heard, indicating that the blood pressure is greater than pressure exerted by the cuff. The pressure at this point (120 mm Hg) is the systolic pressure, which is read from the scale. Systolic pressure is exerted by ventricles contracting.

c The cuff is loosened until bl flows freely through the art and the sounds below the c disappear. The pressure at this point (80 mm Hg) is the diastolic pressure, the press between heart contractions.

▪ *Cultural Perspective* ▪

Portia B. Gordon, Ph.D.
(African-American)
Dr. Gordon works at Albert Einstein College's Montefiore Medical Center, where she studies some of the cellular changes that take place during arteriosclerosis. Because she feels it is important to expose young people to

science, Dr. Gordon has students from nearby high schools working in her laboratory each summer.

Diseases of the Heart and Blood Vessels

Diseases of the heart and blood vessels are referred to as **cardiovascular diseases**. Cardiovascular diseases are the leading cause of death in the United States, claiming about 1 million lives every year. An estimated 63 million Americans have some form of cardiovascular disease.

Heart attacks can kill

Heart attacks are the most common cause of death from cardiovascular disease. A heart attack results from an insufficient supply of blood to an area of heart muscle. Without the blood, heart muscle cells are starved for oxygen and die. A heart attack can result from blockage of a blood vessel due to atherosclerosis (*ath uhr oh skluh ROH sihs*), a condition in which fatty deposits form on the insides of arteries, as shown in **Figure 31.11 b**. Recovery from a heart attack depends on how much heart tissue has been damaged, where the damage occurred, and whether other blood vessels can do the work of the damaged blood vessels.

Scale and Structure

How can atherosclerosis increase the risk of heart attack?

Figure 31.11

a **In a normal artery, the passageway is clear for blood to pass through. However, . . .**

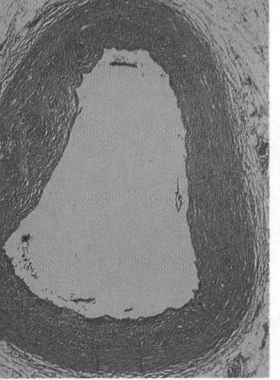
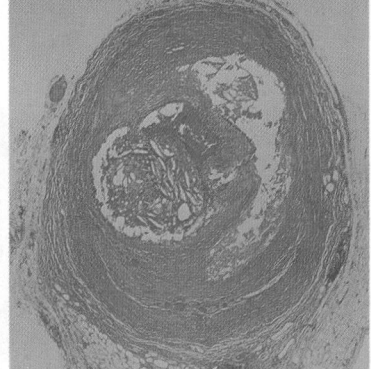

Hypertension is high blood pressure

When the pressure of the blood against the artery wall is continually higher than normal, a condition called **hypertension**, or high blood pressure, results. Hypertension is dangerous because of the damage it can do to the heart, brain, and kidneys if not controlled. Because it has no warning symptoms, hypertension is often called the silent killer.

A healthy lifestyle can reduce your risk of cardiovascular disease

Certain factors can influence an individual's risk of cardiovascular disease. For example, a tendency toward some cardiovascular diseases appears to be hereditary. Nicotine, the drug in cigarette smoke, has been strongly linked to cardiovascular disease in many clinical studies. Scientists have also discovered that obesity, stress, and a lack of regular exercise can greatly increase the risk of high blood pressure, a heart attack, or a stroke. Reducing the amount of animal fat, cholesterol, and salt in your diet and exercising moderately at least three times a week will help maintain resistance to these and many other diseases. Not surprisingly, avoiding smoking greatly lowers your chances of cardiovascular disease.

b **. . . in an artery that is partially blocked because of atherosclerosis, fat deposits decrease the amount of blood that can flow through the artery.**

Section Review

❶ **Describe the roles played by the left and right halves of the heart.**

❷ **Explain the two routes blood can take through the body.**

❸ **What does a blood pressure of 120/80 mean?**

❹ **What steps could you take to reduce your chances of developing cardiovascular disease?**

■ Section Review Answers ■

1. The right half pumps blood to the lungs to pick up oxygen and release carbon dioxide. The left half of the heart pumps oxygen-rich blood to the rest of the body.

2. See Figure 31.8 on page 720 and Figure 31.9 on page 721 for an explanation of the pulmonary system and systemic circulation, respectively.

3. Students should suggest typical blood pressure in a healthy adult. Blood is pushing against the wall of the artery with a pressure of 120 mm Hg as the heart contracts and 80 mm Hg as the heart rests.

4. Students should suggest a healthful lifestyle that includes reducing the amount of animal fat, cholesterol, and salt in the diet; exercising more; and avoiding stress and drugs.

Theme Answer
Scale and Structure
A heart attack results from an insufficient supply of oxygenated blood to an area of heart muscles. Arteries blocked by atherosclerosis decrease the amount of blood that can flow through an artery.

Health Connection

Cholesterol and Heart Attacks
Tell students that the fatty deposits often form from cholesterol that collects on the inner lining of the arteries. Have students explain which artery in Figure 31.11 is more likely to be found in a person with a high cholesterol level. Point out that if atherosclerosis occurs in the coronary artery, a coronary bypass operation may be required. During this procedure, a surgeon inserts segments of a leg vein into the aorta and coronary vessels, in order to bypass areas that are clogged or blocked. In some cases, a double or triple bypass is required since several locations are blocked.

Phase 3

ASSESSMENT OPTIONS

Closure Strategy

The Highs and Lows of Blood Pressure
Have students hypothesize where blood pressure would be the highest (the aorta leaving the left ventricle) and where it would be the lowest (the vena cavae returning to the right atrium). Have them justify their choices.

Section Review

Assign the Section Review.

Reteaching

Assign students to cooperative work groups of four. Have each group list features they would incorporate into the design of an artificial heart that could be used to replace a defective one.

Using the Feature

- You can use this feature to emphasize cholesterol's importance in the human body. Students may inaccurately conclude that cholesterol is "bad" because of its role in atherosclerosis. Remind them that it is an essential part of cell membranes and the raw material of vitamin D and steroid hormones, which are vital to homeostasis. For example, when the body is deprived of cortisol and aldosterone, death results.

- Have students research the diets of cultures around the world and collect information on the staples that comprise them. Afterwards, discuss the components of each of the diets and the effects they most likely have on the individuals in that society.

Discussion

Guide the discussion by posing the following questions.

1. Why should teens be concerned with cholesterol levels in their blood?
   According to studies, monitoring cholesterol levels in childhood and during the teenage years can prevent atherosclerosis and lead to a healthier, longer life.

2. Discuss five guidelines for lowering blood cholesterol levels.
   See the table on page 726.

Cholesterol and Your Health

Cholesterol and Teens

You've read the claims "Cholesterol Free" and "Low Cholesterol" splashed across food packages. Clearly, food makers are hoping to tempt customers concerned about their health. But did you know that these messages are important not only for adults, but also for you as a teenager?

Studies show that a condition called atherosclerosis (a buildup of fat deposits in the arteries) can begin in childhood. The process progresses into adulthood and can lead to coronary heart disease, a major cause of death in the United States. According to many studies, preventing atherosclerosis in childhood or in your teenage years could extend your lifetime. By learning the facts about cholesterol and how it affects the human body, you can take the first step toward living a healthier lifestyle.

What Is Cholesterol?

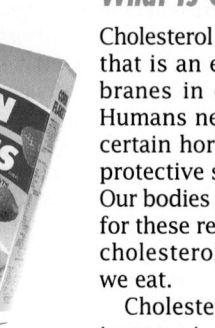

Cholesterol is a fatty, waxlike substance that is an essential part of the membranes in cells of the human body. Humans need cholesterol to produce certain hormones, vitamin D, and the protective sheath around nerve fibers. Our bodies produce enough cholesterol for these requirements. We also take in cholesterol from some of the foods we eat.

Cholesterol is made in the liver and transported to all of the body's cells through the bloodstream. It is carried by lipoproteins, molecules containing fats and proteins. One type of lipoprotein is known as low-density lipoprotein (LDL). A second type is called high-density lipoprotein (HDL). LDL is often called "bad cholesterol" because it deposits cholesterol in body tissues, especially in the walls of arteries. As more and more cholesterol is deposited, the arteries become narrower, making it difficult for blood to flow through. The reduced blood flow may cause heart disease, heart attacks, or strokes. Often referred to as "good cholesterol," HDL removes cholesterol from the LDL and body tissues and transports it back to the liver for removal from the body. In general, a high LDL cholesterol level, or a low HDL cholesterol level, increases the risk for developing coronary heart disease.

Watching your cholesterol intake now could prevent fat deposits from clogging your blood vessels and can maintain your cardiovascular health later in life.

Guidelines for Lowering Blood Cholesterol Levels

- Eat fewer foods high in saturated fat (butter, cheese, whole milk, ice cream, meat, coconut and palm oil)
- Replace part of your saturated fat intake with unsaturated fat
- Eat fewer high-cholesterol foods (eggs, dairy products, liver)
- Choose foods high in complex carbohydrates (starch and fiber)
- Maintain a healthful diet and exercise regularly

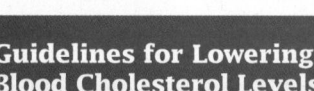

Action ▪ *Science in Action* ▪ *Science in Action* ▪ *Science*

How to Modify Cholesterol Levels

Researchers believe that genetic factors may influence cholesterol levels in the blood. Although these factors cannot be controlled, a person can reduce blood cholesterol levels. One way is to eat foods that are low in cholesterol. Foods that come from animals, such as eggs, cream, and meat, are high in cholesterol. Fruits, vegetables, and grains contain no cholesterol. That's because animals produce cholesterol, whereas plants do not.

The kinds of fats you eat are important too. Saturated fats, such as butter and animal fat, tend to raise blood cholesterol levels. But unsaturated fats may actually help lower the amount of cholesterol carried in the blood. They do this by increasing the levels of HDL, or "good" cholesterol. Unsaturated fats include most of the liquid fats such as vegetable, olive, and fish oils, as well as margarine. Many of the saturated fats in your diet can be substituted with unsaturated fats. People who need to increase their HDL levels are encouraged to eat polyunsaturated fats. These fats are found mainly in safflower, corn, soybean, sesame, and sunflower oils.

Another way to reduce your cholesterol level is by exercising. Some research suggests that HDL levels are higher in people who exercise regularly. And being overweight seems to deplete the amount of HDL in the blood, as does smoking. So, by exercising regularly, not smoking, and eating a diet that is low in saturated fats and low in cholesterol, you can decrease your risk of developing heart disease.

Corn oil, a polyunsaturated fat, and olive oil, an unsaturated fat, should replace saturated fats, such as butter, in your diet.

Food labels should be read carefully. They provide a good source of information about the fat and cholesterol contents of food products.

How to Be a Smart Consumer

Eating to lower your blood cholesterol level means learning to select foods that are low in saturated fats and cholesterol. The packages in the supermarket with "cholesterol free" claims do contain food without cholesterol, but sometimes these statements can be misleading. A food made with highly saturated oil as the only fat source may be cholesterol-free but still be a poor choice if you're trying to reduce your blood cholesterol level.

Food labels can provide important information about the type and amount of fats and cholesterol. Ingredients are presented on the label according to their weight in the product. The ingredient found in the greatest amount is listed first, while the ingredient found in the least amount is listed last. One way to be sure that a certain food fits a low-fat, low-cholesterol diet is to limit your selection of foods in which ingredients high in saturated fats or cholesterol are among the first five on the list. Also, choose sparingly from foods that list many fats or oil.

The listing of fats and cholesterol is optional on food labels. However, under FDA regulations, any food product with a nutritional claim must have the nutritional content listed on the label.

Lesson Plan 31.3

Phase 1

PREPARATION

Key Concepts

- Lungs take in air containing the oxygen needed for cellular respiration.
- Air enters the nose and then takes the following passage: pharynx → larynx → trachea → alveoli.
- During inhalation, the diaphragm contracts and the rib cage moves up, forcing air into the lungs. During exhalation, the diaphragm relaxes and the rib cage moves down, forcing air out of the lungs.
- Oxygen diffuses from the alveoli into the blood, where it is transported by hemoglobin in red blood cells.
- Carbon dioxide diffuses from cells into the blood, where it is mainly transported as bicarbonate ions by the plasma. Upon reaching the lungs, carbon dioxide is again formed and diffuses from the blood into the alveoli to be exhaled.
- Receptors in the brain, sensitive to the concentration of carbon dioxide in the blood, regulate the rate of breathing.
- Respiratory disorders include asthma, emphysema, and lung cancer.

Reading Strategy

The proper environment is important when reading. Suggest that students read this section under circumstances in which distractions are minimal, either in an appropriate place, such as a local library, or at an appropriate time, such as a time of day when they will not be interrupted by others in their home.

HOW GASES ARE EXCHANGED IN ANIMALS DEPENDS ON THE SIZE OF THE ANIMAL. VERY SMALL ANIMALS CAN OBTAIN OXYGEN AND GIVE OFF CARBON DIOXIDE DIRECTLY THROUGH THE PLASMA MEMBRANES OF THEIR OUTERMOST LAYER OF CELLS. IN CONTRAST, LARGER ANIMALS HAVE EVOLVED SPECIAL RESPIRATORY SYSTEMS THAT PROVIDE A LARGE SURFACE AREA TO EXCHANGE GASES.

31.3   The Respiratory System

Objectives

❶ Explain how the diaphragm and rib muscles work to move air into and out of the lungs.

❷ Describe the pathway by which oxygen from the air travels to a cell in the body.

❸ Explain how your body regulates your breathing rate.

❹ Describe three diseases of the respiratory system.

Lungs and Breathing

Figure 31.12
On or off the soccer field, humans must obtain oxygen from the atmosphere and release excess carbon dioxide in the process of gas exchange.

Throughout this book, you have learned that most living organisms, like the soccer players in **Figure 31.12**, need to obtain oxygen from their environment and need to remove carbon dioxide from their bodies. As you read in Chapter 5, the oxygen is needed for cellular respiration, the chemical process that allows organisms to obtain energy from substances such as glucose. During aerobic respiration, carbon dioxide is produced as a waste product and must be removed from the cell. As you learned in Chapter 22, the evolution of lungs enables terrestrial animals to obtain oxygen from the atmosphere and release excess carbon dioxide.

Figure 31.13
Air taken in through the nose or mouth passes through the pharynx, larynx, and trachea into the many branches within the lungs. It soon reaches the alveoli, tiny sacs surrounded by capillaries through which gases are exchanged.

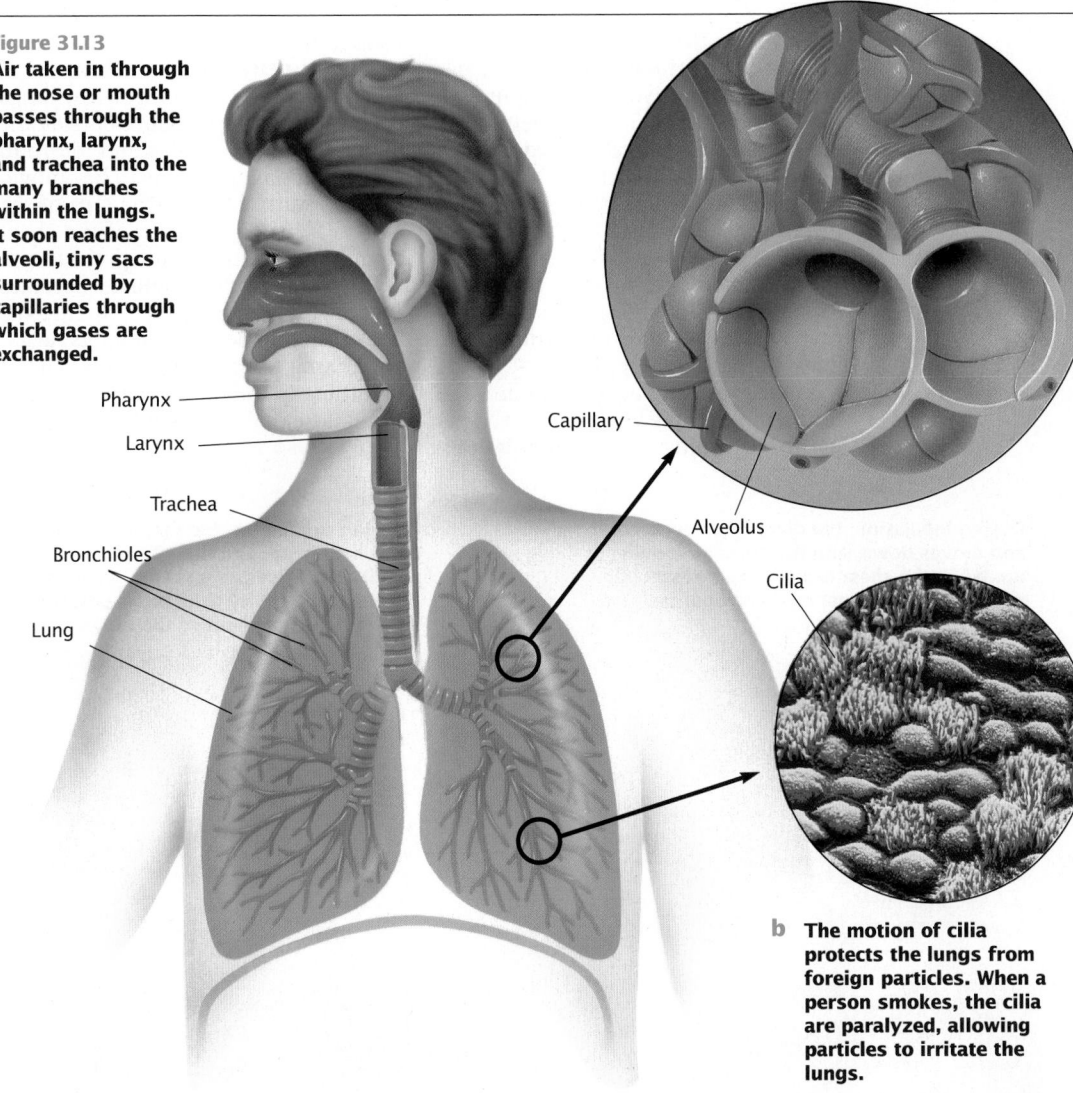

Pharynx

Larynx

Trachea

Bronchioles

Lung

Capillary

Alveolus

Cilia

b The motion of cilia protects the lungs from foreign particles. When a person smokes, the cilia are paralyzed, allowing particles to irritate the lungs.

Lungs contain branched tubes

Follow the passage of air through the respiratory system in **Figure 31.13**. Air enters the body through the nose or mouth and travels to the pharynx (*FAIR ihnks*), a tube at the back of the nose and mouth. From the pharynx, air enters the voice box, or larynx (*LAR ihnks*), and the windpipe, or trachea (*TRAY kee uh*). These passageways are lined with tissues that warm and moisten incoming air. The lower end of the trachea divides into two branches, which divide many times into smaller branches of tubes called bronchioles (*BRAHNG kee ohlz*). This branching network of tubes is lined with cilia, shown in **Figure 31.13b**, and a layer of protective mucus. The smallest of these tubes lead into **alveoli** (*al VEE uh leye*), clusters of tiny air sacs. Gases enter and leave the circulatory system through the alveoli. Each alveolus is surrounded by a network of capillaries. Blood in these capillaries picks up oxygen from the alveoli and releases carbon dioxide to be exhaled. Each lung contains about 150 million alveoli, providing a surface area larger than a small house for gas exchange.

Phase 2

TEACHING STRATEGIES

Visual Strategy

Figure 31.13

Discuss with students where the vocal cords would be found in this figure. Point out that sound is produced when air is forced through these cords. Also inform students that laryngitis is simply an inflammation of the vocal cords in which they lose their ability to vibrate. Hence, people "lose their voice."

Demonstration 1

The Trachea

Have students gently rub their fingers up and down their throats to feel the rings of cartilage that prevent the trachea from collapsing.

Connection: Chapter 4

Diffusion

Point out that gases are exchanged between the air in the alveoli and the blood in the capillaries entirely by diffusion. Then have students explain why a person breathes deeper and faster at higher altitudes, where there is decreased air pressure.

Demonstration 2

Inhalation and Exhalation

Demonstrate how breathing works by making a model of the lung and diaphragm. Attach a small, deflated balloon to a short glass tube that is inserted through a one-holed rubber stopper. Use the stopper to plug the mouth of a bell jar, allowing the balloon to hang inside the jar. Use a rubber band to fasten a thin sheet of rubber to the large opening of the jar. Pull down on the rubber sheet and have students observe what happens to the balloon. Have students relate their observations to inhalation. Release the rubber sheet. Have students note what happens to the balloon. Have them relate this to exhalation.

Visual Strategy

Figure 31.14

Use this figure to discuss lung capacity. Together the lungs can hold between 5 L and 6 L (about 1.5 gal.) of air, which is the total lung capacity. However, a person normally inhales only about 0.5 L (0.13 gal.), which is the same volume normally exhaled. Strenuous exercise can increase these values—it is possible to increase the capacity of a single breath to about 4.5 L (1.13 gal).

Inhalation involves muscle contraction in the chest

Breathing begins when the diaphragm, the dome-shaped muscle below the chest cavity, contracts and moves downward. The muscles between the ribs also contract, causing the rib cage to move up and out. Together, these muscle contractions cause the chest cavity to enlarge. When the chest expands, the air pressure in the chest cavity drops. Air pressure outside the body is then greater than that inside the chest, causing air to rush into the lungs to equalize the pressure. This part of breathing is called **inhalation**.

Muscles return to their relaxed position during exhalation

When the air pressure inside the lungs is equal to the air pressure outside the lungs, the muscles relax and return to their original positions. This movement, in turn, reduces the size of the chest cavity. As the size of the chest cavity decreases, the air pressure inside the chest cavity gradually becomes greater than air pressure outside the body. Air then leaves the lungs, again equalizing the pressure. This part of breathing is called **exhalation**. Inhalation and exhalation are illustrated in **Figure 31.14**.

Figure 31.14

a **During inhalation, the diaphragm contracts and moves down, and the rib cage moves up. When the chest expands as a result, air pressure in the chest cavity drops, causing air to rush into the lungs.**

b **During exhalation, the diaphragm relaxes and moves up, and the size of the chest cavity decreases as a result. Air pressure in the chest cavity thus increases, forcing air to be exhaled out of the lungs .**

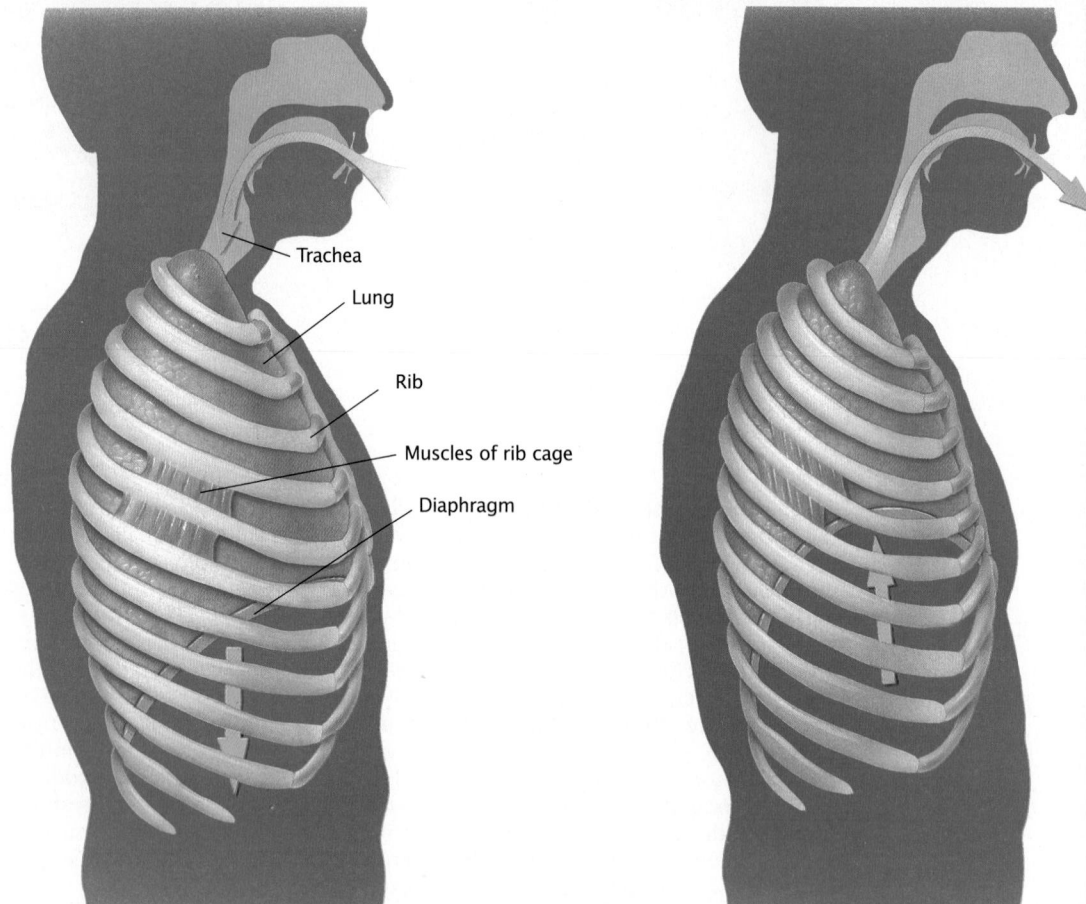

Trachea

Lung

Rib

Muscles of rib cage

Diaphragm

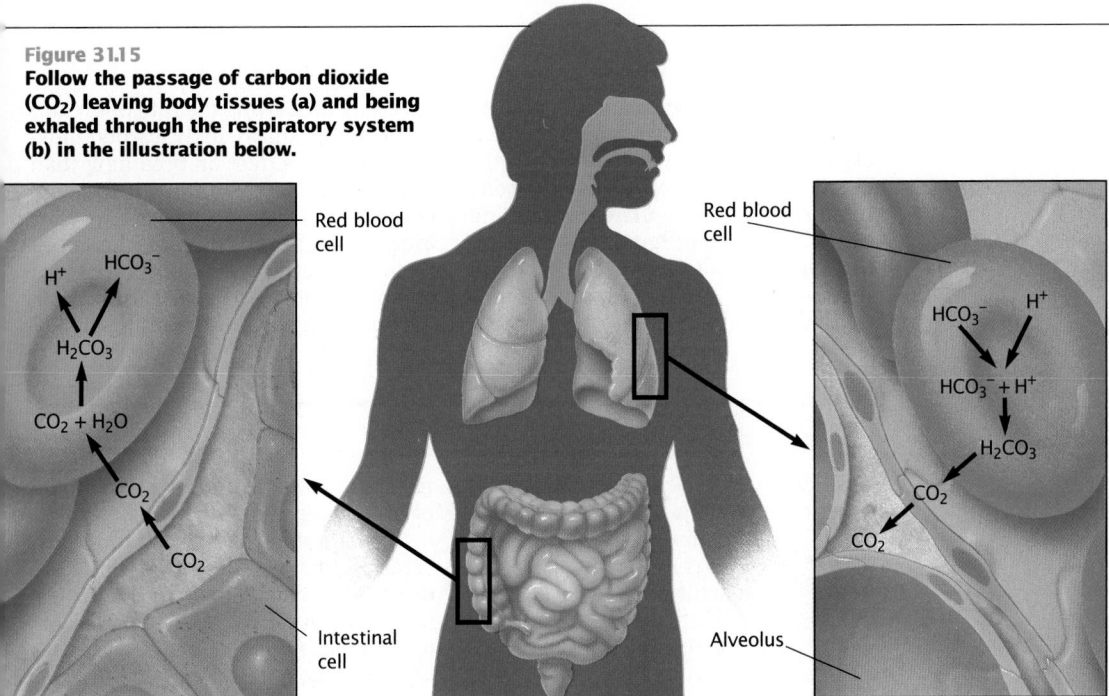

Figure 31.15
Follow the passage of carbon dioxide (CO_2) leaving body tissues (a) and being exhaled through the respiratory system (b) in the illustration below.

Red blood cell

H^+ HCO_3^-

H_2CO_3

$CO_2 + H_2O$

CO_2

CO_2

Intestinal cell

Red blood cell

HCO_3^- H^+

$HCO_3^- + H^+$

H_2CO_3

CO_2

CO_2

Alveolus

a When CO_2 leaves cells of the intestines, it can enter red blood cells and combine with hemoglobin. Or, it may combine with water to form carbonic acid, H_2CO_3, which breaks down to form hydrogen ions and bicarbonate ions.

b When blood reaches the lungs, the process is reversed and CO_2 is released. The CO_2 diffuses from the blood into the alveoli in the lungs. From there, it is exhaled along with water vapor.

Gas Exchange

The destination of the inhaled air is the alveoli, which are surrounded by capillaries. The tissue forming the walls of the alveoli and capillaries is only one cell thick. Gas exchange occurs when oxygen in the alveoli diffuses into the blood in the capillaries. In turn, the carbon dioxide in the blood diffuses into the air of the alveoli.

Oxygen binds to hemoglobin

In the blood, oxygen quickly binds with hemoglobin, the protein in red blood cells. Hemoglobin soaks up oxygen extremely effectively, which causes still more oxygen to enter red blood cells. The red blood cells then give up their oxygen to the cells of body tissues, where it is used in metabolism, the chemical activities of cells. As a result of metabolism, oxygen concentration in the body's cells is low, but carbon dioxide concentration is high.

Carbon dioxide is transported to the lungs to be exhaled

Carbon dioxide is a waste product that must be eliminated from cells. It is transported in the blood in three ways. About 5 percent of the carbon dioxide in the body dissolves in the plasma in blood. Another 25 percent enters the red blood cells and combines with hemoglobin. With help of an enzyme, the remaining 70 percent combines with water in the red blood cells to form carbonic acid, H_2CO_3. Because carbonic acid is unstable, hydrogen ions and bicarbonate ions quickly form, as shown in **Figure 31.15a**.

When blood reaches the lungs, chemical reactions occur that reverse the process, releasing carbon dioxide. As shown in **Figure 31.15b**, the carbon dioxide diffuses from the blood into the alveoli in the lungs. The carbon dioxide is exhaled with water vapor.

Regulation of Breathing

Receptors in the brain and circulatory system continuously monitor the levels of oxygen and carbon dioxide in the blood. These receptors enable the body to automatically regulate oxygen and carbon dioxide concentrations by sending signals to the brain. The brain responds by sending nerve signals to the diaphragm and rib muscles, speeding or slowing the rate of breathing. Perhaps surprisingly, carbon dioxide has more effect on breathing than does oxygen. For example, if the concentration of carbon dioxide in your blood increases, you breathe more deeply, ridding your body of excess carbon dioxide. When the carbon dioxide level drops, your breathing slows.

Diseases of the Respiratory System

Respiratory diseases affect millions of Americans. **Asthma** (*AZ muh*) is a respiratory disease in which certain airways in the lungs become constricted because of sensitivity to certain stimuli. The narrowing of the airways reduces the efficiency of respiration, which decreases the amount of oxygen reaching body cells.

Cigarette smoking is linked to emphysema and lung cancer, two respiratory diseases that claim millions of lives annually. In people who have **emphysema** (*em fuh SEE muh*), the lung tissue loses its elasticity, greatly reducing the efficiency of gas exchange. In **lung cancer**, carcinogens present in tobacco smoke trigger the growth of cancerous cells in lung tissue. More than 90 percent of lung cancer patients are smokers, as suggested by the antismoking posters in **Figure 31.16**. Lung cancer has an extremely low rate of cure. Fewer than 10 percent of its victims live more than five years after diagnosis. In Chapter 30 you learned about the effects of cigarette smoking on the body.

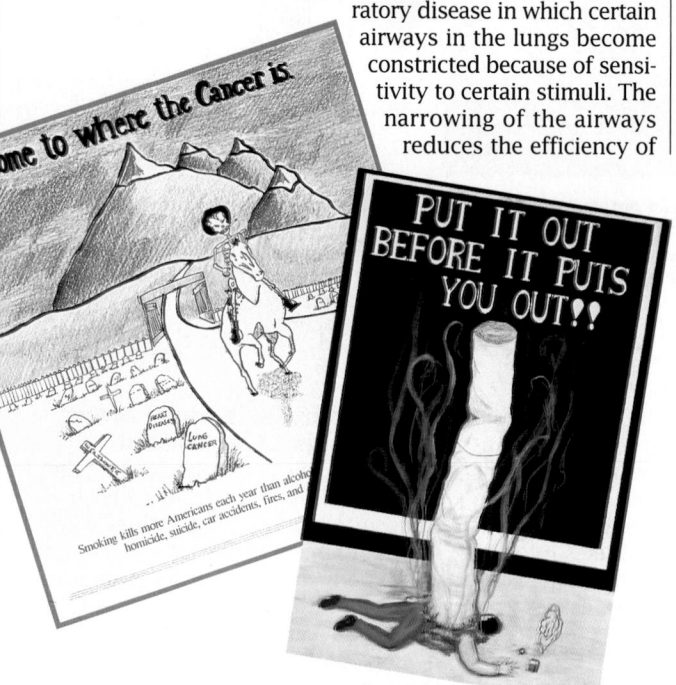

Figure 31.16
These posters, warning about some of the life-threatening illnesses associated with cigarette smoking, were designed by high school students.

Section Review

❶ **What role do the diaphragm and rib muscles play in the processes of inhalation and exhalation?**

❷ **How does the exchange of gases occur in the lungs?**

❸ **When you exercise, you automatically begin to breathe faster. Explain how and why this occurs.**

❹ **How do asthma and emphysema affect respiration?**

■ Section Review Answers ■

1. See Figure 31.14 on page 730.
2. See Figure 31.15 on page 731.

3. Students should suggest that receptors in the brain and circulatory system continuously monitor the levels of oxygen and carbon dioxide in the blood. When exercising, the brain signals the diaphragm and rib muscles to breathe more deeply, ridding the body of excess carbon dioxide.

4. Asthma constricts or narrows airways, whereas emphysema causes the lung tissue to lose its elasticity. Both reduce the efficiency of respiration, decreasing the amount of oxygen in the blood.

Chapter **31** *Highlights*

Strands of a protein called fibrin trap red and white blood cells. Soon a blood clot will form.

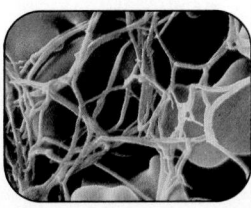

Key Terms	Summary

31.1 *Circulation*

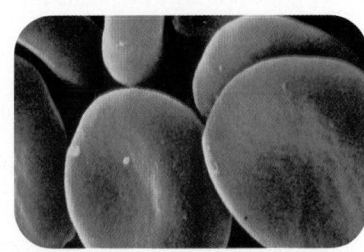

Red blood cells carry oxygen from the lungs to all cells of the body and carry carbon dioxide back to the lungs to be exhaled.

Key Terms:
plasma (p. 712)
red blood cell (p. 712)
anemia (p. 712)
white blood cell (p. 713)
platelet (p. 713)
blood vessel (p. 714)
artery (p. 714)
vein (p. 714)
capillary (p. 714)
blood type (p. 716)
lymphatic system (p. 717)

Summary:
- The circulatory system transports materials to and from cells.
- Blood consists of plasma, red blood cells that transport oxygen, white blood cells that protect against infection, and platelets that help clotting.
- Blood types are defined by proteins on the surface of red blood cells.
- Blood vessels include arteries that carry blood away from the heart, veins that carry blood back to the heart, and capillaries that connect the arteries to the veins.
- The lymphatic system returns fluids back to the blood vessels.

31.2 *How Blood Flows*

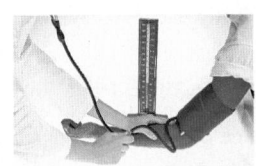

Blood pressure in a healthy adult is typically about 120/80 mm Hg.

Key Terms:
atrium (p. 718)
ventricle (p. 718)
pacemaker (p. 719)
blood pressure (p. 724)
cardiovascular disease (p. 725)
hypertension (p. 725)

Summary:
- The heart has two sides: the right side moves blood to the lungs, and the left side moves blood to the rest of the body.
- Blood pressure consists of a high value (systolic) when the heart contracts and a lower value (diastolic) when the heart rests.
- Hypertension, or high blood pressure, is an example of a cardiovascular disease.

31.3 *The Respiratory System*

The walls of alveoli and the capillaries around them provide a large surface area for gas exchange.

Key Terms:
alveolus (p. 729)
inhalation (p. 730)
exhalation (p. 730)
asthma (p. 732)
emphysema (p. 732)
lung cancer (p. 732)

Summary:
- When the diaphragm and the rib muscles contract, enlarging the chest cavity, inhalation occurs. When the muscles relax, air is forced out and exhalation occurs.
- Gas exchange occurs when oxygen in the alveoli diffuses into the blood in the capillaries. Carbon dioxide in the blood diffuses into the air of the alveoli.
- Breathing is regulated mainly by response to the level of carbon dioxide detected in the blood.
- Cigarette smoking is linked to emphysema and lung cancer.

Chapter Review Answers

Understanding Vocabulary

1. a. Inhalation is the phase of breathing that involves chest expansion and a drop in air pressure that causes air to rush into the lungs. The chest cavity decreases during exhalation and causes the air to leave the lungs.
   b. Red blood cells lack a nucleus and transport oxygen and carbon dioxide. White blood cells have a nucleus and defend the body against invading bacteria, viruses, or other foreign matter.
   c. Arteries carry oxygenated blood away from the heart and to the cells of the body. Veins carry deoxygenated blood toward the heart and away from the cells of the body.
   d. Systolic pressure indicates how much pressure is exerted when the heart contracts and blood flows through the arteries. Diastolic pressure tells how much pressure is exerted when the heart relaxes.

Relating Concepts

2. Map answer is shown on page 709D.

Understanding Concepts

Multiple Choice

3. d	8. c
4. a	9. a
5. c	10. c
6. b	11. b
7. a	12. d

Completion

13. lymph nodes
14. heartbeat, a small bundle of cells, atrium
15. veins, arteries
16. hemoglobin, anemia
17. hypertension

Short Answer

18. The diaphragm divides the abdominal and thoracic cavities. It contracts during inhalation and relaxes during exhalation.
19. The sounds are made by the closing of the valves. Specifically, the "lubb" sound is produced by the valves

---

Chapter 31 Review

Understanding Vocabulary

1. For each pair of terms, explain the differences in their meanings.
   a. inhalation, exhalation
   b. red blood cell, white blood cell
   c. arteries, veins
   d. systolic pressure, diastolic pressure

Relating Concepts

2. Copy the unfinished concept map below onto a sheet of paper. Then complete the concept map by writing the correct word or phrase in each oval containing a question mark.

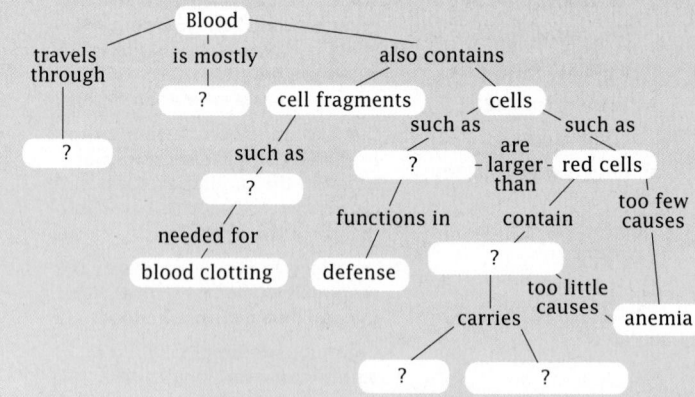

Understanding Concepts

Multiple Choice

3. When blood exits the right ventricle, it
   a. is not under pressure.
   b. is oxygenated.
   c. has more white cells than red cells.
   d. enters the pulmonary artery.

4. The maximum force exerted against the arterial walls occurs during
   a. ventricular contraction.
   b. systolic pressure.
   c. diastolic pressure.
   d. arterial relaxation.

5. By volume, blood is mostly
   a. white cells.   c. plasma.
   b. red cells.   d. platelets.

6. Most of the oxygen transported in the blood is
   a. dissolved in the plasma.
   b. bound to hemoglobin.
   c. in the form of carbonic acid.
   d. in the form of water.

7. Platelets act in
   a. blood clotting.
   b. red cell development.
   c. fighting infection.
   d. transporting oxygen.

8. Fluid balance in the blood is maintained by the
   a. lymphatic system.
   b. plasma.
   c. arteries and veins.
   d. enzymes.

9. Which chambers of the heart contract simultaneously?
   a. all four chambers
   b. right atrium and left atrium
   c. right atrium and right ventricle
   d. left atrium and left ventricle

10. A high count of white cells in the blood is a sign of
    a. bone marrow damage.
    b. infection.
    c. a bleeding wound.
    d. not enough hemoglobin.

11. If the amount of carbon dioxide in the blood increases, breathing
    a. stops.
    b. speeds up.
    c. becomes more infrequent.
    d. is controlled by hormones.

12. The site of gas exchange in the respiratory system is the
    a. lungs.   c. capillaries.
    b. bronchioles.   d. alveoli.

Completion

13. Bacteria and cancer cells that enter the body are destroyed in the _____ .

14. The pacemaker starts each _____ . It is _____ located at the entrance of the right _____ .

15. Blood is carried to the heart by the _____ and away from the heart by the _____ .

16. Red blood cells contain _____ that gives the cells their red color. A person who has too few red blood cells suffers from _____ .

17. A person with a blood pressure of 190/120 suffers from _____ , or high blood pressure.

Short Answer

18. Where is your diaphragm located? How does it function in breathing?

---

between the atria and ventricles and the "dup" sound is made by the valves between the ventricles and the main arteries that carry blood away from the heart.

20. Approximately 5 percent of carbon dioxide is transported in the blood plasma; 25 percent enters the red blood cells and combines with hemoglobin. The remaining 70 percent, with the help of an enzyme, combines with water and blood cells to form carbonic acid.

21. New red blood cells are found in the marrow of leg and arm bones. Old red blood cells are scavenged by macrophages that clean up the bloodstream in the liver.

19. How are the "lubb" and "dup" sounds of the heart and the movement of the heart valves related?

20. Describe the three ways that carbon dioxide is transported in the blood.

21. Where do new red blood cells come from? What happens to old ones?

22. The circulatory system transports oxygen, carbon dioxide, and hormones. What else does it do?

Interpreting Graphics

23. Look at the cross sections of the two coronary arteries below and answer the questions that follow.

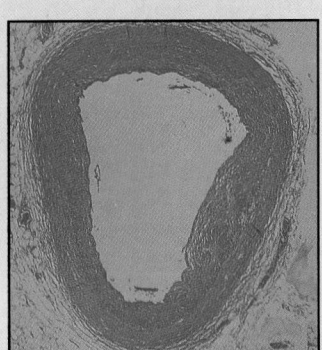

a

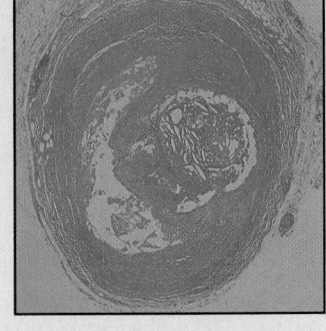

b

- How are the two arteries different?
- How can you explain the differences observed?

Reviewing Themes

24. *Stability*
How do white blood cells help the body maintain homeostasis?

25. *Scale and Structure*
Why can the four-chambered heart of mammals be thought of as two hearts in one?

26. *Evolution*
Explain why the vertebrate lung is considered an adaptation to terrestrial life.

Thinking Critically

27. *Comparing and Contrasting*
How are the pulmonary circulation and systemic circulation different?

28. *Comparing and Contrasting*
How do red blood cells and white blood cells differ in terms of function, size, and number?

29. *Building on What You Have Learned*
In Chapter 25 you learned about the open circulatory system of arthropods. How does the human circulatory system differ from that found in arthropods?

Cross-Discipline Connection

30. *Biology and Health*
Asthma attacks are brought on by something ingested or inhaled. Do library research to discover what substances trigger asthma attacks.

Discovering Through Reading

31. Read the article "Working Out Under Pressure," in *Health*, July/August 1992, pages 98–99. What was wrong with Marc Cohen? What did Dr. Siegel prescribe for Marc? What information serves to justify Dr. Siegel's prescription?

25. In terms of blood flow, the right atrium and ventricle are separate from the left atrium and ventricle. The right side pumps deoxygenated blood to the lungs, while the left side pumps oxygenated blood to the body.

26. The lung solves the problem of keeping the respiratory membrane moist by being located inside the body, where it is sheltered from the drying air.

Thinking Critically

27. Pulmonary circulation moves deoxygenated blood from the heart to the lungs, where it is oxygenated. Systemic circulation moves oxygenated blood from the heart to all parts of the body.

28. Red blood cells transport oxygen and carbon dioxide, while white blood cells defend the body against invading pathogens. Red blood cells are smaller and greater in number than white blood cells.

29. The human circulatory system is closed; blood remains in vessels while materials enter and leave the blood through capillaries. The blood does not enter open sinuses or body cavities, as occurs in arthropods.

Cross-Discipline Connection

30. Most asthma attacks are triggered by pet hairs, feathers in pillows or quilts, pollen, smoke, or dust.

Discovering Through Reading

31. Marc suffered from hypertension, so Dr. Siegel prescribed exercise. Even though Dr. Siegel's study showed no relationship between exercise and lower blood pressure, his decision was based on the fact that exercise improves fitness, contributes to weight loss, and lowers resting heart rate, and seems to reduce the body's response to everyday stress. While no one is certain why, these improvements are linked to lower blood pressure.

22. It maintains homeostasis by controlling body temperature. It carries white blood cells that help the body fight disease and absorbs and distributes heat to maintain body temperature.

Interpreting Graphics

23. • One is a normal artery (a). The other artery (b) is partially blocked.
    • In one artery (a), the passageway is clear and blood can flow. In the other artery (b), deposits decrease the amount of blood flow through the arteries.

Reviewing Themes

24. White blood cells rid the body of cancer and disease-causing cells—maintaining homeostasis.

Procedural Note

1. Review the answers to the Prelab Preparation questions. Be sure that students have a thorough understanding of pulse rate.
2. Have students practice taking each other's pulse rates before beginning the lab.
3. Have students check each other to assure that pulse rate is being taken correctly. Be sure that students do not use their thumbs to check pulse rate because the thumb has a strong pulse rate of its own, which may lead to errors.

Prelab Preparation Answers

• Answers may vary but should explain that the pulse results from the contraction of the left ventricle, which creates a surge of blood that causes the arteries to expand.
• Students should suggest that the respiratory system is responsible for gas exchange, whereas the circulatory system distributes the gases throughout the body via red blood cells.

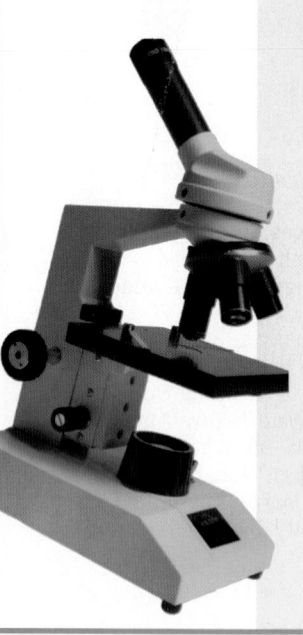

Investigation

How Does Exercise Affect Pulse Rate?

Objectives

In this investigation you will:
• *formulate* a hypothesis stating the effect that exercise has on pulse rate
• *design* an experiment that will compare pulse rate while at rest and immediately after exercise
• *collect* and *record* data

Materials

• unlined paper
• pencil
• watch with a second hand
• stable stool or chair about 30 cm (12 in.) high
• calculator

Prelab Preparation

Review what you have learned about circulation and respiration by answering the following questions:
• What is it that you are feeling when you take your pulse?
• What is the connection between the respiratory system and the circulatory system?

Procedure: Determining the Effect of Exercise on Pulse Rate

1. Form a cooperative team of two students. Work with your partner to complete steps 2–6.
2. Practice finding your partner's pulse by placing your index and middle fingers on the inner part of his or her wrist just below the base of the thumb. Use a watch with a second hand to determine the pulse rate for a 60-second interval.
3. Prepare a data table like the one shown below.

Pulse rate/minute

Subject	Resting rate		After exercise	
	1	2	1	2
Trial 1				
Trial 2				
Trial 3				
Individual average				
Team average				
Class average				

4. Discuss the question "How does exercise affect pulse rate?" with your partner. Formulate a hypothesis that answers the question.
5. Use the following guidelines to design an experiment to test your hypothesis.
   a. One member of the team will be the subject while the other member observes and records data. Team members will then switch roles and repeat the experiment.
   b. Resting pulse rate will be taken while the subject is sitting quietly.

c. **CAUTION: Do not perform the following exercise if you have a health problem that prohibits vigorous exercise. Stop the test immediately if you feel pain, become dizzy, or become extremely tired.**

d. The subject will exercise by stepping onto and off of a stool that is about 30 cm (12 in.) high. The subject should step at a rate of about 30 times a minute for three minutes. Pulse rate is taken immediately afterward.

e. Data should be collected for more than one trial for each subject. Calculate the subject's average pulse rate.

f. Collect data from other teams. Calculate the average pulse rate for the class when appropriate.

g. All data should be organized in a table.

6. After your design is approved by your teacher, conduct your experiment with your partner.

Analysis

1. *Summarizing Data*
   Summarize your data. Explain whether the data support your hypothesis.

2. *Analyzing Data*
   State your conclusion about how pulse rate changes after exercise.

3. *Evaluating Methods*
   Why is it best to average data from many trials and from a number of subjects?

4. *Predicting Outcomes*
   What changes might occur in pulse rate after a person completes an eight-week physical fitness course? Explain the reasons for your answer.

Thinking Critically

Why does your breathing and pulse rate increase when you exercise?

Analysis Answers

1. Answers will vary depending on students' hypotheses.

2. Answers will vary. Students should note that pulse rate climbs sharply after exercise. Also, students may discover that the post-exercise pulse rate of a person with a low resting pulse rate returns to normal more quickly than that of a person with a high resting pulse rate.

3. Students should suggest that these are ways to reduce error.

4. Average pulse rate would probably drop after an individual completed an eight-week physical fitness course. A regular fitness program increases heartbeat efficiency.

Thinking Critically Answer

Students should suggest that carbon dioxide has a greater effect on breathing than does oxygen. You breathe more deeply to rid the body of excess carbon dioxide.

Chapter 32

The Immune System

Planning Guide

	Objectives/Themes	Classwork Resources	Homework Resources
32.1	**1.** Identify two ways in which skin repels pathogens. **2.** Recognize the role of mucous membranes in defending the body. **3.** Summarize the reactions that make up the inflammatory response. **4.** Describe how your body distinguishes its own cells from invading pathogens. **Themes:** Interacting Systems, Patterns of Change, Scale and Structure, Stability	**Teacher's Resource Binder** Focus Activity 32 　*Relating Cell Structure* 　*to Function* **Other Resources** Transparency 145	**Text** Section Review, p. 741 **Teacher's Resource Binder** Directed Reading Worksheet 　32.1 **Other Resources** Audiocassette 32.1*
32.2	**1.** Identify the functions of helper T cells, killer T cells, and B cells. **2.** Explain how fever helps to defeat pathogens. **3.** Describe how the immune system protects against a pathogen's second attack. **4.** Relate vaccines to the functioning of the immune system. **Themes:** Interacting Systems, Patterns of Change, Scale and Structure, Stability	**Teacher's Resource Binder** Extension Worksheet 　*Contagious Diseases* **Other Resources** Transparencies 146–147	**Text** Section Review, p. 747 **Teacher's Resource Binder** Directed Reading Worksheet 　32.2 Reteaching Worksheet* 　*The Immune System* **Other Resources** Audiocassette 32.2*
32.3	**1.** Describe the events of an allergic reaction. **2.** Contrast cancer cells with normal cells. **3.** Relate the symptoms of AIDS to the action of HIV. **4.** Describe how HIV is transmitted. **Themes:** Interacting Systems, Patterns of Change, Scale and Structure	**Text** Investigation 　*How Do Antibody-Antigen* 　*Reactions Work?* pp. 758–759 Science, Technology, and Society 　*Transplant Technology: Saving* 　*Lives and Facing Dilemmas* 　pp. 760–761 **Teacher's Resource Binder** Lab Investigation 32.1 　*Disease Transmission* 　*Simulation* **Other Resources** Transparencies 148–149	**Text** Section Review, p. 754 **Teacher's Resource Binder** Directed Reading Worksheet 　32.3 Vocabulary Review Worksheet* **Other Resources** Audiocassette 32.3*

*Reteaching Options

Demonstrations	Assessment
32.1: p. 739 **32.2:** pp. 743, 747 **32.3:** pp. 748, 752, 753	Chapter Review pp. 756–757 Portfolio Assessment p. 737D Chapter Test—Teacher's Resource Binder Test Generator

Research Notes

Connection to Medicine: Overruling the Immune System

When an organ transplant is performed, frequently the patient's body will recognize the donor organ as being foreign to the body, and will initiate an immune system attack on the organ, leading to rejection.

This happens in as many as two-thirds of heart transplant patients. For a long time the only strategy was to try to find a donor that might be somewhat genetically similar to the patient. A new study may have uncovered a way to use two sets of antibodies to "trick" the body into thinking that any new organ belongs to the body.

A team led by Mitsuaki Isobe of the University of Tokyo treated mice given heart transplants from unmatched donor mice with two sets of antibodies. One antibody binds to a protein called leukocyte function-associated antigen-1 (LFA-1). This protein usually triggers the attack of white blood cells on foreign cells.

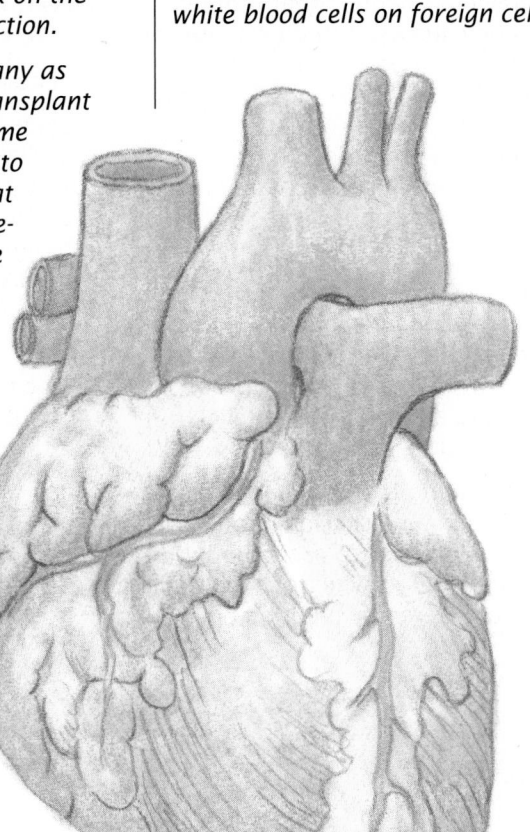

The other antibody binds to intercellular adhesion molecule-1 (ICAM-1), a signaling protein that summons the white blood cells. ICAM-1 is secreted by cells of the body when they are exposed to foreign cells. Each of the antibodies had been tested separately in transplants in the 1980s, but the results were inconsistent.

In Isobe's study, six control mice receiving no treatment died within 10 days. The two sets of six mice receiving only one of the antibodies died within one month. The nine mice that were treated with both antibodies were still alive after six months, and had no indications of tissue rejection. In a follow-up study, the mice accepted skin grafts from their heart donors, but rejected skin grafts of other mice.

According to Isobe, these findings indicate that the treatment caused the immune system to view the heart donor's tissue as its own. Isobe and his team are beginning studies with larger animals to examine safety considerations. If all goes well, the first human trials of this treatment could begin within two years.

Investigation Notes

How Do Antibody-Antigen Reactions Work?

pages 758–759

Purpose: This investigation gives students the opportunity to observe the antibody-antigen reaction common to blood typing. The emphasis of this investigation is placed on using the reaction as a tool in transfusions and other applications.

Prelab Preparation

A blood-typing kit can be obtained from WARD'S, including simulated anti-A and anti-B sera and vials of simulated blood, in compliance with the newly enacted OSHA blood-borne pathogen standard (WARD'S 36 M 0022).

Answers will be found on pages 758–759.

Meeting Individual Needs

Objectives

1. Students will demonstrate appropriate use of core vocabulary for the chapter (Vocabulary File).

2. Students will relate the different defenses of the body against pathogens (Teaching Strategy).

Vocabulary File

(Developing Vocabulary/ Limited English Proficiency)

If you are not already using the Vocabulary File, refer to Chapter 1 for its preparation. See the Chapter Highlights on page 755 for a list of suggested words.

Teaching Strategy

Defense Against Pathogens
for pages 739–746
(Developing Organizational Skills/Visual Learners)

Materials: overhead projector and projector pens

Procedure: Discuss with students the concept of "health." Most of us feel "good" physically much of the time, because our body has lines of defense to protect us against pathogens.

Challenge students to recall different types of tissue discussed in previous chapters that protect the body against pathogens. Students should be able to relate protection to the skin, as studied in Chapter 28 and to the white blood cells, as studied in Chapter 31. These are just two parts of the lines of defense.

Direct students to read silently pages 739 through 746. Have students take notes about the main points of the selection as they read. The points will be incorporated into a visual outline developed through class discussion.

When reading is completed, lead students in a discussion about the lines of defense of the body and the order in which they will be encountered by a pathogen.

Describe the different ways each line of defense can defend against pathogens and describe the types of pathogens it will and will not be able to stop. Incorporate the lines of defense into the outline.

Students are to copy the outline and use it for study. A suggested outline format follows.

I. First line of defense
  A. Skin
    1. Physical barrier to pathogens
    2. Chemical defenses against bacteria: oils and sweat
      a. Acidification
      b. Enzymes

  B. Mucous membrane
    1. Trap pathogens
    2. Sweep pathogens to stomach for destruction

  C. Inflammatory response
    1. Heat suppresses bacterial growth
    2. Attracts white blood cells such as phagocytes

  D. Phagocytes
    1. Consume and digest bacteria
    2. Consume cells infected with viruses
    3. Consume dead cells and foreign particles

Continue this format for the second line of defense, being sure students include the roles of primary and secondary immune responses and antibodies.

Additional Strategies

Visual Strategies

Pages 740, 741, 743, 744, 745, 749, 750, 751, 752, and 754

Auditory Learners

Use *Biology: Visualizing Life* Audiocassettes for Sections 32.1, 32.2, and 32.3.

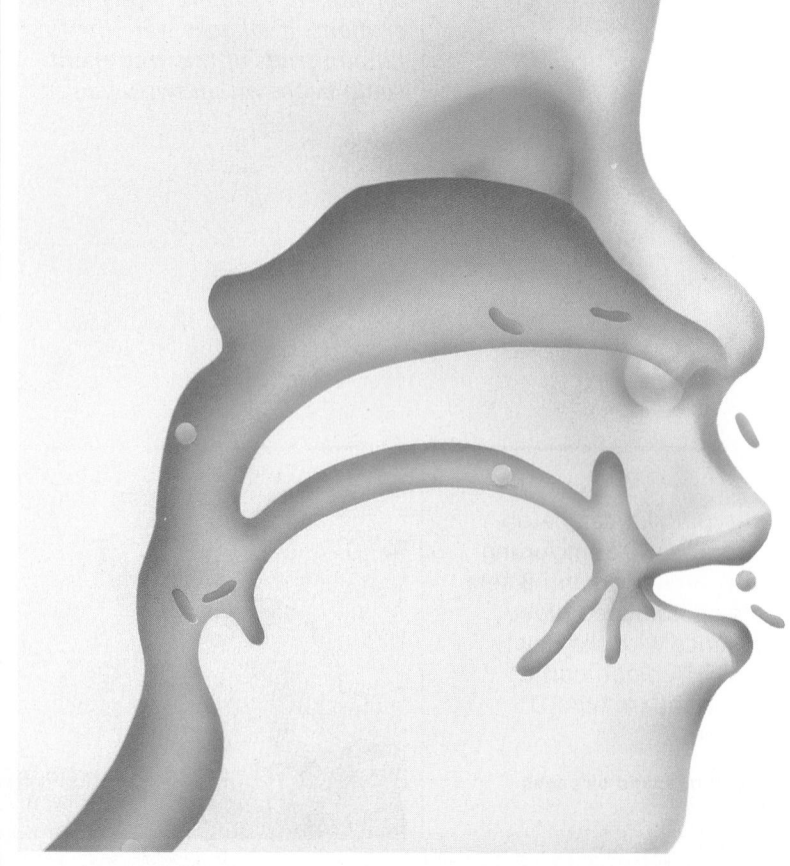

Meeting Individual Needs (cont.)

Cooperative Learning
AIDS and HIV

Timing: Use this activity to introduce Section 32.3.

Group Size: 4 students

Outcome: Students will be able to describe how HIV is transmitted, how it affects the body, and what statistics indicate about the number of people infected with HIV.

Individual Accountability: Each group member is responsible for finding information about AIDS and HIV.

Positive Interdependence: Each group will prepare a report about AIDS and HIV.

Have group members use the library to research AIDS to find out how HIV is transmitted, how HIV affects the body, and to find statistics about the number of people infected with HIV and the number with AIDS. The group members should each take notes on this information, and then share the notes with the others in their group by summarizing them aloud.

Have the group members combine their information into a single group report about HIV and AIDS. The reports should either be given orally or submitted in writing. Oral reports can lead to class discussions about AIDS.

Portfolio Assessment

Students should select their best work and provide a self-reflective rationale for their selections. Students can make selections in the following areas.

1. *Content* — One concept map from the chapter (See page 812 for evaluation criteria.)

2. *Reading Comprehension* — One Directed Reading Worksheet from the Teacher's Resource Binder (Use the answer key to evaluate for accuracy.)

3. *Writing* — Using the Vee Form, summarize a magazine or newspaper article relating to the immune system or tissue rejection. (See page 22T for evaluation criteria.)

   Or: Select a writing task or project from the Chapter Review.

4. *Performance Assessment* — One VEE form from a chapter investigation or lab manual investigation (See page 22T for evaluation criteria.)

Teacher makes selections in the following areas.

1. *Formal Assessment* — Chapter test (Test A, B, or the Test Generator) The teacher-scored test should be reviewed by the student. Incorrect responses should be corrected by the student before the test becomes part of the portfolio.

2. *Informal Assessment* — Use the Direct Observations Checklist, page 33T, during a laboratory or other cooperative learning experience.

3. *Performance Assessment* — Have students portray the story of the body's immune system responding to a pathogen through writing, pictures, or acting.

Concept Map Answer

The following is one possible answer to the Relating Concepts exercise on page 756.

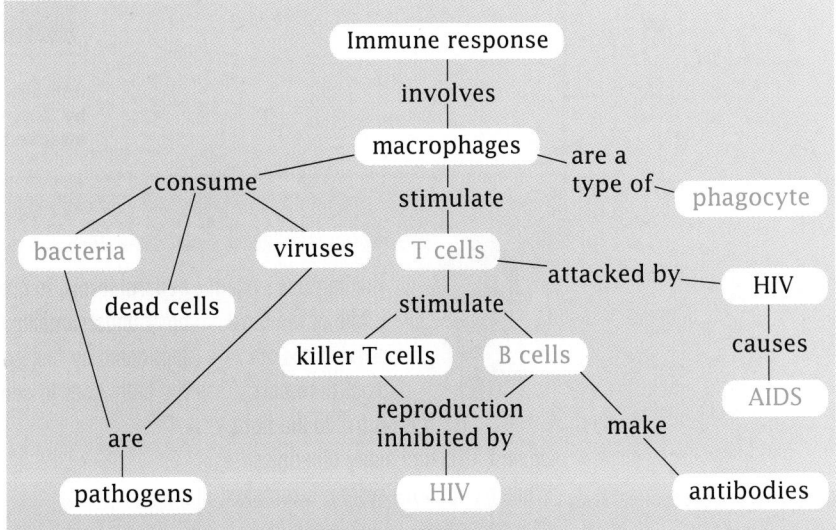

Chapter 32

Determining Prior Knowledge

- Have students use a stereomicroscope to examine the skin on their hands. Have them explain how the skin functions to prevent pathogens from entering the body.
- Have students raise their hands if they have had chickenpox. Have them keep their hands raised if they have had it more than once. Have them explain why most, if not all, hands went down in response to the second question.
- Have students raise their hands if they have an allergy. Have them identify what they are allergic to and describe their symptoms. Have the class explain what causes an allergic reaction.
- Write "AIDS" in large letters on the chalkboard and then ask students what they know about this disease.

Chapter 32

The Immune System

Review

- **membrane marker protein (Section 3.2)**
- **viral reproduction (Section 16.3)**
- **the term *pathogen* (Glossary)**

32.1 First Line of Defense

- **Keeping Pathogens Out**
- **Fighting Off a Local Infection**
- **Recognizing Pathogens**

32.2 The Immune Response

- **Second Line of Defense**
- **T Cells Command and Attack**
- **B Cells: Chemical Warfare**
- **Additional Defenses**
- **Shutting Off the Immune Response**
- **The Immune System "Remembers"**
- **Antigens and Blood Types**

32.3 Immune System Failure

- **Immune Overreaction**
- **Cancer: Unrestrained Cell Division**
- **AIDS: Immune System Collapse**
- **AIDS Is a Worldwide Disease**

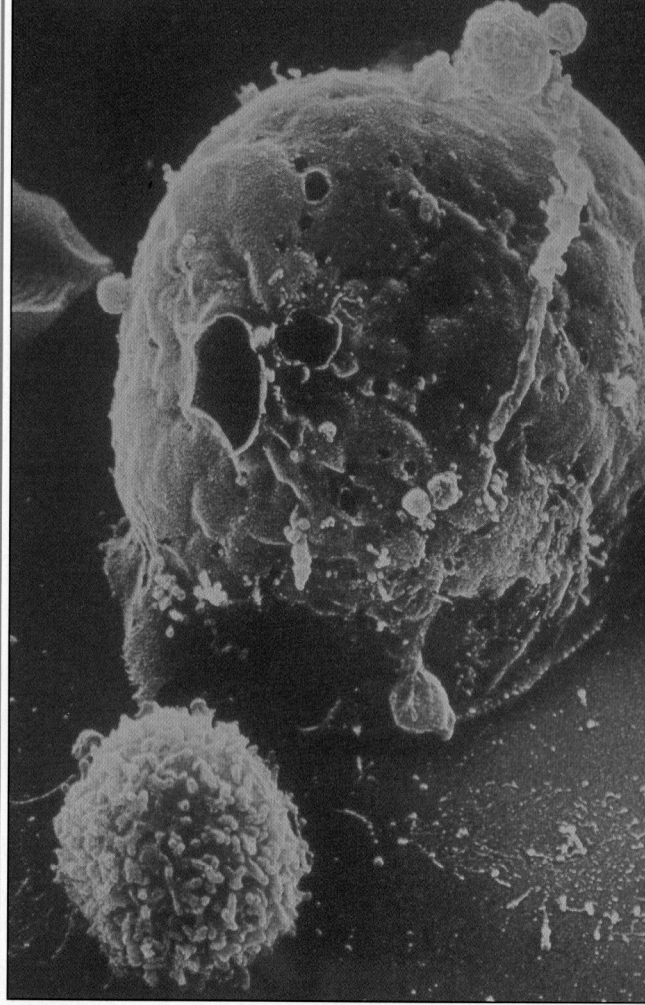

Immune system cells called killer T cells protect the body by destroying tumors. The large tumor cell above is being attacked by the smaller killer T cell.

■ Author's Rationale ■

The immune system is one of the most active areas of biological research. It is central to the fight against AIDS and many other diseases. While the immune system is complex, its workings can be explained in general terms, to give students a basic understanding. Students are fascinated by the lengths to which the body goes to defend itself.

LIKE A CITY UNDER SIEGE, YOUR BODY IS SURROUNDED BY LEGIONS OF WOULD-BE INVADERS: BACTERIA, VIRUSES, PROTISTS, PARASITIC ANIMALS, AND PARASITIC FUNGI. YET YOU ARE USUALLY WELL, BECAUSE YOUR BODY REPELS THESE PATHOGENS. IT ALSO DEFEATS THE FEW PATHOGENS THAT DO MANAGE TO ENTER YOUR BODY, USING A POWERFUL DEFENSE CALLED THE IMMUNE SYSTEM.

32.1   First Line of Defense

Objectives

Figure 32.1
Every time you breathe, you inhale a variety of bacteria and viruses that could make you sick.

① Identify two ways in which skin repels pathogens.

② Recognize the role of mucous membranes in defending the body.

③ Summarize the reactions that make up the inflammatory response.

④ Describe how your body distinguishes its own cells from invading pathogens.

Keeping Pathogens Out

Most pathogens must enter the body to cause disease. Your skin functions as a wall to keep out foreign organisms and viruses. As you learned in Chapter 27, skin is the body's dry, flexible covering. Cells of the outer layers of epidermis are continually being worn away. These lost cells are quickly replaced by cells moving up from the lower layers of epidermis, where cell division is occurring. In only 40 minutes, your body loses and replaces approximately 1 million skin cells. Such rapid replacement of body cells keeps the skin from disintegrating and enables punctures or cuts to be sealed very quickly.

Skin not only acts as a barrier to pathogens, it also engages in chemical warfare against them. Oils and sweat secreted by glands in the skin acidify its surface, creating an unfavorable environment for bacterial growth. In addition, sweat contains enzymes that digest the cell walls of certain bacteria.

As shown in **Figure 32.1**, each time you inhale or eat, bacteria and viruses can bypass the protection provided by your skin. Openings in your skin—such as the mouth, nostrils, eyes, and anus—are essential to allow food, oxygen, and sensory input to enter the body, and to allow wastes such as carbon dioxide, urine, and feces to be eliminated. Your body guards these openings with another series of defenses. For instance, tears and saliva contain the same antibacterial enzyme found in sweat. And internal body surfaces that come into contact with the environment are covered with **mucous membranes**.

A mucous membrane is a moist epithelial layer that is impermeable to most pathogens. Mucous membranes line the nasal passages, mouth, lungs, digestive tract, urethra, and vagina. Mucous membranes contain glands that secrete mucus, a sticky fluid that traps pathogens. Pathogens inhaled into the respiratory tract become lodged in mucus and are swept into the mouth by cilia. They are then swallowed and pass into the stomach. Digestive enzymes and strong acids in the stomach destroy most pathogens that are swallowed.

Lesson Plan 32.1

Phase 1
PREPARATION

Key Concepts
- The skin acts as both a physical and chemical barrier to prevent pathogens from entering the body.
- Tears, saliva, and mucous membranes also help keep pathogens from entering the body.
- An inflammatory response involves a sequence of events designed to fight off an infection.
- A macrophage, a type of white blood cell known as a phagocyte, plays a crucial role in destroying pathogens and dead cells.
- Marker proteins on the cell surface provide the mechanism by which macrophages recognize pathogens and other foreign substances.

Reading Strategy

This section makes several references to material covered in previous chapters. Have students use the index as they read this section to identify specific pages in previous chapters where related material has been discussed.

Phase 2
TEACHING STRATEGIES

Demonstration 1

Bacteria
Obtain two sterile petri dishes containing nutrient agar. Uncover both for 10 minutes, in the classroom, to collect bacteria present in the air. Next, use a sterile cotton applicator to wipe the oils and sweat from a student's arm, and then gently brush the surface of one of the exposed petri dishes. Cover both dishes and have the class check them over the next several days for signs of bacterial growth. Be sure to sterilize the

dishes by using a pressure cooker before disposing of them in the garbage.

Connection: Chapter 27

Epithelial Tissue

Remind students that epithelial tissue is made of tightly connected cells arranged in flat sheets.

Visual Strategy

Figure 32.2

Lead students through the phases of an inflammatory response, as illustrated in this figure. Point out that a mosquito bite also produces an inflammatory response. Before drawing blood, the mosquito injects a small amount of saliva that contains a substance to keep the victim's blood from clotting. This chemical substance makes the skin swell, turn red, and itch.

Connection: Chapter 8

Genetic Engineering

Tell students that a couple of decades ago, the only source of interferon was the small amount obtained from tissue culture cells that had been infected with a virus. In the late 1970s, one-millionth of an ounce of interferon cost $1,500. Today, as a result of genetic engineering, interferon is produced by bacteria at a fraction of this cost.

Fighting Off a Local Infection

A splinter in your fingertip is only a minor injury, even though it punches a hole in your skin through which pathogens can enter. Your body quickly closes the puncture and activates a set of chemical and cellular defenses to destroy any invaders. Your finger swells, becomes red and painful, and feels hot. These are signs that your body is attacking pathogens. The redness and swelling are part of the **inflammatory response** to infection or injury. You can follow the events of an inflammatory response in **Figure 32.2**.

White blood cells destroy pathogens

White blood cells are the "soldiers" of the immune system and are crucial for the inflammatory response. **Phagocytes** (*FAG oh seyets*) are white blood cells that launch direct attacks on pathogens, ingesting them by phagocytosis.

Phagocytes consume bacteria, cells infected by viruses, dead cells, and foreign particles. **Macrophages** (*MAK roh fayjz*) are the most common phagocytes. Another type of phagocyte, called a neutrophil (*NOO troh fihl*), releases the same chemical found in household bleach, killing itself and any nearby bacteria. Phagocytes, dead cells, and pathogens are the components of pus, a yellowish or whitish fluid that often accumulates in a wound. The presence of pus signals that phagocytes are combating pathogens. The inflammatory response is sufficiently powerful to repel most small-scale infections.

Cells infected by viruses release a chemical messenger, a protein called interferon (*in tuhr FIHR ahn*). Interferon activates white blood cells that kill pathogens, and so increases the ability of noninfected cells to resist infection. Interferon may have medical uses.

Figure 32.2
When a splinter enters your finger, bacteria and viruses can gain access to your body. Your body initiates the inflammatory response to combat these pathogens.

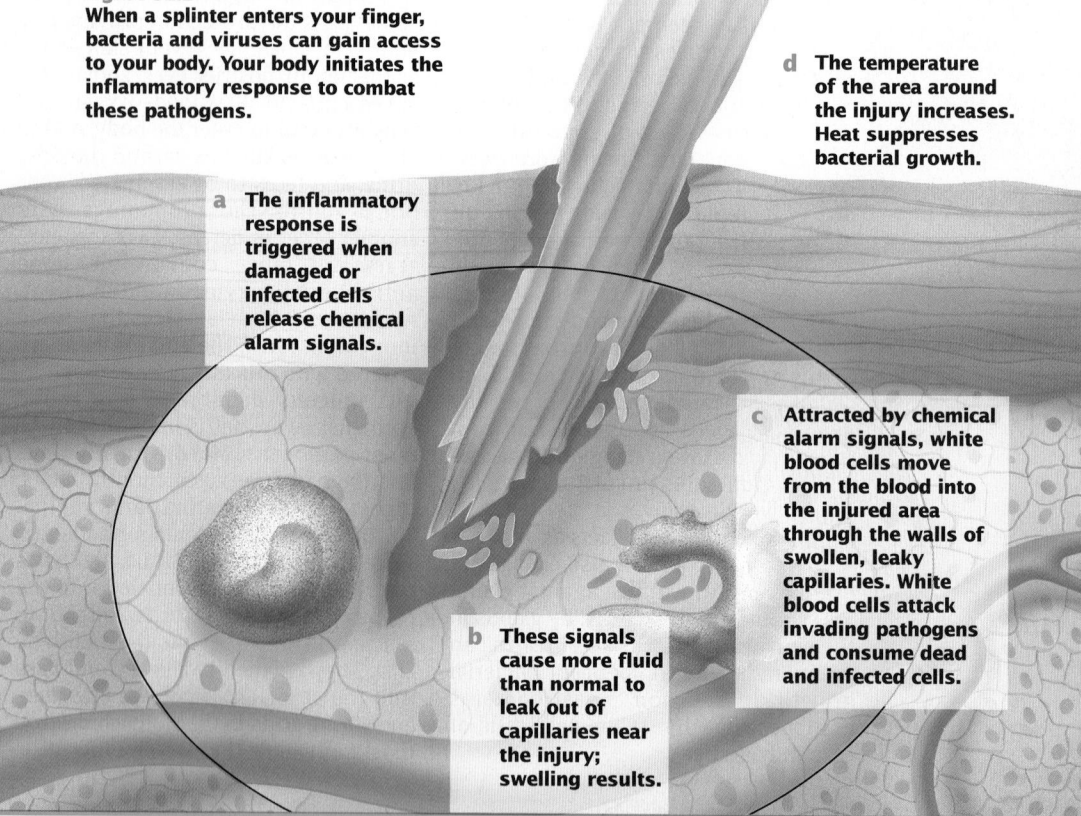

a The inflammatory response is triggered when damaged or infected cells release chemical alarm signals.

b These signals cause more fluid than normal to leak out of capillaries near the injury; swelling results.

c Attracted by chemical alarm signals, white blood cells move from the blood into the injured area through the walls of swollen, leaky capillaries. White blood cells attack invading pathogens and consume dead and infected cells.

d The temperature of the area around the injury increases. Heat suppresses bacterial growth.

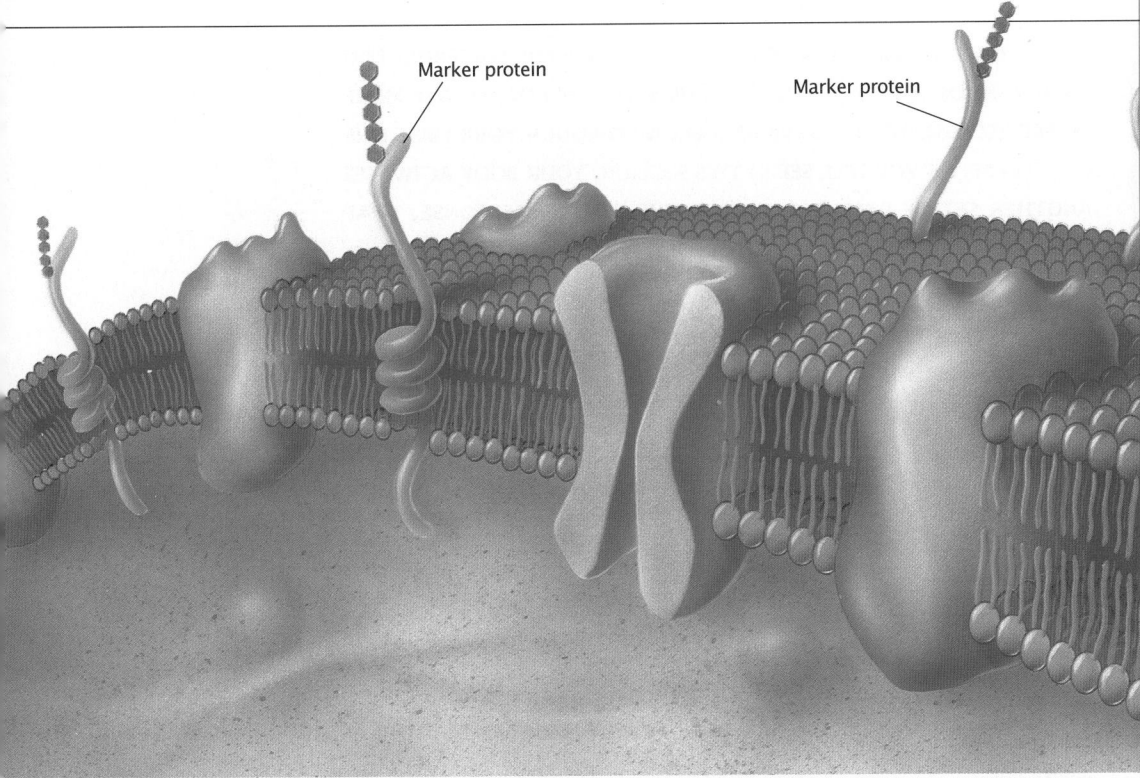

Marker protein

Marker protein

Figure 32.3
Because no two organisms have the same marker proteins, cells of the immune system are able to identify and destroy invading pathogens.

Recognizing Pathogens

How does a phagocyte recognize pathogens and infected cells? Recall from Chapter 3 that channel, receptor, and marker proteins occur in the membrane of a cell. Certain marker proteins serve as identification tags that enable white blood cells to distinguish your cells from foreign cells. As shown in Figure 32.3, each cell in your body carries "self" marker proteins that identify it as your cell. Your marker proteins have a unique amino acid sequence and therefore a unique shape. The shape of your marker proteins differs from the shape of marker proteins on cells of other humans and on the cells of other species, such as pathogenic bacteria. Your marker proteins are also different from viral surface proteins. White blood cells have receptor proteins in their membranes that enable them to recognize proteins on the surface of a cell or virus. Your white blood cells recognize proteins shaped differently from your own as foreign, or "non-self." Molecules that can be recognized by white blood cells and that can trigger a defensive response are known as **antigens**.

Section Review

| ❶ Describe the chemical defenses of skin. | ❷ Explain two ways that your mucous membranes protect you against pathogenic bacteria. | ❸ Describe the events that occur after a splinter enters your finger. | ❹ Explain how white blood cells can distinguish your cells from those of invading pathogens. |

▪ Section Review Answers ▪

1. Students should indicate that oil and sweat acidify the skin's surface, creating an unfavorable environment for bacteria. In addition, sweat contains enzymes that digest the cell walls of certain bacteria.

2. Mucous membranes are impermeable to most bacteria; some contain antibacterial enzymes; others trap pathogens and transport them to areas for destruction.

3. See Figure 32.2 on page 740.

4. Students should note that body cells carry certain marker proteins that enable the body to distinguish its own cells from invading pathogens.

Lesson Plan 32.2

Key Concepts

- Macrophages engulf and consume invading pathogens that are recognized as being foreign because of their surface antigens.
- The immune response involves two kinds of white blood cells: T cells and B cells.
- Helper T cells bind to macrophages that display the antigens from pathogens they have consumed. Once bound, the helper T cells release proteins that promote the rapid division of killer T cells.
- Killer T cells destroy pathogens by pinching holes in their cell membranes.
- Suppressor T cells quell the immune response once the pathogens have been defeated.
- B cells produce antibodies that bind the pathogens so that they can be more easily destroyed by the macrophages.
- Memory B cells remain long after the pathogen has been destroyed, in case they are needed if the same pathogen again invades the body.
- Vaccines promote the development of memory B cells, thus arming the body against possible future invasion by the pathogen.
- Protein antigens present on the membranes of red blood cells account for the various blood types.

Reading Strategy

Analogies can often be useful in understanding what is read. Have students list as many analogies as they can make while reading this section. Examples might include comparing macrophages to a security system that responds to an intruder, comparing killer T cells to a missile that hones in on and destroys a specific target, and comparing antibodies to a lasso that engulfs and entraps several people at the same time.

THE PAIN OF A SPLINTER DOES NOT COMPARE TO THE SUFFERING YOU ENDURE WHEN YOU HAVE THE FLU. THE SYMPTOMS OF FLU ARE MORE SEVERE BECAUSE INFLUENZA VIRUSES BREAK THROUGH YOUR FIRST LINE OF DEFENSE. AS YOU WILL SEE IN THIS SECTION, YOUR BODY ACTIVATES ANOTHER SET OF DEFENSES, CALLED THE IMMUNE RESPONSE, THAT COMBATS AND DEFEATS THE FLU VIRUSES.

32.2 The Immune Response

Objectives

❶ Identify the functions of helper T cells, killer T cells, and B cells.

❷ Explain how fever helps to defeat pathogens.

❸ Describe how the immune system protects against a pathogen's second attack.

❹ Relate vaccines to the functioning of the immune system.

Figure 32.4
Macrophages attack foreign pathogens, such as viruses or the *E. coli* bacteria shown here.

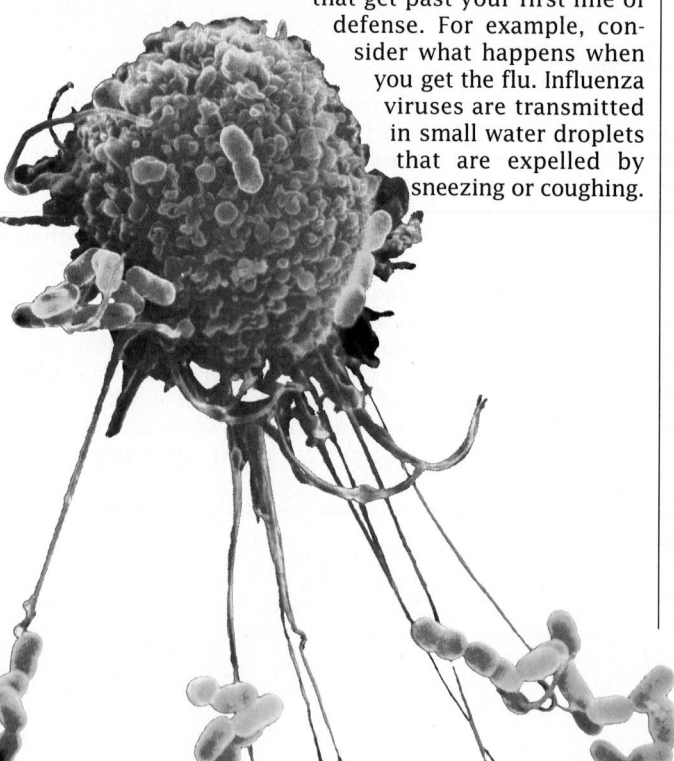

Second Line of Defense

The immune system is the defenses your body uses to attack pathogens that get past your first line of defense. For example, consider what happens when you get the flu. Influenza viruses are transmitted in small water droplets that are expelled by sneezing or coughing.

If you inhale some of these droplets, influenza viruses can enter cells of the mucous membrane lining your respiratory tract. As you recall from Chapter 16, viruses seize control of their host cells and transform them into virus-producing factories. Influenza viruses take over and kill mucous membrane cells. You feel sick because large numbers of the cells lining your respiratory tract are dying.

At this point during a case of flu, the viruses have the upper hand, but your body's defenses are beginning to fight back. The first stages of this counterattack are carried out by macrophages, which consume any influenza viruses they encounter. Macrophages also attack other invaders, including the bacteria shown in **Figure 32.4**. Cells infected by viruses have viral antigens on their membranes. Patrolling macrophages consume these cells as well. Macrophages then display viral antigens on their cell surface, like victory banners. This display stimulates the immune system to carry out a full-blown **immune response**. The immune response is the immune system's attack on a specific pathogen.

T Cells: Command and Attack

White blood cells called **T cells** control the immune response and attack infected cells. T cells are so named because they mature in the thymus, a small gland above the heart. T cells circulate in your blood and lymph, and occur in your spleen and lymph nodes. The three important classes of T cells are helper T cells, killer T cells, and suppressor T cells. Each class carries out a different task during the immune response.

An individual T cell carries specifically shaped receptor proteins on its cell membrane. These receptor proteins enable the T cell to recognize and bind to one particular antigen. Your body can respond to millions of different antigens because it manufactures millions of different types of T cells, each type bearing uniquely shaped proteins. When you have the flu, T cells with receptors that match the antigens of the influenza viruses will be "called up" to fight the viruses.

Helper T cells command the immune response

Figure 32.5 describes the roles of T cells in the immune response. In order for T cells to begin combating influenza viruses, helper T cells, the commanders of the immune response, must be activated. Activation occurs when a helper T cell with a receptor matching the influenza antigen meets a macrophage displaying this antigen (as a "trophy" of its recent encounter with the virus) on its cell membrane.

Figure 32.5
The immune response is activated when a macrophage that has consumed a pathogen, such as an influenza virus, comes into contact with a helper T cell.

Macrophage

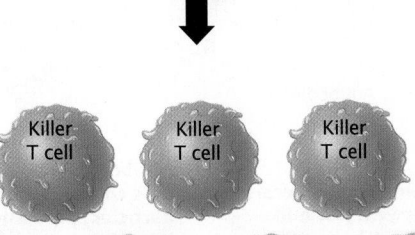

Interleukin-1

Interleukin-2

Interleukin-2

a The macrophage releases a protein called interleukin-1, which stimulates the helper T cell.

Helper T cell

Suppressor T cell

b The helper T cell then releases a slightly different form of interleukin, interleukin-2. In response to interleukin-2, killer T cells and suppressor T cells begin to reproduce.

Suppressor T cell

Suppressor T cell

c Suppressor T cells divide slowly. Their role is to shut down the immune response, as you will learn later.

Killer T cell

Killer T cell Killer T cell Killer T cell

Killer T cell Killer T cell Killer T cell Killer T cell

Killer T cell

d Killer T cells destroy infected cells by puncturing their cell membranes. The action of killer T cells helps to eliminate infected cells but does not attack the viruses themselves.

■ *Cultural Perspective* ■

Jerry Guyden, Ph.D.
African-American
Dr. Guyden is a faculty member at the City College of the City University of New York, where he studies how genes produce and regulate T cells.

Phase 2

TEACHING STRATEGIES

Demonstration 1
Macrophages
Use a microprojector to show the class an *Amoeba*. Have students compare the *Amoeba* to the macrophage illustrated in Figure 32.5. Point out that a macrophage uses amoeboid movements to surround and consume foreign cells and viruses. While some macrophages circulate, others remain stationary (for instance, in the spleen and lymph nodes), simply engulfing anything suspicious that passes by.

Demonstration 2
Thymus Gland
Show students an overhead transparency or other visual that illustrates the size and location of the thymus gland. Point out that the gland is much larger in children than in adults, since it helps establish the immune system early in life. In fact, the thymus gland can be removed from an adult with virtually no physiological consequences.

Visual Strategy

Figure 32.5
Use this figure to explain the difference between helper T and killer T cells. The receptor sites on the helper T cells shown in Figure 32.5a are so specific that helper T cells can recognize the difference between healthy and diseased body cells. Emphasize that special proteins are produced once a helper T cell's receptors have combined with antigens presented by a macrophage. These proteins stimulate the division of killer T cells, which students should locate in Figure 32.5b.

Visual Strategy

Figure 32.6
Emphasize that helper T cells simultaneously stimulate division of killer T cells and B cells. Have students identify the B cells and the antibodies they produce, in Figure 32.6b. Emphasize that these antibodies bind to infected cells that display the foreign antigens and also bind to the same antigens found on the pathogens themselves, as shown in Figure 32.6c and 32.6e.

B Cells: Chemical Warfare

White blood cells known as **B cells** are also important in the immune response. B cells produce and release defensive proteins known as **antibodies**. Antibodies recognize and bind to antigens on the surface of a virus or cell, or free-floating in the blood or lymph. Binding by antibodies does not destroy an antigen. Rather, it marks the antigen for consumption by macrophages.

Like T cells, B cells have receptor proteins scattered over their cell membranes. Your body makes millions of different types of B cells. Each type shows a unique receptor protein that will bind to one particular antigen.

The protein receptors a B cell carries on its cell membrane are identical to the antibodies it produces.

At any one time, millions of B cells are circulating in your blood and lymph but are not producing antibodies. Antibody production is turned on when B cells interact with antigen-displaying macrophages and helper T cells, as illustrated in **Figure 32.6**.

Your body is capable of making millions of different antibodies because of a unique process that occurs in the genes of maturing B cells. Long before contacting any antigen, a B cell scrambles its antibody-coding genes—like shuffling a deck of cards—and becomes specialized to manufacture only one randomly chosen antibody of the millions possible. In this way, your body prepares defenses against a variety of potential invaders, many of which will never be encountered.

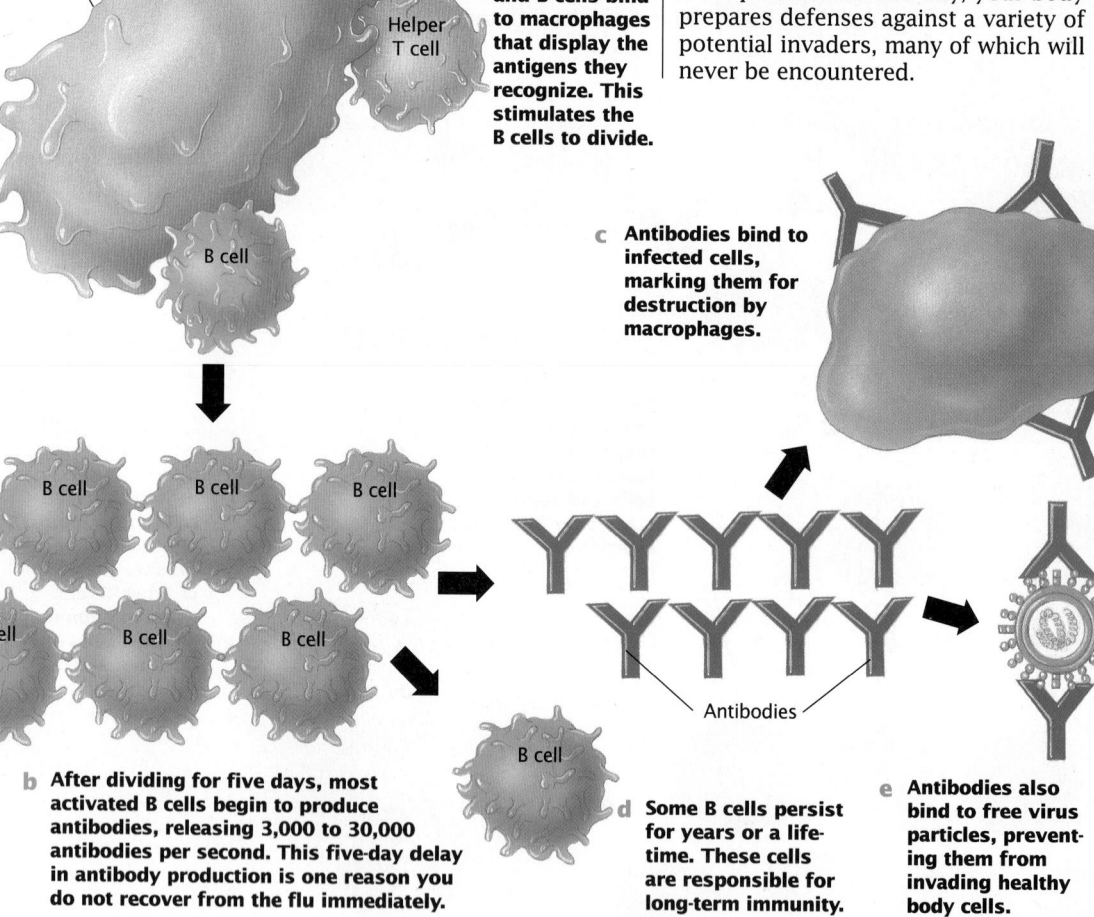

Figure 32.6
A B cell that encounters a macrophage showing the viral antigen will begin to manufacture antibodies.

Macrophage

Helper T cell

B cell

a Helper T cells and B cells bind to macrophages that display the antigens they recognize. This stimulates the B cells to divide.

c Antibodies bind to infected cells, marking them for destruction by macrophages.

B cell

B cell · B cell · B cell

cell · B cell · B cell

Antibodies

B cell

b After dividing for five days, most activated B cells begin to produce antibodies, releasing 3,000 to 30,000 antibodies per second. This five-day delay in antibody production is one reason you do not recover from the flu immediately.

d Some B cells persist for years or a lifetime. These cells are responsible for long-term immunity.

e Antibodies also bind to free virus particles, preventing them from invading healthy body cells.

Additional Defenses

*Scale and
Structure*

*Why is the immune
response
considered a
specific defense
against pathogens,
while the inflam-
matory response is
considered a non-
specific defense?*

A variety of defenses aid cells of the immune system in their fight against infection. When you are sick you often have a fever. Fever is the result of protein messengers sent out by macrophages that have contacted antigens. These protein messengers stimulate the hypothalamus, the area of the brain that controls body temperature, to raise body temperature above the normal 37°C (98.6°F). Moderate fever has beneficial effects: it inhibits the growth of pathogens and stimulates macrophage action. Very high fevers (above 39°C [103°F]) can damage essential proteins, however.

Shutting Off the Immune Response

The immune response is fairly slow in activating to meet an infection. But it usually defeats invading pathogens rapidly once it gets going. After pathogens have been overcome, suppressor T cells shut down the immune response. Recall that these cells begin to divide when stimulated by helper T cells. Suppressor T cells divide slowly, and so only become abundant one to two weeks after infection. Suppressor T cells inhibit reproduction of killer T cells and B cells. Although the immune response slows under the influence of suppressor T cells, it never shuts off completely.

The Immune System "Remembers"

Figure 32.7
The secondary immune response to a pathogen occurs much more quickly and produces more antibodies than the primary immune response.

You may have had mumps when you were younger. Once you have recovered from mumps, you usually cannot catch this illness again, even if exposed to the virus that causes it. Your immune system "remembers" its battles. It maintains a low level of defense that can be quickly mobilized to destroy previously defeated pathogens, should they return.

Recall that B cells begin to divide when stimulated by helper T cells and antigens. Most B cells become short-lived antibody producers. But a few B cells become **memory cells**. Memory cells continue to patrol your body's tissues, circulating through your blood and lymph for long periods of time—sometimes for the rest of your life. If a pathogen you have already encountered enters your body, memory cells recognize it and start to divide, quickly forming a new generation of antibody-producing cells. The antibodies generated by the new cells destroy the pathogen before you become ill; you are not even aware of the battle going on within your body. An immune response to a previously encountered pathogen is known as a **secondary immune response**. A secondary immune response is much faster and produces many more antibodies than the first response, as shown in **Figure 32.7**. The immune response that occurs after the first exposure to a pathogen is known as the **primary immune response**.

Comparison of Primary and Secondary Immune Responses

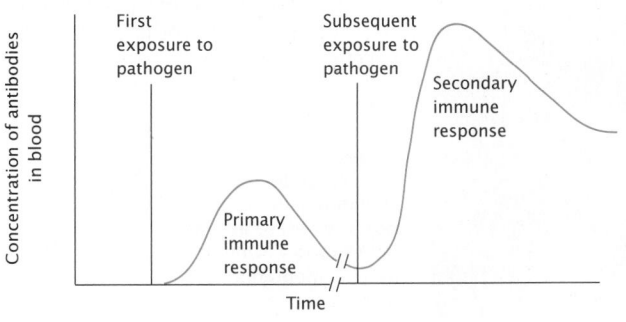

Theme Answer
Scale and Structure
The immune response is targeted toward pathogens it remembers. Memory cells continue to patrol the body's tissues, circulating through the blood and lymph for long periods of time, responding quickly against pathogens they recognize. The inflammatory response can inhibit the growth of pathogens and stimulate macrophage action, but the inflammatory response is not selective.

Health Connection

Monoclonal Antibodies
Monoclonal antibodies are produced by cloning a single cell obtained by fusing a B cell with a tumor cell. Known as a hybridoma, this new cell has properties of both parents—it divides forever (like a cancer cell) and it produces a specific antibody (like a B cell). The use of monoclonal antibodies is currently being studied as a way of focusing a massive attack on a particular pathogen, perhaps even killing cancer cells without harming normal cells.

Visual Strategy

Figure 32.7
Have students explain why there are more antibodies present during the secondary immune response, as compared with the number involved in the primary response. Also have them explain why the variety of memory cells increases as one gets older.

Connection: Chapter 8

Making Vaccines

Remind students that genetic engineering techniques are being used to insert genes from pathogens into harmless bacteria or viruses that act as vectors. These vectors then carry the genes "piggyback" into the human body, where the genes direct the synthesis of protein antigens. In turn, the body produces antibodies. Ideally, enough memory cells remain to be effective even against a rather large-scale, subsequent invasion by the same pathogen.

Theme Connection

Evolution

Natural selection has favored viruses with the ability to mutate rapidly, since these mutations have enabled them to overcome the challenges presented by the immune system.

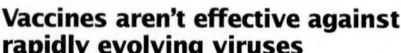

Figure 32.8
This Vietnamese-American child is receiving the vaccinations required before entering school.

Vaccination makes use of the immune system's "memory"

Many serious diseases can now be prevented through vaccination, which you read about in Chapter 16. In the United States and other industrialized countries, children, like the girl in **Figure 32.8**, are routinely immunized against polio, tetanus, diphtheria, whooping cough, and measles. Although these diseases remain killers in the less industrialized countries, groups like the World Health Organization have begun mass immunization programs in these nations. **Figure 32.9** shows a volunteer telling an Ethiopian woman about the vaccination procedure.

Vaccination triggers an immune response against a particular pathogen without causing the disease itself.

Recall that the immune response is triggered by antigens. Vaccines are effective because they contain antigens that have been stripped of their disease-causing abilities. For instance, vaccines for bacterial diseases usually contain bacteria killed by heat or chemical treatment. B cells, T cells, and macrophages respond to the antigens on the dead bacteria as if encountering live, harmful bacteria. Similarly, viral vaccines stimulate the immune response with viruses made harmless by chemicals or genetic engineering. As you learned in Chapter 8, scientists can now produce "piggyback" vaccines through genetic engineering. These vaccines contain harmless viruses that have been altered to express the surface proteins of a pathogen. When injected into the body, the harmless viruses serve as antigens, stimulating production of antibodies and memory cells against the pathogen.

Vaccines aren't effective against rapidly evolving viruses

You can catch the flu more than once, unlike measles or mumps. Furthermore, although you can be immunized against influenza, vaccination does not provide long-term protection against infection. The viruses that cause flu have evolved a way to evade the immune system. Genes coding for the surface proteins of flu viruses mutate, or change, rapidly. Thus, the shapes of these surface proteins alter swiftly. Your body does not recognize viruses with altered surface proteins as the same viruses it has already successfully defeated or been vaccinated against. When these viruses invade your body, they provoke a new primary immune response and you get sick again. HIV (the AIDS virus) also mutates rapidly and has evaded scientists' attempts to produce successful vaccines against it.

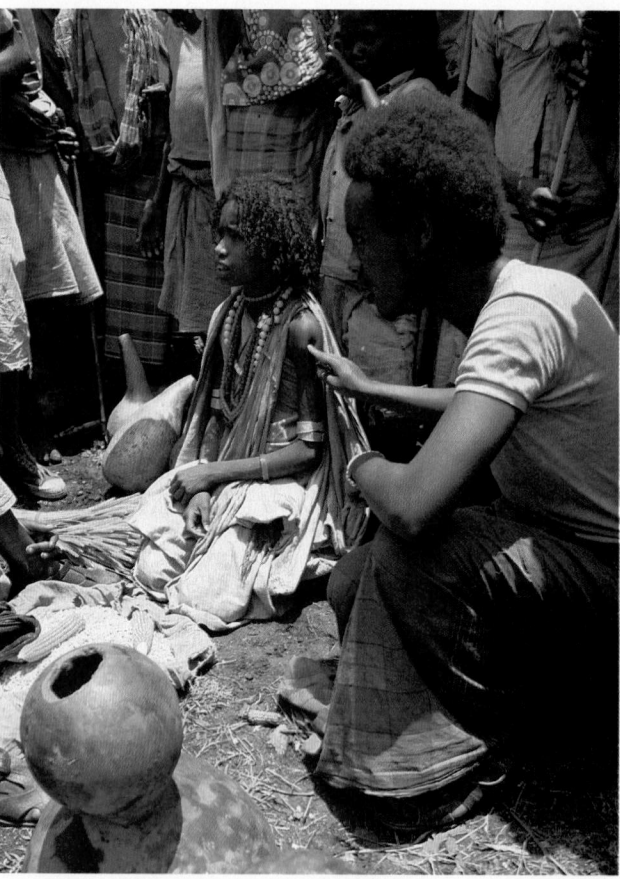

Figure 32.9
This rural doctor is explaining the vaccination procedure and its benefits to a woman at an Ethiopian street market.

Antigens and Blood Types

You learned about the different blood types in Chapter 31. Transfusions between people of different blood types usually are not compatible because the immune system of the recipient attacks the transfused blood. People with type A blood have a marker protein known as the A antigen on the surface of their red blood cells. Red blood cells from individuals with type B blood have a slightly different marker protein called the B antigen. People with type AB blood have both A and B antigens on their red blood cells. Individuals with type O blood have neither A nor B antigens on their red blood cells. Individuals produce antibodies to the marker proteins not found on their own cells. For instance, individuals with type A blood produce antibodies against the B antigen, even if they have never been exposed to it. **Table 32.1** summarizes the antigens and antibodies found in people of each blood type.

If type A blood is transfused into a person with type B or type O blood, antibodies against the A antigen will attack the foreign red blood cells, causing them to clump together, as shown in **Figure 32.10**. Clumps of red blood cells can block capillaries and cut off blood flow, which can be fatal. Clumping also occurs if type B blood is transfused into individuals with type A or type O blood, or if type AB blood is given to people with type O blood.

Figure 32.10
Some blood types are not compatible. When incompatible blood types are mixed, they form clumps (shown below) that prevent blood from circulating properly.

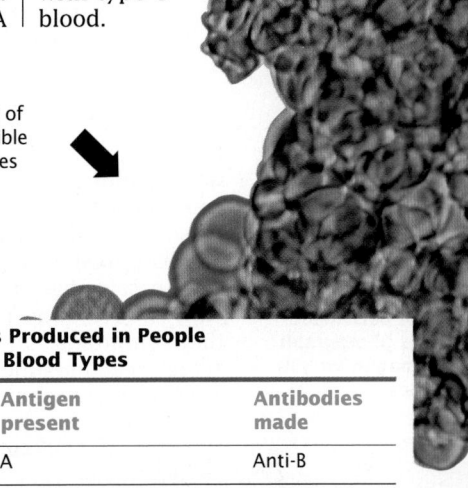

Clumping of incompatible blood types

Table 32.1 Antibodies Produced in People of Various Blood Types

Blood type	Antigen present	Antibodies made
A	A	Anti-B
B	B	Anti-A
AB	A,B	Neither
O	None	Anti-A, Anti-B

Section Review

1 Explain how destruction of helper T cells would affect the immune response.

2 Contrast the roles of B cells and T cells in the immune response.

3 Explain the role of memory cells in the immune system.

4 What can you conclude about the surface proteins of viruses for which effective vaccines exist?

■ *Section Review Answers* ■

1. The destruction of helper T cells would shut down the body's defense against invading microorganisms.

2. See Figure 32.5 on page 743 and Figure 32.6 on page 744.

3. Memory cells patrol the body's tissues, circulating through the blood and lymph, destroying pathogens they have battled before.

4. These surface proteins are not able to change (mutate) quickly enough to escape detection.

Demonstration 3
Antigens and Blood Types
Assign students to cooperative groups of four. Have each group cut out four circles from paper, using a different color for each circle. Have them label the circles so that each color represents a different blood type. Then have students cut out small pieces of white paper in two different shapes to represent the A and B antigens. Tell them to tape pieces of one shape to the periphery of one circle to represent blood type A, and pieces of the second shape to another circle to represent blood type B. Have them explain what they must do to have the other two circles correctly depict blood types AB and O. Next, have them cut out pieces of paper to represent anti-A and anti-B antibodies. Each must fit with its corresponding antigen, just like two pieces of a puzzle. Finally, have them simulate what would happen if two different blood types were mixed. Have them refer to the information listed in Table 32.1 for help when doing their simulations.

Phase 3

ASSESSMENT OPTIONS

Closure Strategy
The Common Cold
Ask students why the older one gets, the fewer number of colds one generally gets. Have them explain why it is not practical to develop a vaccine against the common cold.

Section Review
Assign the *Section Review*.

Reteaching
Assign students to cooperative work groups of four. Have each group write a comic sketch or draw cartoons to show how the second line of defense operates in protecting the body against pathogens.

Lesson Plan 32.3

Phase 1

PREPARATION

Key Concepts

- An allergy involves the response by the immune system to a harmless antigen.
- In an allergic reaction, most cells (a type of white blood cell) release histamines, which cause capillaries to swell and release fluid.
- An autoimmune disease results when the immune system starts to destroy its own body cells.
- The immune system sometimes fails to destroy cancer cells that arise from mutations or as a result of carcinogens.
- HIV destroys T cells, resulting in AIDS.

Reading Strategy

This is a good time to make students aware of how much knowledge still remains to be discovered in biology, especially in the areas of autoimmune diseases, cancer, and AIDS. To do so, tell students to use the school or local library to locate an article that discusses how biologists are attempting to unravel the mysteries concerning one of the diseases discussed in this section. Have each student read the article, provide a short written summary, and then report his or her findings to the rest of the class.

Phase 2

TEACHING STRATEGIES

Demonstration 1

Asthma

Exhibit an inhaler such as those used by asthmatics when breathing becomes difficult. Explain to the class that the inhaler works by dilating the air passages leading to the lungs.

TYPE 1 DIABETES IS A DISEASE IN WHICH CELLS ARE UNABLE TO TAKE IN GLUCOSE BECAUSE THE PANCREAS FAILS TO PRODUCE INSULIN. TYPE 1 DIABETES IS THOUGHT TO BE THE RESULT OF AN IMMUNE SYSTEM DEFECT. INSTEAD OF ATTACKING INVADING PATHOGENS, THE IMMUNE SYSTEM ATTACKS INSULIN-MANUFACTURING CELLS OF THE PANCREAS. TYPE 1 DIABETES IS ONE EXAMPLE OF AN IMMUNE SYSTEM FAILURE.

32.3   Immune System Failure

Objectives

❶ **Describe the events of an allergic reaction.**

❷ **Contrast cancer cells with normal cells.**

❸ **Relate the symptoms of AIDS to the action of HIV.**

❹ **Describe how HIV is transmitted.**

Immune Overreaction

Figure 32.11
People with allergies to dust are actually allergic to the feces of the house-dust mite, which lives on dust particles. This photograph is magnified 300 times.

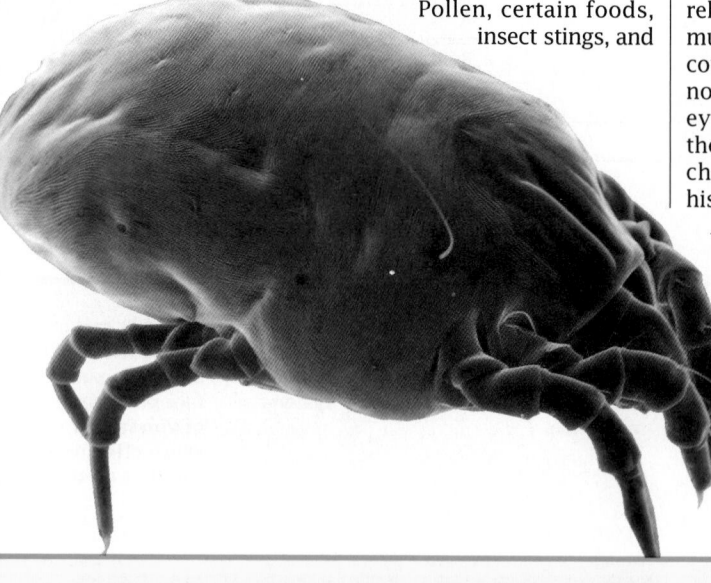

Petting a cat, smelling some flowers, or walking into a dusty room can be miserable experiences for some people. In these individuals, such activities rapidly lead to sneezing, itchy nose and eyes, nasal congestion, and even difficulty in breathing. These are some of the symptoms of an **allergy**. An allergy is an immune system response against a harmless antigen. A variety of substances trigger allergies. Pollen, certain foods, insect stings, and dust can all cause allergies. Allergies to dust are responses to proteins in the feces of tiny mites, such as the one in **Figure 32.11**, that live on dust particles.

If you inhale pollen you are allergic to, cells in the nasal passages release a set of chemical messengers that includes **histamine** (*HIHS tuh meen*), as shown in **Figure 32.12** on page 749. Histamine and the other messengers stimulate nearby capillaries to swell and release fluid. Histamines also increase mucus production by cells of the mucous membranes, resulting in a runny nose or nasal congestion, and watery eyes. Many allergy medicines relieve these symptoms with **antihistamines**, chemicals that block the action of histamines.

Asthma is a form of allergic response that takes place in the lungs. Besides the reactions already described, histamines cause the narrowing of air passages in the lungs of people who have asthma. These individuals will have trouble breathing when exposed to antigens to which they are allergic.

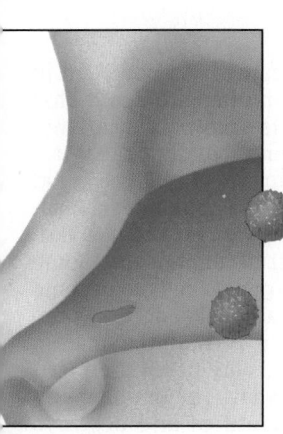
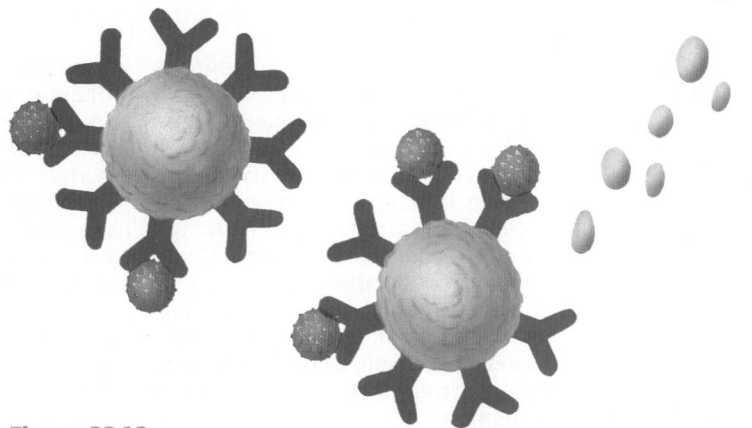

Figure 32.12
Phase 1
When pollen from a plant enters the nose of a person who is allergic to it, pollen antigens attach to antibodies on the surfaces of the cells lining the nasal passages.

Phase 2
Binding of antibodies stimulates cells to release histamine and other chemicals that cause symptoms including sneezing, runny nose, and itchy eyes.

Autoimmune diseases result when the body manufactures "anti-self" antibodies

Distinguishing self from non-self is the key ability of the cells of the immune system. In certain diseases, this ability breaks down, and the body attacks its own tissues. Such diseases are called **autoimmune diseases**. Multiple sclerosis is an autoimmune disease that usually strikes people between the ages of 20 and 40. In multiple sclerosis, the immune system attacks and destroys the insulating myelin sheath that covers motor nerves. Degeneration of the myelin sheath interferes with transmission of nerve impulses, which eventually cannot travel at all. Voluntary functions, such as movement of the limbs, and involuntary functions, such as bladder control, are lost. Multiple sclerosis usually leads to paralysis and death. Scientists do not know what stimulates the immune system to attack myelin. Some other autoimmune diseases, the tissues or organs they affect, and their symptoms are listed in **Table 32.2**.

Table 32.2 Some Autoimmune Diseases

Disease	Areas affected	Symptoms
Systemic lupus erythematosus	Connective tissue, joints, kidney	Facial skin rash, painful joints, fever, fatigue, kidney problems, weight loss
Type 1 diabetes	Insulin-producing cells in pancreas	Excessive urine production, blurred vision, weight loss, fatigue, irritability
Graves' disease	Thyroid	Weakness, irritability, heat intolerance, increased sweating, weight loss, insomnia
Rheumatoid arthritis	Joints	Crippling inflammation of the joints

Connection: Chapter 28

Myelin and Nerve Impulse Conduction
Review how myelin wraps around the axon membrane, forming an insulating sheath. The nerve impulse then "hops" between nodes found between the myelin sheaths, thereby speeding the transmission of the impulse.

Theme Connection

Stability
Autoimmune diseases serve as a perfect example of what happens when internal control mechanisms fail. The stability of the organism becomes seriously threatened as the immune system, designed to fight invaders, now begins to destroy its own body cells.

Visual Strategy

Table 32.2
Lead the class through the various autoimmune diseases listed in this table. When discussing the areas affected, be sure to review the functions of the structures listed:
Chapter 27—connective tissue and joints
Chapter 33—kidneys
Chapter 29—pancreas and thyroid

Connection: Chapter 4

How Cells Divide

Review how cells divide in mitosis. Remind students that cells appear to be programmed for a certain number of divisions, after which they die. Such is not the case with cancer cells.

Health Connection

Transplants and Cancer

Have students explain why transplant patients, who are given drugs to suppress their immune system, are more likely to develop cancer.

Visual Strategy

Figure 32.13

As a way of introducing the next section concerning AIDS, inform students that HIV selectively destroys T cells. From what they understand from this figure, students should then be able to explain why cancer kills more AIDS patients than does any other disease.

Theme Answer

Evolution

Changes in the surface proteins of viruses prevent the immune system from recognizing the virus as one that has already been defeated or vaccinated against.

Evolution

The protists that cause African sleeping sickness can rapidly change the antigens on their cell surfaces. Why is this an effective way to evade the immune system?

Cancer: Unrestrained Cell Division

A major function of your immune system is to ward off cancer. Normally, cells in your body reproduce at a controlled rate. Cancer is a condition in which cells lose the ability to stop dividing. In benign cancers, cells divide rapidly but spread little, forming a mass of cells called a **tumor**. Unless tumors grow so large that they actually crush the internal organs, they are not usually fatal. In malignant cancers, however, the cells aggressively penetrate surrounding tissues. Cells of malignant cancers can even spread throughout the body via the lymph system and bloodstream. Because of the damage they cause to tissues they invade, malignant cancers are often fatal unless detected early.

What causes normal cells to become cancer cells? Scientists think that mutation of growth-regulating genes is what sets off cancer. As you read in Chapter 14, a mutagenic agent that causes cancer is known as a carcinogen. Carcinogens that can transform normal cells into cancer cells are found in cigarette smoke (smoking is a leading cause of lung cancer) and in a variety of industrial chemicals. Some viruses can also cause cells to become cancerous. Mutation also occurs spontaneously in your body all the time, so some cancer cells are being produced "naturally" each day. The immune system usually identifies and destroys these occasional cancer cells before they spread. **Figure 32.13** shows a T cell killing a cancer cell. Surveillance by the immune system is your body's primary defense against cancer. Among AIDS patients, who lack effective immune systems, cancer is the leading cause of death.

Figure 32.13

a   **When a killer T cell comes into contact with the cells of a tumor, it is able to recognize the tumor cells as enemies. Below you can see a killer T cell (upper right) that has bound to a tumor cell.**

b   **The killer T cell releases proteins that disrupt the tumor cell's membrane. The tumor cell tries to defend itself by forming blisters. Eventually, the tumor cell's membrane breaks and the integrity of the cell is lost.**

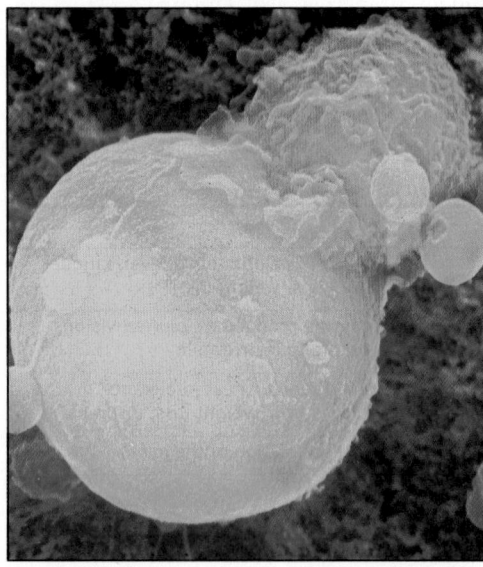

AIDS: Immune System Collapse

When this book was published, more than 150,000 Americans had died of AIDS (Acquired Immune Deficiency Syndrome) and more than 250,000 had the disease. More than 1.5 million Americans were thought to be infected by HIV (Human Immunodeficiency Virus), the virus that causes AIDS.

HIV attacks and cripples the immune system. It invades macrophages and helper T cells, like the one shown in Figure 32.14. HIV transforms these cells into virus factories, killing large numbers of helper T cells. When the number of helper T cells in the blood has fallen to very low levels, a person is said to have AIDS. Without helper T cells to stimulate and direct B cells and killer T cells, the immune response cannot occur. The body is overwhelmed by pathogens and cancers that it normally would defeat. Scientists think that practically everyone infected with HIV will eventually develop AIDS. The time between infection and onset of AIDS can be 10 years or longer. During this time, an infected person can still transmit the virus to others.

HIV is transmitted in body fluids

There is no cure for AIDS. You can protect yourself from AIDS by avoiding exposure to HIV. HIV is a fragile virus that cannot exist for long outside cells. You can get HIV only by contacting the HIV-infected blood cells or body fluids of infected individuals. Because HIV is found in both semen and vaginal secretions, you can contract HIV through sexual intercourse with an infected person; worldwide, most HIV infections are spread this way. The virus can be transmitted in either direction during intercourse. Use of a condom during sex greatly reduces but does not eliminate the risk of getting HIV.

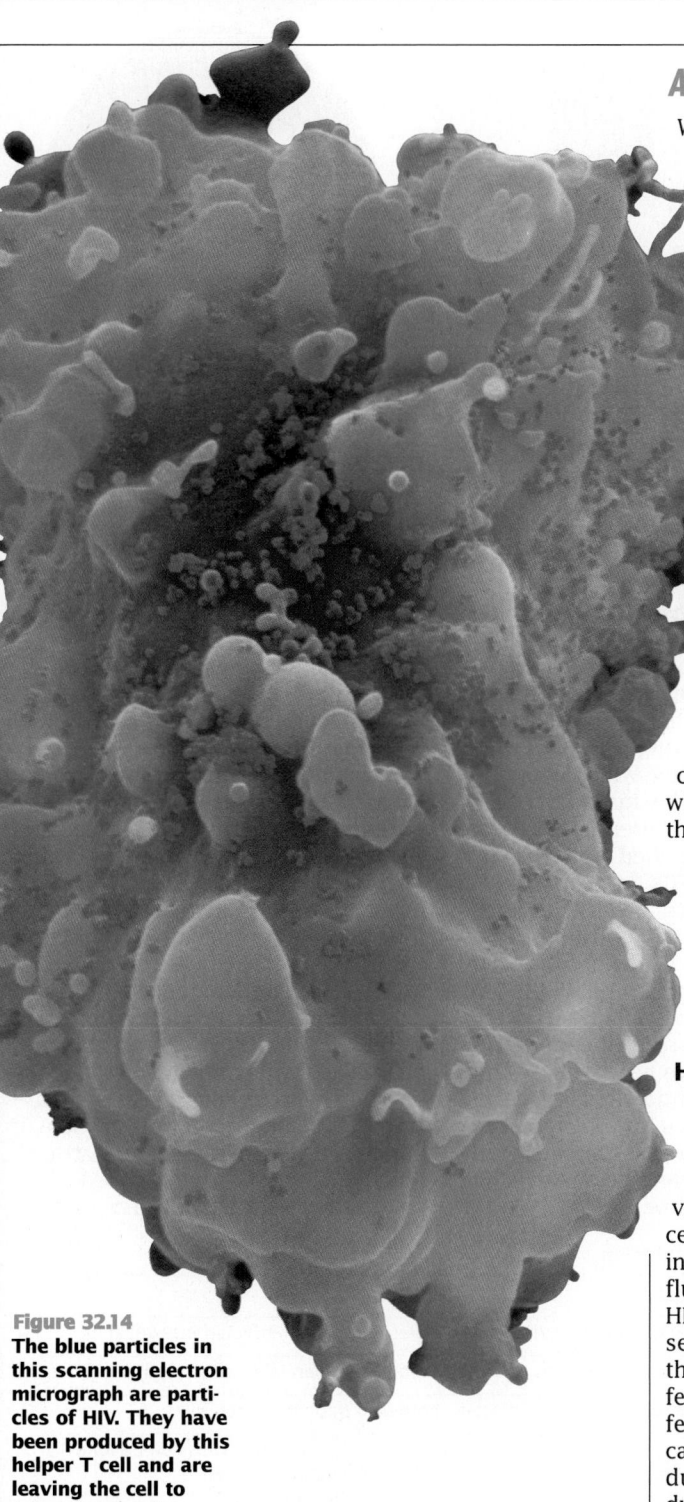

Figure 32.14
The blue particles in this scanning electron micrograph are particles of HIV. They have been produced by this helper T cell and are leaving the cell to infect other cells.

Visual Strategy

Figure 32.14
Point out that what they see happening to the T cell shown in this figure may be the result of an HIV infection that occurred several years earlier. One of the many puzzles about HIV concerns the dormancy of the virus. HIV can remain dormant, perhaps for as long as 10 years, before an infected T cell breaks open to release additional HIV particles.

Visual Strategy

Figure 32.15
Use this figure to emphasize that anyone can get AIDS. You may also wish to use this figure to bring up some bioethical issues. For example, should a physician or dentist with AIDS continue practicing? Should a person with AIDS engage in sexual activity with a noninfected person? If the non-infected person later develops AIDS, should he or she have legal recourse? Point out that a blood test is used to determine whether a person has been infected with HIV. The test assays for the presence of antibodies to HIV, signaling that the person has been infected with the viral antigen.

Demonstration 2
SCIDS
Show students a photograph of a child who lives in a "plastic bubble" because of severe combined immune deficiency syndrome, which is an AIDS-like disease, except that this condition is inherited. Lacking an enzyme known as adenosine deaminase (ADA), which is crucial for a functioning immune system, these children must live in such a protected environment because the slightest infection can lead to death.

Figure 32.15
AIDS is an equal opportunity disease. Anyone can be infected with HIV, regardless of age, sex, or ethnic background.

Ryan White was infected with HIV by blood clotting factor. He died of AIDS at the age of 19.

Arthur Ashe, a professional tennis player, was infected with HIV through a blood transfusion.

HIV can be transmitted when blood from an infected person is transferred to an uninfected person. People who inject intravenous drugs can be infected with HIV if they share or reuse needles or hypodermic syringes, both of which can be contaminated with blood carrying the virus. The majority of HIV infections in the United States in late 1980s were transmitted in this manner.

In the late 1970s and early 1980s, many people contracted HIV by receiving blood transfusions from infected individuals. Many hemophiliacs, including Ryan White (one of the HIV-infected people shown in **Figure 32.15**), were infected by receiving blood clotting factor that had been isolated from blood of infected individuals. Donated blood is now tested for the presence of HIV, so the likelihood of contracting HIV through blood transfusions or blood products is very low.

If a mother is infected, her child has about a one-in-three chance of being infected during pregnancy. HIV can also be transmitted from mother to child in breast milk. **Table 32.3** summarizes the ways that HIV can be transmitted.

HIV is not transmitted through the air or on toilet seats. It cannot be contracted through casual contact, such as shaking hands, sharing food, or drinking from the same water fountain as an infected person. Although HIV is found in saliva, tears, and urine, it occurs there in very low concentrations. One drop of infected blood contains as many viruses as one quart of infected saliva. Scientists think that a large quantity of virus must be received for infection to occur. You cannot catch HIV from the small amount of saliva exchanged when kissing. Biting arthropods, such as mosquitoes, bedbugs, fleas, and ticks, do not transmit HIV.

Table 32.3 Known Routes of HIV Transmission

- Vaginal, oral, or anal intercourse with an infected person
- Injecting drugs or other substances with hypodermic syringes or needles used by an infected individual
- Use of skin piercing equipment, such as tatooing needles, that has been used by an infected person
- From infected mother to fetus through the placenta; from infected mother to baby in breast milk
- Transfusions or injections of blood or blood products drawn from an infected person (Transmission no longer occurs by this route in the United States and other developed nations because blood is tested for the presence of HIV. It is still a transmission route in the less developed countries, where such blood tests are often unavailable.)

Elizabeth Glaser, who is infected with HIV, spoke at the 1992 Democratic National Convention.

Actor Rock Hudson died of AIDS on October 2, 1985.

AIDS Is a Worldwide Disease

AIDS was first recognized as a disease in 1981. Where did this fatal disease come from? HIV probably evolved from a very similar virus, simian immunodeficiency virus (SIV), which occurs in monkeys and apes in Africa. SIV causes an AIDS-like disease in some primates. Scientists are not sure when HIV evolved from SIV or what events allowed the virus to spread from its point of origin.

AIDS is now a worldwide disease, as you can see by looking at the map in **Figure 32.16**. The World Health Organization estimates that 40 million people will be infected with HIV by the year 2000. Most of these cases will occur in the less-developed countries, where AIDS is now spreading rapidly. Indeed, childbirth and AIDS are currently the biggest killers of women in Africa.

Demonstration 3
Spread of HIV

Show the class photographs of people holding hands, embracing, kissing, and other forms of physical contact. Point out that HIV is not spread by casual contact, nor is it spread by insects such as mosquitoes.

Figure 32.16
This map shows the estimated numbers of HIV-infected people throughout the world.

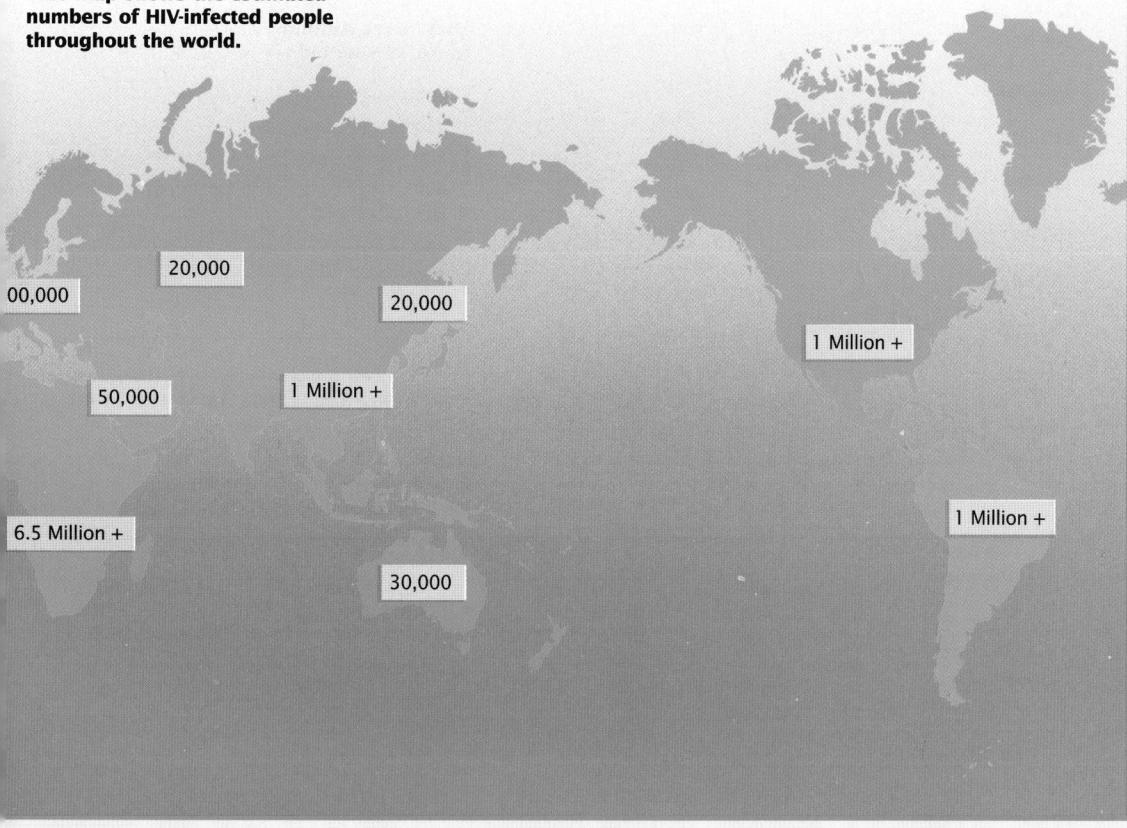

Figure 32.17

Have students examine the increase in the numbers of AIDS cases and AIDS deaths between 1986 and 1992. For example, the number of AIDS cases rose from 13,351 to 226,281, an increase of almost twentyfold.

Visual Strategy

Figure 32.19

Have students calculate the percentage increase in AIDS cases among 13- to 19-year-olds between 1987 and 1992.

Phase 3

ASSESSMENT OPTIONS

Closure Strategy

Public Service

Assign students to cooperative work groups of four. Have each group prepare a poster that dramatizes the impact of AIDS and highlights ways to avoid contracting HIV. Arrange for the posters to be displayed for the rest of the school to see.

Section Review

Assign the *Section Review*.

Reteaching

Invite a local physician or the school nurse to speak to the class about AIDS.

The number of AIDS cases is also increasing in the industrialized countries. As you can see in **Figure 32.17**, the number of AIDS cases and the number of deaths from AIDS in the United States have risen dramatically each year since 1982. American teenagers, such as those in **Figure 32.18**, are also increasingly at risk of contracting HIV, as shown in **Figure 32.19**. More than 20 percent of all reported AIDS cases occur among people in their 20s. Given the average 10-year period before symptoms appear, the majority of these people were probably infected during their teenage years. In 1992, AIDS was the sixth leading cause of death among Americans between the ages of 15 and 24.

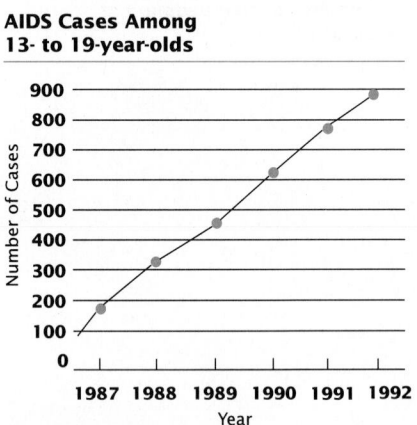

Figure 32.18
Anyone can get AIDS—even you.

Figure 32.17
This graph shows the cumulative numbers of AIDS cases and deaths in the United States.

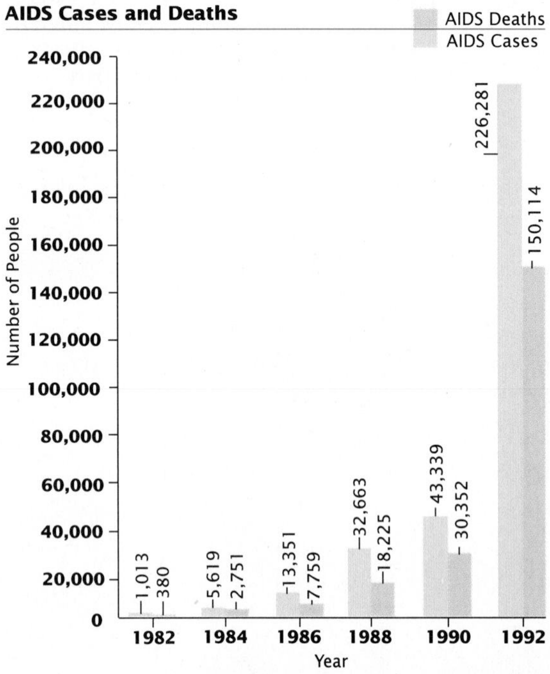

AIDS Cases and Deaths

Legend: AIDS Deaths / AIDS Cases

(Bar graph, Number of People vs. Year)
- 1982: 1,013 / 380
- 1984: 5,619 / 2,751
- 1986: 13,351 / 7,759
- 1988: 32,663 / 18,225
- 1990: 43,339 / 30,352
- 1992: 226,281 / 150,114

AIDS Cases Among 13- to 19-year-olds

(Line graph, Number of Cases vs. Year, 1987–1992)

Figure 32.19
The incidence of AIDS among 13- to 19-year-olds is rising.

Section Review

❶ **Describe what happens when you are exposed to something to which you are allergic.**

❷ **How are cancer cells different from normal cells?**

❸ **Explain why AIDS patients are unable to resist infections.**

❹ **List four ways HIV is transmitted.**

▪ Section Review Answers ▪

1. See Figure 32.12 on page 749.
2. Cancer cells lose their ability to stop dividing, whereas normal cells divide at a controlled rate.

3. AIDS patients have very low levels of helper T cells in their blood, making the patient unable to resist infections and cancers that healthy immune systems normally defeat.

4. See Table 32.3 on page 752.

Chapter **32** *Highlights*

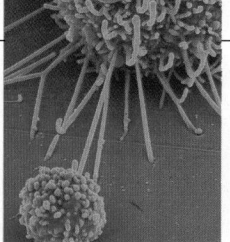

Killer T cells protect the body by destroying tumors, like the one at the left of the photograph.

	Key Terms	Summary
32.1 First Line of Defense 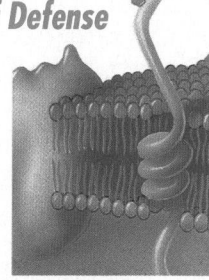 The immune system depends on marker proteins to distinguish self from foreign invaders.	mucous membrane (p. 739) inflammatory response (p. 740) phagocyte (p. 740) macrophage (p. 740) antigen (p. 741)	• When pathogens enter the body through a wound they trigger an inflammatory response. • White blood cells called phagocytes are able to recognize pathogens because the proteins of pathogens differ from the marker proteins of body cells. • A macrophage is a type of phagocyte.
32.2 The Immune Response 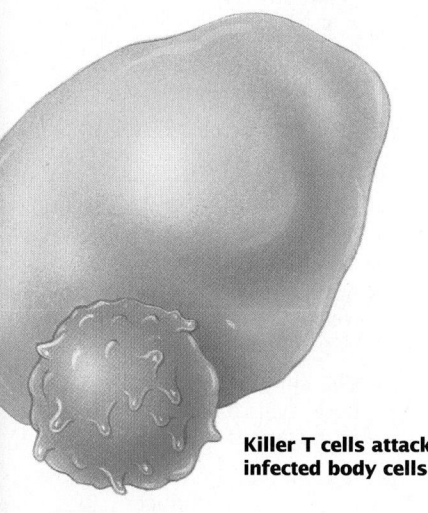 Killer T cells attack infected body cells.	immune response (p. 742) T cell (p. 743) B cell (p. 744) antibody (p. 744) memory cell (p. 745) secondary immune response (p. 745) primary immune response (p. 745)	• Macrophages carrying antigens stimulate an immune response. Defensive white blood cells called T cells begin to divide. Helper T cells stimulate division of killer T cells. • Helper T cells stimulate B cells to divide. B cells are white blood cells that produce antibodies. Macrophages destroy pathogens marked by antibodies. • Some B cells become memory cells. If a pathogen that has already been defeated is encountered again, memory cells produce antibodies against it. • Vaccination stimulates the production of memory cells by exposing B cells to antigens stripped of their ability to cause disease. • Viruses that cause colds, flu, and AIDS mutate rapidly. Your immune system fails to recognize the mutated viruses.
32.3 Immune System Failure 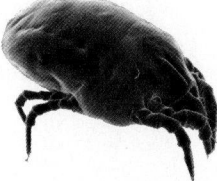 This house-dust mite is a source of misery for thousands of Americans allergic to dust.	allergy (p. 748) histamine (p. 748) antihistamine (p. 748) autoimmune disease (p. 749) tumor (p. 750)	• An allergy is a response to a harmless antigen. • Cancer is uncontrolled cell division. The immune system normally destroys cancer cells before they spread. • HIV causes the immune system to fail by invading helper T cells and macrophages.

Chapter Review Answers

Understanding Vocabulary

1. **a.** T cells are of two types. Helper T cells activate and control the immune response, while killer T cells destroy cells showing viral antigens. Helper T cells are activated by macrophages. Helper T cells, in turn, activate killer T cells and B cells. B cells produce and release antibodies.

b. Antigens are molecules that trigger an immune response; antibodies are protein molecules that mark an antigen for destruction.

c. HIV stands for human immunodeficiency virus; it is the virus that disables the immune system. AIDS stands for acquired immune deficiency syndrome; it is the label used for infections that weaken and eventually kill a person who is infected with HIV.

d. The inflammatory response, part of the first line of defense, is triggered by damaged or infected cells; its symptoms include swelling, red coloration of the tissue, and increased temperature. The immune response is the second line of defense; it involves activation of T cells and B cells.

e. Antihistamine is a chemical that blocks the actions of histamines. A histamine is a chemical messenger that responds to stimulation from the substance to which the person is allergic.

Relating Concepts

2. Map answer is shown on page 737D.

Understanding Concepts

Multiple Choice

3. d   8. a
4. b   9. c
5. a   10. a
6. d   11. c
7. c

Completion

12. vaccination
13. protects
14. faster, more
15. suppressor T

Chapter 32 Review

Understanding Vocabulary

1. For each pair of terms, explain the differences in their meanings.
   a. T cells, B cells
   b. antigens, antibodies
   c. HIV, AIDS
   d. inflammatory response, immune response
   e. antihistamine, histamine

Relating Concepts

2. Copy the unfinished concept map below onto a sheet of paper. Then complete the concept map by writing the correct word or phrase in each oval containing a question mark.

Understanding Concepts

Multiple Choice

3. The skin repels pathogens by
   a. launching phagocytes.
   b. B cell production.
   c. mucous membranes.
   d. secreting oils and sweat.

4. What enables phagocytes to distinguish between pathogens and normal cells?
   a. macrophages   c. lymphatic system
   b. marker proteins   d. enzymes

5. What cells display antigens on their cell membranes but are not attacked by killer T cells?
   a. macrophages   c. B cells
   b. helper T cells   d. pathogens

6. B cells manufacture
   a. antigens.
   b. macrophages.
   c. killer T cells.
   d. antibodies.

7. A successful, long-term vaccine for influenza has not been produced because
   a. the virus that causes influenza has not been isolated.
   b. production of the vaccine is too expensive.
   c. the genetic code for the viral surface protein mutates often.
   d. it is caused by the same virus as AIDS.

8. Cancer cells are different from normal cells in that
   a. the growth regulating genes of cancer cells have mutated.
   b. cancer cells occur in adults, but not in children.
   c. cancer cells are found in the lymph; normal cells are not.
   d. carcinogens change cancer cells to normal cells.

9. The immune cells most affected by HIV are
   a. B cells.
   b. killer T cells.
   c. helper T cells.
   d. neutrophils.

10. Sneezing, nasal congestion, and itchy nose and eyes are symptoms of
    a. an allergy.
    b. AIDS.
    c. multiple sclerosis.
    d. an autoimmune disease.

11. AIDS may be contracted by
    a. kissing.
    b. sharing food.
    c. sexual intercourse.
    d. shaking hands.

Completion

12. A(n) _____ is an injection of a weakened form of a pathogen to produce immunity.

13. The immune system _____ the body against threats from pathogens.

14. A secondary immune response is _____ and produces _____ antibodies than a primary immune response.

15. After invading pathogens are defeated, _____ cells shut down the immune response.

Short Answer

16. What is the function of antibodies in the immune response?

Short Answer

16. Antibodies, produced by B cells, bind to complementary antigens on cell membranes, thus marking them for destruction by macrophages.

17. Macrophages engulf and destroy dead and dying cells and viruses; they also attract helper and killer T cells.

18. Abstain from sexual intercourse, or use a condom to reduce the risk of becoming infected. Don't abuse drugs.

19. HIV suppresses the immune system. The abnormal nature of cancer cells will likely go undetected by an immune system suppressed by HIV infection, and they will reproduce rapidly.

20. In an autoimmune disease, the body loses its ability to distinguish self from nonself. As a result, the immune system begins to attack and destroy healthy body cells.

17. How do macrophages contribute to the defense of the body?

18. What measures should you take to prevent HIV infection?

19. Why is the risk of cancer greater among people infected with HIV than among people not infected?

20. What happens to the immune system's ability to distinguish self from non-self in an autoimmune disease like multiple sclerosis?

Interpreting Graphics

21. Study the diagram below.

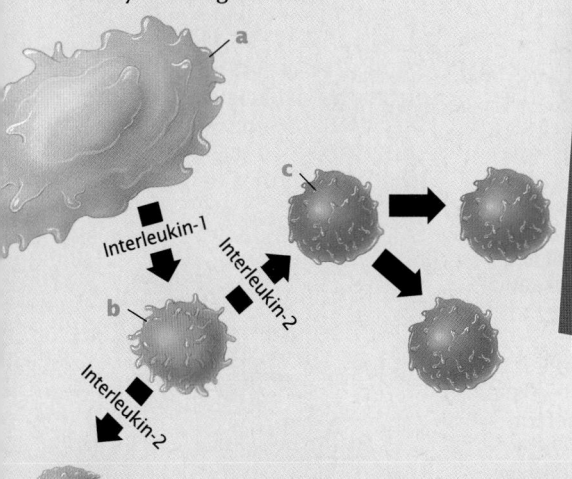

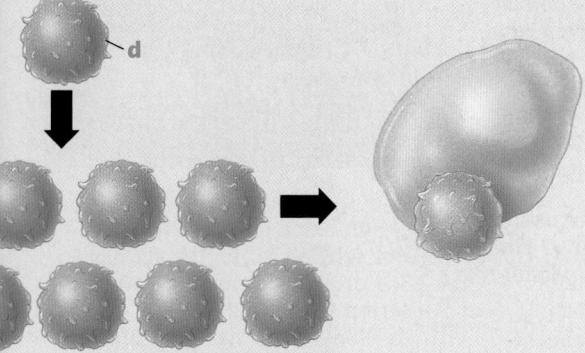

- Identify the cell labeled a. What is its function in the immune response?
- What is the function of the cells labeled d?
- Describe how HIV infection affects the process illustrated above.

Reviewing Themes

22. *Patterns of Change*
What symptoms are associated with an asthma attack? What effect would taking medicines that contain antihistamines have on the symptoms?

23. *Interacting Systems*
Long term smoking causes paralysis of cilia lining the treachea. How would this affect the body's ability to repel pathogens?

24. *Scale and Structure*
Relate the symptoms of an inflammatory response to the events occurring at the cellular level.

Thinking Critically

25. *Inferring Conclusions*
What would be your chances of contracting German measles for a second time if memory cells lived only three months?

26. *Building on What You Have Learned*
In Chapter 28 you learned about the function of myelin in the transmission of nerve impulses. How is myelin affected by the disease multiple sclerosis?

Cross-Discipline Connection

27. *Biology and Geography*
According to the World Health Organization (WHO), the majority of future AIDS cases will occur in Africa, Asia, and South America. Look for information in your library about living conditions and education in these areas, and about the projected population growth. Do you agree with the WHO's prediction about the spread of AIDS?

Discovering Through Reading

28. Read the article "AIDS or Chronic Fatigue?" in *Newsweek*, September 7, 1992, pages 66 and 69. What is ICL and what are its symptoms? How do doctors plan to treat ICL patients like Rosemary Stevens in the future?

24. The swelling is due to more fluid than normal leaking from nearby capillaries, while the redness and heat are a result of enlarged blood vessels in the area. The fluid and blood provide access for phagocytes, T cells, and B cells. The heat serves to suppress bacterial growth.

Thinking Critically

25. If exposed to German measles at least three months after your first bout with the disease, the chances of contracting the disease a second time are great.

26. In multiple sclerosis, the immune system destroys the myelin, which affects the transmission of nerve impulses. The disruption of nerve impulses leads to the loss of voluntary and involuntary muscle functions.

Cross-Discipline Connection

27. In many areas of Africa, Asia, and South America, the living conditions for most people are deplorable and few people are highly educated. In addition, world population growth is expected to be the greatest in Africa, Asia, and South America. Rapid population growth coupled with poor living conditions and little education means the World Health Organization is most likely on target with its prediction.

Discovering Through Reading

28. ICL stands for idiopathic CD4+ T-lymphocytopenia. Its primary symptom is an abnormally low CD4 count (300 or less). Other symptoms include the flu, pneumonia, and those symptoms associated with AIDS and chronic fatigue syndrome. Because the symptoms of ICL resemble those associated with AIDS and chronic fatigue syndrome, doctors are studying ICL in light of what is known about these two diseases. Doctors plan to begin sorting out what they now know about ICL and to develop a profile of the persons who suffer from the illness. It is hoped that the profile will provide information about what causes the illness.

Interpreting Graphics
21. • Macrophages activate an immune response.
• Killer T cells destroy infected cells.
• HIV invades macrophages and helper T cells, transforming them into virus factories.

Reviewing Themes
22. Symptoms of an asthma attack include narrowing of the bronchioles in the lungs which causes difficulty breathing, wheezing, and shortness of breath. Taking medicines that contain antihistamines would block the actions of the histamines, allowing the bronchioles to return to normal.

The wheezing and other symptoms would cease.
23. Without properly functioning cilia, pathogens will not be swept into the mouth, swallowed, and destroyed by digestive enzymes of the stomach. Thus, they will have a greater chance to invade the body through the mucous membranes.

Procedural Note

1. Review the Prelab Preparation questions.
2. Discuss the Procedure. Point out that simulated blood is being used in the lab.
3. Remind students of the dangers involved in handling real blood.

Prelab Preparation Answers

1. • antigen present: A, antibodies made: anti-B
   • antigen present: B, antibodies made: anti-A
   • antigen present: A, antibodies made: neither
   • antigen present: none, antibodies made: anti-A and anti-B

Procedure Answers

9. red blood cells
10. red blood cells

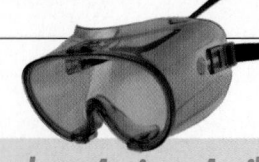

How Do Antibody-Antigen Reactions Work?

Objectives

In this investigation you will:
• *observe* evidence of antibody-antigen reactions
• *relate* these reactions to blood type

Materials

• blood typing slides
• wax pencil
• vials of simulated blood (blood type unknown)
• simulated anti-A blood typing serum
• simulated anti-B blood typing serum
• toothpicks
• microscope slides
• coverslips
• compound light microscope

Prelab Preparation

1. Review what you have learned about antigens, antibodies, and blood types by answering the following questions.
   • What antigens does a person with type A blood have? What antibodies does this person produce?
   • What antigens does a person with type B blood have? What antibodies does this person produce?
   • What antigens does a person with type AB blood have? What antibodies does this person produce?
   • What antigens does a person with type O blood have? What antibodies does this person produce?
2. Review the procedures in the Appendix for using a compound microscope.

Procedure: Antigen-Antibody Reactions

1. Form a cooperative team with another student to complete steps 2–12.
2. Make a table like the one shown at the right.
3. Use a wax pencil to label four blood typing slides "1," "2," "3," and "4."
4. Thoroughly shake each of the four vials of blood. Place 3–4 drops of blood from the vial labeled "Mr. Green" in each well on Slide 1. Place 3–4 drops of blood from the vial labeled "Mr. Smith" in each well on Slide 2. Place 3–4 drops of blood from the vial labeled "Ms. Jones" in each well of slide 3, and place 3–4 drops of blood from the vial labeled "Ms. Brown" in each well on Slide 4.

5. Choose one slide. Place 3–4 drops of simulated anti-A serum in the well labeled "A" on the slide. Place 3–4 drops of simulated anti-B serum in the well labeled "B" on the slide. Use a new toothpick to stir each sample of blood and serum. Dispose of the toothpicks as directed by your teacher.

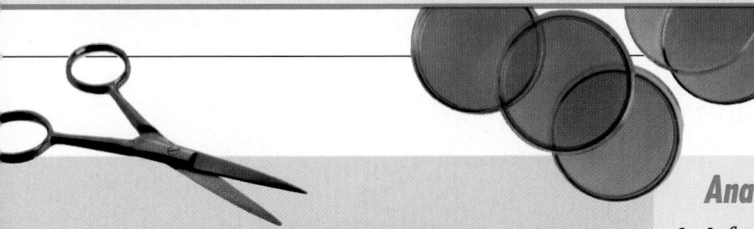

Person	Reaction to Anti-A Serum	Reaction to Anti-B Serum	Blood Type
Mr. Green			
Ms. Jones			
Ms. Brown			
Mr. Smith			

6. Observe each well for two minutes. Look for evidence of clumping. Clumping is evidence that the antibodies in the serum are reacting with the antigens on the blood cells. Record your observations in your table.

7. Repeat steps 5 and 6 for each of the three remaining slides.

8. Thoroughly shake one of the vials of simulated blood. Place a small drop of the blood on a microscope slide. Cover the blood with a coverslip, trying not to trap any air bubbles beneath it.

9. Observe the slide at high power under the microscope. Red blood cells will appear red, while white blood cells will appear blue. *Which kind of blood cell is more common?*

10. Select one vial of blood and place a small drop of blood on a second microscope slide. Now add a drop of blood typing serum that will react to this blood, and mix it with a clean toothpick. Place a coverslip over the blood-serum mixture.

11. Observe the slide at high power under the microscope. *What kind of blood cell has clumped?*

12. Clean up your materials and wash your hands before leaving the lab.

Analysis

1. *Inferring Relationships*
   What antigens are found in Mr. Green's blood? What evidence supports your conclusion?

2. *Inferring Relationships*
   What antibodies would be found in Ms. Brown's blood? What evidence supports your conclusion?

3. *Making Predictions*
   During surgery Ms. Jones is given type O blood. Will her immune system react to the transfusion?

4. *Making Inferences*
   Do white blood cells have A or B antigens? What evidence supports your conclusion?

5. *Making Inferences*
   What might cause a patient to have fewer white blood cells than normal?

Thinking Critically

A wounded soldier with type O blood needs an emergency transfusion, but no type O blood is available. Could this soldier safely receive any other type of blood? Explain your answer.

Analysis Answers

1. antigens A, B; no reaction to either serum
2. anti-A, anti-B; reaction to both serums
3. Ms. Jones can safely receive type O blood without any response from her immune system.
4. No, students should suggest they do not attack healthy blood cells, only infected cells marked by antibodies for destruction.
5. Answers may vary. You might suggest stress and drugs that contain cortisol-related hormones.

Thinking Critically Answer

Yes, the soldier could safely receive type A, B, or AB without any response from his/her immune system.

Using the Feature

- Assign one or more of the activities from the "Acting on the Issue" section at the end of the feature.
- Have students research some of the recent attempts at organ tranplants. Which organ was involved? Was it successful? If not, why? If so, what is the current condition of the organ recipient?

Discussion

Guide the discussion by posing the following questions.

1. List three ethical issues organ transplant technology raises.

   Should a child be conceived for the sole sake of serving as a organ donor for another person? Should the sale of organs be legalized? Who should receive rare organs for transplant? Who should not?

2. Explain the rationale for legalizing the sale of organs.

   Those who take this stance feel that legalizing the sale of organs would increase the supply of otherwise rare organs needed for transplant. If blood, semen, and eggs can be sold, then why not kidneys, hearts, livers, or other organs?

3. Explain the rationale for not legalizing the sale of organs.

   People with this attitude argue that it would be impossible to regulate organ sales. They argue that questions about ownership of an organ are difficult to resolve. For example, who makes the decision to sell an organ in the case of a child?

In a Los Angeles suburb, a couple decides to conceive a child who can serve as a bone marrow donor for their daughter. In a Turkish city, a man sells one of his kidneys so that his daughter can have the operation that will save her life. In an Illinois medical center, a physician asks an accident victim's grieving parents for permission to remove their child's liver so that it can be transplanted to a youngster suffering from liver disease.

This human heart has just been removed from a donor.

TRANSPLANT TECHNOLOGY

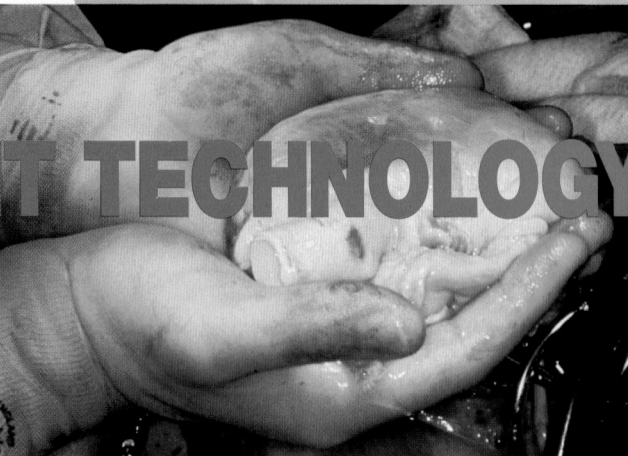

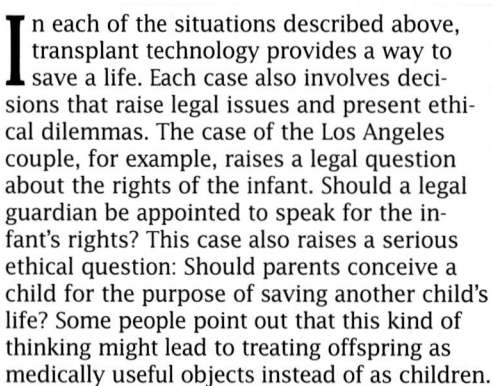

In a Texas hospital, physicians tell a man who needs a heart transplant that 60 other people are on their list waiting to be matched to a suitable heart donor.

In each of the situations described above, transplant technology provides a way to save a life. Each case also involves decisions that raise legal issues and present ethical dilemmas. The case of the Los Angeles couple, for example, raises a legal question about the rights of the infant. Should a legal guardian be appointed to speak for the infant's rights? This case also raises a serious ethical question: Should parents conceive a child for the purpose of saving another child's life? Some people point out that this kind of thinking might lead to treating offspring as medically useful objects instead of as children.

The buying and selling of kidneys, livers, and hearts raises other serious issues. In the United States and most other industrialized nations, the buying and selling of these organs is illegal and subject to heavy penalties. In these countries, organs for transplant are donated.

So far, the supply of donated organs has not kept up with the demand for them. For example, the number of patients on the national waiting list for transplants rose from about 12,000 in 1987 to more than 20,000 in 1989. During that time, the number of donors has remained around 4,000. Many people who are on hospital waiting lists do not survive long enough for a donated organ to become available to them. As the demand for organs continues to exceed the supply, difficult decisions must be made. Who will receive organs for transplant and who will not? How will these decisions be made and who will make them?

The legal aspects of these questions have not yet been resolved in many states. Many times, the guidelines for arriving at decisions are established by the individual hospitals and medical facilities that perform transplant operations. These guidelines often vary from place to place.

The demand for organs has led some people to suggest that the sale of organs should be legalized in the United States. The people who promote this view argue that legalizing the sale of organs would increase the supply of otherwise rare organs needed for transplants. They suggest that citizens who cannot obtain donated organs in the United States may seek them in countries where the buying and selling of organs is legal. They also point out that if blood, semen, and eggs can be sold—then why not kidneys, hearts, livers, or other organs?

This human liver is about to be transplanted into a recipient.

Other people resist legalization of organ sales under any circumstances. These people suggest that it would be impossible to regulate organ sales. They argue that questions about ownership of an organ are difficult to resolve. An adult, for example, may be able to make a decision about whether to sell one of his or her kidneys. But who makes the decision in the case of a child or the victim of a fatal accident? The issues raised by the increasing demand for tissues and organs for transplant present challenges that could not have been imagined a few decades ago.

SAVING LIVES AND
FACING DILEMMAS

Thinking Critically

❶ Do parents have the right to create a life for the purpose of saving another life? In the case of the Los Angeles couple, who owns the infant's bone marrow? Support your view.

❷ What are the similarities and differences between a person selling his or her blood and a person selling one of his or her kidneys? What problems might arise if organs such as kidneys, hearts, and livers could be legally bought and sold in the United States?

❸ When demand for an organ exceeds the supply, how should decisions be made about who receives a transplant? Who should make these decisions?

Acting on the Issue

❶ Contact a local hospital, clinic, or blood bank and find out the criteria for becoming an organ donor.

❷ Write a set of guidelines that prioritize criteria for recipients of scarce organs.

❸ Propose possible ways for organizations to encourage organ donation.

❹ Find out how your state allows a licensed driver to indicate his or her willingness to be an organ donor. What other options are available to people who

want to indicate their willingness to be an organ donor?

❺ Do library research to find information about medical cases in which the organ of an animal was transplanted into the body of a human.

Thinking Critically Answers

1. Students' responses to these two questions should address the ethical and legal dilemmas involved in determining the rights of the donor child. Answers should be supported by sound logic. Have students with opposing attitudes debate their viewpoints.

2. One similarity is that without these organs in their entirety, an individual can still lead a healthy life (new blood cells will be produced to maintain a constant volume—5 to 6 L in adult males and 4 to 5 L in adult females—and one kidney will cleanse the blood just as well as two). However, the difference lies in the legality. In the United States and other industrialized nations, selling blood is legal, whereas selling kidneys, hearts, and livers is illegal and subject to heavy penalty. For these organs, organ donation is encouraged. Many people feel that legalizing the selling of organs would lead to problems in regulating these sales and also raise difficult questions about organ ownership.

3. Students may prioritize patients based on the severity of their medical conditions. Challenge students to consider other factors such as patient's age, economic status, and chances for survival. Students may say that these decisions should be made in line with a carefully considered set of guidelines.

761

Chapter 33 Digestion and Excretion

Planning Guide

	Objectives/Themes	Classwork Resources	Homework Resources
33.1	**1.** List the five nutrients required to maintain good health. **2.** List two ways the body uses lipids, proteins, carbohydrates, vitamins, and minerals. **3.** Discuss two diseases that result from nutrient deficiencies. **4.** Discuss the effects of the eating disorders anorexia nervosa and bulimia. **Themes:** Energy and Life, Interacting Systems, Patterns of Change, Stability	**Teacher's Resource Binder** Focus Activity 33 *Reading Labels: Nutritional Information* **Other Resources** Transparency 150	**Text** Section Review, p. 767 **Teacher's Resource Binder** Directed Reading Worksheet 33.1 **Other Resources** Audiocassette 33.1*

33.2	**1.** List the main organs of the digestive system.
	
	2. Identify the sites of digestion for each of the three nutrients.
	
	3. Describe the absorption of the food from the digestive tract.
	
	4. Explain the roles of the pancreas and liver in the digestive system.
	
	Themes: Energy and Life, Interacting Systems, Patterns of Change, Scale and Structure
		Text
		Science in Action
		  Cheryl Coldwater: Pediatrician
		  pp. 772–773
		
		Teacher's Resource Binder
		Lab Investigation 33.1
		  Enzyme Action in Digestion
		Lab Investigation 33.2
		  Lactose Digestion
		
		Other Resources
		Transparencies 151–152
			Text
			Section Review, p. 771
			
			Teacher's Resource Binder
			Directed Reading Worksheet 33.2
			
			Other Resources
			Audiocassette 33.2*

33.3	**1.** Explain how the kidneys determine what leaves the blood and what remains.
	
	2. Describe the structure of a kidney nephron.
	
	3. Compare and contrast the two stages of urine formation.
	
	4. Explain how medical technology has helped people with kidney disorders.
	
	Themes: Interacting Systems, Patterns of Change, Scale and Structure, Stability
		Text
		Investigation
		  How Do You Test Foods for Nutrients? pp. 780–781
		
		Teacher's Resource Binder
		Extension Worksheet
		  Adaptations of the Respiratory System
		
		Other Resources
		Transparencies 153–154
			Text
			Section Review, p. 776
			
			Teacher's Resource Binder
			Directed Reading Worksheet 33.3
			Vocabulary Review Worksheet*
			Reaching Worksheet*
			  Digestion and Excretion
			
			Other Resources
			Audiocassette 33.3*

*Reteaching Options

Demonstrations	**Assessment**
33.1: pp. 764, 765, 767	Chapter Review pp. 778–779
33.2: pp. 769, 770, 771	Portfolio Assessment p. 761D
33.3: pp. 775, 776	Chapter Test—Teacher's Resource Binder
	Test Generator

Research Notes

Connection to Nutrition: Diet and Colorectal Cancer

Cancer is a complicated disease that frequently does not have a single cause, but rather several different genetic and environmental risk factors associated with it. Many researchers are devoting time and effort identifying these risk factors for different types of cancer. Through a variety of studies, colorectal cancer, which kills more than 50,000 people in the United States each year, was found to be affected by diet.

In 1990, a long-term study of female nurses disclosed that women who ate red meat such as beef and pork daily were more than twice as likely to develop colorectal cancer than those who ate meat less than once a month. Those consuming processed meats such as bologna also faced an additional risk.

The researchers found that those who ate fish or skinless chicken two to four times a week had a reduced risk of cancer, as did those who ate fruit containing fiber. No additional risk or benefit was found to be related to calorie consumption, dairy fats, vegetable fats, calcium, carotene, or vitamins A, C, D, and E.

Researchers working with this study theorized that the findings for meat are related to the amounts and kinds of animal fat in the diet. Red meat has a different proportion of unsaturated and saturated fat than chicken or fish. Other scientists suggest that food preparation can also play a role in cancer risk. Broiled lean red meat could be just as good as, or even better for you than, fried chicken or fish.

In 1991, another long-term study of diet and health trends suggested that modest changes in fat intake had little effect on cancer risk. However, this study's researchers noted that all of their subjects, even those who ate less fat, were eating more than the amount recommended by several health groups, such as the American Heart Association. Even though less radical adjustments did not seem to affect cancer risk, such small adjustments in fat intake can help the body fight heart disease and obesity.

These scientists suggest that to decrease the risk of cancer, people should cut their fat intake substantially, from 38 percent of their calories to 20 percent of their calories. Such a diet would include lots of fruits and vegetables, no cooking oil, and only small servings of lean meat.

Investigation Notes

How Do You Test Foods for Nutrients?

pages 780–781

Purpose: This investigation will introduce the students to techniques used for the identification of lipids, monosaccharides, starches, and proteins.

Note: Wear protective equipment such as gloves, aprons, and goggles during preparation and the investigation. If you prepare the iodine yourself, work under a fume hood.

Solution Preparation

1. Benedict's reagent is available ready-made from WARD'S. To make it yourself, dissolve 17.3 g of sodium citrate and 10 g of sodium carbonate (Na_2CO_3) in 80 mL of distilled water in a beaker. In a smaller beaker, dissolve 1.7 g of copper sulfate ($CuSO_4 \cdot 5H_2O$) in about 10 mL of distilled water. While stirring the first solution, slowly pour the copper sulfate solution into it. Pour the combined solutions into a 100-mL graduated cylinder and dilute to 100 mL. This should be enough for several classes. It can be stored for long periods of time.

2. Lugol's iodine (or "iodine-iodide solution") is available ready-made from WARD'S.

NOTE: Iodine reacts with metal, skin, and many other substances.

To make Lugol's iodine, dissolve 1.0 g of potassium iodide (KI) in about 15 mL of distilled water. Using a porcelain spatula, place 0.7 g of iodine (I_2) in a 100-mL volumetric flask with the KI solution. Stopper and shake the flask. Additional crystals of KI may help dissolve the I_2. Dilute to 100 mL. This should provide enough for several classes. It should be stored in a glass-stoppered bottle in a cool, dark place. Sunlight or strong lights can cause the iodine to react.

3. Biuret reagent is available ready-made from WARD'S.

NOTE: Biuret reagent is a caustic solution. Avoid spilling it on yourself or others.

Dissolve 8.0 g of sodium hydroxide (NaOH) in 100 mL of water in a flask. Add 1.0 g of copper sulfate. Stopper and shake the flask. This should provide enough for several classes. It can be stored for long periods of time.

4. For instructions on preparing starch solution, see p. 67B.

Prelab Preparation

Use colorless liquids and substances that can be easily dissolved for consistent test results.

Answers will be found on pages 780–781.

Meeting Individual Needs

Objectives

1. Students will demonstrate appropriate use of core vocabulary for the chapter (Vocabulary File).

2. Students will explain the different steps of digestion of a meal (Teaching Strategy A).

3. Students will explain how the waste products from the digestion of a meal are handled by the excretory system (Teaching Strategy B).

Vocabulary File
(Developing Vocabulary/ Limited English Proficiency)

If you are not already using the Vocabulary File, refer to Chapter 1 for its preparation. See the Chapter Highlights on page 777 for a list of suggested words.

Teaching Strategy A
Through the Digestive System
for pages 768–771
(Developing Organizational Skills/Verbal Learners)

Materials: overhead projector, transparency, and projector pens

Procedure: Have students silently read Section 33.2, making notes about the digestive system's parts and their functions. Lead a discussion about the order in which food passes through the digestive system, and the aspects of digestion that occur at each step along the way. Create a visual outline describing this on the transparency, using the following as a model:

Digestion

   I. Mouth

      A. Teeth—break food apart

      B. Saliva and enzymes
        1. Moisten food
        2. Begin carbohydrate digestion

Continue the outline with the esophagus and stomach, small intestine, pancreas, liver, and the large intestine. Have an artistically talented student draw the components of this system on the chalkboard.

Ask students to describe their "favorite" meal. For each item, challenge students to break it down into its parts (proteins, sugars, etc.) Then, challenge students to give a "play-by-play"

description of what will happen to all of the different parts of the meal as it passes through the digestive system. Use the drawing of the system and the outline as references when needed.

Also challenge students to discuss the nutritional balance of the sample meal suggested by the students. Be certain students realize that the choices they make about food can affect their health and strength.

Teaching Strategy B
Through the Excretory System
for pages 774–776
(Developing Organizational Skills/Verbal Learners)

A strategy similar to the one described for the digestive system can be used for the excretory system. Be sure students realize that the wastes come from the digestion of the sample "meal" that was described in the digestive system strategy. For example, urea is formed when amino acids are broken down. Food with protein in it is a source for these amino acids. Relate this to proteins in the sample "meal."

Additional Strategies
Visual Strategies
Pages 763, 765, 766, 769, 770, 771, 774, 775, and 776

Auditory Learners
Use *Biology: Visualizing Life* Audiocassettes for Sections 33.1, 33.2, and 33.3.

Meeting Individual Needs (cont.)

Cooperative Learning
Aspects of Digestion

Timing: Use this activity to introduce Section 33.2.

Group Size: 4 students

Outcome: Students will be able to describe several aspects of the process of digestion.

Individual Accountability: Each group member is responsible for describing the organs and processes involved in one aspect of digestion.

Positive Interdependence: Each group will describe the complete process of digestion.

Each group member should write a short description of the processes and organs involved in the following types of digestion: (1) mechanical digestion, (2) chemical digestion, (3) absorption of nutrients, (4) absorption of water.

Have each group member read his or her description aloud so that the group as a whole can change the description as necessary. Then, have group members compile their section descriptions into a single description of what happens in the entire digestive tract.

Portfolio Assessment

Students should select their best work and provide a self-reflective rationale for their selections. Students can make selections in the following areas.

1. *Content* One concept map from the chapter (See page 812 for evaluation criteria.)

2. *Reading Comprehension* One Directed Reading Worksheet from the Teacher's Resource Binder (Use the answer key to evaluate for accuracy.)

3. *Writing* Using the Vee Form, summarize a magazine or newspaper article relating to digestion or excretion, or diseases involving these systems. (See page 22T for evaluation criteria.)

   Or: Select a writing task or project from the Chapter Review.

4. *Performance Assessment* One Vee form from a chapter investigation or lab manual investigation (See page 22T for evaluation criteria.)

Teacher makes selections in the following areas.

1. *Formal Assessment* Chapter test (Test A, B, or the Test Generator) The teacher-scored test should be reviewed by the student. Incorrect responses should be corrected by the student before the test becomes part of the portfolio.

2. *Informal Assessment* Use the Direct Observations Checklist, page 33T, during a laboratory or other cooperative learning experience.

3. *Performance Assessment* Have students determine the nutritional content in several school cafeteria meals, referring to the dietary guidelines for healthy Americans from the United States Food and Drug Administration.

Concept Map Answer

The following is one possible answer to the Relating Concepts exercise on page 778.

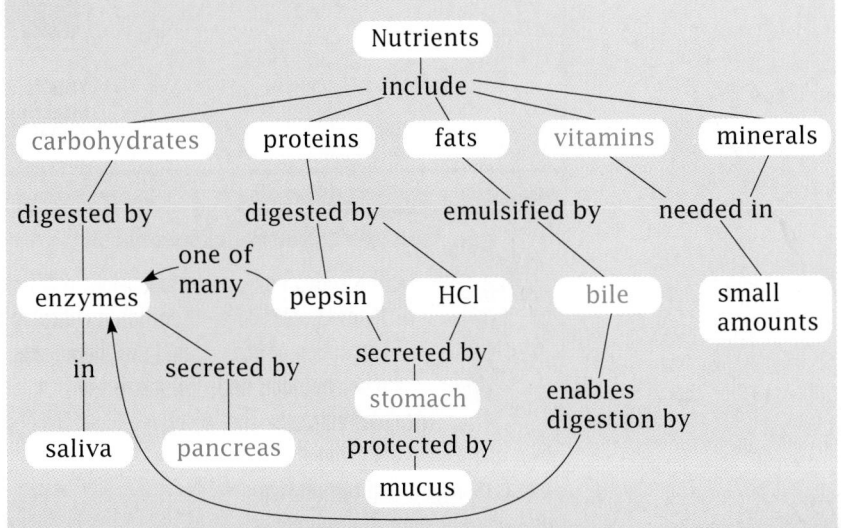

Chapter 33

Determining Prior Knowledge

- Display a potato, a piece of meat, and a stick of butter. Ask what organic compound makes up the bulk of each of these foods. Have students discuss what functions each of these compounds serve in the human body.
- Exhibit a model or overhead transparency of the human digestive system. Have students identify as many organs as possible and explain the role of each in the digestive process.
- Exhibit a model or overhead transparency of the human excretory system.

Chapter 33

Review
- **organic molecules (Section 2.3)**
- **enzymes (Section 5.1)**
- **water conservation (Section 22.2)**

Digestion and Excretion

33.1 Nutrition: What You Eat and Why
- **Carbohydrates, Proteins, and Lipids**
- **Vitamins and Minerals**
- **Nutrition and Health**

33.2 The Digestive System
- **Digestion**
- **Activity in the Intestines**

33.3 The Excretory System
- **Kidney Form and Function**
- **Urine Formation**
- **Kidney Disorders and Treatment**

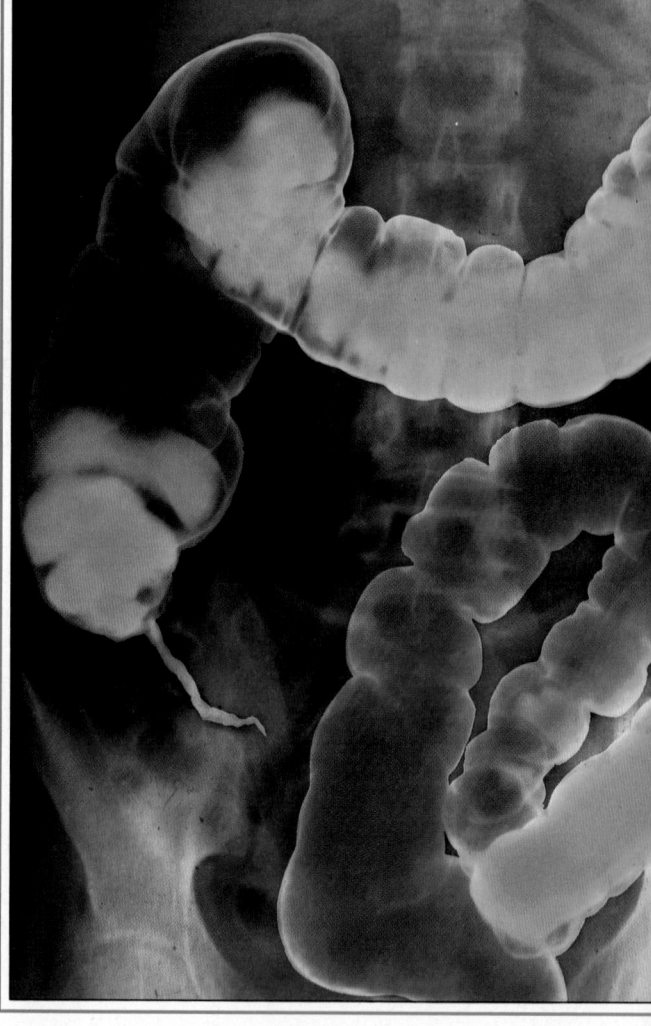

This is an X-ray photo of a human large intestine. The large intestine is an important last stop for food in the digestive system.

▪ Author's Rationale ▪

This chapter extends Chapter 31's description of how the body obtains and circulates nutrients. The liver and kidney are given strong emphasis as organs that control sugar and water balance. The concept of homeostatic regulation of body function is critical to this chapter, and this theme should be the focus of instruction.

ALL ANIMALS MUST GET THE NUTRIENTS THEY NEED FROM A SOURCE OTHER THAN THE SUN. ANIMALS CANNOT MAKE THEIR OWN FOOD AS PLANTS DO, SO THEY MUST EAT OTHER ORGANISMS. SOME ANIMALS, CALLED CARNIVORES, EAT ONLY OTHER ANIMALS; OTHERS, CALLED HERBIVORES, EAT ONLY PLANTS. BUT HUMANS, LIKE BEARS, ARE OMNIVORES, EATING BOTH PLANTS AND ANIMALS.

33.1 *Nutrition: What You Eat and Why*

Objectives

❶ **List the five nutrients required to maintain good health.**

❷ **List two ways the body uses lipids, proteins, carbohydrates, vitamins, and minerals.**

❸ **Discuss two diseases that result from nutrient deficiencies.**

❹ **Discuss the effects of the eating disorders anorexia nervosa and bulimia.**

Carbohydrates, Proteins, and Lipids

Figure 33.1
The USDA food pyramid recommends the foods at the top of the pyramid be eaten sparingly, while the other foods may be eaten more often.

Fats, oils, and sweets

Milk products, meat, poultry, fish, beans, eggs, and nuts (2–3 servings daily)

Vegetables (3–5 servings daily) and fruit (2–4 servings daily)

Bread, cereal, rice, and pasta (6–11 servings daily)

Food contains nutrients, the substances that provide your body with the energy and materials it needs for growth, maintenance, and repair. You need several kinds of nutrients in your diet. These include the carbohydrates, lipids, and proteins that you learned about in Chapter 2, as well as vitamins, minerals, and water.

Carbohydrates, lipids, and proteins make up the bulk of what you eat. Vitamins and minerals, while crucial for health, are required in smaller amounts. Water makes up about two-thirds of your body weight. As shown in **Figure 33.1**, most of the food you eat provides a combination of these nutrients.

Carbohydrates are the body's main source of fuel

Most of your body's energy needs come from carbohydrates, also known as sugars and starches. Carbohydrates have other roles as well. One type of sugar, deoxyribose, is a building block of DNA. Another sugar, glucose, is assembled into long chains called cellulose, the main structural component of plant tissues. Cellulose is not digested by humans but provides roughage, or fiber, in our diets. Fiber aids the passage of food through the digestive system.

The carbohydrate molecule ultimately used by your body cells is also glucose. It is used to make ATP. Brain cells and red blood cells rely almost entirely on glucose to supply their energy needs. Even a temporary shortage of blood glucose can severely depress brain function and lead to the death of neurons. When blood glucose is present in excess amounts, it is often converted to fat and stored.

Lesson Plan 33.1

Phase 1

PREPARATION

Key Concepts

- Five nutrients essential to good health are carbohydrates, lipids, proteins, vitamins, and minerals.
- Carbohydrates serve primarily as the main source of energy for cells.
- Proteins serve primarily as the building blocks for growth and tissue repair.
- Lipids mainly store energy and serve as a major structural material for cells.
- Vitamins are organic compounds needed in small amounts to help regulate body functions.
- Two major eating disorders are anorexia nervosa and bulimia.

Reading Strategy

To help students organize the information they glean from reading this section, have them prepare a table with the headings *Carbohydrates, Proteins, Lipids, Vitamins,* and *Minerals.* Down the left side, have them write *Functions in the human body, Food sources,* and *Additional comments.* Have students complete the table as they read this section.

Phase 2

TEACHING STRATEGIES

Visual Strategy

Figure 33.1
Have students identify whether each of the foods shown in this figure is a carbohydrate, lipid, or protein. Be sure they recognize that the base of the pyramid consists of foods that are complex carbohydrates, while those at the top are lipids and simple carbohydrates.

Demonstration 1
Endurance Events
Invite a marathon runner, cyclist, or some other person who has participated in an endurance event to speak to the class about the carbohydrate-rich foods they consume before competing.

Connection: Chapter 2
Cellulose
Remind students that animals that graze on grass have bacteria that break down cellulose in their intestines. Such bacteria are not present in the human intestine.

Connection: Chapter 5
Respiration
Review how glucose is metabolized in aerobic respiration to produce ATP, which serves as the energy currency of the cell.

Connection: Chapter 27
Collagen
Remind students that collagen provides the strength and support for tendons and ligaments, both of which are types of connective tissues.

Connection: Chapter 28
Myelin
Remind the class that myelin is a lipid layer surrounding the cell membranes of axons. Have students explain what role this lipid layer serves.

Demonstration 2
Fast Foods
Obtain a list from a fast-food establishment that describes the nutritional content of its various food products. Have students describe a typical meal they might have at this restaurant. Then have them compare the nutritional content of this meal with some of the recommended guidelines given in Table 33.1.

Proteins are the major building blocks of body tissue
Most of the material your body uses for growth and tissue repair is devoted to making proteins. The protein collagen provides strength and support in connective tissues. The protein keratin makes your hair and skin waterproof. Proteins are also necessary for physiological processes. For example, proteins in the form of enzymes speed up chemical reactions. Like carbohydrates, excess protein in the bloodstream is converted to fat. Eggs, milk, and meat contain the highest-quality proteins—that is, those with the greatest amount of essential amino acids. The essential amino acids are nine amino acids that must be available in food in order for protein synthesis to occur, because our bodies cannot make them. Beans and other legumes are protein-rich but are low in one or more of the essential amino acids. Vegetarians must carefully plan their diets to obtain all the essential amino acids. For instance, a meal with both beans and cereal grains provides all of the essential amino acids. Nutritious food combinations are staples in many cultures, such as the rice and beans that are a part of many Mexican meals. **Table 33.1** lists some dietary guidelines for staying healthy like the family in **Figure 33.2**.

Lipids store energy
Lipids such as fats and oils are concentrated energy sources. They provide twice the energy per gram as carbohydrates or proteins, possessing many energy-rich carbon-hydrogen bonds. Lipids also form the bilayer foundation of cell membranes and organelles, making lipids one of the most widespread structural materials your body. Excess lipid is stored as fat.

Table 33.1 Dietary Guidelines for Good Health

One gram of fat yields 9 calories. Total fat intake should provide no more than 30 percent of the total calories consumed.

Cholesterol intake should not exceed 300 milligrams per day.

One gram of protein yields 4 calories. Protein should provide approximately 15 percent of total calories consumed.

One gram of carbohydrate yields 4 calories. Carbohydrates should provide between 50 to 55 percent of total calories consumed, and there should be an emphasis on foods containing complex carbohydrates.

Sodium intake should not exceed 3 grams (3,000 milligrams) per day.

Total calories consumed should be sufficient to maintain a person at an appropriate body weight.

A healthful diet includes a variety of foods.

Figure 33.2
These guidelines reflect the recommendations of groups such as the American Heart Association and the American Cancer Society. All families should consider incorporating these guidelines into their daily diets.

Vitamins and Minerals

Vitamins are organic substances needed by the body in very small amounts. Unlike carbohydrates and lipids, vitamins do not provide energy. Instead, many serve as substances that help activate enzymes.

There are two classes of vitamins: those that are fat-soluble and those that are water-soluble. "Fat-soluble" means that the vitamins dissolve in fat. Fat-soluble vitamins, including vitamins A, D, E, and K, can accumulate in body fat. Excessive amounts of these vitamins can be harmful.

Water-soluble vitamins, like vitamins B and C, are not stored in body tissue. Because these vitamins easily dissolve in water, any excess is excreted in urine. Therefore, regular sources of water-soluble vitamins must be available to prevent deficiencies. However, it is best to obtain vitamins from food sources rather than vitamin supplements. A healthful, balanced diet ensures that you will get all the vitamins you need. **Table 33.2** lists some common vitamins, and their sources, and explains their various functions.

Bananas: vitamin A

Spinach: vitamins A, B₂, B₁₂, C, E, K

Tangerine: vitamin C

Yellow pepper: vitamin A

Cheese: vitamins A, B₁₂

Table 33.2 Some Important Vitamins and Their Functions

Vitamin	Function	Sources	Deficiency
Vitamin A (fat-soluble)	Maintains healthy eyes, skin, bones, teeth; keeps lining of digestive tract resistant to infection	Milk products, liver, yellow and leafy dark-green fruits and vegetables	Night blindness, impaired growth
Vitamin B₁ Thiamine (water-soluble)	Assists with conversion of carbohydrates to energy; normal appetite and digestion; nervous system function	Pork products, liver, legumes, enriched and whole-grain breads, cereals, nuts	Beriberi (inflamed nerves, muscle weakness, heart problems)
Vitamin B₂ Riboflavin (water-soluble)	Assists with nerve cell function; healthy appetite; release of energy from carbohydrates, protein, and fats	Milk, eggs, whole-grain products, leafy green vegetables, dried beans, enriched breads, cereals, and pasta	Cheilosis (skin sores on nose and lips, sensitive eyes)
Vitamin B₃ Niacin (water-soluble)	Maintenance of normal metabolism, digestion, nerve function, energy release	Red meats, organ meats, fish, enriched breads and cereals, green vegetables	Pellagra (soreness of tongue and lips, diarrhea, irritability, and depression)
Vitamin B₁₂ (water-soluble)	Necessary for formation of red blood cells; normal cell function	Lean meats, liver, egg products, milk, cheese	Pernicious anemia, stunted growth
Vitamin C (water-soluble)	Needed for normal wound healing and the development of connective tissues, including those holding teeth	Citrus fruits, melons, green vegetables, potatoes	Scurvy (slow healing of wounds, bleeding gums, and loose teeth)
Vitamin D (fat-soluble)	Promotes normal growth; assists with calcium and phosphorus use in building bones and teeth	Fish-liver oils, fortified milk, liver, egg yolk, salmon, tuna	Rickets (inadequate growth of bones and teeth)
Vitamin E (fat-soluble)	Prevents destruction of red blood cells; needed for certain enzymes	Wheat germ, vegetable oils, legumes, nuts, dark-green vegetables	Red blood cell rupture causing anemia
Vitamin K (fat-soluble)	Assists with blood clotting	Leafy green vegetables and vegetable oils, tomatoes, potatoes	Slow blood clotting, hemorrhage

■ Cultural Perspective ■

Myrtle Thierry-Palmer, Ph.D.

African-American

Dr. Thierry-Palmer researches vitamin D metabolism at Morehouse School of Medicine in Atlanta. She is an associate professor of biochemistry.

Visual Strategy

Table 33.3

Use the information in this table as a way of reviewing some points made about the human body in previous chapters. For example, a deficiency of calcium in later life can lead to osteoporosis (Chapter 27); a lack of iodine can cause goiter (Chapter 29); iron deficiency can result in anemia (Chapter 31); phosphorus is needed for bone growth (Chapter 27); and both potassium and sodium play key roles in nerve impulse conduction (Chapter 28).

Social Studies Connection

The Western World

Inform students that Americans and Western Europeans suffer nutritional problems stemming from improper diets. For example, people in developed countries, on average, consume about 20 times more salt than their bodies need. Such excess salt has been correlated with high blood pressure.

Iodized table salt: sodium, iodine

Corn: calcium

Tomatoes: potassium

Peas: iron

Minerals are needed for the growth and maintenance of body structures

Unlike vitamins, carbohydrates, fats, and proteins, **minerals** are inorganic compounds. This means they are not made by living things and must be obtained elsewhere. Plants absorb minerals from the water or soil, and animals get minerals by eating plants or plant-eating animals. Unlike other nutrients, minerals are not broken down within the body.

Like vitamins, minerals do not supply energy but help regulate body functions. Calcium, phosphorus, and magnesium are essential parts of the bones and teeth. Milk and milk products are the richest sources of calcium. Cereals and meats provide phosphorus. Whole-grain cereals, nuts, legumes, and leafy green vegetables are good sources of magnesium. Other minerals such as iron, copper, iodine, and zinc are called trace elements, because they are needed only in extremely small amounts. Iron is an important part of hemoglobin, the oxygen-carrying molecule in red blood cells. Copper helps the body make use of iron to build hemoglobin. Zinc and manganese are required for the normal action of various enzymes. Leafy green vegetables, whole-grain breads and cereals, seafood, liver, and kidney are good sources of most trace elements. **Table 33.3** lists some minerals and their dietary functions.

Table 33.3 Important Minerals and Their Functions

Mineral	Function	Sources
Calcium	Necessary for normal growth of bones and teeth; transmission of nerve cell impulses, muscle contraction	Milk and dairy products, leafy dark-green vegetables
Iodine	Essential for production of thyroid hormone	Iodized salt, seafood
Iron	Found in hemoglobin; needed for some enzymes	Organ meats, whole grains, dark-green vegetables, legumes
Phosphorus	Necessary for normal structure of bones and teeth; plays a role in metabolism	Meats, poultry, fish, cheese, whole-grain products, legumes
Potassium	Helps maintain normal metabolism, nerve and muscle function	Meats, poultry, fish, fruits, vegetables
Sodium	Essential for proper water balance in cells and tissues; nerve cell conduction	Table salt, high-salt meats (ham), most cheeses, crackers

Nutrition and Health

An improper or inadequate diet can lead to a number of health problems. Overeating and inactivity are two factors that lead to **obesity**, a condition in which a high proportion of body weight (over 20 percent) is stored fat. Obesity is associated with a number of disorders, including high blood pressure, heart disease, diabetes, and cancer. A common form of heart disease can lead to heart attacks. Two of the major risk factors for this disease are high blood pressure and high levels of blood cholesterol. Often, these risk factors can be reduced by maintaining a healthful diet. High blood pressure can be reduced by limiting salt and caloric intake. Similarly, blood cholesterol levels can be lowered by reducing the amounts of cholesterol, fat, and calories in the diet.

The fear of obesity has produced two eating disorders

Millions of people are on diets. Unfortunately, in some cases obsessive dieting leads to eating disorders. **Anorexia nervosa** (*an oh RECK see ah ner VOH sah*) is an eating disorder in which afflicted people starve themselves. In contrast, **bulimia** (*buh LEEM ee uh*) is a disorder in which sufferers engage in binge-purge cycles of eating followed by vomiting. Both disorders primarily afflict young middle-class women, but men are also victims.

Individuals with anorexia nervosa (such as the woman in **Figure 33.3**) have a distorted perception of their body weight. With an overwhelming fear of being fat, they starve themselves. They may take laxatives or may vomit to lose weight, and they often exercise obsessively. Because of metabolic changes, many female anorexics cease menstruating. An anorexic may lose 25 percent of normal body weight. The person can become so clever about avoiding food that even friends and family members do not notice that the person

is skipping meals and eating very little at mealtimes. This eating disorder can lead to severe malnutrition, heart irregularities, and even death.

Bulimia is a less obvious disorder that is harder to recognize, because the bulimic usually looks healthy. Secretly, however, the sufferer engages in eating binges, consuming as much as 5,000 calories during one meal. This is usually followed by vomiting or purging of the body with laxatives. The binge-purge routine may occur once a month or several times a day.

Repeated purgings can do severe damage to the body. Vomiting brings stomach acids into the mouth, which can erode teeth. At its most extreme, bulimia can lead to death by heart failure, stomach rupturing, or kidney failure. Anorexia and bulimia both are difficult to detect during the early stages. However, each can be managed with a combination of medical treatment and family support.

Figure 33.3
Roxanne is a recovering anorexic. Today, she looks and feels healthy.

"I got to be very thin, but I thought I looked normal. The disease persists to this day as a constant battle between my perception of my body and my common sense."
–Roxanne A.

Section Review

① **What are the five nutrients the body requires for good health?**

② **What roles do each of the five nutrients play in the body?**

③ **How can scurvy be prevented?**

④ **What kind of damage can an eating disorder do to the body?**

■ Section Review Answers ■

1. carbohydrates, proteins, lipids, vitamins, and minerals

2. Carbohydrates serve as the body's main source of energy; proteins serve as the building blocks for tissue growth and repair; lipids store energy and serve as a major structural material for cells;

vitamins activate enzymes; minerals regulate body functions.

3. by consuming vitamin C

4. Students should discuss anorexia nervosa and bulimia and the damage they do to the body such as severe malnutrition, heart irregularities, stomach rupturing, kidney failure and even death.

Demonstration 4

Thin Is Beautiful

Show the class an advertisement that conveys the message that being thin is glamorous. Have students bring in additional examples of such ads. Invite the school nurse or a local psychologist to discuss anorexia and bulimia with the class.

Phase 3

ASSESSMENT OPTIONS

Closure Strategy

Fad Diets

Have students describe any fad diets they may have read about or perhaps have even tried. Have them analyze these diets in terms of how they may cause problems by failing to provide certain nutritional requirements.

Section Review

Assign the *Section Review*.

Reteaching

Have students bring in a label from a cereal box. Prepare a table on the chalkboard, listing all the brands and the vitamins found in each. Have students also note what percentage of the RDA for each vitamin is provided by one serving of each cereal.

Lesson Plan 32.2

Phase 1

PREPARATION

Key Concepts

- Digestion begins in the mouth as the teeth increase the surface area of the food and as the enzymes in saliva begin to break down complex carbohydrates.
- Food moves down the esophagus by peristalsis.
- Digestion continues in the stomach as pepsin and hydrochloric acid begin the breakdown of proteins.
- Digestion is completed in the small intestine as carbohydrates are broken down into simple sugars, proteins into amino acids, and fats into short chains of carbon and hydrogen molecules.
- Bile, produced by the liver and stored in the gall bladder, and enzymes secreted by the pancreas aid in the digestive process.
- Digested nutrients are absorbed by the villi lining the small intestine and are stored by the liver.
- Undigested materials are compacted and eventually eliminated by the large intestine, which also absorbs water, salts, and vitamin K.

Reading Strategy

Previewing reading material can help students organize their information. Have students skim through this section, noting only the section heads and those main ideas contained in sentences that include boldfaced terms. This previewing of the reading should result in a basic framework of what is contained in this section. Students should then discover that their reading subsequently fills in the details.

ENERGY SUPPLIES AND BUILDING MATERIALS FOR THE BODY EXIST ONLY IN POTENTIAL FORMS IN FOOD. WHATEVER WE EAT MUST BE PROCESSED INTO SMALLER PIECES BEFORE IT CAN BE USED BY THE BODY. FOOD UNDERGOES THIS TRANSFORMATION IN THE DIGESTIVE SYSTEM, A HOLLOW MUSCULAR TUBE THAT BREAKS DOWN FOOD AND MOVES IT THROUGH YOUR BODY.

33.2   The Digestive System

Objectives

① List the main organs of the digestive system.

② Identify the sites of digestion for each of the three nutrients.

③ Describe the absorption of food from the digestive tract.

④ Explain the roles of the pancreas and liver in the digestive system.

Digestion

Suppose you eat a taco for lunch. The taco is rich in the three kinds of large molecules found in food. Proteins occur in the meat, cheese, and beans. Carbohydrates make up the tortilla and beans. Lipids are found in the meat and cheese. The taco also contains essential vitamins and minerals. When you eat the taco, however, the nutrients are in forms that your body cannot absorb and are combined with molecules that your body cannot use. As the taco travels through your digestive system, a journey of more than 8 m (26.24 ft.), it is crushed and churned, and is assaulted by various chemicals in order to release its store of nutrients. In short, it is digested.

Digestion begins in the mouth

Taking a bite of the taco, as shown in **Figure 33.4**, begins the process of digestion. As the food is chewed, it is broken apart so that it will be accessible to digestive enzymes. You have probably noticed that your mouth "waters" when you are hungry and smell or see food (or sometimes just think of food). Salivary glands lying in and near the mouth increase their production of saliva before you eat, and saliva plays an important role in digestion. It moistens and lubricates food, making it easier to swallow. Saliva also contains enzymes that begin to break down starches and other complex carbohydrates present in the food.

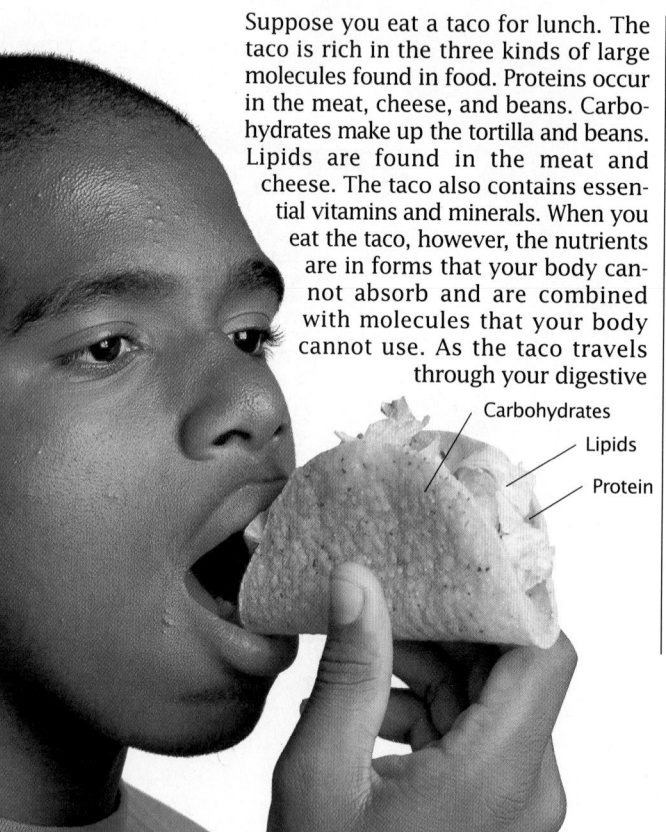

Carbohydrates

Lipids

Protein

Figure 33.4
Biting into a taco and chewing it begins its physical breakdown. Chewing not only makes the taco easier to swallow, but also increases the surface area of the food so that digestive enzymes can come into contact with more of the food.

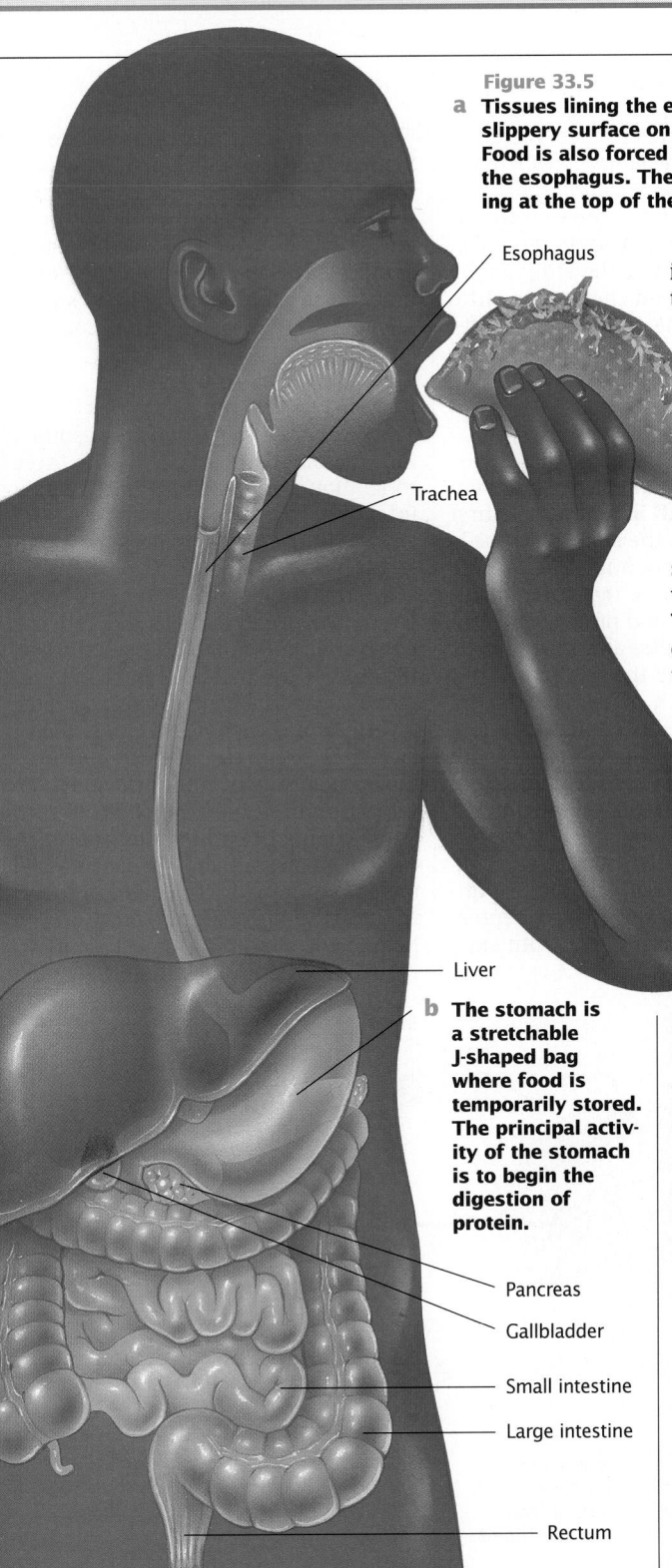

Figure 33.5

a Tissues lining the esophagus secrete mucus, forming a slippery surface on which food slides toward the stomach. Food is also forced downward by smooth muscles ringing the esophagus. These muscles contract in sequence, starting at the top of the esophagus and proceeding downward.

Esophagus

Trachea

Liver

b The stomach is a stretchable J-shaped bag where food is temporarily stored. The principal activity of the stomach is to begin the digestion of protein.

Pancreas

Gallbladder

Small intestine

Large intestine

Rectum

After being thoroughly chewed, food is forced into the rest of the digestive tract by swallowing. The tongue and floor of the mouth rise, squeezing food into a muscular tube called the **esophagus** (*ih SAHF uh guhs*), shown in **Figure 33.5a**. A second tube, the trachea, leads to the lungs. What prevents food from going down the wrong tube into the lungs? As you swallow, touch the front of your neck just under your chin. You should feel the larynx (*LAR ihnks*), or voicebox, slide up and then back down. The upward movement of the larynx causes a flap of tissue overhanging the trachea to swing down like a trapdoor and close off the trachea.

Food slides down the esophagus with the help of mucus and muscle contractions. **Peristalsis** (*pehr uh STAHL sis*) is the name given to this wavelike contraction, which also occurs in the stomach and the rest of the digestive tract. You can swallow food while standing on your head because of peristalsis.

Digestion continues in the stomach

After moving through the 25-cm (9.85-in.) length of the esophagus, the bite of food reaches the stomach, as shown in **Figure 33.5b**. The stomach can expand to hold up to 4 L (4.2 qts.) of food, more than 50 times its empty volume. Cells in the stomach lining release concentrated hydrochloric acid (HCl). Consequently, stomach pH is between 1.5 and 2.5, easily strong enough to damage the stomach lining were it not protected by a thick coating of mucus. Hydrochloric acid and **pepsin**, an enzyme present in the stomach during digestion, break proteins into small chains of amino acids. Agitating this mixture of food, acid, and enzymes are peristaltic waves that pulsate through the stomach's muscular walls.

■ *Matter of Fact* ■

Absorption in the Stomach
Although partially digested, food is not absorbed by the stomach. However, both alcohol and aspirin are absorbed, accounting for their fast-acting effects on the body.

Demonstration 4

Stomach Acid

To dramatize the protective role of the mucous layer, have students observe what happens when iron filings are placed in concentrated hydrochloric acid. Point out that about 500,000 of the mucus-secreting cells are shed every minute. As a result, the mucous layer is replaced every three days. Occasionally, too little mucus or too much acid is present. If this occurs over a long enough period of time, an ulcer may develop. Tell students that "heartburn" is caused when stomach acid "backs up" into the esophagus.

Demonstration 5

Length of the Small Intestine

Show the class a 6 m (19 ft.) long piece of string to demonstrate the average length of the small intestine.

Connection: Chapter 4

Moving Into Cells

Tell students that digestion has resulted in the breakdown of large compounds into ones small enough to enter the cells of the small intestine by diffusion and active transport.

Visual Strategy

Figure 33.6

Have students locate the capillaries and lymph vessels that extend into each villus. Point out that the villi move vigorously to help mix the food and enzymes. The resulting friction wears away millions of small intestinal cells each day. Replacement cells are continuously produced at the base of each villus.

Connection: Chapter 5

Enzyme Activity and pH

Have students explain why the alkaline fluid secreted by the pancreas inactivates the enzymes that were produced by stomach cells.

Activity in the Intestines

The swallowed bites of taco spend up to four hours in the stomach, where the food is reduced to a thin liquid. Most of the proteins and carbohydrates in the taco are broken down in the stomach. Squeezed out of the stomach by peristalsis, the food then moves into the small intestine. Six meters (20 ft.) of small intestine are tightly coiled within your abdomen.

The small intestine absorbs nutrients

The rest of digestion occurs in the entrance to the **small intestine**. The first 25 cm (9.85 in.) of the small intestine is known as the duodenum (*doo uh DEE nuhm*), and it is here that the remaining carbohydrates and proteins are broken down. In adults, almost all lipids are also digested in the small intestine. As in the stomach, peristalsis in the duodenum churns its contents.

The remaining 90 percent of the small intestine has a critical function: it absorbs the nutrients released by digestion and transfers them to the blood. Water is also absorbed. The small intestine is well adapted for absorption. In a cutaway view, the intestine looks as if it were lined with shag carpeting, as shown in **Figure 33.6c**.

The numerous projections of the intestinal lining are **villi** (singular, villus). Each villus is coated with hairlike projections known as **microvilli**. With villi and microvilli, the total surface area available for absorption in the small intestine is about 300 m^2 (3,229 sq. ft.) —an area larger than a tennis court.

The pancreas and liver secrete digestive enzymes

The products of digestion are absorbed into the bloodstream during the three to six hours food spends in the small intestine. Enzymes and chemicals secreted by the upper end of the small intestine cause additional food breakdown. Most of these secretions come from two organs, the **liver** and the **pancreas**. In Chapter 29 you studied the role of the pancreas in regulating blood sugar. The pancreas also secretes a variety of digestive enzymes. These enzymes break down carbohydrates into simple sugars. They also split proteins into amino acids, and fats into short chains of carbon and hydrogen molecules. In addition, the pancreas secretes bicarbonate (the chemical found in antacids and baking soda), which helps neutralize the stomach acids so that they do not digest the wall of the

Figure 33.6

a **The small intestine is located in the lower abdomen.**

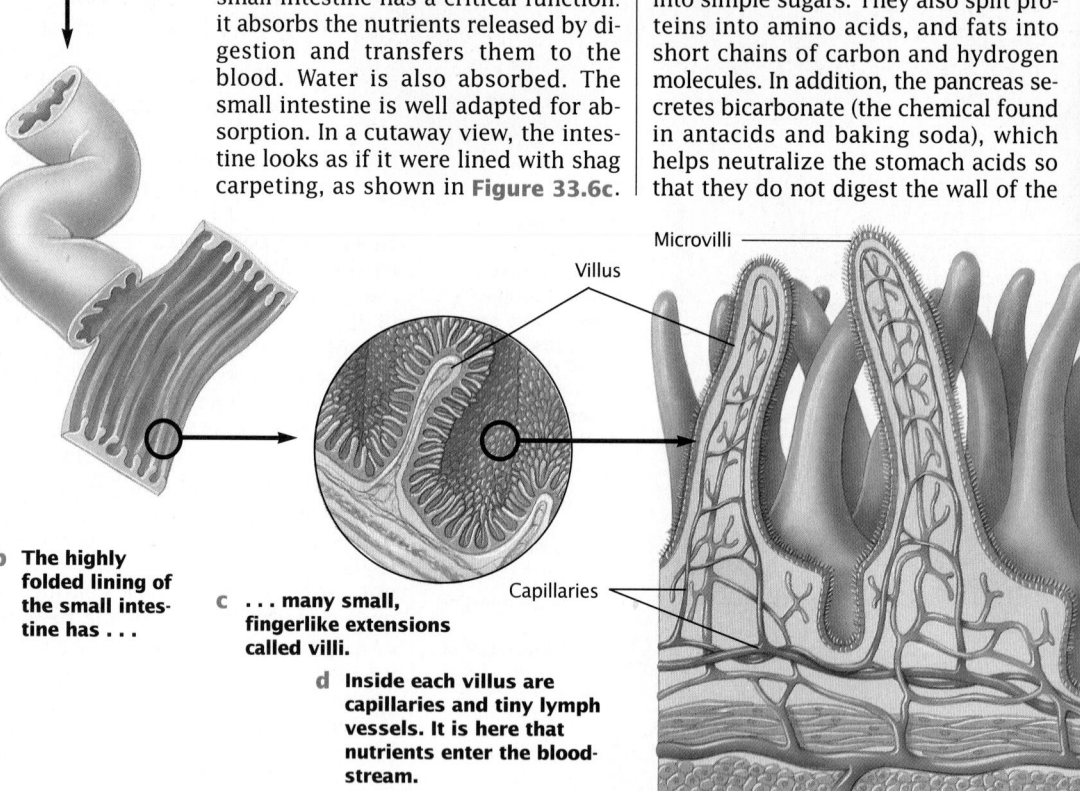

b **The highly folded lining of the small intestine has . . .**

c **. . . many small, fingerlike extensions called villi.**

d **Inside each villus are capillaries and tiny lymph vessels. It is here that nutrients enter the bloodstream.**

Microvilli

Villus

Capillaries

Figure 33.7

a The liver, a football-sized organ, is located above and to the right of the stomach, just below the diaphragm.

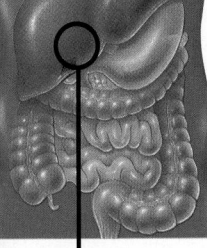

small intestine. When too much stomach acid is produced, the pancreas cannot form enough bicarbonate to compensate and acid begins to eat away the lining of the small intestine or stomach. This painful condition is known as an **ulcer**. Stress and certain foods or medications (alcohol, coffee, aspirin) can cause the stomach to secrete excess acid.

Recall from Chapter 3 that fats are not soluble in water. Fat molecules avoid water and form small clusters or globules in the stomach. Your liver secretes bile, a greenish fluid that emulsifies fat globules so they can be broken down and absorbed. The gallbladder is a green muscular sac attached to the liver that stores bile until it is needed in the small intestine.

Gallbladder

Nutrients are collected in the liver

After digested food molecules are absorbed into the bloodstream, they go first to the liver. The liver, shown in **Figure 33.7**, then sends these molecules into the rest of the bloodstream as needed. For example, when you eat a meal, the liver removes excess glucose from the blood and stores it. When the level of glucose in the blood falls, a hormone causes the liver to release some of this glucose back into the blood. The liver also detoxifies many substances. For instance, it produces enzymes that break down drugs and alcohol.

The large intestine compacts wastes

Your lunch-time taco now is largely undigestible material such as cellulose. From the small intestine, what is left of the food now moves into the **large intestine**. The large intestine's role is to store and compact this undigestible material, and then eliminate it. In addition, some water, sodium, and vitamin K are absorbed there. This vitamin K is produced by the billions of *Escherichia coli* bacteria living in the large intestine.

After traveling the 1.5-m (5-ft.) length of the large intestine, what remains of the bite of taco passes out through the rectum and anus as feces. Feces contain undigested food such as cellulose and bacteria.

b The liver produces bile, stores nutrients, regulates the level of sugar in the blood, and breaks down drugs and alcohol. Blood going into the liver gets cleansed of its dead or damaged red blood cells, debris, and pathogens. Blood exiting the liver contains plasma proteins and nutrients.

Section Review

❶ What organs are needed to digest food?

❷ Which nutrient is digested mainly in the stomach? In the small intestine?

❸ How is the small intestine specialized for absorption?

❹ How would digestion be affected if the pancreas or liver were damaged?

■ *Section Review Answers* ■

1. the mouth, stomach, and small intestines.

2. proteins, carbohydrates

3. Students should suggest that the small intestine has numerous projections in its lining called villi and microvilli, which carry out absorption.

4. Students should indicate that the pancreas and liver secrete digestive enzymes and without them, proteins, fats, and carbohydrates could not be broken down.

Using the Feature

- As a pediatrician, Dr. Coldwater specializes in preventative medicine. Ask students what kind of health advice they would give to young children. Ask them to be as specific as possible.

Discussion

Guide the discussion by posing the following questions.

1. How did Cheryl become interested in pediatrics?

   Because of a birth defect, Cheryl spent a lot of time in hospitals as a child. She was exposed to the medical profession at a very young age. As a high school student, she volunteered in hospitals and clinics.

2. What kind of satisfaction does Cheryl derive out of being a pediatrician?

   She gets to do a lot of preventative medicine—advising children on health matters and watching them grow throughout the years.

3. What personal traits have enabled Cheryl to become a successful pediatrician?

   Besides her caring nature, perseverance, and patience, she likes dealing with people.

Cheryl Coldwater: Pediatrician

How I Became Interested in Pediatrics

"I was born with a birth defect; a broken leg that had healed itself in the womb caused it to be shorter. I got a brace at nine months of age in order to learn how to walk. I wore the brace until I was six, at which point my leg was amputated. That way I could wear a prosthesis and move around as much as possible. As a child I had to spend quite a bit of time in hospitals. And so I was exposed to the medical profession at a very young age.

"In high school I worked as a volunteer both in a clinic and in a hospital. I had a lot of fun and got to see what people did. I worked mostly in the pediatric clinic measuring and weighing the babies and helping the nurses.

"My parents knew before I did that I would go into medicine, because I always had that interest. When I told my mother, she said, 'Well, I thought you would.' They had already figured it out but were going to let me make my own decision. They were very supportive.

"In college I majored in psychology and minored in pre-med. At my college, you couldn't actually major in pre-med. You had to major in something else and fulfill your pre-med requirements at the same time.

"Medical school was very hard—a lot of new things all at once. I didn't pass everything the first time. I had to take two subjects over the first year. It was a lot of work, every day, all day. Some other students had more trouble when it came to working with patients. That was not a big problem for me. That's why I like my specialty so much—you get to deal with people."

Dr. Coldwater is married and has a three-year-old son named Devon. "When I'm not on call, I'm really on my own. That's a new trend in medicine today, particularly with more women going into the field. "

Name:	**Cheryl Coldwater**
Home:	**Austin, Texas**
Employer:	**Austin Regional Clinic**
Personal Traits:	▪ **Perseverance**
	▪ **Good writing skills**
	▪ **Caring**
	▪ **Patience**
	▪ **Sense of humor**

Action ▪ *Science in Action* ▪ *Science in Action* ▪ *Science*

The Satisfaction of Pediatrics

Career Path

High School:
▪ Science
▪ English
• *College:*
▪ Biology
▪ Psychology
▪ Chemistry
Medical School
▪ Biochemistry
▪ Physiology
▪ Anatomy
▪ Pathology

"As a pediatrician, I get to do a lot of preventive medicine. I can help kids with decisions before health becomes a problem. I like to talk to them about good nutrition beginning when they're four years old. I make sure they're eating a fairly balanced diet and keeping junk food for only special occasions. If they learn to eat healthy from the beginning, it's not a big sacrifice. They won't have to go on a diet later on.

"I like talking to kids, questioning them rather than their parents. I talk to them about drugs and alcohol, too. When they get advice only from their parents, they tend not to listen. But when they hear it from their doctor, they pay more attention.

"I like to talk to kids about safety a lot—wearing seat belts and bicycle helmets. It's nice to see that what you're doing is having some effect. When you talk to a child one year about wearing a bicycle helmet, and he comes back the next year and says, 'Oh yeah, I got one,' that feels good.

"I also talk to them about exercise. Many kids aren't interested in organized sports and may not be getting any exercise at all. We talk about ways of getting some movement into their lives. If it's something they're interested in, it will have a long-lasting effect, even into adulthood.

"For our medical group, I produce many patient-education pamphlets and handouts. It's helpful for people to have something they can take home and read later. That way, they can remember the things we talk about in the office."

One of the many aspects of her job that Dr. Coldwater enjoys is working with children. She sees many of the same children year after year and is happy knowing that she is helping them stay healthy.

Research Focus

Sometimes Dr. Coldwater helps pharmaceutical researchers do short-term studies for new pediatric medications. The researchers talk with Dr. Coldwater and her colleagues about using their patients to answer certain questions: Does the drug have any major side effects? Does the medicine work? Do the patients like or dislike the taste? These are things that cannot be learned in the laboratory.

Dr. Coldwater recently helped test a new kind of eye ointment. Newborns are treated with eye ointment to protect them from eye infections caused by bacteria they encounter during the birth process. If there is an infection, it may not show up for a few days. With the parents' permission, Dr. Coldwater helped the researchers find out which eye ointment was most effective and had the fewest side effects.

Lesson Plan 33.3

Key Concepts

- The liver combines ammonia with carbon dioxide to produce urea, a nitrogen-containing waste eliminated mainly by the kidneys.
- Nephrons in the kidney filter the blood by removing urea and other waste products.
- Urine formation involves both the filtration of materials from blood to the nephron and the reabsorption of certain substances from the nephron back into the blood.
- ADH, a hormone produced by the hypothalamus, causes the excretion of a more concentrated urine.
- Kidney disorders include kidney stones, which block the passage of urine, and kidney failure, which can be treated by hemodialysis.

Reading Strategy

Have students practice their note-taking skills while reading this section. Remind them that they should always use their own words, and write their notes in words and phrases (not in complete sentences) in order to save time.

Phase 2
TEACHING STRATEGIES

Visual Strategy

Figure 33.8
Have students identify the two body systems shown in this figure, so that they see the close association between the circulatory and excretory systems.

REGULATING THE CONCENTRATIONS OF SUBSTANCES IN YOUR BODY'S FLUIDS IS ESSENTIAL. A SLIGHT RISE IN POTASSIUM LEVELS IN THE BLOOD CAN CAUSE HEART FAILURE. BUT THIS RARELY HAPPENS BECAUSE THE KIDNEYS MAINTAIN THE LEVEL OF POTASSIUM AND OTHER SUBSTANCES WITHIN NARROW LIMITS. KIDNEYS ALSO SERVE AS FILTERS TO REMOVE NITROGEN-CONTAINING WASTES FROM THE BLOOD.

33.3 The Excretory System

Objectives

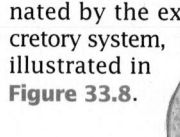

❶ Explain how the kidneys determine what leaves the blood and what remains.

❷ Describe the structure of a kidney nephron.

❸ Compare and contrast the two stages of urine formation.

❹ Explain how medical technology has helped people with kidney disorders.

Kidney Form and Function

When the body breaks down excess amino acids, other metabolic wastes—especially nitrogen compounds in the form of ammonia—are released. In Chapter 22 you read that land animals must rid their bodies of ammonia and other nitrogen-containing wastes. The body is able to make ammonia less poisonous by combining it with carbon dioxide in the liver to form a less toxic compound called urea. Urea enters the bloodstream and circulates throughout the body. Some urea is eliminated from the body through the skin as perspiration, which is a mixture of water, minerals, and urea. Most of the urea, however, is eliminated by the excretory system, illustrated in **Figure 33.8**.

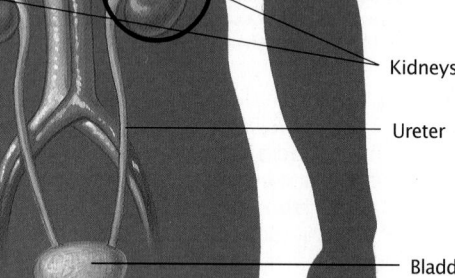

Kidneys

Ureter

Bladder

Urethra

Figure 33.8
The kidneys filter out toxins, urea, water, and mineral salts from the blood. These substances are then stored in the bladder as a liquid called urine until they are eliminated.

The kidneys clean the blood

Your **kidneys** are fist-sized organs located toward the back of your abdomen, about even with the bottom of the rib cage. The kidneys play a vital role in maintaining homeostasis; they remove urea and other wastes, regulate the amount of water in the blood, and adjust the concentrations of various substances in the blood. Because of their crucial function, the kidneys receive large amounts of blood. About one-fourth of the blood your heart pumps every minute travels to the kidneys for cleansing.

Nephrons carry out the processes that form urine

Each kidney contains over 1 million blood-cleaning units called **nephrons** (*NEHF rahns*). A nephron, shown in **Figure 33.9**, is a tiny tube with a cup-shaped capsule at one end. The capsule surrounds a tight ball of capillaries. Substances in the blood inside the capillaries filter into the capsule. The substances then travel through the twists and loops in the nephron, forming urine, an amber-colored fluid that contains nitrogenous waste products and excess amounts of water and solutes.

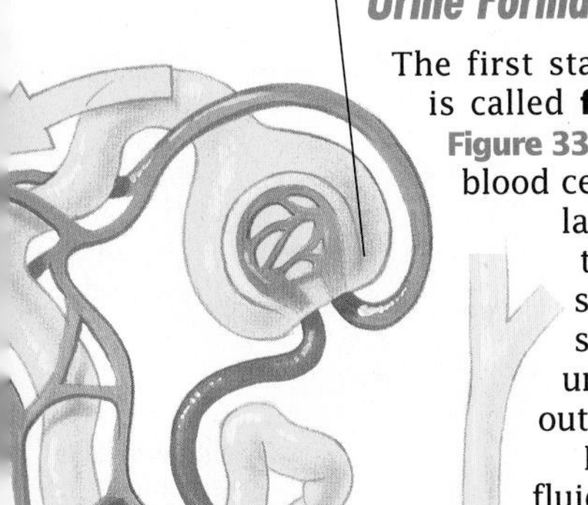

Figure 33.9

a **During filtration, blood pressure in the capillaries forces water, glucose, amino acids, and various salts through small pores in the capillaries and into the capsule of the nephron.**

b **During reabsorption, water and solutes move out of the nephron and into the capillaries.**

Urine Formation

The first stage of urine formation is called **filtration**, as shown in **Figure 33.9a**. During filtration, the blood cells, proteins, and other large solutes remain in the blood, while smaller solutes (such as glucose, salts, amino acids, and urea) and water are forced out of the blood.

Each day, about 180 L of fluid filter out of the blood into the nephrons. Most of this fluid is absorbed back into the blood, leaving about 1 L of urine every day. The amount and content of urine varies, depending on what has been taken into the body.

Reabsorption returns important substances to the blood

From the nephron capsule, the filtered fluid passes into the nephron, beginning **reabsorption**, the second process of urine formation, shown in **Figure 33.9b**. About 99 percent of the water, all of the glucose and amino acids, and many of the salts are reclaimed and sent back into your bloodstream. Water reabsorption is controlled by your hypothalamus. When your body needs to conserve water, the hypothalamus triggers the secretion of a hormone called ADH (antidiuretic hormone). ADH makes the ends of nephrons more permeable to water. As a result, more water is reabsorbed and the urine leaving your body is more concentrated. When your body must get rid of excess water, the secretion of ADH is inhibited, and your urine is more dilute.

After reabsorption, the fluid remaining in the nephron tubes is urine. It is made up of water, urea, and various salts. Urine flows from each kidney into a slender tube called a ureter. It then flows into the urinary bladder, where it is stored. It leaves the body through a tube called the urethra, which leads to the outside of your body.

Matter of Fact

The First Organ Transplant

The first successful organ transplant was that of a kidney, performed in 1954. Since the donor and recipient were identical twins, rejection was not a problem.

Demonstration 1

The Kidneys

Obtain a pig's or sheep's kidney from a local butcher. Cut the kidney in half lengthwise. Have students identify the ammonia smell.

Mathematics Connection

How Much Blood?

The heart pumps about 5 L (2.6 gal.) of blood per minute. Keeping in mind that one-fourth of this volume travels to the kidneys, have students calculate the volume of blood that is filtered each day.

Visual Strategy

Figure 33.9

To be sure that students understand the close association between the circulatory and excretory systems, have them locate the vessels that carry blood to and from the kidney. Emphasize that kidneys clean the blood by removing urea. Have students locate where the blood urea content would be the highest (in the small artery supplying the nephron) and where it would be the lowest (in the small vein leaving the nephron).

Connection: Chapter 4

Diffusion and Osmosis

Review how these two processes account for the filtration of blood in the nephron. Have students locate where filtration occurs in the nephron shown in Figure 33.9.

Connection: Chapter 4

Active Transport

Review how this process accounts for the reabsorption of materials as urine passes through the nephron. Have students locate where reabsorption occurs in the nephron shown in Figure 33.9. Point out that because of active transport, the kidney has an extremely high energy requirement, even higher than that of the heart, when comparing equivalent weights, for example, on a per-gram basis.

Demonstration 2

Kidney Stones

Check with a chemistry teacher on how to prepare a supersaturated salt solution to demonstrate how crystals can form. Point out that kidney stones form when substances crystallize out of the urine in either the kidney or urinary tract. Tell students that passing a kidney stone is extremely painful.

Visual Strategy

Figure 33.10

Have students explain how diffusion is involved in hemodialysis, as illustrated in this figure.

Phase 3

ASSESSMENT OPTIONS

Closure Strategy

Affecting Urine Output

Have students explain why high blood pressure, drinking plenty of fluids, and cold weather promote increased urine output, while low blood pressure, reduced intake of fluids, and hot weather favor decreased urine output.

Section Review

Assign the *Section Review.*

Reteaching

Have students use the notes they were encouraged to take in the Reading Strategy to write a report on the excretory system, either in class or as a homework assignment.

Kidney Disorders and Treatment

An estimated 13 million people in the United States suffer from kidney disorders. One common disorder is the development of kidney stones. Kidney stones are deposits of uric acid, calcium salts, and other substances that have collected inside the kidney. The stones may become lodged in the urethra, the tube through which urine leaves the body, where they interfere with urine flow and cause pain. Kidney stones usually pass naturally from the body, or they can be eliminated by medical or surgical procedures.

When kidneys are damaged by disease or injury and are unable to function, the blood must be artificially filtered. Filtering of the blood is called **hemodialysis**. The term "dialysis" refers to the use of a semipermeable membrane to separate large particles from smaller ones. The kidney machine shown in **Figure 33.10** can be used twice a week to filter the blood of patients with kidney disorders. A recent development in dialysis uses an abdominal cavity membrane as the dialyzing membrane. A dialyzing solution is fed into the abdomen through a catheter. The fluid is changed four times a day.

a Blood in a solution is passed through a membrane where the wastes are removed.

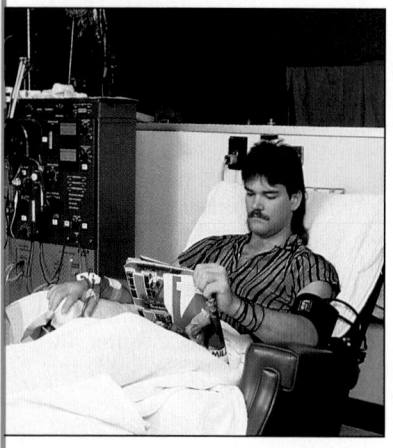

Figure 33.10
The kidney machine is an efficient device for filtering blood.

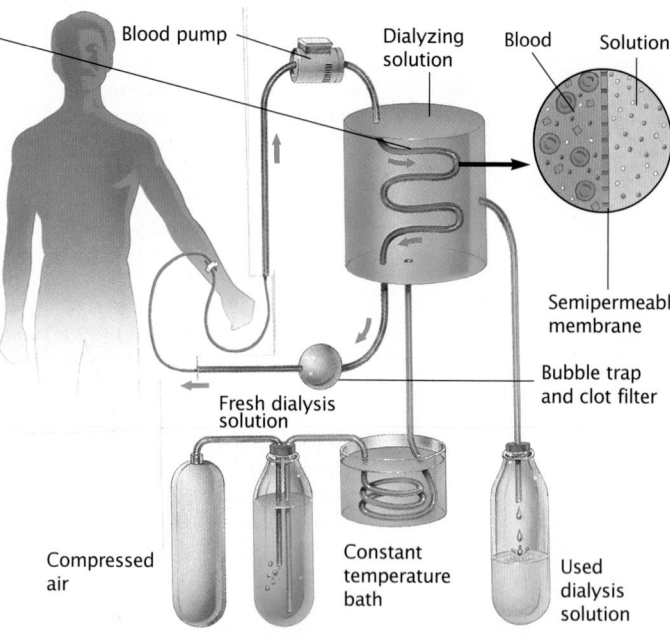

b The blood is maintained at a constant temperature and pH so that it can be pumped back into the patient after it is filtered.

Section Review

❶ How do the kidneys help maintain homeostasis?

❷ How are nephrons specialized for urine formation?

❸ Which substances do nephrons filter out of the blood? Which are reabsorbed?

❹ How does the process of hemodialysis work?

▪ Section Review Answers ▪

1. Students should indicate that kidneys remove urea and other wastes, regulate the amount of water in blood, and adjust the concentration of various substances in the blood.

2. See Figure 33.9 on page 775.

3. Students should indicate that the kidneys filter out toxins, urea, water, and mineral salts from the blood. About 99 percent of the water, all the glucose and amino acids, and many of the salts are reclaimed and sent back into the bloodstream.

4. See Figure 33.10 on page 776.

Chapter 33 *Highlights*

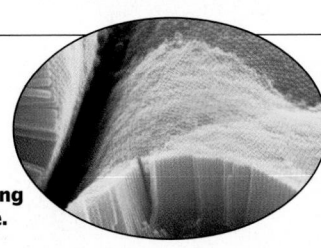

Microvilli can be seen using a scanning electron microscope.

Alternative Assessment

Commercial diet programs are big business, involving millions of people who pay billions of dollars each year with the hope of gaining healthy eating habits while losing unwanted pounds. Have students research these commercial diet programs and report on how each claims to bring about weight loss. Encourage students to check for any information with respect to how these programs might affect either the digestive or excretory systems. For example, a high protein diet places added stress on the liver, which must produce more urea as a result of increased nitrogen intake.

	Key Terms	Summary
33.1 Diet: What You Eat and Why  Eating healthful foods is important for maintaining a healthy body.	vitamin (p. 765) mineral (p. 766) obesity (p. 766) anorexia nervosa (p. 767) bulimia (p. 767)	• A healthful diet includes carbohydrates and lipids (for fuel), proteins (for building tissues), vitamins, and minerals. • Diseases such as scurvy and rickets result from diets deficient in certain nutrients. • Anorexia nervosa and bulimia are eating disorders that can cause serious health problems.
33.2 The Digestive System Digestion is the process of breaking down food so that the body can absorb its nutrients.	esophagus (p. 769) peristalsis (p. 769) pepsin (p. 769) small intestine (p. 770) villus (p. 770) microvillus (p. 770) liver (p. 770) pancreas (p. 770) ulcer (p. 771) large intestine (p. 771)	• In the first stage of digestion, the teeth and saliva break down food. • In the stomach, proteins are broken down by pepsin. • Digestion is completed in the small intestine, where nutrients are absorbed. • The large intestine reabsorbs most of the water from the mass of undigested material left after the removal of nutrients in the small intestine. • The liver stores and regulates the levels of food molecules in the blood.
33.3 The Excretory System 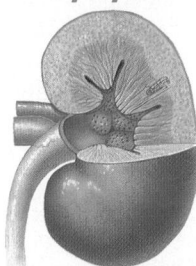 The kidneys are the filtering organs in the body, removing urea and other wastes from the body.	kidney (p. 775) nephron (p. 775) filtration (p. 775) reabsorption (p. 775) hemodialysis (p. 776)	• The functional unit of the kidney is the nephron. • During filtration, water, glucose, amino acids, salts, and urea move out of the capillary and into the nephron. During reabsorption, water, glucose, amino acids, and many salts move out of the nephron and back into the bloodstream. • Some kidney disorders can be managed with hemodialysis, a process that uses a semipermeable membrane to simulate the filtering action of the kidney.

Chapter Review Answers

Understanding Vocabulary
1. **a.** bulimia
   b. liver
   c. reabsorption

Relating Concepts
2. Map answer is shown on page 761D.

Understanding Concepts
Multiple Choice

3. c	8. a
4. c	9. a
5. c	10. b
6. b	11. b
7. b	12. a

Completion
13. digestion
14. duodenum
15. ulcer
16. dialysis

Short Answer
17. Fiber is really cellulose, a carbohydrate that is not digested by humans. It serves to stimulate the muscles of the digestive tract, aiding digestion.
18. Food is moved through the digestive tract by the wave-like contractions of smooth muscles that ring the esophagus, stomach, and intestine.
19. The bicarbonate secreted by the pancreas neutralizes stomach acids before the acids damage the stomach lining.
20. The liver removes glucose from the blood and stores it in an alternate form until it is needed by the body. In addition, the liver produces bile that emulsifies fats.

Chapter 33 Review

Understanding Vocabulary

1. For each set of terms, complete the analogy.
   a. loss of body weight:anorexia nervosa::binge-purge cycle:
   _____
   b. bicarbonate: pancreas::bile:
   _____
   c. movement into nephron:filtration::movement out of nephron:
   _____

Relating Concepts

2. Copy the unfinished concept map below onto a sheet of paper. Then complete the concept map by writing the correct word or phrase in each oval containing a question mark.

Understanding Concepts

Multiple Choice

3. What substances obtained from food regulate body functions but do not supply energy?
   a. proteins   c. vitamins
   b. carbohydrates   d. lipids

4. Vegetarians must take greater care in planning their diets than people who eat meat because
   a. vegetables do not provide proteins.
   b. scurvy can result.
   c. their diets may lack some amino acids.
   d. beans and corn have no fiber.

5. Most carbohydrates come from
   a. dairy products.   c. plants.
   b. meat.   d. fish and poultry.

6. The tube that carries food from the mouth to the stomach is the
   a. epiglottis.   c. duodenum.
   b. esophagus.   d. trachea.

7. Reabsorption of water and sodium is the function of the
   a. stomach.
   b. kidneys.
   c. large intestine.
   d. microvilli.

8. Humans can live for several weeks without eating because the energy needed by the body is stored as
   a. fat.
   b. amino acids.
   c. carbohydrates.
   d. proteins.

9. The functional unit of the kidney is the
   a. nephron.   c. ureter.
   b. capillaries.   d. urinary bladder.

10. Emulsifying fats is a function performed by
    a. mucus.   c. HCl.
    b. bile.   d. bicarbonate.

11. The fingerlike villi of the small intestine function in
    a. enzyme secretion.
    b. nutrient absorption.
    c. bile production.
    d. ulcer formation.

12. ADH secretion increases when
    a. water intake is less than needed.
    b. a lot of fluids are consumed in a short time.
    c. the nephrons are not permeable to water.
    d. your body needs to get rid of excess water.

Completion

13. The breakdown of food to supply the body with amino acids, sugars, lipids, and other nutrients is called _____ .

14. The first section of the small intestine into which bile and pancreatic juice are secreted is called the _____ .

15. When cells lining the stomach or small intestine are attacked by stomach acids, a(n) _____ may form.

16. When kidneys fail, _____ can be used to maintain a proper balance of solutions in the body.

Interpreting Graphics

21. The mouth (a) is where food is crushed and broken into small pieces and acted on by juices secreted by the salivary glands. In the esophagus (b), food is moved to the stomach (c) by the action of smooth muscles. The stomach holds food while it is mixed by muscle action and broken down by HCl and pepsin secreted by the stomach lining. The liver (d) secretes bile into the intestine; bile breaks up fat globules. The pancreas (e) secretes enzymes that break proteins into amino acids and bicarbonate to neutralize stomach acids. Absorption of nutrients occurs in the small intestine (f), and water absorption and waste compaction occur in the large intestine (g).

Short Answer

17. Most medical specialists recommend a high-fiber diet. What is fiber? Why is fiber in the diet important?

18. Astronauts in space are often seen eating while upside down. How is it possible for them to swallow and digest food in this inverted position?

19. The pH of the stomach is very acidic. What prevents damage to the stomach lining?

20. Describe two functions of the liver that are related to digestion.

Interpreting Graphics

21. Name the identified parts of the digestive tract and describe the function of each.

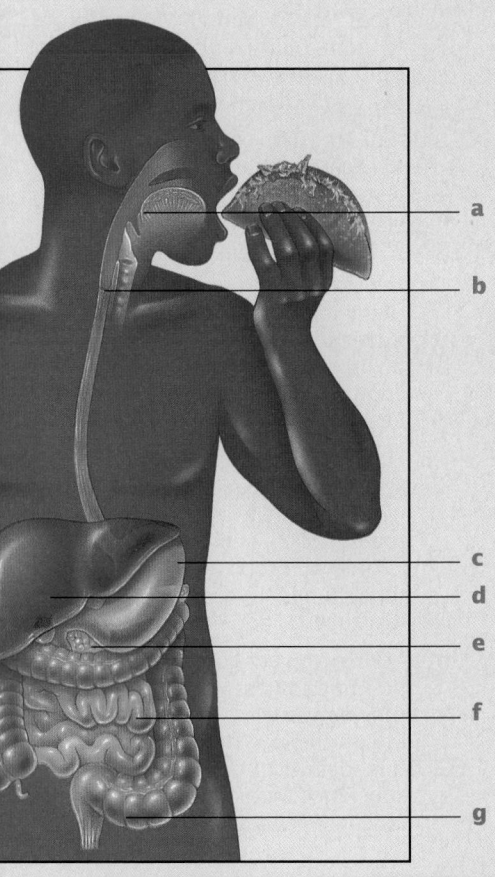

Reviewing Themes

22. *Scale and Structure*
    Both chemical and mechanical processes are involved in digestion in the stomach. Identify one chemical and one mechanical process that occur in the stomach and describe how each aids digestion.

23. *Stability*
    How do the kidneys function to regulate water balance in the human body?

Thinking Critically

24. *Inferring Conclusions*
    A man was diagnosed with cancer of the small intestine, and about a third of his small intestine was removed to save his life. What are the consequences of this life-saving action on the functioning of the small intestine?

25. *Building on What You Have Learned*
    In Chapter 22 you learned about the different forms of nitrogen-containing wastes excreted by organisms. Why would it be a problem for humans to excrete ammonia directly rather than excreting urea?

Cross-Discipline Connection

26. *Biology and Health*
    The "best if used by" date and the "use by" date on products are of interest to consumers. Do library research to learn why product dates are on the labels of packaged foods.

Discovering Through Reading

27. Read the article "There's Always Room for . . . ," in *Health*, May/June 1992, pages 20 and 24–25. What is the source of most of the gelatin eaten today? Besides some desserts, what are some other foods that contain gelatin?

Thinking Critically

24. The removal of about a third of his small intestine would greatly reduce the amount of food that can be absorbed.

25. Ammonia is very toxic to cells of the body. Ammonia is a safe form of waste only for organisms living in a watery environment and those able to excrete their wastes directly through the plasma membrane or skin. Humans are not able to remove waste from the body in this way.

Cross-Discipline Connection

26. From a nutritional perspective, the freshness of packaged foods is very important to consumers. In essence, the date tells the consumer about the freshness of the product and indicates the manufacturer's estimate of how long the product is safe to eat. Food eaten after the product date may present health risks.

Discovering Through Reading

27. Collagen from pork skins and cattle bones and hides is the source of the gelatin eaten today. Gummy candies, ice cream, and throat lozenges are just a few of the foods that contain gelatin.

Reviewing Themes

22. One chemical process is the release of pepsin by the stomach lining; this enzyme breaks proteins into small chains of amino acids. One mechanical process is the contraction of stomach muscles that produce peristaltic waves; the muscle action causes food to be mixed, enabling it to be acted on by acids and enzymes.

23. Water is regulated through reabsorption that is triggered by the hypothalamus. When the body needs to retain water, the nephron becomes more permeable to water. As a result, more water is retained and the urine produced is more concentrated. When the body has too much water, the opposite occurs. The nephron becomes less permeable to water and less water is retained. The urine produced is diluted.

Prelab Preparation Answers

1. • Nutrients provide your body with the energy and materials it needs for growth, maintenance, and repair.
   • carbohydrates, lipids, proteins, vitamins, and minerals

Procedure Answers

4. 1, orange-yellow to red
6. 2, blue-black
7. A translucent spot remains on the paper rubbed with oil but the water leaves no spot.
9. 1, the protein solution turns a faint violet color

Chapter 33

Investigation

How Do You Test Foods for Nutrients?

Objectives

In this investigation you will:
• *perform* chemical identification tests
• *record*, *interpret*, and *evaluate* data
• *compare* the nutrient contents of samples of unknown foods

Materials

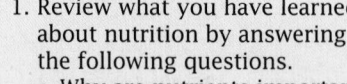

• safety goggles
• lab apron
• disposable gloves
• test tubes
• test-tube holder and rack
• wax pencil
• glucose solution
• starch solution
• Benedict's solution
• water bath
• hot plate
• Lugol's iodine
• brown paper
• cooking oil
• albumin solution
• Biuret reagent
• unknown food substances

Prelab Preparation

1. Review what you have learned about nutrition by answering the following questions.
   • Why are nutrients important to your health?
   • List the five nutrients that you need in your diet in order to maintain good health.
2. Make a data table like the one shown on the next page.
3. In the first row of the table, briefly describe the nutritional role of the substances in each column.

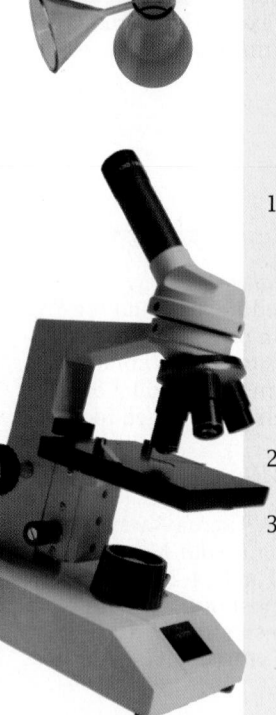

Procedure: Testing Substances for Nutrients

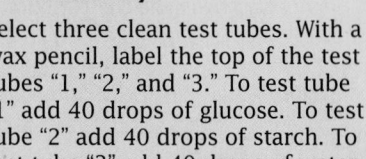

1. Form a cooperative group of four students. Work with another member of your group to complete steps 2–11.

2. **Caution: Put on safety goggles, a lab apron, and disposable gloves. Leave them on for the entire investigation.**

3. Benedict's Test
   Caution: If you get Benedict's solution on your skin, wash it off immediately.

   Select three clean test tubes. With a wax pencil, label the top of the test tubes "1," "2," and "3." To test tube "1" add 40 drops of glucose. To test tube "2" add 40 drops of starch. To test tube "3" add 40 drops of water.

4. Add 10 drops of Benedict's solution to each test tube. Heat the test tubes in a hot water bath for five minutes.

   In which test tube do you see a positive reaction? In your table, write the name of this identification test under the substance that showed a positive reaction. *What color indicates the presence of this substance?* Record this result in your table.

5. Lugol's Iodine Test
   Select three clean test tubes. With a wax pencil, label the tops of the test tubes "1," "2," and "3." To test tube "1" add 40 drops of glucose. To test tube "2" add 40 drops of starch. To test tube "3" add 40 drops of water.

6. Add 2 drops of Lugol's iodine solution to each test tube. *In which test tube do you see a positive reaction?* In your table, write the name of this identification test under the substance that showed a positive reaction. *What color indicates the*

presence of this substance? Record this result in your table.

7. Brown Paper Test.
Rub a few drops of cooking oil (a lipid) on a small piece of brown paper. On a second piece of brown paper, rub an equal amount of water. After 10 minutes, observe both pieces of paper. *On which piece of paper do you see a reaction?* In your table, write the name of the test that identifies this substance. *How does the appearance of the brown paper change in the presence of this substance?* Record this result in your table.

8. Biuret Reagent Test
Caution: Do not get Biuret solution on your skin. If you do, wash it off immediately.

Select two clean test tubes. With a wax pencil, label the top of the test tubes "1" and "2." To test tube "1" add 40 drops of albumin solution (a protein). To test tube "2" add 40 drops of water.

9. Add 3 drops of Biuret reagent to each test tube. *In which test tube do you see a positive reaction?* In your table, write the name of this identification test under the substance that showed a positive reaction. *What color indicates the presence of this substance?* Record this result in your table.

10. Your team will be assigned two unknown food substances for testing. Conduct each of the above tests to identify the presence of nutrients and other substances. For the Benedict's solution test, the Lugol's iodine test, and the Biuret reagent test, use a 5-mL sample of each unknown food substance. In your table, record whether the nutrient is present.

Substance	Glucose	Starch	Lipid	Protein
Nutritional role				
Identification test				
Positive result				
Unknown food substance A				
Unknown food substance B				

11. Clean up your materials and wash your hands before leaving the lab.

Analysis

1. *Summarizing Methods*
Summarize the results of each test you conducted.

2. *Communicating Results*
Compare your results with other members of your group. Offer an explanation for any differences in the test results among the teams.

3. *Analyzing Data*
Which food contains the greatest variety of nutrients? Which food contains the fewest nutrients?

4. *Applying Methods*
Describe an experiment for determining whether milk contains glucose.

5. *Making Inferences*
No color change occurs when table sugar is tested by the Benedict's solution test or the Lugol's iodine test. What conclusion can you draw from these results?

6. *Evaluating Methods*
What was the purpose of conducting the Benedict's solution test on starch and on water?

Thinking Critically

1. How would tests such as these help a nutritionist plan a balanced diet?

2. What additional information would a nutritionist require about food to plan a balanced diet?

Analysis Answers

1. Summaries will vary but should indicate the results that students listed in their data tables.
2. Students should indicate some degree of new information gained from other teams. Variability in results is most likely due to differences in conducting the test, differences among samples, or error.
3. Answers will vary depending on the selection of foods.
4. Students should suggest the use of Benedict's solution to test for glucose.
5. Negative results for both the Benedict's solution and Lugol's iodine tests indicate that the carbohydrate is neither glucose nor starch.
6. Students should suggest that Benedict's solution reacts with glucose, and not with water or starch.

Thinking Critically Answers

1. To plan a diet, a nutritionist needs to know which nutrients are present in each food.
2. A nutritionist would also need to know the caloric value and the quantity of each nutrient in the food.

Chapter 34

Reproduction and Development

Planning Guide

	Objectives/Themes	Classwork Resources	Homework Resources

34.1	1. Identify two features of sperm cells.
	
	2. Draw the pathway of sperm from the testes to the outside of the body.
	
	3. Explain the functions of the two testes.
	
	4. Describe how the hypothalamus regulates the functioning of the testes.
	
	Themes: Interacting Systems, Scale and Structure
		Teacher's Resource Binder
		Focus Activity 34
		    Culturing Frog Embryos
		
		Other Resources
		Transparencies 155–156
		Science in Action Video
		    *Cell Structure and Function/
		    Cell Division**
		
		Text
		Science in Action
		    No Alcohol During Pregnancy
		    pp. 792–793
			Text
			Section Review, p. 786
			
			Teacher's Resource Binder
			Directed Reading Worksheet
			    34.1
			
			Other Resources
			Audiocassette 34.1*

34.2	1. Identify two differences between the male and female reproductive systems.
	
	2. Compare and contrast sperm production with egg production.
	
	3. Relate the events of the ovarian cycle to the levels of LH, FSH, and estrogen.
	
	4. Describe the events of the menstrual cycle.
	
	Themes: Evolution, Interacting Systems, Patterns of Change, Scale and Structure, Stability
		Teacher's Resource Binder
		Science Skills Worksheet
		    *Interpreting Graphs/
		    Critical Thinking*
		
		Other Resources
		Transparencies 157–159
			Text
			Section Review, p. 791
			
			Teacher's Resource Binder
			Directed Reading Worksheet
			    34.2
			
			Other Resources
			Audiocassette 34.2*

34.3	1. Describe the process of fertilization.
	
	2. Explain the process of implantation.
	
	3. List three substances that can pass through the placenta.
	
	4. Describe three changes that occur as the fertilized egg develops into a new individual.
	
	Themes: Patterns of Change, Interacting Systems, Stability
		Teacher's Resource Binder
		Lab Investigation 34.1
		    Embryonic Development
		
		Other Resources
		Transparency 160
			Text
			Section Review, p. 800
			
			Teacher's Resource Binder
			Directed Reading Worksheet
			    34.3
			Reteaching Worksheet*
			    The Reproductive Systems
			
			Other Resources
			Audiocassette 34.3*

34.4	1. Identify three ways HIV is transmitted.
	
	2. Identify three ways to avoid HIV infection.
	
	3. List two sexually transmitted diseases that are caused by bacteria.
	
	4. Describe the cause and results of pelvic inflammatory disease.
	
	Themes: Interacting Systems, Patterns of Change
		Text
		Investigation
		    *How Are Sperm and Eggs
		    Different?* pp. 808–809
			Text
			Section Review, p. 804
			
			Teacher's Resource Binder
			Directed Reading Worksheet
			    34.4
			Vocabulary Review Worksheet*
			
			Other Resources
			Audiocassette 34.4*

*Reteaching Options

Demonstrations
34.3: pp. 794, 797, 800

Assessment
Chapter Review pp. 806–807
Portfolio Assessment p. 781D
Chapter Test—Teacher's Resource Binder
Test Generator

Research Notes

Connection to Genetics: Homeobox Genes and Early Development

One of the greatest mysteries of biology is development. How does a rapidly dividing embryo know to make one end eventually form a head, and the other end legs and feet and toes? How do cells that all have the same genes gradually differentiate to produce the wide variety of cell types found in a body, such as muscles, nerves, and blood cells?

Geneticists and molecular biologists have discovered several clues about the genes and other regulators that control development. Apparently, development of a basic blueprint marking head and tail occurs on different levels all at once.

On one level, it has been shown in fruit flies that the genes that are closer to the "head" end of a cell are those that code for the development of the features of the head. The genes that are closer to the "tail" are active in the cells that become the tail.

In addition, diffusion gradients of certain signaling molecules are set up from head to tail, front to back, and from the sides to the center. In other words, there may be a lot of a certain molecule near the head, a moderate amount in the midsection, and much less near the tail, and so on.

The signaling molecule in turn triggers the expression of a set of genes known as "homeobox" genes. Where the signal molecule is in abundance, all of the homeobox genes are expressed, causing a different body pattern than in places with lower concentrations of signal molecules, where only the first few of the genes are expressed.

Scientists have tested these theories of genetic influence and diffusion gradients. By carefully mutating a front-end fruit fly gene, scientists in the early 1980s succeeded in making a fruit fly with legs on its head instead of antennae. In a later experiment, scientists altered the diffusion gradient of a likely signaling molecule, retinoic acid, within a chicken embryo. The results, including a chicken with extra toes and fingers, confirm this model of development.

The homeobox genes and their signaling molecules seem to be remarkably similar in a number of animals, from the simple animal hydra to fruit flies to humans. Researchers hope that by figuring out such pathways in detail, they will gain insight into how human embryos develop. Such research could help in diagnosing and perhaps healing congenital birth defects.

Investigation Notes

How Does Life Begin?
pages 808–809

Purpose: In this activity, the students will observe sperm and egg cells. They will make comparisons of these structures and relate structural differences to the unique function of each cell. The students will also study a number of early stages of development and attempt to analyze the significance of the changes that occur.

Prelab Preparation

Prepared slides of sperm, egg, and early developmental stages are available from WARD'S (92 M 8238, 92 M 8241, 92 M 8255). Sea star developmental stages (up to gastrula) are recommended because they show a greater degree of similarity to those of mammals than do those of frogs. While you should be aware of this difference, it should not preclude the use of frog developmental stages in this activity.

Answers will be found on pages 808–809.

Meeting Individual Needs

Objectives

1. Students will demonstrate appropriate use of core vocabulary for the chapter (Vocabulary File).

2. Students will relate the roles of hormones in both the male and female reproductive systems (Teaching Strategy).

3. Students will compare infant mortality rates for different regions and plan and produce informational messages about ways to reduce infant mortality (Multicultural Lesson Plan).

Vocabulary File
(Developing Vocabulary/ Limited English Proficiency)

If you are not already using the Vocabulary File, refer to Chapter 1 for its preparation. See the Chapter Highlights on page 805 for a list of suggested words.

Teaching Strategy
Shared Hormones
for pages 783–791
(Developing Organizational Skills/Visual Learners)

Materials: overhead projector, transparency, transparency pens

Procedure: Have each student silently read Sections 34.1 and 34.2, making notes about the roles of hormones in the male and female reproductive systems.

Begin a discussion about hormones, recalling information from Chapter 29. Challenge students to describe the cycle of hormones involved in the male reproductive system. As the discussion continues, sketch the relevant points on the transparency, in a form similar to the flowchart in Figure 34.5 on page 786.

Continue the discussion by turning to the female reproductive system. Create a similar flowchart for the different cycles of hormones and their effects. Be sure students notice the number of hormones that are the same for males and females.

Multicultural Lesson Plan
Infant Mortality Preparation (Dealing with Ethnic Diversity)
Preparation:
1. Identify sources where students can find statistics for the countries with the 10 highest infant mortality rates and the 10 lowest, and the United States cities with the 10 highest and lowest infant mortality rates, as well as the rate for the school's city or town. Possible sources include almanacs, and local or state health departments.

2. Identify possible guest speakers, such as doctors or social workers, who could discuss the causes and prevention of infant mortality.

Teaching Strategies:
1. Have students compare infant mortality rates for nations and for United States cities. Have them note how the rate in their community compares to these rates.

2. Have the guest speaker describe the causes and prevention of infant mortality. Ask the speaker to emphasize what is being done in the community today to combat this problem, and what possibilities there are for improving the infant mortality rate.

Assessment: Students will prepare media messages designed to inform expectant parents and the public about ways that they can reduce the infant mortality rate. Depending on your school's location, you may want to suggest that the students target specific groups or regions. Students may create pamphlets using computers, posters, or even a short video or talk show that highlights methods to help people have healthy babies. The results of their project could be given to a community library or to a local group that reaches out to young prospective mothers.

Additional Strategies
Visual Strategies
Pages 783, 784, 785, 786, 787, 788, 789, 790, 791, 795, 796, 798, 801, and 804

Auditory Learners
Use *Biology: Visualizing Life* Audiocassettes for Sections 34.1, 34.2, 34.3, and 34.4.

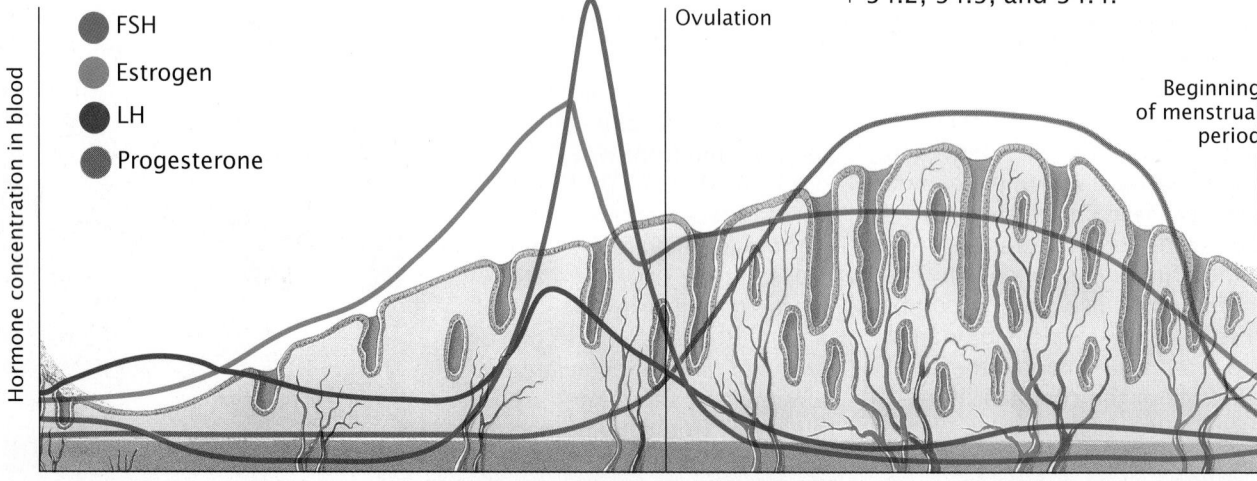

FSH
Estrogen
LH
Progesterone

Hormone concentration in blood

Ovulation

Beginning of menstrual period

Length of one menstrual cycle (about 28 days)

Meeting Individual Needs (cont.)

Cooperative Learning
Multiple Births

Timing: Use this activity to introduce Section 34.3.

Group Size: 3 students

Outcome: Students will be able to list different types of multiple births and make diagrams showing how each type occurs.

Individual Accountability: Each group member is responsible for researching one type of multiple birth.

Positive Interdependence: Each group will construct a poster showing diagrams of how different types of multiple births occur.

Assign one of the following types of multiple births to each group member: (1) identical twins, (2) fraternal twins, and (3) births of more than two infants. Each group member should research this type of multiple birth, diagramming the possible variations of eggs and sperm uniting to form males or females, and the process of cell division that follows.

Each group should compile the information and diagrams of the group members to construct a poster illustrating each type of multiple birth and the possible variations for each occurrence.

Portfolio Assessment

Students should select their best work and provide a self-reflective rationale for their selections. Students can make selections in the following areas.

1. *Content* — One concept map from the chapter (See page 812 for evaluation criteria.)

2. *Reading Comprehension* — One Directed Reading Worksheet from the Teacher's Resource Binder (Use the answer key to evaluate for accuracy.)

3. *Writing* — Using the Vee Form, summarize a magazine or newspaper article relating to the reproductive system or the development of infants. (See page 22T for evaluation criteria.)

   Or: Select a writing task or project from the Chapter Review.

4. *Performance Assessment* — One Vee form from a chapter investigation or lab manual investigation (See page 22T for evaluation criteria.)

Teacher makes selections in the following areas.

1. *Formal Assessment* — Chapter test (Test A, B, or the Test Generator) The teacher-scored test should be reviewed by the student. Incorrect responses should be corrected by the student before the test becomes part of the portfolio.

2. *Informal Assessment* — Use the Direct Observations Checklist, page 33T, during a laboratory or other cooperative learning experience.

3. *Performance Assessment* — Have students write a newspaper editorial about reproduction and education. At what age or grade level do they believe students should learn about reproduction? Why?

Concept Map Answer

The following is one possible answer to the Relating Concepts exercise on page 806.

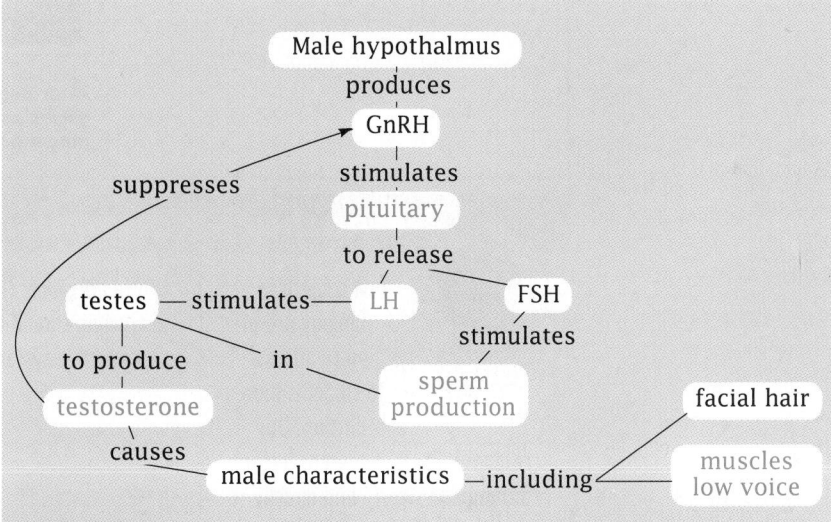

Chapter 34

Determining Prior Knowledge

- Display a model, photographs, or overhead transparency of the male and female reproductive systems. Have students identify as many organs as possible and describe their functions.
- Display photographs of both identical and fraternal twins. Have students explain the differences between the ways each underwent fertilization and development.
- Have students discuss how people generally react to individuals with various communicable diseases. For example, ask how people react to someone with chickenpox. Compare this reaction to what people would likely say about someone with AIDS, syphilis, or gonorrhea. Have the class discuss how these reactions affect the treatment and spread of these diseases.

Chapter 34

Review

- meiosis (Section 6.2)
- placenta (Section 22.3)
- estrogen and testosterone (Section 29.3)

Reproduction and Development

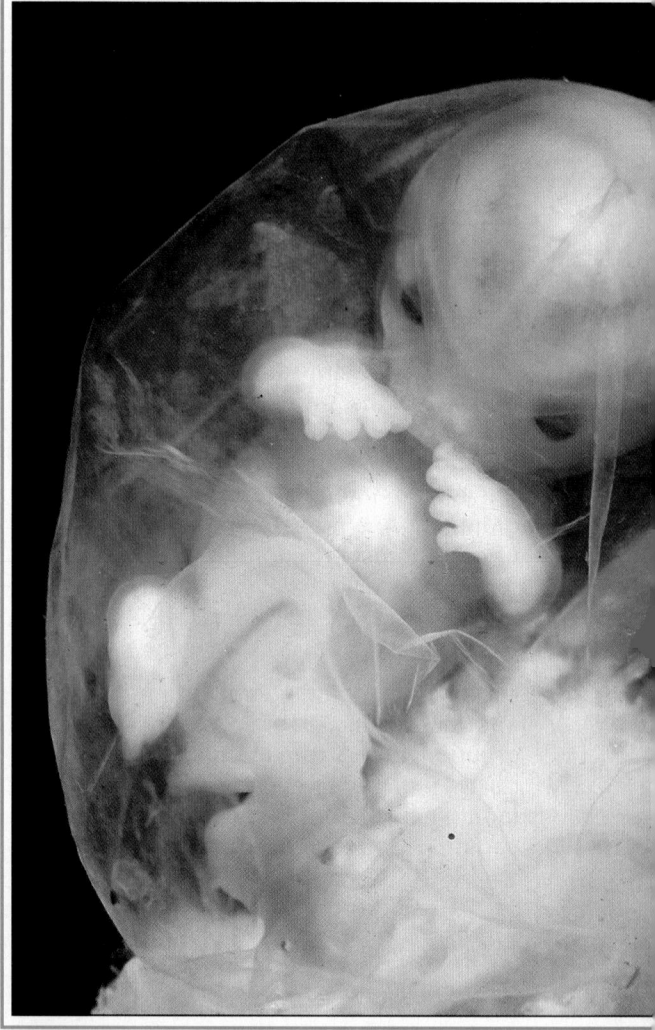

After beginning as a fertilized egg, a human fetus grows and develops inside its mother's uterus for nine months.

■ Author's Rationale ■

Few subjects in biology are of more interest to students than reproduction. Our responsibility is to present factual information clearly and directly so that students can make informed future decisions. The focus of the instruction should be on how the reproductive system works.

BOTH MALES AND FEMALES PRODUCE SPECIALIZED SEX CELLS, CALLED GAMETES, WHICH CONTAIN ONE-HALF OF THE GENETIC INFORMATION NEEDED TO FORM A NEW INDIVIDUAL. THE ROLE OF THE MALE REPRODUCTIVE SYSTEM IS TO PRODUCE SPERM, THE MALE GAMETES, AND DELIVER THEM TO THE FEMALE GAMETE SO THAT FERTILIZATION CAN OCCUR.

34.1 *The Male Reproductive System*

Objectives

❶ **Identify two features of sperm cells.**

❷ **Draw the pathway of sperm from the testes to the outside of the body.**

❸ **Explain the two functions of the testes.**

❹ **Describe how the hypothalamus regulates the functioning of the testes.**

Structure of Sperm

A male produces millions of sperm each day. How are sperm cells different from other cells of the body? Recall from Chapter 6 that sperm are haploid. They have only one set of chromosomes instead of the two sets found in the body's other cells. The cells that give rise to sperm are initially diploid. These cells undergo meiosis, which reduces the number of chromosomes in the developing sperm cell by half.

After meiosis, a sperm matures for several weeks, becoming a highly specialized DNA delivery cell. As you can see in **Figure 34.1**, a sperm has a long tail and has discarded most of its cytoplasm and organelles. Mitochondria remain in the sperm cell, clustering around the top part of the tail. Mitochondria are "motors" for the sperm, providing energy for the tail to whip back and forth. A cap of digestive enzymes forms at the tip of the sperm's head. These enzymes enable the sperm to penetrate the egg.

Figure 34.1

a **A sperm cell consists of a tail used for locomotion and a head that contains DNA.**

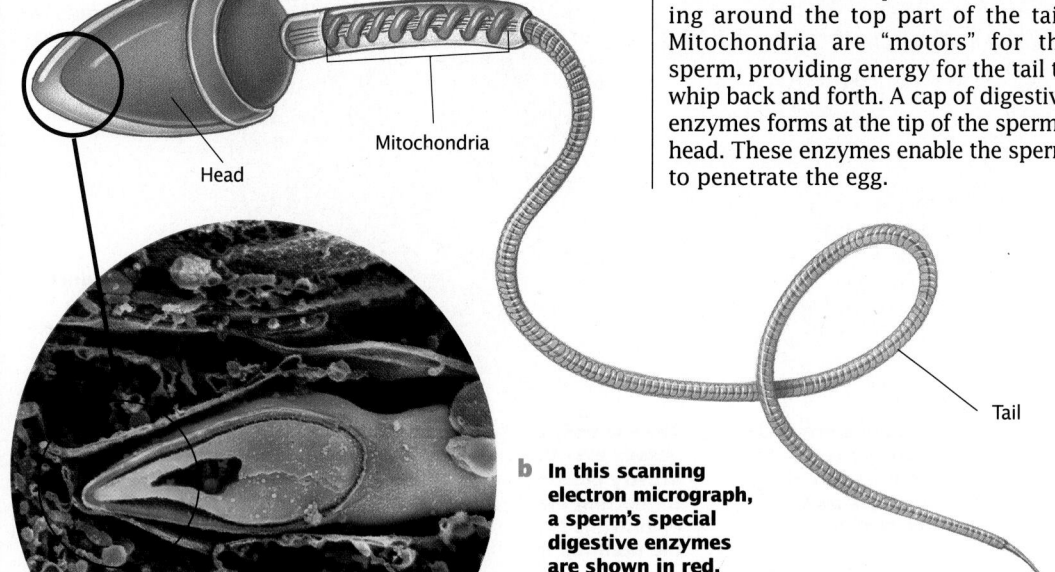

Head

Mitochondria

Tail

b **In this scanning electron micrograph, a sperm's special digestive enzymes are shown in red.**

Lesson Plan 34.1

Phase 1

PREPARATION

Key Concepts

- Sperm consists of a head region, containing DNA found in the 23 chromosomes, and a tail region, filled with mitochondria producing the energy needed for locomotion.
- Sperm, produced in the testes, travel out of the body by passing through the epididymis, vas deferens, urethra, and finally the penis.
- Sperm and the secretions from the seminal vesicles and the prostate gland make up the semen.
- The level of testosterone, the male hormone responsible for secondary sex characteristics, is controlled by negative feedback.

Reading Strategy

Understanding this section requires students not only to read the text but also to study the accompanying figures. Tell students that the illustrations, including the captions, will help them visualize, and better understand, the male reproductive system.

Phase 2

TEACHING STRATEGIES

Visual Strategy

Figure 34.1

Tell students that a normal human male usually produces several hundred million sperm each day. Have students identify what part of the sperm shown in this figure must penetrate the egg. They should recognize that the head region with its DNA, representing paternal genes, must be delivered into the egg.

Figure 34.2

Be sure to spend time with the class discussing the major points made in this illustration:

- The path traveled by sperm is: testes → epididymis → vas deferens → urethra → penis. Have students trace this path in the figure.
- Secretions from the seminal vesicle nourish the sperm, while those from the bulbourethral and prostate glands neutralize the acidic environment present in the female reproductive tract. Have students locate these glands in this figure.
- The testes lie outside the body cavity, since a lower temperature is needed for sperm to mature (undergo meiosis) properly. Have students note the location of the testes in this figure.
- The urethra serves as a common exit for both urine and sperm. Have students trace how urine exits the body. Point out that during ejaculation, small muscles close off the urethra so that urine cannot exit at the same time as sperm.

Circumcision

The penis is covered by a loose-fitting flap of skin, known as the foreskin. The foreskin is often removed after birth; the operation is called circumcision. While sometimes done for religious reasons, circumcision is also performed for health reasons. Glands under the foreskin secrete a thick, white fluid that may accumulate, resulting in irritation and infection. Removing the foreskin prevents this problem.

The Path Traveled by Sperm

Sperm are produced in two oval-shaped organs called **testes** (*TEHS tees*), or testicles. The testes are located in the scrotum, a loose sac hanging below the base of the penis. After sperm form, they mature and are stored temporarily in the **epididymis** (*ehp uh DIHD ih mihs*), shown in **Figure 34.2**. Some sperm are stored in the lower region of the **vas deferens** (*vas DEHF uh REHNZ*).

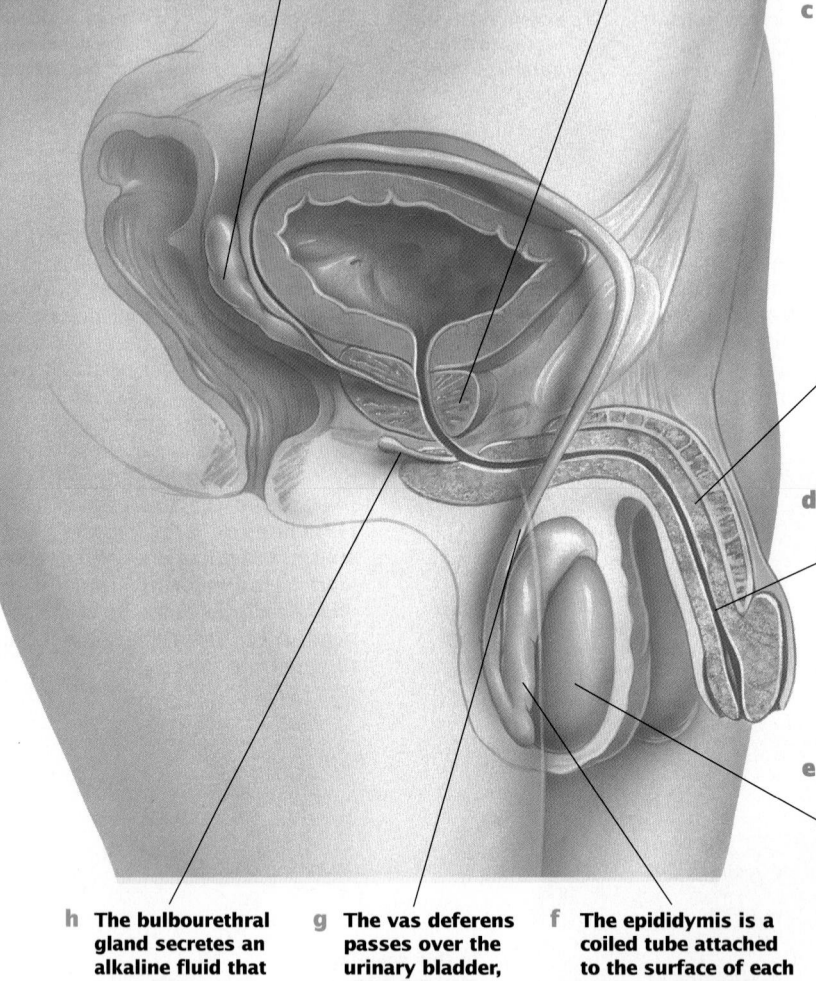

Figure 34.2
The male reproductive system is a set of organs that produces and delivers sperm.

a The seminal vesicles produce fluid that nourishes sperm.

b The prostate gland secretes an alkaline fluid that counteracts the acids found in the male urethra and in the vagina of the female.

c When a male becomes sexually excited, blood accumulates in the penis, causing it to become firm and erect. The firmness of the erect penis makes it possible for the penis to be inserted into the vagina of the female.

d The urethra is the tube that carries urine during urination and semen during ejaculation.

e Sperm production begins in the testes at puberty and continues throughout life.

f The epididymis is a coiled tube attached to the surface of each testis. If uncoiled, an epididymis would be about 6 m (20 ft.) long.

g The vas deferens passes over the urinary bladder, connecting the epididymis to the urethra.

h The bulbourethral gland secretes an alkaline fluid that becomes part of the semen.

Figure 34.3
After sperm are produced in the testes, they are stored in the epididymis and vas deferens. As they pass out of a male's body during ejaculation, they pass through the vas deferens and urethra.

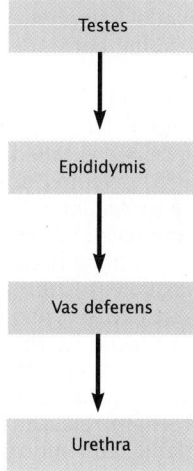

For fertilization to occur, sperm first must be transferred from a male to a female. The penis is the male reproductive organ that makes it possible for sperm to be delivered to the body of the female. **Ejaculation** (*ee jak yoo LAY shuhn*) is the process by which sperm leave the male's body. The walls of the vas deferens contain a thick layer of smooth muscle. During ejaculation, rhythmic contractions of this smooth muscle propel sperm through the vas deferens. Sperm move from the vas deferens into the urethra. Smooth muscles surrounding the urethra near the opening to the urinary bladder contract during ejaculation. Contraction of these muscles prevents urination during ejaculation. The path that sperm travel from the time they are produced to the time they are expelled during ejaculation is outlined in **Figure 34.3**.

Several glands add fluids to the sperm as they travel through the vas deferens and urethra. The seminal vesicles (*SEHM uh nuhl VEHS ih kuhlz*) are two small glands located near the bladder. The fluid produced by these glands nourishes the sperm. The bulbourethral (*buhl boh yoo REE thruhl*) gland is located just past the prostate (*PRAH stayt*) gland. Both glands secrete a thick, clear, alkaline fluid. These

secretions mix with sperm to form **semen**. Semen is expelled from the penis during ejaculation. **Figure 34.4** shows sperm that have reached the uterus of a female.

If sperm cannot reach the egg, fertilization cannot occur. The surest way to prevent a pregnancy is to not have sexual intercourse. Blocking the path travelled by sperm can prevent fertilization. One way to accomplish this is for a male to wear a latex rubber sheath called a condom during sexual intercourse. If the condom is used properly and no semen leaks out, fertilization cannot occur, because no sperm are deposited in the body of the female. If a male chooses to have no more children, he can have an operation called a vasectomy. During a vasectomy, a small incision is made in the scrotum. Each vas deferens is then cut and the ends are tied. A male who has had a vasectomy is sterile, because the sperm's path to the outside of the body is blocked. Sperm are still produced, but they disintegrate in the epididymis.

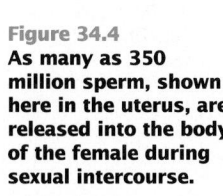

Figure 34.4
As many as 350 million sperm, shown here in the uterus, are released into the body of the female during sexual intercourse.

Health Connection

Prostate Problems
At birth, the prostate is about the size of an almond. During puberty, it will double in size. Then it will stop growing. However, in most men 45 or older, the prostate again starts to enlarge as a result of hormonal activity. As the prostate continues to grow, it can squeeze the urethra, sometimes blocking the passage of urine out of the body.

Health Connection

Vasectomy
A vasectomy does not have any significant effect on ejaculation, since sperm account for only 1 percent of the total volume of the semen. Neither does a vasectomy affect secondary sex characteristics, since male hormones are still produced.

Visual Strategy

Figure 34.4
Tell students to identify the sperm (both the head and tail regions) in this figure. Have them hypothesize why a single ejaculation contains so many sperm.

Connection: Chapter 29

Negative Feedback

Have students explain the principle of negative feedback. Remind them that all hormonal levels, not only those of testosterone, are controlled through negative feedback.

Visual Strategy

Figure 34.5

Have students explain how this figure illustrates the negative feedback control of testosterone. Point out that the level of testosterone is subject to factors other than GnRH. One such factor is stress, which has been shown in several studies to reduce testosterone levels. In vertebrates other than humans, the level of testosterone can be responsive to daylight hours and temperature.

Phase 3

ASSESSMENT OPTIONS

Closure Strategy

Sperm Survival

Have students explain why sperm can survive for only about 48 hours once they enter the female reproductive tract (where the acidic fluid poses a hostile environment), yet survive much longer in the testes (where specialized cells and fluid nourish and protect them).

Section Review

Assign the *Section Review.*

Reteaching

Show the class an overhead transparency of the male reproductive system. Have students identify the structure and function of the various parts.

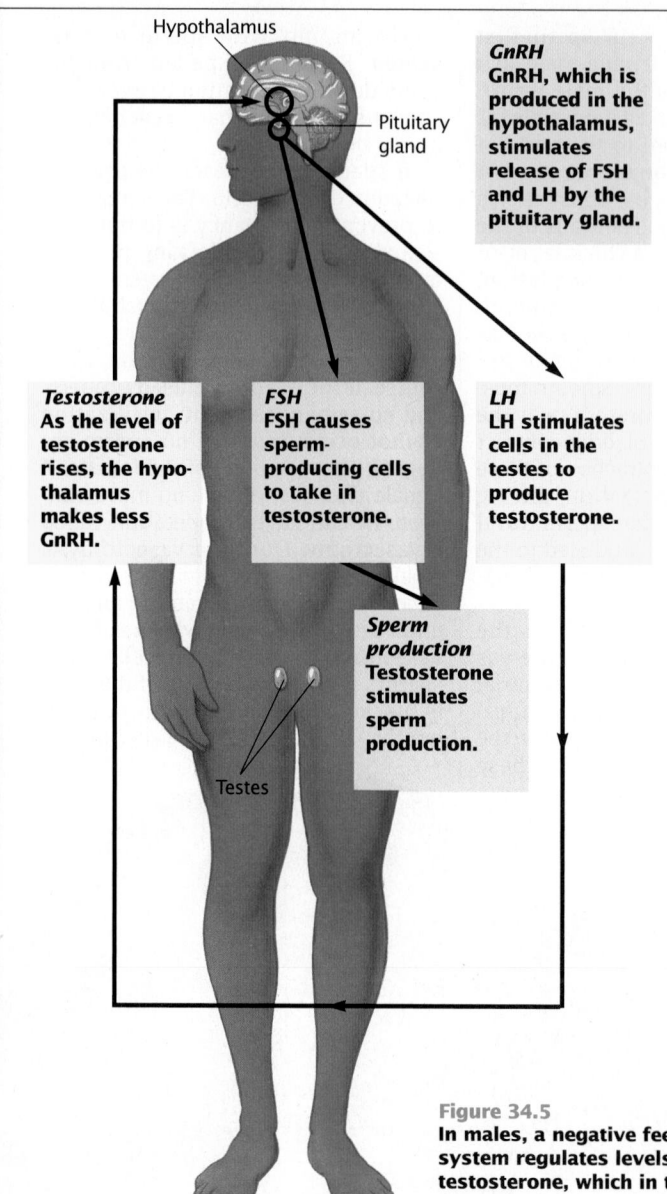

Hypothalamus

Pituitary gland

GnRH
GnRH, which is produced in the hypothalamus, stimulates release of FSH and LH by the pituitary gland.

Testosterone
As the level of testosterone rises, the hypothalamus makes less GnRH.

FSH
FSH causes sperm-producing cells to take in testosterone.

LH
LH stimulates cells in the testes to produce testosterone.

Sperm production
Testosterone stimulates sperm production.

Testes

Figure 34.5
In males, a negative feedback system regulates levels of testosterone, which in turn controls sperm production.

Male Hormones and Reproduction

In addition to producing sperm, the testes also produce the male hormone testosterone. A male begins to make testosterone before birth. Testosterone causes an embryo to develop a penis and scrotum rather than female reproductive structures. At puberty, testosterone levels rise. These higher levels of testosterone cause the development of adult male characteristics, such as a beard, a low voice, and large muscles. Testosterone also stimulates sperm production. Production of sperm and testosterone are regulated by two pituitary hormones: luteinizing hormone (LH) and follicle-stimulating hormone (FSH).

The release of LH and FSH is controlled by a hormone from the hypothalamus. This hormone is called gonadotropin-releasing hormone, or GnRH for short. The hypothalamus is sensitive to testosterone levels in the blood. If the level of testosterone rises, the hypothalamus makes less GnRH. Thus, less LH and FSH are released, and less testosterone is produced by the testes. As the levels of testosterone fall, the hypothalamus begins to make more GnRH. More LH and FSH are released from the pituitary gland, and more testosterone is then produced by the testes. This negative feedback system maintains a nearly constant level of testosterone in the blood. The steps of this negative feedback system are illustrated in **Figure 34.5**.

Section Review

1 In what ways are sperm different from the body's other cells?

2 What is the function of the vas deferens in sperm delivery?

3 Describe two functions of the testes.

4 What role does the hypothalamus play in sperm production?

▪ *Section Review Answers* ▪

1. Answers will vary. Possible answers might include: sperm are haploid; they are highly specialized DNA delivery cells; they have a flagellum; they have discarded most of their organelles and cytoplasm; their mitochondria cluster around the top part of the tail; a cap of digestive enzymes forms at the tips of their heads.

2. The vas deferens connects the epididymis with the urethra.

3. The testes produce sperm and the male hormone testosterone.

4. The hypothalamus produces a hormone called GnRH, which controls the release of two other hormones, LH and FSH. LH and FSH stimulate the production and uptake of testosterone, which stimulates sperm production.

THE MALE'S DIRECT CONTRIBUTION TO THE FERTILIZED EGG IS THE DONATION OF ONE SET OF CHROMOSOMES. THIS CONTRIBUTION IS COMPLETE ONCE FERTILIZATION HAS OCCURRED. IN CONTRAST, THE FEMALE CARRIES THE DEVELOPING FETUS FOR ABOUT NINE MONTHS UNTIL BIRTH. THE FEMALE REPRODUCTIVE SYSTEM THUS MUST PLAY TWO ROLES: MAKING GAMETES AND NOURISHING THE FETUS.

34.2 The Female Reproductive System

Objectives

❶ **Identify two differences between the male and female reproductive systems.**

❷ **Compare and contrast sperm production with egg production.**

❸ **Relate the events of the ovarian cycle to the levels of LH, FSH, and estrogen.**

❹ **Describe the events of the menstrual cycle.**

Structure of the Female Reproductive System

Female gametes are called **ova** (singular, ovum), or eggs. The **ovaries** are the organs responsible for producing eggs. Ovaries are located in the lower part of the abdomen. Eggs released from the ovaries move into the **Fallopian tubes**, which carry the egg into the uterus, as you can see in **Figure 34.6**. The **uterus** is a hollow, muscular organ about the size of your fist. During pregnancy, the uterus expands to many times its usual size.

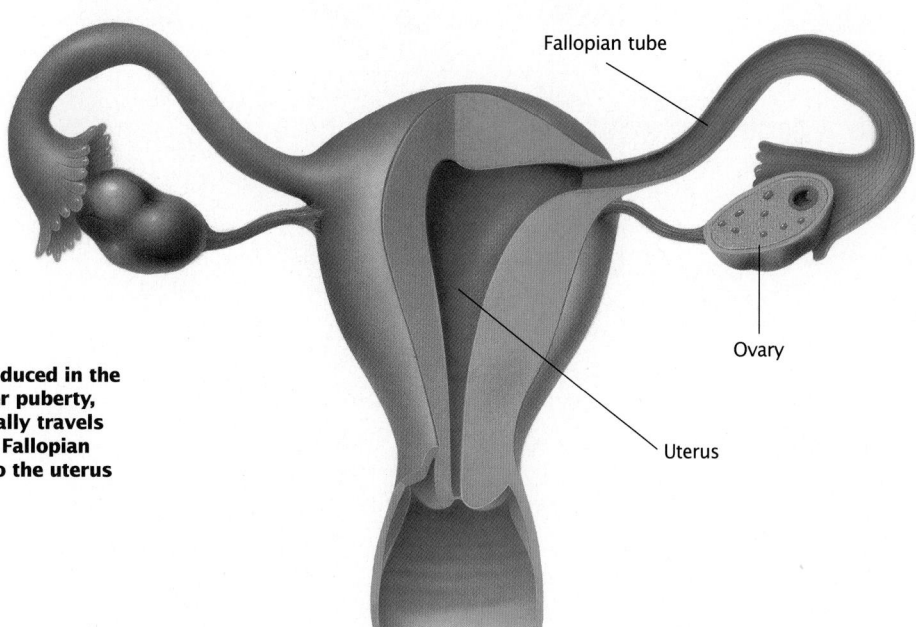

Figure 34.6
Eggs are produced in the ovaries. After puberty, one egg usually travels through one Fallopian tube and into the uterus each month.

Fallopian tube

Ovary

Uterus

Lesson Plan 34.2

Phase 1
PREPARATION

Key Concepts
- Eggs, produced in the ovaries, travel down the Fallopian tubes to the uterus.
- The uterus is separated from the vagina by a ring of muscles known as the cervix.
- During the ovarian cycle that occurs each month, an egg matures and is subsequently released from the ovary during ovulation.
- Both follicle stimulating hormone (FSH) and luteinizing hormone (LH) stimulate the maturation and ovulation of an egg.
- Estrogen, produced by the ovary in response to FSH, prepares the uterus for possible implantation by a fertilized egg.
- The menstrual cycle is the series of changes that occur in the uterus each month.
- Menopause is the shutdown of the menstrual and ovarian cycles.

Reading Strategy

Once again, student understanding will be made easier if they study the figures accompanying the text. Tell students that the illustrations and captions will help them visualize, and better understand, the female reproductive system.

Phase 2
TEACHING STRATEGIES

Visual Strategy

Figure 34.6
Have students locate where sperm enter the female reproductive system illustrated in this figure.

Visual Strategy

Figure 34.7
Be sure to spend time with the class in discussing the major points made in this illustration:

- The path traveled by the egg is: ovary → Fallopian tube → uterus → vagina. Have students trace this path in the figure.

- Unlike males, the urethra in females does not serve as a common exit for both urine and gametes. Have students trace how urine exits the body in a female. Point out that early in development, the embryonic female has a common opening for both the urinary and reproductive tracts. However, during fetal development, a separate opening for the reproductive system (the vagina) is formed.

Language Connection

Cervix and Cervical Vertebrae
The word "cervix" means "neck" in Latin. Have students explain why this is an appropriate term for the ring of strong muscles located in the lower part of the uterus. Ask them where cervical vertebrae would be found.

The lower part of the uterus is a ring of strong muscles known as the **cervix**. As shown in **Figure 34.7**, the **vagina** (*vuh JEYE nuh*) is a muscular tube that receives sperm ejaculated by a male during intercourse. The lower end of the vagina opens to the outside of the body. During birth, the baby must pass through the cervix and the vagina.

Pregnancy can be prevented by not allowing sperm to enter the opening in the cervix. A diaphragm is a shallow, round cup made of rubber that is placed inside the vagina to block the entrance to the cervix. A cervical cap is similar to a diaphragm but fits tightly over the cervix. A diaphragm or a cervical cap must be obtained from a doctor or health-care professional. Spermicidal creams, spermicidal jellies, spermicide-filled sponges, vaginal suppositories, and spermicidal foams work by killing sperm in the vagina so that no living sperm can enter the opening in the cervix ("spermicidal" means "sperm-killing"). Usually spermicides are used with a diaphragm or a cervical cap. A diaphragm and spermicide used together are much more effective than either used by itself.

Figure 34.7
The female reproductive system releases eggs and provides nourishment to a developing fetus.

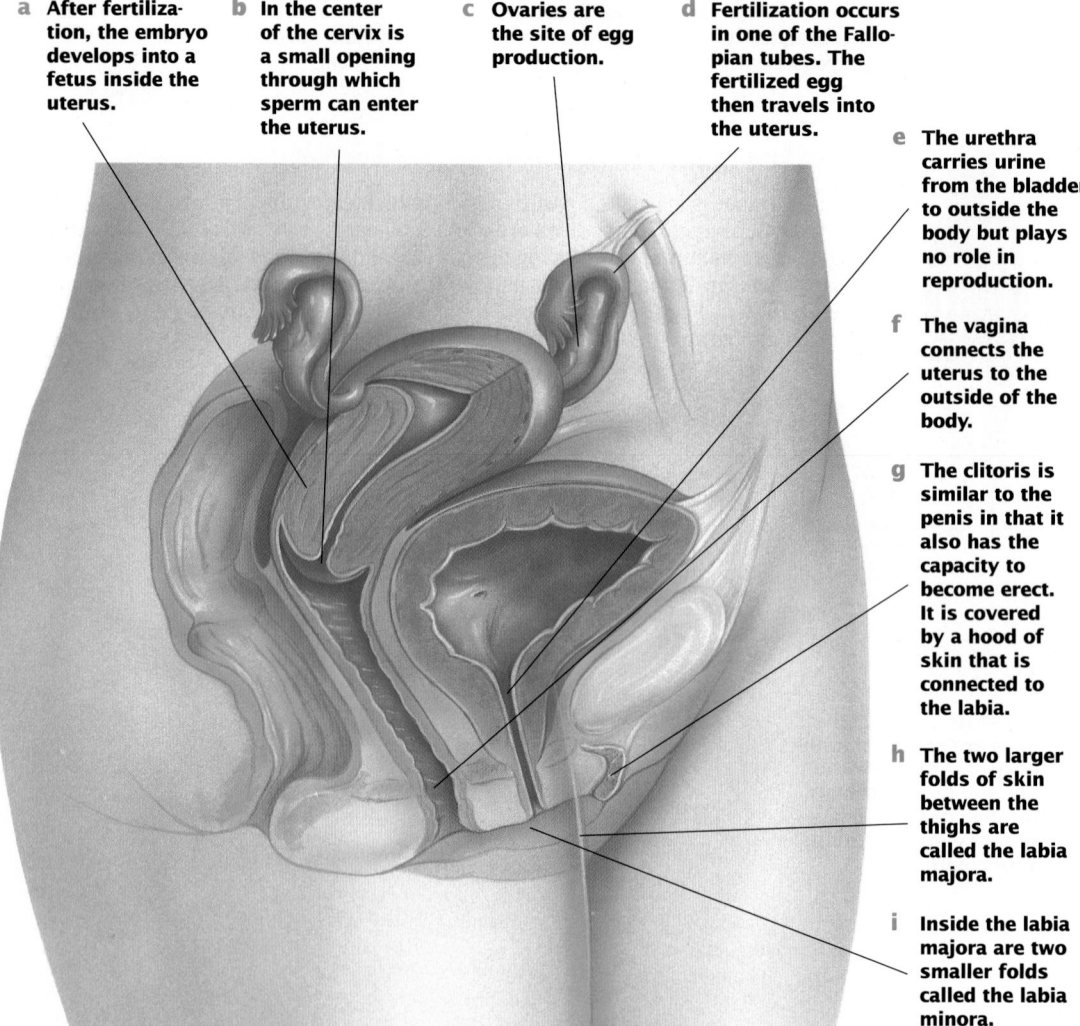

a After fertilization, the embryo develops into a fetus inside the uterus.

b In the center of the cervix is a small opening through which sperm can enter the uterus.

c Ovaries are the site of egg production.

d Fertilization occurs in one of the Fallopian tubes. The fertilized egg then travels into the uterus.

e The urethra carries urine from the bladder to outside the body but plays no role in reproduction.

f The vagina connects the uterus to the outside of the body.

g The clitoris is similar to the penis in that it also has the capacity to become erect. It is covered by a hood of skin that is connected to the labia.

h The two larger folds of skin between the thighs are called the labia majora.

i Inside the labia majora are two smaller folds called the labia minora.

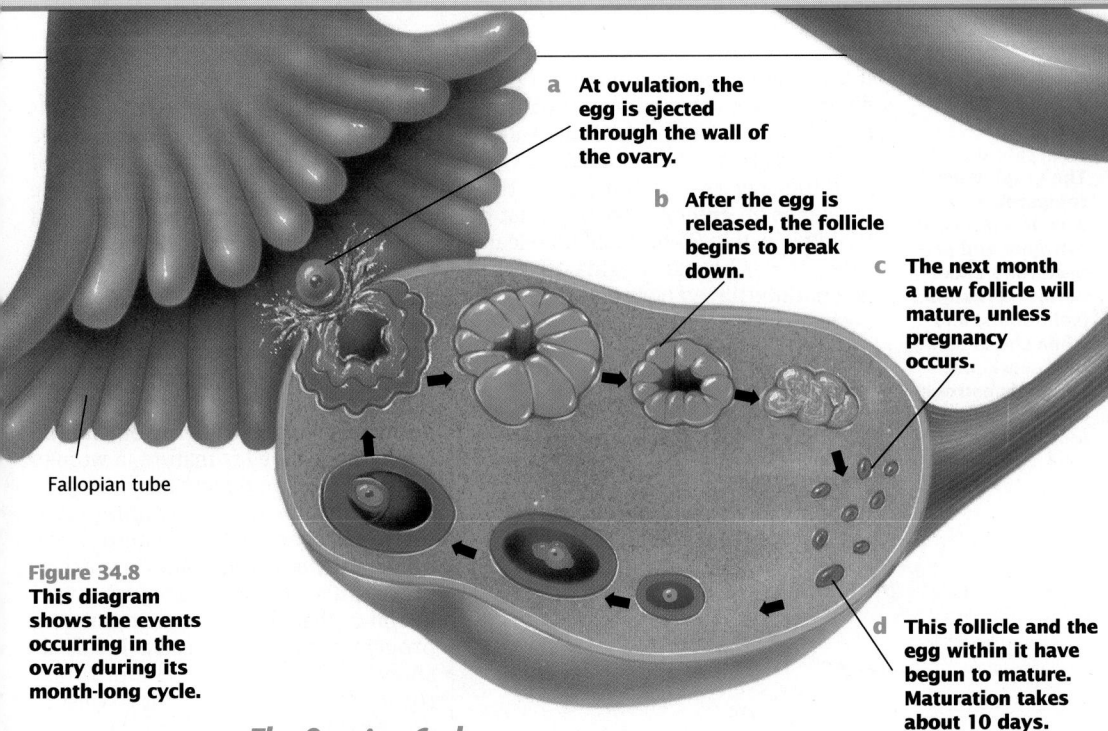

a At ovulation, the egg is ejected through the wall of the ovary.

b After the egg is released, the follicle begins to break down.

c The next month a new follicle will mature, unless pregnancy occurs.

d This follicle and the egg within it have begun to mature. Maturation takes about 10 days.

Fallopian tube

Figure 34.8
This diagram shows the events occurring in the ovary during its month-long cycle.

The Ovarian Cycle

Scale and Structure

How are mature sperm and eggs different in size and structure?

Once a male has reached puberty, his testes continuously produce and release millions of sperm. In a female, however, egg release occurs only at a specific time each month. The release of an egg from the ovary is called **ovulation** (*ahv yoo LAY shun*). A female usually releases only one egg each month. If this egg is not fertilized, the ovaries prepare another egg for release the next month. Thus, the activities of the ovaries occur in a cycle, which is called the ovarian cycle.

Eggs begin to mature in the ovaries of a female while she is still in her mother's uterus. Thousands of her eggs grow and begin to undergo meiosis. Growth and meiosis stop before the eggs are completely mature, however. All the eggs in the ovaries remain in this immature state until puberty.

After puberty, follicle-stimulating hormone (FSH) released by the pituitary signals several eggs to resume the process of maturation each month. As in males, FSH release is stimulated by gonadotropin-releasing hormone (GnRH) produced by the hypothalamus. The maturing eggs become large, highly complex cells, growing to be nearly 200,000 times larger than a sperm. Although many eggs begin to mature each month, usually only one egg completes the process. If two eggs mature, fraternal, or nonidentical, twins may result.

The egg matures within the follicle

Eggs grow inside a fluid-filled chamber called a **follicle** (*FAHL ih kuhl*), which is located in the ovary. The follicle supports growing eggs and produces the hormone estrogen. As you read in Chapter 29, estrogen is the female sex hormone. At puberty, estrogen causes females to develop breasts and wider hips than males. Cells of the follicle produce the greatest amount of estrogen when the egg nears maturity. This high level of estrogen triggers the pituitary to release a burst of luteinizing hormone (LH). LH triggers the egg to resume meiosis. Stimulated by high LH levels, the fluid-filled follicle ruptures, as shown in **Figure 34.8**. After ovulation, currents of fluid sweep the egg into one of the Fallopian tubes. The Fallopian tubes provide a way for an egg to travel from the ovary to the uterus.

Visual Strategy

Figure 34.9

Have students explain how this figure illustrates that negative feedback is controlling the levels of both estrogen and progesterone. Have them explain how this figure demonstrates that prescription pills containing synthetic estrogen and progesterone "override" this negative feedback system.

Figure 34.9

The graph below compares a woman's levels of estrogen and progesterone under normal conditions (solid lines) and while she is using pills containing synthetic estrogen and progesterone (dotted lines).

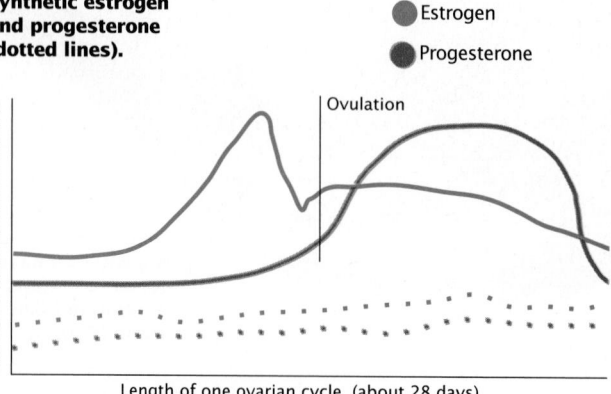

If a woman chooses to have no more children, she can undergo tubal ligation, an operation in which the Fallopian tubes are severed or tied.

If the egg is fertilized it is then known as an **embryo**. An egg must be fertilized within 24 hours of its release. After that, the egg begins to break down. Unfertilized eggs dissolve in the uterus.

Hormone levels change during the ovarian cycle

Levels of estrogen and progesterone are high after ovulation. High levels of estrogen and progesterone cause the pituitary gland to stop releasing LH and FSH. If no embryo reaches the uterus, estrogen and progesterone levels fall, and FSH and LH levels rise again. Another egg will mature and be released. During pregnancy, however, estrogen and progesterone levels remain high. High levels of these hormones suppress release of FSH and LH. Because no eggs mature, a woman does not ovulate during pregnancy.

Prescription pills containing synthetic estrogen and progesterone can prevent the egg from maturing. Use of these pills prevents pregnancy by keeping the levels of estrogen and progesterone at a steady level, as shown in **Figure 34.9**. The pituitary responds as if a pregnancy were in progress. It does not release FSH or LH. Since ovulation does not occur, pregnancy cannot take place.

The Menstrual Cycle

Like the ovaries, the uterus follows a monthly cycle. Each month, the uterus prepares to receive and nourish an embryo. The **menstrual** (*MEHN struhl*) **cycle** is the series of changes that occur in the uterus each month.

During the first part of the menstrual cycle, as the egg matures, the lining of the uterus grows thicker. Many tiny blood vessels grow into this thickened lining. The lining will provide a safe environment for an embryo. After ovulation, the follicle begins to produce progesterone. This hormone causes blood vessels in the lining of the uterus to enlarge even more. The lining is prepared to receive the embryo four or five days after the egg is released from the ovary. It takes an embryo about this long to travel to the uterus. An embryo that settles into the rich lining of the uterus releases hormones. One of these hormones causes the uterus to maintain its thickened lining. Most of the

time, no embryo arrives. The follicle begins to break down and produces less and less estrogen and progesterone. As the levels of estrogen and progesterone fall, the blood vessels in the uterine lining begin to close and then to break. The cells of the uterine lining do not receive adequate blood supply and come loose from the inside of the uterus. Blood from the broken blood vessels helps wash these cells out of the uterus. This mixture of blood and the cells that made up the lining of the uterus is called menstrual fluid. The passage of this fluid through the vagina and out of the body is called a menstrual period. It usually lasts from three to seven days. The steps of the menstrual cycle are shown in **Figure 34.10**.

The average menstrual cycle is 28 days long. Almost all women start their menstrual period 14 days after ovulation occurs. The length of the first stage of the cycle, the period when the follicle is growing, differs from woman to woman. For example, women with regular 28-day cycles probably ovulate 14 days after the first day of their last menstrual period. Women with regular 33-day cycles probably ovulate 19 days after the first day of their last period. Women with irregular menstrual cycles have a difficult time determining when ovulation occurs.

Some couples try to prevent pregnancy by abstaining from sexual intercourse around the time of ovulation. This approach, usually called the rhythm method, is not an effective means of preventing pregnancy. Since very few women know exactly when

their next period will begin, it is difficult to know when ovulation will occur. Also, sperm can survive in the female reproductive tract for two days or longer, so there are usually three or four days each month when fertilization is possible.

The menstrual and ovarian cycles stop in middle age

Between the ages of 45 and 55, women usually stop having menstrual periods and their ovaries stop releasing eggs. The shutdown of menstrual and ovarian cycles is known as **menopause**. Once a woman has undergone menopause, she is no longer able to have children.

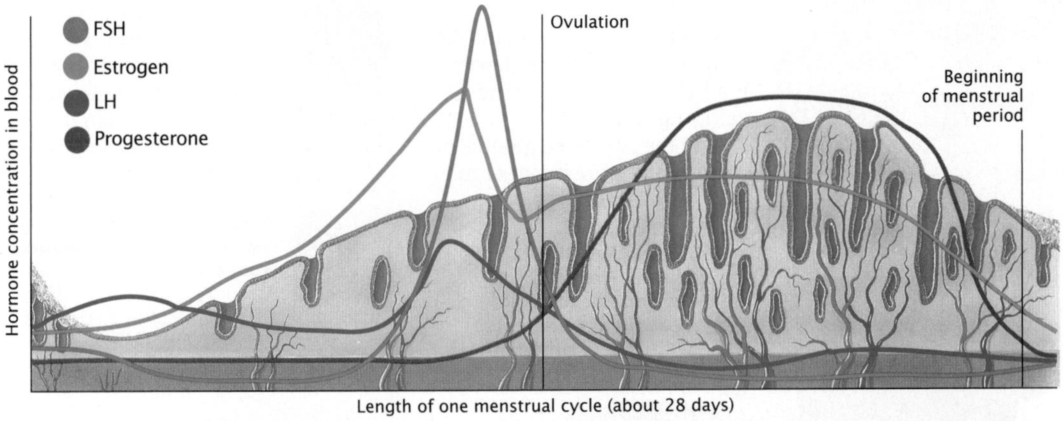

Figure 34.10
This graph shows the changes in uterine lining during the menstrual cycle.

- **FSH**
- **Estrogen**
- **LH**
- **Progesterone**

Ovulation

Beginning of menstrual period

Hormone concentration in blood

Length of one menstrual cycle (about 28 days)

a During the first part of the menstrual cycle, blood vessels grow in the uterine lining as it thickens.

b The lining is ready to receive an embryo four or five days after the egg is released from the ovary.

c If no embryo arrives, the blood vessels in the uterine lining begin to break. Blood and cells from the uterine lining pass through the vagina during the menstrual period. After the menstrual period, menstrual and ovarian cycles begin anew.

Section Review

❶ Trace the path of an egg through the female reproductive system.

❷ Identify two differences between a sperm and an egg.

❸ What role does LH play in the uterine cycle?

❹ Explain the role of progesterone in the menstrual cycle.

▪ *Section Review Answers* ▪

1. See Figure 34.6 on page 787.
2. Answers may vary. The most obvious answers might indicate that sperm have long tails for movement and an oval-shaped head, whereas eggs are much larger and lack a tail.

3. LH causes the follicle to rupture and release the mature egg from the ovary surface—ovulation.
4. Progesterone causes blood vessels in the uterus to enlarge, preparing the uterus to receive the embryo after the egg is released from the ovary.

Using the Feature

Alcohol is a teratogen, an agent that causes abnormal development. Radiation, the rubella (German measles) virus, herpes simplex viruses, mercury, thalidomide (a banned sedative), and the antibiotic tetracycline are other known teratogens. Fetal alcohol syndrome was first described in 1973 by researchers at the University of Washington.

- Have students research the teratogen thalidomide and its effects. Why did so few birth defects caused by thalidomide occur in the United States?
  Thalidomide was never approved by the Food and Drug Administration. Nevertheless, a few women obtained the drug outside the country, and some of their offspring were affected.

- DES (diethylstilbestrol) is a synthetic hormone that was prescribed to millions of women to prevent miscarriages. The daughters of women who received DES have a higher-than-average incidence of certain kinds of cancers. Have students research the uses and effects of DES.

Discussion

1. By the end of the first trimester of pregnancy, all major organ systems have been established. Why is it still unwise to drink alcohol after this time?
   Alcohol can cause brain damage throughout pregnancy, since the brain develops over the entire nine months.

2. What are three symptoms of fetal alcohol syndrome?
   Possible answers include mental retardation, hyperactivity, stunted growth, and facial deformities.

No Alcohol During Pregnancy

Alcohol and Fetal Development

New signs may already be up or may be going up in restaurants and bars in your state. The signs are similar to the caution labels found on containers for alcoholic beverages. These labels warn women about the dangers of drinking alcohol during pregnancy. Alcohol use by pregnant women is one of the leading causes of birth defects.

When a woman drinks alcohol during pregnancy, the alcohol in her blood travels through the placenta and into the blood of the fetus. What may not seem like much alcohol to the mother can be a harmful dose for the developing fetus.

In the first three months of pregnancy, the major organs and the circulatory system of a fetus are forming. The last six months of a fetus's development centers on growth and maturation. Ingesting alcohol during the first three months of pregnancy can result in severe physical damage—organs can be malformed, the spine can be misshapen, and brain development can be impaired. Drinking later in pregnancy often results in inhibited physical growth, hyperactivity, and behavioral problems after the child is born. Because the brain develops throughout the nine months of pregnancy, brain damage due to alcohol can occur at any time.

The mother of this healthy fetus has decided not to drink alcoholic beverages while she is pregnant.

Warning labels now appear on all alcoholic beverage containers.

GOVERNMENT WARNING: ACCORDING TO THE SURGEON GENERAL WOMEN SHOULD NOT DRINK ALCOHOLIC BEVERAGES DURING PREGNANCY BECAUSE OF THE RISK OF BIRTH DEFECTS.

Substance:	Alcohol
Related Illnesses:	Fetal Alcohol Syndrome Fetal Alcohol Effect
Symptoms:	Brain damage, facial deformities, behavioral problems, hyperactivity
Causes:	Drinking during pregnancy

Fetal Alcohol Syndrome

A baby born with severe impairment due to his or her mother's drinking has a condition called fetal alcohol syndrome (FAS). FAS babies often have deformed faces, with improperly formed eyes and lips. They often are mentally retarded and tend to have retarded growth. Most cannot advance beyond the second- to fourth-grade level. These children often have trouble with simple functions such as telling time or counting. The average IQ of children with FAS is 30–40 points below normal.

The damage to these babies is non-treatable and permanent. When FAS children become adults, they cannot take care of themselves and therefore always require care from others.

Some babies may be born with less severe, and less obvious, forms of impairment due to their mother's drinking. These babies have a condition called fetal alcohol effect (FAE). FAE babies usually have fewer physical problems but often have trouble learning and have behavioral problems including a short attention span, consistently poor judgment, and social withdrawal. These behavioral problems, though most obvious during childhood, do seem to continue into adulthood.

This child was born with Fetal Alcohol Syndrome (FAS). The symptoms of FAS include a small head; low, prominent ears; poorly developed cheek bones; and a long, smooth upper lip. In the United States, about 1 in 750 newborns are affected with FAS.

How Much Is Too Much?

The severity of FAS or FAE is directly related to the amount of alcohol the mother consumed during pregnancy—the more alcohol, the worse the impairment. Also, binge drinking, where a large amount of alcohol is consumed at one time, is especially harmful. No one knows if there is any amount of alcohol a pregnant woman can safely drink. For this reason, a "better-safe-than-sorry" approach is recommended. The Surgeon General of the United States, the American Medical Association, and the March of Dimes Birth Defects Foundation all recommend that women consume no alcohol during pregnancy.

Warnings have appeared on every bottle of beer, wine, and liquor sold in the United States since 1989. Despite the warnings, 3,800 to 7,600 FAS babies and an estimated 15,000 to 30,000 FAE babies are born in the United States each year. Getting the word out to women is an important goal of the signs in restaurants and bars.

There is hope for women who drink alcohol. If a woman drinks prior to but not during pregnancy, there is no risk of an FAS or FAE baby. Unfortunately, many women do not realize they are pregnant during the early weeks of pregnancy. If a woman stops drinking as soon as she knows she is pregnant, the risk of FAS and FAE greatly decreases.

The message is clear: Don't drink alcohol while you are pregnant. Give your baby a fair start in life.

How Maternal Consumption of Alcohol May Affect the Fetus and Newborn

Interferes with delivery of maternal nutrients to the fetus

Reduces the supply of oxygen to the fetus

Inhibits protein synthesis and metabolism in the fetus

Causes low birth weight

Leads to alcohol withdrawal in newborns—tremors, abnormal muscle tension, inconsolable crying, reflex abnormalities, restlessness

3. Although all containers of liquor now carry warnings to pregnant women, these cautions have only been mandatory since 1989. If a woman had a FAS baby before 1989, should she be able to collect damages from alcohol manufacturers? Justify your answer.
Answers should be logical. Issues to consider include the manufacturer's responsibility to warn of dangers from its products and the mother's responsibility to identify and refrain from behaviors widely known to be harmful to her fetus.

Lesson Plan 34.3

Phase 1

PREPARATION

Key Concepts

- Sperm deposited in the vagina pass the cervix and uterus and then travel up the Fallopian tube, where one will fertilize the egg released by the ovary.
- The fertilized egg travels down the Fallopian tube and implants itself in the uterus, where it will develop until birth.
- A placenta forms to allow for the exchange of nutrients and wastes between the fetus and mother.
- Viruses and drugs can pass from mother to fetus by way of the placenta.
- Development is divided into three 3-month periods known as trimesters. During the first, structures are formed. During the second, structures become more developed. During the third, additional growth takes place.
- About 270 days following fertilization, labor starts and birth follows.

Reading Strategy

Have students use mapping to organize the ideas and information they will get from reading this section. Each main idea should be written and circled in the center of the page. Each supporting detail should be placed on a line that is connected to the circle around the main idea.

Phase 2
TEACHING STRATEGIES

Demonstration 1

Flagella

Use a microprojector to show the class *Euglena* or other organisms propelled by a flagellum. Point out that the tail of a sperm functions in the same manner.

BEFORE FERTILIZATION, A SPERM AND AN EGG ARE SEPARATE CELLS, EACH CONTAINING ONE-HALF OF THE INFORMATION NEEDED FOR THE DEVELOPMENT OF AN EMBRYO. DURING FERTILIZATION, GAMETES FUSE INTO ONE CELL CONTAINING ALL THE INSTRUCTIONS NEEDED TO FORM A NEW INDIVIDUAL. AFTER FERTILIZATION, THE EMBRYO WILL GROW AND DEVELOP FOR ABOUT NINE MONTHS INSIDE ITS MOTHER'S UTERUS.

34.3   Fertilization and Development

Objectives

❶ **Describe the process of fertilization.**

❷ **Explain the process of implantation.**

❸ **List three substances that can pass through the placenta.**

❹ **Describe three changes that occur as the fertilized egg develops into a new individual.**

Fertilization: Fusion of Gametes

During ejaculation, 150 million to 350 million sperm are deposited just a few inches away from the Fallopian tubes, in which fertilization can occur. Now a long and difficult journey begins. To reach the egg, a sperm must first pass through the small opening in the cervix. Then it must cross the uterus and swim up one of the Fallopian tubes. The vast majority of ejaculated sperm never leave the vagina. Of the sperm that find the small opening to the cervix, many will be trapped in mucus that covers this opening. Over the next two days, sperm will gradually free themselves from this mucus and start their journey across the uterus. Even if a mature egg is waiting in one of the Fallopian tubes, half of the sperm will swim up the wrong tube.

Although only one sperm can fuse with the egg, many sperm can arrive at the egg at about the same time. The egg is surrounded by a protective layer of smaller cells. Enzymes in the heads of the sperm cells help loosen this layer of cells. Once a sperm reaches the surface of the egg, as shown in **Figure 34.11**, it leaves its tail behind. The genetic information inside the head of the sperm enters the egg and fuses with the genetic information inside the egg. A protective shield then forms around the egg to prevent other sperm from entering.

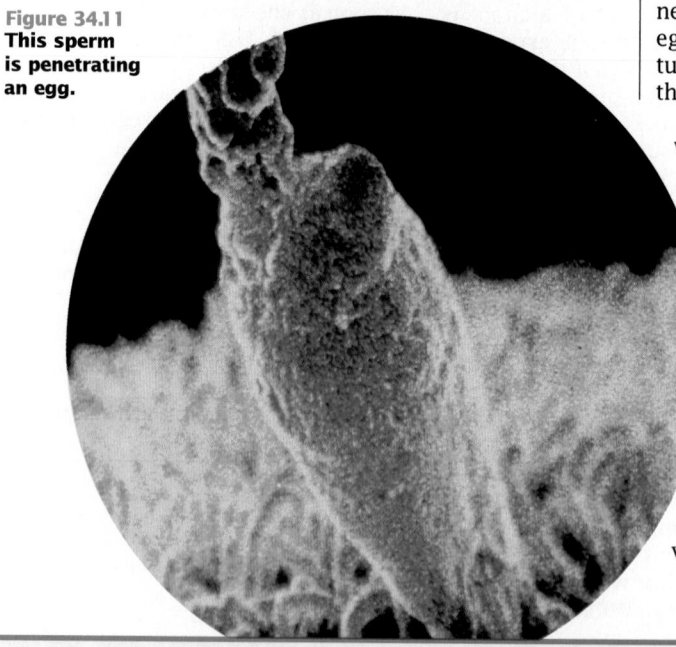

**Figure 34.11
This sperm is penetrating an egg.**

■ Cultural Perspective ■

**Joseph C. Hall, Ph.D.
(African-American)**
Dr. Hall is a molecular biochemist at Pennsylvania State University, where he studies a sperm cell membrane protein that is responsible for recognizing eggs.

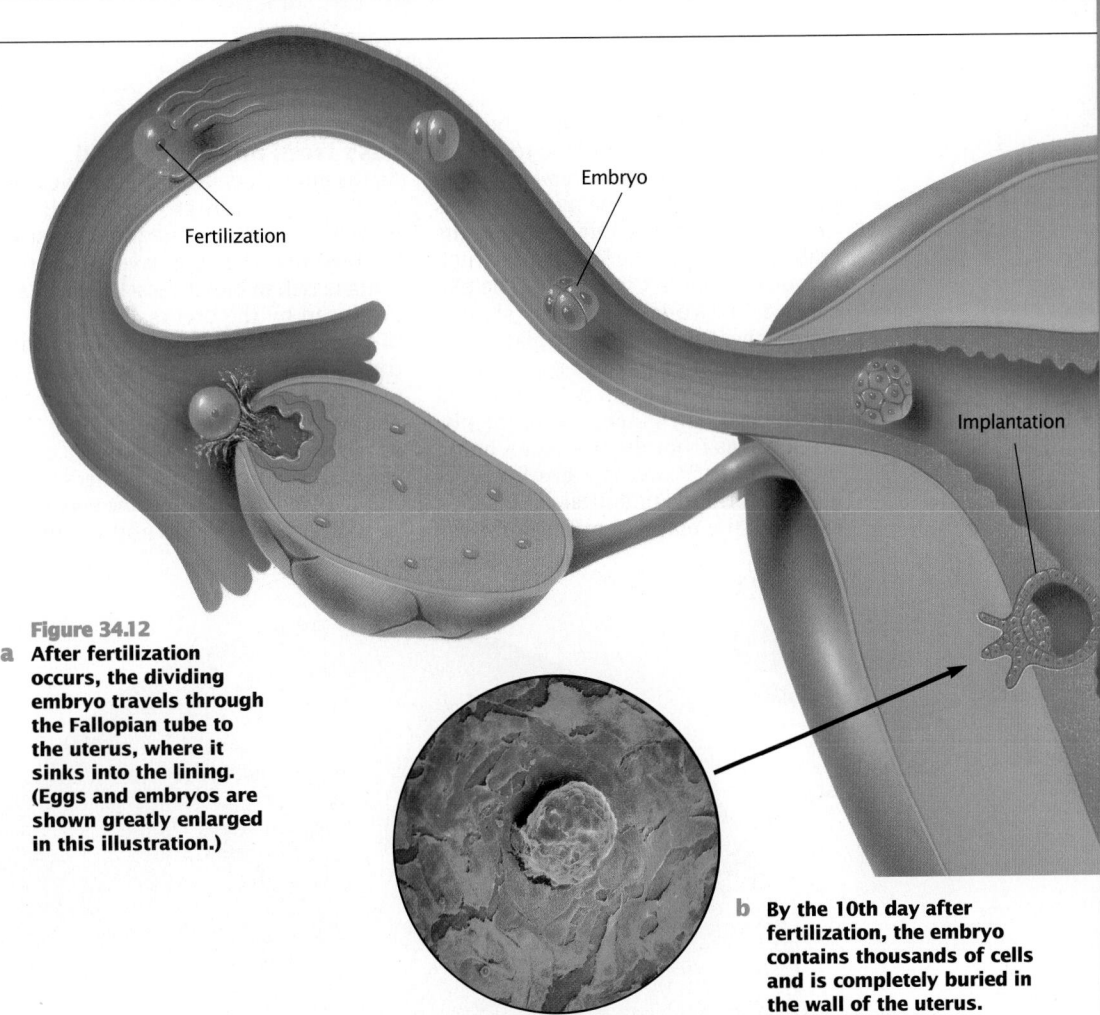

Fertilization

Embryo

Implantation

Figure 34.12

a After fertilization occurs, the dividing embryo travels through the Fallopian tube to the uterus, where it sinks into the lining. (Eggs and embryos are shown greatly enlarged in this illustration.)

b By the 10th day after fertilization, the embryo contains thousands of cells and is completely buried in the wall of the uterus.

The Embryo Enters the Lining of the Uterus

After the egg is fertilized, it continues to travel through the Fallopian tube toward the uterus, as shown in Figure 34.12. After about 30 hours the fertilized egg cell divides into two cells and is called an embryo. The cells of the embryo continue to divide by mitosis. (If the young embryo splits in two, identical twins will result.) From the ninth week of development until birth, it is called a **fetus**.

By the time the embryo reaches the uterus, it is made up of a hundred or more cells. The embryo embeds itself in the soft, thickened uterine lining. The embryo then releases hormones to ensure that the thick lining of the uterus remains healthy and is not washed away by a menstrual period. The embryo's entry into the uterine wall is known as **implantation**.

There are several ways for a female to determine whether she is pregnant. A missed menstrual period is one possible sign of pregnancy. A more accurate way of determining pregnancy is to test her blood or urine for the presence of the hormone produced by the embryo. By three weeks after fertilization, a pregnant woman will probably have enough of this hormone in her blood and urine to produce a positive test result. Doctors and health clinics are equipped to administer these tests.

▪ *Matter of Fact* ▪

In Vitro Fertilization
The male may have a low sperm count. On the other hand, the female may not have enough hormones to ovulate or may have a blocked Fallopian tube. In such cases, the events that ordinarily happen in the oviduct can be caused in a culture dish in a process known as *in vitro* fertilization. In the process, hormones are given to the woman to ensure precise timing of ovulation. Within hours of ovulation, one or more eggs are surgically removed from the ovary and placed in a culture dish. Sperm taken from the male are then added. If successful, one or more fertilized eggs are then implanted in the woman's uterus to complete development.

Figure 34.13
Have students locate the chorion, uterus, and umbilical cord in the inset figure that shows the embryo. Also, have students trace the fetal blood vessels shown in the magnified view so that they see that fetal and maternal bloods do not mix.

Connection: Chapter 4

Diffusion
Remind students that diffusion is the process by which molecules move from a region of high concentration to a region of lower concentration. Have students follow the passage of nutrients and wastes from the fetus to the mother in Figure 34.13.

Connection: Chapter 32

The Immune System
The embryo, although a "foreign" entity in the body, is not attacked and rejected by the immune system of the mother. Somehow, in ways that are poorly understood, the embryo protects itself from immunological attack.

The Placenta

As the embryo implants into the wall of the uterus, it is beginning to use up the raw materials that were stored in the egg. The body of the mother begins to transfer nutrients and oxygen to the embryo through the placenta, which you read about in Chapter 22. The placenta is a two-layered, disk-shaped membrane. One layer of the placenta is derived from the chorion (*KAWR ee ahn*), the outermost membrane around the embryo. The other layer of the placenta forms from the thickened lining of the mother's uterus. Cells from the embryo form the umbilical cord, which connects the placenta to the embryo.

Nutrients and oxygen pass from mother to child

As the embryo develops, blood vessels from the embryo grow through the umbilical cord into the inner layer of the placenta. The outer layer of the placenta is rich in blood vessels, which are produced by the uterus. The blood of the mother does not flow into the embryo, however. As shown in **Figure 34.13**, the mother's part of the placenta forms pools of blood that bathe the outside of the embryo's blood vessels. Even though the mother's blood and the embryo's blood come very close to each other, they do not mix in the

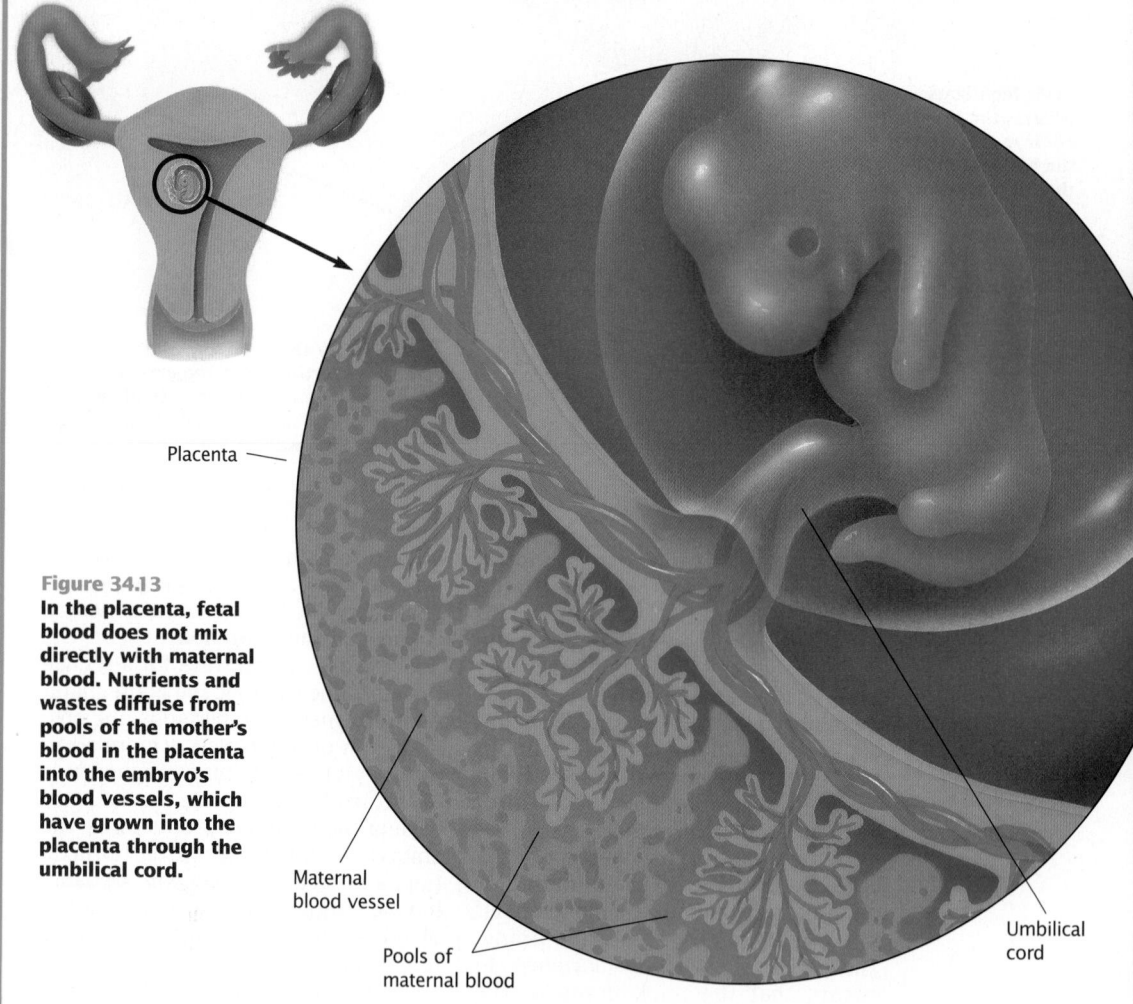

Placenta

Figure 34.13
In the placenta, fetal blood does not mix directly with maternal blood. Nutrients and wastes diffuse from pools of the mother's blood in the placenta into the embryo's blood vessels, which have grown into the placenta through the umbilical cord.

Maternal blood vessel

Pools of maternal blood

Umbilical cord

Drugs
Babies of mothers who use drugs during pregnancy are often born underweight. These babies can face a variety of medical complications. For this reason, no pregnant woman should take any drug without checking with her doctor first.

Nutrition and Exercise
Nutrition and exercise are important during pregnancy. A mother must provide all the nutrients a fetus needs to develop normally. An undernourished mother will give birth to a smaller, less healthy baby.

Smoking
Smoking cigarettes decreases the amount of oxygen available to the growing fetus. Babies born to mothers who smoke heavily tend to be much smaller than the babies of non-smokers. Babies of heavy smokers also face a much higher risk of complications after birth than the babies of nonsmokers.

Alcohol
Women who drink alcohol during pregnancy may have babies with severe facial deformities and mental retardation. Most doctors think that even a small amount of alcohol may damage a fetus, and recommend that women do not drink at all during pregnancy.

Demonstration 2
Harmful Effects on the Fetus
Invite the school nurse or health teacher to talk to the class about the effects of substance abuse during pregnancy. Have the speaker emphasize that in many cases, doctors do not know how much of a drug is necessary to harm a fetus. Consequently, a pregnant woman is best advised to avoid all unnecessary drugs, including alcohol and nicotine.

Figure 34.14
This baby was born healthy because his mother was careful during pregnancy. She ate healthy foods, exercised regularly, and avoided drugs and alcohol.

placenta. Oxygen and nutrients from the mother's blood diffuse through the walls of the embryo's blood vessels in the placenta. The oxygen and nutrients then travel through the blood vessels in the umbilical cord to the embryo.

Since the fetus is completely enclosed in the mother's body, it has no way to get rid of waste products on its own. Waste products such as carbon dioxide and urea diffuse through the placenta into the mother's bloodstream and are eliminated by the mother's body.

The behavior of a pregnant woman can affect her child

Although an embryo contains all the genetic information necessary for development, the health and activities of its mother influence whether it develops normally. As **Figure 34.14** indicates, the diet and behavior of a woman during pregnancy can affect the mental and physical development of the child, both before and after birth. For example, viruses can pass from the mother's blood, through the placenta, into the blood of the fetus. Viruses, such as those that cause rubella (German measles) and herpes, can cause a fetus to develop abnormally.

Drugs can also pass through the placenta. Some drugs can cause birth defects. A growing problem in the United States is the increasing number of babies born to mothers who use crack cocaine during pregnancy. These "crack babies" have a very high risk of mental retardation, learning disabilities, and personality disorders. Other illegal drugs also affect the fetus. Babies of mothers addicted to heroin may themselves be addicted when born.

Alcohol is a legal drug that is also very dangerous when consumed during pregnancy. Nicotine, which is found in cigarettes, is also harmful to babies before birth.

Growth and Development

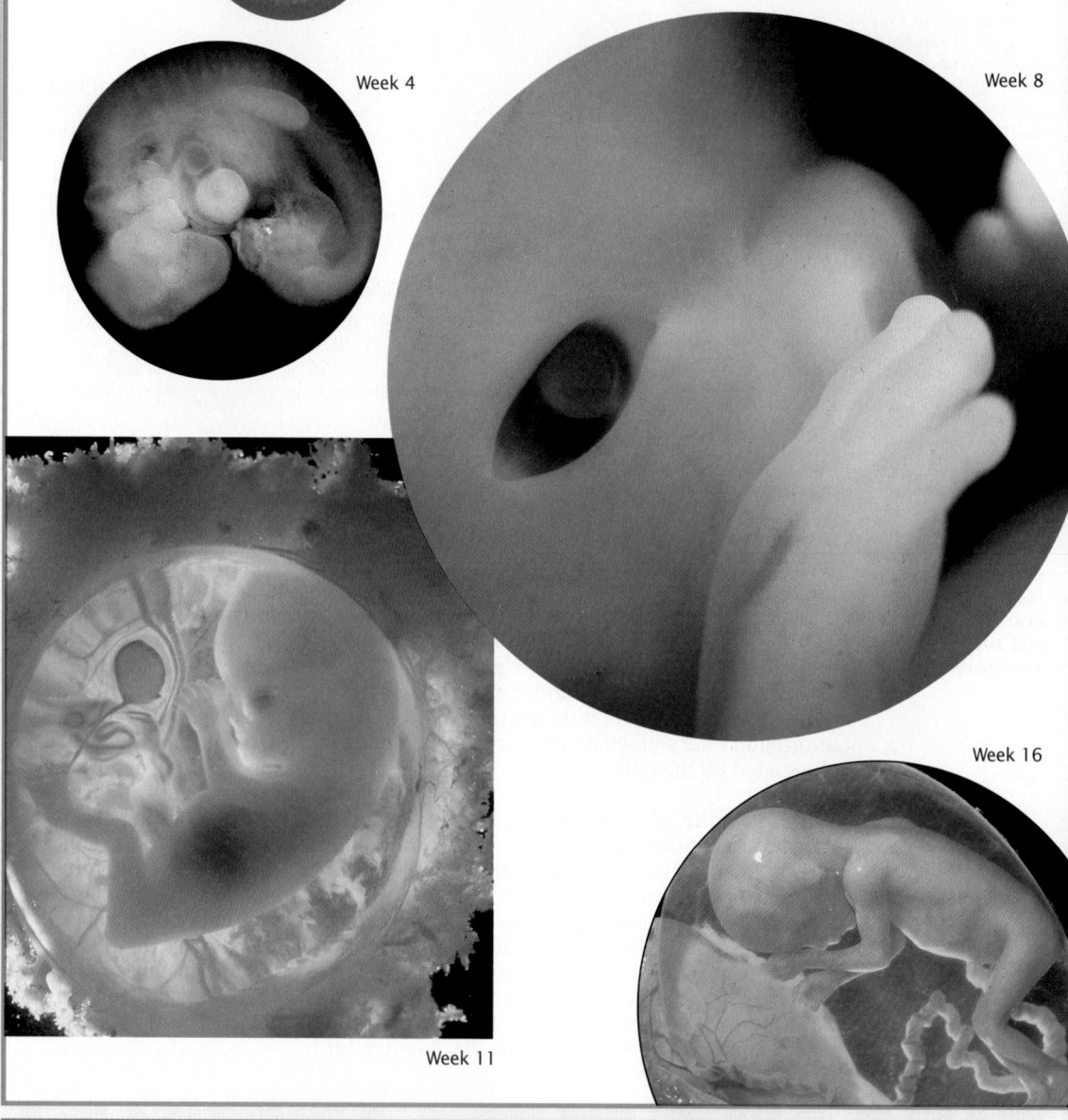

Figure 34.15
Profound changes take place during the nine months of development.

24 hours

Once the embryo has implanted in the uterus, it grows rapidly. A heart and brain begin to form, eyes appear, the face takes shape, small buds on the sides of the body become arms and legs, and internal organs such as the lungs, stomach, and liver develop.

Eight weeks after fertilization, the embryo is only 30 mm (a little over 1 in.) long. However, all the major organ systems of its body have begun to form and are growing, and its heart is beating. The nine months of pregnancy are often divided into three 3-month periods known as **trimesters**. Most of the fetus's development is complete by the end of the second trimester. Follow the changes that occur during development in **Table 34.1** and **Figure 34.15**.

Week 4

Week 8

Week 16

Week 11

■ *Cultural Perspective* ■

Jose E. Garcia-Arraras, Ph.D.
(Puerto Rican)
Dr. Garcia-Arraras studies the differentiation of vertebrate nerve cells in chick embryos at the University of Puerto Rico at Rio Piedras.

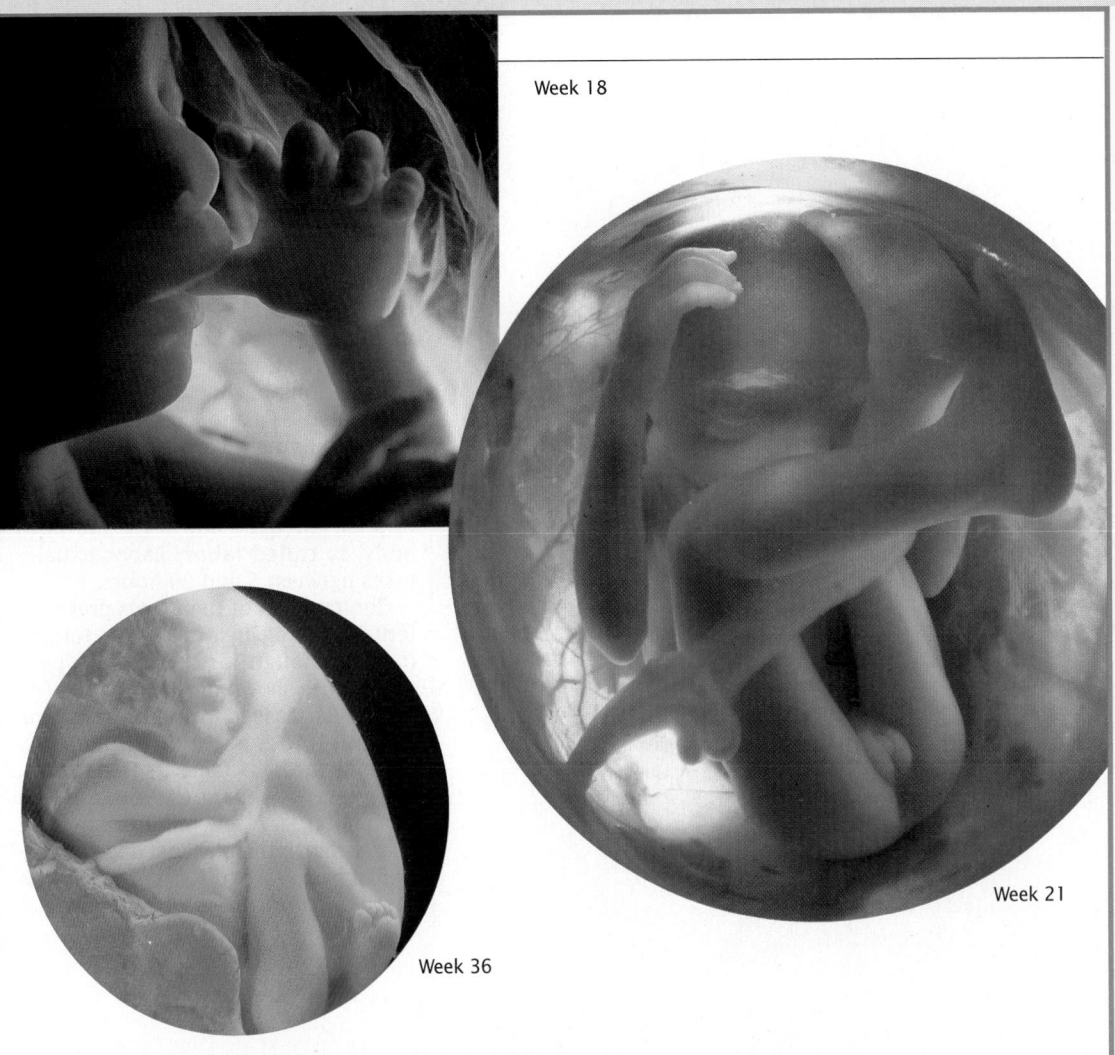

Week 18

Week 21

Week 36

Theme Connection

Evolution

During the first two to three weeks, the human embryo resembles the embryos of many other animals in their development. If possible, show the class photographs or overhead transparencies of fish, amphibian, reptilian, avian, and human embryos at comparable stages of development. Have students identify similarities among the different embryos. Point out that such similarities are evidence suggesting a common ancestor.

Table 34.1 Stages of Embryonic and Fetal Development

Stage	Major Changes
0–4 weeks	Fertilization occurs; embryo travels through Fallopian tubes and implants in uterine wall; brain, ears, and arms begin to form; heart forms and begins to beat
5–8 weeks	Nostrils, eyelids, nose, hands, fingers, legs, feet, toes, and bones begin to form; females develop ovaries, males develop testes; head is nearly as large as body; cardiovascular system is fully functional; about 22 mm (less than 1 in.) long
9–12 weeks	Embryo becomes a fetus; penis of males is distinct; growth of chin and other facial structures give fetus a human face and profile; head still dominant, but body is lengthening; about 36 mm (1.5 in.) long
13–16 weeks	Blinking of eyes and sucking of lips occurs; body begins to outgrow head; mother can feel muscular activity of fetus; about 140 mm (5.5 in.) long
17–20 weeks	Limbs achieve final proportions; eyelashes and eyebrows are present; about 190 mm (6.5 in.) long
21–30 weeks	Substantial increase in weight; may survive if born at this stage; skin is wrinkled and red; about 280 mm (13 in.)
30–40 weeks	Fingernails and toenails are present; about 360 mm (14.5 in.) long

Demonstration 3

The Navel

Have students put their fingers on their navels. Point out that the only evidence left of the umbilical cord following birth is the navel. Once removed from its mother, the baby's umbilical cord is clamped, tied off, and then cut. What remains on the baby slowly shrivels and falls off, leaving a scar known as the navel.

Phase 3

ASSESSMENT OPTIONS

Closure Strategy

Fetal Abnormalities

About 20 years ago, a number of pregnant women took a drug called thalidomide. Those who took this drug during the first trimester of pregnancy had babies with serious birth defects, including severely shortened arms and legs. Those who took this drug during the last trimester had normal babies. Have students explain the reason for this difference.

Section Review

Assign the *Section Review*.

Reteaching

Show the class a videotape that describes the sequence of events from fertilization to birth.

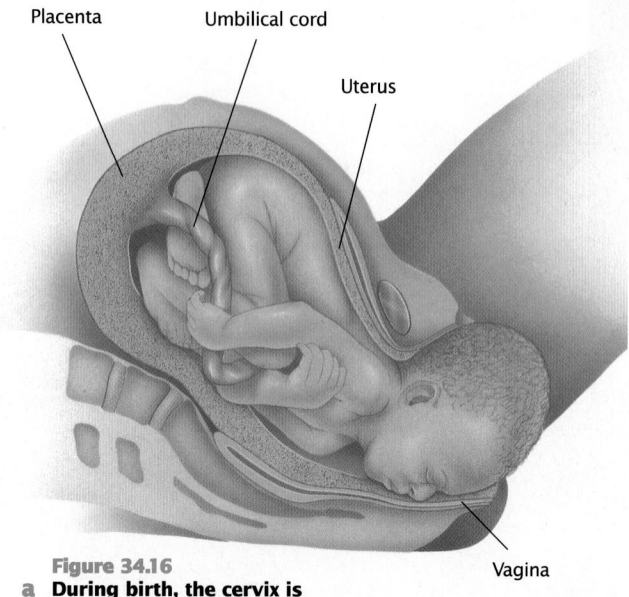

Figure 34.16

a **During birth, the cervix is forced open and the baby is pushed out through the vagina.**

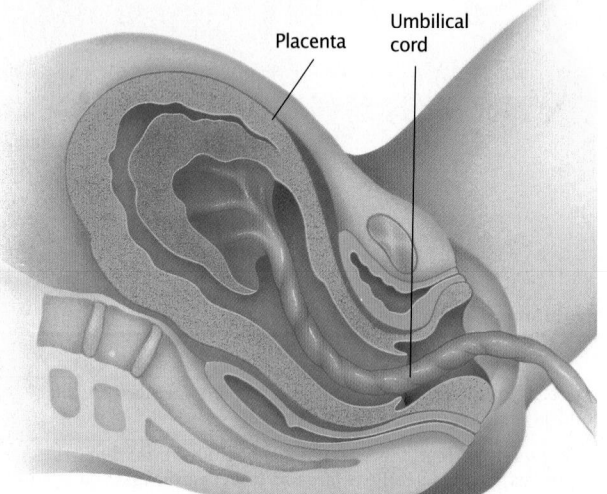

b **Shortly after a child is born, fluid, blood, and the placenta are expelled from the uterus. This material is the afterbirth.**

Birth

About nine months after fertilization, the fetus is ready to be born. By this time, it has usually moved so that its head is against the cervix. Near the end of pregnancy, hormones cause muscles in the walls of the uterus to begin contracting. Weak, irregular contractions may occur for several weeks before birth. As the time for birth approaches, contractions become much stronger. They also start to come at regular intervals and are closer together. The final stage of pregnancy, when the uterus is contracting strongly enough to push the baby out of the mother's body, is called **labor**. Labor usually takes between 5 and 20 hours.

Contractions of the uterus press the fetus's head against the cervix, forcing it open enough for the baby to pass through, as shown in **Figure 34.16**. The baby then passes through the vagina and out into the world. Within a few seconds the baby takes its first breath. The placenta separates from the wall of the uterus. A few minutes after the baby is born, the placenta and the umbilical cord are pushed out of the uterus. The process of childbirth is complete.

Section Review

1 **Explain why only one sperm fertilizes an egg.**

2 **Describe the events of implantation.**

3 **Why can it be harmful to a fetus for its mother to drink alcohol during pregnancy?**

4 **Describe two changes that occur in the fertilized egg before it implants in the uterus.**

■ *Section Review Answers* ■

1. After the head of one sperm enters and fuses with the egg, a protective shield forms around the egg to prevent other sperm from entering.

2. See Figure 34.12 on page 795.

3. Drugs such as alcohol can pass through the placenta and affect the mental and physical development of the child before and after birth.

4. First, a protective shield forms around the egg. Then, after 30 hours, the fertilized egg cell divides into two cells and is called an embryo.

DISEASE-CAUSING BACTERIA AND VIRUSES TRAVEL FROM ONE HOST TO ANOTHER IN CHARACTERISTIC WAYS. SOME ARE CARRIED IN WATER, AIR, OR BY INSECTS, WHILE OTHERS ARE TRANSMITTED BY SEXUAL CONTACT. DISEASES THAT CAN BE SPREAD FROM ONE PERSON TO ANOTHER BY SEXUAL CONTACT ARE CALLED SEXUALLY TRANSMITTED DISEASES, OR STDS.

34.4   *Sexually Transmitted Diseases*

Objectives

❶ Identify three ways HIV is transmitted.

❷ Identify three ways to avoid HIV infection.

❸ List two sexually transmitted diseases that are caused by bacteria.

❹ Describe the cause and results of pelvic inflammatory disease.

Sexually Transmitted Viral Diseases

As you learned in Chapter 16, viruses cause many human diseases, including rubella, measles, mumps, and influenza. In addition, viruses cause the sexually transmitted diseases AIDS and genital herpes.

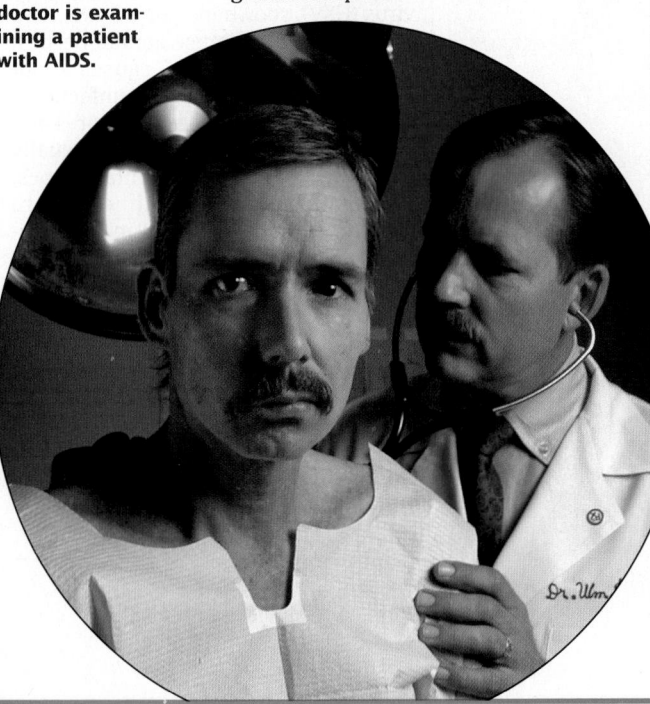

Figure 34.17
This California doctor is examining a patient with AIDS.

AIDS is a fatal viral disease

As you learned in Chapter 32, Acquired Immune Deficiency Syndrome (AIDS) is caused by HIV, Human Immuno-deficiency Virus. HIV destroys the immune system of infected individuals by destroying helper T cells. Helper T cells are essential for the immune system to defend the body against pathogens. AIDS patients, such as the man being examined by his doctor in **Figure 34.17**, die from infections and cancers that the immune system normally defeats.

Researchers now know that HIV is found at high levels within white blood cells called macrophages that are present in blood, semen, and vaginal fluids of infected individuals. HIV and HIV-containing cells can be transmitted from one person to another during sexual intercourse. Anal intercourse, oral sex, and vaginal intercourse are all capable of transmitting HIV, as all permit macrophages to pass from one person to another—*in either direction*. During vaginal intercourse, a woman is at a somewhat higher risk of getting the disease from an infected man than vice versa. However, female-to-male

Genital Herpes

Symptoms of genital herpes include painful, itching sores in or around the genitals. These sores usually appear 2 to 20 days after exposure to the virus. They may last as long as three weeks. Other symptoms include a burning sensation during urination and a fever. Since the virus can spread to other parts of the body, an infected person is advised not to rub the affected area and to keep it dry.

transmission also occurs easily. Basketball star Earvin "Magic" Johnson contracted HIV in this way. In Africa, where AIDS is most widespread, the numbers of HIV-infected males and females are equal.

Sexual intercourse is not the only way in which AIDS is spread. Since HIV is found in blood, it can be transmitted from one person to another by sharing hypodermic syringes and needles used for injecting drugs, and by needles used for tattooing. In a doctor's office or hospital, a new, disposable syringe is used for each injection. Users of illegal drugs sometimes reuse and share syringes and needles, both of which can be contaminated with HIV.

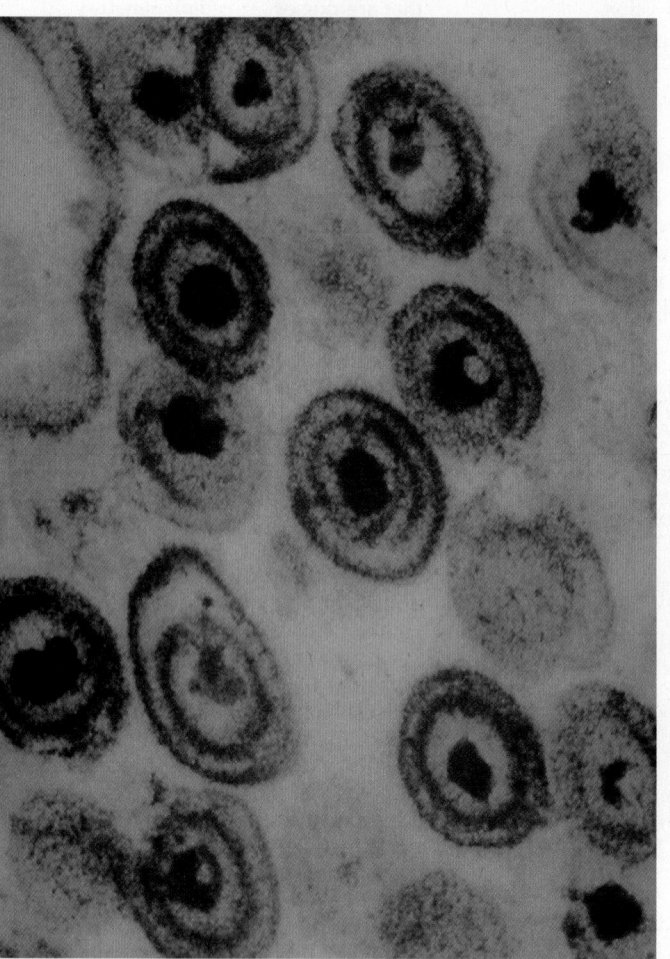

Like the viruses that cause rubella and herpes, HIV can be transmitted from mother to child across the placenta. About one-third of the babies born to HIV-infected mothers are infected with the virus.

Since it takes up to 10 years for HIV to damage the immune system enough to cause illness, a person can be infected with HIV for many years without being aware of it. During this time, the infected person can be spreading the disease through sexual contact or shared needles. Laboratory tests, however, can reveal whether an individual has been infected with HIV.

There is no cure for AIDS. HIV infection can be prevented by avoiding behaviors that allow HIV to be transmitted. Abstaining from sexual intercourse and not sharing needles are the only ways to completely avoid infection with HIV. Using a condom during sexual intercourse greatly reduces but does not eliminate the risk of being infected with HIV.

Genital herpes is also caused by a virus

Genital herpes is a sexually transmitted disease caused by herpes simplex virus (HSV), shown in **Figure 34.18**. Two types of HSV can cause genital herpes: HSV-1 and HSV-2. HSV-2 causes about 80 percent of genital herpes infections. HSV-1 commonly causes cold sores and fever blisters but can also be sexually transmitted and cause genital herpes. The symptoms of genital herpes are painful, recurring blisters in the genital area. Antiviral drugs can temporarily eliminate the blisters, but these drugs cannot eliminate the virus from the body.

HSV can be passed from mother to fetus during pregnancy or birth. A baby infected with HSV can suffer severe damage to its nervous system or even die as a result of the infection.

Figure 34.18
Herpes virus 2 causes the majority of genital herpes infections.

Sexually Transmitted Bacterial Diseases

Evolution

Strains of

penicillin-resistant

gonorrhea bacteria

are becoming more

common. How does

penicillin resistance

represent an

adaptation by

these bacteria?

Like viruses, bacteria cause many human diseases. Among these are sexually transmitted diseases such as syphilis, gonorrhea, and chlamydia. These sexually transmitted diseases, unlike those caused by viruses, can be treated with antibiotics.

Syphilis often goes unnoticed
Unlike HIV and herpes, syphilis (*SIHF uh lihs*) is caused by a bacterium. **Figure 34.19** shows the incidence of syphilis among 15- to 19-year-olds. Even though it can be cured with antibiotics, syphilis often is not diagnosed or treated because its early symptoms are mild or go unnoticed. About two to three weeks after infection, syphilis usually causes a small painless ulcer called a chancre (*SHAHNG kuhr*) on the genitals. In males, the chancre usually appears on the penis. Because it is painless, the chancre may not be noticed or may be ignored. In females, the chancre

may form inside the vagina or on the cervix, making it even less likely to be detected.

If syphilis is not treated, a few weeks later it may cause a rash, fever, or swollen lymph glands. These symptoms also disappear without treatment. Years later, however, the infection may cause serious damage to the nervous system, blood vessels, or skin.

Untreated syphilis can be transmitted from an infected mother to her fetus. Babies infected in the uterus may be stillborn or have serious complications involving damage to major organ systems.

Gonorrhea can damage the reproductive organs
Gonorrhea (*gahn uh REE uh*) is another sexually transmitted disease that is caused by a bacterium. **Figure 34.19** also shows the incidence of gonorrhea among 15- to 19-year-olds. Gonorrhea can be cured with antibiotics, although some strains of gonorrhea are resistant to the more commonly used antibiotics, such as penicillin. In males, gonorrhea usually causes painful urination and pus discharge from the penis. In females, it sometimes causes vaginal discharge but more often causes no symptoms.

In males, untreated gonorrhea can spread to the vas deferens, epididymis, or testes. In females, it can spread to the Fallopian tubes. Infection of the Fallopian tubes can cause scarring that results in infertility.

Cases of Syphilis in 15- to 19-year-olds in the United States

Number of reported cases

- Females
- Males

3,500 / 3,000 / 2,500 / 2,000 / 1,500 / 1,000 / 500

1981 1982 1983 1984 1985 1986 1987 1988 1989 1990 1991
Year

Figure 34.19
The incidence of both syphilis and gonorrhea is quite high among 15- to 19-year-olds. As you can see in the graphs, the number of reported cases of syphilis and gonorrhea is greater for females than for males.

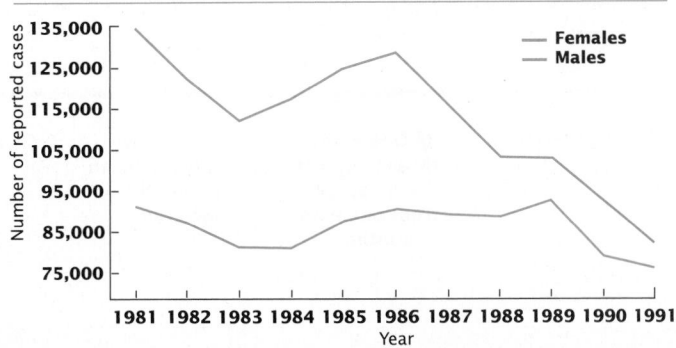

Cases of Gonorrhea in 15- to 19-year-olds in the United States

Number of reported cases

- Females
- Males

135,000 / 125,000 / 115,000 / 105,000 / 95,000 / 85,000 / 75,000

1981 1982 1983 1984 1985 1986 1987 1988 1989 1990 1991
Year

Theme Answer
Evolution
Because penicillin resistance is naturally selected, penicillin-resistant gonorrhea adapts to survive and reproduce. As a result of this selection process the antibiotic becomes less effective each time it is given. Over a period of time, greater numbers of resistant bacteria will become present in an infected individual.

Health Connection

Syphilis
The body cannot mount an immune response against the bacterium that causes syphilis. Consequently, a person can be reinfected at any time. In addition, repeated use of an antibiotic, usually penicillin, may result in decreased effectiveness, since bacteria resistant to the antibiotic will be selected to survive and reproduce. As a result of this selection process, the antibiotic becomes less effective each time it is given, since greater numbers of resistant bacteria are present within the infected individual. A different antibiotic may have to be administered.

Social Studies Connection

State Laws
Tell students that many states require a blood test for syphilis before they will issue a marriage license. Have students check for such a law in their state.

Social Studies Connection

Sixteenth-Century Italy
In sixteenth-century Italy, the spread of sexually transmitted diseases was of particular concern. The Italian anatomist Gabriel Fallopius (who first described the tubes that bear his name) developed a condom, not as a means of birth control but rather as an attempt to prevent the spread of syphilis. His device was made from linen treated with certain drugs.

Figure 34.20
Have students locate the bacteria shown in this figure. Point out that in the past, chlamydia bacteria were difficult to detect. Recently, a new and accurate laboratory test has been developed.

Phase 3

ASSESSMENT OPTIONS

Closure Strategy
Cure and Prevention
Have students name the sexually transmitted diseases that can be cured and those that cannot. Emphasize that all of them, however, are preventable. Yet they continue to spread in epidemic proportions. Conduct a class discussion about the reasons for this alarming rise. Have students list steps that can be taken to combat the spread of these diseases.

Section Review
Assign the *Section Review.*

Reteaching
Contact a local radio station that might be willing to sponsor a competition where students are asked to develop a one-minute public service announcement on sexually transmitted diseases. Arrange for the station to broadcast the winning announcement selected by a panel of judges that may include the school nurse, the health teacher, and a local physician.

Chlamydia is the most common sexually transmitted disease

Chlamydia (*kluh MIHD ee ah*) is the most common sexually transmitted disease in the United States. There are between 3 million and 10 million new cases each year. Like gonorrhea, chlamydia is caused by a bacterium and can be cured with antibiotics. The *Chlamydia* bacterium shown in **Figure 34.20** lives within an infected person's cells. The symptoms of chlamydia are similar to those of a mild case of gonorrhea: painful urination in males and vaginal discharge in females. Also like gonorrhea, chlamydia often produces no symptoms. In people with mild symptoms or no symptoms, chlamydia is often not diagnosed. Untreated chlamydia can spread to the Fallopian tubes and can cause infertility. A baby born to a mother with chlamydia can be infected as it passes through the birth canal and may suffer eye or lung damage.

Although gonorrhea and chlamydia often have no symptoms, routine laboratory tests available at hospitals and clinics can detect infection by either bacterium.

Untreated gonorrhea or chlamydia can cause pelvic inflammatory disease

Infection of the uterus, ovaries, pelvic cavity, or Fallopian tubes is called pelvic inflammatory disease, or PID. PID is usually caused by gonorrhea or chlamydia. When the Fallopian tubes are infected, scar tissue may form and close the tubes. If the tubes are closed, an egg can no longer be fertilized. PID is one of the most common causes of infertility in women.

Fallopian tubes that are damaged, but not completely closed, may allow sperm to reach the egg, but not allow the embryo to reach the uterus. This can cause an ectopic pregnancy. An **ectopic pregnancy** occurs when the embryo grows somewhere other than the uterus. An ectopic pregnancy in a Fallopian tube can be life-threatening to a woman. As the embryo grows, it can cause the Fallopian tube to rupture.

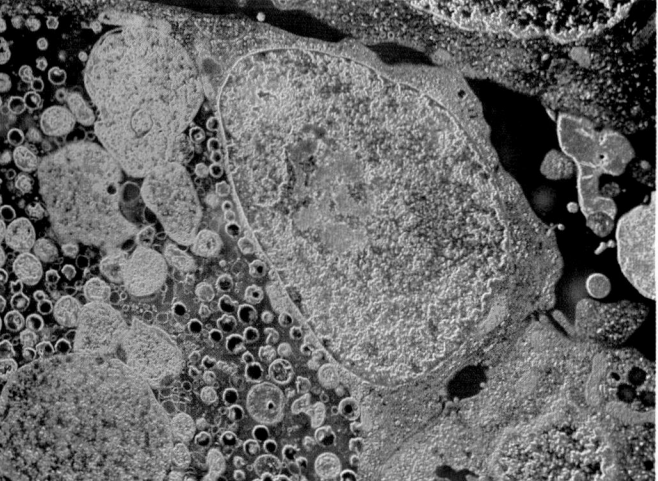

Figure 34.20
The bacterium that causes chlamydia, shown here in brown, grows within cells of its human host.

Section Review

① **Describe three ways HIV can be spread from one person to another.**

② **List two ways you can avoid HIV infection.**

③ **List two sexually transmitted diseases that are caused by bacteria.**

④ **What are the possible results of untreated gonorrhea or chlamydia?**

■ *Section Review Answers* ■

1. HIV can be transmitted by sexual intercourse, by sharing hypodermic needles, through body fluids, and from mother to child across the placenta.

2. abstain from sexual intercourse, do not share needles

3. gonorrhea and chlamydia

4. Untreated gonorrhea or chlamydia can cause pelvic inflammatory disease, which may cause infertility or ectopic pregnancies.

Chapter 34 *Highlights*

This young lady, not yet two years old, has much to learn before she can wear daddy's hat.

	Key Terms	Summary
34.1 The Male Reproductive System 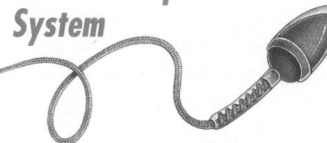 Millions of sperm are expelled by the male during ejaculation.	testis (p. 784) epididymis (p. 784) vas deferens (p. 784) ejaculation (p. 785) semen (p. 785)	• Sperm are produced in the testes. They are stored in the epididymis and vas deferens, and are expelled during ejaculation. • Luteinizing hormone stimulates the testes to produce testosterone. Follicle-stimulating hormone stimulates sperm production.
34.2 The Female Reproductive System 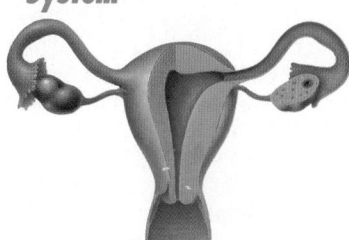 Usually, one egg is released from an ovary each month. If it is fertilized by a sperm, an embryo will develop.	ovum (p. 787) ovary (p. 787) Fallopian tubes (p. 787) uterus (p. 787) cervix (p. 788) vagina (p. 788) ovulation (p. 789) follicle (p. 789) embryo (p. 790) menstrual cycle (p. 790) menopause (p. 791)	• Eggs are produced in the ovaries. Ovulation occurs about 14 days after the egg begins to develop. • Follicle-stimulating hormone causes eggs to mature. Luteinizing hormone causes release of eggs. • Each month, the uterus prepares to receive an embryo. If no embryo arrives, the lining of the uterus is washed out of the body by blood. This bleeding is known as the menstrual period.
34.3 Fertilization and Development 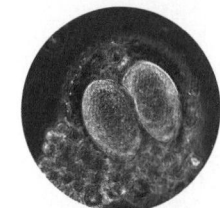 The fetus will grow and develop for about nine months, until it is ready to be born.	fetus (p. 795) implantation (p. 795) trimester (p. 798) labor (p. 800)	• A fertilized egg drifts down the Fallopian tube into the uterus. • After the embryo settles into the uterine lining, the placenta forms. Nutrients, oxygen, wastes, and carbon dioxide are transferred through the placenta. • After eight weeks, the embryo is known as a fetus. • Birth occurs after about nine months of pregnancy.
34.4 Sexually Transmitted Diseases 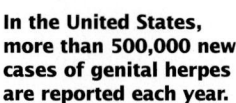 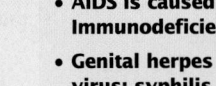 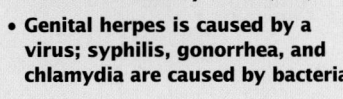 In the United States, more than 500,000 new cases of genital herpes are reported each year.	ectopic pregnancy (p. 804)	• AIDS is caused by the Human Immunodeficiency Virus (HIV). • Genital herpes is caused by a virus; syphilis, gonorrhea, and chlamydia are caused by bacteria.

Chapter Review Answers

Understanding Vocabulary

1. **a.** Epididymis does not fit the pattern because it is a structure of the male, not the female, reproductive system.
b. Genital herpes is caused by a virus; the others are caused by bacteria and can be cured with antibiotics.
c. Menstrual cycle does not fit the pattern because it is a series of changes, and not a single event.
d. FSH does not fit the pattern because FSH is responsible for the production of testosterone and estrogen.

Relating Concepts

2. Map answer is shown on page 781D.

Understanding Concepts

Multiple Choice

3.	a	8.	a
4.	b	9.	c
5.	a	10.	a
6.	a	11.	c
7.	c		

Completion

12. gonorrhea, chlamydia, ectopic
13. fetus
14. menstrual
15. nine, trimester

Short Answer

16. Drinking alcohol can cause fetal alcohol syndrome, the symptoms of which include heart defects, severe facial deformities, and depressed IQ.
17. Prior to ejaculation, the prostate gland discharges a few drops of colorless liquid. This liquid passes through the urethra and neutralizes the acid.
18. Possible answers might include: sexual intercourse, anal intercourse, oral sex, vaginal intercourse, the transmittal of body fluids, shared needles, or the transmittal from mother to child across the placenta.
19. The diaphragm prevents pregnancy by blocking the entrance to the cervix, so that sperm cannot enter the uterus.
20. It dissolves in the uterus.

Chapter 34 Review

Understanding Vocabulary

1. For each set of terms, choose the term that does not fit the pattern and explain why it does not fit.
   a. uterus, ovaries, epididymis, cervix
   b. genital herpes, syphilis, gonorrhea, chlamydia
   c. ovulation, fertilization, ejaculation, menstrual period
   d. estrogen, FSH, progesterone, testosterone

Relating Concepts

2. Copy the unfinished concept map below onto a sheet of paper. Then complete the concept map by writing the correct word or phrase in each blank oval.

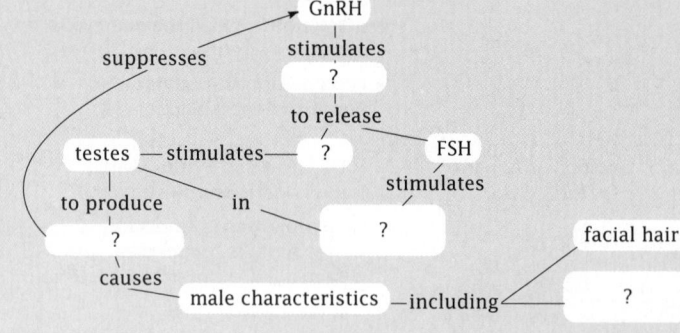

Understanding Concepts

Multiple Choice

3. Testosterone and sperm are produced by the
   a. testes.
   b. epididymis.
   c. vas deferens.
   d. semen.

4. Sperm are stored in the
   a. urethra.   c. kidneys.
   b. epididymis.   d. urinary bladder.

5. Adult male characteristics such as facial hair and a deep voice are caused by
   a. high levels of testosterone.
   b. the onset of sperm production.
   c. low levels of LH and FSH.
   d. ejaculation.

6. The muscular tube in females that serves as a birth canal is the
   a. vagina.   c. ovum.
   b. cervix.   d. uterus.

7. In females, FSH and LH are at their lowest levels during
   a. puberty.
   b. ovulation.
   c. menstruation.
   d. sexual intercourse.

8. An egg travels from the ovary to the uterus through the
   a. Fallopian tube.
   b. urethra.
   c. vas deferens.
   d. follicle.

9. Rupture of the fluid-filled follicle, which signals the beginning of ovulation, is stimulated by
   a. fertilization.
   b. the release of FSH.
   c. high LH levels.
   d. high levels of estrogen.

10. Which substance *cannot* be passed from mother to fetus through the placenta?
    a. blood   c. oxygen
    b. heroin   d. nutrients

11. Which sexually transmitted disease can be cured with antibiotics?
    a. genital herpes   c. gonorrhea
    b. HIV   d. measles

Completion

12. Pelvic inflammatory disease is caused by untreated _____ or _____ and may result in blocked Fallopian tubes or a(n) _____ pregnancy.

13. After eight weeks of development, an embryo is called a(n) _____ .

14. The sequence of events during which one or more eggs mature and are released is known as the _____ cycle.

15. Pregnancy in the human female lasts _____ months. This time is divided into three-month periods called _____ .

Interpreting Graphics

21. • c
    • b
    • c, blockage developed by scarring from sexually transmitted diseases

Reviewing Themes

22. Both follow a monthly cycle that ensures that a fertilized egg will implant in the uterus. At the same time that the ovum is maturing, the lining of the uterus is thickening. The two cycles are further synchronized by the release of the hormone progesterone from the follicle; progesterone causes blood vessels in the lining of the uterus to further enlarge.

Short Answer

16. How can alcohol consumption by a pregnant woman affect the fetus she carries?

17. Urine is acidic. After urine passes through the urethra, it too becomes acidic. How does the body prevent sperm cells from being killed as they pass through the urethra?

18. List behaviors that enable HIV to be transmitted.

19. Describe how the diaphragm prevents pregnancy.

20. What happens to an egg if it is not fertilized within 24 hours after its release?

Interpreting Graphics

21. Examine the diagram below.

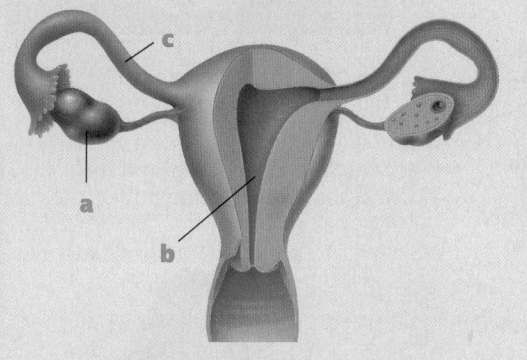

- In which structure does fertilization occur?
- Where does the embryo implant?
- In which structure could an ectopic pregnancy occur? What could cause an ectopic pregnancy?

Reviewing Themes

22. *Patterns of Change*
Describe how the ovarian and menstrual cycles are related.

23. *Interacting Systems*
Why does a woman who has undergone tubal ligation not get pregnant?

24. *Evolution*
What characteristics of sexual reproduction are responsible for the similarities between parents and their children?

Thinking Critically

25. *Applying Concepts*
Why is it important for men who have contracted gonorrhea to inform their female sex partners?

26. *Applying Concepts*
A 48-year-old woman stops having menstrual periods. She believes she is pregnant. What is another possible explanation?

27. *Building on What You Have Learned*
In Chapter 28 you learned about the hypothalamus. Describe how the hypothalamus regulates the levels of testosterone and sperm production in males.

Cross-Discipline Connection

28. *Biology and Health*
The Lamaze method of childbirth stresses muscle relaxation, controlled breathing, and the involvement of a birthing coach. Look for information about the advantages of Lamaze.

Discovering Through Reading

29. Read the article "The Mysterious Origin of AIDS," in *Natural History*, September 1992, pages 24–29. Summarize the three hypotheses to account for the appearance of the AIDS epidemic. Explain the weaknesses in each hypothesis.

26. Another explanation is possible; she could be experiencing menopause.

27. The hypothalamus regulates the levels of testosterone and sperm production in the male through its release of GnRH. GnRH controls the release of LH and ESH. LH stimulates cells of the testes to produce testosterone, and FSH causes sperm-producing cells of the testes to take in testosterone, which stimulates sperm production.

Cross-Discipline Connection

28. The method tends to make childbirth a less stressful ordeal.

Discovering Through Reading

29. All three theories agree that both forms of HIV (HIV-1 and HIV-2) evolved from SIV, and that humans have contracted SIV for many years. The question is how SIV transformed so recently into a human pathogen. One theory is that a new mutation increased the virulence of the virus. A weakness in this theory is that the mutation would have had to occur almost simultaneously in both forms of HIV. The second theory holds that HIV was present in isolated, sedentary populations in Africa. Deaths from AIDS went unnoticed against the high background of mortality for other diseases. AIDS only became widespread when changes in social and political conditions caused these local populations to move. The weakness in this theory is that social upheaval and migration have been occurring for centuries in Africa. The third theory is that HIV was transmitted to humans in medical preparations, such as vaccines, made using monkey tissues. The flaw in this theory is that use of these preparations either preceded the appearance of AIDS by several decades, or came too close to the first deaths to allow for the long latent period of HIV.

23. The pathway that the egg travels is blocked. Eggs cannot travel through the Fallopian tubes, and sperm cannot enter. Fertilization is blocked.

24. In sexual reproduction, the genetic material from each parent contributes to the child's characteristics.

Thinking Critically

25. Women do not experience symptoms of gonorrhea as early as men, and some never show symptoms. However, infection can spread to the uterus and into the Fallopian tubes.

Procedural Note

1. Review the Prelab Preparation questions and clarify the nature of sperm and egg cells.
2. Review the proper use of the microscope if the students have not used the microscope recently.

Prelab Preparation Answers

1. • Sperm cells are produced by the male in the testes.
   • Egg cells are produced by the female in the ovary.
   • A sperm and egg unite during fertilization to form an embryo.
   • meiosis
2. A protective shield forms around the egg to prevent other sperm from entering.

Procedure Answers

3. Answers will vary but should describe the egg as being a relatively large, spherical cell with a centrally located nucleus. The egg contains a very large amount of cytoplasm compared to the nucleus. There appears to be 3 to 5 times more cytoplasmic volume than nuclear volume.
6. There is very little cytoplasm, compared with the nucleus, in the sperm cell. The cytoplasmic volume appears to be only one-fifth to one-fourth that of the nuclear volume.

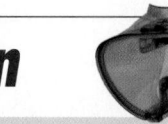

Chapter 34 **Investigation**

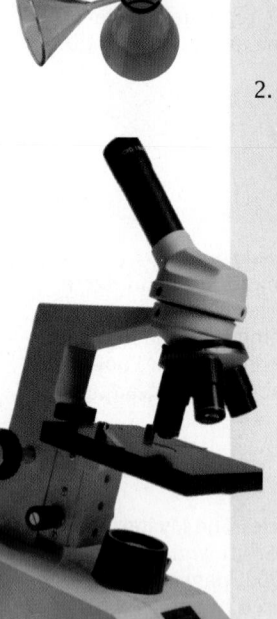

How Are Sperm and Eggs Different?

Objectives

In this investigation you will:
- *observe* and *compare* sperm and eggs
- *observe* early stages of development

Materials

- prepared slides of sea-star eggs and sperm
- compound light microscope
- prepared slide of early developmental stages of a sea star

Prelab Preparation

1. Review what you have learned about eggs and sperm by answering the following questions:
   • Where are sperm produced?
   • Where are eggs produced?
   • What purpose do sperm and eggs serve?
   • What process forms both sperm and eggs?
2. Review what you have learned about early development by answering the following question:
   • What are the first changes to occur in an egg after fertilization?

Procedure

1. Form a cooperative group with another student to complete steps 2–7.
2. Place the slide of the sea-star egg on the microscope stage. Observe the egg cell using low power. Make a labeled drawing of the egg cell.

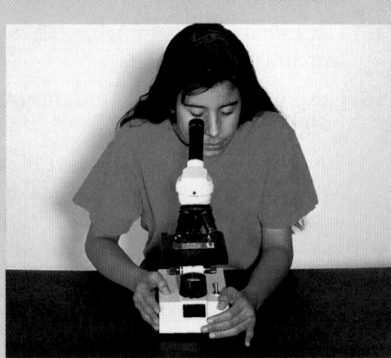

3. Describe the appearance of the egg. *Which region is larger, the nucleus or the cytoplasm?*
4. Switch to high power and make a second labeled drawing of the egg cell.

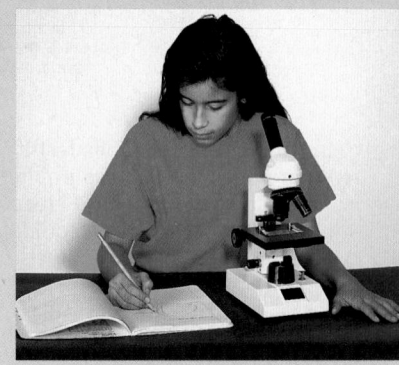

5. Place the slide of sea-star sperm on the microscope stage. Observe the sperm using low power. Make a labeled drawing of the sperm.

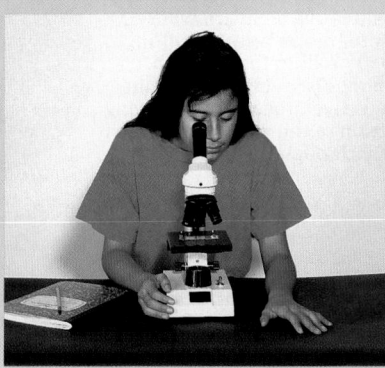

6. Switch to high power and make a second labeled drawing of an individual sperm. *Which region of the sperm is larger, the nucleus or the cytoplasm?*

7. Now examine the slide of early developmental stages of the starfish. Make a drawing of each stage in development. Note the number of cells and the size of the cells at each developmental stage.

Analysis

1. *Analyzing Observations* How do egg and sperm cells compare in size?

2. *Analyzing Observations* How does the amount of cytoplasm in the sperm compare with the amount in the egg?

3. *Making Inferences* Suggest an explanation for the difference in size between sperm and egg cells.

4. *Analyzing Observations* Why are sperm cells capable of locomotion while egg cells are not?

5. *Relating Structure to Function* How does the shape of a sperm cell reflect its function?

6. *Analyzing Observations* Of the developmental stages of the sea star you observed, which stage was the latest in development? Explain.

7. *Comparing Structures* Which developmental stage of the sea star was the earliest in development? Where would you find a human embryo at the same developmental stage?

Thinking Critically

A sperm and an egg each contribute one-half of an embryo's genes. Yet there are some genes that can only be inherited from the mother. Where in the genome are these genes located?

809

Using the Feature

- Discuss with students ways that computers have affected our daily lives. Focus on how computers have made many tasks easier and faster to complete. Have students brainstorm about how advances in computer technology have led to the uses of computers in biomedical engineering. Encourage students to make predictions about how computers will be used in biomedical engineering in the future.

- Explain that the human body is often compared with a complex machine, but that it differs in several respects. Machines can be shut off, new parts can be ordered, and repairs made when they break down, whereas human bodies cannot be shut off or new parts ordered. However, with new technology, ways to treat human disorders and replace some body parts are increasing. Have students use their imagination as researchers given the task of developing an artifical body part. NOTE: You could have students work in cooperative groups. Encourage students to identify and describe all the necessary components their artificial body part would require to function. Have students make and label a drawing of their design and explain how it functions to the class.

- Explain that animals are frequently used instead of humans in medical experiments. Have students write a report on the pros and cons of this practice. Also have them express their feelings on the subject in a class discussion.

- To make a connection with Chapter 8, explain that the promises of genetic engineering were first revealed in the 1970s and that public concern grew because of possible dangers posed by new life forms being released into the environment. Yet genetic engineering offers hope for a better future. Drugs such as human insulin and interferon, which were previously difficult to produce, are being produced by genetically engineered bacteria. Have students discuss the question of whether genetic engineering would expand resources or seriously endanger humans.

UNCOVERING THE SECRETS OF THE BODY

1625

1628 British physician **William Harvey** shows that blood circulates around the body through blood vessels. Harvey bases his conclusions on his own observations and experiments, not the untested ideas of authority. The experimental method used by Harvey provides a model that continues to be used by scientists.

1822 and years following American physician **William Beaumont** begins treating a patient whose injury leaves an opening through the body wall to the stomach. Beaumont discovers much about the digestive process by observing the action of the patient's stomach directly through the wound.

William Beaumont

1796 After developing a vaccine for smallpox, English physician **Edward Jenner** gives the first vaccination.

Edward Jenner

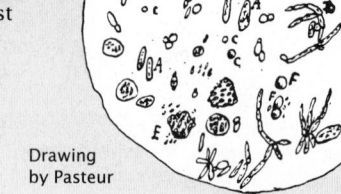

Drawing by Pasteur

1850 and years following **Louis Pasteur** performs experiments showing that bacteria cause fermentation and decay. Pasteur's work leads to the Germ Theory of Disease, which states that bacteria, not "bad humors," cause diseases.

1900

Emil Behring

1901 German scientist **Emil Behring** shows that immunity to a disease can be provided by inoculating animals with the blood serum of other animals that have recovered from the disease.

1921 Canadians **Frederick Banting**, **Charles Best**, and **J.J.R. MacCleod** discover the hormone insulin. Their work links inadequate insulin production with the disease diabetes and shows that insulin injections can control this disease.

1928 British bacteriologist **Alexander Fleming** notices that the mold *Penicillium notatum* prevents the growth of other microorganisms. Fleming's observation leads to the development of the first antibiotic, penicillin, which proves remarkably successful against bacterial infections.

Penicillin mold

Jonas Salk giving polio vaccine

1954 American physician **Jonas Salk** develops an effective vaccine against the virus that causes polio, a crippling disease of children and young adults. The Salk vaccine reduces the number of polio cases in the United States by 90 percent between 1954 and 1962.

ience ▪ *Discoveries in Science* ▪ *Discoveries in Science*

1858 German physician **Rudolf Virchow's** work shows that different kinds of cells in the body perform different tasks that contribute to the smooth functioning of the body.

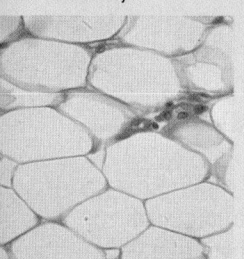

Human body cells

1865 Armed with his knowledge of Pasteur's work, English surgeon **Joseph Lister** introduces a system of antiseptic surgery. Lister's methods reduce the number of fatal infections resulting from surgery.

Anopheles mosquito

1877-1897 Scottish physician **Sir Patrick Manson** shows how insects may carry diseases. Manson's work leads other scientists to search for insect carriers of human diseases. English physician **Sir Ronald Ross** finds the malaria parasite in the stomach of the *Anopheles* mosquito. Observations made by American physician **Carlos Finlay** in Cuba help Finlay hypothesize that the *Aedes* mosquito transmits yellow fever to humans.

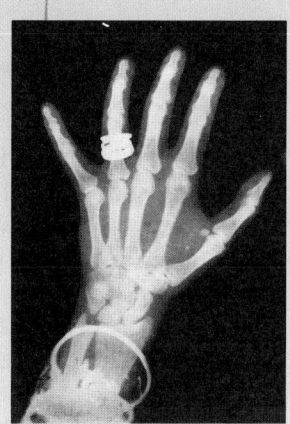

X ray of a hand

1895 German physicist **Wilhelm Roentgen** discovers X rays. X rays revolutionize medicine by allowing physicians to "see" into the body without performing surgery.

1900

1900 Austrian-born American scientist **Karl Landsteiner** defines four major human blood groups—A, B, AB, and O—based on antigens found on red blood cell surfaces.

Red blood cells

1967 South African surgeon **Christiaan Barnard** performs the first successful human heart transplant.

Christiaan Barnard

1972 The **CAT (Computerized Axial Tomography) scanner** is developed in Britain. The CAT scanner takes a series of X rays through an organ. When viewed as a set, these X-ray images allow physicians to pinpoint the exact size and location of diseased tissue in the body.

CAT scanner

1981 The United States Centers for Disease Control officially recognizes **AIDS (Acquired Immune Deficiency Syndrome)** as a disease.

1982 The first commercial product of **genetic engineering** is approved for sale by the United States Food and Drug Administration. The product, humulin, is human insulin produced by bacteria that have had the human gene for insulin production spliced into their DNA.

1990

1990 Researchers use **gene therapy** on a human for the first time in an attempt to cure a patient of severe combined immune deficiency, a fatal hereditary disease of the immune system. Gene therapy is a technique in which healthy genes are introduced into a patient in order to cure or control an inborn error of metabolism.

DNA sequencing

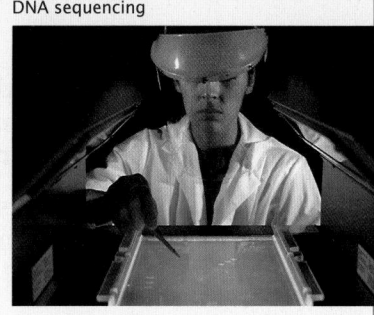

Discussion

Guide the discussion by posing the following questions.

1. In what ways are animals valid substitutes for humans in experiments? In what ways are animals not valid substitutes?
   Answers will vary. You might suggest that many experiments with humans would be impossible or unethical. Animals are valid substitutes for humans only to a certain extent. They are similar enough to humans so that experiments on them suggest results that might occur if humans were used in the experiments. However, the results are not necessarily those that would be obtained if human subjects were used. The knowledge derived is knowledge about animals and can be applied only with caution to humans.

2. Why is the identification of defective genes only one step in eliminating genetic disease?
   Answers will vary. You might suggest that ways must be found to neutralize the effect of the defective genes, or to replace them with normal genes.

3. Would you be interested in viewing your own genetic map and replacing any defective genes present? Why or why not?
   Answers will vary.

DO YOU EVER FIND YOURSELF REREADING A PARAGRAPH OVER AND OVER AND STILL NOT GETTING IT? WHAT DO YOU DO TO STUDY? DO YOU READ OVER THE CHAPTER THE NIGHT BEFORE A TEST AND HOPE FOR THE BEST? DO YOU TRY TO MAKE UP FLASH-CARD QUESTIONS AND GIVE YOURSELF A QUIZ? IF YOU ANSWER "YES" TO ANY OF THESE QUESTIONS, YOU ARE LEARNING IN A WAY THAT ONLY LASTS FOR A SHORT TIME.

Study Skill: Concept Mapping

What's the Best Way to Study?

If you want to learn information you can remember longer and use better, you need to move that information into your long-term memory.

Connecting Ideas

Concept mapping may help you understand the meanings of ideas by showing you their connections to other ideas. If you define soccer as a team sport, you have already related the idea or concept of "soccer" to the ideas of "team" and "sport." Making an idea map may help you see the information more clearly.

This is different from note-taking or making an outline. Your map not only identifies the major ideas of interest from a chapter or your class notes, but it also shows you the relationships among the ideas in much the same way that a road map illustrates how highways and other roads link cities.

Examine these words:

pool	water	biking
grass	sky	raining
tree	playing	thinking

Every one of these words is a concept. Concepts usually form a picture in your mind.

Look at the following words. Are they concepts?

the	when	with
to	was	can
has	be	

No. They don't cause a picture to form in your mind. They are linking words. They connect ideas or concepts.

Organizing Concepts

From the concept map in **Figure A**, on the next page, you can see what a concept map looks like, but we will continue to discuss the fine points of mapping and you will do some maps on your own.

Another feature of mapping ideas is that some concepts or ideas are more main, or general, and include other concepts. What is the main idea for this list of concepts: Tokyo, Mexico City, Seoul, New York, Bombay? Cities, of course. What is the main idea for this list: car, bus, train, bicycle, truck? Each of these concepts is an example of a vehicle.

Try to remember the words below after reading them once or twice:

racket	net
ellipse	helmet
starfish	mitt
tuna	bat
square	diamond
sea horse	star

You probably couldn't remember all of them. Even if you did remember almost all of them, it's a good bet you were already categorizing them.

Try this list and you won't need as much time to memorize the words.

square	**touchdown**	**necklace**
circle	**goal**	**bracelet**
triangle	**basket**	**earring**

Easier, wasn't it? That second list was already categorized. If you learn in a way that carries more meaning, it's easier to remember! That's how concept mapping works, too.

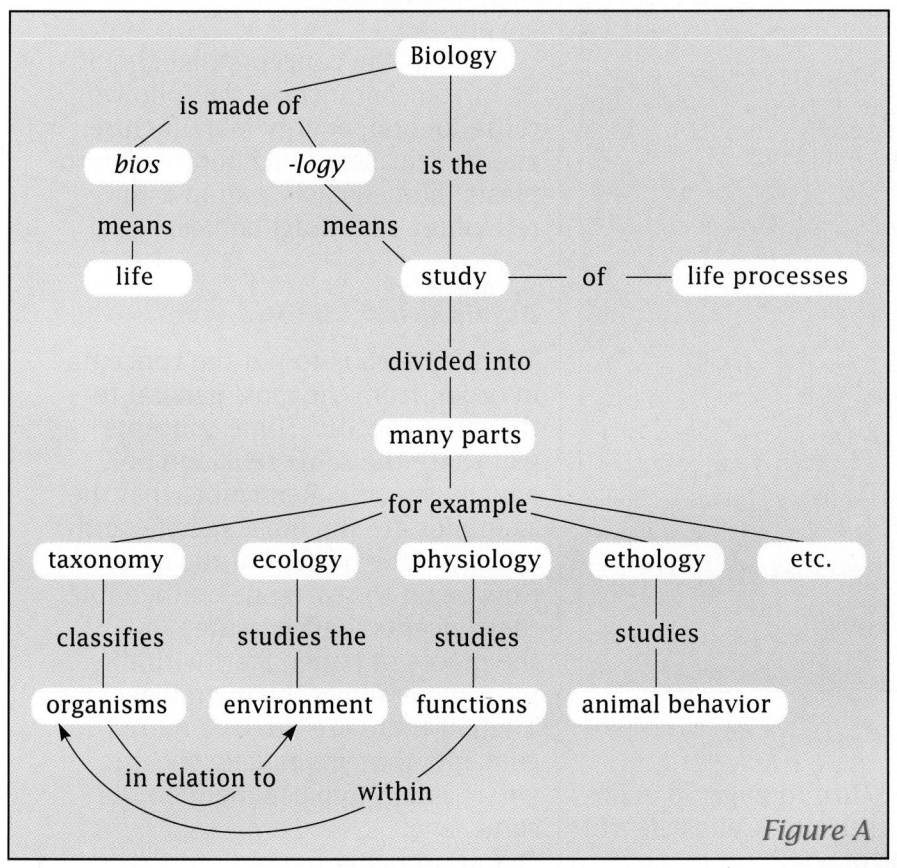

Figure A

Making Learning Faster

You can see in **Figure A** that every concept map should have at its top the most general and all-inclusive concept. The smaller, more specific concepts and examples should go below. It might help you to capitalize the first letter of the main concept or idea. You can write the other, more specific, concepts using all lowercase letters. Your maps should show a pattern of going from most general to most specific, from the top of the map to the bottom of the map. Try to understand how the concept map in **Figure B** was made before you try one yourself. First read the following paragraph and underline any important words you think should be learned.

What is life? What is the difference between living and nonliving things? If you were in a natural wilderness area, it would be easy for you to pick out the living things and nonliving things. The animals and plants are the living organisms. Organisms are made up of many substances organized into living systems. The rocks, air, water, and soil you see are nonliving. They contribute substances to the living organisms.

Connecting Ideas

In the preceding paragraph, there seem to be two general ideas: living and nonliving. We either can make two separate maps, the chief concept of one being "living" and the other being "nonliving," or we can make one map and use "natural things" as our main idea or main concept. We'll capitalize the first letter and put it at the top of our map. "Natural things" did not actually appear in the paragraph, but frequently the main idea does. Since natural things can be living or nonliving, we can split off these subconcepts as parts of the main one. As we read the paragraph, we learned that living things are organisms, such as plants and animals, and that they are made up of substances that are organized into living systems.

Our paragraph also told us that living things are different from nonliving things, but that they are related to nonliving things. So we make that connection up near the top of the map. And the rest of our map should explain how that happens, so we follow what we learn in the paragraph. Nonliving things such as soil, water, and rocks all contribute substances which are organized into living systems.

Now we have the entire map shown in **Figure B**. If you had a choice of studying all those sentences, or looking at this map, you might agree that the map shows the concepts more clearly. This map gives you the main idea more quickly, and it's easier to understand everything because you've now put this in a two-dimensional form that shows how all these things are related.

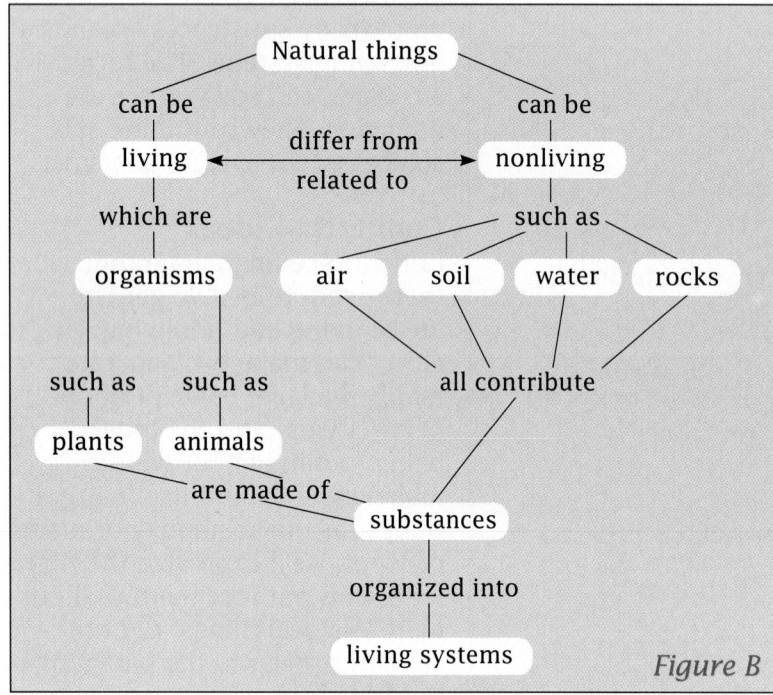

Natural things
can be — living — differ from / related to — nonliving — can be
which are
living → organisms
such as — air, soil, water, rocks — such as
organisms → plants (such as), animals (such as)
air, soil, water, rocks → all contribute
plants, animals — are made of — substances
substances — organized into — living systems

Figure B

Planning the Concept Map

Now you try it. Turn to page 58 and read the major titles in "Cells Perform Basic Functions of Life."

You should have seen "Cells reproduce," "Cells manufacture and release energy," and "Cells maintain homeostasis." Of course, within those titles there was quite a lot of information. Our job is to turn that information from pages 58–59 into a concept map.

Identifying the main idea is easy— it's in the main title. It could be "Basic cell functions" or "Life functions of cells," or some such shortened version of the title. This book is organized and written around concepts. In some other texts the main idea may be buried in the paragraphs.

So now we can put that main idea at the top of our map in a box.

Next it's time to look at the other major important concepts and list them in the margin of your paper, or on small piece of paper. Use one tiny piece of paper for each concept. Remember, your list shows how the concepts appeared in the reading, but this may not necessarily represent how the concepts are related to each other. Making those connections is your job.

Some of the concepts that should be in your list include the following: reproduction, energy manufacture, energy release, mitochondria, chloroplasts, homeostasis, endoplasmic reticulum, and Golgi bodies.

Building the Concept Map

The next step is to put the concepts in order, from the most general to the most specific. Some concepts will share the same rank and be equally specific. Remember that the examples are the most specific and will be at the bottom of the map. Now begin to rearrange on the table the concepts you've written on the pieces of paper. Start with the most general; get the main idea. Then, if there are three or more concepts that are equally specific, you're going to place them on the same level.

You probably figured out that reproduction, energy manufacture, energy release, and homeostasis are all the basic functions described by the title. They should be placed below the main idea so that they are equally spread out across the map.

Now continue to lay out all the concepts under the subconcepts on the first row or layer until you have used them all. You can rearrange your pieces of paper at any time, so keep pushing them around as if they were the pieces of a jigsaw puzzle, until you've arranged it the way you think it belongs.

Remember not to use only the words given here. You're going to have to read the text material and add concepts to explain the major ones. For instance, under "reproduction" you need to have "cell division" and maybe "growth." Under "energy manufacture" you will have "photosynthesis," "food," and "sugar," and "green plants."

Concept mapping takes some time at first, it will take about six times longer the first few times than after a week or two of practice.

Now you're going to make the connections among the concepts. Use lines to connect the concepts, and write linkage words on the lines that show or tell why they are connected. Do this for all the lines connecting all the concepts. Glue or tape down your concept papers if you want to make this permanent, or sketch on paper the way you've arranged the concepts.

Features of Good Concept Maps

Remember, practice is the key to good concept mapping and you'll get better at this as you go along. Here are some things to remember that will help you get started.

- A concept map does not have to be symmetrical. It can have more concepts on one side than the other.
- There are no perfectly correct concept maps, only maps that come closer to the meanings you have for those concepts. As the map-maker, you must make it correct for you.
- Do not put more than three words in a concept box.
- Do not string out more than three boxes in a row or line without branching out.
- Write linkage words connecting every two concepts with as few words as it takes to make sense between them.

If the relationships you have made between any two concepts are wrong, your teacher will help you sort out the misconception. Even if you are absolutely correct in your relationships, you may see a fellow student's map that has differences. They could be equally correct even though your maps may look nothing alike. Everyone thinks a little bit differently, and, as a result, we'll see different relationships between certain concepts.

As you get practice in making concept maps, your teacher will examine your line connection statements more closely. Since these lines represent the relationships between concepts, whatever you write on the line will tell if you really understand how those two concepts are connected. After you have practiced, your maps should always:

- be two dimensional—not just a list of concepts and lines. It goes from top to bottom and from general to specific with lines which show how the concepts are related.
- show which concepts are more important by their placement on the map and by what concepts come off of them.
- have many branches with no more than three concept boxes in a row and no more than three words in a concept box.
- have only concepts in the boxes and only linkage words on the lines.

Evaluating Your Skills

For the first map you make on your own, think about something you know very well. Do you play a team sport or an individual sport? Do you have a hobby? Do you enjoy a particular kind of music? Whatever you choose, this will be the major main concept for your next concept map! This will be more fun and easier since you know this area so well. Good luck!

Safety

Your biology laboratory is a unique place where you can learn by doing things that you couldn't do elsewhere. It also involves some dangers that can be controlled if you follow these special safety notes and all instructions from your teacher.

It is your responsibility to protect yourself and other students by conducting yourself in a safe manner while in the laboratory. Familiarize yourself with the printed safety symbols—they indicate additional measures that you must take.

While in the Laboratory, at All Times...

Familiarize yourself with the investigation—especially safety issues—before entering the lab. Know the potential hazards of the materials, equipment, and the procedures required. Ask the teacher to explain any parts you do not understand before you start.

Never perform any experiment not specifically assigned by your teacher.

Never work alone in the laboratory.

Know the location of all safety/emergency equipment used in the laboratory. Examples include eyewash stations, safety blankets, safety shower, fire extinguisher, first aid kit, chemical spill kit(s).

Know the location of the closest telephone. Be sure there is a posted list of emergency phone numbers, including poison control center, fire department, police, and ambulance.

Before beginning work: tie back long hair, roll up loose sleeves, and put on any required personal protective equipment as required by your teacher. Avoid or confine loose clothing that could knock things over, ignite from flame, or soak up chemical solutions.

Report any accident, incident, or hazard—no matter how trivial— to your teacher immediately. Any incident involving bleeding, burns, fainting, chemical exposure, or ingestion should also be reported to your school nurse and/or school physician.

In case of fire, alert the teacher and leave the laboratory.

Never consume food or drink, nor apply cosmetics, in the laboratory. Never store food in the laboratory. Keep hands away from faces. Wash your hands at the conclusion of each laboratory investigation and whenever leaving the laboratory. Remember that some hair products are highly flammable, even after application.

Keep your work area neat and uncluttered. Only bring books and other materials that are needed to conduct the experiment.

Clean your work area at the conclusion of the experiment as your teacher directs.

When Called for, Use These Safety Procedures:

Eye Safety

Wear approved chemical safety goggles as directed. Goggles should always be worn when you are working with a chemical or solution, heating substances, or any mechanical device, or physical process.

In case of eye contact:
Go to an eyewash station and flush eyes (including under the eyelids) with running water for at least 15 minutes. Notify your teacher or other adult in charge.

Wearing of contact lenses for cosmetic reasons is prohibited in the laboratory. Liquids or gases can be drawn up under the contact lens and onto the eyeball. If you must wear contacts prescribed by a physician, inform your teacher. You must wear approved eye-cup safety goggles.

Never look directly at the sun through any optical device or lens system, nor gather direct sunlight to illuminate a microscope. Such actions will concentrate light rays that will severely burn your retina, possibly causing blindness!

Electrical Supply

Never use equipment with frayed cords.

Assure that electrical cords are taped to work surfaces. This will prevent falls, and equipment can't be pulled off the table.

Never use electrical equipment around water, nor with wet hands or clothing.

Clothing Protection

Wear an apron or laboratory coat when working in the laboratory, to prevent chemicals or chemical solutions from contacting skin or street clothes.

Animal Care

Do not touch or approach any animal in the wild. Be aware of poisonous or dangerous animals in any area where you will be doing outside fieldwork.

Always obtain your teacher's permission before bringing any animal (or pet) into the school building.

Handle any animal only as your teacher directs. Mishandling or abuse of any animal will not be tolerated!

Sharp Object Safety

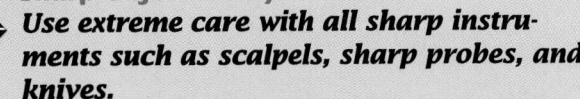

Use extreme care with all sharp instruments such as scalpels, sharp probes, and knives.

Never use double-edged razors in the laboratory.

Never cut objects while holding them in your hand. Place objects on a suitable work surface.

Chemical Safety

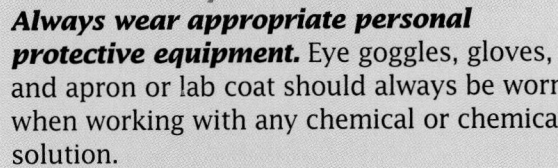

Always wear appropriate personal protective equipment. Eye goggles, gloves, and apron or lab coat should always be worn when working with any chemical or chemical solution.

Never taste, touch, or smell any substance, nor bring it close to your eyes, unless specifically told to do so by your teacher. If you are directed by your teacher to note the odor of a substance, do so by waving the fumes towards you with your hand. Never pipet any substance by mouth; use a suction bulb as directed by your teacher.

Always handle any chemical or chemical solution with care. Check the label on the bottle and observe safe-use procedures. Never return unused chemicals or solutions to their containers. Return unused reagent bottles or containers to your teacher.

Never mix any chemical unless specifically told to do so by your teacher.

Never pour water into a strong acid or base. The mixture can produce heat and splatter. Remember this rhyme:

> Do as you oughta—
> Add acid (or base) to water.

Report any spill immediately to your teacher. Handle spills only as your teacher directs.

Check for the presence of any source of flames, sparks, or heat (open flame, electric heating coils, etc.) before working with flammable liquids or gases.

Plant Safety

Do not ingest any plant part used in the laboratory (especially seeds sold commercially). Do not rub any sap or plant juice on your eyes, skin, or mucous membranes.

Wear protective gloves (disposable polyethylene gloves) when handling any wild plant.

Wash hands thoroughly after handling any plant or plant part (particularly seeds). Avoid touching your face and eyes.

Do not inhale or expose yourself to the smoke of any burning plant.

Do not pick wildflowers or other plants unless directed to do so by your teacher.

Proper Waste Disposal

Clean and decontaminate all work surfaces and personal protective equipment as directed by your teacher.

Dispose of all sharps (broken glass and sharp objects) and other contaminated materials (biological/chemical) in special containers as directed by your teacher.

Hygienic Care

Keep your hands away from your face and mouth.

Wash your hands thoroughly before leaving the laboratory.

Remove contaminated clothing immediately; launder contaminated clothing separately. Use the proper technique demonstrated by your teacher when handling bacteria or similar microorganisms. Examine microorganism cultures (such as petri dishes) without opening them.

Return all stock and experimental cultures to your teacher for proper disposal.

Heating Safety

When heating chemicals or reagents in a test tube, never point the test tube toward anyone.

Use hot plates, not open flames. Be sure hot plates have an "On-Off" switch and indicator light. Never leave hot plates unattended, even for a minute. Never use alcohol lamps.

Know the location of laboratory fire extinguishers and fire blankets. Have ice readily available in case of burns or scalds.

Use tongs or appropriate insulated holders when heating objects. Heated objects often do not have the appearance of being hot. Never pick up an object with your hand unless you are certain it is cold.

Keep combustibles away from heat and other ignition sources.

Hand Safety

Never cut objects while holding them in your hand.

Wear protective gloves when working with stains, chemicals, chemical solutions, or with wild (unknown) plants.

Glassware Safety

Inspect glassware before use; never use chipped or cracked glassware. Use borosilicate glass for heating.

Do not attempt to insert glass tubing into a rubber stopper without specific instruction from your teacher.

Always clean up broken glass by using tongs and/or a brush and dustpan. Discard the pieces in an appropriately labeled "sharps" container.

Safety With Gases

Never directly inhale any gas or vapor.

Handle materials prone to emit vapors or gases in a well-ventilated area. This work should be done in an approved chemical fume hood.

THE METRIC SYSTEM IS USED FOR MAKING MEASUREMENTS IN SCIENCE.
THE OFFICIAL NAME OF THIS SYSTEM IS THE SYSTÈME INTERNATIONAL
D'UNITÉS, OR INTERNATIONAL SYSTEM OF MEASUREMENTS (SI).

SI Conversions

This girl's length is measured in meters.

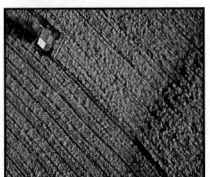

The area of this field is measured in hectares.

The volume of these spoons is measured in milliliters.

The mass of this nickel is measured in grams.

Internation System of Measurements (SI) Conversion Table

SI Units	From SI to English	From English to SI
Length		
kilometer (km) = 1,000 m	1 km = 0.62 mile	1 mile = 1.609 km
meter (m) = 100 cm	1 m = 3.28 feet	1 foot = 0.305 m
centimeter (cm) = 0.01 m	1 cm = 0.394 inch	1 foot = 30.5 cm
millimeter (mm) = 0.001 m	1 mm = 0.039 inch	1 inch = 2.54 cm
micrometer (μm) = 0.000 001 m		
nanometer (nm) = 0.000 000 001 m		
Area		
square kilometer (km^2) = 100 hectares	1 km^2 = 0.3861 square mile	1 square mile = 2.590 km^2
hectare (ha) = 10,000 m^2	1 ha = 2.471 acres	1 acre = 0.4047 ha
square meter (m^2) = 10,000 cm^2	1 m^2 = 10.765 square feet	1 square foot = 0.0929 m^2
square centimeter (cm^2) = 100 mm^2	1 cm^2 = 0.155 square inch	1 square inch = 6.4516 cm^2
Volume		
liter (L) = 1,000 mL = 1 dm^3	1 L = 1.06 fluid quarts	1 fluid quart = 0.946 L
milliliter (mL) = 0.001 L = 1 cm^3	1 mL = 0.034 fluid ounce	1 fluid ounce = 29.57 mL
microliter (μL) = 0.000 001 L		
Mass		
kilogram (kg) = 1,000 g	1 kg = 2.205 pounds	1 pound = 0.4536 kg
gram (g) = 1,000 mg	1 g = 0.0353 ounce	1 ounce = 28.35 g
milligram (mg) = 0.001 g		
microgram (μg) = 0.000 001 g		

Temperature

°F 0 20 40 60 80 100 120 140 160 180 200 220

°C −20 −10 0 10 20 30 40 50 60 70 80 90 100

Freezing point of water   Normal human body temperature
            Room temperature

The top of the thermometer is marked off in degrees Fahrenheit (°F). To read the corresponding temperature in degrees Celsius (°C), look at the bottom side of the thermometer. For example, 50°F is the same temperature as 10°C. You may also use the formulas at the right for conversions.

Conversion of Fahrenheit to Celsius:
$°C = \frac{5}{9}(°F - 32)$

Conversion of Celsius to Fahrenheit:
$°F = (\frac{9}{5}°C) + 32$

Laboratory Skills

Using a Compound Light Microscope

Parts of the Compound Light Microscope

- The *eyepiece* magnifies the image, usually 10X.

- The *low-power objective* magnifies the image more, often 10X.

- The *high-power objective* magnifies the image even more, such as 43X.

- The *revolving nosepiece* holds the objectives and can be turned to change from one magnification to another.

- The *body tube* maintains the correct distance between eyepiece and objectives. This is usually about 25 cm (10 in.) the normal distance for reading and viewing objects with the naked eye.

- The *coarse adjustment* moves the body tube up and down in large increments to allow gross positioning and focusing of the objective lens.

- The *fine adjustment* moves the body tube slightly to bring the image into sharp focus.

- The *stage* supports a slide.

- *Stage clips* secure the slide in position for viewing.

- The *diaphragm* (or *iris diaphragm*) controls the amount of light allowed to pass through the object being viewed.

- The *light source* provides light for viewing the image. It can either be a light reflected with a mirror or incandescent light from a small lamp.

  NEVER use reflected direct sunlight as a light source.

- The *arm* supports the body tube.

- The *base* supports the microscope.

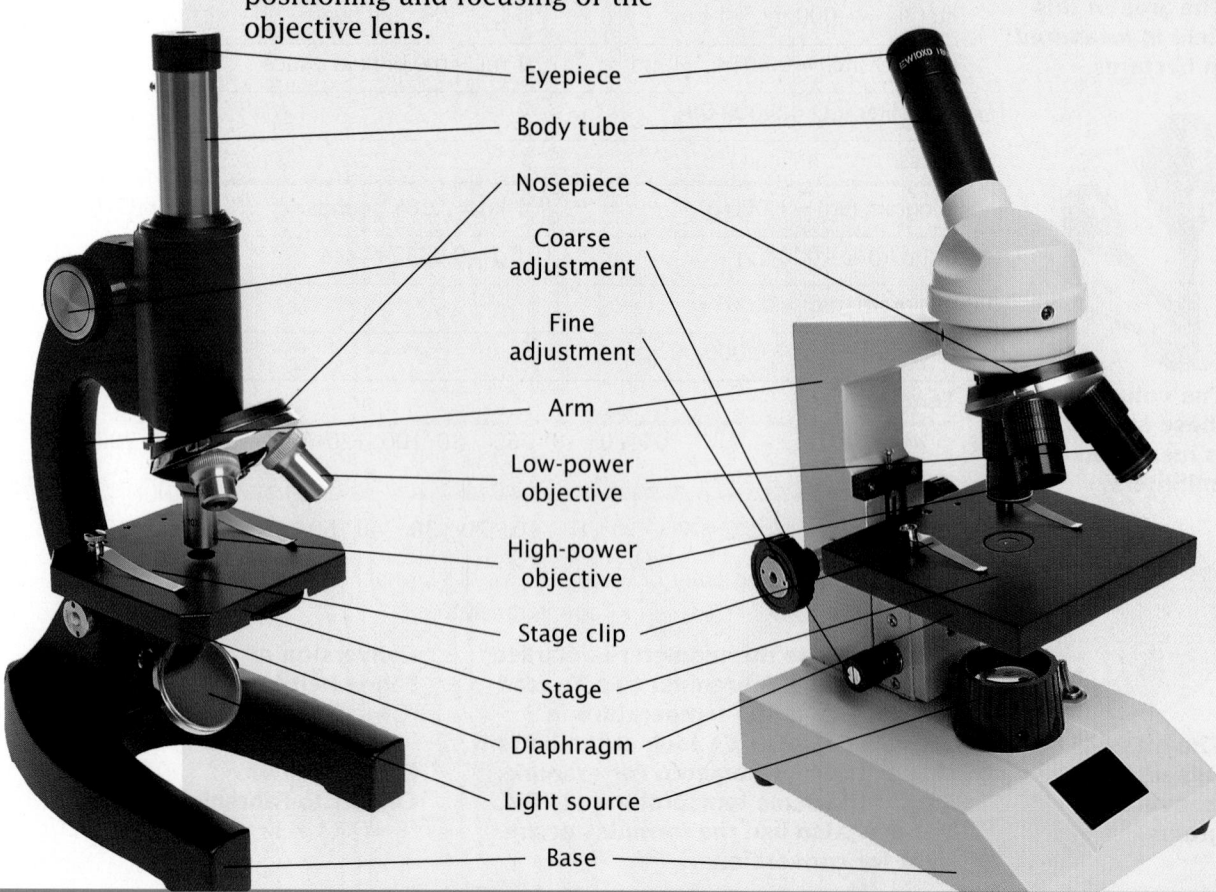

Eyepiece

Body tube

Nosepiece

Coarse adjustment

Fine adjustment

Arm

Low-power objective

High-power objective

Stage clip

Stage

Diaphragm

Light source

Base

Proper Handling and Use of the Compound Light Microscope

1. Carry the microscope to your lab table using both hands, one beneath the base and the other hand holding the arm of the microscope. Hold the microscope close to your body.

2. Place the microscope on the lab table, at least 5 cm (2 in.) in from the edge of the table.

3. Check to see what type of light source the microscope has. If the microscope has a lamp, plug it in, making sure that the cord is out of the way. If the microscope has a mirror, adjust it to reflect light through the hole in the stage.
   CAUTION: If your microscope has a mirror, do not use direct sunlight as a light source. Direct sunlight can damage your eyes.

4. Adjust the revolving nosepiece so that the low-power objective is in line with the body tube.

5. Place a prepared slide over the hole in the stage and secure the slide with stage clips.

6. Look through the eyepiece and move the diaphragm to adjust the amount of light coming through the specimen.

7. Now, look at the stage from eye level, and slowly turn the coarse adjustment to lower the objective until it almost touches the slide. Do not allow the objective to touch the slide.

8. While looking through the eyepiece, turn the coarse adjustment to raise the objective until the image is in focus. *Never focus objectives downward.* Use the fine adjustment to achieve a sharply-focused image. Keep both eyes open while viewing a slide.

9. Make sure that the image is exactly in the center of your field of vision. Then, switch to the high-power objective. Focus the image with the fine adjustment. *Never use the coarse adjustment at high power.*

10. When you are finished using the microscope, remove the slide. Clean the eyepiece and objectives with lens paper and return the microscope to its storage area.

Making a Wet Mount

1. Use lens paper to clean a glass slide and coverslip.

2. Place the specimen you wish to observe in the center of the slide.

3. Using a medicine dropper, place one drop of water on the specimen.

4. Position the coverslip so that it is at the edge of the drop of water and at a 45° angle to the slide. Make sure that the water runs along the edge of the coverslip.

5. Lower the coverslip slowly to avoid trapping air bubbles.

6. If a stain or solution is to be added to a wet mount, place a drop of the staining solution on the microscope slide along one side of the coverslip. Place a small piece of paper towel on the opposite side of the coverslip.

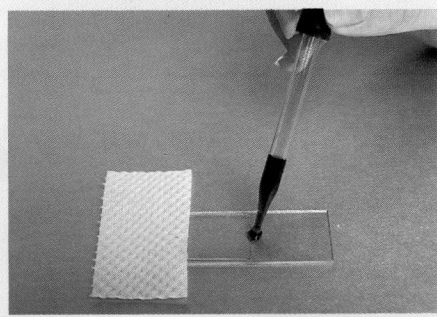

7. As the water evaporates from the slide, add another drop of water by placing the tip of the medicine dropper next to the edge of the coverslip, just as you would do if adding stains or solutions to a wet mount. If you have added too much water, remove the excess by using the corner of a paper towel as a blotter. Do not lift the coverslip to add or remove water.

The Five Kingdom System

Kingdom	Number of Species*	Main Characteristics	Examples
Monera	5,000 +	Prokaryotic; unicellular; lack nucleus or other internal compartments; most primitive form of life; bacteria existed for more than 2 billion years before protists appeared	*Escherichia coli*, *Vibrio cholerae*, *Salmonella typhimurium*
Protista	43,000	Eukaryotic; all but algae are unicellular; aquatic; all possess nucleus and internal compartments; almost all possess mitochondria and many also possess chloroplasts; most protists possess cell walls; aquatic; most diverse kingdom	*Volvox*, *Paramecium*, red algae
Fungi	77,000	Eukaryotic; multicellular; almost all terrestrial; cell walls of chitin; nonmotile cells; exclusively heterotrophic with nutrition by absorption; mitosis occurs within nuclear envelope	Mushrooms, yeasts
Plantae	265,350	Eukaryotic; multicellular; terrestrial; cell walls of cellulose; nonmotile cells (except gametes of a few forms); photosynthetic, with chloroplasts	Oak trees, mosses, roses, ferns,
Animalia	1,203,000	Eukaryotic; multicellular; both aquatic and terrestrial; no cell walls or chloroplasts; heterotrophic, with nutrition by ingestion; reproduction is predominately sexual	Humans, sea stars, earthworms flies

* Approximate number identified

Escherichia coli

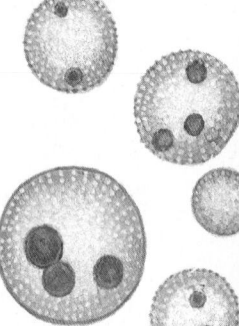

Volvox

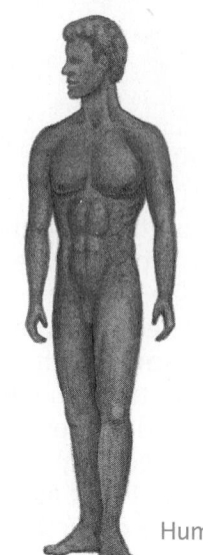

Human

Oak tree

Mushrooms

Kingdom Monera

Phylum*	Main Characteristics	Examples
Archaebacteria	Anaerobic methane-producing bacteria (some reduce sulfur); among the oldest forms of life on earth; many scientists classify them as a separate kingdom	Methanogens, halobacteria
Cyanobacteria	Photosynthetic bacteria; gooey coating that is deeply pigmented; common on land and in the ocean; responsible for "blooms" in polluted waters	*Anabaena, Spirulina, Oscillatoria*
Chemoautotrophs	Ancient group of bacteria that can grow without sunlight or other organisms; derive energy from reduced gases—ammonia (NH_3), methane (CH_4), hydrogen sulfide (H_2S); play critical role in the Earth's nitrogen cycle	Nitrobacteria, *Nitrosomonas*, sulfur bacteria
Enterobacteria	Typically rigid, rod-shaped, heterotrophic bacteria; usually aerobic; flagellated; some, called *Vibrio*, are comma-shaped; responsible for many important diseases of animals and plants	*Salmonella typhimurium, Escherichia coli, Vibrio cholerae*
Pseudomonads	Straight or curved rods with flagella at one end; very common in soil; many are serious plant pathogens	*Pseudomonas aeruginosa*
Spirochaetes	Long spiral-shaped cells; flagella originating at each end; responsible for several serious diseases	*Borrelia* (Lyme disease), *Treponema* (syphilis)
Actinomycetes	Filamentous (and thus sometimes mistaken for fungi); spore-producing; sources of antibiotics including streptomycin, tetracycline, and chloramphenicol	*Myco bacterium*, dental plaques
Rickettsias	Obligatory parasites within the cells of vertebrates and arthropods; responsible for Rocky Mountain spotted fever	*Rickettsia rickettsii*
Gliding and budding bacteria	Long rods that secrete slimy mass of polysaccharides; often aggregate into gliding masses; some members form upright spore-bearing structures of unusual complexity for bacteria	Myxobacteria

* The monerans are one of the least studied kingdoms. Since thousands of species are yet to be discovered and named, numbers of species identified are not provided in this table.

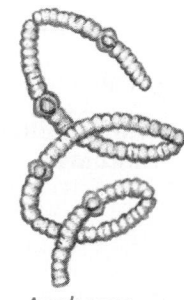

Anabaena

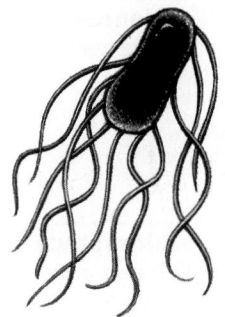

Salmonella typhimurium

Borrelia

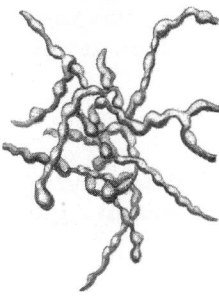

Mycobacteria

Kingdom Protista

Phylum	Number of Species*	Main Characteristics	Examples
Bacillariophyta "Diatoms"	11,500 +	Unicellular; photosynthetic; cells within a unique shell made of opaline silica—resembles a small box with a lid; some radially symmetrical; chloroplasts resemble those of brown algae; shells often form very thick deposits called "diatomaceous earth"	Diatoms
Ciliophyta "Ciliates"	8,000	Very complex single cells; heterotrophic; possess rows of cilia and two types of cell nuclei; genetic code differs in several respects from other organisms; some scientists argue that ciliates should be assigned to a separate kingdom of their own	*Vorticella, Paramecium*
Chlorophyta "Green algae"	7,000	Unicellular, colonial, and multicellular species; all photosynthetic; chloroplasts very similar to those of green plants; most scientists believe that plants evolved from green algae	*Chlamydomonas, Chlorella, Volvox, Ulva, Spirogyra*
Rhodophyta "Red algae"	4,000	Almost all multicellular; photosynthetic; chloroplasts evolved symbiotic cyanobacteria; no cilia or centrioles; one of the most ancient groups of protists	Nori (edible algae)
Sporozoa "Sporozoans"	3,900	Unicellular; heterotrophic; nonmotile; spore-forming parasites of animals; complex life cycles, often involving alternation between different hosts; *Plasmodium* is responsible for malaria, which kills more than 1 million people each year	*Plasmodium*
Zoomastigina "Zoomastigotes"	3,000	Mostly unicellular; heterotrophic; flagellated; free-living or parasitic; a highly diverse phylum; similar cells exist in sponges, suggesting that animals evolved from a zoomastigote	Choanoflagellates
Phaeophyta "Brown algae"	1,500	Multicellular; photosynthetic; marine; some are the longest, fastest growing, and most photosynthetically productive species on earth	Kelps, *Sargassum*

* Approximate number identified

Diatom

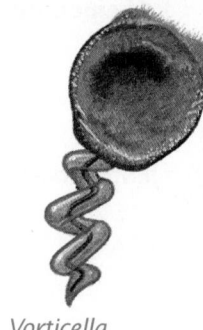

Vorticella

Chlamydomonas

Kelp

Phylum	Number of Species*	Main Characteristics	Examples
Dinoflagellata "Dinoflagellates"	1,000 +	Unicellular; photosynthetic; body enclosed within cellulose plates; two unequal flagella beating in grooves that encircle the body; often symbiotic with corals	*Gonyaulax*, "Red tides"
Euglenophyta "Euglenoids"	800	Unicellular; both photosynthetic and heterotrophic species; asexual; most live in fresh water; very similar to zoomastigotes; chloroplasts are similar to those of green algae, and are thought to have evolved from the same symbiotic bacteria	*Euglena*
Oomycota "Water molds"	475	Unicellular parasites on living or dead tissue; cell walls of cellulose (not chitin, like fungi); spores have two unequal flagella, one directed forward, the other backward; responsible for Irish potato famine of 1845–1847	Rusts, mildews, *Saprolegnia*
Myxomycota "Plasmodial slime molds"	450	Heterotrophic; individuals stream along as a nonwalled multinucleate mass of "slime" cytoplasm; no centrioles; all nuclei undergo mitosis synchronously; when they are dry or starving, they form spores that start a new individual in a more favorable environment	*Arcyria*
Foraminifera "Forams"	300	Unicellular; heterotrophic; marine; possess shells of organic material with many pores through which thin cytoplasmic threads project; shells are reinforced with grains of calcium carbonate or sand; the White Cliffs of Dover and other limestone deposits are made almost entirely of foram shells	Forams
Rhizopoda "Amoebas"	300	Unicellular; heterotrophic; amorphously shaped cells that move using cytoplasmic extensions called pseudopods; asexual reproduction (fission) only	Amoeba
Acrasiomycota "Cellular slime molds"	65	Heterotrophic; amoeba-shaped cells that aggregate into a moving mass called a slug when deprived of food; the slug produces spores that form new amoebas elsewhere	*Discoideum, Dictyostelium*
Caryoblastea "Pelomyxa"	1	Unicellular; lacks mitochondria and chloroplasts; has two kinds of bacterial symbionts; does not undergo mitosis, possibly an early stage in the evolution of eukaryotes	*Pelomyxa palustris*

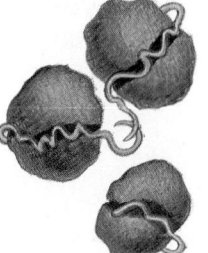

Gonyaulax

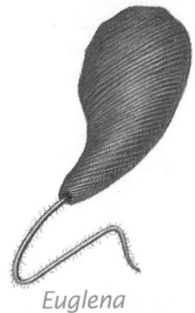

Euglena

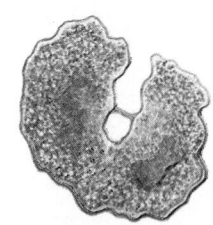

Foram

Amoeba

825

Kingdom Fungi

Division	Number of Species*	Main Characteristics	Examples
Ascomycota "Sac fungi"	30,000	Fusion of hyphae leads to the formation of reproductive structures of densely interwoven hyphae (visible part) that contain characteristic microscopic structures called asci (singular, ascus); two nuclei fuse within the ascus, forming a diploid zygote that undergoes meiosis; asexual reproduction is also common	Morels, yeasts, truffles, cup fungi
Deuteromycetes "Fungi imperfecti"	17,000	Sexual life cycle is not observed; mostly ascomycetes that have lost the ability to reproduce sexually	*Penicillium*, *Aspergillus*
Basidiomycota "Club fungi"	16,000	Fusion of hyphae leads to the formation of a densely interwoven reproductive structure (mushroom) with characteristic microscopic, club-shaped structures called basidia (singular, basidium); only the terminal cell of a basidium produces spores (all cells of an ascus do); reproduction is typically sexual	Puffballs, mushrooms, toadstools, rusts
Zygomycota "Molds"	665	Multinucleate hyphae without septa except for reproductive structures; fusion of hyphae leads directly to the formation of a zygote, which divides by meiosis when it germinates; asexual reproduction is also common	*Rhizopus* (black bread mold)

Morel

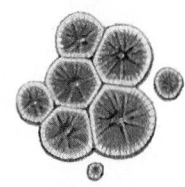

Penicillium

Puffball

Symbiotic Associations**

Lichens	15,000	Fungi (almost always ascomycetes) in the cells of which are cyanobacteria or green algae or both; derives its energy from its photosynthetic partner and cannot survive without it; able to survive freezing or drying out; can invade the harshest of habits; breaks down rocks and sets the stage for invasion by other organisms	
Mycorrhizae	5,000	Symbiotic associations between fungi and plants; 80 percent of all plants have mycorrhizae in their roots; by far the most common are endomycorrhizae, in which zygomycete fungal hyphae penetrate the cells of the plant root; some plants of temperate regions have ectomycorrhizae, in which basidiomycete hyphae (or, rarely, ascomycete) surround but do not penetrate the root	

Rhizopus

* Approximate number identified
** Fungi can form mutualistic associations with algae to make lichens or with plants to make mycorrhizae.

Kingdom Plantae

Division	Number of Species*	Main Characteristics	Examples	
Anthophyta "Flowering plants"	235,000	Angiosperms; ovules are fully enclosed by carpel; after fertilization, carpel and seeds mature to become fruit; flowers are reproductive structures; fertilization involves two sperm nuclei—one forms the gamete, the other fuses with polar bodies to form endosperm for the seed	Roses, oak trees, wheat	 Rose bush
Bryophyta "Mosses"	16,600	No vascular tissue; lack true roots, stem, and leaves; obtain nutrients by osmosis and diffusion; found chiefly in moist habitats; seedless; sperm must swim to eggs; two other divisions: Hepaticophyta (liverworts) and Anthero-cerophyta (hornworts) are also considered bryophytes—they make up 40 percent of bryophyte species	*Polytrichum*, *Sphagnum* (peat moss)	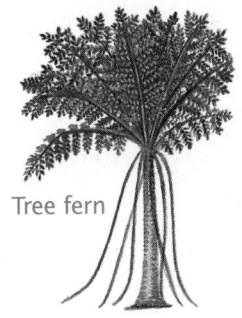 Tree fern
Pterophyta "Ferns"	12,000	Seedless vascular plants; motile sperm; haploid spores germinate into free-living haploid individuals; two minor divisions: Sphenophyta (horsetails) and Psilophyta (whisk ferns) contain 21 additional species	*Sphaer-opteris* (tree ferns), *Azolla*	
Lycophyta "Lycopods"	1,000	Seedless vascular plants; motile sperm; similar in appearance to mosses, but diploid; found chiefly in moist habitats	*Lycopodium*, club mosses	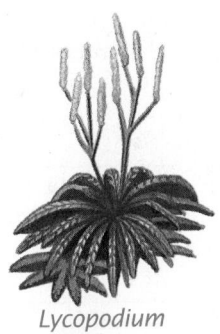 *Lycopodium*
Coniferophyta "Conifers"	550	Gymnosperms; ovules are exposed at time of pollination; pollen is dispersed by wind; sperm lack flagella; leaves are needle-like or scale-like; most species are evergreens	Fir trees, pine trees, spruce trees, redwood trees	
Cycadophyta "Cycads"	100	Gymnosperms; sperm flagellated and motile, but reach vicinity of egg by pollen tube; palmlike trees; very slow growing	Sago palms, cycads	
Gnetophyta "Shrub teas"	70	Gymnosperms; sperm not motile; shrubs and vines	*Welwitschia*, Mormon teas	
Ginkgophyta "Ginkgo"	1	Gymnosperms; like cycads, motile sperm reach vicinity of egg by a pollen tube; deciduous (drop leaves in winter); fan-like leaves; seeds fleshy and ill-scented	Ginkgos	Fir tree

* Approximate number identified

Kingdom Animalia

Phylum	Number of Species*	Main Characteristics	Examples
Arthropoda "Arthropods"	1,000,000	Segmented bodies with paired, jointed appendages; chitinous exoskeletons; open circulatory system; many groups possess wings; most successful of all animal phyla	Beetles, spiders, flies, lobsters
Mollusca "Mollusks"	110,000	Soft-bodied animals with a true coelom; three-part body plan consisting of head-foot, visceral mass, and mantle; many have shells; 35,000 species are terrestrial; almost all possess a unique rasping tongue called a radula	Oysters, snails, squids, octopuses
Chordata "Chordates"	42,500	Possess a notochord, a dorsal nerve cord, pharyngeal slits, and a tail at some stage of life; notochord is replaced during development by the spinal column in vertebrates; pharyngeal slits are lost during development of terrestrial vertebrates; 20,000 species are terrestrial	Fishes, dinosaurs, mammals
Platyhelminthes "Flatworms"	12,000	Solid worms; bilaterally symmetrical; three germ layers; digestive cavity has only one opening; no coelom or pseudo-coelom; simplest animals in which organs occur	Planaria, liver flukes, tapeworms
Nematoda "Roundworms"	12,000+	Tiny, parasitic, unsegmented worms; tubular; bilaterally symmetrical; complete digestive tract with mouth and anus; pseudocoelom; no cilia	*Ascaris*, pinworms, *Filaria*, hook worms
Annelida "Segmented worms"	12,000	Serially segmented worms; complete digestive tract with coelom; some worms possess chitinous bristles (setae) on each segment that anchor them during crawling	Leeches, earthworms, polychaetes
Cnidaria "Jellyfish"	10,100	Radially symmetrical; gelatinous; digestive cavity with only one opening; tentacles armed with stinging cells called nematocysts	Man-of-wars, hydras
Porifera "Sponges"	5,000	Asymmetrical animals; lack distinct tissues and organs; sessile; mostly marine; body consists of two layers that are supported by a stiffening skeleton with many pores; internal cavity is lined with unique food-filtering cells called choanocytes	Barrel sponges

* Approximate number identified

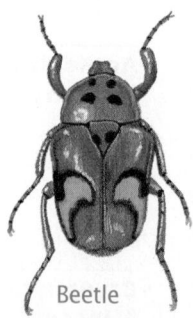

Beetle

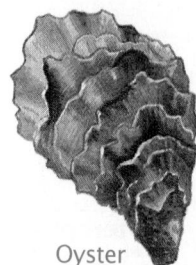

Oyster

Planarian

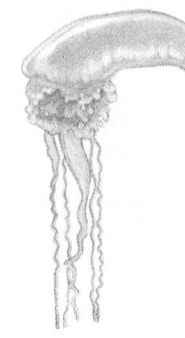

Man-of-war

Phylum	Number of Species*	Main Characteristics	Examples
Echinodermata "Echinoderms"	6,000	Deuterostomes; adults radially symmetrical; endoskeleton of calcium plates; water vascular system with tube feet; five-part body plan, able to regenerate lost body parts	Sea stars, sea urchins, sand dollars
Bryozoa "Moss animals"	4,000	Deuterostomes; microscopic; aquatic organisms that form branching colonies; U-shaped row of ciliated tentacles for feeding (the "lophophore"); this phylum is often called "Ectoprocta"; anus or proct is external to the lophophore	*Bowerbankia*, *Plumatella*
Rotifera "Wheel animals"	2,000	Small, wormlike or spherical animals; have digestive tract; crown of cilia around mouth resembling a wheel; pseudocoelomates; almost all live in fresh water	Rotifers
Minor worms	980	**Rhynchocoela** (ribbon worm): bilaterally symmetrical; acoelomates; marine; long, extendible proboscis **Pognophora** (tube worm): deep-sea worms that live within chitinous tubes in ocean floor; sessile; long, beardlike tentacles **Hemichordata** (acorn worm): marine worms with dorsal and ventral nerve cords **Onychophora** (velvet worms): protostomes; chitinous exoskeleton; evolutionary relicts **Chaetognatha** (arrow worms): deuterostomes; coelomates; bilaterally symmetrical; marine worms with large eyes and powerful jaws	Acorn worms, giant tube worms, velvet worms
Brachiopoda "Lamp shells"	250	Like bryozoans, possess a lophophore, but within two clamlike shells; more than 30,000 species known as fossils	*Lingula*
Ctenophora "Comb jellies"	100	Transparent, gelatinous marine animals; eight bands of cilia; largest animals that use cilia for locomotion; often bioluminescent; digestive tract with anal pore	Comb jellies, sea walnuts
Phoronida "Lophophores"	12	Lophophorate tube worm; U-shaped gut, instead of the straight digestive tube of other tube worms; protostomes; often live in dense populations	*Phoronis*
Loricifera "Sand animals"	6	Minute; bilaterally symmetrical; pseudocoelomates; live in the spaces between grains of sand; mouthparts exhibit a unique flexible tube; discovered in 1983	*Nanoloricus mysticus*

Sea star

Bowerbankia

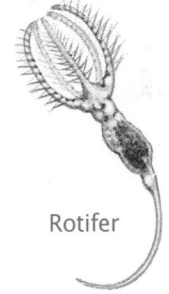

Rotifer

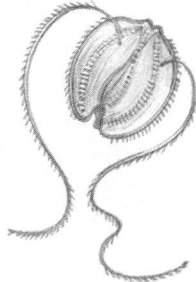

Comb jelly

Glossary

Pronunciation Key

Sound	As In	Phonetic Respelling
a	b**a**t	(BAT)
ay	f**a**ce	(FAYS)
ah	l**o**ck	(LAHK)
	argue	(AHR gyoo)
ow	**ou**t	(OWT)
ch	**ch**apel	(CHAP uhl)
eh	t**e**st	(TEHST)
ai	r**a**re	(RAIR)
ee	**ea**t	(EET)
	f**ee**t	(FEET)
	sk**i**	(SKEE)
ih	b**i**t	(BIHT)
eye	**i**dea	(eye DEE uh)
y	r**i**pe	(RYP)

Sound	As In	Phonetic Respelling
ihng	go**ing**	(GOH ihng)
k	**c**ard	(KAHRD)
	kite	(KYT)
ng	a**n**ger	(ANG guhr)
oh	**o**ver	(OH vuhr)
aw	d**o**g	(DAWG)
	h**o**rn	(HAWRN)
oy	f**oi**l	(FOYL)
u	p**u**ll	(PUL)
oo	p**oo**l	(POOL)
s	**c**ell	(SEHL)
	sit	(SIHT)
sh	**sh**eep	(SHEEP)

Sound	As In	Phonetic Respelling
th	**th**at	(THAT)
	thin	(THIHN)
uh	c**u**t	(CUHT)
ur	f**er**n	(FURN)
y	**y**es	(YEHS)
yoo	gl**o**bule	(GLAHB yool)
yu	c**u**re	(KYUR)
z	bag**s**	(BAGZ)
zh	trea**s**ure	(TREHZH uhr)
uh	med**a**l	(MEHD uhl)
	tal**e**nt	(TAHL uhnt)
	penc**i**l	(PEHN suhl)
	onion	(UHN yuhn)
	playf**u**l	(PLAY ful)
	d**u**ll	(DUHL)
uhr	pap**er**	(PAY puhr)

A

acid: compound capable of donating a hydrogen ion to a water molecule; from the Latin *acere*, meaning "to be sour" (p. 295)

acid rain: rain or any precipitation with unusually high acidity (p. 295)

acoelomate (ay SEEL oh mayt): term meaning without a body cavity; from the Greek *a + koilia*, meaning "not + hollow" (p. 464)

action potential: sudden local reversal of electrical voltage that is carried along the membrane of a neuron (p. 641)

activation energy: amount of energy needed to start a chemical reaction (p. 90)

active site: pocket formed in the folds of an enzyme; reaction catalyzed by the enzyme occurs at the active site (p. 92)

active transport: one-way transport of substances across a membrane; requires energy (p. 77)

adaptation: process by which a species becomes better suited to its environment (p. 188)

addiction: a cell membrane's physiological response to the presence of psychoactive drug molecules (p. 694)

adrenal gland: either of a pair of endocrine organs lying atop the kidney and producing over two dozen hormones; from the Latin *ad + renes*, meaning "at + kidneys" (p. 677)

aerobic exercise (ehr OH bihk): steady, low-intensity exercise; from the Greek *aër* and *bias*, meaning "air" and "life" (p. 631)

AIDS: **A**cquired **I**mmune **D**eficiency **S**yndrome, a condition in which the immune system has been disabled; caused by HIV (p. 353)

allele (uh LEEL): one of two or more alternate forms of a gene; from the Greek *allēlōn*, meaning "of one another" (p. 119)

allergy: immune system response to harmless antigens such as pollen; from the Greek *all + ergon*, meaning "different + work" (p. 748)

alternation of generations: a life cycle in plants, in which a haploid form alternates with a diploid form (p. 391)

alveolus (al VEE oh luhs): tiny air sac within the lungs, through which gases enter and leave the circulatory system; from the Latin *alveus*, meaning "hollow cavity" (p. 729)

amnion (AM nee uhn): thin membrane forming a closed sac around the embryo; from the Greek *amnos*, meaning "lamb" (p. 495)

amniotic egg: the egg of reptiles, birds, and mammals; the embryo develops within two protective membranes, the chorion and amnion (p. 495)

anaerobic exercise (an uh ROH bihk): exercise that requires muscles to work at high intensity for a very brief period; from the Greek *an + aër* and *bios*, meaning "not + air" and "life" (p. 630)

analogous characters: features of organisms that are similar but evolved independently; from the Greek *analogos*, meaning "proportionate" (p. 326)

anemia (uh NEE mee uh): condition in which the body has too few red blood cells or not enough hemoglobin; from the Greek *anaimia*, meaning "bloodless" (p. 712)

angiosperm (AN jee uh sperm): flowering plant having seeds produced within an ovary; from the Greek *angeion* and *speirein*, meaning "case" and "to sow" (p. 396)

anorexia nervosa (an oh RECK see ah ner VOH sah): eating disorder in which sufferers starve themselves; from the Greek *an* + *oregein*, meaning "without + appetite" (p. 767)

anther: part of the stamen that produces and releases pollen; from the Greek *anthēros*, meaning "blooming" (p. 415)

anthropoid (AN thruh poyd): member of the primate suborder that contains monkeys, apes, and humans; from the Greek *anthrōpos*, meaning "man" (p. 223)

antibiotic: chemical substance with capacity to destroy or inhibit growth of bacteria; from the Greek *anti* + *bios*, meaning "against + life" (p. 352)

antibody: defensive protein produced by white blood cells in response to a foreign substance, released by B cells (pp. 166, 744)

anticodon: nucleotide sequence of transfer RNA (tRNA) molecule, which is complementary to an mRNA codon; from the Latin *anti-* + *code*, meaning "against + book" (p. 144)

antigen: molecule that can be recognized by white blood cells and that can trigger a

defensive response (p. 741)

antihistamine: chemical that blocks the action of histamines (p. 748)

anuran (uh NUR uhn): member of the order Anura, which contains frogs and toads; from the Greek *an* + *oura*, meaning "without + tail" (p. 564)

anus: posterior opening of the digestive tract; from the Latin *anus*, meaning "ring" (p. 465)

arachnid (uh RAK nihd): any of the animals of the class Arachnida, such as spiders, scorpions, mites, and ticks; from the Greek *arachnē*, meaning "spider" (p. 525)

archaebacteria (AHR kee bahk TIHR ee ah): distinctive group of bacteria thought to resemble early bacteria; from the Greek *archaikos* + *baktērion*, meaning "ancient + small staff" (p. 203)

artery: blood vessel that carries blood away from the heart (p. 714)

asthma (AZ muh): respiratory disease in which airways in the lungs become restricted because of sensitivity to stimuli; from the Greek *asthma*, meaning "panting" (p. 732)

atom: smallest unit of matter that cannot be broken down by chemical means; from the Greek *a* + *tomos*, meaning "not + cut" (p. 30)

ATP: adenosine triphosphate, used by cells to carry energy (p. 95)

atrium (AY tree uhm): chamber of the heart that receives blood from the body; from the Latin *atrium*, meaning "large room" (p. 718)

autoimmune disease: disease in which the ability of the immune system to distinguish self from non-self breaks down and the body attacks its own cells (p. 749)

autonomic nervous system: component of the nervous system that carries messages to muscles and glands, and that usually works without our noticing (p. 654)

autotroph (AWT oh trahf): organism that makes its own food; from the Greek *auto-* + *trophē*, meaning "self + food" (p. 251)

axon (AK sahn): long extension of a nerve cell that carries impulses away from the cell body; from the Greek *axōn*, meaning "axis" (p. 639)

balancing selection: situation in which natural selection for and against an allele balance out, so the allele is maintained (p. 189)

B cell: white blood cell that produces and releases antibodies (p. 744)

bilateral symmetry: characteristic of an animal that can be separated into nearly mirror-image halves by cutting it lengthwise along the midline of the body; from the Latin *bi-* + *latus* and *symmetria*, meaning "two + side" and "measured together" (p. 462)

binocular vision (beye NAHK yuh luhr VIHZH uhn): vision in which both eyes simultaneously view the same object; from the Latin *bi-* + *ocularis* and *videre*, meaning "double + of the eye" and "to see" (p. 222)

biochemical pathway: linked chemical reactions in which the product of one serves as the reactant in the next (p. 95)

biology: study of living things; from the Greek *bios* and *logos*, meaning "life" and "study of" (p. 5)

biome (BEYE ohm): major type of terrestrial ecological community, such as grassland and desert; from the Greek

bi- and the Latin *oma*, meaning "life" and "mass" (p. 258)

bipedal (beye PEHD uhl): walking on two legs; from the Latin *bi-* + *pedalis*, meaning "double + of the foot" (p. 228)

blood pressure: force exerted on artery walls when ventricles of the heart contract (p. 724)

blood type: classification of blood according to the presence or absence of specific marker proteins found on the surface of red blood cells (p. 716)

blood vessel: any of the tubes that carry blood from the heart to the organs of the body and back again; from the Latin *vas*, meaning "vessel" (p. 714)

body plan: overall organization of an animal's body (p. 455)

book lung: highly folded sacs used for gas exchange that are located inside the body of most arachnids and that open to the outside through a spiracle (p. 489)

brain stem: stalklike structure that connects the cerebrum to the spinal cord (p. 649)

bulimia (buh LEEM ee uh): eating disorder in which sufferers engage in binge-purge cycles of eating followed by vomiting; from the Greek *boulimia*, meaning "ox hunger" (p. 767)

cancer: a condition in which normal restraints on cell reproduction break down, resulting in rapid, uncontrolled cell division; from the Greek *karkinōma*, meaning "crab" or "cancer" (p. 82)

capillary: one of the network of tiny blood vessels from which substances in the blood move into surrounding cells; from the Latin *capillus*, meaning "hair" (p. 714)

capillary action: force of attraction that causes liquid to move up narrow tubes (p. 410)

carbohydrate: compound composed of carbon, hydrogen, and oxygen, used by living things as a source of energy; from the Latin *carbon* and Greek *hydōr*, meaning "coal" and "water" (p. 34)

carcinogen (kahr SIN uh juhn): substance that causes cancer; from the Greek *karkinōma*, meaning "crab" or "cancer" (p. 293)

cardiac muscle: muscle found in the heart; from the Greek *kardia* and the Latin *musculus*, meaning "heart" and "little mouse (muscle)" (p. 626)

cardiovascular disease: disease of the heart or blood vessels (p. 725)

carnivore (KAHR nuh vawr): flesh-eating organism; from the Latin *caro* and *vorare*, meaning "flesh" and "to devour" (p. 251)

catalyst (KAT uh lihst): material that speeds up a chemical reaction without being used itself; from the Greek *katalyein*, meaning "to dissolve" (p. 91)

cell: smallest unit that can perform all the life processes; from the Latin *cella*, meaning "small room" (p. 45)

cell membrane: bilipid layer surrounding the cytoplasm, essential to the cell's metabolism, also called the plasma membrane; from the Latin *membrana*, meaning "skin" (p. 45)

cell surface marker: membrane protein of a cell that distinguishes it from other cells and foreign matter (p. 71)

cellular respiration: process by which living things obtain energy from the bonds of food molecules; from the Latin *respirare*, meaning "to breathe" (p. 96)

cell wall: firm, nonliving covering that encloses and supports most plant, fungal, and bacterial cells and some protist cells (p. 60)

central nervous system: brain and spinal cord (p. 647)

cephalization (sehf uh lih ZAY shuhn): evolution of a definite head end; from the Greek *kephalikos*, meaning "head" (p. 462)

cerebellum (sehr uh BEHL uhm): small structure in the rear part of the brain that controls balance, posture, and coordination; from the Latin *cerebellum*, diminutive for "brain" (p. 649)

cerebrum (suhr REE bruhm): large rounded area of the brain divided into left and right hemispheres; from the Latin *cerebrum*, meaning "brain" (p. 648)

cervix: ring of strong muscles forming the lower part of the uterus; from the Latin *cervix*, meaning "neck" (p. 788)

chelicera (kuh LIHS uh ruh): one of the first pair of appendages of arachnids and their relatives; used for feeding, and in some species as poison-delivering fangs; from the Greek *chēlē*, meaning "claw" (p. 525)

chemical bond: attractive force that holds two atoms together in a molecule; from the Latin *alchymia*, meaning "alchemy" (p. 31)

chemiosmosis (kehm ee ahz MOH sihs): the mechanism of energy release by the action of proton pumps; from the Greek *othein*, meaning "to push" (p. 78)

chemosynthesis: production of food using the energy contained in inorganic molecules; from the Greek *syn-* + *tithenai*, meaning "together + to place" (p. 343)

chlorofluorocarbon (KLAWR oh FLOH roh KAHR buhn): any of a group of compounds that contain carbon, chlorine, and fluorine, often used as coolants, propellants, or foaming agents (p. 296)

chloroplast (KLAWR uh plast): carbohydrate-producing organelles of plants and protists; from the Greek *chlōros*, meaning "pale green" (p. 58)

chorion (KAWR ee ahn): watertight protective membrane surrounding the embryo of reptiles, birds, and mammals; from the Greek *chorion*, meaning "fetal membrane" (p. 495)

chromosome (KROH muh sohm): threadlike structure formed in the cell during prophase of mitosis that contains DNA; from the Greek *chrōma*, meaning "color" (pp. 37, 80)

cilia (SIL ee uh): hairlike projections on certain cells and mobile organisms; from the Latin *celare*, meaning "to conceal" (p. 368)

circulatory system: network of blood-carrying vessels that transports nutrients, oxygen, wastes and carbon dioxide; from the Latin *circulari*, meaning "to form a circle" (p. 466)

cirrhosis (suh ROH sihs): liver disease in which the liver swells with fat deposits and cannot function properly, usually caused by excessive alcohol consumption; from the Greek *kirrhos*, meaning "tawny" (p. 702)

cladogram (KLAD uh gram): branching tree diagram showing evolutionary relationships among a set of organisms; from the Greek *klados*, meaning "branch" (p. 327)

class: collection of similar orders (p. 324)

cloning (KLOHN ihng): growing a large number of genetically identical cells from one cell; from the Greek *klōn*, meaning "twig" (p. 158)

coacervate (KOH ah SUHR vayt): spherical droplet surrounded by layer of lipids; may resemble early stage of evolution of cells; from the Latin *coacerverare*, meaning "to heap up" (p. 202)

codon (KOH dahn): basic unit of genetic code, triplet nucleotides in mRNA specifying a particular amino acid; from the Latin *code*, meaning "book" (p. 143)

coelacanth (SEE luh kanth): any member of an order of bony fishes formerly believed to be extinct; from the Greek *koilos* and *akantha*, meaning "hollow" and "thorn" (p. 559)

coelom (SEE luhm): fluid-filled body cavity that separates the muscles of the body wall from the muscles that surround the gut (p. 466)

coevolution (KOH ev uh LOO shuhn): situation in which two or more species evolve in response to each other; from the Latin *co-* + *evolvere*, meaning "with + to roll" (p. 267)

commensalism (kuh MEN suhl ihz uhm): relationship between two organisms in which one benefits and the other neither benefits nor is harmed; from the Latin *com-* + *mensa*, meaning "with + table" (p. 271)

community: all the organisms that live in an ecosystem (p. 248)

competition: situation in which two or more organisms attempt to use the same resource (p. 273)

competitive exclusion: process in which one species is outcompeted and dies out within an ecosystem (p. 273)

complete metamorphosis: series of abrupt changes in form of young insects as they develop into adults; stages include larva, nymph, pupa (or chrysalis), adult (p. 534)

cone: receptor cell in the retina of the eye that detects color, produces sharp images, and aids in seeing in bright light; from the Greek *kōnos*, meaning "cone" (p. 661)

connective tissue: bones, blood, ligaments, and tendons; from the Latin *texere*, meaning "to weave" (p. 612)

consumer: organism that obtains its energy by eating other organisms; from the Latin *consumere*, meaning "to take up" (p. 250)

contour feather: feathers that cover the body of a bird giving shape to its wings and tail (p. 585)

control experiment: a comparison of an experimental group with a control group (p. 13)

convergent evolution: process whereby organisms that are not related to each other evolve similar features (p. 326)

copepod (KOH puh pahd): very small, sometimes parasitic, crustacean living in either salt water or fresh water; from the Greek *kōpē*, meaning "oar" (p. 542)

cotyledon (kaht uh LEED uhn): leaflike structure in a plant embryo that stores food or helps absorb food stored elsewhere; from the Greek *kotylēdon*, meaning "cavity" (p. 394)

covalent bond: nonionic chemical bond formed by shared electrons; from the Latin *co-* + *valere*, meaning "with + to be strong" (p. 32)

crop rotation: alternating cultivation of two or more crops in the same field (p. 441)

crossing over: exchange of genetic material between paired homologous chromosomes during meiosis (p. 126)

cross-pollination: transfer of pollen to another plant of the same species (p. 416)

cuticle (KYOOT ih kuhl): surface layer of a vascular plant; from the Latin *cutis*, meaning "skin" (p. 391)

cyanobacteria (SY ahn oh bak TIHR ee ah): group of photosynthetic bacteria thought to be the first organisms to release oxygen; from the Greek *cyano-* + *bactērion*, meaning "blue + small staff" (p. 204)

cytokinesis (syt oh kuh NEE sihs): division of the cytoplasm following nuclear division; from the Greek *kytos* and *kinēsis*, meaning "a hollow" and "motion" (p. 80)

cytoplasm (SYT uh plaz uhm): material inside the cell membrane containing the necessary components for cell life; from the Greek *plassein*, meaning "to form" (p. 56)

 D

decapod (DEHK uh pahd): a crustacean such as a lobster or a shrimp with 10 legs; from the Greek *deka* and *pous*, meaning "ten" and "foot" (p. 541)

decomposer: organism that obtains its energy by consuming organic wastes; from the Latin *de-* + *componere*, meaning "from + to put together" (p. 250)

dendrite (DEHN dryt): short branch of a nerve cell that carries impulses toward the cell body; from the Greek *dendron*, meaning "tree" (p. 639)

derived character: a unique characteristic, such as hair, that defines a group of organisms (p. 327)

dermal tissue: layer of tightly packed, flattened cells found chiefly on plant surfaces; from the Greek *derma* and the Latin *texere*, meaning "skin" and "to weave" (p. 406)

dermis: layer of skin below the epidermis (p. 616)

deuterostome (DOOT uhr oh stohm): animal in which the first opening in the embryo does not form the mouth, but often becomes the anus; from the Greek *deuteros* and *stoma*, meaning "second" and "mouth" (p. 471)

development: process of growth and cell specialization (p. 19)

diabetes mellitus (deye uh BEET eez muh LYT uhs): disease in which cells are unable to absorb glucose from blood and tissue fluids; from the Latin *diabetes* and *mellitus*, meaning "siphon" and "honeyed" (p. 682)

diffusion (dif FYOO zhuhn): movement of molecules from a region of higher concentration to one of lower concentration; from the Latin *diffusus*, meaning "poured in different directions" (p. 74)

diploid (DIHP loyd): condition in which a chromosome set is present in duplicate; from the Greek *diploos*, meaning "double" (p. 124)

direct development: development pattern in arthropods in which the young are miniature copies of adults (p. 534)

directional selection: situation in which selection for or against an allele in a population is unopposed (p. 189)

divergence: accumulation of differences between two or more species or populations; from the Latin *divergere*, meaning "to turn apart" (p. 191)

diversity: measure of the number of species living in an ecosystem scaled by how common each species is (p. 248)

division: substitute for phylum in classification of plants, bacteria, and fungi (p. 324)

DNA fingerprint: a photograph with a pattern of dark bands that reflects the composition of an organism's DNA (p. 168)

DNA profiling: the technique used to identify the base sequence in a sample of DNA (p. 168)

dominant: describes the member of a pair of alleles in heterozygous individuals that is expressed (p. 119)

double fertilization: an event in angiosperms wherein one sperm fuses with an egg and a second sperm fuses with two nuclei (p. 417)

double helix: spiral of two strands, as in DNA structure; from the Greek *helissein*, meaning "to turn around" (p. 139)

down feather: fine feathers that grow underneath contour feathers and that are specialized for insulation (p. 585)

ecological race: a population of a species that differs genetically from other populations of the same species because they have adapted to different environments (p. 190)

ecology: study of relationships of organisms to their environment; from the Greek *oikos* and *logos*, meaning "house" and "study of" (p. 248)

ecosystem: self-sustaining collection of organisms and their physical environment (p. 248)

ectoderm: outer layer of cells of an embryo; from the Greek *ektos* and *derma*, meaning "outside" and "skin" (p. 459)

ectopic pregnancy: pregnancy in which the embryo grows somewhere other than in the uterus; from the Greek *ektopos*, meaning "out of place" (p. 804)

ectotherm (EHK tuh thurm): animal that uses heat from the environment to control its body temperature; from the Greek *thermē*, meaning "heat" (p. 214)

ejaculation (ee jak yoo LAY shuhn): expulsion of sperm from the male's body; from the Latin *ejaculari*, meaning "to throw out" (p. 785)

electron: elementary particle with negative electric charge; from the Greek *elektron*, meaning "amber" (p. 30)

element: matter that is made of only one kind of atom; from the Latin *elementum*, meaning "first principle" (p. 30)

embryo: fertilized egg that has begun to divide; from the Greek *en-* + *bryein*, meaning "in + to swell" (p. 790)

emphysema (em fuh SEE muh): condition of the lung in which it has lost its elasticity, greatly reducing the efficiency of gas exchange, often caused by smoking; from the Greek *emphysēma*, meaning "inflation" (p. 732)

endergonic (ehn duhr GAHN ihk) **reaction:** biochemical reaction requiring the absorption of energy; from the Greek *end-* + *ergon*, meaning "within + work" (p. 90)

endocrine (EN duh krihn) **gland:** an organ that is part of the endocrine system, which produces most of the body's hormones; from the Greek *krinein*, meaning "to separate" (p. 669)

endocytosis (ehn doh seye TOH sihs): process in which outside materials are engulfed by the cytoplasm of cells; from the Greek *kytos*, meaning and "hollow" (p. 78)

endoderm: inner layer of cells of the embryo; from the Greek *derma*, meaning "skin" (p. 459)

endoplasmic reticulum (ehn duh PLAZ mihk rih TIK yuh luhm): network of membranes in the cytoplasm that transports substances made by the cell; from the Greek *plassein* and Latin *reticulum*, meaning "to form" and "network" (p. 59)

endoskeleton: internal supporting structure, such as your bones; from the Greek *skeletos*, meaning "dried up" (p. 472)

endotherm (EHN doh thurm): animal that uses heat produced by metabolism to control its body temperature; from the Greek *thermē*, meaning "heat" (p. 214)

enkephalin (ihn KEHF uh luhn): neuromodulator that helps block pain messages from traveling to the brain; a type of endorphin; from the Greek *en-* + *kephalē*, meaning "in + head" (p. 696)

epidermis: outermost layer of the skin, consisting of several layers of cells; from the Greek *epi-* + *derma*, meaning "upon + skin" (p. 618)

epididymis (ehp uh DIHD ih mihs): duct on top of and behind the testis that temporarily stores sperm; from the Greek *didymos*, meaning "testicle" (p. 784)

epithelial (ehp uh THEE lee uhl) **tissue:** tightly connected cells, arranged in flat sheets, such as the skin (p. 611)

esophagus (ih SAHF uh guhs): muscular tube from the pharynx that permits food to descend to the stomach; from the Greek *oisein* and *phagein*, meaning "to be going to carry" and "to eat" (p. 769)

eubacteria (YOO bak TIHR ee ah): group of bacteria thought to have evolved more recently than the archaebacteria; from the Greek *eu-* + *baktērion*, meaning "true + small staff" (p. 203)

eukaryote (yoo KAR ee oht): organisms whose cells contain a membrane-bound nucleus and membrane-bound organelles, such as plants, animals, fungi, and protists; from the Greek *karyon* and *-otes*, meaning "nut (kernel)" and "inhabitant" (p. 56)

evolution (ehv uh LOO shuhn): genetic change in a species over time (p. 17)

exergonic (ehk suhr GAHN ihk) **reaction:** chemical reaction that releases energy; from the Greek *ex-* + *ergon*, meaning "from + work" (p. 90)

exhalation: exit of air from the lungs during breathing, equalizing air pressure (p. 730)

exocrine gland: cell or group of cells that produces and releases secretions onto a body surface; from the Greek *krinein*, meaning "to separate" (p. 611)

exocytosis (ehk soh sye TOH sihs): extrusion of material from cell; from the Greek *kytos*, meaning "hollow" (p. 78)

exon: sequence in the genetic code that supplies the information for protein formation (p. 150)

exoskeleton: supporting structure that surrounds the body; from the Greek *skeletos*, meaning "dried up" (p. 471)

external fertilization: fertilization that takes place outside the body of either parent (p. 495)

facilitated diffusion: transport of substances through channels in either direction through the cell membrane; from the Latin *facere* and *diffusus*, meaning "to do or make" and "poured in different directions" (p. 76)

Fallopian tube: tube that carries eggs released from the ovaries to the uterus; named for Gabriel Fallopius, Italian anatomist (p. 787)

family: group of closely related genera (p. 324)

feedback inhibition: slowing or stopping of an early reaction in a biochemical pathway when levels of the end product become high (p. 106)

fermentation: breakdown of organic compounds such as glucose; from the Latin *fermentum*, meaning "leavening, yeast" (p. 103)

fetal alcohol syndrome (FAS): physical and mental disabilities in children born of women who drink during pregnancy; from the Latin *fetus* and the Greek *syn-* + *dramein*, meaning "offspring" and "with + to run" (p. 702)

fetus: the embryo from the ninth week of development until birth (p. 795)

filtration: first stage of urine formation, in which smaller solutes are forced out of the blood (p. 775)

flagellum (fluh JEHL uhm): whiplike structure of some cells, used for locomotion; from the Latin *flagellum*, meaning "whip" (p. 341)

follicle (FAHL ih kuhl): fluid-filled chamber within the ovary in which an egg matures; from the Latin *folliculus*, meaning "small bag" (p. 789)

fracture: broken bone; from the Latin *fractus*, meaning "broken" (p. 622)

fruit: enlarged ovary of a flowering plant that contains seeds; from the Latin *fructus*, meaning "enjoyment" (p. 417)

fundamental niche: total niche that an organism could potentially use within an ecosystem (p. 272)

fungus (FUHN guhs): members of the kingdom Fungi, heterotrophs with chitin cell walls (p. 385)

gametophyte (guh MEET uh fyt): haploid form of a plant's life cycle during which gametes are produced; from the Greek *gamein* and *phyein*, meaning "to marry" and "to grow" (p. 391)

gastrulation (gas troo LAY shuhn): process of forming layers of cells in the embryo; from the Greek *gastēr*, meaning "stomach" (p. 459)

gated ion channel: a protein channel in a plasma membrane enabling ions to pass in and out, creating an electrical current for the transmission of information (p. 640)

gene: section of chromosome that contains information coding for a trait (pp. 18, 123)

gene expression: use of genetic information in DNA to make proteins (p. 141)

genetic code: the correspondence between nucleotide triplets in DNA and the amino acids in proteins; from the Greek *gignesthai* and the Latin *code*, meaning "to be born" and "book" (p. 143)

genetic engineering: intentional production of new genes and alteration of genomes by the substitution or addition of new genetic material (p. 157)

genetics (juh NEHT ihks): science of biological inheritance (p. 117)

genotype (JEE nuh typ): constitution of an organism in terms of its genes; from the Greek *genos*, meaning "race, kind" (p. 119)

genus (JEE nuhs): category in biological classification in which closely related species are grouped; from the Latin *genus*, meaning "race, kind" (p. 320)

germination: to sprout or start to grow; from the Latin *germinare*, meaning "to sprout" (p. 419)

gill: one of the structures used by aquatic animals for gas exchange (p. 486)

glycolysis (gly KAHL uh sihs): process of enzymatic breakdown and the release of energy in the cell's cytoplasm; from the Greek *glykys* and *lysis*, meaning "sweet" and "dissolving" (p. 102)

goiter (GOY tuhr): swollen thyroid gland; from the Latin *guttur*, meaning "throat" (p. 680)

Golgi (GOHL jee) **body**: organelle that delivers substances released by the endoplasmic reticulum of the cell; named after Camillo Golgi, Italian neurologist (p. 59)

gradualism: hypothesis that evolution occurs at a slow, constant rate (p. 192)

grain: small seedlike fruit of cereal plants; from the Latin *granum*, meaning "seed, kernel" (p. 427)

Gram staining: use of dyes to stain bacterial cell walls to aid in identification and diagnosis; named after Hans Christian Gram, Danish physician (p. 340)

greenhouse effect: warming of Earth's atmosphere resulting from heat trapped by carbon dioxide and other gases (p. 256)

Green Revolution: intensive plant breeding program to produce high-yield crops, started by Norman Borlaug (p. 442)

ground tissue: bulk of a plant's body, some parts of which are for storage, the rest for strength (p. 406)

guard hair: long, thick outer hair responsible for the color of an animal's coat (p. 595)

gut: internal passage through which food passes while being digested (p. 458)

gymnosperm (JIHM nuh sperm): plant that produces its seeds in cones; from the Greek *gymnos* and *speirein*, meaning "naked" and "to sow" (p. 395)

habitat: physical location where an organism lives in an ecosystem; from the Latin *habitare*, meaning "to inhabit" (p. 248)

half-life: the time it takes for one-half of the radioactive atoms in a sample to decay (p. 181)

hallucinogen (huh LOOS uh nuh jehn): drug that distorts the way the brain translates signals from the sensory organs; from the Latin *hallucinari*, meaning "to wonder" (p. 704)

haploid (HAP loyd): condition in which there is only one chromosome of each pair in a nucleus; from the Greek *haploos*, meaning "single" (p. 124)

hemodialysis (hee moh deye AL uh sis): process of removing blood from an artery, purifying it by dialysis, and returning it to a vein; from the Greek *haima* and *dialyein*, meaning "flowing blood" and "to dissolve" (p. 776)

herbivore (HUR buh vawr): consumer that eats only plants; from the Latin *herba* and *vorare*, meaning "grass" and "to devour" (p. 251)

heredity (huh RED ih tee): transmission of characteristics from parent to offspring by means of genes on the chromosomes (p. 27)

hermaphrodite (huhr MAHF roh deyet): animal with both female and male sexual organs; from the Greek myth in which Hermaphroditus shared a body with a nymph (p. 507)

heterotroph (HET uhr oh trohf): organism that cannot make its own food and feeds on other organisms or organic wastes; from the Greek *hetero-* + *trophē*, meaning "different + food" (p. 251)

heterozygous (heht uh roh ZY guhs): having two different alleles for a trait; from the Greek *zygon*, meaning "yoke" (p. 119)

histamine (HIHS tuh meen): chemical messenger that can increase mucus production and stimulate capillaries to swell and release fluids (p. 748)

HIV: **H**uman **I**mmunodeficiency **V**irus, the RNA virus that causes AIDS (p. 353)

homeostasis (hoh mee oh STAY sihs): tendency to maintain stability in an organism amid environmental change; from the Greek *homo-* + *histanai*, meaning "similar + to stand" (pp. 17, 27)

hominid (HAHM ih nihd): two-legged primates such as humans and australopithecines; from the Latin *homo*, meaning "man" (p. 226)

homologous structures (hoh MAHL uh guhs): structures that have a common ancestry; from the Greek *homologos*, meaning "agreeing" (p. 182)

homozygous (hoh muh ZY guhs): having identical alleles for a trait; from the Greek *zygon*, meaning "yoke" (p. 119)

hormone: chemical signal that travels through the bloodstream and regulates the body's activities; from the Greek *hormeinō*, meaning "to stimulate" (p. 669)

human genome: entire collection of genes within human cells (p. 168)

Human Genome Project: effort to determine the nucleotide sequence of every human gene (p. 168)

hybrid: offspring that results from inter-breeding between members of different species (p. 329)

hydrogen bond: linkage formed when a hydrogen atom in one molecule makes an additional bond to an atom in another molecule; from the Greek *hydōr* + *genēs*, meaning "water + born" (p. 48)

hydroponics: cultivation of plants in nutrient solutions instead of soil; from the Greek *ponein*, meaning "to toil" (p. 444)

hypertension: condition in which the blood pressure is continually higher than normal; from the Greek *hyper* + the Latin *tensus*, meaning "above, over + stretched" (p. 725)

hyperthyroidism: overproduction of thyroxine (p. 679)

hypha (HY fuh): one of the slender filaments that make up the body of a fungus; from the Greek *hyphē*, meaning "web" (p. 385)

hypothalamus (heye poh THAL uh muhs): part of the upper end of the brain stem that regulates activities in the nervous and endocrine systems; from the Greek *hypo-* + *thalamos*, meaning "under + inner chamber" (p. 649)

hypothesis: a testable explanation for an observation; from the Greek *hypotithenai*, meaning "to put under" (p. 12)

hypothyroidism: condition in which insufficient thyroxine is produced (p. 680)

immune response: immune system's attack on a specific pathogen; from the Latin *in* + *munus*, meaning "in + service" (p. 742)

implantation: attachment of the embryo to the uterine wall at the beginning of pregnancy (p. 795)

incomplete metamorphosis: the gradual change of a body from young to adult stage (p. 534)

inducer: molecule that binds to the repressor protein, allowing for transcription of DNA; from the Latin *inducere*, meaning "to lead into" (p. 149)

inflammatory response: the body's response to injury, indicated by redness and swelling; from the Latin *flammare*, meaning "to flame" (p. 740)

inhalation: entry of air into the lungs during breathing, equalizing air pressure (p. 730)

insulin: hormone that enables cells to take in glucose molecules; from the Latin *insula*, meaning "island" (p. 681)

integration: interaction of signals from both excitatory and inhibitory synapses, which tend to reinforce or cancel one another; from the Latin *integrare*, meaning "to make whole" (p. 645)

internal fertilization: fertilization that occurs inside the female's body (p. 495)

introns: noncoding portions of the DNA (p. 150)

ion: an atom that has lost or gained one or more electrons and therefore has a net electric charge; from the Greek *ienai*, meaning "to go" (p. 31)

ionic bond: force of attraction between oppositely charged ions (p. 31)

joint: place where two or more bones connect; from the Latin *junctus*, meaning "joined" (p. 625)

keystone species: species whose niche affects many other species in the ecosystem so it cannot be readily replaced if lost (p. 275)

kidney: one of two organs located toward the back of the abdomen that maintains homeostasis, removes wastes, regulates water in the blood, and adjusts concentrations of other substances in the blood (p. 775)

kingdom: collection of similar phyla (p. 324)

labor: final stage of pregnancy, when the uterus contracts strongly enough to push the baby out of the mother's body (p. 800)

large intestine: organ of the digestive system that stores, compacts, and then eliminates indigestible material; from the Latin *intestinus*, meaning "inward" (p. 771)

lateral line: row of pressure-sensitive cells within a fluid-filled canal on each side of a fish; from the Latin *latus* meaning "side" (p. 484)

legume (LEHG yoom): member of an order of plants that includes peas and beans; from the Latin *legumen*, meaning "that which can be gathered" (p. 430)

lichen (LEYE kihn): organism consisting of a fungus and an alga living in a mutualistic relationship; from the Greek *leichein*, meaning "to lick" (p. 388)

ligament: strong, elastic bands of connective tissue joining bones together; from the Latin *ligare*, meaning "to bind" (p. 625)

lipid: organic compound containing fats, alcohol, and waxes that organisms may use to store energy; from the Greek *lipos*, meaning "fat" (p. 35)

lipid bilayer: the double layer of phospholipids that makes up organelle and cell membranes (p. 50)

liver: very large glandular organ that secretes bile and modifies substances contained in the blood (p. 771)

lobe-finned fishes: any of a group of fishes having fleshy fins supported by a series of bones (p. 559)

lung cancer: cancer in the lung, often caused by carcinogens in tobacco smoke (p. 732)

lymphatic system: vessels that return to the circulatory system excess fluids that bathe the cells; from the Latin *lympha*, meaning "spring water" (p. 717)

macromolecule: large molecule composed of several simple structural units, such as DNA or proteins; from the Greek *makros* and the Latin *moles*, meaning "long" and "mass" (p. 33)

macrophage (MAK roh fayj): type of phagocyte that fights pathogens; from the Greek *phagein*, meaning "to eat" (p. 740)

Malpighian (mal PIHG ee uhn) **tubule**: organ responsible for filtering blood in insects and spiders; named after Italian physiologist Marcello Malpighi (p. 493)

mammary gland: gland of female mammals that secretes milk for suckling the young; from the Latin *mamma*, meaning "breast" (p. 591)

mandible: jaw of certain animals; from the Latin *mandibulum*, meaning "jaw" (p. 532)

marrow: spongelike center of long bones, that produces blood cells (p. 621)

marsupial (mahr SOO pee uhl): mammal that lacks a placenta and has an external pouch for young; from the Greek *marsypos*, meaning "pouch" (p. 592)

mass extinction: an episode during which large numbers of species became extinct (p. 205)

medusa: one of the two life stages of cnidarians, the jellyfish stage; from Greek mythology, a snake-haired beast-woman called Medousa (p. 506)

meiosis (my OH sihs): cell division that reduces the number of chromosomes in a cell by half; from the Greek *meioun*, meaning "to make smaller" (p. 124)

melanin (MEHL uh nihn): brown-colored product of skin cells; from the Greek *melas*, meaning "black" (p. 618)

memory cell: B cell that remains after infection and continues to patrol the body's tissues (p. 745)

menopause: shutdown of menstrual and ovarian cycles; from the Latin *mens*, meaning "month" (p. 791)

menstrual (MEHN struhl) **cycle**: the series of changes that occur in the uterus each month (p. 790)

meristem (MEHR uh stehm): plant tissue consisting of actively growing and dividing cells that give rise to permanent tissues; from the Greek *meristos*, meaning "divided" (p. 406)

mesoderm: layer of tissue between the endoderm and ectoderm that develops into muscle, reproductive organs, and circulatory vessels; from the Greek *mesos* and *derma*, meaning "middle" and "skin" (p. 464)

metabolism (muh TAB uh liz uhm): physical and chemical changes in an organism in which energy is released or used; from the Greek *metabolē*, meaning "change" (pp. 27, 90)

metamorphosis (meht uh MAWR fuh sihs): change in form, structure, or function of an animal as a result of development; from the Greek *meta-morphē*, meaning "change" (p. 534)

microvillus: any of the hairlike projections coating the villi of the small intestine; from the Greek *mikros* and the Latin *villus*, meaning "small" and "tuft of hair" (p. 770)

mineral: inorganic compound, one not made from living things; from the Latin *minera*, meaning "ore" (p. 766)

mitochondrion (myt uh KAHN dree uhn): energy-producing cellular organelles in eukaryotes; from the Greek *mitos* and *chondros*, meaning "thread" and "grain" (p. 58)

mitosis (meye TOH sihs): nuclear division involving duplication and separation of the chromosomes; from the Greek *mitos*, meaning "thread" (p. 80)

molecule: group of atoms that form the smallest unit of a substance that can exist and retain its chemical properties; from the Latin *moles*, meaning "mass" (p. 32)

monotreme (MAHN oh treem): mammal that lays eggs and has a single opening for the digestive and urinary tracts and for reproduction; from the Greek *monos* and *trēma*, meaning "single" and "hole" (p. 592)

motor neuron: nerve cell carrying information from the central nervous system to a muscle or gland (p. 652)

mucous (MYOO kuhs) **membrane**: epithelial layer that releases mucus and is impenetrable to most pathogens; from the Latin *mucus*, meaning "slimy secretion" (p. 739)

multicellularity (MUHL tee sehl yoo LAR uh tee): having more than one cell (p. 204)

muscle tissue: tissue made of cells that contract and return to normal length, allowing for movement of parts of the body; from the Latin *musculum* and *texere*, meaning "little mouse (muscle)" and "to weave" (p. 612)

mutagen (MYOOT uh juhn): environmental agent that can alter the structure of DNA; from the Latin *mutare*, meaning "to change" (p. 140)

mutation (myoo TAY shuhn): abrupt change in genotype of an organism (p. 127)

mutualism (MYOO choo uhl ihz ohm): ecological relationship in which all organisms involved benefit (pp. 207, 270)

mycorrhiza (MY koh REYE zuh): mutualistic association of certain fungi with root cells of some vascular plants; from the Greek *mykes* and *rhiza*, meaning "fungus" and "root" (p. 207)

myelin (MEYE uh lihn) **sheath**: a fatty, segmented covering wrapped around axons, which increases the speed of nerve impulse transmissions (p. 642)

narcotic: drug used to relieve pain and induce sleep; from the Greek *narkoun*, meaning "to make numb" (p. 695)

natural selection: process by which organisms best-suited to their environmental conditions are most likely to survive and reproduce (p. 179)

negative feedback: a process in which an end product inhibits the first step of a system (p. 672)

nematocyst (NEHM uh toh sihst): stinging, harpoon-like structure on the tentacles and outer body surface of cnidarians; from the Greek *nēma* and *kystis*, meaning "thread" and "sac" (p. 506)

nephridium (nee FRIHD ee uhm): tubule that collects wastes from the coelom and discharges them from the body; from the Greek *nephros*, meaning "kidney" (p. 511)

nephron (NEHF rahn): blood-cleaning unit of the kidney (p. 775)

nerve tissue: tissue made up of cells called neurons that generate and transfer electrical impulses to other cells; from the Latin *nervus* and *texere*, meaning "sinew, string" and "to weave" (p. 612)

neuroendocrine system: term covering the closely linked nervous system and endocrine system; from the Greek *neuron* and *endon + krinein*, meaning "nerve" and "within + to separate" (p. 670)

neuromodulator: long-lived chemical that enables the body to prolong the transmission of a signal across a synapse by slowing the destruction of neurotransmitters (p. 693)

neuron (NOO rahn): basic cell of the nervous system, specialized to carry nerve impulses; from the Greek *neuron*, meaning "nerve" (p. 639)

neurotransmitter: chemical released by a neuron into a synapse that may stimulate or inhibit other neurons (p. 644)

niche (NIHCH): an organism's "profession," the sum of its interactions with its environment (p. 272)

nitrogen fixation: conversion of atmospheric nitrogen to ammonia and other compounds by bacteria (pp. 162, 254)

nonpolar molecule: molecule with an even charge distribution (p. 49)

nonrenewable resource: natural resource such as iron ore that is not regenerated by natural processes (p. 299)

notochord (NOHT uh kawrd): rod of cartilage that extends along the back in chordates; from the Greek *nōton* and Latin *chorda*, meaning "back" and "cord" (p. 211)

nucleic acid: DNA or RNA; large chainlike molecule containing phosphoric acid, sugar, and purine and pyrimidine bases; from the Latin *nucleus*, meaning "nut or kernel" (p. 37)

nucleotide (noo KLAY oh tyd): structural unit of a nucleic acid (subunit of DNA or RNA) consisting of three parts: a sugar, a phosphate group, and a base (pp. 37, 139)

nucleus (NOO klee uhs): the organelle in eukaryotic cells that contains most of the DNA; from the Latin *nucleus*, meaning "nut or kernel" (p. 56)

nymph (NIHMF): stage in the development of an insect, which is a small but incomplete version of the adult; from the Greek *nymphe*, meaning "a young goddess" (p. 534)

obesity (oh BEES ih tee): condition in which a high proportion of the body weight (over 20 percent) is stored fat; from the Latin *obesus*, meaning "devoured" (p. 766)

omnivore (AHM nih vawr): organism that eats both plants and animals; from the Latin *omni- + vorare*, meaning "all + to devour" (p. 251)

opposable thumb: digit that can be bent toward other digits (p. 221)

order: collection of similar families of organisms (p. 324)

organ: collection of different kinds of tissue that are dedicated to one function; from the Latin *organum*, meaning "instrument" or "tool" (pp. 463, 613)

organelle (awr guh NEHL): subcellular part such as a ribosome that has a special cellular function (p. 57)

organic compound: compound containing the element carbon (p. 33)

organ system: group of interrelated organs that carry out one essential body function (pp. 463, 613)

osmosis (ahz MOH sihs): movement of water through a membrane from an area of low solute concentration to an area of high solute concentration; from the Greek *othein*, meaning "to push" (p. 75)

osmotic (ahz MAH tihk) **pressure**: force required to prevent the flow of water through a membrane (p. 75)

osteoporosis (ahs tee oh puh ROH sihs): extreme loss of bone density, causing brittleness; from the Greek *osteon* and Latin *porosis*, meaning "bone" and "porous condition" (p. 622)

ovary: one of the two organs that produce eggs (p. 787)

ovulation (ahv yoo LAY shun): release of an egg from the ovary; from the Latin *ovum*, meaning "egg" (p. 789)

ovule (AHV yool): small egg or seed in early stage of development (p. 415)

ovum: female gamete, or egg (p. 787)

oxidation (ahks ih DAY shuhn) **reduction reaction**: chemical reaction in which electrons are transferred from one molecule to another; from the Greek *oxys*, meaning "acid" (p. 99)

oxidative respiration (AHKS ih day tihv rehs puh RAY shuhn): biochemical pathway by which cells get most of their energy; from the Latin *respirare*, meaning "to breathe" (p. 102)

ozone: unstable pale-blue gas that absorbs ultraviolet radiation; from the Greek *ozein*, meaning "to smell" (p. 206)

pacemaker: small bundle of cells near the entrance to the right atrium of the heart that emits an electrical signal to start the heartbeat (p. 719)

pancreas (PAN kree uhs): large gland situated behind the stomach that secretes digestive juices into the small intestine and produces insulin; from the Greek *pan- + kreas*, meaning "all + flesh" (p. 770)

parasitism: ecological relationship in which one organism lives on or in another organism and absorbs nutrients from it (p. 270)

pasteurization (PAS tuhr ih zay shuhn): method of destroying disease-producing bacteria with heat; named after Louis Pasteur, French chemist (p. 351)

pathogen: disease-causing agent; from the Greek *pathos*, meaning "suffering" (p. 348)

pedigree (PEHD uh gree): record covering several generations that indicates how a trait is inherited (p. 130)

pedipalp: second pair of appendages on spiders and other arachnids, used to capture and manipulate prey and for courtship displays and mating; from the Latin *pes* and *palpus*, meaning "foot" and "soft palm of the hand" (p. 525)

penicillin (PEN ih sihl ihn): antibiotic compound obtained from the mold *Penicillium* (p. 352)

pepsin: enzyme present in the stomach that, with hydrochloric acid, breaks proteins into small chains of amino acids; from the Greek *pepsis*, meaning "digestion" (p. 769)

peptide hormone: hormone made of amino acids; from the Greek *peptein* and *hormeinō*, meaning "to digest" and "to stimulate" (p. 676)

peripheral nervous system: the nervous system, excluding the spinal cord and brain (p. 652)

peristalsis (pehr uh STAHL sihs): wavelike muscle contractions that propel food through the digestive system; from the Greek *peristaltikos*, meaning "surrounding" (p. 769)

pH: term used to describe the acidity of a solution (p. 295)

phagocyte (FAG oh seyet): type of white blood cell that consumes pathogens; from the Greek *phagein*, meaning "to eat" (p. 740)

phenotype (FEE nuh typ): apparent characteristic of an organism, resulting from both its heredity and environment; from the Greek *phainein*, meaning "to show" (p. 119)

phloem (FLOH ehm): vascular tissue of plant that transports sugars and other dissolved nutrients; from the Greek *phloos*, meaning "bark" (p. 406)

phospholipid (fahs foh LIHP ihd): a molecule found in the cell membrane that is made of two nonpolar fatty acid chains joined to a short polar phosphorus-containing head (p. 50)

photon (FOH tahn): tiny packet of light energy with both particle and wave properties; from the Greek *phōs*, meaning "light" (p. 98)

photoperiodism (foh toh PIHR ee uhd ihzuhm): behavioral reaction of an organism to variations in day length (p. 414)

photosynthesis (foh toh SIHN thuh sihs): process by which organisms use the energy of light to produce food; from the Greek *synthenai*, meaning "to put together" (p. 96)

phylum (FEYE luhm): any of the broad, principal divisions of the kingdoms Monera, Animalia or Protista, coined by Georges Cuvier, French anatomist; from the Greek *phylon*, meaning "tribe" (pp. 205, 324)

phytoplankton (feye tuh PLANK tuhn): photosynthetic plankton, usually found near the surface of a body of water; from the Greek *phyein* and *planktos*, meaning "to grow" and "wandering" (p. 372)

pigment: molecule that absorbs and reflects different colors of light; from the Latin *pingere*, meaning "to paint" (p. 98)

pistil: seed-bearing organ of a flowering plant; from the Latin *pistillum*, meaning "pestle" (p. 415)

pituitary (puh TOO uh tehr ee) **gland**: small endocrine gland that produces at least six different hormones and is attached by a stalk to the base of the brain; from the Latin *pituita*, meaning "phlegm" (p. 671)

placenta (pluh SEHN tuh): structure through which nutrients and oxygen are transferred to the fetus; from the Greek *plakous*, meaning "flat cake" (p. 496)

placental (pluh SENT uhl) **mammal**: mammal with a placenta, through which offspring are nourished until birth (p. 592)

plankton: passively floating or weakly mobile aquatic algae, bacteria, and animals; from the Greek *planktos*,

meaning "wandering" (p. 257)

plasma (PLAZ muh): protein-rich fluid that makes up 55 percent of the blood; from the Greek *plassein*, meaning "to form" (p. 712)

plasmid (PLAZ mihd): bacterial genetic element not a part of the chromosome (p. 159)

platelet: fragment of a cell that is formed within bone marrow and helps form blood clots; from the Greek *platys*, meaning "broad, flat" (p. 713)

polar molecule: molecule in which charge is unevenly distributed (p. 48)

pollen grain: male gametophyte resulting from evolutionary development in gymnosperms; from the Greek *palē*, meaning "dust" (p. 395)

pollination: the transfer of pollen from male to female flower structures (p. 395)

pollution: addition of harmful substances to the environment; from the Latin *polluere*, meaning "to soil" (p. 293)

polyp: one of the two life stages of cnidarians, generally attached to a hard surface; from the Greek *poly- + pous*, meaning "many + foot" (p. 506)

prediction: expected outcome if a hypothesis is accurate; from the Latin *praedicere*, meaning "to foretell" (p. 12)

primary immune response: immune response that occurs after the first exposure to a pathogen (p. 745)

primary succession: regular pattern of species replacement occurring where nothing has lived before (p. 274)

primate: order of mammals to which people, apes, monkeys, and prosimians belong; from the Latin *primus*, meaning "first" (p. 221)

producer: organism that is able to make its own food (p. 250)

prokaryote (pro KAR ee oht): organism lacking a true nucleus and reproducing by splitting; from the Greek *pro-* and *karyon* and *otes*, meaning "before" and "nut" or "kernel" and "inhabitant" (p. 56)

prosimian (proh SIHM ee uhn): member of a group of tree-dwelling primates; from the Latin *pro + simia*, meaning "before + ape" (p. 222)

prostaglandins (pras tuh GLAN dihnz): a group of lipids produced in small quantities by specialized cells that function in the body as hormones; from the Greek *proïstanai*, meaning "to put in front of" (p. 684)

protein: any of the polymers of amino acids; from the Greek *prōtos*, meaning "first" (p. 36)

proton pump: mechanism in cells that expels protons in the cell's metabolism process (p. 78)

protostome (PROHT uh stohm): animal in which the initial depression that starts gastrulation becomes the mouth of the adult organism; from the Greek *prōtos* and *stoma*, meaning "first" and "mouth" (p. 471)

pseudocoelom (SOO duh see luhm): body cavity between mesoderm and endoderm of roundworms (nematodes); from the Greek *koilia*, meaning hollow" (p. 465)

pseudopod (SOO doh pahd): projection of the cell body used for movement or feeding; from the Greek *podos*, meaning "foot" (p. 368)

psychoactive drug: drug that affects the nervous system, such as alcohol, heroin, and nicotine; from the Greek *psychē*, meaning "breath, spirit, soul" (p. 691)

punctuated equilibrium: hypothesis that evolution occurs at an irregular rate; from the Latin *punctus* and *aequus* and *libra*, meaning "point" and "equal" and "balance" (p. 192)

purine (PYUR een): one of the larger two of four different nucleotide bases, namely adenine and guanine (p. 139)

pyrimidine (py RIHM uh deen): one of the smaller bases in nucleotides, namely cytosine, thymine (in DNA), and uracil (in RNA) (p. 139)

pyruvate (py ROO vayt): one of two identical molecules that result from splitting a molecule of glucose during glycolysis (p. 103)

radial symmetry: wheel-like symmetry in which body parts radiate from a central "hub"; from the Latin *radius* and *symmetria*, meaning "spoke, ray" and "measured together" (p. 459)

radioactive dating: technique for measuring the age of an object by determining the ratio of the concentrations of a radioisotope to that of a stable isotope in it (p. 181)

radula (RAJ oo lah): flexible toothy structure in the mouth of mollusks that is used for feeding; from the Latin *radere*, meaning "to scrape" (p. 512)

ray-finned fish: bony fish whose fins are fan-shaped and supported by bony rays (p. 559)

reabsorption: second process of urine formation, in which filtered fluid passes into the nephron, returning important substances to the blood (p. 775)

realized niche: that part of a fundamental niche that an organism actually occupies as a result of competition or predation (p. 273)

receptor protein: protein of cell membranes that senses chemical signals (p. 70)

recessive: describes an allele that is not expressed in heterozygous individuals (p. 119)

recombinant (ree KAHM buh nuhnt) **DNA**: molecule formed when fragments of DNA from two or more different organisms are spliced together; (p. 157)

red blood cell: cell that carries oxygen from the lungs to all cells of the body and carries carbon dioxide back to the lungs to be exhaled (p. 712)

reflex: sudden, involuntary response to a stimulus; from the Latin *reflexus*, meaning "reflected" (p. 653)

renewable resource: natural resource that can be replaced by a natural process, such as water and timber (p. 299)

replication: duplication of DNA chromosomes before cell division (p. 140)

repressor protein: protein that blocks transcription by preventing RNA polymerase from moving along the gene; from the Latin *repressus*, meaning "held back" (p. 148)

resistance exercise: exercise that requires exerting a great deal of force against an immovable object (p. 631)

resting potential: starting point for the transmission of a nerve impulse, when the outside of the neuron is more positive than the inside; from the Latin *potere*, meaning "to be powerful" (p. 640)

restriction enzyme: enzyme that recognizes and binds to specific short sequences of DNA, then cuts the DNA at a specific site within that sequence; from the Greek *en-* + *zyme*, meaning "in + leaven" (p. 158)

rhizoid (REYE zoyd): any of the rootlike structures that attach plants or fungi to a substrate; from the Greek *rhiza*, meaning "root" (p. 391)

ribonucleic acid (RNA): a single-stranded nucleic acid containing the bases adenine, guanine, cytosine, and uracil (p. 141)

ribosome (RY buh sohm): cytoplasmic body that contains RNA and produces proteins for the cell (p. 56)

rod: receptor cell in the retina of the eye that is extremely sensitive to light and can detect shades of gray (p. 661)

science: is a way of investigating the world and observing nature in order to form general rules about what causes things to happen; from the Latin *scire*, meaning "to know" (p. 5)

scientific name: the unique two-word name assigned to each kind of organism on earth (p. 319)

secondary immune response: immune response to a previously encountered pathogen; usually occurs much faster and produces more antibodies than the first response (p. 745)

secondary succession: regular pattern of species replacement occurring where there has been previous growth (p. 274)

second messenger: molecule that carries information into the cytoplasm from the cell membrane (p. 676)

seed: specialized structure that develops from the fertilized egg of certain plants (p. 394)

segmentation: characteristic of animals that are composed of repeated body units; from the Latin *segmentum*, meaning "part" (p. 468)

selective transport: process by which only one type of molecule is transported through membrane protein channels (p. 76)

self-pollination: pollination that occurs when pollen grains fall from the anther to the stigma of the same flower (p. 416)

semen: fluid consisting of the secretions of several glands mixed with sperm, expelled from the penis during ejaculation; from the Latin *semen*, meaning "seed" (p. 785)

sense organ: a specialized organ of the body that responds to stimuli by producing nerve impulses in a sensory neuron (p. 658)

sensitivity: the responsiveness of an organ or an organism (p. 25)

sensory neuron: nerve cell that carries nerve impulses from sense organs to the central nervous system (p. 652)

sex chromosome: either member of a pair of chromosomes responsible for sex determination (p. 126)

sex hormone: hormone responsible for the development and maintenance of sexual characteristics; from the Greek *hormeinō*, meaning "to stimulate" (p. 683)

sex-linked traits: traits whose genes are located on the X chromosome (p. 129)

skeletal muscle: muscle attached to the bones; from the Latin *musculus*, meaning "little mouse (muscle)" (p. 626)

small intestine: digestive organ that receives food from the stomach, where it is further broken down and where nutrients are transferred to the blood; from the Latin *intestinus*, meaning "inward" (p. 770)

smooth muscle: muscle that is not under conscious control; from the Latin *musculus*, meaning "little mouse (muscle)" (p. 626)

sodium potassium pump: mechanism that enables cells to admit ions important to biological processes such as the conduction of nerve impulses (p. 77)

species (SPEE seez): a group of organisms that can interbreed with each other but not with members of any other group; from the Latin *species*, meaning "appearance, kind" (pp. 18, 320)

spicule (SPIHK yool): needlelike, bony structure that forms the skeleton of a sponge; from the Latin *spica*, meaning "point" (p. 503)

spinneret: nozzle-like structure on the abdomen of spiders that directs the flow of silk from glands in the abdomen (p. 526)

spore: haploid reproductive cell capable of developing into a new organism; from the Greek *speirein*, meaning "to sow" (p. 386)

sporophyte (SPOR uh fyt): in plants, the diploid form that is the result of the fusing of two gametes; from the Greek *phyein*, meaning "to grow" (p. 391)

sprain: injury resulting from a ligament that is overstretched (p. 625)

stability: ability of an ecosystem to resist change in the face of disturbance (p. 275)

stamen (STAY mehn): male part of a flower that produces pollen grains; from the Latin *stamen*, meaning "thread" (p. 415)

steroid hormone: hormone synthesized from cholesterol, such as testosterone or estrogen; from the

Greek *hormeinō*, meaning "to stimulate" (p. 675)

stoma (STOH muh): tiny pore in a plant leaf that allows carbon dioxide to enter the leaf and water and oxygen to exit; from the Greek *stoma*, meaning "mouth" (p. 404)

substrate: molecule on which an enzyme acts; from the Latin *substernere*, meaning "to strew about" (p. 92)

succession: regular progression of species replacement in a developing ecosystem; from the Latin *succedere*, meaning "to follow" (p. 274)

swim bladder: gas-filled sac in bony fishes that provides buoyancy (p. 557)

symbiosis (sihm beye OH sihs): close, long-term association between two or more species; from the Greek *syn + bios*, meaning "together + life" (p. 270)

synapse (SIHN ahps): junction of a neuron with another cell; from the Greek *synaptein*, meaning "to fasten together" (p. 644)

 T

target: in the endocrine system, the specific destination (tissue or organ) of a hormone (p. 669)

taxonomy: science of classifying living things (p. 323)

T cell: white blood cell that controls the immune response or attacks infected cells (p. 743)

tendon: inelastic, tough, fibrous connective tissue attaching muscles to bones; from the Latin *tendere*, meaning "to stretch" (p. 628)

tendonitis: inflammation of a tendon caused by friction or stress (p. 632)

testis (TEHS tis): either of the oval-shaped organs of the male that produce

sperm; from the Latin *testis*, meaning "witness" (p. 784)

theory: explanation of several phenomena based on hypotheses that have been tested many times; from the Greek *thēorein*, meaning "to look at" (p. 13)

tissue: group of similar cells that are organized into a functional unit; from the Latin *texere*, meaning "to weave" (pp. 458, 611)

trachea (TRAY kee uh): the cartilaginous, membranous tube by which air passes to and from the lungs in humans and many vertebrates, also one of the tubes leading into an insect's body cavity from pores (spiracles) in the surface of the exoskeleton (pp. 489, 729)

transcription: first stage of gene expression, in which an RNA copy of DNA is made (p. 141)

transformation (trans fawr MAY shuhn): the transfer of genetic material from one organism to another; first observed by Griffith (p. 137)

translation: second stage of gene expression, in which three kinds of RNA work together to assemble amino acids into a protein molecule; (p. 141)

translocation: transport of sugars made by photosynthesis from the leaves to the rest of the plant (p. 411)

transpiration: loss of water by the leaves and stem of a plant; from the Latin *trans-* + *spirare*, meaning "across + to breathe" (p. 411)

transposon (tranz POH zahn): "jumping gene," a gene with the ability to move from one location to another in the chromosomes; from the Latin *transponere*, meaning "to transfer" (p. 150)

trimester: one of the three 3-month periods

of the 9-month pregnancy; from the Latin *trimestris*, meaning "of three months" (p. 798)

trophic (TROHF ihk) **level**: group of organisms whose energy source is the same number of steps away from the sun; from the Greek *trophē*, meaning "food" (p. 251)

tuber: enlarged end of a rhizome that grows underground and can be used for food; from the Latin *tuber*, meaning "swelling" (p. 430)

tumor: formation of a mass of cells as a result of abnormally rapid cell division (p. 750)

 U

ulcer: painful condition in which acid eats away the lining of the stomach or small intestine; from the Greek *helkos*, meaning "abscess, wound" (p. 771)

underhair: shorter, thinner hairs under the guard hair coat of animals, which reduce the amount of body-heat loss (p. 595)

urea: nitrogenous waste that is excreted by mammals, and requires some water for elimination; from the Greek *ouron*, meaning "urine" (p. 492)

uric acid: nontoxic solid waste eliminated by reptiles and birds (p. 492)

uterus: hollow muscular organ of the female where the fetus develops during pregnancy; from the Latin *uterus*, meaning "womb" (p. 787)

 V

vaccine (vak SEEN): a preparation of killed or weakened pathogens introduced into the body to produce immunity; from the Latin *vacca*, meaning "cow" (pp. 166, 351)

vacuole (VAK yoo ohl): membrane-bound cavity in a cell that serves as a storage area and may function in digestion, secretion,

or excretion; from the Latin *vacuus*, meaning "empty" (p. 60)

vagina (vuh JEYE nuh): muscular tube in female reproductive system that receives sperm ejaculated by a male during intercourse; from the Latin *vagina*, meaning "sheath" (p. 788)

vascular tissue: tubes and vessels of a plant's internal system that transport water and nutrients; from the Latin *vasculum*, meaning "small vessel" (p. 391)

vas deferens (vas DEHF uh REHNZ): duct that connects the epididymis to the urethra and carries sperm during ejaculation; from the Latin *vas deferentia*, meaning "deferent vessel" (p. 784)

vector: agent used to transfer genes in genetic engineering; from the Latin *vectus*, meaning "carried" (p. 159)

vein: blood vessel that returns blood to the heart (p. 714)

ventricle (VEHN trih kuhl): chamber of the heart that pumps blood out; from the Latin *venter*, meaning "belly" (p. 718)

vertebrate: animal with a backbone; from the Latin *vertebratus*, meaning "jointed" (p. 211)

vestigial (ves TIHJ ee uhl) **structure**: structure that is a remnant of an organism's evolutionary past and has no function; from the Latin *vestigium*, meaning "footprint" (p. 182)

villus: any of the projections lining the small intestine; from the Latin *villus*, meaning "tuft of hair" (p. 770)

virus: any of the microscopic particles that invade cells of plants, animals, fungi, and bacteria, often destroying the cells; from the Latin *virus*, meaning "poison" (p. 353)

vitamin: organic substance needed by the body in small amounts (p. 765)

voltage-sensitive channel: protein channel of cells that respond to electrical signals (p. 69)

 W

water vascular system: complex arrangement of water-filled tubes that function in the locomotion, feeding, and gas exchange in echinoderms; from the Latin *vasculum*, meaning "small vessel" (p. 517)

white blood cell: colorless blood cell that protects the body against invading bacteria, viruses, or other foreign cells (p. 713)

wood: the secondary xylem produced in plants that live more than one growing season (p. 436)

 X

xylem (ZEYE luhm): vascular tissue of plants that transports water and minerals; from the Greek *xylon*, meaning "wood" (p. 406)

Index

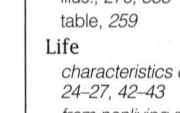

Credits